Auger Electron Spectroscopy

A BIBLIOGRAPHY: 1925–1975

Auger Electron Spectroscopy

A BIBLIOGRAPHY: 1925–1975

Compiled by

Donald T. Hawkins

Bell Laboratories
Murray Hill, New Jersey

With a Foreword by

C. C. Chang

Bell Laboratories
Murray Hill, New Jersey

IFI/PLENUM · NEW YORK - WASHINGTON - LONDON

Library of Congress Cataloging in Publication Data

Hawkins, Donald T
  Auger electron spectroscopy.

  Includes indexes.
  1. Auger effect—Bibliography. 2. Photoelectron spectroscopy—Bibliography. I. Title.
Z7144.S7H37 [QC793.5.E627]   016.5391'52   76-55815
ISBN-13: 978-1-4684-1389-2   e-ISBN-13: 978-1-4684-1387-8
DOI: 10.1007/978-1-4684-1387-8

Foreword

Auger electron spectroscopy is rapidly developing into the single most powerful analytical technique in basic and applied science for investigating the chemical and structural properties of solids. Its explosive growth beginning in 1967 was triggered by the development of Auger analyzers capable of detecting one atom layer of material in a fraction of a second. Continued growth was guaranteed firstly by the commercial availability of apparatus which combined the capabilities of scanning electron microscopy and ion-mill depth profiling with Auger analysis, and secondly by the increasing need to know the atomistics of many processes in fundamental research and engineering applications.

The expanding use of Auger analysis was accompanied by an increase in the number of publications dealing with it. Because of the developing nature of Auger spectroscopy, the articles have appeared in many different sources covering diverse disciplines, so that it is extremely difficult to discover just what has or has not been subjected to Auger analysis. In this situation, a comprehensive bibliography is obviously useful to those both inside and outside the field. For those in the field, this bibliography should be a wonderful time saver for locating certain references, in researching a particular topic, or when considering various aspects of instrumentation or data analysis.

This bibliography not only provides the most complete listing of references pertinent to surface Auger analysis available today, but it is also a basis for extrapolating from past trends to future expectations. The first outstanding trend is the increasing diversity of fields to which Auger analysis has been successfully applied. These now include materials science, communications, transportation, energy research, and manufacturing. This trend will continue with the advancing technological sophistication of each field. Another trend is the rapid development of data interpretation from the initial, modest efforts at chemical identification of surface atoms to quantitative analyses, complex geometrical analyses of structures in three dimensions, and chemical bonding and electronic band structures. It is already evident that the basic analysis of Auger intensities is relatively simple compared, for example, to intensity analyses of electron microscope data. A third trend is the improvement in instrumentation which is still continuing. For instance, computerized data acquisition and reduction is no longer a dream in many laboratories and the flexibility of such systems should make Auger spectroscopy about as commonplace as electron microscopy or gas chromatography. It is evident from these trends that in the very near future comprehensive bibliographies such as this one will be too large and cumbersome to compile and use.

In summary, Auger spectroscopy provides information on the atomistics of physical and chemical processes. Since matter is composed of atoms, Auger analysis should prove useful to every advanced technology and the past explosive growth of this field should continue for some time. This bibliography is a timely work to which those in and out of this field can look for a picture of the initial developments.

C. C. Chang

Murray Hill, New Jersey
September 29, 1976

Acknowledgment

It is a pleasure to thank Mrs. Marion A. Marasco for her careful editing, proofreading, and attention to detail, which have contributed substantially to the appearance of this work.

Introduction

Coverage

In 1925, Auger published a description (Reference 2-0249) of the effect that now bears his name. He was studying particle tracks in a cloud chamber which had been irradiated with X-rays and observed that electrons of constant energy were emitted, independent of the X-ray energy. The significance of Auger's work seems to have been largely unnoticed until 1953, when Lander (Reference 4-0926) pointed out the potential usefulness of the technique in surface analysis. Although Harrower (Reference 4-0874) published a paper in 1956 dealing with secondary (Auger) emission in the spectra of molybdenum and tungsten, little attention seems to have been paid to this application of Auger's work until 1967 and 1968, when Harris (Reference 5-1401) and Weber and Peria (Reference 5-1959) showed that Auger electron spectroscopy (AES) was a useful tool for surface analysis. Harris first used the differentiation of the energy curve to obtain the spectrum so familiar today, and Weber and Peria showed that LEED equipment was suitable for Auger work. The use of the cylindrical mirror analyzer by Palmberg, Bohn, and Tracy (Reference 3-0672) paved the way for the commercial development of Auger spectrometers, which are becoming quite commonplace in laboratories dealing with surface analysis. Since that time, AES has become increasingly popular, and it is rapidly taking its place as a standard analytical tool beside such well-established techniques as NMR, infrared spectroscopy, and so on. In fact, many of the recent papers describing experiments using AES do not mention the term in their title or abstract, showing that AES is becoming one of the recognized common analytical techniques in use today. Such papers are becoming more frequent, making the preparation of a comprehensive bibliography considerably more difficult.

With the increasing popularity of AES, the literature has, of course, also grown rapidly. New applications of AES are described, and new techniques are continually being devised. One of the newer applications is the use of AES in metallurgy, to study the temper embrittlement of steels, surface defects, grain boundary reactions, and so on. Rapid progress is also being made in the development of quantitative AES.

This bibliography of 2146 references represents an effort to cover the AES literature comprehensively, especially from 1967 to the present. Older papers, including Auger's original work, are included when applicable. Many papers on the theory and physics of the Auger effect are included (although this area has not been covered comprehensively before 1967). Particular attention was paid to the use of AES for surface analysis. In such a rapidly growing field, keeping abreast of the literature becomes more and more difficult, especially as its volume grows and it appears in increasingly diverse sources. Several excellent reviews are included here; they have been very helpful in searching out the numerous earlier references to AES. The relevant references in these reviews have been included.

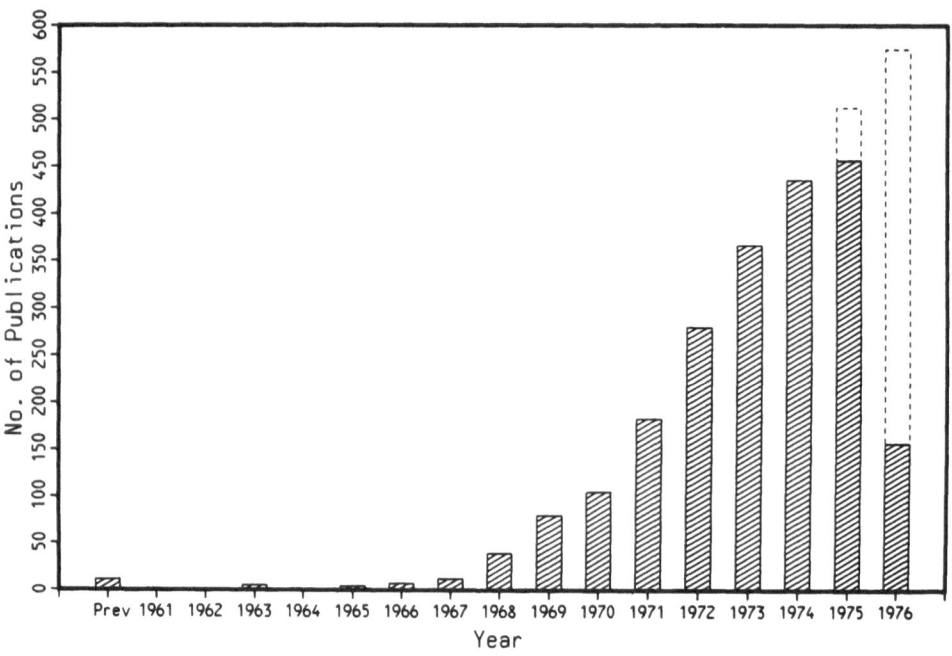

Figure 1. Yearly growth of the AES literature.

Organization

The bibliography is organized into six sections as shown in the Contents. In each section, the title, authors, and source are listed for each reference. Many of the references were obtained from *Chemical Abstracts* (CA), and *Science Abstracts A* (SAA). The latest issues of these two abstract journals which were searched are CA Volume 85(8) and SAA 1976(13). When available, the abstract number of the reference is listed on the last line of the entry. Each unique reference bears a number consisting of the section number followed by a running serial number. Within each section, entries are listed alphabetically by author.

Permuted title and author indexes follow the bibliographic listing. The permuted title index is particularly useful as a subject guide to the bibliography and provides many additional entry points beside the section groupings. The terms AUGER and SPECTROSCOP... are not indexed. Terms prefixed by common prefixes such as DI..., TRI..., etc., are indexed under the root word. Minimal editing was done in either index (particularly in the case of transliterations from Russian) to remove inconsistencies. Parenthetical notes were appended to many of the titles in order to enrich the permuted title index.

This volume has been computer produced and formatted by the Bell Laboratories proprietary system BELDEX.*

Statistical Information

Figure 1 shows the rapid and dramatic growth of the AES literature. From seven papers in 1966, the annual volume of publications has grown to 436 in 1974. The data for 1975 and 1976 are incomplete due to the time lag in the indexing and abstracting services. Extrapolating the 1967–1974 growth

*The BELDEX system has been copyrighted by Bell Laboratories. For further information, see W. K. Lowry, Use of computers in information systems, *Science,* **175**: 841-846 (1972).

TABLE 1. Authorship of Papers on AES

Number of papers, P	Number of authors' names appearing on P papers	Number of papers, P	Number of authors' names appearing on P papers
1	1206	13	3
2	342	14	9
3	128	15	6
4	92	16	3
5	59	17	2
6	42	18	2
7	24	19	0
8	17	20	1
9	20	more	10
10	11		
11	15		
12	8	Total	2000

rate to 1975 and 1976, one would expect about 513 publications for 1975 and 575 for 1976. (Of course, a large and increasing annual growth rate cannot be maintained for more than a few years, especially as the field becomes "saturated.")

A total of 2000 authors' names appear in this bibliography. There are 4717 entries in the author index, for an average of just over two authors per publication. Table 1 shows the number of names associated with one paper, two papers, and so on. The 11 authors with 20 or more papers each are listed in Table 2. Together, these authors account for 300 papers (14% of the total).

Table 3 is an analysis of the sources of the AES literature. Each type of literature is listed, and the journal literature is analyzed by journals contributing one, two, three, four, five, or more papers. Although journal articles predominate, conference proceedings contribute 254 entries (12% of the total). The 61 journals contributing five or more papers are listed in Table 4. It is not surprising to find that the three most prolific journals are *Surface Science* with 335 entries, *Journal of Vacuum*

TABLE 2. Authors Publishing 20 or More Papers on AES

Author	Number of papers
Matthew, J.A.D.	20
McGuire, E.J.	21
Mehlhorn, W.	21
Stolterfoht, N.	21
Gallon, T.E.	23
Riviere, J.C.	23
Stein, D.F.	24
Joshi, A.	25
Palmberg, P.W.	32
Grant, J.T.	41
Haas, T.W.	49
Total	300

TABLE 3. Sources of Literature on AES

Source	Number of papers
Books	4
Conference proceedings	254
Government reports	70
Patents	9
Theses	29
Journals:	
One paper each (140 journals)	140
Two papers each (38 journals)	76
Three papers each (24 journals)	72
Four papers each (15 journals)	60
Five papers each (9 journals)	45
More than five papers each (49 journals)	1387
Total	2146

TABLE 4. Journals Publishing Five or More Papers on AES

Journal	Number of papers	Journal	Number of papers
Surface Sci.	335	*Sov. Phys. Solid State*	11
J. Vacuum Sci. Technol.	132	*J. Phys. D*	11
Phys. Rev. (all sections)	103	*Bull. Amer. Ceram. Soc.*	11
J. Appl. Phys.	68	*J. Catal.*	10
Appl. Phys. Lett.	52	*Anal. Chem.*	10
Phys. Lett.	50	*Proc. Roy. Soc. A*	9
Rev. Sci. Instrum.	37	*Appl. Phys.*	9
Phys. Rev. Lett.	34	*Vide*	8
Solid State Commun.	30	*J. Metals*	8
Thin Solid Films	28	*J. Amer. Ceram. Soc.*	8
J. Phys. C	28	*Sov. Phys. Semicond.*	7
J. Phys. B	27	*Phys. Status Solidi B*	7
J. Electron Spectrosc. Relat. Phenomena	25	*Phys. Scr.*	7
Bull. Acad. Sci. USSR, Phys. Ser.	23	*Nucl. Instrum. Meth.*	7
Z. Phys.	22	*Z. Naturforsch.*	6
J. Chem. Phys.	22	*Oyo Butsuri*	6
J. Phys. F	19	*Kratk. Soderzh. Dokl.*	6
Compt. Rend.	19	*J. Chem. Soc., Faraday Trans.*	6
J. Electrochem. Soc.	18	*Appl. Spectrosc.*	6
Vacuum	17	*Vak. Tech.*	5
Jap. J. Appl. Phys.	17	*Sov. Phys. Tech. Phys.*	5
J. Phys. E	16	*Sov. Phys. JETP*	5
Phys. Fenn.	15	*Nucl. Phys. A*	5
Met. Trans.	15	*Mikrochim. Acta*	5
Shinku	14	*J. Lumin.*	5
Scr. Met.	14	*J. Crystal Grow.*	5
Chem. Phys. Lett.	14	*IEEE Trans. Parts, Hybrids Packag.*	5
Bull. Amer. Phys. Soc.	14	*Hyomen*	5
Phys. Status Solidi A	13		
J. Phys. (Paris)	13	Total	1432

Science and Technology with 132, and *Physical Review* with 103. The 570 entries from these three journals comprise 26.6% of this bibliography.

The number of references in each section is listed in the Table of Contents. Considering the rapid growth of AES as an analytical technique, it is not surprising to find that the section on surface analysis has by far the largest number of references.

Contents

		Number of References	Page
1	General, Reviews, Miscellaneous.	222	3
2	Theory, Physics of the Auger Effect.	360	11
3	Apparatus, Technique	140	25
4	Auger Spectra of Materials, Spectral Line Energies.	404	31
5	Surface Analysis by Auger Spectroscopy	864	45
6	Metallurgical Applications of Auger Spectroscopy	156	79
	Permuted Title Index		89
	Author Index		245

Bibliography

1. GENERAL, REVIEWS, MISCELLANEOUS

1-0001 ELECTRON SPECTROSCOPY. (PHOTOELECTRON, AUGER, AND IONIZATION SPECTROSCOPY, REVIEW, 41 REFS)
AKSELA S
SUOM KEMISTILEHTI A 45(2): 44-54 (1972)
CA 77: 26585J

1-0002 QUANTITATIVE AUGER ELECTRON SPECTROSCOPY USING COADSORPTION.
ARGILE C + RHEAD GE
J PHYS C 7(14): L261-L264 (1974)
CA 81: 114124

1-0003 COMBINED AUGER ELECTRON SPECTROSCOPY AND SCANNING ELECTRON MICROSCOPY.
ASHWELL GWB + TODD CJ + HECKINGBOTTOM R
J PHYS E 6(5): 435-8 (1973)
CA 79: 024532

1-0004 AUGER EFFECT. (REVIEW, 16 REFS)
AUGER P
SURFACE SCI 48(1): 1-8 (1975)
CA 82: 161949

1-0005 HIGH TEMPERATURE AUGER ELECTRON SPECTROSCOPY.
BAS EB + BAENNINGER U
SURFACE SCI 41(1): 1-10 (1974)
CA 80: 032206

1-0006 PRESENT STATE OF AUGER ELECTRON SPECTROSCOPY AND LEED.
BAUER E
Z METALLK 63: 437-47 (1972) (IN GERMAN)

1-0007 LOW ENERGY ELECTRON DIFFRACTION AND AUGER METHODS. (REVIEW)
BAUER E
P255-74 OF INTERACTIONS ON METAL SURFACES, GOMER R (ED), SPRINGER, 1975. (TOPICS APPL PHYS, VOL 4)
CA 84: 36794

1-0008 COMBINED LOW ENERGY ELECTRON DIFFRACTION, AUGER ELECTRON, AND FLASH DESORPTION SPECTROSCOPY OF METALS ON SINGLE CRYSTAL SURFACES.
BAUER E + POPPA H
VIDE 28(164): 73-4 (1973)
CA 79: 139984

1-0009 SECONDARY ION MASS SPECTROMETRY (AND AUGER ELECTRON SPECTROSCOPY): A USEFUL COMPLEMENT TO OTHER DEPTH PROFILING TECHNIQUES.
BAUN WL + MCDEVITT NT
J VACUUM SCI TECHNOL 12(1): 504 (1975)
SAA 1976: 07699

1-0010 ELECTRON BEAM MICROANALYSIS. PART-1. (REVIEW OF ELECTRON PROBE ANALYZER, SCANNING ELECTRON MICROSCOPE, TRANSMISSION ELECTRON MICROSCOPE, ENERGY DISPERSIVE SPECTROMETER, ION MASS ANALYSIS, AUGER ELECTRON MICROSCOPY)
BEAMAN DR + ISASI JA
MATER RES STAND 11(11): 8-35,45,48-68,70-1,73-5,77-8 (1971)
CA 76: 20853C

1-0011 NEW DEVELOPMENTS IN THE SURFACE ANALYSIS OF SOLIDS.
BENNINGHOVEN A
APPL PHYS 1(1): 3-16 (1973)
CA 81: 044835

1-0012 THE AUGER EFFECT.
BERGSTROM J + NORDLING C
P1532-43 OF ALPHA-, BETA-, AND GAMMA-RAY SPECTROSCOPY, VOL 2, SIEGBAHN K (ED), NORTH-HOLLAND, 1965.

1-0013 (AUGER) ELECTRON SPECTROSCOPY, A METHOD FOR STUDYING THE SOLID GAS INTERFACE.
BILOEN P + BOUWMAN R
NED TIJDSCHR NATUURKD 41(13): 196-204 (1975) (IN DUTCH)
CA 84: 36757

1-0014 CHEMICAL EFFECTS IN AUGER ELECTRON SPECTROSCOPY.
BRAUN P + BETZ G + FAERBER W
MIKROCHIM ACTA SUPPL 5: 365-75 (1973) (IN GERMAN)
CA 80: 150620

1-0015 LASER WINDOW SURFACE FINISHING AND COATING TECHNOLOGY. (AUGER ELECTRON SPECTROSCOPY)
BRAUNSTEIN M + BRAUNSTEIN AI + ALLEN SD + GENTILE AL + GIULIANO CR
HUGHES RES LABS, MALIBU, CALIF, 31 JAN 1975, 103P. AD-A007791/7ST

1-0016 SURFACE ANALYSIS, METHODS OF STUDYING THE OUTER ATOMIC LAYERS OF SOLIDS.
BRONGERSMA HH + MEIJER F + WERNER HW
PHILIPS TECH REV 34(11-12): 357-69 (1974)
CA 83: 187813

1-0017 ELECTRON SPECTROSCOPY OF SURFACES.
BRUNDLE CR
SURFACE SCI 27(3): 681-5 (1971)

1-0018 APPLICATION OF ELECTRON SPECTROSCOPY TO SURFACE STUDIES. (REVIEW, 85 REFS)
BRUNDLE CR
SURF DEFECT PROP SOLIDS 1: 171-204 (1972); J VACUUM SCI TECHNOL 11(1): 212-24 (1974)
CA 78: 49905Y; 81: 016948

1-0019 ELECTRON SPECTROSCOPY STUDIES OF ADSORPTION AND OXIDATION PROCESSES AT METAL SURFACES. (REVIEW, 50 REFS)
BRUNDLE CR
J ELECTRON SPECTROSC RELAT PHENOMENA 5: 291-319
CA 82: 78536

1-0020 AUGER SPECTROSCOPY. (REVIEW)
BRUNDLE CR
PROC SOC ANAL CHEM 10(8): 194-5 (1973)
CA 83: 37026

1-0021 TECHNIQUES RELATED TO PHOTOELECTRON SPECTROSCOPY. (REVIEW)
BRUNDLE CR + BAKER AD
P425-47 OF ELECTRONIC STATES OF INORGANIC COMPOUNDS: NEW EXPERIMENTAL TECHNIQUES, DAY P (ED), 1974, NATO, OXFORD, ENGLAND, REIDEL, DORDRECHT, NETHERLANDS, 1975.
SAA 1976: 14656

1-0022 THE AUGER EFFECT AND OTHER RADIATIONLESS TRANSITIONS.
BURHOP EHS
CAMBRIDGE UNIV PRESS, 1952, 188P.

1-0023 AUGER EFFECT.
BURHOP EHS + ASAAD WN
P163-284 OF ADVANCES IN ATOMIC AND MOLECULAR PHYSICS, VOL-8, BATES DR + ESTERMANN I (EDS), ACADEMIC PRESS, 1972.
CA 80: 075985

1-0024 GATED ELECTRON BEAM TECHNIQUE FOR AUGER ELECTRON SPECTROSCOPY.
BYRUM BW
REV SCI INSTRUM 45(5): 707-8 (1974)
CA 81: 031717

1-0025 PRESENT AND FUTURE APPLICATIONS OF AUGER SPECTROSCOPY.
CARLSON TA
P2274-304 OF INT CONF INNER SHELL IONIZATION PHENOMENA AND FUTURE APPLICATIONS, VOL 4, 1972, FINK RW (ED), NTIS, 1973.
CA 78: 129973

1-0026 PHOTOELECTRON AND AUGER SPECTROSCOPY.
CARLSON TA
PLENUM, 1975, 417P.

1-0027 X-RAY PHOTOELECTRON (AND AUGER ELECTRON)
MICROANALYSIS AND MICROSCOPY. PRINCIPLES AND
EXPECTED PERFORMANCES.
CAZAUX J
REV PHYS APPL 10(5): 263-80 (1975) (IN
FRENCH)
      CA 84: 98796

1-0028 ELECTRON PROBE MICROANALYSIS AND ELECTRON
SPECTROSCOPY.
CHAMBERS WJ
CHEM ENG (LONDON) 297: 313-15 (1975)
      CA 83: 172044

1-0029 AUGER ELECTRON SPECTROSCOPY FOR CHEMICAL
ANALYSIS.
CHANG CC
ACS ANN MEET 1970, 160TH PROC ABSTR PAPERS NO
PHYS-7.

1-0030 AUGER ELECTRON SPECTROSCOPY. (REVIEW)
CHANG CC
SURFACE SCI 25: 53-79 (1971)

1-0031 ANALYTICAL AUGER ELECTRON SPECTROSCOPY.
CHANG CC
P509-75 OF CHARACTERIZATION OF SOLID
SURFACES. KANE PF & LARRABEE GB (EDS), NY,
PLENUM PRESS, 1974, 670P.
      CA 81: 098855

1-0032 GENERAL FORMALISM FOR QUANTITATIVE AUGER
ANALYSIS.
CHANG CC
SURFACE SCI 48: 9-21 (1975)

1-0033 THEORY OF AUGER TRANSITIONS.
CHATTARJI D
ACADEMIC PRESS, 1976, 256P.

1-0034 APPLICATION OF ION, ELECTRON, AND X-RAY BEAMS
IN MODERN MATERIAL TESTING.
CHRISTIAN H + WOITSCHECK A
MASCHINENSCHADEN 45(4): 151-9 (1972) (IN
GERMAN)
      CA 78: 113108

1-0035 ELECTRON SPECTROMETRY. (REVIEW)
CIOMARTAN D
IND USOARA 22(8-9): 384-5 (1975) (IN
ROMANIAN)
      CA 84: 51547

1-0036 APPLICATION OF AUGER ELECTRON SPECTROSCOPY TO
REACTOR MATERIALS RESEARCH.
CLAUSING RE
J VACUUM SCI TECHNOL 7: 5124 (1970)

1-0037 CHARACTERIZATION OF A CHROMIUM- GOLD THIN
FILM DEPOSITION PROCESS BY AUGER ANALYSIS.
CLAY FA
BENDIX CORP, 1974, 18P. BDX-613-1173
      CA 82: 115018

1-0038 SURFACE ANALYSIS TODAY. (REVIEW OF METHODS,
INCLUDING AES)
COBURN JW + KAY E
RES DEVELOP 23(12): 37-44 (1972)

1-0039 COMPUTER PROGRAM TO CALCULATE AUGER
TRANSITIONS FOR A SURFACE CONTAINING SEVERAL
KINDS OF ATOMS.
COGHLAN WA
OAK RIDGE NAT LAB, 1971, 50P. ORNL-TM-3575
      CA 76: 147089P

1-0040 CATALOG OF CALCULATED AUGER TRANSITIONS FOR
THE ELEMENTS. (REVIEW)
COGHLAN WA + CLAUSING RE
SURFACE SCI 33(2): 411-13 (1972); J VACUUM
SCI TECHNOL 10(1): 325 (1973); OAK RIDGE NAT
LAB 1971, 570P. ORNL-TM-3576
      CA 77: 170611G; 76: 147088N

1-0041 AUGER CATALOG CALCULATED TRANSITION ENERGIES
LISTED BY ENERGY AND ELEMENT.
COGHLAN WA + CLAUSING RE
ATOMIC DATA 5(4): 317-476 (1973)
      CA 79: 098773

1-0042 USE OF PHOTOEMISSION, AUGER, AND THERMAL
DESORPTION TECHNIQUES TO STUDY THE MECHANISM
OF A CATALYTIC REACTION.
COLLINS DM + LEE JB + SPICER WE
PHYS REV LETT 35(9): 592-4 (1975)
      CA 83: 137439

1-0043 AUGER ELECTRON SPECTROSCOPY.
CONNELL GL + GUPTA YP
MATER RES STAND 11(1): 8-13 (1971)

1-0044 AUGER RECOMBINATION IN SEMICONDUCTORS.
(REVIEW, 26 REFS)
CONRADT R
FESTKOERPERPROBLEME 12: 449-64 (1972)
      CA 77: 145629M

1-0045 A NEW TYPE OF AUGER EFFECT AND ITS INFLUENCE
ON THE X-RAY SPECTRUM.
COSTER D + KRONIG RL
PHYSICA 2: 13-24 (1921)

1-0046 SPECTROSCOPY OF AUGER ELECTRONS. (REVIEW, 24
REFS)
DEVILLE JP
REV PHYS APPL, SUPPL J PHYS 3: 351-5 (1968)
(IN FRENCH)

1-0047 SPECTROGRAPHY BY AUGER ELECTRONS.
DEVILLE JP
J MICROSC 14: 231 (1972) (IN FRENCH)

1-0048 APPLICATIONS OF AUGER ELECTRON SPECTROSCOPY
IN ELECTROPLATING.
DINI JOW + MUSKET RG
PLATING 60(8): 811-14 (1973)
      CA 79: 099752

1-0049 RECENT ADVANCES IN AES.
DOOLEY GJ + HAAS TW
BULL AMER CERAM SOC 51: 722 (1972) (ABSTRACT)

1-0050 STRUCTURE ANALYSIS WITH X-RAY PHOTOELECTRON
SPECTROSCOPY AND AUGER ELECTRON SPECTROSCOPY.
EBEL H + FALK H
FRESENIUS Z ANAL CHEM 273(5): 368-73 (1975)
(IN GERMAN)
      CA 82: 177281

1-0051 RECENT ADVANCES IN THE STUDY OF SOLID
SURFACES.
ELLIS WP
P79-96 OF PHYSICAL CHEMISTRY, SER 1, VOL 7,
SURFACE CHEM COLLOIDS. KERKER M(ED),
UNIVERSITY PARK, 1972, 306P.
      CA 78: 152291

1-0052 COMPARISON OF THE TECHNIQUES FOR SILICON
SURFACE ANALYSIS. (AES, ESCA, SIMS, AND
OTHERS)
EVANS CA
P219-32 OF SEMICONDUCTOR MEASUREMENT
TECHNOLOGY. PART-4: ARPA/NBS NAT BUR STAND
SPEC PUB 400-23, 1976.

1-0053 SURFACE AND THIN FILM COMPOSITIONAL ANALYSIS.
DESCRIPTION AND COMPARISON OF TECHNIQUES.
(REVIEW)
EVANS CA
ANAL CHEM 47(9):
818A-820A,822A,824A,826A-829A (1975)
      CA 83: 125579

1-0054 THIN FILM COMPOSITIONAL ANALYSIS. COMPARISON
OF TECHNIQUES. (AUGER SPECTROSCOPY)
EVANS CA
J VACUUM SCI TECHNOL 12(1): 144-50 (1975)
      CA 83: 21348

1-0055 MICRODOSIMETERY OF AUGER ELECTRONS. (REVIEW)
FEIGE Y + GAVRON A
P557-69 OF INT CONGR RADIAT RES, 5TH PROC,
1974, NYGAARD OF + ADLER HI + SINCLAIR
WK(EDS), ACADEMIC PRESS, 1975.
      CA 85: 30134

1-0056 BIOLOGICAL DAMAGE FROM THE AUGER EFFECT,
POSSIBLE BENEFITS. (DNA)
FEINENDEGEN LE
RADIAT ENVIRON BIOPHYS 12(2): 85-99 (1975)
CA 83: 143712

1-0057 USE OF LOW ENERGY ELECTRON DIFFRACTION, AUGER
EMISSION SPECTROSCOPY AND FIELD ION
MICROSCOPY IN MICROSTRUCTURAL STUDIES.
FERRANTE J + BUCKLEY DH + PEPPER SV
+ BRAINARD WA
P241-79 OF INT MICROSTRUCT ANAL SOC, TOOLS
TECH, 1972 SYMP.
CA 81: 042398

1-0058 COMPARISON OF X-RAY AND ELECTRON IMPACT
IONIZATION IN AUGER SPECTROSCOPY.
GALLON TE + MATTHEW JAD
J PHYS D 5: L69-72 (AUG 1972)

1-0059 AUGER ELECTRON SPECTROSCOPY AND ITS
APPLICATION TO SURFACE STUDIES. (REVIEW, 130
REFS)
GALLON TE + MATTHEW JAD
REV PHYS TECHNOL 3(1): 31-64 (1972)
CA 77: 144882B

1-0060 THE K-SHELL AUGER ELECTRON SPECTRA. REVIEW.
GEIGER JS
P523-47 OF INT CONF INNER SHELL IONIZATION
PHENOMENA AND FUTURE APPLICATIONS, VOL 1,
1972, FINK RW (ED), NTIS, 1973.
CA 78: 129962

1-0061 AUGER ELECTRON SPECTROSCOPY.
GERGELY GY
FIZ SZ 23(8): 237-244 (AUG 1973) (IN
HUNGARIAN)
SAA 1973: 80124

1-0062 APPLICATIONS OF ELECTRON SPECTROSCOPY.
(RELATIONSHIP OF IONIZATION SPECTROSCOPY TO
PHOTOELECTRON AND AES)
GERLACH RL
P885-94 OF ELECTRON SPECTROSCOPY, SHIRLEY DA
(ED), AMER ELSEVIER, 1972.

1-0063 MODERN TECHNIQUES FOR SURFACE STUDIES.
GJOSTEIN NA + BONZEL HP + CHAUKA NG
RES DEVELOP 21(10): 24-30 (1970)

1-0064 TECHNOLOGICAL APPLICATION OF AUGER ELECTRON
SPECTROSCOPY.
GJOSTEIN NA + CHAUKA NG
J TEST EVAL 1(3): 183-93 (1973)
CA 79: 059388

1-0065 ROLE OF ULTRAHIGH VACUUM TECHNIQUES IN BASIC
RESEARCH. (REVIEW, AUGER, MASS SPECTROMETRY)
GOPALARAMAN CP
P37-65 OF NUCL PHYS SOLID STATE PHYS SYMP,
PROC, VOL 16A, 1973.
CA 83: 117647

1-0066 QUANTITATIVE AUGER ANALYSIS USING INTEGRATION
TECHNIQUES.
GRANT JT + HAAS TW + HOUSTON JE
PHYS LETT A 45(4): 309-10 (1973)
CA 80: 033427

1-0067 APPLICATION OF DYNAMIC BACKGROUND SUBTRACTION
TO QUANTIFICATION OF AUGER ELECTRON AND SOFT
X-RAY APPEARANCE POTENTIAL SPECTROSCOPIES.
GRANT JT + HAAS TW + HOUSTON JE
P811-14 OF INT VACUUM CONGRESS, 6TH PROC, INT
CONF ON SOLID SURFACES, 2ND, 1974, KYOTO.
(JAP J APPL PHYS, SUPPL 2, PART 2, 1974)
CA 82: 177694

1-0068 SPECTRUM SUBTRACTION TECHNIQUES IN AUGER
ELECTRON SPECTROSCOPY.
GRANT JT + HOOKER MP + HAAS TW
SURFACE SCI 51(1): 318-22 (1975)
CA 83: 187816

1-0069 USE OF ANALOG INTEGRATION IN DYNAMIC
BACKGROUND SUBTRACTION FOR QUANTITATIVE AUGER
ELECTRON SPECTROSCOPY. (STUDY OF CARBON
MONOXIDE ON MOLYBDENUM (110))
GRANT JT + HOOKER MP + HAAS TW
SURFACE SCI 46(2): 672-5 (DEC 1974)
SAA 1975: 23597

1-0070 SURFACE ANALYSIS WITH HEAVY ION INDUCED AUGER
ELECTRONS. BASIC PROPERTIES OF A NEW METHOD
FOR MATERIALS SCIENCES.
GROENEVELD KO + MANN R + MECKBACH W + SPOHR R
VACUUM 25(1): 9-12 (1975)
CA 83: 90227

1-0071 A BIBLIOGRAPHY OF LOW ENERGY ELECTRON
DIFFRACTION AND AUGER ELECTRON SPECTROSCOPY.
(766 REFS)
HAAS TW + DOOLEY GJ + GRANT JT + JACKSON AG
+ HOOKER MP
PROGR SURF SCI 1(2): 155-236 (1971)

1-0072 AUGER ELECTRON SPECTROSCOPY FOR SURFACE
ANALYSIS.
HARRIS LA
J METALS 20(12): A16 (1968)

1-0073 AUGER ELECTRON EMISSION ANALYSIS.
HARRIS LA
ANAL CHEM A40(14): 24-34 (1968)

1-0074 SECONDARY ELECTRON SPECTROSCOPY.
HARRIS LA
IND RES 10: 52 (1968)

1-0075 MISCELLANEOUS TOPICS IN AUGER ELECTRON
SPECTROSCOPY.
HARRIS LA
J VACUUM SCI TECHNOL 11(1): 23-8 (1974)
CA 80: 138891

1-0076 MICROANALYSIS BY AUGER ELECTRON SPECTROSCOPY.
(REVIEW, 15 REFS)
HAYAKAWA K + KAWASE S + OKANO H
HYOMEN 12(9): 518-25 (1974) (IN JAPANESE)
CA 82: 67627

1-0077 THREE DIMENSIONAL CHARACTERIZATION OF METALS
AND SEMICONDUCTORS BY THE AUGER ELECTRON
MICROANALYTIC METHOD. (REVIEW)
HAYAKAWA K + KAWASE S + OKANO H
SHINKU 18(12): 450-8 (1975) (IN JAPANESE)
CA 84: 158400

1-0078 PHYSICAL METHODS OF SURFACE ANALYSIS.
HECKINGBOTTOM R
PHYS TECHNOL 6(2): 47-53 (1975)
CA 83: 157223

1-0079 SECONDARY ION MASS SPECTROMETRY.
HEINRICH KJF + NEWBURY DE
NAT BUR STANDARDS SPEC PUB 427, OCT 1975,
227P.
SAA 1976: 16109

1-0080 LOW ENERGY PHOTO- AUGER ELECTRON
SPECTROSCOPY.
HENKE BL
APPL SPECTROSC 22: 372 (1968)

1-0081 ELECTRON SPECTROSCOPY: X-RAY AND ELECTRON
EXCITATION. (REVIEW, ESCA, AES)
HERCULES DM
ANAL CHEM 48(5): 294R-313R (1976)
CA 84: 173288

1-0082 ELECTRON SPECTROSCOPY. X-RAY AND ELECTRON
EXCITATION.
HERCULES DM + CARVER JC
ANAL CHEM 46(5): 133R-150R (1974)
CA 80: 152354

1-0083 ELECTRONS AND X-RAYS IN NONDESTRUCTIVE
ANALYSIS. (AUGER, PHOTOELECTRON SPECTRA,
REVIEW, 6 REFS)
HERRMANN KH
METALL 26(8): 785-91 (1972) (IN GERMAN)
CA 77: 107492D

1-0084　ELECTRON SPECTROSCOPY. A NEW TECHNIQUE FOR
MATERIAL AND STRUCTURAL ANALYSIS AND A REVIEW
OF ITS POSSIBLE APPLICATIONS TO PROBLEMS IN
GLASS RESEARCH. (PHOTOELECTRON AND AES)
HICKSON K
GLASTECH BER 44(12): 537-42 (1971)
　　　SAA 1972: 22141

1-0085　QUANTITATIVE DEPTH PROFILING BY AUGER ION
SPUTTERING TECHNIQUE.
HO PS + LEWIS JE + HOWARD JK + WILDMAN HS
P245 OF ELECTROCHEM SOC FALL MEET ABSTR,
1975.
　　　SAA 1976: 33729

1-0086　THE RELATIONSHIP OF SOLID SURFACES TO VACUUM
SCIENCE AND TECHNOLOGY. (REVIEW)
HOBSON JP
P317-25 OF INT VACUUM CONGRESS, 6TH PROC, INT
CONF ON SOLID SURFACES, 2ND, 1974, KYOTO.
(JAP J APPL PHYS, SUPPL 2, PART 1, 1974)

1-0087　APPLICATIONS OF DEPTH PROFILING BY AUGER-
SPUTTER TECHNIQUES.
HOLLOWAY DM
J VACUUM SCI TECHNOL 12(1): 392-9 (1975)
　　　CA 82: 145492

1-0088　PHOTO- AND AUGER ELECTRON SPECTROSCOPY (ESCA)
AND SECONDARY ION MASS SPECTROSCOPY.
COMPARISON OF TWO SURFACE ANALYSIS METHODS.
HOLM R
METALLOBERFLAECHE - ANGEW ELETROCHEM 27(6):
199-207 (1973) (IN GERMAN)
　　　CA 79: 086996

1-0089　SURFACE AND THIN FILM ANALYSIS OF
SEMICONDUCTOR MATERIALS. (SCANNING AUGER
MICROPROBE)
HONIG RE
THIN SOLID FILMS 31(1-2): 89-122 (1976)
　　　SAA 1976: 26319

1-0090　RETRIEVAL OF ELECTRON EXCITED AUGER STRUCTURE
BY DYNAMIC BACKGROUND SUBTRACTION.
HOUSTON JE
APPL PHYS LETT 24(1): 42-4 (1 JAN 1974)
　　　SAA 1974: 37714

1-0091　THE SECONDARY ELECTRON BACKGROUND PROBLEM IN
ELECTRON EXCITED AUGER ELECTRON SPECTROSCOPY.
HOUSTON JE
APPL PHYS 6(2): 281-2 (MAR 1975)
　　　SAA 1975: 26208

1-0092　COMPARISON OF SOFT X-RAY AND AUGER ELECTRON
APPEARANCE POTENTIAL SPECTROSCOPY.
HOUSTON JE + PARK RL
BULL AMER PHYS SOC 17: 662 (1972) (ABSTRACT)

1-0093　TRACE ANALYSIS TECHNIQUES FOR SOLIDS.
(ELECTRON PROBE ANALYSIS, AUGER EFFECT,
REVIEW)
HUGGINS RA
P33-68 OF ANNU REV MATER SCI, HUGGINS RA(ED),
VOL-2, 1972.
　　　SAA 1973: 22555

1-0094　SECONDARY ELECTRON SPECTROSCOPY. SURFACE
SENSITIVE TOOL. (REVIEW, 22 REFS)
IGNATIEV A + RHODIN TN
AMER LAB 4(11): 8-17 (1972)
　　　CA 78: 49895V

1-0095　SECONDARY ELECTRON SPECTROSCOPY. SURFACE
SENSITIVE TOOL. (AES)
IGNATIEV A + RHODIN TN
INT LAB 1973(1): 18-25
　　　CA 83: 90231

1-0096　LEED AND AES. (REVIEW, 53 REFS)
INO S
NIPPON KINZOKU GAKKAI KAIHO 11(11): 773-80
(1972) (IN JAPANESE)
　　　CA 78: 76653W

1-0097　CHARACTERIZING MATERIALS WITH THE SEM TODAY.
(AUGER SPECTROSCOPY)
JOHARI O
MICROSCOP ACTA 76(5): 393-404 (1975) (IN
GERMAN); INDIA 10(1): 12-19 (1975)
　　　SAA 1975: 63785; 1976: 264

1-0098　AUGER ELECTRON SPECTROSCOPY. (REVIEW, 18
REFS)
JOHNSON WC + STEIN DF + JOSHI A
CAN J SPECTROSC 17(3): 88-92 (1972)
　　　CA 77: 145640H

1-0099　LOW ENERGY ELECTRON DIFFRACTION, CIRCA 1968.
(REVIEW OF LEED AND AUGER)
JONA F
HELV PHYS ACTA 41: 960-4 (1968)

1-0100　THIN FILMS. LOW ENERGY ELECTRON DIFFRACTION
AND AUGER ELECTRON SPECTROSCOPY.
JONA F + STROZIER JA
P309-29 OF EPITAXIAL GROWTH, SER A, MATTHEWS
JW(ED), ACADEMIC PRESS, 1975.
　　　CA 83: 106284

1-0101　AUGER ELECTRON SPECTROSCOPY. (REVIEW, 154
REFS)
JOSHI A + DAVIS LE + PALMBERG PW
P159-222 OF METHODS AND PHENOMENA VOL 1,
METHODS OF SURFACE ANALYSIS, CZANDERNA AW
(ED), ELSEVIER, 1975, 481P.

1-0102　AUGER ELECTRON SPECTROSCOPY AND ITS
APPLICATIONS IN SURFACE CHEMISTRY. (REVIEW)
JOYNER RW + ROBERTS MW
SURF DEFECT PROP SOLIDS 4: 68-102 (1975)
　　　CA 84: 23781

1-0103　SURFACE ANALYSIS BY AUGER SPECTROSCOPY.
KARASEK FW
RES DEVELOP 25(10): 48-50, 52, 54, 56, 58, 60
(1974)
　　　CA 82: 010758

1-0104　ADHESION, FRICTION, WEAR, AND LUBRICATION
RESEARCH BY MODERN SURFACE SCIENCE
TECHNIQUES. (AES, ELLIPSOMETRY, OTHER
TECHNIQUES)
KELLER DV
J VACUUM SCI TECHNOL 9: 133-42 (1972)
　　　SAA 1972: 39848

1-0105　AUGER ELECTRON SPECTROSCOPY. (REVIEW, 39
REFS)
KOMIYA S + LYO K
SHINKU 15(2): 35-46 (1972) (IN JAPANESE)
　　　CA 76: 160176J

1-0106　SOME TOPICS ON AES. (SURFACE ANALYSIS)
KOMIYA S + LYO K
J VACUUM SOC JAP 15(2): 35-46 (1972) (IN
JAPANESE)
　　　SAA 1972: 57264

1-0107　ELECTRONIC AUGER SPECTROSCOPY: TEXTBOOK FOR
A COURSE OF LECTURES ON PHYSICAL ELECTRONICS.
KORABLEV VV
LENINGRAD POLITEKH INST 1973, 62P. (IN
RUSSIAN)
　　　CA 80: 076577

1-0108　ELECTRON SPECTROMETRY.
KUCHITSU K + ONISHI T
SHOKUBAI 13(6): 209-21 (1971) (IN JAPANESE)
　　　CA 80: 138673

1-0109　(AUGER) TRANSITION ENERGIES. (REVIEW)
LARKINS FP
P377-409 OF ATOMIC INNER SHELL PROCESSES, VOL
1, IONIZATION AND TRANSITION PROBABILITIES,
CRASEMANN B (ED), ACADEMIC PRESS, 1975.
　　　CA 83: 87483

1-0110　THE AUGER EFFECT (REVIEW, 116 REFS)
LISTENGARTEN MA
BULL ACAD SCI USSR, PHYS SER 24: 1050-83
(1962)

1-0111 PERTURBING EFFECTS OF SPUTTERING DURING THE
AUGER STUDY OF INSULATORS.
LOSCH WHP
J VACUUM SCI TECHNOL 13(3): 737 (1976)
CA 85: 55298

1-0112 BEAM FOIL AUGER SPECTROSCOPY.
LUCAS MW + HARRISON EG
J PHYS B 5(2): L20-622 (1972)
CA 76: 119410T

1-0113 POTENTIAL MAPPING USING AUGER ELECTRON
SPECTROSCOPY.
MACDONALD NC
P481-7 OF SCANNING ELECTRON MICROSCOPE SYMP,
3RD PROC, 1970, JOHARI O(ED), ILL INST
TECHNOL RES INST, 1970, 534P.

1-0114 AUGER ELECTRON SPECTROSCOPY FOR SCANNING
ELECTRON MICROSCOPY.
MACDONALD NC
P89-96 OF SCANNING ELECTRON MICROSCOPE SYMP,
4TH PROC, JOHARI O(ED), ILL INST TECHNOL RES
INST, 1971, 618P.
SAA 1971: 52003

1-0115 HIGH SPATIAL RESOLUTION AES.
MACDONALD NC
PROC ELECTRON MICROSC SOC AMER 30: 366-7
(1972); J VACUUM SCI TECHNOL 10(1): 275
(1973) (ABSTRACT)
CA 78: 21121R

1-0116 SCANNING AUGER MICROANALYSIS.
MACDONALD NC
ABSTR 46A OF ANN CONF MICROBEAM ANAL SOC, 9TH
PROC SUMMARIES, LEHIGH UNIV, 1974.
SAA 1974: 77306

1-0117 THIRD DIMENSION IN SCANNING ELECTRON
MICROSCOPY. SCANNING AUGER MICROSCOPY.
MACDONALD NC
P431-57 OF PHYS ASPECTS ELECTRON MICROSC
MICROBEAM ANAL, SIEGEL BM + BEAMAN DR(EDS),
WILEY, NY, 1975.
CA 83: 187778

1-0118 THIN FILM ANALYSIS FOR PROCESS EVALUATION.
(AES)
MACDONALD NC + RIACH GE
ELECTRON PACKAG PROD 13(4): 50-60 (APR 1973)

1-0119 APPLICATIONS OF SCANNING AUGER MICROANALYSIS.
MACDONALD NC + RIACH GE + GERLACH RL
RES DEVELOP 27(8): 42-50 (AUG 1976)

1-0120 AUGER ELECTRON SPECTROSCOPY. (REVIEW)
MAITREPIERRE PH
BULL CERCLE ETUD MET 12(11): 721-51 (1974)
(IN FRENCH); CIRC INF TECH CENT DOC SIDER
32(7-8): 1749-71 (1975) (IN FRENCH)
CA 82: 161936; 83: 199542

1-0121 REVIEW OF FLUORESCENCE YIELDS AND AUGER
ELECTRON MISSION.
MARK H
P154-68 OF INT CONF PHYS ELECTRON ATOMIC
COLLISIONS, INV PAP PROGR REP, 7TH, GOVERS TR
(ED), NORTH-HOLLAND, 1972.
CA 77: 132553G

1-0122 DIRECTIONAL CORRELATION BETWEEN
PHOTOELECTRONS AND AUGER ELECTRONS.
MCFARLANE SC
J PHYS B 8(6): 895-904 (1975)
CA 83: 35387

1-0123 AUGER AND COSTER-KRONIG TRANSITIONS. (REVIEW)
MCGUIRE EJ
P293-330 OF ATOMIC INNER SHELL PROCESSES, VOL
1, IONIZATION AND TRANSITION PROBABILITIES,
CRASEMANN B (ED), ACADEMIC PRESS, 1975.
CA 83: 105487

1-0124 LEED AND AUGER ELECTRON SPECTROSCOPY.
(REVIEW)
MCKEE CS + ROBERTS MW
CHEM BRIT 6: 106-10 (1970)

1-0125 AUGER ELECTRON SPECTROSCOPY. (REVIEW, 44
REFS)
MEHLHORN W
P169-80 OF INT CONF PHYS ELECTRON ATOMIC
COLLISIONS, INV PAP PROGR REP, 7TH, GOVERS TR
(ED), NORTH-HOLLAND, 1972.
CA 77: 132555J

1-0126 RECENT TRENDS IN AUGER ELECTRON SPECTROSCOPY.
MEHLHORN W
P437-54 OF INT CONF INNER SHELL IONIZATION
PHENOMENA AND FUTURE APPLICATIONS, VOL 1,
1972, FINK RW(ED), NTIS, 1973.
CA 78: 141886

1-0127 APPLICATION OF (AUGER) ELECTRON SPECTROSCOPY
TO CHEMISTRY. (REVIEW)
MEI JY + ZHOW LD
HUA HSUEH TUNG PAO 6: 360-9 (1974) (IN
CHINESE)
CA 82: 147181

1-0128 ELLIPSOMETRY AS A TOOL IN SURFACE STUDIES.
MEYER F
NED TIJDSCHR VACUUMTECH 11(6): 77-83 (1973)
CA 80: 075937

1-0129 QUANTITATIVE ASPECTS OF AUGER ELECTRON
SPECTROSCOPY.
MEYER F + VRAKKING JJ
SURFACE SCI 33(2): 271-94 (1972)
CA 77: 169119W

1-0130 IN DEPTH INFORMATION FROM AUGER ELECTRON
SPECTROSCOPY. (SILICON MONOLAYERS)
MEYER F + VRAKKING JJ
SURFACE SCI 45(2): 409-18 (1974)
CA 81: 180787

1-0131 LOW ENERGY ELECTRON DIFFRACTION.
MITCHELL KAR
CONTEMP PHYS 14(3): 251-71 (1973)
CA 79: 058434

1-0132 ELECTRON AUGER SPECTROSCOPY: A METHOD OF
INVESTIGATING SURFACE PHENOMENA (REVIEW).
MITYAGIN AYU + CHEREVATSKII NYA
+ DVORYANKIN VF
INORG MATER 7: 1890-9 (DEC 1971)

1-0133 SAMPLES HEATED DIRECTLY BY ELECTRIC CURRENT
OBSERVED BY AUGER ELECTRON SPECTROSCOPY.
MIURA T + TUZI Y + ASANO K
SEISAN-KENKYU 23(6): 230-2 (1971) (IN
JAPANESE)
CA 76: 8706Z

1-0134 CHARACTERIZATION OF SOLID SURFACES. (REVIEW,
AUGER SPECTROSCOPY, ESCA)
MODDEMAN WE + COTHERN CR + BLACK JN
J ENVIRON SCI 18(5): 27-33 (1975)
CA 84: 35658

1-0135 COMPARISON OF AUGER ELECTRON SPECTROSCOPY AND
SECONDARY ION MASS SPECTROMETRY.
MORABITO JM
P191-224 OF SECONDARY ION MASS SPECTROMETRY
AND ION MICROPROBE MASS ANALYSIS WORKSHOP,
HEINRICH KFJ + NEWBURY DE (EDS), NAT BUR
STANDARDS, 1975.
SAA 1976: 33724

1-0136 USE OF AUGER ELECTRON SPECTROSCOPY AND
SECONDARY ION MASS SPECTROMETRY IN THE
MICROELECTRONIC TECHNOLOGY. (REVIEW, 78 REFS)
MORABITO JM + LEWIS RK
P279-328 OF METHODS AND PHENOMENA, VOL 1,
METHODS OF SURFACE ANALYSIS, CZANDERNA A(ED),
ELSEVIER, 1975, 481P.
CA 85: 56064

1-0137 RECENT ADVANCES IN SURFACE STUDIES. ION BEAM
ANALYSIS. (AUGER SPECTROSCOPY, REVIEW)
MORGAN DV
CONTEMP PHYS 16(3): 221-41 (1975)
CA 83: 201405

1-0138 DECONVOLUTION TECHNIQUES IN AUGER ELECTRON
SPECTROSCOPY.
MULARIE WM + PERIA WT
SURFACE SCI 26: 125-41 (1971); J VACUUM SCI
TECHNOL 8: 90 (1971)

1-0139 INELASTIC EFFECTS IN AUGER ELECTRON
SPECTROSCOPY.
MULARIE WM + RUSCH TW
SURFACE SCI 19: 469-74 (1970)

1-0140 AUGER ELECTRON SPECTROSCOPY.
MURATA Y
KAGAKU (TOKYO) 44(2): 82-9 (1974) (IN
JAPANESE)
        CA 81: 055874

1-0141 AUGER ELECTRON SPECTROSCOPY AND ITS
APPLICATION.
MUROTANI T + FUJIWARA K + OTANI M
+ NISHIJIMA M
MITSUBISHI DENKI GIHO 47(6): 667-74 (1973)
(IN JAPANESE)
        CA 79: 109127

1-0142 DIRECT COMPARISON OF AUGER ELECTRON
SPECTROSCOPY WITH APPEARANCE POTENTIAL
SPECTROSCOPY.
MUSKET RG
J VACUUM SCI TECHNOL 9(2): 603-7 (1972)

1-0143 SURFACE ANALYSIS USING PROTON BEAMS.
MUSKET RG + BAUER W
THIN SOLID FILMS 19(1): 69-80 (1973)
        CA 80: 066347

1-0144 SURFACE STUDIES BY AUGER ELECTRON
SPECTROSCOPY. (REVIEW, 44 REFS)
NAKAYAMA K
HYOMEN 9(12): 724-36 (1971) (IN JAPANESE)
        CA 77: 94797E

1-0145 AUGER ELECTRON SPECTROSCOPY.
NAKAYAMA K
BOSHOKU GIJUTSU 20(6): 255-63 (1971) (IN
JAPANESE)
        CA 79: 130773

1-0146 AUGER ELECTRON SPECTROSCOPY.
NISHIO M
SUMITOMO KEIKINZOKU GIHO 15(1): 41-8 (1974)
(IN JAPANESE)
        CA 81: 070322

1-0147 PRESENT STATUS AND PROBLEMS ON THE
APPLICATION OF AUGER ELECTRON SPECTROSCOPY.
OKADA H + OGAWA H + OMATA H
HYOMEN 12(3): 148-56 (1974) (IN JAPANESE)
        CA 81: 031270

1-0148 BIBLIOGRAPHY ON AUGER ELECTRON SPECTROSCOPY.
ONO M + TUZI Y
SHINKU 16(1): 9-21 (1973) (IN JAPANESE)
        CA 78: 166198

1-0149 PHOTOELECTRON SPECTROSCOPY AND ALLIED
TECHNIQUES: GENERAL INTRODUCTION.
ORCHARD AF
P267-304 OF ELECTRONIC STATES OF INORGANIC
COMPOUNDS: NEW EXPERIMENTAL TECHNIQUES, DAY
P(ED), 1974, NATO, OXFORD, ENGLAND, REIDEL,
DORDRECHT, NETHERLANDS, 1975.
        SAA 1976: 14652

1-0150 AUGER SPECTROSCOPY FOR THE CHEMICAL ANALYSIS
OF SURFACES.
OSWALD RC
UNIV OF MINNESOTA, ELECTRICAL ENG DEPT, MS
THESIS, 1969.

1-0151 TECHNIQUE AND APPLICATIONS OF AUGER ELECTRON
SPECTROSCOPY.
PALMBERG PW
J VACUUM SCI TECHNOL 7: 76 (1970); 6: 903
(1969)

1-0152 OPTIMIZATION OF AUGER ELECTRON SPECTROSCOPY
IN LEED SYSTEMS.
PALMBERG PW
APPL PHYS LETT 13: 183-5 (1968)

1-0153 AUGER ELECTRON SPECTROSCOPY IN LEED SYSTEMS.
PALMBERG PW
P29-1 TO 29-18 OF STRUCTURE AND CHEMISTRY OF
SOLID SURFACES, SOMORJAI GA (ED), WILEY,
1969.

1-0154 AUGER ELECTRON SPECTROSCOPY. (REVIEW, 37
REFS)
PALMBERG PW
P835-59 OF INT CONF ELECTRON SPECTROSCOPY,
1972 PROC, AMSTERDAM, NORTH HOLLAND, 1972.
        CA 77: 132557M

1-0155 RECENT ADVANCES IN COMPOSITION PROFILING BY
SIMULTANEOUS SPUTTERING AND AUGER ANALYSIS.
PALMBERG PW
J VACUUM SCI TECHNOL 10(1): 274 (1973)
(ABSTRACT)

1-0156 USE OF AUGER ELECTRON SPECTROSCOPY AND INERT
GAS SPUTTERING FOR OBTAINING CHEMICAL
PROFILES.
PALMBERG PW
J VACUUM SCI TECHNOL 9(1): 160-3 (1972)

1-0157 QUANTITATIVE AUGER ELECTRON SPECTROSCOPY
USING ELEMENTAL SENSITIVITY FACTORS.
PALMBERG PW
J VACUUM SCI TECHNOL 13(1): 214-18 (1976)
        SAA 1976: 47654

1-0158 AUGER ELECTRON SPECTROSCOPY IN LEED SYSTEMS.
PALMBERG PW
P29-1 TO 29-18 OF STRUCTURE AND CHEMISTRY OF
SOLID SURFACES, SOMORJAI GA (ED), WILEY,
1969. (BERKELEY INT MATERIALS CONF, 4TH PROC,
1968)

1-0159 HANDBOOK OF AUGER ELECTRON SPECTROSCOPY.
(EXPERIMENTAL SPECTRA)
PALMBERG PW + RIACH GE + WEBER RE
+ MACDONALD NC
PHYSICAL ELECTRONICS INDUSTRIES, EDINA, MINN,
1972.

1-0160 INNER SHELL SPECTROSCOPY. (REVIEW)
PARK RL
PHYS TODAY 28(4): 52-7, 59 (1975)
        CA 83: 17812

1-0161 CHARACTERIZATION OF CHEMISORPTION BY LEED.
PARK RL + HOUSTON JE
J METALS 21(3): A87 (1969)

1-0162 AUGER SPECTROMETRY AS A TOOL FOR SURFACE
CHEMISTRY. (REVIEW)
PENTENERO A
P199-228 OF CATAL REV, VOL 5, HEINEMANN
H(ED), MARCEL DEKKER, NY, 1972.

1-0163 QUANTITATIVE USE OF AUGER SPECTROSCOPY:
STANDARDIZATION OF THE METHOD.
PERDEREAU M
SURFACE SCI 24: 239-47 (1971)

1-0164 LOW ENERGY ELECTRON SPECTROSCOPY. (REVIEW, 16
REFS)
PESSA M
ANN UNIV TURKU, SER A1, NO 152, 1971, 22P.
        CA 78: 64318C

1-0165 SEMICONDUCTOR SURFACE SPECTROSCOPY. (REVIEW,
21 REFS)
PHILLIPS JC
J VACUUM SCI TECHNOL 13(1): 178-82 (1976)
        SAA 1976: 49186

1-0166 HIGH SPATIAL RESOLUTION AUGER SPECTROSCOPY
AND AUGER INTEGRATION APPLICATIONS.
POCKER DJ + HAAS TW
J VACUUM SCI TECHNOL 12(1): 370-4 (1975)
        CA 82: 162650

1-0167 SECONDARY (AUGER) ELECTRON SPECTROSCOPY.
RAMSEY JA
VACUUM 21: 115-19 (1971)

1-0168 AUGER ELECTRON SPECTROSCOPY. (REVIEW)
RAMSEY JA
P J4 OF AUSTRAL SPECTROSC CONF 1971, CLAYTON,
AUSTRALIA, 8TH CONF, ABSTRACTS ONLY.
SAA 1972: 10670

1-0169 ION- ATOM COLLISIONS. (AUGER EMISSION,
REVIEW)
RICHARD P
P73-158 OF ATOMIC INNER SHELL PROCESSES, VOL
1, IONIZATION AND TRANSITION PROBABILITIES,
CRASEMANN B (ED), ACADEMIC PRESS, 1975.
CA 83: 105485

1-0170 CHARACTERISTIC AUGER ELECTRON EMISSION AS A
TOOL FOR THE ANALYSIS OF SURFACE COMPOSITION.
RIVIERE JC
PHYS BULL 20: 85 (1969)

1-0171 AUGER ELECTRON SPECTROSCOPY.
RIVIERE JC
CONTEMP PHYS 14(6): 513-39 (1973)
CA 80: 075962

1-0172 AUGER SPECTROSCOPY. (REVIEW)
RIVIERE JC
P187-205 OF MOD PHYS TECH MATER TECHNOL,
MULVEY T + WEBSTER RK(EDS), OXFORD UNIV
PRESS, 1974.
CA 83: 70913

1-0173 IONIZATION (CHARACTERISTIC LOSS) SPECTROSCOPY
AND APPEARANCE POTENTIAL SPECTROSCOPY.
COMPARISON WITH AUGER SPECTROSCOPY AND X-RAY
PHOTOELECTRON SPECTROSCOPY.
RIVIERE JC
DECHEMA MONOGR 78(1537-1548): 83-115 (1975)
CA 83: 17798

1-0174 AUGER ELECTRON SPECTROSCOPY. IONIZATION LOSS
SPECTROSCOPY. APPEARANCE POTENTIAL
SPECTROSCOPY.
RIWAN R
CEN SACLAY, COMM ENERG AT, FRANCE, 1973, 47P.
REP CEA-CONF-2471 (IN FRENCH)
CA 81: 129057

1-0175 AUGER ELECTRON SPECTROSCOPY: NEW TOOL IN THE
CHARACTERIZATION OF GLASS FIBER SURFACES.
RYND JP + RASTOGI AK
BULL AMER CERAM SOC 53(9): 631-4, 637 (1974)
CA 82: 006997

1-0176 INTERFACE PHENOMENA IN ENGINEERING MATERIALS.
SCHEIBNER EJ
GEORGIA INST TECH REPORT, 1972, AD 753099.

1-0177 QUANTITATIVE AUGER ELECTRON SPECTROSCOPY AND
ELECTRON RANGES.
SEAH MP
SURFACE SCI 32(3): 703-28 (1972); VACUUM
22(10): 475-6 (1972)

1-0178 QUANTITATIVE AUGER ELECTRON SPECTROSCOPY:
COMPARISON OF TECHNIQUES FOR ADSORBED TIN ON
IRON.
SEAH MP
SURFACE SCI 40(3): 595-608 (1973)
CA 80: 064174

1-0179 SECONDARY ELECTRON EMISSION. (REVIEW, 86
REFS)
SEILER H
Z ANGEW PHYS 22: 249-63 (1967)

1-0180 THE AUGER, COSTER-KRONIG, AND AUTOIONIZATION
EFFECTS. (REVIEW, 477 REFS)
SEVIER KD
P51-178 OF LOW ENERGY ELECTRON SPECTROMETRY,
CHAP-3, SEVIER KD, WILEY, 1972.

1-0181 PRESENT STATUS OF BEAM ANALYSIS AND ITS
FUTURE. (REVIEW, AUGER, MASS SPECTROSCOPY)
SHINODA G
SHOKUBAI 17(2): 2-9 (1975) (IN JAPANESE)
CA 83: 90223

1-0182 SPUTTERING MEASUREMENTS ON CTR MATERIALS
USING AUGER ELECTRON SPECTROSCOPY
SMITH JN + MEYER CH + LAYTON JK
GEN ATOMICS CO, 1975, 13P. GA-A-13665
CA 84: 172664

1-0183 DIGITIZING AUGER DATA. THREEFOLD APPROACH TO
ENHANCE ANALYTICAL CAPABILITIES.
SOLOMON JS + BAUN WL
J VACUUM SCI TECHNOL 12(1): 375-8 (1975)
CA 83: 21350

1-0184 AUGER PROFILE ANOMALIES. (REVIEW)
SOLOMON JS + MEYERS V
AMER LAB 8(3): 31-2, 34, 36, 38, 40 (1976)
CA 84: 157306

1-0185 LOW ENERGY ELECTRON DIFFRACTION AND SURFACE
TOPOGRAPHY. (REVIEW, AUGER SPECTROSCOPY)
SOMORJAI GA
P173-263 OF SURFACE SCIENCE LECTURES, 1974,
INT AT ENERGY AGENCY, VIENNA, 1975.
SAA 1976: 30958

1-0186 AUGER ELECTRON SPECTROSCOPY AS A TOOL FOR
MEASURING THE DIFFUSION OF FOREIGN ATOMS IN
SOLIDS NEAR THEIR SURFACE.
SPARNAAY MJ + VAN BOMMEL AJ + VAN TOOREN A
SURFACE SCI 39(1): 251-4 (1973)
CA 79: 108208

1-0187 INTEGRAL AUGER INFORMATION VIA TAILORED
MODULATION TECHNIQUES.
SPRINGER RW + POCKER DJ + HAAS TW
APPL PHYS LETT 27(6): 368-70 (1975)
CA 83: 155122

1-0188 A MEASURING METHOD TO DISTINGUISH AUGER FROM
ENERGY LOSS PEAKS OF THE PRIMARY BEAM IN
SECONDARY ELECTRON ANALYSIS.
STADLER K
VACUUM 22(11): 553-5 (NOV 1972)
SAA 1973: 48571

1-0189 HIGHLY SENSITIVE AUGER ELECTRON SPECTROSCOPY.
STAIB P
SURFACE SCI 53(1): 582 (1975)
SAA 1976: 26965

1-0190 APPLICATIONS OF AUGER SPECTROSCOPY TO
MATERIALS RESEARCH. (REVIEW, 19 REFS)
STEIN DF
J VACUUM SCI TECHNOL 12(1): 268-75 (1975)
CA 82: 161943

1-0191 AUGER ELECTRON SPECTROMETRY.
STUPIAN GW
P57-82 OF SYSTEMATIC MATERIALS ANALYSIS,
VOL-1, RICHARDSON JH + PETERSON RV(EDS),
ACADEMIC PRESS, 1974, 363P.
CA 81: 072120

1-0192 AUGER ELECTRON SPECTROSCOPY.
SZYMERSKA I
PR OSRODKA BADAW ROZWOJOWEGO ELEKTRON
PROZNIOWEJ 3(5): 35-49 (1975) (IN POLISH)
SAA 1976: 24489

1-0193 MECHANISM OF FRICTION BY AES. (REVIEW)
TAKAHASHI N
SHINKU 18(12): 425-31 (1975) (IN JAPANESE)
CA 84: 169745

1-0194 ROLE OF AUGER ELECTRON SPECTROSCOPY IN
SURFACE ELEMENTAL ANALYSIS.
TAYLOR NJ
VACUUM 19: 575-8 (1969)

1-0195 AES FOR SURFACE ANALYSIS. (REVIEW, 24 REFS)
TAYLOR NJ
CESK CAS FIS A 21(5): 511-21 (1971) (IN
CZECH)
SAA 1972: 6303

1-0196 TECHNIQUE OF AUGER ELECTRON SPECTROSCOPY IN
SURFACE ANALYSIS. (REVIEW, 76 REFS)
TAYLOR NJ
P117-59 OF TECHNIQUES OF METALS RESEARCH, VOL
7, PART-1, BUNSHAH RF(ED), INTERSCIENCE,
1972, 527; P339-83 OF INT SUMMER SCHOOL ON
LEED, SURFACE STRUCT SOLIDS, 1972, 1ST PROC.
CA 82: 79918; 79: 098675

1-0197 TECHNIQUE OF PREPARING POWDER SAMPLES FOR AES
AND/OR ESCA ANALYSIS.
THERIAULT GE + BARRY TL + THOMAS MJB
ANAL CHEM 47(8): 1492-3 (1975)
CA 83: 107622

1-0198 ANALYTICAL ASPECTS OF AUGER SPECTROSCOPY.
(REVIEW)
THOMPSON M
PROC SOC ANAL CHEM 9(5): 118-20 (1972)
CA 83: 21311

1-0199 QUANTIFICATION OF AUGER ELECTRON
SPECTROSCOPY.
TOKUTAKA H + NISHIMORI K + TAKASHIMA K
NIPPON BUTSURI GAKKAISHI 30(9): 673-5 (1975)
(IN JAPANESE)
CA 83: 185561

1-0200 AUGER ELECTRON SPECTROSCOPY FOR SURFACE
ANALYSIS.
TRACY JC
P295-372 OF ELECTRON EMISSION SPECTROSCOPY,
DEKEYSER W (ED), REIDEL, 1973, 507P.

1-0201 RECENT ADVANCES IN SURFACE CHARACTERIZATION
USING LOW ENERGY ELECTRON DIFFRACTION AND
AUGER ELECTRON SPECTROSCOPY.
TRACY JC + BURKSTRAND JM
CRIT REV SOLID STATE SCI 4(3): 381-93 (1974)
CA 81: 112925

1-0202 AUGER ELECTRON ANALYSIS AND ITS APPLICATIONS.
TSUJI Y
KINZOKU HYOMEN GIJUTSU 24(2): 111-17 (1973)
(IN JAPANESE)
CA 79: 013068

1-0203 X-RAY SPECTROSCOPY. (REVIEW, AUGER AND
PHOTOELECTRON SPECTROSCOPY)
URCH DS
P449-93 OF ELECTRONIC STATES OF INORGANIC
COMPOUNDS: NEW EXPERIMENTAL TECHNIQUES, DAY
P(ED), 1974, NATO, OXFORD, ENGLAND, REIDEL,
DORDRECHT, NETHERLANDS, 1975.
SAA 1976: 14613

1-0204 AUGER ELECTRON SPECTROSCOPY.
VENNIK J + FIERMANS L
SILICATES IND 39(6): 173-80 (1974)
CA 81: 085449

1-0205 AUGER ELECTRON SPECTROSCOPY MADE QUANTITATIVE
BY ELLIPSOMETRIC CALIBRATION.
VRAKKING JJ + MEYER F
APPL PHYS LETT 18: 226-8 (1971)

1-0206 QUANTITATIVE ASPECTS OF AUGER ELECTRON
SPECTROSCOPY. IMPORTANCE OF ATTENUATION OF
THE PRIMARY ELECTRON BEAM.
VRAKKING JJ + MEYER F
SURFACE SCI 35(1): 34-41 (1973)
CA 78: 77585N

1-0207 QUANTITATIVE AUGER ELECTRON SPECTROSCOPY.
VRAKKING JJ + MEYER F
NED TIJDSCHR VACUUMTECH 11(1): 6-7 (1973) (IN
DUTCH)
CA 78: 164514

1-0208 AUGER LINES IN X-RAY PHOTOELECTRON
SPECTROMETRY.
WAGNER CD
ANAL CHEM 44(6): 967-73 (1972)
CA 76: 160608B

1-0209 AUGER SPECTRA IN X-RAY PHOTOELECTRON
SPECTROSCOPY.
WAGNER CD
P861-84 OF ELECTRON SPECTROSCOPY, SHIRLEY DA
(ED), AMER ELSEVIER, 1972.
CA 77: 132995C

1-0210 THIN FILM ANALYSIS BY AUGER ELECTRON
SPECTROSCOPY. (REVIEW)
WEBER RE
SOLID STATE TECHNOL 13(12): 49-53 (1970)

1-0211 AUGER ELECTRON SPECTROSCOPY FOR THIN FILM
ANALYSIS.
WEBER RE
J CRYST GROW 17: 342-53 (1972); RES DEVELOP
23(10): 22-8 (OCT 1972); CHIMIA 27(6): 335-43
(1973) (IN GERMAN)
SAA 1973: 1946; CA 78: 51996X; CA 79: 03

1-0212 AUGER ELECTRON SPECTROSCOPY FOR THIN FILM
ANALYSIS.
WEBER RE
J CRYST GROW 17: 342-53 (1972)

1-0213 BASIC INVESTIGATION IN THE AREA OF SCANNING
ELECTRON MICROSCOPY. PART-1: APPLICATION OF
AUGER ELECTRON SPECTROSCOPY TO ELECTRON PROBE
MICROANALYSIS.
WEBER U
P1-41 OF BUNDESMINIST FORSCH TECHNOL,
FORSCHUNGSBER TECHNOL FORSCH ENTWICKL,
BMFT-FBT 74-09, 1974. (IN GERMAN)
CA 84: 11749

1-0214 SPUTTERING IN THIN FILM ANALYSIS METHODS.
(ESCA, AUGER SPECTROSCOPY)
WEHNER GK
P495-500 OF INT VACUUM CONGRESS, 6TH PROC,
INT CONF ON SOLID SURFACES, 2ND, 1974, KYOTO.
(JAP J APPL PHYS, SUPPL 2, PART 1, 1974)

1-0215 ASPECTS OF SPUTTERING IN SURFACE ANALYSIS
METHODS. (REVIEW AUGER SPECTROSCOPY AND ESCA)
WEHNER GK
P5-37 OF METHODS AND PHENOMENA, VOL 1,
METHODS OF SURFACE ANALYSIS, CZANDERNA
AW(ED), ELSEVIER, 1975.
CA 85: 56060

1-0216 DIRECT COMPARISON OF AUGER, SECONDARY ION
MASS SPECTROSCOPY, AND PROTON RESONANCE
PROFILING FOR RELIABILITY STUDIES.
WEISENBERGER WH + GRAY H + HUBLER GK
+ DUNNING KL + COMAS J
P7-15 OF RELIAB PHYS SYMP, ANNU PROC, 12TH
PROC, 1974, IEEE, 1974.
CA 82: 67640

1-0217 USE OF AUGER AND CONVERSION ELECTRONS FOR THE
DETECTION OF GAMMA EMITTING RADIONUCLIDES:
CHROMATOGRAPHIC QUALITY CONTROL OF
RADIOPHARMACEUTICALS.
WENZEL M + WAHID AA
INT J APPL RADIAT ISOTOPES 26(3): 119-23
(1975)
SAA 1975: 59960

1-0218 AUGER TRANSITIONS.
WILSON JM + PACKER ME
DAWSON PUBL, FOLKESTONE, ENGLAND, 1971, 458P.
CA 80: 008926

1-0219 AUGER ELECTRON DOSIMETRY.
WRENN ME + HOWELLS GP + HAIRR LM + PASCHOA AS
HEALTH PHYS 24(6): 645-53 (1973)
CA 79: 037771

1-0220 FUTURE GROWTH OF SEM: TOWARD SUPER EFFICIENT
MICROSCOPY. (AUGER SPECTROSCOPY)
YAKOWITZ H
P1-10 OF SCANNING ELECTRON MICROSCOPY CONF,
JOHARI O + CORVIN I(EDS), ILLINOIS INST
TECHNOL RES INST, 1975.
SAA 1976: 3998

1-0221 AUGER ELECTRON SPECTROSCOPY. (REVIEW, 8 REFS)
YAMASHINA T
KAGAKU NO RYOIKI 26(5): 409 (1972) (IN
JAPANESE)
    CA 77: 67888D

1-0222 SOME PARAMETERS AFFECTING AUGER AND
PHOTOELECTRON SPECTROSCOPY AS AN ANALYTICAL
TECHNIQUE.
YIN LI + ADLER I + LAMOTHE R
APPL SPECTROSC 23: 41-50 (1969)

2. THEORY, PHYSICS OF THE AUGER EFFECT

2-0223 EVIDENCE FOR A RADIATIVE AUGER EFFECT IN
X-RAY PHOTON EMISSION.
ABERG T
PHYS REV LETT 22: 1346-8 (1969)

2-0224 THEORY OF THE RADIATIVE AUGER EFFECT.
ABERG T
PHYS REV A 4: 1735-40 (NOV 1971)

2-0225 TWO-PHOTON EMISSION, THE RADIATIVE AUGER
EFFECT, AND THE DOUBLE AUGER PROCESS.
ABERG T
P353-75 OF ATOMIC INNER SHELL PROCESSES, VOL
1, IONIZATION AND TRANSITION PROBABILITIES,
CRASEMANN B(ED), ACADEMIC PRESS, 1975.
    CA 83: 87482

2-0226 INFLUENCE OF CHEMICAL BONDING ON THE LOW
ENERGY K-ALPHA SPECTRUM OF SILICON.
(RADIATIVE AUGER TRANSITIONS)
ABERG T + UTRIAINEN J
J PHYS (PARIS) 32, COLLOQ NO 4: 295-300
(1971)
    CA 77: 26815J

2-0227 RADIATIVE AUGER RECOMBINATION OF TWO INDIRECT
EXCITONS.
ABLOVA LA + KHADZHI PI
P158-67 OF NOVYE POLUPROVDN SOEDIN I IKH
SVOISTVA, 1975. (IN RUSSIAN)
    CA 83: 186876

2-0228 AUGER EFFECT IN THE MULTIPLE IONIZATION OF
MANGANESE AND CADMIUM BY ELECTRON IMPACT.
ABOUAF R
J PHYS (PARIS) 31: 277-83 (1970) (IN FRENCH)

2-0229 MULTIPLE IONIZATION AND AUGER TRANSITIONS IN
INDIUM AND SILVER BY ELECTRON IMPACT.
ABOUAF R
J PHYS (PARIS) 32: 603-8 (1971) (IN FRENCH)

2-0230 AUGER TRANSITION AND SUPPLEMENTARY ELECTRON
EJECTION IN THE IONIZATION OF ALUMINUM BY
ELECTRON IMPACT.
ABOUAF R
PHYS LETT A 35: 319-20 (28 JUN 1971)

2-0231 MECHANISMS OF SIMPLE AND MULTIPLE IONIZATION
INDUCED BY ELECTRON IMPACT IN SOME METAL
VAPORS. (AUGER PROCESS)
ABOUAF R
J PHYS (PARIS) 32, COLLOQ NO 4: 128-31 (1971)
    CA 77: 27174M

2-0232 DOUBLE AND TRIPLE IONIZATION IN ALUMINUM,
GALLIUM, AND SCANDIUM BY ELECTRON IMPACT.
(AUGER SPECTRA)
ABOUAF R
P1667-72, VOL-3 OF INT CONF INNER SHELL
IONIZATION PHENOMENA AND FUTURE APPLICATIONS,
1972, FINK RW(ED), NTIS, 1973.
    CA 78: 130096

2-0233 ENERGY LOSS AND IONIZATION IN FAST ION-ATOM
COLLISIONS.
AFROSIMOV VV
P1297-319, VOL-2 OF INT CONF INNER SHELL
IONIZATION PHENOMENA AND FUTURE APPLICATIONS,
1972, FINK RW(ED), NTIS, 1973.
    CA 78: 115283

2-0234 DETECTION OF A NEW TYPE OF AUGER TRANSITIONS
IN ATOMS WITH TWO INTERNAL VACANCIES.
AFROSIMOV VV + GORDEEV YS + ZINOVEV AN
+ RASULOV DK + SHERGIN AP
JETP LETT 21(9): 249-51 (1975)
    CA 83: 88245

2-0235   OBSERVATION OF THREE ELECTRON AUGER
          TRANSITIONS IN ATOMS WITH TWO INNER
          VACANCIES.
          AFROSIMOV VV + GORDEEV YS + ZINOVEV AN
          + RASULOV DK + SHERGIN AP
          P1068-9 OF INT CONF ON PHYS ELECTRON ATOMIC
          COLLISIONS, (EXTENDED ABSTRACT) JULY 1975,
          RISLEY JS + GEBALLE R (EDS), WASHINGTON UNIV
          PRESS, SEATTLE, 1975.
          SAA 1976: 25491

2-0236   EFFECT OF SURFACE PLASMON OSCILLATIONS ON
          AUGER ELECTRON EMISSION.
          ALLIE G + BLANC E + DUFAYARD D + HAYMANN P
          SURFACE SCI 47(2): 635-44 (1975)
          CA 82: 147630

2-0237   EFFECT OF THE INCIDENCE ANGLE OF PRIMARY
          ELECTRONS ON THE AUGER EMISSION YIELD.
          ALLIE G + BLANC E + DUFAYARD D + STERN RM
          SURFACE SCI 46(1): 188-96 (1974) (IN FRENCH)
          CA 82: 009496

2-0238   INCOMPLETE RELAXATION IN A MODEL PROBLEM FOR
          THE AUGER PROCESS.
          ALMBLADH CO
          NUOVO CIMENTO B 23(1): 75-89 (1974)
          CA 81: 143719

2-0239   INTERPRETATION OF HIGH ENERGY X-RAY
          SATELLITES OF L2,3 EMISSION BANDS OF SODIUM,
          MAGNESIUM, ALUMINUM, AND SILICON.
          ALMBLADH CO + VON BARTH ULF
          J PHYS C 8(23): 4117-23 (1975)
          CA 84: 67215

2-0240   AUGER ELECTRON SPECTROSCOPY AND SECONDARY
          EMISSION IN SEMICONDUCTORS. (SILICON,
          GERMANIUM, GRAPHITE, ELECTRON DIFFUSION
          THEORY)
          AMELIO GF
          GEORGIA INST TECHNOL, THESIS 1968.

2-0241   THEORY FOR ENERGY DISTRIBUTION OF SECONDARY
          ELECTRONS.
          AMELIO GF
          J VACUUM SCI TECHNOL 7: 593-604 (1970)

2-0242   TRUE SECONDARY ELECTRON ENERGY DISTRIBUTIONS.
          AMELIO GF + SCHEIBNER EJ
          P11-1 TO 11-15 OF STRUCTURE AND CHEMISTRY OF
          SOLID SURFACES, SOMORJAI GA (ED), NY, WILEY,
          1969.

2-0243   DIFFRACTION EFFECTS IN BACKSCATTERING AND
          AUGER PRODUCTION NEAR CRYSTAL SURFACES.
          ANDERSEN SK + KJAER HA
          SURFACE SCI 50(1): 197-214 (1975)
          CA 83: 69555

2-0244   FORMATION OF A PLASMA SHEATH WITH SECONDARY
          ELECTRON EMISSION.
          ANDERSON N
          INT J ELECTRON 32: 425-33 (APR 1972)

2-0245   ANISOTROPY OF SECONDARY ELECTRON EMISSION
          FROM PASSAGE OF LITHIUM(+) IONS THROUGH
          COPPER SINGLE CRYSTALS.
          ARIFOV UA + DZHEMILEV NKH + RADZHABOV TD
          BULL ACAD SCI USSR, PHYS SER 35(2): 231-3
          (1971)

2-0246   ENERGY SPECTRUM OF ELECTRONS EMITTED BY
          ALKALI HALIDE CRYSTALS UNDER IMPACT OF ION
          AND ATOM BOMBARDMENT. (AUGER EFFECT)
          ARIFOV UA + GAIPOV S + RAKHIMOV RR
          RADIAT EFF 19(3): 151-4 (1973)
          SAA 1973: 80132

2-0247   AUGER NEUTRALIZATION OF MULTIPLY CHARGED IONS
          ON A METAL SURFACE.
          ARIFOV UA + KISHINEVSKII LM + MUKHAMADIEV ES
          + PARILIS ES
          SOV PHYS TECH PHYS 18(1): 118-22 (1973)
          CA 78: 116498

2-0248   THE L1,L2,3M4,5 COSTER-KRONIG TRANSITIONS.
          ASAAD WN
          NUCL PHYS 63: 337-48 (1965)

2-0249   ON THE COMPOUND PHOTOELECTRIC EFFECT.
          AUGER P
          J PHYS RADIUM 6: 205-8 (1925) (IN FRENCH)

2-0250   PICOSECOND OPTICAL MEASUREMENTS OF
          BAND-TO-BAND AUGER RECOMBINATION OF HIGH
          DENSITY PLASMAS IN GERMANIUM.
          AUSTON DH + SHANK CV + LEFUR P
          PHYS REV LETT 35(15): 1022-5 (1975)
          CA 83: 185640

2-0251   KLL AUGER TRANSITION ABSOLUTE PROBABILITIES.
          (ATOMIC NUMBERS 30-94)
          BABENKOV MI + BOBYKIN BV + PETUKHOV VK
          PHYS LETT A 40(4): 347-8 (1972)
          CA 77: 107129R

2-0252   ABSOLUTE PROBABILITIES FOR AUGER- KLL
          TRANSITIONS DEPENDENT ON THE ATOMIC NUMBER
          Z=30-94. PART-2.
          BABENKOV MI + BOBYKIN BV + PETUKHOV VK
          BULL ACAD SCI USSR, PHYS SER 36(10): 1890-4
          (1972)
          CA 78: 64791B

2-0253   RELATIVE PROBABILITIES FOR AUGER KLL
          TRANSITIONS DEPENDENT ON THE ATOMIC NUMBER 29
          TO 94.
          BABENKOV MI + BOBYKIN BV + PETUKHOV VK
          BULL ACAD SCI USSR, PHYS SER 36(10): 1883-9
          (1972)
          CA 78: 77611T

2-0254   CRYSTALLINE EFFECTS IN BACKSCATTERING AND
          AUGER PRODUCTION. (ALUMINUM)
          BAINES M + HOWIE A + ANDERSEN SK
          SURFACE SCI 53(1): 546-53 (1975)
          CA 84: 67564

2-0255   RADIATIONLESS DEEXCITATION OF EXCITED HELIUM
          ATOMS AT SURFACES. (AUGER PROCESS)
          BAIRD WE + ZIVITZ M + LARSEN J + THOMAS EW
          PHYS REV A 10(6): 2063-9 (1974)
          CA 82: 105083

2-0256   RESONANCE IONIZATION OF ATOMIC HELIUM BY FAST
          ELECTRONS. (ANGULAR DISTRIBUTION OF AUGER
          ELECTRONS)
          BALASHOV VV + LIPOVETSKII SS
          + PAVLICHENKOV AV + POLYUDOV AN
          OPT SPECTROSC 32(1): 4-7 (1972)
          CA 76: 119662B

2-0257   POSSIBILITY OF INVESTIGATING ATOMIC
          AUTOIONIZING STATES IN COINCIDENCE
          EXPERIMENTS. (AUGER SPECTRUM OF HELIUM)
          BALASHOV VV + LIPOVETSKII SS + SENASHENKO VS
          PHYS LETT A 39(2): 103-4 (1972)
          CA77: 106969C

2-0258   TEMPERATURE DEPENDENCE OF THE EDGE EMISSION
          OF GALLIUM SELENIDE SINGLE CRYSTALS AT HIGH
          EXCITATION LEVELS. (AUGER RECOMBINATION)
          BALTRAMIEJUNAS R + GUSEINOV G
          + NARKEVICIUS V + NIUNKA V + VAITKUS J
          + VISCAKAS J
          OPT COMMUN 11(3): 274-6 (JUL 1974)
          SAA 1974: 67645

2-0259   X-RAY FLUORESCENCE YIELDS, AUGER, AND
          COSTER-KRONIG TRANSITION PROBABILITIES.
          BAMBYNEK W + CRASEMANN B + FINK RW
          + FREUND HU + MARK H + SWIFT CD
          REV MOD PHYS 44(4): 716-813 (1972)
          CA 77: 158013F

2-0260   EVIDENCES FOR AUGER RECOMBINATION IN ELECTRON
          HOLE DROPLETS. (SILICON)
          BARRAU J + BRABANT JC + BROUSSEAU M
          + COLLET J + HECKMANN M + MAAREFF H
          SOLID STATE COMMUN 16(9): 1079-82 (1975)
          CA 83: 51654

2-0261   COMPARISON OF IONIZATION CROSS SECTIONS AND
          OTHER FACTORS IN X-RAY EXCITED AND ELECTRON
          EXCITED AUGER SPECTROSCOPY.
          BARRIE A
          J ELECTRON SPECTROSC RELAT PHENOMENA 7(1):
          75-80 (1975)
          CA 83: 123532

2-0262 AUGER LIMITED CARRIER LIFETIMES IN MERCURY CADMIUM TELLURIDE AT HIGH EXCESS CARRIER CONCENTRATIONS.
BARTOLI F + ALLEN R + ESTEROWITZ L + KRUER M
J APPL PHYS 45(5): 2150-4 (1974)
CA 81: 019077

2-0263 NATURE OF ANNEALED SEMICONDUCTOR SURFACES. (AUGER)
BAUER E
PHYS LETT A 26: 530-1 (1968)

2-0264 MOLECULAR ADSORPTION: INCLUSION OF BULK DIFFUSION AND ITS EFFECT ON AUGER INTENSITIES.
BECK DE + MIYAZAKI E
SURFACE SCI 39(1): 37-50 (1973)
CA 79: 097275

2-0265 ACCURATE ONE ELECTRON BINDING ENERGIES AND AUGER ENERGIES IN MANY ELECTRON ATOMS.
BECK DR + NICOLAIDES CA
PHYS FENN 9, SUPPL S1: 244-6 (1974)
CA 82: 147741

2-0266 AUGER RECOMBINATION IN SILICON.
BECK JD + CONRADT R
SOLID STATE COMMUN 13(1): 93-5 (1973) (IN GERMAN)
CA 79: 071478

2-0267 IONIZATION CROSS SECTION OF THE L2,3 SHELL OF ALUMINUM IN A NEON(+) - ALUMINUM COLLISION.
BENAZETH C + VIEL L
COMPT REND B 279(14): 349-52 (30 SEP 1974) (IN FRENCH)
SAA 1975: 22423

2-0268 INTERPRETATION OF ENERGY LOSS BY AUGER ELECTRONS EMITTED DURING ION BOMBARDMENT.
BENAZETH C + VIEL L + COLOMBIE N
COMPT REND B 276(23): 863-6 (1973) (IN FRENCH)
CA 79: 072082

2-0269 ENERGY DISTRIBUTION OF SECONDARY ELECTRONS EMITTED BY INDIUM PHOSPHIDE AND GALLIUM PHOSPHIDE SUBMITTED TO A BOMBARDMENT BY 40 KEV ELECTRONS. (AUGER EFFECT)
BENAZETH C + VIEL L + COTOMBIE N
SURFACE SCI 32: 618-22 (1972) (IN FRENCH)

2-0270 X-RAY EDGE ANOMALIES IN LITHIUM.
BERGERSEN B + JENA P + MCMULLEN T
J PHYS F 4(10): L219-2 (1974)
CA 82: 009619

2-0271 AUGER ELECTRONS FROM THE L-SHELL IN MERCURY.
BERGSTROM I + HILL RD
ARKIV FYS 8: 21-6 (1954)

2-0272 AUGER SIGNALS OBTAINED WITH A PHOTOELECTRON SPECTROMETER.
BERTHOU H + JOERGENSEN CK
J ELECTRON SPECTROSC RELAT PHENOMENA 5: 935-61 (1974)
CA 82: 66207

2-0273 CHEMICAL SHIFTS AND VARIATIONS OF WIDTH AND INTENSITY OF AUGER SIGNALS OBTAINED WITH A PHOTOELECTRON SPECTROMETER.
BERTHOU H + JCRGENSEN CK
PHYS FENN 9, SUPPL S1: 321-3 (JUL 1974)
SAA 1975: 19049

2-0274 RELAXATION OF AUGER EXCITED CARRIERS IN SILICON.
BETZLER K
SOLID STATE CCMMUN 15(11-12): 1837-40 (15 DEC 1974)
SAA 1975: 22361

2-0275 TWO ELECTRON BAND-TO-BAND TRANSITIONS IN SOLIDS. (AUGER RECOMBINATION IN SILICON)
BETZLER K + WELLER T + CONRADT R
PHYS REV B 6(4): 1394-9 (1972)
CA 77: 81516N

2-0276 ELECTRON PHOTOEMISSION INDUCED BY LASER-PRODUCED PLASMA X-RAYS. (PLASMA ELECTRONS, AUGER PROCESS)
BEVERLY RE
J APPL PHYS 46(2): 624-6 (1975)
CA 82: 131747

2-0277 NONRELATIVISTIC FLUORESCENCE YIELDS FOR THE 3P AND THE 3D SHELLS.
BHALLA CP
PHYS REV A 6(4): 1409-13 (1972)
CA 77: 143889X

2-0278 AUGER SPECTRA AND FLUORESCENCE YIELDS FOR THE 3P AND THE 3D SHELL.
BHALLA CP
P59-79, VOL-1 OF INT CONF INNER SHELL IONIZATION PHENOMENA AND FUTURE APPLICATIONS, 1972, FINK RW(ED), NTIS, 1973.
CA 78: 130138

2-0279 EFFECTS OF THE CONFIGURATION INTERACTION ON THE K-SHELL AUGER SPECTRUM OF NEON.
BHALLA CP
PHYS LETT A 44(2): 103-4 (1973)
CA 79: 059789

2-0280 K-SHELL AUGER RATES, TRANSITION ENERGIES, AND FLUORESCENCE YIELDS OF VARIOUSLY IONIZED STATES OF ARGON.
BHALLA CP
PHYS REV A 8(6): 2877-82 (1973)
CA 80: 054236

2-0281 DEEXCITATION OF MULTIPLY IONIZED ATOMS. (AUGER EFFECT)
BHALLA CP
PHYS FENN 9, SUPPL S1: 435-7 (JUL 1974)
SAA 1975: 17799

2-0282 K-SHELL AUGER RATES FOR MULTIPLY IONIZED ATOMS. PART-1: NEON.
BHALLA CP
J ELECTRON SPECTROSC RELAT PHENOMENA 7(4): 28-95 (1975)
CA 83: 199772

2-0283 THEORETICAL FLUORESCENCE YIELDS FOR NEON. (AUGER RATE)
BHALLA CP
PHYS REV A 12(1): 122-8 (1975)
CA 83: 105742

2-0284 THEORETICAL K-SHELL AUGER RATES, TRANSITION ENERGIES, AND FLUORESCENCE YIELDS FOR MULTIPLY IONIZED NEON. (AUGER EFFECT)
BHALLA CP + FOLLAND NO + HEIN MA
PHYS REV A 8(2): 649-57 (1973)
CA 79: 109992

2-0285 EXCITATION OF MULTIPLET STATES IN NEON.
BHALLA CP + MATTHEWS DL + MOORE CF
PHYS LETT A 46(5): 336-8 (1974)
CA 80: 112960

2-0286 ANGULAR DEPENDENCIES IN ELECTRON EXCITED AUGER EMISSION. (COMMENTS)
BISHOP HE + RIVIERE JC
SURFACE SCI 17: 446-7 (1969)

2-0287 CHARACTERISTIC IONIZATION LOSSES OBSERVED IN AUGER EMISSION SPECTROSCOPY.
BISHOP HE + RIVIERE JC
APPL PHYS LETT 16: 21-3 (1970)

2-0288 ESTIMATES OF EFFICIENCIES OF PRODUCTION AND DETECTION OF ELECTRON EXCITED AUGER EMISSION.
BISHOP HE + RIVIERE JC
J APPL PHYS 40: 1740-4 (1969)

2-0289 PHOTOEMISSION STUDIES OF THE DECAY OF CORE EXCITONS IN ALKALI HALIDES. (AUGER PROCESS IN VALENCE BAND)
BLECHSCHMIDT D + SKIBOWSKI M + STEINMANN W
IN INT CONF ON VACUUM ULTRAVIOLET RADIATION PHYS, PHYS SOC JAPAN, TOKYO, SEP 1971, 3RD PROC, 4P.
SAA 1972: 19802

2-0290 NATURE OF SOME X-RAY SATELLITES (RADIATIVE
AUGER TRANSITIONS).
BLOKHIN MA + DEMJOKHIN VF + SACHENKO VP
PHYS FENN 9, SUPPL S1: 231-8 (JUL 1974)
SAA 1975: 28670

2-0291 PLASMON COUPLING TO CORE HOLE EXCITATIONS IN
CARBON. (AUGER APPEARANCE SPECTROSCOPY)
BRADSHAW AM + CEDERBAUM SL + DOMCKE W
+ KRAUSE U
J PHYS C 7(24): 4503-12 (1974)
CA 82: 91814

2-0292 G-R NOISE FOR AUGER BAND-TO-BAND PROCESSES.
BREITZER D + SARD E + PEYTON B + MCELROY J
INFRARED PHYS 11(4): 237-9 (1971)
SAA 1972: 19781

2-0293 NEW SATELLITE BANDS CORRESPONDING TO DOUBLE L
IONIZATION. (OF AUGER ELECTRONS)
BRIAND JP + CHEVALLIER P + TAVERNIER M
J PHYS (PARIS) COLLOQ NO 4: 165-71 (1971)
CA 77: 81452P; 11680N

2-0294 (AUGER) NEUTRALIZATION BEHAVIOR IN SCATTERING
OF LOW ENERGY IONS FROM SOLID SURFACES.
BRONGERSMA HH + BUCK TM
NUCL INSTRUM METH 132: 559-64 (1976)
SAA 1976: 54370

2-0295 METASTABLE AUTOIONIZING THREE- AND FOUR
ELECTRON STATES IN BERYLLIUM. (AUGER DECAY)
BRUCH R + PAUL G + ANDRAE J
J PHYS B 8(11): L253-8 (1975)
CA 83: 154965

2-0296 AUGER TRANSITIONS IN LITHIUM-LIKE
BERYLLIUM-LIKE IONS.
BRUCH R + PAUL G + ANDRAE J + FRICKE B
PHYS LETT A 53(4): 293-4 (1975)
CA 83: 105939

2-0297 ANGLE DISTRIBUTION OF SODIUM CHLORIDE SINGLE
CRYSTALS. (FHCTC- AND AUGER ELECTRON
EMISSION)
BRUNNER J + SIEGRIST M + ZOGG H
HELV PHYS ACTA 45(6): 896-7 (31 DEC 1972) (IN
GERMAN)
SAA 1973: 53165

2-0298 OUTER SHELL EXCITATION OF THE RARE GASES BY
30 MEV OXYGEN IONS. (INNER SHELL AUGER
TRANSITIONS)
BURCH D + BICKEL WS + WIEMAN H
PHYS LETT A 46(6): 405-6 (28 JAN 1974)
SAA 1974: 24009

2-0299 EFFECT OF MULTIPLE IONIZATION ON THE
FLUORESCENCE YIELD OF NEON. (K-AUGER AND K
X-RAY YIELDS)
BURCH D + INGALLS WB + RISLEY JS + HEFFNER R
PHYS REV LETT 29: 1719-22 (1972)
CA 78: 50090K

2-0300 PROJECTILE CHARGE STATE DEPENDENCE OF NEON
K-SHELL IONIZATION AND FLUORESCENCE YIELD IN
50 MEV CHLORINE(N+) ION- NECN COLLISIONS.
BURCH D + STOLTERFOHT N + SCHNEIDER D
+ WIEMAN H + RISLEY JS
PHYS REV LETT 32(21): 1151-4 (1974)
CA 81: 031392

2-0301 SIMPLE MODEL FOR K X-RAY AND AUGER ELECTRON
ENERGY SHIFTS IN HEAVY ION COLLISIONS.
COMMENTS.
BURCH D + WILETS L + MEYERHOF WE
PHYS REV A 9(2): 1007-10 (1974)
CA 80: 114338

2-0302 OPTICALLY STIMULATED SECONDARY ELECTRON
EMISSION AND PHOTOELECTRON EMISSION STUDY OF
ADSORPTION PROCESSES.
BURIBAEV I + SHISHKIN BB
BULL ACAD SCI USSR, PHYS SER 35(2): 296-8
(1971)

2-0303 SENSITIVITY OF L1,L2,3M4,5 COSTER-KRONIG
TRANSITION RATES TO VARIATIONS IN ENERGY
LEVELS AND SCREENING PARAMETERS.
CALLAN EJ
REV MOD PHYS 35: 524-8 (1963)

2-0304 KLL AUGER TRANSITION PROBABILITIES.
CALLAN EJ
PHYS REV 124: 793-9 (1961)

2-0305 MEASUREMENT OF THE AUGER EFFECT IN THE MU-4
HELIUM(2S) (+) IONIC SYSTEM.
CARBONI G + PLACCI A + ZAVATTINI E
+ GASTALDI U + GORINI G + PITZURRA O
+ NERI G + POLACCO E + TORELLI G + DUCLOS J
+ PICARD J + VITALE A
LETT NUOVO CIMENTO SOC ITAL FIS 6(7): 233-7
(1973)
CA 78: 117960

2-0306 PHOTOELECTRIC PROPERTIES OF MAGNESIUM(2)
SILICIDE, MAGNESIUM(2) GERMANIUM, AND
MAGNESIUM(2) STANNIDE. PART-1: X-RAY
EXCITATION PART-2: ULTRAVIOLET EXCITATION.
(AUGER ELECTRONS)
CARDONA M + TEJEDA J + SHEVCHIK NJ
+ LANGER DW
PHYS STATUS SOLIDI B 58(1): 188-200; 58(2):
483-91 (1973)
CA 79: 098313

2-0307 NATURE OF SECONDARY ELECTRONS CREATED AS THE
RESULT OF ELECTRON SHAKE-OFF AND (SUGAR)
VACANCY CASCADES.
CARLSON TA
RADIAT RES 64(1): 53-69 (1975)
CA 83: 197913

2-0308 RELATIVE ABUNDANCES AND RECOIL ENERGIES OF
FRAGMENT IONS FORMED FROM THE X-RAY
PHOTOIONIZATION OF NITROGEN, OXYGEN, CARBON
MONOXIDE, NITRIC OXIDE, CARBON DIOXIDE, AND
CARBON TETRAFLUORIDE. (KLL AUGER PROCESS)
CARLSON TA + KRAUSE MO
J CHEM PHYS 56: 3206-9 (1972)
SAA 1972: 28310

2-0309 SECONDARY ELECTRON EMISSION FROM SOLID
SURFACES BOMBARDED BY MEDIUM ENERGY IONS
(PLATINUM, NICKEL, TANTALUM, GRAPHITE).
CAWTHRON ER
AUSTRAL J PHYS 24: 859-69 (DEC 1971)

2-0310 DOUBLE PHOTOIONIZATION OF NEON. (AUGER
TRANSITIONS)
CHANG TN + POE RT
PHYS REV A 12(4): 1432-9 (1975)
SAA 1976: 16936

2-0311 APPLICATION OF MANY BODY PERTURBATION THEORY
TO THE CALCULATION OF AUGER RATES,
CORRELATION ENERGIES, AND PHOTOIONIZATION
CROSS SECTIONS.
CHASE RL
UNIV VIRGINIA, THESIS 1972, 87P.
CA 78: 102178

2-0312 CORRELATION ENERGIES AND AUGER RATES IN ATOMS
WITH INNER SHELL VACANCIES (NEON WITH 1S OR
2S VACANCY).
CHASE RL + KELLY HP + KOHLER HS
PHYS REV A 3: 1550-7 (MAY 1971)

2-0313 THEORETICAL L2 SUBSHELL FLUORESCENCE YIELDS
AND COSTER-KRONIG TRANSITION PROBABILITIES
F23 BASED ON THE GREEN-SELLIN-ZACHOR
INDEPENDENT PARTICLE MODEL.
CHEN MH + CRASEMANN B
P43-58 OF INT CONF INNER SHELL IONIZATION
PHENOMENA AND FUTURE APPLICATIONS, VOL 1,
1972, FINK RW(ED), NTIS, 1973.
CA 78: 130137

2-0314 MULTIPLET EFFECT IN FLUORESCENCE YIELDS OF
MULTIPLY IONIZED ATOMS. (AUGER EFFECT)
CHEN MH + CRASEMANN B
PHYS FENN 9, SUPPL S1: 250-2 (1974)
CA 82: 147742

2-0315 AUGER AND RADIATIVE DEEXCITATION OF MULTIPLY
IONIZED NEON.
CHEN MH + CRASEMANN B
PHYS REV A 12(3): 959-67 (1975)
CA 83: 200000

2-0316 THEORETICAL L2 AND L3 SUBSHELL FLUORESCENCE
YIELDS AND L2-L3(X) COSTER-KRONIG TRANSITION
PROBABILITIES.
CHEN MH + CRASEMANN B + KOSTROUN VO
PHYS REV A 4: 1-7 (1971)

2-0317 INTERATOMIC AUGER PROCESS: EFFECTS ON
LIFETIMES OF CORE HOLE STATES.
CITRIN PH
PHYS REV LETT 31(19): 1164-7 (1973)
CA 80: 008045

2-0318 INTERATOMIC AUGER TRANSITIONS. SURVEY. (36
REFS)
CITRIN PH
J ELECTRON SPECTROSC RELAT PHENOMENA 5:
273-87 (1974)
CA 82: 78554

2-0319 AUGER ELECTRON EMISSION SPECTRA: A SIMPLE
TREATMENT OF ENERGIES AND INTENSITIES.
CLARKE TA + MASON R + RANDACCIO L + THOMAS JM
J CHEM A, INORG, PHYS, THEOR 9: 1156-60
(1971)

2-0320 ANGULAR DISTRIBUTION OF AUGER ELECTRONS.
CLEFF B + MEHLHORN W
PHYS LETT A 37: 3-4 (25 OCT 1971)

2-0321 ANGULAR DISTRIBUTION OF AUGER ELECTRONS
FOLLOWING IMPACT IONIZATION.
CLEFF B + MEHLHORN W
J PHYS B 7(5): 593-604 (1974)
CA 80: 138882

2-0322 ANGULAR DISTRIBUTION OF L3M2,3M2,3(1S0) AUGER
ELECTRONS OF ARGON.
CLEFF B + MEHLHORN W
J PHYS B 7(5): 605-11 (1974)
CA 80: 138883

2-0323 ARGON K-AUGER CROSS SECTIONS IN COLLISION
WITH 30 MEV FLUORINE IONS.
COCKE CL + RANDALL RR + CURNUTTE B
P933-4 OF INT CONF ON PHYS ELECTRON ATOMIC
COLLISIONS, (EXTENDED ABSTRACTS) JULY 1975,
WASHINGTON UNIV PRESS, SEATTLE.
SAA 1976: 21199

2-0324 AUGER EFFECT AND RADIATIVE RECOMBINATION IN
N-GALLIUM ARSENIDE- P-ALUMINUM GALLIUM
ARSENIDE HETEROJUNCTIONS.
CONSTANTINESCU C + GOLDENBLUM A + SCSTARICH M
REV ROUM PHYS 16(9): 969-77 (1971)
CA 76: 119516G

2-0325 EMISSION SPECTRA OF SILICON COMPENSATED
ALUMINUM(X) GALLIUM(1-X) ARSENIDE
ELECTROLUMINESCENT DIODES. (AUGER
RECOMBINATION MECHANISM)
CONSTANTINESCU L + GEORGESCU S
+ GOLDENBLUM A + IVAN E
PHYS STATUS SCLIDI A 24(1): 367-74 (16 JUL
1974)
SAA 1974: 67662

2-0326 "SEMI-AUGER" PROCESSES IN L2,3 EMISSION IN
ARGON AND POTASSIUM CHLORIDE.
COOPER JW + LAVILLA RE
PHYS REV LETT 25: 1745-7 (1970)

2-0327 INNER SHELL DOUBLE IONIZATION AND CHEMICAL
BONDING. PART-1: AUGER ELECTRONS. (KLL
TRANSITION ENERGIES FOR SULFUR, ALUMINUM,
SILICON)
COULSON CA + GIANTURCO FA
PROC PHYS SOC AT MOLEC PHYS (J PHYS B) 1:
605-13 (1968)

2-0328 AUGER AND RADIATIVE TRANSITION PROBABILITIES.
CRASEMANN B
P9-42 OF INT CONF INNER SHELL IONIZATION
PHENOMENA AND FUTURE APPLICATIONS, VOL 1,
1972, FINK RW(ED), NTIS, 1973.
CA 78: 129967

2-0329 AUGER AND COSTER-KRONIG TRANSITION
PROBABILITIES TO THE ATOMIC 2S STATE AND
THEORETICAL L1 FLUORESCENCE YIELDS.
CRASEMANN B + CHEN MH + KOSTROUN VO
PHYS REV A 4: 2161-3 (1971)

2-0330 EFFICIENCY OF THE AUGER CASCADE IN SOLID
STATE.
CRUSET A
RADIOCHEM RADIOANAL LETT 16(6): 309-14 (1974)
CA 80: 151630

2-0331 ENERGY DISTRIBUTION OF (AUGER) ELECTRONS
RESULTING FROM ION- ATOM COLLISIONS WITH A
FIXED IMPACT PARAMETER.
DAHL P + LORENTS DC
P395-7 OF INT CONF PHYS ELECTRON ATOMIC
COLLISIONS, AMSTERDAM, 1971, 7TH PROC,
ABSTRACTS, NORTH-HOLLAND, 1971.
SAA 1972: 693

2-0332 CARRIER LIFETIMES IN EPITAXIAL INDIUM
ARSENIDE. (AUGER RECOMBINATION)
DALAL VL + HICINBOTHEM WA + KRESSEL H
APPL PHYS LETT 24(4): 184-5 (1974)
CA 80: 101472

2-0333 ADSORBATE EFFECTS IN ELECTRON EJECTION BY
RARE GAS METASTABLE ATOMS. (AUGER PROCESS)
DELCHAR TA + MACLENNAN DA + LANDERS AM
J CHEM PHYS 50: 1779-87 (1969)

2-0334 MERCURY AND GOLD ATOMIC CLOUDS REARRANGEMENT
FOLLOWING INTERNAL CONVERSION AND ELECTRON
CAPTURE. (K-AUGER TRANSITION INTENSITIES)
DIONISIO JS + VIEW C
J ELECTRON SPECTROSC RELAT PHENOMENA 5:
173-91 (1974)
CA 82: 78069

2-0335 INNER SHELL ALIGNMENT OF ATOMS IN ELECTRON
IMPACT IONIZATION. (AUGER ELECTRONS)
DOBELIN E + SANDNER W + MEHLHORN W
PHYS LETT A 49(1): 7-8 (12 AUG 1974)
SAA 1974: 74897

2-0336 DECAY OF HOLE STATES IN OXYGEN-16 VIA A
"NUCLEAR AUGER EFFECT".
DOLL P + FORD JLC + GOLDSCHMIDT M
+ KNOPFLE KT + MAIRLE G + THORNTON ST
+ WAGNER GJ + WIEDNER CA
P181 OF INT CONF ON NUCLEAR STRUCTURE AND
SPECTROSCOPY, ABSTR, SCHOLAR'S PRESS,
AMSTERDAM, 1974.
SAA 1975: 32371

2-0337 K X-RAY EMISSION EDGE SHAPES OF FREE ELECTRON
METALS. (AUGER BROADENING)
DOW JD + WATSON LN + FABIAN DJ
J PHYS F 4(4): L76-9 (APR 1974)
SAA 1974: 45383

2-0338 ROLE OF COSTER-KRONIG TRANSITIONS IN THE
IONIZATION OF SURFACE ATOMS.
DUCHARME AR
J VACUUM SCI TECHNOL 11(1): 281-3 (JAN-FEB
1974)
SAA 1974: 76346

2-0339 L-SHELL IONIZATION CROSS SECTIONS IN AUGER
ELECTRON SPECTROSCOPY.
DUCHARME AR + GERLACH RL
J VACUUM SCI TECHNOL 10(1): 188-91 (1973)
CA 78: 103405

2-0340 SECONDARY ELECTRON EJECTION FROM METAL
SURFACES BY METASTABLE ATOMS. PART-1:
MEASUREMENTS OF SECONDARY EMISSION
COEFFICIENTS USING A CROSSED BEAM METHOD.
PART-2: MEASUREMENTS OF SECONDARY EMISSION
COEFFICIENTS USING A GAS CELL METHOD.
DUNNING FB + SMITH ACH + STEBBINGS RF
J PHYS B 4: 1683-95, 1696-1710 (DEC 1971)

2-0341 VALENCE BAND AND CORE LEVEL STUDIES OF INDIUM
ANTIMONIDE VIA PHOTOEMISSION MEASUREMENTS IN
THE 20-70 EV RANGE. (AUGER PROCESSES)
EASTMAN DE + FREEOUF J
SOLID STATE CCMMUN 13(11): 1815-18 (1 DEC
1973)
SAA 1974: 10101

2-0342 AUGER TRANSITION RATE IN MIXED COUPLING
SCHEME.
EL IBYARI SN + ASAAD WN + MCGUIRE EJ
PHYS REV A 5: 1048-51 (MAR 1972)

2-0343 WARM ELECTRON EFFECTS IN NARROW GAP
MERCURY(1-X) CADMIUM(X) TELLURIDE AMBIENT
TEMPERATURES. (AUGER RECOMBINATION)
ELLIOTT CT + SPAIN IL
J PHYS C 7(4): 727-35 (21 FEB 1974)
SAA 1974: 32960

2-0344 RADIATIVE AUGER TRANSITIONS IN SOFT X-RAY
PLASMA EMISSION.
ELTON RC + PALUMBO LJ
PHYS REV A 9(5): 1873-84 (1974)
CA 81: 018736

2-0345 K-VACANCY SHARING STUDIED FROM AUGER ELECTRON
YIELDS.
FASTRUP B + AAGAARD B + BOVING E
+ SCHNEIDER D + ZIEM P + STOLTERFOHT N
P1058-9 OF INT CONF ON PHYS ELECTRON ATOMIC
COLLISIONS, (EXTENDED ABSTRACT), JULY 1975,
RISLEY JS + GEBALLE R(EDS), WASHINGTON UNIV
PRESS, SEATTLE.
SAA 1976: 25486

2-0346 DYNAMIC CHARGE STATE DEPENDENCE OF THE
K-VACANCY PRODUCTION YIELD IN HEAVY ION ATOM
COLLISIONS AT KEV ENERGIES.
FASTRUP B + BOVING E + LARSEN GA + CAHL P
J PHYS B 7(7): L206-L209 (1974)
CA 81: 031358

2-0347 AUGER PLASMON SATELLITE INTENSITIES VERSUS
DEPTH: A MEANS FOR DETERMINING ADATOM
CONCENTRATION PROFILES.
FEIBELMAN PJ
PHYS REV B 7(6): 2305-17 (15 MAR 1973)
SAA 1973: 30904

2-0348 EFFECTS OF LOW ENERGY ARGON- ION BOMBARDMENT
ON MOLYBDENUM DISULFIDE. (AUGER MEASUREMENTS)
FENG HC + CHEN JM
J PHYS C 7(5): L75-8 (7 MAR 1974)
SAA 1974: 33537

2-0349 THE L(II) SUBSHELL ATOMIC YIELDS OF ELEMENTS
WITH Z APPRCXIMATELY 90.
FERREIRA JG
P233-43 OF INT CONF INNER SHELL IONIZATION
PHENOMENA AND FUTURE APPLICATIONS, VOL 1,
1972, FINK RW(ED), NTIS, 1973.
CA 78: 129954

2-0350 USE OF CHARACTERISTIC IONIZATION LOSSES IN
AUGER ELECTRON EMISSION SPECTROSCOPY.
FIERMANS L + VENNIK J
SURFACE SCI 38(1): 257-60 (1973)
CA 79: 036916

2-0351 ATOMIC FLUORESCENCE YIELDS. (L-SHELL
COSTER-KRONIG YIELDS, K-, L-, AND M-SHELL
FLUORESCENCE YIELDS)
FINK RW + JOPSON RC + MARK H + SWIFT CD
REV MOD PHYS 38(3): 513-40 (1966)

2-0352 ANGULAR DISTRIBUTION OF AUGER ELECTRONS
FOLLOWING PHOTOIONIZATION.
FLUGGE S + MEHLHORN W + SCHMIDT V
PHYS REV LETT 29(1): 7-9 (1973)
CA 79: 059794

2-0353 AUGER EFFECT AND CORE BROADENING IN LOW-Z
ATOMS, IONS, AND METALS.
FRANCESCHETTI DR
P663-4 OF INT CONF VAC ULTRAVIOLET RADIAT
PHYS, 4TH PROC, 1974, KOCH EE + HAENSEL R +
KUNZ C(EDS), PERGAMON, 1974.
CA 83: 68741

2-0354 EFFECTS OF DIFFUSION CURRENT ON
GALVANOMAGNETIC PROPERTIES IN THIN INTRINSIC
INDIUM ANTIMONIDE AT ROOM TEMPERATUPE. (AUGER
PROCESS)
FUJISADA H
J APPL PHYS 45(8): 3530-40 (AUG 1974)
SAA 1974: 82499

2-0355 SIMPLE MODEL FOR DEPENDENCE OF AUGER
INTENSITIES ON SPECIMEN THICKNESS.
GALLON TE
SURFACE SCI 17: 486-9 (1969)

2-0356 SECONDARY ELECTRON EMISSION AT SUPERHIGH
FREQUENCIES. (COPPER- GERMANIUM JUNCTION).
GANICHEV DA + STANSKII VA + FRIDRIKHOV SA
BULL ACAD SCI USSR, PHYS SER 35(2): 245-6
(1971)

2-0357 INFLUENCE OF AN ELECTRIC FIELD ON THE AUGER
RECOMBINATION IN SEMICONDUCTORS.
GEBRANZIG U + HAUG A + ROSENTHAL W
J LUMIN 12-13: 547-50 (1976)
CA 84: 172808

2-0358 DIFFERENTIAL CROSS SECTIONS FOR K-SHELL
IONIZATION OF SURFACE ATOMS BY ELECTRON
IMPACT.
GERLACH RL + DUCHARME AR
PHYS REV LETT 27: 290-2 (1971)

2-0359 INNER SHELL IONIZATION OF SURFACE ATOMS BY
ELECTRON IMPACT.
GEFLACH RL + DUCHARME AR
SURFACE SCI 29: 317-23 (1972)

2-0360 L-SHELL IONIZATION THRESHOLD STRUCTURE FROM
AUGER ELECTRON SPECTRA.
GEFLACH RL + DUCHARME AR
P675-8 OF INT VACUUM CONGRESS, 6TH PROC, INT
CONF ON SOLID SURFACES, 2ND, 1974, KYOTO.
(JAP J APPL PHYS, SUPPL 2, PART 1, 1974)
CA 83: 18148

2-0361 MANY BODY EFFECTS IN AUGER DEEXCITATION OF
ATOMS NEAR SOLIDS.
GERSTEN JI + TZOAR N
PHYS REV B 9(10): 4038-41 (1974)
CA 81: 056356

2-0362 MULTIPLE EXCITATION PROCESSES IN TRANSITION
METAL ATOMS. PART-2: INNER AUGER ELECTRONS.
GIANTURCO FA + SEMPRINI E + STEFANI F
PHYSICA B AND C 80(5-6): 627-35 (1975)
CA 84: 10735

2-0363 ABSOLUTE ELECTRON IMPACT IONIZATION CROSS
SECTIONS OF ATOMIC NITROGEN, ATOMIC OXYGEN,
AND NEON. (AUGER TRANSITIONS)
GLUPE G + MEHLHORN W
J PHYS (PARIS) COLLOQ NO 4: 40-3 (1971)
CA 77: 27181M

2-0364 PARITY NONCONSERVATION IN RADIATIONLESS
(AUGER) TRANSITIONS IN MU MESIC ATOMS.
GORSHKOV VG + MIKHAILOV AI + MOSKALEV AN
ZH EKSP TEOR FIZ 69(5): 1507-16 (1975) (IN
RUSSIAN)
SAA 1976: 12883

2-0365 AUGER AND SECONDARY ELECTRONS EXCITED BY
BACKSCATTERED ELECTRONS.
GOTO K + ISHIKAWA K + KOSHIKAWA T + SHIMIZU R
APPL PHYS LETT 24(8): 358-9 (1974); P633-5 OF
INT VACUUM CONGRESS, 6TH, PROC INT CONF ON
SOLID SURFACES, 2ND, 1974, KYOTO. (JAP J APPL
PHYS, SUPPL 2, PART 1, 1974)
CA 80: 150846

2-0366 AUGER AND SECONDARY ELECTRONS EXCITED BY
BACKSCATTERED ELECTRONS. APPROACH TO
QUANTITATIVE ANALYSIS.
GOTO K + ISHIKAWA K + KOSHIKAWA T + SHIMIZU R
SURFACE SCI 47(2): 477-94 (1975)
CA 82: 147628

2-0367 EFFECT OF MODULATION AMPLITUDE ON ELECTRON
EXCITED AUGER DATA FROM TITANIUM.
GRANT JT + HAAS TW + HOUSTON JE
SURFACE SCI 42(1): 1-11 (1974)
        CA 80: 089041

2-0368 AUGER RECOMBINATION IN SEMICONDUCTORS DURING
INTENSE EXCITATION.
GRIBKOVSKII VP + YASKEVICH GN
ZH PRIKL SPEKTROSK 20(3): 406-11 (1974) (IN
RUSSIAN)
        CA 81: 097303

2-0369 RECOMBINATION IN DONOR ACCEPTOR COMPLEXES IN
CADMIUM SULFIDE SINGLE CRYSTALS. (AUGER)
GRIN VF + LYUBCHENKO AV + SAL'KOV EA
+ SHEINKMAN MK
SOV PHYS SEMICOND 9(2): 197-201 (1975)
        SAA 1976: 14571

2-0370 AUGER ELECTRONS FROM FOIL EXCITED HEAVY ION
BEAMS. (OF NEON)
GRONEVELD KO + MANN R + NOLTE G
+ SCHUMANN S + SPOHR R
PHYS FENN 9, SUPPL S1: 31-2 (JUL 1974)
        SAA 1975: 17788

2-0371 AUGER TRANSITIONS LEADING TO DEFINED FINAL
CHARGE STATES BY A COINCIDENCE TECHNIQUE.
GRONEVELD KO + NOLTE G + SCHUMANN S
+ SEVIER KD
PHYS LETT A 56(1): 29-30 (1976)
        CA 84: 142778

2-0372 EXCITON SPECTRA OF HEAVILY DOPED CADMIUM
SULFIDE SINGLE CRYSTALS. (AUGER PROCESS)
GROSS EF + PERMOGOROV SA + REZNITSKII AN
+ USAROV EN
SOV PHYS SEMICOND 7(7): 844-8 (JAN 1974)
        SAA 1974: 33448

2-0373 SECONDARY ELECTRON EMISSION OF SOLID ALUMINUM
AND MAGNESIUM TARGETS.
GUENNOU H + DUFOUR G + BONNELLE C
SURFACE SCI 41(2): 547-54 (1974) (IN FRENCH)
        CA 80: 076056

2-0374 ELECTRON EMISSION FROM SOLIDS BY PARTICLES
CARRYING POTENTIAL ENERGY. (AUGER
NEUTRALIZATION, REVIEW, 16 REFS)
HAGSTRUM HD
P193-204 OF INT CONF PHENOMENA IONIZ GASES,
10TH PROC, DAVENPORT PA (ED), DONALD PARSONS
CO, LTD, OXFORD, ENGLAND, 1971.
        CA 76: 146240U

2-0375 RESONANCE, AUGER, AND AUTOIONIZATION
PROCESSES INVOLVING HELIUM(+) (2S) AND
HELIUM(2+) NEAR SOLID SURFACES.
HAGSTRUM HD + BECKER GE
PHYS REV B 8(1): 107-21 (1973)
        CA 79: 035219

2-0376 CROSS SECTIONS FOR THE PRODUCTION OF
NITROGEN(2)(X), NITROGEN(X), AND
NITROGEN(2)(2X) FROM NITROGEN BY ELECTRONS IN
THE ENERGY RANGE 16-600 EV.
HALAS ST + ADAMCZYK B
INT J MASS SPECTROM ION PHYS 10(2): 157-60
(1972) (IN POLISH)
        CA 77: 170208F

2-0377 RADIATION CHEMICAL RESONANCE EFFECT IN SOLID
5-BROMO DEOXYURIDINE: CHEMICAL CONSEQUENCES
OF THE AUGER EFFECT.
HALPERN A + STOECKLIN G
RADIAT RES 58(3): 329-37 (1974)
        CA 81: 031781

2-0378 ANGULAR DEPENDENCIES OF ELECTRON EXCITED
AUGER EMISSION.
HARRIS LA
SURFACE SCI 15: 77-93 (1969)

2-0379 INFLUENCE OF SCREENING EFFECTS ON THE AUGER
RECOMBINATION IN SEMICONDUCTORS. (SILICON,
GERMANIUM)
HAUG A + EKARDT W
SOLID STATE COMMUN 17(3): 267-8 (1975)
        SAA 1975: 77013

2-0380 METHOD FOR PACKET OF WAVES IN AUGER
ELECTRON-ELECTRON PROCESSES.
HAYMANN P
COMPT REND B 272: 1029-32 (1971)

2-0381 DYNAMIC THEORY OF AUGER EFFECT.
HAYMANN P
SURFACE SCI 30(3): 536-52 (1972) (IN FRENCH);
VIDE 28(164): 56-7 (1973)
        CA 79: 119599

2-0382 LOW ENERGY ION SCATTERING: ELASTIC AND
INELASTIC EFFECTS. (AUGER NEUTRALIZATION)
HEILAND W + TAGLAUER E
NUCL INSTRUM METH 132: 535-45 (1976)
        SAA 1976: 54369

2-0383 EMISSION OF AUGER ELECTRONS BY ATOMS IN A
METALLIC TARGET SUBJECTED TO AN IONIC
BOMBARDMENT.
HENNEQUIN JF
J PHYS (PARIS) 29: 1053-65 (1968) (IN FRENCH)

2-0384 EFFECTS OF CESIATION ON SECONDARY ELECTRON
EMISSION FROM MAGNESIUM OXIDE- GOLD CERMETS.
HENRICH VE + FAN JCC
J APPL PHYS 45(12): 5484-6 (1974)
        CA 82: 105417

2-0385 A FORMALISM FOR THE INDIRECT AUGER EFFECT.
PART-1. PART-2.
HILL D + LANDSBERG PT
PROC ROY SOC A 347(1651): 547-64, 565-73
(1976)
        CA 84: 142864, 157524

2-0386 MANY BODY PERTURBATION THEORY APPLIED TO
ATOMIC RADIATIONLESS TRANSITIONS.
HOLMES CP + KOSTROUN VO
P475-89 OF INT CONF INNER SHELL IONIZATION
PHENOMENA AND FUTURE APPLICATIONS, VOL 1,
1972, FINK RW(ED), NTIS, 1973.
        CA 78: 129993

2-0387 MEASUREMENTS OF THE L-SHELL FLUORESCENCE
YIELDS OF ARGON AND CHLORINE BY ELECTRON
IMPACT. (AUGER EMISSION)
HOOGKAMER TP + SARIS FW + DEHEER FJ
J PHYS B 8(7): L105-8 (1975)
        SAA 1975: 49575

2-0388 CONCENTRATION-DEPENDENT NONRADIATIVE
PROCESSES IN ZINC SULFIDE PHOSPHORS. (AUGER
RECOMBINATIONS)
HOSHINA T + KAWAI H
J LUMIN 12-13: 453-9 (1976)
        SAA 1976: 54299

2-0389 EXACT CORRECTIONS FOR POTENTIAL MODULATION
DISTORTION IN AUGER YIELD MEASUREMENTS.
HOUSTON JE
SURFACE SCI 38(2): 283-91 (1973)
        CA 79: 047380

2-0390 AUGER EXCITATION BY INTERNAL SECONDARY
ELECTRONS.
HOUSTON JE + PARK RL
APPL PHYS LETT 14: 358-60 (1969)

2-0391 CROSS CORRELATION TECHNIQUES IN AUGER
SPECTROSCOPY.
HOUSTON JE + PARK RL
BULL AMER PHYS SOC 14: 793 (1969)

2-0392 DIRECT COMPARISON OF CORE ELECTRON BINDING
ENERGIES OF SURFACE AND BULK ATOMS OF
TITANIUM, CHROMIUM, AND NICKEL. (AUGER
EFFECT)
HOUSTON JE + PARK RL + LARAMORE GE
PHYS REV LETT 30(18): 846-9 (30 APR 1973)
        SAA 1973: 39210

2-0393 K X-RAY AND AUGER ELECTRON ENERGIES FOR NEON
BY A TRANSITION OPERATOR METHOD.
HOWAT G + GOSCINSKI O + ABERG T
PHYS FENN 9, SUPPL S1: 241-3 (1974)
        CA 82: 147740

2-0394   BAND-TO-BAND AUGER RECOMBINATION IN INDIRECT
         GAP SEMICONDUCTORS.
         HULDT L
         PHYS STATUS SOLIDI A 8(1): 173-87 (1972)
              SAA 1972: 7089

2-0395   AUGER RECOMBINATION IN GERMANIUM.
         HULDT L
         PHYS STATUS SOLIDI A 24(1): 221-9 (1974)
              CA 81: 083254

2-0396   AUGER RECOMBINATION AND IMPACT IONIZATION IN
         INDIRECT GAP SEMICONDUCTORS.
         HULDT L
         P1097-104 OF INT CONF PHYS SEMICOND 1972,
         11TH PROC, VOL-2, AMER ELSEVIER, 1972.
              CA 81: 179630

2-0397   PHONON ASSISTED AUGER RECOMBINATION IN
         GERMANIUM.
         HULDT L
         PHYS STATUS SCLIDI A 33(2): 607-14 (1976)
              CA 84: 157638

2-0398   MATRIX ELEMENTS DEPENDENCE OF OPTICAL
         EXCITATION AND AUGER DECAY OF 5D CORE HOLES
         IN BISMUTH(III) TELLURIDE.
         HURYCH Z + SHAFFER JC + DAVIS DL + KNECHT TA
         + LAPEYRE GJ + GOBBY PL + KNAPP JA + OLSON C
         PHYS REV LETT 33(14): 830-3 (1974)
              CA 81: 161618

2-0399   MODEL FOR THE AUGER ELECTRON SPECTROSCOPY OF
         SYSTEMS EXHIBITING LAYER GROWTH, AND ITS
         APPLICATION TO THE DEPOSITION OF SILVER ON
         NICKEL.
         JACKSON DC + GALLON TE + CHAMBERS A
         SURFACE SCI 36(2): 381-94 (1973)
              CA 78: 128759

2-0400   OXYGEN BEAM GAS AUGER ELECTRON EMISSION.
         JOHNSON BM
         UNIV TEXAS PHD THESIS, 1975, 110P.
              CA 85: 26999

2-0401   DETERMINATION OF THE ESCAPE LENGTH OF LOW
         ENERGY ELECTRCNS BY AUGER AND LOSS
         SPECTROSCOPY.
         KANAJI T + KAGOTANI T + NAGATA S
         THIN SOLID FILMS 32(2): 362 (1976)
              SAA 1976: 45932

2-0402   CONCENTRATION DEPENDENCE OF GREEN- COPPER
         LUMINESCENCE IN COPPER AND ALUMINUM DOPED
         ZINC SULFIDE. (AUGER EFFECT, COLOR CENTERS)
         KAWAI H + KUBONIWA S + HOSHINA T
         JAP J APPL PHYS 13(10): 1593-603 (OCT 1974)
              SAA 1975: 01637

2-0403   EVIDENCE OF RADIATIVE AUGER TRANSITIONS IN
         THE FREE ARGON ATOM.
         KESKI-RAHKONEN O + UTRIAINEN J
         COMMENT PHYS-MATH, SOC SCI FENN 42(4): 297-8
         (1972); J PHYS B 7(1): 55-8 (1974)
              CA 78: 117086; 80: 065223

2-0404   EVIDENCE OF RADIATIVE AUGER TRANSITIONS IN
         THE FREE ARGON ATOM.
         KESKI-RAHKONEN O + UTRIAINEN J
         J PHYS B 7(1): 55-8 (1974)

2-0405   PRODUCTION OF INNER SHELL VACANCIES IN HEAVY
         ION- ATOM COLLISIONS. (AUGER EFFECT)
         KESSEL QC + FASTRUP B
         CASE STUD AT PHYS 3(3): 137-213 (FEB 1973)
              SAA 1973: 46192

2-0406   RECOMBINATION MECHANISMS IN 8-14 MICRON
         MERCURY(1-X) CADMIUM(X) TELLURIDE.
         (AUGER-LIMITED)
         KINCH MA + BRAU MJ + SIMMONS A
         J APPL PHYS 44(4): 1649-63 (APR 1973)
              SAA 1973: 36634

2-0407   AUGER IONIZATION OF ATOMS BY MULTIPLY CHARGED
         IONS. MODEL OF TWO COULCMB CENTERS.
         KISHINEVSKII LM + PARILIS ES
         P87-93 OF SECCNDARY EMISSION AND STRUCTURAL
         PROPERTIES OF SOLIDS, ARIFOV US (ED), NY,
         CONSULTANTS BUREAU, 1971.

2-0408   ELECTRON ESCAPE DEPTH IN SILICON. (AUGER
         ENERGIES)
         KLASSON M + BERNDTSSON A + HEDMAN J
         + NILSSON R + NYHOLM R + NORDLING C
         J ELECTRON SPECTROSC RELAT PHENOMENA 3(6):
         427-34 (JUN 1974)
              SAA 1974: 67688

2-0409   ESCAPE DEPTHS OF X-RAY EXCITED (AUGER)
         ELECTRONS.
         KLASSON M + HEDMAN J + BERNDTSSON A
         + NILSSON R + NORDLING C + MELNIK PV
         PHYS SCR 5(1-2): 93-5 (1972)
              CA 77: 68339N

2-0410   EXOEMISSION PROPERTIES OF ZIRCONIUM DIOXIDE.
         (AUGER EFFECT)
         KORTOV VS + SHIFRIN VP
         PHYS STATUS SOLIDI A 25(2): 377-85 (16 OCT
         1974)
              SAA 1975: 16036

2-0411   BACKGROUND INFORMATION IN AUGER ELECTRON
         SPECTROSCOPY AND SECONDARY ELECTRON EMISSION
         DUE TO CASCADE PROCESS.
         KOSHIKAWA T + SHIMIZU R
         P637-40 OF INT VACUUM CONGRESS, 6TH PROC, INT
         CONF ON SOLID SURFACES, 2ND, 1974, KYOTO.
         (JAP J APPL PHYS, SUPPL 2, PART 1, 1974)
              CA 83: 20275

2-0412   RELATIVE EFFECT OF EXTRAATOMIC RELAXATION ON
         AUGER AND BINDING ENERGY SHIFTS IN TRANSITION
         METALS AND SALTS.
         KOWALCZYK SP + LEY L + MCFEELY FR
         + POLLAK RA + SHIRLEY DA
         PHYS REV B 9(2): 381-91 (1974)
              CA 81: 043768

2-0413   REARRANGEMENT OF INNER SHELL IONIZED ATOMS.
         (AUGER SPECTRA)
         KRAUSE MO
         J PHYS (PARIS) COLLOQ NO 4: 67-75 (1971)
              CA 77: 26856Y; 81894J

2-0414   ARGON KLL AUGER SPECTRUM. TEST OF THEORY.
         KRAUSE MO
         PHYS REV LETT 34(11): 633-5 (1975)
              CA 83: 35368

2-0415   AUGER RECOMBINATION IN SILICON RECTIFIERS AND
         THYRISTORS LOADED IN FORWARD DIRECTION IN THE
         MIDDLE REGION.
         KRAUSSE J
         SOLID STATE ELECTRON 17(5): 427-9 (1974) (IN
         GERMAN)
              CA 81: 018319

2-0416   STIMULATED PHOTOLUMINESCENCE IN GALLIUM
         SELENIDE. (AUGER PROCESS)
         KURODA N + NISHINA Y
         J LUMIN 12-13: 623-8 (1976)
              CA 84: 187064

2-0417   DETAILED BALANCE BETWEEN AUGER RECOMBINATION
         AND IMPACT IONIZATION IN SEMICONDUCTORS.
         LANDSBERG PT
         PROC ROY SOC A 331(1584): 103-8 (1972)
              CA 77: 170229P

2-0418   IMPACT IONIZATION AND AUGER RECOMBINATION IN
         BANDS.
         LANDSBERG PT
         SOLID STATE COMMUN 10: 479-82 (5 MAR 1972)

2-0419   RADIATIVE AND AUGER PROCESSES IN
         SEMICONDUCTORS.
         LANDSBERG PT + ADAMS MJ
         J LUMIN 7: 3-34 (1973)
              CA 79: 084702

2-0420   THEORY OF DONOR- ACCEPTOR RADIATIVE AND AUGER
         RECOMBINATION IN SIMPLE SEMICONDUCTORS.
         LANDSBERG PT + ADAMS MJ
         PROC ROY SOC A 334(1599): 523-39 (1973)
              CA 79: 109526

2-0421 RECOMBINATION-INDUCED NONEQUILIBRIUM PHASE
TRANSITIONS IN SEMICONDUCTORS. (AUGER
RECOMBINATION)
LANDSBERG PT + PIMPALE A
J PHYS C 9(7): 1243-52 (1976)
SAA 1976: 45855

2-0422 PHOTOEMISSION STUDIES OF CORE EXCITON DECAY
IN POTASSIUM IODIDE.
LAPEYRE GJ + BAER AD + HERMANSON J
+ ANDERSON J + KNAPP JA + GOBBY PL
SOLID STATE COMMUN 15(10): 1601-5 (1974)
CA 82: 024101

2-0423 FAST ELECTRON- SLOW ELECTRON COINCIDENCE
STUDY OF VIOLENT ARGON(+)- ARGON COLLISIONS.
COMMENTS. (AUGER ELECTRON ENERGY)
LARKINS FP
PHYS REV A 5: 2288-9 (1972)
SAA 1972: 35130

2-0424 SEMI-AUGER ELECTRON TRANSITIONS.
LAVILLA RE
P509-22 OF INT CONF INNER SHELL IONIZATION
PHENOMENA AND FUTURE APPLICATIONS, VOL 1,
1972, FINK RW(ED), NTIS, 1973.
CA 78: 130140

2-0425 QUANTITATIVE ASPECTS OF A SOFT X-RAY LASER.
(AUGER EFFECT)
LAX B + GUENTHER AH
APPL PHYS LETT 21(8): 361-3 (1972)
CA 77: 171035J

2-0426 STRUCTURAL AND ELECTRONIC MODEL OF NEGATIVE
ELECTRON AFFINITY ON THE SILICON- CESIUM-
OXYGEN SURFACE. (AUGER EFFECT)
LEVINE JD
SURFACE SCI 34(1): 90-107 (JAN 1973)
SAA 1973: 24339

2-0427 MANY BODY EFFECTS IN X-RAY PHOTOEMISSION FROM
MAGNESIUM. (AUGER EFFECT)
LEY L + MCFEELY FR + KOWALCZYK SP
+ JENKIN JG + SHIRLEY DA
PHYS REV B 11(2): 600-12 (1975)
CA 82: 148073

2-0428 PROBING DEPTH IN PHOTOEMISSION AND AUGER
ELECTRON SPECTROSCOPY.
LINDAU I + SPICER WE
J ELECTRON SPECTROSC RELAT PHENOMENA 3(5):
409-13 (1974)
CA 81: 007784

2-0429 A SIMPLE ANALYTICAL CALCULATION FOR X-RAY
PHOTOEMISSION. (AUGER PROCESS)
LOPEZ O
J APPL PHYS 46(6): 2504-8 (1975)
SAA 1975: 66518

2-0430 AUGER EMISSION AND COLOR CENTERS IN LITHIUM
FLUORIDE.
LORD DG + GALION TE
SURFACE SCI 36(2): 606-21 (1973)
CA 78: 130197

2-0431 SCINTILLATION EFFICIENCY FOR LOW VELOCITY
HEAVY IONS. (AUGER EMISSION)
LUNTZ M + FIGUEROA J
PHYS REV B 9(1): 87-94 (1 JAN 1974)
SAA 1974: 49412

2-0432 PHOTOCONDUCTIVITY OF HIGHLY EXCITED CADMIUM
SELENIDE AND CADMIUM SULFIDE(0.48)
SELENIDE(0.52) CRYSTALS. (AUGER
RECOMBINATION)
MARINOVA K + OPANOWICZ A + RUPPEL W
PHYS STATUS SOLIDI A 33(1): 195-204 (1976)
SAA 1976: 35865

2-0433 INELASTIC SCATTERING EFFECTS IN QUANTITATIVE
AUGER ELECTRON SPECTROSCOPY.
MARTIN AD
J PHYS D 8(17): 2074-8 (1975)
CA 84: 10622

2-0434 A TEMPERATURE DEPENDENT CONTRIBUTION TO AUGER
ELECTRON ENERGY DISTRIBUTIONS.
MATTHEW JAD
SURFACE SCI 20: 183-6 (1970)

2-0435 EFFECT OF POLARIZATION OF THE ELECTRON GAS ON
THE ENERGIES OF AUGER TRANSITION IN METALS.
MATTHEW JAD
SURFACE SCI 40(2): 451-5 (1973)
CA 80: 008532

2-0436 DISTINCTION BETWEEN WEAK AUGER TRANSITIONS
AND INTERNAL X-RAY PHOTOEMISSION.
MATTHEW JAD
SOLID STATE COMMUN 13(8): 1203-5 (1973)
CA 80: 042010

2-0437 POSSIBLE LOCALISATION OF CONDUCTION ELECTRONS
IN SOFT X-RAY DOUBLE IONIZATION SATELLITE
(AUGER) TRANSITION.
MATTHEW JAD
PHYS LETT A 50(6): 401-2 (13 JAN 1975)
SAA 1975: 25612

2-0438 TRANSITION RATES FOR INTERATOMIC AUGER
PROCESSES.
MATTHEW JAD + KOMNINOS Y
SURFACE SCI 53(1): 716-25 (1975)
CA 84: 114051

2-0439 ENHANCEMENT OF COSTER-KRONIG RATES IN SOLID
STATE ENVIRONMENTS. (XENON, CADMIUM, ZINC)
MATTHEW JAD + NUTTALL JD + GALLON TE
J PHYS C 9(5): 883-8 (1976)
SAA 1976: 41005

2-0440 MECHANISM OF PLASMON GAIN IN AUGER SPECTRA.
MATTHEW JAD + WATTS CMK
PHYS LETT A 37: 239-40 (22 NOV 1971)

2-0441 SEQUENTIAL STRIPPING IN HIGHLY IONIZED NEON.
(AUGER TRANSITIONS)
MATTHEWS DL + FORTNER RJ + CHEN MH
+ CRASEMANN B
P941-2 OF INT CONF ON PHYS ELECTRON ATOMIC
COLLISIONS, (EXTENDED ABSTRACTS) JULY 1975,
RISLEY JS + GEBALLE R(EDS), WASHINGTON UNIV
PRESS, SEATTLE.
SAA 1976: 25377

2-0442 INTERATOMIC AUGER EFFECT AT HIGH VELOCITIES.
MATVEEV VI
PISMA ZH TEKH FIZ 2(3): 101-4 (1976) (IN
RUSSIAN)
SAA 1976: 39381

2-0443 SOFT X-RAY AMPLIFIED SPONTANEOUS EMISSION VIA
THE AUGER EFFECT.
MCGUIRE EJ
PHYS REV LETT 35(13): 844-8 (1975)
CA 83: 170679

2-0444 AUGER TRANSITION RATES IN J-J COUPLING FOR D
HOLES AND F ELECTRONS.
MCGUIRE EJ
NUCL PHYS A 172: 127-38 (1971)

2-0445 MULTIPLET EFFECTS ON THE WIDTHS OF
PHOTOELECTRON PEAKS. (AUGER EFFECT)
MCGUIRE GE
PHYS REV A 10(1): 32-50 (JUL 1974)
SAA 1974: 74862

2-0446 CHARGE DEPENDENCE OF DEPENDENCE OF TOTAL
K-SHELL VACANCY PRODUCTION CROSS SECTIONS IN
ARGON BY THE IMPACT OF HEAVY IONS. (AUGER
EFFECT)
MCGUIRE JH
PHYS REV A 8(5): 2760-2 (NOV 1973)
SAA 1974: 24015

2-0447 PLASMON SATELLITES IN AUGER SPECTRA.
MCMULLEN T + BERGERSEN B
CAN J PHYS 52(7): 624-38 (1 APR 1974); PHYS
FENN 9, SUPPL S1: 370-2 (JUL 1974)
SAA 1974: 53556; 1975: 19050

2-0448   INELASTIC INTERACTIONS OF SLOW ELECTRONS WITH ABSORBED PARTICLES. (AUGER SPECTROSCOPY)
MENZEL D
ANGEW CHEM INTERNAT ED 9: 255-66 (1970)

2-0449   EFFECT OF THE ELECTRON EJECTION DEPTH ON QUANTITATIVE ANALYSIS BY AUGER SPECTROSCOPY.
MITYAGIN AYU + PANTELEEV VV + CHEREVATSKII NYA
RADIOTEKH ELEKTRON 21(3): 589-93 (1976) (IN RUSSIAN)
CA 85: 27149

2-0450   ELECTRONIC SUPERPOSITION OF SAMPLE CURRENT AND SECONDARY ELECTRON IMAGES IN AUGER ELECTRON SPECTROSCOPY.
MORABITO JM + MUNRO DF
APPL PHYS LETT 21(12): 572-5 (1972)
CA 78: 64568J

2-0451   LUMINESCENCE DUE TO EXCITON-EXCITON COLLISION IN GALLIUM ARSENIDE.
MORIYA T + KUSHIDA T
SOLID STATE COMMUN 12(6): 495-8 (1973)
CA 78: 153266

2-0452   AUGER ELECTRON SPECTROSCOPY STUDY OF ELECTRON IMPACT DESORPTION.
MUSKET RG + FERRANTE J
SURFACE SCI 21: 440-2 (1970)

2-0453   PHOTOCONDUCTIVITY ASSOCIATED WITH AUGER RECOMBINATION OF EXCITONS BOUND TO NEUTRAL DONORS IN TELLURIUM DOPED GALLIUM PHOSPHIDE.
NAKAMURA A + MORIGAKI K
J PHYS SOC JAP 34(3): 672-6 (1973)
CA 78: 116738

2-0454   EXCITON LUMINESCENCE AND PHOTOCONDUCTIVITY IN HIGHLY EXCITED GALLIUM PHOSPHIDE.
NAKAMURA A + MORIGAKI K
P144-8 OF INT CONF LUMIN CRYST MOL SOLUTIONS, 1972 PROC, WILLIAMS F (ED), PLENUM, 1973.
CA 79: 047438

2-0455   DEPENDENCE OF AUGER ELECTRON YIELD ON PRIMARY BEAM ENERGY AT NORMAL AND GLANCING INCIDENCE.
NEAVE JH + FOKON CT + JOYCE BA
SURFACE SCI 29(2): 411-23 (1972)

2-0456   DETERMINATION OF AUGER ELECTRON EXCITATION CROSS SECTIONS AND RANGES USING PROTON EXCITED X-RAYS.
NEEDHAM PB + DRISCOLL TJ
J VACUUM SCI TECHNOL 11(1): 278 (JAN-FEB 1974)
SAA 1974: 73224

2-0457   AUGER THEORY AT DEFECTS. APPLICATION TO STATES WITH TWO BOUND PARTICLES IN GALLIUM PHOSPHIDE.
NEUMARK GF
PHYS REV B 7(8): 3802-10 (1973)
CA 79: 024755

2-0458   BAND-TO-BAND AUGER RECOMBINATION AND CARRIER-CARRIER SCATTERING IN POWER RECTIFIERS.
NILSSON NG
ELECTRON LETT 8(23): 580-2 (1972)
CA 78: 35364J

2-0459   BAND-TO-BAND AUGER RECOMBINATION IN SILICON AND GERMANIUM.
NILSSON NG
PHYS SCR 8(4): 165-76 (1973)
CA 79: 151232

2-0460   SPECTRUM AND DECAY OF THE RECOMBINATION RADIATION FROM STRONGLY EXCITED SILICON. (AUGER RECOMBINATION PROCESS)
NILSSON NG + SVANTESSON KG
SOLID STATE COMMUN 11(1): 155-9 (1972)
CA 77: 95212X

2-0461   TRANSIENT CARRIER DECAY AND TRANSPORT PROPERTIES IN MERCURY CADMIUM TELLURIDE. (AUGER RECOMBINATION)
NIMTZ G + BAUER G + DORNHAUS R + MUELLER KH
PHYS REV B 10(8): 3302-10 (1974)
CA 82: 50321

2-0462   STIMULATED EMISSION OF LIGHT FROM A NONIDEAL EXCITON GAS IN A SEMICONDUCTOR. (AUGER PROCESS)
NOLLE EL
SOV PHYS SEMICOND 8(8): 954-8 (1975)
SAA 1975: 42744

2-0463   DEPENDENCE OF ATOMIC PHOTOEFFECT ON A BOUND ELECTRON MAGNETIC SUBSTATE. (AUGER ELECTRON DISTRIBUTION)
OH SD + PRATT RH
PHYS REV A 10(4): 1198-203 (1974)
CA 82: 009605

2-0464   ELECTRON SPRAYING ATOMS. (AUGER CASCADES)
ORE A
FRA FYS VERDEN 35(2): 30-3 (1973) (IN NORWEGIAN)
CA 80: 054036

2-0465   AUGER DISSOCIATION OF MOLECULES AS A RESULT OF INNER SHELL IONIZATION.
OXENGENDLER BL + PARILIS ES
P1773-4 OF INT CONF INNER SHELL IONIZATION PHENOMENA AND FUTURE APPLICATIONS, VOL 3, 1972, FINK RW (ED), NTIS, 1973.

2-0466   IMPURITY IONIZATION IN N-TYPE GERMANIUM. (AUGER RECOMBINATION PARAMETER)
PALMIER JF
PHYS REV B 6(12): 4557-71 (15 DEC 1972)
SAA 1973: 08192

2-0467   IDENTIFICATION OF AUGER ELECTRONS IN GALLIUM ARSENIDE.
PANKOV JI + TOMASETTA L + WILLIAMS BF
PHYS REV LETT 27: 29-32 (5 JUL 1971)

2-0468   ELECTRONIC STRUCTURE OF SOLID SURFACES: CORE LEVEL EXCITATION TECHNIQUES. (AUGER EFFECT)
PARK RL + HOUSTON JE
J VACUUM SCI TECHNOL 10(1): 176-82 (JAN-FEB 1973)
SAA 1973: 38768

2-0469   THE AUGER EFFECT.
PARILIS ES
P129-63 OF ISSLED VO FAN UZBEKH SSR, TASHKENT, 1969. (IN RUSSIAN)

2-0470   CARRIER TEMPERATURE EFFECTS AND ENERGY TRANSFERS IN NONRADIATIVE (AUGER) RECOMBINATION.
PARROTT JE
SOLID STATE ELECTRON 19(3): 229-33 (1976)
SAA 1976: 26500

2-0471   STATIC RELAXATION OF GERMANE AND THE ESTIMATION OF RELAXATION ENERGY DIFFERENCES FROM AUGER AND CORE BINDING ENERGIES OF GERMANIUM COMPOUNDS.
PERRY WB + JOLLY WL
CHEM PHYS LETT 23(4): 529-32 (1973)
CA 80: 089309

2-0472   ELECTRON DECAY-IN-FLIGHT SPECTRA FROM AUTOIONIZING STATES OF HIGHLY STRIPPED OXYGEN, FLUORINE, CHLORINE, AND ARGON IONS. (AUGER EFFECT)
PESS DJ + SELLIN IA + PETERSON R + MOWAT JR + SMITH WW + BROWN MD + MACDONALD JR
PHYS REV A 8(3): 1350-64 (SEP 1973)
SAA 1974: 09410

2-0473   NEW FINE STRUCTURE IN ELECTRON EXCITED AUGER SPECTRA FROM SOLID SURFACES.
PESSA M + AKSELA S + KARRAS M
PHYS LETT A 31: 382-3 (1970)

2-0474 COINCIDENCE METHODS OF STANDARDIZATION FOR
CESIUM-131 AND MEASUREMENT OF DECAY
PARAMETERS. (DETECTION CF L AUGER ELECTRONS)
PICH J + ZDERADICKA J + KOKTA L
INT J APPL RADIAT ISOTOPES 25(10): 433-44
(1974)
CA 82: 65473

2-0475 CROSS SECTICNS FOR IONIZATION OF INNER SHELL
ELECTRONS BY ELECTRON IMPACT. (AUGER
SPECTROSCOPY)
POWELL CJ
J VACUUM SCI TECHNOL 13(1): 219-20 (1976)
SAA 1976: 48467

2-0476 RADIATIVE AUGER EFFECT IN HEAVY ATOMS.
PRESSER G
PHYS LETT A 56(4): 273-4 (1976)
CA 85: 12135

2-0477 DIFFERENTIAL CROSS SECTION FOR K-SHELL
IONIZATION OF COPPER AND SILVER BY ELECTRON
BOMBARDMENT. (AUGER ELECTRONS)
QUARLES CA + FAULK JD
PHYS REV LETT 31(14): 859-62 (1 OCT 1973)
SAA 1973: 71411

2-0478 PENETRATION EFFECTS IN CONVERTED MUONIC
TRANSITIONS. (AUGER COEFFICIENTS)
RAFF U + VIOLLIER RD + ALDER K
NUCL PHYS A 223(3): 429-41 (1974)
CA 81: 008497

2-0479 X-RAY PRODUCTION BY HEAVY IONS. (OXYGEN
BOMBARDMENT, AUGER SPECTRUM)
RICHARD P
P1641-65 OF INT CONF INNER SHELL IONIZATION
PHENOMENA AND FUTURE APPLICATIONS, VCL 3,
1972, FINK RW(ED), NTIS, 1973.
CA 78: 130131

2-0480 ION- ATOM COLLISIONS AT HIGH ENERGIES. (AUGER
EFFECT)
RICHARD P
PHYS FENN 9, SUPPL S1: 3-10 (JUL 1974)
SAA 1975: 17812

2-0481 RADIATIVE AUGER EFFECT IN ION- ATOM
COLLISIONS.
RICHARD P + OLTJEN J + JAMISON KA
+ KAUFFMAN RL + WOODS CW + HALL JM
PHYS LETT A 54(2): 169-70 (1975)
CA 83: 185746

2-0482 MEASUREMENT OF THE PRODUCTION OF LL VACANCY
STATES BY THE L X-RAY- L X-RAY COINCIDENCE
METHOD. (AUGER TRANSITIONS)
RISLEY JS + GEBALLE R
P945 OF INT CONF ON PHYS ELECTRON ATOMIC
COLLISIONS, (EXTENDED ABSTRACTS) JULY 1975,
RISLEY JS + GEBALLE R(EDS), WASHINGTON UNIV
PRESS, SEATTLE.
SAA 1976: 25378

2-0483 BAND-TO-BAND AUGER PROCESSES AT HIGH CARRIER
CONCENTRATION.
ROSENTHAL W
SOLID STATE COMMUN 13(8): 1215-18 (1973)
CA 80: 031686

2-0484 PHONON ASSISTED AUGER PROCESSES IN GROUP
III-V COMPOUNDS.
ROSENTHAL W
Z NATURFORSCH A28(8): 1233-46 (1973) (IN
GERMAN)
CA 80: 031709

2-0485 RELATIVISTIC CALCULATIONS OF ATOMIC X-RAY AND
AUGER TRANSITION RATES.
ROSNER HR
KANSAS STATE UNIV PHD THESIS, 1970, 228P.
CA 76: 52319Z

2-0486 THEORETICAL STUDY OF THE AUGER EFFECT IN THE
LIGHT TO MEDIUM RANGE OF ATOMIC NUMBER.
RUBENSTEIN RA
UNIV OF ILLINCIS PHD THESIS, 1955.
DISSERTATION ABST 15: 851 (1955)

2-0487 MECHANISMS OF INNER SHELL EXCITATION AND
DEEXCITATION IN MULTIPLY IONIZED ATOMS.
RUDD ME
P1485-508 OF INT CONF INNER SHELL IONIZATION
PHENOMENA AND FUTURE APPLICATIONS, VOL 3,
1972, FINK RW(ED), NTIS, 1973.
CA 78: 129965

2-0488 MECHANISMS OF ELECTRON PRODUCTION IN ION-
ATOM COLLISIONS. (AUGER EFFECT)
RUDD ME
RADIAT RES 64(1): 153-80 (1975)
SAA 1976: 44242

2-0489 PRODUCTION AND DECAY OF DOUBLE L VACANCIES IN
ARGON AND PHOSPHORUS.
RUDD ME + FASTRUP B + DAHL P
+ SCHOWENGERDT FD
PHYS REV A 8(1): 220-5 (1973)
CA 79: 085122

2-0490 MECHANISMS OF ELECTRON PRODUCTION IN ION-
ATOM COLLISIONS. (AUGER EFFECT, REVIEWS)
RUDD ME + MACEK JH
CASE STUD AT PHYS 3(2): 47-136 (NOV 1972)
SAA 1973: 26601

2-0491 INTERATOMIC AUGER PROCESSES AND THE DENSITY
OF STATES.
SALMERON M + BARO AM + ROJO JM
SURFACE SCI 53(1): 689-97 (1975)
CA 84: 82036

2-0492 MULTIIONIZATION OF NEON, ARGON, XENON AND
THEIR IONS BY HIGH ENERGY ELECTRON IMPACT.
(AUGER EMISSION)
SALOP A
PHYS REV A 9(6): 2496-504 (JUN 1974)
SAA 1974: 72087

2-0493 MULTIIONIZATION OF KRYPTON AND ITS IONS BY
HIGH ENERGY ELECTRON IMPACT. (AUGER EFFECT)
SALOP A
PHYS REV A 8(6): 3032-44 (DEC 1973)
SAA 1974: 31738

2-0494 RADIATIVE AUGER EFFECT WITH AND WITHOUT
ELECTRON EMISSION.
SAWADA M + TANIGUCHI K + NAKAMURA H
J PHYS SOC JAP 33(5): 1496 (1972)
CA 78: 64711A

2-0495 INELASTIC SCATTERING OF LOW ENERGY ELECTRONS
FROM SURFACES.
SCHEIBNER EJ + THARP LN
SURFACE SCI 8: 247-65 (1967)

2-0496 HIGH RESOLUTION ARGON K-AUGER SPECTRUM
PRODUCED IN 4 MEV PROTON ON ARGON COLLISIONS.
SCHNEIDER D + ROBERTS K + HODGE W + MOORE CF
PHYS REV LETT 36(17): 1065-7 (1976)
CA 85: 11930

2-0497 RELATIVE SATELLITE LINE INTENSITIES IN ARGON
K AUGER ELECTRON SPECTRA PRODUCED BY 0.4 TO 4
MEV PROTON IMPACT.
SCHNEIDER D + SMITH LE + ROBERTS K + HODGE W
+ WHITENTON J + MOORE CF
PHYS LETT A 56(3): 189-91 (1976)
CA 84: 157651

2-0498 THEORY OF THE AUGER EFFECT IN III-V
SEMICONDUCTORS.
SCHRENE D
Z NATURFORSCH A24: 1752-9 (1969)

2-0499 INVESTIGATION OF THE CHARACTERISTIC AUGER
ENERGY LOSSES OF PLATINUM (111).
SCHRODER W + PETERS E + HOLZI J
APPL PHYS 3(2): 135-40 (FEB 1974)
SAA 1974: 21117

2-0500 SLOW ELECTRON SCATTERING FROM METALS. PART-1:
EMISSION OF TRUE SECONDARY ELECTRONS. PART-2:
INELASTICALLY SCATTERED PRIMARY ELECTRONS.
SEAH MP
SURFACE SCI 17: 132-60 (1969); 161-80 (1969)

2-0501 MEASUREMENT OF MOMENTUM ACCOMMODATION
COEFFICIENTS CN SURFACES CHARACTERIZED BY
AUGER SPECTROSCOPY, SECONDARY ION MASS
SPECTROSCOPY AND LOW ENERGY ELECTRON
DIFFRACTION.
SEIDL M + STEINHEIL E
P E9/1-12 OF INT SYMP ON RAREFIED GAS
DYNAMICS, 9TH PROC, US OFF NAVAL RES, 1974.
SAA 1975: 16039

2-0502 THE K-FAR-M2 RADIATIVE AUGER EFFECT IN
TRANSITION METALS.
SERVOMAA A + KESKI-RAHKONEN O
J PHYS C 8(23): 4124-30 (1975)
CA 84: 51797

2-0503 OPTICAL GAIN IN LIGHTLY DOPED GALLIUM
ARSENIDE. (AUGER RECOMBINATION)
SHAKLESS KL + LEHENY RF + NAHORY RE
APPL PHYS LETT 19(8): 302-5 (1971)
CA 76: 8730C

2-0504 ROLE OF THE AUGER EFFECT IN THE FORMATION OF
AN ELECTRON SHOWER SPECTRUM.
SHEVTSOVA IN
IZV VYSSH UCHEB ZAVED, FIZ 16(4): 122-3
(1973) (IN RUSSIAN)
CA 79: 036700

2-0505 RELAXATION EFFECTS ON AUGER ENERGIES.
SHIRLEY DA
CHEM PHYS LETT 17(3): 312-5 (1972)
CA 78: 50043X

2-0506 THEORY OF KLL AUGER ENERGIES INCLUDING STATIC
RELAXATION.
SHIRLEY DA
PHYS REV A 7(5): 1520-8 (1973)
CA 78: 166277

2-0507 THEORY OF AUGER SATELLITE ENERGY SHIFTS.
SHIRLEY DA
PHYS REV A 9(4): 1549-54 (1974)
CA 80: 150587

2-0508 SECONDARY EMISSION AND ELASTIC REFLECTION OF
ELECTRONS FROM MONOCRYSTALS.
SHULMAN AR + KORABLEV VV + MOROZOV YA
BULL ACAD SCI USSR, PHYS SER 35(2): 199-205
(1971)

2-0509 SELF-CONSISTENT ANALYSIS OF CORE BAND-TO-BAND
EXCITATION: AUGER FINE STRUCTURE.
SICKAFUS EN
SURFACE SCI 36(2): 472-7 (1973)
CA 78: 130230

2-0510 TRANSITION POTENTIAL MODELS FOR AUGER
ELECTRON SPECTRA.
SIEGBAHN H + GOSCINSKI O
PHYS SCR 13(4): 225-8 (1976)
CA 85: 39104

2-0511 SURFACE MADELUNG POTENTIALS IN ELECTRON
SPECTROSCOPY.
SLATER RR
SURFACE SCI 23: 403-8 (1970)

2-0512 SPREADING FUNCTIONS IN SECONDARY ELECTRON
ANALYSIS.
SLOCOMB CA + ELLIS WP
J APPL PHYS 45(8): 3332-6 (1974)
CA 81: 129545

2-0513 AUGER EMISSION FROM SOLIDS. ESTIMATION OF
BACKSCATTERING EFFECTS AND IONIZATICN CROSS
SECTIONS.
SMITH DM + GALLON TE
J PHYS D 7(1): 151-61 (1974)
CA 80: 076073

2-0514 EMISSION AND RECOMBINATION INVOLVING INNER
SHELLS. (AUGER PROCESS, AES, PHOTOELECTRON
SPECTROSCOPY, CHEMICAL SHIFT)
SOMORJAI GA
P175-84 OF PRINCIPLES OF SURFACE CHEMISTRY,
SOMORJAI GA(ED), PRENTICE-HALL, 1972.

2-0515 ELECTRON SPECTROSCOPIC INVESTIGATION OF AUGER
PROCESSES IN BROMINE-SUBSTITUTED METHANES AND
SOME HYDROCARBONS.
SPOHR R + BERGMARK T + MAGNUSSON N
+ WERME LO + NORDLING C + SIEGBAHN K
PHYS SCR 2: 31-7 (1970)

2-0516 MOSSBAUER STUDIES OF THE AFTEREFFECTS OF
AUGER IONIZATION FOLLOWING ELECTRON CAPTURE
IN TRIS (1,10-PHENANTHROLINE) COBALT(III)-57
PERCHLORATE DIHYDRATE.
SRIVASTAVA TS + NATH A
RADIOCHEM RADIOANAL LETT 16(3): 103-7 (1974)
CA 80: 102011

2-0517 CHARGE STATE DEPENDENCE OF THE ARGON L-SHELL
FLUORESCENCE YIELD STUDIED BY ATOMIC
HYDROGEN(+), MOLECULAR HYDROGEN(+), ATOMIC
HELIUM(+) IMPACT.
STOLTERFOHT N + DE HEER FJ + VAN ECK J
PHYS REV LETT 30(23): 1159-62 (1973)
CA 79: 036827

2-0518 K-SHELL IONIZATION AND FLUORESCENCE YIELD IN
50 MEV CHLORINE(+)- NEON COLLISIONS. (AUGER
ELECTRON PRODUCTION)
STOLTERFOHT N + SCHNEIDER D + BURCH D
+ WIEMAN H + RISLEY JS
PHYS FENN 9, SUPPL S1: 44-5 (JUL 1974)
SAA 1975: 17822

2-0519 K-SHELL IONIZATION OF MOLECULAR NITROGEN AND
METHANE BY 50- TO 600 KEV HYDROGEN(+),
DEUTERIUM(+), MOLECULAR HYDROGEN(+), AND
HELIUM(+) ION IMPACT.
STOLTERFOHT N + SCHNEIDER D + HARRISON KG
PHYS REV A 8(5): 2363-71 (1973)
CA 80: 032183

2-0520 CHARGE STATE DEPENDENCE OF THE NEON K
FLUORESCENCE YIELD DEDUCED FROM HIGH
RESOLUTION EMISSION (AUGER) SPECTRA.
STOLTERFOHT N + SCHNEIDER D + RICHARD P
+ KAUFFMAN RL
PHYS REV LETT 33(24): 1418-22 (1974)
CA 82: 104979

2-0521 ARGON L-SHELL IONIZATION BY 50 TO 600 KEV
PROTON, DEUTERON, MOLECULAR HYDROGEN, AND
HELIUM(+) IMPACT.
STOLTERFOHT N + SCHNEIDER D + ZIEM P
PHYS REV A 10(1): 81-91 (1974)
CA 81: 070462

2-0522 MOLECULAR EFFECTS ON K-VAANCY SHARING IN
ASYMMETRIC ION- ATOM COLLISIONS. (AUGER
SPECTRA)
STOLTERFOHT N + SCHNEIDER D + RIDDER D
P1060-1 OF INT CONF ON PHYS ELECTRON ATOMIC
COLLISIONS, (EXTENDED ABSTRACT), JULY 1975,
RISLEY JS + GEBALLE R(EDS), WASHINGTON UNIV
PRESS, SEATTLE.
SAA 1976: 25487

2-0523 GAS TARGET EXPERIMENTS OF K-SHELL VACANCY
SHARING IN ASYMMETRIC ION- ATOM COLLISIONS.
(AUGER EFFECT)
STOLTERFOHT N + ZIEM P + RIDDER D
J PHYS B 7(14): L409-13 (1 OCT 1974)
SAA 1974: 74910

2-0524 DEPENDENCE OF ATOMIC PHOTOEFFECT ON BOUND
ELECTRON MAGNETIC SUBSTATE (AUGER EFFECT).
SUNG DO + PRATT RH
PHYS FENN 9, SUPPL S1: 247-9 (JUL 1974)
SAA 1975: 28690

2-0525 ENERGY AND ANGULAR DISTRIBUTION OF ELECTRONS
RESULTING FROM DIRECT IONIZING COLLISICNS OF
ELECTRONS WITH ATOMS. (AUGER PROCESS)
TAHIRA S
MEM FAC SCI KYOTO UNIV SER PHYS ASTROPHYS
GECPHYS CHEM 34(1): 29-40 (DEC 1972)
SAA 1973: 57610

2-0526 APPLICATION OF THE ZETA EFFECTIVE VALUES OF
HYDROGENIC WAVE FUNCTIONS IN THE CALCULATION
OF THE KLL AUGER ENERGIES.
TANKHIWALE AV + KHARE PL + MANDE C
J PHYS B 4: 1250-4 (OCT 1971)

2-0527 ESCAPE LENGTH OF AUGER ELECTRONS.
TARNG ML + WEHNER GK
J APPL PHYS 44(4): 1534-40 (1973)
CA 78: 142041

2-0528 X-RAY EMISSION CROSS SECTIONS AND
FLUORESCENCE YIELDS FOR LIGHT ATOMS AND
MOLECULES BY ELECTRON IMPACT. (AUGER ELECTRON
EJECTION CROSS SECTIONS)
TAWARA H + HARRISON KG + DEHEER FJ
PHYSICA 63(2): 351-67 (15 JAN 1973)
SAA 1973: 17468

2-0529 ELECTRON ION COINCIDENCE STUDIES OF HIGHLY
IONIZING ARGON(+)- ARGON COLLISIONS. (AUGER
EFFECT)
THOMSON GM
P1168-9 OF INT CONF ON PHYS ELECTRON ATOMIC
COLLISIONS, (EXTENDED ABSTRACTS) JULY 1975,
RISLEY JS + GEBALLE R(EDS), WASHINGTON UNIV
PRESS, SEATTLE.
SAA 1976: 25499

2-0530 EXTENDED ION- ELECTRON COINCIDENCE
MEASUREMENTS OF THE VIOLENT ARGON(+)- ARGON
COLLISION.
THOMSON GM + SMITH WW + RUSSEK A
PHYS REV A 7(1): 168-74 (1973)
CA 78: 64679W

2-0531 BIOLOGICAL RADIATION EFFECT ON INCLUSION OF
MODERATELY HEAVY NUCLEI IN THE TISSUE: USE
OF SOFT X-RAYS. (AUGER ELECTRON IRRADIATION)
TISLJAR-LENTULIS G + FEINENDEGEN LE + BOND VP
STRAHLENTHERAPIE 145(6): 656-62 (1973) (IN
GERMAN)
CA 79: 088846

2-0532 K-SHELL IONIZATION OF CARBON BY FAST PROTONS.
(AUGER ELECTRON YIELDS OF METHANE, ETHANE,
ETHYLENE, ACETYLENE)
TOBUREN LH
PHYS REV A 5(6): 2482-7 (1972)
CA 77: 27023M

2-0533 K-SHELL IONIZATION BY FAST PROTONS.
TOBUREN LH
P979-92 OF INT CONF INNER SHELL IONIZATION
PHENOMENA AND FUTURE APPLICATIONS, VOL 2,
1972, FINK RW(ED), NTIS, 1973.
CA 78: 130160

2-0534 DISTRIBUTION IN ENERGY AND ANGLE OF ELECTRONS
EJECTED FROM XENON BY 0.3 TO 2.0 MEV PROTONS.
(AUGER PEAKS)
TOBUREN LH
PHYS REV A 9(6): 2505-17 (JUN 1974)
SAA 1974: 72097

2-0535 AUGER ELECTRON SPECTROSCOPY APPLIED TO INNER
SHELL IONIZATION BY FAST CHARGED PARTICLES.
TOBUREN LH
P120-31 OF CONF ON APPL SMALL ACCEL, 3RD
PROC, 1974, DUGGAN JL + MORGAN IL (EDS),
NTIS, CONF-741040-P1, 1975.
CA 85: 54175

2-0536 NONRADIATIVE AUGER RECOMBINATION OF ELECTRONS
ON DONOR- ACCEPTOR PAIRS.
TOLPYGO EI + TOLPYGO KB + SHEINKMAN MK
SOV PHYS SEMICOND 8(3): 326-8 (1974)
CA 81: 018091

2-0537 ELECTRON ESCAPE DEPTHS IN ALUMINUM. (AUGER
ELECTRONS)
TRACY JC
J VACUUM SCI TECHNOL 11(1): 280 (JAN-FEB
1974)
SAA 1974: 76345

2-0538 IONIZATION LOSSES IN (AUGER) SPECTRA OF
ELECTRON INELASTIC SCATTERING FROM RARE EARTH
METAL SURFACES.
TSVEIMAN EV + ZASHKVARA VV + KORSUNSKII MI
+ REDKIN VS + TAZHIBAEVA S
SOV PHYS SOLID STATE 14(5): 1333-5 (1972)
CA 77: 107478D

2-0539 ROLE OF AUGER EFFECT IN THERMOSTIMULATED
PHENOMENA IN IONIC CRYSTALS.
TUAN BT + VELICKY B + BOHOUN A
Z PHYS 251(4): 289-99 (1972)

2-0540 ELECTRON- HOLE INTERACTION EFFECTS IN THE
KL2,3L2,3 RADIATIVE AUGER SPECTRUM OF
INSULATORS. (MAGNESIUM OXIDE, MAGNESIUM
FLUORIDE)
UTRIAINEN J + ABERG T
PHYS LETT A 50(4): 263-4 (11 DEC 1974)
SAA 1975: 22430

2-0541 EXCITATION OF INNER SHELL ELECTRONS BY
PHOTONS AND FAST ELECTRONS.
VAN DER WIEL MJ
P140-53 OF INT CONF PHYS ELECTRON ATOMIC
COLLISIONS 1972, 7TH INV PAP, PROGR REP.
CA 77: 132554H

2-0542 ANGLE OF INCIDENCE EFFECTS IN AUGER ELECTRON
EMISSION FROM CLEAN MOLYBDENUM.
VANCE DW
BULL AMER PHYS SOC 13: 947 (1968)

2-0543 AUGER ELECTRON EMISSION FROM CLEAN AND CARBON
CONTAMINATED MOLYBDENUM BOMBARDED BY POSITIVE
IONS. PART-1. PART-2: EFFECT OF ANGLE OF
INCIDENCE. PART-3: EFFECT OF ELECTRONICALLY
EXCITED IONS.
VANCE DW
PHYS REV 164: 372-80 (1967); 169: 252-72
(1968)

2-0544 ATOMIC VACANCY DISTRIBUTIONS PRODUCED BY
INNER SHELL IONIZATION (AUGER AND RADIATIVE
TRANSITIONS).
VENTGOPALA RP + CHEN MH + CRASEMANN B
PHYS REV A 5: 997-1012 (MAR 1972)

2-0545 CHARGE EXCHANGE OF LOW ENERGY HELIUM IONS AND
ATOMS SCATTERED FROM A COPPER SINGLE CRYSTAL.
(AUGER NEUTRALIZATION)
VERHEY LK + POELSEMA B + BOERS AL
NUCL INSTRUM METH 132: 565-70 (1976)
SAA 1976: 54371

2-0546 CHARGE EXCHANGE OF LOW ENERGY HELIUM(+) IONS
(LESS THAN 10 KEV) SCATTERED FROM A (100)
FACE OF A COPPER SINGLE CRYSTAL. (AUGER
NEUTRALIZATION)
VERHEY LK + POELSEMA B + BOERS AL
RADIAT EFF 27(1-2): 47-52 (1975)
CA 85: 10592

2-0547 ARGON AUGER EMISSION BY ARGON(+) ION
BOMBARDMENT OF SOME METALS.
VIARIS DE LESEGNO P + HENNEQUIN JF
J PHYS (PARIS) 35(10): 759-71 (1974) (IN
FRENCH)
CA 81: 161663

2-0548 RELATIVE EMISSION INTENSITIES OF AUGER
ELECTRONS FROM LIGHT METALS (LITHIUM, SODIUM)
SUBJECTED TO ION BOMBARDMENT.
VIARIS DE LESEGNO P + JOYES P + HENNEQUIN JF
COMPT REND C 275: 93-6 (1972) (IN FRENCH)

2-0549 ORIGIN OF ION INDUCED AUGER ELECTRON EMISSION
FROM METALS.
VIARIS DE LESEGNO P + RIVAIS G + HENNEQUIN JF
PHYS LETT A 49(3): 265-6 (1974)
CA 81: 179487

2-0550 EMISSION OF AUGER ELECTRONS BY ION
BOMBARDMENT.
VIEL L + BENAZETH C + FAGOT B + LOUCHET F
+ COLOMBIE N
COMPT REND B 273: 30-2 (5 JUN 1971) (IN
FRENCH)

2-0551 OBSERVATION OF AUGER ELECTRONS IN THE ENERGY
SPECTRUM OF SECONDARY ELECTRONS EMITTED BY
COPPER UNDER BOMBARDMENT BY RARE GAS IONS.
VIEL L + FAGOT B + COLOMBIE N
COMPT REND B 272: 623-8 (1971)

2-0552 AUGER ELECTRONS FROM ARGON WITH ENERGIES
150-210 EV PRODUCED BY HYDROGEN(+) IMPACTS.
VOLZ DJ + RUDD ME
PHYS REV A 2(4): 1395-1403 (OCT 1970)

2-0553 ELECTRON IMPACT IONIZATION CROSS SECTIONS OF
INNER SHELLS MEASURED BY AUGER ELECTRON
SPECTROSCOPY.
VRAKKING JJ + MEYER F
PHYS REV A 9(5): 1932-7 (1974)
CA 81: 031573

2-0554 MEASUREMENT OF IONIZATION CROSS SECTIONS AND
BACKSCATTERING FACTORS FOR USE IN
QUANTITATIVE AES.
VRAKKING JJ + MEYER F
SURFACE SCI 47(1): 50-6 (1975)
CA 83: 21333

2-0555 POTENTIAL PROFILING ACROSS SEMICONDUCTOR
JUNCTIONS BY AUGER ELECTRON SPECTROSCOPY IN
THE SCANNING ELECTRON MICROSCOPE.
WALDROP JR + HARRIS JS
J APPL PHYS 46(12): 5214-7 (1975)
CA 84: 129467

2-0556 NONRELATIVISTIC AUGER RATES, X-RAY RATES, AND
FLUORESCENCE YIELDS FOR THE K-SHELL.
WALTERS DL + BALLA CP
PHYS REV A 3: 1919-27 (1971)

2-0557 Z DEPENDANCE OF THE KLL AUGER RATES.
WALTERS DL + BALLA CP
PHYS REV A 3: 519-20 (1971)

2-0558 NONRELATIVISTIC AUGER RATES, X-RAY RATES, AND
FLUORESCENCE YIELDS FOR THE 2P SHELL.
WALTERS DL + BHALLA CP
PHYS REV A 4: 2164-70 (1971)

2-0559 RELAXATION PROCESSES ACCOMPANYING CADMIUM
M4,5N4,5N4,5 AUGER TRANSITIONS.
WEIGHTMAN P
J PHYS C 9(6): 1117-28 (1976)
CA 85: 27150

2-0560 L2,3MM AUGER PROCESSES IN SELENIUM.
WEIGHTMAN P + ROBERTS ED + JOHNSON CE
J PHYS C 8(4): 550-66 (21 FEB 1975)
SAA 1975: 26235

2-0561 AUGER RECOMBINATION IN GALLIUM ARSENIDE.
WEISBERG LR
J APPL PHYS 39: 6096-8 (1968)

2-0562 X-RAY IONIZATION CROSS SECTIONS, AND
IONIZATION EQUILIBRIUM EQUATIONS MODIFIED BY
AUGER TRANSITIONS.
WEISHEIT JC
ASTROPHYS J 190(3, PT-1): 735-40 (1974)
CA 81: 056482

2-0563 THEORY OF AUGER EJECTION OF ELECTRONS FROM
METALS BY IONS.
WENAAS EP + HOWSMAN AJ
P13-1 TO 13-22 OF STRUCTURE AND CHEMISTRY OF
SOLID SURFACES, SOMORJAI GA(ED), WILEY, 1969.

2-0564 CHEMICAL EFFECTS OF NUCLEAR TRANSFORMATIONS
IN MOSSBAUER SPECTROSCOPY. (AUGER
AFTEREFFECTS)
WERTHEIM GK
ACCOUNTS CHEM RES 4(11): 373-9 (1971)
CA 76: 8290J

2-0565 SIZE EFFECT IN IONIC CHARGE RELAXATION
FOLLOWING AUGER EFFECT.
WERTHEIM GK + GUGGENHEIM HJ + BUCHANAN DN
J CHEM PHYS 51: 1931-4 (1969)

2-0566 IDENTIFICATION AND MEASUREMENT OF LOW ENERGY
ELECTRONS BY CLOUD CHAMBER TECHNIQUES:
OXYGEN AND NEON AUGER ELECTRON PROBLEM.
WESTON WF
UNIV CALIF, LIVERMORE LAB, 1973, 16P. REP
UCRL-51360
CA 79: 110984

2-0567 MINORITY CARRIER LIFETIME IN INDIUM ARSENIDE
EPILAYERS. (AUGER RECOMBINATION PROCESS)
WIEDER HH + COLLINS DA
APPL PHYS LETT 25(12): 742-3 (15 DEC 1974)
SAA 1975: 25783

2-0568 GRAPHITE CONDUCTION BAND STATES FROM
SECONDARY ELECTRON EMISSION SPECTRA.
WILLIS RF + FEUERBACHER B + FITTON B
PHYS LETT A 34: 231-3 (8 MAR 1971)

2-0569 BAND STRUCTURE OF GRAPHITE STUDIED BY
SECONDARY ELECTRON EMISSION.
WILLIS RF + FITTON B
J VACUUM SCI TECHNOL 9(2): 651-6 (1972)

2-0570 K-SHELL AUGER ELECTRON PRODUCTION CROSS
SECTIONS FROM ION BOMBARDMENT.
WOODS CW
KANSAS STATE UNIV PHD THESIS, 1975, 116P.
CA 84: 157508

2-0571 NEON K-AUGER CROSS SECTIONS FROM 1.5 MEV PER
AMU AND ATOMIC HYDROGEN, OXYGEN, AND
FLUORINE.
WOODS CW + KAUFFMAN RL + JAMISON KA
+ COCKE CL + RICHARD P
J PHYS B 7(17): L474-7 (1974)
CA 82: 49685

2-0572 NEON K-SHELL VACANCY PRODUCTION BY
INTERMEDIATE ENERGY FLUORINE IONS. (AUGER
ELECTRON EMISSION)
WOODS CW + KAUFFMAN RL + JAMISON KA
+ STOLTERFOHT N + RICHARD P
J PHYS B 8(4): L61-4 (1975)
CA 82: 148107

2-0573 AUGER ELECTRONS FROM DOUBLE K-SHELL VACANCY
STATES IN NEON.
WOODS CW + KAUFFMAN RL + JAMISON KA
+ STOLTERFOHT N + RICHARD P
P117-8 OF INT CONF ON PHYS ELECTRONIC
COLLISIONS, (EXTENDED ABSTRACTS) JULY 1975,
RISLEY JS + GEBALLE R(EDS), WASHINGTON UNIV
PRESS, SEATTLE.
SAA 1976: 17100

2-0574 K-SHELL AUGER ELECTRON PRODUCTION CROSS
SECTIONS FROM ION BOMBARDMENT. (NEON,
FLUORINE)
WOODS CW + KAUFFMAN RL + JAMISON KA
+ STOLTERFOHT N + RICHARD P
PHYS REV A 13(4): 1358-69 (1976)
CA 84: 169930

2-0575 DETERMINATION OF OSCILLATOR STRENGTH FOR THE
FIRST DISCRETE TRANSITION IN NITROGEN FROM
THE AUGER DECAY.
WUILLEUMIER F + KRAUSE MO
P773-91 OF INT CONF INNER SHELL IONIZATION
PHENOMENA AND FUTURE APPLICATIONS, VOL 1,
1972, FINK RW(ED), NTIS, 1973.
CA 78: 130130

2-0576 BACKSCATTER MOSSBAUER STUDIES USING
CONVERSION ELECTRONS. TIN-119.
YAGNIK CM + MAZAK RA + COLLINS RL
NUCL INSTRUM METH 114(1): 1-4 (1974)
CA 80: 065287

2-0577 ATOMIC N-SHELL VACANCY WIDTHS IN SOLIDS.
YIN LI + TSANG T + ADLER I
PHYS FENN 9, SUPPL S1: 256-7 (JUL 1974)
SAA 1975: 30212

2-0578 INVESTIGATION OF LOW-Z COSTER-KRONIG
TRANSITIONS BY AUGER AND PHOTOELECTRON
SPECTROSCOPY.
YIN LI + TSANG T + ADLER I
P694-708 OF INT CONF INNER SHELL IONIZATION
PHENOMENA AND FUTURE APPLICATIONS, VOL 1,
1972, FINK RW(ED), NTIS, 1973.
CA 78: 130157

2-0579   FREE ATOM BEHAVIOR OF AUGER ELECTRONS IN
          SOLIDS.
          YIN LI + TSANG T + ADLER I
          P146-53 OF CONF ON APPL SMALL ACCEL, 3RD
          PROC, 1974, DUGGAN JL + MORGAN IL(EDS), NTIS,
          CONF-741040-P2, 1975.
               CA 85: 12038

2-0580   INVESTIGATION OF MECHANISM OF S IONIZATION OF
          NOBLE GAS ATOMS BY THE ULTRASOFT X-RAY
          SPECTROSCOPY TECHNIQUE. (AUGER DECAYS)
          ZAPESOCHNYI IP + ZHUKOV IG
          SOV PHYS JETP 38(4): 675-9 (1974)
               SAA 1975: 84581

2-0581   AUGER TRANSITIONS INITIATED FROM SINGLY AND
          DOUBLY IONIZED STATES IN LITHIUM METAL.
          ZEHNER DM + CLAUSING RE + MCGUIRE GE
          + JENKINS LH
          SOLID STATE COMMUN 13(6): 681-4 (1973)
               CA 79: 130895

2-0582   K-SHELL VACANCY SHARING IN 300 TO 500 KEV
          NEON(+)- SODIUM COLLISIONS. (QUASIMOLECULE
          AUGER SPECTRA)
          ZIEM P + STOLTERFOHT N + RIDDER D
          PHYS FENN 9, SUPPL S1: 39-40 (JULY 1974)
               SAA 1975: 17820

3.   APPARATUS, TECHNIQUE

3-0583   IMPROVED SPHERICAL GRID RETARDING POTENTIAL
          ANALYZER. (FOR USE IN AUGER ELECTRON
          SPECTROSCOPY)
          ABBATI A + BRAICOVICH L + DEMICHELIS B
          J PHYS E 5: 627-9 (1972)

3-0584   SECONDARY EMISSION ANALOG FOR IMPROVED AUGER
          SPECTROSCOPY WITH RETARDING POTENTIAL
          ANALYZERS. COMMENTS.
          ABBATI I + BRAICOVICH L
          REV SCI INSTRUM 43(7): 1054-5 (1972)
               CA 77: 40984V

3-0585   SPECTROMETER FOR INDUCED ELECTRON EMISSION.
          (AUGER ELECTRONS)
          ANDERSON WA
          GER PAT 2,223,086, 14P. VARIAN ASSOC, 30 NOV
          1972.
               CA 78: 50485T

3-0586   DEVELOPMENT AND APPLICATION OF AN "AUGER
          ELECTRON SPECTROMETER" FOR HIGH TEMPERATURE
          ADSORPTION STUDIES.
          ARAMATI VS
          MASS INST TECHNOL, MS THESIS, 1971.

3-0587   CALIBRATION IN AUGER ELECTRON SPECTROSCOPY BY
          MEANS OF COADSORPTION.
          ARGILE C + RHEAD GE
          SURFACE SCI 53(1): 659-75 (1975)
               CA 84: 113977

3-0588   CONSTRUCTION AND OPERATION OF A CYLINDRICAL
          MIRROR TYPE OF AUGER SPECTROMETER.
          ARNOTT DR + RAMSEY JA
          VACUUM 22(9): 355-8 (1972)
               CA 78: 50463J

3-0589   ELECTRON BEAM CURRENT CONTROLLER FOR LEED-
          AUGER STUDIES.
          BAINES M + DEAN EM + WILSON JM
          J PHYS E 8(4): 305-7 (1975)
               SAA 1975: 46039, SAB 1975: 21698, SAC 19

3-0590   CYLINDRICAL MIRROR ANALYZER INCORPORATING
          PRERETARDATION. (TUNGSTEN AUGER ELECTRON)
          BAKER JH + WILLIAMS EM
          VACUUM 25(8): 373-6 (1975)
               CA 84: 24304

3-0591   IMPROVED AUGER ELECTRON SPECTROMETER USING
          CONCENTRIC HEMISPHERES.
          BASSETT PJ
          J PHYS E 7(6): 461-3 (1974)
               CA 81: 019133

3-0592   HIGH ENERGY RESOLUTION AUGER ELECTRON
          SPECTROMETER USING CONCENTRIC HEMISPHERES.
          BASSETT PJ + GALLON TE + PRUTTON M
          J PHYS E 5(10): 1008-13 (OCT 1972)

3-0593   A SCANNING ELECTRON MICROSCOPE WITH
          FACILITIES FOR AUGER EMISSION SPECTROSCOPY.
          BISHOP HE
          P41-2 OF UKAEA DIFFRACTION ANALYSIS CONF ON
          IRRADIATION DAMAGE, 1973, UKAEA, 1974.
               SAA 1975: 11710

3-0594   DESIGN FOR A CYLINDRICAL MIRROR ANALYZER FOR
          USE IN AUGER SPECTROSCOPY.
          BISHOP HE + COAD JP + RIVIERE JC
          J ELECTRON SPECTROSC RELAT PHENOMENA 1(4):
          389-401 (1973)
               CA 78: 090874

3-0595   APPARATUS FOR ANALYSIS BY AUGER ELECTRON
          SPECTROSCOPY.
          BOHN GK + WEBER RE
          PHYSICAL ELECTRONICS INDUSTRIES, US PAT
          3739170, 6P, 1973.
               CA 79: 048969

3-0596 CONTAMINATION IN THE SCANNING ELECTRON
MICROSCOPE. (AUGER ELECTRON SPECTROSCOPY,
SURFACE ELEMENTAL COMPOSITON)
BOTTOMS WR + JOHARI O + CORVIN I
P181-8 OF SCANNING ELECTRON MICROSC SYMP,
1973, 6TH PROC, JOHARI O (ED), ILL INST
TECHNOL, RES INST, 1973.
SAA 1973: 70426

3-0597 HIGH SPATIAL RESOLUTION AUGER ELECTRON
SPECTROSCOPY IN AN ORDINARY DIFFUSION PUMPED
SCANNING ELECTRON MICROSCOPE.
BRANDIS EK
P141-8 OF SCANNING ELECTRON MICROSCOPY CONF
1975, JOHARI C + CORVIN I (EDS), ILL INST
TECHNOL RES INST, 1975
SAA 1976: 3938

3-0598 AUGER ELECTRON SPECTROSCOPY IN A DIFFUSION
PUMPED SCANNING ELECTRON MICROSCOPE.
BRANDIS EK + HOOVER RA
P84A-84B OF NAT CONF ELECTRON PROBE ANAL
1972, 7TH PROC, ELECTRON PROBE ANAL SOC AMER,
1972.
CA 80: 114487

3-0599 X-RAY ELECTRON SPECTROMETERS AND AUGER
ANALYZERS.
BRYTOV IA + KOMYAK NI + KAVROV VP
APP METODY RENTGENOVSK ANAL NO 11: 219-36
(1972) (IN RUSSIAN)
CA 80: 077853

3-0600 HIGH RESOLUTICN ELECTRON SPECTROMETER FOR
PHOTOELECTRON AND AUGER ELECTRON STUDIES.
CARLSON TA + PULLEN BP + MODDEMAN WE
+ KRAUSE MO + WARD FW
ACS ANN MEET 1969, 158TH PROC ABSTR PAPERS NO
PHYS-179.

3-0601 BEAM EFFECTS IN AUGER ELECTRON SPECTROSCOPY
REVEALED BY X-RAY PHOTOELECTRON SPECTROSCOPY.
COAD JP + GETTINGS M + RIVIERE JC
FARADAY DISCUSS CHEM SOC 60: 269-78 (1976)
CA 84: 171874

3-0602 ANALYSIS OF HIGH CHARGE STATE OUTPUT FROM A
PENNING ION SCURCE.
DARLING RL + DAVIS RH
REV SCI INSTRUM 44(4): 375-8 (1973)
CA 78: 152670

3-0603 A MULTIPLE SAMPLE HOLDER FOR USE IN ULTRAHIGH
VACUUM.
DAVIS LE + BRYSON CE + MELLES JJ + HENSON JE
+ LEVENSON LL
J VACUUM SCI TECHNOL 10(4): 564-5 (1973)

3-0604 AUGER ELECTRON SPECTROSCOPY IN THE ELECTRON
EMISSION MICRCSCOPE.
DELONG A + DRAHOS V + KOLARIK V + LENC M
P2 OF CZECH CONF ON ELECTRONICS AND VACUUM
PHYSICS, 5TH PROC, EXTENDED ABSTR, 1972.
SAA 1973: 13470

3-0605 POSSIBILITY OF AUGER ELECTRON SPECTROSCOPY IN
THE ELECTRON EMISSION MICROSCOPE.
DELONG A + DRAHOS V + KOLARIK V + LENC M
P IC-6 OF CZECH CONF ELECTRON VAC PHYS, 5TH
PROC, CESK AKAD VED, 1972.
CA 83: 51763

3-0606 ENERGY RESOLUTION OF THE PHOTOEMISSION
ANALYZER.
DISTEFANO TH + PIERCE DT
REV SCI INSTRUM 41: 180-8 (1970)

3-0607 COMBINATION OF AUGER SPECTROSCOPY WITH
IMAGING OF STUDIED SURFACE.
DRAHOS V
CESK CAS FIS A 25(1): 82-3 (1975) (IN CZECH)
SAA 1975: 70815

3-0608 IRON-55 AS AN AUGER ELECTRON EMITTER. NOVEL
SOURCE FOR GAS CHROMATOGRAPHY DETECTORS.
DWIGHT DJ + LCRCH EA + IOVELOCK JE
J CHROMATOGR 116(2): 257-61 (1976)
CA 84: 159331

3-0609 SURFACE AND THIN FILM ANALYSIS. (AUGER
SPECTROMETERS)
EVANS CA
ANAL CHEM 47(9): 855A-6,858-62,864-6 (1975)
SAA 1975: 83545

3-0610 AVERAGING, CURVE FITTING, AND RELATED
TECHNIQUES IN AUGER ELECTRON EMISSION
SPECTROSCOPY USING RETARDING FIELD
SPECTROMETERS.
FIERMANS L + VENNIK J
SURFACE SCI 38(1): 237-44 (1973)
CA 79: 036804

3-0611 MEASUREMENT OF WORK FUNCTION CHANGE IN A
DISPLAY TYPE LOW ENERGY ELECTRON DIFFRACTION
AUGER APPARATUS.
FRITZ JH + HAQUE CA
REV SCI INSTRUM 44(4): 394-5 (1973)
CA 79: 024654

3-0612 IMPROVED APPARATUS FOR MEASUREMENT OF AUGER
ELECTRON SPECTRA.
GALLON TE + HIGGINBOTHAM IG + PRUTTON M
J PHYS E 2: 894-6 (1969)

3-0613 DISTORTION OF DIFFERENTIATED DEFLECTION
ANALYZER CURRENT. (USED IN AES)
GERLACH RL
J VACUUM SCI TECHNOL 9(2): 1043-4 (1972)
CA 77: 171148Y

3-0614 RETARDING FIELD CYLINDRICAL MIRROR ANALYZER.
(FOR AUGER SPECTROSCOPY)
GERLACH RL
J VACUUM SCI TECHNOL 10(1): 122-5 (1973)

3-0615 IONIZATION SPECTROMETER FOR ELEMENTAL
ANALYSIS OF SURFACES.
GERLACH RL + TIPPING DW
REV SCI INSTRUM 42: 151-4 (1971)

3-0616 AN ENERGY MODULATED (CYLINDRICAL) HIGH ENERGY
RESOLUTION ELECTRON SPECTROMETER.
GOLDEN DE + ZECCA A
REV SCI INSTRUM 42: 210-16 (1971)

3-0617 DRYING OF BUTADIENE- NITRILE RUBBERS IN AN
AUGER MACHINE.
GOLOVANOVA AN + LOSIEVSKAYA NV
+ PERELYGINA SA + GRIGOREV VB + FILINOV GP
+ TITOV AP + YADREEV FI + LASKIN A
PROM SIN KAUCH NAUCH-TEKH SB NO 1: 12-14
(1973) (IN RUSSIAN)
CA 80: 134505

3-0618 SENSITIVITY VARIATIONS IN BAYARD- ALPERT
GAUGES CAUSED BY AUGER EMISSION AT COLLECTOR.
GOPALARAMAN CP + ARMSTRONG RA + REDHEAD PA
J VACUUM SCI TECHNOL 6: 910 (1969)

3-0619 COMPACT, INEXPENSIVE HIGH RESOLUTION
RETARDING FIELD ENERGY ANALYZER. (AUGER
ELECTRONS)
GOTO K + ISHIKAWA K
REV SCI INSTRUM 43(3): 427-30 (1972)

3-0620 CORRECTIONS OF AUGER ELECTRON SIGNAL
STRENGTHS FOR MODULATION AMPLITUDE DISTORTION
IN A 4-GRID RETARDING POTENTIAL ENERGY
ANALYZER.
GRANT JT + HAAS TW
SURFACE SCI 44(2): 617-23 (1974)
CA 81: 129541

3-0621 ULTRAHIGH VACUUM SCANNING ELECTRON MICROSCOPE
WITH A FIELD EMISSION CATHODE AND AUGER
ANALYZER.
GRIFFITHS BW + HENRICH K + POWELL BD
+ WOODRUFF DP
MESSTECHNIK 82(6): 135-41 (1974) (IN GERMAN)
CA 81: 145074

3-0622 ULTRAHIGH VACUUM SCANNING ELECTRON MICROSCOPE
WITH AUGER ANALYSIS FACILITIES.
GRIFFITHS BW + JONES AV + WARDELL IRM
P42-5 OF SCANNING ELECTRON MICROSC SYMP 1973,
6TH PROC, JOHARI O (ED), ILL INST TECHNOL RES
INST, 1973.
CA 80: 113664

3-0623 COMPARISON OF SPHERICAL DEFLECTOR AND
CYLINDRICAL MIRROR ANALYZERS.
HAFNER H + SIMPSON JA + KUYATT CE
REV SCI INSTRUM 39: 33-5 (1968)

3-0624 SOME PROPERTIES OF RETARDING FIELD
SPECTROMETERS USING POST MONOCHROMATORS. (FOR
AUGER ELECTRON SPECTROSCOPY)
HARRIS FM + BASSETT PJ + PRUTTON M
J PHYS E 8(1): 11-13 (JAN 1975)
SAA 1975: 20175

3-0625 INVESTIGATION OF HIGH RESOLUTION SCANNING
TECHNIQUES FOR AUGER ELECTRON SPECTROSCOPY.
HART RK + THARP LN + WOODWARD RP
GEORGIA INST TECHNOL, ATLANTA, JULY 1971,
20P. ORO-4069-1

3-0626 A COMPARISON OF THE EXTENDURE OF ELECTRON
SPECTROMETERS. (USED IN AUGER SPECTROSCOPY)
HEDDLE DWO
J PHYS E 4: 589-92 (1971)

3-0627 INTRODUCTION TO LOW ENERGY X-RAY AND ELECTRON
ANALYSIS. (ANALYZER FOR PHOTO AUGER ELECTRON
SPECTROSCOPY)
HENKE BL
P1-25 OF ADVAN X-RAY ANALYSIS, MUELLER
WM(ED), VOL-13, PLENUM, 1969. (ANN CONF
APPLIC X-RAY ANALYSIS, 13TH PROC)

3-0628 USE OF CYLINDRICAL AUGER SPECTROMETERS FOR
RETARDING POTENTIAL SECONDARY ELECTRON YIELD
MEASUREMENTS.
HENRICH VE
REV SCI INSTRUM 45(6): 861-2 (1974)
CA 81: 055515

3-0629 FAST, ACCURATE SECONDARY ELECTRON YIELD
MEASUREMENTS AT LOW PRIMARY ENERGIES. (LEED-
AUGER SYSTEM)
HENRICH VE
REV SCI INSTRUM 44(4): 456-62 (APR 1973)
SAA 1973: 35496

3-0630 DECONVOLUTION METHOD FOR COMPOSITION
PROFILING BY AUGER SPUTTERING TECHNIQUE.
HO PS + LEWIS JE
SURFACE SCI 55(1): 335-48 (1976)
CA 84: 185322

3-0631 A VERSATILE SUBSTRATE DESIGN FOR LEED AND AES
STUDIES IN ULTRAHIGH VACUUM.
HOLLOWAY PH + HUDSON JB
REV SCI INSTRUM 43: 1045-7 (1972)
SAA 1972: 60617

3-0632 AN ANALYTICAL SYSTEM FOR SECONDARY ION MASS
SPECTROMETRY IN ULTRAHIGH VACUUM. (AES)
HUBER WK + SELHOFER H + BENNINGHOVEN A
J VACUUM SCI TECHNOL 9: 482-6 (1972)
SAA 1972: 38786

3-0633 HIGH SENSITIVITY ELECTRON SPECTROMETER.
HUCHITAL DA + RIGDEN JD
APPL PHYS LETT 16: 348-51 (1970)

3-0634 RESOLUTION AND SENSITIVITY OF THE SPHERICAL
GRID RETARDING POTENTIAL ANALYZER.
APPLICATION TO AUGER ELECTRON SPECTROSCOPY
AND X-RAY PHOTOELECTRON SPECTROSCOPY.
HUCHITAL DA + RIGDEN JD
J APPL PHYS 43: 2291-302 (1972); P79-104 OF
ELECTRON SPECTROSCOPY, SHIRLEY DA(ED), AMER
ELSEVIER, 1972, 916P.
CA 77: 158669Z

3-0635 ANALOG TECHNIQUE FOR THE MEASUREMENT OF AUGER
ELECTRON CURRENTS. APPLICATION TO SULFUR ON
NICKEL (110).
ISETT LC + BLAKELY JM
REV SCI INSTRUM 45(11): 1382-5 (1974)
CA 82: 009760

3-0636 NEW SEM AUGER MICROANALYZER WITH ULTRAHIGH
VACUUM CAPABILITY.
ISHIDA T + UCHIYAMA M + ODA Z + HASHIMOTO H
J VACUUM SCI TECHNOL 13(3): 711-5 (1976)
CA 85: 39217

3-0637 A CRYSTAL CLEAVAGE DEVICE FOR USE IN LOW
ENERGY ELECTRON DIFFRACTION- AUGER STUDIES.
JANSSEN AP + CHAMBERS A
J PHYS E 7(6): 425-6 (JUN 1974)
SAA 1974: 52665

3-0638 SECONDARY ELECTRON EMISSION FROM THIN FILMS.
(OF GOLD, APPARATUS)
KADLEC J + ECKERTORA L
Z ANGEW PHYS 30: 142-5 (1970) (IN GERMAN)

3-0639 ELECTRODES AND CIRCUITS OF A LEED- AES
SYSTEM.
KANAJI T + ASANO K + NAGATA S
MEM FAC ENG KOBE UNIV 18: 203-11 (1972)
SAA 1972: 74612

3-0640 DIGITAL METHODS FOR THIN FILM ANALYSIS USING
A COMPUTER COUPLED AUGER SPECTROMETER.
KEENAN JA + MCGUIRE GE
P212-13 OF ELECTROCHEM SOC FALL MEET ABSTR,
1975.
SAA 1976: 33643

3-0641 DESIGN OF AN X-RAY DOUBLE CRYSTAL
SPECTROMETER INSTALLED IN VACUUM AND THE
MEASUREMENT OF EXCITED ELECTRONS UNDER THE
X-RAY DIFFRACTION PROCESS. (AUGER SPECTRA OF
SILICON)
KIKUTA S + TAKAHASHI T + TUZI Y + FUKUDOME R
SEISAN KENKYU 26(10): 386-90 (1974) (IN
JAPANESE)
CA 82: 178043

3-0642 SIMPLIFIED RETARDING FIELD PHOTOELECTRON AND
AUGER ELECTRON SPECTROGRAPH.
KLEIN W
REV SCI INSTRUM 42: 1292-5 (SEP 1971)

3-0643 HIGH SENSITIVITY APPEARANCE POTENTIAL
SPECTROMETER FOR THE DETECTION OF PHOTONS,
ELECTRONS AND IONS.
KLUGE A
REV SCI INSTRUM 46(9): 1179-81 (1975)
SAA 1976: 16148

3-0644 SMALL RETARDING FIELD ANALYZER FOR AUGER
ELECTRON SPECTROSCOPY. (PALLADIUM- NICKEL
ALLOYS, ADSORBED OXYGEN)
KOCH J
REV SCI INSTRUM 45(10): 1212-13 (1974)
CA 81: 144082

3-0645 SCANNING AUGER MICROSCOPE FOR THIN FILM
ANALYSIS.
KROYER JM
AIR FORCE INST TECHNOL 1973, 74P. AD-777849
CA 81: 085486

3-0646 X-RAYS AND ELECTRONS IN ANALYTICAL CHEMISTRY
WITH EMPHASIS ON INSTRUMENTATIION.
LIEBHAFSKY HA + PFEIFFER HG
J CHEM EDUC 50(2): A73-4, A78-82 (1973)
CA 78: 105562

3-0647 DETECTION OF SODIUM ON SILICON AND SILICON
DIOXIDE SURFACES BY CYLINDRICAL MIRROR AUGER
ANALYZER.
LYO K + KOMIYA S
SHINKU 16(1): 41-4 (1973) (IN JAPANESE)
CA 79: 024766

3-0648 AUGER ELECTRON SPECTROSCOPY IN SCANNING
ELECTRON MICROSCOPY: POTENTIAL MEASUREMENTS.
MACDONALD NC
APPL PHYS LETT 16: 76-80 (1970)

3-0649 AUGER ELECTRON IMAGES IN A SCANNING ELECTRON
MICROSCOPE.
MACDONALD NC
ABSTR 45A OF ANN CONF MICROBEAM ANAL SOC, 9TH
PROC SUMMARIES, LEHIGH UNIV, 1974.
SAA 1974: 77308

3-0650 AUGER ELECTRON SPECTROSCOPY IN THE SCANNING
ELECTRON MICROSCOPE: AUGER ELECTRON IMAGES.
MACDONALD NC + WALDROP JR
APPL PHYS LETT 19: 315-18 (1 NOV 1971)

3-0651 LABORATORY COMPUTER SCANNING ELECTRON
MICROSCOPE SYSTEM.
MACDONALD NC + WALDROP JR
PROC ELECTRON MICROSC SOC AMER 29: 86-7
(1971)
CA 80: 089089

3-0652 A NEW PHOTOELECTRON SPECTROMETER: COMBINATION
OF CYLINDRICAL MIRROR ANALYZER WITH SOFT
X-RAY SOURCE.
MAEDA K + IHARA T
REV SCI INSTRUM 42: 1480-4 (1971)

3-0653 SPECTROMETER OF AUGER ELECTRONS.
MASYAGIN VE + IL'IN AI + REDKIN VS
+ ZASHKVARA VV
INSTRUM EXP TECH 17(2, PT-1): 345-7 (1974)
CA 81: 043972

3-0654 SURFACE CLEANLINESS OF 316 L-N STAINLESS
STEEL STUDIES BY SECONDARY ION MASS
SPECTROMETRY AND AUGER ELECTRON SPECTROSCOPY.
COMBINED ESCA- AUGER SYSTEM BASED ON THE
DOUBLE PASS CYLINDRICAL MIRROR ANALYZER.
MATHEWSON AG + PALMBERG PW
J ELECTRON SPECTROSC RELAT PHENOMENA 5:
691-703 (1974)
CA 82: 118114

3-0655 A SMALL RETARDING FIELD ENERGY ANALYZER WITH
CEM. (AUGER SPECTROMETER)
MATSUDAIRA T + ONCHI M
SHINKU 18(12): 468-72 (1975) (IN JAPANESE)
CA 84: 128675

3-0656 AUGER ELECTRON SPECTROSCOPY USING THE
TWO-GRID LEED SYSTEM AND PHOTOELECTRON
MULTIPLIER.
MATSUDAIRA T + WATANABE M + ONCHI M
JAP J APPL PHYS 11(12): 1853-4 (1972); SHINKU
15(12): 435-40 (1972) (IN JAPANESE)
CA 78: 65078M; 128754

3-0657 ULTRAHIGH VACUUM LOW TEMPERATURE STAGE FOR
IN-SITU FILM DEPOSITION AND CHARACTERIZATION.
(LEED- AUGER SYSTEM)
MILLER DL + ARNS T + HRABAK G
REV SCI INSTRUM 46(12): 1642-5 (1975)
SAA 1976: 31682

3-0658 AUGER ELECTRON SPECTROMETER INTEGRATED WITH A
LOW ENERGY ELECTRON DIFFRACTION SYSTEM.
MITYAGIN AYU + CHEREVATSKII NYA
+ ALEKSANDROV AL + BALANDIN GD + KOROLKOV NS
INSTRUM EXP TECH 15(1): 217-20 (1972)
CA 76: 146196J

3-0659 MECHANICAL TESTING: IN-SITU FRACTURE DEVICE
FOR AUGER ELECTRON SPECTROSCOPY.
MOORHEAD RD
REV SCI INSTRUM 47(4): 455-9 (1976)
CA 85: 39205

3-0660 FIRST ORDER APPROXIMATION TO QUANTITATIVE
AUGER ANALYSIS IN THE RANGE 100 TO 1000 EV
USING THE CYLINDRICAL MIRROR ANALYZER.
MORABITO JM
SURFACE SCI 49(1): 318-24 (1975)
CA 83: 125646

3-0661 APPLICATION OF TRIPLE GRID LEED SYSTEM TO
AUGER SPECTRUM ANALYSES.
MORRISON J + LANDER JJ
J VACUUM SCI TECHNOL 6: 338-42 (1969)

3-0662 LOW NOISE DETECTION SYSTEM FOR THE
MEASUREMENT OF AUGER SPECTRA.
NATHAN R + HOPKINS BJ
J PHYS E 6(10): 1040-2 (1973)
CA 79: 131248

3-0663 AN AC RETARDING POTENTIAL TECHNIQUE FOR THE
CONTINUOUS MEASUREMENT OF CHANGES IN WORK
FUNCTION. (LEED- AUGER SYSTEM)
NATHAN R + HOPKINS BJ
J PHYS E 7(10): 851-4 (OCT 1974)
SAA 1974: 77122

3-0664 APPARATUS FOR SURFACE ANALYSIS BY ELECTRON
SPECTROSCOPY.
NEAVE JH + BOUDRY MR
GER OFFEN PAT, 242317, 19 DEC 1974, 21P.; NV
PHILIPS GLOELAMPENFABRIKEN
CA 83: 52903

3-0665 LOW ENERGY ION BACKSCATTERING SPECTROSCOPY
WITH A COMMERCIAL AUGER CYLINDRICAL MIRROR
ANALYZER.
NIEHUS H + BAUER E
REV SCI INSTRUM 46(9): 1275-7 (1975)
CA 83: 170750

3-0666 QUANTITATIVE ASPECTS OF ION SCATTERING
SPECTROSCOPY. (CYLINDRICAL MIRROR ANALYZER)
NIEHUS H + BAUER E
SURFACE SCI 47(1): 222-33 (1975)
SAA 1975: 47341

3-0667 HIGH RESOLUTION ELECTRON SPECTROMETER FOR
AUGER ELECTRON SPECTROSCOPY.
NOELLER HG + POLASCHEGG HD + SCHILLALIES H
P343-6 OF INT VACUUM CONGRESS, 6TH PROC, INT
CONF ON SOLID SURFACES, 2ND, 1974, KYOTO.
(JAP J APPL PHYS, SUPPL 2, PART 2, 1974)
CA 83: 18806

3-0668 PRODUCTION OF A SPHERICAL GRID FOR LOW ENERGY
ELECTRON DIFFRACTION- AUGER ELECTRON
SPECTROSCOPY.
ONO M + BAKU J
SHINKU 15(12): 450-2 (1972) (IN JAPANESE)
CA 78: 141332

3-0669 ULTRAHIGH VACUUM AIRLOCK FOR LEED AND AUGER
SPECTROSCOPY.
ONO M + SHIMIZU H + NAKAYAMA K
J VACUUM SCI TECHNOL 10(4): 566-7 (JUL-AUG
1973)
SAA 1974: 26495

3-0670 INSTRUMENTAL PROBLEMS FOR ELECTRON MICROPROBE
AUGER SPECTROSCOPY.
ONO M + SHIMIZU H + NAKAYAMA K
P359-62 OF INT VACUUM CONGRESS, 6TH PROC, INT
CONF ON SOLID SURFACES, 2ND, 1974, KYOTO.
(JAP J APPL PHYS, SUPPL 2, PART 1, 1974)

3-0671 COMBINED ESCA AND AUGER SPECTROMETER.
PALMBERG PW
J VACUUM SCI TECHNOL 12(1): 379-84 (1975)
CA 82: 162857

3-0672 HIGH SENSITIVITY AUGER ELECTRON SPECTROMETER.
PALMBERG PW + BOHN GK + TRACY JC
APPL PHYS LETT 15: 254-5 (1969)

3-0673 A NOVEL DETECTION CIRCUIT FOR AUGER
SPECTROSCOPY.
PATERSON PJK + DENNE DR + LECKEY RCG
J PHYS E 9(4): 304-7 (1976)
SAA 1976: 43106

3-0674 SPECTROSCOPY IN THE SCANNING ELECTRON
MICROSCOPE NOW INCLUDES SECONDARY ION MASS
ANALYSIS. (AUGER)
PEASE DE
CAN RES DEVELOP 7(5): 36-8, 40-1 (1974)
CA 82: 50910

3-0675 NEW TECHNIQUE FOR AUGER ANALYSIS OF SURFACE
SPECIES SUBJECT TO ELECTRON INDUCED
DESORPTION.
PEPPER SV
REV SCI INSTRUM 44(7): 826-9 (1973)
CA 79: 046092

3-0676 ELECTRON GUN FOR AUGER AND ULTRASOFT X-RAY
APPARATUS.
PESSA M
ACTA POLYTECH SCAND PHYS NUCL SER NO 84: 1-16
(1971)
CA 77: 120673H

3-0677 DOUBLY RESONANT CURRENT-TO-VOLTAGE CONVERTER
PREAMPLIFIER FOR SURFACE SPECTROSCOPY.
POCKER DJ
REV SCI INSTRUM 45(10): 1271-4 (OCT 1974)
SAA 1975: 17121

3-0678 TOWARD OPTIMUM UTILIZATION OF THE CYLINDRICAL
MIRROR ANALYZER IN AUGER ELECTRON
SPECTROSCOPY.
POCKER DJ
REV SCI INSTRUM 46(1): 105-6 (1975)
CA 82: 148321

3-0679 SCANNING AUGER ELECTRON MICROSCOPE FOR
SURFACE STUDIES.
POWELL BD + WOODRUFF DP + GRIFFITHS BW
J PHYS E 8(7): 548-52 (1975)
CA 83: 68967

3-0680 SEMIAUTOMATED DATA RECORDING AND CONTROL
SYSTEM FOR AN ELECTRON ENERGY ANALYZER.
POWELL CJ
REV SCI INSTRUM 44(8): 1031-3 (1973)
CA 79: 099109

3-0681 GAIN FATIGUE MECHANISM IN CHANNEL ELECTRON
MULTIPLIERS. (SURFACE OXYGEN ON LEAD OXIDE BY
AES)
PRINCE RH + CROSS JA
REV SCI INSTRUM 42: 66-71 (1971)

3-0682 AUGER CYLINDRICAL MIRROR SPECTROMETER FOR THE
ANALYSIS OF SURFACE CONSTITUTENTS AND SOME
RESULTS.
PROJAHN H
P351 OF COLLOQ INT APPLIC TECH, VIDE IND,
SEMICOND COMPOSANTS ELECTRON MICROELECTRON,
1971, 3RD PROC. (IN FRENCH)
CA 81: 072137

3-0683 ULTRAHIGH VACUUM EVAPORATOR FOR LEED AND
AUGER EMISSION STUDIES.
PRUTTON M + TOKUTAKA H
THIN SOLID FILMS 3: 411-16 (1969)

3-0684 FOCUSING OF CHARGED PARTICLES BY A SPHERICAL
CONDENSER.
PURCELL EM
PHYS REV 54: 818-26 (1938)

3-0685 SPECTROSCOPIC METHODS OF MICROANALYSIS USING
THE SCANNING ELECTRON MICROSCOPE. (REVIEW,
AUGER SCANNING)
REHME H + WOLFGANG E
SIEMENS Z 49(11): 732-9 (1975) (IN GERMAN)
CA 84: 173267

3-0686 RETARDING FIELD ELECTRON SPECTROMETER.
RIGDEN JD + HUCHITAL DA
US PAT 3681600, 13P. 1 AUG 1972, PERKIN-ELMER
CORP.
CA 77: 121870G

3-0687 DESIGN PARAMETERS FOR THE CYLINDRICAL MIRROR
ENERGY ANALYZER.
RISLEY JS
REV SCI INSTRUM 43(1): 95-103 (1972)

3-0688 EXPERIMENTAL ASSEMBLY FOR OBTAINING THE
SPECTRUM OF SECONDARY ELECTRON EMISSION FROM
METALS AND SEMICONDUCTORS. (AUGER ENERGIES)
SALMERON M + BARO AM + PRADAL F
ANN FIS 67(9-10): 405-7 (1971) (IN FRENCH)
SAA 1972: 33094

3-0689 CYLINDRICAL CAPACITOR AS AN ANALYZER. PART-1:
NONRELATIVISTIC PART.
SAR-EL HZ
REV SCI INSTRUM 38: 1210-16 (1967)

3-0690 REALIZATION OF A MOSSBAUER SPECTROMETER FOR
SURFACES AND THIN FILMS STUDY. (AUGER)
SCHUNCK JP
CENT RECH NUCL, 1975, 117P. CRN-CNPA-74-15
(IN FRENCH)
CA 83: 155556

3-0691 FARADAY CUP LEED APPARATUS WITH FACILITY FOR
INVESTIGATING ENERGY AND ANGULAR
DISTRIBUTIONS OF INELASTICALLY SCATTERED OR
PHOTOEMITTED ELECTRONS.
SEAH MP + FORTY AJ
J PHYS E 3: 833-41 (1970)

3-0692 ELECTRON CHANNELING PATTERNS IN AN AUGER
ELECTRON SPECTROMETER WITH SCANNING SAMPLE
POSITIONER.
SEILER H + KUHNLE G + BAUER H
APPL PHYS 6(2): 167-71 (1975) (IN GERMAN)
CA 82: 131994

3-0693 ORIENTATION DEPENDENCE OF OVERLAYER
ATTENUATION OF ELECTRONS FOR THE CYLINDRICAL
MIRROR ANALYZER AND A RETARDING FIELD
ANALYZER.
SHELTON JC
J ELECTRON SPECTROSC RELAT PHENOMENA 3:
417-25 (1974)

3-0694 SECONDARY EMISSION ANALOG FOR IMPROVED AUGER
SPECTROSCOPY WITH RETARDING POTENTIAL
ANALYZERS. (REPLY TO COMMENTS)
SICKAFUS EN
REV SCI INSTRUM 42: 933-41 (1971); 43(7):
1055-6 (1972)
CA 77: 40985W

3-0695 A MULTICHANNEL MONITOR FOR REPETITIVE AUGER
ELECTRON SPECTROSCOPY WITH APPLICATION TO
SURFACE COMPOSITION CHANGES.
SICKAFUS EN + COLVIN AD
REV SCI INSTRUM 41: 1349-54 (1970)

3-0696 SPECIMEN POSITION EFFECTS ON ENERGY SHIFTS
AND SIGNAL INTENSITY IN A SINGLE STAGE
CYLINDRICAL MIRROR ANALYZER.
SICKAFUS EN + HOLLOWAY DM
SURFACE SCI 51(1): 131-9 (1975)
CA 83: 155557

3-0697 HIGH RESOLUTION, LOW ENERGY ELECTRON
SPECTROMETER.
SIMPSON JA
REV SCI INSTRUM 35: 1698-704 (1964)

3-0698 REFINEMENTS TO A STANDARD LEED- AUGER SYSTEM
FOR THE ANALYSIS OF ELECTRON EMISSION AT LOW
PRIMARY BEAM ENERGIES.
SKINNER DK + WILLIS RF
REV SCI INSTRUM 43(5): 731-4 (1972)
CA 76: 159783Y

3-0699 NEW, STRIATED, HIGH EFFICIENCY PHOTOGRAPHIC
PLATES FOR RECORDING CONVERSION AND/OR AUGER
ELECTRON SPECTRA.
SLATIS H
PHYS SCR 7(6): 298-302 (1973)
CA 79: 059566

3-0700 EASY METHOD TO ACCURATELY ALIGN ION
BOMBARDMENT GUNS FOR DEPTH PROFILING IN AUGER
ELECTRON SPECTROSCOPY.
SPRINGER RW + HAAS TW + GRANT JT + HOOKER MP
REV SCI INSTRUM 45(9): 1113-14 (1974)
CA 81: 143711

3-0701 IMPROVED RETARDING FIELD ANALYZER. (FOR AUGER
AND PHOTOELECTRON SPECTROSCOPY).
STAIB P
J PHYS E 5: 484-7 (MAY 1972)

3-0702 NEW COMPACT ENERGY ANALYZER FOR AUGER AND
ESCA SPECTROSCOPY.
STAIB P
VACUUM 22(10): 481-4 (1972)
CA 78: 77947G

3-0703 HIGH RESOLUTION ELECTRON SPECTROGRAPHS FOR
SOLID STATE RESEARCH. (REVIEW, 12 REFS)
SUONINEN EJ
P653-63 OF INT SEMINAR ON SELEC PROBL THEORY
IMPURITY CENT CRYST, 1970 PROC, ZAVT GS (ED),
AKAD NAUK EST SSR, TALLIN, USSR, 1972.
CA 78: 166234

3-0704 SORBAS: AN APPARATUS FOR INVESTIGATING ION
SCATTERING FROM SURFACES AT ENERGIES 100-2000
EV. (AES)
TAGLAUER E + MELCHIOR W + SCHUSTER F
+ HEILAND W
J PHYS E 8(9): 768-72 (1975) (IN GERMAN)
SAA 1975: 81284

3-0705   RESOLUTION AND SENSITIVITY CONSIDERATIONS OF
AN AUGER ELECTRON SPECTROMETER BASED ON LEED
DISPLAY OPTICS.
TAYLOR NJ
REV SCI INSTRUM 40: 792-804 (1969)

3-0706   AUGER ELECTRON SPECTROMETER AS TOOL FOR
SURFACE ANALYSIS. (CONTAMINATION MONITOR)
TAYLOR NJ
J VACUUM SCI TECHNOL 6: 241-5 (1969)

3-0707   ULTRAHIGH VACUUM EVAPORATOR AND MULTIPLE
CLEAVAGE DEVICE FOR LOW ENERGY ELECTRON
DIFFRACTION AND AUGER EMISSION STUDIES.
TOKUTAKA H + PRUTTON M
SHINKU 15(12): 427-34 (1972) (IN JAPANESE)
     CA 79: 010611

3-0708   AUGER ELECTRON SPECTROMETER PREAMPLIFIER.
TRACY JC + BOHN GK
REV SCI INSTRUM 41: 591-2 (1970)

3-0709   SECONDARY ELECTRON SPECTROMETER.
VASINA P
CESK CAS FIS A 25(1): 53-6 (1975)
     CA 83: 18797

3-0710   AUGER ELECTRON SPECTROSCOPY IN THE SCANNING
ELECTRON MICROSCOPE.
WALDROP JR + MARCUS HL
J TEST EVAL 1(3): 194-201 (1973)
     CA 79: 047242

3-0711   ELECTRON BEAM MICROANALYZER WITH AUGER
ELECTRON DETECTION.
WEBER U
GER PAT 2151167, SIEMENS AG, 10 MAY 1973,
22P.
     CA 79: 026872

3-0712   ELECTRON BEAM ANALYZER FOR DETECTING AUGER
ELECTRONS.
WEBER U
GER PAT 2202347, SIEMENS AG, 1973, 25P.
     CA 79: 098460

3-0713   ANALYSIS APPARATUS FOR INVESTIGATING A SAMPLE
BY MEANS OF EMITTED ELECTRONS.
WEBER U
GER PAT 2216821, SIEMENS AG, 27 SEP 1973,
10P.
     CA 80: 055563

3-0714   ELECTRON BEAM MICROPROBES USED IN THE
ANALYSIS AND DEVELOPMENT OF MATERIALS. (AUGER
ELECTRON SIGNALS)
WEBER U
SIEMENS FORSCH-ENTWICKLUNGSBER 3(5): 294-8
(1974)
     CA 82: 50913

3-0715   ELECTRON SPECTROMETER FOR SURFACE ANALYSIS.
WEI PSP + SMITH AW
BOEING SCI RES LABS, SEATTLE, WASH, NOV 1970,
39P. AD-717280

3-0716   RECENT ADVANCES IN INSTRUMENTATION FOR
MICROPROBE ANALYSIS. (AUGER ELECTRON
SPECTROSCOPY AND ION MICROPROBE MASS
SPECTROMETRY)
WITTRY DB
P206-18 OF INT CONG ON X-RAY OPTICS AND
MICROANALYSIS, 5TH PROC, TUBINGEN, GERMANY,
1968, 612P.

3-0717   HIGH RESOLUTION ELECTROSTATIC SPECTROMETER
USED IN ELECTRON SPECTROMETRY FOR CHEMICAL
ANALYSIS. (AUTOMATIC RECORDING OF PHOTO AND
AUGER ELECTRONS)
YAVOR SY + PETROV IA + DENISOV EP
SOV PHYS TECH PHYS 16(9): 1455-60 (1972)
     CA 76: 8792Z

3-0718   WIDTHS OF ATOMIC M-SHELL VACANCY STATES AND
QUASIATOMIC ASPECTS OF RADIATIONLESS
TRANSITIONS IN SOLIDS.
YIN LI + ADLER I + TSANG T + CHEN MH
+ RINGERS DA + CRASEMANN B
PHYS REV A 9(3): 1070-80 (1974)
     CA 80: 139209

3-0719   INSTRUMENTATION FOR THE CHEMICAL ANALYSIS OF
MANUFACTURED SURFACES. (REVIEW, SPECTROSCOPY)
YOUNG RD
CIRP ANN 24(2): 549-54 (1975)
     CA 85: 13256

3-0720   FOCUSING A CHARGED PARTICLE BEAM OF FINITE
ANGULAR SPREAD IN AN ELECTROSTATIC ANALYZER
OF THE CYLINDRICAL MIRROR TYPE. (AUGER
SPECTROMETER)
ZASHKVARA VV
SOV PHYS TECH PHYS 16: 146-9 (1971)

3-0721   FOCUSING PROPERTIES OF AN ELECTROSTATIC
MIRROR WITH CYLINDRICAL FIELD.
ZASHKVARA VV + KORSUNSKII MI + KOSMACHEV OS
SOV PHYS TECH PHYS 11(1): 96-9 (1966)

3-0722   DIGITAL CONTROL UNIT IN TTL LOGIC FOR THE
AUTOMATIC EXECUTION OF AUGER AND ESCA TEST
PROGRAMS.
ZINSER J
MAX PLANCK INST, MUNICH 1973, 33P. IPP9/10
(IN GERMAN)
     CA 80: 102187

4. AUGER SPECTRA OF MATERIALS, SPECTRAL LINE ENERGIES

4-0723 INTERPRETATION OF KL2,3L2,3 RADIATIVE AUGER THRESHOLDS IN METALS AND SEMICONDUCTORS. (SILICON)
ABERG T + UTRIAINEN J
SOLID STATE COMMUN 16(5): 571-3 (1975)
CA 82: 162639

4-0724 SCF AND LIMITED CI CALCULATIONS FOR ASSIGNMENT OF THE AUGER SPECTRUM AND OF THE SATELLITES IN THE SOFT X-RAY SPECTRUM OF WATER.
AGREN H + SVENSSON S + WAHLGREN UI
CHEM PHYS LETT 35(3): 336-44 (1975)
CA 83: 185603

4-0725 KLL AUGER ELECTRON SPECTRUM OF SAMARIUM.
AKHMETOV KM + BASIN LA + BOBYKIN BV
+ PETUKHOV VK + CHEBOTINKOV VA
P170-1 OF PROGRAMMA TEZISY DOKL SOVESHCH YAD SPEKTROSK STRUKT AT YADRA, 25TH, LEBEDEV PP + PODKOPAEV YUN (EDS), NAUKA, LENINGRAD, 1975. (IN RUSSIAN)
CA 83: 139294

4-0726 SPECTRUM OF AUGER KLL ELECTRONS OF SAMARIUM.
AKHMETOV KM + BOBYKIN BV + PETUKHOV VK
IZV AKAD NAUK SSSR SER FIZ 39(8): 1727-31 (1975) (IN RUSSIAN)
CA 83: 10886

4-0727 CHEMICAL EFFECTS IN THE AUGER SPECTRA OF IRON DUE TO OXYGEN ADSORPTION.
AKIMOTO K + ONCHI M
SHINKU 18(12): 459-67 (1975) (IN JAPANESE)
CA 84: 128353

4-0728 M4,5N4,5N4,5 AND M4,5O1 AUGER ELECTRON SPECTRA FROM CADMIUM VAPOR.
AKSELA H + AKSELA S
J PHYS B 7(11): 1262-9 (1974)
CA 81: 083847

4-0729 AUGER ELECTRON STUDIES OF METAL VAPORS.
AKSELA H + AKSELA S + LAHTEENKORVA E
+ VAYRYNEN J
P56 OF ANNU CONF FINNISH PHYS SOC, 1975 PROC.
SAA 1975: 36560

4-0730 HIGH RESOLUTION MNN AUGER SPECTRA OF SILVER, CADMIUM, INDIUM, ANTIMONY, TELLURIUM AND IODINE.
AKSELA S
Z PHYS 244: 268-74 (1971)

4-0731 L2,3MM AUGER ELECTRON SPECTRUM FROM FREE ZINC ATOMS.
AKSELA S + AKSELA H
PHYS LETT A 48(1): 19-20 (1974)
CA 81: 056058

4-0732 HIGH RESOLUTION LMM AUGER SPECTRA OF LOW ENERGY FROM SOLID SURFACES.
AKSELA S + PESSA M + KARRAS M
Z PHYS 237: 381-7 (1970)

4-0733 HIGH RESOLUTION L2,3M4,5M4,5 AUGER SPECTRUM OF FREE ZINC ATOMS.
AKSELA S + VAYRYNEN J + AKSELA H
PHYS REV LETT 33(17): 999-1002 (1974)
CA 81: 161855

4-0734 KLL AUGER SPECTRUM OF FLUORINE. (IN FLUORIDE SALTS)
ALBRIDGE RG + HAMRIN K + JOHANSSON G
+ FAHLMAN A
Z PHYS 209: 419-27 (1968) (IN GERMAN)

4-0735 SPECTRAL LINE EXCITATION FUNCTIONS FOR SODIUM, POTASSIUM, RUBIDIUM, AND CESIUM IONS IN THE VACUUM ULTRAVIOLET SPECTRAL RANGE. (AUGER DECAYS)
ALEKSAKHIN IS + VUKSTICH VS + ZAPESOCHNYI IP
SOV PHYS JETP 39(6): 971-4 (1974)
SAA 1974: 78353

4-0736 RECORDING OF COMPONENTS ON THE SURFACE OF FUSED EMITTERS. (AUGER SPECTRA)
ALEKSEEV YUV + BORISOV VL + GNUCHEV NM
+ KANICHEVA I + KIRSANOVA TS
+ LEPESHINSKAYA VN + MUSIKHIN VA + SEN AP
+ SMOLYAKOV A
BULL ACAD SCI USSR, PHYS SER 37(12): 54-8 (1973)
CA 80: 101951

4-0737 CHEMICAL SHIFTS IN URANIUM AUGER TRANSITIONS.
ALLEN GC + WILD RK
CHEM PHYS LETT 15(2): 279-81 (1972)
SAA 1972: 67901; CA 77: 120194C

4-0738 AUGER SPECTROSCOPY OF URANIUM.
ALLEN GC + WILD RK
J CHEM SOC DALTON TRANS 1974(5): 493-8
CA 80: 150636

4-0739 ENERGY ANALYSIS OF BACKSCATTERED ELECTRONS FROM A SILICON SINGLE CRYSTAL UNDER A LOW ENERGY BOMBARDMENT. (AUGER SPECTRA OF SILICON, SILICON CARBIDE AND SILICON DIOXIDE)
ALLIE G + GERVAIS A
COMPT REND B 273: 395-8 (3 SEP 1971) (IN FRENCH)

4-0740 BAND STRUCTURE OF SILICON BY CHARACTERISTIC AUGER SPECTRUM ANALYSIS.
AMELIO GF
SURFACE SCI 22: 301-18 (1970)

4-0741 AUGER SPECTROSCOPY OF GRAPHITE SINGLE CRYSTALS WITH LOW ENERGY ELECTRONS.
AMELIO GF + SCHEIBNER EJ
SURFACE SCI 11: 242-54 (1968)

4-0742 SECONDARY ELECTRON EMISSION FROM TUNGSTEN AND MOLYBDENUM SINGLE CRYSTALS.
ARIFOV UA + VAKHIDOV UV + KASYMOV AKH
BULL ACAD SCI USSR, PHYS SER 35(3): 513-15 (1971)

4-0743 AUGER L-SERIES ELECTRONIC SPECTRA ACCOMPANYING THE DECAY OF ERBIUM-160, YTTERBIUM-166, YTTERBIUM-169, AND LUTETIUM-171.
ARTAMONOVA KP + VORONKOV AA + GRIGOREV EP
+ ZOLOTAVIN AV + SERGEEV VO
P115-16 OF PROGRAMMA TEZISY DOKL SOVESHCH YAD SPEKTROSK STRUKT AT YADRA, 24TH, LEBEDEV PP + PODKOPAEV YUN (EDS), NAUKA, LENINGRAD OTD, 1974. (IN RUSSIAN)
CA 82: 130246

4-0744 L-SERIES AUGER ELECTRON SPECTRA FOR ELEMENTS WITH Z = 66-70.
ARTAMONOVA KP + VORONKOV AA + GRIGOREV EP
+ ZOLOTAVIN AV + SERGEEV VO
P43-8 OF TEOR AT AT SPEKTROV MATER VSES SEMIN, VOL 1, 1973, REZ-IZD OTD LATV GOS UNIV, RIGA, USSR (IN RUSSIAN)
CA 85: 26187

4-0745 A NEW TREATMENT OF THE KLL AUGER SPECTRUM.
ASAAD WN
NUCL PHYS 66: 494-512 (1965).

4-0746 ENERGIES OF THE KLL AUGER LINES.
ASAAD WN
J PHYS (PARIS) COLLOQ NO 4: 132-8 (1971)
CA 77: 26864Z

4-0747 POSITIONS OF THE KLL AUGER LINES.
ASAAD WN
P455-74, VOL-1 OF INT CONF INNER SHELL IONIZATION PHENOMENA AND FUTURE APPLICATIONS, 1972, FINK RW (ED), NTIS, 1973.
CA 78: 129995

4-0748 INTENSITIES OF KLL AUGER LINES.
ASAAD WN
NUCL PHYS 44: 399-414 (1963)

4-0749 CALCULATION OF AUGER SPECTRA IN J-J COUPLING.
ASAAD WN
NUCL PHYS 44: 415-30 (1963)

4-0750 RELATIVISTIC CALCULATION OF THE KLL AUGER
SPECTRUM. (OF MERCURY)
ASAAD WN
PROC ROY SOC A 249: 555-73 (1959)

4-0751 THE K-AUGER SPECTRUM.
ASAAD WN + BURHOP EHS
PROC PHYS SOC 71: 369-82 (1958).

4-0752 METAL- DIELECTRIC TRANSITION IN GERMANIUM AND
SILICON STUDIED BY A MICROWAVE METHOD.
(PRODUCTION OF AUGER ELECTRONS)
ASHKINADZE BM + SULTANOV FK
JETP LETT 16(5): 190-4 (1972)
CA 78: 152899

4-0753 KLL AUGER ELECTRON SPECTRA OF THULIUM AND
LUTETIUM.
BABENKOV MI + BOBYKIN BV
BULL ACAD SCI USSR, PHYS SER 32: 1840-3
(1968)

4-0754 KLL AUGER ELECTRONS OF CADMIUM AND ANTIMONY.
BABENKOV MI + BOBYKIN BV + NORGORODOV AF
+ TASKARIN BT
BULL ACAD SCI USSR, PHYS SER 33: 1205-9
(1969)

4-0755 SPECTRUM OF AUGER KLL ELECTRONS OF TELLURIUM
AND XENON.
BABENKOV MI + BOBYKIN BV + PETUKHOV VK
BULL ACAD SCI USSR, PHYS SER 35(8): 1446-53
(1971)
CA 76: 53237H

4-0756 SPECTRUM OF AUGER K,L,M,N,O ELECTRONS OF
SAMARIUM.
BABENKOV MI + BOBYKIN BV + ZHDANOV VS
+ PETUKHOV VK
IZV AKAD NAUK SSSR, SER FIZ 38(10): 2197-200
(1974) (IN RUSSIAN)
CA 82: 77980

4-0757 KLM SPECTRUM OF SAMARIUM AUGER ELECTRONS.
BABENKOV MI + BOBYKIN BV + ZHDANOV VS
+ PETUKHOV VK
P254-5 OF PROGRAMMA TEZISY DOKL SOVESHCH YAD
SPEKTROSK STRUKT AT YADRA, 24TH, LEBEDEV PP +
PODKOPAEV YUN (EDS), NAUKA, LENINGRAD OTD,
1974. (IN RUSSIAN)
CA 82: 117961

4-0758 KMM AUGER TRANSITIONS IN XENON-131.
BABENKOV MI + OBBYKIN BV + ZHDANOV VS
+ PETUKHOV VK
PHYS LETT A 56(5): 363-5 (1976)
CA 85: 12018

4-0759 SECONDARY ELECTRON EMISSION FROM SODIUM
CHLORINE SINGLE CRYSTAL BOMBARDED BY RARE GAS
IONS.
BABOUX JC + PERDRIX M + GOUTTE R + GUILLAUD C
J PHYS D 4: 1617-23 (1971) (IN FRENCH)

4-0760 AUGER STUDIES OF ELECTRCLESS NICKEL FILMS.
BAETZOLD RC + MACK RE
J CATAL 37(2): 251-7 (1975)
CA 82: 160742

4-0761 CARRIER RECOMBINATION IN N-INDIUM ANTIMONIDE.
BAEV IA
COMPT REND ACAD BULG SCI 28(5): 615-17 (1975)
SAA 1975: 77132

4-0762 X-RAY FLUORESCENCE L SPECTRA OF NIOBIUM IN
NIOBIUM METAL AND SOME OF ITS COMPOUNDS.
BAHL MK + KCSTER AS
J PHYS C 7(24): 4537-50 (21 DEC 1974)
SAA 1975: 19014

4-0763 RELAXATION DURING PHOTOEMISSION AND LMM AUGER
DECAY IN ARSENIC AND SOME OF ITS COMPOUNDS.
BAHL MK + WOODALL RO + WATSON RL + IRGOLIC KJ
J CHEM PHYS 64(3): 1210-18 (1976)
CA 84: 97563

4-0764 SECONDARY ELECTRON EMISSION ROCKING CURVE
FROM COPPER.
BARO AM + SALMERON M
PHYS STATUS SOLIDI B 49(2): K135-7 (1972)

4-0765 A STUDY OF THE INTENSITY OF THE L2,3VV AUGER
TRANSITION IN MAGNESIUM AND SILICON FROM THE
CROSS SECTION MODEL OF GRYZINSKI. (AUGER
EFFECT)
BARO AM + SALMERON M
ELECTRON FIS APPL 17(1-2): 166 (1974)
SAA 1975: 03978

4-0766 MVV AUGER SPECTRA OF COPPER AND NICKEL.
BARO AM + SALMERON M + ROJO JM
J PHYS F 5(4): 826-35 (1975)
CA 83: 35385

4-0767 THRESHOLD EFFECTS ON RELATIVE INTENSITIES IN
LMM AUGER SPECTRA OF GERMANIUM AND SELENIUM.
BARRIE A
APPL PHYS LETT 26(11): 621-2 (1975)
CA 83: 50317

4-0768 COMPARISON OF X-RAY INDUCED AND ELECTRON
IMPACT INDUCED PHOTOELECTRON AND AUGER
SPECTRA. CARBON MONOXIDE ADSORPTION ON
MOLYBDENUM.
BARRIE A + BRUNDLE CR
J ELECTRON SPECTROSC RELAT PHENOMENA 5: 321-8
(1974)
CA 82: 90365

4-0769 EXPERIMENTAL DETERMINATION OF THE INTERNAL
PHOTOEMISSION CONTRIBUTION TO KLM AUGER
SPECTRA OF SODIUM, MAGNESIUM, AND ALUMINUM.
BARRIE A + STREET FJ
SOLID STATE COMMUN 15(6): 1017-19 (1974)
CA 81: 161849

4-0770 AUGER AND X-RAY PHOTOELECTRON SPECTROSCOPIC
STUDY OF SODIUM METAL AND SODIUM OXIDE.
BARRIE A + STREET FJ
J ELECTRON SPECTROSC RELAT PHENOMENA 7(1):
1-31 (1975)
CA 83: 105954

4-0771 QUASIATOMIC FINE STRUCTURE IN THE AUGER
SPECTRA OF SOLID SILVER AND INDIUM.
BASSETT PB + GALLON TE + MATTHEW JAD
+ PRUTTON M
SURFACE SCI 35(1): 63-74 (1973)
CA 78: 77594Q

4-0772 HIGH RESOLUTION AUGER ELECTRON SPECTRUM OF
MAGNESIUM OXIDE (100).
BASSETT PJ + GALLON TE + PRUTTON M
+ MATTHEW JAD
SURFACE SCI 33(1): 213-8 (1972)

4-0773 PHOTOELECTRON DETERMINATION OF THE
ATTENUATION OF THE LOW ENERGY ELECTRONS IN
ALUMINUM OXIDE. (AUGER ELECTRON LINES)
BATTYE FL + JENKIN JG + LIESEGANG JL
+ LECKEY RCG
PHYS REV B 9(7): 2887-93 (1 APR 1974)
SAA 1974: 57669

4-0774 INNER ELECTRON BINDING ENERGIES AND AUGER
ENERGIES IN FREE ATOMS FOR USE IN X-RAY
SPECTROSCOPY.
BECK DR + NICOLAIDES CA
J ELECTRON SPECTROSC RELAT PHENOMENA 8(3):
249-54 (1976)
CA 84: 171888

4-0775 DIFFRACTION PEAKS IN SECONDARY ELECTRON
ENERGY SPECTRA. (NICKEL (100))
BECKER GE + HAGSTRUM HD
J VACUUM SCI TECHNOL 11(1): 284-7 (JAN-FEB
1974)
SAA 1974: 76347

4-0776 ELECTRONIC EMISSION FROM SOLID TARGETS
BOMBARDED BY NOBLE GAS IONS (10-100 KEV):
ENERGETIC AND SPATIAL DISTRIBUTIONS. (AUGER
ELECTRON EMISSION)
BENAZETH N + AGUSTI J + BENAZETH C
+ MISCHLER J + VIEL L
NUCL INSTRUM METH 132: 477-82 (1976)
CA 85: 11987

4-0777 ELECTRON HOLE DROPS IN GERMANIUM- SILICON
ALLOYS. (AUGER RECOMBINATION)
BENOITALAGUILIAUME C + VOOS M + PETROFF Y
PHYS REV B 10(12): 4995-5002 (1974)
SAA 1975: 38212

4-0778 AUGER RECOMBINATION IN GALLIUM ANTIMONIDE.
BENZ G + CONRADT R
P1262-6 OF INT CONF PHYS SEMICOND, 12TH PROC,
PILKUHN MH(ED), TEUBNER, STUTTGART, 1974.
CA 83: 18374

4-0779 L1,L2,3V AUGER TRANSITION IN SILICON.
BERRER S + BARO AM + SALMERON M
SOLID STATE COMMUN 16(5): 651-3 (1975)
CA 82: 162641

4-0780 MAGNETIC FIELD DEPENDENT INTENSITY
OSCILLATIONS OF THE ELECTRON HOLE
LUMINESCENCE IN PURE GERMANIUM. (AUGER
RECOMBINATION)
BETZLER K + ZHURKIN BG + KARUZSKII AL
SOLID STATE COMMUN 17(5): 577-9 (1975)
SAA 1975: 77664

4-0781 FLUORESCENCE YIELDS AND AUGER RATES FOR
MULTIPLY IONIZED NITROGEN.
BHALLA CP
J PHYS B 8(17): 2792-7 (1975)
CA 84: 51757

4-0782 RELATIVISTIC KLL AUGER TRANSITION RATES.
BHALLA CP + RAMSDALE DJ
Z PHYS 239: 95-102 (1970)

4-0783 AUGER SPECTROSCOPY OF TITANIUM.
BISHOP HE + RIVIERE JC
SURFACE SCI 24: 1-17 (1971)

4-0784 AUGER SPECTROSCOPY OF SILICON.
BISHOP HE + RIVIERE JC + TAYLOR NJ
SURFACE SCI 17: 462-5 (1969)

4-0785 RECOMBINATION OF CARRIERS IN N-TYPE INDIUM
ARSENIDE AT 77 DEGREES K. (AUGER)
BLAUTBLACHEV AN + BALAGUROV LA + KARATAEV VV
+ OMELYANOVSKII EM
SOV PHYS SEMICOND 9(4): 515-6 (1975)
SAA 1976: 22389

4-0786 SECONDARY ELECTRON YIELDS FROM COPPER
BERYLLIUM OXYGEN SURFACE BY THERMAL CARBON
OXYGEN NITROGEN(2), AND NOBLE GAS
METASTABLES.
BORST WL
REV SCI INSTRUM 42: 1543-4 (OCT 1971)

4-0787 LIFETIME EFFECT ON THE AUGER SPECTRA OF
METALS. (SODIUM)
BOSE SM + FITCHEK J
PHYS LETT A 48(6): 443-4 (1974)
CA 81: 143788

4-0788 CHEMISORPTION REACTIONS ON HIGH INDEX ZINC
SULFIDE SURFACES. (AUGER EFFECT)
BOTTOMS WR + LIDOW DB
J ELECTROCHEM SOC 122(1): 119-21 (JAN 1975)
SAA 1975: 25550

4-0789 THE K-AUGER SPECTRUM OF THE FREE MAGNESIUM
ATOM.
BREUCKMAN B + SCHMIDT V
Z PHYS 268(2): 235-9 (1974)
CA 81: 084132

4-0790 EMISSION LINES OF THE M SERIES OF RARE EARTH
METALS. (AUGER EFFECT)
BRYTOV IA + MSTIBOVSKAYA LE
OPT SPECTROSC 39(2): 229-30 (1975)
SAA 1976: 34633

4-0791 IDENTIFICATION OF HIGH ENERGY LINES IN THE
KLL AUGER SPECTRUM OF NITROGEN.
CARLSON TA + MODDEMAN WE + PULLEN BP
+ KRAUSE MO
CHEM PHYS LETT 5: 390-2 (1970)

4-0792 STUDY OF MUSCOVITE AND SILICON BY AUGER
ELECTRON SPECTROSCOPY.
CARRIERE B + DEVILLE JP + GOLDSZTAUB S
COMPT REND B 271: 796-8 (1970) (IN FRENCH)

4-0793 AUGER ELECTRON SPECTROSCOPY STUDY OF SOME
OXYGENATED SILICON COMPOUNDS.
CARRIERE B + DEVILLE JP + GOLDSZTAUB S
COMPT REND B 272: 951-4 (1971) (IN FRENCH)

4-0794 AUGER SPECTROSCOPY OF SILICON IN SILICATES.
CARRIERE B + DEVILLE JP + GOLDSZTAUB S
COMPT REND B 274: 415-8 (7 FEB 1972) (IN
FRENCH)

4-0795 AUGER ELECTRON SPECTROSCOPY OF INSULATING
SILICON COMPOUNDS.
CARRIERE B + DEVILLE JP + GOLDSZTAUB S
VACUUM 22(10): 485-7 (1972)
CA 78: 77695Y

4-0796 STUDY OF AMORPHOUS AND CRYSTALLINE SILICATES
BY AUGER ELECTRON SPECTROSCOPY.
CARRIERE B + DEVILLE JP + GOLDSZTAUB S
SILICATES IND 39(11): 313-21 (1974)
CA 82: 102914

4-0797 CHEMICAL SHIFTS IN PHOTOEXCITED AUGER
SPECTRA.
CASTLE JE + EPLER D
PROC ROY SOC LONDON A 339(1616): 49-72 (1974)
CA 81: 070628

4-0798 ULTRAVIOLET EMISSIONS IN CALCITE. (AUGER
EFFECT)
CEVA T + MARTI C
J LUMIN 10(3): 205-9 (1975)
SAA 1975: 58755

4-0799 FLUORESCENCE PHOTON AND AUGER ELECTRON
SPECTRA.
CHARLTON DE
RADIAT RES 50(3): 455-63 (1972)
CA 77: 41663B

4-0800 EFFECT OF ELECTRONIC CORRELATIONS ON KLL
AUGER LINES IN ELEMENTS MAGNESIUM TO
SCANDIUM.
CHATTARJI D + TALUKDAR B
PHYS LETT A 39: 399-401 (5 JUN 1972)

4-0801 EFFECT OF ATOMIC CORRELATIONS ON KLL AUGER
TRANSITIONS.
CHATTARJI D + TALUKDAR B
PHYS REV A 7(1): 42-7 (1973); P490-508 OF INT
CONF INNER SHELL IONIZATION PHENOMENA AND
FUTURE APPLICATIONS, 1972, VOL 1, FINK
RW(ED), NTIS, 1973.
CA 78: 64369V

4-0802 RELATIVISTIC STUDY OF KLL AUGER TRANSITION
PROBABILITIES.
CHATTARJI D + TALUKDAR B
PHYS REV 174: 44-7 (1968)

4-0803 KLL AUGER TRANSITION PROBABILITIES FOR
ELEMENTS WITH LOW AND INTERMEDIATE ATOMIC
NUMBERS.
CHEN MH + CRASEMANN B
PHYS REV A 8(1): 7-13 (1973)
CA 79: 085120

4-0804 LLM VERSUS LMM AUGER TRANSITIONS FOR THE
LIGHT ELEMENTS.
CHUNG MF + JENKINS LH
SURFACE SCI 28: 637-44 (1971)

4-0805 AUGER ELECTRON SPECTROSCOPY OF THE OUTER
SHELL ELECTRONS.
CHUNG MF + JENKINS LH
SURFACE SCI 22: 479-85 (1970)

4-0806 AUGER SPECTRA OF LITHIUM METAL AND LITHIUM
OXIDE.
CLAUSING RE + EASTON DS + POWELL GL
SURFACE SCI 36(1): 377-9 (1973)
CA 78: 104007

4-0807 LEED- AES ELECTRON INDUCED EMISSION STUDIES
OF LITHIUM HYDRIDE.
CLAUSING RE + POWELL GL + HOLCOMBE CE
UNION CARBIDE CORP, 1970, 14P. Y-DA-4246
CA 76: 146144R

4-0808 KLL AUGER SPECTRUM OF CHLORINE. (IN CARBON
TETRACHLORIDE)
CLEFF B + MEHLHCRN W
Z PHYS 219: 311-24 (1969) (IN GERMAN)

4-0809 CHEMICAL SHIFTS IN THE AUGER SPECTRA FROM
OXIDIZED CHROMIUM AND VANADIUM.
COAD JP
PHYS LETT A 35: 185-6 (1971)

4-0810 LMM AUGER SPECTRA OF SOME TRANSITION METALS
OF THE FIRST SERIES.
COAD JP
Z PHYS 244: 19-30 (1971)

4-0811 LLM VERSUS LMM AUGER TRANSITIONS FOR LIGHT
ELEMENTS.
COAD JP
SURFACE SCI 30(3): 687-91 (1972)

4-0812 UNEXPLAINED PHENOMENA IN THE AUGER SPECTRA OF
THE FIRST TRANSITION SERIES. (COMMENTS ON
PAPER BY CHUNG AND JENKINS)
COAD JP
PHYS LETT A 37: 437-8 (20 DEC 1971)

4-0813 COMBINATION OF AUGER SPECTRCSCOPY AND
CHARACTERISTIC LOSS SPECTROSCOPY FOR THE
ELEMENTS VANADIUM TO COBALT.
COAD JP + RIVIERE JC
PHYS STATUS SOLIDI A 7: 571-5 (1971)

4-0814 EXPERIMENTAL AND THEORETICAL STUDY OF THE
AUGER SPECTRUM OF NITROUS OXIDE.
CONNOR JA + HILLIER IH + KENDRICK J
+ BARBER M + BARRIE A
J CHEM PHYS 64(8): 3325-9 (1976)
CA 85: 12132

4-0815 SEMIEMPIRICAL AUGER ELECTRON ENERGIES.
PART-2: THE ARGON ATOM.
CROTTY JM + LARKINS FP
J PHYS B 9(6): 881-5 (1976)
CA 85: 38912

4-0816 CHEMICAL SHIFT IN AUGER SPECTRA OF TUNGSTEN
AND MOLYBDENUM DURING THE ADSORPTION OF
OXYGEN.
DADAYAN KA + KRIGER YUG + TAPILIN VM
+ SAVCHENKO VI
BULL ACAD SCI USSR, PHYS SER 38(2): 83-7
(1974)
CA 80: 150706

4-0817 AUGER SPECTRA OF LITHIUM AND SODIUM UNDER ION
BOMBARDMENT.
DE LESEGNO PV + JOYES P + HENNEQUIN JF
COMPT REND B 275: 93-6 (JUL 1972) (IN FRENCH)

4-0818 K-SHELL FLUORESCENCE YIELDS FOR LIGHT
ELEMENTS.
DICK CE + LUCAS AC
PHYS REV A 2: 580-6 (1971)

4-0819 CHEMICAL EFFECTS ON KLL AUGER ELECTRON
SPECTRUM FROM OXYGEN.
DOOLEY GJ + GRANT JT + HAAS TW
BULL AMER PHYS SOC 15: 1507 (1970)

4-0820 AUGER SPECTRA OF CARBON. TANTALUM CARBIDES,
ETHYLENE AND CARBON MONOXIDE ADSORPTION ON
TANTALUM AND POLYCRYSTALLINE RHENIUM.
DUCROS R + PICUARD G + WEBER B + CASSUTO A
SURFACE SCI 54(2): 513-8 (1976) (IN FRENCH)
CA 84: 128324

4-0821 N4,5 AUGER SPECTRA OF GADOLINIUM.
DUFOUR G + BONNELLE C
J PHYS (PARIS) LETT 35(12): L255-6 (1974) (IN
FRENCH)
CA 82: 66235

4-0822 AUGER SPECTRA OF MAGNESIUM AND ALUMINUM.
DUFOUR G + GUENNOU H + BONNELLE C
SURFACE SCI 32(3): 731-4 (1972)

4-0823 4D AUGER TRANSITIONS FROM EXCITED STATES IN
SAMARIUM METAL.
DUFOUR G + KARNATAK RC + MARIOT JM
+ BONNELLE C
J PHYS LETT 37(5): 119-21 (1976)
CA 85: 38888

4-0824 AUGER ELECTRON ANALYSIS OF ELECTROPOLISHED
HIGH PURITY ALUMINUM.
DUNN CG + HARRIS LA
J ELECTROCHEM SOC 117: 81-2 (1970)

4-0825 CHEMICAL SHIFTS IN AUGER ELECTRON SPECTRA.
FAHLMAN A
P757-72 OF INT CONF INNER SHELL IONIZATION
PHENOMENA AND FUTURE APPLICATIONS, VOL 1,
1972, FINK RW(ED), NTIS, 1973.
CA 78: 129968

4-0826 CHEMICAL SHIFT IN AUGER SPECTRA.
FAHLMAN A + HAMRIN K + NORDBERG R
+ NORDLING C + SIEGBAHN K
PHYS LETT 20: 159-60 (1966)

4-0827 AUGER SPECTRA FOR ELEMENTS OF LOW ATOMIC
NUMBER.
FAHLMAN A + NORDBERG R + NORDLING C
+ SIEGBAHN K
Z PHYS 192: 476-83 (1966)

4-0828 WORK FUNCTION VARIATION WITH ALLOY
COMPOSITION SILVER- GOLD. (AUGER ELECTRON
SPECTROSCOPY)
FAIN SC + MCDAVID JM
PHYS REV B 9(12): 5099-107 (15 JUN 1974)
SAA 1974: 75946

4-0829 LMM AUGER SPECTRA OF SULFUR IN SODIUM SULFATE
AND SODIUM PYROSULFITE.
FARRELL HH
SURFACE SCI 34(2): 465-70 (1973)
CA 78: 64517S

4-0830 EMISSION OF KLL AUGER ELECTRONS PRODUCED IN
Z(1)- NITROGEN AND Z(1)- NEON COLLISIONS AT
KEV ENERGIES.
FASTRUP B + LARSEN GA
P392-4 OF INT CONF PHYS ELECTRON ATOMIC
COLLISIONS, AMSTERDAM, 1971, 7TH PROC,
ABSTRACTS.
SAA 1972: 663

4-0831 TIGHT BINDING CALCULATION OF A CORE VALENCE
VALENCE AUGER LINE SHAPE: SILICON (111).
FEIBELMAN PJ + MCGUIRE EJ + PANDEY KC
PHYS REV LETT 36(19): 1154-7 (1976)
CA 85: 53987

4-0832 LIFETIME BROADENING IN X-RAY INDUCED
PHOTOEMISSION SPECTROSCOPY AND AUGER SPECTRA
OF THE ZINC CHALCOGENIDES.
FIERMANS L + VENNIK J
NED TIJDSCHR VACUUMTECH 11(1): 1-5 (1973) (IN
DUTCH)
CA 78: 166747

4-0833 LIFETIME AND SURFACE EFFECTS ON THE L23 VV
AUGER SPECTRA OF ALUMINUM.
FITCHEK J + BOSE SM
PHYS LETT A 54(6): 460-2 (1975)
CA 84: 51794

4-0834 CHEMICAL EFFECTS ON THE AUGER ELECTRON
SPECTRA OF BERYLLIUM.
FORTNER RJ + MUSKET RG
SURFACE SCI 28: 339-43 (1971)

4-0835 AUGER RATES FOR SOFT X-RAY TRANSITIONS IN
IONIC, ATOMIC, AND METALLIC LITHIUM.
FRANCESCHETTI DR + DOW JD
J PHYS F 4(6): L151-5 (1974)
CA 81: 070785

4-0836 THE L-AUGER SPECTRA OF PLUTONIUM, AMERICIUM,
CALIFORNIUM, AND FERMIUM.
FREEDMAN MS + PORTER FT
P680-93 OF INT CONF INNER SHELL IONIZATION
PHENOMENA AND FUTURE APPLICATIONS, VOL 1,
1972, FINK RW(ED), NTIS, 1973.
CA 78: 130735

4-0837 X-RAY EXCITED AUGER AND PHOTOELECTRON SPECTRA
OF MAGNESIUM, SOME ALLOYS OF MAGNESIUM, AND
ITS OXIDE.
FUGGLE JC + WATSON LM + FABIAN DJ
+ AFFROSSMAN S
J PHYS F 5(2): 375-83 (FEB 1975)
SAA 1975: 26239

4-0838 COMPARISON OF THE SOFT X-RAY PHOTOEMISSION,
AND AUGER ELECTRON SPECTRA OF THE VALENCE
BANDS OF MAGNESIUM, MAGNESIUM(2) COPPER AND
MAGNESIUM(3) GOLD.
FUGGLE JC + WATSON LM + NORRIS PR + FABIAN DJ
J PHYS F 5(3): 590-6 (1975)
CA 83: 18581

4-0839 VALENCE BAND AUGER ELECTRON SPECTRA FOR
ALUMINUM: COMMENTS.
GADZUK JW
PHYS REV B 9(4): 1978-80 (1974)
CA 81: 008139

4-0840 (AUGER) RECOMBINATION OF NONEQUILIBRIUM
CHARGE CARRIERS OF INDIUM ARSENIDE AT HIGH
PHOTOEXCITATION LEVELS.
GALKIN GN + KHARAKHORIN FF + SHATKOVSKII EV
LIET FIZ RINKINYS 15(6): 1001-7 (1975) (IN
RUSSIAN)
CA 84: 143656

4-0841 ESTIMATION OF BACKSCATTERING EFFECTS IN
ELECTRON INDUCED AUGER SPECTRA (SILVER AND
SILICON, AUGER EFFECT ENHANCED BY
BACKSCATTERING).
GALLON TE
J PHYS D 5: 822-32 (APR 1972)

4-0842 LOW ENERGY AUGER EMISSION FROM LITHIUM
FLUORIDE.
GALLON TE + MATTHEW JAD
PHYS STATUS SOLIDI B 41: 343-51 (1970)

4-0843 SOLID STATE BROADENING IN AUGER SPECTRA.
(CADMIUM, RARE GASES)
GALLON TE + NUTTALL JD
SURFACE SCI 53(1): 698-715 (1975)
CA 84: 113703

4-0844 INFLUENCE OF AN ELECTRIC FIELD ON THE AUGER
RECOMBINATION IN SEMICONDUCTORS.
GEBRANZIG U + HAUG A + ROSENTHAL W
PHYS STATUS SOLIDI B 68(2): 749-60 (1975)
SAA 1975: 47270

4-0845 (AUGER) ELECTRON SPECTROSCOPY OF TUNGSTEN.
GERGELY GY
KOZL MAGY TUD ADAD MUSZ FIZ KUT INTEZ NO 0-8:
151-90 (1973) (IN HUNGARIAN)
SAA 1973: 72256

4-0846 ELECTRON BINDING ENERGIES OF BARIUM FROM THE
SECONDARY ELECTRON YIELD SPECTRUM.
GERLACH RL
SURFACE SCI 28: 648-50 (1971)

4-0847 TOTAL ELECTRON IMPACT IONIZATION CROSS
SECTIONS ON K-SHELLS OF SURFACE ATOMS. (AUGER
INTENSITIES OF TUNGSTEN (100))
GERLACH RL + DUCHARME AR
SURFACE SCI 32: 329-40 (1972)

4-0848 A NEW METHOD FOR MEASURING ELECTRON IMPACT
IONIZATION CROSS SECTIONS OF INNER SHELLS.
GLUPE G + MEHLHORN W
PHYS LETT A 25: 274-5 (1967)

4-0849 AUGER SPECTROSCOPY OF A BARIUM- METAL FILM
SYSTEM.
GNUCHEV NM + KIRSANOVA TS
SOV PHYS SOLID STATE 16(5): 866-8 (1974)
CA 81: 043500

4-0850 SECONDARY ELECTRON EMISSION FROM TUNGSTEN AND
MOLYBDENUM.
GOMOYUNOVA MV + ALIEV BZ
BULL ACAD SCI USSR, PHYS SER 35(2): 211-15
(1971)

4-0851 X-RAY PHOTOELECTRON SPECTRUM OF DIAMOND.
(AUGER TRANSITIONS)
GORA T + STALEY R + RIMSTIDT JD + SHARMA J
PHYS REV B 5(6): 2309-14 (1972)
CA 76: 119650E

4-0852 METHOD FOR DETECTING FINE STRUCTURE IN THE
SECONDARY ELECTRON EMISSION YIELD AND THE
APPLICATION TO SILICON (111).
GOTO K + ISHIKAWA K
J APPL PHYS 43: 1559-62 (APR 1972)

4-0853 FINE STRUCTURES AND ENERGY DISTRIBUTION OF
SECONDARY ELECTRON EMISSION FROM SILICON
(111). (AUGER ELECTRON EXCITATION)
GOTO K + ISHIKAWA K
J APPL PHYS 44(1): 132-7 (1973)
CA 78: 63850H

4-0854 AUGER ELECTRON SPECTROSCOPY OF SILICON.
GRANT JT + HAAS TW
SURFACE SCI 23: 347-62 (1970)

4-0855 COMPARISON OF MAGNESIUM L2,3MM AUGER CURRENTS
USING ELECTRON AND ION EXCITATION.
GRANT JT + HOOKER MP
PHYS LETT A 54(2): 167-8 (1975)
CA 83: 185986

4-0856 AUGER CURRENT MEASUREMENTS FOR QUANTITATIVE
AUGER ELECTRON SPECTROSCOPY OF SOLIDS.
GRANT JT + HOOKER MP + HAAS TW
J COLLOID INTERFACE SCI 55(2): 370-6 (1976)
CA 85: 13441

4-0857 COMPARISON OF AUGER SPECTRA OF MAGNESIUM,
ALUMINUM, AND SILICON EXCITED BY LOW ENERGY
ELECTRON AND LOW ENERGY ARGON ION
BOMBARDMENT.
GRANT JT + HOOKER MP + SPRINGER RW + HAAS TW
J VACUUM SCI TECHNOL 12(1): 481-4 (1975)
CA 83: 18571

4-0858 AUGER ELECTRON SPECTROSCOPY OF LUNAR SAMPLES.
GRANT RW + HOUSLEY RM + SZALKOWSKI FJ
+ MARCUS HL
P2423-39 OF LUNAR SCI CONF, 1974, 5TH PROC,
GOSE WA(ED), PERGAMON, 1974.
CA 83: 167121

4-0859 DOPPLER SHIFTED AUGER ELECTRONS FROM FOIL
EXCITED HEAVY ION BEAMS.
GROENEVELD KO + MANN R + NOLTE G
+ SCHUMANN S + SPOHR R
P197 OF INT CONF REACT COMPLEX NUCL, 1ST
PROC, ROBINSON RL + MCGOWAN FK + BALL
JB(EDS), NORTH-HOLLAND, 1974.
CA 83: 50483

4-0860 BEAM FOIL EXCITED AUGER TRANSITIONS IN NEON.
GROENEVELD KO + MANN R + NOLTE G
+ SCHUMANN S + SPOHR R + FRICKE B
Z PHYS A 274(3): 191-4 (1975)
CA 83: 199125

4-0861 SECONDARY ELECTRON EMISSION FROM GALLIUM
ARSENIDE.
GUTIERREZ WA + POMMERRENIG HD + HOLT SL
APPL PHYS LETT 21: 249-50 (5 SEP 1972)

4-0862 CHEMICAL EFFECTS ON THE KLL AUGER ELECTRON
SPECTRUM FROM SURFACE CARBON.
HAAS TW + GRANT JT
APPL PHYS LETT 16: 172-3 (1970)

4-0863   CHEMICAL SHIFTS IN AUGER ELECTRON
          SPECTROSCOPY.
          HAAS TW + GRANT JT + DOOLEY GJ
          BULL AMER PHYS SOC 14: 948 (1969)

4-0864   AUGER ELECTRON SPECTROSCOPY OF TRANSITION
          METALS.
          HAAS TW + GRANT JT + DOOLEY GJ
          PHYS REV B 1: 1449-59 (1970)

4-0865   AUGER ELECTRON SPECTROSCOPY OF SOME
          REFRACTORY METALS.
          HAAS TW + GRANT JT + DOOLEY GJ
          J VACUUM SCI TECHNOL 6: 903 (1969)

4-0866   SOME PROBLEMS IN ANALYSIS OF AUGER ELECTRON
          SPECTRA.
          HAAS TW + GRANT JT + DOOLEY GJ
          J VACUUM SCI TECHNOL 7: 43-5 (1970)

4-0867   CHEMICAL EFFECTS IN AUGER ELECTRON
          SPECTROSCOPY (STATES OF CARBON).
          HAAS TW + GRANT JT + DOOLEY GJ
          J APPL PHYS 43: 1853-60 (APR 1972)

4-0868   IDENTIFICATION OF THE CHEMICAL STATE OF
          ADSORBED SPECIES WITH AUGER ELECTRON
          SPECTROSCOPY.
          HAAS TW + GRANT JT + DOOLEY GJ
          P359-68 OF INT CONF ADSORPTION-DESORPTION
          PHENOMENA, 2ND PROC, RICCA F (ED), 1971,
          ACADEMIC PRESS, 1972.
              CA 77: 169117U

4-0869   ION EXCITED AUGER SPECTRA OF ALUMINUM.
          HAAS TW + SPRINGER RW + HOOKER MP + GRANT JT
          PHYS LETT A 47(4): 317-18 (1974)
              CA 80: 150686

4-0870   SOFT X-RAY APPEARANCE POTENTIAL SPECTROSCOPY
          IN A DISPLAY LEED SYSTEM.
          HAAS TW + THOMAS S + DOOLEY GJ
          SURFACE SCI 28: 645-7 (1971)

4-0871   PLASMON SATELLITE IN THE AUGER EMISSION
          SPECTRUM OF METALS. (ALUMINUM)
          HAGEN AL + GLICK AJ
          PHYS REV B 13(4): 1580-6 (1976)
              CA 84: 128384

4-0872   SEARCH FOR CORRELATION EFFECTS IN THE
          M4,5N4,5N4,5 AUGER SPECTRUM OF XENON.
          HAGMANN S + HERMANN G + MEHLHORN W
          Z PHYS 266(3): 189-94 (1974)
              CA 80: 126568

4-0873   AUGER ELECTRON SPECTRA OF MAGNESIUM THIN
          FILMS.
          HALDER NC + ALONSO J + SWARTZ WE
          PHYS REV B 13(6): 2418-25 (1976)
              CA 84: 186982

4-0874   AUGER ELECTRON EMISSION IN THE ENERGY SPECTRA
          OF SECONDARY ELECTRONS FROM MOLYBDENUM AND
          TUNGSTEN.
          HARROWER GA
          PHYS REV 102: 340-7 (1956)

4-0875   PHENOMENOLOGICAL SYSTEMATICS OF L- AND
          M-AUGER SPECTRA.
          HAYNES SK
          P559-661, VOL-1 OF INT CONF INNER SHELL
          IONIZATION PHENOMENA AND FUTURE APPLICATIONS,
          1972, FINK RW (ED), NTIS, 1973.
              CA 78: 129966

4-0876   ENERGY DISTRIBUTION OF AUGER ELECTRONS
          EMITTED FROM BERYLLIUM, MAGNESIUM, ALUMINUM
          AND SILICON UNDER ION BOMBARDMENT.
          HENNEQUIN JF + DE LESEGNO PV
          COMPT REND B 272: 1259-62 (2 JUN 1971) (IN
          FRENCH)

4-0877   AUGER ELECTRON EMISSION BY ION BOMBARDMENT OF
          LIGHT METALS.
          HENNEQUIN JF + VIARIS DE LESEGNO P
          SURFACE SCI 42(1): 50-60 (1974)
              CA 80: 088541

4-0878   VACUUM ULTRAVIOLET EMISSION OF XENON(III)
          LEVELS EXCITED BY ELECTRON IMPACT VIA N4,5
          AUGER TRANSITIONS.
          HERTZ H
          Z PHYS A 274(4): 289-91 (1975)
              SAA 1976: 12863

4-0879   CONFIGURATION INTERACTION CALCULATIONS OF THE
          AUGER SPECTRUM OF METHANE, HYDROGEN FLUORIDE,
          WATER, AND CARBON MONOXIDE.
          HILLIER IN + KENDRICK J
          MOL PHYS 31(3): 849-53 (1976)
              SAA 1976: 48586

4-0880   K-AUGER TRANSITIONS OF THE FREE SODIUM ATOM.
          HILLIG H + CLEFF B + MEHLHORN W + SCHMITZ W
          Z PHYS 268(2): 225-33 (1974)
              CA 81: 084125

4-0881   CHEMICAL SHIFTS IN AUGER ELECTRON SPECTRA
          FROM SILICON IN SILICON NITRIDE.
          HOLLOWAY PH
          SURFACE SCI 54(2): 506-8 (1976)
              CA 84: 113865

4-0882   RELAXATION PHENOMENA INVOLVED IN THE
          L3M4,5M4,.5M4,.5:1G4 AUGER PROCESS IN ZINC
          METAL.
          HOOGEWIJS R + FIERMANS L + VENNIK J
          CHEM PHYS LETT 38(3): 471-4 (1976)
              CA 84: 186986

4-0883   ATOMIC AND EXTRAATOMIC STATIC RELAXATION
          ENERGIES IN THE L3M4,5M4,5 AUGER PROCESS IN
          ARSENIC AND SELENIUM.
          HOOGEWIJS R + FIERMANS L + VENNIK J
          J PHYS C 9(4): L103-5 (1976)
              CA 84: 171702

4-0884   DETAILED ANALYSIS OF THE L3M4,5M4,5: 1G4
          AUGER TRANSITION IN ATOMIC ZINC: IMPROVED
          AUGER ENERGY CALCULATIONS.
          HOOGEWIJS R + FIERMANS L + VENNIK J
          CHEM PHYS LETT 38(1): 192-6 (1976)
              CA 84: 157554

4-0885   CHEMICAL EFFECTS IN THE M4,5NN AUGER SPECTRUM
          OF MOLYBDENUM (110) DUE TO ADSORPTION OF
          MOLECULAR OXYGEN AND CARBON MONOXIDE.
          HOOKER MP + GRANT JT + HAAS TW
          J VACUUM SCI TECHNOL 12(1): 325-8 (1975)
              CA 82: 162649

4-0886   Z DEPENDENCE OF ENERGIES AND RELATIVE
          INTENSITIES IN THE KLL AUGER SPECTRUM.
          HORNFELDT O
          ARKIV FYS 23(23): 235-45 (1963)

4-0887   AUGER SPECTRA IN THE INTERMEDIATE COUPLING
          REGION.
          HORNFELDT O + FAHLINAN A + NORDLING C
          ARKIV FYS 23(16): 155-63 (1963)

4-0888   CLUSTER CALCULATIONS OF THE SOFT X-RAY L3
          EMISSION SPECTRA OF NICKEL AND COPPER.
          (AUGER-TYPE PROCESSES)
          HOUSE D + SMITH PV + GYORFFY BL
          J PHYS F 3(4): 745-52 (APR 1973)
              SAA 1973: 36954

4-0889   VALENCE BAND STRUCTURE IN THE AUGER SPECTRUM
          OF ALUMINUM.
          HOUSTON JE
          J VACUUM SCI TECHNOL 12(1): 255-9 (1975)
              CA 82: 162642

4-0890   DIFFERENCE IN THE CHROMIUM L3/L2 INTENSITY
          RATIO MEASURED BY SOFT X-RAY AND AUGER
          ELECTRON APPEARANCE POTENTIAL SPECTROSCOPY.
          HOUSTON JE + PARK RL
          PHYS REV B 5(10): 3808-9 (1972)
              CA 77: 26979R

4-0891 USE OF X-RAY SPECTROSCOPIC DATA TO EXPLAIN
THE FINE STRUCTURE OF THE DEPENDENCE OF THE
COEFFICIENT OF SECONDARY ELECTRON EMISSION ON
ENERGY OF PRIMARY ELECTRONS (POTASSIUM
IODIDE).
IVANOV AV + TYUTIKOV AM
BULL ACAD SCI USSR, PHYS SER 36(2): 246-50
(1972)

4-0892 THE KVV AUGER SPECTRUM OF LITHIUM METAL.
JACKSON AJ + TATE C + GALLON TE + BASSETT PJ
+ MATTHEW JAD
J PHYS F 5(2): 363-74 (FEB 1975)
SAA 1975: 26217

4-0893 AUGER ELECTRON EMISSION OF THIN CARBON FOILS
IN REFLECTION AND TRANSMISSION.
JACOBI K
SURFACE SCI 26: 54 (1971) (IN GERMAN)

4-0894 SECONDARY ELECTRON EMISSION FROM THIN FILMS.
JAHRREISS H
THIN SOLID FILMS 12: 187-200 (SEP 1972)

4-0895 LOW ENERGY AUGER AND LOSS ELECTRON SPECTRA
FROM MAGNESIUM AND ITS OXIDE.
JANSSEN AP + SCHOONMAKER RC + CHAMBERS A
+ PRUTTON M
SURFACE SCI 45(1): 45-60 (1974)
CA 81: 129544

4-0896 AUGER AND OTHER CHARACTERISTIC ENERGIES IN
SECONDARY ELECTRON SPECTRA FROM ALUMINUM
SURFACES.
JENKINS LH + CHUNG MF
SURFACE SCI 28: 409-22 (1971)

4-0897 ENERGY SPECTRUM OF BACK SCATTERED ELECTRONS
AND CHARACTERISTIC LOSS AND GAIN PHENOMENA OF
COPPER (111).
JENKINS LH + CHUNG MF
SURFACE SCI 26: 151-64 (1971)

4-0898 AUGER SATELLITE AND OTHER CHARACTERISTIC
EVENTS IN MAGNESIUM SECONDARY ELECTRON
SPECTRA.
JENKINS LH + CHUNG MF
SURFACE SCI 33(1): 159-71 (1972)

4-0899 PLASMON ENERGY GAIN AND DOUBLE IONIZATION
SATELLITES IN THE BERYLLIUM AUGER SPECTRUM.
JENKINS LH + ZEHNER DM
SOLID STATE COMMUN 12(11): 1149-51 (1973)
CA 79: 047504

4-0900 HIGHER ENERGY SATELLITES IN AUGER SPECTRA
FROM METALS.
JENKINS LH + ZEHNER DM
P33 OF EUROP PHYS SOC STUDY CONF ON METAL
SURFACES, ABSTR, 1973.
SAA 1974: 33545

4-0901 CHARACTERISTIC ENERGY GAIN AND LOSS, DOUBLE
IONIZATION, AND IONIZATION LOSS EVENTS IN
BERYLLIUM AND BERYLLIUM OXIDE SECONDARY
ELECTRON SPECTRA.
JENKINS LH + ZEHNER DM + CHUNG MF
SURFACE SCI 38(2): 327-40 (1973)
CA 79: 047622

4-0902 PHOTOELECTRON AND AUGER SPECTRA OF XENON IN
SOLID PERXENATES.
JOERGENSEN CK + BERTHOU H
CHEM PHYS LETT 36(4): 432-5 (1975)
CA 84: 51923

4-0903 HIGH RESOLUTION STUDY OF ARGON LMM AUGER
ELECTRONS PRODUCED BY ION BOMBARDMENT.
JOHNSON BM + MATTHEWS DL + SMITH LE
+ MOORE CF
J PHYS B 6(12): L369-71 (1973)
CA 80: 054124

4-0904 HIGH RESOLUTION BEAM GAS AUGER ELECTRON
SPECTRA FOR OXYGEN IONS EXCITED BY COLLISIONS
WITH HELIUM, NEON, AND ARGON.
JOHNSON BM + SCHNEIDER D + ROBERTS KS
+ BOLGER JE + MOORE CF
PHYS LETT A 53(3): 254-6 (1975)
CA 83: 87906

4-0905 LINEAR LIGAND FIELDS INFLUENCING AUGER
SPECTRA OF FLUORIDES, COPPER(I) AND
SILVER(I).
JORGENSEN CK + BERTHOU H
CHEM PHYS LETT 25(1): 21-5 (1974)
CA 80: 139222

4-0906 AUGER SPECTRA OF HIGHLY IONIZED OXYGEN AND
FLUORINE.
JUNKER BR + BARDSLEY JN
PHYS REV A 8(3): 1345-9 (1973)
CA 79: 131016

4-0907 SECONDARY ELECTRON EMISSION DUE TO PRIMARY
AND BACKSCATTERED ELECTRONS (TABLE FOR 60
ELEMENTS).
KANAYA K + KAWAKATSU H
J PHYS D 5: 1727-42 (SEP 1972)

4-0908 N5,4 AND M5,4 APPEARANCE POTENTIAL SPECTRA
FROM PRASEODYMIUM.
KANSKI J + NILSSON PO
PHYS FENN 9, SUPPL S1: 68-70 (JUL 1974)
SAA 1975: 19018

4-0909 K-AUGER RATES CALCULATED FOR NEON(+).
KELLY HP
PHYS REV A 11(2): 556-65 (1975)
CA 82: 147590

4-0910 SULFUR KLL AUGER SPECTRA IN HYDROGEN SULFIDE,
SULFUR DIOXIDE, AND SULFUR HEXAFLUORIDE.
KESKI-RAHKONEN O
P57 OF ANNU CONF FINNISH PHYS SOC, 1975 PROC.
SAA 1975: 36732

4-0911 TOTAL AND PARTIAL ATOMIC LEVEL WIDTHS.
KESKI-RAHKONEN O + KRAUSE MO
ATOMIC DATA NUCL DATA TABLES 14(2): 139-46
(1974)
CA 82: 023906

4-0912 HIGH RESOLUTION L2,3M2,3M2,3 AUGER ELECTRON
SPECTRA OF ARGON.
KONDOW T + KAWAI T + KUNIMORI K + ONISHI T
+ TAMARU K
J PHYS B 6(6): L156-8 (1973)
CA 79: 059664

4-0913 SECONDARY ELECTRON EMISSION FROM ION
IMPLANTED SILICON (CESIUM).
KONRAD GT
J APPL PHYS 43: 1996-8 (APR 1972)

4-0914 K-AUGER SPECTRUM OF NEON.
KORBER H + MEHLHORN W
Z PHYS 191: 217-30 (1966) (IN GERMAN)

4-0915 EXOELECTRON SPECTROMETRY AS A METHOD FOR
DETERMINATION OF THE ENERGY DEPTH OF LEVELS
OF SURFACE ELECTRON CENTERS. (AUGER EFFECT,
ZIRCONIA, SILICA)
KORTOV VS + SHIFRIN VP
SOV PHYS SOLID STATE 17(7): 1404-5 (1975)
SAA 1976: 45416

4-0916 SECONDARY AND AUGER ELECTRON SPECTRA OF A
COPPER- BERYLLIUM ALLOY.
KOSHIKAWA T + SHIMIZU R
PHYS LETT A 44(2): 112-14 (1973)
CA 79: 059788

4-0917 SECONDARY ELECTRON EMISSION FROM IRON SINGLE
AND POLYCRYSTALS.
KOSHIKAWA T + UEDA K + SHIMIZU R
TECHNOL REP OSAKA UNIV FAC ENG 22: 51-76 (FEB
1972)

4-0918 ATOMIC RADIATION TRANSITION PROBABILITIES TO
THE 1S STATE AND THEORETICAL K-SHELL
FLUORESCENCE YIELDS.
KOSTROUN VO + CHEN MH + CRASEMANN B
PHYS REV A 3: 533-45 (1971)

4-0919 L2,3M4,5M4,5 AUGER SPECTRA OF METALLIC COPPER
AND ZINC: THEORY AND EXPERIMENT.
KOWALCZYK SP + POLLAK RA + MCFEELY FR
+ LEY L + SHIRLEY DA
PHYS REV B 8(6): 2387-91 (1973)
CA 80: 008711

4-0920 MANIFESTATION OF ATOMIC DYNAMICS THROUGH THE
AUGER EFFECT. (NEON AUGER SPECTRUM)
KRAUSE MO + CARLSON TA + MODDEMAN WE
J PHYS (PARIS) COLLOQ NO 4: 139-44 (1971)
CA 77: 26854W; 94977P

4-0921 EXOELECTRON EMISSION FROM ZINC OXIDE. (OXYGEN
ADSORPTION EFFECT, AUGER ELECTRON
SPECTROSCOPY)
KRIEGSEIS W + SCHARMANN A
P96-9 OF INT SYMP ON EXOELECTRON EMISSION AND
DOSIMETRY, 4TH PROC, CZECH ACAD SCI, 1974.
SAA 1974: 49544

4-0922 CHEMICAL SHIFT IN THE AUGER SPECTRUM OF
TUNGSTEN FROM OXYGEN ADSORPTION.
KRIGER YUG + TAPILIN VM
KRATK SODERZH DOKL (1): 68-9 (1973) (IN
RUSSIAN)
CA 83: 155000

4-0923 FILM STRUCTURE EFFECT ON THE SECONDARY
ELECTRON EMISSION CHARACTERISTICS (GERMANIUM
AND BERYLLIUM).
KRYNKO YUN + KOVAL IF + MELNIK PV
UKR FIZ ZH 17: 302-5 (FEB 1972) (IN RUSSIAN)

4-0924 PHOTOELECTRON CHEMICAL SHIFTS IN COORDINATION
COMPOUNDS. X-RAY EXCITED LMM AUGER SPECTRA OF
GERMANIUM.
KUMAR G
VANDERBILT UNIV, THESIS 1973, 151P.
CA 81: 097446

4-0925 CHEMICAL EFFECTS OF THE NITROGEN KLL AUGER
SPECTRA OF CHEMISORBED NITROGEN ON IRON AND
MOLYBDENUM SURFACES.
KUNIMORI K + KAWAI T + KONDOW T + ONISHI T
+ TAMARU K
SURFACE SCI 54(2): 525-7 (1976)
CA 84: 113866

4-0926 AUGER PEAKS IN THE ENERGY SPECTRA OF
SECONDARY ELECTRONS FROM VARIOUS MATERIALS.
LANDER JJ
PHYS REV 91: 1382-7 (1953)

4-0927 CHARGE STATE DEPENDENCE OF X-RAY AND AUGER
ELECTRON EMISSION SPECTRA FOR RARE GAS ATOMS.
PART-1. THE ARGON ATOM. PART-2. KRYPTON ATOM
M-SHELL. PART-3.
LARKINS FP
J PHYS B 4: 1-19 (JAN 1971); 6(11): 2450-63
(1973)

4-0928 KRYPTON M-SHELL AUGER AND COSTER-KRONIG
ELECTRON ENERGIES.
LARKINS FP
J PHYS B 6(8): 1556-61 (1973)
CA 79: 109919

4-0929 L1,L2,3M COSTER-KRONIG TRANSITION ENERGIES.
LARKINS FP
J PHYS B 7(1): 37-46 (11 JAN 1974)
SAA 1974: 12478

4-0930 SEMI-EMPIRICAL AUGER ELECTRON ENERGIES.
PART-1: GENERAL METHOD AND KLL LINE ENERGIES.
LARKINS FP
J PHYS B 9(1): 47-58 (1976)
CA 84: 81876

4-0931 AUGER ELECTRON AND CHARACTERISTIC ENERGY LOSS
SPECTRA OF PLUTONIUM.
LARSON DT + ADAMS RO
SURFACE SCI 47(2): 413-21 (1975)
CA 82: 147627

4-0932 X-RAY INDUCED PHOTOELECTRON AND AUGER SPECTRA
OF COPPER, COPPER(II) OXIDE, COPPER(I) OXIDE,
AND COPPER(I) SULFIDE THIN FILMS.
LARSON PE
J ELECTRON SPECTROSC RELAT PHENOMENA 4(3):
213-18 (1974)
CA 81: 161837

4-0933 ABSORPTION BANDS IN SECONDARY ELECTRON
CONTINUOUS SPECTRUM.
LEGRESSUS C + MASSIGNON D
COMPT REND B 280(13): 401-4 (1975) (IN
FRENCH)
CA 83: 36442

4-0934 NEW METHOD FOR ANALYZING THE SECONDARY
ELECTRON SPECTRUM.
LEGRESSUS C + MASSIGNON D + SOPIZET R
COMPT REND B 280(14): 439-42 (1975) (IN
FRENCH)
CA 83: 51769

4-0935 INTERPRETATION OF THE AUGER ELECTRON SPECTRUM
FROM OXIDIZED BERYLLIUM.
LEJEUNE EJ + DIXON RD
J APPL PHYS 43: 1998-9 (APR 1972)

4-0936 HIGH RESOLUTION LMM AUGER SPECTRA OF
GERMANIUM.
LINDAU I + RIBBING CG
PHYS LETT A 44(7): 509-10 (1973)
CA 79: 120098

4-0937 FINE STRUCTURE DETAILS IN THE LMM AUGER
SPECTRA OF GERMANIUM.
LINDAU I + RIBBING CG
PHYS STATUS SOLIDI B 59(1): 259-67 (1973)
CA 79: 120198

4-0938 KLL AUGER SPECTRUM OF MANGANESE.
LIU YY + ALBRIDGE RG
NUCL PHYS A 92: 139-144 (1967)

4-0939 WIDTH OF ATOMIC L2 AND L3 VACANCY STATES NEAR
Z = 30. (AUGER EFFECT)
LO IY + ADLER I + MAU HC + CRASEMANN B
PHYS REV A 7(3): 897-903 (MAR 1973)
SAA 1973: 29512

4-0940 HIGH ENERGY SATELLITES IN AUGER SPECTRA FROM
METALS.
LOFGREN H + WALLDEN L
SOLID STATE COMMUN 12(1): 19-21 (1973)
CA 78: 090649

4-0941 NUCLEAR AUGER EFFECTS IN MUONIC ATOMS. (LEAD,
BISMUTH)
LOHS KP + WOLSCHIN GW
NUCL PHYS A 236(2): 457-68 (1974)
CA 82: 91054

4-0942 SECONDARY ELECTRON SPECTRA OF SOME SOLID
TARGETS UNDER IONIC BOMBARDMENT.
LOUCHET F + VIEL L + BENAZETH C + FAGOT B
+ COLOMBIE N
RADIAT EFF 14(1-2) 123-30 (1972)
CA 77: 41155N

4-0943 HIGH RESOLUTION MEASUREMENTS OF P(+) ARGON
KLL AND KLM AUGER ELECTRONS.
MACKEY JJ + SMITH LE + JOHNSON BM + MOORE CF
+ MATTHEWS DL
J PHYS B 7(16): L447-L449 (1974)
CA 82: 009778

4-0944 INFLUENCE OF THE BAND STRUCTURE ON THE AUGER
ELECTRON SPECTRUM OF SILICON.
MAGUIRE HG + AUGUSTUS PD
J PHYS C 4: L174-9 (1971)

4-0945 CHARACTERISTIC K-SHELL LOW ENERGY ELECTRON
LOSS SPECTRA FROM DIAMOND, GRAPHITE, AND
EVAPORATED CARBON SURFACES.
MAGUIRE HG + CILLIE GG
J PHYS C 9(5): L135-8 (1976)
SAA 1976: 41004

4-0946 HIGH FREQUENCY EXCITON BREAKDOWN AND FREE
CARRIER AND EXCITON KINETICS IN GERMANIUM IN
THE PRESENCE OF ELECTRON HOLE DROPS. (AUGER
RECOMBINATION)
MANENKOV AA + MILYAEV VA + MIKHAILOVA GN
+ SANINA VA + SEFEROV AS
ZH EKSP TEOR FIZ 70(2): 695-701 (1976) (IN
RUSSIAN)
SAA 1976: 45378

4-0947 CHEMICAL ANALYSIS OF ELECTRODEPOSITED NICKEL-
NICKEL BONDS BY AES.
MARCUS HL + WALDROP JR + SCHULER FT
+ CAIN EFC
J ELECTROCHEM SOC 119(10): 1348-9 (1972)
CA 78: 23125U

4-0948 STRUCTURE OF LOW ENERGY ELECTRON SPECTRUM.
MASSIGNON D + LE GRESSUS C + SOPIZET R
+ DURAUD JP
COMPT REND D 281(23): 579-82 (1975) (IN
FRENCH)
CA 84: 82251

4-0949 AUGER ELECTRON FINE PROFILES IN IONIC
CRYSTALS.
MATTHEW JAD
PHYS LETT A 32: 261-2 (1970)

4-0950 EFFECTS OF EXTRA CLOSED CORE RELAXATION ON
AUGER AND BINDING ENERGY SHIFTS.
MATTHEW JAD
J PHYS C 8(8): L136-8 (1975)
CA 83: 35221

4-0951 MEASUREMENTS OF NEON K-AUGER ELECTRONS
PRODUCED BY ENERGETIC HEAVY ION BOMBARDMENT.
MATTHEWS DL
UNIV TEXAS, THESIS 1974, 145P.
CA 81: 113376

4-0952 HIGH RESOLUTION K-AUGER SPECTRA FOR MULTIPLY
IONIZED NEON.
MATTHEWS DL + JOHNSON BM + MACKEY JJ
+ MOORE CF
PHYS REV LETT 31(22): 1331-4 (1973)
CA 80: 021022

4-0953 HIGH RESOLUTION NEON KLL AUGER SPECTRA
PRODUCED BY HYDROGEN(+) AND HELIUM(+) ION
BOMBARDMENT.
MATTHEWS DL + JOHNSON BM + MACKEY JJ
+ MOORE CF
PHYS LETT A 45(6): 447-8 (1973)
CA 80: 042507

4-0954 AUGER DECAY OF NEON FOLLOWING ENERGETIC ION
BOMBARDMENT.
MATTHEWS DL + JOHNSON BM + MACKEY JJ
+ SMITH LE + HODGE W + MOORE CF
PHYS REV A 10(4): 1177-87 (1974)
CA 82: 009661

4-0955 CALCULATED K-AUGER ELECTRON AND K X-RAY
TRANSITION ENERGIES FOR THE MULTIPLY IONIZED
NEON ATOM.
MATTHEWS DL + JOHNSON BM + MOORE CF
ATOMIC DATA NUCL DATA TABLES 15(1): 41-7
(1975)
CA 83: 87909

4-0956 MULTIPLET EFFECTS ON THE L2,3 FLUORESCENCE
YIELD OF MULTIPLY IONIZED ARGON. (AUGER
EFFECT)
MAU HC + CRASEMANN B
PHYS REV A 10(6): 2232-9 (DEC 1974); PHYS
FENN 9, SUPPL S1: 250-2 (1974)
SAA 1975: 20865

4-0957 STRIPPING MECHANISM DEPENDENCE OF THE ARGON
2S FLUORESCENCE YIELD. (AUGER TRANSITION)
MAUHSIUNG C + CRASEMANN B
PHYS REV A 12(2): 710-11 (1975)
SAA 1976: 1072

4-0958 ANGULAR DEPENDENCE OF AUGER ELECTRON EMISSION
FROM A SINGLE CRYSTAL SPECIMEN.
MCDONNELL L + WOODRUFF DP
VACUUM 22(10): 477-80 (1972)
CA 78: 77606V

4-0959 L- AND M-SHELL AND SUBSHELL FLUORESCENCE AND
COSTER-KRONIG YIELDS.
MCGEORGE JC
P175-96 OF INT CONF INNER SHELL IONIZATION
PHENOMENA AND FUTURE APPLICATIONS, VOL 1,
1972, FINK REV (ED), NTIS, 1973.

4-0960 L-SHELL AUGER AND COSTER-KRONIG ELECTRON
SPECTRA.
MCGUIRE EJ
PHYS REV A 3: 1801-10 (1971)

4-0961 K-SHELL AUGER TRANSITION RATES AND
FLUORESCENT YIELDS FOR ELEMENTS ARGON TO
XENON.
MCGUIRE EJ
PHYS REV A 2: 273-8 (1970)

4-0962 K-SHELL AUGER TRANSITION RATES AND
FLUORESCENT YIELDS FOR ELEMENTS BERYLLIUM TO
ARGON.
MCGUIRE EJ
PHYS REV 185: 1-6 (1969)

4-0963 ATOMIC L-SHELL COSTER-KRONIG, AUGER, AND
RADIATIVE RATES AND FLUORESCENCE YIELDS FOR
SODIUM-THORIUM.
MCGUIRE EJ
PHYS REV A 3: 587-94 (1971)

4-0964 ATOMIC M-SHELL COSTER-KRONIG, AUGER, AND
RADIATIVE RATES, AND FLUORESCENCE YIELDS FOR
CALCIUM TO THORIUM.
MCGUIRE EJ
PHYS REV A 5: 1043-7 (MAR 1972)

4-0965 M-SHELL AUGER AND COSTER-KRONIG ELECTRON
SPECTRA.
MCGUIRE EJ
PHYS REV A 5: 1052-9 (MAR 1972)

4-0966 K- AND L-SHELL FLUORESCENCE AND AUGER YIELDS
AND AUGER ELECTRON SPECTROSCOPY.
MCGUIRE EJ
J PHYS (PARIS) COLLOQ NO 4: 124-7 (1971)
CA 77: 26852U

4-0967 M-SHELL AUGER, COSTER-KRONIG, AND RADIATIVE
MATRIX ELEMENTS, AND AUGER AND COSTER-KRONIG
TRANSITION RATES IN J-J COUPLING.
MCGUIRE EJ
SANDIA LAB, 1972, 98P. SC-RR-710835
CA 77: 158293X

4-0968 L- AND M-SHELL YIELDS AND ELECTRON SPECTRA
FOR THE TRANSURANIC ELEMENTS.
MCGUIRE EJ
P662-79 OF INT CONF INNER SHELL IONIZATION
PHENOMENA AND FUTURE APPLICATIONS, VOL 1,
1972, FINK RW(ED), NTIS, 1973.
CA 78: 130129

4-0969 ATOMIC N-SHELL COSTER-KRONIG, AUGER, AND
RADIATIVE RATES, AND FLUORESCENCE YIELDS FOR
Z BETWEEN 38 AND 103.
MCGUIRE EJ
PHYS REV A 9(5): 1840-51 (1974)
CA 81: 018723

4-0970 WIDTHS OF 4S AND 5S MULTIPLET COMPONENTS IN
THE RARE EARTHS.
MCGUIRE EJ
PHYS REV A 10(1): 13-20 (1974)
CA 81: 070333

4-0971 AUGER SPECTRA OF THE NOBLE GASES. KRYPTON.
MCGUIRE EJ
PHYS REV A 11(1): 17-25 (1975)
CA 82: 177463

4-0972 AUGER TRANSITION RATE FOR THE NOBLE GAS
SATELLITE SPECTRA.
MCGUIRE EJ
PHYS REV A 11(1): 10-16 (1975)
CA 82: 177462

4-0973 AUGER CASCADES IN ALUMINUM.
MCGUIRE EJ
PHYS REV A 11(6): 1889-98 (1975)
CA 83: 88250

4-0974 AUGER SPECTRUM OF THE NOBLE GASES. PART-2:
ARGON.
MCGUIRE EJ
PHYS REV A 11(6): 1880-8 (1975)
CA 83: 68763

4-0975　K-SHELL FLUORESCENCE YIELD OF IONIZED
　　　　ALUMINUM. (AUGER RATE)
　　　　MCGUIRE EJ
　　　　SANDIA LAB, 1975, 15P. SAND-74-0324
　　　　　　CA 83: 35173

4-0976　DOUBLE IONIZATION EFFECTS ON RADIATIVE RATES,
　　　　AUGER RATES, AND FLUORESCENCE YIELDS.
　　　　MCGUIRE JH + MITTLEMAN MH
　　　　PHYS REV A 5(4): 1971-3 (1972)
　　　　　　SAA 1972: 31682; CA 76: 119317T

4-0977　COSTER-KRONIG AND AUGER SPECTRUM OF THE L1
　　　　SHELL OF ARGON.
　　　　MEHLHORN W
　　　　Z PHYS 208: 1-27 (1968) (IN GERMAN).

4-0978　AUGER ELECTRON SPECTROMETRY OF SINGLY AND
　　　　MULTIPLY IONIZED ATOMS.
　　　　MEHLHORN W
　　　　PHYS FENN 9, SUPPL S1: 223-30 (1974)
　　　　　　CA 82: 147185

4-0979　AUGER AND COSTER-KRONIG STUDIES OF THE
　　　　M-SHELL OF KRYPTON.
　　　　MEHLHORN W
　　　　Z PHYS 187: 21-40 (1965) (IN GERMAN)

4-0980　THE KLL AUGER SPECTRUM FOR ATOMIC NUMBERS
　　　　BETWEEN 10 AND 36.
　　　　MEHLHORN W + ASAAD WN
　　　　Z PHYS 191: 231-47 (1966) (IN GERMAN).

4-0981　CORRELATION EFFECTS IN THE AUGER SPECTRA OF
　　　　NOBLE GASES.
　　　　MEHLHORN W + SCHMITZ W + STAHLHERM D
　　　　Z PHYS 252(5): 399-411 (1972) (IN GERMAN)
　　　　　　CA 77: 68149A

4-0982　ENERGY WIDTHS OF ROENTGEN LEVELS BY AUGER
　　　　ELECTRON SPECTROSCOPY.
　　　　MEHLHORN W + STAHLHERM D + VERBEEK H
　　　　Z NATURFORSCH A23: 287-94 (1968)

4-0983　AUGER SPECTRUM OF L2 AND L3 SHELLS OF ARGON.
　　　　MEHLHORN W + STAHLHERM D
　　　　Z PHYS 217: 294-300 (1968) (IN GERMAN)

4-0984　AUGER SATELLITES ON THE L23 AUGER EMISSION
　　　　BANDS OF ALUMINUM, SILICON, AND PHOSPHORUS.
　　　　MELLES JJ
　　　　UNIV MISSOURI PHD THESIS, 1973, 80P.
　　　　　　CA 83: 155091

4-0985　AUGER SATELLITES OF THE L2,3 AUGER EMISSION
　　　　BANDS OF ALUMINUM, SILICON, AND PHOSPHORUS.
　　　　MELLES JJ + DAVIS LE + LEVENSON LL
　　　　PHYS REV B 9(11): 4618-27 (1974)
　　　　　　CA 81: 055959

4-0986　AUGER AND OTHER CHARACTERISTIC ENERGIES IN
　　　　SECONDARY ELECTRON SPECTRA FROM THIN
　　　　EVAPORATED PHOSPHORUS FILMS.
　　　　MELLES JJ + DAVIS LE + LEVENSON LL
　　　　J VACUUM SCI TECHNOL 10(1): 140-3 (1973)
　　　　　　CA 78: 104084

4-0987　MEASUREMENT OF THE IONIZATION CROSS SECTION
　　　　OF THE L2,3 SHELL OF CHLORINE USING AUGER
　　　　ELECTRON SPECTROSCOPY.
　　　　MEYER F + VRAKKING JJ
　　　　PHYS LETT A 44(7): 511-12 (1973)
　　　　　　CA 79: 118347

4-0988　ENERGY SPECTRUM AND ANGULAR DISTRIBUTION OF
　　　　SECONDARY ELECTRONS EMITTED BY IONIC
　　　　BOMBARDMENT.
　　　　MISCHLER J + BENAZETH C + COLOMBIE N + VIEL L
　　　　VIDE 28(164): 63-4 (1973) (IN FRENCH)
　　　　　　CA 79: 120279

4-0989　AUGER SPECTROSCOPY OF SIMPLE GASEOUS
　　　　MOLECULES.
　　　　MODDEMAN WE
　　　　UNIV TENNESSEE PHD THESIS, 1970, 96P.
　　　　　　CA 76: 8549A

4-0990　DETERMINATION OF KLL AUGER SPECTRA OF
　　　　NITROGEN, OXYGEN, CARBON DIOXIDE, NITRIC
　　　　OXIDE, WATER, AND CARBON MONOXIDE.
　　　　MODDEMAN WE + CARLSON TA + KRAUSE MO
　　　　+ PULLEN BP + BULL WE + SCHWEITZ G
　　　　J CHEM PHYS 55: 2317 (1971)

4-0991　AUGER SPECTRA OF SIMPLE MOLECULES.
　　　　MODDEMAN WE + CARLSON TA + PULLEN BP
　　　　+ KRAUSE MO
　　　　ACS ANN MEET 1969, 158TH PROC ABSTR PAPER NO
　　　　PHYS-178

4-0992　AUGER SPECTRA OF HYDROGEN CHLORIDE VAPOR
　　　　ETCHED N+ GALLIUM ARSENIDE (100) SUBSTRATES.
　　　　MOON RL + JAMES LW
　　　　J ELECTROCHEM SOC 120(4): 581-3 (1973)
　　　　　　CA 78: 141622

4-0993　OBSERVATION OF A PROJECTILE CHARGE DEPENDENCE
　　　　FOR THE NEON K-AUGER ELECTRON SPECTRUM.
　　　　MOORE CF + MACKEY JJ + SMITH LE + BOLGER JE
　　　　+ JOHNSON BM + MATTHEWS DL
　　　　J PHYS B 7(9): L302-4 (1974)
　　　　　　CA 81: 113127

4-0994　K-SHELL (AUGER) FLUORESCENT YIELDS FOR
　　　　MULTIPLY IONIZED NEON.
　　　　MOORE CF + MATTHEWS DL
　　　　P419-26 OF CONF ON APPL SMALL ACCEL, 3RD
　　　　PROC, 1974, DUGGAN JL + MORGAN IL (EDS),
　　　　NTIS, CONF-741040-P1, 1975.
　　　　　　CA 85: 54177

4-0995　IN-DEPTH PROFILES BY AUGER SPECTROSCOPY AND
　　　　SECONDARY ION EMISSION.
　　　　MORABITO JM
　　　　J VACUUM SCI TECHNOL 10(1): 278 (1973)
　　　　（ABSTRACT)

4-0996　IN-DEPTH PROFILES OF PHOSPHORUS ION IMPLANTED
　　　　SILICON BY AUGER SPECTROSCOPY AND SECONDARY
　　　　ION EMISSION.
　　　　MORABITO JM + TSAI JC
　　　　SURFACE SCI 33(2): 422-6 (1972)
　　　　　　CA 78: 9340D

4-0997　OBSERVATION AND INTERPRETATION OF THE AUGER
　　　　ELECTRON SPECTRUM FROM CLEAN BERYLLIUM.
　　　　MUSKET RG + FARTNER RJ
　　　　PHYS REV LETT 26: 80-2 (1971)

4-0998　SINGULARITIES IN AUGER EMISSION SPECTRUM OF
　　　　METALS.
　　　　NATTA M + JOYES P
　　　　J PHYS CHEM SOLIDS 31: 447-52 (1970)

4-0999　TEMPERATURE DEPENDENCE OF THE LOW ENERGY
　　　　AUGER SPECTRUM OF NICKEL(II) OXIDE (100).
　　　　NETZER FP + PRUTTON M
　　　　SOLID STATE COMMUN 15(2): 341-3 (1974)
　　　　　　CA 81: 113198

4-1000　THEORY OF KLL AUGER ENERGIES.
　　　　NICOLAIDES CA + BECK DR
　　　　CHEM PHYS LETT 27(2): 269-75 (1974)
　　　　　　CA 81: 113171

4-1001　EFFECT OF THE POLARITY OF INDIUM ANTIMONIDE
　　　　ON THE EXTERNAL PHOTOEFFECT IN THE X-RAY
　　　　SPECTRAL REGION. (EFFECT OF IMPURITIES ON THE
　　　　AUGER SPECTRUM.)
　　　　NIKOLAENYA AZ + NEKRASHEVICH IG
　　　　+ SEMERENKO VV
　　　　IZV VYSSH UCHEB ZAVED, FIZ 12(3): 68-73
　　　　(1969) (FOR TRANSLATION, SEE SOV PHYS J)

4-1002　NEW RADIATIVE TRANSITION IN GERMANIUM.
　　　　NILSSON NG + SVANTESSON KG
　　　　P1105-12 OF INT CONF PHYS SEMICOND 1972, 11TH
　　　　PROC, VOL-2, AMER ELSEVIER, 1972.
　　　　　　CA 81: 179414

4-1003　ANGULAR DISTRIBUTION OF M4,5NN AUGER
　　　　ELECTRONS EJECTED FROM KRYPTON BY ELECTRON
　　　　IMPACT.
　　　　NISHIMURA F + TAHIRA S + ODA N
　　　　J PHYS SOC JAP 35(3): 861-5 (SEP 1973)
　　　　　　SAA 1973: 69112

4-1004 ORIGIN OF THE PEAK SHAPES IN THE AUGER
SPECTRUM OF OXYGEN (KL2,3L2,3) ON MOLYBDENUM.
NOZOYE H + MATSUMOTO Y + ONISHI T + KONSOW T
+ TAMARU K
J PHYS C 8(23): 4131-4 (1975)
CA 84: 51798

4-1005 SOLID STATE BROADENING EFFECTS IN THE AUGER
SPECTRUM OF XENON.
NUTTALL JD + GALLON TE
SOLID STATE COMMUN 15(2): 329-32 (1974)
CA 81: 097487

4-1006 SOLID STATE BROADENING IN THE AUGER SPECTRA
OF RARE GASES. (ARGON, KRYPTON, XENON)
NUTTALL JD + GALLON TE
PHYS STATUS SOLIDI B 71(1): 259-68 (1975)
CA 84: 10659

4-1007 ENERGY SPECTRA OF ELECTRONS EJECTED IN
ARGON(+)- ARGON COLLISIONS. (AUGER AND
COSTER-KRONIG TRANSITIONS)
OGURTSOV GN
P400-3 OF INT CONF PHYS ELECTRON ATOMIC
COLLISIONS, 7TH, AMSTERDAM, 1971, NORTH
HOLLAND, 1971.
SAA 1972: 660

4-1008 ENERGY DISTRIBUTION OF ELECTRONS EMITTED BY
ARGON ATOMS UNDER ELECTRON IMPACT.
OGURTSOV GN
SOV PHYS JETP 37(4): 584-6 (1973)
CA 79: 011085

4-1009 INFORMATION OBTAINED FROM THE WIDTH OF
SPECTRAL LINES IN THE ELECTRON SPECTROSCOPY
FOR CHEMICAL ANALYSIS.
OHTA T
KAGAKU NO RYOIKI 27(6): 516 (1973) (IN
JAPANESE)
CA 79: 084976

4-1010 AUGER ELECTRON EJECTION FROM XENON N4,5OO AND
KRYPTON M4,5NN PROCESSES BY ELECTRON IMPACT
NEAR THRESHOLD.
OHTANI S + NISHIMURA H + SUZUKI HG + WAKIYA K
PHYS REV LETT 36(15): 863-6 (1976)
CA 84: 186964

4-1011 ARGON CHARGE STATES DUE TO IMPACT OF
HYDROGEN(+) AND HELIUM(+). (AUGER SPECTRA)
OONA H
PHYS REV LETT 32(11): 571-4 (18 MAR 1974)
SAA 1974: 36444

4-1012 THEORETICAL ANALYSIS OF THE AUGER SPECTRA OF
METHANE.
ORTENBURGER IB + BAGUS PS
PHYS REV A 11(5): 1501-3 (1975)
CA 83: 50314

4-1013 AUGER TRANSITIONS BETWEEN 0 AND 1500 EV FOR
ELEMENTS UP TO Z EQUALS 88.
PACKER ME + WILSON JM
INST PHYS, LONDON, 1971, 458P.
CA 78: 50471K

4-1014 K-SHELL FLUORESCENCE YIELDS OF ARGON,
CHLORINE AND SULFUR. (AUGER TRANSITION)
PAHOR J + MOLJK A + KODRE A + RUPNIK T
+ HRIBAR M
FIZIKA 4, SUPPL: 19-21 (1971)
SAA 1973: 26570

4-1015 SECONDARY ELECTRON EMISSION FROM THIN FILM
GOLD METAL- OXIDE CERMETS.
PAKSWER S + STEVENS MT
J APPL PHYS 42: 5853-4 (DEC 1971)

4-1016 SECONDARY ELECTRON SPECTROSCOPY OF
POLYCRYSTALLINE SILVER.
PATTINSON EB + HARRIS PR
J PHYS D 5(7): L59-65 (1972)

4-1017 STUDY OF INTENSITIES OF AUGER LINES EXCITED
BY ELECTRON BOMBARDMENT AND ALUMINUM K-ALPHA
X-RAY IRRADIATION.
PESSA M
J APPL PHYS 42: 5831-6 (1971)

4-1018 L-SHELL AUGER ENERGIES AND INTENSITIES IN 3D
TRANSITION METALS.
PHILLIPS JD
GEORGIA INST TECHNOL, THESIS, 1973, 207P.
CA 78: 166476

4-1019 PROTON STATES, ANOMALOUS CONVERSION, AND
K-AUGER SPECTRA IN AMERICIUM-241 FROM
CURIUM-241 (ELECTRON CAPTURE) AND
BERKELIUM-245 ALPHA DECAYS.
PORTER FT + AHMAD I + FREEDMAN MS
+ MILSTED J + FRIEDMAN AM
PHYS REV C 10(2): 803-24 (1974)
CA 81: 129998

4-1020 PLASMON EFFECTS IN ELECTRON ENERGY LOSS AND
GAIN SPECTRA IN ALUMINUM.
POWELL BD + WOODRUFF DP
SURFACE SCI 33(3): 437-44 (1972)
CA 78: 9657N

4-1021 STRUCTURE ON THE HIGH ENERGY SIDE OF THE
KL2,3M AUGER PEAK FROM SOLID ALUMINUM:
INTERNAL PHOTOEMISSION.
POWELL CJ
APPL PHYS LETT 20: 335-7 (1 MAY 1972)

4-1022 CONTRASTING VALENCE BAND AUGER ELECTRON
SPECTRA FOR SILVER AND ALUMINUM.
POWELL CJ
PHYS REV LETT 30(23): 1179-82 (1973)
CA 79: 036768

4-1023 HIGH RESOLUTION MEASUREMENTS OF AUGER
ELECTRON AND PHOTOELECTRON STRUCTURE IN THE
SECONDARY ELECTRON ENERGY DISTRIBUTIONS OF
ALUMINUM, NICKEL, AND COPPER.
POWELL CJ + MANDL A
PHYS REV B 6: 4418-29 (1972); P743-56 OF INT
CONF INNER SHELL IONIZATION PHENOMENA AND
FUTURE APPLICATIONS, VOL 1, 1972, FINK
RW(ED), NTIS, 1973.
CA 78: 130132

4-1024 HIGH RESOLUTION MEASUREMENTS OF THE
L3M2,3M4,5 AUGER TRANSITONS IN NICKEL AND
COPPER.
POWELL CJ + MANDL A
PHYS REV LETT 29: 1153-6 (1972)

4-1025 AUGER SPECTRA OF LITHIUM HYDRIDE.
POWELL GL + MCGUIRE GE + EASTON DS
+ CLAUSING RE
SURFACE SCI 46(2): 345-57 (1974)
CA 82: 77430

4-1026 ELECTROCHEMISTRY AT THIN SOLID FILMS. (AUGER
SPECTRA)
QUINN RK + ARMSTRONG NR + VAN DER BORGH NE
J VACUUM SCI TECHNOL 12(1): 160-8 (1975)
SAA 1975: 55324

4-1027 IDENTIFICATION OF AUGER SPECTRA FROM
ALUMINUM.
QUINTO DT + ROBERTSON WD
SURFACE SCI 27(3): 645-8 (1971)

4-1028 AUGER SPECTRA OF COPPER- NICKEL ALLOYS.
QUINTO DT + SUNDARAM VS + ROBERTSON WD
SURFACE SCI 28: 504-16 (1971)

4-1029 SECONDARY ELECTRON EMISSION OF METALS
BOMBARDED WITH 120-EV TO 5-KEV PROTONS.
RAY JA + BARNETT CF
J APPL PHYS 42: 3260 (JUL 1971)

4-1030 AUGER ELECTRON SPECTRUM OF OSMIUM AT ENERGIES
UP TO 300 EV.
REDKIN VS + ZASHKVARA VV + KORSUNSKII MI
+ ISVEIMAN EV
SOV PHYS SOLID STATE 13: 1269-70 (1971)

4-1031 ORIENTATION ANISOTROPY OF BACKSCATTERING
COEFFICIENT AND SECONDARY ELECTRON EMISSION
FOR 10-100 KEV ELECTRONS.
REIMER L + BADDE HG + SEIDEL H + BUHRING W
Z ANGEW PHYS 31(3): 145-51 (1971) (IN GERMAN)

4-1032 WIDTHS AND INTENSITIES OF HIGHLY RESOLVED
LINES IN ARGON L2,3 AUGER SPECTRA.
RIDDER D + DIERINGER J + STOLTERFOHT N
P419-20 OF INT CONF ON PHYS ELECTRONIC
COLLISIONS, (EXTENDED ABSTRACTS) JULY 1975,
RISLEY JS + GEBALLE R(EDS), WASHINGTON UNIV
PRESS, SEATTLE.
SAA 1976: 21171

4-1033 TRANSITION PROBABILITIES FOR THE L2,3MM AUGER
SPECTRUM OF SELENIUM.
ROBERTS ED + WEIGHTMAN P + JOHNSON CE
J PHYS C 8(14): 2336-42 (1975)
CA 83: 123675

4-1034 AUGER VACANCY SATELLITE STRUCTURE IN THE
L3,M4,5M4,5 AUGER SPECTRA OF COPPER.
ROBERTS ED + WEIGHTMAN P + JOHNSON CE
J PHYS C 8(13): L301-4 (1975)
CA 83: 105736

4-1035 PHOTOELECTRON AND L2,3MM AUGER ELECTRON
ENERGIES FOR ARSENIC.
ROBERTS ED + WEIGHTMAN P + JOHNSON CE
J PHYS C 8(8): 1301-9 (1975)
CA 83: 35396

4-1036 HIGH ENERGY FINE STRUCTURE IN THE AUGER
SPECTRA OF SILICON AND SILICON CARBIDE.
ROWE JE + CHRISTMAN SB
J VACUUM SCI TECHNOL 10(1): 276 (1973)
(ABSTRACT)

4-1037 ABSENCE OF PLASMON GAIN SATELLITES IN THE
AUGER SPECTRUM OF SILICON.
ROWE JE + CHRISTMAN SB
SOLID STATE COMMUN 13(3): 315-18 (1973)
CA 79: 109962

4-1038 COMPARISON OF SILICON L3,2 ELECTRON LOSS
SPECTRA AND OPTICAL ABSORPTION. (AUGER
EFFECT)
ROWE JE + CHRISTMAN SB
PHYS LETT A 43(4): 377-8 (26 MAR 1973)
SAA 1973: 34449

4-1039 SURFACE STATE TRANSITIONS OF SILICON IN
ELECTRON ENERGY LOSS SPECTRA. (AUGER EFFECT)
ROWE JE + IBACH H
PHYS REV LETT 31(2): 102-5 (9 JUL 1973)
SAA 1973: 52887

4-1040 CROSS SECTIONS FOR L2,3 AUGER EMISSION IN 22
TO 300 KEV PROTON- ARGON COLLISIONS.
RUDD ME
PHYS REV A 10(2): 518-21 (1974)
CA 81: 113180

4-1041 SHIFT ENERGIES OF CHARACTERISTIC X-RAYS AND
AUGER ELECTRONS FOR IONIZED ATOMS.
SAKISAKA M + KOBAYASHI N + MAEDA N + HORI H
MEM FAC ENG 37 PART-2: 77-87 (1975)
CA 84: 36927

4-1042 INTERATOMIC TRANSITIONS AND RELAXATION
EFFECTS IN AUGER SPECTRA OF SEVERAL GAS
ADSORBATES ON TRANSITION METALS.
SALMERON M + BARO AM + ROJO JM
PHYS REV B 13(10): 4348-63 (1976)
CA 85: 54309

4-1043 HIGH ENERGY SATELLITES OF THE M2,3VV AND M1VV
AUGER PEAKS OF COPPER.
SALMERON M
SURFACE SCI 41(2): 584-6 (1974)
CA 80: 076131

4-1044 EXPERIMENTAL OBSERVATION OF CHEMICAL SHIFTS
IN AUGER SPECTRUM FROM SURFACE LAYERS OF
SILICON DIOXIDE DURING ELECTRON BOMBARDMENT.
SALMERON M + BARO AM
SURFACE SCI 29: 300-2 (1972)

4-1045 INTERATOMIC AUGER PROCESSES IN SOME
ADSORBATES ON TRANSITION METALS.
SALMERON M + BARO AM
SURFACE SCI 49(1): 356-62 (1975)
CA 83: 35027

4-1046 HIGH ENERGY SATELLITES IN AUGER PEAKS OF
MAGNESIUM AND SILICON.
SALMERON M + BARO AM + ROJO JM
SURFACE SCI 41(1): 11-20 (1974)
CA 80: 032207

4-1047 ENERGY SPECTRA OF SECONDARY ELECTRONS EMITTED
BY MAGNESIUM AND SILICON WHEN BOMBARDED BY
PRIMARY ELECTRONS. PART-1: ENHANCED PLASMON
EXCITATION ON INNER LEVEL IONIZATION. PART-2:
DOUBLE IONIZATION ORIGIN OF THE HIGH ENERGY
SATELLITE OF AUGER PEAK.
SALMERON M + BARO AM + ROJO JM
P36 OF EUROP PHYS SOC STUDY CONF ON METAL
SURFACES, ABSTR, 1973.
SAA 1974: 33546

4-1048 EXCITATION SPECTRUM OF OXYGEN DOMINATED
COMPOUNDS IN THE 3-21 EV REGION OF THE
SPECTRUM. (INTERBAND AUGER TRANSITIONS,
RADIATIONLESS RECOMBINATIONS)
SAVIKHINA TI
EESTI NSV TEAD AKAD FUUS AND ASTRON INST
UURIM NO 40: 24-52 (1972) (IN RUSSIAN)
SAA 1973: 50611

4-1049 TRANSITION PROBABILITIES OF KL- LLL AUGER
SATELLITES IN NEON.
SCHMIDT V
P548-58 OF INT CONF INNER SHELL IONIZATION
PHENOMENA AND FUTURE APPLICATIONS, VOL 1,
1972, FINK RW(ED), NTIS, 1973.
CA 78: 142075

4-1050 AUGER ELECTRON EMISSION SPECTRA FROM FOIL AND
GAS EXCITED CARBON BEAMS.
SCHNEIDER D + GODGE W + JOHNSON BM
+ SMITH LE + MOORE CF
PHYS LETT A 54(2): 174-6 (1975)
CA 83: 185987

4-1051 HIGH RESOLUTION NEON K-AUGER ELECTRON
SPECTRUM PRODUCED BY 45 MEV CHLORINE(12+) ION
IMPACT.
SCHNEIDER D + MOORE CF + JOHNSON BM
J PHYS B 9(6): L153-6 (1976)
CA 85: 54036

4-1052 HIGH RESOLUTION AUGER ELECTRON SPECTROSCOPY
OF METALLIC COPPER.
SCHON G
J ELECTRON SPECTROSC RELAT PHENOMENA 1(4):
377-87 (1973)
CA 78: 090522

4-1053 SPIN SPLITTING IN COPPER AUGER ELECTRON
SPECTRUM.
SCHON G
PHYS LETT A 42(5): 381-2 (1973)
CA 78: 103910

4-1054 AUGER AND DIRECT ELECTRON SPECTRA IN X-RAY
PHOTOELECTRON STUDIES OF ZINC, ZINC OXIDE,
GALLIUM, AND GALLIUM OXIDE.
SChON G
J ELECTRON SPECTROSC RELAT PHENOMENA 2(1):
75-86 (1973)
CA 79: 059792

4-1055 ESCA STUDIES OF SILVER, DISILVER OXIDE AND
SILVER OXIDE.
SCHON G
ACTA CHEM SCAND 27(7): 2623-33 (1973)
CA 80: 032329

4-1056 CHEMICAL SHIFTS IN THE AUGER SPECTRA OF
PASSIVE FILMS.
SEO M + LUMSDEN JB + STAEHLE RW
SURFACE SCI 42(1): 337-9 (1974)
CA 80: 101988

4-1057 RADIATIVE KM2 AUGER EFFECT IN SOME TRANSITION
METALS.
SERVOMAA A + KESKI-RAHKONEN O
P59 OF ANNU CONF FINNISH PHYS SOC, 1975 PROC.
SAA 1975: 38262

4-1058 AUGER ELECTRON SPECTRUM AND IONIZATION
POTENTIALS OF THE HYDROGEN FLUORIDE MOLECULE.
SHAW RW + THOMAS TD
PHYS REV A 11(5): 1491-7 (1975)
CA 83: 50313

4-1059 INELASTIC MEAN FREE PATH FOR ELECTRONS IN
BULK JELLIUM. (AUGER, LEED, ENERGY LOSS
SPECTROSCOPY)
SHELTON JC
SURFACE SCI 44: 305-9 (1974)

4-1060 INTERFACIAL AND PLASMON DECAY PEAKS IN THE
AUGER SPECTRA OF TITANIUM AND IRON.
SHIH HD + LEGG KO + JONA F
SURFACE SCI 54(2): 355-64 (1976)
CA 84: 113864

4-1061 CHANGES IN THE FINE STRUCTURE OF THE ANGULAR
DEPENDENCE OF THE SECONDARY ELECTRON EMISSION
COEFFICIENT OF SINGLE CRYSTALS IN THE 2-10
KEV RANGE OF PRIMARY ELECTRON ENERGIES.
SHULMAN AR + KORABLEV VV
RADIO ENG ELECTRON PHYS 15: 2154-6 (NOV 1970)

4-1062 SECONDARY ELECTRON EMISSION OF SILICON
DIOXIDE SINGLE CRYSTALS.
SHULMAN AR + KORABLEV VV + MOROZOV YA
SOV PHYS SOLID STATE 12: 519-20 (1970)

4-1063 SECONDARY ELECTRON EMISSION OF MOLYBDENUM
CRYSTALS.
SHULMAN AR + KORABLEV VV + MOROZOV YA
SOV PHYS SOLID STATE 12: 586-9 (1970)

4-1064 ENERGY SPECTRA OF INELASTICALLY SCATTERED AND
AUGER ELECTRONS FROM SINGLE CRYSTALS.
SHULMAN AR + KORABLEV VV + MOROZOV YA
SOV PHYS SOLID STATE 12: 1487-8 (1970)

4-1065 ELECTRON EXCITED AUGER ELECTRON SPECTRUM OF A
NICKEL (110)- C(2X2) SULFUR SURFACE: LINE
SHAPE ANALYSIS AND CORRELATION WITH ION
NEUTRALIZATION SPECTROSCOPY.
SICKAFUS EN
PHYS REV B 7(12): 5100-14 (1973)
CA 79: 036627

4-1066 AUGER ELECTRON SPECTRUM OF WATER VAPOR.
SIEGBAHN H + ASPLUND L + KELFVE P
CHEM PHYS LETT 35(3): 330-5 (1975)
CA 83: 185826

4-1067 ANOMALOUS THRESHOLD BEHAVIOR FOR ELECTRON
INDUCED INNER SHELL IONIZATION OF THE N6,7
LEVEL IN GOLD, BISMUTH, AND LEAD MONITORED BY
AUGER SPECTROSCOPY.
SMITH DM + GALLON TE + MATTHEW JAD
J PHYS B 7(11): 1255-61 (1974)
CA 81: 083846

4-1068 A1 KLL AUGER PROFILE ARTIFACT AT THE OXIDE-
METAL INTERFACE.
SOLOMON JS
APPL SPECTROSC 30(1): 46-9 (1976)
CA 84: 113978

4-1069 MOLECULAR ORBITAL EFFECTS ON THE TITANIUM LMV
AUGER SPECTRA OF TITANIUM(II) AND
TITANIUM(IV) OXIDES.
SOLOMON JS + BAUN WL
SURFACE SCI 51(1): 228-36 (1975)
CA 83: 155171

4-1070 INTENSITIES OF SPECTRA- AUGER AND
PHOTOELECTRON.
STADNIKOV CG + NIKOLSHII AP
IZV AKAD NAUK SSSR, FIZ ESKA 35: 330 (1971)
(IN RUSSIAN)

4-1071 ENERGIES OF EXCITED STATES OF DOUBLY IONIZED
MOLECULES BY MEANS OF AUGER ELECTRON
SPECTROSCOPY. PART-1: ELECTRONIC STATES OF
NITROGEN.
STAHLHERM D + CLEFF B + HILLIG H + MEHLHORN W
Z NATURFORSCH A24: 1728-33 (1969)

4-1072 PREDICTION OF THE RELATIVE INTENSITIES OF
AUGER LINES IN MICA.
STAIB P
PHYS LETT A 41: 3-4 (28 AUG 1972)

4-1073 OXIDE FILMS: ELEMENT PROFILES BY AUGER
ELECTRON SPECTROSCOPY.
STODDART CTH + HONDROS ED
NATURE PHYS SCI 237: 90-1 (5 JUN 1972)

4-1074 AUGER ELECTRON SPECTROSCOPY STUDY OF BETA
ALUMINA ELECTROLYTE.
STODDART CTH + HONDROS ED
TRANS J BRIT CERAM SOC 73(2): 61-4 (1974)
CA 81: 015418

4-1075 ANGULAR AND ENERGY DISTRIBUTION OF ELECTRONS
PRODUCED BY 200-500 KEV PROTONS IN GASES.
PART-2: NITROGEN. (AUGER SPECTRUM)
STOLTERFOHT N
Z PHYS 248(1): 92-100 (1971)
CA 76: 8458V

4-1076 PRODUCTION OF INTENSE SATELLITE LINES IN
AUGER SPECTRA OF NITROGEN BY SLOW PROTON
IMPACT.
STOLTERFOHT N
PHYS LETT A 41(5): 400-2 (1972)
CA 78: 9673Q

4-1077 CROSS SECTIONS FOR INNER SHELL IONIZATION OF
GASEOUS MOLECULES BY 50-500 KEV PROTONS.
STOLTERFOHT N
P1043-56 OF INT CONF INNER SHELL IONIZATION
PHENOMENA AND FUTURE APPLICATIONS, VOL 2,
1972, FINK RW(ED), NTIS, 1973.
CA 78: 128496

4-1078 HIGH RESOLUTION NEON AUGER SPECTRUM PRODUCED
IN 4.2 MEV ATOMIC HYDROGEN(+)- NEON
COLLISIONS.
STOLTERFOHT N + GABLER H + LEITHAEUSER U
PHYS LETT A 45(5): 351-2 (1973)
CA 80: 032104

4-1079 AUGER ELECTRON AND X-RAY PRODUCTION IN 50- TO
200 KEV NEON- NEON COLLISIONS.
STOLTERFOHT N + SCHNEIDER D + BURCH D
+ AAGAARD B + BOEVING E
PHYS REV A 12(4): 1313-29 (1975)
CA 83: 210974

4-1080 ARGON L-SHELL AUGER SPECTRA PRODUCED IN
ARGON(+) ION- ARGON COLLISIONS.
STOLTERFOHT N + SCHNEIDER D + GABLER H
PHYS LETT A 47(4): 271-2 (1974)
CA 80: 150552

4-1081 AUGER ELECTRON SPECTROSCOPY STUDY OF SILICON
SPECTRA FROM SILICON MONOXIDE, SILICON
DIOXIDE, AND SILICON NITRIDE.
STRAUSSER YE + JOHANNESSEN JS
P125-38 OF SEMICONDUCTOR MEASUREMENT
TECHNOLOGY. PART-4: ARPA/NBS WORKSHOP ON
SURFACE ANALYSIS FOR SILICON DEVICES,
LIEBERMAN AG(ED), NAT BUR STAND SPEC PUB
400-23, 1976.
CA 84: 187272

4-1082 VARIAN CHART OF AUGER ELECTRON ENERGIES.
STRAUSSER YE + UEBBING JJ
VARIAN CORP, PALO ALTO, CALIF, 1971.

4-1083 AUGER ELECTRON AND CHARACTERISTIC ENERGY LOSS
SPECTROSCOPY OF SOME METALS.
SULEMAN M
UNIV KEELE, ENGLAND, PHD THESIS, 1971.
SAA 1973: 34452

4-1084 AUGER SPECTROSCOPY OF BERYLLIUM.
SULEMAN M + PATTINSON EB
J PHYS F 1: 124-7 (1971)

4-1085 OBSERVATION OF A PLASMON GAIN IN THE FINE
STRUCTURE OF THE ALUMINUM AUGER SPECTRUM.
SULEMAN M + PATTINSON EB
J PHYS F 1: L21-4 (1971)

4-1086 CHANGES IN AUGER SPECTRA OF MAGNESIUM AND
IRON DUE TO OXIDATION.
SULEMAN M + PATTINSON EB
SURFACE SCI 35(1): 75-81 (1973)
CA 78: 77593P

4-1087 INTERPRETATION OF THE AUGER SPECTRUM OF CLEAN
AND OXIDIZED BERYLLIUM.
SULEMAN M + PATTINSON EB
J PHYS F 3(3): 497-504 (1973)
CA 78: 140694

4-1088 DETERMINATION OF ELECTRON ATTENUATION LENGTHS
IN METALS: TRANSMISSION THROUGH THIN ADSORBED
FILMS. (AUGER EFFECT)
SUNJIC M + SOKCEVIC D + GADZUK JW
JAP J APPL PHYS SUPPL 2 PART-2: 753-6 (1974)
SAA 1975: 46598

4-1089 AUGER ELECTRON SPECTROSCOPY OF THE SURFACE
CHEMICAL COMPOSITION OF VANADIUM OXIDES, AND
OXIDIZED VANADIUM: CHEMICAL SHIFT AND PEAK
INTENSITY ANALYSIS.
SZALKOWSKI FJ + SOMORJAI GA
J CHEM PHYS 56: 6097-6103 (5 JUN 1972)

4-1090 AUGER ELECTRON SPECTROSCOPIC INVESTIGATIONS
OF CHEMICAL SHIFTS IN SOME VANADIUM
COMPOUNDS. (VANADIUM OXIDE, NITRIDE, CARBIDE,
SILICIDE)
SZALKOWSKI FJ + SOMORJAI GA
J CHEM PHYS 61(5): 2064-70 (1974)
CA 81: 143710

4-1091 AUGER RECOMBINATION IN INDIUM ARSENIDE,
GALLIUM ANTIMONIDE, INDIUM PHOSPHIDE AND
GALLIUM ARSENIDE.
TAKESHIMA M
J APPL PHYS 43: 4114-9 (OCT 1972)

4-1092 AUGER RECOMBINATION IN A SEMICONDUCTOR
(MERCURY(1-X) CADMIUM(X) TELLURIDE) UNDER A
MAGNETIC FIELD.
TAKESHIMA M
J APPL PHYS 44(10): 4717-23 (1973)
CA 79: 150564

4-1093 EFFECT OF ELECTRON HOLE INTERACTION ON THE
AUGER RECOMBINATION PROCESS IN A
SEMICONDUCTOR.
TAKESHIMA M
J APPL PHYS 46(7): 3082-8 (1975)
CA 83: 87951

4-1094 ENERGY SPECTRA OF INELASTICALLY SCATTERED
ELECTRONS AND LEED STUDIES OF TUNGSTEN.
THARP LN + SCHEIBNER EJ
J APPL PHYS 38: 3320-30 (1967)

4-1095 PLASMON ENERGY GAIN IN THE BERYLLIUM AUGER
SPECTRUM.
THOMAS S
SOLID STATE COMMUN 13(10): 1593-4 (1973)
CA 80: 042451

4-1096 ELECTRON IRRADIATION EFFECT IN THE AUGER
ANALYSIS OF SILICON DIOXIDE.
THOMAS S
J APPL PHYS 45(1): 161-6 (1974)
CA 80: 089102

4-1097 K-, L-, AND M-AUGER AND L- COSTER-KRONIG
SPECTRA OF PLATINUM.
TOBUREN LH + ALBRIDGE RG
NUCL PHYS A 90: 529-44 (1967)

4-1098 INFLUENCE OF COSTER-KRONIG PROCESSES ON
L-AUGER YIELDS.
TRACY JC
SURFACE SCI 38(1): 265-8 (1973)
CA 79: 036802

4-1099 SEPARATION OF THE SPECTRA OF PHOTO AND AUGER
ELECTRONS.
TRAPEZNIKOV VA
USSR PAT 322704, 30 NOV 1971.
CA 76: 119829M

4-1100 AUGER ELECTRON SPECTRA OF SOME RARE EARTH
METALS. (LANTHANUM, PRASEODYMIUM, NEODYMIUM,
GADOLINIUM, DYSPROSIUM, YTTERBIUM, HAFNIUM).
TSVEIMAN EV + KORSUNSKII MI + ZASHKVARA VV
+ REDKIN VS
DOKL AKAD NAUK SSSR 204(4): 828-30 (1972) (IN
RUSSIAN)

4-1101 INTERPRETATION OF THE LOW ENERGY K-BETA
SPECTRUM IN SOLIDS.
UTRIAINEN J + LINKOAHO M + ABERG T
COMMENT PHYS-MATH, SOC SCI FENN 42(4): 296-7
(1972)
CA 78: 130253

4-1102 SCANDIUM L X-RAY EMISSION FROM PURE METAL AND
SCANDIUM OXIDE.
UTRIAINEN J + VALJAKKA J + LINKOAHO M
J PHYS C 8(21): 3710-4 (1975)
SAA 1976: 14608

4-1103 ELECTRON HOLE SCATTERING AND (AUGER)
RECOMBINATION OF NONEQUILIBRIUM CHARGE
CARRIERS IN SILICON AT A HIGH EXCITATION
LEVEL.
VAITKUS J + GRIVICKAS V + STORASTA J
SOV PHYS SEMICOND 9(7): 883-6 (1975)
CA 83: 124817

4-1104 AUGER TRANSITIONS AND SHAPE OF X-RAY
SPECTRUM.
VELRINSKII RV + KOLESNIKOV VV
BULL ACAD SCI USSR, PHYS SER 31: 904-10
(1967)

4-1105 L-SHELL FLUORESCENCE YIELDS OF DOUBLE VACANCY
STATES IN LEAD.
VELURI VR + WOOD RE + PALMS JM + RAO PV
P251-6 OF INT CONF INNER SHELL IONIZATION
PHENOMENA AND FUTURE APPLICATIONS, VOL 1,
1972, FINK RW(ED), NTIS, 1973.
CA 78: 130721

4-1106 AUGER SPECTRA OF SOME FREE ELEMENTS AND
BINARY COMPOUNDS BOMBARDED BY 60 KEV ARGON(+)
IONS.
VIEL L + BENAZETH C + BENAZETH N
SURFACE SCI 54(3): 635-46 (1976) (IN FRENCH)
CA 85: 27146

4-1107 CHEMICAL SHIFTS OF AUGER LINES, AND THE AUGER
PARAMETER.
WAGNER CD
FARADAY DISCUSS CHEM SOC 60: 291-300 (1976)
CA 84: 171667

4-1108 AUGER ELECTRON SPECTROSCOPY OF A STABLE
GERMANIUM OXIDE.
WANG KL + JOSHI A
J VACUUM SCI TECHNOL 12(4): 927-32 (1975)
CA 83: 105732

4-1109 AUGER SPECTRA OF CARBON AND ARGON FOLLOWING
IONIZATION BY EQUAL VELOCITY ALPHA PARTICLES
AND DEUTERONS.
WATSON RL + TOBUREN LH
PHYS REV A 7(6): 1853-63 (1973)
CA 79: 036727

4-1110 DYNAMICAL SCREENING EFFECTS IN THE AUGER
SPECTRUM OF METALS.
WATTS CMK
J PHYS F 2: 574-83 (MAY 1972)

4-1111 AUGER ELECTRON SPECTROSCOPY OF HIGH
TEMPERATURE OXYGEN- RHENIUM INTERACTIONS.
WEBER B + BIGEARD B + MIHE JP + CASSUTO A
J VACUUM SCI TECHNOL 12(1): 338-40 (1975)
CA 83: 16086

4-1112 X-RAY SPECTRA, L-SUBSHELL FLUORESCENCE AND
COSTER-KRONIG YIELDS IN BISMUTH AND
NEPTUNIUM.
WEKSLER M + DE PINHO AG
REV BRAS FIS 3(2): 291-309 (OCT 1973)
SAA 1974: 47848

4-1113 HIGH RESOLUTICN L2,3MM AND M4,5NN AUGER
SPECTRA FROM KRYPTON AND M4, 5NN AND N4,5OO
AUGER SPECTRA FROM XENON.
WERME LO + BERGMARK T + SIEGBAHN K
PHYS SCR 6(2-3): 141-50 (1972)
CA 78: 090616

4-1114 L2,3MM AUGER SPECTRUM OF ARGON.
WERME LO + BERGMARK T + SIEGBAHN K
PHYS SCR 8(4): 149-58 (1973)
CA 79: 151231

4-1115 CHEMICAL SHIFTS IN AUGER SPECTRA (MONOLAYER
ADSORPTION, ESCA SIMPLE CHARGE TRANSFER
MODEL).
WHITAKER MAB
J PHYS C 5: L102-5 (1972)

4-1116 LINE WIDTH AND AUGER EFFECT IN CALCIUM,
STRONTIUM, AND BARIUM FLUORIDES.
WIESNER H + HCENERLAGE B
Z PHYS 256(1): 43-8 (1972)
CA 78: 22057M

4-1117 K-SHELL AUGER ELECTRON HYPERSATELLITES OF
NEON.
WOODS CW + KAUFFMAN RL + JAMISON KA
+ STOLTERFOHT N + RICHARD P
PHYS REV A 12(4): 1393-8 (1975)
CA 83: 210977

4-1118 OXIDATION EFFECTS ON THE SPECTRUM OF THE SLOW
SECONDARY ELECTRON EMISSION PEAK IN
POLYCRYSTALLINE MAGNESIUM.
WRIGHT B + PATTINSON EB
J PHYS F 4(1): 176-82 (1974)
CA 80: 089312

4-1119 AUGER ELECTRON ENERGIES (0-2000 EV) FOR
ELEMENTS OF ATOMIC NUMBER 5-103.
YASKO RN + WHITMOYER RD
J VACUUM SCI TECHNOL 8: 733-7 (1971)

4-1120 CHARACTERISTICS OF SHAPING SECONDARY ELECTRON
EMISSION FRCM POROUS POTASSIUM CHLORIDE
FILMS.
YASNOPOLSKII NL + BALASHOVA AP
RADIO ENG ELECTRON PHYS 15: 1141-3 (JUN 1970)

4-1121 QUASIATOMIC LMM AUGER SPECTRA OF SOLID COPPER
AND ZINC.
YIN LI + ADLER I + TSANG T + CHEN MH
+ CRASEMANN B
PHYS LETT A 46(2): 113-14 (1973)
CA 80: 076192

4-1122 L-S COUPLING INTERPRETATION OF HIGH
RESOLUTION LMM AUGER SPECTRA OF COPPER AND
ZINC.
YIN LI + TSANG T + ADLER I + YELLIN E
J APPL PHYS 43(8): 3464-7 (1972)

4-1123 X-RAY EXCITED LMM AUGER SPECTRA OF COPPER,
NICKEL AND IRCN.
YIN LI + YELLIN E + ADLER I
J APPL PHYS 42: 3595-600 (1971)

4-1124 PLASMA SATELLITES IN THE AUGER ELECTRON
SPECTRUM OF MCLYBDENUM.
ZASHKVARA VV + REDKIN VS + KORSUNSKII MI
+ SOTNIKOV VT + CHAIKOVSKII EF
UKR FIZ ZH 17(8): 1364-6 (1972) (IN RUSSIAN)
CA 77: 132779K

4-1125 AUGER SPECTRA OF URANIUM.
ZENDER MJ
Z PHYS 218: 245-59 (1969)

4-1126 DETECTION OF EXCITONS NEAR THE L2,3 EDGE OF
THE CHLORIDE ION IN SODIUM CHLORIDE BY MEANS
OF THE DECOMPOSITION PRCDUCTS.
ZHURAKOVSKII AP + KADCHENKO VN + SAAR A
+ ELANGO M
IZV AKAD NAUK SSSR SER FIZ 40(2): 267-9
(1976) (IN RUSSIAN)
CA 84: 186864

5. SURFACE ANALYSIS BY AUGER SPECTROSCOPY

5-1127 APPLICATION OF AUGER ELECTRON SPECTROSCOPY TO
THE STUDY OF DISCOLORING OF ALUMINUM SURFACES
IN BOILING WATER.
ABE T + AIZAWA K + UCHIYAMA T + ISOYAMA E
KEIKINZOKU 24(6): 254-62 (1974) (IN JAPANESE)
CA 82: 006647

5-1128 ELECTRON BEAM EFFECTS IN DEPTH PROFILING
MEASUREMENTS WITH AUGER ELECTRON
SPECTROSCOPY.
AHN J + PERLEBERG CR + WILCOX DL + COBURN JW
+ WINTERS HF
J APPL PHYS 46(10): 4581-3 (1975)
CA 84: 24926

5-1129 CHARACTERISTICS OF SURFACE OF THE PALLADIUM
BARIUM ALLOY CATHODE IN RELATION TO ITS FILM
SYSTEM. (AUGER EMISSION)
ALEKSEEV YUV + KANICHEVA I + SMOLYAKOV A
KRATK SODERZH DOKL 2: 115-16 (1973) (IN
RUSSIAN)
CA 83: 187068

5-1130 EFFECT OF IMPURITIES ON INTRINSIC STRESS IN
THIN NICKEL FILMS. (AUGER SPECTROSCOPY)
ALEXANDER PM
CASE WESTERN RESERVE UNIV, CLEVELAND, PHD
THESIS, 1975, 97P.
SAA 1976: 13892

5-1131 EFFECT OF IMPURITIES ON INTRINSIC STRESS IN
THIN NICKEL FILMS. (AUGER EFFECT)
ALEXANDER PM + HOFFMAN RW
J VACUUM SCI TECHNOL 13(1): 96-8 (1976)
SAA 1976: 49107

5-1132 STUDY OF ALUMINUM WITH A COMBINED AUGER
ELECTRON SPECTROMETER ELLIPSOMETER SYSTEM.
ALLEN TH
J VACUUM SCI TECHNOL 13(1): 112-5 (1976)
CA 84: 141024

5-1133 SECONDARY ELECTRON EJECTION FROM METAL
SURFACES BY METASTABLE ATOMS. PART-3: ENERGY
AND ANGULAR DISTRIBUTIONS OF THE EJECTED
ELECTRONS. (SURFACE CONTAMINATION)
ALLISON W + DUNNING FB + SMITH ACH
J PHYS B 5: 1175-85 (JUN 1972)

5-1134 CLEAN TELLURIUM SURFACES STUDIED BY LEED.
ANDERSSON S + ANDERSSON D + MARKLUND I
SURFACE SCI 12: 284-98 (1968)

5-1135 MEASUREMENTS OF CONDENSATION COEFFICIENTS OF
GOLD ON ROCK SALT (100) CLEAVAGE SURFACES AND
ON AMORPHOUS CARBON BY MEANS OF QUANTITATIVE
AUGER ELECTRON SPECTROSCOPY.
ANTON R
VAK TECH 23(6): 172-7 (1974) (IN GERMAN)
CA 82: 78831

5-1136 QUANTITATIVE AUGER ELECTRON SPECTROSCOPY OF
GOLD CONDENSED ON ROCK SALT AND ON AMORPHOUS
CARBON.
ANTON R + HARSDORFF M
THIN SOLID FILMS 22(2): 523-26 (1974)
CA 81: 083905

5-1137 HETEROGENEOUS NUCLEATION AND GROWTH OF GOLD
ON SUBSTRATES. (AUGER SPECTRA)
ANTON R + HARSDORFF M + PAUNOV M
+ SCHMEISSER H
JAP J APPL PHYS SUPPL 2 PART-1: 563-6 (1974)
SAA 1975: 37627, SAC 1975: 14481

5-1138 INCREASING THE ACCURACY OF MATERIAL TESTS BY
THE APPLICATION OF ELECTRON SPECTROSCOPY.
(AUGER EFFECT)
ANTONIEWICZ J
PRZEGL ELEKTROTECH 50(5): 197-202 (MAY 1974)
(IN POLISH)
SAA 1974: 70597

5-1139 INVESTIGATION OF REORDERED (001) GOLD
SURFACES BY PCSITIVE ION CHANNELING
SPECTROSCOPY, LEED AND AES.
APPLETON BR + NOGGLE TS + MILLER JW
+ SCHOW OE + ZEHNER DM + JENKINS LH
+ BARRETT JH
P86-96 OF CONF ON APPL SMALL ACCEL, 3RD PROC,
1974, DUGGAN JL + MORGAN IL (EDS), NTIS,
CONF-741040-P1, 1975.
CA 85: 52144

5-1140 AUGER ELECTRON SPECTROSCOPY OF LAYERED GROWTH
OF TITANIUM ON TUNGSTEN.
ARMSTRONG RA
SURFACE SCI 50(2): 615-20 (1975)
CA 83: 139948

5-1141 SURFACE VALENCE BAND AND PLASMON FEATURES ON
CLEAN CLEAVED SILICON.
ARNOTT DR + HANEMAN D
SURFACE SCI 45(1): 128-40 (1974)
CA 81: 129535

5-1142 USE OF SCANNING AUGER MICROSCOPY IN MOLECULAR
BEAM EPITAXY OF GALLIUM ARSENIDE AND GALLIUM
PHOSPHIDE.
ARTHUR JR
J VACUUM SCI TECHNOL 10(1): 136-9 (1973)
CA 78: 102839

5-1143 SURFACE STOICHICMETRY AND STRUCTURE OF
GALLIUM ARSENIDE.
ARTHUR JR
SURFACE SCI 43(2): 449-61 (JUN 1974)
SAA 1974: 56947

5-1144 ADSORPTION OF ZINC ON GALLIUM ARSENIDE.
(AUGER SPECTROSCOPY)
ARTHUR JR
SURFACE SCI 38(2): 394-412 (JUL 1973)
SAA 1973: 55150

5-1145 LEED- AUGER, AND WORK FUNCTION STUDY OF
IODINE ADSORBED ON TUNGSTEN (110).
AVERY NR
SURFACE SCI 43(1): 101-22 (1974)
CA 81: 016988

5-1146 AUGER OBSERVATION OF A SURFACE STATE ON
TUNGSTEN.
AVERY NR
PHYS REV LETT 32(22): 1248-50 (1974)
CA 81: 043485

5-1147 INVESTIGATIONS OF THE GAS- SOLID INTERFACE
INTERACTIONS AT HIGH TEMPERATURES BY AUGER
ELECTRON SPECTROSCOPY.
BAENNINGER U + BAS EB
HELV PHYS ACTA 45(6): 977-9 (31 DEC 1972) (IN
GERMAN)
SAA 1973: 52749

5-1148 ADSORPTION ISCTHERMS OF OXYGEN ON TUNGSTEN
(100), AND (111) SURFACES IN THE PRESSURE
RANGE 10(-9) TO 10(-4) TORR AND TEMPERATURE
RANGE 1500 TO 2600 DEGREES K AS MEASURED BY
AES.
BAENNINGER U + BAS EB
SURFACE SCI 50(2): 279-95 (1975)
CA 83: 137303

5-1149 SURFACE IMPURITIES IN SILVER HALIDE FILMS.
BAETZOLD RC
APPL PHYS LETT 26(12): 709-11 (1975)
CA 83: 87776

5-1150 QUANTITATIVE AUGER SPECTROSCOPY OF PHYSICALLY
ADSORBED XENON ON NICKEL.
BAKER BG + SEXTON BA
P275-8 OF INT VACUUM CONGRESS, 6TH, PROC INT
CONF ON SOLID SURFACES, 2ND, 1974, KYOTO.
(JAP J APPL PHYS, SUPPL 2, PART 2, 1974)
CA 82: 160618

5-1151 ELECTRON BEAM EFFECTS IN AUGER ANALYSIS OF
PHYSISORBED XENON.
BAKER BG + SEXTON BA
SURFACE SCI 52(2): 353-64 (1975)
CA 84: 22559

5-1152 PHOTOEMISSION SPECTRA FROM ADSORBED OXYGEN ON
TUNGSTEN (110) AND CARBON MONOXIDE ON
TUNGSTEN (100). (BY LEED AND AES)
BAKER JM + EASTMAN DE
J VACUUM SCI TECHNOL 10(1): 223-6 (1973)

5-1153 CHEMICAL SHIFTS IN THE AUGER SPECTRUM OF
YTTRIUM ON OXYGEN ADSORPTION.
BAKER JM + MCNATT JL
J VACUUM SCI TECHNOL 9(2): 792-6 (1972)
CA 77: 81485B

5-1154 HOW TO ASSURE ONESELF OF THE CLEANLINESS OF A
SURFACE. (USE OF AES TO DETERMINE TRACE
IMPURITIES)
BALIBAR F
RECHERCHE 3(22): 370-1 (1972) (IN FRENCH)
CA 77: 80488T

5-1155 ANGULAR DISTRIBUTIONS OF HYDROGEN DESORBED
FROM THE (100), (110), AND (111) FACES OF
COPPER CRYSTALS. (AUGER ELECTRON
SPECTROSCOPY)
BALOOCH M + STICKNEY RE
SURFACE SCI 44(2): 310-20 (AUG 1974)
SAA 1974: 75709

5-1156 INVESTIGATION OF OXYGEN ADSORPTION ON
TUNGSTEN AT HIGH TEMPERATURES BY AUGER
ELECTRON SPECTROSCOPY.
BAS EB + BAENNINGER U
P197-200 OF INT VACUUM CONGRESS, 6TH, PROC
INT CONF ON SOLID SURFACES, 2ND, 1974, KYOTO.
(JAP J APPL PHYS, SUPPL 2, PART 2, 1974)
CA 82: 175599

5-1157 INVESTIGATION OF EUROPIUM(II) OXIDE (100)
CLEAVAGE SURFACE BY LOW ENERGY ELECTRON
DIFFRACTION AND AUGER ELECTRON SPECTROSCOPY.
BAS EB + BAENNINGER U + MUEHLETHALER H
P671-3 OF INT VACUUM CONGRESS, 6TH, PROC INT
CONF ON SOLID SURFACES, 2ND, 1974, KYOTO.
(JAP J APPL PHYS, SUPPL 2, PART 2, 1974)
CA 82: 178706

5-1158 EFFECT OF OXIDATION ON THE HIGH RESOLUTION
AUGER SPECTRUM OF TITANIUM.
BASSETT PJ + GALLON TE
J ELECTRON SPECTROSC RELAT PHENOMENA 2(1):
101-4 (1973)
CA 79: 071953

5-1159 METHODS OF SURFACE STUDIES DEPENDING ON
INELASTIC SCATTERING OF ELECTRONS. (AUGER
ELECTRON SPECTROSCOPY)
BAUER E
VACUUM 22(11): 539-52 (NOV 1972)
SAA 1973: 48570

5-1160 ADSORPTION AND CONDENSATION OF COPPER ON
TUNGSTEN SINGLE CRYSTAL SURFACES.
BAUER E + POPPA H + TODD G + BONCZEK F
J APPL PHYS 45(12): 5164-75 (DEC 1974)
SAA 1975: 21700

5-1161 SURFACE STRUCTURE AND COMPOSITION OF VARIOUS
ELEMENTAL AND III-V SEMICONDUCTORS.
BAYLISS CR
UNIV NOTTINGHAM, ENGLAND, PHD THESIS, 1975
SAA 1976: 05640

5-1162 DEPARTURES FROM STOICHIOMETRY IN (100)
SURFACES OF GALLIUM PHOSPHIDE, INDUCED BY
BOMBARDMENT WITH 2 KV ELECTRONS. (AUGER
ELECTRON DAMAGE)
BAYLISS CR + KIRK DL
THIN SOLID FILMS 29(2): L35-8 (1975)
CA 84: 24611

5-1163 COMPOSITIONAL AND STRUCTURAL CHANGES THAT
ACCOMPANY THE THERMAL ANNEALING OF (100)
SURFACES OF GALLIUM ARSENIDE, INDIUM
PHOSPHIDE, AND GALLIUM PHOSPHIDE IN VACUUM.
(AUGER SPECTROSCOPY)
BAYLISS CR + KIRK DL
J PHYS D 9(2): 233-44 (1976)
SAA 1976: 26311

5-1164 EFFECT OF TEMPERATURE ON THE SURFACE
STRUCTURE AND STOICHIOMETRY OF (100) INDIUM
PHOSPHIDE SURFACES.
BAYLISS CR + KIRK DL
THIN SOLID FILMS 29(1): 97-106 (1975)
SAA 1976: 09570

5 1165 DETECTION OF THIN CONTAMINATED LAYERS ON
CONTACT SURFACES. (BY AUGER SPECTROSCOPY)
BECK K + SCHIFF KL
Z WERKSTOFFTECH 8(4): 117-21 (1975) (IN
GERMAN)
CA 83: 141312

5-1166 SURFACE COMPOSITION AND STRUCTURE ANALYSIS
WITH A ONE APPARATUS ION PROBE TECHNIQUE.
(COMPARISON BETWEEN ION PROBE AND AUGER- LEED
TECHNIQUES)
BEGEMANN SHA + BOERS AL
SURFACE SCI 32(3): 607-17 (1972)
SAA 1972: 71297

5-1167 ANALYSIS OF SURFACE LAYERS BY LIGHT ION
BACKSCATTERING AND SPUTTERING COMBINED WITH
AUGER ELECTRON SPECTROSCOPY.
BEHRISCH R + SCHERZER BMU + STAIB P
THIN SOLID FILMS 19(1): 57-67 (1973)
CA 80: 066471

5-1168 AN ORDERED BORON STRUCTURE AFTER DEPOSITION
ON SILICON (111). (AUGER ELECTRON
SPECTROSCOPY)
BELLINA JJ
JAP J APPL PHYS 13(10): 1659-60 (1974)
SAA 1975: 01261

5-1169 EFFECTS OF PREADSORBED OXYGEN CONTAMINANT ON
THE ADSORPTION OF ALUMINUM ON SILICON (111).
BELLINA JJ
J VACUUM SCI TECHNOL 11(6): 1133-40 (1974)
SAA 1975: 46465

5-1170 PHASE EQUILIBRIA IN TWO DIMENSIONAL
CHEMISORBED LAYERS. (AUGER SPECTROSCOPY,
SULFIDE SYSTEMS)
BENARD J
ELECTRON FIS APPL 17(1-2): 216-20 (1974)
SAA 1975: 06647

5-1171 ISOMORPHISM AND PURITY OF COPPER VAPOR
DEPOSITED ON A COPPER (111) SURFACE. (AUGER)
BENNDORF C + KESSLER J + THIEME F
SURFACE SCI 51(2): 553-6 (1975) (IN GERMAN)
SAA 1976: 5694

5-1172 AUGER ELECTRON SPECTROSCOPY STUDY OF THE
SURFACE COMPOSITION OF THE LEAD- INDIUM
SYSTEM.
BERGLUND S + SOMORJAI GA
J CHEM PHYS 59(10): 5537-45 (1973)
CA 80: 074488

5-1173 SMALL MOLECULE REACTIONS ON STEPPED SINGLE
CRYSTAL PLATINUM SURFACES.
BERNASEK SL + SOMORJAI GA
SURFACE SCI 48(1): 204-13 (1975)
SAA 1975: 38730

5-1174 CHARACTERIZATION OF CERAMIC SURFACES. (AUGER
SPECTROSCOPY)
BERRIN L + SUNDAHL RC
P583-623 OF CHARACTERIZATION OF CERAMICS,
HENCH LL + GOULD RW (EDS), DEKKER, 1972.

5-1175 LOW ENERGY ELECTRON DIFFRACTION AND AUGER
SPECTROSCOPY OF THE ADSORPTION OF SULFUR ON
PLATINUM.
BERTHIER Y + PERDEREAU M + OUDAR J
SURFACE SCI 36(1): 225-41 (1973) (IN FRENCH)
CA 78: 088861

5-1176 AUGER ELECTRON SPECTROSCOPIC STUDY OF THE
SURFACE POISONING OF COPPER CATALYSTS.
BHASIN MM
J CATAL 34(3): 356-9 (1974)
CA 82: 007885

5-1177 AUGER SPECTROSCOPIC STUDY OF THE POISONING OF
A COMMERCIAL PALLADIUM ALUMINA HYDROGENATION
CATALYST.
BHASIN MM
J CATAL 38(1): 218-22 (1975)
CA 83: 48832

5-1178 SPONTANEOUS ALLOYING OF A GOLD SUBSTRATE WITH
LEAD MONOLAYERS. (INTERFACE AUGER ELECTRON
SPECTROSCOPY)
BIBERIAN JP + RHEAD GE
J PHYS F 3(4): 675-82 (APR 1973)
SAA 1973: 36463

5-1179 SURFACE ABSORPTION BY POLYPEPTIDE FILMS
EXAMINED BY AUGER EMISSION SPECTROSCOPY.
BISHOP HE + RIVIERE JC
J COLLOID INTERFACE SCI 33: 272-7 (1970)

5-1180 SEGREGATION OF GOLD TO SILICON (111) SURFACE
OBSERVED BY AUGER EMISSION SPECTROSCOPY AND
BY LEED.
BISHOP HE + RIVIERE JC
J PHYS D 2: 1635-42 (1969)

5-1181 SURFACE SEGREGATION IN BORON DOPED IRON
OBSERVED BY AUGER EMISSION SPECTROSCOPY.
BISHOP HE + RIVIERE JC
ACTA MET 18: 813-17 (1970)

5-1182 EFFECT OF OXIDIZING AMBIENTS ON PLATINUM
SILICIDE FORMATION. PART-2: AUGER AND
BACKSCATTERING ANALYSES.
BLATTNER RJ + EVANS CA + LAU SS + MAYER JW
+ ULLRICH BM
J ELECTROCHEM SOC 122(12): 1732-6 (1975)
CA 84: 57375

5-1183 AUGER ELECTRON SPECTROSCOPY OF A SULFUR-
OXYGEN SURFACE REACTION ON A COPPER (110)
CRYSTAL.
BONZEL HP
SURFACE SCI 27(3): 387-410 (1971)

5-1184 MEASUREMENT OF EQUILIBRIUM SURFACE
SEGREGATION USING AUGER ELECTRON
SPECTROSCOPY.
BONZEL HP + AARON HB
SCR MET 5(12): 1057-60 (1971)
CA 76: 62430T

5-1185 DETECTION OF IMPURITIES ON COPPER SURFACES BY
LEED- AUGER ELECTRON SPECTROSCOPY AND THEIR
INFLUENCE ON SURFACE SELF DIFFUSION OF
COPPER.
BONZEL HP + GJOSTEIN NA
J METALS 21(3): A88 (1969)

5-1186 CHANGE IN THE SURFACE ELECTRONIC STRUCTURE OF
PLATINUM (100) INDUCED BY RECONSTRUCTION.
BONZEL HP + HELMS CR + KELEMEN S
PHYS REV LETT 35(18): 1237-40 (1975)
SAA 1976: 14080

5-1187 CARBON MONOXIDE OXIDATION ON A PLATINUM (110)
SINGLE CRYSTAL SURFACE.
BONZEL HP + KU R
J VACUUM SCI TECHNOL 9(2): 663-7 (1972)

5-1188 MECHANISMS OF THE CATALYTIC CARBON MONOXIDE
OXIDATION ON PLATINUM (110).
BONZEL HP + KU R
SURFACE SCI 33(1): 91-106 (1972)

5-1189 KINETICS OF OXYGEN ADSORPTION ON A PLATINUM
(111) SURFACE. (AUGER ELECTRON SPECTROSCOPY,
STICKING COEFFICIENT)
BONZEL HP + KU R
SURFACE SCI 40(1): 85-101 (OCT 1973)
SAA 1974: 01167

5-1190 ADSORBATE INTERACTIONS ON A PLATINUM (110)
SURFACE. PART-1: SULFUR AND CARBON MONOXIDE
LEED PATTERNS. (AUGER ELECTRON SPECTROSCOPY,
STICKING COEFFICIENT) PART-2: EFFECT OF
SULFUR ON THE CATALYTIC CARBON MONOXIDE
OXIDATION. (AUGER ELECTRON SPECTROSCOPY)
BONZEL HP + KU R
J CHEM PHYS 58(10): 4617-24 (1973); 59(4):
1641-51 (1974)
SAA 1973: 41869, 73166

5-1191 STRUCTURE AND SECONDARY EMISSION OF MAGNESIUM
OXIDE LAYERS APPLIED ON A MOLYBDENUM SINGLE
CRYSTAL.
BORISOV VL + DOSYAGAEV AV
RADIOENG ELECTRON PHYS 18(5): 803-5 (1973)
CA 79: 046803

5-1192 AUGER SPECTROSCOPIC STUDY OF THE SURFACES OF
ALLOY EMITTERS.
BORISOV VL + LEPESHINSKAYA VN + MUSIKHIN VA
KRATK SODERZH DOKL 2: 111-13 (1973) (IN
RUSSIAN)
CA 83: 187066

5-1193 AUGER ELECTRON SPECTROSCOPY FOR THE ANALYSIS
OF SOLID SURFACES.
BOUWMAN R
NED TIJDSCHR VACUUMTECH 11(3): 37-48 (1973)
CA 79: 059381

5-1194 ELECTRON SPECTROSCOPY AND SURFACE STUDIES.
BOUWMAN R
CHEM WEEKBLAD 70(40): V28-9 (1974) (IN DUTCH)
CA 82: 147164

5-1195 SURFACE COMPOSITION AND DEPTH CONCENTRATION
PROFILE OF PLATINUM- TIN ALLOYS FROM COMBINED
X-RAY PHOTOELECTRON AND AUGER SPECTROSCOPIC
DATA.
BOUWMAN R + BILOEN P
SURFACE SCI 41(2): 348-58 (1974)
CA 81: 040565

5-1196 AUGER SPECTROSCOPIC STUDY OF THE SURFACE
COMPOSITION OF PLATINUM- TIN IN ULTRAHIGH
VACUUM AND IN THE PRESENCE OF OXYGEN AND
HYDROGEN.
BOUWMAN R + TCNEMAN LH + HOLSCHER AA
SURFACE SCI 35: 8-33 (MAR 1973)
SAA 1973: 30318

5-1197 MEASUREMENTS OF THE SPATIAL DISTRIBUTION OF
HYDROGEN DESORBED FROM NICKEL SURFACES:
EFFECTS OF SURFACE COMPOSITION AND CRYSTAL
ORIENTATION. (SURFACE COMPOSITION DETERMINED
BY AES).
BRADLEY TL + CABIRI AE + STICKNEY RE
SURFACE SCI 29(2): 590-602 (1972)
SAA 1972: 25321

5-1198 ADHESION AND FRICTION OF POLYTETRAFLUORO
ETHYLENE IN CCNTACT WITH METALS AS STUDIED BY
AUGER SPECTROSCOPY, FIELD ION AND SCANNING
ELECTRON MICRCSCOPY.
BRAINARD WA + BUCKLEY DH
WEAR 26(1): 75-93 (1973)
CA 80: 027583

5-1199 AES STUDIES OF SURFACE COMPOSITION OF SILVER-
COPPER ALLOYS.
BRAUN P + FAERBER W
SURFACE SCI 47(1): 57-63 (1975); 51(1): 342-3
(1975)
CA 83: 21619; SAA 1975: 81937

5-1200 AUGER ELECTRON SPECTROSCOPY STUDIES OF
SURFACE COMPOSITION OF SILVER- COPPER ALLOYS.
QUANTITATIVE AUGER ANALYSIS OF COPPER- NICKEL
ALLOY SURFACES AFTER ARGON ION BOMBARDMENT.
REPLY TO COMMENTS.
BRAUN P + FAERBER W
SURFACE SCI 51(1): 342-3 (1975)
CA 84: 77959

5-1201 SURFACE PHOTOVOLTAGE AND AUGER SPECTROSCOPY
STUDIES OF (1120) CADMIUM SULFIDE SURFACE.
BRILLSON LJ
J VACUUM SCI TECHNOL 12(1): 249-52 (1975)
CA 83: 20434

5-1202 OBSERVATION OF EXTRINSIC SURFACE STATES ON
(1120) CADMIUM SULFIDE BY AUGER SPECTROSCOPY.
BRILLSON LJ
SURFACE SCI 51(1): 45-60 (1975)
SAA 1975: 81505

5-1203 X-RAY PHOTOELECTRON AND VACUUM ULTRAVIOLET
PHOTOELECTRON STUDIES OF ADSORPTION ON CLEAN
METAL FILM SURFACES. (AUGER SPECTRA)
BRUNDLE CR
P48-9 EUROP PHYS SOC STUDY CONF ON METAL
SURFACES, ABSTR, 1973.
SAA 1974: 37097

5-1204 ELECTRON SPECTROSCOPY AND ADSORPTION STUDIES.
CHEMICAL SHIFTS IN X-RAY PHOTOELECTRON AND
AUGER SPECTRA FOR MONOLAYER ADSORPTION.
BRUNDLE CR + CARLEY AF
CHEM PHYS LETT 33(1): 41-5 (1975)
CA 83: 87793

5-1205 USE OF X-RAY PHOTOELECTRON AND AUGER
SPECTROSCOPY TO DETERMINE THE SURFACE
COMPOSITION OF OXIDIZED VANADIUM.
BRUNDLE CR + SZALKOWSKI FJ + SOMORJAI GA
SURFACE SCI 52(2): 426-33 (1975)
CA 84: 37122

5-1206 AUGER ANALYSIS OF OXYGEN AND SULFUR
INTERACTIONS WITH VARIOUS METALS AND THE
EFFECT OF SLIDING ON THESE INTERACTIONS.
BUCKLEY DH
LEWIS RES CENTER 1973, 21P. NASA TN D-7340
CA 79: 058003

5-1207 ABSORPTION OF ETHYLENE OXIDE AND VINYL
CHLORIDE ON AN IRON (011) SURFACE AND EFFECT
OF THESE FILMS ON ADHESION.
BUCKLEY DH
NASA, LEWIS RES CENTER, CLEVELAND, OHIO, SEPT
1970, 17P. NASA-TN-D-5999

5-1208 INTERACTION OF METHANE, ETHANE, ETHYLENE, AND
ACETYLENE WITH AN IRON (001) SURFACE AND
THEIR INFLUENCE ON ADHESION STUDIED WITH LOW
ENERGY ELECTRON DIFFRACTION AND AUGER.
BUCKLEY DH
NASA, LEWIS RES CENTER, CLEVELAND, OHIO, MAY
1970, 25P. NASA-TN-D-5822

5-1209 (AUGER STUDIES OF) ADHESION, FRICTION,
LUBRICATION IN VACUUM.
BUCKLEY DH
P297-304 OF INT VACUUM CONGRESS, 6TH, PROC
INT CONF ON SOLID SURFACES, 2ND, 1974, KYOTO.
(JAP J APPL PHYS, SUPPL 2, PART 1, 1974)

5-1210 WEAR AND INTERFACIAL TRANSPORT OF MATERIAL.
(AES)
BUCKLEY DH
J VACUUM SCI TECHNOL 13(1): 88-95 (1976)
SAA 1976: 53279

5-1211 EFFECT OF SULFUR, OXYGEN, AND HYDROGEN
SULFIDE SURFACE FILMS ON THE ADHESION OF
CLEAN IRON. (USING LEED AND AES)
BUCKLEY DH
INT J NONDESTRUCT TEST 2(2): 171-88 (1970)

5-1212 ATOMIC NATURE OF POLYMER- METAL INTERACTIONS
IN ADHESION, FRICTION, AND WEAR.
BUCKLEY DH + BRAINARD WA
AMER CHEM SOC, DIV ORG COAT PLAST CHEM PAP
34(1): 298-306 (1974)
CA 84: 17905

5-1213 STUDY OF SURFACE PROCESSES ON COPPER-
BERYLLIUM BY COMBINED SECONDARY ION MASS
SPECTROMETRY AND AUGER ELECTRON SPECTROSCOPY.
BUHL R + HUBER WK + LOEBACH E
P807-10 OF INT VACUUM CONGRESS, 6TH, PROC INT
CONF ON SOLID SURFACES, 2ND, 1974, KYOTO.
(JAP J APPL PHYS, SUPPL 2, PART 2, 1974)
CA 82: 175612

5-1214 COMBINATION OF SECONDARY ION MASS
SPECTROMETRY AND AUGER ELECTRON SPECTROSCOPY,
FOR THE ANALYSIS OF THIN FILMS.
BUHL R + HUBER WK + LOEBACH E
P665-8 OF INT VACUUM CONGRESS, 6TH, PROC INT
CONF ON SOLID SURFACES, 2ND, 1974, KYOTO.
(JAP J APPL PHYS, SUPPL 2, PART 2, 1974)
CA 83: 107607

5-1215 EFFECT OF ICN BCMBARDMENT ON THE COMPOSITION
AND STRUCTURE OF TUNGSTEN AND MOLYBDENUM
SINGLE CRYSTALS STUDIED BY AUGER ELECTRON
SPECTROSCOPY AND SLOW ELECTRON DIFFRACTION.
BURMAKA DS + IVASHCHENKC YN + CHEREPIN VT
P136-9 OF VZAIMODEISTVIE ATOM CHASTIIS S
TVERD TELOV, 1974. (IN RUSSIAN)
CA 83: 150992

5-1216 SURFACE SEGREGATION IN ALLOYS: EQUILIBRIUM
SEGREGATION OF GOLD TO THE (111) SURFACE OF
NICKEL. (AES)
BURTON JJ + HELMS CR + POLIZZOTTI RS
J VACUUM SCI TECHNOL 13(1): 204-8 (1976)
SAA 1976: 49065

5-1217 AUGER ELECTRON STUDIES OF SURFACES: URANIUM
DIOXIDE, URANIUM, GRAPHITE, 300 SERIES
STAINLESS STEEL, AND NICBIUM.
CAMPBELL BD
LOS ALAMOS SCI LABS 1969. REPT NO LA-4010

5-1218 AUGER ELECTRON STUDIES OF URANIUM DIOXIDE
SURFACES.
CAMPBELL BD + ELLIS WP
J CHEM PHYS 52: 3303-4 (1970)

5-1219 IN-DEPTH AUGER ANALYSIS OF ALUMINUM- SILICON
INTERFACIAL REACTIONS.
CARD HC + SINGER KE
THIN SOLID FILMS 28(2): 265-8 (1975)
CA 83: 140836

5-1220 OPTICAL PROPERTIES AND AUGER CHARACTERISTICS
OF MAGNESIUM SINGLE CRYSTALS AND MAGNESIUM
OXIDE AND MAGNESIUM NITRIDE SURFACE FILMS.
CARDEN JL + FERRANDINO F + LARSEN J + WALL W
+ STEVENSON JR
P572-3 OF INT CONF VAC ULTRAVIOLET RADIAT
PHYS, 4TH PROC, KOCH EE + HAENSEL R + KUNZ
C(EDS), PERGAMON, 1974.
CA 83: 50172

5-1221 ADSORPTION OF KRYPTON ON THE (111) FACE OF
COPPER SINGLE CRYSTALS. (AUGER ELECTRON
SPECTROSCOPY)
CARDEN JL + PIEROTTI RA
J COLLOID INTERFACE SCI 47(2): 379-94 (MAY
1974)
SAA 1974: 56970

5-1222 INJECTION OF IONS INTO GLASS FROM A GLOW
DISCHARGE. (AUGER SPECTROSCOPY)
CARLSON DE + TRACY CE
J APPL PHYS 46(4): 1575-80 (1975)
SAA 1975: 51275

5-1223 STUDY OF THE INTERACTION OF OXYGEN WITH
PLATINUM SINGLE CRYSTAL SURFACES BY LOW
ENERGY ELECTRON DIFFRACTION AND AUGER
ELECTRON SPECTROSCOPY.
CARRIERE B + LEGARE P + MAIRE G
J CHIM PHYS, PHYS CHIM BIOL 71(3): 355-65
(1974) (IN FRENCH)
CA 80: 150828

5-1224 OXIDATION OF CUPRONICKEL ALLOYS. PART-1: XPS
STUDY OF INTERDIFFUSION. (AND AES)
CASTLE JE + NASSERIAN-RIABI M
CORROS SCI 15(9): 537-43 (1975)
SAA 1976: 35701

5-1225 AUGER ELECTRON SPECTROSCOPY ANALYSIS OF ION
PLATED GOLD FILMS ON SILICON.
CAVALERU A + LINCA G + DELCEA B
STUD CERCET FIZ 28(2): 113-9 (1976) (IN
ROMANIAN)
CA 84: 183531

5-1226 DIFFUSION OF SULFUR AND PHOSPHORUS IMPURITY
ATOMS TO A TUNGSTEN SURFACE.
CHAIKOVSKII EF + SOTNIKOV VT
SOV PHYS SOLID STATE 15(9): 1864 (1973)
CA 79: 140707

5-1227 GROWTH AND STRUCTURE OF THIN COPPER FILMS ON
(001) SURFACES OF NICKEL.
CHAMBERS A + JACKSON DC
PHIL MAG 31(6): 1357-71 (1975)
CA 83: 124165

5-1228 AUGER ANALYSIS OF SILICON THIN FILMS
DEPOSITED ON CARBON AT HIGH TEMPERATURES.
CHANG CA + SIEKHAUS WJ
J APPL PHYS 46(8): 3402-7 (1975)
CA 83: 166856

5-1229 CONTAMINANTS ON CHEMICALLY ETCHED SILICON
SURFACES. LEED- AUGER METHOD.
CHANG CC
SURFACE SCI 23: 283-98 (1970)

5-1230 SILICON ON SAPPHIRE EPITAXY BY VACUUM
SUBLIMATION: LEED- AUGER STUDIES AND
ELECTRONIC PROPERTIES OF FILMS.
CHANG CC
J VACUUM SCI TECHNOL 8: 500-11 (1971)

5-1231 PHOSPHORUS CONCENTRATION PROFILES IN P-DOPED
SILICON DIOXIDE MEASURED USING AUGER
SPECTROSCOPY.
CHANG CC + ADAMS AC + QUINTANA G + SHENG TT
J APPL PHYS 45(1): 252-6 (1974)
CA 80: 087995

5-1232 INTERDIFFUSIONS IN THIN FILM GOLD ON PLATINUM
ON GALLIUM ARSENIDE (100) STUDIED WITH AUGER
SPECTROSCOPY.
CHANG CC + MURARKA SP + KUMAR V + QUINTANA G
J APPL PHYS 46(10): 4237-43 (1975)
CA 84: 35605

5-1233 STRUCTURE AND CHEMISTRY OF SILICON SURFACES
AFTER PRE- AND BACKSPUTTERING, STUDIED WITH
AUGER SPECTROSCOPY, ELLIPSOMETRY, AND
REFLECTION HIGH ENERGY ELECTRON DIFFRACTION.
CHANG CC + PETROFF P + QUINTANA G + SOSNIAK J
SURFACE SCI 38(2): 341-56 (1973)
CA 79: 046076

5-1234 SILICON DIFFUSION THROUGH THIN TUNGSTEN FILMS
ON SILICON, STUDIED WITH AUGER SPECTROSCOPY.
CHANG CC + QUINTANA G
J ELECTRON SPECTROS RELAT PHENOMENA 2(4):
363-76 (1973)
CA 80: 019723

5-1235 DIFFUSION KINETICS OF GOLD THROUGH PLATINUM
FILMS ABOUT 2000 AND 600 ANGSTROMS THICK
STUDIED WITH AUGER SPECTROSCOPY.
CHANG CC + QUINTANA G
THIN SOLID FILMS 31(3): 265-73 (1976)
CA 84: 155835

5-1236 AUGER ELECTRON SPECTROSCOPY STUDY OF FLOAT
GLASS SURFACES.
CHAPPELL RA + STODDART CTH
PHYS CHEM GLASSES 15(5): 130-6 (1974)
CA 83: 167779

5-1237 CARBON CONTAMINATION OF SILICON (111)
SURFACES.
CHARIG JM + SKINNER DK
SURFACE SCI 15: 277-85 (1969)

5-1238 AUGER ELECTRON SPECTROSCOPY OF NICKEL
DEPOSITS ON THE SILICON (111) SURFACE.
CHARIG JM + SKINNER DK
SURFACE SCI 19: 283-90 (1970)

5-1239 SECONDARY ELECTRON EMISSION OF AMORPHOUS
SEMICONDUCTOR THIN FILMS.
CHEN ACM + NORTON JF + WANG JM
APPL PHYS LETT 18: 443-4 (5 MAY 1971)

5-1240 LEED, AUGER, AND WORK FUNCTION STUDIES OF
CLEAN AND SODIUM- COVERED SURFACES OF GALLIUM
ARSENIDE.
CHEN JM
SURFACE SCI 25: 305-14 (1971)

5-1241 DITUNGSTEN CARBIDE OVERLAYER ON TUNGSTEN
(112).
CHEN JM + PAPGEORGOPOULOS CA
SURFACE SCI 20: 195-200 (1970)

5-1242 ELECTRON AUGER SPECTROSCOPIC STUDY OF THE
SURFACE OF CARBIDE CATHODES.
CHEREVATSKII NYA + KUL'VARSKAYA BS
KRATK SODERZH DOKI 2: 118 (1973) (IN RUSSIAN)
CA 83: 187070

5-1243 CHARACTERIZATION OF ADSORBED SPECIES USING
AUGER PEAK SHAPES. ETHYLENE AND CARBON
MONOXIDE ON TUNGSTEN (100).
CHESTERS MA + HOPKINS BJ + JONES AR
+ NATHAN R
SURFACE SCI 45(2): 740-4 (1974)
CA 82: 022146

5-1244 ELECTRON BEAM EFFECTS ON CARBON MONOXIDE
SATURATED TUNGSTEN (100), ETHYLENE SATURATED
TUNGSTEN (100), AND TUNGSTEN CARBIDE.
CHESTERS MA + HOPKINS BJ + JONES AR
+ NATHAN R
J PHYS C 7(24): 4486-92 (1974)
CA 82: 77439

5-1245 CHEMISORPTION ON TANTALUM (100). (AUGER
SPECTROSCOPY)
CHESTERS MA + HOPKINS BJ + LEGGETT MR
SURFACE SCI 43(1): 1-10 (MAY 1974)
SAA 1974: 48856

5-1246 LEED AND SURFACE POTENTIAL STUDY OF CARBON
MONOXIDE AND XENON ADSORBED ON COPPER (100).
CHESTERS MA + PRITCHARD J
SURFACE SCI 28: 460-8 (1971)

5-1247 MAGNESIUM DOPED GALLIUM ARSENIDE AND
ALUMINUM(X) GALLIUM(1-X) ARSENIDE BY
MOLECULAR BEAM EPITAXY. (AUGER SPECTROMETRY)
CHO AY + PANISH MB
J APPL PHYS 43(12): 5118-23 (DEC 1972)
SAA 1973: 18553

5-1248 AUGER AND ELLIPSOMETRIC STUDIES OF ULTRATHIN
LEAD(II) OXIDE GROWTH ON LEAD.
CHOU NJ + ELDRIDGE JM + HAMMER R + DONG D
J ELECTRON MATER 2(1): 115-26 (1973)
CA 78: 129155

5-1249 AUGER ANALYSIS OF THIN OXIDE FILMS ON LEAD-
INDIUM ALLOYS.
CHOU NJ + LAHIRI SK + HAMMER R + KOMAREK KL
J CHEM PHYS 63(6): 2758-64 (1975)
CA 83: 184117

5-1250 AUGER AND ELLIPSOMETRIC STUDY OF PHOSPHORUS
SEGREGATION IN OXIDIZED DEGENERATE SILICON.
CHOU NJ + VAN DER MEULEN YJ + HAMMER R
+ CAHILL J
APPL PHYS LETT 24(4): 200-2 (1974)
CA 80: 101471

5-1251 AUGER ANALYSIS OF CHLORINE IN HYDROGEN
CHLORIDE OR CHLORINE GROWN SILICON DIOXIDE
FILMS.
CHOU NJ + VAN DER MEULEN YJ + HAMMER R
APPL PHYS LETT 22: 380-1 (1973)

5-1252 INTERACTIONS OF CARBON MONOXIDE AND OXYGEN
WITH IRIDIUM (110) SURFACES. (AUGER
SPECTROSCOPY)
CHRISTMANN K + ERTL G
Z NATURFORSCH A28: 1144-8 (1973)

5-1253 SURFACE STUDIES WITH EPITAXIALLY GROWN METAL
FILMS. (AUGER EFFECT)
CHRISTMANN K + ERTL G
THIN SOLID FILMS 28(1): 3-18 (1975)
SAA 1975: 72722

5-1254 ADSORPTION OF CARBON MONOXIDE ON SILVER-
PALLADIUM ALLOYS.
CHRISTMANN K + ERTL G
SURFACE SCI 33: 254-70 (1972)

5-1255 ADSORPTION OF HYDROGEN ON A PLATINUM (111)
SURFACE.
CHRISTMANN K + ERTL G + PIGNET TP
SURFACE SCI 54(2): 365-92 (1976)
SAA 1976: 35687

5-1256 ADSORPTION OF COBALT ON A NICKEL (111)
SURFACE. (AUGER EFFECT, AUGER SPECTROSCOPY)
CHRISTMANN K + SCHOBER O + ERTL G
J CHEM PHYS 60(12): 4719-24 (15 JUN 1974)
SAA 1974: 75692

5-1257 ADSORPTION OF HYDROGEN ON NICKEL SINGLE
CRYSTAL SURFACES. (AUGER ELECTRON
SPECTROSCOPY)
CHRISTMANN K + SCHOBER O + ERTL G + NEUMANN M
J CHEM PHYS 60(11): 4528-40 (1 JUN 1974)
SAA 1974: 67097

5-1258 AUGER SPECTROSCOPY OF SOLID SURFACES IN A DRY
PUMPED HIGH RESOLUTION SCANNING ELECTRON
MICROSCOPE.
CHRISTOU A
P1419-56 OF SCANNING ELECTRON MICROSCOPY CONF
1975, JOHARI O + CORVIN I (EDS), ILL INST
TECHNOL RES 1975.
SAA 1976: 3939

5-1259 SCANNING MICROSPOT AUGER SPECTROSCOPY STUDY
OF INTERDIFFUSION AND EUTECTIC FORMATION IN
TUNGSTEN PLATINUM- TUNGSTEN GOLD THIN FILMS.
CHRISTOU A + DAY HM
IEEE TRANS PARTS HYBRIDS PACKAG PHP-11(3):
229-35 (1975)
CA 84: 24493

5-1260 SILICIDE FORMATION AND INTERDIFFUSION EFFECTS
IN SILICON- TANTALUM, SILICON DIOXIDE-
TANTALUM, AND SILICON- PLATINUM SILICIDE-
TANTALUM THIN FILM STRUCTURES. (AES)
CHRISTOU A + DAY HM
J ELECTRON MATER 5(1): 1-12 (1976)
SAA 1976: 45427

5-1261 SCANNING ELECTRON MICROSCOPY, AUGER
SPECTROSCOPY, AND ION BACKSCATTERING
TECHNIQUES APPLIED TO ANALYSIS OF GOLD-
REFRACTORY METALLIZATIONS.
CHRISTOU A + JARVIS L + WEISENBERGER WH
+ HIRVONEN JK
J ELECTRON MATER 4(2): 329-45 (1975)
CA 82: 164134

5-1262 COMBINED SCANNING ELECTRON MICROSCOPY- AUGER
SPECTROSCOPY FOR MICROSPOT SURFACE AND
IN-DEPTH ANALYSIS OF SILICON AND TRANSISTOR
METALLIZATIONS.
CHRISTOU A + WEISENBERGER W + DAY HM
P143-50 OF SEMICONDUCTOR MEASUREMENT
TECHNOLOGY. PART-4: ARPA/NBS WORKSHOP ON
SURFACE ANALYSIS FOR SILICON DEVICES,
LIEBERMAN AG(ED), NAT BUR STAND SPEC PUB
400-23, 1976.

5-1263 CHARACTERISTIC ENERGIES IN SECONDARY ELECTRON
SPECTRA FROM SILICON (111) SURFACES.
CHUNG MF + JENKINS LH
SURFACE SCI 26: 649-63 (1971)

5-1264 AUGER SPECTROSCOPIC ANALYSIS OF BIOGLASS
CORROSION FILMS.
CLARK AE + PANTANO CG + HENCH LL
J AMER CERAM SOC 59(1-2): 37-9 (1976)
CA 84: 140694

5-1265 AQUEOUS CORROSION OF SODA- SILICA AND SODA-
LIME- SILICA GLASS. (AES)
CLARK DE + DILMORE MF + ETHRIDGE EC
+ HENCH LL
J AMER CERAM SOC 59(1-2): 62-5 (1976)
SAA 1976: 41236

5-1266 AUGER ANALYSIS OF BRASS- ENAMEL AND STAINLESS STEEL- ENAMEL INTERFACES.
CLARK DE + PANTANO CG + ONODA GY
J AMER CERAM SOC 58(7-8): 336-7 (1975)
CA 83: 182969

5-1267 AUGER SPECTROSCOPY AND LOW ENERGY ELECTRON DIFFRACTION STUDIES OF THE CHEMISORPTION OF UNSATURATED MOLECULES BY THE (100) SURFACE OF PLATINUM.
CLARKE TA + MASON R + TESCARI M
PROC ROY SOC A 331(1586): 321-33 (1972)
SAA 1973: 11808

5-1268 X-RAY PHOTOELECTRON AND AUGER SPECTROSCOPIC ANALYSES OF STEEL SURFACES.
COAD JP
UK AT ENERGY RES ESTAB, 1975, 7P. AERE-R-7944
CA 83: 182631

5-1269 ELECTRON BEAM ASSISTED ADSORPTION ON SILICON (111) SURFACE.
COAD JP + BISHOP HE + RIVIERE JC
SURFACE SCI 21: 253-64 (1970)

5-1270 AUGER SPECTROSCOPY OF CARBON ON NICKEL.
COAD JP + RIVIERE JC
SURFACE SCI 25: 609-24 (1971)

5-1271 ORIGIN OF FINE STRUCTURE IN THE AUGER SPECTRUM OF SULFUR ON A NICKEL SURFACE.
COAD JP + RIVIERE JC
PROC ROY SOC A 331: 403-15 (1972)
CA 78: 50104T

5-1272 SURFACE ANALYSES BY COMBINED X-RAY PHOTOELECTRON SPECTROSCOPY AND AUGER ELECTRON EMISSION SPECTROSCOPY.
COAD JP + RIVIERE JC
VIDE 28(167): 176-8 (1973)
CA 80: 140775

5-1273 TECHNIQUES FOR ELEMENTAL COMPOSITION PROFILING IN THIN FILMS. (AUGER SPECTROSCOPY)
COBURN JW + KAY E
CRIT REV SOLID STATE SCI 4(4): 561-90 (1974)
SAA 1975: 39984

5-1274 STUDY OF MULBERRY SURFACES BY AUGER AND IEE SPECTROSCOPY.
COLMENARES CA
UNIV CALIF, LAWRENCE RADIATION LAB 1970. REPT UCID-15599

5-1275 APPLICATION OF AUGER ELECTRON SPECTROSCOPY TO SURFACE CHEMISTRY STUDIES OF CERAMIC SUBSTRATES.
CONLEY DK
BULL AMER CERAM SOC 52(4): 375 (APR 1973)
SAA 1974: 07262

5-1276 AUGER ELECTRON SPECTROSCOPY OF GRAPHITE FIBER SURFACES.
CONNELL GL
NATURE 230: 377 (1971)

5-1277 AUGER ELECTRON SPECTROSCOPY OF LUNAR MATERIAL.
CONNELL GL + GUPTA YP
IN 2ND NASA LUNAR SCIENCE CONF, 1971 PROC.

5-1278 INTERACTION OF NITRIC OXIDE WITH A NICKEL (111) SURFACE.
CONRAD H + ERTL G + KUPPERS J + LATTA EE
SURFACE SCI 50(2): 269-310 (1975)
SAA 1975: 69109

5-1279 STUDY OF SURFACE DISSOCIATION IN POTASSIUM CHLORIDE AND LITHIUM FLUORIDE BY OBSERVATION OF PLASMON LOSSES ASSOCIATED WITH POTASSIUM AND LITHIUM. (AUGER SPECTROSCOPY)
COTAARAIZA IS + POWELL BD
SURFACE SCI 51(2): 504-12 (1975)
SAA 1976: 6337

5-1280 USE OF SCANNING AUGER SPECTROSCOPY FOR SURFACE ANALYSIS.
COUNTS RL
AIR FORCE INST TECHNOL WRIGHT-PATTERSON AFB, OHIO, MS THESIS, MAR 1973, 63P. AD-760550

5-1281 MECHANISM OF SURFACE ALIGNMENT IN NEMATIC LIQUID CRYSTALS. (SUBSTRATE SURFACE EFFECTS, AUGER SPECTROMETRY)
CREAGH LT + KMETZ AR
MOL CRYST LIQ CRYST 24(1-2): 59-68 (1973)
SAA 1974: 78902

5-1282 AUGER ELECTRON SPECTROSCOPY MEASUREMENTS OF ADSORPTION ISOBARS FOR OXYGEN ON TUNGSTEN AT LOW PRESSURE AND HIGH TEMPERATURE.
DABIRI AE + ARAMATI VS + STICKNEY RE
SURFACE SCI 40(2): 205-15 (1973)
CA 79: 149722

5-1283 AUGER SPECTROSCOPY OF CARBON ON NICKEL. COMMENTS.
DALMAI-IMELIK G + BERTOLINI JC + ROUSSEAU J
SURFACE SCI 27: 379 (1971)

5-1284 STUDY OF THE ADSORPTION OF BENZENE ON THE (100) FACE OF NICKEL. INFLUENCE OF SULFUR AT THE SURFACE.
DALMAI-IMELIK G + ROUSSEAU J + BERTOLINI K
VIDE 27(158, SUPPL): 36-46 (1972)

5-1285 PLATINUM SILICIDE FORMATION. ELECTRON SPECTROSCOPY OF THE PLATINUM- PLATINUM SILICIDE INTERFACE.
DANYLUK S + MCGUIRE GE
J APPL PHYS 45(12): 5141-4 (1974)
CA 82: 79509

5-1286 DIFFUSION STUDIES IN CHROMIUM- PLATINUM THIN FILMS USING AUGER ELECTRON SPECTROSCOPY.
DANYLUK S + MCGUIRE GE + KOLIWAD KM + YANG MG
THIN SOLID FILMS 25(2): 483-9 (1975)
CA 82: 160423

5-1287 SOLID PHASE EPITAXIAL STUDIES USING VACUUM DEPOSITION ON HEATED SILICON SUBSTRATES. (SCANNING AUGER SPECTROSCOPY)
DAVEY JE + CHRISTOU A + DAY HM
APPL PHYS LETT 28(7): 365-7 (1976)
SAA 1976: 49507

5-1288 A STUDY OF PHOSPHORUS ADSORPTION AND DESORPTION KINETICS ON SILICON (111) SURFACES. (BY LEED AND AES)
DAVIS LE + LEVENSON LL + MELLES JJ
J CRYST GROW 17: 354-6 (1972)
SAA 1973: 1883

5-1289 SILICON DEVICE APPLICATIONS USING A COMBINED ESCA- AUGER ANALYSIS SYSTEM.
DAVIS LE + RIACH GE
P183-7 OF SEMICONDUCTOR MEASUREMENT TECHNOLOGY. PART-4: ARPA/NBS WORKSHOP ON SURFACE ANALYSIS FOR SILICON DEVICES, LIEBERMAN AG(ED), NAT BUR STAND SPEC PUB 400-23, 1976.
CA 84: 188450

5-1290 CHEMISORPTION STUDIES OF NICKEL SURFACES UTILIZING WORK FUNCTION MEASUREMENTS, AUGER, AND LOW ENERGY ELECTRON DIFFRACTION SPECTROSCOPY.
DEMUTH JE
CORNELL UNIV, THESIS 1973, 272P.
CA 81: 040518

5-1291 CHEMISORPTION ON (001), (110), AND (111) NICKEL SURFACES. CORRELATED STUDY USING LOW ENERGY ELECTRON DIFFRACTION SPECTRA, AUGER SPECTRA, AND WORK FUNCTION CHANGE MEASUREMENTS.
DEMUTH JE + RHODIN TN
SURFACE SCI 45(1): 249-307 (1974)
CA 81: 159316

5-1292 ELASTIC LEED INTENSITY ENERGY STUDIES OF CLEAN (001), (110) AND ELASTIC (111) NICKEL SURFACES. (AUGER SPECTROSCOPY)
DEMUTH JE + RHODIN TN
SURFACE SCI 42(1): 261-98 (MAR 1974)
SAA 1974: 28264

5-1293 KINETICS OF THERMAL DESORPTION USING AUGER
ELECTRON SPECTROSCOPY. APPLICATION TO
CESIUM-COVERED (110) GALLIUM ARSENIDE.
DERRIEN J + DAVITAYA FA
REV PHYS APPL 11(3): 377-85 (1976)
CA 84: 185337

5-1294 LOW ENERGY ELECTRON DIFFRACTION, AUGER
ELECTRON SPECTRCSCOPY, AND WORK FUNCTION
MEASUREMENTS ON CLEAN AND CESIUM COVERED
POLAR FACES OF GALLIUM ARSENIDE.
DERRIEN J + D'AVITAYA FA + GLACHANT A
SURFACE SCI 47(1): 162-6 (1975)
CA 82: 129724

5-1295 AUGER ELECTRON SPECTROSCOPY OF CESIUM
ADSORBED ON CLEAN AND OXYGEN- COVERED (100)
TUNGSTEN.
DESPLAT JL
SOLID STATE COMMUN 13(6): 689-91 (1973)
CA 79: 130893

5-1296 CESIUM- OXYGEN COADSORPTION ON (100)
TUNGSTEN.
DESPLAT JL
VIDE 28(167): 184-6 (1973) (IN FRENCH)
CA 80: 113063

5-1297 LOW ENERGY ELECTRON DIFFRACTION AUGER, WORK
FUNCTION STUDY OF OXYGEN- CESIUM ADSORPTION
AND COADSORPTION ON (100) TUNGSTEN.
DESPLAT JL
P177-80 OF INT VACUUM CONGRESS, 6TH PROC, INT
CONF ON SOLID SURFACES, 2ND, 1974, KYOTO.
(JAP J APPL PHYS, SUPPL 2, PART 2, 1974)
CA 82: 175598

5-1298 MUSCOVITE CLEAVAGE STUDIED BY AUGER ELECTRON
SPECTROSCOPY.
DEVILLE JP + GOLDSZTAUB S
COMPT REND B 268: 629-30 (1969) (IN FRENCH)

5-1299 PURITY AND MORPHOLOGY OF ALUMINUM FILMS.
(AUGER SPECTRCSCOPY)
DHERE NG + ARSENIO TP + PATNAIK BK
THIN SOLID FILMS 30(2): 267-79 (1975)
SAA 1976: 26387

5-1300 STRUCTURE DEPENDENT PROPERTIES OF THIN FILM
SUPERCONDUCTORS.
DICKEY JM
CITY UNIV NY RES FOUNDATION, AUG 1974, 9P.
AD-785424/3; GOVT REP ANNOUNCE 74(23): 174
(1974)
CA 82: 163961

5-1301 INTERACTION OF NITROGEN WITH NIOBIUM. (AUGER
EFFECT)
DICKEY JM
SURFACE SCI 50(2): 515-26 (1975)
SAA 1975: 72693

5-1302 PHASE TRANSITIONS IN TWO DIMENSIONAL ABSORBED
SULFUR LAYERS ON COPPER INVESTIGATED BY LEED
AND AUGER SPECTROSCOPY.
DOMANGE JL
J VACUUM SCI TECHNOL 9(2): 682-4 (1972)

5-1303 CHEMISORPTION ON SINGLE CRYSTAL MOLYBDENUM
(112) SURFACES.
DOOLEY GJ + HAAS TW
J VACUUM SCI TECHNOL 7: 49-52 (1970)

5-1304 SOME PROPERTIES OF THE RHENIUM (0001)
SURFACE.
DOOLEY GJ + HAAS TW
SURFACE SCI 19: 1-8 (1970)

5-1305 FURTHER STUDIES OF GAS ADSORPTION ON
MOLYBDENUM (100) SURFACE.
DOOLEY GJ + HAAS TW
J CHEM PHYS 52: 461-2 (1970)

5-1306 ADSORPTION OF OXYGEN ON CLEAN CLEAVED (110)
GALLIUM ARSENIDE SURFACES.
DORN R + LUTH H + RUSSELL GJ
PHYS REV B 10(12): 5049-56 (1974)
SAA 1975: 37544

5-1307 USE OF AUGER ELECTRON SPECTROSCOPY TO
DETERMINE THE COMPOSITION OF FRACTURE
SURFACES OF MINERALS.
EADINGTON P
TRANS INST MIN MET C 83: C223-7 (1974)
CA 83: 135423

5-1308 PHOTOEMISSION STUDIES OF CHEMISORPTION ON
NICKEL AND TUNGSTEN SURFACES.
EASTMAN DE + BAKER JM + DEMUTH JE
P45 OF EUROP PHYS SOC STUDY CONF ON METAL
SURFACES, ABSTR, 1973.
SAA 1974: 37112

5-1309 (AUGER ELECTRON AND MOSSBAUER)
SPECTROSCOPICAL STUDIES OF THE COMPOSITION OF
PASSIVE STATE FILMS.
EBIKO H
HYOMEN 12(11): 668-77 (1974) (IN JAPANESE)
CA 82: 147171

5-1310 AUGER SPECTRA FROM THE (100) SURFACE OF
SILICON DURING OXIDATION AND NITRIDATION.
EDELMAN FL + DADAYAN KA + SAVCHENKO VI
P252-6 OF ELEKTRON PROTESSY NA POVERKHNOSTI
POLUPROVOD I NA GRANITSE VOZDELA POLUPROVOD
DIELEKTRIK, 1974. (IN RUSSIAN)
CA 83: 105661

5-1311 PLATINUM (100) STRUCTURES: IMPURITY CAUSATION
AND PROPERTIES. (AUGER EFFECT)
EDMONDS T + MCCARROLL JJ + PITKETHLY RC
P53 OF EUROP PHYS SOC STUDY CONF ON METAL
SURFACES, ABSTR, 1973.
SAA 1974: 37067

5-1312 CONDENSATION OF GOLD ONTO TANTALUM (100)
SINGLE CRYSTAL SURFACES. LEED AND AES
ANALYSIS.
ELLIOT AG
SURFACE SCI 51(2): 489-503 (1975)
CA 83: 200365

5-1313 SEGREGATION OF IMPURITIES ON SURFACES OF
THORIUM. THERMODYNAMICS AND KINETICS.
(DETERMINATION OF SURFACES BY AUGER METHOD)
ELLIS WP
USAEC, 1970, 7P. LA-DC-12348
CA 76: 29410P

5-1314 IMPURITIES ON THORIUM METAL: AUGER STUDY OF
BULK TO SURFACE EQUILIBRIA.
ELLIS WP
J VACUUM SCI TECHNOL 9(2): 1027-32 (1972)
SAA 1972: 50465

5-1315 ARSENIC (0001) SURFACES: AUGER, LOSS, AND
PHOTOELECTRON SPECTROSCOPIC STUDIES.
ELLIS WP
SURFACE SCI 41(1): 125-41 (1974)
CA 80: 032204

5-1316 SECONDARY ELECTRON ENERGY DISTRIBUTION
STUDIES OF URANIUM DIOXIDE SURFACES.
ELLIS WP + CAMPBELL BD
J APPL PHYS 41: 1858-61 (1970)

5-1317 GROWTH OF THIN METALLIC FILMS ON INSULATING
SUBSTRATES IN THE PRESENCE OF ELECTRON
IRRADIATION. (AUGER SPECTROSCOPY)
ELOVICOV SS + DUBININA EM + NOVOJILOV VP
THIN SOLID FILMS 32(2): 220 (1976)
SAA 1976: 40353

5-1318 ADSORPTION OF OXYGEN ON THE (110) PLANE OF
TUNGSTEN. (AUGER ELECTRON SPECTROSCOPY)
ENGEL T + NIEHUS H + BAUER E
VIDE 28(164): 82-3 (MAR-APR 1973); P46-7 OF
EUROP PHYS SOC STUDY CONF ON METAL SURFACES,
ABSTR, 1973.
SAA 1973: 79564; 1974: 37113

5-1319 AN ELECTRICAL, CHEMICAL, AND STRUCTURAL STUDY
OF THE GERMANIUM (100)- CESIUM- OXYGEN
PHOTOSURFACE.
ERICSON RE
UNIV MINNESOTA PHD THESIS, 1975.
SAA 1976: 49489

5-1320 ELECTRON SPECTROSCOPY AND LEED STUDIES OF
GAS- METAL INTERFACES. (AUGER EFFECT)
ERTL G
SURFACE SCI 47(1): 86-8 (1975)
SAA 1975: 46467

5-1321 INTERACTION OF CARBON MONOXIDE AND OXYGEN
WITH A PALLADIUM (100) SURFACE.
(CHEMISORPTION AND CATALYTIC REACTIONS, LEED,
AUGER, CONTACT POTENTIAL METHOD AND MASS
SPECTROMETRY)
ERTL G + KOCH J
Z PHYS CHEM (FRANKFURT) 69: 323-7 (1970) (IN
GERMAN)

5-1322 ADSORPTION ON SINGLE CRYSTAL SURFACES OF
COPPER- NICKEL ALLOYS. PART-1.
ERTL G + KUPPERS J
SURFACE SCI 24: 104-24 (1971) (IN GERMAN)

5-1323 ELECTRON SPECTROSCOPIC STUDIES OF CLEAN AND
OXIDIZED IRON. (AUGER STUDY OF SURFACE)
ERTL G + WANDELT K
SURFACE SCI 50(2): 479-92 (1975)
CA 83: 139378

5-1324 COMPUTER-CONTROLLED MOLECULAR BEAM EPITAXY.
(AUGER SURFACE ANALYSIS, GALLIUM(X)
ALUMINUM(1-X) ARSENIDE)
ESAKI L
P821-8 OF INT VACUUM CONGRESS, 6TH PROC, INT
CONF ON SOLID SURFACES, 2ND, 1974, KYOTO.
(JAP J APPL PHYS, SUPPL 2, PART 1, 1974)

5-1325 THERMALLY STIMULATED EXOELECTRON EMISSION OF
BERYLLIUM OXIDE LAYERS. AUGER ELECTRON
SPECTROSCOPY INVESTIGATIONS.
EULER M + SCHARMANN A
PHYS STATUS SOLIDI A 34(1): 297-306 (1976)

5-1326 TECHNIQUES FOR THE CHEMICAL CHARACTERIZATIONS
OF CERAMIC SURFACES. (ICN SCATTERING
SPECTROMETRY, AUGER ELECTRON SPECTROMETRY)
EVANS CA
BULL AMER CERAM SOC 52(4): 437 (APR 1973)
SAA 1973: 80418

5-1327 COMPARISON OF THE TECHNIQUES FOR SILICON
SURFACE ANALYSIS.
EVANS CA
P219-32 OF NAT BUR STAND SPEC PUBL 400-23,
1976
CA 85: 40383

5-1328 AUGER ANALYSIS OF THREE ENHANCED MULTILAYER
DIELECTRIC MIRROR DESIGNS. (ZINC SELENIDE,
THORIUM FLUORIDE)
EWING WS
US AIR FORCE CAMBRIDGE RES LAB PHYS SCI RES
PAP 637, 1975, 16P.
CA 84: 23952

5-1329 INVESTIGATIONS OF THE OXIDE FORMATION ON
METALS WITH AUGER ELECTRON SPECTROSCOPY.
FAERBER W + BRAUN P
MIKROCHIM ACTA SUPPL 6: 391-402 (1975) (IN
GERMAN)
CA 83: 209766

5-1330 WORK FUNCTION VARIATION WITH ALLOY
COMPOSITION: COPPER- GOLD. (AUGER
SPECTROSCOPY)
FAIN SC + MCDAVID JM
PHYS REV B 13(4): 1853-4 (1976)
SAA 1976: 49190

5-1331 SPUTTERING OF THE ALLOY SYSTEMS SILVER- GOLD,
GOLD- COPPER, AND SILVER- COPPER STUDIED BY
AUGER ELECTRON SPECTROSCOPY.
FARBER W + BETZ G + BRAUN P
NUCL INSTRUM METH 132: 351-4 (1976)
SAA 1976: 54347

5-1332 OXYGEN EXPOSURE OF SAMARIUM, GADOLINIUM, AND
TERBIUM STUDIED BY AUGER ELECTRON
SPECTROSCOPY.
FARBER W + BRAUN P
SURFACE SCI 41(1): 195-204 (1974)
CA 80: 032205

5-1333 INTERACTION OF OXYGEN AND NITROGEN WITH THE
NIOBIUM (100) SURFACE. PART-2: REACTION
KINETICS. (AUGER EFFECT)
FARRELL HH + ISAACS HS + STRONGIN M
SURFACE SCI 38(1): 31-52 (JUL 1973)
SAA 1973: 55166

5-1334 MOLECULAR BEAM STUDIES OF INDIUM PHOSPHIDE
EVAPORATION AND GROWTH. (AES)
FARROW RFC
P547-8 OF ELECTROCHEM SOC FALL MEET ABSTR,
1975.
SAA 1976: 36514

5-1335 INTERACTION BETWEEN CHLORINE AND THE (100)
SURFACE OF GOLD.
FEDAK DG + FLORIO JV + ROBERTSON WD
P74-1 TO 74-18 OF STRUCTURE AND CHEMISTRY OF
SOLID SURFACES, SOMORJAI GA (ED), NY, WILEY,
1969.

5-1336 STUDY OF PALLADIUM SILICIDE FILMS ON SILICON
USING AUGER ELECTRON SPECTROSCOPY.
FERTIG DJ + ROBINSON GY
SOLID STATE ELECTRON 19(5): 407-13 (1976)
CA 85: 12431

5-1337 COMBINED LOW ENERGY ELECTRON DIFFRACTION,
AUGER ELECTRON SPECTROSCOPY, AND X-RAY
PHOTOELECTRON SPECTROSCOPY STUDY OF THE ZINC
OXIDE (0001) POLAR SURFACES.
FIERMANS L + ARJIS E + VENNIK J
+ MAENHOUT-VAN DER VORST W
SURFACE SCI 39(2): 357-67 (1973)
CA 79: 108368

5-1338 AUGER ELECTRON EMISSION SPECTROSCOPY (AES)
AND LOW ENERGY ELECTRON DIFFRACTION (LEED) OF
THE ZINC OXIDE (0001) POLAR SURFACES.
FIERMANS L + ARJIS E + VENNIK J
+ MAENHOUT-VAN DER VORST W
PHYS STATUS SOLIDI A 6(1): K59-63 (1971)

5-1339 (AUGER) ELECTRON SPECTROSCOPY OF TRANSITION
METAL OXIDE SURFACES.
FIERMANS L + HOOGEWIJS R + VENNIK J
SURFACE SCI 47(1): 1-40 (1975)
CA 82: 147163

5-1340 INELASTIC EFFECTS AND STRUCTURE IN THE AUGER
ELECTRON EMISSION SPECTRA OF VANADIUM
PENTOXIDE (010) AND VANADIUM (100) SURFACES.
FIERMANS L + VENNIK J
SURFACE SCI 35(1): 42-62 (1973)
CA 78: 77588R

5-1341 STRUCTURE AND COMPOSITION OF METAL SURFACES:
A REVIEW OF RECENT LEED, AES, AND XPS
RESULTS.
FIERMANS L + VENNIK J
REV M 21(1): 65-70 (1975)
CA 85: 25652

5-1342 DETERMINATION OF SURFACE STRUCTURES BY LEED
(AND AES): THE SILICON (111) SURFACE.
FLORIO JV
P42 OF AMER CRYSTALLOGR ASSOC SUMMER MEETING,
AMES, IOWA, 1971, AMER CRYSTALLOGR ASSOC,
1971.
SAA 1972: 3502

5-1343 PHASE TRANSFORMATIONS OF SILICON (111)
SURFACE.
FLORIO JV + ROBERTSON WD
SURFACE SCI 22: 459-64 (1970)

5-1344 CHLORINE REACTIONS ON THE SILICON (111)
SURFACE.
FLORIO JV + ROBERTSON WD
SURFACE SCI 18: 398-427 (1970)

5-1345 INVESTIGATIONS OF CARBON RESIDUES ON SURFACES
OF SILICON INTEGRATED CIRCUITS. (BY AUGER
ELECTRON SPECTROSCOPY)
FONTANA PV + DECOSTERD JP + WEGMANN L
J ELECTROCHEM SOC 121(1): 146-50 (1974)

5-1346 SURFACE STRUCTURES ON IRON (110) ULTRAHIGH
VACUUM.
GAFNER G + FEDER R
ACTA CRYSTALLOGR A 31(S3): S288 (1975)
SAA 1976: 09559

5-1347 (100) SURFACES OF ALKALI HALIDES. PART-1: AIR
AND VACUUM CLEANED SURFACES.
GALLON TE + HIGGINBOTHAM IG + PRUTTON M
+ TOKUTAKA H
SURFACE SCI 21: 224-32 (1970)

5-1348 GROWTH OF SILVER ON POTASSIUM CHLORIDE
OBSERVED BY LEED AND AUGER EMISSION
SPECTROSCOPY.
GALLON TE + HIGGINBOTHAM IG + PRUTTON M
+ TOKUTAKA H
THIN SOLID FILMS 2: 369-73 (1968)

5-1349 THE (111) SURFACE OF N-TYPE SILICON.
GALLON TE + PRUTTON M + WRAY L
J VACUUM SCI TECHNOL 9(2): 911-14 (1972)
CA 77: 26206M

5-1350 EFFECT OF YTTRIA ADDITIONS ON HOT PRESSED
SILICON NITRIDE. (AUGER ANALYSIS)
GAZZA GE
BULL AMER CERAM SOC 54(9): 778-81 (1975)
SAA 1976: 31943

5-1351 IONIZATION SPECTROSCOPY OF CONTAMINATED METAL
SURFACES.
GERLACH RL
J VACUUM SCI TECHNOL 8: 599-604 (1971)

5-1352 IONIZATION SPECTROSCOPY OF SURFACES.
GERLACH RL + HOUSTON JE + PARK RL
APPL PHYS LETT 16: 179-81 (1970)

5-1353 A PRELIMINARY STUDY OF PURE METAL SURFACES
USING AUGER ELECTRON SPECTROSCOPY X-RAY
PHOTOELECTRON SPECTROSCOPY AND SECONDARY ION
MASS SPECTROSCOPY.
GETTINGS M + COAD JP
SURFACE SCI 53(1): 636-48 (1975)
CA 84: 80125

5-1354 OBSERVATIONS OF THE EPITAXIAL GROWTH OF
NICKEL AND COPPER ON (111) SILVER. (BY AUGER
SPECTROSCOPY)
GIBSON MJ + DOBSON PJ
J PHYS F 5(5): 864-72 (1975)
SAA 1975: 54431

5-1355 AUGER SPECTROMETRIC AND MASS SPECTROMETRIC
STUDY OF THE CHEMISORPTION OF CARBON MONOXIDE
ON POLYCRYSTALLINE MOLYBDENUM.
GILLET E + CHIARENA JC + GILLET M
SURFACE SCI 54(3): 601-16 (1976) (IN FRENCH)
CA 84: 141045

5-1356 DIFFUSION OF SILVER THROUGH EPITAXIAL FILMS
OF COPPER AND NICKEL ON SILVER (111):
FORMATION OF AN EQUILIBRIUM LAYER OF SILVER.
(AUGER SPECTROSCOPY)
GISON MJ + DOBSON PJ
J PHYS F 5(10): 1828-35 (1975)
SAA 1976: 5703

5-1357 AUGER SPECTROSCOPY OF THE BARIUM- PLATINUM
FILM SYSTEM.
GNUCHEV NM + KIRSANOVA TS + LITVINOV OK
SOV PHYS SOLID STATE 17(3): 600-1 (1975)
CA 83: 35811

5-1358 RECORDING THE COMPONENTS ON THE SURFACE OF A
BINARY ALLOY. (AUGER SPECTROSCOPY)
GNUCHEV NM + KIRSANOVA TS + SEN AP
KRATK SODERZH DOKL 2: 113-15 (1973) (IN
RUSSIAN)
CA 83: 187067

5-1359 EFFECT OF STRUCTURE ON THE AUGER SPECTRA OF
BARIUM FILMS ON PLATINUM.
GNUCHEV NM + KIRSANOVA TS
TR LENINGR POLITEKHN IN-TA NO 345: 27-9
(1975) (IN RUSSIAN)
CA 85: 11931

5-1360 DETECTION OF IMPURITIES ON A SILICON (111)
SURFACE BY AUGER- LEED ANALYSIS.
GOFF RF + JACOBSON RL
BULL AMER PHYS SOC 13: 944 (1968)

5-1361 COMPOSITION OF THE SURFACE OF COPPER-
MAGNESIUM OXIDE CATALYSTS STUDIED BY AUGER
ELECTRON SPECTROSCOPY.
GOLDOBIN AN + SAVCHENKO VI
KINET KATAL 15(5): 1363-5 (1974) (IN RUSSIAN)
CA 82: 48063

5-1362 ELECTRON BEAM INDUCED REDUCTION OF CARBON
CONCENTRATION ON BERYLLIUM OXIDE AND ITS
EFFECTS ON SECONDARY ELECTRON EMISSION.
(AUGER SPECTROSCOPY)
GOLDSTEIN B
SURFACE SCI 39(2): 261-71 (SEP 1973)
SAA 1973: 75674

5-1363 LEED, AUGER AND PLASMON STUDIES OF NEGATIVE
ELECTRON AFFINITY ON SILICON PRODUCED BY THE
ADSORPTION OF CESIUM AND OXYGEN.
GOLDSTEIN B
SURFACE SCI 35: 227-45 (MAR 1973)
SAA 1973: 30536

5-1364 LOW ENERGY ELECTRON DIFFRACTION- AUGER
CHARACTERIZATION OF GALLIUM ARSENIDE DURING
ACTIVATION TO NEGATIVE ELECTRON AFFINITY BY
THE ADSORPTION OF CESIUM AND OXYGEN.
GOLDSTEIN B
SURFACE SCI 47(1): 143-61 (1975)
CA 82: 145530

5-1365 DETERMINATION OF THE COMPOSITION OF GLASS
SURFACES BY AUGER SPECTROSCOPY.
GOLDSTEIN B + CARLSON DE
J AMER CERAM SOC 55: 51-2 (JAN 1972)

5-1366 PREFERENTIAL EVAPORATION OF INDIUM FROM
GALLIUM(X) INDIUM(1-X) ARSENIDE. (AUGER
EFFECT)
GOLDSTEIN B + SZOSTAK D
APPL PHYS LETT 26(12): 685-7 (1975)
SAA 1975: 73290, SAB 1975: 37385

5-1367 INTERACTION OF SULFUR DIOXIDE WITH TUNGSTEN.
(AUGER ANALYSIS)
GOLUB S + FEDAK DG
SURFACE SCI 45(1): 213-26 (SEP 1974)
SAA 1975: 03649

5-1368 STUDY OF BARIUM FILMS ON A (110) TUNGSTEN
FACE BY THE DIFFRACTION OF SLOW ELECTRONS AND
AUGER SPECTROSCOPY.
GORODETSKII DA + MAKSIMENKO VI + MELNIK YP
KRATK SODERZH DOKL 1: 9-10 (1973) (IN
RUSSIAN)
CA 83: 152925

5-1369 EQUILIBRIUM SURFACE SEGREGATION OF CARBON ON
IRON (100) FACES. (AUGER EFFECT)
GRABKE HJ + TAUBER G + VIEFHAUS H
SCR MET 9(11): 1181-4 (1975)
SAA 1976: 27105

5-1370 LEED STUDY OF IRIDIUM (100) SURFACE.
GRANT JT
SURFACE SCI 18: 228-38 (1969)

5-1371 STUDY OF IRIDIUM (100) SURFACE USING LEED AND
AUGER ELECTRON SPECTROSCOPY.
GRANT JT
BULL AMER PHYS SOC 14: 794 (1969)

5-1372 STUDIES ON THE IRIDIUM (111) SURFACE USING
LEED AND AUGER ELECTRON SPECTROSCOPY.
GRANT JT
SURFACE SCI 25: 451-6 (1971)

5-1373 COMBINED LEED AND AES STUDIES OF INDIUM
ARSENIDE SURFACES.
GRANT JT
J VACUUM SCI TECHNOL 9: 911 (1972) (ABSTRACT)
SAA 1972: 50449

5-1374 STRUCTURE OF THE PLATINUM (100) SURFACE.
GRANT JT + HAAS TW
SURFACE SCI 18: 457-61 (1969)

5-1375 AUGER STUDIES OF (111) SILICON SURFACES.
GRANT JT + HAAS TW
J VACUUM SCI TECHNOL 6: 903 (1969) (ABSTRACT)

5-1376 AUGER ELECTRON SPECTROSCOPY STUDIES OF CARBON
OVERLAYERS ON METAL SURFACES.
GRANT JT + HAAS TW
SURFACE SCI 24: 332-4 (1971)

5-1377 LEED STUDY OF PLATINUM (100) SURFACE.
GRANT JT + HAAS TW
BULL AMER PHYS SOC 14: 948 (1969)

5-1378 IDENTIFICATION OF THE FORM OF CARBON AT A
SILICON (100) SURFACE USING AUGER ELECTRON
SPECTROSCOPY.
GRANT JT + HAAS TW
PHYS LETT A 33: 386-7 (1970)

5-1379 COMBINED LEED AND AUGER ELECTRON SPECTROSCOPY
STUDIES OF SILICON, GERMANIUM, GALLIUM
ARSENIDE, AND INDIUM ANTIMONIDE SURFACES.
GRANT JT + HAAS TW
J VACUUM SCI TECHNOL 8(1): 94-7 (1971)

5-1380 STUDY OF RUTHENIUM (0001) AND RHODIUM (111)
SURFACES USING LEED AND AUGER ELECTRON
SPECTROSCOPY.
GRANT JT + HAAS TW
SURFACE SCI 21: 76-85 (1970)

5-1381 STUDY OF INDIUM ARSENIDE (111) AND (-1-1-1)
SURFACES USING LEED AND AUGER ELECTRON
SPECTROSCOPY.
GRANT JT + HAAS TW
SURFACE SCI 26(2): 669-76 (1971)

5-1382 AUGER STUDIES OF CLEAVED SILICON (111)
SURFACES.
GRANT JT + HAAS TW
J VACUUM SCI TECHNOL 7: 77-9 (1970)

5-1383 NATURE OF SILICON (111) SURFACES.
GRANT JT + HAAS TW
APPL PHYS LETT 15: 140-1 (1969)

5-1384 QUANTITATIVE COMPARISON OF TITANIUM AND
TITANIUM MONOXIDE SURFACES USING AUGER
ELECTRON AND SOFT X-RAY APPEARANCE POTENTIAL
SPECTROSCOPIES.
GRANT JT + HAAS TW + HOUSTON JE
J VACUUM SCI TECHNOL 11(1): 227-30 (1974)
CA 80: 140872

5-1385 MORPHOLOGICAL AND ELECTRICAL PROPERTIES OF RF
SPUTTERED YTTRIA-DOPED ZIRCONIA THIN FILMS.
(AUGER EFFECT)
GREENE JE + WICKERSHAM CE + ZILDO JL
+ WELSH LB + SZOFRAN FR
J VACUUM SCI TECHNOL 13(1): 72-5 (1976)
SAA 1976: 53508

5-1386 AUGER ANALYSIS OF SURFACE FILMS ON SILVER-
TIN.
GRENGA HE + CARDEN JL + OKABE T + HOCHMAN RF
J BIOMED MATER RES 9(2): 207-11 (1975)
CA 83: 135893

5-1387 SURFACE ANALYSIS WITH HEAVY ION INDUCED AUGER
ELECTRONS.
GROENEVELD KO + SPOHR R
VAK TECH 23(8): 225-9 (1974) (IN GERMAN)
CA 82: 179859

5-1388 ANODIC OXIDATION OF GALLIUM ARSENIDE
PHOSPHIDE. (AES)
GUENTHNER CS + RAO PVS + THOMAS S + PAULSON W
P420-2 OF ELECTROCHEM SOC FALL MEET ABSTR,
1975.
SAA 1976: 36499

5-1389 STRUCTURAL STUDIES OF THE ADSORPTION OF
CESIUM AND OXYGEN ON SILICON (100).
GUNDRY PM + HOLTON R + LEVERETT V
SURFACE SCI 43(2): 647-52 (JUN 1974)
SAA 1974: 56982

5-1390 SURFACE ANALYSIS USING COMPLEMENTARY
ELECTRONIC AND CHEMICAL MEASUREMENTS. (AUGER)
HAAS GA
J VACUUM SCI TECHNOL 13(1): 479-86 (1976)
CA 84: 127087

5-1391 EFFECTS OF SULFUR, BARIUM, AND CALCIUM ON
IMPREGNATED CATHODE SURFACES. (AES)
HAAS GA + GRAY HF + THOMAS RE
J APPL PHYS 46(8): 3293-301 (1975)
SAA 1975: 86337

5-1392 X-RAY PHOTOELECTRON AND AUGER ELECTRON
SPECTRA OF MAGNESIUM THIN FILMS.
HALDER NC + ALONSO J + SWARTZ WE
THIN SOLID FILMS 34(1): 99-102 (1976)
CA 85: 27167

5-1393 DIFFUSION MECHANISMS IN THE PALLADIUM- GOLD
THIN FILM SYSTEM AND THE CORRELATION OF
RESISTIVITY CHANGES WITH AUGER ELECTRON
SPECTROSCOPY AND RUTHERFORD BACKSCATTERING
PROFILES.
HALL PM + MORABITO JM + POATE JM
THIN SOLID FILMS 33(1): 107-34 (1976)
CA 84: 185073

5-1394 SURFACE CHEMISTRY OF CONTACT MATERIALS. (BY
AUGER ELECTRON SPECTROSCOPY)
HAQUE CA
P287-97 OF HOLM SEMINAR ELECTRICAL CONTACT
PHENOMENA 1972, ILL INST TECHNOL, 1972.

5-1395 WORK FUNCTION CHANGES ON CONTACT MATERIALS.
(AUGER ELECTRON SPECTROSCOPY)
HAQUE CA + FRITZ JH
IEEE TRANS PARTS HYBRIDS PACKAG PHP-10(1):
27-31 (MAR 1974)
SAA 1974: 49053

5-1396 AUGER SPECTROSCOPY AND LEED OF SILICON
SURFACES.
HARMAN R + LIDAY J + VESELY M
P2 OF CZECH CONF ON ELECTRONICS AND VACUUM
PHYSICS, 5TH PROC, 1972.
SAA 1973: 15061

5-1397 AUGER ELECTRON SPECTROSCOPY OF SILICON
SURFACES.
HARMAN R + LIDAY J + VESELY M
P823-6 OF INT VACUUM CONGRESS 6TH PROC, INT
CONF ON SOLID SURFACES, 2ND, 1974, KYOTO.
(JAP J APPL PHYS, SUPPL 2, PART 2, 1974)
CA 82: 177696

5-1398 FLUORINE ION IMPLANTATION PROFILES IN GALLIUM
ARSENIDE AS DETERMINED BY AES.
HARRIS JS + HARRIS JM + MARCUS HL
APPL PHYS LETT 21(12): 598-60 (1972)

5-1399 ANALYSIS OF MATERIALS BY ELECTRON EXCITED
AUGER ELECTRONS.
HARRIS LA
J APPL PHYS 39: 1419-27 (1968)

5-1400 CARBON EVAPORATION FROM A THORIUM DISPENSER
CATHODE OBSERVED BY AUGER ELECTRON EMISSION.
HARRIS LA
J APPL PHYS 39: 4862 (1968)

5-1401 SOME OBSERVATIONS OF SURFACE SEGREGATION BY
AUGER ELECTRON EMISSION.
HARRIS LA
J APPL PHYS 39: 1428-31 (1968)

5-1402 SURFACE ANALYSIS BY AUGER ELECTRON
SPECTROSCOPY.
HARRIS LA
J ELECTROCHEM SOC 115: C250 (1968)

5-1403 SECONDARY ELECTRON SPECTROSCOPY. (SURFACE
ANALYSIS OF MATERIALS BY ELECTRON EXCITED
AUGER ELECTRON EMISSION)
HARRIS LA
APPL SPECTROSC 22: 372 (1968)

5-1404 AUGER SPECTROSCOPY OF SEMICONDUCTOR AND METAL
SURFACES.
HARRIS LA
P259-60 OF ELECTROCHEM SOC SPRING MEET,
EXTENDED ABSTR, WASHINGTON DC, MAY 1971.
ELECTROCHEM SOC, 1971.
SAA 1972: 1503

5-1405 ELECTRON SPECTROSCOPY ON THIN FILMS.
(SCANNING AUGER MICROSCOPE)
HARRIS LA
GENERAL ELEC CO, 1971, 70P. AD891470

5-1406 ELECTRON INDUCED SECONDARY ELECTRON EMISSION
SPECTROSCOPY OF SURFACES.
HARRIS PR
UNIV KEELE, ENGLAND, PHD THESIS, 1973.
SAA 1974: 25203

5-1407 SECONDARY ELECTRON EMISSION OF A ZIRCONIUM-
ALUMINUM BULK GETTER. (AUGER SPECTROSCOPY)
HARRIS PR + PATTINSON EB
VACUUM 22(9): 405-7 (SEP 1972)
SAA 1972: 19182

5-1408 EXPERIMENTAL PROBLEMS OF STUDYING THIN FILM
NUCLEATION. (ELECTRON MICROSCOPY, AUGER
EFFECT)
HARSDORFF M
THIN SOLID FILMS 32(1): 103-15 (1976)
SAA 1976: 45178

5-1409 (AUGER ELECTRON SPECTROSCOPY OF) PRECISION
SINGLE SIDEBAND CRYSTAL UNITS.
HART RK + HICKLIN WH
GEORGIA INST TECHNOL, 1973, FINAL REPT
ECOM-0087F, 1974. AD 786559

5-1410 METHODS OF CLEANING CONTAMINANTS FROM QUARTZ
SURFACES DURING RESONATOR FABRICATION. (AUGER
SPECTRA)
HART RK + HICKLIN WH + PHILLIPS LA
P89-95 OF SYMP ON FREQUENCY CONTROL, 28TH
PROC, ARMY ELECTRON COMMAND, 1974.

5-1411 IN-PROCESS CONTROL TECHNIQUES FOR COMPLEX
SEMICONDUCTOR STRUCTURES. TASK-II.
APPLICATION OF SECONDARY ELECTRON
SPECTROSCOPY (AUGER ELECTRON ANALYSIS) TO
SURFACE CONTROL PROGRAM IN THE MANUFACTURING
OF SILICON DEVICES.
HARTMANN DK + HARRIS LA + AFFLECK JH
GEN ELEC CO, SEMICOND PRODUCT DEPT, 1968,
1969. AD-845596, AD-845997, AD-847775,
AD-852179, AD-856920, AD-861921, AD-882887

5-1412 THREE-DIMENSIONAL ELEMENT ANALYSIS OF
SURFACES AND INTERFACES OF SOLIDS BY THE
AUGER ELECTRON MICROANALYTIC METHOD. (REVIEW)
HAYAKAWA K + KAWASE S + ICHIKAWA M + OKANO H
OYO BUTSURI 44(12): 1294-302 (1975) (IN
JAPANESE)
CA 84: 98791

5-1413 AUGER ELECTRON EMISSION MICROGRAPHIC STUDIES
OF THE CLEAVAGE SURFACE OF GRAPHITE SINGLE
CRYSTAL.
HAYAKAWA K + OKANO H + KAWASE S + YAMAMOTO S
J APPL PHYS 44(6): 2575-9 (1973)
CA 79: 048852

5-1414 AUGER ELECTRON SPECTROSCOPY AND MICROANALYSIS
OF SOLID SURFACES.
HAYAKAWA K + OKANO H + KAWASE S + ICHIKAWA M
+ GOTOH Y
SHOKUBAI 17(2): 9-21 (1975) (IN JAPANESE)
CA 83: 90222

5-1415 AUGER ELECTRON EMISSION MICROGRAPHY AND
MICROANALYSIS OF A STAINED SURFACE OF PURE
IRON PLATE.
HAYAKAWA K + OKANO H + KAWASE S + YAMAMOTO S
PROC ELECTRON MICROSC SOC AMER 31: 142-3
(1973)
CA 80: 087806

5-1416 AUGER ELECTRON EMISSION MICROGRAPHY AND
MICROANALYSIS OF SOLID SURFACES.
HAYAKAWA K + OKANO H + KAWASE S + YAMAMOTO S
P803-6 OF INT VACUUM CONGRESS, 6TH PROC, INT
CONF ON SOLID SURFACES, 2ND, 1974, KYOTO.
(JAP J APPL PHYS, SUPPL 2, PART 1, 1974)
P498-508 OF ADVAN X-RAY ANAL, VOL 17, MUELLER
WM(ED), UNIV DENVER, 1974.
CA 83: 21392, 21777

5-1417 STUDY OF THE NITRIDATION OF SILICON SURFACES
BY LOW ENERGY ELECTRON DIFFRACTION AND AUGER
ELECTRON SPECTROSCOPY.
HECKINGBOTTOM R + WOOD PR
SURFACE SCI 36(2): 594-605 (1973)
CA 78: 128741

5-1418 AUGER ELECTRON SPECTROSCOPY AND ITS
APPLICATION IN STUDIES OF SOME SILICON
SURFACES.
HECKINGBOTTOM R + WOOD PR
UK POST OFFICE RES DEPT, 1973, 16P. REPT NO
330.
CA 84: 23591

5-1419 SURFACE PHYSICS OF SOLIDS FROM THE VIEWPOINT
OF ULTRAVACUUM. (AUGER SPECTROSCOPY, SURFACE
PREPARATION AND CLEANLINESS)
HEDBAVNY P
SLABOPROUDY OBZ 37(3): 107-116 (1976) (IN
CZECH)
SAA 1976: 53467

5-1420 SULFUR ON SINGLE CRYSTAL PLATINUM PLANES. LOW
ENERGY ELECTRON DIFFRACTION AND AUGER
INVESTIGATIONS.
HEEGEMANN W + BECHTOLD E + HAYEK K
P165-7 OF INT VACUUM CONGRESS, 6TH PROC, INT
CONF ON SOLID SURFACES, 2ND, 1974, KYOTO.
(JAP J APPL PHYS, SUPPL 2, PART 2, 1974)
CA 82: 160688

5-1421 ADSORPTION OF SULFUR ON THE (100) AND (111)
FACES OF PLATINUM. LOW ENERGY ELECTRON
DIFFRACTION AND AUGER ELECTRON SPECTROSCOPY
STUDY.
HEEGEMANN W + MEISTER KH + BECHTOLD E
+ HAYEK K
SURFACE SCI 49(1): 161-80 (1975)
CA 82: 175713

5-1422 OXYGEN ADSORPTION ON (110) SILVER.
HEILAND W + IBERL F + TAGLAUER E + MENZEL D
SURFACE SCI 53(1): 383-92 (1975)
SAA 1976: 22113

5-1423 SURFACE ANALYTICAL STUDIES USING ION
SCATTERING SPECTROMETRY, AUGER ELECTRON
SPECTROSCOPY AND SECONDARY ION MASS
SPECTROMETRY.
HEILAND W + TAGLAUER E
J VACUUM SCI TECHNOL 12(1): 352-3 (1975)
SAA 1975: 69105

5-1424 DETERMINATION OF THE SURFACE COMPOSITION OF
COPPER- NICKEL ALLOYS FOR CLEAN AND ADSORBATE
COVERED SURFACES.
HELMS CR + YU KY
J VACUUM SCI TECHNOL 12(1): 276-8 (1975)
CA 82: 145567

5-1425 CORROSION AND STRUCTURAL CHANGES IN GLASS.
(AUGER ELECTRON SPECTROSCOPY)
HENCH LL + GOULD RW + REED-HILL RE
+ PERSON WB + SANDERS DM
UNIV FLA, OCT 1974, 15P. AD/A-000432/5ST

5-1426 SILICON CLEANING WITH HYDROGEN PEROXIDE
SOLUTIONS: HIGH ENERGY ELECTRON DIFFRACTION
AND AUGER ELECTRON SPECTROSCOPY STUDY.
HENDERSON RC
J ELECTROCHEM SOC 119: 772-5 (JUN 1972)

5-1427 HEED AND AES STUDY OF SILICON EPITAXY BY
SILANE PYROLYSIS.
HENDERSON RC + HELM RF
J VACUUM SCI TECHNOL 9(1): 164 (1972)

5-1428 SILICON HOMOEPITAXIAL THIN FILMS VIA SILANE
PYROLYSIS: HEED AND AUGER ELECTRON
SPECTROSCOPY STUDY.
HENDERSON RC + HELM RF
SURFACE SCI 30(2): 310-34 (1972)

5-1429 SURFACE CHARACTERIZATION BY LOW ENERGY PHOTO
AUGER ELECTRON SPECTROSCOPY.
HENKE BL + ANDERSON MW
P367-84 OF INT CONF X-RAY OPT MICROANAL 1971,
6TH PROC, MOLLENSTADT G + GANKLER KH(EDS),
SPRINGER, 1969, 612P.
CA 81: 029946

5-1430 DIFFERENTIAL SPUTTERING OF MAGNESIUM OXIDE-
GOLD CERMET FILMS AND ITS APPLICATION TO HIGH
YIELD SECONDARY ELECTRON EMITTERS. (AUGER
ELECTRON SPECTROSCOPY)
HENRICH VE + FAN JCC
SURFACE SCI 42(1): 139-56 (MAR 1974)
SAA 1974: 28982

5-1431 AUGER SPECTROSCOPY STUDIES OF THE OXIDATION
OF AMORPHOUS AND CRYSTALLINE GERMANIUM.
HENRICH VE + FAN JCC
J APPL PHYS 46(3): 1206-13 (1975)
CA 83: 21243

5-1432 ADSORPTION OF WATER VAPOR ON CLEAN CLEAVED
GERMANIUM. PART-1: STRUCTURAL AND KINETIC
PROPERTIES. (AUGER ELECTRON SPECTROSCOPY)
HENZLER M + TOPLER J
SURFACE SCI 40(2): 388-96 (NOV 1973)
SAA 1974: 28289

5-1433 USE OF ADVANCED ANALYTICAL TECHNIQUES FOR
CHARACTERIZATION OF THE CHEMICAL, PHYSICAL,
AND CRYSTALLOGRAPHIC NATURE OF A SURFACE.
HERFERT RE
PROC ELECTRON MICROSC SOC AMER 31: 132-3
(1973)
CA 81: 040522

5-1434 ADHESION AT GOLD MICA INTERFACES. (LEED, AES)
HIGGINBOTHAM IG + WILLIAMS RH + MCEVOY AJ
SURFACE SCI 53(1): 581 (1975)
SAA 1976: 26384

5-1435 ELECTRONIC STRUCTURE OF A THIN GOLD FILM
DEPOSITED ON A SILICON SUBSTRATE STUDIED BY
AUGER ELECTRON AND X-RAY PHOTOELECTRON
SPECTROSCOPIES.
HIRAKI A + IWAMI M
P749-52 OF INT VACUUM CONGRESS, 6TH PROC, INT
CONF ON SOLID SURFACES, 2ND, 1974, KYOTO.
(JAP J APPL PHYS, SUPPL 2, PART 2, 1974)
CA 82: 177693

5-1436 DIFFUSE INTERFACE STRUCTURES IN SOME
(SUBSTRATE- VACUUM EVAPORATED FILM) SYSTEMS
STUDIED BY AUGER ELECTRON SPECTROSCOPY.
HIRAKI A + IWAMI M + NARUSAWA T + KOMIYA S
P219-22 OF INT CONF ON VACUUM METALLURGY, 4TH
PROC, TOKYO, 1973. IRON STEEL INST JAP, 1974.
CA 82: 020607

5-1437 POSSIBLE ORIGIN OF LOW TEMPERATURE SILICON
MIGRATION IN SILICON- METAL SYSTEMS. (AUGER
SPECTROSCOPY)
HIRAKI A + SHIMIZU A + IWAMI M
P265-6 OF ELECTROCHEM SOC SPRING MEET
EXTENDED ABSTR, 1975.
SAA 1975: 85591

5-1438 METALLIC STATE OF SILICON IN SILICON- NOBLE
METAL VAPOR QUENCHED ALLOYS STUDIED BY AUGER
ELECTRON SPECTROSCOPY.
HIRAKI A + SHIMIZU A + IWAMI M + NARUSAWA T
+ KOMIYA S
APPL PHYS LETT 26(2): 57-60 (1975)
CA 82: 162612

5-1439 DIFFUSION MEASUREMENT IN THIN FILMS BY AUGER
SPUTTER ETCHING TECHNIQUE.
HO PS + LEWIS JE + HOWARD JK + WILDMAN HS
P243-4 OF ELECTROCHEM SOC SPRING MEET
EXTENDED ABSTR, 1975.
SAA 1975: 85587

5-1440 INTERDIFFUSION IN BIMETALLIC POLYCRYSTALLINE
FILMS MEASURED BY AUGER ION SPUTTERING
TECHNIQUE.
HO PS + LEWIS JE + HOWARD JK + WILDMAN HS
P246 OF ELECTROCHEM SOC FALL MEET ABSTR,
1975.
SAA 1976: 35746

5-1441 CORRELATION OF THE WETTING BEHAVIOR OF SODIUM
ON NIMONIC-PE16 ALLOY WITH THE SURFACE
OXYGEN- CHROMIUM RATIO DETERMINED BY AUGER
SPECTROSCOPY.
HODKIN EN + MORTIMER DA + NICHOLAS MG
+ RIVIERE JC
J NUCL MATER 52(2): 131-41 (1974)
CA 82: 35280

5-1442 EVALUATION OF CONCENTRATION DEPTH PROFILES BY
SPUTTERING IN SIMS AND AES.
HOFMANN S
APPL PHYS 9(1): 59-66 (1976)
CA 84: 98300

5-1443 SURFACE SEGREGATION OF OXYGEN IN NIOBIUM.
(AES)
HOFMANN S + BLANK G + SCHULTZ H
Z METALLK 67(3): 189-94 (1976) (IN GERMAN)
SAA 1976: 53446

5-1444 STUDY OF FRACTURE SURFACES OF HOT PRESSED
SILICON NITRIDE BY AUGER ELECTRON
SPECTROSCOPY.
HOFMANN S + GAUCKLER LJ
POWDER MET INT 6(2): 90-2 (1974)
CA 81: 056351

5-1445 AUGER ELECTRON SPECTROSCOPY ON THE FRACTURE
OF NICKEL TUNGSTEN MATERIALS AND SILICON
NITRIDE.
HOFMANN S + GAUCKLER LJ + TILLMANN L
MIKROCHIM ACTA SUPPL 6: 373-82 (1975) (IN
GERMAN)
CA 84: 38279

5-1446 LEED AND AES STUDIES OF LITHIUM HYDRIDE (100)
SURFACE.
HOLCOMBE CE + POWELL GL + CLAUSING RE
SURFACE SCI 30(3): 561-72 (1972)
CA 77: 10754C; SAA 1972: 47078

5-1447 ANGULAR DEPENDENCE OF AUGER ELECTRON EMISSION
FROM SOLID SURFACES.
HOLLAND BW + MCDONNELL L + WOODRUFF DP
SOLID STATE COMMUN 11(8): 991-3 (1972)
CA 77: 170796W

5-1448 EFFECT OF SURFACE ROUGHNESS OF AUGER ELECTRON
SPECTROSCOPY.
HOLLOWAY PH
J ELECTRON SPECTROSC RELAT PHENOMENA 7(3):
215-32 (1975)
CA 83: 185785

5-1449 THICKNESS DETERMINATION OF ULTRATHIN FILMS BY
AUGER ELECTRON SPECTROSCOPY.
HOLLOWAY PH
J VACUUM SCI TECHNOL 12(6): 1418-22 (1976)
SAA 1976: 37970

5-1450 DETECTION BY AUGER ELECTRON SPECTROSCOPY AND
REMOVAL BY OZONIZATION OF PHOTORESIST
RESIDUES.
HOLLOWAY PH + BUSHMIRE DW
P180-6 OF ANNU PROC RELIAB PHYS SYMP, 12TH
PROC, 1974, IEEE; SANDIA LAB 1974, 26P. REP
SLA-74-5238.
CA 81: 178854; 82: 92199

5-1451 KINETICS OF THE REACTION BETWEEN OXYGEN AND
SULFUR ON A NICKEL (111) SURFACE. (STUDIED BY
LEED AND AES)
HOLLOWAY PH + HUDSON JB
SURFACE SCI 33: 56-68 (1972)

5-1452 KINETICS OF THE REACTION OF OXYGEN WITH CLEAN
NICKEL SINGLE CRYSTAL SURFACES. PART-1:
NICKEL (100) SURFACE. PART-2: NICKEL (111)
SURFACE. (AUGER ELECTRON SPECTROSCOPY)
HOLLOWAY PH + HUDSON JB
SURFACE SCI 43(1): 123-40, 141-9 (MAY 1974)
SAA 1974: 48869-70

5-1453 CHEMICAL CHANGES IN SECONDARY ELECTRON
EMISSION DURING OXIDATION OF NICKEL (100) AND
(111) CRYSTAL SURFACES.
HOLLOWAY PH + HUDSON JB
J VACUUM SCI TECHNOL 12(2): 647-9 (1975)
SAA 1975: 46498

5-1454 CHEMICAL CLEANING FOR THERMOCOMPRESSION
BONDING. (REMOVAL OF OXIDE FILMS, AUGER
SPECTROSCOPY OF SURFACE)
HOLLOWAY PH + LONG RL
IEEE TRANS PARTS HYBRIDS PACKAG 11(2): 83-8
(1975)

5-1455 AUGER ELECTRON SPECTROSCOPIC ANALYSIS OF
SILICON NITRIDE ON SILICON.
HOLLOWAY PH + STEIN HJ
P218-20 OF ELECTROCHEM SOC FALL MEET ABSTR,
1975.
SAA 1976: 35743

5-1456 EFFECTS OF INTERFACIAL OXIDES ON THE CONTACT
RESISTANCE OF TUNGSTEN ON HEAVILY DOPED
SILICON- GERMANIUM ALLOY. (AUGER
SPECTROSCOPY)
HOLLOWAY PH + SWEET JN
P292-3 OF ELECTROCHEM SOC SPRING MEET
EXTENDED ABST, 1975.
SAA 1975: 85839

5-1457 FACTORS AFFECTING DEPTH PROFILING
MEASUREMENTS USING AUGER ELECTRON
SPECTROSCOPY.
HOOKER MP + GRANT JT
SURFACE SCI 51(1): 328-32 (1975)
CA 83: 155173

5-1458 AUGER ELECTRON SPECTROSCOPY STUDIES OF CARBON
MONOXIDE ON NICKEL. SPECTRAL LINE SHAPES AND
QUANTITATIVE ASPECTS.
HOOKER MP + GRANT JT
SURFACE SCI 55(2): 741-6 (1976)
CA 85: 52119

5-1459 SOME ASPECTS OF AN AES AND XPS STUDY OF THE
ADSORPTION OF OXYGEN ON NICKEL.
HOOKER MP + GRANT JT + HAAS TW
J VACUUM SCI TECHNOL 13(1): 296-300 (1976)
CA 84: 141026

5-1460 AUGER ELECTRON SPECTROSCOPY OF MATERIALS
EXPOSED TO FLOWING SODIUM.
HOOPER AJ + WILD RK
BERKELEY NUCLEAR LAB, ENGLAND, 1973, 26P.
RD/B/N-2683
CA 81: 019845

5-1461 FLASH DESORPTION- AUGER ELECTRON
SPECTROSCOPY. DISCREPANCY FOR CARBON DIOXIDE
ON TUNGSTEN (100).
HOPKINS BJ + JONES AR + WINTON RI
SURFACE SCI 47(2): 677-80 (1975)
CA 82: 162624

5-1462 ROLE OF IMPURITIES IN THE STABILITY OF ZINC
OXIDE.
HOPKINS BJ + LEYSEN R + TAYLOR PA
SURFACE SCI 48(2): 486-96 (1975)
SAA 1975: 42058

5-1463 INTERACTION OF CHLORINE WITH THE ZINC OXIDE
(0001)- ZINC SURFACE: A LEED, AUGER AND WORK
FUNCTION STUDY.
HOPKINS BJ + TAYLOR PA
J PHYS C 9(4): 571-8 (1976)
CA 84: 156111

5-1464 COMBINED LEED- RHEED AUGER STUDY OF OXYGEN
ADSORPTION ON THE TUNGSTEN (112) SURFACE.
HOPKINS BJ + WATTS GD
SURFACE SCI 44(1): 237-46 (1974)
CA 81: 176636

5-1465 AUGER SPECTROSCOPIC STUDY OF THE COADSORPTION
OF MOLECULAR OXYGEN AND CARBON MONOXIDE, AND
MOLECULAR OXYGEN AND CARBON DIOXIDE ON A
TUNGSTEN (111) SURFACE.
HOPKINS BJ + WATTS GD
SURFACE SCI 55(2): 729-34 (1976)
CA 85: 52118

5-1466 STUDY OF THE ADSORPTION OF OXYGEN ON
NICKEL(III) USING AUGER ELECTRON
SPECTROSCOPY: CHEMICAL SHIFTS AND VALENCE
SPECTRA.
HORGAN AM + DALINS I
SURFACE SCI 36(2): 526-43 (1973)
CA 78: 128758

5-1467 EFFECTIVENESS OF THE STANDARD SURFACE
CLEANING TECHNIQUES AS APPLIED TO NICKEL
(111), NICKEL (100), AND NICKEL SHEET USING
AUGER ELECTRON SPECTROSCOPY.
HORGAN AM + DALINS I
J VACUUM SCI TECHNOL 10(4): 523-32 (1973)
CA 79: 082485

5-1468 STICKING PROBABILITIES FOR THE ADSORPTION OF
HYDROGEN ON CLEAN NICKEL (111), NICKEL (100),
SHEET AND EVAPORATED NICKEL FILMS. (AUGER
ELECTRON SPECTROSCOPY)
HORGAN AM + DALINS I
SURFACE SCI 41(2): 624-8 (FEB 1974)
SAA 1974: 24643

5-1469 STRUCTURAL STUDY OF OXYGEN- IRON (001) BY LOW
ENERGY ELECTRON DIFFRACTION AND AUGER
ELECTRON SPECTROSCOPY. PART-1.
HORIGUCHI T + NAKANISHI S
P89-92 OF INT VACUUM CONGRESS, 6TH PROC, INT
CONF ON SOLID SURFACES, 2ND, 1974, KYOTO.
(JAP J APPL PHYS, SUPPL 2, PART 2, 1974)
CA 82: 178936

5-1470 ANALYSIS OF THE SURFACE STRUCTURE WITH
ANTIPHASE DOMAINS OF SULFUR ON IRON (001) BY
LEED- AES.
HORIGUCHI T + NAKANISHI S
ACTA CRYSTALLOGR A 31(S3): S288 (1975)
SAA 1976: 09558

5-1471 SURFACE CONDITION AND THERMIONIC EMISSION OF
LANTHANUM HEXABORIDE. (AUGER SPECTROSCOPY)
HOSOKI S+ YAMAMOTO S + HAYAKAWA K + OKANO H
P285-8 OF INT VACUUM CONGRESS, 6TH PROC, INT
CONF ON SOLID SURFACES, 2ND, 1974, KYOTO.
(JAP J APPL PHYS, SUPPL 2, PART 1, 1974)

5-1472 NIOBIUM SURFACES FOR RF SUPERCONDUCTORS.
(AUGER SURFACE ANALYSIS)
HOYT EW
J VACUUM SCI TECHNOL 9: 144 (1972) (ABSTRACT)
SAA 1972: 39462

5-1473 EFFECT OF LIGHT ELEMENTS NITROGEN, CARBON AND
OXYGEN ON THE PHYSICAL PROPERTIES OF
SPUTTERED TANTALUM FILMS. (AUGER ELECTRON
SPECTROSCOPY)
HUTTEMANN RD + MORABITO JM + STEIDEL CA
+ GERSTENBERG D
P513-16 OF INT VACUUM CONGRESS, 6TH PROC, INT
CONF ON SOLID SURFACES, 2ND, 1974, KYOTO.
(JAP J APPL PHYS, SUPPL 2, PART 1, 1974)

5-1474 ADSORPTION OF OXYGEN ON SILICON (111)
SURFACES. PART-1. (AUGER ELECTRON
SPECTROSCOPY)
IBACH H + HORN K + DORN R + LUTH H
SURFACE SCI 38(2): 433-54 (JUL 1973)
SAA 1973: 55160

5-1475 ELECTRON ORBITAL ENERGIES OF OXYGEN ADSORBED
ON SILICON SURFACES AND OF SILICON DIOXIDE.
IBACH H + ROWE JE
PHYS REV B 10(2): 710-18 (1974)
CA 81: 097478

5-1476 ELECTRONIC STATES OF OXYGEN ADSORBED ON CLEAN
SILICON (111) AND (100) SURFACES. (AUGER
SPECTROSCOPY)
IBACH H + ROWE JE
J VACUUM SCI TECHNOL 11(1): 279 (JAN-FEB
1974)
        SAA 1974: 75699

5-1477 SURFACE DEBYE TEMPERATURE OF THE SILICON
(001) - 2X2 STRUCTURE. (AUGER EFFECT)
IGNATIEV A + JONA F
SURFACE SCI 42(2): 605-8 (APR 1974)
        SAA 1974: 40957

5-1478 ATOMIC ARRANGEMENT IN THE 1 K 1 STRUCTURE OF
A SILICON ORDERED MONOLAYER ON MOLYBDENUM
(001).
IGNATIEV A + JONA F + JEPSEN DW + MARCUS PM
J VACUUM SCI TECHNOL 12(1): 226-9 (1975)
        SAA 1975: 61897

5-1479 LEED INVESTIGATIONS OF XENON SINGLE CRYSTAL
FILMS AND THEIR USE IN STUDYING THE IRIDIUM
(100) SURFACE. (AUGER SPECTRA OF IRIDIUM)
IGNATIEV A + JONES AV + RHODIN TN
SURFACE SCI 30: 573-91 (1972)

5-1480 AUGER ANALYSIS OF TANTALUM SURFACES.
INGREY SJ + WESTWOOD WD
P99 OF INT CONF ON ATOMIC SPECTROSCOPY, 4TH
PROC, AND CANAD SPECTROSC SYMP 20TH PROC, NAT
RES COUNC, TORONTO, 1973.
        SAA 1974: 17156

5-1481 AUGER STUDIES OF DIFFUSED IMPURITY PROFILES
IN SILICON DIOXIDE- SILICON AND POLYSILICON-
SILICON DIOXIDE- SILICON STRUCTURES.
INOUE T + HORIUCHI S + YOSHII S + TANUMA Y
P329-37 OF CONF SOLID STATE DEVICES, 6TH
PROC, 1974, JAP SOC APPL PHYS, 1974, 343P.
        CA 83: 171573

5-1482 OBSERVATION OF IMPURITY PROFILE IN ION
IMPLANTED SILICON BY AUGER ELECTRON
SPECTROSCOPY.
INOUE T + OHMURA Y + KOIKE K + YOSHIKOKA T
P819-22 OF INT VACUUM CONGRESS, 6TH PROC, INT
CONF ON SOLID SURFACES, 2ND, 1974, KYOTO.
(JAP J APPL PHYS, SUPPL 2, PART 1, 1974)
        CA 83: 20594

5-1483 DEVELOPMENT OF STEPPED SURFACE REGIONS ON
POLYCRYSTALLINE GOLD. LOW ENERGY ELECTRON
DIFFRACTIONS AND AUGER STUDIES.
ISA SA + JOYNER RW + ROBERTS MW
J CHEM SOC FARADAY TRANS 72(2): 540-2 (1976)
        CA 84: 82241

5-1484 BINDING ENERGIES OF CARBON TO NICKEL (100)
FROM EQUILIBRIUM SEGREGATION STUDIES. (AUGER
EFFECT)
ISETT LC + BLAKELY JM
SURFACE SCI 47(2): 645-9 (FEB 1975)
        SAA 1975: 33380

5-1485 BINDING OF CARBON ATOMS AT A STEPPED NICKEL
SURFACE. (AUGER SPECTROSCOPY)
ISETT LC + BLAKELY JM
J VACUUM SCI TECHNOL 12(1): 237-41 (1975)
        SAA 1975: 58270

5-1486 STUDY OF PLATINUM ELECTRODES BY MEANS OF THIN
LAYER ELECTROCHEMISTRY AND LOW ENERGY
ELECTRON DIFFRACTION. PART-1: ELECTRODE
SURFACE STRUCTURE AFTER EXPOSURE TO WATER AND
AQUEOUS ELECTROLYTES.
ISHIKAWA RM + HUBBARD AT
J ELECTROANAL CHEM INTERFACIAL ELECTROCHEM
69(3): 317-38 (1976)
        CA 84: 186572

5-1487 DETECTION OF ORGANIC CONTAMINANTS ON ALUMINUM
SURFACES BY AUGER ELECTRON SPECTROSCOPY.
ISOYAMA E + ABE T + UCHIYAMA T + HAYASHI Y
+ MURAKAWA T
KINZOKU HYOMEN GIJUTSU 25(6): 353-5 (1974)
(IN JAPANESE)
        CA 81: 181019

5-1488 EPITAXY OF ULTRATHIN METAL FILMS ON BODY
CENTERED CUBIC SUBSTRATES USING LEED- AUGER
TECHNIQUES.
JACKSON AG
J VACUUM SCI TECHNOL 8: 23 (1971) (ABSTRACT)

5-1489 AUGER- LEED INVESTIGATION OF THE DEPOSITION
OF ALUMINUM ONTO THE MOLYBDENUM (110)
SURFACE.
JACKSON AG + HOOKER MP
SURFACE SCI 28: 373-94 (1971)

5-1490 AUGER- LEED INVESTIGATION OF TIN ON
MOLYBDENUM (100).
JACKSON AG + HOOKER MP
SURFACE SCI 27: 197-210 (1971)

5-1491 LEED AUGER STUDY OF GROWTH OF ALUMINUM ON
NIOBIUM (110).
JACKSON AG + HOOKER MP
J VACUUM SCI TECHNOL 9: 784 (1972) (ABSTRACT)
        SAA 1972: 50484

5-1492 A CLOSE LOOK AT SURFACES. (SURFACE
STRUCTURES, SPECIES, TOPOGRAPHY, PHENOMENA,
FIELD ION MICROSCOPY, LEED, HEED, AUGER
SPECTROSCOPY, ION BOMBARDMENT,
MICROELECTRONICS)
JACKSON KA
BELL LAB REC 50: 180-8 (1972)

5-1493 OXIDATION AND ANNEALING OF GALLIUM PHOSPHIDE
AND GALLIUM ARSENIDE (111) FACES STUDIED BY
AES AND UPS.
JACOBI K + RANKE W
J ELECTRON SPECTROSC RELAT PHENOMENA 8(3):
225-38 (1976)
        CA 85: 25805

5-1494 ELECTRON STIMULATED DESORPTION OF SODIUM AND
MEASUREMENT OF SURFACE DIFFUSION ON MAGNESIUM
OXIDE (001) BY AUGER SPECTROSCOPY.
JANSSEN AP
SURFACE SCI 52(1): 230-6 (1975)
        CA 84: 9299

5-1495 COMPARATIVE STUDY OF SINGLE CRYSTAL MAGNESIUM
OXIDE AND OXIDIZED MAGNESIUM BY AUGER
ELECTRON SPECTROSCOPY.
JANSSEN AP + SCHOONMAKER RC + CHAMBERS A
SURFACE SCI 47(1): 41-7 (1975)
        CA 82: 147631

5-1496 STUDY OF THE EPITAXIAL GROWTH OF MAGNESIUM ON
MAGNESIUM OXIDE (001) USING REFLECTION
DIFFRACTION, LEED AND AUGER SPECTROSCOPY.
JANSSEN AP + SCHOONMAKER RC + CHAMBERS A
SURFACE SCI 49(1): 143-60 (1975)
        CA 82: 178411

5-1497 INTERFACIAL AUGER TRANSITIONS IN OXIDIZED
SODIUM AND MAGNESIUM.
JANSSEN AP + SCHOONMAKER RC + MATTHEW JAD
+ CHAMBERS A
SOLID STATE COMMUN 14(11): 1263-7 (1974)
        CA 81: 082732

5-1498 AUGER AND REFLECTION ELECTRON DIFFRACTION
INVESTIGATION OF OXYGEN AND HYDROGEN
ADSORPTION ON STRONTIUM SURFACES.
JANSSEN AP + SCHOONMAKER RC
SURFACE SCI 55(1): 109-25 (1976)
        CA 84: 185315

5-1499 LEED AND AUGER INVESTIGATIONS OF COPPER (111)
SURFACE.
JENKINS LH + CHUNG MF
SURFACE SCI 24: 125-39 (1971)

5-1500 CHEMISORPTION OF OXYGEN ON PLATINUM (111)
SURFACES. (AUGER SPECTRA)
JOEBSTL JA
J VACUUM SCI TECHNOL 12(1): 347-50 (1975)
        SAA 1975: 65815

5-1501 USE OF AUGER ELECTRON SPECTROSCOPY TO
DETERMINE THE STRUCTURE OF SILICON OXIDE
FILMS.
JOHANNESSEN JS + SPICER WE
P119-23 OF SEMICONDUCTOR MEASUREMENT
TECHNOLOGY. PART-4: ARPA/NBS WORKSHOP ON
SURFACE ANALYSIS FOR SILICON DEVICES,
LIEBERMAN AG(ED), NAT BUR STAND SPEC PUB
400-23, 1976.

5-1502 PHASE SEPARATION IN SILICON OXIDES AS SEEN BY
AUGER ELECTRON SPECTROSCOPY.
JOHANNESSEN JS + SPICER WE + STRAUSSER YE
APPL PHYS LETT 27(8): 452-4 (1975)
      CA 83: 186619

5-1503 STUDY OF THE CHEMICAL COMPOSITION OF MOS AND
MNOS STRUCTURES BY AUGER ELECTRON
SPECTROSCOPY.
JOHANNESSEN JS + SPICER WE + STRAUSSER YE
THIN SOLID FILMS 32(2): 311-4 (1976)
      CA 85: 39831

5-1504 COMPOSITION OF PLATINUM ANODE SURFACES BY
AUGER SPECTROSCOPY.
JOHNSON WC + HELDT LA
J ELECTROCHEM SOC 121(1): 34-6 (1974)
      CA 80: 055202

5-1505 ADDITIVE AND IMPURITY DISTRIBUTIONS AT GRAIN
BOUNDARIES IN SINTERED ALUMINA. (SCANNING
AUGER MICROPROBE)
JOHNSON WC + STEIN DF
J AMER CERAM SOC 58(11-12): 485-8 (1975)
      SAA 1976: 27220

5-1506 ANALYSIS OF GRAIN BOUNDARY IMPURITIES AND
FLUORIDE ADDITIVES IN HOT PRESSED OXIDES BY
AUGER ELECTRON SPECTROSCOPY.
JOHNSON WC + STEIN DF + RICE RW
J AMER CERAM SOC 57(8): 342-4 (1974)
      CA 81: 157720

5-1507 REEXAMINATION OF ALUMINUM (110) WITH LOW
ENERGY DIFFRACTION AND AUGER SPECTROSCOPY.
JONA F + STROZIER JA + WONG C
SURFACE SCI 30(1): 225-31 (1972)

5-1508 SURFACE SEGREGATION AND INTERACTION OF OXYGEN
AND NITROGEN WITH ZIRCONIUM IN NIOBIUM- 1%
ZIRCONIUM.
JOSHI A + VARMA MN + STRONGIN M
MET TRANS 5(4): 861-3 (APR 1974)

5-1509 GROWTH AND STRUCTURE OF SEMICONDUCTING THIN
FILMS. (AUGER ELECTRON SPECTROSCOPY)
JOYCE BA
REP PROG PHYS 37(3): 363-420 (1974)
      SAA 1974: 37778

5-1510 SOME ASPECTS OF THE SURFACE BEHAVIOR OF
SILICON. (AUGER ELECTRON SPECTROSCOPY)
JOYCE BA
SURFACE SCI 35: 1-7 (MAR 1973)
      SAA 1973: 30317

5-1511 SILICON- OXYGEN INTERACTIONS USING AUGER
ELECTRON SPECTROSCOPY.
JOYCE BA + NEAVE JH
SURFACE SCI 27(3): 499-515 (1971)

5-1512 INTERACTION OF OXYGEN WITH SILICON (111)
SURFACES. (REPLY TO COMMENT)
JOYCE BA + NEAVE JH
SURFACE SCI 30: 710-11 (1972)

5-1513 ELECTRON BEAM ADSORBATE INTERACTIONS ON
SILICON SURFACES. (AUGER SPECTROSCOPY)
JOYCE BA + NEAVE JH
SURFACE SCI 34(2 PART-2): 401-19 (JAN 1973)
      SAA 1973: 20939

5-1514 INFLUENCE OF SUBSTRATE CONDITIONS ON THE
NUCLEATION AND GROWTH OF EPITAXIAL SILICON
FILMS.
JOYCE BA + NEAVE JH + WATTS BE
SURFACE SCI 15: 1-13 (1969)

5-1515 ADSORPTION OF CARBON MONOXIDE ON COPPER
(001). LEED AND AUGER EMISSION STUDIES.
JOYNER RW + MCKEE CS + ROBERTS MW
SURFACE SCI 26: 303-9 (1971)

5-1516 INTERACTION OF HYDROGEN SULFIDE WITH COPPER
(001).
JOYNER RW + MCKEE CS + ROBERTS MW
SURFACE SCI 27: 279-85 (1971)

5-1517 STUDY OF THE PREPARATION OF ATOMICALLY CLEAN
TUNGSTEN SURFACES BY AUGER ELECTRON
SPECTROSCOPY.
JOYNER RW + RICKMAN J + ROBERTS MW
SURFACE SCI 39(2): 445-9 (1973)
      CA 79: 108375

5-1518 CHEMISORPTION OF NITROGEN ON TUNGSTEN STUDIED
BY AUGER ELECTRON SPECTROSCOPY.
JOYNER RW + RICKMAN J + ROBERTS MW
J CHEM SOC FARADAY TRANS 1 70(10): 1825-39
(1974)
      CA 82: 64769

5-1519 AUGER ELECTRON SPECTROSCOPY STUDIES OF CLEAN
POLYCRYSTALLINE GOLD AND OF THE ADSORPTION OF
MERCURY ON GOLD.
JOYNER RW + ROBERTS MW
J CHEM SOC FARADAY TRANS 1 69(PT-7): 1242-50
(1973)
      CA 79: 072080

5-1520 STABILITY OF FIELD ELECTRON EMISSION AND
VACUUM BREAKDOWN. (FROM SURFACES)
INVESTIGATIONS WITH FIELD EMISSION MICROSCOPY
AND AUGER SPECTROSCOPY.
JUTTNER B + WOLFF H + ALTRICHTER B
PHYS STATUS SOLIDI A 27(2): 403-12 (16 FEB
1975)
      SAA 1975: 26248

5-1521 MEASUREMENTS OF SURFACE MATERIAL STATES.
KAMADA H + NIKEI Y + NISHIJIMA A + HASEGAWA Y
KAGAKU KOGYO 24(5): 617-27 (1973) (IN
JAPANESE)
      CA 79: 083745

5-1522 LOW ENERGY ELECTRON DIFFRACTION- AUGER
ELECTRON SPECTROSCOPY STUDY OF IRON FILM
DEPOSITED ON MAGNESIUM OXIDE.
KANAJI T + ASANO K + NAGATA S
SHINKU 16(1): 22-7 (1973) (IN JAPANESE)
      CA 79: 095372

5-1523 BEHAVIOR OF IMPURITY ATOMS AND ADSORBED
OXYGEN ATOMS ON THE (001) FACE OF IRON
EPITAXIAL FILM. (AUGER EFFECT)
KANAJI T + ASANO K + NAGATA S
VACUUM 23(2): 55-9 (FEB 1973)
      SAA 1973: 55132

5-1524 AUGER AND LOSS SPECTROSCOPY STUDY OF SURFACE
CONTAMINATION EFFECT ON THE GROWTH MODE OF
IRON EPITAXIAL FILMS ON MAGNESIUM OXIDE
(001).
KANAJI T + KANOTANI T + NAGATA S
THIN SOLID FILMS 32(2): 217-9 (1976)
      CA 84: 143255

5-1525 LEED- AES OBSERVATION OF GROWTH MECHANISM OF
DEPOSITED IRON FILM ON MAGNESIUM OXIDE (001)
SURFACE.
KANAJI T + MAMEDA J + KAGOTANI T + NAGATA S
J VACUUM SOC JAP 18(2): 45-53 (1975)
      SAA 1975: 85622

5-1526 LOW ENERGY ELECTRON DIFFRACTION- AUGER
ELECTRON SPECTROSCOPY STUDY OF AN EPITAXIAL
IRON FILM.
KANAJI T + MAMEDA J + NAGATA S
P223-7 OF INT CONF VACUUM METALLURGY, 4TH
PROC, 1973, JAPAN IRON STEEL INST, 1974.
      CA 82: 46661

5-1527   LOW ENERGY ELECTRON DIFFRACTION- AUGER
          ELECTRON SPECTROSCOPY STUDY OF CLEAN (001)
          IRON SURFACE.
          KANAJI T + MAMEDA J + NAGATA S
          P860 OF INT VACUUM CONGRESS, 6TH PROC, INT
          CONF ON SOLID SURFACES, 2ND, 1974, KYOTO.
          (JAP J APPL PHYS, SUPPL 2, PART 1, 1974)

5-1528   GROWTH MODE OF IRON FILM DEPOSITED ON
          MAGNESIUM OXIDE (001) SURFACE. (AUGER
          SPECTROSCOPY)
          KANAJI T + MAMEDA J + NAGATA S
          MEM FAC ENG KOBE UNIV (21): 173-6 (1975)
                SAA 1976: 6373

5-1529   AUGER ELECTRON SPECTRUM OF CARBON SUBOXIDE.
          KARLSSON L + WERME LO + BERGMARK T
          + SIEGBAHN K
          J ELECTRON SPECTROSC RELAT PHENOMENA 3(3):
          181-9 (1974)
                CA 80: 114397

5-1530   STUDIES OF THE MECHANISM OF THE REACTION
          BETWEEN SULFUR AND OXYGEN ON A MOLYBDENUM
          SURFACE BY AUGER ELECTRON SPECTROSCOPY:
          WORKING STATE OF THE CATALYST SURFACE.
          KAWAI T + KUNIMORI K + KONDOW T + ONISHI T
          + TAMARU K
          CHEM LETT 1973(10): 1101-4
                CA 79: 140048

5-1531   STUDY OF THE OXIDATION OF MOLYBDENUM SURFACES
          BY ENERGY LOSS SPECTROSCOPY COMBINED WITH
          AUGER ELECTRON SPECTROSCOPY.
          KAWAI T + KUNIMORI K + KONDOW T + ONISHI T
          + TAMARU K
          J CHEM SOC FARADAY TRANS 1 70(1): 137-44
          (1974)
                CA 80: 089212

5-1532   HIGH RESOLUTION MEASUREMENTS OF THE NITROGEN
          KLL AUGER TRANSITIONS OF CHEMISORBED AMMONIA
          ON A MOLYBDENUM SURFACE.
          KAWAI T + KUNIMORI K + KONDOW T + ONISHI T
          + TAMARU K
          PHYS REV LETT 33(9): 533-6 (1974)
                CA 81: 113104

5-1533   APPLICATIONS OF AUGER ELECTRON AND ENERGY
          LOSS SPECTROSCOPIES TO CATALYTIC PROCESSES.
          CHEMISORPTION OF AMMONIA, NITROGEN, AND
          NITRIDE FORMATION ON A MOLYBDENUM.
          KAWAI T + KUNIMORI K + KONDOW T + ONISHI T
          + TAMARU K
          P513-19 OF INT VACUUM CONGRESS, 6TH PROC, INT
          CONF ON SOLID SURFACES, 2ND, 1974, KYOTO.
          (JAP J APPL PHYS, SUPPL 2, PART 2, 1974)
                CA 82: 175760

5-1534   DYNAMIC INVESTIGATION OF THE MECHANISM OF
          REACTION BETWEEN SULFUR AND OXYGEN ON A
          MOLYBDENUM SURFACE BY MEANS OF AUGER ELECTRON
          SPECTROSCOPY.
          KAWAI T + KUNIMORI K + KONDOW T + ONISHI T
          + TAMARU K
          J CHEM SOC FARADAY TRANS 1 72(4): 833-41
          (1976)
                CA 84: 170233

5-1535   DYNAMIC ELUCIDATION OF REACTION STEPS ON A
          SOLID SURFACE BY ELECTRON SPECTROSCOPY.
          KAWAI T + TAMARU K
          SHOKUBAI 17(2): 21-32 (1975) (IN JAPANESE)
                CA 83: 65992

5-1536   INITIAL OXIDATION OF MOLYBDENUM. PART-3:
          OXIDE NUCLEATION RATES ON (100) AND (110)
          MOLYBDENUM SURFACES. (AUGER SPECTRA)
          KENNETT HM + LEE AE
          SURFACE SCI 48(2): 617-23 (1975)
                SAA 1975: 58278

5-1537   LOW TEMPERATURE EPITAXY OF GERMANIUM FILMS BY
          SPUTTER DEPOSITION. (HEED, LEED, AUGER
          ELECTRON SPECTROSCOPY)
          KHAN IH
          J APPL PHYS 44(1): 14-19 (JAN 1973)
                SAA 1973: 18554

5-1538   SURFACE STRUCTURE OF TITANIUM AND ITS
          INTERACTION WITH BROMINE AND CHLORINE. (AUGER
          SPECTROSCOPY)
          KHAN IH
          SURFACE SCI 48(2): 537-48 (1975)
                SAA 1975: 58249

5-1539   A STUDY OF ACTIVATED SILICON SURFACES FOR
          ELECTROLESS NICKEL PLATING BY AES.
          KHERA RP + THOMAS S
          P267-8 OF ELECTROCHEM SOC FALL MEET, EXTENDED
          ABSTR, 1974.
                SAA 1975: 69615, SAB 1975: 33058

5-1540   GERMANIUM SURFACE CLEANING: AUGER ANALYSIS.
          KIEWIT DA + D'HAENENS IJ + ROTH JA
          J ELECTROCHEM SOC 121(2): 310-11 (1974)
                CA 80: 113988

5-1541   SUPERCONDUCTING TRANSITION TEMPERATURES OF
          REACTIVELY SPUTTERED FILMS OF TANTALUM
          NITRIDE AND TUNGSTEN NITRIDE. (AUGER EFFECT)
          KILBANE FM + HABIG PS
          J VACUUM SCI TECHNOL 12(1): 107-9 (1975)
                SAA 1975: 54607

5-1542   ELECTRON BEAM INDUCED EFFECTS ON GAS
          ADSORPTION STUDIES UTILIZING AUGER ELECTRON
          SPECTROSCOPY: CARBON MONOXIDE AND OXYGEN ON
          SILICON.
          KIRBY RE
          UNIV WISCONSIN, THESIS 1973, 130P.
                CA 79: 149742

5-1543   ELECTRON BEAM INDUCED EFFECTS ON GAS
          ADSORPTION UTILIZING AUGER ELECTRON
          SPECTROSCOPY. CARBON MONOXIDE AND MOLECULAR
          OXYGEN ON SILICON. PART-1: ADSORPTION
          STUDIES. PART-2: STRUCTURAL EFFECTS.
          KIRBY RE + LICHTMAN D + DIEBALL JW
          SURFACE SCI 41(2): 447-66 (1974); 41(2):
          467-74 (1974)
                CA 80: 074520; 074521

5-1544   EVALUATION OF THE FREQUENCY OF AUGER
          NEUTRALIZATION OF INERT GAS IONS ON THE
          SURFACE OF ALKALI HALIDE CRYSTALS.
          KISHINEVSKII LM
          P24-7, CH 2 OF VZAIMODEISTVIE ATOM CHASTITS S
          TVERD TELOM, 1974. (IN RUSSIAN)
                CA 83: 152515

5-1545   SURFACE ANALYSIS BY APPEARANCE POTENTIAL
          SPECTROSCOPY.
          KLUGE A
          VAK TECH 23(4): 104-8 (1974) (IN GERMAN)
                CA 81: 128382

5-1546   (AUGER) ADSORPTION STUDIES OF OXYGEN AND
          CARBON MONOXIDE ON A PLATINUM (100) SURFACE.
          KNERINGER G + NETZER FP
          SURFACE SCI 49(1): 125-42 (1975)
                SAA 1975: 42090

5-1547   IN-SITU TRANSPORT MEASUREMENTS IN ULTRAHIGH
          VACUUM GROWN AMORPHOUS GERMANIUM. (DEPTH
          PROFILES OF OXYGEN AND CARBON BY AUGER
          SPECTROSCOPY)
          KNOTEK ML
          P297-310 OF INT CONF ON TETRAHEDRALLY BONDED
          AMORPHOUS SEMICONDUCTORS, BRODSKY MH +
          KIRKPATRICK S + WEARIE D(EDS), AIP, 1974.
          (AIP CONF PROC NO 20)
                CA 82: 50259

5-1548   TRANSPORT OF OXYGEN IN AMORPHOUS GERMANIUM
          THIN FILMS DURING ANNEALING. (AUGER
          SPECTROSCOPY)
          KNOTEK ML
          J VACUUM SCI TECHNOL 12(1): 117-20 (1975)
                SAA 1975: 54358

5-1549   ADSORPTION AND SOLUTION OF HYDROGEN,
          DEUTERIUM, NITROGEN, OXYGEN AND CARBON
          MONOXIDE BY (100) TANTALUM. (AUGER EFFECT)
          KO SM + SCHMIDT LD
          SURFACE SCI 47(2): 557-68 (FEB 1975)
                SAA 1975: 29504

5-1550 COMPOSITION PROFILES OF SEVERAL CONTAMINATED
AND CLEANED SURFACES OF GOLD THICK FILMS ON
COPPER PLATES BY AUGER ELECTRON AND SECONDARY
ION MASS SPECTROSCOPY.
KOMIYA A + MIZUNO M + NARUSAWA T + MAEDA H
+ YOSHIKAWA M
P363-6 OF INT VACUUM CONGRESS, 6TH PROC, INT
CONF ON SOLID SURFACES, 2ND, 1974, KYOTO.
(JAP J APPL PHYS, SUPPL 2, PART 1, 1974)
CA 83: 49720

5-1551 EFFECT OF GOLD COATING ON SORPTION
CHARACTERISTICS OF CARBON MONOXIDE, SULFUR,
AND OTHER GASES ON STAINLESS STEEL SHEET.
(AES OF GOLD COATING)
KOMIYA S + NARUSAWA T + HAYASHI C
J VACUUM SCI TECHNOL 9: 302-5 (1972)
SAA 1972: 39558

5-1552 SPUTTER ETCHING AND AUGER ELECTRON
SPECTROSCOPY OF METAL SURFACES.
KOMIYA S + NARUSAWA T + HAYASHI C + SAITO S
P2-20 OF METHODES USINAGE NON TRADII, JOURN
PRINTEMPS MEC IND, 4TH, SER 3, 1974. (IN
FRENCH)
CA 83: 151053

5-1553 SIMULTANEOUS OBSERVATIONS OF PARTIALLY
OXIDIZED SURFACES BY AUGER ELECTRON
SPECTROSCOPY AND SECONDARY ION MASS
SPECTROMETRY FOR ALUMINUM, SILICON, TITANIUM,
VANADIUM, AND CHROMIUM.
KOMIYA S + NARUSAWA T + SATAKE T
J VACUUM SCI TECHNOL 12(1): 361-5 (1975)
CA 83: 14437

5-1554 SPUTTER ETCHING AND AUGER ELECTRON
SPECTROSCOPY OF METAL SURFACES.
KOMIYA S + NARUSAWA T + HAYASHI C + SAITO S
MECH MATER ELECTR 306-307: 40-4 (1975)
CA 84: 172787

5-1555 SECONDARY ELECTRON EMISSION IN THE STUDY OF
SOLID SURFACES.
KOSHIKAWA T + SHIMIZU R
OYO BUTURI 44(3): 215-30 (1975) (IN JAPANESE)
SAA 1976: 2653

5-1556 MICROSTRUCTURE OF HOT PRESSED SILICON
NITRIDE. (AUGER ANALYSIS, MICROPROBE
ANALYSIS)
KOSSOWSKY R
J MATER SCI 8(11): 1603-15 (NOV 1973)
SAA 1974: 04192

5-1557 IMPURITIES AND INCLUSIONS IN HOT PRESSED
SILICON NITRIDE. (AUGER SPECTROGRAPHIC
ANALYSIS)
KOSSOWSKY R
BULL AMER CERAM SOC 52(4): 426 (APR 1973)
SAA 1973: 76284

5-1558 ADSORPTION OF SULFUR ON GOLD BY LOW ENERGY
ELECTRON DIFFRACTION AND AUGER SPECTROSCOPY.
KOSTELITZ M + DOMANGE JL + OUDAR J
SURFACE SCI 34(2 PART-2): 431-49 (JAN 1973)
SAA 1973: 20940

5-1559 ADSORPTION OF KRYPTON ON THE BASAL PLANE OF
GRAPHITE: LEED AND AUGER MEASUREMENTS.
KRAMER HM + SUZANNE J
SURFACE SCI 54(3): 659-69 (1976)
CA 84: 141047

5-1560 OBSERVATIONS OF BETA- SILICON CARBIDE
FORMATION ON RECONSTRUCTED SILICON SURFACES.
KRAUSE MO
PHYS STATUS SOLIDI A 3: 899-906 (1970)

5-1561 ELECTRON AUGER SPECTROSCOPIC STUDY OF THE
SURFACE OF CARBIDE CATHODES.
KUL'VARSKAYA BS + CHEREVATSKII NYA
BULL ACAD SCI USSR, PHYS SER 38(2): 176
(1974)
CA 81: 018170

5-1562 SURFACE OF THERMIONIC EMITTERS MADE OF
CARBIDE COMPOUNDS STUDIED BY AN ELECTRONIC
AUGER SPECTROSCOPIC METHOD.
KUL'VARSKAYA BS + CHEREVATSKII NYA
SOV PHYS TECH PHYS 18(6): 827-9 (1973)
CA 79: 119602

5-1563 REACTION OF SPUTTERED PLATINUM FILMS ON
GALLIUM ARSENIDE. (AUGER SPECTROSCOPY)
KUMAR V
J PHYS CHEM SOLIDS 36(6): 535-41 (1975)
SAA 1975: 50677, SAB 1975: 24311

5-1564 SEGREGATION OF SULFUR ON A MOLYBDENUM SURFACE
STUDIED BY AUGER ELECTRON SPECTROSCOPY.
KUNIMORI K + KAWAI T + KONDOW T + ONISHI T
+ TAMARU K
SURFACE SCI 46(2): 567-76 (1974)
CA 82: 64847

5-1565 HIGH RESOLUTION AUGER ELECTRON SPECTROSCOPY
OF NITRIC OXIDE ADSORBED ON A TUNGSTEN
SURFACE.
KUNIMORI K + KAWAI T + KONDOW T + ONISHI T
+ TAMARU K
CHEM LETT 1975(12): 1303-4
CA 84: 50203

5-1566 SURFACE SEGREGATION: COMPARISON BETWEEN
THEORY AND EXPERIMENT.
LAGUES M + DOMANGE JL
SURFACE SCI 47: 77-85 (1975)

5-1567 ROLE OF PRIMARY AND SECONDARY ELECTRONS IN
ELECTRON INDUCED DESORPTION AND DISSOCIATION:
CARBON MONOXIDE ON PLATINUM (111). (AUGER
SPECTRUM)
LAMBERT RM + COMRIE CM
SURFACE SCI 38(1): 197-209 (JUL 1973)
SAA 1973: 55152

5-1568 ADSORPTION OF OXYGEN ON MOLYBDENUM (111):
EFFECT OF TRACE IMPURITIES.
LAMBERT RM + LINNETT JW + SCHWARZ JA
SURFACE SCI 26: 572-86 (1971)

5-1569 LEED- AUGER INVESTIGATION OF A STABLE CARBIDE
OVERLAYER ON A PLATINUM (111) SURFACE.
LAMBERT RM + WEINBERG WH + COMRIE CM
+ LINNETT JW
SURFACE SCI 27: 653-8 (1971)

5-1570 (AUGER) ADSORPTION STUDIES ON THE PLATINUM
(100) SURFACE.
LANG B + LEGARE P + MAIRE G
SURFACE SCI 47(1): 89-97 (1975)
SAA 1975: 46468

5-1571 FLUORESCENCE YIELD OF CARBON. (AUGER
MEASUREMENTS)
LANGENBERG A + VANECK J
PHYS REV LETT 31(2): 71-3 (9 JUL 1973)
SAA 1973: 52271

5-1572 ESTABLISHING DETECTION AND DETERMINATION
LIMITS IN SURFACE AND THIN FILM ANALYSIS.
(AUGER SPECTROSCOPY)
LARRABEE GB + DOBROTT RD + KEENAN JA
+ MCGUIRE GE
P234-6 OF ELECTROCHEM SOC FALL MEET ABSTR,
1975.
SAA 1976: 33726

5-1573 SURFACE CHARACTERIZATION OF NICKEL- COPPER
ALLOY DISPLAYING POOR WETTABILITY DURING
BRAZING. (MONEL, AUGER SPECTRA)
LARSON DT + KORBITZ FW
J VACUUM SCI TECHNOL 12(3): 721-2 (1975)
CA 83: 182787

5-1574 DETECTION OF CONTAMINANTS ON POLYCRYSTALLINE
SILVER BY AES AND WORK FUNCTION MEASUREMENTS.
LASSITER WS
J VACUUM SCI TECHNOL 9: 310 (1972) (ABSTRACT)
SAA 1972: 39552

5-1575 INTERACTION OF SULFUR DIOXIDE AND CARBON
DIOXIDE WITH CLEAN SILVER IN ULTRAHIGH
VACUUM. (AES)
LASSITER WS
J PHYS CHEM 76(9): 1289-92 (1972)
CA 76: 158697M

5-1576 AUGER SPECTROSCOPY OF HYDROCHLORIC ACID
INTERACTION WITH ALUMINUM AT LOW PRESSURES.
LASSITER WS
SURFACE SCI 47(2): 669-76 (1975)
CA 82: 145529

5-1577 ANTIMONY DOPING OF SILICON LAYERS GROWN BY
SOLID PHASE EPITAXY. (AES MEASUREMENTS)
LAU SS + CANALI C + LIAU ZL + MAKAMURA K
+ NICOLET MA + MAYER JW + BLATTNER RJ
+ EVANS CA
APPL PHYS LETT 28(3): 148-50 (1976)
SAA 1976: 45983

5-1578 SLOW ELECTRON SPECTROSCOPY ON MOLYBDENUM
(100).
LECANTE J
P56 OF EUROP PHYS SOC STUDY CONF ON METAL
SURFACES, ABSTR, 1973.
SAA 1974: 37101

5-1579 CARBON MONOXIDE ADSORPTION KINETICS ON
MOLYBDENUM (100). (AUGER SPECTROSCOPY)
LECANTE J + RIWAN R + GUILLOT C
SURFACE SCI 35: 271-87 (MAR 1973)
SAA 1973: 30348

5-1580 A LEED STUDY OF MAGNESIUM OXIDE (100).
PART-1: EXPERIMENT.
LEGG K + PRUTTON M + KINNIBURGH C
J PHYS C 7(23): 4236-46 (7 DEC 1974)
SAA 1975: 12890

5-1581 LEED- AUGER ANALYSIS OF THE BERYLLIUM (0001)
SURFACE.
LEJEUNE EJ
J VACUUM SCI TECHNOL 8(1): 9 (1971)

5-1582 STUDY OF METAL SURFACES BY IONIC
MICROANALYSIS AND AUGER SPECTROMETRY.
LEROY V + SERVAIS JP + GRAAS H + HABRAKEN L
SILICATES IND 40(2): 37-47 (1975) (IN FRENCH)
CA 83: 101561

5-1583 QUANTITATIVE STUDY OF AUGER ELECTRON SIGNALS
OF PHOSPHORUS ON SILICON USING A QUARTZ
CRYSTAL MICROBALANCE.
LEVENSON LL + DAVIS LE + BRYSON CE
+ MELLES JJ + KOU WH
J VACUUM SCI TECHNOL 9(2): 608-11 (1972)

5-1584 CORRELATION OF ELECTRONIC, LEED, AND AUGER
DIAGNOSTICS ON ZINC OXIDE SURFACES.
LEVINE JD + WILLIS A + BOTTOMS WR + MARK P
SURFACE SCI 29: 144-64 (1972)

5-1585 LOW ENERGY ELECTRON DIFFRACTION- AUGER
ELECTRON SPECTROSCOPY STUDY OF THE OXIDATION
OF IRON (110) AND IRON (100).
LEYGRAF C + EKELUND S
SURFACE SCI 40(3): 609-35 (1973)
CA 80: 031286

5-1586 A LOW ENERGY ELECTRON DIFFRACTION- AUGER
ELECTRON SPECTROSCOPY STUDY OF THE OXIDATION
OF IRON (100) AND IRON (110).
LEYGRAF C + EKELUND S
J VACUUM SCI TECHNOL 11(1): 189 (JAN-FEB
1974)
SAA 1974: 75720

5-1587 LEED- AES STUDY OF THE OXIDATION OF IRON-
CHROMIUM (100) AND (110).
LEYGRAF C + HULTQUIST G + EKELUND S
SURFACE SCI 51(2): 409-32 (1975)
CA 83: 200477

5-1588 INTERACTION OF CESIUM WITH CLEAN ZINC OXIDE
SURFACES. (AUGER SPECTROSCOPY)
LEYSEN R + HOPKINS BJ + TAYLOR PA
J PHYS C 8(7): 907-16 (1975)
SAA 1975: 46456

5-1589 CHEMICAL IDENTIFICATION OF SURFACE
MONOLAYERS. (BY AUGER EMISSION)
LICHTMAN D
PHYS TEACH 14(1): 50-1 (1976)
SAA 1976: 47659

5-1590 LIQUID PHASE EPITAXY OF III-V COMPOUNDS ON
INDIUM PHOSPHIDE SUBSTRATES. (AUGER
SPECTROSCOPY)
LINH NT + HOMBROUCK JP + SOL N
J CRYST GROW 31(1): 204-9 (1975)
SAA 1976: 22959

5-1591 SURFACE COMPOSITION STUDIES OF THE (001) AND
(110) FACES OF MONOCRYSTALLINE IRON(0.84)
CHROMIUM(0.16). (AUGER EFFECT)
LLYGRAF C + HULTQUIST G + EKELUND S
+ ERIKSSON JC
SURFACE SCI 46(1): 157-76 (1974)
SAA 1975: 42053

5-1592 NEW METHODS FOR THE DETECTION AND ELIMINATION
OF COATING FLAWS.
LOHMEYER S
DEFAZET - DEUT FARBEN-Z 28(9): 406-7 (1974)
(IN GERMAN)
CA 82: 018660

5-1593 EVALUATION OF LEAD(0.8) TIN(0.2) TELLURIDE
DETECTOR FABRICATION USING SURFACE ANALYSIS.
(AUGER)
LONGSHORE R + JASPER M + SUMNER B
+ LOVECCHIO P
INFRARED PHYS 15(4): 311-5 (1975)
SAA 1976: 18891

5-1594 EQUILIBRATION OF HYDROGEN AND DEUTERIUM ON
SINGLE CRYSTAL SURFACES OF PLATINUM. (AUGER
SPECTROSCOPY)
LU KE + RYE RR
J VACUUM SCI TECHNOL 12(1): 334-7 (1975)
SAA 1975: 65807

5-1595 (AUGER) ELECTRON SPECTROSCOPY OF GALLIUM
ARSENIDE AND ALUMINUM ARSENIDE SURFACES.
LUDEKE R + ESAKI L
SURFACE SCI 47(1): 132-42 (1975)
CA 82: 148079

5-1596 GALLIUM(1-X) ALUMINUM(X) ARSENIDE
SUPERLATTICES PROFILED BY AUGER ELECTRON
SPECTROSCOPY.
LUDEKE R + ESAKI L + CHANG LL
APPL PHYS LETT 24(9): 417-19 (1974)
CA 81: 007153

5-1597 THREE DIMENSIONAL ELEMENTAL ANALYSIS OF
MATERIALS BY AUGER ELECTRON SPECTROSCOPY.
MACDONALD NC
P58-63 OF SYMP ON ELECTROCATALYSIS, 1974,
BREITER MW(ED), ELECTROCHEM SOC, 1974.
CA 82: 67623

5-1598 MICROSCOPIC AUGER ELECTRON ANALYSIS OF
FRACTURE SURFACES.
MACDONALD NC + MARCUS HL + PALMBERG PW
P25-31 OF SCANNING ELECTRON MICROSCOPE SYMP,
3RD PROC, JOHARI O(ED), ILL INST TECHNOL RES
INST, 1970, 534P.

5-1599 AUGER EJECTION OF ELECTRONS FROM TUNGSTEN BY
OXYGEN CHEMISORPTION.
MACLENNAN DA
BULL AMER PHYS SOC 13: 197 (1968)

5-1600 ROLE OF WORK FUNCTION IN ELECTRON EJECTION BY
METASTABLE ATOMS: HELIUM AND ARGON ON (111)
AND (110) TUNGSTEN. (AUGER)
MACLENNAN DA + DELCHAR TA
J CHEM PHYS 50: 1772-8 (1969)

5-1601 WORK FUNCTION EFFECTS ON AUGER ELECTRON
EJECTION BY NOBLE GAS METASTABLE ATOMS.
MACLENNAN DA + DELCHAR TA
BULL AMER PHYS SOC 13: 197 (1968)

5-1602 LOW ENERGY ELECTRON DIFFRACTION STUDY OF THE
POLAR (111) SURFACES OF GALLIUM ARSENIDE AND
GALLIUM ANTIMONIDE.
MACRAE AU
SURFACE SCI 4: 247-64 (1966)

5-1603 AN ELECTRON DIFFRACTION STUDY OF CESIUM
ADSORPTION ON TUNGSTEN.
MACRAE AU + MULLER K + LANDER JJ + MORRISON J
SURFACE SCI 15: 483-97 (1969)

5-1604 ELECTRONIC AND LATTICE STRUCTURE OF CESIUM
FILMS ADSORBED ON TUNGSTEN.
MACRAE AU + MULLER K + LANDER JJ
+ MORRISON J + PHILLIPS JC
PHYS REV LETT 22: 1048-51 (1969)

5-1605 ELECTRON STIMULATED PROCESSES AT SOLID
SURFACES. (AUGER SPECTRCSCOPY)
MADDEN HH
J VACUUM SCI TECHNOL 13(1): 228-32 (1976)
SAA 1976: 49473

5-1606 DECOMPOSITION OF CARBON MONOXIDE ON A (110)
NICKEL SURFACE. (AUGER ELECTRON SPECTROSCOPY)
MADDEN HH + ERTL G
SURFACE SCI 35: 211-26 (MAR 1973)
SAA 1973: 30346

5-1607 INTERACTION OF CARBON MONOXIDE WITH (110)
NICKEL SURFACES. (AUGER EFFECT)
MADDEN HH + KUPPERS J + ERTL G
J CHEM PHYS 58(8): 3401-10 (15 APR 1973)
SAA 1973: 38716

5-1608 ADSORPTION CF OXYGEN AND OXIDATION CF COBALT
ON THE RUTHENIUM (001) SURFACE. (AUGER
SPECTRA)
MADEY TE + ENGELHARDT HA + MENZEL D
SURFACE SCI 48(2): 304-28 (1975)
SAA 1975: 42088

5-1609 ADSORPTION OF COBALT ON (001) RUTHENIUM AT
TEMPERATURES BELOW 300 DEGREES K.
MADEY TE + MENZEL D
JAP J APPL PHYS SUPPL 2 PART-2: 229-35 (1974)
SAA 1975: 37531

5-1610 QUANTITATIVE DETERMINATION OF CARBON ON
SILVER BY AUGER SPECTROSCOPY.
MAGLIETTA M + PRATESI F + ROVIDA G
CHEM PHYS LETT 36(4): 436-40 (1975)
CA 84: 35742

5-1611 DETECTION OF SILICON OXYNITRIDE LAYERS ON THE
SURFACES OF SILICON NITRIDE FILMS BY AUGER
ELECTRON EMISSION.
MAGUIRE HG + AUGUSTUS PD
J ELECTROCHEM SOC 119: 791-3 (JUN 1972)

5-1612 STUDY OF THE OXIDATION AND REDUCTION BY WATER
VAPOR OF SILICON (111) SURFACES USING AUGER
ELECTRON EMISSION.
MAGUIRE HG + AUGUSTUS PD
PHYS STATUS SCLIDI A 17(1): 101-8 (1973)
CA 79: 023780

5-1613 AUGER ELECTRON EMISSION FROM BERYLLIUM
SURFACES.
MAGUIRE HG + AUGUSTUS PD
PHIL MAG 30(1): 95-103 (1974)
CA 81: 179417

5-1614 AUGER ELECTRON SPECTROSCOPIC ANALYSIS OF
ANODIC FILMS CN ALUMINUM.
MAITRA S + VERINK ED
P216-17 OF ELECTROCHEM SOC FALL MEET ABSTR,
1975.
SAA 1976: 35720

5-1615 GRAIN BOUNDARY SEGREGATION IN MAGNESIA DOPED
ALUMINA. (HIGH VACUUM AUGER ELECTRON
SPECTROMETER, FRACTURED ALUMINA)
MARCUS HL + FINE ME
J AMER CERAM SOC 55(11): 568-70 (NOV 1972)
SAA 1973: 27739

5-1616 AUGER SPECTROSCOPY OF FRACTURE SURFACES OF
CERAMICS.
MARCUS HL + HARRIS JM + SZALKOWSKI FJ
P387-98 OF SYMP FRACT MECH CERAM, 1973 PROC,
VOL 1, BRADT RC(ED), PLENUM PRESS, 1974, NY.
CA 81: 157709

5-1617 AUGER ELECTRON SPECTROSCOPY STUDY OF ELECTRON
BEAM STIMULATED DESORPTION OF OXYGEN FROM A
SEMICONDUCTOR SURFACE.
MARGONINSKI Y
PHYS LETT A 54(5): 391-2 (1975)
CA 84: 9261

5-1618 THICKNESS MEASUREMENTS OF ADSORBED LAYERS BY
AUGER ELECTRON SPECTROSCOPY.
MARGONINSKI Y
SOLID STATE COMMUN 17(3): 373-5 (1975)
CA 83: 152775

5-1619 LEED AND AES OBSERVATIONS ON THE POLAR
SURFACES OF ZINC OXIDE.
MARGONINSKI Y + KIRBY RE
J PHYS C 8(10): 1516-23 (1975)
CA 83: 51233

5-1620 INTERFERENCE OF AN ELECTRON BEAM WITH THE
SURFACE REACTION BETWEEN OXYGEN AND
GERMANIUM. (AES)
MARGONINSKI Y + SEGAL D + KIRBY RE
SURFACE SCI 53(1): 488-99 (1975)
SAA 1976: 26340

5-1621 SCATTERING OF LOW ENERGY ELECTRONS FROM A
COPPER (111) SURFACE.
MARKLUND I + ANDERSSON S + MARTINSON J
ARKIV FYS 37: 127-39 (1967)

5-1622 ADSORPTION AND DECOMPOSITION OF CARBON
MONOXIDE ON PLATINUM (111). (AUGER
SPECTROMETRY)
MARTINEZ JM + HUDSON JB
J VACUUM SCI TECHNOL 10(1): 35-8 (JAN-FEB
1973)
SAA 1973: 38722

5-1623 ANGULAR DISTRIBUTION OF AUGER ELECTRON
EMISSION FROM CLEAN AND GAS-COVERED IRON
(100) SURFACES.
MATSUDAIRA T + WATANABE M + ONCHI M
P181-4 OF INT VACUUM CONGRESS, 6TH PROC, INT
CONF ON SOLID SURFACES, 2ND, 1974, KYOTO.
(JAP J APPL PHYS, SUPPL 2, PART 2, 1974)
CA 82: 177849

5-1624 GROWTH MODE DISCRIMINATION OF THIN FILMS BY
AUGER ELECTRON SPECTROSCOPY.
MATSUSHITA Y + YAGI K + NARUSAWA T + HONJO G
P567-70 OF INT VACUUM CONGRESS, 6TH PROC, INT
CONF ON SOLID SURFACES, 2ND, 1974, KYOTO.
(JAP J APPL PHYS, SUPPL 2, PART 2, 1974)
CA 82: 174310

5-1625 CLEAN SURFACES.
MATTHEW JAD
PHYS EDUC 8(5): 315-20 (1973)
CA 79: 083744

5-1626 AUGER EFFECT IN SURFACE SCIENCE.
MATTHEW JAD
ENDEAVOUR 33(119): 86-91 (1974)
CA 80: 138677

5-1627 COMPARISON OF SURFACE LAYER ANALYSIS
TECHNIQUES. (AUGER ELECTRON SPECTROSCOPY,
SECONDARY ION MASS SPECTROSCOPY, ELECTRON
MICROPROBE)
MAYER JW + TUROS A
THIN SOLID FILMS 19(1): 1-10 (DEC 1973)
SAA 1974: 08579

5-1628 DETECTION OF LEAD, BY AUGER SPECTROSCOPY, ON
IRON INHIBITED IN LEAD AZELATE SOLUTION.
MAYNE JEO + TURGOOSE S + WILSON JM
BRIT CORROS J 8(5): 236-7 (1973)
CA 80: 148009

5-1629 STUDY OF THE KINETICS AND MECHANISM OF THE DECOMPOSITION OF FORMIC ACID ON CARBURIZED AND GRAPHITIZED NICKEL (110) USING AES, LEED, AND FLASH DESORPTION.
MCCARTY JG + MADIX RJ
J CATAL 38(1): 402-17 (1975)
CA 83: 146882

5-1630 FORMIC ACID DESORPTION FROM GRAPHITIZED NICKEL (110). (LEED- AES)
MCCARTY JG + MADIX RJ
SURFACE SCI 54(2): 210-28 (1976)
SAA 1976: 35694

5-1631 SEGREGATION AT COPPER- GOLD ALLOY SURFACES. (AUGER SPECTROSCOPY)
MCDAVID JM + FAIN SC
SURFACE SCI 52(1): 161-73 (1975); J VACUUM SCI TECHNOL 12(1): 351 (1974)
SAA 1975: 69133

5-1632 ANGULAR DEPENDENCE OF AUGER ELECTRON EMISSION FROM COPPER (111) AND (100) SURFACES.
MCDONNELL I + WOODRUFF DP + HOLLAND BW
SURFACE SCI 51(1): 249-69 (1975)
CA 83: 155172

5-1633 ADSORPTION OF XENON AND COBALT ON SILVER (111).
MCELHINEY G + PAPP H + PRITCHARD J
SURFACE SCI 54(3): 617-34 (1976)
SAA 1976: 40311

5-1634 ELECTRONIC PROPERTIES OF CLEAVED MOLYBDENUM DISULFIDE SURFACES.
MCGOVERN IT + WILLIAMS RH + MEE CHB
SURFACE SCI 46(2): 427-40 (DEC 1974)
SAA 1975: 21934

5-1635 CHOOSING BETWEEN ESCA AND AUGER FOR SURFACE ANALYSIS.
MCGUIRE GE
P175-82 OF SEMICONDUCTOR MEASUREMENT TECHNOLOGY. PART-4: ARPA/NBS WORKSHOP ON SURFACE ANALYSIS FOR SILICON DEVICES, LIEBERMAN AG(ED), NAT BUR STAND SPEC PUB 400-23, 1976.

5-1636 DIFFUSION STUDIES IN GOLD- PLATINUM THIN FILMS BY AUGER ELECTRON SPECTROSCOPY.
MCGUIRE GE + WISSEMAN WR
P221-3 OF ELECTROCHEM SOC FALL MEET ABSTR, 1975.
SAA 1976: 35744

5-1637 ION SELECTIVE SPUTTERING OF III-V COMPOUND SEMICONDUCTOR SURFACES. (AUGER EFFECT)
MCGUIRE GE + YANG MG
P247-9 OF ELECTROCHEM SOC FALL MEET ABSTR, 1975.
SAA 1976: 36416

5-1638 ADSORPTION OF OXYGEN ON A CLEAN SILICON SURFACE. (AUGER ELECTRON SPECTROSCOPY)
MEYER F + VRAKKING JJ
SURFACE SCI 38(1): 275-81 (JUL 1973)
SAA 1973: 55149

5-1639 INVESTIGATION OF THE PURE AND CESIUM COATED (100) SURFACE OF GALLIUM ARSENIDE BY SLOW ELECTRON DIFFRACTION. (AUGER SPECTROSCOPY)
MITYAGIN AYU + ORLOV VP
SOV PHYS CRYSTALLOGR 18(4): 554-5 (1973)
SAA 1974: 48822

5-1640 (111) SURFACE OF GALLIUM PHOSPHIDE STUDIED BY AUGER SPECTROSCOPY AND DIFFRACTION OF SLOW ELECTRONS.
MITYAGIN AYU + ORLOV VP + CHEREVATSKII NYA
SOV PHYS CRYSTALLOGR 18(2): 268-9 (1973) (IN RUSSIAN)
CA 78: 165410

5-1641 PROCESSES OF ADSORPTION OF OXYGEN ON GALLIUM ARSENIDE METHOD OF LOW ENERGY ELECTRON SPECTROSCOPY. (AUGER PEAK AMPLITUDE CALIBRATION)
MITYAGIN AYU + ORLOV VP + KHRONOPULO KA
BULL ACAD SCI USSR, PHYS SER 38(2): 88-91 (1974)
SAA 1974: 32805

5-1642 AUGER ELECTRON SPECTRUM DURING THE INTERACTION BETWEEN OXYGEN AND THE (111) SURFACE OF GALLIUM PHOSPHIDE.
MITYAGIN AYU + ORLOV VP + KHRONOPULO KA + CHEREVATSKII NYA
SOV PHYS JETP 36(5): 906-8 (1973)
CA 78: 34839P

5-1643 SOME FEATURES OF THE AUGER ELECTRON SPECTRUM IN THE INTERACTION BETWEEN OXYGEN AND THE (111) SURFACE OF GALLIUM PHOSPHIDE.
MITYAGIN AYU + ORLOV VP + KHRONOPULO KA + CHEREVATSKII NYA
SOV PHYS JETP 36(5): 906-8 (MAY 1973)
SAA 1974: 10512

5-1644 INVESTIGATION OF CESIUM ADSORPTION ONTO A GALLIUM ARSENIDE (110) SURFACE. (AUGER ELECTRON SPECTROSCOPY)
MITYAGIN AYU + ORLOV VP + PANTELEEV VV + KHRONOPULO KA + CHEREVATSKII NYA
SOV PHYS SOLID STATE 14(7): 1623-6 (JAN 1973)
SAA 1973: 20941

5-1645 QUANTITATIVE ESTIMATES OF THE CHEMICAL COMPOSITION NEAR THE SURFACE BASED ON THE ELECTRON AUGER SPECTROSCOPY METHOD.
MITYAGIN AYU + ORLOVA GA
SOV PHYS SOLID STATE 16(9): 1823-4 (1974)
SAA 1975: 51061

5-1646 GROWTH OF GALLIUM ARSENIDE ON (100) GERMANIUM STUDIED BY LOW ENERGY ELECTRON DIFFRACTION AND AUGER SPECTROSCOPIC METHODS.
MITYAGIN AYU + PANTELEEV VV
SOV PHYS CRYSTALLOGR 20(3): 379-81 (1975)
CA 83: 155820

5-1647 DETERMINATION OF SULFUR CONTAMINATION ON MOLYBDENUM SURFACES. (AUGER EFFECT)
MIURA T
JAP J APPL PHYS 15(2): 403-4 (1976)
CA 84: 188975

5-1648 CARBON CONTAMINATION PRODUCED BY ELECTRON BOMBARDMENT. POLYCRYSTALLINE MOLYBDENUM- CARBON MONOXIDE SYSTEM. (AUGER SPECTROSCOPY)
MIURA T + TUZI Y
SEISAN KENKYU 24(5): 210-13 (1972) (IN JAPANESE)
CA 78: 48322U

5-1649 CLEANING PROCEDURES OF MOLYBDENUM (110) SURFACE.
MIURA T + TUZI Y
SHINKU 16(1): 28-32 (1973) (IN JAPANESE)
CA 78: 162787

5-1650 INITIAL OXIDATION OF THE MOLYBDENUM (110) SURFACE OBSERVED BY AES AND LEED.
MIURA T + TUZI Y
P85-8 OF INT VACUUM CONGRESS, 6TH PROC, INT CONF ON SOLID SURFACES, 2ND, 1974, KYOTO. (JAP J APPL PHYS, SUPPL 2, PART 1, 1974)
CA 83: 16397

5-1651 EVALUATION OF THE TRANSITION DENSITY FUNCTION OF CARBON ON VARIOUS SUBSTRATES FROM THE KLL AUGER SPECTRUM. APPLICATION OF AUGER ELECTRON SPECTROSCOPY TO THE BULK TO SURFACE PRECIPITATION AND SURFACE DIFFUSION OF CARBON ON POLYCRYSTALLINE NICKEL.
MOJICA JF
UNIV MISSOURI PHD THESIS, 1975, 151P.
CA 84: 127019

5-1652   QUANTITATIVE ANALYSIS OF LIGHT ELEMENTS
          (NITROGEN, CARBON, AND OXYGEN) IN SPUTTERED
          TANTALUM FILMS BY AUGER ELECTRON SPECTROSCOPY
          AND SECONDARY ION MASS SPECTROMETRY.
          MORABITO JM
          ANAL CHEM 46(2): 189-96 (1974)
              CA 80: 066411

5-1653   SELECTED AREA AND IN-DEPTH AUGER ANALYSIS OF
          THIN FILMS.
          MORABITO JM
          THIN SOLID FILMS 19(1): 21-41 (1973)
              CA 80: 103545

5-1654   ELEMENTAL ANALYSIS WITH AUGER ELECTRON
          SPECTROSCOPY AND SECONDARY ION MASS
          SPECTROMETRY.
          MORABITO JM
          P99-111 OF SYMP MATER SCI ASPECTS THIN FILM
          SYST SOLAR ENERGY COVERS, PROC, NTIS, 1974.
          PB-239270; P139-52 OF INT CONF ELECTRON ION
          BEAM SCI TECHNOL, 6TH PROC, 1974, BAKISH
          R(ED), ELECTROCHEM SOC, 1974.
              CA 83: 157230; 107743

5-1655   APPLICATIONS OF SCANNING AUGER SPECTROSCOPY
          TO THE SILICON INTEGRATED CIRCUIT TECHNOLOGY.
          MORABITO JM
          P105-18 OF SEMICONDUCTOR MEASUREMENT
          TECHNOLOGY. PART-4: ARPA/NBS WORKSHOP ON
          SURFACE ANALYSIS FOR SILICON DEVICES,
          LIEBERMAN AG(ED), NAT BUR STAND SPEC PUB
          400-23, 1976.
              CA 84: 188448

5-1656   COMPOSITION PROFILES OF CHEMICALLY VAPOR
          DEPOSITED PLATINUM AND PLATINUM SILICIDE BY
          AUGER ELECTRON SPECTROSCOPY AND SECONDARY ION
          MASS SPECTROMETRY.
          MORABITO JM + RAND MJ
          THIN SOLID FILMS 22(3): 293-303 (1974)
              CA 81: 085534

5-1657   CHARACTERISTIC ENERGY LOSS AND AUGER ELECTRON
          SPECTRA OF GALLIUM PHOSPHIDE (110).
          MORGAN AE + VAN VELZEN WJM
          SURFACE SCI 40(2): 360-74 (1973)
              CA 79: 151162

5-1658   NEW METHOD OF THIN FILM THICKNESS MEASUREMENT
          USING RADIOISOTOPE AUGER ELECTRONS.
          MORI C + KOIKE J + WATANABE T
          NUCL INSTRUM METH 121(2): 253-8 (1974)
              CA 82: 48949

5-1659   ADSORPTION OF IONIC SALTS ON A TUNGSTEN (100)
          SURFACE.
          MORRISON J + LANDER JJ
          SURFACE SCI 18: 428-30 (1969)

5-1660   EPITAXIAL GROWTH OF COPPER ON (110) SURFACE
          OF A TUNGSTEN SINGLE CRYSTAL STUDIED BY LEED,
          AUGER ELECTRON AND WORK FUNCTION TECHNIQUES.
          MOSS ARL + BLOTT BH
          SURFACE SCI 17: 240-61 (1969)

5-1661   CESIUM ADSORPTION ON TUNGSTEN STUDIED BY LEED
          AND SECONDARY ELECTRON SPECTROSCOPY.
          MULLER K
          P1-5 OF SURFACE PHENOMENA OF THERMIONIC
          EMITTERS, 1969, JULICH NUCL RES ESTAB,
          JULICH, GERMANY, 1969.
              SAA 1971: 23789

5-1662   AUGER ELECTRONS, INDICATORS IN ANALYSIS OF
          SOLID BODY SURFACES.
          MULLER K
          MIKROCHIM ACTA SUPPL 4: 1-9 (1969)

5-1663   HOW MUCH CAN AUGER ELECTRONS TELL US ABOUT
          SOLID SURFACES.
          MULLER K
          P97-125 OF SURFACE PHYSICS, HOHLER G(ED),
          SPRINGER TRACTS IN MODERN PHYSICS NO 77,
          SPRINGER-VERLAG, 1975.
              CA 84: 128089

5-1664   ROOM TEMPERATURE ADSORPTION OF OXYGEN ON
          TUNGSTEN SURFACES. (AUGER MEASUREMENT
          METHODS)
          MUSKET RG
          J LESS COMMON METALS 22: 175-91 (1970)

5-1665   ENERGY DEPENDENCE OF THE SENSITIVITY OF
          IONIZATION SPECTROSCOPY: SILICON ON
          BERYLLIUM. (AUGER ELECTRON SPECTROSCOPY)
          MUSKET RG
          SURFACE SCI 44(2): 629-34 (AUG 1974)
              SAA 1974: 76354

5-1666   PROTON INDUCED ELECTRON EMISSION FROM
          CHARACTERIZED NIOBIUM SURFACES. (AUGER
          SPECTROSCOPY)
          MUSKET RG
          J VACUUM SCI TECHNOL 12(1): 444-7 (1975)
              SAA 1976: 2652

5-1667   PROTON AND ELECTRON EXCITED AUGER ELECTRON
          SPECTRA OF MOLYBDENUM SURFACES.
          MUSKET RG + BAUER W
          APPL PHYS LETT 20: 455-6 (1 JUN 1972)

5-1668   AUGER ELECTRON SPECTROSCOPY STUDY OF OXYGEN
          ADSORPTION ON TUNGSTEN (110).
          MUSKET RG + FERRANTE J
          J VACUUM SCI TECHNOL 7: 14-7 (1970)

5-1669   LEED AND AES STUDY OF OXYGEN ABSORPTION ON
          TUNGSTEN (110).
          MUSKET RG + FERRANTE J
          LEWIS RES CENTER, CLEVELAND, 1971, 35P. NASA
          TN-D-6399.

5-1670   A STUDY OF THE IRON (111) SURFACE USING LEED
          AND AUGER EMISSION SPECTROSCOPY.
          NAKAJIMA K + ISOGAI A + TAGA Y
          ACTA CRYSTALLOGR A 28 (PART-4) SUPPL: S150
          (15 JUL 1972)
              SAA 1973: 15054

5-1671   CHEMICAL ANALYSIS OF SEMICONDUCTORS.
          (IMPURITIES AND COMPOSITION)
          NAKAJIMA S + KOHARA R
          OYO BUTURI 43(5): 422-42 (MAY 1974) (IN
          JAPANESE)
              SAA 1975: 01713

5-1672   INTERACTION OF ALUMINUM LAYERS WITH
          POLYCRYSTALLINE SILICON. (AUGER SPECTROSCOPY)
          NAKAMURA K + NICOLET MA + MAYER JW
          + BLATTNER RJ + EVANS CA
          J APPL PHYS 46(11): 4678-84 (1975)
              SAA 1976: 22163

5-1673   LOW TEMPERATURE DIFFUSION OF GOLD INTO
          SILICON IN THE SILICON (SUBSTRATE)-GOLD
          (FILM) SYSTEM.
          NAKASHIMA K + IWAMI M + HIRAKI A
          THIN SOLID FILMS 25(2): 423-30 (1975)
              SAA 1975: 42018

5-1674   ANALYSIS OF IMPURITIES ON SILICON SURFACES BY
          AUGER ELECTRON SPECTROSCOPY.
          NAKAYAMA K
          BUSSEI 12(11): 647-53 (1971) (IN JAPANESE)
              CA 76: 91067K

5-1675   AUGER SPECTROSCOPY STUDY ON THE SURFACE
          COMPOSITION OF COPPER- NICKEL ALLOYS AFTER
          ANNEALING AND SPUTTERING.
          NAKAYAMA K + ONO M + SHIMIZU H
          J VACUUM SCI TECHNOL 9(2): 749-51 (1972)

5-1676   AUGER ELECTRON EMISSION FROM GOLD DEPOSITED
          ON SILICON (111) SURFACE.
          NARUSAWA T
          JAP J APPL PHYS 10: 280-1 (1971)

5-1677   LEED AND AUGER ELECTRON SPECTROSCOPIC STUDIES
          ON TUNGSTEN (110) SURFACE.
          NARUSAWA T + HAYAKAWA K
          OYO BUTURI 40(6): 616-22 (1971) (IN JAPANESE)
              SAA 1972: 29011

5-1678 LEED AND AUGER ELECTRON SPECTROSCOPIC
OBSERVATIONS OF THE SUBSTRATE SURFACE DURING
SPUTTERING AND VACUUM VAPOR DEPOSITION.
NARUSAWA T + KOMIYA S
VACUUM 22(9): 349-53 (1972)
CA 78: 49051S

5-1679 COMPOSITION PROFILE OF ION PLATED GOLD FILM
ON COPPER ANALYZED BY AUGER ELECTRON
SPECTROSCOPY AND SECONDARY ION MASS SPECTRA
DURING XENON ION BOMBARDMENT.
NARUSAWA T + KOMIYA S
J VACUUM SCI TECHNOL 11(1): 312-16 (1974)
CA 80: 140873

5-1680 AUGER SPECTROSCOPIC OBSERVATION OF SILICON-
GOLD MIXED PHASE FORMATION AT LOW
TEMPERATURES.
NARUSAWA T + KOMIYA S + HIRAKI A
APPL PHYS LETT 21: 272-3 (5 SEP 1972)

5-1681 DIFFUSE INTERFACE IN SILICON (SUBSTRATE)-
GOLD (EVAPORATED FILM) SYSTEM. (AUGER
ELECTRON SPECTROSCOPY)
NARUSAWA T + KOMIYA S + HIRAKI A
APPL PHYS LETT 22(8): 389-90 (15 APR 1973)
SAA 1973: 38751

5-1682 CORRELATION OF FRACTIONAL MONOLAYER OXYGEN
DETERMINATIONS OBTAINED BY PROTON EXCITED
X-RAY, AES ANALYSIS OF IRON SURFACES.
NEEDHAM PB + CRISCOLL TJ + RAO NG
APPL PHYS LETT 21: 502-5 (1972)
SAA 1973: 1879

5-1683 PREPARATION AND SURFACE OXIDATION OF A (110)
FACE OF VANADIUM.
NEMEH EAK + CINTI RC
SURFACE SCI 40: 583-94 (1973)

5-1684 ADSORPTION OF CYANOGEN ON PLATINUM (100).
NETZER FP
SURFACE SCI 52(3): 709-14 (1975)
SAA 1976: 22123

5-1685 ADSORPTION OF HYDROGEN AND THE REACTION OF
HYDROGEN WITH OXYGEN ON PLATINUM (100).
(AUGER SPECTROSCOPY)
NETZER FP + KNERINGER G
SURFACE SCI 51(2): 526-38 (1975)
SAA 1976: 5665

5-1686 LOW ENERGY ELECTRON SPECTROSCOPY OF PLATINUM
(100) AND PLATINUM (100)- CARBON MONOXIDE.
NETZER FP + MATTHEW JAD
SURFACE SCI 51(2): 352-64 (1975)
CA 83: 211115

5-1687 LEED AND ELECTRON SPECTROSCOPIC OBSERVATIONS
ON NICKEL MONOXIDE (100).
NETZER FP + PRUTTON M
J PHYS C 8(15): 2401-12 (1975)
CA 83: 140592

5-1688 ADSORPTION AND DECOMPOSITION OF ACETYLENE ON
POLYCRYSTALLINE RHENIUM FILMS AND RIBBONS.
(AUGER SPECTROSCOPY)
NICLOU Y + EHRHARDT JJ + DUCROS R + CASSUTO A
JAP J APPL PHYS SUPPL 2 PART-2: 521-4 (1974)
SAA 1975: 47802

5-1689 COMPARISON BETWEEN AUGER ELECTRON
SPECTROSCOPY AND SECONDARY ION MASS
SPECTROSCOPY ON WELL CHARACTERIZED SINGLE
CRYSTAL SURFACES. (SILVER AND COPPER ON
TUNGSTEN)
NIEHUS H + BAUER E
ELECTRON FIS APPL 17(1-2): 53-6 (1974)
CA 81: 145303

5-1690 X-RAY EXCITED PHOTOELECTRON AND AUGER
ELECTRON SPECTRA OF OXIDIZED STATES ON SOME
METAL SURFACES.
NISHIJIMA A + KUDO M + NIHEI Y + KAMADA H
PHYS FENN 9, SUPPL S1: 324-6 (JUL 1974)
SAA 1975: 19071

5-1691 COMBINED AUGER ELECTRON SPECTROSCOPY AND
ELECTRON IMPACT DESORPTION STUDIES OF THE
INTERACTIONS OF GASES WITH SILICON SURFACES.
NISHIJIMA M + FUJIWARA K + MUROTANI T
P303-6 OF INT VACUUM CONGRESS, 6TH PROC, INT
CONF ON SOLID SURFACES, 2ND, 1974, KYOTO.
(JAP J APPL PHYS, SUPPL 2, PART 2, 1974)
CA 82: 175608

5-1692 IDENTIFICATION OF ION SPECIES IN THE COMBINED
AUGER ELECTRON SPECTROSCOPY AND ELECTRON
IMPACT DESORPTION STUDIES.
NISHIJIMA M + MUROTANI T
JAP J APPL PHYS 12(5): 777-8 (1973)
CA 79: 084667

5-1693 INTERACTION OF OXYGEN WITH SILICON (111)
SURFACES. (AUGER ELECTRON SPECTROSCOPY)
NISHIJIMA M + MUROTANI T
SURFACE SCI 39(2): 441-4 (SEP 1973)
SAA 1973: 75670

5-1694 COMBINED AUGER ELECTRON SPECTROSCOPY AND
ELECTRON IMPACT DESORPTION STUDIES OF SILICON
SURFACES.
NISHIJIMA M + MUROTANI T
MITSUBISHI DENKI LAB REP 13(1-4): 19-25 (OCT
1972); SURFACE SCI 32(2): 459-65 (1972)
SAA 1973: 68505

5-1695 COMBINED AUGER ELECTRON SPECTROSCOPY AND
ELECTRON IMPACT DESORPTION STUDIES OF SILICON
SURFACES.
NISHIJIMA M + MUROTANI T
SURFACE SCI 32: 459-65 (1972)

5-1696 EPITAXIAL GROWTH OF SILVER ON POTASSIUM
CHLORIDE (100) SURFACE OBSERVED BY LOW ENERGY
ELECTRON DIFFRACTION AND AUGER ELECTRON
SPECTROSCOPY.
NISHIMORI K + TOKUTAKA H + ISHIHARA N
+ TAKASHIMA K
SHINKU 17(5): 161-6 (1974) (IN JAPANESE)
CA 81: 160244

5-1697 ANGULAR RESOLVED AUGER EMISSION SPECTRA FROM
A CLEAN COPPER (100) SURFACE.
NOONAN JR + ZEHNER DM + JENKINS LH
J VACUUM SCI TECHNOL 13(1): 183-7 (1976)
CA 84: 142790

5-1698 INITIAL OXIDATION OF MOLYBDENUM STUDIED BY A
HIGH RESOLUTION AUGER PHOTOELECTRON
SPECTROMETER.
NOZOYE H + MATSUMOTO Y + ONISHI T + TAMARU K
J CHEM SOC FARADAY TRANS 72(2): 389-98 (1976)
CA 84: 83566

5-1699 STUDIES ON SURFACE PLASMA OXIDIZED FILMS OF
III-V COMPOUND SEMICONDUCTORS BY AUGER
ELECTRON SPECTROSCOPY.
ODA T + SUGANO T
ANNU REP ENG RES INST FAC 33: 153-9 (1974)
(IN JAPANESE)
SAA 1975: 38283

5-1700 RESIDUAL GAS AND THE OPTICAL PROPERTIES OF
SILVER FILMS. (AUGER SPECTROSCOPY)
OHANDLEY RC + BURGE DK + JASPERSON SN
+ ASHLEY EJ
SURFACE SCI 50(2): 407-33 (1975)
SAA 1975: 69539

5-1701 MATERIAL ANALYSIS USING THE SCANNING ELECTRON
MICROSCOPE. (AUGER SPECTROSCOPY)
OHNSORGE J + HOLM R
SCR MED FAC MED UNIV 47(3): 143-9 (1974) (IN
GERMAN)
CA 83: 37042

5-1702 QUANTITATIVE AUGER ANALYSIS OF COPPER- NICKEL
ALLOY SURFACES AFTER ARGON ION BOMBARDMENT.
REPLY TO COMMENTS.
ONO M + SHIMIZU H + NAKAYAMA K
SURFACE SCI 52(3): 681-4 (1975)
CA 85: 50371

5-1703 AUGER SPECTROSCOPY ON COPPER- NICKEL ALLOY
SURFACES RELATED TO CATALYSIS.
ONO M + TAKASU Y + NAKAYAMA K + YAMASHINA T
SURFACE SCI 26: 313-6 (1971)

5-1704 ANALYSIS OF GLASS CONTAINER COATINGS BY AUGER
ELECTRON SPECTROSCOPY.
ONODA GY + DOVE DB + HENCH LL
BULL AMER CERAM SOC 52(4): 379 (APR 1973)
SAA 1974: 07264

5-1705 APPLICATIONS OF SURFACE CHARACTERIZATION TO
GLASS SURFACES.
ONODA GY + DOVE DB + PANTANO CG
P39-55 OF INT VACUUM CONGRESS, 6TH PROC, INT
CONF ON SOLID SURFACES, 2ND, 1974, KYOTO.
(JAP J APPL PHYS, SUPPL 2, PART 1, 1974)

5-1706 AUGER ELECTRON SPECTROSCOPY STUDY OF
OXIDATION ON LANTHANUM HEXABORIDE.
OSHIMA C + KAWAI S
APPL PHYS LETT 23(5): 215-16 (1973)
CA 80: 042023

5-1707 SURFACES TEXTURES: AN INVESTIGATION OF
SURFACES CONTAINING IMPURITIES. (AUGER
ELECTRON SPECTROSCOPY)
OUDAR J
SCI PROG DECOUV NO 3446: 33-40 (JUL-AUG 1972)
SAA 1973: 01864

5-1708 GOLD-INDUCED SUPERSTRUCTURES ON SILICON (100)
SURFACES AS OBSERVED BY LEED AES.
OURA K + MAKINO Y + HANAWA T
JAP J APPL PHYS 15(4): 737-8 (1976)
CA 84: 187791

5-1709 AUGER ELECTRON SPECTROSCOPY OF ALLOY
SURFACES. (GOLD- SILVER, LEAD- INDIUM)
OVERBURY SH + SOMORJAI GA
UNIV CALIF, 1975, 25P. LBL-3742
CA 83: 170459

5-1710 SURFACE COMPOSITION OF THE SILVER- GOLD
SYSTEM BY AUGER ELECTRON SPECTROSCOPY.
OVERBURY SH + SOMORJAI GA
SURFACE SCI 55(1): 209-26 (1976)
CA 84: 185377

5-1711 PHYSICAL ADSORPTION OF XENON ON PALLADIUM
(100).
PALMBERG PW
SURFACE SCI 25: 598-608 (1971)

5-1712 CHEMICAL ANALYSIS OF PLATINUM (100) AND GOLD
(100) SURFACES BY AUGER ELECTRON
SPECTROSCOPY.
PALMBERG PW
P29-1 TO 29-18 OF STRUCTURE AND CHEMISTRY OF
SOLID SURFACES, SOMORJAI GA (ED), WILEY,
1969.

5-1713 STRUCTURE TRANSFORMATIONS ON CLEAVED AND
ANNEALED GERMANIUM (111) SURFACES.
PALMBERG PW
SURFACE SCI 11: 153-8 (1958)

5-1714 SECONDARY EMISSION STUDIES ON GERMANIUM AND
SODIUM- COVERED GERMANIUM.
PALMBERG PW
J APPL PHYS 38: 2137-47 (1967)

5-1715 QUANTITATIVE ANALYSIS OF SOLID SURFACES BY
AUGER ELECTRON SPECTROSCOPY.
PALMBERG PW
ANAL CHEM 45: A549-56 (1973)

5-1716 ATOMIC ARRANGEMENT OF GOLD (100) AND RELATED
METAL OVERLAYER SURFACE STRUCTURES.
PALMBERG PW + RHODIN TN
J CHEM PHYS 49: 134-46 (1968)

5-1717 AUGER ELECTRON SPECTROSCOPY OF FACE CENTERED
CUBIC METAL SURFACES. (GOLD, SILVER,
PALLADIUM, COPPER, NICKEL)
PALMBERG PW + RHODIN TN
J APPL PHYS 39: 2425-32 (1968)

5-1718 ATOMIC ARRANGEMENT OF GOLD (100),
GOLD-COVERED (100), AND SILVER-COVERED COPPER
(100) SURFACES.
PALMBERG PW + RHODIN TN
BULL AMER PHYS SOC 13: 944 (1968)

5-1719 SURFACE DISSOCIATION OF POTASSIUM CHLORIDE BY
LOW ENERGY ELECTRON BOMBARDMENT.
PALMBERG PW + RHODIN TN
J PHYS CHEM SOLIDS 29: 1917-24 (1968)

5-1720 GLASS SURFACE ANALYSIS BY AUGER ELECTRON
SPECTROSCOPY.
PANTANO CG + DOVE DB + ONODA GY
J NONCRYST SOLIDS 19: 41-53 (1975)
CA 84: 125724

5-1721 AES ANALYSIS OF SODIUM IN A CORRODED BIOGLASS
USING A LOW TEMPERATURE TECHNIQUE.
PANTANO CG + DOVE DB + ONODA GY
APPL PHYS LETT 26(11): 601-2 (1975)
SAA 1975: 63818

5-1722 AES COMPOSITIONAL PROFILES OF MOBILE IONS IN
THE SURFACE REGION OF GLASS.
PANTANO CG + DOVE DB + ONODA BY
J VACUUM SCI TECHNOL 13(1): 414-8 (1976)
CA 84: 168561

5-1723 COMPARISON OF MANY BODY EFFECTS IN CORE LEVEL
SURFACE SPECTROSCOPIES.
PARK RL + HOUSTON JE + LARAMORE GE
P757-65 OF INT VACUUM CONGRESS, 6TH PROC, INT
CONF ON SOLID SURFACES, 2ND, 1974, KYOTO.
(JAP J APPL PHYS, SUPPL 2, PART 1, 1974)
CA 83: 18149

5-1724 CHARACTERISTIC GAIN PHENOMENA IN
POLYCRYSTALLINE COPPER, SILVER, AND ALUMINUM.
(AUGER SPECTROSCOPY)
PATTINSON EB + HARRIS PR
J ELECTRON SPECTROSC RELAT PHENOMENA 1: 500-5
(1972/1973)

5-1725 CHARACTERIZATION OF ALUMINUM ADHEREND
SURFACES (BY AUGER ELECTRON SPECTROSCOPY).
PATTNAIK A + MEAKIN JD
FRANKLIN INST RES LABS, PHILA, PA, PICATINNY
ARSENAL, DOVER, NJ, JULY 1974, 94P.
AD-786597/5ST

5-1726 ANGULAR DEPENDENCE OF ELECTRON EMISSION FROM
SURFACES. (AUGER EFFECT)
PENDTRY JB
J PHYS C 8(15): 2413-22 (1975)
SAA 1975: 81945

5-1727 SLIDING OF POLY(VINYL CHLORIDE) ON METALS
STUDIED BY AUGER ELECTRON SPECTROSCOPY.
PEPPER SV
LEWIS RES CENTER, 1974, 22P. NASA TN D-7533.
CA 81: 064409

5-1728 AUGER ANALYSIS OF FILMS FORMED ON METALS IN
SLIDING CONTACT WITH HALOGENATED POLYMERS.
PEPPER SV
J APPL PHYS 45(7): 2947-56 (1974)
CA 81: 153360

5-1729 METALLIC TRANSFER BETWEEN METALS IN SLIDING
CONTACT EXAMINED BY AUGER EMISSION
SPECTROSCOPY.
PEPPER SV
NASA LEWIS RES CENTER, CLEVELAND, MAR 1972,
16P. NASA-TN-D-6716

5-1730 ADHESION AND TRANSFER OF PLATINUM IRON TO
METALS STUDIED BY AUGER EMISSION
SPECTROSCOPY.
PEPPER SV + BUCKLEY DH
NASA LEWIS RES CENTER, CLEVELAND, 1972, 22P.
NASA TM S-68076

5-1731 ADHESION AND TRANSFER OF POLYTETRAFLUORO
ETHYLENE TO METALS STUDIED BY AUGER EMISSION
SPECTROSCOPY.
PEPPER SV + BUCKLEY DH
NASA LEWIS RES CENTER, CLEVELAND, 1972, 21P.
NASA TN-D-6983
CA 78: 17724X

5-1732 ADSORPTION OF SULFUR ON THE (100) FACE OF
MOLYBDENUM.
PERALTA L + BERTHIER Y + OUDAR J
SURFACE SCI 55(1): 199-208 (1976)
SAA 1976: 49085

5-1733 ADSORPTION AND SURFACE ALLOYING OF LEAD
MONOLAYERS ON (111) AND (110) FACES OF GOLD.
(AUGER EFFECT)
PERDEREAU J + BIBERIAN JP + RHEAD GE
J PHYS F 4(5): 798-806 (MAY 1974)
SAA 1974: 52992

5-1734 DISCRIMINATION BY AUGER SPECTROSCOPY BETWEEN
ADSORBED SULFUR AND SULFIDES OF COPPER AND
NICKEL.
PERDEREAU M
COMPT REND C 274: 448-50 (1972) (IN FRENCH)

5-1735 AUGER MEASUREMENTS ON T-027 SAMPLES EXPOSED
DURING THE SKYLAB-2 MISSION.
PETERS PN
NASA, MARSHALL SPACE FLIGHT CENTER,
HUNTSVILLE, ALA, FEB 1974, 57P.
NASA-TM-X-64834

5-1736 INTERACTION OF LOW ENERGY ATMOSPHERIC IONS
WITH CONTROLLED SURFACES. (AUGER
NEUTRALIZATION AT (100) FACE OF TUNGSTEN,
POLYCRYSTALLINE MOLYBDENUM).
PIERCE RH + FRENCH JB
UNIV TORONTO REPT AD682373, 1968; AD675206,
1969.
CA 70: 81182; 71: 16846

5-1737 SURFACE COMPOSITION OF PLATINUM HEATED IN
OXYGEN. (AUGER SPECTRA)
PIGNET TP + SCHMIDT LD + JARVIS NL
J CATAL 31(1): 145-7 (1973)

5-1738 SURFACE COMPOSITION AND CHEMISTRY OF
EVAPORATED PERMALLOY FILMS OBSERVED BY X-RAY
PHOTOEMISSION SPECTROSCOPY AND BY AUGER
ELECTRON SPECTROSCOPY.
POLLAK RA + BAJOREK CH
J APPL PHYS 46(3): 1382-8 (1975)
CA 83: 36789

5-1739 A CORRELATION OF AUGER SPECTROSCOPY, LEED AND
WORK FUNCTION MEASUREMENTS FOR EPITAXIAL
GROWTH OF THORIUM ON A TUNGSTEN (100)
SUBSTRATE.
POLLARD JH
SURFACE SCI 20: 269-84 (1970)

5-1740 AUGER ELECTRON SPECTROSCOPY STUDIES OF
SURFACE COMPOSITION OF SILVER- COPPER ALLOYS.
QUANTITATIVE AUGER ANALYSIS OF COPPER- NICKEL
ALLOY SURFACES AFTER ARGON ION BOMBARDMENT.
COMMENTS.
PONS F + LE HERICY J + LANGERON JP
SURFACE SCI 51(1): 336-41 (1975)
CA 84: 98819

5-1741 INTEGRATED CIRCUIT PROCESS CHARACTERIZATION
USING AES.
POPE M + HURD J
BULL AMER CERAM SOC 51: 722 (1972) (ABSTRACT)

5-1742 SURFACE COMPOSITION OF MICA SUBSTRATES.
POPPA H + ELLIOT AG
SURFACE SCI 24: 149-63 (1971)

5-1743 CHANGE OF SURFACE PROPERTIES OF MICA AFTER
CLEAVAGE. (OUTGASSING AND DECORATION STUDIES,
AUGER SPECTRA)
POPPA H + LEE EH
THIN SOLID FILMS 32(2): 223-8 (1976)
SAA 1976: 40319

5-1744 SODIUM ADSORPTION ON ALUMINUM (111) SURFACES:
ELECTRIC LOW ENERGY ELECTRON DIFFRACTION,
AUGER, AND CONTACT POTENTIAL MEASUREMENTS.
PORTEUS JO
SURFACE SCI 41(2): 515-32 (1974); J VACUUM
SCI TECHNOL 11(1): 191 (1974)
CA 80: 074547

5-1745 SURFACE PLASMON STUDIES ON ALUMINUM (111) AND
ALUMINUM (001) SINGLE CRYSTAL SURFACES USING
INELASTIC LOW ENERGY ELECTRON DIFFRACTION.
(AND AUGER ANALYSIS)
PORTEUS JO + FAITH WN
PHYS REV B 12(6): 2097-110 (1975)
SAA 1976: 18244

5-1746 SURFACE CHARACTERISTICS RELATED TO LASER
DAMAGE OF LITHIUM NIOBATE AND POTASSIUM
CHLORIDE SURFACES. (AUGER SURFACE PROFILE)
PORTEUS JO + TEPPO EA + DANCY JH
P135-40 OF SYMP ON LASER DAMAGE IN OPTICAL
MATERIALS, 6TH, 1974, GLASS AJ + GUENTHER AH
(EDS), NAT BUR STAND SPEC PUB 414, 1974.
CA 83: 155380

5-1747 CHARACTERIZATION OF COPPER- GOLD ALLOY SINGLE
SURFACES.
POTTER HC
CORNELL UNIV PHD THESIS, 1970.

5-1748 LEED, AUGER SPECTROSCOPY, AND CONTACT
POTENTIAL STUDIES OF COPPER- GOLD ALLOY
SINGLE CRYSTAL SURFACES.
POTTER HC + BLAKELY JM
J VACUUM SCI TECHNOL 12(2): 635-42 (1975)
CA 83: 51203

5-1749 IDENTIFICATION OF A GRAIN BOUNDARY PHASE IN
HOT PRESSED SILICON NITRIDE BY AUGER ELECTRON
SPECTROSCOPY.
POWELL BD + DREW P
J MATER SCI 9(11): 1867-70 (1974)
CA 82: 89489

5-1750 SODIUM SEGREGATION ONTO A LITHIUM METAL
SURFACE. (AUGER SPECTRA)
POWELL GL + CLAUSING RE + MCGUIRE GE
SURFACE SCI 49(1): 310-14 (1975)
SAA 1975: 37510

5-1751 SPECTROSCOPY OF A METAL- GAS INTERFACE.
PRITCHARD J
ANN REP PROGR CHEM SECT A66: 65 (1969)

5-1752 LOW ENERGY ELECTRON DIFFRACTION FROM CLEAN
(100) NICKEL OXIDE.
PRUTTON M + DOYLE WD + LEGG K
P1430-3 OF CONF ON MAGNETISM MAGN MATER,
PART-2, 1972, AMER INST PHYS, NY, 1972. (AIP
CONF PROC NO 5)
CA 77: 53402D

5-1753 AUGER SCANNING ELECTRON MICROSCOPIC STUDY OF
THE PASSIVE FILM FORMED ON AN IRON- CHROMIUM
ALLOY.
QUDAR J + LEGRESSUS C
MEM SCI REV MET 72(12): 929-31 (1975) (IN
FRENCH)
CA 85: 50022

5-1754 ELLIPSOMETRIC STUDY OF ADSORPTION: ADSORPTION
OF XENON ON GRAPHITE (001).
QUENTEL G + RICKARD JM + KERN R
SURFACE SCI 50(2): 343-59 (1975) (IN FRENCH)
SAA 1975: 69110

5-1755 LOW ENERGY ELECTRON DIFFRACTION AND AUGER
SPECTROSCOPY STUDIES OF THE (100) SURFACES OF
ALUMINUM AND ALUMINUM- COPPER ALLOYS.
QUINTO DT
YALE UNIV THESIS, 1972, 181P.
CA 78: 129373

5-1756 ELECTRON STIMULATED OXIDATION OF GALLIUM
ARSENIDE, STUDIED BY QUANTITATIVE AUGER
ELECTRON SPECTROSCOPY.
RANKE W + JACOBI K
SURFACE SCI 47(2): 525-42 (1975)
CA 82: 129808

5-1757 AUGER SPECTROSCOPY OF OXYGEN AND OTHER
"CONTAMINANTS" ON SILVER.
RAWLINGS KJ + DOBSON PJ
CHEM PHYS LETT 37(2): 353-4 (1976)
CA 84: 112075

5-1758　ROLE OF SURFACE CONTAMINANTS IN ELECTRON
DEVICE TECHNOLOGY.
REDHEAD PA
P35-45 OF SYMP ON ELECTRONICS (ELECTRON TUBE
SEMICOND DEVICES), 1972, 1ST PROC, BHABHA AT
RES CENTER, 1972.
CA 78: 152648

5-1759　DECREASE OF CARBON CONCENTRATION ON THE
SURFACE OF MOLYBDENUM CARBIDE DURING HEATING
ACCORDING TO AUGER ELECTRON SPECTRUM DATA.
REDKIN VS + ZASHKVARA VV + KORSUNSKII MI
+ CHAIKOVSKII EF + VLASENKO VA + SOTNIKOV VT
MONOKRIST TEKH NO 6: 126-9 (1972) (IN
RUSSIAN)
CA 78: 102348

5-1760　AUGER ELECTRON SPECTROSCOPY OF CLEAN
POLYCRYSTALLINE GOLD FILMS.
REID RJ
P815-17 OF INT VACUUM CONGRESS, 6TH PROC, INT
CONF ON SOLID SURFACES, 2ND, 1974, KYOTO.
(JAP J APPL PHYS, SUPPL 2, PART 2, 1974)
CA 82: 177695

5-1761　PASSIVE FILM ON IRON. APPLICATION OF AUGER
ELECTRON SPECTROSCOPY.
REVIE RW + BAKER BG + BOCKRIS JOM
J ELECTROCHEM SOC 122(11): 1460-6 (1975)
CA 84: 10403

5-1762　DEGRADED PASSIVE FILM ON IRON. APPLICATION OF
AUGER ELECTRON SPECTROSCOPY.
REVIE RW + BOCKRIS JOM + BAKER BG
SURFACE SCI 52(3): 664-9 (1975)
CA 84: 168014

5-1763　BINDING, VALENCY AND CHARGE TRANSFER IN
CHEMICAL REACTIONS ON METAL SURFACES.
RHODIN TN
CORNELL UNIV, ITHACA, NY, JUN 1974, 9P.
AD-780934/6

5-1764　CHEMISORPTION ON (001), (110) AND (111)
NICKEL SURFACES. (AUGER SPECTRA)
RHODIN TN + DEMUTH JE
P41-2 EUROP PHYS SOC STUDY CONF ON METAL
SURFACES, ABSTR, 1973.
SAA 1974: 37109

5-1765　ELECTRONS OR IONS. (IN SURFACE ANALYSIS)
RIACH GE + GCFF RF
IND RES 16(6): 84-92 (1974)
CA 81: 043309

5-1766　CORROSION, SURFACE, AND MAGNETIC PROPERTIES
OF NICKEL- IRON- CHROMIUM THIN FILMS. (AES)
RICE DW + SUITS JC + LEWIS SJ
P145-7 OF ELECTROCHEM SOC FALL MEET ABSTR,
1975.
SAA 1976: 36118

5-1767　SILICON (111)-7 STRUCTURE OBTAINED BY
CLEAVAGE AT ROOM TEMPERATURE. (AUGER SURFACE
SPECTRA)
RIDGWAY JWT + HANEMAN D
APPL PHYS LETT 17: 130-1 (1970)

5-1768　AUGER SPECTRA AND LEED PATTERNS FROM NICKEL
DEPOSITS ON CLEAVED SILICON.
RIDGWAY JWT + HANEMAN D
SURFACE SCI 26: 683-7 (1971)

5-1769　AUGER SPECTRA AND LEED PATTERNS FROM VACUUM
CLEANED SILICON CRYSTALS WITH CALIBRATED
DEPOSITS OF IRON.
RIDGWAY JWT + HANEMAN D
SURFACE SCI 24: 451-8 (1971)

5-1770　AUGER SPECTROSCOPY OF OSMIUM, AND THE
SEGREGATION OF PLATINUM TO THE OSMIUM
SURFACE.
RIVIERE JC
J LESS COMMON METALS 38(2-3): 193-200 (1974)
CA 81: 176896

5-1771　DIFFUSION OF SULFUR TOWARD THE (110) FACE OF
NICKEL.
RIWAN R
SURFACE SCI 27: 267-72 (1971)

5-1772　OXYGEN ADSORPTION ON CLEAN MOLYBDENUM (100)
SURFACES.
RIWAN R + GUILLOT C + PAIGNE J
SURFACE SCI 47(1): 183-90 (1975)
CA 82: 129674

5-1773　PALLADIUM SILICIDE FORMATION OBSERVED BY
AUGER ELECTRON SPECTROSCOPY.
ROBINSON GY
APPL PHYS LETT 25(3): 158-60 (1974)
CA 81: 083497

5-1774　AUGER ELECTRON SPECTROSCOPY STUDIES OF
GALLIUM ARSENIDE AND SILICON METAL-
SEMICONDUCTOR STRUCTURES.
ROBINSON GY + FERTIG DJ
CRIT REV SOLID STATE SCI 5(3): 291-6 (1975)
CA 84: 52650

5-1775　AES AND SPUTTER ETCHING OF NICKEL- GOLD-
GERMANIUM ON N-GALLIUM ARSENIDE.
ROBINSON GY + JARVIS NL
APPL PHYS LETT 21(11): 507-10 (1972)
CA 78: 21615E

5-1776　CHARACTERISTIC ENERGY LOSS AND AUGER ELECTRON
SPECTROSCOPY APPLIED TO THE STUDY OF
ADSORPTION PHENOMENA AT METAL SURFACES.
ROUSSEAU J + PRALIAND H
J CHIM PHYS 67: 1493-505 (1970) (IN FRENCH)

5-1777　CHLORINE MONOLAYERS ON THE LOW INDEX FACES OF
SILVER. (AUGER EFFECT)
ROVIDA G + PRATESI F
SURFACE SCI 51(1): 270-82 (1975)
SAA 1975: 81315

5-1778　CHEMISORPTION OF OXYGEN ON THE SILVER (110)
SURFACE.
ROVIDA G + PRATESI F
SURFACE SCI 52(3): 542-55 (1975)
SAA 1976: 22138

5-1779　LEED (AUGER) STUDY OF CHLORINE CHEMISORPTION
ON THE SILVER (111) SURFACE.
ROVIDA G + PRATESI F + MAGLIETTA M
+ FERRONI E
JAP J APPL PHYS SUPPL 2 PART-2: 117-20 (1974)
SAA 1975: 42099

5-1780　AMORPHOUS TO CRYSTALLINE TRANSITION OF
SILICON (111) SURFACES. (AUGER ELECTRON
SPECTROSCOPY)
ROWE JE
P60-4 OF INT CONF ON TETRAHEDRALLY BONDED
AMORPHOUS SEMICONDUCTORS, 1974 PROC, BRODSKY
MH + KIRKPATRICK S + WEAIRE D (EDS), AIP,
1974. (AIP CONF PROC NO 20)
SAA 1975: 01239

5-1781　PHOTOEMISSION MEASUREMENTS OF BULK, SURFACE,
AND HYDROGEN INDUCED STATES ON CLEAVED
GERMANIUM (111). (AUGER SPECTROSCOPY)
ROWE JE
SOLID STATE COMMUN 17(6): 673-6 (1975)
CA 83: 200003

5-1782　INVESTIGATION OF SILVER CESIUM OXIDE
PHOTOEMISSIVE SURFACES USING AUGER ELECTRON
SPECTROSCOPY.
RUSCH TW
UNIV MINNESOTA THEISIS, 1973, 271P.
CA 79: 150653

5-1783　KIKUCHI CORRELATIONS IN AUGER ELECTRON
SPECTROSCOPY.
RUSCH TW + BERTINO JP + ELLIS WP
APPL PHYS LETT 23(7): 359-60 (1973)
CA 80: 031461

5-1784　HIGH ANGULAR RESOLUTION SECONDARY ELECTRON
SPECTROSCOPY. KIKUCHI CORRELATIONS FOR
ARSENIC (00001).
RUSCH TW + ELLIS WP
APPL PHYS LETT 26(2): 44-6 (1975)
CA 82: 162611

5-1785 METHODS FOR SURFACE ANALYSIS OF PRECIOUS
METALS CONTACT MATERIALS.
RUTHARDT R
METALL 28(8): 798-803 (1974) (IN GERMAN)
CA 81: 145031

5-1786 LOW TEMPERATURE OXIDATION OF SILICON STUDIED
BY PHOTOSENSITIVE ESR AND AUGER ELECTRON
SPECTROSCOPY.
RUZYLLO J + SHIOTA I + MIYAMOTO N
+ NISHIZAWA J
J ELECTROCHEM SOC 123(1): 26-9 (1976)
CA 84: 68533

5-1787 CHARACTERIZATION OF GLASS SURFACES BY
ELECTRON SPECTROSCOPY.
RYND JP + PIASTOGI AK
SURFACE SCI 48: 22-43 (1974)

5-1788 SURFACE COMPOSITION OF (PALLADIUM- SILVER)
ALLOYS IN EQUILIBRIUM.
SACHTLER WMH
VIDE 28(164): 67-71 (MAR-APR 1973)
SAA 1973: 79577

5-1789 CHEMISORPTION OF ATOMIC HYDROGEN ON THE
SILICON (111) 7*7 SURFACE. (AUGER EFFECT)
SAKURAI T + HAGSTRUM HD
PHYS REV B 12(12): 5349-54 (1975)
SAA 1976: 30986

5-1790 PROPERTIES OF THE LEAD SULFIDE- LEAD OXIDE
CRYSTALLINE FILMS. PART-1: PREPARATION AND
PHYSICAL STRUCTURE. (AUGER SPECTROSCOPY)
SAMEH-SAID M + ZEMEL JN
J APPL PHYS 47(3): 866-70 (1976)
SAA 1976: 53503

5-1791 AUGER SPECTROSCOPIC STUDY OF PLATINUM,
NICKEL, AND IRON SURFACES.
SAVCHENKO VI
KINET KATAL 13(6): 1583-91 (1972) (IN
RUSSIAN)
CA 78: 114464

5-1792 CHEMICAL SHIFT IN THE AUGER SPECTRUM OF IRON
DURING OXYGEN ADSORPTION.
SAVCHENKO VI
DOKL PHYS CHEM 208(5): 163-5 (1973)
CA 79: 010184

5-1793 PHYSICAL AND CHEMICAL CHARACTERIZATION OF
PLATINUM- RHODIUM GAUZE CATALYSTS. (AUGER
SPECTROSCOPY)
SCHMIDT LD + LUSS D
J CATAL 22(2): 269-79 (1971)

5-1794 ESCA (AND AUGER) STUDIES OF COPPER, COPPER(I)
OXIDE, AND COPPER(II) OXIDE.
SCHON G
SURFACE SCI 35(1): 96-108 (1973)
CA 78: 77848A

5-1795 ULTRAHIGH VACUUM TECHNIQUES IN THE
MEASUREMENT OF CONTACT ANGLES. PART-4: WATER
ON GRAPHITE (0001). (AES)
SCHRADER ME
J PHYS CHEM 79(23): 2508-15 (1975)
SAA 1976: 35673

5-1796 MOSSBAUER SPECTROSCOPY OF IRON-57 AND TIN-119
BY DETECTION OF CONVERSION AND AUGER
ELECTRONS. APPLICATION TO SURFACE STUDIES.
SCHUNCK JP + FRIEDT JM + LLABADOR Y
REV PHYS APPL 10(3): 121-6 (1975) (IN FRENCH)
CA 83: 35559

5-1797 DISTINCTION BETWEEN ADSORBED MONOLAYERS AND
THICKER LAYERS IN AUGER ELECTRON
SPECTROSCOPY.
SEAH MP
J PHYS F 3(8): 1538-47 (1973)
CA 79: 109354

5-1798 MATERIAL ANALYSIS OF THIN FOILS BY AUGER
ELECTRON SPECTROSCOPY.
SEILER H + WIEDMANN P
OPTIK 44(4): 505-7 (1976)
SAA 1976: 28553

5-1799 AUGER ELECTRON SPECTROSCOPY, SCANNING
ELECTRON MICROSCOPY AND NONDISPERSIVE X-RAY
ANALYSIS TECHNIQUES FOR EXAMINATION OF
SILVER- CADMIUM OXIDE CONTACT MATERIALS.
SELZER A + DIEBALL JW + LICHTMAN D
IEEE TRANS PARTS HYBRIDS PACKAG PHP11(2):
139-47 (1975); P224-37 OF HOLM SEMINAR ON
ELECTRICAL CONTACTS, 20TH PROC, ILL INST
TECHNOL, CHICAGO, 1974.
SAA 1975: 34062

5-1800 AES ANALYSIS OF OXIDE FILMS ON IRON.
SEO M + LUMSDEN JB + STAEHLE RW
SURFACE SCI 50(2): 541-52 (1975)
CA 83: 139379

5-1801 CHEMISORPTION, PHOTODESORPTION AND
CONDUCTIVITY MEASUREMENTS ON ZINC OXIDE
SURFACES. (AUGER ANALYSIS)
SHAPIRA Y + COX SM + LICHTMAN D
SURFACE SCI 54(1): 43-59 (1976)
SAA 1976: 26352

5-1802 QUANTITATIVE AUGER SPECTROSCOPIC ANALYSIS OF
SEGREGATION OF PHOSPHORUS IN IRON.
SHELL CA + RIVIERE JC
SURFACE SCI 40(1): 149-56 (1973)
CA 79: 118356

5-1803 EQUILIBRIUM SEGREGATION OF CARBON TO A NICKEL
(111) SURFACE: A SURFACE PHASE TRANSITION.
(AUGER SPECTROSCOPY)
SHELTON JC + PATIL HR + BLAKELY JM
SURFACE SCI 43(2): 493-520 (JUN 1974)
SAA 1974: 56948

5-1804 DETECTION OF SURFACE NITROGEN IN HIGH
CRITICAL TEMPERATURE NIOBIUM(3) GERMANIUM
FILMS BY AUGER ANALYSIS.
SHEN LYL
J APPL PHYS 46(8): 3614-18 (1975)
CA 83: 187815

5-1805 OBSERVATION OF SUPERSTRUCTURES ON CARBON
COVERED GERMANIUM (100) SURFACE BY HIGH
ENERGY ELECTRON DIFFRACTION.
SHEN LYL
SURFACE SCI 47(2): 685-91 (1975)
SAA 1975: 37485

5-1806 FINAL STATE EFFECTS IN THE PHOTOEMISSION CORE
LEVEL AND AUGER SPECTRA OF NICKEL AND NICKEL
ALLOYED WITH ZINC.
SHEVCHIK NJ + PENCHINA CM
J PHYS F 5(10): 2008-15 (1975)
CA 84: 10629

5-1807 QUANTITATIVE AUGER ANALYSIS OF COPPER- NICKEL
ALLOY SURFACES AFTER ARGON ION BOMBARDMENT.
SHIMIZU H + ONO M + NAKAYAMA K
SURFACE SCI 36(3): 817-21 (1973)
CA 78: 140707

5-1808 EFFECT OF TARGET TEMPERATURE ON SURFACE
COMPOSITION CHANGES OF COPPER- NICKEL ALLOYS
DURING ARGON BOMBARDMENT.
SHIMIZU H + ONO M + NAKAYAMA K
J APPL PHYS 46: 460-2 (1975)

5-1809 SURVEYING SILICON SURFACES BY MEANS OF THE
LEED- AES METHOD.
SHIMOJO T + MORITA I
OYO BUTURI 40(9): 1029-32 (1971) (IN
JAPANESE)
SAA 75: 39470

5-1810 BEAM ANALYSIS METHOD OF FILM AND SURFACE
COMPOSITION DETERMINATION.
SHINODA G
OYO BUTSURI 43(4): 310-23 (1974) (IN
JAPANESE)
CA 81: 098847

5-1811 X-RAY PHOTOEMISSION AND SURFACE STRUCTURE.
(AUGER DISPLACEMENTS)
SHIRLEY DA
J VACUUM SCI TECHNOL 12(1): 280-5 (1975)
SAA 1975: 66520

5-1812 AUGER ELECTRON SPECTROSCOPY APPLIED TO
SURFACE COMPOSITION PROBLEMS.
SICKAFUS EN
J METALS 21(3): A72 (1969)

5-1813 SULFUR AND CARBON ON (110) SURFACE OF NICKEL.
SICKAFUS EN
SURFACE SCI 19: 181-97 (1970)

5-1814 LEED AND AUGER ELECTRON SPECTROSCOPY STUDY OF
NICKEL (110) SURFACE EFFECTS DUE TO CARBON
AND SULFUR.
SICKAFUS EN
BULL AMER PHYS SOC 14: 793 (1969)

5-1815 SURFACE CHARACTERIZATION BY AUGER ELECTRON
SPECTROMETRY.
SICKAFUS EN
J VACUUM SCI TECHNOL 11(1): 299-311 (1974)
CA 81: 006404

5-1816 AUGER LINE SHAPE COMPARISON OF NITROGEN AND
SULFUR IN TWO DIFFERENT CHEMICAL
ENVIRONMENTS.
SICKAFUS EN + STEINRISSER F
J VACUUM SCI TECHNOL 10(1): 43-6 (1973)
CA 78: 102318

5-1817 SURFACE COMPOSITION OF PROMOTED IRON
CATALYSTS BY AUGER ELECTRON SPECTROSCOPY.
SILVERMAN DC
STANFORD UNIV PHD THESIS, 1976, 228P.
CA 85: 52251

5-1818 AUGER ELECTRON SPECTROSCOPY AND INELASTIC
ELECTRON SCATTERING IN THE STUDIES OF METAL
SURFACES.
SIMMONS GW
J COLLOID INTERFACE SCI 34: 343-56 (1970)

5-1819 ORDER- DISORDER PHENOMENA AT THE SURFACE OF
ALPHA TITANIUM- OXYGEN SOLID SOLUTIONS.
SIMMONS GW + SCHEIBNER EJ
J MATER 5: 933-49 (1970)

5-1820 BINDING STATES OF WATER VAPOR ON CLEAVED
GERMANIUM. (LEED, AES)
SINHAROY S + HENZLER M
SURFACE SCI 51(1): 75-88 (1975)
SAA 1975: 81314

5-1821 ADSORPTION KINETICS OF CESIUM ON GALLIUM
ARSENIDE. (USING AES AND X-RAY PHOTOELECTRON
SPECTROSCOPY)
SMITH DL + HUCHITAL DA
J APPL PHYS 43: 2624-8 (1972)

5-1822 SURFACE ANALYSIS BY LOW ENERGY ION
REFLECTION.
SMITH DP
APPL SPECTROSC 25: 147 (1971)

5-1823 ANALYSIS OF SURFACE COMPOSITION WITH LOW
ENERGY BACKSCATTERED IONS.
SMITH DP
SURFACE SCI 25: 171-91 (1971)

5-1824 SURFACE CONCENTRATION OF MAGNESIUM IN 6061
ALUMINUM BY AUGER ELECTRON SPECTROSCOPY.
SMITH JA + MARTINSEN WE
BULL AMER CERAM SOC 52(11): 855-6 (1973)
CA 80: 062776

5-1825 AUGER ELECTRON SPECTROSCOPY IN SPUTTERING
MEASUREMENTS. APPLICATION TO LOW ENERGY
ARGON(+) ION SPUTTERING OF SILVER AND
NIOBIUM.
SMITH JN + MEYER CH + LAYTON JK
J APPL PHYS 46(10): 4291-3 (1975)
CA 84: 52768

5-1826 SPUTTERING MEASUREMENTS ON CONTROLLED
THERMONUCLEAR REACTOR MATERIALS USING AUGER
ELECTRON SPECTROSCOPY.
SMITH JN + MEYER CH + LAYTON JK
NUCL TECHNOL 29(3): 318-21 (1976)
CA 85: 38326

5-1827 AUGER ELECTRON SPECTROSCOPY DETERMINATION OF
THE OXYGEN- SILICON RATIO IN SPIN ON GLASS
FILMS.
SMITH JN + THOMAS S + RITCHIE K
J ELECTROCHEM SOC 121(6): 827-9 (1974)
CA 81: 095466

5-1828 SPUTTER CLEANING AND ETCHING OF CRYSTAL
SURFACES (TITANIUM, TUNGSTEN, SILICON)
MONITORED BY AUGER SPECTROSCOPY, ELLIPSOMETRY
AND WORK FUNCTION CHANGE.
SMITH T
SURFACE SCI 27: 45-59 (1971)

5-1829 ELLIPSOMETRY, AUGER SPECTROSCOPY, AND WORK
FUNCTION OF CESIUM ON TUNGSTEN AND TITANIUM
SINGLE CRYSTALS IN THE SUBMONOLAYER REGION.
SMITH T
P427-33 OF IEEE CONF REC THERMION CONVERS
SPEC CONF, 9TH PROC, IEEE, 1970.
CA 76: 38744C

5-1830 CESIUM ADSORPTION ON TUNGSTEN: ELLIPSOMETRY,
AUGER SPECTROSCOPY AND SURFACE POTENTIAL
DIFFERENCE STUDIES.
SMITH T
J APPL PHYS 43: 2964-70 (JUL 1972)

5-1831 CHLORINE ADSORPTION ON CLEAN ALUMINUM. (AUGER
SPECTOSCOPY)
SMITH T
SURFACE SCI 32: 527-42 (1972)

5-1832 MARKER EXPERIMENTS WITH AUGER SPECTROSCOPY
AND ELLIPSOMETRY.
SMITH T
SURFACE SCI 36(1): 373-6 (1973)
CA 78: 102428

5-1833 CHLORINE ADSORPTION AT A TITANIUM SURFACE.
SEGREGATION FROM THE BULK CRYSTAL INTERACTION
WITH CHLORINE GAS. (AUGER SPECTROSCOPY,
ELLIPSOMETRY)
SMITH T
J ELECTROCHEM SOC 119(10): 1398-406 (OCT
1972)
SAA 1973: 01882

5-1834 CHANGES IN THE TITANIUM DIOXIDE AUGER AND
APPEARANCE POTENTIAL SPECTRA INDUCED BY
INCIDENT ELECTRON BEAM INTERACTION.
SOLOMON JS + CHAMBERLAIN MB + BAUN WL
ABSTR 46A OF ANNU CONF MICROBEAM ANAL SOC,
9TH PROC SUMMARIES, LEHIGH UNIV, 1974.
SAA 1974: 77309

5-1835 LEED AND AES STUDIES OF THE STRUCTURE OF
ADSORBED GASES ON SOLID SURFACES.
SOMORJAI GA
SURFACE SCI 34(1): 156-76 (1973)

5-1836 STUDIES OF CRYSTAL SURFACES BY LOW ENERGY
ELECTRON DIFFRACTION AND AUGER ELECTRON
SPECTROSCOPY.
SOMORJAI GA
ACTA CRYSTALLOGR A 28 (PART-4) SUPPL: S149
(15 JUL 1972)
SAA 1973: 18008

5-1837 SURFACE STRUCTURE OF AND CATALYSIS BY
PLATINUM SINGLE CRYSTAL SURFACES.
SOMORJAI GA
P87-120 OF CATALYSIS REVIEW, VOL 7, HEINEMANN
H (ED), MARCEL DEKKER, 1973.

5-1838 REACTIVITY OF LOW INDEX ((111) AND (100)) AND
STEPPED PLATINUM SINGLE CRYSTAL SURFACES.
(LEED, AUGER ELECTRON SPECTROSCOPY)
SOMORJAI GA + JOYNER RW + LANG B
PROC ROY SOC A 331(1586): 335-46 (1972)
SAA 1973: 11787

5-1839 AUGER ELECTRON SPECTROSCOPY OF (SILVER- GOLD
AND INDIUM- LEAD) ALLOY SURFACES.
SOMORJAI GA + OVERBURY SH
FARADAY DISCUSS CHEM SOC 60: 279-90 (1976)
CA 84: 171666

5-1840 AUGER ELECTRON SPECTROSCOPY ON SURFACES.
SOMORJAI GA + SZALKOWSKI FJ
P137-60 OF ADVAN HIGH TEMP CHEM, EYRING
L(ED), ACADEMIC PRESS, 1971.
CA 76: 159338G

5-1841 AUGER ELECTRON SPECTROSCOPY STUDY OF CATHODE
SURFACES DURING ACTIVATION AND POISONING.
PART-1: BARIUM ON OXYGEN ON TUNGSTEN
DISPENSER CATHODE.
SPRINGER RW + HAAS TW
J APPL PHYS 45(12): 5260-3 (1974)
CA 82: 66939

5-1842 DEPTH ANALYSIS OF CLEAVED MICA SURFACES
MONITORED BY AUGER SPECTROSCOPY.
STAIB P
RADIAT EFF 18(3-4): 217-19 (1973)
CA 79: 109055

5-1843 ABSOLUTE ATOMIC DENSITIES DETERMINED BY AUGER
ELECTRON SPECTROSCOPY.
STAIB P + KIRSCHNER J
APPL PHYS 3(5): 421-7 (1974)
CA 81: 113380

5-1844 ATTENUATION OF LOW ENERGY ELECTRONS FROM
SOLIDS: RESULTS FROM X-RAY PHOTOELECTRON (AND
AUGER) SPECTROSCOPY.
STEINHARDT RG + HUDIS J + PERLMAN ML
PHYS REV B 5: 1016-20 (1972)

5-1845 AUGER ELECTRON SPECTROSCOPY AND LOW ENERGY
ELECTRON DIFFRACTION STUDY OF THE ACTIVATION
OF GALLIUM ARSENIDE- CESIUM- OXYGEN NEGATIVE
ELECTRON AFFINITY SURFACES.
STOCKER BJ
SURFACE SCI 47(2): 501-13 (1975)
CA 82: 163996

5-1846 DETERMINATION OF THE SURFACE COMPOSITION OF
PALLADIUM- NICKEL ALLOY FILM CATALYSTS USING
AUGER ELECTRON SPECTROSCOPY.
STODDART CTH + MOSS RL + POPE D
SURFACE SCI 53(1): 241-56 (1975)
CA 84: 188964

5-1847 AES CHARACTERIZATION OF OXIDIZED FILMS OF
MAGNESIUM, ALUMINUM, AND SILICON.
STRAUSSER YE + JOHANNESSEN JS
J VACUUM SCI TECHNOL 13(1): 48-9 (1976)
SAA 1976: 49474

5-1848 SOME PROPERTIES OF TRANSITION METAL
SUPERCONDUCTORS AT SURFACES AND IN THIN
FILMS. (LEED- AES STUDIES OF NIOBIUM
SURFACES)
STRONGIN M
P223-36 OF SUPERCONDUCTIVITY IN D- AND F-BAND
METALS, DOUGLASS DH (ED), METALS, 1971, AMER
INST PHYS, 1972. (AIP CONF PROC NO 4)
SAA 1972: 32943

5-1849 DIFFERENTIAL AUGER SPECTROMETRY. (NIOBIUM-
ZIRCONIUM ALLOY ANALYSIS)
STRONGIN M + VARMA MN + JOSHI A
US PAT 519,324, 30 OCT 1974, 20P.
CA 84: 53576

5-1850 AUGER SPECTROSCOPY OF SILICONES.
STUPIAN GW
J APPL PHYS 45(12): 5278-82 (1974)
CA 82: 105010

5-1851 RELATIONSHIPS BETWEEN SUBSTRATE SURFACE
CHEMISTRY AND ADHESION OF THIN FILMS. (AUGER
ELECTRON SPECTROSCOPY OF ALUMINA SUBSTRATES)
SUNDAHL RC
J VACUUM SCI TECHNOL 9: 181-5 (1972)

5-1852 CRYSTALLOGRAPHY AND CHEMISTRY OF CERAMIC
SURFACES. (AUGER SPECTROSCOPY, LEED, ION
MICROPROBE ANALYSIS)
SUNDAHL RC
BULL AMER CERAM SOC 52(4): 437 (APR 1973)
SAA 1974: 01614

5-1853 ANALYSIS AND CHARACTERIZATION OF CERAMIC
SURFACES FOR ELECTRONIC APPLICATIONS.
SUNDAHL RC + BERRIN L
P270-89 OF SCI CERAMIC MACHINING AND SURF
FINISHING, SCHNEIDER SJ + RICE RW (EDS), NBS
SPEC PUB 348, 1972.

5-1854 LOW ENERGY ELECTRON DIFFRACTION AND AUGER
INVESTIGATION OF THE (100) SURFACE OF
COPPER(3) GOLD. TEMPERATURE DEPENDENCE OF
LONG RANGE ORDER AT A CRYSTAL SURFACE.
SUNDARAM VS
YALE UNIV THESIS, 1973, 123P.
CA 80: 075198

5-1855 ORDER- DISORDER TRANSFORMATION AT A (100)
SURFACE OF COPPER(3) GOLD. THEORY,
EXPERIMENT.
SUNDARAM VS + ALBEN RS + ROBERTSON WD
SURFACE SCI 46: 653-71 (1974)

5-1856 AUGER ELECTRON SPECTROSCOPY AND LOW ENERGY
ELECTRON DIFFRACTION STUDIES OF ADSORPTION
ISOTHERMS: XENON ON (0001) GRAPHITE.
SUZANNE J + COULOMB JP + BIENFAIT M
SURFACE SCI 40(2): 414-18 (1973)
CA 80: 007358

5-1857 FIRST ORDER TWO DIMENSIONAL TRANSITION: XENON
ADSORBED ON THE (0001) FACE OF GRAPHITE.
(AUGER ELECTRON SPECTROSCOPY)
SUZANNE J + COULOMB JP + BIENFAIT M
SURFACE SCI 44(1): 141-56 (JUL 1974) (IN
FRENCH)
SAA 1974: 63981

5-1858 TWO-DIMENSIONAL PHASE TRANSITION IN XENON
SUBMONOLAYER FILMS ADSORBED ON (0001)
GRAPHITE. (AUGER SPECTROSCOPY, LOW ENERGY
ELECTRON DIFFRACTION)
SUZANNE J + COULOMB JP + BIENFAIT M
SURFACE SCI 47(1): 204-6 (1975)
SAA 1975: 46472

5-1859 THERMODYNAMICS AND KINETICS OF THE FIRST
MONOLAYER ADSORPTION OF XENON ON THE (0001)
GRAPHITE FACE.
SUZANNE J + COULOMB JP + BIENFAIT M
J CRYST GROW 31(1): 87-91 (1975)
SAA 1976: 22101

5-1860 ENTROPY OF ADSORPTION OF XENON IN EPITAXY ON
THE (0001) FACE OF GRAPHITE. (AUGER
SPECTROSCOPY)
SUZANNE J + MASRI P + BIENFAIT M
SURFACE SCI 43(2): 441-8 (JUN 1974)
SAA 1974: 56977

5-1861 CORRELATIONS BETWEEN INTERFACIAL OXIDES AND
RESISTANCE CHANGES FOR TUNGSTEN CONTACTS ON
HEAVILY DOPED SILICON- GERMANIUM. (AUGER
SPECTROSCOPY)
SWEET JN + HOLLOWAY PH
THIN SOLID FILMS 27(2): 263-71 (1975)
SAA 1975: 58418, SAB 1975: 28724

5-1862 AUGER ELECTRON SPECTROSCOPY ANALYSIS OF
VANADIUM AND VANADIUM COMPOUND SURFACES.
SZALKOWSKI FJ
UNIV CALIF, BERKELEY LAB 1973, 256P. REP
LBL-888
CA 81: 007992

5-1863 USE OF X-RAY PHOTOELECTRON AND AUGER
SPECTROSCOPY TO DETERMINE THE SURFACE
COMPOSITION OF OXIDIZED VANADIUM. (COMMENTS)
SZALKOWSKI FJ + SOMORJAI GA
SURFACE SCI 52(2): 431-3 (1975)
SAA 1976: 13863

5-1864 CHARACTERIZATION OF SOME VANADIUM (100)
SURFACE STRUCTURES USING LEED AND AES.
SZALKOWSKI FJ + SOMORJAI GA
J CHEM PHYS 64(7): 2985-9 (1976)
CA 84: 156121

5-1865 LEED- AES STUDIES OF CHEMISORPTION INDUCED
SULFUR SEGREGATION FROM THE BULK TO THE
SURFACE OF PALLADIUM (100).
SZYMERSKA I + LIPSKI M
J CATAL 41(2): 197-201 (1976)
CA 84: 156122

5-1866 LOW ENERGY ELECTRON DIFFRACTION- AUGER
ELECTRON SPECTROSCOPY STUDY OF IRON SINGLE
CRYSTAL SURFACES.
TAGA Y + ISOGAI A + NAKAJIMA K
J JAP INST MET 37(12): 1300-4 (DEC 1973) (IN
JAPANESE)
SAA 1974: 40945

5-1867 CONCENTRATION CHANGE OF TIN IN THE SURFACE
LAYER OF COPPER- TIN ALLOY SYSTEM DUE TO
FRICTION. (AUGER SPECTRCSCOPY)
TAGA Y + ISOGAI A + NAKAJIMA K
J JAP INST MET 38(8): 716-19 (AUG 1974) (IN
JAPANESE)
SAA 1975: 09905

5-1868 EFFECT OF PHOSPHORUS ON THE FRICTION AND WEAR
CHARACTERISTICS OF COPPER- TIN- PHOSPHORUS
ALLOYS. (AUGER SPECTROSCOPY)
TAGA Y + NAKAJIMA K
J JAP INST MET 40(4): 382-6 (1976) (IN
JAPANESE)
SAA 1976: 54610

5-1869 DIRECT COMPARISON OF LOW ENERGY ION
BACKSCATTERING WITH AUGER ELECTRON
SPECTROSCOPY IN THE ANALYSIS OF SULFUR
ABSORBED ON NICKEL.
TAGLAUER E + HEILAND W
APPL PHYS LETT 24(9): 437-9 (1974)
CA 81: 009375

5-1870 LOW ENERGY ION SCATTERING AND AUGER ELECTRON
SPECTROSCOPY STUDIES OF CLEAN NICKEL SURFACES
AND ADSORBED LAYERS.
TAGLAUER E + HEILAND W
SURFACE SCI 47(1): 234-43 (1975)
CA 83: 21620

5-1871 SURFACE ANALYSIS WITH LOW ENERGY ION
SCATTERING.
TAGLAUER E + HEILAND W
APPL PHYS 9(4): 261-75 (1976)
SAA 1976: 43094

5-1872 (AUGER) MICROSCOPICAL STUDY OF FRICTIONAL
PROPERTIES OF MOLYBDENUM DISULFIDE.
TAKAHASHI N + OKADA K
WEAR 33(1): 153-67 (1975)
SAA 1975: 76753

5-1873 WORK FUNCTION OF WELL-DEFINED SURFACE OF
COPPER- NICKEL ALLOY PLATES. (AUGER ELECTRON
SPECTROSCOPY)
TAKASU Y + KONNO H + YAMASHINA T
SURFACE SCI 45(1): 321-4 (SEP 1974)
SAA 1975: 03987

5-1874 ANALYSIS OF THE SURFACE COMPOSITION OF
PASSIVE FILMS CN COPPER- NICKEL ALLOYS BY
AUGER ELECTRON SPECTROSCOPY AND X-RAY
PHOTOELECTRON SPECTROSCOPY.
TAKASU Y + MARU S + MATSUDA Y + SHIMIZU H
BULL CHEM SOC JAP 48(1): 219-21 (1975)
CA 82: 117789

5-1875 RELATION BETWEEN CATALYTIC ACTIVITY PATTERN
AND SURFACE COMPOSITION OF COPPER- NICKEL
ALLOY. APPLICATION OF AUGER ELECTRON
SPECTROSCOPY TO THE STUDY ON CATALYSTS.
TAKASU Y + SHIMIZU H
J CATAL 29(3): 479-85 (1973)
CA 79: 046146

5-1876 GROWTH OF GOLD FILM ON GALLIUM ARSENIDE (110)
OBSERVED BY LCW ENERGY ELECTRON DIFFRACTION
AND AUGER ELECTRON SPECTROSCOPY.
TAKEDA K + HANAWA T + SHIMOJO T
P589-92 OF INT VACUUM CONGR, 6TH PRCC, 1974
(JAP J APPL PHYS SUPPL 2, PART 1, 1974)
CA 82: 178407

5-1877 ALLOY SPUTTERING STUDIES WITH IN-SITU AUGER
ELECTRON SPECTROSCOPY.
TARNG ML + WEHNER GK
J APPL PHYS 42: 2449-52 (1971); J VACUUM SCI
TECHNOL 8: 23 (1971)

5-1878 AUGER ELECTRON SPECTROSCOPY STUDIES OF
SPUTTER DEPOSITION AND SPUTTER REMOVAL OF
MOLYBDENUM FROM VARIOUS METAL SURFACES
(COPPER, GOLD, ALUMINUM, TUNGSTEN).
TARNG ML + WEHNER GK
J APPL PHYS 43: 2268-72 (1972); J VACUUM SCI
TECHNOL 9: 625 (1972)

5-1879 THIN REACTION LAYERS AND SURFACE STRUCTURE OF
SILICON (111).
TAYLOR NJ
SURFACE SCI 15: 169-74 (1969)

5-1880 AUGER SPECTROSCOPY OF SILICON. (REPLY TO
COMMENTS)
TAYLOR NJ
SURFACE SCI 17: 466-8 (1969)

5-1881 SOME PROPERTIES OF CLEAN AND ALKALI METAL
COVERED ZINC OXIDE SURFACES. (LEED, AES)
TAYLOR PA + LEYSEN R + HOPKINS BJ
SURFACE SCI 53(1): 500 (1975)
SAA 1976: 26315

5-1882 STABILITY OF CESIUM COVERED ZINC OXIDE
SURFACES.
TAYLOR PA + LEYSEN R + HOPKINS BJ
SOLID STATE COMMUN 17(8): 983-6 (1975)
SAA 1976: 09566

5-1883 THORIUM (100) CRYSTAL FACE. A STUDY OF ITS
CLEANING, SURFACE STRUCTURE, AND INTERACTION
WITH OXYGEN AND CARBON MONOXIDE.
TAYLOR TN + COLMENARES CA + SMITH RL
+ SOMORJAI GA
SURFACE SCI 54(2): 317-39 (1976)
SAA 1976: 35666

5-1884 CARBON MONOXIDE ADSORPTION ON NICKEL (110).
(AUGER EFFECT)
TAYLOR TN + ESTRUP PJ
J VACUUM SCI TECHNOL 10(1): 26-30 (JAN-FEB
1973)
SAA 1973: 38721

5-1885 VALENCE BAND OF MAGNESIUM STANNIDE DETERMINED
BY AUGER AND PHOTOEMISSION SPECTROSCOPY.
TEJEDA J + SHEVCHIK NJ + LANGER DW
+ CARDONA M
PHYS REV LETT 30(9): 370-3 (1973)
CA 78: 103526

5-1886 CARBIDE FORMATION ON NIOBIUM SURFACES DURING
HIGH TEMPERATURE PROTON IRRADIATION.
THOMAS GJ + BAUER W
J VACUUM SCI TECHNOL 12(1): 490-5 (1975)
SAA 1976: 09319

5-1887 DETECTABILITY LIMITS FOR BORON AND PHOSPHORUS
IN SILICON BY AUGER ELECTRON SPECTROSCOPY.
THOMAS JH + MORABITO JM
SURFACE SCI 40(2): 629-33 (1974)
CA 80: 090849

5-1888 AUGER SPECTROSCOPY OF SUBMONOLAYER GOLD
DEPOSITIONS ON SILICON.
THOMAS MT + STYRIS DL
PHYS STATUS SOLIDI B 57(1): K83-K85 (1973)
CA 79: 011574

5-1889 DIFFUSION MEASUREMENTS IN THIN FILMS
UTILIZING WORK FUNCTION CHANGES: CHROMIUM
INTO GOLD. (AUGER ELECTRON SPECTROSCOPY)
THOMAS RE + HAAS GA
J APPL PHYS 43(12): 4900-7 (DEC 1972)
SAA 1973: 15030

5-1890 AUGER EXAMINATION OF CONTAMINANTS IN THIN
FILM METALLIZATIONS.
THOMAS RE + HAAS GA
J APPL PHYS 46(2): 963-5 (1975)
CA 82: 143715

5-1891 SURFACE COMPOSITIONAL CHANGES WITH ELECTRON
BOMBARDMENT OBSERVED BY AUGER ELECTRON
SPECTROSCOPY.
THOMAS S
P139-41 OF SEMICONDUCTOR MEASUREMENT
TECHNOLOGY. PART-4: ARPA/NBS WORKSHOP ON
SURFACE ANALYSIS FOR SILICON DEVICES,
LIEBERMAN AG(ED), NAT BUR STAND SPEC PUB
400-23, 1976.

5-1892 SURFACE ENRICHMENT OF INDIUM IN EVAPORATED
GOLD- INDIUM FILMS. (AUGER ELECTRON
SPECTROSCOPY)
THOMAS S
APPL PHYS LETT 24(1): 1-3 (1 JAN 1974)
SAA 1974: 37125

5-1893 AES ION SPUTTERING ANALYSIS AND THE SURFACE
COMPOSITION OF TITANIUM(IV) OXIDE.
THOMAS S
SURFACE SCI 55(2): 754-8 (1976)
CA 85: 37483

5-1894 AUGER SPECTROSCOPY STUDY OF THE ADSORPTION OF
RUBIDIUM ON MOLYBDENUM (100).
THOMAS S + HAAS TW
SURFACE SCI 28: 632-6 (1971)

5-1895 QUANTITATIVE AUGER ELECTRON SPECTROSCOPY AND
LOW ENERGY ELECTRON DIFFRACTION STUDY OF
ALKALI METAL OVERLAYERS ON TUNGSTEN (100).
THOMAS S + HAAS TW
J VACUUM SCI TECHNOL 10(1): 218-22 (1973)
CA 78: 102324

5-1896 LEED- AUGER SPECTROSCOPY STUDY OF THE
ADSORPTION OF ALKALI METALS ON MOLYBDENUM
(110).
THOMAS S + HAAS TW
J VACUUM SCI TECHNOL 9(2): 840-3 (1972)

5-1897 ELECTRON SPECTROSCOPY OF METAL BLACKS.
THOMAS S + SULEMAN M + PATTINSON EB
J PHYS D 3: L77-80 (1970)

5-1898 AUGER SPECTROSCOPY ANALYSIS OF PALLADIUM
SILICIDE FILMS.
THOMAS S + TERRY LE
APPL PHYS LETT 26(8): 433-5 (1975)
CA 82: 177654

5-1899 AES STUDY OF THE SILICIDE FORMATION IN 85%
NICKEL- 15% PLATINUM ALLOY FILMS.
THOMAS S + TERRY LE
J VACUUM SCI TECHNOL 13(1): 156 (1976)
SAA 1976: 49185

5-1900 COMPOSITION PROFILES AND SCHOTTKY BARRIER
HEIGHTS OF SILICIDES FORMED IN NICKEL-
PLATINUM ALLOY FILMS.
THOMAS S + TERRY LE
J APPL PHYS 47(1): 301-7 (1976)
CA 84: 172774

5-1901 ELECTRON SPECTROSCOPY FROM SOLID SURFACES IN
ULTRAHIGH VACUUM. CURRENT PROGRESS WITH
TECHNIQUES INVOLVING ELECTRON, AUGER ELECTRON
SPECTROSCOPY, PHOTON, ELECTRON SPECTROSCOPY
FOR CHEMICAL ANALYSIS, AND FIELD STIMULATED
ELECTRON EMISSION.
TODD CJ
VACUUM 23(6): 195-203 (1973)
CA 80: 008497

5-1902 THERMALLY INDUCED PROCESSES AT GOLD- GALLIUM
ARSENIDE INTERFACES: AN ASSESSMENT BY
RUTHERFORD BACKSCATTERING, SCANNING ELECTRON
MICROSCOPY, AND DEPTH PROFILING AUGER
ELECTRON SPECTROSCOPY.
TODD CJ + ASHWELL GWB + SPEIGHT JD
+ HECKINGBOTTOM R
P171-83 OF CONF ON METAL-SEMICONDUCTOR
CONTACTS, INST PHYS, LONDON, 1974.
SAA 1975: 19093

5-1903 RELATIVE ESCAPE DEPTH OF AUGER AND
PHOTOELECTRONS FROM A SOLID SURFACE.
TODD CJ + HECKINGBOTTOM R
PHYS LETT A 42(6): 455-6 (1973)
CA 78: 089997

5-1904 AUGER ELECTRON SPECTROSCOPY AT HIGH SPATIAL
RESOLUTION AND PRIMARY BEAM CURRENTS. (CARBON
ON SILVER FILM)
TODD G + POPPA H + MOORHEAD D + BALES M
J VACUUM SCI TECHNOL 12(4): 953-5 (1975)
SAA 1976: 11929

5-1905 SCATTERING OF MOLECULAR BEAMS FROM SURFACES.
(AUGER SPECTROSCOPY)
TOENNIES JP
APPL PHYS 3(2): 91-144 (FEB 1974)
SAA 1974: 21116

5-1906 STRUCTURES OF SUBSTRATE CRYSTALS AND
EPITAXIAL GROWTH OF METAL ON THEM OBSERVED BY
LOW ENERGY ELECTRON DIFFRACTION- AUGER
ELECTRON SPECTROSCOPY AND MASS FILTER.
TOKUTAKA H
NIPPON KESSHO GAKKAISHI 16(1): 74-86 (1974)
(IN JAPANESE)
CA 80: 149446

5-1907 CRYSTAL SURFACES OF ALKALI HALIDES AND THIN
FILM GROWTH OBSERVED BY LEED AND AES.
TOKUTAKA H
NIPPON BUTSURI GAKKAISHI 30(7): 527-35 (1975)
(IN JAPANESE)
CA 83: 106291

5-1908 (100) SURFACES OF ALKALI HALIDES. PART-2:
ELECTRON STIMULATED DISSOCIATION.
TOKUTAKA H + PRUTTON M + HIGGINBOTHAM IG
+ GALLON TE
SURFACE SCI 21: 233-40 (1970)

5-1909 LOW INTRINSIC STRESS AND GOOD ADHESION ALLOY
THIN FILMS. (TITANIUM AND CHROMIUM ALLOYS,
AES)
TONG HC + LO CM + TRABER WF
J VACUUM SCI TECHNOL 13(1): 99-101 (1976)
SAA 1976: 53509

5-1910 SURFACE CHEMICAL ANALYSIS BY AUGER ELECTRON
SPECTROSCOPY AND APPEARANCE POTENTIAL
SPECTROSCOPY: A COMPARISON.
TRACY JC
APPL PHYS LETT 19: 353-6 (1971)

5-1911 STRUCTURAL INFLUENCES ON ADSORPTION ENERGY.
PART-2: CARBON MONOXIDE ON NICKEL (100).
PART-3: CARBON MONOXIDE ON COPPER (100).
(AUGER SPECTRUM OF NICKEL AND COPPER)
TRACY JC
J CHEM PHYS 56: 2736-47; 2748-54 (1972)

5-1912 THE KINETICS OF OXYGEN ADSORPTION ON THE
(112) AND (110) PLANES OF TUNGSTEN.
TRACY JC + BLAKELY JM
SURFACE SCI 15: 257-76 (1969)

5-1913 STRUCTURAL INFLUENCES ON ADSORBATE BINDING
ENERGY. PART-1: CARBON MONOXIDE ON (100)
PALLADIUM.
TRACY JC + PALMBERG PW
J CHEM PHYS 51: 4852-62 (1969)

5-1914 AUGER ELECTRON SPECTROSCOPY STUDIES OF
ARSENIC VACANCIES IN (-1-1-1) GALLIUM
ARSENIDE ANNEALED SURFACES.
TRUEBA A + MUNOZ E + PIQUERAS J
SOLID STATE COMMUN 15(2): 199-202 (1974)
CA 81: 112100

5-1915 SURFACE CARRIER CONCENTRATION AND AS
EVAPORATION IN HEAT TREATED GALLIUM ARSENIDE.
(AUGER SPECTROSCOPY)
TRUEBA A + MUNOZ E + PIQUERAS J
ELECTRON FIS APPL 17(1-2): 215 (1974)
SAA 1975: 06874

5-1916 IN-DEPTH PROFILE DETECTION LIMITS OF NITROGEN
IN GALLIUM PHOSPHIDE (AND) NITROGEN, OXYGEN,
AND FLUORINE IN SILICON BY SECONDARY ION MASS
SPECTROMETRY AND AUGER ELECTRON SPECTROSCOPY.
TSAI JC + MORABITO JM
P115-24 OF INT CONF ON ION IMPLANTATION IN
SEMICONDUCTORS, 4TH PROC, 1974, NAMBA S(ED),
PLENUM, 1975.
CA 84: 25438

5-1917 STRUCTURE AND GROWTH KINETICS OF NICKEL(2) SILICIDE ON SILICON. (AUGER SPECTROSCOPY)
TU KN + CHU WK + MAYER JW
THIN SOLID FILMS 25(2): 403-13 (1975)
  SAA 1975: 46520

5-1918 SEGREGATION OF CADMIUM TO A (111) SILVER CHLORIDE SURFACE. (AUGER SPECTROSCOPY)
TU Y + BLAKELY JM
PHOTOGR SCI ENG 20(2): 59-60 (1976)
  CA 84: 157993

5-1919 USE OF AUGER ELECTRON SPECTROSCOPY IN DETERMINING THE EFFECT OF CARBON AND OTHER SURFACE CONTAMINANTS ON GALLIUM ARSENIDE- CESIUM- OXYGEN PHOTOCATHODES.
UEBBING JJ
J APPL PHYS 41: 802-4 (1970)

5-1920 AUGER ELECTRON SPECTROSCOPY OF CONTAMINATED GALLIUM ARSENIDE SURFACES.
UEBBING JJ + JAMES LW
J VACUUM SCI TECHNOL 7: 81-3 (1970)

5-1921 AUGER ELECTRON SPECTROSCOPY OF CLEAN GALLIUM ARSENIDE.
UEBBING JJ + TAYLOR NJ
J APPL PHYS 41: 804-8 (1970)

5-1922 AUGER ELECTRON SPECTROSCOPY OF GALLIUM ARSENIDE PHOTOSURFACES.
UEBBING JJ + TAYLOR NJ
BULL AMER PHYS SOC 14: 792 (1969)

5-1923 IMPURITY CONCENTRATIONS ON METAL SURFACES BY AUGER ELECTRON SPECTROSCOPY.
UEDA K + SHIMIZU R
SURFACE SCI 36(3): 789-96 (1973)
  CA 78: 143498

5-1924 STUDIES OF INITIAL OXIDATION ON SILICON- IRON ALLOY (100) BY MEANS OF WORK FUNCTION AND AUGER ELECTRON SPECTROSCOPY.
UEDA K + SHIMIZU R
APPL PHYS LETT 22(8): 393-5 (1973)
  CA 78: 151905

5-1925 PHOTOELECTRIC WORK FUNCTION AND AUGER ELECTRON SPECTROSCOPY OF 5.59 ATOMIC PERCENT SILICON- IRON ALLOY (100) SINGLE CRYSTALS.
UEDA K + SHIMIZU R
PHYS STATUS SOLIDI A 18(1): 329-36 (1973)
  CA 79: 084660

5-1926 PHOTOELECTRIC WORK FUNCTION STUDY ON IRON (100) SURFACE COMBINED WITH AUGER ELECTRON SPECTROSCOPY.
UEDA K + SHIMIZU R
JAP J APPL PHYS 12(12): 1869-73 (1973)
  CA 80: 075521

5-1927 INITIAL OXIDATION STUDIES ON IRON (100) BY MEANS OF PHOTOELECTRIC WORK FUNCTION MEASUREMENTS COMBINED WITH AUGER ELECTRON SPECTROSCOPY.
UEDA K + SHIMIZU R
SURFACE SCI 43(1): 77-87 (1974)
  CA 80: 139204

5-1928 INITIAL OXIDATION STUDIES BY MEANS OF PHOTOELECTRIC WORK FUNCTION MEASUREMENTS AND AUGER ELECTRON SPECTROSCOPY ON 5.59% SILICON- IRON (100) SINGLE CRYSTALS.
UEDA K + SHIMIZU R + HASHIMOTO H
TECHNOL REP OSAKA UNIV 25(1230-1253): 35-42 (1975)
  CA 83: 150812

5-1929 ADSORPTION OF NITRIC OXIDE ON TUNGSTEN.
USAMI S + NAKAGIMA T
JAP J APPL PHYS SUPPL 2 PART-2: 237-40 (1974)
  SAA 1975: 42081

5-1930 LOW ENERGY ELECTRON DIFFRACTION AND AUGER ELECTRON SPECTROSCOPY STUDY OF PHOSPHINE ADSORPTION ON CLEAN SILICON (111).
VAN BOMMEL AJ + CROMBEEN JE
SURFACE SCI 36(3): 773-7 (1973)
  CA 78: 140705

5-1931 LOW ENERGY ELECTRON DIFFRACTION, AUGER ELECTRON SPECTROSCOPY, AND PHOTOEMISSION STUDIES OF THE CESIUM COVERED GALLIUM ARSENIDE (110) SURFACE.
VAN BOMMEL AJ + CROMBEEN JE
SURFACE SCI 45(1): 308-13 (1974)
  CA 81: 160981

5-1932 LOW ENERGY ELECTRON DIFFRACTION AND AUGER ELECTRON OBSERVATIONS OF THE SILICON CARBIDE (0001) SURFACE.
VAN BOMMEL AJ + CROMBEEN JE
SURFACE SCI 48(2): 463-72 (1975)
  CA 82: 160683

5-1933 APPLICATION OF AES TO THE STUDY OF SELECTIVE SPUTTERING OF THIN FILMS.
VAN OOSTROM A
J VACUUM SCI TECHNOL 13(1): 224-7 (1976)
  CA 84: 143494

5-1934 SURFACE ENRICHMENT IN COPPER(3) GOLD AND GOLD(3) COPPER ALLOYS.
VAN SANTEN RA + TONEMAN LH + BONSAAN R
SURFACE SCI 47: 64-76 (1975)

5-1935 SOME RECENT STUDIES OF VERY THIN SILICON DIOXIDE FILMS.
VAN DER MEULEN YJ
J VACUUM SCI TECHNOL 11(6): 985-89 (1974)
  SAA 1975: 46790, SAB 1975: 24258

5-1936 INVESTIGATION OF VACUUM BREAKDOWN USING AUGER SPECTROSCOPY.
VAN OOSTROM A
P795-801 OF INT VACUUM CONGRESS, 6TH PROC, INT CONF ON SOLID SURFACES, 2ND, 1974, KYOTO. (JAP J APPL PHYS, SUPPL 2 PART 1, 1974)
  CA 83: 20210

5-1937 DIFFERENCE AUGER SPECTROSCOPY FOR STUDYING SMALL QUANTITIES OF ELEMENTS ON METALLIC SURFACES.
VARMA MN + JOSHI A + STRONGIN M + RADEKA V
REV SCI INSTRUM 44(11): 1643-5 (1973)
  CA 80: 077876

5-1938 EFFECT OF HIGH TEMPERATURE HEAT TREATMENT ON PENETRATION DEPTH OF SUPERCONDUCTING NIOBIUM. (AUGER SPECTROSCOPIC MEASUREMENTS)
VARMAZIS C + JOSHI A + LUHMAN T + STRONGIN M
APPL PHYS LETT 24(8): 394-5 (15 APR 1974)
  SAA 1974: 57211

5-1939 AUGER SPECTROSCOPIC STUDY OF THE SURFACE OF IRON AFTER IONIC BOMBARDMENT AND AIR OXIDATION.
VASILEV MA + BURMAKE LS + IVASHCHENKO YN + KOSYACHKOV AA + CHEREPIN VT
P134-5 OF VZAIMODEISTVIE ATOM CHASTITS S TVERD TELOM, 1974. (IN RUSSIAN)
  CA 83: 196900

5-1940 CHEMICAL AND STRUCTURAL ANALYSIS OF IRON SURFACES BY MEANS OF AUGER ELECTRONIC SPECTROSCOPY AND LOW ENERGY ELECTRON DIFFRACTION.
VIEFHAUS H + TAUBER G + GRABKE HJ
MIKROCHIM ACTA 6: 383-90 (1975) (IN GERMAN)
  CA 84: 80195

5-1941 AUGER EMISSION SPECTROSCOPY VANADIUM PENTOXIDE (010) AND VANADIUM (100) SURFACES.
VIERMANS L + VENNIK J
SURFACE SCI 24: 541-54 (1971)

5-1942 SURFACE STUDIES FOR QUARTZ RESONATORS. (CLEANING, AUGER SPECTROSCOPY)
VIG JR + COOK CF + SCHWIDTAL K + LEBUS JW + HAFNER E
P96-108 OF SYMP ON FREQUENCY CONTROL, 28TH PROC, ARMY ELECTRON COMMAND, 1974.

5-1943 SURFACE PREPARATION AND CHARACTERIZATION TECHNIQUES (AUGER, CLEANING) FOR QUARTZ RESONATORS.
VIG JR + WASSHAUSEN H + COOK CF + KATZ M + HAFNER E
P98-112 OF SYMP ON FREQUENCY CONTROL, 27TH PROC, ARMY ELECTRON COMMAND, 1973.

5-1944 OXIDATION OF CARBON ON TUNGSTEN (100) AND
MOLYBDENUM (100). (AUGER ELECTRON
SPECTROSCOPY)
VISWANATH Y + SCHMIDT LD
J CHEM PHYS 59(8): 4184-91 (15 OCT 1973)
SAA 1974: 20480

5-1945 AUGER ELECTRON SPECTROSCOPY OF SILICON
SURFACES.
VLACHOVA B
CZECH J PHYS 23(9): 931-46 (1973)
CA 79: 139978

5-1946 AUGER SPECTROSCOPY OF SILICON.
VLACHOVA B
P2 OF CZECH CONF ON ELECTRONICS AND VACUUM
PHYSICS, 5TH PROC, EXTENDED ABSTR, 1972.
SAA 1973: 15062

5-1947 UPS, ESCA, AND AUGER SPECTROSCOPY STUDY OF
VACUUM DEPOSITED SILVER HALIDE FILMS.
VON BACHO PS + CEASAR GP + SALTSBURG HM
+ BAETZOLD RC
J VACUUM SCI TECHNOL 13(1): 107-11 (1976)
CA 84: 157742

5-1948 AUGER PARAMETER IN ELECTRON SPECTROSCOPY FOR
THE IDENTIFICATION OF CHEMICAL SPECIES.
WAGNER CD
ANAL CHEM 47(7): 1201-3 (1975)
CA 83: 37097

5-1949 X-RAY EXCITED AUGER AND PHOTOELECTRON SPECTRA
OF PARTIALLY OXIDIZED MAGNESIUM SURFACES.
ABNORMAL CHEMICAL SHIFTS.
WAGNER CD + BILCEN P
SURFACE SCI 35(1): 82-95 (1973)
CA 78: 77583K

5-1950 DEPLETION EFFECTS IN SEMIINSULATING GALLIUM
ARSENIDE. (AUGER SPECTROSCOPY)
WALDROP JR + ZUCCA R + WEN CP
APPL PHYS LETT 26(6): 322-4 (1975)
CA 82: 164087

5-1951 COMBINED FIELD ION MICROSCOPY. AUGER ELECTRON
SPECTROSCOPY, AND LEED STUDY OF THE STRUCTURE
AND COMPOSITION OF ION BOMBARDED TUNGSTEN
SURFACES.
WALLS JM + MARTIN AD + SOUTHWORTH HN
SURFACE SCI 50(2): 360-78 (1975)
CA 83: 140002

5-1952 SOFT X-RAY APPEARANCE POTENTIAL SPECTROSCOPY
STUDIES WITH COPPER- NICKEL AND IRON- NICKEL
ALLOYS. (AES)
WANDELT K + ERTL G
BER BUNSENGES PHYSIK CHEM 79(11): 1101-7
(1975)
SAA 1976: 40988

5-1953 ELECTRON SPECTROSCOPIC STUDIES OF THE
OXIDATION OF NICKEL- PALLADIUM ALLOYS.
WANDELT K + ERTL G
Z NATURFORSCH A 31(2): 205-10 (1976)
SAA 1976: 40331

5-1954 AUGER ELECTRON SPECTROSCOPY STUDY OF THE
ELECTRON STIMULATED DESORPTION OF OXYGEN FROM
A TUNGSTEN (100) SURFACE.
WATTS GD + JONES AR + HOPKINS BJ
SURFACE SCI 45(2): 705-12 (1974)
CA 81: 176715

5-1955 APPLICATION OF AUGER ELECTRON SPECTROSCOPY TO
THE PLATINUM- OXYGEN SYSTEM AT HIGH
TEMPERATURE.
WEBER B + FUSY J + CASSUTO A
J CHIM PHYS, PHYS CHIM BIOL 7(11-12): 1551-3
(1974) (IN FRENCH)
CA 82: 103633

5-1956 DETERMINATION OF SURFACE STRUCTURES USING
LEED AND ENERGY ANALYSIS OF SCATTERED
ELECTRONS.
WEBER RE + JOHNSON AL
J APPL PHYS 40: 314-8 (1969)

5-1957 DETERMINATION OF SURFACE STRUCTURES BY LEED
AND AUGER ELECTRON SPECTROSCOPY.
WEBER RE + JOHNSON AL
BULL AMER PHYS SOC 13: 945 (1968)

5-1958 WORK FUNCTION AND STRUCTURAL STUDIES OF
ALKALI- COVERED SEMICONDUCTORS.
WEBER RE + PERIA WT
SURFACE SCI 14: 13-38 (1969)

5-1959 USE OF LEED APPARATUS FOR THE DETECTION AND
IDENTIFICATION OF SURFACE CONTAMINANTS.
WEBER RE + PERIA WT
J APPL PHYS 38: 4355-8 (1967)

5-1960 AUGER THIN FILM ANALYSIS AS APPLIED TO
VARIOUS ASPECTS IN SOLAR ENERGY CONVERSION.
WEHNER GK
P135-44 OF SYMP MATER SCI ASPECTS THIN FILM
SYST SOLAR ENERGY CONVERS, NTIS, 1974.
PB-239270
CA 83: 181998

5-1961 HIGH SPATIAL RESOLUTION SCANNING AUGER
SPECTROSCOPY APPLIED TO ANALYSIS OF X-BAND
DIODE BURNOUT.
WEISENBERGER WH + CHRISTOU A + ANAND Y
J VACUUM SCI TECHNOL 12(6): 1365-8 (1975)
CA 84: 68559

5-1962 DEPOSITION AND AUGER ANALYSIS OF DEPOSITED
SILICON DIOXIDE ON ALUMINUM(X) GALLIUM(1-X)
ARSENIDE.
WEISSMAN I + ANTYPAS G + GERLACH RL
+ HOFFMAN V + TAYLOR NJ
VACUUM 24(2): 81-8 (1974)
CA 81: 018273

5-1963 STUDIES OF GAS- SURFACE INTERACTIONS AT THE
(100) CRYSTAL FACE OF PLATINUM BY MOLECULAR
BEAM SCATTERING, LOW ENERGY ELECTRON
DIFFRACTION, AND AUGER ELECTRON SPECTROSCOPY.
WEST LA
UNIV CALIF, LAWRENCE BERKELEY LAB, NOV 1971,
277P. LBL-107

5-1964 RELATIVE SPUTTERING YIELDS AND QUANTITATIVE
SURFACE ANALYSIS BY AUGER SPECTROSCOPY.
WEST LA
J VACUUM SCI TECHNOL 13(1): 198-203 (1976)
SAA 1976: 49472

5-1965 POROSITY IN SPUTTERED PLATINUM FILMS. (AUGER
ELECTRON SPECTROSCOPY)
WESTWOOD WD
J VACUUM SCI TECHNOL 11(1): 466-71 (JAN-FEB
1974)
SAA 1974: 67133

5-1966 SURFACE COMPOSITION OF NICKEL- GOLD ALLOYS.
WILLIAMS FL + BOUDART M
J CATAL 30(3): 438-43 (1973)

5-1967 ELECTRON EMISSION STUDIES FROM GALLIUM
SELENIDE SURFACES.
WILLIAMS RH + MCEVOY AJ
J VACUUM SCI TECHNOL 9(2): 867-70 (1972)
CA 77: 26057P

5-1968 SURFACE CHARACTERIZATION OF INDIUM PHOSPHIDE.
(BY AUGER SPECTROSCOPY)
WILLIAMS RH + MCGOVERN IT
SURFACE SCI 51(1): 14-28 (1975)
SAA 1975: 81287, SAB 1975: 41866

5-1969 SURFACE STUDIES OF SOME TRANSITION METAL
DICHALCOGENIDES.
WILLIAMS RH + MCGOVERN IT
JAP J APPL PHYS SUPPL 2 PART-2: 413-16 (1974)
SAA 1975: 37481

5-1970 STUDY OF CARBON FIBER SURFACES USING AUGER
AND SECONDARY ELECTRON EMISSION SPECTROSCOPY.
WILLIS RF + FITTON B + SKINNER DK
J APPL PHYS 43(11): 4412-19 (1972)
CA 78: 21488R

5-1971   CORRELATION BETWEEN THE COMPOSITIONAL PROFILE
          AND ELECTRICAL CONDUCTIVITY OF THE THERMAL
          AND ANODIC OXIDES OF INDIUM ANTIMONIDE. (AES)
          WILMSEN CW
          J VACUUM SCI TECHNOL 13(1): 64-7 (1976)
              SAA 1976: 53780

5-1972   STUDY OF DIFFUSION IN THIN COPPER- LEAD FILMS
          USING AUGER ELECTRON SPECTROSCOPY.
          WILSON JM
          PHIL MAG 27(6): 1467-74 (1973)
              CA 79: 057861

5-1973   LEED AND AES STUDY OF THE INTERACTION OF
          HYDROGEN SULFIDE AND MOLYBDENUM (100).
          WILSON JM
          SURFACE SCI 53(1): 330-40 (1975)
              CA 84: 80338

5-1974   INVESTIGATION OF HYDROGEN CHEMISORPTION ON
          THE TUNGSTEN (100) SURFACE. (AUGER ELECTRON
          SPECTROSCOPY)
          WITHROW SP
          ARMY ELECTRON COMMAND, FT MONMOUTH, NJ, MAR
          1975, 201P. AD-A007434/4ST

5-1975   RECENT ASPECTS OF GLASS FIBER- RESIN
          INTERFACES. (AUGER SPECTROSCOPY OF GLASS
          FIBERS)
          WONG R
          J ADHES 4(2): 171-9 (1972)
              CA 77: 127398

5-1976   SURFACE COMPOSITION OF PALLADIUM- GOLD AND
          PALLADIUM- SILVER CATALYSTS BY AUGER ELECTRON
          SPECTROSCOPY.
          WOOD BJ + WISE H
          SURFACE SCI 52(1): 151-60 (1975)
              CA 84: 35767

5-1977   SURFACE STRUCTURE FROM ANGULAR DEPENDENCE OF
          AUGER ELECTRON EMISSION.
          WOODRUFF DP
          SURFACE SCI 53(1): 538-45 (1975)
              CA 84: 80202

5-1978   AUGER ELECTRON SPECTROSCOPY FOR THE STUDY OF
          SURFACE CONTAMINATION IN SEMICONDUCTOR
          DEVICES.
          WOODWARD RP
          GEORGIA INST TECHNOL THESIS, 1973, 305P.
              CA 80: 075646

5-1979   SECONDARY ELECTRON EMISSION FROM PLATINUM
          BLACK COATED SURFACES. (AUGER SPECTRA)
          WRIGHT B + PATTINSON EB
          J PHYS D 7(11): 1560-5 (21 JUL 1974)
              SAA 1974: 64460

5-1980   GROWTH AND STRUCTURE OF TITANIUM OXIDE THIN
          FILMS. PART-1. (AUGER SPECTRA)
          YAMADA Y + YOSHIDA K
          JAP J APPL PHYS 15(2): 249-56 (1976)
              SAA 1976: 31011

5-1981   MEASUREMENTS ON SURFACES OF ELECTRICAL SOLID
          MATERIALS BY THE LEED METHOD.
          YAMAGUCHI K
          KANAGAWA-KEN KOGYO SHIKENSHO KENKYU HOKOKU
          (33): 15-22 (1972) (IN JAPANESE)
              CA 77: 131762N

5-1982   OBSERVATION OF SURFACE COMPOSITION OF COPPER-
          NICKEL ALLOY BY AUGER SPECTRA IN THE LOWER
          ENERGY REGION AT AROUND 100EV.
          YAMASHINA T + WATANABE K + FUKUDA Y
          + HASHIBA M
          SURFACE SCI 50(2): 591-6 (1975)
              CA 83: 212172

5-1983   AUGER ELECTRON SPECTROSCOPY OF
          CLEANUP-RELATED CONTAMINATION ON SILICON
          SURFACES.
          YANG MG + KOLIWAD KM + MCGUIRE GE
          J ELECTROCHEM SOC 122(5): 675-8 (1975)
              CA 83: 20608

5-1984   DETERMINATION OF FILM THICKNESS OR ION ETCH
          RATE USING AUGER ELECTRON SPECTROSCOPY.
          YASKO RN + FRIED LJ
          REV SCI INSTRUM 43: 335-7 (FEB 1972)

5-1985   CHEMICAL COMPOSITION OF THE SURFACE OF
          MOLYBDENUM AND STEEL BASED ON AUGER ELECTRON
          SPECTRAL DATA.
          ZASHKVARA VV + REDKIN VS + KORSUNSKII MI
          IZV AKAD NAUK KAZ SSR SER FIZ MAT 10(2): 46-8
          (1972) (IN RUSSIAN)
              CA 77: 53399H

5-1986   CHARACTERIZATION OF REORDERED (001) GOLD
          SURFACES BY POSITIVE ION CHANNELING
          SPECTROSCOPY, LOW ENERGY ELECTRON DIFFRACTION
          AND AUGER ELECTRON SPECTROSCOPY.
          ZEHNER DM + APPLETON BR + NOGGLE TS
          + MILLER JW + BARRETT JH + JENKINS LH
          + SCHOW OE
          J VACUUM SCI TECHNOL 12(1): 454-7 (1975)
              CA 82: 160588

5-1987   AUGER AND IONIZATION LOSS SPECTRA FROM
          BERYLLIUM SURFACES.
          ZEHNER DM + BARBULESCO N
          SURFACE SCI 34(2 PART-2): 385-93 (JAN 1973)
              SAA 1973: 21508

5-1988   ANGULAR EFFECTS IN AUGER ELECTRON EMISSION
          FROM COPPER (110).
          ZEHNER DM + NOONAN JR + JENKINS LH
          SOLID STATE COMMUN 18(4): 483-6 (1976)
              CA 84: 157564

5-1989   A CARBON STRUCTURE ON THE RHENIUM (0001)
          SURFACE.
          ZIMMER RS + ROBERTSON WD
          SURFACE SCI 29: 230-6 (1972)

5-1990   MANY PARTICLE RECOMBINATION PROCESSES (AUGER)
          ON THE SURFACES AND IN THIN FILMS OF SILICON
          AND GERMANIUM SUBJECTED TO LASER EXCITATION.
          ZUEV VA + LITOVCHENKO VG + SUKACH GA
          SOV PHYS SEMICOND 9(9): 1083-7 (1975)
              SAA 1976: 54277

6. METALLURGICAL APPLICATIONS OF AUGER
SPECTROSCOPY

6-1991   LEED- AUGER INVESTIGATION ON A (100) SURFACE
OF AUSTENITIC CHROMIUM- NICKEL STEEL.
ADOLPHI B + MUESSIG HJ + SCHOEPE H
KRIST TECH 8(10): 1181-7 (1973) (IN GERMAN)
CA 80: 039722

6-1992   PREPARATION AND SURFACE OXIDATION OF A
VANADIUM (100) PLANE. (AUGER EFFECT)
ALKHOURY NE + CINTI RC
SURFACE SCI 40(3): 583-94 (DEC 1973)
SAA 1974: 10053

6-1993   AUGER ELECTRON SPECTROSCOPY STUDY OF 18-8
STAINLESS STEEL OXIDATION.
ALLEN GC + WILD RK
J ELECTRON SPECTROSC RELAT PHENOMENA 5:
409-15 (1974)
SAA 1975: 47334

6-1994   OXIDATION OF TUNGSTEN (110). PART-2: LEED-
AUGER STUDY OF CXIDE FORMATION AT 650-1100
DEGREES K.
AVERY NR
SURFACE SCI 41(2): 533-46 (1974)
CA 80: 075039

6-1995   SURFACE CONCENTRATION OF MOLYBDENUM IN TYPES
316 AND 304 STAINLESS STEEL BY AUGER ELECTRON
SPECTROSCOPY.
BARNES GJ + ALDAG AW + JERNER RC
J ELECTROCHEM SOC 119(6): 684-6 (1972)
CA 77: 65103T

6-1996   CHEMICAL PROFILE ANALYSIS OF TYPE 316
STAINLESS STEEL SHEET BY AUGER SPECTROSCOPY.
BARNES GJ + ALDAG AW + JERNER RC
PROC OKLA ACAD SCI 53: 78-80 (1973)
CA 79: 142640

6-1997   DETERMINATION OF THE INFLUENCE GRAIN BOUNDARY
IMPURITIES ON THE PROPERTIES OF SINTERED
TUNGSTEN. (BY AUGER SPECTROSCOPY)
BENESOVSKY F + LASSNER E + PETTER H
+ TILES B + BRAUN P + FRABER W + VIEHBOCK FP
PLANSEEBER PULVERMET 23(2): 101-20 (1975)
SAA 1976: 6527

6-1998   COMPOSITION VERSUS DEPTH PROFILES OBTAINED
WITH AUGER ELECTRON SPECTROSCOPY OF AIR
OXIDIZED STAINLESS STEEL SURFACES.
BETZ G + WEHNER GK + TOTH L + JOSHI A
J APPL PHYS 45(12): 5312-16 (1974)
CA 82: 63095

6-1999   COMPARISON OF SURFACE CHEMISTRIES OBTAINED BY
USING ION SCATTERING SPECTROMETRY AND AUGER
ELECTRON SPECTROSCOPY. (STEEL SURFACES)
BINGLE WD
ABSTR 47A OF ANNU CONV MICROBEAM ANAL SOC,
9TH PROC SUMMARIES, 3P, LEHIGH UNIV, 1974.
SAA 1974: 77310

6-2000   SEGREGATION OF CARBON TO (100) SURFACE OF
NICKEL.
BLAKELY JM + KIM JS + POTTER HC
J APPL PHYS 41: 2693-7 (1970)

6-2001   AUGER ELECTRON SPECTROSCOPY OF FRACTURE
SURFACES IN IRRADIATED TYPE 304 STAINLESS
STEEL.
BLOOM EE + CLAUSING RE
OAK RIDGE NAT LAB, 1974, 22P. REP
ORNL-TM-4495.
CA 81: 084931

6-2002   AUGER SPECTROSCOPY OF FRACTURE SURFACES OF
IRRADIATED STAINLESS STEEL.
BLOOM EE + CLAUSING RE + MCGUIRE GE
TRANS AMER NUCL SOC 16: 96-7 (JUN 1973)
SAA 1973: 69248

6-2003   A TECHNIQUE FOR INVESTIGATING THE CATALYTIC
ACTIVITY OF ALLOY SYSTEMS. (SURFACE
COMPOSITION, INTERDIFFUSION, AUGER ELECTRON
SPECTROSCOPY)
BONZEL HP + WYNBLATT P
SURFACE SCI 36(2): 822-8 (MAY 1973)
SAA 1973: 44743

6-2004   AUGER SPECTRA OF ACTIVATED SILVER- MAGNESIUM
AND SILVER- LITHIUM ALLOYS.
BORISOV VL + REDOZUBOV VD
RADIO ENG ELECTRON PHYS 19(1): 151-3 (1974)
CA 80: 125909

6-2005   QUANTITATIVE STUDY OF THE SURFACE COMPOSITION
OF BINARY ALLOYS BY AUGER SPECTROSCOPY.
BOUWMAN R + TONEMAN LH + HOLSCHER AA
VACUUM 23(5): 163-4 (1973)
CA 80: 010059

6-2006   SPATIAL DISTRIBUTIONS OF HYDROGEN DESORBED
FROM IRON, PLATINUM, COPPER, NIOBIUM,
STAINLESS STEEL SURFACES. (AUGER ELECTRON
SPECTROSCOPY)
BRADLEY TL + STICKNEY RE
SURFACE SCI 38(2): 313-26 (JUL 1973)
SAA 1973: 55154

6-2007   AUGER ELECTRON SPECTROSCOPY OF METALS DURING
OXIDATION.
BRAUN P + FAERBER W
VAK TECH 23(1): 7-11 (1974) (IN GERMAN)
CA 81: 054800

6-2008   ADHESION OF METALS TO A CLEAN IRON SURFACE
STUDIED WITH LEED AND AUGER EMISSION
SPECTROSCOPY.
BUCKLEY DH
WEAR 20: 89-103 (MAY 1972)

6-2009   ADHESION, FRICTION AND AUGER SPECTROSCOPY
ANALYSIS OF A COMMERCIAL COBALT BASE AIRCRAFT
TURBINE SHROUD ALLOY.
BUCKLEY DH
NASA, LEWIS RES CENTER, CLEVELAND, OHIO,
1974, 14P. NASA-TM-X-71609

6-2010   LEED AND AUGER STUDIES OF EFFECT OF OXYGEN ON
ADHESION OF CLEAN IRON (001) AND (011)
SURFACES.
BUCKLEY DH
NASA, LEWIS RES CENTER, CLEVELAND, OHIO, APR
1970, 18P. NASA-TN-D-5756

6-2011   INFLUENCE OF SILICON ON FRICTION AND WEAR OF
IRON- COBALT ALLOYS. (AUGER ANALYSIS)
BUCKLEY DH + BRAINARD WA
NASA, 1972, 15P. NASA-TN-D-6769
SAA 1973: 07846

6-2012   AUGER ELECTRON SPECTROSCOPY STUDY OF SURFACE
SEGREGATION IN COPPER- ALUMINUM ALLOYS.
BUCKLEY DH + FERRANTE J
NASA, LEWIS RES CENTER, CLEVELAND, OHIO, NOV
1970, 16P. NASA-TN-D-6095

6-2013   ELEMENTAL ANALYSIS OF FRICTION AND WEAR
SURFACE DURING SLIDING USING AUGER
SPECTROSCOPY.
BUCKLEY DH + PEPPER SV
AMER SOC LUBRIC ENG TRANS 15(4): 252-60
(1972); NASA 1971, 18P. TECH NOTE
NASA-TN-D-6497
CA 78: 31914R; 76: 6142P

6-2014   MASS TRANSPORT OF STAINLESS STEEL CORROSION
PRODUCTS IN FLOWING LIQUID SODIUM. (SCANNING
ELECTRON MICROSCOPY, AUGER SPECTROSCOPY, ION
BEAM ANALYSIS)
CLAXTON KT + COLLIER JG
J BRIT NUCL ENERGY SOC 12(1): 63-75 (JAN
1973)
SAA 1973: 14252

6-2015   STUDY OF SULFUR EMBRITTLED OXYGEN FREE
COPPER. (AES, GRAIN BOUNDARY SEGREGATION)
CLOUGH SP + STEIN DF
SCR MET 9(11): 1163-6 (1975)
SAA 1976: 27102

6-2016 ANALYSIS PROFILES OF OXIDE FILMS ON CHROMIUM
STEEL BY AUGER EMISSION AND X-RAY
PHOTOELECTRON SPECTROSCOPIES.
COAD JP + CUNNINGHAM JG
J ELECTRON SPECTROSC RELAT PHENOMENA 3(6):
435-48 (1974)
CA 81: 057855

6-2017 SPUTTERING OF IONS FROM STAINLESS STEEL BY
HYDROGEN BOMBARDMENT. (AES)
COHEN SA
TRANS AMER NUCL SOC 22: 31-2 (1975)
SAA 1976: 36413

6-2018 IDENTIFICATION OF ATOMIC BORON IN STEEL BY
AUGER ELECTRON SPECTROSCOPY.
COLDREN AP + JOSHI A + STEIN DF
MET TRANS A 6(12): 2304-5 (1975)
CA 84: 109209

6-2019 SURFACE SEGREGATION STUDIES IN ALLOYS USING
AUGER ELECTRON SPECTROSCOPY.
DOOLEY GJ
J VACUUM SCI TECHNOL 9(1): 145-9 (1972)

6-2020 BEHAVIOR OF REFRACTORY METAL SURFACES IN
ULTRAHIGH VACUUM AS OBSERVED BY LEED AND
AUGER ELECTRON SPECTROSCOPY.
DOOLEY GJ + HAAS TW
J VACUUM SCI TECHNOL 7: S90-100 (1970)

6-2021 AUGER ELECTRON SPECTROSCOPY: METALLURGICAL
APPLICATIONS.
DOOLEY GJ + HAAS TW
J METALS 22(11): 17-24 (1970)

6-2022 LOW ENERGY ELECTRON DIFFRACTION- AUGER
ELECTRON SPECTROSCOPY STUDY OF THE OXIDATION
OF CHROMIUM (110) AND CHROMIUM (100).
EKELUND S + LEYGRAF C
SURFACE SCI 40(1): 179-99 (1973)
CA 79: 118624

6-2023 OXIDATION STUDIES OF AMORPHOUS AND
CRYSTALLINE GERMANIUM FILMS BY AUGER
SPECTROSCOPY.
FAN JCC + HENRICH VE
APPL PHYS LETT 25(7): 401-3 (1974)
CA 81: 144955

6-2024 AUGER ELECTRON SPECTROSCOPY OF SILVER- GOLD
ALLOYS.
FARBER W + BRAUN P
VAK TECH 23(8): 239-42 (1974) (IN GERMAN)
CA 82: 143835

6-2025 AUGER SPECTROSCOPY AND LEED STUDY OF
EQUILIBRIUM SURFACE SEGREGATION IN COPPER-
ALUMINUM ALLOYS.
FERRANTE J
ACTA MET 19: 743-8 (1971)

6-2026 MEASUREMENT OF EQUILIBRIUM SURFACE
SEGREGATION USING AUGER ELECTRON
SPECTROSCOPY. (REPLY TO COMMENTS)
FERRANTE J
SCR MET 5: 1129-34 (1971)
SAA 1972: 25627

6-2027 AUGER ELECTRON SPECTROSCOPY STUDY OF SURFACE
SEGREGATION IN THE BINARY ALLOYS COPPER- 1
ATOMIC PERCENT INDIUM, COPPER- 2 ATOMIC
PERCENT TIN, AND IRON- 6.55 ATOMIC PERCENT
SILICON.
FERRANTE J
LEWIS RES CENTER 1973, 20P. NASA TN D-6982
CA 78: 127594

6-2028 AUGER ELECTRON SPECTROSCOPY STUDY OF INITIAL
STAGES OF OXIDATION IN A COPPER- 19.6 ATOMIC
PERCENT ALUMINUM ALLOY.
FERRANTE J
LEWIS RES CENTER, 1973, 18P. NASA TN D-7479.
CA 80: 073402

6-2029 AUGER ELECTRON SPECTROSCOPY AND DEPTH PROFILE
STUDY OF OXIDATION MODIFIED 440C STEEL.
FERRANTE J
NASA, LEWIS RES CENTER, CLEVELAND, OHIO, OCT
1974, 24P. NASA-TN-D-7789

6-2030 THERMAL EFFECTS IN EQUILIBRIUM SURFACE
SEGREGATION IN A COPPER- 10 ATOMIC PERCENT
ALUMINUM ALLOY USING AUGER ELECTRON
SPECTROSCOPY.
FERRANTE J
NASA, LEWIS RES CENTER, CLEVELAND, OHIO, APR
1972, 13P. NASA-TM-X-2543

6-2031 REACTION OF SULFUR DIOXIDE WITH MODIFIED 440C
STEEL STUDIED BY AUGER ELECTRON SPECTROSCOPY
AND DEPTH PROFILING.
FERRANTE J
NASA TECH NOTE, 1974, 21P. NASA-TN-D-7933
CA 83: 167435

6-2032 AUGER ELECTRON SPECTROSCOPY AND DEPTH PROFILE
STUDY OF OXIDATION OF MODIFIED 440C STEEL.
FERRANTE J
NASA TECH NOTE, 1974, 21P. NASA-TN-D-7789
CA 83: 46419

6-2033 A REVIEW OF SURFACE SEGREGATION, ADHESION,
AND FRICTION STUDIES PERFORMED ON COPPER-
ALUMINUM, COPPER- TIN, AND IRON- ALUMINUM
ALLOYS. (BY LEED AND AES)
FERRANTE J + BUCKLEY DH
P35-40 OF ASLE-ASME LUBRICATION CONF,
PITTSBURGH, 1971, AMER SOC LUBRICATION
ENGINEERS, 1971.
SAA 1972: 22303

6-2034 FAILURE ANALYSIS APPLICATIONS OF AUGER
ELECTRON SPECTROSCOPY.
GONZALES AJ
P185-8 OF REL PHYS SYMP 1973, 11TH PROC,
IEEE, 1973.
CA 80: 051545

6-2035 NEW METHODS IN SURFACE ANALYSIS AND THEIR
APPLICATION FOR STUDIES ON IRON AND STEEL.
(AES, ESCA, SIMS)
GRABKE HJ + VIEFHAUS H
ARCH EISENHUTTENW 46(11): 689-94 (1975) (IN
GERMAN)
SAA 1976: 45111

6-2036 STRESS CORROSION CRACKING OF AN ALUMINUM- 5%
MAGNESIUM TERNARY AND VARIOUS QUATERNARY
ALLOYS. (AES)
GREEN JAS + MONTAGUE WG
CORROSION 31(6): 209-13 (1975)
SAA 1976: 41140

6-2037 EFFECT OF A THIN DIFFUSED SOLUTE LAYER ON THE
NUCLEATION OF FATIGUE CRACKS. (AUGER ELECTRON
SPECTROSCOPY)
GREENFIELD IG + SWAIN M + SNYDER DM
UNIV DELAWARE, JAN 1975, 23P. AD/A-005722/4ST

6-2038 CHEMICAL SHIFTS IN AUGER ELECTRON
SPECTROSCOPY FROM INITIAL OXIDATION OF
TANTALUM (110).
HAAS TW + GRANT JT
PHYS LETT A 30: 272 (1969)

6-2039 APPLICATIONS OF AUGER ELECTRON SPECTROSCOPY
TO METALLURGICAL PROBLEMS.
HAAS TW + POCKER DJ
J VACUUM SCI TECHNOL 11(6): 1087-92 (1974)
CA 82: 114847

6-2040 OBSERVATIONS ON THE SURFACE OXIDATION OF
STEEL DURING ANNEALING IN HYDROGEN-FREE
PROTECTIVE GASES. (USING AES)
HABEL L + STEIGER J
SCHWEIZ ARCH 38(7): 235-41 (JUL 1972) (IN
GERMAN)
SAA 1972: 67899

6-2041 SURFACE SEGREGATION ON PALLADIUM- SILVER 40
ATOMIC PERCENT, AND PALLADIUM- GOLD 70 ATOMIC
PERCENT CONTACT METAL ALLOYS. (BY AUGER
ELECTRON SPECTROSCOPY)
HAQUE CA
P41-9 OF HOLM SEMINAR ELECTRICAL CONTACT
PHENOMENA 1971, ILL INST TECHNOL, 1971.

6-2042 COMBINED MASS SPECTROMETRIC AND AUGER
ELECTRON SPECTROSCOPIC TECHNIQUES FOR METAL
CONTACTS.
HAQUE CA
IEEE TRANS PARTS, HYBRIDS PACKAG 9(1): 58-64
(1973)
CA 78: 141641

6-2043 EFFECTS OF COPPER AND PHOSPHORUS ON TEMPER
EMBRITTLEMENT OF MANGANESE- MOLYBDENUM-
NICKEL LOW ALLOY STEEL. (AES)
HASEGAWA M + NAKAJIMA N + KUSUNOKI N
+ SUZUKI K
TRANS JAP INST METALS 16(10): 641-6 (1975)
SAA 1976: 36711

6-2044 OBSERVATION OF THE SEGREGATION OF COPPER TO
THE SURFACE OF A CLEAN ANNEALED 50-50 NICKEL
ALLOY BY AUGER ELECTRON SPECTROSCOPY.
HELMS CR
J CATAL 36(1): 114-17 (1975)
CA 82: 62918

6-2045 VERIFICATION OF THE METAL-RICH SURFACE MODEL
FOR THE OXIDATION OF STRONTIUM, BY AUGER
ELECTRON SPECTROSCOPY.
HELMS CR + SPICER WE
PHYS REV LETT 32(5): 228-32 (4 FEB 1974)
SAA 1974: 24654

6-2046 RECRYSTALLIZATION OF TUNGSTEN FIBERS IN
NICKEL MATRIX COMPOSITES. (AUGER ELECTRON
SPECTROSCOPIC TECHNIQUES)
HOFFMANN J + HOFMANN S + TILLMANN L
Z METALLK 65(11): 721-6 (NOV 1974) (IN
GERMAN)
SAA 1975: 07624

6-2047 AUGER ELECTRON SPECTROSCOPY OF CONTRAST
FORMING LAYERS ON METALS.
HOFMANN S + EXNER HE
Z METALLK 65(12): 778-81 (DEC 1974) (IN
GERMAN)
SAA 1975: 16125

6-2048 QUANTITATIVE DETERMINATION OF SURFACE OXIDE
THICKNESS ON DEPOSITED METAL FILMS BY
COMBINATION AUGER SPECTROSCOPY AND INERT GAS
ION BOMBARDMENT.
HOLLOWAY DM
APPL SPECTROSC 27(2): 95-8 (1973)
CA 78: 151881

6-2049 GRAIN BOUNDARY ACTIVITY MEASUREMENTS BY AUGER
ELECTRON SPECTROSCOPY.
HONDROS ED + SEAH MP
SCR MET 6(10): 1007-12 (OCT 1972)

6-2050 APPLICATION OF PHYSICAL EXAMINATION
TECHNIQUES IN THE STUDY OF CORROSION AND
DEPOSITION IN LIQUID METALS. (AUGER
SPECTROSCOPY, X-RAY PHOTOELECTRON
SPECTROSCOPY, MOSSBAUER SPECTROSCOPY)
HOOPER AJ
P201-7 OF INT CONF ON LIQUID ALKALI METALS,
1973 PROC, BRIT NUCL ENERGY SOC, LONDON,
1973.
SAA 1974: 27462

6-2051 APPLICATION OF MICROAUTORADIOGRAPHY BY
BETA-RAY AUGER ELECTRONS TO THE STUDIES OF
THE STRUCTURE OF ALUMINUM ALLOYS AND
CORROSION OF THE ALLOYS.
INOUE T
KEIKINZOKU 22(12): 731-44 (1972) (IN
JAPANESE)
CA 79: 008715

6-2052 NEW MICROAUTORADIOGRAPHY OF METAL STRUCTURE
USING AUGER ELECTRONS.
INOUE T + SATO O + KATO M
RADIOISOTOPES 20(7): 338-9 (1971) (IN
JAPANESE)
CA 76: 48996A

6-2053 CONFIRMATION OF SULFUR EMBRITTLEMENT IN
NICKEL ALLOYS. (AUGER ELECTRON SPECTROSCOPY)
JOHNSON WC + DOHERTY JE + KEAR BH + GIAMEI AF
SCR MET 8(8): 971-4 (AUG 1974)
SAA 1974: 76505

6-2054 CONFIRMATION OF IMPURITY ADSORPTION AT FLAKE
IRON INTERFACES IN GRAY CAST IRON. (SCANNING
AUGER MICROPROBE)
JOHNSON WC + SMARTT HB
SCR MET 9(11): 1205-10 (1975)
SAA 1976: 27108

6-2055 STUDY OF GRAIN BOUNDARY SEGREGATION IN
THERMALLY EMBRITTLED MARAGING STEEL. (AUGER
ELECTRON SPECTROSCOPY)
JOHNSON WC + STEIN DF
MET TRANS 5(3): 549-54 (MAR 1974)
SAA 1975: 07526

6-2056 GRAIN BOUNDARY SEGREGATION IN SINTERED
ALUMINA. (AUGER ELECTRON SPECTROSCOPY)
JOHNSON WC + STEIN DF
BULL AMER CERAM SOC 52(4): 341 (APR 1973)
SAA 1973: 69343

6-2057 SEGREGATION AT SELECTIVE GRAIN BOUNDARIES AND
ITS ROLE IN TEMPER EMBRITTLEMENT OF LOW ALLOY
STEELS. (AUGER SPECTROSCOPY)
JOSHI A
SCR MET 9(3): 251-60 (1975)
SAA 1975: 43001

6-2058 GRAIN BOUNDARY EMBRITTLEMENT IN SOME METALS
AND ALLOYS. (AUGER EMISSION ANALYSIS,
TUNGSTEN, COPPER, STEELS)
JOSHI A
UNIV MINNESOTA PHD THESIS, 1970, 99P.

6-2059 ROLE OF MANGANESE AND SILICON IN TEMPER
EMBRITTLEMENT OF LOW ALLOY STEELS. (AUGER
SPECTROSCOPY)
JOSHI A + PALMBERG PW + STEIN DF
MET TRANS A 6(11): 2160-1 (1975)
SAA 1976: 41185

6-2060 AUGER SPECTROSCOPIC ANALYSIS OF BISMUTH
SEGREGATED TO GRAIN BOUNDARIES IN COPPER.
JOSHI A + STEIN DF
J INST METALS 99: 178-81 (1971)

6-2061 INTERANGULAR BRITTLENESS STUDIES IN TUNGSTEN
USING AUGER SPECTROSCOPY.
JOSHI A + STEIN DF
MET TRANS 1: 2543-6 (1970)

6-2062 TEMPER EMBRITTLEMENT OF LOW ALLOY STEELS.
(AES OF FRACTURE SURFACES)
JOSHI A + STEIN DF
P59-89 OF AMER SOC TEST MATER, SPEC TECH PUBL
499, 1972.
CA 77: 91514Z

6-2063 IMPURITY SEGREGATION TO GRAIN BOUNDARIES.
JOSHI A + STEIN DF
J TEST EVAL 1(3): 202-8 (1973)
CA 79: 082173

6-2064 CHEMISTRY OF GRAIN BOUNDARIES AND ITS
RELATION TO INTERGRANULAR CORROSION OF
AUSTENITIC STAINLESS STEEL. (AUGER
SPECTROSCOPY)
JOSHI A + STEIN DF
CORROSION 28(9): 321-30 (1972)

6-2065 SURFACE SEGREGATION OF OXYGEN IN NIOBIUM-
OXYGEN AND TANTALUM- OXYGEN ALLOYS. (AUGER
ELECTRON SPECTROSCOPY)
JOSHI A + STRONGIN M
SCR MET 8(4): 413-24 (APR 1974)
SAA 1974: 49730

6-2066 NEW MICRORADIOGRAPHY USING AUGER ELECTRONS IN
THE STUDY OF THE STRUCTURE AND PITTING
CORROSION OF ALUMINUM.
KATO M + INOUE T + SATO O
P541-5 OF NUCL TECH BASIC MET IND SYMP, 1972,
IAEA, VIENNA, 1973.
CA 82: 89356

6-2067 SURFACE SEGREGATION IN TITANIUM AS MONITORED
BY AUGER ELECTRON SPECTROSCOPY.
KHAN IH
SURFACE SCI 40(3): 723-7 (1973)
CA 80: 136425

6-2068 DISAPPEARANCE POTENTIAL SPECTROSCOPY
(VANADIUM OXIDATION).
KIRSCHNER J + STAIB P
APPL PHYS 6(1): 99-109 (FEB 1975)
    SAA 1975: 527494

6-2069 IMPURITIES SEGREGATION TO THE (001) NICKEL
SURFACE DURING THERMAL TREATMENT. WORK
FUNCTION CHANGES AND AUGER ELECTRON
SPECTROSCOPY USING THE LEED CAMERA.
KOLACZKIEWICZ J + KOZIOL C + MROZ S
ACTA PHYS POL A 41(6): 783-4 (1972)
    CA 77: 80599E

6-2070 SEGREGATION OF MANGANESE AND ANTIMONY TO THE
GRAIN BOUNDARIES OF TEMPER EMBRITTLED STEEL.
(AUGER ELECTRON SPECTROSCOPY)
KRAHE PR + GUTTMAN M
SCR MET 7(4): 387-94 (APR 1973)
    SAA 1973: 39346

6-2071 SECONDARY EMISSION AND CONTAMINATION OF METAL
SURFACES.
KULOV SK + SHERTNEV LG
INSTRUM EXP TECH 4: 917-8 (1967)

6-2072 INTERGRANULAR EMBRITTLEMENT OF NICKEL BY
HYDROGEN: EFFECT OF GRAIN BOUNDARY
SEGREGATION. (AUGER ELECTRON SPECTROSCOPY)
LATANISION RM + OPPERHAUSER H
MET TRANS 5(2): 483-92 (FEB 1974)
    SAA 1974: 82905

6-2073 SURFACE SEGREGATION AS A GUIDE TO GRAIN
BOUNDARY SEGREGATION. (AUGER SPECTROSCOPY)
LEA C + SEAH MP
SCR MET 9(6): 583-6 (1975)
    SAA 1975: 69630

6-2074 SITE COMPETITION IN SURFACE SEGREGATION.
(AUGER EFFECT)
LEA C + SEAH MP
SURFACE SCI 53(1): 272-85 (1975)
    SAA 1976: 22109

6-2075 POSSIBILITIES OF THE ION MICROPROBE MASS
ANALYZER AND AUGER SPECTROMETER IN PHYSICAL
METALLURGY.
LEROY V + SERVAIS JP + HAERAKEN L
P69-83 OF MET REP CENT RECH MET, BRUSSELS, NO
35, 1973.
    CA 79: 100050

6-2076 AUGER ELECTRON SPECTROSCOPY AND ITS
APPLICATION IN METALLURGY.
LOSCH WHP
MET ASSOC BRASIL METALS 29(189): 493-9 (1973)
(IN SPANISH)
    CA 79: 110051

6-2077 SPECTROSCOPY OF AUGER ELECTRONS AND ITS USE
IN METALLURGY.
LOSCH WHP
UNIV RIO DE JANEIRO, 1972, 21P. COPPE-TP-6/72
(IN PORTUGESE)
    CA 82: 148102

6-2078 IMPURITIES, INTERFACES, AND BRITTLE FRACTURE.
(REVIEW OF AUGER ELECTRON SPECTROSCOPY AS
APPLIED TO INTERFACES)
LOW JR
TRANS MET SOC AIME 245: 2481-94 (1969)

6-2079 GRAIN BOUNDARY SEGREGATION OF IMPURITIES IN
METALS AND INTERGRANULAR BRITTLE FRACTURE.
(STUDIED BY AES)
LOW JR + SMITH CL + VERDI FW
METALS RES LAB, CAMBRIDGE, MASS, 1972, 38P.
AD742072
    CA 77: 117010J

6-2080 APPLICATON OF AUGER ELECTRON SPECTROSCOPY TO
THE DETERMINATION OF THE COMPOSITION OF
PASSIVE FILMS ON TYPE 316 STAINLESS STEEL.
LUMSDEN JB + STAEHLE RW
SCR MET 6(12): 1205-8 (1972)
    CA 78: 101140

6-2081 AES AND TEMPER EMBRITTLEMENT OF STEELS.
MAGEE CL + GJOSTEIN NA
P261-2 OF ELECTROCHEM SOC SPRING MEETING
EXTENDED ABSTRACTS, WASHINGTON DC, 1971.
    SAA 1972: 3897

6-2082 GRAIN BOUNDARY SEGREGATION RELATING TO
INTERGRANULAR FRACTURE. (AUGER ELECTRON
SPECTROSCOPY)
MARCUS HL
ROCKWELL INT CORP, THOUSAND OAKS, CALIF,
1975, 15P. AD/A-006422/0ST

6-2083 EFFECT OF SOLUTE ELEMENTS ON TEMPER
EMBRITTLEMENT OF LOW ALLOY STEELS. (IMPURITY
GRAIN BOUNDARY SEGREGATION, AES OF FRACTURE
SURFACES)
MARCUS HL + HACKETT LH + PALMBERG PW
P90-103 OF AMER SOC TEST MATER SPEC TECH PUBL
499, 1972.
    CA 77: 91515A

6-2084 FRACTURE SURFACE ANALYSIS OF TEMPER
EMBRITTLED STEEL BY AUGER ELECTRON
SPECTROSCOPY.
MARCUS HL + PALMBERG PW
J METALS 211(3): A96 (1969)

6-2085 AUGER FRACTURE SURFACE ANALYSIS OF A TEMPER
EMBRITTLED 3340- STAINLESS STEEL.
MARCUS HL + PALMBERG PW
TRANS MET SOC AIME 245: 1664-6 (1969)

6-2086 METASTABLE GRAIN BOUNDARY SEGREGATION IN
COPPER CONTAINING TELLURIUM. (AUGER
SPECTROSCOPY)
MARCUS HL + PATON NE
MET TRANS 5(10): 2135-8 (1974)
    SAA 1975: 42985

6-2087 SURFACE CLEANLINESS OF 316 L-N STAINLESS
STEEL STUDIED BY SECONDARY ION MASS
SPECTROMETRY AND AUGER ELECTRON SPECTROSCOPY.
MATHEWSON AG
VACUUM 24(10): 505-9 (1974)

6-2088 QUANTITATIVE AUGER ELECTRON SPECTROSCOPY
ANALYSIS OF SILVER- PALLADIUM AND NICKEL-
PALLADIUM ALLOYS.
MATHIEU HJ + LANDOLT D
SURFACE SCI 53(1): 228-40 (1975)
    CA 85: 56129

6-2089 A LOW ENERGY ELECTRON DIFFRACTION- AUGER
ELECTRON SPECTROSCOPY STUDY OF THE ADSORPTION
OF OXYGEN ON COPPER (100) AND (111) SURFACES.
MCDONNELL L + WOODRUFF DP
SURFACE SCI 46(2): 505-36 (DEC 1974); P47 OF
EUROP PHYS SOC STUDY CONF ON METAL SURFACES,
ABSTR, 1973.
    SAA 1975: 21706; 1974: 37096

6-2090 COMPOSITION OF BINARY ALLOYS BY SIMULTANEOUS
SIMS AND AES MEASUREMENTS.
NARUSAWA T + SATAKE T + KOMIYA S
J VACUUM SCI TECHNOL 13(1): 514-8 (1976)
    CA 84: 144289

6-2091 SEGREGATION OF ALLOYING ELEMENTS TO FREE
SURFACES DURING IRRADIATION. (AUGER
SPECTROSCOPY ANALYSIS, STAINLESS STEEL)
OKAMOTO PR + WIEDERSICH H
J NUCL MATER 53(1): 336-45 (SEP 1974)
    SAA 1974: 81649

6-2092 IRRADIATION INDUCED SOLUTE SEGREGATION (IN
STEEL, AUGER SPECTROSCOPY).
OKAMOTO PR + WIEDERSICH H
P231-8 OF CONSULTANT SYMP ON PHYS OF
IRRADIATION PRODUCED VOIDS, UKAEA, 1975.
    SAA 1975: 33204

6-2093 PLASMA CLEANING OF METAL SURFACES. (AUGER
ELECTRON SPECTROSCOPY)
OKANE DF + MITTAL KL
J VACUUM SCI TECHNOL 11(3): 567-9 (MAY-JUN
1974)
    SAA 1974: 56996

6-2094 AUGER SPECTROSCOPIC ANALYSIS OF THE EXTENT OF
GRAIN BOUNDARY SEGREGATION.
PALMBERG PW + MARCUS HL
TRANS AMER SOC METALS 62: 1016-8 (1969)

6-2095 STRESS CORROSION CRACKING OF ALPHA BRASS IN A
TARNISHING AMMONIACAL ENVIRONMENT:
FRACTOGRAPHY AND CHEMICAL ANALYSIS. (AUGER
SPECTROSCOPY)
PINCHBACK TR + CLOUGH SP + HELDT LA
MET TRANS A 6(8): 1479-83 (1975)
SAA 1976: 31892

6-2096 SEGREGATION OF BISMUTH TO GRAIN BOUNDARIES IN
COPPER- BISMUTH ALLOYS.
POWELL BD + MYKURA H
ACTA MET 21(8): 1151-6 (1973)
CA 79: 095742

6-2097 STUDY OF INTERGRANULAR FRACTURE IN IRON USING
AUGER SPECTROSCOPY.
POWELL BD + WESTWOOD HJ + TAPLIN DMR
+ MYKURA M
MET TRANS 4(10): 2357-61 (1973)
CA 80: 006172

6-2098 GRAIN BOUNDARY EMBRITTLEMENT IN IRON-
PHOSPHORUS, IRON- PHOSPHORUS- SULFUR AND
IRON- ANTIMONY- SULFUR ALLOYS. (AUGER
ELECTRON EMISSION SPECTROSCOPY)
RAMASUBRAMANIAN PV + STEIN DF
MET TRANS 4(7): 1735-42 (JUL 1973)
SAA 1974: 41581

6-2099 EFFECT OF TELLURIUM ON INTERGRANULAR COHESION
OF IRON.
RELLICK JR + MCMAHON CJ + MARCUS HL
+ PALMBERG PW
MET TRANS 2: 1492-4 (1971)

6-2100 AUGER SPECTROSCOPY AND ITS APPLICATIONS TO
SOME TECHNICAL PROBLEMS. (EMBRITTLEMENT,
SEGREGATION OF STEELS AND ALLOYS)
RIVIERE JC
BULL SOC FR MINERAL CRISTALLOGR 94(3): 187-94
(1971)
CA 76: 38486V

6-2101 METALLURGICAL AND ELECTRICAL PROPERTIES OF
ALLOYED NICKEL- GOLD- GERMANIUM FILMS ON
N-TYPE GALLIUM ARSENIDE. (BY AUGER
SPECTROSCOPY)
ROBINSON GY
SOLID STATE ELECTRON 18(4): 331-42 (APR 1975)
SAA 1975: 25759

6-2102 AUGER SPECTROMETRY STUDY OF PASSIVE FILMS
FORMED ON STAINLESS STEELS.
RONDOT B + PONS F + LEHERICY J
+ DACUNHABELO M + LANGERON JP
VIDE 30(176): 70-5 (1975) (IN FRENCH)
SAA 1976: 6519

6-2103 GRAIN BOUNDARY EMBRITTLEMENT OF STEELS
STUDIED BY AUGER ELECTRON SPECTROSCOPY.
SEAH MP + HONDROS ED
MET METAL FORM 39(3): 100-3 (1972)
CA 77: 51455Z

6-2104 GRAIN BOUNDARY SEGREGATION. (AUGER ELECTRON
SPECTROSCOPY)
SEAH MP + HONDROS ED
PROC ROY SOC A 335(1601): 191-212 (1973)
SAA 1974: 04101

6-2105 SCANNING ELECTRON AUGER MICROSCOPY IN
METALLURGY.
SEAH MP + LEA C
P276-81 OF SCANNING ELECTRON MICROSC SYMP 6TH
PROC, JOHARI C(ED), ILL INST TECHNOL RES
INST, 1973, 789P.
CA 80: 136424

6-2106 SURFACE SEGREGATION AND ITS RELATION TO GRAIN
BOUNDARY SEGREGATION (AUGER ELECTRON
SPECTROSCOPY).
SEAH MP + LEA C
PHIL MAG 31(3): 627-45 (MAR 1975)
SAA 1975: 33940

6-2107 IDENTIFICATION OF BUBBLE FORMING IMPURITIES
IN DOPED TUNGSTEN. (AUGER SPECTROSCOPY)
SELL HG + STEIN DF + STICKLER R + JOSHI A
+ BERKEY E
J INST METALS 100: 275-88 (1972)

6-2108 EFFECTS OF TEN YEARS EXPERIMENTAL BREEDER
REACTOR II SERVICE ON CHROMIUM- MOLYBDENUM
STEEL.
SHIELDS JA + LONGUA KJ
NUCL TECHNOL 28: 471-81 (1976)
SAA 1976: 34556

6-2109 SURFACE AND GRAIN BOUNDARY SEGREGATION OF
STEEL IMPURITIES. (TEMPER EMBRITTLEMENT BY
AUGER SPECTROSCOPY)
SHIMIZU H + ONO M + NAKAYAMA K + YAMADA M
P351-4 OF INT VACUUM CONGRESS, 6TH PROC, INT
CONF ON SOLID SURFACES, 2ND, 1974, KYOTO.
(JAP J APPL PHYS, SUPPL 2, PART 1, 1974)

6-2110 LOW ENERGY ELECTRON DIFFRACTION- AUGER
ELECTRON SPECTROSCOPY STUDY OF THE INITIAL
STAGES OF OXIDATION OF IRON (001).
SIMMONS GW + DWYER DJ
SURFACE SCI 48(2): 373-92 (1975)
CA 82: 160593

6-2111 SOLID SOLUTION SOFTENING (RHENIUM DUCTILIZING
EFFECT AND BUBBLE STRENGTHENING IN TUNGSTEN-
RHENIUM ALLOYS, AUGER ELECTRON MICROSCOPY).
SIMPSON RP
AIR FORCE MATERIALS LAB, WRIGHT-PATTERSON AFB
OHIO, SEPT 1973, 165P. AD-769343/5

6-2112 STUDY OF GRAIN BOUNDARY FRACTURE SURFACES IN
DOPED TUNGSTEN- RHENIUM ALLOYS. (AUGER
ELECTRON SPECTROSCOPY)
SIMPSON RP + DOOLEY GJ + HAAS TW
MET TRANS 5(3): 585-91 (MAR 1974)
SAA 1975: 07527

6-2113 AUGER ELECTRON SPECTROSCOPY ANALYSIS OF THE
FRACTURE SURFACES OF IRRADIATED PRESSURE
VESSEL STEELS.
SMIDT FA + STEIN DF + JOSHI A
NAVAL RES LAB 1973, 12P. REP NRL-7660
CA 81: 109594

6-2114 EFFECT OF PRIOR AUSTENITIC GRAIN BOUNDARY
COMPOSITION ON TEMPER BRITTLENESS IN A
NICKEL- CHROMIUM- ANTIMONY STEEL. (AUGER
ANALYSIS)
SMITH CL + LOW JR
MET TRANS 5(1): 279-87 (JAN 1974)
SAA 1975: 10653

6-2115 OXIDATION OF TITANIUM BETWEEN 25 DEGREES C
AND 400 DEGREES C.
SMITH T
SURFACE SCI 38(2): 292-312 (JUL 1973)
SAA 1973: 55167

6-2116 CHARACTERIZING OF ALUMINUM FATIGUE DAMAGE BY
PHOTOEMISSION, ELLIPSOMETRY, AND AUGER
SPECTROSCOPY.
SMITH T + THOMPSON DO
P212-14 OF INT SYMP EXOELECTRON EMISS DOSIM,
4TH PROC, BOHUN A(ED), CZECH ACAD SCI,
PRAGUE, 1974.
CA 82: 143856

6-2117 STUDY OF GRAIN BOUNDARY SEGREGATION USING
AUGER ELECTRON EMISSION SPECTROSCOPY.
STEIN DF
J METALS 21(3): A88 (1969)

6-2118 GRAIN BOUNDARY STRUCTURE AND SEGREGATION.
AUGER SPECTROGRAPHIC ANALYSIS OF FRACTURE
SURFACES.
STEIN DF + JOSHI A
METAL SCI J 6: 67-8 (MAR 1972)

6-2119 STUDIES USING AUGER ELECTRON EMISSION
SPECTROSCOPY ON TEMPER EMBRITTLEMENT IN LOW
ALLOY STEELS.
STEIN DF + JOSHI A + LAFORCE RP
TRANS AMER SOC METALS 62: 776-83 (1969)

6-2120 RELATION OF GRAIN BOUNDARY SEGREGATION TO
REDUCED STRENGTH OF METALS AND ALLOYS.
(SEGREGATION IN STEEL, TUNGSTEN, COPPER-
BISMUTH ALLOYS STUDIED BY AUGER SPECTROSCOPY)
STEIN DF + JOSHI A + RAMASUBRAMANIAN PV
P317-21 OF INT CONF ON STRENGTH OF METALS AND
ALLOYS, 2ND PROC, 1970, VOL 1, AMER SOC
METALS, 1970.

6-2121 STUDY OF GRAIN BOUNDARY SEGREGATION USING
AUGER EMISSION SPECTRA.
STEIN DF + RAMASUBRAMANIAN PV
UNIV OF MINNESOTA ANN TECH PROG REPT I, 1968,
N69-26102.

6-2122 AUGER ELECTRON SPECTROSCOPY OF METAL
SURFACES.
STEIN DF + WEBER RE + PALMBERG PW
J METALS 23(2): 39-44 (1971)

6-2123 METALLURGICAL APPLICATION OF AUGER ELECTRON
SPECTROSCOPY.
STEIN DF + WEBER RE + PALMBERG PW
UNIV MINNESOTA, MINNEAPOLIS, 1970, 24P.
COO-1778-5

6-2124 LEAD MONOLAYER LUBRICATION IN STEEL MACHINING
STUDIED BY AUGER ELECTRON SPECTROSCOPY.
STODDART CTH + SEAH MP + MCLEAN M + MILLS B
NATURE (LONDON) 253(5488): 187-9 (1975)
CA 83: 64065

6-2125 (AUGER) ANALYSIS OF CHROMATE CONVERSION
COATINGS ON ALUMINUM.
SUNDERLAND RJ
P347-50 OF INT VACUUM CONGRESS, 6TH PROC, INT
CONF ON SOLID SURFACES, 2ND, 1974, KYOTO.
(JAP J APPL PHYS, SUPPL 2, PART 1, 1974)

6-2126 APPLICATION OF AUGER ELECTRON SPECTROSCOPY TO
THE STUDY OF INTERFACE IN STEELS.
SUZUKI HG
NIPPON KINZOKU GAKKAI KAIHO 14(3): 173-85
(1975) (IN JAPANESE)
CA 83: 101367

6-2127 INTERGRANULAR FRACTURE AND IMPURITY
SEGREGATION IN STEEL. STUDIES BY AUGER
ELECTRON SPECTROSCOPY.
SUZUKI HG
HYOMEN 13(6): 309-27 (1975) (IN JAPANESE)
CA 84: 47808

6-2128 TEMPER EMBRITTLEMENT OF TIN DOPED STEELS BY
AUGER ELECTRON SPECTROSCOPIC METHOD.
SUZUKI HG + ONO M
TRANS IRON STEEL INST JAP 12: 251 (1972)

6-2129 AES STUDIES ON SURFACES OF 18-8 STAINLESS
STEEL FINISHED MECHANICALLY AND
ELECTROLYTICALLY.
TAKAHASHI N + OKADA K
JAP J APPL PHYS 11: 1580 (1972)
CA 78: 32742V

6-2130 AUGER ELECTRON SPECTROSCOPIC STUDIES OF THE
SURFACES OF 18-8 STAINLESS STEEL
ELECTROPOLISHED OR ABRADED BY EMERY PAPERS.
TAKAHASHI N + OKADA K
SHINKU 15(12): 441-4 (1972) (IN JAPANESE)
CA 79: 008387

6-2131 AUGER ELECTRON SPECTROSCOPY STUDIES ON
ABRADED SURFACE OF ALLOYS.
TAKAHASHI N + OKADA K
SHINKU 16(4): 141-5 (1973) (IN JAPANESE)
CA 79: 117357

6-2132 SURFACE DEFORMATION CAUSED ON NATURAL
MOLYBDENITE BY ABRASION. (AUGER ELECTRON
SPECTROSCOPY)
TAKAHASHI N + OKADA K
J VACCUM SOC JAP 16(12): 438-42 (1973) (IN
JAPANESE)

6-2133 STUDY OF THE INTERMITTENT MOTION OF FRICTION
IN STAINLESS STEEL BY AN AUGER MICROPROBE.
TAKAHASHI N + OKADA K + HOTTA M
COMPT REND B 280(4): 73-6 (1975) (IN FRENCH)
CA 83: 63983

6-2134 GRAIN BOUNDARY SEGREGATION IN ALUMINUM OXIDE.
(AUGER SPECTROSCOPY)
TAYLOR RI + COAD JP + BROOK RJ
J AMER CERAM SOC 57(12): 539-40 (1974)
SAA 1975: 58948

6-2135 DIFFUSION OF COBALT OUT OF COBALT HARDENED
GOLD MEASURED WITH AUGER ELECTRON
SPECTROSCOPY.
TOMPKINS HG
J ELECTROCHEM SOC 122(7): 983-7 (1975)
CA 83: 103454

6-2136 SURFACE DIFFUSION OF TIN OVER A GOLD SURFACE
OBSERVED WITH AUGER ELECTRON SPECTROSCOPY.
TOMPKINS HG
J VACUUM SCI TECHNOL 12(2): 650-1 (1975)
CA 83: 48576

6-2137 EFFECT OF SURFACE METALLURGY ON THE
PENETRATION DEPTH AND RF BREAKDOWN FIELD OF
SUPERCONDUCTING NIOBIUM. (AUGER EFFECT)
VARMAZIS C + LUHMAN TS + JOSHI A
+ KAMMERER OF + GIORDANO S + STRONGIN M
IEEE TRANS MAGN MAG-11(2): 423-6 (1975)
SAA 1975: 62144; SAB 1975: 28521

6-2138 TEMPER EMBRITTLEMENT IN A NICKEL- CHROMIUM
STEEL CONTAINING PHOSPHORUS AS AN IMPURITY.
(AUGER EMISSION ANALYSIS)
VISWANATHAN R
MET TRANS 2: 809-15 (1971).

6-2139 EFFECT OF MICROSTRUCTURE ON GRAIN BOUNDARY
SEGREGATION OF PHOSPHORUS IN A CHROMIUM-
MOLYBDENUM- VANADIUM STEEL. (AUGER
SPECTROSCOPY)
VISWANATHAN R + JOSHI A
SCR MET 9(5): 475-7 (1975)
SAA 1975: 55068

6-2140 LONG TIME ISOTHERMAL TEMPER EMBRITTLEMENT IN
NICKEL- CHROMIUM- MOLYBDENUM- VANADIUM.
VISWANATHAN R + SHERLOCK TP
MET TRANS 3(2): 459-68 (FEB 1972)

6-2141 STUDY ON THE INITIAL OXIDATION AND REDUCTION
PROCESSES OF IRON (100) SURFACE BY MEANS OF
COMBINED LOW ENERGY ELECTRON DIFFRACTION-
AUGER ELECTRON SPECTROSCOPY- MASS
SPECTROSCOPY SYSTEM.
WATANABE M + MIYAMURA M + MATSUDAIRA T
+ ONCHI M
P501-4 OF INT VACUUM CONGRESS, 6TH PROC, INT
CONF ON SOLID SURFACES, 2ND, 1974, KYOTO.
(JAP J APPL PHYS, SUPPL 2, PART 2, 1974)
CA 82: 160730

6-2142 A TECHNIQUE FOR DETERMINING THE ELEMENTAL
COMPOSITION OF FRACTURE SURFACES PRODUCED BY
CRACK GROWTH IN HYDROGEN AND IN WATER VAPOR.
(AUGER SPECTROMETRY, STEEL)
WEI RP + SIMMONS GW
SCR MET 10(2): 153-7 (1976)
CA 84: 188990

6-2143 EFFECT OF ADSORBED CHLORINE AND OXYGEN ON
SHEAR STRENGTH OF IRON AND COPPER JUNCTIONS.
(SPECTROSCOPY)
WHEELER DR
NASA, LEWIS RES CENTER, CLEVELAND, OHIO, JAN
1975, 26P. NASA-TN-D-7894

6-2144 OXIDATION OF NIMONIC-80A ALLOY AT 800 DEGREES
C IN LOW OXYGEN PRESSURES. (AUGER ELECTRON
SPECTROSCOPY, SURFACE CONTAMINANTS)
WILD RK
CORROS SCI 13(2): 105-12 (FEB 1973)
SAA 1973: 30358

6-2145 VACUUM ANNEALING OF STAINLESS STEEL AT
TEMPERATURES BETWEEN 770 AND 1470 K.
WILD RK
CORROS SCI 14(10): 575-86 (NOV 1974)
SAA 1975: 34073

6-2146   ANALYSIS OF GRAIN BOUNDARY DIFFUSION IN
           BIMETALLIC THIN FILM STRUCTURES USING AUGER
           ELECTRON SPECTROSCOPY.
           WILDMAN HS + HOWARD JK + HO PS
           J VACUUM SCI TECHNOL 12(1): 75-8 (1975)
                 CA 83: 48314

Permuted Title Index

A

ECTRON SPECTRA OF PARTIALLY OXIDIZED MAGNESIUM SURFACES. /ITED AUGER AND PHOTOEL 5-1949
F THE SURFACES OF 18-8 STAINLESS STEEL ELECTROPOLISHED OR ABRADED BY EMERY PAPERS. /SPECTROSCOPIC STUDIES O 6-2130
AUGER ELECTRON SPECTROSCOPY STUDIES ON ABRADED SURFACE OF ALLOYS. 6-2131
SURFACE DEFORMATION CAUSED ON NATURAL MOLYBDENITE BY ABRASION. (AUGER ELECTRON SPECTROSCOPY) 6-2132
PECTRUM OF SILICON. ABSENCE OF PLASMON GAIN SATELLITES IN THE AUGER S 4-1037
CTRON SPECTROSCOPY. ABSOLUTE ATOMIC DENSITIES DETERMINED BY AUGER ELE 5-1843
S OF ATOMIC NITROGEN, ATOMIC OXYGEN, AND NEON. (AUGER TR/ ABSOLUTE ELECTRON IMPACT IONIZATION CROSS SECTION 2-0363
KLL AUGER TRANSITION ABSOLUTE PROBABILITIES. (ATOMIC NUMBERS 30-94) 2-0251
DEPENDENT ON THE ATOMIC NUMBER Z=30-94. PART-2. ABSOLUTE PROBABILITIES FOR AUGER- KLL TRANSITIONS 2-0252
ITH AUGER ELECTRON SPECTROSCOPY IN THE ANALYSIS OF SULFUR ABSORBED ON NICKEL. / ENERGY ION BACKSCATTERING W 5-1869
INELASTIC INTERACTIONS OF SLOW ELECTRONS WITH ABSORBED PARTICLES. (AUGER SPECTROSCOPY) 2-0448
LEED AND AUGER SPEC/ PHASE TRANSITIONS IN TWO DIMENSIONAL ABSORBED SULFUR LAYERS ON COPPER INVESTIGATED BY 5-1302
PARISON OF SILICON L3,2 ELECTRON LOSS SPECTRA AND OPTICAL ABSORPTION. (AUGER EFFECT) COM 4-1038
SPECTRUM. ABSORPTION BANDS IN SECONDARY ELECTRON CONTINUOUS 4-0933
EMISSION SPECTROSCOPY. SURFACE ABSORPTION BY POLYPEPTIDE FILMS EXAMINED BY AUGER 5-1179
N AN IRON (011) SURFACE AND EFFECT OF THESE FILMS ON ADH/ ABSORPTION OF ETHYLENE OXIDE AND VINYL CHLORIDE O 5-1207
LEED AND AES STUDY OF OXYGEN ABSORPTION ON TUNGSTEN (110). 5-1669
ORMED FROM THE X-RAY PHOTOIONIZATION OF NITROGE/ RELATIVE ABUNDANCES AND RECOIL ENERGIES OF FRAGMENT IONS F 2-0308
AC SEE ALTERNATING CURRENT
TALS. (AUGER) RECOMBINATION IN DONOR ACCEPTOR COMPLEXES IN CADMIUM SULFIDE SINGLE CRYS 2-0369
NONRADIATIVE AUGER RECOMBINATION OF ELECTRONS ON DONOR- ACCEPTOR PAIRS. 2-0536
PLE SEMICONDUCTORS. THEORY OF DONOR- ACCEPTOR RADIATIVE AND AUGER RECOMBINATION IN SIM 2-0420
ZED BY AUGER SPECTROSCOPY, SECON/ MEASUREMENT OF MOMENTUM ACCOMMODATION COEFFICIENTS ON SURFACES CHARACTERI 2-0501
OF GALLIUM AR/ COMPOSITIONAL AND STRUCTURAL CHANGES THAT ACCOMPANY THE THERMAL ANNEALING OF (100) SURFACES 5-1163
ELECTRON SPECTROSCOPY. (AUGER EFFECT) INCREASING THE ACCURACY OF MATERIAL TESTS BY THE APPLICATION OF 5-1138
ENERGIES IN MANY ELECTRON ATOMS. ACCURATE ONE ELECTRON BINDING ENERGIES AND AUGER 2-0265
LOW PRIMARY ENERGIES. (LEED- AUGER SYSTEM) FAST, ACCURATE SECONDARY ELECTRON YIELD MEASUREMENTS AT 3-0629
ROFILING IN AUGER ELECTRON SPECTROSCOPY. EASY METHOD TO ACCURATELY ALIGN ION BOMBARDMENT GUNS FOR DEPTH P 3-0700
ONS. (AUGER ELECTRON YIELDS OF METHANE, ETHANE, ETHYLENE, ACETYLENE) /ELL IONIZATION OF CARBON BY FAST PROT 2-0532
BBONS. (AUGER SPECTROSCO/ ADSORPTION AND DECOMPOSITION OF ACETYLENE ON POLYCRYSTALLINE RHENIUM FILMS AND RI 5-1688
FLUENCE ON/ INTERACTION OF METHANE, ETHANE, ETHYLENE, AND ACETYLENE WITH AN IRON (001) SURFACE AND THEIR IN 5-1208
EED- AES) FORMIC ACID DESORPTION FROM GRAPHITIZED NICKEL (110). (L 5-1630
AUGER SPECTROSCOPY OF HYDROCHLORIC ACID INTERACTION WITH ALUMINUM AT LOW PRESSURES. 5-1576
/HE KINETICS AND MECHANISM OF THE DECOMPOSITION OF FORMIC ACID ON CARBURIZED AND GRAPHITIZED NICKEL (110) / 5-1629
PLATING BY AES. A STUDY OF ACTIVATED SILICON SURFACES FOR ELECTROLESS NICKEL 5-1539
LLOYS. AUGER SPECTRA OF ACTIVATED SILVER- MAGNESIUM AND SILVER- LITHIUM A 6-2004
/R ELECTRON SPECTROSCOPY STUDY OF CATHODE SURFACES DURING ACTIVATION AND POISONING. PART-1: BARIUM ON OXYG/ 5-1841
/ROSCOPY AND LOW ENERGY ELECTRON DIFFRACTION STUDY OF THE ACTIVATION OF GALLIUM ARSENIDE- CESIUM- OXYGEN N/ 5-1845
/CTION- AUGER CHARACTERIZATION OF GALLIUM ARSENIDE DURING ACTIVATION TO NEGATIVE ELECTRON AFFINITY BY THE / 5-1364
OPY. GRAIN BOUNDARY ACTIVITY MEASUREMENTS BY AUGER ELECTRON SPECTROSC 6-2049
INTERDIFFUSI/ A TECHNIQUE FOR INVESTIGATING THE CATALYTIC ACTIVITY OF ALLOY SYSTEMS. (SURFACE COMPOSITION, 6-2003
R- NICKEL ALLOY. APPLICATION/ RELATION BETWEEN CATALYTIC ACTIVITY PATTERN AND SURFACE COMPOSITION OF COPPE 5-1350
TELLITE INTENSITIES VERSUS DEPTH: A MEANS FOR DETERMINING ADATOM CONCENTRATION PROFILES. AUGER PLASMON SA 2-0347
ANALYSIS) EFFECT OF YTTRIA ADDITIONS ON HOT PRESSED SILICON NITRIDE. (AUGER 5-1350
DARIES IN SINTERED ALUMINA. (SCANNING AUGER MICROPROBE) ADDITIVE AND IMPURITY DISTRIBUTIONS AT GRAIN BOUN 5-1505
SPEC/ ANALYSIS OF GRAIN BOUNDARY IMPURITIES AND FLUORIDE ADDITIVES IN HOT PRESSED OXIDES BY AUGER ELECTRON 5-1506
). CHARACTERIZATION OF ALUMINUM ADHEREND SURFACES (BY AUGER ELECTRON SPECTROSCOPY 5-1725
IDE ON AN IRON (011) SURFACE AND EFFECT OF THESE FILMS ON ADHESION. /TION OF ETHYLENE OXIDE AND VINYL CHLOR 5-1207
ALLOYS, AES) LOW INTRINSIC STRESS AND GOOD ADHESION ALLOY THIN FILMS. (TITANIUM AND CHROMIUM 5-1909
IN CONTACT WITH METALS AS STUDIED BY AUGER SPECTROSCOPY/ ADHESION AND FRICTION OF POLYTETRAFLUORO ETHYLENE 5-1198
R- ALUMINUM, COPPER- TI/ A REVIEW OF SURFACE SEGREGATION, ADHESION, AND FRICTION STUDIES PERFORMED ON COPPE 6-2033
STUDIED BY AUGER EMISSION SPECTROSCOPY. ADHESION AND TRANSFER OF PLATINUM IRON TO METALS 5-1730
TO METALS STUDIED BY AUGER EMISSION SPECTROSCOPY. ADHESION AND TRANSFER OF POLYTETRAFLUORO ETHYLENE 5-1731
ADHESION AT GOLD MICA INTERFACES. (LEED, AES) 5-1434
S OF A COMMERCIAL COBALT BASE AIRCRAFT TURBINE SHROUD AL/ ADHESION, FRICTION AND AUGER SPECTROSCOPY ANALYSI 6-2009
ATOMIC NATURE OF POLYMER- METAL INTERACTIONS IN ADHESION, FRICTION, AND WEAR. 5-1212
(AUGER STUDIES OF) ADHESION, FRICTION, LUBRICATION IN VACUUM. 5-1209
H BY MODERN SURFACE SCIENCE TECHNIQUES. (AES, ELLIPSOMET/ ADHESION, FRICTION, WEAR, AND LUBRICATION RESEARC 1-0104
SULFUR, OXYGEN, AND HYDROGEN SULFIDE SURFACE FILMS ON THE ADHESION OF CLEAN IRON. (USING LEED AND AES) /OF 5-1211
LEED AND AUGER STUDIES OF EFFECT OF OXYGEN ON ADHESION OF CLEAN IRON (001) AND (011) SURFACES. 6-2010
D WITH LEED AND AUGER EMISSION SPECTROSCOPY. ADHESION OF METALS TO A CLEAN IRON SURFACE STUDIE 6-2008
OP/ RELATIONSHIPS BETWEEN SUBSTRATE SURFACE CHEMISTRY AND ADHESION OF THIN FILMS. (AUGER ELECTRON SPECTROSC 5-1851
/TYLENE WITH AN IRON (001) SURFACE AND THEIR INFLUENCE ON ADHESION STUDIED WITH LOW ENERGY ELECTRON DIFFRA/ 5-1208
ON (100) PALLADIUM. STRUCTURAL INFLUENCES ON ADSORBATE BINDING ENERGY. PART-1: CARBON MONOXIDE 5-1913
URFACE COMPOSITION OF COPPER- NICKEL ALLOYS FOR CLEAN AND ADSORBATE COVERED SURFACES. /TERMINATION OF THE S 5-1424
S METASTABLE ATOMS. (AUGER PROCESS) ADSORBATE EFFECTS IN ELECTRON EJECTION BY RARE GA 2-0333
E. PART-1: SULFUR AND CARBON MONOXIDE LEED PATTERNS. (AU/ ADSORBATE INTERACTIONS ON A PLATINUM (110) SURFAC 5-1190
R SPECTROSCOPY) ELECTRON BEAM ADSORBATE INTERACTIONS ON SILICON SURFACES. (AUGE 5-1513
INTERATOMIC AUGER PROCESSES IN SOME ADSORBATES ON TRANSITION METALS. 4-1045
NS AND RELAXATION EFFECTS IN AUGER SPECTRA OF SEVERAL GAS ADSORBATES ON TRANSITION METALS. /TOMIC TRANSITIO 4-1042
IRON AND COPPER JUNCTIONS. (SPECTROSCOPY) EFFECT OF ADSORBED CHLORINE AND OXYGEN ON SHEAR STRENGTH OF 6-2143
ATTENUATION LENGTHS IN METALS: TRANSMISSION THROUGH THIN ADSORBED FILMS. (AUGER EFFECT) /ATION OF ELECTRON 4-1088
LEED AND AES STUDIES OF THE STRUCTURE OF ADSORBED GASES ON SOLID SURFACES. 5-1835
LECTRON SPECTROSCOPY STUDIES OF CLEAN NICKEL SURFACES AND ADSORBED LAYERS. /ERGY ION SCATTERING AND AUGER E 5-1870
THICKNESS MEASUREMENTS OF ADSORBED LAYERS BY AUGER ELECTRON SPECTROSCOPY. 5-1618
LECTRON SPECTROSCOPY. DISTINCTION BETWEEN ADSORBED MONOLAYERS AND THICKER LAYERS IN AUGER E 5-1797
GH RESOLUTION AUGER ELECTRON SPECTROSCOPY OF NITRIC OXIDE ADSORBED ON A TUNGSTEN SURFACE. HI 5-1565
TEN. AUGER ELECTRON SPECTROSCOPY OF CESIUM ADSORBED ON CLEAN AND OXYGEN- COVERED (100) TUNGS 5-1295
S. (AUGER SPECTROSCOPY) ELECTRONIC STATES OF OXYGEN ADSORBED ON CLEAN SILICON (111) AND (100) SURFACE 5-1476
AND SURFACE POTENTIAL STUDY OF CARBON MONOXIDE AND XENON ADSORBED ON COPPER (100). LEED 5-1246
DE. ELECTRON ORBITAL ENERGIES OF OXYGEN ADSORBED ON SILICON SURFACES AND OF SILICON DIOXI 5-1475
LECTRON SP/ FIRST ORDER TWO DIMENSIONAL TRANSITION: XENON ADSORBED ON THE (0001) FACE OF GRAPHITE. (AUGER E 5-1857
ELECTRONIC AND LATTICE STRUCTURE OF CESIUM FILMS ADSORBED ON TUNGSTEN. 5-1604
LEED- AUGER, AND WORK FUNCTION STUDY OF IODINE ADSORBED ON TUNGSTEN (110). 5-1145
/DIMENSIONAL PHASE TRANSITION IN XENON SUBMONOLAYER FILMS ADSORBED ON (0001) GRAPHITE. (AUGER SPECTROSCOPY/ 5-1858
R AUGER ELECTRON SPECTROSCOPY. (PALLADIUM- NICKEL ALLOYS, ADSORBED OXYGEN) /ALL RETARDING FIELD ANALYZER FO 3-0644
PITAXIAL FILM. (AUGER EFF/ BEHAVIOR OF IMPURITY ATOMS AND ADSORBED OXYGEN ATOMS ON THE (001) FACE OF IRON E 5-1523

ALUMINUM ON SILICON (111).   EFFECTS OF PREADSORBED OXYGEN CONTAMINANT ON THE ADSORPTION OF 5-1169
XIDE ON TUNGSTEN (100). (BY L/ PHOTOEMISSION SPECTRA FROM ADSORBED OXYGEN ON TUNGSTEN (110) AND CARBON MONO 5-1152
E AND CARBON MONOXIDE ON TUNGSTEN (1/ CHARACTERIZATION OF ADSORBED SPECIES USING AUGER PEAK SHAPES. ETHYLEN 5-1243
.   IDENTIFICATION OF THE CHEMICAL STATE OF ADSORBED SPECIES WITH AUGER ELECTRON SPECTROSCOPY 4-0868
.   DISCRIMINATION BY AUGER SPECTROSCOPY BETWEEN ADSORBED SULFUR AND SULFIDES OF COPPER AND NICKEL 5-1734
AUGER ELECTRON SPECTROSCOPY: COMPARISON OF TECHNIQUES FOR ADSORBED TIN ON IRON.   QUANTITATIVE 1-0178
       QUANTITATIVE AUGER SPECTROSCOPY OF PHYSICALLY ADSORBED XENON ON NICKEL.   5-1150
    CALIBRATION IN AUGER ELECTRON SPECTROSCOPY BY MEANS OF COADSORPTION.   3-0587
EMICAL EFFECTS IN THE AUGER SPECTRA OF IRON DUE TO OXYGEN ADSORPTION.   CH 4-0727
MICAL SHIFT IN THE AUGER SPECTRUM OF TUNGSTEN FROM OXYGEN ADSORPTION.   CHE 4-0922
HEMICAL SHIFT IN THE AUGER SPECTRUM OF IRON DURING OXYGEN ADSORPTION.   C 5-1792
HEM/CAL SHIFTS IN THE AUGER SPECTRUM OF YTTRIUM ON OXYGEN ADSORPTION.   C 5-1153
       QUANTITATIVE AUGER ELECTRON SPECTROSCOPY USING COADSORPTION.   1-0002
.   ELLIPSOMETRIC STUDY OF ADSORPTION: ADSORPTION OF XENON ON GRAPHITE (001) 5-1754
  DIFFRACTION AUGER, WORK FUNCTION STUDY OF OXYGEN- CESIUM ADSORPTION AND COADSORPTION ON (100) TUNGSTEN. /N 5-1297
SINGLE CRYSTAL SURFACES.   ADSORPTION AND CONDENSATION OF COPPER ON TUNGSTEN 5-1160
CRYSTALLINE RHENIUM FILMS AND RIBBONS. (AUGER SPECTROSCO/ ADSORPTION AND DECOMPOSITION OF ACETYLENE ON POLY 5-1688
N PLATINUM (111). (AUGER SPECTROMETRY)   ADSORPTION AND DECOMPOSITION OF CARBON MONOXIDE O 5-1622
1) SURFACES. (BY LEED AND AES)   A STUDY OF PHOSPHORUS ADSORPTION AND DESORPTION KINETICS ON SILICON (11 5-1288
CES. (REVIEW, 50 REFS)   ELECTRON SPECTROSCOPY STUDIES OF ADSORPTION AND OXIDATION PROCESSES AT METAL SURFA 1-0019
ITROGEN, OXYGEN AND CARBON MONOXIDE BY (100) TANTALUM. (/ ADSORPTION AND SOLUTION OF HYDROGEN, DEUTERIUM, N 5-1549
S ON (111) AND (110) FACES OF GOLD. (AUGER EFFECT)   ADSORPTION AND SURFACE ALLOYING OF LEAD MONOLAYER 5-1733
M THE BULK CRYSTAL INTERACTION WITH CHLORINE GA/ CHLORINE ADSORPTION AT A TITANIUM SURFACE. SEGREGATION FRO 5-1833
IRON. (SCANNING AUGER MICRCFROB/ CONFIRMATICN OF IMPURITY ADSORPTION AT FLAKE IRON INTERFACES IN GRAY CAST 6-2054
      EXOELECTRON EMISSION FROM ZINC OXIDE. (OXYGEN ADSORPTION EFFECT, AUGER ELECTRON SPECTROSCOPY) 4-0921
KEL (100). PART-3: CARBON MONOX/ STRUCTURAL INFLUENCES ON ADSORPTION ENERGY. PART-2: CARBON MONOXIDE ON NIC 5-1911
      CHEMICAL SHIFTS IN AUGER SPECTRA (MONOLAYER ADSORPTION, ESCA SIMPLE CHARGE TRANSFER MODEL). 4-1115
EFFECT ON AUGER INTENSITIES.   MOLECULAR ADSORPTION: INCLUSION OF BULK DIFFUSION AND ITS 2-0264
PRESSURE AND/ AUGER ELECTRON SPECTROSCOPY MEASUREMENTS OF ADSORPTION ISOBARS FOR OXYGEN ON TUNGSTEN AT LOW 5-1282
AND (111) SURFACES IN THE PRESSURE RANGE 10(-9) TO 10(-/ ADSORPTION ISOTHERMS OF OXYGEN ON TUNGSTEN (100). 5-1148
/CTROSCOPY AND LOW ENERGY ELECTRON DIFFRACTICN STUDIES OF ADSORPTION ISOTHERMS: XENON ON (0001) GRAPHITE. 5-1856
. (USING AES AND X-RAY PHOTOELECTRON SPECTROSCOPY)   ADSORPTION KINETICS OF CESIUM ON GALLIUM ARSENIDE 5-1821
PECTROSCOPY)   CARBON MONOXIDE ADSORPTION KINETICS ON MOLYBDENUM (100). (AUGER S 5-1579
       LEED- AUGER SPECTROSCOPY STUDY OF THE ADSORPTION OF ALKALI METALS ON MOLYBDENUM (110). 5-1896
      EFFECTS OF PREADSORBED OXYGEN CONTAMINANT ON THE ADSORPTION OF ALUMINUM ON SILICON (111). 5-1169
. INFLUENCE OF SULFUR AT THE SURFACE.   STUDY OF THE ADSORPTION OF BENZENE ON THE (100) FACE OF NICKEL 5-1284
ED AND AUGER EMISSION STUDIES.   ADSORPTION OF CARBON MONOXIDE ON COPPER (001). LE 5-1515
M ALLOYS.   ADSORPTION OF CARBON MONOXIDE ON SILVER- PALLADIU 5-1254
OF NEGATIVE ELECTRON AFFINITY ON SILICON PRODUCED BY THE ADSORPTION OF CESIUM AND OXYGEN. /PLASMON STUDIES 5-1363
DE DURING ACTIVATION TC NEGATIVE ELECTRON AFFINITY BY THE ADSORPTION OF CESIUM AND OXYGEN. / GALLIUM ARSENI 5-1364
      STRUCTURAL STUDIES OF THE ADSORPTION OF CESIUM AND OXYGEN ON SILICON (100). 5-1389
AUGER EFFECT, AUGER SPECTROSCOPY)   ADSORPTION OF COBALT ON A NICKEL (111) SURFACE. (5-1256
ATURES BELOW 300 DEGREES K.   ADSORPTION OF COBALT ON (001) RUTHENIUM AT TEMPER 5-1609
                                                          ADSORPTION OF CYANOGEN ON PLATINUM (100). 5-1684
EN WITH OXYGEN ON PLATINUM (100). (AUGER SPECTROSCOPY)   ADSORPTION OF HYDROGEN AND THE REACTION OF HYDROG 5-1685
E.   ADSORPTION OF HYDROGEN ON A PLATINUM (111) SURFAC 5-1255
KEL (100), SHEET AND EVAP/ STICKING PROBABILITIES FOR THE ADSORPTION OF HYDROGEN ON CLEAN NICKEL (111), NIC 5-1468
URFACES. (AUGER ELECTRCN SPECTROSCOPY)   ADSORPTION OF HYDROGEN ON NICKEL SINGLE CRYSTAL S 5-1257
FACE.   ADSORPTION OF IONIC SALTS ON A TUNGSTEN (100) SUR 5-1659
ITE: LEED AND AUGER MEASUREMENTS.   ADSORPTION OF KRYPTON ON THE BASAL PLANE OF GRAPH 5-1559
SINGLE CRYSTALS. (AUGER ELECTRON SPECTROSCOFY)   ADSORPTION OF KRYPTON ON THE (111) FACE OF COPPER 5-1221
TROSCOPY STUDIES OF CLEAN POLYCRYSTALLINE GOLD AND OF THE ADSORPTION OF MERCURY ON GOLD. /GER ELECTRON SPEC 5-1519
/ IN THE M4,5NN AUGER SPECTRUM OF MOLYBDENUM (110) DUE TO ADSORPTION OF MOLECULAR OXYGEN AND CARBON MONOXI/ 4-0885
E, AND MOLECULAR OXYG/ AUGER SPECTROSCOPIC STUDY OF THE COADSORPTION OF MOLECULAR OXYGEN AND CARBON MONOXID 5-1465
                                                          ADSORPTION OF NITRIC OXIDE ON TUNGSTEN. 5-1929
FT IN AUGER SPECTRA OF TUNGSTEN AND MOLYBDENUM DURING THE ADSORPTION OF CXYGEN.   CHEMICAL SHI 4-0816
HE RUTHENIUM (001) SURFACE. (AUGER SPECTRA)   ADSORPTION OF OXYGEN AND OXIDATION OF COBALT ON T 5-1608
(AUGER ELECTRON SPECTRCSCOPY)   ADSORPTION OF OXYGEN ON A CLEAN SILICON SURFACE. 5-1638
UM ARSENIDE SURFACES.   ADSORPTION OF OXYGEN ON CLEAN CLEAVED (110) GALLI 5-1306
/ON DIFFRACTION- AUGER ELECTRON SPECTROSCOPY STUDY OF THE ADSORPTION OF OXYGEN ON COPPER (100) AND (111) S/ 6-2089
F LOW ENERGY ELECTRON SPECTROSCOPY. (AUGER / PROCESSES OF ADSORPTION OF OXYGEN ON GALLIUM ARSENIDE METHOD O 5-1641
OF TRACE IMPURITIES.   ADSORPTION OF OXYGEN ON MOLYBDENUM (111): EFFECT 5-1568
      SOME ASPECTS OF AN AES AND XPS STUDY OF THE ADSORPTION OF OXYGEN ON NICKEL. 5-1459
LECTRON SPECTROSCOPY: CHEMICAL SHIFTS AND / STUDY OF THE ADSORPTION OF OXYGEN ON NICKEL(III) USING AUGER E 5-1466
ART-1. (AUGER ELECTRCN SPECTROSCOPY)   ADSORPTION OF OXYGEN ON SILICON (111) SURFACES. P 5-1474
EN. (AUGER ELECTRON SPECTROSCOPY)   ADSORPTION OF OXYGEN ON THE (110) PLANE OF TUNGST 5-1318
MEASUREMENT METHODS)   ROOM TEMPERATURE ADSORPTION OF OXYGEN ON TUNGSTEN SURFACES. (AUGER 5-1664
      AUGER SPECTROSCOPY STUDY OF THE ADSORPTION OF RUBIDIUM ON MOLYBDENUM (100). 5-1894
ON DIFFRACTION AND AUGER SPECTROSCOPY.   ADSORPTION OF SULFUR ON GOLD BY LOW ENERGY ELECTR 5-1558
ENERGY ELECTRON DIFFRACTION AND AUGER SPECTROSCOPY OF THE ADSORPTION OF SULFUR ON PLATINUM.   LOW 5-1175
OF PLATINUM. LOW ENERGY ELECTRON DIFFRACTION AND AUGER / ADSORPTION OF SULFUR ON THE (100) AND (111) FACES 5-1421
NUM.   ADSORPTION OF SULFUR ON THE (100) FACE OF MOLYBDE 5-1732
IUM. PART-1: STRUCTURAL AND KINETIC PROPERTIES. (AUGER E/ ADSORPTION OF WATER VAPOR ON CLEAN CLEAVED GERMAN 5-1432
                                                          ADSORPTION OF XENON AND COBALT ON SILVER (111). 5-1633
OF GRAPHITE. (AUGER SPECTROSCOPY)   ENTROPY OF ADSORPTION OF XENON IN EPITAXY ON THE (0001) FACE 5-1860
                     PHYSICAL ADSORPTION OF XENON ON PALLADIUM (100). 5-1711
       THERMODYNAMICS AND KINETICS OF THE FIRST MONOLAYER ADSORPTION OF XENON ON THE (0001) GRAPHITE FACE. 5-1859
ECTROSCOPY)   ADSORPTION OF ZINC ON GALLIUM ARSENIDE. (AUGER SP 5-1144
ECTRON SPECTROSCOPY, STICKING COEFFIC/ KINETICS OF OXYGEN ADSORPTION ON A PLATINUM (111) SURFACE. (AUGER EL 5-1189
OW ENERGY ELECTRON DIFFRACTION, AUGER, AND CONTAC/ SODIUM ADSORPTION ON ALUMINUM (111) SURFACES: ELECTRIC L 5-1744
      CHLORINE ADSORPTION ON CLEAN ALUMINUM. (AUGER SPECTOSCOPY) 5-1831
/ELECTRON AND VACUUM ULTRAVIOLET PHOTOELECTRON STUDIES OF ADSORPTION ON CLEAN METAL FILM SURFACES. (AUGER / 5-1203
      OXYGEN ADSORPTION ON CLEAN MOLYBDENUM (100) SURFACES. 5-1772
ACTION AND AUGER ELECTRON SPECTROSCOPY STUDY OF PHOSPHINE ADSORPTION ON CLEAN SILICON (111). /LECTRON DIFFR 5-1930
INDUCED PHOTOELECTRON AND AUGER SPECTRA. CARBON MONOXIDE ADSORPTION ON MOLYBDENUM. /ED AND ELECTRON IMPACT 4-0768
      FURTHER STUDIES OF GAS ADSORPTION ON MOLYBDENUM (100) SURFACE. 5-1305
      CARBON MONOXIDE ADSORPTION ON NICKEL (110). (AUGER EFFECT) 5-1884
      ELECTRON BEAM ASSISTED ADSORPTION ON SILICON (111) SURFACE. 5-1269
NICKEL ALLOYS. PART-1.   ADSORPTION ON SINGLE CRYSTAL SURFACES OF COPPER- 5-1322
ELECTRON DIFFRACTION INVESTIGATICN OF OXYGEN AND HYDROGEN ADSORPTION ON STRONTIUM SURFACES. /ND REFLECTION 5-1498

/ CARBON. TANTALUM CARBIDES, ETHYLENE AND CARBON MONOXIDE ADSORPTION ON TANTALUM AND POLYCRYSTALLINE RHENI/ 4-0820
COMBINED LEED- RHEED AUGER STUDY OF OXYGEN ADSORPTION ON THE TUNGSTEN (112) SURFACE. 5-1464
TEN. THE KINETICS OF OXYGEN ADSORPTION ON THE (112) AND (110) PLANES OF TUNGS 5-1912
AN ELECTRON DIFFRACTION STUDY OF CESIUM ADSORPTION ON TUNGSTEN. 5-1603
GER ELECTRON SPECTROSCOPY. INVESTIGATION OF OXYGEN ADSORPTION ON TUNGSTEN AT HIGH TEMPERATURES BY AU 5-1156
ROSCOPY AND SURFACE POTENTIAL DIFFERENCE STUDIES. CESIUM ADSORPTION ON TUNGSTEN: ELLIPSOMETRY, AUGER SPECT 5-1830
ARY ELECTRON SPECTROSCOPY. CESIUM ADSORPTION ON TUNGSTEN STUDIED BY LEED AND SECOND 5-1661
AUGER ELECTRON SPECTROSCOPY STUDY OF OXYGEN ADSORPTION ON TUNGSTEN (110). 5-1660
CESIUM- OXYGEN COADSORPTION ON (100) TUNGSTEN. 5-1296
R, WORK FUNCTION STUDY OF OXYGEN- CESIUM ADSORPTION AND COADSORPTION ON (100) TUNGSTEN. /N DIFFRACTION AUGE 5-1297
OXYGEN ADSORPTION ON (110) SILVER. 5-1422
(AUGER ELECTRON SPECTROSCOPY) INVESTIGATION OF CESIUM ADSORPTION ONTO A GALLIUM ARSENIDE (110) SURFACE. 5-1644
S AND AUGER ELECTRON SPECTROSCOPY APPLIED TO THE STUDY OF ADSORPTION PHENOMENA AT METAL SURFACES. /ERGY LOS 5-1776
ARY ELECTRON EMISSION AND PHOTOELECTRON EMISSION STUDY OF ADSORPTION PROCESSES. OPTICALLY STIMULATED SECOND 2-0302
OF AN "AUGER ELECTRON SPECTROMETER" FOR HIGH TEMPERATURE ADSORPTION STUDIES. DEVELOPMENT AND APPLICATION 3-0586
OELECTRON AND AUGER SPECTRA FO/ ELECTRON SPECTROSCOPY AND ADSORPTION STUDIES. CHEMICAL SHIFTS IN X-RAY PHOT 5-1204
ON A PLATINUM (100) SURFACE. (AUGER) ADSORPTION STUDIES OF OXYGEN AND CARBON MONOXIDE 5-1546
(AUGER) ADSORPTION STUDIES ON THE PLATINUM (100) SURFACE. 5-1570
ROSCOPY: CARBON MO/ ELECTRON BEAM INDUCED EFFECTS ON GAS ADSORPTION STUDIES UTILIZING AUGER ELECTRON SPECT 5-1542
CARBON MONOXIDE A/ ELECTRON BEAM INDUCED EFFECTS ON GAS ADSORPTION UTILIZING AUGER ELECTRON SPECTROSCOPY. 5-1543
RECENT ADVANCES IN AES. 1-0049
SPUTTERING AND AUGER ANALYSIS. RECENT ADVANCES IN COMPOSITION PROFILING BY SIMULTANEOUS 1-0155
IS. (AUGER ELECTRON SPECTROSCOPY AND ION MICROPRO/ RECENT ADVANCES IN INSTRUMENTATION FOR MICROPROBE ANALYS 3-0716
ERGY ELECTRON DIFFRACTION AND AUGER ELECTRON SPEC/ RECENT ADVANCES IN SURFACE CHARACTERIZATION USING LOW EN 1-0201
AUGER SPECTROSCOPY, REVIEW) RECENT ADVANCES IN SURFACE STUDIES. ION BEAM ANALYSIS. (1-0137
RECENT ADVANCES IN THE STUDY OF SOLID SURFACES. 1-0051
ON OF THE CHEMICAL, PHYSICAL, AND CRYSTALLOGRAPHI/ USE OF ADVANCED ANALYTICAL TECHNIQUES FOR CHARACTERIZATI 5-1433
AES IS NOT INDEXED.
/ GALLIUM ARSENIDE DURING ACTIVATION TO NEGATIVE ELECTRON AFFINITY BY THE ADSORPTION OF CESIUM AND OXYGEN. 5-1364
CE/ LEED, AUGER AND PLASMON STUDIES OF NEGATIVE ELECTRON AFFINITY ON SILICON PRODUCED BY THE ADSORPTION OF 5-1363
(AU/ STRUCTURAL AND ELECTRONIC MODEL OF NEGATIVE ELECTRON AFFINITY ON THE SILICON- CESIUM- OXYGEN SURFACE. 2-0426
ION OF GALLIUM ARSENIDE- CESIUM- OXYGEN NEGATIVE ELECTRON AFFINITY SURFACES. /FRACTION STUDY OF THE ACTIVAT 5-1845
NUCLEAR TRANSFORMATIONS IN MOSSBAUER SPECTROSCOPY. (AUGER AFTEREFFECTS) CHEMICAL EFFECTS OF 2-0564
ON CAPTURE IN TRIS (1,10-PHENAN/ MOSSBAUER STUDIES OF THE AFTEREFFECTS OF AUGER IONIZATION FOLLOWING ELECTR 2-0516
(100) SURFACES OF ALKALI HALIDES. PART-1: AIR AND VACUUM CLEANED SURFACES. 5-1347
STUDY OF THE SURFACE OF IRON AFTER IONIC BOMBARDMENT AND AIR OXIDATION. AUGER SPECTROSCOPIC 5-1939
PTH PROFILES OBTAINED WITH AUGER ELECTRON SPECTROSCOPY OF AIR OXIDIZED STAINLESS STEEL SURFACES. /VERSUS DE 6-1998
D AUGER SPECTROSCOPY ANALYSIS OF A COMMERCIAL COBALT BASE AIRCRAFT TURBINE SHROUD ALLOY. /SION, FRICTION AN 6-2009
ULTRAHIGH VACUUM AIRLOCK FOR LEED AND AUGER SPECTROSCOPY. 3-0669
AUGER ELECTRON SPECTROSCOPY. EASY METHOD TO ACCURATELY ALIGN ION BOMBARDMENT GUNS FOR DEPTH PROFILING IN 3-0700
SURFACE EFFECTS, AUGER SPECTROMETRY) MECHANISM OF SURFACE ALIGNMENT IN NEMATIC LIQUID CRYSTALS. (SUBSTRATE 5-1212
(AUGER ELECTRONS) INNER SHELL ALIGNMENT OF ATOMS IN ELECTRON IMPACT IONIZATION. 2-0335
AUGER NEUTRALIZATION OF INERT GAS IONS ON THE SURFACE OF ALKALI HALIDE CRYSTALS. /TION OF THE FREQUENCY OF 5-1544
OM BOMBARDMENT. / ENERGY SPECTRUM OF ELECTRONS EMITTED BY ALKALI HALIDE CRYSTALS UNDER IMPACT OF ION AND AT 2-0246
EED AND AES. CRYSTAL SURFACES OF ALKALI HALIDES AND THIN FILM GROWTH OBSERVED BY L 5-1907
PHOTOEMISSION STUDIES OF THE DECAY OF CORE EXCITONS IN ALKALI HALIDES. (AUGER PROCESS IN VALENCE BAND) 2-0289
RFACES. (100) SURFACES OF ALKALI HALIDES. PART-1: AIR AND VACUUM CLEANED SU 5-1347
CIATION. (100) SURFACES OF ALKALI HALIDES. PART-2: ELECTRON STIMULATED DISSO 5-1908
AES) SOME PROPERTIES OF CLEAN AND ALKALI METAL COVERED ZINC OXIDE SURFACES. (LEED, 5-1881
SPECTROSCOPY AND LOW ENERGY ELECTRON DIFFRACTION STUDY OF ALKALI METAL OVERLAYERS ON TUNGSTEN (100). /TRON 5-1895
LEED- AUGER SPECTROSCOPY STUDY OF THE ADSORPTION OF ALKALI METALS ON MOLYBDENUM (110). 5-1896
WORK FUNCTION AND STRUCTURAL STUDIES OF ALKALI- COVERED SEMICONDUCTORS. 5-1958
PHOTOELECTRON SPECTROSCOPY AND ALLIED TECHNIQUES: GENERAL INTRODUCTION. 1-0149
LYSIS OF A COMMERCIAL COBALT BASE AIRCRAFT TURBINE SHROUD ALLOY. /SION, FRICTION AND AUGER SPECTROSCOPY ANA 6-2009
ES OF OXIDATION IN A COPPER- 19.6 ATOMIC PERCENT ALUMINUM ALLOY. /ECTRON SPECTROSCOPY STUDY OF INITIAL STAG 6-2028
PIC STUDY OF THE PASSIVE FILM FORMED ON AN IRON- CHROMIUM ALLOY. AUGER SCANNING ELECTRON MICROSCO 5-1753
CONDARY AND AUGER ELECTRON SPECTRA OF A COPPER- BERYLLIUM ALLOY. SE 4-0916
DIFFERENTIAL AUGER SPECTROMETRY. (NIOBIUM- ZIRCONIUM ALLOY ANALYSIS) 5-1849
/TIVITY PATTERN AND SURFACE COMPOSITION OF COPPER- NICKEL ALLOY. APPLICATION OF AUGER ELECTRON SPECTROSCO/ 5-1875
AUGER ELECTRON SPECTROSCOPY, SU/ OXIDATION OF NIMONIC-80A ALLOY AT 800 DEGREES C IN LOW OXYGEN PRESSURES. (6-2144
ESISTANCE OF TUNGSTEN ON HEAVILY DOPED SILICON- GERMANIUM ALLOY. (AUGER SPECTROSCOPY) /DES ON THE CONTACT R 5-1456
RECORDING THE COMPONENTS ON THE SURFACE OF A BINARY ALLOY. (AUGER SPECTROSCOPY) 5-1358
OF COPPER TO THE SURFACE OF A CLEAN ANNEALED 50-50 NICKEL ALLOY BY AUGER ELECTRON SPECTROSCOPY. /GREGATION 6-2044
AT/ OBSERVATION OF SURFACE COMPOSITION OF COPPER- NICKEL ALLOY BY AUGER SPECTRA IN THE LOWER ENERGY REGION 5-1982
GER E/ CHARACTERISTICS OF SURFACE OF THE PALLADIUM BARIUM ALLOY CATHODE IN RELATION TO ITS FILM SYSTEM. (AU 5-1129
OPY) WORK FUNCTION VARIATION WITH ALLOY COMPOSITION: COPPER- GOLD. (AUGER SPECTROSC 5-1330
PECTROSCOPY) WORK FUNCTION VARIATION WITH ALLOY COMPOSITION SILVER- GOLD. (AUGER ELECTRON S 4-0828
(MONEL, AUGE/ SURFACE CHARACTERIZATION OF NICKEL- COPPER ALLOY DISPLAYING POOR WETTABILITY DURING BRAZING. 5-1573
AUGER SPECTROSCOPIC STUDY OF THE SURFACES OF ALLOY EMITTERS. 5-1192
/MINATION OF THE SURFACE COMPOSITION OF PALLADIUM- NICKEL ALLOY FILM CATALYSTS USING AUGER ELECTRON SPECTR/ 5-1846
UDY OF THE SILICIDE FORMATION IN 85% NICKEL- 15% PLATINUM ALLOY FILMS. AES ST 5-1899
Y BARRIER HEIGHTS OF SILICIDES FORMED IN NICKEL- PLATINUM ALLOY FILMS. COMPOSITION PROFILES AND SCHOTTK 5-1900
WORK FUNCTION OF WELL-DEFINED SURFACE OF COPPER- NICKEL ALLOY PLATES. (AUGER ELECTRON SPECTROSCOPY) 5-1873
ECTROSCOPY, AND CONTACT POTENTIAL STUDIES OF COPPER- GOLD ALLOY SINGLE CRYSTAL SURFACES. LEED, AUGER SP 5-1748
CHARACTERIZATION OF COPPER- GOLD ALLOY SINGLE SURFACES. 5-1747
RON SPECTROSCOPY. ALLOY SPUTTERING STUDIES WITH IN-SITU AUGER ELECT 5-1877
TEMPER EMBRITTLEMENT OF MANGANESE- MOLYBDENUM- NICKEL LOW ALLOY STEEL. (AES) / OF COPPER AND PHOSPHORUS ON 6-2043
TRON EMISSION SPECTROSCOPY ON TEMPER EMBRITTLEMENT IN LOW ALLOY STEELS. STUDIES USING AUGER ELEC 6-2119
TEMPER EMBRITTLEMENT IN LOW ALLOY STEELS. (AES OF FRACTURE SURFACES) 6-2062
E OF MANGANESE AND SILICON IN TEMPER EMBRITTLEMENT OF LOW ALLOY STEELS. (AUGER SPECTROSCOPY) ROL 6-2059
IN BOUNDARIES AND ITS ROLE IN TEMPER EMBRITTLEMENT OF LOW ALLOY STEELS. (AUGER SPECTROSCOPY) /SELECTIVE GRA 6-2057
EFFECT OF SOLUTE ELEMENTS ON TEMPER EMBRITTLEMENT IN LOW ALLOY STEELS. (IMPURITY GRAIN BOUNDARY SEGREGATI/ 6-2083
ELECTRON SPECTROSCOPY OF (SILVER- GOLD AND INDIUM- LEAD) ALLOY SURFACES. AUGER 5-1839
QUANTITATIVE AUGER ANALYSIS OF COPPER- NICKEL ALLOY SURFACES AFTER ARGON ION BOMBARDMENT. 5-1807
/ER ALLOYS. QUANTITATIVE AUGER ANALYSIS OF COPPER- NICKEL ALLOY SURFACES AFTER ARGON ION BOMBARDMENT. COMM/ 5-1740
/ER ALLOYS. QUANTITATIVE AUGER ANALYSIS OF COPPER- NICKEL ALLOY SURFACES AFTER ARGON ION BOMBARDMENT. REPL/ 5-1200
TO COMMEN/ QUANTITATIVE AUGER ANALYSIS OF COPPER- NICKEL ALLOY SURFACES AFTER ARGON ION BOMBARDMENT. REPLY 5-1702
SEGREGATION AT COPPER- GOLD ALLOY SURFACES. (AUGER SPECTROSCOPY) 5-1631
AUGER ELECTRON SPECTROSCOPY OF ALLOY SURFACES. (GOLD- SILVER, LEAD- INDIUM) 5-1709
AUGER SPECTROSCOPY ON COPPER- NICKEL ALLOY SURFACES RELATED TO CATALYSIS. 5-1703

/RATION CHANGE OF TIN IN THE SURFACE LAYER OF COPPER- TIN ALLOY SYSTEM DUE TO FRICTION. (AUGER SPECTROSCOP/ 5-1867
VER- COPPER STUDIED BY AUGER ELECTRON / SPUTTERING OF THE ALLOY SYSTEMS SILVER- GOLD, GOLD- COPPER, AND SIL 5-1331
/ A TECHNIQUE FOR INVESTIGATING THE CATALYTIC ACTIVITY OF ALLOY SYSTEMS. (SURFACE COMPOSITION, INTERDIFFUSI 6-2003
AES)   LOW INTRINSIC STRESS AND GOOD ADHESION ALLOY THIN FILMS. (TITANIUM AND CHROMIUM ALLOYS, 5-1909
RFACE SEGREGATION IN A CCPPER- 10 ATCMIC PERCENT ALUMINUM ALLOY USING AUGER ELECTRON SPECTROSCOPY. /RIUM SU 6-2030
/LATION OF THE WETTING BEHAVIOR OF SODIUM ON NIMONIC-PE16 ALLOY WITH THE SURFACE OXYGEN- CHROMIUM RATIO DE/ 5-1441
LECTRON SP/ STUDIES OF INITIAL OXIDATION ON SILICON- IRON ALLOY (100) BY MEANS OF WORK FUNCTION AND AUGER E 5-1924
LECTRON SPECTROSCOPY OF 5.59 ATOMIC PERCENT SILICON- IRON ALLOY (100) SINGLE CRYSTALS. /UNCTION AND AUGER E 5-1925
ADSORPTION OF CARBON MONOXIDE ON SILVER- PALLADIUM ALLOYS. 5-1254
AES STUDIES OF SURFACE COMPCSITION OF SILVER- COPPER ALLOYS. 5-1199
AUGER ANALYSIS OF THIN OXIDE FILMS ON LEAD- INDIUM ALLOYS. 5-1249
AUGER ELECTRON SPECTROSCOPY OF SILVER- GOLD ALLOYS. 6-2024
AUGER ELECTRON SPECTRCSCOPY STUDIES ON ABRADED SURFACE OF ALLOYS.   AUGER ELECTRON SPEC 6-2131
TROSCOPY STUDY OF SURFACE SEGREGATION IN COPPER- ALUMINUM ALLOYS.   AUGER ELECTRON SPEC 6-2012
PECTRA OF ACTIVATED SILVER- MAGNESIUM AND SILVER- LITHIUM ALLOYS.   AUGER S 6-2004
AUGER SPECTRA OF COPPER- NICKEL ALLOYS. 4-1028
NICAL PROBLEMS. (EMBRITTLEMENT, SEGREGATION OF STEELS AND ALLOYS) /OSCOPY AND ITS APPLICATIONS TO SOME TECH 6-2100
DY OF EQUILIBRIUM SURFACE SEGREGATION IN COPPER- ALUMINUM ALLOYS.   AUGER SPECTROSCOPY AND LEED STU 6-2025
ECTROSCOPIC STUDIES OF THE OXIDATION OF NICKEL- PALLADIUM ALLOYS.   ELECTRON SP 5-1953
ES OF THE (100) SURFACES OF ALUMINUM AND ALUMINUM- COPPER ALLOYS. /DIFFRACTION AND AUGER SPECTROSCOPY STUDI 5-1755
SCOPY ANALYSIS OF SILVER- PALLADIUM AND NICKEL- PALLADIUM ALLOYS.   QUANTITATIVE AUGER ELECTRON SPECTRO 6-2088
EGATION OF BISMUTH TO GRAIN BOUNDARIES IN COPPER- BISMUTH ALLOYS.   SEGR 6-2096
SURFACE COMPOSITION OF NICKEL- GOLD ALLOYS. 5-1966
SURFACE ENRICHMENT IN COPPER(3) GOLD AND GOLD(3) COPPER ALLOYS. 5-1934
LYZER FOR AUGER ELECTRON SPECTROSCOPY. (PALLADIUM- NICKEL ALLOYS, ADSORBED OXYGEN) /ALL RETARDING FIELD ANA 3-0644
ND GOOD ADHESION ALLCY THIN FILMS. (TITANIUM AND CHROMIUM ALLOYS, AES)   LOW INTRINSIC STRESS A 5-1909
SPECTROSCOPY STUDIES WITH COPPER- NICKEL AND IRON- NICKEL ALLOYS. (AES)   SOFT X-RAY APPEARANCE POTENTIAL 5-1952
AN ALUMINUM- 5% MAGNESIUM TERNARY AND VARIOUS QUATERNARY ALLOYS. (AES)   STRESS CORROSION CRACKING OF 6-2036
OSCOPY STUDY ON THE SURFACE COMPCSITION CF CCPPER- NICKEL ALLOYS AFTER ANNEALING AND SPUTTERING. /ER SPECTR 5-1675
GER ELECTRONS TO THE STUDIES OF THE STRUCTURE OF ALUMINUM ALLOYS AND CORROSION OF THE ALLOYS. / BETA-RAY AU 6-2051
INFLUENCE OF SILICON ON FRICTION AND WEAR OF IRON- COBALT ALLOYS. (AUGER ANALYSIS) 6-2011
ORUS, IRON- PHOSPHORUS- SULFUR AND IRON- ANTIMONY- SULFUR ALLOYS. (AUGER ELECTRON EMISSION SPECTROSCOPY) /H 6-2098
ZING EFFECT AND BUBBLE STRENGTHENING IN TUNGSTEN- RHENIUM ALLOYS, AUGER ELECTRON MICROSCOPY). /NIUM DUCTILI 6-2111
CONFIRMATION OF SULFUR EMBRITTLEMENT IN NICKEL ALLOYS. (AUGER ELECTRON SPECTROSCOPY) 6-2053
AIN BOUNDARY FRACTURE SURFACES IN DOPED TUNGSTEN- RHENIUM ALLOYS. (AUGER ELECTRON SPECTROSCOPY) STUDY OF GR 6-2112
EGATION OF OXYGEN IN NIOBIUM- OXYGEN AND TANTALUM- OXYGEN ALLOYS. (AUGER ELECTRON SPECTROSCOPY) /RFACE SEGR 6-2065
R, STEEL/ GRAIN BOUNDARY EMBRITTLEMENT IN SOME METALS AND ALLOYS. (AUGER EMISSION ANALYSIS, TUNGSTEN, COPPE 6-2058
ELECTRON HOLE DROPS IN GERMANIUM- SILICON ALLOYS. (AUGER RECOMBINATION) 4-0777
CTION AND WEAR CHARACTERISTICS OF COPPER- TIN- PHOSPHORUS ALLOYS. (AUGER SPECTROSCOPY) /OSPHORUS ON THE FRI 5-1868
CENT, AND PALLADIUM- GOLD 70 ATOMIC PERCENT CONTACT METAL ALLOYS. (BY AUGER ELECTRON SPECTROSCOPY) /MIC PER 6-2041
/E SURFACE COMPOSITION OF PASSIVE FILMS CN CCPPER- NICKEL ALLOYS BY AUGER ELECTRON SPECTROSCOPY AND X-RAY / 5-1874
QUANTITATIVE STUDY OF THE SURFACE COMPOSITION OF BINARY ALLOYS BY AUGER SPECTROSCOPY. 6-2005
RMED CN COPPER- ALUMINUM, COPPER- TIN, AND IRON- ALUMINUM ALLOYS. (BY LEED AND AES) /FRICTION STUDIES PERFO 6-2033
COMPOSITION OF BINARY ALLOYS BY SIMULTANEOUS SIMS AND AES MEASUREMENTS. 6-2090
/ SPECTROSCOPY STUDY OF SURFACE SEGREGATION IN THE BINARY ALLOYS COPPER- 1 ATOMIC PERCENT INDIUM, COPPER- / 6-2027
PERATURE ON SURFACE COMPOSITION CHANGES CF COPPER- NICKEL ALLOYS DURING ARGON BOMBARDMENT. /T OF TARGET TEM 5-1808
11) SURFACE OF NICKEL. (AES)   SURFACE SEGREGATION IN ALLOYS: EQUILIBRIUM SEGREGATION OF GOLD TO THE (1 5-1216
/TERMINATION OF THE SURFACE COMPOSITION CF CCPPER- NICKEL ALLOYS FOR CLEAN AND ADSORBATE COVERED SURFACES. 5-1424
/OSITION AND DEPTH CONCENTRATION PROFILE OF PLATINUM- TIN ALLOYS FROM COMBINED X-RAY PHOTOELECTRON AND AUG/ 5-1195
SURFACE COMPOSITICN OF (PALLADIUM- SILVER) ALLOYS IN EQUILIBRIUM. 5-1788
XCITED AUGER AND PHOTOELECTRON SPECTRA OF MAGNESIUM, SOME ALLOYS OF MAGNESIUM, AND ITS OXIDE.   X-RAY E 4-0837
ADSORPTION ON SINGLE CRYSTAL SURFACES CF CCPPER- NICKEL ALLOYS. PART-1. 5-1322
AES)   OXIDATION OF CUPRONICKEL ALLOYS. PART-1: XPS STUDY OF INTERDIFFUSION. (AND 5-1224
/ROSCOPY STUDIES OF SURFACE COMPOSITION OF SILVER- COPPER ALLOYS. QUANTITATIVE AUGER ANALYSIS OF COPPER- N/ 5-1200
/ROSCOPY STUDIES OF SURFACE COMPCSITION OF SILVER- COPPER ALLOYS. QUANTITATIVE AUGER ANALYSIS OF COPPER- N/ 5-1740
/N BOUNDARY SEGREGATION TO REDUCED STRENGTH OF METALS AND ALLOYS. (SEGREGATION IN STEEL, TUNGSTEN, COPPER-/ 6-2120
C STATE OF SILICON IN SILICON- NOBLE METAL VAPOR QUENCHED ALLOYS STUDIED BY AUGER ELECTRON SPECTROSCOPY. /I 5-1438
SURFACE SEGREGATION STUDIES IN ALLOYS USING AUGER ELECTRON SPECTROSCOPY. 6-2019
ALLIUM ARSENI/ METALLURGICAL AND ELECTRICAL PROPERTIES OF ALLOYED NICKEL- GOLD- GERMANIUM FILMS ON N-TYPE G 6-2101
MISSION CORE LEVEL AND AUGER SPECTRA OF NICKEL AND NICKEL ALLOYED WITH ZINC. /L STATE EFFECTS IN THE PHOTOE 5-1806
TION. (AUGER SPECTROSCCPY ANALYSIS, STAIN/ SEGREGATION OF ALLOYING ELEMENTS TO FREE SURFACES DURING IRRADIA 6-2091
. (INTERFACE AUGER ELECTRON SPECTROSCOPY)   SPONTANEOUS ALLOYING OF A GOLD SUBSTRATE WITH LEAD MONOLAYERS 5-1178
CES OF GOLD. (AUGER EFFECT)   ADSORPTION AND SURFACE ALLOYING OF LEAD MONOLAYERS ON (111) AND (110) FA 5-1733
OR.   SENSITIVITY VARIATIONS IN BAYARD- ALPERT GAUGES CAUSED BY AUGER EMISSION AT COLLECT 3-0618
FOR THE CONTINUOUS MEASUREMENT OF CHANGES IN WORK FU/ AN ALTERNATING CURRENT RETARDING POTENTIAL TECHNIQUE 3-0663
GRAIN BOUNDARY SEGREGATION IN SINTERED ALUMINA. (AUGER ELECTRON SPECTROSCOPY) 6-2056
AUGER ELECTRON SPECTROSCOPY STUDY OF BETA ALUMINA ELECTROLYTE. 4-1074
, FRACTURED/ GRAIN BOUNDARY SEGREGATION IN MAGNESIA DOPED ALUMINA. (HIGH VACUUM AUGER ELECTRON SPECTROMETER 5-1615
ROSCOPIC STUDY OF THE POISONING OF A COMMERCIAL PALLADIUM ALUMINA HYDROGENATION CATALYST.   AUGER SPECT 5-1177
ND IMPURITY DISTRIBUTIONS AT GRAIN BOUNDARIES IN SINTERED ALUMINA. (SCANNING AUGER MICROPROBE)   ADDITIVE A 5-1505
D ADHESION OF THIN FILMS. (AUGER ELECTRON SPECTROSCOPY OF ALUMINA SUBSTRATES) /BSTRATE SURFACE CHEMISTRY AN 5-1851
(AUGER) ANALYSIS OF CHROMATE CONVERSION COATINGS ON ALUMINUM. 6-2125
AUGER CASCADES IN ALUMINUM. 4-0973
AUGER ELECTRON ANALYSIS OF ELECTROPOLISHED HIGH PURITY ALUMINUM. 4-0824
AUGER ELECTRON SPECTROSCOPIC ANALYSIS OF ANODIC FILMS ON ALUMINUM. 5-1614
AUGER SPECTRA OF MAGNESIUM AND ALUMINUM. 4-0822
ASTING VALENCE BAND AUGER ELECTRON SPECTRA FCR SILVER AND ALUMINUM.   CONTR 4-1022
STALLINE EFFECTS IN BACKSCATTERING AND AUGER PRODUCTION. (ALUMINUM)   CRY 2-0254
NTRIBUTION TO KLM AUGER SPECTRA OF SODIUM, MAGNESIUM, AND ALUMINUM. /ATION OF THE INTERNAL PHOTOEMISSION CO 4-0769
IDENTIFICATION OF AUGER SPECTRA FROM ALUMINUM. 4-1027
ION EXCITED AUGER SPECTRA OF ALUMINUM. 4-0869
FETIME AND SURFACE EFFECTS ON THE L23 VV AUGER SPECTRA OF ALUMINUM.   LI 4-0833
NS IN THE STUDY OF THE STRUCTURE AND PITTING CORRCSION OF ALUMINUM. /W MICRORADIOGRAPHY USING AUGER ELECTRO 6-2066
ASMON EFFECTS IN ELECTRON ENERGY LOSS AND GAIN SPECTRA IN ALUMINUM.   PL 4-1020
SMON SATELLITE IN THE AUGER EMISSION SPECTRUM OF METALS. (ALUMINUM)   PLA 4-0871
VALENCE BAND STRUCTURE IN THE AUGER SPECTRUM OF ALUMINUM. 4-0889
CTRCSCOPY).   CHARACTERIZATION OF ALUMINUM ADHEREND SURFACES (BY AUGER ELECTRON SPE 5-1725
TIAL STAGES OF OXIDATICN IN A COPPER- 19.6 ATCMIC PERCENT ALUMINUM ALLOY. /ECTRON SPECTROSCOPY STUDY OF INI 6-2028
/BRIUM SURFACE SEGREGATION IN A COPPER- 10 ATCMIC PERCENT ALUMINUM ALLOY USING AUGER ELECTRON SPECTROSCOPY. 6-2030
TRON SPECTROSCOPY STUDY OF SURFACE SEGREGATION IN COPPER- ALUMINUM ALLOYS.   AUGER ELEC 6-2012

LEED STUDY OF EQUILIBRIUM SURFACE SEGREGATION IN COPPER- ALUMINUM ALLOYS.   AUGER SPECTROSCOPY AND 6-2025
TA-RAY AUGER ELECTRONS TO THE STUDIES OF THE STRUCTURE OF ALUMINUM ALLOYS AND CORROSION OF THE ALLOYS. / BE 6-2051
N AND AUGER SPECTROSCOPY STUDIES OF THE (100) SURFACES OF ALUMINUM AND ALUMINUM- COPPER ALLOYS. /DIFFRACTIO 5-1755
               SECONDARY ELECTRON EMISSION OF SOLID ALUMINUM AND MAGNESIUM TARGETS.   2-0373
     AES CHARACTERIZATION OF OXIDIZED FILMS CF MAGNESIUM, ALUMINUM, AND SILICON.   5-1847
Y SATELLITES OF L2,3 EMISSION BANDS OF SODIUM, MAGNESIUM, ALUMINUM, AND SILICON. /ATION OF HIGH ENERGY X-RA 2-0239
RON AND LOW EN/ COMPARISON OF AUGER SPECTRA CF MAGNESIUM, ALUMINUM, AND SILICON EXCITED BY LOW ENERGY ELECT 4-0857
ION OF AUGER ELECTRONS EMITTED FROM BERYLLIUM, MAGNESIUM, ALUMINUM AND SILICON UNDER ION BOMBARDMENT. /IBUT 4-0876
       (AUGER) ELECTRON SPECTROSCOPY OF GALLIUM ARSENIDE AND ALUMINUM ARSENIDE SURFACES.   5-1595
       AUGER SPECTROSCOPY OF HYDROCHLORIC ACID INTERACTION WITH ALUMINUM AT LOW PRESSURES.   5-1576
                            ELECTRON ESCAPE DEPTHS IN ALUMINUM. (AUGER ELECTRONS)   2-0537
          K-SHELL FLUORESCENCE YIELD OF IONIZED ALUMINUM. (AUGER RATE)   4-0975
                  CHLORINE ADSORPTION ON CLEAN ALUMINUM. (AUGER SPECTOSCOPY)   5-1831
TIC GAIN PHENOMENA IN POLYCRYSTALLINE COPPER, SILVER, AND ALUMINUM. (AUGER SPECTROSCOPY)   CHARACTERIS 5-1724
BSERVATION OF A PLASMON GAIN IN THE FINE STRUCTURE OF THE ALUMINUM AUGER SPECTRUM.   O 4-1085
     SECONDARY ELECTRON EMISSION OF A ZIRCONIUM- ALUMINUM BULK GETTER. (AUGER SPECTROSCOPY)   5-1407
        SURFACE CONCENTRATION OF MAGNESIUM IN 6061 ALUMINUM BY AUGER ELECTRON SPECTROSCOPY.   5-1824
AND SUPPLEMENTARY ELECTRON EJECTION IN THE IONIZATION OF ALUMINUM BY ELECTRON IMPACT.   AUGER TRANSITION 2-0230
        VALENCE BAND AUGER ELECTRON SPECTRA FOR ALUMINUM: COMMENTS.   4-0839
/ION, ADHESION, AND FRICTION STUDIES PERFORMED CN COPPER- ALUMINUM, COPPER- TIN, AND IRON- ALUMINUM ALLOYS/ 6-2033
/N DEPENDENCE OF GREEN- COPPER LUMINESCENCE IN COPPER AND ALUMINUM DOPED ZINC SULFIDE. (AUGER EFFECT, COLO/   2-0402
METRY, AND AUGER SPECTROSCOPY.   CHARACTERIZING OF ALUMINUM FATIGUE DAMAGE BY PHOTOEMISSION, ELLIPSO 6-2116
            PURITY AND MORPHOLOGY OF ALUMINUM FILMS. (AUGER SPECTROSCOPY)   5-1299
T. (AUGER SPECTRA)   DOUBLE AND TRIPLE IONIZATION IN ALUMINUM, GALLIUM, AND SCANDIUM BY ELECTRON IMPAC 2-0232
FECT AND RADIATIVE RECOMBINATION IN N-GALLIUM ARSENIDE- P-ALUMINUM GALLIUM ARSENIDE HETEROJUNCTIONS. /ER EF 2-0324
        IONIZATION CROSS SECTION OF THE L2,3 SHELL OF ALUMINUM IN A NEON (+)- ALUMINUM COLLISION.   2-0267
THE HIGH ENERGY SIDE CF THE KL2,3M AUGER PEAK FROM SOLID ALUMINUM: INTERNAL PHOTOEMISSION.   STRUCTURE ON 4-1021
SITIES OF AUGER LINES EXCITED BY ELECTRON BOMBARDMENT AND ALUMINUM K-ALPHA X-RAY IRRADIATION. /UDY OF INTEN 4-1017
GER SPECTROSCOPY)   INTERACTION OF ALUMINUM LAYERS WITH POLYCRYSTALLINE SILICON. (AU 5-1672
RUCTURE IN THE SECONDARY ELECTRON ENERGY DISTRIBUTIONS OF ALUMINUM, NICKEL, AND COPPER. /D PHOTOELECTRON ST 4-1023
              LEED AUGER STUDY CF GROWTH OF ALUMINUM ON NIOBIUM (110).   5-1491
TS OF PREADSORBED OXYGEN CONTAMINANT ON THE ADSORPTION OF ALUMINUM ON SILICON (111).   EFFEC 5-1169
          AUGER- LEED INVESTIGATION OF THE DEPOSITION OF ALUMINUM ONTO THE MOLYBDENUM (110) SURFACE.   5-1489
INATION OF THE ATTENUATION OF THE LOW ENERGY ELECTRONS IN ALUMINUM OXIDE. (AUGER ELECTRON LINES) /ON DETERM 4-0773
                GRAIN BOUNDARY SEGREGATION IN ALUMINUM OXIDE. (AUGER SPECTROSCOPY)   6-2134
-1: AUGER ELECTRONS. (KLL TRANSITION ENERGIES FOR SULFUR, ALUMINUM, SILICON) /ON AND CHEMICAL BONDING. PART 2-0327
         AUGER SATELLITES ON THE L23 AUGER EMISSION BANDS OF ALUMINUM, SILICON, AND PHOSPHORUS.   4-0984
         AUGER SATELLITES OF THE L2,3 AUGER EMISSION BANDS OF ALUMINUM, SILICON, AND PHOSPHORUS.   4-0985
/RON SPECTROSCOPY AND SECONDARY ION MASS SPECTROMETRY FOR ALUMINUM, SILICON, TITANIUM, VANADIUM, AND CHROM/ 5-1553
HARACTERISTIC ENERGIES IN SECONDARY ELECTRON SPECTRA FROM ALUMINUM SURFACES.   AUGER AND OTHER C 4-0896
               DETECTION OF ORGANIC CONTAMINANTS ON ALUMINUM SURFACES BY AUGER ELECTRON SPECTROSCOPY. 5-1487
UGER ELECTRON SPECTROSCOPY TO THE STUDY OF DISCOLORING OF ALUMINUM SURFACES IN BOILING WATER. /ICATION OF A 5-1127
OF MOLYBDENUM FROM VARIOUS METAL SURFACES (COPPER, GOLD, ALUMINUM, TUNGSTEN). /OSITION AND SPUTTER REMOVAL 5-1878
TER ELLIPSOMETER SYSTEM.   STUDY OF ALUMINUM WITH A COMBINED AUGER ELECTRON SPECTROME 5-1132
GER ELECTRON SPECTROSCOPY.   GALLIUM(1-X) ALUMINUM(X) ARSENIDE SUPERLATTICES PROFILED BY AU 5-1596
SITION AND AUGER ANALYSIS CF DEPOSITED SILICON DIOXIDE ON ALUMINUM(X) GALLIUM(1-X) ARSENIDE.   DEPO 5-1962
AM EPITAXY. (AUGER / MAGNESIUM DOPED GALLIUM ARSENIDE AND ALUMINUM(X) GALLIUM(1-X) ARSENIDE BY MOLECULAR BE 5-1247
ENT DIODES. (AUG/ EMISSION SPECTRA OF SILICON COMPENSATED ALUMINUM(X) GALLIUM(1-X) ARSENIDE ELECTROLUMINESC 2-0325
LECULAR BEAM EPITAXY. (AUGER SURFACE ANALYSIS, GALLIUM(X) ALUMINUM(1-X) ARSENIDE)   COMPUTER-CONTROLLED MO 5-1324
GER SPECTROSCOPY.   REEXAMINATION OF ALUMINUM (110) WITH LOW ENERGY DIFFRACTION AND AU 5-1507
SURFACES USING INELASTIC LOW / SURFACE PLASMON STUDIES ON ALUMINUM (111) AND ALUMINUM (001) SINGLE CRYSTAL 5-1745
TRON DIFFRACTION, AUGER, AND CONTAC/ SODIUM ADSORPTION ON ALUMINUM (111) SURFACES: ELECTRIC LOW ENERGY ELEC 5-1744
PECTROSCOPY STUDIES OF THE (100) SURFACES OF ALUMINUM AND ALUMINUM- COPPER ALLOYS. /DIFFRACTION AND AUGER S 5-1755
                 IN-DEPTH AUGER ANALYSIS OF ALUMINUM- SILICON INTERFACIAL REACTIONS.   5-1219
NARY ALLOYS. (AES)   STRESS CORROSION CRACKING OF AN ALUMINUM- 5% MAGNESIUM TERNARY AND VARIOUS QUATER 6-2036
N EFFECTS IN NARROW GAP MERCURY(1-X) CADMIUM(X) TELLURIDE AMBIENT TEMPERATURES. (AUGER RECOMBINATION) /CTRO 2-0343
AUGER AND BACKSCATTERING ANALYSES.   EFFECT OF OXIDIZING AMBIENTS ON PLATINUM SILICIDE FORMATION. PART-2: 5-1182
        THE L-AUGER SPECTRA OF PLUTONIUM, AMERICIUM, CALIFORNIUM, AND FERMIUM.   4-0836
/TON STATES, ANOMALOUS CONVERSION, AND K-AUGER SPECTRA IN AMERICIUM-241 FROM CURIUM-241 (ELECTRON CAPTURE)/ 4-1019
/ SPECTROSCOPIES TO CATALYTIC PROCESSES. CHEMISORPTION OF AMMONIA, NITROGEN, AND NITRIDE FORMATION ON A MO/ 5-1533
ENTS OF THE NITROGEN KLL AUGER TRANSITIONS OF CHEMISORBED AMMONIA ON A MOLYBDENUM SURFACE. /LUTION MEASUREM 5-1532
STRESS CORROSION CRACKING OF ALPHA BRASS IN A TARNISHING AMMONIACAL ENVIRONMENT: FRACTOGRAPHY AND CHEMICA/ 6-2095
       AUGER SPECTROSCOPY STUDIES OF THE OXIDATION OF AMORPHOUS AND CRYSTALLINE GERMANIUM.   5-1431
R SPECTROSCOPY.   OXIDATION STUDIES OF AMORPHOUS AND CRYSTALLINE GERMANIUM FILMS BY AUGE 6-2023
TRON SPECTROSCOPY.   STUDY OF AMORPHOUS AND CRYSTALLINE SILICATES BY AUGER ELEC 4-0796
ECTRON SPECTROSCOPY OF GOLD CONDENSED ON ROCK SALT AND ON AMORPHOUS CARBON.   QUANTITATIVE AUGER EL 5-1136
/ENTS OF GOLD ON ROCK SALT (100) CLEAVAGE SURFACES AND ON AMORPHOUS CARBON BY MEANS OF QUANTITATIVE AUGER / 5-1135
IN-SITU TRANSPORT MEASUREMENTS IN ULTRAHIGH VACUUM GROWN AMORPHOUS GERMANIUM. (DEPTH PROFILES OF OXYGEN A/ 5-1547
(AUGER SPECTROSCOPY)   TRANSPORT OF OXYGEN IN AMORPHOUS GERMANIUM THIN FILMS DURING ANNEALING. 5-1548
        SECONDARY ELECTRON EMISSION OF AMORPHOUS SEMICONDUCTOR THIN FILMS.   5-1239
11) SURFACES. (AUGER ELECTRON SPECTROSCOPY) AMORPHOUS TO CRYSTALLINE TRANSITION OF SILICON (1 5-1780
CT.   SOFT X-RAY AMPLIFIED SPONTANEOUS EMISSION VIA THE AUGER EFFE 2-0443
E METHOD OF LOW ENERGY ELECTRON SPECTROSCOPY. (AUGER PEAK AMPLITUDE CALIBRATION) /OXYGEN ON GALLIUM ARSENID 5-1641
/CTIONS OF AUGER ELECTRON SIGNAL STRENGTHS BY MODULATION AMPLITUDE DISTORTION IN A 4-GRID RETARDING POTEN/ 3-0620
ANIUM.   EFFECT OF MODULATION AMPLITUDE ON ELECTRON EXCITED AUGER DATA FROM TIT 2-0367
DING POTENTIAL ANALYZERS. COMMENTS.   SECONDARY EMISSION ANALOG FOR IMPROVED AUGER SPECTROSCOPY WITH RETAR 3-0584
DING POTENTIAL ANALYZERS. (REPLY TO C/ SECONDARY EMISSION ANALOG FOR IMPROVED AUGER SPECTROSCOPY WITH RETAR 3-0694
ION FOR QUANTITATIVE AUGER ELECTRON SPECTROSCOPY./ USE OF ANALOG INTEGRATION IN DYNAMIC BACKGROUND SUBTRACT 1-0069
CTRON CURRENTS. APPLICATION TO SULFUR ON NICKEL (110).   ANALOG TECHNIQUE FOR THE MEASUREMENT OF AUGER ELE 3-0635
                 APPLICATION OF TRIPLE GRID LEED SYSTEM TO AUGER SPECTRUM ANALYSES.   3-0661
INUM SILICIDE FORMATION. PART-2: AUGER AND BACKSCATTERING ANALYSES.   EFFECT OF OXIDIZING AMBIENTS ON PLAT 5-1182
COPY AND AUGER ELECTRON EMISSION SPECTROSCOPY.   SURFACE ANALYSES BY COMBINED X-RAY PHOTOELECTRON SPECTROS 5-1272
       X-RAY PHOTOELECTRON AND AUGER SPECTROSCOPIC ANALYSES OF STEEL SURFACES.   5-1268
ERGY LOSS PEAKS OF THE PRIMARY BEAM IN SECONDARY ELECTRON ANALYSIS. /NG METHOD TO DISTINGUISH AUGER FROM EN 1-0188
AUGER SPECTROSCOPY AND ELECTRON SPECTROSCOPY FOR CHEMICAL ANALYSIS / IN SURFACE ANALYSIS METHODS. (REVIEW 1-0215
ITED BY BACKSCATTERED ELECTRONS. APPROACH TO QUANTITATIVE ANALYSIS.   AUGER AND SECONDARY ELECTRONS EXC 2-0366
              AUGER ELECTRON EMISSION ANALYSIS.   1-0073
      AUGER ELECTRON SPECTROSCOPY FOR CHEMICAL ANALYSIS.   1-0029
        AUGER ELECTRON SPECTROSCOPY FOR SURFACE ANALYSIS.   1-0072
        AUGER ELECTRON SPECTROSCOPY FOR SURFACE ANALYSIS.   1-0200

AUGER ELECTRON SPECTROSCOPY FOR THIN FILM ANALYSIS. 1-0211
AUGER ELECTRON SPECTROSCOPY FOR THIN FILM ANALYSIS. 1-0212
AND OXIDIZED VANADIUM: CHEMICAL SHIFT AND PEAK INTENSITY ANALYSIS. /EMICAL COMPOSITION OF VANADIUM OXIDES, 4-1089
AND STRUCTURE OF SILICON BY CHARACTERISTIC AUGER SPECTRUM ANALYSIS. B 4-0740
OF A CHROMIUM- GOLD THIN FILM DEPOSITION PROCESS BY AUGER ANALYSIS. CHARACTERIZATION 1-0037
W, AUGER SPECTROSCOPY, ELECTRON SPECTROSCOPY FOR CHEMICAL ANALYSIS /ACTERIZATION OF SOLID SURFACES. (REVIE 1-0134
CONDUCTIVITY MEASUREMENTS ON ZINC OXIDE SURFACES. (AUGER ANALYSIS) CHEMISORPTION, PHOTODESORPTION AND 5-1801
CHOOSING BETWEEN ESCA AND AUGER FOR SURFACE ANALYSIS. 5-1635
COMPARISON OF THE TECHNIQUES FOR SILICON SURFACE ANALYSIS. 5-1327
ONTO TANTALUM (100) SINGLE CRYSTAL SURFACES. LEED AND AES ANALYSIS. CONDENSATION OF GOLD 5-1312
RAMIC SURFACES. (AUGER SPECTROSCCPY, LEED, ION MICROPROBE ANALYSIS) CRYSTALLOGRAPHY AND CHEMISTRY OF CE 5-1852
OF IMPURITIES ON A SILICON (111) SURFACE BY AUGER- LEED ANALYSIS. DETECTION 5-1360
CRITICAL TEMPERATURE NIOBIUM(3) GERMANIUM FILMS BY AUGER ANALYSIS. DETECTION OF SURFACE NITROGEN IN HIGH 5-1804
IFFERENTIAL AUGER SPECTRCMETRY. (NIOBIUM- ZIRCONIUM ALLOY ANALYSIS) D 5-1849
BRITTLENESS IN A NICKEL- CHROMIUM- ANTIMONY STEEL. (AUGER ANALYSIS) / GRAIN BOUNDARY COMPOSITION ON TEMPER 6-2114
F YTTRIA ADDITIONS ON HOT PRESSED SILICON NITRIDE. (AUGER ANALYSIS) EFFECT O 5-1350
ELECTRON SPECTROMETER FOR SURFACE ANALYSIS. 3-0715
ELECTRONS OR IONS. (IN SURFACE ANALYSIS) 5-1765
GENERAL FORMALISM FOR QUANTITATIVE AUGER ANALYSIS. 1-0032
GERMANIUM SURFACE CLEANING: AUGER ANALYSIS. 5-1540
ONS IN HOT PRESSED SILICON NITRIDE. (AUGER SPECTROGRAPHIC ANALYSIS) IMPURITIES AND INCLUSI 5-1557
LICON ON FRICTION AND WEAR OF IRON- COBALT ALLOYS. (AUGER ANALYSIS) INFLUENCE OF SI 6-2011
SPECTRAL LINES IN THE ELECTRON SPECTROSCOPY FOR CHEMICAL ANALYSIS. INFORMATION OBTAINED FROM THE WIDTH OF 4-1009
INTERACTION OF SULFUR DIOXIDE WITH TUNGSTEN. (AUGER ANALYSIS) 5-1367
CANNING ELECTRON MICROSCOPY, AUGER SPECTROSCOPY, ION BEAM ANALYSIS) / PRODUCTS IN FLOWING LIQUID SODIUM. (S 6-2014
NIOBIUM SURFACES FOR RF SUPERCONDUCTORS. (AUGER SURFACE ANALYSIS) 5-1472
PHYSICAL METHODS OF SURFACE ANALYSIS. 1-0078
OMPOSITION PROFILING BY SIMULTANEOUS SPUTTERING AND AUGER ANALYSIS. RECENT ADVANCES IN C 1-0155
ROLE OF AUGER ELECTRON SPECTROSCOPY IN SURFACE ELEMENTAL ANALYSIS. 1-0194
SCANNING AUGER MICROSCOPE FOR THIN FILM ANALYSIS. 3-0645
SOME TOPICS ON AES. (SURFACE ANALYSIS) 1-0106
SPREADING FUNCTIONS IN SECONDARY ELECTRON ANALYSIS. 2-0512
ING INELASTIC LOW ENERGY ELECTRON DIFFRACTION. (AND AUGER ANALYSIS) /MINUM (001) SINGLE CRYSTAL SURFACES US 5-1745
TECHNIQUE OF PREPARING POWDER SAMPLES FOR AES AND/OR ESCA ANALYSIS. 1-0197
EEL CONTAINING PHOSPHORUS AS AN IMPURITY. (AUGER EMISSION ANALYSIS) /EMBRITTLEMENT IN A NICKEL- CHROMIUM ST 6-2138
USE OF SCANNING AUGER SPECTROSCOPY FOR SURFACE ANALYSIS. 5-1280
N EXCITATION. (REVIEW, ELECTRON SPECTROSCOPY FOR CHEMICAL ANALYSIS AES) /N SPECTROSCOPY: X-RAY AND ELECTRO 1-0081
COMPARISON OF THE TECHNIQUES FOR SILICON SURFACE ANALYSIS. (AES, ESCA, SIMS, AND OTHERS) 1-0052
SAMPLES FOR AES AND/OR ELECTRON SPECTROSCOPY FOR CHEMICAL ANALYSIS. TECHNIQUE OF PREPARING POWDER 1-0197
TROSCOPY) INTRODUCTION TO LOW ENERGY X-RAY AND ELECTRON ANALYSIS. (ANALYZER FOR PHOTO AUGER ELECTRON SPEC 3-0627
/PECTROSCOPY. A NEW TECHNIQUE FOR MATERIAL AND STRUCTURAL ANALYSIS AND A REVIEW OF ITS POSSIBLE APPLICATIO/ 1-0084
CHOOSING BETWEEN ELECTRON SPECTROSCOPY FOR CHEMICAL ANALYSIS AND AUGER FOR SURFACE ANALYSIS. 5-1635
COMBINED ELECTRON SPECTROSCOPY FOR CHEMICAL ANALYSIS AND AUGER SPECTROMETER. 3-0671
DEPOSITED SILVER/ UPS, ELECTRON SPECTROSCOPY FOR CHEMICAL ANALYSIS AND AUGER SPECTROSCOPY STUDY OF VACUUM 5-1947
OXIDE, AND COPPER(II/ ELECTRON SPECTROSCOPY FOR CHEMICAL ANALYSIS (AND AUGER) STUDIES OF COPPER, COPPER(I) 5-1794
FOR ELECTRONIC APPLICATIONS. ANALYSIS AND CHARACTERIZATION OF CERAMIC SURFACES 5-1853
/UM OF A NICKEL (110)- C(2X2) SULFUR SURFACE: LINE SHAPE ANALYSIS AND CORRELATION WITH ION NEUTRALIZATIO/ 4-1065
CTRON SIGNALS) ELECTRON BEAM MICROPROBES USED IN THE ANALYSIS AND DEVELOPMENT OF MATERIALS. (AUGER ELE 3-0714
/SPECTROSCOPY, PHOTON, ELECTRON SPECTROSCOPY FOR CHEMICAL ANALYSIS, AND FIELD STIMULATED ELECTRON EMISSION. 5-1901
AUGER ELECTRON ANALYSIS AND ITS APPLICATIONS. 1-0202
CTROSCOPY) PRESENT STATUS OF BEAM ANALYSIS AND ITS FUTURE. (REVIEW, AUGER, MASS SPE 1-0181
/LECTRON SPECTROSCOPY (ELECTRON SPECTROSCOPY FOR CHEMICAL ANALYSIS AND SECONDARY ION MASS SPECTROSCOPY. / 1-0088
IV) OXIDE. AES ION SPUTTERING ANALYSIS AND THE SURFACE COMPOSITION OF TITANIUM(5-1893
N AND STEEL. (AES, ESCA, SIMS) NEW METHODS IN SURFACE ANALYSIS AND THEIR APPLICATION FOR STUDIES ON IRO 6-2035
MEANS OF EMITTED ELECTRONS. ANALYSIS APPARATUS FOR INVESTIGATING A SAMPLE BY 3-0713
OPY. FAILURE ANALYSIS APPLICATIONS OF AUGER ELECTRON SPECTROSC 6-2034
NERGY CONVERSION. AUGER THIN FILM ANALYSIS AS APPLIED TO VARIOUS ASPECTS IN SOLAR E 5-1960
.8) TIN(0.2) TELLURIDE DETECTOR FABRICATION USING SURFACE ANALYSIS. (AUGER) EVALUATION OF LEAD(0 5-1593
NNING ELECTRON MICRCSCOPE NOW INCLUDES SECONDARY ION MASS ANALYSIS. (AUGER) SPECTROSCOPY IN THE SCA 3-0674
TIONS USING A COMBINED ELECTRON SPECTROSCOPY FOR CHEMICAL ANALYSIS AUGER ANALYSIS SYSTEM. /N DEVICE APPLICA 5-1289
TRON MICROSCOPE, ENERGY DISPERSIVE SPECTROMETER, ION MASS ANALYSIS, AUGER ELECTRON MICROSCOPY) /ISSION ELEC 1-0010
CROPRO/ RECENT ADVANCES IN INSTRUMENTATION FCR MICROPROBE ANALYSIS. (AUGER ELECTRON SPECTROSCOPY AND ION MI 3-0716
6 REFS) ELECTRONS AND X-RAYS IN NONDESTRUCTIVE ANALYSIS. (AUGER, PHOTOELECTRON SPECTRA, REVIEW, 1-0083
SURFACE AND THIN FILM ANALYSIS. (AUGER SPECTROMETERS) 3-0609
TECTION AND DETERMINATION LIMITS IN SURFACE AND THIN FILM ANALYSIS. (AUGER SPECTROSCOPY) ESTABLISHING DE 5-1572
ILM ANALYSIS METHODS. (ELECTRON SPECTROSCOPY FOR CHEMICAL ANALYSIS AUGER SPECTROSCOPY) /UTTERING IN THIN F 1-0214
NISHING AMMONIACAL ENVIRONMENT: FRACTOGRAPHY AND CHEMICAL ANALYSIS. (AUGER SPECTROSCOPY) /HA BRASS IN A TAR 6-2095
RECENT ADVANCES IN SURFACE STUDIES. ION BEAM ANALYSIS. (AUGER SPECTROSCOPY, REVIEW) 1-0137
/PECTROSCOPY. COMBINED ELECTRON SPECTROSCOPY FOR CHEMICAL ANALYSIS AUGER SYSTEM BASED ON THE DOUBLE PASS C/ 3-0654
/ SPECTROMETER USED IN ELECTRON SPECTROMETRY FOR CHEMICAL ANALYSIS. (AUTOMATIC RECORDING OF PHOTO AND AUGE/ 3-0717
SURFACE ANALYSIS BY APPEARANCE POTENTIAL SPECTROSCOPY. 5-1545
APPARATUS FOR ANALYSIS BY AUGER ELECTRON SPECTROSCOPY. 3-0595
GLASS SURFACE ANALYSIS BY AUGER ELECTRON SPECTROSCOPY. 5-1720
SURFACE ANALYSIS BY AUGER ELECTRON SPECTROSCOPY. 5-1402
RANCE POTENTIAL SPECTROSCOPY: A COMPAR/ SURFACE CHEMICAL ANALYSIS BY AUGER ELECTRON SPECTROSCOPY AND APPEA 5-1910
THIN FILM ANALYSIS BY AUGER ELECTRON SPECTROSCOPY. (REVIEW) 1-0210
EFFECT OF THE ELECTRON EJECTION DEPTH ON QUANTITATIVE ANALYSIS BY AUGER SPECTROSCOPY. 2-0449
RELATIVE SPUTTERING YIELDS AND QUANTITATIVE SURFACE ANALYSIS BY AUGER SPECTROSCOPY. 5-1964
SURFACE ANALYSIS BY AUGER SPECTROSCOPY. 1-0103
APPARATUS FOR SURFACE ANALYSIS BY ELECTRON SPECTROSCOPY. 3-0666
SURFACE ANALYSIS BY LOW ENERGY ION REFLECTION. 5-1822
OSCOPY) THIN FILM COMPOSITIONAL ANALYSIS. COMPARISON OF TECHNIQUES. (AUGER SPECTR 1-0054
AUGER ELECTRCN SPECTROMETER AS TOOL FOR SURFACE ANALYSIS. (CONTAMINATION MONITOR) 3-0706
S. (REVIEW) SURFACE AND THIN FILM COMPOSITIONAL ANALYSIS. DESCRIPTION AND COMPARISON OF TECHNIQUE 1-0053
ULTRAHIGH VACUUM SCANNING ELECTRON MICROSCOPE WITH AUGER ANALYSIS FACILITIES. 3-0622
THIN FILM ANALYSIS FOR PROCESS EVALUATION. (AES) 1-0118
OMPUTER-CONTROLLED MOLECULAR BEAM EPITAXY. (AUGER SURFACE ANALYSIS, GALLIUM(X) ALUMINUM(1-X) ARSENIDE) C 5-1324
LINDRICA/ FIRST ORDER APPROXIMATICN TO QUANTITATIVE AUGER ANALYSIS IN THE RANGE 100 TO 1000 EV USING THE CY 3-0660
ETERMINATION. BEAM ANALYSIS METHOD OF FILM AND SURFACE COMPOSITION D 5-1810
CONDARY ION MASS SPECTROSCOPY. COMPARISON OF TWO SURFACE ANALYSIS METHODS. /RON SPECTROSCOPY (ESCA) AND SE 1-0088

	SPUTTERING IN THIN FILM ANALYSIS METHODS. (ESCA, AUGER SPECTROSCOPY)	1-0214
YERS OF SOLIDS.	SURFACE ANALYSIS, METHODS OF STUDYING THE OUTER ATOMIC LA	1-0016
ESCA)	ASPECTS OF SPUTTERING IN SURFACE ANALYSIS METHODS. (REVIEW AUGER SPECTROSCOPY AND	1-0215
MICROSTRUCTURE OF HOT PRESSED SILICON NITRIDE. (AUGER ANALYSIS, MICROPROBE ANALYSIS)		5-1556
BINE SHROUD AL/ ADHESION, FRICTION AND AUGER SPECTROSCOPY ANALYSIS OF A COMMERCIAL COBALT BASE AIRCRAFT TUR		6-2009
TEEL.	AUGER FRACTURE SURFACE ANALYSIS OF A TEMPER EMBRITTLED 3340- STAINLESS S	6-2085
NS.	IN-DEPTH AUGER ANALYSIS OF ALUMINUM- SILICON INTERFACIAL REACTIO	5-1219
	AUGER ELECTRON SPECTROSCOPIC ANALYSIS OF ANODIC FILMS ON ALUMINUM.	5-1614
	SOME PROBLEMS IN ANALYSIS OF AUGER ELECTRON SPECTRA.	4-0866
N SINGLE CRYSTAL UNDER A LOW ENERGY BOMBARDMENT. / ENERGY ANALYSIS OF BACKSCATTERED ELECTRONS FROM A SILICO		4-0739
	AUGER SPECTROSCOPIC ANALYSIS OF BIOGLASS CORROSION FILMS.	5-1264
S IN COPPER.	AUGER SPECTROSCOPIC ANALYSIS OF BISMUTH SEGREGATED TO GRAIN BOUNDARIE	6-2060
AMEL INTERFACES.	AUGER ANALYSIS OF BRASS- ENAMEL AND STAINLESS STEEL- EN	5-1266
RINE GROWN SILICON DIOXIDE FILMS.	AUGER ANALYSIS OF CHLORINE IN HYDROGEN CHLORIDE OR CHLO	5-1251
NUM.	(AUGER) ANALYSIS OF CHROMATE CONVERSION COATINGS ON ALUMI	6-2125
GER SPECTROSCOPY.	DEPTH ANALYSIS OF CLEAVED MICA SURFACES MONITORED BY AU	5-1842
/COMPOSITION OF SILVER- COPPER ALLOYS. QUANTITATIVE AUGER ANALYSIS OF COPPER- NICKEL ALLOY SURFACES AFTER /		5-1200
RGON ION BOMBARDMENT. REPLY TO CCMMEN/ QUANTITATIVE AUGER ANALYSIS OF COPPER- NICKEL ALLOY SURFACES AFTER A		5-1702
/COMPOSITION OF SILVER- COPPER ALLOYS. QUANTITATIVE AUGER ANALYSIS OF COPPER- NICKEL ALLOY SURFACES AFTER /		5-1740
RGON ION BCMBARDMENT.	QUANTITATIVE AUGER ANALYSIS OF COPPER- NICKEL ALLOY SURFACES AFTER A	5-1807
INE STRUCTURE.	SELF-CONSISTENT ANALYSIS OF CORE BAND-TO-BAND EXCITATION: AUGER F	2-0509
(X) GALLIUM(1-X) ARSENIDE.	DEPOSITION AND AUGER ANALYSIS OF DEPOSITED SILICON DIOXIDE ON ALUMINUM	5-1962
BY AES.	CHEMICAL ANALYSIS OF ELECTRODEPOSITED NICKEL- NICKEL BONDS	4-0947
EN/ REFINEMENTS TO A STANDARD LEED- AUGER SYSTEM FOR THE ANALYSIS OF ELECTRON EMISSION AT LOW PRIMARY BEAM		3-0698
	AUGER ELECTRON ANALYSIS OF ELECTROPOLISHED HIGH PURITY ALUMINUM.	4-0824
TACT WITH HALOGENATED POLYMERS.	AUGER ANALYSIS OF FILMS FORMED ON METALS IN SLIDING CON	5-1728
BOUNDARY STRUCTURE AND SEGREGATION. AUGER SPECTROGRAPHIC ANALYSIS OF FRACTURE SURFACES. GRAIN		6-2118
MICROSCOPIC AUGER ELECTRON ANALYSIS OF FRACTURE SURFACES.		5-1598
ING USING AUGER SPECTROSCOPY.	ELEMENTAL ANALYSIS OF FRICTION AND WEAR SURFACE DURING SLID	6-2013
CTRON SPECTROSCOPY.	ANALYSIS OF GLASS CONTAINER COATINGS BY AUGER ELE	5-1704
PECTROSCOPY, AND ION BACKSCATTERING TECHNIQUES APPLIED TO ANALYSIS OF GOLD- REFRACTORY METALLIZATIONS. /R S		5-1261
C THIN FILM STRUCTURES USING AUGER ELECTRON SPECTROSCOPY. ANALYSIS OF GRAIN BOUNDARY DIFFUSION IN BIMETALLI		6-2146
E ADDITIVES IN HOT PRESSED OXIDES BY AUGER ELECTRON SPEC/ ANALYSIS OF GRAIN BOUNDARY IMPURITIES AND FLUORID		5-1506
NG ION SOURCE.	ANALYSIS OF HIGH CHARGE STATE OUTPUT FROM A PENNI	3-0602
ER ELECTRON SPECTROSCOPY.	ANALYSIS OF IMPURITIES ON SILICON SURFACES BY AUG	5-1674
	AUGER ELECTRON SPECTROSCOPY ANALYSIS OF ION PLATED GOLD FILMS ON SILICON.	5-1225
YGEN DETERMINATIONS OBTAINED BY PROTON EXCITED X-RAY, AES ANALYSIS OF IRON SURFACES. /ACTIONAL MONOLAYER OX		5-1682
RONIC SPECTROSCOPY AND LOW ENERG/ CHEMICAL AND STRUCTURAL ANALYSIS OF IRON SURFACES BY MEANS OF AUGER ELECT		5-1940
OXYGEN) IN SPUTTERED TANTALUM FILMS BY AUG/ QUANTITATIVE ANALYSIS OF LIGHT ELEMENTS (NITROGEN, CARBON, AND		5-1652
ROSCOPY)	INSTRUMENTATION FOR THE CHEMICAL ANALYSIS OF MANUFACTURED SURFACES. (REVIEW, SPECT	3-0719
OPY.	THREE DIMENSIONAL ELEMENTAL ANALYSIS OF MATERIALS BY AUGER ELECTRON SPECTROSC	5-1597
LECTRONS.	ANALYSIS OF MATERIALS BY ELECTRON EXCITED AUGER E	5-1399
LECTRON EMISSI/ SECONDARY ELECTRON SPECTROSCOPY. (SURFACE ANALYSIS OF MATERIALS BY ELECTRON EXCITED AUGER E		5-1403
AES	ANALYSIS OF OXIDE FILMS ON IRON.	5-1800
ARIOUS METALS AND THE EFFECT OF SLIDING ON THESE I/ AUGER ANALYSIS OF OXYGEN AND SULFUR INTERACTIONS WITH V		5-1206
	AUGER SPECTROSCOPY ANALYSIS OF PALLADIUM SILICIDE FILMS.	5-1898
ELECTRON BEAM EFFECTS IN AUGER ANALYSIS OF PHYSISORBED XENON.		5-1151
S BY AUGER ELECTRON SPECTROSCOPY.	CHEMICAL ANALYSIS OF PLATINUM (100) AND GOLD (100) SURFACE	5-1712
	METHODS FOR SURFACE ANALYSIS OF PRECIOUS METALS CONTACT MATERIALS.	5-1785
DETERMINATION OF SURFACE STRUCTURES USING LEED AND ENERGY ANALYSIS OF SCATTERED ELECTRONS.		5-1956
QUANTITATIVE AUGER SPECTROSCOPIC ANALYSIS OF SEGREGATION OF PHOSPHORUS IN IRON.		5-1802
GER MICROPROBE)	SURFACE AND THIN FILM ANALYSIS OF SEMICONDUCTOR MATERIALS. (SCANNING AU	1-0089
SITION)	CHEMICAL ANALYSIS OF SEMICONDUCTORS. (IMPURITIES AND COMPO	5-1671
/Y- AUGER SPECTROSCOPY FCR MICROSPOT SURFACE AND IN-DEPTH ANALYSIS OF SILICON AND TRANSISTOR METALLIZATION/		5-1262
ELECTRON IRRADIATION EFFECT IN THE AUGER ANALYSIS OF SILICON DIOXIDE.		4-1096
AUGER ELECTRON SPECTROSCOPIC ANALYSIS OF SILICON NITRIDE ON SILICON.		5-1455
N AT HIGH TEMPERATURES.	AUGER ANALYSIS OF SILICON THIN FILMS DEPOSITED ON CARBO	5-1228
UM ALLOYS. QUANTITATIVE AUGER ELECTRON SPECTROSCOPY ANALYSIS OF SILVER- PALLADIUM AND NICKEL- PALLADI		6-2088
LOW TEMPERATURE TECHNIQUE.	AES ANALYSIS OF SODIUM IN A CORRODED BIOGLASS USING A	5-1721
AUGER ELECTRONS, INDICATORS IN ANALYSIS OF SOLID BODY SURFACES.		5-1662
AUGER ELECTRON SPECTROSCOPY FOR THE ANALYSIS OF SOLID SURFACES.		5-1193
TROSCOPY.	QUANTITATIVE ANALYSIS OF SOLID SURFACES BY AUGER ELECTRON SPEC	5-1715
NEW DEVELOPMENTS IN THE SURFACE ANALYSIS OF SOLIDS.		1-0011
ON BACKSCATTERING WITH AUGER ELECTRON SPECTRCSCOPY IN THE ANALYSIS OF SULFUR ABSORBED ON NICKEL. / ENERGY I		5-1869
CHARACTERISTIC AUGER ELECTRON EMISSION AS A TOOL FOR THE ANALYSIS OF SURFACE COMPOSITION.		1-0170
	ANALYSIS OF SURFACE COMPOSITION WITH LOW ENERGY B	5-1823
ACKSCATTERED IONS.		
S. AUGER CYLINDRICAL MIRROR SPECTROMETER FOR THE ANALYSIS OF SURFACE CONSTITUTENTS AND SOME RESULT		3-0682
AUGER ANALYSIS OF SURFACE FILMS ON SILVER- TIN.		5-1386
ERING AND SPUTTERING CCMBINED WITH AUGER ELECTRON SPECTR/ ANALYSIS OF SURFACE LAYERS BY LIGHT ION BACKSCATT		5-1167
NDUCED DESORPTION.	NEW TECHNIQUE FOR AUGER ANALYSIS OF SURFACE SPECIES SUBJECT TO ELECTRON I	3-0675
AUGER SPECTROSCOPY FOR THE CHEMICAL ANALYSIS OF SURFACES.		1-0150
IONIZATION SPECTROMETER FCR ELEMENTAL ANALYSIS OF SURFACES.		3-0615
THE AUGER ELECTRON MICROANALYT/ THREE-DIMENSIONAL ELEMENT ANALYSIS OF SURFACES AND INTERFACES OF SOLIDS BY		5-1412
AUGER ANALYSIS OF TANTALUM SURFACES.		5-1480
TRON SPECTROSCOPY.	FRACTURE SURFACE ANALYSIS OF TEMPER EMBRITTLED STEEL BY AUGER ELEC	6-2084
THEORETICAL ANALYSIS OF THE AUGER SPECTRA OF METHANE.		4-1012
LEED- AUGER ANALYSIS OF THE BERYLLIUM (0001) SURFACE.		5-1581
ION.	AUGER SPECTROSCOPY ANALYSIS OF THE EXTENT OF GRAIN BOUNDARY SEGREGAT	6-2094
RESSURE VESSEL STEELS. AUGER ELECTRON SPECTROSCOPY ANALYSIS OF THE FRACTURE SURFACES OF IRRADIATED P		6-2113
IN ATOMIC ZINC: IMPROVED AUGER ENERGY CALCULATI/ DETAILED ANALYSIS OF THE L3M4,5M4,5: 1G4 AUGER TRANSITION		4-0884
LMS ON COPPER- NICKEL ALLOYS BY AUGER ELECTRON SPECTROSC/ ANALYSIS OF THE SURFACE COMPOSITION OF PASSIVE FI		5-1874
DOMAINS OF SULFUR ON IRON (001) BY LEED- AES.	ANALYSIS OF THE SURFACE STRUCTURE WITH ANTIPHASE	5-1470
ASS SPECTROMETRY AND AUGER ELECTRON SPECTROSCOPY, FOR THE ANALYSIS OF THIN FILMS. /ATION OF SECONDARY ION M		5-1214
SELECTED AREA AND IN-DEPTH AUGER ANALYSIS OF THIN FILMS.		5-1653
COPY.	MATERIAL ANALYSIS OF THIN FOILS BY AUGER ELECTRON SPECTROS	5-1798
YS.	AUGER ANALYSIS OF THIN OXIDE FILMS ON LEAD- INDIUM ALLO	5-1249
MIRROR DESIGNS. (ZINC SELENIDE, THORIUM FLUCRIDE) AUGER ANALYSIS OF THREE ENHANCED MULTILAYER DIELECTRIC		5-1328
ER SPECTROSCOPY.	CHEMICAL PROFILE ANALYSIS OF TYPE 316 STAINLESS STEEL SHEET BY AUG	6-1996
ES.	AUGER ELECTRON SPECTROSCOPY ANALYSIS OF VANADIUM AND VANADIUM COMPOUND SURFAC	5-1862
SPATIAL RESOLUTION SCANNING AUGER SPECTROSCOPY APPLIED TO ANALYSIS OF X-BAND DIODE BURNOUT. HIGH		5-1961

L BY AUGER EMISSION AND X-RAY PHOTOELECTRON SPECTROSCOPI/ ANALYSIS PROFILES OF OXIDE FILMS ON CHROMIUM STEE 6-2016
AES FOR SURFACE ANALYSIS. (REVIEW, 24 REFS) 1-0195
TECHNIQUE OF AUGER ELECTRON SPECTROSCOPY IN SURFACE.ANALYSIS. (REVIEW, 76 REFS) 1-0196
IRON AND STEEL. (AES, ELECTRON SPECTROSCOPY FOR CHEMICAL ANALYSIS SIMS) /THEIR APPLICATION FOR STUDIES ON 6-2035
URFACE ANALYSIS. (AES, ELECTRON SPECTROSCOPY FOR CHEMICAL ANALYSIS SIMS, AND OTHERS) /NIQUES FOR SILICON S 1-0052
(MONOLAYER ADSORPTION, ELECTRON SPECTROSCOPY FOR CHEMICAL ANALYSIS SIMPLE CHARGE TRANSFER MODEL). /SPECTRA 4-1115
ANALYZER FOR AUGER AND ELECTRON SPECTROSCOPY FOR CHEMICAL ANALYSIS SPECTROSCOPY. NEW COMPACT ENERGY 3-0702
TO FREE SURFACES DURING IRRADIATION. (AUGER SPECTROSCOPY ANALYSIS, STAINLESS STEEL) / OF ALLOYING ELEMENTS 6-2091
LVER OXIDE. ELECTRON SPECTROSCOPY FOR CHEMICAL ANALYSIS STUDIES OF SILVER, DISILVER OXIDE AND SI 4-1055
SILICON DEVICE APPLICATIONS USING A COMBINED ESCA- AUGER ANALYSIS SYSTEM. 5-1289
, SECONDARY ION MASS SPECTRO/ COMPARISON OF SURFACE LAYER ANALYSIS TECHNIQUES. (AUGER ELECTRON SPECTROSCOPY 5-1627
/PY, SCANNING ELECTRON MICROSCOPY AND NONDISPERSIVE X-RAY ANALYSIS TECHNIQUES FOR EXAMINATION OF SILVER- C/ 5-1799
NALYSIS, AUGER EFFECT, REVIEW) TRACE ANALYSIS TECHNIQUES FOR SOLIDS. (ELECTRON PROBE A 1-0093
EXECUTION OF AUGER AND ELECTRON SPECTROSCOPY FOR CHEMICAL ANALYSIS TEST PROGRAMS. /LOGIC FOR THE AUTOMATIC 3-0722
/ATION OF SECONDARY ELECTRON SPECTROSCOPY (AUGER ELECTRON ANALYSIS) TO SURFACE CONTROL PROGRAM IN THE MANU/ 5-1411
) SURFACE ANALYSIS TODAY. (REVIEW OF METHODS, INCLUDING AES 1-0038
EMBRITTLEMENT IN SOME METALS AND ALLOYS. (AUGER EMISSION ANALYSIS, TUNGSTEN, COPPER, STEELS) /AIN BOUNDARY 6-2058
TER. DIGITAL METHODS FOR THIN FILM ANALYSIS USING A COMPUTER COUPLED AUGER SPECTROME 3-0640
CAL MEASUREMENTS. (AUGER) SURFACE ANALYSIS USING COMPLEMENTARY ELECTRONIC AND CHEMI 5-1390
QUANTITATIVE AUGER ANALYSIS USING INTEGRATION TECHNIQUES. 1-0066
SURFACE ANALYSIS USING PROTON BEAMS. 1-0143
(AUGER SPECTROSCOPY) MATERIAL ANALYSIS USING THE SCANNING ELECTRON MICROSCOPE. 5-1701
. (COMPARISON BETWEEN / SURFACE COMPOSITION AND STRUCTURE ANALYSIS WITH A ONE APPARATUS ION PROBE TECHNIQUE 5-1166
ONDARY ION MASS SPECTROMETRY. ELEMENTAL ANALYSIS WITH AUGER ELECTRON SPECTROSCOPY AND SEC 5-1654
BASIC PROPERTIES OF A NEW METHOD FOR MATERIALS S/ SURFACE ANALYSIS WITH HEAVY ION INDUCED AUGER ELECTRONS. 5-1387
SURFACE ANALYSIS WITH HEAVY ION INDUCED AUGER ELECTRONS. 1-0070
SURFACE ANALYSIS WITH LOW ENERGY ION SCATTERING. 5-1871
D AUGER ELECTRON SPECTROSCOPY. STRUCTURE ANALYSIS WITH X-RAY PHOTOELECTRON SPECTROSCOPY AN 1-0050
) ANALYTICAL ASPECTS OF AUGER SPECTROSCOPY. (REVIEW 1-0198
ANALYTICAL AUGER ELECTRON SPECTROSCOPY. 1-0031
AUGER PROCESS) A SIMPLE ANALYTICAL CALCULATION FOR X-RAY PHOTOEMISSION. (2-0429
DIGITIZING AUGER DATA. THREEFOLD APPROACH TO ENHANCE ANALYTICAL CAPABILITIES. 1-0183
TIION. X-RAYS AND ELECTRONS IN ANALYTICAL CHEMISTRY WITH EMPHASIS ON INSTRUMENTA 3-0646
TRY, AUGER ELECTRON SPECTROSCOPY AND SECONDARY I/ SURFACE ANALYTICAL STUDIES USING ION SCATTERING SPECTROME 5-1423
ETRY IN ULTRAHIGH VACUUM. (AES) AN ANALYTICAL SYSTEM FOR SECONDARY ION MASS SPECTROM 3-0632
TERS AFFECTING AUGER AND PHOTOELECTRON SPECTROSCOPY AS AN ANALYTICAL TECHNIQUE. SOME PARAME 1-0222
CHEMICAL, PHYSICAL, AND CRYSTALLOGRAPHI/ USE OF ADVANCED ANALYTICAL TECHNIQUES FOR CHARACTERIZATION OF THE 5-1433
DA/ COMPOSITION PROFILE OF ION PLATED GOLD FILM ON COPPER ANALYZED BY AUGER ELECTRON SPECTROSCOPY AND SECON 5-1679
PLITUDE DISTORTION IN A 4-GRID RETARDING POTENTIAL ENERGY ANALYZER. /RON SIGNAL STRENGTHS FOR MODULATION AM 3-0620
DESIGN PARAMETERS FOR THE CYLINDRICAL MIRROR ENERGY ANALYZER. 3-0687
AND SILICON DIOXIDE SURFACES BY CYLINDRICAL MIRROR AUGER ANALYZER. DETECTION OF SODIUM ON SILICON 3-0647
ENERGY RESOLUTION OF THE PHOTOEMISSION ANALYZER. 3-0606
IN THE RANGE 100 TO 1000 EV USING THE CYLINDRICAL MIRROR ANALYZER. /IMATION TO QUANTITATIVE AUGER ANALYSIS 3-0660
G SPECTROSCOPY WITH A COMMERCIAL AUGER CYLINDRICAL MIRROR ANALYZER. LOW ENERGY ION BACKSCATTERIN 3-0665
PECTS OF ION SCATTERING SPECTROSCOPY. (CYLINDRICAL MIRROR ANALYZER) QUANTITATIVE AS 3-0666
DATA RECORDING AND CONTROL SYSTEM FOR AN ELECTRON ENERGY ANALYZER. SEMIAUTOMATED 3-0680
AND SIGNAL INTENSITY IN A SINGLE STAGE CYLINDRICAL MIRROR ANALYZER. /MEN POSITION EFFECTS ON ENERGY SHIFTS 3-0696
AUGER SYSTEM BASED ON THE DOUBLE PASS CYLINDRICAL MIRROR ANALYZER. / ELECTRON SPECTROSCOPY. COMBINED ESCA- 3-0654
ECTRON MICROSCOPE WITH A FIELD EMISSION CATHODE AND AUGER ANALYZER. ULTRAHIGH VACUUM SCANNING EL 3-0621
LAYER ATTENUATION OF ELECTRONS FOR THE CYLINDRICAL MIRROR ANALYZER AND A RETARDING FIELD ANALYZER. /OF OVER 3-0693
LURGY. POSSIBILITIES OF THE ION MICROPROBE MASS ANALYZER AND AUGER SPECTROMETER IN PHYSICAL METAL 6-2075
/ND SENSITIVITY OF THE SPHERICAL GRID RETARDING POTENTIAL ANALYZER. APPLICATION TO AUGER ELECTRON SPECTROS/ 3-0634
MPACT, INEXPENSIVE HIGH RESOLUTION RETARDING FIELD ENERGY ANALYZER. (AUGER ELECTRONS) CO 3-0619
DISTORTION OF DIFFERENTIATED DEFLECTION ANALYZER CURRENT. (USED IN AES) 3-0613
NEW COMPACT ENERGY ANALYZER FOR AUGER AND ESCA SPECTROSCOPY. 3-0702
PY). IMPROVED RETARDING FIELD ANALYZER. (FOR AUGER AND PHOTOELECTRON SPECTROSCO 3-0701
IUM- NICKEL ALLOYS, ADSORBED OXYGE/ SMALL RETARDING FIELD ANALYZER FOR AUGER ELECTRON SPECTROSCOPY. (PALLAD 3-0644
RETARDING FIELD CYLINDRICAL MIRROR ANALYZER. (FOR AUGER SPECTROSCOPY) 3-0614
ELECTRON BEAM ANALYZER FOR DETECTING AUGER ELECTRONS. 3-0712
INTRODUCTION TO LOW ENERGY X-RAY AND ELECTRON ANALYSIS. (ANALYZER FOR PHOTO AUGER ELECTRON SPECTROSCOPY) 3-0627
) IMPROVED SPHERICAL GRID RETARDING POTENTIAL ANALYZER. (FOR USE IN AUGER ELECTRON SPECTROSCOPY 3-0583
DESIGN FOR A CYLINDRICAL MIRROR ANALYZER FOR USE IN AUGER SPECTROSCOPY. 3-0594
TOWARD OPTIMUM UTILIZATION OF THE CYLINDRICAL MIRROR ANALYZER IN AUGER ELECTRON SPECTROSCOPY. 3-0678
AUGER ELECTRON) CYLINDRICAL MIRROR ANALYZER INCORPORATING PRERETARDATION. (TUNGSTEN 3-0590
/RTICLE BEAM OF FINITE ANGULAR SPREAD IN AN ELECTROSTATIC ANALYZER OF THE CYLINDRICAL MIRROR TYPE. (AUGER / 3-0720
CYLINDRICAL CAPACITOR AS AN ANALYZER. PART-1: NONRELATIVISTIC PART. 3-0689
/ON BEAM MICROANALYSIS. PART-1. (REVIEW OF ELECTRON PROBE ANALYZER, SCANNING ELECTRON MICROSCOPE, TRANSMIS/ 1-0010
A SMALL RETARDING FIELD ENERGY ANALYZER WITH CEM. (AUGER SPECTROMETER) 3-0655
OELECTRON SPECTROMETER: COMBINATION OF CYLINDRICAL MIRROR ANALYZER WITH SOFT X-RAY SOURCE. A NEW PHOT 3-0652
COMPARISON OF SPHERICAL DEFLECTOR AND CYLINDRICAL MIRROR ANALYZERS. 3-0623
X-RAY ELECTRON SPECTROMETERS AND AUGER ANALYZERS. 3-0599
FOR IMPROVED AUGER SPECTROSCOPY WITH RETARDING POTENTIAL ANALYZERS. COMMENTS. SECONDARY EMISSION ANALOG 3-0584
FOR IMPROVED AUGER SPECTROSCOPY WITH RETARDING POTENTIAL ANALYZERS. (REPLY TO COMMENTS) /Y EMISSION ANALOG 3-0694
NEW METHOD FOR ANALYZING THE SECONDARY ELECTRON SPECTRUM. 4-0934
TALS. (PHOTO- AND AUGER ELECTRON EMISSION) ANGLE DISTRIBUTION OF SODIUM CHLORIDE SINGLE CRYS 2-0297
.0 MEV PROTONS. (AUGER PEAKS) DISTRIBUTION IN ENERGY AND ANGLE OF ELECTRONS EJECTED FROM XENON BY 0.3 TO 2 2-0534
SION FROM CLEAN MOLYBDENUM. ANGLE OF INCIDENCE EFFECTS IN AUGER ELECTRON EMIS 2-0542
/UM BOMBARDED BY POSITIVE IONS. PART-1. PART-2: EFFECT OF ANGLE OF INCIDENCE. PART-3: EFFECT OF ELECTRONIC/ 2-0543
YIELD. EFFECT OF THE INCIDENCE ANGLE OF PRIMARY ELECTRONS ON THE AUGER EMISSION 2-0237
ULTRAHIGH VACUUM TECHNIQUES IN THE MEASUREMENT OF CONTACT ANGLES. PART-4: WATER ON GRAPHITE (0001). (AES) 5-1795
UCED BY 200-500 KEV PROTONS IN GASES. PART-2: NITROGEN. / ANGULAR AND ENERGY DISTRIBUTION OF ELECTRONS PROD 4-1075
SURFACE STRUCTURE FROM ANGULAR DEPENDENCE OF AUGER ELECTRON EMISSION. 5-1977
M A SINGLE CRYSTAL SPECIMEN. ANGULAR DEPENDENCE OF AUGER ELECTRON EMISSION FRO 4-0958
M SOLID SURFACES. ANGULAR DEPENDENCE OF AUGER ELECTRON EMISSION FRO 5-1447
M COPPER (111) AND (100) SURFACES. ANGULAR DEPENDENCE OF AUGER ELECTRON EMISSION FRO 5-1632
ACES. (AUGER EFFECT) ANGULAR DEPENDENCE OF ELECTRON EMISSION FROM SURF 5-1726
SION COEFFICIENT OF/ CHANGES IN THE FINE STRUCTURE OF THE ANGULAR DEPENDENCE OF THE SECONDARY ELECTRON EMIS 4-1061
ISSION. (COMMENTS) ANGULAR DEPENDENCIES IN ELECTRON EXCITED AUGER EM 2-0286
ISSION. ANGULAR DEPENDENCIES OF ELECTRON EXCITED AUGER EM 2-0378
ROM CLEAN AND GAS-COVERED IRON (100) SURFACES. ANGULAR DISTRIBUTION OF AUGER ELECTRON EMISSION F 5-1623

	ANGULAR DISTRIBUTION OF AUGER ELECTRONS. 2-0320
RESONANCE IONIZATION OF ATOMIC HELIUM BY FAST ELECTRONS.	(ANGULAR DISTRIBUTION OF AUGER ELECTRONS) 2-0256
IMPACT IONIZATION.	ANGULAR DISTRIBUTION OF AUGER ELECTRONS FOLLOWING 2-0321
PHOTOIONIZATION.	ANGULAR DISTRIBUTION OF AUGER ELECTRONS FOLLOWING 2-0352
DIRECT IONIZING COLLISIONS CF ELECTRONS WITH / ENERGY AND	ANGULAR DISTRIBUTION OF ELECTRONS RESULTING FROM 2-0525
CTRONS OF ARGON.	ANGULAR DISTRIBUTION OF L3M2,3M2,3(1SO) AUGER ELE 2-0322
ECTED FROM KRYPTON BY ELECTRON IMPACT.	ANGULAR DISTRIBUTION OF M4,5NN AUGER ELECTRONS EJ 4-1003
ED BY IONIC BOMBARDMENT. ENERGY SPECTRUM AND	ANGULAR DISTRIBUTION OF SECONDARY ELECTRONS EMITT 4-0988
HE (100), (110), AND (111) FACES OF COPPER CRYSTALS. (AU/	ANGULAR DISTRIBUTIONS OF HYDROGEN DESORBED FROM T 5-1155
/EED APPARATUS WITH FACILITY FOR INVESTIGATING ENERGY AND	ANGULAR DISTRIBUTIONS OF INELASTICALLY SCATTERED/ 3-0691
/M METAL SURFACES BY METASTABLE ATOMS. PART-3: ENERGY AND	ANGULAR DISTRIBUTIONS OF THE EJECTED ELECTRONS. / 5-1133
OPPER (110).	ANGULAR EFFECTS IN AUGER ELECTRON EMISSION FROM C 5-1988
Y. KIKUCHI CORRELATIONS FOR ARSENIC (00001). HIGH	ANGULAR RESOLUTION SECONDARY ELECTRON SPECTROSCOP 5-1784
EAN COPPER (100) SURFACE.	ANGULAR RESOLVED AUGER EMISSION SPECTRA FROM A CL 5-1697
E CYLINDRICAL/ FOCUSING A CHARGED PARTICLE BEAM OF FINITE	ANGULAR SPREAD IN AN ELECTROSTATIC ANALYZER OF TH 3-0720
NDARY ELECTRON EMISSION FOR 10-100 KEV ELECT/ ORIENTATION	ANISOTROPY OF BACKSCATTERING COEFFICIENT AND SECO 4-1031
SSAGE OF LITHIUM(+) IONS THROUGH COPPER SINGLE CRYSTALS.	ANISOTROPY OF SECONDARY ELECTRON EMISSION FROM PA 2-0245
STRUCTURE TRANSFORMATIONS ON CLEAVED AND	ANNEALED GERMANIUM (111) SURFACES. 5-1713
NATURE OF	ANNEALED SEMICONDUCTOR SURFACES. (AUGER) 2-0263
STUDIES OF ARSENIC VACANCIES IN (-1-1-1) GALLIUM ARSENIDE	ANNEALED SURFACES. AUGER ELECTRON SPECTROSCOPY 5-1914
/N OF THE SEGREGATION CF COPPER TO THE SURFACE OF A CLEAN	ANNEALED 50-50 NICKEL ALLOY BY AUGER ELECTRON SP/ 6-2044
ON THE SURFACE COMPOSITICN CF COPPER- NICKEL ALLOYS AFTER	ANNEALING AND SPUTTERING. /ER SPECTROSCOPY STUDY 5-1675
NSPORT OF OXYGEN IN AMCRPHOUS GERMANIUM THIN FILMS DURING	ANNEALING. (AUGER SPECTROSCOPY) TRA 5-1548
NG/ OBSERVATIONS ON THE SURFACE OXIDATION OF STEEL DURING	ANNEALING IN HYDROGEN-FREE PROTECTIVE GASES. (USI 6-2040
DE (111) FACES STUDIED BY AES AND UPS. OXIDATION AND	ANNEALING OF GALLIUM PHOSPHIDE AND GALLIUM ARSENI 5-1493
EEN 770 AND 1470 K.	ANNEALING OF STAINLESS STEEL AT TEMPERATURES BETW 6-2145
/TIONAL AND STRUCTURAL CHANGES THAT ACCOMPANY THE THERMAL	ANNEALING OF (100) SURFACES OF GALLIUM ARSENIDE,/ 5-1163
COMPOSITION OF PLATINUM	ANODE SURFACES BY AUGER SPECTROSCOPY. 5-1504
AUGER ELECTRON SPECTROSCOPIC ANALYSIS OF	ANODIC FILMS ON ALUMINUM. 5-1614
AES)	ANODIC OXIDATION OF GALLIUM ARSENIDE PHOSPHIDE. (5-1388
AL PROFILE AND ELECTRICAL CONDUCTIVITY OF THE THERMAL AND	ANODIC OXIDES OF INDIUM ANTIMONIDE. (AES) /SITION 5-1971
X-RAY EDGE	ANOMALIES IN LITHIUM. 2-0270
AUGER PROFILE	ANOMALIES. (REVIEW) 1-0184
ICIUM-241 FROM CURIUM-241 (ELECTRON CAPTU/ PROTON STATES,	ANOMALOUS CONVERSION, AND K-AUGER SPECTRA IN AMER 4-1019
INNER SHELL IONIZATION OF THE N6,7 LEVEL IN GOLD, BISMU/	ANOMALOUS THRESHOLD BEHAVIOR FOR ELECTRON INDUCED 4-1067
AUGER RECOMBINATION IN GALLIUM	ANTIMONIDE. 4-0778
CARRIER RECOMBINATION IN N-INDIUM	ANTIMONIDE. 4-0761
THE POLAR (111) SURFACES OF GALLIUM ARSENIDE AND GALLIUM	ANTIMONIDE. /ENERGY ELECTRON DIFFRACTION STUDY OF 5-1602
L CONDUCTIVITY OF THE THERMAL AND ANODIC OXIDES OF INDIUM	ANTIMONIDE. (AES) /SITIONAL PROFILE AND ELECTRICA 5-1971
NT ON GALVANOMAGNETIC PROPERTIES IN THIN INTRINSIC INDIUM	ANTIMONIDE AT ROOM TEMPERATURE. (AUGER PROCESS) / 2-0354
. AUGER RECOMBINATION IN INDIUM ARSENIDE, GALLIUM	ANTIMONIDE, INDIUM PHOSPHIDE AND GALLIUM ARSENIDE 4-1091
AY SPECTRAL REGION. (EF/ EFFECT CF THE POLARITY OF INDIUM	ANTIMONIDE ON THE EXTERNAL PHOTOEFFECT IN THE X-R 4-1001
UDIES OF SILICON, GERMANIUM, GALLIUM ARSENIDE, AND INDIUM	ANTIMONIDE SURFACES. /ER ELECTRON SPECTROSCOPY ST 5-1379
20-70 EV R/ VALENCE BAND AND CORE LEVEL STUDIES CF INDIUM	ANTIMONIDE VIA PHOTOEMISSION MEASUREMENTS IN THE 2-0341
KLL AUGER ELECTRONS OF CADMIUM AND	ANTIMONY. 4-0754
PHASE EPITAXY. (AES MEASUREMENTS)	ANTIMONY DOPING OF SILICON LAYERS GROWN BY SOLID 5-1577
COMPOSITION ON TEMPER BRITTLENESS IN A NICKEL- CHROMIUM-	ANTIMONY STEEL. (AUGER ANALYSIS) / GRAIN BOUNDARY 6-2114
RESOLUTION MNN AUGER SPECTRA OF SILVER, CADMIUM, INDIUM,	ANTIMONY, TELLURIUM AND IODINE. HIGH 4-0730
TLED STEEL. (AUGER ELECTRON/ SEGREGATION OF MANGANESE AND	ANTIMONY TO THE GRAIN BOUNDARIES OF TEMPER EMBRIT 6-2070
/ IN IRON- PHOSPHORUS, IRON- PHOSPHORUS- SULFUR AND IRON-	ANTIMONY- SULFUR ALLOYS. (AUGER ELECTRON EMISSIO/ 6-2098
- AES. ANALYSIS OF THE SURFACE STRUCTURE WITH	ANTIPHASE DOMAINS OF SULFUR ON IRON (001) BY LEED 5-1470
N5,4 AND M5,4	APPEARANCE POTENTIAL SPECTRA FROM PRASEODYMIUM. 4-0908
ELECTRON BEAM / CHANGES IN THE TITANIUM DIOXIDE AUGER AND	APPEARANCE POTENTIAL SPECTRA INDUCED BY INCIDENT 5-1834
ON OF PHOTONS, ELECTRONS AND IONS. HIGH SENSITIVITY	APPEARANCE POTENTIAL SPECTROMETER FOR THE DETECTI 3-0643
ACTION TO QUANTIFICATION OF AUGER ELECTRON AND SOFT X-RAY	APPEARANCE POTENTIAL SPECTROSCOPIES. /ROUND SUBTR 1-0067
IUM MONOXIDE SURFACES USING AUGER ELECTRON AND SOFT X-RAY	APPEARANCE POTENTIAL SPECTROSCOPIES. /M AND TITAN 5-1384
GER ELECTRON SPECTROSCOPY. IONIZATION LOSS SPECTROSCOPY.	APPEARANCE POTENTIAL SPECTROSCOPY. AU 1-0174
COMPARISON OF SOFT X-RAY AND AUGER ELECTRON	APPEARANCE POTENTIAL SPECTROSCOPY. 1-0092
INTENSITY RATIO MEASURED BY SOFT X-RAY AND AUGER ELECTRON	APPEARANCE POTENTIAL SPECTROSCOPY. /ROMIUM L3/L2 4-0890
DIRECT COMPARISON CF AUGER ELECTRON SPECTROSCOPY WITH	APPEARANCE POTENTIAL SPECTROSCOPY. 1-0142
SURFACE ANALYSIS BY	APPEARANCE POTENTIAL SPECTROSCOPY. 5-1545
/ACE CHEMICAL ANALYSIS BY AUGER ELECTRON SPECTROSCOPY AND	APPEARANCE POTENTIAL SPECTROSCOPY: A COMPARISON. 5-1910
H AUGE/ IONIZATION (CHARACTERISTIC LCSS) SPECTROSCOPY AND	APPEARANCE POTENTIAL SPECTROSCOPY. COMPARISON WIT 1-0173
ED SYSTEM. SOFT X-RAY	APPEARANCE POTENTIAL SPECTROSCOPY IN A DISPLAY LE 4-0870
PPER- NICKEL AND IRON- NICKEL ALLOYS. (AES) SOFT X-RAY	APPEARANCE POTENTIAL SPECTROSCOPY STUDIES WITH CO 5-1952
ASMON COUPLING TO CORE HOLE EXCITATIONS IN CARBON. (AUGER	APPEARANCE SPECTROSCOPY) PL 2-0291
DETERMINATION OF THE COMPOSITION OF PASSIVE FILMS ON TYP/	APPLICATON OF AUGER ELECTRON SPECTROSCOPY TO THE 6-2080
HE RANGE 100 TO 1000 EV USING THE CYLINDRICA/ FIRST ORDER	APPROXIMATION TO QUANTITATIVE AUGER ANALYSIS IN T 3-0660
SILICA GLASS. (AES)	AQUEOUS CORROSION OF SODA- SILICA AND SODA- LIME- 5-1265
: ELECTRODE SURFACE STRUCTURE AFTER EXPOSURE TO WATER AND	AQUEOUS ELECTROLYTES. /ECTRON DIFFRACTION. PART-1 5-1486
SELECTED	AREA AND IN-DEPTH AUGER ANALYSIS OF THIN FILMS. 5-1653
LICATION OF AUGER ELECTRON SP/ BASIC INVESTIGATION IN THE	AREA OF SCANNING ELECTRON MICROSCOPY. PART-1: APP 1-0213
NGULAR DISTRIBUTION OF L3M2,3M2,3(1SO) AUGER ELECTRONS OF	ARGON. A 2-0322
AUGER SPECTRUM OF L2 AND L3 SHELLS OF	ARGON. 4-0983
AUGER SPECTRUM OF THE NOBLE GASES. PART-2:	ARGON. 4-0974
COSTER-KRONIG AND AUGER SPECTRUM OF THE L1 SHELL OF	ARGON. 4-0977
OXYGEN IONS EXCITED BY COLLISIONS WITH HELIUM, NEON, AND	ARGON. /UTION BEAM GAS AUGER ELECTRON SPECTRA FOR 4-0904
HIGH RESOLUTION L2,3M2,3M2,3 AUGER ELECTRON SPECTRA OF	ARGON. 4-0912
S, AND FLUORESCENCE YIELDS OF VARIOUSLY IONIZED STATES OF	ARGON. K-SHELL AUGER RATES, TRANSITION ENERGIE 2-0280
ON RATES AND FLUORESCENT YIELDS FOR ELEMENTS BERYLLIUM TO	ARGON. K-SHELL AUGER TRANSITI 4-0962
L2,3MM AUGER SPECTRUM OF	ARGON. 4-1114
DED ION- ELECTRON CCINCIDENCE MEASUREMENTS OF THE VIOLENT	ARGON(+)- ARGON COLLISION. EXTEN 2-0530
NIG TRANSITIONS) ENERGY SPECTRA OF ELECTRONS EJECTED IN	ARGON(+)- ARGON COLLISIONS. (AUGER AND COSTER-KRO 4-1007
ELECTRON ION CCINCIDENCE STUDIES OF HIGHLY IONIZING	ARGON(+)- ARGON COLLISIONS. (AUGER EFFECT) 2-0529
/AST ELECTRON- SLOW ELECTRON COINCIDENCE STUDY OF VIOLENT	ARGON(+)- ARGON COLLISIONS. COMMENTS. (AUGER ELE/ 2-0423
ARGON AUGER EMISSION BY	ARGON(+) ION BOMBARDMENT OF SOME METALS. 2-0547
PY IN SPUTTERING MEASUREMENTS. APPLICATION TO LOW ENERGY	ARGON(+) ION SPUTTERING OF SILVER AND NIOBIUM. /O 5-1825
ME FREE ELEMENTS AND BINARY COMPOUNDS BOMBARDED BY 60 KEV	ARGON(+) IONS. AUGER SPECTRA OF SO 4-1106
ARGON L-SHELL AUGER SPECTRA PRODUCED IN	ARGON(+) ION- ARGON COLLISIONS. 4-1080
SSION) MEASUREMENTS OF THE L-SHELL FLUORESCENCE YIELDS OF	ARGON AND CHLORINE BY ELECTRON IMPACT. (AUGER EMI 2-0387

PRODUCTION AND DECAY OF DOUBLE L VACANCIES IN ARGON AND PHOSPHORUS. 2-0489
"SEMI-AUGER" PROCESSES IN L2,3 EMISSION IN ARGON AND POTASSIUM CHLORIDE. 2-0326
EVIDENCE OF RADIATIVE AUGER TRANSITIONS IN THE FREE ARGON ATOM. 2-0403
EVIDENCE OF RADIATIVE AUGER TRANSITIONS IN THE FREE ARGON ATOM. 2-0404
SEMIEMPIRICAL AUGER ELECTRON ENERGIES. PART-2: THE ARGON ATOM. 4-0815
/LECTRON EMISSION SPECTRA FOR RARE GAS ATOMS. PART-1. THE ARGON ATOM. PART-2. KRYPTON ATOM M-SHELL. PART-3. 4-0927
ENERGY DISTRIBUTION CF ELECTRCNS EMITTED BY ARGON ATOMS UNDER ELECTRON IMPACT. 4-1008
FFECTS ON THE L2,3 FLUCRESCENCE YIELD OF MULTIPLY IONIZED ARGON. (AUGER EFFECT)   MULTIPLET E 4-0956
OF SOME METALS.   ARGON AUGER EMISSION BY ARGON(+) ION BOMBARDMENT 2-0547
RFACE COMPOSITION CHANGES CF COPPER- NICKEL ALLOYS DURING ARGON BOMBARDMENT. /T OF TARGET TEMPERATURE ON SU 5-1808
/CE OF TOTAL K-SHELL VACANCY PRODUCTION CROSS SECTIONS IN ARGON BY THE IMPACT OF HEAVY IONS. (AUGER EFFECT) 2-0446
AND HELIUM(+). (AUGER SPECTRA)   ARGON CHARGE STATES DUE TO IMPACT OF HYDROGEN(+) 4-1011
K-SHELL FLUORESCENCE YIELDS OF ARGON, CHLORINE AND SULFUR. (AUGER TRANSITION) 4-1014
LECTRON COINCIDENCE MEASUREMENTS OF THE VIOLENT ARGON(+)- ARGON COLLISION.   EXTENDED ION- E 2-0530
SECTIONS FOR L2,3 AUGER EMISSION IN 22 TO 300 KEV PROTON- ARGON COLLISIONS.   CROSS 4-1040
TIONS)   ENERGY SPECTRA OF ELECTRONS EJECTED IN ARGON(+)- ARGON COLLISIONS. (AUGER AND COSTER-KRONIG TRANSI 4-1007
TRON ION COINCIDENCE STUDIES OF HIGHLY IONIZING ARGON(+)- ARGON COLLISIONS. (AUGER EFFECT)   ELEC 2-0529
/ON- SLOW ELECTRON COINCIDENCE STUDY OF VIOLENT ARGON(+)- ARGON COLLISIONS. COMMENTS. (AUGER ELECTRON ENER/ 2-0423
A PARTICLES AND DEUTERONS.   AUGER SPECTRA OF CARBON AND ARGON FOLLOWING IONIZATION BY EQUAL VELOCITY ALPH 4-1109
AND SILICON EXCITED BY LCW ENERGY ELECTRON AND LOW ENERGY ARGON ION BOMBARDMENT. / OF MAGNESIUM, ALUMINUM, 4-0857
IVE AUGER ANALYSIS OF COPPER- NICKEL ALLOY SURFACES AFTER ARGON ION BOMBARDMENT.   QUANTITAT 5-1807
IVE AUGER ANALYSIS OF COPPER- NICKEL ALLOY SURFACES AFTER ARGON ION BOMBARDMENT. COMMENTS. /LOYS. QUANTITAT 5-1740
IVE AUGER ANALYSIS OF COPPER- NICKEL ALLOY SURFACES AFTER ARGON ION BOMBARDMENT. REPLY TO COMMENTS. /NTITAT 5-1200
IVE AUGER ANALYSIS OF COPPER- NICKEL ALLOY SURFACES AFTER ARGON ION BOMBARDMENT. REPLY TO COMMENTS. /NTITAT 5-1702
STATES OF HIGHLY STRIPPED OXYGEN, FLUORINE, CHLORINE, AND ARGON IONS. (AUGER EFFECT) /RA FROM AUTOIONIZING 2-0472
4 MEV PROTON IMP/ RELATIVE SATELLITE LINE INTENSITIES IN ARGON K AUGER ELECTRON SPECTRA PRODUCED BY 0.4 TO 2-0497
MEV FLUORINE IONS.   ARGON K-AUGER CROSS SECTIONS IN COLLISION WITH 30 2-0323
N ARGON COLLISIONS.   HIGH RESOLUTION ARGON K-AUGER SPECTRUM PRODUCED IN 4 MEV PROTON O 2-0496
HIGH RESOLUTION MEASUREMENTS OF P(+) ARGON KLL AND KLM AUGER ELECTRONS. 4-0943
ARGON KLL AUGER SPECTRUM. TEST OF THEORY. 2-0414
LID STATE BROADENING IN THE AUGER SPECTRA OF RARE GASES. (ARGON, KRYPTON, XENON)   SO 4-1006
ION- ARGON COLLISIONS.   ARGON L-SHELL AUGER SPECTRA PRODUCED IN ARGON(+) 4-1080
C HYDROGEN(+), MOLECULAR / CHARGE STATE DEPENDENCE OF THE ARGON L-SHELL FLUORESCENCE YIELD STUDIED BY ATOMI 2-0517
DEUTERON, MOLECULAR HYDROGEN, AND HELIUM(+) IMPACT. ARGON L-SHELL IONIZATION BY 50 TO 600 KEV PROTON, 2-0521
MENT.   HIGH RESOLUTION STUDY OF ARGON LMM AUGER ELECTRONS PRODUCED BY ION BOMBARD 4-0903
WIDTHS AND INTENSITIES OF HIGHLY RESOLVED LINES IN ARGON L2,3 AUGER SPECTRA. 4-1032
TION IN ELECTRON EJECTION BY METASTABLE ATOMS: HELIUM AND ARGON ON (111) AND (110) TUNGSTEN. (AUGER) / FUNC 5-1600
UGER TRANSITION RATES AND FLUORESCENT YIELDS FOR ELEMENTS ARGON TO XENON.   K-SHELL A 4-0961
EN(+) IMPACTS.   AUGER ELECTRONS FROM ARGON WITH ENERGIES 150-210 EV PRODUCED BY HYDROG 2-0552
ON IMPACT. (AUGER EMISSION)   MULTIIONIZATION OF NEON, ARGON, XENON AND THEIR IONS BY HIGH ENERGY ELECTR 2-0492
STRIPPING MECHANISM DEPENDENCE OF THE ARGON 2S FLUORESCENCE YIELD. (AUGER TRANSITION) 4-0957
AUGER MEASUREMENTS)   EFFECTS OF LOW ENERGY ARGON- ION BOMBARDMENT ON MOLYBDENUM DISULFIDE. (2-0348
RDERED MONOLAYER ON MOLYBDENUM (001).   ATOMIC ARRANGEMENT IN THE 1 K 1 STRUCTURE OF A SILICON O 5-1478
AYER SURFACE STRUCTURES.   ATOMIC ARRANGEMENT OF GOLD (100) AND RELATED METAL OVERL 5-1716
D SILVER-COVERED COPPER (100) SURFACES.   ATOMIC ARRANGEMENT OF GOLD (100), GOLD-COVERED (100), AN 5-1718
PHOTOELECTRON AND L2,3MM AUGER ELECTRON ENERGIES FOR ARSENIC. 4-1035
IC RELAXATION ENERGIES IN THE L3M4,5M4,5 AUGER PROCESS IN ARSENIC AND SELENIUM. ATOMIC AND EXTRAATOMIC STAT 4-0883
RELAXATION DURING PHOTOEMISSION AND LMM AUGER DECAY IN ARSENIC AND SOME OF ITS COMPOUNDS. 4-0763
NEALED SURFACES.   AUGER ELECTRON SPECTROSCOPY STUDIES OF ARSENIC VACANCIES IN (-1-1-1) GALLIUM ARSENIDE AN 5-1914
SECONDARY ELECTRON SPECTROSCOPY. KIKUCHI CORRELATIONS FOR ARSENIC (00001).   HIGH ANGULAR RESOLUTION 5-1784
LECTRON SPECTROSCOPIC STUDIES.   ARSENIC (0001) SURFACES: AUGER, LOSS, AND PHOTOE 5-1315
D SPUTTER ETCHING OF NICKEL- GOLD- GERMANIUM ON N-GALLIUM ARSENIDE.   AES AN 5-1775
AUGER ELECTRON SPECTROSCOPY OF CLEAN GALLIUM ARSENIDE. 5-1921
AUGER RECOMBINATION IN GALLIUM ARSENIDE. 2-0561
PITAXY. (AUGER SURFACE ANALYSIS, GALLIUM(X) ALUMINUM(1-X) ARSENIDE)   COMPUTER-CONTROLLED MOLECULAR BEAM E 5-1324
OF DEPOSITED SILICON DIOXIDE ON ALUMINUM(X) GALLIUM(1-X) ARSENIDE.   DEPOSITION AND AUGER ANALYSIS 5-1962
IDENTIFICATION OF AUGER ELECTRONS IN GALLIUM ARSENIDE. 2-0467
SPECTROSCOPY. APPLICATION TO CESIUM-COVERED (110) GALLIUM ARSENIDE. /ERMAL DESORPTION USING AUGER ELECTRON 5-1293
STUDIES OF CLEAN AND SODIUM- COVERED SURFACES OF GALLIUM ARSENIDE.   LEED, AUGER, AND WORK FUNCTION 5-1240
EMENTS ON CLEAN AND CESIUM COVERED POLAR FACES OF GALLIUM ARSENIDE. /SPECTROSCOPY, AND WORK FUNCTION MEASUR 5-1294
LUMINESCENCE DUE TO EXCITON-EXCITON COLLISION IN GALLIUM ARSENIDE. 2-0541
SECONDARY ELECTRON EMISSION FROM GALLIUM ARSENIDE. 4-0861
SURFACE STOICHIOMETRY AND STRUCTURE OF GALLIUM ARSENIDE. 5-1143
(AUGER) ELECTRON SPECTROSCOPY OF GALLIUM ARSENIDE AND ALUMINUM ARSENIDE SURFACES. 5-1595
MOLECULAR BEAM EPITAXY. (AUGER / MAGNESIUM DOPED GALLIUM ARSENIDE AND ALUMINUM(X) GALLIUM(1-X) ARSENIDE BY 5-1247
DIFFRACTION STUDY OF THE POLAR (111) SURFACES OF GALLIUM ARSENIDE AND GALLIUM ANTIMONIDE. /ENERGY ELECTRON 5-1602
ING AUGER MICROSCOPY IN MOLECULAR BEAM EPITAXY OF GALLIUM ARSENIDE AND GALLIUM PHOSPHIDE. USE OF SCANN 5-1142
CTRON SPECTROSCOPY STUDIES OF SILICON, GERMANIUM, GALLIUM ARSENIDE, AND INDIUM ANTIMONIDE SURFACES. /ER ELE 5-1379
RES.   AUGER ELECTRON SPECTROSCOPY STUDIES OF GALLIUM ARSENIDE AND SILICON METAL- SEMICONDUCTOR STRUCTU 5-1774
TROSCOPY STUDIES OF ARSENIC VACANCIES IN (-1-1-1) GALLIUM ARSENIDE ANNEALED SURFACES.   AUGER ELECTRON SPEC 5-1914
FLUORINE ION IMPLANTATION PROFILES IN GALLIUM ARSENIDE AS DETERMINED BY AES. 5-1398
RECOMBINATION OF NONEQUILIBRIUM CHARGE CARRIERS OF INDIUM ARSENIDE AT HIGH PHOTOEXCITATION LEVELS. (AUGER) 4-0840
RECOMBINATION OF CARRIERS IN N-TYPE INDIUM ARSENIDE AT 77 DEGREES K. (AUGER) 4-0785
RENTIAL EVAPORATION CF INDIUM FROM GALLIUM(X) INDIUM(1-X) ARSENIDE. (AUGER EFFECT)   PREFE 5-1366
CARRIER LIFETIMES IN EPITAXIAL INDIUM ARSENIDE. (AUGER RECOMBINATION) 2-0332
OPTICAL GAIN IN LIGHTLY DOPED GALLIUM ARSENIDE. (AUGER RECOMBINATION) 2-0503
ADSORPTION OF ZINC ON GALLIUM ARSENIDE. (AUGER SPECTROSCOPY) 5-1144
DEPLETION EFFECTS IN SEMIINSULATING GALLIUM ARSENIDE. (AUGER SPECTROSCOPY) 5-1950
REACTION OF SPUTTERED PLATINUM FILMS ON GALLIUM ARSENIDE. (AUGER SPECTROSCOPY) 5-1563
CONCENTRATION AND AS EVAPORATION IN HEAT TREATED GALLIUM ARSENIDE. (AUGER SPECTROSCOPY)   SURFACE CARRIER 5-1915
F ALLOYED NICKEL- GOLD- GERMANIUM FILMS ON N-TYPE GALLIUM ARSENIDE. (BY AUGER SPECTROSCOPY) /L PROPERTIES O 6-2101
/N OF THE PURE AND CESIUM COATED (100) SURFACE OF GALLIUM ARSENIDE BY SLOW ELECTRON DIFFRACTION. (AUGER SP/ 5-1639
/ ELECTRON DIFFRACTION- AUGER CHARACTERIZATION OF GALLIUM ARSENIDE DURING ACTIVATION TO NEGATIVE ELECTRON / 5-1364
/ SPECTRA OF SILICON CCMPENSATED ALUMINUM(X) GALLIUM(1-X) ARSENIDE ELECTROLUMINESCENT DIODES. (AUGER RECOM/ 2-0325
MINORITY CARRIER LIFETIME IN INDIUM ARSENIDE EPILAYERS. (AUGER RECOMBINATION PROCESS) 2-0567
D GALLIUM ARSENIDE.   AUGER RECOMBINATION IN INDIUM ARSENIDE, GALLIUM ANTIMONIDE, INDIUM PHOSPHIDE AN 4-1091
E RECOMBINATION IN N-GALLIUM ARSENIDE- P-ALUMINUM GALLIUM ARSENIDE HETEROJUNCTIONS. /ER EFFECT AND RADIATIV 2-0324
/MPANY THE THERMAL ANNEALING OF (100) SURFACES OF GALLIUM ARSENIDE, INDIUM PHOSPHIDE, AND GALLIUM PHOSPHID/ 5-1163
BACKSCATTER/ THERMALLY INDUCED PROCESSES AT GOLD- GALLIUM ARSENIDE INTERFACES: AN ASSESSMENT BY RUTHERFORD 5-1902
PY. (AUGER / PROCESSES OF ADSORPTION OF OXYGEN ON GALLIUM ARSENIDE METHOD OF LOW ENERGY ELECTRON SPECTROSCO 5-1641

ELECTRON DIFFRACTION AND AUGER SPECTR/ GROWTH OF GALLIUM ARSENIDE ON (100) GERMANIUM STUDIED BY LOW ENERGY 5-1646
ANODIC OXIDATION OF GALLIUM ARSENIDE PHOSPHIDE. (AES) 5-1388
AUGER ELECTRON SPECTROSCOPY OF GALLIUM ARSENIDE PHOTOSURFACES. 5-1922
SPECTROSCOPY. ELECTRON STIMULATED OXIDATION OF GALLIUM ARSENIDE, STUDIED BY QUANTITATIVE AUGER ELECTRON 5-1756
SPECTROSCOPY. GALLIUM(1-X) ALUMINUM(X) ARSENIDE SUPERLATTICES PROFILED BY AUGER ELECTRON 5-1596
ADSORPTION OF OXYGEN ON CLEAN CLEAVED (110) GALLIUM ARSENIDE SURFACES. 5-1306
AUGER ELECTRON SPECTROSCOPY OF CONTAMINATED GALLIUM ARSENIDE SURFACES. 5-1920
COMBINED LEED AND AES STUDIES OF INDIUM ARSENIDE SURFACES. 5-1373
TROSCOPY) ADSORPTION KINETICS OF CESIUM ON GALLIUM ARSENIDE. (USING AES AND X-RAY PHOTOELECTRON SPEC 5-1821
INTERDIFFUSIONS IN THIN FILM GOLD ON PLATINUM ON GALLIUM ARSENIDE (100) STUDIED WITH AUGER SPECTROSCOPY. 5-1232
UGER SPECTRA OF HYDROGEN CHLORIDE VAPOR ETCHED N+ GALLIUM ARSENIDE (100) SUBSTRATES. A 4-0992
FFRACTION AND AUGER ELECT/ GROWTH OF GOLD FILM ON GALLIUM ARSENIDE (110) OBSERVED BY LOW ENERGY ELECTRON DI 5-1876
, AND PHOTOEMISSION STUDIES OF THE CESIUM COVERED GALLIUM ARSENIDE (110) SURFACE. /ER ELECTRON SPECTROSCOPY 5-1931
OPY) INVESTIGATION OF CESIUM ADSORPTION ONTO A GALLIUM ARSENIDE (110) SURFACE. (AUGER ELECTRON SPECTROSC 5-1644
ND AUGER ELECTRON SPECTROSCOPY. STUDY OF INDIUM ARSENIDE (111) AND (-1-1-1) SURFACES USING LEED A 5-1381
OXIDATION AND ANNEALING OF GALLIUM PHOSPHIDE AND GALLIUM ARSENIDE (111) FACES STUDIED BY AES AND UPS. 5-1493
/ ELECTRON DIFFRACTION STUDY OF THE ACTIVATION OF GALLIUM ARSENIDE- CESIUM- OXYGEN NEGATIVE ELECTRON AFFIN/ 5-1845
FFECT OF CARBON AND OTHER SURFACE CONTAMINANTS ON GALLIUM ARSENIDE- CESIUM- OXYGEN PHOTOCATHODES. /NG THE E 5-1919
IO/ AUGER EFFECT AND RADIATIVE RECOMBINATION IN N-GALLIUM ARSENIDE- P-ALUMINUM GALLIUM ARSENIDE HETEROJUNCT 2-0324
A1 KLL AUGER PROFILE ARTIFACT AT THE OXIDE- METAL INTERFACE. 4-1068
ELECTRON EMISSION FROM METALS AND SEMICONDU/ EXPERIMENTAL ASSEMBLY FOR OBTAINING THE SPECTRUM OF SECONDARY 3-0688
/DUCED PROCESSES AT GOLD- GALLIUM ARSENIDE INTERFACES: AN ASSESSMENT BY RUTHERFORD BACKSCATTERING, SCANNIN/ 5-1902
LITES IN THE SOFT X-/ SCF AND LIMITED CI CALCULATIONS FOR ASSIGNMENT OF THE AUGER SPECTRUM AND OF THE SATEL 4-0724
ELECTRON BEAM ASSISTED ADSORPTION ON SILICON (111) SURFACE. 5-1269
PHONON ASSISTED AUGER PROCESSES IN GROUP III-V COMPOUNDS 2-0484
PHONON ASSISTED AUGER RECOMBINATION IN GERMANIUM. 2-0397
USE OF AES TO DETERMINE TRACE IMPURITIES) HOW TO ASSURE ONESELF OF THE CLEANLINESS OF A SURFACE. (5-1154
GAS TARGET EXPERIMENTS OF K-SHELL VACANCY SHARING IN ASYMMETRIC ION- ATOM COLLISIONS. (AUGER EFFECT) 2-0523
MOLECULAR EFFECTS ON K-VAANCY SHARING IN ASYMMETRIC ION- ATOM COLLISIONS. (AUGER SPECTRA) 2-0522
NEUTRALIZATION AT (100) FACE / INTERACTION OF LOW ENERGY ATMOSPHERIC IONS WITH CONTROLLED SURFACES. (AUGER 5-1736
K X-RAY TRANSITION ENERGIES FOR THE MULTIPLY IONIZED NEON ATOM. CALCULATED K-AUGER ELECTRON AND 4-0955
EVIDENCE OF RADIATIVE AUGER TRANSITIONS IN THE FREE ARGON ATOM. 2-0403
EVIDENCE OF RADIATIVE AUGER TRANSITIONS IN THE FREE ARGON ATOM. 2-0404
K-AUGER TRANSITIONS OF THE FREE SODIUM ATOM. 4-0880
SEMIEMPIRICAL AUGER ELECTRON ENERGIES. PART-2: THE ARGON ATOM. 4-0815
THE K-AUGER SPECTRUM OF THE FREE MAGNESIUM ATOM. 4-0789
FREE ATOM BEHAVIOR OF AUGER ELECTRONS IN SOLIDS. 2-0579
EMITTED BY ALKALI HALIDE CRYSTALS UNDER IMPACT OF ION AND ATOM BOMBARDMENT. (AUGER EFFECT) /M OF ELECTRONS 2-0246
ENERGY LOSS AND IONIZATION IN FAST ION- ATOM COLLISIONS. 2-0233
RADIATIVE AUGER EFFECT IN ICN- ATOM COLLISIONS. 2-0481
ION- ATOM COLLISIONS AT HIGH ENERGIES. (AUGER EFFECT) 2-0480
DEPENDENCE OF THE K-VACANCY PRODUCTION YIELD IN HEAVY ION ATOM COLLISIONS AT KEV ENERGIES. /C CHARGE STATE 2-0346
EXPERIMENTS OF K-SHELL VACANCY SHARING IN ASYMMETRIC ION- ATOM COLLISIONS. (AUGER EFFECT) GAS TARGET 2-0523
MECHANISMS OF ELECTRON PRODUCTION IN ION- ATOM COLLISIONS. (AUGER EFFECT) 2-0488
PRODUCTION OF INNER SHELL VACANCIES IN HEAVY ION- ATOM COLLISIONS. (AUGER EFFECT) 2-0405
MECHANISMS OF ELECTRON PRODUCTION IN ION- ATOM COLLISIONS. (AUGER EFFECT, REVIEWS) 2-0490
ION- ATOM COLLISIONS. (AUGER EMISSION, REVIEW) 1-0169
MOLECULAR EFFECTS ON K-VAANCY SHARING IN ASYMMETRIC ION- ATOM COLLISIONS. (AUGER SPECTRA) 2-0522
RGY DISTRIBUTION OF (AUGER) ELECTRONS RESULTING FROM ION- ATOM COLLISIONS WITH A FIXED IMPACT PARAMETER. /E 2-0331
ON EMISSION SPECTRA FOR RARE GAS ATOMS. PART-1. THE ARGON ATOM. PART-2. KRYPTON ATOM M-SHELL. PART-3. /ECTR 4-0927
TRON BINDING ENERGIES AND AUGER ENERGIES IN MANY ELECTRON ATOMS. ACCURATE ONE ELEC 4-0978
UGER ELECTRON SPECTROMETRY OF SINGLY AND MULTIPLY IONIZED ATOMS. A 4-0265
GER TRANSITIONS FOR A SURFACE CONTAINING SEVERAL KINDS OF ATOMS. COMPUTER PROGRAM TO CALCULATE AU 1-0039
HIGH RESOLUTION L2,3M4,5M4,5 AUGER SPECTRUM OF FREE ZINC ATOMS. 4-0733
L2,3MM AUGER ELECTRON SPECTRUM FROM FREE ZINC ATOMS. 4-0731
NER SHELL EXCITATION AND DEEXCITATION IN MULTIPLY IONIZED ATOMS. MECHANISMS OF IN 2-0487
ERVATION IN RADIATIONLESS (AUGER) TRANSITIONS IN MU MESIC ATOMS. PARITY NONCONS 2-0364
RADIATIVE AUGER EFFECT IN HEAVY ATOMS. 2-0476
OF COSTER-KRONIG TRANSITIONS IN THE IONIZATION OF SURFACE ATOMS. ROLE 2-0338
OF CHARACTERISTIC X-RAYS AND AUGER ELECTRONS FOR IONIZED ATOMS. SHIFT ENERGIES 4-1041
FFECTS ON AUGER ELECTRON EJECTION BY NOBLE GAS METASTABLE ATOMS. WORK FUNCTION E 5-1601
OF IRON EPITAXIAL FILM. (AUGER EFF/ BEHAVIOR OF IMPURITY ATOMS AND ADSORBED OXYGEN ATOMS ON THE (001) FACE 5-1523
/MISSION CROSS SECTIONS AND FLUORESCENCE YIELDS FOR LIGHT ATOMS AND MOLECULES BY ELECTRON IMPACT. (AUGER E/ 2-0528
SCOPY) BINDING OF CARBON ATOMS AT A STEPPED NICKEL SURFACE. (AUGER SPECTRO 5-1485
RADIATIONLESS DEEXCITATION OF EXCITED HELIUM ATOMS AT SURFACES. (AUGER PROCESS) 2-0255
ELECTRON SPRAYING ATOMS. (AUGER CASCADES) 2-0464
DEEXCITATION OF MULTIPLY IONIZED ATOMS. (AUGER EFFECT) 2-0281
LTIPLET EFFECT IN FLUORESCENCE YIELDS OF MULTIPLY IONIZED ATOMS. (AUGER EFFECT) MU 2-0314
N IMPACT IONIZATION CROSS SECTIONS ON K-SHELLS OF SURFACE ATOMS. (AUGER INTENSITIES OF TUNGSTEN (100)) /TRO 4-0847
RBATE EFFECTS IN ELECTRON EJECTION BY RARE GAS METASTABLE ATOMS. (AUGER PROCESS) ADSO 2-0333
SULTING FROM DIRECT IONIZING COLLISIONS OF ELECTRONS WITH ATOMS. (AUGER PROCESS) /TRIBUTION OF ELECTRONS RE 2-0525
REARRANGEMENT OF INNER SHELL IONIZED ATOMS. (AUGER SPECTRA) 2-0413
ERENTIAL CROSS SECTIONS FOR K-SHELL IONIZATION OF SURFACE ATOMS BY ELECTRON IMPACT. DIFF 2-0358
INNER SHELL IONIZATION OF SURFACE ATOMS BY ELECTRON IMPACT. 2-0359
OMB CENTERS. AUGER IONIZATION OF ATOMS BY MULTIPLY CHARGED IONS. MODEL OF TWO COUL 2-0407
/ INVESTIGATION OF MECHANISM OF S IONIZATION OF NOBLE GAS ATOMS BY THE ULTRASOFT X-RAY SPECTROSCOPY TECHNIQ 2-0580
NNER ELECTRON BINDING ENERGIES AND AUGER ENERGIES IN FREE ATOMS FOR USE IN X-RAY SPECTROSCOPY. I 4-0774
ROLE OF WORK FUNCTION IN ELECTRON EJECTION BY METASTABLE ATOMS: HELIUM AND ARGON ON (111) AND (110) TUNGS/ 5-1600
BOMBARDMENT. EMISSION OF AUGER ELECTRONS BY ATOMS IN A METALLIC TARGET SUBJECTED TO AN IONIC 2-0383
RONS) INNER SHELL ALIGNMENT OF ATOMS IN ELECTRON IMPACT IONIZATION. (AUGER ELECT 2-0335
TROSCOPY AS A TOOL FOR MEASURING THE DIFFUSION OF FOREIGN ATOMS IN SOLIDS NEAR THEIR SURFACE. /LECTRON SPEC 1-0186
AUGER EFFECT AND CORE BROADENING IN LOW-Z ATOMS, IONS, AND METALS. 2-0353
NUCLEAR AUGER EFFECTS IN MUONIC ATOMS. (LEAD, BISMUTH) 4-0941
MANY BODY EFFECTS IN AUGER DEEXCITATION OF ATOMS NEAR SOLIDS. 2-0361
/ON OF CORE ELECTRON BINDING ENERGIES OF SURFACE AND BULK ATOMS OF TITANIUM, CHROMIUM, AND NICKEL. (AUGER / 2-0392
/DARY ELECTRON EJECTION FROM METAL SURFACES BY METASTABLE ATOMS. PART-1: MEASUREMENTS OF SECONDARY EMISSIO/ 2-0340
K-SHELL AUGER RATES FOR MULTIPLY IONIZED ATOMS. PART-1: NEON. 2-0282
/F X-RAY AND AUGER ELECTRON EMISSION SPECTRA FOR RARE GAS ATOMS. PART-1. THE ARGON ATOM. PART-2. KRYPTON A/ 4-0927
MULTIPLE EXCITATION PROCESSES IN TRANSITION METAL ATOMS. PART-2: INNER AUGER ELECTRONS. 2-0362
/DARY ELECTRON EJECTION FROM METAL SURFACES BY METASTABLE ATOMS. PART-3: ENERGY AND ANGULAR DISTRIBUTIONS / 5-1133

GER NEUTRA/ CHARGE EXCHANGE OF LOW ENERGY HELIUM IONS AND ATOMS SCATTERED FROM A COPPER SINGLE CRYSTAL. (AU 2-0545
DIFFUSION OF SULFUR AND PHOSPHORUS IMPURITY ATOMS TO A TUNGSTEN SURFACE. 5-1226
ENERGY DISTRIBUTION OF ELECTRONS EMITTED BY ARGON ATOMS UNDER ELECTRON IMPACT. 4-1008
2S VACANCY). CORRELATION ENERGIES AND AUGER RATES IN ATOMS WITH INNER SHELL VACANCIES (NEON WITH 1S OR 2-0312
OBSERVATION OF THREE ELECTRON AUGER TRANSITIONS IN ATOMS WITH TWO INNER VACANCIES. 2-0235
DETECTION CF A NEW TYPE OF AUGER TRANSITIONS IN ATOMS WITH TWO INTERNAL VACANCIES. 2-0234
IN THE L3M4,5M4,5 AUGER PROCESS IN ARSENIC AND SELENIUM. ATOMIC AND EXTRAATOMIC STATIC RELAXATION ENERGIES 4-0883
AUGER RATES FOR SOFT X-RAY TRANSITIONS IN IONIC, ATOMIC, AND METALLIC LITHIUM. 4-0835
LICON ORDERED MONOLAYER ON MOLYBDENUM (001). ATOMIC ARRANGEMENT IN THE 1 K 1 STRUCTURE OF A SI 5-1478
L OVERLAYER SURFACE STRUCTURES. ATOMIC ARRANGEMENT OF GOLD (100) AND RELATED META 5-1716
00), AND SILVER-COVERED COPPER (100) SURFACES. ATOMIC ARRANGEMENT OF GOLD (100), GOLD-COVERED (1 5-1718
ENTS. (AUGER SPECTRUM CF HE/ POSSIBILITY OF INVESTIGATING ATOMIC AUTOIONIZING STATES IN COINCIDENCE EXPERIM 2-0257
OPY. IDENTIFICATION OF ATOMIC BORON IN STEEL BY AUGER ELECTRON SPECTROSC 6-2018
NVERSION AND ELECTRON CAPTURE. (K-AUGER/ MERCURY AND GCLD ATOMIC CLOUDS REARRANGEMENT FOLLOWING INTERNAL CO 2-0334
EFFECT OF ATOMIC CORRELATIONS ON KLL AUGER TRANSITIONS. 4-0801
CTRCSCOPY. ABSOLUTE ATOMIC DENSITIES DETERMINED BY AUGER ELECTRON SPE 1-1843
UGER SPECTRUM) MANIFESTATION OF ATOMIC DYNAMICS THROUGH THE AUGER EFFECT. (NEON A 4-0920
G YIELDS, K-, L-, AND M-SHELL FLUORESCENCE YIELDS) ATOMIC FLUORESCENCE YIELDS. (L-SHELL COSTER-KRONI 2-0351
UTION OF AUGER ELECTRONS) RESONANCE IONIZATION OF ATOMIC HELIUM BY FAST ELECTRONS. (ANGULAR DISTRIB 2-0256
HIGH RESOLUTION NEON AUGER SPECTRUM PRODUCED IN 4.2 MEV ATOMIC HYDROGEN(+)- NEON COLLISIONS. 4-1078
/DENCE OF THE ARGON I-SHELL FLUORESCENCE YIELD STUDIED BY ATOMIC HYDROGEN(+), MOLECULAR HYDROGEN(+), ATOMI/ 2-0517
(AUGER EFFECT) CHEMISORPTION OF ATOMIC HYDROGEN ON THE SILICON (111) 7*7 SURFACE. 5-1789
NEON K-AUGER CROSS SECTIONS FROM 1.5 MEV PER AMU AND ATOMIC HYDROGEN, OXYGEN, AND FLUORINE. 2-0571
E RATES AND FLUORESCENCE YIELDS FOR SODIUM- THORIUM. ATOMIC L-SHELL COSTER-KRONIG, AUGER, AND RADIATIV 4-0963
SURFACE ANALYSIS, METHCDS OF STUDYING THE OUTER ATOMIC LAYERS OF SOLIDS. 1-0016
TOTAL AND PARTIAL ATOMIC LEVEL WIDTHS. 4-0911
ER EFFECT) WIDTH OF ATOMIC L2 AND L3 VACANCY STATES NEAR Z = 30. (AUG 4-0939
E RATES, AND FLUORESCENCE YIELDS FOR CALCIUM TO THORIUM. ATOMIC M-SHELL COSTER-KRONIG, AUGER, AND RADIATIV 4-0964
ECTS OF RADIATIONLESS TRANSITIONS IN SOLIDS. WIDTHS OF ATOMIC M-SHELL VACANCY STATES AND QUASIATOMIC ASP 3-0718
E RATES, AND FLUORESCENCE YIELDS FOR Z BETWEEN 38 AND 10/ ATOMIC N-SHELL COSTER-KRONIG, AUGER, AND RADIATIV 4-0969
ATOMIC N-SHELL VACANCY WIDTHS IN SOLIDS. 2-0577
DHESION, FRICTION, AND WEAR. ATOMIC NATURE OF POLYMER- METAL INTERACTIONS IN A 5-1212
TR/ ABSOLUTE ELECTRON IMPACT IONIZATION CROSS SECTIONS OF ATOMIC NITROGEN, ATOMIC OXYGEN, AND NEON. (AUGER 2-0363
AUGER SPECTRA FOR ELEMENTS OF LOW ATOMIC NUMBER. 4-0827
STUDY OF THE AUGER EFFECT IN THE LIGHT TO MEDIUM RANGE OF ATOMIC NUMBER. THEORETICAL 2-0486
PROBABILITIES FOR AUGER- KLL TRANSITIONS DEPENDENT ON THE ATOMIC NUMBER Z=30-94. PART-2. ABSOLUTE 2-0252
PROBABILITIES FOR AUGER KLL TRANSITIONS DEPENDENT ON THE ATOMIC NUMBER 29 TO 94. RELATIVE 2-0253
AUGER ELECTRON ENERGIES (0-2000 EV) FOR ELEMENTS OF ATOMIC NUMBER 5-103. 4-1119
TION PROBABILITIES FOR ELEMENTS WITH LOW AND INTERMEDIATE ATOMIC NUMBERS. KLL AUGER TRANSI 4-0803
THE KLL AUGER SPECTRUM FOR ATOMIC NUMBERS BETWEEN 10 AND 36. 4-0980
KLL AUGER TRANSITION ABSOLUTE PROBABILITIES. (ATOMIC NUMBERS 30-94) 2-0251
PY STUDY OF INITIAL STAGES OF OXIDATION IN A COPPER- 19.6 ATOMIC PERCENT ALUMINUM ALLOY. /ECTRON SPECTROSCO 6-2028
/DY OF SURFACE SEGREGATION IN THE BINARY ALLOYS COPPER- 1 ATOMIC PERCENT INDIUM, COPPER- 2 ATOMIC PERCENT / 6-2027
UBSTATE. (AUGER ELECTRON DISTRIBUTION) DEPENDENCE OF ATOMIC PHOTOEFFECT ON A BOUND ELECTRON MAGNETIC S 2-0463
STATE (AUGER EFFECT). DEPENDENCE OF ATOMIC PHOTOEFFECT ON BOUND ELECTRON MAGNETIC SUB 2-0524
1S STATE AND THEORETICAL K-SHELL FLUCRESCENCE YIELDS. ATOMIC RADIATION TRANSITION PROBABILITIES TO THE 4-0918
MANY BODY PERTURBATION THEORY APPLIED TO ATOMIC RADIATIONLESS TRANSITIONS. 2-0386
ELL IONIZATION (AUGER AND RADIATIVE TRANSITICNS). ATOMIC VACANCY DISTRIBUTIONS PRODUCED BY INNER SH 2-0544
RELATIVISTIC CALCULATIONS OF ATOMIC X-RAY AND AUGER TRANSITION RATES. 2-0485
THE L(II) SUBSHELL ATOMIC YIELDS OF ELEMENTS WITH Z APPROXIMATELY 90 2-0349
/ILED ANALYSIS OF THE L3M4,5M4,5: 1G4 AUGER TRANSITION IN ATOMIC ZINC: IMPROVED AUGER ENERGY CALCULATIONS. 4-0884
/ AUGER AND COSTER-KRONIG TRANSITION PROBABILITIES TO THE ATOMIC 2S STATE AND THEORETICAL L1 FLUORESCENCE Y 2-0329
RON SPECTROSCOPY. STUDY OF THE PREPARATION OF ATOMICALLY CLEAN TUNGSTEN SURFACES BY AUGER ELECT 5-1517
GH THIN ADSORBED FILMS. (AUGER/ DETERMINATION OF ELECTRON ATTENUATION LENGTHS IN METALS: TRANSMISSION THROU 4-1088
OR ANALYZER AND A RE/ CRIENTATION DEPENDENCE OF OVERLAYER ATTENUATION OF ELECTRONS FOR THE CYLINDRICAL MIRR 3-0693
RESULTS FROM X-RAY PHOTOELECTRON (AND AUGER) SPECTROSCOP/ ATTENUATION OF LOW ENERGY ELECTRONS FROM SOLIDS: 5-1844
UM OXIDE. (AUGER ELEC/ PHOTCELECTRON DETERMINATION OF THE ATTENUATION OF THE LOW ENERGY ELECTRONS IN ALUMIN 4-0773
IVE ASPECTS OF AUGER ELECTRON SPECTROSCOPY. IMPORTANCE OF ATTENUATION OF THE PRIMARY ELECTRON BEAM. /NTITAT 1-0206
AUGER IS NOT INDEXED.
ANISMS IN 8-14 MICRON MERCURY(1-X) CADMIUM(X) TELLURIDE. (AUGER-LIMITED) RECOMBINATION MECH 2-0406
THE SOFT X-RAY L3 EMISSION SPECTRA OF NICKEL AND COPPER. (AUGER-TYPE PROCESSES) CLUSTER CALCULATIONS OF 4-0888
LEED- AUGER INVESTIGATION ON A (100) SURFACE OF AUSTENITIC CHROMIUM- NICKEL STEEL. 6-1991
RITTLENESS IN A NICKEL- CHROMIUM- ANTIMO/ EFFECT OF PRIOR AUSTENITIC GRAIN BOUNDARY COMPOSITION ON TEMPER B 6-2114
/OUNDARIES AND ITS RELATION TO INTERGRANULAR CORROSION OF AUSTENITIC STAINLESS STEEL. (AUGER SPECTROSCOPY) 6-2064
THE AUGER, COSTER-KRONIG, AND AUTOIONIZATION EFFECTS. (REVIEW, 477 REFS) 1-0180
AND HELIUM(2+) NEAR SOLID SURFACES. RESONANCE, AUGER, AND AUTOIONIZATION PROCESSES INVOLVING HELIUM(+) (2S) 2-0375
AUGER SPECTRUM OF HE/ POSSIBILITY OF INVESTIGATING ATOMIC AUTOIONIZING STATES IN COINCIDENCE EXPERIMENTS. (2-0257
UORINE, CHLORINE, / ELECTRON DECAY-IN-FLIGHT SPECTRA FROM AUTOIONIZING STATES OF HIGHLY STRIPPED OXYGEN, FL 2-0472
ERYLLIUM. (AUGER DECAY) METASTABLE AUTOIONIZING THREE- AND FOUR ELECTRON STATES IN B 2-0295
MS. DIGITAL CONTROL UNIT IN TTL LOGIC FOR THE AUTOMATIC EXECUTION OF AUGER AND ESCA TEST PROGRA 3-0722
/ER USED IN ELECTRON SPECTROMETRY FOR CHEMICAL ANALYSIS. (AUTOMATIC RECORDING OF PHOTO AND AUGER ELECTRONS) 3-0717
IN AUGER ELECTRON EMISSION SPECTROSCOPY USING RETARDING / AVERAGING, CURVE FITTING, AND RELATED TECHNIQUES 3-0610
OF LEAD, BY AUGER SPECTRCSCOPY, ON IRON INHIBITED IN LEAD AZELATE SOLUTION. DETECTION 5-1628
INTERFACE. A1 KLL AUGER PROFILE ARTIFACT AT THE OXIDE- METAL 4-1068

B

AND GAIN PHENOMENA OF COPPER (111). ENERGY SPECTRUM OF BACK SCATTERED ELECTRONS AND CHARACTERISTIC LOSS 4-0897
COPY AND SECONDARY ELECTRON EMISSICN DUE TO CASCADE PROC/ BACKGROUND INFORMATION IN AUGER ELECTRON SPECTROS 2-0411
TRON SPECTROSCOPY. THE SECONDARY ELECTRON BACKGROUND PROBLEM IN ELECTRON EXCITED AUGER ELEC 1-0091
RETRIEVAL OF ELECTRON EXCITED AUGER STRUCTURE BY DYNAMIC BACKGROUND SUBTRACTION. 1-0090
CTRON SPECTROSCOPY./ USE OF ANALOG INTEGRATION AS DYNAMIC BACKGROUND SUBTRACTION FOR QUANTITATIVE AUGER ELE 1-0069
ELECTRON AND SOFT X-RAY APPEARAN/ APPLICATICN OF DYNAMIC BACKGROUND SUBTRACTION TO QUANTIFICATION OF AUGER 1-0067
ECTRONS. TIN-119. BACKSCATTER MOSSBAUER STUDIES USING CONVERSION EL 2-0576
AUGER AND SECONDARY ELECTRONS EXCITED BY BACKSCATTERED ELECTRONS. 2-0365
ANALYSIS. AUGER AND SECONDARY ELECTRONS EXCITED BY BACKSCATTERED ELECTRONS. APPROACH TO QUANTITATIVE 2-0366
STAL UNDER A LOW ENERGY BOMBARDMENT. / ENERGY ANALYSIS OF BACKSCATTERED ELECTRONS FROM A SILICON SINGLE CRY 4-0739
SECONDARY ELECTRON EMISSION DUE TO PRIMARY AND BACKSCATTERED ELECTRONS (TABLE FOR 60 ELEMENTS). 4-0907
ANALYSIS OF SURFACE COMPOSITION WITH LOW ENERGY BACKSCATTERED IONS. 5-1823
MBIENTS ON PLATINUM SILICIDE FORMATION. PART-2: AUGER AND BACKSCATTERING ANALYSES. EFFECT OF OXIDIZING A 5-1182

CRYSTALLINE EFFECTS IN BACKSCATTERING AND AUGER PRODUCTION. (ALUMINUM) 2-0254
SURFACES. DIFFRACTION EFFECTS IN BACKSCATTERING AND AUGER PRODUCTION NEAR CRYSTAL 2-0243
ELECTRON SPECTR/ ANALYSIS OF SURFACE LAYERS BY LIGHT ION BACKSCATTERING AND SPUTTERING COMBINED WITH AUGER 5-1167
EMISSION FOR 10-100 KEV ELECT/ ORIENTATION ANISOTROPY OF BACKSCATTERING COEFFICIENT AND SECONDARY ELECTRON 4-1031
ONS. AUGER EMISSION FROM SOLIDS. ESTIMATION OF BACKSCATTERING EFFECTS AND IONIZATION CROSS SECTI 2-0513
SPECTRA (SILVER AND SILICON, AUGER EFFECT / ESTIMATION OF BACKSCATTERING EFFECTS IN ELECTRON INDUCED AUGER 4-0841
S. MEASUREMENT OF IONIZATION CROSS SECTIONS AND BACKSCATTERING FACTORS FOR USE IN QUANTITATIVE AE 2-0554
Y CHANGES WITH AUGER ELECTRON SPECTROSCOPY AND RUTHERFORD BACKSCATTERING PROFILES. /RRELATION OF RESISTIVIT 5-1393
/GALLIUM ARSENIDE INTERFACES: AN ASSESSMENT BY RUTHERFORD BACKSCATTERING, SCANNING ELECTRON MICROSCOPY, AN/ 5-1902
ER CYLINDRICAL MIRROR ANALYZER. LOW ENERGY ION BACKSCATTERING SPECTROSCOPY WITH A COMMERCIAL AUG 3-0665
/CANNING ELECTRON MICROSCOPY, AUGER SPECTROSCOPY, AND ION BACKSCATTERING TECHNIQUES APPLIED TO ANALYSIS OF/ 5-1261
N THE ANALYSIS OF SU/ DIRECT COMPARISON OF LOW ENERGY ION BACKSCATTERING WITH AUGER ELECTRON SPECTROSCOPY I 5-1869
/RUCTURE AND CHEMISTRY OF SILICON SURFACES AFTER PRE- AND BACKSPUTTERING, STUDIED WITH AUGER SPECTROSCOPY,/ 5-1233
NIZATION IN SEMICONDUCTORS. DETAILED BALANCE BETWEEN AUGER RECOMBINATION AND IMPACT IO 2-0417
ORE EXCITONS IN ALKALI HALIDES. (AUGER PROCESS IN VALENCE BAND) PHOTOEMISSION STUDIES OF THE DECAY OF C 2-0289
VIA PHOTOEMISSION MEASUREMENTS IN THE 20-70 EV R/ VALENCE BAND AND CORE LEVEL STUDIES OF INDIUM ANTIMONIDE 2-0341
N. SURFACE VALENCE BAND AND PLASMON FEATURES ON CLEAN CLEAVED SILICO 5-1141
S. VALENCE BAND AUGER ELECTRON SPECTRA FOR ALUMINUM: COMMENT 4-0839
UM. CONTRASTING VALENCE BAND AUGER ELECTRON SPECTRA FOR SILVER AND ALUMIN 4-1022
TION SCANNING AUGER SPECTROSCOPY APPLIED TO ANALYSIS OF X-BAND DIODE BURNOUT. HIGH SPATIAL RESOLU 5-1961
D PHOTOEMISSION SPECTROSCOPY. VALENCE BAND OF MAGNESIUM STANNIDE DETERMINED BY AUGER AN 5-1885
TRA. GRAPHITE CONDUCTION BAND STATES FROM SECONDARY ELECTRON EMISSION SPEC 2-0568
VALENCE BAND STRUCTURE IN THE AUGER SPECTRUM OF ALUMINUM. 4-0889
LECTRON EMISSION. BAND STRUCTURE OF GRAPHITE STUDIED BY SECONDARY E 2-0569
SPECTRUM ANALYSIS. BAND STRUCTURE OF SILICON BY CHARACTERISTIC AUGER 4-0740
SILICON. INFLUENCE OF THE BAND STRUCTURE ON THE AUGER ELECTRON SPECTRUM OF 4-0944
IMPACT IONIZATION AND AUGER RECOMBINATION IN BANDS. 2-0418
UGER ELECTRONS) NEW SATELLITE BANDS CORRESPONDING TO DOUBLE L IONIZATION. (OF A 2-0293
ABSORPTION BANDS IN SECONDARY ELECTRON CONTINUOUS SPECTRUM. 4-0933
AUGER SATELLITES ON THE L23 AUGER EMISSION BANDS OF ALUMINUM, SILICON, AND PHOSPHORUS. 4-0984
AUGER SATELLITES OF THE L2,3 AUGER EMISSION BANDS OF ALUMINUM, SILICON, AND PHOSPHORUS. 4-0985
/PHOTOEMISSION, AND AUGER ELECTRON SPECTRA OF THE VALENCE BANDS OF MAGNESIUM, MAGNESIUM(2) COPPER AND MAGN/ 4-0838
/ETATION OF HIGH ENERGY X-RAY SATELLITES OF L2,3 EMISSION BANDS OF SODIUM, MAGNESIUM, ALUMINUM, AND SILICO/ 2-0239
ENTRATION. BAND-TO-BAND AUGER PROCESSES AT HIGH CARRIER CONC 2-0483
IER SCATTERING IN POWER RECTIFIERS. BAND-TO-BAND AUGER RECOMBINATION AND CARRIER-CARR 2-0458
SEMICONDUCTORS. BAND-TO-BAND AUGER RECOMBINATION IN INDIRECT GAP 2-0394
ERMANIUM. BAND-TO-BAND AUGER RECOMBINATION IN SILICON AND G 2-0459
PLASMAS IN GERMANIUM. PICOSECOND OPTICAL MEASUREMENTS OF BAND-TO-BAND AUGER RECOMBINATION OF HIGH DENSITY 2-0250
SELF-CONSISTENT ANALYSIS OF CORE BAND-TO-BAND EXCITATION: AUGER FINE STRUCTURE. 2-0509
G-R NOISE FOR AUGER BAND-TO-BAND PROCESSES. 2-0292
INATION IN SILICON) TWO ELECTRON BAND-TO-BAND TRANSITIONS IN SOLIDS. (AUGER RECOMB 2-0275
EM. (AUGER E/ CHARACTERISTICS OF SURFACE OF THE PALLADIUM BARIUM ALLOY CATHODE IN RELATION TO ITS FILM SYST 5-1129
ES. (AES) EFFECTS OF SULFUR, BARIUM, AND CALCIUM ON IMPREGNATED CATHODE SURFAC 5-1391
RACTION OF SLOW ELECTRONS AND AUGER SPECTROSCO/ STUDY OF BARIUM FILMS ON A (110) TUNGSTEN FACE BY THE DIFF 5-1368
EFFECT OF STRUCTURE ON THE AUGER SPECTRA OF BARIUM FILMS ON PLATINUM. 5-1359
LINE WIDTH AND AUGER EFFECT IN CALCIUM, STRONTIUM, AND BARIUM FLUORIDES. 4-1116
ELECTRON BINDING ENERGIES OF BARIUM FROM THE SECONDARY ELECTRON YIELD SPECTRUM 4-0846
CATHODE SURFACES DURING ACTIVATION AND POISONING. PART-1: BARIUM ON OXYGEN ON TUNGSTEN DISPENSER CATHODE. / 5-1841
AUGER SPECTROSCOPY OF A BARIUM- METAL FILM SYSTEM. 4-0849
AUGER SPECTROSCOPY OF THE BARIUM- PLATINUM FILM SYSTEM. 5-1357
ATINUM ALLOY FILMS. COMPOSITION PROFILES AND SCHOTTKY BARRIER HEIGHTS OF SILICIDES FORMED IN NICKEL- PL 5-1900
NTS. ADSORPTION OF KRYPTON ON THE BASAL PLANE OF GRAPHITE: LEED AND AUGER MEASUREME 5-1559
ON AND AUGER SPECTROSCOPY ANALYSIS OF A COMMERCIAL COBALT BASE AIRCRAFT TURBINE SHROUD ALLOY. /SION, FRICTI 6-2009
COLLECTOR. SENSITIVITY VARIATIONS IN BAYARD- ALPERT GAUGES CAUSED BY AUGER EMISSION A 3-0618
OSCOPY. IMPORTANCE OF ATTENUATION OF THE PRIMARY ELECTRON BEAM. /NTITATIVE ASPECTS OF AUGER ELECTRON SPECTR 1-0206
(AUGER SPECTROSCOPY) ELECTRON BEAM ADSORBATE INTERACTIONS ON SILICON SURFACES. 5-1513
M. (SCANNING ELECTRON MICROSCOPY, AUGER SPECTROSCOPY, ION BEAM ANALYSIS) / PRODUCTS IN FLOWING LIQUID SODIU 6-2014
S SPECTROSCOPY) PRESENT STATUS OF BEAM ANALYSIS AND ITS FUTURE. (REVIEW, AUGER, MAS 1-0181
RECENT ADVANCES IN SURFACE STUDIES. ION BEAM ANALYSIS. (AUGER SPECTROSCOPY, REVIEW) 1-0137
ION DETERMINATION. BEAM ANALYSIS METHOD OF FILM AND SURFACE COMPOSIT 5-1810
ELECTRON BEAM ANALYZER FOR DETECTING AUGER ELECTRONS. 3-0712
ELECTRON BEAM ASSISTED ADSORPTION ON SILICON (111) SURFACE 5-1269
ELECTRON BEAM CURRENT CONTROLLER FOR LEED- AUGER STUDIES. 3-0589
CTRON SPECTROSCOPY AT HIGH SPATIAL RESOLUTION AND PRIMARY BEAM CURRENTS. (CARBON ON SILVER FILM) AUGER ELE 5-1904
ON. ELECTRON BEAM EFFECTS IN AUGER ANALYSIS OF PHYSISORBED XEN 5-1151
LED BY X-RAY PHOTOELECTRON SPECTROSCOPY. BEAM EFFECTS IN AUGER ELECTRON SPECTROSCOPY REVEA 3-0601
AUGER ELECTRON SPECTROSCOPY. ELECTRON BEAM EFFECTS IN DEPTH PROFILING MEASUREMENTS WITH 5-1128
N (100), ETHYLENE SATURATED TUNGSTEN (100), AND/ ELECTRON BEAM EFFECTS ON CARBON MONOXIDE SATURATED TUNGSTE 5-1244
STEM FOR THE ANALYSIS OF ELECTRON EMISSION AT LOW PRIMARY BEAM ENERGIES. /ENTS TO A STANDARD LEED- AUGER SY 3-0698
DEPENDENCE OF AUGER ELECTRON YIELD ON THE PRIMARY BEAM ENERGY AT NORMAL AND GLANCING INCIDENCE. 2-0455
SENIDE AND ALUMINUM(X) GALLIUM(1-X) ARSENIDE BY MOLECULAR BEAM EPITAXY. (AUGER SPECTROMETRY) /ED GALLIUM AR 5-1247
ALUMINUM(1-X) ARSENIDE) COMPUTER-CONTROLLED MOLECULAR BEAM EPITAXY. (AUGER SURFACE ANALYSIS, GALLIUM(X) 5-1324
PHIDE. USE OF SCANNING AUGER MICROSCOPY IN MOLECULAR BEAM EPITAXY OF GALLIUM ARSENIDE AND GALLIUM PHOS 5-1142
BEAM FOIL AUGER SPECTROSCOPY. 1-0112
BEAM FOIL EXCITED AUGER TRANSITIONS IN NEON. 4-0860
OXYGEN BEAM GAS AUGER ELECTRON EMISSION. 2-0400
XCITED BY COLLISIONS WITH HELIUM, NEON, / HIGH RESOLUTION BEAM GAS AUGER ELECTRON SPECTRA FOR OXYGEN IONS E 4-0904
O DISTINGUISH AUGER FROM ENERGY LOSS PEAKS OF THE PRIMARY BEAM IN SECONDARY ELECTRON ANALYSIS. /NG METHOD T 1-0188
ILIZING AUGER ELECTRON SPECTROSCOPY: CARBON MO/ ELECTRON BEAM INDUCED EFFECTS ON GAS ADSORPTION STUDIES UT 5-1542
AUGER ELECTRON SPECTROSCOPY. CARBON MONOXIDE A/ ELECTRON BEAM INDUCED EFFECTS ON GAS ADSORPTION UTILIZING 5-1543
BERYLLIUM OXIDE AND ITS EFFECTS ON SECONDARY E/ ELECTRON BEAM INDUCED REDUCTION OF CARBON CONCENTRATION ON 5-1362
APPEARANCE POTENTIAL SPECTRA INDUCED BY INCIDENT ELECTRON BEAM INTERACTION. /HE TITANIUM DIOXIDE AUGER AND 5-1834
/MENTS OF SECONDARY EMISSION COEFFICIENTS USING A CROSSED BEAM METHOD. PART-2: MEASUREMENTS OF SECONDARY E/ 2-0340
ROBE ANALYZER, SCANNING ELECTRON MICROSCOPE, TR/ ELECTRON BEAM MICROANALYSIS. PART-1. (REVIEW OF ELECTRON P 1-0010
ELECTRON BEAM MICROANALYZER WITH AUGER ELECTRON DETECTION. 3-0711
MENT OF MATERIALS. (AUGER ELECTRON SIGNALS) ELECTRON BEAM MICROPROBES USED IN THE ANALYSIS AND DEVELOP 3-0714
ANALYZER OF THE CYLINDRICAL/ FOCUSING A CHARGED PARTICLE BEAM OF FINITE ANGULAR SPREAD IN AN ELECTROSTATIC 3-0720
/TIONS AT THE (100) CRYSTAL FACE OF PLATINUM BY MOLECULAR BEAM SCATTERING, LOW ENERGY ELECTRON DIFFRACTION/ 5-1963
ONDUCTOR S/ AUGER ELECTRON SPECTROSCOPY STUDY OF ELECTRON BEAM STIMULATED DESORPTION OF OXYGEN FROM A SEMIC 5-1617
GROWTH. (AES) MOLECULAR BEAM STUDIES OF INDIUM PHOSPHIDE EVAPORATION AND 5-1334

101

GATED ELECTRON BEAM TECHNIQUE FOR AUGER ELECTRON SPECTROSCOPY. 1-0024
GERMANIUM. (AES)   INTERFERENCE OF AN ELECTRON BEAM WITH THE SURFACE REACTION BETWEEN OXYGEN AND 5-1620
LECTRON EMISSION SPECTRA FROM FOIL AND GAS EXCITED CARBON BEAMS.   AUGER E 4-1050
PPLER SHIFTED AUGER ELECTRONS FROM FOIL EXCITED HEAVY ION BEAMS.   DO 4-0859
SURFACE ANALYSIS USING PROTON BEAMS.   1-0143
SCATTERING CF MOLECULAR BEAMS FROM SURFACES. (AUGER SPECTROSCOPY) 5-1905
APPLICATION OF ION, ELECTRON, AND X-RAY BEAMS IN MODERN MATERIAL TESTING. 1-0034
AUGER ELECTRONS FRCM FOIL EXCITED HEAVY ION BEAMS. (OF NEON) 2-0370
BIOLOGICAL DAMAGE FROM THE AUGER EFFECT, POSSIBLE BENEFITS. (DNA) 1-0056
SULFUR AT THE SURFACE.   STUDY OF THE ADSORPTION OF BENZENE ON THE (100) FACE OF NICKEL. INFLUENCE OF 5-1284
A IN AMERICIUM-241 FROM CURIUM-241 (ELECTRON CAPTURE) AND BERKELIUM-245 ALPHA DECAYS. /, AND K-AUGER SPECTR 4-1019
AUGER SPECTROSCOPY OF BERYLLIUM.   4-1084
CHEMICAL EFFECTS ON THE AUGER ELECTRON SPECTRA OF BERYLLIUM.   4-0834
ECONDARY ELECTRON EMISSION CHARACTERISTICS (GERMANIUM AND BERYLLIUM).   FILM STRUCTURE EFFECT ON THE S 4-0923
TERPRETATION OF THE AUGER ELECTRON SPECTRUM FROM OXIDIZED BERYLLIUM.   IN 4-0935
NTERPRETATION OF THE AUGER SPECTRUM OF CLEAN AND OXIDIZED BERYLLIUM.   I 4-1087
INTERPRETATION OF THE AUGER ELECTRON SPECTRUM FROM CLEAN BERYLLIUM.   OBSERVATION AND 4-0997
SECONDARY ANC AUGER ELECTRON SPECTRA OF A CCPPER- BERYLLIUM ALLOY.   4-0916
/D LOSS, DOUBLE IONIZATION, AND IONIZATION LCSS EVENTS IN BERYLLIUM AND BERYLLIUM OXIDE SECONDARY ELECTRON/ 4-0901
ETASTABLE AUTOIONIZING THREE- AND FOUR ELECTRON STATES IN BERYLLIUM. (AUGER DECAY)   M 2-0295
OF THE SENSITIVITY CF IONIZATION SPECTROSCOPY: SILICON ON BERYLLIUM. (AUGER ELECTRON SPECTROSCOPY) /NDENCE 5-1665
ASMON ENERGY GAIN AND DOUBLE IONIZATION SATELLITES IN THE BERYLLIUM AUGER SPECTRUM.   PL 4-0899
PLASMON ENERGY GAIN IN THE BERYLLIUM AUGER SPECTRUM.   4-1095
ETRY AND AUGER ELE/ STUDY OF SURFACE PROCESSES ON COPPER- BERYLLIUM BY COMBINED SECONDARY ION MASS SPECTROM 5-1213
ION / ENERGY DISTRIBUTION OF AUGER ELECTRONS EMITTED FROM BERYLLIUM, MAGNESIUM, ALUMINUM AND SILICON UNDER 4-0876
/ECTRON BEAM INDUCED REDUCTION OF CARBON CONCENTRATION ON BERYLLIUM OXIDE AND ITS EFFECTS ON SECONDARY ELE/ 5-1362
PY INVESTIG/ THERMALLY STIMULATED EXOELECTRON EMISSION OF BERYLLIUM OXIDE LAYERS. AUGER ELECTRON SPECTROSCO 5-1325
NITROGEN(2), AND / SECONDARY ELECTRON YIELDS FROM COPPER BERYLLIUM OXYGEN SURFACE BY THERMAL CARBON OXYGEN 4-0786
AUGER AND IONIZATION LOSS SPECTRA FROM BERYLLIUM SURFACES.   5-1987
AUGER ELECTRON EMISSION FROM BERYLLIUM SURFACES.   5-1613
UGER TRANSITION RATES AND FLUORESCENT YIELDS FOR ELEMENTS BERYLLIUM TO ARGON.   K-SHELL A 4-0962
LEED- AUGER ANALYSIS OF THE BERYLLIUM (0001) SURFACE.   5-1581
AUGER TRANSITIONS IN LITHIUM-LIKE BERYLLIUM-LIKE IONS.   2-0296
RUCTURE OF ALUMIN/ APPLICATION OF MICROAUTORADIOGRAPHY BY BETA-RAY AUGER ELECTRONS TO THE STUDIES OF THE ST 6-2051
ND AUGER ELECTRON SPECTROSCOPY. (766 REFS)   A BIBLIOGRAPHY OF LOW ENERGY ELECTRON DIFFRACTION A 1-0071
R ION SPUTTERING TECHNIQUE.   BIBLIOGRAPHY ON AUGER ELECTRON SPECTROSCOPY. 1-0148
RON SPECTROSCOPY. ANALYSIS OF GRAIN BOUNDARY DIFFUSION IN BIMETALLIC POLYCRYSTALLINE FILMS MEASURED BY AUGE 5-1440
RECORDING THE COMPONENTS ON THE SURFACE OF A BIMETALLIC THIN FILM STRUCTURES USING AUGER ELECT 6-2146
QUANTITATIVE STUDY OF THE SURFACE COMPOSITION OF BINARY ALLOY. (AUGER SPECTROSCOPY) 5-1358
EMENTS.   COMPOSITION OF BINARY ALLOYS BY AUGER SPECTROSCOPY. 6-2005
/LECTRON SPECTROSCOPY STUDY OF SURFACE SEGREGATION IN THE BINARY ALLOYS BY SIMULTANEOUS SIMS AND AES MEASUR 6-2090
S.   AUGER SPECTRA OF SOME FREE ELEMENTS AND BINARY ALLOYS COPPER- 1 ATOMIC PERCENT INDIUM, C/ 6-2027
ER LINE SHAPE: SILICON (111).   TIGHT BINARY COMPOUNDS BOMBARDED BY 60 KEV ARGON(+) ION 4-1106
FOR USE IN X-RAY SPECTROSCOPY.   INNER ELECTRON BINDING CALCULATION OF A CORE VALENCE VALENCE AUG 4-0831
RON ATOMS.   ACCURATE ONE ELECTRON BINDING ENERGIES AND AUGER ENERGIES IN FREE ATOMS 4-0774
CTRON YIELD SPECTRUM.   ELECTRON BINDING ENERGIES AND AUGER ENERGIES IN MANY ELECT 2-0265
QUILIBRIUM SEGREGATION STUDIES. (AUGER EFFECT)   BINDING ENERGIES OF BARIUM FROM THE SECONDARY ELE 4-0846
TION OF RELAXATION ENERGY DIFFERENCES FROM AUGER AND CORE BINDING ENERGIES OF CARBON TO NICKEL (100) FROM E 5-1484
ANIUM, CHROMIUM, AND / DIRECT COMPARISON CF CCRE ELECTRON BINDING ENERGIES OF GERMANIUM COMPOUNDS. / ESTIMA 2-0471
PALLADIUM.   STRUCTURAL INFLUENCES ON ADSORBATE BINDING ENERGIES OF SURFACE AND BULK ATOMS OF TIT 2-0392
EFFECTS OF EXTRA CLOSED CORE RELAXATION ON AUGER AND BINDING ENERGY. PART-1: CARBON MONOXIDE ON (100) 5-1913
L/ RELATIVE EFFECT OF EXTRAATOMIC RELAXATION CN AUGER AND BINDING ENERGY SHIFTS.   4-0950
CE. (AUGER SPECTROSCOPY)   BINDING ENERGY SHIFTS IN TRANSITION METALS AND SA 2-0412
M. (LEED, AES)   BINDING OF CAREON ATOMS AT A STEPPED NICKEL SURFA 5-1485
REACTIONS ON METAL SURFACES.   BINDING STATES OF WATER VAPOR ON CLEAVED GERMANIU 5-1820
AUGER SPECTROSCOPIC ANALYSIS OF BINDING, VALENCY AND CHARGE TRANSFER IN CHEMICAL 5-1763
AES ANALYSIS OF SODIUM IN A CORRODED BIOGLASS CORROSION FILMS.   5-1264
BENEFITS. (DNA)   BIOGLASS USING A LOW TEMPERATURE TECHNIQUE. 5-1721
ATELY HEAVY NUCLEI IN THE TISSUE: USE OF SOFT X-RAYS. (/ BIOLOGICAL DAMAGE FROM THE AUGER EFFECT, POSSIBLE 1-0056
NUCLEAR AUGER EFFECTS IN MUONIC ATOMS. (LEED, AES)   BIOLOGICAL RADIATION EFFECT ON INCLUSION OF MODER 2-0531
TALS AND ALLOYS. (SEGREGATION IN STEEL, TUNGSTEN, COPPER- BISMUTH)   4-0941
/NDUCED INNER SHELL IONIZATION OF THE N6,7 LEVEL IN GOLD, BISMUTH ALLOYS STUDIED BY AUGER SPECTROSCOPY) /ME 6-2120
CTRA, L-SUBSHELL FLUCRESCENCE AND COSTER-KRONIG YIELDS IN BISMUTH, AND LEAD MONITORED BY AUGER SPECTROSCOP/ 4-1067
OF OPTICAL EXCITATION AND AUGER DECAY OF 5D CORE HOLES IN BISMUTH AND NEPTUNIUM.   X-RAY SPE 4-1112
AUGER SPECTROSCOPIC ANALYSIS OF BISMUTH(III) TELLURIDE. /RIX ELEMENTS DEPENDENCE 2-0398
LOYS.   SEGREGATION OF BISMUTH SEGREGATED TO GRAIN BOUNDARIES IN COPPER. 6-2060
SECONDARY ELECTRON EMISSION FROM PLATINUM BISMUTH TO GRAIN BOUNDARIES IN COPPER- BISMUTH AL 6-2096
ELECTRON SPECTROSCOPY OF METAL BLACK COATED SURFACES. (AUGER SPECTRA) 5-1979
TECHNIQUES.   EPITAXY OF ULTRATHIN METAL FILMS ON BLACKS.   5-1897
SOLIDS.   MANY BODY CENTERED CUBIC SUBSTRATES USING LEED- AUGER 5-1488
.   COMPARISON OF MANY BODY EFFECTS IN AUGER DEEXCITATION OF ATOMS NEAR 2-0361
M. (AUGER EFFECT)   MANY BODY EFFECTS IN CORE LEVEL SURFACE SPECTROSCOPIES 5-1723
IONLESS TRANSITIONS.   MANY BODY EFFECTS IN X-RAY PHOTOEMISSION FROM MAGNESIU 2-0427
GER RATES, CORRELATICN ENERGIES, AND/ APPLICATION OF MANY BODY PERTURBATION THEORY APPLIED TO ATOMIC RADIAT 2-0386
AUGER ELECTRONS, INDICATORS IN ANALYSIS OF SOLID BODY PERTURBATION THEORY TO THE CALCULATION OF AU 2-0311
SCOPY TO THE STUDY OF DISCCLORING OF ALUMINUM SURFACES IN BODY SURFACES.   5-1662
, TANTAL/ SECONDARY ELECTRON EMISSION FROM SCLID SURFACES BOILING WATER. /ICATION OF AUGER ELECTRON SPECTRO 5-1127
IC AND SPATIAL DI/ ELECTRONIC EMISSICN FRCM SCLID TARGETS BOMBARDED BY MEDIUM ENERGY IONS (PLATINUM, NICKEL 2-0309
/N EMISSION FROM CLEAN AND CARBON CONTAMINATED MOLYBDENUM BOMBARDED BY NOBLE GAS IONS (10-100 KEV): ENERGET 4-0776
/ECONDARY ELECTRONS EMITTED BY MAGNESIUM AND SILICON WHEN BOMBARDED BY POSITIVE IONS. PART-2: EFFE/ 4-0543
ARY ELECTRON EMISSION FROM SODIUM CHLORINE SINGLE CRYSTAL BOMBARDED BY PRIMARY ELECTRONS. PART-1: ENHANCED/ 4-1047
AUGER SPECTRA OF SOME FREE ELEMENTS AND BINARY COMPOUNDS BOMBARDED BY RARE GAS IONS.   SECOND 4-0759
Y, AND LEED STUDY OF THE STRUCTURE AND COMPOSITION OF ION BOMBARDED BY 60 KEV ARGON(+) IONS.   4-1106
SECONDARY ELECTRON EMISSION OF METALS BOMBARDED TUNGSTEN SURFACES. /LECTRON SPECTROSCOP 5-1951
AUGER DECAY OF NEON FOLLOWING ENERGETIC ION BOMBARDED WITH 120-EV TO 5-KEV PROTONS. 4-1029
AUGER SPECTRA OF LITHIUM AND SODIUM UNDER ION BOMBARDMENT.   4-0954
N EXCITED BY LOW ENERGY ELECTRON AND LOW ENERGY ARGON ION BOMBARDMENT.   4-0817
CTROSCOPY AND SECONDARY ION MASS SPECTRA DURING XENON ICN BOMBARDMENT. / OF MAGNESIUM, ALUMINUM, AND SILICO 4-0857
COMPOSITION CHANGES CF COPPER- NICKEL ALLOYS DURING ARGON BOMBARDMENT. /PPER ANALYZED BY AUGER ELECTRON SPE 5-1679
BOMBARDMENT. /1 OF TARGET TEMPERATURE ON SURFACE 5-1808

TRONS BY ATOMS IN A METALLIC TARGET SUBJECTED TO AN IONIC BOMBARDMENT.   EMISSION OF AUGER ELEC 2-0383
                         EMISSION OF AUGER ELECTRONS BY ION BOMBARDMENT.   2-0550
FROM BERYLLIUM, MAGNESIUM, ALUMINUM AND SILICON UNDER ION BOMBARDMENT. /IBUTION OF AUGER ELECTRONS EMITTED 4-0876
ULAR DISTRIBUTION OF SECONDARY ELECTRONS EMITTED BY IONIC BOMBARDMENT.   ENERGY SPECTRUM AND ANG 4-0988
UM FROM SURFACE LAYERS OF SILICON DIOXIDE DURING ELECTRON BOMBARDMENT. / OF CHEMICAL SHIFTS IN AUGER SPECTR 4-1044
L AUGER SPECTRA PRODUCED BY HYDROGEN(+) AND HELIUM(+) ICN BOMBARDMENT.   HIGH RESOLUTION NEON KL 4-0953
LUTION STUDY OF ARGON LMM AUGER ELECTRONS PRODUCED BY ION BOMBARDMENT.   HIGH RESO 4-0903
TION OF ENERGY LOSS BY AUGER ELECTRONS EMITTED DURING ION BOMBARDMENT.   INTERPRETA 2-0268
K-SHELL AUGER ELECTRON PRODUCTION CROSS SECTIONS FROM ION BOMBARDMENT.   2-0570
OF NEON K-AUGER ELECTRONS PRODUCED BY ENERGETIC HEAVY ION BOMBARDMENT.   MEASUREMENTS 4-0951
ANALYSIS OF COPPER- NICKEL ALLOY SURFACES AFTER ARGON ION BOMBARDMENT. QUANTITATIVE AUGER 5-1807
FILMS BY COMBINATION AUGER SPECTROSCOPY AND INERT GAS ION BOMBARDMENT. /OXIDE THICKNESS ON DEPOSITED METAL 6-2048
RONS FROM LIGHT METALS (LITHIUM, SODIUM) SUBJECTED TO ION BOMBARDMENT. /EMISSION INTENSITIES OF AUGER ELECT 2-0548
ONDARY ELECTRON SPECTRA OF SOME SOLID TARGETS UNDER IONIC BOMBARDMENT.   SEC 4-0942
DISSOCIATION OF POTASSIUM CHLORIDE BY LOW ENERGY ELECTRON BOMBARDMENT.   SURFACE 5-1719
                SPUTTERING OF IONS FROM STAINLESS STEEL BY HYDROGEN BOMBARDMENT. (AES)   6-2017
ER SPECTROSCOPIC STUDY OF THE SURFACE OF IRON AFTER IONIC BOMBARDMENT AND AIR OXIDATION.   AUG 5-1939
/ STUDY OF INTENSITIES OF AUGER LINES EXCITED BY ELECTRON BOMBARDMENT AND ALUMINUM K-ALPHA X-RAY IRRADIATIO 4-1017
ED BY ALKALI HALIDE CRYSTALS UNDER IMPACT OF ION AND ATOM BOMBARDMENT. (AUGER EFFECT) /M OF ELECTRONS EMITT 2-0246
N FOR K-SHELL IONIZATION OF COPPER AND SILVER BY ELECTRON BOMBARDMENT. (AUGER ELECTRONS) /TIAL CROSS SECTIO 2-0477
/ECTRONS FROM A SILICON SINGLE CRYSTAL UNDER A LOW ENERGY BOMBARDMENT. (AUGER SPECTRA OF SILICON, SILICON / 4-0739
                     X-RAY PRODUCTION BY HEAVY IONS. (OXYGEN BOMBARDMENT, AUGER SPECTRUM)   2-0479
Y SPECTRUM OF SECONDARY ELECTRONS EMITTED BY COPPER UNDER BOMBARDMENT BY RARE GAS IONS. /TRONS IN THE ENERG 2-0551
BY INDIUM PHOSPHIDE AND GALLIUM PHOSPHIDE SUBMITTED TO A BOMBARDMENT BY 40 KEV ELECTRONS. (AUGER EFFECT) / 2-0269
ANALYSIS OF COPPER- NICKEL ALLOY SURFACES AFTER ARGON ION BOMBARDMENT. COMMENTS. /LOYS. QUANTITATIVE AUGER 5-1740
CTRON SPECTROSCOPY.   EASY METHOD TO ACCURATELY ALIGN ION BOMBARDMENT GUNS FOR DEPTH PROFILING IN AUGER ELE 3-0700
FIELD ION MICROSCOPY, LEED, HEED, AUGER SPECTROSCOPY, ION BOMBARDMENT, MICROELECTRONICS) /APHY, PHENOMENA, 5-1492
K-SHELL AUGER ELECTRON PRODUCTION CROSS SECTIONS FROM ION BOMBARDMENT. (NEON, FLUORINE)   2-0574
PY.   SURFACE COMPOSITIONAL CHANGES WITH ELECTRON BOMBARDMENT OBSERVED BY AUGER ELECTRON SPECTROSCO 5-1891
                     AUGER ELECTRON EMISSION BY ION BOMBARDMENT OF LIGHT METALS.   4-0877
                     ARGON AUGER EMISSION BY ARGON(+) ION BOMBARDMENT OF SOME METALS.   2-0547
REMENTS)   EFFECTS OF LOW ENERGY ARGON- ION BOMBARDMENT ON MOLYBDENUM DISULFIDE. (AUGER MEASU 2-0348
UNGSTEN AND MOLYBDENUM SINGLE CRYSTALS STU/ EFFECT OF ION BOMBARDMENT ON THE COMPOSITION AND STRUCTURE OF T 5-1215
ONOXIDE SYSTEM/ CARBON CONTAMINATION PRODUCED BY ELECTRON BOMBARDMENT. POLYCRYSTALLINE MOLYBDENUM- CARBON M 5-1648
ANALYSIS OF COPPER- NICKEL ALLOY SURFACES AFTER ARGON ION BOMBARDMENT. REPLY TO COMMENTS. /NTITATIVE AUGER 5-1200
ANALYSIS OF COPPER- NICKEL ALLOY SURFACES AFTER ARGON ION BOMBARDMENT. REPLY TO COMMENTS. /NTITATIVE AUGER 5-1702
/METRY IN (100) SURFACES OF GALLIUM PHOSPHIDE, INDUCED BY BOMBARDMENT WITH 2 KV ELECTRONS. (AUGER ELECTRON/ 5-1162
     CHEMICAL ANALYSIS OF ELECTRODEPOSITED NICKEL- NICKEL BONDS BY AES.   4-0947
ICON. (RADIATIVE AUGER TRANSITIONS) INFLUENCE OF CHEMICAL BONDING ON THE LOW ENERGY K-ALPHA SPECTRUM OF SIL 2-0226
ENERGIES FOR/ INNER SHELL DOUBLE IONIZATION AND CHEMICAL BONDING. PART-1: AUGER ELECTRONS. (KLL TRANSITION 2-0327
OPY OF SURFACE)   CHEMICAL CLEANING FOR THERMOCOMPRESSION BONDING. (REMOVAL OF OXIDE FILMS, AUGER SPECTROSC 5-1454
ELECTRON SPECTROSCOPY STUDY OF OXIDATION ON LANTHANUM HEXABORIDE.   AUGER 5-1706
URFACE CONDITION AND THERMIONIC EMISSION OF LANTHANUM HEXABORIDE. (AUGER SPECTROSCOPY)   S 5-1471
SPECTROSCOPY.   DETECTABILITY LIMITS FOR BORON AND PHOSPHORUS IN SILICON BY AUGER ELECTRON 5-1887
ROSCOPY.   SURFACE SEGREGATION IN BORON DOPED IRON OBSERVED BY AUGER EMISSION SPECT 5-1181
             IDENTIFICATION OF ATOMIC BORON IN STEEL BY AUGER ELECTRON SPECTROSCOPY.   6-2018
. (AUGER ELECTRON SPECTROSCOPY)   AN ORDERED BORON STRUCTURE AFTER DEPOSITION ON SILICON (111) 5-1168
          DEPENDENCE OF ATOMIC PHOTOEFFECT ON BOUND ELECTRON MAGNETIC SUBSTATE (AUGER EFFECT).   2-0524
DISTRIBUTION)   DEPENDENCE OF ATOMIC PHOTOEFFECT ON A BOUND ELECTRON MAGNETIC SUBSTATE. (AUGER ELECTRON 2-0463
AUGER THEORY AT DEFECTS.   APPLICATION TO STATES WITH TWO BOUND PARTICLES IN GALLIUM PHOSPHIDE.   2-0457
/UCTIVITY ASSOCIATED WITH AUGER RECOMBINATION OF EXCITONS BOUND TO NEUTRAL DONORS IN TELLURIUM DOPED GALLI/ 2-0453
     IMPURITY SEGREGATION TO GRAIN BOUNDARIES.   6-2063
OSION OF AUSTENITIC STAINLESS STEEL. / CHEMISTRY OF GRAIN BOUNDARIES AND ITS RELATION TO INTERGRANULAR CORR 6-2064
F LOW ALLOY STEELS. (AUGE/ SEGREGATION AT SELECTIVE GRAIN BOUNDARIES AND ITS ROLE IN TEMPER EMBRITTLEMENT O 6-2057
GER SPECTROSCOPIC ANALYSIS OF BISMUTH SEGREGATED TO GRAIN BOUNDARIES IN COPPER.   AU 6-2060
             SEGREGATION OF BISMUTH TO GRAIN BOUNDARIES IN COPPER- BISMUTH ALLOYS.   6-2096
ICROPROBE)   ADDITIVE AND IMPURITY DISTRIBUTIONS AT GRAIN BOUNDARIES IN SINTERED ALUMINA. (SCANNING AUGER M 5-1505
CTRON/ SEGREGATION OF MANGANESE AND ANTIMONY TO THE GRAIN BOUNDARIES OF TEMPER EMBRITTLED STEEL. (AUGER ELE 6-2070
SPECTROSCOPY.   GRAIN BOUNDARY ACTIVITY MEASUREMENTS BY AUGER ELECTRON 6-2049
ICKEL- CHROMIUM- ANTIMO/ EFFECT OF PRIOR AUSTENITIC GRAIN BOUNDARY COMPOSITION ON TEMPER BRITTLENESS IN A N 6-2114
URES USING AUGER ELECTRON SPECTROSCOPY. ANALYSIS OF GRAIN BOUNDARY DIFFUSION IN BIMETALLIC THIN FILM STRUCT 6-2146
PHOSPHORUS- SULFUR AND IRON- ANTIMONY- SULFUR ALL/ GRAIN BOUNDARY EMBRITTLEMENT IN IRON- PHOSPHORUS, IRON- 6-2098
(AUGER EMISSION ANALYSIS, TUNGSTEN, COPPER, STEEL/ GRAIN BOUNDARY EMBRITTLEMENT IN SOME METALS AND ALLOYS. 6-2058
ELECTRON SPECTROSCOPY.   GRAIN BOUNDARY EMBRITTLEMENT OF STEELS STUDIED BY AUGER 6-2103
NIUM ALLOYS. (AUGER ELECTRON SPECTROSCOPY) STUDY OF GRAIN BOUNDARY FRACTURE SURFACES IN DOPED TUNGSTEN- RHE 6-2112
PRESSED OXIDES BY AUGER ELECTRON SPEC/ ANALYSIS OF GRAIN BOUNDARY IMPURITIES AND FLUORIDE ADDITIVES IN HOT 5-1506
TUNGSTEN. (BY AUGE/ DETERMINATION OF THE INFLUENCE GRAIN BOUNDARY IMPURITIES ON THE PROPERTIES OF SINTERED 5-1997
AUGER ELECTRON SPECTROSCOPY.   IDENTIFICATION OF A GRAIN BOUNDARY PHASE IN HOT PRESSED SILICON NITRIDE BY 5-1749
ING STEEL. (AUGER ELECTRON SPECTROSCOPY)   STUDY OF GRAIN BOUNDARY SEGREGANTS IN THERMALLY EMBRITTLED MARAG 6-2055
   AUGER SPECTROSCOPIC ANALYSIS OF THE EXTENT OF GRAIN BOUNDARY SEGREGATION.   6-2094
TUDY OF SULFUR EMBRITTLED OXYGEN FREE COPPER. (AES, GRAIN BOUNDARY SEGREGATION)   S 6-2015
TEMPER EMBRITTLEMENT OF LOW ALLOY STEELS. (IMPURITY GRAIN BOUNDARY SEGREGATION, AES OF FRACTURE SURFACES) / 6-2083
/LAR EMBRITTLEMENT OF NICKEL BY HYDROGEN: EFFECT OF GRAIN BOUNDARY SEGREGATION. (AUGER ELECTRON SPECTROSCO/ 6-2072
Y)   GRAIN BOUNDARY SEGREGATION. (AUGER ELECTRON SPECTROSCOP 6-2104
).   SURFACE SEGREGATION AND ITS RELATION TO GRAIN BOUNDARY SEGREGATION (AUGER ELECTRON SPECTROSCOPY) 6-2106
     SURFACE SEGREGATION AS A GUIDE TO GRAIN BOUNDARY SEGREGATION. (AUGER SPECTROSCOPY) 6-2073
ECTROSCOPY)   GRAIN BOUNDARY SEGREGATION IN ALUMINUM OXIDE. (AUGER SP 6-2134
UM. (AUGER SPECTROSCOPY)   METASTABLE GRAIN BOUNDARY SEGREGATION IN COPPER CONTAINING TELLURI 6-2086
HIGH VACUUM AUGER ELECTRON SPECTROMETER, FRACTURED/ GRAIN BOUNDARY SEGREGATION IN MAGNESIA DOPED ALUMINA. (5-1615
ELECTRON SPECTROSCOPY)   GRAIN BOUNDARY SEGREGATION IN SINTERED ALUMINA. (AUGER 6-2056
INTERGRANULAR BRITTLE FRACTURE. (STUDIED BY AES) GRAIN BOUNDARY SEGREGATION OF IMPURITIES IN METALS AND 6-2079
MOLYBDENUM- VANADIUM / EFFECT OF MICROSTRUCTURE ON GRAIN BOUNDARY SEGREGATION OF PHOSPHORUS IN A CHROMIUM- 6-2139
EMBRITTLEMENT BY AUGER SPECTROSCOPY)   SURFACE AND GRAIN BOUNDARY SEGREGATION OF STEEL IMPURITIES. (TEMPER 6-2109
ACTURE. (AUGER ELECTRON SPECTROSCOPY)   GRAIN BOUNDARY SEGREGATION RELATING TO INTERGRANULAR FR 6-2082
S AND ALLOYS. (SEGREGATION IN STEEL, T/ RELATION OF GRAIN BOUNDARY SEGREGATION TO REDUCED STRENGTH OF METAL 6-2120
N SPECTROSCOPY.   STUDY OF GRAIN BOUNDARY SEGREGATION USING AUGER ELECTRON EMISSIO 6-2117
.   STUDY OF GRAIN BOUNDARY SEGREGATION USING AUGER EMISSION SPECTRA 6-2121
             GRAIN BOUNDARY STRUCTURE AND SEGREGATION.   AUGER SPECTR 6-2118
OGRAPHIC ANALYSIS OF FRACTURE SURFACES.   GRAIN BOUNDARY STRUCTURE AND SEGREGATION.   AUGER SPECTR 6-2118
CTOGRAPHY AND CHEMICA/ STRESS CORROSION CRACKING OF ALPHA BRASS IN A TARNISHING AMMONIACAL ENVIRONMENT: FRA 6-2095
CES.   AUGER ANALYSIS OF BRASS- ENAMEL AND STAINLESS STEEL- ENAMEL INTERFA 5-1266

F NICKEL- COPPER ALLOY DISPLAYING POOR WETTABILITY DURING BRAZING. (MONEL, AUGER SPECTRA) /ARACTERIZATION O 5-1573
N GERMANIUM IN THE PRESENCE OF EL/ HIGH FREQUENCY EXCITON BREAKDOWN AND FREE CARRIER AND EXCITON KINETICS I 4-0946
/CT OF SURFACE METALLURGY ON THE PENETRATION DEPTH AND RF BREAKDOWN FIELD OF SUPERCONDUCTING NIOBIUM. (AUG/ 6-2137
ELD EMIS/ STABILITY OF FIELD ELECTRON EMISSION AND VACUUM BREAKDOWN. (FROM SURFACES) INVESTIGATIONS WITH FI 5-1520
INVESTIGATION OF VACUUM BREAKDOWN USING AUGER SPECTROSCOPY. 5-1936
M STEEL.   EFFECTS OF TEN YEARS EXPERIMENTAL BREEDER REACTOR II SERVICE ON CHROMIUM- MOLYBDENU 6-2108
ROSCOPY AS APPLIED TO INTERF/ IMPURITIES, INTERFACES, AND BRITTLE FRACTURE. (REVIEW OF AUGER ELECTRON SPECT 6-2078
ARY SEGREGATION OF IMPURITIES IN METALS AND INTERGRANULAR BRITTLE FRACTURE. (STUDIED BY AES)   GRAIN BOUND 6-2079
/OF PRIOR AUSTENITIC GRAIN BOUNDARY COMPOSITION ON TEMPER BRITTLENESS IN A NICKEL- CHROMIUM- ANTIMONY STEE/ 6-2114
ROSCOPY.   INTERANGULAR BRITTLENESS STUDIES IN TUNGSTEN USING AUGER SPECT 6-2061
-RAY EMISSION EDGE SHAPES OF FREE ELECTRON METALS. (AUGER BROADENING)   K X 2-0337
.   SOLID STATE BROADENING EFFECTS IN THE AUGER SPECTRUM OF XENON 4-1005
)   SOLID STATE BROADENING IN AUGER SPECTRA. (CADMIUM, RARE GASES 4-0843
AUGER EFFECT AND CORE BROADENING IN LOW-Z ATOMS, IONS, AND METALS. 2-0353
RGON, KRYPTON, XENON)   SOLID STATE BROADENING IN THE AUGER SPECTRA OF RARE GASES. (A 4-1006
SCOPY AND AUGER SPECTRA OF THE ZINC CHALCOGENID/ LIFETIME BROADENING IN X-RAY INDUCED PHOTOEMISSION SPECTRO 4-0832
SURFACE STRUCTURE OF TITANIUM AND ITS INTERACTION WITH BROMINE AND CHLORINE. (AUGER SPECTROSCOPY) 5-1538
/ECTRON SPECTROSCOPIC INVESTIGATION OF AUGER PROCESSES IN BROMINE-SUBSTITUTED METHANES AND SOME HYDROCARBO/ 2-0515
AUGER EFF/ RADIATION CHEMICAL RESONANCE EFFECT IN SOLID 5-BROMO DEOXYURIDINE: CHEMICAL CONSEQUENCES OF THE 2-0377
ER SPECTROSCOPY)   IDENTIFICATION OF BUBBLE FORMING IMPURITIES IN DOPED TUNGSTEN. (AUG 6-2107
SOLID SOLUTION SOFTENING (RHENIUM DUCTILIZING EFFECT AND BUBBLE STRENGTHENING IN TUNGSTEN- RHENIUM ALLOYS/ 6-2111
/PARISON OF CORE ELECTRON BINDING ENERGIES OF SURFACE AND BULK ATOMS OF TITANIUM, CHROMIUM, AND NICKEL. (A/ 2-0392
/E ADSORPTION AT A TITANIUM SURFACE. SEGREGATION FROM THE BULK CRYSTAL INTERACTION WITH CHLORINE GAS. (AUG/ 5-1833
S.   MOLECULAR ADSORPTION: INCLUSION OF BULK DIFFUSION AND ITS EFFECT ON AUGER INTENSITIE 2-0264
SECONDARY ELECTRON EMISSION OF A ZIRCONIUM- ALUMINUM BULK GETTER. (AUGER SPECTROSCOPY) 5-1407
OPY)   INELASTIC MEAN FREE PATH FOR ELECTRONS IN BULK JELLIUM. (AUGER, LEED, ENERGY LOSS SPECTROSC 4-1059
AVED GERMANIUM (111). (AUG/ PHOTOEMISSION MEASUREMENTS OF BULK, SURFACE, AND HYDROGEN INDUCED STATES ON CLE 5-1781
IMPURITIES ON THORIUM METAL: AUGER STUDY OF BULK TO SURFACE EQUILIBRIA. 5-1314
/CTRUM. APPLICATION OF AUGER ELECTRON SPECTROSCOPY TO THE BULK TO SURFACE PRECIPITATION AND SURFACE DIFFUS/ 5-1651
DIES OF CHEMISORPTION INDUCED SULFUR SEGREGATION FROM THE BULK TO THE SURFACE OF PALLADIUM (100). / AES STU 5-1865
NG AUGER SPECTROSCOPY APPLIED TO ANALYSIS OF X-BAND DIODE BURNOUT.   HIGH SPATIAL RESOLUTION SCANNI 5-1961
DRYING OF BUTADIENE- NITRILE RUBBERS IN AN AUGER MACHINE. 3-0617

C

/CTRON EXCITED AUGER ELECTRON SPECTRUM OF A NICKEL (110)- C(2X2) SULFUR SURFACE: LINE SHAPE ANALYSIS AND / 4-1065
KLL AUGER ELECTRONS OF CADMIUM AND ANTIMONY. 4-0754
AUGER EFFECT IN THE MULTIPLE IONIZATION OF MANGANESE AND CADMIUM BY ELECTRON IMPACT. 2-0228
HIGH RESOLUTION MNN AUGER SPECTRA OF SILVER, CADMIUM, INDIUM, ANTIMONY, TELLURIUM AND IODINE. 4-0730
RELAXATION PROCESSES ACCOMPANYING CADMIUM M4,5N4,5N4,5 AUGER TRANSITIONS. 2-0559
SIVE X-RAY ANALYSIS TECHNIQUES FOR EXAMINATION OF SILVER- CADMIUM OXIDE CONTACT MATERIALS. /Y AND NONDISPER 5-1799
SOLID STATE BROADENING IN AUGER SPECTRA. (CADMIUM, RARE GASES) 4-0843
DE(0.52) CRYSTALS. (/ PHOTOCONDUCTIVITY OF HIGHLY EXCITED CADMIUM SELENIDE AND CADMIUM SULFIDE(0.48) SELENI 2-0432
OBSERVATION OF EXTRINSIC SURFACE STATES ON (1120) CADMIUM SULFIDE BY AUGER SPECTROSCOPY. 5-1202
RECOMBINATION IN DONOR ACCEPTOR COMPLEXES IN CADMIUM SULFIDE SINGLE CRYSTALS. (AUGER) 2-0369
EXCITON SPECTRA OF HEAVILY DOPED CADMIUM SULFIDE SINGLE CRYSTALS. (AUGER PROCESS) 2-0372
ACE PHOTOVOLTAGE AND AUGER SPECTROSCOPY STUDIES OF (1120) CADMIUM SULFIDE SURFACE.   SURF 5-1201
ATIONS.   AUGER LIMITED CARRIER LIFETIMES IN MERCURY CADMIUM TELLURIDE AT HIGH EXCESS CARRIER CONCENTR 2-0262
ANSIENT CARRIER DECAY AND TRANSPORT PROPERTIES IN MERCURY CADMIUM TELLURIDE. (AUGER RECOMBINATION)   TR 2-0461
R SPECTROSCOPY)   SEGREGATION OF CADMIUM TO A (111) SILVER CHLORIDE SURFACE. (AUGE 5-1918
M4,5N4,5N4,5 AND M4,5O1 AUGER ELECTRON SPECTRA FROM CADMIUM VAPOR. 4-0728
RECOMB/ WARM ELECTRON EFFECTS IN NARROW GAP MERCURY(1-X) CADMIUM(X) TELLURIDE AMBIENT TEMPERATURES. (AUGER 2-0343
RECOMBINATION MECHANISMS IN 8-14 MICRON MERCURY(1-X) CADMIUM(X) TELLURIDE. (AUGER-LIMITED) 2-0406
AUGER RECOMBINATION IN A SEMICONDUCTOR (MERCURY(1-X) CADMIUM(X) TELLURIDE) UNDER A MAGNETIC FIELD. 4-1092
COSTER-KRONIG RATES IN SOLID STATE ENVIRONMENTS. (XENON, CADMIUM, ZINC)   ENHANCEMENT OF 2-0439
ULTRAVIOLET EMISSIONS IN CALCITE. (AUGER EFFECT) 4-0798
EFFECTS OF SULFUR, BARIUM, AND CALCIUM ON IMPREGNATED CATHODE SURFACES. (AES) 5-1391
LINE WIDTH AND AUGER EFFECT IN CALCIUM, STRONTIUM, AND BARIUM FLUORIDES. 4-1116
, AUGER, AND RADIATIVE RATES, AND FLUORESCENCE YIELDS FOR CALCIUM TO THORIUM. ATOMIC M-SHELL COSTER-KRONIG 4-0964
ING SEVERAL KINDS OF ATOMS.   COMPUTER PROGRAM TO CALCULATE AUGER TRANSITIONS FOR A SURFACE CONTAIN 4-0039
EVIEW)   CATALOG OF CALCULATED AUGER TRANSITIONS FOR THE ELEMENTS. (R 1-0040
K-AUGER RATES CALCULATED FOR NEON(+). 4-0909
N ENERGIES FOR THE MULTIPLY IONIZED NEON ATOM. CALCULATED K-AUGER ELECTRON AND K X-RAY TRANSITIO 4-0955
ND ELEMENT.   AUGER CATALOG CALCULATED TRANSITION ENERGIES LISTED BY ENERGY A 1-0041
SS)   A SIMPLE ANALYTICAL CALCULATION FOR X-RAY PHOTOEMISSION. (AUGER PROCE 2-0429
SHAPE: SILICON (111).   TIGHT BINDING CALCULATION OF A CORE VALENCE VALENCE AUGER LINE 4-0831
AND/ APPLICATION OF MANY BODY PERTURBATION THEORY TO THE CALCULATION OF AUGER RATES, CORRELATION ENERGIES, 2-0311
CALCULATION OF AUGER SPECTRA IN J-J COUPLING. 4-0749
ZETA EFFECTIVE VALUES OF HYDROGENIC WAVE FUNCTIONS IN THE CALCULATION OF THE KLL AUGER ENERGIES. /N OF THE 2-0526
Y)   RELATIVISTIC CALCULATION OF THE KLL AUGER SPECTRUM. (OF MERCUR 4-0750
G4 AUGER TRANSITION IN ATOMIC ZINC: IMPROVED AUGER ENERGY CALCULATIONS. /ILED ANALYSIS OF THE L3M4,5M4,5: 1 4-0884
AND OF THE SATELLITES IN THE SOFT X-/ SCF AND LIMITED CI CALCULATIONS FOR ASSIGNMENT OF THE AUGER SPECTRUM 4-0724
RATES.   RELATIVISTIC CALCULATIONS OF ATOMIC X-RAY AND AUGER TRANSITION 2-0485
DROGEN FLUORIDE, WATER, AND CA/ CONFIGURATION INTERACTION CALCULATIONS OF THE AUGER SPECTRUM OF METHANE, HY 4-0879
A OF NICKEL AND COPPER. (AUGER-TYPE PROCESSES) CLUSTER CALCULATIONS OF THE SOFT X-RAY L3 EMISSION SPECTR 4-0888
D LEED PATTERNS FROM VACUUM CLEANED SILICON CRYSTALS WITH CALIBRATED DEPOSITS OF IRON.   AUGER SPECTRA AN 5-1769
ELECTRON SPECTROSCOPY MADE QUANTITATIVE BY ELLIPSOMETRIC CALIBRATION.   AUGER 1-0205
F LOW ENERGY ELECTRON SPECTROSCOPY. (AUGER PEAK AMPLITUDE CALIBRATION) /OXYGEN ON GALLIUM ARSENIDE METHOD O 5-1641
NS OF COADSORPTION.   CALIBRATION IN AUGER ELECTRON SPECTROSCOPY BY MEA 3-0587
THE L-AUGER SPECTRA OF PLUTONIUM, AMERICIUM, CALIFORNIUM, AND FERMIUM. 4-0836
ON CHANGES AND AUGER ELECTRON SPECTROSCOPY USING THE LEED CAMERA. /CE DURING THERMAL TREATMENT. WORK FUNCTI 6-2069
ZING AUGER DATA. THREEFOLD APPROACH TO ENHANCE ANALYTICAL CAPABILITIES.   DIGITI 1-0183
PART.   CYLINDRICAL CAPACITOR AS AN ANALYZER. PART-1: NONRELATIVISTIC 3-0689
-AUGER SPECTRA IN AMERICIUM-241 FROM CURIUM-241 (ELECTRON CAPTURE) AND BERKELIUM-245 ALPHA DECAYS. /, AND K 4-1019
/ THE AFTEREFFECTS OF AUGER IONIZATION FOLLOWING ELECTRON CAPTURE IN TRIS (1,10-PHENANTHROLINE) COBALT(III 2-0516
REARRANGEMENT FOLLOWING INTERNAL CONVERSION AND ELECTRON CAPTURE. (K-AUGER TRANSITION INTENSITIES) /CLOUDS 2-0334
EN (100), ETHYLENE SATURATED TUNGSTEN (100), AND TUNGSTEN CARBIDE. /CTS ON CARBON MONOXIDE SATURATED TUNGST 5-1244
INE STRUCTURE IN THE AUGER SPECTRA OF SILICON AND SILICON CARBIDE.   HIGH ENERGY F 4-1036
OW ENERGY BOMBARDMENT. (AUGER SPECTRA OF SILICON, SILICON CARBIDE AND SILICON DIOXIDE) /E CRYSTAL UNDER A L 4-0739
ELECTRON AUGER SPECTROSCOPIC STUDY OF THE SURFACE OF CARBIDE CATHODES. 5-1242

ELECTRON AUGER SPECTROSCOPIC STUDY OF THE SURFACE OF CARBIDE CATHODES.   5-1561
SPECTROSCOPIC MET/ SURFACE OF THERMIONIC EMITTERS MADE OF CARBIDE COMPOUNDS STUDIED BY AN ELECTRONIC AUGER   5-1562
/ASE OF CARBON CONCENTRATION ON THE SURFACE OF MOLYBDENUM CARBIDE DURING HEATING ACCORDING TO AUGER ELECTR/   5-1759
TEMPERATURE PROTON IRRADIATION.   CARBIDE FORMATION ON NIOBIUM SURFACES DURING HIGH   5-1886
ES.   OBSERVATIONS OF BETA- SILICON   CARBIDE FORMATION ON RECONSTRUCTED SILICON SURFAC   5-1560
          LEED- AUGER INVESTIGATION OF A STABLE   CARBIDE OVERLAYER ON A PLATINUM (111) SURFACE.   5-1569
                                       DITUNGSTEN   CARBIDE OVERLAYER ON TUNGSTEN (112).   5-1241
FTS IN SOME VANADIUM COMPOUNDS. (VANADIUM OXIDE, NITRIDE, CARBIDE, SILICIDE) /NVESTIGATIONS OF CHEMICAL SHI   4-1090
IFFRACTION AND AUGER ELECTRON OBSERVATIONS OF THE SILICON CARBIDE (0001) SURFACE.   LOW ENERGY ELECTRON D   5-1932
ON TANTALUM AND POLYC/ AUGER SPECTRA OF CARBON. TANTALUM CARBIDES, ETHYLENE AND CARBON MONOXIDE ADSORPTION   4-0820
HEMICAL EFFECTS IN AUGER ELECTRON SPECTROSCOPY (STATES OF CARBON).   C   4-0867
L EFFECTS ON THE KLL AUGER ELECTRON SPECTRUM FROM SURFACE CARBON.   CHEMICA   4-0862
CTROSCOPY OF GOLD CONDENSED ON ROCK SALT AND ON AMORPHOUS CARBON.   QUANTITATIVE AUGER ELECTRON SPE   5-1136
LOCITY ALPHA PARTICLES AND DEUTERONS.   AUGER SPECTRA OF CARBON AND ARGON FOLLOWING IONIZATION BY EQUAL VE   4-1109
/AUGER ELECTRON SPECTROSCOPY IN DETERMINING THE EFFECT OF CARBON AND OTHER SURFACE CONTAMINANTS ON GALLIUM/   5-1919
Y AUG/ QUANTITATIVE ANALYSIS OF LIGHT ELEMENTS (NITROGEN, CARBON, AND OXYGEN) IN SPUTTERED TANTALUM FILMS B   5-1652
PUTTERED TANTALUM FIL/ EFFECT OF LIGHT ELEMENTS NITROGEN, CARBON AND OXYGEN ON THE PHYSICAL PROPERTIES OF S   5-1473
SPECTROSCOPY STUDY OF NICKEL (110) SURFACE EFFECTS DUE TO CARBON AND SULFUR.   LEED AND AUGER ELECTRON   5-1814
CTRON SPECTROSCOPY.   IDENTIFICATION OF THE FORM OF   CARBON AT A SILICON (100) SURFACE USING AUGER ELE   5-1378
AUGER ANALYSIS OF SILICON THIN FILMS DEPOSITED ON   CARBON AT HIGH TEMPERATURES.   5-1228
SPECTROSCOPY)   BINDING OF   CARBON ATOMS AT A STEPPED NICKEL SURFACE. (AUGER   5-1485
        PLASMON COUPLING TO CORE HOLE EXCITATIONS IN   CARBON. (AUGER APPEARANCE SPECTROSCOPY)   2-0291
                        FLUORESCENCE YIELD OF   CARBON. (AUGER MEASUREMENTS)   5-1571
AUGER ELECTRON EMISSION SPECTRA FROM FOIL AND GAS EXCITED CARBON BEAMS.   4-1050
GROWN AMORPHOUS GERMANIUM. (DEPTH PROFILES OF OXYGEN AND CARBON BY AUGER SPECTROSCOPY) /N ULTRAHIGH VACUUM   5-1547
METHANE, ETHANE, ETHYLENE, ACETYL/ K-SHELL IONIZATION OF CARBON BY FAST PROTONS. (AUGER ELECTRON YIELDS OF   2-0532
/LD ON ROCK SALT (100) CLEAVAGE SURFACES AND ON AMORPHOUS CARBON BY MEANS OF QUANTITATIVE AUGER ELECTRON S/   5-1135
FFECTS ON SECONDARY E/ ELECTRON BEAM INDUCED REDUCTION OF CARBON CONCENTRATION ON BERYLLIUM OXIDE AND ITS E   5-1362
CARBIDE DURING HEATING ACCORDING TO AUGER E/ DECREASE OF CARBON CONCENTRATION ON THE SURFACE OF MOLYBDENUM   5-1759
IVE IONS. PART-1./ AUGER ELECTRON EMISSION FROM CLEAN AND CARBON CONTAMINATED MOLYBDENUM BOMBARDED BY POSIT   2-0543
                                                          CARBON CONTAMINATION OF SILICON (111) SURFACES.   5-1237
MENT. POLYCRYSTALLINE MOLYBDENUM- CARBON MONOXIDE SYSTEM/ CARBON CONTAMINATION PRODUCED BY ELECTRON BOMBARD   5-1648
ERGY ELECTRON DIFFRACT/ OBSERVATION OF SUPERSTRUCTURES ON CARBON COVERED GERMANIUM (100) SURFACE BY HIGH EN   5-1805
/ DETERMINATION OF KLL AUGER SPECTRA OF NITROGEN, OXYGEN, CARBON DIOXIDE, NITRIC OXIDE, WATER, AND CARBON M   4-0990
DESORPTION- AUGER ELECTRON SPECTROSCOPY. DISCREPANCY FOR CARBON DIOXIDE ON TUNGSTEN (100).   FLASH   5-1461
UUM. (AES)   INTERACTION OF SULFUR DIOXIDE AND   CARBON DIOXIDE WITH CLEAN SILVER IN ULTRAHIGH VAC   5-1575
DE OBSERVED BY AUGER ELECTRON EMISSION.   CARBON EVAPORATION FROM A THORIUM DISPENSER CATHO   5-1400
LECTRON EMISSION SPECTROSCOPY.   STUDY OF   CARBON FIBER SURFACES USING AUGER AND SECONDARY E   5-1970
            AUGER ELECTRON EMISSION OF THIN   CARBON FOILS IN REFLECTION AND TRANSMISSION.   4-0893
OLYBDENUM (110) DUE TO ADSORPTION OF MOLECULAR OXYGEN AND CARBON MONOXIDE. / THE M4,5NN AUGER SPECTRUM OF M   4-0885
AUGER SPECTRUM OF METHANE, HYDROGEN FLUORIDE, WATER, AND CARBON MONOXIDE. /INTERACTION CALCULATIONS OF THE   4-0879
ECTRON SPECTROSCOPY OF PLATINUM (100) AND PLATINUM (100)- CARBON MONOXIDE.   LOW ENERGY EL   5-1686
ANING, SURFACE STRUCTURE, AND INTERACTION WITH OXYGEN AND CARBON MONOXIDE. /RYSTAL FACE. A STUDY OF ITS CLE   5-1883
(100). (AUGER SPECTROSCOPY)   CARBON MONOXIDE ADSORPTION KINETICS ON MOLYBDENUM   5-1579
ELECTRON IMPACT INDUCED PHOTOELECTRON AND AUGER SPECTRA. CARBON MONOXIDE ADSORPTION ON MOLYBDENUM. /ED AND   4-0768
R EFFECT)   CARBON MONOXIDE ADSORPTION ON NICKEL (110). (AUGE   5-1884
/SCOPIC STUDY OF THE COADSORPTION OF MOLECULAR OXYGEN AND CARBON MONOXIDE, AND MOLECULAR OXYGEN AND CARBON/   5-1465
/N GAS ADSORPTION UTILIZING AUGER ELECTRON SPECTROSCOPY. CARBON MONOXIDE AND MOLECULAR OXYGEN ON SILICON./   5-1543
DSORPTION STUDIES UTILIZING AUGER ELECTRON SPECTROSCOPY: CARBON MONOXIDE AND OXYGEN ON SILICON. / ON GAS A   5-1542
SURFACE. (CHEMISORPTION AND CATALYTIC RE/ INTERACTION OF CARBON MONOXIDE AND OXYGEN WITH A PALLADIUM (100)   5-1321
FACES. (AUGER SPECTROSCOPY)   INTERACTIONS OF   CARBON MONOXIDE AND OXYGEN WITH IRIDIUM (110) SUR   5-1252
).   LEED AND SURFACE POTENTIAL STUDY OF   CARBON MONOXIDE AND XENON ADSORBED ON COPPER (100   5-1246
/ND SOLUTION OF HYDROGEN, DEUTERIUM, NITROGEN, OXYGEN AND CARBON MONOXIDE BY (100) TANTALUM. (AUGER EFFECT)   5-1549
/RACTIONS ON A PLATINUM (110) SURFACE. PART-1: SULFUR AND CARBON MONOXIDE LEED PATTERNS. (AUGER ELECTRON S/   5-1190
/RMED FROM THE X-RAY PHOTOIONIZATION OF NITROGEN, OXYGEN, CARBON MONOXIDE, NITRIC OXIDE, CARBON DIOXIDE, A/   2-0308
        (AUGER) ADSORPTION STUDIES OF OXYGEN AND   CARBON MONOXIDE ON A PLATINUM (100) SURFACE.   5-1546
ELECTRON SPECTROSCOPY)   DECOMPOSITION OF   CARBON MONOXIDE ON A (110) NICKEL SURFACE. (AUGER   5-1606
MISSION STUDIES.   ADSORPTION OF   CARBON MONOXIDE ON COPPER (001). LEED AND AUGER E   5-1515
N FOR QUANTITATIVE AUGER ELECTRON SPECTROSCOPY. (STUDY OF CARBON MONOXIDE ON MOLYBDENUM (110)) / SUBTRACTIO   1-0069
ND QUANTITATIVE A/ AUGER ELECTRON SPECTROSCOPY STUDIES OF CARBON MONOXIDE ON NICKEL. SPECTRAL LINE SHAPES A   5-1458
ONOX/ STRUCTURAL INFLUENCES ON ADSORPTION ENERGY. PART-2: CARBON MONOXIDE ON NICKEL (100). PART-3: CARBON M   5-1911
METRY)   ADSORPTION AND DECOMPOSITION OF   CARBON MONOXIDE ON PLATINUM (111). (AUGER SPECTRO   5-1622
/ECTRONS IN ELECTRON INDUCED DESORPTION AND DISSOCIATION: CARBON MONOXIDE ON PLATINUM (111). (AUGER SPECTR/   5-1567
TRIC AND MASS SPECTROMETRIC STUDY OF THE CHEMISORPTION OF CARBON MONOXIDE ON POLYCRYSTALLINE MOLYBDENUM. /E   5-1355
                            ADSORPTION OF   CARBON MONOXIDE ON SILVER- PALLADIUM ALLOYS.   5-1254
OF ADSORBED SPECIES USING AUGER PEAK SHAPES. ETHYLENE AND CARBON MONOXIDE ON TUNGSTEN (100). /ACTERIZATION   5-1243
/SSION SPECTRA FROM ADSORBED OXYGEN ON TUNGSTEN (110) AND CARBON MONOXIDE ON TUNGSTEN (100). (BY LEED AND /   5-1152
TRUCTURAL INFLUENCES ON ADSORBATE BINDING ENERGY. PART-1: CARBON MONOXIDE ON (100) PALLADIUM.   S   5-1913
GLE CRYSTAL SURFACE.   CARBON MONOXIDE OXIDATION ON A PLATINUM (110) SIN   5-1187
             MECHANISMS OF THE CATALYTIC   CARBON MONOXIDE OXIDATION ON PLATINUM (110).   5-1188
E SATURATED TUNGSTEN (100), AND/ ELECTRON BEAM EFFECTS ON CARBON MONOXIDE SATURATED TUNGSTEN (100), ETHYLEN   5-1244
LE/ EFFECT OF GOLD COATING ON SORPTION CHARACTERISTICS OF CARBON MONOXIDE, SULFUR, AND OTHER GASES ON STAIN   5-1551
R EFFECT)   INTERACTION OF   CARBON MONOXIDE WITH (110) NICKEL SURFACES. (AUGE   5-1607
        EQUILIBRIUM SURFACE SEGREGATION OF   CARBON ON IRON (100) FACES. (AUGER EFFECT)   5-1369
                 AUGER SPECTROSCOPY OF   CARBON ON NICKEL.   5-1270
                 AUGER SPECTROSCOPY OF   CARBON ON NICKEL. COMMENTS.   5-1283
        QUANTITATIVE DETERMINATION OF   CARBON ON SILVER BY AUGER SPECTROSCOPY.   5-1610
PY AT HIGH SPATIAL RESOLUTION AND PRIMARY BEAM CURRENTS. (CARBON ON SILVER FILM) AUGER ELECTRON SPECTROSCO   5-1904
UGER ELECTRON SPECTROSCOPY)   OXIDATION OF   CARBON ON TUNGSTEN (100) AND MOLYBDENUM (100). (A   5-1944
PECTRUM/ EVALUATION OF THE TRANSITION DENSITY FUNCTION OF CARBON ON VARIOUS SUBSTRATES FROM THE KLL AUGER S   5-1651
                    SULFUR AND   CARBON ON (110) SURFACE OF NICKEL.   5-1813
        AUGER ELECTRON SPECTROSCOPY STUDIES OF   CARBON OVERLAYERS ON METAL SURFACES.   5-1376
/N YIELDS FROM COPPER BERYLLIUM OXYGEN SURFACE BY THERMAL CARBON OXYGEN NITROGEN(2), AND NOBLE GAS METASTA/   4-0786
CIRCUITS. (BY AUGER ELECTRON SPECTROS/ INVESTIGATIONS OF CARBON RESIDUES ON SURFACES OF SILICON INTEGRATED   5-1345
                              A   CARBON STRUCTURE ON THE RHENIUM (0001) SURFACE.   5-1989
        AUGER ELECTRON SPECTRUM OF   CARBON SUBOXIDE.   5-1529
CTRON LOSS SPECTRA FROM DIAMOND, GRAPHITE, AND EVAPORATED CARBON SURFACES. /TERISTIC K-SHELL LOW ENERGY ELE   4-0945
NOXIDE ADSORPTION ON TANTALUM AND POLYC/ AUGER SPECTRA OF CARBON. TANTALUM CARBIDES, ETHYLENE AND CARBON MO   4-0820
        KLL AUGER SPECTRUM OF CHLORINE. (IN   CARBON TETRACHLORIDE)   4-0808
TRANSITION. (AUGER SPECTROSC/ EQUILIBRIUM SEGREGATION OF CARBON TO A NICKEL (111) SURFACE: A SURFACE PHASE   5-1803

ON STUDIES. (AUGER EFFECT) BINDING ENERGIES OF CARBON TO NICKEL (100) FROM EQUILIBRIUM SEGREGATI 5-1484
SEGREGATION OF CARBON TO (100) SURFACE OF NICKEL. 6-2000
/ICS AND MECHANISM OF THE DECOMPOSITION OF FORMIC ACID ON CARBURIZED AND GRAPHITIZED NICKEL (110) USING AE/ 5-1629
PRESENCE OF EL/ HIGH FREQUENCY EXCITON BREAKDOWN AND FREE CARRIER AND EXCITON KINETICS IN GERMANIUM IN THE 4-0946
BAND-TO-BAND AUGER PROCESSES AT HIGH CARRIER CONCENTRATION. 2-0483
TREATED GALLIUM ARSENIDE. (AUGER SPECTROSCOPY) SURFACE CARRIER CONCENTRATION AND AS EVAPORATION IN HEAT 5-1915
CADMIUM TELLURIDE. (AUGER RECOMBINATION) TRANSIENT CARRIER DECAY AND TRANSPORT PROPERTIES IN MERCURY 2-0461
UGER RECOMBINATION PROCESS) MINORITY CARRIER LIFETIME IN INDIUM ARSENIDE EPILAYERS. (A 2-0567
CARRIER LIFETIMES IN EPITAXIAL INDIUM ARSENIDE. (2-0332
HIGH EXCESS CARRIER CONCENTRATIONS. AUGER LIMITED CARRIER LIFETIMES IN MERCURY CADMIUM TELLURIDE AT 2-0262
CARRIER RECOMBINATION IN N-INDIUM ANTIMONIDE. 4-0761
IN NONRADIATIVE (AUGER) RECOMBINATION. CARRIER TEMPERATURE EFFECTS AND ENERGY TRANSFERS 2-0470
K. (AUGER) RECOMBINATION OF CARRIERS IN N-TYPE INDIUM ARSENIDE AT 77 DEGREES 4-0785
RELAXATION OF AUGER EXCITED CARRIERS IN SILICON. 2-0274
TERING AND (AUGER) RECOMBINATION OF NONEQUILIBRIUM CHARGE CARRIERS IN SILICON AT A HIGH EXCITATION LEVEL. / 4-1103
ON LEVELS. (AUGER) RECOMBINATION OF NONEQUILIBRIUM CHARGE CARRIERS OF INDIUM ARSENIDE AT HIGH PHOTOEXCITATI 4-0840
BAND-TO-BAND AUGER RECOMBINATION AND CARRIER-CARRIER SCATTERING IN POWER RECTIFIERS. 2-0458
REVIEW, 16 R/ ELECTRON EMISSION FROM SOLIDS BY PARTICLES CARRYING POTENTIAL ENERGY. (AUGER NEUTRALIZATION, 2-0374
EFFICIENCY OF THE AUGER CASCADE IN SOLID STATE. 2-0330
CTRON SPECTROSCOPY AND SECONDARY ELECTRON EMISSION DUE TO CASCADE PROCESS. /GROUND INFORMATION IN AUGER ELE 2-0411
ELECTRON SPRAYING ATOMS. (AUGER CASCADES) 2-0464
D AS THE RESULT OF ELECTRON SHAKE-OFF AND (SUGAR) VACANCY CASCADES. NATURE OF SECONDARY ELECTRONS CREATE 2-0307
AUGER CASCADES IN ALUMINUM. 4-0973
N OF IMPURITY ADSORPTION AT FLAKE IRON INTERFACES IN GRAY CAST IRON. (SCANNING AUGER MICROPROBE) /NFIRMATIO 6-2054
ENERGY AND ELEMENT. AUGER CATALOG CALCULATED TRANSITION ENERGIES LISTED BY 1-0041
LEMENTS. (REVIEW) CATALOG OF CALCULATED AUGER TRANSITIONS FOR THE E 1-0040
SPECTROSCOPY ON COPPER- NICKEL ALLOY SURFACES RELATED TO CATALYSIS. AUGER 5-1703
SURFACE STRUCTURE OF AND CATALYSIS BY PLATINUM SINGLE CRYSTAL SURFACES. 5-1837
POISONING OF A COMMERCIAL PALLADIUM ALUMINA HYDROGENATION CATALYST. AUGER SPECTROSCOPIC STUDY OF THE 5-1177
ACE BY AUGER ELECTRON SPECTROSCOPY: WORKING STATE OF THE CATALYST SURFACE. /ND OXYGEN ON A MOLYBDENUM SURF 5-1530
ON SPECTROSCOPIC STUDY OF THE SURFACE POISONING OF COPPER CATALYSTS. AUGER ELECTR 5-1176
PPLICATION OF AUGER ELECTRON SPECTROSCOPY TO THE STUDY ON CATALYSTS. /MPOSITION OF COPPER- NICKEL ALLOY. A 5-1875
AND CHEMICAL CHARACTERIZATION OF PLATINUM- RHODIUM GAUZE CATALYSTS. (AUGER SPECTROSCOPY) PHYSICAL 5-1793
SURFACE COMPOSITION OF PROMOTED IRON CATALYSTS BY AUGER ELECTRON SPECTROSCOPY. 5-1817
FACE COMPOSITION OF PALLADIUM- GOLD AND PALLADIUM- SILVER CATALYSTS BY AUGER ELECTRON SPECTROSCOPY. SUR 5-1976
COMPOSITION OF THE SURFACE OF COPPER- MAGNESIUM OXIDE CATALYSTS STUDIED BY AUGER ELECTRON SPECTROSCOPY. 5-1361
F THE SURFACE COMPOSITION OF PALLADIUM- NICKEL ALLOY FILM CATALYSTS USING AUGER ELECTRON SPECTROSCOPY. /N O 5-1846
POSITION, INTERDIFFUSI/ A TECHNIQUE FOR INVESTIGATING THE CATALYTIC ACTIVITY OF ALLOY SYSTEMS. (SURFACE COM 6-2003
N OF COPPER- NICKEL ALLOY. APPLICATION/ RELATION BETWEEN CATALYTIC ACTIVITY PATTERN AND SURFACE COMPOSITIO 5-1875
/Y, STICKING COEFFICIENT) PART-2: EFFECT OF SULFUR ON THE CATALYTIC CARBON MONOXIDE OXIDATION. (AUGER ELEC/ 5-1190
110). MECHANISMS OF THE CATALYTIC CARBON MONOXIDE OXIDATION ON PLATINUM (5-1188
/IONS OF AUGER ELECTRON AND ENERGY LOSS SPECTROSCOPIES TO CATALYTIC PROCESSES. CHEMISORPTION OF AMMONIA, N/ 5-1533
THERMAL DESORPTION TECHNIQUES TO STUDY THE MECHANISM OF A CATALYTIC REACTION. /F PHOTOEMISSION, AUGER, AND 1-0042
/XYGEN WITH A PALLADIUM (100) SURFACE. (CHEMISORPTION AND CATALYTIC REACTIONS, LEED, AUGER, CONTACT POTENT/ 5-1321
VACUUM SCANNING ELECTRON MICROSCOPE WITH A FIELD EMISSION CATHODE AND AUGER ANALYZER. ULTRAHIGH 3-0621
CHARACTERISTICS OF SURFACE OF THE PALLADIUM BARIUM ALLOY CATHODE IN RELATION TO ITS FILM SYSTEM. (AUGER E/ 5-1129
CARBON EVAPORATION FROM A THORIUM DISPENSER CATHODE OBSERVED BY AUGER ELECTRON EMISSION. 5-1400
EFFECTS OF SULFUR, BARIUM, AND CALCIUM ON IMPREGNATED CATHODE SURFACES. (AES) 5-1391
PART-1: BARIUM ON / AUGER ELECTRON SPECTROSCOPY STUDY OF CATHODE SURFACES DURING ACTIVATION AND POISONING. 5-1841
CTRON AUGER SPECTROSCOPIC STUDY OF THE SURFACE OF CARBIDE CATHODES. ELE 5-1242
CTRON AUGER SPECTROSCOPIC STUDY OF THE SURFACE OF CARBIDE CATHODES. ELE 5-1561
PLATINUM (100) STRUCTURES: IMPURITY CAUSATION AND PROPERTIES. (AUGER EFFECT) 5-1311
ASUREMENTS OF SECONDARY EMISSION COEFFICIENTS USING A GAS CELL METHOD. /G A CROSSED BEAM METHOD. PART-2: ME 2-0340
A SMALL RETARDING FIELD ENERGY ANALYZER WITH CEM. (AUGER SPECTROMETER) 3-0655
N OF ATOMS BY MULTIPLY CHARGED ICNS. MODEL OF TWO COULOMB CENTERS. AUGER IONIZATIO 2-0407
PER AND ALUMINUM DOPED ZINC SULFIDE. (AUGER EFFECT, COLOR CENTERS) /CE OF GREEN- COPPER LUMINESCENCE IN COP 2-0402
INATION OF THE ENERGY DEPTH OF LEVELS OF SURFACE ELECTRON CENTERS. (AUGER EFFECT, ZIRCONIA, SILICA) /DETERM 4-0915
AUGER EMISSION AND COLOR CENTERS IN LITHIUM FLUORIDE. 2-0430
LADIUM, COPPER, NICK/ AUGER ELECTRON SPECTROSCOPY OF FACE CENTERED CUBIC METAL SURFACES. (GOLD, SILVER, PAL 2-1717
IQUES. EPITAXY OF ULTRATHIN METAL FILMS ON BODY CENTERED CUBIC SUBSTRATES USING LEED- AUGER TECHN 5-1488
GER ELECTRON SPECTROSCOPY TO SURFACE CHEMISTRY STUDIES OF CERAMIC SUBSTRATES. APPLICATION OF AU 5-1275
CHARACTERIZATION OF CERAMIC SURFACES. (AUGER SPECTROSCOPY) 5-1174
MICROPROBE ANALYSIS) CRYSTALLOGRAPHY AND CHEMISTRY OF CERAMIC SURFACES. (AUGER SPECTROSCOPY, LEED, ION 5-1852
ANALYSIS AND CHARACTERIZATION OF CERAMIC SURFACES FOR ELECTRONIC APPLICATIONS. 5-1853
UGER EL/ TECHNIQUES FOR THE CHEMICAL CHARACTERIZATIONS OF CERAMIC SURFACES. (ION SCATTERING SPECTROMETRY, A 5-1326
AUGER SPECTROSCOPY OF FRACTURE SURFACES OF CERAMICS. 5-1616
CONDARY/ DIFFERENTIAL SPUTTERING OF MAGNESIUM OXIDE- GOLD CERMET FILMS AND ITS APPLICATION TO HIGH YIELD SE 5-1430
ON SECONDARY ELECTRON EMISSION FROM MAGNESIUM OXIDE- GOLD CERMETS. EFFECTS OF CESIATION 2-0384
ONDARY ELECTRON EMISSION FROM THIN FILM GOLD METAL- OXIDE CERMETS. SEC 4-1015
NESIUM OXIDE- GOLD CERMETS. EFFECTS OF CESIATION ON SECONDARY ELECTRON EMISSION FROM MAG 2-0384
SECONDARY ELECTRON EMISSION FROM ION IMPLANTED SILICON (CESIUM). 4-0913
) TUNGSTEN. AUGER ELECTRON SPECTROSCOPY OF CESIUM ADSORBED ON CLEAN AND OXYGEN- COVERED (100 5-1295
/ECTRON DIFFRACTION AUGER, WORK FUNCTION STUDY OF OXYGEN- CESIUM ADSORPTION AND COADSORPTION ON (100) TUNG/ 5-1297
AN ELECTRON DIFFRACTION STUDY OF CESIUM ADSORPTION ON TUNGSTEN. 5-1603
R SPECTROSCOPY AND SURFACE POTENTIAL DIFFERENCE STUDIES. CESIUM ADSORPTION ON TUNGSTEN: ELLIPSOMETRY, AUGE 5-1830
SECONDARY ELECTRON SPECTROSCOPY. CESIUM ADSORPTION ON TUNGSTEN STUDIED BY LEED AND 5-1661
URFACE. (AUGER ELECTRON SPECTROSCOPY) INVESTIGATION OF CESIUM ADSORPTION ONTO A GALLIUM ARSENIDE (110) S 5-1644
LECTRON AFFINITY ON SILICON PRODUCED BY THE ADSORPTION OF CESIUM AND OXYGEN. /PLASMON STUDIES OF NEGATIVE E 5-1363
VATION TO NEGATIVE ELECTRON AFFINITY BY THE ADSORPTION OF CESIUM AND OXYGEN. / GALLIUM ARSENIDE DURING ACTI 5-1364
STRUCTURAL STUDIES OF THE ADSORPTION OF CESIUM AND OXYGEN ON SILICON (100). 5-1389
Y SLOW ELECTRON DIFFRACTIO/ INVESTIGATION OF THE PURE AND CESIUM COATED (100) SURFACE OF GALLIUM ARSENIDE B 5-1639
R ELECTRON SPECTROSCOPY, AND PHOTOEMISSION STUDIES OF THE CESIUM COVERED GALLIUM ARSENIDE (110) SURFACE. /E 5-1931
SPECTROSCOPY, AND WORK FUNCTION MEASUREMENTS ON CLEAN AND CESIUM COVERED POLAR FACES OF GALLIUM ARSENIDE. / 5-1294
STABILITY OF CESIUM COVERED ZINC OXIDE SURFACES. 5-1882
ELECTRONIC AND LATTICE STRUCTURE OF CESIUM FILMS ADSORBED ON TUNGSTEN. 5-1604
/XCITATION FUNCTIONS FCR SODIUM, POTASSIUM, RUBIDIUM, AND CESIUM IONS IN THE VACUUM ULTRAVIOLET SPECTRAL R/ 4-0735
PHOTOELECTRON SPECTROSCOPY) ADSORPTION KINETICS OF CESIUM ON GALLIUM ARSENIDE. (USING AES AND X-RAY 5-1821
N/ ELLIPSOMETRY, AUGER SPECTROSCOPY, AND WORK FUNCTION OF CESIUM ON TUNGSTEN AND TITANIUM SINGLE CRYSTALS I 5-1829
LECTRON SPECTROSCOPY. INVESTIGATION CF SILVER CESIUM OXIDE PHOTOEMISSIVE SURFACES USING AUGER E 5-1782
CTRCSCOPY) INTERACTION OF CESIUM WITH CLEAN ZINC OXIDE SURFACES. (AUGER SPE 5-1588

/DIFFRACTION STUDY OF THE ACTIVATION OF GALLIUM ARSENIDE-
ARBON AND OTHER SURFACE CONTAMINANTS ON GALLIUM ARSENIDE-
L, CHEMICAL, AND STRUCTURAL STUDY OF THE GERMANIUM (100)-
RONIC MODEL OF NEGATIVE ELECTRON AFFINITY ON THE SILICON-
ORPTION USING AUGER ELECTRON SPECTROSCOPY. AFFLICATION TO
DETECTION OF / COINCIDENCE METHODS OF STANDARDIZATION FOR
PHOTOEMISSION SPECTROSCOPY AND AUGER SPECTRA OF THE ZINC
/ICATION AND MEASUREMENT OF LOW ENERGY ELECTRONS BY CLOUD
LEAD OXIDE BY AES)   GAIN FATIGUE MECHANISM IN
ETER WITH SCANNING SAMPLE POSITICNER.   ELECTRON
TIGATION OF REORDERED (001) GOLD SURFACES BY POSITIVE ION
/IZATION OF REORDERED (001) GOLD SURFACES BY POSITIVE ION
                                                   SURFACE
ON SPECTROSCOPY.   SURFACE
UUM LOW TEMPERATURE STAGE FOR IN-SITU FILM DEFOSITION AND
POSITION PROCESS BY AUGER ANALYSIS.
PEAK SHAPES. ETHYLENE AND CARBON MONOXIDE ON TUNGSTEN (1/
Y AUGER ELECTRON SPECTRCSCOPY).
TROSCOPY)
IC APPLICATIONS.   ANALYSIS AND

FACES.
ATION TO NEGATIVE/ LOW ENERGY ELECTRON DIFFRACTION- AUGER
           AUGER ELECTRON SPECTROSCOPY: NEW TOOL IN THE
ECTROSCOPY.
PECTROSCOPY)
THE AUGER ELECTRON MICROANALYTIC METHO/ THREE DIMENSIONAL
NG POOR WETTABILITY DURING BRAZING. (MONEL, AUGE/ SURFACE
ALUMINUM, AND SILICON.   AES
YSTS. (AUGER SPECTROSCCPY)   PHYSICAL AND CHEMICAL
BY POSITIVE ION CHANNELING SPECTRCSCOPY, LOW ENERGY ELE/
R SPECTROSCOPY, ESCA)
TRUCTURES USING LEED AND AES.
RYSTALLOGRAPHI/ USE OF ADVANCED ANALYTICAL TECHNIQUES FOR
QUARTZ RESONATORS.
                        SURFACE PREPARATION AND
                        APPLICATIONS OF SURFACE
                        INTEGRATED CIRCUIT PROCESS
CTION AND AUGER ELECTRCN SPEC/ RECENT ADVANCES IN SURFACE
ERING SPECTROMETRY, AUGER EL/ TECHNIQUES FOR THE CHEMICAL
/E SCATTERING AND (AUGER) RECOMBINATION OF NONEQUILIBRIUM
XCITATION LEVELS. (AUGER) RECOMBINATION OF NONEQUILIBRIUM
PECTRUM.   OBSERVATION OF A PROJECTILE
VACANCY PRODUCTION CROSS SECTIONS IN ARGON BY THE IMPACT/
S THAN 10 KEV) SCATTERED FROM A (100) FACE OF A COPPER S/
MS SCATTERED FROM A COFPER SINGLE CRYSTAL. (AUGER NEUTRA/
                        SIZE EFFECT IN IONIC
N AND FLUORESCENCE YIELD IN 50 MEV CHLORINE(N/ PROJECTILE
RESCENCE YIELD STUDIED BY ATOMIC HYDROGEN(+), MOLECULAR /
ON YIELD IN HEAVY ION ATOM COLLISIONS AT KEV ENE/ DYNAMIC
E YIELD DEDUCED FROM HIGH RESOLUTION EMISSION (AUGER) SP/
ON EMISSION SPECTRA FOR RARE GAS ATOMS. PART-1. THE ARGO/
                        ANALYSIS OF HIGH
          AUGER TRANSITIONS LEADING TO DEFINED FINAL
LIUM(+). (AUGER SPECTRA)   ARGON
RFACES.
                        BINDING, VALENCY AND
HIFTS IN AUGER SPECTRA (MONOLAYER ADSORPTION, ESCA SIMPLE
          AUGER IONIZATION OF ATOMS BY MULTIPLY
          AUGER NEUTRALIZATION OF MULTIPLY
AN ELECTROSTATIC ANALYZER OF THE CYLINDRICAL/ FOCUSING A
ON SPECTROSCOPY APPLIED TO INNER SHELL IONIZATION BY FAST
                        FOCUSING OF
                                                   VARIAN
(REVIEW AUGER SPECTROSCOPY AND ELECTRON SPECTROSCOPY FOR
          AUGER ELECTRON SPECTRCSCOPY FOR
S. (REVIEW, AUGER SPECTROSCOPY, ELECTRON SPECTRCSCOPY FOR
WIDTH OF SPECTRAL LINES IN THE ELECTRON SPECTRCSCOPY FOR
D ELECTRON EXCITATION. (REVIEW, ELECTRON SPECTROSCOPY FOR
G POWDER SAMPLES FOR AES AND/OR ELECTRON SPECTROSCOPY FOR
CHOOSING BETWEEN ELECTRON SPECTROSCOPY FOR
          CCMBINED ELECTRON SPECTROSCOPY FOR
F VACUUM DEPOSITED SILVER/ UPS, ELECTRON SPECTROSCOPY FOR
COPPER(I) OXIDE, AND CCPPER(II/ ELECTRON SPECTROSCOPY FOR
/ELECTRON SPECTROSCOPY, PHOTON, ELECTRON SPECTROSCOPY FOR
/D AUGER ELECTRON SPECTROSCOPY (ELECTRON SPECTROSCOPY FOR
E CCMPLICATIONS USING A CCMBINED ELECTRON SPECTROSCOPY FOR
IN THIN FILM ANALYSIS METHODS. (ELECTRON SPECTROSCOPY FOR
IN A TARNISHING AMMONIACAL ENVIRONMENT: FRACTOGRAPHY AND
/LECTRON SPECTROSCOPY. CCMBINED ELECTRON SPECTROSCOPY FOR
/TROSTATIC SPECTROMETER USED IN ELECTRON SPECTROMETRY FOR
AND APPEARANCE POTENTIAL SPECTROSCOPY: A COMPAR/ SURFACE
KEL BONDS BY AES.
EW, SPECTROSCOPY)   INSTRUMENTATION FOR THE
) SURFACES BY AUGER ELECTRON SPECTROSCOPY.
AND COMPOSITION)
                      AUGER SPECTROSCOPY FOR THE
TUDIES ON IRON AND STEEL. (AES, ELECTRON SPECTROSCOPY FOR
SILICON SURFACE ANALYSIS. (AES, ELECTRON SPECTROSCOPY FOR
/SPECTRA (MONOLAYER ADSORPTION, ELECTRON SPECTROSCOPY FOR
T ENERGY ANALYZER FOR AUGER AND ELECTRON SPECTROSCOPY FOR

CESIUM- OXYGEN COADSORPTION ON (100) TUNGSTEN.   5-1296
CESIUM- OXYGEN NEGATIVE ELECTRON AFFINITY SURFAC/   5-1845
CESIUM- OXYGEN PHOTOCATHODES. /NG THE EFFECT OF C   5-1919
CESIUM- OXYGEN PHOTOSURFACE.   AN ELECTRICA   5-1319
CESIUM- OXYGEN SURFACE. (AUGER EFFECT) /AND ELECT   2-0426
CESIUM-COVERED (110) GALLIUM ARSENIDE. /ERMAL DES   5-1293
CESIUM-131 AND MEASUREMENT OF DECAY PARAMETERS. (2-0474
CHALCOGENIDES. /ETIME BROADENING IN X-RAY INDUCED   4-0932
CHAMBER TECHNIQUES: OXYGEN AND NEON AUGER ELECT/   2-0566
CHANNEL ELECTRON MULTIPLIERS. (SURFACE OXYGEN ON   3-0681
CHANNELING PATTERNS IN AN AUGER ELECTRON SPECTROM   3-0692
CHANNELING SPECTROSCOPY, LEED AND AES.   INVES   5-1139
CHANNELING SPECTROSCOPY, LOW ENERGY ELECTRON DIF/   5-1986
CHARACTERIZATION BY AUGER ELECTRON SPECTROMETRY.   5-1815
CHARACTERIZATION BY LOW ENERGY PHOTO AUGER ELECTR   5-1429
CHARACTERIZATION. (LEED- AUGER SYSTEM) /AHIGH VAC   3-0657
CHARACTERIZATION OF A CHROMIUM- GOLD THIN FILM DE   1-0037
CHARACTERIZATION OF ADSORBED SPECIES USING AUGER   5-1243
CHARACTERIZATION OF ALUMINUM ADHEREND SURFACES (B   5-1725
CHARACTERIZATION OF CERAMIC SURFACES. (AUGER SPEC   5-1174
CHARACTERIZATION OF CERAMIC SURFACES FOR ELECTRON   5-1853
CHARACTERIZATION OF CHEMISORPTION BY LEED.   1-0161
CHARACTERIZATION OF COPPER- GOLD ALLOY SINGLE SUR   5-1747
CHARACTERIZATION OF GALLIUM ARSENIDE DURING ACTIV   5-1364
CHARACTERIZATION OF GLASS FIBER SURFACES.   1-0175
CHARACTERIZATION OF GLASS SURFACES BY ELECTRON SP   5-1787
CHARACTERIZATION OF INDIUM PHOSPHIDE. (BY AUGER S   5-1968
CHARACTERIZATION OF METALS AND SEMICONDUCTORS BY   1-0077
CHARACTERIZATION OF NICKEL- COPPER ALLOY DISPLAYI   5-1573
CHARACTERIZATION OF OXIDIZED FILMS OF MAGNESIUM,   5-1847
CHARACTERIZATION OF PLATINUM- RHODIUM GAUZE CATAL   5-1793
CHARACTERIZATION OF REORDERED (001) GOLD SURFACES   5-1986
CHARACTERIZATION OF SOLID SURFACES. (REVIEW, AUGE   1-0134
CHARACTERIZATION OF SOME VANADIUM (100) SURFACE S   5-1864
CHARACTERIZATION OF THE CHEMICAL, PHYSICAL, AND C   5-1433
CHARACTERIZATION TECHNIQUES (AUGER, CLEANING) FOR   5-1943
CHARACTERIZATION TO GLASS SURFACES.   5-1705
CHARACTERIZATION USING AES.   5-1741
CHARACTERIZATION USING LOW ENERGY ELECTRON DIFFRA   1-0201
CHARACTERIZATIONS OF CERAMIC SURFACES. (ION SCATT   5-1326
CHARGE CARRIERS IN SILICON AT A HIGH EXCITATION   4-1103
CHARGE CARRIERS OF INDIUM ARSENIDE AT HIGH PHOTOE   4-0840
CHARGE DEPENDENCE FOR THE NEON K-AUGER ELECTRON S   4-0993
CHARGE DEPENDENCE OF DEPENDENCE OF TOTAL K-SHELL   2-0446
CHARGE EXCHANGE OF LOW ENERGY HELIUM(+) IONS (LES   2-0546
CHARGE EXCHANGE OF LOW ENERGY HELIUM IONS AND ATO   2-0545
CHARGE RELAXATION FOLLOWING AUGER EFFECT.   2-0565
CHARGE STATE DEPENDENCE OF NEON K-SHELL IONIZATIO   2-0300
CHARGE STATE DEPENDENCE OF THE ARGON L-SHELL FLUO   2-0517
CHARGE STATE DEPENDENCE OF THE K-VACANCY PRODUCTI   2-0346
CHARGE STATE DEPENDENCE OF THE NEON K FLUORESCENC   2-0520
CHARGE STATE DEPENDENCE OF X-RAY AND AUGER ELECTR   4-0927
CHARGE STATE OUTPUT FROM A PENNING ION SOURCE.   3-0602
CHARGE STATES BY A COINCIDENCE TECHNIQUE.   2-0371
CHARGE STATES DUE TO IMPACT OF HYDROGEN(+) AND HE   4-1011
CHARGE TRANSFER IN CHEMICAL REACTIONS ON METAL SU   5-1763
CHARGE TRANSFER MODEL).   CHEMICAL S   4-1115
CHARGED IONS. MODEL OF TWO COULOMB CENTERS.   2-0407
CHARGED IONS ON A METAL SURFACE.   2-0247
CHARGED PARTICLE BEAM OF FINITE ANGULAR SPREAD IN   3-0720
CHARGED PARTICLES.   AUGER ELECTR   2-0535
CHARGED PARTICLES BY A SPHERICAL CONDENSER.   3-0684
CHART OF AUGER ELECTRON ENERGIES.   4-1082
CHEMICAL ANALYSIS / IN SURFACE ANALYSIS METHODS.   1-0215
CHEMICAL ANALYSIS.   1-0029
CHEMICAL ANALYSIS /ACTERIZATION OF SOLID SURFACE   1-0134
CHEMICAL ANALYSIS. INFORMATION OBTAINED FROM THE   4-1009
CHEMICAL ANALYSIS AES) /N SPECTROSCOPY: X-RAY AN   1-0081
CHEMICAL ANALYSIS ANALYSIS. TECHNIQUE OF PREPARIN   1-0197
CHEMICAL ANALYSIS AND AUGER FOR SURFACE ANALYSIS.   5-1635
CHEMICAL ANALYSIS AND AUGER SPECTROMETER.   3-0671
CHEMICAL ANALYSIS AND AUGER SPECTROSCOPY STUDY O   5-1947
CHEMICAL ANALYSIS (AND AUGER) STUDIES OF COPPER,   5-1794
CHEMICAL ANALYSIS AND FIELD STIMULATED ELECTRON/   5-1901
CHEMICAL ANALYSIS AND SECONDARY ION MASS SPECTR/   1-0088
CHEMICAL ANALYSIS AUGER ANALYSIS SYSTEM. /N DEVIC   5-1289
CHEMICAL ANALYSIS AUGER SPECTROSCOPY) /UTTERING   1-0214
CHEMICAL ANALYSIS. (AUGER SPECTROSCOPY) /HA BRASS   6-2095
CHEMICAL ANALYSIS AUGER SYSTEM BASED ON THE DOUB/   3-0654
CHEMICAL ANALYSIS. (AUTOMATIC RECORDING OF PHOTO/   3-0717
CHEMICAL ANALYSIS BY AUGER ELECTRON SPECTROSCOPY   5-1910
CHEMICAL ANALYSIS OF ELECTRODEPOSITED NICKEL- NIC   4-0947
CHEMICAL ANALYSIS OF MANUFACTURED SURFACES. (REVI   3-0719
CHEMICAL ANALYSIS OF PLATINUM (100) AND GOLD (100   5-1712
CHEMICAL ANALYSIS OF SEMICONDUCTORS. (IMPURITIES   5-1671
CHEMICAL ANALYSIS OF SURFACES.   1-0150
CHEMICAL ANALYSIS SIMS) /THEIR APPLICATION FOR S   6-2035
CHEMICAL ANALYSIS SIMS, AND OTHERS) /NIQUES FOR   1-0052
CHEMICAL ANALYSIS SIMPLE CHARGE TRANSFER MODEL).   4-1115
CHEMICAL ANALYSIS SPECTROSCOPY.   NEW COMPAC   3-0702

DE AND SILVER OXIDE.   ELECTRON SPECTROSCOPY FOR CHEMICAL ANALYSIS STUDIES OF SILVER, DISILVER OXI 4-1055
UTOMATIC EXECUTION OF AUGER AND ELECTRON SPECTROSCOPY FOR CHEMICAL ANALYSIS TEST PROGRAMS. /LOGIC FOR THE A 3-0722
BY MEANS OF AUGER ELECTRONIC SPECTROSCOPY AND LOW ENERG/ CHEMICAL AND STRUCTURAL ANALYSIS OF IRON SURFACES 5-1940
100)- CESIUM- OXYGEN PHOTOSURFACE.   AN ELECTRICAL, CHEMICAL, AND STRUCTURAL STUDY OF THE GERMANIUM (5-1319
UM OF SILICON. (RADIATIVE AUGER TRANSITIONS) INFLUENCE OF CHEMICAL BONDING ON THE LOW ENERGY K-ALPHA SPECTR 2-0226
RANSITION ENERGIES FOR/ INNER SHELL DOUBLE IONIZATION AND CHEMICAL BONDING. PART-1: AUGER ELECTRONS. (KLL T 2-0327
URING OXIDATION OF NICKEL (100) AND (111) CRYSTAL SURFAC/ CHEMICAL CHANGES IN SECONDARY ELECTRON EMISSION D 5-1453
UZE CATALYSTS. (AUGER SPECTROSCOPY)   PHYSICAL AND CHEMICAL CHARACTERIZATION OF PLATINUM- RHODIUM GA 5-1793
ION SCATTERING SPECTROMETRY, AUGER EL/ TECHNIQUES FOR THE CHEMICAL CHARACTERIZATIONS OF CERAMIC SURFACES. (5-1326
(REMOVAL OF OXIDE FILMS, AUGER SPECTROSCOPY OF SURFACE) CHEMICAL CLEANING FOR THERMOCOMPRESSION BONDING. 5-1454
E ELECTRON AUGER SPECTROSC/ QUANTITATIVE ESTIMATES OF THE CHEMICAL COMPOSITION NEAR THE SURFACE BASED ON TH 5-1645
Y AUGER ELECTRON SPECTROSCOPY.   STUDY OF THE CHEMICAL COMPOSITION OF MOS AND MNOS STRUCTURES B 5-1503
AND STEEL BASED ON AUGER ELECTRON SPECTRAL DATA. CHEMICAL COMPOSITION OF THE SURFACE OF MOLYBDENUM 5-1985
IZED VANADIUM/ AUGER ELECTRON SPECTROSCOPY OF THE SURFACE CHEMICAL COMPOSITION OF VANADIUM OXIDES, AND OXID 4-1089
                                                        CHEMICAL EFFECTS IN AUGER ELECTRON SPECTROSCOPY. 1-0014
STATES OF CARBON).   CHEMICAL EFFECTS IN AUGER ELECTRON SPECTROSCOPY (4-0867
TO OXYGEN ADSORPTION.   CHEMICAL EFFECTS IN THE AUGER SPECTRA OF IRON DUE 4-0727
MOLYBDENUM (110) DUE TO ADSORPTION OF MOLECULAR OXYGEN A/ CHEMICAL EFFECTS IN THE M4,5NN AUGER SPECTRUM OF 4-0885
SSBAUER SPECTROSCOPY. (AUGER AFTEREFFECTS) CHEMICAL EFFECTS OF NUCLEAR TRANSFORMATIONS IN MO 2-0564
A OF CHEMISORBED NITROGEN ON IRON AND MOLYBDENUM SURFACE/ CHEMICAL EFFECTS OF THE NITROGEN KLL AUGER SPECTR 4-0925
ROM OXYGEN. CHEMICAL EFFECTS ON KLL AUGER ELECTRON SPECTRUM F 4-0819
BERYLLIUM. CHEMICAL EFFECTS ON THE AUGER ELECTRON SPECTRA OF 4-0834
UM FROM SURFACE CARBON. CHEMICAL EFFECTS ON THE KLL AUGER ELECTRON SPECTR 4-0862
SHAPE COMPARISON OF NITROGEN AND SULFUR IN TWO DIFFERENT CHEMICAL ENVIRONMENTS.   AUGER LINE 5-1816
Y AUGER EMISSION. CHEMICAL IDENTIFICATION OF SURFACE MONOLAYERS. (B 5-1589
        SURFACE ANALYSIS USING COMPLEMENTARY ELECTRONIC AND CHEMICAL MEASUREMENTS. (AUGER) 5-1390
/VANCED ANALYTICAL TECHNIQUES FOR CHARACTERIZATION OF THE CHEMICAL, PHYSICAL, AND CRYSTALLOGRAPHIC NATURE / 5-1433
TEEL SHEET BY AUGER SPECTROSCOPY. CHEMICAL PROFILE ANALYSIS OF TYPE 316 STAINLESS S 6-1996
CTRON SPECTROSCOPY AND INERT GAS SPUTTERING FOR OBTAINING CHEMICAL PROFILES.   USE OF AUGER ELE 1-0156
        BINDING, VALENCY AND CHARGE TRANSFER IN CHEMICAL REACTIONS ON METAL SURFACES. 5-1763
RIDINE: CHEMICAL CONSEQUENCES OF THE AUGER EFF/ RADIATION CHEMICAL RESONANCE EFFECT IN SOLID 5-BROMO DEOXYU 2-0377
SHELLS. (AUGER PROCESS, AES, PHOTOELECTRON SPECTROSCOPY, CHEMICAL SHIFT) /ND RECOMBINATION INVOLVING INNER 2-0514
                                                        CHEMICAL SHIFT IN AUGER SPECTRA. 4-0826
OLYBDENUM DURING THE ADSORPTION OF OXYGEN. CHEMICAL SHIFT IN AUGER SPECTRA OF TUNGSTEN AND M 4-0816
NG OXYGEN ADSORPTION. CHEMICAL SHIFT IN THE AUGER SPECTRUM OF IRON DURI 5-1792
FROM OXYGEN ADSORPTION. CHEMICAL SHIFT IN THE AUGER SPECTRUM OF TUNGSTEN 4-0922
ECTRA OF PARTIALLY OXIDIZED MAGNESIUM SURFACES. ABNORMAL CHEMICAL SHIFTS. /ITED AUGER AND PHOTOELECTRON SP 5-1949
OXYGEN ON NICKEL(III) USING AUGER ELECTRON SPECTROSCOPY: CHEMICAL SHIFTS AND VALENCE SPECTRA. /ORPTION OF 5-1466
SITY OF AUGER SIGNALS OBTAINED WITH A PHOTOELECTRON SPEC/ CHEMICAL SHIFTS AND VARIATIONS OF WIDTH AND INTEN 2-0273
                                                        CHEMICAL SHIFTS IN AUGER ELECTRON SPECTRA. 4-0825
LICON IN SILICON NITRIDE. CHEMICAL SHIFTS IN AUGER ELECTRON SPECTRA FROM SI 4-0881
                                                        CHEMICAL SHIFTS IN AUGER ELECTRON SPECTROSCOPY. 4-0863
OM INITIAL OXIDATION OF TANTALUM (110). CHEMICAL SHIFTS IN AUGER ELECTRON SPECTROSCOPY FR 6-2038
PTION, ESCA SIMPLE CHARGE TRANSFER MODEL). CHEMICAL SHIFTS IN AUGER SPECTRA (MONOLAYER ADSOR 4-1115
YERS OF SILICON DIOXIDE DURI/ EXPERIMENTAL OBSERVATION OF CHEMICAL SHIFTS IN AUGER SPECTRUM FROM SURFACE LA 4-1044
EXCITED LMM AUGER SPECTRA OF GERMANIUM.   PHOTOELECTRON CHEMICAL SHIFTS IN COORDINATION COMPOUNDS. X-RAY 4-0924
                                                        CHEMICAL SHIFTS IN PHOTOEXCITED AUGER SPECTRA. 4-0797
DIUM OXID/ AUGER ELECTRON SPECTROSCOPIC INVESTIGATIONS OF CHEMICAL SHIFTS IN SOME VANADIUM COMPOUNDS. (VANA 4-1090
D CHROMIUM AND VANADIUM. CHEMICAL SHIFTS IN THE AUGER SPECTRA FROM OXIDIZE 4-0809
ILMS. CHEMICAL SHIFTS IN THE AUGER SPECTRA OF PASSIVE F 4-1056
ON OXYGEN ADSORPTION. CHEMICAL SHIFTS IN THE AUGER SPECTRUM OF YTTRIUM 5-1153
SPECTRA FO/ ELECTRON SPECTROSCOPY AND ADSORPTION STUDIES. CHEMICAL SHIFTS IN URANIUM AUGER TRANSITIONS. 4-0737
AMETER. CHEMICAL SHIFTS IN X-RAY PHOTOELECTRON AND AUGER 5-1204
AMETER IN ELECTRON SPECTROSCOPY FOR THE IDENTIFICATION OF CHEMICAL SHIFTS OF AUGER LINES, AND THE AUGER PAR 4-1107
CTRON SPECTROSCOPY.   IDENTIFICATION OF THE CHEMICAL SPECIES.   AUGER PAR 5-1948
ETHOD.   CONTAMINANTS ON CHEMICAL STATE OF ADSORBED SPECIES WITH AUGER ELE 4-0868
SILICIDE BY AUGER ELECTRON SPECT/ COMPOSITION PROFILES OF CHEMICALLY ETCHED SILICON SURFACES. LEED- AUGER M 5-1229
ION MEASUREMENTS OF THE NITROGEN KLL AUGER TRANSITIONS OF CHEMICALLY VAPOR DEPOSITED PLATINUM AND PLATINUM 5-1656
SYSTEMS)   PHASE EQUILIBRIA IN TWO DIMENSIONAL CHEMISORBED AMMONIA ON A MOLYBDENUM SURFACE. /LUT 5-1532
CE/ CHEMICAL EFFECTS OF THE NITROGEN KLL AUGER SPECTRA OF CHEMISORBED LAYERS. (AUGER SPECTROSCOPY, SULFIDE 5-1170
        AUGER EJECTION OF ELECTRONS FROM TUNGSTEN BY OXYGEN CHEMISORBED NITROGEN ON IRON AND MOLYBDENUM SURFA 4-0925
/BON MONOXIDE AND OXYGEN WITH A PALLADIUM (100) SURFACE. CHEMISORPTION. 5-1599
        CHARACTERIZATION OF CHEMISORPTION AND CATALYTIC REACTIONS, LEED, AUG/ 5-1321
BULK TO THE SURFACE OF PALLADIUM (/ LEED- AES STUDIES OF CHEMISORPTION BY LEED. 1-0161
/N AND ENERGY LOSS SPECTROSCOPIES TO CATALYTIC PROCESSES. CHEMISORPTION INDUCED SULFUR SEGREGATION FROM THE 5-1865
111) 7*7 SURFACE. (AUGER EFFECT) CHEMISORPTION OF AMMONIA, NITROGEN, AND NITRIDE / 5-1533
/ AUGER SPECTROMETRIC AND MASS SPECTROMETRIC STUDY OF THE CHEMISORPTION OF ATOMIC HYDROGEN ON THE SILICON (5-1789
AUGER ELECTRON SPECTROSCOPY. CHEMISORPTION OF CARBON MONOXIDE ON POLYCRYSTALLI 5-1355
S. (AUGER SPECTRA) CHEMISORPTION OF NITROGEN ON TUNGSTEN STUDIED BY 5-1518
CE. CHEMISORPTION OF OXYGEN ON PLATINUM (111) SURFACE 5-1500
/SCOPY AND LOW ENERGY ELECTRON DIFFRACTION STUDIES OF THE CHEMISORPTION OF OXYGEN ON THE SILVER (110) SURFA 5-1778
        PHOTOEMISSION STUDIES OF CHEMISORPTION OF UNSATURATED MOLECULES BY THE (1/ 5-1267
SURFACES. CHEMISORPTION ON NICKEL AND TUNGSTEN SURFACES. 5-1308
OPY) CHEMISORPTION ON SINGLE CRYSTAL MOLYBDENUM (112) 5-1303
        LEED (AUGER) STUDY OF CHLORINE CHEMISORPTION ON TANTALUM (100). (AUGER SPECTROSC 5-1245
ER ELECTRON SPECTROSCOPY)   INVESTIGATION OF HYDROGEN CHEMISORPTION ON THE SILVER (111) SURFACE. 5-1779
URFACES. CORRELATED STUDY USING LOW ENERGY ELECTRON DIFF/ CHEMISORPTION ON THE TUNGSTEN (100) SURFACE. (AUG 5-1974
RFACES. (AUGER SPECTRA) CHEMISORPTION ON (001), (110), AND (111) NICKEL S 5-1291
EASUREMENTS ON ZINC OXIDE SURFACES. (AUGER ANALYSIS) CHEMISORPTION ON (001), (110) AND (111) NICKEL SU 5-1764
E SURFACES. (AUGER EFFECT) CHEMISORPTION, PHOTODESORPTION AND CONDUCTIVITY M 5-1801
G WORK FUNCTION MEASUREMENTS, AUGER, AND LOW ENERGY ELEC/ CHEMISORPTION REACTIONS ON HIGH INDEX ZINC SULFID 4-0788
TROMETRY AND AUGER ELECTRON SPECTR/ COMPARISON OF SURFACE CHEMISORPTION STUDIES OF NICKEL SURFACES UTILIZIN 5-1290
TRON SPECTROSCOP/ RELATIONSHIPS BETWEEN SUBSTRATE SURFACE CHEMISTRIES OBTAINED BY USING ION SCATTERING SPEC 6-1999
Y, LEED, ION MICROPROBE ANALYSIS)   CRYSTALLOGRAPHY AND CHEMISTRY AND ADHESION OF THIN FILMS. (AUGER ELEC 5-1851
N SPECTROSCOPY) SURFACE CHEMISTRY OF CERAMIC SURFACES. (AUGER SPECTROSCOP 5-1852
BY X-RAY PHOTOEMISSION SPECTROSC/ SURFACE COMPOSITION AND CHEMISTRY OF CONTACT MATERIALS. (BY AUGER ELECTRO 5-1394
INTERGRANULAR CORROSION OF AUSTENITIC STAINLESS STEEL. / CHEMISTRY OF EVAPORATED PERMALLOY FILMS OBSERVED 5-1738
SPUTTERING, STUDIED WITH AUGER SPECTROSCOP/ STRUCTURE AND CHEMISTRY OF GRAIN BOUNDARIES AND ITS RELATION TO 6-2064
        APPLICATION OF (AUGER) ELECTRON SPECTROSCOPY TO CHEMISTRY OF SILICON SURFACES AFTER PRE- AND BACK 5-1233
                                                        CHEMISTRY. (REVIEW) 1-0127

GER ELECTRON SPECTROSCOPY AND ITS APPLICATIONS IN SURFACE CHEMISTRY. (REVIEW) AU 1-0102
AUGER SPECTROMETRY AS A TOOL FOR SURFACE CHEMISTRY. (REVIEW) 1-0162
APPLICATION OF AUGER ELECTRON SPECTROSCOPY TO SURFACE CHEMISTRY STUDIES OF CERAMIC SUBSTRATES. 5-1275
X-RAYS AND ELECTRONS IN ANALYTICAL CHEMISTRY WITH EMPHASIS ON INSTRUMENTATIION. 3-0646
KLL AUGER SPECTRUM OF CHLORINE. (IN CARBON TETRACHLORIDE) 4-0808
-AUGER" PROCESSES IN L2,3 EMISSION IN ARGON AND POTASSIUM CHLORIDE. "SEMI 2-0326
LASMON LOSSES/ STUDY OF SURFACE DISSOCIATION IN POTASSIUM CHLORIDE AND LITHIUM FLUORIDE BY OBSERVATION OF P 5-1279
SURFACE DISSOCIATION OF POTASSIUM CHLORIDE BY LOW ENERGY ELECTRON BOMBARDMENT. 5-1719
SHAPING SECONDARY ELECTRON EMISSION FROM POROUS POTASSIUM CHLORIDE FILMS. CHARACTERISTICS OF 4-1120
ECOMPOSI/ DETECTION OF EXCITONS NEAR THE L2,3 EDGE OF THE CHLORIDE ION IN SODIUM CHLORIDE BY MEANS OF THE D 4-1126
TROSCOPY. GROWTH OF SILVER ON POTASSIUM CHLORIDE OBSERVED BY LEED AND AUGER EMISSION SPEC 5-1348
HESE FILMS ON ADH/ ABSORPTION OF ETHYLENE OXIDE AND VINYL CHLORIDE ON AN IRON (011) SURFACE AND EFFECT OF T 5-1207
CTROSCOPY. SLIDING OF POLY(VINYL CHLORIDE) ON METALS STUDIED BY AUGER ELECTRON SPE 5-1727
AUGER ANALYSIS OF CHLORINE IN HYDROGEN CHLORIDE OR CHLORINE GROWN SILICON DIOXIDE FILMS. 5-1251
RON EMISSION) ANGLE DISTRIBUTION OF SODIUM CHLORIDE SINGLE CRYSTALS. (PHOTO- AND AUGER ELECT 2-0297
SEGREGATION OF CADMIUM TO A (111) SILVER CHLORIDE SURFACE. (AUGER SPECTROSCOPY) 5-1918
RELATED TO LASER DAMAGE OF LITHIUM NIOBATE AND POTASSIUM CHLORIDE SURFACES. (AUGER SURFACE PROFILE) /STICS 5-1746
UBSTRATES. AUGER SPECTRA OF HYDROGEN CHLORIDE VAPOR ETCHED N+ GALLIUM ARSENIDE (100) S 4-0992
CTRON DIFFRACTIO/ EPITAXIAL GROWTH OF SILVER ON POTASSIUM CHLORIDE (100) SURFACE OBSERVED BY LOW ENERGY ELE 5-1696
DUCT/ K-SHELL IONIZATION AND FLUORESCENCE YIELD IN 50 MEV CHLORINE(+)- NEON COLLISIONS. (AUGER ELECTRON PRO 2-0518
ATION FROM THE BULK CRYSTAL INTERACTION WITH CHLORINE GA/ CHLORINE ADSORPTION AT A TITANIUM SURFACE. SEGREG 5-1833
CTOSCOPY) CHLORINE ADSORPTION ON CLEAN ALUMINUM. (AUGER SPE 5-1831
AUTOIONIZING STATES OF HIGHLY STRIPPED OXYGEN, FLUORINE, CHLORINE, AND ARGON IONS. (AUGER EFFECT) /RA FROM 2-0472
COPPER JUNCTIONS. (SPECTROSCOPY) EFFECT OF ADSORBED CHLORINE AND OXYGEN ON SHEAR STRENGTH OF IRON AND 6-2143
K-SHELL FLUORESCENCE YIELDS OF ARGON, CHLORINE AND SULFUR. (AUGER TRANSITION) 4-1014
INTERACTION BETWEEN CHLORINE AND THE (100) SURFACE OF GOLD. 5-1538
TRUCTURE OF TITANIUM AND ITS INTERACTION WITH BROMINE AND CHLORINE. (AUGER SPECTROSCOPY) SURFACE S 5-1538
SUREMENTS OF THE L-SHELL FLUORESCENCE YIELDS OF ARGON AND CHLORINE BY ELECTRON IMPACT. (AUGER EMISSION) MEA 2-0387
E. LEED (AUGER) STUDY OF CHLORINE CHEMISORPTION ON THE SILVER (111) SURFAC 5-1779
KLL AUGER SPECTRUM OF CHLORINE. (IN CARBON TETRACHLORIDE) 4-0808
ILICON DIOXIDE FILMS. AUGER ANALYSIS OF CHLORINE IN HYDROGEN CHLORIDE OR CHLORINE GROWN S 5-1251
VER. (AUGER EFFECT) CHLORINE MONOLAYERS ON THE LOW INDEX FACES OF SIL 5-1777
NEON K-SHELL IONIZATION AND FLUORESCENCE YIELD IN 50 MEV CHLORINE(N+) ION- NEON COLLISIONS. /DEPENDENCE OF 2-0300
CHLORINE REACTIONS ON THE SILICON (111) SURFACE. 5-1344
S. SECONDARY ELECTRON EMISSION FROM SODIUM CHLORINE SINGLE CRYSTAL BOMBARDED BY RARE GAS ION 4-0759
MENT OF THE IONIZATION CROSS SECTION OF THE L2,3 SHELL OF CHLORINE USING AUGER ELECTRON SPECTROSCOPY. /SURE 4-0997
: A LEED, AUGER AND WORK FUNCTION STUDY. INTERACTION OF CHLORINE WITH THE ZINC OXIDE (0001)- ZINC SURFACE 5-1463
OLUTION NEON K-AUGER ELECTRON SPECTRUM PRODUCED BY 45 MEV CHLORINE(12+) ION IMPACT. HIGH RES 4-1051
SIS. CHOOSING BETWEEN ESCA AND AUGER FOR SURFACE ANALY 5-1921
(AUGER) ANALYSIS OF CHROMATE CONVERSION COATINGS ON ALUMINUM. 6-2125
/TRONS FOR THE DETECTION OF GAMMA EMITTING RADIONUCLIDES: CHROMATOGRAPHIC QUALITY CONTROL OF RADIOPHARMACE/ 1-0217
RON-55 AS AN AUGER ELECTRON EMITTER. NOVEL SOURCE FOR GAS CHROMATOGRAPHY DETECTORS. I 3-0608
ECTROMETRY FOR ALUMINUM, SILICON, TITANIUM, VANADIUM, AND CHROMIUM. /SPECTROSCOPY AND SECONDARY ION MASS SP 5-1553
MICROSCOPIC STUDY OF THE PASSIVE FILM FORMED ON AN IRON- CHROMIUM ALLOY. AUGER SCANNING ELECTRON 5-1753
STRESS AND GOOD ADHESION ALLOY THIN FILMS. (TITANIUM AND CHROMIUM ALLOYS, AES) LOW INTRINSIC 5-1909
N BINDING ENERGIES OF SURFACE AND BULK ATOMS OF TITANIUM, CHROMIUM, AND NICKEL. (AUGER EFFECT) /ORE ELECTRO 2-0392
CHEMICAL SHIFTS IN THE AUGER SPECTRA FROM OXIDIZED CHROMIUM AND VANADIUM. 4-0809
/SUREMENTS IN THIN FILMS UTILIZING WORK FUNCTION CHANGES: CHROMIUM INTO GOLD. (AUGER ELECTRON SPECTROSCOPY) 5-1889
-RAY AND AUGER ELECTRON APPEARANCE POT/ DIFFERENCE IN THE CHROMIUM L3/L2 INTENSITY RATIO MEASURED BY SOFT X 4-0890
/OF SODIUM ON NIMONIC-PE16 ALLOY WITH THE SURFACE OXYGEN- CHROMIUM RATIO DETERMINED BY AUGER SPECTROSCOPY. 5-1441
TY. (AUGER EMISSION AN/ TEMPER EMBRITTLEMENT IN A NICKEL- CHROMIUM STEEL BY AUGER EMISSION AND X-RAY PHOTOE 6-2016
LECTRON SPECTROSCOPI/ ANALYSIS PROFILES OF OXIDE FILMS ON CHROMIUM STEEL CONTAINING PHOSPHORUS AS AN IMPURI 6-2138
ROSION, SURFACE, AND MAGNETIC PROPERTIES OF NICKEL- IRON- CHROMIUM THIN FILMS. (AES) COR 5-1766
F THE (001) AND (110) FACES OF MONOCRYSTALLINE IRON(0.84) CHROMIUM(0.16). (AUGER EFFECT) /OSITION STUDIES O 5-1591
LEED- AES STUDY OF THE OXIDATION OF IRON- CHROMIUM (100) AND (110). 5-1587
ON- AUGER ELECTRON SPECTROSCOPY STUDY OF THE OXIDATION OF CHROMIUM (110) AND CHROMIUM (100). /RON DIFFRACTI 6-2022
N BOUNDARY COMPOSITION ON TEMPER BRITTLENESS IN A NICKEL- CHROMIUM- ANTIMONY STEEL. (AUGER ANALYSIS) / GRAI 6-2114
GER ANALYSIS. CHARACTERIZATION OF A CHROMIUM- GOLD THIN FILM DEPOSITION PROCESS BY AU 1-0037
S OF TEN YEARS EXPERIMENTAL BREEDER REACTOR II SERVICE ON CHROMIUM- MOLYBDENUM STEEL. EFFECT 6-2108
LONG TIME ISOTHERMAL TEMPER EMBRITTLEMENT IN NICKEL- CHROMIUM- MOLYBDENUM- VANADIUM. 6-2140
/RUCTURE ON GRAIN BOUNDARY SEGREGATION OF PHOSPHORUS IN A CHROMIUM- MOLYBDENUM- VANADIUM STEEL. (AUGER SPE/ 6-2139
EED- AUGER INVESTIGATION ON A (100) SURFACE OF AUSTENITIC CHROMIUM- NICKEL STEEL. L 6-1991
N SPECTROSCOPY. DIFFUSION STUDIES IN CHROMIUM- PLATINUM THIN FILMS USING AUGER ELECTRO 5-1286
RUM AND OF THE SATELLITES IN THE SOFT X-/ SCF AND LIMITED CI CALCULATIONS FOR ASSIGNMENT OF THE AUGER SPECT 4-0724
A NOVEL DETECTION CIRCUIT FOR AUGER SPECTROSCOPY. 3-0673
INTEGRATED CIRCUIT PROCESS CHARACTERIZATION USING AES. 5-1741
OF SCANNING AUGER SPECTROSCOPY TO THE SILICON INTEGRATED CIRCUIT TECHNOLOGY. APPLICATIONS 5-1655
IONS OF CARBON RESIDUES ON SURFACES OF SILICON INTEGRATED CIRCUITS. (BY AUGER ELECTRON SPECTROSCOPY) /TIGAT 5-1345
ELECTRODES AND CIRCUITS OF A LEED- AES SYSTEM. 3-0639
CHLORINE ADSORPTION ON CLEAN ALUMINUM. (AUGER SPECTOSCOPY) 5-1831
N OF THE SURFACE COMPOSITION OF COPPER- NICKEL ALLOYS FOR CLEAN AND ADSORBATE COVERED SURFACES. /TERMINATIO 5-1424
S. (LEED, AES) SOME PROPERTIES OF CLEAN AND ALKALI METAL COVERED ZINC OXIDE SURFACE 5-1881
D BY POSITIVE IONS. PART-1./ AUGER ELECTRON EMISSION FROM CLEAN AND CARBON CONTAMINATED MOLYBDENUM BOMBARDE 2-0543
/ELECTRON SPECTROSCOPY, AND WORK FUNCTION MEASUREMENTS ON CLEAN AND CESIUM COVERED POLAR FACES OF GALLIUM / 5-1294
ANGULAR DISTRIBUTION OF AUGER ELECTRON EMISSION FROM CLEAN AND GAS-COVERED IRON (100) SURFACES. 5-1623
INTERPRETATION OF THE AUGER SPECTRUM OF CLEAN AND OXIDIZED BERYLLIUM. 4-1087
ELECTRON SPECTROSCOPIC STUDIES OF CLEAN AND OXIDIZED IRON. (AUGER STUDY OF SURFACE) 5-1323
AUGER ELECTRON SPECTROSCOPY OF CESIUM ADSORBED ON CLEAN AND OXYGEN- COVERED (100) TUNGSTEN. 5-1295
ENIDE. LEED, AUGER, AND WORK FUNCTION STUDIES OF CLEAN AND SODIUM- COVERED SURFACES OF GALLIUM ARS 5-1240
/RVATION OF THE SEGREGATION OF COPPER TO THE SURFACE OF A CLEAN ANNEALED 50-50 NICKEL ALLOY BY AUGER ELECT/ 6-2044
ON AND INTERPRETATION OF THE AUGER ELECTRON SPECTRUM FROM CLEAN BERYLLIUM. OBSERVATI 4-0997
INETIC PROPERTIES. (AUGER E/ ADSORPTION OF WATER VAPOR ON CLEAN CLEAVED GERMANIUM. PART-1: STRUCTURAL AND K 5-1432
SURFACE VALENCE BAND AND PLASMON FEATURES OF CLEAN CLEAVED SILICON. 5-1141
ADSORPTION OF OXYGEN ON CLEAN CLEAVED (110) GALLIUM ARSENIDE SURFACES. 5-1306
ANGULAR RESOLVED AUGER EMISSION SPECTRA FROM A CLEAN COPPER (100) SURFACE. 5-1697
AUGER ELECTRON SPECTROSCOPY OF CLEAN GALLIUM ARSENIDE. 5-1921
ISSION SPECTROSCOPY. ADHESION OF METALS TO A CLEAN IRON SURFACE STUDIED WITH LEED AND AUGER EM 6-2008
EN, AND HYDROGEN SULFIDE SURFACE FILMS ON THE ADHESION OF CLEAN IRON. (USING LEED AND AES) /OF SULFUR, OXYG 5-1211
LEED AND AUGER STUDIES OF EFFECT OF OXYGEN ON ADHESION OF CLEAN IRON (001) AND (011) SURFACES. 6-2010
VACUUM ULTRAVIOLET PHOTOELECTRON STUDIES OF ADSORPTION ON CLEAN METAL FILM SURFACES. (AUGER SPECTRA) / AND 5-1203

NGLE OF INCIDENCE EFFECTS IN AUGER ELECTRON EMISSION FROM CLEAN MOLYBDENUM. A 2-0542
OXYGEN ADSORPTION ON CLEAN MOLYBDENUM (100) SURFACES. 5-1772
KEL (100) SURFAC/ KINETICS OF THE REACTION OF OXYGEN WITH CLEAN NICKEL SINGLE CRYSTAL SURFACES. PART-1: NIC 5-1452
ION SCATTERING AND AUGER ELECTRON SPECTROSCOPY STUDIES OF CLEAN NICKEL SURFACES AND ADSORBED LAYERS. /ERGY 5-1870
STICKING PROBABILITIES FOR THE ADSORPTION OF HYDROGEN ON CLEAN NICKEL (111), NICKEL (100), SHEET AND EVAP/ 5-1468
OF MERCURY CN GOL/ AUGER ELECTRON SPECTROSCOPY STUDIES OF CLEAN POLYCRYSTALLINE GOLD AND OF THE ADSORPTION 5-1519
AUGER ELECTRON SPECTROSCOPY OF CLEAN POLYCRYSTALLINE GOLD FILMS. 5-1760
PY) ADSORPTION OF OXYGEN ON A CLEAN SILICON SURFACE. (AUGER ELECTRON SPECTROSCO 5-1638
ER ELECTRON SPECTROSCOPY STUDY OF PHOSPHINE ADSORPTION ON CLEAN SILICON (111). /LECTRON DIFFRACTION AND AUG 5-1930
ECTROSCOPY) ELECTRONIC STATES OF OXYGEN ADSORBED ON CLEAN SILICON (111) AND (100) SURFACES. (AUGER SP 5-1476
INTERACTION OF SULFUR DIOXIDE AND CARBON DIOXIDE WITH CLEAN SILVER IN ULTRAHIGH VACUUM. (AES) 5-1575
CLEAN SURFACES. 5-1625
CLEAN TELLURIUM SURFACES STUDIED BY LEED. 5-1134
SCOPY. STUDY OF THE PREPARATION OF ATOMICALLY CLEAN TUNGSTEN SURFACES BY AUGER ELECTRON SPECTRO 5-1517
INTERACTION OF CESIUM WITH CLEAN ZINC OXIDE SURFACES. (AUGER SPECTROSCOPY) 5-1588
LECTRON DIFFRACTION- AUGER ELECTRON SPECTROSCOPY STUDY OF CLEAN (001) IRON SURFACE. LOW ENERGY E 5-1527
CES. (AUGER SPE/ ELASTIC LEED INTENSITY ENERGY STUDIES OF CLEAN (001), (110) AND ELASTIC (111) NICKEL SURFA 5-1292
LOW ENERGY ELECTRON DIFFRACTION FROM CLEAN (100) NICKEL OXIDE. 5-1752
OF IRON. AUGER SPECTRA AND LEED PATTERNS FROM VACUUM CLEANED SILICON CRYSTALS WITH CALIBRATED DEPOSITS 5-1769
(100) SURFACES OF ALKALI HALIDES. PART-1: AIR AND VACUUM CLEANED SURFACES. 5-1347
ATES BY/ COMPOSITION PROFILES OF SEVERAL CONTAMINATED AND CLEANED SURFACES OF GOLD THICK FILMS ON COPPER PL 5-1550
M, TUNGSTEN, SILICON) MONITORED BY AUGER SPECTRO/ SPUTTER CLEANING AND ETCHING OF CRYSTAL SURFACES (TITANIU 5-1828
GERMANIUM SURFACE CLEANING: AUGER ANALYSIS. 5-1540
SURFACE STUDIES FOR QUARTZ RESONATORS. (CLEANING, AUGER SPECTROSCOPY) 5-1942
RESONATOR FABRICATION. (AUGER SPECTRA) METHODS OF CLEANING CONTAMINANTS FROM QUARTZ SURFACES DURING 5-1410
RFACE PREPARATION AND CHARACTERIZATION TECHNIQUES (AUGER, CLEANING) FOR QUARTZ RESONATORS. SU 5-1943
OF OXIDE FILMS, AUGER SPECTROSCOPY OF SURFACE) CHEMICAL CLEANING FOR THERMOCOMPRESSION BONDING, (REMOVAL 5-1454
ROSCOPY) PLASMA CLEANING OF METAL SURFACES. (AUGER ELECTRON SPECT 6-2093
CLEANING PROCEDURES OF MOLYBDENUM (110) SURFACE. 5-1649
OXYGEN AND C/ THORIUM (100) CRYSTAL FACE. A STUDY OF ITS CLEANING, SURFACE STRUCTURE, AND INTERACTION WITH 5-1883
ICKEL (100), AND N/ EFFECTIVENESS OF THE STANDARD SURFACE CLEANING TECHNIQUES AS APPLIED TO NICKEL (111), N 5-1467
ENERGY ELECTRON DIFFRACTION AND AUGER ELECTRON / SILICON CLEANING WITH HYDROGEN PEROXIDE SOLUTIONS: HIGH 5-1426
ULTRAVACUUM. (AUGER SPECTROSCOPY, SURFACE PREPARATION AND CLEANLINESS) /CS OF SOLIDS FROM THE VIEWPOINT OF 5-1419
E TRACE IMPURITIES) HOW TO ASSURE ONESELF OF THE CLEANLINESS OF A SURFACE. (USE OF AES TO DETERMIN 5-1154
SECONDARY ION MASS SPECTROMETRY AND AUGER ELECT/ SURFACE CLEANLINESS OF 316 L-N STAINLESS STEEL STUDIES BY 3-0654
SECONDARY ION MASS SPECTROMETRY AND AUGER ELECT/ SURFACE CLEANLINESS OF 316 L-N STAINLESS STEEL STUDIED BY 6-2087
. AUGER ELECTRON SPECTROSCOPY OF CLEANUP-RELATED CONTAMINATION ON SILICON SURFACES 5-1983
TRA) SILICON (111)-7 STRUCTURE OBTAINED BY CLEAVAGE AT ROOM TEMPERATURE. (AUGER SURFACE SPEC 5-1767
ON AND AUGER EM/ ULTRAHIGH VACUUM EVAPORATOR AND MULTIPLE CLEAVAGE DEVICE FOR LOW ENERGY ELECTRON DIFFRACTI 3-0707
FFRACTION- AUGER STUDIES. A CRYSTAL CLEAVAGE DEVICE FOR USE IN LOW ENERGY ELECTRON DI 3-0637
ER SPECTRA) CHANGE OF SURFACE PROPERTIES OF MICA AFTER CLEAVAGE. (OUTGASSING AND DECORATION STUDIES, AUG 5-1743
MUSCOVITE CLEAVAGE STUDIED BY AUGER ELECTRON SPECTROSCOPY. 5-1298
ON AND AUGER E/ INVESTIGATION OF EUROPIUM(II) OXIDE (100) CLEAVAGE SURFACE BY LOW ENERGY ELECTRON DIFFRACTI 5-1157
AUGER ELECTRON EMISSION MICROGRAPHIC STUDIES OF THE CLEAVAGE SURFACE OF GRAPHITE SINGLE CRYSTAL. 5-1413
/ OF CONDENSATION COEFFICIENTS OF GOLD ON ROCK SALT (100) CLEAVAGE SURFACES AND ON AMORPHOUS CARBON BY MEA/ 5-1135
STRUCTURE TRANSFORMATIONS ON CLEAVED AND ANNEALED GERMANIUM (111) SURFACES. 5-1713
BINDING STATES OF WATER VAPOR ON CLEAVED GERMANIUM. (LEED, AES) 5-1820
PROPERTIES. (AUGER E/ ADSORPTION OF WATER VAPOR ON CLEAN CLEAVED GERMANIUM. PART-1: STRUCTURAL AND KINETIC 5-1432
UREMENTS OF BULK, SURFACE, AND HYDROGEN INDUCED STATES ON CLEAVED GERMANIUM (111). (AUGER SPECTROSCOPY) /AS 5-1781
COPY. DEPTH ANALYSIS OF CLEAVED MICA SURFACES MONITORED BY AUGER SPECTROS 5-1842
ELECTRONIC PROPERTIES OF CLEAVED MOLYBDENUM DISULFIDE SURFACES. 5-1634
AUGER SPECTRA AND LEED PATTERNS FROM NICKEL DEPOSITS ON CLEAVED SILICON. 5-1768
SURFACE VALENCE BAND AND PLASMON FEATURES ON CLEAN CLEAVED SILICON. 5-1141
AUGER STUDIES OF CLEAVED SILICON (111) SURFACES. 5-1382
ADSORPTION OF OXYGEN ON CLEAN CLEAVED (110) GALLIUM ARSENIDE SURFACES. 5-1306
IES, TOPOGRAPHY, PHENOMENA, FIELD ION MICROSCOPY, LEED/ A CLOSE LOOK AT SURFACES. (SURFACE STRUCTURES, SPEC 5-1492
Y SHIFTS. EFFECTS OF EXTRA CLOSED CORE RELAXATION ON AUGER AND BINDING ENERG 4-0950
/DENTIFICATION AND MEASUREMENT OF LOW ENERGY ELECTRONS BY CLOUD CHAMBER TECHNIQUES: OXYGEN AND NEON AUGER/ 2-0566
N AND ELECTRON CAPTURE. (K-AUGER/ MERCURY AND GOLD ATOMIC CLOUDS REARRANGEMENT FOLLOWING INTERNAL CONVERSIO 2-0334
N SPECTRA OF NICKEL AND COPPER. (AUGER-TYPE PROCESSES) CLUSTER CALCULATIONS OF THE SOFT X-RAY L3 EMISSIO 4-0888
SECONDARY ELECTRON EMISSION FROM PLATINUM BLACK COATED SURFACES. (AUGER SPECTRA) 5-1979
ELECTRON DIFFRACTIO/ INVESTIGATION OF THE PURE AND CESIUM COATED (100) SURFACE OF GALLIUM ARSENIDE BY SLOW 5-1639
NEW METHODS FOR THE DETECTION AND ELIMINATION OF COATING FLAWS. 5-1592
OXIDE, SULFUR, AND OTHER GASES ON STAINLE/ EFFECT OF GOLD COATING ON SORPTION CHARACTERISTICS OF CARBON MON 5-1551
LASER WINDOW SURFACE FINISHING AND COATING TECHNOLOGY. (AUGER ELECTRON SPECTROSCOPY) 1-0015
ANALYSIS OF GLASS CONTAINER COATINGS BY AUGER ELECTRON SPECTROSCOPY. 5-1704
(AUGER) ANALYSIS OF CHROMATE CONVERSION COATINGS ON ALUMINUM. 6-2125
ACTERISTIC LOSS SPECTROSCOPY FOR THE ELEMENTS VANADIUM TO COBALT. /MBINATION OF AUGER SPECTROSCOPY AND CHAR 4-0813
INFLUENCE OF SILICON ON FRICTION AND WEAR OF IRON- COBALT ALLOYS. (AUGER ANALYSIS) 6-2011
FRICTION AND AUGER SPECTROSCOPY ANALYSIS OF A COMMERCIAL COBALT BASE AIRCRAFT TURBINE SHROUD ALLOY. /SION, 6-2009
FOLLOWING ELECTRON CAPTURE IN TRIS (1,10-PHENANTHROLINE) COBALT(III)-57 PERCHLORATE DIHYDRATE. /IONIZATION 2-0516
AUGER SPECTROSCOPY) ADSORPTION OF COBALT ON A NICKEL (111) SURFACE. (AUGER EFFECT, 5-1256
CTRA) ADSORPTION OF XENON AND COBALT ON SILVER (111). 5-1633
00 DEGREES K. ADSORPTION OF OXYGEN AND OXIDATION OF COBALT ON THE RUTHENIUM (001) SURFACE. (AUGER SPE 5-1608
AUGER ELECTRON SPECTROSCOPY. ADSORPTION OF COBALT ON (001) RUTHENIUM AT TEMPERATURES BELOW 3 5-1609
NUM (111) SURFACE. (AUGER ELECTRON SPECTROSCOPY, STICKING DIFFUSION OF COBALT OUT OF COBALT HARDENED GOLD MEASURED WITH 6-2135
0-100 KEV ELECT/ ORIENTATION ANISOTROPY OF BACKSCATTERING COEFFICIENT) /ICS OF OXYGEN ADSORPTION ON A PLATI 5-1189
/A TO EXPLAIN THE FINE STRUCTURE OF THE DEPENDENCE OF THE COEFFICIENT AND SECONDARY ELECTRON EMISSION FOR 1 4-1031
/HE ANGULAR DEPENDENCE OF THE SECONDARY ELECTRON EMISSION COEFFICIENT OF SECONDARY ELECTRON EMISSION ON EN/ 4-0891
/DE LEED PATTERNS. (AUGER ELECTRON SPECTROSCOPY, STICKING COEFFICIENT OF SINGLE CRYSTALS IN THE 2-10 KEV R/ 4-1061
NETRATION EFFECTS IN CONVERTED MUONIC TRANSITIONS. (AUGER COEFFICIENT) PART-2: EFFECT OF SULFUR ON THE CAT/ 5-1190
SURFACES AND ON AMORPHOUS C/ MEASUREMENTS OF CONDENSATION COEFFICIENTS) PE 2-0478
PECTROSCOPY, SECON/ MEASUREMENT OF MOMENTUM ACCOMMODATION COEFFICIENTS OF GOLD ON ROCK SALT (100) CLEAVAGE 5-1135
/STABLE ATOMS. PART-1: MEASUREMENTS OF SECONDARY EMISSION COEFFICIENTS ON SURFACES CHARACTERIZED BY AUGER S 5-0501
EFFECT OF TELLURIUM ON INTERGRANULAR COEFFICIENTS USING A CROSSED BEAM METHOD. PART-2/ 2-0340
/SSIBILITY OF INVESTIGATING ATOMIC AUTOIONIZING STATES IN COHESION OF IRON. 6-2099
ARGON COLLISION. EXTENDED ICN- ELECTRON COINCIDENCE EXPERIMENTS. (AUGER SPECTRUM OF HELI/ 2-0530
E PRODUCTION OF LL VACANCY STATES BY THE L X-RAY- L X-RAY COINCIDENCE MEASUREMENTS OF THE VIOLENT ARGON (+)- 2-0482
-131 AND MEASUREMENT OF DECAY PARAMETERS. (DETECTION OF / COINCIDENCE METHOD. (AUGER TRANSITIONS) /NT OF TH 2-0474
COINCIDENCE METHODS OF STANDARDIZATION FOR CESIUM 2-0474

ARGON COLLISIONS. (AUGER EFFECT)   ELECTRON ION COINCIDENCE STUDIES OF HIGHLY IONIZING ARGON(+)-   2-0529
ISIONS. COMMENTS. (AUGER EL/ FAST ELECTRON- SLOW ELECTRON COINCIDENCE STUDY OF VIOLENT ARGON(+)- ARGON COLL   2-0423
R TRANSITIONS LEADING TO DEFINED FINAL CHARGE STATES BY A COINCIDENCE TECHNIQUE.   AUGE   2-0371
IONS IN BAYARD- ALPERT GAUGES CAUSED BY AUGER EMISSION AT COLLECTOR.   SENSITIVITY VARIAT   3-0618
N COINCIDENCE MEASUREMENTS OF THE VIOLENT ARGON(+)- ARGON COLLISION.   EXTENDED ION- ELECTRO   2-0530
TION OF THE L2,3 SHELL OF ALUMINUM IN A NEON(+)- ALUMINUM COLLISION.   IONIZATION CROSS SEC   2-0267
                 LUMINESCENCE DUE TO EXCITON-EXCITON COLLISION IN GALLIUM ARSENIDE.   2-0451
              ARGON K-AUGER CROSS SECTIONS IN COLLISION WITH 30 MEV FLUORINE IONS.   2-0323
GON L-SHELL AUGER SPECTRA PRODUCED IN ARGON- ARGCN COLLISIONS.   AR   4-1080
LECTRON AND X-RAY PRODUCTION IN 50- TO 200 KEV NEON- NEON COLLISIONS.   AUGER E   4-1079
NS FOR L2,3 AUGER EMISSION IN 22 TO 300 KEV PROTON- ARGON COLLISIONS.   CROSS SECTIO 4-1040
                ENERGY LOSS AND IONIZATION IN FAST ION- ATOM COLLISIONS.   2-0233
ARGON K-AUGER SPECTRUM PRODUCED IN 4 MEV PROTON ON ARGON COLLISIONS.   HIGH RESOLUTION 2-0496
GER SPECTRUM PRODUCED IN 4.2 MEV ATOMIC HYDROGEN(+)- NEON COLLISIONS.   HIGH RESOLUTION NEON AU 4-1078
N AND FLUORESCENCE YIELD IN 50 MEV CHLORINE(N+) ION- NEON COLLISIONS. /DEPENDENCE OF NEON K-SHELL IONIZATIO   2-0300
                RADIATIVE AUGER EFFECT IN ION- ATOM COLLISIONS.   2-0481
                                    ION- ATOM COLLISIONS AT HIGH ENERGIES. (AUGER EFFECT)   2-0480
DENCE OF THE K-VACANCY PRODUCTION YIELD IN HEAVY ION ATOM COLLISIONS AT KEV ENERGIES. /C CHARGE STATE DEPEN   2-0346
AUGER ELECTRONS PRODUCED IN Z(1)- NITROGEN AND Z(1)- NEON COLLISIONS AT KEV ENERGIES.   EMISSION OF KLL   4-0830
ENERGY SPECTRA OF ELECTRONS EJECTED IN ARGCN(+)- ARGON COLLISIONS. (AUGER AND COSTER-KRONIG TRANSITIONS)   4-1007
ON COINCIDENCE STUDIES OF HIGHLY IONIZING ARGON(+)- ARGON COLLISIONS. (AUGER EFFECT)   ELECTRON I 2-0529
IMENTS OF K-SHELL VACANCY SHARING IN ASYMMETRIC ICN- ATCM COLLISIONS. (AUGER EFFECT)   GAS TARGET EXPER   2-0523
       MECHANISMS OF ELECTRON PRODUCTION IN ION- ATOM COLLISIONS. (AUGER EFFECT)   2-0488
     PRODUCTION OF INNER SHELL VACANCIES IN HEAVY ION- ATOM COLLISIONS. (AUGER EFFECT)   2-0405
       MECHANISMS OF ELECTRON PRODUCTION IN ION- ATOM COLLISIONS. (AUGER EFFECT, REVIEWS)   2-0490
ZATION AND FLUORESCENCE YIELD IN 50 MEV CHLORINE(+)- NEON COLLISIONS. (AUGER ELECTRON PRODUCTION) /ELL IONI   2-0518
                            ION- ATOM COLLISIONS. (AUGER EMISSION, REVIEW)   1-0169
CULAR EFFECTS ON K-VAANCY SHARING IN ASYMMETRIC ION- ATCM COLLISIONS. (AUGER SPECTRA)   MOLE 2-0522
FOR K X-RAY AND AUGER ELECTRON ENERGY SHIFTS IN HEAVY ICN COLLISIONS. COMMENTS.   SIMPLE MODEL   2-0301
LOW ELECTRON COINCIDENCE STUDY CF VIOLENT ARGCN(+)- ARGON COLLISIONS. COMMENTS. (AUGER ELECTRON ENERGY) / S   2-0423
/DISTRIBUTION OF ELECTRONS RESULTING FROM DIRECT IONIZING COLLISIONS OF ELECTRONS WITH ATOMS. (AUGER PROCE/   2-0525
K-SHELL VACANCY SHARING IN 300 TO 500 KEV NECN(+)- SODIUM COLLISIONS. (QUASIMOLECULE AUGER SPECTRA)   2-0582
ISTRIBUTION OF (AUGER) ELECTRONS RESULTING FROM ION- ATOM COLLISIONS WITH A FIXED IMPACT PARAMETER. /ERGY D   2-0331
EAM GAS AUGER ELECTRCN SPECTRA FOR OXYGEN IONS EXCITED BY COLLISIONS WITH HELIUM, NEON, AND ARGON. /UTION B   4-0904
IN COPPER AND ALUMINUM DOPED ZINC SULFIDE. (AUGER EFFECT, COLOR CENTERS) /CE OF GREEN- COPPER LUMINESCENCE   2-0402
                       AUGER EMISSION AND COLOR CENTERS IN LITHIUM FLUORIDE.   2-0430
        LOW ENERGY ION BACKSCATTERING SPECTRCSCOPY WITH A COMMERCIAL AUGER CYLINDRICAL MIRROR ANALYZER.   3-0665
/ ADHESION, FRICTION AND AUGER SPECTROSCOPY ANALYSIS OF A COMMERCIAL COBALT BASE AIRCRAFT TURBINE SHROUD AL   6-2090
ST.   AUGER SPECTRCSCOPIC STUDY OF THE PCISONING OF A COMMERCIAL PALLADIUM ALUMINA HYDROGENATION CATALY   5-1177
OSCOPY.   NEW COMPACT ENERGY ANALYZER FOR AUGER AND ESCA SPECTR   3-0702
ELD ENERGY ANALYZER. (AUGER ELECTRONS)   COMPACT, INEXPENSIVE HIGH RESOLUTION RETARDING FI   3-0619
CTROLUMINESCENT DIODES. (AUG/ EMISSION SPECTRA OF SILICON COMPENSATED ALUMINUM(X) GALLIUM(1-X) ARSENIDE ELE   2-0325
)   SITE COMPETITION IN SURFACE SEGREGATION. (AUGER EFFECT   6-2074
SPECTROMETRY (AND AUGER ELECTRON SPECTROSCOPY): A USEFUL COMPLEMENT TO OTHER DEPTH PROFILING TECHNIQUES. /   1-0009
S. (AUGER)   SURFACE ANALYSIS USING COMPLEMENTARY ELECTRONIC AND CHEMICAL MEASUREMENT   5-1390
ATION OF SECONDARY ELE/ IN-PROCESS CONTROL TECHNIQUES FOR COMPLEX SEMICONDUCTOR STRUCTURES. TASK-II. APPLIC   5-1411
GER)   RECOMBINATION IN DONOR ACCEPTOR COMPLEXES IN CADMIUM SULFIDE SINGLE CRYSTALS. (AU   2-0369
         WIDTHS OF 4S AND 5S MULTIPLET COMPONENTS IN THE RARE EARTHS.   4-0970
ER SPECTROSCOPY)   RECORDING THE COMPONENTS ON THE SURFACE OF A BINARY ALLOY. (AUG   5-1358
ER SPECTRA)   RECORDING OF COMPONENTS ON THE SURFACE OF FUSED EMITTERS. (AUG   4-0736
UE/ RECRYSTALLIZATION OF TUNGSTEN FIBERS IN NICKEL MATRIX COMPOSITES. (AUGER ELECTRON SPECTROSCOPIC TECHNIQ   6-2046
R ELECTRON EMISSION AS A TOOL FOR THE ANALYSIS OF SURFACE COMPOSITION.   CHARACTERISTIC AUGE   1-0170
CHEMICAL ANALYSIS OF SEMICONDUCTORS. (IMPURITIES AND COMPOSITION)   5-1671
FILMS OBSERVED BY X-RAY PHOTOEMISSION SPECTROSC/ SURFACE COMPOSITION AND CHEMISTRY OF EVAPORATED PERMALLOY   5-1738
/DROGEN DESORBED FROM NICKEL SURFACES: EFFECTS OF SURFACE COMPOSITION AND CRYSTAL ORIENTATION. (SURFACE CO/   5-1197
ATINUM- TIN ALLOYS FROM COMBINED X-RAY PHOTOELEC/ SURFACE COMPOSITION AND DEPTH CONCENTRATION PROFILE OF PL   5-1195
ARATUS ION PROBE TECHNIQUE. (COMPARISON BETWEEN / SURFACE COMPOSITION AND STRUCTURE ANALYSIS WITH A ONE APP   5-1166
NUM SINGLE CRYSTALS STU/ EFFECT CF ION BOMBARDMENT ON THE COMPOSITION AND STRUCTURE OF TUNGSTEN AND MOLYBDE   5-1215
E AUGER ELECTRON SPECTROSCOPY WITH APPLICATICN TO SURFACE COMPOSITION CHANGES. /ANNEL MONITOR FOR REPETITIV   3-0695
NG ARGON BOMBARD/ EFFECT OF TARGET TEMPERATURE ON SURFACE COMPOSITION CHANGES OF COPPER- NICKEL ALLOYS DURI   5-1808
         WORK FUNCTION VARIATION WITH ALLOY COMPOSITION: COPPER- GOLD. (AUGER SPECTROSCOPY)   5-1330
       BEAM ANALYSIS METHOD OF FILM AND SURFACE COMPOSITION DETERMINATION.   5-1810
/GATING THE CATALYTIC ACTIVITY OF ALLOY SYSTEMS. (SURFACE COMPOSITION, INTERDIFFUSION, AUGER ELECTRON SPEC/   6-2003
N AUGER SPECTROSC/ QUANTITATIVE ESTIMATES OF THE CHEMICAL COMPOSITION NEAR THE SURFACE BASED ON THE ELECTRO   6-1645
Y.   QUANTITATIVE STUDY OF THE SURFACE COMPOSITION OF BINARY ALLOYS BY AUGER SPECTROSCOP   6-2005
AND AES MEASUREMENTS.   COMPOSITION OF BINARY ALLOYS BY SIMULTANEOUS SIMS   6-2090
/ RELATION BETWEEN CATALYTIC ACTIVITY PATTERN AND SURFACE COMPOSITION OF COPPER- NICKEL ALLOY. APPLICATION   5-1875
TRA IN THE LOWER ENERGY REGION AT/ OBSERVATICN OF SURFACE COMPOSITION OF COPPER- NICKEL ALLOY BY AUGER SPEC   5-1982
ING AND SPUTTERI/ AUGER SPECTROSCOPY STUDY ON THE SURFACE COMPOSITION OF COPPER- NICKEL ALLOYS AFTER ANNEAL   5-1675
D ADSORBATE COVERED SURFACE/ DETERMINATION OF THE SURFACE COMPOSITION OF COPPER- NICKEL ALLOYS FOR CLEAN AN   5-1424
USE OF AUGER ELECTRON SPECTROSCOPY TO DETERMINE THE COMPOSITION CF FRACTURE SURFACES OF MINERALS.   5-1307
K GROWTH IN HY/ A TECHNIQUE FOR DETERMINING THE ELEMENTAL COMPOSITION OF FRACTURE SURFACES PRODUCED BY CRAC   6-2142
PY.   DETERMINATION OF THE COMPOSITION OF GLASS SURFACES BY AUGER SPECTROSCO   5-1365
LECTRON SPECTROSCOPY, AND LEED STUDY OF THE STRUCTURE AND COMPOSITION OF ION BOMBARDED TUNGSTEN SURFACES. /   5-1951
LEED, AES, AND XPS RESULTS.   STRUCTURE AND COMPOSITION OF METAL SURFACES: A REVIEW OF RECENT   5-1341
                     SURFACE COMPOSITION OF MICA SUBSTRATES.   5-1742
LECTRON SPECTROSCOPY.   STUDY OF THE CHEMICAL COMPOSITION OF MOS AND MNOS STRUCTURES BY AUGER E   5-1503
                     SURFACE COMPOSITION OF NICKEL- GOLD ALLOYS.   5-1966
OELECTRON AND AUGER SPECTROSCOPY TO DETERMINE THE SURFACE COMPOSITION OF OXIDIZED VANADIUM. / OF X-RAY PHOT   5-1205
OELECTRON AND AUGER SPECTROSCOPY TO DETERMINE THE SURFACE COMPOSITION OF OXIDIZED VANADIUM. (COMMENTS) /HOT   5-1863
VER CATALYSTS BY AUGER ELECTRON SPECTROSCOPY.   SURFACE COMPOSITION OF PALLADIUM- GOLD AND PALLADIUM- SIL   5-1976
YSTS USING AUGER ELECTRON S/ DETERMINATION OF THE SURFACE COMPOSITION OF PALLADIUM- NICKEL ALLOY FILM CATAL   5-1846
LIBRIUM.   SURFACE COMPOSITION OF (PALLADIUM- SILVER) ALLOYS IN EQUI   5-1788
LOYS BY AUGER ELECTRON SPECTROSC/ ANALYSIS OF THE SURFACE COMPOSITION OF PASSIVE FILMS ON COPPER- NICKEL AL   5-1874
/ AUGER ELECTRON SPECTROSCOPY TO THE DETERMINATION OF THE COMPOSITION OF PASSIVE FILMS ON TYPE 316 STAINLE/   6-2080
ER ELECTRON AND MOSSBAUER) SPECTROSCOPICAL STUDIES OF THE COMPOSITION OF PASSIVE STATE FILMS.   (AUG   5-1309
PECTROSCOPY.   COMPOSITION OF PLATINUM ANODE SURFACES BY AUGER S   5-1504
SPECTRA)   SURFACE COMPOSITION OF PLATINUM HEATED IN OXYGEN. (AUGER   5-1737
AND IN THE PRES/ AUGER SPECTROSCOPIC STUDY OF THE SURFACE COMPOSITION OF PLATINUM- TIN IN ULTRAHIGH VACUUM   5-1196
LECTRON SPECTROSCOPY.   SURFACE COMPOSITION OF PROMOTED IRON CATALYSTS BY AUGER E   5-1817

AES STUDIES OF SURFACE COMPOSITION OF SILVER- COPPER ALLOYS. 5-1199
E AUGER A/ AUGER ELECTRON SPECTROSCOPY STUDIES OF SURFACE COMPOSITION OF SILVER- COPPER ALLOYS. QUANTITATIV 5-1200
E AUGER A/ AUGER ELECTRON SPECTROSCOPY STUDIES OF SURFACE COMPOSITION OF SILVER- COPPER ALLOYS. QUANTITATIV 5-1740
AUGER ELECTRON SPECTROSCOPY STUDY OF THE SURFACE COMPOSITION OF THE LEAD- INDIUM SYSTEM. 5-1172
LECTRON SPECTROSCOPY. SURFACE COMPOSITION OF THE SILVER- GOLD SYSTEM BY AUGER E 5-1710
XIDE CATALYSTS STUDIED BY AUGER ELECTRON SPECTROSCOPY. COMPOSITION OF THE SURFACE OF COPPER- MAGNESIUM O 5-1361
L BASED ON AUGER ELECTRON SPECTRAL DATA. CHEMICAL COMPOSITION OF THE SURFACE OF MOLYBDENUM AND STEE 5-1985
AES ION SPUTTERING ANALYSIS AND THE SURFACE COMPOSITION OF TITANIUM(IV) OXIDE. 5-1893
DIUM/ AUGER ELECTRON SPECTROSCOPY OF THE SURFACE CHEMICAL COMPOSITION OF VANADIUM OXIDES, AND OXIDIZED VANA 4-1089
NDUCTORS. SURFACE STRUCTURE AND COMPOSITION OF VARIOUS ELEMENTAL AND III-V SEMICO 5-1161
ROMIUM- ANTIMO/ EFFECT OF PRIOR AUSTENITIC GRAIN BOUNDARY COMPOSITION ON TEMPER BRITTLENESS IN A NICKEL- CH 6-2114
AUGER ELECTRON SPECTROSCOPY APPLIED TO SURFACE COMPOSITION PROBLEMS. 5-1812
PPER ANALYZED BY AUGER ELECTRON SPECTROSCOPY AND SECONDA/ COMPOSITION PROFILE OF ION PLATED GOLD FILM ON CO 5-1679
OF SILICIDES FORMED IN NICKEL- PLATINUM ALLOY FILMS. COMPOSITION PROFILES AND SCHOTTKY BARRIER HEIGHTS 5-1900
D PLATINUM AND PLATINUM SILICIDE BY AUGER ELECTRON SPECT/ COMPOSITION PROFILES OF CHEMICALLY VAPOR DEPOSITE 5-1656
CLEANED SURFACES OF GOLD THICK FILMS ON COPPER PLATES BY/ COMPOSITION PROFILES OF SEVERAL CONTAMINATED AND 5-1550
UE. DECONVOLUTION METHOD FOR COMPOSITION PROFILING BY AUGER SPUTTERING TECHNIQ 3-0630
AND AUGER ANALYSIS. RECENT ADVANCES IN COMPOSITION PROFILING BY SIMULTANEOUS SPUTTERING 1-0155
ROSCOPY) TECHNIQUES FOR ELEMENTAL COMPOSITION PROFILING IN THIN FILMS. (AUGER SPECT 5-1273
SCOPY) WORK FUNCTION VARIATION WITH ALLOY COMPOSITION SILVER- GOLD. (AUGER ELECTRON SPECTRO 4-0828
OF MONOCRYSTALLINE IRON(0.84) CHROMIUM(0.16). (A/ SURFACE COMPOSITION STUDIES OF THE (001) AND (110) FACES 5-1591
UGER ELECTRON SPECTROSCOPY OF AIR OXIDIZED STAINLESS STE/ COMPOSITION VERSUS DEPTH PROFILES OBTAINED WITH A 6-1998
ANALYSIS OF SURFACE COMPOSITION WITH LOW ENERGY BACKSCATTERED IONS. 5-1823
(AUGER SPECTROSCOPY) THIN FILM COMPOSITIONAL ANALYSIS. COMPARISON OF TECHNIQUES. 1-0054
N OF TECHNIQUES. (REVIEW) SURFACE AND THIN FILM COMPOSITIONAL ANALYSIS. DESCRIPTION AND COMPARISO 1-0053
NY THE THERMAL ANNEALING OF (100) SURFACES OF GALLIUM AR/ COMPOSITIONAL AND STRUCTURAL CHANGES THAT ACCOMPA 5-1163
BSERVED BY AUGER ELECTRON SPECTROSCOPY. SURFACE COMPOSITIONAL CHANGES WITH ELECTRON BOMBARDMENT O 5-1891
OF THE THERMAL AND ANODIC OXIDE/ CORRELATION BETWEEN THE COMPOSITIONAL PROFILE AND ELECTRICAL CONDUCTIVITY 5-1971
ACE REGION OF GLASS. AES COMPOSITIONAL PROFILES OF MOBILE IONS IN THE SURF 5-1722
CROSCOPE. (AUGER ELECTRON SPECTROSCOPY, SURFACE ELEMENTAL COMPOSITON) /MINATION IN THE SCANNING ELECTRON MI 3-0596
ON THE COMPOUND PHOTOELECTRIC EFFECT. 2-0249
ION SELECTIVE SPUTTERING OF III-V COMPOUND SEMICONDUCTOR SURFACES. (AUGER EFFECT) 5-1637
SCOPY. STUDIES ON SURFACE PLASMA OXIDIZED FILMS OF III-V COMPOUND SEMICONDUCTORS BY AUGER ELECTRON SPECTRO 5-1699
R ELECTRON SPECTROSCOPY ANALYSIS OF VANADIUM AND VANADIUM COMPOUND SURFACES. AUGE 5-1862
AUGER ELECTRON SPECTROSCOPY OF INSULATING SILICON COMPOUNDS. 4-0795
ER ELECTRON SPECTROSCOPY STUDY OF SOME OXYGENATED SILICON COMPOUNDS. AUG 4-0793
PHONON ASSISTED AUGER PROCESSES IN GROUP III-V COMPOUNDS. 2-0484
TOEMISSION AND LMM AUGER DECAY IN ARSENIC AND SOME OF ITS COMPOUNDS. RELAXATION DURING PHO 4-0763
ERENCES FROM AUGER AND CORE BINDING ENERGIES OF GERMANIUM COMPOUNDS. / ESTIMATION OF RELAXATION ENERGY DIFF 2-0471
NCE L SPECTRA OF NIOBIUM IN NIOBIUM METAL AND SOME OF ITS COMPOUNDS. X-RAY FLUORESCE 4-0762
AUGER SPECTRA OF SOME FREE ELEMENTS AND BINARY COMPOUNDS BOMBARDED BY 60 KEV ARGON(+) IONS. 4-1106
(INTERBAND AUGER/ EXCITATION SPECTRUM OF OXYGEN DOMINATED COMPOUNDS IN THE 3-21 EV REGION OF THE SPECTRUM. 4-1048
SPECTROSCOPY) LIQUID PHASE EPITAXY OF III-V COMPOUNDS ON INDIUM PHOSPHIDE SUBSTRATES. (AUGER 5-1590
COPIC MET/ SURFACE OF THERMIONIC EMITTERS MADE OF CARBIDE COMPOUNDS STUDIED BY AN ELECTRONIC AUGER SPECTROS 5-1562
/COPIC INVESTIGATIONS OF CHEMICAL SHIFTS IN SOME VANADIUM COMPOUNDS. (VANADIUM OXIDE, NITRIDE, CARBIDE, SI/ 4-1090
MANIUM. PHOTOELECTRON CHEMICAL SHIFTS IN COORDINATION COMPOUNDS. X-RAY EXCITED LMM AUGER SPECTRA OF GER 4-0924
DIGITAL METHODS FOR THIN FILM ANALYSIS USING A COMPUTER COUPLED AUGER SPECTROMETER. 3-0640
OR A SURFACE CONTAINING SEVERAL KINDS OF ATOMS. COMPUTER PROGRAM TO CALCULATE AUGER TRANSITIONS F 1-0039
LABORATORY COMPUTER SCANNING ELECTRON MICROSCOPE SYSTEM. 3-0651
R SURFACE ANALYSIS, GALLIUM(X) ALUMINUM(1-X) ARSENIDE/ COMPUTER-CONTROLLED MOLECULAR BEAM EPITAXY. (AUGE 5-1324
BAND-TO-BAND AUGER PROCESSES AT HIGH CARRIER CONCENTRATION. 2-0483
GALLIUM ARSENIDE. (AUGER SPECTROSCOPY) SURFACE CARRIER CONCENTRATION AND AS EVAPORATION IN HEAT TREATED 5-1915
OF COPPER- TIN ALLOY SYSTEM DUE TO FRICTION. (AUGER SPEC/ CONCENTRATION CHANGE OF TIN IN THE SURFACE LAYER 5-1867
CENCE IN COPPER AND ALUMINUM DOPED ZINC SULFIDE. (AUGER / CONCENTRATION DEPENDENCE OF GREEN- COPPER LUMINES 2-0402
S AND AES. EVALUATION OF CONCENTRATION DEPTH PROFILES BY SPUTTERING IN SIM 5-1442
GER ELECTRON SPECTROSCOPY. SURFACE CONCENTRATION OF MAGNESIUM IN 6061 ALUMINUM BY AU 5-1824
STAINLESS STEEL BY AUGER ELECTRON SPECTROSCOPY. SURFACE CONCENTRATION OF MOLYBDENUM IN TYPES 316 AND 304 6-1995
ON SECONDARY E/ ELECTRON BEAM INDUCED REDUCTION OF CARBON CONCENTRATION ON BERYLLIUM OXIDE AND ITS EFFECTS 5-1362
E DURING HEATING ACCORDING TO AUGER E/ DECREASE OF CARBON CONCENTRATION ON THE SURFACE OF MOLYBDENUM CARBID 5-1759
M COMBINED X-RAY PHOTOELEC/ SURFACE COMPOSITION AND DEPTH CONCENTRATION PROFILE OF PLATINUM- TIN ALLOYS FRO 5-1195
INTENSITIES VERSUS DEPTH: A MEANS FOR DETERMINING ADATOM CONCENTRATION PROFILES. AUGER PLASMON SATELLITE 2-0347
MEASURED USING AUGER SPECTROSCOPY. PHOSPHORUS CONCENTRATION PROFILES IN P-DOPED SILICON DIOXIDE 5-1231
TIMES IN MERCURY CADMIUM TELLURIDE AT HIGH EXCESS CARRIER CONCENTRATIONS. AUGER LIMITED CARRIER LIFE 2-0262
N SPECTROSCOPY. IMPURITY CONCENTRATIONS ON METAL SURFACES BY AUGER ELECTRO 5-1923
ZINC SULFIDE PHOSPHORS. (AUGER RECOMBINATIONS) CONCENTRATION-DEPENDENT NONRADIATIVE PROCESSES IN 2-0388
HIGH ENERGY RESOLUTION AUGER ELECTRON SPECTROMETER USING CONCENTRIC HEMISPHERES. 3-0592
IMPROVED AUGER ELECTRON SPECTROMETER USING CONCENTRIC HEMISPHERES. 3-0591
00) CLEAVAGE SURFACES AND ON AMORPHOUS C/ MEASUREMENTS OF CONDENSATION COEFFICIENTS OF GOLD ON ROCK SALT (1 5-1135
SURFACES. ADSORPTION AND CONDENSATION OF COPPER ON TUNGSTEN SINGLE CRYSTAL 5-1160
RYSTAL SURFACES. LEED AND AES ANALYSIS. CONDENSATION OF GOLD ONTO TANTALUM (100) SINGLE C 5-1312
QUANTITATIVE AUGER ELECTRON SPECTROSCOPY OF GOLD CONDENSED ON ROCK SALT AND ON AMORPHOUS CARBON. 5-1136
FOCUSING OF CHARGED PARTICLES BY A SPHERICAL CONDENSER. 3-0684
ISSION SPECTRA. GRAPHITE CONDUCTION BAND STATES FROM SECONDARY ELECTRON EM 2-0568
ION SATELLITE (AUGER) TRANSITIO/ POSSIBLE LOCALISATION OF CONDUCTION ELECTRONS IN SOFT X-RAY DOUBLE IONIZAT 2-0437
(AUGER ANALYSIS) CHEMISORPTION, PHOTODESORPTION AND CONDUCTIVITY MEASUREMENTS ON ZINC OXIDE SURFACES. 5-1801
/ELATION BETWEEN THE COMPOSITIONAL PROFILE AND ELECTRICAL CONDUCTIVITY OF THE THERMAL AND ANODIC OXIDES OF/ 5-1971
ER SPECTRUM OF METHANE, HYDROGEN FLUORIDE, WATER, AND CA/ CONFIGURATION INTERACTION CALCULATIONS OF THE AUG 4-0879
ECTRUM OF NEON. EFFECTS OF THE CONFIGURATION INTERACTION ON THE K-SHELL AUGER SP 2-0279
INTERFACES IN GRAY CAST IRON. (SCANNING AUGER MICROPROB/ CONFIRMATION OF IMPURITY ADSORPTION AT FLAKE IRON 6-2054
LOYS. (AUGER ELECTRON SPECTROSCOPY) CONFIRMATION OF SULFUR EMBRITTLEMENT IN NICKEL AL 6-2053
IN MU MESIC ATOMS. PARITY NONCONSERVATION IN RADIATIONLESS (AUGER) TRANSITIONS 2-0364
BASED ON LEED DISPLAY OPTICS. RESOLUTION AND SENSITIVITY CONSIDERATIONS OF AN AUGER ELECTRON SPECTROMETER 3-0705
LINDRICAL MIRROR SPECTROMETER FOR THE ANALYSIS OF SURFACE CONSTITUTENTS AND SOME RESULTS. AUGER CY 3-0682
R TYPE OF AUGER SPECTROMETER. CONSTRUCTION AND OPERATION OF A CYLINDRICAL MIRRO 3-0588
(AES) ULTRAHIGH VACUUM TECHNIQUES IN THE MEASUREMENT OF CONTACT ANGLES. PART-4: WATER ON GRAPHITE (0001). 5-1795
METALLIC TRANSFER BETWEEN METALS IN SLIDING CONTACT EXAMINED BY AUGER EMISSION SPECTROSCOPY. 5-1729
LYSIS TECHNIQUES FOR EXAMINATION OF SILVER- CADMIUM OXIDE CONTACT MATERIALS. /Y AND NONDISPERSIVE X-RAY ANA 5-1799
METHODS FOR SURFACE ANALYSIS OF PRECIOUS METALS CONTACT MATERIALS. 5-1785
WORK FUNCTION CHANGES ON CONTACT MATERIALS. (AUGER ELECTRON SPECTROSCOPY) 5-1395
Y) SURFACE CHEMISTRY OF CONTACT MATERIALS. (BY AUGER ELECTRON SPECTROSCOP 5-1394

/40 ATOMIC PERCENT, AND PALLADIUM- GOLD 70 ATOMIC PERCENT CONTACT METAL ALLOYS. (BY AUGER ELECTRON SPECTRO/ 6-2041
CES: ELECTRIC LOW ENERGY ELECTRON DIFFRACTION, AUGER, AND CONTACT POTENTIAL MEASUREMENTS. /INUM (111) SURFA 5-1744
ACE. (CHEMISORPTION AND CATALYTIC REACTIONS, LEED, AUGER, CONTACT POTENTIAL METHOD AND MASS SPECTROMETRY) / 5-1321
INGLE CRYSTAL SURFACES.   LEED, AUGER SPECTROSCOPY, AND CONTACT POTENTIAL STUDIES OF COPPER- GOLD ALLCY S 5-1748
ILICON- GERMANIUM A/ EFFECTS OF INTERFACIAL OXIDES ON THE CONTACT RESISTANCE OF TUNGSTEN ON HEAVILY DOPED S 5-1456
DETECTION OF THIN CONTAMINATED LAYERS ON CONTACT SURFACES. (BY AUGER SPECTROSCOPY) 5-1165
AUGER ANALYSIS OF FILMS FORMED ON METALS IN SLIDING CONTACT WITH HALOGENATED POLYMERS. 5-1728
OPY/ ADHESION AND FRICTION OF POLYTETRAFLUORO ETHYLENE IN CONTACT WITH METALS AS STUDIED BY AUGER SPECTROSC 5-1198
RIC AND AUGER ELECTRON SPECTROSCOPIC TECHNIQUES FCR METAL CONTACTS.   COMBINED MASS SPECTROMET 6-2042
/N INTERFACIAL OXIDES AND RESISTANCE CHANGES FOR TUNGSTEN CONTACTS ON HEAVILY DOPED SILICON- GERMANIUM. (A/ 5-1861
.   ANALYSIS OF GLASS CONTAINER COATINGS BY AUGER ELECTRON SPECTROSCOPY 5-1704
CON (111).   EFFECTS OF PREADSORBED OXYGEN CONTAMINANT ON THE ADSORPTION OF ALUMINUM ON SILI 5-1169
W OXYGEN PRESSURES. (AUGER ELECTRON SPECTROSCCPY, SURFACE CONTAMINANTS) /C-80A ALLOY AT 800 DEGREES C IN LO 6-2144
APPARATUS FOR THE DETECTION AND IDENTIFICATION OF SURFACE CONTAMINANTS.   USE OF LEED 5-1959
R FABRICATION. (AUGER SPECTRA)   METHODS OF CLEANING CONTAMINANTS FROM QUARTZ SURFACES DURING RESONATO 5-1410
ROLE OF SURFACE CONTAMINANTS IN ELECTRON DEVICE TECHNOLOGY. 5-1758
AUGER EXAMINATION OF CONTAMINANTS IN THIN FILM METALLIZATIONS. 5-1890
ON SPECTROSCOPY.   DETECTION OF ORGANIC CONTAMINANTS ON ALUMINUM SURFACES BY AUGER ELECTR 5-1487
S. LEED- AUGER METHOD.   CONTAMINANTS ON CHEMICALLY ETCHED SILICON SURFACE 5-1229
/PY IN DETERMINING THE EFFECT OF CARBON AND OTHER SURFACE CONTAMINANTS ON GALLIUM ARSENIDE- CESIUM- OXYGEN/ 5-1919
WORK FUNCTION MEASUREMENTS.   DETECTION OF CONTAMINANTS ON POLYCRYSTALLINE SILVER BY AES AND 5-1574
AUGER SPECTROSCOPY OF OXYGEN AND OTHER "CONTAMINANTS" ON SILVER. 5-1757
ILMS ON COPPER PLATES BY/ COMPOSITION PROFILES OF SEVERAL CONTAMINATED AND CLEANED SURFACES OF GOLD THICK F 5-1550
AUGER ELECTRON SPECTROSCOPY OF CONTAMINATED GALLIUM ARSENIDE SURFACES. 5-1920
R SPECTROSCOPY)   DETECTION OF THIN CONTAMINATED LAYERS ON CONTACT SURFACES. (BY AUGE 5-1165
IONIZATION SPECTROSCOPY OF CONTAMINATED METAL SURFACES. 5-1351
S. PART-1./ AUGER ELECTRON EMISSION FROM CLEAN AND CARBON CONTAMINATION) /ASTABLE ATOMS. PART-3: ENERGY AND 2-0543
ANGULAR DISTRIBUTICNS OF THE EJECTED ELECTRCNS. (SURFACE CONTAMINATION EFFECT ON THE GROWTH MCDE OF IRON E 5-1133
PITAXIAL FI/ AUGER AND LOSS SPECTROSCOPY STUDY OF SURFACE CONTAMINATION IN SEMICONDUCTOR DEVICES. 5-1524
AUGER ELECTRON SPECTROSCOPY FOR THE STUDY OF SURFACE CONTAMINATION IN THE SCANNING ELECTRON MICROSCOPE 5-1978
. (AUGER ELECTRON SPECTROSCOPY, SURFACE ELEMENTAL COMPOS/ CONTAMINATION IN THE SCANNING ELECTRON MICROSCOPE 3-0596
UGER ELECTRON SPECTROMEIER AS TOCL FOR SURFACE ANALYSIS. (CONTAMINATION MONITOR)   A 3-0706
SECONDARY EMISSION AND CONTAMINATION OF METAL SURFACES. 6-2071
CARBON CONTAMINATION OF SILICON (111) SURFACES. 5-1237
CT)   DETERMINATION OF SULFUR CONTAMINATION ON MOLYBDENUM SURFACES. (AUGER EFFE 5-1647
AUGER ELECTRON SPECTRCSCOPY OF CLEANUP-RELATED CONTAMINATION ON SILICON SURFACES. 5-1983
OLYCRYSTALLINE MOLYBDENUM- CARBON MONOXIDE SYSTEM/ CARBON CONTAMINATION PRODUCED BY ELECTRON BOMBARDMENT. P 5-1648
N. (LEED- AU/ AN AC RETARDING POTENTIAL TECHNIQUE FOR THE CONTINUOUS MEASUREMENT OF CHANGES IN WORK FUNCTIO 3-0663
ABSORPTION BANDS IN SECONDARY ELECTRON CONTINUOUS SPECTRUM. 4-0933
AUGER ELECTRON SPECTROSCOPY OF CONTRAST FORMING LAYERS ON METALS. 6-2047
OR SILVER AND ALUMINUM.   CONTRASTING VALENCE BAND AUGER ELECTRON SPECTRA F 4-1022
OF GAMMA EMITTING RADIONUCLIDES: CHROMATOGRAPHIC QUALITY CONTROL OF RADIOPHARMACEUTICALS. /R THE DETECTION 1-0217
SEMIAUTOMATED DATA RECORDING AND CONTROL SYSTEM FOR AN ELECTRON ENERGY ANALYZER. 3-0680
CTURES. TASK-II. APPLICATION OF SECONDARY ELE/ IN-PROCESS CONTROL TECHNIQUES FOR COMPLEX SEMICONDUCTOR STRU 5-1411
TION OF AUGER AND ESCA TEST PROGRAMS.   DIGITAL CONTROL UNIT IN TTL LOGIC FOR THE AUTOMATIC EXECU 3-0722
0) FACE / INTERACTION OF LOW ENERGY ATMOSPHERIC IONS WITH CONTROLLED SURFACES. (AUGER NEUTRALIZATION AT (10 5-1736
AUGER ELECTRON SPECTROSCOPY.   SPUTTERING MEASUREMENTS ON CONTROLLED THERMONUCLEAR REACTOR MATERIALS USING 5-1826
ELECTRON BEAM CURRENT CONTROLLER FOR LEED- AUGER STUDIES. 3-0589
LM ANALYSIS AS APPLIED TO VARIOUS ASPECTS IN SOLAR ENERGY CONVERSION.   AUGER THIN FI 5-1960
/AUER SPECTROSCOPY OF IRON-57 AND TIN-119 BY DETECTION OF CONVERSION AND AUGER ELECTRONS. APPLICATION TO S/ 5-1796
/ AND GOLD ATOMIC CLOUDS REARRANGEMENT FOLLOWING INTERNAL CONVERSION AND ELECTRON CAPTURE. (K-AUGER TRANSI/ 2-0334
FROM CURIUM-241 (ELECTRON CAPTU/ PROTON STATES, ANOMALOUS CONVERSION, AND K-AUGER SPECTRA IN AMERICIUM-241 4-1019
RIATED, HIGH EFFICIENCY FHOTOGRAPHIC PLATES FOR RECORDING CONVERSION AND/OR AUGER ELECTRON SPECTRA. NEW, ST 3-0699
(AUGER) ANALYSIS OF CHROMATE CONVERSION COATINGS ON ALUMINUM. 6-2125
MITTING RADIONUCLIDES: CHRCMATOGRAPHIC / USE OF AUGER AND CONVERSION ELECTRONS FOR THE DETECTION OF GAMMA E 1-0217
BACKSCATTER MOSSBAUER STUDIES USING CONVERSION ELECTRONS. TIN-119. 2-0576
)   PENETRATICN EFFECTS IN CONVERTED MUONIC TRANSITIONS. (AUGER COEFFICIENTS 2-0478
DOUBLY RESONANT CURRENT-TO-VOLTAGE CONVERTER PREAMPLIFIER FOR SURFACE SPECTROSCOPY. 3-0677
PECTRA OF GERMANIUM.   PHOTOELECTRON CHEMICAL SHIFTS IN COORDINATION COMPOUNDS. X-RAY EXCITED LMM AUGER S 4-0924
PIC ANALYSIS OF BISMUTH SEGREGATED TO GRAIN BOUNDARIES IN COPPER.   AUGER SPECTROSCO 6-2060
Y SATELLITE STRUCTURE IN THE L3,M4,5M4,5 AUGER SPECTRA OF COPPER.   AUGER VACANC 4-1034
H ENERGY SATELLITES OF THE M2,3VV AND M1VV AUGER PEAKS OF COPPER.   HIG 4-1043
HIGH RESOLUTION AUGER ELECTRON SPECTROSCOPY OF METALLIC COPPER. 4-1052
RY ELECTRON ENERGY DISTRIBUTIONS OF ALUMINUM, NICKEL, AND COPPER. /D PHOTOELECTRON STRUCTURE IN THE SECONDA 4-1023
UREMENTS OF THE L3M2,3M4,5 AUGER TRANSITONS IN NICKEL AND COPPER.   HIGH RESOLUTION MEAS 4-1024
SECONDARY ELECTRON EMISSION ROCKING CURVE FROM COPPER. 4-0764
STUDY OF SULFUR EMBRITTLED OXYGEN FREE COPPER. (AES, GRAIN BOUNDARY SEGREGATION) 6-2015
RAZING. (MONEL, AUGE/ SURFACE CHARACTERIZATION OF NICKEL- COPPER ALLOY DISPLAYING POOR WETTABILITY DURING B 5-1573
AES STUDIES OF SURFACE COMPOSITION CF SILVER- COPPER ALLOYS. 5-1199
Y STUDIES OF THE (100) SURFACES OF ALUMINUM AND ALUMINUM- COPPER ALLOYS. /DIFFRACTION AND AUGER SPECTROSCOP 5-1755
SURFACE ENRICHMENT IN COPPER(3) GOLD AND GOLD(3) COPPER ALLOYS. 5-1934
/N SPECTROSCOPY STUDIES OF SURFACE COMPOSITION OF SILVER- COPPER ALLOYS. QUANTITATIVE AUGER ANALYSIS OF CO/ 5-1200
/N SPECTROSCOPY STUDIES OF SURFACE COMPOSITION OF SILVER- COPPER ALLOYS. QUANTITATIVE AUGER ANALYSIS OF CO/ 5-1740
D SECONDA/ COMPOSITION PROFILE OF ION PLATED GOLD FILM ON COPPER ANALYZED BY AUGER ELECTRON SPECTROSCOPY AN 5-1679
N SPECTRA OF THE VALENCE BANDS OF MAGNESIUM, MAGNESIUM(2) COPPER AND MAGNESIUM(3) GOLD. / AND AUGER ELECTRO 4-0838
UGER SPECTROSCOPY BETWEEN ADSORBED SULFUR AND SULFIDES OF COPPER AND NICKEL.   DISCRIMINATION BY A 5-1734
MVV AUGER SPECTRA OF COPPER AND NICKEL. 4-0766
N EQUILIB/ DIFFUSION OF SILVER THROUGH EPITAXIAL FILMS CF COPPER AND NICKEL ON SILVER (111): FORMATION OF A 5-1356
MANGANESE- MOLYBDENUM- NICKEL LOW ALLOY STEEL/ EFFECTS OF COPPER AND PHOSPHORUS ON TEMPER EMBRITTLEMENT OF 6-2043
EL/ DIFFERENTIAL CROSS SECTION FOR K-SHELL IONIZATION OF COPPER AND SILVER BY ELECTRON BOMBARDMENT. (AUGER 2-0477
RON / SPUTTERING OF THE ALLOY SYSTEMS SILVER- GOLD, GOLD- COPPER, AND SILVER- COPPER STUDIED BY AUGER ELECT 5-1331
NG INTERPRETATION OF HIGH RESOLUTION LMM AUGER SPECTRA OF COPPER AND ZINC.   L-S COUPLI 4-1122
QUASIATOMIC LMM AUGER SPECTRA OF SOLID COPPER AND ZINC. 4-1121
L2,3M4,5M4,5 AUGER SPECTRA OF METALLIC COPPER AND ZINC: THEORY AND EXPERIMENT. 4-0919
SPIN SPLITTING IN COPPER AUGER ELECTRON SPECTRUM. 4-1053
TIONS OF THE SOFT X-RAY L3 EMISSION SPECTRA OF NICKEL AND COPPER. (AUGER-TYPE PROCESSES)   CLUSTER CALCULA 4-0888
OXYGEN NITROGEN(2), AND / SECONDARY ELECTRON YIELDS FROM COPPER BERYLLIUM OXYGEN SURFACE BY THERMAL CARBON 4-0786
ELECTRON SPECTROSCOPIC STUDY OF THE SURFACE POISONING OF COPPER CATALYSTS.   AUGER 5-1176
METASTABLE GRAIN BOUNDARY SEGREGATION IN COPPER CONTAINING TELLURIUM. (AUGER SPECTROSCOPY) 6-2086
ESCA (AND AUGER) STUDIES CF COPPER, COPPER(I) OXIDE, AND COPPER(II) OXIDE. 5-1794

PPER(I)/ X-RAY INDUCED PHOTOELECTRON AND AUGER SPECTRA OF COPPER, COPPER(II) OXIDE, COPPER(I) OXIDE, AND CO 4-0932
DROGEN DESORBED FROM THE (100), (110), AND (111) FACES OF COPPER CRYSTALS. (AUGER ELECTRON SPECTROSCOPY) /Y 5-1155
GROWTH AND STRUCTURE OF THIN COPPER FILMS ON (001) SURFACES OF NICKEL. 5-1227
PUTTER REMOVAL OF MOLYBDENUM FROM VARIOUS METAL SURFACES (COPPER, GOLD, ALUMINUM, TUNGSTEN). /OSITION AND S 5-1878
EAR LIGAND FIELDS INFLUENCING AUGER SPECTRA CF FLUORIDES, COPPER(I) AND SILVER(I). LIN 4-0905
/OELECTRON AND AUGER SPECTRA OF COPPER, CCPPER(II) OXIDE, COPPER(I) OXIDE, AND COPPER(I) SULFIDE THIN FILM/ 4-0932
ESCA (AND AUGER) STUDIES CF COPPER, COPPER(I) OXIDE, AND COPPER(II) OXIDE. 5-1794
ESCA (AND AUGER) STUDIES CF COPPER, COPPER(I) OXIDE, AND COPPER(II) OXIDE. 5-1794
X-RAY INDUCED PHOTOELECTRON AND AUGER SPECTRA OF COPPER, COPPER(II) OXIDE, COPPER(I) OXIDE, AND COPPER(I)/ 4-0932
/TRANSITIONS IN TWO DIMENSIONAL ABSORBED SULFUR LAYERS ON COPPER INVESTIGATED BY LEED AND AUGER SPECTROSCO/ 5-1302
DSORBED CHLORINE AND OXYGEN ON SHEAR STRENGTH OF IRON AND COPPER JUNCTIONS. (SPECTROSCOPY) EFFECT OF A 6-2143
ZINC SULFIDE. (AUGER / CONCENTRATION DEPENDENCE OF GREEN- COPPER LUMINESCENCE IN COPPER AND ALUMINUM DOPED 2-0402
CENTERED CUBIC METAL SURFACES. (GOLD, SILVER, PALLADIUM, COPPER, NICKEL) /ER ELECTRON SPECTROSCOPY OF FACE 5-1717
X-RAY EXCITED LMM AUGER SPECTRA CF COPPER, NICKEL AND IRON. 4-1123
/ DISTRIBUTIONS OF HYDROGEN DESORBED FRCM IRCN, PLATINUM, COPPER, NIOBIUM, STAINLESS STEEL SURFACES. (AUGE/ 6-2006
N WELL CHARACTERIZED SINGLE CRYSTAL SURFACES. (SILVER AND COPPER ON TUNGSTEN /DARY ION MASS SPECTROSCOPY O 5-1689
ADSORPTION AND CONDENSATION OF COPPER ON TUNGSTEN SINGLE CRYSTAL SURFACES. 5-1160
TAL STUDIED BY LEED, AUGER ELECTRON / EPITAXIAL GROWTH CF COPPER ON (110) SURFACE OF A TUNGSTEN SINGLE CRYS 5-1660
OBSERVATIONS OF THE EPITAXIAL GROWTH CF NICKEL AND COPPER ON (111) SILVER. (BY AUGER SPECTROSCOPY) 5-1354
/CONTAMINATED AND CLEANED SURFACES OF GOLD THICK FILMS CN COPPER PLATES BY AUGER ELECTRON AND SECONDARY IO/ 5-1550
) CHARACTERISTIC GAIN PHENOMENA IN PCLYCRYSTALLINE COPPER, SILVER, AND ALUMINUM. (AUGER SPECTROSCOPY 5-1724
ANGE OF LOW ENERGY HELIUM IONS AND ATOMS SCATTERED FROM A COPPER SINGLE CRYSTAL. (AUGER NEUTRALIZATION) /CH 2-0545
IONS (LESS THAN 10 KEV) SCATTERED FROM A (100) FACE OF A COPPER SINGLE CRYSTAL. (AUGER NEUTRALIZATION) /+) 2-0546
ELECTRON EMISSION FROM PASSAGE OF LITHIUM(+) IONS THROUGH CCPPER SINGLE CRYSTALS. ANISOTROPY OF SECONDARY 2-0245
OPY) ADSORPTION CF KRYPTON ON THE (111) FACE OF COPPER SINGLE CRYSTALS. (AUGER ELECTRON SPECTROSC 5-1221
ME METALS AND ALLOYS. (AUGER EMISSION ANALYSIS, TUNGSTEN, COPPER, STEELS) /AIN BOUNDARY EMBRITTLEMENT IN SO 6-2058
OPY AND THEIR INFLUENCE ON SU/ DETECTION OF IMPURITIES ON COPPER SURFACES BY LEED- AUGER ELECTRON SPECTROSC 5-1185
ICKEL ALLOY BY AUGER E/ OBSERVATION OF THE SEGREGATION OF COPPER TO THE SURFACE OF A CLEAN ANNEALED 50-50 N 6-2044
IN THE ENERGY SPECTRUM OF SECONDARY ELECTRONS EMITTED BY COPPER UNDER BOMBARDMENT BY RARE GAS IONS. /TRONS 2-0551
(AUGER) ISOMORPHISM AND PURITY OF COPPER VAPOR DEPOSITED ON A COPPER (111) SURFACE. 5-1171
INTERACTION OF HYDROGEN SULFIDE WITH COPPER (001). 5-1516
ADSORPTION OF CARBON MONOXIDE ON COPPER (001). LEED AND AUGER EMISSION STUDIES. 5-1515
POTENTIAL STUDY OF CARBON MONOXIDE AND XENON ADSORBED ON COPPER (100). LEED AND SURFACE 5-1246
LECTRON SPECTROSCOPY STUDY OF THE ADSORPTION OF OXYGEN ON COPPER (100) AND (111) SURFACES. /ACTION- AUGER E 6-2089
/BON MONOXIDE ON NICKEL (100). PART-3: CARBON MONOXIDE ON COPPER (100). (AUGER SPECTRUM OF NICKEL AND COPP/ 5-1911
ANGULAR RESOLVED AUGER EMISSION SPECTRA FROM A CLEAN COPPER (100) SURFACE. 5-1697
ENT OF GOLD (100), GCLD-COVERED (100), AND SILVER-COVERED COPPER (100) SURFACES. ATOMIC ARRANGEM 5-1718
ANGULAR EFFECTS IN AUGER ELECTRON EMISSION FROM COPPER (110). 5-1988
ON SPECTROSCOPY OF A SULFUR- OXYGEN SURFACE REACTION ON A COPPER (110) CRYSTAL. AUGER ELECTR 5-1183
D ELECTRONS AND CHARACTERISTIC LOSS AND GAIN PHENOMENA OF COPPER (111). ENERGY SPECTRUM OF BACK SCATTERE 4-0897
ANGULAR DEPENDENCE OF AUGER ELECTRON EMISSION FRCM COPPER (111) AND (100) SURFACES. 5-1632
LEED AND AUGER INVESTIGATIONS OF COPPER (111) SURFACE. 5-1499
SCATTERING OF LOW ENERGY ELECTRONS FROM A COPPER (111) SURFACE. 5-1621
SURFACE ENRICHMENT IN COPPER(3) GOLD AND GOLD(3) COPPER ALLOYS. 5-1934
/FRACTION AND AUGER INVESTIGATION OF THE (100) SURFACE CF COPPER(3) GOLD. TEMPERATURE DEPENDENCE OF LONG R/ 5-1854
ORDER- DISORDER TRANSFORMATION AT A (100) SURFACE CF COPPER(3) GOLD, THEORY, EXPERIMENT. 5-1855
GER ELECTRON SPECTROSCCPY STUDY OF SURFACE SEGREGATION IN COPPER- ALUMINUM ALLOYS. AU 6-2012
COPY AND LEED STUDY CF EQUILIBRIUM SURFACE SEGREGATION IN CCPPER- ALUMINUM ALLOYS. AUGER SPECTROS 6-2025
/SEGREGATION, ADHESICN, AND FRICTION STUDIES PERFORMED CN COPPER- ALUMINUM, COPPER- TIN, AND IRON- ALUMINU/ 6-2033
SECONDARY AND AUGER ELECTRON SPECTRA OF A COPPER- BERYLLIUM ALLOY. 4-0916
SPECTROMETRY AND AUGER ELE/ STUDY OF SURFACE PROCESSES ON COPPER- BERYLLIUM BY COMBINED SECONDARY ION MASS 5-1213
SEGREGATION OF BISMUTH TO GRAIN BOUNDARIES IN COPPER- BISMUTH ALLOYS. 6-2096
/H OF METALS AND ALLCYS. (SEGREGATION IN STEEL, TUNGSTEN, COPPER- BISMUTH ALLOYS STUDIED BY AUGER SPECTROS/ 6-2120
SECONDARY ELECTRON EMISSION AT SUPERHIGH FREQUENCIES. (COPPER- GERMANIUM JUNCTION). 2-0356
EED, AUGER SPECTROSCOPY, AND CONTACT POTENTIAL STUDIES OF COPPER- GOLD ALLOY SINGLE CRYSTAL SURFACES. L 5-1748
CHARACTERIZATION OF COPPER- GOLD ALLOY SINGLE SURFACES. 5-1747
SEGREGATION AT COPPER- GOLD ALLOY SURFACES. (AUGER SPECTROSCOPY) 5-1631
WORK FUNCTION VARIATION WITH ALLOY COMPOSITION: COPPER- GOLD. (AUGER SPECTROSCOPY) 5-1330
OPY. STUDY OF DIFFUSION IN THIN COPPER- LEAD FILMS USING AUGER ELECTRON SPECTROSC 5-1972
R ELECTRON SPECTROSCOPY. COMPOSITION OF THE SURFACE CF COPPER- MAGNESIUM OXIDE CATALYSTS STUDIED BY AUGE 5-1361
/EN CATALYTIC ACTIVITY PATTERN AND SURFACE COMPOSITION OF COPPER- NICKEL ALLOY. APPLICATION OF AUGER ELEC/ 5-1875
R ENERGY REGION AT/ OBSERVATION OF SURFACE COMPOSITION OF COPPER- NICKEL ALLOY BY AUGER SPECTRA IN THE LOWE 5-1982
TROSCOPY) WORK FUNCTION OF WELL-DEFINED SURFACE OF COPPER- NICKEL ALLOY PLATES. (AUGER ELECTRON SPEC 5-1873
/OF SILVER- COPPER ALLCYS. QUANTITATIVE AUGER ANALYSIS CF COPPER- NICKEL ALLOY SURFACES AFTER ARGON ION BO/ 5-1200
BARDMENT. REPLY TO CCMMEN/ QUANTITATIVE AUGER ANALYSIS CF COPPER- NICKEL ALLOY SURFACES AFTER ARGON ION BOM 5-1702
/OF SILVER- COPPER ALLCYS. QUANTITATIVE AUGER ANALYSIS OF COPPER- NICKEL ALLOY SURFACES AFTER ARGON ION BO/ 5-1740
BARDMENT. QUANTITATIVE AUGER ANALYSIS OF COPPER- NICKEL ALLOY SURFACES AFTER ARGON ION BOM 5-1807
S. AUGER SPECTROSCOPY CN COPPER- NICKEL ALLOY SURFACES RELATED TO CATALYSI 5-1703
AUGER SPECTRA CF COPPER- NICKEL ALLOYS. 4-1028
I/ AUGER SPECTROSCOPY STUDY ON THE SURFACE COMPOSITION CF COPPER- NICKEL ALLOYS AFTER ANNEALING AND SPUTTER 5-1675
/ ANALYSIS OF THE SURFACE COMPOSITION OF PASSIVE FILMS CN COPPER- NICKEL ALLOYS BY AUGER ELECTRON SPECTROSC 5-1874
T OF TARGET TEMPERATURE ON SURFACE COMPOSITION CHANGES OF COPPER- NICKEL ALLOYS DURING ARGON BOMBARDMENT. / 5-1808
ERED SURFACE/ DETERMINATION OF THE SURFACE COMPOSITION OF COPPER- NICKEL ALLOYS FOR CLEAN AND ADSORBATE COV 5-1424
ADSORPTION ON SINGLE CRYSTAL SURFACES CF COPPER- NICKEL ALLOYS. PART-1. 5-1322
SOFT X-RAY APPEARANCE POTENTIAL SPECTROSCCPY STUDIES WITH COPPER- NICKEL AND IRON- NICKEL ALLOYS. (AES) 5-1952
SPEC/ CONCENTRATION CHANGE OF TIN IN THE SURFACE LAYER OF COPPER- TIN ALLOY SYSTEM DUE TO FRICTION. (AUGER 5-1867
/F PHOSPHORUS ON THE FRICTION AND WEAR CHARACTERISTICS OF COPPER- TIN- PHOSPHORUS ALLOYS. (AUGER SPECTROSC/ 5-1868
/OSCOPY STUDY OF SURFACE SEGREGATION IN THE BINARY ALLCYS COPPER- 1 ATOMIC PERCENT INDIUM, COPPER- 2 ATOMI/ 6-2027
/ THERMAL EFFECTS IN EQUILIBRIUM SURFACE SEGREGATION IN A COPPER- 10 ATOMIC PERCENT ALUMINUM ALLOY USING AU 6-2030
ON SPECTROSCOPY STUDY CF INITIAL STAGES OF OXIDATION IN A COPPER- 19.6 ATOMIC PERCENT ALUMINUM ALLOY. /ECTR 6-2028
E. SELF-CONSISTENT ANALYSIS OF CORE BAND-TO-BAND EXCITATION: AUGER FINE STRUCTUR 2-0509
STIMATION OF RELAXATION ENERGY DIFFERENCES FROM AUGER AND CORE BINDING ENERGIES OF GERMANIUM COMPOUNDS. / E 2-0471
AUGER EFFECT AND CORE BROADENING IN LOW-Z ATOMS, IONS, AND METALS. 2-0353
K ATOMS OF TITANIUM, CHROMIUM, AND / DIRECT CCMPARISON CF CORE ELECTRON BINDING ENERGIES OF SURFACE AND BUL 2-0392
PHOTOEMISSION STUDIES OF CORE EXCITON DECAY IN POTASSIUM IODIDE. 2-0422
N VALENCE BAND) FHCTOEMISSION STUDIES OF THE DECAY OF CORE EXCITONS IN ALKALI HALIDES. (AUGER PROCESS I 2-0289
E SPECTROSCOPY) PLASMCN COUPLING TO CORE HOLE EXCITATIONS IN CARBON. (AUGER APPEARANC 2-0291
INTERATOMIC AUGER PROCESS: EFFECTS ON LIFETIMES OF CORE HOLE STATES. 2-0317
TS DEPENDENCE OF OPTICAL EXCITATION AND AUGER DECAY OF 5D CORE HOLES IN BISMUTH(III) TELLURIDE. /RIX ELEMEN 2-0398
ALLOYED WITH Z/ FINAL STATE EFFECTS IN THE PHOTOEMISSION CORE LEVEL AND AUGER SPECTRA OF NICKEL AND NICKEL 5-1806

ELECTRONIC STRUCTURE OF SOLID SURFACES: CORE LEVEL EXCITATION TECHNIQUES. (AUGER EFFECT) 2-0468
EMISSION MEASUREMENTS IN THE 20-70 EV R/ VALENCE BAND AND CORE LEVEL STUDIES OF INDIUM ANTIMONIDE VIA PHOTO 2-0341
CCMPARISON OF MANY BODY EFFECTS IN CORE LEVEL SURFACE SPECTROSCOPIES. 5-1723
S. EFFECTS OF EXTRA CLOSED CORE RELAXATION ON AUGER AND BINDING ENERGY SHIFT 4-0950
11). TIGHT BINDING CALCULATION OF A CORE VALENCE VALENCE AUGER LINE SHAPE: SILICON (1 4-0831
N AUGER YIELD MEASUREMENTS. EXACT CORRECTIONS FOR POTENTIAL MODULATION DISTORTION I 2-0389
R MODULATION AMPLITUDE DISTORTION IN A 4-GRID RETARDING / CORRECTIONS OF AUGER ELECTRON SIGNAL STRENGTHS FO 3-0620
/HEMISORPTION ON (001), (110), AND (111) NICKEL SURFACES. CORRELATED STUDY USING LOW ENERGY ELECTRON DIFFR/ 5-1391
TRONS. DIRECTIONAL CORRELATION BETWEEN PHOTOELECTRONS AND AUGER ELEC 1-0122
ELECTRICAL CONDUCTIVITY OF THE THERMAL AND ANODIC OXIDE/ CORRELATION BETWEEN THE COMPOSITIONAL PROFILE AND 5-1971
GASES. CORRELATION EFFECTS IN THE AUGER SPECTRA OF NOBLE 4-0981
CTRUM OF XENON. SEARCH FOR CORRELATION EFFECTS IN THE M4,5N4,5N4,5 AUGER SPE 4-0872
H INNER SHELL VACANCIES (NEON WITH 1S OR 2S VACANCY). CORRELATION ENERGIES AND AUGER RATES IN ATOMS WIT 2-0312
/Y PERTURBATION THEORY TO THE CALCULATION OF AUGER RATES, CORRELATION ENERGIES, AND PHOTOIONIZATION CROSS / 2-0311
FUNCTION MEASUREMENTS FOR EPITAXIAL GROWTH OF THORIUM / A CORRELATION OF AUGER SPECTROSCOPY, LEED AND WORK 5-1739
STICS ON ZINC OXIDE SURFACES. CORRELATION OF ELECTRONIC, LEED, AND AUGER DIAGNO 5-1584
INATIONS OBTAINED BY PROTON EXCITED X-RAY, AES ANALYSIS / CORRELATION OF FRACTIONAL MONOLAYER OXYGEN DETERM 5-1682
/CHANISMS IN THE PALLADIUM- GOLD THIN FILM SYSTEM AND THE CORRELATION OF RESISTIVITY CHANGES WITH AUGER EL/ 5-1393
NIMONIC-PE16 ALLOY WITH THE SURFACE OXYGEN- CHROMIUM RAT/ CORRELATION OF THE WETTING BEHAVIOR OF SODIUM ON 5-1441
CROSS CORRELATION TECHNIQUES IN AUGER SPECTROSCOPY. 2-0391
/L (110)- C(2X2) SULFUR SURFACE: LINE SHAPE ANALYSIS AND CORRELATION WITH ION NEUTRALIZATION SPECTROSCOP/ 4-1065
TANCE CHANGES FOR TUNGSTEN CONTACTS ON HEAVILY DOPED SIL/ CORRELATIONS BETWEEN INTERFACIAL OXIDES AND RESIS 5-1861
GULAR RESOLUTION SECONDARY ELECTRON SPECTROSCOPY. KIKUCHI CORRELATIONS FOR ARSENIC (00001). HIGH AN 5-1784
KIKUCHI CORRELATIONS IN AUGER ELECTRON SPECTROSCOPY. 5-1783
SIUM TO SCANDIUM. EFFECT OF ELECTRONIC CORRELATIONS ON KLL AUGER LINES IN ELEMENTS MAGNE 4-0800
EFFECT OF ATOMIC CORRELATIONS ON KLL AUGER TRANSITIONS. 4-0801
LECTRONS) NEW SATELLITE BANDS CORRESPONDING TO DOUBLE L IONIZATION. (OF AUGER E 2-0293
UE. AES ANALYSIS OF SODIUM IN A CORRODED BIOGLASS USING A LOW TEMPERATURE TECHNIQ 5-1721
/ATION OF PHYSICAL EXAMINATION TECHNIQUES IN THE STUDY OF CORROSION AND DEPOSITION IN LIQUID METALS. (AUGE/ 6-2050
ELECTRON SPECTROSCOPY) CORROSION AND STRUCTURAL CHANGES IN GLASS. (AUGER 5-1425
AMMONIACAL ENVIRONMENT: FRACTOGRAPHY AND CHEMICA/ STRESS CORROSION CRACKING OF ALPHA BRASS IN A TARNISHING 6-2095
ERNARY AND VARIOUS QUATERNARY ALLOYS. (AES) STRESS CORROSION CRACKING OF AN ALUMINUM- 5% MAGNESIUM T 6-2036
AUGER SPECTROSCOPIC ANALYSIS OF BIOGLASS CORROSION FILMS. 5-1264
AUGER ELECTRONS IN THE STUDY OF THE STRUCTURE AND PITTING CORROSION OF ALUMINUM. /W MICRORADIOGRAPHY USING 6-2066
/RY OF GRAIN BOUNDARIES AND ITS RELATION TO INTERGRANULAR CORROSION OF AUSTENITIC STAINLESS STEEL. (AUGER / 6-2064
GLASS. (AES) AQUEOUS CORROSION OF SODA- SILICA AND SODA- LIME- SILICA 5-1265
NS TO THE STUDIES OF THE STRUCTURE OF ALUMINUM ALLOYS AND CORROSION OF THE ALLOYS. / BETA-RAY AUGER ELECTRO 6-2051
NNING ELECTRON MICROSC/ MASS TRANSPORT OF STAINLESS STEEL CORROSION PRODUCTS IN FLOWING LIQUID SODIUM. (SCA 6-2014
CKEL- IRON- CHROMIUM THIN FILMS. (AES) CORROSION, SURFACE, AND MAGNETIC PROPERTIES OF NI 5-1766
OF ARGON. COSTER-KRONIG AND AUGER SPECTRUM OF THE L1 SHELL 4-0977
W, 477 REFS) THE AUGER, COSTER-KRONIG, AND AUTOIONIZATION EFFECTS. (REVIE 1-0180
AUGER AND COSTER-KRONIG TRANSITION RATES/ M-SHELL AUGER, COSTER-KRONIG, AND RADIATIVE MATRIX ELEMENTS, AND 4-0967
ORESCENCE YIELDS FOR SODIUM- THORIUM. ATOMIC L-SHELL COSTER-KRONIG, AUGER, AND RADIATIVE RATES AND FLU 4-0963
UORESCENCE YIELDS FOR CALCIUM TO THORIUM. ATOMIC M-SHELL COSTER-KRONIG, AUGER, AND RADIATIVE RATES, AND FL 4-0964
UORESCENCE YIELDS FOR Z BETWEEN 38 AND 10/ ATOMIC N-SHELL COSTER-KRONIG, AUGER, AND RADIATIVE RATES, AND FL 4-0969
KRYPTON M-SHELL AUGER AND COSTER-KRONIG ELECTRON ENERGIES. 4-0928
L-SHELL AUGER AND COSTER-KRONIG ELECTRON SPECTRA. 4-0960
M-SHELL AUGER AND COSTER-KRONIG ELECTRON SPECTRA. 4-0965
INFLUENCE OF COSTER-KRONIG PROCESSES ON L-AUGER YIELDS. 4-1098
(XENON, CADMIUM, ZINC) ENHANCEMENT OF COSTER-KRONIG RATES IN SOLID STATE ENVIRONMENTS. 2-0439
K-, L-, AND M-AUGER AND L- COSTER-KRONIG SPECTRA OF PLATINUM. 4-1097
AUGER AND COSTER-KRONIG STUDIES OF THE M-SHELL OF KRYPTON. 4-0979
L1,L2,3M COSTER-KRONIG TRANSITION ENERGIES. 4-0929
TICAL L2 AND L3 SUBSHELL FLUORESCENCE YIELDS AND L2-L3(X) COSTER-KRONIG TRANSITION PROBABILITIES. THEORE 2-0316
X-RAY FLUORESCENCE YIELDS, AUGER, AND COSTER-KRONIG TRANSITION PROBABILITIES. 2-0259
ON THE G/ THEORETICAL L2 SUBSHELL FLUORESCENCE YIELDS AND COSTER-KRONIG TRANSITION PROBABILITIES F23 BASED 2-0313
MIC 2S STATE AND THEORETICAL L1 FLUORESCENCE Y/ AUGER AND COSTER-KRONIG TRANSITION PROBABILITIES TO THE ATO 2-0329
NERGY LEVELS AND SCREENING PA/ SENSITIVITY OF L1,L2,3M4,5 COSTER-KRONIG TRANSITION RATES TO VARIATIONS IN E 2-0303
ECTRONS EJECTED IN ARGON(+)- ARGON COLLISIONS. (AUGER AND COSTER-KRONIG TRANSITIONS) ENERGY SPECTRA OF EL 4-1007
THE L1,L2,3M4,5 COSTER-KRONIG TRANSITIONS. 2-0248
RON SPECTROSCOPY. INVESTIGATION OF LOW-Z COSTER-KRONIG TRANSITIONS BY AUGER AND PHOTOELECT 2-0578
RFACE ATOMS. ROLE OF COSTER-KRONIG TRANSITIONS IN THE IONIZATION OF SU 2-0338
AUGER AND COSTER-KRONIG TRANSITIONS. (REVIEW) 1-0123
L- AND M-SHELL AND SUBSHELL FLUORESCENCE AND COSTER-KRONIG YIELDS. 4-0959
X-RAY SPECTRA, L-SUBSHELL FLUORESCENCE AND COSTER-KRONIG YIELDS IN BISMUTH AND NEPTUNIUM. 4-1112
CENCE YIELDS) ATOMIC FLUORESCENCE YIELDS. (L-SHELL COSTER-KRONIG YIELDS, K-, L-, AND M-SHELL FLUORES 2-0351
ONIZATION OF ATOMS BY MULTIPLY CHARGED IONS. MODEL OF TWO COULOMB CENTERS. AUGER I 2-0407
DIGITAL METHODS FOR THIN FILM ANALYSIS USING A COMPUTER COUPLED AUGER SPECTROMETER. 3-0640
CALCULATION OF AUGER SPECTRA IN J-J COUPLING. 4-0749
ENTS, AND AUGER AND COSTER-KRONIG TRANSITION RATES IN J-J COUPLING. /STER-KRONIG, AND RADIATIVE MATRIX ELEM 4-0967
AUGER TRANSITION RATES IN J-J COUPLING FOR D HOLES AND F ELECTRONS. 2-0444
GER SPECTRA OF COPPER AND ZINC. L-S COUPLING INTERPRETATION OF HIGH RESOLUTION LMM AU 4-1122
AUGER SPECTRA IN THE INTERMEDIATE COUPLING REGION. 4-0887
AUGER TRANSITION RATE IN MIXED COUPLING SCHEME. 2-0342
ER APPEARANCE SPECTROSCOPY) PLASMON COUPLING TO CORE HOLE EXCITATIONS IN CARBON. (AUG 2-0291
ELECTRONIC AUGER SPECTROSCOPY: TEXTBOOK FOR A COURSE OF LECTURES ON PHYSICAL ELECTRONICS. 1-0107
RON SPECTROSCOPY, AND PHOTOEMISSION STUDIES OF THE CESIUM COVERED GALLIUM ARSENIDE (110) SURFACE. /ER ELECT 5-1931
SECONDARY EMISSION STUDIES ON GERMANIUM AND SODIUM- COVERED GERMANIUM. 5-1714
ECTRON DIFFRACT/ OBSERVATION OF SUPERSTRUCTURES ON CARBON COVERED GERMANIUM (100) SURFACE BY HIGH ENERGY EL 5-1805
SCOPY, AND WORK FUNCTION MEASUREMENTS ON CLEAN AND CESIUM COVERED POLAR FACES OF GALLIUM ARSENIDE. /SPECTRO 5-1294
WORK FUNCTION AND STRUCTURAL STUDIES OF ALKALI- COVERED SEMICONDUCTORS. 5-1958
POSITION OF COPPER- NICKEL ALLOYS FOR CLEAN AND ADSORBATE COVERED SURFACES. /TERMINATION OF THE SURFACE COM 5-1424
ED, AUGER, AND WORK FUNCTION STUDIES OF CLEAN AND SODIUM- COVERED SURFACES OF GALLIUM ARSENIDE. LE 5-1240
STABILITY OF CESIUM COVERED ZINC OXIDE SURFACES. 5-1882
SOME PROPERTIES OF CLEAN AND ALKALI METAL COVERED ZINC OXIDE SURFACES. (LEED, AES) 5-1881
TRON SPECTROSCOPY OF CESIUM ADSORBED ON CLEAN AND OXYGEN- COVERED (100) TUNGSTEN. AUGER ELEC 5-1295
/E ELEMENTAL COMPOSITION OF FRACTURE SURFACES PRODUCED BY CRACK GROWTH IN HYDROGEN AND IN WATER VAPOR. (AU/ 6-2142
A THIN DIFFUSED SOLUTE LAYER ON THE NUCLEATION OF FATIGUE CRACKS. (AUGER ELECTRON SPECTROSCOPY) EFFECT OF 6-2037
L ENVIRONMENT: FRACTOGRAPHY AND CHEMICA/ STRESS CORROSION CRACKING OF ALPHA BRASS IN A TARNISHING AMMONIACA 6-2095
VARIOUS QUATERNARY ALLOYS. (AES) STRESS CORROSION CRACKING OF AN ALUMINUM- 5% MAGNESIUM TERNARY AND 6-2036

SUGAR) VACANCY CASCADES.   NATURE OF SECONDARY ELECTRONS CREATED AS THE RESULT OF ELECTRON SHAKE-OFF AND (2-0307
Y AUGER ANALYSIS.   DETECTION OF SURFACE NITROGEN IN HIGH CRITICAL TEMPERATURE NIOBIUM(3) GERMANIUM FILMS B 5-1804
Y.   CROSS CORRELATION TECHNIQUES IN AUGER SPECTROSCOP 2-0391
D SILVER BY ELECTRON BOMBARDMENT. (AUGER EL/ DIFFERENTIAL CROSS SECTION FOR K-SHELL IONIZATION OF COPPER AN 2-0477
/2,3VV AUGER TRANSITION IN MAGNESIUM AND SILICON FROM THE CROSS SECTION MODEL OF GRYZINSKI. (AUGER EFFECT) 4-0765
NEON(+)- ALUMINUM COLLISION.   IONIZATION CROSS SECTION OF THE L2,3 SHELL OF ALUMINUM IN A 2-0267
AUGER ELECTRON SPECTROSCO/ MEASUREMENT OF THE IONIZATION CROSS SECTION OF THE L2,3 SHELL OF CHLORINE USING 4-0987
OF AUGER RATES, CORRELATION ENERGIES, AND PHOTOIONIZATION CROSS SECTIONS. /ATION THEORY TO THE CALCULATION 2-0311
IDS. ESTIMATION OF BACKSCATTERING EFFECTS AND IONIZATION CROSS SECTIONS.   AUGER EMISSION FROM SOL 2-0513
IN QUANTITATIVE AES.   MEASUREMENT OF IONIZATION CROSS SECTIONS AND BACKSCATTERING FACTORS FOR USE 2-0554
ATOMS AND MOLECULES BY ELECTRON IMPACT. (/ X-RAY EMISSION CROSS SECTIONS AND FLUORESCENCE YIELDS FOR LIGHT 2-0528
ONS MODIFIED BY AUGER TRANSITIONS.   X-RAY IONIZATION CROSS SECTIONS, AND IONIZATION EQUILIBRIUM EQUATI 2-0562
AND ELECTRON EXCITED AUGER SPE/ COMPARISON OF IONIZATION CROSS SECTIONS AND OTHER FACTORS IN X-RAY EXCITED 2-0261
RAYS.   DETERMINATION OF AUGER ELECTRON EXCITATION CROSS SECTIONS AND RANGES USING PROTON EXCITED X- 2-0456
OUS MOLECULES BY 50-500 KEV PROTONS.   CROSS SECTIONS FOR INNER SHELL IONIZATION OF GASE 4-1077
TRONS BY ELECTRON IMPACT. (AUGER SPECTROSCOPY)   CROSS SECTIONS FOR IONIZATION OF INNER SHELL ELEC 2-0475
ATOMS BY ELECTRON IMPACT.   DIFFERENTIAL CROSS SECTIONS FOR K-SHELL IONIZATION OF SURFACE 2-0358
00 KEV PROTON- ARGON COLLISIONS.   CROSS SECTIONS FOR L2,3 AUGER EMISSION IN 22 TO 3 4-1040
X), NITROGEN(X), AND NITROGEN(2)(2X) FROM NITROGEN BY EL/ CROSS SECTIONS FOR THE PRODUCTION OF NITROGEN(2) (2-0376
K-SHELL AUGER ELECTRON PRODUCTION CROSS SECTIONS FROM ION BOMBARDMENT.   2-0570
INE)   K-SHELL AUGER ELECTRON PRODUCTION CROSS SECTIONS FROM ION BOMBARDMENT. (NEON, FLUOR 2-0574
DROGEN, OXYGEN, AND FLUORINE.   NEON K-AUGER CROSS SECTIONS FROM 1.5 MEV PER AMU AND ATOMIC HY 2-0571
/NDENCE OF DEPENDENCE OF TOTAL K-SHELL VACANCY PRODUCTION CROSS SECTIONS IN ARGON BY THE IMPACT OF HEAVY I/ 2-0446
L-SHELL IONIZATION CROSS SECTIONS IN AUGER ELECTRON SPECTROSCOPY. 2-0339
IONS.   ARGON K-AUGER CROSS SECTIONS IN COLLISION WITH 30 MEV FLUORINE 2-0323
AND NEON. (AUGER TR/ ABSOLUTE ELECTRON IMPACT IONIZATION CROSS SECTIONS OF ATOMIC NITROGEN, ATOMIC OXYGEN, 2-0363
A NEW METHOD FOR MEASURING ELECTRON IMPACT IONIZATION CROSS SECTIONS OF INNER SHELLS. 4-0848
ELECTRON SPECTROSCOPY.   ELECTRON IMPACT IONIZATION CROSS SECTIONS OF INNER SHELLS MEASURED BY AUGER 2-0553
ER INTENSITIES OF TUNGS/ TOTAL ELECTRON IMPACT IONIZATION CROSS SECTIONS ON K-SHELLS OF SURFACE ATOMS. (AUG 4-0847
/ MEASUREMENTS OF SECONDARY EMISSION COEFFICIENTS USING A CROSSED BEAM METHOD. PART-2: MEASUREMENTS OF SEC/ 2-0340
RAPHIC STUDIES OF THE CLEAVAGE SURFACE OF GRAPHITE SINGLE CRYSTAL.   AUGER ELECTRON EMISSION MICROG 5-1413
PY OF A SULFUR- OXYGEN SURFACE REACTION ON A COPPER (110) CRYSTAL.   AUGER ELECTRON SPECTROSCO 5-1183
OF MAGNESIUM OXIDE LAYERS APPLIED ON A MOLYBDENUM SINGLE CRYSTAL.   STRUCTURE AND SECONDARY EMISSION 5-1191
ERGY HELIUM IONS AND ATOMS SCATTERED FROM A COPPER SINGLE CRYSTAL. (AUGER NEUTRALIZATION) /CHANGE OF LOW EN 2-0545
AN 10 KEV) SCATTERED FROM A (100) FACE OF A COPPER SINGLE CRYSTAL. (AUGER NEUTRALIZATION) /+) IONS (LESS TH 2-0546
SECONDARY ELECTRON EMISSION FROM SODIUM CHLORINE SINGLE CRYSTAL BOMBARDED BY RARE GAS IONS.   4-0759
CTRON DIFFRACTION- AUGER STUDIES.   A CRYSTAL CLEAVAGE DEVICE FOR USE IN LOW ENERGY ELE 3-0637
RUCTURE, AND INTERACTION WITH OXYGEN AND C/ THORIUM (100) CRYSTAL FACE. A STUDY OF ITS CLEANING, SURFACE ST 5-1883
RING, / STUDIES OF GAS- SURFACE INTERACTIONS AT THE (100) CRYSTAL FACE OF PLATINUM BY MOLECULAR BEAM SCATTE 5-1963
UM (100) SURFACE. (A/ LEED INVESTIGATIONS OF XENON SINGLE CRYSTAL FILMS AND THEIR USE IN STUDYING THE IRIDI 5-1479
/ORPTION AT A TITANIUM SURFACE. SEGREGATION FROM THE BULK CRYSTAL INTERACTION WITH CHLORINE GAS. (AUGER SP/ 5-1833
AUGER ELECTRON SPECTROSCOPY. COMPARATIVE STUDY OF SINGLE CRYSTAL MAGNESIUM OXIDE AND OXIDIZED MAGNESIUM BY 5-1495
ELECTRON SIGNALS OF PHOSPHORUS ON SILICON USING A QUARTZ CRYSTAL MICROBALANCE. QUANTITATIVE STUDY OF AUGER 5-1583
CHEMISORPTION ON SINGLE CRYSTAL MOLYBDENUM (112) SURFACES.   5-1303
/FROM NICKEL SURFACES: EFFECTS OF SURFACE COMPOSITION AND CRYSTAL ORIENTATION. (SURFACE COMPOSITION DETERM/ 5-1197
RACTION AND AUGER INVESTIGATIONS.   SULFUR ON SINGLE CRYSTAL PLATINUM PLANES. LOW ENERGY ELECTRON DIFF 5-1420
SMALL MOLECULE REACTIONS ON STEPPED SINGLE CRYSTAL PLATINUM SURFACES.   5-1173
GULAR DEPENDENCE OF AUGER ELECTRON EMISSION FROM A SINGLE CRYSTAL SPECIMEN.   AN 4-0958
MEASUREMENT OF EXCITED ELECTRO/ DESIGN OF AN X-RAY DOUBLE CRYSTAL SPECTROMETER INSTALLED IN VACUUM AND THE 3-0641
/L GROWTH OF COPPER ON (110) SURFACE OF A TUNGSTEN SINGLE CRYSTAL STUDIED BY LEED, AUGER ELECTRON AND WORK/ 5-1660
CARBON MONOXIDE OXIDATION ON A PLATINUM (110) SINGLE CRYSTAL SURFACE.   5-1187
(3) GOLD. TEMPERATURE DEPENDENCE OF LONG RANGE ORDER AT A CRYSTAL SURFACE. / OF THE (100) SURFACE OF COPPER 5-1854
ADSORPTION AND CONDENSATION OF COPPER ON TUNGSTEN SINGLE CRYSTAL SURFACES.   5-1160
CTRON EMISSION DURING OXIDATION OF NICKEL (100) AND (111) CRYSTAL SURFACES. /MICAL CHANGES IN SECONDARY ELE 5-1453
ON, AND FLASH DESORPTION SPECTROSCOPY OF METALS ON SINGLE CRYSTAL SURFACES. /TRON DIFFRACTION, AUGER ELECTR 1-0008
CTION EFFECTS IN BACKSCATTERING AND AUGER PRODUCTION NEAR CRYSTAL SURFACES.   DIFFRA 2-0243
ND CONTACT POTENTIAL STUDIES OF COPPER- GOLD ALLOY SINGLE CRYSTAL SURFACES.   LEED, AUGER SPECTROSCOPY, A 5-1748
RACTION- AUGER ELECTRON SPECTROSCOPY STUDY OF IRON SINGLE CRYSTAL SURFACES.   LOW ENERGY ELECTRON DIFF 5-1866
SURFACE STRUCTURE OF AND CATALYSIS BY PLATINUM SINGLE CRYSTAL SURFACES.   5-1837
ADSORPTION OF HYDROGEN ON NICKEL SINGLE CRYSTAL SURFACES. (AUGER ELECTRON SPECTROSCOPY) 5-1257
/ STUDY OF THE INTERACTION OF OXYGEN WITH PLATINUM SINGLE CRYSTAL SURFACES BY LOW ENERGY ELECTRON DIFFRACTI 5-1223
ON AND AUGER ELECTRON SPECTROSCOPY.   STUDIES OF CRYSTAL SURFACES BY LOW ENERGY ELECTRON DIFFRACTI 5-1836
CONDENSATION OF GOLD ONTO TANTALUM (100) SINGLE CRYSTAL SURFACES. LEED AND AES ANALYSIS.   5-1212
/ LOW INDEX ((111) AND (100)) AND STEPPED PLATINUM SINGLE CRYSTAL SURFACES. (LEED, AUGER ELECTRON SPECTROS/ 5-1838
GROWTH OBSERVED BY LEED AND AES.   CRYSTAL SURFACES OF ALKALI HALIDES AND THIN FILM 5-1907
.   ADSORPTION ON SINGLE CRYSTAL SURFACES OF COPPER- NICKEL ALLOYS. PART-1 5-1322
)   EQUILIBRATION OF HYDROGEN AND DEUTERIUM ON SINGLE CRYSTAL SURFACES OF PLATINUM. (AUGER SPECTROSCOPY 5-1594
/ETICS OF THE REACTION OF OXYGEN WITH CLEAN NICKEL SINGLE CRYSTAL SURFACES. PART-1: NICKEL (100) SURFACE. / 5-1452
/NDARY ION MASS SPECTROSCOPY ON WELL CHARACTERIZED SINGLE CRYSTAL SURFACES. (SILVER AND COPPER ON TUNGSTEN) 5-1689
NITORED BY AUGER SPECTRO/ SPUTTER CLEANING AND ETCHING OF CRYSTAL SURFACES (TITANIUM, TUNGSTEN, SILICON) MO 5-1828
/SMON STUDIES ON ALUMINUM (111) AND ALUMINUM (001) SINGLE CRYSTAL SURFACES USING INELASTIC LOW ENERGY ELEC/ 5-1745
/NALYSIS OF BACKSCATTERED ELECTRONS FROM A SILICON SINGLE CRYSTAL UNDER A LOW ENERGY BOMBARDMENT. (AUGER S/ 4-0739
AUGER ELECTRON SPECTROSCOPY OF) PRECISION SINGLE SIDEBAND CRYSTAL UNITS.   (5-1409
ION FROM PASSAGE OF LITHIUM(+) IONS THROUGH COPPER SINGLE CRYSTALS.   ANISOTROPY OF SECONDARY ELECTRON EMISS 2-0245
AUGER ELECTRON FINE PROFILES IN IONIC CRYSTALS.   4-0949
F INELASTICALLY SCATTERED AND AUGER ELECTRONS FROM SINGLE CRYSTALS.   ENERGY SPECTRA O 4-1064
IZATION OF INERT GAS IONS ON THE SURFACE OF ALKALI HALIDE CRYSTALS. /TION OF THE FREQUENCY OF AUGER NEUTRAL 5-1544
ELECTRON SPECTROSCOPY ON 5.59% SILICON- IRON (100) SINGLE CRYSTALS. / WORK FUNCTION MEASUREMENTS AND AUGER 5-1928
Y OF 5.59 ATOMIC PERCENT SILICON- IRON ALLOY (100) SINGLE CRYSTALS. /UNCTION AND AUGER ELECTRON SPECTROSCOP 5-1925
LE OF AUGER EFFECT IN THERMOSTIMULATED PHENOMENA IN IONIC CRYSTALS.   RO 2-0539
ARY ELECTRON EMISSION FROM TUNGSTEN AND MOLYBDENUM SINGLE CRYSTALS.   SECOND 4-0742
SECONDARY ELECTRON EMISSION OF MOLYBDENUM CRYSTALS.   4-1063
SECONDARY ELECTRON EMISSION OF SILICON DIOXIDE SINGLE CRYSTALS.   4-1062
SERVED BY LOW ENERGY ELECTRON DI/ STRUCTURES OF SUBSTRATE CRYSTALS AND EPITAXIAL GROWTH OF METAL ON THEM OB 5-1906
/PROPERTIES AND AUGER CHARACTERISTICS OF MAGNESIUM SINGLE CRYSTALS AND MAGNESIUM OXIDE AND MAGNESIUM NITRI/ 5-1220
/PENDENCE OF THE EDGE EMISSION OF GALLIUM SELENIDE SINGLE CRYSTALS AT HIGH EXCITATION LEVELS. (AUGER RECOM/ 2-0258
ION IN DONOR ACCEPTOR COMPLEXES IN CADMIUM SULFIDE SINGLE CRYSTALS. (AUGER)   RECOMBINAT 2-0369
ADSORPTION OF KRYPTON ON THE (111) FACE OF COPPER SINGLE CRYSTALS. (AUGER ELECTRON SPECTROSCOPY) 5-1221
DESORBED FROM THE (100), (110), AND (111) FACES OF COPPER CRYSTALS. (AUGER ELECTRON SPECTROSCOPY) /YDROGEN 5-1155
EXCITON SPECTRA OF HEAVILY DOPED CADMIUM SULFIDE SINGLE CRYSTALS. (AUGER PROCESS)   2-0372

CADMIUM SELENIDE AND CADMIUM SULFIDE(0.48) SELENIDE(0.52) CRYSTALS. (AUGER RECOMBINATION) / HIGHLY EXCITED 2-0432
D WORK FUNCTION OF CESIUM ON TUNGSTEN AND TITANIUM SINGLE CRYSTALS IN THE SUBMONOLAYER REGION. /ROSCOPY, AN 5-1829
/OF THE SECONDARY ELECTRON EMISSION COEFFICIENT OF SINGLE CRYSTALS IN THE 2-10 KEV RANGE OF PRIMARY ELECTR/ 4-1061
ANGLE DISTRIBUTION OF SODIUM CHLORIDE SINGLE CRYSTALS. (PHOTO- AND AUGER ELECTRON EMISSION) 2-0297
/POSITION AND STRUCTURE OF TUNGSTEN AND MOLYBDENUM SINGLE CRYSTALS STUDIED BY AUGER ELECTRON SPECTROSCOPY / 5-1215
ROMETRY) MECHANISM OF SURFACE ALIGNMENT IN NEMATIC LIQUID CRYSTALS. (SUBSTRATE SURFACE EFFECTS, AUGER SPECT 5-1281
. / ENERGY SPECTRUM OF ELECTRONS EMITTED BY ALKALI HALIDE CRYSTALS UNDER IMPACT OF ION AND ATOM BOMBARDMENT 2-0246
GER SPECTRA AND LEED PATTERNS FROM VACUUM CLEANED SILICON CRYSTALS WITH CALIBRATED DEPOSITS OF IRON.   AU 5-1769
AUGER SPECTROSCOPY OF GRAPHITE SINGLE CRYSTALS WITH LOW ENERGY ELECTRONS. 4-0741
RODUCTION. (ALUMINUM) CRYSTALLINE EFFECTS IN BACKSCATTERING AND AUGER P 2-0254
AL STRUCTURE./ PROPERTIES OF THE LEAD SULFIDE- LEAD OXIDE CRYSTALLINE FILMS. PART-1: PREPARATION AND PHYSIC 5-1790
ER SPECTROSCOPY STUDIES OF THE OXIDATION OF AMORPHOUS AND CRYSTALLINE GERMANIUM.   AUG 5-1431
.   OXIDATION STUDIES OF AMORPHOUS AND CRYSTALLINE GERMANIUM FILMS BY AUGER SPECTROSCOPY 6-2023
OPY.   STUDY OF AMORPHOUS AND CRYSTALLINE SILICATES BY AUGER ELECTRON SPECTROSC 4-0796
(AUGER ELECTRON SPECTROSCOPY)   AMORPHOUS TO CRYSTALLINE TRANSITION OF SILICON (111) SURFACES. 5-1780
IQUES FOR CHARACTERIZATION OF THE CHEMICAL, PHYSICAL, AND CRYSTALLOGRAPHIC NATURE OF A SURFACE. /ICAL TECHN 5-1433
. (AUGER SPECTROSCOPY, LEED, ION MICROPROBE ANALYSIS) CRYSTALLOGRAPHY AND CHEMISTRY OF CERAMIC SURFACES 5-1852
SPUTTERING MEASUREMENTS ON CTR MATERIALS USING AUGER ELECTRON SPECTROSCOPY 1-0182
OPPER, NICK/ AUGER ELECTRON SPECTROSCOPY OF FACE CENTERED CUBIC METAL SURFACES. (GOLD, SILVER, PALLADIUM, C 5-1717
EPITAXY OF ULTRATHIN METAL FILMS CN BCDY CENTERED CUBIC SUBSTRATES USING LEED- AUGER TECHNIQUES. 5-1488
G ENERGY AND ANGULAR DISTRIBUTIONS OF INELASTICA/ FARADAY CUP LEED APPARATUS WITH FACILITY FOR INVESTIGATIN 3-0691
FUSION. (AND AES)   OXIDATION OF CUPRONICKEL ALLOYS. PART-1: XPS STUDY OF INTERDIF 5-1224
/US CONVERSION, AND K-AUGER SPECTRA IN AMERICIUM-241 FROM CURIUM-241 (ELECTRON CAPTURE) AND BERKELIUM-245 / 4-1019
ECTRON SPECTROSCOPY.   ELECTRONIC SUPERPOSITION OF SAMPLE CURRENT AND SECONDARY ELECTRON IMAGES IN AUGER EL 2-0450
ELECTRON BEAM CURRENT CONTROLLER FOR LEED- AUGER STUDIES. 3-0589
RON SPECTROSCOPY OF SOLIDS.   AUGER CURRENT MEASUREMENTS FOR QUANTITATIVE AUGER ELECT 4-0856
SAMPLES HEATED DIRECTLY BY ELECTRIC CURRENT OBSERVED BY AUGER ELECTRON SPECTROSCOPY. 1-0133
RINSIC INDIUM ANTIMONIDE AT ROOM TE/ EFFECTS OF DIFFUSION CURRENT ON GALVANOMAGNETIC PROPERTIES IN THIN INT 2-0354
/ON SPECTROSCOPY FROM SOLID SURFACES IN ULTRAHIGH VACUUM. CURRENT PROGRESS WITH TECHNIQUES INVOLVING ELECT/ 5-1901
TINUOUS MEASUREMENT OF CHANGES IN WORK FU/ AN ALTERNATING CURRENT RETARDING POTENTIAL TECHNIQUE FOR THE CON 3-0663
DISTORTION OF DIFFERENTIATED DEFLECTION ANALYZER CURRENT. (USED IN AES) 3-0613
ANALOG TECHNIQUE FOR THE MEASUREMENT OF AUGER ELECTRON CURRENTS. APPLICATION TO SULFUR ON NICKEL (110). 3-0635
SPECTROSCOPY AT HIGH SPATIAL RESOLUTION AND PRIMARY BEAM CURRENTS. (CARBON ON SILVER FILM) AUGER ELECTRON 5-1904
CCMPARISON OF MAGNESIUM L2,3MM AUGER CURRENTS USING ELECTRON AND ION EXCITATION. 4-0855
FACE SPECTROSCOPY.   DOUBLY RESONANT CURRENT-TO-VOLTAGE CONVERTER PREAMPLIFIER FOR SUR 3-0677
ECTRON EMISSION SPECTROSCOPY USING RETARDING / AVERAGING, CURVE FITTING, AND RELATED TECHNIQUES IN AUGER EL 3-0610
SECONDARY ELECTRON EMISSION ROCKING CURVE FROM COPPER. 4-0764
ADSORPTION OF CYANOGEN ON PLATINUM (100). 5-1684
ENTIAL SECONDARY ELECTRON YIELD MEASUREMENTS.   USE OF CYLINDRICAL AUGER SPECTROMETERS FOR RETARDING POT 3-0628
RELATIVISTIC PART. CYLINDRICAL CAPACITOR AS AN ANALYZER. PART-1: NON 3-0689
FOCUSING PROPERTIES OF AN ELECTROSTATIC MIRROR WITH CYLINDRICAL FIELD. 3-0721
TROMETER.   AN ENERGY MODULATED (CYLINDRICAL) HIGH ENERGY RESOLUTION ELECTRON SPEC 3-0616
TIVE AUGER ANALYSIS IN THE RANGE 100 TO 1000 EV USING THE CYLINDRICAL MIRROR ANALYZER. /IMATION TO QUANTITA 3-0660
Y ION BACKSCATTERING SPECTROSCOPY WITH A COMMERCIAL AUGER CYLINDRICAL MIRROR ANALYZER.   LOW ENERG 3-0665
QUANTITATIVE ASPECTS OF ION SCATTERING SPECTROSCOPY. (CYLINDRICAL MIRROR ANALYZER) 3-0666
S ON ENERGY SHIFTS AND SIGNAL INTENSITY IN A SINGLE STAGE CYLINDRICAL MIRROR ANALYZER. /MEN POSITION EFFECT 3-0696
OPY. COMBINED ESCA- AUGER SYSTEM BASED ON THE DOUBLE PASS CYLINDRICAL MIRROR ANALYZER. / ELECTRON SPECTROSC 3-0654
/DEPENDENCE OF OVERLAYER ATTENUATION OF ELECTRONS FOR THE CYLINDRICAL MIRROR ANALYZER AND A RETARDING FIEL/ 3-0693
OPY)   RETARDING FIELD CYLINDRICAL MIRROR ANALYZER. (FOR AUGER SPECTROSC 3-0614
TROSCOPY.   DESIGN FOR A CYLINDRICAL MIRROR ANALYZER FOR USE IN AUGER SPEC 3-0594
CTROSCOPY.   TOWARD OPTIMUM UTILIZATION OF THE CYLINDRICAL MIRROR ANALYZER IN AUGER ELECTRON SPE 3-0678
RDATION. (TUNGSTEN AUGER ELECTRON) CYLINDRICAL MIRROR ANALYZER INCORPORATING PRERETA 3-0590
E.   A NEW PHOTOELECTRON SPECTROMETER: CCMBINATION OF CYLINDRICAL MIRROR ANALYZER WITH SOFT X-RAY SOURC 3-0652
COMPARISON OF SPHERICAL DEFLECTOR AND CYLINDRICAL MIRROR ANALYZERS. 3-0623
TION OF SODIUM ON SILICON AND SILICON DIOXIDE SURFACES BY CYLINDRICAL MIRROR AUGER ANALYZER.   DETEC 3-0647
DESIGN PARAMETERS FOR THE CYLINDRICAL MIRROR ENERGY ANALYZER. 3-0687
OF SURFACE CONSTITUENTS AND SOME RESULTS.   AUGER CYLINDRICAL MIRROR SPECTROMETER FOR THE ANALYSIS 3-0682
FINITE ANGULAR SPREAD IN AN ELECTROSTATIC ANALYZER OF THE CYLINDRICAL MIRROR TYPE. (AUGER SPECTROMETER) /F 3-0720
CONSTRUCTION AND OPERATION OF A CYLINDRICAL MIRROR TYPE OF AUGER SPECTROMETER. 3-0588

D

AUGER TRANSITION RATES IN J-J COUPLING FOR D HOLES AND F ELECTRONS. 2-0444
DUCED BY BOMBARDMENT WITH 2 KV ELECTRONS. (AUGER ELECTRON DAMAGE) / (100) SURFACES OF GALLIUM PHOSPHIDE, IN 6-1162
SPECTROSCOPY.   CHARACTERIZING OF ALUMINUM FATIGUE DAMAGE BY PHOTOEMISSION, ELLIPSOMETRY, AND AUGER 6-2116
(DNA)   BIOLOGICAL DAMAGE FROM THE AUGER EFFECT, POSSIBLE BENEFITS. 1-0056
SURFACES. (AUGE/ SURFACE CHARACTERISTICS RELATED TO LASER DAMAGE OF LITHIUM NIOBATE AND POTASSIUM CHLORIDE 5-1746
OF MOLYBDENUM AND STEEL BASED ON AUGER ELECTRON SPECTRAL DATA.   CHEMICAL COMPOSITION OF THE SURFACE 5-1985
RBIDE DURING HEATING ACCORDING TO AUGER ELECTRON SPECTRUM DATA. /CENTRATION ON THE SURFACE OF MOLYBDENUM CA 5-1759
FROM COMBINED X-RAY PHOTOELECTRON AND AUGER SPECTROSCOPIC DATA. /ENTRATION PROFILE OF PLATINUM- TIN ALLOYS 5-1195
EFFECT OF MODULATION AMPLITUDE ON ELECTRON EXCITED AUGER DATA FROM TITANIUM. 2-0367
ENERGY ANALYZER.   SEMIAUTOMATED DATA RECORDING AND CONTROL SYSTEM FOR AN ELECTRON 3-0680
PABILITIES.   DIGITIZING AUGER DATA. THREEFOLD APPROACH TO ENHANCE ANALYTICAL CA 1-0183
NCE OF THE COEFFICIENT OF SEC/ USE OF X-RAY SPECTROSCOPIC DATA TO EXPLAIN THE FINE STRUCTURE OF THE DEPENDE 4-0891
CTURE. (AUGER EFFECT)   SURFACE DEBYE TEMPERATURE OF THE SILICON (001) - 2X2 STRU 5-1477
THE FIRST DISCRETE TRANSITION IN NITROGEN FROM THE AUGER DECAY.   DETERMINATION OF OSCILLATOR STRENGTH FOR 2-0575
ZING THREE- AND FOUR ELECTRON STATES IN BERYLLIUM. (AUGER DECAY) METASTABLE AUTOIONI 2-0295
TELLURIDE. (AUGER RECCMBINATION)   TRANSIENT CARRIER DECAY AND TRANSPORT PROPERTIES IN MERCURY CADMIUM 2-0461
RELAXATION DURING PHCTOEMISSICN AND LMM AUGER DECAY IN ARSENIC AND SOME OF ITS COMPOUNDS. 4-0763
PHCTOEMISSION STUDIES OF CORE EXCITON DECAY IN POTASSIUM IODIDE. 2-0422
PROCESS IN VALENCE BAND)   PHOTOEMISSION STUDIES OF THE DECAY OF CORE EXCITONS IN ALKALI HALIDES. (AUGER 2-0289
US.   PRODUCTION AND DECAY OF DOUBLE L VACANCIES IN ARGON AND PHOSPHOR 2-0489
, AND/ AUGER L-SERIES ELECTRONIC SPECTRA ACCOMPANYING THE DECAY OF ERBIUM-160, YTTERBIUM-166, YTTERBIUM-169 4-0743
AUGER EFFECT".   DECAY OF HOLE STATES IN OXYGEN-16 VIA A "NUCLEAR 2-0336
.   AUGER DECAY OF NEON FOLLOWING ENERGETIC ION BOMBARDMENT 4-0954
Y EXCITED SILICON. (AUGER RECOMBINATION PRO/ SPECTRUM AND DECAY OF THE RECOMBINATION RADIATION FROM STRONGL 2-0460
/TRIX ELEMENTS DEPENDENCE OF OPTICAL EXCITATION AND AUGER DECAY OF 5D CORE HOLES IN BISMUTH(III) TELLURIDE. 2-0398
/ODS OF STANDARDIZATION FOR CESIUM-131 AND MEASUREMENT OF DECAY PARAMETERS. (DETECTION OF L AUGER ELECTRON/ 2-0474
IRON.   INTERFACIAL AND PLASMON DECAY PEAKS IN THE AUGER SPECTRA OF TITANIUM AND 4-1060
OMS BY THE ULTRASOFT X-RAY SPECTROSCOPY TECHNIQUE. (AUGER DECAYS) /ECHANISM OF S IONIZATION OF NOBLE GAS AT 2-0580

117

ROM CURIUM-241 (ELECTRCN CAPTURE) AND BERKELIUM-245 ALPHA DECAYS. /, AND K-AUGER SPECTRA IN AMERICIUM-241 F 4-1019
IUM IONS IN THE VACUUM ULTRAVIOLET SPECTRAL RANGE. (AUGER DECAYS) /FOR SODIUM, POTASSIUM, RUBIDIUM, AND CES 4-0735
OF HIGHLY STRIPPED OXYGEN, FLUORINE, CHLORINE, / ELECTRON DECAY-IN-FLIGHT SPECTRA FROM AUTOIONIZING STATES 2-0472
NIUM FILMS AND RIBBONS. (AUGER SPECTROSCO/ ADSORPTION AND DECOMPOSITION OF ACETYLENE ON POLYCRYSTALLINE RHE 5-1688
L SURFACE. (AUGER ELECTRON SPECTROSCOPY) DECOMPOSITION OF CARBON MONOXIDE ON A (110) NICKE 5-1606
). (AUGER SPECTROMETRY)   ADSORPTION AND DECOMPOSITION OF CARBON MONOXIDE ON PLATINUM (111 5-1622
APHITIZED NIC/ STUDY OF THE KINETICS AND MECHANISM OF THE DECOMPOSITION OF FORMIC ACID ON CARBURIZED AND GR 5-1629
GE OF THE CHLORIDE ION IN SODIUM CHLORIDE BY MEANS OF THE DECOMPOSITION PRODUCTS. /XCITONS NEAR THE L2,3 ED 4-1126
AUGER SPUTTERING TECHNIQUE.   DECONVOLUTION METHOD FOR COMPOSITION PROFILING BY 3-0630
OSCOPY.   DECONVOLUTION TECHNIQUES IN AUGER ELECTRON SPECTR 1-0138
URFACE PROPERTIES OF MICA AFTER CLEAVAGE. (OUTGASSING AND DECORATION STUDIES, AUGER SPECTRA)   CHANGE OF S 5-1743
F MOLYBDENUM CARBIDE DURING HEATING ACCORDING TO AUGER E/ DECREASE OF CARBON CONCENTRATION ON THE SURFACE O 5-1759
CHARGE STATE DEPENDENCE OF THE NEON K FLUORESCENCE YIELD DEDUCED FROM HIGH RESOLUTION EMISSION (AUGER) SP/ 2-0520
MECHANISMS OF INNER SHELL EXCITATION AND DEEXCITATION IN MULTIPLY IONIZED ATOMS. 2-0487
MANY BODY EFFECTS IN AUGER DEEXCITATION OF ATOMS NEAR SOLIDS. 2-0361
(AUGER PROCESS)   RADIATIONLESS DEEXCITATION OF EXCITED HELIUM ATOMS AT SURFACES. 2-0255
FECT)   DEEXCITATION OF MULTIPLY IONIZED ATOMS. (AUGER EF 2-0281
AUGER AND RADIATIVE DEEXCITATION OF MULTIPLY IONIZED NEON. 2-0315
RTICLES IN GALLIUM PHOSPHIDE.   AUGER THEORY AT DEFECTS.   APPLICATION TO STATES WITH TWO BOUND PA 2-0457
NIQUE.   AUGER TRANSITIONS LEADING TO DEFINED FINAL CHARGE STATES BY A COINCIDENCE TECH 2-0371
DISTORTION OF DIFFERENTIATED DEFLECTION ANALYZER CURRENT. (USED IN AES) 3-0613
COMPARISON CF SPHERICAL DEFLECTOR AND CYLINDRICAL MIRROR ANALYZERS. 3-0623
SION. (AUGER ELECTRON SPECTROSCOPY)   SURFACE DEFORMATION CAUSED ON NATURAL MOLYBDENITE BY ABRA 6-2132
ELLIPSOMETRIC STUDY OF PHOSPHORUS SEGREGATION IN OXIDIZED DEGENERATE SILICON.   AUGER AND 5-1250
ER ELECTRON SPECTROSCOPY. DEGRADED PASSIVE FILM ON IRON. APPLICATION OF AUG 5-1762
PY.   ABSOLUTE ATOMIC DENSITIES DETERMINED BY AUGER ELECTRON SPECTROSCO 5-1843
FROM THE KLL AUGER SPECTRUM/ EVALUATION OF THE TRANSITION DENSITY FUNCTION OF CARBON ON VARIOUS SUBSTRATES 5-1651
INTERATOMIC AUGER PROCESSES AND THE DENSITY OF STATES. 2-0491
MEASUREMENTS OF BAND-TO-BAND AUGER RECOMBINATION OF HIGH DENSITY PLASMAS IN GERMANIUM.   PICOSECOND OPTICAL 2-0250
EFF/ RADIATION CHEMICAL RESONANCE EFFECT IN SCLID 5-BROMO DEOXYURIDINE: CHEMICAL CONSEQUENCES OF THE AUGER 2-0377
F GALLIUM PHOSPHIDE, INDUCED BY BOMBARDMENT WITH 2 KV EL/ DEPARTURES FROM STOICHIOMETRY IN (100) SURFACES O 5-1162
Z DEPENDANCE OF THE KLL AUGER RATES. 2-0557
(COMMENTS)   ANGULAR DEPENDENCIES IN ELECTRON EXCITED AUGER EMISSION. 2-0286
ANGULAR DEPENDENCIES OF ELECTRON EXCITED AUGER EMISSION. 2-0378
ISTRIBUTIONS.   A TEMPERATURE DEPENDENT CONTRIBUTION TO AUGER ELECTRON ENERGY D 2-0434
HOLE LUMINESCENCE IN PURE GERMANIUM. (AUG/ MAGNETIC FIELD DEPENDENT INTENSITY OSCILLATIONS OF THE ELECTRON 4-0780
ABSOLUTE PROBABILITIES FCR AUGER- KLL TRANSITIONS DEPENDENT ON THE ATOMIC NUMBER Z=30-94. PART-2. 2-0252
RELATIVE PROBABILITIES FOR AUGER KLL TRANSITIONS DEPENDENT ON THE ATOMIC NUMBER 29 TO 94. 2-0253
STRUCTURE DEPENDENT PROPERTIES OF THIN FILM SUPERCONDUCTORS 5-1300
IDE. (AUGER SPECTROSCOPY) DEPLETION EFFECTS IN SEMIINSULATING GALLIUM ARSEN 5-1950
ERNS FROM VACUUM CLEANED SILICON CRYSTALS WITH CALIBRATED DEPOSITS OF IRON.   AUGER SPECTRA AND LEED PATT 5-1769
AUGER SPECTRA AND LEED PATTERNS FROM NICKEL DEPOSITS ON CLEAVED SILICON. 5-1768
AUGER ELECTRON SPECTROSCCPY OF NICKEL DEPOSITS ON THE SILICON (111) SURFACE. 5-1238
ACE.   LEED- AES OBSERVATION OF GROWTH MECHANISM OF DEPOSITED IRON FILM ON MAGNESIUM OXIDE (001) SURF 5-1525
QUANTITATIVE DETERMINATION OF SURFACE OXIDE THICKNESS ON DEPOSITED METAL FILMS BY COMBINATION AUGER SPECT/ 6-2048
ISOMORPHISM AND PURITY OF COPPER VAPOR DEPOSITED ON A COPPER (111) SURFACE. (AUGER) 5-1171
ELECTRON AND X/ ELECTRONIC STRUCTURE OF A THIN GOLD FILM DEPOSITED ON A SILICON SUBSTRATE STUDIED BY AUGER 5-1435
AUGER ANALYSIS OF SILICON THIN FILMS DEPOSITED ON CARBON AT HIGH TEMPERATURES. 5-1228
FFRACTION- AUGER ELECTRON SPECTRCSCOPY STUDY OF IRON FILM DEPOSITED ON MAGNESIUM OXIDE. /ENERGY ELECTRON DI 5-1522
R SPECTROSCOPY)   GROWTH MODE OF IRON FILM DEPOSITED ON MAGNESIUM OXIDE (001) SURFACE. (AUGE 5-1528
AUGER ELECTRON EMISSICN FROM GOLD DEPOSITED ON SILICON (111) SURFACE. 5-1676
ELECTRON SPECT/ COMPOSITION PROFILES OF CHEMICALLY VAPOR DEPOSITED PLATINUM AND PLATINUM SILICIDE BY AUGER 5-1656
1-X) ARSENIDE.   DEPOSITION AND AUGER ANALYSIS OF DEPOSITED SILICON DIOXIDE ON ALUMINUM(X) GALLIUM(5-1962
UPS, ESCA, AND AUGER SPECTRCSCOPY STUDY OF VACUUM DEPOSITED SILVER HALIDE FILMS. 5-1947
THE SUBSTRATE SURFACE DURING SPUTTERING AND VACUUM VAPOR DEPOSITION. /ECTRON SPECTROSCOPIC OBSERVATIONS OF 5-1678
N DIOXIDE ON ALUMINUM(X) GALLIUM(1-X) ARSENIDE. DEPOSITION AND AUGER ANALYSIS OF DEPOSITED SILICO 5-1962
/ ULTRAHIGH VACUUM LOW TEMPERATURE STAGE FOR IN-SITU FILM DEPOSITION AND CHARACTERIZATION. (LEED- AUGER SYS 3-0657
VARIOUS / AUGER ELECTRON SPECTRCSCOPY STUDIES OF SPUTTER DEPOSITION AND SPUTTER REMOVAL OF MOLYBDENUM FROM 5-1878
OP/ LOW TEMPERATURE EPITAXY OF GERMANIUM FILMS BY SPUTTER DEPOSITION. (HEED, LEED, AUGER ELECTRON SPECTROSC 5-1537
/CAL EXAMINATION TECHNIQUES IN THE STUDY OF CORROSION AND DEPOSITION IN LIQUID METALS. (AUGER SPECTROSCOPY/ 6-2050
SURFACE.   AUGER- LEED INVESTIGATION OF THE DEPOSITION OF ALUMINUM ONTO THE MOLYBDENUM (110) 5-1489
STEMS EXHIBITING LAYER GROWTH, AND ITS APPLICATION TO THE DEPOSITION OF SILVER ON NICKEL. /ECTROSCOPY OF SY 5-0399
G AUGER SPECT/ SOLID PHASE EPITAXIAL STUDIES USING VACUUM DEPOSITION ON HEATED SILICON SUBSTRATES. (SCANNIN 5-1287
TROSCOPY)   AN ORDERED BORON STRUCTURE AFTER DEPOSITION ON SILICON (111). (AUGER ELECTRON SPEC 5-1168
CHARACTERIZATION OF A CHROMIUM- GCLD THIN FILM DEPOSITION PROCESS BY AUGER ANALYSIS. 1-0037
AUGER SPECTROSCOPY OF SUBMCNOLAYER GOLD DEPOSITIONS ON SILICON. 5-1888
ON PROFILES.   AUGER PLASMON SATELLITE INTENSITIES VERSUS DEPTH: A MEANS FOR DETERMINING ADATOM CONCENTRATI 2-0347
BY AUGER SPECTROSCOPY. DEPTH ANALYSIS OF CLEAVED MICA SURFACES MONITORED 5-1842
IOBIUM. / EFFECT OF SURFACE METALLURGY ON THE PENETRATION DEPTH AND RF BREAKDOWN FIELD OF SUPERCONDUCTING N 6-2137
YS FROM COMBINED X-RAY PHOTOELEC/ SURFACE COMPOSITION AND DEPTH CONCENTRATION PROFILE OF PLATINUM- TIN ALLO 5-1195
SCOPY.   PROBING DEPTH IN PHOTOEMISSION AND AUGER ELECTRON SPECTRO 2-0428
ELECTRON ESCAPE DEPTH IN SILICON. (AUGER ENERGIES) 2-0408
Y. (SILICON MONOLAYERS)   IN DEPTH INFORMATION FROM AUGER ELECTRON SPECTROSCOP 1-0130
RFACE.   RELATIVE ESCAPE DEPTH OF AUGER AND PHOTOELECTRONS FROM A SOLID SU 1-0903
/SPECTROMETRY AS A METHOD FOR DETERMINATION OF THE ENERGY DEPTH OF LEVELS OF SURFACE ELECTRON CENTERS. (AU/ 4-0915
EFFECT OF HIGH TEMPERATURE HEAT TREATMENT ON PENETRATION DEPTH OF SUPERCONDUCTING NIOBIUM. (AUGER SPECTRO/ 5-1938
OPY.   EFFECT OF THE ELECTRON EJECTION DEPTH ON QUANTITATIVE ANALYSIS BY AUGER SPECTROSC 2-0449
EEL.   AUGER ELECTRON SPECTROSCOPY AND DEPTH PROFILE STUDY OF OXIDATION MODIFIED 440C ST 6-2029
STEEL.   AUGER ELECTRON SPECTROSCOPY AND DEPTH PROFILE STUDY OF OXIDATION OF MODIFIED 440C 6-2032
EVALUATION OF CONCENTRATION DEPTH PROFILES BY SPUTTERING IN SIMS AND AES. 5-1442
ROSCOPY OF AIR OXIDIZED STAINLESS STE/ COMPOSITION VERSUS DEPTH PROFILES OBTAINED WITH AUGER ELECTRON SPECT 6-1998
/UREMENTS IN ULTRAHIGH VACUUM GROWN AMORPHOUS GERMANIUM. (DEPTH PROFILES OF OXYGEN AND CARBON BY AUGER SPE/ 5-1547
IED 440C STEEL STUDIED BY AUGER ELECTRON SPECTROSCOPY AND DEPTH PROFILING. /ON OF SULFUR DIOXIDE WITH MODIF 6-2031
HERFORD BACKSCATTERING, SCANNING ELECTRON MICROSCOPY, AND DEPTH PROFILING AUGER ELECTRON SPECTROSCOPY. /RUT 5-1902
QUANTITATIVE DEPTH PROFILING BY AUGER ION SPUTTERING TECHNIQUE 1-0085
APPLICATIONS CF DEPTH PROFILING BY AUGER- SPUTTER TECHNIQUES. 1-0087
EASY METHOD TO ACCURATELY ALIGN ICN BOMBARDMENT GUNS FOR DEPTH PROFILING IN AUGER ELECTRON SPECTROSCOPY. 3-0700
SPECTROSCOPY.   FACTORS AFFECTING DEPTH PROFILING MEASUREMENTS USING AUGER ELECTRON 5-1457
SPECTROSCOPY.   ELECTRON BEAM EFFECTS IN DEPTH PROFILING MEASUREMENTS WITH AUGER ELECTRON 5-1128
UGER ELECTRON SPECTRCSCOPY): A USEFUL COMPLEMENT TO OTHER DEPTH PROFILING TECHNIQUES. / SPECTROMETRY (AND A 1-0009

118

ELECTRON ESCAPE	DEPTHS IN ALUMINUM. (AUGER ELECTRONS)	2-0537
ESCAPE	DEPTHS OF X-RAY EXCITED (AUGER) ELECTRONS.	2-0409
AINLESS STEEL SURFACES/ SPATIAL DISTRIBUTIONS OF HYDROGEN	DESORBED FROM IRON, PLATINUM, COPPER, NIOBIUM, ST	6-2006
CO/ MEASUREMENTS OF THE SPATIAL DISTRIBUTION OF HYDROGEN	DESORBED FROM NICKEL SURFACES: EFFECTS OF SURFACE	5-1197
F COPPER CRYSTALS. (AU/ ANGULAR DISTRIBUTIONS OF HYDROGEN	DESORBED FROM THE (100), (110), AND (111) FACES O	5-1155
AUGER ELECTRON SPECTROSCOPY STUDY OF ELECTRON IMPACT	DESORPTION.	2-0452
R ANALYSIS OF SURFACE SPECIES SUBJECT TO ELECTRON INDUCED	DESORPTION. NEW TECHNIQUE FOR AUGE	3-0675
D AND GRAPHITIZED NICKEL (110) USING AES, LEED, AND FLASH	DESORPTION. /POSITION OF FORMIC ACID ON CARBURIZE	5-1629
/E CF PRIMARY AND SECONDARY ELECTRONS IN ELECTRON INDUCED	DESORPTION AND DISSOCIATION: CARBON MONOXIDE ON /	5-1567
AES) FORMIC ACID	DESORPTION FROM GRAPHITIZED NICKEL (110). (LEED-	5-1630
Y LEED AND AES) A STUDY OF PHOSPHORUS ADSORPTION AND	DESORPTION KINETICS ON SILICON (111) SURFACES. (B	5-1288
/ ELECTRON SPECTROSCOPY STUDY OF ELECTRON BEAM STIMULATED	DESORPTION OF OXYGEN FROM A SEMICONDUCTOR SURFAC/	5-1617
/R ELECTRON SPECTROSCOPY STUDY OF THE ELECTRON STIMULATED	DESORPTION OF OXYGEN FROM A TUNGSTEN (100) SURFA/	5-1954
IFFUSION ON MAGNESIUM OXIDE (001) BY/ ELECTRON STIMULATED	DESORPTION OF SODIUM AND MEASUREMENT OF SURFACE D	5-1494
/W ENERGY ELECTRON DIFFRACTION, AUGER ELECTRON, AND FLASH	DESORPTION SPECTROSCOPY OF METALS ON SINGLE CRYS/	1-0008
COMBINED AUGER ELECTRON SPECTROSCOPY AND ELECTRON IMPACT	DESORPTION STUDIES. /CATION OF ION SPECIES IN THE	5-1692
COMBINED AUGER ELECTRON SPECTROSCOPY AND ELECTRON IMPACT	DESORPTION STUDIES OF SILICON SURFACES.	5-1694
COMBINED AUGER ELECTRON SPECTROSCOPY AND ELECTRON IMPACT	DESORPTION STUDIES OF SILICON SURFACES.	5-1695
COMBINED AUGER ELECTRON SPECTROSCOPY AND ELECTRON IMPACT	DESORPTION STUDIES OF THE INTERACTIONS OF GASES /	5-1691
CATALYTIC REAC/ USE OF PHOTOEMISSION, AUGER, AND THERMAL	DESORPTION TECHNIQUES TO STUDY THE MECHANISM OF A	1-0042
LICATION TO CESIUM-COVERED (110) GAL/ KINETICS OF THERMAL	DESORPTION USING AUGER ELECTRON SPECTROSCOPY. APP	5-1293
NCY FOR CARBON DIOXIDE ON TUNGSTEN (100). FLASH	DESORPTION- AUGER ELECTRON SPECTROSCOPY. DISCREPA	5-1461
RA, REVIEW, 6 REFS) ELECTRONS AND X-RAYS IN NONDESTRUCTIVE ANALYSIS. (AUGER, PHOTOELECTRON SPECT		1-0083
SILICON BY AUGER ELECTRON SPECTROSCOPY.	DETECTABILITY LIMITS FOR BORON AND PHOSPHORUS IN	5-1887
ELECTRON BEAM ANALYZER FOR	DETECTING AUGER ELECTRONS.	3-0712
N EMISSION YIELD AND THE APPLICATION TO SILIC/ METHOD FOR	DETECTING FINE STRUCTURE IN THE SECONDARY ELECTRO	4-0852
ELECTRON BEAM MICROANALYZER WITH AUGER ELECTRON	DETECTION.	3-0711
THIN FILM ANALYSIS. (AUGER SPECTROSCOPY) ESTABLISHING	DETECTION AND DETERMINATION LIMITS IN SURFACE AND	5-1572
NEW METHODS FOR THE	DETECTION AND ELIMINATION OF COATING FLAWS.	5-1592
NTS. USE OF LEED APPARATUS FOR THE	DETECTION AND IDENTIFICATION OF SURFACE CONTAMINA	5-1959
VAL BY OZONIZATION OF PHOTORESIST RESIDUES.	DETECTION BY AUGER ELECTRON SPECTROSCOPY AND REMO	5-1450
A NOVEL	DETECTION CIRCUIT FOR AUGER SPECTROSCOPY.	3-0673
(AND) NITROGEN, OXYGEN, AND FLUORINE I/ IN-DEPTH PROFILE	DETECTION LIMITS OF NITROGEN IN GALLIUM PHOSPHIDE	5-1916
TOMS WITH TWO INTERNAL VACANCIES.	DETECTION OF A NEW TYPE OF AUGER TRANSITIONS IN A	2-0234
ER BY AES AND WORK FUNCTION MEASUREMENTS.	DETECTION OF CONTAMINANTS ON POLYCRYSTALLINE SILV	5-1574
ICATION/ MOSSBAUER SPECTROSCOPY CF IRON-57 AND TIN-119 BY	DETECTION OF CONVERSION AND AUGER ELECTRONS. APPL	5-1796
ESTIMATES OF EFFICIENCIES OF PRODUCTION AND	DETECTION OF ELECTRON EXCITED AUGER EMISSION.	2-0288
HLORIDE ION IN SODIUM CHLORIDE BY MEANS OF THE DECOMPOSI/	DETECTION OF EXCITONS NEAR THE L2,3 EDGE OF THE C	4-1126
TOGRAPHIC / USE OF AUGER AND CONVERSION ELECTRONS FOR THE	DETECTION OF GAMMA EMITTING RADIONUCLIDES: CHROMA	1-0217
E BY AUGER- LEED ANALYSIS.	DETECTION OF IMPURITIES ON A SILICON (111) SURFAC	5-1360
D- AUGER ELECTRON SPECTROSCOPY AND THEIR INFLUENCE ON SU/	DETECTION OF IMPURITIES ON COPPER SURFACES BY LEE	5-1185
TION FOR CESIUM-131 AND MEASUREMENT OF DECAY PARAMETERS.	(DETECTION OF L AUGER ELECTRONS) /S OF STANDARDIZA	2-0474
INHIBITED IN LEAD AZELATE SOLUTION.	DETECTION OF LEAD, BY AUGER SPECTROSCOPY, ON IRON	5-1628
FACES BY AUGER ELECTRON SPECTROSCOPY.	DETECTION OF ORGANIC CONTAMINANTS ON ALUMINUM SUR	5-1487
IGH SENSITIVITY APPEARANCE POTENTIAL SPECTROMETER FOR THE	DETECTION OF PHOTONS, ELECTRONS AND IONS. H	3-0643
FACES OF SILICON NITRIDE FILMS BY AUGER ELECTRON EMISSIO/	DETECTION OF SILICON OXYNITRIDE LAYERS ON THE SUR	5-1611
E SURFACES BY CYLINDRICAL MIRROR AUGER ANALYZER.	DETECTION OF SODIUM ON SILICON AND SILICON DIOXID	3-0647
MPERATURE NIOBIUM(3) GERMANIUM FILMS BY AUGER ANALYSIS.	DETECTION OF SURFACE NITROGEN IN HIGH CRITICAL TE	5-1804
SURFACES. (BY AUGER SPECTROSCOPY)	DETECTION OF THIN CONTAMINATED LAYERS ON CONTACT	5-1165
CTRA. LOW NOISE	DETECTION SYSTEM FOR THE MEASUREMENT OF AUGER SPE	3-0662
ER) EVALUATION OF LEAD(0.8) TIN(0.2) TELLURIDE	DETECTOR FABRICATION USING SURFACE ANALYSIS. (AUG	5-1593
GER ELECTRON EMITTER. NOVEL SOURCE FOR GAS CHROMATOGRAPHY	DETECTORS. IRON-55 AS AN AU	3-0608
AES ANALYSIS / CORRELATION OF FRACTIONAL MONOLAYER OXYGEN	DETERMINATIONS OBTAINED BY PROTON EXCITED X-RAY,	5-1682
/ULAR NITROGEN AND METHANE BY 50- TO 600 KEV HYDROGEN(+),	DEUTERIUM(+), MOLECULAR HYDROGEN (+), AND HELIUM(/	2-0519
Y (100) TANTALUM. (/ ADSORPTION AND SOLUTION CF HYDROGEN,	DEUTERIUM, NITROGEN, OXYGEN AND CARBON MONOXIDE B	5-1549
(AUGER SPECTROSCOPY) EQUILIBRATION OF HYDROGEN AND	DEUTERIUM ON SINGLE CRYSTAL SURFACES OF PLATINUM.	5-1594
T. ARGON L-SHELL IONIZATION BY 50 TO 600 KEV PROTON,	DEUTERON, MOLECULAR HYDROGEN, AND HELIUM(+) IMPAC	5-1521
OLLOWING IONIZATION BY EQUAL VELOCITY ALPHA PARTICLES AND	DEUTERONS. AUGER SPECTRA OF CARBON AND ARGON F	4-1109
	DI..., SEE THE PARENT WORD.	
CORRELATION OF ELECTRONIC, LEED, AND AUGER	DIAGNOSTICS ON ZINC OXIDE SURFACES.	5-1584
X-RAY PHOTOELECTRON SPECTRUM OF	DIAMOND. (AUGER TRANSITIONS)	4-0851
/ACTERISTIC K-SHELL LOW ENERGY ELECTRON LOSS SPECTRA FROM	DIAMOND, GRAPHITE, AND EVAPORATED CARBON SURFACE/	4-0945
SURFACE STUDIES OF SOME TRANSITION METAL	DICHALOCOGENIDES.	5-1969
M FLUORIDE) AUGER ANALYSIS OF THREE ENHANCED MULTILAYER	DIELECTRIC MIRROR DESIGNS. (ZINC SELENIDE, THORIU	5-1328
UDIED BY A MICROWAVE METHOD. (PRODUCTION OF AUGER/ METAL-	DIELECTRIC TRANSITION IN GERMANIUM AND SILICON ST	4-0752
QUANTITIES OF ELEMENTS ON METALLIC SURFACES.	DIFFERENCE AUGER SPECTROSCOPY FOR STUDYING SMALL	4-1937
MEASURED BY SOFT X-RAY AND AUGER ELECTRON APPEARANCE POT/	DIFFERENCE IN THE CHROMIUM L3/L2 INTENSITY RATIO	4-0890
N: ELLIPSOMETRY, AUGER SPECTROSCOPY AND SURFACE POTENTIAL	DIFFERENCE STUDIES. CESIUM ADSORPTION ON TUNGSTE	5-1830
/ATION OF GERMANE AND THE ESTIMATION OF RELAXATION ENERGY	DIFFERENCES FROM AUGER AND CORE BINDING ENERGIES/	2-0471
IUM ALLOY ANALYSIS)	DIFFERENTIAL AUGER SPECTROMETRY. (NIOBIUM- ZIRCON	5-1849
OF COPPER AND SILVER BY ELECTRON BOMBARDMENT. (AUGER EL/	DIFFERENTIAL CROSS SECTION FOR K-SHELL IONIZATION	2-0477
N OF SURFACE ATOMS BY ELECTRON IMPACT.	DIFFERENTIAL CROSS SECTIONS FOR K-SHELL IONIZATIO	2-0358
CERMET FILMS AND ITS APPLICATION TO HIGH YIELD SECONDARY/	DIFFERENTIAL SPUTTERING OF MAGNESIUM OXIDE- GOLD	5-1430
IN AES) DISTORTION CF	DIFFERENTIATED DEFLECTION ANALYZER CURRENT. (USED	3-0613
CHARACTERIZATION OF CHEMISORPTION BY LOW ENERGY ELECTRON	DIFFRACTION	1-0161
OF ELECTRONIC SPECTROSCOPY AND LOW ENERGY ELECTRON	DIFFRACTION. / ANALYSIS OF IRON SURFACES BY MEANS	1-0940
CLEAN TELLURIUM SURFACES STUDIED BY LOW ENERGY ELECTRON	DIFFRACTION	5-1134
STUDIED BY AUGER ELECTRON SPECTROSCOPY AND SLOW ELECTRON	DIFFRACTION. /STEN AND MOLYBDENUM SINGLE CRYSTALS	5-1215
NUM (110) SURFACE OBSERVED BY AES AND LOW ENERGY ELECTRON	DIFFRACTION INITIAL OXIDATION OF THE MOLYBDE	5-1650
LOW ENERGY ELECTRON	DIFFRACTION.	1-0131
, SECONDARY ION MASS SPECTROSCOPY AND LOW ENERGY ELECTRON	DIFFRACTION. /CHARACTERIZED BY AUGER SPECTROSCOPY	2-0501
N COVERED GERMANIUM (100) SURFACE BY HIGH ENERGY ELECTRON	DIFFRACTION. /RVATION OF SUPERSTRUCTURES ON CARBO	5-1805
TE OF AUGER ELECTRON SPECTROSCOPY AND LOW ENERGY ELECTRON	DIFFRACTION PRESENT STA	1-0006
BY AUGER EMISSION SPECTROSCOPY AND BY LOW ENERGY ELECTRON	DIFFRACTION / TO SILICON (111) SURFACE OBSERVED	5-1180
OSCOPY, ELLIPSOMETRY, AND REFLECTION HIGH ENERGY ELECTRON	DIFFRACTION. /UTTERING, STUDIED WITH AUGER SPECTR	5-1233
(0001) GRAPHITE. (AUGER SPECTROSCOPY, LOW ENERGY ELECTRON	DIFFRACTION) /NON SUBMONOLAYER FILMS ADSORBED ON	5-1858
ADHESION AT GOLD MICA INTERFACES. (LOW ENERGY ELECTRON	DIFFRACTION AES)	5-1434
SE DOMAINS OF SULFUR ON IRON (001) BY LOW ENERGY ELECTRON	DIFFRACTION AES. / SURFACE STRUCTURE WITH ANTIPHA	5-1470
OF WATER VAPOR ON CLEAVED GERMANIUM. (LOW ENERGY ELECTRON	DIFFRACTION AES) BINDING STATES	5-1820
PTION FROM GRAPHITIZED NICKEL (110). (LOW ENERGY ELECTRON	DIFFRACTION AES) FORMIC ACID DESOR	5-1630

SILICON (100) SURFACES AS OBSERVED BY LOW ENERGY ELECTRON DIFFRACTION AES. GOLD-INDUCED SUPERSTRUCTURES ON   5-1708
I METAL COVERED ZINC OXIDE SURFACES. (LOW ENERGY ELECTRON DIFFRACTION AES) / PROPERTIES OF CLEAN AND ALKAL   5-1881
OF METAL SURFACES: A REVIEW OF RECENT LOW ENERGY ELECTRON DIFFRACTION AES, AND XPS RESULTS. / COMPOSITION   5-1341
OF LITHIUM HYDRIDE.   LOW ENERGY ELECTRON DIFFRACTION AES ELECTRON INDUCED EMISSION STUDIES   4-0807
YING SILICON SURFACES BY MEANS OF THE LOW ENERGY ELECTRON DIFFRACTION AES METHOD.   SURVE   5-1809
F DEPOSITED IRON FILM CN MAGNESIUM O/ LOW ENERGY ELECTRON DIFFRACTION AES OBSERVATION OF GROWTH MECHANISM O   5-1525
SULFUR SEGREGATION FROM THE BULK TO / LOW ENERGY ELECTRON DIFFRACTION AES STUDIES OF CHEMISORPTION INDUCED   5-1865
CTORS AT SURFACES AND IN THIN FILMS. (LOW ENERGY ELECTRON DIFFRACTION AES STUDIES OF NIOBIUM SURFACES) /NDU   5-1848
HROMIUM (100) AND (110).   LOW ENERGY ELECTRON DIFFRACTION AES STUDY OF THE OXIDATION OF IRON- C   5-1587
        ELECTRODES ANC CIRCUITS OF A LOW ENERGY ELECTRON DIFFRACTION AES SYSTEM.   3-0639
ON A SILICON (111) SURFACE BY AUGER- LOW ENERGY ELECTRON DIFFRACTION ANALYSIS.   DETECTION OF IMPURITIES   5-1360
- TIN, AND IRON- ALUMINUM ALLOYS. (BY LOW ENERGY ELECTRON DIFFRACTION AND AES) /ON COPPER- ALUMINUM, COPPER   6-2033
NETICS ON SILICON (111) SURFACES. (BY LOW ENERGY ELECTRON DIFFRACTION AND AES) /DSORPTION AND DESORPTION KI   5-1288
NADIUM (100) SURFACE STRUCTURES USING LOW ENERGY ELECTRON DIFFRACTION AND AES. CHARACTERIZATION OF SOME VA   5-1864
IDES AND THIN FILM GROWTH OBSERVED BY LOW ENERGY ELECTRON DIFFRACTION AND AES. /STAL SURFACES OF ALKALI HAL   5-1907
ON THE ADHESION OF CLEAN IRON. (USING LOW ENERGY ELECTRON DIFFRACTION AND AES) /OGEN SULFIDE SURFACE FILMS   5-1211
POSITIVE ION CHANNELING SPECTROSCOPY, LOW ENERGY ELECTRON DIFFRACTION AND AES. /RED (001) GOLD SURFACES BY   5-1139
N A NICKEL (111) SURFACE. (STUDIED WITH LOW ENERGY ELECTRON DIFFRACTION AND AES) /BETWEEN OXYGEN AND SULFUR O   5-1451
ARBCN MONOXIDE ON TUNGSTEN (100). (BY LOW ENERGY ELECTRON DIFFRACTION AND AES) /GEN ON TUNGSTEN (110) AND C   5-1152
NTALUM (100) SINGLE CRYSTAL SURFACES. LOW ENERGY ELECTRON DIFFRACTION AND AES ANALYSIS. /ON OF GOLD ONTO TA   5-1312
FACES OF ZINC OXIDE.   LOW ENERGY ELECTRON DIFFRACTION AND AES OBSERVATIONS ON THE POLAR SUR   5-1619
                              LOW ENERGY ELECTRON DIFFRACTION AND AES. (REVIEW, 53 REFS)   1-0096
    A VERSATILE SUBSTRATE DESIGN FOR LOW ENERGY ELECTRON DIFFRACTION AND AES STUDIES IN ULTRAHIGH VACUUM.   3-0631
RFACES.   COMBINED LOW ENERGY ELECTRON DIFFRACTION AND AES STUDIES OF INDIUM ARSENIDE SU   5-1373
00) SURFACE.   LOW ENERGY ELECTRON DIFFRACTION AND AES STUDIES OF LITHIUM HYDRIDE (1   5-1446
DSORBED GASES ON SOLID SURFACES.   LOW ENERGY ELECTRON DIFFRACTION AND AES STUDIES OF THE STRUCTURE OF A   5-1835
  TUNGSTEN (110).   LOW ENERGY ELECTRON DIFFRACTION AND AES STUDY OF OXYGEN ABSORPTION ON   5-1669
ILANE PYROLYSIS.   HIGH ENERGY ELECTRON DIFFRACTION AND AES STUDY OF SILICON EPITAXY BY S   5-1427
YDROGEN SULFIDE AND MOLYBDENUM (100)/ LOW ENERGY ELECTRON DIFFRACTION AND AES STUDY OF THE INTERACTION OF H   5-1973
/TERMINATION OF SURFACE STRUCTURES BY LOW ENERGY ELECTRON DIFFRACTION (AND AES): THE SILICON (111) SURFACE.   5-1342
IR INFLUENCE ON ADHESION STUDIED WITH LOW ENERGY ELECTRON DIFFRACTION AND AUGER. /RON (001) SURFACE AND THE   5-1208
N DIFFRACTION, CIRCA 1968. (REVIEW OF LOW ENERGY ELECTRON DIFFRACTION AND AUGER)   LOW ENERGY ELECTRO   1-0099
NGLE CRYSTAL SURFACES USING INELASTIC LOW ENERGY ELECTRON DIFFRACTION. (AND AUGER ANALYSIS) /MINUM (001) SI   5-1745
SURFACES.   CORRELATICN OF ELECTRONIC, LOW ENERGY ELECTRON DIFFRACTION AND AUGER DIAGNOSTICS ON ZINC OXIDE   5-1584
E SILICON CARBIDE (0001) SURFACE.   LOW ENERGY ELECTRON DIFFRACTION AND AUGER ELECTRON OBSERVATIONS OF TH   5-1932
RVATIONS OF THE SUBSTRATE SURFACE DU/ LOW ENERGY ELECTRON DIFFRACTION AND AUGER ELECTRON SPECTROSCOPIC OBSE   5-1678
IES ON TUNGSTEN (110) SURFACE.   LOW ENERGY ELECTRON DIFFRACTION AND AUGER ELECTRON SPECTROSCOPIC STUD   5-1677
ES IN ULTRAHIGH VACUUM AS OBSERVED BY LOW ENERGY ELECTRON DIFFRACTION AND AUGER ELECTRON SPECTROSCOPY. /FAC   6-2020
POSITIVE ION CHANNELING SPECTROSCOPY, LOW ENERGY ELECTRON DIFFRACTION AND AUGER ELECTRON SPECTROSCOPY. /BY   5-1986
ETERMINATION OF SURFACE STRUCTURES BY LOW ENERGY ELECTRON DIFFRACTION AND AUGER ELECTRON SPECTROSCOPY.   D   5-1957
UM CHLORIDE (100) SURFACE OBSERVED BY LOW ENERGY ELECTRON DIFFRACTION AND AUGER ELECTRON SPECTROSCOPY. /SSI   5-1696
ON GALLIUM ARSENIDE (110) OBSERVED BY LOW ENERGY ELECTRON DIFFRACTION AND AUGER ELECTRON SPECTROSCOPY. /LM   5-1876
M(II) OXIDE (100) CLEAVAGE SURFACE BY LOW ENERGY ELECTRON DIFFRACTION AND AUGER ELECTRON SPECTROSCOPY. /PIU   5-1157
CES IN SURFACE CHARACTERIZATION USING LOW ENERGY ELECTRON DIFFRACTION AND AUGER ELECTRON SPECTROSCOPY. /VAN   1-0201
H PLATINUM SINGLE CRYSTAL SURFACES BY LOW ENERGY ELECTRON DIFFRACTION AND AUGER ELECTRON SPECTROSCOPY. /WIT   5-1223
    STUDY OF IRIDIUM (100) SURFACE USING LOW ENERGY ELECTRON DIFFRACTION AND AUGER ELECTRON SPECTROSCOPY.   5-1371
ES CN THE IRIDIUM (111) SURFACE USING LOW ENERGY ELECTRON DIFFRACTION AND AUGER ELECTRON SPECTROSCOPY. /UDI   5-1372
001) AND RHODIUM (111) SURFACES USING LOW ENERGY ELECTRON DIFFRACTION AND AUGER ELECTRON SPECTROSCOPY. / (0   5-1380
IDE (111) AND (-1-1-1) SURFACES USING LOW ENERGY ELECTRON DIFFRACTION AND AUGER ELECTRON SPECTROSCOPY. /SEN   5-1381
HE NITRIDATION OF SILICON SURFACES BY LOW ENERGY ELECTRON DIFFRACTION AND AUGER ELECTRON SPECTROSCOPY. /F T   5-1417
    STUDIES OF CRYSTAL SURFACES BY LOW ENERGY ELECTRON DIFFRACTION AND AUGER ELECTRON SPECTROSCOPY.   5-1836
LATINUM BY MOLECULAR BEAM SCATTERING, LOW ENERGY ELECTRON DIFFRACTION, AND AUGER ELECTRON SPECTROSCOPY. / P   5-1963
    THIN FILMS. LCW ENERGY ELECTRON DIFFRACTION AND AUGER ELECTRON SPECTROSCOPY.   1-0100
/TURAL STUDY OF OXYGEN- IRON (001) BY LOW ENERGY ELECTRON DIFFRACTION AND AUGER ELECTRON SPECTROSCOPY. PAR/   5-1469
IEW)   LOW ENERGY ELECTRON DIFFRACTION AND AUGER ELECTRON SPECTROSCOPY. (REV   1-0124
ES OF SILICON, GERMANIUM, G/ COMBINED LOW ENERGY ELECTRON DIFFRACTION AND AUGER ELECTRON SPECTROSCOPY STUDI   5-1379
/E (100) AND (111) FACES OF PLATINUM. LOW ENERGY ELECTRON DIFFRACTION AND AUGER ELECTRON SPECTROSCOPY STUD/   5-1421
/WITH HYDROGEN PEROXIDE SOLUTIONS: HIGH ENERGY ELECTRON DIFFRACTION AND AUGER ELECTRON SPECTROSCOPY STUD/   5-1426
/AL THIN FILMS VIA SILANE PYROLYSIS: HIGH ENERGY ELECTRON DIFFRACTION AND AUGER ELECTRON SPECTROSCOPY STUD/   5-1428
OF NICKEL (110) SURFACE EFFECTS DUE/ LOW ENERGY ELECTRON DIFFRACTION AND AUGER ELECTRON SPECTROSCOPY STUDY   5-1814
OF PHOSPHINE ADSORPTICN ON CLEAN SI/ LOW ENERGY ELECTRON DIFFRACTION AND AUGER ELECTRON SPECTROSCOPY STUDY   5-1930
REFS)   A BIBLIOGRAPHY OF LOW ENERGY ELECTRON DIFFRACTION AND AUGER ELECTRON SPECTROSCOPY. (766   1-0071
STUDY OF THE IRON (111) SURFACE USING LOW ENERGY ELECTRON DIFFRACTION AND AUGER EMISSION SPECTROSCOPY.   A   5-1670
TO A CLEAN IRON SURFACE STUDIED WITH LOW ENERGY ELECTRON DIFFRACTION AND AUGER EMISSION SPECTROSCOPY. /ALS   6-2008
VER ON POTASSIUM CHLCRIDE OBSERVED BY LOW ENERGY ELECTRON DIFFRACTION AND AUGER EMISSION SPECTROSCOPY. /SIL   5-1348
N OF CARBON MONOXIDE ON COPPER (001). LOW ENERGY ELECTRON DIFFRACTION AND AUGER EMISSION STUDIES. ADSORPTIO   5-1515
    ULTRAHIGH VACUUM EVAPORATOR FOR LOW ENERGY ELECTRON DIFFRACTION AND AUGER EMISSION STUDIES.   3-0683
ATOR AND MULTIPLE CLEAVAGE DEVICE FOR LOW ENERGY ELECTRON DIFFRACTION AND AUGER EMISSION STUDIES. /M EVAPOR   3-0707
SURFACE OF COPPER(3) GOLD. TEMPERATU/ LOW ENERGY ELECTRON DIFFRACTION AND AUGER INVESTIGATION OF THE (100)   5-1854
UR ON SINGLE CRYSTAL PLATINUM PLANES. LOW ENERGY ELECTRON DIFFRACTION AND AUGER INVESTIGATIONS.   SULF   5-1420
11) SURFACE.   LOW ENERGY ELECTRON DIFFRACTION AND AUGER INVESTIGATIONS OF COPPER (1   5-1499
YPTON ON THE BASAL PLANE OF GRAPHITE: LOW ENERGY ELECTRON DIFFRACTION AND AUGER MEASUREMENTS. /RPTION OF KR   5-1559
                         LOW ENERGY ELECTRON DIFFRACTION AND AUGER METHODS. (REVIEW)   1-0007
RSENIDE ON (100) GERMANIUM STUDIED BY LOW ENERGY ELECTRON DIFFRACTION AND AUGER SPECTROSCOPIC METHODS. /M A   5-1646
    ADSORPTION OF SULFUR ON GOLD BY LOW ENERGY ELECTRON DIFFRACTION AND AUGER SPECTROSCOPY.   5-1558
LFUR LAYERS ON COPPER INVESTIGATED BY LOW ENERGY ELECTRON DIFFRACTION AND AUGER SPECTROSCOPY. / ABSORBED SU   5-1302
    REEXAMINATION OF ALUMINUM (110) WITH LOW ENERGY DIFFRACTION AND AUGER SPECTROSCOPY.   5-1507
E (001) USING REFLECTICN DIFFRACTION, LOW ENERGY ELECTRON DIFFRACTION AND AUGER SPECTROSCOPY. /GNESIUM OXID   5-1496
    ULTRAHIGH VACUUM AIRLOCK FOR LOW ENERGY ELECTRON DIFFRACTION AND AUGER SPECTROSCOPY.   3-0669
ION OF SULFUR ON PLATINUM.   LOW ENERGY ELECTRON DIFFRACTION AND AUGER SPECTROSCOPY OF THE ADSORPT   5-1175
(100) SURFACES OF ALUMINUM AND ALUM/ LOW ENERGY ELECTRON DIFFRACTION AND AUGER SPECTROSCOPY STUDIES OF THE   5-1755
ON ADHESION OF CLEAN IRON (001) AND/ LOW ENERGY ELECTRON DIFFRACTION AND AUGER STUDIES OF EFFECT OF OXYGEN   6-2010
NS ON NICKEL MONOXIDE (100).   LOW ENERGY ELECTRON DIFFRACTION AND ELECTRON SPECTROSCOPIC OBSERVATIO   5-1687
/MINATION OF SURFACE STRUCTURES USING LOW ENERGY ELECTRON DIFFRACTION AND ENERGY ANALYSIS OF SCATTERED ELE/   5-1956
D GRAPHITIZED NICKEL (110) USING AES, LOW ENERGY ELECTRON DIFFRACTION AND FLASH DESORPTION. /CARBURIZED AN   5-1629
/UM ADSORPTION ON TUNGSTEN STUDIED BY LOW ENERGY ELECTRON DIFFRACTION AND SECONDARY ELECTRON SPECTROSCOPY.   5-1661
MONOXIDE AND XENON ADSORBED ON COPP/ LOW ENERGY ELECTRON DIFFRACTION AND SURFACE POTENTIAL STUDY OF CARBON   5-1246
R SPECTROSCOPY)   LOW ENERGY ELECTRON DIFFRACTION AND SURFACE TOPOGRAPHY. (REVIEW, AUGE   1-0185
A CORRELATION OF AUGER SPECTROSCOPY, LOW ENERGY ELECTRON DIFFRACTION AND WORK FUNCTION MEASUREMENTS FOR E/   5-1739
IFICATION OF SURFACE CONTAMINA USE OF LOW ENERGY ELECTRON DIFFRACTION APPARATUS FOR THE DETECTION AND IDENT   5-1959

TING ENERGY AND ANGULAR / FARADAY CUP LOW ENERGY ELECTRON DIFFRACTION APPARATUS WITH FACILITY FOR INVESTIGA 3-0691
) SURFACE.   LOW ENERGY ELECTRON DIFFRACTION AUGER ANALYSIS OF THE BERYLLIUM (0001 5-1581
/ON ALUMINUM (111) SURFACES: ELECTRIC LOW ENERGY ELECTRON DIFFRACTION, AUGER, AND CONTACT POTENTIAL MEASUR/ 5-1744
E ELECTRON AFFINITY ON SILICON PRODU/ LOW ENERGY ELECTRON DIFFRACTION AUGER AND PLASMON STUDIES OF NEGATIV 5-1363
CLEAN AND SODIUM- COVERED SURFACES O/ LOW ENERGY ELECTRON DIFFRACTION AUGER, AND WORK FUNCTION STUDIES OF 5-1240
HE ZINC OXIDE (0001)- ZINC SURFACE: A LOW ENERGY ELECTRON DIFFRACTION AUGER AND WORK FUNCTION STUDY. /TH T 5-1463
INE ADSORBED ON TUNGSTEN (110).   LOW ENERGY ELECTRON DIFFRACTION AUGER, AND WORK FUNCTION STUDY OF IOD 5-1145
ORK FUNCTION CHANGE IN A DISPLAY TYPE LOW ENERGY ELECTRON DIFFRACTION AUGER APPARATUS.   MEASUREMENT OF W 3-0611
/EMISORPTION AND CATALYTIC REACTIONS, LOW ENERGY ELECTRON DIFFRACTION AUGER, CONTACT POTENTIAL METHOD AND/ 5-1321
SPECTROSCOPY OF METALS ON / COMBINED LOW ENERGY ELECTRON DIFFRACTION, AUGER ELECTRON, AND FLASH DESORPTION 1-0008
/A TUNGSTEN SINGLE CRYSTAL STUDIED BY LOW ENERGY ELECTRON DIFFRACTION AUGER ELECTRON AND WORK FUNCTION TE/ 5-1660
M FILMS BY SPUTTER DEPOSITION. (HEED, LOW ENERGY ELECTRON DIFFRACTION AUGER ELECTRON SPECTROSCOPY) /RMANIU 5-1537
ED PLATINUM SINGLE CRYSTAL SURFACES. (LOW ENERGY ELECTRON DIFFRACTION AUGER ELECTRON SPECTROSCOPY) / STEPP 5-1838
TOEMISSION STUDIES OF THE CESIUM COV/ LOW ENERGY ELECTRON DIFFRACTION, AUGER ELECTRON SPECTROSCOPY, AND PHO 5-1931
/ OF IMPURITIES ON COPPER SURFACES BY LOW ENERGY ELECTRON DIFFRACTION AUGER ELECTRON SPECTROSCOPY AND THEI/ 5-1185
K FUNCTION MEASUREMENTS ON CLEAN AND/ LOW ENERGY ELECTRON DIFFRACTION, AUGER ELECTRON SPECTROSCOPY, AND WOR 5-1294
AY PHOTOELECTRON SPECTROSCO/ COMBINED LOW ENERGY ELECTRON DIFFRACTION, AUGER ELECTRON SPECTROSCOPY, AND X-R 5-1337
D ION MICROSCOPY IN MICROSTRU/ USE OF LOW ENERGY ELECTRON DIFFRACTION, AUGER EMISSION SPECTROSCOPY AND FIEL 1-0057
DE OVERLAYER ON A PLATINUM (111) SUR/ LOW ENERGY ELECTRON DIFFRACTION AUGER INVESTIGATION OF A STABLE CARBI 5-1569
E OF AUSTENITIC CHROMIUM- NICKEL STE/ LOW ENERGY ELECTRON DIFFRACTION AUGER INVESTIGATION ON A (100) SURFAC 6-1991
N CHEMICALLY ETCHED SILICON SURFACES. LOW ENERGY ELECTRON DIFFRACTION AUGER METHOD.   CONTAMINANTS O 5-1229
COATED (100) SURFACE OF GALLIUM ARSENIDE BY SLOW ELECTRON DIFFRACTION. (AUGER SPECTROSCOPY) /RE AND CESIUM 5-1639
NTIAL STUDIES OF COPPER- GOLD ALLOY / LOW ENERGY ELECTRON DIFFRACTION AUGER SPECTROSCOPY, AND CONTACT POTE 5-1748
/NOMENA, FIELD ION MICROSCOPY, LEED, HIGH ENERGY ELECTRON DIFFRACTION AUGER SPECTROSCOPY, ION BOMBARDMENT/ 5-1492
PTION OF ALKALI METALS ON MOLYBDENUM/ LOW ENERGY ELECTRON DIFFRACTION AUGER SPECTROSCOPY STUDY OF THE ADSOR 5-1896
ELECTRON BEAM CURRENT CONTROLLER FOR LOW ENERGY ELECTRON DIFFRACTION AUGER STUDIES.   3-0589
/PHIRE EPITAXY BY VACUUM SUBLIMATION: LOW ENERGY ELECTRON DIFFRACTION AUGER STUDIES AND ELECTRONIC PROPERT/ 5-1230
ON ON THE SILVER (111) SURFACE.   LOW ENERGY ELECTRON DIFFRACTION (AUGER) STUDY OF CHLORINE CHEMISORPTI 5-1779
NIOBIUM (110).   LOW ENERGY ELECTRON DIFFRACTION AUGER STUDY OF GROWTH OF ALUMINUM ON 5-1491
/IDATION OF TUNGSTEN (110). PART-2:   LOW ENERGY ELECTRON DIFFRACTION AUGER STUDY OF OXIDE FORMATION AT 65/ 6-1994
HE TUNGST/ COMBINED LEED- REFLECTION HIGH ENERGY ELECTRON DIFFRACTION AUGER STUDY OF OXYGEN ADSORPTION ON T 5-1464
UREMENT OF CHANGES IN WORK FUNCTION. (LOW ENERGY ELECTRON DIFFRACTION AUGER SYSTEM) /OR THE CONTINUOUS MEAS 3-0663
EASUREMENTS AT LOW PRIMARY ENERGIES. (LOW ENERGY ELECTRON DIFFRACTION AUGER SYSTEM) /NDARY ELECTRON YIELD M 3-0629
ILM DEPOSITION AND CHARACTERIZATION. (LOW ENERGY ELECTRON DIFFRACTION AUGER SYSTEM) /RE STAGE FOR IN-SITU F 3-0657
TRON EMISS/ REFINEMENTS TO A STANDARD LOW ENERGY ELECTRON DIFFRACTION AUGER SYSTEM FOR THE ANALYSIS OF ELEC 3-0698
BODY CENTERED CUBIC SUBSTRATES USING LOW ENERGY ELECTRON DIFFRACTION AUGER TECHNIQUES. /HIN METAL FILMS ON 5-1289
CESIUM ADSORPTION AND COADSORPTION / LOW ENERGY ELECTRON DIFFRACTION AUGER, WORK FUNCTION STUDY OF OXYGEN- 5-1297
AUGER ELECTRON SPECTROSCOPY USING THE LOW ENERGY ELECTRON DIFFRACTION CAMERA. /. WORK FUNCTION CHANGES AND 6-2069
R)   LOW ENERGY ELECTRON DIFFRACTION, CIRCA 1968. (REVIEW OF LEED AND AUGE 1-0099
AUGER ELECTRON SPECTROMETER BASED ON LOW ENERGY ELECTRON DIFFRACTION DISPLAY OPTICS. /CONSIDERATIONS OF AN 3-0705
RODUCTION NEAR CRYSTAL SURFACES.   DIFFRACTION EFFECTS IN BACKSCATTERING AND AUGER P 2-0243
OR ELECTRONS IN BULK JELLIUM. (AUGER, LOW ENERGY ELECTRON DIFFRACTION ENERGY LOSS SPECTROSCOPY) /EE PATH F 4-1059
                                      LOW ENERGY ELECTRON DIFFRACTION FROM CLEAN (100) NICKEL OXIDE.   5-1752
/HY, PHENOMENA, FIELD ION MICROSCOPY, LOW ENERGY ELECTRON DIFFRACTION HEED, AUGER SPECTROSCOPY, ION BOMBA/ 5-1492
1), (110) AND ELASTIC (111) / ELASTIC LOW ENERGY ELECTRON DIFFRACTION INTENSITY ENERGY STUDIES OF CLEAN (00 5-1292
ADSORPTION ON STRONTIUM SU/ AUGER AND REFLECTION ELECTRON DIFFRACTION INVESTIGATION OF OXYGEN AND HYDROGEN 5-1498
UMINUM ONTO THE MOLYBDENUM (1/ AUGER- LOW ENERGY ELECTRON DIFFRACTION INVESTIGATION OF THE DEPOSITION OF AL 5-1489
00).   AUGER- LOW ENERGY ELECTRON DIFFRACTION INVESTIGATION OF TIN ON MOLYBDENUM (1 5-1490
L FILMS AND THEIR USE IN STUDYING TH/ LOW ENERGY ELECTRON DIFFRACTION INVESTIGATIONS OF XENON SINGLE CRYSTA 5-1479
ERAMIC SURFACES. (AUGER SPECTROSCOPY, LOW ENERGY ELECTRON DIFFRACTION ION MICROPROBE ANALYSIS) /ISTRY OF C 5-1852
TH OF MAGNESIUM ON MAGNESIUM OXIDE (001) USING REFLECTION DIFFRACTION, LEED AND AUGER SPECTROSCOPY. /L GROW 5-1496
MANIUM FILMS BY SPUTTER DEPOSITION. (HIGH ENERGY ELECTRON DIFFRACTION LEED, AUGER ELECTRON SPECTROSCOPY) / 5-1537
/TRON EMISSION SPECTROSCOPY (AES) AND LOW ENERGY ELECTRON DIFFRACTION (LEED) OF THE ZINC OXIDE (0001) POLA 5-1338
OF ELECTRICAL SOLID MATERIALS BY THE LOW ENERGY ELECTRON DIFFRACTION METHOD.   MEASUREMENTS ON SURFACES 5-1981
AUGER SPECTROSCOPY AND LOW ENERGY ELECTRON DIFFRACTION OF SILICON SURFACES.   5-1396
CE OF GALLIUM PHOSPHIDE STUDIED BY AUGER SPECTROSCOPY AND DIFFRACTION OF SLOW ELECTRONS.   (111) SURFA 5-1640
OP/ STUDY OF BARIUM FILMS ON A (110) TUNGSTEN FACE BY THE DIFFRACTION OF SLOW ELECTRONS AND AUGER SPECTROSC 5-1368
/AND LOW ENERGY ELECTRON DIFFRACTION (LOW ENERGY ELECTRON DIFFRACTION OF THE ZINC OXIDE (0001) POLAR SURF/ 5-1338
/S OF THIN LAYER ELECTROCHEMISTRY AND LOW ENERGY ELECTRON DIFFRACTION. PART-1: ELECTRODE SURFACE STRUCTURE/ 5-1486
/. PART-1: SULFUR AND CARBON MONOXIDE LOW ENERGY ELECTRON DIFFRACTION PATTERNS. (AUGER ELECTRON SPECTROSCO/ 5-1190
VED SILICON.   AUGER SPECTRA AND LOW ENERGY ELECTRON DIFFRACTION PATTERNS FROM NICKEL DEPOSITS ON CLEA 5-1768
CRYSTALS WITH CALI/ AUGER SPECTRA AND LOW ENERGY ELECTRON DIFFRACTION PATTERNS FROM VACUUM CLEANED SILICON 5-1769
ECTRA. (NICKEL (100))   DIFFRACTION PEAKS IN SECONDARY ELECTRON ENERGY SP 4-0775
AND THE MEASUREMENT OF EXCITED ELECTRONS UNDER THE X-RAY DIFFRACTION PROCESS. (AUGER SPECTRA OF SILICON) / 3-0641
N ON THE TUNGSTEN (112) SUR/ COMBINED LOW ENERGY ELECTRON DIFFRACTION RHEED AUGER STUDY OF OXYGEN ADSORPTIO 5-1464
/KEL SURFACES. CORRELATED STUDY USING LOW ENERGY ELECTRON DIFFRACTION SPECTRA, AUGER SPECTRA, AND WORK FUN/ 5-1291
ORK FUNCTION MEASUREMENTS, AUGER, AND LOW ENERGY ELECTRON DIFFRACTION SPECTROSCOPY. /L SURFACES UTILIZING W 5-1290
ON O/ AUGER ELECTRON SPECTROSCOPY AND LOW ENERGY ELECTRON DIFFRACTION STUDIES OF ADSORPTION ISOTHERMS: XEN 5-1856
GER EFFECT) ELECTRON SPECTROSCOPY AND LOW ENERGY ELECTRON DIFFRACTION STUDIES OF GAS- METAL INTERFACES. (AU 5-1320
URATED MOLECU/ AUGER SPECTROSCOPY AND LOW ENERGY ELECTRON DIFFRACTION STUDIES OF THE CHEMISORPTION OF UNSAT 5-1267
INELASTICALLY SCATTERED ELECTRONS AND LOW ENERGY ELECTRON DIFFRACTION STUDIES OF TUNGSTEN. /RGY SPECTRA OF 4-1094
/TIVE AUGER SPECTROSCOPY AND LOW ENERGY ELECTRON DIFFRACTION STUDY OF ALKALI METAL OVERLAYERS ON / 5-1895
N.   AN ELECTRON DIFFRACTION STUDY OF CESIUM ADSORPTION ON TUNGSTE 5-1603
ION IN COPPER/ AUGER SPECTROSCOPY AND LOW ENERGY ELECTRON DIFFRACTION STUDY OF EQUILIBRIUM SURFACE SEGREGAT 6-2025
                                      LOW ENERGY ELECTRON DIFFRACTION STUDY OF IRIDIUM (100) SURFACE.   5-1370
1: EXPERIMENT.   A LOW ENERGY ELECTRON DIFFRACTION STUDY OF MAGNESIUM OXIDE (100). PART- 5-1580
                                      LOW ENERGY ELECTRON DIFFRACTION STUDY OF PLATINUM (100) SURFACE.   5-1377
SENI/ AUGER ELECTRON SPECTROSCOPY AND LOW ENERGY ELECTRON DIFFRACTION STUDY OF THE ACTIVATION OF GALLIUM AR 5-1845
GALLIUM ARSENIDE AND GALLIUM ANTIMON/ LOW ENERGY ELECTRON DIFFRACTION STUDY OF THE POLAR (111) SURFACES OF 5-1602
/PY. AUGER ELECTRON SPECTROSCOPY, AND LOW ENERGY ELECTRON DIFFRACTION STUDY OF THE STRUCTURE AND COMPOSITI/ 5-1951
ECTRON SPECTROMETER INTEGRATED WITH A LOW ENERGY ELECTRON DIFFRACTION SYSTEM.   AUGER EL 3-0658
E POTENTIAL SPECTROSCOPY IN A DISPLAY LOW ENERGY ELECTRON DIFFRACTION SYSTEM.   SOFT X-RAY APPEARANC 4-0870
/TRON SPECTROSCOPY USING THE TWO-GRID LOW ENERGY ELECTRON DIFFRACTION SYSTEM AND PHOTOELECTRON MULTIPLIER. 3-0656
APPLICATION OF TRIPLE GRID LOW ENERGY ELECTRON DIFFRACTION SYSTEM TO AUGER SPECTRUM ANALYSES. 3-0661
AUGER ELECTRON SPECTROSCOPY IN LOW ENERGY ELECTRON DIFFRACTION SYSTEMS.   1-0153
AUGER ELECTRON SPECTROSCOPY IN LOW ENERGY ELECTRON DIFFRACTION SYSTEMS.   1-0158
ION OF AUGER ELECTRON SPECTROSCOPY IN LOW ENERGY ELECTRON DIFFRACTION SYSTEMS.   OPTIMIZAT 1-0152
MPARISON BETWEEN ION PROBE AND AUGER- LOW ENERGY ELECTRON DIFFRACTION TECHNIQUES) /ION PROBE TECHNIQUE. (CO 1-0166
FACE REGIONS ON POLYCRYSTALLINE GOLD. LOW ENERGY ELECTRON DIFFRACTIONS AND AUGER STUDIES. /T OF STEPPED SUR 5-1483
SENIDE DURING ACTIVATION TO NEGATIVE/ LOW ENERGY ELECTRON DIFFRACTION- AUGER CHARACTERIZATION OF GALLIUM AR 5-1364

PRODUCTION OF A SPHERICAL GRID FOR LOW ENERGY ELECTRON DIFFRACTION- AUGER ELECTRON SPECTROSCOPY. 3-0668
/ GROWTH OF METAL ON THEM OBSERVED BY LOW ENERGY ELECTRON DIFFRACTION- AUGER ELECTRON SPECTROSCOPY AND MAS/ 5-1906
IRON FILM DEPOSITED ON MAGNESIUM OX/ LOW ENERGY ELECTRON DIFFRACTION- AUGER ELECTRON SPECTROSCOPY STUDY OF 5-1522
AN EPITAXIAL IRON FILM.   LOW ENERGY ELECTRON DIFFRACTION- AUGER ELECTRON SPECTROSCOPY STUDY OF 5-1526
CLEAN (001) IRON SURFACE.   LOW ENERGY ELECTRON DIFFRACTION- AUGER ELECTRON SPECTROSCOPY STUDY OF 5-1527
THE OXIDATION OF IRCN (110) AND IRO/ LOW ENERGY ELECTRON DIFFRACTION- AUGER ELECTRON SPECTROSCOPY STUDY OF 5-1585
THE OXIDATION OF IRON (100) AND I/ A LOW ENERGY ELECTRON DIFFRACTION- AUGER ELECTRON SPECTROSCOPY STUDY OF 5-1586
IRON SINGLE CRYSTAL SURFACES.   LOW ENERGY ELECTRON DIFFRACTION- AUGER ELECTRON SPECTROSCOPY STUDY OF 5-1866
THE OXIDATION OF CHROMIUM (110) AND/ LOW ENERGY ELECTRON DIFFRACTION- AUGER ELECTRON SPECTROSCOPY STUDY OF 6-2022
THE ADSORPTION OF OXYGEN CN COPPE/ A LOW ENERGY ELECTRON DIFFRACTION- AUGER ELECTRON SPECTROSCOPY STUDY OF 6-2089
THE INITIAL STAGES OF OXIDATION OF / LOW ENERGY ELECTRON DIFFRACTION- AUGER ELECTRON SPECTROSCOPY STUDY OF 6-2110
/N (100) SURFACE BY MEANS OF COMBINED LOW ENERGY ELECTRON DIFFRACTION- AUGER ELECTRON SPECTROSCOPY- MASS S/ 6-2141
A CRYSTAL CLEAVAGE DEVICE FOR USE IN LOW ENERGY ELECTRON DIFFRACTION- AUGER ELECTRON STUDIES. 3-0637
VAPORATED FILM) SYSTEM. (AUGER ELECTRON SPECTROSCOPY) DIFFUSE INTERFACE IN SILICON (SUBSTRATE)- GOLD (E 5-1681
VACUUM EVAPORATED FILM) SYSTEMS STUDIED BY AUGER ELECTRO/ DIFFUSE INTERFACE STRUCTURES IN SOME (SUBSTRATE- 5-1436
LICON AND POLYSILICON- SILICON DIOXIDE-/ AUGER STUDIES OF DIFFUSED IMPURITY PROFILES IN SILICON DIOXIDE- SI 5-1481
E CRACKS. (AUGER ELECTRON SPECTROSCOPY) EFFECT OF A THIN DIFFUSED SOLUTE LAYER ON THE NUCLEATION OF FATIGU 6-2037
MOLECULAR ADSORPTION: INCLUSION OF BULK DIFFUSION AND ITS EFFECT ON AUGER INTENSITIES. 2-0264
N THIN INTRINSIC INDIUM ANTIMONIDE AT ROOM TE/ EFFECTS OF DIFFUSION CURRENT ON GALVANOMAGNETIC PROPERTIES I 2-0354
G AUGER ELECTRON SPECTROSCOPY. ANALYSIS OF GRAIN BOUNDARY DIFFUSION IN BIMETALLIC THIN FILM STRUCTURES USIN 6-2146
ELECTRON SPECTROSCOPY.   STUDY OF DIFFUSION IN THIN COPPER- LEAD FILMS USING AUGER 5-1972
ABOUT 2000 AND 600 ANGTROMS THICK STUDIED WITH AUGER SP/ DIFFUSION KINETICS OF GOLD THROUGH PLATINUM FILMS 5-1235
TER ETCHING TECHNIQUE.   DIFFUSION MEASUREMENT IN THIN FILMS BY AUGER SPUT 5-1439
RK FUNCTION CHANGES: CHROMIUM INTO GOLD. (AUGER ELECTRON/ DIFFUSION MEASUREMENTS IN THIN FILMS UTILIZING WO 5-1889
FILM SYSTEM AND THE CORRELATION OF RESISTIVITY CHANGES W/ DIFFUSION MECHANISMS IN THE PALLADIUM- GOLD THIN 5-1393
TROSCOPY TO THE BULK TC SURFACE PRECIPITATION AND SURFACE DIFFUSION OF CARBON ON POLYCRYSTALLINE NICKEL. /C 5-1651
EASURED WITH AUGER ELECTRON SPECTROSCOPY.   DIFFUSION OF COBALT OUT OF COBALT HARDENED GOLD M 6-2135
ELECTRON SPECTROSCOPY AND THEIR INFLUENCE ON SURFACE SELF DIFFUSION CF COPPER. /ER SURFACES BY LEED- AUGER 5-1185
/ AUGER ELECTRON SPECTROSCOPY AS A TOOL FOR MEASURING THE DIFFUSION OF FOREIGN ATOMS IN SOLIDS NEAR THEIR S 1-0186
BSTRATE)- GOLD (FILM) SYSTEM.   LOW TEMPERATURE DIFFUSION OF GOLD INTO SILICON IN THE SILICON (SU 5-1673
PPER AND NICKEL ON SILVER (111): FORMATION OF AN EQUILIB/ DIFFUSION OF SILVER THROUGH EPITAXIAL FILMS OF CO 5-1356
TO A TUNGSTEN SURFACE.   DIFFUSION OF SULFUR AND PHOSPHORUS IMPURITY ATOMS 5-1226
EL.   DIFFUSION OF SULFUR TOWARD THE (110) FACE OF NICK 5-1771
H AUGER ELECTRON SPECTROSCOPY.   SURFACE DIFFUSION OF TIN OVER A GOLD SURFACE OBSERVED WIT 6-2136
/IMULATED DESORPTION OF SODIUM AND MEASUREMENT OF SURFACE DIFFUSION ON MAGNESIUM OXIDE (001) BY AUGER SPEC/ 5-1494
IAL RESOLUTION AUGER ELECTRON SPECTROSCOPY IN AN ORDINARY DIFFUSION PUMPED SCANNING ELECTRON MICROSCOPE. /T 3-0597
AUGER ELECTRON SPECTROSCOPY IN A DIFFUSION PUMPED SCANNING ELECTRON MICROSCOPE. 3-0598
S USING AUGER ELECTRON SPECTROSCOPY. DIFFUSION STUDIES IN CHROMIUM- PLATINUM THIN FILM 5-1286
AUGER ELECTRON SPECTROSCOPY.   DIFFUSION STUDIES IN GOLD- PLATINUM THIN FILMS BY 5-1636
N SEMICONDUCTORS. (SILICON, GERMANIUM, GRAPHITE, ELECTRON DIFFUSION THEORY) /SCOPY AND SECONDARY EMISSION I 2-0240
STUDIED WITH AUGER SPECTROSCOPY.   SILICON DIFFUSION THROUGH THIN TUNGSTEN FILMS ON SILICON, 5-1234
IC EXECUTION OF AUGER AND ESCA TEST PROGRAMS.   DIGITAL CONTROL UNIT IN TTL LOGIC FOR THE AUTOMAT 3-0722
MPUTER COUPLED AUGER SPECTROMETER.   DIGITAL METHODS FOR THIN FILM ANALYSIS USING A CO 3-0640
NCE ANALYTICAL CAPABILITIES.   DIGITIZING AUGER DATA. THREEFOLD APPROACH TO ENHA 1-0183
STIGATED BY LEED AND AUGER SPEC/ PHASE TRANSITIONS IN TWO DIMENSIONAL ABSORBED SULFUR LAYERS ON COPPER INVE 5-1302
NDUCTORS BY THE AUGER ELECTRON MICROANALYTIC METHO/ THREE DIMENSIONAL CHARACTERIZATION OF METALS AND SEMICO 1-0077
PY, SULFIDE SYSTEMS)   PHASE EQUILIBRIA IN TWO DIMENSIONAL CHEMISORBED LAYERS. (AUGER SPECTROSCO 5-1170
GER ELECTRON SPECTROSCCPY.   THREE DIMENSIONAL ELEMENTAL ANALYSIS OF MATERIALS BY AU 5-1597
01) FACE OF GRAPHITE. (AUGER ELECTRON SP/ FIRST ORDER TWO DIMENSIONAL TRANSITION: XENON ADSORBED ON THE (00 5-1857
SCANNING AUGER SPECTROSCOPY APPLIED TO ANALYSIS OF X-BAND DIODE BURNOUT.   HIGH SPATIAL RESOLUTION 5-1961
ATED ALUMINUM(X) GALLIUM(1-X) ARSENIDE ELECTROLUMINESCENT DIODES. (AUGER RECOMBINATION MECHANISM) / COMPENS 2-0325
WITH APPEARANCE POTENTIAL SPECTROSCOPY.   DIRECT COMPARISON OF AUGER ELECTRON SPECTROSCOPY 1-0142
ECTROSCOPY, AND PROTON RESONANCE PROFILING FOR RELIABILI/ DIRECT COMPARISON OF AUGER, SECONDARY ION MASS SP 1-0216
ES OF SURFACE AND BULK ATOMS OF TITANIUM, CHROMIUM, AND / DIRECT COMPARISON OF CORE ELECTRON BINDING ENERGI 2-0392
G WITH AUGER ELECTRON SPECTROSCOPY IN THE ANALYSIS OF SU/ DIRECT COMPARISON OF LOW ENERGY ION BACKSCATTERIN 5-1869
UDIES OF ZINC, ZINC OXIDE, GALLIUM, AND GALLIU/ AUGER AND DIRECT ELECTRON SPECTRA IN X-RAY PHOTOELECTRON ST 4-1054
/RGY AND ANGULAR DISTRIBUTION OF ELECTRONS RESULTING FROM DIRECT IONIZING COLLISIONS OF ELECTRONS WITH ATO/ 2-0525
ON IN SILICON RECTIFIERS AND THYRISTORS LOADED IN FORWARD DIRECTION IN THE MIDDLE REGION. AUGER RECOMBINATI 2-0415
D AUGER ELECTRONS.   DIRECTIONAL CORRELATION BETWEEN PHOTOELECTRONS AN 1-0122
ECTRON SPECTROSCOPY.   SAMPLES HEATED DIRECTLY BY ELECTRIC CURRENT OBSERVED BY AUGER EL 1-0133
IDATION).   DISAPPEARANCE POTENTIAL SPECTROSCOPY (VANADIUM OX 2-0068
INJECTION OF IONS INTO GLASS FROM A GLOW DISCHARGE. (AUGER SPECTROSCOPY) 5-1222
/PLICATION OF AUGER ELECTRON SPECTROSCOPY TO THE STUDY OF DISCOLORING OF ALUMINUM SURFACES IN BOILING WATE/ 5-1127
FLASH DESORPTION- AUGER ELECTRON SPECTROSCOPY. DISCREPANCY FOR CARBON DIOXIDE ON TUNGSTEN (100). 5-1461
CAY.   DETERMINATION OF OSCILLATOR STRENGTH FOR THE FIRST DISCRETE TRANSITION IN NITROGEN FROM THE AUGER DE 2-0575
RBED SULFUR AND SULFIDES OF COPPER AND NICKEL. DISCRIMINATION BY AUGER SPECTROSCOPY BETWEEN ADSO 5-1734
ECTROSCOPY.   GROWTH MODE DISCRIMINATION OF THIN FILMS BY AUGER ELECTRON SP 5-1624
UM- OXYGEN SOLID SOLUTIONS.   ORDER- DISORDER PHENOMENA AT THE SURFACE OF ALPHA TITANI 5-1819
PER(3) GOLD, THEORY, EXPERIMENT.   ORDER- DISORDER TRANSFORMATION AT A (100) SURFACE OF COP 5-1855
ATION AND POISONING. PART-1: BARIUM ON OXYGEN ON TUNGSTEN DISPENSER CATHODE. /CATHODE SURFACES DURING ACTIV 5-1841
SION.   CARBON EVAPORATION FRCM A THORIUM DISPENSER CATHODE OBSERVED BY AUGER ELECTRON EMIS 5-1400
/RON MICROSCOPE, TRANSMISSION ELECTRON MICROSCOPE, ENERGY DISPERSIVE SPECTROMETER, ION MASS ANALYSIS, AUGE/ 1-0010
/ECTRON SPECTROSCOPY, SCANNING ELECTRON MICRCSCOPY AND NONDISPERSIVE X-RAY ANALYSIS TECHNIQUES FOR EXAMINA/ 5-1799
X-RAY PHOTOEMISSION AND SURFACE STRUCTURE. (AUGER DISPLACEMENTS) 5-1811
SOFT X-RAY APPEARANCE POTENTIAL SPECTROSCOPY IN A DISPLAY LEED SYSTEM. 4-0870
DERATIONS OF AN AUGER ELECTRON SPECTROMETER BASED ON LEED DISPLAY OPTICS. RESOLUTION AND SENSITIVITY CONSI 3-0705
R APPARATUS.   MEASUREMENT OF WORK FUNCTION CHANGE IN A DISPLAY TYPE LOW ENERGY ELECTRON DIFFRACTION AUGE 3-0611
L, AUGE/ SURFACE CHARACTERIZATION OF NICKEL- COPPER ALLOY DISPLAYING POOR WETTABILITY DURING BRAZING. (MONE 5-1573
) SURFACES OF ALKALI HALIDES. PART-2: ELECTRON STIMULATED DISSOCIATION.   (100 5-1908
/D SECONDARY ELECTRONS IN ELECTRON INDUCED DESORPTION AND DISSOCIATION: CARBON MONOXIDE ON PLATINUM (111)./ 5-1567
UORIDE BY CBSERVATION OF PLASMON LOSSES/ STUDY OF SURFACE DISSOCIATION IN POTASSIUM CHLORIDE AND LITHIUM FL 5-1279
ELL IONIZATION.   AUGER DISSOCIATION OF MOLECULES AS A RESULT OF INNER SH 2-0465
ELECTRON BOMBARDMENT.   SURFACE DISSOCIATION OF POTASSIUM CHLORIDE BY LOW ENERGY 5-1719
ER LAYERS IN AUGER ELECTRON SPECTROSCOPY. DISTINCTION BETWEEN ADSORBED MONOLAYERS AND THICK 5-1797
TERNAL X-RAY PHOTOEMISSION.   DISTINCTION BETWEEN WEAK AUGER TRANSITIONS AND IN 2-0436
RIMARY BEAM IN SECONDARY ELECTRON / A MEASURING METHOD TO DISTINGUISH AUGER FROM ENERGY LOSS PEAKS OF THE P 1-0188
/AUGER ELECTRON SIGNAL STRENGTHS FOR MODULATICN AMPLITUDE DISTORTION IN A 4-GRID RETARDING POTENTIAL ENERG/ 3-0620
EXACT CORRECTIONS FOR POTENTIAL MODULATICN DISTORTION IN AUGER YIELD MEASUREMENTS. 2-0389
CURRENT. (USED IN AES)   DISTORTION OF DIFFERENTIATED DEFLECTION ANALYZER 3-0613
CT ON A BOUND ELECTRON MAGNETIC SUBSTATE. (AUGER ELECTRON DISTRIBUTION)   DEPENDENCE OF ATOMIC PHOTOEFFE 2-0463

CTED FROM XENON BY 0.3 TO 2.0 MEV PROTONS. (AUGER PEAKS) DISTRIBUTION IN ENERGY AND ANGLE OF ELECTRONS EJE 2-0534
N AND GAS-COVERED IRON (100) SURFACES. ANGULAR DISTRIBUTION OF AUGER ELECTRON EMISSION FROM CLEA 5-1623
ANGULAR DISTRIBUTION OF AUGER ELECTRONS. 2-0320
E IONIZATION OF ATOMIC HELIUM BY FAST ELECTRCNS. (ANGULAR DISTRIBUTION OF AUGER ELECTRONS) RESONANC 2-0256
LLIUM, MAGNESIUM, ALUMINUM AND SILICON UNDER ION / ENERGY DISTRIBUTION OF AUGER ELECTRONS EMITTED FROM BERY 4-0876
IONIZATION. ANGULAR DISTRIBUTION OF AUGER ELECTRONS FOLLOWING IMPACT 2-0321
NIZATION. ANGULAR DISTRIBUTION OF AUGER ELECTRONS FOLLOWING PHOTOIO 2-0352
ION- ATOM COLLISIONS WITH A FIXED IMPACT PARAMETE/ ENERGY DISTRIBUTION OF (AUGER) ELECTRONS RESULTING FROM 2-0331
UNDER ELECTRON IMPACT. ENERGY DISTRIBUTION OF ELECTRONS EMITTED BY ARGON ATOMS 4-1008
PROTONS IN GASES. PART-2: NITROGEN. / ANGULAR AND ENERGY DISTRIBUTION OF ELECTRONS PRODUCED BY 200-500 KEV 4-1075
ONIZING COLLISIONS OF ELECTRONS WITH / ENERGY AND ANGULAR DISTRIBUTION OF ELECTRONS RESULTING FROM DIRECT I 2-0525
FACES: EFFECTS OF SURFACE CO/ MEASUREMENTS OF THE SPATIAL DISTRIBUTION OF HYDROGEN DESORBED FROM NICKEL SUR 5-1197
F ARGON. ANGULAR DISTRIBUTION OF L3M2,3M2,3(1SO) AUGER ELECTRONS O 2-0322
OM KRYPTON BY ELECTRON IMPACT. ANGULAR DISTRIBUTION OF M4,5NN AUGER ELECTRONS EJECTED FR 4-1003
SILICON (111). (AUGER ELECTRO/ FINE STRUCTURES AND ENERGY DISTRIBUTION OF SECONDARY ELECTRON EMISSION FROM 4-0853
THEORY FOR ENERGY DISTRIBUTION OF SECONDARY ELECTRONS. 2-0241
DIUM PHOSPHIDE AND GALLIUM PHOSPHIDE SUBMITTED TO/ ENERGY DISTRIBUTION OF SECONDARY ELECTRONS EMITTED BY IN 2-0269
NIC BOMBARDMENT. ENERGY SPECTRUM AND ANGULAR DISTRIBUTION OF SECONDARY ELECTRONS EMITTED BY IO 4-0988
(PHOTO- AND AUGER ELECTRON EMISSION) ANGLE DISTRIBUTION OF SODIUM CHLORIDE SINGLE CRYSTALS. 2-0297
SECONDARY ELECTRON ENERGY DISTRIBUTION STUDIES OF URANIUM DIOXIDE SURFACES. 5-1316
MPERATURE DEPENDENT CONTRIBUTION TO AUGER ELECTRON ENERGY DISTRIBUTIONS. A TE 2-0434
TRUE SECONDARY ELECTRON ENERGY DISTRIBUTIONS. 2-0242
MINA. (SCANNING AUGER MICRCPROBE) ADDITIVE AND IMPURITY DISTRIBUTIONS AT GRAIN BOUNDARIES IN SINTERED ALU 5-1505
DED BY NOBLE GAS IONS (10-100 KEV): ENERGETIC AND SPATIAL DISTRIBUTIONS. (AUGER ELECTRON EMISSION) / BOMBAR 4-0776
PHOTOELECTRON STRUCTURE IN THE SECONDARY ELECTRON ENERGY DISTRIBUTIONS OF ALUMINUM, NICKEL, AND COPPER. /D 4-1023
TINUM, COPPER, NIOBIUM, STAINLESS STEEL SURFACES/ SPATIAL DISTRIBUTIONS OF HYDROGEN DESORBED FROM IRON, PLA 6-2006
, (110), AND (111) FACES OF COPPER CRYSTALS. (AU/ ANGULAR DISTRIBUTIONS OF HYDROGEN DESORBED FROM THE (100) 5-1155
/RATUS WITH FACILITY FCR INVESTIGATING ENERGY AND ANGULAR DISTRIBUTIONS OF INELASTICALLY SCATTERED OR PHOT/ 3-0691
/SURFACES BY METASTABLE ATOMS. PART-3: ENERGY AND ANGULAR DISTRIBUTIONS OF THE EJECTED ELECTRONS. (SURFACE/ 5-1133
(AUGER AND RADIATIVE TRANSITIONS). ATOMIC VACANCY DISTRIBUTIONS PRODUCED BY INNER SHELL IONIZATION 2-0544
LOGICAL DAMAGE FROM THE AUGER EFFECT, POSSIBLE BENEFITS. (DNA) BIO 1-0056
ANALYSIS OF THE SURFACE STRUCTURE WITH ANTIPHASE DOMAINS OF SULFUR ON IRON (001) BY LEED- AES. 5-1470
SPECTRUM. (INTERBAND AUGER/ EXCITATION SPECTRUM OF OXYGEN DOMINATED COMPOUNDS IN THE 3-21 EV REGION OF THE 4-1048
E CRYSTALS. (AUGER) RECCMBINATION IN DONOR ACCEPTOR COMPLEXES IN CADMIUM SULFIDE SINGL 2-0369
NONRADIATIVE AUGER RECOMBINATION OF ELECTRONS IN DONOR- ACCEPTOR PAIRS. 2-0536
IN SIMPLE SEMICONDUCTORS. THEORY OF DONOR- ACCEPTOR RADIATIVE AND AUGER RECOMBINATION 2-0420
TED WITH AUGER RECOMBINATION OF EXCITONS BOUND TO NEUTRAL DONORS IN TELLURIUM DOPED GALLIUM PHOSPHIDE. /CIA 2-0453
OMETER, FRACTURED/ GRAIN BOUNDARY SEGREGATION IN MAGNESIA DOPED ALUMINA. (HIGH VACUUM AUGER ELECTRON SPECTR 5-1615
CESS) EXCITON SPECTRA OF HEAVILY DOPED CADMIUM SULFIDE SINGLE CRYSTALS. (AUGER PRO 2-0372
X) ARSENIDE BY MOLECULAR BEAM EPITAXY. (AUGER / MAGNESIUM DOPED GALLIUM ARSENIDE AND ALUMINUM(X) GALLIUM(1- 5-1247
OPTICAL GAIN IN LIGHTLY DOPED GALLIUM ARSENIDE. (AUGER RECOMBINATION) 2-0503
BINATION OF EXCITONS BCUND TO NEUTRAL DONORS IN TELLURIUM DOPED GALLIUM PHOSPHIDE. /CIATED WITH AUGER RECOM 2-0453
Y. SURFACE SEGREGATION IN BORON DOPED IRON OBSERVED BY AUGER EMISSION SPECTROSCOP 5-1181
/ OXIDES ON THE CONTACT RESISTANCE OF TUNGSTEN ON HEAVILY DOPED SILICON- GERMANIUM ALLOY. (AUGER SPECTROSC/ 5-1456
S AND RESISTANCE CHANGES FOR TUNGSTEN CONTACTS ON HEAVILY DOPED SILICON- GERMANIUM. (AUGER SPECTROSCOPY) /E 5-1861
OD. TEMPER EMBRITTLEMENT OF TIN DOPED STEELS BY AUGER ELECTRON SPECTROSCOPIC METH 6-2128
IDENTIFICATION OF BUBBLE FORMING IMPURITIES IN DOPED TUNGSTEN. (AUGER SPECTROSCOPY) 6-2107
PECTROSCOPY) STUDY OF GRAIN BOUNDARY FRACTURE SURFACES IN DOPED TUNGSTEN- RHENIUM ALLOYS. (AUGER ELECTRON S 6-2112
/NCE OF GREEN- COPPER LUMINESCENCE IN COPPER AND ALUMINUM DOPED ZINC SULFIDE. (AUGER EFFECT, COLOR CENTERS) 2-0402
TAXY. (AES MEASUREMENTS) ANTIMONY DOPING OF SILICON LAYERS GROWN BY SOLID PHASE EPI 5-1577
HEAVY ION BEAMS. DOPPLER SHIFTED AUGER ELECTRONS FROM FOIL EXCITED 4-0859
AUGER ELECTRON DOSIMETRY. 1-0219
AND SCANDIUM BY ELECTRON IMPACT. (AUGER SPECTRA) DOUBLE AND TRIPLE IONIZATION IN ALUMINUM, GALLIUM 2-0232
TWO-PHOTON EMISSION, THE RADIATIVE AUGER EFFECT, AND THE DOUBLE AUGER PROCESS. 2-0225
ND THE MEASUREMENT OF EXCITED ELECTRO/ DESIGN OF AN X-RAY DOUBLE CRYSTAL SPECTROMETER INSTALLED IN VACUUM A 3-0641
UGER ELECTRONS. (KLL TRANSITION ENERGIES FOR/ INNER SHELL DOUBLE IONIZATION AND CHEMICAL BONDING. PART-1: A 2-0327
BERYLLIUM AND BERYL/ CHARACTERISTIC ENERGY GAIN AND LOSS, DOUBLE IONIZATION, AND IONIZATION LOSS EVENTS IN 4-0901
ER RATES, AND FLUORESCENCE YIELDS. DOUBLE IONIZATION EFFECTS ON RADIATIVE RATES, AUG 4-0976
/ED PLASMON EXCITATION ON INNER LEVEL IONIZATION. PART-2: DOUBLE IONIZATION ORIGIN OF THE HIGH ENERGY SATE/ 4-1047
SSIBLE LOCALISATION OF CONDUCTION ELECTRONS IN SOFT X-RAY DOUBLE IONIZATION SATELLITE (AUGER) TRANSITION. / 2-0437
ER SPECTRUM. PLASMON ENERGY GAIN AND DOUBLE IONIZATION SATELLITES IN THE BERYLLIUM AUG 4-0899
AUGER ELECTRONS FRCM DOUBLE K-SHELL VACANCY STATES IN NEON. 2-0573
NEW SATELLITE BANDS CORRESPONDING TO DOUBLE L IONIZATION. (OF AUGER ELECTRONS) 2-0293
PRODUCTION AND DECAY OF DOUBLE L VACANCIES IN ARGON AND PHOSPHORUS. 2-0489
ON SPECTROSCOPY. COMBINED ESCA- AUGER SYSTEM BASED ON THE DOUBLE PASS CYLINDRICAL MIRROR ANALYZER. / ELECTR 3-0654
S) DOUBLE PHOTOIONIZATION OF NEON. (AUGER TRANSITION 2-0310
L-SHELL FLUORESCENCE YIELDS OF DOUBLE VACANCY STATES IN LEAD. 4-1105
ON SPECTROSCOPY. PART-1: E/ ENERGIES OF EXCITED STATES OF DOUBLY IONIZED MOLECULES BY MEANS OF AUGER ELECTR 4-1071
AUGER TRANSITIONS INITIATED FROM SINGLY AND DOUBLY IONIZED STATES IN LITHIUM METAL. 2-0581
MPLIFIER FOR SURFACE SPECTROSCOPY. DOUBLY RESONANT CURRENT-TO-VOLTAGE CONVERTER PREA 3-0677
ON KINETICS IN GERMANIUM IN THE PRESENCE OF ELECTRON HOLE DROPS. (AUGER RECOMBINATION) /E CARRIER AND EXCIT 4-0946
INATION) ELECTRON HOLE DROPS IN GERMANIUM- SILICON ALLOYS. (AUGER RECOMB 4-0777
EVIDENCES FOR AUGER RECOMBINATION IN ELECTRON HOLE DROPLETS. (SILICON) 2-0260
OSCOPE. AUGER SPECTROSCOFY OF SOLID SURFACES IN A DRY PUMPED HIGH RESOLUTION SCANNING ELECTRON MICR 5-1258
MACHINE. DRYING OF BUTADIENE- NITRILE RUBBERS IN AN AUGER 3-0617
NGSTEN- RHENIUM ALLOYS/ SOLID SOLUTION SOFTENING (RHENIUM DUCTILIZING EFFECT AND BUBBLE STRENGTHENING IN TU 6-2111
RETRIEVAL CF ELECTRON EXCITED AUGER STRUCTURE BY DYNAMIC BACKGROUND SUBTRACTION. 1-0090
UGER ELECTRON SPECTRCSCOPY./ USE OF ANALOG INTEGRATION IN DYNAMIC BACKGROUND SUBTRACTION FOR QUANTITATIVE A 1-0069
OF AUGER ELECTRON AND SOFT X-RAY APPEARAN/ APPLICATION OF DYNAMIC BACKGROUND SUBTRACTION TO QUANTIFICATION 1-0067
PRODUCTION YIELD IN HEAVY ION ATOM COLLISIONS AT KEV ENE/ DYNAMIC CHARGE STATE DEPENDENCE OF THE K-VACANCY 2-0346
SURFACE BY ELECTRON SPECTROSCOPY. DYNAMIC ELUCIDATION OF REACTION STEPS ON A SOLID 5-1535
N BETWEEN SULFUR AND OXYGEN ON A MOLYBDENUM SURFACE BY M/ DYNAMIC INVESTIGATION OF THE MECHANISM OF REACTIO 5-1534
DYNAMIC THEORY OF AUGER EFFECT. 2-0381
ECTRUM) MANIFESTATION OF ATOMIC DYNAMICS THROUGH THE AUGER EFFECT. (NEON AUGER SP 4-0920
OF METALS. DYNAMICAL SCREENING EFFECTS IN THE AUGER SPECTRUM 4-1110
METALS. (LANTHANUM, PRASEODYMIUM, NEODYMIUM, GADOLINIUM, DYSPROSIUM, YTTERBIUM, HAFNIUM). /SOME RARE EARTH 4-1100

E

AUGER) SPECTRA OF ELECTRON INELASTIC SCATTERING FROM RARE EARTH METAL SURFACES.   IONIZATION LOSSES IN (2-0538
EMISSION LINES OF THE M SERIES OF RARE EARTH METALS. (AUGER EFFECT) 4-0790
, GADOLINIUM, DYSPRO/ AUGER ELECTRON SPECTRA CF SOME RARE EARTH METALS. (LANTHANUM, PRASEODYMIUM, NEODYMIUM 4-1100
WIDTHS OF 4S AND 5S MULTIPLET COMPONENTS IN THE RARE EARTHS. 4-0970
UNS FOR DEPTH PROFILING IN AUGER ELECTRON SPECTROSCOPY. EASY METHOD TO ACCURATELY ALIGN ION BOMBARDMENT G 3-0700
X-RAY EDGE ANOMALIES IN LITHIUM. 2-0270
AT HIGH EXCITATION LEVELS/ TEMPERATURE DEPENDENCE OF THE EDGE EMISSION OF GALLIUM SELENIDE SINGLE CRYSTALS 2-0258
ANS OF THE DECOMPOSI/ DETECTION OF EXCITONS NEAR THE L2,3 EDGE OF THE CHLORIDE ION IN SODIUM CHLORIDE BY ME 4-1126
K X-RAY EMISSION EDGE SHAPES OF FREE ELECTRON METALS. (AUGER BROAD 2-0337
THE CALCULATION OF THE KLL AUGER/ APPLICATION OF THE ZETA EFFECTIVE VALUES OF HYDROGENIC WAVE FUNCTIONS IN 2-0526
CHNIQUES AS APPLIED TO NICKEL (111), NICKEL (100), AND N/ EFFECTIVENESS OF THE STANDARD SURFACE CLEANING TE 5-1467
LE ATOMS. PART-3: ENERGY AND ANGULAR DISTRIBUTIONS OF THE EJECTED ELECTRONS. (SURFACE CONTAMINATION) /ASTAB 5-1133
ANGULAR DISTRIBUTION OF M4,5NN AUGER ELECTRONS EJECTED FROM KRYPTON BY ELECTRON IMPACT. 4-1003
GER PEAKS) DISTRIBUTION IN ENERGY AND ANGLE CF ELECTRONS EJECTED FROM XENON BY 0.3 TO 2.0 MEV PROTONS. (AU 2-0534
COSTER-KRONIG TRANSITIONS) ENERGY SPECTRA OF ELECTRONS EJECTED IN ARGON(+)- ARGON COLLISIONS. (AUGER AND 4-1007
(111) AND (110) TUNGS/ ROLE OF WORK FUNCTION IN ELECTRON EJECTION BY METASTABLE ATOMS: HELIUM AND ARGON ON 5-1600
WORK FUNCTION EFFECTS ON AUGER ELECTRON EJECTION BY NOBLE GAS METASTABLE ATOMS. 5-1601
CESS) ADSORBATE EFFECTS IN ELECTRON EJECTION BY RARE GAS METASTABLE ATOMS. (AUGER PRO 2-0333
T ATOMS AND MOLECULES BY ELECTRON IMPACT. (AUGER ELECTRON EJECTION CROSS SECTIONS) /ESCENCE YIELDS FOR LIGH 2-0528
SPECTROSCOPY. EFFECT OF THE ELECTRON EJECTION DEPTH ON QUANTITATIVE ANALYSIS BY AUGER 2-0449
PART-1: MEASUREMENTS CF SECONDARY EM/ SECONDARY ELECTRON EJECTION FROM METAL SURFACES BY METASTABLE ATOMS. 2-0340
PART-3: ENERGY AND ANGULAR DISTRIBUT/ SECONDARY ELECTRON EJECTION FROM METAL SURFACES BY METASTABLE ATOMS. 5-1133
CESSES BY ELECTRON IMPACT NEAR THRESHOLD. AUGER ELECTRON EJECTION FROM XENON N4,5OO AND KRYPTON M4,5NN PRO 4-1010
N IMPACT. AUGER TRANSITION AND SUPPLEMENTARY ELECTRON EJECTION IN THE IONIZATION OF ALUMINUM BY ELECTRO 2-0230
THEORY OF AUGER EJECTION OF ELECTRONS FROM METALS BY IONS. 2-0563
MISORPTION. AUGER EJECTION OF ELECTRONS FROM TUNGSTEN BY OXYGEN CHE 5-1599
ION) LOW ENERGY ION SCATTERING: ELASTIC AND INELASTIC EFFECTS. (AUGER NEUTRALIZAT 2-0382
O1), (110) AND ELASTIC (111) NICKEL SURFACES. (AUGER SPE/ ELASTIC LEED INTENSITY ENERGY STUDIES OF CLEAN (0 5-1292
. SECONDARY EMISSION AND ELASTIC REFLECTION OF ELECTRONS FROM MONOCRYSTALS 2-0508
ROSCOPY. SAMPLES HEATED DIRECTLY BY ELECTRIC CURRENT OBSERVED BY AUGER ELECTRON SPECT 1-0133
CONDUCTORS. INFLUENCE OF AN ELECTRIC FIELD ON THE AUGER RECOMBINATION IN SEMI 2-0357
CONDUCTORS. INFLUENCE OF AN ELECTRIC FIELD ON THE AUGER RECOMBINATION IN SEMI 4-0844
AND CONTAC/ SODIUM ADSORPTION ON ALUMINUM (111) SURFACES: ELECTRIC LOW ENERGY ELECTRON DIFFRACTION, AUGER, 5-1744
GERMANIUM (100)- CESIUM- OXYGEN PHOTOSURFACE. AN ELECTRICAL, CHEMICAL, AND STRUCTURAL STUDY OF THE 5-1319
OXIDE/ CORRELATION BETWEEN THE COMPOSITIONAL PROFILE AND ELECTRICAL CONDUCTIVITY OF THE THERMAL AND ANODIC 5-1971
RMANIUM FILMS ON N-TYPE GALLIUM ARSENI/ METALLURGICAL AND ELECTRICAL PROPERTIES OF ALLOYED NICKEL- GOLD- GE 6-2101
D ZIRCONIA THIN FILMS. (AUGER EFFECT) MORPHOLOGICAL AND ELECTRICAL PROPERTIES OF RF SPUTTERED YTTRIA-DOPE 5-1385
MEASUREMENTS ON SURFACES OF ELECTRICAL SOLID MATERIALS BY THE LEED METHOD. 5-1981
ION./ STUDY OF PLATINUM ELECTRODES BY MEANS CF THIN LAYER ELECTROCHEMISTRY AND LOW ENERGY ELECTRON DIFFRACT 5-1486
TRA) ELECTROCHEMISTRY AT THIN SOLID FILMS. (AUGER SPEC 4-1026
/ROCHEMISTRY AND LOW ENERGY ELECTRON DIFFRACTION. PART-1: ELECTRODE SURFACE STRUCTURE AFTER EXPOSURE TO WA/ 5-1486
ELECTRODES AND CIRCUITS OF A LEED- AES SYSTEM. 3-0639
Y AND LOW ENERGY ELECTRON DIFFRACTION./ STUDY OF PLATINUM ELECTRODES BY MEANS OF THIN LAYER ELECTROCHEMISTR 5-1486
CHEMICAL ANALYSIS OF ELECTRODEPOSITED NICKEL- NICKEL BONDS BY AES. 4-0947
AUGER STUDIES OF ELECTROLESS NICKEL FILMS. 4-0760
A STUDY OF ACTIVATED SILICON SURFACES FOR ELECTROLESS NICKEL PLATING BY AES. 5-1539
/OF SILICON COMPENSATED ALUMINUM(X) GALLIUM(1-X) ARSENIDE ELECTROLUMINESCENT DIODES. (AUGER RECOMBINATION / 2-0325
AUGER ELECTRON SPECTROSCOPY STUDY OF BETA ALUMINA ELECTROLYTE. 4-1074
ODE SURFACE STRUCTURE AFTER EXPOSURE TO WATER AND AQUEOUS ELECTROLYTES. /ECTRON DIFFRACTION. PART-1: ELECTR 5-1486
URFACES OF 18-8 STAINLESS STEEL FINISHED MECHANICALLY AND ELECTROLYTICALLY. AES STUDIES ON S 6-2129
OR ANALYZER INCORPORATING PRERETARDATION. (TUNGSTEN AUGER ELECTRON) CYLINDRICAL MIRR 3-0590
RPTION CF CE/ LEED, AUGER AND PLASMON STUDIES OF NEGATIVE ELECTRON AFFINITY ON SILICON PRODUCED BY THE ADSO 5-1363
SURFACE. (AU/ STRUCTURAL AND ELECTRONIC MODEL OF NEGATIVE ELECTRON AFFINITY ON THE SILICON- CESIUM- OXYGEN 2-0426
R FROM ENERGY LOSS PEAKS OF THE PRIMARY BEAM IN SECONDARY ELECTRON ANALYSIS. /NG METHOD TO DISTINGUISH AUGE 4-0188
SPREADING FUNCTIONS IN SECONDARY ELECTRON ANALYSIS. 2-0512
TRON SPECTROSCOPY) INTRODUCTION TO LOW ENERGY X-RAY AND ELECTRON ANALYSIS. (ANALYZER FOR PHOTO AUGER ELEC 3-0627
AUGER ELECTRON ANALYSIS AND ITS APPLICATIONS. 1-0202
ALUMINUM. AUGER ELECTRON ANALYSIS OF ELECTROPOLISHED HIGH PURITY 4-0824
MICROSCOPIC AUGER ELECTRON ANALYSIS OF FRACTURE SURFACES. 5-1598
F PLUTONIUM. AUGER ELECTRON AND CHARACTERISTIC ENERGY LOSS SPECTRA O 4-0931
OPY OF SOME METALS. AUGER ELECTRON AND CHARACTERISTIC ENERGY LOSS SPECTROSC 4-1083
TIC PROCESSES. CHEMISORPTION OF AM/ APPLICATIONS OF AUGER ELECTRON AND ENERGY LOSS SPECTROSCOPIES TO CATALY 5-1533
COMPARISON OF MAGNESIUM L2,3MM AUGER CURRENTS USING ELECTRON AND ION EXCITATION. 4-0855
MULTIPLY IONIZED NEON ATOM. CALCULATED K-AUGER ELECTRON AND K X-RAY TRANSITION ENERGIES FOR THE 4-0955
OF MAGNESIUM, ALUMINUM, AND SILICON EXCITED BY LOW ENERGY ELECTRON AND LOW ENERGY ARGON ION BOMBARDMENT. / 4-0857
F THE COMPOSITION OF PASSIVE STATE FILMS. (AUGER ELECTRON AND MOSSBAUER) SPECTROSCOPICAL STUDIES O 5-1309
DARY ELECTRON ENER/ HIGH RESOLUTION MEASUREMENTS OF AUGER ELECTRON AND PHOTOELECTRON STRUCTURE IN THE SECON 4-1023
ED SURFACES OF GOLD THICK FILMS ON COPPER PLATES BY AUGER ELECTRON AND SECONDARY ION MASS SPECTROSCOPY. /AN 5-1550
/YNAMIC BACKGROUND SUBTRACTION TO QUANTIFICATION OF AUGER ELECTRON AND SOFT X-RAY APPEARANCE POTENTIAL SPE/ 1-0067
/N OF TITANIUM AND TITANIUM MONOXIDE SURFACES USING AUGER ELECTRON AND SOFT X-RAY APPEARANCE POTENTIAL SPE/ 5-1384
RFACE OF A TUNGSTEN SINGLE CRYSTAL STUDIED BY LEED, AUGER ELECTRON AND WORK FUNCTION TECHNIQUES. / (110) SU 5-1660
ING. APPLICATION OF ION, ELECTRON, AND X-RAY BEAMS IN MODERN MATERIAL TEST 5-1550
/D FILM DEPOSITED ON A SILICON SUBSTRATE STUDIED BY AUGER ELECTRON AND X-RAY PHOTOELECTRON SPECTROSCOPIES. 5-1435
EON- NEON COLLISIONS. AUGER ELECTRON AND X-RAY PRODUCTION IN 50- TO 200 KEV N 4-1079
COMPARISON OF SOFT X-RAY AND AUGER ELECTRON APPEARANCE POTENTIAL SPECTROSCOPY. 1-0092
UM L3/L2 INTENSITY RATIO MEASURED BY SOFT X-RAY AND AUGER ELECTRON APPEARANCE POTENTIAL SPECTROSCOPY. /ROMI 4-0890
ION THROUGH THIN ADSORBED FILMS. (AUGER/ DETERMINATION OF ELECTRON ATTENUATION LENGTHS IN METALS: TRANSMISS 4-1088
SCANNING ELECTRON AUGER MICROSCOPY IN METALLURGY. 6-2105
ELECTRON AUGER SPECTROSCOPIC STUDY OF THE SURFACE 5-1242
OF CARBIDE CATHODES. ELECTRON AUGER SPECTROSCOPIC STUDY OF THE SURFACE 5-1561
OF CARBIDE CATHODES. ELECTRON AUGER SPECTROSCOPY: A METHOD OF INVESTIG 1-0132
ATING SURFACE PHENOMENA (REVIEW). ELECTRON AUGER SPECTROSCOPY METHOD. /E ESTIMATES 5-1645
OF THE CHEMICAL COMPOSITION NEAR THE SURFACE BASED ON THE ELECTRON AUGER TRANSITIONS IN ATOMS WITH TWO INNE 2-0235
R VACANCIES. OBSERVATION OF THREE ELECTRON BACKGROUND PROBLEM IN ELECTRON EXCITED A 1-0091
UGER ELECTRON SPECTROSCOPY. THE SECONDARY ELECTRON BAND-TO-BAND TRANSITIONS IN SOLIDS. (AUG 2-0275
ER RECOMBINATION IN SILICON) TWO ELECTRON BEAM ADSORBATE INTERACTIONS ON SILICON S 5-1513
URFACES. (AUGER SPECTROSCOPY) ELECTRON BEAM ANALYZER FOR DETECTING AUGER ELECTR 3-0712
ONS.

) SURFACE.
STUDIES.
ORBED XENON.
ENTS WITH AUGER ELECTRON SPECTROSCOPY.
D TUNGSTEN (100), ETHYLENE SATURATED TUNGSTEN (100), AND/
TUDIES UTILIZING AUGER ELECTRON SPECTROSCOPY: CARBON MC/
TILIZING AUGER ELECTRON SPECTROSCOPY.   CARBON MONOXIDE A/
RATION ON BERYLLIUM OXIDE AND ITS EFFECTS ON SECONDARY E/
UGER AND APPEARANCE POTENTIAL SPECTRA INDUCED BY INCIDENT
LECTRON PROBE ANALYZER, SCANNING ELECTRON MICROSCOPE, TR/
ETECTION.
D DEVELOPMENT OF MATERIALS. (AUGER ELECTRON SIGNALS)
OSCOPY.   GATED
XYGEN AND GERMANIUM. (AES)   INTERFERENCE OF AN
REE ATOMS FOR USE IN X-RAY SPECTROSCOPY.   INNER
ANY ELECTRON ATOMS.   ACCURATE ONE
NDARY ELECTRON YIELD SPECTRUM.
MS OF TITANIUM, CHROMIUM, AND / DIRECT COMPARISON OF CORE
ER SPECTRUM FROM SURFACE LAYERS OF SILICON DIOXIDE DURING
SURFACE DISSOCIATION OF POTASSIUM CHLORIDE BY LOW ENERGY
RRADIATIO/ STUDY OF INTENSITIES OF AUGER LINES EXCITED BY
SS SECTION FOR K-SHELL IONIZATION OF COPPER AND SILVER BY
PECTROSCOPY.   SURFACE COMPOSITIONAL CHANGES WITH
CARBON MONOXIDE SYSTEM/ CARBON CONTAMINATION PRODUCED BY
/N, AND K-AUGER SPECTRA IN AMERICIUM-241 FROM CURIUM-241 (
/TUDIES OF THE AFTEREFFECTS OF AUGER IONIZATION FOLLOWING
/C CLOUDS REARRANGEMENT FOLLOWING INTERNAL CONVERSION AND
/R DETERMINATION OF THE ENERGY DEPTH OF LEVELS OF SURFACE
SPECTROMETER WITH SCANNING SAMPLE POSITIONER.
ARGON (+) - ARGON COLLISION.   EXTENDED ION-
RGON COLLISIONS. COMMENTS. (AUGER EL/ FAST ELECTRON- SLOW
ABSORPTION BANDS IN SECONDARY
L (110).   ANALOG TECHNIQUE FOR THE MEASUREMENT OF AUGER
PHIDE, INDUCED BY BOMBARDMENT WITH 2 KV ELECTRONS. (AUGER
G STATES OF HIGHLY STRIPPED OXYGEN, FLUORINE, CHLORINE, /
ROLE OF SURFACE CONTAMINANTS IN
CHARACTERIZATION OF CHEMISORPTION BY LOW ENERGY
BY MEANS OF AUGER ELECTRONIC SPECTROSCOPY AND LOW ENERGY
CLEAN TELLURIUM SURFACES STUDIED BY LOW ENERGY
E MOLYBDENUM (110) SURFACE OBSERVED BY AES AND LOW ENERGY
LOW ENERGY
CTROSCOPY, SECONDARY ION MASS SPECTROSCOPY AND LOW ENERGY
ON CARBON COVERED GERMANIUM (100) SURFACE BY HIGH ENERGY
ESENT STATE OF AUGER ELECTRON SPECTROSCOPY AND LOW ENERGY
OBSERVED BY AUGER EMISSION SPECTROSCOPY AND BY LOW ENERGY
ER SPECTROSCOPY, ELLIPSOMETRY, AND REFLECTION HIGH ENERGY
ORBED ON (0001) GRAPHITE. (AUGER SPECTROSCOPY, LOW ENERGY
ADHESION AT GOLD MICA INTERFACES. (LOW ENERGY
H ANTIPHASE DOMAINS OF SULFUR ON IRON (001) BY LOW ENERGY
G STATES OF WATER VAPOR ON CLEAVED GERMANIUM. (LOW ENERGY
CID DESORPTION FROM GRAPHITIZED NICKEL (110). (LOW ENERGY
TURES ON SILICON (100) SURFACES AS OBSERVED BY LOW ENERGY
AND ALKALI METAL COVERED ZINC OXIDE SURFACES. (LOW ENERGY
POSITION OF METAL SURFACES: A REVIEW OF RECENT LOW ENERGY
N STUDIES OF LITHIUM HYDRIDE.   LOW ENERGY
SURVEYING SILICON SURFACES BY MEANS OF THE LOW ENERGY
CHANISM OF DEPOSITED IRON FILM ON MAGNESIUM C/ LOW ENERGY
INDUCED SULFUR SEGREGATION FROM THE BULK TO / LOW ENERGY
/PERCONDUCTORS AT SURFACES AND IN THIN FILMS. (LOW ENERGY
F IRON- CHROMIUM (100) AND (110) .   LOW ENERGY
ELECTRODES AND CIRCUITS OF A LOW ENERGY
PURITIES ON A SILICON (111) SURFACE BY   AUGER- LOW ENERGY
M, COPPER- TIN, AND IRON- ALUMINUM ALLOYS. (BY LOW ENERGY
RPTION KINETICS ON SILICON (111) SURFACES. (BY LOW ENERGY
F SOME VANADIUM (100) SURFACE STRUCTURES USING LOW ENERGY
LKALI HALIDES AND THIN FILM GROWTH OBSERVED BY LOW ENERGY
CE FILMS ON THE ADHESION OF CLEAN IRON. (USING LOW ENERGY
FACES BY POSITIVE ION CHANNELING SPECTROSCOPY, LOW ENERGY
SULFUR ON A NICKEL (111) SURFACE. (STUDIED BY LOW ENERGY
10) AND CARBON MONOXIDE ON TUNGSTEN (100). (BY LOW ENERGY
D ONTO TANTALUM (100) SINGLE CRYSTAL SURFACES. LOW ENERGY
POLAR SURFACES OF ZINC OXIDE.   LOW ENERGY
LOW ENERGY
VACUUM.   A VERSATILE SUBSTRATE DESIGN FOR LOW ENERGY
SENIDE SURFACES.   COMBINED LOW ENERGY
YDRIDE (100) SURFACE.   LOW ENERGY
TURE OF ADSORBED GASES ON SOLID SURFACES.   LOW ENERGY
RPTION ON TUNGSTEN (110) .   LOW ENERGY
TAXY BY SILANE PYROLYSIS.   HIGH ENERGY
TION OF HYDROGEN SULFIDE AND MOLYBDENUM (100)/ LOW ENERGY
SURFA/ DETERMINATION OF SURFACE STRUCTURES BY LOW ENERGY
E AND THEIR INFLUENCE ON ADHESION STUDIED WITH LOW ENERGY
Y ELECTRON DIFFRACTION, CIRCA 1968. (REVIEW OF LOW ENERGY
(001) SINGLE CRYSTAL SURFACES USING INELASTIC LOW ENERGY
NC OXIDE SURFACES. CORRELATION OF ELECTRONIC, LOW ENERGY
ONS OF THE SILICON CARBIDE (0001) SURFACE.   LOW ENERGY
OPY. (766 REFS)   A BIBLIOGRAPHY OF LOW ENERGY
OPY.   THIN FILMS. LOW ENERGY
OPY. (REVIEW)   LOW ENERGY
/NT ADVANCES IN SURFACE CHARACTERIZATION USING LOW ENERGY

ELECTRON BEAM ASSISTED ADSORPTION ON SILICON (111   5-1269
ELECTRON BEAM CURRENT CONTROLLER FOR LEED- AUGER   3-0589
ELECTRON BEAM EFFECTS IN AUGER ANALYSIS OF PHYSIS   5-1151
ELECTRON BEAM EFFECTS IN DEPTH PROFILING MEASUREM   5-1128
ELECTRON BEAM EFFECTS ON CARBON MONOXIDE SATURATE   5-1244
ELECTRON BEAM INDUCED EFFECTS ON GAS ADSORPTION S   5-1542
ELECTRON BEAM INDUCED EFFECTS ON GAS ADSORPTION U   5-1543
ELECTRON BEAM INDUCED REDUCTION OF CARBON CONCENT   5-1362
ELECTRON BEAM INTERACTION. /HE TITANIUM DIOXIDE A   5-1834
ELECTRON BEAM MICROANALYSIS. PART-1. (REVIEW OF E   1-0010
ELECTRON BEAM MICROANALYZER WITH AUGER ELECTRON D   3-0711
ELECTRON BEAM MICROPROBES USED IN THE ANALYSIS AN   3-0714
ELECTRON BEAM TECHNIQUE FOR AUGER ELECTRON SPECTR   1-0024
ELECTRON BEAM WITH THE SURFACE REACTION BETWEEN O   5-1620
ELECTRON BINDING ENERGIES AND AUGER ENERGIES IN F   4-0774
ELECTRON BINDING ENERGIES AND AUGER ENERGIES IN M   2-0265
ELECTRON BINDING ENERGIES OF BARIUM FROM THE SECO   4-0846
ELECTRON BINDING ENERGIES OF SURFACE AND BULK ATO   2-0392
ELECTRON BOMBARDMENT. / OF CHEMICAL SHIFTS IN AUG   4-1044
ELECTRON BOMBARDMENT.   5-1719
ELECTRON BOMBARDMENT AND ALUMINUM K-ALPHA X-RAY I   4-1017
ELECTRON BOMBARDMENT. (AUGER ELECTRONS) /TIAL CRO   2-0477
ELECTRON BOMBARDMENT OBSERVED BY AUGER ELECTRON S   5-1891
ELECTRON BOMBARDMENT. POLYCRYSTALLINE MOLYBDENUM-   5-1648
ELECTRON CAPTURE) AND BERKELIUM-245 ALPHA DECAYS.   4-1019
ELECTRON CAPTURE IN TRIS (1,10-PHENANTHROLINE) C/   2-0516
ELECTRON CAPTURE. (K-AUGER TRANSITION INTENSITIE/   2-0334
ELECTRON CENTERS. (AUGER EFFECT, ZIRCONIA, SILIC/   4-0915
ELECTRON CHANNELING PATTERNS IN AN AUGER ELECTRON   3-0692
ELECTRON COINCIDENCE MEASUREMENTS OF THE VIOLENT   2-0530
ELECTRON COINCIDENCE STUDY OF VIOLENT ARGON (+)- A   2-0423
ELECTRON CONTINUOUS SPECTRUM.   4-0933
ELECTRON CURRENTS. APPLICATION TO SULFUR ON NICKE   3-0635
ELECTRON DAMAGE) / (100) SURFACES OF GALLIUM PHOS   5-1162
ELECTRON DECAY-IN-FLIGHT SPECTRA FROM AUTOIONIZIN   2-0472
ELECTRON DEVICE TECHNOLOGY.   5-1758
ELECTRON DIFFRACTION   1-0161
ELECTRON DIFFRACTION. / ANALYSIS OF IRON SURFACES   5-1940
ELECTRON DIFFRACTION   5-1134
ELECTRON DIFFRACTION   INITIAL OXIDATION OF TH 5-1650
ELECTRON DIFFRACTION.   1-0131
ELECTRON DIFFRACTION. /CHARACTERIZED BY AUGER SPE   2-0501
ELECTRON DIFFRACTION. /RVATION OF SUPERSTRUCTURES   5-1805
ELECTRON DIFFRACTION   PR   1-0006
ELECTRON DIFFRACTION. / TO SILICON (111) SURFACE   5-1180
ELECTRON DIFFRACTION. /UTTERING, STUDIED WITH AUG   5-1233
ELECTRON DIFFRACTION) /NON SUBMONOLAYER FILMS ADS   5-1858
ELECTRON DIFFRACTION AES)   5-1434
ELECTRON DIFFRACTION AES. / SURFACE STRUCTURE WIT   5-1470
ELECTRON DIFFRACTION AES)   BINDIN   5-1820
ELECTRON DIFFRACTION AES)   FORMIC A 5-1630
ELECTRON DIFFRACTION AES. GOLD-INDUCED SUPERSTRUC   5-1708
ELECTRON DIFFRACTION AES) / PROPERTIES OF CLEAN   5-1881
ELECTRON DIFFRACTION AES, AND XPS RESULTS. / COM   5-1341
ELECTRON DIFFRACTION AES ELECTRON INDUCED EMISSIO   4-0807
ELECTRON DIFFRACTION AES METHOD.   5-1809
ELECTRON DIFFRACTION AES OBSERVATION OF GROWTH ME   5-1525
ELECTRON DIFFRACTION AES STUDIES OF CHEMISORPTION   5-1865
ELECTRON DIFFRACTION AES STUDIES OF NIOBIUM SURF/   5-1848
ELECTRON DIFFRACTION AES STUDY OF THE OXIDATION O   5-1587
ELECTRON DIFFRACTION AES SYSTEM.   3-0639
ELECTRON DIFFRACTION ANALYSIS.   DETECTION OF IM   5-1360
ELECTRON DIFFRACTION AND AES) /ON COPPER- ALUMINU   6-2033
ELECTRON DIFFRACTION AND AES) /DSORPTION AND DESO   5-1288
ELECTRON DIFFRACTION AND AES. CHARACTERIZATION O   5-1864
ELECTRON DIFFRACTION AND AES. /STAL SURFACES OF A   5-1907
ELECTRON DIFFRACTION AND AES) /OGEN SULFIDE SURFA   5-1211
ELECTRON DIFFRACTION AND AES) /RED (001) GOLD SUR   5-1139
ELECTRON DIFFRACTION AND AES) /BETWEEN OXYGEN AND   5-1451
ELECTRON DIFFRACTION AND AES) /GEN ON TUNGSTEN (1   5-1152
ELECTRON DIFFRACTION AND AES ANALYSIS. /ON OF GOL   5-1312
ELECTRON DIFFRACTION AND AES OBSERVATIONS ON THE   5-1619
ELECTRON DIFFRACTION AND AES. (REVIEW, 53 REFS)   1-0096
ELECTRON DIFFRACTION AND AES STUDIES IN ULTRAHIGH   3-0631
ELECTRON DIFFRACTION AND AES STUDIES OF INDIUM AR   5-1373
ELECTRON DIFFRACTION AND AES STUDIES OF LITHIUM H   5-1446
ELECTRON DIFFRACTION AND AES STUDIES OF THE STRUC   5-1835
ELECTRON DIFFRACTION AND AES STUDY OF OXYGEN ABSO   5-1669
ELECTRON DIFFRACTION AND AES STUDY OF SILICON EPI   5-1427
ELECTRON DIFFRACTION AND AES STUDY OF THE INTERAC   5-1973
ELECTRON DIFFRACTION (AND AES): THE SILICON (111)   5-1342
ELECTRON DIFFRACTION AND AUGER. /RON (001) SURFAC   5-1208
ELECTRON DIFFRACTION AND AUGER)   LOW ENERG   1-0099
ELECTRON DIFFRACTION. (AND AUGER ANALYSIS) /MINUM   5-1745
ELECTRON DIFFRACTION   AND AUGER DIAGNOSTICS ON ZI   5-1584
ELECTRON DIFFRACTION AND AUGER ELECTRON OBSERVATI   5-1932
ELECTRON DIFFRACTION AND AUGER ELECTRON SPECTROSC   1-0071
ELECTRON DIFFRACTION AND AUGER ELECTRON SPECTROSC   1-0100
ELECTRON DIFFRACTION AND AUGER ELECTRON SPECTROSC   1-0124
ELECTRON DIFFRACTION AND AUGER ELECTRON SPECTROS/   1-0201

```
/ EUROPIUM(II) OXIDE (100) CLEAVAGE SURFACE BY LOW ENERGY ELECTRON DIFFRACTION AND AUGER ELECTRON SPECTROS/ 5-1157
/YGEN WITH PLATINUM SINGLE CRYSTAL SURFACES BY LOW ENERGY ELECTRON DIFFRACTION AND AUGER ELECTRON SPECTROS/ 5-1223
OPY.        STUDY OF IRIDIUM (100) SURFACE USING LOW ENERGY ELECTRON DIFFRACTION AND AUGER ELECTRON SPECTROSC  5-1371
OP/ STUDIES ON THE IRIDIUM (111) SURFACE USING LOW ENERGY ELECTRON DIFFRACTION AND AUGER ELECTRON SPECTROSC  5-1372
OPY STUDIES OF SILICON, GERMANIUM, G/ COMBINED LOW ENERGY ELECTRON DIFFRACTION AND AUGER ELECTRON SPECTROSC  5-1379
/ENIUM (0001) AND RHODIUM (111) SURFACES USING LOW ENERGY ELECTRON DIFFRACTION AND AUGER ELECTRON SPECTROS/ 5-1380
/UM ARSENIDE (111) AND (-1-1-1) SURFACES USING LOW ENERGY ELECTRON DIFFRACTION AND AUGER ELECTRON SPECTROS/ 5-1381
/UDY OF THE NITRIDATION OF SILICON SURFACES BY LOW ENERGY ELECTRON DIFFRACTION AND AUGER ELECTRON SPECTROS/ 5-1417
/FUR ON THE (100) AND (111) FACES OF PLATINUM. LOW ENERGY ELECTRON DIFFRACTION AND AUGER ELECTRON SPECTROS/ 5-1421
/CLEANING WITH HYDROGEN PEROXIDE SOLUTIONS:   HIGH ENERGY ELECTRON DIFFRACTION AND AUGER ELECTRON SPECTROS/ 5-1426
/MOEPITAXIAL THIN FILMS VIA SILANE PYROLYSIS: HIGH ENERGY ELECTRON DIFFRACTION AND AUGER ELECTRON SPECTROS/ 5-1428
OPY/ STRUCTURAL STUDY OF OXYGEN- IRON (001) BY LOW ENERGY ELECTRON DIFFRACTION AND AUGER ELECTRON SPECTROSC  5-1469
OPIC STUDIES ON TUNGSTEN (110) SURFACE.        LOW ENERGY ELECTRON DIFFRACTION AND AUGER ELECTRON SPECTROSC  5-1677
OPIC OBSERVATIONS OF THE SUBSTRATE SURFACE DU/ LOW ENERGY ELECTRON DIFFRACTION AND AUGER ELECTRON SPECTROSC  5-1678
/ POTASSIUM CHLORIDE (100) SURFACE OBSERVED BY LOW ENERGY ELECTRON DIFFRACTION AND AUGER ELECTRON SPECTROS/ 5-1696
OPY STUDY OF NICKEL (110) SURFACE EFFECTS DUE/ LOW ENERGY ELECTRON DIFFRACTION AND AUGER ELECTRON SPECTROS/ 5-1814
OPY.        STUDIES OF CRYSTAL SURFACES BY LOW  LOW ENERGY ELECTRON DIFFRACTION AND AUGER ELECTRON SPECTROSC  5-1836
/LD FILM ON GALLIUM ARSENIDE (110) OBSERVED BY LOW ENERGY ELECTRON DIFFRACTION AND AUGER ELECTRON SPECTROS/ 5-1876
OPY STUDY OF PHOSPHINE ADSORPTION ON CLEAN SI/ LOW ENERGY ELECTRON DIFFRACTION AND AUGER ELECTRON SPECTROSC  5-1930
OPY.   DETERMINATION OF SURFACE STRUCTURES BY LOW ENERGY ELECTRON DIFFRACTION AND AUGER ELECTRON SPECTROS/ 5-1957
/ACE OF PLATINUM BY MOLECULAR BEAM SCATTERING, LOW ENERGY ELECTRON DIFFRACTION, AND AUGER ELECTRON SPECTRO/ 5-1963
/ACES BY POSITIVE ION CHANNELING SPECTROSCOPY, LOW ENERGY ELECTRON DIFFRACTION AND AUGER ELECTRON SPECTROS/ 5-1986
/L SURFACES IN ULTRAHIGH VACUUM AS OBSERVED BY LOW ENERGY ELECTRON DIFFRACTION AND AUGER ELECTRON SPECTROS/ 6-2020
/H OF SILVER ON POTASSIUM CHLORIDE OBSERVED BY LOW ENERGY ELECTRON DIFFRACTION AND AUGER EMISSION SPECTROS/ 5-1348
OPY.   A STUDY OF THE IRON (111) SURFACE USING LOW ENERGY ELECTRON DIFFRACTION AND AUGER EMISSION SPECTROSC  5-1670
/F METALS TO A CLEAN IRON SURFACE STUDIED WITH LOW ENERGY ELECTRON DIFFRACTION AND AUGER EMISSION SPECTROS/ 6-2008
              ULTRAHIGH VACUUM EVAPORATOR FOR LOW ENERGY ELECTRON DIFFRACTION AND AUGER EMISSION STUDIES.   3-0683
/M EVAPORATOR AND MULTIPLE CLEAVAGE DEVICE FOR LOW ENERGY ELECTRON DIFFRACTION AND AUGER EMISSION STUDIES.   3-0707
ADSORPTION OF CARBON MONOXIDE ON COPPER (001). LOW ENERGY ELECTRON DIFFRACTION AND AUGER EMISSION STUDIES.   5-1515
HE (100) SURFACE OF COPPER(3) GOLD. TEMPERATU/ LOW ENERGY ELECTRON DIFFRACTION AND AUGER INVESTIGATION OF T 5-1854
         SULFUR ON SINGLE CRYSTAL PLATINUM PLANES. LOW ENERGY ELECTRON DIFFRACTION AND AUGER INVESTIGATIONS.    5-1420
COPPER (111) SURFACE.                          LOW ENERGY ELECTRON DIFFRACTION AND AUGER INVESTIGATIONS OF   5-1499
ION OF KRYPTON ON THE BASAL PLANE OF GRAPHITE: LOW ENERGY ELECTRON DIFFRACTION AND AUGER MEASUREMENTS. /RPT 5-1559
                                               LOW ENERGY ELECTRON DIFFRACTION AND AUGER METHODS. (REVIEW)   1-0007
/ALLIUM ARSENIDE ON (100) GERMANIUM STUDIED BY LOW ENERGY ELECTRON DIFFRACTION AND AUGER SPECTROSCOPIC MET/ 5-1646
              ADSORPTION OF SULFUR ON GOLD BY LOW ENERGY ELECTRON DIFFRACTION AND AUGER SPECTROSCOPY.       5-1558
SORBED SULFUR LAYERS ON COPPER INVESTIGATED BY LOW ENERGY ELECTRON DIFFRACTION AND AUGER SPECTROSCOPY. / AB 5-1302
SIUM OXIDE (001) USING REFLECTION DIFFRACTION, LOW ENERGY ELECTRON DIFFRACTION AND AUGER SPECTROSCOPY. /GNE 5-1496
        ULTRAHIGH VACUUM AIRLOCK FOR LOW ENERGY ELECTRON DIFFRACTION AND AUGER SPECTROSCOPY.       3-0669
E ADSORPTION OF SULFUR ON PLATINUM.            LOW ENERGY ELECTRON DIFFRACTION AND AUGER SPECTROSCOPY OF TH 5-1175
ES OF THE (100) SURFACES OF ALUMINUM AND ALUM/ LOW ENERGY ELECTRON DIFFRACTION AND AUGER SPECTROSCOPY STUDI 5-1755
OF OXYGEN ON ADHESION OF CLEAN IRON (001) AND/ LOW ENERGY ELECTRON DIFFRACTION AND AUGER STUDIES OF EFFECT  6-2010
BSERVATIONS ON NICKEL MONOXIDE (100).          LOW ENERGY ELECTRON DIFFRACTION AND ELECTRON SPECTROSCOPIC O 5-1687
ERE/ DETERMINATION OF SURFACE STRUCTURES USING LOW ENERGY ELECTRON DIFFRACTION AND ENERGY ANALYSIS OF SCATT 5-1956
URIZED AND GRAPHITIZED NICKEL (110) USING AES, LOW ENERGY ELECTRON DIFFRACTION  AND FLASH DESORPTION. /CARB 5-1629
ROSC/ CESIUM ADSORPTION ON TUNGSTEN STUDIED BY LOW ENERGY ELECTRON DIFFRACTION AND SECONDARY ELECTRON SPECT 5-1661
OF CARBON MONOXIDE AND XENON ADSORBED ON COPP/ LOW ENERGY ELECTRON DIFFRACTION AND SURFACE POTENTIAL STUDY  5-1246
IEW, AUGER SPECTROSCOPY)                        LOW ENERGY ELECTRON DIFFRACTION AND SURFACE TOPOGRAPHY. (REV 1-0185
TS FOR E/ A CORRELATION OF AUGER SPECTROSCOPY, LOW ENERGY ELECTRON DIFFRACTION AND WORK FUNCTION MEASUREMEN 5-1739
AND IDENTIFICATION OF SURFACE CONTAMINA USE OF LOW ENERGY ELECTRON DIFFRACTION APPARATUS FOR THE DETECTION  5-1959
INVESTIGATING ENERGY AND ANGULAR / FARADAY CUP LOW ENERGY ELECTRON DIFFRACTION APPARATUS WITH FACILITY FOR  3-0691
IUM (0001) SURFACE.                            LOW ENERGY ELECTRON DIFFRACTION AUGER ANALYSIS OF THE BERYLL 5-1581
/SORPTION ON ALUMINUM (111) SURFACES: ELECTRIC LOW ENERGY ELECTRON DIFFRACTION, AUGER, AND CONTACT POTENTI/ 5-1744
F NEGATIVE ELECTRON AFFINITY ON SILICON PRODU/ LOW ENERGY ELECTRON DIFFRACTION  AUGER AND PLASMON STUDIES O 5-1363
DY OF IODINE ADSORBED ON TUNGSTEN (110).       LOW ENERGY ELECTRON DIFFRACTION AUGER, AND WORK FUNCTION STU 5-1145
UDIES OF CLEAN AND SODIUM- COVERED SURFACES C/ LOW ENERGY ELECTRON DIFFRACTION  AUGER, AND WORK FUNCTION ST 5-1240
/E WITH THE ZINC OXIDE (0001)- ZINC SURFACE: A LOW ENERGY ELECTRON DIFFRACTION  AUGER AND WORK FUNCTION ST/ 5-1463
MENT OF WORK FUNCTION CHANGE IN A DISPLAY TYPE LOW ENERGY ELECTRON DIFFRACTION AUGER APPARATUS.      MEASURE 3-0611
/FACE. (CHEMISORPTION AND CATALYTIC REACTIONS, LOW ENERGY ELECTRON DIFFRACTION  AUGER, CONTACT POTENTIAL M/ 5-1321
ESORPTION SPECTROSCOPY OF METALS ON / COMBINED LOW ENERGY ELECTRON DIFFRACTION, AUGER ELECTRON, AND FLASH D 1-0008
/RFACE OF A TUNGSTEN SINGLE CRYSTAL STUDIED BY LOW ENERGY ELECTRON DIFFRACTION  AUGER ELECTRON AND WORK FU/ 5-1660
DETECTION OF IMPURITIES ON COPPER SURFACES BY LOW ENERGY ELECTRON DIFFRACTION AUGER ELECTRON SPECTROSCOPY/ 5-1185
, AND WORK FUNCTION MEASUREMENTS ON CLEAN AND/ LOW ENERGY ELECTRON DIFFRACTION, AUGER ELECTRON SPECTROSCOPY 5-1294
, AND X-RAY PHOTOELECTRON SPECTROSCO/ COMBINED LOW ENERGY ELECTRON DIFFRACTION  AUGER ELECTRON SPECTROSCOPY 5-1337
/GERMANIUM FILMS BY SPUTTER DEPOSITION. (HEED, LOW ENERGY ELECTRON DIFFRACTION  AUGER ELECTRON SPECTROSCOP/ 5-1537
/ND STEPPED PLATINUM SINGLE CRYSTAL SURFACES. (LOW ENERGY ELECTRON DIFFRACTION  AUGER ELECTRON SPECTROSCOP/ 5-1838
, AND PHOTOEMISSION STUDIES OF THE CESIUM COV/ LOW ENERGY ELECTRON DIFFRACTION, AUGER ELECTRON SPECTROSCOP/ 5-1931
AND FIELD ION MICROSCOPY IN MICROSTRU/ USE OF LOW ENERGY ELECTRON DIFFRACTION, AUGER EMISSION SPECTROSCOPY 1-0057
BLE CARBIDE OVERLAYER ON A PLATINUM (111) SUR/ LOW ENERGY ELECTRON DIFFRACTION AUGER INVESTIGATION OF A STA 5-1569
0) SURFACE OF AUSTENITIC CHROMIUM- NICKEL STE/ LOW ENERGY ELECTRON DIFFRACTION AUGER INVESTIGATION ON A 10 6-1991
MINANTS ON CHEMICALLY ETCHED SILICON SURFACES. LOW ENERGY ELECTRON DIFFRACTION AUGER METHOD.      CONTA 5-1229
D CESIUM COATED (100) SURFACE OF GALLIUM ARSENIDE BY SLOW ELECTRON DIFFRACTION. (AUGER SPECTROSCOPY) /RE AN 5-1639
TACT POTENTIAL STUDIES OF COPPER- GOLD ALLOY / LOW ENERGY ELECTRON DIFFRACTION  AUGER SPECTROSCOPY, AND CON 5-1748
/APHY, PHENOMENA, FIELD ION MICROSCOPY, LEED, HIGH ENERGY ELECTRON DIFFRACTION  AUGER SPECTROSCOPY, ION BO/ 5-1492
THE ADSORPTION OF ALKALI METALS ON MOLYBDENUM/ LOW ENERGY ELECTRON DIFFRACTION  AUGER SPECTROSCOPY STUDY OF 5-1896
        ELECTRON BEAM CURRENT CONTROLLER FOR LOW ENERGY ELECTRON DIFFRACTION AUGER STUDIES.        3-0589
/ON ON SAPPHIRE EPITAXY BY VACUUM SUBLIMATION: LOW ENERGY ELECTRON DIFFRACTION AUGER STUDIES AND ELECTRONI/ 5-1230
EMISORPTION ON THE SILVER (111) SURFACE.       LOW ENERGY ELECTRON DIFFRACTION (AUGER) STUDY OF CHLORINE CH 5-1779
MINUM ON NIOBIUM (110).                         LOW ENERGY ELECTRON DIFFRACTION AUGER STUDY OF GROWTH OF ALU 5-1491
ON AT / OXIDATION OF TUNGSTEN (110). PART-2:   LOW ENERGY ELECTRON DIFFRACTION AUGER STUDY OF OXIDE FORMATI 6-1994
TION ON THE TUNGST/ COMBINED LEED- REFLECTION HIGH ENERGY ELECTRON DIFFRACTION AUGER STUDY OF OXYGEN ADSORP 5-1464
UOUS MEASUREMENT OF CHANGES IN WORK FUNCTION. (LOW ENERGY ELECTRON DIFFRACTION AUGER SYSTEM) /OR THE CONTIN 3-0663
N YIELD MEASUREMENTS AT LOW PRIMARY ENERGIES. (LOW ENERGY ELECTRON DIFFRACTION AUGER SYSTEM) /NDARY ELECTRO 3-0629
IN-SITU FILM DEPOSITION AND CHARACTERIZATION. (LOW ENERGY ELECTRON DIFFRACTION AUGER SYSTEM) /RE STAGE FOR  3-0657
S OF ELECTRON EMISS/ REFINEMENTS TO A STANDARD LOW ENERGY ELECTRON DIFFRACTION AUGER SYSTEM FOR THE ANALYSI 3-0698
FILMS ON BODY CENTERED CUBIC SUBSTRATES USING LOW ENERGY ELECTRON DIFFRACTION AUGER TECHNIQUES. /HIN METAL 5-1488
F OXYGEN- CESIUM ADSORPTION AND COADSORPTION / LOW ENERGY ELECTRON DIFFRACTION AUGER, WORK FUNCTION STUDY O 5-1297
NGES AND AUGER ELECTRON SPECTROSCOPY USING THE LOW ENERGY ELECTRON DIFFRACTION CAMERA. /. WORK FUNCTION CHA 6-2069
AND AUGER)                                     LOW ENERGY ELECTRON DIFFRACTION, CIRCA 1968. (REVIEW OF LEED 1-0099
ONS OF AN AUGER ELECTRON SPECTROMETER BASED ON LOW ENERGY ELECTRON DIFFRACTION DISPLAY OPTICS. /CONSIDERATI 3-0705
```

126

EE PATH FOR ELECTRONS IN BULK JELLIUM. (AUGER, LOW ENERGY ELECTRON DIFFRACTION ENERGY LOSS SPECTROSCOPY) / 4-1059
E.   LOW ENERGY ELECTRON DIFFRACTION FROM CLEAN (100) NICKEL OXID 5-1752
/ TOPOGRAPHY, PHENOMENA, FIELD ION MICROSCOPY, LOW ENERGY ELECTRON DIFFRACTION HEED, AUGER SPECTROSCOPY, / 5-1492
CLEAN (001), (110) AND ELASTIC (111) / ELASTIC LOW ENERGY ELECTRON DIFFRACTION INTENSITY ENERGY STUDIES OF 5-1292
HYDROGEN ADSORPTION ON STRONTIUM SU/ AUGER AND REFLECTION ELECTRON DIFFRACTION INVESTIGATION OF OXYGEN AND 5-1498
ION OF ALUMINUM ONTO THE MOLYBDENUM (1/ AUGER- LOW ENERGY ELECTRON DIFFRACTION INVESTIGATION OF THE DEPOSIT 5-1489
BDENUM (100).   AUGER- LOW ENERGY ELECTRON DIFFRACTION INVESTIGATION OF TIN ON MOLY 5-1490
LD CRYSTAL FILMS AND THEIR USE IN STUDYING TH/ LOW ENERGY ELECTRON DIFFRACTION INVESTIGATIONS OF XENON SING 5-1479
STRY OF CERAMIC SURFACES. (AUGER SPECTROSCOPY, LOW ENERGY ELECTRON DIFFRACTION ION MICROPROBE ANALYSIS) /I 5-1852
/Y OF GERMANIUM FILMS BY SPUTTER DEPOSITION. (HIGH ENERGY ELECTRON DIFFRACTION LEED, AUGER ELECTRON SPECT/ 5-1537
SURFACES OF ELECTRICAL SOLID MATERIALS BY THE LOW ENERGY ELECTRON DIFFRACTION METHOD.   MEASUREMENTS ON 5-1981
AUGER SPECTROSCOPY AND LOW ENERGY ELECTRON DIFFRACTION OF SILICON SURFACES.   (5-1396
/PY (AES) AND LOW ENERGY ELECTRON DIFFRACTION (LOW ENERGY ELECTRON DIFFRACTION OF THE ZINC OXIDE (0001) P/ 5-1338
/S BY MEANS OF THIN LAYER ELECTROCHEMISTRY AND LOW ENERGY ELECTRON DIFFRACTION. PART-1: ELECTRODE SURFACE / 5-1486
/) SURFACE. PART-1: SULFUR AND CARBON MONOXIDE LOW ENERGY ELECTRON DIFFRACTION PATTERNS. (AUGER ELECTRON S/ 5-1190
S ON CLEAVED SILICON.   AUGER SPECTRA AND LOW ENERGY ELECTRON DIFFRACTION PATTERNS FROM NICKEL DEPOSIT 5-1768
SILICON CRYSTALS WITH CALI/ AUGER SPECTRA AND LOW ENERGY ELECTRON DIFFRACTION PATTERNS FROM VACUUM CLEANED 5-1769
ADSORPTION ON THE TUNGSTEN (112) SUR/ COMBINED LOW ENERGY ELECTRON DIFFRACTION RHEED AUGER STUDY OF OXYGEN 5-1464
/(111) NICKEL SURFACES. CORRELATED STUDY USING LOW ENERGY ELECTRON DIFFRACTION SPECTRA, AUGER SPECTRA, AND/ 5-1291
ILIZING WORK FUNCTION MEASUREMENTS, AUGER, AND LOW ENERGY ELECTRON DIFFRACTION SPECTROSCOPY. /L SURFACES UT 5-1290
ACES. (AUGER EFFECT) ELECTRON SPECTROSCOPY AND LOW ENERGY ELECTRON DIFFRACTION STUDIES OF GAS- METAL INTERF 5-1320
OF UNSATURATED MOLECU/ AUGER SPECTROSCOPY AND LOW ENERGY ELECTRON DIFFRACTION STUDIES OF THE CHEMISORPTION 5-1267
ECTRA OF INELASTICALLY SCATTERED ELECTRONS AND LOW ENERGY ELECTRON DIFFRACTION STUDIES OF TUNGSTEN. /RGY SP 4-1094
N TUNGSTEN.   AN ELECTRON DIFFRACTION STUDY OF CESIUM ADSORPTION O 5-1603
SEGREGATION IN COPPER/ AUGER SPECTROSCOPY AND LOW ENERGY ELECTRON DIFFRACTION STUDY OF EQUILIBRIUM SURFACE 6-2025
CE.   LOW ENERGY ELECTRON DIFFRACTION STUDY OF IRIDIUM (100) SURFA 5-1370
0). PART-1: EXPERIMENT.   A LOW ENERGY ELECTRON DIFFRACTION STUDY OF MAGNESIUM OXIDE (10 5-1580
ACE.   LOW ENERGY ELECTRON DIFFRACTION STUDY OF PLATINUM (100) SURF 5-1377
FACES OF GALLIUM ARSENIDE AND GALLIUM ANTIMON/ LOW ENERGY ELECTRON DIFFRACTION STUDY OF THE POLAR (111) SUR 5-1602
/ MICROSCOPY. AUGER ELECTRON SPECTROSCOPY, AND LOW ENERGY ELECTRON DIFFRACTION STUDY OF THE STRUCTURE AND / 5-1951
APPEARANCE POTENTIAL SPECTROSCOPY IN A DISPLAY LOW ENERGY ELECTRON DIFFRACTION SYSTEM.   SOFT X-RAY 4-0870
/UGER ELECTRON SPECTROSCOPY USING THE TWO-GRID LOW ENERGY ELECTRON DIFFRACTION SYSTEM AND PHOTOELECTRON MU/ 3-0656
LYSES.   APPLICATION OF TRIPLE GRID LOW ENERGY ELECTRON DIFFRACTION SYSTEM TO AUGER SPECTRUM ANA 3-0661
AUGER ELECTRON SPECTROSCOPY IN LOW ENERGY ELECTRON DIFFRACTION SYSTEMS.   1-0153
AUGER ELECTRON SPECTROSCOPY IN LOW ENERGY ELECTRON DIFFRACTION SYSTEMS.   1-0158
OPTIMIZATION OF AUGER ELECTRON SPECTROSCOPY IN LOW ENERGY ELECTRON DIFFRACTION SYSTEMS.   1-0152
IQUE. (COMPARISON BETWEEN ION PROBE AND AUGER- LOW ENERGY ELECTRON DIFFRACTION TECHNIQUES) /ION PROBE TECHN 5-1166
EPPED SURFACE REGIONS ON POLYCRYSTALLINE GOLD. LOW ENERGY ELECTRON DIFFRACTIONS AND AUGER STUDIES. /T OF ST 5-1166
ALLIUM ARSENIDE DURING ACTIVATION TO NEGATIVE/ LOW ENERGY ELECTRON DIFFRACTION- AUGER CHARACTERIZATION OF G 5-1364
PRODUCTION OF A SPHERICAL GRID FOR LOW ENERGY ELECTRON DIFFRACTION- AUGER ELECTRON SPECTROSCOPY 3-0668
STUDY OF IRON FILM DEPOSITED ON MAGNESIUM OX/ LOW ENERGY ELECTRON DIFFRACTION- AUGER ELECTRON SPECTROSCOPY 5-1522
STUDY OF AN EPITAXIAL IRON FILM.   LOW ENERGY ELECTRON DIFFRACTION- AUGER ELECTRON SPECTROSCOPY 5-1526
STUDY OF CLEAN (001) IRON SURFACE.   LOW ENERGY ELECTRON DIFFRACTION- AUGER ELECTRON SPECTROSCOPY 5-1527
STUDY OF THE OXIDATION OF IRON (110) AND IRO/ LOW ENERGY ELECTRON DIFFRACTION- AUGER ELECTRON SPECTROSCOPY 5-1585
STUDY OF THE OXIDATION OF IRON (100) AND I/ A LOW ENERGY ELECTRON DIFFRACTION- AUGER ELECTRON SPECTROSCOPY 5-1586
STUDY OF IRON SINGLE CRYSTAL SURFACES.   LOW ENERGY ELECTRON DIFFRACTION- AUGER ELECTRON SPECTROSCOPY 5-1866
/EPITAXIAL GROWTH OF METAL ON THEM OBSERVED BY LOW ENERGY ELECTRON DIFFRACTION- AUGER ELECTRON SPECTROSCOPY 5-1906
STUDY OF THE OXIDATION OF CHROMIUM (110) AND/ LOW ENERGY ELECTRON DIFFRACTION- AUGER ELECTRON SPECTROSCOPY 6-2022
STUDY OF THE ADSORPTION OF OXYGEN ON COPPE/ A LOW ENERGY ELECTRON DIFFRACTION- AUGER ELECTRON SPECTROSCOPY 6-2089
STUDY OF THE INITIAL STAGES OF OXIDATION OF / LOW ENERGY ELECTRON DIFFRACTION- AUGER ELECTRON SPECTROSCOPY 6-2110
/ES OF IRON (100) SURFACE BY MEANS OF COMBINED LOW ENERGY ELECTRON DIFFRACTION- AUGER ELECTRON SPECTROSCOP/ 6-2141
A CRYSTAL CLEAVAGE DEVICE FOR USE IN LOW ENERGY ELECTRON DIFFRACTION- AUGER STUDIES.   3-0637
AUGER ELECTRON DOSIMETRY.   1-0219
UM(X) TELLURIDE AMBIENT TEMPERATURES. (AUGER RECOMB/ WARM ELECTRON EFFECTS IN NARROW GAP MERCURY(1-X) CADMI 2-0343
ARGON ON (111) AND (110) TUNGS/ ROLE OF WORK FUNCTION IN ELECTRON EJECTION BY METASTABLE ATOMS: HELIUM AND 5-1600
WORK FUNCTION EFFECTS ON AUGER ELECTRON EJECTION BY NOBLE GAS METASTABLE ATOMS.   5-1601
AUGER PROCESS)   ADSORBATE EFFECTS IN ELECTRON EJECTION BY RARE GAS METASTABLE ATOMS. (2-0333
BY AUGER SPECTROSCOPY.   EFFECT OF THE ELECTRON EJECTION DEPTH ON QUANTITATIVE ANALYSIS 2-0449
LE ATOMS. PART-1: MEASUREMENTS OF SECONDARY EM/ SECONDARY ELECTRON EJECTION FROM METAL SURFACES BY METASTAB 2-0340
LE ATOMS. PART-3: ENERGY AND ANGULAR DISTRIBU/ SECONDARY ELECTRON EJECTION FROM METAL SURFACES BY METASTAB 5-1133
4,5NN PROCESSES BY ELECTRON IMPACT NEAR THRESHOLD.   AUGER ELECTRON EJECTION FROM XENON N4,5OO AND KRYPTON M 4-1010
Y ELECTRON IMPACT.   AUGER TRANSITION AND SUPPLEMENTARY ELECTRON EJECTION IN THE IONIZATION OF ALUMINUM B 2-0230
ION OF SODIUM CHLORIDE SINGLE CRYSTALS. (PHOTO- AND AUGER ELECTRON EMISSION)   ANGLE DISTRIBUT 2-0207
BAND STRUCTURE OF GRAPHITE STUDIED BY SECONDARY ELECTRON EMISSION.   2-0569
RATION FROM A THORIUM DISPENSER CATHODE OBSERVED BY AUGER ELECTRON EMISSION.   CARBON EVAPO 5-1400
LAYERS ON THE SURFACES OF SILICON NITRIDE FILMS BY AUGER ELECTRON EMISSION. /TECTION OF SILICON OXYNITRIDE 5-1611
EFFECT OF SURFACE PLASMON OSCILLATIONS ON AUGER ELECTRON EMISSION.   2-0236
(10-100 KEV): ENERGETIC AND SPATIAL DISTRIBUTIONS. (AUGER ELECTRON EMISSION) / BOMBARDED BY NOBLE GAS IONS 4-0776
FORMATION OF A PLASMA SHEATH WITH SECONDARY ELECTRON EMISSION.   2-0244
Y PRODUCTION BY INTERMEDIATE ENERGY FLUORINE IONS. (AUGER ELECTRON EMISSION)   NEON K-SHELL VACANC 2-0572
OXYGEN BEAM GAS AUGER ELECTRON EMISSION.   2-0400
RADIATIVE AUGER EFFECT WITH AND WITHOUT ELECTRON EMISSION.   2-0494
SOME OBSERVATIONS OF SURFACE SEGREGATION BY AUGER ELECTRON EMISSION.   5-1401
TION BY WATER VAPOR OF SILICON (111) SURFACES USING AUGER ELECTRON EMISSION. /DY OF THE OXIDATION AND REDUC 5-1612
SURFACE STRUCTURE FROM ANGULAR DEPENDENCE OF AUGER ELECTRON EMISSION.   5-1977
AUGER ELECTRON EMISSION ANALYSIS.   1-0073
Y OF ADSORPTION PROCESSES. OPTICALLY STIMULATED SECONDARY ELECTRON EMISSION AND PHOTOELECTRON EMISSION STUD 2-0302
FACES) INVESTIGATIONS WITH FIELD EMIS/ STABILITY OF FIELD ELECTRON EMISSION AND VACUUM BREAKDOWN. (FROM SUR 5-1520
URFACE COMPOSITION.   CHARACTERISTIC AUGER ELECTRON EMISSION AS A TOOL FOR THE ANALYSIS OF S 1-0170
ENTS TO A STANDARD LEED- AUGER SYSTEM FOR THE ANALYSIS OF ELECTRON EMISSION AT LOW PRIMARY BEAM ENERGIES. / 3-0698
ER- GERMANIUM JUNCTION).   SECONDARY ELECTRON EMISSION AT SUPERHIGH FREQUENCIES. (COPP 2-0356
SPECTROMETER FOR INDUCED ELECTRON EMISSION. (AUGER ELECTRONS)   3-0585
ALS.   AUGER ELECTRON EMISSION BY ION BOMBARDMENT OF LIGHT MET 4-0877
BERYLLIUM).   FILM STRUCTURE EFFECT ON THE SECONDARY ELECTRON EMISSION CHARACTERISTICS (GERMANIUM AND 4-0923
/INE STRUCTURE OF THE ANGULAR DEPENDENCE OF THE SECONDARY ELECTRON EMISSION COEFFICIENT OF SINGLE CRYSTALS/ 4-1061
D ELECTRONS (TABLE FOR 60 ELEMENTS).   SECONDARY ELECTRON EMISSION DUE TO PRIMARY AND BACKSCATTERE 4-0907
) AND (111) CRYSTAL SURFAC/ CHEMICAL CHANGES IN SECONDARY ELECTRON EMISSION DURING OXIDATION OF NICKEL (100 5-1453
ON ANISOTROPY OF BACKSCATTERING COEFFICIENT AND SECONDARY ELECTRON EMISSION FOR 10-100 KEV ELECTRONS. /TATI 4-1031
ANGULAR DEPENDENCE OF AUGER ELECTRON EMISSION FROM A SINGLE CRYSTAL SPECIMEN. 4-0958
AUGER ELECTRON EMISSION FROM BERYLLIUM SURFACES.   5-1613

ACES. (AUGER SPECTROSCOPY)   PROTON INDUCED ELECTRON EMISSION FROM CHARACTERIZED NIOBIUM SURF 5-1666
TED MOLYBDENUM BOMBARDED BY POSITIVE IONS. PART-1./ AUGER ELECTRON EMISSION FROM CLEAN AND CARBON CONTAMINA 2-0543
   (100) SURFACES.   ANGULAR DISTRIBUTION OF AUGER ELECTRON EMISSION FROM CLEAN AND GAS-COVERED IRON 5-1623
              ANGLE OF INCIDENCE EFFECTS IN AUGER ELECTRON EMISSION FROM CLEAN MOLYBDENUM.   2-0542
                    ANGULAR EFFECTS IN AUGER ELECTRON EMISSION FROM COPPER (110).   5-1988
FACES.   ANGULAR DEPENDENCE OF AUGER ELECTRON EMISSION FROM COPPER (111) AND (100) SUR 5-1632
                              SECONDARY ELECTRON EMISSION FROM GALLIUM ARSENIDE.   4-0861
(111) SURFACE.   AUGER ELECTRON EMISSION FROM GOLD DEPOSITED ON SILICON 5-1676
IUM).   SECONDARY ELECTRON EMISSION FROM ION IMPLANTED SILICON (CES 4-0913
LS.   SECONDARY ELECTRON EMISSION FROM IRON SINGLE AND POLYCRYSTA 4-0917
ETS.   EFFECTS OF CESIATION ON SECONDARY ELECTRON EMISSION FROM MAGNESIUM OXIDE- GOLD CERM 2-0384
                    ORIGIN OF ION INDUCED AUGER ELECTRON EMISSION FROM METALS.   2-0549
/IMENTAL ASSEMBLY FOR OBTAINING THE SPECTRUM OF SECONDARY ELECTRON EMISSION FROM METALS AND SEMICONDUCTORS/ 3-0688
   THROUGH COPPER SINGLE CRYSTALS. ANISOTROPY OF SECONDARY ELECTRON EMISSION FROM PASSAGE OF LITHIUM(+) IONS 2-0245
ACES. (AUGER SPECTRA)   SECONDARY ELECTRON EMISSION FROM PLATINUM BLACK COATED SURF 5-1979
FILMS.   CHARACTERISTICS OF SHAPING SECONDARY ELECTRON EMISSION FROM POROUS POTASSIUM CHLORIDE 4-1120
TRO/ FINE STRUCTURES AND ENERGY DISTRIBUTION OF SECONDARY ELECTRON EMISSION FROM SILICON (111). (AUGER ELEC 4-0853
STAL BOMBARDED BY RARE GAS IONS.   SECONDARY ELECTRON EMISSION FROM SODIUM CHLORINE SINGLE CRY 4-0759
                    ANGULAR DEPENDENCE OF AUGER ELECTRON EMISSION FROM SOLID SURFACES.   5-1447
Y MEDIUM ENERGY IONS (PLATINUM, NICKEL, TANTAL/ SECONDARY ELECTRON EMISSION FROM SOLID SURFACES BOMBARDED B 2-0309
NG POTENTIAL ENERGY. (AUGER NEUTRALIZATION, REVIEW, 16 R/ ELECTRON EMISSION FROM SOLIDS BY PARTICLES CARRYI 2-0374
                    ANGULAR DEPENDENCE OF ELECTRON EMISSION FROM SURFACES. (AUGER EFFECT)   5-1726
E CERMETS.   SECONDARY ELECTRON EMISSION FROM THIN FILM GOLD METAL- OXID 4-1015
                              SECONDARY ELECTRON EMISSION FROM THIN FILMS.   4-0894
RATUS)   SECONDARY ELECTRON EMISSION FROM THIN FILMS. (OF GOLD, APPA 3-0638
                              SECONDARY ELECTRON EMISSION FROM TUNGSTEN AND MOLYBDENUM.   4-0850
NGLE CRYSTALS.   SECONDARY ELECTRON EMISSION FROM TUNGSTEN AND MOLYBDENUM SI 4-0742
ARY ELECTRONS FROM MOLYBDENUM AND TUNGSTEN.   AUGER ELECTRON EMISSION IN THE ENERGY SPECTRA OF SECOND 4-0874
                              SECONDARY ELECTRON EMISSION IN THE STUDY OF SOLID SURFACES. 5-1555
AVAGE SURFACE OF GRAPHITE SINGLE CRYSTAL.   AUGER ELECTRON EMISSION MICROGRAPHIC STUDIES OF THE CLE 5-1413
F A STAINED SURFACE OF PURE IRON PLATE.   AUGER ELECTRON EMISSION MICROGRAPHY AND MICROANALYSIS O 5-1415
F SOLID SURFACES.   AUGER ELECTRON EMISSION MICROGRAPHY AND MICROANALYSIS O 5-1416
ETTER. (AUGER SPECTROSCOPY)   SECONDARY ELECTRON EMISSION OF A ZIRCONIUM- ALUMINUM BULK G 5-1407
   FILMS.   SECONDARY ELECTRON EMISSION OF AMORPHOUS SEMICONDUCTOR THIN 5-1239
   TO 5-KEV PROTONS.   SECONDARY ELECTRON EMISSION OF METALS BOMBARDED WITH 120-EV 4-1029
                              SECONDARY ELECTRON EMISSION OF MOLYBDENUM CRYSTALS.   4-1063
ALS.   SECONDARY ELECTRON EMISSION OF SILICON DIOXIDE SINGLE CRYST 4-1062
   TARGETS.   SECONDARY ELECTRON EMISSION OF SOLID ALUMINUM AND MAGNESIUM 2-0373
ION AND TRANSMISSION.   AUGER ELECTRON EMISSION OF THIN CARBON FOILS IN REFLECT 4-0893
/UCTURE OF THE DEPENDENCE OF THE COEFFICIENT OF SECONDARY ELECTRON EMISSION ON ENERGY OF PRIMARY ELECTRONS/ 4-0891
/ OXIDATION EFFECTS ON THE SPECTRUM OF THE SLOW SECONDARY ELECTRON EMISSION PEAK IN POLYCRYSTALLINE MAGNESI 4-1118
                              SECONDARY ELECTRON EMISSION. (REVIEW, 86 REFS)   1-0179
                              SECONDARY ELECTRON EMISSION ROCKING CURVE FROM COPPER.   4-0764
              GRAPHITE CONDUCTION BAND STATES FROM SECONDARY ELECTRON EMISSION SPECTRA.   2-0568
ENERGIES AND INTENSITIES.   AUGER ELECTRON EMISSION SPECTRA: A SIMPLE TREATMENT OF 2-0319
T-1. THE ARGO/ CHARGE STATE DEPENDENCE OF X-RAY AND AUGER ELECTRON EMISSION SPECTRA FOR RARE GAS ATOMS. PAR 4-0927
ED CARBON BEAMS.   AUGER ELECTRON EMISSION SPECTRA FROM FOIL AND GAS EXCIT 4-1050
010) AND VA/ INELASTIC EFFECTS AND STRUCTURE IN THE AUGER ELECTRON EMISSION SPECTRA OF VANADIUM PENTOXIDE (5-1340
SPHORUS- SULFUR AND IRON- ANTIMONY- SULFUR ALLOYS. (AUGER ELECTRON EMISSION SPECTROSCOPY) /HORUS, IRON- PHO 6-2098
STUDY OF CARBON FIBER SURFACES USING AUGER AND SECONDARY ELECTRON EMISSION SPECTROSCOPY.   5-1970
          STUDY OF GRAIN BOUNDARY SEGREGATION USING AUGER ELECTRON EMISSION SPECTROSCOPY.   6-2117
ES BY COMBINED X-RAY PHOTOELECTRON SPECTROSCOPY AND AUGER ELECTRON EMISSION SPECTROSCOPY.   SURFACE ANALYS 5-1272
   USE OF CHARACTERISTIC IONIZATION LOSSES IN AUGER ELECTRON EMISSION SPECTROSCOPY.   2-0350
GY ELECTRON DIFFRACTION (LEED) OF THE ZINC OXIDE (/ AUGER ELECTRON EMISSION SPECTROSCOPY (AES) AND LOW ENER 5-1338
EMENT IN LOW ALLOY STEELS.   STUDIES USING AUGER ELECTRON EMISSION SPECTROSCOPY ON TEMPER EMBRITTL 6-2119
/VERAGING, CURVE FITTING, AND RELATED TECHNIQUES IN AUGER ELECTRON EMISSION SPECTROSCOPY USING RETARDING F/ 3-0610
URFACES.   ELECTRON EMISSION STUDIES FROM GALLIUM SELENIDE S 5-1967
LIC/ METHOD FOR DETECTING FINE STRUCTURE IN THE SECONDARY ELECTRON EMISSION YIELD AND THE APPLICATION TO SI 4-0852
APHY DETECTORS.   IRON-55 AS AN AUGER ELECTRON EMITTER. NOVEL SOURCE FOR GAS CHROMATOGR 3-0608
/CERMET FILMS AND ITS APPLICATION TO HIGH YIELD SECONDARY ELECTRON EMITTERS. (AUGER ELECTRON SPECTROSCOPY) 5-1430
   KRYPTON M-SHELL AUGER AND COSTER-KRONIG ELECTRON ENERGIES.   4-0928
              VARIAN CHART OF AUGER ELECTRON ENERGIES.   4-1082
              PHOTOELECTRON AND L2,3MM AUGER ELECTRON ENERGIES.   4-1035
OR METHOD.   K X-RAY AND AUGER ELECTRON ENERGIES FOR ARSENIC.   2-0393
   LINE ENERGIES.   SEMI-EMPIRICAL AUGER ELECTRON ENERGIES. PART-1: GENERAL METHOD AND KLL 4-0930
              SEMIEMPIRICAL AUGER ELECTRON ENERGIES. PART-2: THE ARGON ATOM.   4-0815
MIC NUMBER 5-103.   AUGER ELECTRON ENERGIES (0-2000 EV) FOR ELEMENTS OF ATO 4-1119
   SEMIAUTOMATED DATA RECORDING AND CONTROL SYSTEM FOR AN ELECTRON ENERGY ANALYZER.   3-0680
IOXIDE SURFACES.   SECONDARY ELECTRON ENERGY DISTRIBUTION STUDIES OF URANIUM D 5-1316
   A TEMPERATURE DEPENDENT CONTRIBUTION TO AUGER ELECTRON ENERGY DISTRIBUTIONS.   2-0434
                    TRUE SECONDARY ELECTRON ENERGY DISTRIBUTIONS.   2-0242
              PLASMON EFFECTS IN ELECTRON ENERGY LOSS AND GAIN SPECTRA IN ALUMINUM 4-1020
   SURFACE STATE TRANSITIONS OF SILICON IN ELECTRON ENERGY LOSS SPECTRA. (AUGER EFFECT)   4-1039
COMMENTS.   SIMPLE MODEL FOR K X-RAY AND AUGER ELECTRON ENERGY SHIFTS IN HEAVY ION COLLISIONS. 2-0301
   DIFFRACTION PEAKS IN SECONDARY ELECTRON ENERGY SPECTRA. (NICKEL (100))   4-0775
)   ELECTRON ESCAPE DEPTH IN SILICON. (AUGER ENERGIES 2-0408
ONS)   ELECTRON ESCAPE DEPTHS IN ALUMINUM. (AUGER ELECTR 2-0537
NG PROTON EXCITED X-RAYS.   DETERMINATION OF AUGER ELECTRON EXCITATION CROSS SECTIONS AND RANGES USI 2-0456
          EFFECT OF MODULATION AMPLITUDE ON ELECTRON EXCITED AUGER DATA FROM TITANIUM.   2-0367
ENUM SURFACES.   PROTON AND ELECTRON EXCITED AUGER ELECTRON SPECTRA OF MOLYBD 5-1667
KEL (110)- C(2X2) SULFUR SURFACE: LINE SHAPE ANALYSIS A/ ELECTRON EXCITED AUGER ELECTRON SPECTRUM OF A NIC 4-1065
          ANALYSIS OF MATERIALS BY ELECTRON EXCITED AUGER ELECTRON SPECTRUM. 5-1399
          ANGULAR DEPENDENCIES OF ELECTRON EXCITED AUGER EMISSION.   2-0378
   ESTIMATES OF EFFICIENCIES OF PRODUCTION AND DETECTION OF ELECTRON EXCITED AUGER EMISSION.   2-0288
          ANGULAR DEPENDENCIES IN ELECTRON EXCITED AUGER EMISSION. (COMMENTS)   2-0286
S.   NEW FINE STRUCTURE IN ELECTRON EXCITED AUGER SPECTRA FROM SOLID SURFACE 2-0473
ION CROSS SECTIONS AND OTHER FACTORS IN X-RAY EXCITED AND ELECTRON EXCITED AUGER SPECTROSCOPY. / OF IONIZAT 2-0261
ROUND SUBTRACTION.   RETRIEVAL OF ELECTRON EXCITED AUGER STRUCTURE BY DYNAMIC BACKG 1-0090
                    AUGER ELECTRON FINE PROFILES IN IONIC CRYSTALS.   4-0949
IN METALS.   EFFECT OF POLARIZATION OF THE ELECTRON GAS ON THE ENERGIES OF AUGER TRANSITION 2-0435

TUS.   ELECTRON GUN FOR AUGER AND ULTRASOFT X-RAY APPARA 3-0676
RIER AND EXCITON KINETICS IN GERMANIUM IN THE PRESENCE OF ELECTRON HOLE DROPS. (AUGER RECOMBINATION) /E CAR 4-0946
(AUGER RECOMBINATICN)   ELECTRON HOLE DROPS IN GERMANIUM- SILICON ALLOYS. 4-0777
              EVIDENCES FOR AUGER RECOMBINATION IN   ELECTRON HOLE DROPLETS. (SILICON)   2-0260
ION PROCESS IN A SEMICCNDUCTOR.   EFFECT OF   ELECTRON HOLE INTERACTION ON THE AUGER RECOMBINAT 4-1093
G/ MAGNETIC FIELD DEPENDENT INTENSITY OSCILLATIONS OF THE ELECTRON HOLE LUMINESCENCE IN PURE GERMANIUM. (AU 4-0780
N OF NONEQUILIBRIUM CHARGE CARRIERS IN SILICON AT A HIGH/ ELECTRON HOLE SCATTERING AND (AUGER) RECOMBINATIO 4-1103
                          K-SHELL AUGER ELECTRON   ELECTRON HYPERSATELLITES OF NEON.   4-1117
.   AUGER ELECTRON   ELECTRON IMAGES IN A SCANNING ELECTRON MICROSCOPE 3-0649
  ELECTRONIC SUPERPOSITION OF SAMPLE CURRENT AND SECONDARY ELECTRON IMAGES IN AUGER ELECTRON SPECTROSCOPY. 2-0450
IBUTION OF M4,5NN AUGER ELECTRONS EJECTED FRCM KRYPTON BY ELECTRON IMPACT.   ANGULAR DISTR   4-1003
CT IN THE MULTIPLE ICNIZATION OF MANGANESE AND CADMIUM BY ELECTRON IMPACT.   AUGER EFFE   2-0228
CROSS SECTIONS FOR K-SHELL IONIZATION OF SURFACE ATOMS BY ELECTRON IMPACT.   DIFFERENTIAL   2-0358
GY DISTRIBUTION OF ELECTRONS EMITTED BY ARGCN ATOMS UNDER ELECTRON IMPACT.   ENER   4-1008
              INNER SHELL ICNIZATION OF SURFACE ATOMS BY ELECTRON IMPACT.   2-0359
  IONIZATION ANC AUGER TRANSITIONS IN INDIUM AND SILVER BY ELECTRON IMPACT.   MULTIPLE   2-0229
  MULTIIONIZATION OF KRYPTON AND ITS IONS BY HIGH ENERGY ELECTRON IMPACT. (AUGER EFFECT)   2-0493
/AND FLUORESCENCE YIELDS FCR LIGHT ATOMS AND MOLECULES BY ELECTRON IMPACT. (AUGER ELECTRON EJECTION CROSS / 2-0528
  THE L-SHELL FLUCRESCENCE YIELDS CF ARGON AND CHLORINE BY ELECTRON IMPACT. (AUGER EMISSION) MEASUREMENTS OF 2-0387
ATION OF NEON, ARGON, XENON AND THEIR IONS BY HIGH ENERGY ELECTRON IMPACT. (AUGER EMISSION)   MULTIIONIZ 2-0492
D TRIPLE IONIZATION IN ALUMINUM, GALLIUM, AND SCANDIUM BY ELECTRON IMPACT. (AUGER SPECTRA)   DOUBLE AN   2-0232
CROSS SECTIONS FOR IONIZATION OF INNER SHELL ELECTRONS BY ELECTRON IMPACT. (AUGER SPECTROSCOPY)   2-0475
/ MECHANISMS OF SIMPLE AND MULTIPLE IONIZATICN INDUCED BY ELECTRON IMPACT IN SOME METAL VAPORS. (AUGER PROC 4-0768
PECTRA. CARBON MONOXIDE / COMPARISON OF X-RAY INDUCED AND ELECTRON IMPACT INDUCED PHOTOELECTRON AND AUGER S 4-0768
              INNER SHELL ALIGNMENT OF ATOMS IN   ELECTRON IMPACT IONIZATION. (AUGER ELECTRONS)   2-0335
IC NITROGEN, ATOMIC OXYGEN, AND NEON. (AUGER TR/ ABSOLUTE ELECTRON IMPACT IONIZATION CROSS SECTIONS OF ATOM 2-0363
R SHELLS MEASURED BY AUGER ELECTRON SPECTROSCCPY.   ELECTRON IMPACT IONIZATION CROSS SECTIONS OF INNE 2-0553
R SHELLS.   A NEW METHOD FOR MEASURING   ELECTRON IMPACT IONIZATION CROSS SECTIONS OF INNE 4-0848
ELLS OF SURFACE ATOMS. (AUGER INTENSITIES OF TUNGS/ TOTAL ELECTRON IMPACT IONIZATION CROSS SECTIONS ON K-SH 4-0847
              COMPARISON CF X-RAY AND   ELECTRON IMPACT IONIZATION IN AUGER SPECTROSCOPY. 1-0058
CUUM ULTRAVIOLET EMISSION OF XENON(III) LEVELS EXCITED BY ELECTRON IMPACT VIA N4,5 AUGER TRANSITIONS.   VA 4-0878
N, AUGER EFFECT / ESTIMATION OF BACKSCATTERING EFFECTS IN ELECTRON INDUCED AUGER SPECTRA (SILVER AND SILICO 4-0841
ECHNIQUE FOR AUGER ANALYSIS OF SURFACE SPECIES SUBJECT TO ELECTRON INDUCED DESORPTION.   NEW T   3-0675
BON MONOXIDE / ROLE CF PRIMARY AND SECONDARY ELECTRONS IN ELECTRON INDUCED DESORPTION AND DISSOCIATION: CAR 5-1567
IDE.   LEED- AES   ELECTRON INDUCED EMISSION STUDIES OF LITHIUM HYDR 4-0807
,7 LEVEL IN GOLD, BISMU/ ANOMALOUS THRESHOLD BEHAVIOR FOR ELECTRON INDUCED INNER SHELL IONIZATION OF THE N6 4-1067
TROSCOPY OF SURFACES.   ELECTRON INDUCED SECONDARY ELECTRON EMISSION SPEC 5-1406
AL SURFACES.   IONIZATION LOSSES IN (AUGER) SPECTRA OF ELECTRON INELASTIC SCATTERING FROM RARE EARTH MET 2-0538
NG ARGON(+)- ARGON COLLISIONS. (AUGER EFFECT)   ELECTRON ION COINCIDENCE STUDIES OF HIGHLY IONIZI 2-0529
Y HEAVY NUCLEI IN THE TISSUE: USE OF SOFT X-RAYS. (AUGER ELECTRON IRRADIATION. / ON INCLUSION OF MODERATEL 2-0531
ETALLIC FILMS ON INSULATING SUBSTRATES IN THE PRESENCE OF ELECTRON IRRADIATION. (AUGER SPECTROSCOPY) /HIN M 5-1317
OF SILICON DIOXIDE.   ELECTRON IRRADIATION EFFECT IN THE AUGER ANALYSIS 4-1096
ION OF THE LOW ENERGY ELECTRONS IN ALUMINUM OXIDE. (AUGER ELECTRON LINES) /ON DETERMINATION OF THE ATTENUAT 4-0773
GER EFFECT)   COMPARISON OF SILICON L3,2   ELECTRON LOSS SPECTRA AND OPTICAL ABSORPTION. (AU 4-1038
EVAPORATED CARBON SUR/ CHARACTERISTIC K-SHELL LOW ENERGY ELECTRON LOSS SPECTRA FROM DIAMOND, GRAPHITE, AND 4-0945
              DEPENDENCE OF ATCMIC PHOTOEFFECT ON BOUND ELECTRON MAGNETIC SUBSTATE (AUGER EFFECT).   2-0524
IBUTION)   DEPENDENCE OF ATOMIC PHOTOEFFECT ON A BOUND   ELECTRON MAGNETIC SUBSTATE. (AUGER ELECTRON DISTR 2-0463
              K X-RAY EMISSION EDGE SHAPES OF FREE   ELECTRON METALS. (AUGER BROADENING)   2-0337
S AND EXPECTED PERFORMANC/ X-RAY PHOTOELECTRCN (AND AUGER ELECTRON) MICROANALYSIS AND MICROSCOPY. PRINCIPLE 1-0027
HARACTERIZATION OF METALS AND SEMICONDUCTCRS BY THE AUGER ELECTRON MICROANALYTIC METHOD. (REVIEW) /SIONAL C 1-0077
NALYSIS OF SURFACES AND INTERFACES OF SOLIDS BY THE AUGER ELECTRON MICROANALYTIC METHOD. (REVIEW) /LEMENT A 5-1412
              INSTRUMENTAL PROBLEMS FOR   ELECTRON MICROPROBE AUGER SPECTROSCOPY.   3-0670
F SOLID SURFACES IN A DRY PUMPED HIGH RESOLUTION SCANNING ELECTRON MICROSCOPE.   AUGER SPECTROSCOPY O 1-1258
, SURFACE ELEMENTAL COMPCS/ CONTAMINATION IN THE SCANNING ELECTRON MICROSCOPE. (AUGER ELECTRON SPECTROSCOPY 3-0596
              MATERIAL ANALYSIS USING THE SCANNING   ELECTRON MICROSCOPE. (AUGER SPECTROSCOPY)   5-1701
              SCANNING AUGER   ELECTRON MICROSCOPE FOR SURFACE STUDIES.   3-0679
SS ANALYSIS. (AUGER)   SPECTROSCOPY IN THE SCANNING   ELECTRON MICROSCOPE NOW INCLUDES SECONDARY ION MA 3-0674
SPECTROSCOPIC METHODS OF MICROANALYSIS USING THE SCANNING ELECTRON MICROSCOPE. (REVIEW, AUGER SCANNING)   3-0685
              LABCRATORY COMPUTER SCANNING   ELECTRON MICROSCOPE SYSTEM.   3-0651
AND AUGER ANALYZER.   ULTRAHIGH VACUUM SCANNING   ELECTRON MICROSCOPE WITH A FIELD EMISSION CATHODE 3-0621
S.   ULTRAHIGH VACUUM SCANNING   ELECTRON MICROSCOPE WITH AUGER ANALYSIS FACILITIE 3-0622
SSION SPECTROSCOPY.   A SCANNING   ELECTRON MICROSCOPE WITH FACILITIES FOR AUGER EMI 3-0593
RMED ON AN IRON- CHROMIUM ALLOY.   AUGER SCANNING   ELECTRON MICROSCOPIC STUDY OF THE PASSIVE FILM FO 5-1753
AS STUDIED BY AUGER SPECTROSCOPY, FIELD ION AND SCANNING ELECTRON MICROSCOPY. /LENE IN CONTACT WITH METALS 5-1198
D BUBBLE STRENGTHENING IN TUNGSTEN- RHENIUM ALLOYS, AUGER ELECTRON MICROSCOPY). /NIUM DUCTILIZING EFFECT AN 6-2111
/ES: AN ASSESSMENT BY RUTHERFORD BACKSCATTERING, SCANNING ELECTRON MICROSCOPY, AND DEPTH PROFILING AUGER E/ 5-1902
EXPERIMENTAL PROBLEMS OF STUDYING THIN FILM NUCLEATION. (ELECTRON MICROSCOPY, AUGER EFFECT)   5-1408
BACKSCATTERING TECHNIQUES APPLIED TO ANALYSIS O/ SCANNING ELECTRON MICROSCOPY, AUGER SPECTROSCOPY, AND ION 5-1261
/L CORROSION PRODUCTS IN FLOWING LIQUID SODIUM. (SCANNING ELECTRON MICROSCOPY, AUGER SPECTROSCOPY, ION BEA/ 6-2014
ELECTRON SP/ BASIC INVESTIGATION IN THE AREA OF SCANNING ELECTRON MICROSCOPY. PART-1: APPLICATION OF AUGER 1-0213
              THIRD DIMENSION IN SCANNING   ELECTRON MICROSCOPY- SCANNING AUGER MICROSCOPY.   1-0117
SPOT SURFACE AND IN-DEPTH ANALYSIS OF / COMBINED SCANNING ELECTRON MICROSCOPY- AUGER SPECTROSCOPY FOR MICRO 1-1262
              REVIEW OF FLUORESCENCE YIELDS AND AUGER   ELECTRON MISSION.   1-0121
DE BY AES)   GAIN FATIGUE MECHANISM IN CHANNEL   ELECTRON MULTIPLIERS. (SURFACE OXYGEN ON LEAD OXI 5-0681
ILICON SURFACES AND OF SILICON DIOXIDE.   ELECTRON ORBITAL ENERGIES OF OXYGEN ADSORBED ON S 5-1475
PLASMA X-RAYS. (PLASMA ELECTRONS, AUGER PROCESS)   ELECTRON PHOTOEMISSION INDUCED BY LASER-PRODUCED 2-0276
              TRACE ANALYSIS TECHNIQUES FOR SOLIDS. (ELECTRON PROBE ANALYSIS, AUGER EFFECT, REVIEW)   1-0093
SCOPY.   ELECTRON PROBE MICROANALYSIS AND ELECTRON SPECTRO 1-0028
TRONS BY CLOUD CHAMBER TECHNIQUES: OXYGEN AND NEON AUGER ELECTRON PROBLEM. /MEASUREMENT OF LOW ENERGY ELEC 2-0566
ENCE YIELD IN 50 MEV CHLORINE(+)- NEON COLLISIONS. (AUGER ELECTRON PRODUCTION) /ELL IONIZATION AND FLUORESC 2-0518
RDMENT.   K-SHELL AUGER   ELECTRON PRODUCTION CROSS SECTIONS FROM ION BOMBA 2-0570
RDMENT. (NEON, FLUORINE)   K-SHELL AUGER   ELECTRON PRODUCTION CROSS SECTIONS FROM ION BOMBA 2-0574
ER EFFECT)   MECHANISMS OF   ELECTRON PRODUCTION IN ION- ATOM COLLISIONS. (AUG 2-0488
ER EFFECT, REVIEWS)   MECHANISMS OF   ELECTRON PRODUCTION IN ION- ATOM COLLISIONS. (AUG 2-0490
OF TRUE SECONDARY ELECTRONS. PART-2: INELASTICALLY/ SLOW ELECTRON SCATTERING FROM METALS. PART-1: EMISSION 2-0500
NATURE OF SECONDARY ELECTRONS CREATED AS THE RESULT OF   ELECTRON SHAKE-OFF AND (SUGAR) VACANCY CASCADES. 2-0307
ROLE OF THE AUGER EFFECT IN THE FORMATION OF AN   ELECTRON SHOWER SPECTRUM.   2-0504
E DISTORTION IN A 4-GRID RETARDING / CORRECTIONS OF AUGER ELECTRON SIGNAL STRENGTHS FOR MODULATION AMPLITUD 3-0620
QUARTZ CRYSTAL MICRCBALANCE. QUANTITATIVE STUDY OF AUGER ELECTRON SIGNALS OF PHOSPHORUS ON SILICON USING A 5-1583
TE AND OTHER CHARACTERISTIC EVENTS IN MAGNESIUM SECONDARY ELECTRON SPECTKA.   AUGER SATELLI   4-0898

ON LOSS EVENTS IN BERYLLIUM AND BERYLLIUM OXIDE SECONDARY ELECTRON SPECTRA. /OUBLE IONIZATION, AND IONIZATI 4-0901
CHEMICAL SHIFTS IN AUGER ELECTRON SPECTRA. 4-0825
FLUORESCENCE PHOTON AND AUGER ELECTRON SPECTRA. 4-0799
IMPROVED APPARATUS FOR MEASUREMENT OF AUGER ELECTRON SPECTRA. 3-0612
L-SHELL AUGER AND COSTER-KRONIG ELECTRON SPECTRA. 4-0960
L-SHELL IONIZATION THRESHOLD STRUCTURE FROM AUGER ELECTRON SPECTRA. 2-0360
M-SHELL AUGER AND COSTER-KRONIG ELECTRON SPECTRA. 4-0965
PHOTOGRAPHIC PLATES FOR RECORDING CONVERSION AND/OR AUGER ELECTRON SPECTRA. NEW, STRIATED, HIGH EFFICIENCY 3-0699
SOME PROBLEMS IN ANALYSIS OF AUGER ELECTRON SPECTRA. 4-0866
TRANSITION POTENTIAL MODELS FOR AUGER ELECTRON SPECTRA. 2-0510
VALENCE BAND AUGER ELECTRON SPECTRA FOR ALUMINUM: COMMENTS. 4-0839
L-SERIES AUGER ELECTRON SPECTRA FOR ELEMENTS WITH Z = 66-70. 4-0744
SIONS WITH HELIUM, NEON, / HIGH RESOLUTION BEAM GAS AUGER ELECTRON SPECTRA FOR OXYGEN IONS EXCITED BY COLLI 4-0904
CONTRASTING VALENCE BAND AUGER ELECTRON SPECTRA FOR SILVER AND ALUMINUM. 4-1022
L- AND M-SHELL YIELDS AND ELECTRON SPECTRA FOR THE TRANSURANIC ELEMENTS. 4-0968
AUGER AND OTHER CHARACTERISTIC ENERGIES IN SECONDARY ELECTRON SPECTRA FROM ALUMINUM SURFACES. 4-0896
M4,5N4,5N4,5 AND M4,5O1 AUGER ELECTRON SPECTRA FROM CADMIUM VAPOR. 4-0728
LOW ENERGY AUGER AND LOSS ELECTRON SPECTRA FROM MAGNESIUM AND ITS OXIDE. 4-0895
CHEMICAL SHIFTS IN AUGER ELECTRON SPECTRA FROM SILICON IN SILICON NITRIDE. 4-0881
CHARACTERISTIC ENERGIES IN SECONDARY ELECTRON SPECTRA FROM SILICON (111) SURFACES. 5-1263
FIL/ AUGER AND OTHER CHARACTERISTIC ENERGIES IN SECONDARY ELECTRON SPECTRA FROM THIN EVAPORATED PHOSPHORUS 4-0986
F ZINC, ZINC OXIDE, GALLIUM, AND GALLIU/ AUGER AND DIRECT ELECTRON SPECTRA IN X-RAY PHOTOELECTRON STUDIES O 4-1054
SECONDARY AND AUGER ELECTRON SPECTRA OF A COPPER- BERYLLIUM ALLOY. 4-0916
HIGH RESOLUTION L2,3M2,3M2,3 AUGER ELECTRON SPECTRA OF ARGON. 4-0912
CHEMICAL EFFECTS ON THE AUGER ELECTRON SPECTRA OF BERYLLIUM. 4-0834
CHARACTERISTIC ENERGY LOSS AND AUGER ELECTRON SPECTRA OF GALLIUM PHOSPHIDE (110). 5-1657
AUGER ELECTRON SPECTRA OF MAGNESIUM THIN FILMS. 4-0873
X-RAY PHOTOELECTRON AND AUGER ELECTRON SPECTRA OF MAGNESIUM THIN FILMS. 5-1392
SURFACES. X-RAY EXCITED PHOTOELECTRON AND AUGER ELECTRON SPECTRA OF OXIDIZED STATES ON SOME METAL 5-1690
HANUM, PRASEODYMIUM, NEODYMIUM, GADOLINIUM, DYSPRO/ AUGER ELECTRON SPECTRA OF SOME RARE EARTH METALS. (LANT 4-1100
C BOMBARDMENT. SECONDARY ELECTRON SPECTRA OF SOME SOLID TARGETS UNDER IONI 4-0942
M,/ COMPARISON OF THE SOFT X-RAY PHOTOEMISSION, AND AUGER ELECTRON SPECTRA OF THE VALENCE BANDS OF MAGNESIU 4-0838
KLL AUGER ELECTRON SPECTRA OF THULIUM AND LUTETIUM. 4-0753
IMP/ RELATIVE SATELLITE LINE INTENSITIES IN ARGON K AUGER ELECTRON SPECTRA PRODUCED BY 0.4 TO 4 MEV PROTON 2-0497
THE K-SHELL AUGER ELECTRON SPECTRA. REVIEW. 1-0060
ION OF THE SURFACE OF MOLYBDENUM AND STEEL BASED ON AUGER ELECTRON SPECTRAL DATA. CHEMICAL COMPOSIT 5-1985
SIMPLIFIED RETARDING FIELD PHOTOELECTRON AND AUGER ELECTRON SPECTROGRAPH. 3-0642
(REVIEW, 12 REFS) HIGH RESOLUTION ELECTRON SPECTROGRAPHS FOR SOLID STATE RESEARCH. 3-0703
AN ENERGY MODULATED (CYLINDRICAL) HIGH ENERGY RESOLUTION ELECTRON SPECTROMETER. 3-0616
HIGH RESOLUTION, LOW ENERGY ELECTRON SPECTROMETER. 3-0697
HIGH SENSITIVITY AUGER ELECTRON SPECTROMETER. 3-0672
HIGH SENSITIVITY ELECTRON SPECTROMETER. 3-0633
RETARDING FIELD ELECTRON SPECTROMETER. 3-0686
SECONDARY ELECTRON SPECTROMETER. 3-0709
S. (CONTAMINATION MONITOR) AUGER ELECTRON SPECTROMETER AS TOOL FOR SURFACE ANALYSI 3-0706
S. RESOLUTION AND SENSITIVITY CONSIDERATIONS OF AN AUGER ELECTRON SPECTROMETER BASED ON LEED DISPLAY OPTIC 3-0705
STUDY OF ALUMINUM WITH A COMBINED AUGER ELECTRON SPECTROMETER ELLIPSOMETER SYSTEM. 5-1132
COPY. HIGH RESOLUTION ELECTRON SPECTROMETER FOR AUGER ELECTRON SPECTROS 3-0667
PTION STUDIES. DEVELOPMENT AND APPLICATION OF AN "AUGER ELECTRON SPECTROMETER" FOR HIGH TEMPERATURE ADSOR 3-0586
ELECTRON STUDIES. HIGH RESOLUTION ELECTRON SPECTROMETER FOR PHOTOELECTRON AND AUGER 3-0600
ELECTRON SPECTROMETER FOR SURFACE ANALYSIS. 3-0715
SEGREGATION IN MAGNESIA DOPED ALUMINA. (HIGH VACUUM AUGER ELECTRON SPECTROMETER, FRACTURED ALUMINA) /NDARY 5-1615
Y ELECTRON DIFFRACTION SYSTEM. AUGER ELECTRON SPECTROMETER INTEGRATED WITH A LOW ENERG 3-0658
AUGER ELECTRON SPECTROMETER PREAMPLIFIER. 3-0708
S. IMPROVED AUGER ELECTRON SPECTROMETER USING CONCENTRIC HEMISPHERE 3-0591
S. HIGH ENERGY RESOLUTION AUGER ELECTRON SPECTROMETER USING CONCENTRIC HEMISPHERE 3-0592
X-RAY ELECTRON SPECTROMETERS AND AUGER ANALYZERS. 3-0599
PY) A COMPARISON OF THE EXTENDURE OF ELECTRON SPECTROMETERS. (USED IN AUGER SPECTROSCO 3-0626
ELECTRON SPECTROMETRY. 1-0108
AUGER ELECTRON SPECTROMETRY. 1-0191
SURFACE CHARACTERIZATION BY AUGER ELECTRON SPECTROMETRY. 5-1815
OF CERAMIC SURFACES. (ION SCATTERING SPECTROMETRY, AUGER ELECTRON SPECTROMETRY) /HEMICAL CHARACTERIZATIONS 5-1326
OMATI/ HIGH RESOLUTION ELECTROSTATIC SPECTROMETER USED IN ELECTRON SPECTROMETRY FOR CHEMICAL ANALYSIS. (AUT 3-0717
ZED ATOMS. AUGER ELECTRON SPECTROMETRY OF SINGLY AND MULTIPLY IONI 4-0978
ELECTRON SPECTROMETRY. (REVIEW) 1-0035
N ALUMINUM. AUGER ELECTRON SPECTROSCOPIC ANALYSIS OF ANODIC FILMS O 5-1614
E ON SILICON. AUGER ELECTRON SPECTROSCOPIC ANALYSIS OF SILICON NITRID 5-1455
CESSES IN BROMINE-SUBSTITUTED METHANES AND SOME HYDROCAR/ ELECTRON SPECTROSCOPIC INVESTIGATION OF AUGER PRO 2-0515
SHIFTS IN SOME VANADIUM COMPOUNDS. (VANADIUM OXID/ AUGER ELECTRON SPECTROSCOPIC INVESTIGATIONS OF CHEMICAL 4-1090
TEMPER EMBRITTLEMENT OF TIN DOPED STEELS BY AUGER ELECTRON SPECTROSCOPIC METHOD. 6-2128
ATE SURFACE DURING SPUTTERING AND VACUUM / LEED AND AUGER ELECTRON SPECTROSCOPIC OBSERVATIONS OF THE SUBSTR 5-1678
OXIDE (100). LEED AND ELECTRON SPECTROSCOPIC OBSERVATIONS ON NICKEL MON 5-1687
ZED IRON. (AUGER STUDY OF SURFACE) ELECTRON SPECTROSCOPIC STUDIES OF CLEAN AND OXIDI 5-1323
F NICKEL- PALLADIUM ALLOYS. ELECTRON SPECTROSCOPIC STUDIES OF THE OXIDATION O 5-1953
18-8 STAINLESS STEEL ELECTROPOLISHED OR ABRADED B/ AUGER ELECTRON SPECTROSCOPIC STUDIES OF THE SURFACES OF 6-2130
SURFACE. LEED AND AUGER ELECTRON SPECTROSCOPIC STUDIES ON TUNGSTEN (110) 5-1677
NING OF COPPER CATALYSTS. ELECTRON SPECTROSCOPIC STUDY OF THE SURFACE POISO 5-1176
ON OF TUNGSTEN FIBERS IN NICKEL MATRIX COMPOSITES. (AUGER ELECTRON SPECTROSCOPIC TECHNIQUES) /CRYSTALLIZATI 6-2046
CTS. COMBINED MASS SPECTROMETRIC AND AUGER ELECTRON SPECTROSCOPIC TECHNIQUES FOR METAL CONTA 6-2042
LLOY SYSTEMS. (SURFACE COMPOSITION, INTERDIFFUSION, AUGER ELECTRON SPECTROSCOPY) /E CATALYTIC ACTIVITY OF A 6-2003
ABSOLUTE ATOMIC DENSITIES DETERMINED BY AUGER ELECTRON SPECTROSCOPY. 5-1843
ION OF HYDROGEN ON NICKEL SINGLE CRYSTAL SURFACES. (AUGER ELECTRON SPECTROSCOPY) ADSORPT 5-1257
YPTON ON THE (111) FACE OF COPPER SINGLE CRYSTALS. (AUGER ELECTRON SPECTROSCOPY) ADSORPTION OF KR 5-1221
SORPTION OF OXYGEN ON THE (110) PLANE OF TUNGSTEN. (AUGER ELECTRON SPECTROSCOPY) AD 5-1318
PTION OF OXYGEN ON SILICON (111) SURFACES. PART-1. (AUGER ELECTRON SPECTROSCOPY) ADSOR 5-1474
ADSORPTION OF OXYGEN ON A CLEAN SILICON SURFACE. (AUGER ELECTRON SPECTROSCOPY) 5-1638
MANIUM. PART-1: STRUCTURAL AND KINETIC PROPERTIES. (AUGER ELECTRON SPECTROSCOPY) /APOR ON CLEAN CLEAVED GER 5-1432
ALLOY SPUTTERING STUDIES WITH IN-SITU AUGER ELECTRON SPECTROSCOPY. 5-1877
CRYSTALLINE TRANSITION OF SILICON (111) SURFACES. (AUGER ELECTRON SPECTROSCOPY) AMORPHOUS TO 5-1780
BORON STRUCTURE AFTER DEPOSITION ON SILICON (111). (AUGER ELECTRON SPECTROSCOPY) AN ORDERED 5-1168

ANALYSIS OF GLASS CONTAINER COATINGS BY AUGER ELECTRON SPECTROSCOPY.   5-1704
IES AND FLUORIDE ADDITIVES IN HOT PRESSED OXIDES BY AUGER ELECTRON SPECTROSCOPY. /OF GRAIN BOUNDARY IMPURIT 5-1506
DIFFUSION IN BIMETALLIC THIN FILM STRUCTURES USING AUGER ELECTRON SPECTROSCOPY. ANALYSIS OF GRAIN BOUNDARY 6-2146
ANALYSIS OF IMPURITIES ON SILICON SURFACES BY AUGER ELECTRON SPECTROSCOPY.   5-1674
GHT ION BACKSCATTERING AND SPUTTERING COMBINED WITH AUGER ELECTRON SPECTROSCOPY. /S OF SURFACE LAYERS BY LI 5-1167
ANALYTICAL AUGER ELECTRON SPECTROSCOPY.   1-0031
(100), (110), AND (111) FACES OF COPPER CRYSTALS. (AUGER ELECTRON SPECTROSCOPY) /YDROGEN DESORBED FROM THE 5-1155
APPARATUS FOR ANALYSIS BY AUGER ELECTRON SPECTROSCOPY.   3-0595
APPARATUS FOR SURFACE ANALYSIS BY ELECTRON SPECTROSCOPY.   3-0664
AUGER ELECTRON SPECTROSCOPY.   1-0043
AUGER ELECTRON SPECTROSCOPY.   1-0061
AUGER ELECTRON SPECTROSCOPY.   1-0140
AUGER ELECTRON SPECTROSCOPY.   1-0145
AUGER ELECTRON SPECTROSCOPY.   1-0146
AUGER ELECTRON SPECTROSCOPY.   1-0171
AUGER ELECTRON SPECTROSCOPY.   1-0192
AUGER ELECTRON SPECTROSCOPY.   1-0204
URFACES IN ULTRAHIGH VACUUM AS OBSERVED BY LEED AND AUGER ELECTRON SPECTROSCOPY. /IOR OF REFRACTORY METAL S 6-2020
BIBLIOGRAPHY ON AUGER ELECTRON SPECTROSCOPY.   1-0148
SIUM ADSORPTION ON TUNGSTEN STUDIED BY LEED AND SECONDARY ELECTRON SPECTROSCOPY. CE 5-1661
CHARACTERIZATION OF ALUMINUM ADHEREND SURFACES (BY AUGER ELECTRON SPECTROSCOPY).   5-1725
CHARACTERIZATION OF GLASS SURFACES BY AUGER ELECTRON SPECTROSCOPY.   5-1787
ALYSIS OF PLATINUM (100) AND GOLD (100) SURFACES BY AUGER ELECTRON SPECTROSCOPY. CHEMICAL AN 5-1712
CHEMICAL EFFECTS IN AUGER ELECTRON SPECTROSCOPY.   1-0014
CHEMICAL SHIFTS IN AUGER ELECTRON SPECTROSCOPY.   4-0863
CHEMISORPTION OF NITROGEN ON TUNGSTEN STUDIED BY AUGER ELECTRON SPECTROSCOPY.   5-1518
E CRYSTAL MAGNESIUM OXIDE AND OXIDIZED MAGNESIUM BY AUGER ELECTRON SPECTROSCOPY. COMPARATIVE STUDY OF SINGL 5-1495
ACE OF COPPER- MAGNESIUM OXIDE CATALYSTS STUDIED BY AUGER ELECTRON SPECTROSCOPY. COMPOSITION OF THE SURF 5-1361
IRMATION OF SULFUR EMBRITTLEMENT IN NICKEL ALLOYS. (AUGER ELECTRON SPECTROSCOPY) CONF 6-2053
CORROSION AND STRUCTURAL CHANGES IN GLASS. (AUGER ELECTRON SPECTROSCOPY)   5-1425
TION OF CARBON MONOXIDE ON A (110) NICKEL SURFACE. (AUGER ELECTRON SPECTROSCOPY) DECOMPOSI 5-1606
DECONVOLUTION TECHNIQUES IN AUGER ELECTRON SPECTROSCOPY.   1-0138
DEGRADED PASSIVE FILM ON IRON. APPLICATION OF AUGER ELECTRON SPECTROSCOPY.   5-1762
ILITY LIMITS FOR BORON AND PHOSPHORUS IN SILICON BY AUGER ELECTRON SPECTROSCOPY. DETECTAB 5-1887
ION OF ORGANIC CONTAMINANTS ON ALUMINUM SURFACES BY AUGER ELECTRON SPECTROSCOPY. DETECT 5-1487
ERMINATION OF FILM THICKNESS OR ION ETCH RATE USING AUGER ELECTRON SPECTROSCOPY. DET 5-1984
DETERMINATION OF SURFACE STRUCTURES BY LEED AND AUGER ELECTRON SPECTROSCOPY.   5-1957
ION OF PALLADIUM- NICKEL ALLOY FILM CATALYSTS USING AUGER ELECTRON SPECTROSCOPY. /N OF THE SURFACE COMPOSIT 5-1846
ILICON (SUBSTRATE)- GOLD (EVAPORATED FILM) SYSTEM. (AUGER ELECTRON SPECTROSCOPY) DIFFUSE INTERFACE IN S 5-1681
BSTRATE- VACUUM EVAPORATED FILM) SYSTEMS STUDIED BY AUGER ELECTRON SPECTROSCOPY. /CE STRUCTURES IN SOME (SU 5-1436
ILIZING WORK FUNCTION CHANGES: CHROMIUM INTO GOLD. (AUGER ELECTRON SPECTROSCOPY) /UREMENTS IN THIN FILMS UT 5-1889
OF COBALT OUT OF COBALT HARDENED GOLD MEASURED WITH AUGER ELECTRON SPECTROSCOPY. DIFFUSION 6-2135
SION STUDIES IN CHROMIUM- PLATINUM THIN FILMS USING AUGER ELECTRON SPECTROSCOPY. DIFFU 5-1286
DIFFUSION STUDIES IN GOLD- PLATINUM THIN FILMS BY AUGER ELECTRON SPECTROSCOPY.   5-1636
N BETWEEN ADSORBED MONOLAYERS AND THICKER LAYERS IN AUGER ELECTRON SPECTROSCOPY. DISTINCTIO 5-1797
NAMIC ELUCIDATION OF REACTION STEPS ON A SOLID SURFACE BY ELECTRON SPECTROSCOPY. DY 5-1535
LFUR AND OXYGEN ON A MOLYBDENUM SURFACE BY MEANS OF AUGER ELECTRON SPECTROSCOPY. /SM OF REACTION BETWEEN SU 5-1534
Y ALIGN ION BOMBARDMENT GUNS FOR DEPTH PROFILING IN AUGER ELECTRON SPECTROSCOPY. EASY METHOD TO ACCURATEL 3-0700
SOLUTE LAYER ON THE NUCLEATION OF FATIGUE CRACKS. (AUGER ELECTRON SPECTROSCOPY) EFFECT OF A THIN DIFFUSED 6-2037
E PHYSICAL PROPERTIES OF SPUTTERED TANTALUM FILMS. (AUGER ELECTRON SPECTROSCOPY) /, CARBON AND OXYGEN ON TH 5-1473
EFFECT OF SURFACE ROUGHNESS OF AUGER ELECTRON SPECTROSCOPY.   5-1448
NICKEL (111), NICKEL (100), AND NICKEL SHEET USING AUGER ELECTRON SPECTROSCOPY. / TECHNIQUES AS APPLIED TO 5-1467
OF IONIZATION SPECTROSCOPY: SILICON ON BERYLLIUM. (AUGER ELECTRON SPECTROSCOPY) /NDENCE OF THE SENSITIVITY 5-1665
ENERGY WIDTHS OF ROENTGEN LEVELS IN AUGER ELECTRON SPECTROSCOPY.   4-0982
MISSION FROM ZINC OXIDE. (OXYGEN ADSORPTION EFFECT, AUGER ELECTRON SPECTROSCOPY) EXOELECTRON E 4-0921
ACTORS AFFECTING DEPTH PROFILING MEASUREMENTS USING AUGER ELECTRON SPECTROSCOPY. F 5-1457
FAILURE ANALYSIS APPLICATIONS OF AUGER ELECTRON SPECTROSCOPY.   5-2034
ON: XENON ADSORBED ON THE (0001) FACE OF GRAPHITE. (AUGER ELECTRON SPECTROSCOPY) / TWO DIMENSIONAL TRANSITI 5-1857
TURE SURFACE ANALYSIS OF TEMPER EMBRITTLED STEEL BY AUGER ELECTRON SPECTROSCOPY. FRAC 6-2084
1-X) ALUMINUM(X) ARSENIDE SUPERLATTICES PROFILED BY AUGER ELECTRON SPECTROSCOPY. GALLIUM(5-1596
GLASS SURFACE ANALYSIS BY AUGER ELECTRON SPECTROSCOPY.   5-1720
GRAIN BOUNDARY ACTIVITY MEASUREMENTS BY AUGER ELECTRON SPECTROSCOPY.   6-2049
GRAIN BOUNDARY EMBRITTLEMENT OF STEELS STUDIED BY AUGER ELECTRON SPECTROSCOPY.   6-2103
GRAIN BOUNDARY SEGREGATION IN SINTERED ALUMINA. (AUGER ELECTRON SPECTROSCOPY)   6-2056
RY SEGREGATION RELATING TO INTERGRANULAR FRACTURE. (AUGER ELECTRON SPECTROSCOPY) GRAIN BOUNDA 6-2082
GRAIN BOUNDARY SEGREGATION. (AUGER ELECTRON SPECTROSCOPY)   6-2104
GROWTH AND STRUCTURE OF SEMICONDUCTING THIN FILMS. (AUGER ELECTRON SPECTROSCOPY)   5-1509
GROWTH MODE DISCRIMINATION OF THIN FILMS BY AUGER ELECTRON SPECTROSCOPY.   5-1624
HIGH TEMPERATURE AUGER ELECTRON SPECTROSCOPY.   1-0005
HIGHLY SENSITIVE AUGER ELECTRON SPECTROSCOPY.   1-0189
IN BOUNDARY PHASE IN HOT PRESSED SILICON NITRIDE BY AUGER ELECTRON SPECTROSCOPY. IDENTIFICATION OF A GRA 5-1749
IDENTIFICATION OF ATOMIC BORON IN STEEL BY AUGER ELECTRON SPECTROSCOPY.   6-2018
TION OF THE CHEMICAL STATE OF ADSORBED SPECIES WITH AUGER ELECTRON SPECTROSCOPY. IDENTIFICA 4-0868
THE FORM OF CARBON AT A SILICON (100) SURFACE USING AUGER ELECTRON SPECTROSCOPY. IDENTIFICATION OF 5-1378
ICAL GRID RETARDING POTENTIAL ANALYZER. (FOR USE IN AUGER ELECTRON SPECTROSCOPY) IMPROVED SPHER 3-0583
IMPURITY CONCENTRATIONS ON METAL SURFACES BY AUGER ELECTRON SPECTROSCOPY.   5-1923
E IN SILICON BY SECONDARY ION MASS SPECTROMETRY AND AUGER ELECTRON SPECTROSCOPY. /OGEN, OXYGEN, AND FLUORIN 5-1916
INELASTIC EFFECTS IN AUGER ELECTRON SPECTROSCOPY.   1-0139
INELASTIC SCATTERING EFFECTS IN QUANTITATIVE AUGER ELECTRON SPECTROSCOPY.   2-0433
TOELECTRIC WORK FUNCTION MEASUREMENTS COMBINED WITH AUGER ELECTRON SPECTROSCOPY. /RON (100) BY MEANS OF PHO 5-1927
INTERACTION OF OXYGEN WITH SILICON (111) SURFACES. (AUGER ELECTRON SPECTROSCOPY)   5-1693
BY HYDROGEN: EFFECT OF GRAIN BOUNDARY SEGREGATION. (AUGER ELECTRON SPECTROSCOPY) / EMBRITTLEMENT OF NICKEL 6-2072
TURE AND IMPURITY SEGREGATION IN STEEL. STUDIES BY AUGER ELECTRON SPECTROSCOPY. INTERGRANULAR FRAC 6-2127
ADSORPTION ONTO A GALLIUM ARSENIDE (110) SURFACE. (AUGER ELECTRON SPECTROSCOPY) INVESTIGATION OF CESIUM 5-1644
TIGATION OF HIGH RESOLUTION SCANNING TECHNIQUES FOR AUGER ELECTRON SPECTROSCOPY. INVES 3-0625
ROGEN CHEMISORPTION ON THE TUNGSTEN (100) SURFACE. (AUGER ELECTRON SPECTROSCOPY) INVESTIGATION OF HYD 5-1974
YGEN ADSORPTION ON TUNGSTEN AT HIGH TEMPERATURES BY AUGER ELECTRON SPECTROSCOPY. INVESTIGATION OF OX 5-1156
OF SILVER CESIUM OXIDE PHOTOEMISSIVE SURFACES USING AUGER ELECTRON SPECTROSCOPY. INVESTIGATION 5-1782
UES ON SURFACES OF SILICON INTEGRATED CIRCUITS. (BY AUGER ELECTRON SPECTROSCOPY) /TIGATIONS OF CARBON RESID 5-1345
OLID INTERFACE INTERACTIONS AT HIGH TEMPERATURES BY AUGER ELECTRON SPECTROSCOPY. /ESTIGATIONS OF THE GAS- S 5-1147

NVESTIGATIONS OF THE OXIDE FORMATION ON METALS WITH AUGER ELECTRON SPECTROSCOPY.   I 5-1329
K- AND L-SHELL FLUORESCENCE AND AUGER YIELDS AND AUGER ELECTRON SPECTROSCOPY.   4-0966
                         KIKUCHI CORRELATIONS IN AUGER ELECTRON SPECTROSCOPY.   5-1783
ICKEL (100) SURFACE. PART-2: NICKEL (111) SURFACE. (AUGER ELECTRON SPECTROSCOPY) /YSTAL SURFACES. PART-1: N 5-1452
         L-SHELL IONIZATION CROSS SECTIONS IN AUGER ELECTRON SPECTROSCOPY.   2-0339
R WINDOW SURFACE FINISHING AND COATING TECHNCLOGY. (AUGER ELECTRON SPECTROSCOPY)   LASE 1-0015
MONOLAYER LUBRICATION IN STEEL MACHINING STUDIED BY AUGER ELECTRON SPECTROSCOPY.   LEAD 6-2124
                              LOW ENERGY PHOTO- AUGER ELECTRON SPECTROSCOPY.   1-0080
GERMANIUM FILMS BY SPUTTER DEPOSITION. (HEED, LEED, AUGER ELECTRON SPECTROSCOPY) /W TEMPERATURE EPITAXY OF 5-1537
DATION OF SILICON STUDIED BY PHOTOSENSITIVE ESR AND AUGER ELECTRON SPECTROSCOPY.   LOW TEMPERATURE OXI 5-1786
            MATERIAL ANALYSIS OF THIN FCILS BY AUGER ELECTRON SPECTROSCOPY.   5-1798
EASUREMENT OF EQUILIBRIUM SURFACE SEGREGATION USING AUGER ELECTRON SPECTROSCOPY.   M 5-1184
N CROSS SECTION OF THE L2,3 SHELL CF CHLORINE USING AUGER ELECTRON SPECTROSCOPY. /SUREMENT OF THE IONIZATIO 4-0987
ES AND ON AMORPHOUS CARBON BY MEANS OF QUANTITATIVE AUGER ELECTRON SPECTROSCOPY. /ALT (100) CLEAVAGE SURFAC 5-1135
     MECHANICAL TESTING: IN-SITU FRACTURE DEVICE FOR AUGER ELECTRON SPECTROSCOPY.   3-0659
LICON- NOBLE METAL VAFCR QUENCHED ALLOYS STUDIED BY AUGER ELECTRON SPECTROSCOPY. /IC STATE OF SILICON IN SI 5-1438
             METALLURGICAL APPLICATION OF AUGER ELECTRON SPECTROSCOPY.   6-2123
ES DEPENDING ON INELASTIC SCATTERING OF ELECTRONS. (AUGER ELECTRON SPECTROSCOPY) METHODS OF SURFACE STUDI 5-1159
              MISCELLANEOUS TOPICS IN AUGER ELECTRON SPECTROSCOPY.   1-0075
          MUSCOVITE CLEAVAGE STUDIED BY AUGER ELECTRON SPECTROSCOPY.   5-1298
ION OF IMPURITY PROFILE IN ION IMPLANTED SILICON BY AUGER ELECTRON SPECTROSCOPY.   OBSERVAT 5-1482
E SURFACE OF A CLEAN ANNEALED 50-50 NICKEL ALLOY BY AUGER ELECTRON SPECTROSCOPY. /GREGATION OF COPPER TO TH 6-2044
OF CARBON ON TUNGSTEN (100) AND MCLYBDENUM (100). (AUGER ELECTRON SPECTROSCOPY)   OXIDATION 5-1944
          OXIDE FILMS: ELEMENT PROFILES BY AUGER ELECTRON SPECTROSCOPY.   4-1073
URE OF SAMARIUM, GADOLINIUM, AND TERBIUM STUDIED BY AUGER ELECTRON SPECTROSCOPY.   OXYGEN EXPOS 5-1332
         PALLADIUM SILICIDE FORMATION OBSERVED BY AUGER ELECTRON SPECTROSCOPY.   5-1773
      PASSIVE FILM ON IRON. APPLICATION OF AUGER ELECTRON SPECTROSCOPY.   5-1761
     PHASE SEPARATION IN SILICON OXIDES AS SEEN BY AUGER ELECTRON SPECTROSCOPY.   5-1502
FUNCTION STUDY ON IRON (100) SURFACE COMBINED WITH AUGER ELECTRON SPECTROSCOPY. PHOTOELECTRIC WORK 5-1926
         PLASMA CLEANING OF METAL SURFACES. (AUGER ELECTRON SPECTROSCOPY)   6-2093
    POROSITY IN SPUTTERED PLATINUM FILMS. (AUGER ELECTRON SPECTROSCOPY)   5-1965
              POTENTIAL MAPPING USING AUGER ELECTRON SPECTROSCOPY.   1-0113
PRESENT STATUS AND PROBLEMS ON THE APPLICATION OF AUGER ELECTRON SPECTROSCOPY.   1-0147
     PROBING DEPTH IN PHOTOEMISSION AND AUGER ELECTRON SPECTROSCOPY.   2-0428
                  QUANTIFICATION OF AUGER ELECTRON SPECTROSCOPY.   1-0199
     QUANTITATIVE ANALYSIS OF SOLID SURFACES BY AUGER ELECTRON SPECTROSCOPY.   5-1715
        QUANTITATIVE ASPECTS OF AUGER ELECTRON SPECTROSCOPY.   1-0129
              QUANTITATIVE AUGER ELECTRON SPECTROSCOPY.   1-0207
ND STEPPED PLATINUM SINGLE CRYSTAL SURFACES. (LEED, AUGER ELECTRON SPECTROSCOPY) /INDEX ((111) AND (100)) A 5-1838
              RECENT TRENDS IN AUGER ELECTRON SPECTROSCOPY.   1-0126
LES HEATED DIRECTLY BY ELECTRIC CURRENT OBSERVED BY AUGER ELECTRON SPECTROSCOPY.   SAMP 1-0133
              SECONDARY (AUGER) ELECTRON SPECTROSCOPY.   1-0167
                    SECONDARY ELECTRON SPECTROSCOPY.   1-0074
O THE GRAIN BOUNDARIES OF TEMPER EMBRITTLED STEEL. (AUGER ELECTRON SPECTROSCOPY) / MANGANESE AND ANTIMONY T 6-2070
GATION OF SULFUR ON A MOLYBDENUM SURFACE STUDIED BY AUGER ELECTRON SPECTROSCOPY.   SEGRE 5-1564
      SILICCN- OXYGEN INTERACTIONS USING AUGER ELECTRON SPECTROSCOPY.   5-1511
LIDING OF POLY(VINYL CHLCRIDE) ON METALS STUDIED BY AUGER ELECTRON SPECTROSCOPY.   S 5-1727
SOME ASPECTS OF THE SURFACE BEHAVIOR OF SILICON. (AUGER ELECTRON SPECTROSCOPY)   5-1510
FIELD SPECTROMETERS USING POST MCNOCHROMATORS. (FOR AUGER ELECTRON SPECTROSCOPY) / PROPERTIES OF RETARDING 3-0624
ATINUM, COPPER, NIOBIUM, STAINLESS STEEL SURFACES. (AUGER ELECTRON SPECTROSCOPY) /EN DESORBED FROM IRON, PL 6-2006
          SPECTRUM SUBTRACTION TECHNIQUES IN AUGER ELECTRON SPECTROSCOPY.   1-0068
F A GOLD SUBSTRATE WITH LEAD MONCLAYERS. (INTERFACE AUGER ELECTRON SPECTROSCOPY) SPONTANEOUS ALLOYING O 5-1178
SPUTTERING MEASUREMENTS ON CTR MATERIALS USING AUGER ELECTRON SPECTROSCOPY   1-0182
ON CONTROLLED THERMONUCLEAR REACTOR MATERIALS USING AUGER ELECTRON SPECTROSCOPY. SPUTTERING MEASUREMENTS 5-1826
- GOLD, GOLD- COPPER, AND SILVER- COPPER STUDIED BY AUGER ELECTRON SPECTROSCOPY. / THE ALLOY SYSTEMS SILVER 5-1331
, NICKEL (100), SHEET AND EVAPORATED NICKEL FILMS. (AUGER ELECTRON SPECTROSCOPY) /GEN ON CLEAN NICKEL (111) 5-1468
              ANALYSIS WITH X-RAY PHOTOELECTRON SPECTROSCOPY AND AUGER ELECTRON SPECTROSCOPY. STRUCTURE 1-0050
CON- IRON ALLOY (100) BY MEANS OF WORK FUNCTION AND AUGER ELECTRON SPECTROSCOPY. /INITIAL OXIDATION ON SILI 5-1924
OXIDIZED FILMS OF III-V COMPOUND SEMICONDUCTORS BY AUGER ELECTRON SPECTROSCOPY. STUDIES ON SURFACE PLASMA 5-1699
STUDIES ON THE IRIDIUM (111) SURFACE USING LEED AND AUGER ELECTRON SPECTROSCOPY.   5-1372
STUDY OF AMORPHOUS AND CRYSTALLINE SILICATES BY AUGER ELECTRON SPECTROSCOPY.   4-0796
STUDY OF DIFFUSION IN THIN COPPER- LEAD FILMS USING AUGER ELECTRON SPECTROSCOPY.   5-1972
FRACTURE SURFACES OF HCT PRESSED SILICON NITRIDE BY AUGER ELECTRON SPECTROSCOPY.   STUDY OF 5-1444
SEGREGANTS IN THERMALLY EMBRITTLED MARAGING STEEL. (AUGER ELECTRON SPECTROSCOPY) STUDY OF GRAIN BOUNDARY 6-2055
ACTURE SURFACES IN DOPED TUNGSTEN- RHENIUM ALLOYS. (AUGER ELECTRON SPECTROSCOPY) STUDY OF GRAIN BOUNDARY FR 6-2112
ARSENIDE (111) AND (-1-1-1) SURFACES USING LEED AND AUGER ELECTRON SPECTROSCOPY.   STUDY OF INDIUM 5-1381
STUDY OF IRIDIUM (100) SURFACE USING LEED AND AUGER ELECTRON SPECTROSCOPY.   5-1371
      STUDY OF MUSCOVITE AND SILICON BY AUGER ELECTRON SPECTROSCOPY.   4-0792
STUDY OF PALLADIUM SILICIDE FILMS ON SILICON USING AUGER ELECTRON SPECTROSCOPY.   5-1336
UM (0001) AND RHODIUM (111) SURFACES USING LEED AND AUGER ELECTRON SPECTROSCOPY.   STUDY OF RUTHENI 5-1380
IUM BY COMBINED SECONDARY ION MASS SPECTROMETRY AND AUGER ELECTRON SPECTROSCOPY. /OCESSES ON COPPER- BERYLL 5-1213
CHEMICAL COMPOSITION CF MOS AND MNOS STRUCTURES BY AUGER ELECTRON SPECTROSCOPY.   STUDY OF THE 5-1503
SURFACES BY ENERGY LOSS SPECTROSCOPY COMBINED WITH AUGER ELECTRON SPECTROSCOPY. /E OXIDATION OF MOLYBDENUM 5-1531
REPARATION OF ATOMICALLY CLEAN TUNGSTEN SURFACES BY AUGER ELECTRON SPECTROSCOPY.   STUDY OF THE P 5-1517
              SURFACE ANALYSIS BY AUGER ELECTRON SPECTROSCOPY.   5-1402
SURFACE CHARACTERIZATION BY LOW ENERGY PHOTO AUGER ELECTRON SPECTROSCOPY.   5-1429
SURFACE CHEMISTRY OF CONTACT MATERIALS. (BY AUGER ELECTRON SPECTROSCOPY)   5-1394
TEEL STUDIED BY SECONDARY ION MASS SPECTROMETRY AND AUGER ELECTRON SPECTROSCOPY. /SS OF 316 L-N STAINLESS S 6-2087
OBSERVED BY X-RAY PHOTOEMISSION SPECTROSCOPY AND AUGER ELECTRON SPECTROSCOPY. /APORATED PERMALLOY FILMS 5-1738
SURFACE COMPOSITION CF PROMOTED IRON CATALYSTS BY AUGER ELECTRON SPECTROSCOPY.   5-1817
PALLADIUM- GOLD AND PALLADIUM- SILVER CATALYSTS BY AUGER ELECTRON SPECTROSCOPY. SURFACE COMPOSITION OF 5-1976
SURFACE COMPOSITION CF THE SILVER- GOLD SYSTEM BY AUGER ELECTRON SPECTROSCOPY.   5-1710
FACE CONCENTRATION OF MAGNESIUM IN 6061 ALUMINUM BY AUGER ELECTRON SPECTROSCOPY.   SUR 5-1824
MOLYBDENUM IN TYPES 316 AND 304 STAINLESS STEEL BY AUGER ELECTRON SPECTROSCOPY. SURFACE CONCENTRATION OF 6-1995
RMATION CAUSED ON NATURAL MOLYBDENITE BY ABRASION. (AUGER ELECTRON SPECTROSCOPY) SURFACE DEFO 6-2132
DIFFUSION OF TIN OVER A GOLD SURFACE OBSERVED WITH AUGER ELECTRON SPECTROSCOPY.   SURFACE 6-2136
CHMENT OF INDIUM IN EVAPORATED GOLD- INDIUM FILMS. (AUGER ELECTRON SPECTROSCOPY) SURFACE ENRI 5-1892
          SURFACE MADELUNG POTENTIALS IN ELECTRON SPECTROSCOPY.   2-0511
ION AND ITS RELATION TO GRAIN BOUNDARY SEGREGATION (AUGER ELECTRON SPECTROSCOPY). SURFACE SEGREGAT 6-2106
SURFACE SEGREGATION IN TITANIUM AS MONITCRED BY AUGER ELECTRON SPECTROSCOPY.   6-2067

EN IN NIOBIUM- OXYGEN AND TANTALUM- OXYGEN ALLOYS. (AUGER ELECTRON SPECTROSCOPY) /RFACE SEGREGATION OF OXYG 6-2065
M- GOLD 70 ATOMIC PERCENT CONTACT METAL ALLOYS. (BY AUGER ELECTRON SPECTROSCOPY) /MIC PERCENT, AND PALLADIU 6-2041
                SURFACE SEGREGATION STUDIES IN ALLOYS USING AUGER ELECTRON SPECTROSCOPY.   6-2019
N INVESTIGATION OF SURFACES CONTAINING IMPURITIES. (AUGER ELECTRON SPECTROSCOPY)   SURFACES TEXTURES: A   5-1707
                        TECHNIQUE AND APPLICATIONS OF AUGER ELECTRON SPECTROSCOPY.   1-0151
                        TECHNOLOGICAL APPLICATION OF AUGER ELECTRON SPECTROSCOPY.   1-0064
IN A COPPER- 10 ATOMIC PERCENT ALUMINUM ALLCY USING AUGER ELECTRON SPECTROSCOPY. /RIUM SURFACE SEGREGATION   6-2030
        THICKNESS DETERMINATION OF ULTRATHIN FILMS BY AUGER ELECTRON SPECTROSCOPY.   5-1449
        THICKNESS MEASUREMENTS OF ADSORBED LAYERS BY AUGER ELECTRON SPECTROSCOPY.   5-1618
HREE DIMENSIONAL ELEMENTAL ANALYSIS OF MATERIALS BY AUGER ELECTRON SPECTROSCOPY.   T   5-1597
M UTILIZATION OF THE CYLINDRICAL MIRROR ANALYZER IN AUGER ELECTRON SPECTROSCOPY.   TOWARD OPTIMU   3-0678
CH SURFACE MODEL FOR THE OXIDATION OF STRONTIUM, BY AUGER ELECTRON SPECTROSCOPY. /IFICATION OF THE METAL-RI   6-2045
                WORK FUNCTION CHANGES ON CONTACT MATERIALS. (AUGER ELECTRON SPECTROSCOPY)   5-1395
LL-DEFINED SURFACE OF COPPER- NICKEL ALLOY PLATES. (AUGER ELECTRON SPECTROSCOPY)   WORK FUNCTION OF WE   5-1873
ION VARIATION WITH ALLCY COMPOSITION SILVER- GOLD. (AUGER ELECTRON SPECTROSCOPY)   WORK FUNCT   4-0828
SOLID GAS INTERFACE.   (AUGER) ELECTRON SPECTROSCOPY, A METHOD FOR STUDYING THE   1-0013
AL AND STRUCTURAL ANALYSIS AND A REVIEW OF ITS POSSIBLE / ELECTRON SPECTROSCOPY. A NEW TECHNIQUE FOR MATERI   1-0084
HER DEPTH PRO/ SECONDARY ION MASS SPECTROMETRY (AND AUGER ELECTRON SPECTROSCOPY): A USEFUL COMPLEMENT TO OT   1-0009
    FILMS ON SILICON.   AUGER ELECTRON SPECTROSCOPY ANALYSIS OF ION PLATED GOLD   5-1225
UM AND NICKEL- PALLADIUM ALLOYS.   QUANTITATIVE AUGER ELECTRON SPECTROSCOPY ANALYSIS OF SILVER- PALLADI   6-2088
RFACES OF IRRADIATED PRESSURE VESSEL STEELS.   AUGER ELECTRON SPECTROSCOPY ANALYSIS OF THE FRACTURE SU   6-2113
NADIUM COMPOUND SURFACES.   AUGER ELECTRON SPECTROSCOPY ANALYSIS OF VANADIUM AND VA   5-1862
MICAL SHIFTS IN X-RAY PHOTOELECTRON AND AUGER SPECTRA FO/ ELECTRON SPECTROSCOPY AND ADSORPTION STUDIES. CHE   5-1204
ECTROSCOPY: A COMPAR/ SURFACE CHEMICAL ANALYSIS BY AUGER ELECTRON SPECTROSCOPY AND APPEARANCE POTENTIAL SP   5-1910
OXIDATION MODIFIED 440C STEEL.   AUGER ELECTRON SPECTROSCOPY AND DEPTH PROFILE STUDY OF   6-2029
OXIDATION OF MODIFIED 440C STEEL.   AUGER ELECTRON SPECTROSCOPY AND DEPTH PROFILE STUDY OF   6-2032
    SULFUR DIOXIDE WITH MODIFIED 440C STEEL STUDIED BY AUGER ELECTRON SPECTROSCOPY AND DEPTH PROFILING. /ON OF 6-2031
ION STUDIES OF THE INTERACTIONS CF GASES / COMBINED AUGER ELECTRON SPECTROSCOPY AND ELECTRON IMPACT DESORPT   5-1691
ION / IDENTIFICATION OF ION SPECIES IN THE COMBINED AUGER ELECTRON SPECTROSCOPY AND ELECTRON IMPACT DESORPT   5-1692
ION STUDIES OF SILICON SURFACES.   COMBINED AUGER ELECTRON SPECTROSCOPY AND ELECTRON IMPACT DESORPT   5-1694
ION STUDIES OF SILICON SURFACES.   COMBINED AUGER ELECTRON SPECTROSCOPY AND ELECTRON IMPACT DESORPT   5-1695
                        QUANTITATIVE AUGER ELECTRON SPECTROSCOPY AND ELECTRON RANGES.   1-0177
TERING IN THE STUDIES CF METAL SURFACES.   AUGER ELECTRON SPECTROSCOPY AND INELASTIC ELECTRON SCAT   5-1818
R OBTAINING CHEMICAL PROFILES.   USE OF AUGER ELECTRON SPECTROSCOPY AND INERT GAS SPUTTERING FO   1-0156
/ANCES IN INSTRUMENTATION FOR MICROPROBE ANALYSIS. (AUGER ELECTRON SPECTROSCOPY AND ION MICROPROBE MASS SP/   3-0716
                                            AUGER ELECTRON SPECTROSCOPY AND ITS APPLICATION.   1-0141
LLURGY.   AUGER ELECTRON SPECTROSCOPY AND ITS APPLICATION IN META   6-2076
IES OF SOME SILICON SURFACES.   AUGER ELECTRON SPECTROSCOPY AND ITS APPLICATION IN STUD   5-1418
ACE STUDIES. (REVIEW, 130 REFS)   AUGER ELECTRON SPECTROSCOPY AND ITS APPLICATION TO SURF   1-0059
FACE CHEMISTRY. (REVIEW)   AUGER ELECTRON SPECTROSCOPY AND ITS APPLICATIONS IN SUR   1-0102
                        PRESENT STATE OF AUGER ELECTRON SPECTROSCOPY AND LEED.   1-0006
TAL INTERFACES. (AUGER EFFECT)   ELECTRON SPECTROSCOPY AND LEED STUDIES OF GAS- ME   5-1320
CTURE AND COMPOSITI/ CCMBINED FIELD ION MICRCSCOPY. AUGER ELECTRON SPECTROSCOPY, AND LEED STUDY OF THE STRU   5-1951
FRACTION STUDY OF THE ACTIVATION OF GALLIUM ARSENI/ AUGER ELECTRON SPECTROSCOPY AND LOW ENERGY ELECTRON DIF   5-1845
FRACTION STUDIES OF ADSORPTION ISOTHERMS: XENCN O/ AUGER ELECTRON SPECTROSCOPY AND LOW ENERGY ELECTRON DIF   5-1856
FRACTION STUDY OF ALKALI METAL OVERLA/ QUANTITATIVE AUGER ELECTRON SPECTROSCOPY AND LOW ENERGY ELECTRON DIF   5-1895
SURFACES.   AUGER ELECTRON SPECTROSCOPY AND MICROANALYSIS OF SOLID   5-1414
OF PHOTORESIST RESIDUES.   DETECTION BY AUGER ELECTRON SPECTROSCOPY AND REMOVAL BY OZONIZATION   5-1450
/EM AND THE CORRELATION OF RESISTIVITY CHANGES WITH AUGER ELECTRON SPECTROSCOPY AND RUTHERFORD BACKSCATTER/   5-1393
SCOPY.   COMBINED AUGER ELECTRON SPECTROSCOPY AND SCANNING ELECTRON MICRO   1-0003
SION DUE TO CASCADE PROC/ BACKGROUND INFORMATION IN AUGER ELECTRON SPECTROSCOPY AND SECONDARY ELECTRON EMIS   2-0411
EMICONDUCTORS. (SILICON, GERMANIUM, GRAPHITE, ELEC/ AUGER ELECTRON SPECTROSCOPY AND SECONDARY EMISSION IN S   2-0240
TROMETRY.   COMPARISON OF AUGER ELECTRON SPECTROSCOPY AND SECONDARY ION MASS SPEC   1-0135
TROMETRY IN THE MICROELECTRONIC TECHNOLOGY./ USE OF AUGER ELECTRON SPECTROSCOPY AND SECONDARY ION MASS SPEC   1-0136
/LYTICAL STUDIES USING ION SCATTERING SPECTROMETRY, AUGER ELECTRON SPECTROSCOPY AND SECONDARY ION MASS SPE/   5-1423
/OUS OBSERVATIONS OF PARTIALLY OXIDIZED SURFACES BY AUGER ELECTRON SPECTROSCOPY AND SECONDARY ION MASS SPE/   5-1553
/CARBON, AND OXYGEN) IN SPUTTERED TANTALUM FILMS BY AUGER ELECTRON SPECTROSCOPY AND SECONDARY ION MASS SPE/   5-1652
TROMETRY.   ELEMENTAL ANALYSIS WITH AUGER ELECTRON SPECTROSCOPY AND SECONDARY ION MASS SPEC   5-1654
/ VAPOR DEPOSITED PLATINUM AND PLATINUM SILICIDE BY AUGER ELECTRON SPECTROSCOPY AND SECONDARY ION MASS SPE/   5-1656
/FILE OF ION PLATED GOLD FILM ON COPPER ANALYZED BY AUGER ELECTRON SPECTROSCOPY AND SECONDARY ION MASS SPE/   5-1679
TROSCOPY ON WELL CHARACTERIZED / COMPARISON BETWEEN AUGER ELECTRON SPECTROSCOPY AND SECONDARY ION MASS SPE/   5-1689
/TUNGSTEN AND MOLYBDENUM SINGLE CRYSTALS STUDIED BY AUGER ELECTRON SPECTROSCOPY AND SLOW ELECTRON DIFFRACT/   5-1215
                                                  ELECTRON SPECTROSCOPY AND SURFACE STUDIES.   5-1194
/ETECTION OF IMPURITIES ON COPPER SURFACES BY LEED- AUGER ELECTRON SPECTROSCOPY AND THEIR INFLUENCE ON SUR/   5-1185
/ GRID RETARDING POTENTIAL ANALYZER. APPLICATION TO AUGER ELECTRON SPECTROSCOPY AND X-RAY PHOTOELECTRON SP/   3-0634
/ITION OF PASSIVE FILMS ON COPPER- NICKEL ALLCYS BY AUGER ELECTRON SPECTROSCOPY AND X-RAY PHOTOELECTRON SP/   5-1874
RED (110) GAL/ KINETICS OF THERMAL DESORPTION USING AUGER ELECTRON SPECTROSCOPY. APPLICATION TO CESIUM-COVE   5-1293
ZATION BY FAST CHARGED PARTICLES.   AUGER ELECTRON SPECTROSCOPY APPLIED TO INNER SHELL IONI   2-0535
ION PROBLEMS.   AUGER ELECTRON SPECTROSCOPY APPLIED TO SURFACE COMPOSIT   5-1812
ORPTION PHENOMENA A/ CHARACTERISTIC ENERGY LCSS AND AUGER ELECTRON SPECTROSCOPY APPLIED TO THE STUDY OF ADS   5-1776
DIFFUSION OF FOREIGN ATOMS IN SOLIDS NEAR THE S/ AUGER ELECTRON SPECTROSCOPY AS A TOOL FOR MEASURING THE   1-0186
ITIES, INTERFACES, AND BRITTLE FRACTURE. (REVIEW OF AUGER ELECTRON SPECTROSCOPY AS APPLIED TO INTERFACES) /   6-2078
AND PRIMARY BEAM CURRENTS. (CARBON ON SILVER FILM) AUGER ELECTRON SPECTROSCOPY AT HIGH SPATIAL RESOLUTION   5-1904
SING THE ACCURACY OF MATERIAL TESTS BY THE APPLICATION OF ELECTRON SPECTROSCOPY. (AUGER EFFECT)   INCREA   5-1138
/ICONDUCTOR STRUCTURES. TASK-II. APPLICATION OF SECONDARY ELECTRON SPECTROSCOPY. (AUGER ELECTRON ANALYSIS) / 5-1411
/PTION OF OXYGEN ON GALLIUM ARSENIDE METHOD OF LOW ENERGY ELECTRON SPECTROSCOPY. (AUGER PEAK AMPLITUDE CAL/   5-1641
                        CALIBRATION IN AUGER ELECTRON SPECTROSCOPY BY MEANS OF COADSORPTION.   3-0587
/Y OF THE ADSORPTION OF OXYGEN ON NICKEL(III) USING AUGER ELECTRON SPECTROSCOPY: CHEMICAL SHIFTS AND VALE/   5-1466
/EEL STUDIES BY SECONDARY ION MASS SPECTROMETRY AND AUGER ELECTRON SPECTROSCOPY. COMBINED ESCA- AUGER SYST/   3-0654
OR ADSORBED TIN ON IRON.   QUANTITATIVE AUGER ELECTRON SPECTROSCOPY. COMPARISON OF TECHNIQUES F   1-0178
- SILICON RATIO IN SPIN ON GLASS FILMS.   AUGER ELECTRON SPECTROSCOPY DETERMINATION OF THE OXYGEN   5-1827
XIDE ON TUNGSTEN (100).   FLASH DESORPTION- AUGER ELECTRON SPECTROSCOPY. DISCREPANCY FOR CARBON DIO   5-1461
SS SPECTROSCOPY.   COMPARISON OF TWO SUR/ PHOTC- AND AUGER ELECTRON SPECTROSCOPY (ESCA) AND SECONDARY ION MA   1-0088
                        HANDBOOK OF AUGER ELECTRON SPECTROSCOPY. (EXPERIMENTAL SPECTRA)   1-0159
SURFACE ANALYSIS METHODS. (REVIEW AUGER SPECTROSCOPY AND ELECTRON SPECTROSCOPY FOR CHEMICAL ANALYSIS / IN   1-0215
                        AUGER ELECTRON SPECTROSCOPY FOR CHEMICAL ANALYSIS.   1-0029
ERIZATION OF SOLID SURFACES. (REVIEW, AUGER SPECTROSCOPY AND ELECTRON SPECTROSCOPY FOR CHEMICAL ANALYSIS /ACT 1-0134
ORMATION OBTAINED FROM THE WIDTH OF SPECTRAL LINES IN THE ELECTRON SPECTROSCOPY FOR CHEMICAL ANALYSIS. INF   4-1009
/ON SPECTROSCOPY: X-RAY AND ELECTRON EXCITATION. (REVIEW, ELECTRON SPECTROSCOPY FOR CHEMICAL ANALYSIS AES)   1-0081
SIS. TECHNIQUE OF PREPARING POWDER SAMPLES FOR AES AND/OR ELECTRON SPECTROSCOPY FOR CHEMICAL ANALYSIS ANALY   1-0197

UGER SPECTROMETER.	COMBINED ELECTRON	SPECTROSCOPY FOR CHEMICAL ANALYSIS AND A	3-0671
UGER FOR SURFACE ANALYSIS.	CHOOSING BETWEEN ELECTRON	SPECTROSCOPY FOR CHEMICAL ANALYSIS AND A	5-1635
AUGER) STUDIES OF COPPER, COPPER(I) OXIDE, AND COPPER(II/	ELECTRON	SPECTROSCOPY FOR CHEMICAL ANALYSIS (AND	5-1794
AUGER SPECTROSCOPY STUDY OF VACUUM DEPOSITED SILVER/ UPS,	ELECTRON	SPECTROSCOPY FOR CHEMICAL ANALYSIS AND	5-1947
SECONDARY ION MA/ PHOTO- AND AUGER ELECTRON SPECTROSCOPY (ELECTRON	SPECTROSCOPY FOR CHEMICAL ANALYSIS AND	1-0088
R SPECTROSCOP/ SPUTTERING IN THIN FILM ANALYSIS METHODS. (ELECTRON	SPECTROSCOPY FOR CHEMICAL ANALYSIS AUGE	1-0214
/S SPECTROMETRY AND AUGER ELECTRON SPECTROSCOPY. COMBINED ELECTRON	SPECTROSCOPY FOR CHEMICAL ANALYSIS AUGE/	3-0654
ANALYSIS S/ SILICON DEVICE APPLICATIONS USING A COMBINED ELECTRON	SPECTROSCOPY FOR CHEMICAL ANALYSIS AUGER	5-1289
/ON OF THE TECHNIQUES FOR SILICON SURFACE ANALYSIS. (AES, ELECTRON	SPECTROSCOPY FOR CHEMICAL ANALYSIS SIM/	1-0052
/D THEIR APPLICATION FOR STUDIES ON IRON AND STEEL. (AES, ELECTRON	SPECTROSCOPY FOR CHEMICAL ANALYSIS SIM/	6-2035
/ CHEMICAL SHIFTS IN AUGER SPECTRA (MONOLAYER ADSORPTION, ELECTRON	SPECTROSCOPY FOR CHEMICAL ANALYSIS SIMPL	4-1115
ROSCOPY. NEW COMPACT ENERGY ANALYZER FOR AUGER AND ELECTRON	SPECTROSCOPY FOR CHEMICAL ANALYSIS SPECT	3-0702
ES OF SILVER, DISILVER OXIDE AND SILVER OXIDE. ELECTRON	SPECTROSCOPY FOR CHEMICAL ANALYSIS STUDI	4-1055
/IT IN TTL LOGIC FOR THE AUTOMATIC EXECUTION OF AUGER AND ELECTRON	SPECTROSCOPY FOR CHEMICAL ANALYSIS TEST/	3-0722
SCOPY.	AUGER ELECTRON	SPECTROSCOPY FOR SCANNING ELECTRON MICRO	1-0114
	AUGER ELECTRON	SPECTROSCOPY FOR SURFACE ANALYSIS.	1-0072
	AUGER ELECTRON	SPECTROSCOPY FOR SURFACE ANALYSIS.	1-0200
URFACES.	AUGER ELECTRON	SPECTROSCOPY FOR THE ANALYSIS OF SOLID S	5-1193
COMBINATION OF SECONDARY ION MASS SPECTROMETRY AND AUGER ELECTRON	SPECTROSCOPY, FOR THE ANALYSIS OF THIN /	5-1214
HEMICAL SPECIES. AUGER PARAMETER IN AUGER ELECTRON	SPECTROSCOPY FOR THE IDENTIFICATION OF C	5-1948
NTAMINATION IN SEMICONDUCTOR DEVICES. AUGER ELECTRON	SPECTROSCOPY FOR THE STUDY OF SURFACE CO	5-1978
	AUGER ELECTRON	SPECTROSCOPY FOR THIN FILM ANALYSIS.	1-0211
	AUGER ELECTRON	SPECTROSCOPY FOR THIN FILM ANALYSIS.	1-0212
ANTALUM (110). CHEMICAL SHIFTS IN AUGER ELECTRON	SPECTROSCOPY FROM INITIAL OXIDATION OF T	6-2038
AHIGH VACUUM. CURRENT PROGRESS WITH TECHNIQUES INVOLVING/ ELECTRON	SPECTROSCOPY FROM SOLID SURFACES IN ULTR	5-1901
OF THE PRIMARY ELECTRON BE/ QUANTITATIVE ASPECTS OF AUGER ELECTRON	SPECTROSCOPY. IMPORTANCE OF ATTENUATION	1-0206
ING ELECTRON MICROSCOPE. AUGER ELECTRON	SPECTROSCOPY IN A DIFFUSION PUMPED SCANN	3-0598
MPED SCANNING ELECTRON MIC/ HIGH SPATIAL RESOLUTION AUGER ELECTRON	SPECTROSCOPY IN AN ORDINARY DIFFUSION PU	3-0597
F CARBON AND OTHER SURFACE CONTAMINANTS ON / USE OF AUGER ELECTRON	SPECTROSCOPY IN DETERMINING THE EFFECT O	5-1919
APPLICATIONS OF AUGER ELECTRON	SPECTROSCOPY IN ELECTROPLATING.	1-0048
	AUGER ELECTRON	SPECTROSCOPY IN LEED SYSTEMS.	1-0153
	AUGER ELECTRON	SPECTROSCOPY IN LEED SYSTEMS.	1-0158
OPTIMIZATION OF AUGER ELECTRON	SPECTROSCOPY IN LEED SYSTEMS.	1-0152
COPY: POTENTIAL MEASUREMENTS. AUGER ELECTRON	SPECTROSCOPY IN SCANNING ELECTRON MICROS	3-0648
APPLICATION TO LOW ENERGY ARGON(+) ION SPUTTERIN/ AUGER ELECTRON	SPECTROSCOPY IN SPUTTERING MEASUREMENTS.	5-1825
W, 76 REFS) TECHNIQUE OF AUGER ELECTRON	SPECTROSCOPY IN SURFACE ANALYSIS. (REVIE	1-0196
IS. RCLE OF AUGER ELECTRON	SPECTROSCOPY IN SURFACE ELEMENTAL ANALYS	1-0194
/T COMPARISON OF LOW ENERGY ION BACKSCATTERING WITH AUGER ELECTRON	SPECTROSCOPY IN THE ANALYSIS OF SULFUR /	5-1869
CROSCOPE. AUGER ELECTRON	SPECTROSCOPY IN THE ELECTRON EMISSION MI	3-0604
CROSCOPE. POSSIBILITY OF AUGER ELECTRON	SPECTROSCOPY IN THE ELECTRON EMISSION MI	3-0605
/ENTIAL PROFILING ACROSS SEMICONDUCTOR JUNCTIONS BY AUGER ELECTRON	SPECTROSCOPY IN THE SCANNING ELECTRON M/	2-0555
CROSCOPE: AUGER ELECTRON IMAGES. AUGER ELECTRON	SPECTROSCOPY IN THE SCANNING ELECTRON MI	3-0650
CROSCOPE. AUGER ELECTRON	SPECTROSCOPY IN THE SCANNING ELECTRON MI	3-0710
TED EXOELECTRON EMISSION OF BERYLLIUM OXIDE LAYERS. AUGER ELECTRON	SPECTROSCOPY INVESTIGATIONS. /LY STIMULA	5-1325
OPY. APPEARANCE POTENTIAL SPECTROSCOPY. AUGER ELECTRON	SPECTROSCOPY. IONIZATION LOSS SPECTROSC	1-0174
RSENIC (00001). HIGH ANGULAR RESOLUTION SECONDARY ELECTRON	SPECTROSCOPY. KIKUCHI CORRELATIONS FOR A	5-1784
OMETRIC CALIBRATION. AUGER ELECTRON	SPECTROSCOPY MADE QUANTITATIVE BY ELLIPS	1-0205
ISOBARS FOR OXYGEN ON TUNGSTEN AT LOW PRESSURE AND/ AUGER ELECTRON	SPECTROSCOPY MEASUREMENTS OF ADSORPTION	5-1282
. AUGER ELECTRON	SPECTROSCOPY: METALLURGICAL APPLICATIONS	6-2021
IZATION OF GLASS FIBER SURFACES. AUGER ELECTRON	SPECTROSCOPY: NEW TOOL IN THE CHARACTER	1-0175
. AUGER ELECTRON	SPECTROSCOPY OF A STABLE GERMANIUM OXIDE	4-1108
REACTION ON A COPPER (110) CRYSTAL. AUGER ELECTRON	SPECTROSCOPY OF A SULFUR- OXYGEN SURFACE	5-1183
TE/ COMPOSITION VERSUS DEPTH PROFILES OBTAINED WITH AUGER ELECTRON	SPECTROSCOPY OF AIR OXIDIZED STAINLESS S	6-1998
ILVER, LEAD- INDIUM) AUGER ELECTRON	SPECTROSCOPY OF ALLOY SURFACES. (GOLD- S	5-1709
RATE SURFACE CHEMISTRY AND ADHESION OF THIN FILMS. (AUGER ELECTRON	SPECTROSCOPY OF ALUMINA SUBSTRATES) /BST	5-1851
AND OXYGEN- COVERED (100) TUNGSTEN. AUGER ELECTRON	SPECTROSCOPY OF CESIUM ADSORBED ON CLEAN	5-1295
	AUGER ELECTRON	SPECTROSCOPY OF CLEAN GALLIUM ARSENIDE.	5-1921
LD FILMS. AUGER ELECTRON	SPECTROSCOPY OF CLEAN POLYCRYSTALLINE GO	5-1760
ATION ON SILICON SURFACES. AUGER ELECTRON	SPECTROSCOPY OF CLEANUP-RELATED CONTAMIN	5-1983
ENIDE SURFACES. AUGER ELECTRON	SPECTROSCOPY OF CONTAMINATED GALLIUM ARS	5-1920
ON METALS. AUGER ELECTRON	SPECTROSCOPY OF CONTRAST FORMING LAYERS	6-2047
L SURFACES. (GOLD, SILVER, PALLADIUM, COPPER, NICK/ AUGER ELECTRON	SPECTROSCOPY OF FACE CENTERED CUBIC META	5-1717
ADIATED TYPE 304 STAINLESS STEEL. AUGER ELECTRON	SPECTROSCOPY OF FRACTURE SURFACES IN IRR	6-2001
MINUM ARSENIDE SURFACES. (AUGER) ELECTRON	SPECTROSCOPY OF GALLIUM ARSENIDE AND ALU	5-1595
RFACES. AUGER ELECTRON	SPECTROSCOPY OF GALLIUM ARSENIDE PHOTOSU	5-1922
ALT AND ON AMORPHOUS CARBON. QUANTITATIVE AUGER ELECTRON	SPECTROSCOPY OF GOLD CONDENSED ON ROCK S	5-1136
	AUGER ELECTRON	SPECTROSCOPY OF GRAPHITE FIBER SURFACES.	5-1276
RHENIUM INTERACTIONS. AUGER ELECTRON	SPECTROSCOPY OF HIGH TEMPERATURE OXYGEN-	4-1111
UNDS. AUGER ELECTRON	SPECTROSCOPY OF INSULATING SILICON COMPO	4-0795
UM ON TUNGSTEN. AUGER ELECTRON	SPECTROSCOPY OF LAYERED GROWTH OF TITANI	5-1140
	AUGER ELECTRON	SPECTROSCOPY OF LUNAR MATERIAL.	5-1277
	AUGER ELECTRON	SPECTROSCOPY OF LUNAR SAMPLES.	4-0858
WING SODIUM. AUGER ELECTRON	SPECTROSCOPY OF MATERIALS EXPOSED TO FLO	5-1460
	ELECTRON	SPECTROSCOPY OF METAL BLACKS.	5-1897
	AUGER ELECTRON	SPECTROSCOPY OF METAL SURFACES.	6-2122
SPUTTER ETCHING AND AUGER ELECTRON	SPECTROSCOPY OF METAL SURFACES.	5-1552
SPUTTER ETCHING AND AUGER ELECTRON	SPECTROSCOPY OF METAL SURFACES.	5-1554
	AUGER ELECTRON	SPECTROSCOPY OF METALS DURING OXIDATION.	6-2007
HIGH RESOLUTION AUGER ELECTRON	SPECTROSCOPY OF METALLIC COPPER.	4-1052
ILICON (111) SURFACE. AUGER ELECTRON	SPECTROSCOPY OF NICKEL DEPOSITS ON THE S	5-1238
A TUNGSTEN SURFACE. HIGH RESOLUTION AUGER ELECTRON	SPECTROSCOPY OF NITRIC OXIDE ADSORBED ON	5-1565
NUM (100)- CARBON MONOXIDE. LOW ENERGY ELECTRON	SPECTROSCOPY OF PLATINUM (100) AND PLATI	5-1686
SECONDARY ELECTRON	SPECTROSCOPY OF POLYCRYSTALLINE SILVER.	4-1016
ND CRYSTAL UNITS. (AUGER ELECTRON	SPECTROSCOPY OF) PRECISION SINGLE SIDEBA	5-1409
	AUGER ELECTRON	SPECTROSCOPY OF SILICON.	4-0854
	AUGER ELECTRON	SPECTROSCOPY OF SILICON SURFACES.	5-1397
	AUGER ELECTRON	SPECTROSCOPY OF SILICON SURFACES.	5-1945
	AUGER ELECTRON	SPECTROSCOPY OF SILVER- GOLD ALLOYS.	6-2024
- LEAD) ALLOY SURFACES. AUGER ELECTRON	SPECTROSCOPY OF (SILVER- GOLD AND INDIUM	5-1839
AUGER CURRENT MEASUREMENTS FOR QUANTITATIVE AUGER ELECTRON	SPECTROSCOPY OF SOLIDS.	4-0856

	AUGER	ELECTRON	SPECTROSCOPY OF SOME REFRACTORY METALS.	4-0865
		ELECTRON	SPECTROSCOPY OF SURFACES.	1-0017
GROWTH, AND ITS APPLICATION TO THE / MODEL FOR THE	AUGER	ELECTRON	SPECTROSCOPY OF SYSTEMS EXHIBITING LAYER	2-0399
S.	AUGER	ELECTRON	SPECTROSCOPY OF THE OUTER SHELL ELECTRON	4-0805
ILICIDE INTERFACE. PLATINUM SILICIDE FORMATION.		ELECTRON	SPECTROSCOPY OF THE PLATINUM- PLATINUM S	5-1285
POSITION OF VANADIUM OXIDES, AND OXIDIZED VANADIUM/	(AUGER)	ELECTRON	SPECTROSCOPY OF THE SURFACE CHEMICAL COM	4-1089
URFACES.	(AUGER)	ELECTRON	SPECTROSCOPY OF TRANSITION METAL OXIDE S	5-1339
	AUGER	ELECTRON	SPECTROSCOPY OF TRANSITION METALS.	4-0864
	(AUGER)	ELECTRON	SPECTROSCOPY OF TUNGSTEN.	4-0845
CON- IRON ALLOY (1/ PHOTOELECTRIC WORK FUNCTION AND	AUGER	ELECTRON	SPECTROSCOPY OF 5.59 ATOMIC PERCENT SILI	5-1925
	SLOW	ELECTRON	SPECTROSCOPY ON MOLYBDENUM (100).	5-1578
	AUGER	ELECTRON	SPECTROSCOPY ON SURFACES.	5-1840
UNGSTEN MATERIALS AND SILICON NITRIDE.	AUGER	ELECTRON	SPECTROSCOPY ON THE FRACTURE OF NICKEL T	5-1445
GER MICROSCOPE)		ELECTRON	SPECTROSCOPY ON THIN FILMS. (SCANNING AU	5-1405
/NS OF PHOTOELECTRIC WORK FUNCTION MEASUREMENTS AND	AUGER	ELECTRON	SPECTROSCOPY ON 5.59% SILICON- IRON (10/	5-1928
ADSORBED OXYGE/ SMALL RETARDING FIELD ANALYZER FOR	AUGER	ELECTRON	SPECTROSCOPY. (PALLADIUM- NICKEL ALLOYS,	3-0644
/TED STATES OF DOUBLY IONIZED MOLECULES BY MEANS OF	AUGER	ELECTRON	SPECTROSCOPY. PART-1: ELECTRONIC STATES/	4-1071
IONIZATION SPECTROSCOPY, REVIEW, 41 REFS)		ELECTRON	SPECTROSCOPY. (PHOTOELECTRON, AUGER, AND	1-0001
N SPECTROSCOPY TO PHOTOELECTRON AND AES) APPLICATIONS OF		ELECTRON	SPECTROSCOPY. (RELATIONSHIP OF IONIZATIO	1-0062
EASUREMENT OF EQUILIBRIUM SURFACE SEGREGATION USING	AUGER	ELECTRON	SPECTROSCOPY. (REPLY TO COMMENTS) M	6-2026
TRON SPECTROSCOPY. BEAM EFFECTS IN	AUGER	ELECTRON	SPECTROSCOPY REVEALED BY X-RAY PHOTOELEC	3-0601
	AUGER	ELECTRON	SPECTROSCOPY. (REVIEW)	1-0030
	AUGER	ELECTRON	SPECTROSCOPY. (REVIEW)	1-0120
	AUGER	ELECTRON	SPECTROSCOPY. (REVIEW)	1-0168
	LEED AND	ELECTRON	SPECTROSCOPY. (REVIEW)	1-0124
THIN FILM ANALYSIS BY	AUGER	ELECTRON	SPECTROSCOPY. (REVIEW)	1-0210
MICROANALYSIS BY	AUGER	ELECTRON	SPECTROSCOPY. (REVIEW, 15 REFS)	1-0076
	AUGER	ELECTRON	SPECTROSCOPY. (REVIEW, 154 REFS)	1-0101
LOW ENERGY	ELECTRON	ELECTRON	SPECTROSCOPY. (REVIEW, 16 REFS)	1-0164
	AUGER	ELECTRON	SPECTROSCOPY. (REVIEW, 18 REFS)	1-0098
	AUGER	ELECTRON	SPECTROSCOPY. (REVIEW, 37 REFS)	1-0154
	AUGER	ELECTRON	SPECTROSCOPY. (REVIEW, 39 REFS)	1-0105
	AUGER	ELECTRON	SPECTROSCOPY. (REVIEW, 44 REFS)	1-0125
SURFACE STUDIES BY	AUGER	ELECTRON	SPECTROSCOPY. (REVIEW, 44 REFS)	1-0144
	AUGER	ELECTRON	SPECTROSCOPY. (REVIEW, 8 REFS)	1-0221
PY AND NONDISPERSIVE X-RAY ANALYSIS TECHNIQUES FOR/	AUGER	ELECTRON	SPECTROSCOPY, SCANNING ELECTRON MICROSCO	5-1799
/ COMPARISON OF SURFACE LAYER ANALYSIS TECHNIQUES. (AUGER		ELECTRON	SPECTROSCOPY, SECONDARY ION MASS SPECTRO	5-1627
IN DEPTH INFORMATION FROM	AUGER	ELECTRON	SPECTROSCOPY. (SILICON MONOLAYERS)	1-0130
CHEMICAL EFFECTS IN	AUGER	ELECTRON	SPECTROSCOPY (STATES OF CARBON).	4-0867
S OBTAINED BY USING ION SCATTERING SPECTROMETRY AND	AUGER	ELECTRON	SPECTROSCOPY. (STEEL SURFACES) /EMISTRIE	6-1999
OF OXYGEN ADSORPTION CN A PLATINUM (111) SURFACE. (AUGER	ELECTRON	SPECTROSCOPY, STICKING COEFFICIENT) /ICS	5-1189
/PART-1: SULFUR AND CARBON MONOXIDE LEED PATTERNS. (AUGER		ELECTRON	SPECTROSCOPY, STICKING COEFFICIENT) PAR/	5-1190
XIDATION PROCESSES AT METAL SURFACES. (REVIEW, 50 REFS)		ELECTRON	SPECTROSCOPY STUDIES OF ADSORPTION AND O	1-0019
S IN (-1-1-1) GALLIUM ARSENIDE ANNEALED SURFACES.	AUGER	ELECTRON	SPECTROSCOPY STUDIES OF ARSENIC VACANCIE	5-1914
ON NICKEL. SPECTRAL LINE SHAPES AND QUANTITATIVE A/	AUGER	ELECTRON	SPECTROSCOPY STUDIES OF CARBON MONOXIDE	5-1458
S ON METAL SURFACES.	AUGER	ELECTRON	SPECTROSCOPY STUDIES OF CARBON OVERLAYER	5-1376
FACES AND ADSORBED L/ LOW ENERGY ION SCATTERING AND	AUGER	ELECTRON	SPECTROSCOPY STUDIES OF CLEAN NICKEL SUR	5-1870
LLINE GOLD AND OF THE ADSORPTION OF MERCURY CN GOL/	AUGER	ELECTRON	SPECTROSCOPY STUDIES OF CLEAN POLYCRYSTA	5-1519
AND SILICON METAL- SEMICONDUCTOR STRUCTURES.	AUGER	ELECTRON	SPECTROSCOPY STUDIES OF GALLIUM ARSENIDE	5-1774
UM, GALLIUM ARSENIDE, AND INDIUM/ COMBINED LEED AND	AUGER	ELECTRON	SPECTROSCOPY STUDIES OF SILICON, GERMANI	5-1379
ON AND SPUTTER REMOVAL OF MOLYBDENUM FROM VARIOUS /	AUGER	ELECTRON	SPECTROSCOPY STUDIES OF SPUTTER DEPOSITI	5-1878
ION OF SILVER- COPPER ALLOYS. QUANTITATIVE AUGER A/	AUGER	ELECTRON	SPECTROSCOPY STUDIES OF SURFACE COMPOSIT	5-1200
ION OF SILVER- COPPER ALLOYS. QUANTITATIVE AUGER A/	AUGER	ELECTRON	SPECTROSCOPY STUDIES OF SURFACE COMPOSIT	5-1740
OF ALLOYS.	AUGER	ELECTRON	SPECTROSCOPY STUDIES ON ABRADED SURFACE	6-2131
EPITAXIAL THIN FILMS VIA SILANE PYROLYSIS: HEED AND	AUGER	ELECTRON	SPECTROSCOPY STUDY. SILICON HOMO	5-1428
ROLYTE.	AUGER	ELECTRON	SPECTROSCOPY STUDY OF BETA ALUMINA ELECT	4-1074
/IN DYNAMIC BACKGROUND SUBTRACTION FOR QUANTITATIVE	AUGER	ELECTRON	SPECTROSCOPY. (STUDY OF CARBON MONOXIDE)	1-0069
URING ACTIVATION AND POISONING. PART-1: BARIUM ON /	AUGER	ELECTRON	SPECTROSCOPY STUDY OF CATHODE SURFACES D	5-1841
ULATED DESORPTION OF OXYGEN FROM A SEMICONDUCTOR S/	AUGER	ELECTRON	SPECTROSCOPY STUDY OF ELECTRON BEAM STIM	5-1617
SORPTION.	AUGER	ELECTRON	SPECTROSCOPY STUDY OF ELECTRON IMPACT DE	2-0452
ES.	AUGER	ELECTRON	SPECTROSCOPY STUDY OF FLOAT GLASS SURFAC	5-1236
OXIDATION IN A COPPER- 19.6 ATOMIC PERCENT ALUMINU/	AUGER	ELECTRON	SPECTROSCOPY STUDY OF INITIAL STAGES OF	6-2028
CE EFFECTS DUE TO CARBON AND SULFUR. LEED AND	AUGER	ELECTRON	SPECTROSCOPY STUDY OF NICKEL (110) SURFA	5-1814
ANUM HEXABORIDE.	AUGER	ELECTRON	SPECTROSCOPY STUDY OF OXIDATION ON LANTH	5-1706
ON TUNGSTEN (110).	AUGER	ELECTRON	SPECTROSCOPY STUDY OF OXYGEN ADSORPTION	5-1668
OM SILICON MONOXIDE, SILICON DIOXIDE, AND SILICON /	AUGER	ELECTRON	SPECTROSCOPY STUDY OF SILICON SPECTRA FR	4-1081
LICON COMPOUNDS.	AUGER	ELECTRON	SPECTROSCOPY STUDY OF SOME OXYGENATED SI	4-0793
N IN COPPER- ALUMINUM ALLOYS.	AUGER	ELECTRON	SPECTROSCOPY STUDY OF SURFACE SEGREGATIO	6-2012
N IN THE BINARY ALLOYS COPPER- 1 ATOMIC PERCENT IN/	AUGER	ELECTRON	SPECTROSCOPY STUDY OF SURFACE SEGREGATIO	6-2027
LATED DESORPTION OF OXYGEN FROM A TUNGSTEN (100) S/	AUGER	ELECTRON	SPECTROSCOPY STUDY OF THE ELECTRON STIMU	5-1954
ITION OF THE LEAD- INDIUM SYSTEM.	AUGER	ELECTRON	SPECTROSCOPY STUDY OF THE SURFACE COMPOS	5-1172
EL OXIDATION.	AUGER	ELECTRON	SPECTROSCOPY STUDY OF 18-8 STAINLESS STE	6-1993
IALS BY ELECTRON EXCITED AUGER ELECTRON EMISSI/ SECONDARY	ELECTRON	SPECTROSCOPY. (SURFACE ANALYSIS OF MATER	5-1403
0A ALLOY AT 800 DEGREES C IN LOW OXYGEN PRESSURES. (AUGER	ELECTRON	SPECTROSCOPY, SURFACE CONTAMINANTS) /C-8	6-2144
ES)	SECONDARY	ELECTRON	SPECTROSCOPY. SURFACE SENSITIVE TOOL. (A	1-0095
EVIEW, 22 REFS)	SECONDARY	ELECTRON	SPECTROSCOPY. SURFACE SENSITIVE TOOL. (R	1-0094
APPLICATION OF (AUGER)	ELECTRON	SPECTROSCOPY TO CHEMISTRY. (REVIEW)	1-0127
N OF FRACTURE SURFACES OF MINERALS. USE OF	AUGER	ELECTRON	SPECTROSCOPY TO DETERMINE THE COMPOSITIO	5-1307
OF SILICON OXIDE FILMS. USE OF	AUGER	ELECTRON	SPECTROSCOPY TO DETERMINE THE STRUCTURE	5-1501
APPLICATIONS OF	AUGER	ELECTRON	SPECTROSCOPY TO METALLURGICAL PROBLEMS.	6-2039
CH. APPLICATION OF	AUGER	ELECTRON	SPECTROSCOPY TO REACTOR MATERIALS RESEAR	1-0036
S OF CERAMIC SUBSTRATES. APPLICATION OF	AUGER	ELECTRON	SPECTROSCOPY TO SURFACE CHEMISTRY STUDIE	5-1275
, 85 REFS) APPLICATION OF		ELECTRON	SPECTROSCOPY TO SURFACE STUDIES. (REVIEW	1-0018
/TRATES FROM THE KLL AUGER SPECTRUM. APPLICATION OF	AUGER	ELECTRON	SPECTROSCOPY TO THE BULK TO SURFACE PRE/	5-1651
COMPOSITION OF PASSIVE FILMS ON TYP/ APPLICATON OF	AUGER	ELECTRON	SPECTROSCOPY TO THE DETERMINATION OF THE	6-2080
TEM AT HIGH TEMPERATURE. APPLICATION OF	AUGER	ELECTRON	SPECTROSCOPY TO THE PLATINUM- OXYGEN SYS	5-1955
OF ALUMINUM SURFACES IN BOILING WA/ APPLICATION OF	AUGER	ELECTRON	SPECTROSCOPY TO THE STUDY OF DISCOLORING	5-1127
N STEELS. APPLICATION OF	AUGER	ELECTRON	SPECTROSCOPY TO THE STUDY OF INTERFACE I	6-2126
/MPOSITION OF COPPER- NICKEL ALLCY. APPLICATION OF	AUGER	ELECTRON	SPECTROSCOPY TO THE STUDY ON CATALYSTS.	5-1875
QUANTITATIVE	AUGER	ELECTRON	SPECTROSCOPY USING COADSORPTION.	1-0002

FACTORS. QUANTITATIVE AUGER ELECTRON	SPECTROSCOPY USING ELEMENTAL SENSITIVITY	1-0157
DURING THERMAL TREATMENT. WORK FUNCTION CHANGES AND AUGER ELECTRON	SPECTROSCOPY USING THE LEED CAMERA. /CE	6-2069
TEM AND PHOTOELECTRON MULTIPLIER. AUGER ELECTRON	SPECTROSCOPY USING THE TWO-GRID LEED SYS	3-0656
PECTROSCOPY. DIRECT COMPARISON OF AUGER ELECTRON	SPECTROSCOPY WITH APPEARANCE POTENTIAL S	1-0142
COMPOSITION/ A MULTICHANNEL MONITOR FOR REPETITIVE AUGER ELECTRON	SPECTROSCOPY WITH APPLICATION TO SURFACE	3-0695
/TWEEN SULFUR AND OXYGEN ON A MOLYBDENUM SURFACE BY AUGER ELECTRON	SPECTROSCOPY: WORKING STATE OF THE CAT/	5-1530
ION. (REVIEW, ESCA, AES) ELECTRON	SPECTROSCOPY: X-RAY AND ELECTRON EXCITAT	1-0081
TION. ELECTRON	SPECTROSCOPY. X-RAY AND ELECTRON EXCITA	1-0082
S/ A PRELIMINARY STUDY OF PURE METAL SURFACES USING AUGER ELECTRON	SPECTROSCOPY X-RAY PHOTOELECTRON SPECTRO	5-1353
NEW METHOD FOR ANALYZING THE SECONDARY ELECTRON	SPECTRUM.	4-0934
ON OF A PROJECTILE CHARGE DEPENDENCE FOR THE NEON K-AUGER ELECTRON	SPECTRUM. OBSERVATI	4-0993
SPIN SPLITTING IN COPPER AUGER ELECTRON	SPECTRUM.	4-1053
STRUCTURE OF LOW ENERGY ELECTRON	SPECTRUM.	4-0948
E HYDROGEN FLUORIDE MOLECULE. AUGER ELECTRON	SPECTRUM AND IONIZATION POTENTIALS OF TH	4-1058
E OF MOLYBDENUM CARBIDE DURING HEATING ACCORDING TO AUGER ELECTRON	SPECTRUM DATA. /CENTRATION ON THE SURFAC	5-1759
OXYGEN AND THE (111) SURFACE OF GALLIUM PHOSPHIDE. AUGER ELECTRON	SPECTRUM DURING THE INTERACTION BETWEEN	5-1642
OBSERVATION AND INTERPRETATION OF THE AUGER ELECTRON	SPECTRUM FROM CLEAN BERYLLIUM.	4-0997
L2,3MM AUGER ELECTRON	SPECTRUM FROM FREE ZINC ATOMS.	4-0731
INTERPRETATION OF THE AUGER ELECTRON	SPECTRUM FROM OXIDIZED BERYLLIUM.	4-0935
CHEMICAL EFFECTS ON KLL AUGER ELECTRON	SPECTRUM FROM OXYGEN.	4-0819
CHEMICAL EFFECTS ON THE KLL AUGER ELECTRON	SPECTRUM FROM SURFACE CARBON.	4-0862
EN AND THE (111) SURFACE OF G/ SOME FEATURES OF THE AUGER ELECTRON	SPECTRUM IN THE INTERACTION BETWEEN OXYG	5-1643
AUGER ELECTRON	SPECTRUM OF CARBON SUBOXIDE.	5-1529
HIGH RESOLUTION AUGER ELECTRON	SPECTRUM OF MAGNESIUM OXIDE (100).	4-0772
PLASMA SATELLITES IN THE AUGER ELECTRON	SPECTRUM OF MOLYBDENUM.	4-1124
EV. AUGER ELECTRON	SPECTRUM OF OSMIUM AT ENERGIES UP TO 300	4-1030
KLL AUGER ELECTRON	SPECTRUM OF SAMARIUM.	4-0725
INFLUENCE OF THE BAND STRUCTURE ON THE AUGER ELECTRON	SPECTRUM OF SILICON.	4-0944
AUGER ELECTRON	SPECTRUM OF WATER VAPOR.	4-1066
) ION IMPACT. HIGH RESOLUTION NEON K-AUGER ELECTRON	SPECTRUM PRODUCED BY 45 MEV CHLORINE(12+	4-1051
ELECTRON	SPRAYING ATOMS. (AUGER CASCADES)	2-0464
METASTABLE AUTOIONIZING THREE- AND FOUR ELECTRON	STATES IN BERYLLIUM. (AUGER DECAY)	2-0295
UREMENT OF SURFACE DIFFUSION ON MAGNESIUM OXIDE (001) BY/ ELECTRON	STIMULATED DESORPTION OF SODIUM AND MEAS	5-1494
(100) SURFACES OF ALKALI HALIDES. PART-2: ELECTRON	STIMULATED DISSOCIATION.	5-1908
, STUDIED BY QUANTITATIVE AUGER ELECTRON SPECTROSCOPY. ELECTRON	STIMULATED OXIDATION OF GALLIUM ARSENIDE	5-1756
(AUGER SPECTROSCOPY) ELECTRON	STIMULATED PROCESSES AT SOLID SURFACES.	5-1605
AUGER ELECTRON	STUDIES OF METAL VAPORS.	4-0729
ANIUM, GRAPHITE, 300 SERIES STAINLESS STEEL, AND N/ AUGER ELECTRON	STUDIES OF SURFACES: URANIUM DIOXIDE, UR	5-1217
AUGER ELECTRON	STUDIES OF URANIUM DIOXIDE SURFACES.	5-1218
SEMI-AUGER ELECTRON	TRANSITIONS.	2-0424
CAL AUGER SPECTROMETERS FOR RETARDING POTENTIAL SECONDARY ELECTRON	YIELD MEASUREMENTS. USE OF CYLINDRI	3-0628
ES. (LEED- AUGER SYSTEM) FAST, ACCURATE SECONDARY ELECTRON	YIELD MEASUREMENTS AT LOW PRIMARY ENERGI	3-0629
ND GLANCING INCIDENCE. DEPENDENCE OF AUGER ELECTRON	YIELD ON PRIMARY BEAM ENERGY AT NORMAL A	2-0455
K-VACANCY SHARING STUDIED FROM AUGER ELECTRON	YIELDS.	2-0345
ACE BY THERMAL CARBON OXYGEN NITROGEN(2), AND / SECONDARY ELECTRON	YIELDS FROM COPPER BERYLLIUM OXYGEN SURF	4-0786
TYL/ K-SHELL IONIZATION OF CARBON BY FAST PROTONS. (AUGER ELECTRON	YIELDS OF METHANE, ETHANE, ETHYLENE, ACE	2-0532
APPARATUS FOR INVESTIGATING A SAMPLE BY MEANS OF EMITTED ELECTRONS.	ANALYSIS	3-0713
ANALYSIS OF MATERIALS BY ELECTRON EXCITED AUGER ELECTRONS.		5-1399
ANGULAR DISTRIBUTION OF AUGER ELECTRONS.		2-0320
AUGER ELECTRON SPECTROSCOPY OF THE OUTER SHELL ELECTRONS.		4-0805
AUGER EXCITATION BY INTERNAL SECONDARY ELECTRONS.		2-0390
SPECTROSCOPY OF GRAPHITE SINGLE CRYSTALS WITH LOW ENERGY ELECTRONS.	AUGER	4-0741
AUGER TRANSITION RATES IN J-J COUPLING FOR D HOLES AND F ELECTRONS.		2-0444
ND MEASUREMENT OF DECAY PARAMETERS. (DETECTION OF L AUGER ELECTRONS)	/S OF STANDARDIZATION FOR CESIUM-131 A	2-0474
E HIGH RESOLUTION RETARDING FIELD ENERGY ANALYZER. (AUGER ELECTRONS)	COMPACT, INEXPENSIV	3-0619
CE STRUCTURES USING LEED AND ENERGY ANALYSIS OF SCATTERED ELECTRONS.	DETERMINATION OF SURFA	5-1956
TION OF COPPER AND SILVER BY ELECTRON BOMBARDMENT. (AUGER ELECTRONS)	/TIAL CROSS SECTION FOR K-SHELL IONIZA	2-0477
DIRECTIONAL CORRELATION BETWEEN PHOTOELECTRONS AND AUGER ELECTRONS.		1-0122
ELECTRON BEAM ANALYZER FOR DETECTING AUGER ELECTRONS.		3-0712
ELECTRON ESCAPE DEPTHS IN ALUMINUM. (AUGER ELECTRONS)		2-0537
ESCAPE DEPTHS OF X-RAY EXCITED (AUGER) ELECTRONS.		2-0409
ESCAPE LENGTH OF AUGER ELECTRONS.		2-0527
DISTRIBUTIONS OF INELASTICALLY SCATTERED OR PHOTOEMITTED ELECTRONS.	/ FOR INVESTIGATING ENERGY AND ANGULAR	3-0691
HEMICAL ANALYSIS. (AUTOMATIC RECORDING OF PHOTO AND AUGER ELECTRONS)	/R USED IN ELECTRON SPECTROMETRY FOR C	3-0717
H RESOLUTION MEASUREMENTS OF P(+) ARGON KLL AND KLM AUGER ELECTRONS.	HIG	4-0943
ALIGNMENT OF ATOMS IN ELECTRON IMPACT IONIZATION. (AUGER ELECTRONS)	INNER SHELL	2-0335
KLM SPECTRUM OF SAMARIUM AUGER ELECTRONS.		4-0757
LICON STUDIED BY A MICROWAVE METHOD. (PRODUCTION OF AUGER ELECTRONS)	/ECTRIC TRANSITION IN GERMANIUM AND SI	4-0752
PROCESSES IN TRANSITION METAL ATOMS. PART-2: INNER AUGER ELECTRONS.	MULTIPLE EXCITATION	2-0362
THIN FILM THICKNESS MEASUREMENT USING RADIOISOTOPE AUGER ELECTRONS.	NEW METHOD OF	5-1658
NEW MICROAUTORADIOGRAPHY OF METAL STRUCTURE USING AUGER ELECTRONS.		6-2052
ITE BANDS CORRESPONDING TO DOUBLE L IONIZATION. (OF AUGER ELECTRONS)	NEW SATELL	2-0293
OEFFICIENT AND SECONDARY ELECTRON EMISSION FOR 10-100 KEV ELECTRONS.	/TATION ANISOTROPY OF BACKSCATTERING C	2-0306
: X-RAY EXCITATION PART-2: ULTRAVIOLET EXCITATION. (AUGER ELECTRONS)	/UM, AND MAGNESIUM(2) STANNIDE. PART-1	2-0306
SEPARATION OF THE SPECTRA OF PHOTO AND AUGER ELECTRONS.		4-1099
SPECTROGRAPHY BY AUGER ELECTRONS.		1-0047
SPECTROMETER FOR INDUCED ELECTRON EMISSION. (AUGER ELECTRONS)		3-0585
SPECTROMETER OF AUGER ELECTRONS.		3-0653
SURFACE ANALYSIS WITH HEAVY ION INDUCED AUGER ELECTRONS.		5-1387
THEORY FOR ENERGY DISTRIBUTION OF SECONDARY ELECTRONS.		2-0241
IDE STUDIED BY AUGER SPECTROSCOPY AND DIFFRACTION OF SLOW ELECTRONS.	(111) SURFACE OF GALLIUM PHOSPH	5-1640
FILMS ON A (110) TUNGSTEN FACE BY THE DIFFRACTION OF SLOW ELECTRONS AND AUGER SPECTROSCOPY.	/UDY OF BARIUM	5-1368
ENA OF COPPER (111). ENERGY SPECTRUM OF BACK SCATTERED ELECTRONS AND CHARACTERISTIC LOSS AND GAIN PHENOM		4-0897
ANCE POTENTIAL SPECTROMETER FOR THE DETECTION OF PHOTONS, ELECTRONS AND IONS. HIGH SENSITIVITY APPEAR		3-0643
SPECTROSCOPY OF AUGER ELECTRONS AND ITS USE IN METALLURGY.		6-2077
ENERGY SPECTRA OF INELASTICALLY SCATTERED ELECTRONS AND LEED STUDIES OF TUNGSTEN.		1-0094
(AUGER, PHOTOELECTRON SPECTRA, REVIEW, 6 REFS) ELECTRONS AND X-RAYS IN NONDESTRUCTIVE ANALYSIS.		1-0083
NS) RESONANCE IONIZATION OF ATOMIC HELIUM BY FAST ELECTRONS. (ANGULAR DISTRIBUTION OF AUGER ELECTRO		2-0256
IRON-57 AND TIN-119 BY DETECTION OF CONVERSION AND AUGER ELECTRONS. APPLICATION TO SURFACE STUDIES. /PY OF		5-1796
ES OF GALLIUM PHOSPHIDE, INDUCED BY BOMBARDMENT WITH 2 KV ELECTRONS. (AUGER ELECTRON DAMAGE) / (100) SURFAC		5-1162

S OF SURFACE STUDIES DEPENDING ON INELASTIC SCATTERING OF ELECTRONS. (AUGER ELECTRON SPECTROSCOPY)   METHOD 5-1159
EMISSION INDUCED BY LASER-PRODUCED PLASMA X-RAYS. (PLASMA ELECTRONS, AUGER PROCESS)   ELECTRON PHOTO 2-0276
ATERIALS S/ SURFACE ANALYSIS WITH HEAVY ION INDUCED AUGER ELECTRONS. BASIC PROPERTIES OF A NEW METHOD FOR M 1-0070
TO AN IONIC BOMBARDMENT.   EMISSION OF AUGER ELECTRONS BY ATOMS IN A METALLIC TARGET SUBJECTED 2-0383
DETERMINATION OF THE ESCAPE LENGTH OF LOW ENERGY ELECTRONS BY AUGER AND LOSS SPECTROSCOPY. 2-0401
D NEON AUGE/ IDENTIFICATION AND MEASUREMENT OF LOW ENERGY ELECTRONS BY CLOUD CHAMBER TECHNIQUES: OXYGEN AN 2-0566
) CROSS SECTIONS FOR IONIZATION OF INNER SHELL ELECTRONS BY ELECTRON IMPACT. (AUGER SPECTROSCOPY 2-0475
EMISSION OF AUGER ELECTRONS BY ION BOMBARDMENT. 2-0550
EXCITATION OF INNER SHELL ELECTRONS BY PHOTONS AND FAST ELECTRONS. 2-0541
-OFF AND (SUGAR) VACANCY CASCADES.   NATURE OF SECONDARY ELECTRONS CREATED AS THE RESULT OF ELECTRON SHAKE 2-0307
.   ANGULAR DISTRIBUTION OF M4,5NN AUGER ELECTRONS EJECTED FROM KRYPTON BY ELECTRON IMPACT 4-1003
OTONS. (AUGER PEAKS) DISTRIBUTION IN ENERGY AND ANGLE OF ELECTRONS EJECTED FROM XENON BY 0.3 TO 2.0 MEV PR 2-0534
(AUGER AND COSTER-KRONIG TRANSITIONS) / ENERGY SPECTRA OF ELECTRONS EJECTED IN ARGON (+)- ARGON COLLISIONS. 4-1007
IMPACT OF ION AND ATOM BOMBARDMENT. / ENERGY SPECTRUM OF ELECTRONS EMITTED BY ALKALI HALIDE CRYSTALS UNDER 2-0246
MPACT.   ENERGY DISTRIBUTION OF ELECTRONS EMITTED BY ARGON ATOMS UNDER ELECTRON I 4-1008
PHOSPHIDE SUBMITTED TO/ ENERGY DISTRIBUTION OF SECONDARY ELECTRONS EMITTED BY INDIUM PHOSPHIDE AND GALLIUM 2-0269
ENERGY SPECTRUM AND ANGULAR DISTRIBUTION OF SECONDARY ELECTRONS EMITTED BY IONIC BOMBARDMENT. 4-0988
OMBARDED BY PRIMARY ELECTRON/ ENERGY SPECTRA OF SECONDARY ELECTRONS EMITTED BY MAGNESIUM AND SILICON WHEN B 4-1047
INTERPRETATION OF ENERGY LOSS BY AUGER ELECTRONS EMITTED DURING ION BOMBARDMENT. 2-0268
INUM AND SILICON UNDER ION / ENERGY DISTRIBUTION OF AUGER ELECTRONS EMITTED FROM BERYLLIUM, MAGNESIUM, ALUM 4-0876
AUGER AND SECONDARY ELECTRONS EXCITED BY BACKSCATTERED ELECTRONS. 2-0365
ROACH TO QUANTITATIVE ANALYSIS.   AUGER AND SECONDARY ELECTRONS EXCITED BY BACKSCATTERED ELECTRONS. APP 2-0366
ANGULAR DISTRIBUTION OF AUGER ELECTRONS FOLLOWING IMPACT IONIZATION. 2-0321
ANGULAR DISTRIBUTION OF AUGER ELECTRONS FOLLOWING PHOTOIONIZATION. 2-0352
SHIFT ENERGIES OF CHARACTERISTIC X-RAYS AND AUGER ELECTRONS FOR IONIZED ATOMS. 4-1041
A RE/ ORIENTATION DEPENDENCE OF OVERLAYER ATTENUATION OF ELECTRONS FOR THE CYLINDRICAL MIRROR ANALYZER AND 3-0693
IONUCLIDES: CHROMATOGRAPHIC / USE OF AUGER AND CONVERSION ELECTRONS FOR THE DETECTION OF GAMMA EMITTING RAD 1-0217
SCATTERING OF LOW ENERGY ELECTRONS FROM A COPPER (111) SURFACE. 5-1621
OW ENERGY BOMBARDMENT. / ENERGY ANALYSIS OF BACKSCATTERED ELECTRONS FROM A SILICON SINGLE CRYSTAL UNDER A L 4-0739
DUCED BY HYDROGEN(+) IMPACTS.   AUGER ELECTRONS FROM ARGON WITH ENERGIES 150-210 EV PRO 2-0552
EON.   AUGER ELECTRONS FROM DOUBLE K-SHELL VACANCY STATES IN N 2-0573
DOPPLER SHIFTED AUGER ELECTRONS FROM FOIL EXCITED HEAVY ION BEAMS. 4-0859
NEON)   AUGER ELECTRONS FROM FOIL EXCITED HEAVY ION BEAMS. (OF 2-0370
JECTED TO ION BOM/ RELATIVE EMISSION INTENSITIES OF AUGER ELECTRONS FROM LIGHT METALS (LITHIUM, SODIUM) SUB 2-0548
THEORY OF AUGER EJECTION OF ELECTRONS FROM METALS BY IONS. 2-0563
UGER ELECTRON EMISSION IN THE ENERGY SPECTRA OF SECONDARY ELECTRONS FROM MOLYBDENUM AND TUNGSTEN.   A 4-0874
SECONDARY EMISSION AND ELASTIC REFLECTION OF ELECTRONS FROM MONOCRYSTALS. 2-0508
ENERGY SPECTRA OF INELASTICALLY SCATTERED AND AUGER ELECTRONS FROM SINGLE CRYSTALS. 4-1064
ECTRON (AND AUGER) SPECTROSCOP/ ATTENUATION OF LOW ENERGY ELECTRONS FROM SOLIDS: RESULTS FROM X-RAY PHOTOEL 5-1844
INELASTIC SCATTERING OF LOW ENERGY ELECTRONS FROM SURFACES. 2-0495
AUGER ELECTRONS FROM THE L-SHELL IN MERCURY. 2-0271
AUGER EJECTION OF ELECTRONS FROM TUNGSTEN BY OXYGEN CHEMISORPTION. 5-1599
AUGER PEAKS IN THE ENERGY SPECTRA OF SECONDARY ELECTRONS FROM VARIOUS MATERIALS. 4-0926
,3 RADIATIVE AUGER SPECTRUM OF INSULATORS. (MAGNESIUM OX/ ELECTRON- HOLE INTERACTION EFFECTS IN THE KL2,3L2 2-0540
/CTRON DETERMINATION OF THE ATTENUATION OF THE LOW ENERGY ELECTRONS IN ALUMINUM OXIDE. (AUGER ELECTRON LIN/ 4-0773
N INSTRUMENTATIION.   X-RAYS AND ELECTRONS IN ANALYTICAL CHEMISTRY WITH EMPHASIS O 3-0646
OSS SPECTROSCOPY)   INELASTIC MEAN FREE PATH FOR ELECTRONS IN BULK JELLIUM. (AUGER, LEED, ENERGY L 1-1059
OCIATION: CARBON MONOXIDE / ROLE OF PRIMARY AND SECONDARY ELECTRONS IN ELECTRON INDUCED DESORPTION AND DISS 5-1567
IDENTIFICATION OF AUGER ELECTRONS IN GALLIUM ARSENIDE. 2-0467
TE (AUGER) TRANSITIO/ POSSIBLE LOCALISATION OF CONDUCTION ELECTRONS IN SOFT X-RAY DOUBLE IONIZATION SATELLI 2-0437
FREE ATOM BEHAVIOR OF AUGER ELECTRONS IN SOLIDS. 2-0579
(2)(X), NITROGEN(X), AND NITROGEN(2)(2X) FROM NITROGEN BY ELECTRONS IN THE ENERGY RANGE 16-600 EV. /ITROGEN 2-0376
CTRONS EMITTED BY COPPER UNDER BOMB/ OBSERVATION OF AUGER ELECTRONS IN THE ENERGY SPECTRUM OF SECONDARY ELE 2-0551
NG CORROSION OF ALUMINU/ NEW MICRORADIOGRAPHY USING AUGER ELECTRONS IN THE STUDY OF THE STRUCTURE AND PITTI 6-2066
URFACES.   AUGER ELECTRONS, INDICATORS IN ANALYSIS OF SOLID BODY S 5-1662
/LL DOUBLE IONIZATION AND CHEMICAL BONDING. PART-1: AUGER ELECTRONS. (KLL TRANSITION ENERGIES FOR SULFUR, / 2-0327
ANGULAR DISTRIBUTION OF L3M2,3M2,3(1S0) AUGER ELECTRONS OF ARGON. 2-0322
KLL AUGER ELECTRONS OF CADMIUM AND ANTIMONY. 4-0754
SPECTRUM OF AUGER K,L,M,N,O ELECTRONS OF SAMARIUM. 4-0756
SPECTRUM OF AUGER KLL ELECTRONS OF SAMARIUM. 4-0726
SPECTRUM OF AUGER KLL ELECTRONS OF TELLURIUM AND XENON. 4-0755
NONRADIATIVE AUGER RECOMBINATION OF ELECTRONS ON DONOR- ACCEPTOR PAIRS. 2-0536
EFFECT OF THE INCIDENCE ANGLE OF PRIMARY ELECTRONS ON THE AUGER EMISSION YIELD. 2-0237
ELECTRONS OR IONS. (IN SURFACE ANALYSIS) 5-1765
/ATTERING FROM METALS. PART-1: EMISSION OF TRUE SECONDARY ELECTRONS. PART-2: INELASTICALLY SCATTERED PRIMA/ 2-0500
CIENT OF SECONDARY ELECTRON EMISSION ON ENERGY OF PRIMARY ELECTRONS (POTASSIUM IODIDE). /ENCE OF THE COEFFI 4-0891
MENT.   MEASUREMENTS OF NEON K-AUGER ELECTRONS PRODUCED BY ENERGETIC HEAVY ION BOMBARD 4-0951
HIGH RESOLUTION STUDY OF ARGON LMM AUGER ELECTRONS PRODUCED BY ION BOMBARDMENT. 4-0903
S. PART-2: NITROGEN. / ANGULAR AND ENERGY DISTRIBUTION OF ELECTRONS PRODUCED BY 200-500 KEV PROTONS IN GASE 4-1075
ON COLLISIONS AT KEV ENERGIES.   EMISSION OF KLL AUGER ELECTRONS PRODUCED IN Z(1)- NITROGEN AND Z(1)- NE 4-0830
NS OF ELECTRONS WITH / ENERGY AND ANGULAR DISTRIBUTION OF ELECTRONS RESULTING FROM DIRECT IONIZING COLLISIO 2-0525
H A FIXED IMPACT PARAMETE/ ENERGY DISTRIBUTION OF (AUGER) ELECTRONS RESULTING FROM ION- ATOM COLLISIONS WIT 2-0331
MICRODOSIMETERY OF AUGER ELECTRONS. (REVIEW) 1-0055
SPECTROSCOPY OF AUGER ELECTRONS. (REVIEW, 24 REFS) 1-0046
ENT ARGON(+)- ARGON COLLISIONS. COMMENTS. (AUGER EL/ FAST ELECTRON- SLOW ELECTRON COINCIDENCE STUDY OF VIOL 2-0423
. PART-3: ENERGY AND ANGULAR DISTRIBUTIONS OF THE EJECTED ELECTRONS. (SURFACE CONTAMINATION) /ASTABLE ATOMS 5-1133
ONDARY ELECTRON EMISSION DUE TO PRIMARY AND BACKSCATTERED ELECTRONS (TABLE FOR 60 ELEMENTS).   SEC 4-0907
HOW MUCH CAN AUGER ELECTRONS TELL US ABOUT SOLID SURFACES. 5-1663
BACKSCATTER MOSSBAUER STUDIES USING CONVERSION ELECTRONS. TIN-119. 2-0576
IN/ APPLICATION OF MICROAUTORADIOGRAPHY BY BETA-RAY AUGER ELECTRONS TO THE STUDIES OF THE STRUCTURE OF ALUM 6-2051
/METER INSTALLED IN VACUUM AND THE MEASUREMENT OF EXCITED ELECTRONS UNDER THE X-RAY DIFFRACTION PROCESS. (/ 3-0641
SCOPY)   INELASTIC INTERACTIONS OF SLOW ELECTRONS WITH ABSORBED PARTICLES. (AUGER SPECTRO 2-0448
METHOD FOR PACKET OF WAVES IN AUGER ELECTRON-ELECTRON PROCESSES. 2-0380
SURFACE ANALYSIS USING COMPLEMENTARY ELECTRONIC AND CHEMICAL MEASUREMENTS. (AUGER) 5-1390
ADSORBED ON TUNGSTEN.   ELECTRONIC AND LATTICE STRUCTURE OF CESIUM FILMS 5-1604
ANALYSIS AND CHARACTERIZATION OF CERAMIC SURFACES FOR ELECTRONIC APPLICATIONS. 5-1853
ERMIONIC EMITTERS MADE OF CARBIDE COMPOUNDS STUDIED BY AN ELECTRONIC AUGER SPECTROSCOPIC METHOD. /ACE OF TH 5-1562
URSE OF LECTURES ON PHYSICAL ELECTRONICS.   ELECTRONIC AUGER SPECTROSCOPY: TEXTBOOK FOR A CO 1-0107
MENTS MAGNESIUM TO SCANDIUM.   EFFECT OF ELECTRONIC CORRELATIONS ON KLL AUGER LINES IN ELE 4-0800
BY NOBLE GAS IONS (10-100 KEV): ENERGETIC AND SPATIAL DI/ ELECTRONIC EMISSION FROM SOLID TARGETS BOMBARDED 4-0776

XIDE SURFACES.   CORRELATION OF ELECTRONIC, LEED, AND AUGER DIAGNOSTICS ON ZINC O 5-1584
    THE SILICON- CESIUM- OXYGEN SURFACE. (AU/ STRUCTURAL AND ELECTRONIC MODEL OF NEGATIVE ELECTRON AFFINITY ON 2-0426
FIDE SURFACES.   ELECTRONIC PROPERTIES OF CLEAVED MOLYBDENUM DISUL 5-1634
RE EPITAXY BY VACUUM SUBLIMATION: LEED- AUGER STUDIES AND ELECTRONIC PROPERTIES OF FILMS. SILICON ON SAPPHI 5-1230
UM-160, YTTERBIUM-166, YTTERBIUM-169, AND/ AUGER L-SERIES ELECTRONIC SPECTRA ACCOMPANYING THE DECAY OF ERBI 4-0743
/D STRUCTURAL ANALYSIS OF IRON SURFACES BY MEANS OF AUGER ELECTRONIC SPECTROSCOPY AND LOW ENERGY ELECTRON / 5-1940
OLECULES BY MEANS OF AUGER ELECTRON SPECTROSCOPY. PART-1: ELECTRONIC STATES OF NITROGEN. / DOUBLY IONIZED M 4-1071
ICON (111) AND (100) SURFACES. (AUGER SPECTRCSCOPY)   ELECTRONIC STATES OF OXYGEN ADSORBED ON CLEAN SIL 5-1476
D ON A SILICON SUBSTRATE STUDIED BY AUGER ELECTRON AND X/ ELECTRONIC STRUCTURE OF A THIN GOLD FILM DEPOSITE 5-1435
    RECONSTRUCTION.   CHANGE IN THE SURFACE ELECTRONIC STRUCTURE OF PLATINUM (100) INDUCED BY 5-1186
L EXCITATION TECHNIQUES. (AUGER EFFECT)   ELECTRONIC STRUCTURE OF SOLID SURFACES: CORE LEVE 2-0468
CONDARY ELECTRON IMAGES IN AUGER ELECTRON SPECTROSCOPY. ELECTRONIC SUPERPOSITION OF SAMPLE CURRENT AND SE 2-0450
CTROSCOPY: TEXTBOOK FCR A COURSE OF LECTURES ON PHYSICAL ELECTRONICS.   ELECTRONIC AUGER SPE 1-0107
    PART-2: EFFECT OF ANGLE OF INCIDENCE. PART-3: EFFECT OF ELECTRONICALLY EXCITED IONS. /SITIVE IONS. PART-1 2-0543
               APPLICATIONS OF AUGER ELECTRON SPECTROSCOPY IN ELECTROPLATING.   1-0048
               AUGER ELECTRON ANALYSIS OF ELECTROPOLISHED HIGH PURITY ALUMINUM.   4-0824
TROSCOPIC STUDIES OF THE SURFACES OF 18-8 STAINLESS STEEL ELECTROPOLISHED OR ABRADED BY EMERY PAPERS. /SPEC 6-2130
/G A CHARGED PARTICLE BEAM OF FINITE ANGULAR SPREAD IN AN ELECTROSTATIC ANALYZER OF THE CYLINDRICAL MIRROR/ 3-0720
               FOCUSING PRCFERTIES OF AN ELECTROSTATIC MIRROR WITH CYLINDRICAL FIELD. 3-0721
ROMETRY FOR CHEMICAL ANALYSIS. (AUTCMATI/ HIGH RESOLUTION ELECTROSTATIC SPECTROMETER USED IN ELECTRON SPECT 3-0717
TALOG CALCULATED TRANSITION ENERGIES LISTED BY ENERGY AND ELEMENT.   AUGER CA 1-0041
LIDS BY THE AUGER ELECTRON MICROANALYT/ THREE-DIMENSIONAL ELEMENT ANALYSIS OF SURFACES AND INTERFACES OF SO 5-1412
               CXIDE FILMS: ELEMENT PROFILES BY AUGER ELECTRON SPECTROSCOPY. 4-1073
               K-SHELL FLUORESCENCE YIELDS FOR LIGHT ELEMENTS.   4-0818
D M-SHELL YIELDS AND ELECTRON SPECTRA FOR THE TRANSURANIC ELEMENTS.   L- AN 4-0968
    LLM VERSUS LMM AUGER TRANSITIONS FCR THE LIGHT ELEMENTS.   4-0804
    LLM VERSUS LMM AUGER TRANSITIONS FOR LIGHT ELEMENTS.   4-0811
DUE TO PRIMARY AND BACKSCATTERED ELECTRONS (TABLE FOR 60 ELEMENTS).   SECONDARY ELECTRON EMISSION 4-0907
RATES/ M-SHELL AUGER, COSTER-KRONIG, AND RADIATIVE MATRIX ELEMENTS, AND AUGER AND COSTER-KRONIG TRANSITION 4-0967
ARGON(+) IONS.   AUGER SPECTRA CF SOME FREE ELEMENTS AND BINARY COMPOUNDS BOMBARDED BY 60 KEV 4-1106
K-SHELL AUGER TRANSITICN RATES AND FLUORESCENT YIELDS FOR ELEMENTS ARGON TO XENON.   4-0961
K-SHELL AUGER TRANSITION RATES AND FLUORESCENT YIELDS FCR ELEMENTS BERYLLIUM TO ARGON.   4-0962
ER DECAY OF 5D CORE HOLES IN BISMUTH(III) TELLURI/ MATRIX ELEMENTS DEPENDENCE OF OPTICAL EXCITATION AND AUG 2-0398
    EFFECT OF ELECTRONIC CORRELATIONS ON KLL AUGER LINES IN ELEMENTS MAGNESIUM TO SCANDIUM.   4-0800
RED TANTALUM FILMS BY AUG/ QUANTITATIVE ANALYSIS OF LIGHT ELEMENTS (NITROGEN, CARBON, AND OXYGEN) IN SPUTTE 5-1652
CAL PROPERTIES OF SPUTTERED TANTALUM FIL/ EFFECT OF LIGHT ELEMENTS NITROGEN, CARBON AND OXYGEN ON THE PHYSI 5-1473
               AUGER ELECTRON ENERGIES (0-2000 EV) FOR ELEMENTS OF ATOMIC NUMBER 5-103.   4-1119
               AUGER SPECTRA FOR ELEMENTS OF LOW ATOMIC NUMBER.   4-0827
RENCE AUGER SPECTROSCOFY FOR STUDYING SMALL QUANTITIES OF ELEMENTS ON METALLIC SURFACES.   DIFFE 5-1937
ELS. (IMPURITY GRAIN BCUNDARY SEGREGATI/ EFFECT OF SOLUTE ELEMENTS ON TEMPER EMBRITTLEMENT OF LOW ALLOY STE 6-2083
               CATALOG OF CALCULATED AUGER TRANSITIONS FOR THE ELEMENTS. (REVIEW)   1-0040
GER SPECTROSCOPY ANALYSIS, STAIN/ SEGREGATION OF ALLOYING ELEMENTS TO FREE SURFACES DURING IRRADIATION. (AU 6-2091
    AUGER TRANSITIONS BETWEEN 0 AND 1500 EV FOR ELEMENTS UP TO Z EQUALS 88.   4-1013
SPECTROSCOPY AND CHARACTERISTIC LOSS SPECTROSCOPY FOR THE ELEMENTS VANADIUM TO COBALT. /MBINATION OF AUGER 4-0813
               KLL AUGER TRANSITION PROBABILITIES FOR ELEMENTS WITH LOW AND INTERMEDIATE ATOMIC NUMBERS 4-0803
               THE L(II) SUBSHELL ATOMIC YIELDS OF ELEMENTS WITH Z APPROXIMATELY 90.   2-0349
               L-SERIES AUGER ELECTRON SPECTRA FOR ELEMENTS WITH Z = 66-70.   4-0744
               ROLE OF AUGER ELECTRON SPECTROSCOPY IN SURFACE ELEMENTAL ANALYSIS.   1-0194
URING SLIDING USING AUGER SPECTROSCOPY.   ELEMENTAL ANALYSIS OF FRICTION AND WEAR SURFACE D 6-2013
    SPECTROSCOPY.   THREE DIMENSIONAL ELEMENTAL ANALYSIS OF MATERIALS BY AUGER ELECTRON 5-1597
               IONIZATION SPECTROMETER FOR ELEMENTAL ANALYSIS OF SURFACES.   3-0615
PY AND SECONDARY ION MASS SPECTRCMETRY.   ELEMENTAL ANALYSIS WITH AUGER ELECTRON SPECTROSCO 5-1654
               SURFACE STRUCTURE AND COMPOSITICN OF VARIOUS ELEMENTAL AND III-V SEMICONDUCTORS.   5-1161
ED BY CRACK GROWTH IN HY/ A TECHNIQUE FOR DETERMINING THE ELEMENTAL COMPOSITION OF FRACTURE SURFACES PRODUC 6-2142
UGER SPECTROSCOPY)   TECHNIQUES FOR ELEMENTAL COMPOSITION PROFILING IN THIN FILMS. (A 5-1273
LECTRON MICROSCOPE. (AUGER ELECTRON SPECTROSCOPY, SURFACE ELEMENTAL COMPOSITON) /MINATION IN THE SCANNING E 3-0596
QUANTITATIVE AUGER ELECTRON SPECTROSCOPY USING ELEMENTAL SENSITIVITY FACTORS.   1-0157
NEW METHODS FOR THE DETECTION AND ELIMINATION OF COATING FLAWS.   5-1592
Y OF ALUMINUM WITH A CCMBINED AUGER ELECTRON SPECTROMETER ELLIPSOMETER SYSTEM.   STUD 5-1132
    AUGER ELECTRON SPECTROSCOPY MADE QUANTITATIVE BY ELLIPSOMETRIC CALIBRATION.   1-0205
               GROWTH ON LEAD.   AUGER AND ELLIPSOMETRIC STUDIES OF ULTRATHIN LEAD(II) OXIDE 5-1248
XENON ON GRAPHITE (001).   ELLIPSOMETRIC STUDY OF ADSORPTION: ADSORPTION OF 5-1754
OXIDIZED DEGENERATE SILICON.   AUGER AND ELLIPSOMETRIC STUDY OF PHOSPHORUS SEGREGATION IN 5-1250
YSTAL INTERACTION WITH CHLORINE GAS. (AUGER SPECTROSCOPY, ELLIPSOMETRY) /FACE. SEGREGATION FROM THE BULK CR 5-1833
               MARKER EXPERIMENTS WITH AUGER SPECTROSCOPY AND ELLIPSOMETRY.   5-1832
ARACTERIZING OF ALUMINUM FATIGUE DAMAGE BY PHOTOEMISSION, ELLIPSOMETRY, AND AUGER SPECTROSCOPY.   CH 6-2116
/RE- AND BACKSPUTTERING, STUDIED WITH AUGER SPECTROSCOPY, ELLIPSOMETRY, AND REFLECTION HIGH ENERGY ELECTRO/ 5-1233
NIUM, TUNGSTEN, SILICON) MONITORED BY AUGER SPECTROSCOPY, ELLIPSOMETRY AND WORK FUNCTION CHANGE. /CES (TITA 5-1828
               ELLIPSOMETRY AS A TOOL IN SURFACE STUDIES.   1-0128
NTIAL DIFFERENCE STUDIES. CESIUM ADSORPTION ON TUNGSTEN: ELLIPSOMETRY, AUGER SPECTROSCOPY AND SURFACE POTE 5-1830
ON OF CESIUM ON TUNGSTEN AND TITANIUM SINGLE CRYSTALS IN/ ELLIPSOMETRY, AUGER SPECTROSCOPY, AND WORK FUNCTI 5-1829
TION RESEARCH BY MODERN SURFACE SCIENCE TECHNIQUES. (AES, ELLIPSOMETRY, OTHER TECHNIQUES) /EAR, AND LUBRICA 1-0104
BY ELECTRON SPECTROSCOPY.   DYNAMIC ELUCIDATION OF REACTION STEPS ON A SOLID SURFACE 5-1535
OSCOPY)   STUDY OF GRAIN BOUNDARY SEGREGANTS IN THERMALLY EMBRITTLED MARAGING STEEL. (AUGER ELECTRON SPECTR 6-2055
RY SEGREGATION)   STUDY OF SULFUR EMBRITTLED OXYGEN FREE COPPER. (AES, GRAIN BOUNDA 6-2015
MANGANESE AND ANTIMONY TO THE GRAIN BOUNDARIES OF TEMPER EMBRITTLED STEEL. (AUGER ELECTRON SPECTROSCOPY) / 6-2070
FRACTURE SURFACE ANALYSIS OF A TEMPER EMBRITTLED STEEL BY AUGER ELECTRON SPECTROSCOPY. 6-2084
AUGER FRACTURE SURFACE ANALYSIS OF A TEMPER EMBRITTLED 3340- STAINLESS STEEL.   6-2085
D GRAIN BOUNDARY SEGREGATION OF STEEL IMPURITIES. (TEMPER EMBRITTLEMENT BY AUGER SPECTROSCOPY)   SURFACE AN 6-2109
ING PHOSPHORUS AS AN IMPURITY. (AUGER EMISSICN AN/ TEMPER EMBRITTLEMENT IN A NICKEL- CHROMIUM STEEL CONTAIN 6-2138
US- SULFUR AND IRON- ANTIMONY- SULFUR ALL/ GRAIN BOUNDARY EMBRITTLEMENT IN IRON- PHOSPHORUS, IRON- PHOSPHOR 6-2098
DIES USING AUGER ELECTRON EMISSION SPECTROSCOPY ON TEMPER EMBRITTLEMENT IN LOW ALLOY STEELS.   STU 6-2119
PECTROSCOPY)   CONFIRMATION OF SULFUR EMBRITTLEMENT IN NICKEL ALLOYS. (AUGER ELECTRON S 6-2053
NADIUM.   LONG TIME ISOTHERMAL TEMPER EMBRITTLEMENT IN NICKEL- CHROMIUM- MOLYBDENUM- VA 6-2140
MISSION ANALYSIS, TUNGSTEN, COPPER, STEEL/ GRAIN BOUNDARY EMBRITTLEMENT IN SOME METALS AND ALLOYS. (AUGER E 6-2058
RE SURFACES)   TEMPER EMBRITTLEMENT OF LOW ALLOY STEELS. (AES OF FRACTU 6-2062
/ION AT SELECTIVE GRAIN BOUNDARIES AND ITS RCLE IN TEMPER EMBRITTLEMENT OF LOW ALLOY STEELS. (AUGER SPECTR/ 6-2057
SCOPY)   ROLE OF MANGANESE AND SILICON IN TEMPER EMBRITTLEMENT OF LOW ALLOY STEELS. (AUGER SPECTRO 6-2059
N BOUNDARY SEGREGATI/ EFFECT OF SOLUTE ELEMENTS ON TEMPER EMBRITTLEMENT OF LOW ALLOY STEELS. (IMPURITY GRAI 6-2083
W ALLOY STEEL/ EFFECTS OF COPPER AND PHOSPHORUS ON TEMPER EMBRITTLEMENT OF MANGANESE- MOLYBDENUM- NICKEL LO 6-2043

AIN BOUNDARY SEGREGATICN. (AUGER ELECTRON / INTERGRANULAR EMBRITTLEMENT OF NICKEL BY HYDROGEN: EFFECT OF GR 6-2072
AES AND TEMPER EMBRITTLEMENT OF STEELS. 6-2081
SPECTROSCOPY. GRAIN BOUNDARY EMBRITTLEMENT OF STEELS STUDIED BY AUGER ELECTRON 6-2103
ON SPECTROSCOPIC METHOD. TEMPER EMBRITTLEMENT OF TIN DOPED STEELS BY AUGER ELECTR 6-2128
/OSCOPY AND ITS APPLICATIONS TO SOME TECHNICAL PROBLEMS. (EMBRITTLEMENT, SEGREGATION OF STEELS AND ALLOYS) 6-2100
CES OF 18-8 STAINLESS STEEL ELECTROPOLISHED OR AERADED BY EMERY PAPERS. /SPECTROSCOPIC STUDIES OF THE SURFA 6-2130
DIUM CHLORIDE SINGLE CRYSTALS. (PHOTO- AND AUGER ELECTRON EMISSION) ANGLE DISTRIBUTION OF SO 2-0297
ANGULAR DEPENDENCIES CF ELECTRON EXCITED AUGER EMISSION. 2-0378
BAND STRUCTURE OF GRAPHITE STUDIED BY SECONDARY ELECTRON EMISSION. 2-0569
OM A THORIUM DISPENSER CATHODE OBSERVED BY AUGER ELECTRON EMISSION. CARBON EVAPORATION FR 5-1400
RIUM ALLOY CATHODE IN RELATION TC ITS FILM SYSTEM. (AUGER EMISSION) /RISTICS OF SURFACE OF THE PALLADIUM BA 5-1129
CHEMICAL IDENTIFICATICN OF SURFACE MONOLAYERS. (BY AUGER EMISSION) 5-1589
N THE SURFACES OF SILICON NITRIDE FILMS BY AUGER ELECTRON EMISSION. /TECTION OF SILICON OXYNITRIDE LAYERS O 5-1611
EFFECT OF SURFACE PLASMON OSCILLATIONS ON AUGER EMISSION. 2-0236
COPY FOR CHEMICAL ANALYSIS, AND FIELD STIMULATED ELECTRON EMISSION. /PECTROSCOPY, PHOTON, ELECTRON SPECTROS 5-1901
IES OF PRODUCTION AND DETECTION OF ELECTRON EXCITED AUGER EMISSION. ESTIMATES OF EFFICIENC 2-0288
EVIDENCE FOR A RADIATIVE AUGER EFFECT IN X-RAY PHOTON EMISSION. 2-0223
FORMATION OF A PLASMA SHEATH WITH SECONDARY ELECTRON EMISSION. 2-0244
IN-DEPTH PROFILES BY AUGER SPECTROSCOPY AND SECONDARY ION EMISSION. 4-0995
IMPLANTED SILICON BY AUGER SPECTROSCOPY AND SECONDARY ION EMISSION. IN-DEPTH PROFILES OF PHOSPHORUS ION 4-0996
E YIELDS OF ARGON AND CHLORINE BY ELECTRON IMPACT. (AUGER EMISSION) MEASUREMENTS OF THE L-SHELL FLUORESCENC 2-0387
NON AND THEIR IONS BY HIGH ENERGY ELECTRON IMPACT. (AUGER EMISSION) MULTIIONIZATION OF NEON, ARGON, XE 2-0492
ION BY INTERMEDIATE ENERGY FLUORINE IONS. (AUGER EMISSION) NEON K-SHELL VACANCY PRODUCT 2-0572
OXYGEN BEAM GAS AUGER ELECTRON EMISSION. 2-0400
RADIATIVE AUGER EFFECT WITH AND WITHOUT ELECTRON EMISSION. 2-0494
RADIATIVE AUGER TRANSITIONS IN SOFT X-RAY PLASMA EMISSION. 2-0344
NTILLATION EFFICIENCY FOR LOW VELOCITY HEAVY IONS. (AUGER EMISSION) SCI 2-0431
ANALYSIS OF MATERIALS BY ELECTRON EXCITED AUGER ELECTRON EMISSION) /ONDARY ELECTRON SPECTROSCOPY. (SURFACE 5-1403
OME OBSERVATIONS OF SURFACE SEGREGATION BY AUGER ELECTRON EMISSION. S 5-1401
ATER VAPOR OF SILICON (111) SURFACES USING AUGER ELECTRON EMISSION. /DY OF THE OXIDATION AND REDUCTION BY W 5-1612
RFACE STRUCTURE FROM ANGULAR DEPENDENCE OF AUGER ELECTRON EMISSION. SU 5-1977
ITH RETARDING POTENTIAL ANALYZERS. COMMENTS. / SECONDARY EMISSION ANALOG FOR IMPROVED AUGER SPECTROSCOPY W 3-0584
ITH RETARDING POTENTIAL ANALYZERS. (REPLY TC C/ SECONDARY EMISSION ANALOG FOR IMPROVED AUGER SPECTROSCOPY W 3-0694
AUGER ELECTRON EMISSION ANALYSIS. 1-0073
ROMIUM STEEL CONTAINING PHOSPHORUS AS AN IMPURITY. (AUGER EMISSION ANALYSIS) /EMBRITTLEMENT IN A NICKEL- CH 6-2138
BOUNDARY EMBRITTLEMENT IN SOME METALS AND ALLOYS. (AUGER EMISSION ANALYSIS, TUNGSTEN, COPPER, STEELS) /AIN 6-2058
AUGER EMISSION AND COLOR CENTERS IN LITHIUM FLUORIDE. 2-0430
SECONDARY EMISSION AND CONTAMINATION OF METAL SURFACES. 6-2071
MONOCRYSTALS. SECONDARY EMISSION AND ELASTIC REFLECTION OF ELECTRONS FROM 2-0508
RPTION PROCESSES. OPTICALLY STIMULATED SECONDARY ELECTRON EMISSION AND PHOTOELECTRON EMISSION STUDY OF ADSO 2-0302
. (AUGER PROCESS, AES, PHOTOELECTRON SPECTROSCOPY, CHEMI/ EMISSION AND RECOMBINATION INVOLVING INNER SHELLS 2-0514
VESTIGATIONS WITH FIELD EMIS/ STABILITY OF FIELD ELECTRON EMISSION AND VACUUM BREAKDOWN. (FROM SURFACES) IN 5-1520
/LYSIS PROFILES OF OXIDE FILMS ON CHROMIUM STEEL BY AUGER EMISSION AND X-RAY PHOTOELECTRON SPECTROSCOPIES. 6-2016
MPOSITION. CHARACTERISTIC AUGER ELECTRON EMISSION AS A TOOL FOR THE ANALYSIS OF SURFACE CO 1-0170
IVITY VARIATIONS IN BAYARD- ALPERT GAUGES CAUSED BY AUGER EMISSION AT COLLECTOR. SENSIT 3-0618
STANDARD LEED- AUGER SYSTEM FOR THE ANALYSIS OF ELECTRON EMISSION AT LOW PRIMARY BEAM ENERGIES. /ENTS TO A 3-0698
NIUM JUNCTION). SECONDARY ELECTRON EMISSION AT SUPERHIGH FREQUENCIES. (COPPER- GERMA 2-0356
SPECTROMETER FOR INDUCED ELECTRON EMISSION. (AUGER ELECTRONS) 3-0585
HE NEON K FLUORESCENCE YIELD DEDUCED FROM HIGH RESOLUTION EMISSION (AUGER) SPECTRA. / STATE DEPENDENCE OF T 2-0520
ON BERYLLIUM OXIDE AND ITS EFFECTS ON SECONDARY ELECTRON EMISSION. (AUGER SPECTROSCOPY) /BON CONCENTRATION 5-1362
US. AUGER SATELLITES ON THE L23 AUGER EMISSION BANDS OF ALUMINUM, SILICON, AND PHOSPHOR 4-0984
US. AUGER SATELLITES OF THE L2,3 AUGER EMISSION BANDS OF ALUMINUM, SILICON, AND PHOSPHOR 4-0985
D/ INTERPRETATION OF HIGH ENERGY X-RAY SATELLITES OF L2,3 EMISSION BANDS OF SODIUM, MAGNESIUM, ALUMINUM, AN 2-0239
LS. ARGON AUGER EMISSION BY ARGON(+) ION BOMBARDMENT OF SOME META 2-0547
AUGER ELECTRON EMISSION BY ION BOMBARDMENT OF LIGHT METALS. 4-0877
LTRAHIGH VACUUM SCANNING ELECTRON MICROSCOPE WITH A FIELD EMISSION CATHODE AND AUGER ANALYZER. U 3-0621
). FILM STRUCTURE EFFECT ON THE SECONDARY ELECTRON EMISSION CHARACTERISTICS (GERMANIUM AND BERYLLIUM 4-0923
/TURE OF THE ANGULAR DEPENDENCE OF THE SECONDARY ELECTRON EMISSION COEFFICIENT OF SINGLE CRYSTALS IN THE 2/ 4-1061
/S BY METASTABLE ATOMS. PART-1: MEASUREMENTS OF SECONDARY EMISSION COEFFICIENTS USING A CROSSED BEAM METHO/ 2-0340
ANGULAR DEPENDENCIES IN ELECTRON EXCITED AUGER EMISSION. (COMMENTS) 2-0286
OR LIGHT ATOMS AND MOLECULES BY ELECTRON IMPACT. (/ X-RAY EMISSION CROSS SECTIONS AND FLUORESCENCE YIELDS F 2-0528
ION IN AUGER ELECTRON SPECTROSCOPY AND SECONDARY ELECTRON EMISSION DUE TO CASCADE PROCESS. /GROUND INFORMAT 2-0411
NS (TABLE FOR 60 ELEMENTS). SECONDARY ELECTRON EMISSION DUE TO PRIMARY AND BACKSCATTERED ELECTRO 4-0907
1) CRYSTAL SURFAC/ CHEMICAL CHANGES IN SECONDARY ELECTRON EMISSION DURING OXIDATION OF NICKEL (100) AND (11 5-1453
GER BROADENING) K X-RAY EMISSION EDGE SHAPES OF FREE ELECTRON METALS. (AU 2-0337
ROPY OF BACKSCATTERING COEFFICIENT AND SECONDARY ELECTRON EMISSION FOR 10-100 KEV ELECTRONS. /TATION ANISOT 4-1031
ANGULAR DEPENDENCE OF AUGER ELECTRON EMISSION FROM A SINGLE CRYSTAL SPECIMEN. 4-0958
AUGER ELECTRON EMISSION FROM BERYLLIUM SURFACES. 5-1613
GER SPECTROSCOPY) PROTON INDUCED ELECTRON EMISSION FROM CHARACTERIZED NIOBIUM SURFACES. (AU 5-1666
DENUM BOMBARDED BY POSITIVE IONS. PART-1./ AUGER ELECTRON EMISSION FROM CLEAN AND CARBON CONTAMINATED MOLYB 2-0543
RFACES. ANGULAR DISTRIBUTION OF AUGER ELECTRON EMISSION FROM CLEAN AND GAS-COVERED IRON (100) SU 5-1623
ANGLE OF INCIDENCE EFFECTS IN AUGER ELECTRON EMISSION FROM CLEAN MOLYBDENUM. 2-0542
ANGULAR EFFECTS IN AUGER ELECTRON EMISSION FROM COPPER (110). 5-1988
ANGULAR DEPENDENCE OF AUGER ELECTRON EMISSION FROM COPPER (111) AND (100) SURFACES. 5-1632
SECONDARY ELECTRON EMISSION FROM GALLIUM ARSENIDE. 4-0861
FACE. AUGER ELECTRON EMISSION FROM GOLD DEPOSITED ON SILICON (111) SUR 5-1676
SECONDARY ELECTRON EMISSION FROM ION IMPLANTED SILICON (CESIUM). 4-0913
SECONDARY ELECTRON EMISSION FROM IRON SINGLE AND POLYCRYSTALS. 4-0917
LOW ENERGY AUGER EMISSION FROM LITHIUM FLUORIDE. 4-0842
EFFECTS OF CESIATION ON SECONDARY ELECTRON EMISSION FROM MAGNESIUM OXIDE- GOLD CERMETS. 2-0384
ORIGIN CF ICN INDUCED AUGER ELECTRON EMISSION FROM METALS. 2-0549
/SSEMBLY FOR OBTAINING THE SPECTRUM OF SECONDARY ELECTRON EMISSION FROM METALS AND SEMICONDUCTORS. (AUGER / 3-0688
COPPER SINGLE CRYSTALS. ANISOTRCPY CF SECONDARY ELECTRON EMISSION FROM PASSAGE OF LITHIUM(+) IONS THROUGH 2-0245
GER SPECTRA) SECONDARY ELECTRON EMISSION FROM PLATINUM BLACK COATED SURFACES. (AU 5-1979
CHARACTERISTICS OF SHAPING SECONDARY ELECTRON EMISSION FROM POROUS POTASSIUM CHLORIDE FILMS. 4-1120
SCANDIUM L X-RAY EMISSION FROM PURE METAL AND SCANDIUM OXIDE. 4-1102
/STRUCTURES AND ENERGY DISTRIBUTION OF SECONDARY ELECTRON EMISSION FROM SILICON (111). (AUGER ELECTRON EXC/ 4-0853
ARDED BY RARE GAS IONS. SECONDARY ELECTRON EMISSION FROM SODIUM CHLORINE SINGLE CRYSTAL BOMB 4-0759
ANGULAR DEPENDENCE OF AUGER ELECTRON EMISSION FROM SOLID SURFACES. 5-1447
ENERGY IONS (PLATINUM, NICKEL, TANTAL/ SECONDARY ELECTRON EMISSION FROM SOLID SURFACES BOMBARDED BY MEDIUM 2-0309

```
S IONS (10-100 KEV): ENERGETIC AND SPATIAL DI/ ELECTRONIC EMISSION FROM SOLID TARGETS BOMBARDED BY NOBLE GA    4-0776
IAL ENERGY. (AUGER NEUTRALIZATION, REVIEW, 16 R/ ELECTRON EMISSION FROM SOLIDS BY PARTICLES CARRYING POTENT     2-0374
NG EFFECTS AND IONIZATION CROSS SECTIONS.            AUGER EMISSION FROM SOLIDS.   ESTIMATION OF BACKSCATTERI    2-0513
              ANGULAR DEPENDENCE OF ELECTRON EMISSION FROM SURFACES, (AUGER EFFECT)                             5-1726
                           SECONDARY ELECTRON EMISSION FROM THIN FILM GOLD METAL- OXIDE CERMETS                 4-1015
                           SECONDARY ELECTRON EMISSION FROM THIN FILMS.                                         4-0894
                           SECONDARY ELECTRON EMISSION FROM THIN FILMS. (OF GOLD, APPARATUS)                    3-0638
                           SECONDARY ELECTRON EMISSION FROM TUNGSTEN AND MOLYBDENUM.                            4-0850
TALS.                      SECONDARY ELECTRON EMISSION FROM TUNGSTEN AND MOLYBDENUM SINGLE CRYS                 4-0742
CT, AUGER ELECTRON SPECTROSCOPY)         EXOELECTRON EMISSION FROM ZINC OXIDE. (OXYGEN ADSORPTION EFFE          4-0921
                "SEMI-AUGER" PROCESSES IN L2,3 EMISSION IN ARGON AND POTASSIUM CHLORIDE.                        2-0326
GRAPHITE, ELEC/ AUGER ELECTRON SPECTROSCOPY AND SECONDARY EMISSION IN SEMICONDUCTORS. (SILICON, GERMANIUM,      2-0240
RONS FROM MOLYBDENUM AND TUNGSTEN.          AUGER EMISSION IN THE ENERGY SPECTRA OF SECONDARY ELECT             4-0874
                           SECONDARY ELECTRON EMISSION IN THE STUDY OF SOLID SURFACES.                          5-1555
S.              CROSS SECTIONS FOR L2,3 AUGER EMISSION IN 22 TO 300 KEV PROTON- ARGON COLLISION                 4-1040
T METALS (LITHIUM, SODIUM) SUBJECTED TO ION BOM/ RELATIVE EMISSION INTENSITIES OF AUGER ELECTRONS FROM LIGH     2-0548
LS. (AUGER EFFECT)                            EMISSION LINES OF THE M SERIES OF RARE EARTH META                 4-0790
FACE OF GRAPHITE SINGLE CRYSTAL.          AUGER ELECTRON EMISSION MICROGRAPHIC STUDIES OF THE CLEAVAGE SUR      5-1413
ED SURFACE OF PURE IRON PLATE.            AUGER ELECTRON EMISSION MICROGRAPHY AND MICROANALYSIS OF A STAIN      5-1415
URFACES.                                  AUGER ELECTRON EMISSION MICROGRAPHY AND MICROANALYSIS OF SOLID S      5-1416
        AUGER ELECTRON SPECTROSCOPY IN THE ELECTRON EMISSION MICROSCOPE.                                        3-0604
OSSIBILITY OF AUGER ELECTRON SPECTROSCOPY IN THE ELECTRON EMISSION MICROSCOPE.                               P  3-0605
UGER SPECTROSCOPY)            SECONDARY ELECTRON EMISSION OF A ZIRCONIUM- ALUMINUM BULK GETTER. (A             5-1407
                           SECONDARY ELECTRON EMISSION OF AMORPHOUS SEMICONDUCTOR THIN FILMS.                   5-1239
C TARGET SUBJECTED TO AN IONIC BOMBARDMENT.          EMISSION OF AUGER ELECTRONS BY ATOMS IN A METALLI          2-0383
                                                     EMISSION OF AUGER ELECTRONS BY ION BOMBARDMENT.            2-0550
N SPECTROSCOPY INVESTIG/ THERMALLY STIMULATED EXOELECTRON EMISSION OF BERYLLIUM OXIDE LAYERS. AUGER ELECTRO    5-1325
IGH EXCITATION LEVELS/ TEMPERATURE DEPENDENCE OF THE EDGE EMISSION OF GALLIUM SELENIDE SINGLE CRYSTALS AT H    4-0258
NITROGEN AND Z(1)- NEON COLLISIONS AT KEV ENERGIES.    EMISSION OF KLL AUGER ELECTRONS PRODUCED IN Z(1)-       4-0830
COPY)              SURFACE CONDITION AND THERMIONIC EMISSION OF LANTHANUM HEXABORIDE. (AUGER SPECTROS          5-1471
A SEMICONDUCTOR. (AUGER PROCESS)          STIMULATED EMISSION OF LIGHT FROM A NONIDEAL EXCITON GAS IN          2-0462
OLYBDENUM SINGLE CRYSTAL.    STRUCTURE AND SECONDARY EMISSION OF MAGNESIUM OXIDE LAYERS APPLIED ON A M          5-1191
PROTONS.                     SECONDARY ELECTRON EMISSION OF METALS BOMBARDED WITH 120-EV TO 5-KEV              4-1029
                           SECONDARY ELECTRON EMISSION OF MOLYBDENUM CRYSTALS.                                 4-1063
                           SECONDARY ELECTRON EMISSION OF SILICON DIOXIDE SINGLE CRYSTALS.                     4-1062
                           SECONDARY ELECTRON EMISSION OF SOLID ALUMINUM AND MAGNESIUM TARGETS.                2-0373
RANSMISSION.                    AUGER ELECTRON EMISSION OF THIN CARBON FOILS IN REFLECTION AND T               4-0893
LASTICALLY/ SLOW ELECTRON SCATTERING FROM METALS. PART-1: EMISSION OF TRUE SECONDARY ELECTRONS. PART-2: INE    2-0500
IMPACT VIA N4,5 AUGER TRANSITIONS.    VACUUM ULTRAVIOLET EMISSION OF XENON(III) LEVELS EXCITED BY ELECTRON      4-0878
/ THE DEPENDENCE OF THE COEFFICIENT OF SECONDARY ELECTRON EMISSION ON ENERGY OF PRIMARY ELECTRONS (POTASSI/    4-0891
ON EFFECTS ON THE SPECTRUM OF THE SLOW ELECTRON EMISSION PEAK IN POLYCRYSTALLINE MAGNESIUM. /DATI              4-1118
                ION- ATOM COLLISIONS. (AUGER EMISSION, REVIEW)                                                 1-0169
                           SECONDARY ELECTRON EMISSION. (REVIEW, 86 REFS)                                      1-0179
                           SECONDARY ELECTRON EMISSION ROCKING CURVE FROM COPPER.                              4-0764
        GRAPHITE CONDUCTION BAND STATES FROM SECONDARY ELECTRON EMISSION SPECTRA.                              2-0568
              STUDY OF GRAIN BOUNDARY SEGREGATION USING AUGER EMISSION SPECTRA.                                6-2121
AND INTENSITIES.                    AUGER ELECTRON EMISSION SPECTRA: A SIMPLE TREATMENT OF ENERGIES            2-0319
ARGO/ CHARGE STATE DEPENDENCE OF X-RAY AND AUGER ELECTRON EMISSION SPECTRA FOR RARE GAS ATOMS. PART-1. THE     4-0927
E.                    ANGULAR RESOLVED AUGER EMISSION SPECTRA FROM A CLEAN COPPER (100) SURFAC                 5-1697
BEAMS.                              AUGER ELECTRON EMISSION SPECTRA FROM FOIL AND GAS EXCITED CARBON           4-1050
E PROCESSES)    CLUSTER CALCULATIONS OF THE SCFT X-RAY L3 EMISSION SPECTRA OF NICKEL AND COPPER. (AUGER-TYP    4-0888
X) GALLIUM(1-X) ARSENIDE ELECTROLUMINESCENT DIODES. (AUG/ EMISSION SPECTRA OF SILICON COMPENSATED ALUMINUM(    2-0325
VA/ INELASTIC EFFECTS AND STRUCTURE IN THE AUGER ELECTRON EMISSION SPECTRA OF VANADIUM PENTOXIDE (010) AND     5-1340
        A SCANNING ELECTRON MICROSCOPE WITH FACILITIES FOR AUGER EMISSION SPECTROSCOPY.                       3-0593
              A STUDY OF THE IRON (111) SURFACE USING LEED AND AUGER EMISSION SPECTROSCOPY.                   5-1670
AND TRANSFER OF PLATINUM IRON TO METALS STUDIED BY AUGER EMISSION SPECTROSCOPY.               ADHESION        5-1730
ER OF POLYTETRAFLUORO ETHYLENE TO METALS STUDIED BY AUGER EMISSION SPECTROSCOPY.         ADHESION AND TRANSF  5-1731
ETALS TO A CLEAN IRON SURFACE STUDIED WITH LEED AND AUGER EMISSION SPECTROSCOPY.           ADHESION OF M      6-2008
        CHARACTERISTIC IONIZATION LOSSES OBSERVED IN AUGER EMISSION SPECTROSCOPY.                             2-0287
SULFUR AND IRON- ANTIMONY- SULFUR ALLOYS. (AUGER ELECTRON EMISSION SPECTROSCOPY) /HORUS, IRON- PHOSPHORUS-    6-2098
F SILVER ON POTASSIUM CHLORIDE OBSERVED BY LEED AND AUGER EMISSION SPECTROSCOPY.             GROWTH O         5-1348
NSFER BETWEEN METALS IN SLIDING CONTACT EXAMINED BY AUGER EMISSION SPECTROSCOPY.        METALLIC TRA          5-1729
        CARBON FIBER SURFACES USING AUGER AND SECONDARY EMISSION SPECTROSCOPY.             STUDY OF           5-1970
        STUDY OF GRAIN BOUNDARY SEGREGATION USING AUGER ELECTRON EMISSION SPECTROSCOPY.                       6-2117
SURFACE ABSORPTION BY POLYPEPTIDE FILMS EXAMINED BY AUGER EMISSION SPECTROSCOPY.                              5-1179
BINED X-RAY PHOTOELECTRON SPECTROSCOPY AND AUGER ELECTRON EMISSION SPECTROSCOPY.      SURFACE ANALYSES BY COM 5-1272
SURFACE SEGREGATION IN BORON DOPED IRON OBSERVED BY AUGER EMISSION SPECTROSCOPY.                              5-1181
USE OF CHARACTERISTIC IONIZATION LOSSES IN AUGER EMISSION SPECTROSCOPY.                                       2-0350
ON DIFFRACTION (LEED) OF THE ZINC OXIDE (/ AUGER ELECTRON EMISSION SPECTROSCOPY (AES) AND LOW ENERGY ELECTR   5-1338
GATION OF GOLD TO SILICON (111) SURFACE OBSERVED BY AUGER EMISSION SPECTROSCOPY AND BY LEED.       SEGRE      5-1180
MICROSTRU/ USE OF LOW ENERGY ELECTRON DIFFRACTION, AUGER EMISSION SPECTROSCOPY AND FIELD ION MICROSCOPY IN    1-0057
        ELECTRON INDUCED SECONDARY ELECTRON EMISSION SPECTROSCOPY OF SURFACES.                                5-1406
              STUDIES USING AUGER EMISSION SPECTROSCOPY ON TEMPER EMBRITTLEMENT IN                            6-2119
/ CURVE FITTING, AND RELATED TECHNIQUES IN AUGER ELECTRON EMISSION SPECTROSCOPY USING RETARDING FIELD SPEC/   3-0610
LOW ALLOY STEELS.                    AUGER EMISSION SPECTROSCOPY VANADIUM PENTOXIDE (010) AN                  5-1941
D VANADIUM (100) SURFACES.          SINGULARITIES IN AUGER EMISSION SPECTRUM OF METALS.                       4-0998
        PLASMON SATELLITE IN THE AUGER EMISSION SPECTRUM OF METALS. (ALUMINUM)                                4-0871
RPTION OF CARBON MONOXIDE ON COPPER (001). LEED AND AUGER EMISSION STUDIES.                  ADSO             5-1515
VAGE DEVICE FOR LOW ENERGY ELECTRON DIFFRACTION AND AUGER EMISSION STUDIES. /M EVAPORATOR AND MULTIPLE CLEA   3-0707
        ULTRAHIGH VACUUM EVAPORATOR FOR LEED AND AUGER EMISSION STUDIES.                                      3-0683
              ELECTRON EMISSION STUDIES FROM GALLIUM SELENIDE SURFACES.                                       5-1967
        LEED- AES ELECTRON INDUCED EMISSION STUDIES OF LITHIUM HYDRIDE.                                       4-0807
GERMANIUM.                    SECONDARY EMISSION STUDIES ON GERMANIUM AND SODIUM- COVERED                     5-1714
BLE AUGER PROCESS.    TWO- PHOTON EMISSION, THE RADIATIVE AUGER EFFECT, AND THE DOU                           2-0225
        SOFT X-RAY AMPLIFIED SPONTANEOUS EMISSION VIA THE AUGER EFFECT.                                       2-0443
OF THE INCIDENCE ANGLE OF PRIMARY ELECTRONS ON THE AUGER EMISSION YIELD.                          EFFECT     2-0237
/D FOR DETECTING FINE STRUCTURE IN THE SECONDARY ELECTRON EMISSION YIELD AND THE APPLICATION TO SILICON (1/   4-0852
        ULTRAVIOLET EMISSIONS IN CALCITE. (AUGER EFFECT)                                                      4-0798
ION AND ATOM BOMBARDMENT. / ENERGY SPECTRUM OF ELECTRONS EMITTED BY ALKALI HALIDE CRYSTALS UNDER IMPACT OF    2-0246
        ENERGY DISTRIBUTION OF ELECTRONS EMITTED BY ARGON ATOMS UNDER ELECTRON IMPACT.                        4-1008
```

/ ELECTRONS IN THE ENERGY SPECTRUM OF SECONDARY ELECTRONS EMITTED BY COPPER UNDER BOMBARDMENT BY RARE GAS / 2-0551
SUBMITTED TO/ ENERGY DISTRIBUTION OF SECONDARY ELECTRONS EMITTED BY INDIUM PHOSPHIDE AND GALLIUM PHOSPHIDE 2-0269
SPECTRUM AND ANGULAR DISTRIBUTION OF SECONDARY ELECTRONS EMITTED BY IONIC BOMBARDMENT.   ENERGY 4-0988
Y PRIMARY ELECTRON/ ENERGY SPECTRA OF SECONDARY ELECTRONS EMITTED BY MAGNESIUM AND SILICON WHEN BOMBARDED B 4-1047
INTERPRETATION OF ENERGY LOSS BY AUGER ELECTRONS EMITTED DURING ION BOMBARDMENT.   2-0268
ANALYSIS APPARATUS FCR INVESTIGATING A SAMPLE BY MEANS OF EMITTED ELECTRONS.   3-0713
ILICON UNDER ION / ENERGY DISTRIBUTION OF AUGER ELECTRONS EMITTED FROM BERYLLIUM, MAGNESIUM, ALUMINUM AND S 4-0876
CTORS.   IRON-55 AS AN AUGER ELECTRON EMITTER. NOVEL SOURCE FOR GAS CHROMATOGRAPHY DETE 3-0608
AUGER SPECTROSCCPIC STUDY OF THE SURFACES OF ALLOY EMITTERS.   5-1192
ILMS AND ITS APPLICATION TO HIGH YIELD SECONDARY ELECTRON EMITTERS. (AUGER ELECTRON SPECTROSCOPY) /CERMET F 5-1430
RECORDING OF COMPONENTS ON THE SURFACE OF FUSED EMITTERS. (AUGER SPECTRA)   4-0736
ELECTRONIC AUGER SPECTROSCOPIC MET/ SURFACE OF THERMIONIC EMITTERS MADE OF CARBIDE COMPOUNDS STUDIED BY AN 5-1562
/UGER AND CONVERSION ELECTRONS FOR THE DETECTION OF GAMMA EMITTING RADIONUCLIDES: CHROMATOGRAPHIC QUALITY / 1-0217
AUGER ANALYSIS OF BRASS- ENAMEL AND STAINLESS STEEL- ENAMEL INTERFACES.   5-1266
/ SOLID TARGETS BOMBARDED BY NOBLE GAS IONS (10-100 KEV): ENERGETIC AND SPATIAL DISTRIBUTIONS. (AUGER ELEC/ 4-0776
MEASUREMENTS OF NEON K-AUGER ELECTRONS PRODUCED BY ENERGETIC HEAVY ION BOMBARDMENT.   4-0951
AUGER DECAY OF NEON FOLLOWING ENERGETIC ION BOMBARDMENT.   4-0954
OGENIC WAVE FUNCTIONS IN THE CALCULATION OF THE KLL AUGER ENERGIES. /N OF THE ZETA EFFECTIVE VALUES OF HYDR 2-0526
SINGLE CRYSTALS IN THE 2-10 KEV RANGE OF PRIMARY ELECTRON ENERGIES. /DARY ELECTRON EMISSION COEFFICIENT OF 4-1061
ANCY PRODUCTION YIELD IN HEAVY ION ATOM COLLISIONS AT KEV ENERGIES. /C CHARGE STATE DEPENDENCE OF THE K-VAC 2-0346
ELECTRON ESCAPE DEPTH IN SILICON. (AUGER ENERGIES) 2-0408
ODUCED IN Z(1)- NITROGEN AND Z(1)- NEON COLLISIONS AT KEV ENERGIES.   EMISSION OF KLL AUGER ELECTRONS PR 4-0830
ELECTRON EMISSION FRCM METALS AND SEMICONDUCTORS. (AUGER ENERGIES) /OR OBTAINING THE SPECTRUM OF SECONDARY 3-0688
KRYPTON M-SHELL AUGER AND COSTER-KRONIG ELECTRON ENERGIES.   4-0928
L1,L2,3M COSTER-KRONIG TRANSITION ENERGIES.   4-0929
FOR THE ANALYSIS OF ELECTRON EMISSION AT LOW PRIMARY BEAM ENERGIES. /ENTS TO A STANDARD LEED- AUGER SYSTEM 3-0698
RELAXATION EFFECTS ON AUGER ENERGIES.   2-0505
THEORY OF KLL AUGER ENERGIES.   4-1000
VARIAN CHART OF AUGER ELECTRON ENERGIES.   4-1082
IN X-RAY SPECTROSCOPY.   INNER ELECTRON BINDING ENERGIES AND AUGER ENERGIES IN FREE ATOMS FOR USE 4-0774
S.   ACCURATE ONE ELECTRON BINDING ENERGIES AND AUGER ENERGIES IN MANY ELECTRON ATOM 2-0265
L VACANCIES (NEON WITH 1S OR 2S VACANCY).   CORRELATION ENERGIES AND AUGER RATES IN ATOMS WITH INNER SHEL 2-0312
NIZED NEON. / THEORETICAL K-SHELL AUGER RATES, TRANSITION ENERGIES, AND FLUORESCENCE YIELDS FOR MULTIPLY IO 2-0284
NIZED STATES OF ARGON.   K-SHELL AUGER RATES, TRANSITION ENERGIES, AND FLUORESCENCE YIELDS OF VARIOUSLY IO 2-0280
AUGER ELECTRON EMISSION SPECTRA: A SIMPLE TREATMENT OF ENERGIES AND INTENSITIES.   2-0319
I-SHELL AUGER ENERGIES AND INTENSITIES IN 3D TRANSITION METALS. 4-1018
ION THEORY TO THE CALCULATION OF AUGER RATES, CORRELATION ENERGIES, AND PHOTOIONIZATION CROSS SECTIONS. /AT 2-0311
R SPECTRUM.   Z DEPENDENCE OF ENERGIES AND RELATIVE INTENSITIES IN THE KLL AUGE 4-0886
ION- ATOM COLLISIONS AT HIGH ENERGIES. (AUGER EFFECT)   2-0480
PHOTOELECTRON AND L2,3MM AUGER ELECTRON ENERGIES FOR ARSENIC.   4-1035
K X-RAY AND AUGER ELECTRON ENERGIES FOR NEON BY A TRANSITION OPERATOR METHOD 2-0393
HEMICAL BONDING. PART-1: AUGER ELECTRONS. (KLL TRANSITION ENERGIES FOR SULFUR, ALUMINUM, SILICON) /ON AND C 2-0327
CALCULATED K-AUGER ELECTRON AND K X-RAY TRANSITION ENERGIES FOR THE MULTIPLY IONIZED NEON ATOM. 4-0955
NUM SURFACES.   AUGER AND OTHER CHARACTERISTIC ENERGIES IN SECONDARY ELECTRON SPECTRA FROM ALUMI 4-0896
ON (111) SURFACES.   CHARACTERISTIC ENERGIES IN SECONDARY ELECTRON SPECTRA FROM SILIC 5-1263
EVAPORATED PHOSPHORUS FIL/ AUGER AND OTHER CHARACTERISTIC ENERGIES IN SECONDARY ELECTRON SPECTRA FROM THIN 4-0986
IC AND SELENIUM. ATOMIC AND EXTRAATOMIC STATIC RELAXATION ENERGIES IN THE L3M4,5M4,5 AUGER PROCESS IN ARSEN 4-0883
THEORY OF KLL AUGER ENERGIES INCLUDING STATIC RELAXATION.   2-0506
RATE SECONDARY ELECTRON YIELD MEASUREMENTS AT LOW PRIMARY ENERGIES. (LEED- AUGER SYSTEM)   FAST, ACCU 4-0629
AUGER CATALOG CALCULATED TRANSITION ENERGIES LISTED BY ENERGY AND ELEMENT.   1-0041
EFFECT OF POLARIZATION OF THE ELECTRON GAS ON THE ENERGIES OF AUGER TRANSITION IN METALS.   2-0435
ELD SPECTRUM.   ELECTRON BINDING ENERGIES OF BARIUM FROM THE SECONDARY ELECTRON YI 4-0846
UM SEGREGATION STUDIES. (AUGER EFFECT)   BINDING ENERGIES OF CARBON TO NICKEL (100) FROM EQUILIBRI 5-1484
RONS FOR IONIZED ATOMS.   SHIFT ENERGIES OF CHARACTERISTIC X-RAYS AND AUGER ELECT 4-1041
CULES BY MEANS OF AUGER ELECTRON SPECTROSCOPY. PART-1: E/ ENERGIES OF EXCITED STATES OF DOUBLY IONIZED MOLE 4-1071
HOTOIONIZATION OF NITROGE/ RELATIVE ABUNDANCES AND RECOIL ENERGIES OF FRAGMENT IONS FORMED FROM THE X-RAY P 2-0308
RELAXATION ENERGY DIFFERENCES FRCM AUGER AND CORE BINDING ENERGIES OF GERMANIUM COMPOUNDS. / ESTIMATION OF 2-0471
ND OF SILICON DIOXIDE.   ELECTRON ORBITAL ENERGIES CF OXYGEN ADSORBED ON SILICON SURFACES A 5-1475
HROMIUM, AND / DIRECT COMPARISON OF CORE ELECTRON BINDING ENERGIES OF SURFACE AND BULK ATOMS OF TITANIUM, C 2-0392
ENERGIES OF THE KLL AUGER LINES.   4-0746
RGIES.   SEMI-EMPIRICAL AUGER ELECTRON ENERGIES. PART-1: GENERAL METHOD AND KLL LINE ENE 4-0930
SEMIEMPIRICAL AUGER ELECTRON ENERGIES. PART-2: THE ARGON ATOM.   4-0815
(AUGER) TRANSITION ENERGIES. (REVIEW)   1-0109
AUGER ELECTRCN SPECTRUM OF OSMIUM AT ENERGIES UP TO 300 EV.   4-1030
R 5-103.   AUGER ELECTRON ENERGIES (0-2000 EV) FOR ELEMENTS OF ATOMIC NUMBE 4-1119
PARATUS FOR INVESTIGATING ION SCATTERING FROM SURFACES AT ENERGIES 100-2000 EV. (AES)   SORBAS: AN AP 3-0704
TS.   AUGER ELECTRONS FRCM ARGON WITH ENERGIES 150-210 EV PRODUCED BY HYDROGEN(+) IMPAC 2-0552
ENT ARGON(+)- ARGON COLLISIONS. COMMENTS. (AUGER ELECTRON ENERGY) / SLOW ELECTRON COINCIDENCE STUDY OF VIOL 4-0423
SILICON SINGLE CRYSTAL UNDER A LOW ENERGY BCMBARDMENT. / ENERGY ANALYSIS OF BACKSCATTERED ELECTRONS FROM A 4-0739
DETERMINATION OF SURFACE STRUCTURES USING LEED AND ENERGY ANALYSIS OF SCATTERED ELECTRONS.   5-1956
TION AMPLITUDE DISTORTION IN A 4-GRID RETARDING POTENTIAL ENERGY ANALYZER. /RON SIGNAL STRENGTHS FOR MODULA 3-0620
DESIGN PARAMETERS FOR THE CYLINDRICAL MIRROR ENERGY ANALYZER.   3-0687
TOMATED DATA RECORDING AND CONTROL SYSTEM FOR AN ELECTRON ENERGY ANALYZER.   SEMIAU 3-0680
COMPACT, INEXPENSIVE HIGH RESOLUTION RETARDING FIELD ENERGY ANALYZER. (AUGER ELECTRONS)   3-0619
NEW COMPACT ENERGY ANALYZER FOR AUGER AND ESCA SPECTROSCOPY. 3-0702
A SMALL RETARDING FIELD ENERGY ANALYZER WITH CEM. (AUGER SPECTROMETER) 3-0655
BY 0.3 TO 2.0 MEV PROTONS. (AUGER PEAKS)   DISTRIBUTION IN ENERGY AND ANGLE OF ELECTRONS EJECTED FROM XENON 2-0534
LTING FROM DIRECT IONIZING COLLISIONS OF ELECTRONS WITH / ENERGY AND ANGULAR DISTRIBUTION OF ELECTRONS RESU 2-0525
/RADAY CUP LEED APPARATUS WITH FACILITY FOR INVESTIGATING ENERGY AND ANGULAR DISTRIBUTIONS OF INELASTICALL/ 3-0691
/JECTION FROM METAL SURFACES BY METASTABLE ATCMS. PART-3: ENERGY AND ANGULAR DISTRIBUTIONS OF THE EJECTED / 5-1133
AUGER CATALOG CALCULATED TRANSITION ENERGIES LISTED BY ENERGY AND ELEMENT.   1-0041
/TROSCOPY IN SPUTTERING MEASUREMENTS.   APPLICATION TO LOW ENERGY ARGON(+) ION SPUTTERING OF SILVER AND NIO/ 5-1825
FIDE. (AUGER MEASUREMENTS)   EFFECTS OF LOW ENERGY ARGON- ION BOMBARDMENT ON MOLYBDENUM DISUL 2-0348
DEPENDENCE OF AUGER ELECTRON YIELD ON PRIMARY BEAM ENERGY AT NORMAL AND GLANCING INCIDENCE.   2-0455
(AUGER NEUTRALIZATION AT (100) FACE / INTERACTION OF LOW ENERGY ATMOSPHERIC IONS WITH CONTROLLED SURFACES. 5-1736
SIUM AND ITS OXIDE.   LOW ENERGY AUGER AND LOSS ELECTRON SPECTRA FROM MAGNE 4-0895
LOW ENERGY AUGER EMISSION FROM LITHIUM FLUORIDE. 4-0842
TRON EMISSION FROM SOLIDS BY PARTICLES CARRYING POTENTIAL ENERGY. (AUGER NEUTRALIZATION, REVIEW, 16 REFS) / 2-0374
TEMPERATURE DEPENDENCE OF THE LOW ENERGY AUGER SPECTRUM OF NICKEL(II) OXIDE (100). 4-0999
ANALYSIS OF SURFACE COMPOSITION WITH LOW ENERGY BACKSCATTERED IONS.   5-1823

M4,5: 1G4 AUGER TRANSITION IN ATOMIC ZINC: IMPROVED AUGER ENERGY CALCULATIONS. /ILED ANALYSIS OF THE L3M4,5 4-0884
THIN FILM ANALYSIS AS APPLIED TO VARIOUS ASPECTS IN SOLAR ENERGY CONVERSION. 5-1960
N SPECTROSCOPY: SILICON ON BERYLLIUM. (AUGER ELECTRON SP/ ENERGY DEPENDENCE OF THE SENSITIVITY OF IONIZATIO 5-1665
/ECTRON SPECTROMETRY AS A METHOD FOR DETERMINATION OF THE ENERGY DEPTH OF LEVELS OF SURFACE ELECTRON CENTE/ 4-0915
/C RELAXATION OF GERMANE AND THE ESTIMATION OF RELAXATION ENERGY DIFFERENCES FROM AUGER AND CORE BINDING E/ 2-0471
REEXAMINATION OF ALUMINUM (110) WITH LOW ENERGY DIFFRACTION AND AUGER SPECTROSCOPY. 5-1507
/G ELECTRON MICROSCOPE, TRANSMISSION ELECTRON MICROSCOPE, ENERGY DISPERSIVE SPECTROMETER, ION MASS ANALYSI/ 1-0010
OM BERYLLIUM, MAGNESIUM, ALUMINUM AND SILICON UNDER ION ENERGY DISTRIBUTION OF AUGER ELECTRONS EMITTED FR 4-0876
G FROM ION- ATOM COLLISIONS WITH A FIXED IMPACT PARAMETE/ ENERGY DISTRIBUTION OF (AUGER) ELECTRONS RESULTIN 2-0331
ATOMS UNDER ELECTRON IMPACT. ENERGY DISTRIBUTION OF ELECTRONS EMITTED BY ARGON 4-1008
500 KEV PROTONS IN GASES. PART-2: NITROGEN. / ANGULAR AND ENERGY DISTRIBUTION OF ELECTRONS PRODUCED BY 200- 4-1075
N FROM SILICON (111). (AUGER ELECTRO/ FINE STRUCTURES AND ENERGY DISTRIBUTION OF SECONDARY ELECTRON EMISSIO 4-0853
THEORY FOR ENERGY DISTRIBUTION OF SECONDARY ELECTRONS. 2-0241
D BY INDIUM PHOSPHIDE AND GALLIUM PHOSPHIDE SUBMITTED TO/ ENERGY DISTRIBUTION OF SECONDARY ELECTRONS EMITTE 2-0269
RFACES. SECONDARY ELECTRON ENERGY DISTRIBUTION STUDIES OF URANIUM DIOXIDE SU 5-1316
A TEMPERATURE DEPENDENT CONTRIBUTION TO AUGER ELECTRON ENERGY DISTRIBUTIONS. 2-0434
TRUE SECONDARY ELECTRON ENERGY DISTRIBUTIONS. 2-0242
/ON AND PHOTOELECTRON STRUCTURE IN THE SECONDARY ELECTRON ENERGY DISTRIBUTIONS OF ALUMINUM, NICKEL, AND CO/ 4-1023
/ECTRA OF MAGNESIUM, ALUMINUM, AND SILICON EXCITED BY LOW ENERGY ELECTRON AND LOW ENERGY ARGON ION BOMBARD/ 4-0857
SURFACE DISSOCIATION OF POTASSIUM CHLORIDE BY LOW ENERGY ELECTRON BOMBARDMENT. 5-1719
CHARACTERIZATION OF CHEMISORPTION BY LOW ENERGY ELECTRON DIFFRACTION 1-0161
URFACES BY MEANS OF AUGER ELECTRONIC SPECTROSCOPY AND LOW ENERGY ELECTRON DIFFRACTION. / ANALYSIS OF IRON S 5-1940
CLEAN TELLURIUM SURFACES STUDIED BY LOW ENERGY ELECTRON DIFFRACTION 5-1134
N OF THE MOLYBDENUM (110) SURFACE OBSERVED BY AES AND LOW ENERGY ELECTRON DIFFRACTION INITIAL OXIDATIO 5-1650
LOW ENERGY ELECTRON DIFFRACTION. 1-0131
GER SPECTROSCOPY, SECONDARY ION MASS SPECTROSCOPY AND LOW ENERGY ELECTRON DIFFRACTION. /CHARACTERIZED BY AU 2-0501
UCTURES ON CARBON COVERED GERMANIUM (100) SURFACE BY HIGH ENERGY ELECTRON DIFFRACTION. /RVATION OF SUPERSTR 5-1805
PRESENT STATE OF AUGER ELECTRON SPECTROSCOPY AND LOW ENERGY ELECTRON DIFFRACTION 1-0006
URFACE OBSERVED BY AUGER EMISSION SPECTROSCOPY AND BY LOW ENERGY ELECTRON DIFFRACTION / TO SILICON (111) S 5-1180
ITH AUGER SPECTROSCOPY, ELLIPSOMETRY, AND REFLECTION HIGH ENERGY ELECTRON DIFFRACTION. /UTTERING, STUDIED W 5-1233
LMS ADSORBED ON (0001) GRAPHITE. (AUGER SPECTROSCOPY, LOW ENERGY ELECTRON DIFFRACTION) /NON SUBMONOLAYER FI 5-1858
ADHESION AT GOLD MICA INTERFACES. (LOW ENERGY ELECTRON DIFFRACTION AES) 5-1434
URE WITH ANTIPHASE DOMAINS OF SULFUR ON IRON (001) BY LOW ENERGY ELECTRON DIFFRACTION AES. / SURFACE STRUCT 5-1470
BINDING STATES OF WATER VAPOR ON CLEAVED GERMANIUM. (LOW ENERGY ELECTRON DIFFRACTION AES) 5-1820
ORMIC ACID DESORPTION FROM GRAPHITIZED NICKEL (110). (LOW ENERGY ELECTRON DIFFRACTION AES) F 5-1630
ERSTRUCTURES ON SILICON (100) SURFACES AS OBSERVED BY LOW ENERGY ELECTRON DIFFRACTION AES. GOLD-INDUCED SUP 5-1708
CLEAN AND ALKALI METAL COVERED ZINC OXIDE SURFACES. (LOW ENERGY ELECTRON DIFFRACTION AES) / PROPERTIES OF 5-1881
/ND COMPOSITION OF METAL SURFACES: A REVIEW OF RECENT LOW ENERGY ELECTRON DIFFRACTION AES, AND XPS RESULT/ 5-1341
EMISSION STUDIES OF LITHIUM HYDRIDE. LOW ENERGY ELECTRON DIFFRACTION AES ELECTRON INDUCED 4-0807
SURVEYING SILICON SURFACES BY MEANS OF THE LOW ENERGY ELECTRON DIFFRACTION AES METHOD. 5-1809
OWTH MECHANISM OF DEPOSITED IRON FILM ON MAGNESIUM O/ LOW ENERGY ELECTRON DIFFRACTION AES OBSERVATION OF GR 5-1525
ORPTION INDUCED SULFUR SEGREGATION FROM THE BULK TO / LOW ENERGY ELECTRON DIFFRACTION AES STUDIES OF CHEMIS 5-1865
/ETAL SUPERCONDUCTORS AT SURFACES AND IN THIN FILMS. (LOW ENERGY ELECTRON DIFFRACTION AES STUDIES OF NIOBI/ 5-1848
ATION OF IRON- CHROMIUM (100) AND (110). LOW ENERGY ELECTRON DIFFRACTION AES STUDY OF THE OXID 5-1587
ELECTRODES AND CIRCUITS OF A LOW ENERGY ELECTRON DIFFRACTION AES SYSTEM. 3-0639
N OF IMPURITIES ON A SILICON (111) SURFACE BY AUGER- LOW ENERGY ELECTRON DIFFRACTION ANALYSIS. DETECTIO 5-1816
ALUMINUM, COPPER- TIN, AND IRON- ALUMINUM ALLOYS. (BY LOW ENERGY ELECTRON DIFFRACTION AND AES) /ON COPPER- 6-2033
ND DESORPTION KINETICS ON SILICON (111) SURFACES. (BY LOW ENERGY ELECTRON DIFFRACTION AND AES) /DSORPTION A 5-1288
ATION OF SOME VANADIUM (100) SURFACE STRUCTURES USING LOW ENERGY ELECTRON DIFFRACTION AND AES. CHARACTERIZ 5-1864
ES OF ALKALI HALIDES AND THIN FILM GROWTH OBSERVED BY LOW ENERGY ELECTRON DIFFRACTION AND AES) /STAL SURFAC 5-1907
E SURFACE FILMS ON THE ADHESION OF CLEAN IRON. (USING LOW ENERGY ELECTRON DIFFRACTION AND AES) /OGEN SULFID 5-1211
OLD SURFACES BY POSITIVE ION CHANNELING SPECTROSCOPY, LOW ENERGY ELECTRON DIFFRACTION AND AES. /RED (001) G 5-1139
GEN AND SULFUR ON A NICKEL (111) SURFACE. (STUDIED BY LOW ENERGY ELECTRON DIFFRACTION AND AES) /BETWEEN OXY 5-1451
STEN (110) AND CARBON MONOXIDE ON TUNGSTEN (100). (BY LOW ENERGY ELECTRON DIFFRACTION AND AES) /GEN ON TUNG 5-1152
OF GOLD ONTO TANTALUM (100) SINGLE CRYSTAL SURFACES. LOW ENERGY ELECTRON DIFFRACTION AND AES ANALYSIS. /ON 5-1312
ON THE POLAR SURFACES OF ZINC OXIDE. LOW ENERGY ELECTRON DIFFRACTION AND AES OBSERVATIONS 5-1619
REFS) LOW ENERGY ELECTRON DIFFRACTION AND AES. (REVIEW, 53 1-0096
TRAHIGH VACUUM. A VERSATILE SUBSTRATE DESIGN FOR LOW ENERGY ELECTRON DIFFRACTION AND AES STUDIES IN UL 3-0631
DIUM ARSENIDE SURFACES. COMBINED LOW ENERGY ELECTRON DIFFRACTION AND AES STUDIES OF IN 5-1373
THIUM HYDRIDE (100) SURFACE. LOW ENERGY ELECTRON DIFFRACTION AND AES STUDIES OF LI 5-1446
E STRUCTURE OF ADSORBED GASES ON SOLID SURFACES. LOW ENERGY ELECTRON DIFFRACTION AND AES STUDIES OF TH 5-1835
EN ABSORPTION ON TUNGSTEN (110). LOW ENERGY ELECTRON DIFFRACTION AND AES STUDIES OF OXYG 5-1669
CON EPITAXY BY SILANE PYROLYSIS. HIGH ENERGY ELECTRON DIFFRACTION AND AES STUDY OF SILI 5-1427
INTERACTION OF HYDROGEN SULFIDE AND MOLYBDENUM (100)/ LOW ENERGY ELECTRON DIFFRACTION AND AES STUDY OF THE 5-1973
N (111) SURFA/ DETERMINATION OF SURFACE STRUCTURES BY LOW ENERGY ELECTRON DIFFRACTION (AND AES): THE SILICO 5-1342
SURFACE AND THEIR INFLUENCE ON ADHESION STUDIED WITH LOW ENERGY ELECTRON DIFFRACTION AND AUGER. /RON (001) 5-1208
W ENERGY ELECTRON DIFFRACTION, CIRCA 1968. (REVIEW OF LOW ENERGY ELECTRON DIFFRACTION AND AUGER) LO 1-0099
/UMINUM (001) SINGLE CRYSTAL SURFACE USING INELASTIC LOW ENERGY ELECTRON DIFFRACTION. (AND AUGER ANALYSIS) 5-1745
S ON ZINC OXIDE SURFACES. CORRELATION OF ELECTRONIC, LOW ENERGY ELECTRON DIFFRACTION AND AUGER DIAGNOSTIC 5-1584
SERVATIONS OF THE SILICON CARBIDE (0001) SURFACE. LOW ENERGY ELECTRON DIFFRACTION AND AUGER ELECTRON OB 5-1932
ECTROSCOPY. (766 REFS) A BIBLIOGRAPHY OF LOW ENERGY ELECTRON DIFFRACTION AND AUGER ELECTRON SP 1-0071
ECTROSCOPY. THIN FILMS. LOW ENERGY ELECTRON DIFFRACTION AND AUGER ELECTRON SP 1-0100
ECTROSCOPY. (REVIEW) LOW ENERGY ELECTRON DIFFRACTION AND AUGER ELECTRON SP 1-0124
EC/ RECENT ADVANCES IN SURFACE CHARACTERIZATION USING LOW ENERGY ELECTRON DIFFRACTION AND AUGER ELECTRON SP 1-0201
/TION OF EUROPIUM(II) OXIDE (100) CLEAVAGE SURFACE BY LOW ENERGY ELECTRON DIFFRACTION AND AUGER ELECTRON S/ 1-0223
/N OF OXYGEN WITH PLATINUM SINGLE CRYSTAL SURFACES BY LOW ENERGY ELECTRON DIFFRACTION AND AUGER ELECTRON S/ 5-1157
ECTROSCOPY. STUDY OF IRIDIUM (100) SURFACE USING LOW ENERGY ELECTRON DIFFRACTION AND AUGER ELECTRON SP 5-1371
ECTROSCOP/ STUDIES ON THE IRIDIUM (111) SURFACE USING LOW ENERGY ELECTRON DIFFRACTION AND AUGER ELECTRON SP 5-1372
ECTROSCOPY STUDIES OF SILICON, GERMANIUM, G/ COMBINED LOW ENERGY ELECTRON DIFFRACTION AND AUGER ELECTRON SP 5-1379
/OF RUTHENIUM (0001) AND RHODIUM (111) SURFACES USING LOW ENERGY ELECTRON DIFFRACTION AND AUGER ELECTRON S/ 5-1380
/OF INDIUM ARSENIDE (111) AND (-1-1-1) SURFACES USING LOW ENERGY ELECTRON DIFFRACTION AND AUGER ELECTRON S/ 5-1381
ECTR/ STUDY OF THE NITRIDATION OF SILICON SURFACES BY LOW ENERGY ELECTRON DIFFRACTION AND AUGER ELECTRON SP 5-1417
/ OF SULFUR ON THE (100) AND (111) FACES OF PLATINUM. LOW ENERGY ELECTRON DIFFRACTION AND AUGER ELECTRON S/ 5-1421
/ILICON CLEANING WITH HYDROGEN PEROXIDE SOLUTIONS: HIGH ENERGY ELECTRON DIFFRACTION AND AUGER ELECTRON S/ 5-1426
/ICON HOMOEPITAXIAL THIN FILMS VIA SILANE PYROLYSIS: HIGH ENERGY ELECTRON DIFFRACTION AND AUGER ELECTRON S/ 5-1428
ECTROSCOPY/ STRUCTURAL STUDY OF OXYGEN- IRON (001) BY LOW ENERGY ELECTRON DIFFRACTION AND AUGER ELECTRON SP 5-1469
ECTROSCOPIC STUDIES ON TUNGSTEN (110) SURFACE. LOW ENERGY ELECTRON DIFFRACTION AND AUGER ELECTRON SP 5-1677
ECTROSCOPIC OBSERVATIONS OF THE SUBSTRATE SURFACE DU/ LOW ENERGY ELECTRON DIFFRACTION AND AUGER ELECTRON SP 5-1678
/LVER ON POTASSIUM CHLORIDE (100) SURFACE OBSERVED BY LOW ENERGY ELECTRON DIFFRACTION AND AUGER ELECTRON S/ 5-1696
ECTROSCOPY STUDY OF NICKEL (110) SURFACE EFFECTS DUE/ LOW ENERGY ELECTRON DIFFRACTION AND AUGER ELECTRON SP 5-1814

ECTROSCOPY.   STUDIES OF CRYSTAL SURFACES BY LOW ENERGY ELECTRON DIFFRACTION AND AUGER ELECTRON SP 5-1836
/H CF GOLD FILM ON GALLIUM ARSENIDE (110) OBSERVED BY LOW ENERGY ELECTRON DIFFRACTION AND AUGER ELECTRON S/ 5-1876
ECTROSCOPY STUDY OF PHCSPHINE ADSORPTION ON CLEAN SI/ LOW ENERGY ELECTRON DIFFRACTION AND AUGER ELECTRON SP 5-1930
ECTROSCOPY.   DETERMINATION OF SURFACE STRUCTURES BY LOW ENERGY ELECTRON DIFFRACTION AND AUGER ELECTRON SP 5-1957
/YSTAL FACE OF PLATINUM BY MOLECULAR BEAM SCATTERING, LOW ENERGY ELECTRON DIFFRACTION, AND AUGER ELECTRON / 5-1963
/LD SURFACES BY POSITIVE ION CHANNELING SPECTROSCOPY, LOW ENERGY ELECTRON DIFFRACTION AND AUGER ELECTRON S/ 5-1986
/RY METAL SURFACES IN ULTRAHIGH VACUUM AS OBSERVED BY LOW ENERGY ELECTRON DIFFRACTION AND AUGER ELECTRON S/ 6-2020
D./ GROWTH OF SILVER ON POTASSIUM CHLORIDE OBSERVED BY LOW ENERGY ELECTRON DIFFRACTION AND AUGER EMISSION OF 5 1840
ECTROSCOPY.   A STUDY OF THE IRON (111) SURFACE USING LOW ENERGY ELECTRON DIFFRACTION AND AUGER EMISSION SP 5-1670
/ESION OF METALS TO A CLEAN IRON SURFACE STUDIED WITH LOW ENERGY ELECTRON DIFFRACTION AND AUGER EMISSION S/ 6-2008
UDIES.   ULTRAHIGH VACUUM EVAPORATOR FOR LOW ENERGY ELECTRON DIFFRACTION AND AUGER EMISSION ST 3-0683
/H VACUUM EVAPORATOR AND MULTIPLE CLEAVAGE DEVICE FOR LOW ENERGY ELECTRON DIFFRACTION AND AUGER EMISSION S/ 3-0707
UDIES. ADSORPTION OF CARBON MONOXIDE ON COPPER (001). LOW ENERGY ELECTRON DIFFRACTION AND AUGER EMISSION ST 5-1515
ONS.   SULFUR ON SINGLE CRYSTAL PLATINUM PLANES. LOW ENERGY ELECTRON DIFFRACTION AND AUGER INVESTIGATI 5-1420
ONS OF COPPER (111) SURFACE.   LOW ENERGY ELECTRON DIFFRACTION AND AUGER INVESTIGATI 5-1499
ON OF THE (100) SURFACE CF COPPER(3) GOLD. TEMPERATU/ LOW ENERGY ELECTRON DIFFRACTION AND AUGER INVESTIGATI 5-1854
/DSORPTION OF KRYPTON ON THE BASAL PLANE OF GRAPHITE: LOW ENERGY ELECTRON DIFFRACTION AND AUGER MEASUREMEN/ 5-1559
EVIEW)   LOW ENERGY ELECTRON DIFFRACTION AND AUGER METHODS. (R 1-0007
Y.   ULTRAHIGH VACUUM AIRLOCK FOR LOW ENERGY ELECTRON DIFFRACTION AND AUGER SPECTROSCOP 3-0669
Y OF THE ADSORPTION CF SULFUR ON PLATINUM.   LOW ENERGY ELECTRON DIFFRACTION AND AUGER SPECTROSCOP 5-1175
/NAL ABSORBED SULFUR LAYERS ON COPPER INVESTIGATED BY LOW ENERGY ELECTRON DIFFRACTION AND AUGER SPECTROSCO/ 5-1302
/ MAGNESIUM OXIDE (001) USING REFLECTION DIFFRACTION, LOW ENERGY ELECTRON DIFFRACTION AND AUGER SPECTROSCO/ 5-1496
Y.   ADSORPTION OF SULFUR ON GOLD BY LOW ENERGY ELECTRON DIFFRACTION AND AUGER SPECTROSCOP 5-1558
/TH OF GALLIUM ARSENIDE ON (100) GERMANIUM STUDIED BY LOW ENERGY ELECTRON DIFFRACTION AND AUGER SPECTROSCO/ 5-1646
Y STUDIES OF THE (100) SURFACES OF ALUMINUM AND ALUM/ LOW ENERGY ELECTRON DIFFRACTION AND AUGER SPECTROSCO/ 5-1755
EFFECT OF OXYGEN ON ADHESION OF CLEAN IRON (001) AND/ LOW ENERGY ELECTRON DIFFRACTION AND AUGER STUDIES OF 6-2010
COPIC OBSERVATIONS ON NICKEL MONCXIDE (100).   LOW ENERGY ELECTRON DIFFRACTION AND ELECTRON SPECTROS 5-1687
F SCATTERE/ DETERMINATION OF SURFACE STRUCTURES USING LOW ENERGY ELECTRON DIFFRACTION AND ENERGY ANALYSIS O 5-1956
/N CARBURIZED AND GRAPHITIZED NICKEL (110) USING AES, LOW ENERGY ELECTRON DIFFRACTION AND FLASH DESORPTIO/ 5-1629
N SPECTROSC/ CESIUM ADSORPTION ON TUNGSTEN STUDIED BY LOW ENERGY ELECTRON DIFFRACTION AND SECONDARY ELECTRO 5-1661
STUDY OF CARBON MONOXIDE AND XENON ADSORBED ON COPP/ LOW ENERGY ELECTRON DIFFRACTION AND SURFACE POTENTIAL 5-1246
Y. (REVIEW, AUGER SPECTROSCOPY)   LOW ENERGY ELECTRON DIFFRACTION AND SURFACE TOPOGRAPH 1-0185
SUREMENTS FOR E/ A CORRELATION OF AUGER SPECTROSCOPY, LOW ENERGY ELECTRON DIFFRACTION AND WORK FUNCTION MEA 5-1739
ECTION AND IDENTIFICATION OF SURFACE CONTAMINA USE OF LOW ENERGY ELECTRON DIFFRACTION APPARATUS FOR THE DET 5-1959
TY FOR INVESTIGATING ENERGY AND ANGULAR / FARADAY CUP LOW ENERGY ELECTRON DIFFRACTION APPARATUS WITH FACILI 3-0691
BERYLLIUM (0001) SURFACE.   LOW ENERGY ELECTRON DIFFRACTION AUGER ANALYSIS OF THE 5-1581
/DIUM ADSORPTION ON ALUMINUM (111) SURFACES: ELECTRIC LOW ENERGY ELECTRON DIFFRACTION, AUGER, AND CONTACT / 5-1744
UDIES OF NEGATIVE ELECTRON AFFINITY ON SILICON PRODU/ LOW ENERGY ELECTRON DIFFRACTION AUGER AND PLASMON ST 5-1363
ION STUDY OF IODINE ADSORBED ON TUNGSTEN (110).   LOW ENERGY ELECTRON DIFFRACTION AUGER, AND WORK FUNCT 5-1145
TION STUDIES OF CLEAN AND SODIUM- COVERED SURFACES O/ LCW ENERGY ELECTRON DIFFRACTION AUGER, AND WORK FUNC 5-1240
/CHLORINE WITH THE ZINC OXIDE (0001)- ZINC SURFACE: A LOW ENERGY ELECTRON DIFFRACTION AUGER AND WORK FUNC/ 5-1463
MEASUREMENT OF WORK FUNCTION CHANGE IN A DISPLAY TYPE LOW ENERGY ELECTRON DIFFRACTION AUGER APPARATUS.   3-0611
/00) SURFACE. (CHEMISORPTION AND CATALYTIC REACTIONS, LOW ENERGY ELECTRON DIFFRACTION AUGER, CONTACT POTE/ 5-1321
FLASH DESORPTION SPECTROSCOPY OF METALS ON / COMBINED LOW ENERGY ELECTRON DIFFRACTION, AUGER ELECTRON, AND 1-0008
/110) SURFACE OF A TUNGSTEN SINGLE CRYSTAL STUDIED BY LOW ENERGY ELECTRON DIFFRACTION AUGER ELECTRON AND / 5-1660
OSCOPY/ DETECTION OF IMPURITIES ON COPPER SURFACES BY LOW ENERGY ELECTRON DIFFRACTION AUGER ELECTRON SPECTR 5-1185
ROSCOPY, AND WORK FUNCTION MEASUREMENTS ON CLEAN AND/ LOW ENERGY ELECTRON DIFFRACTION, AUGER ELECTRON SPECT 5-1294
ROSCOPY, AND X-RAY PHOTOELECTRON SPECTROSCO/ COMBINED LOW ENERGY ELECTRON DIFFRACTION, AUGER ELECTRON SPECT 5-1337
/AXY OF GERMANIUM FILMS BY SPUTTER DEPOSITION. (HEED, LOW ENERGY ELECTRON DIFFRACTION AUGER ELECTRON SPEC/ 5-1537
/100)) AND STEPPED PLATINUM SINGLE CRYSTAL SURFACES. (LOW ENERGY ELECTRON DIFFRACTION, AUGER ELECTRON SPEC/ 5-1838
ROSCOPY, AND PHOTOEMISSION STUDIES OF THE CESIUM COV/ LOW ENERGY ELECTRON DIFFRACTION, AUGER ELECTRON SPECT 5-1931
ROSCOPY AND FIELD ION MICROSCOPY IN MICROSTRU/ USE OF LOW ENERGY ELECTRON DIFFRACTION, AUGER EMISSION SPECT 1-0057
F A STABLE CARBIDE OVERLAYER ON A PLATINUM (111) SUR/ LOW ENERGY ELECTRON DIFFRACTION AUGER INVESTIGATION O 5-1569
N A (100) SURFACE OF AUSTENITIC CHROMIUM- NICKEL STE/ LOW ENERGY ELECTRON DIFFRACTION AUGER INVESTIGATION O 6-1991
CONTAMINANTS ON CHEMICALLY ETCHED SILICON SURFACES.   LOW ENERGY ELECTRON DIFFRACTION AUGER METHOD.   5-1229
AND CONTACT POTENTIAL STUDIES OF COPPER- GOLD ALLOY / LOW ENERGY ELECTRON DIFFRACTION AUGER SPECTROSCOPY, 5-1748
/ TOPOGRAPHY, PHENOMENA, FIELD ION MICROSCOPY, LEED, HIGH ENERGY ELECTRON DIFFRACTION AUGER SPECTROSCOPY/ 5-1492
UDY OF THE ADSORPTION CF ALKALI METALS ON MOLYBDENUM/ LCW ENERGY ELECTRON DIFFRACTION AUGER SPECTROSCOPY ST 5-1896
ELECTRON BEAM CURRENT CONTRCLLER FOR LOW ENERGY ELECTRON DIFFRACTION AUGER STUDIES.   3-0589
C/ SILICON ON SAPPHIRE EPITAXY BY VACUUM SUBLIMATION: LOW ENERGY ELECTRON DIFFRACTION AUGER STUDIES AND ELE 5-1230
RINE CHEMISORPTION ON THE SILVER (111) SURFACE.   LOW ENERGY ELECTRON DIFFRACTION (AUGER) STUDY OF CHLO 5-1779
OF ALUMINUM ON NIOBIUM (110).   LOW ENERGY ELECTRON DIFFRACTION AUGER STUDY OF GROWTH 5-1491
FORMATION AT / OXIDATION OF TUNGSTEN (110). PART-2:   LOW ENERGY ELECTRON DIFFRACTION AUGER STUDY OF OXIDE 6-1994
ADSORPTION ON THE TUNGST/ COMBINED LEED- REFLECTION HIGH ENERGY ELECTRON DIFFRACTION AUGER STUDY OF OXYGEN 5-1464
CONTINUOUS MEASUREMENT OF CHANGES IN WORK FUNCTION. (LOW ENERGY ELECTRON DIFFRACTION AUGER SYSTEM) /OR THE 3-0663
ELECTRON YIELD MEASUREMENTS AT LCW PRIMARY ENERGIES. (LOW ENERGY ELECTRON DIFFRACTION AUGER SYSTEM) /NDARY 3-0629
GE FOR IN-SITU FILM DEPOSITION AND CHARACTERIZATION. (LOW ENERGY ELECTRON DIFFRACTION AUGER SYSTEM) /RE STA 3-0657
ANALYSIS OF ELECTRON EMISS/ REFINEMENTS TO A STANDARD LOW ENERGY ELECTRON DIFFRACTION AUGER SYSTEM FOR THE 3-0698
N METAL FILMS ON BODY CENTERED CUBIC SUBSTRATES USING LOW ENERGY ELECTRON DIFFRACTION AUGER TECHNIQUES. /HI 5-1488
STUDY OF OXYGEN- CESIUM ADSORPTION AND COADSORPTION / LOW ENERGY ELECTRON DIFFRACTION AUGER, WORK FUNCTION 5-1297
ION CHANGES AND AUGER ELECTRON SPECTROSCOPY USING THE LOW ENERGY ELECTRON DIFFRACTION CAMERA. /. WORK FUNCT 6-2069
OF LEED AND AUGER)   LOW ENERGY ELECTRON DIFFRACTION, CIRCA 1968. (REVIEW 1-0099
IDERATIONS OF AN AUGER ELECTRON SPECTROMETER BASED ON LOW ENERGY ELECTRON DIFFRACTION DISPLAY OPTICS. /CONS 3-0705
/EAN FREE PATH FOR ELECTRONS IN BULK JELLIUM. (AUGER, LOW ENERGY ELECTRON DIFFRACTION ENERGY LOSS SPECTRO/ 4-1059
EL OXIDE.   LOW ENERGY ELECTRON DIFFRACTION FROM CLEAN (100) NICK 5-1752
/PECIES, TOPOGRAPHY, PHENOMENA, FIELD ION MICROSCOPY, LOW ENERGY ELECTRON DIFFRACTION HEED, AUGER SPECTRO/ 5-1492
IES OF CLEAN (001), (110) AND ELASTIC (111) / ELASTIC LOW ENERGY ELECTRON DIFFRACTION INTENSITY ENERGY STUD 5-1292
DEPOSITION OF ALUMINUM ONTO THE MOLYBDENUM (1/ AUGER- LOW ENERGY ELECTRON DIFFRACTION INVESTIGATION OF THE 5-1489
ON MOLYBDENUM (100).   AUGER- LOW ENERGY ELECTRON DIFFRACTION INVESTIGATION OF TIN 5-1490
ON SINGLE CRYSTAL FILMS AND THEIR USE IN STUDYING TH/ LOW ENERGY ELECTRON DIFFRACTION INVESTIGATIONS OF XEN 5-1479
/ CHEMISTRY OF CERAMIC SURFACES. (AUGER SPECTROSCOPY, LOW ENERGY ELECTRON DIFFRACTION ION MICROPROBE ANAL/ 5-1852
/ EPITAXY OF GERMANIUM FILMS BY SPUTTER DEPOSITION. (HIGH ENERGY ELECTRON DIFFRACTION LEED, AUGER ELECTRO/ 5-1537
IDE (/ AUGER ELECTRON EMISSION SPECTROSCOPY (AES) AND LOW ENERGY ELECTRON DIFFRACTION (LEED) OF THE ZINC OX 5-1338
ENTS ON SURFACES OF ELECTRICAL SOLID MATERIALS BY THE LOW ENERGY ELECTRON DIFFRACTION METHOD.   MEASUREM 5-1981
AUGER SPECTROSCOPY AND LOW ENERGY ELECTRON DIFFRACTION OF SILICON SURFACES.   5-1396
/CTROSCOPY (AES) AND LOW ENERGY ELECTRON DIFFRACTION (LOW ENERGY ELECTRON DIFFRACTION OF THE ZINC OXIDE (/ 5-1338
/ECTRODES BY MEANS OF THIN LAYER ELECTROCHEMISTRY AND LOW ENERGY ELECTRON DIFFRACTION. PART-1: ELECTRODE S/ 5-1486
/UM (110) SURFACE. PART-1: SULFUR AND CARBON MONOXIDE LCW ENERGY ELECTRON DIFFRACTION PATTERNS. (AUGER ELE/ 5-1190
DEPOSITS ON CLEAVED SILICON.   AUGER SPECTRA AND LOW ENERGY ELECTRON DIFFRACTION PATTERNS FROM NICKEL 5-1768
CLEANED SILICON CRYSTALS WITH CALI/ AUGER SPECTRA AND LOW ENERGY ELECTRON DIFFRACTION PATTERNS FROM VACUUM 5-1769

OXYGEN ADSORPTION ON THE TUNGSTEN (112) SUR/ COMBINED LOW ENERGY ELECTRON DIFFRACTION RHEED AUGER STUDY OF 5-1464
/), AND (111) NICKEL SURFACES. CORRELATED STUDY USING LOW ENERGY ELECTRON DIFFRACTION SPECTRA, AUGER SPECT/ 5-1291
ACES UTILIZING WORK FUNCTION MEASUREMENTS, AUGER, AND LOW ENERGY ELECTRON DIFFRACTION SPECTROSCOPY. /L SURF 5-1290
ISOTHERMS: XENON O/ AUGER ELECTRON SPECTROSCOPY AND LOW ENERGY ELECTRON DIFFRACTION STUDIES OF ADSORPTION 5-1856
INTERFACES. (AUGER EFFECT) ELECTRON SPECTROSCOPY AND LOW ENERGY ELECTRON DIFFRACTION STUDIES OF GAS- METAL 5-1320
ORPTION OF UNSATURATED MOLECU/ AUGER SPECTROSCOPY AND LOW ENERGY ELECTRON DIFFRACTION STUDIES OF THE CHEMIS 5-1267
/RGY SPECTRA OF INELASTICALLY SCATTERED ELECTRONS AND LOW ENERGY ELECTRON DIFFRACTION STUDIES OF TUNGSTEN. 4-1094
OVERLA/ QUANTITATIVE AUGER ELECTRON SPECTROSCOPY AND LOW ENERGY ELECTRON DIFFRACTION STUDY OF ALKALI METAL 5-1895
SURFACE SEGREGATION IN COPPER/ AUGER SPECTROSCOPY AND LOW ENERGY ELECTRON DIFFRACTION STUDY OF EQUILIBRIUM 6-2025
) SURFACE. LOW ENERGY ELECTRON DIFFRACTION STUDY OF IRIDIUM (100 5-1370
IDE (100). PART-1: EXPERIMENT. A LOW ENERGY ELECTRON DIFFRACTION STUDY OF MAGNESIUM OX 5-1580
0) SURFACE. LOW ENERGY ELECTRON DIFFRACTION STUDY OF PLATINUM (10 5-1377
ON OF GALLIUM ARSENI/ AUGER ELECTRON SPECTROSCOPY AND LOW ENERGY ELECTRON DIFFRACTION STUDY OF THE ACTIVATI 5-1845
11) SURFACES OF GALLIUM ARSENIDE AND GALLIUM ANTIMON/ LOW ENERGY ELECTRON DIFFRACTION STUDY OF THE POLAR (1 5-1602
/ELD ION MICROSCOPY. AUGER ELECTRON SPECTROSCOPY, AND LOW ENERGY ELECTRON DIFFRACTION STUDY OF THE STRUCTU/ 5-1951
AUGER ELECTRON SPECTROMETER INTEGRATED WITH A LOW ENERGY ELECTRON DIFFRACTION SYSTEM. 3-0658
X-RAY APPEARANCE POTENTIAL SPECTROSCOPY IN A DISPLAY LOW ENERGY ELECTRON DIFFRACTION SYSTEM. SOFT 4-0870
RON M/ AUGER ELECTRON SPECTROSCOPY USING THE TWO-GRID LOW ENERGY ELECTRON DIFFRACTION SYSTEM AND PHOTOELECT 3-0656
RUM ANALYSES. APPLICATION OF TRIPLE GRID LOW ENERGY ELECTRON DIFFRACTION SYSTEM TO AUGER SPECT 3-0661
AUGER ELECTRON SPECTROSCOPY IN LOW ENERGY ELECTRON DIFFRACTION SYSTEMS. 1-0153
AUGER ELECTRON SPECTROSCOPY IN LOW ENERGY ELECTRON DIFFRACTION SYSTEMS. 1-0158
OPTIMIZATION OF AUGER ELECTRON SPECTROSCOPY IN LOW ENERGY ELECTRON DIFFRACTION SYSTEMS. 1-0152
E TECHNIQUE. (COMPARISON BETWEEN ION PROBE AND AUGER- LOW ENERGY ELECTRON DIFFRACTION TECHNIQUES) /ION PROB 5-1166
T OF STEPPED SURFACE REGIONS ON POLYCRYSTALLINE GOLD. LOW ENERGY ELECTRON DIFFRACTIONS AND AUGER STUDIES. / 5-1483
ON OF GALLIUM ARSENIDE DURING ACTIVATION TO NEGATIVE/ LOW ENERGY ELECTRON DIFFRACTION- AUGER CHARACTERIZATI 5-1364
ROSCOPY. PRODUCTION OF A SPHERICAL GRID FOR LOW ENERGY ELECTRON DIFFRACTION- AUGER ELECTRON SPECT 3-0668
ROSCOPY STUDY OF IRON FILM DEPOSITED ON MAGNESIUM OX/ LOW ENERGY ELECTRON DIFFRACTION- AUGER ELECTRON SPECT 5-1522
ROSCOPY STUDY OF AN EPITAXIAL IRON FILM. LOW ENERGY ELECTRON DIFFRACTION- AUGER ELECTRON SPECT 5-1526
ROSCOPY STUDY OF CLEAN (001) IRON SURFACE. LOW ENERGY ELECTRON DIFFRACTION- AUGER ELECTRON SPECT 5-1527
ROSCOPY STUDY OF THE OXIDATION OF IRON (110) AND IRO/ LOW ENERGY ELECTRON DIFFRACTION- AUGER ELECTRON SPECT 5-1585
ROSCOPY STUDY OF THE OXIDATION OF IRON (100) AND I/ A LOW ENERGY ELECTRON DIFFRACTION- AUGER ELECTRON SPECT 5-1586
ROSCOPY STUDY OF IRON SINGLE CRYSTAL SURFACES. LOW ENERGY ELECTRON DIFFRACTION- AUGER ELECTRON SPECT 5-1866
/LS AND EPITAXIAL GROWTH OF METAL ON THEM OBSERVED BY LOW ENERGY ELECTRON DIFFRACTION- AUGER ELECTRON SPEC/ 5-1906
ROSCOPY STUDY OF THE OXIDATION OF CHROMIUM (110) AND/ LOW ENERGY ELECTRON DIFFRACTION- AUGER ELECTRON SPECT 6-2022
ROSCOPY STUDY OF THE ADSORPTION OF OXYGEN ON COPPE/ A LOW ENERGY ELECTRON DIFFRACTION- AUGER ELECTRON SPECT 6-2089
ROSCOPY STUDY OF THE INITIAL STAGES OF OXIDATION OF / LOW ENERGY ELECTRON DIFFRACTION- AUGER ELECTRON SPECT 6-2110
/PROCESSES OF IRON (100) SURFACE BY MEANS OF COMBINED LOW ENERGY ELECTRON DIFFRACTION- AUGER ELECTRON SPEC/ 6-2141
A CRYSTAL CLEAVAGE DEVICE FOR USE IN LOW ENERGY ELECTRON DIFFRACTION- AUGER STUDIES. 3-0637
MULTIIONIZATION OF KRYPTON AND ITS IONS BY HIGH ENERGY ELECTRON IMPACT. (AUGER EFFECT) 2-0493
TIIONIZATION OF NEON, ARGON, XENON AND THEIR IONS BY HIGH ENERGY ELECTRON IMPACT. (AUGER EMISSION) MUL 2-0492
TE, AND EVAPORATED CARBON SUR/ CHARACTERISTIC K-SHELL LOW ENERGY ELECTRON LOSS SPECTRA FROM DIAMOND, GRAPHI 4-0945
HIGH RESOLUTION, LOW ENERGY ELECTRON SPECTROMETER. 3-0697
/F ADSORPTION OF OXYGEN ON GALLIUM ARSENIDE METHOD OF LOW ENERGY ELECTRON SPECTROSCOPY. (AUGER PEAK AMPLIT/ 5-1641
D PLATINUM (100)- CARBON MONOXIDE. LOW ENERGY ELECTRON SPECTROSCOPY OF PLATINUM (100) AN 5-1686
LOW ENERGY ELECTRON SPECTROSCOPY. (REVIEW, 16 REFS) 1-0164
STRUCTURE OF LOW ENERGY ELECTRON SPECTRUM. 4-0948
AUGER SPECTROSCOPY OF GRAPHITE SINGLE CRYSTALS WITH LOW ENERGY ELECTRONS. 4-0741
DETERMINATION OF THE ESCAPE LENGTH OF LOW ENERGY ELECTRONS BY AUGER AND LOSS SPECTROSCOPY. 2-0401
YGEN AND NEON AUGE/ IDENTIFICATION AND MEASUREMENT OF LOW ENERGY ELECTRONS BY CLOUD CHAMBER TECHNIQUES: OX 2-0566
SCATTERING OF LOW ENERGY ELECTRONS FROM A COPPER (111) SURFACE. 5-1621
PHOTOELECTRON (AND AUGER) SPECTROSCOP/ ATTENUATION OF LOW ENERGY ELECTRONS FROM SOLIDS: RESULTS FROM X-RAY 5-1844
INELASTIC SCATTERING OF LOW ENERGY ELECTRONS FROM SURFACES. 2-0495
/HOTOELECTRON DETERMINATION OF THE ATTENUATION OF THE LOW ENERGY ELECTRONS IN ALUMINUM OXIDE. (AUGER ELECT/ 4-0773
ICON AND SILICON CARBIDE. HIGH ENERGY FINE STRUCTURE IN THE AUGER SPECTRA OF SIL 4-1036
NEON K-SHELL VACANCY PRODUCTION BY INTERMEDIATE ENERGY FLUORINE IONS. (AUGER ELECTRON EMISSION) 2-0572
HIGH RESOLUTION LMM AUGER SPECTRA OF LOW ENERGY FROM SOLID SURFACES. 4-0732
HE BERYLLIUM AUGER SPECTRUM. PLASMON ENERGY GAIN AND DOUBLE IONIZATION SATELLITES IN T 4-0899
ZATION LOSS EVENTS IN BERYLLIUM AND BERYL/ CHARACTERISTIC ENERGY GAIN AND LOSS, DOUBLE IONIZATION, AND IONI 4-0901
PLASMON ENERGY GAIN IN THE BERYLLIUM AUGER SPECTRUM. 4-1095
D FROM A (100) FACE OF A COPPER S/ CHARGE EXCHANGE OF LOW ENERGY HELIUM(+) IONS (LESS THAN 10 KEV) SCATTERE 2-0546
PER SINGLE CRYSTAL. (AUGER NEUTRA/ CHARGE EXCHANGE OF LOW ENERGY HELIUM IONS AND ATOMS SCATTERED FROM A COP 2-0545
MERCIAL AUGER CYLINDRICAL MIRROR ANALYZER. LOW ENERGY ION BACKSCATTERING SPECTROSCOPY WITH A COM 3-0665
CTROSCOPY IN THE ANALYSIS OF SU/ DIRECT COMPARISON OF LOW ENERGY ION BACKSCATTERING WITH AUGER ELECTRON SPE 5-1869
SURFACE ANALYSIS BY LOW ENERGY ION REFLECTION. 5-1822
SURFACE ANALYSIS WITH LOW ENERGY ION SCATTERING. 5-1871
COPY STUDIES OF CLEAN NICKEL SURFACES AND ADSORBED L/ LOW ENERGY ION SCATTERING AND AUGER ELECTRON SPECTROS 5-1870
CTS. (AUGER NEUTRALIZATION) LOW ENERGY ION SCATTERING: ELASTIC AND INELASTIC EFFE 2-0382
(AUGER) NEUTRALIZATION BEHAVIOR IN SCATTERING OF LOW ENERGY IONS FROM SOLID SURFACES. 2-0294
/LECTRON EMISSION FROM SOLID SURFACES BOMBARDED BY MEDIUM ENERGY IONS (PLATINUM, NICKEL, TANTALUM, GRAPHIT/ 2-0309
GER TRANSITIONS) INFLUENCE OF CHEMICAL BONDING ON THE LOW ENERGY K-ALPHA SPECTRUM OF SILICON. (RADIATIVE AU 2-0226
INTERPRETATION OF THE LOW ENERGY K-BETA SPECTRUM IN SOLIDS. 4-1101
,L2,3M4,5 COSTER-KRONIG TRANSITION RATES TO VARIATIONS IN ENERGY LEVELS AND SCREENING PARAMETERS. /TY OF L1 2-0303
N. IDENTIFICATION OF HIGH ENERGY LINES IN THE KLL AUGER SPECTRUM OF NITROGE 4-0791
PHOSPHIDE (110). CHARACTERISTIC ENERGY LOSS AND AUGER ELECTRON SPECTRA OF GALLIUM 5-1657
ED TO THE STUDY OF ADSORPTION PHENOMENA A/ CHARACTERISTIC ENERGY LOSS AND AUGER ELECTRON SPECTROSCOPY APPLI 5-1776
PLASMON EFFECTS IN ELECTRON ENERGY LOSS AND GAIN SPECTRA IN ALUMINUM. 4-1020
ISIONS. ENERGY LOSS AND IONIZATION IN FAST ION- ATOM COLL 2-0233
BOMBARDMENT. INTERPRETATION OF ENERGY LOSS BY AUGER ELECTRONS EMITTED DURING ION 2-0268
Y ELECTRON / A MEASURING METHOD TO DISTINGUISH AUGER FROM ENERGY LOSS PEAKS OF THE PRIMARY BEAM IN SECONDAR 1-0188
SURFACE STATE TRANSITIONS OF SILICON IN ELECTRON ENERGY LOSS SPECTRA. (AUGER EFFECT) 4-1039
AUGER ELECTRON AND CHARACTERISTIC ENERGY LOSS SPECTRA OF PLUTONIUM. 4-0931
. CHEMISORPTION OF AM/ APPLICATIONS OF AUGER ELECTRON AND ENERGY LOSS SPECTROSCOPIES TO CATALYTIC PROCESSES 5-1533
AN FREE PATH FOR ELECTRONS IN BULK JELLIUM. (AUGER, LEED, ENERGY LOSS SPECTROSCOPY) INELASTIC ME 4-1059
TRON SP/ STUDY OF THE OXIDATION OF MOLYBDENUM SURFACES BY ENERGY LOSS SPECTROSCOPY COMBINED WITH AUGER ELEC 5-1531
AUGER ELECTRON AND CHARACTERISTIC ENERGY LOSS SPECTROSCOPY OF SOME METALS. 4-1083
INVESTIGATION OF THE CHARACTERISTIC AUGER ENERGY LOSSES OF PLATINUM (111). 2-0499
TION ELECTRON SPECTROMETER. AN ENERGY MODULATED (CYLINDRICAL) HIGH ENERGY RESOLU 3-0616
ENCE OF THE COEFFICIENT OF SECONDARY ELECTRON EMISSION ON ENERGY OF PRIMARY ELECTRONS (POTASSIUM IODIDE). / 4-0891
M. STRUCTURAL INFLUENCES ON ADSORBATE BINDING ENERGY. PART-1: CARBON MONOXIDE ON (100) PALLADIU 5-1913
PART-3: CARBON MONOX/ STRUCTURAL INFLUENCES ON ADSORPTION ENERGY. PART-2: CARBON MONOXIDE ON NICKEL (100). 5-1911

144

SURFACE CHARACTERIZATION BY LOW ENERGY PHOTO AUGER ELECTRON SPECTROSCOPY. 5-1429
LOW ENERGY PHOTO- AUGER ELECTRON SPECTROSCOPY. 1-0080
X), AND NITROGEN(2) (2X) FROM NITROGEN BY ELECTRONS IN THE ENERGY RANGE 16-600 EV. /ITROGEN(2)(X), NITROGEN(2-0376
ION OF COPPER- NICKEL ALLOY BY AUGER SPECTRA IN THE LOWER ENERGY REGION AT AROUND 100EV. / SURFACE COMPOSIT 5-1982
NG CONCENTRIC HEMISPHERES. HIGH ENERGY RESOLUTION AUGER ELECTRON SPECTROMETER USI 3-0592
ENERGY RESOLUTION OF THE PHOTOEMISSION ANALYZER. 3-0606
SILICON. HIGH ENERGY SATELLITES IN AUGER PEAKS OF MAGNESIUM AND 4-1046
HIGHER ENERGY SATELLITES IN AUGER SPECTRA FROM METALS. 4-0900
HIGH ENERGY SATELLITES IN AUGER SPECTRA FROM METALS. 4-0940
AKS OF COPPER. HIGH ENERGY SATELLITES OF THE M2,3VV AND M1VV AUGER PE 4-1043
ECTS OF EXTRA CLOSED CORE RELAXATION ON AUGER AND BINDING ENERGY SHIFTS. EFF 4-0950
THEORY OF AUGER SATELLITE ENERGY SHIFTS. 2-0507
AGE CYLINDRICAL MIRROR ANAL/ SPECIMEN POSITICN EFFECTS ON ENERGY SHIFTS AND SIGNAL INTENSITY IN A SINGLE ST 3-0606
SIMPLE MCDEL FOR K X-RAY AND AUGER ELECTRON ENERGY SHIFTS IN HEAVY ION COLLISIONS. COMMENTS. 2-0301
IVE EFFECT OF EXTRAATOMIC RELAXATION ON AUGER AND BINDING ENERGY SHIFTS IN TRANSITION METALS AND SALTS. /AT 2-0412
LUMINUM: INTERNAL PHOTOEMISSION. STRUCTURE ON THE HIGH ENERGY SIDE OF THE KL2,3M AUGER PEAK FROM SOLID A 4-1021
DIFFRACTION PEAKS IN SECONDARY ELECTRON ENERGY SPECTRA. (NICKEL (100)) 4-0775
ARGON COLLISIONS. (AUGER AND COSTER-KRONIG TRANSITIONS) ENERGY SPECTRA OF ELECTRONS EJECTED IN ARGON(+)- 4-1007
ER ELECTRONS FROM SINGLE CRYSTALS. ENERGY SPECTRA OF INELASTICALLY SCATTERED AND AUG 4-1064
NS AND LEED STUDIES CF TUNGSTEN. ENERGY SPECTRA OF INELASTICALLY SCATTERED ELECTRO 4-1094
MAGNESIUM AND SILICON WHEN BOMBARDED BY PRIMARY ELECTRON/ ENERGY SPECTRA OF SECONDARY ELECTRONS EMITTED BY 4-1047
ENUM AND TUNGSTEN. AUGER ELECTRON EMISSION IN THE ENERGY SPECTRA OF SECONDARY ELECTRONS FROM MOLYBD 4-0874
S MATERIALS. AUGER PEAKS IN THE ENERGY SPECTRA OF SECONDARY ELECTRONS FROM VARIOU 4-0926
DARY ELECTRONS EMITTED BY IONIC BOMBARDMENT. ENERGY SPECTRUM AND ANGULAR DISTRIBUTION OF SECON 4-0988
HARACTERISTIC LOSS AND GAIN PHENOMENA OF COPPER (111). ENERGY SPECTRUM OF BACK SCATTERED ELECTRONS AND C 4-0897
LIDE CRYSTALS UNDER IMPACT OF ION AND ATOM BCMBARDMENT. / ENERGY SPECTRUM OF ELECTRONS EMITTED BY ALKALI HA 4-0246
COPPER UNDER BOMB/ OBSERVATION OF AUGER ELECTRONS IN THE ENERGY SPECTRUM OF SECONDARY ELECTRONS EMITTED BY 2-0551
(111) NICKEL SURFACES. (AUGER SPE/ ELASTIC LEED INTENSITY ENERGY STUDIES OF CLEAN (001), (110) AND ELASTIC 5-1292
ATION. CARRIER TEMPERATURE EFFECTS AND ENERGY TRANSFERS IN NONRADIATIVE (AUGER) RECOMBIN 2-0470
N SPECTROSCOPY. ENERGY WIDTHS OF ROENTGEN LEVELS BY AUGER ELECTRO 4-0982
PHOTO AUGER ELECTRON SPECTROSCOPY) INTRODUCTION TO LOW ENERGY X-RAY AND ELECTRON ANALYSIS. (ANALYZER FOR 3-0627
SODIUM, MAGNESIUM, ALUMINUM, AND/ INTERPRETATION OF HIGH ENERGY X-RAY SATELLITES OF L2,3 EMISSION BANDS OF 2-0239
INTERFACE PHENOMENA IN ENGINEERING MATERIALS. 1-0176
DIGITIZING AUGER DATA. THREEFOLD APPROACH TO ENHANCE ANALYTICAL CAPABILITIES. 1-0183
N INDUCED AUGER SPECTRA (SILVER AND SILICON, AUGER EFFECT ENHANCED BY BACKSCATTERING). / EFFECTS IN ELECTRO 4-0841
INC SELENIDE, THORIUM FLUORIDE) AUGER ANALYSIS OF THREE ENHANCED MULTILAYER DIELECTRIC MIRROR DESIGNS. (Z 5-1328
/AND SILICON WHEN BOMBARDED BY PRIMARY ELECTRCNS. PART-1: ENHANCED PLASMON EXCITATION ON INNER LEVEL IONIZ/ 4-1047
ENVIRONMENTS. (XENON, CADMIUM, ZINC) ENHANCEMENT OF COSTER-KRONIG RATES IN SOLID STATE 2-0439
LLOYS. SURFACE ENRICHMENT IN COPPER(3) GOLD AND GOLD(3) COPPER A 5-1934
ILMS. (AUGER ELECTRON SPECTROSCOPY) SURFACE ENRICHMENT OF INDIUM IN EVAPORATED GOLD- INDIUM F 5-1892
(0001) FACE OF GRAPHITE. (AUGER SPECTROSCOPY) ENTROPY OF ADSORPTION OF XENON IN EPITAXY ON THE 5-1860
/OSION CRACKING OF ALPHA BRASS IN A TARNISHING AMMONIACAL ENVIRONMENT: FRACTOGRAPHY AND CHEMICAL ANALYSIS./ 6-2095
MPARISON OF NITROGEN AND SULFUR IN TWO DIFFERENT CHEMICAL ENVIRONMENTS. AUGER LINE SHAPE CO 2-1816
ENHANCEMENT CF COSTER-KRONIG RATES IN SOLID STATE ENVIRONMENTS. (XENON, CADMIUM, ZINC) 2-0439
MINORITY CARRIER LIFETIME IN INDIUM ARSENIDE EPILAYERS. (AUGER RECOMBINATION PROCESS) 2-0567
ATOMS AND ADSORBED OXYGEN ATCMS ON THE (001) FACE OF IRON EPITAXIAL FILM. (AUGER EFFECT) /VIOR OF IMPURITY 5-1523
11): FORMATION OF AN EQUILIB/ DIFFUSION OF SILVER THROUGH EPITAXIAL FILMS CF COPPER AND NICKEL ON SILVER (1 5-1356
F SURFACE CONTAMINATION EFFECT ON THE GROWTH MCDE OF IRON EPITAXIAL FILMS ON MAGNESIUM OXIDE (001). /TUDY O 5-1524
TUNGSTEN SINGLE CRYSTAL STUDIED BY LEED, AUGER ELECTRON / EPITAXIAL GROWTH OF COPPER ON (110) SURFACE OF A 5-1660
(001) USING REFLECTION DIFFRACTION, LEED AN/ STUDY OF THE EPITAXIAL GROWTH OF MAGNESIUM ON MAGNESIUM OXIDE 5-1496
ENERGY ELECTRON DI/ STRUCTURES OF SUBSTRATE CRYSTALS AND EPITAXIAL GROWTH OF METAL ON THEM OBSERVED BY LOW 5-1906
LVER. (BY AUGER SPECTROSCOPY) OBSERVATIONS OF THE EPITAXIAL GROWTH OF NICKEL AND COPPER ON (111) SI 5-1354
(100) SURFACE OBSERVED BY LOW ENERGY ELECTRON DIFFRACTIO/ EPITAXIAL GROWTH OF SILVER ON POTASSIUM CHLORIDE 5-1696
/ER SPECTROSCOPY, LEED AND WORK FUNCTION MEASUREMENTS FOR EPITAXIAL GROWTH OF THORIUM ON A TUNGSTEN (100) / 5-1739
CARRIER LIFETIMES IN EPITAXIAL INDIUM ARSENIDE. (AUGER RECOMBINATION) 2-0332
TRON DIFFRACTION- AUGER ELECTRON SPECTROSCOPY STUDY OF AN EPITAXIAL IRON FILM. LOW ENERGY ELEC 5-1526
E OF SUBSTRATE CONDITICNS ON THE NUCLEATION AND GROWTH OF EPITAXIAL SILICON FILMS. INFLUENC 5-1514
ED SILICON SUBSTRATES. (SCANNING AUGER SPECT/ SOLID PHASE EPITAXIAL STUDIES USING VACUUM DEPOSITION ON HEAT 5-1287
SURFACE STUDIES WITH EPITAXIALLY GROWN METAL FILMS. (AUGER EFFECT) 5-1253
ANTIMONY DOPING OF SILICON LAYERS GROWN BY SOLID PHASE EPITAXY. (AES MEASUREMENTS) 5-1577
E AND ALUMINUM(X) GALLIUM(1-X) ARSENIDE BY MCLECULAR BEAM EPITAXY. (AUGER SPECTROMETRY) /ED GALLIUM ARSENID 5-1247
INUM(1-X) ARSENIDE) COMPUTER-CONTROLLED MCLECULAR BEAM EPITAXY. (AUGER SURFACE ANALYSIS, GALLIUM(X) ALUM 5-1324
HEED AND AES STUDY OF SILICON EPITAXY BY SILANE PYROLYSIS. 5-1427
S AND ELECTRONIC PROPERTIES OF FILMS. SILICCN ON SAPPHIRE EPITAXY BY VACUUM SUBLIMATION: LEED- AUGER STUDIE 5-1230
. USE OF SCANNING AUGER MICROSCOPY IN MOLECULAR BEAM EPITAXY OF GALLIUM ARSENIDE AND GALLIUM PHOSPHIDE 5-1142
(HEED, LEED, AUGER ELECTRON SPECTROSCOP/ LOW TEMPERATURE EPITAXY OF GERMANIUM FILMS BY SPUTTER DEPOSITION. 5-1537
BSTRATES. (AUGER SPECTROSCOPY) LIQUID PHASE EPITAXY OF III-V COMPOUNDS ON INDIUM PHOSPHIDE SU 5-1590
CUBIC SUBSTRATES USING LEED- AUGER TECHNIQUES. EPITAXY OF ULTRATHIN METAL FILMS ON BODY CENTERED 5-1488
ECTROSCOPY) ENTROPY OF ADSORPTION OF XENON IN EPITAXY ON THE (0001) FACE OF GRAPHITE. (AUGER SP 5-1860
RAY IONIZATION CROSS SECTIONS, AND IONIZATION EQUILIBRIUM EQUATIONS MODIFIED BY AUGER TRANSITIONS. X- 2-0562
CRYSTAL SURFACES OF PLATINUM. (AUGER SPECTRCSCOPY) EQUILIBRATION OF HYDROGEN AND DEUTERIUM ON SINGLE 5-1594
PURITIES ON THORIUM METAL: AUGER STUDY OF BULK TO SURFACE EQUILIBRIA. IM 5-1314
(AUGER SPECTROSCOPY, SULFIDE SYSTEMS) PHASE EQUILIBRIA IN TWO DIMENSIONAL CHEMISORBED LAYERS. 5-1170
SURFACE COMPOSITICN OF (PALLADIUM- SILVER) ALLOYS IN EQUILIBRIUM. 5-1788
ELECTRON HOLE SCATTERING AND (AUGER) RECOMBINATION OF NONEQUILIBRIUM CHARGE CARRIERS IN SILICON AT A HIGH/ 4-1103
HIGH PHOTOEXCITATION LEVELS. (AUGER) RECOMBINATION OF NONEQUILIBRIUM CHARGE CARRIERS OF INDIUM ARSENIDE AT 4-0840
NS. X-RAY IONIZATION CROSS SECTIONS, AND IONIZATION EQUILIBRIUM EQUATIONS MODIFIED BY AUGER TRANSITIO 2-0562
/MS OF COPPER AND NICKEL ON SILVER (111): FORMATION OF AN EQUILIBRIUM LAYER OF SILVER. (AUGER SPECTROSCOPY) 5-1356
(AUGER RECOMBINATION) RECOMBINATICN-INDUCED NONEQUILIBRIUM PHASE TRANSITIONS IN SEMICONDUCTORS. 2-0421
1) SURFACE: A SURFACE PHASE TRANSITION. (AUGER SPECTROSC/ EQUILIBRIUM SEGREGATION OF CARBON TO A NICKEL (11 5-1803
ACE OF NICKEL. (AES) SURFACE SEGREGATION IN ALLOYS: EQUILIBRIUM SEGREGATION OF GOLD TO THE (111) SURF 5-1216
BINDING ENERGIES OF CARBON TO NICKEL (100) FROM EQUILIBRIUM SEGREGATION STUDIES. (AUGER EFFECT) 5-1484
TOMIC PERCENT ALUMINUM ALLOY USING AU/ THERMAL EFFECTS IN EQUILIBRIUM SURFACE SEGREGATION IN A COPPER- 10 A 6-2030
UM ALLOYS. AUGER SPECTROSCOPY AND LEED STUDY OF EQUILIBRIUM SURFACE SEGREGATION IN COPPER- ALUMIN 6-2025
(100) FACES. (AUGER EFFECT) EQUILIBRIUM SURFACE SEGREGATION OF CARBON ON IRON 5-1369
RON SPECTROSCOPY. MEASUREMENT OF EQUILIBRIUM SURFACE SEGREGATION USING AUGER ELECT 5-1184
RON SPECTROSCOPY. (REPLY TO COMMENTS) MEASUREMENT OF EQUILIBRIUM SURFACE SEGREGATION USING AUGER ELECT 6-2026
/ER L-SERIES ELECTRONIC SPECTRA ACCOMPANYING THE DECAY OF ERBIUM-160, YTTERBIUM-166, YTTERBIUM-169, AND LU/ 4-0743
SIS ESCA SEE ELECTRON SPECTROSCOPY FOR CHEMICAL ANALY
YSIS ESCA- SEE ELECTRON SPECTROSCOPY FOR CHEMICAL ANAL

OLID SURFACE. ELECTRON ESCAPE DEPTH IN SILICON. (AUGER ENERGIES) 2-0408
RELATIVE ESCAPE DEPTH OF AUGER AND PHOTOELECTRONS FROM A S 5-1903
ELECTRON ESCAPE DEPTHS IN ALUMINUM. (AUGER ELECTRONS) 2-0537
ESCAPE DEPTHS OF X-RAY EXCITED (AUGER) ELECTRONS. 2-0409
ESCAPE LENGTH OF AUGER ELECTRONS. 2-0527
D LOSS SPECTROSCOPY. DETERMINATION OF THE ESCAPE LENGTH OF LOW ENERGY ELECTRONS BY AUGER AN 2-0401
EMPERATURE OXIDATION OF SILICON STUDIED BY PHOTOSENSITIVE ESR AND AUGER ELECTRON SPECTROSCOPY. LOW T 5-1786
DETERMINATION OF FILM THICKNESS OR ION ETCH RATE USING AUGER ELECTRON SPECTROSCOPY. 5-1984
AUGER SPECTRA OF HYDROGEN CHLORIDE VAPOR ETCHED N+ GALLIUM ARSENIDE (100) SUBSTRATES. 4-0992
CONTAMINANTS ON CHEMICALLY ETCHED SILICON SURFACES. LEED- AUGER METHOD. 5-1229
SURFACES. SPUTTER ETCHING AND AUGER ELECTRON SPECTROSCOPY OF METAL 5-1552
SURFACES. SPUTTER ETCHING AND AUGER ELECTRON SPECTROSCOPY OF METAL 5-1554
SILICON) MONITORED BY AUGER SPECTRO/ SPUTTER CLEANING AND ETCHING OF CRYSTAL SURFACES (TITANIUM, TUNGSTEN, 5-1828
RSENIDE. AES AND SPUTTER ETCHING OF NICKEL- GOLD- GERMANIUM ON N-GALLIUM A 5-1775
DIFFUSION MEASUREMENT IN THIN FILMS BY AUGER SPUTTER ETCHING TECHNIQUE. 5-1439
ARBON BY FAST PROTONS. (AUGER ELECTRON YIELDS OF METHANE, ETHANE, ETHYLENE, ACETYLENE) /ELL IONIZATION OF C 2-0532
) SURFACE AND THEIR INFLUENCE ON/ INTERACTION OF METHANE, ETHANE, ETHYLENE, AND ACETYLENE WITH AN IRON (001 5-1208
FAST PROTONS. (AUGER ELECTRON YIELDS OF METHANE, ETHANE, ETHYLENE, ACETYLENE) /ELL IONIZATION OF CARBON BY 2-0532
E AND THEIR INFLUENCE CN/ INTERACTION OF METHANE, ETHANE, ETHYLENE, AND ACETYLENE WITH AN IRON (001) SURFAC 5-1208
UM AND POLYC/ AUGER SPECTRA OF CARBON. TANTALUM CARBIDES, ETHYLENE AND CARBON MONOXIDE ADSORPTION ON TANTAL 4-0820
ACTERIZATION OF ADSORBED SPECIES USING AUGER PEAK SHAPES. ETHYLENE AND CARBON MONOXIDE ON TUNGSTEN (100). / 5-1243
ER SPECTROSCOPY/ ADHESION AND FRICTION OF POLYTETRAFLUORO ETHYLENE IN CONTACT WITH METALS AS STUDIED BY AUG 5-1198
) SURFACE AND EFFECT OF THESE FILMS ON ADH/ ABSORPTION OF ETHYLENE OXIDE AND VINYL CHLORIDE ON AN IRON (011 5-1207
/EAM EFFECTS ON CARBON MONOXIDE SATURATED TUNGSTEN (100), ETHYLENE SATURATED TUNGSTEN (100), AND TUNGSTEN / 5-1244
TROSCOPY. ADHESION AND TRANSFER OF POLYTETRAFLUORO ETHYLENE TO METALS STUDIED BY AUGER EMISSION SPEC 5-1731
ENERGY ELECTRON DIFFRACTION AND AUGER E/ INVESTIGATION OF EUROPIUM(II) OXIDE (100) CLEAVAGE SURFACE BY LOW 5-1157
/MICROSPOT AUGER SPECTROSCOPY STUDY OF INTERDIFFUSION AND EUTECTIC FORMATION IN TUNGSTEN PLATINUM- TUNGSTE/ 5-1259
ENERGY ELECTRON LOSS SPECTRA FROM DIAMOND, GRAPHITE, AND EVAPORATED CARBON SURFACES. /TERISTIC K-SHELL LOW 4-0945
COPY) DIFFUSE INTERFACE IN SILICON (SUBSTRATE)- GOLD (EVAPORATED FILM) SYSTEM. (AUGER ELECTRON SPECTROS 5-1681
/ DIFFUSE INTERFACE STRUCTURES IN SOME (SUBSTRATE- VACUUM EVAPORATED FILM) SYSTEMS STUDIED BY AUGER ELECTRO 5-1436
ECTROSCOPY) SURFACE ENRICHMENT OF INDIUM IN EVAPORATED GOLD- INDIUM FILMS. (AUGER ELECTRON SP 5-1892
/ HYDROGEN ON CLEAN NICKEL (111), NICKEL (100), SHEET AND EVAPORATED NICKEL FILMS. (AUGER ELECTRON SPECTRO/ 5-1468
OEMISSION SPECTROSC/ SURFACE COMPOSITION AND CHEMISTRY CF EVAPORATED PERMALLOY FILMS OBSERVED BY X-RAY PHOT 5-1738
TERISTIC ENERGIES IN SECONDARY ELECTRON SPECTRA FROM THIN EVAPORATED PHOSPHORUS FILMS. /ER AND OTHER CHARAC 4-0986
MOLECULAR BEAM STUDIES OF INDIUM PHOSPHIDE EVAPORATION AND GROWTH. (AES) 5-1334
RVED BY AUGER ELECTRON EMISSION. CARBON EVAPORATION FROM A THORIUM DISPENSER CATHODE OBSE 5-1400
GER SPECTROSCOPY) SURFACE CARRIER CONCENTRATION AND AS EVAPORATION IN HEAT TREATED GALLIUM ARSENIDE. (AU 5-1915
ARSENIDE. (AUGER EFFECT) PREFERENTIAL EVAPORATION OF INDIUM FROM GALLIUM(X) INDIUM(1-X) 5-1366
NERGY ELECTRON DIFFRACTION AND AUGER EM/ ULTRAHIGH VACUUM EVAPORATOR AND MULTIPLE CLEAVAGE DEVICE FOR LOW E 3-0707
ULTRAHIGH VACUUM EVAPORATOR FOR LEED AND AUGER EMISSION STUDIES. 3-0683
/GY GAIN AND LOSS, DOUBLE IONIZATION, AND IONIZATION LOSS EVENTS IN BERYLLIUM AND BERYLLIUM OXIDE SECONDAR/ 4-0901
AUGER SATELLITE AND OTHER CHARACTERISTIC EVENTS IN MAGNESIUM SECONDARY ELECTRON SPECTRA. 4-0898
OTON EMISSION. EVIDENCE FOR A RADIATIVE AUGER EFFECT IN X-RAY PH 2-0223
EE ARGON ATOM. EVIDENCE OF RADIATIVE AUGER TRANSITIONS IN THE FR 2-0403
EE ARGON ATOM. EVIDENCE OF RADIATIVE AUGER TRANSITIONS IN THE FR 2-0404
E DROPLETS. (SILICON) EVIDENCES FOR AUGER RECOMBINATION IN ELECTRON HOL 2-0260
TION IN AUGER YIELD MEASUREMENTS. EXACT CORRECTIONS FOR POTENTIAL MODULATION DISTOR 2-0389
ATIONS. AUGER EXAMINATION OF CONTAMINANTS IN THIN FILM METALLIZ 5-1890
/CROSCOPY AND NONDISPERSIVE X-RAY ANALYSIS TECHNIQUES FOR EXAMINATION OF SILVER- CADMIUM OXIDE CONTACT MAT/ 5-1799
AND DEPOSITION IN LIQUID METALS./ APPLICATION OF PHYSICAL EXAMINATION TECHNIQUES IN THE STUDY OF CORROSION 6-2050
METALLIC TRANSFER EETWEEN METALS IN SLIDING CONTACT EXAMINED BY AUGER EMISSION SPECTROSCOPY. 5-1729
SURFACE ABSORPTION BY POLYPEPTIDE FILMS EXAMINED BY AUGER EMISSION SPECTROSCOPY. 5-1179
ED CARRIER LIFETIMES IN MERCURY CADMIUM TELLURIDE AT HIGH EXCESS CARRIER CONCENTRATIONS. AUGER LIMIT 2-0262
10 KEV) SCATTERED FROM A (100) FACE CF A COPPER S/ CHARGE EXCHANGE OF LOW ENERGY HELIUM(+) IONS (LESS THAN 2-0546
TERED FROM A COPPER SINGLE CRYSTAL. (AUGER NEUTRA/ CHARGE EXCHANGE OF LOW ENERGY HELIUM IONS AND ATOMS SCAT 2-0545
AUGER RECOMBINATION IN SEMICONDUCTORS DURING INTENSE EXCITATION. 2-0368
OF MAGNESIUM L2,3MM AUGER CURRENTS USING ELECTRON AND ION EXCITATION. COMPARISON 4-0855
ELECTRON SPECTROSCOPY. X-RAY AND ELECTRON EXCITATION. 1-0082
ARY ELECTRON EMISSION FROM SILICON (111). (AUGER ELECTRON EXCITATION) /ES AND ENERGY DISTRIBUTION OF SECOND 4-0853
IN THIN FILMS OF SILICON AND GERMANIUM SUBJECTED TO LASER EXCITATION. /OCESSES (AUGER) ON THE SURFACES AND 5-1990
SMUTH(III) TELLURI/ MATRIX ELEMENTS DEPENDENCE OF OPTICAL EXCITATION AND AUGER DECAY OF 5D CORE HOLES IN BI 2-0398
TOMS. MECHANISMS OF INNER SHELL EXCITATION AND DEEXCITATION IN MULTIPLY IONIZED A 2-0487
SELF-CONSISTENT ANALYSIS OF CORE BAND-TO-BAND EXCITATION: AUGER FINE STRUCTURE. 2-0509
AUGER EXCITATION BY INTERNAL SECONDARY ELECTRONS. 2-0390
EXCITED X-RAYS. DETERMINATION OF AUGER ELECTRON EXCITATION CROSS SECTIONS AND RANGES USING PROTON 2-0456
IUM, AND CESIUM IONS IN THE VACUUM ULTRAVI/ SPECTRAL LINE EXCITATION FUNCTIONS FOR SODIUM, POTASSIUM, RUBID 4-0735
ON CF NONEQUILIBRIUM CHARGE CARRIERS IN SILICCN AT A HIGH EXCITATION LEVEL. /TERING AND (AUGER) RECOMBINATI 4-1103
EDGE EMISSION OF GALLIUM SELENIDE SINGLE CRYSTALS AT HIGH EXCITATION LEVELS. (AUGER RECOMBINATION) /OF THE 2-0258
EXCITATION OF INNER SHELL ELECTRONS BY PHOTONS AN 2-0541
D FAST ELECTRONS. EXCITATION OF MULTIPLET STATES IN NEON. 2-0285
EXCITATION OF THE RARE GASES BY 30 MEV OXYGEN ION 2-0298
S. (INNER SHELL AUGER TRANSITIONS) OUTER SHELL EXCITATION ON INNER LEVEL IONIZATION. PART-2: DO/ 4-1047
/BOMBARDED BY PRIMARY ELECTRONS. PART-1: ENHANCED PLASMON EXCITATION PART-2: ULTRAVIOLET EXCITATION. (AUGE/ 2-0306
/M(2) GERMANIUM, AND MAGNESIUM(2) STANNIDE. PART-1: X-RAY EXCITATION PROCESSES IN TRANSITION METAL ATOMS. P 2-0362
ART-2: INNER AUGER ELECTRONS. MULTIPLE EXCITATION. (REVIEW, ESCA, AES) 1-0081
ELECTRON SPECTROSCOPY: X-RAY AND ELECTRON EXCITATION SPECTRUM OF OXYGEN DOMINATED COMPOUNDS 4-1048
IN THE 3-21 EV REGION OF THE SPECTRUM. (INTERBAND AUGER/ EXCITATION TECHNIQUES. (AUGER EFFECT) 2-0468
ELECTRONIC STRUCTURE OF SOLID SURFACES: CORE LEVEL EXCITATIONS IN CARBON. (AUGER APPEARANCE SPECTROS 2-0291
COPY) PLASMON COUPLING TO CORE HOLE EXCITED AND ELECTRON EXCITED AUGER SPECTROSCOPY. 2-0261
/ OF IONIZATION CROSS SECTIONS AND OTHER FACTORS IN X-RAY EXCITED AUGER AND PHOTOELECTRON SPECTRA OF MAGNES 4-0837
IUM, SOME ALLOYS OF MAGNESIUM, AND ITS OXIDE. X-RAY EXCITED AUGER AND PHOTOELECTRON SPECTRA OF PARTIA 5-1949
LLY OXIDIZED MAGNESIUM SURFACES. ABNORMAL CHEMICA/ X-RAY EXCITED AUGER DATA FROM TITANIUM. 2-0367
EFFECT CF MODULATION AMPLITUDE ON ELECTRON EXCITED AUGER ELECTRON EMISSION) /ONDARY ELECTRON 5-1403
SPECTROSCOPY. (SURFACE ANALYSIS OF MATERIALS BY ELECTRON EXCITED AUGER ELECTRON SPECTRA OF MOLYBDENUM SURF 5-1667
ACES. PROTON AND ELECTRON EXCITED AUGER ELECTRON SPECTROSCOPY. 1-0091
THE SECONDARY ELECTRON BACKGROUND PROBLEM IN ELECTRON EXCITED AUGER ELECTRON SPECTRUM OF A NICKEL (110) 4-1065
- C(2X2) SULFUR SURFACE: LINE SHAPE ANALYSIS A/ ELECTRON EXCITED AUGER ELECTRONS. 5-1399
ANALYSIS OF MATERIALS BY ELECTRON EXCITED (AUGER) ELECTRONS. 2-0409
ESCAPE DEPTHS OF X-RAY EXCITED AUGER EMISSION. 2-0378
S OF EFFICIENCIES OF PRODUCTION AND DETECTICN OF ELECTRON EXCITED AUGER EMISSION. ESTIMATE 2-0288

ANGULAR DEPENDENCIES IN ELECTRON EXCITED AUGER EMISSION. (COMMENTS) 2-0286
NEW FINE STRUCTURE IN ELECTRON EXCITED AUGER SPECTRA FROM SOLID SURFACES. 2-0473
ION EXCITED AUGER SPECTRA OF ALUMINUM. 4-0869
TRACTION. RETRIEVAL OF ELECTRON EXCITED AUGER STRUCTURE BY DYNAMIC BACKGROUND SUB 1-0090
BEAM FOIL EXCITED AUGER TRANSITIONS IN NEON. 4-0860
AUGER AND SECONDARY ELECTRONS EXCITED BY BACKSCATTERED ELECTRONS. 2-0365
UANTITATIVE ANALYSIS. AUGER AND SECONDARY ELECTRONS EXCITED BY BACKSCATTERED ELECTRONS. APPROACH TO Q 2-0366
/SOLUTION BEAM GAS AUGER ELECTRON SPECTRA FOR OXYGEN IONS EXCITED BY COLLISIONS WITH HELIUM, NEON, AND ARG/ 4-0904
PHA X-RAY IRRADIATIO/ STUDY OF INTENSITIES OF AUGER LINES EXCITED BY ELECTRON BOMBARDMENT AND ALUMINUM K-AL 4-1017
IONS. VACUUM ULTRAVIOLET EMISSION OF XENON(III) LEVELS EXCITED BY ELECTRON IMPACT VIA N4,5 AUGER TRANSIT 4-0878
/SON OF AUGER SPECTRA OF MAGNESIUM, ALUMINUM, AND SILICON EXCITED BY LOW ENERGY ELECTRON AND LOW ENERGY AR/ 4-0857
) SELENIDE(0.52) CRYSTALS. (/ PHOTOCONDUCTIVITY OF HIGHLY EXCITED CADMIUM SELENIDE AND CADMIUM SULFIDE(0.48 2-0432
AUGER ELECTRON EMISSION SPECTRA FROM FOIL AND GAS EXCITED CARBON BEAMS. 4-1050
RELAXATION OF AUGER EXCITED CARRIERS IN SILICON. 2-0274
/ SPECTROMETER INSTALLED IN VACUUM AND THE MEASUREMENT OF EXCITED ELECTRONS UNDER THE X-RAY DIFFRACTION PR/ 3-0641
EXCITON LUMINESCENCE AND PHOTOCONDUCTIVITY IN HIGHLY EXCITED GALLIUM PHOSPHIDE. 2-0454
DOPPLER SHIFTED AUGER ELECTRONS FROM FOIL EXCITED HEAVY ION BEAMS. 4-0859
AUGER ELECTRONS FROM FOIL EXCITED HEAVY ION BEAMS. (OF NEON) 2-0370
RADIATIONLESS DEEXCITATION OF EXCITED HELIUM ATOMS AT SURFACES. (AUGER PROCESS) 2-0255
T OF ANGLE OF INCIDENCE. PART-3: EFFECT OF ELECTRONICALLY EXCITED IONS. /SITIVE IONS. PART-1. PART-2: EFFEC 2-0543
RON. X-RAY EXCITED LMM AUGER SPECTRA OF COPPER, NICKEL AND I 4-1123
ELECTRON CHEMICAL SHIFTS IN COORDINATION COMPOUNDS. X-RAY EXCITED LMM AUGER SPECTRA OF GERMANIUM. PHOTO 4-0924
OF OXIDIZED STATES ON SOME METAL SURFACES. X-RAY EXCITED PHOTOELECTRON AND AUGER ELECTRON SPECTRA 5-1690
UM AND DECAY OF THE RECOMBINATION RADIATION FROM STRONGLY EXCITED SILICON. (AUGER RECOMBINATION PROCESS) /R 2-0460
4D AUGER TRANSITIONS FROM EXCITED STATES IN SAMARIUM METAL. 4-0823
NS OF AUGER ELECTRON SPECTROSCOPY. PART-1: E/ ENERGIES OF EXCITED STATES OF DOUBLY IONIZED MOLECULES BY MEA 4-1071
TIONAL MONOLAYER OXYGEN DETERMINATIONS OBTAINED BY PROTON EXCITED X-RAY, AES ANALYSIS OF IRON SURFACES. /AC 5-1682
LECTRON EXCITATION CROSS SECTIONS AND RANGES USING PROTON EXCITED X-RAYS. DETERMINATION OF AUGER E 2-0456
NETICS IN GERMANIUM IN THE PRESENCE OF EL/ HIGH FREQUENCY EXCITON BREAKDOWN AND FREE CARRIER AND EXCITON KI 4-0946
PHOTOEMISSION STUDIES OF CORE EXCITON DECAY IN POTASSIUM IODIDE. 2-0422
STIMULATED EMISSION OF LIGHT FROM A NONIDEAL EXCITON GAS IN A SEMICONDUCTOR. (AUGER PROCESS) 2-0462
HLY EXCITED GALLIUM PHOSPHIDE. EXCITON LUMINESCENCE AND PHOTOCONDUCTIVITY IN HIG 2-0454
SINGLE CRYSTALS. (AUGER PROCESS) EXCITON SPECTRA OF HEAVILY DOPED CADMIUM SULFIDE 2-0372
RADIATIVE AUGER RECOMBINATION OF TWO INDIRECT EXCITONS. 2-0227
PHOTOCONDUCTIVITY ASSOCIATED WITH AUGER RECOMBINATION OF EXCITONS BOUND TO NEUTRAL DONORS IN TELLURIUM DO/ 2-0453
ENCE BAND) PHOTOEMISSION STUDIES OF THE DECAY OF CORE EXCITONS IN ALKALI HALIDES. (AUGER PROCESS IN VAL 2-0289
N SODIUM CHLORIDE BY MEANS OF THE DECOMPOSI/ DETECTION OF EXCITONS NEAR THE L2,3 EDGE OF THE CHLORIDE ION I 4-1126
LUMINESCENCE DUE TO EXCITON-EXCITON COLLISION IN GALLIUM ARSENIDE. 2-0451
DIGITAL CONTROL UNIT IN TTL LOGIC FOR THE AUTOMATIC EXECUTION OF AUGER AND ESCA TEST PROGRAMS. 3-0722
HE / MODEL FOR THE AUGER ELECTRON SPECTROSCOPY OF SYSTEMS EXHIBITING LAYER GROWTH, AND ITS APPLICATION TO T 2-0399
ORPTION EFFECT, AUGER ELECTRON SPECTROSCOPY) EXOELECTRON EMISSION FROM ZINC OXIDE. (OXYGEN ADS 5-0921
UGER ELECTRON SPECTROSCOPY INVESTIG/ THERMALLY STIMULATED EXOELECTRON EMISSION OF BERYLLIUM OXIDE LAYERS. A 5-1325
ATION OF THE ENERGY DEPTH OF LEVELS OF SURFACE ELECTRON / EXOELECTRON SPECTROMETRY AS A METHOD FOR DETERMIN 4-0915
ER EFFECT) EXOEMISSION PROPERTIES OF ZIRCONIUM DIOXIDE. (AUG 4-0410
AUGER MEASUREMENTS ON T-027 SAMPLES EXPOSED DURING THE SKYLAB-2 MISSION. 5-1735
AUGER ELECTRON SPECTROSCOPY OF MATERIALS EXPOSED TO FLOWING SODIUM. 5-1460
DIED BY AUGER ELECTRON SPECTROSCOPY. OXYGEN EXPOSURE OF SAMARIUM, GADOLINIUM, AND TERBIUM STU 5-1332
ON DIFFRACTION. PART-1: ELECTRODE SURFACE STRUCTURE AFTER EXPOSURE TO WATER AND AQUEOUS ELECTROLYTES. /ECTR 5-1486
F THE VIOLENT ARGON(+) - ARGON COLLISION. EXTENDED ION- ELECTRON COINCIDENCE MEASUREMENTS O 2-0530
ER SPECTROSCOPY) A COMPARISON OF THE EXTENDURE OF ELECTRON SPECTROMETERS. (USED IN AUG 3-0626
AUGER SPECTROSCOPIC ANALYSIS OF THE EXTENT OF GRAIN BOUNDARY SEGREGATION. 6-2094
. (EF/ EFFECT OF THE POLARITY OF INDIUM ANTIMONIDE ON THE EXTERNAL PHOTOEFFECT IN THE X-RAY SPECTRAL REGION 4-1001
ENERGY SHIFTS. EFFECTS OF EXTRA CLOSED CORE RELAXATION ON AUGER AND BINDING 4-0950
Y SHIFTS IN TRANSITION METALS AND SAL/ RELATIVE EFFECT OF EXTRAATOMIC RELAXATION ON AUGER AND BINDING ENERG 2-0412
4,5M4,5 AUGER PROCESS IN ARSENIC AND SELENIUM. ATOMIC AND EXTRAATOMIC STATIC RELAXATION ENERGIES IN THE L3M 4-0883
E BY AUGER SPECTROSCOPY. OBSERVATION OF EXTRINSIC SURFACE STATES ON (1120) CADMIUM SULFID 5-1202

F

AUGER TRANSITION RATES IN J-J COUPLING FOR D HOLES AND F ELECTRONS. 2-0444
EANING CONTAMINANTS FROM QUARTZ SURFACES DURING RESONATOR FABRICATION. (AUGER SPECTRA) METHODS OF CL 5-1410
EVALUATION OF LEAD(0.8) TIN(0.2) TELLURIDE DETECTOR FABRICATION USING SURFACE ANALYSIS. (AUGER) 5-1593
IRST MONOLAYER ADSORPTION OF XENON ON THE (0001) GRAPHITE FACE. THERMODYNAMICS AND KINETICS OF THE F 5-1859
AND INTERACTION WITH OXYGEN AND C/ THORIUM (100) CRYSTAL FACE. A STUDY OF ITS CLEANING, SURFACE STRUCTURE, 5-1883
ER SPECTROSCOP/ STUDY OF BARIUM FILMS ON A (110) TUNGSTEN FACE BY THE DIFFRACTION OF SLOW ELECTRONS AND AUG 5-1368
, PALLADIUM, COPPER, NICK/ AUGER ELECTRON SPECTROSCOPY OF FACE CENTERED CUBIC METAL SURFACES. (GOLD, SILVER 5-1717
/HELIUM(+) IONS (LESS THAN 10 KEV) SCATTERED FROM A (100) FACE OF A COPPER SINGLE CRYSTAL. (AUGER NEUTRALI/ 2-0546
PECTROSCOPY) ADSORPTION OF KRYPTON ON THE (111) FACE OF COPPER SINGLE CRYSTALS. (AUGER ELECTRON S 5-1221
TWO DIMENSIONAL TRANSITION: XENON ADSORBED ON THE (0001) FACE OF GRAPHITE. (AUGER ELECTRON SPECTROSCOPY) / 5-1857
ENTROPY OF ADSORPTION OF XENON IN EPITAXY ON THE (0001) FACE OF GRAPHITE. (AUGER SPECTROSCOPY) 5-1860
OF IMPURITY ATOMS AND ADSORBED OXYGEN ATOMS ON THE (001) FACE OF IRON EPITAXIAL FILM. (AUGER EFFECT) /VIOR 5-1523
ADSORPTION OF SULFUR ON THE (100) FACE OF MOLYBDENUM. 5-1732
DIFFUSION OF SULFUR TOWARD THE (110) FACE OF NICKEL. 5-1771
E. STUDY OF THE ADSORPTION OF BENZENE ON THE (100) FACE OF NICKEL. INFLUENCE OF SULFUR AT THE SURFAC 5-1284
/TUDIES OF GAS- SURFACE INTERACTIONS AT THE (100) CRYSTAL FACE OF PLATINUM BY MOLECULAR BEAM SCATTERING, L/ 5-1963
WITH CONTROLLED SURFACES. (AUGER NEUTRALIZATION AT (100) FACE OF TUNGSTEN, POLYCRYSTALLINE MOLYBDENUM). /S 5-1736
PREPARATION AND SURFACE OXIDATION OF A (110) FACE OF VANADIUM. 5-1683
EQUILIBRIUM SURFACE SEGREGATION OF CARBON ON IRON (100) FACES. (AUGER EFFECT) 5-1369
/NS OF HYDROGEN DESORBED FROM THE (100), (110), AND (111) FACES OF COPPER CRYSTALS. (AUGER ELECTRON SPECTR/ 5-1155
K FUNCTION MEASUREMENTS ON CLEAN AND CESIUM COVERED POLAR FACES OF GALLIUM ARSENIDE. /SPECTROSCOPY, AND WOR 5-1294
ND SURFACE ALLOYING OF LEAD MONOLAYERS ON (111) AND (110) FACES OF GOLD. (AUGER EFFECT) ADSORPTION A 5-1733
). (A/ SURFACE COMPOSITION STUDIES OF THE (001) AND (110) FACES OF MONOCRYSTALLINE IRON(0.84) CHROMIUM(0.16 5-1591
N AND AUGER / ADSORPTION OF SULFUR ON THE (100) AND (111) FACES OF PLATINUM. LOW ENERGY ELECTRON DIFFRACTIO 5-1421
CHLORINE MONOLAYERS ON THE LOW INDEX FACES OF SILVER. (AUGER EFFECT) 5-1777
ANNEALING OF GALLIUM PHOSPHIDE AND GALLIUM ARSENIDE (111) FACES STUDIED BY AES AND UPS. OXIDATION AND 5-1493
PECTROSCOPY. FAILURE ANALYSIS APPLICATIONS OF AUGER ELECTRON S 6-2034
STIGATING ENERGY AND ANGULAR DISTRIBUTIONS OF INELASTICA/ FARADAY CUP LEED APPARATUS WITH FACILITY FOR INVE 3-0691
FECT OF A THIN DIFFUSED SOLUTE LAYER ON THE NUCLEATION OF FATIGUE CRACKS. (AUGER ELECTRON SPECTROSCOPY) EF 6-2037
D AUGER SPECTROSCOPY. CHARACTERIZING OF ALUMINUM FATIGUE DAMAGE BY PHOTOEMISSION, ELLIPSOMETRY, AN 6-2116
. (SURFACE OXYGEN ON LEAD OXIDE BY AES) GAIN FATIGUE MECHANISM IN CHANNEL ELECTRON MULTIPLIERS 3-0681

L-AUGER SPECTRA OF PLUTONIUM, AMERICIUM, CALIFORNIUM, AND FERMIUM.   THE 4-0836
SPECTROSCOPY: NEW TOOL IN THE CHARACTERIZATION OF GLASS FIBER SURFACES.   AUGER ELECTRON 1-0175
                     AUGER ELECTRON SPECTROSCOPY OF GRAPHITE FIBER SURFACES.   5-1276
EMISSION SPECTROSCOPY.   STUDY OF CARBON FIBER SURFACES USING AUGER AND SECONDARY ELECTRON 5-1970
ASS FIBER- RESIN INTERFACES. (AUGER SPECTROSCOPY OF GLASS FIBERS)   RECENT ASPECTS OF GL 5-1975
ON SPECTROSCOPIC TECHNIQUE/ RECRYSTALLIZATION OF TUNGSTEN FIBERS IN NICKEL MATRIX COMPOSITES. (AUGER ELECTR 6-2046
LASS FIBERS)   RECENT ASPECTS OF GLASS FIBER- RESIN INTERFACES. (AUGER SPECTROSCOPY OF G 5-1975
CTOR (MERCURY(1-X) CADMIUM(X) TELLURIDE) UNDER A MAGNETIC FIELD.   AUGER RECOMBINATION IN A SEMICONDU 4-1092
NG PROPERTIES OF AN ELECTROSTATIC MIRROR WITH CYLINDRICAL FIELD.   FOCUSI 3-0721
TRONS FOR THE CYLINDRICAL MIRROR ANALYZER AND A RETARDING FIELD ANALYZER. /OF OVERLAYER ATTENUATION OF ELEC 3-0693
TROSCOPY).   IMPROVED RETARDING FIELD ANALYZER. (FOR AUGER AND PHOTOELECTRON SPEC 3-0701
PALLADIUM- NICKEL ALLOYS, ADSORBED OXYGE/ SMALL RETARDING FIELD ANALYZER FOR AUGER ELECTRON SPECTROSCOPY. (3-0644
CTROSCOPY)   RETARDING FIELD CYLINDRICAL MIRROR ANALYZER. (FOR AUGER SPE 3-0614
CTRON HOLE LUMINESCENCE IN PURE GERMANIUM. (AUG/ MAGNETIC FIELD DEPENDENT INTENSITY OSCILLATIONS OF THE ELE 4-0780
OM SURFACES) INVESTIGATIONS WITH FIELD EMIS/ STABILITY OF FIELD ELECTRON EMISSION AND VACUUM BREAKDOWN. (FR 5-1520
                                    RETARDING FIELD ELECTRON SPECTROMETER.   5-0686
   ULTRAHIGH VACUUM SCANNING ELECTRON MICROSCOPE WITH A FIELD EMISSION CATHODE AND AUGER ANALYZER.   3-0621
     COMPACT, INEXPENSIVE HIGH RESOLUTION RETARDING FIELD ENERGY ANALYZER. (AUGER ELECTRONS)   3-0619
ER)   A SMALL RETARDING FIELD ENERGY ANALYZER WITH CEM. (AUGER SPECTROMET 3-0655
   IN CONTACT WITH METALS AS STUDIED BY AUGER SPECTROSCOPY, FIELD ION AND SCANNING ELECTRON MICROSCOPY. /LENE 5-1198
, AND LEED STUDY OF THE STRUCTURE AND COMPOSITI/ COMBINED FIELD ION MICROSCOPY. AUGER ELECTRON SPECTROSCOPY 5-1951
/GY ELECTRON DIFFRACTION, AUGER EMISSION SPECTROSCOPY AND FIELD ION MICROSCOPY IN MICROSTRUCTURAL STUDIES. 1-0057
/ES. (SURFACE STRUCTURES, SPECIES, TOPOGRAPHY, PHENOMENA, FIELD ION MICROSCOPY, LEED, HEED, AUGER SPECTROS/ 5-1492
/ACE METALLURGY ON THE PENETRATION DEPTH AND RF BREAKDOWN FIELD OF SUPERCONDUCTING NIOBIUM. (AUGER EFFECT) 6-2137
S.   INFLUENCE OF AN ELECTRIC FIELD ON THE AUGER RECOMBINATION IN SEMICONDUCTOR 2-0357
S.   INFLUENCE OF AN ELECTRIC FIELD ON THE AUGER RECOMBINATION IN SEMICONDUCTOR 4-0844
PH.   SIMPLIFIED RETARDING FIELD PHOTOELECTRON AND AUGER ELECTRON SPECTROGRA 3-0642
S IN AUGER ELECTRON EMISSION SPECTROSCOPY USING RETARDING FIELD SPECTROMETERS. /TING, AND RELATED TECHNIQUE 3-0610
OR AUGER ELECTRON SPECTROSC/ SOME PROPERTIES OF RETARDING FIELD SPECTROMETERS USING POST MONOCHROMATORS. (F 3-0624
   PHOTON, ELECTRON SPECTROSCOPY FOR CHEMICAL ANALYSIS, AND FIELD STIMULATED ELECTRON EMISSION. /PECTROSCOPY, 5-1901
PPER(I) AND SILVER(I)   LINEAR LIGAND FIELDS INFLUENCING AUGER SPECTRA OF FLUORIDES, CO 4-0905
L RESOLUTION AND PRIMARY BEAM CURRENTS. (CARBON ON SILVER FILM)   AUGER ELECTRON SPECTROSCOPY AT HIGH SPATIA 5-1904
N- AUGER ELECTRON SPECTROSCOPY STUDY OF AN EPITAXIAL IRON FILM.   LOW ENERGY ELECTRON DIFFRACTIO 5-1526
           AUGER ELECTRON SPECTROSCOPY FOR THIN FILM ANALYSIS.   1-0211
           AUGER ELECTRON SPECTROSCOPY FOR THIN FILM ANALYSIS.   1-0212
           SCANNING AUGER MICROSCOPE FOR THIN FILM ANALYSIS.   3-0645
LAR ENERGY CONVERSION.   AUGER THIN FILM ANALYSIS AS APPLIED TO VARIOUS ASPECTS IN SO 5-1960
           SURFACE AND THIN FILM ANALYSIS. (AUGER SPECTROMETERS)   3-0609
NG DETECTION AND DETERMINATION LIMITS IN SURFACE AND THIN FILM ANALYSIS. (AUGER SPECTROSCOPY)   ESTABLISHI 5-1572
VIEW)   THIN FILM ANALYSIS BY AUGER ELECTRON SPECTROSCOPY. (RE 1-0210
           THIN FILM ANALYSIS FOR PROCESS EVALUATION. (AES)   1-0118
       SPUTTERING IN THIN FILM ANALYSIS METHODS. (ESCA, AUGER SPECTROSCOPY) 1-0214
NG AUGER MICROPROBE)   SURFACE AND THIN FILM ANALYSIS OF SEMICONDUCTOR MATERIALS. (SCANNI 1-0089
TROMETER.   DIGITAL METHODS FOR THIN FILM ANALYSIS USING A COMPUTER COUPLED AUGER SPEC 3-0640
           BEAM ANALYSIS METHOD OF FILM AND SURFACE COMPOSITION DETERMINATION.   5-1810
ADSORBED OXYGEN ATOMS ON THE (001) FACE OF IRON EPITAXIAL FILM. (AUGER EFFECT) /VIOR OF IMPURITY ATOMS AND 5-1523
/ON OF THE SURFACE COMPOSITION OF PALLADIUM- NICKEL ALLOY FILM CATALYSTS USING AUGER ELECTRON SPECTROSCOPY. 5-1846
QUES. (AUGER SPECTROSCOPY)   THIN FILM COMPOSITIONAL ANALYSIS. COMPARISON OF TECHNI 1-0054
ARISON OF TECHNIQUES. (REVIEW)   SURFACE AND THIN FILM COMPOSITIONAL ANALYSIS. DESCRIPTION AND COMP 1-0053
AUGER ELECTRON AND X/ ELECTRONIC STRUCTURE OF A THIN GOLD FILM DEPOSITED ON A SILICON SUBSTRATE STUDIED BY 5-1435
ON DIFFRACTION- AUGER ELECTRON SPECTROSCOPY STUDY OF IRON FILM DEPOSITED ON MAGNESIUM OXIDE. /ENERGY ELECTR 5-1522
(AUGER SPECTROSCOPY)   GROWTH MODE OF IRON FILM DEPOSITED ON MAGNESIUM OXIDE (001) SURFACE. 5-1528
R SYS/ ULTRAHIGH VACUUM LOW TEMPERATURE STAGE FOR IN-SITU FILM DEPOSITION AND CHARACTERIZATION. (LEED- AUGE 3-0657
     CHARACTERIZATION OF A CHROMIUM- GOLD THIN FILM DEPOSITION PROCESS BY AUGER ANALYSIS.   1-0037
AUGER SCANNING ELECTRON MICROSCOPIC STUDY OF THE PASSIVE FILM FORMED ON AN IRON- CHROMIUM ALLOY.   5-1753
     SECONDARY ELECTRON EMISSION FROM THIN FILM GOLD METAL- OXIDE CERMETS.   4-1015
TUDIED WITH AUGER SPECTROSCOPY.   INTERDIFFUSIONS IN THIN FILM GOLD ON PLATINUM ON GALLIUM ARSENIDE (100) S 5-1232
CRYSTAL SURFACES OF ALKALI HALIDES AND THIN FILM GROWTH OBSERVED BY LEED AND AES.   5-1907
     AUGER EXAMINATION OF CONTAMINANTS IN THIN FILM METALLIZATIONS.   5-1890
CT)   EXPERIMENTAL PROBLEMS OF STUDYING THIN FILM NUCLEATION. (ELECTRON MICROSCOPY, AUGER EFFE 5-1408
SCOPY AND SECONDA/ COMPOSITION PROFILE OF ION PLATED GOLD FILM ON COPPER ANALYZED BY AUGER ELECTRON SPECTRO 5-1679
ERGY ELECTRON DIFFRACTION AND AUGER ELECT/ GROWTH OF GOLD FILM ON GALLIUM ARSENIDE (110) OBSERVED BY LOW EN 5-1876
ROSCOPY.   PASSIVE FILM ON IRON. APPLICATION OF AUGER ELECTRON SPECT 5-1761
ROSCOPY.   DEGRADED PASSIVE FILM ON IRON. APPLICATION OF AUGER ELECTRON SPECT 5-1762
ED- AES OBSERVATION OF GROWTH MECHANISM OF DEPOSITED IRON FILM ON MAGNESIUM OXIDE (001) SURFACE.   LE 5-1525
MISSION CHARACTERISTICS (GERMANIUM AND BERYLLIUM).   FILM STRUCTURE EFFECT ON THE SECONDARY ELECTRON E 4-0923
- TANTALUM, AND SILICON- PLATINUM SILICIDE- TANTALUM THIN FILM STRUCTURES. (AES) /TANTALUM, SILICON DIOXIDE 5-1260
. ANALYSIS OF GRAIN BOUNDARY DIFFUSION IN BIMETALLIC THIN FILM STRUCTURES USING AUGER ELECTRON SPECTROSCOPY 6-2146
     STRUCTURE DEPENDENT PROPERTIES OF THIN FILM SUPERCONDUCTORS.   5-1300
VIOLET PHOTOELECTRON STUDIES OF ADSORPTION ON CLEAN METAL FILM SURFACES. (AUGER SPECTRA) / AND VACUUM ULTRA 5-1203
     AUGER SPECTROSCOPY OF A BARIUM- METAL FILM SYSTEM.   4-0849
     AUGER SPECTROSCOPY OF THE BARIUM- PLATINUM FILM SYSTEM.   5-1357
ON OF GOLD INTO SILICON IN THE SILICON (SUBSTRATE)- GOLD (FILM) SYSTEM.   LOW TEMPERATURE DIFFUSI 5-1673
ANGES W/ DIFFUSION MECHANISMS IN THE PALLADIUM- GOLD THIN FILM SYSTEM AND THE CORRELATION OF RESISTIVITY CH 5-1393
IFFUSE INTERFACE IN SILICON (SUBSTRATE)- GOLD (EVAPORATED FILM) SYSTEM. (AUGER ELECTRON SPECTROSCOPY)   D 5-1681
OF THE PALLADIUM BARIUM ALLOY CATHODE IN RELATION TO ITS FILM SYSTEM. (AUGER EMISSION) /RISTICS OF SURFACE 5-1129
/TERFACE STRUCTURES IN SOME (SUBSTRATE- VACUUM EVAPORATED FILM) SYSTEMS STUDIED BY AUGER ELECTRON SPECTROS/ 5-1436
ER ELECTRONS.   NEW METHOD OF THIN FILM THICKNESS MEASUREMENT USING RADIOISOTOPE AUG 5-1658
RON SPECTROSCOPY.   DETERMINATION OF THIN FILM THICKNESS OR ION ETCH RATE USING AUGER ELECT 5-1984
THE SILICIDE FORMATION IN 85% NICKEL- 15% PLATINUM ALLOY FILMS.   AES STUDY OF 5-1899
ATION OF AES TO THE STUDY OF SELECTIVE SPUTTERING OF THIN FILMS.   APPLIC 5-1933
NE IN HYDROGEN CHLORIDE OR CHLORINE GROWN SILICON DIOXIDE FILMS.   AUGER ANALYSIS OF CHLORI 5-1251
ECONDARY ELECTRON SPECTRA FROM THIN EVAPORATED PHOSPHORUS FILMS. /ER AND OTHER CHARACTERISTIC ENERGIES IN S 4-0986
ECTROSCOPICAL STUDIES OF THE COMPOSITION OF PASSIVE STATE FILMS. (AUGER ELECTRON AND MOSSBAUER) SP 5-1309
     AUGER ELECTRON SPECTRA OF MAGNESIUM THIN FILMS.   4-0873
TERMINATION OF THE OXYGEN- SILICON RATIO IN SPIN ON GLASS FILMS.   AUGER ELECTRON SPECTROSCOPY DE 5-1827
AUGER ELECTRON SPECTROSCOPY OF CLEAN POLYCRYSTALLINE GOLD FILMS.   5-1760
     AUGER SPECTROSCOPIC ANALYSIS OF BIOGLASS CORROSION FILMS.   5-1264
     AUGER SPECTROSCOPY ANALYSIS OF PALLADIUM SILICIDE FILMS.   5-1898
     AUGER STUDIES OF ELECTROLESS NICKEL FILMS.   4-0760

ECONDARY ELECTRON EMISSION FROM POROUS POTASSIUM CHLORIDE FILMS.   CHARACTERISTICS OF SHAPING S 4-1120
CHEMICAL SHIFTS IN THE AUGER SPECTRA OF PASSIVE FILMS.   4-1056
AND AUGER ELECTRON SPECTROSCOPY, FOR THE ANALYSIS OF THIN FILMS. /ATION OF SECONDARY ION MASS SPECTROMETRY 5-1214
IER HEIGHTS OF SILICIDES FORMED IN NICKEL- PLATINUM ALLOY FILMS.   COMPOSITION PROFILES AND SCHOTTKY BARR 5-1900
DITIONS ON THE NUCLEATION AND GROWTH OF EPITAXIAL SILICON FILMS.   INFLUENCE OF SUBSTRATE CON 5-1514
RYSTALS AND MAGNESIUM OXIDE AND MAGNESIUM NITRIDE SURFACE FILMS. /GER CHARACTERISTICS OF MAGNESIUM SINGLE C 5-1220
TECTIC FORMATION IN TUNGSTEN PLATINUM- TUNGSTEN GOLD FILMS. /ECTROSCOPY STUDY OF INTERDIFFUSION AND FII 5-1259
SECONDARY ELECTRON EMISSION FROM THIN FILMS.   4-0894
CONDARY ELECTRON EMISSION OF AMORPHOUS SEMICONDUCTOR THIN FILMS.   SE 5-1239
SELECTED AREA AND IN-DEPTH AUGER ANALYSIS OF THIN FILMS.   5-1653
IMATION: LEED- AUGER STUDIES AND ELECTRONIC PROPERTIES OF FILMS. SILICON ON SAPPHIRE EPITAXY BY VACUUM SUBL 5-1230
SOME RECENT STUDIES OF VERY THIN SILICON DIOXIDE FILMS.   5-1935
SURFACE IMPURITIES IN SILVER HALIDE FILMS.   5-1149
UGER SPECTROSCOPY STUDY OF VACUUM DEPOSITED SILVER HALIDE FILMS.   UPS, ESCA, AND A 5-1947
SPECTROSCOPY TO DETERMINE THE STRUCTURE OF SILICON OXIDE FILMS.   USE OF AUGER ELECTRON 5-1501
ER(II) OXIDE, COPPER(I) OXIDE, AND COPPER(I) SULFIDE THIN FILMS. /LECTRON AND AUGER SPECTRA OF COPPER, COPP 4-0932
HOTOELECTRON AND AUGER ELECTRON SPECTRA OF MAGNESIUM THIN FILMS.   X-RAY P 5-1392
ITH AUGER SP/ DIFFUSION KINETICS OF GOLD THROUGH PLATINUM FILMS ABOUT 2000 AND 600 ANGTROMS THICK STUDIED W 5-1235
ELECTRONIC AND LATTICE STRUCTURE OF CESIUM FILMS ADSORBED ON TUNGSTEN.   5-1604
S/ TWO-DIMENSIONAL PHASE TRANSITION IN XENON SUBMONOLAYER FILMS ADSORBED ON (0001) GRAPHITE. (AUGER SPECTRO 5-1858
E, AND MAGNETIC PROPERTIES OF NICKEL- IRON- CHROMIUM THIN FILMS. (AES)   CORROSION, SURFAC 5-1766
/ DIFFERENTIAL SPUTTERING OF MAGNESIUM OXIDE- GOLD CERMET FILMS AND ITS APPLICATION TO HIGH YIELD SECONDARY 5-1430
AND DECOMPOSITION OF ACETYLENE ON POLYCRYSTALLINE RHENIUM FILMS AND RIBBONS. (AUGER SPECTROSCOPY) /ORPTION 5-1588
SURFACE. (A/ LEED INVESTIGATIONS OF XENON SINGLE CRYSTAL FILMS AND THEIR USE IN STUDYING THE IRIDIUM (100) 5-1479
ION LENGTHS IN METALS: TRANSMISSION THROUGH THIN ADSORBED FILMS. (AUGER EFFECT) /ATION OF ELECTRON ATTENUAT 4-1088
EFFECT OF IMPURITIES ON INTRINSIC STRESS IN THIN NICKEL FILMS. (AUGER EFFECT)   5-1131
CAL PROPERTIES OF RF SPUTTERED YTTRIA-DOPED ZIRCONIA THIN FILMS. (AUGER EFFECT)   MORPHOLOGICAL AND ELECTRI 5-1385
SURFACE STUDIES WITH EPITAXIALLY GROWN METAL FILMS. (AUGER EFFECT)   5-1253
D OXYGEN ON THE PHYSICAL PROPERTIES OF SPUTTERED TANTALUM FILMS. (AUGER ELECTRON SPECTROSCOPY) /, CARBON AN 5-1473
GROWTH AND STRUCTURE OF SEMICONDUCTING THIN FILMS. (AUGER ELECTRON SPECTROSCOPY) 5-1509
POROSITY IN SPUTTERED PLATINUM FILMS. (AUGER ELECTRON SPECTROSCOPY) 5-1965
N NICKEL (111), NICKEL (100), SHEET AND EVAPORATED NICKEL FILMS. (AUGER ELECTRON SPECTROSCOPY) /GEN ON CLEA 5-1468
SURFACE ENRICHMENT OF INDIUM IN EVAPORATED GOLD- INDIUM FILMS. (AUGER ELECTRON SPECTROSCOPY) 5-1892
/BETWEEN SUBSTRATE SURFACE CHEMISTRY AND ADHESION OF THIN FILMS. (AUGER ELECTRON SPECTROSCOPY OF ALUMINA S/ 5-1851
ELECTROCHEMISTRY AT THIN SOLID FILMS. (AUGER SPECTRA) 4-1026
EFFECT OF IMPURITIES ON INTRINSIC STRESS IN THIN NICKEL FILMS. (AUGER SPECTROSCOPY) 5-1130
PURITY AND MORPHOLOGY OF ALUMINUM FILMS. (AUGER SPECTROSCOPY) 5-1299
RESIDUAL GAS AND THE OPTICAL PROPERTIES OF SILVER FILMS. (AUGER SPECTROSCOPY) 5-1700
TECHNIQUES FOR ELEMENTAL COMPOSITION PROFILING IN THIN FILMS. (AUGER SPECTROSCOPY) 5-1273
CLEANING FOR THERMOCOMPRESSION BONDING. (REMOVAL OF OXIDE FILMS, AUGER SPECTROSCOPY OF SURFACE) CHEMICAL 5-1454
ITROGEN IN HIGH CRITICAL TEMPERATURE NIOBIUM (3) GERMANIUM FILMS BY AUGER ANALYSIS. DETECTION OF SURFACE N 5-1804
ICON OXYNITRIDE LAYERS ON THE SURFACES OF SILICON NITRIDE FILMS BY AUGER ELECTRON EMISSION. /TECTION OF SIL 5-1611
DIFFUSION STUDIES IN GOLD- PLATINUM THIN FILMS BY AUGER ELECTRON SPECTROSCOPY. 5-1636
GROWTH MODE DISCRIMINATION OF THIN FILMS BY AUGER ELECTRON SPECTROSCOPY. 5-1624
THICKNESS DETERMINATION OF ULTRATHIN FILMS BY AUGER ELECTRON SPECTROSCOPY. 5-1449
/NTS (NITROGEN, CARBON, AND OXYGEN) IN SPUTTERED TANTALUM FILMS BY AUGER ELECTRON SPECTROSCOPY AND SECONDA/ 5-1652
OXIDATION STUDIES OF AMORPHOUS AND CRYSTALLINE GERMANIUM FILMS BY AUGER SPECTROSCOPY. 6-2023
DIFFUSION MEASUREMENT IN THIN FILMS BY AUGER SPUTTER ETCHING TECHNIQUE. 5-1439
/ERMINATION OF SURFACE OXIDE THICKNESS ON DEPOSITED METAL FILMS BY COMBINATION AUGER SPECTROSCOPY AND INER/ 6-2048
LECTRON SPECTROSCOP/ LOW TEMPERATURE EPITAXY OF GERMANIUM FILMS BY SPUTTER DEPOSITION. (HEED, LEED, AUGER E 5-1537
AUGER ANALYSIS OF SILICON THIN FILMS DEPOSITED ON CARBON AT HIGH TEMPERATURES. 5-1228
TRANSPORT OF OXYGEN IN AMORPHOUS GERMANIUM THIN FILMS DURING ANNEALING. (AUGER SPECTROSCOPY) 5-1548
SCOPY.   OXIDE FILMS: ELEMENT PROFILES BY AUGER ELECTRON SPECTRO 4-1073
SURFACE ABSORPTION BY POLYPEPTIDE FILMS EXAMINED BY AUGER EMISSION SPECTROSCOPY. 5-1179
LOGENATED POLYMERS.   AUGER ANALYSIS OF FILMS FORMED ON METALS IN SLIDING CONTACT WITH HA 5-1728
AUGER SPECTROMETRY STUDY OF PASSIVE FILMS FORMED ON STAINLESS STEELS. 6-2102
TRANSITION METAL SUPERCONDUCTORS AT SURFACES AND IN THIN FILMS. (LEED- AES STUDIES OF NIOBIUM SURFACES) /F 5-1848
ELECTRON SPECTROSCOPY.   THIN FILMS. LOW ENERGY ELECTRON DIFFRACTION AND AUGER 1-0100
INTERDIFFUSION IN BIMETALLIC POLYCRYSTALLINE FILMS MEASURED BY AUGER ION SPUTTERING TECHNIQUE. 5-1440
/URFACE COMPOSITION AND CHEMISTRY OF EVAPORATED PERMALLOY FILMS OBSERVED BY X-RAY PHOTOEMISSION SPECTROSCO/ 5-1738
TION OF AN EQUILIB/ DIFFUSION OF SILVER THROUGH EPITAXIAL FILMS OF COPPER AND NICKEL ON SILVER (111): FORMA 5-1356
SECONDARY ELECTRON EMISSION FROM THIN FILMS. (OF GOLD, APPARATUS) 3-0638
LECTRON SPECTROSCOPY. STUDIES ON SURFACE PLASMA OXIDIZED FILMS OF III-V COMPOUND SEMICONDUCTORS BY AUGER E 5-1699
AES CHARACTERIZATION OF OXIDIZED FILMS OF MAGNESIUM, ALUMINUM, AND SILICON. 5-1847
/OMBINATION PROCESSES (AUGER) ON THE SURFACES AND IN THIN FILMS OF SILICON AND GERMANIUM SUBJECTED TO LASE/ 5-1990
/NDUCTING TRANSITION TEMPERATURES OF REACTIVELY SPUTTERED FILMS OF TANTALUM NITRIDE AND TUNGSTEN NITRIDE. / 5-1541
OF SLOW ELECTRONS AND AUGER SPECTROSCOP/ STUDY OF BARIUM FILMS ON A (110) TUNGSTEN FACE BY THE DIFFRACTION 5-1368
NYL CHLORIDE ON AN IRON (011) SURFACE AND EFFECT OF THESE FILMS ON ADHESION. /TION OF ETHYLENE OXIDE AND VI 5-1207
AUGER ELECTRON SPECTROSCOPIC ANALYSIS OF ANODIC FILMS ON ALUMINUM. 5-1614
D- AUGER TECHNIQUES.   EPITAXY OF ULTRATHIN METAL FILMS ON BODY CENTERED CUBIC SUBSTRATES USING LEE 5-1488
AY PHOTOELECTRON SPECTROSCOPI/ ANALYSIS PROFILES OF OXIDE FILMS ON CHROMIUM STEEL BY AUGER EMISSION AND X-R 6-2016
/ SEVERAL CONTAMINATED AND CLEANED SURFACES OF GOLD THICK FILMS ON COPPER PLATES BY AUGER ELECTRON AND SEC/ 5-1550
SPECTROSC/ ANALYSIS OF THE SURFACE COMPOSITION OF PASSIVE FILMS ON COPPER- NICKEL ALLOYS BY AUGER ELECTRON 5-1874
REACTION OF SPUTTERED PLATINUM FILMS ON GALLIUM ARSENIDE. (AUGER SPECTROSCOPY) 5-1563
ELECTRON IRRADIATION. (AUGER SP/ GROWTH OF THIN METALLIC FILMS ON INSULATING SUBSTRATES IN THE PRESENCE OF 5-1317
AES ANALYSIS OF OXIDE FILMS ON IRON. 5-1800
AUGER ANALYSIS OF THICK OXIDE FILMS ON LEAD- INDIUM ALLOYS. 5-1249
CONTAMINATION EFFECT ON THE GROWTH MODE OF IRON EPITAXIAL FILMS ON MAGNESIUM OXIDE (001). /TUDY OF SURFACE 5-1524
/ELECTRICAL PROPERTIES OF ALLOYED NICKEL- GOLD- GERMANIUM FILMS ON N-TYPE GALLIUM ARSENIDE. (BY AUGER SPEC/ 6-2101
EFFECT OF STRUCTURE ON THE AUGER SPECTRA OF BARIUM FILMS ON PLATINUM. 5-1359
AUGER ELECTRON SPECTROSCOPY ANALYSIS OF ION PLATED GOLD FILMS ON SILICON. 5-1225
.   SILICON DIFFUSION THROUGH THIN TUNGSTEN FILMS ON SILICON, STUDIED WITH AUGER SPECTROSCOPY 5-1234
Y.   STUDY OF PALLADIUM SILICIDE FILMS ON SILICON USING AUGER ELECTRON SPECTROSCOP 5-1336
AUGER ANALYSIS OF SURFACE FILMS ON SILVER- TIN. 5-1386
A/ EFFECT OF SULFUR, OXYGEN, AND HYDROGEN SULFIDE SURFACE FILMS ON THE ADHESION OF CLEAN IRON. (USING LEED 5-1211
OSCOPY TO THE DETERMINATION OF THE COMPOSITION OF PASSIVE FILMS ON TYPE 316 STAINLESS STEEL. /ECTRON SPECTR 6-2080
GROWTH AND STRUCTURE OF THIN COPPER FILMS ON (001) SURFACES OF NICKEL. 5-1227
GROWTH AND STRUCTURE OF TITANIUM OXIDE THIN FILMS. PART-1. (AUGER SPECTRA) 5-1980
./ PROPERTIES OF THE LEAD SULFIDE- LEAD OXIDE CRYSTALLINE FILMS. PART-1: PREPARATION AND PHYSICAL STRUCTURE 5-1790
ELECTRON SPECTROSCOPY ON THIN FILMS. (SCANNING AUGER MICROSCOPE) 5-1405

IZATION OF A MOSSBAUER SPECTROMETER FOR SURFACES AND THIN FILMS STUDY. (AUGER)   REAL 3-0690
LOW INTRINSIC STRESS AND GOOD ADHESION ALLOY THIN FILMS. (TITANIUM AND CHROMIUM ALLOYS, AES) 5-1909
DIFFUSION STUDIES IN CHROMIUM- PLATINUM THIN FILMS USING AUGER ELECTRON SPECTROSCOPY. 5-1286
STUDY OF DIFFUSION IN THIN COPPER- LEAD FILMS USING AUGER ELECTRON SPECTROSCOPY. 5-1972
NTO GOLD. (AUGER ELECTRON/ DIFFUSION MEASUREMENTS IN THIN FILMS UTILIZING WORK FUNCTION CHANGES: CHROMIUM I 5-1889
ON SPECTROSCOPY STUDY.   SILICON HOMOEPITAXIAL THIN FILMS VIA SILANE PYROLYSIS: HEED AND AUGER ELECTR 5-1428
LECTRON DIFFRACTION- AUGER ELECTRON SPECTROSCPPY AND MASS FILTER. /F METAL ON THEM OBSERVED BY LOW ENERGY E 5-1906
AUGER ELECTRON FINE PROFILES IN IONIC CRYSTALS. 4-0949
ONSISTENT ANALYSIS OF CORE BAND-TO-BAND EXCITATION: AUGER FINE STRUCTURE.   SELF-C 2-0509
F GERMANIUM.   FINE STRUCTURE DETAILS IN THE LMM AUGER SPECTRA O 4-0937
FROM SOLID SURFACES.   NEW FINE STRUCTURE IN ELECTRON EXCITED AUGER SPECTRA 2-0473
D SILICON CARBIDE.   HIGH ENERGY FINE STRUCTURE IN THE AUGER SPECTRA OF SILICON AN 4-1036
ER AND INDIUM.   QUASIATOMIC FINE STRUCTURE IN THE AUGER SPECTRA OF SOLID SILV 4-0771
A NICKEL SURFACE.   ORIGIN OF FINE STRUCTURE IN THE AUGER SPECTRUM OF SULFUR ON 5-1271
YIELD AND THE APPLICATION TO SILIC/ METHOD FOR DETECTING FINE STRUCTURE IN THE SECONDARY ELECTRON EMISSION 4-0852
OBSERVATION OF A PLASMON GAIN IN THE FINE STRUCTURE OF THE ALUMINUM AUGER SPECTRUM. 4-1085
ECONDARY ELECTRON EMISSION COEFFICIENT OF/ CHANGES IN THE FINE STRUCTURE OF THE ANGULAR DEPENDENCE OF THE S 4-1061
NT OF SEC/ USE OF X-RAY SPECTROSCOPIC DATA TO EXPLAIN THE FINE STRUCTURE OF THE DEPENDENCE OF THE COEFFICIE 4-0891
ARY ELECTRON EMISSION FRCM SILICON (111). (AUGER ELECTRO/ FINE STRUCTURES AND ENERGY DISTRIBUTION OF SECOND 4-0853
AES STUDIES ON SURFACES OF 18-8 STAINLESS STEEL FINISHED MECHANICALLY AND ELECTROLYTICALLY. 6-2129
SPECTROSCOPY)   LASER WINDOW SURFACE FINISHING AND COATING TECHNOLOGY. (AUGER ELECTRON 1-0015
R OF THE CYLINDRICAL/ FOCUSING A CHARGED PARTICLE BEAM OF FINITE ANGULAR SPREAD IN AN ELECTROSTATIC ANALYZE 3-0720
GER DECAY.   DETERMINATION OF OSCILLATOR STRENGTH FOR THE FIRST DISCRETE TRANSITION IN NITROGEN FROM THE AU 2-0575
GRAPHITE FACE.   THERMODYNAMICS AND KINETICS OF THE FIRST MONOLAYER ADSORPTION OF XENON ON THE (0001) 5-1859
NALYSIS IN THE RANGE 100 TO 1000 EV USING THE CYLINDRICA/ FIRST ORDER APPROXIMATION TO QUANTITATIVE AUGER A 3-0660
ORBED ON THE (0001) FACE OF GRAPHITE. (AUGER ELECTRON SP/ FIRST ORDER TWO DIMENSIONAL TRANSITION: XENON ADS 5-1857
LMM AUGER SPECTRA OF SOME TRANSITION METALS OF THE FIRST SERIES. 4-0810
UNG AN/ UNEXPLAINED PHENOMENA IN THE AUGER SPECTRA OF THE FIRST TRANSITION SERIES. (COMMENTS ON PAPER BY CH 4-0812
EMISSION SPECTROSCOPY USING RETARDING / AVERAGING, CURVE FITTING, AND RELATED TECHNIQUES IN AUGER ELECTRON 3-0610
GER) ELECTRONS RESULTING FROM ION- ATOM COLLISIONS WITH A FIXED IMPACT PARAMETER. /ERGY DISTRIBUTION OF (AU 4-0331
G AUGER MICROPROB/ CONFIRMATION OF IMPURITY ADSORPTION AT FLAKE IRON INTERFACES IN GRAY CAST IRON. (SCANNIN 6-2054
BURIZED AND GRAPHITIZED NICKEL (110) USING AES, LEED, AND FLASH DESORPTION. /POSITION OF FORMIC ACID ON CAR 5-1629
/NED LOW ENERGY ELECTRON DIFFRACTION, AUGER ELECTRON, AND FLASH DESORPTION SPECTROSCOPY OF METALS ON SINGL/ 1-0008
SCREPANCY FOR CARBON DIOXIDE ON TUNGSTEN (100).   FLASH DESORPTION- AUGER ELECTRON SPECTROSCOPY. DI 5-1461
NEW METHODS FOR THE DETECTION AND ELIMINATICN OF COATING FLAWS. 5-1592
AUGER ELECTRON SPECTROSCOPY STUDY OF FLOAT GLASS SURFACES. 5-1236
/ MASS TRANSPORT OF STAINLESS STEEL CORROSION PRODUCTS IN FLOWING LIQUID SODIUM. (SCANNING ELECTRON MICROSC 6-2014
AUGER ELECTRON SPECTROSCOPY OF MATERIAIS EXPOSED TO FLOWING SODIUM. 5-1460
SPECTROSCOPY.   K- AND L-SHELL FLUORESCENCE AND AUGER YIELDS AND AUGER ELECTRON 4-0966
L- AND M-SHELL AND SUBSHELL FLUORESCENCE AND COSTER-KRONIG YIELDS. 4-0959
AND NEPTUNIUM.   X-RAY SPECTRA, L-SUBSHELL FLUORESCENCE AND COSTER-KRONIG YIELDS IN BISMUTH 4-1112
L AND SOME OF ITS COMPOUNDS.   X-RAY FLUORESCENCE L SPECTRA OF NIOBIUM L META 4-0762
FLUORESCENCE PHOTON AND AUGER ELECTRON SPECTRA. 4-0799
STRIPPING MECHANISM DEPENDENCE OF THE ARGON 2S FLUORESCENCE YIELD. (AUGER TRANSITION) 4-0957
MISSION (AUGER) SP/ CHARGE STATE DEPENDENCE OF THE NEON K FLUORESCENCE YIELD DEDUCED FROM HIGH RESOLUTION E 2-0520
LLISIONS. (AUGER ELECTRON PRODUCT/ K-SHELL IONIZATION AND FLUORESCENCE YIELD IN 50 MEV CHLORINE(+)- NEON CO 2-0518
/E CHARGE STATE DEPENDENCE OF NEON K-SHELL IONIZATION AND FLUORESCENCE YIELD IN 50 MEV CHLORINE(N+) ION- N/ 2-0300
)   FLUORESCENCE YIELD OF CARBON. (AUGER MEASUREMENTS 5-1571
TE)   K-SHELL FLUORESCENCE YIELD OF IONIZED ALUMINUM. (AUGER RA 4-0975
GER EFFECT)   MULTIPLET EFFECTS ON THE L2,3 FLUORESCENCE YIELD OF MULTIPLY IONIZED ARGON. (AU 4-0956
YIELDS)   EFFECT OF MULTIPLE IONIZATION ON THE FLUORESCENCE YIELD OF NEON. (K-AUGER AND K X-RAY 2-0299
MOLECULAR / CHARGE STATE DEPENDENCE OF THE ARGON L-SHELL FLUORESCENCE YIELD STUDIED BY ATOMIC HYDROGEN(+), 2-0517
ION PROBABILITIES TO THE 1S STATE AND THEORETICAL K-SHELL FLUORESCENCE YIELDS.   ATOMIC RADIATION TRANSIT 4-0918
N PROBABILITIES TO THE ATOMIC 2S STATE AND THEORETICAL l1 FLUORESCENCE YIELDS. /AND COSTER-KRONIG TRANSITIO 2-0329
E IONIZATION EFFECTS ON RADIATIVE RATES, AUGER RATES, AND FLUORESCENCE YIELDS.   DOUBL 4-0976
REVIEW OF FLUORESCENCE YIELDS AND AUGER ELECTRON MISSION. 1-0121
IONIZED NITROGEN.   FLUORESCENCE YIELDS AND AUGER RATES FOR MULTIPLY 4-0781
PROBABILITIES F23 BASED ON THE G/ THEORETICAL L2 SUBSHELL FLUORESCENCE YIELDS AND COSTER-KRONIG TRANSITION 2-0313
ANSITION PROBABILITIES.   THEORETICAL L2 AND L3 SUBSHELL FLUORESCENCE YIELDS AND L2-L3(X) COSTER-KRONIG TR 2-0316
NSITION PROBABILITIES.   X-RAY FLUORESCENCE YIELDS, AUGER, AND COSTER-KRONIG TRA 2-0259
IC M-SHELL COSTER-KRONIG, AUGER, AND RADIATIVE RATES, AND FLUORESCENCE YIELDS FOR CALCIUM TO THORIUM. ATOM 4-0964
BY ELECTRON IMPACT. (/ X-RAY EMISSION CROSS SECTIONS AND FLUORESCENCE YIELDS FOR LIGHT ATOMS AND MOLECULES 2-0528
K-SHELL FLUORESCENCE YIELDS FOR LIGHT ELEMENTS. 4-0818
/HEORETICAL K-SHELL AUGER RATES, TRANSITION ENERGIES, AND FLUORESCENCE YIELDS FOR MULTIPLY IONIZED NEON. (/ 2-0284
THEORETICAL FLUORESCENCE YIELDS FOR NEON. (AUGER RATE) 2-0283
MIC L-SHELL COSTER-KRONIG, AUGER, AND RADIATIVE RATES AND FLUORESCENCE YIELDS FOR SODIUM- THORIUM. ATO 4-0963
NONRELATIVISTIC AUGER RATES, X-RAY RATES, AND FLUORESCENCE YIELDS FOR THE K-SHELL. 2-0556
NONRELATIVISTIC AUGEF RATES, X-RAY RATES, AND FLUORESCENCE YIELDS FOR THE 2P SHELL. 2-0558
AUGER SPECTRA AND FLUORESCENCE YIELDS FOR THE 3P AND THE 3D SHELL. 2-0278
NONFELATIVISTIC FLUORESCENCE YIELDS FOR THE 3P AND THE 3D SHELLS. 2-0277
IC N-SHELL COSTER-KRCNIG, AUGER, AND RADIATIVE RATES, AND FLUORESCENCE YIELDS FOR Z BETWEEN 38 AND 103. /OM 4-0969
S, K-, L-, AND M-SHELL FLUORESCENCE YIELDS)   ATOMIC FLUORESCENCE YIELDS. (L-SHELL COSTER-KRONIG YIELD 2-0351
TRON IMPACT. (AUGER EMISSION) MEASUREMENTS OF THE L-SHELL FLUORESCENCE YIELDS OF ARGON AND CHLORINE BY ELEC 2-0387
. (AUGER TRANSITION)   K-SHELL FLUORESCENCE YIELDS OF ARGON, CHLORINE AND SULFUR 4-1014
EAD.   L-SHELL FLUORESCENCE YIELDS OF DOUBLE VACANCY STATES IN L 4-1105
UGER EFFECT)   MULTIPLET EFFECT IN FLUORESCENCE YIELDS OF MULTIPLY IONIZED ATOMS. (A 2-0314
F ARGON.   K-SHELL AUGER RATES, TRANSITION ENERGIES, AND FLUORESCENCE YIELDS OF VARIOUSLY IONIZED STATES O 2-0280
K-SHELL AUGER TRANSITION RATES AND FLUORESCENT YIELDS FOR ELEMENTS ARGON TO XENON. 4-0961
N.   K-SHELL AUGER TRANSITICN RATES AND FLUORESCENT YIELDS FOR ELEMENTS BERYLLIUM TO ARGO 4-0962
K-SHELL (AUGER) FLUORESCENT YIELDS FOR MULTIPLY IONIZED NEON. 4-0994
ILAYER DIELECTRIC MIRRCR DESIGNS. (ZINC SELENIDE, THORIUM FLUORIDE)   AUGER ANALYSIS OF THREE ENHANCED MULT 5-1328
AUGER EMISSION AND COLOR CENTERS IN LITHIUM FLUORIDE. 2-0430
AUGER SPECTRUM OF INSULATORS. (MAGNESIUM OXIDE, MAGNESIUM FLUORIDE) /ON EFFECTS IN THE KL2,3L2,3 RADIATIVE 2-0540
LCW ENERGY AUGER EMISSION FROM LITHIUM FLUORIDE. 4-0842
ECTRA IN HYDROGEN SULFIDE, SULFUR DICXIDE, AND SULFUR HEXAFLUORIDE.   SULFUR KLL AUGER SP 4-0910
ELECTRON SPEC/ ANALYSIS OF GRAIN BOUNDARY IMPURITIES AND FLUORIDE ADDITIVES IN HOT PRESSED OXIDES BY AUGER 5-1506
/F SURFACE CISSOCIATION IN POTASSIUM CHLCRIDE AND LITHIUM FLUORIDE BY OBSERVATION OF PLASMON LOSSES ASSOCI/ 5-1279
N MONOXIDE, NITRIC OXIDE, CARBON DIOXIDE, ANC CARBON TETRAFLUORIDE. (KLL AUGER PROCESS) /GEN, OXYGEN, CARBO 2-0308
ECTRON SPECTRUM AND IONIZATION POTENTIALS OF THE HYDROGEN FLUORIDE MOLECULE.   AUGER EL 4-1058
KLL AUGER SPECTRUM OF FLUORINE. (IN FLUORIDE SALTS) 4-0734

N CALCULATIONS OF THE AUGER SPECTRUM OF METHANE, HYDROGEN FLUORIDE, WATER, AND CARBON MONOXIDE. /INTERACTIO 4-0879
WIDTH AND AUGER EFFECT IN CALCIUM, STRONTIUM, AND BARIUM FLUORIDES.   LINE 4-1116
LINEAR LIGAND FIELDS INFLUENCING AUGER SPECTRA OF FLUORIDES, COPPER(I) AND SILVER(I).   4-0905
AUGER SPECTRA OF HIGHLY IONIZED OXYGEN AND FLUORINE.   4-0906
ON PRODUCTION CROSS SECTIONS FROM ION BOMBARDMENT. (NEON, FLUORINE)   K-SHELL AUGER ELECTR 2-0574
ONS FROM 1.5 MEV PER AMU AND ATOMIC HYDROGEN, OXYGEN, AND FLUORINE.   NEON K-AUGER CROSS SECTI 2-0571
/CTRA FROM AUTOIONIZING STATES OF HIGHLY STRIPPED OXYGEN, FLUORINE, CHLORINE, AND ARGON IONS. (AUGER EFFEC/ 2-0472
KLL AUGER SPECTRUM OF FLUORINE. (IN FLUORIDE SALTS)   4-0734
/ITROGEN IN GALLIUM PHOSPHIDE (AND) NITROGEN, OXYGEN, AND FLUORINE IN SILICON BY SECONDARY ION MASS SPECTR/ 5-1916
ENIDE AS DETERMINED BY AES.   FLUORINE ION IMPLANTATION PROFILES IN GALLIUM ARS 5-1398
ARGON K-AUGER CROSS SECTIONS IN COLLISION WITH 30 MEV FLUORINE IONS.   2-0323
NEON K-SHELL VACANCY PRODUCTION BY INTERMEDIATE ENERGY FLUORINE IONS. (AUGER ELECTRON EMISSION)   2-0572
R SPREAD IN AN ELECTROSTATIC ANALYZER OF THE CYLINDRICAL/ FOCUSING A CHARGED PARTICLE BEAM OF FINITE ANGULA 3-0720
ENSER.   FOCUSING OF CHARGED PARTICLES BY A SPHERICAL COND 3-0684
TH CYLINDRICAL FIELD.   FOCUSING PROPERTIES OF AN ELECTROSTATIC MIRROR WI 3-0721
AUGER ELECTRON EMISSION SPECTRA FROM FOIL AND GAS EXCITED CARBON BEAMS.   4-1050
BEAM FOIL AUGER SPECTROSCOPY.   1-0112
BEAM FOIL EXCITED AUGER TRANSITIONS IN NEON.   4-0860
DOPPLER SHIFTED AUGER ELECTRONS FROM FOIL EXCITED HEAVY ION BEAMS.   4-0859
AUGER ELECTRONS FROM FOIL EXCITED HEAVY ION BEAMS. (OF NEON)   2-0370
MATERIAL ANALYSIS OF THIN FOILS BY AUGER ELECTRON SPECTROSCOPY.   5-1798
AUGER ELECTRON EMISSION OF THIN CARBON FOILS IN REFLECTION AND TRANSMISSION.   4-0893
RON SPECTROSCOPY AS A TOOL FOR MEASURING THE DIFFUSION OF FOREIGN ATOMS IN SOLIDS NEAR THEIR SURFACE. /LECT 1-0186
UGER ELECTRON SPECTROSCOPY.   IDENTIFICATION OF THE FORM OF CARBON AT A SILICON (100) SURFACE USING A 5-1378
GENERAL FORMALISM FOR QUANTITATIVE AUGER ANALYSIS.   1-0032
PART-2.   A FORMALISM FOR THE INDIRECT AUGER EFFECT. PART-1.   2-0385
TANTALUM, SILICON DIOXIDE- TANTALUM, AND SILICO/ SILICIDE FORMATION AND INTERDIFFUSION EFFECTS IN SILICON- 5-1260
ER SPECTROSCOPIC OBSERVATION OF SILICON- GOLD MIXED PHASE FORMATION AT LOW TEMPERATURES.   AUG 5-1680
N OF TUNGSTEN (110).   PART-2: LEED- AUGER STUDY OF OXIDE FORMATION AT 650-1100 DEGREES K.   OXIDATIO 6-1994
PLATINUM SILICIDE INTERFACE.   PLATINUM SILICIDE FORMATION. ELECTRON SPECTROSCOPY OF THE PLATINUM- 5-1285
/ AUGER SPECTROSCOPY STUDY OF INTERDIFFUSION AND EUTECTIC FORMATION IN TUNGSTEN PLATINUM- TUNGSTEN GOLD TH/ 5-1259
AES STUDY OF THE SILICIDE FORMATION IN 85% NICKEL- 15% PLATINUM ALLOY FILMS 5-1899
PALLADIUM SILICIDE FORMATION OBSERVED BY AUGER ELECTRON SPECTROSCOPY 5-1773
RON EMISSION.   FORMATION OF A PLASMA SHEATH WITH SECONDARY ELECT 2-0244
ROLE OF THE AUGER EFFECT IN THE FORMATION OF AN ELECTRON SHOWER SPECTRUM.   2-0504
/GH EPITAXIAL FILMS OF COPPER AND NICKEL ON SILVER (111): FORMATION OF AN EQUILIBRIUM LAYER OF SILVER. (AU/ 5-1356
ROCESSES. CHEMISORPTION OF AMMONIA, NITROGEN, AND NITRIDE FORMATION ON A MOLYBDENUM. /COPIES TO CATALYTIC P 5-1533
OPY.   INVESTIGATIONS OF THE OXIDE FORMATION ON METALS WITH AUGER ELECTRON SPECTROSC 5-1329
TURE PROTON IRRADIATION.   CARBIDE FORMATION ON NIOBIUM SURFACES DURING HIGH TEMPERA 5-1886
OBSERVATIONS OF BETA- SILICON CARBIDE FORMATION ON RECONSTRUCTED SILICON SURFACES.   5-1560
SES.   EFFECT OF OXIDIZING AMBIENTS ON PLATINUM SILICIDE FORMATION. PART-2: AUGER AND BACKSCATTERING ANALY 5-1182
RELATIVE ABUNDANCES AND RECOIL ENERGIES OF FRAGMENT IONS FORMED FROM THE X-RAY PHOTOIONIZATION OF NITROGE/ 2-0308
SITION PROFILES AND SCHOTTKY BARRIER HEIGHTS OF SILICIDES FORMED IN NICKEL- PLATINUM ALLOY FILMS.   COMPO 5-1900
R SCANNING ELECTRON MICROSCOPIC STUDY OF THE PASSIVE FILM FORMED ON AN IRON- CHROMIUM ALLOY.   AUGE 5-1753
TED POLYMERS.   AUGER ANALYSIS OF FILMS FORMED ON METALS IN SLIDING CONTACT WITH HALOGENA 5-1728
AUGER SPECTROMETRY STUDY OF PASSIVE FILMS FORMED ON STAINLESS STEELS.   6-2102
10). (LEED- AES)   FORMIC ACID DESORPTION FROM GRAPHITIZED NICKEL (1 5-1630
/DY OF THE KINETICS AND MECHANISM OF THE DECOMPOSITION OF FORMIC ACID ON CARBURIZED AND GRAPHITIZED NICKEL/ 5-1629
TROSCOPY)   IDENTIFICATION OF BUBBLE FORMING IMPURITIES IN DOPED TUNGSTEN. (AUGER SPEC 6-2107
AUGER ELECTRON SPECTROSCOPY OF CONTRAST FORMING LAYERS ON METALS.   6-2047
OMBINATION IN SILICON RECTIFIERS AND THYRISTORS LOADED IN FORWARD DIRECTION IN THE MIDDLE REGION. AUGER REC 2-0415
ED BY PROTON EXCITED X-RAY, AES ANALYSIS / CORRELATION OF FRACTIONAL MONOLAYER OXYGEN DETERMINATIONS OBTAIN 5-1682
/G OF ALPHA BRASS IN A TARNISHING AMMONIACAL ENVIRONMENT: FRACTOGRAPHY AND CHEMICAL ANALYSIS. (AUGER SPECT/ 6-2095
IES BY AUGER ELECTRON SPECTROSCOPY.   INTERGRANULAR FRACTURE AND IMPURITY SEGREGATION IN STEEL.   STUD 6-2127
GRAIN BOUNDARY SEGREGATION RELATING TO INTERGRANULAR FRACTURE. (AUGER ELECTRON SPECTROSCOPY)   6-2082
MECHANICAL TESTING: IN-SITU FRACTURE DEVICE FOR AUGER ELECTRON SPECTROSCOPY.   3-0659
STUDY OF INTERGRANULAR FRACTURE IN IRON USING AUGER SPECTROSCOPY.   6-2097
NITRIDE.   AUGER ELECTRON SPECTROSCOPY ON THE FRACTURE OF NICKEL TUNGSTEN MATERIALS AND SILICON 6-1445
AS APPLIED TO INTERF/ IMPURITIES, INTERFACES, AND BRITTLE FRACTURE. (REVIEW OF AUGER ELECTRON SPECTROSCOPY 6-2078
EGATION OF IMPURITIES IN METALS AND INTERGRANULAR BRITTLE FRACTURE. (STUDIED BY AES)   GRAIN BOUNDARY SEGR 6-2079
3340- STAINLESS STEEL.   AUGER FRACTURE SURFACE ANALYSIS OF A TEMPER EMBRITTLED 6-2085
EEL BY AUGER ELECTRON SPECTROSCOPY.   FRACTURE SURFACE ANALYSIS OF TEMPER EMBRITTLED ST 6-2084
LLOY STEELS. (IMPURITY GRAIN BOUNDARY SEGREGATION, AES OF FRACTURE SURFACES) /TEMPER EMBRITTLEMENT OF LOW A 6-2083
UCTURE AND SEGREGATION. AUGER SPECTROGRAPHIC ANALYSIS OF FRACTURE SURFACES.   GRAIN BOUNDARY STR 6-2118
MICROSCOPIC AUGER ELECTRON ANALYSIS OF FRACTURE SURFACES.   5-1598
TEMPER EMBRITTLEMENT OF LOW ALLOY STEELS. (AES OF FRACTURE SURFACES)   6-2062
YS. (AUGER ELECTRON SPECTROSCOPY) STUDY OF GRAIN BOUNDARY FRACTURE SURFACES IN DOPED TUNGSTEN- RHENIUM ALLO 6-2112
S STEEL.   AUGER ELECTRON SPECTROSCOPY OF FRACTURE SURFACES IN IRRADIATED TYPE 304 STAINLES 6-2001
AUGER SPECTROSCOPY OF FRACTURE SURFACES OF CERAMICS.   5-1616
BY AUGER ELECTRON SPECTROSCOPY.   STUDY OF FRACTURE SURFACES OF HOT PRESSED SILICON NITRIDE 5-1444
TEELS.   AUGER ELECTRON SPECTROSCOPY ANALYSIS OF THE FRACTURE SURFACES OF IRRADIATED PRESSURE VESSEL S 6-2113
AUGER SPECTROSCOPY OF FRACTURE SURFACES OF IRRADIATED STAINLESS STEEL.   6-2002
GER ELECTRON SPECTROSCOPY TO DETERMINE THE COMPOSITION OF FRACTURE SURFACES OF MINERALS.   USE OF AU 5-1307
A TECHNIQUE FOR DETERMINING THE ELEMENTAL COMPOSITION OF FRACTURE SURFACES PRODUCED BY CRACK GROWTH IN HY/ 6-2142
DOPED ALUMINA. (HIGH VACUUM AUGER ELECTRON SPECTROMETER, FRACTURED ALUMINA) /NDARY SEGREGATION IN MAGNESIA 5-1615
ON OF NITROGE/ RELATIVE ABUNDANCES AND RECOIL ENERGIES OF FRAGMENT IONS FORMED FROM THE X-RAY PHOTOIONIZATI 2-0308
EVIDENCE OF RADIATIVE AUGER TRANSITIONS IN THE FREE ARGON ATOM.   2-0403
EVIDENCE OF RADIATIVE AUGER TRANSITIONS IN THE FREE ARGON ATOM.   2-0404
FREE ATOM BEHAVIOR OF AUGER ELECTRONS IN SOLIDS.   2-0579
INNER ELECTRON BINDING ENERGIES AND AUGER ENERGIES IN FREE ATOMS FOR USE IN X-RAY SPECTROSCOPY.   4-0774
THE PRESENCE OF EL/ HIGH FREQUENCY EXCITON BREAKDOWN AND FREE CARRIER AND EXCITON KINETICS IN GERMANIUM IN 4-0946
STUDY OF SULFUR EMBRITTLED OXYGEN FREE COPPER. (AES, GRAIN BOUNDARY SEGREGATION) 6-2015
K X-RAY EMISSION EDGE SHAPES OF FREE ELECTRON METALS. (AUGER BROADENING)   2-0337
0 KEV ARGON(+) IONS.   AUGER SPECTRA OF SOME FREE ELEMENTS AND BINARY COMPOUNDS BOMBARDED BY 6 4-1106
THE K-AUGER SPECTRUM OF THE FREE MAGNESIUM ATOM.   4-0789
LEED, ENERGY LOSS SPECTROSCOPY)   INELASTIC MEAN FREE PATH FOR ELECTRONS IN BULK JELLIUM. (AUGER, 4-1059
K-AUGER TRANSITIONS OF THE FREE SODIUM ATOM.   4-0880
COPY ANALYSIS, STAIN/ SEGREGATION OF ALLOYING ELEMENTS TO FREE SURFACES DURING IRRADIATION. (AUGER SPECTROS 6-2091
HIGH RESOLUTION L2,3M4,5M4,5 AUGER SPECTRUM OF FREE ZINC ATOMS.   4-0733
L2,3MM AUGER ELECTRON SPECTRUM FROM FREE ZINC ATOMS.   4-0731

SECONDARY ELECTRON EMISSION AT SUPERHIGH FREQUENCIES. (COPPER- GERMANIUM JUNCTION). 2-0356
EXCITON KINETICS IN GERMANIUM IN THE PRESENCE OF EL/ HIGH FREQUENCY EXCITON BREAKDOWN AND FREE CARRIER AND 4-0946
NS ON THE SURFACE OF ALKALI HALIDE CRY/ EVALUATION OF THE FREQUENCY OF AUGER NEUTRALIZATION OF INERT GAS IO 6-1544
MERCIAL COBALT BASE AIRCRAFT TURBINE SHROUD AL/ ADHESION, FRICTION AND AUGER SPECTROSCOPY ANALYSIS OF A COM 6-2009
ATOMIC NATURE OF POLYMER- METAL INTERACTIONS IN ADHESION, FRICTION, AND WEAR. 5-1212
PHOSPHORUS ALLOYS. (AUGER S/ EFFECT OF PHOSPHORUS ON THE FRICTION AND WEAR CHARACTERISTICS OF COPPER- TIN- 5-1868
ANALYSIS) INFLUENCE OF SILICCN ON FRICTION AND WEAR OF IRON- COBALT ALLOYS. (AUGER 6-2011
GER SPECTROSCOPY. ELEMENTAL ANALYSIS OF FRICTION AND WEAR SURFACE DURING SLIDING USING AU 6-2013
N IN THE SURFACE LAYER OF COPPER- TIN ALLOY SYSTEM DUE TO FRICTION. (AUGER SPECTROSCOPY) /TION CHANGE OF TI 5-1867
MECHANISM OF FRICTION BY AES. (REVIEW) 1-0193
E. STUDY OF THE INTERMITTENT MOTION OF FRICTION IN STAINLESS STEEL BY AN AUGER MICROPROB 6-2133
(AUGER STUDIES OF) ADHESION, FRICTION, LUBRICATION IN VACUUM. 5-1209
ITH METALS AS STUDIED BY AUGER SPECTROSCOPY/ ADHESION AND FRICTION OF POLYTETRAFLUORO ETHYLENE IN CONTACT W 5-1198
OPPER- TI/ A REVIEW OF SURFACE SEGREGATION, ADHESION, AND FRICTION STUDIES PERFORMED ON COPPER- ALUMINUM, C 6-2033
N SURFACE SCIENCE TECHNIQUES. (AES, ELLIPSOMET/ ADHESION, FRICTION, WEAR, AND LUBRICATION RESEARCH BY MODER 1-0104
(AUGER) MICROSCOPICAL STUDY OF FRICTIONAL PROPERTIES OF MOLYBDENUM DISULFIDE. 5-1872
L OXIDATION ON SILICON- IRON ALLOY (100) BY MEANS OF WORK FUNCTION AND AUGER ELECTRON SPECTROSCOPY. /INITIA 5-1924
ATOMIC PERCENT SILICON- IRON ALLOY (1/ PHOTOELECTRIC WORK FUNCTION AND AUGER ELECTRON SPECTROSCOPY OF 5.59 5-1925
D SEMICONDUCTORS. WORK FUNCTION AND STRUCTURAL STUDIES OF ALKALI- COVERE 5-1958
N) MONITORED BY AUGER SPECTROSCOPY, ELLIPSOMETRY AND WORK FUNCTION CHANGE. /CES (TITANIUM, TUNGSTEN, SILICO 5-1828
TRON DIFFRACTION AUGER APPARATUS. MEASUREMENT OF WORK FUNCTION CHANGE IN A DISPLAY TYPE LOW ENERGY ELEC 3-0611
RGY ELECTRON DIFFRACTICN SPECTRA, AUGER SPECTRA, AND WORK FUNCTION CHANGE MEASUREMENTS. /TUDY USING LOW ENE 5-1291
/ THE (001) NICKEL SURFACE DURING THERMAL TREATMENT. WORK FUNCTION CHANGES AND AUGER ELECTRON SPECTROSCOPY/ 6-2069
TRON/ DIFFUSION MEASUREMENTS IN THIN FILMS UTILIZING WORK FUNCTION CHANGES: CHROMIUM INTO GOLD. (AUGER ELEC 5-1889
CTRON SPECTROSCOPY) WORK FUNCTION CHANGES ON CONTACT MATERIALS. (AUGER ELE 5-1395
BLE GAS METASTABLE ATOMS. WORK FUNCTION EFFECTS ON AUGER ELECTRON EJECTION BY NO 5-1601
: HELIUM AND ARGON ON (111) AND (110) TUNGS/ ROLE OF WORK FUNCTION IN ELECTRON EJECTION BY METASTABLE ATOMS 5-1600
CHNIQUE FOR THE CONTINUOUS MEASUREMENT OF CHANGES IN WORK FUNCTION. (LEED- AUGER SYSTEM) /DING POTENTIAL TE 3-0663
OF CONTAMINANTS ON POLYCRYSTALLINE SILVER BY AES AND WORK FUNCTION MEASUREMENTS. DETECTION 5-1574
INITIAL OXIDATION STUDIES BY MEANS OF PHOTOELECTRIC WORK FUNCTION MEASUREMENTS AND AUGER ELECTRON SPECTRO/ 5-1928
/ CHEMISORPTION STUDIES OF NICKEL SURFACES UTILIZING WORK FUNCTION MEASUREMENTS, AUGER, AND LOW ENERGY ELEC 5-1290
/ION STUDIES ON IRON (100) BY MEANS OF PHOTOELECTRIC WORK FUNCTION MEASUREMENTS COMBINED WITH AUGER ELECTR/ 5-1927
RIUM / A CORRELATION OF AUGER SPECTRCSCOPY, LEED AND WORK FUNCTION MEASUREMENTS FOR EPITAXIAL GROWTH OF THO 5-1739
/CTRON DIFFRACTION, AUGER ELECTRON SPECTROSCCPY, AND WORK FUNCTION MEASUREMENTS ON CLEAN AND CESIUM COVERE/ 5-1294
KLL AUGER SPECTRUM/ EVALUATION OF THE TRANSITION DENSITY FUNCTION OF CARBON ON VARIOUS SUBSTRATES FROM THE 5-1651
E CRYSTALS IN/ ELLIPSOMETRY, AUGER SPECTROSCOPY, AND WORK FUNCTION OF CESIUM ON TUNGSTEN AND TITANIUM SINGL 5-1829
L ALLOY PLATES. (AUGER ELECTRON SPECTROSCOPY) WORK FUNCTION OF WELL-DEFINED SURFACE OF COPPER- NICKE 5-1873
FACES OF GALLIUM ARSENIDE. LEED, AUGER, AND WORK FUNCTION STUDIES OF CLEAN AND SODIUM- COVERED SUR 5-1240
E ZINC OXIDE (0001)- ZINC SURFACE: A LEED, AUGER AND WORK FUNCTION STUDY. INTERACTION OF CHLORINE WITH TH 5-1463
0). LEED- AUGER, AND WORK FUNCTION STUDY OF IODINE ADSORBED ON TUNGSTEN (11 5-1145
OADSORPTION / LOW ENERGY ELECTRON DIFFRACTION AUGER, WORK FUNCTION STUDY OF OXYGEN- CESIUM ADSORPTION AND C 5-1297
H AUGER ELECTRON SPECTROSCOPY. PHOTOELECTRIC WORK FUNCTION STUDY ON IRON (100) SURFACE COMBINED WIT 5-1926
N SINGLE CRYSTAL STUDIED BY LEED, AUGER ELECTRON AND WORK FUNCTION TECHNIQUES. / (110) SURFACE OF A TUNGSTE 5-1660
- GOLD. (AUGER SPECTROSCOPY) WORK FUNCTION VARIATION WITH ALLOY COMPOSITION: COPPER 5-1330
GOLD. (AUGER ELECTRON SPECTROSCOPY) WORK FUNCTION VARIATION WITH ALLOY COMPOSITION SILVER- 4-0828
SIUM IONS IN THE VACUUM ULTRAVI/ SPECTRAL LINE EXCITATION FUNCTIONS FOR SODIUM, POTASSIUM, RUBIDIUM, AND CE 4-0735
SPREADING FUNCTIONS IN SECONDARY ELECTRON ANALYSIS. 2-0512
/LICATION OF THE ZETA EFFECTIVE VALUES OF HYDROGENIC WAVE FUNCTIONS IN THE CALCULATION OF THE KLL AUGER EN/ 2-0526
RECORDING OF COMPONENTS ON THE SURFACE OF FUSED EMITTERS. (AUGER SPECTRA) 4-0736
PRESENT AND FUTURE APPLICATIONS OF AUGER SPECTROSCOPY. 1-0025
OSCOPY. (AUGER SPECTROSCOPY) FUTURE GROWTH OF SEM: TOWARD SUPER EFFICIENT MICR 4-0220
PRESENT STATUS OF BEAM ANALYSIS AND ITS FUTURE. (REVIEW, AUGER, MASS SPECTROSCOPY) 1-0181
/SCENCE YIELDS AND COSTER- KRONIG TRANSITION PROBABILITIES F23 BASED ON THE GREEN-SELLIN-ZACHOR INDEPENDENT/ 2-0313

G

G-R NOISE FOR AUGER BAND-TO-BAND PROCESSES. 2-0292
N4,5 AUGER SPECTRA OF GADOLINIUM. 4-0821
SPECTROSCOPY. OXYGEN EXPOSURE OF SAMARIUM, GADOLINIUM, AND TERBIUM STUDIED BY AUGER ELECTRON 5-1332
E RARE EARTH METALS. (LANTHANUM, PRASEODYMIUM, NEODYMIUM, GADOLINIUM, DYSPROSIUM, YTTERBIUM, HAFNIUM). /SOM 4-1100
LLIUM AUGER SPECTRUM. PLASMON ENERGY GAIN AND DOUBLE IONIZATION SATELLITES IN THE BERY 4-0899
LOSS EVENTS IN BERYLLIUM AND BERYL/ CHARACTERISTIC ENERGY GAIN AND LOSS, DOUBLE IONIZATION, AND IONIZATION 4-0901
LIERS. (SURFACE OXYGEN ON LEAD OXIDE BY AES) GAIN FATIGUE MECHANISM IN CHANNEL ELECTRON MULTIP 3-0681
MECHANISM OF PLASMON GAIN IN AUGER SPECTRA. 2-0440
COMBINATION) OPTICAL GAIN IN LIGHTLY DOPED GALLIUM ARSENIDE. (AUGER RE 2-0503
PLASMON ENERGY GAIN IN THE BERYLLIUM AUGER SPECTRUM. 4-1095
SPECTRUM. OBSERVATION OF A PLASMON GAIN IN THE FINE STRUCTURE OF THE ALUMINUM AUGER 4-1085
AND ALUMINUM. (AUGER SPECTROSCOPY) CHARACTERISTIC GAIN PHENOMENA IN POLYCRYSTALLINE COPPER, SILVER, 5-1724
M OF BACK SCATTERED ELECTRONS AND CHARACTERISTIC LOSS GAIN PHENOMENA OF COPPER (111). ENERGY SPECTRU 4-0897
ABSENCE OF PLASMON GAIN SATELLITES IN THE AUGER SPECTRUM OF SILICON. 4-1037
PLASMON EFFECTS IN ELECTRON ENERGY LOSS AND GAIN SPECTRA IN ALUMINUM. 4-1020
ECTRA IN X-RAY PHOTOELECTRON STUDIES OF ZINC, ZINC OXIDE, GALLIUM, AND GALLIUM OXIDE. /D DIRECT ELECTRON SP 4-1054
SPECTRA) DOUBLE AND TRIPLE IONIZATION IN ALUMINUM, GALLIUM, AND SCANDIUM BY ELECTRON IMPACT. (AUGER 2-0232
AUGER RECCMBINATION IN GALLIUM ANTIMONIDE. 4-0778
ARSENIDE. AUGER RECOMBINATION IN INDIUM ARSENIDE, GALLIUM ANTIMONIDE, INDIUM PHOSPHIDE AND GALLIUM 4-1091
AES AND SPUTTER ETCHING OF NICKEL- GOLD- GERMANIUM ON N-GALLIUM ARSENIDE. 5-1775
AUGER ELECTRON SPECTROSCCPY OF CLEAN GALLIUM ARSENIDE. 5-1921
AUGER RECCMBINATION IN GALLIUM ARSENIDE. 2-0561
IDENTIFICATION CF AUGER ELECTRONS IN GALLIUM ARSENIDE. 2-0467
LECTRON SPECTROSCOPY. APPLICATION TO CESIUM-COVERED (110) GALLIUM ARSENIDE. /ERMAL DESORPTION USING AUGER E 5-1293
FUNCTION STUDIES OF CLEAN AND SODIUM- COVERED SURFACES OF GALLIUM ARSENIDE. LEED, AUGER, AND WORK 5-1240
N MEASUREMENTS ON CLEAN AND CESIUM COVERED POLAR FACES OF GALLIUM ARSENIDE. /SPECTROSCOPY, AND WORK FUNCTIO 5-1294
LUMINESCENCE DUE TO EXCITON-EXCITON COLLISION IN GALLIUM ARSENIDE. 2-0451
SECONDARY ELECTRON EMISSION FROM GALLIUM ARSENIDE. 4-0861
SURFACE STOICHIOMETRY AND STRUCTURE OF GALLIUM ARSENIDE. 5-1143
(AUGER) ELECTRON SPECTROSCOPY OF GALLIUM ARSENIDE AND ALUMINUM ARSENIDE SURFACES. 5-1595
ENIDE BY MOLECULAR BEAM EPITAXY. (AUGER / MAGNESIUM DOPED GALLIUM ARSENIDE AND ALUMINUM(X) GALLIUM(1-X) ARS 5-1247
ELECTRON DIFFRACTION STUDY OF THE POLAR (111) SURFACES OF GALLIUM ARSENIDE AND GALLIUM ANTIMONIDE. /ENERGY 5-1602
OF SCANNING AUGER MICRCSCOPY IN MOLECULAR BEAM EPITAXY OF GALLIUM ARSENIDE AND GALLIUM PHOSPHIDE. USE 5-1142
/GER ELECTRON SPECTROSCOPY STUDIES OF SILICON, GERMANIUM, GALLIUM ARSENIDE, AND INDIUM ANTIMONIDE SURFACES, 5-1379

STRUCTURES.   AUGER ELECTRON SPECTROSCOPY STUDIES OF GALLIUM ARSENIDE AND SILICON METAL- SEMICONDUCTOR 5-1774
RON SPECTROSCOPY STUDIES OF ARSENIC VACANCIES IN (-1-1-1) GALLIUM ARSENIDE ANNEALED SURFACES.   AUGER ELECT 5-1914
            FLUORINE ION IMPLANTATION PROFILES IN GALLIUM ARSENIDE AS DETERMINED BY AES.   5-1398
               OPTICAL GAIN IN LIGHTLY DOPED GALLIUM ARSENIDE. (AUGER RECOMBINATION)   2-0503
                    ADSORPTION OF ZINC ON GALLIUM ARSENIDE. (AUGER SPECTROSCOPY)   5-1144
         DEPLETION EFFECTS IN SEMIINSULATING GALLIUM ARSENIDE. (AUGER SPECTROSCOPY)   5-1950
          REACTION OF SPUTTERED PLATINUM FILMS ON GALLIUM ARSENIDE. (AUGER SPECTROSCOPY)   5-1563
CARRIER CONCENTRATION AND AS EVAPORATION IN HEAT TREATED GALLIUM ARSENIDE. (AUGER SPECTROSCOPY)   SURFACE 5-1915
ERTIES OF ALLOYED NICKEL- GOLD- GERMANIUM FILMS ON N-TYPE GALLIUM ARSENIDE. (BY AUGER SPECTROSCOPY) /L PROP 6-2101
/STIGATION OF THE PURE AND CESIUM COATED (100) SURFACE OF GALLIUM ARSENIDE BY SLOW ELECTRON DIFFRACTION. (/ 5-1639
/W ENERGY ELECTRON DIFFRACTION- AUGER CHARACTERIZATION OF GALLIUM ARSENIDE DURING ACTIVATION TO NEGATIVE E/ 5-1364
RADIATIVE RECOMBINATION IN N-GALLIUM ARSENIDE- P-ALUMINUM GALLIUM ARSENIDE HETEROJUNCTIONS. /ER EFFECT AND 2-0324
/HAT ACCOMPANY THE THERMAL ANNEALING OF (100) SURFACES OF GALLIUM ARSENIDE, INDIUM PHOSPHIDE, AND GALLIUM / 5-1163
HERFORD BACKSCATTER/ THERMALLY INDUCED PROCESSES AT GOLD- GALLIUM ARSENIDE INTERFACES: AN ASSESSMENT BY RUT 5-1902
ECTROSCOPY. (AUGER / PROCESSES OF ADSORPTION OF OXYGEN ON GALLIUM ARSENIDE METHOD OF LOW ENERGY ELECTRON SP 5-1641
W ENERGY ELECTRON DIFFRACTION AND AUGER SPECTR/ GROWTH OF GALLIUM ARSENIDE ON (100) GERMANIUM STUDIED BY LO 5-1646
                  ANODIC OXIDATION OF GALLIUM ARSENIDE PHOSPHIDE. (AES)   5-1388
          AUGER ELECTRON SPECTROSCOPY OF GALLIUM ARSENIDE PHOTOSURFACES.   5-1922
LECTRON SPECTROSCOPY.   ELECTRON STIMULATED OXIDATION OF GALLIUM ARSENIDE, STUDIED BY QUANTITATIVE AUGER E 5-1756
       ADSORPTION OF OXYGEN ON CLEAN CLEAVED (110) GALLIUM ARSENIDE SURFACES.   5-1306
      AUGER ELECTRON SPECTROSCOPY OF CONTAMINATED GALLIUM ARSENIDE SURFACES.   5-1920
RON SPECTROSCOPY)   ADSORPTION KINETICS OF CESIUM ON GALLIUM ARSENIDE. (USING AES AND X-RAY PHOTOELECT 5-1821
SCOPY.   INTERDIFFUSIONS IN THIN FILM GOLD ON PLATINUM ON GALLIUM ARSENIDE (100) STUDIED WITH AUGER SPECTRO 5-1232
   AUGER SPECTRA OF HYDROGEN CHLORIDE VAPOR ETCHED N+ GALLIUM ARSENIDE (100) SUBSTRATES.   4-0992
CTRON DIFFRACTION AND AUGER ELECT/ GROWTH OF GOLD FILM ON GALLIUM ARSENIDE (110) OBSERVED BY LOW ENERGY ELE 5-1876
TROSCOPY, AND PHOTOEMISSION STUDIES OF THE CESIUM COVERED GALLIUM ARSENIDE (110) SURFACE. /ER ELECTRON SPEC 5-1931
PECTROSCOPY)   INVESTIGATION OF CESIUM ADSORPTION ONTO A GALLIUM ARSENIDE (110) SURFACE. (AUGER ELECTRON S 5-1644
/W ENERGY ELECTRON DIFFRACTION STUDY OF THE ACTIVATION OF GALLIUM ARSENIDE- CESIUM- OXYGEN NEGATIVE ELECTR/ 5-1845
NG THE EFFECT OF CARBON AND OTHER SURFACE CONTAMINANTS ON GALLIUM ARSENIDE- CESIUM- OXYGEN PHOTOCATHODES. / 5-1919
EROJUNCTIO/ AUGER EFFECT AND RADIATIVE RECOMBINATION IN N-GALLIUM ARSENIDE- P-ALUMINUM GALLIUM ARSENIDE HET 2-0324
G THE INTERACTION BETWEEN OXYGEN AND THE (111) SURFACE OF GALLIUM PHOSPHIDE.   AUGER ELECTRON SPECTRUM DURIN 5-1642
FECTS.   APPLICATION TO STATES WITH TWO BOUND PARTICLES IN GALLIUM PHOSPHIDE.   AUGER THEORY AT DE 2-0457
ITON LUMINESCENCE AND PHOTOCONDUCTIVITY IN HIGHLY EXCITED GALLIUM PHOSPHIDE.   EXC 2-0454
ON OF EXCITONS BOUND TO NEUTRAL DONORS IN TELLURIUM DOPED GALLIUM PHOSPHIDE. /CIATED WITH AUGER RECOMBINATI 2-0453
N THE INTERACTION BETWEEN OXYGEN AND THE (111) SURFACE OF GALLIUM PHOSPHIDE. /THE AUGER ELECTRON SPECTRUM I 5-1643
S STUDIED BY AES AND UES.   OXIDATION AND ANNEALING OF GALLIUM PHOSPHIDE AND GALLIUM ARSENIDE (111) FACE 5-1493
ORINE I/ IN-DEPTH PROFILE DETECTION LIMITS OF NITROGEN IN GALLIUM PHOSPHIDE (AND) NITROGEN, OXYGEN, AND FLU 5-1916
KV EL/ DEPARTURES FROM STOICHIOMETRY IN (100) SURFACES OF GALLIUM PHOSPHIDE, INDUCED BY BOMBARDMENT WITH 2 5-1162
ND DIFFRACTION OF SLOW ELECTRONS.   (111) SURFACE OF GALLIUM PHOSPHIDE STUDIED BY AUGER SPECTROSCOPY A 5-1640
/N OF SECONDARY ELECTRONS EMITTED BY INDIUM PHOSPHIDE AND GALLIUM PHOSPHIDE SUBMITTED TO A BOMBARDMENT BY / 2-0269
CHARACTERISTIC ENERGY LOSS AND AUGER ELECTRON SPECTRA OF GALLIUM PHOSPHIDE (110).   5-1657
             STIMULATED PHOTOLUMINESCENCE IN GALLIUM SELENIDE. (AUGER PROCESS)   2-0416
ON LEVELS/ TEMPERATURE DEPENDENCE OF THE EDGE EMISSION OF GALLIUM SELENIDE SINGLE CRYSTALS AT HIGH EXCITATI 2-0258
          ELECTRON EMISSION STUDIES FROM GALLIUM SELENIDE SURFACES.   5-1967
NTROLLED MOLECULAR BEAM EPITAXY. (AUGER SURFACE ANALYSIS, GALLIUM(X) ALUMINUM(1-X) ARSENIDE   COMPUTER-CO 5-1324
          PREFERENTIAL EVAPORATION OF INDIUM FROM GALLIUM(X) INDIUM(1-X) ARSENIDE. (AUGER EFFECT)   5-1366
ROFILED BY AUGER ELECTRON SPECTROSCOPY.   GALLIUM(1-X) ALUMINUM(X) ARSENIDE SUPERLATTICES P 5-1596
UGER ANALYSIS OF DEPOSITED SILICON DIOXIDE ON ALUMINUM(X) GALLIUM(1-X) ARSENIDE.   DEPOSITION AND A 5-1962
(AUGER / MAGNESIUM DOPED GALLIUM ARSENIDE AND ALUMINUM(X) GALLIUM(1-X) ARSENIDE BY MOLECULAR BEAM EPITAXY. 5-1247
(AUG/ EMISSION SPECTRA OF SILICON COMPENSATED ALUMINUM(X) GALLIUM(1-X) ARSENIDE ELECTROLUMINESCENT DIODES. 2-0325
UM ANTIMONIDE AT ROOM TE/ EFFECTS OF DIFFUSION CURRENT ON GALVANOMAGNETIC PROPERTIES IN THIN INTRINSIC INDI 2-0354
/E OF AUGER AND CONVERSION ELECTRONS FOR THE DETECTION OF GAMMA EMITTING RADIONUCLIDES: CHROMATOGRAPHIC QU/ 1-0217
PERATURES. (AUGER RECOMB/ WARM ELECTRON EFFECTS IN NARROW GAP MERCURY(1-X) CADMIUM(X) TELLURIDE AMBIENT TEM 2-0343
     AUGER RECOMBINATION AND IMPACT IONIZATION IN INDIRECT GAP SEMICONDUCTORS.   2-0396
          BAND-TO-BAND AUGER RECOMBINATION IN INDIRECT GAP SEMICONDUCTORS.   2-0394
ITIONS AND RELAXATION EFFECTS IN AUGER SPECTRA OF SEVERAL GAS ADSORBATES ON TRANSITION METALS. /TOMIC TRANS 4-1042
               FURTHER STUDIES OF GAS ADSORPTION ON MOLYBDENUM (100) SURFACE.   5-1305
PECTROSCOPY: CARBON MO/ ELECTRON BEAM INDUCED EFFECTS ON GAS ADSORPTION STUDIES UTILIZING AUGER ELECTRON S 5-1542
OPY.   CARBON MONOXIDE A/ ELECTRON BEAM INDUCED EFFECTS ON GAS ADSORPTION UTILIZING AUGER ELECTRON SPECTROSC 5-1543
AUGER SPECTROSCOPY)   RESIDUAL GAS AND THE OPTICAL PROPERTIES OF SILVER FILMS. (5-1700
HNIQ/ INVESTIGATION OF MECHANISM OF S IONIZATION OF NOBLE GAS ATOMS BY THE ULTRASOFT X-RAY SPECTROSCOPY TEC 2-0580
/CE OF X-RAY AND AUGER ELECTRON EMISSION SPECTRA FOR RARE GAS ATOMS. PART-1. THE ARGON ATOM. PART-2. KRYPT/ 4-0927
               OXYGEN BEAM GAS AUGER ELECTRON EMISSION.   2-0400
D BY COLLISIONS WITH HELIUM, NEON, / HIGH RESOLUTION BEAM GAS AUGER ELECTRON SPECTRA FOR OXYGEN IONS EXCITE 4-0904
GREGATION FROM THE BULK CRYSTAL INTERACTION WITH CHLORINE GAS. (AUGER SPECTROSCOPY, ELLIPSOMETRY) /FACE. SE 5-1833
: MEASUREMENTS OF SECONDARY EMISSION COEFFICIENTS USING A GAS CELL METHOD. /G A CROSSED BEAM METHOD. PART-2 2-0340
IRON-55 AS AN AUGER ELECTRON EMITTER. NOVEL SOURCE FOR GAS CHROMATOGRAPHY DETECTORS.   3-0608
     AUGER ELECTRON EMISSION SPECTRA FROM FOIL AND GAS EXCITED CARBON BEAMS.   4-1050
STIMULATED EMISSION OF LIGHT FROM A NONIDEAL EXCITON GAS IN A SEMICONDUCTOR. (AUGER PROCESS)   2-0462
R) ELECTRON SPECTROSCOPY, A METHOD FOR STUDYING THE SOLID GAS INTERFACE.   (AUGE 1-0013
   SPECTROSCOPY OF A METAL-GAS INTERFACE.   5-1751
D METAL FILMS BY COMBINATION AUGER SPECTROSCOPY AND INERT GAS ION BOMBARDMENT. /OXIDE THICKNESS ON DEPOSITE 6-2048
ARY ELECTRONS EMITTED BY COPPER UNDER BOMBARDMENT BY RARE GAS IONS. /TRONS IN THE ENERGY SPECTRUM OF SECOND 2-0551
ION FROM SODIUM CHLORINE SINGLE CRYSTAL BOMBARDED BY RARE GAS IONS.   SECONDARY ELECTRON EMISS 4-0759
/UATION OF THE FREQUENCY OF AUGER NEUTRALIZATION OF INERT GAS IONS ON THE SURFACE OF ALKALI HALIDE CRYSTAL/ 5-1544
/LECTRONIC EMISSION FROM SOLID TARGETS BOMBARDED BY NOBLE GAS IONS (10-100 KEV): ENERGETIC AND SPATIAL DIS/ 4-0776
WORK FUNCTION EFFECTS ON AUGER ELECTRON EJECTION BY NOBLE GAS METASTABLE ATOMS.   5-1601
      ADSORBATE EFFECTS IN ELECTRON EJECTION BY RARE GAS METASTABLE ATOMS. (AUGER PROCESS)   2-0333
N SURFACE BY THERMAL CARBON OXYGEN NITROGEN(2), AND NOBLE GAS METASTABLES. /LDS FROM COPPER BERYLLIUM OXYGE 4-0786
  .   EFFECT OF POLARIZATION OF THE ELECTRON GAS ON THE ENERGIES OF AUGER TRANSITION IN METALS 2-0435
          AUGER TRANSITION RATE FOR THE NOBLE GAS SATELLITE SPECTRA.   4-0972
     USE OF AUGER ELECTRON SPECTROSCOPY AND INERT GAS SPUTTERING FOR OBTAINING CHEMICAL PROFILES. 1-0156
IN ASYMMETRIC ION- ATOM COLLISIONS. (AUGER EFFECT)   GAS TARGET EXPERIMENTS OF K-SHELL VACANCY SHARING 2-0523
     ELECTRON SPECTROSCOPY AND LEED STUDIES OF GAS- METAL INTERFACES. (AUGER EFFECT)   5-1320
TURES BY AUGER ELECTRON SPECTROSCO/ INVESTIGATIONS OF THE GAS- SOLID INTERFACE INTERACTIONS AT HIGH TEMPERA 5-1147
CE OF PLATINUM BY MOLECULAR BEAM SCATTERING, / STUDIES OF GAS- SURFACE INTERACTIONS AT THE (100) CRYSTAL FA 5-1963
AR DISTRIBUTION OF AUGER ELECTRON EMISSION FROM CLEAN AND GAS-COVERED IRON (100) SURFACES.   ANGUL 5-1623
     CORRELATION EFFECTS IN THE AUGER SPECTRA OF NOBLE GASES.   4-0981
   SOLID STATE BROADENING IN AUGER SPECTRA. (CADMIUM, RARE GASES)   4-0843
   SOLID STATE BROADENING IN THE AUGER SPECTRA OF RARE GASES. (ARGON, KRYPTON, XENON)   4-1006

RANSITIONS)   OUTER SHELL EXCITATION OF THE RARE GASES BY 30 MEV OXYGEN IONS. (INNER SHELL AUGER T 2-0298
                              AUGER SPECTRA OF THE NOBLE GASES. KRYPTON.   4-0971
      LEED AND AES STUDIES OF THE STRUCTURE OF ADSORBED GASES ON SOLID SURFACES.   5-1835
/ON CHARACTERISTICS OF CARBCN MONOXIDE, SULFUR, AND OTHER GASES ON STAINLESS STEEL SHEET. (AES OF GOLD COA/   5-1551
                          AUGER SPECTRUM CF THE NOBLE GASES. PART-2: ARGON.   4-0974
TRIBUTION OF ELECTRONS PROLUCED BY 200-500 KEV PROTONS IN GASES. PART-2: NITROGEN. (AUGER SPECTRUM) /GY DIS   4-1075
ION OF STEEL DURING ANNEALING IN HYDROGEN-FREE PROTECTIVE GASES. (USING AES) /VATIONS ON THE SURFACE OXIDAT   6-2040
ELECTRON IMPACT DESORPTION STUDIES OF THE INTERACTIONS OF GASES WITH SILICON SURFACES. /N SPECTROSCOPY AND   5-1691
                      AUGER SPECTROSCOPY OF SIMPLE GASEOUS MOLECULES.   4-0989
      CROSS SECTIONS FOR INNER SHELL ICNIZATION OF GASEOUS MOLECULES BY 50-500 KEV PROTONS.   4-1077
SPECTROSCOPY.   GATED ELECTRON BEAM TECHNIQUE FOR AUGER ELECTRON   1-0024
      SENSITIVITY VARIATIONS IN BAYARD- ALPERT GAUGES CAUSED BY AUGER EMISSION AT COLLECTOR.   3-0618
YSICAL ANC CHEMICAL CHARACTERIZATION OF PLATINUM- RHODIUM GAUZE CATALYSTS. (AUGER SPECTROSCOPY)   PH   5-1793
IFFERENCES FROM AUGER ANC CORE BIND/ STATIC RELAXATION OF GERMANE AND THE ESTIMATION OF RELAXATION ENERGY D   2-0471
                              AUGER RECCMBINATICN IN GERMANIUM.   2-0395
OPY STUDIES OF THE OXICATION OF AMORPHOUS ANC CRYSTALLINE GERMANIUM.   AUGER SPECTROSC   5-1431
      BAND-TO-BANC AUGER RECOMBINATION IN SILICON AND GERMANIUM.   2-0459
      FINE STRUCTURE CETAILS IN THE LMM AUGER SPECTRA OF GERMANIUM.   4-0937
      HIGH RESOLUTICN LMM AUGER SPECTRA OF GERMANIUM.   4-0936
S ON THE AUGER RECOMBINATION IN SEMICONDUCTCRS. (SILICON, GERMANIUM)   INFLUENCE OF SCREENING EFFECT   2-0379
      NEW RADIATIVE TRANSITION IN GERMANIUM.   4-1002
      PHONON ASSISTED AUGER RECCMBINATION IN GERMANIUM.   2-0397
OORDINATION COMPOUNDS. X-RAY EXCITED LMM AUGER SPECTRA OF GERMANIUM.   PHOTOELECTRON CHEMICAL SHIFTS IN C   4-0924
ND-TO-BAND AUGER RECOMBINATION OF HIGH DENSITY PLASMAS IN GERMANIUM. PICOSECOND OPTICAL MEASUREMENTS OF BA   2-0250
LECTRON BEAM WITH THE SURFACE REACTION RETWEEN OXYGEN AND GERMANIUM. (AES)   INTERFERENCE OF AN E   5-1620
CONTACT RESISTANCE CF TUNGSTEN CN HEAVILY DCPED SILICON- GERMANIUM ALLOY. (AUGER SPECTROSCOPY) /DES ON THE   5-1456
FFECT ON THE SECONDARY ELECTFON EMISSION CHAFACTERISTICS (GERMANIUM AND BERYLLIUM).   FILM STRUCTURE E   4-0923
/ECTRIC PROPERTIES OF MAGNESIUM(2) SILICIDE, MAGNESIUM(2) GERMANIUM, AND MAGNESIUM(2) STANNIDE. PART-1: X-/   2-0306
D EFFECTS ON RELATIVE INTENSITIES IN LMM AUGER SPECTRA OF GERMANIUM AND SELENIUM.   THRESHOL   4-0767
OD. (PRODUCTION OF AUGER/ METAL- DIELECTRIC TRANSITION IN GERMANIUM AND SILICON STUDIED BY A MICROWAVE METH   4-0752
      SECONDARY EMISSICN STUDIES ON GERMANIUM AND SODIUM- COVERED GERMANIUM.   5-1714
TY OSCILLATIONS OF THE ELECTRON HOLE LUMINESCENCE IN PURE GERMANIUM. (AUGER RECOMBINATION) /PENDENT INTENSI   4-0780
      IMPURITY IONIZATION IN N-TYPE GERMANIUM. (AUGER RECOMBINATION PARAMETER)   2-0466
E CHANGES FOR TUNGSTEN CONTACTS ON HEAVILY DOPED SILICON- GERMANIUM. (AUGER SPECTROSCOPY) /ES AND RESISTANC   5-1861
NERGY DIFFERENCES FROM AUGER AND CORE BINDING ENERGIES OF GERMANIUM COMPOUNDS. / ESTIMATION OF RELAXATION E   2-0471
/ANSPORT MEASUREMENTS IN ULTRAHIGH VACUUM GROWN AMORPHOUS GERMANIUM. (DEPTH PROFILES OF OXYGEN AND CARBON /   5-1547
SURFACE NITROGEN IN HIGH CRITICAL TEMPERATURE NIOBIUM(3) GERMANIUM FILMS BY AUGER ANALYSIS.   DETECTION OF   5-1804
      OXIDATION STUDIES OF AMORPHOUS AND CRYSTALLINE GERMANIUM FILMS BY AUGER SPECTROSCOPY.   6-2023
D, AUGER ELECTRON SPECTROSCO/ LOW TEMPERATURE EPITAXY OF GERMANIUM FILMS BY SPUTTER DEPOSITION. (HEED, LEE   5-1537
/GICAL AND ELECTRICAL PRCPERTIES OF ALLOYED NICKEL- GOLD- GERMANIUM FILMS ON N-TYPE GALLIUM ARSENIDE. (BY /   6-2101
/LEED AND AUGER ELECTRON SPECTROSCOPY STUDIES OF SILICON, GERMANIUM, GALLIUM ARSENIDE, AND INDIUM ANTIMONI/   5-1379
SCOPY AND SECONDARY EMISSION IN SEMICONDUCTORS. (SILICON, GERMANIUM, GRAPHITE, ELECTRON DIFFUSION THEORY) /   2-0240
/CITON BREAKDOWN AND FREE CARRIER AND EXCITON KINETICS IN GERMANIUM IN THE PRESENCE OF ELECTRON HOLE DROPS/   4-0946
DARY ELECTRON EMISSION AT SUPERHIGH FREQUENCIES. (COPPER- GERMANIUM JUNCTION).   SECON   2-0356
      BINDING STATES OF WATER VAPOR ON CLEAVED GERMANIUM. (LEED, AES)   5-1820
      AES AND SPUTTER ETCHING OF NICKEL- GOLD- GERMANIUM ON N-GALLIUM ARSENIDE.   5-1775
      AUGER ELECTRON SPECTROSCOPY OF A STABLE GERMANIUM OXIDE.   4-1108
IES. (AUGER E/ ADSORPTION OF WATER VAPOR ON CLEAN CLEAVED GERMANIUM. PART-1: STRUCTURAL AND KINETIC PROPERT   5-1432
ION AND AUGER SPECTR/ GROWTH OF GALLIUM ARSENIDE ON (100) GERMANIUM STUDIED BY LOW ENERGY ELECTRON DIFFRACT   5-1646
(AUGER) ON THE SURFACES AND IN THIN FILMS OF SILICON AND GERMANIUM SUBJECTED TO LASER EXCITATION. /OCESSES   5-1990
                                                  GERMANIUM SURFACE CLEANING: AUGER ANALYSIS.   5-1540
CTROSCOPY)   TRANSPORT OF OXYGEN IN AMORPHOUS GERMANIUM THIN FILMS DURING ANNEALING. (AUGER SPE   5-1548
AN ELECTRICAL, CHEMICAL, AND STRUCTURAL STUDY OF THE GERMANIUM (100)- CESIUM- OXYGEN PHOTOSURFACE.   5-1319
IFFRACT/ OBSERVATION OF SUPERSTRUCTURES ON CARBON COVERED GERMANIUM (100) SURFACE BY HIGH ENERGY ELECTRON D   5-1805
OF BULK, SURFACE, AND HYDROGEN INDUCED STATES ON CLEAVED GERMANIUM (111). (AUGER SPECTROSCOPY) /ASUREMENTS   5-1781
      STRUCTURE TRANSFORMATIONS ON CLEAVED AND ANNEALED GERMANIUM (111) SURFACES.   5-1713
      ELECTRON HOLE DROPS IN GERMANIUM- SILICON ALLOYS. (AUGER RECOMBINATION)   4-0777
SECONDARY ELECTRON EMISSION OF A ZIRCONIUM- ALUMINUM BULK GETTER. (AUGER SPECTROSCOPY)   5-1407
AUGER ELECTRON YIELD ON PRIMARY BEAM ENERGY AT NORMAL AND GLANCING INCIDENCE.   DEPENDENCE OF   2-0455
SITIONAL PROFILES OF MOBILE IONS IN THE SURFACE REGION OF GLASS.   AES COMPO   5-1722
AQUEOUS CORROSION CF SODA- SILICA AND SODA- LIME- SILICA GLASS. (AES)   5-1265
      CORROSION AND STRUCTURAL CHANGES IN GLASS. (AUGER ELECTRON SPECTROSCOPY)   5-1425
OSCOPY.   ANALYSIS OF GLASS CONTAINER COATINGS BY AUGER ELECTRON SPECTR   5-1704
ECTRON SPECTROSCOPY: NEW TOOL IN THE CHARACTERIZATION OF GLASS FIBER SURFACES.   AUGER EL   1-0175
Y OF GLASS FIBERS)   RECENT ASPECTS OF GLASS FIBER- RESIN INTERFACES. (AUGER SPECTROSCOP   5-1975
OPY DETERMINATION OF THE OXYGEN- SILICON RATIO IN SPIN ON GLASS FILMS.   AUGER ELECTRON SPECTROSC   5-1827
      INJECTION OF IONS INTO GLASS FROM A GLOW DISCHARGE. (AUGER SPECTROSCOPY)   5-1222
AND A REVIEW OF ITS PCSSIELE APPLICATIONS TC PROBLEMS IN GLASS RESEARCH. (PHOTOELECTRON AND AES) /ANALYSIS   1-0084
COPY.   GLASS SURFACE ANALYSIS BY AUGER ELECTRON SPECTROS   5-1720
      APPLICATIONS OF SURFACE CHARACTERIZATICN TO GLASS SURFACES.   5-1705
      AUGER ELECTRON SPECTROSCOPY STUDY OF FLOAT GLASS SURFACES.   5-1236
      DETERMINATION CF THE CCMPOSITION OF GLASS SURFACES BY AUGER SPECTROSCOPY.   5-1365
      CHARACTERIZATION OF GLASS SURFACES BY ELECTRON SPECTROSCOPY.   5-1787
      INJECTION CF IONS INTC GLASS FROM A GLOW DISCHARGE. (AUGER SPECTROSCOPY)   5-1222
BANDS OF MAGNESIUM, MAGNESIUM(2) COPPER AND MAGNESIUM(3) GOLD. / AND AUGER ELECTRON SPECTRA OF THE VALENCE   4-0838
INTERACTION BETWEEN CHLORINE AND THE (100) SURFACE OF GOLD.   5-1335
ER SPECTROSCOPY, AND CCNTACT POTENTIAL STUDIES CF COPPER- GOLD ALLOY SINGLE CRYSTAL SURFACES.   LEED, AUG   5-1748
      CHARACTERIZATION OF COPPER- GOLD ALLOY SINGLE SURFACES.   5-1747
      SEGREGATION AT COPPER- GOLD ALLOY SURFACES. (AUGER SPECTROSCOPY)   5-1631
      AUGER ELECTRON SPECTROSCOPY OF SILVER- GOLD ALLOYS.   6-2024
      SURFACE COMPOSITION OF NICKEL- GOLD ALLOYS.   5-1966
EMOVAL OF MOLYBDENUM FROM VARIOUS METAL SURFACES (COPPER, GOLD, ALUMINUM, TUNGSTEN). /OSITION AND SPUTTER R   5-1878
      SURFACE ENRICHMENT IN COPPER(3) GOLD AND GOLD(3) COPPER ALLOYS.   5-1934
      AUGER ELECTRON SPECTROSCOPY OF (SILVER- GOLD AND INDIUM- LEAD) ALLOY SURFACES.   5-1519
ER ELECTRON SPECTROSCOPY STUDIES OF CLEAN PCLYCRYSTALLINE GOLD AND OF THE ADSORPTION OF MERCURY ON GOLD. /G   5-1976
CTRON SPECTROSCOPY.   SURFACE COMPOSITION OF PALLADIUM- GOLD AND PALLADIUM- SILVER CATALYSTS BY AUGER ELE   3-0638
      SECONDARY ELECTRON EMISSION FROM THIN FILMS. (OF GOLD, APPARATUS)   0334
AL CONVERSION AND ELECTRON CAPTURE. (K-AUGER/ MERCURY AND GOLD ATOMIC CLOUDS REARRANGEMENT FOLLOWING INTERN   5-0334
E ALLOYING OF LEAD MONOLAYERS ON (111) AND (110) FACES OF GOLD. (AUGER EFFECT)   ADSORPTION AND SURFAC   5-1733
THIN FILMS UTILIZING WORK FUNCTION CHANGES: CHROMIUM INTO GOLD. (AUGER ELECTRON SPECTROSCOPY) /UREMENTS IN   5-1889

WORK FUNCTION VARIATION WITH ALLOY COMPOSITION SILVER- GOLD. (AUGER ELECTRON SPECTROSCOPY) 4-0828
WORK FUNCTION VARIATION WITH ALLOY COMPOSITION: COPPER- GOLD. (AUGER SPECTROSCOPY) 5-1330
/TRON INDUCED INNER SHELL IONIZATION OF THE N6,7 LEVEL IN GOLD, BISMUTH, AND LEAD MONITORED BY AUGER SPECT/ 4-1067
SPECTROSCOPY. ADSORPTION OF SULFUR ON GOLD BY LOW ENERGY ELECTRON DIFFRACTION AND AUGER 5-1558
LD SECONDARY/ DIFFERENTIAL SPUTTERING OF MAGNESIUM OXIDE- GOLD CERMET FILMS AND ITS APPLICATION TO HIGH YIE 5-1430
TION ON SECONDARY ELECTRON EMISSION FROM MAGNESIUM OXIDE- GOLD CERMETS. EFFECTS OF CESIA 2-0384
N MONOXIDE, SULFUR, AND OTHER GASES ON STAINLE/ EFFECT OF GOLD COATING ON SORPTION CHARACTERISTICS OF CARBO 5-1551
ON. QUANTITATIVE AUGER ELECTRON SPECTROSCOPY OF GOLD CONDENSED ON ROCK SALT AND ON AMORPHOUS CARB 5-1136
AUGER ELECTRON EMISSION FROM GOLD DEPOSITED ON SILICON (111) SURFACE. 5-1676
AUGER SPECTROSCOPY OF SUBMONOLAYER GOLD DEPOSITIONS ON SILICON. 5-1888
ECTROSCOPY) DIFFUSE INTERFACE IN SILICON (SUBSTRATE)- GOLD (EVAPORATED FILM) SYSTEM. (AUGER ELECTRON SP 5-1681
D BY AUGER ELECTRON AND X/ ELECTRONIC STRUCTURE OF A THIN GOLD FILM DEPOSITED ON A SILICON SUBSTRATE STUDIE 5-1435
ECTROSCOPY AND SECONDA/ COMPOSITION PROFILE OF ION PLATED GOLD FILM ON COPPER ANALYZED BY AUGER ELECTRON SP 5-1679
OW ENERGY ELECTRON DIFFRACTION AND AUGER ELECT/ GROWTH OF GOLD FILM ON GALLIUM ARSENIDE (110) OBSERVED BY L 5-1876
AUGER ELECTRON SPECTROSCOPY OF CLEAN POLYCRYSTALLINE GOLD FILMS. 5-1760
AUGER ELECTRON SPECTROSCOPY ANALYSIS OF ION PLATED GOLD FILMS ON SILICON. 5-1225
AUGER ELECTRON / SPUTTERING OF THE ALLOY SYSTEMS SILVER- GOLD, GOLD- COPPER, AND SILVER- COPPER STUDIED BY 5-1331
D (FILM) SYSTEM. LOW TEMPERATURE DIFFUSION OF GOLD INTO SILICON IN THE SILICON (SUBSTRATE)- GOL 5-1673
/EVELOPMENT OF STEPPED SURFACE REGIONS ON POLYCRYSTALLINE GOLD. LOW ENERGY ELECTRON DIFFRACTIONS AND AUGER/ 5-1483
DIFFUSION OF COBALT OUT OF COBALT HARDENED GOLD MEASURED WITH AUGER ELECTRON SPECTROSCOPY. 6-2135
SECONDARY ELECTRON EMISSION FROM THIN FILM GOLD METAL- OXIDE CERMETS. 4-1015
ADHESION AT GOLD MICA INTERFACES. (LEED, AES) 5-1434
AUGER SPECTROSCOPIC OBSERVATION OF SILICON- GOLD MIXED PHASE FORMATION AT LOW TEMPERATURES. 5-1680
D WITH AUGER SPECTROSCOPY. INTERDIFFUSIONS IN THIN FILM GOLD ON PLATINUM ON GALLIUM ARSENIDE (100) STUDIE 5-1232
AMORPHOUS C/ MEASUREMENTS OF CONDENSATION COEFFICIENTS OF GOLD ON ROCK SALT (100) CLEAVAGE SURFACES AND ON 5-1135
HETEROGENEOUS NUCLEATION AND GROWTH OF GOLD ON SUBSTRATES. (AUGER SPECTRA) 5-1137
LEED AND AES ANALYSIS. CONDENSATION OF GOLD ONTO TANTALUM (100) SINGLE CRYSTAL SURFACES. 5-1312
TRON SPECTROSCOPY OF FACE CENTERED CUBIC METAL SURFACES. (GOLD, SILVER, PALLADIUM, COPPER, NICKEL) /ER ELEC 5-1717
UGER ELECTRON SPECTROSCOPY. SPONTANEOUS ALLOYING OF A GOLD SUBSTRATE WITH LEAD MONOLAYERS. (INTERFACE A 5-1178
SCOPY. SURFACE DIFFUSION OF TIN OVER A GOLD SURFACE OBSERVED WITH AUGER ELECTRON SPECTRO 6-2136
COPY, LEED AND AES. INVESTIGATION OF REORDERED (001) GOLD SURFACES BY POSITIVE ION CHANNELING SPECTROS 5-1139
COPY, LOW ENERGY ELE/ CHARACTERIZATION OF REORDERED (001) GOLD SURFACES BY POSITIVE ION CHANNELING SPECTROS 5-1986
SURFACE COMPOSITION OF THE SILVER- GOLD SYSTEM BY AUGER ELECTRON SPECTROSCOPY. 5-1710
/ND AUGER INVESTIGATION OF THE (100) SURFACE OF COPPER(3) GOLD. TEMPERATURE DEPENDENCE OF LONG RANGE ORDER/ 5-1854
- DISORDER TRANSFORMATION AT A (100) SURFACE OF COPPER(3) GOLD, THEORY, EXPERIMENT. ORDER 5-1855
/PROFILES OF SEVERAL CONTAMINATED AND CLEANED SURFACES OF GOLD THICK FILMS ON COPPER PLATES BY AUGER ELECT/ 5-1550
IS. CHARACTERIZATION OF A CHROMIUM- GOLD THIN FILM DEPOSITION PROCESS BY AUGER ANALYS 1-0037
STIVITY CHANGES W/ DIFFUSION MECHANISMS IN THE PALLADIUM- GOLD THIN FILM SYSTEM AND THE CORRELATION OF RESI 5-1393
ION AND EUTECTIC FORMATION IN TUNGSTEN PLATINUM- TUNGSTEN GOLD THIN FILMS. /ECTROSCOPY STUDY OF INTERDIFFUS 5-1259
GTROMS THICK STUDIED WITH AUGER SP/ DIFFUSION KINETICS OF GOLD THROUGH PLATINUM FILMS ABOUT 2000 AND 600 AN 5-1235
MISSION SPECTROSCOPY AND BY LEED. SEGREGATION OF GOLD TO SILICON (111) SURFACE OBSERVED BY AUGER E 5-1180
SURFACE SEGREGATION IN ALLOYS: EQUILIBRIUM SEGREGATION OF GOLD TO THE (111) SURFACE OF NICKEL. (AES) 5-1216
RUCTURES. ATOMIC ARRANGEMENT OF GOLD (100) AND RELATED METAL OVERLAYER SURFACE ST 5-1716
D COPPER (100) SURFACES. ATOMIC ARRANGEMENT OF GOLD (100), GOLD-COVERED (100), AND SILVER-COVERE 5-1718
Y. CHEMICAL ANALYSIS OF PLATINUM (100) AND GOLD (100) SURFACES BY AUGER ELECTRON SPECTROSCOP 5-1712
SURFACE ENRICHMENT IN COPPER(3) GOLD AND GOLD(3) COPPER ALLOYS. 5-1934
/N ON PALLADIUM- SILVER 40 ATOMIC PERCENT, AND PALLADIUM- GOLD 70 ATOMIC PERCENT CONTACT METAL ALLOYS. (BY/ 6-2041
ELECTRON / SPUTTERING OF THE ALLOY SYSTEMS SILVER- GOLD, GOLD- COPPER, AND SILVER- COPPER STUDIED BY AUGER 5-1331
BY RUTHERFORD BACKSCATTER/ THERMALLY INDUCED PROCESSES AT GOLD- GALLIUM ARSENIDE INTERFACES: AN ASSESSMENT 5-1902
/TALLURGICAL AND ELECTRICAL PROPERTIES OF ALLOYED NICKEL- GOLD- GERMANIUM FILMS ON N-TYPE GALLIUM ARSENIDE/ 6-2101
AES AND SPUTTER ETCHING OF NICKEL- GOLD- GERMANIUM ON N-GALLIUM ARSENIDE. 5-1775
SURFACE ENRICHMENT OF INDIUM IN EVAPORATED GOLD- INDIUM FILMS. (AUGER ELECTRON SPECTROSCOPY) 5-1892
ROSCOPY. DIFFUSION STUDIES IN GOLD- PLATINUM THIN FILMS BY AUGER ELECTRON SPECT 5-1636
AND ION BACKSCATTERING TECHNIQUES APPLIED TO ANALYSIS OF GOLD- REFRACTORY METALLIZATIONS. /R SPECTROSCOPY, 5-1261
AUGER ELECTRON SPECTROSCOPY OF ALLOY SURFACES. (GOLD- SILVER, LEAD- INDIUM) 5-1709
0) SURFACES. ATOMIC ARRANGEMENT OF GOLD (100), GOLD-COVERED (100), AND SILVER-COVERED COPPER (10 5-1718
FACES AS OBSERVED BY LEED AES. GOLD-INDUCED SUPERSTRUCTURES ON SILICON (100) SUR 5-1708
IMPURITY SEGREGATION TO GRAIN BOUNDARIES. 6-2063
R CORROSION OF AUSTENITIC STAINLESS STEEL. / CHEMISTRY OF GRAIN BOUNDARIES AND ITS RELATION TO INTERGRANULA 6-2064
MENT OF LOW ALLOY STEELS. (AUGE/ SEGREGATION AT SELECTIVE GRAIN BOUNDARIES AND ITS ROLE IN TEMPER EMBRITTLE 6-2057
AUGER SPECTROSCOPIC ANALYSIS OF BISMUTH SEGREGATED TO GRAIN BOUNDARIES IN COPPER. 6-2060
SEGREGATION OF BISMUTH TO GRAIN BOUNDARIES IN COPPER- BISMUTH ALLOYS. 6-2096
UGER MICROPROBE) ADDITIVE AND IMPURITY DISTRIBUTIONS IN GRAIN BOUNDARIES IN SINTERED ALUMINA. (SCANNING A 5-1505
ER ELECTRON/ SEGREGATION OF MANGANESE AND ANTIMONY TO THE GRAIN BOUNDARIES OF TEMPER EMBRITTLED STEEL. (AUG 6-2070
CTRON SPECTROSCOPY. GRAIN BOUNDARY ACTIVITY MEASUREMENTS BY AUGER ELE 6-2049
IN A NICKEL- CHROMIUM- ANTIMO/ EFFECT OF PRIOR AUSTENITIC GRAIN BOUNDARY COMPOSITION ON TEMPER BRITTLENESS 6-2114
STRUCTURES USING AUGER ELECTRON SPECTROSCOPY. ANALYSIS OF GRAIN BOUNDARY DIFFUSION IN BIMETALLIC THIN FILM 6-2146
IRON- PHOSPHORUS- SULFUR AND IRON- ANTIMONY- SULFUR ALL/ GRAIN BOUNDARY EMBRITTLEMENT IN IRON- PHOSPHORUS, 6-2098
LLOYS. (AUGER EMISSION ANALYSIS, TUNGSTEN, COPPER, STEEL/ GRAIN BOUNDARY EMBRITTLEMENT IN SOME METALS AND A 6-2058
AUGER ELECTRON SPECTROSCOPY. GRAIN BOUNDARY EMBRITTLEMENT OF STEELS STUDIED BY 6-2103
N- RHENIUM ALLOYS. (AUGER ELECTRON SPECTROSCOPY) STUDY OF GRAIN BOUNDARY FRACTURE SURFACES IN DOPED TUNGSTE 6-2112
IN HOT PRESSED OXIDES BY AUGER ELECTRON SPEC/ ANALYSIS OF GRAIN BOUNDARY IMPURITIES AND FLUORIDE ADDITIVES 5-1506
NTERED TUNGSTEN. (BY AUGE/ DETERMINATION OF THE INFLUENCE GRAIN BOUNDARY IMPURITIES ON THE PROPERTIES OF SI 6-1997
DE BY AUGER ELECTRON SPECTROSCOPY. IDENTIFICATION OF A GRAIN BOUNDARY PHASE IN HOT PRESSED SILICON NITRI 5-1749
MARAGING STEEL. (AUGER ELECTRON SPECTROSCOPY) STUDY OF GRAIN BOUNDARY SEGREGANTS IN THERMALLY EMBRITTLED 6-2055
AUGER SPECTROSCOPIC ANALYSIS OF THE EXTENT OF GRAIN BOUNDARY SEGREGATION. 6-2094
STUDY OF SULFUR EMBRITTLED OXYGEN FREE COPPER. (AES, GRAIN BOUNDARY SEGREGATION) 6-2015
/S ON TEMPER EMBRITTLEMENT OF LOW ALLOY STEELS. (IMPURITY GRAIN BOUNDARY SEGREGATION, AES OF FRACTURE SURF/ 6-2083
/RGRANULAR EMBRITTLEMENT OF NICKEL BY HYDROGEN: EFFECT OF GRAIN BOUNDARY SEGREGATION. (AUGER ELECTRON SPEC/ 6-2072
ROSCOPY) GRAIN BOUNDARY SEGREGATION. (AUGER ELECTRON SPECT 6-2104
OSCOPY). SURFACE SEGREGATION AND ITS RELATION TO GRAIN BOUNDARY SEGREGATION (AUGER ELECTRON SPECTR 6-2106
SURFACE SEGREGATION AS A GUIDE TO GRAIN BOUNDARY SEGREGATION. (AUGER SPECTROSCOPY) 6-2073
GER SPECTROSCOPY) GRAIN BOUNDARY SEGREGATION IN ALUMINUM OXIDE. (AU 6-2134
ELLURIUM. (AUGER SPECTROSCOPY) METASTABLE GRAIN BOUNDARY SEGREGATION IN COPPER CONTAINING T 6-2086
INA. (HIGH VACUUM AUGER ELECTRON SPECTROMETER, FRACTURED/ GRAIN BOUNDARY SEGREGATION IN MAGNESIA DOPED ALUM 5-1615
AUGER ELECTRON SPECTROSCOPY) GRAIN BOUNDARY SEGREGATION IN SINTERED ALUMINA. (6-2056
S AND INTERGRANULAR BRITTLE FRACTURE. (STUDIED BY AES) GRAIN BOUNDARY SEGREGATION OF IMPURITIES IN METAL 6-2079
OMIUM- MOLYBDENUM- VANADIUM / EFFECT OF MICROSTRUCTURE ON GRAIN BOUNDARY SEGREGATION OF PHOSPHORUS IN A CHR 6-2139
TEMPER EMBRITTLEMENT BY AUGER SPECTROSCOPY) SURFACE AND GRAIN BOUNDARY SEGREGATION OF STEEL IMPURITIES. (6-2109
LAR FRACTURE. (AUGER ELECTRON SPECTROSCOPY) GRAIN BOUNDARY SEGREGATION RELATING TO INTERGRANU 6-2082

METALS AND ALLOYS. (SEGREGATION IN STEEL, T/ RELATION OF GRAIN BOUNDARY SEGREGATION TO REDUCED STRENGTH OF 6-2120
MISSION SPECTROSCOPY. STUDY OF GRAIN BOUNDARY SEGREGATION USING AUGER ELECTRON E 6-2117
PECTRA. STUDY OF GRAIN BOUNDARY SEGREGATION USING AUGER EMISSION S 6-2121
SPECTROGRAPHIC ANALYSIS OF FRACTURE SURFACES. GRAIN BOUNDARY STRUCTURE AND SEGREGATION. AUGER 6-2118
RACTION STUDIES OF ADSORPTION ISCTHERMS: XENCN ON (0001) GRAPHITE. /CTROSCOPY AND LOW ENERGY ELECTRON DIFF 5-1856
BARDED BY MEDIUM ENERGY IONS (PLATINUM, NICKEL, TANTALUM, GRAPHITE). /TRON EMISSION FROM SOLID SURFACES BOM 2-0309
IC K-SHELL LOW ENERGY ELECTRON LCSS SPECTRA FROM DIAMOND, GRAPHITE, AND EVAPORATED CARBON SURFACES. /TERIST 4-0945
ENSIONAL TRANSITION: XENON ADSORBED ON THE (0001) FACE OF GRAPHITE. (AUGER ELECTRON SPECTROSCOPY) / TWO DIM 5-1857
Y OF ADSORPTION OF XENON IN EPITAXY ON THE (0001) FACE OF GRAPHITE. (AUGER SPECTROSCOPY) ENTROP 5-1860
/RANSITION IN XENON SUBMONOLAYER FILMS ADSORBED ON (0001) GRAPHITE. (AUGER SPECTROSCOPY, LOW ENERGY ELECTR/ 5-1858
ECTRON EMISSION SPECTRA. GRAPHITE CONDUCTION BAND STATES FROM SECONDARY EL 2-0568
ECONDARY EMISSION IN SEMICONDUCTORS. (SILICON, GERMANIUM, GRAPHITE, ELECTRON DIFFUSION THEORY) /SCOPY AND S 2-0240
OF THE FIRST MONOLAYER ADSORPTION OF XENON ON THE (0001) GRAPHITE FACE. THERMODYNAMICS AND KINETICS 5-1859
AUGER ELECTRON SPECTROSCOPY GRAPHITE FIBER SURFACES. 5-1276
ADSORPTION OF KRYPTON ON THE BASAL PLANE OF GRAPHITE: LEED AND AUGER MEASUREMENTS. 5-1559
EMISSION MICROGRAPHIC STUDIES OF THE CLEAVAGE SURFACE OF GRAPHITE SINGLE CRYSTAL. AUGER ELECTRON 5-1413
S. AUGER SPECTROSCOPY OF GRAPHITE SINGLE CRYSTALS WITH LOW ENERGY ELECTRON 4-0741
ES IN THE MEASUREMENT CF CONTACT ANGLES. PART-4: WATER ON GRAPHITE STUDIED BY SECONDARY ELECTRON EMISSION. 2-0569
ELLIPSOMETRIC STUDY OF ADSORPTION: ADSORPTION OF XENON ON GRAPHITE (0001). (AES) ULTRAHIGH VACUUM TECHNIQU 5-1795
/ ELECTRON STUDIES OF SURFACES: URANIUM DICXIDE, URANIUM, GRAPHITE (001). 5-1754
FORMIC ACID DESCRPTION FRCM GRAPHITE, 300 SERIES STAINLESS STEEL, AND NIOBIU/ 5-1217
/SM OF THE DECOMPOSITICN OF FORMIC ACID ON CARBURIZED AND GRAPHITIZED NICKEL (110). (LEED- AES) 5-1630
MATION OF IMPURITY ADSCRPTION AT FLAKE IRON INTERFACES IN GRAPHITIZED NICKEL (110) USING AES, LEED, AND FL/ 5-1629
DOPED ZINC SULFIDE. (AUGER / CONCENTRATION DEPENDENCE OF GRAY CAST IRON. (SCANNING AUGER MICROPROBE) /NFIR 6-2054
D COSTER-KRONIG TRANSITION PROBABILITIES F23 BASED ON THE GREEN- COPPER LUMINESCENCE IN COPPER AND ALUMINUM 2-0402
LECTRON SPECTROSCOPY. GREEN-SELLIN-ZACHOR INDEPENDENT PARTICLE MODEL. / 2-0313
PRODUCTION OF A SPHERICAL GRID FOR LOW ENERGY ELECTRON DIFFRACTION- AUGER E 3-0668
APPLICATION OF TRIPLE GRID LEED SYSTEM TO AUGER SPECTRUM ANALYSES. 3-0661
AUGER ELECT/ RESOLUTICN AND SENSITIVITY OF THE SPHERICAL GRID RETARDING POTENTIAL ANALYZER. APPLICATION TO 3-0634
GER ELECTRON SPECTROSCCPY) IMPROVED SPHERICAL GRID RETARDING POTENTIAL ANALYZER. (FOR USE IN AU 3-0583
GNAL STRENGTHS FOR MCDULATION AMPLITUDE DISTORTION IN A 4-GRID RETARDING POTENTIAL ENERGY ANALYZER. /RON SI 3-0620
PHONON ASSISTED AUGER PROCESSES IN GROUP III-V COMPOUNDS. 2-0484
GEN A/ IN-SITU TRANSPORT MEASUREMENTS IN ULTRAHIGH VACUUM GROWN AMORPHOUS GERMANIUM. (DEPTH PROFILES OF OXY 5-1547
ANTIMONY DOPING OF SILICON LAYERS GROWN BY SOLID PHASE EPITAXY. (AES MEASUREMENTS) 5-1577
SURFACE STUDIES WITH EPITAXIALLY GROWN METAL FILMS. (AUGER EFFECT) 5-1253
GER ANALYSIS OF CHLORINE IN HYDRCGEN CHLORIDE OR CHLORINE GROWN SILICON DIOXIDE FILMS. AU 5-1251
OLECULAR BEAM STUDIES CF INDIUM FHCSPHIDE EVAPORATION AND GROWTH. (AES) M 5-1334
/ AUGER ELECTRON SPECTROSCOPY OF SYSTEMS EXHIBITING LAYER GROWTH, AND ITS APPLICATION TO THE DEPOSITION OF/ 2-0399
. (AUGER ELECTRON SPECTRCSCOPY) GROWTH AND STRUCTURE OF SEMICONDUCTING THIN FILMS 5-1509
) SURFACES OF NICKEL. GROWTH AND STRUCTURE OF THIN COPPER FILMS ON (001 5-1227
. PART-1. (AUGER SPECTRA) GROWTH AND STRUCTURE OF TITANIUM OXIDE THIN FILMS 5-1980
/ENTAL COMPOSITION OF FRACTURE SURFACES PRODUCED BY CRACK GROWTH IN HYDROGEN AND IN WATER VAPOR. (AUGER SP/ 6-2142
(AUGER SPECTROSCOPY) STRUCTURE AND GROWTH KINETICS OF NICKEL(2) SILICIDE ON SILICON. 5-1917
IUM OXIDE (001) SURFACE. LEED- AES OBSERVATION OF GROWTH MECHANISM OF DEPOSITED IRON FILM ON MAGNES 5-1525
ELECTRON SPECTROSCOPY. GROWTH MODE DISCRIMINATION OF THIN FILMS BY AUGER 5-1624
/PECTROSCOPY STUDY OF SURFACE CONTAMINATION EFFECT ON THE GROWTH MODE OF IRON EPITAXIAL FILMS ON MAGNESIUM/ 5-1524
XIDE (001) SURFACE. (AUGER SPECTROSCOPY) GROWTH MODE OF IRON FILM DEPOSITED ON MAGNESIUM O 5-1528
CRYSTAL SURFACES OF ALKALI HALIDES AND THIN FILM GROWTH OBSERVED BY LEED AND AES. 5-1907
LEED AUGER STUDY OF GROWTH OF ALUMINUM ON NIOBIUM (110). 5-1491
INGLE CRYSTAL STUDIED BY LEED, AUGER ELECTRON / EPITAXIAL GROWTH OF COPPER ON (110) SURFACE OF A TUNGSTEN S 5-1660
INFLUENCE OF SUBSTRATE CONDITIONS ON THE NUCLEATION AND GROWTH OF EPITAXIAL SILICON FILMS. 5-1514
DIED BY LOW ENERGY ELECTRON DIFFRACTION AND AUGER SPECTR/ GROWTH OF GALLIUM ARSENIDE ON (100) GERMANIUM STU 5-1646
ERVED BY LOW ENERGY ELECTRON DIFFRACTION AND AUGER ELECT/ GROWTH OF GOLD FILM ON GALLIUM ARSENIDE (110) OBS 5-1876
HETEROGENEOUS NUCLEATION AND GROWTH OF GOLD ON SUBSTRATES. (AUGER SPECTRA) 5-1137
G REFLECTION DIFFRACTICN, LEED AN/ STUDY OF THE EPITAXIAL GROWTH OF MAGNESIUM ON MAGNESIUM OXIDE (001) USIN 5-1496
ECTRON DI/ STRUCTURES CF SUBSTRATE CRYSTALS AND EPITAXIAL GROWTH OF METAL ON THEM OBSERVED BY LOW ENERGY EL 5-1906
AUGER SPECTROSCOPY) OBSERVATIONS OF THE EPITAXIAL GROWTH OF NICKEL AND COPPER ON (111) SILVER. (BY 5-1354
(AUGER SPECTROSCOPY) FUTURE GROWTH OF SEM: TOWARD SUPER EFFICIENT MICROSCOPY. 1-0220
Y LEED AND AUGER EMISSION SPECTROSCOPY. GROWTH OF SILVER ON POTASSIUM CHLORIDE OBSERVED B 5-1348
ACE OBSERVED BY LOW ENERGY ELECTRON DIFFRACTIO/ EPITAXIAL GROWTH OF SILVER ON POTASSIUM CHLORIDE (100) SURF 5-1696
RATES IN THE PRESENCE OF ELECTRON IRRADIATION. (AUGER SP/ GROWTH OF THIN METALLIC FILMS ON INSULATING SUBST 5-1317
/SCOPY, LEED AND WORK FUNCTION MEASUREMENTS FOR EPITAXIAL GROWTH OF THORIUM ON A TUNGSTEN (100) SUBSTRATE. 5-1739
AUGER ELECTRON SPECTRCSCOPY OF LAYERED GROWTH OF TITANIUM ON TUNGSTEN. 5-1140
GER AND ELLIPSOMETRIC STUDIES OF ULTRATHIN LEAD(II) OXIDE GROWTH ON LEAD. AU 5-1248
IN MAGNESIUM AND SILICON FROM THE CROSS SECTION MODEL OF GRYZINSKI. (AUGER EFFECT) /2,3VV AUGER TRANSITION 4-0765
ROSCOPY) SURFACE SEGREGATION AS A GUIDE TO GRAIN BOUNDARY SEGREGATION. (AUGER SPECT 6-2073
ELECTRON GUN FOR AUGER AND ULTRASOFT X-RAY APPARATUS. 3-0676
OSCOPY. EASY METHOD TO ACCURATELY ALIGN ICN BCMBARDMENT GUNS FOR DEPTH PROFILING IN AUGER ELECTRON SPECTR 3-0700

H

ASEODYMIUM, NEODYMIUM, GADOLINIUM, DYSPROSIUM, YTTERBIUM, HAFNIUM). /SOME RARE EARTH METALS. (LANTHANUM, PR 4-1100
NEUTRALIZATION OF INERT GAS IONS ON THE SURFACE OF ALKALI HALIDE CRYSTALS. /TION OF THE FREQUENCY OF AUGER 5-1544
ARDMENT. / ENERGY SPECTRUM OF ELECTRONS EMITTED BY ALKALI HALIDE CRYSTALS UNDER IMPACT OF ION AND ATOM BOMB 2-0246
SURFACE IMPURITIES IN SILVER HALIDE FILMS. 5-1149
, AND AUGER SPECTRCSCCFY STUDY OF VACUUM DEPOSITED SILVER HALIDE FILMS. UPS, ESCA 5-1947
AES. CRYSTAL SURFACES OF ALKALI HALIDES AND THIN FILM GROWTH OBSERVED BY LEED AND 5-1907
OEMISSION STUDIES OF THE DECAY OF CORE EXCITONS IN ALKALI HALIDES. (AUGER PROCESS IN VALENCE BAND) PHOT 2-0289
(100) SURFACES OF ALKALI HALIDES. PART-1: AIR AND VACUUM CLEANED SURFACES. 5-1347
(100) SURFACES OF ALKALI HALIDES. PART-2: ELECTRON STIMULATED DISSOCIATION 5-1908
NALYSIS OF FILMS FORMED ON METALS IN SLIDING CONTACT WITH HALOGENATED POLYMERS. AUGER A 5-1728
ENTAL SPECTRA) HANDBOOK OF AUGER ELECTRON SPECTROSCOPY. (EXPERIM 1-0159
OSCOPY. DIFFUSION OF COBALT CUT OF COBALT HARDENED GOLD MEASURED WITH AUGER ELECTRON SPECTR 6-2135
Y) SURFACE CARRIER CONCENTRATION AND AS EVAPORATION IN HEAT TREATED GALLIUM ARSENIDE. (AUGER SPECTROSCOP 5-1915
CTING NIOBIUM. (AUGER SPECTRO/ EFFECT OF HIGH TEMPERATURE HEAT TREATMENT ON PENETRATION DEPTH OF SUPERCONDU 5-1938
UGER ELECTRON SPECTRCSCOPY. SAMPLES HEATED DIRECTLY BY ELECTRIC CURRENT OBSERVED BY A 5-1737
SURFACE COMPOSITION OF PLATINUM HEATED IN OXYGEN. (AUGER SPECTRA) 5-1737
SOLID PHASE EPITAXIAL STUDIES USING VACUUM DEPOSITION ON HEATED SILICON SUBSTRATES. (SCANNING AUGER SPECT/ 5-1287
/ONCENTRATION ON THE SURFACE OF MOLYBDENUM CARBIDE DURING HEATING ACCORDING TO AUGER ELECTRON SPECTRUM DAT/ 5-1759
UGER PROCESS) EXCITCN SPECTRA OF HEAVILY DOPED CADMIUM SULFIDE SINGLE CRYSTALS. (A 2-0372

/ERFACIAL OXIDES ON THE CONTACT RESISTANCE OF TUNGSTEN ON HEAVILY DOPED SILICON- GERMANIUM ALLOY. (AUGER S/ 5-1456
/L OXIDES AND RESISTANCE CHANGES FOR TUNGSTEN CONTACTS ON HEAVILY DOPED SILICON- GERMANIUM. (AUGER SPECTRO/ 5-1861
                              RADIATIVE AUGER EFFECT IN HEAVY ATOMS. 2-0476
RGE STATE DEPENDENCE OF THE K-VACANCY PRODUCTION YIELD IN HEAVY ION ATOM COLLISIONS AT KEV ENERGIES. /C CHA 2-0346
            DOPPLER SHIFTED AUGER ELECTRONS FROM FOIL EXCITED HEAVY ION BEAMS. /C CHA 4-0859
                       AUGER ELECTRONS FROM FOIL EXCITED HEAVY ION BEAMS. (OF NEON) 2-0370
SUREMENTS OF NEON K-AUGER ELECTRONS PRODUCED BY ENERGETIC HEAVY ION BOMBARDMENT.   MEA 4-0951
PLE MODEL FOR K X-RAY AND AUGER ELECTRON ENERGY SHIFTS IN HEAVY ION COLLISIONS. COMMENTS.   SIM 2-0301
                            SURFACE ANALYSIS WITH HEAVY ION INDUCED AUGER ELECTRONS. 5-1387
ES OF A NEW METHOD FOR MATERIALS S/ SURFACE ANALYSIS WITH HEAVY ION INDUCED AUGER ELECTRONS. BASIC PROPERTI 1-0070
               PRODUCTION OF INNER SHELL VACANCIES IN HEAVY ION- ATOM COLLISIONS. (AUGER EFFECT) 2-0405
CANCY PRODUCTION CROSS SECTIONS IN ARGON BY THE IMPACT OF HEAVY IONS. (AUGER EFFECT) /E OF TOTAL K-SHELL VA 2-0446
           SCINTILLATION EFFICIENCY FOR LOW VELOCITY HEAVY IONS. (AUGER EMISSION) 2-0431
                            X-RAY PRODUCTION BY HEAVY IONS. (OXYGEN BOMBARDMENT, AUGER SPECTRUM) 2-0479
(/ BIOLOGICAL RADIATION EFFECT ON INCLUSION OF MODERATELY HEAVY NUCLEI IN THE TISSUE: USE OF SOFT X-RAYS. 2-0531
                                HEED SEE HIGH ENERGY ELECTRON DIFFRACTION
LLOY FILMS.   COMPOSITION PROFILES AND SCHOTTKY BARRIER HEIGHTS OF SILICIDES FORMED IN NICKEL- PLATINUM A 5-1900
ING STATES IN COINCIDENCE EXPERIMENTS. (AUGER SPECTRUM OF HELIUM) /BILITY OF INVESTIGATING ATOMIC AUTOIONIZ 2-0257
RESONANCE, AUGER, AND AUTOIONIZATION PROCESSES INVOLVING HELIUM(+) (2S) AND HELIUM(2+) NEAR SOLID SURFACES. 2-0375
ARGON CHARGE STATES DUE TO IMPACT OF HYDROGEN(+) AND HELIUM(+). (AUGER SPECTRA) 4-1011
Y 50 TO 600 KEV PROTON, DEUTERON, MOLECULAR HYDROGEN, AND HELIUM(+) IMPACT.   ARGON L-SHELL IONIZATION B 2-0521
DIED BY ATOMIC HYDROGEN(+), MOLECULAR HYDROGEN(+), ATOMIC HELIUM(+) IMPACT. /L-SHELL FLUORESCENCE YIELD STU 2-0517
LUTION NEON KLL AUGER SPECTRA PRODUCED BY HYDROGEN(+) AND HELIUM(+) ION BOMBARDMENT.   HIGH RESO 4-0953
KEV HYDROGEN(+), DEUTERIUM(+), MOLECULAR HYDROGEN(+), AND HELIUM(+) ION IMPACT. /AND METHANE BY 50- TO 600 2-0519
A (100) FACE OF A COPPER S/ CHARGE EXCHANGE OF LOW ENERGY HELIUM(+) IONS (LESS THAN 10 KEV) SCATTERED FROM 2-0546
/ WORK FUNCTION IN ELECTRON EJECTION BY METASTABLE ATOMS: HELIUM AND ARGON ON (111) AND (110) TUNGSTEN. (A/ 5-1600
                RADIATIONLESS DEEXCITATION OF EXCITED HELIUM ATOMS AT SURFACES. (AUGER PROCESS) 2-0255
F AUGER ELECTRONS)   RESONANCE IONIZATION OF ATOMIC HELIUM BY FAST ELECTRONS. (ANGULAR DISTRIBUTION O 2-0256
GLE CRYSTAL. (AUGER NEUTRA/ CHARGE EXCHANGE OF LOW ENERGY HELIUM IONS AND ATOMS SCATTERED FROM A COPPER SIN 2-0545
ECTRON SPECTRA FOR OXYGEN IONS EXCITED BY COLLISIONS WITH HELIUM, NEON, AND ARGON. /UTION BEAM GAS AUGER EL 4-0904
AND AUTOIONIZATION PROCESSES INVOLVING HELIUM(+) (2S) AND HELIUM(2+) NEAR SOLID SURFACES. RESONANCE, AUGER, 2-0375
         MEASUREMENT OF THE AUGER EFFECT IN THE MU-4 HELIUM(2S) (+) IONIC SYSTEM. 2-0305
Y RESOLUTION AUGER ELECTRON SPECTROMETER USING CONCENTRIC HEMISPHERES.   HIGH ENERG 3-0592
IMPROVED AUGER ELECTRON SPECTROMETER USING CONCENTRIC HEMISPHERES. 3-0591
BSTRATES. (AUGER SPECTRA)   HETEROGENEOUS NUCLEATION AND GROWTH OF GOLD ON SU 5-1137
NATION IN N-GALLIUM ARSENIDE- P-ALUMINUM GALLIUM ARSENIDE HETEROJUNCTIONS. /ER EFFECT AND RADIATIVE RECOMBI 2-0324
                                HEXA..., SEE THE PARENT WORD.
                  SILICON EPITAXY BY SILANE PYROLYSIS. HIGH ENERGY ELECTRON DIFFRACTION AND AES STUDY OF 5-1427
O/ SILICON HOMOEPITAXIAL THIN FILMS VIA SILANE PYROLYSIS: HIGH ENERGY ELECTRON DIFFRACTION AND AUGER ELECTR 5-1428
/CIES, TOPOGRAPHY, PHENOMENA, FIELD ION MICROSCOPY, LEED, HIGH ENERGY ELECTRON DIFFRACTION AUGER SPECTROS/ 5-1492
XYGEN ADSORPTION ON THE TUNGST/ COMBINED LEED- REFLECTION HIGH ENERGY ELECTRON DIFFRACTION AUGER STUDY OF O 5-1464
/ATURE EPITAXY OF GERMANIUM FILMS BY SPUTTER DEPOSITION. (HIGH ENERGY ELECTRON DIFFRACTION LEED, AUGER EL/ 5-1537
DE(0.48) SELENIDE(0.52) CRYSTALS. (/ PHOTOCONDUCTIVITY OF HIGHLY EXCITED CADMIUM SELENIDE AND CADMIUM SULFI 2-0432
EXCITON LUMINESCENCE AND PHOTOCONDUCTIVITY IN HIGHLY EXCITED GALLIUM PHOSPHIDE. 2-0454
        SEQUENTIAL STRIPPING OF HIGHLY IONIZED NEON. (AUGER TRANSITIONS) 2-0441
                       AUGER SPECTRA OF HIGHLY IONIZED OXYGEN AND FLUORINE. 4-0906
R EFFECT)   ELECTRON ION COINCIDENCE STUDIES OF HIGHLY IONIZING ARGON(+)- ARGON COLLISIONS. (AUGE 2-0529
.   WIDTHS AND INTENSITIES OF HIGHLY RESOLVED LINES IN ARGON L2,3 AUGER SPECTRA 4-1032
                                HIGHLY SENSITIVE AUGER ELECTRON SPECTROSCOPY. 1-0189
/TRON DECAY-IN-FLIGHT SPECTRA FROM AUTOIONIZING STATES OF HIGHLY STRIPPED OXYGEN, FLUORINE, CHLORINE, AND / 2-0472
                   A MULTIPLE SAMPLE HOLDER FOR USE IN ULTRAHIGH VACUUM. 3-0603
EXCITON KINETICS IN GERMANIUM IN THE PRESENCE OF ELECTRON HOLE DROPS. (AUGER RECOMBINATION) /E CARRIER AND 4-0946
ECOMBINATION)   ELECTRON HOLE DROPS IN GERMANIUM- SILICON ALLOYS. (AUGER R 4-0777
        EVIDENCES FOR AUGER RECOMBINATION IN ELECTRON HOLE DROPLETS. (SILICON) 2-0260
CTROSCOPY)   PLASMON COUPLING TO CORE HOLE EXCITATIONS IN CARBON. (AUGER APPEARANCE SPE 2-0291
VE AUGER SPECTRUM OF INSULATORS. (MAGNESIUM OX/ ELECTRON- HOLE INTERACTION EFFECTS IN THE KL2,3L2,3 RADIATI 2-0540
SS IN A SEMICONDUCTOR.   EFFECT OF ELECTRON HOLE INTERACTION ON THE AUGER RECOMBINATION PROCE 4-1093
/C FIELD DEPENDENT INTENSITY OSCILLATIONS OF THE ELECTRON HOLE LUMINESCENCE IN PURE GERMANIUM. (AUGER RECO/ 4-0780
QUILIBRIUM CHARGE CARRIERS IN SILICON AT A HIGH/ ELECTRON HOLE SCATTERING AND (AUGER) RECOMBINATION OF NONE 4-1103
INTERATOMIC AUGER PROCESS: EFFECTS ON LIFETIMES OF CORE HOLE STATES. 2-0317
ECT".   DECAY OF HOLE STATES IN OXYGEN-16 VIA A "NUCLEAR AUGER EFF 2-0336
       AUGER TRANSITION RATES IN J-J COUPLING FOR D HOLES AND F ELECTRONS. 2-0444
PENDENCE OF OPTICAL EXCITATION AND AUGER DECAY OF 5D CORE HOLES IN BISMUTH(III) TELLURIDE. /RIX ELEMENTS DE 2-0398
ED AND AUGER ELECTRON SPECTROSCOPY STUDY.   SILICON HOMOEPITAXIAL THIN FILMS VIA SILANE PYROLYSIS: HE 5-1428
/S OF GRAIN BOUNDARY IMPURITIES AND FLUORIDE ADDITIVES IN HOT PRESSED OXIDES BY AUGER ELECTRON SPECTROSCOP/ 5-1506
        EFFECT OF YTTRIA ADDITIONS ON HOT PRESSED SILICON NITRIDE. (AUGER ANALYSIS) 5-1350
ROPROBE ANALYSIS)   MICROSTRUCTURE OF HOT PRESSED SILICON NITRIDE. (AUGER ANALYSIS, MIC 5-1556
C ANALYSIS)   IMPURITIES AND INCLUSIONS IN HOT PRESSED SILICON NITRIDE. (AUGER SPECTROGRAPHI 5-1557
CTROSCOPY.   STUDY OF FRACTURE SURFACES OF HOT PRESSED SILICON NITRIDE BY AUGER ELECTRON SPE 5-1444
CTROSCOPY.   IDENTIFICATION OF A GRAIN BOUNDARY PHASE IN HOT PRESSED SILICON NITRIDE BY AUGER ELECTRON SPE 5-1749
N TRIS (1,10-PHENANTHROLINE) COBALT(III)-57 PERCHLORATE DIHYDRATE. /IONIZATION FOLLOWING ELECTRON CAPTURE I 2-0516
                    AUGER SPECTRA OF LITHIUM HYDRIDE. 4-1025
  LEED- AES ELECTRON INDUCED EMISSION STUDIES OF LITHIUM HYDRIDE. 4-0807
        LEED AND AES STUDIES OF LITHIUM HYDRIDE (100) SURFACE. 5-1446
AUGER PROCESSES IN BROMINE-SUBSTITUTED METHANES AND SOME HYDROCARBONS. /RON SPECTROSCOPIC INVESTIGATION OF 2-0515
W PRESSURES.   AUGER SPECTROSCOPY OF HYDROCHLORIC ACID INTERACTION WITH ALUMINUM AT LO 5-1576
TIN IN ULTRAHIGH VACUUM AND IN THE PRESENCE OF OXYGEN AND HYDROGEN. / THE SURFACE COMPOSITION OF PLATINUM- 5-1196
RESOLUTION NEON AUGER SPECTRUM PRODUCED IN 4.2 MEV ATOMIC HYDROGEN(+)- NEON COLLISIONS.   HIGH 4-1078
ARGON CHARGE STATES DUE TO IMPACT OF HYDROGEN(+) AND HELIUM(+). (AUGER SPECTRA) 4-1011
HIGH RESOLUTION NEON KLL AUGER SPECTRA PRODUCED BY HYDROGEN(+) AND HELIUM(+) ION BOMBARDMENT. 4-0953
/TION OF MOLECULAR NITROGEN AND METHANE BY 50- TO 600 KEV HYDROGEN(+), DEUTERIUM(+), MOLECULAR HYDROGEN(+)/ 2-0519
ELECTRONS FROM ARGON WITH ENERGIES 150-210 EV PRODUCED BY HYDROGEN(+) IMPACTS.   AUGER 2-0552
/F THE ARGON L-SHELL FLUORESCENCE YIELD STUDIED BY ATOMIC HYDROGEN(+), MOLECULAR HYDROGEN(+), ATOMIC HELIU/ 2-0517
FLECTION ELECTRON DIFFRACTION INVESTIGATION OF OXYGEN AND HYDROGEN ADSORPTION ON STRONTIUM SURFACES. /ND RE 5-1498
OF PLATINUM. (AUGER SPECTROSCOPY)   EQUILIBRATION OF HYDROGEN AND DEUTERIUM ON SINGLE CRYSTAL SURFACES 5-1594
L IONIZATION BY 50 TO 600 KEV PROTON, DEUTERON, MOLECULAR HYDROGEN, AND HELIUM(+) IMPACT.   ARGON L-SHEL 2-0521
/OSITION OF FRACTURE SURFACES PRODUCED BY CRACK GROWTH IN HYDROGEN AND IN WATER VAPOR. (AUGER SPECTROMETRY/ 6-2142
ON PLATINUM (100). (AUGER SPECTROSCOPY)   ADSORPTION OF HYDROGEN AND THE REACTION OF HYDROGEN WITH OXYGEN 5-1685
        SPUTTERING OF IONS FROM STAINLESS STEEL BY HYDROGEN BOMBARDMENT. (AES) 6-2017
ACE. (AUGER ELECTRON SPECTROSCOPY)   INVESTIGATION OF HYDROGEN CHEMISORPTION ON THE TUNGSTEN (100) SURF 5-1974

DE FILMS.   AUGER ANALYSIS OF CHLORINE IN HYDROGEN CHLORIDE OR CHLORINE GROWN SILICON DIOXI 5-1251
E (100) SUBSTRATES.   AUGER SPECTRA OF HYDROGEN CHLORIDE VAPOR ETCHED N+ GALLIUM ARSENID 4-0992
OBIUM, STAINLESS STEEL SURFACES/ SPATIAL DISTRIBUTIONS OF HYDROGEN DESORBED FROM IRON, PLATINUM, COPPER, NI 6-2006
F SURFACE CO/ MEASUREMENTS OF THE SPATIAL DISTRIBUTION OF HYDROGEN DESORBED FROM NICKEL SURFACES: EFFECTS O 5-1197
) FACES OF COPPER CRYSTALS. (AU/ ANGULAR DISTRIBUTIONS OF HYDROGEN DESORBED FROM THE (100), (110), AND (111 5-1155
MONOXIDE BY (100) TANTALUM. (/ ADSORPTION AND SOLUTION OF HYDROGEN, DEUTERIUM, NITROGEN, OXYGEN AND CARBON 5-1549
AUGER ELECTRON / INTERGRANULAR EMBRITTLEMENT OF NICKEL BY HYDROGEN: EFFECT OF GRAIN BOUNDARY SEGREGATION. (6-2072
AUGER ELECTRON SPECTRUM AND IONIZATION POTENTIALS OF THE HYDROGEN FLUORIDE MOLECULE.   4-1058
NTERACTION CALCULATIONS OF THE AUGER SPECTRUM OF METHANE, HYDROGEN FLUORIDE, WATER, AND CARBON MONOXIDE. /I 4-0879
). (AUG/ PHOTOEMISSION MEASUREMENTS OF BULK, SURFACE, AND HYDROGEN INDUCED STATES ON CLEAVED GERMANIUM (111 5-1781
ADSORPTION OF HYDROGEN ON A PLATINUM (111) SURFACE.   5-1255
ET AND EVAP/ STICKING PROBABILITIES FOR THE ADSORPTION OF HYDROGEN ON CLEAN NICKEL (111), NICKEL (100), SHE 5-1468
R ELECTRON SPECTROSCOPY)   ADSORPTION OF HYDROGEN ON NICKEL SINGLE CRYSTAL SURFACES. (AUGE 5-1257
EFFECT)   CHEMISORPTION OF ATOMIC HYDROGEN ON THE SILICON (111) 7*7 SURFACE. (AUGER 5-1789
ON K-AUGER CROSS SECTIONS FROM 1.5 MEV PER AMU AND ATOMIC HYDROGEN, OXYGEN, AND FLUORINE.   NE 2-0571
ON DIFFRACTION AND AUGER ELECTRON / SILICON CLEANING WITH HYDROGEN PEROXIDE SOLUTIONS: HIGH ENERGY ELECTR 5-1426
LEED AND AES STUDY OF THE INTERACTION OF HYDROGEN SULFIDE AND MOLYBDENUM (100).   5-1973
FLUORIDE.   SULFUR KLL AUGER SPECTRA IN HYDROGEN SULFIDE, SULFUR DIOXIDE, AND SULFUR HEXA 4-0910
CLEAN IRON. (USING LEED A/ EFFECT OF SULFUR, OXYGEN, AND HYDROGEN SULFIDE SURFACE FILMS ON THE ADHESION OF 5-1211
INTERACTION OF HYDROGEN SULFIDE WITH COPPER (001).   5-1516
ONS ON THE SURFACE OXIDATION OF STEEL DURING ANNEALING IN HYDROGEN-FREE PROTECTIVE GASES. (USING AES) /VATI 6-2040
STUDY OF THE POISONING OF A COMMERCIAL PALLADIUM ALUMINA HYDROGENATION CATALYST.   AUGER SPECTROSCOPIC 5-1177
HE KLL AUGER/ APPLICATION OF THE ZETA EFFECTIVE VALUES OF HYDROGENIC WAVE FUNCTIONS IN THE CALCULATION OF T 2-0526
K-SHELL AUGER ELECTRON HYPERSATELLITES OF NEON.   4-1117

                                    I

ESS)   STIMULATED EMISSION OF LIGHT FROM A NONIDEAL EXCITON GAS IN A SEMICONDUCTOR. (AUGER PROC 2-0462
TRONS BY CLOUD CHAMBER TECHNIQUES: OXYGEN AND NEON AUGE/ IDENTIFICATION AND MEASUREMENT OF LOW ENERGY ELEC 2-0566
RESSED SILICON NITRIDE BY AUGER ELECTRON SPECTROSCOPY.   IDENTIFICATION OF A GRAIN BOUNDARY PHASE IN HOT P 5-1749
ELECTRON SPECTROSCOPY.   IDENTIFICATION OF ATOMIC BORON IN STEEL BY AUGER 6-2018
NIDE.   IDENTIFICATION OF AUGER ELECTRONS IN GALLIUM ARSE 2-0467
IDENTIFICATION OF AUGER SPECTRA FROM ALUMINUM.   4-1027
PED TUNGSTEN. (AUGER SPECTROSCOPY)   IDENTIFICATION OF BUBBLE FORMING IMPURITIES IN DO 6-2107
AUGER PARAMETER IN ELECTRON SPECTROSCOPY FOR THE IDENTIFICATION OF CHEMICAL SPECIES.   5-1948
GER SPECTRUM OF NITROGEN.   IDENTIFICATION OF HIGH ENERGY LINES IN THE KLL AU 4-0791
ER ELECTRON SPECTROSCOPY AND ELECTRON IMPACT DESORPTION / IDENTIFICATION OF ION SPECIES IN THE COMBINED AUG 5-1692
USE OF LEED APPARATUS FOR THE DETECTION AND IDENTIFICATION OF SURFACE CONTAMINANTS.   5-1959
MISSION)   CHEMICAL IDENTIFICATION OF SURFACE MONOLAYERS. (BY AUGER E 5-1589
SPECIES WITH AUGER ELECTRON SPECTROSCOPY.   IDENTIFICATION OF THE CHEMICAL STATE OF ADSORBED 4-0868
(100) SURFACE USING AUGER ELECTRON SPECTROSCOPY.   IDENTIFICATION OF THE FORM OF CARBON AT A SILICON 5-1378
STUDY OF MULBERRY SURFACES BY AUGER AND IEE SPECTROSCOPY.   5-1274
ECT)   ION SELECTIVE SPUTTERING OF III-V COMPOUND SEMICONDUCTOR SURFACES. (AUGER EFF 5-1637
PECTROSCOPY. STUDIES ON SURFACE PLASMA OXIDIZED FILMS OF III-V COMPOUND SEMICONDUCTORS BY AUGER ELECTRON S 5-1699
PHONON ASSISTED AUGER PROCESSES IN GROUP III-V COMPOUNDS.   2-0484
AUGER SPECTROSCOPY)   LIQUID PHASE EPITAXY OF III-V COMPOUNDS ON INDIUM PHOSPHIDE SUBSTRATES. (5-1590
URFACE STRUCTURE AND COMPOSITION OF VARIOUS ELEMENTAL AND III-V SEMICONDUCTORS.   S 5-1161
THEORY OF THE AUGER EFFECT IN III-V SEMICONDUCTORS.   2-0498
COPY IN THE SCANNING ELECTRON MICROSCOPE: AUGER ELECTRON IMAGES.   AUGER ELECTRON SPECTROS 3-0650
AUGER ELECTRON IMAGES IN A SCANNING ELECTRON MICROSCOPE.   3-0649
IC SUPERPOSITION OF SAMPLE CURRENT AND SECONDARY ELECTRON IMAGES IN AUGER ELECTRON SPECTROSCOPY.   ELECTRON 2-0450
COMBINATION OF AUGER SPECTROSCOPY WITH IMAGING OF STUDIED SURFACE.   3-0607
F M4,5NN AUGER ELECTRONS EJECTED FROM KRYPTON BY ELECTRON IMPACT.   ANGULAR DISTRIBUTION O 4-1003
O KEV PROTON, DEUTERON, MOLECULAR HYDROGEN, AND HELIUM(+) IMPACT.   ARGON L-SHELL IONIZATION BY 50 TO 60 2-0521
MULTIPLE IONIZATION OF MANGANESE AND CADMIUM BY ELECTRON IMPACT.   AUGER EFFECT IN THE 2-0228
ECTRON EJECTION IN THE IONIZATION OF ALUMINUM BY ELECTRON IMPACT.   AUGER TRANSITION AND SUPPLEMENTARY EL 2-0230
OMIC HYDROGEN(+), MOLECULAR HYDROGEN(+), ATOMIC HELIUM(+) IMPACT. /L-SHELL FLUORESCENCE YIELD STUDIED BY AT 2-0517
TIONS FOR K-SHELL IONIZATION OF SURFACE ATOMS BY ELECTRON IMPACT.   DIFFERENTIAL CROSS SEC 2-0358
BUTION OF ELECTRONS EMITTED BY ARGON ATOMS UNDER ELECTRON IMPACT.   ENERGY DISTRI 4-1008
ER ELECTRON SPECTRUM PRODUCED BY 45 MEV CHLORINE(12+) ION IMPACT.   HIGH RESOLUTION NEON K-AUG 4-1051
INNER SHELL IONIZATION OF SURFACE ATOMS BY ELECTRON IMPACT.   2-0359
), DEUTERIUM(+), MOLECULAR HYDROGEN(+), AND HELIUM(+) ION IMPACT. /AND METHANE BY 50- TO 600 KEV HYDROGEN(+ 2-0519
ON AND AUGER TRANSITIONS IN INDIUM AND SILVER BY ELECTRON IMPACT.   MULTIPLE IONIZATI 2-0229
TELLITE LINES IN AUGER SPECTRA OF NITROGEN BY SLOW PROTON IMPACT.   PRODUCTION OF INTENSE SA 4-1076
K AUGER ELECTRON SPECTRA PRODUCED BY 0.4 TO 4 MEV PROTON IMPACT. /TIVE SATELLITE LINE INTENSITIES IN ARGON 2-0497
ONIZATION OF KRYPTON AND ITS IONS BY HIGH ENERGY ELECTRON IMPACT. (AUGER EFFECT)   MULTII 2-0493
/ESCENCE YIELDS FOR LIGHT ATOMS AND MOLECULES BY ELECTRON IMPACT. (AUGER ELECTRON EJECTION CROSS SECTIONS) 2-0528
ELL FLUORESCENCE YIELDS OF ARGON AND CHLORINE BY ELECTRON IMPACT. (AUGER EMISSION) MEASUREMENTS OF THE L-SH 2-0387
NEON, ARGON, XENON AND THEIR IONS BY HIGH ENERGY ELECTRON IMPACT. (AUGER EMISSION)   MULTIIONIZATION OF 2-0492
IONIZATION IN ALUMINUM, GALLIUM, AND SCANDIUM BY ELECTRON IMPACT. (AUGER SPECTRA)   DOUBLE AND TRIPLE 2-0232
TIONS FOR IONIZATION OF INNER SHELL ELECTRONS BY ELECTRON IMPACT. (AUGER SPECTROSCOPY)   CROSS SEC 2-0475
AUGER ELECTRON SPECTROSCOPY STUDY OF ELECTRON IMPACT DESORPTION.   2-0452
IN THE COMBINED AUGER ELECTRON SPECTROSCOPY AND ELECTRON IMPACT DESORPTION STUDIES. /CATION OF ION SPECIES 5-1692
COMBINED AUGER ELECTRON SPECTROSCOPY AND ELECTRON IMPACT DESORPTION STUDIES OF SILICON SURFACES.   5-1694
COMBINED AUGER ELECTRON SPECTROSCOPY AND ELECTRON IMPACT DESORPTION STUDIES OF SILICON SURFACES.   5-1695
GASES / COMBINED AUGER ELECTRON SPECTROSCOPY AND ELECTRON IMPACT DESORPTION STUDIES OF THE INTERACTIONS OF 5-1691
SMS OF SIMPLE AND MULTIPLE IONIZATION INDUCED BY ELECTRON IMPACT IN SOME METAL VAPORS. (AUGER PROCESS) /ANI 4-0768
ARBON MONOXIDE / COMPARISON OF X-RAY INDUCED AND ELECTRON IMPACT INDUCED PHOTOELECTRON AND AUGER SPECTRA. C 2-0321
ANGULAR DISTRIBUTION OF AUGER ELECTRONS FOLLOWING IMPACT IONIZATION.   2-0418
S.   IMPACT IONIZATION AND AUGER RECOMBINATION IN BAND 2-0335
INNER SHELL ALIGNMENT OF ATOMS IN ELECTRON IMPACT IONIZATION. (AUGER ELECTRONS)   2-0363
EN, ATOMIC OXYGEN, AND NEON. (AUGER TR/ ABSOLUTE ELECTRON IMPACT IONIZATION CROSS SECTIONS OF ATOMIC NITROG 2-0553
MEASURED BY AUGER ELECTRON SPECTROSCOPY.   ELECTRON IMPACT IONIZATION CROSS SECTIONS OF INNER SHELLS 4-0848
A NEW METHOD FOR MEASURING ELECTRON IMPACT IONIZATION CROSS SECTIONS OF INNER SHELLS. 4-0847
URFACE ATOMS. (AUGER INTENSITIES OF TUNGS/ TOTAL ELECTRON IMPACT IONIZATION CROSS SECTIONS ON K-SHELLS OF S 1-0058
COMPARISON OF X-RAY AND ELECTRON IMPACT IONIZATION IN AUGER SPECTROSCOPY.   2-0396
AUGER RECOMBINATION AND IMPACT IONIZATION IN INDIRECT GAP SEMICONDUCTORS.   2-0417
DETAILED BALANCE BETWEEN AUGER RECOMBINATION AND IMPACT IONIZATION IN SEMICONDUCTORS.   4-1010
ROM XENON N4,500 AND KRYPTON M4,5NN PROCESSES BY ELECTRON IMPACT NEAR THRESHOLD. AUGER ELECTRON EJECTION F 2-0446
K-SHELL VACANCY PRODUCTION CROSS SECTIONS IN ARGON BY THE IMPACT OF HEAVY IONS. (AUGER EFFECT) /E OF TOTAL 2-0446

RA)   ARGON CHARGE STATES DUE TO IMPACT OF HYDROGEN(+) AND HELIUM(+). (AUGER SPECT 4-1011
/RUM OF ELECTRONS EMITTED BY ALKALI HALIDE CRYSTALS UNDER IMPACT OF ION AND ATOM BOMBARDMENT. (AUGER EFFEC/ 2-0246
LECTRONS RESULTING FROM ION- ATOM COLLISIONS WITH A FIXED IMPACT PARAMETER. /ERGY DISTRIBUTION OF (AUGER) E 2-0331
AVIOLET EMISSION OF XENON(III) LEVELS EXCITED BY ELECTRON IMPACT VIA N4,5 AUGER TRANSITIONS.   VACUUM ULTR 4-0878
OM ARGON WITH ENERGIES 150-210 EV PRODUCED BY HYDROGEN(+) IMPACTS.   AUGER ELECTRONS FR 2-0552
RMINED BY AES.   FLUORINE ION IMPLANTATION PROFILES IN GALLIUM ARSENIDE AS DETE 5-1398
          OBSERVATION OF IMPURITY PROFILE IN ION IMPLANTED SILICON BY AUGER ELECTRON SPECTROSCOPY. 5-1482
DARY ION EMISSION.   IN-DEPTH PROFILES OF PHOSPHORUS ION IMPLANTED SILICON BY AUGER SPECTROSCOPY AND SECON 4-0996
             SECONDARY ELECTRON EMISSION FROM ION IMPLANTED SILICON (CESIUM).   4-0913
       EFFECTS OF SULFUR, BARIUM, AND CALCIUM ON IMPREGNATED CATHODE SURFACES. (AES)   5-1391
DSORPTION OF OXYGEN ON MOLYBDENUM (111): EFFECT OF TRACE IMPURITIES.   A 5-1568
  CLEANLINESS OF A SURFACE. (USE OF AES TO DETERMINE TRACE IMPURITIES)   HOW TO ASSURE ONESELF OF THE 5-1154
                    CHEMICAL ANALYSIS OF SEMICONDUCTORS. (IMPURITIES AND COMPOSITION)   5-1671
OXIDES BY AUGER ELECTRON SPEC/ ANALYSIS OF GRAIN BOUNDARY IMPURITIES AND FLUORIDE ADDITIVES IN HOT PRESSED 5-1506
NITRIDE. (AUGER SPECTROGRAPHIC ANALYSIS)   IMPURITIES AND INCLUSIONS IN HOT PRESSED SILICON 5-1557
URFACES TEXTURES: AN INVESTIGATION OF SURFACES CONTAINING IMPURITIES. (AUGER ELECTRON SPECTROSCOPY)   S 5-1707
)   IDENTIFICATION OF BUBBLE FORMING IMPURITIES IN DOPED TUNGSTEN. (AUGER SPECTROSCOPY 6-2107
ACTURE. (STUDIED BY AES)   GRAIN BOUNDARY SEGREGATION OF IMPURITIES IN METALS AND INTERGRANULAR BRITTLE FR 6-2079
                        SURFACE IMPURITIES IN SILVER HALIDE FILMS.   5-1149
                        ROLE OF IMPURITIES IN THE STABILITY OF ZINC OXIDE.   5-1462
VIEW OF AUGER ELECTRON SPECTROSCOPY AS APPLIED TO INTERF/ IMPURITIES, INTERFACES, AND BRITTLE FRACTURE. (RE 6-2078
LEED ANALYSIS.   DETECTION OF IMPURITIES ON A SILICON (111) SURFACE BY AUGER- 5-1360
TRON SPECTROSCOPY AND THEIR INFLUENCE ON SU/ DETECTION OF IMPURITIES ON COPPER SURFACES BY LEED- AUGER ELEC 5-1185
MS. (AUGER SPECTROSCOPY)   EFFECT OF IMPURITIES ON INTRINSIC STRESS IN THIN NICKEL FIL 5-1130
MS. (AUGER EFFECT)   EFFECT OF IMPURITIES ON INTRINSIC STRESS IN THIN NICKEL FIL 5-1131
SPECTROSCOPY.   ANALYSIS OF IMPURITIES ON SILICON SURFACES BY AUGER ELECTRON 5-1674
  AND KINETICS. (DETERMINATION OF SURFACES/ SEGREGATION OF IMPURITIES ON SURFACES OF THORIUM. THERMODYNAMICS 5-1313
RNAL PHOTOEFFECT IN THE X-RAY SPECTRAL REGION. (EFFECT OF IMPURITIES ON THE AUGER SPECTRUM.) /E ON THE EXTE 4-1001
. (BY AUGE/ DETERMINATION OF THE INFLUENCE GRAIN BOUNDARY IMPURITIES ON THE PROPERTIES OF SINTERED TUNGSTEN 6-1997
TO SURFACE EQUILIBRIA.   IMPURITIES ON THORIUM METAL: AUGER STUDY OF BULK 5-1314
E DURING THERMAL TREATMENT. WORK FUNCTION CHANGES AND AU/ IMPURITIES SEGREGATION TO THE (001) NICKEL SURFAC 6-2069
OSCOPY)   SURFACE AND GRAIN BOUNDARY SEGREGATION OF STEEL IMPURITIES. (TEMPER EMBRITTLEMENT BY AUGER SPECTR 6-2109
RAY CAST IRON. (SCANNING AUGER MICROPROB/ CONFIRMATION OF IMPURITY ADSORPTION AT FLAKE IRON INTERFACES IN G 6-2054
001) FACE OF IRON EPITAXIAL FILM. (AUGER EFF/ BEHAVIOR OF IMPURITY ATOMS AND ADSORBED OXYGEN ATOMS ON THE (5-1523
            DIFFUSION OF SULFUR AND PHOSPHORUS IMPURITY ATOMS TO A TUNGSTEN SURFACE.   5-1226
T IN A NICKEL- CHROMIUM STEEL CONTAINING PHOSPHORUS AS AN IMPURITY. (AUGER EMISSION ANALYSIS) /EMBRITTLEMEN 6-2138
           PLATINUM (100) STRUCTURES: IMPURITY CAUSATION AND PROPERTIES. (AUGER EFFECT) 5-1311
R ELECTRON SPECTROSCOPY.   IMPURITY CONCENTRATIONS ON METAL SURFACES BY AUGE 5-1923
TERED ALUMINA. (SCANNING AUGER MICROPROBE)   ADDITIVE AND IMPURITY DISTRIBUTIONS AT GRAIN BOUNDARIES IN SIN 5-1505
/E ELEMENTS ON TEMPER EMBRITTLEMENT OF LOW ALLOY STEELS. (IMPURITY GRAIN BOUNDARY SEGREGATION, AES OF FRAC/ 6-2083
ECOMBINATION PARAMETER)   IMPURITY IONIZATION IN N-TYPE GERMANIUM. (AUGER R 2-0466
R ELECTRON SPECTROSCOPY.   OBSERVATION OF IMPURITY PROFILE IN ION IMPLANTED SILICON BY AUGE 5-1482
  POLYSILICON- SILICON DIOXIDE-/ AUGER STUDIES OF DIFFUSED IMPURITY PROFILES IN SILICON DIOXIDE- SILICON AND 5-1481
ELECTRON SPECTROSCOPY.   INTERGRANULAR FRACTURE AND IMPURITY SEGREGATION IN STEEL. STUDIES BY AUGER 6-2127
                        IMPURITY SEGREGATION TO GRAIN BOUNDARIES.   6-2063
/MICROSCOPY- AUGER SPECTROSCOPY FOR MICROSPOT SURFACE AND IN-DEPTH ANALYSIS OF SILICON AND TRANSISTOR META/ 5-1262
RFACIAL REACTIONS.   IN-DEPTH AUGER ANALYSIS OF ALUMINUM- SILICON INTE 5-1219
                   SELECTED AREA AND IN-DEPTH AUGER ANALYSIS OF THIN FILMS.   5-1653
GALLIUM PHOSPHIDE (AND) NITROGEN, OXYGEN, AND FLUORINE I/ IN-DEPTH PROFILE DETECTION LIMITS OF NITROGEN IN 5-1916
DARY ION EMISSION.   IN-DEPTH PROFILES BY AUGER SPECTROSCOPY AND SECON 4-0995
ICON BY AUGER SPECTROSCOPY AND SECONDARY ION EMISSION. IN-DEPTH PROFILES OF PHOSPHORUS ION IMPLANTED SIL 4-0996
DUCTOR STRUCTURES. TASK-II. APPLICATION OF SECONDARY ELE/ IN-PROCESS CONTROL TECHNIQUES FOR COMPLEX SEMICON 5-1411
             ALLOY SPUTTERING STUDIES WITH IN-SITU AUGER ELECTRON SPECTROSCOPY.   5-1877
ED- AUGER SYS/ ULTRAHIGH VACUUM LOW TEMPERATURE STAGE FOR IN-SITU FILM DEPOSITION AND CHARACTERIZATION. (LE 3-0657
OSCOPY.   MECHANICAL TESTING: IN-SITU FRACTURE DEVICE FOR AUGER ELECTRON SPECTR 3-0659
M GROWN AMORPHOUS GERMANIUM. (DEPTH PROFILES OF OXYGEN A/ IN-SITU TRANSPORT MEASUREMENTS IN ULTRAHIGH VACUU 5-1547
CTRON YIELD ON PRIMARY BEAM ENERGY AT NORMAL AND GLANCING INCIDENCE.   DEPENDENCE OF AUGER ELE 2-0455
  EMISSION YIELD.   EFFECT OF THE INCIDENCE ANGLE OF PRIMARY ELECTRONS ON THE AUGER 2-0237
  CLEAN MOLYBDENUM.   ANGLE OF INCIDENCE EFFECTS IN AUGER ELECTRON EMISSION FROM 2-0542
/DED BY POSITIVE IONS. PART-1. PART-2: EFFECT OF ANGLE OF INCIDENCE. PART-3: EFFECT OF ELECTRONICALLY EXCI/ 2-0543
DIOXIDE AUGER AND APPEARANCE POTENTIAL SPECTRA INDUCED BY INCIDENT ELECTRON BEAM INTERACTION. /HE TITANIUM 5-1834
ER INTENSITIES.   MOLECULAR ADSORPTION: INCLUSION OF BULK DIFFUSION AND ITS EFFECT ON AUG 2-0264
E: USE OF SOFT X-RAYS. (/ BIOLOGICAL RADIATION EFFECT ON INCLUSION OF MODERATELY HEAVY NUCLEI IN THE TISSU 2-0531
SPECTROGRAPHIC ANALYSIS)   IMPURITIES AND INCLUSIONS IN HOT PRESSED SILICON NITRIDE. (AUGER 5-1557
SITION PROBABILITIES F23 BASED ON THE GREEN-SELLIN-ZACHOR INDEPENDENT PARTICLE MODEL. /D COSTER-KRONIG TRAN 2-0313
              CHLORINE MONOLAYERS ON THE LOW INDEX FACES OF SILVER. (AUGER EFFECT)   5-1777
             CHEMISORPTION REACTIONS ON HIGH INDEX ZINC SULFIDE SURFACES. (AUGER EFFECT)   4-0788
LE CRYSTAL SURFACES. (LEED, AUGER ELEC/ REACTIVITY OF LOW INDEX ((111) AND (100)) AND STEPPED PLATINUM SING 5-1838
                     AUGER ELECTRONS, INDICATORS IN ANALYSIS OF SOLID BODY SURFACES. 5-1662
            A FORMALISM FOR THE INDIRECT AUGER EFFECT. PART-1. PART-2.   2-0385
       RADIATIVE AUGER RECOMBINATION OF TWO INDIRECT EXCITONS.   2-0227
  AUGER RECOMBINATION AND IMPACT IONIZATION IN INDIRECT GAP SEMICONDUCTORS.   2-0396
    BAND-TO-BAND AUGER RECOMBINATION IN INDIRECT GAP SEMICONDUCTORS.   2-0394
TRON SPECTROSCOPY OF ALLOY SURFACES. (GOLD- SILVER, LEAD- INDIUM)   AUGER ELEC 5-1709
C FINE STRUCTURE IN THE AUGER SPECTRA OF SOLID SILVER AND INDIUM.   QUASIATOMI 4-0771
  AUGER ANALYSIS OF THIN OXIDE FILMS ON LEAD- INDIUM ALLOYS.   5-1249
  MULTIPLE IONIZATION AND AUGER TRANSITIONS IN INDIUM AND SILVER BY ELECTRON IMPACT.   2-0229
        CARRIER RECOMBINATION IN N-INDIUM ANTIMONIDE.   4-0761
ECTRICAL CONDUCTIVITY OF THE THERMAL AND ANODIC OXIDES OF INDIUM ANTIMONIDE. (AES) /SITIONAL PROFILE AND EL 5-1971
/ CURRENT ON GALVANOMAGNETIC PROPERTIES IN THIN INTRINSIC INDIUM ANTIMONIDE AT ROOM TEMPERATURE. (AUGER PR/ 2-0354
THE X-RAY SPECTRAL REGION. (EF/ EFFECT OF THE POLARITY OF INDIUM ANTIMONIDE ON THE EXTERNAL PHOTOEFFECT IN 4-1001
COPY STUDIES OF SILICON, GERMANIUM, GALLIUM ARSENIDE, AND INDIUM ANTIMONIDE SURFACES. /ER ELECTRON SPECTROS 5-1379
IN THE 20-70 EV R/ VALENCE BAND AND CORE LEVEL STUDIES OF INDIUM ANTIMONIDE VIA PHOTOEMISSION MEASUREMENTS 2-0341
    HIGH RESOLUTION MNN AUGER SPECTRA OF SILVER, CADMIUM, INDIUM, ANTIMONY, TELLURIUM AND IODINE. 4-0730
AUGER) RECOMBINATION OF NONEQUILIBRIUM CHARGE CARRIERS OF INDIUM ARSENIDE AT HIGH PHOTOEXCITATION LEVELS. (4-0840
      RECOMBINATION OF CARRIERS IN N-TYPE INDIUM ARSENIDE AT 77 DEGREES K. (AUGER)   4-0785
     CARRIER LIFETIMES IN EPITAXIAL INDIUM ARSENIDE. (AUGER RECOMBINATION)   2-0332
ROCESS)   MINORITY CARRIER LIFETIME IN INDIUM ARSENIDE EPILAYERS. (AUGER RECOMBINATION P 2-0567
HIDE AND GALLIUM ARSENIDE.   AUGER RECOMBINATION IN INDIUM ARSENIDE, GALLIUM ANTIMONIDE, INDIUM PHOSP 4-1091
         COMBINED LEED AND AES STUDIES OF INDIUM ARSENIDE SURFACES.   5-1373

LEED AND AUGER ELECTRON SPECTROSCOPY.   STUDY OF INDIUM ARSENIDE (111) AND (-1-1-1) SURFACES USING 5-1381
/EGREGATION IN THE BINARY ALLOYS COPPER- 1 ATOMIC PERCENT INDIUM, COPPER- 2 ATOMIC PERCENT TIN, AND IRON- / 6-2027
ER EFFECT)   PREFERENTIAL EVAPORATION OF INDIUM FROM GALLIUM(X) INDIUM(1-X) ARSENIDE. (AUG 5-1366
LECTRON SPECTROSCOPY)   SURFACE ENRICHMENT OF INDIUM IN EVAPORATED GOLD- INDIUM FILMS. (AUGER E 5-1892
/THERMAL ANNEALING OF (100) SURFACES OF GALLIUM ARSENIDE, INDIUM PHOSPHIDE, AND GALLIUM PHOSPHIDE IN VACUU/ 5-1163
TO/ ENERGY DISTRIBUTION OF SECONDARY ELECTRONS EMITTED BY INDIUM PHOSPHIDE AND GALLIUM PHOSPHIDE SUBMITTED 2-0269
SURFACE CHARACTERIZATION OF INDIUM PHOSPHIDE. (BY AUGER SPECTROSCOPY) 5-1968
MOLECULAR BEAM STUDIES OF INDIUM PHOSPHIDE EVAPORATION AND GROWTH. (AES) 5-1334
LIQUID PHASE EPITAXY OF III-V COMPOUNDS ON INDIUM PHOSPHIDE SUBSTRATES. (AUGER SPECTROSCOPY) 5-1590
ATURE ON THE SURFACE STRUCTURE AND STOICHIOMETRY OF (100) INDIUM PHOSPHIDE SURFACES.   EFFECT OF TEMPER 5-1164
PECTROSCOPY STUDY OF THE SURFACE COMPOSITION CF THE LEAD- INDIUM SYSTEM.   AUGER ELECTRON S 5-1172
PREFERENTIAL EVAPORATION OF INDIUM FROM GALLIUM(X) INDIUM(1-X) ARSENIDE. (AUGER EFFECT) 5-1366
AUGER ELECTRON SPECTROSCOPY OF (SILVER- GOLD AND INDIUM- LEAD) ALLOY SURFACES. 5-1839
TRON EMISSION SPECTRA CF VANADIUM PENTOXIDE (010) AND VA/ INELASTIC EFFECTS AND STRUCTURE IN THE AUGER ELEC 5-1340
LOW ENERGY ION SCATTERING: ELASTIC AND INELASTIC EFFECTS. (AUGER NEUTRALIZATION) 2-0382
INELASTIC EFFECTS IN AUGER ELECTRON SPECTROSCOPY. 1-0139
ETAL SURFACES.   AUGER ELECTRON SPECTROSCOPY AND INELASTIC ELECTRON SCATTERING IN THE STUDIES OF M 5-1818
ORBED PARTICLES. (AUGER SPECTROSCOPY) INELASTIC INTERACTIONS OF SLOW ELECTRONS WITH ABS 2-0448
/M (111) AND ALUMINUM (001) SINGLE CRYSTAL SURFACES USING INELASTIC LOW ENERGY ELECTRON DIFFRACTION. (AND / 5-1745
LLIUM. (AUGER, LEED, ENERGY LOSS SPECTROSCOPY) INELASTIC MEAN FREE PATH FOR ELECTRONS IN BULK JE 4-1059
R ELECTRON SPECTROSCOPY. INELASTIC SCATTERING EFFECTS IN QUANTITATIVE AUGE 2-0433
ES.   IONIZATION LOSSES IN (AUGER) SPECTRA OF ELECTRON INELASTIC SCATTERING FROM RARE EARTH METAL SURFAC 2-0538
N SPECTROSCOPY)   METHCDS CF SURFACE STUDIES DEPENDING ON INELASTIC SCATTERING OF ELECTRONS. (AUGER ELECTRO 5-1159
SURFACES. INELASTIC SCATTERING OF LOW ENERGY ELECTRONS FROM 2-0495
SINGLE CRYSTALS.   ENERGY SPECTRA OF INELASTICALLY SCATTERED AND AUGER ELECTRONS FROM 4-1064
S OF TUNGSTEN.   ENERGY SPECTRA OF INELASTICALLY SCATTERED ELECTRONS AND LEED STUDIE 4-1094
/TY FOR INVESTIGATING ENERGY AND ANGULAR DISTRIBUTIONS OF INELASTICALLY SCATTERED OR PHOTOEMITTED ELECTRON/ 3-0691
LS. PART-1: EMISSION OF TRUE SECONDARY ELECTRONS. PART-2: INELASTICALLY SCATTERED PRIMARY ELECTRONS. / META 2-0500
POSITED METAL FILMS BY COMBINATION AUGER SPECTROSCOPY AND INERT GAS ION BOMBARDMENT. /OXIDE THICKNESS ON DE 6-2048
Y/ EVALUATION OF THE FREQUENCY OF AUGER NEUTRALIZATION OF INERT GAS IONS ON THE SURFACE OF ALKALI HALIDE CR 5-1544
LES.   USE OF AUGER ELECTRON SPECTROSCOPY AND INERT GAS SPUTTERING FOR OBTAINING CHEMICAL PROFI 1-0156
Y ANALYZER. (AUGER ELECTRONS)   COMPACT, INEXPENSIVE HIGH RESOLUTION RETARDING FIELD ENERG 3-0619
ARBON MONOXIDE ON (100) PALLADIUM.   STRUCTURAL INFLUENCES ON ADSORBATE BINDING ENERGY. PART-1: C 5-1913
ONOXIDE ON NICKEL (100). PART-3: CARBON MONOX/ STRUCTURAL INFLUENCES ON ADSORPTION ENERGY. PART-2: CARBON M 5-1911
LICON MONOLAYERS)   IN DEPTH INFORMATION FROM AUGER ELECTRON SPECTROSCOPY. (SI 1-0130
CONDARY ELECTRON EMISSION DUE TO CASCADE PROC/ BACKGROUND INFORMATION IN AUGER ELECTRON SPECTROSCOPY AND SE 2-0411
INES IN THE ELECTRON SPECTROSCOPY FOR CHEMICAL ANALYSIS. INFORMATION OBTAINED FROM THE WIDTH OF SPECTRAL L 4-1009
INTEGRAL AUGER INFORMATION VIA TAILORED MODULATION TECHNIQUES. 1-0187
DETECTION OF LEAD, BY AUGER SPECTROSCOPY, ON IRON INHIBITED IN LEAD AZELATE SOLUTION. 5-1628
(100) SURFACE BY MEANS OF COMBINED LOW ENE/ STUDY ON THE INITIAL OXIDATION AND REDUCTION PROCESSES OF IRON 6-2141
CLEATION RATES ON (100) AND (110) MOLYBDENUM SURFACES. (/ INITIAL OXIDATION OF MOLYBDENUM. PART-3: OXIDE NU 5-1536
RESOLUTION AUGER PHOTCELECTRON SPECTROMETER. INITIAL OXIDATION OF MOLYBDENUM STUDIED BY A HIGH 5-1698
CHEMICAL SHIFTS IN AUGER ELECTRON SPECTROSCOPY FROM INITIAL OXIDATION OF TANTALUM (110). 6-2038
OBSERVED BY AES AND LEED. INITIAL OXIDATION OF THE MOLYBDENUM (110) SURFACE 5-1650
MEANS OF WORK FUNCTION AND AUGER ELECTRON SP/ STUDIES OF INITIAL OXIDATION ON SILICON- IRON ALLOY (100) BY 5-1924
IC WORK FUNCTION MEASUREMENTS AND AUGER ELECTRON SPECTRO/ INITIAL OXIDATION STUDIES BY MEANS OF PHOTOELECTR 5-1928
OF PHOTOELECTRIC WORK FUNCTION MEASUREMENTS COMBINED WIT/ INITIAL OXIDATION STUDIES ON IRON (100) BY MEANS 5-1927
MIC PERCENT ALUMINU/ AUGER ELECTRON SPECTROSCOPY STUDY OF INITIAL STAGES OF OXIDATION IN A COPPER- 19.6 ATO 6-2028
RON DIFFRACTION- AUGER ELECTRON SPECTROSCOPY STUDY OF THE INITIAL STAGES OF OXIDATION OF IRON (001). /ELECT 6-2110
N LITHIUM METAL.   AUGER TRANSITIONS INITIATED FROM SINGLY AND DOUBLY IONIZED STATES I 2-0581
E. (AUGER SPECTROSCCPY) INJECTION OF IONS INTO GLASS FROM A GLOW DISCHARG 5-1222
E EXCITATION PROCESSES IN TRANSITION METAL ATOMS. PART-2: INNER AUGER ELECTRONS.   MULTIPL 2-0362
S IN FREE ATOMS FOR USE IN X-RAY SPECTROSCOPY. INNER ELECTRON BINDING ENERGIES AND AUGER ENERGIE 4-0774
/RIMARY ELECTRONS. PART-1: ENHANCED PLASMON EXCITATION ON INNER LEVEL IONIZATION. PART-2: DOUBLE IONIZATIO/ 4-1047
IONIZATION. (AUGER ELECTRONS) INNER SHELL ALIGNMENT OF ATOMS IN ELECTRON IMPACT 2-0335
HELL EXCITATION OF THE RARE GASES BY 30 MEV CXYGEN IONS. (INNER SHELL AUGER TRANSITIONS)   OUTER S 2-0298
G. PART-1: AUGER ELECTRONS. (KLL TRANSITION ENERGIES FOR/ INNER SHELL DOUBLE IONIZATION AND CHEMICAL BONDIN 2-0327
SPECTROSCOPY)   CROSS SECTIONS FOR IONIZATION OF INNER SHELL ELECTRONS BY ELECTRON IMPACT. (AUGER 2-0475
NS.   EXCITATION OF INNER SHELL ELECTRONS BY PHOTONS AND FAST ELECTRO 2-0541
LY IONIZED ATOMS.   MECHANISMS OF INNER SHELL EXCITATION AND DEEXCITATION IN MULTIP 2-0487
AUGER DISSOCIATION OF MOLECULES AS A RESULT OF INNER SHELL IONIZATION. 2-0465
ITIONS).   ATOMIC VACANCY DISTRIBUTIONS PRODUCED BY INNER SHELL IONIZATION (AUGER AND RADIATIVE TRANS 2-0544
AUGER ELECTRON SPECTROSCOPY APPLIED TO INNER SHELL IONIZATION BY FAST CHARGED PARTICLES. 2-0535
-500 KEV PROTONS.   CROSS SECTIONS FOR INNER SHELL IONIZATION OF GASEOUS MOLECULES BY 50 4-1077
ON IMPACT. INNER SHELL IONIZATION OF SURFACE ATOMS BY ELECTR 2-0359
BISMU/ ANOMALOUS THRESHOLD BEHAVIOR FOR ELECTRON INDUCED INNER SHELL IONIZATION OF THE N6,7 LEVEL IN GOLD, 4-1067
REARRANGEMENT OF INNER SHELL IONIZED ATOMS. (AUGER SPECTRA) 2-0413
NS. (AUGER EFFECT)   PRODUCTION OF INNER SHELL SPECTROSCOPY. (REVIEW) 1-0160
).   CORRELATION ENERGIES AND AUGER RATES IN ATOMS WITH INNER SHELL VACANCIES IN HEAVY ION- ATOM COLLISIO 2-0405
OR MEASURING ELECTRON IMPACT IONIZATION CROSS SECTIONS OF INNER SHELL VACANCIES (NEON WITH 1S OR 2S VACANCY 2-0312
SPECTROSCOPY, CHEMI/ EMISSION AND RECOMBINATION INVOLVING INNER SHELLS.   A NEW METHOD F 4-0848
OPY.   ELECTRON IMPACT IONIZATION CROSS SECTIONS OF INNER SHELLS. (AUGER PROCESS, AES, PHOTOELECTRON 2-0514
ION OF THREE ELECTRON AUGER TRANSITIONS IN ATOMS WITH TWO INNER SHELLS MEASURED BY AUGER ELECTRON SPECTROSC 2-0553
D ELECTRO/ DESIGN OF AN X-RAY DOUBLE CRYSTAL SPECTROMETER INNER VACANCIES.   OBSERVAT 2-0235
ER SPECTROSCOPY. INSTALLED IN VACUUM AND THE MEASUREMENT OF EXCITE 3-0641
YS AND ELECTRONS IN ANALYTICAL CHEMISTRY WITH EMPHASIS ON INSTRUMENTAL PROBLEMS FOR ELECTRON MICROPROBE AUG 3-0670
LECTRON SPECTROSCOPY AND ION MICROPRO/ RECENT ADVANCES IN INSTRUMENTATIION.   X-RA 3-0646
FACTURED SURFACES. (REVIEW, SPECTROSCOPY) INSTRUMENTATION FOR MICROPROBE ANALYSIS. (AUGER E 3-0716
AUGER ELECTRON SPECTROSCOPY OF INSTRUMENTATION FOR THE CHEMICAL ANALYSIS OF MANU 3-0719
IRRADIATION. (AUGER SP/ GROWTH OF THIN METALLIC FILMS ON INSULATING SILICON COMPOUNDS. 4-0795
ERTURBING EFFECTS OF SPUTTERING DURING THE AUGER STUDY OF INSULATING SUBSTRATES IN THE PRESENCE OF ELECTRON 5-1317
/ION EFFECTS IN THE KL2,3L2,3 RADIATIVE AUGER SPECTRUM OF INSULATORS.   P 1-0111
N TECHNIQUES. INSULATORS. (MAGNESIUM OXIDE, MAGNESIUM FLUORIDE) 2-0540
AES. INTEGRAL AUGER INFORMATION VIA TAILORED MODULATIO 1-0187
PPLICATIONS OF SCANNING AUGER SPECTROSCOPY TO THE SILICON INTEGRATED CIRCUIT PROCESS CHARACTERIZATION USING 5-1741
INVESTIGATIONS OF CARBON RESIDUES ON SURFACES OF SILICON INTEGRATED CIRCUIT TECHNOLOGY.   A 5-1655
SYSTEM.   AUGER ELECTRON SPECTROMETER INTEGRATED CIRCUITS. (BY AUGER ELECTRON SPECTROS/ 5-1345
HIGH SPATIAL RESOLUTION AUGER SPECTROSCCPY AND AUGER INTEGRATED WITH A LOW ENERGY ELECTRON DIFFRACTION 3-0658
QUANTITATIVE AUGER ELECTRON SPECTROSCOPY./ USE OF ANALOG INTEGRATION APPLICATIONS. 1-0166
INTEGRATION IN DYNAMIC BACKGROUND SUBTRACTION FOR 1-0069

QUANTITATIVE AUGER ANALYSIS USING INTEGRATION TECHNIQUES.   1-0066
AUGER RECOMBINATION IN SEMICONDUCTORS DURING INTENSE EXCITATION.   2-0368
GEN BY SLOW PROTON IMPACT.   PRODUCTION OF INTENSE SATELLITE LINES IN AUGER SPECTRA OF NITRO 4-1076
RON EMISSION SPECTRA: A SIMPLE TREATMENT OF ENERGIES AND INTENSITIES.   AUGER ELECT 2-0319
RNAL CONVERSION AND ELECTRON CAPTURE. (K-AUGER TRANSITION INTENSITIES) /CLOUDS REARRANGEMENT FOLLOWING INTE 2-0334
ION: INCLUSION OF BULK DIFFUSION AND ITS EFFECT ON AUGER INTENSITIES.   MOLECULAR ADSORPT 2-0264
DUCED BY 0.4 TO 4 MEV PROTON IMP/ RELATIVE SATELLITE LINE INTENSITIES IN ARGON K AUGER ELECTRON SPECTRA PRO 2-0497
SELENIUM.   THRESHOLD EFFECTS ON RELATIVE INTENSITIES IN LMM AUGER SPECTRA OF GERMANIUM AND 4-0767
              Z DEPENDENCE OF ENERGIES AND RELATIVE INTENSITIES IN THE KLL AUGER SPECTRUM.   4-0886
                  L-SHELL AUGER ENERGIES AND INTENSITIES IN 3D TRANSITION METALS.   4-1018
(LITHIUM, SODIUM) SUBJECTED TO ICN BCM/ RELATIVE EMISSION INTENSITIES OF AUGER ELECTRONS FROM LIGHT METALS 2-0548
MBARDMENT AND ALUMINUM K-ALPHA X-RAY IRRADIATIO/ STUDY OF INTENSITIES OF AUGER LINES EXCITED BY ELECTRON BO 4-1017
              PREDICTION OF THE RELATIVE INTENSITIES OF AUGER LINES IN MICA.   4-1072
3 AUGER SPECTRA.   WIDTHS AND INTENSITIES OF HIGHLY RESOLVED LINES IN ARGON L2, 4-1032
                                          INTENSITIES OF KLL AUGER LINES.   4-0748
                                          INTENSITIES OF SPECTRA- AUGER AND PHOTOELECTRON. 4-1070
ATION CROSS SECTIONS ON K-SHELLS OF SURFACE ATOMS. (AUGER INTENSITIES OF TUNGSTEN (100) /TRON IMPACT IONIZ 4-0847
              SIMPLE MODEL FOR DEPENDENCE OF AUGER INTENSITIES ON SPECIMEN THICKNESS.   2-0355
ADATOM CONCENTRATION PROFILES.   AUGER PLASMON SATELLITE INTENSITIES VERSUS DEPTH: A MEANS FOR DETERMINING 2-0347
UM OXIDES, AND OXIDIZED VANADIUM: CHEMICAL SHIFT AND PEAK INTENSITY ANALYSIS. /EMICAL COMPOSITION OF VANADI 4-1089
D ELASTIC (111) NICKEL SURFACES. (AUGER SPE/ ELASTIC LEED INTENSITY ENERGY STUDIES OF CLEAN (001), (110) AN 5-1292
AL/ SPECIMEN POSITION EFFECTS ON ENERGY SHIFTS AND SIGNAL INTENSITY IN A SINGLE STAGE CYLINDRICAL MIRROR AN 3-0696
LECTRON SPEC/ CHEMICAL SHIFTS AND VARIATIONS OF WIDTH AND INTENSITY OF AUGER SIGNALS OBTAINED WITH A PHOTOE 2-0273
SIUM AND SILICON FROM THE CROSS SECTION M/ A STUDY OF THE INTENSITY OF THE L2,3VV AUGER TRANSITION IN MAGNE 4-0765
ESCENCE IN PURE GERMANIUM. (AUG/ MAGNETIC FIELD DEPENDENT INTENSITY OSCILLATIONS OF THE ELECTRON HOLE LUMIN 4-0780
ELECTRON APPEARANCE POT/ DIFFERENCE IN THE CHROMIUM L3/L2 INTENSITY RATIO MEASURED BY SOFT X-RAY AND AUGER 4-0890
RANCE POTENTIAL SPECTRA INDUCED BY INCIDENT ELECTRON BEAM INTERACTION. /HE TITANIUM DIOXIDE AUGER AND APPEA 5-1834
E OF GOLD.   INTERACTION BETWEEN CHLORINE AND THE (100) SURFAC 5-1335
OF GALLIUM PHOSPHIDE.   AUGER ELECTRON SPECTRUM DURING THE INTERACTION BETWEEN OXYGEN AND THE (111) SURFACE 5-1642
OF G/ SOME FEATURES OF THE AUGER ELECTRON SPECTRUM IN THE INTERACTION BETWEEN OXYGEN AND THE (111) SURFACE 5-1643
METHANE, HYDROGEN FLUORIDE, WATER, AND CA/ CCNFIGURATION INTERACTION CALCULATIONS OF THE AUGER SPECTRUM OF 4-0879
NE SILICON. (AUGER SPECTROSCOPY)   ELECTRON- HOLE INTERACTION EFFECTS IN THE KL2,3L2,3 RADIATIVE AU 2-0540
PALLADIUM (100) SURFACE. (CHEMISORPTION AND CATALYTIC RE/ INTERACTION OF ALUMINUM LAYERS WITH POLYCRYSTALLI 5-1672
SURFACES. (AUGER EFFECT)   INTERACTION OF CARBON MONOXIDE AND OXYGEN WITH A 5-1321
CES. (AUGER SPECTROSCOPY)   INTERACTION OF CARBON MONOXIDE WITH (110) NICKEL 5-1607
)- ZINC SURFACE: A LEED, AUGER AND WORK FUNCTION STUDY.   INTERACTION OF CESIUM WITH ZINC OXIDE SURFA 5-1588
00).   LEED AND AES STUDY OF THE INTERACTION OF CHLORINE WITH THE ZINC OXIDE (0001 5-1463
.   INTERACTION OF HYDROGEN SULFIDE AND MOLYBDENUM (1 5-1973
ONTROLLED SURFACES. (AUGER NEUTRALIZATION AT (100) FACE / INTERACTION OF HYDROGEN SULFIDE WITH COPPER (100) 5-1516
TYLENE WITH AN IRON (001) SURFACE AND THEIR INFLUENCE ON/ INTERACTION OF LOW ENERGY ATMOSPHERIC IONS WITH C 5-1736
URFACE.   INTERACTION OF METHANE, ETHANE, ETHYLENE, AND ACE 5-1208
CT)   INTERACTION OF NITRIC OXIDE WITH A NICKEL (111) S 5-1278
UM (100) SURFACE. PART-2: REACTION KINETICS. (AUGER EFFE/ INTERACTION OF NITROGEN WITH NIOBIUM. (AUGER EFFE 5-1301
IN NIOBIUM- 1% ZIRCONIUM.   SURFACE SEGREGATION AND INTERACTION OF OXYGEN AND NITROGEN WITH THE NIOBI 5-1333
L SURFACES BY LOW ENERGY ELECTRON DIFFRACTI/ STUDY OF THE INTERACTION OF OXYGEN AND NITROGEN WITH ZIRCONIUM 5-1508
. (AUGER ELECTRON SPECTROSCOPY)   INTERACTION OF OXYGEN WITH PLATINUM SINGLE CRYSTA 5-1223
. (REPLY TO COMMENT)   INTERACTION OF OXYGEN WITH SILICON (111) SURFACES 5-1693
WITH CLEAN SILVER IN ULTRAHIGH VACUUM. (AES)   INTERACTION OF OXYGEN WITH SILICON (111) SURFACES 5-1512
ER ANALYSIS)   INTERACTION OF SULFUR DIOXIDE AND CARBON DIOXIDE 5-1575
A SEMICONDUCTOR.   EFFECT OF ELECTRON HOLE INTERACTION OF SULFUR DIOXIDE WITH TUNGSTEN. (AUG 5-1367
.   EFFECTS OF THE CONFIGURATION INTERACTION ON THE AUGER RECOMBINATION PROCESS IN 4-1093
       AUGER SPECTROSCOPY OF HYDROCHLORIC ACID INTERACTION ON THE K-SHELL AUGER SPECTRUM OF NEON 2-0279
CTROSCOPY)   SURFACE STRUCTURE OF TITANIUM AND ITS INTERACTION WITH ALUMINUM AT LOW PRESSURES.   5-1576
/AT A TITANIUM SURFACE. SEGREGATION FROM THE BULK CRYSTAL INTERACTION WITH BROMINE AND CHLORINE. (AUGER SPE 5-1538
TAL FACE. A STUDY OF ITS CLEANING, SURFACE STRUCTURE, AND INTERACTION WITH CHLORINE GAS. (AUGER SPECTROSCO/ 5-1833
ELECTRON SPECTROSCOPY CF HIGH TEMPERATURE OXYGEN- RHENIUM INTERACTION WITH OXYGEN AND CARBON MONOXIDE. /RYS 5-1883
ON SPECTROSCO/ INVESTIGATIONS OF THE GAS- SOLID INTERFACE INTERACTIONS.   AUGER 4-1111
M BY MOLECULAR BEAM SCATTERING, / STUDIES OF GAS- SURFACE INTERACTIONS AT HIGH TEMPERATURES BY AUGER ELECTR 5-1147
              ATOMIC NATURE OF POLYMER- METAL INTERACTIONS AT THE (100) CRYSTAL FACE OF PLATINU 5-1963
RIDIUM (110) SURFACES. (AUGER SPECTROSCOPY)   INTERACTIONS IN ADHESION, FRICTION, AND WEAR.   5-1212
PECTROSCOPY AND ELECTRON IMPACT DESORPTION STUDIES OF THE INTERACTIONS OF CARBON MONOXIDE AND OXYGEN WITH I 5-1252
ICLES. (AUGER SPECTROSCOPY)   INELASTIC INTERACTIONS OF GASES WITH SILICON SURFACES. /N S 5-1691
SULFUR AND CARBON MONOXIDE LEED PATTERNS. (AU/ ADSORBATE INTERACTIONS OF SLOW ELECTRONS WITH ABSORBED PART 2-0448
COPY)   ELECTRON BEAM ADSORBATE INTERACTIONS ON A PLATINUM (110) SURFACE. PART-1: 5-1190
       SILICON- OXYGEN INTERACTIONS ON SILICON SURFACES. (AUGER SPECTROS 5-1513
F SLIDING ON THESE I/ AUGER ANALYSIS OF OXYGEN AND SULFUR INTERACTIONS USING AUGER ELECTRON SPECTROSCOPY. 5-1511
G AUGER SPECTROSCOPY.   INTERACTIONS WITH VARIOUS METALS AND THE EFFECT O 5-1206
                                          INTERANGULAR BRITTLENESS STUDIES IN TUNGSTEN USIN 6-2061
OF CORE HOLE STATES.   INTERATOMIC AUGER EFFECT AT HIGH VELOCITIES.   2-0442
                    TRANSITION RATES FOR INTERATOMIC AUGER PROCESS: EFFECTS ON LIFETIMES 2-0317
ATES.   INTERATOMIC AUGER PROCESSES.   2-0438
TRANSITION METALS.   INTERATOMIC AUGER PROCESSES AND THE DENSITY OF ST 2-0491
                                          INTERATOMIC AUGER PROCESSES IN SOME ADSORBATES ON 4-1045
AUGER SPECTRA OF SEVERAL GAS ADSORBATES ON TRANSITION M/ INTERATOMIC AUGER TRANSITIONS. SURVEY. (36 REFS) 2-0318
/INATED COMPOUNDS IN THE 3-21 EV REGION OF THE SPECTRUM. (INTERBAND AUGER TRANSITIONS, RADIATIONLESS RECOM/ 4-1048
       OXIDATION OF CUPRONICKEL ALLOYS. PART-1: XPS STUDY OF INTERDIFFUSION. (AND AES)   5-1224
PLATINUM/ SCANNING MICROSPOT AUGER SPECTROSCOPY STUDY OF INTERDIFFUSION AND EUTECTIC FORMATION IN TUNGSTEN 5-1259
ATALYTIC ACTIVITY OF ALLCY SYSTEMS. (SURFACE COMPOSITION, INTERDIFFUSION, AUGER ELECTRON SPECTROSCOPY) /E C 6-2003
CON DIOXIDE- TANTALUM, AND SILICO/ SILICIDE FORMATION AND INTERDIFFUSION EFFECTS IN SILICON- TANTALUM, SILI 5-1260
S MEASURED BY AUGER ICN SPUTTERING TECHNIQUE.   INTERDIFFUSION IN BIMETALLIC POLYCRYSTALLINE FILM 5-1440
GALLIUM ARSENIDE (100) STUDIED WITH AUGER SPECTROSCOPY.   INTERDIFFUSIONS IN THIN FILM GOLD ON PLATINUM ON 5-1232
LECTRON SPECTROSCOPY, A METHOD FCR STUDYING THE SOLID GAS INTERFACE.   (AUGER) E 1-0013
A1 KLL AUGER PROFILE ARTIFACT AT THE OXIDE- METAL INTERFACE.   4-1068
ELECTRON SPECTROSCOPY OF THE PLATINUM- PLATINUM SILICIDE INTERFACE.   PLATINUM SILICIDE FORMATION. 5-1285
       SPECTROSCOPY OF A METAL- GAS INTERFACE.   5-1751
NEOUS ALLOYING OF A GOLD SUBSTRATE WITH LEAD MONOLAYERS. (INTERFACE AUGER ELECTRON SPECTROSCOPY)   SPONTA 5-1178
D FILM) SYSTEM. (AUGER ELECTRON SPECTROSCOPY)   DIFFUSE INTERFACE IN SILICON (SUBSTRATE)- GOLD (EVAPORATE 5-1681
PPLICATION OF AUGER ELECTRON SPECTROSCOPY TO THE STUDY OF INTERFACE IN STEELS.   A 6-2126
GER ELECTRON SPECTROSCO/ INVESTIGATIONS OF THE GAS- SOLID INTERFACE INTERACTIONS AT HIGH TEMPERATURES BY AU 5-1147

INTERFACE PHENOMENA IN ENGINEERING MATERIALS.   1-0176
VAPORATED FILM) SYSTEMS STUDIED BY AUGER ELECTRO/ DIFFUSE INTERFACE STRUCTURES IN SOME (SUBSTRATE- VACUUM E 5-1436
GER ANALYSIS OF BRASS- ENAMEL AND STAINLESS STEEL- ENAMEL INTERFACES.   AU 5-1266
ER/ THERMALLY INDUCED PROCESSES AT GOLD- GALLIUM ARSENIDE INTERFACES: AN ASSESSMENT BY RUTHERFORD BACKSCATT 5-1902
R ELECTRON SPECTROSCOPY AS APPLIED TO INTERF/ IMPURITIES, INTERFACES, AND BRITTLE FRACTURE. (REVIEW OF AUGE 6-2078
ELECTRON SPECTROSCOPY AND LEED STUDIES OF GAS- METAL INTERFACES. (AUGER EFFECT)   5-1320
RECENT ASPECTS OF GLASS FIBER- RESIN INTERFACES. (AUGER SPECTROSCOPY OF GLASS FIBERS) 5-1975
ROPROB/ CONFIRMATION OF IMPURITY ADSORPTION AT FLAKE IRON INTERFACES IN GRAY CAST IRON. (SCANNING AUGER MIC 6-2054
ADHESION AT GOLD MICA INTERFACES. (LEED, AES)   5-1434
NALYT/ THREE-DIMENSIONAL ELEMENT ANALYSIS OF SURFACES AND INTERFACES OF SOLIDS BY THE AUGER ELECTRON MICROA 5-1412
SPECTRA OF TITANIUM AND IRON.   INTERFACIAL AND PLASMON DECAY PEAKS IN THE AUGER 4-1060
AND MAGNESIUM.   INTERFACIAL AUGER TRANSITIONS IN OXIDIZED SODIUM 5-1497
GSTEN CONTACTS ON HEAVILY DOPED SIL/ CORRELATIONS BETWEEN INTERFACIAL OXIDES AND RESISTANCE CHANGES FOR TUN 5-1861
UNGSTEN ON HEAVILY DOPED SILICON- GERMANIUM A/ EFFECTS OF INTERFACIAL OXIDES ON THE CONTACT RESISTANCE OF T 5-1456
IN-DEPTH AUGER ANALYSIS OF ALUMINUM- SILICON INTERFACIAL REACTIONS.   5-1219
WEAR AND INTERFACIAL TRANSPORT OF MATERIAL. (AES)   5-1210
REACTION BETWEEN OXYGEN AND GERMANIUM. (AES) INTERFERENCE OF AN ELECTRON BEAM WITH THE SURFACE 5-1620
GRAIN BOUNDARY SEGREGATION OF IMPURITIES IN METALS AND INTERGRANULAR BRITTLE FRACTURE. (STUDIED BY AES) 6-2079
EFFECT OF TELLURIUM ON INTERGRANULAR COHESION OF IRON.   6-2099
TEEL. / CHEMISTRY OF GRAIN BOUNDARIES AND ITS RELATION TO INTERGRANULAR CORROSION OF AUSTENITIC STAINLESS S 6-2064
: EFFECT OF GRAIN BOUNDARY SEGREGATION. (AUGER ELECTRON / INTERGRANULAR EMBRITTLEMENT OF NICKEL BY HYDROGEN 6-2072
N STEEL.   STUDIES BY AUGER ELECTRON SPECTROSCOPY. INTERGRANULAR FRACTURE AND IMPURITY SEGREGATION I 6-2127
OPY)   GRAIN BOUNDARY SEGREGATION RELATING TO INTERGRANULAR FRACTURE. (AUGER ELECTRON SPECTROSC 6-2082
OSCOPY.   STUDY OF INTERGRANULAR FRACTURE IN IRON USING AUGER SPECTR 6-2097
AUGER TRANSITION PROBABILITIES FOR ELEMENTS WITH LOW AND INTERMEDIATE ATOMIC NUMBERS.   KLL 4-0803
AUGER SPECTRA IN THE INTERMEDIATE COUPLING REGION.   4-0887
N EMISSION)   NEON K-SHELL VACANCY PRODUCTION BY INTERMEDIATE ENERGY FLUORINE IONS. (AUGER ELECTRO 2-0572
L BY AN AUGER MICROPROBE.   STUDY OF THE INTERMITTENT MOTION OF FRICTION IN STAINLESS STEE 6-2133
R/ MERCURY AND GOLD ATOMIC CLOUDS REARRANGEMENT FOLLOWING INTERNAL CONVERSION AND ELECTRON CAPTURE. (K-AUGE 2-0334
ENERGY SIDE OF THE KL2,3M AUGER PEAK FROM SOLID ALUMINUM: INTERNAL PHOTOEMISSION.   STRUCTURE ON THE HIGH 4-1021
SPECTRA OF SODIUM, MAG/ EXPERIMENTAL DETERMINATION OF THE INTERNAL PHOTOEMISSION CONTRIBUTION TO KLM AUGER 4-0769
AUGER EXCITATION BY INTERNAL SECONDARY ELECTRONS.   2-0390
TION OF A NEW TYPE OF AUGER TRANSITIONS IN ATOMS WITH TWO INTERNAL VACANCIES.   DETEC 2-0234
DISTINCTION BETWEEN WEAK AUGER TRANSITIONS AND INTERNAL X-RAY PHOTOEMISSION.   2-0436
EMITTED DURING ION BOMBARDMENT.   INTERPRETATION OF ENERGY LOSS BY AUGER ELECTRONS 2-0268
L2,3 EMISSION BANDS OF SODIUM, MAGNESIUM, ALUMINUM, AND/ INTERPRETATION OF HIGH ENERGY X-RAY SATELLITES 2-0239
RA OF COPPER AND ZINC.   L-S COUPLING INTERPRETATION OF HIGH RESOLUTION LMM AUGER SPECT 4-1122
HOLDS IN METALS AND SEMICONDUCTORS. (SILICON) INTERPRETATION OF KL2,3L2,3 RADIATIVE AUGER THRES 4-0723
M OXIDIZED BERYLLIUM.   INTERPRETATION OF THE AUGER ELECTRON SPECTRUM FRO 4-0935
M CLEAN BERYLLIUM.   OBSERVATION AND INTERPRETATION OF THE AUGER ELECTRON SPECTRUM FRO 4-0997
OXIDIZED BERYLLIUM.   INTERPRETATION OF THE AUGER SPECTRUM OF CLEAN AND 4-1087
IN SOLIDS.   INTERPRETATION OF THE LOW ENERGY K-BETA SPECTRUM 4-1101
/ DIFFUSION CURRENT ON GALVANOMAGNETIC PROPERTIES IN THIN INTRINSIC INDIUM ANTIMONIDE AT ROOM TEMPERATURE./ 2-0354
MS. (TITANIUM AND CHROMIUM ALLOYS, AES)   LOW INTRINSIC STRESS AND GOOD ADHESION ALLOY THIN FIL 5-1909
ECT)   EFFECT OF IMPURITIES ON INTRINSIC STRESS IN THIN NICKEL FILMS. (AUGER EFF 5-1131
CTROSCOPY)   EFFECT OF IMPURITIES ON INTRINSIC STRESS IN THIN NICKEL FILMS. (AUGER SPE 5-1130
TIONS IN TWO DIMENSIONAL ABSORBED SULFUR LAYERS ON COPPER INVESTIGATED BY LEED AND AUGER SPECTROSCOPY. /NSI 5-1302
ONS.   ANALYSIS APPARATUS FOR INVESTIGATING A SAMPLE BY MEANS OF EMITTED ELECTR 3-0713
IDENCE EXPERIMENTS. (AUGER SPECTRUM OF HE/ POSSIBILITY OF INVESTIGATING ATOMIC AUTOIONIZING STATES IN COINC 2-0257
INELASTICA/ FARADAY CUP LEED APPARATUS WITH FACILITY FOR INVESTIGATING ENERGY AND ANGULAR DISTRIBUTIONS OF 3-0691
RGIES 100-2000 EV. (AES)   SORBAS: AN APPARATUS FOR INVESTIGATING ION SCATTERING FROM SURFACES AT ENE 3-0704
ELECTRON AUGER SPECTROSCPY: A METHOD OF INVESTIGATING SURFACE PHENOMENA (REVIEW).   1-0132
TEMS. (SURFACE COMPOSITION, INTERDIFFUSI/ A TECHNIQUE FOR INVESTIGATING THE CATALYTIC ACTIVITY OF ALLOY SYS 6-2003
PHOTOEMISSION STUDIES OF CORE EXCITON DECAY IN POTASSIUM IODIDE.   2-0422
ECTRON EMISSION ON ENERGY OF PRIMARY ELECTRONS (POTASSIUM IODIDE). /ENCE OF THE COEFFICIENT OF SECONDARY EL 4-0891
ECTRA OF SILVER, CADMIUM, INDIUM, ANTIMONY, TELLURIUM AND IODINE.   HIGH RESOLUTION MNN AUGER SP 4-0730
LEED- AUGER, AND WORK FUNCTION STUDY OF IODINE ADSORBED ON TUNGSTEN (110).   5-1145
ECTRONS EMITTED BY ALKALI HALIDE CRYSTALS UNDER IMPACT OF ION AND ATOM BOMBARDMENT. (AUGER EFFECT) /M OF EL 2-0246
NTACT WITH METALS AS STUDIED BY AUGER SPECTROSCOPY, FIELD ION AND SCANNING ELECTRON MICROSCOPY. /LENE IN CO 5-1198
ATE DEPENDENCE OF THE K-VACANCY PRODUCTION YIELD IN HEAVY ION ATOM COLLISIONS AT KEV ENERGIES. /C CHARGE ST 2-0346
UGER ELECTRON SPECTR/ ANALYSIS OF SURFACE LAYERS BY LIGHT ION BACKSCATTERING AND SPUTTERING COMBINED WITH A 5-1167
AUGER CYLINDRICAL MIRROR ANALYZER.   LOW ENERGY ION BACKSCATTERING SPECTROSCOPY WITH A COMMERCIAL 3-0665
O/ SCANNING ELECTRON MICROSCOPY, AUGER SPECTROSCOPY, AND ION BACKSCATTERING TECHNIQUES APPLIED TO ANALYSIS 5-1261
PY IN THE ANALYSIS OF SU/ DIRECT COMPARISON OF LOW ENERGY ION BACKSCATTERING WITH AUGER ELECTRON SPECTROSCO 5-1869
ODIUM. (SCANNING ELECTRON MICROSCOPY, AUGER SPECTROSCOPY, ION BEAM ANALYSIS) / PRODUCTS IN FLOWING LIQUID S 6-2014
RECENT ADVANCES IN SURFACE STUDIES. ION BEAM ANALYSIS. (AUGER SPECTROSCOPY, REVIEW) 1-0137
DOPPLER SHIFTED AUGER ELECTRONS FROM FOIL EXCITED HEAVY ION BEAMS.   4-0859
AUGER ELECTRONS FROM FOIL EXCITED HEAVY ION BEAMS. (OF NEON)   2-0370
AUGER DECAY OF NEON FOLLOWING ENERGETIC ION BOMBARDMENT.   4-0954
AUGER SPECTRA OF LITHIUM AND SODIUM UNDER ION BOMBARDMENT.   4-0817
LICON EXCITED BY LOW ENERGY ELECTRON AND LOW ENERGY ARGON ION BOMBARDMENT. / OF MAGNESIUM, ALUMINUM, AND SI 4-0857
EMISSION OF AUGER ELECTRONS BY ION BOMBARDMENT.   2-0550
TED FROM BERYLLIUM, MAGNESIUM, ALUMINUM AND SILICON UNDER ION BOMBARDMENT. /IBUTION OF AUGER ELECTRONS EMIT 4-0876
N KLL AUGER SPECTRA PRODUCED BY HYDROGEN(+) AND HELIUM(+) ION BOMBARDMENT.   HIGH RESOLUTION NEO 4-0953
RESOLUTION STUDY OF ARGON LMM AUGER ELECTRONS PRODUCED BY ION BOMBARDMENT.   HIGH 4-0903
RETATION OF ENERGY LOSS BY AUGER ELECTRONS EMITTED DURING ION BOMBARDMENT.   INTERP 2-0268
K-SHELL AUGER ELECTRON PRODUCTION CROSS SECTIONS FROM ION BOMBARDMENT.   2-0570
NTS OF NEON K-AUGER ELECTRONS PRODUCED BY ENERGETIC HEAVY ION BOMBARDMENT.   MEASUREME 4-0951
GER ANALYSIS OF COPPER- NICKEL ALLOY SURFACES AFTER ARGON ION BOMBARDMENT.   QUANTITATIVE AU 5-1807
TAL FILMS BY COMBINATION AUGER SPECTROSCOPY AND INERT GAS ION BOMBARDMENT. /OXIDE THICKNESS ON DEPOSITED ME 6-2048
LECTRONS FROM LIGHT METALS (LITHIUM, SODIUM) SUBJECTED TO ION BOMBARDMENT. /EMISSION INTENSITIES OF AUGER E 6-2048
GER ANALYSIS OF COPPER- NICKEL ALLOY SURFACES AFTER ARGON ION BOMBARDMENT. COMMENTS. /LOYS. QUANTITATIVE AU 5-1740
ELECTRON SPECTROSCOPY.   EASY METHOD TO ACCURATELY ALIGN ION BOMBARDMENT GUNS FOR DEPTH PROFILING IN AUGER 3-0700
K-SHELL AUGER ELECTRON PRODUCTION CROSS SECTIONS FROM ION BOMBARDMENT. (NEON, FLUORINE)   2-0574
AUGER ELECTRON EMISSION BY ION BOMBARDMENT OF LIGHT METALS.   4-0877
ARGON AUGER EMISSION BY ARGON(+) ION BOMBARDMENT OF SOME METALS.   2-0547
EASUREMENTS)   EFFECTS OF LOW ENERGY ARGON- ION BOMBARDMENT ON MOLYBDENUM DISULFIDE. (AUGER M 2-0348
OF TUNGSTEN AND MOLYBDENUM SINGLE CRYSTALS STU/ EFFECT OF ION BOMBARDMENT ON THE COMPOSITION AND STRUCTURE 2-1215
GER ANALYSIS OF COPPER- NICKEL ALLOY SURFACES AFTER ARGON ION BOMBARDMENT. REPLY TO COMMENTS. /NTITATIVE AU 5-1200
GER ANALYSIS OF COPPER- NICKEL ALLOY SURFACES AFTER ARGON ION BOMBARDMENT. REPLY TO COMMENTS. /NTITATIVE AU 5-1702

NVESTIGATION OF REORDERED (001) GOLD SURFACES BY POSITIVE ION CHANNELING SPECTROSCOPY, LEED AND AES.   I 5-1139
/CTERIZATION OF REORDERED (001) GOLD SURFACES BY POSITIVE ION CHANNELING SPECTROSCOPY, LOW ENERGY ELECTRON/ 5-1986
+)- ARGON COLLISIONS. (AUGER EFFECT)   ELECTRON ION CCINCIDENCE STUDIES OF HIGHLY IONIZING ARGON/ 2-0529
DEL FOR K X-RAY AND AUGER ELECTRON ENERGY SHIFTS IN HEAVY ION COLLISIONS. COMMENTS.   SIMPLE MO 2-0301
TESTING.   APPLICATION OF ION ELECTRON, AND X-RAY BEAMS IN MODERN MATERIAL 1-0034
IN-DEPTH PROFILES BY AUGER SPECTROSCOPY AND SECONDARY ION EMISSION.   4-0995
OON OF MAGNESIUM L2,3MM AUGER CURRENTS USING ELECTRON AND ION EXCITATION.   COMPARI 4-0866
ION EXCITED AUGER SPECTRA OF ALUMINUM.   4-0869
-AUGER ELECTRON SPECTRUM PRODUCED BY 45 MEV CHLORINE(12+) ION IMPACT.   HIGH RESOLUTION NEON K 4-1051
EN(+), DEUTERIUM(+), MCLECULAR HYDROGEN(+), AND HELIUM(+) ION IMPACT. /AND METHANE BY 50- TO 600 KEV HYDROG 2-0519
DETERMINED BY AES.   FLUORINE ION IMPLANTATION PROFILES IN GALLIUM ARSENIDE AS 5-1398
OPY.   OBSERVATION OF IMPURITY PROFILE IN ION IMPLANTED SILICON BY AUGER ELECTRON SPECTROSC 5-1482
ECONDARY ION EMISSION.   IN-DEPTH PROFILES CF PHOSPHORUS ION IMPLANTED SILICON BY AUGER SPECTROSCOPY AND S 4-0996
SECONDARY ELECTRON EMISSION FROM ION IMPLANTED SILICON (CESIUM).   4-0913
DETECTION OF EXCITONS NEAR THE L2,3 EDGE OF THE CHLORIDE ION IN SODIUM CHLORIDE BY MEANS OF THE DECOMPOSI/ 4-1126
ORIGIN OF ION INDUCED AUGER ELECTRON EMISSION FROM METALS. 2-0549
SURFACE ANALYSIS WITH HEAVY ION INDUCED AUGER ELECTRONS.   5-1387
A NEW METHOD FOR MATERIALS S/ SURFACE ANALYSIS WITH HEAVY ION INDUCED AUGER ELECTRONS. BASIC PROPERTIES OF 1-0070
N THE SCANNING ELECTRON MICROSCOPE NOW INCLUDES SECONDARY ION MASS ANALYSIS. (AUGER)   SPECTROSCOPY I 3-0674
COMPARISON OF AUGER ELECTRON SPECTROSCOPY AND SECONDARY ION MASS ANALYSIS, AUGER ELECTRON MICROSCOPY) /IS 1-0010
NUM SILICIDE BY AUGER ELECTRON SPECTROSCOPY AND SECONDARY ION MASS SPECTROMETRY.   1-0135
L ANALYSIS WITH AUGER ELECTRON SPECTROSCOPY AND SECONDARY ION MASS SPECTROMETRY. /OSITED PLATINUM AND PLATI 5-1656
NTALUM FILMS BY AUGER ELECTRON SPECTROSCOPY AND SECONDARY ION MASS SPECTROMETRY.   ELEMENTA 5-1654
SECONDARY ION MASS SPECTROMETRY. /D OXYGEN) IN SPUTTERED TA 5-1652
SCOPY): A USEFUL COMPLEMENT TO OTHER DEPTH PRO/ SECONDARY ION MASS SPECTROMETRY.   1-0079
/ANLINESS OF 316 L-N STAINLESS STEEL STUDIES BY SECONDARY ION MASS SPECTROMETRY (AND AUGER ELECTRON SPECTRO 1-0009
/ACE PROCESSES ON COPPER- BERYLLIUM BY COMBINED SECONDARY ION MASS SPECTROMETRY AND AUGER ELECTRON SPECTRO/ 3-0654
COPY, FOR THE ANALYSIS OF THIN / COMBINATION CF SECONDARY ION MASS SPECTROMETRY AND AUGER ELECTRON SPECTROS 5-1213
/) NITROGEN, OXYGEN, AND FLUORINE IN SILICON BY SECONDARY ION MASS SPECTROMETRY AND AUGER ELECTRON SPECTROS 5-1214
/ANLINESS OF 316 L-N STAINLESS STEEL STUDIED BY SECONDARY ION MASS SPECTROMETRY AND AUGER ELECTRON SPECTRO/ 5-1916
/ED SURFACES BY AUGER ELECTRON SPECTROSCOPY AND SECONDARY ION MASS SPECTROMETRY AND AUGER ELECTRON SPECTRO/ 6-2087
NOLOGY./ USE OF AUGER ELECTRON SPECTROSCOPY AND SECONDARY ION MASS SPECTROMETRY FOR ALUMINUM, SILICON, TIT/ 5-1553
AN ANALYTICAL SYSTEM FOR SECONDARY ION MASS SPECTROMETRY IN THE MICROELECTRONIC TECH 1-0136
ECTROSCOPY X-RAY PHOTOELECTRON SPECTROSCOPY AND SECONDARY ION MASS SPECTROMETRY IN ULTRAHIGH VACUUM. (AES) 3-0632
CK FILMS ON COPPER PLATES BY AUGER ELECTRON AND SECONDARY ION MASS SPECTROSCOPY. /S USING AUGER ELECTRON SP 5-1353
/ SURFACES CHARACTERIZED BY AUGER SPECTROSCOPY, SECONDARY ION MASS SPECTROSCOPY. /ANED SURFACES OF GOLD THI 5-1550
LING FOR RELIABILI/ DIRECT COMPARISON OF AUGER, SECONDARY ION MASS SPECTROSCOPY AND LOW ENERGY ELECTRON DI/ 2-0501
/TO- AND AUGER ELECTRON SPECTROSCOPY (ESCA) AND SECONDARY ION MASS SPECTROSCOPY, AND PROTON RESONANCE PROFI 1-0216
LYSIS TECHNIQUES. (AUGER ELECTRON SPECTROSCOPY, SECONDARY ION MASS SPECTROSCOPY. COMPARISON OF TWO SURFAC/ 1-0088
/ARISON BETWEEN AUGER ELECTRON SPECTROSCOPY AND SECONDARY ION MASS SPECTROSCOPY, ELECTRON MICROPROBE) / ANA 5-1627
CHEMISTRY OF CERAMIC SURFACES. (AUGER SPECTROSCOPY, LEED, ION MASS SPECTROSCOPY ON WELL CHARACTERIZED SING/ 5-1689
ER IN PHYSICAL METALLURGY.   POSSIBILITIES OF THE ION MICROPROBE ANALYSIS)   CRYSTALLOGRAPHY AND 5-1852
FOR MICROPROBE ANALYSIS. (AUGER ELECTRON SPECTROSCOPY AND ION MICROPROBE MASS ANALYZER AND AUGER SPECTROMET 6-2075
LEED STUDY OF THE STRUCTURE AND COMPOSITI/ CCMBINED FIELD ION MICROPROBE MASS SPECTROMETRY) /STRUMENTATION 3-0716
ECTRON DIFFRACTION, AUGER EMISSION SPECTROSCOPY AND FIELD ION MICROSCOPY. AUGER ELECTRON SPECTROSCOPY, AND 5-1951
/URFACE STRUCTURES, SPECIES, TOPOGRAPHY, PHENOMENA, FIELD ION MICROSCOPY IN MICROSTRUCTURAL STUDIES. /GY EL 1-0057
SULFUR SURFACE: LINE SHAPE ANALYSIS AND CORRELATION WITH ION MICROSCOPY, LEED, HEED, AUGER SPECTROSCOPY, / 5-1492
ELECTRON SPECTROSCOPY AND SECONDA/ COMPOSITICN PROFILE OF ION NEUTRALIZATION SPECTROSCOPY. /(110)- C(2X2) 4-1065
AUGER ELECTRON SPECTROSCOPY ANALYSIS OF ION PLATED GOLL FILM ON COPPER ANALYZED BY AUGER 5-1679
/ COMPOSITION AND STRUCTURE ANALYSIS WITH A CNE APPARATUS ION PLATED GOLD FILMS ON SILICON.   5-1225
SURFACE ANALYSIS BY LOW ENERGY ION PROBE TECHNIQUE. (COMPARISON BETWEEN ION PRO/ 5-1166
SURFACE ANALYSIS WITH LOW ENERGY ION REFLECTION.   5-1822
UDIES OF CLEAN NICKEL SURFACES AND ADSORBED L/ LOW ENERGY ION SCATTERING.   5-1871
UGER NEUTRALIZATION)   LOW ENERGY ION SCATTERING AND AUGER ELECTRON SPECTROSCOPY ST 5-1870
EV. (AES)   SORBAS: AN APPARATUS FCR INVESTIGATING ION SCATTERING: ELASTIC AND INELASTIC EFFECTS. (A 2-0382
ECTR/ COMPARISON OF SURFACE CHEMISTRIES OBTAINED BY USING ION SCATTERING FROM SURFACES AT ENERGIES 100-2000 3-0704
/FOR THE CHEMICAL CHARACTERIZATICNS OF CERAMIC SURFACES. (ION SCATTERING SPECTROMETRY AND AUGER ELECTRON SP 6-1999
ROSCOPY AND SECONDARY I/ SURFACE ANALYTICAL STUDIES USING ION SCATTERING SPECTROMETRY, AUGER ELECTRON SPEC/ 5-1326
ANALYZER)   QUANTITATIVE ASPECTS OF ION SCATTERING SPECTROMETRY, AUGER ELECTRON SPECT/ 5-1423
NDUCTOR SURFACES. (AUGER EFFECT) ION SCATTERING SPECTROSCOPY. (CYLINDRICAL MIRROR 3-0666
ANALYSIS OF HIGH CHARGE STATE OUTPUT FROM A PENNING ION SELECTIVE SPUTTERING OF III-V COMPOUND SEMICO 5-1637
OSCOPY AND ELECTRON IMPACT DESORPTION / IDENTIFICATION CF ION SOURCE.   3-0602
ON OF TITANIUM(IV) OXIDE.   AES ION SPECIES IN THE COMBINED AUGER ELECTRON SPECTR 5-1692
TTERING MEASUREMENTS. APPLICATION TO LOW ENERGY ARGON(+) ION SPUTTERING ANALYSIS AND THE SURFACE COMPOSITI 5-1893
ION IN BIMETALLIC POLYCRYSTALLINE FILMS MEASURED BY AUGER ION SPUTTERING OF SILVER AND NIOBIUM. /OPY IN SPU 5-1825
QUANTITATIVE DEPTH PROFILING BY AUGER ION SPUTTERING TECHNIQUE.   INTERDIFFUS 5-1440
YSIS OF SURFACE COMPOSITION WITH LOW ENERGY BACKSCATTERED ION SPUTTERING TECHNIQUE.   1-0085
IONS.   ANAL 5-1823
K-AUGER CROSS SECTIONS IN COLLISION WITH 30 MEV FLUORINE IONS.   ARGON 2-0323
LEMENTS AND BINARY CCMPOUNDS BOMBARDED BY 60 KEV ARGON(+) IONS.   AUGER SPECTRA OF SOME FREE E 4-1106
AUGER TRANSITIONS IN LITHIUM-LIKE BERYLLIUM-LIKE IONS.   2-0296
SPECTROMETER FOR THE DETECTION OF PHOTONS, ELECTRONS AND IONS.   HIGH SENSITIVITY APPEARANCE POTENTIAL 3-0643
ELECTRONS EMITTED BY COPPER UNDER BOMBARDMENT BY RARE GAS IONS. /TRONS IN THE ENERGY SPECTRUM OF SECONDARY 2-0551
FROM SODIUM CHLORINE SINGLE CRYSTAL BOMBARDED BY RARE GAS IONS.   SECONDARY ELECTRON EMISSION 4-0759
THEORY OF AUGER EJECTION OF ELECTRONS FRCM METALS BY IONS.   2-0563
STAL. (AUGER NEUTRA/ CHARGE EXCHANGE OF LOW ENERGY HELIUM IONS AND ATOMS SCATTERED FROM A COPPER SINGLE CRY 2-0545
AUGER EFFECT AND CORE BROADENING IN LOW-Z ATOMS, IONS, AND METALS.   2-0353
ARGON L-SHELL AUGER SPECTRA PRODUCED IN ARGON(+) ION- ARGON COLLISIONS.   4-1080
ENERGY LOSS AND IONIZATION IN FAST ION- ATOM COLLISIONS.   2-0233
RADIATIVE AUGER EFFECT IN ION- ATOM COLLISIONS.   2-0481
ECT)   ION- ATOM COLLISIONS AT HIGH ENERGIES. (AUGER EFF 2-0480
RGET EXPERIMENTS OF K-SHELL VACANCY SHARING IN ASYMMETRIC ION- ATOM COLLISIONS. (AUGER EFFECT)   GAS TA 2-0523
MECHANISMS OF ELECTRON PRODUCTION IN ION- ATOM COLLISIONS. (AUGER EFFECT)   2-0488
PRODUCTION OF INNER SHELL VACANCIES IN HEAVY ION- ATOM COLLISIONS. (AUGER EFFECT)   2-0405
MECHANISMS OF ELECTRON PRODUCTION IN ION- ATOM COLLISIONS. (AUGER EFFECT, REVIEWS)   2-0490
ION- ATOM COLLISIONS. (AUGER EMISSION, REVIEW)   1-0169
MOLECULAR EFFECTS ON K-VAANCY SHARING IN ASYMMETRIC ION- ATOM COLLISIONS. (AUGER SPECTRA)   2-0522
/ ENERGY DISTRIBUTION OF (AUGER) ELECTRONS RESULTING FROM ION- ATOM COLLISIONS WITH A FIXED IMPACT PARAMETE 2-0331
PRODUCTION CROSS SECTIONS IN ARGON BY THE IMPACT OF HEAVY IONS. (AUGER EFFECT) /E OF TOTAL K-SHELL VACANCY 2-0446

OF HIGHLY STRIPPED CXYGEN, FLUORINE, CHLORINE, AND ARGON IONS. (AUGER EFFECT) /RA FROM AUTOIONIZING STATES 2-0472
-SHELL VACANCY PRODUCTION BY INTERMEDIATE ENERGY FLUORINE IONS. (AUGER ELECTRON EMISSION) NEON K 2-0572
SCINTILLATION EFFICIENCY FCR LOW VELOCITY HEAVY IONS. (AUGER EMISSION) 2-0431
T) MULTIIONIZATION OF KRYPTON AND ITS IONS BY HIGH ENERGY ELECTRON IMPACT. (AUGER EFFEC 2-0493
ION) MULTIIONIZATION OF NEON, ARGON, XENCN AND THEIR IONS BY HIGH ENERGY ELECTRON IMPACT. (AUGER EMISS 2-0492
LENT ARGON(+)- ARGON CCLIISION. EXTENDED ION- ELECTRON COINCIDENCE MEASUREMENTS OF THE VIO 2-0530
/GH RESOLUTION BEAM GAS AUGER ELECTRON SPECTRA FOR OXYGEN IONS EXCITED BY COLLISIONS WITH HELIUM, NEON, AN/ 4-0904
ROGE/ RELATIVE ABUNDANCES AND RECOIL ENERGIES OF FRAGMENT IONS FORMED FROM THE X-RAY PHOTOIONIZATION OF NIT 2-0308
UGER) NEUTRALIZATION BEHAVIOR IN SCATTERING CF LOW ENERGY IONS FROM SOLID SURFACES. (A 2-0294
. (AES) SPUTTERING OF IONS FROM STAINLESS STEEL BY HYDROGEN BOMBARDMENT 6-2017
ELECTRONS OR IONS. (IN SURFACE ANALYSIS) 5-1765
AES COMPOSITIONAL PROFILES OF MOBILE IONS IN THE SURFACE REGION OF GLASS. 5-1722
/ON FUNCTIONS FOR SODIUM, FOTASSIUM, RUBIDIUM, AND CESIUM IONS IN THE VACUUM ULTRAVIOLET SPECTRAL RANGE. (/ 4-0735
OUTER SHELL EXCITATION OF THE RARE GASES BY 30 MEV OXYGEN IONS. (INNER SHELL AUGER TRANSITIONS) 2-0298
CTROSCOPY) INJECTION OF IONS INTO GLASS FROM A GLOW DISCHARGE. (AUGER SPE 5-1222
CE OF A COPPER S/ CHARGE EXCHANGE OF LOW ENERGY HELIUM(+) IONS (LESS THAN 10 KEV) SCATTERED FROM A (100) FA 2-0546
AUGER IONIZATION OF ATOMS BY MULTIPLY CHARGED IONS. MODEL OF TWO COULOMB CENTERS. 2-0407
IONIZATION AND FLUORESCENCE YIELD IN 50 MEV CHLORINE(N+) ION- NEON COLLISIONS. /DEPENDENCE OF NEON K-SHELL 2-0300
AUGER NEUTRALIZATION OF MULTIPLY CHARGED IONS ON A METAL SURFACE. 2-0247
ION OF THE FREQUENCY OF AUGER NEUTRALIZATION OF INERT GAS IONS ON THE SURFACE OF ALKALI HALIDE CRYSTALS. /T 5-1544
X-RAY PRODUCTION BY HEAVY IONS. (OXYGEN BOMBARDMENT, AUGER SPECTRUM) 2-0479
/AND CARBON CONTAMINATED MOLYBDENUM BOMBARDED BY POSITIVE IONS. PART-1. PART-2: EFFECT OF ANGLE OF INCIDEN/ 2-0543
N EMISSION FROM SOLID SURFACES BOMBARDED BY MEDIUM ENERGY IONS (PLATINUM, NICKEL, TANTALUM, GRAPHITE). /TRO 2-0309
OF SECONDARY ELECTRON EMISSICN FROM FASSAGE CF LITHIUM(+) IONS THROUGH COPPER SINGLE CRYSTALS. ANISOTROPY 2-0245
ION AT (100) FACE / INTERACTION CF LOW ENERGY ATMOSPHERIC IONS WITH CONTROLLED SURFACES. (AUGER NEUTRALIZAT 5-1736
/RONIC EMISSION FROM SCLID TARGETS BCMBARDED BY NOBLE GAS IONS (10-100 KEV): ENERGETIC AND SPATIAL DISTRIB/ 4-0776
AUGER RATES FOR SOFT X-RAY TRANSITIONS IN IONIC, ATOMIC, AND METALLIC LITHIUM. 4-0835
R ELECTRONS BY ATOMS IN A METALLIC TARGET SUBJECTED TO AN IONIC BOMBARDMENT. EMISSION OF AUGE 2-0383
ND ANGULAR DISTRIBUTION CF SECONDARY ELECTRONS EMITTED BY IONIC BOMBARDMENT. ENERGY SPECTRUM A 4-0988
SECONDARY ELECTRON SPECTRA OF SOME SOLID TARGETS UNDER IONIC BOMBARDMENT. 4-0942
AUGER SPECTROSCOPIC STUDY OF THE SURFACE OF IRON AFTER IONIC BOMBARDMENT AND AIR OXIDATION. 5-1939
SIZE EFFECT IN IONIC CHARGE RELAXATICN FOLLOWING AUGER EFFECT. 2-0565
AUGER ELECTRON FINE PROFILES IN IONIC CRYSTALS. 4-0949
ROLE OF AUGER EFFECT IN THERMOSTIMULATED PHENOMENA IN IONIC CRYSTALS. 2-0539
STUDY OF METAL SURFACES BY IONIC MICROANALYSIS AND AUGER SPECTROMETRY. 5-1582
ADSORPTION OF IONIC SALTS ON A TUNGSTEN (100) SURFACE. 5-1659
MEASUREMENT OF THE AUGER EFFECT IN THE MU-4 HELIUM(2S) (+) IONIC SYSTEM. 2-0305
ANGULAR DISTRIBUTION OF AUGER ELECTRONS FOLLOWING IMPACT IONIZATION. 2-0321
UGER DISSOCIATION OF MOLECULES AS A RESULT OF INNER SHELL IONIZATION. A 2-0465
IMPACT IONIZATION AND AUGER RECOMBINATION IN BANDS. 2-0418
LVER BY ELECTRON IMPACT. MULTIPLE IONIZATION AND AUGER TRANSITIONS IN INDIUM AND SI 2-0229
ECTRONS. (KLL TRANSITION ENERGIES FOR/ INNER SHELL DOUBLE IONIZATION AND CHEMICAL BONDING. PART-1: AUGER EL 2-0327
INE(N/ PROJECTILE CHARGE STATE DEPENDENCE OF NEON K-SHELL IONIZATION AND FLUORESCENCE YIELD IN 50 MEV CHLOR 2-0300
INE(+)- NEON COLLISICNS. (AUGER ELECTRON PRODUCT/ K-SHELL IONIZATION AND FLUORESCENCE YIELD IN 50 MEV CHLOR 2-0518
UM AND BERYL/ CHARACTERISTIC ENERGY GAIN ANC LOSS, DOUBLE IONIZATION, AND IONIZATION LOSS EVENTS IN BERYLLI 2-0901
ATOMIC VACANCY DISTRIBUTIONS FRODUCED BY INNER SHELL IONIZATION (AUGER AND RADIATIVE TRANSITIONS). 2-0544
INNER SHELL ALIGNMENT OF ATOMS IN ELECTRON IMPACT IONIZATION. (AUGER ELECTRONS) 2-0335
DEUTERONS. AUGER SPECTRA OF CARBON AND ARGCN FOLLCWING IONIZATION BY EQUAL VELOCITY ALPHA PARTICLES AND 4-1109
AUGER ELECTRON SPECTRCSCOFY APPLIED TC INNER SHELL IONIZATION BY FAST CHARGED PARTICLES. 2-0535
K-SHELL IONIZATION BY FAST PROTONS. 2-0533
ECULAR HYDROGEN, AND HELIUM(+) IMPACT. ARGON L-SHELL IONIZATION BY 50 TO 600 KEV PROTON, DEUTERON, MOL 2-0521
APPEARANCE POTENTIAL SPECTROSCOPY. COMPARISON WITH AUGE/ IONIZATION (CHARACTERISTIC LOSS) SPECTROSCOPY AND 1-0173
MINUM IN A NEON(+)- ALUMINUM COLLISICN. IONIZATION CROSS SECTION OF THE L2,3 SHELL OF ALU 2-0267
ORINE USING AUGER ELECTRON SPECTROSCO/ MEASUREMENT OF THE IONIZATION CROSS SECTION OF THE L2,3 SHELL OF CHL 4-0987
ON FROM SOLIDS. ESTIMATION OF BACKSCATTERING EFFECTS AND IONIZATION CROSS SECTIONS. AUGER EMISSI 2-0513
ORS FOR USE IN QUANTITATIVE AES. MEASUREMENT OF IONIZATION CROSS SECTIONS AND BACKSCATTERING FACT 2-0554
RIUM EQUATIONS MODIFIED BY AUGER TRANSITIONS. X-RAY IONIZATION CROSS SECTIONS, AND IONIZATION EQUILIB 2-0562
RAY EXCITED AND ELECTRON EXCITED AUGER SPE/ CCMPARISON OF IONIZATION CROSS SECTIONS AND OTHER FACTORS IN X- 2-0261
ROSCOPY. L-SHELL IONIZATION CROSS SECTIONS IN AUGER ELECTRON SPECT 2-0339
MIC OXYGEN, AND NEON. (AUGER TR/ ABSOLUTE ELECTRON IMPACT IONIZATION CROSS SECTIONS OF ATOMIC NITROGEN, ATO 2-0363
A NEW METHOD FOR MEASURING ELECTRON IMPACT IONIZATION CROSS SECTIONS OF INNER SHELLS. 4-0848
D BY AUGER ELECTRON SPECTROSCOPY. ELECTRON IMPACT IONIZATION CROSS SECTIONS OF INNER SHELLS MEASURE 2-0553
ATOMS. (AUGER INTENSITIES OF TUNGS/ TOTAL ELECTRON IMPACT IONIZATION CROSS SECTIONS ON K-SHELLS OF SURFACE 4-0847
S, AND FLUORESCENCE YIELDS. DOUBLE IONIZATION EFFECTS ON RADIATIVE RATES, AUGER RATE 4-0976
10-PHENAN/ MOSSBAUER STUDIES OF THE AFTEREFFECTS OF AUGER IONIZATION FOLLOWING ELECTRON CAPTURE IN TRIS (1, 2-0516
ELECTRON IMPACT. (AUGER SPECTRA) DOUBLE AND TRIPLE IONIZATION IN ALUMINUM, GALLIUM, AND SCANDIUM BY 2-0232
COMPARISON OF X-RAY AND ELECTRON IMPACT IONIZATION IN AUGER SPECTROSCOPY. 1-0058
ENERGY LOSS AND IONIZATION IN FAST ION- ATOM COLLISIONS. 2-0233
AUGER RECOMBINATICN AND IMPACT IONIZATION IN INDIRECT GAP SEMICONDUCTORS. 2-0396
ION PARAMETER) IMPURITY IONIZATION IN N-TYPE GERMANIUM. (AUGER RECOMBINAT 2-0466
DETAILED BALANCE BETWEEN AUGER RECOMBINATICN AND IMPACT IONIZATION IN SEMICONDUCTORS. 2-0417
AL VAPORS. (AUGER PROC/ MECHANISMS OF SIMPLE AND MULTIPLE IONIZATION INDUCED BY ELECTRON IMPACT IN SOME MET 2-0231
AUGER AND IONIZATION LOSS SPECTRA FROM BERYLLIUM SURFACES. 1-0174
L SPECTROSCOPY. AUGER ELECTRON SFECTROSCOPY. IONIZATION LOSS SPECTROSCOPY. APPEARANCE POTENTIA 2-0350
TROSCOPY. USE OF CHARACTERISTIC IONIZATION LOSSES IN AUGER ELECTRON EMISSION SPEC 2-0538
INELASTIC SCATTERING FROM RARE EARTH METAL SURFACES. IONIZATION LOSSES IN (AUGER) SPECTRA OF ELECTRON 2-0287
TROSCOPY. CHARACTERISTIC IONIZATION LOSSES OBSERVED IN AUGER EMISSION SPEC 2-0230
GER TRANSITION AND SUPPLEMENTARY ELECTRON EJECTION IN THE IONIZATION OF ALUMINUM BY ELECTRON IMPACT. AU 2-0407
EL OF TWO COULOMB CENTERS. AUGER IONIZATION OF ATOMS BY MULTIPLY CHARGED IONS. MOD 2-0256
NGULAR DISTRIBUTION OF AUGER ELECTRONS) RESONANCE IONIZATION OF ATOMIC HELIUM BY FAST ELECTRONS. (A 2-0293
NEW SATELLITE BANDS CORRESPONDING TO DOUBLE L IONIZATION. (OF AUGER ELECTRONS) 2-0532
TRON YIELDS OF METHANE, ETHANE, ETHYLENE, ACETYL/ K-SHELL IONIZATION OF CARBON BY FAST PROTONS. (AUGER ELEC 2-0477
RDMENT. (AUGER EL/ DIFFERENTIAL CROSS SECTION FOR K-SHELL IONIZATION CF COPPER AND SILVER BY ELECTRON BOMBA 4-1077
TONS. CROSS SECTIONS FOR INNER SHELL IONIZATION OF GASEOUS MOLECULES BY 50-500 KEV PRO 2-0475
MPACT. (AUGER SPECTROSCOPY) CROSS SECTIONS FOR IONIZATION OF INNER SHELL ELECTRONS BY ELECTRON I 2-0228
MPACT. AUGER EFFECT IN THE MULTIPLE IONIZATION OF MANGANESE AND CADMIUM BY ELECTRON I 2-0519
0- TO 600 KEV HYDROGEN(+), DEUTERIUM(+), MOLECUL/ K-SHELL IONIZATION OF MOLECULAR NITROGEN AND METHANE BY 5 2-0580
RAY SPECTROSCOPY TECHNIQ/ INVESTIGATION OF MECHANISM OF S IONIZATION OF NOBLE GAS ATOMS BY THE ULTRASOFT X- 2-0338
ROLE OF COSTER-KRONIG TRANSITIONS IN THE IONIZATION OF SURFACE ATOMS. 2-0358
DIFFERENTIAL CROSS SECTIONS FOR K-SHELL IONIZATION OF SURFACE ATOMS BY ELECTRON IMPACT.

INNER SHELL IONIZATION OF SURFACE ATOMS BY ELECTRON IMPACT. 2-0359
/LOUS THRESHOLD BEHAVIOR FOR ELECTRON INDUCED INNER SHELL IONIZATION OF THE N6,7 LEVEL IN GOLD, BISMUTH, A/ 4-1067
AUGER AND K X-RAY YIELDS) EFFECT OF MULTIPLE IONIZATION ON THE FLUORESCENCE YIELD OF NEON. (K- 2-0299
/RONS. PART-1: ENHANCED PLASMON EXCITATION ON INNER LEVEL IONIZATION. PART-2: DOUBLE IONIZATION ORIGIN OF / 4-1047
LECULE. AUGER ELECTRON SPECTRUM AND IONIZATION POTENTIALS OF THE HYDROGEN FLUORIDE MO 4-1058
LOCALISATION OF CONDUCTION ELECTRONS IN SOFT X-RAY DOUBLE IONIZATION SATELLITE (AUGER) TRANSITION. /SSIBLE 2-0437
TRUM. PLASMON ENERGY GAIN AND DOUBLE IONIZATION SATELLITES IN THE BERYLLIUM AUGER SPEC 4-0899
SURFACES. IONIZATION SPECTROMETER FOR ELEMENTAL ANALYSIS OF 3-0615
FACES. IONIZATION SPECTROSCOPY OF CONTAMINATED METAL SUR 5-1351
IONIZATION SPECTROSCOPY OF SURFACES. 5-1352
ELECTRON SPECTROSCOPY. (PHOTOELECTRON, AUGER, AND IONIZATION SPECTROSCOPY, REVIEW, 41 REFS) 1-0001
UGER ELECTRON SP/ ENERGY DEPENDENCE OF THE SENSITIVITY OF IONIZATION SPECTROSCOPY: SILICON ON BERYLLIUM. (A 5-1665
APPLICATIONS OF ELECTRON SPECTROSCOPY. (RELATIONSHIP OF IONIZATION SPECTROSCOPY TO PHOTOELECTRON AND AES) 1-0062
N SPECTRA. L-SHELL IONIZATION THRESHOLD STRUCTURE FROM AUGER ELECTRO 2-0360
K-SHELL FLUORESCENCE YIELD OF IONIZED ALUMINUM. (AUGER RATE) 4-0975
TIPLET EFFECTS ON THE L2,3 FLUORESCENCE YIELD OF MULTIPLY IONIZED ARGON. (AUGER EFFECT) MUL 4-0956
AUGER ELECTRON SPECTROMETRY OF SINGLY AND MULTIPLY IONIZED ATOMS. 4-0978
MS OF INNER SHELL EXCITATION AND DEEXCITATION IN MULTIPLY IONIZED ATOMS. MECHANIS 2-0487
ENERGIES OF CHARACTERISTIC X-RAYS AND AUGER ELECTRONS FOR IONIZED ATOMS. SHIFT 4-1041
DEEXCITATION OF MULTIPLY IONIZED ATOMS. (AUGER EFFECT) 2-0281
MULTIPLET EFFECT IN FLUORESCENCE YIELDS OF MULTIPLY IONIZED ATOMS. (AUGER EFFECT) 2-0314
REARRANGEMENT OF INNER SHELL IONIZED ATOMS. (AUGER SPECTRA) 2-0413
K-SHELL AUGER RATES FOR MULTIPLY IONIZED ATOMS. PART-1: NEON. 2-0282
TROSCOPY. PART-1: E/ ENERGIES OF EXCITED STATES OF DOUBLY IONIZED MOLECULES BY MEANS OF AUGER ELECTRON SPEC 4-1071
AUGER AND RADIATIVE DEEXCITATION OF MULTIPLY IONIZED NEON. 2-0315
HIGH RESOLUTION K-AUGER SPECTRA FOR MULTIPLY IONIZED NEON. 4-0952
K-SHELL (AUGER) FLUORESCENT YIELDS FOR MULTIPLY IONIZED NEON. 4-0994
ELECTRON AND K X-RAY TRANSITION ENERGIES FOR THE MULTIPLY IONIZED NEON ATOM. CALCULATED K-AUGER 4-0955
TRANSITION ENERGIES, AND FLUORESCENCE YIELDS FOR MULTIPLY IONIZED NEON. (AUGER EFFECT) /SHELL AUGER RATES, 2-0284
SEQUENTIAL STRIPPING IN HIGHLY IONIZED NEON. (AUGER TRANSITIONS) 2-0441
FLUORESCENCE YIELDS AND AUGER RATES FOR MULTIPLY IONIZED NITROGEN. 4-0781
AUGER SPECTRA OF HIGHLY IONIZED OXYGEN AND FLUORINE. 4-0906
AUGER TRANSITIONS INITIATED FROM SINGLY AND DOUBLY IONIZED STATES IN LITHIUM METAL. 2-0581
TRANSITION ENERGIES, AND FLUORESCENCE YIELDS OF VARIOUSLY IONIZED STATES OF ARGON. K-SHELL AUGER RATES, 2-0280
T) ELECTRON ION COINCIDENCE STUDIES OF HIGHLY IONIZING ARGON(+)- ARGON COLLISIONS. (AUGER EFFEC 2-0529
/ ANGULAR DISTRIBUTION OF ELECTRONS RESULTING FROM DIRECT IONIZING COLLISIONS OF ELECTRONS WITH ATOMS. (AU/ 2-0525
LEED STUDY OF IRIDIUM (100) SURFACE. 5-1370
/XENON SINGLE CRYSTAL FILMS AND THEIR USE IN STUDYING THE IRIDIUM (100) SURFACE. (AUGER SPECTRA OF IRIDIUM) 5-1479
ON SPECTROSCOPY. STUDY OF IRIDIUM (100) SURFACE USING LEED AND AUGER ELECTR 5-1371
INTERACTIONS OF CARBON MONOXIDE AND OXYGEN WITH IRIDIUM (110) SURFACES. (AUGER SPECTROSCOPY) 5-1252
ON SPECTROSCOPY. STUDIES ON THE IRIDIUM (111) SURFACE USING LEED AND AUGER ELECTR 5-1372
AES ANALYSIS OF OXIDE FILMS ON IRON. 5-1800
CUUM CLEANED SILICON CRYSTALS WITH CALIBRATED DEPOSITS OF IRON. AUGER SPECTRA AND LEED PATTERNS FROM VA 1-0099
EFFECT OF TELLURIUM ON INTERGRANULAR COHESION OF IRON. 6-2099
PLASMON DECAY PEAKS IN THE AUGER SPECTRA OF TITANIUM AND IRON. INTERFACIAL AND 4-1060
PECTROSCOPY: COMPARISON OF TECHNIQUES FOR ADSORBED TIN ON IRON. QUANTITATIVE AUGER ELECTRON S 1-0178
ER SPECTROSCOPIC ANALYSIS OF SEGREGATION OF PHOSPHORUS IN IRON. QUANTITATIVE AUG 5-1802
X-RAY EXCITED LMM AUGER SPECTRA OF COPPER, NICKEL AND IRON. 4-1123
AUGER SPECTROSCOPIC STUDY OF THE SURFACE OF IRON AFTER IONIC BOMBARDMENT AND AIR OXIDATION. 5-1939
GER ELECTRON SP/ STUDIES OF INITIAL OXIDATION ON SILICON- IRON ALLOY (100) BY MEANS OF WORK FUNCTION AND AU 5-1924
GER ELECTRON SPECTROSCOPY OF 5.59 ATOMIC PERCENT SILICON- IRON ALLOY (100) SINGLE CRYSTALS. /UNCTION AND AU 5-1925
FECT OF ADSORBED CHLORINE AND OXYGEN ON SHEAR STRENGTH OF IRON AND COPPER JUNCTIONS. (SPECTROSCOPY) EF 6-2143
THE NITROGEN KLL AUGER SPECTRA OF CHEMISORBED NITROGEN ON IRON AND MOLYBDENUM SURFACES. /EMICAL EFFECTS OF 4-0925
IN SURFACE ANALYSIS AND THEIR APPLICATION FOR STUDIES ON IRON AND STEEL. (AES, ESCA, SIMS) NEW METHODS 6-2035
PASSIVE FILM ON IRON. APPLICATION OF AUGER ELECTRON SPECTROSCOPY. 5-1761
DEGRADED PASSIVE FILM ON IRON. APPLICATION OF AUGER ELECTRON SPECTROSCOPY. 5-1762
ELECTRON SPECTROSCOPIC STUDIES OF CLEAN AND OXIDIZED IRON. (AUGER STUDY OF SURFACE) 5-1323
SURFACE COMPOSITION OF PROMOTED IRON CATALYSTS BY AUGER ELECTRON SPECTROSCOPY. 5-1817
CHANGES IN AUGER SPECTRA OF MAGNESIUM AND IRON DUE TO OXIDATION. 4-1086
CHEMICAL EFFECTS IN THE AUGER SPECTRA OF IRON DUE TO OXYGEN ADSORPTION. 4-0727
CHEMICAL SHIFT IN THE AUGER SPECTRUM OF IRON DURING OXYGEN ADSORPTION. 5-1792
RITY ATOMS AND ADSORBED OXYGEN ATOMS ON THE (001) FACE OF IRON EPITAXIAL FILM. (AUGER EFFECT) /VIOR OF IMPU 5-1523
UDY OF SURFACE CONTAMINATION EFFECT ON THE GROWTH MODE OF IRON EPITAXIAL FILMS ON MAGNESIUM OXIDE (001). /T 5-1524
ACTION- AUGER ELECTRON SPECTROSCOPY STUDY OF AN EPITAXIAL IRON FILM. LOW ENERGY ELECTRON DIFFR 5-1526
LECTRON DIFFRACTION- AUGER ELECTRON SPECTROSCOPY STUDY OF IRON FILM DEPOSITED ON MAGNESIUM OXIDE. /ENERGY E 5-1522
ACE. (AUGER SPECTROSCOPY) GROWTH MODE OF IRON FILM DEPOSITED ON MAGNESIUM OXIDE (001) SURF 5-1528
LEED- AES OBSERVATION OF GROWTH MECHANISM OF DEPOSITED IRON FILM ON MAGNESIUM OXIDE (001) SURFACE. 5-1525
DETECTION OF LEAD, BY AUGER SPECTROSCOPY, ON IRON INHIBITED IN LEAD AZELATE SOLUTION. 5-1628
R MICROPROB/ CONFIRMATION OF IMPURITY ADSORPTION AT FLAKE IRON INTERFACES IN GRAY CAST IRON. (SCANNING AUGE 6-2054
SURFACE SEGREGATION IN BORON DOPED IRON OBSERVED BY AUGER EMISSION SPECTROSCOPY. 5-1181
ICROGRAPHY AND MICROANALYSIS OF A STAINED SURFACE OF PURE IRON PLATE. AUGER ELECTRON EMISSION M 5-1415
SURFACES/ SPATIAL DISTRIBUTIONS OF HYDROGEN DESORBED FROM IRON, PLATINUM, COPPER, NIOBIUM, STAINLESS STEEL 6-2006
SECONDARY ELECTRON EMISSION FROM IRON SINGLE AND POLYCRYSTALS. 4-0917
LECTRON DIFFRACTION- AUGER ELECTRON SPECTROSCOPY STUDY OF IRON SINGLE CRYSTAL SURFACES. LOW ENERGY E 5-1866
RACTION- AUGER ELECTRON SPECTROSCOPY STUDY OF CLEAN (001) IRON SURFACE. LOW ENERGY ELECTRON DIFF 5-1527
SPECTROSCOPY. ADHESION OF METALS TO A CLEAN IRON SURFACE STUDIED WITH LEED AND AUGER EMISSION 6-2008
AUGER SPECTROSCOPIC STUDY OF PLATINUM, NICKEL, AND IRON SURFACES. 5-1791
NATIONS OBTAINED BY PROTON EXCITED X-RAY, AES ANALYSIS OF IRON SURFACES. /ACTIONAL MONOLAYER OXYGEN DETERMI 5-1682
OSCOPY AND LOW ENERG/ CHEMICAL AND STRUCTURAL ANALYSIS OF IRON SURFACES BY MEANS OF AUGER ELECTRONIC SPECTR 5-1940
COPY. ADHESION AND TRANSFER OF PLATINUM IRON TO METALS STUDIED BY AUGER EMISSION SPECTROS 5-1730
STUDY OF INTERGRANULAR FRACTURE IN IRON USING AUGER SPECTROSCOPY. 6-2097
D HYDROGEN SULFIDE SURFACE FILMS ON THE ADHESION OF CLEAN IRON. (USING LEED AND AES) /OF SULFUR, OXYGEN, AN 5-1211
N STUDIES OF THE (001) AND (110) FACES OF MONOCRYSTALLINE IRON(0.84) CHROMIUM(0.16). (AUGER EFFECT) /OSITIO 5-1591
SPECTROSCOPY STUDY OF THE INITIAL STAGES OF OXIDATION OF IRON (001). /ELECTRON DIFFRACTION- AUGER ELECTRON 6-2110
ND AUGER STUDIES OF EFFECT OF OXYGEN ON ADHESION OF CLEAN IRON (001). AND (011) SURFACES. LEED A 6-2010
THE SURFACE STRUCTURE WITH ANTIPHASE DOMAINS OF SULFUR ON IRON (001) BY LEED- AES. ANALYSIS OF 5-1470
AUGER ELECTRON SPECTROSCOPY/ STRUCTURAL STUDY OF OXYGEN- IRON (001) BY LOW ENERGY ELECTRON DIFFRACTION AND 5-1469
/TION OF METHANE, ETHANE, ETHYLENE, AND ACETYLENE WITH AN IRON (001) SURFACE AND THEIR INFLUENCE ON ADHESI/ 5-1208
DH/ ABSORPTION OF ETHYLENE OXIDE AND VINYL CHLORIDE ON AN IRON (011) SURFACE AND EFFECT OF THESE FILMS ON A 5-1207
ON- AUGER ELECTRON SPECTROSCOPY STUDY OF THE OXIDATION OF IRON (100) AND IRON (110). /GY ELECTRON DIFFRACTI 5-1586

N MEASUREMENTS COMBINED WIT/ INITIAL OXIDATION STUDIES ON IRON (100) BY MEANS OF PHOTOELECTRIC WORK FUNCTIO 5-1927
EQUILIBRIUM SURFACE SEGREGATION OF CARBON ON IRON (100) FACES. (AUGER EFFECT) 5-1369
REMENTS AND AUGER ELECTRON SPECTROSCOPY ON 5.59% SILICON- IRON (100) SINGLE CRYSTALS. / WORK FUNCTION MEASU 5-1928
/TUDY ON THE INITIAL OXIDATION AND REDUCTION PROCESSES OF IRON (100) SURFACE BY MEANS OF COMBINED LOW ENER/ 6-2141
PECTROSCOPY. PHOTOELECTRIC WORK FUNCTION STUDY ON IRON (100) SURFACE COMBINED WITH AUGER ELECTRON S 5-1926
ION OF AUGER ELECTRON EMISSION FROM CLEAN AND GAS-COVERED IRON (100) SURFACES. ANGULAR DISTRIBUT 5-1623
ON- AUGER ELECTRON SPECTROSCOPY STUDY OF THE OXIDATION OF IRON (110) AND IRON (100). /GY ELECTRON DIFFRACTI 5-1585
SURFACE STRUCTURES ON IRON (110) ULTRAHIGH VACUUM. 5-1346
SPECTROSCOPY. A STUDY OF THE IRON (111) SURFACE USING LEED AND AUGER EMISSION 5-1670
N STUDIES PERFORMED ON COPPER- ALUMINUM, COPPER- TIN, AND IRON- ALUMINUM ALLOYS. (BY LEED AND AES) /FRICTIO 6-2033
ECTRON MICROSCOPIC STUDY OF THE PASSIVE FILM FORMED ON THE IRON- CHROMIUM ALLOY. AUGER SCANNING EL 5-1753
CORROSION, SURFACE, AND MAGNETIC PROPERTIES OF NICKEL- IRON- CHROMIUM THIN FILMS. (AES) 5-1766
LEED- AES STUDY OF THE OXIDATION OF IRON- CHROMIUM (100) AND (110). 5-1587
INFLUENCE OF SILICON ON FRICTION AND WEAR OF IRON- COBALT ALLOYS. (AUGER ANALYSIS) 6-2011
CE POTENTIAL SPECTROSCOPY STUDIES WITH COPPER- NICKEL AND IRON- NICKEL ALLOYS. (AES) SOFT X-RAY APPEARAN 5-1952
ON- ANTIMONY- SULFUR ALL/ GRAIN BOUNDARY EMBRITTLEMENT IN IRON- PHOSPHORUS, IRON- PHOSPHORUS- SULFUR AND IR 6-2098
ATOMIC PERCENT INDIUM, COPPER- 2 ATOMIC PERCENT TIN, AND IRON- 6.55 ATOMIC PERCENT SILICON. /OYS COPPER- 1 6-2027
E FOR GAS CHROMATOGRAPHY DETECTORS. IRON-55 AS AN AUGER ELECTRON EMITTER. NOVEL SOURC 3-0608
D AUGER ELECTRONS. APPLICATION/ MOSSBAUER SPECTROSCOPY OF IRON-57 AND TIN-119 BY DETECTION OF CONVERSION AN 5-1796
LECTRON SPECTROSCOPY ANALYSIS OF THE FRACTURE SURFACES OF IRRADIATED PRESSURE VESSEL STEELS. AUGER E 6-2113
AUGER SPECTROSCOPY OF FRACTURE SURFACES OF IRRADIATED STAINLESS STEEL. 6-2002
AUGER ELECTRON SPECTROSCOPY OF FRACTURE SURFACES IN IRRADIATED TYPE 304 STAINLESS STEEL. 6-2001
UCLEI IN THE TISSUE: USE OF SOFT X-RAYS. (AUGER ELECTRON IRRADIATION) / ON INCLUSION OF MODERATELY HEAVY N 2-0531
MATION ON NIOBIUM SURFACES DURING HIGH TEMPERATURE PROTON IRRADIATION. CARBIDE FOR 5-1886
XCITED BY ELECTRON BOMBARDMENT AND ALUMINUM K-ALPHA X-RAY IRRADIATION. /UDY OF INTENSITIES OF AUGER LINES E 4-1017
ILMS ON INSULATING SUBSTRATES IN THE PRESENCE OF ELECTRON IRRADIATION. (AUGER SPECTROSCOPY) /HIN METALLIC F 5-1317
SEGREGATION OF ALLOYING ELEMENTS TO FREE SURFACES DURING IRRADIATION. (AUGER SPECTROSCOPY ANALYSIS, STAIN/ 6-2091
ON DIOXIDE. ELECTRON IRRADIATION EFFECT IN THE AUGER ANALYSIS OF SILIC 4-1096
AUGER SPECTROSCOPY). IRRADIATION INDUCED SOLUTE SEGREGATION (IN STEEL, 6-2092
D/ AUGER ELECTRON SPECTROSCOPY MEASUREMENTS OF ADSORPTION ISOBARS FOR OXYGEN ON TUNGSTEN AT LOW PRESSURE AN 5-1282
ON A COPPER (111) SURFACE. (AUGER) ISOMORPHISM AND PURITY OF COPPER VAPOR DEPOSITED 5-1171
SURFACES IN THE PRESSURE RANGE 10(-9) TO 10(-/ ADSORPTION ISOTHERMS OF OXYGEN ON TUNGSTEN (100), AND (111) 5-1148
AND LOW ENERGY ELECTRON DIFFRACTION STUDIES OF ADSORPTION ISOTHERMS: XENON ON (0001) GRAPHITE. /CTROSCOPY 5-1856
UM- MOLYBDENUM- VANADIUM. LONG TIME ISOTHERMAL TEMPER EMBRITTLEMENT IN NICKEL- CHROMI 6-2140

J

CALCULATION OF AUGER SPECTRA IN J-J COUPLING. 4-0749
ELEMENTS, AND AUGER AND COSTER-KRONIG TRANSITION RATES IN J-J COUPLING. /STER-KRONIG, AND RADIATIVE MATRIX 4-0967
AUGER TRANSITION RATES IN J-J COUPLING FOR D HOLES AND F ELECTRONS. 2-0444
INELASTIC MEAN FREE PATH FOR ELECTRONS IN BULK JELLIUM. (AUGER, LEED, ENERGY LOSS SPECTROSCOPY) 4-1059
RON EMISSION AT SUPERHIGH FREQUENCIES. (COPPER- GERMANIUM JUNCTION). SECONDARY ELECT 2-0356
CANNING ELECTRO/ POTENTIAL PROFILING ACROSS SEMICONDUCTOR JUNCTIONS BY AUGER ELECTRON SPECTROSCOPY IN THE S 2-0555
CHLORINE AND OXYGEN ON SHEAR STRENGTH OF IRON AND COPPER JUNCTIONS. (SPECTROSCOPY) EFFECT OF ADSORBED 6-2143

K

PROTON IMP/ RELATIVE SATELLITE LINE INTENSITIES IN ARGON K AUGER ELECTRON SPECTRA PRODUCED BY 0.4 TO 4 MEV 2-0497
EMISSION (AUGER) SP/ CHARGE STATE DEPENDENCE OF THE NEON K FLUORESCENCE YIELD DEDUCED FROM HIGH RESOLUTION 2-0520
SPECTRUM OF AUGER K,L,M,N,O ELECTRONS OF SAMARIUM. 4-0756
TRANSITION OPERATOR METHOD. K X-RAY AND AUGER ELECTRON ENERGIES FOR NEON BY A 2-0393
ION COLLISIONS. COMMENTS. SIMPLE MODEL FOR K X-RAY AND AUGER ELECTRON ENERGY SHIFTS IN HEAVY 2-0301
ALS. (AUGER BROADENING) K X-RAY EMISSION EDGE SHAPES OF FREE ELECTRON MET 2-0337
ZED NEON ATOM. CALCULATED K-AUGER ELECTRON AND K X-RAY TRANSITION ENERGIES FOR THE MULTIPLY IONI 4-0955
ONIZATION ON THE FLUORESCENCE YIELD OF NEON. (K-AUGER AND K X-RAY YIELDS) EFFECT OF MULTIPLE I 2-0299
AUGER ELECTRON SPECTROSCOPY. K- AND L-SHELL FLUORESCENCE AND AUGER YIELDS AND 4-0966
OF PLATINUM. K-, L-, AND M-AUGER AND L- COSTER-KRONIG SPECTRA 4-1097
TOMIC FLUORESCENCE YIELDS. (L-SHELL COSTER-KRONIG YIELDS, K-, L-, AND M-SHELL FLUORESCENCE YIELDS) A 2-0351
NSITIONS) INFLUENCE OF CHEMICAL BONDING ON THE LOW ENERGY K-ALPHA SPECTRUM OF SILICON. (RADIATIVE AUGER TRA 2-0226
AUGER LINES EXCITED BY ELECTRON BOMBARDMENT AND ALUMINUM K-ALPHA X-RAY IRRADIATION. /UDY OF INTENSITIES OF 4-1017
F MULTIPLE IONIZATION ON THE FLUORESCENCE YIELD OF NEON. (K-AUGER AND K X-RAY YIELDS) EFFECT O 2-0299
TOMIC HYDROGEN, OXYGEN, AND FLUORINE. NEON K-AUGER CROSS SECTIONS FROM 1.5 MEV PER AMU AND A 2-0571
LUORINE IONS. ARGON K-AUGER CROSS SECTIONS IN COLLISION WITH 30 MEV F 2-0323
FOR THE MULTIPLY IONIZED NEON ATOM. CALCULATED K-AUGER ELECTRON AND K X-RAY TRANSITION ENERGIES 4-0955
BSERVATION OF A PROJECTILE CHARGE DEPENDENCE FOR THE NEON K-AUGER ELECTRON SPECTRUM. O 4-0993
RINE(12+) ION IMPACT. HIGH RESOLUTION NEON K-AUGER ELECTRON SPECTRUM PRODUCED BY 45 MEV CHLO 4-1051
BOMBARDMENT. MEASUREMENTS OF NEON K-AUGER ELECTRONS PRODUCED BY ENERGETIC HEAVY ION 4-0951
K-AUGER RATES CALCULATED FOR NEON(+). 4-0909
HIGH RESOLUTION K-AUGER SPECTRA FOR MULTIPLY IONIZED NEON. 4-0952
(ELECTRON CAPTU/ PROTON STATES, ANOMALOUS CONVERSION, AND K-AUGER SPECTRA IN AMERICIUM-241 FROM CURIUM-241 4-1019
THE K-AUGER SPECTRUM. 4-0751
K-AUGER SPECTRUM OF NEON. 4-0914
THE K-AUGER SPECTRUM OF THE FREE MAGNESIUM ATOM. 4-0789
N COLLISIONS. HIGH RESOLUTION ARGON K-AUGER SPECTRUM PRODUCED IN 4 MEV PROTON ON ARGO 2-0496
MENT FOLLOWING INTERNAL CONVERSION AND ELECTRON CAPTURE. (K-AUGER TRANSITION INTENSITIES) /CLOUDS REARRANGE 2-0334
K-AUGER TRANSITIONS OF THE FREE SODIUM ATOM. 4-0880
INTERPRETATION OF THE LOW ENERGY K-BETA SPECTRUM IN SOLIDS. 4-1101
ALS. THE K-FAR-M2 RADIATIVE AUGER EFFECT IN TRANSITION MET 2-0502
NS. (AUGER SPECTRA) MOLECULAR EFFECTS ON K-VACANCY SHARING IN ASYMMETRIC ION- ATOM COLLISIO 2-0522
HIGH ANGULAR RESOLUTION SECONDARY ELECTRON SPECTROSCOPY. KIKUCHI CORRELATIONS FOR ARSENIC (00001). 5-1784
PY. KIKUCHI CORRELATIONS IN AUGER ELECTRON SPECTROSCO 5-1783
CULATE AUGER TRANSITIONS FOR A SURFACE CONTAINING SEVERAL KINDS OF ATOMS. COMPUTER PROGRAM TO CAL 1-0039
/VAPOR ON CLEAN CLEAVED GERMANIUM. PART-1: STRUCTURAL AND KINETIC PROPERTIES. (AUGER ELECTRON SPECTROSCOPY) 5-1432
RMIC ACID ON CARBURIZED AND GRAPHITIZED NIC/ STUDY OF THE KINETICS AND MECHANISM OF THE DECOMPOSITION OF FO 5-1629
NITROGEN WITH THE NIOBIUM (100) SURFACE. PART-2: REACTION KINETICS. (AUGER EFFECT) /ERACTION OF OXYGEN AND 5-1333
/OF IMPURITIES ON SURFACES OF THORIUM. THERMODYNAMICS AND KINETICS. (DETERMINATION OF SURFACES BY AUGER ME/ 5-1313
/FREQUENCY EXCITON BREAKDOWN AND FREE CARRIER AND EXCITON KINETICS IN GERMANIUM IN THE PRESENCE OF ELECTRO/ 4-0946
S AND X-RAY PHOTOELECTRON SPECTROSCOPY) ADSORPTION KINETICS OF CESIUM ON GALLIUM ARSENIDE. (USING AE 5-1821
0 AND 600 ANGSTROMS THICK STUDIED WITH AUGER SP/ DIFFUSION KINETICS OF GOLD THROUGH PLATINUM FILMS ABOUT 200 5-1235
SPECTROSCOPY) STRUCTURE AND GROWTH KINETICS OF NICKEL(2) SILICIDE ON SILICON. (AUGER 5-1917

SURFACE. (AUGER ELECTRON SPECTROSCOPY, STICKING COEFFIC/ KINETICS OF OXYGEN ADSORPTION ON A PLATINUM (111) 5-1189
10) PLANES OF TUNGSTEN.   THE KINETICS OF OXYGEN ADSORPTION ON THE (112) AND (1 5-1912
ON ON THE (0001) GRAPHITE FACE.   THERMODYNAMICS AND KINETICS OF THE FIRST MONOLAYER ADSORPTION OF XEN 5-1859
R ON A NICKEL (111) SURFACE. (STUDIED BY LEED AND AES)   KINETICS OF THE REACTION BETWEEN OXYGEN AND SULFU 5-1451
KEL SINGLE CRYSTAL SURFACES. PART-1: NICKEL (100) SURFAC/ KINETICS OF THE REACTION OF OXYGEN WITH CLEAN NIC 5-1452
ON SPECTROSCOPY. APPLICATION TO CESIUM-COVERED (110) GAL/ KINETICS OF THERMAL DESORPTION USING AUGER ELECTR 5-1293
)   CARBON MONOXIDE ADSORPTION KINETICS ON MOLYBDENUM (100). (AUGER SPECTROSCOPY 5-1579
AES)   A STUDY OF PHOSPHORUS ADSORPTION AND DESORPTION KINETICS ON SILICON (111) SURFACES. (BY LEED AND 5-1288
                         TRANSITION PROBABILITIES OF KL- LLL AUGER SATELLITES IN NEON.   4-1049
        HIGH RESOLUTION MEASUREMENTS OF P(+) ARGON KLL AND KLM AUGER ELECTRONS.   4-0943
M.   KLL AUGER ELECTRON SPECTRA OF THULIUM AND LUTETIU 4-0753
                        CHEMICAL EFFECTS ON KLL AUGER ELECTRON SPECTRUM FROM OXYGEN.   4-0819
                    CHEMICAL EFFECTS ON THE KLL AUGER ELECTRON SPECTRUM FROM SURFACE CARBON. 4-0862
                                                    KLL AUGER ELECTRON SPECTRUM OF SAMARIUM.   4-0725
                                                    KLL AUGER ELECTRONS OF CADMIUM AND ANTIMONY.   4-0754
D Z(1)- NEON COLLISIONS AT KEV ENERGIES.   EMISSION OF KLL AUGER ELECTRONS PRODUCED IN Z(1)- NITROGEN AN 4-0830
ES OF HYDROGENIC WAVE FUNCTIONS IN THE CALCULATION OF THE KLL AUGER ENERGIES. /N OF THE ZETA EFFECTIVE VALU 2-0526
                                       THEORY OF KLL AUGER ENERGIES.   4-1000
                                       THEORY OF KLL AUGER ENERGIES INCLUDING STATIC RELAXATION.   2-0506
                          ENERGIES OF THE KLL AUGER LINES.   4-0746
                         INTENSITIES OF KLL AUGER LINES.   4-0748
                          POSITIONS OF THE KLL AUGER LINES.   4-0747
          EFFECT OF ELECTRONIC CORRELATIONS ON KLL AUGER LINES IN ELEMENTS MAGNESIUM TO SCANDIUM 4-0800
NITRIC OXIDE, CARBON DIOXIDE, AND CARBON TETRAFLUORIDE. (KLL AUGER PROCESS) /GEN, OXYGEN, CARBON MONOXIDE, 2-0308
TERFACE.   A1 KLL AUGER PROFILE ARTIFACT AT THE OXIDE- METAL IN 4-1068
                       Z DEPENDANCE OF THE KLL AUGER RATES.   4-0557
XIDE, AND SULFUR HEXAFLUORIDE.   SULFUR KLL AUGER SPECTRA IN HYDROGEN SULFIDE, SULFUR DIO 4-0910
AND MOLYBDENUM SURFACE/ CHEMICAL EFFECTS OF THE NITROGEN KLL AUGER SPECTRA OF CHEMISORBED NITROGEN ON IRON 4-0925
XIDE, NITRIC OXIDE, WATER, AND CARBON M/ DETERMINATION OF KLL AUGER SPECTRA OF NITROGEN, OXYGEN, CARBON DIO 4-0990
IUM(+) ION BOMBARDMENT.   HIGH RESOLUTION NEON KLL AUGER SPECTRA PRODUCED BY HYDROGEN(+) AND HEL 4-0953
                A NEW TREATMENT OF THE KLL AUGER SPECTRUM.   4-0745
Z DEPENDENCE OF ENERGIES AND RELATIVE INTENSITIES IN THE KLL AUGER SPECTRUM.   4-0886
/ENSITY FUNCTION OF CARBON ON VARIOUS SUBSTRATES FROM THE KLL AUGER SPECTRUM. APPLICATION OF AUGER ELECTRO/ 5-1651
AND 36.   THE KLL AUGER SPECTRUM FOR ATOMIC NUMBERS BETWEEN 10 4-0980
HLORIDE)   KLL AUGER SPECTRUM OF CHLORINE. (IN CARBON TETRAC 4-0808
S)   KLL AUGER SPECTRUM OF FLUORINE. (IN FLUORIDE SALT 4-0734
                                                    KLL AUGER SPECTRUM OF MANGANESE.   4-0938
               RELATIVISTIC CALCULATION OF THE KLL AUGER SPECTRUM. (OF MERCURY)   4-0750
        IDENTIFICATION OF HIGH ENERGY LINES IN THE KLL AUGER SPECTRUM OF NITROGEN.   4-0791
                                       ARGON KLL AUGER SPECTRUM. TEST OF THEORY.   2-0414
MIC NUMBERS 30-94)   KLL AUGER TRANSITION ABSOLUTE PROBABILITIES. (ATO 2-0251
                                                    KLL AUGER TRANSITION PROBABILITIES.   2-0304
                  RELATIVISTIC STUDY OF KLL AUGER TRANSITION PROBABILITIES.   4-0802
ITH LOW AND INTERMEDIATE ATOMIC NUMBERS.   KLL AUGER TRANSITION PROBABILITIES FOR ELEMENTS W 4-0803
                            RELATIVISTIC KLL AUGER TRANSITION RATES.   4-0782
           EFFECT OF ATOMIC CORRELATIONS ON KLL AUGER TRANSITIONS.   4-0801
MOLYBDENUM/ HIGH RESOLUTION MEASUREMENTS OF THE NITROGEN KLL AUGER TRANSITIONS OF CHEMISORBED AMMONIA ON A 5-1532
                        SPECTRUM OF AUGER KLL ELECTRONS OF SAMARIUM.   4-0726
                        SPECTRUM OF AUGER KLL ELECTRONS OF TELLURIUM AND XENON.   4-0755
RICAL AUGER ELECTRON ENERGIES. PART-1: GENERAL METHOD AND KLL LINE ENERGIES.   SEMI-EMPI 4-0930
/NIZATION AND CHEMICAL BONDING. PART-1: AUGER ELECTRONS. (KLL TRANSITION ENERGIES FOR SULFUR, ALUMINUM, SI/ 4-0327
30-94. PART-2.   ABSOLUTE PROBABILITIES FOR AUGER- KLL TRANSITIONS DEPENDENT ON THE ATOMIC NUMBER Z= 2-0252
TO 94.   RELATIVE PROBABILITIES FOR AUGER KLL TRANSITIONS DEPENDENT ON THE ATOMIC NUMBER 29 2-0253
        HIGH RESOLUTION MEASUREMENTS OF P(+) ARGON KLL AND KLM AUGER ELECTRONS.   4-0943
/ERMINATION OF THE INTERNAL PHOTOEMISSION CONTRIBUTION TO KLM AUGER SPECTRA OF SODIUM, MAGNESIUM, AND ALUM/ 4-0769
                                                    KLM SPECTRUM OF SAMARIUM AUGER ELECTRONS.   4-0757
RIGIN OF THE PEAK SHAPES IN THE AUGER SPECTRUM OF OXYGEN (KL2,3L2,3) ON MOLYBDENUM.   O 4-1004
(MAGNESIUM OX/ ELECTRON- HOLE INTERACTION EFFECTS IN THE KL2,3L2,3 RADIATIVE AUGER SPECTRUM OF INSULATORS. 2-0540
D SEMICONDUCTORS. (SILICON)   INTERPRETATION OF KL2,3L2,3 RADIATIVE AUGER THRESHOLDS IN METALS AN 4-0723
HOTOEMISSION.   STRUCTURE ON THE HIGH ENERGY SIDE OF THE KL2,3M AUGER PEAK FROM SOLID ALUMINUM: INTERNAL P 4-1021
                                                    KMM AUGER TRANSITIONS IN XENON-131.   4-0758
                                RADIATIVE KM2 AUGER EFFECT IN SOME TRANSITION METALS.   4-1057
        AUGER AND COSTER-KRONIG STUDIES OF THE M-SHELL OF KRYPTON.   4-0979
                AUGER SPECTRA OF THE NOBLE GASES. KRYPTON.   4-0971
CT. (AUGER EFFECT)   MULTIICNIZATION OF KRYPTON AND ITS IONS BY HIGH ENERGY ELECTRON IMPA 2-0493
XE/ HIGH RESOLUTION L2,3MM AND M4,5NN AUGER SPECTRA FROM KRYPTON AND M4, 5NN AND N4,5OO AUGER SPECTRA FROM 4-1113
ECTRA FOR RARE GAS ATOMS. PART-1. THE ARGON ATOM. PART-2. KRYPTON ATOM M-SHELL. PART-3. /ECTRON EMISSION SP 4-0927
GULAR DISTRIBUTION OF M4,5NN AUGER ELECTRONS EJECTED FROM KRYPTON BY ELECTRON IMPACT.   AN 4-1003
ENERGIES.   KRYPTON M-SHELL AUGER AND COSTER-KRONIG ELECTRON 4-0928
THRESHOLD.   AUGER ELECTRON EJECTION FROM XENON N4,5OO AND KRYPTON M4,5NN PROCESSES BY ELECTRON IMPACT NEAR 4-1010
AUGER MEASUREMENTS.   ADSORPTION OF KRYPTON ON THE BASAL PLANE OF GRAPHITE: LEED AND 5-1559
LS. (AUGER ELECTRON SPECTROSCOPY)   ADSORPTION OF KRYPTON ON THE (111) FACE OF COPPER SINGLE CRYSTA 5-1221
TE BROADENING IN THE AUGER SPECTRA OF RARE GASES. (ARGON, KRYPTON, XENON)   SOLID STA 4-1006
                            THE KVV AUGER SPECTRUM OF LITHIUM METAL.   4-0892

                                          L

UM-131 AND MEASUREMENT OF DECAY PARAMETERS. (DETECTION OF L AUGER ELECTRONS) /S OF STANDARDIZATION FOR CESI 2-0474
PPROXIMATELY 90.   THE L(II) SUBSHELL ATOMIC YIELDS OF ELEMENTS WITH Z A 2-0349
        NEW SATELLITE BANDS CORRESPONDING TO DOUBLE L IONIZATION. (OF AUGER ELECTRONS)   2-0293
                 SPECTRUM OF AUGER K,L,M,N,O ELECTRONS OF SAMARIUM.   4-0756
ITS COMPOUNDS.   X-RAY FLUORESCENCE L SPECTRA OF NIOBIUM IN NIOBIUM METAL AND SOME OF 4-0762
          PRODUCTION AND DECAY OF DOUBLE L VACANCIES IN ARGON AND PHOSPHORUS.   2-0489
DE.   SCANDIUM L X-RAY EMISSION FROM PURE METAL AND SCANDIUM OXI 4-1102
/EASUREMENT OF THE PRODUCTION OF LL VACANCY STATES BY THE L X-RAY- L X-RAY COINCIDENCE METHOD. (AUGER TRAN/ 2-0482
LATINUM.   K-, L-, AND M-AUGER AND L- COSTER-KRONIG SPECTRA OF P 4-1097
          PHENOMENOLOGICAL SYSTEMATICS OF L- AND M-AUGER SPECTRA.   4-0875
ER-KRONIG YIELDS.   L- AND M-SHELL AND SUBSHELL FLUORESCENCE AND COST 4-0959
C FLUORESCENCE YIELDS. (L-SHELL COSTER-KRONIG YIELDS, K-, L-, AND M-SHELL FLUORESCENCE YIELDS)   ATOMI 2-0351
E TRANSURANIC ELEMENTS.   L- AND M-SHELL YIELDS AND ELECTRON SPECTRA FOR TH 4-0968

IUM, AND FERMIUM.   THE L-AUGER SPECTRA OF PLUTONIUM, AMERICIUM, CALIFORN 4-0836
                  INFLUENCE OF COSTER-KRONIG PROCESSES ON L-AUGER YIELDS.   4-1098
SPECTROMETRY AND AUGER ELECT/ SURFACE CLEANLINESS OF 316 L-N STAINLESS STEEL STUDIES BY SECONDARY ION MASS   3-0654
SPECTROMETRY AND AUGER ELECT/ SURFACE CLEANLINESS OF 316 L-N STAINLESS STEEL STUDIED BY SECONDARY ION MASS   6-2087
M AUGER SPECTRA OF COPPER AND ZINC.   L-S COUPLING INTERPRETATION OF HIGH RESOLUTION LM   4-1122
Z = 66-70.   L-SERIES AUGER ELECTRON SPECTRA FOR ELEMENTS WITH   4-0744
Y OF ERBIUM-160, YTTERBIUM-166, YTTERBIUM-169, AND/ AUGER L-SERIES ELECTRONIC SPECTRA ACCOMPANYING THE DECA   4-0743
SYSTEM.   LABORATORY COMPUTER SCANNING ELECTRON MICROSCOPE   3-0651
         AUGER ELECTRON SPECTROSCOPY STUDY OF OXIDATION ON LANTHANUM HEXABORIDE.   5-1706
          SURFACE CONDITION AND THERMIONIC EMISSION OF LANTHANUM HEXABORIDE. (AUGER SPECTROSCOPY)   5-1471
YSPRO/ AUGER ELECTRON SPECTRA OF SOME RARE EARTH METALS. (LANTHANUM, PRASEODYMIUM, NEODYMIUM, GADOLINIUM, D   4-1100
             QUANTITATIVE ASPECTS OF A SOFT X-RAY LASER. (AUGER EFFECT)   2-0425
ORIDE SURFACES. (AUGE/ SURFACE CHARACTERISTICS RELATED TO LASER DAMAGE OF LITHIUM NIOBATE AND POTASSIUM CHL   5-1746
S AND IN THIN FILMS OF SILICON AND GERMANIUM SUBJECTED TO LASER EXCITATION. /OCESSES (AUGER) ON THE SURFACE   5-1990
LOGY. (AUGER ELECTRON SPECTROSCOPY)   LASER WINDOW SURFACE FINISHING AND COATING TECHNO   1-0015
AUGER PROCESS)   ELECTRON PHOTOEMISSION INDUCED BY LASER-PRODUCED PLASMA X-RAYS. (PLASMA ELECTRONS,   2-0276
GSTEN.   ELECTRONIC AND LATTICE STRUCTURE OF CESIUM FILMS ADSORBED ON TUN   5-1604
OSCOPY, SECONDARY ION MASS SPECTRO/ COMPARISON OF SURFACE LAYER ANALYSIS TECHNIQUES. (AUGER ELECTRON SPECTR   5-1627
FFRACTION./ STUDY OF PLATINUM ELECTRODES BY MEANS OF THIN LAYER ELECTROCHEMISTRY AND LOW ENERGY ELECTRON DI   5-1486
/OR THE AUGER ELECTRON SPECTROSCOPY OF SYSTEMS EXHIBITING LAYER GROWTH, AND ITS APPLICATION TO THE DEPOSIT/   2-0399
. (AUGER SPEC/ CONCENTRATION CHANGE OF TIN IN THE SURFACE LAYER OF COPPER- TIN ALLOY SYSTEM DUE TO FRICTION   5-1867
R AND NICKEL ON SILVER (111): FORMATION OF AN EQUILIBRIUM LAYER OF SILVER. (AUGER SPECTROSCOPY) /S OF COPPE   5-1356
ELECTRON SPECTROSCOPY) EFFECT OF A THIN DIFFUSED SOLUTE LAYER ON THE NUCLEATION OF FATIGUE CRACKS. (AUGER   6-2037
PECTROSCOPY STUDIES OF CLEAN NICKEL SURFACES AND ADSORBED LAYERS. /ERGY ION SCATTERING AND AUGER ELECTRON S   5-1870
                                       THIN REACTION LAYERS AND SURFACE STRUCTURE OF SILICON (111).   5-1879
           STRUCTURE AND SECONDARY EMISSION OF MAGNESIUM OXIDE LAYERS APPLIED ON A MOLYBDENUM SINGLE CRYSTAL.   5-1191
/MALLY STIMULATED EXOELECTRON EMISSION OF BERYLLIUM OXIDE LAYERS. AUGER ELECTRON SPECTROSCOPY INVESTIGATIO/   5-1325
          PHASE EQUILIBRIA IN TWO DIMENSIONAL CHEMISORBED LAYERS. (AUGER SPECTROSCOPY, SULFIDE SYSTEMS)   5-1170
             THICKNESS MEASUREMENTS OF ADSORBED LAYERS BY AUGER ELECTRON SPECTROSCOPY.   5-1618
COMBINED WITH AUGER ELECTRON SPECTR/ ANALYSIS OF SURFACE LAYERS BY LIGHT ION BACKSCATTERING AND SPUTTERING   5-1167
MENTS)   ANTIMONY DOPING OF SILICON LAYERS GROWN BY SOLID PHASE EPITAXY. (AES MEASURE   5-1577
         DISTINCTION BETWEEN ADSORBED MONOLAYERS AND THICKER LAYERS IN AUGER ELECTRON SPECTROSCOPY.   5-1797
/VATION OF CHEMICAL SHIFTS IN AUGER SPECTRUM FROM SURFACE LAYERS OF SILICON DIOXIDE DURING ELECTRON BOMBAR/   4-1044
    SURFACE ANALYSIS, METHODS OF STUDYING THE OUTER ATOMIC LAYERS OF SOLIDS.   1-0016
Y)   DETECTION OF THIN CONTAMINATED LAYERS ON CONTACT SURFACES. (BY AUGER SPECTROSCOP   5-1165
PEC/ PHASE TRANSITIONS IN TWO DIMENSIONAL ABSORBED SULFUR LAYERS ON COPPER INVESTIGATED BY LEED AND AUGER S   5-1302
           AUGER ELECTRON SPECTROSCOPY OF CONTRAST FORMING LAYERS ON METALS.   6-2047
Y AUGER ELECTRON EMISSIO/ DETECTION OF SILICON OXYNITRIDE LAYERS ON THE SURFACES OF SILICON NITRIDE FILMS B   5-1611
ROSCOPY)   INTERACTION OF ALUMINUM LAYERS WITH POLYCRYSTALLINE SILICON. (AUGER SPECT   5-1672
        AUGER ELECTRON SPECTROSCOPY OF LAYERED GROWTH OF TITANIUM ON TUNGSTEN.   5-1140
LIPSOMETRIC STUDIES OF ULTRATHIN LEAD(II) OXIDE GROWTH ON LEAD.   AUGER AND EL   5-1248
L-SHELL FLUORESCENCE YIELDS OF DOUBLE VACANCY STATES IN LEAD.   4-1105
AUGER ELECTRON SPECTROSCOPY OF (SILVER- GOLD AND INDIUM- LEAD) ALLOY SURFACES.   5-1839
        NUCLEAR AUGER EFFECTS IN MUONIC ATOMS. (LEAD, BISMUTH)   4-0941
LEAD AZELATE SOLUTION.   DETECTION OF LEAD, BY AUGER SPECTROSCOPY, ON IRON INHIBITED IN   5-1628
        STUDY OF DIFFUSION IN THIN COPPER- LEAD FILMS USING AUGER ELECTRON SPECTROSCOPY.   5-1972
SHELL IONIZATION OF THE N6,7 LEVEL IN GOLD, BISMUTH, AND LEAD(II) OXIDE GROWTH ON LEAD.   5-1248
DIED BY AUGER ELECTRON SPECTROSCOPY.   LEAD MONITORED BY AUGER SPECTROSCOPY. /UCED INNER   4-1067
OSCOPY)   SPONTANEOUS ALLOYING OF A GOLD SUBSTRATE WITH LEAD MONOLAYER LUBRICATION IN STEEL MACHINING STU   6-2124
(AUGER EFFECT)   ADSORPTION AND SURFACE ALLOYING OF LEAD MONOLAYERS. (INTERFACE AUGER ELECTRON SPECTR   5-1178
ANISM IN CHANNEL ELECTRON MULTIPLIERS. (SURFACE OXYGEN ON LEAD MONOLAYERS ON (111) AND (110) FACES OF GOLD.   5-1733
1: PREPARATION AND PHYSICAL STRUCTURE./ PROPERTIES OF THE LEAD OXIDE BY AES)   GAIN FATIGUE MECH   3-0681
USING SURFACE ANALYSIS. (AUGER)   EVALUATION OF LEAD SULFIDE- LEAD OXIDE CRYSTALLINE FILMS. PART-   5-1790
R ELECTRON SPECTROSCOPY OF ALLOY SURFACES. (GOLD- SILVER, LEAD(0.8) TIN(0.2) TELLURIDE DETECTOR FABRICATION   5-1593
         AUGER ANALYSIS OF THIN OXIDE FILMS ON LEAD- INDIUM)   AUGE   5-1709
TRON SPECTROSCOPY STUDY OF THE SURFACE COMPOSITION OF THE LEAD- INDIUM ALLOYS.   5-1249
IDENCE TECHNIQUE.   AUGER TRANSITIONS LEAD- INDIUM SYSTEM.   AUGER ELEC   5-1172
ELECTRONIC AUGER SPECTROSCOPY: TEXTBOOK FOR A COURSE OF LECTURES ON PHYSICAL ELECTRONICS.   1-0107
                  LEED SEE LOW ENERGY ELECTRON DIFFRACTION
                  LEED- SEE LOW ENERGY ELECTRON DIFFRACTION
                           ESCAPE LENGTH OF AUGER ELECTRONS.   2-0527
SPECTROSCOPY.   DETERMINATION OF THE ESCAPE LENGTH OF LOW ENERGY ELECTRONS BY AUGER AND LOSS   2-0401
RBED FILMS. (AUGER/ DETERMINATION OF ELECTRON ATTENUATION LENGTHS IN METALS: TRANSMISSION THROUGH THIN ADSO   4-1088
UILIBRIUM CHARGE CARRIERS IN SILICON AT A HIGH EXCITATION LEVEL. /TERING AND (AUGER) RECOMBINATION OF NONEQ   4-1103
YED WITH Z/ FINAL STATE EFFECTS IN THE PHOTOEMISSION CORE LEVEL AND AUGER SPECTRA OF NICKEL AND NICKEL ALLO   5-1806
ELECTRONIC STRUCTURE OF SOLID SURFACES: CORE LEVEL EXCITATION TECHNIQUES. (AUGER EFFECT)   2-0468
/ FOR ELECTRON INDUCED INNER SHELL IONIZATION OF THE N6,7 LEVEL IN GOLD, BISMUTH, AND LEAD MONITORED BY AU/   4-1067
/ ELECTRONS. PART-1: ENHANCED PLASMON EXCITATION ON INNER LEVEL IONIZATION. PART-2: DOUBLE IONIZATION ORIG/   4-1047
ION MEASUREMENTS IN THE 20-70 EV R/ VALENCE BAND AND CORE LEVEL STUDIES OF INDIUM ANTIMONIDE VIA PHOTOEMISS   2-0341
           COMPARISON OF MANY BODY EFFECTS IN CORE LEVEL SURFACE SPECTROSCOPIES.   5-1723
         TOTAL AND PARTIAL ATOMIC LEVEL WIDTHS.   4-0911
HARGE CARRIERS OF INDIUM ARSENIDE AT HIGH PHOTOEXCITATION LEVELS. (AUGER) RECOMBINATION OF NONEQUILIBRIUM C   4-0840
,5 COSTER-KRONIG TRANSITION RATES TO VARIATIONS IN ENERGY LEVELS AND SCREENING PARAMETERS. /TY OF L1,L2,3M4   2-0303
ON OF GALLIUM SELENIDE SINGLE CRYSTALS AT HIGH EXCITATION LEVELS. (AUGER RECOMBINATION) /OF THE EDGE EMISSI   2-0258
           ENERGY WIDTHS OF ROENTGEN LEVELS BY AUGER ELECTRON SPECTROSCOPY.   4-0982
TRANSITIONS.   VACUUM ULTRAVIOLET EMISSION OF XENON(III) LEVELS EXCITED BY ELECTRON IMPACT VIA N4,5 AUGER   4-0878
/TRY AS A METHOD FOR DETERMINATION OF THE ENERGY DEPTH OF LEVELS OF SURFACE ELECTRON CENTERS. (AUGER EFFEC/   4-0915
SPECTRA OF ALUMINUM.   LIFETIME AND SURFACE EFFECTS ON THE L23 VV AUGER   4-0833
N SPECTROSCOPY AND AUGER SPECTRA OF THE ZINC CHALCOGENID/ LIFETIME BROADENING IN X-RAY INDUCED PHOTOEMISSIO   4-0832
SODIUM)   LIFETIME EFFECT ON THE AUGER SPECTRA OF METALS. (4-0787
OMBINATION PROCESS)   MINORITY CARRIER LIFETIME IN INDIUM ARSENIDE EPILAYERS. (AUGER REC   2-0567
COMBINATION)   CARRIER LIFETIMES IN EPITAXIAL INDIUM ARSENIDE. (AUGER RE   2-0332
CESS CARRIER CONCENTRATIONS.   AUGER LIMITED CARRIER LIFETIMES IN MERCURY CADMIUM TELLURIDE AT HIGH EX   2-0262
INTERATOMIC AUGER PROCESS: EFFECTS ON LIFETIMES OF CORE HOLE STATES.   2-0317
DES, COPPER(I) AND SILVER(I).   LINEAR LIGAND FIELDS INFLUENCING AUGER SPECTRA OF FLUORI   4-0905
/-RAY EMISSION CROSS SECTIONS AND FLUORESCENCE YIELDS FOR LIGHT ATOMS AND MOLECULES BY ELECTRON IMPACT. (A/   2-0528
           K-SHELL FLUORESCENCE YIELDS FOR LIGHT ELEMENTS.   4-0818
           LLM VERSUS LMM AUGER TRANSITIONS FOR THE LIGHT ELEMENTS.   4-0804
           LLM VERSUS LMM AUGER TRANSITIONS FOR LIGHT ELEMENTS.   4-0811

SPUTTERED TANTALUM FILMS BY AUG/ QUANTITATIVE ANALYSIS OF LIGHT ELEMENTS (NITROGEN, CARBON, AND OXYGEN) IN 5-1652
PHYSICAL PROPERTIES OF SPUTTERED TANTALUM FIL/ EFFECT OF LIGHT ELEMENTS NITROGEN, CARBON AND OXYGEN ON THE 5-1473
TOR. (AUGER PROCESS)   STIMULATED EMISSION OF LIGHT FROM A NONIDEAL EXCITON GAS IN A SEMICONDUC 2-0462
WITH AUGER ELECTRON SPECTR/ ANALYSIS OF SURFACE LAYERS BY LIGHT ION BACKSCATTERING AND SPUTTERING COMBINED 5-1167
AUGER ELECTRON EMISSION BY ION BOMBARDMENT OF LIGHT METALS. 4-0877
OM/ RELATIVE EMISSION INTENSITIES OF AUGER ELECTRONS FROM LIGHT METALS (LITHIUM, SODIUM) SUBJECTED TO ION B 2-0548
THEORETICAL STUDY OF THE AUGER EFFECT IN THE LIGHT TO MEDIUM RANGE OF ATOMIC NUMBER. 2-0486
ION)   OPTICAL GAIN IN LIGHTLY DOPED GALLIUM ARSENIDE. (AUGER RECOMBINAT 2-0503
AQUEOUS CORROSION OF SODA- SILICA AND SODA- LIME- SILICA GLASS. (AES) 5-1265
URIDE AT HIGH EXCESS CARRIER CONCENTRATIONS.   AUGER LIMITED CARRIER LIFETIMES IN MERCURY CADMIUM TELL 2-0262
ER SPECTRUM AND OF THE SATELLITES IN THE SOFT X-/ SCF AND LIMITED CI CALCULATIONS FOR ASSIGNMENT OF THE AUG 4-0724
L AUGER ELECTRON ENERGIES. PART-1: GENERAL METHOD AND KLL LINE ENERGIES.   SEMI-EMPIRICA 4-0930
RUBIDIUM, AND CESIUM IONS IN THE VACUUM ULTRAVI/ SPECTRAL LINE EXCITATION FUNCTIONS FOR SODIUM, POTASSIUM, 4-0735
A PRODUCED BY 0.4 TO 4 MEV PROTON IMP/ RELATIVE SATELLITE LINE INTENSITIES IN ARGON K AUGER ELECTRON SPECTR 2-0497
/TRON SPECTRUM OF A NICKEL (110) - C(2X2) SULFUR SURFACE: LINE SHAPE ANALYSIS AND CORRELATION WITH ION NE/ 4-1065
WO DIFFERENT CHEMICAL ENVIRONMENTS.   AUGER LINE SHAPE COMPARISON OF NITROGEN AND SULFUR IN T 5-1816
TIGHT BINDING CALCULATION OF A CORE VALENCE VALENCE AUGER LINE SHAPE: SILICON (111). 4-0831
ECTROSCOPY STUDIES OF CARBON MONOXIDE ON NICKEL. SPECTRAL LINE SHAPES AND QUANTITATIVE ASPECTS. /LECTRON SP 5-1458
, AND BARIUM FLUORIDES.   LINE WIDTH AND AUGER EFFECT IN CALCIUM, STRONTIUM 4-1116
ENERGIES OF THE KLL AUGER LINES. 4-0746
INTENSITIES OF KLL AUGER LINES. 4-0748
E LOW ENERGY ELECTRONS IN ALUMINUM OXIDE. (AUGER ELECTRON LINES) /ON DETERMINATION OF THE ATTENUATION OF TH 4-0773
POSITIONS OF THE KLL AUGER LINES. 4-0747
CHEMICAL SHIFTS OF AUGER LINES, AND THE AUGER PARAMETER. 4-1107
M K-ALPHA X-RAY IRRADIATIO/ STUDY OF INTENSITIES OF AUGER LINES EXCITED BY ELECTRON BOMBARDMENT AND ALUMINU 4-1017
WIDTHS AND INTENSITIES OF HIGHLY RESOLVED LINES IN ARGON L2,3 AUGER SPECTRA. 4-1032
IMPACT.   PRODUCTION OF INTENSE SATELLITE LINES IN AUGER SPECTRA OF NITROGEN BY SLOW PROTON 4-1076
EFFECT OF ELECTRONIC CORRELATIONS ON KLL AUGER LINES IN ELEMENTS MAGNESIUM TO SCANDIUM. 4-0800
PREDICTION OF THE RELATIVE INTENSITIES OF AUGER LINES IN MICA. 4-1072
'NALYSIS.   INFORMATION OBTAINED FROM THE WIDTH OF SPECTRAL LINES IN THE ELECTRON SPECTROSCOPY FOR CHEMICAL A 4-1009
IDENTIFICATION OF HIGH ENERGY LINES IN THE KLL AUGER SPECTRUM OF NITROGEN. 4-0791
AUGER LINES IN X-RAY PHOTOELECTRON SPECTROMETRY. 1-0208
R EFFECT)   EMISSION LINES OF THE M SERIES OF RARE EARTH METALS. (AUGE 4-0790
FLUORIDES, COPPER(I) AND SILVER(I).   LINEAR LIGAND FIELDS INFLUENCING AUGER SPECTRA OF 4-0905
R SPECTROMETRY) MECHANISM OF SURFACE ALIGNMENT IN NEMATIC LIQUID CRYSTALS. (SUBSTRATE SURFACE EFFECTS, AUGE 5-1281
/N TECHNIQUES IN THE STUDY OF CORROSION AND DEPOSITION IN LIQUID METALS. (AUGER SPECTROSCOPY, X-RAY PHOTOE/ 6-2050
PHOSPHIDE SUBSTRATES. (AUGER SPECTROSCOPY) LIQUID PHASE EPITAXY OF III-V COMPOUNDS ON INDIUM 5-1590
/ANSPORT OF STAINLESS STEEL CORROSION PRODUCTS IN FLOWING LIQUID SODIUM. (SCANNING ELECTRON MICROSCOPY, AU/ 6-2014
AUGER CATALOG CALCULATED TRANSITION ENERGIES LISTED BY ENERGY AND ELEMENT. 1-0041
FOR SOFT X-RAY TRANSITIONS IN IONIC, ATOMIC, AND METALLIC LITHIUM.   AUGER RATES 4-0835
X-RAY EDGE ANOMALIES IN LITHIUM. 2-0270
ANISOTROPY OF SECONDARY ELECTRON EMISSION FROM PASSAGE OF LITHIUM(+) IONS THROUGH COPPER SINGLE CRYSTALS. 2-0245
AUGER SPECTRA OF ACTIVATED SILVER- MAGNESIUM AND SILVER- LITHIUM ALLOYS. 6-2004
AUGER SPECTRA OF LITHIUM AND SODIUM UNDER ION BOMBARDMENT. 4-0817
AUGER EMISSION AND COLOR CENTERS IN LITHIUM FLUORIDE. 2-0430
LOW ENERGY AUGER EMISSION FROM LITHIUM FLUORIDE. 4-0842
/ STUDY OF SURFACE DISSOCIATION IN POTASSIUM CHLORIDE AND LITHIUM FLUORIDE BY OBSERVATION OF PLASMON LOSSES 5-1279
AUGER SPECTRA OF LITHIUM HYDRIDE. 4-1025
LEED- AES ELECTRON INDUCED EMISSION STUDIES OF LITHIUM HYDRIDE. 4-0807
LEED AND AES STUDIES OF LITHIUM HYDRIDE (100) SURFACE. 5-1446
ITIONS INITIATED FROM SINGLY AND DOUBLY IONIZED STATES IN LITHIUM METAL.   AUGER TRANS 2-0581
THE KVV AUGER SPECTRUM OF LITHIUM METAL. 4-0892
AUGER SPECTRA OF LITHIUM METAL AND LITHIUM OXIDE. 4-0806
SODIUM SEGREGATION ONTO A LITHIUM METAL SURFACE. (AUGER SPECTRA) 5-1750
(AUGE/ SURFACE CHARACTERISTICS RELATED TO LASER DAMAGE OF LITHIUM NIOBATE AND POTASSIUM CHLORIDE SURFACES. 5-1746
MISSION INTENSITIES OF AUGER ELECTRONS FROM LIGHT METALS (LITHIUM, SODIUM) SUBJECTED TO ION BOMBARDMENT. /E 2-0548
AUGER TRANSITIONS IN LITHIUM-LIKE BERYLLIUM-LIKE IONS. 2-0296
ENCE METHOD. (AUGER TRA/ MEASUREMENT OF THE PRODUCTION OF LL VACANCY STATES BY THE L X-RAY- L X-RAY COINCID 2-0482
TRANSITION PROBABILITIES OF KL- LLL AUGER SATELLITES IN NEON. 4-1049
TS.   LLM VERSUS LMM AUGER TRANSITIONS FOR LIGHT ELEMEN 4-0811
EMENTS.   LLM VERSUS LMM AUGER TRANSITIONS FOR THE LIGHT EL 4-0804
NDS.   RELAXATION DURING PHOTOEMISSION AND LMM AUGER DECAY IN ARSENIC AND SOME OF ITS COMPOU 4-0763
HIGH RESOLUTION STUDY OF ARGON LMM AUGER ELECTRONS PRODUCED BY ION BOMBARDMENT. 4-0903
L-S COUPLING INTERPRETATION OF HIGH RESOLUTION LMM AUGER SPECTRA OF COPPER AND ZINC. 4-1122
X-RAY EXCITED LMM AUGER SPECTRA OF COPPER, NICKEL AND IRON. 4-1123
FINE STRUCTURE DETAILS IN THE LMM AUGER SPECTRA OF GERMANIUM. 4-0937
HIGH RESOLUTION LMM AUGER SPECTRA OF GERMANIUM. 4-0936
CHEMICAL SHIFTS IN COORDINATION COMPOUNDS. X-RAY EXCITED LMM AUGER SPECTRA OF GERMANIUM.   PHOTOELECTRON 4-0924
THRESHOLD EFFECTS ON RELATIVE INTENSITIES IN LMM AUGER SPECTRA OF GERMANIUM AND SELENIUM. 4-0767
ES.   HIGH RESOLUTION LMM AUGER SPECTRA OF LOW ENERGY FROM SOLID SURFAC 4-0732
QUASIATOMIC LMM AUGER SPECTRA OF SOLID COPPER AND ZINC. 4-1121
E FIRST SERIES.   LMM AUGER SPECTRA OF SOME TRANSITION METALS OF TH 4-0810
SODIUM PYROSULFITE.   LMM AUGER SPECTRA OF SULFUR IN SODIUM SULFATE AND 4-0829
LLM VERSUS LMM AUGER TRANSITIONS FOR LIGHT ELEMENTS. 4-0811
LLM VERSUS LMM AUGER TRANSITIONS FOR THE LIGHT ELEMENTS. 4-0804
) OXIDES.   MOLECULAR ORBITAL EFFECTS ON THE TITANIUM LMV AUGER SPECTRA OF TITANIUM(II) AND TITANIUM(IV 4-1069
AUGER RECOMBINATION IN SILICON RECTIFIERS AND THYRISTORS LOADED IN FORWARD DIRECTION IN THE MIDDLE REGION. 2-0415
Y DOUBLE IONIZATION SATELLITE (AUGER) TRANSITIO/ POSSIBLE LOCALISATION OF CONDUCTION ELECTRONS IN SOFT X-RA 2-0437
CA TEST PROGRAMS.   DIGITAL CONTROL UNIT IN TTL LOGIC FOR THE AUTOMATIC EXECUTION OF AUGER AND ES 3-0722
100) SURFACE OF COPPER(3) GOLD. TEMPERATURE DEPENDENCE OF LONG RANGE ORDER AT A CRYSTAL SURFACE. / OF THE (5-1854
EL- CHROMIUM- MOLYBDENUM- VANADIUM.   LONG TIME ISOTHERMAL TEMPER EMBRITTLEMENT IN NICK 6-2140
OPOGRAPHY, PHENOMENA, FIELD ION MICROSCOPY, LEED/ A CLOSE LOOK AT SURFACES. (SURFACE STRUCTURES, SPECIES, T 5-1492
IDE (110).   CHARACTERISTIC ENERGY LOSS AND AUGER ELECTRON SPECTRA OF GALLIUM PHOSPH 5-1657
HE STUDY OF ADSORPTION PHENOMENA A/ CHARACTERISTIC ENERGY LOSS AND AUGER ELECTRON SPECTROSCOPY APPLIED TO T 5-1776
Y SPECTRUM OF BACK SCATTERED ELECTRONS AND CHARACTERISTIC LOSS AND GAIN PHENOMENA OF COPPER (111).   ENERG 4-0897
PLASMON EFFECTS IN ELECTRON ENERGY LOSS AND GAIN SPECTRA IN ALUMINUM. 4-1020
ENERGY LOSS AND IONIZATION IN FAST ION- ATOM COLLISIONS. 2-0233
ARSENIC (0001) SURFACES: AUGER, LOSS, AND PHOTOELECTRON SPECTROSCOPIC STUDIES. 5-1315
DMENT.   INTERPRETATION OF ENERGY LOSS BY AUGER ELECTRONS EMITTED DURING ION BOMBAR 2-0268
TS IN BERYLLIUM AND BERYL/ CHARACTERISTIC ENERGY GAIN AND LOSS, DOUBLE IONIZATION, AND IONIZATION LOSS EVEN 4-0901

```
E.                                      LOW ENERGY AUGER AND LOSS ELECTRON SPECTRA FROM MAGNESIUM AND ITS OXID 4-0895
RON / A MEASURING METHOD TO DISTINGUISH AUGER FROM ENERGY LOSS PEAKS OF THE PRIMARY BEAM IN SECONDARY ELECT 1-0188
T)                   COMPARISON OF SILICON L3,2 ELECTRON LOSS SPECTRA AND OPTICAL ABSORPTION. (AUGER EFFEC 4-1038
    SURFACE STATE TRANSITIONS OF SILICON IN ELECTRON ENERGY LOSS SPECTRA. (AUGER EFFECT)                   4-1039
                           AUGER AND IONIZATION LOSS SPECTRA FROM BERYLLIUM SURFACES.             5-1987
ED CARBON SUR/ CHARACTERISTIC K-SHELL LOW ENERGY ELECTRON LOSS SPECTRA FROM DIAMOND, GRAPHITE, AND EVAPORAT 4-0945
          AUGER ELECTRON AND CHARACTERISTIC ENERGY LOSS SPECTRA OF PLUTONIUM.                     4-0931
SORPTION OF AM/ APPLICATIONS OF AUGER ELECTRON AND ENERGY LOSS SPECTROSCOPIES TO CATALYTIC PROCESSES. CHEMI 5-1533
OF THE ESCAPE LENGTH CF LOW ENERGY ELECTRONS BY AUGER AND LOSS SPECTROSCOPY.            DETERMINATION    2-0401
    PATH FOR ELECTRONS IN BULK JELLIUM. (AUGER, LEED, ENERGY LOSS SPECTROSCOPY)       INELASTIC MEAN FREE  4-1059
ROSCOPY. COMPARISON WITH AUGE/ IONIZATION (CHARACTERISTIC LOSS) SPECTROSCOPY AND APPEARANCE POTENTIAL SPECT 1-0173
OPY.              AUGER ELECTRON SPECTROSCOPY.  IONIZATION LOSS SPECTROSCOPY. APPEARANCE POTENTIAL SPECTROSC 1-0174
/ STUDY OF THE OXIDATION OF MOLYBDENUM SURFACES BY ENERGY LOSS SPECTROSCOPY COMBINED WITH AUGER ELECTRON SP 5-1531
BAL/ COMBINATION OF AUGER SPECTROSCOPY AND CHARACTERISTIC LOSS SPECTROSCOPY FOR THE ELEMENTS VANADIUM TO CO 4-0813
            AUGER ELECTRON AND CHARACTERISTIC ENERGY LOSS SPECTROSCOPY OF SOME METALS.            4-1083
EFFECT ON THE GROWTH MODE OF IRON EPITAXIAL FI/ AUGER AND LOSS SPECTROSCOPY STUDY OF SURFACE CONTAMINATION  5-1524
/ CHLORIDE AND LITHIUM FLUORIDE BY OBSERVATION OF PLASMON LOSSES ASSOCIATED WITH POTASSIUM AND LITHIUM. (A/ 5-1279
         USE OF CHARACTERISTIC IONIZATION LOSSES IN AUGER ELECTRON EMISSION SPECTROSCOPY.       2-0350
CATTERING FROM RARE EARTH METAL SURFACES.       IONIZATION LOSSES IN (AUGER) SPECTRA OF ELECTRON INELASTIC S 2-0538
         CHARACTERISTIC IONIZATION LOSSES OBSERVED IN AUGER EMISSION SPECTROSCOPY.      2-0287
    INVESTIGATION OF THE CHARACTERISTIC AUGER ENERGY LOSSES OF PLATINUM (111).                  2-0499
         CHARACTERIZATION OF CHEMISORPTION BY LOW ENERGY ELECTRON DIFFRACTION                    1-0161
                CLEAN TELLURIUM SURFACES STUDIED BY LOW ENERGY ELECTRON DIFFRACTION             5-1134
ATION OF THE MOLYBDENUM (110) SURFACE OBSERVED BY AES AND LOW ENERGY ELECTRON DIFFRACTION    INITIAL OXID 5-1650
    PRESENT STATE OF AUGER ELECTRON SPECTROSCOPY AND LOW ENERGY ELECTRON DIFFRACTION              1-0006
1) SURFACE OBSERVED BY AUGER EMISSION SPECTROSCOPY AND BY LOW ENERGY ELECTRON DIFFRACTION   / TO SILICON (11 5-1180
        ADHESION AT GOLD MICA INTERFACES. (LOW ENERGY ELECTRON DIFFRACTION AES)              5-1434
RUCTURE WITH ANTIPHASE DOMAINS OF SULFUR ON IRON (001) BY LOW ENERGY ELECTRON DIFFRACTION AES. / SURFACE ST 5-1470
    BINDING STATES OF WATER VAPOR ON CLEAVED GERMANIUM. (LOW ENERGY ELECTRON DIFFRACTION AES)     5-1820
    FORMIC ACID DESORPTION FROM GRAPHITIZED NICKEL (110). (LOW ENERGY ELECTRON DIFFRACTION AES)     5-1630
    SUPERSTRUCTURES ON SILICON (100) SURFACES AS OBSERVED BY LOW ENERGY ELECTRON DIFFRACTION AES. GOLD-INDUCED 5-1708
S OF CLEAN AND ALKALI METAL COVERED ZINC OXIDE SURFACES. (LOW ENERGY ELECTRON DIFFRACTION AES) / PROPERTIE 5-1881
/RE AND COMPOSITION OF METAL SURFACES: A REVIEW OF RECENT LOW ENERGY ELECTRON DIFFRACTION AES, AND XPS RE/ 5-1341
CED EMISSION STUDIES OF LITHIUM HYDRIDE.         LOW ENERGY ELECTRON DIFFRACTION AES ELECTRON INDU 4-0807
            SURVEYING SILICON SURFACES BY MEANS OF THE LOW ENERGY ELECTRON DIFFRACTION AES METHOD.  5-1809
F GROWTH MECHANISM OF DEPOSITED IRON FILM ON MAGNESIUM C/ LOW ENERGY ELECTRON DIFFRACTION AES OBSERVATION O 5-1525
EMISORPTION INDUCED SULFUR SEGREGATION FROM THE BULK TO / LOW ENERGY ELECTRON DIFFRACTION AES STUDIES OF CH 5-1865
/ON METAL SUPERCONDUCTORS AT SURFACES AND IN THIN FILMS. (LOW ENERGY ELECTRON DIFFRACTION AES STUDIES OF N/ 5-1848
OXIDATION OF IRON- CHROMIUM (100) AND (110).      LOW ENERGY ELECTRON DIFFRACTION AES STUDY OF THE 5-1587
            ELECTRODES AND CIRCUITS OF A LOW ENERGY ELECTRON DIFFRACTION AES SYSTEM.      3-0639
CTION OF IMPURITIES ON A SILICON (111) SURFACE BY  AUGER- LOW ENERGY ELECTRON DIFFRACTION ANALYSIS.   DETE 5-1360
ER- ALUMINUM, COPPER- TIN, AND IRON- ALUMINUM ALLOYS. (BY LOW ENERGY ELECTRON DIFFRACTION AND AES) /ON COPP 6-2033
ON AND DESORPTION KINETICS ON SILICON (111) SURFACES. (BY LOW ENERGY ELECTRON DIFFRACTION AND AES) /DSORPTI 5-1288
ERIZATION OF SOME VANADIUM (100) SURFACE STRUCTURES USING LOW ENERGY ELECTRON DIFFRACTION AND AES.  CHARACT 5-1864
RFACES OF ALKALI HALIDES AND THIN FILM GROWTH OBSERVED BY LOW ENERGY ELECTRON DIFFRACTION AND AES. /STAL SU 5-1907
LFIDE SURFACE FILMS ON THE ADHESION OF CLEAN IRON. (USING LOW ENERGY ELECTRON DIFFRACTION AND AES) /OGEN SU 5-1211
1) GOLD SURFACES BY POSITIVE ION CHANNELING SPECTROSCOPY, LOW ENERGY ELECTRON DIFFRACTION AND AES. /RED (00 5-1139
    OXYGEN AND SULFUR ON A NICKEL (111) SURFACE. (STUDIED BY LOW ENERGY ELECTRON DIFFRACTION AND AES) /BETWEEN 5-1451
    TUNGSTEN (110) AND CARBON MONOXIDE ON TUNGSTEN (100). (BY LOW ENERGY ELECTRON DIFFRACTION AND AES) /GEN ON 5-1152
/ION OF GOLD ONTO TANTALUM (100) SINGLE CRYSTAL SURFACES. LOW ENERGY ELECTRON DIFFRACTION AND AES ANALYSIS. 5-1312
ONS ON THE POLAR SURFACES OF ZINC OXIDE.         LOW ENERGY ELECTRON DIFFRACTION AND AES OBSERVATI 5-1619
53 REFS)                          LOW ENERGY ELECTRON DIFFRACTION AND AES. (REVIEW, 5-1096
N ULTRAHIGH VACUUM.     A VERSATILE SUBSTRATE DESIGN FOR LOW ENERGY ELECTRON DIFFRACTION AND AES STUDIES I 3-0631
F INDIUM ARSENIDE SURFACES.            COMBINED LOW ENERGY ELECTRON DIFFRACTION AND AES STUDIES O 5-1373
F LITHIUM HYDRIDE (100) SURFACE.            LOW ENERGY ELECTRON DIFFRACTION AND AES STUDIES O 5-1446
F THE STRUCTURE OF ADSORBED GASES ON SOLID SURFACES.    LOW ENERGY ELECTRON DIFFRACTION AND AES STUDIES O 5-1835
OXYGEN ABSORPTION ON TUNGSTEN (110).          LOW ENERGY ELECTRON DIFFRACTION AND AES STUDY OF 5-1669
THE INTERACTION OF HYDROGEN SULFIDE AND MOLYBDENUM (100)/ LOW ENERGY ELECTRON DIFFRACTION AND AES STUDY OF 5-1973
LICON (111) SURFA/ DETERMINATION OF SURFACE STRUCTURES BY LOW ENERGY ELECTRON DIFFRACTION (AND AES): THE SI 5-1342
    LOW ENERGY ELECTRON DIFFRACTION, CIRCA 1968. (REVIEW OF LOW ENERGY ELECTRON DIFFRACTION AND AUGER)   1-0099
STICS ON ZINC OXIDE SURFACES.  CORRELATION OF ELECTRONIC, LOW ENERGY ELECTRON DIFFRACTION  AND AUGER DIAGNO 5-1584
N SPECTROSCOPY. (REVIEW)                 LOW ENERGY ELECTRON DIFFRACTION AND AUGER ELECTRO 1-0124
N SPECTROSCOPY.     STUDY OF IRIDIUM (100) SURFACE USING LOW ENERGY ELECTRON DIFFRACTION AND AUGER ELECTRO 5-1371
N SPECTROSCOP/ STUDIES ON THE IRIDIUM (111) SURFACE USING LOW ENERGY ELECTRON DIFFRACTION AND AUGER ELECTRO 5-1372
N SPECTROSCOPY STUDIES OF SILICON, GERMANIUM, G/ COMBINED LOW ENERGY ELECTRON DIFFRACTION AND AUGER ELECTRO 5-1379
/UDY OF RUTHENIUM (0001) AND RHODIUM (111) SURFACES USING LOW ENERGY ELECTRON DIFFRACTION AND AUGER ELECTR/ 5-1380
/UDY OF INDIUM ARSENIDE (111) AND (-1-1-1) SURFACES USING LOW ENERGY ELECTRON DIFFRACTION AND AUGER ELECTR/ 5-1381
N SPECTROSCOPIC STUDIES ON TUNGSTEN (110) SURFACE.     LOW ENERGY ELECTRON DIFFRACTION AND AUGER ELECTRO 5-1677
N SPECTROSCOPIC OBSERVATIONS OF THE SUBSTRATE SURFACE DU/ LOW ENERGY ELECTRON DIFFRACTION AND AUGER ELECTRO 5-1678
N SPECTROSCOPY STUDY OF NICKEL (110) SURFACE EFFECTS DUE/ LOW ENERGY ELECTRON DIFFRACTION AND AUGER ELECTRO 5-1814
N SPECTROSCOPY.    DETERMINATION OF SURFACE STRUCTURES BY LOW ENERGY ELECTRON DIFFRACTION AND AUGER ELECTRO 5-1957
/ACTORY METAL SURFACES IN ULTRAHIGH VACUUM AS OBSERVED BY LOW ENERGY ELECTRON DIFFRACTION AND AUGER ELECTR/ 6-2020
N STUDIES.          ULTRAHIGH VACUUM EVAPORATOR FOR LOW ENERGY ELECTRON DIFFRACTION AND AUGER EMISSIO 3-0683
N SPE/ GROWTH OF SILVER ON POTASSIUM CHLORIDE OBSERVED BY LOW ENERGY ELECTRON DIFFRACTION AND AUGER EMISSIO 5-1348
N STUDIES. ADSORPTION OF CARBON MONOXIDE ON COPPER (001). LOW ENERGY ELECTRON DIFFRACTION AND AUGER EMISSIO 5-1515
N SPECTROSCOPY.   A STUDY OF THE IRON (111) SURFACE USING LOW ENERGY ELECTRON DIFFRACTION AND AUGER EMISSIO 5-1670
/ ADHESION OF METALS TO A CLEAN IRON SURFACE STUDIED WITH LOW ENERGY ELECTRON DIFFRACTION AND AUGER EMISSIO 6-2008
GATIONS OF COPPER (111) SURFACE.           LOW ENERGY ELECTRON DIFFRACTION AND AUGER INVESTI 5-1499
ME/ ADSORPTION OF KRYPTON ON THE BASAL PLANE OF GRAPHITE: LOW ENERGY ELECTRON DIFFRACTION AND AUGER MEASURE 5-1559
SCOPY.            ULTRAHIGH VACUUM AIRLOCK FOR LOW ENERGY ELECTRON DIFFRACTION AND AUGER SPECTRO 3-0669
/NSIONAL ABSORBED SULFUR LAYERS ON COPPER INVESTIGATED BY LOW ENERGY ELECTRON DIFFRACTION AND AUGER SPECTR/ 5-1302
/M ON MAGNESIUM OXIDE (001) USING REFLECTION DIFFRACTION, LOW ENERGY ELECTRON DIFFRACTION AND AUGER SPECTR/ 5-1496
OF EFFECT OF OXYGEN ON ADHESION OF CLEAN IRON (001) AND/ LOW ENERGY ELECTRON DIFFRACTION AND AUGER STUDIES 6-2010
TROSCOPIC OBSERVATIONS ON NICKEL MONOXIDE (100).      LOW ENERGY ELECTRON DIFFRACTION AND ELECTRON SPEC 5-1687
IS OF SCATTERE/ DETERMINATION OF SURFACE STRUCTURES USING LOW ENERGY ELECTRON DIFFRACTION AND ENERGY ANALYS 5-1956
/ID ON CARBURIZED AND GRAPHITIZED NICKEL (110) USING AES, LOW ENERGY ELECTRON DIFFRACTION  AND FLASH DESOR/ 5-1629
CTRON SPECTROSC/ CESIUM ADSORPTION ON TUNGSTEN STUDIED BY LOW ENERGY ELECTRON DIFFRACTION AND SECONDARY ELE 5-1661
TIAL STUDY OF CARBON MONOXIDE AND XENON ADSORBED ON COPP/ LOW ENERGY ELECTRON DIFFRACTION AND SURFACE POTEN 5-1246
    MEASUREMENTS FOR E/ A CORRELATION OF AUGER SPECTROSCOPY, LOW ENERGY ELECTRON DIFFRACTION AND WORK FUNCTION 5-1739
    DETECTION AND IDENTIFICATION OF SURFACE CONTAMINA USE OF LOW ENERGY ELECTRON DIFFRACTION APPARATUS FOR THE 5-1959
```

CILITY FOR INVESTIGATING ENERGY AND ANGULAR / FARADAY CUP LOW ENERGY ELECTRON DIFFRACTION APPARATUS WITH FA 3-0691
THE BERYLLIUM (0001) SURFACE. LOW ENERGY ELECTRON DIFFRACTION AUGER ANALYSIS OF 5-1581
N STUDIES OF NEGATIVE ELECTRON AFFINITY ON SILICON PRODU/ LOW ENERGY ELECTRON DIFFRACTION AUGER AND PLASMO 5-1363
UNCTION STUDY OF IODINE ADSORBED ON TUNGSTEN (110). LOW ENERGY ELECTRON DIFFRACTION AUGER, AND WORK F 5-1145
FUNCTION STUDIES OF CLEAN AND SODIUM- COVERED SURFACES O/ LOW ENERGY ELECTRON DIFFRACTION AUGER, AND WORK / 5-1240
/ OF CHLORINE WITH THE ZINC OXIDE (0001)- ZINC SURFACE: A LOW ENERGY ELECTRON DIFFRACTION AUGER AND WORK / 5-1463
/M (100) SURFACE. (CHEMISORPTION AND CATALYTIC REACTIONS, LOW ENERGY ELECTRON DIFFRACTION AUGER, CONTACT / 5-1321
/ON (110) SURFACE OF A TUNGSTEN SINGLE CRYSTAL STUDIED BY LOW ENERGY ELECTRON DIFFRACTION AUGER ELECTRON / 5-1660
ECTROSCOPY/ DETECTION OF IMPURITIES ON COPPER SURFACES BY LOW ENERGY ELECTRON DIFFRACTION AUGER ELECTRON SP 5-1185
/EPITAXY OF GERMANIUM FILMS BY SPUTTER DEPOSITION. (HEED, LOW ENERGY ELECTRON DIFFRACTION AUGER ELECTRON / 5-1537
/ND (100)) AND STEPPED PLATINUM SINGLE CRYSTAL SURFACES. (LOW ENERGY ELECTRON DIFFRACTION AUGER ELECTRON / 5-1838
ON OF A STABLE CARBIDE OVERLAYER ON A PLATINUM (111) SUR/ LOW ENERGY ELECTRON DIFFRACTION AUGER INVESTIGATI 5-1569
ON ON A (100) SURFACE OF AUSTENITIC CHROMIUM- NICKEL STE/ LOW ENERGY ELECTRON DIFFRACTION AUGER INVESTIGATI 6-1991
CONTAMINANTS ON CHEMICALLY ETCHED SILICON SURFACES. LOW ENERGY ELECTRON DIFFRACTION AUGER METHOD. 5-1229
PY, AND CONTACT POTENTIAL STUDIES OF COPPER- GOLD ALLOY / LOW ENERGY ELECTRON DIFFRACTION AUGER SPECTROSCO 5-1748
Y STUDY OF THE ADSORPTION OF ALKALI METALS ON MOLYBDENUM/ LOW ENERGY ELECTRON DIFFRACTION AUGER SPECTROSCOP 5-1896
ELECTRON BEAM CURRENT CONTROLLER FOR LOW ENERGY ELECTRON DIFFRACTION AUGER STUDIES. 3-0589
ELEC/ SILICON ON SAPPHIRE EPITAXY BY VACUUM SUBLIMATION: LOW ENERGY ELECTRON DIFFRACTION AUGER STUDIES AND 5-1230
CHLORINE CHEMISORPTION ON THE SILVER (111) SURFACE. LOW ENERGY ELECTRON DIFFRACTION (AUGER) STUDY OF 5-1779
OWTH OF ALUMINUM ON NIOBIUM (110). LOW ENERGY ELECTRON DIFFRACTION AUGER STUDY OF GR 5-1491
IDE FORMATION AT / OXIDATION OF TUNGSTEN (110). PART-2: LOW ENERGY ELECTRON DIFFRACTION AUGER STUDY OF OX 6-1994
THE CONTINUOUS MEASUREMENT OF CHANGES IN WORK FUNCTION. (LOW ENERGY ELECTRON DIFFRACTION AUGER SYSTEM) /OR 3-0663
ARY ELECTRON YIELD MEASUREMENTS AT LOW PRIMARY ENERGIES. (LOW ENERGY ELECTRON DIFFRACTION AUGER SYSTEM) /ND 3-0629
STAGE FOR IN-SITU FILM DEPOSITION AND CHARACTERIZATION. (LOW ENERGY ELECTRON DIFFRACTION AUGER SYSTEM) /RE 3-0657
THE ANALYSIS OF ELECTRON EMISS/ REFINEMENTS TO A STANDARD LOW ENERGY ELECTRON DIFFRACTION AUGER SYSTEM FOR 3-0698
/THIN METAL FILMS ON BODY CENTERED CUBIC SUBSTRATES USING LOW ENERGY ELECTRON DIFFRACTION AUGER TECHNIQUES. 5-1488
UNCTION CHANGES AND AUGER ELECTRON SPECTROSCOPY USING THE LOW ENERGY ELECTRON DIFFRACTION CAMERA. /. WORK F 6-2069
CONSIDERATIONS OF AN AUGER ELECTRON SPECTROMETER BASED ON LOW ENERGY ELECTRON DIFFRACTION DISPLAY OPTICS. / 3-0705
/IC MEAN FREE PATH FOR ELECTRONS IN BULK JELLIUM. (AUGER, LOW ENERGY ELECTRON DIFFRACTION ENERGY LOSS SPE/ 4-1059
/S, SPECIES, TOPOGRAPHY, PHENOMENA, FIELD ION MICROSCOPY, LOW ENERGY ELECTRON DIFFRACTION HEED, AUGER SPE/ 5-1492
STUDIES OF CLEAN (001), (110) AND ELASTIC (111) / ELASTIC LOW ENERGY ELECTRON DIFFRACTION INTENSITY ENERGY 5-1292
THE DEPOSITION OF ALUMINUM ONTO THE MOLYBDENUM (1/ AUGER- LOW ENERGY ELECTRON DIFFRACTION INVESTIGATION OF 5-1489
TIN ON MOLYBDENUM (100). AUGER- LOW ENERGY ELECTRON DIFFRACTION INVESTIGATION OF 5-1490
XENON SINGLE CRYSTAL FILMS AND THEIR USE IN STUDYING TH/ LOW ENERGY ELECTRON DIFFRACTION INVESTIGATIONS OF 5-1479
/ AND CHEMISTRY OF CERAMIC SURFACES. (AUGER SPECTROSCOPY, LOW ENERGY ELECTRON DIFFRACTION ION MICROPROBE / 5-1852
UREMENTS ON SURFACES OF ELECTRICAL SOLID MATERIALS BY THE LOW ENERGY ELECTRON DIFFRACTION METHOD. MEAS 5-1981
ES. AUGER SPECTROSCOPY AND LOW ENERGY ELECTRON DIFFRACTION OF SILICON SURFAC 5-1396
/ SPECTROSCOPY (AES) AND LOW ENERGY ELECTRON DIFFRACTION (LOW ENERGY ELECTRON DIFFRACTION OF THE ZINC OXI/ 5-1338
/ATINUM (110) SURFACE. PART-1: SULFUR AND CARBON MONOXIDE LOW ENERGY ELECTRON DIFFRACTION PATTERNS. (AUGER/ 5-1190
KEL DEPOSITS ON CLEAVED SILICON. AUGER SPECTRA AND LOW ENERGY ELECTRON DIFFRACTION PATTERNS FROM NIC 5-1768
UUM CLEANED SILICON CRYSTALS WITH CALI/ AUGER SPECTRA AND LOW ENERGY ELECTRON DIFFRACTION PATTERNS FROM VAC 5-1770
OF OXYGEN ADSORPTION ON THE TUNGSTEN (112) SUR/ COMBINED LOW ENERGY ELECTRON DIFFRACTION RHEED AUGER STUDY 5-1464
ETAL INTERFACES. (AUGER EFFECT) ELECTRON SPECTROSCOPY AND LOW ENERGY ELECTRON DIFFRACTION STUDIES OF GAS- M 5-1320
/ ENERGY SPECTRA OF INELASTICALLY SCATTERED ELECTRONS AND LOW ENERGY ELECTRON DIFFRACTION STUDIES OF TUNGST 4-1094
IUM SURFACE SEGREGATION IN COPPER/ AUGER SPECTROSCOPY AND LOW ENERGY ELECTRON DIFFRACTION STUDY OF EQUILIBR 6-2025
(100) SURFACE. LOW ENERGY ELECTRON DIFFRACTION STUDY OF IRIDIUM 5-1370
M OXIDE (100). PART-1: EXPERIMENT. A LOW ENERGY ELECTRON DIFFRACTION STUDY OF MAGNESIU 5-1580
(100) SURFACE. LOW ENERGY ELECTRON DIFFRACTION STUDY OF PLATINUM 5-1377
/D FIELD ION MICROSCOPY. AUGER ELECTRON SPECTROSCOPY, AND LOW ENERGY ELECTRON DIFFRACTION STUDY OF THE STR/ 5-1951
SOFT X-RAY APPEARANCE POTENTIAL SPECTROSCOPY IN A DISPLAY LOW ENERGY ELECTRON DIFFRACTION SYSTEM. 4-0870
LECTRON M/ AUGER ELECTRON SPECTROSCOPY USING THE TWO-GRID LOW ENERGY ELECTRON DIFFRACTION SYSTEM AND PHOTOE 3-0656
PECTRUM ANALYSES. APPLICATION OF TRIPLE GRID LOW ENERGY ELECTRON DIFFRACTION SYSTEM TO AUGER S 3-0661
AUGER ELECTRON SPECTROSCOPY IN LOW ENERGY ELECTRON DIFFRACTION SYSTEMS. 1-0153
AUGER ELECTRON SPECTROSCOPY IN LOW ENERGY ELECTRON DIFFRACTION SYSTEMS. 1-0158
OPTIMIZATION OF AUGER ELECTRON SPECTROSCOPY IN LOW ENERGY ELECTRON DIFFRACTION SYSTEMS. 1-0152
PROBE TECHNIQUE. (COMPARISON BETWEEN ION PROBE AND AUGER- LOW ENERGY ELECTRON DIFFRACTION TECHNIQUES) /ION 5-1166
AUGER EFFECT AND CORE BROADENING IN LOW-Z ATOMS, IONS, AND METALS. 2-0353
OELECTRON SPECTROSCOPY. INVESTIGATION OF LOW-Z COSTER-KRONIG TRANSITIONS BY AUGER AND PHOT 2-0578
LECTRON SPECTROSCOPY. LEAD MONOLAYER LUBRICATION IN STEEL MACHINING STUDIED BY AUGER E 6-2124
(AUGER STUDIES OF) ADHESION, FRICTION, LUBRICATION IN VACUUM. 5-1209
CHNIQUES. (AES, ELLIPSOMET/ ADHESION, FRICTION, WEAR, AND LUBRICATION RESEARCH BY MODERN SURFACE SCIENCE TE 1-0104
TED GALLIUM PHOSPHIDE. EXCITON LUMINESCENCE AND PHOTOCONDUCTIVITY IN HIGHLY EXCI 2-0454
GALLIUM ARSENIDE. LUMINESCENCE DUE TO EXCITON-EXCITON COLLISION IN 2-0451
LFIDE. (AUGER / CONCENTRATION DEPENDENCE OF GREEN- COPPER LUMINESCENCE IN COPPER AND ALUMINUM DOPED ZINC SU 2-0402
/LD DEPENDENT INTENSITY OSCILLATIONS OF THE ELECTRON HOLE LUMINESCENCE IN PURE GERMANIUM. (AUGER RECOMBINA/ 4-0780
AUGER ELECTRON SPECTROSCOPY OF LUNAR MATERIAL. 5-1277
AUGER ELECTRON SPECTROSCOPY OF LUNAR SAMPLES. 4-0858
KLL AUGER ELECTRON SPECTRA OF THULIUM AND LUTETIUM. 4-0753
HE DECAY OF ERBIUM-160, YTTERBIUM-166, YTTERBIUM-169, AND LUTETIUM-171. / ELECTRONIC SPECTRA ACCOMPANYING T 4-0743
TION PROBABILITIES TO THE ATOMIC 2S STATE AND THEORETICAL L1 FLUORESCENCE YIELDS. /AND COSTER-KRONIG TRANSI 2-0329
L1,L2,3M COSTER-KRONIG TRANSITION ENERGIES. 4-0929
IATIONS IN ENERGY LEVELS AND SCREENING PA/ SENSITIVITY OF L1,L2,3M4,5 COSTER-KRONIG TRANSITION RATES TO VAR 2-0303
THE L1,L2,3M4,5 COSTER-KRONIG TRANSITIONS. 2-0248
L1,L2,3V AUGER TRANSITION IN SILICON. 4-0779
COSTER-KRONIG AND AUGER SPECTRUM OF THE L1 SHELL OF ARGON. 4-0977
AUGER SPECTRUM OF L2 AND L3 SHELLS OF ARGON. 4-0983
X) COSTER-KRONIG TRANSITION PROBABILITIES. THEORETICAL L2 AND L3 SUBSHELL FLUORESCENCE YIELDS AND L2-L3(2-0316
CT) WIDTH OF ATOMIC L2 AND L3 VACANCY STATES NEAR Z = 30. (AUGER EFFE 4-0939
TRANSITION PROBABILITIES F23 BASED ON THE G/ THEORETICAL L2 SUBSHELL FLUORESCENCE YIELDS AND COSTER-KRONIG 2-0313
ND PHOSPHORUS. AUGER SATELLITES OF THE L2,3 AUGER EMISSION BANDS OF ALUMINUM, SILICON, A 4-0985
N COLLISIONS. CROSS SECTIONS FOR L2,3 AUGER EMISSION IN 22 TO 300 KEV PROTON- ARGO 4-1040
WIDTHS AND INTENSITIES OF HIGHLY RESOLVED LINES IN ARGON L2,3 AUGER SPECTRA. 4-1032
BY MEANS OF THE DECOMPOSI/ DETECTION OF EXCITONS NEAR THE L2,3 EDGE OF THE CHLORIDE ION IN SODIUM CHLORIDE 4-1126
M, AND/ INTERPRETATION OF HIGH ENERGY X-RAY SATELLITES OF L2,3 EMISSION BANDS OF SODIUM, MAGNESIUM, ALUMINU 2-0239
"SEMI-AUGER" PROCESSES IN L2,3 EMISSION IN ARGON AND POTASSIUM CHLORIDE. 2-0326
. (AUGER EFFECT) MULTIPLET EFFECTS ON THE L2,3 FLUORESCENCE YIELD OF MULTIPLY IONIZED ARGON 4-0956
LISION. IONIZATION CROSS SECTION OF THE L2,3 SHELL OF ALUMINUM IN A NEON (+)- ALUMINUM COL 2-0267
ROSCO/ MEASUREMENT OF THE IONIZATION CROSS SECTION OF THE L2,3 SHELL OF CHLORINE USING AUGER ELECTRON SPECT 4-0987
L1,L2,3M COSTER-KRONIG TRANSITION ENERGIES. 4-0929
M4, 5NN AND N4,500 AUGER SPECTRA FROM XE/ HIGH RESOLUTION L2,3MM AND M4,5NN AUGER SPECTRA FROM KRYPTON AND 4-1113

TATION.   COMPARISON CF MAGNESIUM L2,3MM AUGER CURRENTS USING ELECTRON AND ION EXCI   4-0855
                            PHOTOELECTRON AND L2,3MM AUGER ELECTRON ENERGIES FOR ARSENIC.   4-1035
MS.   L2,3MM AUGER ELECTRON SPECTRUM FROM FREE ZINC ATO   4-0731
                                                   L2,3MM AUGER PROCESSES IN SELENIUM.   2-0560
                                                   L2,3MM AUGER SPECTRUM OF ARGON.   4-1114
                 TRANSITION PROBABILITIES FOR THE L2,3MM AUGER SPECTRUM OF SELENIUM.   4-1033
                            HIGH RESOLUTION L2,3M2,3M2,3 AUGER ELECTRON SPECTRA OF ARGON.   4-0912
IONS IN ENERGY LEVELS AND SCREENING PA/ SENSITIVITY OF L1,L2,3M4,5 COSTER-KRONIG TRANSITION RATES TO VARIAT   2-0303
                          THE L1,L2,3M4,5 COSTER-KRONIG TRANSITIONS.   2-0248
ZINC: THEORY AND EXPERIMENT.   L2,3M4,5M4,5 AUGER SPECTRA OF METALLIC COPPER AND   4-0919
                            HIGH RESOLUTION L2,3M4,5M4,5 AUGER SPECTRUM OF FREE ZINC ATOMS.   4-0733
                                                   L1,L2,3V AUGER TRANSITION IN SILICON.   4-0779
FROM THE CROSS SECTION M/ A STUDY OF THE INTENSITY OF THE L2,3VV AUGER TRANSITION IN MAGNESIUM AND SILICON   4-0765
     THEORETICAL L2 AND L3 SUBSHELL FLUORESCENCE YIELDS AND L2-L3(X) COSTER-KRONIG TRANSITION PROBABILITIES.   2-0316
D PHOSPHORUS.   AUGER SATELLITES ON THE L23 AUGER EMISSION BANDS OF ALUMINUM, SILICON, AN 4-0984
                         LIFETIME AND SURFACE EFFECTS ON THE L23 VV AUGER SPECTRA OF ALUMINUM.   4-0833
TYPE PROCESSES)   CLUSTER CALCULATIONS OF THE SOFT X-RAY L3 EMISSION SPECTRA OF NICKEL AND COPPER. (AUGER-   4-0888
AUGER ELECTRON APPEARANCE POT/ DIFFERENCE IN THE CHROMIUM L3/L2 INTENSITY RATIO MEASURED BY SOFT X-RAY AND   4-0890
                AUGER VACANCY SATELLITE STRUCTURE IN THE L3,M4,5M4,5 AUGER SPECTRA OF COPPER.   4-1034
                                  AUGER SPECTRUM OF L2 AND L3 SHELLS OF ARGON.   4-0983
ER-KRONIG TRANSITION PROBABILITIES.   THEORETICAL L2 AND L3 SUBSHELL FLUORESCENCE YIELDS AND L2-L3(X) COST   2-0316
. (AUGER EFFECT)   WIDTH OF ATOMIC L2 AND L3 VACANCY STATES NEAR Z = 30. (AUGER EFFECT)   4-0939
                           COMPARISON OF SILICON L3,2 ELECTRON LOSS SPECTRA AND OPTICAL ABSORPTION   4-1038
                       ANGULAR DISTRIBUTION OF L3M2,3M2,3(1S0) AUGER ELECTRONS OF ARGON.   2-0322
             HIGH RESOLUTION MEASUREMENTS OF THE L3M2,3M4,5 AUGER TRANSITONS IN NICKEL AND COPPER.   4-1024
ATOMIC AND EXTRAATOMIC STATIC RELAXATION ENERGIES IN THE L3M4,5M4,5 AUGER PROCESS IN ARSENIC AND SELENIUM.   4-0883
IMPROVED AUGER ENERGY CALCULATI/ DETAILED ANALYSIS OF THE L3M4,5M4,5: 1G4 AUGER TRANSITION IN ATOMIC ZINC: 4-0884
         RELAXATION PHENOMENA INVOLVED IN THE L3M4,5M4,.5M4,.5:1G4 AUGER PROCESS IN ZINC METAL.   4-0882

                                               M

                      SPECTRUM OF AUGER K,L,M,N,O ELECTRONS OF SAMARIUM.   4-0756
                      EMISSION LINES OF THE M SERIES OF RARE EARTH METALS. (AUGER EFFECT)   4-0790
              K-, L-, AND M-AUGER AND L- COSTER-KRONIG SPECTRA OF PLATINUM.   4-1097
     PHENOMENOLOGICAL SYSTEMATICS OF L- AND M-AUGER SPECTRA.   4-0875
DRYING OF BUTADIENE- NITRILE RUBBERS IN AN AUGER MACHINE.   3-0617
        LEAD MONOLAYER LUBRICATION IN STEEL MACHINING STUDIED BY AUGER ELECTRON SPECTROSCOPY.   6-2124
                      SURFACE MADELUNG POTENTIALS IN ELECTRON SPECTROSCOPY.   2-0511
ON SPECTROMETER, FRACTURED/ GRAIN BOUNDARY SEGREGATION IN MAGNESIA DOPED ALUMINA. (HIGH VACUUM AUGER ELECTR   5-1615
     INTERFACIAL AUGER TRANSITIONS IN OXIDIZED SODIUM AND MAGNESIUM.   5-1497
SLOW SECONDARY ELECTRON EMISSION PEAK IN POLYCRYSTALLINE MAGNESIUM. /DATION EFFECTS ON THE SPECTRUM OF THE   4-1118
     AES CHARACTERIZATION OF OXIDIZED FILMS OF MAGNESIUM, ALUMINUM, AND SILICON.   5-1847
ENERGY X-RAY SATELLITES OF L2,3 EMISSION BANDS OF SODIUM, MAGNESIUM, ALUMINUM, AND SILICON. /ATION OF HIGH   2-0239
NERGY ELECTRON AND LOW EN/ COMPARISON OF AUGER SPECTRA OF MAGNESIUM, ALUMINUM, AND SILICON EXCITED BY LOW E   4-0857
/ DISTRIBUTION OF AUGER ELECTRONS EMITTED FROM BERYLLIUM, MAGNESIUM, ALUMINUM AND SILICON UNDER ION BOMBAR/   4-0876
                      AUGER SPECTRA OF MAGNESIUM AND ALUMINUM.   4-0822
HOTOEMISSION CONTRIBUTION TO KLM AUGER SPECTRA OF SODIUM, MAGNESIUM, AND ALUMINUM. /ATION OF THE INTERNAL P   4-0769
             CHANGES IN AUGER SPECTRA OF MAGNESIUM AND IRON DUE TO OXIDATION.   4-1086
LOW ENERGY AUGER AND LOSS ELECTRON SPECTRA FROM MAGNESIUM AND ITS OXIDE.   4-0895
            HIGH ENERGY SATELLITES IN AUGER PEAKS OF MAGNESIUM AND SILICON.   4-1046
/STUDY OF THE INTENSITY OF THE L2,3VV AUGER TRANSITION IN MAGNESIUM AND SILICON FROM THE CROSS SECTION MOD/   4-0765
LECTRON/ ENERGY SPECTRA OF SECONDARY ELECTRONS EMITTED BY MAGNESIUM AND SILICON WHEN BOMBARDED BY PRIMARY E   4-1047
          AUGER SPECTRA OF ACTIVATED SILVER- MAGNESIUM AND SILVER- LITHIUM ALLOYS.   6-2004
              THE K-AUGER SPECTRUM OF THE FREE MAGNESIUM ATOM.   4-0789
        MANY BODY EFFECTS IN X-RAY PHOTOEMISSION FROM MAGNESIUM. (AUGER EFFECT)   2-0427
GALLIUM(1-X) ARSENIDE BY MOLECULAR BEAM EPITAXY. (AUGER / MAGNESIUM DOPED GALLIUM ARSENIDE AND ALUMINUM(X)   5-1247
TROSCOPY.   SURFACE CONCENTRATION OF MAGNESIUM IN 6061 ALUMINUM BY AUGER ELECTRON SPEC   5-1824
D ION EXCITATION.   COMPARISON OF MAGNESIUM L2,3MM AUGER CURRENTS USING ELECTRON AN   4-0855
/SION, AND AUGER ELECTRON SPECTRA OF THE VALENCE BANDS OF MAGNESIUM, MAGNESIUM(2) COPPER AND MAGNESIUM(3) /   4-0838
ON DIFFRACTION, LEED AN/ STUDY OF THE EPITAXIAL GROWTH OF MAGNESIUM ON MAGNESIUM OXIDE (001) USING REFLECTI   5-1496
GER ELECTRON SPECTROSCOPY STUDY OF IRON FILM DEPOSITED ON MAGNESIUM OXIDE. /ENERGY ELECTRON DIFFRACTION- AU   5-1522
LECTRON SPECTROSCOPY. COMPARATIVE STUDY OF SINGLE CRYSTAL MAGNESIUM OXIDE AND OXIDIZED MAGNESIUM BY AUGER E   5-1495
ON SPECTROSCOPY.   COMPOSITION CF THE SURFACE OF COPPER- MAGNESIUM OXIDE CATALYSTS STUDIED BY AUGER ELECTR   5-1361
NGLE CRYSTAL.   STRUCTURE AND SECONDARY EMISSION OF MAGNESIUM OXIDE LAYERS APPLIED ON A MOLYBDENUM SI   5-1191
IN THE KL2,3L2,3 RADIATIVE AUGER SPECTRUM OF INSULATORS. (MAGNESIUM OXIDE, MAGNESIUM FLUORIDE) /ON EFFECTS   2-0540
TION EFFECT ON THE GROWTH MODE OF IRON EPITAXIAL FILMS ON MAGNESIUM OXIDE (001). /TUDY OF SURFACE CONTAMINA   5-1524
ORPTION OF SODIUM AND MEASUREMENT OF SURFACE DIFFUSION ON MAGNESIUM OXIDE (001) BY AUGER SPECTROSCOPY. /DES   5-1494
OBSERVATION OF GROWTH MECHANISM OF DEPOSITED IRON FILM ON MAGNESIUM OXIDE (001) SURFACE.   LEED- AES   5-1525
Y)   GROWTH MODE OF IRON FILM DEPOSITED ON MAGNESIUM OXIDE (001) SURFACE. (AUGER SPECTROSCOP   5-1528
            HIGH RESOLUTION AUGER ELECTRON SPECTRUM OF MAGNESIUM OXIDE (100).   4-0772
                      A LEED STUDY OF MAGNESIUM OXIDE (100). PART-1: EXPERIMENT.   5-1580
ATION TO HIGH YIELD SECONDARY/ DIFFERENTIAL SPUTTERING OF MAGNESIUM OXIDE- GOLD CERMET FILMS AND ITS APPLIC   5-1430
EFFECTS OF CESIATION ON SECONDARY ELECTRON EMISSION FROM MAGNESIUM OXIDE- GOLD CERMETS.   2-0384
     AUGER SATELLITE AND OTHER CHARACTERISTIC EVENTS IN MAGNESIUM SECONDARY ELECTRON SPECTRA.   4-0898
MAGNESI/ OPTICAL PROPERTIES AND AUGER CHARACTERISTICS OF MAGNESIUM SINGLE CRYSTALS AND MAGNESIUM OXIDE AND   5-1220
E.   X-RAY EXCITED AUGER AND PHOTOELECTRON SPECTRA OF MAGNESIUM, SOME ALLOYS OF MAGNESIUM, AND ITS OXID   4-0837
MISSION SPECTROSCOPY.   VALENCE BAND OF MAGNESIUM STANNIDE DETERMINED BY AUGER AND PHOTOE   5-1885
TED AUGER AND PHOTOELECTRON SPECTRA OF PARTIALLY OXIDIZED MAGNESIUM SURFACES. ABNORMAL CHEMICAL SHIFTS. /I   5-1949
     SECONDARY ELECTRON EMISSION OF SOLID ALUMINUM AND MAGNESIUM TARGETS.   2-0373
(AES)   STRESS CORROSION CRACKING OF AN ALUMINUM- 5% MAGNESIUM TERNARY AND VARIOUS QUATERNARY ALLOYS.   6-2036
                      AUGER ELECTRON SPECTRA OF MAGNESIUM THIN FILMS.   4-0873
     X-RAY PHOTOELECTRON AND AUGER ELECTRON SPECTRA OF MAGNESIUM THIN FILMS.   5-1392
OF ELECTRONIC CORRELATIONS ON KLL AUGER LINES IN ELEMENTS MAGNESIUM TO SCANDIUM.   EFFECT   4-0800
AUGER ELECTRON SPECTRA OF THE VALENCE BANDS OF MAGNESIUM, MAGNESIUM(2) COPPER AND MAGNESIUM(3) GOLD. / AND   4-0838
D MAGNESIUM(2) STANNIDE. PAR/ PHOTOELECTRIC PROPERTIES OF MAGNESIUM(2) SILICIDE, MAGNESIUM(2) GERMANIUM, AN   2-0306
F THE VALENCE BANDS OF MAGNESIUM, MAGNESIUM(2) COPPER AND MAGNESIUM(3) GOLD. / AND AUGER ELECTRON SPECTRA O   4-0838
SEMICONDUCTOR (MERCURY(1-X) CADMIUM(X) TELLURIDE) UNDER A MAGNETIC FIELD.   AUGER RECOMBINATION IN A   4-1092
F THE ELECTRON HOLE LUMINESCENCE IN PURE GERMANIUM. (AUG/ MAGNETIC FIELD DEPENDENT INTENSITY OSCILLATIONS O   4-0780
N FILMS. (AES)   CORROSION, SURFACE, AND MAGNETIC PROPERTIES OF NICKEL- IRON- CHROMIUM THI   5-1766
     DEPENDENCE OF ATOMIC PHOTOEFFECT ON BOUND ELECTRON MAGNETIC SUBSTATE (AUGER EFFECT).   2-0524

DEPENDENCE OF ATOMIC PHOTOEFFECT ON A BOUND ELECTRON MAGNETIC SUBSTATE. (AUGER ELECTRON DISTRIBUTION) 2-0463
KLL AUGER SPECTRUM OF MANGANESE. 4-0938
TEMPER EMBRITTLED STEEL. (AUGER ELECTRON/ SEGREGATION OF MANGANESE AND ANTIMONY TO THE GRAIN BOUNDARIES OF 6-2070
AUGER EFFECT IN THE MULTIPLE IONIZATION OF MANGANESE AND CADMIUM BY ELECTRON IMPACT. 2-0228
LOW ALLOY STEELS. (AUGER SPECTROSCOPY) ROLE OF MANGANESE AND SILICON IN TEMPER EMBRITTLEMENT OF 6-2059
/ECTS OF COPPER AND PHOSPHORUS ON TEMPER EMBRITTLEMENT OF MANGANESE- MOLYBDENUM- NICKEL LOW ALLOY STEEL. (/ 6-2043
R EFFECT. (NEON AUGER SPECTRUM) MANIFESTATION OF ATOMIC DYNAMICS THROUGH THE AUGE 4-0920
INSTRUMENTATION FOR THE CHEMICAL ANALYSIS OF MANUFACTURED SURFACES. (REVIEW, SPECTROSCOPY) 3-0719
UGER ELECTRON ANALYSIS) TO SURFACE CONTROL PROGRAM IN THE MANUFACTURING OF SILICON DEVICES. /PECTROSCOPY (A 5-1411
POTENTIAL MAPPING USING AUGER ELECTRON SPECTROSCOPY. 1-0113
TUDY OF GRAIN BOUNDARY SEGREGANTS IN THERMALLY EMBRITTLED MARAGING STEEL. (AUGER ELECTRON SPECTROSCOPY) S 6-2055
LIPSOMETRY. MARKER EXPERIMENTS WITH AUGER SPECTROSCOPY AND EL 5-1832
E SCANNING ELECTRON MICROSCOPE NOW INCLUDES SECONDARY ION MASS ANALYSIS. (AUGER) SPECTROSCOPY IN TH 3-0674
ELECTRON MICROSCOPE, ENERGY DISPERSIVE SPECTROMETER, ION MASS ANALYSIS, AUGER ELECTRON MICROSCOPY) /ISSION 1-0010
METALLURGY. POSSIBILITIES OF THE ION MICROPROBE MASS ANALYZER AND AUGER SPECTROMETER IN PHYSICAL 6-2075
RGY ELECTRON DIFFRACTION- AUGER ELECTRON SPECTROSCOPY AND MASS FILTER. /F METAL ON THEM OBSERVED BY LOW ENE 5-1906
ANALYZED BY AUGER ELECTRON SPECTROSCOPY AND SECONDARY ION MASS SPECTRA DURING XENON ION BOMBARDMENT. /PPER 5-1679
IC TECHNIQUES FOR METAL CONTACTS. COMBINED MASS SPECTROMETRIC AND AUGER ELECTRON SPECTROSCOP 6-2042
CARBON MONOXIDE ON POLYCRYSTALLI/ AUGER SPECTROMETRIC AND MASS SPECTROMETRIC STUDY OF THE CHEMISORPTION OF 5-1355
MPARISON OF AUGER ELECTRON SPECTROSCOPY AND SECONDARY ION MASS SPECTROMETRY. CO 1-0135
SILICIDE BY AUGER ELECTRON SPECTROSCOPY AND SECONDARY ION MASS SPECTROMETRY. /OSITED PLATINUM AND PLATINUM 5-1656
ALYSIS WITH AUGER ELECTRON SPECTROSCOPY AND SECONDARY ION MASS SPECTROMETRY. ELEMENTAL AN 5-1654
YTIC REACTIONS, LEED, AUGER, CONTACT POTENTIAL METHOD AND MASS SPECTROMETRY) /ACE. (CHEMISORPTION AND CATAL 5-1321
UM FILMS BY AUGER ELECTRON SPECTROSCOPY AND SECONDARY ION MASS SPECTROMETRY. /D OXYGEN) IN SPUTTERED TANTAL 5-1652
ANALYSIS. (AUGER ELECTRON SPECTROSCOPY AND ION MICROPROBE MASS SPECTROMETRY) /STRUMENTATION FOR MICROPROBE 3-0716
HIGH VACUUM TECHNIQUES IN BASIC RESEARCH. (REVIEW, AUGER, MASS SPECTROMETRY) ROLE OF ULTRA 1-0065
SECONDARY ION MASS SPECTROMETRY. 1-0079
ECTROMETRY, AUGER ELECTRON SPECTROSCOPY AND SECONDARY ION MASS SPECTROMETRY. /UDIES USING ION SCATTERING SP 5-1423
Y): A USEFUL COMPLEMENT TO OTHER DEPTH PRO/ SECONDARY ION MASS SPECTROMETRY (AND AUGER ELECTRON SPECTROSCOP 1-0009
/NESS OF 316 L-N STAINLESS STEEL STUDIES BY SECONDA/ MASS SPECTROMETRY AND AUGER ELECTRON SPECTROSCOP/ 3-0654
/PROCESSES ON COPPER- BERYLLIUM BY COMBINED SECONDARY ION MASS SPECTROMETRY AND AUGER ELECTRON SPECTROSCOP/ 5-1213
, FOR THE ANALYSIS OF THIN / COMBINATION OF SECONDARY ION MASS SPECTROMETRY AND AUGER ELECTRON SPECTROSCOPY 5-1214
/TROGEN, OXYGEN, AND FLUORINE IN SILICON BY SECONDARY ION MASS SPECTROMETRY AND AUGER ELECTRON SPECTROSCOP/ 5-1916
/NESS OF 316 L-N STAINLESS STEEL STUDIED BY SECONDARY ION MASS SPECTROMETRY AND AUGER ELECTRON SPECTROSCOP/ 6-2087
/URFACES BY AUGER ELECTRON SPECTROSCOPY AND SECONDARY ION MASS SPECTROMETRY FOR ALUMINUM, SILICON, TITANIU/ 5-1553
GY./ USE OF AUGER ELECTRON SPECTROSCOPY AND SECONDARY ION MASS SPECTROMETRY IN THE MICROELECTRONIC TECHNOLO 1-0136
AN ANALYTICAL SYSTEM FOR SECONDARY ION MASS SPECTROMETRY IN ULTRAHIGH VACUUM. (AES) 3-0632
OSCOPY X-RAY PHOTOELECTRON SPECTROSCOPY AND SECONDARY ION MASS SPECTROSCOPY. /S USING AUGER ELECTRON SPECTR 5-1353
ILMS ON COPPER PLATES BY AUGER ELECTRON AND SECONDARY ION MASS SPECTROSCOPY. /ANED SURFACES OF GOLD THICK F 5-1550
T STATUS OF BEAM ANALYSIS AND ITS FUTURE. (REVIEW, AUGER, MASS SPECTROSCOPY) PRESEN 1-0181
/FACES CHARACTERIZED BY AUGER SPECTROSCOPY, SECONDARY ION MASS SPECTROSCOPY AND LOW ENERGY ELECTRON DIFFRA/ 2-0501
FOR RELIABILI/ DIRECT COMPARISON OF AUGER, SECONDARY ION MASS SPECTROSCOPY, AND PROTON RESONANCE PROFILING 1-0216
/AND AUGER ELECTRON SPECTROSCOPY (ESCA) AND SECONDARY ION MASS SPECTROSCOPY. COMPARISON OF TWO SURFACE AN/ 1-0088
S TECHNIQUES. (AUGER ELECTRON SPECTROSCOPY, SECONDARY ION MASS SPECTROSCOPY, ELECTRON MICROPROBE) / ANALYSI 1-1627
/ON BETWEEN AUGER ELECTRON SPECTROSCOPY AND SECONDARY ION MASS SPECTROSCOPY ON WELL CHARACTERIZED SINGLE C/ 5-1689
ENERGY ELECTRON DIFFRACTION- AUGER ELECTRON SPECTROSCOPY/ MASS SPECTROSCOPY SYSTEM. /MEANS OF COMBINED LOW 6-2141
CTS IN FLOWING LIQUID SODIUM. (SCANNING ELECTRON MICROSC/ MASS TRANSPORT OF STAINLESS STEEL CORROSION PRODU 6-2014
TECHNIQUE/ RECRYSTALLIZATION OF TUNGSTEN FIBERS IN NICKEL MATRIX COMPOSITES. (AUGER ELECTRON SPECTROSCOPIC 6-2046
SITION RATES/ M-SHELL AUGER, COSTER-KRONIG, AND RADIATIVE MATRIX ELEMENTS, AND AUGER AND COSTER-KRONIG TRAN 4-0967
AND AUGER DECAY OF 5D CORE HOLES IN BISMUTH(III) TELLURI/ MATRIX ELEMENTS DEPENDENCE OF OPTICAL EXCITATION 2-0398
GER, LEED, ENERGY LOSS SPECTROSCOPY) INELASTIC MEAN FREE PATH FOR ELECTRONS IN BULK JELLIUM. (AU 4-1059
(-4) TORR AND TEMPERATURE RANGE 1500 TO 2600 DEGREES K AS MEASURED BY AES. /THE PRESSURE RANGE 10(-9) TO 10 5-1148
ELECTRON IMPACT IONIZATION CROSS SECTIONS OF INNER SHELLS MEASURED BY AUGER ELECTRON SPECTROSCOPY. 2-0553
INTERDIFFUSION IN BIMETALLIC POLYCRYSTALLINE FILMS MEASURED BY AUGER ION SPUTTERING TECHNIQUE. 5-1440
NCE POT/ DIFFERENCE IN THE CHROMIUM L3/L2 INTENSITY RATIO MEASURED BY SOFT X-RAY AND AUGER ELECTRON APPEARA 4-0890
SPHORUS CONCENTRATION PROFILES IN P-DOPED SILICON DIOXIDE MEASURED USING AUGER SPECTROSCOPY. PHO 5-1231
DIFFUSION OF COBALT OUT OF COBALT HARDENED GOLD MEASURED WITH AUGER ELECTRON SPECTROSCOPY. 6-2135
G TECHNIQUE. DIFFUSION MEASUREMENT IN THIN FILMS BY AUGER SPUTTER ETCHIN 5-1439
ERATURE ADSORPTION OF OXYGEN ON TUNGSTEN SURFACES. (AUGER MEASUREMENT METHODS) ROOM TEMP 5-1664
ON TO SULFUR ON NICKEL (110). ANALOG TECHNIQUE FOR THE MEASUREMENT OF AUGER ELECTRON CURRENTS. APPLICATI 3-0635
IMPROVED APPARATUS FOR MEASUREMENT OF AUGER ELECTRON SPECTRA. 3-0612
LOW NOISE DETECTION SYSTEM FOR THE MEASUREMENT OF AUGER SPECTRA. 3-0662
U/ AN AC RETARDING POTENTIAL TECHNIQUE FOR THE CONTINUOUS MEASUREMENT OF CHANGES IN WORK FUNCTION. (LEED- A 3-0663
RAPHITE (0001). (AES) ULTRAHIGH VACUUM TECHNIQUES IN THE MEASUREMENT OF CONTACT ANGLES. PART-4: WATER ON G 5-1795
/OINCIDENCE METHODS OF STANDARDIZATION FOR CESIUM-131 AND MEASUREMENT OF DECAY PARAMETERS. (DETECTION OF L/ 2-0474
ING AUGER ELECTRON SPECTROSCOPY. MEASUREMENT OF EQUILIBRIUM SURFACE SEGREGATION US 5-1184
ING AUGER ELECTRON SPECTROSCOPY. (REPLY TO COMMENTS) MEASUREMENT OF EQUILIBRIUM SURFACE SEGREGATION US 6-2026
/ DOUBLE CRYSTAL SPECTROMETER INSTALLED IN VACUUM AND THE MEASUREMENT OF EXCITED ELECTRONS UNDER THE X-RAY/ 3-0641
SCATTERING FACTORS FOR USE IN QUANTITATIVE AES. MEASUREMENT OF IONIZATION CROSS SECTIONS AND BACK 2-0554
BER TECHNIQUES: OXYGEN AND NEON AUGE/ IDENTIFICATION AND MEASUREMENT OF LOW ENERGY ELECTRONS BY CLOUD CHAM 2-0566
S ON SURFACES CHARACTERIZED BY AUGER SPECTROSCOPY, SECON/ MEASUREMENT OF MOMENTUM ACCOMMODATION COEFFICIENT 2-0501
DE (001) BY/ ELECTRON STIMULATED DESORPTION OF SODIUM AND MEASUREMENT OF SURFACE DIFFUSION ON MAGNESIUM OXI 5-1494
M(2S) (+) IONIC SYSTEM. MEASUREMENT OF THE AUGER EFFECT IN THE MU-4 HELIU 2-0305
E L2,3 SHELL OF CHLORINE USING AUGER ELECTRON SPECTROSCO/ MEASUREMENT OF THE IONIZATION CROSS SECTION OF TH 4-0987
S BY THE L X-RAY- L X-RAY COINCIDENCE METHOD. (AUGER TRA/ MEASUREMENT OF THE PRODUCTION OF LL VACANCY STATE 2-0482
TYPE LOW ENERGY ELECTRON DIFFRACTION AUGER APPARATUS. MEASUREMENT OF WORK FUNCTION CHANGE IN A DISPLAY 3-0611
NEW METHOD OF THIN FILM THICKNESS MEASUREMENT USING RADIOISOTOPE AUGER ELECTRONS. 5-1658
OF KRYPTON ON THE BASAL PLANE OF GRAPHITE: LEED AND AUGER MEASUREMENTS. ADSORPTION 5-1559
PING OF SILICON LAYERS GROWN BY SOLID PHASE EPITAXY. (AES MEASUREMENTS) ANTIMONY DO 5-1577
N SPECTROSCOPY IN SCANNING ELECTRON MICROSCOPY: POTENTIAL MEASUREMENTS. AUGER ELECTRO 3-0648
FRACTION SPECTRA, AUGER SPECTRA, AND WORK FUNCTION CHANGE MEASUREMENTS. /TUDY USING LOW ENERGY ELECTRON DIF 5-1291
COMPOSITION OF BINARY ALLOYS BY SIMULTANEOUS SIMS AND AES MEASUREMENTS. 6-2090
INANTS ON POLYCRYSTALLINE SILVER BY AES AND WORK FUNCTION MEASUREMENTS. DETECTION OF CONTAM 5-1574
ON DEPTH OF SUPERCONDUCTING NIOBIUM. (AUGER SPECTROSCOPIC MEASUREMENTS) /RATURE HEAT TREATMENT ON PENETRATI 5-1938
GY ARGON- ION BOMBARDMENT ON MOLYBDENUM DISULFIDE. (AUGER MEASUREMENTS) EFFECTS OF LOW ENER 2-0348
CTIONS FOR POTENTIAL MODULATION DISTORTION IN AUGER YIELD MEASUREMENTS. EXACT CORRE 2-0389
FLUORESCENCE YIELD OF CARBON. (AUGER MEASUREMENTS) 5-1571
ENERGY ELECTRON DIFFRACTION, AUGER, AND CONTACT POTENTIAL MEASUREMENTS. /INUM (111) SURFACES: ELECTRIC LOW 5-1744
ROMETERS FOR RETARDING POTENTIAL SECONDARY ELECTRON YIELD MEASUREMENTS. USE OF CYLINDRICAL AUGER SPECT 3-0628
/XIDATION STUDIES BY MEANS OF PHOTOELECTRIC WORK FUNCTION MEASUREMENTS AND AUGER ELECTRON SPECTROSCOPY ON / 5-1928

ION SPUTTERIN/ AUGER ELECTRON SPECTROSCOPY IN SPUTTERING MEASUREMENTS. APPLICATION TO LOW ENERGY ARGON(+) 5-1825
R SYSTEM)   FAST, ACCURATE SECONDARY ELECTRON YIELD MEASUREMENTS AT LOW PRIMARY ENERGIES. (LEED- AUGE 3-0629
FACE ANALYSIS USING COMPLEMENTARY ELECTRONIC AND CHEMICAL MEASUREMENTS. (AUGER)   SUR 5-1390
/PTION STUDIES OF NICKEL SURFACES UTILIZING WORK FUNCTION MEASUREMENTS, AUGER, AND LOW ENERGY ELECTRON DIF/ 5-1290
GRAIN BOUNDARY ACTIVITY MEASUREMENTS BY AUGER ELECTRON SPECTROSCOPY. 6-2049
/ES ON IRON (100) BY MEANS OF PHOTOELECTRIC WORK FUNCTION MEASUREMENTS COMBINED WITH AUGER ELECTRON SPECTR/ 5-1927
/ORRELATION OF AUGER SPECTROSCOPY, LEED AND WORK FUNCTION MEASUREMENTS FOR EPITAXIAL GROWTH OF THORIUM ON / 5-1739
TROSCOPY OF SOLIDS.   AUGER CURRENT MEASUREMENTS FOR QUANTITATIVE AUGER ELECTRON SPEC 4-0856
/ORE LEVEL STUDIES OF INDIUM ANTIMONIDE VIA PHOTOEMISSION MEASUREMENTS IN THE 20-70 EV RANGE. (AUGER PROCE/ 2-0341
N CHANGES: CHROMIUM INTO GOLD. (AUGER ELECTRON/ DIFFUSION MEASUREMENTS IN THIN FILMS UTILIZING WORK FUNCTIO 5-1889
GERMANIUM. (DEPTH PROFILES OF OXYGEN A/ IN-SITU TRANSPORT MEASUREMENTS IN ULTRAHIGH VACUUM GROWN AMORPHOUS 5-1547
SPECTROSCOPY.   THICKNESS MEASUREMENTS OF ADSORBED LAYERS BY AUGER ELECTRON 5-1618
TUNGSTEN AT LOW PRESSURE AND/ AUGER ELECTRON SPECTROSCOPY MEASUREMENTS OF ADSORPTION ISOBARS FOR OXYGEN ON 5-1282
STRUCTURE IN THE SECONDARY ELECTRON ENER/ HIGH RESOLUTION MEASUREMENTS OF AUGER ELECTRON AND PHOTOELECTRON 4-1023
OF HIGH DENSITY PLASMAS IN GERMANIUM.   PICOSECOND OPTICAL MEASUREMENTS OF BAND-TO-BAND AUGER RECOMBINATION 2-0250
ED STATES ON CLEAVED GERMANIUM (111). (AUG/ PHOTOEMISSION MEASUREMENTS OF BULK, SURFACE, AND HYDROGEN INDUC 5-1781
ON ROCK SALT (100) CLEAVAGE SURFACES AND ON AMORPHOUS C/ MEASUREMENTS OF CONDENSATION COEFFICIENTS OF GOLD 5-1135
Y ENERGETIC HEAVY ION BOMBARDMENT.   MEASUREMENTS OF NEON K-AUGER ELECTRONS PRODUCED B 4-0951
TRONS.   HIGH RESOLUTION MEASUREMENTS OF P(+) ARGON KLL AND KLM AUGER ELEC 4-0943
/JECTION FROM METAL SURFACES BY METASTABLE ATOMS. PART-1: MEASUREMENTS OF SECONDARY EMISSION COEFFICIENTS / 2-0340
MEASUREMENTS OF SURFACE MATERIAL STATES. 5-1521
F ARGON AND CHLORINE BY ELECTRON IMPACT. (AUGER EMISSION) MEASUREMENTS OF THE L-SHELL FLUORESCENCE YIELDS O 2-0387
N NICKEL AND COPPER.   HIGH RESOLUTION MEASUREMENTS OF THE L3M2,3M4,5 AUGER TRANSITONS I 4-1024
S OF CHEMISORBED AMMONIA ON A MOLYBDENUM/ HIGH RESOLUTION MEASUREMENTS OF THE NITROGEN KLL AUGER TRANSITION 5-1532
GEN DESORBED FROM NICKEL SURFACES: EFFECTS OF SURFACE CO/ MEASUREMENTS OF THE SPATIAL DISTRIBUTION OF HYDRO 5-1197
SION.   EXTENDED ION- ELECTRON COINCIDENCE MEASUREMENTS OF THE VIOLENT ARGON(+)- ARGON COLLI 2-0530
/FRACTION, AUGER ELECTRON SPECTROSCOPY, AND WORK FUNCTION MEASUREMENTS ON CLEAN AND CESIUM COVERED POLAR F/ 5-1294
MATERIALS USING AUGER ELECTRON SPECTROSCOPY.   SPUTTERING MEASUREMENTS ON CONTROLLED THERMONUCLEAR REACTOR 5-1826
N SPECTROSCOPY   SPUTTERING MEASUREMENTS ON CTR MATERIALS USING AUGER ELECTRO 1-0182
RIALS BY THE LEED METHOD.   MEASUREMENTS ON SURFACES OF ELECTRICAL SOLID MATE 5-1981
SKYLAB-2 MISSION.   AUGER MEASUREMENTS ON T-027 SAMPLES EXPOSED DURING THE 5-1735
SIS)   CHEMISORPTION, PHOTODESORPTION AND CONDUCTIVITY MEASUREMENTS ON ZINC OXIDE SURFACES. (AUGER ANALY 5-1801
FACTORS AFFECTING DEPTH PROFILING MEASUREMENTS USING AUGER ELECTRON SPECTROSCOPY. 5-1457
ELECTRON BEAM EFFECTS IN DEPTH PROFILING MEASUREMENTS WITH AUGER ELECTRON SPECTROSCOPY. 5-1128
NS OF INNER SHELLS.   A NEW METHOD FOR MEASURING ELECTRON IMPACT IONIZATION CROSS SECTIO 4-0848
LOSS PEAKS OF THE PRIMARY BEAM IN SECONDARY ELECTRON / A MEASURING METHOD TO DISTINGUISH AUGER FROM ENERGY 1-0188
S NEAR THEIR S/ AUGER ELECTRON SPECTROSCOPY AS A TOOL FOR MEASURING THE DIFFUSION OF FOREIGN ATOMS IN SOLID 1-0186
UGER ELECTRON SPECTROSCOPY.   MECHANICAL TESTING: IN-SITU FRACTURE DEVICE FOR A 3-0659
AES STUDIES ON SURFACES OF 18-8 STAINLESS STEEL FINISHED MECHANICALLY AND ELECTROLYTICALLY. 6-2124
ARSENIDE ELECTROLUMINESCENT DIODES. (AUGER RECOMBINATION MECHANISM) / COMPENSATED ALUMINUM(X) GALLIUM(1-X) 2-0325
YIELD. (AUGER TRANSITION)   STRIPPING MECHANISM DEPENDENCE OF THE ARGON 2S FLUORESCENCE 4-0957
CE OXYGEN ON LEAD OXIDE BY AES)   GAIN FATIGUE MECHANISM IN CHANNEL ELECTRON MULTIPLIERS. (SURFA 3-0681
ON, AUGER, AND THERMAL DESORPTION TECHNIQUES TO STUDY THE MECHANISM OF A CATALYTIC REACTION. /F PHOTOEMISSI 1-0042
DE (001) SURFACE.   LEED- AES OBSERVATION OF GROWTH MECHANISM OF DEPOSITED IRON FILM ON MAGNESIUM OXI 5-1525
MECHANISM OF FRICTION BY AES. (REVIEW) 1-0193
MECHANISM OF PLASMON GAIN IN AUGER SPECTRA. 2-0440
N A MOLYBDENUM SURFACE BY M/ DYNAMIC INVESTIGATION OF THE MECHANISM OF REACTION BETWEEN SULFUR AND OXYGEN O 5-1534
HE ULTRASOFT X-RAY SPECTROSCOPY TECHNIQ/ INVESTIGATION OF MECHANISM OF S IONIZATION OF NOBLE GAS ATOMS BY T 2-0580
CRYSTALS. (SUBSTRATE SURFACE EFFECTS, AUGER SPECTROMETRY) MECHANISM OF SURFACE ALIGNMENT IN NEMATIC LIQUID 5-1281
CARBURIZED AND GRAPHITIZED NIC/ STUDY OF THE KINETICS AND MECHANISM OF THE DECOMPOSITION OF FORMIC ACID ON 5-1629
EN ON A MOLYBDENUM SURFACE BY AUGER ELECT/ STUDIES OF THE MECHANISM OF THE REACTION BETWEEN SULFUR AND OXYG 5-1530
M AND THE CORRELATION OF RESISTIVITY CHANGES W/ DIFFUSION MECHANISMS IN THE PALLADIUM- GOLD THIN FILM SYSTE 5-1393
TELLURIDE. (AUGER-LIMITED)   RECOMBINATION MECHANISMS IN 8-14 MICRON MERCURY(1-X) CADMIUM(X) 2-0406
LLISIONS. (AUGER EFFECT)   MECHANISMS OF ELECTRON PRODUCTION IN ION- ATOM CO 2-0488
LLISIONS. (AUGER EFFECT, REVIEWS)   MECHANISMS OF ELECTRON PRODUCTION IN ION- ATOM CO 2-0490
TION IN MULTIPLY IONIZED ATOMS.   MECHANISMS OF INNER SHELL EXCITATION AND DEEXCITA 2-0487
CED BY ELECTRON IMPACT IN SOME METAL VAPORS. (AUGER PROC/ MECHANISMS OF SIMPLE AND MULTIPLE IONIZATION INDU 2-0231
TION ON PLATINUM (110).   MECHANISMS OF THE CATALYTIC CARBON MONOXIDE OXIDA 5-1188
/NDARY ELECTRON EMISSION FROM SOLID SURFACES BOMBARDED BY MEDIUM ENERGY IONS (PLATINUM, NICKEL, TANTALUM, / 2-0309
THEORETICAL STUDY OF THE AUGER EFFECT IN THE LIGHT TO MEDIUM RANGE OF ATOMIC NUMBER. 2-0486
AUGER ELECTRONS FROM THE L-SHELL IN MERCURY. 2-0271
RELATIVISTIC CALCULATION OF THE KLL AUGER SPECTRUM. (OF MERCURY) 4-0750
OWING INTERNAL CONVERSION AND ELECTRON CAPTURE. (K-AUGER/ MERCURY AND GOLD ATOMIC CLOUDS REARRANGEMENT FOLL 2-0334
CONCENTRATIONS.   AUGER LIMITED CARRIER LIFETIMES IN MERCURY CADMIUM TELLURIDE AT HIGH EXCESS CARRIER 2-0262
TRANSIENT CARRIER DECAY AND TRANSPORT PROPERTIES IN MERCURY CADMIUM TELLURIDE. (AUGER RECOMBINATION) 2-0461
ES OF CLEAN POLYCRYSTALLINE GOLD AND OF THE ADSORPTION OF MERCURY ON GOLD. /GER ELECTRON SPECTROSCOPY STUDI 5-1519
TURES. (AUGER RECOMB/ WARM ELECTRON EFFECTS IN NARROW GAP MERCURY(1-X) CADMIUM(X) TELLURIDE AMBIENT TEMPERA 2-0343
)   RECOMBINATION MECHANISMS IN 8-14 MICRON MERCURY(1-X) CADMIUM(X) TELLURIDE. (AUGER-LIMITED 2-0406
IC FIELD.   AUGER RECOMBINATION IN A SEMICONDUCTOR (MERCURY(1-X) CADMIUM(X) TELLURIDE) UNDER A MAGNET 4-1092
ONCONSERVATION IN RADIATIONLESS (AUGER) TRANSITIONS IN MU MESIC ATOMS.   PARITY N 2-0364
NITIATED FROM SINGLY AND DOUBLY IONIZED STATES IN LITHIUM METAL.   AUGER TRANSITIONS I 2-0581
NVOLVED IN THE L3M4,5M4,.5M4,.5:1G4 AUGER PROCESS IN ZINC METAL.   RELAXATION PHENOMENA I 4-0882
THE KVV AUGER SPECTRUM OF LITHIUM METAL. 4-0892
4D AUGER TRANSITIONS FROM EXCITED STATES IN SAMARIUM METAL. 4-0823
IC PERCENT, AND PALLADIUM- GOLD 70 ATOMIC PERCENT CONTACT METAL ALLOYS. (BY AUGER ELECTRON SPECTROSCOPY) /M 6-2041
AUGER SPECTRA OF LITHIUM METAL AND LITHIUM OXIDE. 4-0806
SCANDIUM L X-RAY EMISSION FROM PURE METAL AND SCANDIUM OXIDE. 4-1102
GER AND X-RAY PHOTOELECTRON SPECTROSCOPIC STUDY OF SODIUM METAL AND SODIUM OXIDE.   AU 4-0770
X-RAY FLUORESCENCE L SPECTRA OF NIOBIUM IN NIOBIUM METAL AND SOME OF ITS COMPOUNDS. 4-0762
MULTIPLE EXCITATION PROCESSES IN TRANSITION METAL ATOMS. PART-2: INNER AUGER ELECTRONS. 2-0362
IMPURITIES ON THORIUM METAL: AUGER STUDY OF BULK TO SURFACE EQUILIBRIA. 5-1314
ELECTRON SPECTROSCOPY OF METAL BLACKS. 5-1897
TROMETRIC AND AUGER ELECTRON SPECTROSCOPIC TECHNIQUES FOR METAL CONTACTS.   COMBINED MASS SPEC 6-2042
SOME PROPERTIES OF CLEAN AND ALKALI METAL COVERED ZINC OXIDE SURFACES. (LEED, AES) 5-1881
SURFACE STUDIES OF SOME TRANSITION METAL DICHALCOGENIDES. 5-1969
ULTRAVIOLET PHOTOELECTRON STUDIES OF ADSORPTION ON CLEAN METAL FILM SURFACES. (AUGER SPECTRA) / AND VACUUM 5-1203
AUGER SPECTROSCOPY OF A BARIUM- METAL FILM SYSTEM. 4-0849
SURFACE STUDIES WITH EPITAXIALLY GROWN METAL FILMS. (AUGER EFFECT) 5-1253
/VE DETERMINATION OF SURFACE OXIDE THICKNESS ON DEPOSITED METAL FILMS BY COMBINATION AUGER SPECTROSCOPY AN/ 6-2048
NG LEED- AUGER TECHNIQUES.   EPITAXY OF ULTRATHIN METAL FILMS ON BODY CENTERED CUBIC SUBSTRATES USI 5-1488

R.   ATOMIC NATURE OF POLYMER- METAL INTERACTIONS IN ADHESION, FRICTION, AND WEA 5-1212
                 A1 KLL AUGER PROFILE ARTIFACT AT THE OXIDE- METAL INTERFACE.   4-1068
              ELECTRON SPECTROSCOPY AND LEED STUDIES OF GAS- METAL INTERFACES. (AUGER EFFECT)   5-1320
STRUCTURES OF SUBSTRATE CRYSTALS AND EPITAXIAL GROWTH OF METAL ON THEM OBSERVED BY LOW ENERGY ELECTRON DI/   5-1906
       ATOMIC ARRANGEMENT CF GOLD (100) AND RELATED METAL OVERLAYER SURFACE STRUCTURES.   5-1716
SCOPY AND LOW ENERGY ELECTRON DIFFRACTION STUDY OF ALKALI METAL OVERLAYERS ON TUNGSTEN (100). /TRON SPECTRO   5-1895
      (AUGER) ELECTRON SPECTROSCOPY CF TRANSITION METAL OXIDE SURFACES.   5-1339
              NEW MICROAUTORADIOGRAPHY OF METAL STRUCTURE USING AUGER ELECTRONS.   6-2052
MS. (LEED- AES STUDIES OF / SOME PROPERTIES CF TRANSITION METAL SUPERCONDUCTORS AT SURFACES AND IN THIN FIL   5-1848
      AUGER NEUTRALIZATION OF MULTIPLY CHARGED IONS ON A METAL SURFACE.   2-0247
            SODIUM SEGREGATION ONTO A LITHIUM METAL SURFACE. (AUGER SPECTRA)   5-1750
SCOPY AND INELASTIC ELECTRON SCATTERING IN THE STUDIES OF METAL SURFACES.   AUGER ELECTRON SPECTRO 5-1818
            AUGER ELECTRON SPECTROSCOPY OF METAL SURFACES.   6-2122
GER ELECTRON SPECTROSCOPY STUDIES OF CARBON OVERLAYERS ON METAL SURFACES.   AU 5-1376
           AUGER SPECTROSCOPY OF SEMICONDUCTOR AND METAL SURFACES.   5-1404
ING, VALENCY AND CHARGE TRANSFER IN CHEMICAL REACTIONS ON METAL SURFACES.   BIND 5-1763
CTROSCOPY APPLIED TO THE STUDY OF ADSORPTION PHENOMENA AT METAL SURFACES. /ERGY LOSS AND AUGER ELECTRON SPE   5-1776
SPECTRA OF ELECTRON INELASTIC SCATTERING FROM RARE EARTH METAL SURFACES.   IONIZATION LOSSES IN (AUGER) 2-0538
         IONIZATION SPECTROSCOPY OF CONTAMINATED METAL SURFACES.   5-1351
            SECONDARY EMISSION AND CONTAMINATION OF METAL SURFACES.   6-2071
       SPUTTER ETCHING AND AUGER ELECTRON SPECTROSCOPY OF METAL SURFACES.   5-1552
       SPUTTER ETCHING AND AUGER ELECTRON SPECTROSCOPY OF METAL SURFACES.   5-1554
RON AND AUGER ELECTRON SPECTRA OF OXIDIZED STATES ON SOME METAL SURFACES.   X-RAY EXCITED PHOTOELECT 5-1690
XPS RESULTS.   STRUCTURE AND COMPOSITION OF METAL SURFACES: A REVIEW OF RECENT LEED, AES, AND   5-1341
           PLASMA CLEANING OF METAL SURFACES. (AUGER ELECTRON SPECTROSCOPY)   6-2093
          IMPURITY CONCENTRATIONS ON METAL SURFACES BY AUGER ELECTRON SPECTROSCOPY.   5-1923
PECTROMETRY.   STUDY OF METAL SURFACES BY IONIC MICROANALYSIS AND AUGER S 5-1582
REMENTS OF SECONDARY EM/ SECONDARY ELECTRON EJECTION FROM METAL SURFACES BY METASTABLE ATOMS. PART-1: MEASU   2-0340
Y AND ANGULAR DISTRIBUT/ SECONDARY ELECTRON EJECTION FROM METAL SURFACES BY METASTABLE ATOMS. PART-3: ENERG   5-1133
/EPOSITION AND SPUTTER REMOVAL OF MOLYBDENUM FROM VARIOUS METAL SURFACES (COPPER, GOLD, ALUMINUM, TUNGSTEN/   5-1878
NICK/ AUGER ELECTRON SPECTROSCOPY OF FACE CENTERED CUBIC METAL SURFACES. (GOLD, SILVER, PALLADIUM, COPPER,   5-1717
LEED AND AUGER ELECTRON SPECTROS/ BEHAVIOR OF REFRACTORY METAL SURFACES IN ULTRAHIGH VACUUM AS OBSERVED BY 6-2020
TROSCOPY STUDIES OF ADSORPTION AND OXIDATION PROCESSES AT METAL SURFACES. (REVIEW, 50 REFS)   ELECTRON SPEC 1-0019
X-RAY PHOTOELECTRON SPECTROS/ A PRELIMINARY STUDY OF PURE METAL SURFACES USING AUGER ELECTRON SPECTROSCOPY   5-1353
E ORIGIN OF LOW TEMPERATURE SILICON MIGRATION IN SILICON- METAL SYSTEMS. (AUGER SPECTROSCOPY)   POSSIBL 5-1437
TRON SPECTRO/ METALLIC STATE OF SILICON IN SILICON- NOBLE METAL VAPOR QUENCHED ALLOYS STUDIED BY AUGER ELEC   5-1438
            AUGER ELECTRON STUDIES OF METAL VAPORS.   4-0729
ND MULTIPLE IONIZATION INDUCED BY ELECTRON IMPACT IN SOME METAL VAPORS. (AUGER PROCESS) /ANISMS OF SIMPLE A 2-0231
ARGON AUGER EMISSION BY ARGON(+) ION BOMBARDMENT OF SOME METALS.   2-0547
UGER EFFECT AND CORE BROADENING IN LOW-Z ATOMS, IONS, AND METALS.   A 2-0353
CTRON AND CHARACTERISTIC ENERGY LOSS SPECTROSCOPY OF SOME METALS.   AUGER ELE 4-1083
          AUGER ELECTRON EMISSION BY ION BOMBARDMENT OF LIGHT METALS.   4-0877
AUGER ELECTRON SPECTROSCOPY OF CONTRAST FORMING LAYERS ON METALS.   6-2047
          AUGER ELECTRON SPECTROSCOPY OF SOME REFRACTORY METALS.   4-0865
          AUGER ELECTRON SPECTROSCOPY OF TRANSITION METALS.   4-0864
          DYNAMICAL SCREENING EFFECTS IN THE AUGER SPECTRUM OF METALS.   4-1110
F THE ELECTRON GAS ON THE ENERGIES OF AUGER TRANSITION IN METALS.   EFFECT OF POLARIZATION O 2-0435
          HIGH ENERGY SATELLITES IN AUGER SPECTRA FROM METALS.   4-0940
          HIGHER ENERGY SATELLITES IN AUGER SPECTRA FROM METALS.   4-0900
ERATOMIC AUGER PROCESSES IN SOME ADSORBATES ON TRANSITION METALS.   INT 4-1045
IN AUGER SPECTRA OF SEVERAL GAS ADSORBATES ON TRANSITION METALS. /TOMIC TRANSITIONS AND RELAXATION EFFECTS 4-1042
L-SHELL AUGER ENERGIES AND INTENSITIES IN 3D TRANSITION METALS.   4-1018
          ORIGIN OF ION INDUCED AUGER ELECTRON EMISSION FROM METALS.   2-0549
          RADIATIVE KM2 AUGER EFFECT IN SOME TRANSITION METALS.   4-1057
          SINGULARITIES IN AUGER EMISSION SPECTRUM OF METALS.   4-0998
          THE K-FAR-M2 RADIATIVE AUGER EFFECT IN TRANSITION METALS.   2-0502
          PLASMON SATELLITE IN THE AUGER EMISSION SPECTRUM OF METALS. (ALUMINUM)   4-0871
STEN, COPPER, STEEL/ GRAIN BOUNDARY EMBRITTLEMENT IN SOME METALS AND ALLOYS. (AUGER EMISSION ANALYSIS, TUNG 6-2058
/ION OF GRAIN BOUNDARY SEGREGATION TO REDUCED STRENGTH OF METALS AND ALLOYS. (SEGREGATION IN STEEL, TUNGST/   6-2120
ED BY AES)   GRAIN BOUNDARY SEGREGATION OF IMPURITIES IN METALS AND INTERGRANULAR BRITTLE FRACTURE. (STUDI 6-2079
LAXATION ON AUGER AND BINDING ENERGY SHIFTS IN TRANSITION METALS AND SALTS. /ATIVE EFFECT OF EXTRAATOMIC RE 2-0412
BTAINING THE SPECTRUM OF SECONDARY ELECTRON EMISSION FROM METALS AND SEMICONDUCTORS. (AUGER ENERGIES) /OR O 3-0688
ICROANALYTIC METHO/ THREE DIMENSIONAL CHARACTERIZATION OF METALS AND SEMICONDUCTORS BY THE AUGER ELECTRON M 1-0077
INTERPRETATION OF KL2,3L2,3 RADIATIVE AUGER THRESHOLDS IN METALS AND SEMICONDUCTORS. (SILICON)   4-0723
/ ANALYSIS OF OXYGEN AND SULFUR INTERACTIONS WITH VARIOUS METALS AND THE EFFECT OF SLIDING ON THESE INTERA/   5-1206
/AND FRICTION OF POLYTETRAFLUORO ETHYLENE IN CONTACT WITH METALS AS STUDIED BY AUGER SPECTROSCOPY, FIELD I/   5-1198
K X-RAY EMISSION EDGE SHAPES OF FREE ELECTRON METALS. (AUGER BROADENING)   2-0337
          EMISSION LINES OF THE M SERIES CF RARE EARTH METALS. (AUGER EFFECT)   4-0790
/IQUES IN THE STUDY OF CORROSION AND DEPOSITION IN LIQUID METALS. (AUGER SPECTROSCOPY, X-RAY PHOTOELECTRON/   6-2050
          SECONDARY ELECTRON EMISSION OF METALS BOMBARDED WITH 120-EV TO 5-KEV PROTONS.   4-1029
          THEORY OF AUGER EJECTION OF ELECTRONS FROM METALS BY IONS.   2-0563
          METHODS FOR SURFACE ANALYSIS OF PRECIOUS METALS CONTACT MATERIALS.   5-1785
ICON STUDIED BY A MICROWAVE METHCD. (PRODUCTION OF AUGER/ METAL- DIELECTRIC TRANSITION IN GERMANIUM AND SIL 4-0752
          AUGER ELECTRON SPECTROSCOPY OF METALS DURING OXIDATION.   6-2007
          SPECTROSCOPY OF A METAL- GAS INTERFACE.   5-1751
ION SPECTROSCOPY.   METALLIC TRANSFER BETWEEN METALS IN SLIDING CONTACT EXAMINED BY AUGER EMISS 5-1729
RS.   AUGER ANALYSIS OF FILMS FORMED ON METALS IN SLIDING CONTACT WITH HALOGENATED POLYME 5-1728
LINIUM, DYSPRO/ AUGER ELECTRON SPECTRA OF SOME RARE EARTH METALS. (LANTHANUM, PRASEODYMIUM, NEODYMIUM, GADO 4-1100
/ATIVE EMISSION INTENSITIES OF AUGER ELECTRONS FROM LIGHT METALS (LITHIUM, SODIUM) SUBJECTED TO ION BOMBAR/ 2-0548
          LMM AUGER SPECTRA OF SOME TRANSITION METALS OF THE FIRST SERIES.   4-0810
EED- AUGER SPECTROSCOPY STUDY OF THE ADSORPTION OF ALKALI METALS ON MOLYBDENUM (110).   L 5-1896
ION, AUGER ELECTRON, AND FLASH DESORPTION SPECTROSCOPY OF METALS ON SINGLE CRYSTAL SURFACES. /TRON DIFFRACT 1-0008
          SECONDARY ELECTRON EMISSION FROM THIN FILM GOLD METAL- OXIDE CERMETS.   4-1015
ONS. PART-2: INELASTICALLY/ SLOW ELECTRON SCATTERING FROM METALS. PART-1: EMISSION OF TRUE SECONDARY ELECTR 2-0500
TRON SPECTROSCOPY STUDIES OF GALLIUM ARSENIDE AND SILICON METAL- SEMICONDUCTOR STRUCTURES.   AUGER ELEC 5-1774
          LIFETIME EFFECT ON THE AUGER SPECTRA OF METALS. (SODIUM)   4-0787
          SLIDING OF POLY(VINYL CHLORIDE) ON METALS STUDIED BY AUGER ELECTRON SPECTROSCOPY.   5-1727
          ADHESION AND TRANSFER OF PLATINUM IRON TO METALS STUDIED BY AUGER EMISSION SPECTROSCOPY.   5-1730
      ADHESION AND TRANSFER OF POLYTETRAFLUORO ETHYLENE TO METALS STUDIED BY AUGER EMISSION SPECTROSCOPY.   5-1731
AND AUGER EMISSION SPECTROSCOPY.   ADHESION OF METALS TO A CLEAN IRON SURFACE STUDIED WITH LEED 6-2008

175

(AUGER/ DETERMINATION OF ELECTRON ATTENUATION LENGTHS IN METALS: TRANSMISSION THROUGH THIN ADSORBED FILMS. 4-1088
INVESTIGATIONS CF THE OXIDE FORMATION ON METALS WITH AUGER ELECTRON SPECTROSCOPY. 5-1329
ONTIUM, BY AUGER ELECTRON SPECTROSCO/ VERIFICATION OF THE METAL-RICH SURFACE MODEL FOR THE OXIDATION OF STR 6-2045
HIGH RESOLUTION AUGER ELECTRON SPECTROSCOPY OF METALLIC COPPER. 4-1052
L2,3M4,5M4,5 AUGER SPECTRA OF METALLIC COPPER AND ZINC: THEORY AND EXPERIMENT. 4-0919
ESENCE OF ELECTRON IRRADIATION. (AUGER SP/ GROWTH OF THIN METALLIC FILMS ON INSULATING SUBSTRATES IN THE PR 5-1317
ER RATES FOR SOFT X-RAY TRANSITIONS IN IONIC, ATOMIC, AND METALLIC LITHIUM. AUG 4-0835
VAPOR QUENCHED ALLOYS STUDIED BY AUGER ELECTRON SPECTRO/ METALLIC STATE OF SILICON IN SILICON- NOBLE METAL 5-1438
SPECTROSCOPY FOR STUDYING SMALL QUANTITIES OF ELEMENTS ON METALLIC SURFACES. DIFFERENCE AUGER 5-1937
. EMISSICN OF AUGER ELECTRONS BY ATOMS IN A METALLIC TARGET SUBJECTED TO AN IONIC BOMBARDMENT 2-0383
CT EXAMINED BY AUGER EMISSION SPECTROSCOPY. METALLIC TRANSFER BETWEEN METALS IN SLIDING CONTA 5-1729
AUGER ELECTRON EXAMINATION OF CONTAMINANTS IN THIN FILM METALLIZATIONS. 5-1890
T SURFACE AND IN-DEPTH ANALYSIS OF SILICON AND TRANSISTOR METALLIZATIONS. / AUGER SPECTROSCOPY FOR MICROSPO 5-1262
TERING TECHNIQUES APPLIED TO ANALYSIS OF GOLD- REFRACTORY METALLIZATIONS. /R SPECTROSCOPY, AND ION BACKSCAT 5-1261
D NICKEL- GOLD- GERMANIUM FILMS ON N-TYPE GALLIUM ARSENI/ METALLURGICAL AND ELECTRICAL PROPERTIES OF ALLOYE 6-2101
ROSCOPY. METALLURGICAL APPLICATION OF AUGER ELECTRON SPECT 6-2123
APPLICATIONS OF AUGER ELECTRON SPECTROSCOPY TO METALLURGICAL APPLICATIONS. 6-2021
AUGER ELECTRON SPECTROSCOPY AND ITS APPLICATION IN METALLURGICAL PROBLEMS. 6-2039
CROPROBE MASS ANALYZER AND AUGER SPECTROMETER IN PHYSICAL METALLURGY. 6-2076
SCANNING ELECTRON AUGER MICROSCOPY IN METALLURGY. POSSIBILITIES OF THE ION MI 6-2075
SPECTROSCCPY OF AUGER ELECTRONS AND ITS USE IN METALLURGY. 6-2105
OWN FIELD OF SUPERCONDUCTING NIOBIUM. / EFFECT OF SURFACE METALLURGY ON THE PENETRATION DEPTH AND RF BREAKD 6-2137
FUNCTION EFFECTS ON AUGER ELECTRON EJECTION BY NOBLE GAS METASTABLE ATOMS. WORK 5-1601
ADSORBATE EFFECTS IN ELECTRON EJECTION BY RARE GAS METASTABLE ATOMS. (AUGER PROCESS) 2-0333
110) TUNGS/ ROLE OF WORK FUNCTION IN ELECTRON EJECTION BY METASTABLE ATOMS: HELIUM AND ARGON ON (111) AND (5-1600
RY EM/ SECONDARY ELECTRON EJECTION FROM METAL SURFACES BY METASTABLE ATOMS. PART-1: MEASUREMENTS OF SECONDA 2-0340
RIBUT/ SECONDARY ELECTRON EJECTION FROM METAL SURFACES BY METASTABLE ATOMS. PART-3: ENERGY AND ANGULAR DIST 5-1133
STATES IN BERYLLIUM. (AUGER DECAY) METASTABLE AUTOIONIZING THREE- AND FOUR ELECTRON 2-0295
ONTAINING TELLURIUM. (AUGER SPECTROSCOPY) METASTABLE GRAIN BOUNDARY SEGREGATION IN COPPER C 6-2086
RFACE BY THERMAL CARBON OXYGEN NITROGEN(2), AND NOBLE GAS METASTABLES. /LDS FROM COPPER BERYLLIUM OXYGEN SU 4-0786
THEORETICAL ANALYSIS OF THE AUGER SPECTRA OF METHANE. 4-1012
+), MOLECUL/ K-SHELL IONIZATION OF MOLECULAR NITROGEN AND METHANE BY 50- TO 600 KEV HYDROGEN(+), DEUTERIUM(2-0519
TION OF CARBON BY FAST PROTONS. (AUGER ELECTRON YIELDS OF METHANE, ETHANE, ETHYLENE, ACETYLENE) /ELL IONIZA 2-0532
IRON (001) SURFACE AND THEIR INFLUENCE ON/ INTERACTION OF METHANE, ETHANE, ETHYLENE, AND ACETYLENE WITH AN 5-1208
/RATION INTERACTION CALCULATIONS OF THE AUGER SPECTRUM OF METHANE, HYDROGEN FLUORIDE, WATER, AND CARBON MO/ 4-0879
C INVESTIGATION OF AUGER PROCESSES IN BROMINE-SUBSTITUTED METHANES AND SOME HYDROCARBONS. /RON SPECTROSCOPI 2-0515
PREDICTION OF THE RELATIVE INTENSITIES OF AUGER LINES IN MICA. 4-1072
TUDIES, AUGER SPECTRA) CHANGE OF SURFACE PROPERTIES OF MICA AFTER CLEAVAGE. (OUTGASSING AND DECORATION S 5-1743
ADHESION AT GOLD MICA INTERFACES. (LEED, AES) 5-1434
SURFACE COMPOSITION OF MICA SUBSTRATES. 5-1742
DEPTH ANALYSIS OF CLEAVED MICA SURFACES MONITORED BY AUGER SPECTROSCOPY. 5-1842
APPLICATIONS OF SCANNING AUGER MICROANALYSIS. 1-0119
LICATION OF AUGER ELECTRON SPECTROSCOPY TO ELECTRON PROBE MICROANALYSIS. / ELECTRON MICROSCOPY. PART-1: APP 1-0213
SCANNING AUGER MICROANALYSIS. 1-0116
STUDY OF METAL SURFACES BY IONIC MICROANALYSIS AND AUGER SPECTROMETRY. 5-1582
ELECTRON PROBE MICROANALYSIS AND ELECTRON SPECTROSCOPY. 1-0028
CTED PERFORMANC/ X-RAY PHOTOELECTRON (AND AUGER ELECTRON) MICROANALYSIS AND MICROSCOPY. PRINCIPLES AND EXPE 1-0027
VIEW, 15 REFS) MICROANALYSIS BY AUGER ELECTRON SPECTROSCOPY. (RE 1-0076
LATE. AUGER ELECTRON EMISSION MICROGRAPHY AND MICROANALYSIS OF A STAINED SURFACE OF PURE IRON P 5-1415
AUGER ELECTRON SPECTROSCOPY AND MICROANALYSIS OF SOLID SURFACES. 5-1414
AUGER ELECTRON EMISSICN MICROGRAPHY AND MICROANALYSIS OF SOLID SURFACES. 5-1416
ANALYZER, SCANNING ELECTRON MICROSCOPE, TR/ ELECTRON BEAM MICROANALYSIS. PART-1. (REVIEW OF ELECTRON PROBE 1-0010
OPE. (REVIEW, AUGER SCANNING) SPECTROSCOPIC METHODS OF MICROANALYSIS USING THE SCANNING ELECTRON MICROSC 3-0685
ZATION OF METALS AND SEMICONDUCTORS BY THE AUGER ELECTRON MICROANALYTIC METHOD. (REVIEW) /SIONAL CHARACTERI 1-0077
F SURFACES AND INTERFACES OF SOLIDS BY THE AUGER ELECTRON MICROANALYTIC METHOD. (REVIEW) /LEMENT ANALYSIS O 5-1414
ELECTRON BEAM MICROANALYZER WITH AUGER ELECTRON DETECTION. 3-0711
NEW SEM AUGER MICROANALYZER WITH ULTRAHIGH VACUUM CAPABILITY. 3-0636
TO THE STUDIES OF THE STRUCTURE OF ALUMIN/ APPLICATION OF MICROAUTORADIOGRAPHY BY BETA-RAY AUGER ELECTRONS 6-2051
ER ELECTRONS. NEW MICROAUTORADIOGRAPHY OF METAL STRUCTURE USING AUG 6-2052
N SIGNALS OF PHOSPHORUS ON SILICON USING A QUARTZ CRYSTAL MICROBALANCE. QUANTITATIVE STUDY OF AUGER ELECTRO 5-1583
MICRODOSIMETERY OF AUGER ELECTRONS. (REVIEW) 1-0055
N SPECTROSCOPY AND SECONDARY ION MASS SPECTROMETRY IN THE MICROELECTRONIC TECHNOLOGY. (REVIEW, 78 REFS) /RO 1-0136
ROSCOPY, LEED, HEED, AUGER SPECTROSCOPY, ION BOMBARDMENT, MICROELECTRONICS /APHY, PHENOMENA, FIELD ION MIC 5-1492
RAPHITE SINGLE CRYSTAL. AUGER ELECTRON EMISSION MICROGRAPHIC STUDIES OF THE CLEAVAGE SURFACE OF G 5-1413
E OF PURE IRON PLATE. AUGER ELECTRON EMISSION MICROGRAPHY AND MICROANALYSIS OF A STAINED SURFAC 5-1415
AUGER ELECTRON EMISSION MICROGRAPHY AND MICROANALYSIS OF SOLID SURFACES. 5-1416
AT GRAIN BOUNDARIES IN SINTERED ALUMINA. (SCANNING AUGER MICROPROBE) ADDITIVE AND IMPURITY DISTRIBUTIONS 5-1505
N SPECTROSCOPY, SECONDARY ION MASS SPECTROSCOPY, ELECTRON MICROPROBE) / ANALYSIS TECHNIQUES. (AUGER ELECTRO 6-1627
FLAKE IRON INTERFACES IN GRAY CAST IRON. (SCANNING AUGER MICROPROBE) /NFIRMATION OF IMPURITY ADSORPTION AT 6-2054
MITTENT MOTION OF FRICTION IN STAINLESS STEEL BY AN AUGER MICROPROBE. STUDY OF THE INTER 6-2133
FILM ANALYSIS OF SEMICONDUCTOR MATERIALS. (SCANNING AUGER MICROPROBE) SURFACE AND THIN 1-0089
ISTRY OF CERAMIC SURFACES. (AUGER SPECTROSCOPY, LEED, ION MICROPROBE ANALYSIS) CRYSTALLOGRAPHY AND CHEM 5-1852
TRUCTURE OF HOT PRESSED SILICON NITRIDE. (AUGER ANALYSIS, MICROPROBE ANALYSIS) MICROS 5-1556
AND ION MICROPRO/ RECENT ADVANCES IN INSTRUMENTATION FOR MICROPROBE ANALYSIS. (AUGER ELECTRON SPECTROSCOPY 3-0716
INSTRUMENTAL PROBLEMS FOR ELECTRON MICROPROBE AUGER SPECTROSCOPY. 3-0670
N PHYSICAL METALLURGY. POSSIBILITIES OF THE ION MICROPROBE MASS ANALYZER AND AUGER SPECTROMETER I 6-2075
OF MATERIALS. (AUGER ELECTRON SIGNALS) ELECTRON BEAM MICROPROBES USED IN THE ANALYSIS AND DEVELOPMENT 3-0714
DY OF THE STRUCTURE AND PITTING CORROSION OF ALUMINU/ NEW MICRORADIOGRAPHY USING AUGER ELECTRONS IN THE STU 6-2066
AUGER ELECTRON IMAGES IN A SCANNING ELECTRON/ 3-0649
TRON SPECTROSCOPY IN A DIFFUSION PUMPED SCANNING ELECTRON MICROSCOPE. AUGER ELEC 3-0598
AUGER ELECTRON SPECTROSCOPY IN THE ELECTRON EMISSION MICROSCOPE. 3-0604
AUGER ELECTRON SPECTROSCOPY IN THE SCANNING ELECTRON MICROSCOPE. 3-0710
URFACES IN A DRY PUMPED HIGH RESOLUTION SCANNING ELECTRON MICROSCOPE. AUGER SPECTROSCOPY OF SOLID S 5-1258
ELECTRON SPECTROSCOPY ON THIN FILMS. (SCANNING AUGER MICROSCOPE) 5-1405
ROSCOPY IN AN ORDINARY DIFFUSION PUMPED SCANNING ELECTRON MICROSCOPE. /TIAL RESOLUTION AUGER ELECTRON SPECT 5-0597
Y OF AUGER ELECTRON SPECTROSCOPY IN THE ELECTRON EMISSION MICROSCOPE. POSSIBILIT 3-0605
S BY AUGER ELECTRON SPECTROSCOPY IN THE SCANNING ELECTRON MICROSCOPE. /FILING ACROSS SEMICONDUCTOR JUNCTION 2-0555
AUGER ELECTRON SPECTROSCOPY IN THE SCANNING ELECTRON MICROSCOPE: AUGER ELECTRON IMAGES. 3-0650
ELEMENTAL COMPOS/ CONTAMINATION IN THE SCANNING ELECTRON MICROSCOPE. (AUGER ELECTRON SPECTROSCOPY, SURFACE 3-0596

MATERIAL ANALYSIS USING THE SCANNING ELECTRON MICROSCOPE. (AUGER SPECTROSCOPY)   5-1701
                          SCANNING AUGER ELECTRON MICROSCOPE FOR SURFACE STUDIES.   3-0679
                          SCANNING AUGER MICROSCOPE FOR THIN FILM ANALYSIS.   3-0645
IS. (AUGER)   SPECTROSCOPY IN THE SCANNING ELECTRON MICROSCOPE NOW INCLUDES SECONDARY ION MASS ANALYS 3-0674
OPIC METHODS OF MICROANALYSIS USING THE SCANNING ELECTRON MICROSCOPE. (REVIEW, AUGER SCANNING)   SPECTROSC 3-0685
        LABORATORY COMPUTER SCANNING ELECTRON MICROSCOPE SYSTEM.   3-0651
/1. (REVIEW OF ELECTRON PROBE ANALYZER, SCANNING ELECTRON MICROSCOPE, TRANSMISSION ELECTRON MICROSCOPE. EN/ 1-0010
R ANALYZER.   ULTRAHIGH VACUUM SCANNING ELECTRON MICROSCOPE WITH A FIELD EMISSION CATHODE AND AUGE 3-0621
             ULTRAHIGH VACUUM SCANNING ELECTRON MICROSCOPE WITH AUGER ANALYSIS FACILITIES.   3-0622
CTROSCOPY.   A SCANNING ELECTRON MICROSCOPE WITH FACILITIES FOR AUGER EMISSION SPE   3-0593
URFACES.   MICROSCOPIC AUGER ELECTRON ANALYSIS OF FRACTURE S   5-1598
N IRON- CHROMIUM ALLCY.   AUGER SCANNING ELECTRON MICROSCOPIC STUDY OF THE PASSIVE FILM FORMED ON A 5-1753
OLYBDENUM DISULFIDE.   (AUGER) MICROSCOPICAL STUDY OF FRICTIONAL PROPERTIES OF M 5-1872
ED BY AUGER SPECTROSCOPY, FIELD ION AND SCANNING ELECTRON MICROSCOPY. /LENE IN CONTACT WITH METALS AS STUDI 5-1198
     AUGER ELECTRON SPECTROSCOPY FOR SCANNING ELECTRON MICROSCOPY.   1-0114
OMBINED AUGER ELECTRON SPECTROSCOPY AND SCANNING ELECTRON MICROSCOPY.   C 1-0003
ISPERSIVE SPECTROMETER, ION MASS ANALYSIS, AUGER ELECTRON MICROSCOPY) /ISSION ELECTRON MICROSCOPE, ENERGY D 1-0010
STRENGTHENING IN TUNGSTEN- RHENIUM ALLOYS, AUGER ELECTRCN MICROSCOPY). /NIUM DUCTILIZING EFFECT AND BUBBLE 6-2111
KDOWN. (FROM SURFACES) INVESTIGATIONS WITH FIELD EMISSION MICROSCOPY AND AUGER SPECTROSCOPY. /D VACUUM BREA 5-1520
/SESSMENT BY RUTHERFCRD BACKSCATTERING, SCANNING ELECTRON MICROSCOPY, AND DEPTH PROFILING AUGER ELECTRON S/ 5-1902
IQUES FOR/ AUGER ELECTRON SPECTROSCOPY, SCANNING ELECTRON MICROSCOPY AND NONDISPERSIVE X-RAY ANALYSIS TECHN 5-1799
NTAL PROBLEMS OF STUDYING THIN FILM NUCLEATION. (ELECTRON MICROSCOPY, AUGER EFFECT)   EXPERIME 5-1408
STUDY OF THE STRUCTURE AND COMPOSITI/ COMBINED FIELD ION MICROSCOPY. AUGER ELECTRON SPECTROSCOPY, AND LEED 5-1951
          FUTURE GROWTH OF SEM: TOWARD SUPER EFFICIENT MICROSCOPY. (AUGER SPECTROSCOPY)   1-0220
ERING TECHNIQUES APPLIED TO ANALYSIS O/ SCANNING ELECTRON MICROSCOPY, AUGER SPECTROSCOPY, AND ION BACKSCATT 5-1261
/ON PRODUCTS IN FLOWING LIQUID SODIUM. (SCANNING ELECTRON MICROSCOPY, AUGER SPECTROSCOPY, ION BEAM ANALYSI/ 6-2014
              SCANNING ELECTRON AUGER MICROSCOPY IN METALLURGY.   6-2105
ON DIFFRACTION, AUGER EMISSION SPECTROSCOPY AND FIELD ION MICROSCOPY IN MICROSTRUCTURAL STUDIES. /GY ELECTR 1-0057
RSENIDE AND GALLIUM PHOSPHIDE.   USE OF SCANNING AUGER MICROSCOPY IN MOLECULAR BEAM EPITAXY OF GALLIUM A 5-1142
/CE STRUCTURES, SPECIES, TOPOGRAPHY, PHENOMENA, FIELD ION MICROSCOPY, LEED, HEED, AUGER SPECTROSCOPY, ION / 5-1492
SP/ BASIC INVESTIGATICN IN THE AREA OF SCANNING ELECTRON MICROSCOPY: APPLICATION OF AUGER ELECTRON 1-0213
    AUGER ELECTRON SPECTROSCOPY IN SCANNING ELECTRON MICROSCOPY: POTENTIAL MEASUREMENTS.   3-0648
/RAY PHOTOELECTRON (AND AUGER ELECTRON) MICROANALYSIS AND MICROSCOPY. PRINCIPLES AND EXPECTED PERFORMANCES. 1-0027
            THIRD DIMENSION IN SCANNING ELECTRON MICROSCOPY. SCANNING AUGER MICROSCOPY.   1-0117
ACE AND IN-DEPTH ANALYSIS OF / CCMBINED SCANNING ELECTRON MICROSCOPY- AUGER SPECTROSCOPY FOR MICROSPOT SURF 5-1262
ION AND EUTECTIC FORMATION IN TUNGSTEN PLATINUM/ SCANNING MICROSPOT AUGER SPECTROSCOPY STUDY OF INTERDIFFUS 5-1259
/NED SCANNING ELECTRCN MICRCSCOPY- AUGER SPECTROSCOPY FCR MICROSPOT SURFACE AND IN-DEPTH ANALYSIS OF SILIC/ 5-1262
, AUGER EMISSION SPECTROSCOPY AND FIELD ION MICROSCOPY IN MICROSTRUCTURAL STUDIES. /GY ELECTRON DIFFRACTION 1-0057
UGER ANALYSIS, MICROPROBE ANALYSIS)   MICROSTRUCTURE OF HOT PRESSED SILICON NITRIDE. (A 5-1556
HOSPHORUS IN A CHROMIUM- MOLYBDENUM- VANADIUM / EFFECT OF MICROSTRUCTURE ON GRAIN BOUNDARY SEGREGATION OF P 6-2139
/LECTRIC TRANSITION IN GERMANIUM AND SILICON STUDIED BY A MICROWAVE METHOD. (PRODUCTION OF AUGER ELECTRONS) 4-0752
TIFIERS AND THYRISTORS LOADED IN FORWARD DIRECTION IN THE MIDDLE REGION. AUGER RECOMBINATION IN SILICON REC 2-0415
ROSCOPY)   POSSIBLE ORIGIN OF LOW TEMPERATURE SILICON MIGRATION IN SILICON- METAL SYSTEMS. (AUGER SPECT 5-1437
COPY TO DETERMINE THE COMPOSITION OF FRACTURE SURFACES OF MINERALS.   USE OF AUGER ELECTRON SPECTROS 5-1307
AYERS. (AUGER RECOMBINATION PROCESS)   MINORITY CARRIER LIFETIME IN INDIUM ARSENIDE EPIL 2-0567
NALYSIS IN THE RANGE 100 TO 1000 EV USING THE CYLINDRICAL MIRROR ANALYZER. /IMATION TO QUANTITATIVE AUGER A 3-0660
ATTERING SPECTROSCOPY WITH A COMMERCIAL AUGER CYLINDRICAL MIRROR ANALYZER.   LOW ENERGY ION BACKSC 3-0665
TIVE ASPECTS OF ION SCATTERING SPECTROSCOPY. (CYLINDRICAL MIRROR ANALYZER)   QUANTITA 3-0666
SHIFTS AND SIGNAL INTENSITY IN A SINGLE STAGE CYLINDRICAL MIRROR ANALYZER. /MEN POSITION EFFECTS ON ENERGY 3-0696
D ESCA- AUGER SYSTEM BASED ON THE DOUBLE PASS CYLINDRICAL MIRROR ANALYZER. / ELECTRON SPECTROSCOPY. COMBINE 3-0654
OF OVERLAYER ATTENUATICN OF ELECTRONS FOR THE CYLINDRICAL MIRROR ANALYZER AND A RETARDING FIELD ANALYZER. / 3-0693
             RETARDING FIELD CYLINDRICAL MIRROR ANALYZER. (FOR AUGER SPECTROSCOPY)   3-0614
          DESIGN FOR A CYLINDRICAL MIRROR ANALYZER FOR USE IN AUGER SPECTROSCOPY.   3-0594
TOWARD OPTIMUM UTILIZATION OF THE CYLINDRICAL MIRROR ANALYZER IN AUGER ELECTRON SPECTROSCOPY.   3-0678
NGSTEN AUGER ELECTRON)   CYLINDRICAL MIRROR ANALYZER INCORPORATING PRERETARDATION. (TU 3-0590
EW PHOTOELECTRON SPECTROMETER: COMBINATION OF CYLINDRICAL MIRROR ANALYZER WITH SOFT X-RAY SOURCE.   A N 3-0652
COMPARISON OF SPHERICAL DEFLECTOR AND CYLINDRICAL MIRROR ANALYZERS.   3-0623
UM ON SILICON AND SILICON DIOXIDE SURFACES BY CYLINDRICAL MIRROR AUGER ANALYZER.   DETECTION OF SODI 3-0647
AUGER ANALYSIS OF THREE ENHANCED MULTILAYER DIELECTRIC MIRROR DESIGNS. (ZINC SELENIDE, THORIUM FLUORIDE) 5-1328
          DESIGN PARAMETERS FOR THE CYLINDRICAL MIRROR ENERGY ANALYZER.   3-0687
ONSTITUTENTS AND SOME RESULTS.   AUGER CYLINDRICAL MIRROR SPECTROMETER FOR THE ANALYSIS OF SURFACE C 3-0682
AR SPREAD IN AN ELECTRCSTATIC ANALYZER OF THE CYLINDRICAL MIRROR TYPE. (AUGER SPECTROMETER) /F FINITE ANGUL 3-0720
CONSTRUCTICN AND OPERATION OF A CYLINDRICAL MIRROR TYPE OF AUGER SPECTROMETER.   3-0588
FOCUSING PROPERTIES OF AN ELECTROSTATIC MIRROR WITH CYLINDRICAL FIELD.   3-0721
PY.   MISCELLANEOUS TOPICS IN AUGER ELECTRON SPECTROSCO 1-0075
MEASUREMENTS ON T-027 SAMPLES EXPOSED DURING THE SKYLAB-2 MISSION.   AUGER 5-1735
REVIEW OF FLUCRESCENCE YIELDS AND AUGER ELECTRON MISSION.   1-0121
             AUGER TRANSITION RATE IN MIXED COUPLING SCHEME.   2-0342
AUGER SPECTROSCOPIC OBSERVATION OF SILICON- GOLD MIXED PHASE FORMATION AT LOW TEMPERATURES.   5-1680
IMONY, TELLURIUM AND IODINE.   HIGH RESOLUTION MNN AUGER SPECTRA OF SILVER, CADMIUM, INDIUM, ANT 4-0730
STUDY OF THE CHEMICAL COMPOSITION OF MOS AND MNOS STRUCTURES BY AUGER ELECTRON SPECTROSCOPY.   5-1503
            AES COMPOSITIONAL PROFILES OF MOBILE IONS IN THE SURFACE REGION OF GLASS.   5-1722
ON SPECTROSCOPY.   GROWTH MODE DISCRIMINATION OF THIN FILMS BY AUGER ELECTR 5-1624
/COPY STUDY OF SURFACE CONTAMINATION EFFECT CN THE GROWTH MODE OF IRON EPITAXIAL FILMS ON MAGNESIUM OXIDE / 5-1524
01) SURFACE. (AUGER SPECTROSCOPY)   GROWTH MODE OF IRON FILM DEPOSITED ON MAGNESIUM OXIDE (0 5-1528
PECTRA (MONOLAYER ADSORPTION, ESCA SIMPLE CHARGE TRANSFER MODEL).   CHEMICAL SHIFTS IN AUGER S 4-1115
F23 BASED ON THE GREEN-SELLIN-ZACHOR INDEPENDENT PARTICLE MODEL. /D COSTER-KRONIG TRANSITION PROBABILITIES 2-0313
IMEN THICKNESS.   SIMFLE MODEL FOR DEPENDENCE OF AUGER INTENSITIES ON SPEC 2-0355
S IN HEAVY ION COLLISICNS.   COMMENTS.   SIMFLE MODEL FOR K X-RAY AND AUGER ELECTRON ENERGY SHIFT 2-0301
EMS EXHIBITING LAYER GROWTH, AND ITS APPLICATION TO THE / MODEL FOR THE AUGER ELECTRON SPECTROSCOPY OF SYST 2-0399
ECTRON SPECTROSCO/ VERIFICATION OF THE METAL-RICH SURFACE MODEL FOR THE OXIDATION OF STRONTIUM, BY AUGER EL 6-2045
RANSITION IN MAGNESIUM AND SILICCN FROM THE CROSS SECTION MODEL OF GRYZINSKI. (AUGER EFFECT) /2,3VV AUGER T 4-0765
N- CESIUM- OXYGEN SURFACE. (AU/ STRUCTURAL AND ELECTRONIC MODEL OF NEGATIVE ELECTRON AFFINITY ON THE SILICO 2-0426
AUGER IONIZATICN OF ATOMS BY MULTIPLY CHARGED IONS. MODEL OF TWO COULOMB CENTERS.   2-0407
INCOMPLETE RELAXATION IN A MODEL PROBLEM FOR THE AUGER PROCESS.   2-0238
TRANSITICN POTENTIAL MODELS FOR AUGER ELECTRON SPECTRA.   2-0510
FT X-RAYS. (/ BIOLOGICAL RADIATION EFFECT ON INCLUSICN OF MODERATELY HEAVY NUCLEI IN THE TISSUE: USE OF SO 2-0531
APPLICATION OF ION, ELECTRON, AND X-RAY BEAMS IN MODERN MATERIAL TESTING.   1-0034
ET/ ADHESION, FRICTION, WEAR, AND LUBRICATION RESEARCH BY MODERN SURFACE SCIENCE TECHNIQUES. (AES, ELLIPSOM 1-0104
                          MODERN TECHNIQUES FOR SURFACE STUDIES.   1-0063

ECTRON SPECTROMETER. AN ENERGY MODULATED (CYLINDRICAL) HIGH ENERGY RESOLUTION EL 3-0616
DING / CORRECTIONS OF AUGER ELECTRON SIGNAL STRENGTHS FOR MODULATION AMPLITUDE DISTORTION IN A 4-GRID RETAR 3-0620
TA FROM TITANIUM. EFFECT OF MODULATION AMPLITUDE ON ELECTRON EXCITED AUGER DA 2-0367
. EXACT CORRECTIONS FOR POTENTIAL MODULATION DISTORTION IN AUGER YIELD MEASUREMENTS 2-0389
INTEGRAL AUGER INFORMATION VIA TAILORED MODULATION TECHNIQUES. 1-0187
N AND ITS EFFECT ON AUGER INTENSITIES. MOLECULAR ADSORPTION: INCLUSION OF BULK DIFFUSIO 2-0264
GALLIUM ARSENIDE AND ALUMINUM(X) GALLIUM(1-X) ARSENIDE BY MOLECULAR BEAM EPITAXY. (AUGER SPECTROMETRY) /ED 5-1247
GALLIUM(X) ALUMINUM(1-X) ARSENIDE) COMPUTER-CONTROLLED MOLECULAR BEAM EPITAXY. (AUGER SURFACE ANALYSIS. 5-1324
LLIUM PHOSPHIDE. USE OF SCANNING AUGER MICROSCOPY IN MOLECULAR BEAM EPITAXY OF GALLIUM ARSENIDE AND GA 5-1142
/CE INTERACTIONS AT THE (100) CRYSTAL FACE OF PLATINUM BY MOLECULAR BEAM SCATTERING, LOW ENERGY ELECTRON D/ 5-1963
ATION AND GROWTH. (AES) MOLECULAR BEAM STUDIES OF INDIUM PHOSPHIDE EVAPOR 5-1334
Y) SCATTERING OF MOLECULAR BEAMS FROM SURFACES. (AUGER SPECTROSCOP 5-1905
IC ION- ATOM COLLISIONS. (AUGER SPECTRA) MOLECULAR EFFECTS ON K-VAANCY SHARING IN ASYMMETR 2-0522
L-SHELL FLUORESCENCE YIELD STUDIED BY ATOMIC HYDROGEN(+), MOLECULAR HYDROGEN(+), ATOMIC HELIUM(+) IMPACT. / 2-0517
GON L-SHELL IONIZATION BY 50 TO 600 KEV PROTON, DEUTERON, MOLECULAR HYDROGEN, AND HELIUM(+) IMPACT. AR 2-0521
HYDROGEN(+), DEUTERIUM(+), MOLECUL/ K-SHELL IONIZATION OF MOLECULAR NITROGEN AND METHANE BY 50- TO 600 KEV 2-0519
ER SPECTRA OF TITANIUM(II) AND TITANIUM(IV) OXIDES. MOLECULAR ORBITAL EFFECTS ON THE TITANIUM LMV AUG 4-1069
N AUGER SPECTRUM OF MOLYBDENUM (110) DUE TO ADSORPTION OF MOLECULAR OXYGEN AND CARBON MONOXIDE. / THE M4,5N 4-0885
AR OXYG/ AUGER SPECTROSCOPIC STUDY OF THE COADSORPTION OF MOLECULAR OXYGEN AND CARBON MONOXIDE, AND MOLECUL 5-1465
/LIZING AUGER ELECTRON SPECTROSCOPY. CARBON MONOXIDE AND MOLECULAR OXYGEN ON SILICON. PART-1: ADSORPTION / 5-1543
ECTRUM AND IONIZATION POTENTIALS OF THE HYDROGEN FLUORIDE MOLECULE. AUGER ELECTRON SP 4-1058
INUM SURFACES. SMALL MOLECULE REACTIONS ON STEPPED SINGLE CRYSTAL PLAT 5-1173
AUGER SPECTRA OF SIMPLE MOLECULES. 4-0991
AUGER SPECTROSCOPY OF SIMPLE GASEOUS MOLECULES. 4-0989
AUGER DISSOCIATION OF MOLECULES AS A RESULT OF INNER SHELL IONIZATION. 2-0465
/OSS SECTIONS AND FLUORESCENCE YIELDS FOR LIGHT ATOMS AND MOLECULES BY ELECTRON IMPACT. (AUGER ELECTRON EJ/ 2-0528
. PART-1: E/ ENERGIES OF EXCITED STATES OF DOUBLY IONIZED MOLECULES BY MEANS OF AUGER ELECTRON SPECTROSCOPY 4-1071
N DIFFRACTION STUDIES OF THE CHEMISORPTION OF UNSATURATED MOLECULES BY THE (100) SURFACE OF PLATINUM. /CTRO 5-1267
CROSS SECTIONS FOR INNER SHELL IONIZATION OF GASEOUS MOLECULES BY 50-500 KEV PROTONS. 4-1077
COPY) SURFACE DEFORMATION CAUSED ON NATURAL MOLYBDENITE BY ABRASION. (AUGER ELECTRON SPECTROS 6-2132
ADSORPTION OF SULFUR ON THE (100) FACE OF MOLYBDENUM. 5-1732
F INCIDENCE EFFECTS IN AUGER ELECTRON EMISSION FROM CLEAN MOLYBDENUM. ANGLE O 2-0542
SORPTION OF AMMONIA, NITROGEN, AND NITRIDE FORMATION ON A MOLYBDENUM. /COPIES TO CATALYTIC PROCESSES. CHEMI 5-1533
F THE CHEMISORPTION OF CARBON MONOXIDE ON POLYCRYSTALLINE MOLYBDENUM. /ETRIC AND MASS SPECTROMETRIC STUDY O 5-1355
ELECTRON AND AUGER SPECTRA. CARBON MONOXIDE ADSORPTION ON MOLYBDENUM. /ED AND ELECTRON IMPACT INDUCED PHOTO 4-0768
NEUTRALIZATION AT (100) FACE OF TUNGSTEN, POLYCRYSTALLINE MOLYBDENUM). /S WITH CONTROLLED SURFACES. (AUGER 5-1736
EAK SHAPES IN THE AUGER SPECTRUM OF OXYGEN (KL2,3L2,3) ON MOLYBDENUM. ORIGIN OF THE P 4-1004
PLASMA SATELLITES IN THE AUGER ELECTRON SPECTRUM OF MOLYBDENUM. 4-1124
SECONDARY ELECTRON EMISSION FROM TUNGSTEN AND MOLYBDENUM. 4-0850
TRAL DATA. CHEMICAL COMPOSITION OF THE SURFACE OF MOLYBDENUM AND STEEL BASED ON AUGER ELECTRON SPEC 5-1985
MISSION IN THE ENERGY SPECTRA OF SECONDARY ELECTRONS FROM MOLYBDENUM AND TUNGSTEN. AUGER ELECTRON E 4-0874
/GER ELECTRON EMISSION FROM CLEAN AND CARBON CONTAMINATED MOLYBDENUM BOMBARDED BY POSITIVE IONS. PART-1. P/ 2-0543
GER E/ DECREASE OF CARBON CONCENTRATION ON THE SURFACE OF MOLYBDENUM CARBIDE DURING HEATING ACCORDING TO AU 5-1759
SECONDARY ELECTRON EMISSION OF MOLYBDENUM CRYSTALS. 4-1063
(AUGER) MICROSCOPICAL STUDY OF FRICTIONAL PROPERTIES OF MOLYBDENUM DISULFIDE. 5-1872
EFFECTS OF LOW ENERGY ARGON- ION BOMBARDMENT ON MOLYBDENUM DISULFIDE. (AUGER MEASUREMENTS) 2-0348
ELECTRONIC PROPERTIES OF CLEAVED MOLYBDENUM DISULFIDE SURFACES. 5-1634
CHEMICAL SHIFT IN AUGER SPECTRA OF TUNGSTEN AND MOLYBDENUM DURING THE ADSORPTION OF OXYGEN. 4-0816
/OPY STUDIES OF SPUTTER DEPOSITION AND SPUTTER REMOVAL OF MOLYBDENUM FROM VARIOUS METAL SURFACES (COPPER, / 5-1878
Y AUGER ELECTRON SPECTROSCOPY. SURFACE CONCENTRATION OF MOLYBDENUM IN TYPES 316 AND 304 STAINLESS STEEL B 6-1995
0) AND (110) MOLYBDENUM SURFACES. (/ INITIAL OXIDATION OF MOLYBDENUM. PART-3: OXIDE NUCLEATION RATES ON (10 5-1536
SECONDARY EMISSION OF MAGNESIUM OXIDE LAYERS APPLIED ON A MOLYBDENUM SINGLE CRYSTAL. STRUCTURE AND 5-1191
SECONDARY ELECTRON EMISSION FROM TUNGSTEN AND MOLYBDENUM SINGLE CRYSTALS. 4-0742
/ARDMENT ON THE COMPOSITION AND STRUCTURE OF TUNGSTEN AND MOLYBDENUM SINGLE CRYSTALS STUDIED BY AUGER ELEC/ 5-1215
EARS EXPERIMENTAL BREEDER REACTOR II SERVICE ON CHROMIUM- MOLYBDENUM STEEL. EFFECTS OF TEN Y 6-2108
TOELECTRON SPECTROMETER. INITIAL OXIDATION OF MOLYBDENUM STUDIED BY A HIGH RESOLUTION AUGER PHO 5-1698
ITROGEN KLL AUGER TRANSITIONS OF CHEMISORBED AMMONIA ON A MOLYBDENUM SURFACE. /LUTION MEASUREMENTS OF THE N 5-1532
/MECHANISM OF THE REACTION BETWEEN SULFUR AND OXYGEN ON A MOLYBDENUM SURFACE BY AUGER ELECTRON SPECTROSCOP/ 5-1530
/THE MECHANISM OF REACTION BETWEEN SULFUR AND OXYGEN ON A MOLYBDENUM SURFACE BY MEANS OF AUGER ELECTRON SP/ 5-1534
TROSCOPY. SEGREGATION OF SULFUR ON A MOLYBDENUM SURFACE STUDIED BY AUGER ELECTRON SPEC 5-1564
GEN KLL AUGER SPECTRA OF CHEMISORBED NITROGEN ON IRON AND MOLYBDENUM SURFACES. /EMICAL EFFECTS OF THE NITRO 4-0925
PROTON AND ELECTRON EXCITED AUGER ELECTRON SPECTRA OF MOLYBDENUM SURFACES. 5-1667
DETERMINATION OF SULFUR CONTAMINATION ON MOLYBDENUM SURFACES. (AUGER EFFECT) 5-1647
OMBINED WITH AUGER ELECTRON SP/ STUDY OF THE OXIDATION OF MOLYBDENUM SURFACES BY ENERGY LOSS SPECTROSCOPY C 5-1531
IN THE 1 K 1 STRUCTURE OF A SILICON ORDERED MONOLAYER ON MOLYBDENUM (001). ATOMIC ARRANGEMENT 5-1478
AUGER SPECTROSCOPY STUDY OF THE ADSORPTION OF RUBIDIUM ON MOLYBDENUM (100). 5-1894
AUGER- LEED INVESTIGATION OF TIN ON MOLYBDENUM (100). 5-1490
AND AES STUDY OF THE INTERACTION OF HYDROGEN SULFIDE AND MOLYBDENUM (100). LEED 5-1973
SLOW ELECTRON SPECTROSCOPY ON MOLYBDENUM (100). 5-1578
OXIDATION OF CARBON ON TUNGSTEN (100) AND MOLYBDENUM (100). (AUGER ELECTRON SPECTROSCOPY) 5-1944
CARBON MONOXIDE ADSORPTION KINETICS ON MOLYBDENUM (100) (AUGER SPECTROSCOPY) 5-1579
FURTHER STUDIES OF GAS ADSORPTION ON MOLYBDENUM (100) SURFACE. 5-1305
OXYGEN ADSORPTION ON CLEAN MOLYBDENUM (100) SURFACES. 5-1772
SPECTROSCOPY STUDY OF THE ADSORPTION OF ALKALI METALS ON MOLYBDENUM (110). LEED- AUGER 5-1896
AUGER ELECTRON SPECTROSCOPY. (STUDY OF CARBON MONOXIDE ON MOLYBDENUM (110)) / SUBTRACTION FOR QUANTITATIVE 1-0069
XYGEN A/ CHEMICAL EFFECTS IN THE M4,5NN AUGER SPECTRUM OF MOLYBDENUM (110) DUE TO ADSORPTION OF MOLECULAR O 4-0885
LEED INVESTIGATION OF THE DEPOSITION OF ALUMINUM ONTO THE MOLYBDENUM (110) SURFACE. AUGER- 5-1489
CLEANING PROCEDURES OF MOLYBDENUM (110) SURFACE. 5-1649
INITIAL OXIDATION OF THE MOLYBDENUM (110) SURFACE OBSERVED BY AES AND LEED 5-1650
. ADSORPTION OF OXYGEN ON MOLYBDENUM (111): EFFECT OF TRACE IMPURITIES. 5-1568
CHEMISORPTION ON SINGLE CRYSTAL MOLYBDENUM (112) SURFACES. 5-1303
/NATION PRODUCED BY ELECTRON BOMBARDMENT. POLYCRYSTALLINE MOLYBDENUM- CARBON MONOXIDE SYSTEM. (AUGER SPECT/ 5-1648
PPER AND PHOSPHORUS ON TEMPER EMBRITTLEMENT OF MANGANESE- MOLYBDENUM- NICKEL LOW ALLOY STEEL. (AES) / OF CO 6-2043
TIME ISOTHERMAL TEMPER EMBRITTLEMENT IN NICKEL- CHROMIUM- MOLYBDENUM- VANADIUM. LONG 6-2140
/ GRAIN BOUNDARY SEGREGATION OF PHOSPHORUS IN A CHROMIUM- MOLYBDENUM- VANADIUM STEEL. (AUGER SPECTROSCOPY) 6-2139
HARACTERIZED BY AUGER SPECTROSCOPY, SECON/ MEASUREMENT OF MOMENTUM ACCOMMODATION COEFFICIENTS ON SURFACES C 2-0501
COPPER ALLOY DISPLAYING POOR WETTABILITY DURING BRAZING. (MONEL, AUGER SPECTRA) /ARACTERIZATION OF NICKEL- 5-1573
SPECTROMETER AS TOOL FOR SURFACE ANALYSIS. (CONTAMINATION MONITOR) AUGER ELECTRON 3-0706
Y WITH APPLICATION TO SURFACE COMPOSITION/ A MULTICHANNEL MONITOR FOR REPETITIVE AUGER ELECTRON SPECTROSCOP 3-0695
SURFACE SEGREGATION IN TITANIUM AS MONITORED BY AUGER ELECTRON SPECTROSCOPY. 6-2067

L IONIZATION OF THE N6,7 LEVEL IN GOLD, BISMUTH, AND LEAD MONITORED BY AUGER SPECTROSCOPY. /UCED INNER SHEL 4-1067
DEPTH ANALYSIS OF CLEAVED MICA SURFACES MONITORED BY AUGER SPECTROSCOPY. 5-1842
/TCHING OF CRYSTAL SURFACES (TITANIUM, TUNGSTEN, SILICON) MONITORED BY AUGER SPECTROSCOPY, ELLIPSOMETRY AN/ 5-1828
/E PROPERTIES OF RETARCING FIELD SPECTROMETERS USING POST MONOCHROMATORS. (FOR AUGER ELECTRON SPECTROSCOPY) 3-0624
CONDARY EMISSION AND ELASTIC REFLECTION OF ELECTRONS FROM MONOCRYSTALS. SE 2-0508
/FACE COMPOSITION STUDIES OF THE (001) AND (110) FACES OF MONOCRYSTALLINE IRON(0.84) CHROMIUM(0.16). (AUGE/ 5-1591
MICAL SHIFTS IN X-RAY PHOTOELECTRON AND AUGER SPECTRA FOR MONOLAYER ADSORPTION. /ND ADSORPTION STUDIES. CHE 5-1204
MODEL). CHEMICAL SHIFTS IN AUGER SPECTRA (MONOLAYER ADSORPTION, ESCA SIMPLE CHARGE TRANSFER 4-1115
ITE FACE. THERMODYNAMICS AND KINETICS OF THE FIRST MONOLAYER ADSORPTION OF XENON ON THE (0001) GRAPH 5-1859
ER SPECTROS/ TWO-DIMENSIONAL PHASE TRANSITION IN XENON SUBMONOLAYER FILMS ADSORBED ON (0001) GRAPHITE. (AUG 5-1858
AUGER SPECTROSCOPY OF SUBMONOLAYER GOLD DEPOSITIONS ON SILICON. 5-1888
BY AUGER ELECTRON SPECTRCSCOPY. LEAD MONOLAYER LUBRICATION IN STEEL MACHINING STUDIED 6-2124
C ARRANGEMENT IN THE 1 K 1 STRUCTURE OF A SILICON ORDERED MONOLAYER ON MOLYBDENUM (001). ATOMI 5-1478
N EXCITED X-RAY, AES ANALYSIS / CORRELATION CF FRACTIONAL MONOLAYER OXYGEN DETERMINATIONS OBTAINED BY PROTO 5-1682
CESIUM ON TUNGSTEN AND TITANIUM SINGLE CRYSTALS IN THE SUBMONOLAYER REGION. /ROSCOPY, AND WORK FUNCTION OF 5-1829
TH INFORMATION FROM AUGER ELECTRON SPECTROSCOPY. (SILICON MONOLAYERS) IN DEP 1-0130
PECTROSCOPY. DISTINCTION BETWEEN ADSORBED MONOLAYERS AND THICKER LAYERS IN AUGER ELECTRON S 5-1797
CHEMICAL IDENTIFICATION OF SURFACE MONOLAYERS. (BY AUGER EMISSION) 5-1589
Y) SPONTANEOUS ALLCYING OF A GOLD SUBSTRATE WITH LEAD MONOLAYERS. (INTERFACE AUGER ELECTRON SPECTROSCOP 5-1178
ER EFFECT) CHLORINE MONOLAYERS ON THE LOW INDEX FACES OF SILVER. (AUG 5-1777
ER EFFECT) ADSORPTION AND SURFACE ALLOYING OF LEAD MONOLAYERS ON (111) AND (110) FACES OF GOLD. (AUG 5-1733
TTERED YTTRIA-DOPED ZIRCONIA THIN FILMS. (AUGER EFFECT) MORPHOLOGICAL AND ELECTRICAL PROPERTIES OF RF SPU 5-1385
) PURITY AND MORPHOLOGY OF ALUMINUM FILMS. (AUGER SPECTROSCOPY 5-1299
SCOPY. STUDY OF THE CHEMICAL CCMPOSITION OF MOS AND MNOS STRUCTURES BY AUGER ELECTRON SPECTRO 5-1503
S STUDY. (AUGER) REALIZATION OF A MOSSBAUER SPECTROMETER FOR SURFACES AND THIN FILM 3-0690
TION OF PASSIVE STATE FILMS. (AUGER ELECTRON AND MOSSBAUER) SPECTROSCOPICAL STUDIES OF THE COMPOSI 5-1309
S. (AUGER SPECTROSCOPY, X-RAY PHOTOELECTRON SPECTROSCOPY, MOSSBAUER SPECTROSCOPY) /POSITION IN LIQUID METAL 6-2050
CHEMICAL EFFECTS OF NUCLEAR TRANSFCRMATIONS IN MOSSBAUER SPECTROSCOPY. (AUGER AFTEREFFECTS) 2-0564
DETECTION OF CONVERSION AND AUGER ELECTRONS. APPLICATION/ MOSSBAUER SPECTROSCOPY OF IRON-57 AND TIN-119 BY 5-1796
NIZATION FOLLOWING ELECTRON CAPTURE IN TRIS (1,10-PHENAN/ MOSSBAUER STUDIES OF THE AFTEREFFECTS OF AUGER IO 2-0516
-119. BACKSCATTER MOSSBAUER STUDIES USING CONVERSION ELECTRONS. TIN 2-0576
MICROPROBE. STUDY OF THE INTERMITTENT MOTION OF FRICTION IN STAINLESS STEEL BY AN AUGER 6-2133
Y NONCONSERVATION IN RADIATIONLESS (AUGER) TRANSITIONS IN MU MESIC ATOMS. PARIT 2-0364
STUDY OF MULBERRY SURFACES BY AUGER AND IEE SPECTROSCOPY. 5-1274
N SPECTROSCOPY WITH APPLICATION TO SURFACE CCMPOSITION/ A MULTICHANNEL MONITOR FOR REPETITIVE AUGER ELECTRO 3-0695
NERGY ELECTRON IMPACT. (AUGER EFFECT) MULTIIONIZATION OF KRYPTON AND ITS IONS BY HIGH E 2-0493
ONS BY HIGH ENERGY ELECTRON IMPACT. (AUGER EMISSION) MULTIIONIZATION OF NEON, ARGON, XENON AND THEIR I 2-0492
IDE, THORIUM FLUORIDE) AUGER ANALYSIS OF THREE ENHANCED MULTILAYER DIELECTRIC MIRROR DESIGNS. (ZINC SELEN 1-0328
DIFFRACTION AND AUGER EM/ ULTRAHIGH VACUUM EVAPORATOR AND MULTIPLE CLEAVAGE DEVICE FOR LOW ENERGY ELECTRON 3-0707
ATOMS. PART-2: INNER AUGER ELECTRONS. MULTIPLE EXCITATION PROCESSES IN TRANSITION METAL 2-0362
UM AND SILVER BY ELECTRON IMPACT. MULTIPLE IONIZATION AND AUGER TRANSITIONS IN INDI 2-0229
SOME METAL VAPORS. (AUGER PROC/ MECHANISMS OF SIMPLE AND MULTIPLE IONIZATION INDUCED BY ELECTRON IMPACT IN 2-0231
LECTRON IMPACT. AUGER EFFECT IN THE MULTIPLE IONIZATION OF MANGANESE AND CADMIUM BY E 2-0228
NEON. (K-AUGER AND K X-RAY YIELDS) EFFECT OF MULTIPLE IONIZATION ON THE FLUORESCENCE YIELD OF 2-0299
M. A MULTIPLE SAMPLE HOLDER FOR USE IN ULTRAHIGH VACUU 3-0603
WIDTHS OF 4S AND 5S MULTIPLET COMPONENTS IN THE RARE EARTHS. 4-0970
LY IONIZED ATOMS. (AUGER EFFECT) MULTIPLET EFFECT IN FLUORESCENCE YIELDS OF MULTIP 2-0314
OF MULTIPLY IONIZED ARGON. (AUGER EFFECT) MULTIPLET EFFECTS ON THE L2,3 FLUORESCENCE YIELD 4-0956
PEAKS. (AUGER EFFECT) MULTIPLET EFFECTS ON THE WIDTHS OF PHOTOELECTRON 2-0445
EXCITATION OF MULTIPLET STATES IN NEON. 2-0285
TROSCOPY USING THE TWO-GRID LEED SYSTEM AND PHOTOELECTRON MULTIPLIER. AUGER ELECTRON SPEC 3-0656
) GAIN FATIGUE MECHANISM IN CHANNEL ELECTRON MULTIPLIERS. (SURFACE OXYGEN ON LEAD OXIDE BY AES 3-0681
RS. AUGER IONIZATION OF ATOMS BY MULTIPLY CHARGED IONS. MODEL OF TWO COULOMB CENTE 2-0407
AUGER NEUTRALIZATION OF MULTIPLY CHARGED IONS ON A METAL SURFACE. 2-0247
MULTIPLET EFFECTS ON THE L2,3 FLUORESCENCE YIELD OF MULTIPLY IONIZED ARGON. (AUGER EFFECT) 4-0956
AUGER ELECTRON SPECTROMETRY OF SINGLY AND MULTIPLY IONIZED ATOMS. 4-0978
MECHANISMS OF INNER SHELL EXCITATION AND DEEXCITATION IN MULTIPLY IONIZED ATOMS. 2-0487
DEEXCITATION OF MULTIPLY IONIZED ATOMS. (AUGER EFFECT) 2-0281
MULTIPLET EFFECT IN FLUORESCENCE YIELDS OF MULTIPLY IONIZED ATOMS. (AUGER EFFECT) 2-0314
K-SHELL AUGER RATES FOR MULTIPLY IONIZED ATOMS. PART-1: NEON. 2-0282
AUGER AND RADIATIVE DEEXCITATION OF MULTIPLY IONIZED NEON. 2-0315
HIGH RESOLUTION K-AUGER SPECTRA FOR MULTIPLY IONIZED NEON. 4-0952
K-SHELL (AUGER) FLUORESCENT YIELDS FOR MULTIPLY IONIZED NEON. 4-0994
K-AUGER ELECTRON AND K X-RAY TRANSITION ENERGIES FOR THE MULTIPLY IONIZED NEON ATOM. CALCULATED 4-0955
R RATES, TRANSITION ENERGIES, AND FLUORESCENCE YIELDS FCR MULTIPLY IONIZED NEON. (AUGER EFFECT) /SHELL AUGE 2-0284
FLUORESCENCE YIELDS AND AUGER RATES FOR MULTIPLY IONIZED NITROGEN. 4-0781
NUCLEAR AUGER EFFECTS IN MUONIC ATOMS. (LEAD, BISMUTH) 4-0941
PENETRATION EFFECTS IN CONVERTED MUONIC TRANSITIONS. (AUGER COEFFICIENTS) 2-0478
OPY. STUDY OF MUSCOVITE AND SILICON BY AUGER ELECTRON SPECTROSC 4-0792
TROSCOPY. MUSCOVITE CLEAVAGE STUDIED BY AUGER ELECTRON SPEC 5-1298
MVV AUGER SPECTRA OF COPPER AND NICKEL. 4-0766
HIGH ENERGY SATELLITES OF THE M2,3VV AND M1VV AUGER PEAKS OF COPPER. 4-1043
HIGH ENERGY SATELLITES OF THE M2,3VV AND M1VV AUGER PEAKS OF COPPER. 4-1043
AUGER VACANCY SATELLITE STRUCTURE IN THE L3,M4,5M4,5 AUGER SPECTRA OF COPPER. 4-1034
ECTRON IMPACT. ANGULAR DISTRIBUTION OF M4,5NN AUGER ELECTRONS EJECTED FROM KRYPTON BY EL 4-1003
N4,500 AUGER SPECTRA FROM XE/ HIGH RESOLUTION L2,3MM AND M4,5NN AUGER SPECTRA FROM KRYPTON AND M4, 5NN AND 4-1113
ADSORPTION OF MOLECULAR OXYGEN A/ CHEMICAL EFFECTS IN THE M4,5NN AUGER SPECTRUM OF MOLYBDENUM (110) DUE TO 4-0885
D. AUGER ELECTRON EJECTION FROM XENON N4,5OO AND KRYPTON M4,5NN PROCESSES BY ELECTRON IMPACT NEAR THRESHOL 4-1010
OM CADMIUM VAPOR. M4,5N4,5 AND M4,501 AUGER ELECTRON SPECTRA FR 4-0728
SEARCH FOR CORRELATION EFFECTS IN THE M4,5N4,5N4,5 AUGER SPECTRUM OF XENON. 4-0872
RELAXATION PROCESSES ACCOMPANYING CADMIUM M4,5N4,5N4,5 AUGER TRANSITIONS. 2-0559
UM. N5,4 AND M5,4 APPEARANCE POTENTIAL SPECTRA FROM PRASEODYMI 4-0908

N

AUGER SPECTRA OF HYDROGEN CHLORIDE VAPOR ETCHED	N+ GALLIUM ARSENIDE (100) SUBSTRATES.	4-0992
SPECTRUM OF AUGER K,L,M,N,O ELECTRONS OF SAMARIUM.		4-0756
/L PROPERTIES OF ALLOYED NICKEL- GOLD- GERMANIUM FILMS ON	N-TYPE GALLIUM ARSENIDE. (BY AUGER SPECTROSCOPY)	6-2101
IMPURITY IONIZATION IN	N-TYPE GERMANIUM. (AUGER RECOMBINATION PARAMETER)	2-0466
RECOMBINATION OF CARRIERS IN	N-TYPE INDIUM ARSENIDE AT 77 DEGREES K. (AUGER)	4-0785
THE (111) SURFACE OF	N-TYPE SILICON.	5-1349
SPECTROSCOPY) SURFACE DEFORMATION CAUSED ON	NATURAL MOLYBDENITE BY ABRASION. (AUGER ELECTRON	6-2132
/HARACTERIZATION OF GALLIUM ARSENIDE DURING ACTIVATION TO	NEGATIVE ELECTRON AFFINITY BY THE ADSORPTION OF /	5-1364
THE ADSORPTION OF CE/ LEED, AUGER AND PLASMON STUDIES OF	NEGATIVE ELECTRON AFFINITY ON SILICON PRODUCED BY	5-1363
- OXYGEN SURFACE. (AU/ STRUCTURAL AND ELECTRONIC MODEL OF	NEGATIVE ELECTRON AFFINITY ON THE SILICON- CESIUM	2-0426
UDY OF THE ACTIVATION OF GALLIUM ARSENIDE- CESIUM- OXYGEN	NEGATIVE ELECTRON AFFINITY SURFACES. /FRACTION ST	5-1845
TS, AUGER SPECTROMETRY) MECHANISM OF SURFACE ALIGNMENT IN	NEMATIC LIQUID CRYSTALS. (SUBSTRATE SURFACE EFFEC	5-1281
/TRA OF SOME RARE EARTH METALS. (LANTHANUM, PRASEODYMIUM,	NEODYMIUM, GADOLINIUM, DYSPROSIUM, YTTERBIUM, HA/	4-1100
K-AUGER RATES CALCULATED FOR	NEON(+).	4-0909
AUGER AND RADIATIVE DEEXCITATION OF MULTIPLY IONIZED	NEON.	2-0315
AUGER ELECTRONS FROM DOUBLE K-SHELL VACANCY STATES IN	NEON.	2-0573
AUGER ELECTRONS FROM FOIL EXCITED HEAVY ION BEAMS. (OF	NEON)	2-0370
BEAM FOIL EXCITED AUGER TRANSITIONS IN	NEON.	4-0860
ONFIGURATION INTERACTION ON THE K-SHELL AUGER SPECTRUM OF	NEON. EFFECTS OF THE C	2-0279
EXCITATION OF MULTIPLET STATES IN	NEON.	2-0285
HIGH RESOLUTION K-AUGER SPECTRA FOR MULTIPLY IONIZED	NEON.	4-0952
K-AUGER SPECTRUM OF	NEON.	4-0914
K-SHELL AUGER ELECTRON HYPERSATELLITES OF	NEON.	4-1117
K-SHELL (AUGER) FLUORESCENT YIELDS FOR MULTIPLY IONIZED	NEON.	4-0994
K-SHELL AUGER RATES FOR MULTIPLY IONIZED ATOMS. PART-1: IN	NEON.	2-0282
TRANSITION PROBABILITIES OF KL- LLL AUGER SATELLITES IN	NEON.	4-1049
NIZATION CROSS SECTION OF THE L2,3 SHELL OF ALUMINUM IN A	NEON(+)- ALUMINUM COLLISION. IO	2-0267
SPECTRA) K-SHELL VACANCY SHARING IN 300 TO 500 KEV	NEON(+)- SODIUM COLLISIONS. (QUASIMOLECULE AUGER	2-0582
PECTRA FOR OXYGEN IONS EXCITED BY COLLISIONS WITH HELIUM,	NEON, AND ARGON. /UTION BEAM GAS AUGER ELECTRON S	4-0904
ELECTRON IMPACT. (AUGER EMISSION) MULTIIONIZATION OF	NEON, ARGON, XENON AND THEIR IONS BY HIGH ENERGY	2-0492
AND K X-RAY TRANSITION ENERGIES FOR THE MULTIPLY IONIZED	NEON ATOM. CALCULATED K-AUGER ELECTRON	4-0955
ON ENERGIES, AND FLUORESCENCE YIELDS FOR MULTIPLY IONIZED	NEON. (AUGER EFFECT) /SHELL AUGER RATES, TRANSITI	2-0284
ENERGY ELECTRONS BY CLOUD CHAMBER TECHNIQUES: OXYGEN AND	NEON AUGER ELECTRON PROBLEM. /MEASUREMENT OF LOW	2-0566
THEORETICAL FLUORESCENCE YIELDS FOR	NEON. (AUGER RATE)	2-0283
NIFESTATION OF ATOMIC DYNAMICS THROUGH THE AUGER EFFECT.	(NEON AUGER SPECTRUM) MA	4-0920
DROGEN(+)- NEON COLLISIONS. HIGH RESOLUTION	NEON AUGER SPECTRUM PRODUCED IN 4.2 MEV ATOMIC HY	4-1078
ION CROSS SECTIONS OF ATOMIC NITROGEN, ATOMIC OXYGEN, AND	NEON. (AUGER TRANSITIONS) /LECTRON IMPACT IONIZAT	2-0363
DOUBLE PHOTOIONIZATION OF	NEON. (AUGER TRANSITIONS)	2-0310
SEQUENTIAL STRIPPING IN HIGHLY IONIZED	NEON. (AUGER TRANSITIONS)	2-0441
K X-RAY AND AUGER ELECTRON ENERGIES FOR	NEON BY A TRANSITION OPERATOR METHOD.	2-0393
GER ELECTRON AND X-RAY PRODUCTION IN 50- TO 200 KEV NEON-	NEON COLLISIONS. AU	4-1079
KLL AUGER ELECTRONS PRODUCED IN Z(1)- NITROGEN AND Z(1)-	NEON COLLISIONS AT KEV ENERGIES. EMISSION OF	4-0830
IONIZATION AND FLUORESCENCE YIELD IN 50 MEV CHLORINE(+)-	NEON COLLISIONS. (AUGER ELECTRON PRODUCTION) /ELL	2-0518
ELECTRON PRODUCTION CROSS SECTIONS FROM ION BOMBARDMENT.	(NEON, FLUORINE) K-SHELL AUGER	2-0574
AUGER DECAY OF	NEON FOLLOWING ENERGETIC ION BOMBARDMENT.	4-0954
UTION EMISSION (AUGER) SP/ CHARGE STATE DEPENDENCE OF THE	NEON K FLUORESCENCE YIELD DEDUCED FROM HIGH RESOL	2-0520
FFECT OF MULTIPLE IONIZATION ON THE FLUORESCENCE YIELD OF	NEON. (K-AUGER AND K X-RAY YIELDS) E	2-0299
AND ATOMIC HYDROGEN, OXYGEN, AND FLUORINE.	NEON K-AUGER CROSS SECTIONS FROM 1.5 MEV PER AMU	2-0571
OBSERVATION OF A PROJECTILE CHARGE DEPENDENCE FOR THE	NEON K-AUGER ELECTRON SPECTRUM.	4-0993
CHLORINE(12+) ION IMPACT. HIGH RESOLUTION	NEON K-AUGER ELECTRON SPECTRUM PRODUCED BY 45 MEV	4-1051
Y ION BOMBARDMENT. MEASUREMENTS OF	NEON K-AUGER ELECTRONS PRODUCED BY ENERGETIC HEAV	4-0951
50 MEV CHLORINE(N/ PROJECTILE CHARGE STATE DEPENDENCE OF	NEON K-SHELL IONIZATION AND FLUORESCENCE YIELD IN	2-0300
NERGY FLUORINE IONS. (AUGER ELECTRON EMISSION)	NEON K-SHELL VACANCY PRODUCTION BY INTERMEDIATE E	2-0572
D HELIUM(+) ION BOMBARDMENT. HIGH RESOLUTION	NEON KLL AUGER SPECTRA PRODUCED BY HYDROGEN(+) AN	4-0953
GIES AND AUGER RATES IN ATOMS WITH INNER SHELL VACANCIES	(NEON WITH 1S OR 2S VACANCY). CORRELATION ENER	2-0312
AUGER ELECTRON AND X-RAY PRODUCTION IN 50- TO 200 KEV NEON-	NEON- NEON COLLISIONS.	4-1079
HELL FLUORESCENCE AND COSTER-KRONIG YIELDS IN BISMUTH AND	NEPTUNIUM. X-RAY SPECTRA, L-SUBS	4-1112
/ASSOCIATED WITH AUGER RECOMBINATION OF EXCITONS BOUND TO	NEUTRAL DONORS IN TELLURIUM DOPED GALLIUM PHOSPH/	2-0453
AND ATOMS SCATTERED FROM A COPPER SINGLE CRYSTAL. (AUGER	NEUTRALIZATION) /CHANGE OF LOW ENERGY HELIUM IONS	2-0545
ERED FROM A (100) FACE OF A COPPER SINGLE CRYSTAL. (AUGER	NEUTRALIZATION) /+) IONS (LESS THAN 10 KEV) SCATT	2-0546
RGY ION SCATTERING: ELASTIC AND INELASTIC EFFECTS. (AUGER	NEUTRALIZATION) LOW ENE	2-0382
/ENERGY ATMOSPHERIC IONS WITH CONTROLLED SURFACES. (AUGER	NEUTRALIZATION AT (100) FACE OF TUNGSTEN, POLYCR/	5-1736
GY IONS FROM SOLID SURFACES. (AUGER	NEUTRALIZATION BEHAVIOR IN SCATTERING OF LOW ENER	2-0294
F ALKALI HALIDE CRY/ EVALUATION OF THE FREQUENCY OF AUGER	NEUTRALIZATION OF INERT GAS IONS ON THE SURFACE O	5-1544
L SURFACE. AUGER	NEUTRALIZATION OF MULTIPLY CHARGED IONS ON A META	2-0247
ROM SOLIDS BY PARTICLES CARRYING POTENTIAL ENERGY. (AUGER	NEUTRALIZATION, REVIEW, 16 REFS) /TRON EMISSION F	2-0374
R SURFACE: LINE SHAPE ANALYSIS AND CORRELATION WITH ION	NEUTRALIZATION SPECTROSCOPY. /(110)- C(2X2) SULFU	4-1065
D CUBIC METAL SURFACES. (GOLD, SILVER, PALLADIUM, COPPER,	NICKEL) /ER ELECTRON SPECTROSCOPY OF FACE CENTERE	5-1717
AUGER SPECTROSCOPY OF CARBON ON	NICKEL.	5-1270
DIFFUSION OF SULFUR TOWARD THE (110) FACE OF	NICKEL.	5-1771
ECTRON SPECTROSCOPY IN THE ANALYSIS OF SULFUR ABSORBED ON	NICKEL. / ENERGY ION BACKSCATTERING WITH AUGER EL	5-1869
OSCOPY BETWEEN ADSORBED SULFUR AND SULFIDES OF COPPER AND	NICKEL. DISCRIMINATION BY AUGER SPECTR	5-1734
TATION AND SURFACE DIFFUSION OF CARBON ON POLYCRYSTALLINE	NICKEL. /CTROSCOPY TO THE BULK TO SURFACE PRECIPI	5-1651
H AND STRUCTURE OF THIN COPPER FILMS ON (001) SURFACES OF	NICKEL. GROWT	5-1227
ROWTH, AND ITS APPLICATION TO THE DEPOSITION OF SILVER ON	NICKEL. /ECTROSCOPY OF SYSTEMS EXHIBITING LAYER G	2-0399
MVV AUGER SPECTRA OF COPPER AND	NICKEL.	4-0766
TATIVE AUGER SPECTROSCOPY OF PHYSICALLY ADSORBED XENON ON	NICKEL. QUANTI	5-1150
SEGREGATION OF CARBON TO (100) SURFACE OF	NICKEL.	6-2000
TS OF AN AES AND XPS STUDY OF THE ADSORPTION OF OXYGEN ON	NICKEL. SOME ASPEC	5-1459
SULFUR AND CARBON ON (110) SURFACE OF	NICKEL.	5-1813
: EQUILIBRIUM SEGREGATION OF GOLD TO THE (111) SURFACE OF	NICKEL. (AES) SURFACE SEGREGATION IN ALLOYS	5-1216
/YTIC ACTIVITY PATTERN AND SURFACE COMPOSITION OF COPPER-	NICKEL ALLOY. APPLICATION OF AUGER ELECTRON SPE/	5-1875
GATION OF COPPER TO THE SURFACE OF A CLEAN ANNEALED 50-50	NICKEL ALLOY BY AUGER ELECTRON SPECTROSCOPY. /GRE	6-2044
REGION AT/ OBSERVATION OF SURFACE COMPOSITION OF COPPER-	NICKEL ALLOY BY AUGER SPECTRA IN THE LOWER ENERGY	5-1982
S/ DETERMINATION OF THE SURFACE COMPOSITION OF PALLADIUM-	NICKEL ALLOY FILM CATALYSTS USING AUGER ELECTRON	5-1846
) WORK FUNCTION OF WELL-DEFINED SURFACE OF COPPER-	NICKEL ALLOY PLATES. (AUGER ELECTRON SPECTROSCOPY	5-1873
/R- COPPER ALLOYS. QUANTITATIVE AUGER ANALYSIS OF COPPER-	NICKEL ALLOY SURFACES AFTER ARGON ION BOMBARDMEN/	5-1200

. REPLY TO COMMEN/ QUANTITATIVE AUGER ANALYSIS OF COPPER- NICKEL ALLOY SURFACES AFTER ARGON ION BOMBARDMENT 5-1702
/R- COPPER ALLOYS. QUANTITATIVE AUGER ANALYSIS CF COPPER- NICKEL ALLOY SURFACES AFTER ARGON ION BOMBARDMEN/ 5-1740
. QUANTITATIVE AUGER ANALYSIS OF COPPER- NICKEL ALLOY SURFACES AFTER ARGON ION BOMBARDMENT 5-1807
AUGER SPECTROSCOPY ON NICKEL ALLOY SURFACES RELATED TO CATALYSIS. 5-1703
AUGER SPECTRA CF COPPER- NICKEL ALLOYS. 4-1028
ELD ANALYZER FOR AUGER ELECTRON SPECTROSCOPY. (PALLADIUM- NICKEL ALLOYS, ADSORBED OXYGEN) /ALL RETARDING FI 3-0644
SPECTROSCOPY STUDY ON THE SURFACE COMPOSITION OF COPPER- NICKEL ALLOYS AFTER ANNEALING AND SPUTTERING. /ER 5-1675
CONFIRMATION OF SULFUR EMBRITTLEMENT IN NICKEL ALLOYS. (AUGER ELECTRON SPECTROSCOPY) 6-2053
/S OF THE SURFACE COMPOSITION OF PASSIVE FILMS ON COPPER- NICKEL ALLOYS BY AUGER ELECTRON SPECTROSCOPY AND/ 5-1874
GET TEMPERATURE ON SURFACE COMPOSITION CHANGES OF COPPER- NICKEL ALLOYS DURING ARGON BOMBARDMENT. /T OF TAR 5-1808
FACE/ DETERMINATION CF THE SURFACE COMPOSITION OF COPPER- NICKEL ALLOYS FOR CLEAN AND ADSORBATE COVERED SUR 5-1424
ADSORPTION ON SINGLE CRYSTAL SURFACES CF COPPER- NICKEL ALLOYS. PART-1. 5-1322
THE SECONDARY ELECTRON ENERGY DISTRIBUTIONS OF ALUMINUM, NICKEL, AND COPPER. /D PHOTOELECTRON STRUCTURE IN 4-1023
LUTION MEASUREMENTS OF THE L3M2,3M4,5 AUGER TRANSITONS IN NICKEL AND COPPER. HIGH RESO 4-1024
TER CALCULATIONS OF THE SOFT X-RAY L3 EMISSION SPECTRA OF NICKEL AND COPPER. (AUGER-TYPE PROCESSES) CLUS 4-0888
TROSCOPY) OBSERVATIONS OF THE EPITAXIAL GROWTH OF NICKEL AND COPPER ON (111) SILVER. (BY AUGER SPEC 5-1354
X-RAY EXCITED LMM AUGER SPECTRA CF COPPER, NICKEL AND IRON. 4-1123
AUGER SPECTROSCOPIC STUDY OF PLATINUM, NICKEL, AND IRON SURFACES. 5-1791
AY APPEARANCE POTENTIAL SPECTROSCOPY STUDIES WITH COPPER- NICKEL AND IRON- NICKEL ALLOYS. (AES) SOFT X-R 5-1952
ECTS IN THE PHOTOEMISSION CORE LEVEL AND AUGER SPECTRA OF NICKEL AND NICKEL ALLOYED WITH ZINC. /L STATE EFF 5-1806
PHOTOEMISSION STUDIES OF CHEMISORPTION ON NICKEL AND TUNGSTEN SURFACES. 5-1308
GIES OF SURFACE AND BULK ATOMS OF TITANIUM, CHROMIUM, AND NICKEL. (AUGER EFFECT) /ORE ELECTRON BINDING ENER 2-0392
CHEMICAL ANALYSIS OF ELECTRODEPOSITED NICKEL- NICKEL BONDS BY AES. 4-0947
EGATION. (AUGER ELECTRON / INTERGRANULAR EMBRITTLEMENT IN NICKEL BY HYDROGEN: EFFECT OF GRAIN BOUNDARY SEGR 6-2072
AUGER SPECTROSCOPY OF CARBON ON NICKEL. COMMENTS. 5-1283
AUGER SPECTRA AND LEED PATTERNS FROM NICKEL DEPOSITS ON CLEAVED SILICON. 5-1768
AUGER ELECTRON SPECTROSCOPY OF NICKEL DEPOSITS ON THE SILICON (111) SURFACE. 5-1238
AUGER STUDIES OF ELECTROLESS NICKEL FILMS. 4-0760
EFFECT OF IMPURITIES ON INTRINSIC STRESS IN THIN NICKEL FILMS. (AUGER EFFECT) 5-1131
EFFECT OF IMPURITIES ON INTRINSIC STRESS IN THIN NICKEL FILMS. (AUGER SPECTROSCOPY) 5-1130
EMPERATURE DEPENDENCE CF THE LOW ENERGY AUGER SPECTRUM OF NICKEL(II) OXIDE (100). T 4-0999
HEMICAL SHIFTS AND / STUDY OF THE ADSORPTION OF OXYGEN ON NICKEL(III) USING AUGER ELECTRON SPECTROSCOPY: C 5-1466
STUDY OF THE ADSORPTION OF BENZENE ON THE (100) FACE OF NICKEL. INFLUENCE OF SULFUR AT THE SURFACE. 5-1284
SPHORUS ON TEMPER EMBRITTLEMENT OF MANGANESE- MOLYBDENUM- NICKEL LOW ALLOY STEEL. (AES) / OF COPPER AND PHO 6-2043
SCOPIC TECHNIQUE/ RECRYSTALLIZATION OF TUNGSTEN FIBERS IN NICKEL MATRIX COMPOSITES. (AUGER ELECTRON SPECTRO 6-2046
LEED AND ELECTRON SPECTROSCOPIC OBSERVATIONS ON NICKEL MONOXIDE (100). 5-1687
/IFFUSION OF SILVER THROUGH EPITAXIAL FILMS CF COPPER AND NICKEL ON SILVER (111): FORMATION OF AN EQUILIBR/ 5-1356
LOW ENERGY ELECTRON DIFFRACTION FROM CLEAN (100) NICKEL OXIDE. 5-1752
A STUDY OF ACTIVATED SILICON SURFACES FOR ELECTROLESS NICKEL PLATING BY AES. 5-1539
PECTROSCOPY) ADSORPTION OF HYDROGEN ON NICKEL SINGLE CRYSTAL SURFACES. (AUGER ELECTRON S 5-1257
00) SURFAC/ KINETICS OF THE REACTION OF OXYGEN WITH CLEAN NICKEL SINGLE CRYSTAL SURFACES. PART-1: NICKEL (1 5-1452
/UGER ELECTRON SPECTROSCOPY STUDIES OF CARBON MONOXIDE ON NICKEL. SPECTRAL LINE SHAPES AND QUANTITATIVE AS/ 5-1458
INVESTIGATION ON A (100) SURFACE OF AUSTENITIC CHROMIUM- NICKEL STEEL. LEED- AUGER 6-1991
IN OF FINE STRUCTURE IN THE AUGER SPECTRUM OF SULFUR ON A NICKEL SURFACE. ORIG 5-1271
DECOMPOSITION OF CARBON MONOXIDE ON A (110) NICKEL SURFACE. (AUGER ELECTRON SPECTROSCOPY) 5-1606
BINDING OF CARBON ATOMS AT A STEPPED NICKEL SURFACE. (AUGER SPECTROSCOPY) 5-1485
CTION CHANGES AND AU/ IMPURITIES SEGREGATION TO THE (001) NICKEL SURFACE DURING THERMAL TREATMENT. WORK FUN 6-2069
ATTERING AND AUGER ELECTRON SPECTROSCOPY STUDIES OF CLEAN NICKEL SURFACES AND ADSORBED LAYERS. /ERGY ION SC 5-1870
INTERACTION OF CARBON MONOXIDE WITH (110) NICKEL SURFACES. (AUGER EFFECT) 5-1607
CHEMISORPTION ON (001), (110) AND (111) NICKEL SURFACES. (AUGER SPECTRA) 5-1764
TY ENERGY STUDIES OF CLEAN (001), (110) AND ELASTIC (111) NICKEL SURFACES. (AUGER SPECTROSCOPY) /ED INTENSI 5-1292
Y ELECTRON DIFF/ CHEMISORPTION ON (001), (110), AND (111) NICKEL SURFACES. CORRELATED STUDY USING LOW ENERG 5-1291
/TS OF THE SPATIAL DISTRIBUTION OF HYDROGEN DESORBED FROM NICKEL SURFACES: EFFECTS OF SURFACE COMPOSITION / 5-1197
NTS, AUGER, AND LOW ENERGY ELEC/ CHEMISORPTION STUDIES OF NICKEL SURFACES UTILIZING WORK FUNCTION MEASUREME 5-1290
SOLID SURFACES BOMBARDED BY MEDIUM ENERGY IONS (PLATINUM, NICKEL, TANTALUM, GRAPHITE). /TRON EMISSION FROM 2-0309
AUGER ELECTRON SPECTROSCOPY ON THE FRACTURE OF NICKEL TUNGSTEN MATERIALS AND SILICON NITRIDE. 5-1445
DIFFRACTION PEAKS IN SECONDARY ELECTRON ENERGY SPECTRA. (NICKEL (100)) 4-0775
HANGES IN SECONDARY ELECTRON EMISSION DURING OXIDATION OF NICKEL (100) AND (111) CRYSTAL SURFACES. /MICAL C 5-1453
. (AUGER EFFECT) BINDING ENERGIES OF CARBON TO NICKEL (100) FROM EQUILIBRIUM SEGREGATION STUDIES 5-1484
/LUENCES ON ADSORPTION ENERGY. PART-2: CARBON MONOXIDE ON NICKEL (100). PART-3: CARBON MONOXIDE ON COPPER / 5-1911
MENT OF AUGER ELECTRON CURRENTS. APPLICATION TO SULFUR ON NICKEL (110). ANALOG TECHNIQUE FOR THE MEASURE 3-0635
ANALYSIS A/ ELECTRON EXCITED AUGER ELECTRON SPECTRUM OF A NICKEL (110)- C(2X2) SULFUR SURFACE: LINE SHAPE 4-1065
CARBON MONOXIDE ADSORPTION ON NICKEL (110). (AUGER EFFECT) 5-1884
FORMIC ACID DESORPTION FROM GRAPHITIZED NICKEL (110). (LEED- AES) 5-1630
LFUR. LEED AND AUGER ELECTRON SPECTROSCOPY STUDY OF NICKEL (110) SURFACE EFFECTS DUE TO CARBON AND SU 5-1814
/COMPOSITION OF FORMIC ACID ON CARBURIZED AND GRAPHITIZED NICKEL (110) USING AES, LEED, AND FLASH DESORPTI/ 5-1629
/F THE STANDARD SURFACE CLEANING TECHNIQUES AS APPLIED TO NICKEL (111), NICKEL (100), AND NICKEL SHEET USI/ 5-1467
/NG PROBABILITIES FOR THE ADSORPTION OF HYDROGEN ON CLEAN NICKEL (111), NICKEL (100), SHEET AND EVAPORATED/ 5-1468
INTERACTION OF NITRIC OXIDE WITH A NICKEL (111) SURFACE. 5-1278
(AUGER SPECTROSC/ EQUILIBRIUM SEGREGATION OF CARBON TO A NICKEL (111) SURFACE: A SURFACE PHASE TRANSITION. 5-1803
OSCOPY) ADSORPTION OF COBALT ON A NICKEL (111) SURFACE. (AUGER EFFECT, AUGER SPECTR 5-1256
KINETICS OF THE REACTION BETWEEN OXYGEN AND SULFUR ON A NICKEL (111) SURFACE. (STUDIED BY LEED AND AES) 5-1451
Y) STRUCTURE AND GROWTH KINETICS OF NICKEL(2) SILICIDE ON SILICON. (AUGER SPECTROSCOP 5-1917
N IMPURITY. (AUGER EMISSION AN/ TEMPER EMBRITTLEMENT IN A NICKEL- CHROMIUM STEEL CONTAINING PHOSPHORUS AS A 6-2138
/IC GRAIN BOUNDARY COMPOSITION ON TEMPER BRITTLENESS IN A NICKEL- CHROMIUM- ANTIMONY STEEL. (AUGER ANALYSI/ 6-2114
LONG TIME ISOTHERMAL TEMPER EMBRITTLEMENT IN NICKEL- CHROMIUM- MOLYBDENUM- VANADIUM. 6-2140
DURING BRAZING. (MONEL, AUGE/ SURFACE CHARACTERIZATION OF NICKEL- COPPER ALLOY DISPLAYING POOR WETTABILITY 5-1573
SURFACE COMPOSITION OF NICKEL- GOLD ALLOYS. 5-1966
RSENI/ METALLURGICAL AND ELECTRICAL PROPERTIES OF ALLOYED NICKEL- GOLD- GERMANIUM FILMS ON N-TYPE GALLIUM A 6-2101
AES AND SPUTTER ETCHING OF NICKEL- GOLD- GERMANIUM ON N-GALLIUM ARSENIDE. 5-1775
CORROSION, SURFACE, AND MAGNETIC PROPERTIES OF NICKEL- IRON- CHROMIUM THIN FILMS. (AES) 5-1766
CHEMICAL ANALYSIS OF ELECTRODEPOSITED NICKEL- NICKEL BONDS BY AES. 4-0947
ELECTRON SPECTROSCOPIC STUDIES OF THE OXIDATION OF NICKEL- PALLADIUM ALLOYS. 5-1953
R ELECTRON SPECTROSCOPY ANALYSIS OF SILVER- PALLADIUM AND NICKEL- PALLADIUM ALLOYS. QUANTITATIVE AUGE 6-2088
FILES AND SCHOTTKY BARRIER HEIGHTS OF SILICIDES FORMED IN NICKEL- PLATINUM ALLOY FILMS. COMPOSITION PRO 5-1900
AES STUDY OF THE SILICIDE FORMATION IN 85% NICKEL- 15% PLATINUM ALLOY FILMS. 5-1899
IUM RAT/ CORRELATION OF THE WETTING BEHAVIOR OF SODIUM WITH NIMONIC-PE16 ALLOY WITH THE SURFACE OXYGEN- CHROM 5-1441
PRESSURES. (AUGER ELECTRON SPECTROSCOPY, SU/ OXIDATION OF NIMONIC-80A ALLOY AT 800 DEGREES C IN LOW OXYGEN 6-2144
/RFACE CHARACTERISTICS RELATED TO LASER DAMAGE OF LITHIUM NIOBATE AND POTASSIUM CHLORIDE SURFACES. (AUGER / 5-1746
ATION TO LOW ENERGY ARGON(+) ION SPUTTERING OF SILVER AND NIOBIUM. /OPY IN SPUTTERING MEASUREMENTS. APPLIC 5-1825

OXIDE, URANIUM, GRAPHITE, 300 SERIES STAINLESS STEEL, AND NIOBIUM. /LECTRON STUDIES OF SURFACES: URANIUM DI 5-1217
SURFACE SEGREGATION OF OXYGEN IN NIOBIUM. (AES) 5-1443
NETRATION DEPTH AND RF BREAKDOWN FIELD OF SUPERCONDUCTING NIOBIUM. (AUGER EFFECT) /ACE METALLURGY ON THE PE 6-2137
INTERACTION OF NITROGEN WITH NIOBIUM. (AUGER EFFECT) 5-1301
RE HEAT TREATMENT ON PENETRATION DEPTH OF SUPERCONDUCTING NIOBIUM. (AUGER SPECTROSCOPIC MEASUREMENTS) /RATU 5-1938
S.   X-RAY FLUORESCENCE L SPECTRA OF NIOBIUM IN NIOBIUM METAL AND SOME OF ITS COMPOUND 4-0762
/UTIONS OF HYDROGEN DESORBED FROM IRON, PLATINUM, COPPER, NIOBIUM, STAINLESS STEEL SURFACES. (AUGER ELECTR/ 6-2006
TORS AT SURFACES AND IN THIN FILMS. (LEED- AES STUDIES OF NIOBIUM SURFACES) /F TRANSITION METAL SUPERCONDUC 5-1848
PROTON INDUCED ELECTRON EMISSION FROM CHARACTERIZED NIOBIUM SURFACES. (AUGER SPECTROSCOPY) 5-1666
RRADIATION.   CARBIDE FORMATION ON NIOBIUM SURFACES DURING HIGH TEMPERATURE PROTON I 5-1886
URFACE ANALYSIS)   NIOBIUM SURFACES FOR RF SUPERCONDUCTORS. (AUGER S 5-1472
(AUGER EFFE/ INTERACTION OF OXYGEN AND NITROGEN WITH THE NIOBIUM (100) SURFACE. PART-2: REACTION KINETICS. 5-1333
LEED AUGER STUDY OF GROWTH OF ALUMINUM ON NIOBIUM (110). 5-1491
ETECTION OF SURFACE NITROGEN IN HIGH CRITICAL TEMPERATURE NIOBIUM(3) GERMANIUM FILMS BY AUGER ANALYSIS.   D 5-1804
ER ELECTRON SPECTROSCOP/ SURFACE SEGREGATION OF OXYGEN IN NIOBIUM- OXYGEN AND TANTALUM- OXYGEN ALLOYS. (AUG 6-2065
DIFFERENTIAL AUGER SPECTROMETRY. (NIOBIUM- ZIRCONIUM ALLOY ANALYSIS) 5-1849
AND INTERACTION OF OXYGEN AND NITROGEN WITH ZIRCONIUM IN NIOBIUM- 1% ZIRCONIUM.   SURFACE SEGREGATION 5-1508
HIGH RESOLUTION AUGER ELECTRON SPECTROSCOPY OF NITRIC OXIDE ADSORBED ON A TUNGSTEN SURFACE. 5-1565
/AY PHOTOIONIZATION OF NITROGEN, OXYGEN, CARBON MONOXIDE, NITRIC OXIDE, CARBON DIOXIDE, AND CARBON TETRAFL/ 2-0308
ADSORPTION OF NITRIC OXIDE ON TUNGSTEN. 5-1929
OF KLL AUGER SPECTRA OF NITROGEN, OXYGEN, CARBON DIOXIDE, NITRIC OXIDE, WATER, AND CARBON MONOXIDE. /ATION 4-0990
INTERACTION OF NITRIC OXIDE WITH A NICKEL (111) SURFACE. 5-1278
RA FROM THE (100) SURFACE OF SILICON DURING OXIDATION AND NITRIDATION.   AUGER SPECT 5-1310
CTRON DIFFRACTION AND AUGER ELECTRON SPECTR/ STUDY OF THE NITRIDATION OF SILICON SURFACES BY LOW ENERGY ELE 5-1417
ON THE FRACTURE OF NICKEL TUNGSTEN MATERIALS AND SILICON NITRIDE.   AUGER ELECTRON SPECTROSCOPY 5-1445
ECTRA FROM SILICON MONOXIDE, SILICON DIOXIDE, AND SILICON NITRIDE. /ECTRON SPECTROSCOPY STUDY OF SILICON SP 4-1081
SHIFTS IN AUGER ELECTRON SPECTRA FROM SILICON NITRIDE.   CHEMICAL 4-0881
ON TEMPERATURES OF REACTIVELY SPUTTERED FILMS OF TANTALUM NITRIDE AND TUNGSTEN NITRIDE. (AUGER EFFECT) /ITI 5-1541
EFFECT OF YTTRIA ADDITIONS ON HOT PRESSED SILICON NITRIDE. (AUGER ANALYSIS) 5-1350
MICROSTRUCTURE OF HOT PRESSED SILICON NITRIDE. (AUGER ANALYSIS, MICROPROBE ANALYSIS) 5-1556
IMPURITIES AND INCLUSIONS IN HOT PRESSED SILICON NITRIDE. (AUGER SPECTROGRAPHIC ANALYSIS) 5-1557
FICATION OF A GRAIN BOUNDARY PHASE IN HOT PRESSED SILICON NITRIDE BY AUGER ELECTRON SPECTROSCOPY.   IDENTI 5-1749
STUDY OF FRACTURE SURFACES OF HOT PRESSED SILICON NITRIDE BY AUGER ELECTRON SPECTROSCOPY. 5-1444
MICAL SHIFTS IN SOME VANADIUM COMPOUNDS. (VANADIUM OXIDE, NITRIDE, CARBIDE, SILICIDE) /NVESTIGATIONS OF CHE 4-1090
N OF SILICON OXYNITRIDE LAYERS ON THE SURFACES OF SILICON NITRIDE FILMS BY AUGER ELECTRON EMISSION. /TECTIO 5-1611
ALYTIC PROCESSES. CHEMISORPTION OF AMMONIA, NITROGEN, AND NITRIDE FORMATION ON A MOLYBDENUM. /COPIES TO CAT 5-1533
AUGER ELECTRON SPECTROSCOPIC ANALYSIS OF SILICON NITRIDE ON SILICON. 5-1455
GNESIUM SINGLE CRYSTALS AND MAGNESIUM OXIDE AND MAGNESIUM NITRIDE SURFACE FILMS. /GER CHARACTERISTICS OF MA 5-1220
DRYING OF BUTADIENE- NITRILE RUBBERS IN AN AUGER MACHINE. 3-0617
AUGER ELECTRON SPECTROSCOPY. PART-1: ELECTRONIC STATES OF NITROGEN. / DOUBLY IONIZED MOLECULES BY MEANS OF 4-1071
FLUORESCENCE YIELDS AND AUGER RATES FOR MULTIPLY IONIZED NITROGEN. 4-0781
ICATION OF HIGH ENERGY LINES IN THE KLL AUGER SPECTRUM OF NITROGEN.   IDENTIF 4-0791
), DEUTERIUM(+), MOLECUL/ K-SHELL IONIZATION OF MOLECULAR NITROGEN AND METHANE BY 50- TO 600 KEV HYDROGEN(+ 2-0519
/COPIES TO CATALYTIC PROCESSES. CHEMISORPTION OF AMMONIA, NITROGEN, AND NITRIDE FORMATION ON A MOLYBDENUM. 5-1533
IRONMENTS.   AUGER LINE SHAPE COMPARISON OF NITROGEN AND SULFUR IN TWO DIFFERENT CHEMICAL ENV 5-1816
S.   EMISSION OF KLL AUGER ELECTRONS PRODUCED IN Z(1)- NITROGEN AND Z(1)- NEON COLLISIONS AT KEV ENERGIE 4-0830
/LUTE ELECTRON IMPACT IONIZATION CROSS SECTIONS OF ATOMIC NITROGEN, ATOMIC OXYGEN, AND NEON. (AUGER TRANSI/ 2-0363
ECTRONS PRODUCED BY 200-500 KEV PROTONS IN GASES. PART-2: NITROGEN. (AUGER SPECTRUM) /GY DISTRIBUTION OF EL 4-1075
/OF NITROGEN(2) (X), NITROGEN(X), AND NITROGEN(2) (2X) FROM NITROGEN BY ELECTRONS IN THE ENERGY RANGE 16-600/ 2-0376
PRODUCTION OF INTENSE SATELLITE LINES IN AUGER SPECTRA OF NITROGEN BY SLOW PROTON IMPACT. 4-1076
UM FILMS BY AUG/ QUANTITATIVE ANALYSIS OF LIGHT ELEMENTS (NITROGEN, CARBON, AND OXYGEN) IN SPUTTERED TANTAL 5-1652
RTIES OF SPUTTERED TANTALUM FIL/ EFFECT OF LIGHT ELEMENTS NITROGEN, CARBON AND OXYGEN ON THE PHYSICAL PROPE 5-1473
OSCILLATOR STRENGTH FOR THE FIRST DISCRETE TRANSITION IN NITROGEN FROM THE AUGER DECAY.   DETERMINATION 2-0575
GEN, AND FLUORINE I/ IN-DEPTH PROFILE DETECTION LIMITS OF NITROGEN IN GALLIUM PHOSPHIDE (AND) NITROGEN, OXY 5-1916
GERMANIUM FILMS BY AUGER ANALYSIS.   DETECTION OF SURFACE NITROGEN IN HIGH CRITICAL TEMPERATURE NIOBIUM(3) 5-1804
N ON IRON AND MOLYBDENUM SURFACE/ CHEMICAL EFFECTS OF THE NITROGEN KLL AUGER SPECTRA OF CHEMISORBED NITROGE 4-0925
ONIA ON A MOLYBDENUM/ HIGH RESOLUTION MEASUREMENTS OF THE NITROGEN KLL AUGER TRANSITIONS OF CHEMISORBED AMM 5-1532
ECTROSCOPY.   CHEMISORPTION OF NITROGEN ON TUNGSTEN STUDIED BY AUGER ELECTRON SP 5-1518
TALUM. (/ ADSORPTION AND SOLUTION OF HYDROGEN, DEUTERIUM, NITROGEN, OXYGEN AND CARBON MONOXIDE BY (100) TAN 5-1549
ATER, AND CARBON M/ DETERMINATION OF KLL AUGER SPECTRA OF NITROGEN, OXYGEN, CARBON DIOXIDE, NITRIC OXIDE, W 4-0990
/F FRAGMENT IONS FORMED FROM THE X-RAY PHOTOIONIZATION OF NITROGEN, OXYGEN, CARBON MONOXIDE, NITRIC OXIDE,/ 2-0308
INTERACTION OF NITROGEN WITH NIOBIUM. (AUGER EFFECT) 5-1301
REACTION KINETICS. (AUGER EFFE/ INTERACTION OF OXYGEN AND NITROGEN WITH THE NIOBIUM (100) SURFACE. PART-2: 5-1333
.   SURFACE SEGREGATION AND INTERACTION OF OXYGEN AND NITROGEN WITH ZIRCONIUM IN NIOBIUM- 1% ZIRCONIUM 5-1508
EL/ CROSS SECTIONS FOR THE PRODUCTION OF NITROGEN(2) (X), NITROGEN(X), AND NITROGEN(2) (2X) FROM NITROGEN 2-0376
FROM NITROGEN BY EL/ CROSS SECTIONS FOR THE PRODUCTION OF NITROGEN(2) (X), NITROGEN(X), AND NITROGEN(2) (2X) 2-0376
/S FOR THE PRODUCTION OF NITROGEN(2) (X), NITROGEN(X), AND NITROGEN(2) (2X) FROM NITROGEN BY ELECTRONS IN TH/ 2-0376
COPPER BERYLLIUM OXYGEN SURFACE BY THERMAL CARBON OXYGEN NITROGEN(2), AND NOBLE GAS METASTABLES. /LDS FROM 4-0786
PERIMENTAL AND THEORETICAL STUDY OF THE AUGER SPECTRUM OF NITROUS OXIDE.   EX 4-0814
PY TECHNIQ/ INVESTIGATION OF MECHANISM OF S IONIZATION OF NOBLE GAS ATOMS BY THE ULTRASOFT X-RAY SPECTROSCO 2-0580
L DI/ ELECTRONIC EMISSION FROM SOLID TARGETS BOMBARDED BY NOBLE GAS IONS (10-100 KEV): ENERGETIC AND SPATIA 4-0776
WORK FUNCTION EFFECTS ON AUGER ELECTRON EJECTION BY NOBLE GAS METASTABLE ATOMS. 5-1601
OXYGEN SURFACE BY THERMAL CARBON OXYGEN NITROGEN(2), AND NOBLE GAS METASTABLES. /LDS FROM COPPER BERYLLIUM 4-0786
AUGER TRANSITION RATE FOR THE NOBLE GAS SATELLITE SPECTRA. 4-0972
CORRELATION EFFECTS IN THE AUGER SPECTRA OF NOBLE GASES. 4-0981
AUGER SPECTRA OF THE NOBLE GASES. KRYPTON. 4-0971
AUGER SPECTRUM OF THE NOBLE GASES. PART-2: ARGON. 4-0974
R ELECTRON SPECTRO/ METALLIC STATE OF SILICON IN SILICON- NOBLE METAL VAPOR QUENCHED ALLOYS STUDIED BY AUGE 5-1438
ER SPECTRA.   LOW NOISE DETECTION SYSTEM FOR THE MEASUREMENT OF AUG 3-0662
G-R NOISE FOR AUGER BAND-TO-BAND PROCESSES. 2-0292
NON..., SEE THE PARENT WORD.
ENDENCE OF AUGER ELECTRON YIELD ON PRIMARY BEAM ENERGY AT NORMAL AND GLANCING INCIDENCE.   DEP 2-0455
A NOVEL DETECTION CIRCUIT FOR AUGER SPECTROSCOPY. 3-0673
IRON-55 AS AN AUGER ELECTRON EMITTER. NOVEL SOURCE FOR GAS CHROMATOGRAPHY DETECTORS. 3-0608
DECAY OF HOLE STATES IN OXYGEN-16 VIA A "NUCLEAR AUGER EFFECT". 2-0336
MUTH)   NUCLEAR AUGER EFFECTS IN MUONIC ATOMS. (LEAD, BIS 4-0941
. (AUGER AFTEREFFECTS)   CHEMICAL EFFECTS OF NUCLEAR TRANSFORMATIONS IN MOSSBAUER SPECTROSCOPY 2-0564
INFLUENCE OF SUBSTRATE CONDITIONS ON THE NUCLEATION AND GROWTH OF EPITAXIAL SILICON FILMS. 5-1514
ER SPECTRA)   HETEROGENEOUS NUCLEATION AND GROWTH OF GOLD ON SUBSTRATES. (AUG 5-1137
EXPERIMENTAL PROBLEMS OF STUDYING THIN FILM NUCLEATION. (ELECTRON MICROSCOPY, AUGER EFFECT) 5-1408

CTROSCOPY) EFFECT OF A THIN DIFFUSED SOLUTE LAYER ON THE NUCLEATION OF FATIGUE CRACKS. (AUGER ELECTRON SPE 6-2037
RFACES. (/ INITIAL OXIDATION OF MOLYBDENUM. PART-3: OXIDE NUCLEATION RATES ON (100) AND (110) MOLYBDENUM SU 5-1536
/OGICAL RADIATION EFFECT ON INCLUSION OF MODERATELY HEAVY NUCLEI IN THE TISSUE: USE OF SOFT X-RAYS. (AUGE/ 2-0531
AUGER SPECTRA FOR ELEMENTS OF LOW ATOMIC NUMBER. 4-0827
F THE AUGER EFFECT IN THE LIGHT TO MEDIUM RANGE OF ATOMIC NUMBER. THEORETICAL STUDY O 2-0486
LITIES FOR AUGER- KLL TRANSITIONS DEPENDENT ON THE ATOMIC NUMBER Z=30-94. PART-2. ABSOLUTE PROBABI 2-0252
ILITIES FOR AUGER KLL TRANSITIONS DEPENDENT ON THE ATOMIC NUMBER 29 TO 94. RELATIVE PROBAB 2-0253
UGER ELECTRON ENERGIES (0-2000 EV) FOR ELEMENTS OF ATOMIC NUMBER 5-103. A 4-1119
OBABILITIES FOR ELEMENTS WITH LOW AND INTERMEDIATE ATOMIC NUMBERS. KLL AUGER TRANSITION PR 4-0803
THE KLL AUGER SPECTRUM FOR ATOMIC NUMBERS BETWEEN 10 AND 36. 4-0980
KLL AUGER TRANSITION ABSOLUTE PROBABILITIES. (ATOMIC NUMBERS 30-94) 2-0251
N4,5 AUGER SPECTRA OF GADOLINIUM. 4-0821
SSION OF XENON(III) LEVELS EXCITED BY ELECTRON IMPACT VIA N4,5 AUGER TRANSITIONS. VACUUM ULTRAVIOLET EMI 4-0878
MPACT NEAR THRESHOLD. AUGER ELECTRON EJECTICN FROM XENON N4,5OO AND KRYPTON M4,5NN PROCESSES BY ELECTRON I 4-1010
3MM AND M4,5NN AUGER SPECTRA FROM KRYPTON AND M4, 5NN AND N4,5OO AUGER SPECTRA FROM XENON. / RESOLUTION L2, 4-1113
RASEODYMIUM. N5,4 AND M5,4 APPEARANCE POTENTIAL SPECTRA FROM P 4-0908
/AVIOR FOR ELECTRON INDUCED INNER SHELL IONIZATION OF THE N6,7 LEVEL IN GOLD, BISMUTH, AND LEAD MONITORED / 4-1067

O

SPECTRUM OF AUGER K,L,M,N,O ELECTRONS OF SAMARIUM. 4-0756
-RAY AND AUGER ELECTRON ENERGIES FOR NEON BY A TRANSITION OPERATOR METHOD. K X 2-0393
S OF AN AUGER ELECTRON SPECTROMETER BASED ON LEED DISPLAY OPTICS. RESOLUTION AND SENSITIVITY CONSIDERATION 3-0705
COMPARISON OF SILICON L3,2 ELECTRON LOSS SPECTRA AND OPTICAL ABSORPTION. (AUGER EFFECT) 4-1038
ES IN BISMUTH(III) TELLURI/ MATRIX ELEMENTS DEPENDENCE OF OPTICAL EXCITATION AND AUGER DECAY OF 5D CORE HOL 2-0398
AUGER RECOMBINATION. OPTICAL GAIN IN LIGHTLY DOPED GALLIUM ARSENIDE. (2-0503
INATION OF HIGH DENSITY PLASMAS IN GERMANIUM. PICOSECOND OPTICAL MEASUREMENTS OF BAND-TO-BAND AUGER RECOMB 2-0250
AGNESIUM SINGLE CRYSTALS AND MAGNESIUM CXIDE AND MAGNESI/ OPTICAL PROPERTIES AND AUGER CHARACTERISTICS OF M 5-1220
OSCOPY) RESIDUAL GAS AND THE OPTICAL PROPERTIES OF SILVER FILMS. (AUGER SPECTR 5-1700
AND PHOTOELECTRON EMISSION STUDY OF ADSORPTICN PROCESSES. OPTICALLY STIMULATED SECONDARY ELECTRON EMISSION 2-0302
ED SYSTEMS. OPTIMIZATION OF AUGER ELECTRON SPECTROSCOPY IN LE 1-0152
OF TITANIUM(II) AND TITANIUM(IV) OXIDES. MOLECULAR ORBITAL EFFECTS ON THE TITANIUM LMV AUGER SPECTRA 4-1069
RFACES AND OF SILICON CIOXIDE. ELECTRON ORBITAL ENERGIES OF OXYGEN ADSORBED ON SILICON SU 5-1475
S IN THE RANGE 100 TO 1000 EV USING THE CYLINDRICA/ FIRST ORDER APPROXIMATION TO QUANTITATIVE AUGER ANALYSI 3-0660
E OF COPPER(3) GOLD. TEMPERATURE DEPENDENCE OF LONG RANGE ORDER AT A CRYSTAL SURFACE. / OF THE (100) SURFAC 5-1854
ON THE (0001) FACE CF GRAPHITE. (AUGER ELECTRON SP/ FIRST ORDER TWO DIMENSIONAL TRANSITION: XENON ADSORBED 5-1857
TITANIUM- OXYGEN SOLID SOLUTIONS. ORDER- DISORDER PHENOMENA AT THE SURFACE OF ALPHA 5-1819
OF COPPER(3) GOLD, THEORY, EXPERIMENT. ORDER- DISORDER TRANSFORMATION AT A (100) SURFACE 5-1855
ON (111). (AUGER ELECTRON SPECTROSCOPY) AN ORDERED BORON STRUCTURE AFTER DEPOSITION ON SILIC 5-1168
ATOMIC ARRANGEMENT IN THE 1 K 1 STRUCTURE OF A SILICON ORDERED MONOLAYER ON MOLYBDENUM (001). 5-1478
/IGH SPATIAL RESOLUTION AUGER ELECTRON SPECTROSCOPY IN AN ORDINARY DIFFUSION PUMPED SCANNING ELECTRON MICR/ 3-0597
R ELECTRON SPECTROSCOPY. DETECTION OF ORGANIC CONTAMINANTS ON ALUMINUM SURFACES BY AUGE 5-1487
ENT AND SECONDARY ELECTRON EMISSION FOR 10-100 KEV ELECT/ ORIENTATION ANISOTROPY OF BACKSCATTERING COEFFICI 4-1031
F ELECTRONS FOR THE CYLINDRICAL MIRRCR ANALYZER AND A RE/ ORIENTATION DEPENDENCE OF OVERLAYER ATTENUATION O 3-0693
/KEL SURFACES: EFFECTS OF SURFACE COMPOSITION AND CRYSTAL ORIENTATION. (SURFACE COMPOSITION DETERMINED BY / 5-1197
SULFUR ON A NICKEL SURFACE. ORIGIN OF FINE STRUCTURE IN THE AUGER SPECTRUM OF 5-1271
M METALS. ORIGIN OF ION INDUCED AUGER ELECTRON EMISSION FRO 2-0549
LICON- METAL SYSTEMS. (AUGER SPECTROSCOPY) POSSIBLE ORIGIN OF LCW TEMPERATURE SILICON MIGRATION IN SI 5-1437
/ION ON INNER LEVEL IONIZATICN. FART-2: DOUBLE IONIZATICN ORIGIN OF THE HIGH ENERGY SATELLITE OF AUGER PEA/ 4-1047
F OXYGEN (KL2,3L2,3) ON MOLYBDENUM. ORIGIN OF THE PEAK SHAPES IN THE AUGER SPECTRUM O 4-1004
PURE GERMANIUM. (AUG/ MAGNETIC FIELD DEPENDENT INTENSITY OSCILLATIONS OF THE ELECTRON HOLE LUMINESCENCE IN 4-0780
EFFECT OF SURFACE PLASMON OSCILLATIONS ON AUGER ELECTRON EMISSION. 2-0236
TION IN NITROGEN FROM THE AUGER DECAY. DETERMINATION OF OSCILLATOR STRENGTH FOR THE FIRST DISCRETE TRANSI 2-0575
MIUM SURFACE. AUGER SPECTROSCOPY OF OSMIUM, AND THE SEGREGATION OF PLATINUM TO THE OS 5-1770
AUGER ELECTRON SPECTRUM OF OSMIUM AT ENERGIES UP TO 300 EV. 4-1030
SURFACE ANALYSIS, METHODS OF STUDYING THE OUTER ATOMIC LAYERS OF SOLIDS. 1-0016
AUGER ELECTRON SPECTROSCOPY OF THE OUTER SHELL ELECTRONS. 4-0805
V OXYGEN IONS. (INNER SHELL AUGER TRANSITIONS) OUTER SHELL EXCITATION OF THE RARE GASES BY 30 ME 2-0298
CHANGE OF SURFACE PROPERTIES OF MICA AFTER CLEAVAGE. (OUTGASSING AND DECORATION STUDIES, AUGER SPECTRA) 5-1743
ANALYSIS OF HIGH CHARGE STATE OUTPUT FROM A PENNING ION SOURCE. 3-0602
RICAL MIRROR ANALYZER AND A RE/ ORIENTATION CEPENDENCE OF OVERLAYER ATTENUATION OF ELECTRONS FOR THE CYLIND 3-0693
LEED- AUGER INVESTIGATION OF A STABLE CARBIDE OVERLAYER ON A PLATINUM (111) SURFACE. 5-1569
DITUNGSTEN CARBIDE OVERLAYER ON TUNGSTEN (112). 5-1241
ATOMIC ARRANGEMENT OF GOLD (100) AND RELATED METAL OVERLAYER SURFACE STRUCTURES. 5-1716
AUGER ELECTRON SPECTROSCOPY STUDIES OF CARBON OVERLAYERS ON METAL SURFACES. 5-1376
AND LOW ENERGY ELECTRON DIFFRACTION STUDY OF ALKALI METAL OVERLAYERS ON TUNGSTEN (100). /TRON SPECTROSCOPY 5-1895
AUGER ELECTRON SPECTROSCOPY OF METALS DURING OXIDATION. 6-2007
AUGER ELECTRON SPECTROSCOPY STUDY OF 18-8 STAINLESS STEEL OXIDATION. 6-1993
DY OF THE SURFACE OF IRON AFTER IONIC BOMBARDMENT AND AIR OXIDATION. AUGER SPECTROSCOPIC STU 5-1939
CHANGES IN AUGER SPECTRA OF MAGNESIUM AND IRON DUE TO OXIDATION. 4-1086
DISAPPEARANCE POTENTIAL SPECTROSCOPY (VANADIUM OXIDATION). 6-2068
GALLIUM ARSENIDE (111) FACES STUDIED BY AES AND UPS. OXIDATION AND ANNEALING OF GALLIUM PHOSPHIDE AND 5-1493
AUGER SPECTRA FROM THE (100) SURFACE OF SILICON DURING OXIDATION AND NITRIDATION. 5-1310
(111) SURFACES USING AUGER ELECTRON EMISSI/ STUDY OF THE OXIDATION AND REDUCTION BY WATER VAPOR OF SILICON 5-1612
URFACE BY MEANS OF CCMBINED LOW ENE/ STUDY ON THE INITIAL OXIDATION AND REDUCTION PROCESSES OF IRON (100) S 6-2141
PART-2: EFFECT OF SULFUR ON THE CATALYTIC CARBON MONOXIDE OXIDATION. (AUGER SPECTROSCOPY) /CIENT 5-1190
ONDARY ELECTRON EMISSICN PEAK IN POLYCRYSTALLINE MAGNESI/ OXIDATION EFFECTS ON THE SPECTRUM OF THE SLOW SEC 4-1118
U/ AUGER ELECTRON SPECTRCSCOPY STUDY OF INITIAL STAGES OF OXIDATION IN A COPPER- 19.6 ATOMIC PERCENT ALUMIN 6-2028
AUGER ELECTRON SPECTROSCOPY AND DEPTH PROFILE STUDY OF OXIDATION MODIFIED 440C STEEL. 6-2029
T) PREPARATION AND SURFACE OXIDATION OF A VANADIUM (100) PLANE. (AUGER EFFEC 6-1992
PREPARATION AND SURFACE OXIDATION OF A (110) FACE OF VANADIUM. 5-1683
AUGER SPECTROSCOPY STUDIES OF THE OXIDATION OF AMORPHOUS AND CRYSTALLINE GERMANIUM. 5-1431
NUM (100). (AUGER ELECTRON SPECTROSCOPY) OXIDATION OF CARBON ON TUNGSTEN (100) AND MOLYBDE 5-1944
RON DIFFRACTION- AUGER ELECTRON SPECTROSCOPY STUDY OF THE OXIDATION OF CHROMIUM (110) AND CHROMIUM (100). / 6-2022
E. (AUGER SPECTRA) ADSORPTION CF OXYGEN AND OXIDATION OF COBALT ON THE RUTHENIUM (001) SURFAC 5-1608
Y OF INTERDIFFUSION. (AND AES) OXIDATION OF CUPRONICKEL ALLOYS. PART-1: XPS STUD 5-1224
ANODIC OXIDATION OF GALLIUM ARSENIDE PHOSPHIDE. (AES) 5-1388
ATIVE AUGER ELECTRON SFECTROSCOPY. ELECTRON STIMULATED OXIDATION OF GALLIUM ARSENIDE, STUDIED BY QUANTIT 5-1756
UGER ELECTRON SPECTROSCOPY STUDY OF THE INITIAL STAGES OF OXIDATION OF IRON (001). /ELECTRON DIFFRACTION- A 6-2110
RON DIFFRACTION- AUGER ELECTRON SPECTROSCOPY STUDY OF THE OXIDATION OF IRON (100) AND IRON (110). /GY ELECT 5-1586

RON DIFFRACTION- AUGER ELECTRON SPECTROSCOPY STUDY OF THE OXIDATION OF IRON (110) AND IRON (100). /GY ELECT 5-1585
LEED- AES STUDY OF THE OXIDATION OF IRON- CHROMIUM (100) AND (110). 5-1587
AUGER ELECTRON SPECTROSCOPY AND DEPTH PROFILE STUDY OF OXIDATION OF MODIFIED 440C STEEL. 6-2032
RATES ON (100) AND (110) MOLYBDENUM SURFACES. (/ INITIAL OXIDATION OF MOLYBDENUM. PART-3: OXIDE NUCLEATION 5-1536
ION AUGER PHOTOELECTRON SPECTROMETER. INITIAL OXIDATION OF MOLYBDENUM STUDIED BY A HIGH RESOLUT 5-1698
PECTROSCOPY COMBINED WITH AUGER ELECTRON SP/ STUDY OF THE OXIDATION OF MOLYBDENUM SURFACES BY ENERGY LOSS S 5-1531
C/ CHEMICAL CHANGES IN SECONDARY ELECTRON EMISSION DURING OXIDATION OF NICKEL (100) AND (111) CRYSTAL SURFA 5-1453
ELECTRON SPECTROSCOPIC STUDIES OF THE OXIDATION OF NICKEL- PALLADIUM ALLOYS. 5-1953
N LOW OXYGEN PRESSURES. (AUGER ELECTRON SPECTROSCOPY, SU/ OXIDATION OF NIMONIC-80A ALLOY AT 800 DEGREES C I 6-2144
R AND AUGER ELECTRON SPECTROSCOPY. LOW TEMPERATURE OXIDATION OF SILICON STUDIED BY PHOTOSENSITIVE ES 5-1786
REE PROTECTIVE GASES. (USING/ OBSERVATIONS ON THE SURFACE OXIDATION OF STEEL DURING ANNEALING IN HYDROGEN-F 6-2040
SCO/ VERIFICATION OF THE METAL-RICH SURFACE MODEL FOR THE OXIDATION OF STRONTIUM, BY AUGER ELECTRON SPECTRO 6-2045
EMICAL SHIFTS IN AUGER ELECTRON SPECTROSCOPY FROM INITIAL OXIDATION OF TANTALUM (110). CH 6-2038
D BY AES AND LEED. INITIAL OXIDATION OF THE MOLYBDENUM (110) SURFACE OBSERVE 5-1650
0 DEGREES C. OXIDATION OF TITANIUM BETWEEN 25 DEGREES C AND 40 6-2115
R STUDY OF OXIDE FORMATION AT 650-1100 DEGREES K. OXIDATION OF TUNGSTEN (110). PART-2: LEED- AUGE 6-1994
ACE. CARBON MONOXIDE OXIDATION ON A PLATINUM (110) SINGLE CRYSTAL SURF 5-1187
AUGER ELECTRON SPECTROSCOPY STUDY OF OXIDATION ON LANTHANUM HEXABORIDE. 5-1706
MECHANISMS OF THE CATALYTIC CARBON MONOXIDE OXIDATION ON PLATINUM (110). 5-1188
F WORK FUNCTION AND AUGER ELECTRON SP/ STUDIES OF INITIAL OXIDATION ON SILICON- IRON ALLOY (100) BY MEANS O 5-1924
F TITANIUM. EFFECT OF OXIDATION ON THE HIGH RESOLUTION AUGER SPECTRUM O 5-1158
0 REFS) ELECTRON SPECTROSCOPY STUDIES OF ADSORPTION AND OXIDATION PROCESSES AT METAL SURFACES. (REVIEW, 5 1-0019
FUNCTION MEASUREMENTS AND AUGER ELECTRON SPECTRO/ INITIAL OXIDATION STUDIES BY MEANS OF PHOTOELECTRIC WORK 5-1928
RMANIUM FILMS BY AUGER SPECTROSCCPY. OXIDATION STUDIES OF AMORPHOUS AND CRYSTALLINE GE 6-2023
ELECTRIC WORK FUNCTION MEASUREMENTS COMBINED WIT/ INITIAL OXIDATION STUDIES ON IRON (100) BY MEANS OF PHOTO 5-1927
RING ANALYSIS AND THE SURFACE COMPOSITION OF TITANIUM(IV) OXIDE. AES ION SPUTTE 5-1893
TOELECTRON SPECTROSCOPIC STUDY OF SODIUM METAL AND SODIUM OXIDE. AUGER AND X-RAY PHO 4-0770
AUGER ELECTRON SPECTROSCOPY OF A STABLE GERMANIUM OXIDE. 4-1108
AUGER SPECTRA OF LITHIUM METAL AND LITHIUM OXIDE. 4-0806
(110) DUE TO ADSORPTION OF MOLECULAR OXYGEN AND CARBON MONOXIDE. / THE M4,5NN AUGER SPECTRUM OF MOLYBDENUM 4-0885
CTRUM OF METHANE, HYDROGEN FLUORIDE, WATER, AND CARBON MONOXIDE. /INTERACTION CALCULATIONS OF THE AUGER SPE 4-0879
XYGEN, CARBON DIOXIDE, NITRIC OXIDE, WATER, AND CARBON MONOXIDE. /ATION OF KLL AUGER SPECTRA OF NITROGEN, O 4-0990
RON IRRADIATION EFFECT IN THE AUGER ANALYSIS OF SILICON DIOXIDE. ELECT 4-1096
S OF OXYGEN ADSORBED ON SILICON SURFACES AND CF SILICON DIOXIDE. ELECTRON ORBITAL ENERGIE 5-1475
(AUGER SPECTRA OF SILICON, SILICCN CARBIDE AND SILICON DIOXIDE) /E CRYSTAL UNDER A LOW ENERGY BOMBARDMENT. 4-0739
AL AND THEORETICAL STUDY OF THE AUGER SPECTRUM OF NITROUS OXIDE. EXPERIMENT 4-0814
LEED AND AES OBSERVATIONS ON THE POLAR SURFACES OF ZINC OXIDE. 5-1619
GY AUGER AND LOSS ELECTRON SPECTRA FROM MAGNESIUM AND ITS OXIDE. LOW ENER 4-0895
LOW ENERGY ELECTRON DIFFRACTION FROM CLEAN (100) NICKEL OXIDE. 5-1752
ON SPECTROSCOPY STUDY CF IRON FILM DEPOSITED ON MAGNESIUM OXIDE. /ENERGY ELECTRON DIFFRACTION- AUGER ELECTR 5-1522
CTROSCOPY OF PLATINUM (100) AND PLATINUM (100)- CARBON MONOXIDE. LOW ENERGY ELECTRON SPE 5-1686
ROLE OF IMPURITIES IN THE STABILITY OF ZINC OXIDE. 5-1462
SCANDIUM L X-RAY EMISSION FROM PURE METAL AND SCANDIUM OXIDE. 4-1102
FACE STRUCTURE, AND INTERACTION WITH OXYGEN AND CARBON MONOXIDE. /RYSTAL FACE. A STUDY OF ITS CLEANING, SUR 5-1883
N SPECTRA OF MAGNESIUM, SOME ALLCYS OF MAGNESIUM, AND ITS OXIDE. X-RAY EXCITED AUGER AND PHOTOELECTRO 4-0837
HIGH RESOLUTION AUGER ELECTRON SPECTROSCOPY OF NITRIC OXIDE ADSORBED ON A TUNGSTEN SURFACE. 5-1565
UGER SPECTROSCOPY) CARBON MONOXIDE ADSORPTION KINETICS ON MOLYBDENUM (100). (A 5-1579
IMPACT INDUCED PHOTOELECTRON AND AUGER SPECTRA. CARBON MONOXIDE ADSORPTION ON MOLYBDENUM. /ED AND ELECTRON 4-0768
CARBON MONOXIDE ADSORPTION ON NICKEL (110). (AUGER EFFECT) 5-1884
/TRA OF CARBON. TANTALUM CARBIDES, ETHYLENE AND CARBON MONOXIDE ADSORPTION ON TANTALUM AND POLYCRYSTALLINE/ 4-0820
RAHIGH VACUUM. (AES) INTERACTION OF SULFUR DIOXIDE AND CARBON DIOXIDE WITH CLEAN SILVER IN ULT 5-1575
/ITROGEN, OXYGEN, CARBCN MONOXIDE, NITRIC OXIDE, CARBON DIOXIDE, AND CARBON TETRAFLUORIDE. (KLL AUGER PROC/ 2-0308
ESCA (AND AUGER) STUDIES OF COPPER, COPPER(I) OXIDE, AND COPPER(II) OXIDE. 5-1794
/M INDUCED REDUCTION OF CARBON CONCENTRATION ON BERYLLIUM OXIDE AND ITS EFFECTS ON SECONDARY ELECTRON EMIS/ 5-1362
HARACTERISTICS OF MAGNESIUM SINGLE CRYSTALS AND MAGNESIUM OXIDE AND MAGNESIUM NITRIDE SURFACE FILMS. /GER C 5-1220
/DY OF THE COADSORPTION OF MOLECULAR OXYGEN AND CARBCN MONOXIDE, AND MOLECULAR OXYGEN AND CARBON DIOXIDE O/ 5-1465
/RPTION UTILIZING AUGER ELECTRON SPECTROSCOPY. CARBON MONOXIDE AND MOLECULAR OXYGEN ON SILICON. PART-1: A/ 5-1543
ECTROSCOPY. COMPARATIVE STUDY OF SINGLE CRYSTAL MAGNESIUM OXIDE AND OXIDIZED MAGNESIUM BY AUGER ELECTRON SP 5-1495
STUDIES UTILIZING AUGER ELECTRON SPECTROSCOPY: CARBON MONOXIDE AND OXYGEN ON SILICON. / ON GAS ADSORPTION 5-1542
(CHEMISORPTION AND CATALYTIC RE/ INTERACTION OF CARBON MONOXIDE AND OXYGEN WITH A PALLADIUM (100) SURFACE. 5-1321
GER SPECTROSCOPY) INTERACTIONS CF CARBON MONOXIDE AND OXYGEN WITH IRIDIUM (110) SURFACES. (AU 5-1252
STUDY OF SILICON SPECTRA FROM SILICON MONOXIDE, SILICON DIOXIDE, AND SILICON NITRIDE. /ECTRON SPECTROSCOPY 4-1081
ESCA STUDIES OF SILVER, DISILVER OXIDE AND SILVER OXIDE. 4-1055
SULFUR KLL AUGER SPECTRA IN HYDROGEN SULFIDE, SULFUR DIOXIDE, AND SULFUR HEXAFLUORIDE. 4-0910
AND EFFECT OF THESE FILMS ON ADH/ ABSORPTION OF ETHYLENE OXIDE AND VINYL CHLORIDE ON AN IRON (011) SURFACE 5-1207
LEED AND SURFACE POTENTIAL STUDY OF CARBON MONOXIDE AND XENON ADSORBED ON COPPER (100). 5-1246
CED BY INCIDENT ELECTRCN BEAM / CHANGES IN THE TITANIUM DIOXIDE AUGER AND APPEARANCE POTENTIAL SPECTRA INDU 5-1834
EXCEMISSION FROPERTIES OF ZIRCONIUM DIOXIDE. (AUGER EFFECT) 2-0410
F THE ATTENUATION OF THE LOW ENERGY ELECTRONS IN ALUMINUM OXIDE. (AUGER ELECTRON LINES) /ON DETERMINATION O 4-0773
GRAIN BCUNDARY SEGREGATION IN ALUMINUM OXIDE. (AUGER SPECTROSCOPY) 6-2134
IN CHANNEL ELECTRON MULTIPLIERS. (SURFACE OXYGEN ON LEAD OXIDE BY AES) GAIN FATIGUE MECHANISM 3-0681
ON OF HYDROGEN, DEUTERIUM, NITROGEN, OXYGEN AND CARBON MONOXIDE BY (100) TANTALUM. (AUGER EFFECT) /D SOLUTI 5-1549
/OIONIZATION OF NITRCGEN, OXYGEN, CARBON MONCXIDE, NITRIC OXIDE, CARBON DIOXIDE, AND CARBON TETRAFLUORIDE./ 2-0308
SCOPY. COMPOSITION CF THE SURFACE OF COPPER- MAGNESIUM OXIDE CATALYSTS STUDIED BY AUGER ELECTRON SPECTRO 5-1361
SECONDARY ELECTRON EMISSION FROM THIN FILM GOLD METAL- OXIDE CERMETS. 4-1015
AY ANALYSIS TECHNIQUES FOR EXAMINATION OF SILVER- CADMIUM OXIDE CONTACT MATERIALS. /Y AND NONDISPERSIVE X-R 5-1799
/ED PHOTOELECTRON ANC AUGER SPECTRA OF COPPER, COPPER(II) OXIDE, COPPER(I) OXIDE, AND COPPER(I) SULFIDE TH/ 4-0932
PHYSICAL STRUCTURE./ PROPERTIES OF THE LEAD SULFIDE- LEAD OXIDE CRYSTALLINE FILMS. PART-1: PREPARATION AND 5-1790
SHIFTS IN AUGER SPECTRUM FRCM SURFACE LAYERS OF SILICON DIOXIDE DURING ELECTRON BOMBARDMENT. / OF CHEMICAL 4-1044
CHLORINE IN HYDROGEN CHLORIDE OR CHLORINE GROWN SILICON DIOXIDE FILMS. AUGER ANALYSIS OF 5-1251
SOME RECENT STUDIES OF VERY THIN SILICON DIOXIDE FILMS. 5-1935
ECTRON SPECTROSCOPY TO DETERMINE THE STRUCTURE OF SILICON OXIDE FILMS. USE OF AUGER EL 5-1501
MICAL CLEANING FOR THERMOCOMPRESSION BONDING. (REMOVAL OF OXIDE FILMS, AUGER SPECTROSCOPY OF SURFACE) CHE 5-1454
PECTROSCOPY. OXIDE FILMS: ELEMENT PROFILES BY AUGER ELECTRON S 4-1073
ND X-RAY PHOTOELECTRON SPECTROSCCPI/ ANALYSIS PROFILES OF OXIDE FILMS ON CHROMIUM STEEL BY AUGER EMISSION A 6-2016
AES ANALYSIS OF OXIDE FILMS ON IRON. 5-1800
AUGER ANALYSIS OF THIN OXIDE FILMS ON LEAD- INDIUM ALLOYS. 5-1249
IDATION OF TUNGSTEN (110). PART-2: LEED- AUGER STUDY OF OXIDE FORMATION AT 650-1100 DEGREES K. OX 6-1994
CTROSCOPY. INVESTIGATIONS OF THE OXIDE FORMATION ON METALS WITH AUGER ELECTRON SPE 5-1329
TRON SPECTRA IN X-RAY FHOTOELECTRON STUDIES OF ZINC, ZINC OXIDE, GALLIUM, AND GALLIUM OXIDE. /D DIRECT ELEC 4-1054

AUGER AND ELLIPSOMETRIC STUDIES OF ULTRATHIN LEAD(II) OXIDE GROWTH ON LEAD. 5-1248
AL. STRUCTURE AND SECONDARY EMISSION OF MAGNESIUM OXIDE LAYERS APPLIED ON A MOLYBDENUM SINGLE CRYST 5-1191
G/ THERMALLY STIMULATED EXOELECTRON EMISSION OF BERYLLIUM OXIDE LAYERS. AUGER ELECTRON SPECTROSCOPY INVESTI 5-1325
/N A PLATINUM (110) SURFACE. PART-1: SULFUR AND CARBON MONOXIDE LEED PATTERNS. (AUGER ELECTRON SPECTROSCOP/ 5-1190
,3L2,3 RADIATIVE AUGER SPECTRUM OF INSULATORS. (MAGNESIUM OXIDE, MAGNESIUM FLUORIDE) /ON EFFECTS IN THE KL2 2-0540
PHOSPHORUS CONCENTRATION PROFILES IN P-DOPED SILICON DIOXIDE MEASURED USING AUGER SPECTROSCOPY. 5-1231
/THE X-RAY PHOTOIONIZATION OF NITROGEN, OXYGEN, CARBON MONOXIDE, NITRIC OXIDE, CARBON DIOXIDE, AND CARBON / 2-0308
/ATION OF KLL AUGER SPECTRA OF NITROGEN, OXYGEN, CARBON DIOXIDE, NITRIC OXIDE, WATER, AND CARBON MONOXIDE. 4-0990
OF CHEMICAL SHIFTS IN SOME VANADIUM COMPOUNDS. (VANADIUM OXIDE, NITRIDE, CARBIDE, SILICIDE) /NVESTIGATIONS 4-1090
NUM SURFACES. (/ INITIAL OXIDATION OF MOLYBDENUM. PART-3: OXIDE NUCLEATION RATES ON (100) AND (110) MOLYBDE 5-1536
(AUGER) ADSORPTION STUDIES OF OXYGEN AND CARBON MONOXIDE ON A PLATINUM (100) SURFACE. 5-1546
EN AND CARBON MONOXIDE, AND MOLECULAR OXYGEN AND CARBON DIOXIDE ON A TUNGSTEN (111) SURFACE. /OLECULAR OXYG 5-1465
SPECTROSCOPY) DECOMPOSITION OF CARBON MONOXIDE ON A (110) NICKEL SURFACE. (AUGER ELECTRON 5-1606
DEPOSITION AND AUGER ANALYSIS OF DEPOSITED SILICON DIOXIDE ON ALUMINUM(X) GALLIUM(1-X) ARSENIDE. 5-1962
UDIES. ADSORPTION OF CARBON MONOXIDE ON COPPER (001). LEED AND AUGER EMISSION ST 5-1515
TITATIVE AUGER ELECTRON SPECTROSCOPY. (STUDY OF CARBON MONOXIDE ON MOLYBDENUM (110)) / SUBTRACTION FOR QUAN 1-0069
ATIVE A/ AUGER ELECTRON SPECTROSCOPY STUDIES OF CARBON MONOXIDE ON NICKEL. SPECTRAL LINE SHAPES AND QUANTIT 5-1458
/TURAL INFLUENCES ON ADSORPTION ENERGY. PART-2: CARBON MONOXIDE ON NICKEL (100). PART-3: CARBON MONOXIDE O/ 5-1911
ADSORPTION AND DECOMPOSITION OF CARBON MONOXIDE ON PLATINUM (111). (AUGER SPECTROMETRY) 5-1622
N ELECTRON INDUCED DESORPTION AND DISSOCIATION: CARBON MONOXIDE ON PLATINUM (111). (AUGER SPECTRUM) /RONS I 5-1567
ASS SPECTROMETRIC STUDY OF THE CHEMISORPTION OF CARBON MONOXIDE ON POLYCRYSTALLINE MOLYBDENUM. /ETRIC AND M 5-1355
ADSORPTION OF CARBON MONOXIDE ON SILVER- PALLADIUM ALLOYS. 5-1254
ADSORPTION OF NITRIC OXIDE ON TUNGSTEN. 5-1929
D SPECIES USING AUGER PEAK SHAPES. ETHYLENE AND CARBON MONOXIDE ON TUNGSTEN (100). /ACTERIZATION OF ADSORBE 5-1243
ON- AUGER ELECTRON SPECTROSCOPY. DISCREPANCY FOR CARBON DIOXIDE ON TUNGSTEN (100). FLASH DESORPTI 5-1461
CTRA FROM ADSORBED OXYGEN ON TUNGSTEN (110) AND CARBON MONOXIDE ON TUNGSTEN (100). (BY LEED AND AES) /N SPE 5-1152
INFLUENCES ON ADSORBATE BINDING ENERGY. PART-1: CARBON MONOXIDE ON (100) PALLADIUM. STRUCTURAL 5-1913
L SURFACE. CARBON MONOXIDE OXIDATION ON A PLATINUM (110) SINGLE CRYSTA 5-1187
MECHANISMS OF THE CATALYTIC CARBON MONOXIDE OXIDATION ON PLATINUM (110). 5-1188
SPECTROSCOPY) EXOELECTRON EMISSION FROM ZINC OXIDE. (OXYGEN ADSORPTION EFFECT, AUGER ELECTRON 4-0921
SPECTROSCOPY. INVESTIGATION OF SILVER CESIUM OXIDE PHOTOEMISSIVE SURFACES USING AUGER ELECTRON 5-1782
D TUNGSTEN (100), AND/ ELECTRON BEAM EFFECTS ON CARBON MONOXIDE SATURATED TUNGSTEN (100), ETHYLENE SATURATE 5-1244
ON, AND IONIZATION LOSS EVENTS IN BERYLLIUM AND BERYLLIUM OXIDE SECONDARY ELECTRON SPECTRA. /OUBLE IONIZATI 4-0901
RON SPECTROSCOPY STUDY OF SILICON SPECTRA FROM SILICON MONOXIDE, SILICON DIOXIDE, AND SILICON NITRIDE. /ECT 4-1081
SECONDARY ELECTRON EMISSION OF SILICON DIOXIDE SINGLE CRYSTALS. 4-1062
/OF GOLD COATING ON SORPTION CHARACTERISTICS OF CARBON MONOXIDE, SULFUR, AND OTHER GASES ON STAINLESS STEE/ 5-1551
(AUGER) ELECTRON SPECTROSCOPY OF TRANSITION METAL OXIDE SURFACES. 5-1339
AUGER ELECTRON STUDIES OF URANIUM DIOXIDE SURFACES. 5-1218
LATION OF ELECTRONIC, LEED, AND AUGER DIAGNOSTICS ON ZINC OXIDE SURFACES. CORRE 5-1584
CONDARY ELECTRON ENERGY DISTRIBUTION STUDIES OF URANIUM DIOXIDE SURFACES. SE 5-1316
STABILITY OF CESIUM COVERED ZINC OXIDE SURFACES. 5-1882
ON, PHOTODESORPTION AND CONDUCTIVITY MEASUREMENTS ON ZINC OXIDE SURFACES. (AUGER ANALYSIS) CHEMISORPTI 5-1801
INTERACTION OF CESIUM WITH CLEAN ZINC OXIDE SURFACES. (AUGER SPECTROSCOPY) 5-1588
ER. DETECTION OF SODIUM ON SILICON AND SILICON DIOXIDE SURFACES BY CYLINDRICAL MIRROR AUGER ANALYZ 3-0647
SOME PROPERTIES OF CLEAN AND ALKALI METAL COVERED ZINC OXIDE SURFACES. (LEED, AES) 5-1881
Y AP/ QUANTITATIVE COMPARISON OF TITANIUM AND TITANIUM MONOXIDE SURFACES USING AUGER ELECTRON AND SOFT X-RA 5-1384
ECTRON BOMBARDMENT. POLYCRYSTALLINE MOLYBDENUM- CARBON MONOXIDE SYSTEM. (AUGER SPECTROSCOPY) /RODUCED BY EL 5-1648
NATION AUGER SPECT/ QUANTITATIVE DETERMINATION OF SURFACE OXIDE THICKNESS ON DEPOSITED METAL FILMS BY COMBI 6-2048
GROWTH AND STRUCTURE OF TITANIUM OXIDE THIN FILMS. PART-1. (AUGER SPECTRA) 5-1980
EEL, AND N/ AUGER ELECTRON STUDIES OF SURFACES: URANIUM DIOXIDE, URANIUM, GRAPHITE, 300 SERIES STAINLESS ST 5-1217
AUGER SPECTRA OF NITROGEN, OXYGEN, CARBON DIOXIDE, NITRIC OXIDE, WATER, AND CARBON MONOXIDE. /ATION OF KLL 4-0990
INTERACTION OF NITRIC OXIDE WITH A NICKEL (111) SURFACE. 5-1278
LECTRON SPECTROSCOPY AND DEPTH PROF/ REACTION OF SULFUR DIOXIDE WITH MODIFIED 440C STEEL STUDIED BY AUGER E 6-2031
INTERACTION OF SULFUR DIOXIDE WITH TUNGSTEN. (AUGER ANALYSIS) 5-1367
INTERACTION OF CARBON MONOXIDE WITH (110) NICKEL SURFACES. (AUGER EFFECT) 5-1607
K FUNCTION STUDY. INTERACTION OF CHLORINE WITH THE ZINC OXIDE (0001)- ZINC SURFACE: A LEED, AUGER AND WOR 5-1463
S) AND LOW ENERGY ELECTRON DIFFRACTION (LEED) OF THE ZINC OXIDE (0001) POLAR SURFACES. /ON SPECTROSCOPY (AE 5-1338
Y, AND X-RAY PHOTOELECTRON SPECTROSCOPY STUDY OF THE ZINC OXIDE (0001) POLAR SURFACES. /LECTRON SPECTROSCO 5-1337
T ON THE GROWTH MODE OF IRON EPITAXIAL FILMS ON MAGNESIUM OXIDE (001). /TUDY OF SURFACE CONTAMINATION EFFEC 5-1524
SODIUM AND MEASUREMENT OF SURFACE DIFFUSION ON MAGNESIUM OXIDE (001) BY AUGER SPECTROSCOPY. /DESORPTION OF 5-1494
N OF GROWTH MECHANISM OF DEPOSITED IRON FILM ON MAGNESIUM OXIDE (001) SURFACE. LEED- AES OBSERVATIO 5-1525
GROWTH MODE OF IRON FILM DEPOSITED ON MAGNESIUM OXIDE (001) SURFACE. (AUGER SPECTROSCOPY) 5-1528
/ STUDY OF THE EPITAXIAL GROWTH OF MAGNESIUM ON MAGNESIUM OXIDE (001) USING REFLECTION DIFFRACTION, LEED AN 5-1496
AUGER EMISSION SPECTROSCOPY VANADIUM PENTOXIDE (010) AND VANADIUM (100) SURFACES. 5-1941
RE IN THE AUGER ELECTRON EMISSION SPECTRA OF VANADIUM PENTOXIDE (010) AND VANADIUM (100) SURFACES. /STRUCTU 5-1340
HIGH RESOLUTION AUGER ELECTRON SPECTRUM OF MAGNESIUM OXIDE (100). 4-0772
LEED AND ELECTRON SPECTROSCOPIC OBSERVATIONS ON NICKEL MONOXIDE (100). 5-1687
DEPENDENCE OF THE LOW ENERGY AUGER SPECTRUM OF NICKEL(II) OXIDE (100). TEMPERATURE 4-0999
ON DIFFRACTION AND AUGER E/ INVESTIGATION OF EUROPIUM(II) OXIDE (100) CLEAVAGE SURFACE BY LOW ENERGY ELECTR 5-1157
A LEED STUDY OF MAGNESIUM OXIDE (100). PART-1: EXPERIMENT. 5-1580
TANIUM LMV AUGER SPECTRA OF TITANIUM(II) AND TITANIUM(IV) OXIDES. MOLECULAR ORBITAL EFFECTS ON THE TI 4-1069
/TROSCOPY OF THE SURFACE CHEMICAL COMPOSITION OF VANADIUM OXIDES, AND OXIDIZED VANADIUM: CHEMICAL SHIFT AN/ 4-1089
TS ON HEAVILY DOPED SIL/ CORRELATIONS BETWEEN INTERFACIAL OXIDES AND RESISTANCE CHANGES FOR TUNGSTEN CONTAC 5-1861
PHASE SEPARATION IN SILICON OXIDES AS SEEN BY AUGER ELECTRON SPECTROSCOPY. 5-1502
BOUNDARY IMPURITIES AND FLUORIDE ADDITIVES IN HOT PRESSED OXIDES BY AUGER ELECTRON SPECTROSCOPY. /OF GRAIN 5-1506
IGH YIELD SECONDARY/ DIFFERENTIAL SPUTTERING OF MAGNESIUM OXIDE- GOLD CERMET FILMS AND ITS APPLICATION TO H 5-1430
F CESIATION ON SECONDARY ELECTRON EMISSION FROM MAGNESIUM OXIDE- GOLD CERMETS. EFFECTS O 2-0384
A1 KLL AUGER PROFILE ARTIFACT AT THE OXIDE- METAL INTERFACE. 4-1068
ILE AND ELECTRICAL CONDUCTIVITY OF THE THERMAL AND ANODIC OXIDES OF INDIUM ANTIMONIDE. (AES) /SITIONAL PROF 5-1971
EAVILY DOPED SILICON- GERMANIUM A/ EFFECTS OF INTERFACIAL OXIDES ON THE CONTACT RESISTANCE OF TUNGSTEN ON H 5-1456
AUGER STUDIES OF DIFFUSED IMPURITY PROFILES IN SILICON DIOXIDE- SILICON AND POLYSILICON- SILICON DIOXIDE-/ 5-1481
/D INTERDIFFUSION EFFECTS IN SILICON- TANTALUM, SILICON DIOXIDE- TANTALUM, AND SILICON- PLATINUM SILICIDE-/ 5-1260
INTERPRETATION OF THE AUGER ELECTRON SPECTRUM FROM OXIDIZED BERYLLIUM. 4-0935
INTERPRETATION OF THE AUGER SPECTRUM OF CLEAN AND OXIDIZED BERYLLIUM. 4-1087
CHEMICAL SHIFTS IN THE AUGER SPECTRA FROM OXIDIZED CHROMIUM AND VANADIUM. 4-0809
UGER AND ELLIPSOMETRIC STUDY OF PHOSPHORUS SEGREGATION IN OXIDIZED DEGENERATE SILICON. A 5-1250
Y AUGER ELECTRON SPECTROSCOPY. STUDIES ON SURFACE PLASMA OXIDIZED FILMS OF III-V COMPOUND SEMICONDUCTORS B 5-1699
N. AES CHARACTERIZATION OF OXIDIZED FILMS OF MAGNESIUM, ALUMINUM, AND SILICO 5-1847
ELECTRON SPECTROSCOPIC STUDIES OF CLEAN AND OXIDIZED IRON. (AUGER STUDY OF SURFACE) 5-1323
. COMPARATIVE STUDY OF SINGLE CRYSTAL MAGNESIUM OXIDE AND OXIDIZED MAGNESIUM BY AUGER ELECTRON SPECTROSCOPY 5-1495

/RAY EXCITED AUGER AND PHOTOELECTRON SPECTRA CF PARTIALLY OXIDIZED MAGNESIUM SURFACES. ABNORMAL CHEMICAL / 5-1949
                    INTERFACIAL AUGER TRANSITIONS IN OXIDIZED SODIUM AND MAGNESIUM.   5-1497
PROFILES OBTAINED WITH AUGER ELECTRON SPECTRCSCOPY OF AIR OXIDIZED STAINLESS STEEL SURFACES. /VERSUS DEPTH 6-1998
X-RAY EXCITED PHOTOELECTRON AND AUGER ELECTRCN SPECTRA CF OXIDIZED STATES ON SOME METAL SURFACES.   5-1690
AND SECONDARY ION/ SIMULTANEOUS CBSERVATICNS OF PARTIALLY CXIDIZED SURFACES BY AUGER ELECTRON SPECTROSCOPY 5-1553
UGER SPECTROSCOPY TC DETERMINE THE SURFACE CCMPOSITION OF OXIDIZED VANADIUM. / OF X-RAY PHOTOELECTRON AND A 5-1205
/THE SURFACE CHEMICAL COMPOSITION OF VANADIUM OXIDES, AND OXIDIZED VANADIUM: CHEMICAL SHIFT AND PEAK INTEN/ 4-1089
UGER SPECTROSCOPY TC DETERMINE THE SURFACE CCMPOSITION OF OXIDIZED VANADIUM. (COMMENTS) /HOTOELECTRON AND A 5-1863
. PART-2: AUGER AND BACKSCATTERING ANALYSES.   EFFECT OF OXIDIZING AMBIENTS ON PLATINUM SILICIDE FORMATION 5-1182
          CHEMICAL EFFECTS CN KLL AUGER ELECTRON SPECTRUM FROM OXYGEN.   4-0819
ECTRA OF TUNGSTEN AND MOLYBDENUM DURING THE ADSORPTION OF OXYGEN.   CHEMICAL SHIFT IN AUGER SP 4-0816
INITY ON SILICON PRODUCED BY THE ADSORPTION OF CESIUM AND OXYGEN. /PLASMON STUDIES OF NEGATIVE ELECTRON AFF 5-1363
EGATIVE ELECTRON AFFINITY BY THE ADSORPTION OF CESIUM AND OXYGEN. / GALLIUM ARSENIDE DURING ACTIVATION TO N 5-1364
LECTRON SPECTROSCOPY. (PALLADIUM- NICKEL ALLOYS, ADSORBED OXYGEN) /ALL RETARDING FIELD ANALYZER FOR AUGER E 3-0644
                      LEED AND AES STUDY OF OXYGEN ABSORPTION ON TUNGSTEN (110).   5-1669
SURFACES. (AUGER SPECTROSCOPY)   ELECTRONIC STATES OF OXYGEN ADSORBED ON CLEAN SILICON (111) AND (100)   5-1476
N DIOXIDE.   ELECTRON ORBITAL ENERGIES OF OXYGEN ADSORBED ON SILICON SURFACES AND OF SILICO   5-1475
       CHEMICAL EFFECTS IN THE AUGER SPECTRA OF IRON DUE TO OXYGEN ADSORPTION.   4-0727
     CHEMICAL SHIFT IN THE AUGER SPECTRUM OF TUNGSTEN FROM OXYGEN ADSORPTION.   4-0922
       CHEMICAL SHIFT IN THE AUGER SPECTRUM OF IRON DURING OXYGEN ADSORPTION.   5-1792
     CHEMICAL SHIFTS IN THE AUGER SPECTRUM CF YTTRIUM ON OXYGEN ADSORPTION.   5-1153
COPY)   EXOELECTRON EMISSION FROM ZINC OXIDE. (OXYGEN ADSORPTION EFFECT, AUGER ELECTRON SPECTROS   4-0921
UGER ELECTRON SPECTROSCOPY, STICKING COEFFIC/ KINETICS OF OXYGEN ADSORPTION ON A PLATINUM (111) SURFACE. (A 5-1189
CES.   OXYGEN ADSORPTION ON CLEAN MOLYBDENUM (100) SURFA   5-1772
       COMBINED LEED- RHEED AUGER STUDY OF OXYGEN ADSORPTION ON THE TUNGSTEN (112) SURFACE.   5-1464
F TUNGSTEN.   THE KINETICS OF OXYGEN ADSORPTION ON THE (112) AND (110) PLANES O   5-1912
S BY AUGER ELECTRON SPECTROSCOPY.   INVESTIGATION OF OXYGEN ADSORPTION ON TUNGSTEN AT HIGH TEMPERATURE   5-1156
        AUGER ELECTRON SPECTROSCCPY STUDY OF OXYGEN ADSORPTION ON TUNGSTEN (110).   5-1668
                                       OXYGEN ADSORPTION ON (110) SILVER.   5-1422
HIGH VACUUM GROWN AMORPHOUS GERMANIUM. (DEPTH PROFILES CF OXYGEN AND CAREON BY AUGER SPECTROSCOPY) /N ULTRA 5-1547
ECTRUM OF MOLYBDENUM (110) DUE TC ADSORPTION CF MOLECULAR OXYGEN AND CARBON MONOXIDE. / THE M4,5NN AUGER SP 4-0885
    OF ITS CLEANING, SURFACE STRUCTURE, AND INTERACTION WITH OXYGEN AND CARBON MONOXIDE. /RYSTAL FACE. A STUDY 5-1883
/GER SPECTROSCOPIC STUDY OF THE COADSORPTION OF MOLECULAR OXYGEN AND CARBON MONOXIDE, AND MOLECULAR OXYGEN/ 5-1465
/DSORPTION AND SOLUTION OF HYDROGEN, DEUTERIUM, NITROGEN, OXYGEN AND CARBON MONOXIDE BY (100) TANTALUM. (A/ 5-1549
RFACE.   (AUGER) ADSORPTION STUDIES OF OXYGEN AND CARBON MONOXIDE ON A PLATINUM (100) SU   5-1546
       AUGER SPECTRA OF HIGHLY ICNIZED OXYGEN AND FLUORINE.   4-0906
CROSS SECTIONS FROM 1.5 MEV PER AMU AND ATOMIC HYDROGEN, OXYGEN, AND FLUORINE.   NEON K-AUGER   2-0571
/ LIMITS OF NITROGEN IN GALLIUM FHOSPHIDE (AND) NITROGEN, OXYGEN, AND FLUORINE IN SILICON BY SECONDARY ION/ 5-1916
NCE OF AN ELECTRON BEAM WITH THE SURFACE REACTION BETWEEN OXYGEN AND GERMANIUM. (AES)   INTERFERE 5-1620
PLATINUM- TIN IN ULTRAHIGH VACUUM AND IN THE PRESENCE OF OXYGEN AND HYDROGEN. / THE SURFACE COMPOSITION OF 5-1196
/GER AND REFLECTION ELECTRON DIFFRACTION INVESTIGATION OF OXYGEN AND HYDROGEN ADSORPTION ON STRONTIUM SURF/ 5-1498
ADHESION OF CLEAN IRON. (USING LEED A/ EFFECT OF SULFUR, OXYGEN, AND HYDROGEN SULFIDE SURFACE FILMS ON THE 5-1211
ENT OF LOW ENERGY ELECTRONS BY CLOUD CHAMBER TECHNIQUES: OXYGEN AND NEON AUGER ELECTRON PROBLEM. /MEASUREM 2-0566
PACT IONIZATION CROSS SECTIONS OF ATOMIC NITROGEN, ATOMIC OXYGEN, AND NEON. (AUGER TRANSITIONS) /LECTRON IM 2-0363
E. PART-2: REACTION KINETICS. (AUGER EFFE/ INTERACTION OF OXYGEN AND NITROGEN WITH THE NIOBIUM (100) SURFAC 5-1333
-% ZIRCONIUM.   SURFACE SEGREGATION AND INTERACTION OF OXYGEN AND NITROGEN WITH ZIRCONIUM IN NIOBIUM- 1 5-1508
                  AUGER SPECTROSCOPY OF OXYGEN AND OTHER "CONTAMINANTS" ON SILVER.   5-1757
001) SURFACE. (AUGER SPECTRA)   ADSORPTION OF OXYGEN AND OXIDATION OF COBALT ON THE RUTHENIUM (5-1608
S AND THE EFFECT OF SLIDING ON THESE I/ AUGER ANALYSIS OF OXYGEN AND SULFUR INTERACTIONS WITH VARIOUS METAL 5-1206
DIED BY LEED AND AES)   KINETICS OF THE REACTION BETWEEN OXYGEN AND SULFUR ON A NICKEL (111) SURFACE. (STU 5-1451
. AUGER ELECTRON SPECTRUM DURING THE INTERACTION BETWEEN OXYGEN AND THE (111) SURFACE OF GALLIUM PHOSPHIDE 5-1642
/F THE AUGER ELECTRON SPECTRUM IN THE INTERACTION BETWEEN OXYGEN AND THE (111) SURFACE OF GALLIUM PHOSPHID/ 5-1643
FILM. (AUGER EFF/ BEHAVIOR OF IMPURITY ATOMS AND ADSORBED OXYGEN ATOMS ON THE (001) FACE OF IRON EPITAXIAL 5-1523
          SURFACE COMPOSITION CF PLATINUM HEATED IN OXYGEN. (AUGER SPECTRA)   5-1737
                              OXYGEN BEAM GAS AUGER ELECTRON EMISSION.   2-0400
          X-RAY PRODUCTION BY HEAVY IONS. (OXYGEN BOMBARDMENT, AUGER SPECTRUM)   2-0479
CARBON M/ DETERMINATION CF KLL AUGER SPECTRA OF NITROGEN, OXYGEN, CARBON DIOXIDE, NITRIC OXIDE, WATER, AND 4-0990
/ IONS FORMED FROM THE X-RAY PHOTOIONIZATION OF NITROGEN, OXYGEN, CARBON MONOXIDE, NITRIC OXIDE, CARBON DI/ 2-0308
          AUGER EJECTION OF ELECTRONS FROM TUNGSTEN BY OXYGEN CHEMISORPTION.   5-1599
                           CESIUM- OXYGEN COADSORPTION ON (100) TUNGSTEN.   5-1296
ON SILICON (111).   EFFECTS OF PREADSORBED OXYGEN CONTAMINANT ON THE ADSORPTION OF ALUMINUM   5-1169
X-RAY, AES ANALYSIS / CCRRELATION OF FRACTIONAL MONOLAYER OXYGEN DETERMINATIONS OBTAINED BY PROTON EXCITED 5-1682
OF THE SPECTRUM. (INTERBAND AUGER/ EXCITATICN SPECTRUM OF OXYGEN DOMINATED COMPOUNDS IN THE 3-21 EV REGION 4-1048
IUM STUDIED BY AUGER ELECTRON SPECTRCSCOPY.   OXYGEN EXPOSURE OF SAMARIUM, GADOLINIUM, AND TERB   5-1332
/IGHT SPECTRA FROM AUTOIONIZING STATES OF HIGHLY STRIPPED OXYGEN, FLUORINE, CHLORINE, AND ARGON IONS. (AUG/ 2-0472
ION)   STUDY OF SULFUR EMBRITTLED OXYGEN FREE COPPER. (AES, GRAIN BOUNDARY SEGREGAT   6-2015
CTROSCOPY STUDY OF ELECTRON BEAM STIMULATED DESORPTION OF OXYGEN FROM A SEMICONDUCTOR SURFACE. /LECTRON SPE 5-1617
ECTROSCOPY STUDY OF THE ELECTRON STIMULATED DESORPTION OF OXYGEN FROM A TUNGSTEN (100) SURFACE. /LECTRON SP 5-1954
NNEALING. (AUGER SPECTROSCOPY)   TRANSPORT OF OXYGEN IN AMORPHOUS GERMANIUM THIN FILMS DURING A   5-1548
          SURFACE SEGREGATION OF OXYGEN IN NIOBIUM. (AES)   5-1443
LOYS. (AUGER ELECTRON SPECTROSCOP/ SURFACE SEGREGATION OF OXYGEN IN NIOBIUM- OXYGEN AND TANTALUM- OXYGEN AL 6-2065
/TATIVE ANALYSIS OF LIGHT ELEMENTS (NITROGEN, CARBON, AND OXYGEN) IN SPUTTERED TANTALUM FILMS BY AUGER ELE/ 5-1652
COPY.   SILICON- OXYGEN INTERACTIONS USING AUGER ELECTRON SPECTROS   5-1511
ON, / HIGH RESOLUTION BEAM GAS AUGER ELECTRON SPECTRA OF OXYGEN IONS EXCITED BY COLLISIONS WITH HELIUM, NE 4-0904
     OUTER SHELL EXCITATION OF THE RARE GASES BY 30 MEV OXYGEN IONS. (INNER SHELL AUGER TRANSITIONS)   2-0298
     ORIGIN OF THE PEAK SHAPES IN THE AUGER SPECTRUM OF OXYGEN (KL2,3L2,3) ON MOLYBDENUM.   4-1004
TION STUDY OF THE ACTIVATION OF GALLIUM ARSENIDE- CESIUM- OXYGEN NEGATIVE ELECTRON AFFINITY SURFACES. /FRAC 5-1845
N SPECTROSCOPY)   ADSORPTION OF OXYGEN ON A CLEAN SILICON SURFACE. (AUGER ELECTRO   5-1638
/DIES OF THE MECHANISM OF THE REACTION BETWEEN SULFUR AND OXYGEN ON A MOLYBDENUM SURFACE BY AUGER ELECTRON/ 5-1530
/TIGATION OF THE MECHANISM OF REACTION BETWEEN SULFUR AND OXYGEN ON A MOLYBDENUM SURFACE BY MEANS OF AUGER/ 5-1534
SURFACES.   LEED AND AUGER STUDIES OF EFFECT OF OXYGEN ON ADHESION OF CLEAN IRON (001) AND (011)   6-2010
RFACES.   ADSORPTION OF OXYGEN ON CLEAN CLEAVED (110) GALLIUM ARSENIDE SU   5-1306
N- AUGER ELECTRON SPECTRCSCCPY STUDY OF THE ADSORPTION OF OXYGEN ON COPPER (100) AND (111) SURFACES. /ACTIO 6-2089
LECTRON SPECTROSCOPY. (AUGER / PROCESSES OF ADSORPTION OF OXYGEN ON GALLIUM ARSENIDE METHOD OF LOW ENERGY E 5-1641
TIGUE MECHANISM IN CHANNEL ELECTRON MULTIPLIERS. (SURFACE OXYGEN ON LEAD OXIDE BY AES)   GAIN FA   3-0681
RITIES.   ADSORPTION OF OXYGEN ON MOLYBDENUM (111): EFFECT OF TRACE IMPU   5-1568
SOME ASPECTS OF AN AES AND XPS STUDY OF THE ADSORPTION OF OXYGEN ON NICKEL.   5-1459
OSCOPY: CHEMICAL SHIFTS AND / STUDY OF THE ADSORPTION OF OXYGEN ON NICKEL(III) USING AUGER ELECTRON SPECTR 5-1466
ADSORPTION OF HYDROGEN AND THE REACTION OF HYDROGEN WITH OXYGEN ON PLATINUM (100). (AUGER SPECTROSCOPY)   5-1685
)   CHEMISORPTION OF OXYGEN ON PLATINUM (111) SURFACES. (AUGER SPECTRA   5-1500

IONS. (SPECTROSCOPY)   EFFECT OF ADSORBED CHLORINE AND OXYGEN ON SHEAR STRENGTH OF IRON AND COPPER JUNCT 6-2143
ILIZING AUGER ELECTRON SPECTROSCOPY: CARBON MONOXIDE AND OXYGEN ON SILICON. / ON GAS ADSORPTION STUDIES UT 5-1542
/ER ELECTRON SPECTROSCCPY.   CARBON MONOXIDE AND MOLECULAR OXYGEN ON SILICON. PART-1: ADSORPTION STUDIES. P/ 5-1543
        STRUCTURAL STUDIES OF THE ADSORPTION OF CESIUM AND OXYGEN ON SILICON (100).   5-1389
ELECTRON SPECTROSCOPY)   ADSORPTION OF OXYGEN ON SILICON (111) SURFACES. PART-1. (AUGER 5-1474
NTALUM FIL/ EFFECT OF LIGHT ELEMENTS NITROGEN, CARBON AND OXYGEN ON THE PHYSICAL PROPERTIES OF SPUTTERED TA 5-1473
                          CHEMISORPTION OF OXYGEN ON THE SILVER (110) SURFACE.   5-1778
CTRON SPECTROSCOPY)   ADSORPTION OF OXYGEN ON THE (110) PLANE OF TUNGSTEN. (AUGER ELE 5-1318
/TRON SPECTROSCOPY MEASUREMENTS OF ADSORPTION ISOBARS FOR OXYGEN ON TUNGSTEN AT LOW PRESSURE AND HIGH TEMP/ 5-1282
RFACES DURING ACTIVATION AND POISONING. PART-1: BARIUM ON OXYGEN ON TUNGSTEN DISPENSER CATHODE. /CATHODE SU 5-1841
ETHODS)   ROOM TEMPERATURE ADSORPTION OF OXYGEN ON TUNGSTEN SURFACES. (AUGER MEASUREMENT M 5-1664
HE PRESSURE RANGE 10(-9) TO 10(-/ ADSORPTION ISOTHERMS OF OXYGEN ON TUNGSTEN (100), AND (111) SURFACES IN T 5-1148
UNGSTEN (100). (BY L/ PHOTOEMISSION SPECTRA FROM ADSORBED OXYGEN ON TUNGSTEN (110) AND CARBON MONOXIDE ON T 5-1152
D OTHER SURFACE CONTAMINANTS ON GALLIUM ARSENIDE- CESIUM- OXYGEN PHOTOCATHODES. /NG THE EFFECT OF CARBON AN 5-1919
CAL, AND STRUCTURAL STUDY OF THE GERMANIUM (100)- CESIUM- OXYGEN PHOTOSURFACE.   AN ELECTRICAL, CHEMI 5-1319
U/ OXIDATION OF NIMONIC-80A ALLOY AT 800 DEGREES C IN LOW OXYGEN PRESSURES. (AUGER ELECTRON SPECTROSCOPY, S 6-2144
DER- DISORDER PHENOMENA AT THE SURFACE OF ALPHA TITANIUM- OXYGEN SOLID SOLUTIONS.   OR 5-1819
DEL OF NEGATIVE ELECTRON AFFINITY ON THE SILICON- CESIUM- OXYGEN SURFACE. (AUGER EFFECT) /AND ELECTRONIC MO 2-0426
2), AND / SECONDARY ELECTRON YIELDS FROM COPPER BERYLLIUM OXYGEN SURFACE BY THERMAL CARBON OXYGEN NITROGEN(4-0786
.   AUGER ELECTRON SPECTROSCCPY OF A SULFUR- OXYGEN SURFACE REACTION ON A COPPER (110) CRYSTAL 5-1183
PLICATION OF AUGER ELECTRON SPECTROSCOPY TO THE PLATINUM- OXYGEN SYSTEM AT HIGH TEMPERATURE.   AP 5-1955
TION AND CATALYTIC RE/ INTERACTION OF CARBON MONOXIDE AND OXYGEN WITH A PALLADIUM (100) SURFACE. (CHEMISORP 5-1321
 PART-1: NICKEL (100) SURFAC/ KINETICS OF THE REACTION OF OXYGEN WITH CLEAN NICKEL SINGLE CRYSTAL SURFACES. 5-1452
OSCOPY)   INTERACTIONS OF CARBON MONOXIDE AND OXYGEN WITH IRIDIUM (110) SURFACES. (AUGER SPECTR 5-1252
OW ENERGY ELECTRON DIFFRACTI/ STUDY OF THE INTERACTION OF OXYGEN WITH PLATINUM SINGLE CRYSTAL SURFACES BY L 5-1223
ON SPECTROSCOPY)   INTERACTION OF OXYGEN WITH SILICON (111) SURFACES. (AUGER ELECTR 5-1693
MENT)   INTERACTION OF OXYGEN WITH SILICON (111) SURFACES. (REPLY TO COM 5-1512
/NERGY ELECTRON DIFFRACTION AUGER, WORK FUNCTION STUDY OF OXYGEN- CESIUM ADSORPTION AND COADSORPTION ON (1/ 5-1297
/EHAVIOR OF SODIUM ON NIMONIC-PE16 ALLOY WITH THE SURFACE OXYGEN- CHROMIUM RATIO DETERMINED BY AUGER SPECT/ 5-1441
GER ELECTRON SPECTROSCCPY CF CESIUM ADSORBED ON CLEAN AND OXYGEN- COVERED (100) TUNGSTEN.   AU 5-1295
TION AND AUGER ELECTRCN SPECTROSCOFY/ STRUCTURAL STUDY OF OXYGEN- IRON (001) BY LOW ENERGY ELECTRON DIFFRAC 5-1297
        AUGER ELECTRON SPECTROSCCPY OF HIGH TEMPERATURE OXYGEN- RHENIUM INTERACTIONS.   4-1111
     AUGER ELECTRON SPECTROSCOPY DETERMINATION OF THE OXYGEN- SILICON RATIO IN SPIN ON GLASS FILMS.   5-1827
              DECAY OF HCLE STATES IN OXYGEN-16 VIA A "NUCLEAR AUGER EFFECT".   2-0336
        AUGER ELECTRON SPECTROSCCPY STUDY OF SOME OXYGENATED SILICON COMPOUNDS.   4-0793
IDE FILMS BY AUGER ELECTRON EMISSIO/ DETECTION OF SILICON OXYNITRIDE LAYERS ON THE SURFACES OF SILICON NITR 5-1611
DETECTION BY AUGER ELECTRON SPECTROSCOPY AND REMOVAL BY OZONIZATION OF PHOTORESIST RESIDUES.   5-1450

P

              HIGH RESOLUTION MEASUREMENTS OF P(+) ARGON KLL AND KLM AUGER ELECTRONS.   4-0943
TROSCOPY.   PHOSPHORUS CONCENTRATION PROFILES IN P-DOPED SILICON DIOXIDE MEASURED USING AUGER SPEC 5-1231
SES.   METHOD FOR PACKET OF WAVES IN AUGER ELECTRON-ELECTRON PROCES 2-0380
ATIVE AUGER RECOMBINATION OF ELECTRONS ON DONOR- ACCEPTOR PAIRS.   NONRADI 2-0536
DSORBATE BINDING ENERGY. PART-1: CARBON MONOXIDE ON (100) PALLADIUM.   STRUCTURAL INFLUENCES ON A 5-1913
          ADSORPTION OF CARBON MONOXIDE ON SILVER- PALLADIUM ALLOYS.   5-1254
LECTRON SPECTROSCOPIC STUDIES OF THE OXIDATION OF NICKEL- PALLADIUM ALLOYS.   E 5-1953
UGER SPECTROSCOPIC STUDY OF THE POISONING OF A COMMERCIAL PALLADIUM ALUMINA HYDROGENATION CATALYST.   A 5-1177
NTITATIVE AUGER ELECTRCN SPECTROSCOPY ANALYSIS OF SILVER- PALLADIUM AND NICKEL- PALLADIUM ALLOYS.   QUA 6-2088
FILM SYSTEM. (AUGER E/ CHARACTERISTICS OF SURFACE OF THE PALLADIUM BARIUM ALLOY CATHODE IN RELATION TO ITS 5-1129
OPY OF FACE CENTERED CUBIC METAL SURFACES. (GOLD, SILVER, PALLADIUM, COPPER, NICKEL) /ER ELECTRON SPECTROSC 5-1717
        AUGER SPECTROSCOPY ANALYSIS OF PALLADIUM SILICIDE FILMS.   5-1898
LECTRON SPECTROSCOPY.   STUDY OF PALLADIUM SILICIDE FILMS ON SILICON USING AUGER E 5-1336
ECTRON SPECTROSCOPY.   PALLADIUM SILICIDE FORMATION OBSERVED BY AUGER EL 5-1773
NDUCED SULFUR SEGREGATION FROM THE BULK TO THE SURFACE OF PALLADIUM (100). / AES STUDIES OF CHEMISORPTION I 5-1865
        PHYSICAL ADSORPTION OF XENON ON PALLADIUM (100).   5-1711
YTIC RE/ INTERACTION OF CARBON MONOXIDE AND OXYGEN WITH A PALLADIUM (100) SURFACE. (CHEMISORPTION AND CATAL 5-1321
Y AUGER ELECTRON SPECTROSCOPY.   SURFACE COMPOSITION OF PALLADIUM- GOLD AND PALLADIUM- SILVER CATALYSTS B 5-1976
ION OF RESISTIVITY CHANGES W/ DIFFUSION MECHANISMS IN THE PALLADIUM- GOLD THIN FILM SYSTEM AND THE CORRELAT 5-1393
R ELECTRON S/ DETERMINATION OF THE SURFACE COMPOSITION OF PALLADIUM- NICKEL ALLOY FILM CATALYSTS USING AUGE 5-1846
ETARDING FIELD ANALYZER FOR AUGER ELECTRON SPECTROSCOPY. (PALLADIUM- NICKEL ALLOYS, ADSORBED OXYGEN) /ALL R 3-0644
        SURFACE COMPOSITION OF (PALLADIUM- SILVER) ALLOYS IN EQUILIBRIUM.   5-1788
M- GOLD 70 ATOMIC PERCENT CONTACT/ SURFACE SEGREGATION ON PALLADIUM- SILVER 40 ATOMIC PERCENT, AND PALLADIU 6-2041
18-8 STAINLESS STEEL ELECTROPOLISHED OR ABRADED BY EMERY PAPERS. /SPECTROSCOPIC STUDIES OF THE SURFACES OF 6-2130
RANSITIONS IN MU MESIC ATOMS.   PARITY NONCONSERVATION IN RADIATIONLESS (AUGER) T 2-0364
TROSTATIC ANALYZER OF THE CYLINDRICAL/ FOCUSING A CHARGED PARTICLE BEAM OF FINITE ANGULAR SPREAD IN AN ELEC 3-0720
BILITIES F23 BASED ON THE GREEN-SELLIN-ZACHOR INDEPENDENT PARTICLE MODEL. /D COSTER-KRONIG TRANSITION PROBA 2-0313
URFACES AND IN THIN FILMS OF SILICON AND GERMANIUM / MANY PARTICLE RECOMBINATION PROCESSES (AUGER) ON THE S 5-1990
ROSCOPY APPLIED TO INNER SHELL IONIZATION BY FAST CHARGED PARTICLES.   AUGER ELECTRON SPECT 2-0535
ON AND ARGON FOLLOWING IONIZATION BY EQUAL VELOCITY ALPHA PARTICLES AND DEUTERONS.   AUGER SPECTRA OF CARB 4-1109
        FOCUSING OF CHARGED PARTICLES. (AUGER SPECTROSCOPY)   2-0448
ALIZATION, REVIEW, 16 R/ ELECTRON EMISSION FROM SOLIDS BY PARTICLES BY A SPHERICAL CONDENSER.   3-0684
THEORY AT DEFECTS.   APPLICATION TO STATES WITH TWO BOUND PARTICLES CARRYING POTENTIAL ENERGY. (AUGER NEUTR 2-0374
TROSCOPY. COMBINED ESCA- AUGER SYSTEM BASED ON THE DOUBLE PARTICLES IN GALLIUM PHOSPHIDE.   AUGER 2-0457
CRYSTALS.   ANISOTROPY OF SECONDARY ELECTRON EMISSION FROM PASS CYLINDRICAL MIRROR ANALYZER. / ELECTRON SPEC 3-0654
        AUGER SCANNING ELECTRON MICROSCOPIC STUDY OF THE PASSAGE OF LITHIUM(+) IONS THROUGH COPPER SINGLE 2-0245
ON SPECTROSCOPY.   PASSIVE FILM FORMED ON AN IRON- CHROMIUM ALLOY.   5-1753
ON SPECTROSCOPY.   PASSIVE FILM ON IRON. APPLICATION OF AUGER ELECTR 5-1761
        DEGRADED PASSIVE FILM ON IRON. APPLICATION OF AUGER ELECTR 5-1762
CHEMICAL SHIFTS IN THE AUGER SPECTRA OF PASSIVE FILMS.   4-1056
AUGER SPECTROMETRY STUDY OF PASSIVE FILMS FORMED ON STAINLESS STEELS.   6-2102
LECTRON SPECTROSC/ ANALYSIS OF THE SURFACE COMPOSITION OF PASSIVE FILMS ON COPPER- NICKEL ALLOYS BY AUGER E 5-1874
N SPECTROSCOPY TO THE DETERMINATION OF THE COMPOSITION OF PASSIVE FILMS ON TYPE 316 STAINLESS STEEL. /ECTRO 6-2080
MOSSBAUER) SPECTROSCOPICAL STUDIES OF THE CCMPOSITION OF PASSIVE STATE FILMS.   (AUGER ELECTRON AND 5-1309
ENERGY LOSS SPECTROSCOPY)   INELASTIC MEAN FREE PATH FOR ELECTRONS IN BULK JELLIUM. (AUGER, LEED, 4-1059
ALLOY.   APPLICATION/ RELATION BETWEEN CATALYTIC ACTIVITY PATTERN AND SURFACE COMPOSITION OF COPPER- NICKEL 5-1875
/M (110) SURFACE. PART-1: SULFUR AND CARBON MONOXIDE LEED PATTERNS. (AUGER ELECTRON SPECTROSCOPY, STICKING/ 5-1190
     AUGER SPECTRA AND LEED PATTERNS FROM NICKEL DEPOSITS ON CLEAVED SILICON. 5-1768
H CALIBRATED DEPOSITS OF IRON.   AUGER SPECTRA AND LEED PATTERNS FROM VACUUM CLEANED SILICON CRYSTALS WIT 5-1769
CANNING SAMPLE POSITIONER.   ELECTRON CHANNELING PATTERNS IN AN AUGER ELECTRON SPECTROMETER WITH S 3-0692

E IONIZATION ORIGIN OF THE HIGH ENERGY SATELLITE OF AUGER PEAK. /N ON INNER LEVEL IONIZATION. PART-2: DOUBL 4-1047
SENIDE METHOD OF LOW ENERGY ELECTRON SPECTROSCOPY. (AUGER PEAK AMPLITUDE CALIBRATION) /OXYGEN ON GALLIUM AR 5-1641
STRUCTURE ON THE HIGH ENERGY SIDE OF THE KL2,3M AUGER PEAK FROM SOLID ALUMINUM: INTERNAL PHOTOEMISSION. 4-1021
S ON THE SPECTRUM OF THE SLOW SECONDARY ELECTRON EMISSION PEAK IN POLYCRYSTALLINE MAGNESIUM. /DATION EFFECT 4-1118
ANADIUM OXIDES, AND OXIDIZED VANADIUM: CHEMICAL SHIFT AND PEAK INTENSITY ANALYSIS. /EMICAL COMPOSITION OF V 4-1089
STEN (1/ CHARACTERIZATION OF ADSORBED SPECIES USING AUGER PEAK SHAPES. ETHYLENE AND CARBON MONOXIDE ON TUNG 5-1243
3L2,3) ON MOLYBDENUM.   ORIGIN OF THE PEAK SHAPES IN THE AUGER SPECTRUM OF OXYGEN (KL2, 4-1004
RONS EJECTED FROM XENON BY 0.3 TO 2.0 MEV PROTONS. (AUGER PEAKS)   DISTRIBUTION IN ENERGY AND ANGLE OF ELECT 2-0534
          MULTIPLET EFFECTS ON THE WIDTHS OF PHOTOELECTRON PEAKS. (AUGER EFFECT) 2-0445
EL (100))   DIFFRACTION PEAKS IN SECONDARY ELECTRON ENERGY SPECTRA. (NICK 4-0775
          INTERFACIAL AND PLASMON DECAY PEAKS IN THE AUGER SPECTRA OF TITANIUM AND IRON. 4-1060
S FROM VARIOUS MATERIALS.   AUGER PEAKS IN THE ENERGY SPECTRA OF SECONDARY ELECTRON 4-0926
HIGH ENERGY SATELLITES OF THE M2,3VV AND M1VV AUGER PEAKS OF COPPER. 4-1043
          HIGH ENERGY SATELLITES IN AUGER PEAKS OF MAGNESIUM AND SILICON. 4-1046
A MEASURING METHOD TO DISTINGUISH AUGER FROM ENERGY LOSS PEAKS OF THE PRIMARY BEAM IN SECONDARY ELECTRON / 1-0188
CONDUCTING NIOBIUM. / EFFECT OF SURFACE METALLURGY ON THE PENETRATION DEPTH AND RF BREAKDOWN FIELD OF SUPER 6-2137
GER SPECTRO/ EFFECT OF HIGH TEMPERATURE HEAT TREATMENT ON PENETRATION DEPTH OF SUPERCONDUCTING NIOBIUM. (AU 5-1938
NS. (AUGER COEFFICIENTS)   PENETRATION EFFECTS IN CONVERTED MUONIC TRANSITIO 2-0478
          ANALYSIS OF HIGH CHARGE STATE OUTPUT FROM A PENNING ION SOURCE. 3-0602
                                        PENT..., SEE THE PARENT WORD.
TRON CAPTURE IN TRIS (1,10-PHENANTHROLINE) COBALT(III)-57 PERCHLORATE DIHYDRATE. /IONIZATION FOLLOWING ELEC 2-0516
ON) MICROANALYSIS AND MICROSCOPY. PRINCIPLES AND EXPECTED PERFORMANCES. /AY PHOTOELECTRON (AND AUGER ELECTR 1-0027
PECTROSC/ SURFACE COMPOSITION AND CHEMISTRY OF EVAPORATED PERMALLOY FILMS OBSERVED BY X-RAY PHOTOEMISSION S 5-1738
CTION AND AUGER ELECTRON / SILICON CLEANING WITH HYDROGEN PEROXIDE SOLUTIONS: HIGH ENERGY ELECTRON DIFFRA 5-1426
SS TRANSITIONS.   MANY BODY PERTURBATION THEORY APPLIED TO ATOMIC RADIATIONLE 2-0386
ATES, CORRELATION ENERGIES, AND/ APPLICATION OF MANY BODY PERTURBATION THEORY TO THE CALCULATION OF AUGER R 2-0311
STUDY OF INSULATORS.   PERTURBING EFFECTS OF SPUTTERING DURING THE AUGER 1-0111
          PHOTOELECTRON AND AUGER SPECTRA OF XENON IN SOLID PERXENATES. 4-0902
N HEATED SILICON SUBSTRATES. (SCANNING AUGER SPECT/ SOLID PHASE EPITAXIAL STUDIES USING VACUUM DEPOSITION O 5-1287
          ANTIMONY DOPING OF SILICON LAYERS GROWN BY SOLID PHASE EPITAXY. (AES MEASUREMENTS) 5-1577
IDE SUBSTRATES. (AUGER SPECTROSCOPY)   LIQUID PHASE EPITAXY OF III-V COMPOUNDS ON INDIUM PHOSPH 5-1590
AYERS. (AUGER SPECTROSCOPY, SULFIDE SYSTEMS)   PHASE EQUILIBRIA IN TWO DIMENSIONAL CHEMISORBED L 5-1170
AUGER SPECTROSCOPIC OBSERVATION OF SILICON- GOLD MIXED PHASE FORMATION AT LOW TEMPERATURES. 5-1680
CTRON SPECTROSCOPY.   IDENTIFICATION OF A GRAIN BOUNDARY PHASE IN HOT PRESSED SILICON NITRIDE BY AUGER ELE 5-1749
ER ELECTRON SPECTROSCOPY.   PHASE SEPARATION IN SILICON OXIDES AS SEEN BY AUG 5-1502
                                        PHASE TRANSFORMATIONS OF SILICON (111) SURFACE. 5-1343
EGREGATION OF CARBON TO A NICKEL (111) SURFACE: A SURFACE PHASE TRANSITION. (AUGER SPECTROSCOPY) /LIBRIUM S 5-1803
RBED ON (0001) GRAPHITE. (AUGER SPECTROS/ TWO-DIMENSIONAL PHASE TRANSITION IN XENON SUBMONOLAYER FILMS ADSO 5-1858
BINATION)   RECOMBINATION-INDUCED NONEQUILIBRIUM PHASE TRANSITIONS IN SEMICONDUCTORS. (AUGER RECOM 2-0421
FUR LAYERS ON COPPER INVESTIGATED BY LEED AND AUGER SPEC/ PHASE TRANSITIONS IN TWO DIMENSIONAL ABSORBED SUL 5-1302
/UGER IONIZATION FOLLOWING ELECTRON CAPTURE IN TRIS (1,10-PHENANTHROLINE) COBALT(III)-57 PERCHLORATE DIHYD/ 2-0516
ELECTRON SPECTROSCOPY APPLIED TO THE STUDY OF ADSORPTION PHENOMENA AT METAL SURFACES. /ERGY LOSS AND AUGER 5-1776
N SOLID SOLUTIONS.   ORDER- DISORDER PHENOMENA AT THE SURFACE OF ALPHA TITANIUM- OXYGE 5-1819
/K AT SURFACES. (SURFACE STRUCTURES, SPECIES, TOPOGRAPHY, PHENOMENA, FIELD ION MICROSCOPY, LEED, HEED, AUG/ 5-1492
                              INTERFACE PHENOMENA IN ENGINEERING MATERIALS. 1-0176
ROLE OF AUGER EFFECT IN THERMOSTIMULATED PHENOMENA IN IONIC CRYSTALS. 2-0539
ALUMINUM. (AUGER SPECTROSCOPY)   CHARACTERISTIC GAIN PHENOMENA IN POLYCRYSTALLINE COPPER, SILVER, AND 5-1724
ITION SERIES. (COMMENTS ON PAPER BY CHUNG AN/ UNEXPLAINED PHENOMENA IN THE AUGER SPECTRA OF THE FIRST TRANS 4-0812
GER PROCESS IN ZINC METAL.   RELAXATION PHENOMENA INVOLVED IN THE L3M4,5M4,.5M4,.5:1G4 AU 4-0882
BACK SCATTERED ELECTRONS AND CHARACTERISTIC LOSS AND GAIN PHENOMENA OF COPPER (111).   ENERGY SPECTRUM OF 4-0897
RON AUGER SPECTROSCOPY: A METHOD OF INVESTIGATING SURFACE PHENOMENA (REVIEW).   ELECT 1-0132
ECTRA.   PHENOMENOLOGICAL SYSTEMATICS OF L- AND M-AUGER SP 4-0875
MPOUNDS.   PHONON ASSISTED AUGER PROCESSES IN GROUP III-V CO 2-0484
                                        PHONON ASSISTED AUGER RECOMBINATION IN GERMANIUM. 2-0397
TERACTION BETWEEN OXYGEN AND THE (111) SURFACE OF GALLIUM PHOSPHIDE.   AUGER ELECTRON SPECTRUM DURING THE IN 5-1642
APPLICATION TO STATES WITH TWO BOUND PARTICLES IN GALLIUM PHOSPHIDE.   AUGER THEORY AT DEFECTS. 2-0457
INESCENCE AND PHOTOCONDUCTIVITY IN HIGHLY EXCITED GALLIUM PHOSPHIDE.   EXCITON LUM 2-0454
CITONS BOUND TO NEUTRAL DONORS IN TELLURIUM DOPED GALLIUM PHOSPHIDE. /CIATED WITH AUGER RECOMBINATION OF EX 2-0453
TERACTION BETWEEN OXYGEN AND THE (111) SURFACE OF GALLIUM PHOSPHIDE. /THE AUGER ELECTRON SPECTRUM IN THE IN 5-1643
IN MOLECULAR BEAM EPITAXY OF GALLIUM ARSENIDE AND GALLIUM PHOSPHIDE.   USE OF SCANNING AUGER MICROSCOPY 5-1142
          ANODIC OXIDATION OF GALLIUM ARSENIDE PHOSPHIDE. (AES) 5-1388
OMBINATION IN INDIUM ARSENIDE, GALLIUM ANTIMONIDE, INDIUM PHOSPHIDE AND GALLIUM ARSENIDE.   AUGER REC 4-1091
D BY AES AND UPS.   OXIDATION AND ANNEALING OF GALLIUM PHOSPHIDE AND GALLIUM ARSENIDE (111) FACES STUDIE 5-1493
/ ANNEALING OF (100) SURFACES OF GALLIUM ARSENIDE, INDIUM PHOSPHIDE, AND GALLIUM PHOSPHIDE IN VACUUM. (AUG/ 5-1163
/GY DISTRIBUTION OF SECONDARY ELECTRONS EMITTED BY INDIUM PHOSPHIDE AND GALLIUM PHOSPHIDE SUBMITTED TO A B/ 2-0269
IN-DEPTH PROFILE DETECTION LIMITS OF NITROGEN IN GALLIUM PHOSPHIDE (AND) NITROGEN, OXYGEN, AND FLUORINE I/ 5-1916
          SURFACE CHARACTERIZATION OF INDIUM PHOSPHIDE. (BY AUGER SPECTROSCOPY) 5-1968
          MOLECULAR BEAM STUDIES OF INDIUM PHOSPHIDE EVAPORATION AND GROWTH. (AES) 5-1334
/PARTURES FROM STOICHIOMETRY IN (100) SURFACES OF GALLIUM PHOSPHIDE, INDUCED BY BOMBARDMENT WITH 2 KV ELEC/ 5-1162
ACTION OF SLOW ELECTRONS.   (111) SURFACE OF GALLIUM PHOSPHIDE STUDIED BY AUGER SPECTROSCOPY AND DIFFR 5-1640
          LIQUID PHASE EPITAXY OF III-V COMPOUNDS ON INDIUM PHOSPHIDE SUBSTRATES. (AUGER SPECTROSCOPY) 5-1590
N THE SURFACE STRUCTURE AND STOICHIOMETRY OF (100) INDIUM PHOSPHIDE SURFACES.   EFFECT OF TEMPERATURE O 5-1164
ERISTIC ENERGY LOSS AND AUGER ELECTRON SPECTRA OF GALLIUM PHOSPHIDE (110).   CHARACT 5-1657
TRON DIFFRACTION AND AUGER ELECTRON SPECTROSCOPY STUDY OF PHOSPHINE ADSORPTION ON CLEAN SILICON (111). /LEC 5-1930
NTRATION-DEPENDENT NONRADIATIVE PROCESSES IN ZINC SULFIDE PHOSPHORS. (AUGER RECOMBINATIONS)   CONCE 2-0388
F THE L2,3 AUGER EMISSION BANDS OF ALUMINUM, SILICON, AND PHOSPHORUS.   AUGER SATELLITES O 4-0985
ON THE L23 AUGER EMISSION BANDS OF ALUMINUM, SILICON, AND PHOSPHORUS.   AUGER SATELLITES 4-0984
PRODUCTION AND DECAY OF DOUBLE L VACANCIES IN ARGON AND PHOSPHORUS. 2-0489
SILICON (111) SURFACES. (BY LEED AND AES)   A STUDY OF PHOSPHORUS ADSORPTION AND DESORPTION KINETICS ON 5-1288
/PER EMBRITTLEMENT IN A NICKEL- CHROMIUM STEEL CONTAINING PHOSPHORUS AS AN IMPURITY. (AUGER EMISSION ANALY/ 6-2138
CON DIOXIDE MEASURED USING AUGER SPECTROSCOPY.   PHOSPHORUS CONCENTRATION PROFILES IN P-DOPED SILI 5-1231
ERGIES IN SECONDARY ELECTRON SPECTRA FROM THIN EVAPORATED PHOSPHORUS FILMS. /ER AND OTHER CHARACTERISTIC EN 4-0986
          DIFFUSION OF SULFUR AND PHOSPHORUS IMPURITY ATOMS TO A TUNGSTEN SURFACE. 5-1226
/FFECT OF MICROSTRUCTURE ON GRAIN BOUNDARY SEGREGATION OF PHOSPHORUS IN A CHROMIUM- MOLYBDENUM- VANADIUM S/ 6-2139
ANTITATIVE AUGER SPECTROSCOPIC ANALYSIS OF SEGREGATION OF PHOSPHORUS IN IRON.   QU 5-1802
OPY.   DETECTABILITY LIMITS FOR BORON AND PHOSPHORUS IN SILICON BY AUGER ELECTRON SPECTROSC 5-1887
SCOPY AND SECONDARY ION EMISSION.   IN-DEPTH PROFILES OF PHOSPHORUS ION IMPLANTED SILICON BY AUGER SPECTRO 4-0996
TIMONY- SULFUR ALL/ GRAIN BOUNDARY EMBRITTLEMENT IN IRON- PHOSPHORUS, IRON- PHOSPHORUS- SULFUR AND IRON- AN 6-2098
OBALANCE. QUANTITATIVE STUDY OF AUGER ELECTRON SIGNALS OF PHOSPHORUS ON SILICON USING A QUARTZ CRYSTAL MICR 5-1583
MOLYBDENUM- NICKEL LOW ALLOY STEEL/ EFFECTS OF COPPER AND PHOSPHORUS ON TEMPER EMBRITTLEMENT OF MANGANESE- 6-2043

CS OF COPPER- TIN- PHOSPHORUS ALLOYS. (AUGER S/ EFFECT OF PHOSPHORUS ON THE FRICTION AND WEAR CHARACTERISTI 5-1868
ICON.   AUGER AND ELLIPSOMETRIC STUDY OF PHOSPHORUS SEGREGATION IN OXIDIZED DEGENERATE SIL 5-1250
/ GRAIN BOUNDARY EMBRITTLEMENT IN IRON- PHOSPHORUS, IRON- PHOSPHORUS- SULFUR AND IRON- ANTIMONY- SULFUR ALL 6-2098
ECTROMETRY FOR CHEMICAL ANALYSIS. (AUTOMATIC RECORDING OF PHOTO AND AUGER ELECTRONS) /R USED IN ELECTRON SP 3-0717
SEPARATION OF THE SPECTRA OF PHOTO AND AUGER ELECTRONS. 4-1099
TO LOW ENERGY X-RAY AND ELECTRON ANALYSIS. (ANALYZER FOR PHOTO AUGER ELECTRON SPECTROSCOPY)   INTRODUCTION 3-0627
SURFACE CHARACTERIZATION BY LOW ENERGY PHOTO AUGER ELECTRON SPECTROSCOPY. 5-1429
ANGLE DISTRIBUTION OF SODIUM CHLORIDE SINGLE CRYSTALS. (PHOTO- AND AUGER ELECTRON EMISSION) 2-0297
SECONDARY ION MASS SPECTROSCOPY. COMPARISON OF TWO SUR/ PHOTO- AND AUGER ELECTRON SPECTROSCOPY (ESCA) AND 1-0088
LOW ENERGY PHOTO- AUGER ELECTRON SPECTROSCOPY. 1-0080
SURFACE CONTAMINANTS ON GALLIUM ARSENIDE- CESIUM- OXYGEN PHOTOCATHODES. /NG THE EFFECT OF CARBON AND OTHER 5-1919
TION OF EXCITONS BOUND TO NEUTRAL DONORS IN TELLURIUM DO/ PHOTOCONDUCTIVITY ASSOCIATED WITH AUGER RECOMBINA 2-0453
HIDE.   EXCITON LUMINESCENCE AND PHOTOCONDUCTIVITY IN HIGHLY EXCITED GALLIUM PHOSP 2-0454
IDE AND CADMIUM SULFIDE(0.48) SELENIDE(0.52) CRYSTALS. (/ PHOTOCONDUCTIVITY OF HIGHLY EXCITED CADMIUM SELEN 2-0432
ZINC OXIDE SURFACES. (AUGER ANALYSIS)   CHEMISORPTION, PHOTODESORPTION AND CONDUCTIVITY MEASUREMENTS ON 5-1801
/ECT OF THE POLARITY OF INDIUM ANTIMONIDE ON THE EXTERNAL PHOTOEFFECT IN THE X-RAY SPECTRAL REGION. (EFFEC/ 4-1001
. (AUGER ELECTRON DISTRIBUTION)   DEPENDENCE OF ATOMIC PHOTOEFFECT ON A BOUND ELECTRON MAGNETIC SUBSTATE 2-0463
AUGER EFFECT).   DEPENDENCE OF ATOMIC PHOTOEFFECT ON BOUND ELECTRON MAGNETIC SUBSTATE (2-0524
ON THE COMPOUND PHOTOELECTRIC EFFECT. 2-0249
, MAGNESIUM(2) GERMANIUM, AND MAGNESIUM(2) STANNIDE. PAR/ PHOTOELECTRIC PROPERTIES OF MAGNESIUM(2) SILICIDE 2-0306
ECTROSCOPY OF 5.59 ATOMIC PERCENT SILICON- IRON ALLOY (1/ PHOTOELECTRIC WORK FUNCTION AND AUGER ELECTRON SP 5-1925
R ELECTRON SPECTRO/ INITIAL OXIDATION STUDIES BY MEANS OF PHOTOELECTRIC WORK FUNCTION MEASUREMENTS AND AUGE 5-1928
WIT/ INITIAL OXIDATION STUDIES ON IRON (100) BY MEANS OF PHOTOELECTRIC WORK FUNCTION MEASUREMENTS COMBINED 5-1927
URFACE COMBINED WITH AUGER ELECTRON SPECTROSCOPY. PHOTOELECTRIC WORK FUNCTION STUDY ON IRON (100) S 5-1926
INTENSITIES OF SPECTRA- AUGER AND PHOTOELECTRON. 4-1070
SPECTROSCOPY. (RELATIONSHIP OF IONIZATION SPECTROSCOPY TO PHOTOELECTRON AND AES)   APPLICATIONS OF ELECTRON 1-0062
ITS POSSIBLE APPLICATIONS TO PROBLEMS IN GLASS RESEARCH. (PHOTOELECTRON AND AES) /ANALYSIS AND A REVIEW OF 1-0084
AND MICROSCOPY. PRINCIPLES AND EXPECTED PERFORMANC/ X-RAY PHOTOELECTRON (AND AUGER ELECTRON) MICROANALYSIS 1-0027
SIUM THIN FILMS.   X-RAY PHOTOELECTRON AND AUGER ELECTRON SPECTRA OF MAGNE 5-1392
ZED STATES ON SOME METAL SURFACES.   X-RAY EXCITED PHOTOELECTRON AND AUGER ELECTRON SPECTRA OF OXIDI 5-1690
SIMPLIFIED RETARDING FIELD PHOTOELECTRON AND AUGER ELECTRON SPECTROGRAPH. 3-0642
HIGH RESOLUTION ELECTRON SPECTROMETER FOR PHOTOELECTRON AND AUGER ELECTRON STUDIES. 3-0600
/ COMPARISON OF X-RAY INDUCED AND ELECTRON IMPACT INDUCED PHOTOELECTRON AND AUGER SPECTRA. CARBON MONOXIDE 4-0768
/ROSCOPY AND ADSORPTION STUDIES. CHEMICAL SHIFTS IN X-RAY PHOTOELECTRON AND AUGER SPECTRA FOR MONOLAYER AD/ 5-1204
(II) OXIDE, COPPER(I) OXIDE, AND COPPER(I)/ X-RAY INDUCED PHOTOELECTRON AND AUGER SPECTRA OF COPPER, COPPER 4-0932
PERXENATES.   PHOTOELECTRON AND AUGER SPECTRA OF XENON IN SOLID 4-0902
STEEL SURFACES.   X-RAY PHOTOELECTRON AND AUGER SPECTROSCOPIC ANALYSES OF 5-1268
ATION PROFILE OF PLATINUM- TIN ALLOYS FROM COMBINED X-RAY PHOTOELECTRON AND AUGER SPECTROSCOPIC DATA. /ENTR 5-1195
PHOTOELECTRON AND AUGER SPECTROSCOPY. 1-0026
N OF LOW ENERGY ELECTRONS FROM SOLIDS: RESULTS FROM X-RAY PHOTOELECTRON (AND AUGER) SPECTROSCOPY. /TENUATIO 5-1844
THE SURFACE COMPOSITION OF OXIDIZED VANADI/ USE OF X-RAY PHOTOELECTRON AND AUGER SPECTROSCOPY TO DETERMINE 5-1205
THE SURFACE COMPOSITION OF OXIDIZED VANADI/ USE OF X-RAY PHOTOELECTRON AND AUGER SPECTROSCOPY TO DETERMINE 5-1863
FOR ARSENIC.   PHOTOELECTRON AND L2,3MM AUGER ELECTRON ENERGIES 4-1035
N STUDIES OF ADSORPTION ON CLEAN METAL FILM SURFAC/ X-RAY PHOTOELECTRON AND VACUUM ULTRAVIOLET PHOTOELECTRO 5-1203
, REVIEW, 41 REFS)   ELECTRON SPECTROSCOPY. (PHOTOELECTRON, AUGER, AND IONIZATION SPECTROSCOPY 1-0001
POUNDS. X-RAY EXCITED LMM AUGER SPECTRA OF GERMANIUM. PHOTOELECTRON CHEMICAL SHIFTS IN COORDINATION COM 4-0924
THE LOW ENERGY ELECTRONS IN ALUMINUM OXIDE. (AUGER ELEC/ PHOTOELECTRON DETERMINATION OF THE ATTENUATION OF 4-0773
SES. OPTICALLY STIMULATED SECONDARY ELECTRON EMISSION AND PHOTOELECTRON EMISSION STUDY OF ADSORPTION PROCES 2-0302
ELECTRON SPECTROSCOPY USING THE TWO-GRID LEED SYSTEM AND PHOTOELECTRON MULTIPLIER.   AUGER 3-0656
MULTIPLET EFFECTS ON THE WIDTHS OF PHOTOELECTRON PEAKS. (AUGER EFFECT) 2-0445
F MAGNESIUM, AND ITS OXIDE.   X-RAY EXCITED AUGER AND PHOTOELECTRON SPECTRA OF MAGNESIUM, SOME ALLOYS O 4-0837
SIUM SURFACES. ABNORMAL CHEMICA/ X-RAY EXCITED AUGER AND PHOTOELECTRON SPECTRA OF PARTIALLY OXIDIZED MAGNE 5-1949
ELECTRONS AND X-RAYS IN NONDESTRUCTIVE ANALYSIS. (AUGER, PHOTOELECTRON SPECTRA, REVIEW, 6 REFS) 1-0083
AUGER SIGNALS OBTAINED WITH A PHOTOELECTRON SPECTROMETER. 2-0272
S OF WIDTH AND INTENSITY OF AUGER SIGNALS OBTAINED WITH A PHOTOELECTRON SPECTROMETER. /SHIFTS AND VARIATION 2-0273
XIDATION OF MOLYBDENUM STUDIED BY A HIGH RESOLUTION AUGER PHOTOELECTRON SPECTROMETER.   INITIAL O 5-1698
RICAL MIRROR ANALYZER WITH SOFT X-RAY SOURCE.   A NEW PHOTOELECTRON SPECTROMETER: COMBINATION OF CYLIND 3-0652
AUGER LINES IN X-RAY PHOTOELECTRON SPECTROMETRY. 1-0208
ARSENIC (0001) SURFACES: AUGER, LOSS, AND PHOTOELECTRON SPECTROSCOPIC STUDIES. 5-1315
AND SODIUM OXIDE.   AUGER AND X-RAY PHOTOELECTRON SPECTROSCOPIC STUDY OF SODIUM METAL 4-0770
OXIDE FILMS ON CHROMIUM STEEL BY AUGER EMISSION AND X-RAY PHOTOELECTRON SPECTROSCOPIES. /LYSIS PROFILES OF 6-2016
N A SILICON SUBSTRATE STUDIED BY AUGER ELECTRON AND X-RAY PHOTOELECTRON SPECTROSCOPIES. /D FILM DEPOSITED O 5-1435
ETICS OF CESIUM ON GALLIUM ARSENIDE. (USING AES AND X-RAY PHOTOELECTRON SPECTROSCOPY)   ADSORPTION KIN 5-1821
R- NICKEL ALLOYS BY AUGER ELECTRON SPECTROSCOPY AND X-RAY PHOTOELECTRON SPECTROSCOPY. /SSIVE FILMS ON COPPE 5-1874
AUGER SPECTRA IN X-RAY PHOTOELECTRON SPECTROSCOPY. 1-0209
EFFECTS IN AUGER ELECTRON SPECTROSCOPY REVEALED BY X-RAY PHOTOELECTRON SPECTROSCOPY.   BEAM 3-0601
IMPROVED RETARDING FIELD ANALYZER. (FOR AUGER AND PHOTOELECTRON SPECTROSCOPY). 3-0701
STIGATION OF LOW-Z COSTER-KRONIG TRANSITIONS BY AUGER AND PHOTOELECTRON SPECTROSCOPY.   INVE 3-0578
PECTROSCOPY. COMPARISON WITH AUGER SPECTROSCOPY AND X-RAY PHOTOELECTRON SPECTROSCOPY. /PEARANCE POTENTIAL S 1-0173
ZER. APPLICATION TO AUGER ELECTRON SPECTROSCOPY AND X-RAY PHOTOELECTRON SPECTROSCOPY. /DING POTENTIAL ANALY 3-0634
X-RAY SPECTROSCOPY. (REVIEW, AUGER AND PHOTOELECTRON SPECTROSCOPY) 1-0203
GENERAL INTRODUCTION.   PHOTOELECTRON SPECTROSCOPY AND ALLIED TECHNIQUES: 1-0149
SSION SPECTROSCOPY.   SURFACE ANALYSES BY COMBINED X-RAY PHOTOELECTRON SPECTROSCOPY AND AUGER ELECTRON EMI 5-1272
CTROSCOPY.   STRUCTURE ANALYSIS WITH X-RAY PHOTOELECTRON SPECTROSCOPY AND AUGER ELECTRON SPE 1-0050
/E METAL SURFACES USING AUGER ELECTRON SPECTROSCOPY X-RAY PHOTOELECTRON SPECTROSCOPY AND SECONDARY ION MAS/ 5-1353
IQUE.   SOME PARAMETERS AFFECTING AUGER AND PHOTOELECTRON SPECTROSCOPY AS AN ANALYTICAL TECHN 1-0222
ECOMBINATION INVOLVING INNER SHELLS. (AUGER PROCESS, AES, PHOTOELECTRON SPECTROSCOPY, CHEMICAL SHIFT) /ND R 2-0514
/ DEPOSITION IN LIQUID METALS. (AUGER SPECTROSCOPY, X-RAY PHOTOELECTRON SPECTROSCOPY, MOSSBAUER SPECTROSCO/ 6-2050
TECHNIQUES RELATED TO PHOTOELECTRON SPECTROSCOPY. (REVIEW) 1-0021
/TRON DIFFRACTION, AUGER ELECTRON SPECTROSCOPY, AND X-RAY PHOTOELECTRON SPECTROSCOPY STUDY OF THE ZINC OXI/ 5-1337
IONS)   X-RAY PHOTOELECTRON SPECTRUM OF DIAMOND. (AUGER TRANSIT 4-0851
ENER/ HIGH RESOLUTION MEASUREMENTS OF AUGER ELECTRON AND PHOTOELECTRON STRUCTURE IN THE SECONDARY ELECTRON 4-1023
M, AND GALLIU/ AUGER AND DIRECT ELECTRON SPECTRA IN X-RAY PHOTOELECTRON STUDIES OF ZINC, ZINC OXIDE, GALLIU 4-1054
DIRECTIONAL CORRELATION BETWEEN PHOTOELECTRONS AND AUGER ELECTRONS. 1-0122
RELATIVE ESCAPE DEPTH OF AUGER AND PHOTOELECTRONS FROM A SOLID SURFACE. 5-1903
INCTION BETWEEN WEAK AUGER TRANSITIONS AND INTERNAL X-RAY PHOTOEMISSION.   DIST 2-0436
DE OF THE KL2,3M AUGER PEAK FROM SOLID ALUMINUM: INTERNAL PHOTOEMISSION.   STRUCTURE ON THE HIGH ENERGY SI 4-1021
ENERGY RESOLUTION OF THE PHOTOEMISSION ANALYZER. 3-0606
VALENCE BANDS OF MAGNESIUM,/ COMPARISON OF THE SOFT X-RAY PHOTOEMISSION, AND AUGER ELECTRON SPECTRA OF THE 4-0838
PROBING DEPTH IN PHOTOEMISSION AND AUGER ELECTRON SPECTROSCOPY. 2-0428

SOME OF ITS COMPOUNDS. RELAXATION DURING	PHOTOEMISSION AND LMM AUGER DECAY IN ARSENIC AND	4-0763
ACEMENTS) X-RAY	PHOTOEMISSION AND SURFACE STRUCTURE. (AUGER DISPL	5-1811
NIQUES TO STUDY THE MECHANISM OF A CATALYTIC REAC/ USE OF	PHOTOEMISSION, AUGER, AND THERMAL DESORPTION TECH	1-0042
A SIMPLE ANALYTICAL CALCULATICN FOR X-RAY	PHOTOEMISSION. (AUGER PROCESS)	2-0429
F SODIUM, MAG/ EXPERIMENTAL DETERMINATION OF THE INTERNAL	PHOTOEMISSION CONTRIBUTION TO KLM AUGER SPECTRA O	4-0769
KEL AND NICKEL ALLOYED WITH Z/ FINAL STATE EFFECTS IN THE	PHOTOEMISSION CORE LEVEL AND AUGER SPECTRA OF NIC	5-1806
PY. CHARACTERIZING OF ALUMINUM FATIGUE DAMAGE BY	PHOTOEMISSION, ELLIPSOMETRY, AND AUGER SPECTROSCO	6-2116
MANY BODY EFFECTS IN X-RAY	PHOTOEMISSION FROM MAGNESIUM. (AUGER EFFECT)	2-0427
RAYS. (PLASMA ELECTRONS, AUGER PROCESS) ELECTRON	PHOTOEMISSION INDUCED BY LASER-PRODUCED PLASMA X-	2-0276
/NCE BAND AND CORE LEVEL STUDIES OF INDIUM ANTIMONIDE VIA	PHOTOEMISSION MEASUREMENTS IN THE 20-70 EV RANGE/	2-0341
HYDROGEN INDUCED STATES ON CLEAVED GERMANIUM (111). (AUG/	PHOTOEMISSION MEASUREMENTS OF BULK, SURFACE, AND	5-1781
GSTEN (110) AND CARBON MONOXIDE ON TUNGSTEN (100). (BY L/	PHOTOEMISSION SPECTRA FROM ADSORBED OXYGEN ON TUN	5-1152
ALENCE BAND OF MAGNESIUM STANNIDE DETERMINED BY AUGER AND	PHOTOEMISSION SPECTROSCOPY. V	5-1885
HE ZINC CHALCOGENID/ LIFETIME BROADENING IN X-RAY INDUCED	PHOTOEMISSION SPECTROSCOPY AND AUGER SPECTRA OF T	4-0832
/HEMISTRY OF EVAPORATED PERMALLOY FILMS OBSERVED BY X-RAY	PHOTOEMISSION SPECTROSCOPY AND BY AUGER ELECTRON/	5-1738
AND TUNGSTEN SURFACES.	PHOTOEMISSION STUDIES OF CHEMISORPTION ON NICKEL	5-1308
TASSIUM IODIDE.	PHOTOEMISSION STUDIES OF CORE EXCITON DECAY IN PO	2-0422
/Y ELECTRON DIFFRACTION, AUGER ELECTRON SPECTROSCOPY, AND	PHOTOEMISSION STUDIES OF THE CESIUM COVERED GALL/	5-1931
NS IN ALKALI HALIDES. (AUGER PROCESS IN VALENCE BAND)	PHOTOEMISSION STUDIES OF THE DECAY OF CORE EXCITO	2-0289
ROSCOPY. INVESTIGATION OF SILVER CESIUM OXIDE	PHOTOEMISSIVE SURFACES USING AUGER ELECTRON SPECT	5-1782
Y AND ANGULAR DISTRIBUTIONS OF INELASTICALLY SCATTERED OR	PHOTOEMITTED ELECTRONS. / FOR INVESTIGATING ENERG	5-0691
NONEQUILIBRIUM CHARGE CARRIERS OF INDIUM ARSENIDE AT HIGH	PHOTOEXCITATION LEVELS. (AUGER) RECOMBINATION OF	4-0840
CHEMICAL SHIFTS IN	PHOTOEXCITED AUGER SPECTRA.	4-0797
OR AUGER ELECTRON SPECTRA. NEW, STRIATED, HIGH EFFICIENCY	PHOTOGRAPHIC PLATES FOR RECORDING CONVERSION AND/	3-0699
ANGULAR DISTRIBUTION OF AUGER ELECTRONS FOLLOWING	PHOTOIONIZATION.	2-0352
THE CALCULATION OF AUGER RATES, CORRELATION ENERGIES, AND	PHOTOIONIZATION CROSS SECTIONS. /ATION THEORY TO	2-0311
DOUBLE	PHOTOIONIZATION OF NEON. (AUGER TRANSITIONS)	2-0310
/D RECOIL ENERGIES OF FRAGMENT IONS FORMED FROM THE X-RAY	PHOTOIONIZATION OF NITROGEN, OXYGEN, CARBON MONO/	2-0308
CESS) STIMULATED	PHOTOLUMINESCENCE IN GALLIUM SELENIDE. (AUGER PRO	2-0416
FLUORESCENCE	PHOTON AND AUGER ELECTRON SPECTRA.	4-0799
/HNIQUES INVOLVING ELECTRON, AUGER ELECTRON SPECTROSCOPY,	PHOTON, ELECTRON SPECTROSCOPY FOR CHEMICAL ANALY/	5-1901
EVIDENCE FOR A RADIATIVE AUGER EFFECT IN X-RAY	PHOTON EMISSION.	2-0223
EXCITATION OF INNER SHELL ELECTRONS BY	PHOTONS AND FAST ELECTRONS.	2-0541
TY APPEARANCE POTENTIAL SPECTROMETER FOR THE DETECTION OF	PHOTONS, ELECTRONS AND IONS. HIGH SENSITIVI	3-0643
AUGER ELECTRON SPECTROSCOPY AND REMOVAL BY OZONIZATION OF	PHOTORESIST RESIDUES. DETECTION BY	5-1450
Y. LOW TEMPERATURE OXIDATION OF SILICON STUDIED BY	PHOTOSENSITIVE ESR AND AUGER ELECTRON SPECTROSCOP	5-1786
D STRUCTURAL STUDY OF THE GERMANIUM (100)- CESIUM- OXYGEN	PHOTOSURFACE. AN ELECTRICAL, CHEMICAL, AN	5-1319
AUGER ELECTRON SPECTROSCOPY OF GALLIUM ARSENIDE	PHOTOSURFACES.	5-1922
120) CADMIUM SULFIDE SURFACE. SURFACE	PHOTOVOLTAGE AND AUGER SPECTROSCOPY STUDIES OF (1	5-1201
UM. (AUGER SPECTROSCOPY, SURFACE PREPARATION AND/ SURFACE	PHYSICS OF SOLIDS FROM THE VIEWPOINT OF ULTRAVACU	5-1711
	PHYSICAL ADSORPTION OF XENON ON PALLADIUM (100).	5-1793
M- RHODIUM GAUZE CATALYSTS. (AUGER SPECTROSCOPY)	PHYSICAL AND CHEMICAL CHARACTERIZATION OF PLATINU	5-1433
/LYTICAL TECHNIQUES FOR CHARACTERIZATION OF THE CHEMICAL,	PHYSICAL, AND CRYSTALLOGRAPHIC NATURE OF A SURFA/	1-0107
AUGER SPECTROSCOPY: TEXTBOOK FOR A COURSE OF LECTURES ON	PHYSICAL ELECTRONICS. ELECTRONIC	6-2050
ORROSION AND DEPOSITION IN LIQUID METALS./ APPLICATION OF	PHYSICAL EXAMINATION TECHNIQUES IN THE STUDY OF C	6-2025
HE ION MICROPROBE MASS ANALYZER AND AUGER SPECTROMETER IN	PHYSICAL METALLURGY. POSSIBILITIES OF T	1-0078
	PHYSICAL METHODS OF SURFACE ANALYSIS.	5-1473
/ECT OF LIGHT ELEMENTS NITROGEN, CARBON AND OXYGEN ON THE	PHYSICAL PROPERTIES OF SPUTTERED TANTALUM FILMS./	5-1790
DE- LEAD OXIDE CRYSTALLINE FILMS. PART-1: PREPARATION AND	PHYSICAL STRUCTURE. (AUGER SPECTROSCOPY) /D SULFI	5-1150
QUANTITATIVE AUGER SPECTROSCOPY OF	PHYSICALLY ADSORBED XENON ON NICKEL.	5-1151
ELECTRON BEAM EFFECTS IN AUGER ANALYSIS OF	PHYSISORBED XENON.	2-0250
UGER RECOMBINATION OF HIGH DENSITY PLASMAS IN GERMANIUM.	PICOSECOND OPTICAL MEASUREMENTS OF BAND-TO-BAND A	6-2066
Y USING AUGER ELECTRONS IN THE STUDY OF THE STRUCTURE AND	PITTING CORROSION OF ALUMINUM. /W MICRORADIOGRAPH	6-1992
PREPARATION AND SURFACE OXIDATION OF A VANADIUM (100)	PLANE. (AUGER EFFECT)	5-1559
ADSORPTION OF KRYPTON CN THE BASAL	PLANE OF GRAPHITE: LEED AND AUGER MEASUREMENTS.	5-1318
ADSORPTION OF OXYGEN ON THE (110)	PLANE OF TUNGSTEN. (AUGER ELECTRON SPECTROSCOPY)	5-1420
INVESTIGATIONS. SULFUR ON SINGLE CRYSTAL PLATINUM	PLANES. LOW ENERGY ELECTRON DIFFRACTION AND AUGER	5-1912
THE KINETICS OF OXYGEN ADSORPTION ON THE (112) AND (110)	PLANES OF TUNGSTEN.	6-2093
N SPECTROSCOPY)	PLASMA CLEANING OF METAL SURFACES. (AUGER ELECTRO	2-0344
RADIATIVE AUGER TRANSITIONS IN SOFT X-RAY	PLASMA EMISSION.	5-1699
CTORS BY AUGER ELECTRON SPECTROSCOPY. STUDIES ON SURFACE	PLASMA OXIDIZED FILMS OF III-V COMPOUND SEMICONDU	4-1124
OF MOLYBDENUM.	PLASMA SATELLITES IN THE AUGER ELECTRON SPECTRUM	2-0244
FORMATION OF A	PLASMA SHEATH WITH SECONDARY ELECTRON EMISSION.	2-0276
ELECTRON PHOTOEMISSION INDUCED BY LASER-PRODUCED	PLASMA X-RAYS. (PLASMA ELECTRONS, AUGER PROCESS)	2-0250
MENTS OF BAND-TO-BAND AUGER RECOMBINATION OF HIGH DENSITY	PLASMAS IN GERMANIUM. PICOSECOND OPTICAL MEASURE	2-0291
ON. (AUGER APPEARANCE SPECTROSCOPY)	PLASMON COUPLING TO CORE HOLE EXCITATIONS IN CARB	4-1060
IUM AND IRON. INTERFACIAL AND	PLASMON DECAY PEAKS IN THE AUGER SPECTRA OF TITAN	4-1020
SPECTRA IN ALUMINUM.	PLASMON EFFECTS IN ELECTRON ENERGY LOSS AND AUGER	4-0899
TES IN THE BERYLLIUM AUGER SPECTRUM.	PLASMON ENERGY GAIN AND DOUBLE IONIZATION SATELLI	4-1095
UM.	PLASMON ENERGY GAIN IN THE BERYLLIUM AUGER SPECTR	4-1047
/ON WHEN BOMBARDED BY PRIMARY ELECTRONS. PART-1: ENHANCED	PLASMON EXCITATION ON INNER LEVEL IONIZATION. PA/	2-0440
SURFACE VALENCE BAND AND	PLASMON FEATURES ON CLEAN CLEAVED SILICON.	2-0440
MECHANISM OF	PLASMON GAIN IN AUGER SPECTRA.	4-1085
M AUGER SPECTRUM. OBSERVATION OF A	PLASMON GAIN IN THE FINE STRUCTURE OF THE ALUMINU	4-1037
SILICON. ABSENCE OF	PLASMON GAIN SATELLITES IN THE AUGER SPECTRUM OF	5-1279
/OTASSIUM CHLORIDE AND LITHIUM FLUORIDE BY OBSERVATION OF	PLASMON LOSSES ASSOCIATED WITH POTASSIUM AND LIT/	2-0236
EFFECT OF SURFACE	PLASMON OSCILLATIONS ON AUGER ELECTRON EMISSION.	4-0871
OF METALS. (ALUMINUM)	PLASMON SATELLITE IN THE AUGER EMISSION SPECTRUM	2-0347
NS FOR DETERMINING ADATOM CONCENTRATION PROFILES. AUGER	PLASMON SATELLITE INTENSITIES VERSUS DEPTH: A MEA	2-0447
	PLASMON SATELLITES IN AUGER SPECTRA.	5-1363
SILICON PRODUCED BY THE ADSORPTION OF CE/ LEED, AUGER AND	PLASMON STUDIES OF NEGATIVE ELECTRON AFFINITY ON	5-1745
01) SINGLE CRYSTAL SURFACES USING INELASTIC LOW / SURFACE	PLASMON STUDIES ON ALUMINUM (111) AND ALUMINUM (0	5-1415
RAPHY AND MICROANALYSIS OF A STAINED SURFACE OF PURE IRON	PLATE. AUGER ELECTRON EMISSION MICROG	5-1873
FUNCTION OF WELL-DEFINED SURFACE OF COPPER- NICKEL ALLOY	PLATES. (AUGER ELECTRON SPECTROSCOPY) WORK	5-1550
/NATED AND CLEANED SURFACES OF GOLD THICK FILMS ON COPPER	PLATES BY AUGER ELECTRON AND SECONDARY ION MASS /	3-0699
TRON SPECTRA. NEW, STRIATED, HIGH EFFICIENCY PHOTOGRAPHIC	PLATES FOR RECORDING CONVERSION AND/OR AUGER ELEC	5-1679
TRON SPECTROSCOPY AND SECONDA/ CCMPOSITION PRCFILE OF ION	PLATED GOLD FILM ON COPPER ANALYZED BY AUGER ELEC	5-1225
AUGER ELECTRON SPECTROSCOPY ANALYSIS OF ION	PLATED GOLD FILMS ON SILICON.	5-1539
TUDY OF ACTIVATED SILICON SURFACES FOR ELECTROLESS NICKEL	PLATING BY AES. A S	5-1267
SORPTION OF UNSATURATED MOLECULES BY THE (100) SURFACE OF	PLATINUM. /CTRON DIFFRACTION STUDIES OF THE CHEMI	

FECT OF STRUCTURE ON THE AUGER SPECTRA OF BARIUM FILMS ON PLATINUM.   EF 5-1359
    K-, L-, AND M-AUGER AND L- COSTER-KRONIG SPECTRA OF PLATINUM.   4-1097
ION AND AUGER SPECTROSCOPY OF THE ADSORPTION OF SULFUR ON PLATINUM.   LOW ENERGY ELECTRON DIFFRACT 5-1175
    AES STUDY OF THE SILICIDE FORMATION IN 85% NICKEL- 15% PLATINUM ALLOY FILMS.   5-1899
D SCHOTTKY BARRIER HEIGHTS OF SILICIDES FORMED IN NICKEL- PLATINUM ALLOY FILMS.   COMPOSITION PROFILES AN 5-1900
SPECT/ COMPOSITION PROFILES OF CHEMICALLY VAPOR DEPOSITED PLATINUM AND PLATINUM SILICIDE BY AUGER ELECTRON 5-1656
                         COMPOSITION OF PLATINUM ANODE SURFACES BY AUGER SPECTROSCOPY.   5-1504
N OF HYDROGEN AND DEUTERIUM ON SINGLE CRYSTAL SURFACES OF PLATINUM. (AUGER SPECTROSCOPY)   EQUILIBRATIO 5-1594
                  SECONDARY ELECTRON EMISSION FROM PLATINUM BLACK COATED SURFACES. (AUGER SPECTRA)   5-1979
/F GAS- SURFACE INTERACTIONS AT THE (100) CRYSTAL FACE OF PLATINUM BY MOLECULAR BEAM SCATTERING, LOW ENERG/ 5-1963
ES/ SPATIAL DISTRIBUTIONS OF HYDROGEN DESORBED FROM IRON, PLATINUM, COPPER, NIOBIUM, STAINLESS STEEL SURFAC 6-2006
OCHEMISTRY AND LOW ENERGY ELECTRON DIFFRACTION./ STUDY OF PLATINUM ELECTRODES BY MEANS OF THIN LAYER ELECTR 5-1486
            AUGER SPECTROSCOPY OF THE BARIUM- PLATINUM FILM SYSTEM.   5-1357
STUDIED WITH AUGER SP/ DIFFUSION KINETICS OF GOLD THROUGH PLATINUM FILMS ABOUT 2000 AND 600 ANGSTROMS THICK 5-1235
              POROSITY IN SPUTTERED PLATINUM FILMS. (AUGER ELECTRON SPECTROSCOPY)   5-1965
OSCOPY)   REACTION OF SPUTTERED PLATINUM FILMS ON GALLIUM ARSENIDE. (AUGER SPECTR 5-1563
          SURFACE COMPOSITION OF PLATINUM HEATED IN OXYGEN. (AUGER SPECTRA)   5-1737
SPECTROSCOPY.   ADHESION AND TRANSFER OF PLATINUM IRON TO METALS STUDIED BY AUGER EMISSION 5-1730
ER / ADSORPTION OF SULFUR ON THE (100) AND (111) FACES OF PLATINUM. LOW ENERGY ELECTRON DIFFRACTION AND AUG 5-1421
          AUGER SPECTROSCOPIC STUDY OF PLATINUM, NICKEL, AND IRON SURFACES.   5-1791
SION FROM SOLID SURFACES BOMBARDED BY MEDIUM ENERGY IONS (PLATINUM, NICKEL, TANTALUM, GRAPHITE). /TRON EMIS 2-0309
UGER SPECTROSCOPY.   INTERDIFFUSIONS IN THIN FILM GOLD ON PLATINUM ON GALLIUM ARSENIDE (100) STUDIED WITH A 5-1232
AND AUGER INVESTIGATIONS.   SULFUR ON SINGLE CRYSTAL PLATINUM PLANES. LOW ENERGY ELECTRON DIFFRACTION 5-1420
Y OF THE PLATINUM- PLATINUM SILICIDE INTERFACE.   PLATINUM SILICIDE FORMATION. ELECTRON SPECTROSCOP 5-1285
CKSCATTERING ANALYSES.   EFFECT OF OXIDIZING AMBIENTS ON PLATINUM SILICIDE FORMATION. PART-2: AUGER AND BA 5-1182
/LICON- TANTALUM, SILICON DIOXIDE- TANTALUM, AND SILICON- PLATINUM SILICIDE- TANTALUM THIN FILM STRUCTURES/ 5-1260
      SURFACE STRUCTURE OF AND CATALYSIS BY PLATINUM SINGLE CRYSTAL SURFACES.   5-1837
ECTRON DIFFRACTI/ STUDY OF THE INTERACTION OF OXYGEN WITH PLATINUM SINGLE CRYSTAL SURFACES BY LOW ENERGY EL 5-1223
EC/ REACTIVITY OF LOW INDEX ((111) AND (100)) AND STEPPED PLATINUM SINGLE CRYSTAL SURFACES. (LEED, AUGER EL 5-1838
      SMALL MOLECULE REACTIONS ON STEPPED SINGLE CRYSTAL PLATINUM SURFACES.   5-1173
Y.   DIFFUSION STUDIES IN GOLD- PLATINUM THIN FILMS BY AUGER ELECTRON SPECTROSCOP 5-1636
COPY.   DIFFUSION STUDIES IN CHROMIUM- PLATINUM THIN FILMS USING AUGER ELECTRON SPECTROS 5-1286
AUGER SPECTROSCOPY OF OSMIUM, AND THE SEGREGATION OF PLATINUM TO THE OSMIUM SURFACE.   5-1770
          ADSORPTION OF CYANOGEN ON PLATINUM (100).   5-1684
LECTRON SPECTROSCOPY.   CHEMICAL ANALYSIS OF PLATINUM (100) AND GOLD (100) SURFACES BY AUGER E 5-1712
E.   LOW ENERGY ELECTRON SPECTROSCOPY OF PLATINUM (100) AND PLATINUM (100)- CARBON MONOXID 5-1686
N OF HYDROGEN AND THE REACTION OF HYDROGEN WITH OXYGEN ON PLATINUM (100). (AUGER SPECTROSCOPY)   ADSORPTIO 5-1685
    CHANGE IN THE SURFACE ELECTRONIC STRUCTURE OF PLATINUM (100) INDUCED BY RECONSTRUCTION.   5-1186
PROPERTIES. (AUGER EFFECT)   PLATINUM (100) STRUCTURES: IMPURITY CAUSATION AND 5-1311
ER) ADSORPTION STUDIES OF OXYGEN AND CARBON MONOXIDE ON A PLATINUM (100) SURFACE.   (AUG 5-1546
          (AUGER) ADSORPTION STUDIES ON THE PLATINUM (100) SURFACE.   5-1570
              LEED STUDY OF PLATINUM (100) SURFACE.   5-1377
          STRUCTURE OF THE PLATINUM (100) SURFACE.   5-1374
MECHANISMS OF THE CATALYTIC CARBON MONOXIDE OXIDATION ON PLATINUM (110).   5-1188
          CARBON MONOXIDE OXIDATION ON A PLATINUM (110) SINGLE CRYSTAL SURFACE.   5-1187
MONOXIDE LEED PATTERNS. (AU/ ADSORBATE INTERACTIONS ON A PLATINUM (110) SURFACE. PART-1: SULFUR AND CARBON 5-1190
NVESTIGATION OF THE CHARACTERISTIC AUGER ENERGY LOSSES OF PLATINUM (111).   I 2-0499
ADSORPTION AND DECOMPOSITION OF CARBON MONOXIDE ON PLATINUM (111). (AUGER SPECTROMETRY)   5-1622
N INDUCED DESORPTION AND DISSOCIATION: CARBON MONOXIDE ON PLATINUM (111). (AUGER SPECTRUM) /RONS IN ELECTRO 5-1567
          ADSORPTION OF HYDROGEN ON A PLATINUM (111) SURFACE.   5-1255
D- AUGER INVESTIGATION OF A STABLE CARBIDE OVERLAYER ON A PLATINUM (111) SURFACE.   LEE 5-1569
OPY, STICKING COEFFIC/ KINETICS OF OXYGEN ADSORPTION ON A PLATINUM (111) SURFACE. (AUGER ELECTRON SPECTROSC 5-1189
          CHEMISORPTION OF OXYGEN ON PLATINUM (111) SURFACES. (AUGER SPECTRA)   5-1500
    APPLICATION OF AUGER ELECTRON SPECTROSCOPY TO THE PLATINUM- OXYGEN SYSTEM AT HIGH TEMPERATURE.   5-1955
PLATINUM SILICIDE FORMATION. ELECTRON SPECTROSCOPY OF THE PLATINUM- PLATINUM SILICIDE INTERFACE.   5-1285
SCOPY)   PHYSICAL AND CHEMICAL CHARACTERIZATION OF PLATINUM- RHODIUM GAUZE CATALYSTS. (AUGER SPECTRO 5-1793
C/ SURFACE COMPOSITION AND DEPTH CONCENTRATION PROFILE OF PLATINUM- TIN ALLOYS FROM COMBINED X-RAY PHOTOELE 5-1195
/ AUGER SPECTROSCOPIC STUDY OF THE SURFACE COMPOSITION OF PLATINUM- TIN IN ULTRAHIGH VACUUM AND IN THE PRES 5-1196
TUDY OF INTERDIFFUSION AND EUTECTIC FORMATION IN TUNGSTEN PLATINUM- TUNGSTEN GOLD THIN FILMS. /ECTROSCOPY S 5-1259
AUGER ELECTRON AND CHARACTERISTIC ENERGY LOSS SPECTRA OF PLUTONIUM.   4-0931
          THE L-AUGER SPECTRA OF PLUTONIUM, AMERICIUM, CALIFORNIUM, AND FERMIUM. 4-0836
GENATION CATALYST.   AUGER SPECTROSCOPIC STUDY OF THE POISONING OF A COMMERCIAL PALLADIUM ALUMINA HYDRO 5-1177
    AUGER ELECTRON SPECTROSCOPIC STUDY OF THE SURFACE POISONING OF COPPER CATALYSTS.   5-1176
/TROSCOPY STUDY OF CATHODE SURFACES DURING ACTIVATION AND POISONING. PART-1: BARIUM ON OXYGEN ON TUNGSTEN / 5-1841
ND WORK FUNCTION MEASUREMENTS ON CLEAN AND CESIUM COVERED POLAR FACES OF GALLIUM ARSENIDE. /SPECTROSCOPY, A 5-1294
ERGY ELECTRON DIFFRACTION (LEED) OF THE ZINC OXIDE (0001) POLAR SURFACES. /ON SPECTROSCOPY (AES) AND LOW EN 5-1338
PHOTOELECTRON SPECTROSCOPY STUDY OF THE ZINC OXIDE (0001) POLAR SURFACES. /LECTRON SPECTROSCOPY, AND X-RAY 5-1337
          LEED AND AES OBSERVATIONS ON THE POLAR SURFACES OF ZINC OXIDE.   5-1619
IUM ANTIMON/ LOW ENERGY ELECTRON DIFFRACTION STUDY OF THE POLAR (111) SURFACES OF GALLIUM ARSENIDE AND GALL 5-1602
TOEFFECT IN THE X-RAY SPECTRAL REGION. (EF/ EFFECT OF THE POLARITY OF INDIUM ANTIMONIDE ON THE EXTERNAL PHO 4-1001
OF AUGER TRANSITION IN METALS.   EFFECT OF POLARIZATION OF THE ELECTRON GAS ON THE ENERGIES 2-0435
LECTRON SPECTROSCOPY.   SLIDING OF POLY (VINYL CHLORIDE) ON METALS STUDIED BY AUGER E 5-1727
SECONDARY ELECTRON EMISSION FROM IRON SINGLE AND POLYCRYSTALS.   4-0917
GER SPECTROSCOPY)   CHARACTERISTIC GAIN PHENOMENA IN POLYCRYSTALLINE COPPER, SILVER, AND ALUMINUM. (AU 5-1724
ERING TECHNIQUE.   INTERDIFFUSION IN BIMETALLIC POLYCRYSTALLINE FILMS MEASURED BY AUGER ION SPUTT 5-1440
CURY ON GOL/ AUGER ELECTRON SPECTROSCOPY STUDIES OF CLEAN POLYCRYSTALLINE GOLD AND OF THE ADSORPTION OF MER 5-1519
          AUGER ELECTRON SPECTROSCOPY OF CLEAN POLYCRYSTALLINE GOLD FILMS.   5-1760
TIONS AND AUGE/ DEVELOPMENT OF STEPPED SURFACE REGIONS ON POLYCRYSTALLINE GOLD. LOW ENERGY ELECTRON DIFFRAC 5-1483
SPECTRUM OF THE SLOW SECONDARY ELECTRON EMISSION PEAK IN POLYCRYSTALLINE MAGNESIUM. /DATION EFFECTS ON THE 4-1118
ROMETRIC STUDY OF THE CHEMISORPTION OF CARBON MONOXIDE ON POLYCRYSTALLINE MOLYBDENUM. /ETRIC AND MASS SPECT 5-1355
URFACES. (AUGER NEUTRALIZATION AT (100) FACE OF TUNGSTEN, POLYCRYSTALLINE MOLYBDENUM). /S WITH CONTROLLED S 5-1736
M/ CARBON CONTAMINATION PRODUCED BY ELECTRON BOMBARDMENT. POLYCRYSTALLINE MOLYBDENUM- CARBON MONOXIDE SYSTE 5-1648
SURFACE PRECIPITATION AND SURFACE DIFFUSION OF CARBON ON POLYCRYSTALLINE NICKEL. /CTROSCOPY TO THE BULK TO 5-1651
, ETHYLENE AND CARBON MONOXIDE ADSORPTION ON TANTALUM AND POLYCRYSTALLINE RHENIUM. /RBON. TANTALUM CARBIDES 4-0820
SPECTROSCO/ ADSORPTION AND DECOMPOSITION OF ACETYLENE ON POLYCRYSTALLINE RHENIUM FILMS AND RIBBONS. (AUGER 5-1688
INTERACTION OF ALUMINUM LAYERS WITH POLYCRYSTALLINE SILICON. (AUGER SPECTROSCOPY)   5-1672
          SECONDARY ELECTRON SPECTROSCOPY OF POLYCRYSTALLINE SILVER.   4-1016
EASUREMENTS.   DETECTION OF CONTAMINANTS ON POLYCRYSTALLINE SILVER BY AES AND WORK FUNCTION M 5-1574
ILMS FORMED ON METALS IN SLIDING CONTACT WITH HALOGENATED POLYMERS.   AUGER ANALYSIS OF F 5-1728
, AND WEAR.   ATOMIC NATURE OF POLYMER- METAL INTERACTIONS IN ADHESION, FRICTION 5-1212

```
TROSCOPY.                              SURFACE ABSORPTION BY POLYPEPTIDE FILMS EXAMINED BY AUGER EMISSION SPEC 5-1179
/FFUSED IMPURITY PROFILES IN SILICON DIOXIDE- SILICON AND POLYSILICON- SILICON DIOXIDE- SILICON STRUCTURES.  5-1481
S STUDIED BY AUGER SPECTROSCOPY/ ADHESION AND FRICTION OF POLYTETRAFLUORO ETHYLENE IN CONTACT WITH METALS A 5-1198
ER EMISSION SPECTROSCOPY.        ADHESION AND TRANSFER OF POLYTETRAFLUORO ETHYLENE TO METALS STUDIED BY AUG 5-1731
/FACE CHARACTERIZATION OF NICKEL- COPPER ALLOY DISPLAYING POOR WETTABILITY DURING BRAZING. (MONEL, AUGER S/ 5-1573
TRON SPECTROSCOPY)                                       POROSITY IN SPUTTERED PLATINUM FILMS. (AUGER ELEC 5-1965
ARACTERISTICS OF SHAPING SECONDARY ELECTRON EMISSION FROM POROUS POTASSIUM CHLORIDE FILMS.              CH 4-1120
NSITY IN A SINGLE STAGE CYLINDRICAL MIRROR ANAL/ SPECIMEN POSITION EFFECTS ON ENERGY SHIFTS AND SIGNAL INTE 3-0696
                                                          POSITIONS OF THE KLL AUGER LINES.             4-0747
NS IN AN AUGER ELECTRON SPECTROMETER WITH SCANNING SAMPLE POSITIONER.              ELECTRON CHANNELING PATTER 3-0692
S.        INVESTIGATION OF REORDERED (001) GOLD SURFACES BY POSITIVE ION CHANNELING SPECTROSCOPY, LEED AND AE 5-1139
ELE/ CHARACTERIZATION CF REORDERED (001) GOLD SURFACES BY POSITIVE ION CHANNELING SPECTROSCOPY, LOW ENERGY 5-1986
/OM CLEAN AND CARBON CONTAMINATED MOLYBDENUM BOMBARDED BY POSITIVE IONS. PART-1. PART-2: EFFECT OF ANGLE O/ 2-0543
AND AUGER SPECTROMETER IN PHYSICAL METALLURGY.           POSSIBILITIES OF THE ION MICROPROBE MASS ANALYZER 6-2075
   "SEMI-AUGER" PROCESSES IN L2,3 EMISSION IN ARGON AND POTASSIUM CHLORIDE.                             2-0326
ATION OF PLASMON LOSSES/ STUDY OF SURFACE DISSOCIATION IN POTASSIUM CHLORIDE AND LITHIUM FLUORIDE BY OBSERV 5-1279
MENT.                              SURFACE DISSOCIATION OF POTASSIUM CHLORIDE BY LOW ENERGY ELECTRON BOMBARD 5-1719
ISTICS OF SHAPING SECONDARY ELECTRON EMISSION FROM POROUS POTASSIUM CHLORIDE FILMS.              CHARACTER 4-1120
SSION SPECTROSCOPY.            GROWTH OF SILVER ON POTASSIUM CHLORIDE OBSERVED BY LEED AND AUGER EMI 5-1348
/TERISTICS RELATED TO LASER DAMAGE OF LITHIUM NIOBATE AND POTASSIUM CHLORIDE SURFACES. (AUGER SURFACE PROF/ 5-1746
ENERGY ELECTRON DIFFRACTIO/ EPITAXIAL GROWTH OF SILVER ON POTASSIUM CHLORIDE (100) SURFACE OBSERVED BY LOW 5-1696
   PHOTOEMISSION STUDIES OF CORE EXCITON DECAY IN POTASSIUM IODIDE.                             2-0422
CONDARY ELECTRON EMISSION ON ENERGY OF PRIMARY ELECTRONS (POTASSIUM IODIDE). /ENCE OF THE COEFFICIENT OF SE 4-0891
M ULTRAVI/ SPECTRAL LINE EXCITATION FUNCTIONS FOR SODIUM, POTASSIUM, RUBIDIUM, AND CESIUM IONS IN THE VACUU 4-0735
/SOLUTION AND SENSITIVITY OF THE SPHERICAL GRID RETARDING POTENTIAL ANALYZER. APPLICATION TO AUGER ELECTRO/ 3-0634
ECTROSCOPY)               IMPROVED SPHERICAL GRID RETARDING POTENTIAL ANALYZER. (FOR USE IN AUGER ELECTRON SP 3-0583
ION ANALOG FOR IMPROVED AUGER SPECTROSCOPY WITH RETARDING POTENTIAL ANALYZERS. COMMENTS.      SECONDARY EMISS 3-0584
ION ANALOG FOR IMPROVED AUGER SPECTROSCOPY WITH RETARDING POTENTIAL ANALYZERS. (REPLY TO COMMENTS) /Y EMISS 3-0694
ON TUNGSTEN: ELLIPSOMETRY, AUGER SPECTROSCOPY AND SURFACE POTENTIAL DIFFERENCE STUDIES.   CESIUM ADSORPTION 5-1830
FOR MODULATION AMPLITUDE DISTORTION IN A 4-GRID RETARDING POTENTIAL ENERGY ANALYZER. /RON SIGNAL STRENGTHS 3-0620
16 R/ ELECTRON EMISSION FROM SOLIDS BY PARTICLES CARRYING POTENTIAL ENERGY. (AUGER NEUTRALIZATION, REVIEW, 2-0374
PY.                                                       POTENTIAL MAPPING USING AUGER ELECTRON SPECTROSCO 1-0113
ER ELECTRON SPECTROSCOPY IN SCANNING ELECTRON MICROSCOPY: POTENTIAL MEASUREMENTS.              AUG 3-0648
CTRIC LOW ENERGY ELECTRON DIFFRACTION, AUGER, AND CONTACT POTENTIAL MEASUREMENTS. /INUM (111) SURFACES: ELE 5-1744
EMI SORPTION AND CATALYTIC REACTIONS, LEED, AUGER, CONTACT POTENTIAL METHOD AND MASS SPECTROMETRY) /ACE. (CH 5-1321
                                            TRANSITION POTENTIAL MODELS FOR AUGER ELECTRON SPECTRA.  2-0510
ASUREMENTS.                           EXACT CORRECTIONS FOR POTENTIAL MODULATION DISTORTION IN AUGER YIELD ME 2-0389
S BY AUGER ELECTRON SPECTROSCOPY IN THE SCANNING ELECTRC/ POTENTIAL PROFILING ACROSS SEMICONDUCTOR JUNCTION 2-0555
   USE OF CYLINDRICAL AUGER SPECTROMETERS FOR RETARDING POTENTIAL SECONDARY ELECTRON YIELD MEASUREMENTS. 3-0628
                N5,4 AND M5,4 APPEARANCE POTENTIAL SPECTRA FROM PRASEODYMIUM.          4-0908
AM / CHANGES IN THE TITANIUM DIOXIDE AUGER AND APPEARANCE POTENTIAL SPECTRA INDUCED BY INCIDENT ELECTRON BE 5-1834
NS, ELECTRONS AND IONS.           HIGH SENSITIVITY APPEARANCE POTENTIAL SPECTROMETER FOR THE DETECTION OF PHOTO 3-0643
UANTIFICATION OF AUGER ELECTRON AND SOFT X-RAY APPEARANCE POTENTIAL SPECTROSCOPIES. /ROUND SUBTRACTION TO Q 1-0067
E SURFACES USING AUGER ELECTRON AND SOFT X-RAY APPEARANCE POTENTIAL SPECTROSCOPIES. /M AND TITANIUM MONOXID 5-1384
N SPECTROSCOPY.   IONIZATION LOSS SPECTROSCOPY. APPEARANCE POTENTIAL SPECTROSCOPY.              AUGER ELECTRO 1-0174
   COMPARISON OF SOFT X-RAY AND AUGER ELECTRON APPEARANCE POTENTIAL SPECTROSCOPY.                   1-0092
ATIO MEASURED BY SOFT X-RAY AND AUGER ELECTRON APPEARANCE POTENTIAL SPECTROSCOPY. /ROMIUM L3/L2 INTENSITY R 4-0890
COMPARISON OF AUGER ELECTRON SPECTROSCOPY WITH APPEARANCE POTENTIAL SPECTROSCOPY.              DIRECT 1-0142
   SURFACE ANALYSIS BY APPEARANCE POTENTIAL SPECTROSCOPY.              5-1545
AL ANALYSIS BY AUGER ELECTRON SPECTROSCOPY AND APPEARANCE POTENTIAL SPECTROSCOPY: A COMPARISON. /CE CHEMIC 5-1910
/ZATION (CHARACTERISTIC LOSS) SPECTROSCOPY AND APPEARANCE POTENTIAL SPECTROSCOPY. COMPARISON WITH AUGER SP/ 1-0173
           SOFT X-RAY APPEARANCE POTENTIAL SPECTROSCOPY IN A DISPLAY LEED SYSTEM. 4-0870
L AND IRON- NICKEL ALLCYS. (AES)   SOFT X-RAY APPEARANCE POTENTIAL SPECTROSCOPY STUDIES WITH COPPER- NICKE 5-1952
           DISAPPEARANCE POTENTIAL SPECTROSCOPY (VANADIUM OXIDATION).  6-2068
YSTAL SURFACES.      LEED, AUGER SPECTROSCOPY, AND CONTACT POTENTIAL STUDIES OF COPPER- GOLD ALLOY SINGLE CR 5-1748
RBED ON COPPER (100).              LEED AND SURFACE POTENTIAL STUDY OF CARBON MONOXIDE AND XENON ADSO 5-1246
T OF CHANGES IN WORK FUNCTION. (LEED- AU/ AN AC RETARDING POTENTIAL TECHNIQUE FOR THE CONTINUOUS MEASUREMEN 3-0663
                       SURFACE MADELUNG POTENTIALS IN ELECTRON SPECTROSCOPY.      2-0511
   AUGER ELECTRON SPECTRUM AND IONIZATION POTENTIALS OF THE HYDROGEN FLUORIDE MOLECULE.  4-1058
   TECHNIQUE OF PREPARING POWDER SAMPLES FOR AES AND/OR ESCA ANALYSIS. 1-0197
AND AUGER RECOMBINATION AND CARRIER-CARRIER SCATTERING IN POWER RECTIFIERS.              BAND-TO-B 2-0458
   N5,4 AND M5,4 APPEARANCE POTENTIAL SPECTRA FROM PRASEODYMIUM.          4-0908
/ ELECTRON SPECTRA OF SOME RARE EARTH METALS. (LANTHANUM, PRASEODYMIUM, NEODYMIUM, GADOLINIUM, DYSPROSIUM,/ 4-1100
   AUGER ELECTRON SPECTROMETER PREAMPLIFIER.              3-0708
   DOUBLY RESONANT CURRENT-TO-VOLTAGE CONVERTER PREAMPLIFIER FOR SURFACE SPECTROSCOPY.  3-0677
   METHODS FOR SURFACE ANALYSIS OF PRECIOUS METALS CONTACT MATERIALS.      5-1785
/ON OF AUGER ELECTRON SPECTROSCOPY TO THE BULK TO SURFACE PRECIPITATION AND SURFACE DIFFUSION OF CARBON ON/ 5-1651
   (AUGER ELECTRON SPECTROSCOPY OF) PRECISION SINGLE SIDEBAND CRYSTAL UNITS.  5-1409
                       PREDICTION OF THE RELATIVE INTENSITIES OF AUGER L 4-1072
INES IN MICA.                       PREFERENTIAL EVAPORATION OF INDIUM FROM GALLIUM(X 5-1366
) INDIUM(1-X) ARSENIDE. (AUGER EFFECT)               SURFACE PREPARATION AND CHARACTERIZATION TECHNIQUES (AUGE 5-1943
R, CLEANING) FOR QUARTZ RESONATORS.               SURFACE PREPARATION AND CLEANLINESS) /CS OF SOLIDS FROM T 5-1419
HE VIEWPOINT OF ULTRAVACUUM. (AUGER SPECTROSCOPY, SURFACE PREPARATION AND PHYSICAL STRUCTURE. (AUGER SPECT/ 5-1790
/ THE LEAD SULFIDE- LEAD OXIDE CRYSTALLINE FILMS. PART-1: PREPARATION AND SURFACE OXIDATION OF A VANADIUM ( 6-1992
100) PLANE. (AUGER EFFECT)                                 PREPARATION AND SURFACE OXIDATION OF A (110) FACE 5-1683
OF VANADIUM.                                              PREPARATION AND SURFACE OXIDATION OF A (110) FACE 5-1683
BY AUGER ELECTRON SPECTROSCOPY.               STUDY OF THE PREPARATION OF ATOMICALLY CLEAN TUNGSTEN SURFACES 5-1517
YSIS.                                         TECHNIQUE OF PREPARING POWDER SAMPLES FOR AES AND/OR ESCA ANAL 1-0197
F GRAIN BOUNDARY IMPURITIES AND FLUORIDE ADDITIVES IN HOT PRESSED OXIDES BY AUGER ELECTRON SPECTROSCOPY. /O 5-1506
           EFFECT OF YTTRIA ADDITIONS ON HOT PRESSED SILICON NITRIDE. (AUGER ANALYSIS)       5-1350
OBE ANALYSIS)               MICROSTRUCTURE OF HOT PRESSED SILICON NITRIDE. (AUGER ANALYSIS, MICROPR 5-1556
ALYSIS)               IMPURITIES AND INCLUSIONS IN HOT PRESSED SILICON NITRIDE. (AUGER SPECTROGRAPHIC AN 5-1557
SCOPY.               STUDY OF FRACTURE SURFACES OF HOT PRESSED SILICON NITRIDE BY AUGER ELECTRON SPECTRO 5-1444
SCOPY.   IDENTIFICATION OF A GRAIN BOUNDARY PHASE IN HOT PRESSED SILICON NITRIDE BY AUGER ELECTRON SPECTRO 5-1749
MENTS OF ADSORPTION ISCBARS FOR OXYGEN ON TUNGSTEN AT LOW PRESSURE AND HIGH TEMPERATURE. /CTROSCOPY MEASURE 5-1282
/S OF OXYGEN ON TUNGSTEN (100), AND (111) SURFACES IN THE PRESSURE RANGE 10(-9) TO 10(-4) TORR AND TEMPERA/ 5-1148
CTROSCOPY ANALYSIS OF THE FRACTURE SURFACES OF IRRADIATED PRESSURE VESSEL STEELS.      AUGER ELECTRON SPE 5-2113
OPY OF HYDROCHLORIC ACID INTERACTION WITH ALUMINUM AT LOW PRESSURES.              AUGER SPECTROSC 5-1576
/TION OF NIMONIC-80A ALLOY AT 800 DEGREES C IN LOW OXYGEN PRESSURES. (AUGER ELECTRON SPECTROSCOPY, SURFACE/ 6-2144
ELEMENTS).               SECONDARY ELECTRON EMISSION DUE TO PRIMARY AND BACKSCATTERED ELECTRONS (TABLE FOR 60 4-0907
ED DESORPTION AND DISSOCIATION: CARBON MONOXIDE / ROLE OF PRIMARY AND SECONDARY ELECTRONS IN ELECTRON INDUC 5-1567
```

UGER ELECTRON SPECTROSCOPY AT HIGH SPATIAL RESOLUTION AND PRIMARY BEAM CURRENTS. (CARBON ON SILVER FILM) A 5-1904
AUGER SYSTEM FOR THE ANALYSIS OF ELECTRON EMISSION AT LOW PRIMARY BEAM ENERGIES. /ENTS TO A STANDARD LEED- 3-0698
NCE. DEPENDENCE OF AUGER ELECTRON YIELD ON PRIMARY BEAM ENERGY AT NORMAL AND GLANCING INCIDE 2-0455
METHOD TO DISTINGUISH AUGER FROM ENERGY LOSS PEAKS OF THE PRIMARY BEAM IN SECONDARY ELECTRON ANALYSIS. /NG 1-0188
R ELECTRON SPECTROSCOPY. IMPORTANCE OF ATTENUATION OF THE PRIMARY ELECTRON BEAM. /NTITATIVE ASPECTS OF AUGE 1-0206
N COEFFICIENT OF SINGLE CRYSTALS IN THE 2-10 KEV RANGE OF PRIMARY ELECTRON ENERGIES. /DARY ELECTRON EMISSIO 4-1061
TRUE SECONDARY ELECTRONS. PART-2: INELASTICALLY SCATTERED PRIMARY ELECTRONS. / METALS. PART-1: EMISSION OF 2-0500
EFFECT OF THE INCIDENCE ANGLE OF PRIMARY ELECTRONS ON THE AUGER EMISSION YIELD. 2-0237
/TRONS EMITTED BY MAGNESIUM AND SILICON WHEN BOMBARDED BY PRIMARY ELECTRONS. PART-1: ENHANCED PLASMON EXCI/ 4-1047
E COEFFICIENT OF SECONDARY ELECTRON EMISSION ON ENERGY OF PRIMARY ELECTRONS (POTASSIUM IODIDE). /ENCE OF TH 4-0891
ST, ACCURATE SECONDARY ELECTRON YIELD MEASUREMENTS AT LOW PRIMARY ENERGIES. (LEED- AUGER SYSTEM) FA 3-0629
MPER BRITTLENESS IN A NICKEL- CHROMIUM- ANTIMO/ EFFECT OF PRIOR AUSTENITIC GRAIN BOUNDARY COMPOSITION ON TE 6-2114
AUGER AND RADIATIVE TRANSITION PROBABILITIES. 2-0328
KLL AUGER TRANSITION PROBABILITIES. 2-0304
RELATIVISTIC STUDY OF KLL AUGER TRANSITION PROBABILITIES. 4-0802
FLUORESCENCE YIELDS AND L2-L3(X) COSTER-KRONIG TRANSITION PROBABILITIES. THEORETICAL L2 AND L3 SUBSHELL 2-0316
FLUORESCENCE YIELDS, AUGER, AND COSTER-KRONIG TRANSITION PROBABILITIES. X-RAY 2-0259
KLL AUGER TRANSITION ABSOLUTE PROBABILITIES. (ATOMIC NUMBERS 30-94) 2-0251
ON THE ATOMIC NUMBER 29 TO 94. RELATIVE PROBABILITIES FOR AUGER KLL TRANSITIONS DEPENDENT 2-0253
T ON THE ATOMIC NUMBER Z=30-94. PART-2. ABSOLUTE PROBABILITIES FOR AUGER- KLL TRANSITIONS DEPENDEN 2-0252
ATE ATOMIC NUMBERS. KLL AUGER TRANSITION PROBABILITIES FOR ELEMENTS WITH LOW AND INTERMEDI 4-0803
LEAN NICKEL (111), NICKEL (100), SHEET AND EVAP/ STICKING PROBABILITIES FOR THE ADSORPTION OF HYDROGEN ON C 5-1468
LENIUM. TRANSITION PROBABILITIES FOR THE L2,3MM AUGER SPECTRUM OF SE 4-1033
/UBSHELL FLUORESCENCE YIELDS AND CCSTER-KRONIG TRANSITION PROBABILITIES F23 BASED ON THE GREEN-SELLIN-ZACH/ 2-0313
. TRANSITICN PROBABILITIES OF KL- LLL AUGER SATELLITES IN NEON 4-1049
CAL L1 FLUORESCENCE Y/ AUGER AND COSTER-KRONIG TRANSITION PROBABILITIES TO THE ATOMIC 2S STATE AND THEORETI 2-0339
HELL FLUORESCENCE YIELDS. ATOMIC RADIATION TRANSITION PROBABILITIES TO THE .1S STATE AND THEORETICAL K-S 4-0918
TRACE ANALYSIS TECHNIQUES FOR SOLIDS. (ELECTRON PROBE ANALYSIS, AUGER EFFECT, REVIEW) 1-0093
ELECTRON BEAM MICROANALYSIS. PART-1. (REVIEW OF ELECTRON PROBE ANALYZER, SCANNING ELECTRON MICROSCOPE, TR/ 1-0010
1: APPLICATION OF AUGER ELECTRON SPECTROSCOPY TO ELECTRON PROBE MICROANALYSIS. / ELECTRON MICROSCOPY. PART- 1-0213
ELECTRON PROBE MICROANALYSIS AND ELECTRON SPECTROSCOPY. 1-0028
/POSITION AND STRUCTURE ANALYSIS WITH A ONE APPARATUS ION PROBE TECHNIQUE. (COMPARISON BETWEEN ION PROBE A/ 5-1166
SPECTROSCOPY. PROBING DEPTH IN PHOTOEMISSION AND AUGER ELECTRON 2-0428
RIDE ION IN SODIUM CHLORIDE BY MEANS OF THE DECOMPOSITION PRODUCTS. /XCITONS NEAR THE L2,3 EDGE OF THE CHLO 4-1126
TRON MICROSC/ MASS TRANSPORT OF STAINLESS STEEL CORROSION PRODUCTS IN FLOWING LIQUID SODIUM. (SCANNING ELEC 6-2014
D IN 50 MEV CHLORINE(+)- NEON COLLISIONS. (AUGER ELECTRON PRODUCTION) /ELL IONIZATION AND FLUORESCENCE YIEL 2-0518
CRYSTALLINE EFFECTS IN BACKSCATTERING AND AUGER PRODUCTION. (ALUMINUM) 2-0254
ON AND PHOSPHORUS. PRODUCTION AND DECAY OF DOUBLE L VACANCIES IN ARG 2-0489
R EMISSION. ESTIMATES OF EFFICIENCIES OF PRODUCTION AND DETECTION OF ELECTRON EXCITED AUGE 2-0288
GER SPECTRUM) X-RAY PRODUCTION BY HEAVY IONS. (OXYGEN BOMBARDMENT, AU 2-0479
(AUGER ELECTRON EMISSION) NEON K-SHELL VACANCY PRODUCTION BY INTERMEDIATE ENERGY FLUORINE IONS. 2-0572
K-SHELL AUGER ELECTRON PRODUCTION CROSS SECTIONS FROM ION BOMBARDMENT. 2-0570
NEON, FLUORINE) K-SHELL AUGER ELECTRON PRODUCTION CROSS SECTIONS FROM ION BOMBARDMENT. (2-0574
CHARGE DEPENDENCE OF DEPENDENCE OF TOTAL K-SHELL VACANCY PRODUCTION CROSS SECTIONS IN ARGON BY THE IMPACT/ 2-0446
) MECHANISMS OF ELECTRON PRODUCTION IN ION- ATOM COLLISIONS. (AUGER EFFECT 2-0488
, REVIEWS) MECHANISMS OF ELECTRON PRODUCTION IN ION- ATOM COLLISIONS. (AUGER EFFECT 2-0490
S. AUGER ELECTRON AND X-RAY PRODUCTION IN 50- TO 200 KEV NEON- NEON COLLISION 4-1079
DIFFRACTION EFFECTS IN BACKSCATTERING AND AUGER PRODUCTION NEAR CRYSTAL SURFACES. 2-0243
CTRON DIFFRACTION- AUGER ELECTRON SPECTROSCOPY. PRODUCTION OF A SPHERICAL GRID FOR LOW ENERGY ELE 3-0668
IN GERMANIUM AND SILICON STUDIED BY A MICROWAVE METHOD. (PRODUCTION OF AUGER ELECTRONS) /ECTRIC TRANSITION 4-0752
ATOM COLLISIONS. (AUGER EFFECT) PRODUCTION OF INNER SHELL VACANCIES IN HEAVY ION- 2-0405
ECTRA OF NITROGEN BY SLOW PROTON IMPACT. PRODUCTION OF INTENSE SATELLITE LINES IN AUGER SP 4-1076
X-RAY COINCIDENCE METHOD. (AUGER TRA/ MEASUREMENT OF THE PRODUCTION OF LL VACANCY STATES BY THE L X-RAY- L 2-0482
TROGEN(2)(2X) FROM NITROGEN BY EL/ CROSS SECTIONS FOR THE PRODUCTION OF NITROGEN(2)(X), NITROGEN(X), AND NI 2-0376
KEV ENE/ DYNAMIC CHARGE STATE DEPENDENCE OF THE K-VACANCY PRODUCTION YIELD IN HEAVY ION ATOM COLLISIONS AT 2-0346
M NIOBATE AND POTASSIUM CHLORIDE SURFACES. (AUGER SURFACE PROFILE) /STICS RELATED TO LASER DAMAGE OF LITHIU 5-1746
T BY AUGER SPECTROSCOPY. CHEMICAL PROFILE ANALYSIS OF TYPE 316 STAINLESS STEEL SHEE 6-1996
L AND ANODIC OXIDE/ CORRELATION BETWEEN THE COMPOSITIONAL PROFILE AND ELECTRICAL CONDUCTIVITY OF THE THERMA 5-1971
AUGER PROFILE ANOMALIES. (REVIEW) 1-0184
A1 KLL AUGER PROFILE ARTIFACT AT THE OXIDE- METAL INTERFACE. 4-1068
HOSPHIDE (AND) NITROGEN, OXYGEN, AND FLUORINE I/ IN-DEPTH PROFILE DETECTION LIMITS OF NITROGEN IN GALLIUM P 5-1916
N SPECTROSCOPY. OBSERVATION OF IMPURITY PROFILE IN ION IMPLANTED SILICON BY AUGER ELECTRO 5-1482
D BY AUGER ELECTRON SPECTROSCOPY AND SECONDA/ COMPOSITION PROFILE OF ION PLATED GOLD FILM ON COPPER ANALYZE 5-1679
AY PHOTOELEC/ SURFACE COMPOSITION AND DEPTH CONCENTRATION PROFILE OF PLATINUM- TIN ALLOYS FROM COMBINED X-R 5-1195
AUGER ELECTRON SPECTROSCOPY AND DEPTH PROFILE STUDY OF OXIDATION MODIFIED 440C STEEL. 6-2029
AUGER ELECTRON SPECTROSCOPY AND DEPTH PROFILE STUDY OF OXIDATION OF MODIFIED 440C STEEL 6-2032
ERSUS DEPTH: A MEANS FOR DETERMINING ADATOM CONCENTRATION PROFILES. AUGER PLASMON SATELLITE INTENSITIES V 2-0347
AUGER ELECTRON SPECTROSCOPY AND RUTHERFORD BACKSCATTERING PROFILES. /RRELATION OF RESISTIVITY CHANGES WITH 5-1393
CTROSCOPY AND INERT GAS SPUTTERING FOR OBTAINING CHEMICAL PROFILES. USE OF AUGER ELECTRON SPE 1-0156
S FORMED IN NICKEL- PLATINUM ALLOY FILMS. COMPOSITION PROFILES AND SCHOTTKY BARRIER HEIGHTS OF SILICIDE 5-1900
OXIDE FILMS: ELEMENT PROFILES BY AUGER ELECTRON SPECTROSCOPY. 4-1073
EMISSION. IN-DEPTH PROFILES BY AUGER SPECTROSCOPY AND SECONDARY ION 4-0995
EVALUATION OF CONCENTRATION DEPTH PROFILES BY SPUTTERING IN SIMS AND AES. 5-1442
FLUORINE ION IMPLANTATION PROFILES IN GALLIUM ARSENIDE AS DETERMINED BY AES 5-1398
AUGER ELECTRON FINE PROFILES IN IONIC CRYSTALS. 4-0949
G AUGER SPECTROSCOPY. PHOSPHORUS CONCENTRATION PROFILES IN P-DOPED SILICON DIOXIDE MEASURED USIN 5-1231
CON- SILICON DIOXIDE-/ AUGER STUDIES OF DIFFUSED IMPURITY PROFILES IN SILICON DIOXIDE- SILICON AND POLYSIL 5-1481
Y OF AIR OXIDIZED STAINLESS STE/ COMPOSITION VERSUS DEPTH PROFILES OBTAINED WITH AUGER ELECTRON SPECTROSCOP 6-1998
ND PLATINUM SILICIDE BY AUGER ELECTRON SPECT/ COMPOSITION PROFILES OF CHEMICALLY VAPOR DEPOSITED PLATINUM A 5-1656
GLASS. AES COMPOSITIONAL PROFILES OF MOBILE IONS IN THE SURFACE REGION OF 5-1722
R EMISSION AND X-RAY PHOTOELECTRON SPECTROSCOPI/ ANALYSIS PROFILES OF OXIDE FILMS ON CHROMIUM STEEL BY AUGE 6-2016
/TS IN ULTRAHIGH VACUUM GROWN AMORPHOUS GERMANIUM. (DEPTH PROFILES OF OXYGEN AND CARBON BY AUGER SPECTROSC/ 5-1547
UGER SPECTROSCOPY AND SECONDARY ION EMISSION. IN-DEPTH PROFILES OF PHOSPHORUS ION IMPLANTED SILICON BY A 4-0996
ACES OF GOLD THICK FILMS ON COPPER PLATES BY/ COMPOSITION PROFILES OF SEVERAL CONTAMINATED AND CLEANED SURF 5-1596
GALLIUM(1-X) ALUMINUM(X) ARSENIDE SUPERLATTICES PROFILED BY AUGER ELECTRON SPECTROSCOPY.
OC STEEL STUDIED BY AUGER ELECTRON SPECTROSCOPY AND DEPTH PROFILING. /ON OF SULFUR DIOXIDE WITH MODIFIED 44 6-2031
ELECTRON SPECTROSCOPY IN THE SCANNING ELECTRO/ POTENTIAL PROFILING ACROSS SEMICONDUCTOR JUNCTIONS BY AUGER 2-0555
D BACKSCATTERING, SCANNING ELECTRON MICROSCOPY, AND DEPTH PROFILING AUGER ELECTRON SPECTROSCOPY. /RUTHERFOR 5-1902
QUANTITATIVE DEPTH PROFILING BY AUGER ION SPUTTERING TECHNIQUE. 1-0085
DECONVOLUTION METHOD FOR COMPOSITION PROFILING BY AUGER SPUTTERING TECHNIQUE. 3-0630

APPLICATIONS OF DEPTH PROFILING BY AUGER- SPUTTER TECHNIQUES. 1-0087
ALYSIS. RECENT ADVANCES IN COMPOSITION PROFILING BY SIMULTANEOUS SPUTTERING AND AUGER AN 1-0155
ER, SECONDARY ION MASS SPECTROSCOPY, AND PROTON RESONANCE PROFILING FOR RELIABILITY STUDIES. /ARISON OF AUG 1-0216
METHOD TO ACCURATELY ALIGN ION BOMBARDMENT GUNS FOR DEPTH PROFILING IN AUGER ELECTRON SPECTROSCOPY. EASY 3-0700
TECHNIQUES FOR ELEMENTAL COMPOSITION PROFILING IN THIN FILMS. (AUGER SPECTROSCOPY) 5-1273
ROSCOPY. FACTORS AFFECTING DEPTH PROFILING MEASUREMENTS USING AUGER ELECTRON SPECT 5-1457
OSCOPY. ELECTRON BEAM EFFECTS IN DEPTH PROFILING MEASUREMENTS WITH AUGER ELECTRON SPECTR 5-1128
LECTRON SPECTROSCOPY): A USEFUL COMPLEMENT TO OTHER DEPTH PROFILING TECHNIQUES. / SPECTROMETRY (AND AUGER E 1-0009
/PECTROSCOPY (AUGER ELECTRON ANALYSIS) TO SURFACE CONTROL PROGRAM IN THE MANUFACTURING OF SILICON DEVICES. 5-1411
ACE CONTAINING SEVERAL KINDS OF ATOMS. COMPUTER PROGRAM TO CALCULATE AUGER TRANSITIONS FOR A SURF 1-0039
LOGIC FOR THE AUTOMATIC EXECUTION OF AUGER AND ESCA TEST PROGRAMS. DIGITAL CONTROL UNIT IN TTL 3-0722
ELECTRON SPECTRUM. OBSERVATION OF A PROJECTILE CHARGE DEPENDENCE FOR THE NEON K-AUGER 4-0993
L IONIZATION AND FLUORESCENCE YIELD IN 50 MEV CHLORINE(N/ PROJECTILE CHARGE STATE DEPENDENCE OF NEON K-SHEL 2-0300
SCOPY. SURFACE COMPOSITION OF PROMOTED IRON CATALYSTS BY AUGER ELECTRON SPECTRO 5-1817
FACE OXIDATION OF STEEL DURING ANNEALING IN HYDROGEN-FREE PROTECTIVE GASES. (USING AES) /VATIONS ON THE SUR 6-2040
A OF MOLYBDENUM SURFACES. PROTON AND ELECTRON EXCITED AUGER ELECTRON SPECTR 5-1667
SURFACE ANALYSIS USING PROTON BEAMS. 1-0143
+) IMPACT. ARGON L-SHELL IONIZATION BY 50 TO 600 KEV PROTON, DEUTERON, MOLECULAR HYDROGEN, AND HELIUM(2-0521
/F FRACTIONAL MONOLAYER OXYGEN DETERMINATIONS OBTAINED BY PROTON EXCITED X-RAY, AES ANALYSIS OF IRON SURFA/ 5-1682
AUGER EXCITATION CROSS SECTIONS AND RANGES USING PROTON EXCITED X-RAYS. DETERMINATION OF 4-0456
ENSE SATELLITE LINES IN AUGER SPECTRA OF NITROGEN BY SLOW PROTON IMPACT. PRODUCTION OF INT 4-1076
N ARGON K AUGER ELECTRON SPECTRA PRODUCED BY 0.4 TO 4 MEV PROTON IMPACT. /TIVE SATELLITE LINE INTENSITIES I 2-0497
ED NIOBIUM SURFACES. (AUGER SPECTROSCOPY) PROTON INDUCED ELECTRON EMISSION FROM CHARACTERIZ 5-1666
IDE FORMATION ON NIOBIUM SURFACES DURING HIGH TEMPERATURE PROTON IRRADIATION. CARB 5-1886
HIGH RESOLUTION ARGON K-AUGER SPECTRUM PRODUCED IN 4 MEV PROTON ON ARGON COLLISIONS. 2-0496
/OMPARISON OF AUGER, SECONDARY ION MASS SPECTROSCOPY, AND PROTON RESONANCE PROFILING FOR RELIABILITY STUDI/ 1-0216
SPECTRA IN AMERICIUM-241 FROM CURIUM-241 (ELECTRON CAPTU/ PROTON STATES, ANOMALOUS CONVERSION, AND K-AUGER 4-1019
INNER SHELL IONIZATION OF GASEOUS MOLECULES BY 50-500 KEV PROTONS. CROSS SECTIONS FOR 4-1077
K-SHELL IONIZATION BY FAST PROTONS. 2-0533
LECTRON EMISSION OF METALS BOMBARDED WITH 120-EV TO 5-KEV PROTONS. SECONDARY E 4-1029
CROSS SECTIONS FOR L2,3 AUGER EMISSION IN 22 TO 300 KEV PROTON- ARGON COLLISIONS. 4-1040
E, ETHYLENE, ACETYL/ K-SHELL IONIZATION OF CARBON BY FAST PROTONS. (AUGER ELECTRON YIELDS OF METHANE, ETHAN 2-0532
D ANGLE OF ELECTRONS EJECTED FROM XENON BY 0.3 TO 2.0 MEV PROTONS. (AUGER PEAKS) DISTRIBUTION IN ENERGY AN 4-1075
/ENERGY DISTRIBUTION OF ELECTRONS PRODUCED BY 200-500 KEV PROTONS IN GASES. PART-2: NITROGEN. (AUGER SPECT/ 4-1075
PE. AUGER SPECTROSCOPY OF SOLID SURFACES IN A DRY PUMPED HIGH RESOLUTION SCANNING ELECTRON MICROSCO 5-1258
AUGER ELECTRON SPECTROSCOPY IN A DIFFUSION PUMPED SCANNING ELECTRON MICROSCOPE. 3-0598
TION AUGER ELECTRON SPECTROSCOPY IN AN ORDINARY DIFFUSION PUMPED SCANNING ELECTRON MICROSCOPE. /TIAL RESOLU 3-0597
RSENIDE BY SLOW ELECTRON DIFFRACTIO/ INVESTIGATION OF THE PURE AND CESIUM COATED (100) SURFACE OF GALLIUM A 5-1639
TENSITY OSCILLATIONS OF THE ELECTRON HOLE LUMINESCENCE IN PURE GERMANIUM. (AUGER RECOMBINATION) /PENDENT IN 4-0780
ION MICROGRAPHY AND MICROANALYSIS OF A STAINED SURFACE OF PURE IRON PLATE. AUGER ELECTRON EMISS 5-1415
SCANDIUM L X-RAY EMISSION FROM PURE METAL AND SCANDIUM OXIDE. 4-1102
COPY X-RAY PHOTOELECTRON SPECTROS/ A PRELIMINARY STUDY OF PURE METAL SURFACES USING AUGER ELECTRON SPECTROS 5-1353
AUGER ELECTRON ANALYSIS OF ELECTROPOLISHED HIGH PURITY ALUMINUM. 4-0824
PECTROSCOPY) PURITY AND MORPHOLOGY OF ALUMINUM FILMS. (AUGER S 5-1299
) SURFACE. (AUGER) ISOMORPHISM AND PURITY OF COPPER VAPOR DEPOSITED ON A COPPER (111 5-1171
HEED AND AES STUDY OF SILICON EPITAXY BY SILANE PYROLYSIS. 5-1427
TUDY. SILICON HOMOEPITAXIAL THIN FILMS VIA SILANE PYROLYSIS: HEED AND AUGER ELECTRON SPECTROSCOPY 5-1428
LMM AUGER SPECTRA OF SULFUR IN SODIUM SULFATE AND SODIUM PYROSULFITE. 4-0829

Q

ETECTION OF GAMMA EMITTING RADIONUCLIDES: CHROMATOGRAPHIC QUALITY CONTROL OF RADIOPHARMACEUTICALS. /R THE D 1-0217
PPEARAN/ APPLICATION OF DYNAMIC BACKGROUND SUBTRACTION TO QUANTIFICATION OF AUGER ELECTRON AND SOFT X-RAY A 1-0067
QUANTIFICATION OF AUGER ELECTRON SPECTROSCOPY. 1-0199
TION CROSS SECTIONS AND BACKSCATTERING FACTORS FOR USE IN QUANTITATIVE AES. MEASUREMENT OF IONIZA 2-0554
ELECTRONS EXCITED BY BACKSCATTERED ELECTRONS. APPROACH TO QUANTITATIVE ANALYSIS. AUGER AND SECONDARY 2-0366
EFFECT OF THE ELECTRON EJECTION DEPTH ON QUANTITATIVE ANALYSIS BY AUGER SPECTROSCOPY. 2-0449
, CARBON, AND OXYGEN) IN SPUTTERED TANTALUM FILMS BY AUG/ QUANTITATIVE ANALYSIS OF LIGHT ELEMENTS (NITROGEN 5-1652
ELECTRON SPECTROSCOPY. QUANTITATIVE ANALYSIS OF SOLID SURFACES BY AUGER 5-1715
ES OF CARBON MONOXIDE ON NICKEL. SPECTRAL LINE SHAPES AND QUANTITATIVE ASPECTS. /LECTRON SPECTROSCOPY STUDI 5-1458
R EFFECT) QUANTITATIVE ASPECTS OF A SOFT X-RAY LASER. (AUGE 2-0425
PY. QUANTITATIVE ASPECTS OF AUGER ELECTRON SPECTROSCO 1-0129
PY. IMPORTANCE OF ATTENUATION OF THE PRIMARY ELECTRON BE/ QUANTITATIVE ASPECTS OF AUGER ELECTRON SPECTROSCO 1-0206
PY. (CYLINDRICAL MIRROR ANALYZER) QUANTITATIVE ASPECTS OF ION SCATTERING SPECTROSCO 3-0666
GENERAL FORMALISM FOR QUANTITATIVE AUGER ANALYSIS. 1-0032
000 EV USING THE CYLINDRICA/ FIRST ORDER APPROXIMATION TO QUANTITATIVE AUGER ANALYSIS IN THE RANGE 100 TO 1 3-0660
/STUDIES OF SURFACE COMPOSITION OF SILVER- COPPER ALLOYS. QUANTITATIVE AUGER ANALYSIS OF COPPER- NICKEL AL/ 5-1200
OY SURFACES AFTER ARGON ION BOMBARDMENT. REPLY TO COMMEN/ QUANTITATIVE AUGER ANALYSIS OF COPPER- NICKEL ALL 5-1702
/STUDIES OF SURFACE COMPOSITION OF SILVER- COPPER ALLOYS. QUANTITATIVE AUGER ANALYSIS OF COPPER- NICKEL AL/ 5-1740
OY SURFACES AFTER ARGON ION BOMBARDMENT. QUANTITATIVE AUGER ANALYSIS OF COPPER- NICKEL ALL 5-1807
HNIQUES. QUANTITATIVE AUGER ANALYSIS USING INTEGRATION TEC 5-0066
QUANTITATIVE AUGER ELECTRON SPECTROSCOPY. 1-0207
TRON STIMULATED OXIDATION OF GALLIUM ARSENIDE, STUDIED BY QUANTITATIVE AUGER ELECTRON SPECTROSCOPY. ELEC 5-1756
INELASTIC SCATTERING EFFECTS IN QUANTITATIVE AUGER ELECTRON SPECTROSCOPY. 2-0433
00) CLEAVAGE SURFACES AND ON AMORPHOUS CARBON BY MEANS OF QUANTITATIVE AUGER ELECTRON SPECTROSCOPY. /ALT (1 5-1135
OF SILVER- PALLADIUM AND NICKEL- PALLADIUM ALLOYS. QUANTITATIVE AUGER ELECTRON SPECTROSCOPY ANALYSIS 6-2088
TRON RANGES. QUANTITATIVE AUGER ELECTRON SPECTROSCOPY AND ELEC 1-0177
ENERGY ELECTRON DIFFRACTION STUDY OF ALKALI METAL OVERLA/ QUANTITATIVE AUGER ELECTRON SPECTROSCOPY AND LOW 5-1895
SON OF TECHNIQUES FOR ADSORBED TIN ON IRON. QUANTITATIVE AUGER ELECTRON SPECTROSCOPY: COMPARI 1-0178
CONDENSED ON ROCK SALT AND ON AMORPHOUS CARBON. QUANTITATIVE AUGER ELECTRON SPECTROSCOPY OF GOLD 5-1136
S. AUGER CURRENT MEASUREMENTS FOR QUANTITATIVE AUGER ELECTRON SPECTROSCOPY OF SOLID 4-0856
/ANALOG INTEGRATION IN DYNAMIC BACKGROUND SUBTRACTION FOR QUANTITATIVE AUGER ELECTRON SPECTROSCOPY. (STUDY/ 1-0069
ADSORPTION. QUANTITATIVE AUGER ELECTRON SPECTROSCOPY USING CO 1-0002
EMENTAL SENSITIVITY FACTORS. QUANTITATIVE AUGER ELECTRON SPECTROSCOPY USING EL 1-0157
EGATION OF PHOSPHORUS IN IRON. QUANTITATIVE AUGER SPECTROSCOPIC ANALYSIS OF SEGR 5-1802
ORBED XENON ON NICKEL. QUANTITATIVE AUGER SPECTROSCOPY OF PHYSICALLY ADS 5-1150
AUGER ELECTRON SPECTROSCOPY MADE QUANTITATIVE BY ELLIPSOMETRIC CALIBRATION. 1-0205
MONOXIDE SURFACES USING AUGER ELECTRON AND SOFT X-RAY AP/ QUANTITATIVE COMPARISON OF TITANIUM AND TITANIUM 5-1384
ING TECHNIQUE. QUANTITATIVE DEPTH PROFILING BY AUGER ION SPUTTER 1-0085
AUGER SPECTROSCOPY. QUANTITATIVE DETERMINATION OF CARBON ON SILVER BY 5-1610

NESS ON DEPOSITED METAL FILMS BY COMBINATION AUGER SPECT/ QUANTITATIVE DETERMINATION OF SURFACE OXIDE THICK 6-2048
N NEAR THE SURFACE BASED ON THE ELECTRON AUGER SPECTROSC/ QUANTITATIVE ESTIMATES OF THE CHEMICAL COMPOSITIO 5-1645
HOSPHORUS ON SILICON USING A QUARTZ CRYSTAL MICROBALANCE. QUANTITATIVE STUDY OF AUGER ELECTRON SIGNALS OF P 5-1583
BINARY ALLOYS BY AUGER SPECTROSCOPY. QUANTITATIVE STUDY OF THE SURFACE COMPOSITION OF 6-2005
PY.   RELATIVE SPUTTERING YIELDS AND QUANTITATIVE SURFACE ANALYSIS BY AUGER SPECTROSCO 5-1964
ZATION OF THE METHOD. QUANTITATIVE USE OF AUGER SPECTROSCOPY: STANDARDI 1-0163
F AUGER ELECTRON SIGNALS OF PHOSPHORUS ON SILICON USING A QUARTZ CRYSTAL MICROBALANCE. QUANTITATIVE STUDY O 5-1583
ION AND CHARACTERIZATION TECHNIQUES (AUGER, CLEANING) FOR QUARTZ RESONATORS.   SURFACE PREPARAT 5-1942
                                        SURFACE STUDIES FOR QUARTZ RESONATORS. (CLEANING, AUGER SPECTROSCOPY) 5-1942
GER SPECTRA)   METHODS OF CLEANING CONTAMINANTS FROM QUARTZ SURFACES DURING RESONATOR FABRICATION. (AU 5-1410
IN SOLIDS.   WIDTHS OF ATOMIC M-SHELL VACANCY STATES AND QUASIATOMIC ASPECTS OF RADIATIONLESS TRANSITIONS. 3-0718
F SOLID SILVER AND INDIUM. QUASIATOMIC FINE STRUCTURE IN THE AUGER SPECTRA O 4-0771
ZINC. QUASIATOMIC LMM AUGER SPECTRA OF SOLID COPPER AND 4-1121
CY SHARING IN 300 TO 500 KEV NEON(+)- SODIUM COLLISIONS. (QUASIMOLECULE AUGER SPECTRA)   K-SHELL VACAN 2-0582
CRACKING OF AN ALUMINUM- 5% MAGNESIUM TERNARY AND VARIOUS QUATERNARY ALLOYS. (AES)   STRESS CORROSION 6-2036
/ METALLIC STATE OF SILICON IN SILICON- NOBLE METAL VAPOR QUENCHED ALLOYS STUDIED BY AUGER ELECTRON SPECTRO 5-1438

R

OMO DEOXYURIDINE: CHEMICAL CONSEQUENCES OF THE AUGER EFF/ RADIATION CHEMICAL RESONANCE EFFECT IN SOLID 5-BR 2-0377
NUCLEI IN THE TISSUE: USE OF SCFT X-RAYS. (/ BIOLOGICAL RADIATION EFFECT ON INCLUSION OF MODERATELY HEAVY 2-0531
ECOMBINATION PRO/ SPECTRUM AND DECAY OF THE RECOMBINATION RADIATION FROM STRONGLY EXCITED SILICON. (AUGER R 2-0460
E AND THEORETICAL K-SHELL FLUORESCENCE YIELDS.   ATOMIC RADIATION TRANSITION PROBABILITIES TO THE 1S STAT 4-0918
MS.   PARITY NONCONSERVATION IN RADIATIONLESS (AUGER) TRANSITIONS IN MU MESIC ATO 2-0364
S AT SURFACES. (AUGER PROCESS) RADIATIONLESS DEEXCITATION OF EXCITED HELIUM ATOM 2-0255
EV REGION OF THE SPECTRUM. (INTERBAND AUGER TRANSITIONS, RADIATIONLESS RECOMBINATIONS) /POUNDS IN THE 3-21 4-1048
        MANY BODY PERTURBATION THEORY APPLIED TO ATOMIC RADIATIONLESS TRANSITIONS. 2-0386
                THE AUGER EFFECT AND OTHER RADIATIONLESS TRANSITIONS. 1-0022
ATOMIC M-SHELL VACANCY STATES AND QUASIATOMIC ASPECTS OF RADIATIONLESS TRANSITIONS IN SOLIDS.   WIDTHS OF 3-0718
                                                RADIATIVE AND AUGER PROCESSES IN SEMICONDUCTORS. 2-0419
ONDUCTORS.   THEORY OF DONOR- ACCEPTOR RADIATIVE AND AUGER RECOMBINATION IN SIMPLE SEMIC 2-0420
                        THEORY OF THE RADIATIVE AUGER EFFECT. 2-0224
ESS.   TWO-PHOTON EMISSION, THE RADIATIVE AUGER EFFECT, AND THE DOUBLE AUGER PROC 2-0225
                                RADIATIVE AUGER EFFECT IN HEAVY ATOMS. 2-0476
                        THE K-FAR-M2 RADIATIVE AUGER EFFECT IN ION- ATOM COLLISIONS. 2-0481
                EVIDENCE FOR A RADIATIVE AUGER EFFECT IN TRANSITION METALS. 2-0502
EMISSION.   RADIATIVE AUGER EFFECT IN X-RAY PHOTON EMISSION. 2-0223
    CARRIER TEMPERATURE EFFECTS AND ENERGY TRANSFERS IN NONRADIATIVE RADIATIVE AUGER EFFECT WITH AND WITHOUT ELECTRON 2-0494
OR- ACCEPTOR PAIRS. (AUGER) RECOMBINATION. 2-0470
ITONS.   NONRADIATIVE AUGER RECOMBINATION OF ELECTRONS ON DON 2-0536
M OX/ ELECTRON- HOLE INTERACTION EFFECTS IN THE KL2,3L2,3 RADIATIVE AUGER RECOMBINATION OF TWO INDIRECT EXC 2-0227
UCTORS. (SILICON)   INTERPRETATION OF KL2,3L2,3 RADIATIVE AUGER SPECTRUM OF INSULATORS. (MAGNESIU 2-0540
L BONDING ON THE LOW ENERGY K-ALPHA SPECTRUM OF SILICON. (RADIATIVE AUGER THRESHOLDS IN METALS AND SEMICOND 4-0723
                NATURE OF SOME X-RAY SATELLITES (RADIATIVE AUGER TRANSITIONS) INFLUENCE OF CHEMICA 2-0226
EMISSION.   RADIATIVE AUGER TRANSITIONS). 2-0290
M.   EVIDENCE OF RADIATIVE AUGER TRANSITIONS IN SOFT X-RAY PLASMA 2-0344
M.   EVIDENCE OF RADIATIVE AUGER TRANSITIONS IN THE FREE ARGON ATO 2-0403
                        AUGER AND RADIATIVE AUGER TRANSITIONS IN THE FREE ARGON ATO 2-0404
ALS.   RADIATIVE DEEXCITATION OF MULTIPLY IONIZED NEON. 2-0315
RONIG TRANSITION RATES/ M-SHELL AUGER, COSTER-KRONIG, AND RADIATIVE KM2 AUGER EFFECT IN SOME TRANSITION MET 4-1057
UGER RECOMBINATIONS)   CONCENTRATION-DEPENDENT NONRADIATIVE RADIATIVE MATRIX ELEMENTS, AND AUGER AND COSTER-K 4-0967
IUM TO THORIUM. ATOMIC M-SHELL COSTER-KRONIG, AUGER, AND RADIATIVE PROCESSES IN ZINC SULFIDE PHOSPHORS. (A 2-0388
M- THORIUM. ATOMIC L-SHELL COSTER-KRONIG, AUGER, AND RADIATIVE RATES, AND FLUORESCENCE YIELDS FOR CALC 4-0964
TWEEN 38 AND 10/ ATOMIC N-SHELL COSTER-KRONIG, AUGER, AND RADIATIVE RATES, AND FLUORESCENCE YIELDS FOR SODIU 4-0963
ELDS.   DOUBLE IONIZATION EFFECTS ON RADIATIVE RATES, AND FLUORESCENCE YIELDS FOR Z BE 4-0969
ALUMINUM GALLIUM ARSENIDE HETEROJUNCTIO/ AUGER EFFECT AND RADIATIVE RATES, AUGER RATES, AND FLUORESCENCE YI 4-0976
                        NEW RADIATIVE RECOMBINATION IN N-GALLIUM ARSENIDE- P- 2-0324
                        AUGER AND RADIATIVE TRANSITION IN GERMANIUM. 4-1002
STRIBUTIONS PRODUCED BY INNER SHELL IONIZATION (AUGER AND RADIATIVE TRANSITION PROBABILITIES. 2-0328
/FFECT OF SURFACE METALLURGY ON THE PENETRATION DEPTH AND RADIATIVE TRANSITIONS).   ATOMIC VACANCY DI 2-0544
IN FILMS. (AU/ MORPHOLOGICAL AND ELECTRICAL PROPERTIES OF RADIOFREQUENCY BREAKDOWN FIELD OF SUPERCONDUCTIN/ 6-2137
ALYSIS)   NIOBIUM SURFACES FOR RADIOFREQUENCY SPUTTERED YTTRIA-DOPED ZIRCONIA TH 5-1385
        NEW METHOD OF THIN FILM THICKNESS MEASUREMENT USING RADIOFREQUENCY SUPERCONDUCTORS. (AUGER SURFACE AN 5-1472
/CONVERSION ELECTRONS FOR THE DETECTION OF GAMMA EMITTING RADIOISOTOPE AUGER ELECTRONS. 5-1658
MITTING RADIONUCLIDES: CHROMATOGRAPHIC QUALITY CONTROL OF RADIONUCLIDES: CHROMATOGRAPHIC QUALITY CONTROL O/ 1-0217
        QUANTITATIVE AUGER ELECTRON SPECTROSCOPY AND ELECTRON RADIOPHARMACEUTICALS. /R THE DETECTION OF GAMMA E 1-0217
RMINATION OF AUGER ELECTRON EXCITATION CROSS SECTIONS AND RANGES. 1-0177
IN (AUGER) SPECTRA OF ELECTRON INELASTIC SCATTERING FROM RANGES USING PROTON EXCITED X-RAYS.   DETE 2-0456
        EMISSION LINES OF THE M SERIES OF RARE EARTH METAL SURFACES.   IONIZATION LOSSES 2-0538
YMIUM, GADOLINIUM, DYSPRO/ AUGER ELECTRON SPECTRA OF SOME RARE EARTH METALS. (AUGER EFFECT) 4-0790
        WIDTHS OF 4S AND 5S MULTIPLET COMPONENTS IN THE RARE EARTH METALS. (LANTHANUM, PRASEODYMIUM, NEOD 4-1100
/ENDENCE OF X-RAY AND AUGER ELECTRON EMISSION SPECTRA FOR RARE EARTHS. 4-0970
ECONDARY ELECTRONS EMITTED BY COPPER UNDER BOMBARDMENT BY RARE GAS ATOMS. PART-1. THE ARGON ATOM. PART-2. / 4-0927
EMISSION FROM SODIUM CHLORINE SINGLE CRYSTAL BOMBARDED BY RARE GAS IONS. /TRONS IN THE ENERGY SPECTRUM OF S 2-0551
        ADSORBATE EFFECTS IN ELECTRON EJECTION BY RARE GAS IONS.   SECONDARY ELECTRON 4-0759
        SOLID STATE BROADENING IN AUGER SPECTRA. (CADMIUM, RARE GAS METASTABLE ATOMS. (AUGER PROCESS) 2-0333
        SOLID STATE BROADENING IN THE AUGER SPECTRA OF RARE GASES. 4-0843
GER TRANSITIONS)   OUTER SHELL EXCITATION OF THE RARE GASES. (ARGON, KRYPTON, XENON) 4-1006
K-SHELL FLUORESCENCE YIELD OF IONIZED ALUMINUM. (AUGER RARE GASES BY 30 MEV OXYGEN IONS. (INNER SHELL AU 2-0298
        THEORETICAL FLUORESCENCE YIELDS FOR NEON. (AUGER RATE) 4-0975
                AUGER TRANSITION RATE) 2-0283
                AUGER TRANSITION RATE FOR THE NOBLE GAS SATELLITE SPECTRA. 4-0972
        DETERMINATION OF FILM THICKNESS OR ION ETCH RATE IN MIXED COUPLING SCHEME. 2-0342
IVISTIC CALCULATIONS OF ATOMIC X-RAY AND AUGER TRANSITION RATE USING AUGER ELECTRON SPECTROSCOPY. 5-1984
        RELATIVISTIC KLL AUGER TRANSITION RATES.   RELAT 2-0485
                Z DEPENDANCE OF THE KLL AUGER RATES. 4-0782
RIUM. ATOMIC M-SHELL COSTER-KRONIG, AUGER, AND RADIATIVE RATES. 2-0557
. ATOMIC L-SHELL COSTER-KRONIG, AUGER, AND RADIATIVE RATES, AND FLUORESCENCE YIELDS FOR CALCIUM TO THO 4-0964
ND 10/ ATOMIC N-SHELL COSTER-KRONIG, AUGER, AND RADIATIVE RATES AND FLUORESCENCE YIELDS FOR SODIUM- THORIUM 4-0963
RATES, AND FLUORESCENCE YIELDS FOR Z BETWEEN 38 A 4-0969

O XENON.   K-SHELL AUGER TRANSITION RATES AND FLUORESCENT YIELDS FOR ELEMENTS ARGON T 4-0961
UM TO ARGON.   K-SHELL AUGER TRANSITION RATES AND FLUORESCENT YIELDS FOR ELEMENTS BERYLLI 4-0962
          DOUBLE IONIZATION EFFECTS ON RADIATIVE RATES, AUGER RATES, AND FLUORESCENCE YIELDS.   4-0976
                                           K-AUGER RATES CALCULATED FOR NEON(+).   4-0909
/ANY BODY PERTURBATION THEORY TO THE CALCULATION OF AUGER RATES, CORRELATION ENERGIES, AND PHOTOIONIZATION/ 2-0311
                                    TRANSITION RATES FOR INTERATOMIC AUGER PROCESSES.   2-0438
                             K-SHELL AUGER RATES FOR MULTIPLY IONIZED ATOMS. PART-1: NEON.   2-0282
               FLUORESCENCE YIELDS AND AUGER RATES FOR MULTIPLY IONIZED NITROGEN.   4-0781
, AND METALLIC LITHIUM.   AUGER RATES FOR SOFT X-RAY TRANSITIONS IN IONIC, ATOMIC   4-0835
ITH 1S OR 2S VACANCY).   CORRELATION ENERGIES AND AUGER RATES IN ATOMS WITH INNER SHELL VACANCIES (NEON W   2-0312
E MATRIX ELEMENTS, AND AUGER AND COSTER-KRONIG TRANSITION RATES IN J-J COUPLING. /STER-KRONIG, AND RADIATIV 4-0967
                         AUGER TRANSITION RATES IN J-J COUPLING FOR D HOLES AND F ELECTRONS   2-0444
M, ZINC)   ENHANCEMENT OF COSTER-KRONIG RATES IN SOLID STATE ENVIRONMENTS. (XENON, CADMIU   5-1536
/NITIAL OXIDATION OF MOLYBDENUM. PART-3: OXIDE NUCLEATION RATES ON (100) AND (110) MOLYBDENUM SURFACES. (A/ 2-0303
G PA/ SENSITIVITY OF L1,L2,3M4,5 COSTER-KRONIG TRANSITION RATES TO VARIATIONS IN ENERGY LEVELS AND SCREENIN 2-0280
DS OF VARIOUSLY IONIZED STATES OF ARGON.   K-SHELL AUGER RATES, TRANSITION ENERGIES, AND FLUORESCENCE YIEL 2-0284
DS FOR MULTIPLY IONIZED NEON. / THEORETICAL K-SHELL AUGER RATES, TRANSITION ENERGIES, AND FLUORESCENCE YIEL 2-0556
HE K-SHELL.   NONRELATIVISTIC AUGER RATES, X-RAY RATES, AND FLUORESCENCE YIELDS FOR T   2-0558
HE 2P SHELL.   NONRELATIVISTIC AUGER RATES, X-RAY RATES, AND FLUORESCENCE YIELDS FOR T   1-0042
SORPTION TECHNIQUES TO STUDY THE MECHANISM OF A CATALYTIC REACTION. /F PHOTOEMISSION, AUGER, AND THERMAL DE 5-1620
INTERFERENCE OF AN ELECTRON BEAM WITH THE SURFACE REACTION BETWEEN OXYGEN AND GERMANIUM. (AES)   5-1451
11) SURFACE. (STUDIED BY LEED AND AES)   KINETICS OF THE REACTION BETWEEN OXYGEN AND SULFUR ON A NICKEL (1 5-1530
M SURFACE BY AUGER ELECT/ STUDIES OF THE MECHANISM OF THE REACTION BETWEEN SULFUR AND OXYGEN ON A MOLYBDENU 5-1534
M SURFACE BY M/ DYNAMIC INVESTIGATION OF THE MECHANISM OF REACTION BETWEEN SULFUR AND OXYGEN ON A MOLYBDENU 5-1333
YGEN AND NITROGEN WITH THE NIOBIUM (100) SURFACE. PART-2: REACTION KINETICS. (AUGER EFFECT) /ERACTION OF OX 5-1879
(111).   THIN REACTION LAYERS AND SURFACE STRUCTURE OF SILICON   5-1685
). (AUGER SPECTROSCOPY)   ADSORPTION OF HYDROGEN AND THE REACTION OF HYDROGEN WITH OXYGEN ON PLATINUM (100 5-1452
AL SURFACES. PART-1: NICKEL (100) SURFAC/ KINETICS OF THE REACTION OF OXYGEN WITH CLEAN NICKEL SINGLE CRYST 5-1563
RSENIDE. (AUGER SPECTROSCOPY)   REACTION OF SPUTTERED PLATINUM FILMS ON GALLIUM A   6-2031
EL STUDIED BY AUGER ELECTRON SPECTROSCOPY AND DEPTH PROF/ REACTION OF SULFUR DIOXIDE WITH MODIFIED 440C STE 5-1183
   AUGER ELECTRON SPECTROSCOPY OF A SULFUR- OXYGEN SURFACE REACTION ON A COPPER (110) CRYSTAL.   5-1535
CTROSCOPY.   DYNAMIC ELUCIDATION OF REACTION STEPS ON A SOLID SURFACE BY ELECTRON SPE   5-1219
IN-DEPTH AUGER ANALYSIS OF ALUMINUM- SILICON INTERFACIAL REACTIONS.   5-1321
/ A PALLADIUM (100) SURFACE. (CHEMISORPTION AND CATALYTIC REACTIONS, LEED, AUGER, CONTACT POTENTIAL METHOD/ 4-0788
UGER EFFECT)   CHEMISORPTION REACTIONS ON HIGH INDEX ZINC SULFIDE SURFACES. (A   5-1763
   BINDING, VALENCY AND CHARGE TRANSFER IN CHEMICAL REACTIONS ON METAL SURFACES.   5-1173
ACES.   SMALL MOLECULE REACTIONS ON STEPPED SINGLE CRYSTAL PLATINUM SURF   5-1344
                                  CHLORINE REACTIONS ON THE SILICON (111) SURFACE.   5-1541
D TUNGSTEN NI/ SUPERCONDUCTING TRANSITION TEMPERATURES OF REACTIVELY SPUTTERED FILMS OF TANTALUM NITRIDE AN 5-1838
PPED PLATINUM SINGLE CRYSTAL SURFACES. (LEED, AUGER ELEC/ REACTIVITY OF LOW INDEX ((111) AND (100)) AND STE 6-2108
EFFECTS OF TEN YEARS EXPERIMENTAL BREEDER REACTOR II SERVICE ON CHROMIUM- MOLYBDENUM STEEL.   1-0036
APPLICATION OF AUGER ELECTRON SPECTROSCOPY TO REACTOR MATERIALS RESEARCH.   5-1826
PY.   SPUTTERING MEASUREMENTS ON CONTROLLED THERMONUCLEAR REACTOR MATERIALS USING AUGER ELECTRON SPECTROSCO 3-0690
CES AND THIN FILMS STUDY. (AUGER)   REALIZATION OF A MOSSBAUER SPECTROMETER FOR SURFA   2-0334
LECTRON CAPTURE. (K-AUGER/ MERCURY AND GOLD ATOMIC CLOUDS REARRANGEMENT FOLLOWING INTERNAL CONVERSION AND E 2-0413
R SPECTRA)   REARRANGEMENT OF INNER SHELL IONIZED ATOMS. (AUGE   2-0308
X-RAY PHOTOIONIZATION OF NITROGE/ RELATIVE ABUNDANCES AND RECOIL ENERGIES OF FRAGMENT IONS FORMED FROM THE 2-0332
CARRIER LIFETIMES IN EPITAXIAL INDIUM ARSENIDE. (AUGER RECOMBINATION).   2-0470
TURE EFFECTS AND ENERGY TRANSFERS IN NONRADIATIVE (AUGER RECOMBINATION)   CARRIER TEMPERA   4-0777
ELECTRON HOLE DROPS IN GERMANIUM- SILICON ALLOYS. (AUGER RECOMBINATION)   4-0946
GERMANIUM IN THE PRESENCE OF ELECTRON HOLE DROPS. (AUGER RECOMBINATION)   /E CARRIER AND EXCITON KINETICS IN 4-0780
THE ELECTRON HOLE LUMINESCENCE IN PURE GERMANIUM. (AUGER RECOMBINATION)   /PENDENT INTENSITY OSCILLATIONS OF 2-0503
OPTICAL GAIN IN LIGHTLY DOPED GALLIUM ARSENIDE. (AUGER RECOMBINATION)   2-0432
AND CADMIUM SULFIDE(0.48) SELENIDE(0.52) CRYSTALS. (AUGER RECOMBINATION)   / HIGHLY EXCITED CADMIUM SELENIDE 2-0421
ONEQUILIBRIUM PHASE TRANSITIONS IN SEMICONDUCTORS. (AUGER RECOMBINATION)   RECOMBINATION-INDUCED N   2-0258
ELENIDE SINGLE CRYSTALS AT HIGH EXCITATION LEVELS. (AUGER RECOMBINATION) /OF THE EDGE EMISSION OF GALLIUM S 2-0461
TRANSPORT PROPERTIES IN MERCURY CADMIUM TELLURIDE. (AUGER RECOMBINATION)   TRANSIENT CARRIER DECAY AND   2-0343
RY(1-X) CADMIUM(X) TELLURIDE AMBIENT TEMPERATURES. (AUGER RECOMBINATION) /CTRON EFFECTS IN NARROW GAP MERCU 2-0458
OWER RECTIFIERS.   BAND-TO-BAND AUGER RECOMBINATION AND CARRIER-CARRIER SCATTERING IN P   2-0396
AP SEMICONDUCTORS.   AUGER RECOMBINATION AND IMPACT IONIZATION IN INDIRECT G   2-0417
TORS.   DETAILED BALANCE BETWEEN AUGER RECOMBINATION AND IMPACT IONIZATION IN SEMICONDUC   4-1092
DMIUM(X) TELLURIDE) UNDER A MAGNETIC FIELD.   AUGER RECOMBINATION IN A SEMICONDUCTOR (MERCURY(1-X) CA   2-0418
IMPACT IONIZATION AND AUGER RECOMBINATION IN BANDS.   2-0369
IUM SULFIDE SINGLE CRYSTALS. (AUGER   EVIDENCES FOR AUGER RECOMBINATION IN DONOR ACCEPTOR COMPLEXES IN CADM 2-0260
)   AUGER RECOMBINATION IN ELECTRON HOLE DROPLETS. (SILICON   4-0778
                                     AUGER RECOMBINATION IN GALLIUM ANTIMONIDE.   2-0561
                                     AUGER RECOMBINATION IN GALLIUM ARSENIDE.   2-0395
                     PHONON ASSISTED AUGER RECOMBINATION IN GERMANIUM.   2-0397
                      BAND-TO-BAND AUGER RECOMBINATION IN GERMANIUM.   2-0394
IDE, INDIUM PHOSPHIDE AND GALLIUM ARSENIDE.   AUGER RECOMBINATION IN INDIRECT GAP SEMICONDUCTORS.   4-1091
ALLIUM ARSENIDE HETEROJUNCTIO/ AUGER EFFECT AND RADIATIVE RECOMBINATION IN INDIUM ARSENIDE, GALLIUM ANTIMON 2-0324
CARRIER RECOMBINATION IN N-GALLIUM ARSENIDE- P-ALUMINUM G   4-0761
INFLUENCE OF AN ELECTRIC FIELD ON THE AUGER RECOMBINATION IN N-INDIUM ANTIMONIDE.   2-0357
INFLUENCE OF AN ELECTRIC FIELD ON THE AUGER RECOMBINATION IN SEMICONDUCTORS.   4-0844
CITATION.   AUGER RECOMBINATION IN SEMICONDUCTORS DURING INTENSE EX   2-0368
)   AUGER RECOMBINATION IN SEMICONDUCTORS. (REVIEW, 26 REFS   1-0044
IUM)   INFLUENCE OF SCREENING EFFECTS ON THE AUGER RECOMBINATION IN SEMICONDUCTORS. (SILICON, GERMAN   2-0379
                                     AUGER RECOMBINATION IN SILICON.   2-0266
TWO ELECTRON BAND-TO-BAND TRANSITIONS IN SOLIDS. (AUGER RECOMBINATION IN SILICON)   2-0275
BAND-TO-BAND AUGER RECOMBINATION IN SILICON AND GERMANIUM.   2-0459
S LOADED IN FORWARD DIRECTION IN THE MIDDLE REGION. (AUGER RECOMBINATION IN SILICON RECTIFIERS AND THYRISTOR 2-0415
THEORY OF DONOR- ACCEPTOR RADIATIVE AND AUGER RECOMBINATION IN SIMPLE SEMICONDUCTORS.   2-0420
ESS, AES, PHOTOELECTRON SPECTROSCOPY, CHEMI/ EMISSION AND RECOMBINATION INVOLVING INNER SHELLS. (AUGER PROC 2-0514
) GALLIUM(1-X) ARSENIDE ELECTROLUMINESCENT DIODES. (AUGER RECOMBINATION MECHANISM) / COMPENSATED ALUMINUM(X 2-0325
-X) CADMIUM(X) TELLURIDE. (AUGER-LIMITED)   RECOMBINATION MECHANISMS IN 8-14 MICRON MERCURY(1 2-0406
DE AT 77 DEGREES K. (AUGER)   RECOMBINATION OF CARRIERS IN N-TYPE INDIUM ARSENI   4-0785
RS.   NONRADIATIVE AUGER RECOMBINATION OF ELECTRONS ON DONOR- ACCEPTOR PAI   2-0536
IN TELLURIUM DO/ PHOTOCONDUCTIVITY ASSOCIATED WITH AUGER RECOMBINATION OF EXCITONS BOUND TO NEUTRAL DONORS 2-0453
M. PICOSECOND OPTICAL MEASUREMENTS OF BAND-TO-BAND AUGER RECOMBINATION OF HIGH DENSITY PLASMAS IN GERMANIU 2-0250

N SILICON AT A HIGH/ ELECTRON HOLE SCATTERING AND (AUGER) RECOMBINATION OF NONEQUILIBRIUM CHARGE CARRIERS I 4-1103
F INDIUM ARSENIDE AT HIGH PHOTOEXCITATION LEVELS. (AUGER) RECOMBINATION OF NONEQUILIBRIUM CHARGE CARRIERS O 4-0840
RADIATIVE AUGER RECOMBINATION OF TWO INDIRECT EXCITONS. 2-0227
IMPURITY IONIZATION IN N-TYPE GERMANIUM. (AUGER RECOMBINATION PARAMETER) 2-0466
ITY CARRIER LIFETIME IN INDIUM ARSENIDE EPILAYERS. (AUGER RECOMBINATION PROCESS) MINOR 2-0567
EFFECT OF ELECTRON HOLE INTERACTION ON THE AUGER RECOMBINATION PROCESS IN A SEMICONDUCTOR. 4-1093
ND IN THIN FILMS OF SILICON AND GERMANIUM / MANY PARTICLE RECOMBINATION PROCESSES (AUGER) ON THE SURFACES A 5-1990
ICON. (AUGER RECOMBINATICN PRO/ SPECTRUM AND DECAY OF THE RECOMBINATION RADIATION FROM STRONGLY EXCITED SIL 2-0460
NONRADIATIVE PROCESSES IN ZINC SULFIDE PHOSPHORS. (AUGER RECOMBINATIONS) CONCENTRATION-DEPENDENT 2-0388
THE SPECTRUM. (INTERBAND AUGER TRANSITIONS, RADIATIONLESS RECOMBINATIONS) /POUNDS IN THE 3-21 EV REGION OF 4-1048
TIONS IN SEMICONDUCTORS. (AUGER RECOMBINATION) RECOMBINATION-INDUCED NONEQUILIBRIUM PHASE TRANSI 2-0421
OBSERVATIONS OF BETA- SILICON CARBIDE FORMATION ON RECONSTRUCTED SILICON SURFACES. 5-1560
SURFACE ELECTRONIC STRUCTURE OF PLATINUM (100) INDUCED BY RECONSTRUCTION. CHANGE IN THE 5-1186
GY ANALYZER. SEMIAUTOMATED DATA RECORDING AND CONTROL SYSTEM FOR AN ELECTRON ENER 3-0680
A. NEW, STRIATED, HIGH EFFICIENCY PHOTOGRAPHIC PLATES FOR RECORDING CONVERSION AND/OR AUGER ELECTRON SPECTR 3-0699
MITTERS. (AUGER SPECTRA) RECORDING OF COMPONENTS ON THE SURFACE OF FUSED E 4-0736
N ELECTRON SPECTROMETRY FOR CHEMICAL ANALYSIS. (AUTOMATIC RECORDING OF PHOTO AND AUGER ELECTRONS) /R USED I 3-0717
RY ALLOY. (AUGER SPECTROSCOPY) RECORDING THE COMPONENTS ON THE SURFACE OF A BINA 5-1358
TRIX COMPOSITES. (AUGER ELECTRON SPECTROSCOPIC TECHNIQUE/ RECRYSTALLIZATION OF TUNGSTEN FIBERS IN NICKEL MA 6-2046
GER RECOMBINATION AND CARRIER-CARRIER SCATTERING IN POWER RECTIFIERS. BAND-TO-BAND AU 2-0458
TION IN THE MIDDLE REGION. AUGER RECOMBINATION IN SILICON RECTIFIERS AND THYRISTORS LOADED IN FORWARD DIREC 2-0415
ON IN STEEL, T/ RELATICN OF GRAIN BOUNDARY SEGREGATION TO REDUCED STRENGTH OF METALS AND ALLOYS. (SEGREGATI 6-2120
S USING AUGER ELECTRON EMISSI/ STUDY OF THE OXIDATION AND REDUCTION BY WATER VAPOR OF SILICON (111) SURFACE 5-1612
IDE AND ITS EFFECTS ON SECONDARY E/ ELECTRON BEAM INDUCED REDUCTION OF CARBON CONCENTRATION ON BERYLLIUM OX 5-1362
S OF COMBINED LOW ENE/ STUDY ON THE INITIAL OXIDATION AND REDUCTION PROCESSES OF IRON (100) SURFACE BY MEAN 6-2141
IFFRACTION AND AUGER SPECTROSCOPY. REEXAMINATION OF ALUMINUM (110) WITH LOW ENERGY D 5-1507
THE ANALYSIS OF ELECTRON EMISSION AT LOW PRIMARY BEAM EN/ REFINEMENTS TO A STANDARD LEED- AUGER SYSTEM FOR 3-0698
SURFACE ANALYSIS BY LOW ENERGY ION REFLECTION. 5-1822
AUGER ELECTRON EMISSION OF THIN CARBON FOILS IN REFLECTION AND TRANSMISSION. 4-0893
/AXIAL GROWTH OF MAGNESIUM ON MAGNESIUM OXIDE (001) USING REFLECTION DIFFRACTION, LEED AND AUGER SPECTROSC/ 5-1496
OXYGEN AND HYDROGEN ADSORPTION ON STRONTIUM SU/ AUGER AND REFLECTION ELECTRON DIFFRACTION INVESTIGATION OF 5-1498
ERING, STUDIED WITH AUGER SPECTRCSCOPY, ELLIPSOMETRY, AND REFLECTION HIGH ENERGY ELECTRON DIFFRACTION. /UTT 5-1233
STUDY OF OXYGEN ADSORPTION ON THE TUNGST/ COMBINED LEED- REFLECTION HIGH ENERGY ELECTRON DIFFRACTION AUGER 5-1464
SECONDARY EMISSION AND ELASTIC REFLECTION OF ELECTRONS FROM MONOCRYSTALS. 2-0508
OBSERVED BY LEED AND AUGER ELECTRON SPECTROS/ BEHAVIOR OF REFRACTORY METAL SURFACES IN ULTRAHIGH VACUUM AS 6-2020
AUGER ELECTRON SPECTROSCOPY OF SOME REFRACTORY METALS. 4-0865
ON BACKSCATTERING TECHNIQUES APPLIED TO ANALYSIS OF GOLD- REFRACTORY METALLIZATIONS. /R SPECTROSCOPY, AND I 5-1261
AND THYRISTORS LOADED IN FORWARD DIRECTION IN THE MIDDLE REGION. AUGER RECOMBINATION IN SILICON RECTIFIERS 2-0415
AUGER SPECTRA IN THE INTERMEDIATE COUPLING REGION. 4-0887
TUNGSTEN AND TITANIUM SINGLE CRYSTALS IN THE SUBMONOLAYER REGION. /ROSCOPY, AND WORK FUNCTION OF CESIUM ON 5-1829
COPPER- NICKEL ALLOY BY AUGER SPECTRA IN THE LOWER ENERGY REGION AT AROUND 100EV. / SURFACE COMPOSITION OF 5-1982
/MONIDE ON THE EXTERNAL PHOTOEFFECT IN THE X-RAY SPECTRAL REGION. (EFFECT OF IMPURITIES ON THE AUGER SPECT/ 4-1001
AES COMPOSITIONAL PROFILES OF MOBILE IONS IN THE SURFACE REGION OF GLASS. 5-1722
/ON SPECTRUM OF OXYGEN DOMINATED COMPOUNDS IN THE 3-21 EV REGION OF THE SPECTRUM. (INTERBAND AUGER TRANSIT/ 4-1048
RON DIFFRACTIONS AND AUGE/ DEVELOPMENT OF STEPPED SURFACE REGIONS ON POLYCRYSTALLINE GOLD. LOW ENERGY ELECT 5-1483
AND ADHESION OF THIN FILMS. (AUGER ELECTRON SPECTROSCOP/ RELATIONSHIPS BETWEEN SUBSTRATE SURFACE CHEMISTRY 5-1851
SCENCE YIELDS FOR THE K-SHELL. NONRELATIVISTIC AUGER RATES, X-RAY RATES, AND FLUORE 2-0556
SCENCE YIELDS FOR THE 2P SHELL. NONRELATIVISTIC AUGER RATES, X-RAY RATES, AND FLUORE 2-0558
M. (OF MERCURY) RELATIVISTIC CALCULATION OF THE KLL AUGER SPECTRU 4-0750
ER TRANSITION RATES. RELATIVISTIC CALCULATIONS OF ATOMIC X-RAY AND AUG 2-0485
HE 3D SHELLS. NONRELATIVISTIC FLUORESCENCE YIELDS FOR THE 3P AND T 2-0277
RELATIVISTIC KLL AUGER TRANSITION RATES. 4-0782
CYLINDRICAL CAPACITOR AS AN ANALYZER. PART-1: NONRELATIVISTIC PART. 3-0689
ILITIES. RELATIVISTIC STUDY OF KLL AUGER TRANSITION PROBAB 4-0802
THEORY OF KLL AUGER ENERGIES INCLUDING STATIC RELAXATION. 2-0506
AY IN ARSENIC AND SOME OF ITS COMPOUNDS. RELAXATION DURING PHOTOEMISSION AND LMM AUGER DEC 4-0763
S ADSORBATES ON TRANSITION M/ INTERATOMIC TRANSITIONS AND RELAXATION EFFECTS IN AUGER SPECTRA OF SEVERAL GA 4-1042
RELAXATION EFFECTS ON AUGER ENERGIES. 2-0505
SS IN ARSENIC AND SELENIUM. ATOMIC AND EXTRAATOMIC STATIC RELAXATION ENERGIES IN THE L3M4,5M4,5 AUGER PROCE 4-0883
SIZE EFFECT IN IONIC CHARGE RELAXATION FOLLOWING AUGER EFFECT. 2-0565
SS. INCOMPLETE RELAXATION IN A MODEL PROBLEM FOR THE AUGER PROCE 2-0238
RELAXATION OF AUGER EXCITED CARRIERS IN SILICON. 2-0274
ATION ENERGY DIFFERENCES FROM AUGER AND CORE BIND/ STATIC RELAXATION OF GERMANE AND THE ESTIMATION OF RELAX 2-0471
EFFECTS OF EXTRA CLOSED CORE RELAXATION ON AUGER AND BINDING ENERGY SHIFTS. 4-0950
TRANSITION METALS AND SAL/ RELATIVE EFFECT OF EXTRAATOMIC RELAXATION ON AUGER AND BINDING ENERGY SHIFTS IN 2-0412
4,.5:1G4 AUGER PROCESS IN ZINC METAL. RELAXATION PHENOMENA INVOLVED IN THE L3M4,5M4,.5M 4-0882
5N4,5 AUGER TRANSITIONS. RELAXATION PROCESSES ACCOMPANYING CADMIUM M4,5N4, 2-0559
ION MASS SPECTROSCOPY, AND PROTON RESONANCE PROFILING FOR RELIABILITY STUDIES. /ARISON OF AUGER, SECONDARY 1-0216
DETECTION BY AUGER ELECTRON SPECTROSCOPY AND REMOVAL BY OZONIZATION OF PHOTORESIST RESIDUES. 5-1450
/N SPECTROSCOPY STUDIES OF SPUTTER DEPOSITION AND SPUTTER REMOVAL OF MOLYBDENUM FROM VARIOUS METAL SURFACE/ 5-1878
FACE) CHEMICAL CLEANING FOR THERMOCOMPRESSION BONDING. (REMOVAL OF OXIDE FILMS, AUGER SPECTROSCOPY OF SUR 5-1454
NNELING SPECTROSCOPY, LEED AND AES. INVESTIGATION OF REORDERED (001) GOLD SURFACES BY POSITIVE ION CHA 5-1139
NNELING SPECTROSCOPY, LOW ENERGY ELE/ CHARACTERIZATION OF REORDERED (001) GOLD SURFACES BY POSITIVE ION CHA 5-1986
CATION TO SURFACE COMPOSITION/ A MULTICHANNEL MONITOR FOR REPETITIVE AUGER ELECTRON SPECTROSCOPY WITH APPLI 3-0695
FILMS. (AUGER SPECTROSCOPY) RESIDUAL GAS AND THE OPTICAL PROPERTIES OF SILVER 5-1700
ON SPECTROSCOPY AND REMOVAL BY OZONIZATION OF PHOTORESIST RESIDUES. DETECTION BY AUGER ELECTR 5-1450
TS. (BY AUGER ELECTRON SPECTROS/ INVESTIGATIONS OF CARBON RESIDUES ON SURFACES OF SILICON INTEGRATED CIRCUI 5-1345
BERS) RECENT ASPECTS OF GLASS FIBER- RESIN INTERFACES. (AUGER SPECTROSCOPY OF GLASS FI 5-1975
LY DOPED SIL/ CORRELATIONS BETWEEN INTERFACIAL OXIDES AND RESISTANCE CHANGES FOR TUNGSTEN CONTACTS ON HEAVI 5-1861
GERMANIUM A/ EFFECTS OF INTERFACIAL OXIDES ON THE CONTACT RESISTANCE OF TUNGSTEN ON HEAVILY DOPED SILICON- 5-1456
/ PALLADIUM- GOLD THIN FILM SYSTEM AND THE CORRELATION OF RESISTIVITY CHANGES WITH AUGER ELECTRON SPECTROS/ 5-1393
HIGH SPATIAL RESOLUTION AES. 1-0115
SILVER FILM) AUGER ELECTRON SPECTROSCOPY AT HIGH SPATIAL RESOLUTION AND PRIMARY BEAM CURRENTS. (CARBON ON 5-1904
UGER ELECTRON SPECTRCMETER BASED ON LEED DISPLAY OPTICS. RESOLUTION AND SENSITIVITY CONSIDERATIONS OF AN A 3-0705
RETARDING POTENTIAL ANALYZER. APPLICATION TO AUGER ELECT/ RESOLUTION AND SENSITIVITY OF THE SPHERICAL GRID 3-0634
EV PROTON ON ARGON COLLISIONS. HIGH RESOLUTION ARGON K-AUGER SPECTRUM PRODUCED IN 4 M 2-0496
ENTRIC HEMISPHERES. HIGH ENERGY RESOLUTION AUGER ELECTRON SPECTROMETER USING CONC 3-0592
NARY DIFFUSION PUMPED SCANNING ELECTRON MIC/ HIGH SPATIAL RESOLUTION AUGER ELECTRON SPECTROSCOPY IN AN ORDI 3-0597
C COPPER. HIGH RESOLUTION AUGER ELECTRON SPECTROSCOPY OF METALLI 4-1052
OXIDE ADSORBED ON A TUNGSTEN SURFACE. HIGH RESOLUTION AUGER ELECTRON SPECTROSCOPY OF NITRIC 5-1565

XIDE (100).   HIGH RESOLUTION AUGER ELECTRON SPECTRUM OF MAGNESIUM O 4-0772
          INITIAL OXIDATION OF MOLYBDENUM STUDIED BY A HIGH RESOLUTION AUGER PHOTOELECTRON SPECTROMETER.   5-1698
ON APPLICATIONS.   HIGH SPATIAL RESOLUTION AUGER SPECTROSCOPY AND AUGER INTEGRATI 1-0166
                    EFFECT OF OXIDATION ON THE HIGH RESOLUTION AUGER SPECTRUM OF TITANIUM.   5-1158
YGEN IONS EXCITED BY COLLISIONS WITH HELIUM, NEON, / HIGH RESOLUTION BEAM GAS AUGER ELECTRON SPECTRA FOR OX 4-0904
    RESEARCH. (REVIEW, 12 REFS)   HIGH RESOLUTION ELECTRON SPECTROGRAPHS FOR SOLID STATE   3-0703
          AN ENERGY MODULATED (CYLINDRICAL) HIGH ENERGY RESOLUTION ELECTRON SPECTROMETER.   3-0616
ON SPECTROSCOPY.   HIGH RESOLUTION ELECTRON SPECTROMETER FOR AUGER ELECTR 3-0667
N AND AUGER ELECTRON STUDIES.   HIGH RESOLUTION ELECTRON SPECTROMETER FOR PHOTOELECTRO 3-0600
CTRON SPECTROMETRY FOR CHEMICAL ANALYSIS. (AUTOMATI/ HIGH RESOLUTION ELECTROSTATIC SPECTROMETER USED IN ELE   3-0717
NDENCE OF THE NEON K FLUORESCENCE YIELD DEDUCED FROM HIGH RESOLUTION EMISSION (AUGER) SPECTRA. / STATE DEPE   2-0520
EON.   HIGH RESOLUTION K-AUGER SPECTRA FOR MULTIPLY IONIZED N 4-0952
          L-S COUPLING INTERPRETATION OF HIGH RESOLUTION LMM AUGER SPECTRA OF COPPER AND ZINC.   4-1122
                                                        HIGH RESOLUTION LMM AUGER SPECTRA OF GERMANIUM.   4-0936
OLID SURFACES.   HIGH RESOLUTION LMM AUGER SPECTRA OF LOW ENERGY FROM S 4-0732
                                                        HIGH RESOLUTION LOW ENERGY ELECTRON SPECTROMETER.   3-0697
RYPTON AND M4, 5NN AND N4,500 AUGER SPECTRA FROM XE/ HIGH RESOLUTION L2,3MM AND M4,5NN AUGER SPECTRA FROM K 4-1113
    ARGON.   HIGH RESOLUTION L2,3M2,3M2,3 AUGER ELECTRON SPECTRA OF 4-0912
NC ATOMS.   HIGH RESOLUTION L2,3M4,5M4,5 AUGER SPECTRUM OF FREE ZI 4-0733
TOELECTRON STRUCTURE IN THE SECONDARY ELECTRON ENER/ HIGH RESOLUTION MEASUREMENTS OF AUGER ELECTRON AND PHO 4-1023
    AUGER ELECTRONS.   HIGH RESOLUTION MEASUREMENTS OF P(+) ARGON KLL AND KLM 4-0943
RANSITONS IN NICKEL AND COPPER.   HIGH RESOLUTION MEASUREMENTS OF THE L3M2,3M4,5 AUGER T 4-1024
    TRANSITIONS OF CHEMISORBED AMMONIA ON A MOLYBDENUM/ HIGH RESOLUTION MEASUREMENTS OF THE NITROGEN KLL AUGER 5-1532
INDIUM, ANTIMONY, TELLURIUM AND IODINE.   HIGH RESOLUTION MNN AUGER SPECTRA OF SILVER, CADMIUM, 4-0730
V ATOMIC HYDROGEN(+)- NEON COLLISIONS.   HIGH RESOLUTION NEON AUGER SPECTRUM PRODUCED IN 4.2 ME 4-1078
D BY 45 MEV CHLORINE(12+) ION IMPACT.   HIGH RESOLUTION NEON K-AUGER ELECTRON SPECTRUM PRODUCE 4-1051
ROGEN(+) AND HELIUM(+) ION BOMBARDMENT.   HIGH RESOLUTION NEON KLL AUGER SPECTRA PRODUCED BY HYD 4-0953
                                                     ENERGY RESOLUTION OF THE PHOTOEMISSION ANALYZER.   3-0606
R ELECTRONS)   COMPACT, INEXPENSIVE HIGH RESOLUTION RETARDING FIELD ENERGY ANALYZER. (AUGE   3-0619
    ANALYSIS OF X-BAND DIODE BURNOUT.   HIGH SPATIAL RESOLUTION SCANNING AUGER SPECTROSCOPY APPLIED TO 5-1961
AUGER SPECTROSCOPY OF SOLID SURFACES IN A DRY PUMPED HIGH RESOLUTION SCANNING ELECTRON MICROSCOPE.   5-1258
    SPECTROSCOPY.   INVESTIGATION OF HIGH RESOLUTION SCANNING TECHNIQUES FOR AUGER ELECTRON   3-0625
HI CORRELATIONS FOR ARSENIC (00001).   HIGH ANGULAR RESOLUTION SECONDARY ELECTRON SPECTROSCOPY. KIKUC   5-1784
DUCED BY ION BOMBARDMENT.   HIGH RESOLUTION STUDY OF ARGON LMM AUGER ELECTRONS PRO   4-0903
ER (100) SURFACE.   ANGULAR RESOLVED AUGER EMISSION SPECTRA FROM A CLEAN COPP   5-1697
          WIDTHS AND INTENSITIES OF HIGHLY RESOLVED LINES IN ARGON L2,3 AUGER SPECTRA.   4-1032
VOLVING HELIUM(+)(2S) AND HELIUM(2+) NEAR SOLID SURFACES. RESONANCE, AUGER, AND AUTOIONIZATION PROCESSES IN   2-0375
HEMICAL CONSEQUENCES OF THE AUGER EFF/ RADIATION CHEMICAL RESONANCE EFFECT IN SOLID 5-BROMO DEOXYURIDINE: C   2-0377
CTRONS. (ANGULAR DISTRIBUTION OF AUGER ELECTRONS) RESONANCE IONIZATION OF ATOMIC HELIUM BY FAST ELE   2-0256
SON OF AUGER, SECONDARY ION MASS SPECTROSCOPY, AND PROTON RESONANCE PROFILING FOR RELIABILITY STUDIES. /ARI 1-0216
R FOR SURFACE SPECTROSCOPY.   DOUBLY RESONANT CURRENT-TO-VOLTAGE CONVERTER PREAMPLIFIE   3-0677
HODS OF CLEANING CONTAMINANTS FROM QUARTZ SURFACES DURING RESONATOR FABRICATION. (AUGER SPECTRA)   MET 5-1410
    CHARACTERIZATION TECHNIQUES (AUGER, CLEANING) FOR QUARTZ RESONATORS.   SURFACE PREPARATION AND 5-1943
          SURFACE STUDIES FOR QUARTZ RESONATORS. (CLEANING, AUGER SPECTROSCOPY)   5-1942
          CYLINDRICAL MIRROR ANALYZER INCORPORATING PRERETARDATION. (TUNGSTEN AUGER ELECTRON)   3-0590
ON OF ELECTRONS FOR THE CYLINDRICAL MIRROR ANALYZER AND A RETARDING FIELD ANALYZER. /OF OVERLAYER ATTENUATI 3-0693
CTRON SPECTROSCOPY).   IMPROVED RETARDING FIELD ANALYZER. (FOR AUGER AND PHOTOELE   3-0701
ROSCOPY. (PALLADIUM- NICKEL ALLOYS, ADSORBED OXYGE/ SMALL RETARDING FIELD ANALYZER FOR AUGER ELECTRON SPECT 3-0644
    AUGER SPECTROSCOPY)   RETARDING FIELD CYLINDRICAL MIRROR ANALYZER. (FOR 3-0614
)   RETARDING FIELD ELECTRON SPECTROMETER.   3-0686
)   COMPACT, INEXPENSIVE HIGH RESOLUTION RETARDING FIELD ENERGY ANALYZER. (AUGER ELECTRONS   3-0619
SPECTROMETER)   A SMALL RETARDING FIELD ENERGY ANALYZER WITH CEM. (AUGER   3-0655
SPECTROGRAPH.   SIMPLIFIED RETARDING FIELD PHOTOELECTRON AND AUGER ELECTRON   3-0642
TECHNIQUES IN AUGER ELECTRON EMISSION SPECTROSCOPY USING RETARDING FIELD SPECTROMETERS. /TING, AND RELATED   3-0610
MATORS. (FOR AUGER ELECTRON SPECTROSC/ SOME PROPERTIES OF RETARDING FIELD SPECTROMETERS USING POST MONOCHRO 3-0624
R ELECT/ RESOLUTION AND SENSITIVITY OF THE SPHERICAL GRID RETARDING POTENTIAL ANALYZER. APPLICATION TO AUGE 3-0634
LECTRON SPECTROSCOPY)   IMPROVED SPHERICAL GRID RETARDING POTENTIAL ANALYZER. (FOR USE IN AUGER E 3-0583
DARY EMISSION ANALOG FOR IMPROVED AUGER SPECTROSCOPY WITH RETARDING POTENTIAL ANALYZERS. COMMENTS.   SECON 3-0584
/ARY EMISSION ANALOG FOR IMPROVED AUGER SPECTROSCOPY WITH RETARDING POTENTIAL ANALYZERS. (REPLY TO COMMENT/ 3-0694
STRENGTHS FOR MODULATION AMPLITUDE DISTORTION IN A 4-GRID RETARDING POTENTIAL ENERGY ANALYZER. /RON SIGNAL   3-0620
UREMENTS.   USE OF CYLINDRICAL AUGER SPECTROMETERS FOR RETARDING POTENTIAL SECONDARY ELECTRON YIELD MEAS 3-0628
MEASUREMENT OF CHANGES IN WORK FUNCTION. (LEED- AU/ AN AC RETARDING POTENTIAL TECHNIQUE FOR THE CONTINUOUS   3-0663
DYNAMIC BACKGROUND SUBTRACTION.   RETRIEVAL OF ELECTRON EXCITED AUGER STRUCTURE BY   1-0090
          BEAM EFFECTS IN AUGER ELECTRON SPECTROSCOPY REVEALED BY X-RAY PHOTOELECTRON SPECTROSCOPY.   3-0601
ECTRON PRODUCTION IN ION- ATOM COLLISIONS. (AUGER EFFECT, REVIEWS)   MECHANISMS OF EL 2-0490
                                                    RF SEE RADIOFREQUENCY
TION   RHEED SEE REFLECTION HIGH ENERGY ELECTRON DIFFRAC
ARBON MONOXIDE ADSORPTION ON TANTALUM AND POLYCRYSTALLINE RHENIUM. /RBON. TANTALUM CARBIDES, ETHYLENE AND C 4-0820
DY OF GRAIN BOUNDARY FRACTURE SURFACES IN DOPED TUNGSTEN- RHENIUM ALLOYS. (AUGER ELECTRON SPECTROSCOPY) STU 6-2112
NG IN TUNGSTEN- RHENIUM ALLOYS/ SOLID SOLUTION SOFTENING (RHENIUM DUCTILIZING EFFECT AND BUBBLE STRENGTHENI 6-2111
ORPTION AND DECOMPOSITION OF ACETYLENE ON POLYCRYSTALLINE RHENIUM FILMS AND RIBBONS. (AUGER SPECTROSCOPY) / 5-1688
    AUGER ELECTRON SPECTROSCOPY OF HIGH TEMPERATURE OXYGEN- RHENIUM INTERACTIONS.   4-1111
          A CARBON STRUCTURE ON THE RHENIUM (0001) SURFACE.   5-1989
          SOME PROPERTIES OF THE RHENIUM (0001) SURFACE.   5-1304
PHYSICAL AND CHEMICAL CHARACTERIZATION OF PLATINUM- RHODIUM GAUZE CATALYSTS. (AUGER SPECTROSCOPY)   5-1793
RON SPECTROSCOPY.   STUDY OF RUTHENIUM (0001) AND RHODIUM (111) SURFACES USING LEED AND AUGER ELECT 5-1380
OSITION OF ACETYLENE ON POLYCRYSTALLINE RHENIUM FILMS AND RIBBONS. (AUGER SPECTROSCOPY) /ORPTION AND DECOMP 5-1688
TITATIVE AUGER ELECTRON SPECTROSCOPY OF GOLD CONDENSED ON ROCK SALT AND ON AMORPHOUS CARBON.   QUAN 5-1136
S C/ MEASUREMENTS OF CONDENSATION COEFFICIENTS OF GOLD ON ROCK SALT (100) CLEAVAGE SURFACES AND ON AMORPHOU 5-1135
          SECONDARY ELECTRON EMISSION ROCKING CURVE FROM COPPER.   4-0764
          ENERGY WIDTHS OF ROENTGEN LEVELS BY AUGER ELECTRON SPECTROSCOPY.   4-0982
SURFACES. (AUGER MEASUREMENT METHODS)   ROOM TEMPERATURE ADSORPTION OF OXYGEN ON TUNGSTEN 5-1664
AGNETIC PROPERTIES IN THIN INTRINSIC INDIUM ANTIMONIDE AT ROOM TEMPERATURE. (AUGER PROCESS) /NT ON GALVANOM 2-0354
    SILICON (111)-7 STRUCTURE OBTAINED BY CLEAVAGE AT ROOM TEMPERATURE. (AUGER SURFACE SPECTRA)   5-1767
          EFFECT OF SURFACE ROUGHNESS OF AUGER ELECTRON SPECTROSCOPY.   5-1448
          DRYING OF BUTADIENE- NITRILE RUBBERS IN AN AUGER MACHINE.   3-0617
/PECTRAL LINE EXCITATION FUNCTIONS FOR SODIUM, POTASSIUM, RUBIDIUM, AND CESIUM IONS IN THE VACUUM ULTRAVIO/ 4-0735
    AUGER SPECTROSCOPY STUDY OF THE ADSORPTION OF RUBIDIUM ON MOLYBDENUM (100).   5-1894
          ADSORPTION OF COBALT ON (001) RUTHENIUM AT TEMPERATURES BELOW 300 DEGREES K.   5-1609
LEED AND AUGER ELECTRON SPECTROSCOPY.   STUDY OF RUTHENIUM (0001) AND RHODIUM (111) SURFACES USING 5-1380

ADSORPTION OF OXYGEN AND OXIDATION OF COBALT ON THE RUTHENIUM (001) SURFACE. (AUGER SPECTRA)   5-1608
RESISTIVITY CHANGES WITH AUGER ELECTRON SPECTROSCOPY AND RUTHERFORD BACKSCATTERING PROFILES. /RRELATION OF 5-1393
/S AT GOLD- GALLIUM ARSENIDE INTERFACES: AN ASSESSMENT BY RUTHERFORD BACKSCATTERING, SCANNING ELECTRON MIC/ 5-1902

S

X-RAY SPECTROSCOPY TECHNIQ/ INVESTIGATION OF MECHANISM OF S IONIZATION OF NOBLE GAS ATOMS BY THE ULTRASOFT 2-0580
IVE AUGER ELECTRON SPECTROSCOPY OF GOLD CONDENSED ON ROCK SALT AND ON AMORPHOUS CARBON.   QUANTITAT 5 1136
/EASUREMENTS OF CONDENSATION COEFFICIENTS OF GOLD ON ROCK SALT (100) CLEAVAGE SURFACES AND ON AMORPHOUS CA/ 5-1135
AUGER AND BINDING ENERGY SHIFTS IN TRANSITION METALS AND SALTS. /ATIVE EFFECT OF EXTRAATOMIC RELAXATION ON 2-0412
ADSORPTION OF IONIC SALTS ON A TUNGSTEN (100) SURFACE.   5-1659
KLL AUGER ELECTRON SPECTRUM OF SAMARIUM.   4-0725
SPECTRUM OF AUGER K,L,M,N,O ELECTRONS OF SAMARIUM.   4-0756
SPECTRUM OF AUGER KLL ELECTRONS OF SAMARIUM.   4-0726
KLM SPECTRUM OF SAMARIUM AUGER ELECTRONS.   4-0757
R ELECTRON SPECTROSCPY.   OXYGEN EXPOSURE OF SAMARIUM, GADOLINIUM, AND TERBIUM STUDIED BY AUGE 4-1332
4D AUGER TRANSITIONS FROM EXCITED STATES IN SAMARIUM METAL.   4-0823
ANALYSIS APPARATUS FOR INVESTIGATING A SAMPLE BY MEANS OF EMITTED ELECTRONS.   3-0713
UGER ELECTRON SPECTRCSCOPY.   ELECTRONIC SUPERPOSITION OF SAMPLE CURRENT AND SECONDARY ELECTRON IMAGES IN A 2-0450
A MULTIPLE SAMPLE HOLDER FOR USE IN ULTRAHIGH VACUUM.   3-0603
PATTERNS IN AN AUGER ELECTRON SPECTROMETER WITH SCANNING SAMPLE POSITIONER.   ELECTRON CHANNELING 3-0692
AUGER ELECTRON SPECTROSCCPY OF LUNAR SAMPLES.   4-0858
AUGER MEASUREMENTS ON T-027 SAMPLES EXPOSED DURING THE SKYLAB-2 MISSION.   5-1735
TECHNIQUE OF PREPARING POWDER SAMPLES FOR AES AND/OR ESCA ANALYSIS.   1-0197
VED BY AUGER ELECTRON SPECTROSCOPY.   SAMPLES HEATED DIRECTLY BY ELECTRIC CURRENT OBSER 1-0133
ER STUDIES AND ELECTRONIC PROPERTIES OF FILMS. SILICON ON SAPPHIRE EPITAXY BY VACUUM SUBLIMATION: LEED- AUG 5-1230
ESIUM SECONDARY ELECTRCN SPECTRA.   AUGER SATELLITE AND OTHER CHARACTERISTIC EVENTS IN MAGN 4-0898
N OF CONDUCTION ELECTRCNS IN SOFT X-RAY DOUBLE IONIZATION SATELLITE (AUGER) TRANSITION. /SSIBLE LOCALISATIO 2-0437
ION. (OF AUGER ELECTRONS)   NEW SATELLITE BANDS CORRESPONDING TO DOUBLE L IONIZAT 2-0293
THEORY OF AUGER SATELLITE ENERGY SHIFTS.   2-0507
S. (ALUMINUM)   PLASMON SATELLITE IN THE AUGER EMISSION SPECTRUM OF METAL 4-0871
ETERMINING ADATOM CONCENTRATION PROFILES.   AUGER PLASMON SATELLITE INTENSITIES VERSUS DEPTH: A MEANS FOR D 2-0347
RON SPECTRA PRODUCED BY 0.4 TO 4 MEV PROTON IMP/ RELATIVE SATELLITE LINE INTENSITIES IN ARGON K AUGER ELECT 2-0497
LOW PROTON IMPACT.   PRODUCTION OF INTENSE SATELLITE LINES IN AUGER SPECTRA OF NITROGEN BY S 4-1076
TION. PART-2: DOUBLE IONIZATION ORIGIN OF THE HIGH ENERGY SATELLITE OF AUGER PEAK. /N ON INNER LEVEL IONIZA 4-1047
AUGER TRANSITION RATE FOR THE NOBLE GAS SATELLITE SPECTRA.   4-0972
TRA OF COPPER.   AUGER VACANCY SATELLITE STRUCTURE IN THE L3,M4,5M4,5 AUGER SPEC 4-1034
N.   HIGH ENERGY SATELLITES IN AUGER PEAKS OF MAGNESIUM AND SILICO 4-1046
PLASMON SATELLITES IN AUGER SPECTRA.   2-0447
HIGH ENERGY SATELLITES IN AUGER SPECTRA FROM METALS.   4-0940
HIGHER ENERGY SATELLITES IN AUGER SPECTRA FROM METALS.   4-0900
TRANSITION PROBABILITIES OF KL- LLL AUGER SATELLITES IN NEON.   4-1049
BDENUM.   PLASMA SATELLITES IN THE AUGER ELECTRON SPECTRUM OF MOLY 4-1124
ABSENCE OF PLASMON GAIN SATELLITES IN THE AUGER SPECTRUM OF SILICON.   4-1037
PLASMON ENERGY GAIN AND DOUBLE IONIZATION SATELLITES IN THE BERYLLIUM AUGER SPECTRUM.   4-0899
CULATIONS FOR ASSIGNMENT OF THE AUGER SPECTRUM AND OF THE SATELLITES IN THE SOFT X-RAY SPECTRUM OF WATER. / 4-0724
ESIUM, ALUMINUM, AND/ INTERPRETATION OF HIGH ENERGY X-RAY SATELLITES OF L2,3 EMISSION BANDS OF SODIUM, MAGN 2-0239
UMINUM, SILICON, AND PHOSPHORUS.   AUGER SATELLITES OF THE L2,3 AUGER EMISSION BANDS OF AL 4-0985
COPPER.   HIGH ENERGY SATELLITES OF THE M2,3VV AND M1VV AUGER PEAKS OF 4-1043
MINUM, SILICON, AND PHOSPHORUS.   AUGER SATELLITES ON THE L23 AUGER EMISSION BANDS OF ALU 4-0984
NATURE OF SOME X-RAY SATELLITES (RADIATIVE AUGER TRANSITIONS).   2-0290
STEN (100), AND/ ELECTRON BEAM EFFECTS ON CARBON MONOXIDE SATURATED TUNGSTEN (100), ETHYLENE SATURATED TUNG 5-1244
CORRELATIONS ON KLL AUGER LINES IN ELEMENTS MAGNESIUM TO SCANDIUM.   EFFECT OF ELECTRONIC 4-0800
DOUBLE AND TRIPLE IONIZATION IN ALUMINUM, GALLIUM, AND SCANDIUM BY ELECTRON IMPACT. (AUGER SPECTRA) 2-0232
NDIUM OXIDE.   SCANDIUM L X-RAY EMISSION FROM PURE METAL AND SCA 4-1102
UDIES.   SCANNING AUGER ELECTRON MICROSCOPE FOR SURFACE ST 3-0679
SCANNING AUGER MICROANALYSIS.   1-0116
APPLICATIONS OF SCANNING AUGER MICROANALYSIS.   1-0119
Y DISTRIBUTIONS AT GRAIN BOUNDARIES IN SINTERED ALUMINA. (SCANNING AUGER MICROPROBE)   ADDITIVE AND IMPURIT 5-1505
Y ADSORPTION AT FLAKE IRON INTERFACES IN GRAY CAST IRON. (SCANNING AUGER MICROPROBE) /NFIRMATION OF IMPURIT 6-2054
RFACE AND THIN FILM ANALYSIS OF SEMICONDUCTOR MATERIALS. (SCANNING AUGER MICROPROBE)   SU 1-0089
ELECTRON SPECTROSCOPY ON THIN FILMS. (SCANNING AUGER MICROSCOPE)   5-1405
SCANNING AUGER MICROSCOPE FOR THIN FILM ANALYSIS. 3-0645
XY OF GALLIUM ARSENIDE AND GALLIUM PHOSPHIDE.   USE OF SCANNING AUGER MICROSCOPY IN MOLECULAR BEAM EPITA 5-1142
ES USING VACUUM DEPOSITION ON HEATED SILICON SUBSTRATES. (SCANNING AUGER SPECTROSCOPY) /ASE EPITAXIAL STUDI 5-1287
F X-BAND DIODE BURNOUT.   HIGH SPATIAL RESOLUTION SCANNING AUGER SPECTROSCOPY APPLIED TO ANALYSIS O 5-1961
USE OF SCANNING AUGER SPECTROSCOPY FOR SURFACE ANALYSIS. 5-1280
ATED CIRCUIT TECHNOLCGY.   APPLICATIONS OF SCANNING AUGER SPECTROSCOPY TO THE SILICON INTEGR 5-1655
SCANNING ELECTRON AUGER MICROSCOPY IN METALLURGY. 6-2105
AUGER ELECTRON IMAGES IN A SCANNING ELECTRON MICROSCOPE.   3-0649
AUGER ELECTRON SPECTROSCOPY IN A DIFFUSION PUMPED SCANNING ELECTRON MICROSCOPE.   3-0598
AUGER ELECTRON SPECTRCSCOPY IN THE SCANNING ELECTRON MICROSCOPE.   3-0710
ROSCOPY OF SOLID SURFACES IN A DRY PUMPED HIGH RESOLUTION SCANNING ELECTRON MICROSCOPE.   AUGER SPECT 5-1258
GER ELECTRON SPECTRCSCCPY IN AN ORDINARY DIFFUSION PUMPED SCANNING ELECTRON MICROSCOPE. /TIAL RESOLUTION AU 2-0597
CONDUCTOR JUNCTIONS BY AUGER ELECTRON SPECTROSCOPY IN THE SCANNING ELECTRON MICROSCOPE. /FILING ACROSS SEMI 2-0555
GES.   AUGER ELECTRON SPECTROSCOPY IN THE SCANNING ELECTRON MICROSCOPE: AUGER ELECTRON IMA 3-0650
CTROSCOPY, SURFACE ELEMENTAL COMPOS/ CONTAMINATION IN THE SCANNING ELECTRON MICROSCOPE. (AUGER ELECTRON SPE 3-0596
)   MATERIAL ANALYSIS USING THE SCANNING ELECTRON MICROSCOPE. (AUGER SPECTROSCOPY 5-1701
RY ION MASS ANALYSIS. (AUGER)   SPECTROSCOPY IN THE SCANNING ELECTRON MICROSCOPE NOW INCLUDES SECONDA 3-0674
NING)   SPECTROSCOPIC METHODS OF MICROANALYSIS USING THE SCANNING ELECTRON MICROSCOPE. (REVIEW, AUGER SCAN 3-0685
LABORATORY COMPUTER SCANNING ELECTRON MICROSCOPE SYSTEM.   3-0651
/CROANALYSIS. PART-1. (REVIEW OF ELECTRON PROBE ANALYZER, SCANNING ELECTRON MICROSCOPE, TRANSMISSION ELECT/ 1-0010
N CATHODE AND AUGER ANALYZER.   ULTRAHIGH VACUUM SCANNING ELECTRON MICROSCOPE WITH A FIELD EMISSIO 3-0621
FACILITIES.   ULTRAHIGH VACUUM SCANNING ELECTRON MICROSCOPE WITH AUGER ANALYSIS 3-0622
AUGER EMISSION SPECTROSCOPY.   A SCANNING ELECTRON MICROSCOPE WITH FACILITIES FOR 3-0593
E FILM FORMED ON AN IRON- CHROMIUM ALLOY.   AUGER SCANNING ELECTRON MICROSCOPIC STUDY OF THE PASSIV 5-1753
TH METALS AS STUDIED BY AUGER SPECTROSCOPY, FIELD ION AND SCANNING ELECTRON MICROSCOPY. /LENE IN CONTACT WI 5-1198
AUGER ELECTRON SPECTROSCOPY FOR SCANNING ELECTRON MICROSCOPY.   1-0114
COMBINED AUGER ELECTRON SPECTROSCOPY AND SCANNING ELECTRON MICROSCOPY.   1-0003
/ INTERFACES: AN ASSESSMENT BY RUTHERFORD BACKSCATTERING, SCANNING ELECTRON MICROSCOPY, AND DEPTH PROFILIN/ 5-1902

RAY ANALYSIS TECHNIQUES FOR/ AUGER ELECTRON SPECTROSCOPY, SCANNING ELECTRON MICROSCOPY AND NONDISPERSIVE X- 5-1799
AND ION BACKSCATTERING TECHNIQUES APPLIED TO ANALYSIS O/ SCANNING ELECTRON MICROSCOPY, AUGER SPECTROSCOPY, 5-1261
/LESS STEEL CORROSICN PRODUCTS IN FLOWING LIQUID SODIUM. (SCANNING ELECTRON MICROSCOPY, AUGER SPECTROSCOPY/ 6-2014
OF AUGER ELECTRON SP/ BASIC INVESTIGATION IN THE AREA OF SCANNING ELECTRON MICROSCOPY. PART-1: APPLICATION 1-0213
NTS.   AUGER ELECTRON SPECTROSCOPY IN SCANNING ELECTRON MICROSCOPY: POTENTIAL MEASUREME 3-0648
OSCOPY.   THIRD DIMENSION IN SCANNING ELECTRON MICROSCOPY. SCANNING AUGER MICR 1-0117
FOR MICROSPOT SURFACE AND IN-DEPTH ANALYSIS OF / COMBINED SCANNING ELECTRON MICROSCOPY- AUGER SPECTROSCOPY 5-1262
TERDIFFUSION AND EUTECTIC FORMATION IN TUNGSTEN PLATINUM/ SCANNING MICROSPOT AUGER SPECTROSCOPY STUDY OF IN 5-1259
HANNELING PATTERNS IN AN AUGER ELECTRON SPECTROMETER WITH SCANNING SAMPLE POSITIONER.   ELECTRON C 3-0692
PY.   INVESTIGATION OF HIGH RESOLUTION SCANNING TECHNIQUES FOR AUGER ELECTRON SPECTROSCO 3-0625
S.   ENERGY SPECTRA OF INELASTICALLY SCATTERED AND AUGER ELECTRONS FROM SINGLE CRYSTAL 4-1064
N OF SURFACE STRUCTURES USING LEED AND ENERGY ANALYSIS OF SCATTERED ELECTRONS.   DETERMINATIO 5-1956
AIN PHENOMENA OF COPPER (111).   ENERGY SPECTRUM OF BACK SCATTERED ELECTRONS AND CHARACTERISTIC LOSS AND G 4-0897
                      ENERGY SPECTRA OF INELASTICALLY SCATTERED ELECTRONS AND LEED STUDIES OF TUNGSTEN. 4-1094
UTRA/ CHARGE EXCHANGE OF LOW ENERGY HELIUM IONS AND ATOMS SCATTERED FROM A COPPER SINGLE CRYSTAL. (AUGER NE 2-0545
/EXCHANGE OF LOW ENERGY HELIUM(+) IONS (LESS THAN 10 KEV) SCATTERED FROM A (100) FACE OF A COPPER SINGLE C/ 2-0546
IGATING ENERGY AND ANGULAR DISTRIBUTIONS OF INELASTICALLY SCATTERED OR PHOTOEMITTED ELECTRONS. / FOR INVEST 3-0691
ISSION OF TRUE SECONDARY ELECTRONS. PART-2: INELASTICALLY SCATTERED PRIMARY ELECTRONS. / METALS. PART-1: EM 2-0500
                      SURFACE ANALYSIS WITH LOW ENERGY ION SCATTERING. 5-1871
S OF CLEAN NICKEL SURFACES AND ADSORBED L/ LOW ENERGY ION SCATTERING AND AUGER ELECTRON SPECTROSCOPY STUDIE 5-1870
BRIUM CHARGE CARRIERS IN SILICON AT A HIGH/ ELECTRON HOLE SCATTERING AND (AUGER) RECOMBINATION OF NONEQUILI 4-1103
SPECTROSCOPY.   INELASTIC SCATTERING EFFECTS IN QUANTITATIVE AUGER ELECTRON 2-0433
NEUTRALIZATION)   LOW ENERGY ION SCATTERING: ELASTIC AND INELASTIC EFFECTS. (AUGER 2-0382
SECONDARY ELECTRONS. PART-2: INELASTICALLY/ SLOW ELECTRON SCATTERING FROM METALS. PART-1: EMISSION OF TRUE 2-0500
ONIZATION LOSSES IN (AUGER) SPECTRA OF ELECTRON INELASTIC SCATTERING FROM RARE EARTH METAL SURFACES.   I 2-0538
(AES)   SORBAS: AN APPARATUS FOR INVESTIGATING ION SCATTERING FROM SURFACES AT ENERGIES 100-2000 EV. 3-0704
BAND-TO-BAND AUGER RECOMBINATION AND CARRIER-CARRIER SCATTERING IN POWER RECTIFIERS. 2-0458
AUGER ELECTRON SPECTRCSCOPY AND INELASTIC ELECTRON SCATTERING IN THE STUDIES OF METAL SURFACES. 5-1818
/ AT THE (100) CRYSTAL FACE OF PLATINUM BY MOLECULAR BEAM SCATTERING, LOW ENERGY ELECTRON DIFFRACTION, AND/ 5-1963
COPY) METHODS OF SURFACE STUDIES DEPENDING ON INELASTIC SCATTERING OF ELECTRONS. (AUGER ELECTRON SPECTROS 5-1159
(111) SURFACE.   SCATTERING OF LOW ENERGY ELECTRONS FROM A COPPER 5-1621
              INELASTIC SCATTERING OF LOW ENERGY ELECTRONS FROM SURFACES. 2-0495
              (AUGER) NEUTRALIZATION BEHAVIOR IN SCATTERING OF LOW ENERGY IONS FROM SOLID SURFACES 2-0294
ER SPECTROSCOPY)   SCATTERING OF MOLECULAR BEAMS FROM SURFACES. (AUG 5-1905
/ COMPARISON OF SURFACE CHEMISTRIES OBTAINED BY USING ION SCATTERING SPECTROMETRY AND AUGER ELECTRON SPECTR 6-1999
/THE CHEMICAL CHARACTERIZATIONS OF CERAMIC SURFACES. (ION SCATTERING SPECTROMETRY, AUGER ELECTRON SPECTROM/ 5-1326
OPY AND SECONDARY I/ SURFACE ANALYTICAL STUDIES USING ION SCATTERING SPECTROMETRY, AUGER ELECTRON SPECTROSC 5-1423
YZER)   QUANTITATIVE ASPECTS OF ION SCATTERING SPECTROSCOPY. (CYLINDRICAL MIRROR ANAL 3-0666
THE AUGER SPECTRUM AND OF THE SATELLITES IN THE SOFT X-/ SCF AND LIMITED CI CALCULATIONS FOR ASSIGNMENT OF 4-0724
AUGER TRANSITION RATE IN MIXED COUPLING SCHEME. 2-0342
ICKEL- PLATINUM ALLOY FILMS.   COMPOSITION PROFILES AND SCHOTTKY BARRIER HEIGHTS OF SILICIDES FORMED IN N 5-1900
AUGER EFFECT IN SURFACE SCIENCE. 5-1626
THE RELATIONSHIP OF SOLID SURFACES TO VACUUM SCIENCE AND TECHNOLOGY. (REVIEW) 1-0086
/ICTION, WEAR, AND LUBRICATION RESEARCH BY MODERN SURFACE SCIENCE TECHNIQUES. (AES, ELLIPSOMETRY, OTHER TE/ 1-0104
ELECTRONS. BASIC PROPERTIES OF A NEW METHOD FOR MATERIALS SCIENCES. /ANALYSIS WITH HEAVY ION INDUCED AUGER 1-0070
ONS. (AUGER EMISSION)   SCINTILLATION EFFICIENCY FOR LOW VELOCITY HEAVY I 2-0431
.   DYNAMICAL SCREENING EFFECTS IN THE AUGER SPECTRUM OF METALS 4-1110
EMICONDUCTORS. (SILICON, GERMANIUM)   INFLUENCE OF SCREENING EFFECTS ON THE AUGER RECOMBINATION IN S 2-0379
RONIG TRANSITION RATES TO VARIATIONS IN ENERGY LEVELS AND SCREENIG PARAMETERS. /TY OF L1,L2,3M4,5 COSTER-K 2-0303
5 AUGER SPECTRUM OF XENON.   SEARCH FOR CORRELATION EFFECTS IN THE M4,5N4,5N4, 4-0872
BERYLLIUM ALLOY.   SECONDARY AND AUGER ELECTRON SPECTRA OF A COPPER- 4-0916
                      SECONDARY (AUGER) ELECTRON SPECTROSCOPY. 1-0167
GUISH AUGER FROM ENERGY LOSS PEAKS OF THE PRIMARY BEAM IN SECONDARY ELECTRON ANALYSIS. /NG METHOD TO DISTIN 1-0188
SPREADING FUNCTIONS IN SECONDARY ELECTRON ANALYSIS. 2-0512
EXCITED AUGER ELECTRON SPECTROSCOPY.   THE SECONDARY ELECTRON BACKGROUND PROBLEM IN ELECTRON 1-0091
ABSORPTION BANDS IN SECONDARY ELECTRON CONTINUOUS SPECTRUM. 4-0933
Y METASTABLE ATOMS. PART-1: MEASUREMENTS OF SECONDARY EM/ SECONDARY ELECTRON EJECTION FROM METAL SURFACES B 2-0340
Y METASTABLE ATOMS. PART-3: ENERGY AND ANGULAR DISTRIBUT/ SECONDARY ELECTRON EJECTION FROM METAL SURFACES B 2-0340
BAND STRUCTURE OF GRAPHITE STUDIED BY SECONDARY ELECTRON EMISSION. 2-0569
FORMATION OF A PLASMA SHEATH WITH SECONDARY ELECTRON EMISSION. 2-0244
SSION STUDY OF ADSORPTION PROCESSES. OPTICALLY STIMULATED SECONDARY ELECTRON EMISSION AND PHOTOELECTRON EMI 2-0302
IES. (COPPER- GERMANIUM JUNCTION).   SECONDARY ELECTRON EMISSION AT SUPERHIGH FREQUENC 2-0356
/RBON CONCENTRATION ON BERYLLIUM OXIDE AND ITS EFFECTS ON SECONDARY ELECTRON EMISSION. (AUGER SPECTROSCOPY) 5-1362
ANIUM AND BERYLLIUM).   FILM STRUCTURE EFFECT ON THE SECONDARY ELECTRON EMISSION CHARACTERISTICS (GERM 4-0923
/S IN THE FINE STRUCTURE OF THE ANGULAR DEPENDENCE OF THE SECONDARY ELECTRON EMISSION COEFFICIENT OF SINGL/ 4-1061
/ACKGROUND INFORMATION IN AUGER ELECTRON SPECTROSCOPY AND SECONDARY ELECTRON EMISSION DUE TO CASCADE PROCE/ 2-0411
CKSCATTERED ELECTRONS (TABLE FOR 60 ELEMENTS).   SECONDARY ELECTRON EMISSION DUE TO PRIMARY AND BA 4-0907
ICKEL (100) AND (111) CRYSTAL SURFAC/ CHEMICAL CHANGES IN SECONDARY ELECTRON EMISSION DURING OXIDATION OF N 5-1453
ORIENTATION ANISOTROPY OF BACKSCATTERING COEFFICIENT AND SECONDARY ELECTRON EMISSION FOR 10-100 KEV ELECT/ 4-1031
                      SECONDARY ELECTRON EMISSION FROM GALLIUM ARSENIDE 4-0861
LICON (CESIUM).   SECONDARY ELECTRON EMISSION FROM ION IMPLANTED SI 4-0913
POLYCRYSTALS.   SECONDARY ELECTRON EMISSION FROM IRON SINGLE AND 4-0917
GOLD CERMETS.   EFFECTS OF CESIATION ON SECONDARY ELECTRON EMISSION FROM MAGNESIUM OXIDE- 2-0384
ONDU/ EXPERIMENTAL ASSEMBLY FOR OBTAINING THE SPECTRUM OF SECONDARY ELECTRON EMISSION FROM METALS AND SEMIC 3-0688
UM(+) IONS THROUGH CCPFER SINGLE CRYSTALS. ANISOTROPY OF SECONDARY ELECTRON EMISSION FROM PASSAGE OF LITHI 2-0245
OATED SURFACES. (AUGER SPECTRA)   SECONDARY ELECTRON EMISSION FROM PLATINUM BLACK C 5-1979
CHLORIDE FILMS.   CHARACTERISTICS OF SHAPING SECONDARY ELECTRON EMISSION FROM POROUS POTASSIUM 4-1120
AUGER ELECTRO/ FINE STRUCTURES AND ENERGY DISTRIBUTION OF SECONDARY ELECTRON EMISSION FROM SILICON (111). (4-0853
SINGLE CRYSTAL BOMBARDED BY RARE GAS IONS.   SECONDARY ELECTRON EMISSION FROM SODIUM CHLORINE 4-0759
OMBARDED BY MEDIUM ENERGY IONS (PLATINUM, NICKEL, TANTAL/ SECONDARY ELECTRON EMISSION FROM SOLID SURFACES B 2-0309
ETAL- OXIDE CERMETS.   SECONDARY ELECTRON EMISSION FROM THIN FILM GOLD M 4-1015
                      SECONDARY ELECTRON EMISSION FROM THIN FILMS. 4-0894
GOLD, APPARATUS)   SECONDARY ELECTRON EMISSION FROM THIN FILMS. (OF 3-0638
YBDENUM SINGLE CRYSTALS.   SECONDARY ELECTRON EMISSION FROM TUNGSTEN AND MOL 4-0742
YBDENUM.   SECONDARY ELECTRON EMISSION FROM TUNGSTEN AND MOL 4-0850
SURFACES.   SECONDARY ELECTRON EMISSION IN THE STUDY OF SOLID 5-1555
NUM BULK GETTER. (AUGER SPECTROSCOPY)   SECONDARY ELECTRON EMISSION OF A ZIRCONIUM- ALUMI 5-1407
UCTOR THIN FILMS.   SECONDARY ELECTRON EMISSION OF AMORPHOUS SEMICOND 5-1239
ITH 120-EV TO 5-KEV PROTONS.   SECONDARY ELECTRON EMISSION OF METALS BOMBARDED W 4-1029
S.   SECONDARY ELECTRON EMISSION OF MOLYBDENUM CRYSTAL 4-1063

NGLE CRYSTALS.   SECONDARY ELECTRON EMISSION OF SILICON DIOXIDE SI 4-1062
    MAGNESIUM TARGETS.   SECONDARY ELECTRON EMISSION OF SOLID ALUMINUM AND 2-0373
/E FINE STRUCTURE OF THE DEPENDENCE OF THE COEFFICIENT OF   SECONDARY ELECTRON EMISSION ON ENERGY OF PRIMARY/ 4-0891
NE MAGNESI/ OXIDATION EFFECTS ON THE SPECTRUM OF THE SLOW   SECONDARY ELECTRON EMISSION PEAK IN POLYCRYSTALLI 4-1118
PPER.   SECONDARY ELECTRON EMISSION. (REVIEW, 86 REFS)   1-0179
                    GRAPHITE CONDUCTION BAND STATES FROM   SECONDARY ELECTRON EMISSION ROCKING CURVE FROM CO 4-0764
    STUDY OF CARBON FIBER SURFACES USING AUGER AND   SECONDARY ELECTRON EMISSION SPECTRA.   2-0568
CES.   ELECTRON INDUCED   SECONDARY ELECTRON EMISSION SPECTROSCOPY.   5-1970
TION TO SILIC/ METHOD FOR DETECTING FINE STRUCTURE IN THE   SECONDARY ELECTRON EMISSION SPECTROSCOPY OF SURFA 5-1406
/IDE- GOLD CERMET FILMS AND ITS APPLICATICN TO HIGH YIELD   SECONDARY ELECTRON EMISSION YIELD AND THE APPLICA 4-0852
URANIUM DICXIDE SURFACES.   SECONDARY ELECTRON EMITTERS. (AUGER ELECTRON SPE/ 5-1430
                                        TRUE   SECONDARY ELECTRON ENERGY DISTRIBUTION STUDIES OF 5-1316
/NTS OF AUGER ELECTRON AND PHOTOELECTRON STRUCTURE IN THE   SECONDARY ELECTRON ENERGY DISTRIBUTIONS.   2-0242
                            DIFFRACTION PEAKS IN   SECONDARY ELECTRON ENERGY DISTRIBUTIONS OF ALUMI/ 4-1023
ROSCOPY.   ELECTRONIC SUPERPOSITION OF SAMPLE CURRENT AND SECONDARY ELECTRON ENERGY SPECTRA. (NICKEL (100)) 4-0775
ER SATELLITE AND OTHER CHARACTERISTIC EVENTS IN MAGNESIUM   SECONDARY ELECTRON IMAGES IN AUGER ELECTRON SPECT 2-0450
D IONIZATION LOSS EVENTS IN BERYLLIUM AND BERYLLIUM OXIDE   SECONDARY ELECTRON SPECTRA.   AUG 4-0898
.   AUGER AND OTHER CHARACTERISTIC ENERGIES IN SECONDARY ELECTRON SPECTRA. /OUBLE IONIZATION, AN 4-0901
FACES.   CHARACTERISTIC ENERGIES IN SECONDARY ELECTRON SPECTRA FROM ALUMINUM SURFACES 4-0896
HOSPHORUS FIL/ AUGER AND OTHER CHARACTERISTIC ENERGIES IN   SECONDARY ELECTRON SPECTRA FROM SILICON (111) SUR 5-1263
UNDER IONIC BOMBARDMENT.   SECONDARY ELECTRON SPECTRA FROM THIN EVAPORATED P 4-0986
                                                            SECONDARY ELECTRON SPECTRA OF SOME SOLID TARGETS 4-0942
                                                            SECONDARY ELECTRON SPECTROMETER.   3-0709
                                                            SECONDARY ELECTRON SPECTROSCOPY.   1-0074
          CESIUM ADSORPTION ON TUNGSTEN STUDIED BY LEED AND SECONDARY ELECTRON SPECTROSCOPY.   5-1661
/OMPLEX SEMICONDUCTOR STRUCTURES. TASK-II. APPLICATION OF   SECONDARY ELECTRON SPECTROSCOPY (AUGER ELECTRON / 5-1411
IONS FOR ARSENIC (00001).   HIGH ANGULAR RESOLUTION   SECONDARY ELECTRON SPECTROSCOPY. KIKUCHI CORRELAT 5-1784
E SILVER.   SECONDARY ELECTRON SPECTROSCOPY OF POLYCRYSTALLIN 4-1016
S OF MATERIALS BY ELECTRON EXCITED AUGER ELECTRON EMISSI/   SECONDARY ELECTRON SPECTROSCOPY. (SURFACE ANALYSI 5-1403
E TOOL. (REVIEW, 22 REFS)   SECONDARY ELECTRON SPECTROSCOPY. SURFACE SENSITIV 1-0094
E TOOL. (AES)   SECONDARY ELECTRON SPECTROSCOPY. SURFACE SENSITIV 1-0095
              NEW METHOD FOR ANALYZING THE   SECONDARY ELECTRON SPECTRUM.   4-0934
F CYLINDRICAL AUGER SPECTROMETERS FOR RETARDING POTENTIAL   SECONDARY ELECTRON YIELD MEASUREMENTS.   USE O 3-0628
ARY ENERGIES. (LEED- AUGER SYSTEM)   FAST, ACCURATE   SECONDARY ELECTRON YIELD MEASUREMENTS AT LOW PRIM 3-0629
         ELECTRON BINDING ENERGIES OF BARIUM FROM THE   SECONDARY ELECTRON YIELD SPECTRUM.   4-0846
XYGEN SURFACE BY THERMAL CARBON OXYGEN NITROGEN (2), AND / SECONDARY ELECTRON YIELDS FROM COPPER BERYLLIUM O 4-0786
          AUGER EXCITATION BY INTERNAL   SECONDARY ELECTRONS.   2-0390
          THEORY FOR ENERGY DISTRIBUTION OF   SECONDARY ELECTRONS.   2-0241
TRON SHAKE-OFF AND (SUGAR) VACANCY CASCADES.   NATURE OF   SECONDARY ELECTRONS CREATED AS THE RESULT OF ELEC 2-0307
     OBSERVATION OF AUGER ELECTRONS IN THE ENERGY SPECTRUM OF SECONDARY ELECTRONS EMITTED BY COPPER UNDER BOMB/ 2-0551
ND GALLIUM PHOSPHIDE SUBMITTED TO/ ENERGY DISTRIBUTION OF   SECONDARY ELECTRONS EMITTED BY INDIUM PHOSPHIDE A 2-0269
    ENERGY SPECTRUM AND ANGULAR DISTRIBUTION OF   SECONDARY ELECTRONS EMITTED BY IONIC BOMBARDMENT. 4-0988
CON WHEN BOMBARDED BY PRIMARY ELECTRON/ ENERGY SPECTRA OF   SECONDARY ELECTRONS EMITTED BY MAGNESIUM AND SILI 4-1047
TRONS.   AUGER AND   SECONDARY ELECTRONS EXCITED BY BACKSCATTERED ELEC 2-0365
TRONS. APPROACH TO QUANTITATIVE ANALYSIS.   AUGER AND   SECONDARY ELECTRONS EXCITED BY BACKSCATTERED ELEC 2-0366
     AUGER ELECTRON EMISSION IN THE ENERGY SPECTRA OF   SECONDARY ELECTRONS FROM MOLYBDENUM AND TUNGSTEN. 4-0874
          AUGER PEAKS IN THE ENERGY SPECTRA OF   SECONDARY ELECTRONS FROM VARIOUS MATERIALS.   4-0926
N AND DISSOCIATION: CARBON MONOXIDE / ROLE OF PRIMARY AND   SECONDARY ELECTRONS IN ELECTRON INDUCED DESORPTIO 5-1567
/LECTRON SCATTERING FRCM METALS. PART-1: EMISSION OF TRUE   SECONDARY ELECTRONS. PART-2: INELASTICALLY SCATT/ 2-0500
TROSCOPY WITH RETARDING POTENTIAL ANALYZERS. COMMENTS.   SECONDARY EMISSION ANALOG FOR IMPROVED AUGER SPEC 3-0584
TROSCOPY WITH RETARDING POTENTIAL ANALYZERS. (REPLY TO C/   SECONDARY EMISSION ANALOG FOR IMPROVED AUGER SPEC 3-0694
FACES.   SECONDARY EMISSION AND CONTAMINATION OF METAL SUR 6-2071
TRONS FROM MONOCRYSTALS.   SECONDARY EMISSION AND ELASTIC REFLECTION OF ELEC 2-0508
ERMANIUM, GRAPHITE, ELEC/ AUGER ELECTRON SPECTROSCOPY AND   SECONDARY EMISSION IN SEMICONDUCTORS. (SILICON, G 2-0240
IED ON A MOLYBDENUM SINGLE CRYSTAL.   STRUCTURE AND   SECONDARY EMISSION OF MAGNESIUM OXIDE LAYERS APPL 5-1191
M- COVERED GERMANIUM.   SECONDARY EMISSION STUDIES ON GERMANIUM AND SODIU 5-1714
     IN-DEPTH PROFILES BY AUGER SPECTROSCOPY AND   SECONDARY ION EMISSION.   4-0995
HOSPHORUS ION IMPLANTED SILICON BY AUGER SPECTROSCOPY AND   SECONDARY ION EMISSION.   IN-DEPTH PROFILES OF P 4-0996
TROSCOPY IN THE SCANNING ELECTRON MICROSCOPE NOW INCLUDES   SECONDARY ION MASS ANALYSIS. (AUGER)   SPEC 3-0674
/LM ON COPPER ANALYZED BY AUGER ELECTRON SPECTROSCOPY AND   SECONDARY ION MASS SPECTRA DURING XENON ION BOMB/ 5-1679
                                                            SECONDARY ION MASS SPECTROMETRY.   1-0079
          COMPARISON OF AUGER ELECTRON SPECTROSCOPY AND   SECONDARY ION MASS SPECTROMETRY.   1-0135
AND PLATINUM SILICIDE BY AUGER ELECTRON SPECTROSCOPY AND   SECONDARY ION MASS SPECTROMETRY. /OSITED PLATINUM 5-1656
    ELEMENTAL ANALYSIS WITH AUGER ELECTRON SPECTROSCOPY AND SECONDARY ION MASS SPECTROMETRY.   5-1654
UTTERED TANTALUM FILMS BY AUGER ELECTRON SPECTROSCOPY AND   SECONDARY ION MASS SPECTROMETRY. /D OXYGEN) IN SP 5-1652
    SCATTERING SPECTROMETFY, AUGER ELECTRON SPECTROSCOPY AND SECONDARY ION MASS SPECTROMETRY. /UDIES USING ION 5-1423
ON SPECTROSCOPY): A USEFUL COMPLEMENT TO OTHER DEPTH PRO/   SECONDARY ION MASS SPECTROMETRY (AND AUGER ELECTR 1-0009
/URFACE CLEANLINESS CF 316 L-N STAINLESS STEEL STUDIES BY   SECONDARY ION MASS SPECTROMETRY AND AUGER ELECTR/ 3-0654
/DY OF SURFACE PROCESSES ON COPPER- BERYLLIUM BY COMBINED   SECONDARY ION MASS SPECTROMETRY AND AUGER ELECTR/ 5-1213
N SPECTROSCOPY, FOR THE ANALYSIS OF THIN / COMBINATION OF   SECONDARY ION MASS SPECTROMETRY AND AUGER ELECTRO 5-1214
/PHIDE (AND) NITROGEN, OXYGEN, AND FLUORINE IN SILICON BY   SECONDARY ION MASS SPECTROMETRY AND AUGER ELECTR/ 5-1916
/URFACE CLEANLINESS OF 316 L-N STAINLESS STEEL STUDIED BY   SECONDARY ION MASS SPECTROMETRY AND AUGER ELECTR/ 6-2087
/LLY OXIDIZED SURFACES BY AUGER ELECTRON SPECTROSCOPY AND   SECONDARY ION MASS SPECTROMETRY FOR ALUMINUM, SI/ 5-1553
RONIC TECHNOLOGY./ USE OF AUGER ELECTRON SPECTROSCOPY AND   SECONDARY ION MASS SPECTROMETRY IN THE MICROELECT 1-0136
UM. (AES)   AN ANALYTICAL SYSTEM FOR SECONDARY ION MASS SPECTROMETRY IN ULTRAHIGH VACU 3-0632
LECTRON SPECTROSCOPY X-RAY PHOTOELECTRON SPECTROSCOPY AND   SECONDARY ION MASS SPECTROSCOPY. /S USING AUGER E 5-1353
F GOLD THICK FILMS ON COPPER PLATES BY AUGER SPECTROSCOPY AND SECONDARY ION MASS SPECTROSCOPY. /ANED SURFACES O 5-1510
/ICIENTS ON SURFACES CHARACTERIZED BY AUGER SPECTROSCOPY,   SECONDARY ION MASS SPECTROSCOPY, AND LOW ENERGY E/ 2-0501
ANCE PROFILING FOR RELIABILI/ DIRECT COMPARISON OF AUGER,   SECONDARY ION MASS SPECTROSCOPY, AND PROTON RESON 1-0216
WO SUR/ PHOTO- AND AUGER ELECTRON SPECTROSCOPY (ESCA) AND   SECONDARY ION MASS SPECTROSCOPY. COMPARISON OF T 1-0088
/LAYER ANALYSIS TECHNIQUES. (AUGER ELECTRON SPECTROSCOPY,   SECONDARY ION MASS SPECTROSCOPY, ELECTRON MICROP/ 1-1627
IZED / COMPARISON BETWEEN AUGER ELECTRON SPECTROSCOPY AND   SECONDARY ION MASS SPECTROSCOPY ON WELL CHARACTER 5-1689
ER BY ELECTRON BOMBARDMENT. (AUGER EL/ DIFFERENTIAL CROSS   SECTION FOR K-SHELL IONIZATION OF COPPER AND SILV 2-0477
     AUGER TRANSITION IN MAGNESIUM AND SILICON FROM THE CROSS SECTION MODEL OF GRYZINSKI. (AUGER EFFECT) /2,3VV 4-0765
)- ALUMINUM COLLISION.   IONIZATION CROSS   SECTION OF THE L2,3 SHELL OF ALUMINUM IN A NEON (+ 2-0267
     ELECTRON SPECTROSCO/ MEASUREMENT OF THE IONIZATION CROSS SECTION OF THE L2,3 SHELL OF CHLORINE USING AUGER 4-0987
ER RATES, CORRELATION ENERGIES, AND PHOTOIONIZATION CROSS   SECTIONS. /ATION THEORY TO THE CALCULATION OF AUG 2-0311
ESTIMATION OF BACKSCATTERING EFFECTS AND IONIZATION CROSS   SECTIONS.   AUGER EMISSION FROM SOLIDS. 2-0513
ANTITATIVE AES.   MEASUREMENT OF IONIZATICN CROSS   SECTIONS AND BACKSCATTERING FACTORS FOR USE IN QU 2-0554
AND MOLECULES BY ELECTRON IMPACT. (/ X-RAY EMISSICN CROSS   SECTIONS AND FLUORESCENCE YIELDS FOR LIGHT ATOMS 2-0528

DIFIED BY AUGER TRANSITIONS.   X-RAY IONIZATION CROSS SECTIONS, AND IONIZATION EQUILIBRIUM EQUATIONS MO 2-0562
LECTRON EXCITED AUGER SPE/ COMPARISON OF IONIZATION CROSS SECTIONS AND OTHER FACTORS IN X-RAY EXCITED AND E 2-0261
          DETERMINATION OF AUGER ELECTRON EXCITATION CROSS SECTIONS AND RANGES USING PROTON EXCITED X-RAYS.   2-0456
LECULES BY 50-500 KEV PROTONS.   CROSS SECTIONS FOR INNER SHELL IONIZATION OF GASEOUS MO 4-1077
BY ELECTRON IMPACT. (AUGER SPECTROSCOPY)   CROSS SECTIONS FOR IONIZATION OF INNER SHELL ELECTRONS 2-0475
BY ELECTRON IMPACT.   DIFFERENTIAL CROSS SECTIONS FOR K-SHELL IONIZATION OF SURFACE ATOMS   2-0358
 PROTON- ARGON COLLISIONS.   CROSS SECTIONS FOR L2,3 AUGER EMISSION IN 22 TO 300 KEV 4-1040
TROGEN(X), AND NITROGEN(2)(2X) FROM NITROGEN BY EL/ CROSS SECTIONS FOR THE PRODUCTION OF NITROGEN(2)(X), NI 2-0376
     K-SHELL AUGER ELECTRON PRODUCTION CROSS CRCSS SECTIONS FROM ION BOMBARDMENT.   2-0570
     K-SHELL AUGER ELECTRON PRODUCTION CRCSS CRCSS SECTIONS FROM ION BOMBARDMENT. (NEON, FLUORINE)   2-0574
, OXYGEN, AND FLUORINE.   NEON K-AUGER CROSS SECTIONS FROM 1.5 MEV PER AMU AND ATOMIC HYDROGEN   2-0571
/ OF DEPENDENCE OF TOTAL K-SHELL VACANCY PRODUCTION CROSS SECTIONS IN ARGON BY THE IMPACT OF HEAVY IONS. (/ 2-0446
          L-SHELL IONIZATION CROSS SECTIONS IN AUGER ELECTRON SPECTROSCOPY.   2-0339
          ARGON K-AUGER CROSS SECTIONS IN COLLISION WITH 30 MEV FLUORINE IONS.   2-0323
EON. (AUGER TR/ ABSOLUTE ELECTRON IMPACT IONIZATION CROSS SECTIONS OF ATOMIC NITROGEN, ATOMIC OXYGEN, AND N 2-0363
NEW METHOD FOR MEASURING ELECTRON IMPACT IONIZATION CROSS SECTIONS OF INNER SHELLS.   A 4-0848
ON SPECTROSCOPY.   ELECTRON IMPACT IONIZATION CROSS SECTIONS OF INNER SHELLS MEASURED BY AUGER ELECTR 2-0553
ENSITIES OF TUNGS/ TOTAL ELECTRON IMPACT IONIZATION CROSS SECTIONS ON K-SHELLS OF SURFACE ATOMS. (AUGER INT 4-0847
. (AUGER ELECTRON SPECTROSCOPY)   STUDY OF GRAIN BOUNDARY SEGREGANTS IN THERMALLY EMBRITTLED MARAGING STEEL 6-2055
          AUGER SPECTROSCOPIC ANALYSIS OF BISMUTH SEGREGATED TO GRAIN BOUNDARIES IN COPPER.   6-2060
ER SPECTROSCOPIC ANALYSIS OF THE EXTENT OF GRAIN BOUNDARY SEGREGATION.   AUG 6-2094
ULFUR EMBRITTLED OXYGEN FREE COPPER. (AES, GRAIN BOUNDARY SEGREGATION)   STUDY OF S 6-2015
RMED ON COPPER- ALUMINUM, COPPER- TI/ A REVIEW OF SURFACE SEGREGATION, ADHESION, AND FRICTION STUDIES PERFO 6-2033
BRITTLEMENT OF LOW ALLOY STEELS. (IMPURITY GRAIN BOUNDARY SEGREGATION, AES OF FRACTURE SURFACES) /TEMPER EM 6-2083
N WITH ZIRCONIUM IN NIOBIUM- 1% ZIRCONIUM.   SURFACE SEGREGATION AND INTERACTION OF OXYGEN AND NITROGE 5-1508
GREGATION (AUGER ELECTRON SPECTROSCOPY).   SURFACE SEGREGATION AND ITS RELATION TO GRAIN BOUNDARY SE 6-2106
ION. (AUGER SPECTROSCOPY)   SURFACE SEGREGATION AS A GUIDE TO GRAIN BOUNDARY SEGREGAT 6-2073
R SPECTROSCOPY)   SEGREGATION AT COPPER- GOLD ALLOY SURFACES. (AUGE 5-1631
ROLE IN TEMPER EMBRITTLEMENT OF LOW ALLOY STEELS. (AUGE/ SEGREGATION AT SELECTIVE GRAIN BOUNDARIES AND ITS 6-2057
          SITE COMPETITION IN SURFACE SEGREGATION. (AUGER EFFECT)   6-2074
          GRAIN BOUNDARY SEGREGATION. (AUGER ELECTRON SPECTROSCOPY)   6-2104
ITTLEMENT OF NICKEL BY HYDROGEN: EFFECT OF GRAIN BOUNDARY SEGREGATION. (AUGER ELECTRON SPECTROSCOPY) / EMBR 6-2072
ACTURE SURFACES.   GRAIN BOUNDARY STRUCTURE AND SEGREGATION. AUGER SPECTROGRAPHIC ANALYSIS OF FR 6-2118
          SOME OBSERVATIONS OF SURFACE SEGREGATION BY AUGER ELECTRON EMISSION.   5-1401
MENT.   SURFACE SEGREGATION: COMPARISON BETWEEN THEORY AND EXPERI 5-1566
H CHLORINE GA/ CHLORINE ADSORPTION AT A TITANIUM SURFACE. SEGREGATION FROM THE BULK CRYSTAL INTERACTION WIT 5-1833
DIUM (/ LEED- AES STUDIES OF CHEMISORPTION INDUCED SULFUR SEGREGATION FROM THE BULK TO THE SURFACE OF PALLA 5-1865
UM ALLOY USING AU/ THERMAL EFFECTS IN EQUILIBRIUM SURFACE SEGREGATION IN A COPPER- 10 ATOMIC PERCENT ALUMIN 6-2030
GOLD TO THE (111) SURFACE OF NICKEL. (AES)   SURFACE SEGREGATION IN ALLOYS: EQUILIBRIUM SEGREGATION OF 5-1216
Y)   GRAIN BOUNDARY SEGREGATION IN ALUMINUM OXIDE. (AUGER SPECTROSCOP 6-2134
EMISSION SPECTROSCOPY.   SURFACE SEGREGATION IN BORON DOPED IRON OBSERVED BY AUGER 5-1181
R SPECTROSCOPY)   METASTABLE GRAIN BOUNDARY SEGREGATION IN COPPER CONTAINING TELLURIUM. (AUGE 6-2086
          AUGER ELECTRON SPECTROSCOPY STUDY OF SURFACE SEGREGATION IN COPPER- ALUMINUM ALLOYS.   6-2012
AUGER SPECTROSCOPY AND LEED STUDY OF EQUILIBRIUM SURFACE SEGREGATION IN COPPER- ALUMINUM ALLOYS.   6-2025
UM AUGER ELECTRON SPECTROMETER, FRACTURED/ GRAIN BOUNDARY SEGREGATION IN MAGNESIA DOPED ALUMINA. (HIGH VACU 5-1615
          AUGER AND ELLIPSOMETRIC STUDY OF PHOSPHORUS SEGREGATION IN OXIDIZED DEGENERATE SILICON.   5-1250
SPECTROSCOPY)   GRAIN BOUNDARY SEGREGATION IN SINTERED ALUMINA. (AUGER ELECTRON 6-2056
          IRRADIATION INDUCED SOLUTE SEGREGATION (IN STEEL, AUGER SPECTROSCOPY).   6-2092
SPECTROSCOPY.   INTERGRANULAR FRACTURE AND IMPURITY SEGREGATION IN STEEL. STUDIES BY AUGER ELECTRON 6-2127
PERCENT IN/ AUGER ELECTRON SPECTROSCOPY STUDY OF SURFACE SEGREGATION IN THE BINARY ALLOYS COPPER- 1 ATOMIC 6-2027
CTRON SPECTROSCOPY.   SURFACE SEGREGATION IN TITANIUM AS MONITORED BY AUGER ELE 6-2067
DURING IRRADIATION. (AUGER SPECTROSCOPY ANALYSIS, STAIN/ SEGREGATION OF ALLOYING ELEMENTS TO FREE SURFACES 6-2091
PER- BISMUTH ALLOYS.   SEGREGATION OF BISMUTH TO GRAIN BOUNDARIES IN COP 6-2096
SURFACE. (AUGER SPECTROSCOPY)   SEGREGATION OF CADMIUM TO A (111) SILVER CHLORIDE 5-1918
EFFECT)   EQUILIBRIUM SURFACE SEGREGATION OF CARBON ON IRON (100) FACES. (AUGER 5-1369
A SURFACE PHASE TRANSITION. (AUGER SPECTROSC/ EQUILIBRIUM SEGREGATION OF CARBON TO A NICKEL (111) SURFACE: 5-1803
          SEGREGATION OF CARBON TO (100) SURFACE OF NICKEL. 6-2000
NNEALED 50-50 NICKEL ALLOY BY AUGER E/ OBSERVATION OF THE SEGREGATION OF COPPER TO THE SURFACE OF A CLEAN A 6-2044
RVED BY AUGER EMISSION SPECTROSCOPY AND BY LEED.   SEGREGATION OF GOLD TO SILICON (111) SURFACE OBSE 5-1180
ULAR BRITTLE FRACTURE. (STUDIED BY AES)   GRAIN BOUNDARY SEGREGATION OF IMPURITIES IN METALS AND INTERGRAN 6-2079
THERMODYNAMICS AND KINETICS. (DETERMINATION OF SURFACES/ SEGREGATION OF IMPURITIES ON SURFACES OF THORIUM. 5-1313
N BOUNDARIES OF TEMPER EMBRITTLED STEEL. (AUGER ELECTRON/ SEGREGATION OF MANGANESE AND ANTIMONY TO THE GRAI 6-2070
          SURFACE SEGREGATION OF OXYGEN IN NIOBIUM. (AES)   5-1443
ALUM- OXYGEN ALLOYS. (AUGER ELECTRON SPECTROSCOP/ SURFACE SEGREGATION OF OXYGEN IN NIOBIUM- OXYGEN AND TANT 6-2065
UM- VANADIUM / EFFECT OF MICROSTRUCTURE ON GRAIN BOUNDARY SEGREGATION OF PHOSPHORUS IN A CHROMIUM- MOLYBDEN 6-2139
          QUANTITATIVE AUGER SPECTROSCOPIC ANALYSIS OF SEGREGATION OF PHOSPHORUS IN IRON.   5-1802
          AUGER SPECTROSCOPY OF OSMIUM, AND THE SEGREGATION OF PLATINUM TO THE OSMIUM SURFACE.   5-1770
EMENT BY AUGER SPECTROSCOPY)   SURFACE AND GRAIN BOUNDARY SEGREGATION OF STEEL IMPURITIES. (TEMPER EMBRITTL 6-2109
APPLICATIONS TO SOME TECHNICAL PROBLEMS. (EMBRITTLEMENT, SEGREGATION OF STEELS AND ALLOYS) /OSCOPY AND ITS 6-2100
DIED BY AUGER ELECTRON SPECTROSCOPY.   SEGREGATION OF SULFUR ON A MOLYBDENUM SURFACE STU 5-1564
T, AND PALLADIUM- GOLD 70 ATOMIC PERCENT CONTACT/ SURFACE SEGREGATION ON PALLADIUM- SILVER 40 ATOMIC PERCEN 6-2041
SPECTRA.   SODIUM SEGREGATION ONTO A LITHIUM METAL SURFACE. (AUGER 5-1750
AUGER ELECTRON SPECTROSCOPY)   GRAIN BOUNDARY SEGREGATION RELATING TO INTERGRANULAR FRACTURE. (6-2082
NDING ENERGIES OF CARBON TO NICKEL (100) FROM EQUILIBRIUM SEGREGATION STUDIES. (AUGER EFFECT)   BI 5-1484
N SPECTROSCOPY.   SURFACE SEGREGATION STUDIES IN ALLOYS USING AUGER ELECTRO 6-2019
          IMPURITY SEGREGATION TO GRAIN BOUNDARIES.   6-2063
OYS. (SEGREGATION IN STEEL, T/ RELATION OF GRAIN BOUNDARY SEGREGATION TO REDUCED STRENGTH OF METALS AND ALL 6-2120
ERMAL TREATMENT. WORK FUNCTION CHANGES AND AU/ IMPURITIES SEGREGATION TO THE (001) NICKEL SURFACE DURING TH 6-2069
SCOPY.   STUDY OF GRAIN BOUNDARY SEGREGATION USING AUGER ELECTRON EMISSION SPECTRO 6-2117
          MEASUREMENT OF EQUILIBRIUM SURFACE SEGREGATION USING AUGER ELECTRON SPECTROSCOPY.   5-1184
EPLY TO COMMENTS)   MEASUREMENT OF EQUILIBRIUM SURFACE SEGREGATION USING AUGER ELECTRON SPECTROSCOPY. (R 6-2026
          STUDY OF GRAIN BOUNDARY SEGREGATION USING AUGER EMISSION SPECTRA.   6-2121
FILMS.   SELECTED AREA AND IN-DEPTH AUGER ANALYSIS OF THIN 5-1653
EMBRITTLEMENT OF LOW ALLOY STEELS. (AUGE/ SEGREGATION AT SELECTIVE GRAIN BOUNDARIES AND ITS ROLE IN TEMPER 6-2057
TOR SURFACES. (AUGER EFFECT)   ION SELECTIVE SPUTTERING OF III-V COMPOUND SEMICONDUC 5-1637
          APPLICATION OF AES TO THE STUDY OF SELECTIVE SPUTTERING OF THIN FILMS.   5-1933
CRYSTALS. (/ PHOTOCONDUCTIVITY OF HIGHLY EXCITED CADMIUM SELENIDE AND CADMIUM SULFIDE(0.48) SELENIDE(0.52) 2-0413
          STIMULATED PHOTOLUMINESCENCE IN GALLIUM SELENIDE. (AUGER PROCESS)   2-0416
S/ TEMPERATURE DEPENDENCE OF THE EDGE EMISSION OF GALLIUM SELENIDE SINGLE CRYSTALS AT HIGH EXCITATION LEVEL 2-0258
          ELECTRON EMISSION STUDIES FROM GALLIUM SELENIDE SURFACES.   5-1967

HREE ENHANCED MULTILAYER DIELECTRIC MIRROR DESIGNS. (ZINC SELENIDE, THORIUM FLUORIDE)   AUGER ANALYSIS OF T 5-1328
HIGHLY EXCITED CADMIUM SELENIDE AND CADMIUM SULFIDE(0.48) SELENIDE(0.52) CRYSTALS. (AUGER RECOMBINATION) / 2-0432
N ENERGIES IN THE L3M4,5M4,5 AUGER PROCESS IN ARSENIC AND SELENIUM. ATOMIC AND EXTRAATOMIC STATIC RELAXATIO 4-0883
                                      L2,3MM AUGER PROCESSES IN SELENIUM.   2-0560
ELATIVE INTENSITIES IN LMM AUGER SPECTRA OF GERMANIUM AND SELENIUM.   THRESHOLD EFFECTS ON R 4-0767
TRANSITION PROBABILITIES FOR THE L2,3MM AUGER SPECTRUM OF SELENIUM.   4-1033
UGER ELECTRON SPECTROSCOPY AND THEIR INFLUENCE ON SURFACE SELF DIFFUSION OF COPPER. /ER SURFACES BY LEED- A 5-1185
ITATION: AUGER FINE STRUCTURE.   SELF-CONSISTENT ANALYSIS OF CORE BAND-TO-BAND EXC 2-0509
ABILITY.   NEW SEM AUGER MICROANALYZER WITH ULTRAHIGH VACUUM CAP 3-0636
              CHARACTERIZING MATERIALS WITH THE SEM TODAY. (AUGER SPECTROSCOPY)   1-0097
                   FUTURE GROWTH OF SEM: TOWARD SUPER EFFICIENT MICROSCOPY. (AUGER SP 1-0220
                                                         SEMI-AUGER ELECTRON TRANSITIONS.   2-0424
ND POTASSIUM CHLORIDE.   "SEMI-AUGER" PROCESSES IN L2,3 EMISSION IN ARGON A 2-0326
ENERAL METHOD AND KLL LINE ENERGIES.   SEMI-EMPIRICAL AUGER ELECTRON ENERGIES. PART-1: G 4-0930
OR AN ELECTRON ENERGY ANALYZER.   SEMIAUTOMATED DATA RECORDING AND CONTROL SYSTEM F 3-0680
OSCOPY)   GROWTH AND STRUCTURE OF SEMICONDUCTING THIN FILMS. (AUGER ELECTRON SPECTR 5-1509
  HOLE INTERACTION ON THE AUGER RECOMBINATION PROCESS IN A SEMICONDUCTOR.   EFFECT OF ELECTRON 4-1093
                        AUGER SPECTROSCOPY OF SEMICONDUCTOR AND METAL SURFACES.   5-1404
ULATED EMISSION OF LIGHT FROM A NONIDEAL EXCITON GAS IN A SEMICONDUCTOR. (AUGER PROCESS)   STIM 2-0462
ON SPECTROSCOPY FOR THE STUDY OF SURFACE CONTAMINATION IN SEMICONDUCTOR DEVICES.   AUGER ELECTR 5-1978
SCOPY IN THE SCANNING ELECTRO/ POTENTIAL PROFILING ACROSS SEMICONDUCTOR JUNCTIONS BY AUGER ELECTRON SPECTRO 2-0555
BE)   SURFACE AND THIN FILM ANALYSIS OF SEMICONDUCTOR MATERIALS. (SCANNING AUGER MICROPRO 1-0089
  UNDER A MAGNETIC FIELD.   AUGER RECOMBINATION IN A SEMICONDUCTOR (MERCURY(1-X) CADMIUM(X) TELLURIDE) 4-1092
ECTROSCOPY STUDIES OF GALLIUM ARSENIDE AND SILICON METAL- SEMICONDUCTOR STRUCTURES.   AUGER ELECTRON SP 5-1774
SECONDARY ELE/ IN-PROCESS CONTROL TECHNIQUES FOR COMPLEX SEMICONDUCTOR STRUCTURES. TASK-II. APPLICATION OF 5-1411
Y OF ELECTRON BEAM STIMULATED DESORPTION OF OXYGEN FROM A SEMICONDUCTOR SURFACE. /LECTRON SPECTROSCOPY STUD 5-1617
EFS)   SEMICONDUCTOR SURFACE SPECTROSCOPY. (REVIEW, 21 R 1-0165
                    NATURE OF ANNEALED SEMICONDUCTOR SURFACES. (AUGER)   2-0263
           ION SELECTIVE SPUTTERING OF III-V COMPOUND SEMICONDUCTOR SURFACES. (AUGER EFFECT)   5-1637
        SECONDARY ELECTRON EMISSION CF AMORPHOUS SEMICONDUCTOR THIN FILMS.   5-1239
AUGER RECOMBINATION AND IMPACT IONIZATION IN INDIRECT GAP SEMICONDUCTORS.   2-0396
BAND-TO-BAND AUGER RECOMBINATION IN INDIRECT GAP SEMICONDUCTORS.   2-0394
ANCE BETWEEN AUGER RECCMBINATION AND IMPACT ICNIZATION IN SEMICONDUCTORS.   DETAILED BAL 2-0417
LUENCE OF AN ELECTRIC FIELD ON THE AUGER RECOMBINATION IN SEMICONDUCTORS.   INF 2-0357
LUENCE OF AN ELECTRIC FIELD ON THE AUGER RECCMBINATION IN SEMICONDUCTORS.   INF 4-0844
             RADIATIVE AND AUGER PROCESSES IN SEMICONDUCTORS.   2-0419
STRUCTURE AND COMPOSITION OF VARIOUS ELEMENTAL AND III-V SEMICONDUCTORS.   SURFACE 5-1161
NOR- ACCEPTOR RADIATIVE AND AUGER RECOMBINATION IN SIMPLE SEMICONDUCTORS.   THEORY OF DO 2-0420
              THEORY OF THE AUGER EFFECT IN III-V SEMICONDUCTORS.   2-0498
    WORK FUNCTION AND STRUCTURAL STUDIES OF ALKALI- COVERED SEMICONDUCTORS.   5-1958
E SPECTRUM OF SECONDARY ELECTRON EMISSION FROM METALS AND SEMICONDUCTORS. (AUGER ENERGIES) /OR OBTAINING TH 3-0688
RECOMBINATION-INDUCED NONEQUILIBRIUM PHASE TRANSITIONS IN SEMICONDUCTORS. (AUGER RECOMBINATION)   2-0421
TUDIES ON SURFACE PLASMA OXIDIZED FILMS OF III-V COMPOUND SEMICONDUCTORS BY AUGER ELECTRON SPECTROSCOPY. S 5-1699
C METHO/ THREE DIMENSIONAL CHARACTERIZATION OF METALS AND SEMICONDUCTORS BY THE AUGER ELECTRON MICROANALYTI 1-0077
                        AUGER RECCMBINATION IN SEMICONDUCTORS DURING INTENSE EXCITATION.   2-0368
                    CHEMICAL ANALYSIS OF SEMICONDUCTORS. (IMPURITIES AND COMPOSITION) 5-1671
                        AUGER RECOMBINATION IN SEMICONDUCTORS. (REVIEW, 26 REFS)   1-0044
ION OF KL2,3L2,3 RADIATIVE AUGER THRESHOLDS IN METALS AND SEMICONDUCTORS. (SILICON)   INTERPRETAT 4-0723
LUENCE OF SCREENING EFFECTS ON THE AUGER RECCMBINATION IN SEMICONDUCTORS. (SILICON, GERMANIUM)   INF 2-0379
EC/ AUGER ELECTRON SPECTROSCOPY AND SECONDARY EMISSION IN SEMICONDUCTORS. (SILICON, GERMANIUM, GRAPHITE, EL 2-0240
E ARGON ATOM.   SEMIEMPIRICAL AUGER ELECTRON ENERGIES. PART-2: TH 4-0815
OPY)   DEPLETION EFFECTS IN SEMIINSULATING GALLIUM ARSENIDE. (AUGER SPECTROSC 5-1950
                                        HIGHLY SENSITIVE AUGER ELECTRON SPECTROSCOPY.   1-0189
        SECONDARY ELECTRON SPECTROSCOPY. SURFACE SENSITIVE TOOL. (AES)   1-0095
        SECONDARY ELECTRON SPECTROSCOPY. SURFACE SENSITIVE TOOL. (REVIEW, 22 REFS)   1-0094
THE DETECTION OF PHOTONS, ELECTRONS AND IONS.   HIGH SENSITIVITY APPEARANCE POTENTIAL SPECTROMETER FOR 3-0643
                                          HIGH SENSITIVITY AUGER ELECTRON SPECTROMETER.   3-0672
PECTROMETER BASED ON LEED DISPLAY OPTICS. RESOLUTION AND SENSITIVITY CONSIDERATIONS OF AN AUGER ELECTRON S 3-0705
                                          HIGH SENSITIVITY ELECTRON SPECTROMETER.   3-0633
QUANTITATIVE AUGER ELECTRON SPECTROSCOPY USING ELEMENTAL SENSITIVITY FACTORS.   1-0157
N BERYLLIUM. (AUGER ELECTRON SP/ ENERGY DEPENDENCE OF THE SENSITIVITY OF IONIZATION SPECTROSCOPY: SILICON O 5-1665
ON RATES TO VARIATIONS IN ENERGY LEVELS AND SCREENING PA/ SENSITIVITY OF L1,L2,3M4,5 COSTER-KRONIG TRANSITI 2-0303
TIAL ANALYZER. APPLICATION TO AUGER ELECT/ RESOLUTION AND SENSITIVITY OF THE SPHERICAL GRID RETARDING POTEN 3-0634
AUSED BY AUGER EMISSION AT COLLECTOR.   SENSITIVITY VARIATIONS IN BAYARD- ALPERT GAUGES C 3-0618
CTRON SPECTROSCOPY.   PHASE SEPARATION IN SILICON OXIDES AS SEEN BY AUGER ELE 5-1502
TRONS.   SEPARATION OF THE SPECTRA OF PHOTO AND AUGER ELEC 4-1099
ER TRANSITIONS)   SEQUENTIAL STRIPPING IN HIGHLY IONIZED NEON. (AUG 2-0441
  LMM AUGER SPECTRA OF SOME TRANSITION METALS OF THE FIRST SERIES.   4-0810
/D PHENOMENA IN THE AUGER SPECTRA OF THE FIRST TRANSITION SERIES. (COMMENTS ON PAPER BY CHUNG AND JENKINS) 4-0812
                        EMISSION LINES OF THE M SERIES OF RARE EARTH METALS. (AUGER EFFECT)   4-0790
DIES OF SURFACES: URANIUM DIOXIDE, URANIUM, GRAPHITE, 300 SERIES STAINLESS STEEL, AND NIOBIUM. /LECTRON STU 5-1217
        EFFECTS OF TEN YEARS EXPERIMENTAL BREEDER REACTOR II SERVICE ON CHROMIUM- MOLYBDENUM STEEL.   6-2108
OF SECONDARY ELECTRONS CREATED AS THE RESULT OF ELECTRON SHAKE-OFF AND (SUGAR) VACANCY CASCADES.   NATURE 2-0307
/SPECTRUM OF A NICKEL (110)- C(2X2) SULFUR SURFACE: LINE SHAPE ANALYSIS AND CORRELATION WITH ION NEUTRAL/ 4-1065
FFERENT CHEMICAL ENVIRONMENTS.   AUGER LINE SHAPE COMPARISON OF NITROGEN AND SULFUR IN TWO DI 5-1816
                        AUGER TRANSITIONS AND SHAPE OF X-RAY SPECTRUM.   4-1104
BINDING CALCULATION OF A CORE VALENCE VALENCE AUGER LINE SHAPE: SILICON (111).   TIGHT 4-0831
SCOPY STUDIES OF CARBON MONOXIDE ON NICKEL. SPECTRAL LINE SHAPES AND QUANTITATIVE ASPECTS. /LECTRON SPECTRO 5-1458
(1/ CHARACTERIZATION OF ADSORBED SPECIES USING AUGER PEAK SHAPES. ETHYLENE AND CARBON MONOXIDE ON TUNGSTEN 5-1243
) ON MOLYBDENUM.   ORIGIN OF THE PEAK SHAPES IN THE AUGER SPECTRUM OF OXYGEN (KL2,3L2,3 4-1004
)   K X-RAY EMISSION EDGE SHAPES OF FREE ELECTRON METALS. (AUGER BROADENING 2-0337
OTASSIUM CHLORIDE FILMS.   CHARACTERISTICS SHAPING SECONDARY ELECTRON EMISSION FROM POROUS P 4-1120
R SPECTRA)   MOLECULAR EFFECTS ON K-VAANCY SHARING IN ASYMMETRIC ION- ATOM COLLISIONS. (AUGE 2-0522
R EFFECT)   GAS TARGET EXPERIMENTS OF K-SHELL VACANCY SHARING IN ASYMMETRIC ION- ATOM COLLISIONS. (AUGE 2-0523
ONS. (QUASIMOLECULE AUGER SPECTRA)   K-SHELL VACANCY SHARING IN 300 TO 500 KEV NEON(+)- SODIUM COLLISI 2-0582
                          K-VACANCY SHARING STUDIED FROM AUGER ELECTRON YIELDS.   2-0345
CTROSCOPY)   EFFECT OF ADSORBED CHLORINE AND OXYGEN ON SHEAR STRENGTH OF IRON AND COPPER JUNCTIONS. (SPE 6-2143
            FORMATION OF A PLASMA SHEATH WITH SECONDARY ELECTRON EMISSION.   2-0244
RBON MONOXIDE, SULFUR, AND OTHER GASES ON STAINLESS STEEL SHEET. (AES OF GOLD COATING) /ARACTERISTICS OF CA 5-1551
/ORPTION OF HYDROGEN ON CLEAN NICKEL (111), NICKEL (100), SHEET AND EVAPORATED NICKEL FILMS. (AUGER ELECTR/ 5-1468

CHEMICAL PROFILE ANALYSIS OF TYPE 316 STAINLESS STEEL SHEET BY AUGER SPECTROSCOPY.   6-1996
QUES AS APPLIED TO NICKEL (111), NICKEL (100), AND NICKEL SHEET USING AUGER ELECTRON SPECTROSCOPY. / TECHNI   5-1467
GER SPECTRA AND FLUORESCENCE YIELDS FOR THE 3P AND THE 3D SHELL.   AU   2-0278
GER RATES, X-RAY RATES, AND FLUORESCENCE YIELDS FOR THE K-SHELL.   NONRELATIVISTIC AU   2-0556
ER RATES, X-RAY RATES, AND FLUORESCENCE YIELDS FOR THE 2P SHELL.   NONRELATIVISTIC AUG   2-0558
ATION. (AUGER ELECTRONS)   INNER SHELL ALIGNMENT OF ATOMS IN ELECTRON IMPACT IONIZ   2-0335
YIELDS.   L- AND M-SHELL AND SUBSHELL FLUORESCENCE AND COSTER-KRONIG   4-0959
ELY 90.   THE L(II) SUBSHELL ATOMIC YIELDS OF ELEMENTS WITH Z APPROXIMAT   2-0349
                                     KRYPTON M-SHELL AUGER AND COSTER-KRONIG ELECTRON ENERGIES.   4-0928
                                             L-SHELL AUGER AND COSTER-KRONIG ELECTRON SPECTRA.   4-0960
                                             M-SHELL AUGER AND COSTER-KRONIG ELECTRON SPECTRA.   4-0965
ELEMENTS, AND AUGER AND COSTER-KRONIG TRANSITION RATES/ M-SHELL AUGER, COSTER-KRONIG, AND RADIATIVE MATRIX   4-0967
                                             K-SHELL AUGER ELECTRON HYPERSATELLITES OF NEON.   4-1117
OM ION BOMBARDMENT.   K-SHELL AUGER ELECTRON PRODUCTION CROSS SECTIONS FR   2-0570
OM ION BOMBARDMENT. (NEON, FLUORINE)   K-SHELL AUGER ELECTRON PRODUCTION CROSS SECTIONS FR   2-0574
                                         THE K-SHELL AUGER ELECTRON SPECTRA. REVIEW.   1-0060
TION METALS.   L-SHELL AUGER ENERGIES AND INTENSITIES IN 3D TRANSI   4-1018
IZED NEON.   K-SHELL (AUGER) FLUORESCENT YIELDS FOR MULTIPLY ION   4-0994
T-1: NEON.   K-SHELL AUGER RATES FOR MULTIPLY IONIZED ATOMS. PAR   2-0282
ESCENCE YIELDS OF VARIOUSLY IONIZED STATES OF ARGON.   K-SHELL AUGER RATES, TRANSITION ENERGIES, AND FLUOR   2-0280
ESCENCE YIELDS FOR MULTIPLY IONIZED NEON. / THEORETICAL K-SHELL AUGER RATES, TRANSITION ENERGIES, AND FLUOR   2-0284
ON COLLISIONS.   ARGON L-SHELL AUGER SPECTRA PRODUCED IN ARGON(+) ION- ARG   4-1080
          EFFECTS OF THE CONFIGURATION INTERACTION ON THE K-SHELL AUGER SPECTRUM OF NEON.   2-0279
DS FOR ELEMENTS ARGON TO XENON.   K-SHELL AUGER TRANSITION RATES AND FLUORESCENT YIEL   4-0961
DS FOR ELEMENTS BERYLLIUM TO ARGON.   K-SHELL AUGER TRANSITION RATES AND FLUORESCENT YIEL   4-0962
ND FLUORESCENCE YIELDS FOR SODIUM- THORIUM.   ATOMIC L-SHELL COSTER-KRONIG, AUGER, AND RADIATIVE RATES A   4-0963
AND FLUORESCENCE YIELDS FOR CALCIUM TO THORIUM. ATOMIC M-SHELL COSTER-KRONIG, AUGER, AND RADIATIVE RATES,   4-0964
AND FLUORESCENCE YIELDS FOR Z BETWEEN 38 AND 10/ ATOMIC N-SHELL COSTER-KRONIG, AUGER, AND RADIATIVE RATES,   4-0969
LUORESCENCE YIELDS)   ATOMIC FLUORESCENCE YIELDS. (L-SHELL COSTER-KRONIG YIELDS, K-, L-, AND M-SHELL F   2-0351
T-1: AUGER ELECTRONS. (KLL TRANSITION ENERGIES FOR/ INNER SHELL DOUBLE IONIZATION AND CHEMICAL BONDING. PAR   2-0327
          AUGER ELECTRON SPECTROSCOPY OF THE OUTER SHELL ELECTRONS.   4-0805
OSCOPY)   CROSS SECTIONS FOR IONIZATION OF INNER SHELL ELECTRONS BY ELECTRON IMPACT. (AUGER SPECTR   2-0475
          EXCITATION OF INNER SHELL ELECTRONS BY PHOTONS AND FAST ELECTRONS.   2-0541
IZED ATOMS.   MECHANISMS OF INNER SHELL EXCITATION AND DEEXCITATION IN MULTIPLY ION   2-0487
EN IONS. (INNER SHELL AUGER TRANSITIONS)   OUTER SHELL EXCITATION OF THE RARE GASES BY 30 MEV OXYG   2-0298
CTRON SPECTROSCOPY.   K- AND L-SHELL FLUORESCENCE AND AUGER YIELDS AND AUGER ELE   4-0966
          L- AND M-SHELL AND SUBSHELL FLUORESCENCE AND COSTER-KRONIG YIELDS.   4-0959
SMUTH AND NEPTUNIUM.   X-RAY SPECTRA, L-SUBSHELL FLUORESCENCE AND COSTER-KRONIG YIELDS IN BI   4-1112
GER RATE)   K-SHELL FLUORESCENCE YIELD OF IONIZED ALUMINUM. (AU   4-0975
EN(+), MOLECULAR / CHARGE STATE DEPENDENCE OF THE ARGON L-SHELL FLUORESCENCE YIELD STUDIED BY ATOMIC HYDROG   2-0517
ENCE YIELDS. (L-SHELL COSTER-KRONIG YIELDS, K-, L-, AND M-SHELL FLUORESCENCE YIELDS)   ATOMIC FLUORESC   2-0351
RANSITION PROBABILITIES TO THE 1S STATE AND THEORETICAL K-SHELL FLUORESCENCE YIELDS.   ATOMIC RADIATION T   4-0918
ITION PROBABILITIES F23 BASED ON THE G/ THEORETICAL L2 SUBSHELL FLUORESCENCE YIELDS AND COSTER-KRONIG TRANS   2-0313
NIG TRANSITION PROBABILITIES.   THEORETICAL L2 AND L3 SUBSHELL FLUORESCENCE YIELDS AND L2-L3(X) COSTER-KRO   2-0316
          K-SHELL FLUORESCENCE YIELDS FOR LIGHT ELEMENTS.   4-0818
Y ELECTRON IMPACT. (AUGER EMISSION) MEASUREMENTS OF THE L-SHELL FLUORESCENCE YIELDS OF ARGON AND CHLORINE B   2-0387
SULFUR. (AUGER TRANSITION)   K-SHELL FLUORESCENCE YIELDS OF ARGON, CHLORINE AND   4-1014
S IN LEAD.   L-SHELL FLUORESCENCE YIELDS OF DOUBLE VACANCY STATE   2-0271
          AUGER ELECTRONS FROM THE L-SHELL IN MERCURY.   2-0465
CHLORINE(N/ PROJECTILE CHARGE STATE DEPENDENCE OF NEON K-SHELL IONIZATION AND FLUORESCENCE YIELD IN 50 MEV   2-0300
CHLORINE(+)- NEON COLLISIONS. (AUGER ELECTRON PRODUCT/ K-SHELL IONIZATION AND FLUORESCENCE YIELD IN 50 MEV   2-0518
).   ATOMIC VACANCY DISTRIBUTIONS PRODUCED BY INNER SHELL IONIZATION (AUGER AND RADIATIVE TRANSITIONS   2-0544
          AUGER ELECTRON SPECTROSCOPY APPLIED TO INNER SHELL IONIZATION BY FAST CHARGED PARTICLES.   2-0535
          K-SHELL IONIZATION BY FAST PROTONS.   2-0533
N, MOLECULAR HYDROGEN, AND HELIUM(+) IMPACT.   ARGON L-SHELL IONIZATION BY 50 TO 600 KEV PROTON, DEUTERO   2-0521
SPECTROSCOPY.   K-SHELL IONIZATION CROSS SECTIONS IN AUGER ELECTRON   2-0339
R ELECTRON YIELDS OF METHANE, ETHANE, ETHYLENE, ACETYL/ K-SHELL IONIZATION OF CARBON BY FAST PROTONS. (AUGE   2-0532
BOMBARDMENT. (AUGER EL/ DIFFERENTIAL CROSS SECTION FOR K-SHELL IONIZATION OF COPPER AND SILVER BY ELECTRON   2-0477
EV PROTONS.   CROSS SECTIONS FOR K-SHELL IONIZATION OF GASEOUS MOLECULES BY 50-500 K   4-1077
E BY 50- TO 600 KEV HYDROGEN(+), DEUTERIUM(+), MOLECUL/ K-SHELL IONIZATION OF MOLECULAR NITROGEN AND METHAN   2-0519
ACT.   DIFFERENTIAL CROSS SECTIONS FOR K-SHELL IONIZATION OF SURFACE ATOMS BY ELECTRON IMP   2-0358
ACT.   INNER SHELL IONIZATION OF SURFACE ATOMS BY ELECTRON IMP   2-0359
/ ANOMALOUS THRESHOLD BEHAVIOR FOR ELECTRON INDUCED INNER SHELL IONIZATION OF THE N6,7 LEVEL IN GOLD, BISMU   4-1067
LECTRON SPECTRA.   L-SHELL IONIZATION THRESHOLD STRUCTURE FROM AUGER E   2-0360
          REARRANGEMENT OF INNER SHELL IONIZED ATOMS. (AUGER SPECTRA)   2-0413
ND, GRAPHITE, AND EVAPORATED CARBON SUR/ CHARACTERISTIC K-SHELL LOW ENERGY ELECTRON LOSS SPECTRA FROM DIAMO   4-0945
N.   IONIZATION CROSS SECTION OF THE L2,3 SHELL OF ALUMINUM IN A NEON(+)- ALUMINUM COLLISIO   2-0267
          COSTER-KRONIG AND AUGER SPECTRUM OF THE L1 SHELL OF ARGON.   4-0977
/ MEASUREMENT OF THE IONIZATION CROSS SECTION OF THE L2,3 SHELL OF CHLORINE USING AUGER ELECTRON SPECTROSCO   4-0987
          AUGER AND COSTER-KRONIG STUDIES OF THE M-SHELL OF KRYPTON.   4-0979
GAS ATOMS. PART-1. THE ARGON ATOM. PART-2. KRYPTON ATOM M-SHELL. PART-3. /ECTRON EMISSION SPECTRA FOR RARE   4-0927
          INNER SHELL SPECTROSCOPY. (REVIEW)   1-0160
UGER EFFECT)   PRODUCTION OF INNER SHELL VACANCIES IN HEAVY ION- ATOM COLLISIONS. (A   2-0405
CORRELATION ENERGIES AND AUGER RATES IN ATOMS WITH INNER SHELL VACANCIES (NEON WITH 1S OR 2S VACANCY).   2-0312
LUORINE IONS. (AUGER ELECTRON EMISSION)   NEON K-SHELL VACANCY PRODUCTION BY INTERMEDIATE ENERGY F   2-0446
BY THE IMPACT/ CHARGE DEPENDENCE OF DEPENDENCE OF TOTAL K-SHELL VACANCY PRODUCTION CROSS SECTIONS IN ARGON   2-0523
LISIONS. (AUGER EFFECT)   GAS TARGET EXPERIMENTS OF K-SHELL VACANCY SHARING IN ASYMMETRIC ION- ATOM COL   2-0582
SODIUM COLLISIONS. (QUASIMOLECULE AUGER SPECTRA)   K-SHELL VACANCY SHARING IN 300 TO 500 KEV NEON(+)-   3-0718
ADIATIONLESS TRANSITIONS IN SOLIDS.   WIDTHS OF ATOMIC M-SHELL VACANCY STATES AND QUASIATOMIC ASPECTS OF R   2-0573
          AUGER ELECTRONS FROM DOUBLE K-SHELL VACANCY STATES IN NEON.   4-0968
          ATOMIC N-SHELL VACANCY WIDTHS IN SOLIDS.   4-0848
ANIC ELEMENTS.   L- AND M-SHELL YIELDS AND ELECTRON SPECTRA FOR THE TRANSUR   2-0277
SURING ELECTRON IMPACT IONIZATION CROSS SECTIONS OF INNER SHELLS.   A NEW METHOD FOR MEA   2-0514
NONRELATIVISTIC FLUORESCENCE YIELDS FOR THE 3P AND THE 3D SHELLS.   2-0553
OSCOPY, CHEMI/ EMISSION AND RECOMBINATION INVOLVING INNER SHELLS. (AUGER PROCESS, AES, PHOTOELECTRON SPECTR   4-0983
ELECTRON IMPACT IONIZATION CROSS SECTIONS OF INNER SHELLS MEASURED BY AUGER ELECTRON SPECTROSCOPY.   4-0847
          AUGER SPECTRUM OF L2 AND L3 SHELLS OF ARGON.   2-0514
NGS/ TOTAL ELECTRON IMPACT IONIZATION CROSS SECTIONS ON K-SHELLS OF SURFACE ATOMS. (AUGER INTENSITIES OF TU
(AUGER PROCESS, AES, PHOTOELECTRON SPECTROSCOPY, CHEMICAL SHIFT) /ND RECOMBINATION INVOLVING INNER SHELLS.

ITION OF VANADIUM OXIDES, AND OXIDIZED VANADIUM: CHEMICAL SHIFT AND PEAK INTENSITY ANALYSIS. /EMICAL COMPOS 4-1089
ELECTRONS FOR IONIZED ATOMS. SHIFT ENERGIES OF CHARACTERISTIC X-RAYS AND AUGER 4-1041
CHEMICAL SHIFT IN AUGER SPECTRA. 4-0826
DURING THE ADSORPTION OF OXYGEN. CHEMICAL SHIFT IN AUGER SPECTRA OF TUNGSTEN AND MOLYBDENUM 4-0816
ADSORPTION. CHEMICAL SHIFT IN THE AUGER SPECTRUM OF IRON DURING OXYGEN 5-1792
EN ADSORPTION. CHEMICAL SHIFT IN THE AUGER SPECTRUM OF TUNGSTEN FROM OXYG 4-0922
EXTRA CLOSED CORE RELAXATION ON AUGER AND BINDING ENERGY SHIFTS. EFFECTS OF 4-0950
THEORY OF AUGER SATELLITE ENERGY SHIFTS. 2-0507
PARTIALLY OXIDIZED MAGNESIUM SURFACES. ABNORMAL CHEMICAL SHIFTS. /ITED AUGER AND PHOTOELECTRON SPECTRA OF 5-1949
INDRICAL MIRROR ANAL/ SPECIMEN POSITION EFFECTS ON ENERGY SHIFTS AND SIGNAL INTENSITY IN A SINGLE STAGE CYL 3-0696
NICKEL(III) USING AUGER ELECTRON SPECTROSCOPY: CHEMICAL SHIFTS AND VALENCE SPECTRA. /ORPTION OF OXYGEN ON 5-1466
UGER SIGNALS OBTAINED WITH A PHOTOELECTRON SPEC/ CHEMICAL SHIFTS AND VARIATIONS OF WIDTH AND INTENSITY OF A 2-0273
CHEMICAL SHIFTS IN AUGER ELECTRON SPECTRA. 4-0825
SILICON NITRIDE. CHEMICAL SHIFTS IN AUGER ELECTRON SPECTRA FROM SILICON IN 4-0881
CHEMICAL SHIFTS IN AUGER ELECTRON SPECTROSCOPY. 4-0863
L OXIDATION OF TANTALUM (110). CHEMICAL SHIFTS IN AUGER ELECTRON SPECTROSCOPY FROM INITIA 6-2038
CA SIMPLE CHARGE TRANSFER MODEL). CHEMICAL SHIFTS IN AUGER SPECTRA (MONOLAYER ADSORPTION, ES 4-1115
ILICON DIOXIDE DURI/ EXPERIMENTAL OBSERVATION OF CHEMICAL SHIFTS IN AUGER SPECTRUM FROM SURFACE LAYERS OF S 4-1044
MM AUGER SPECTRA OF GERMANIUM. PHOTOELECTRON CHEMICAL SHIFTS IN COORDINATION COMPOUNDS. X-RAY EXCITED L 4-0924
SIMPLE MODEL FOR K X-RAY AND AUGER ELECTRON ENERGY SHIFTS IN HEAVY ION COLLISIONS. COMMENTS. 2-0301
CHEMICAL SHIFTS IN PHOTOEXCITED AUGER SPECTRA. 4-0797
/ AUGER ELECTRON SPECTROSCOPIC INVESTIGATIONS OF CHEMICAL SHIFTS IN SOME VANADIUM COMPOUNDS. (VANADIUM OXID 4-1090
M AND VANADIUM. CHEMICAL SHIFTS IN THE AUGER SPECTRA FROM OXIDIZED CHROMIU 4-0809
CHEMICAL SHIFTS IN THE AUGER SPECTRA OF PASSIVE FILMS. 4-1056
ADSORPTION. CHEMICAL SHIFTS IN THE AUGER SPECTRUM OF YTTRIUM ON OXYGEN 5-1153
ECT OF EXTRAATOMIC RELAXATION ON AUGER AND BINDING ENERGY SHIFTS IN TRANSITION METALS AND SALTS. /ATIVE EFF 2-0412
CHEMICAL SHIFTS IN URANIUM AUGER TRANSITIONS. 4-0737
O/ ELECTRON SPECTROSCOPY AND ADSORPTION STUDIES. CHEMICAL SHIFTS IN X-RAY PHOTOELECTRON AND AUGER SPECTRA F 5-1204
CHEMICAL SHIFTS OF AUGER LINES, AND THE AUGER PARAMETER. 4-1107
ON BEAMS. DOPPLER SHIFTED AUGER ELECTRONS FROM FOIL EXCITED HEAVY I 4-0859
ROLE OF THE AUGER EFFECT IN THE FORMATION OF AN ELECTRON SHOWER SPECTRUM. 2-0504
OPY ANALYSIS OF A COMMERCIAL COBALT BASE AIRCRAFT TURBINE SHROUD ALLOY. /SION, FRICTION AND AUGER SPECTROSC 6-2009
: INTERNAL PHOTOEMISSION. STRUCTURE ON THE HIGH ENERGY SIDE OF THE KL2,3M AUGER PEAK FROM SOLID ALUMINUM 4-1021
(AUGER ELECTRON SPECTROSCOPY OF) PRECISION SINGLE SIDEBAND CRYSTAL UNITS. 5-1409
RROR ANAL/ SPECIMEN POSITION EFFECTS ON ENERGY SHIFTS AND SIGNAL INTENSITY IN A SINGLE STAGE CYLINDRICAL MI 3-0696
ION IN A 4-GRID RETARDING / CORRECTIONS OF AUGER ELECTRON SIGNAL STRENGTHS FOR MODULATION AMPLITUDE DISTORT 4-0620
HE ANALYSIS AND DEVELOPMENT OF MATERIALS. (AUGER ELECTRON SIGNALS) ELECTRON BEAM MICROPROBES USED IN T 3-0714
R. AUGER SIGNALS OBTAINED WITH A PHOTOELECTRON SPECTROMETE 2-0272
/AL SHIFTS AND VARIATIONS OF WIDTH AND INTENSITY OF AUGER SIGNALS OBTAINED WITH A PHOTOELECTRON SPECTROMET/ 2-0273
RYSTAL MICROBALANCE. QUANTITATIVE STUDY OF AUGER ELECTRON SIGNALS OF PHOSPHORUS ON SILICON USING A QUARTZ C 5-1583
HEED AND AES STUDY OF SILICON EPITAXY BY SILANE PYROLYSIS. 5-1427
SCOPY STUDY. SILICON HOMOEPITAXIAL THIN FILMS VIA SILANE PYROLYSIS: HEED AND AUGER ELECTRON SPECTRO 5-1428
ELS OF SURFACE ELECTRON CENTERS. (AUGER EFFECT, ZIRCONIA, SILICA) /DETERMINATION OF THE ENERGY DEPTH OF LEV 4-0915
AQUEOUS CORROSION OF SODA- SILICA AND SODA- LIME- SILICA GLASS. (AES) 5-1265
AUGER SPECTROSCOPY OF SILICON IN SILICATES. 4-0794
STUDY OF AMORPHOUS AND CRYSTALLINE SILICATES BY AUGER ELECTRON SPECTROSCOPY. 4-0796
ME VANADIUM COMPOUNDS. (VANADIUM OXIDE, NITRIDE, CARBIDE, SILICIDE) /NVESTIGATIONS OF CHEMICAL SHIFTS IN SO 4-1090
/ILES OF CHEMICALLY VAPOR DEPOSITED PLATINUM AND PLATINUM SILICIDE BY AUGER ELECTRON SPECTROSCOPY AND SECO/ 5-1656
AUGER SPECTROSCOPY ANALYSIS OF PALLADIUM SILICIDE FILMS. 5-1898
ECTROSCOPY. STUDY OF PALLADIUM SILICIDE FILMS ON SILICON USING AUGER ELECTRON SP 5-1336
SILICON- TANTALUM, SILICON DIOXIDE- TANTALUM, AND SILICO/ SILICIDE FORMATION AND INTERDIFFUSION EFFECTS IN 5-1260
PLATINUM- PLATINUM SILICIDE INTERFACE. PLATINUM SILICIDE FORMATION. ELECTRON SPECTROSCOPY OF THE 5-1285
LOY FILMS. AES STUDY OF THE SILICIDE FORMATION IN 85% NICKEL- 15% PLATINUM AL 5-1899
CTROSCOPY. PALLADIUM SILICIDE FORMATION OBSERVED BY AUGER ELECTRON SPE 5-1773
ING ANALYSES. EFFECT OF OXIDIZING AMBIENTS ON PLATINUM SILICIDE FORMATION. PART-2: AUGER AND BACKSCATTER 5-1182
) STANNIDE. PAR/ PHOTOELECTRIC PROPERTIES OF MAGNESIUM(2) SILICIDE, MAGNESIUM(2) GERMANIUM, AND MAGNESIUM(2 2-0306
STRUCTURE AND GROWTH KINETICS OF NICKEL(2) SILICIDE ON SILICON. (AUGER SPECTROSCOPY) 5-1917
COMPOSITION PROFILES AND SCHOTTKY BARRIER HEIGHTS OF SILICIDES FORMED IN NICKEL- PLATINUM ALLOY FILMS. 5-1900
ANTALUM, SILICON DIOXIDE- TANTALUM, AND SILICON- PLATINUM SILICIDE- TANTALUM THIN FILM STRUCTURES. (AES) /T 5-1260
SENCE OF PLASMON GAIN SATELLITES IN THE AUGER SPECTRUM OF SILICON. AB 4-1037
CTERIZATION OF OXIDIZED FILMS OF MAGNESIUM, ALUMINUM, AND SILICON. AES CHARA 5-1847
IC STUDY OF PHOSPHORUS SEGREGATION IN OXIDIZED DEGENERATE SILICON. AUGER AND ELLIPSOMETR 5-1250
LECTRON SPECTROSCOPY ANALYSIS OF ION PLATED GOLD FILMS ON SILICON. AUGER E 5-1225
AUGER ELECTRON SPECTROSCOPY OF SILICON. 4-0854
PPER- 2 ATOMIC PERCENT TIN, AND IRON- 6.55 ATOMIC PERCENT SILICON. /OYS COPPER- 1 ATOMIC PERCENT INDIUM, CO 6-2027
AUGER RECOMBINATION IN SILICON. 2-0266
SPECTRA AND LEED PATTERNS FROM NICKEL DEPOSITS ON CLEAVED SILICON. AUGER 5-1768
AUGER SPECTROSCOPY OF SILICON. 4-0784
AUGER SPECTROSCOPY OF SILICON. 5-1946
AUGER SPECTROSCOPY OF SUBMONOLAYER GOLD DEPOSITIONS ON SILICON. 5-1888
NS UNDER THE X-RAY DIFFRACTION PROCESS. (AUGER SPECTRA OF SILICON) / AND THE MEASUREMENT OF EXCITED ELECTRO 3-0641
GER ELECTRON SPECTROSCOPY: CARBON MONOXIDE AND OXYGEN ON SILICON. / ON GAS ADSORPTION STUDIES UTILIZING AU 5-1542
ENCES FOR AUGER RECOMBINATION IN ELECTRON HOLE DROPLETS. (SILICON) EVID 2-0260
HIGH ENERGY SATELLITES IN AUGER PEAKS OF MAGNESIUM AND SILICON. 4-1046
E OF THE BAND STRUCTURE ON THE AUGER ELECTRON SPECTRUM OF SILICON. INFLUENC 4-0944
ELECTRONS. (KLL TRANSITION ENERGIES FOR SULFUR, ALUMINUM, SILICON) /ON AND CHEMICAL BONDING. PART-1: AUGER 2-0327
F L2,3 EMISSION BANDS OF SODIUM, MAGNESIUM, ALUMINUM, AND SILICON. /ATION OF HIGH ENERGY X-RAY SATELLITES O 2-0239
RADIATIVE AUGER THRESHOLDS IN METALS AND SEMICONDUCTORS. (SILICON) INTERPRETATION OF KL2,3L2,3 4-0723
L1,L2,3V AUGER TRANSITION IN SILICON. 4-0779
RELAXATION OF AUGER EXCITED CARRIERS IN SILICON. 2-0274
URFACE VALENCE BAND AND PLASMON FEATURES ON CLEAN CLEAVED SILICON. S 5-1141
THE (111) SURFACE OF N-TYPE SILICON. 5-1349
ND-TO-BAND TRANSITIONS IN SOLIDS. (AUGER RECOMBINATION IN SILICON) TWO ELECTRON BA 2-0275
ELECTRON HOLE DROPS IN GERMANIUM- SILICON ALLOYS. (AUGER RECOMBINATION) 4-0777
BAND-TO-BAND AUGER RECOMBINATION IN SILICON AND GERMANIUM. 2-0459
/N PROCESSES (AUGER) ON THE SURFACES AND IN THIN FILMS OF SILICON AND GERMANIUM SUBJECTED TO LASER EXCITAT/ 5-1990
SATELLITES OF THE L2,3 AUGER EMISSION BANDS OF ALUMINUM, SILICON, AND PHOSPHORUS. AUGER 4-0985
R SATELLITES ON THE L23 AUGER EMISSION BANDS OF ALUMINUM, SILICON, AND PHOSPHORUS. AUGE 4-0984
HIGH ENERGY FINE STRUCTURE IN THE AUGER SPECTRA OF SILICON AND SILICON CARBIDE. 4-1036
AL MIRROR AUGER ANALYZER. DETECTION OF SODIUM ON SILICON AND SILICON DIOXIDE SURFACES BY CYLINDRIC 3-0647

ECTROSCOPY FOR MICROSPCT SURFACE AND IN-DEPTH ANALYSIS OF SILICON AND TRANSISTOR METALLIZATIONS. / AUGER SP 5-1262
AUGER) RECOMBINATION OF NONEQUILIBRIUM CHARGE CARRIERS IN SILICON AT A HIGH EXCITATION LEVEL. /TERING AND (4-1103
/NG EFFECTS IN ELECTRON INDUCED AUGER SPECTRA (SILVER AND SILICON, AUGER EFFECT ENHANCED BY BACKSCATTERING/ 4-0841
SOME ASPECTS OF THE SURFACE BEHAVIOR OF SILICON. (AUGER ELECTRON SPECTROSCOPY) 5-1510
ELECTRON ESCAPE DEPTH IN SILICON. (AUGER ENERGIES) 2-0408
ECAY OF THE RECOMBINATION RADIATION FROM STRONGLY EXCITED SILICON. (AUGER RECOMBINATION PROCESS) /RUM AND D 2-0460
INTERACTION OF ALUMINUM LAYERS WITH POLYCRYSTALLINE SILICON. (AUGER SPECTROSCOPY) 5-1672
STRUCTURE AND GROWTH KINETICS OF NICKEL(2) SILICIDE ON SILICON. (AUGER SPECTROSCOPY) 5-1917
DETECTABILITY LIMITS FOR BORON AND PHOSPHORUS IN SILICON BY AUGER ELECTRON SPECTROSCOPY. 5-1887
OBSERVATION OF IMPURITY PROFILE IN ICN IMPLANTED SILICON BY AUGER ELECTRON SPECTROSCOPY. 5-1482
STUDY OF MUSCOVITE AND SILICON BY AUGER ELECTRON SPECTROSCOPY. 4-0792
MISSION. IN-DEPTH PROFILES OF PHOSPHORUS ICN IMPLANTED SILICON BY AUGER SPECTROSCOPY AND SECONDARY ION E 4-0996
. BAND STRUCTURE OF SILICON BY CHARACTERISTIC AUGER SPECTRUM ANALYSIS 4-0740
/ALLIUM PHOSPHIDE (AND) NITROGEN, OXYGEN, AND FLUORINE IN SILICON BY SECONDARY ION MASS SPECTROMETRY AND A/ 5-1916
N SURFACES. OBSERVATICNS OF BETA- SILICON CARBIDE FORMATION ON RECONSTRUCTED SILICO 5-1560
ECTRON DIFFRACTION AND AUGER ELECTRON OBSERVATIONS OF THE SILICON CARBIDE (0001) SURFACE. LOW ENERGY EL 5-1932
SECONDARY ELECTRON EMISSION FROM ION IMPLANTED SILICON (CESIUM). 4-0913
: HIGH ENERGY ELECTRON DIFFRACTION AND AUGER ELECTRON / SJLICON CLEANING WITH HYDROGEN PEROXIDE SOLUTIONS 5-1426
NIDE ELECTROLUMINESCENT DIODES. (AUG/ EMISSION SPECTRA OF SILICON COMPENSATED ALUMINUM(X) GALLIUM(1-X) ARSE 2-0325
AUGER ELECTRON SPECTROSCOPY STUDY OF SOME OXYGENATED SILICON COMPOUNDS. 4-0793
AUGER ELECTRON SPECTROSCOPY OF INSULATING SILICON COMPOUNDS. 4-0795
. AUGER SPECTRA AND LEED PATTERNS FROM VACUUM CLEANED SILICON CRYSTALS WITH CALIBRATED DEPOSITS OF IRON 5-1769
- AUGER ANALYSIS SYSTEM. SILICON DEVICE APPLICATIONS USING A COMBINED ESCA 5-1289
LYSIS) TO SURFACE CONTROL PROGRAM IN THE MANUFACTURING OF SILICON DEVICES. /PECTROSCOPY (AUGER ELECTRON ANA 5-1411
SILICON, STUDIED WITH AUGER SPECTROSCOPY. SILICON DIFFUSION THROUGH THIN TUNGSTEN FILMS ON 5-1234
ELECTRON IRRADIATION EFFECT IN THE AUGER ANALYSIS OF SILICON DIOXIDE. 4-1096
CHEMICAL SHIFTS IN AUGER SPECTRUM FROM SURFACE LAYERS OF SILICON DIOXIDE DURING ELECTRON BOMBARDMENT. / OF 4-1044
ALYSIS OF CHLORINE IN HYDROGEN CHLORIDE OR CHLORINE GROWN SILICON DIOXIDE FILMS. AUGER AN 5-1251
SOME RECENT STUDIES OF VERY THIN SILICON DIOXIDE FILMS. 5-1935
. PHOSPHORUS CONCENTRATION PROFILES IN P-DOPED SILICON DIOXIDE MEASURED USING AUGER SPECTROSCOPY 5-1231
IDE. DEPOSITION AND AUGER ANALYSIS OF DEPOSITED SILICON DIOXIDE ON ALUMINUM(X) GALLIUM(1-X) ARSEN 5-1962
SECONDARY ELECTRON EMISSION OF SILICON DIOXIDE SINGLE CRYSTALS. 4-1062
DIOXIDE-/ AUGER STUDIES OF DIFFUSED IMPURITY PROFILES IN SILICON DIOXIDE- SILICON AND POLYSILICON- SILICON 5-1481
/RMATION AND INTERDIFFUSION EFFECTS IN SILICON- TANTALUM, SILICON DIOXIDE- TANTALUM, AND SILICON- PLATINUM/ 5-1260
AUGER SPECTRA FROM THE (100) SURFACE OF SILICON DURING OXIDATION AND NITRIDATION. 5-1310
HEED AND AES STUDY OF SILICON EPITAXY BY SILANE PYROLYSIS. 5-1427
/ COMPARISON OF AUGER SPECTRA OF MAGNESIUM, ALUMINUM, AND SILICON EXCITED BY LOW ENERGY ELECTRON AND LOW EN 4-0857
RATE CONDITIONS ON THE NUCLEATION AND GROWTH OF EPITAXIAL SILICON FILMS. INFLUENCE OF SUBST 5-1514
/NTENSITY OF THE L2,3VV AUGER TRANSITION IN MAGNESIUM AND SILICON FROM THE CROSS SECTION MODEL OF GRYZINSK/ 4-0765
NG EFFECTS ON THE AUGER RECOMBINATION IN SEMICONDUCTORS. (SILICON, GERMANIUM) INFLUENCE OF SCREENI 2-0379
COMBINED LEED AND AUGER ELECTRON SPECTROSCOPY STUDIES OF SILICON, GERMANIUM, GALLIUM ARSENIDE, AND INDIUM/ 5-1379
/ SPECTROSCOPY AND SECCNDARY EMISSION IN SEMICONDUCTORS. (SILICON, GERMANIUM, GRAPHITE, ELECTRON DIFFUSION/ 2-0240
YSIS: HEED AND AUGER ELECTRON SPECTROSCOPY STUDY. SILICON HOMOEPITAXIAL THIN FILMS VIA SILANE PYROL 5-1428
FFECT) SURFACE STATE TRANSITIONS OF SILICON IN ELECTRON ENERGY LOSS SPECTRA. (AUGER E 4-1039
AUGER SPECTROSCOPY OF SILICON IN SILICATES. 4-0794
CHEMICAL SHIFTS IN AUGER ELECTRON SPECTRA FROM SILICON IN SILICON NITRIDE. 4-0881
LOYS STUDIED BY AUGER ELECTRON SPECTRO/ METALLIC STATE OF SILICON IN SILICON- NOBLE METAL VAPOR QUENCHED AL 5-1438
LS. (AUGER SPECTROSCOPY) ROLE OF MANGANESE AND SILICON IN TEMPER EMBRITTLEMENT OF LOW ALLOY STEE 6-2059
YSTEM. LOW TEMPERATURE DIFFUSION CF GOLD INTO SILICON IN THE SILICON (SUBSTRATE)- GOLD (FILM) S 5-1673
APPLICATIONS OF SCANNING AUGER SPECTROSCOPY TO THE SILICON INTEGRATED CIRCUIT TECHNOLOGY. 5-1655
PECTROS/ INVESTIGATIONS OF CARBON RESIDUES ON SURFACES OF SILICON INTEGRATED CIRCUITS. (BY AUGER ELECTRON S 5-1345
IN-DEPTH AUGER ANALYSIS OF ALUMINUM- SILICON INTERFACIAL REACTIONS. 5-1219
MEASUREMENTS) ANTIMONY DOPING OF SILICON LAYERS GROWN BY SOLID PHASE EPITAXY. (AES 5-1577
SORPTION. (AUGER EFFECT) COMPARISON OF SILICON L3,2 ELECTRON LOSS SPECTRA AND OPTICAL AB 4-1038
GER ELECTRON SPECTROSCOPY STUDIES OF GALLIUM ARSENIDE AND SILICON METAL- SEMICONDUCTOR STRUCTURES. AU 5-1774
ER SPECTROSCOPY) POSSIBLE ORIGIN OF LCW TEMPERATURE SILICON MIGRATION IN SILICON- METAL SYSTEMS. (AUG 5-1437
/ING AND ETCHING OF CRYSTAL SURFACES (TITANIUM, TUNGSTEN, SILICON) MONITORED BY AUGER SPECTROSCOPY, ELLIPS/ 5-1828
IN DEPTH INFORMATICN FROM AUGER ELECTRON SPECTROSCOPY. (SILICON MONOLAYERS) 1-0130
TROSCOPY ON THE FRACTURE OF NICKEL TUNGSTEN MATERIALS AND SILICON NITRIDE. AUGER ELECTRON SPEC 5-1445
EFFECT OF YTTRIA ADDITIONS ON HOT PRESSED SILICON NITRIDE. (AUGER ANALYSIS) 5-1350
YSIS) MICROSTRUCTURE OF HOT PRESSED SILICON NITRIDE. (AUGER ANALYSIS, MICROPROBE ANAL 5-1556
IMPURITIES AND INCLUSIONS IN HOT PRESSED SILICON NITRIDE. (AUGER SPECTROGRAPHIC ANALYSIS) 5-1557
STUDY OF FRACTURE SURFACES OF HOT PRESSED SILICON NITRIDE BY AUGER ELECTRON SPECTROSCOPY. 5-1444
IDENTIFICATION OF A GRAIN BOUNDARY PHASE IN HOT PRESSED SILICON NITRIDE BY AUGER ELECTRON SPECTROSCOPY. 5-1749
AUGER ELECTRON SPECTROSCOPIC ANALYSIS OF SILICON NITRIDE ON SILICON. 5-1455
/EPENDENCE OF THE SENSITIVITY OF IONIZATION SPECTROSCOPY: SILICON ON BERYLLIUM. (AUGER ELECTRON SPECTROSCO/ 5-1665
YS. (AUGER ANALYSIS) INFLUENCE OF SILICON ON FRICTION AND WEAR OF IRON- COBALT ALLO 6-2011
: LEED- AUGER STUDIES AND ELECTRONIC PROPERTIES OF FILMS. SILICON ON SAPPHIRE EPITAXY BY VACUUM SUBLIMATION 5-1230
ATOMIC ARRANGEMENT IN THE 1 K 1 STRUCTURE OF A SILICON ORDERED MONOLAYER ON MOLYBDENUM (001). 5-1478
AUGER ELECTRON SPECTROSCOPY TO DETERMINE THE STRUCTURE OF SILICON OXIDE FILMS. USE OF 5-1501
COPY. PHASE SEPARATION IN SILICON OXIDES AS SEEN BY AUGER ELECTRON SPECTROS 5-1502
CON NITRIDE FILMS BY AUGER ELECTRON EMISSIO/ DETECTION OF SILICON OXYNITRIDE LAYERS ON THE SURFACES OF SILI 5-1611
/N SPECTROSCOPY. CARBON MONOXIDE AND MOLECULAR OXYGEN ON SILICON. PART-1: ADSORPTION STUDIES. PART-2: STR/ 5-1543
/GER AND PLASMON STUDIES OF NEGATIVE ELECTRON AFFINITY ON SILICON PRODUCED BY THE ADSORPTION OF CESIUM AND/ 5-1363
OF CHEMICAL BONDING ON THE LOW ENERGY K-ALPHA SPECTRUM OF SILICON. (RADIATIVE AUGER TRANSITIONS) INFLUENCE 2-0226
AUGER ELECTRON SPECTROSCOPY DETERMINATION OF THE OXYGEN- SILICON RATIO IN SPIN ON GLASS FILMS. 5-1827
RD DIRECTION IN THE MIDDLE REGION. AUGER RECOMBINATION IN SILICON RECTIFIERS AND THYRISTORS LOADED IN FORWA 2-0415
AUGER SPECTROSCOPY OF SILICON. (REPLY TO COMMENTS) 5-1880
MENT. / ENERGY ANALYSIS OF BACKSCATTERED ELECTRONS FROM A SILICON SINGLE CRYSTAL UNDER A LOW ENERGY BOMBARD 4-0739
OXIDE, AND SILICON / AUGER ELECTRON SPECTROSCOPY STUDY OF SILICON SPECTRA FROM SILICON MONOXIDE, SILICON DI 4-1081
N OF AUGER/ METAL- CIELECTRIC TRANSITION IN GERMANIUM AND SILICON STUDIED BY A MICROWAVE METHOD. (PRODUCTIO 4-0752
LECTRON SPECTROSCOPY. LOW TEMPERATURE OXIDATION OF SILICON STUDIED BY PHOTOSENSITIVE ESR AND AUGER E 5-1786
M. (AUGER ELECTRON SPECTROSCOPY) DIFFUSE INTERFACE IN SILICON (SUBSTRATE)- GOLD (EVAPORATED FILM) SYSTE 5-1681
/ ELECTRONIC STRUCTURE OF A THIN GOLD FILM DEPOSITED ON A SILICON SUBSTRATE STUDIED BY AUGER ELECTRON AND X 5-1435
/HASE EPITAXIAL STUDIES USING VACUUM DEPOSITION ON HEATED SILICON SUBSTRATES. (SCANNING AUGER SPECTROSCOPY) 5-1287
COMPARISON OF THE TECHNIQUES FOR SILICON SURFACE ANALYSIS. 5-1327
THERS) COMPARISON OF THE TECHNIQUES FOR SILICON SURFACE ANALYSIS. (AES, ESCA, SIMS, AND O 1-0052
ADSORPTION OF OXYGEN ON A CLEAN SILICON SURFACE. (AUGER ELECTRON SPECTROSCOPY) 5-1638
CTRON SPECTROSCOPY AND ITS APPLICATION IN STUDIES OF SOME SILICON SURFACES. AUGER ELE 5-1418
AUGER ELECTRON SPECTROSCOPY OF SILICON SURFACES. 5-1397

AUGER ELECTRON SPECTROSCOPY OF SILICON SURFACES. 5-1945
ELECTRON SPECTROSCOPY OF CLEANUP-RELATED CONTAMINATION ON SILICON SURFACES. AUGER 5-1983
AUGER SPECTROSCOPY AND LEED OF SILICON SURFACES. 5-1396
AUGER STUDIES OF (111) SILICON SURFACES. 5-1375
PACT DESORPTION STUDIES OF THE INTERACTIONS OF GASES WITH SILICON SURFACES. /N SPECTROSCOPY AND ELECTRON IM 5-1691
ON SPECTROSCOPY AND ELECTRON IMPACT DESORPTION STUDIES OF SILICON SURFACES. COMBINED AUGER ELECTR 5-1694
ON SPECTROSCOPY AND ELECTRON IMPACT DESORPTION STUDIES OF SILICON SURFACES. COMBINED AUGER ELECTR 5-1695
TUDIED WITH AUGER SPECTROSCOP/ STRUCTURE AND CHEMISTRY OF SILICON SURFACES AFTER PRE- AND BACKSPUTTERING. S 5-1233
ELECTRON ORBITAL ENERGIES OF OXYGEN ADSORBED ON SILICON SURFACES AND OF SILICON DIOXIDE. 5-1475
ELECTRON BEAM ADSORBATE INTERACTIONS ON SILICON SURFACES. (AUGER SPECTROSCOPY) 5-1513
ANALYSIS OF IMPURITIES ON SILICON SURFACES BY AUGER ELECTRON SPECTROSCOPY. 5-1674
ON AND AUGER ELECTRON SPECTR/ STUDY OF THE NITRIDATION OF SILICON SURFACES BY LOW ENERGY ELECTRON DIFFRACTI 5-1417
. SURVEYING SILICON SURFACES BY MEANS OF THE LEED- AES METHOD 5-1809
Y AES. A STUDY OF ACTIVATED SILICON SURFACES FOR ELECTROLESS NICKEL PLATING B 5-1539
CONTAMINANTS ON CHEMICALLY ETCHED SILICON SURFACES. LEED- AUGER METHOD. 5-1229
MPERATURES. AUGER ANALYSIS OF SILICON THIN FILMS DEPOSITED ON CARBON AT HIGH TE 5-1228
ROSCOPY AND SECONDARY ION MASS SPECTROMETRY FOR ALUMINUM, SILICON, TITANIUM, VANADIUM, AND CHROMIUM. /SPECT 5-1553
ELECTRONS EMITTED FROM BERYLLIUM, MAGNESIUM, ALUMINUM AND SILICON UNDER ION BOMBARDMENT. /IBUTION OF AUGER 4-0876
TITATIVE STUDY OF AUGER ELECTRON SIGNALS OF PHOSPHORUS ON SILICON USING A QUARTZ CRYSTAL MICROBALANCE. QUAN 5-1583
STUDY OF PALLADIUM SILICIDE FILMS ON SILICON USING AUGER ELECTRON SPECTROSCOPY. 5-1336
/ SPECTRA OF SECONDARY ELECTRONS EMITTED BY MAGNESIUM AND SILICON WHEN BOMBARDED BY PRIMARY ELECTRONS. PAR/ 4-1047
SURFACE DEBYE TEMPERATURE OF THE SILICON (001) - 2X2 STRUCTURE. (AUGER EFFECT) 5-1477
UCTURAL STUDIES OF THE ADSORPTION OF CESIUM AND OXYGEN ON SILICON (100). STR 5-1389
OSCOPY. IDENTIFICATION OF THE FORM OF CARBON AT A SILICON (100) SURFACE USING AUGER ELECTRON SPECTR 5-1378
GOLD-INDUCED SUPERSTRUCTURES ON SILICON (100) SURFACES AS OBSERVED BY LEED AES. 5-1708
ORBED OXYGEN CONTAMINANT ON THE ADSORPTION OF ALUMINUM ON SILICON (111). EFFECTS OF PREADS 5-1169
CTRON SPECTROSCOPY STUDY OF PHOSPHINE ADSORPTION ON CLEAN SILICON (111). /LECTRON DIFFRACTION AND AUGER ELE 5-1930
SECONDARY ELECTRON EMISSION YIELD AND THE APPLICATION TO SILICON (111). /R DETECTING FINE STRUCTURE IN THE 4-0852
THIN REACTION LAYERS AND SURFACE STRUCTURE OF SILICON (111). 5-1879
G CALCULATION OF A CORE VALENCE VALENCE AUGER LINE SHAPE: SILICON (111). TIGHT BINDIN 4-0831
ROOM TEMPERATURE. (AUGER SURFACE SPECTRA) SILICON (111)-7 STRUCTURE OBTAINED BY CLEAVAGE AT 5-1767
COPY) ELECTRONIC STATES OF OXYGEN ADSORBED ON CLEAN SILICON (111) AND (100) SURFACES. (AUGER SPECTROS 5-1476
D ENERGY DISTRIBUTION OF SECONDARY ELECTRON EMISSION FROM SILICON (111). (AUGER ELECTRON EXCITATION) /ES AN 4-0853
AN ORDERED BORON STRUCTURE AFTER DEPOSITION ON SILICON (111). (AUGER ELECTRON SPECTROSCOPY) 5-1168
AUGER ELECTRON EMISSION FROM GOLD DEPOSITED ON SILICON (111) SURFACE. 5-1676
AUGER ELECTRON SPECTROSCOPY OF NICKEL DEPOSITS ON THE SILICON (111) SURFACE. 5-1238
CHLORINE REACTIONS ON THE SILICON (111) SURFACE. 5-1344
ETERMINATION OF SURFACE STRUCTURES BY LEED (AND AES): THE SILICON (111) SURFACE. D 5-1342
ELECTRON BEAM ASSISTED ADSORPTION ON SILICON (111) SURFACE. 5-1269
PHASE TRANSFORMATIONS OF SILICON (111) SURFACE. 5-1343
DETECTION OF IMPURITIES ON A SILICON (111) SURFACE BY AUGER- LEED ANALYSIS. 5-1360
SPECTROSCOPY AND BY LEED. SEGREGATION OF GOLD TO SILICON (111) SURFACE OBSERVED BY AUGER EMISSION 5-1180
AUGER STUDIES OF CLEAVED SILICON (111) SURFACES. 5-1382
CARBON CONTAMINATION OF SILICON (111) SURFACES. 5-1237
HARACTERISTIC ENERGIES IN SECONDARY ELECTRON SPECTRA FROM SILICON (111) SURFACES. C 5-1263
NATURE OF SILICON (111) SURFACES. 5-1383
OPY) INTERACTION OF OXYGEN WITH SILICON (111) SURFACES. (AUGER ELECTRON SPECTROSC 5-1693
OPY) AMORPHOUS TO CRYSTALLINE TRANSITION OF SILICON (111) SURFACES. (AUGER ELECTRON SPECTROSC 5-1780
STUDY OF PHOSPHORUS ADSORPTION AND DESORPTION KINETICS ON SILICON (111) SURFACES. (BY LEED AND AES) A 5-1288
PECTROSCOPY) ADSORPTION OF OXYGEN ON SILICON (111) SURFACES. PART-1. (AUGER ELECTRON S 5-1474
INTERACTION OF OXYGEN WITH SILICON (111) SURFACES. (REPLY TO COMMENT) 5-1512
I/ STUDY OF THE OXIDATION AND REDUCTION BY WATER VAPOR OF SILICON (111) SURFACES USING AUGER ELECTRON EMISS 5-1612
CHEMISORPTION OF ATOMIC HYDROGEN ON THE SILICON (111) 7*7 SURFACE. (AUGER EFFECT) 5-1789
AND ELECTRONIC MODEL OF NEGATIVE ELECTRON AFFINITY ON THE SILICON- CESIUM- OXYGEN SURFACE. (AUGER EFFECT) / 2-0426
ES ON THE CONTACT RESISTANCE OF TUNGSTEN ON HEAVILY DOPED SILICON- GERMANIUM ALLOY. (AUGER SPECTROSCOPY) /D 5-1456
RESISTANCE CHANGES FOR TUNGSTEN CONTACTS ON HEAVILY DOPED SILICON- GERMANIUM. (AUGER SPECTROSCOPY) /ES AND 5-1861
ATURES. AUGER SPECTROSCOPIC OBSERVATION OF SILICON- GOLD MIXED PHASE FORMATION AT LOW TEMPER 5-1680
ON AND AUGER ELECTRON SP/ STUDIES OF INITIAL OXIDATION ON SILICON- IRON ALLOY (100) BY MEANS OF WORK FUNCTI 5-1924
ON AND AUGER ELECTRON SPECTROSCOPY OF 5.59 ATOMIC PERCENT SILICON- IRON ALLOY (100) SINGLE CRYSTALS. /UNCTI 5-1925
ION MEASUREMENTS AND AUGER ELECTRON SPECTROSCOPY ON 5.59% SILICON- IRON (100) SINGLE CRYSTALS. / WORK FUNCT 5-1928
POSSIBLE ORIGIN OF LOW TEMPERATURE SILICON MIGRATION IN SILICON- METAL SYSTEMS. (AUGER EFFECT) 5-1437
D BY AUGER ELECTRON SPECTRO/ METALLIC STATE OF SILICON IN SILICON- NOBLE METAL VAPOR QUENCHED ALLOYS STUDIE 5-1438
SPECTROSCOPY. SILICON- OXYGEN INTERACTIONS USING AUGER ELECTRON 5-1511
SILICO/ SILICIDE FORMATION AND INTERDIFFUSION EFFECTS IN SILICON- TANTALUM, SILICON DIOXIDE- TANTALUM, AND 5-1260
AUGER SPECTROSCOPY OF SILICONES. 5-1850
AUGER SPECTROSCOPY OF OXYGEN AND OTHER "CONTAMINANTS" ON SILVER. 5-1757
OXYGEN ADSORPTION ON (110) SILVER. 5-1422
SECONDARY ELECTRON SPECTROSCOPY OF POLYCRYSTALLINE SILVER. 4-1016
SURFACE COMPOSITION OF (PALLADIUM- SILVER) ALLOYS IN EQUILIBRIUM. 5-1788
CONTRASTING VALENCE BAND AUGER ELECTRON SPECTRA FOR SILVER AND ALUMINUM. 4-1022
CHARACTERISTIC GAIN PHENOMENA IN POLYCRYSTALLINE COPPER, SILVER, AND ALUMINUM. (AUGER SPECTROSCOPY) 5-1724
CTROSCOPY ON WELL CHARACTERIZED SINGLE CRYSTAL SURFACES. (SILVER AND COPPER ON TUNGSTEN) /DARY ION MASS SPE 5-1689
QUASIATOMIC FINE STRUCTURE IN THE AUGER SPECTRA OF SOLID SILVER AND INDIUM. 4-0771
TS. APPLICATION TO LOW ENERGY ARGON(+) ION SPUTTERING OF SILVER AND NIOBIUM. /OPY IN SPUTTERING MEASUREMEN 5-1825
/ACKSCATTERING EFFECTS IN ELECTRON INDUCED AUGER SPECTRA (SILVER AND SILICON, AUGER EFFECT ENHANCED BY BAC/ 4-0841
CHLORINE MONOLAYERS ON THE LOW INDEX FACES OF SILVER. (AUGER EFFECT) 5-1777
DETECTION OF CONTAMINANTS ON POLYCRYSTALLINE SILVER BY AES AND WORK FUNCTION MEASUREMENTS. 5-1574
ONS OF THE EPITAXIAL GROWTH OF NICKEL AND COPPER ON (111) SILVER. (BY AUGER SPECTROSCOPY) OBSERVATI 5-1354
QUANTITATIVE DETERMINATION OF CARBON ON SILVER BY AUGER SPECTROSCOPY. 5-1610
/NTIAL CROSS SECTION FOR K-SHELL IONIZATION OF COPPER AND SILVER BY ELECTRON BOMBARDMENT. (AUGER ELECTRONS) 2-0477
MULTIPLE IONIZATION AND AUGER TRANSITIONS IN INDIUM AND SILVER BY ELECTRON IMPACT. 2-0229
IODINE. HIGH RESOLUTION MNN AUGER SPECTRA OF SILVER, CADMIUM, INDIUM, ANTIMONY, TELLURIUM AND 4-0730
SURFACE COMPOSITION OF PALLADIUM- GOLD AND PALLADIUM- SILVER CATALYSTS BY AUGER ELECTRON SPECTROSCOPY. 5-1976
AUGER ELECTRON SPECTROSCOPY. INVESTIGATION OF SILVER CESIUM OXIDE PHOTOEMISSIVE SURFACES USING 5-1782
SEGREGATION OF CADMIUM TO A (111) SILVER CHLORIDE SURFACE. (AUGER SPECTROSCOPY) 5-1918
ESCA STUDIES OF SILVER, DISILVER OXIDE AND SILVER OXIDE. 4-1055
SPATIAL RESOLUTION AND PRIMARY BEAM CURRENTS. (CARBON ON SILVER FILM) AUGER ELECTRON SPECTROSCOPY AT HIGH 5-1904
RESIDUAL GAS AND THE OPTICAL PROPERTIES OF SILVER FILMS. (AUGER SPECTROSCOPY) 5-1700
SURFACE IMPURITIES IN SILVER HALIDE FILMS. 5-1149
S, ESCA, AND AUGER SPECTROSCOPY STUDY OF VACUUM DEPOSITED SILVER HALIDE FILMS. UP 5-1947

LDS INFLUENCING AUGER SPECTRA OF FLUORIDES, COPPER(I) AND SILVER(I).   LINEAR LIGAND FIE 4-0905
TERACTION OF SULFUR DIOXIDE AND CARBON DIOXIDE WITH CLEAN SILVER IN ULTRAHIGH VACUUM. (AES)   IN 5-1575
     AUGER ELECTRON SPECTROSCOPY OF ALLOY SURFACES. (GOLD- SILVER, LEAD- INDIUM)   5-1709
NG LAYER GROWTH, AND ITS APPLICATION TO THE DEPOSITION OF SILVER ON NICKEL. /ECTROSCOPY OF SYSTEMS EXHIBITI 2-0399
     AUGER EMISSION SPECTROSCOPY.   GROWTH OF SILVER ON POTASSIUM CHLORIDE OBSERVED BY LEED AND 5-1348
ED BY LOW ENERGY ELECTRON DIFFRACTIO/ EPITAXIAL GROWTH OF SILVER ON POTASSIUM CHLORIDE (100) SURFACE OBSERV 5-1696
            ESCA STUDIES OF SILVER, DISILVER OXIDE AND SILVER OXIDE.   4-1055
PECTROSCOPY OF FACE CENTERED CUBIC METAL SURFACES. (GOLD, SILVER, PALLADIUM, COPPER, NICKEL) /ER ELECTRON S 5-1717
EL ON SILVER (111): FORMATION OF AN EQUILIB/ DIFFUSION OF SILVER THROUGH EPITAXIAL FILMS OF COPPER AND NICK 5-1356
                    CHEMISORPTION OF OXYGEN ON THE SILVER (110) SURFACE.   5-1778
              ADSORPTION OF XENON AND COBALT ON SILVER (111).   5-1633
     LEED (AUGER) STUDY OF CHLORINE CHEMISORPTION ON THE SILVER (111) SURFACE.   5-1779
ATOMIC PERCENT CONTACT/ SURFACE SEGREGATION ON PALLADIUM- SILVER 40 ATOMIC PERCENT, AND PALLADIUM- GOLD 70 6-2041
ONDISPERSIVE X-RAY ANALYSIS TECHNIQUES FOR EXAMINATION OF SILVER- CADMIUM OXIDE CONTACT MATERIALS. /Y AND N 5-1799
         AES STUDIES OF SURFACE COMPOSITION OF SILVER- COPPER ALLOYS.   5-1199
/ ELECTRON SPECTROSCOPY STUDIES OF SURFACE COMPOSITION OF SILVER- COPPER ALLOYS. QUANTITATIVE AUGER ANALYS/ 5-1200
/ ELECTRON SPECTROSCOPY STUDIES OF SURFACE COMPOSITION OF SILVER- COPPER ALLOYS. QUANTITATIVE AUGER ANALYS/ 5-1740
                  AUGER ELECTRON SPECTROSCOPY OF SILVER- GOLD ALLOYS.   6-2024
           AUGER ELECTRON SPECTROSCOPY OF (SILVER- GOLD AND INDIUM- LEAD) ALLOY SURFACES.   5-1839
     WORK FUNCTION VARIATION WITH ALLOY COMPOSITION SILVER- GOLD. (AUGER ELECTRON SPECTROSCOPY)   4-0828
UDIED BY AUGER ELECTRON / SPUTTERING OF THE ALLOY SYSTEMS SILVER- GOLD, GOLD- COPPER, AND SILVER- COPPER ST 5-1331
Y.   SURFACE COMPOSITION OF THE SILVER- GOLD SYSTEM BY AUGER ELECTRON SPECTROSCOP 5-1710
        AUGER SPECTRA OF ACTIVATED SILVER- MAGNESIUM AND SILVER- LITHIUM ALLOYS.   6-2004
       ADSORPTION OF CARBON MONOXIDE ON SILVER- PALLADIUM ALLOYS.   5-1254
QUANTITATIVE AUGER ELECTRON SPECTROSCOPY ANALYSIS OF SILVER- PALLADIUM AND NICKEL- PALLADIUM ALLOYS.   6-2088
             AUGER ANALYSIS OF SURFACE FILMS ON SILVER- TIN.   5-1386
ATOMIC ARRANGEMENT OF GOLD (100), GOLD-COVERED (100), AND SILVER-COVERED COPPER (100) SURFACES.   5-1718
IR APPLICATION FOR STUDIES ON IRON AND STEEL. (AES, ESCA, SIMS)   NEW METHODS IN SURFACE ANALYSIS AND THE 6-2035
ALUATION OF CONCENTRATION DEPTH PROFILES BY SPUTTERING IN SIMS AND AES.   EV 5-1442
         COMPOSITION OF BINARY ALLOYS BY SIMULTANEOUS SIMS AND AES MEASUREMENTS.   6-2090
THE TECHNIQUES FOR SILICON SURFACE ANALYSIS. (AES, ESCA, SIMS, AND OTHERS)   COMPARISON OF 1-0052
SION. (AUGER PROCESS)   A SIMPLE ANALYTICAL CALCULATION FOR X-RAY PHOTOEMIS 2-0429
N IMPACT IN SOME METAL VAPORS. (AUGER PROC/ MECHANISMS OF SIMPLE AND MULTIPLE IONIZATION INDUCED BY ELECTRO 2-0231
MICAL SHIFTS IN AUGER SPECTRA (MONOLAYER ADSORPTION, ESCA SIMPLE CHARGE TRANSFER MODEL).   CHE 4-1115
                  AUGER SPECTROSCOPY OF SIMPLE GASEOUS MOLECULES.   4-0989
ON SPECIMEN THICKNESS.   SIMPLE MODEL FOR DEPENDENCE OF AUGER INTENSITIES 2-0355
Y SHIFTS IN HEAVY ION COLLISIONS. COMMENTS.   SIMPLE MODEL FOR K X-RAY AND AUGER ELECTRON ENERG 2-0301
                          AUGER SPECTRA OF SIMPLE MOLECULES.   4-0991
Y OF DONOR- ACCEPTOR RADIATIVE AND AUGER RECOMBINATION IN SIMPLE SEMICONDUCTORS.   THEOR 2-0420
              AUGER ELECTRON EMISSION SPECTRA: A SIMPLE TREATMENT OF ENERGIES AND INTENSITIES.   2-0319
R ELECTRON SPECTROGRAPH.   SIMPLIFIED RETARDING FIELD PHOTOELECTRON AND AUGE 3-0642
URFACES BY AUGER ELECTRON SPECTROSCOPY AND SECONDARY ION/ SIMULTANEOUS OBSERVATIONS OF PARTIALLY OXIDIZED S 5-1553
          COMPOSITION OF BINARY ALLOYS BY SIMULTANEOUS SIMS AND AES MEASUREMENTS.   6-2090
       RECENT ADVANCES IN COMPOSITION PROFILING BY SIMULTANEOUS SPUTTERING AND AUGER ANALYSIS.   1-0155
          SECONDARY ELECTRON EMISSION FROM IRON SINGLE AND POLYCRYSTALS.   4-0917
MICROGRAPHIC STUDIES OF THE CLEAVAGE SURFACE OF GRAPHITE SINGLE CRYSTAL.   AUGER ELECTRON EMISSION 5-1413
MISSION OF MAGNESIUM OXIDE LAYERS APPLIED ON A MOLYBDENUM SINGLE CRYSTAL.   STRUCTURE AND SECONDARY E 5-1191
LOW ENERGY HELIUM IONS AND ATOMS SCATTERED FROM A COPPER SINGLE CRYSTAL. (AUGER NEUTRALIZATION) /CHANGE OF 2-0545
LESS THAN 10 KEV) SCATTERED FROM A (100) FACE OF A COPPER SINGLE CRYSTAL. (AUGER NEUTRALIZATION) /+) IONS (2-0546
          SECONDARY ELECTRON EMISSION FROM SODIUM CHLORINE SINGLE CRYSTAL BOMBARDED BY RARE GAS IONS.   4-0759
E IRIDIUM (100) SURFACE. (A/ LEED INVESTIGATIONS OF XENON SINGLE CRYSTAL FILMS AND THEIR USE IN STUDYING TH 5-1479
SIUM BY AUGER ELECTRON SPECTROSCOPY. COMPARATIVE STUDY OF SINGLE CRYSTAL MAGNESIUM OXIDE AND OXIDIZED MAGNE 5-1495
               CHEMISORPTION ON SINGLE CRYSTAL MOLYBDENUM (112) SURFACES.   5-1303
ON DIFFRACTION AND AUGER INVESTIGATIONS.   SULFUR ON SINGLE CRYSTAL PLATINUM PLANES. LOW ENERGY ELECTR 5-1420
        SMALL MOLECULE REACTIONS ON STEPPED SINGLE CRYSTAL PLATINUM SURFACES.   5-1173
       ANGULAR DEPENDENCE OF AUGER ELECTRON EMISSION FROM A SINGLE CRYSTAL SPECIMEN.   4-0958
/PITAXIAL GROWTH OF COPPER ON (110) SURFACE OF A TUNGSTEN SINGLE CRYSTAL STUDIED BY LEED, AUGER ELECTRON A/ 5-1660
       CARBON MONOXIDE OXIDATION ON A PLATINUM (110) SINGLE CRYSTAL SURFACE.   5-1187
       ADSORPTION AND CONDENSATION OF COPPER ON TUNGSTEN SINGLE CRYSTAL SURFACES.   5-1160
ELECTRON, AND FLASH DESORPTION SPECTROSCOPY OF METALS ON SINGLE CRYSTAL SURFACES. /TRON DIFFRACTION, AUGER 1-0008
COPY, AND CONTACT POTENTIAL STUDIES OF COPPER- GOLD ALLOY SINGLE CRYSTAL SURFACES.   LEED, AUGER SPECTROS 5-1748
ON DIFFRACTION- AUGER ELECTRON SPECTROSCOPY STUDY OF IRON SINGLE CRYSTAL SURFACES.   LOW ENERGY ELECTR 5-1866
      SURFACE STRUCTURE OF AND CATALYSIS BY PLATINUM SINGLE CRYSTAL SURFACES.   5-1837
COPY)   ADSORPTION OF HYDROGEN ON NICKEL SINGLE CRYSTAL SURFACES. (AUGER ELECTRON SPECTROS 5-1257
FFRACTI/ STUDY OF THE INTERACTION OF OXYGEN WITH PLATINUM SINGLE CRYSTAL SURFACES BY LOW ENERGY ELECTRON DI 5-1223
          CONDENSATION OF GOLD ONTO TANTALUM (100) SINGLE CRYSTAL SURFACES. LEED AND AES ANALYSIS.   5-1312
/VITY OF LOW INDEX ((111) AND (100)) AND STEPPED PLATINUM SINGLE CRYSTAL SURFACES. (LEED, AUGER ELECTRON S/ 5-1838
PART-1.   ADSORPTION ON SINGLE CRYSTAL SURFACES OF COPPER- NICKEL ALLOYS.   5-1322
ROSCOPY)   EQUILIBRATION OF HYDROGEN AND DEUTERIUM ON SINGLE CRYSTAL SURFACES OF PLATINUM. (AUGER SPECT 5-1594
FAC/ KINETICS OF THE REACTION OF OXYGEN WITH CLEAN NICKEL SINGLE CRYSTAL SURFACES. PART-1: NICKEL (100) SUR 5-1452
/ND SECONDARY ION MASS SPECTROSCOPY ON WELL CHARACTERIZED SINGLE CRYSTAL SURFACES. (SILVER AND COPPER ON T/ 5-1689
/ACE PLASMON STUDIES ON ALUMINUM (111) AND ALUMINUM (001) SINGLE CRYSTAL SURFACES USING INELASTIC LOW ENER/ 5-1745
/NERGY ANALYSIS OF BACKSCATTERED ELECTRONS FROM A SILICON SINGLE CRYSTAL UNDER A LOW ENERGY BOMBARDMENT. (/ 4-0739
N EMISSION FROM PASSAGE OF LITHIUM(+) IONS THROUGH COPPER SINGLE CRYSTALS.   ANISOTROPY OF SECONDARY ELECTRO 2-0245
ECTRA OF INELASTICALLY SCATTERED AND AUGER ELECTRONS FROM SINGLE CRYSTALS.   ENERGY SP 4-1064
     AUGER ELECTRON SPECTROSCOPY ON 5.59% SILICON- IRON (100) SINGLE CRYSTALS. / WORK FUNCTION MEASUREMENTS AND 5-1928
TROSCOPY OF 5.59 ATOMIC PERCENT SILICON- IRON ALLOY (100) SINGLE CRYSTALS. /UNCTION AND AUGER ELECTRON SPEC 5-1925
     SECONDARY ELECTRON EMISSION FROM TUNGSTEN AND MOLYBDENUM SINGLE CRYSTALS.   4-0742
          SECONDARY ELECTRON EMISSION OF SILICON DIOXIDE SINGLE CRYSTALS.   4-1062
/PTICAL PROPERTIES AND AUGER CHARACTERISTICS OF MAGNESIUM SINGLE CRYSTALS AND MAGNESIUM OXIDE AND MAGNESIU/ 5-1220
/TURE DEPENDENCE OF THE EDGE EMISSION OF GALLIUM SELENIDE SINGLE CRYSTALS AT HIGH EXCITATION LEVELS. (AUGE/ 2-0258
OMBINATION IN DONOR ACCEPTOR COMPLEXES IN CADMIUM SULFIDE SINGLE CRYSTALS. (AUGER)   REC 2-0369
       ADSORPTION OF KRYPTON ON THE (111) FACE OF COPPER SINGLE CRYSTALS. (AUGER ELECTRON SPECTROSCOPY)   5-1221
     EXCITON SPECTRA OF HEAVILY DOPED CADMIUM SULFIDE SINGLE CRYSTALS. (AUGER PROCESS)   2-0372
OPY, AND WORK FUNCTION OF CESIUM ON TUNGSTEN AND TITANIUM SINGLE CRYSTALS IN THE SUBMONOLAYER REGION. /ROSC 5-1829
/NDENCE OF THE SECONDARY ELECTRON EMISSION COEFFICIENT OF SINGLE CRYSTALS IN THE 2-10 KEV RANGE OF PRIMARY/ 4-1061
ION)   ANGLE DISTRIBUTION OF SODIUM CHLORIDE SINGLE CRYSTALS. (PHOTO- AND AUGER ELECTRON EMISS 2-0297
/THE COMPOSITION AND STRUCTURE OF TUNGSTEN AND MOLYBDENUM SINGLE CRYSTALS STUDIED BY AUGER ELECTRON SPECTR/ 5-1215
                 AUGER SPECTROSCOPY OF GRAPHITE SINGLE CRYSTALS WITH LOW ENERGY ELECTRONS.   4-0741
     (AUGER ELECTRON SPECTROSCOPY OF) PRECISION SINGLE SIDEBAND CRYSTAL UNITS.   5-1409

SITION EFFECTS ON ENERGY SHIFTS AND SIGNAL INTENSITY IN A SINGLE STAGE CYLINDRICAL MIRROR ANALYZER. /MEN PO 3-0696
CHARACTERIZATION CF COPPER- GOLD ALLOY SINGLE SURFACES. 5-1747
. AUGER TRANSITIONS INITIATED FROM SINGLY AND DOUBLY IONIZED STATES IN LITHIUM METAL 2-0581
AUGER ELECTRON SPECTROMETRY OF SINGLY AND MULTIPLY IONIZED ATOMS. 4-0978
S. SINGULARITIES IN AUGER EMISSION SPECTRUM OF METAL 4-0998
GRAIN BOUNDARY SEGREGATION IN SINTERED ALUMINA. (AUGER ELECTRON SPECTROSCOPY) 6-2056
DDITIVE AND IMPURITY DISTRIBUTIONS AT GRAIN BOUNDARIES IN SINTERED ALUMINA. (SCANNING AUGER MICROPROBE)   A 5-1505
INFLUENCE GRAIN BOUNDARY IMPURITIES ON THE PROPERTIES OF SINTERED TUNGSTEN. (BY AUGER SPECTROSCOPY) /F THE 6-1997
FFECT) SITE COMPETITION IN SURFACE SEGREGATION. (AUGER E 6-2074
AUGER EFFECT. SIZE EFFECT IN IONIC CHARGE RELAXATION FOLLOWING 2-0565
AUGER MEASUREMENTS ON T-027 SAMPLES EXPOSED DURING THE SKYLAB-2 MISSION. 5-1735
OSCOPY. METALLIC TRANSFER BETWEEN METALS IN SLIDING CONTACT EXAMINED BY AUGER EMISSION SPECTR 5-1729
AUGER ANALYSIS OF FILMS FORMED ON METALS IN SLIDING CONTACT WITH HALOGENATED POLYMERS. 5-1728
BY AUGER ELECTRON SPECTROSCOPY. SLIDING OF POLY(VINYL CHLORIDE) ON METALS STUDIED 5-1727
SULFUR INTERACTIONS WITH VARIOUS METALS AND THE EFFECT OF SLIDING ON THESE INTERACTIONS. /IS OF OXYGEN AND 5-1206
ELEMENTAL ANALYSIS OF FRICTION AND WEAR SURFACE DURING SLIDING USING AUGER SPECTROSCOPY. 6-2013
AQUEOUS CORROSION OF SODA- SILICA AND SODA- LIME- SILICA GLASS. (AES) 5-1265
GER ELECTRON SPECTROSCOPY OF MATERIALS EXPOSED TO FLOWING SODIUM. AU 5-1460
LIFETIME EFFECT ON THE AUGER SPECTRA OF METALS. (SODIUM) 4-0787
CTRIC LOW ENERGY ELECTRON DIFFRACTION, AUGER, AND CONTAC/ SODIUM ADSORPTION ON ALUMINUM (111) SURFACES: ELE 5-1744
INTERFACIAL AUGER TRANSITIONS IN OXIDIZED SODIUM AND MAGNESIUM. 5-1497
GNESIUM OXIDE (001) BY/ ELECTRON STIMULATED DESORPTION OF SODIUM AND MEASUREMENT OF SURFACE DIFFUSION ON MA 5-1494
K-AUGER TRANSITIONS OF THE FREE SODIUM ATOM. 4-0880
/ON OF EXCITONS NEAR THE L2,3 EDGE OF THE CHLORIDE ION IN SODIUM CHLORIDE BY MEANS OF THE DECOMPOSITION PR/ 4-1126
R ELECTRON EMISSION) ANGLE DISTRIBUTION OF SODIUM CHLORIDE SINGLE CRYSTALS. (PHOTO- AND AUGE 2-0297
GAS IONS. SECONDARY ELECTRON EMISSION FRCM SODIUM CHLORINE SINGLE CRYSTAL BOMBARDED BY RARE 4-0759
K-SHELL VACANCY SHARING IN 300 TO 500 KEV NEON(+)- SODIUM COLLISIONS. (QUASIMOLECULE AUGER SPECTRA) 2-0582
TURE TECHNIQUE. AES ANALYSIS OF SODIUM IN A CORRODED BIOGLASS USING A LOW TEMPERA 5-1721
OF HIGH ENERGY X-RAY SATELLITES OF L2,3 EMISSION BANDS OF SODIUM, MAGNESIUM, ALUMINUM, AND SILICON. /ATION 2-0239
TERNAL PHOTOEMISSION CONTRIBUTION TO KLM AUGER SPECTRA OF SODIUM, MAGNESIUM, AND ALUMINUM. /ATION OF THE IN 4-0769
AUGER AND X-RAY PHOTOELECTRON SPECTROSCOPIC STUDY OF SODIUM METAL AND SODIUM OXIDE. 4-0770
GEN- CHROMIUM RAT/ CORRELATION OF THE WETTING BEHAVIOR OF SODIUM ON NIMONIC-PE16 ALLOY WITH THE SURFACE OXY 5-1441
CYLINDRICAL MIRROR AUGER ANALYZER. DETECTION OF SODIUM ON SILICON AND SILICON DIOXIDE SURFACES BY 3-0647
HE VACUUM ULTRAVI/ SPECTRAL LINE EXCITATION FUNCTIONS FCR SODIUM, POTASSIUM, RUBIDIUM, AND CESIUM IONS IN T 4-0735
/ OF STAINLESS STEEL CORROSION PRODUCTS IN FLOWING LIQUID SODIUM. (SCANNING ELECTRON MICROSCOPY, AUGER SPE/ 6-2014
(AUGER SPECTRA) SODIUM SEGREGATION ONTO A LITHIUM METAL SURFACE. 5-1750
NTENSITIES OF AUGER ELECTRONS FRCM LIGHT METALS (LITHIUM, SODIUM) SUBJECTED TO ION BOMBARDMENT. /EMISSION I 2-0548
LMM AUGER SPECTRA CF SULFUR IN SODIUM SULFATE AND SODIUM PYROSULFITE. 4-0829
AUGER SPECTRA OF LITHIUM AND SODIUM UNDER ION BOMBARDMENT. 4-0817
SECONDARY EMISSION STUDIES ON GERMANIUM AND SODIUM- COVERED GERMANIUM. 5-1714
LEED, AUGER, AND WORK FUNCTION STUDIES CF CLEAN AND SODIUM- COVERED SURFACES OF GALLIUM ARSENIDE. 5-1240
G, AUGER, AND RADIATIVE RATES AND FLUORESCENCE YIELDS FCR SODIUM- THORIUM. ATOMIC L-SHELL COSTER-KRONI 4-0963
AUGER EFFECT. SOFT X-RAY AMPLIFIED SPONTANEOUS EMISSION VIA THE 2-0443
L SPECTROSCOPY. COMPARISON OF SOFT X-RAY AND AUGER ELECTRON APPEARANCE POTENTIA 1-0092
/ERENCE IN THE CHROMIUM L3/L2 INTENSITY RATIO MEASURED BY SOFT X-RAY AND AUGER ELECTRON APPEARANCE POTENTI/ 4-0890
ROUND SUBTRACTION TO QUANTIFICATION OF AUGER ELECTRON AND SOFT X-RAY APPEARANCE POTENTIAL SPECTROSCOPIES. / 1-0067
M AND TITANIUM MONOXIDE SURFACES USING AUGER ELECTRON AND SOFT X-RAY APPEARANCE POTENTIAL SPECTROSCOPIES. / 5-1384
DISPLAY LEED SYSTEM. SOFT X-RAY APPEARANCE POTENTIAL SPECTROSCOPY IN A 4-0870
IES WITH COPPER- NICKEL AND IRON- NICKEL ALLOYS. (AES) SOFT X-RAY APPEARANCE POTENTIAL SPECTROSCOPY STUD 1-1952
ANSITIO/ POSSIBLE LOCALISATION OF CONDUCTION ELECTRONS IN SOFT X-RAY DOUBLE IONIZATION SATELLITE (AUGER) TR 2-0437
QUANTITATIVE ASPECTS OF A SOFT X-RAY LASER. (AUGER EFFECT) 2-0425
ER. (AUGER-TYPE PROCESSES) CLUSTER CALCULATIONS OF THE SOFT X-RAY L3 EMISSION SPECTRA OF NICKEL AND COPP 4-0888
TRA OF THE VALENCE BANDS OF MAGNESIUM,/ COMPARISON OF THE SOFT X-RAY PHOTOEMISSION, AND AUGER ELECTRON SPEC 4-0838
RADIATIVE AUGER TRANSITIONS IN SOFT X-RAY PLASMA EMISSION. 2-0344
TROMETER: COMBINATION OF CYLINDRICAL MIRROR ANALYZER WITH SOFT X-RAY SOURCE. A NEW PHOTOELECTRON SPEC 3-0652
GNMENT OF THE AUGER SPECTRUM AND OF THE SATELLITES IN THE SOFT X-RAY SPECTRUM OF WATER. /CULATIONS FOR ASSI 4-0724
LLIC LITHIUM. AUGER RATES FOR SOFT X-RAY TRANSITIONS IN IONIC, ATOMIC, AND META 4-0835
CLUSION OF MODERATELY HEAVY NUCLEI IN THE TISSUE: USE OF SOFT X-RAYS. (AUGER ELECTRON IRRADIATION) / ON IN 2-0531
STRENGTHENING IN TUNGSTEN- RHENIUM ALLOYS/ SOLID SOLUTION SOFTENING (RHENIUM DUCTILIZING EFFECT AND BUBBLE 6-2111
AUGER THIN FILM ANALYSIS AS APPLIED TO VARIOUS ASPECTS IN SOLAR ENERGY CONVERSION. 5-1960
SECONDARY ELECTRON EMISSION OF SOLID ALUMINUM AND MAGNESIUM TARGETS. 2-0373
URE ON THE HIGH ENERGY SIDE OF THE KL2,3M AUGER PEAK FROM SOLID ALUMINUM: INTERNAL PHOTOEMISSION. STRUCT 4-1021
AUGER ELECTRONS, INDICATORS IN ANALYSIS OF SOLID BODY SURFACES. 5-1662
QUASIATOMIC LMM AUGER SPECTRA OF SOLID COPPER AND ZINC. 4-1121
ELECTROCHEMISTRY AT THIN SOLID FILMS. (AUGER SPECTRA) 4-1026
(AUGER) ELECTRON SPECTROSCOPY, A METHOD FOR STUDYING THE SOLID GAS INTERFACE. 1-0013
BY AUGER ELECTRON SPECTROSCO/ INVESTIGATIONS OF THE GAS- SOLID INTERFACE INTERACTIONS AT HIGH TEMPERATURES 5-1147
MEASUREMENTS ON SURFACES CF ELECTRICAL SOLID MATERIALS BY THE LEED METHOD. 5-1981
PHOTOELECTRON AND AUGER SPECTRA OF XENON IN SOLID PERXENATES. 4-0902
TION ON HEATED SILICON SUBSTRATES. (SCANNING AUGER SPECT/ SOLID PHASE EPITAXIAL STUDIES USING VACUUM DEPOSI 5-1287
ANTIMONY DOPING OF SILICON LAYERS GROWN BY SOLID PHASE EPITAXY. (AES MEASUREMENTS) 5-1577
QUASIATOMIC FINE STRUCTURE IN THE AUGER SPECTRA OF SOLID SILVER AND INDIUM. 4-0771
ECT AND BUBBLE STRENGTHENING IN TUNGSTEN- RHENIUM ALLOYS/ SOLID SOLUTION SOFTENING (RHENIUM DUCTILIZING EFF 6-2111
SORDER PHENOMENA AT THE SURFACE OF ALPHA TITANIUM- OXYGEN SOLID SOLUTIONS. ORDER- DI 5-1819
EFFICIENCY OF THE AUGER CASCADE IN SOLID STATE. 2-0330
RUM OF XENON. SOLID STATE BROADENING EFFECTS IN THE AUGER SPECT 4-1005
, RARE GASES. SOLID STATE BROADENING IN AUGER SPECTRA. (CADMIUM 4-0843
RE GASES. (ARGON, KRYPTON, XENON) SOLID STATE BROADENING IN THE AUGER SPECTRA OF RA 4-1006
ENHANCEMENT CF COSTER-KRONIG RATES IN SOLID STATE ENVIRONMENTS. (XENON, CADMIUM, ZINC) 2-0439
HIGH RESOLUTION ELECTRON SPECTROGRAPHS FOR SOLID STATE RESEARCH. (REVIEW, 12 REFS) 3-0703
RELATIVE ESCAPE DEPTH OF AUGER AND PHOTOELECTRONS FROM A SOLID SURFACE. 5-1903
DYNAMIC ELUCIDATION OF REACTION STEPS ON A SOLID SURFACE BY ELECTRON SPECTROSCOPY. 5-1535
ANGULAR DEPENDENCE OF AUGER ELECTRON EMISSION FROM SOLID SURFACES. 5-1447
AUGER ELECTRON EMISSION MICROGRAPHY AND MICRANALYSIS OF SOLID SURFACES. 5-1416
AUGER ELECTRON SPECTROSCOPY AND MICROANALYSIS OF SOLID SURFACES. 5-1414
AUGER ELECTRON SPECTROSCOPY FOR THE ANALYSIS OF SOLID SURFACES. 5-1193
RALIZATION BEHAVIOR IN SCATTERING CF LOW ENERGY IONS FROM SOLID SURFACES. (AUGER) NEUT 2-0294
HIGH RESOLUTION LMM AUGER SPECTRA OF LOW ENERGY FROM SOLID SURFACES. 4-0732
HOW MUCH CAN AUGER ELECTRONS TELL US ABOUT SOLID SURFACES. 5-1663
EED AND AES STUDIES OF THE STRUCTURE OF ADSORBED GASES ON SOLID SURFACES. L 5-1835

NEW FINE STRUCTURE IN ELECTRON EXCITED AUGER SPECTRA FROM SOLID SURFACES.   2-0473
                              RECENT ADVANCES IN THE STUDY OF SOLID SURFACES.   1-0051
ION PROCESSES INVOLVING HELIUM(+)(2S) AND HELIUM(2+) NEAR SOLID SURFACES. RESONANCE, AUGER, AND AUTOIONIZAT   2-0375
        SECONDARY ELECTRON EMISSION IN THE STUDY OF SOLID SURFACES.   5-1555
                      ELECTRON STIMULATED PROCESSES AT SOLID SURFACES. (AUGER SPECTROSCOPY)   5-1605
LATINUM, NICKEL, TANTAL/ SECONDARY ELECTRON EMISSION FROM SOLID SURFACES BOMBARDED BY MEDIUM ENERGY IONS (P   2-0309
                      QUANTITATIVE ANALYSIS OF SOLID SURFACES BY AUGER ELECTRON SPECTROSCOPY.   5-1715
(AUGER EFFECT)   ELECTRONIC STRUCTURE OF SOLID SURFACES: CORE LEVEL EXCITATION TECHNIQUES.   2-0468
ANNING ELECTRON MICROSCOPE.   AUGER SPECTROSCOPY OF SOLID SURFACES IN A DRY PUMPED HIGH RESOLUTION SC   5-1258
ESS WITH TECHNIQUES INVOLVING/ ELECTRON SPECTROSCOPY FROM SOLID SURFACES IN ULTRAHIGH VACUUM. CURRENT PROGR   5-1901
)   CHARACTERIZATION OF SOLID SURFACES. (REVIEW, AUGER SPECTROSCOPY, ESCA   1-0134
(REVIEW)   THE RELATIONSHIP OF SOLID SURFACES TO VACUUM SCIENCE AND TECHNOLOGY.   1-0086
KEV): ENERGETIC AND SPATIAL DI/ ELECTRONIC EMISSION FROM SOLID TARGETS BOMBARDED BY NOBLE GAS IONS (10-100   4-0776
                SECONDARY ELECTRON SPECTRA OF SOME SOLID TARGETS UNDER IONIC BOMBARDMENT.   4-0942
OF THE AUGER EFF/ RADIATION CHEMICAL RESONANCE EFFECT IN SOLID 5-BROMO DEOXYURIDINE: CHEMICAL CONSEQUENCES   2-0377
            ATOMIC N-SHELL VACANCY WIDTHS IN SOLIDS.   2-0577
SUREMENTS FOR QUANTITATIVE AUGER ELECTRON SPECTROSCOPY OF SOLIDS.   AUGER CURRENT MEA   4-0856
                FREE ATOM BEHAVIOR OF AUGER ELECTRONS IN SOLIDS.   2-0579
        INTERPRETATION OF THE LOW ENERGY K-BETA SPECTRUM IN SOLIDS.   4-1101
      MANY BODY EFFECTS IN AUGER DEEXCITATION OF ATOMS NEAR SOLIDS.   2-0361
            NEW DEVELOPMENTS IN THE SURFACE ANALYSIS OF SOLIDS.   1-0011
ANALYSIS, METHODS OF STUDYING THE OUTER ATOMIC LAYERS OF SOLIDS.   SURFACE   1-0016
S AND QUASIATOMIC ASPECTS OF RADIATIONLESS TRANSITIONS IN SOLIDS.   WIDTHS OF ATOMIC M-SHELL VACANCY STATE   3-0718
            TWO ELECTRON BAND-TO-BAND TRANSITIONS IN SOLIDS. (AUGER RECOMBINATION IN SILICON)   2-0275
UGER NEUTRALIZATION, REVIEW, 16 R/ ELECTRON EMISSION FROM SOLIDS BY PARTICLES CARRYING POTENTIAL ENERGY. (A   2-0374
/MENSIONAL ELEMENT ANALYSIS OF SURFACES AND INTERFACES OF SOLIDS BY THE AUGER ELECTRON MICROANALYTIC METHO/   5-1412
EVIEW)   TRACE ANALYSIS TECHNIQUES FOR SOLIDS. (ELECTRON PROBE ANALYSIS, AUGER EFFECT, R   1-0093
IONIZATION CROSS SECTIONS.   AUGER EMISSION FROM SOLIDS.   ESTIMATION OF BACKSCATTERING EFFECTS AND   2-0513
SPECTROSCOPY, SURFACE PREPARATION AND/ SURFACE PHYSICS OF SOLIDS FROM THE VIEWPOINT OF ULTRAVACUUM. (AUGER   5-1419
AS A TOOL FOR MEASURING THE DIFFUSION OF FOREIGN ATOMS IN SOLIDS NEAR THEIR SURFACE. /LECTRON SPECTROSCOPY   1-0186
ER) SPECTROSCOP/ ATTENUATION OF LOW ENERGY ELECTRONS FROM SOLIDS: RESULTS FROM X-RAY PHOTOELECTRON (AND AUG   5-1844
LOY STEELS. (IMPURITY GRAIN BOUNDARY SEGREGATI/ EFFECT OF SOLUTE ELEMENTS ON TEMPER EMBRITTLEMENT OF LOW AL   6-2083
(AUGER ELECTRON SPECTROSCOPY) EFFECT OF A THIN DIFFUSED SOLUTE LAYER ON THE NUCLEATION OF FATIGUE CRACKS.   6-2037
.   IRRADIATION INDUCED SOLUTE SEGREGATION (IN STEEL, AUGER SPECTROSCOPY)   6-2092
BY AUGER SPECTROSCOPY, ON IRON INHIBITED IN LEAD AZELATE SOLUTION.   DETECTION OF LEAD,   5-1628
AND CARBON MONOXIDE BY (100) TANTALUM. (/ ADSORPTION OF SOLUTION OF HYDROGEN, DEUTERIUM, NITROGEN, OXYGEN   5-1549
D BUBBLE STRENGTHENING IN TUNGSTEN- RHENIUM ALLOYS/ SOLID SOLUTION SOFTENING (RHENIUM DUCTILIZING EFFECT AN   6-2111
PHENOMENA AT THE SURFACE OF ALPHA TITANIUM- OXYGEN SOLID SOLUTIONS.   ORDER- DISORDER   5-1819
AUGER ELECTRON / SILICON CLEANING WITH HYDROGEN PEROXIDE SOLUTIONS.   HIGH ENERGY ELECTRON DIFFRACTION AND   5-1426
RING FROM SURFACES AT ENERGIES 100-2000 EV. (AES) SORBAS: AN APPARATUS FOR INVESTIGATING ION SCATTE   3-0704
UR, AND OTHER GASES ON STAINLE/ EFFECT OF GOLD COATING ON SORPTION CHARACTERISTICS OF CARBON MONOXIDE, SULF   5-1551
OMBINATION OF CYLINDRICAL MIRROR ANALYZER WITH SOFT X-RAY SOURCE.   A NEW PHOTOELECTRON SPECTROMETER: C   3-0652
ANALYSIS OF HIGH CHARGE STATE OUTPUT FROM A PENNING ION SOURCE.   3-0602
IRON-55 AS AN AUGER ELECTRON EMITTER. NOVEL SOURCE FOR GAS CHROMATOGRAPHY DETECTORS.   3-0608
CKEL SURFACES: EFFECTS OF SURFACE CO/ MEASUREMENTS OF THE SPATIAL DISTRIBUTION OF HYDROGEN DESORBED FROM NI   5-1197
/ BOMBARDED BY NOBLE GAS IONS (10-100 KEV): ENERGETIC AND SPATIAL DISTRIBUTIONS. (AUGER ELECTRON EMISSION)   4-0776
RON, PLATINUM, COPPER, NIOBIUM, STAINLESS STEEL SURFACES/ SPATIAL DISTRIBUTIONS OF HYDROGEN DESORBED FROM I   6-2006
                              HIGH SPATIAL RESOLUTION AES.   1-0115
RBON ON SILVER FILM) AUGER ELECTRON SPECTROSCOPY AT HIGH SPATIAL RESOLUTION AND PRIMARY BEAM CURRENTS. (CA   5-1904
AN ORDINARY DIFFUSION PUMPED SCANNING ELECTRON MIC/ HIGH SPATIAL RESOLUTION AUGER ELECTRON SPECTROSCOPY IN   3-0597
NTEGRATION APPLICATIONS.   HIGH SPATIAL RESOLUTION AUGER SPECTROSCOPY AND AUGER I   1-0166
PLIED TO ANALYSIS OF X-BAND DIODE BURNOUT.   HIGH SPATIAL RESOLUTION SCANNING AUGER SPECTROSCOPY AP   5-1961
ELECTRON SPECTROSCOPY FOR THE IDENTIFICATION OF CHEMICAL SPECIES.   AUGER PARAMETER IN   5-1948
PY AND ELECTRON IMPACT DESORPTION / IDENTIFICATION OF ION SPECIES IN THE COMBINED AUGER ELECTRON SPECTROSCO   5-1692
NEW TECHNIQUE FOR AUGER ANALYSIS OF SURFACE SPECIES SUBJECT TO ELECTRON INDUCED DESORPTION.   3-0675
OPY, LEED/ A CLOSE LOOK AT SURFACES. (SURFACE STRUCTURES, SPECIES, TOPOGRAPHY, PHENOMENA, FIELD ION MICROSC   5-1492
BON MONOXIDE ON TUNGSTEN (1/ CHARACTERIZATION OF ADSORBED SPECIES USING AUGER PEAK SHAPES. ETHYLENE AND CAR   5-1243
IDENTIFICATION OF THE CHEMICAL STATE OF ADSORBED SPECIES WITH AUGER ELECTRON SPECTROSCOPY.   4-0868
PENDENCE OF AUGER ELECTRON EMISSION FROM A SINGLE CRYSTAL SPECIMEN.   ANGULAR DE   4-0958
GNAL INTENSITY IN A SINGLE STAGE CYLINDRICAL MIRROR ANAL/ SPECIMEN POSITION EFFECTS ON ENERGY SHIFTS AND SI   3-0696
SIMPLE MODEL FOR DEPENDENCE OF AUGER INTENSITIES ON SPECIMEN THICKNESS.   2-0355
CHLORINE ADSORPTION ON CLEAN ALUMINUM. (AUGER SPECTOSCOPY)   5-1831
XIDATION OF COBALT ON THE RUTHENIUM (001) SURFACE. (AUGER SPECTRA)   ADSORPTION OF OXYGEN AND O   5-1608
STATES DUE TO IMPACT OF HYDROGEN(+) AND HELIUM(+). (AUGER SPECTRA)   ARGON CHARGE   4-1011
HER CHARACTERISTIC EVENTS IN MAGNESIUM SECONDARY ELECTRON SPECTRA.   AUGER SATELLITE AND OT   4-0898
AUGER TRANSITION RATE FOR THE NOBLE GAS SATELLITE SPECTRA.   4-0972
AFTER CLEAVAGE. (OUTGASSING AND DECORATION STUDIES, AUGER SPECTRA)   CHANGE OF SURFACE PROPERTIES OF MICA   5-1743
VENTS IN BERYLLIUM AND BERYLLIUM OXIDE SECONDARY ELECTRON SPECTRA. /OUBLE IONIZATION, AND IONIZATION LOSS E   4-0901
CENCE YIELD DEDUCED FROM HIGH RESOLUTION EMISSION (AUGER) SPECTRA. / STATE DEPENDENCE OF THE NEON K FLUORES   4-0520
                      CHEMICAL SHIFT IN AUGER SPECTRA.   4-0826
                      CHEMICAL SHIFTS IN AUGER ELECTRON SPECTRA.   4-0825
                      CHEMICAL SHIFTS IN PHOTOEXCITED AUGER SPECTRA.   4-0797
HEMISORPTION OF OXYGEN ON PLATINUM (111) SURFACES. (AUGER SPECTRA)   C   5-1500
ORPTION ON (001), (110) AND (111) NICKEL SURFACES. (AUGER SPECTRA)   CHEMIS   5-1764
LUMINUM, GALLIUM, AND SCANDIUM BY ELECTRON IMPACT. (AUGER SPECTRA)   DOUBLE AND TRIPLE IONIZATION IN A   5-0232
ELECTROCHEMISTRY AT THIN SOLID FILMS. (AUGER SPECTRA)   4-1026
FLUORESCENCE PHOTON AND AUGER ELECTRON SPECTRA.   4-0799
E CONDUCTION BAND STATES FROM SECONDARY ELECTRON EMISSION SPECTRA.   GRAPHIT   2-0568
ND STRUCTURE OF TITANIUM OXIDE THIN FILMS. PART-1. (AUGER SPECTRA)   GROWTH A   5-1980
HANDBOOK OF AUGER ELECTRON SPECTROSCOPY. (EXPERIMENTAL SPECTRA)   1-0159
NEOUS NUCLEATION AND GROWTH OF GOLD ON SUBSTRATES. (AUGER SPECTRA)   HETEROGE   5-1137
IMPROVED APPARATUS FOR MEASUREMENT OF AUGER ELECTRON SPECTRA.   3-0612
TION RATES ON (100) AND (110) MOLYBDENUM SURFACES. (AUGER SPECTRA) /ION OF MOLYBDENUM. PART-3: OXIDE NUCLEA   5-1536
500 KEV NEON(+)- SODIUM COLLISIONS. (QUASIMOLECULE AUGER SPECTRA)   K-SHELL VACANCY SHARING IN 300 TO   5-0582
L-SHELL AUGER AND COSTER-KRONIG ELECTRON SPECTRA.   4-0960
-SHELL IONIZATION THRESHOLD STRUCTURE FROM AUGER ELECTRON SPECTRA.   L   2-0360
LOW NOISE DETECTION SYSTEM FOR THE MEASUREMENT OF AUGER SPECTRA.   3-0662
M-SHELL AUGER AND COSTER-KRONIG ELECTRON SPECTRA.   4-0965
MECHANISM OF PLASMON GAIN IN AUGER SPECTRA.   2-0440
FROM QUARTZ SURFACES DURING RESONATOR FABRICATION. (AUGER SPECTRA)   METHODS OF CLEANING CONTAMINANTS   5-1410

VAANCY SHARING IN ASYMMETRIC ION- ATOM COLLISIONS. (AUGER SPECTRA)   MOLECULAR EFFECTS ON K-   2-0522
HIC PLATES FOR RECORDING CONVERSION AND/OR AUGER ELECTRON SPECTRA. NEW, STRIATED, HIGH EFFICIENCY PHOTOGRAP 3-0699
PHENOMENOLOGICAL SYSTEMATICS OF L- AND M-AUGER SPECTRA.   2-0875
PLASMON SATELLITES IN AUGER SPECTRA.   2-0447
REARRANGEMENT OF INNER SHELL IONIZED ATOMS. (AUGER SPECTRA)   2-0413
NG OF COMPONENTS ON THE SURFACE OF FUSED EMITTERS. (AUGER SPECTRA)   RECORDI 4-0736
TRON EMISSION FROM PLATINUM BLACK COATED SURFACES. (AUGER SPECTRA)   SECONDARY ELEC 5-1979
OBTAINED BY CLEAVAGE AT ROOM TEMPERATURE. (AUGER SURFACE SPECTRA)   SILICON (111)-7 STRUCTURE 5-1767
SODIUM SEGREGATION ONTO A LITHIUM METAL SURFACE. (AUGER SPECTRA)   5-1750
SOME PROBLEMS IN ANALYSIS OF AUGER ELECTRON SPECTRA.   4-0866
STUDY OF GRAIN BOUNDARY SEGREGATION USING AUGER EMISSION SPECTRA.   6-2121
AUGER ELECTRON SPECTROSCOPY: CHEMICAL SHIFTS AND VALENCE SPECTRA. /ORPTION OF OXYGEN ON NICKEL(III) USING 5-1466
DISPLAYING POOR WETTABILITY DURING BRAZING. (MONEL, AUGER SPECTRA) /ARACTERIZATION OF NICKEL- COPPER ALLOY 5-1573
SURFACE COMPOSITION OF PLATINUM HEATED IN OXYGEN. (AUGER SPECTRA)   5-1737
TRANSITION POTENTIAL MODELS FOR AUGER ELECTRON SPECTRA.   2-0510
INTENSITIES OF HIGHLY RESOLVED LINES IN ARGON L2,3 AUGER SPECTRA.   WIDTHS AND 4-1032
TUDIES OF ADSORPTION ON CLEAN METAL FILM SURFACES. (AUGER SPECTRA) / AND VACUUM ULTRAVIOLET PHOTOELECTRON S 5-1203
NSITIES.   AUGER ELECTRON EMISSION SPECTRA: A SIMPLE TREATMENT OF ENERGIES AND INTE   2-0319
ERBIUM-166, YTTERBIUM-169, AND/ AUGER L-SERIES ELECTRONIC SPECTRA ACCOMPANYING THE DECAY OF ERBIUM-160, YTT 4-0743
E 3D SHELL.   AUGER SPECTRA AND FLUORESCENCE YIELDS FOR THE 3P AND TH 2-0278
CLEAVED SILICON.   AUGER SPECTRA AND LEED PATTERNS FROM NICKEL DEPOSITS ON 5-1768
ICON CRYSTALS WITH CALIBRATED DEPOSITS OF IRON.   AUGER SPECTRA AND LEED PATTERNS FROM VACUUM CLEANED SIL 5-1769
COMPARISON OF SILICON L3,2 ELECTRON LOSS SPECTRA AND OPTICAL ABSORPTION. (AUGER EFFECT)   4-1038
FACE STATE TRANSITIONS OF SILICON IN ELECTRON ENERGY LOSS SPECTRA. (AUGER EFFECT)   SUR 4-1039
/. CORRELATED STUDY USING LOW ENERGY ELECTRON DIFFRACTION SPECTRA, AUGER SPECTRA, AND WORK FUNCTION CHANGE/ 5-1291
SOLID STATE BROADENING IN AUGER SPECTRA. (CADMIUM, RARE GASES)   4-0843
/UCED AND ELECTRON IMPACT INDUCED PHOTOELECTRON AND AUGER SPECTRA. CARBON MONOXIDE ADSORPTION ON MOLYBDENU/ 4-0768
ZED BY AUGER ELECTRON SPECTROSCOPY AND SECONDARY ION MASS SPECTRA DURING XENON ION BOMBARDMENT. /PPER ANALY 5-1679
VALENCE BAND AUGER ELECTRON SPECTRA FOR ALUMINUM: COMMENTS.   4-0839
AUGER SPECTRA FOR ELEMENTS OF LOW ATOMIC NUMBER.   4-0827
L-SERIES AUGER ELECTRON SPECTRA FOR ELEMENTS WITH Z = 66-70.   4-0744
STUDIES. CHEMICAL SHIFTS IN X-RAY PHOTOELECTRON AND AUGER SPECTRA FOR MONOLAYER ADSORPTION. /ND ADSORPTION 5-1204
HIGH RESOLUTION K-AUGER SPECTRA FOR MULTIPLY IONIZED NEON.   4-0952
H HELIUM, NEON, / HIGH RESOLUTION BEAM GAS AUGER ELECTRON SPECTRA FOR OXYGEN IONS EXCITED BY COLLISIONS WIT 4-0904
/GE STATE DEPENDENCE OF X-RAY AND AUGER ELECTRON EMISSION SPECTRA FOR RARE GAS ATOMS. PART-1. THE ARGON AT/ 4-0927
CONTRASTING VALENCE BAND AUGER ELECTRON SPECTRA FOR SILVER AND ALUMINUM.   4-1022
L- AND M-SHELL YIELDS AND ELECTRON SPECTRA FOR THE TRANSURANIC ELEMENTS.   4-0968
ANGULAR RESOLVED AUGER EMISSION SPECTRA FROM A CLEAN COPPER (100) SURFACE.   4-1697
D CARBON MONOXIDE ON TUNGSTEN (100). (BY L/ PHOTOEMISSION SPECTRA FROM ADSORBED OXYGEN ON TUNGSTEN (110) AN 5-1152
IDENTIFICATION OF AUGER SPECTRA FROM ALUMINUM.   4-1027
R AND OTHER CHARACTERISTIC ENERGIES IN SECONDARY ELECTRON SPECTRA FROM ALUMINUM SURFACES.   AUGE 4-0896
ED OXYGEN, FLUORINE, CHLORINE, / ELECTRON DECAY-IN-FLIGHT SPECTRA FROM AUTOIONIZING STATES OF HIGHLY STRIPP 2-0472
AUGER AND IONIZATION LOSS SPECTRA FROM BERYLLIUM SURFACES.   5-1987
M4,5N4,5N4,5 AND M4,5O1 AUGER ELECTRON SPECTRA FROM CADMIUM VAPOR.   4-0728
RBON SUR/ CHARACTERISTIC K-SHELL LOW ENERGY ELECTRON LOSS SPECTRA FROM DIAMOND, GRAPHITE, AND EVAPORATED CA 4-0945
AUGER ELECTRON EMISSION SPECTRA FROM FOIL AND GAS EXCITED CARBON BEAMS.   4-1050
SPECTRA FROM XE/ HIGH RESOLUTION L2,3MM AND M4,5NN AUGER SPECTRA FROM KRYPTON AND M4, 5NN AND N4,5OO AUGER 4-1113
LOW ENERGY AUGER AND LOSS ELECTRON SPECTRA FROM MAGNESIUM AND ITS OXIDE.   4-0895
HIGH ENERGY SATELLITES IN AUGER SPECTRA FROM METALS.   4-0940
HIGHER ENERGY SATELLITES IN AUGER SPECTRA FROM METALS.   4-0900
CHEMICAL SHIFTS IN THE AUGER SPECTRA FROM OXIDIZED CHROMIUM AND VANADIUM.   4-0809
N5,4 AND M5,4 APPEARANCE POTENTIAL SPECTRA FROM PRASEODYMIUM.   4-0908
CHEMICAL SHIFTS IN AUGER ELECTRON SPECTRA FROM SILICON IN SILICON NITRIDE.   4-0881
ND SILICON / AUGER ELECTRON SPECTROSCOPY STUDY OF SILICON SPECTRA FROM SILICON MONOXIDE, SILICON DIOXIDE, A 4-1081
CHARACTERISTIC ENERGIES IN SECONDARY ELECTRON SPECTRA FROM SILICON (111) SURFACES.   5-1263
NEW FINE STRUCTURE IN ELECTRON EXCITED AUGER SPECTRA FROM SOLID SURFACES.   2-0473
OXIDATION AND NITRIDATION.   AUGER SPECTRA FROM THE (100) SURFACE OF SILICON DURING 5-1310
R AND OTHER CHARACTERISTIC ENERGIES IN SECONDARY ELECTRON SPECTRA FROM THIN EVAPORATED PHOSPHORUS FILMS. /E 4-0986
PLASMON EFFECTS IN ELECTRON ENERGY LOSS AND GAIN SPECTRA IN ALUMINUM.   4-1020
N CAPTU/ PROTON STATES, ANOMALOUS CONVERSION, AND K-AUGER SPECTRA IN AMERICIUM-241 FROM CURIUM-241 (ELECTRO 4-1019
SULFUR HEXAFLUORIDE.   SULFUR KLL AUGER SPECTRA IN HYDROGEN SULFIDE, SULFUR DIOXIDE, AND   4-0910
CALCULATION OF AUGER SPECTRA IN J-J COUPLING.   4-0749
AUGER SPECTRA IN THE INTERMEDIATE COUPLING REGION.   4-0887
/ OF SURFACE COMPOSITION OF COPPER- NICKEL ALLOY BY AUGER SPECTRA IN THE LOWER ENERGY REGION AT AROUND 100/ 5-1982
AUGER SPECTRA IN X-RAY PHOTOELECTRON SPECTROSCOPY.   1-0209
INC OXIDE, GALLIUM, AND GALLIU/ AUGER AND DIRECT ELECTRON SPECTRA IN X-RAY PHOTOELECTRON STUDIES OF ZINC, Z 4-1054
/S IN THE TITANIUM DIOXIDE AUGER AND APPEARANCE POTENTIAL SPECTRA INDUCED BY INCIDENT ELECTRON BEAM INTERA/ 5-1834
G YIELDS IN BISMUTH AND NEPTUNIUM.   X-RAY SPECTRA, L-SUBSHELL FLUORESCENCE AND COSTER-KRONI   4-1112
TRANSFER MODEL).   CHEMICAL SHIFTS IN AUGER SPECTRA (MONOLAYER ADSORPTION, ESCA SIMPLE CHARGE   4-1115
DIFFRACTION PEAKS IN SECONDARY ELECTRON ENERGY SPECTRA. (NICKEL (100))   4-0775
SECONDARY AND AUGER ELECTRON SPECTRA OF A COPPER- BERYLLIUM ALLOY.   4-0916
- LITHIUM ALLOYS.   AUGER SPECTRA OF ACTIVATED SILVER- MAGNESIUM AND SILVER   6-2004
ION EXCITED AUGER SPECTRA OF ALUMINUM.   4-0869
LIFETIME AND SURFACE EFFECTS ON THE L23 VV AUGER SPECTRA OF ALUMINUM.   4-0833
HIGH RESOLUTION L2,3M2,3M2,3 AUGER ELECTRON SPECTRA OF ARGON.   4-0912
EFFECT OF STRUCTURE ON THE AUGER SPECTRA OF BARIUM FILMS ON PLATINUM.   5-1359
CHEMICAL EFFECTS ON THE AUGER ELECTRON SPECTRA OF BERYLLIUM.   4-0834
BY EQUAL VELOCITY ALPHA PARTICLES AND DEUTERONS.   AUGER SPECTRA OF CARBON AND ARGON FOLLOWING IONIZATION 4-1109
D CARBON MONOXIDE ADSORPTION ON TANTALUM AND POLYC/ AUGER SPECTRA OF CARBON. TANTALUM CARBIDES, ETHYLENE AN 4-0820
DENUM SURFACE/ CHEMICAL EFFECTS OF THE NITROGEN KLL AUGER SPECTRA OF CHEMISORBED NITROGEN ON IRON AND MOLYB 4-0925
UGER VACANCY SATELLITE STRUCTURE IN THE L3,M4,5M4,5 AUGER SPECTRA OF COPPER.   A 4-1034
MVV AUGER SPECTRA OF COPPER AND NICKEL.   4-0766
L-S COUPLING INTERPRETATION OF HIGH RESOLUTION LMM AUGER SPECTRA OF COPPER AND ZINC.   4-1122
IDE, AND COPPER(I)/ X-RAY INDUCED PHOTOELECTRON AND AUGER SPECTRA OF COPPER, COPPER(II) OXIDE, COPPER(I) OX 4-0932
X-RAY EXCITED LMM AUGER SPECTRA OF COPPER, NICKEL AND IRON.   4-1123
AUGER SPECTRA OF COPPER- NICKEL ALLOYS.   4-1028
E EARTH METAL SURFACES.   IONIZATION LOSSES IN (AUGER) SPECTRA OF ELECTRON INELASTIC SCATTERING FROM RAR 2-0538
OLLISIONS. (AUGER AND COSTER-KRONIG TRANSITIONS)   ENERGY SPECTRA OF ELECTRONS EJECTED IN ARGON(+)- ARGON C 4-1007
LINEAR LIGAND FIELDS INFLUENCING AUGER SPECTRA OF FLUORIDES, COPPER(I) AND SILVER(I).   4-0905
N4,5 AUGER SPECTRA OF GADOLINIUM.   4-0821

211

CHARACTERISTIC ENERGY LOSS AND AUGER ELECTRON SPECTRA OF GALLIUM PHOSPHIDE (110).   5-1657
                     FINE STRUCTURE DETAILS IN THE LMM AUGER SPECTRA OF GERMANIUM.   4-0937
                        HIGH RESOLUTION LMM AUGER SPECTRA OF GERMANIUM.   4-0936
SHIFTS IN COORDINATION COMPOUNDS. X-RAY EXCITED LMM AUGER SPECTRA OF GERMANIUM.   PHOTOELECTRON CHEMICAL   4-0924
   THRESHOLD EFFECTS ON RELATIVE INTENSITIES IN LMM AUGER SPECTRA OF GERMANIUM AND SELENIUM.   4-0767
RYSTALS. (AUGER PROCESS)   EXCITON SPECTRA OF HEAVILY DOPED CADMIUM SULFIDE SINGLE C 2-0372
                                                    AUGER SPECTRA OF HIGHLY IONIZED OXYGEN AND FLUORINE.   4-0906
IUM ARSENIDE (100) SUBSTRATES.   AUGER SPECTRA OF HYDROGEN CHLORIDE VAPOR ETCHED N+ GALL 4-0992
TRONS FROM SINGLE CRYSTALS.   ENERGY SPECTRA OF INELASTICALLY SCATTERED AND AUGER ELEC 4-1064
LEED STUDIES OF TUNGSTEN.   ENERGY SPECTRA OF INELASTICALLY SCATTERED ELECTRONS AND 4-1094
D THEIR USE IN STUDYING THE IRIDIUM (100) SURFACE. (AUGER SPECTRA OF IRIDIUM) /ENON SINGLE CRYSTAL FILMS AN 5-1479
         CHEMICAL EFFECTS IN THE AUGER SPECTRA OF IRON DUE TO OXYGEN ADSORPTION.   4-0727
NT.   AUGER SPECTRA OF LITHIUM AND SODIUM UNDER ION BOMBARDME 4-0817
                                                    AUGER SPECTRA OF LITHIUM HYDRIDE.   4-1025
                                                    AUGER SPECTRA OF LITHIUM METAL AND LITHIUM OXIDE.   4-0806
                         HIGH RESOLUTION LMM AUGER SPECTRA OF LOW ENERGY FROM SOLID SURFACES.   4-0732
ED BY LOW ENERGY ELECTRON AND LOW EN/ COMPARISON OF AUGER SPECTRA OF MAGNESIUM, ALUMINUM, AND SILICON EXCIT 4-0857
                                                    AUGER SPECTRA OF MAGNESIUM AND ALUMINUM.   4-0822
                          CHANGES IN AUGER SPECTRA OF MAGNESIUM AND IRON DUE TO OXIDATION.   4-1086
ND ITS OXIDE.   X-RAY EXCITED AUGER AND PHOTOELECTRON SPECTRA OF MAGNESIUM, SOME ALLOYS OF MAGNESIUM, A 4-0837
           AUGER ELECTRON SPECTRA OF MAGNESIUM THIN FILMS.   4-0873
           X-RAY PHOTOELECTRON AND AUGER SPECTRA OF MAGNESIUM THIN FILMS.   5-1392
                 LIFETIME EFFECT CN THE AUGER SPECTRA OF METALS. (SODIUM)   4-0787
XPERIMENT.   L2,3M4,5M4,5 AUGER SPECTRA OF METALLIC COPPER AND ZINC: THEORY AND E 4-0919
                THEORETICAL ANALYSIS OF THE AUGER SPECTRA OF METHANE.   4-1012
    PROTON ANC ELECTRON EXCITED AUGER ELECTRON SPECTRA OF MOLYBDENUM SURFACES.   5-1667
ES)   CLUSTER CALCULATIONS OF THE SOFT X-RAY L3 EMISSION SPECTRA OF NICKEL AND COPPER. (AUGER-TYPE PROCESS 4-0888
L STATE EFFECTS IN THE PHOTOEMISSION CORE LEVEL AND AUGER SPECTRA OF NICKEL AND NICKEL ALLOYED WITH ZINC. / 5-1806
TS COMPOUNDS.   X-RAY FLUORESCENCE L SPECTRA OF NIOBIUM IN NIOBIUM METAL AND SOME OF I 4-0762
      PRODUCTION OF INTENSE SATELLITE LINES IN AUGER SPECTRA OF NITROGEN BY SLOW PROTON IMPACT.   4-1076
IC OXIDE, WATER, AND CARBON M/ DETERMINATION OF KLL AUGER SPECTRA OF NITROGEN, OXYGEN, CARBON DIOXIDE, NITR 4-0990
            CORRELATION EFFECTS IN THE AUGER SPECTRA OF NOBLE GASES.   4-0981
.   X-RAY EXCITED PHOTOELECTRON AND AUGER ELECTRON SPECTRA OF OXIDIZED STATES ON SOME METAL SURFACES 5-1690
   ABNORMAL CHEMICA/ X-RAY EXCITED AUGER AND PHOTOELECTRON SPECTRA OF PARTIALLY OXIDIZED MAGNESIUM SURFACES. 5-1949
           CHEMICAL SHIFTS IN THE AUGER SPECTRA OF PASSIVE FILMS.   4-1056
                 SEPARATION OF THE SPECTRA OF PHOTO AND AUGER ELECTRONS.   4-1099
       K-, L-, AND M-AUGER AND L- COSTER-KRONIG SPECTRA OF PLATINUM.   4-1097
       AUGER ELECTRON AND CHARACTERISTIC ENERGY LOSS SPECTRA OF PLUTONIUM.   4-0931
FERMIUM.   THE L-AUGER SPECTRA OF PLUTONIUM, AMERICIUM, CALIFORNIUM, AND 4-0836
           SOLID STATE BROADENING IN THE AUGER SPECTRA OF RARE GASES. (ARGON, KRYPTON, XENON)   4-1006
UM AND SILICON WHEN BOMBARDED BY PRIMARY ELECTRON/ ENERGY SPECTRA OF SECONDARY ELECTRONS EMITTED BY MAGNESI 4-1047
D TUNGSTEN.   AUGER ELECTRON EMISSION IN THE ENERGY SPECTRA OF SECONDARY ELECTRONS FROM MOLYBDENUM AN 4-0874
IALS.   AUGER PEAKS IN THE ENERGY SPECTRA OF SECONDARY ELECTRONS FROM VARIOUS MATER 4-0926
/ INTERATOMIC TRANSITICNS AND RELAXATION EFFECTS IN AUGER SPECTRA OF SEVERAL GAS ADSORBATES ON TRANSITION M 4-1042
TED ELECTRONS UNDER THE X-RAY DIFFRACTION PROCESS. (AUGER SPECTRA OF SILICON) / AND THE MEASUREMENT OF EXCI 3-0641
      HIGH ENERGY FINE STRUCTURE IN THE AUGER SPECTRA OF SILICON AND SILICON CARBIDE.   4-1036
M(1-X) ARSENIDE ELECTROLUMINESCENT DIODES. (AUG/ EMISSION SPECTRA OF SILICON COMPENSATED ALUMINUM(X) GALLIU 2-0325
/ON SINGLE CRYSTAL UNDER A LOW ENERGY BOMBARDMENT. (AUGER SPECTRA OF SILICON, SILICON CARBIDE AND SILICON / 4-0739
LURIUM AND IODINE.   HIGH RESOLUTION MNN AUGER SPECTRA OF SILVER, CADMIUM, INDIUM, ANTIMONY, TEL 4-0730
                                                    AUGER SPECTRA OF SIMPLE MOLECULES.   4-0991
N OF THE INTERNAL PHOTOEMISSION CONTRIBUTION TO KLM AUGER SPECTRA OF SODIUM, MAGNESIUM, AND ALUMINUM. /ATIO 4-0769
               QUASIATOMIC LMM AUGER SPECTRA OF SOLID COPPER AND ZINC.   4-1121
       QUASIATOMIC FINE STRUCTURE IN THE AUGER SPECTRA OF SOLID SILVER AND INDIUM.   4-0771
S BOMBARDED BY 60 KEV ARGON(+) IONS.   AUGER SPECTRA OF SOME FREE ELEMENTS AND BINARY COMPOUND 4-1106
ASEODYMIUM, NEODYMIUM, GADOLINIUM, DYSPRO/ AUGER ELECTRON SPECTRA OF SOME RARE EARTH METALS. (LANTHANUM, PR 4-1100
MENT.   SECONDARY ELECTRON SPECTRA OF SOME SOLID TARGETS UNDER IONIC BOMBARD 4-0942
RIES.   LMM AUGER SPECTRA OF SOME TRANSITION METALS OF THE FIRST SE 4-0810
ROSULFITE.   LMM AUGER SPECTRA OF SULFUR IN SODIUM SULFATE AND SODIUM PY 4-0829
  ON PAPER BY CHUNG AN/ UNEXPLAINED PHENOMENA IN THE AUGER SPECTRA OF THE FIRST TRANSITION SERIES. (COMMENTS 4-0812
                                                    AUGER SPECTRA OF THE NOBLE GASES. KRYPTON.   4-0971
/ISON OF THE SOFT X-RAY PHOTOEMISSION, AND AUGER ELECTRON SPECTRA OF THE VALENCE BANDS OF MAGNESIUM, MAGNE/ 4-0838
ING IN X-RAY INDUCED PHOTOEMISSION SPECTROSCOPY AND AUGER SPECTRA OF THE ZINC CHALCOGENIDES. /ETIME BROADEN 4-0832
               KLL AUGER ELECTRON SPECTRA OF THULIUM AND LUTETIUM.   4-0753
    INTERFACIAL AND PLASMON DECAY PEAKS IN THE AUGER SPECTRA OF TITANIUM AND IRON.   4-1060
     MOLECULAR ORBITAL EFFECTS ON THE TITANIUM LMV AUGER SPECTRA OF TITANIUM(II) AND TITANIUM(IV) OXIDES. 4-1069
ORPTION OF OXYGEN.   CHEMICAL SHIFT IN THE AUGER SPECTRA OF TUNGSTEN AND MOLYBDENUM DURING THE ADS 4-0816
                                                    AUGER SPECTRA OF URANIUM.   4-1125
/TIC EFFECTS AND STRUCTURE IN THE AUGER ELECTRON EMISSION SPECTRA OF VANADIUM PENTOXIDE (010) AND VANADIUM/ 5-1340
             PHOTOELECTRON AND AUGER SPECTRA OF XENON IN SOLID PERXENATES.   4-0902
BOMBARDMENT.   HIGH RESOLUTION NEON KLL AUGER SPECTRA PRODUCED BY HYDROGEN(+) AND HELIUM(+) ION 4-0953
TIVE SATELLITE LINE INTENSITIES IN ARGON K AUGER ELECTRON SPECTRA PRODUCED BY 0.4 TO 4 MEV PROTON IMPACT. / 2-0497
S.   ARGON L-SHELL AUGER SPECTRA PRODUCED IN ARGON(+) ION- ARGON COLLISION 4-1080
          THE K-SHELL AUGER SPECTRA.   REVIEW.   4-1060
X-RAYS IN NONDESTRUCTIVE ANALYSIS. (AUGER, PHOTOELECTRON SPECTRA, REVIEW, 6 REFS)   ELECTRONS AND 1-0083
/TION OF BACKSCATTERING EFFECTS IN ELECTRON INDUCED AUGER SPECTRA (SILVER AND SILICON, AUGER EFFECT ENHANC/ 4-0841
          INTENSITIES OF SPECTRA- AUGER EFFECT AND PHOTOELECTRON.   4-1070
E SURFACE OF MOLYBDENUM AND STEEL BASED ON AUGER ELECTRON SPECTRAL DATA.   CHEMICAL COMPOSITION OF TH 5-1985
TASSIUM, RUBIDIUM, AND CESIUM IONS IN THE VACUUM ULTRAVI/ SPECTRAL LINE EXCITATION FUNCTIONS FOR SODIUM, PO 4-0735
ECTRON SPECTROSCOPY STUDIES OF CARBON MONOXIDE ON NICKEL. SPECTRAL LINE SHAPES AND QUANTITATIVE ASPECTS. /L 4-1458
HEMICAL ANALYSIS.   INFORMATION OBTAINED FROM THE WIDTH OF SPECTRAL LINES IN THE ELECTRON SPECTROSCOPY FOR C 4-1009
/DIUM ANTIMONIDE ON THE EXTERNAL PHOTOEFFECT IN THE X-RAY SPECTRAL REGION. (EFFECT OF IMPURITIES ON THE AU/ 4-1001
MPLIFIED RETARDING FIELD PHOTOELECTRON AND AUGER ELECTRON SPECTROGRAPH.   SI 3-0642
12 REFS)   HIGH RESOLUTION ELECTRON SPECTROGRAPHS FOR SOLID STATE RESEARCH. (REVIEW, 3-0703
IES AND INCLUSIONS IN HOT PRESSED SILICON NITRIDE. (AUGER SPECTROGRAPHIC ANALYSIS)   IMPURIT 5-1557
   GRAIN BOUNDARY STRUCTURE AND SEGREGATION.   AUGER SPECTROGRAPHIC ANALYSIS OF FRACTURE SURFACES.   6-2118
          SPECTROGRAPHY BY AUGER ELECTRONS.   1-0047
A SMALL RETARDING FIELD ENERGY ANALYZER WITH CEM. (AUGER SPECTROMETER)   3-0655
Y MODULATED (CYLINDRICAL) HIGH ENERGY RESOLUTION ELECTRON SPECTROMETER.   AN ENERG 3-0616
       AUGER SIGNALS OBTAINED WITH A PHOTOELECTRON SPECTROMETER.   2-0272
INTENSITY OF AUGER SIGNALS OBTAINED WITH A PHOTOELECTRON SPECTROMETER. /SHIFTS AND VARIATIONS OF WIDTH AND 2-0273

212

COMBINED ESCA AND AUGER SPECTROMETER. 3-0671
CTION AND OPERATION OF A CYLINDRICAL MIRROR TYPE OF AUGER SPECTROMETER. CONSTRU 3-0588
ODS FOR THIN FILM ANALYSIS USING A COMPUTER COUPLED AUGER SPECTROMETER. DIGITAL METH 3-0640
TROSTATIC ANALYZER OF THE CYLINDRICAL MIRROR TYPE. (AUGER SPECTROMETER) /F FINITE ANGULAR SPREAD IN AN ELEC 3-0720
HIGH RESOLUTION, LOW ENERGY ELECTRON SPECTROMETER. 3-0697
HIGH SENSITIVITY AUGER ELECTRON SPECTROMETER. 3-0672
HIGH SENSITIVITY ELECTRON SPECTROMETER. 3-0633
LYBDENUM STUDIED BY A HIGH RESOLUTION AUGER PHOTOELECTRON SPECTROMETER. INITIAL OXIDATION OF MO 5-1698
RETARDING FIELD ELECTRON SPECTROMETER. 3-0686
SECONDARY ELECTRON SPECTROMETER. 3-0709
MINATION MONITOR) AUGER ELECTRON SPECTROMETER AS TOOL FOR SURFACE ANALYSIS. (CONTA 3-0706
UTION AND SENSITIVITY CONSIDERATIONS OF AN AUGER ELECTRON SPECTROMETER BASED ON LEED DISPLAY OPTICS. RESOL 3-0705
NALYZER WITH SOFT X-RAY SOURCE. A NEW PHOTOELECTRON SPECTROMETER: COMBINATION OF CYLINDRICAL MIRROR A 3-0652
STUDY OF ALUMINUM WITH A COMBINED AUGER ELECTRON SPECTROMETER ELLIPSOMETER SYSTEM. 5-1132
HIGH RESOLUTION ELECTRON SPECTROMETER FOR AUGER ELECTRON SPECTROSCOPY. 3-0667
IONIZATION SPECTROMETER FOR ELEMENTAL ANALYSIS OF SURFACES. 3-0615
DIES. DEVELOPMENT AND APPLICATION OF AN "AUGER ELECTRON SPECTROMETER" FOR HIGH TEMPERATURE ADSORPTION STU 3-0586
R ELECTRONS) SPECTROMETER FOR INDUCED ELECTRON EMISSION. (AUGE 3-0585
STUDIES. HIGH RESOLUTION ELECTRON SPECTROMETER FOR PHOTOELECTRON AND AUGER ELECTRON 3-0600
ELECTRON SPECTROMETER FOR SURFACE ANALYSIS. 3-0715
AUGER) REALIZATION OF A MOSSBAUER SPECTROMETER FOR SURFACES AND THIN FILMS STUDY. (3-0690
TENTS AND SOME RESULTS. AUGER CYLINDRICAL MIRROR SPECTROMETER FOR THE ANALYSIS OF SURFACE CONSTITU 3-0682
ONS AND IONS. HIGH SENSITIVITY APPEARANCE POTENTIAL SPECTROMETER FOR THE DETECTION OF PHOTONS, ELECTR 3-0643
ON IN MAGNESIA DOPED ALUMINA. (HIGH VACUUM AUGER ELECTRON SPECTROMETER, FRACTURED ALUMINA) /NDARY SEGREGATI 5-1615
SSIBILITIES OF THE ION MICROPROBE MASS ANALYZER AND AUGER SPECTROMETER IN PHYSICAL METALLURGY. PO 6-2075
ENT OF EXCITED ELECTRO/ DESIGN OF AN X-RAY DOUBLE CRYSTAL SPECTROMETER INSTALLED IN VACUUM AND THE MEASUREM 3-0641
N DIFFRACTION SYSTEM. AUGER SPECTROMETER INTEGRATED WITH A LOW ENERGY ELECTRO 3-0658
/OPE, TRANSMISSION ELECTRON MICROSCOPE, ENERGY DISPERSIVE SPECTROMETER, ION MASS ANALYSIS, AUGER ELECTRON / 1-0010
SPECTROMETER OF AUGER ELECTRONS. 3-0653
AUGER ELECTRON SPECTROMETER PREAMPLIFIER. 3-0708
EMICAL ANALYSIS. (AUTOMATI/ HIGH RESOLUTION ELECTROSTATIC SPECTROMETER USED IN ELECTRON SPECTROMETRY FOR CH 3-0717
HIGH ENERGY RESOLUTION AUGER ELECTRON SPECTROMETER USING CONCENTRIC HEMISPHERES. 3-0592
IMPROVED AUGER ELECTRON SPECTROMETER USING CONCENTRIC HEMISPHERES. 3-0591
ELECTRON CHANNELING PATTERNS IN AN AUGER ELECTRON SPECTROMETER WITH SCANNING SAMPLE POSITIONER. 3-0692
UGER ELECTRON EMISSION SPECTROSCOPY USING RETARDING FIELD SPECTROMETERS. /TING, AND RELATED TECHNIQUES IN A 3-0610
SURFACE AND THIN FILM ANALYSIS. (AUGER SPECTROMETERS) 3-0609
X-RAY ELECTRON SPECTROMETERS AND AUGER ANALYZERS. 3-0599
LECTRON YIELD MEASUREMENTS. USE OF CYLINDRICAL AUGER SPECTROMETERS FOR RETARDING POTENTIAL SECONDARY E 3-0628
A COMPARISON OF THE EXTENDURE OF ELECTRON SPECTROMETERS. (USED IN AUGER SPECTROSCOPY) 3-0626
ER ELECTRON SPECTROSC/ SOME PROPERTIES OF RETARDING FIELD SPECTROMETERS USING POST MONOCHROMATORS. (FOR AUG 3-0624
CHNIQUES FOR METAL CONTACTS. COMBINED MASS SPECTROMETRIC AND AUGER ELECTRON SPECTROSCOPIC TE 6-2042
CHEMISORPTION OF CARBON MONOXIDE ON POLYCRYSTALLI/ AUGER SPECTROMETRIC AND MASS SPECTROMETRIC STUDY OF THE 5-1355
ECOMPOSITION OF CARBON MONOXIDE ON PLATINUM (111). (AUGER SPECTROMETRY) ADSORPTION AND D 5-1622
AUGER ELECTRON SPECTROMETRY. 1-0191
AUGER LINES IN X-RAY PHOTOELECTRON SPECTROMETRY. 1-0208
SON OF AUGER ELECTRON SPECTROSCOPY AND SECONDARY ION MASS SPECTROMETRY. COMPARI 1-0135
IDE BY AUGER ELECTRON SPECTROSCOPY AND SECONDARY ION MASS SPECTROMETRY. /OSITED PLATINUM AND PLATINUM SILIC 5-1656
ELECTRON SPECTROMETRY. 1-0108
S WITH AUGER ELECTRON SPECTROSCOPY AND SECONDARY ION MASS SPECTROMETRY. ELEMENTAL ANALYSI 5-1654
REACTIONS, LEED, AUGER, CONTACT POTENTIAL METHOD AND MASS SPECTROMETRY) /ACE. (CHEMISORPTION AND CATALYTIC 5-1321
) GALLIUM(1-X) ARSENIDE BY MOLECULAR BEAM EPITAXY. (AUGER SPECTROMETRY) /ED GALLIUM ARSENIDE AND ALUMINUM(X 5-1247
EMATIC LIQUID CRYSTALS. (SUBSTRATE SURFACE EFFECTS, AUGER SPECTROMETRY) MECHANISM OF SURFACE ALIGNMENT IN N 5-1281
LMS BY AUGER ELECTRON SPECTROSCOPY AND SECONDARY ION MASS SPECTROMETRY. /D OXYGEN) IN SPUTTERED TANTALUM FI 5-1652
SIS. (AUGER ELECTRON SPECTROSCOPY AND ION MICROPROBE MASS SPECTROMETRY) /STRUMENTATION FOR MICROPROBE ANALY 3-0716
VACUUM TECHNIQUES IN BASIC RESEARCH. (REVIEW, AUGER, MASS SPECTROMETRY) ROLE OF ULTRAHIGH 1-0065
SECONDARY ION MASS SPECTROMETRY. 1-0079
STUDY OF METAL SURFACES BY IONIC MICROANALYSIS AND AUGER SPECTROMETRY. 5-1582
SURFACE CHARACTERIZATION BY AUGER ELECTRON SPECTROMETRY. 5-1815
EN, OXYGEN, AND FLUORINE IN SILICON BY SECONDARY ION MASS SPECTROMETRY AND AUGER ELECTRON SPECTROSCOPY. /OG 5-1916
ESSES ON COPPER- BERYLLIUM BY COMBINED SECONDARY ION MASS SPECTROMETRY AND AUGER ELECTRON SPECTROSCOPY. /OC 5-1213
OF 316 L-N STAINLESS STEEL STUDIED BY SECONDARY ION MASS SPECTROMETRY AND AUGER ELECTRON SPECTROSCOPY. /SS 6-2087
USEFUL COMPLEMENT TO OTHER DEPTH PRO/ SECONDARY ION MASS SPECTROMETRY (AND AUGER ELECTRON SPECTROSCOPY): A 1-0009
/OF 316 L-N STAINLESS STEEL STUDIES BY SECONDARY ION MASS SPECTROMETRY AND AUGER ELECTRON SPECTROSCOPY. CO/ 3-0654
THE ANALYSIS OF THIN / COMBINATION OF SECONDARY ION MASS SPECTROMETRY AND AUGER ELECTRON SPECTROSCOPY, FOR 5-1214
/ OF SURFACE CHEMISTRIES OBTAINED BY USING ION SCATTERING SPECTROMETRY AND AUGER ELECTRON SPECTROSCOPY. (S/ 6-1999
ENERGY DEPTH OF LEVELS OF SURFACE ELECTRON / EXOELECTRON SPECTROMETRY AS A METHOD FOR DETERMINATION OF THE 4-0915
VIEW) AUGER SPECTROMETRY AS A TOOL FOR SURFACE CHEMISTRY. (RE 1-0162
AL CHARACTERIZATIONS OF CERAMIC SURFACES. (ION SCATTERING SPECTROMETRY, AUGER ELECTRON SPECTROMETRY) /HEMIC 5-1326
ONDARY I/ SURFACE ANALYTICAL STUDIES USING ION SCATTERING SPECTROMETRY, AUGER ELECTRON SPECTROSCOPY AND SEC 5-1423
/ES BY AUGER ELECTRON SPECTROSCOPY AND SECONDARY ION MASS SPECTROMETRY FOR ALUMINUM, SILICON, TITANIUM, VA/ 5-1553
/H RESOLUTION ELECTROSTATIC SPECTROMETER USED IN ELECTRON SPECTROMETRY FOR CHEMICAL ANALYSIS. (AUTOMATIC R/ 3-0717
/SE OF AUGER ELECTRON SPECTROSCOPY AND SECONDARY ION MASS SPECTROMETRY IN THE MICROELECTRONIC TECHNOLOGY. / 1-0136
AN ANALYTICAL SYSTEM FOR SECONDARY ION MASS SPECTROMETRY IN ULTRAHIGH VACUUM. (AES) 3-0632
DIFFERENTIAL AUGER SPECTROMETRY. (NIOBIUM- ZIRCONIUM ALLOY ANALYSIS) 5-1849
AUGER ELECTRON SPECTROMETRY OF SINGLY AND MULTIPLY IONIZED ATOMS 4-0978
ELECTRON SPECTROMETRY. (REVIEW) 1-0035
ED BY CRACK GROWTH IN HYDROGEN AND IN WATER VAPOR. (AUGER SPECTROMETRY, STEEL) /OF FRACTURE SURFACES PRODUC 6-2142
INLESS STEELS. AUGER SPECTROMETRY STUDY OF PASSIVE FILMS FORMED ON STA 6-2102
SPECTROSCOP... IS NOT INDEXED.
A NEW TREATMENT OF THE KLL AUGER SPECTRUM. 4-0745
A NEW TYPE OF AUGER EFFECT AND ITS INFLUENCE ON THE X-RAY SPECTRUM. 1-0045
ABSORPTION BANDS IN SECONDARY ELECTRON CONTINUOUS SPECTRUM. 4-0933
BY 200-500 KEV PROTONS IN GASES. PART-2: NITROGEN. (AUGER SPECTRUM) /GY DISTRIBUTION OF ELECTRONS PRODUCED 4-1075
AUGER TRANSITIONS AND SHAPE OF X-RAY SPECTRUM. 4-1104
X-RAY SPECTRAL REGION. (EFFECT OF IMPURITIES ON THE AUGER SPECTRUM.) /E ON THE EXTERNAL PHOTOEFFECT IN THE 4-1001
DING ENERGIES OF BARIUM FROM THE SECONDARY ELECTRON YIELD SPECTRUM. ELECTRON BIN 4-0846
OF ATOMIC DYNAMICS THROUGH THE AUGER EFFECT. (NEON AUGER SPECTRUM) MANIFESTATION 4-0920
NEW METHOD FOR ANALYZING THE SECONDARY ELECTRON SPECTRUM. 4-0934
PLASMON GAIN IN THE FINE STRUCTURE OF THE ALUMINUM AUGER SPECTRUM. OBSERVATION OF A 4-1085
ROJECTILE CHARGE DEPENDENCE FOR THE NEON K-AUGER ELECTRON SPECTRUM. OBSERVATION OF A P 4-0993

N AND DOUBLE IONIZATION SATELLITES IN THE BERYLLIUM AUGER SPECTRUM.   PLASMON ENERGY GAI 4-0899
                    PLASMON ENERGY GAIN IN THE BERYLLIUM AUGER SPECTRUM.   4-1095
D DISSOCIATION: CARBON MONOXIDE ON PLATINUM (111). (AUGER SPECTRUM) /RONS IN ELECTRON INDUCED DESORPTION AN 5-1567
F THE AUGER EFFECT IN THE FORMATION OF AN ELECTRON SHOWER SPECTRUM.   ROLE O 2-0504
                     SPIN SPLITTING IN COPPER AUGER ELECTRON SPECTRUM.   4-1053
                     STRUCTURE OF LOW ENERGY ELECTRON SPECTRUM.   4-0948
                                    THE K-AUGER SPECTRUM.   4-0751
-RAY PRODUCTION BY HEAVY IONS. (OXYGEN BOMBARDMENT, AUGER SPECTRUM)   X 2-0479
NCE OF ENERGIES AND RELATIVE INTENSITIES IN THE KLL AUGER SPECTRUM.   Z DEPENDE 4-0886
            APPLICATION OF TRIPLE GRID LEED SYSTEM TO AUGER SPECTRUM ANALYSES.   3-0661
          BAND STRUCTURE OF SILICON BY CHARACTERISTIC AUGER SPECTRUM ANALYSIS.   4-0740
ECTRONS EMITTED BY IONIC BOMBARDMENT.   ENERGY SPECTRUM AND ANGULAR DISTRIBUTION OF SECONDARY EL 4-0988
   FROM STRONGLY EXCITED SILICON. (AUGER RECOMBINATION PRO/ SPECTRUM AND DECAY OF THE RECOMBINATION RADIATION 2-0460
                                   AUGER ELECTRON SPECTRUM AND IONIZATION POTENTIALS OF THE HYDROGE 4-1058
N FLUORIDE MOLECULE.
/ AND LIMITED CI CALCULATIONS FOR ASSIGNMENT OF THE AUGER SPECTRUM AND OF THE SATELLITES IN THE SOFT X-RAY/ 4-0724
/CTION OF CARBON ON VARIOUS SUBSTRATES FROM THE KLL AUGER SPECTRUM. APPLICATION OF AUGER ELECTRON SPECTROS/ 5-1651
BDENUM CARBIDE DURING HEATING ACCORDING TO AUGER ELECTRON SPECTRUM DATA. /CENTRATION ON THE SURFACE OF MOLY 5-1759
D THE (111) SURFACE OF GALLIUM PHOSPHIDE. AUGER ELECTRON SPECTRUM DURING THE INTERACTION BETWEEN OXYGEN AN 5-1642
                             THE KLL AUGER SPECTRUM FOR ATOMIC NUMBERS BETWEEN 10 AND 36.   4-0980
OBSERVATION AND INTERPRETATION OF THE AUGER ELECTRON SPECTRUM FROM CLEAN BERYLLIUM.   4-0997
                  L2,3MM AUGER ELECTRON SPECTRUM FROM FREE ZINC ATOMS.   4-0731
           INTERPRETATION OF THE AUGER ELECTRON SPECTRUM FROM OXIDIZED BERYLLIUM.   4-0935
       CHEMICAL EFFECTS ON KLL AUGER ELECTRON SPECTRUM FROM OXYGEN.   4-0819
       CHEMICAL EFFECTS ON THE KLL AUGER ELECTRON SPECTRUM FROM SURFACE CARBON.   4-0862
URI/ EXPERIMENTAL OBSERVATION OF CHEMICAL SHIFTS IN AUGER SPECTRUM FROM SURFACE LAYERS OF SILICON DIOXIDE D 4-1044
       INTERPRETATION OF THE LOW ENERGY K-BETA SPECTRUM IN SOLIDS.   4-1101
E (111) SURFACE OF G/ SOME FEATURES OF THE AUGER ELECTRON SPECTRUM IN THE INTERACTION BETWEEN OXYGEN AND TH 5-1643
: LINE SHAPE ANALYSIS A/ ELECTRON EXCITED AUGER ELECTRON SPECTRUM OF A NICKEL (110)- C(2X2) SULFUR SURFACE 4-1065
            VALENCE BAND STRUCTURE IN THE AUGER SPECTRUM OF ALUMINUM.   4-0889
                         L2,3MM AUGER SPECTRUM OF ARGON.   4-1114
.   SPECTRUM OF AUGER K,L,M,N,O ELECTRONS OF SAMARIUM 4-0756
                                           SPECTRUM OF AUGER KLL ELECTRONS OF SAMARIUM.   4-0726
XENON.   SPECTRUM OF AUGER KLL ELECTRONS OF TELLURIUM AND 4-0755
RISTIC LOSS AND GAIN PHENOMENA OF COPPER (111).   ENERGY SPECTRUM OF BACK SCATTERED ELECTRONS AND CHARACTE 4-0897
                        AUGER ELECTRON SPECTRUM OF CARBON SUBOXIDE.   5-1529
                        KLL AUGER SPECTRUM OF CHLORINE. (IN CARBON TETRACHLORIDE)   4-0808
            INTERPRETATION OF THE AUGER SPECTRUM OF CLEAN AND OXIDIZED BERYLLIUM.   4-1087
                X-RAY PHOTOELECTRON SPECTRUM OF DIAMOND. (AUGER TRANSITIONS)   4-0851
YSTALS UNDER IMPACT OF ION AND ATOM BOMBARDMENT. / ENERGY SPECTRUM OF ELECTRONS EMITTED BY ALKALI HALIDE CR 2-0246
                        KLL AUGER SPECTRUM OF FLUORINE. (IN FLUORIDE SALTS)   4-0734
      HIGH RESOLUTION L2,3M4,5M4,5 AUGER SPECTRUM OF FREE ZINC ATOMS.   4-0733
IC AUTOIONIZING STATES IN COINCIDENCE EXPERIMENTS. (AUGER SPECTRUM OF HELIUM) /BILITY OF INVESTIGATING ATOM 2-0257
/OLE INTERACTION EFFECTS IN THE KL2,3L2,3 RADIATIVE AUGER SPECTRUM OF INSULATORS. (MAGNESIUM OXIDE, MAGNES/ 2-0540
       CHEMICAL SHIFT IN THE AUGER SPECTRUM OF IRON DURING OXYGEN ADSORPTION.   5-1792
              THE KVV AUGER SPECTRUM OF LITHIUM METAL.   4-0892
                        AUGER SPECTRUM OF L2 AND L3 SHELLS OF ARGON.   4-0983
   HIGH RESOLUTION AUGER ELECTRON SPECTRUM OF MAGNESIUM OXIDE (100).   4-0772
                        KLL AUGER SPECTRUM OF MANGANESE.   4-0938
    RELATIVISTIC CALCULATION OF THE KLL AUGER SPECTRUM. (OF MERCURY)   4-0750
    DYNAMICAL SCREENING EFFECTS IN THE AUGER SPECTRUM OF METALS.   4-1110
         SINGULARITIES IN AUGER EMISSION SPECTRUM OF METALS.   4-0998
      PLASMON SATELLITE IN THE AUGER SPECTRUM OF METALS. (ALUMINUM)   4-0871
D CA/ CONFIGURATION INTERACTION CALCULATIONS OF THE AUGER SPECTRUM OF METHANE, HYDROGEN FLUORIDE, WATER, AN 4-0879
      PLASMA SATELLITES IN THE AUGER ELECTRON SPECTRUM OF MOLYBDENUM.   4-1124
MOLECULAR OXYGEN A/ CHEMICAL EFFECTS IN THE M4,5NN AUGER SPECTRUM OF MOLYBDENUM (110) DUE TO ADSORPTION OF 4-0885
CTS OF THE CONFIGURATION INTERACTION ON THE K-SHELL AUGER SPECTRUM OF NEON.   EFFE 2-0279
              K-AUGER SPECTRUM OF NEON.   4-0914
EL (100). PART-3: CARBON MONOXIDE ON COPPER (100). (AUGER SPECTRUM OF NICKEL AND COPPER) / MONOXIDE ON NICK 5-1911
    TEMPERATURE DEPENDENCE OF THE LOW ENERGY AUGER SPECTRUM OF NICKEL(II) OXIDE (100).   4-0999
IDENTIFICATION OF HIGH ENERGY LINES IN THE KLL AUGER SPECTRUM OF NITROGEN.   4-0791
EXPERIMENTAL AND THEORETICAL STUDY OF THE AUGER SPECTRUM OF NITROUS OXIDE.   4-0814
                        AUGER ELECTRON SPECTRUM OF OSMIUM AT ENERGIES UP TO 300 EV.   4-1030
1 EV REGION OF THE SPECTRUM. (INTERBAND AUGER/ EXCITATION SPECTRUM OF OXYGEN DOMINATED COMPOUNDS IN THE 3-2 4-1048
       ORIGIN OF THE PEAK SHAPES IN THE AUGER SPECTRUM OF OXYGEN (KL2,3L2,3) ON MOLYBDENUM. 4-1004
                KLL AUGER ELECTRON SPECTRUM OF SAMARIUM.   4-0725
                            KLM SPECTRUM OF SAMARIUM AUGER ELECTRONS.   4-0757
LS AND SEMICONDU/ EXPERIMENTAL ASSEMBLY FOR OBTAINING THE SPECTRUM OF SECONDARY ELECTRON EMISSION FROM META 3-0688
UNDER BOMB/ OBSERVATION OF AUGER ELECTRONS IN THE ENERGY SPECTRUM OF SECONDARY ELECTRONS EMITTED BY COPPER 2-0551
   TRANSITION PROBABILITIES FOR THE L2,3MM AUGER SPECTRUM OF SELENIUM.   4-1033
   ABSENCE OF PLASMON GAIN SATELLITES IN THE AUGER SPECTRUM OF SILICON.   4-1037
INFLUENCE OF THE BAND STRUCTURE ON THE AUGER ELECTRON SPECTRUM OF SILICON.   4-0944
) INFLUENCE OF CHEMICAL BONDING ON THE LOW ENERGY K-ALPHA SPECTRUM OF SILICON. (RADIATIVE AUGER TRANSITIONS 2-0226
       ORIGIN OF FINE STRUCTURE IN THE AUGER SPECTRUM OF SULFUR ON A NICKEL SURFACE.   5-1271
              THE AUGER SPECTRUM OF THE FREE MAGNESIUM ATOM.   4-0789
   COSTER-KRONIG AND AUGER SPECTRUM OF THE L1 SHELL OF ARGON.   4-0977
              AUGER SPECTRUM OF THE NOBLE GASES. PART-2: ARGON.   4-0974
PEAK IN POLYCRYSTALLINE MAGNESI/ OXIDATION EFFECTS ON THE SPECTRUM OF THE SLOW SECONDARY ELECTRON EMISSION 4-1118
EFFECT OF OXIDATION ON THE HIGH RESOLUTION AUGER SPECTRUM OF TITANIUM.   5-1158
   CHEMICAL SHIFT IN THE AUGER SPECTRUM OF TUNGSTEN FROM OXYGEN ADSORPTION.   4-0922
              AUGER ELECTRON SPECTRUM OF WATER VAPOR.   4-1066
SEARCH FOR CORRELATION EFFECTS IN THE M4,5N4,5N4,5 AUGER SPECTRUM OF XENON.   4-0872
   SOLID STATE BROADENING EFFECTS IN THE AUGER SPECTRUM OF XENON.   4-1005
       CHEMICAL SHIFTS IN THE AUGER SPECTRUM OF YTTRIUM ON OXYGEN ADSORPTION.   5-1153
ACT.   HIGH RESOLUTION NEON K-AUGER ELECTRON SPECTRUM PRODUCED BY 45 MEV CHLORINE (12+) ION IMP 4-1051
IONS.   HIGH RESOLUTION ARGON K-AUGER SPECTRUM PRODUCED IN 4 MEV PROTON ON ARGON COLLIS 2-0496
NEON COLLISIONS.   HIGH RESOLUTION NEON AUGER SPECTRUM PRODUCED IN 4.2 MEV ATOMIC HYDROGEN(+)- 4-1078
SPECTROSCOPY.   SPECTRUM SUBTRACTION TECHNIQUES IN AUGER ELECTRON 1-0068
              ARGON KLL AUGER SPECTRUM. TEST OF THEORY.   2-0414
       FOCUSING OF CHARGED PARTICLES BY A SPHERICAL CONDENSER.   3-0684
ERS.   COMPARISON OF SPHERICAL DEFLECTOR AND CYLINDRICAL MIRROR ANALYZ 3-0623

N- AUGER ELECTRON SPECTROSCOPY.   PRODUCTION OF A SPHERICAL GRID FOR LOW ENERGY ELECTRON DIFFRACTIO 3-0668
ICATION TO AUGER ELECT/ RESOLUTION AND SENSITIVITY OF THE SPHERICAL GRID RETARDING POTENTIAL ANALYZER. APPL 3-0634
USE IN AUGER ELECTRON SPECTROSCOPY)   IMPROVED SPHERICAL GRID RETARDING POTENTIAL ANALYZER. (FOR 3-0583
PECTROSCOPY DETERMINATION OF THE OXYGEN- SILICON RATIO IN SPIN ON GLASS FILMS.   AUGER ELECTRON S 5-1827
                                                           SPIN SPLITTING IN COPPER AUGER ELECTRON SPECTRUM. 4-1053
                                          SPIN SPLITTING IN COPPER AUGER ELECTRON SPECTRUM.   4-1053
D MONOLAYERS. (INTERFACE AUGER ELECTRON SPECTROSCOPY) SPONTANEOUS ALLOYING OF A GOLD SUBSTRATE WITH LEA 5-1178
          SOFT X-RAY AMPLIFIED SPONTANEOUS EMISSION VIA THE AUGER EFFECT.   2-0443
                               ELECTRON SPRAYING ATOMS. (AUGER CASCADES)   2-0464
RICAL/ FOCUSING A CHARGED PARTICLE BEAM OF FINITE ANGULAR SPREAD IN AN ELECTROSTATIC ANALYZER OF THE CYLIND 3-0720
S.   SPREADING FUNCTIONS IN SECONDARY ELECTRON ANALYSI 2-0512
(TITANIUM, TUNGSTEN, SILICON) MONITORED BY AUGER SPECTRO/ SPUTTER CLEANING AND ETCHING OF CRYSTAL SURFACES. 5-1828
NUM FROM VARIOUS / AUGER ELECTRON SPECTROSCOPY STUDIES OF SPUTTER DEPOSITION AND SPUTTER REMOVAL OF MOLYBDE 5-1878
PECTROSCOP/ LOW TEMPERATURE EPITAXY OF GERMANIUM FILMS BY SPUTTER DEPOSITION. (HEED, LEED, AUGER ELECTRON S 5-1537
F METAL SURFACES.   SPUTTER ETCHING AND AUGER ELECTRON SPECTROSCOPY O 5-1552
F METAL SURFACES.   SPUTTER ETCHING AND AUGER ELECTRON SPECTROSCOPY O 5-1554
ALLIUM ARSENIDE.   AES AND SPUTTER ETCHING OF NICKEL- GOLD- GERMANIUM ON N-G 5-1775
          DIFFUSION MEASUREMENT IN THIN FILMS BY AUGER SPUTTER ETCHING TECHNIQUE.   5-1439
          APPLICATIONS OF DEPTH PROFILING BY AUGER- SPUTTER TECHNIQUES.   1-0087
NI/ SUPERCONDUCTING TRANSITION TEMPERATURES OF REACTIVELY SPUTTERED FILMS OF TANTALUM NITRIDE AND TUNGSTEN 5-1541
SCOPY)   POROSITY IN SPUTTERED PLATINUM FILMS. (AUGER ELECTRON SPECTRO 5-1965
GER SPECTROSCOPY)   REACTION OF SPUTTERED PLATINUM FILMS ON GALLIUM ARSENIDE. (AU 5-1563
/ITROGEN, CARBON AND OXYGEN ON THE PHYSICAL PROPERTIES OF SPUTTERED TANTALUM FILMS. (AUGER ELECTRON SPECTR/ 5-1473
/YSIS OF LIGHT ELEMENTS (NITROGEN, CARBON, AND OXYGEN) IN SPUTTERED TANTALUM FILMS BY AUGER ELECTRON SPECT/ 5-1652
R EFFECT)   MORPHOLOGICAL AND ELECTRICAL PROPERTIES OF SPUTTERED YTTRIA-DOPED ZIRCONIA THIN FILMS. (AUGE 5-1385
COMPOSITION OF COPPER- NICKEL ALLOYS AFTER ANNEALING AND SPUTTERING. /ER SPECTROSCOPY STUDY ON THE SURFACE 5-1675
F TITANIUM(IV) OXIDE.   AES ION SPUTTERING ANALYSIS AND THE SURFACE COMPOSITION O 5-1893
RECENT ADVANCES IN COMPOSITION PROFILING BY SIMULTANEOUS SPUTTERING AND AUGER ANALYSIS.   1-0155
PECTROSCOPIC OBSERVATIONS OF THE SUBSTRATE SURFACE DURING SPUTTERING AND VACUUM VAPOR DEPOSITION. /ECTRON S 5-1678
/ALYSIS OF SURFACE LAYERS BY LIGHT ION BACKSCATTERING AND SPUTTERING COMBINED WITH AUGER ELECTRON SPECTROS 5-1167
          PERTURBING EFFECTS OF SPUTTERING DURING THE AUGER STUDY OF INSULATORS. 1-0111
USE OF AUGER ELECTRON SPECTROSCOPY AND INERT GAS SPUTTERING FOR OBTAINING CHEMICAL PROFILES. 1-0156
EVALUATION OF CONCENTRATION DEPTH PROFILES BY SPUTTERING IN SIMS AND AES.   5-1442
UGER SPECTROSCOPY AND ESCA)   ASPECTS OF SPUTTERING IN SURFACE ANALYSIS METHODS. (REVIEW A 1-0215
AUGER SPECTROSCOPY)   SPUTTERING IN THIN FILM ANALYSIS METHODS. (ESCA, 1-0214
GY ARGON(+) ION SPUTTERIN/ AUGER ELECTRON SPECTROSCOPY IN SPUTTERING MEASUREMENTS. APPLICATION TO LOW ENER 5-1825
AR REACTOR MATERIALS USING AUGER ELECTRON SPECTROSCOPY. SPUTTERING MEASUREMENTS ON CONTROLLED THERMONUCLE 5-1826
GER ELECTRON SPECTROSCOPY   SPUTTERING MEASUREMENTS ON CTR MATERIALS USING AU 1-0182
ES. (AUGER EFFECT)   ION SELECTIVE SPUTTERING OF III-V COMPOUND SEMICONDUCTOR SURFAC 5-1637
EN BOMBARDMENT. (AES)   SPUTTERING OF IONS FROM STAINLESS STEEL BY HYDROG 6-2017
AND ITS APPLICATION TO HIGH YIELD SECONDARY/ DIFFERENTIAL SPUTTERING OF MAGNESIUM OXIDE- GOLD CERMET FILMS 5-1430
D- COPPER, AND SILVER- COPPER STUDIED BY AUGER ELECTRON / SPUTTERING OF THE ALLOY SYSTEMS SILVER- GOLD, GOL 5-1331
APPLICATION OF AES TO THE STUDY OF SELECTIVE SPUTTERING OF THIN FILMS.   5-1933
ECTROSCOPY.   ALLOY SPUTTERING STUDIES WITH IN-SITU AUGER ELECTRON SP 5-1877
DECONVOLUTION METHOD FOR COMPOSITION PROFILING BY AUGER SPUTTERING TECHNIQUE.   3-0630
IN BIMETALLIC POLYCRYSTALLINE FILMS MEASURED BY AUGER ION SPUTTERING TECHNIQUE.   INTERDIFFUSION 5-1440
QUANTITATIVE DEPTH PROFILING BY AUGER ION SPUTTERING TECHNIQUE.   1-0085
IS BY AUGER SPECTROSCOPY.   RELATIVE SPUTTERING YIELDS AND QUANTITATIVE SURFACE ANALYS 5-1964
                                         STABILITY OF CESIUM COVERED ZINC OXIDE SURFACES. 5-1882
REAKDOWN. (FROM SURFACES) INVESTIGATIONS WITH FIELD EMIS/ STABILITY OF FIELD ELECTRON EMISSION AND VACUUM B 5-1520
          ROLE OF IMPURITIES IN THE STABILITY OF ZINC OXIDE.   5-1462
ACE.   LEED- AUGER INVESTIGATION OF A STABLE CARBIDE OVERLAYER ON A PLATINUM (111) SURF 5-1569
          AUGER ELECTRON SPECTROSCOPY OF A STABLE GERMANIUM OXIDE.   4-1108
EFFECTS ON ENERGY SHIFTS AND SIGNAL INTENSITY IN A SINGLE STAGE CYLINDRICAL MIRROR ANALYZER. /MEN POSITION 3-0696
ATION. (LEED- AUGER SYS/ ULTRAHIGH VACUUM LOW TEMPERATURE STAGE FOR IN-SITU FILM DEPOSITION AND CHARACTERIZ 3-0657
ENT ALUMINU/ AUGER ELECTRON SPECTROSCOPY STUDY OF INITIAL STAGES OF OXIDATION IN A COPPER- 19.6 ATOMIC PERC 6-2028
RACTION- AUGER ELECTRON SPECTROSCOPY STUDY OF THE INITIAL STAGES OF OXIDATION OF IRON (001). /ELECTRON DIFF 6-2110
UGER ELECTRON EMISSION MICROGRAPHY AND MICROANALYSIS OF A STAINED SURFACE OF PURE IRON PLATE.   A 5-1415
RMINATION OF THE COMPOSITION OF PASSIVE FILMS ON TYPE 316 STAINLESS STEEL. /ECTRON SPECTROSCOPY TO THE DETE 6-2080
SPECTROSCOPY OF FRACTURE SURFACES IN IRRADIATED TYPE 304 STAINLESS STEEL.   AUGER ELECTRON 6-2001
ER FRACTURE SURFACE ANALYSIS OF A TEMPER EMBRITTLED 3340- STAINLESS STEEL.   AUG 6-2085
AUGER SPECTROSCOPY OF FRACTURE SURFACES OF IRRADIATED STAINLESS STEEL.   6-2002
URFACES DURING IRRADIATION. (AUGER SPECTROSCOPY ANALYSIS, STAINLESS STEEL) / OF ALLOYING ELEMENTS TO FREE S 6-2091
SURFACES: URANIUM DIOXIDE, URANIUM, GRAPHITE, 300 SERIES STAINLESS STEEL, AND NIOBIUM. /LECTRON STUDIES OF 5-1217
470 K.   VACUUM ANNEALING OF STAINLESS STEEL AT TEMPERATURES BETWEEN 770 AND 1 6-2145
AND ITS RELATION TO INTERGRANULAR CORROSION OF AUSTENITIC STAINLESS STEEL. (AUGER SPECTROSCOPY) /OUNDARIES 6-2064
STUDY OF THE INTERMITTENT MOTION OF FRICTION IN STAINLESS STEEL BY AN AUGER MICROPROBE.   6-2133
SURFACE CONCENTRATION OF MOLYBDENUM IN TYPES 316 AND 304 STAINLESS STEEL BY AUGER ELECTRON SPECTROSCOPY. 6-1995
          SPUTTERING OF IONS FROM STAINLESS STEEL BY HYDROGEN BOMBARDMENT. (AES) 6-2017
UID SODIUM. (SCANNING ELECTRON MICROSC/ MASS TRANSPORT OF STAINLESS STEEL CORROSION PRODUCTS IN FLOWING LIQ 6-2014
/R ELECTRON SPECTROSCOPIC STUDIES OF THE SURFACES OF 18-8 STAINLESS STEEL ELECTROPOLISHED OR ABRADED BY EM/ 6-2130
LYTICALLY.   AES STUDIES ON SURFACES OF 18-8 STAINLESS STEEL FINISHED MECHANICALLY AND ELECTRO 6-2129
AUGER ELECTRON SPECTROSCOPY STUDY OF 18-8 STAINLESS STEEL OXIDATION.   6-1993
CTERISTICS OF CARBON MONOXIDE, SULFUR, AND OTHER GASES ON STAINLESS STEEL SHEET. (AES OF GOLD COATING) /ARA 5-1551
CHEMICAL PROFILE ANALYSIS OF TYPE 316 STAINLESS STEEL SHEET BY AUGER SPECTROSCOPY.   6-1996
CTROMETRY AND AUGER ELECT/ SURFACE CLEANLINESS OF 316 L-N STAINLESS STEEL STUDIES BY SECONDARY ION MASS SPE 3-0654
CTROMETRY AND AUGER ELECT/ SURFACE CLEANLINESS OF 316 L-N STAINLESS STEEL STUDIED BY SECONDARY ION MASS SPE 6-2087
OBTAINED WITH AUGER ELECTRON SPECTROSCOPY OF AIR OXIDIZED STAINLESS STEEL SURFACES. /VERSUS DEPTH PROFILES 6-1998
/ HYDROGEN DESORBED FROM IRON, PLATINUM, COPPER, NIOBIUM, STAINLESS STEEL SURFACES. (AUGER ELECTRON SPECTR/ 6-2006
AUGER SPECTROMETRY STUDY OF PASSIVE FILMS FORMED ON STAINLESS STEELS.   6-2102
AUGER ANALYSIS OF BRASS- ENAMEL AND STAINLESS STEEL- ENAMEL INTERFACES.   5-1266
LECTRON EMISSION AT LOW PRIMARY BEAM EN/ REFINEMENTS TO A STANDARD LEED- AUGER SYSTEM FOR THE ANALYSIS OF E 3-0698
O NICKEL (111), NICKEL (100), AND N/ EFFECTIVENESS OF THE STANDARD SURFACE CLEANING TECHNIQUES AS APPLIED T 5-1467
DECAY PARAMETERS. (DETECTION OF / COINCIDENCE METHODS OF STANDARDIZATION FOR CESIUM-131 AND MEASUREMENT OF 2-0474
QUANTITATIVE USE OF AUGER SPECTROSCOPY: STANDARDIZATION OF THE METHOD.   1-0163
ECTROSCOPY.   VALENCE BAND OF MAGNESIUM STANNIDE DETERMINED BY AUGER AND PHOTOEMISSION SP 5-1885
/UM(2) SILICIDE, MAGNESIUM(2) GERMANIUM, AND MAGNESIUM(2) STANNIDE. PART-1: X-RAY EXCITATION PART-2: ULTRA/ 2-0306
THEORY OF KLL AUGER ENERGIES INCLUDING STATIC RELAXATION.   2-0506
R PROCESS IN ARSENIC AND SELENIUM. ATOMIC AND EXTRAATOMIC STATIC RELAXATION ENERGIES IN THE L3M4,5M4,5 AUGE 4-0883
F RELAXATION ENERGY DIFFERENCES FROM AUGER AND CORE BIND/ STATIC RELAXATION OF GERMANE AND THE ESTIMATION O 2-0471

215

OWTH IN HYDROGEN AND IN WATER VAPOR. (AUGER SPECTROMETRY, STEEL) /OF FRACTURE SURFACES PRODUCED BY CRACK GR 6-2142
OF THE COMPOSITION OF PASSIVE FILMS ON TYPE 316 STAINLESS STEEL. /ECTRON SPECTROSCOPY TO THE DETERMINATION 6-2080
OSCOPY AND DEPTH PROFILE STUDY OF OXIDATION MODIFIED 440C STEEL. AUGER ELECTRON SPECTR 6-2029
OPY AND DEPTH PROFILE STUDY OF OXIDATION OF MODIFIED 440C STEEL. AUGER ELECTRON SPECTROSC 6-2032
OPY OF FRACTURE SURFACES IN IRRADIATED TYPE 304 STAINLESS STEEL. AUGER ELECTRON SPECTROSC 6-2001
E SURFACE ANALYSIS OF A TEMPER EMBRITTLED 3340- STAINLESS STEEL. AUGER FRACTUR 6-2085
SPECTROSCOPY OF FRACTURE SURFACES OF IRRADIATED STAINLESS STEEL. AUGER 6-2002
MENTAL BREEDER REACTOR II SERVICE ON CHROMIUM- MOLYBDENUM STEEL. EFFECTS OF TEN YEARS EXPERI 6-2108
IGATION ON A (100) SURFACE OF AUSTENITIC CHROMIUM- NICKEL STEEL. LEED- AUGER INVEST 6-1991
RING IRRADIATION. (AUGER SPECTROSCOPY ANALYSIS, STAINLESS STEEL) / OF ALLOYING ELEMENTS TO FREE SURFACES DU 6-2091
EMBRITTLEMENT OF MANGANESE- MOLYBDENUM- NICKEL LOW ALLOY STEEL. (AES) / OF COPPER AND PHOSPHORUS ON TEMPER 6-2043
CE ANALYSIS AND THEIR APPLICATION FOR STUDIES ON IRON AND STEEL. (AES, ESCA, SIMS)   NEW METHODS IN SURFA 6-2035
URANIUM DIOXIDE, URANIUM, GRAPHITE, 300 SERIES STEEL, AND NIOBIUM. /LECTRON STUDIES OF SURFACES: 5-1217
VACUUM ANNEALING OF STAINLESS STEEL AT TEMPERATURES BETWEEN 770 AND 1470 K. 6-2145
ION ON TEMPER BRITTLENESS IN A NICKEL- CHROMIUM- ANTIMONY STEEL. (AUGER ANALYSIS) / GRAIN BOUNDARY COMPOSIT 6-2114
AND ANTIMONY TO THE GRAIN BOUNDARIES OF TEMPER EMBRITTLED STEEL. (AUGER ELECTRON SPECTROSCOPY) / MANGANESE 6-2070
RAIN BOUNDARY SEGREGANTS IN THERMALLY EMBRITTLED MARAGING STEEL. (AUGER ELECTRON SPECTROSCOPY)   STUDY OF G 6-2055
LATION TO INTERGRANULAR CORROSION OF AUSTENITIC STAINLESS STEEL. (AUGER SPECTROSCOPY) /OUNDARIES AND ITS RE 6-2064
EGATION OF PHOSPHORUS IN A CHROMIUM- MOLYBDENUM- VANADIUM STEEL. (AUGER SPECTROSCOPY) / GRAIN BOUNDARY SEGR 6-2139
IRRADIATION INDUCED SOLUTE SEGREGATION (IN STEEL, AUGER SPECTROSCOPY). 6-2092
CHEMICAL COMPOSITION OF THE SURFACE OF MOLYBDENUM AND STEEL BASED ON AUGER ELECTRON SPECTRAL DATA. 5-1985
STUDY OF THE INTERMITTENT MOTION OF FRICTION IN STAINLESS STEEL BY AN AUGER MICROPROBE. 6-2133
FRACTURE SURFACE ANALYSIS OF TEMPER EMBRITTLED STEEL BY AUGER ELECTRON SPECTROSCOPY. 6-2084
IDENTIFICATION OF ATOMIC BORON IN STEEL BY AUGER ELECTRON SPECTROSCOPY. 6-2018
ONCENTRATION OF MOLYBDENUM IN TYPES 316 AND 304 STAINLESS STEEL BY AUGER ELECTRON SPECTROSCOPY.   SURFACE C 6-1995
PECTROSCOPI/ ANALYSIS PROFILES OF OXIDE FILMS ON CHROMIUM STEEL BY AUGER EMISSION AND X-RAY PHOTOELECTRON S 6-2016
SPUTTERING OF IONS FROM STAINLESS STEEL BY HYDROGEN BOMBARDMENT. (AES) 6-2017
R EMISSION AN/ TEMPER EMBRITTLEMENT IN A NICKEL- CHROMIUM STEEL CONTAINING PHOSPHORUS AS AN IMPURITY. (AUGE 6-2138
. (SCANNING ELECTRON MICROSC/ MASS TRANSPORT OF STAINLESS STEEL CORROSION PRODUCTS IN FLOWING LIQUID SODIUM 6-2014
E GASES. (USING/ OBSERVATIONS ON THE SURFACE OXIDATION OF STEEL DURING ANNEALING IN HYDROGEN-FREE PROTECTIV 6-2040
/ SPECTROSCOPIC STUDIES OF THE SURFACES OF 18-8 STAINLESS STEEL ELECTROPOLISHED OR ABRADED BY EMERY PAPERS. 6-2130
AES STUDIES ON SURFACES OF 18-8 STAINLESS STEEL FINISHED MECHANICALLY AND ELECTROLYTICALLY. 6-2129
SPECTROSCOPY)   SURFACE AND GRAIN BOUNDARY SEGREGATION OF STEEL IMPURITIES. (TEMPER EMBRITTLEMENT BY AUGER 6-2109
SCOPY.   LEAD MONOLAYER LUBRICATION IN STEEL MACHINING STUDIED BY AUGER ELECTRON SPECTRO 6-2124
AUGER ELECTRON SPECTROSCOPY STUDY OF 18-8 STAINLESS STEEL OXIDATION. 6-1993
OF CARBON MONOXIDE, SULFUR, AND OTHER GASES ON STAINLESS STEEL SHEET. (AES OF GOLD COATING) /ARACTERISTICS 5-1551
CHEMICAL PROFILE ANALYSIS OF TYPE 316 STAINLESS STEEL SHEET BY AUGER SPECTROSCOPY. 6-1996
INTERGRANULAR FRACTURE AND IMPURITY SEGREGATION IN STEEL.   STUDIES BY AUGER ELECTRON SPECTROSCOPY. 6-2127
AND AUGER ELECT/ SURFACE CLEANLINESS OF 316 L-N STAINLESS STEEL STUDIES BY SECONDARY ION MASS SPECTROMETRY 3-0654
DEPTH PROF/ REACTION OF SULFUR DIOXIDE WITH MODIFIED 440C STEEL STUDIED BY AUGER ELECTRON SPECTROSCOPY AND 6-2031
AND AUGER ELECT/ SURFACE CLEANLINESS OF 316 L-N STAINLESS STEEL STUDIED BY SECONDARY ION MASS SPECTROMETRY 6-2087
SCATTERING SPECTROMETRY AND AUGER ELECTRON SPECTROSCOPY. (STEEL SURFACES) /EMISTRIES OBTAINED BY USING ION 6-1999
ITH AUGER ELECTRON SPECTROSCOPY OF AIR OXIDIZED STAINLESS STEEL SURFACES. /VERSUS DEPTH PROFILES OBTAINED W 6-1998
X-RAY PHOTOELECTRON AND AUGER SPECTROSCOPIC ANALYSES OF STEEL SURFACES. 5-1268
DESORBED FROM IRON, PLATINUM, COPPER, NIOBIUM, STAINLESS STEEL SURFACES. (AUGER ELECTRON SPECTROSCOPY) /EN 6-2006
/O REDUCED STRENGTH OF METALS AND ALLOYS. (SEGREGATION IN STEEL, TUNGSTEN, COPPER- BISMUTH ALLOYS STUDIED / 6-2120
AES AND TEMPER EMBRITTLEMENT OF STEELS. 6-2081
AUGER ELECTRON SPECTROSCOPY TO THE STUDY OF INTERFACE IN STEELS.   APPLICATION OF 6-2126
IS OF THE FRACTURE SURFACES OF IRRADIATED PRESSURE VESSEL STEELS.   AUGER ELECTRON SPECTROSCOPY ANALYS 6-2113
R SPECTROMETRY STUDY OF PASSIVE FILMS FORMED ON STAINLESS STEELS.   AUGE 6-2102
S AND ALLOYS. (AUGER EMISSION ANALYSIS, TUNGSTEN, COPPER, STEELS) /AIN BOUNDARY EMBRITTLEMENT IN SOME METAL 6-2058
MISSION SPECTROSCOPY ON TEMPER EMBRITTLEMENT IN LOW ALLOY STEELS.   STUDIES USING AUGER ELECTRON E 6-2119
TEMPER EMBRITTLEMENT OF LOW ALLOY STEELS. (AES OF FRACTURE SURFACES) 6-2062
O SOME TECHNICAL PROBLEMS. (EMBRITTLEMENT, SEGREGATION OF STEELS AND ALLOYS. /OSCOPY AND ITS APPLICATIONS T 6-2100
ANGANESE AND SILICON IN TEMPER EMBRITTLEMENT OF LOW ALLOY STEELS. (AUGER SPECTROSCOPY)   ROLE OF M 6-2059
NDARIES AND ITS ROLE IN TEMPER EMBRITTLEMENT OF LOW ALLOY STEELS. (AUGER SPECTROSCOPY) /SELECTIVE GRAIN BOU 6-2057
TEMPER EMBRITTLEMENT OF TIN DOPED STEELS BY AUGER ELECTRON SPECTROSCOPIC METHOD. 6-2128
AUGER ANALYSIS OF BRASS- ENAMEL AND STAINLESS STEEL- ENAMEL INTERFACES. 5-1266
/ OF SOLUTE ELEMENTS ON TEMPER EMBRITTLEMENT OF LOW ALLOY STEELS. (IMPURITY GRAIN BOUNDARY SEGREGATION, AE/ 6-2083
GRAIN BOUNDARY EMBRITTLEMENT OF STEELS STUDIED BY AUGER ELECTRON SPECTROSCOPY. 6-2103
.   DYNAMIC ELUCIDATION OF REACTION STEPS ON A SOLID SURFACE BY ELECTRON SPECTROSCOPY 5-1535
BINDING OF CARBON ATOMS AT A STEPPED NICKEL SURFACE. (AUGER SPECTROSCOPY) 5-1485
AUGER ELEC/ REACTIVITY OF LOW INDEX ((111) AND (100)) AND STEPPED PLATINUM SINGLE CRYSTAL SURFACES. (LEED, 5-1838
SMALL MOLECULE REACTIONS ON STEPPED SINGLE CRYSTAL PLATINUM SURFACES. 5-1173
LOW ENERGY ELECTRON DIFFRACTIONS AND AUGE/ DEVELOPMENT OF STEPPED SURFACE REGIONS ON POLYCRYSTALLINE GOLD. 5-1483
N A PLATINUM (111) SURFACE. (AUGER ELECTRON SPECTROSCOPY, STICKING COEFFICIENT) /ICS OF OXYGEN ADSORPTION O 5-1189
/ON MONOXIDE LEED PATTERNS. (AUGER ELECTRON SPECTROSCOPY, STICKING COEFFICIENT) PART-2: EFFECT OF SULFUR O/ 5-1190
OGEN ON CLEAN NICKEL (111), NICKEL (100), SHEET AND EVAP/ STICKING PROBABILITIES FOR THE ADSORPTION OF HYDR 5-1468
TOR S/ AUGER ELECTRON SPECTROSCOPY STUDY OF ELECTRON BEAM STIMULATED DESORPTION OF OXYGEN FROM A SEMICONDUC 5-1617
100) S/ AUGER ELECTRON SPECTROSCOPY STUDY OF THE ELECTRON STIMULATED DESORPTION OF OXYGEN FROM A TUNGSTEN (5-1954
F SURFACE DIFFUSION ON MAGNESIUM OXIDE (001) BY/ ELECTRON STIMULATED DESORPTION OF SODIUM AND MEASUREMENT O 5-1494
(100) SURFACES OF ALKALI HALIDES. PART-2: ELECTRON STIMULATED DISSOCIATION. 5-1908
N, ELECTRON SPECTROSCOPY FOR CHEMICAL ANALYSIS, AND FIELD STIMULATED ELECTRON EMISSION. /PECTROSCOPY, PHOTO 5-1901
TON GAS IN A SEMICONDUCTOR. (AUGER PROCESS) STIMULATED EMISSION OF LIGHT FROM A NONIDEAL EXCI 2-0462
E LAYERS. AUGER ELECTRON SPECTROSCOPY INVESTIG/ THERMALLY STIMULATED EXOELECTRON EMISSION OF BERYLLIUM OXID 5-1325
BY QUANTITATIVE AUGER ELECTRON SPECTROSCOPY.   ELECTRON STIMULATED OXIDATION OF GALLIUM ARSENIDE, STUDIED 5-1756
(AUGER PROCESS) STIMULATED PHOTOLUMINESCENCE IN GALLIUM SELENIDE. 2-0416
ECTROSCOPY)   ELECTRON STIMULATED PROCESSES AT SOLID SURFACES. (AUGER SP 5-1605
LECTRON EMISSION STUDY OF ADSORPTION PROCESSES. OPTICALLY STIMULATED SECONDARY ELECTRON EMISSION AND PHOTOE 2-0302
SURFACE STOICHIOMETRY AND STRUCTURE OF GALLIUM ARSENIDE. 5-1143
IDE, INDUCED BY BOMBARDMENT WITH 2 KV EL/ DEPARTURES FROM STOICHIOMETRY IN (100) SURFACES OF GALLIUM PHOSPH 5-1162
EFFECT OF TEMPERATURE ON THE SURFACE STRUCTURE AND STOICHIOMETRY OF (100) INDIUM PHOSPHIDE SURFACES. 5-1164
ROGEN FROM THE AUGER DECAY.   DETERMINATION OF OSCILLATOR STRENGTH FOR THE FIRST DISCRETE TRANSITION IN NIT 2-0575
OPY)   EFFECT OF ADSORBED CHLORINE AND OXYGEN ON SHEAR STRENGTH OF IRON AND COPPER JUNCTIONS. (SPECTROSC 6-2143
EEL, T/ RELATION OF GRAIN BOUNDARY SEGREGATION TO REDUCED STRENGTH OF METALS AND ALLOYS. (SEGREGATION IN 6-2120
A 4-GRID RETARDING / CORRECTIONS OF AUGER ELECTRON SIGNAL STRENGTHS FOR MODULATION AMPLITUDE DISTORTION IN 3-0620
/OLUTION SOFTENING (RHENIUM DUCTILIZING EFFECT AND BUBBLE STRENGTHENING IN TUNGSTEN- RHENIUM ALLOYS, AUGER/ 6-2111
IUM AND CHROMIUM ALLOYS, AES)   LOW INTRINSIC STRESS AND GOOD ADHESION ALLOY THIN FILMS. (TITAN 5-1909
NISHING AMMONIACAL ENVIRONMENT: FRACTOGRAPHY AND CHEMICA/ STRESS CORROSION CRACKING OF ALPHA BRASS IN A TAR 6-2095
ESIUM TERNARY AND VARIOUS QUATERNARY ALLOYS. (AES) STRESS CORROSION CRACKING OF AN ALUMINUM- 5% MAGN 6-2036

EFFECT OF IMPURITIES ON INTRINSIC STRESS IN THIN NICKEL FILMS. (AUGER EFFECT) 5-1131
EFFECT OF IMPURITIES ON INTRINSIC STRESS IN THIN NICKEL FILMS. (AUGER SPECTROSCOPY) 5-1130
RECORDING CONVERSION AND/OR AUGER ELECTRON SPECTRA. NEW, STRIATED, HIGH EFFICIENCY PHOTOGRAPHIC PLATES FOR 3-0699
/CAY-IN-FLIGHT SPECTRA FROM AUTOIONIZING STATES OF HIGHLY STRIPPED OXYGEN, FLUORINE, CHLORINE, AND ARGON I/ 2-0472
ONS) SEQUENTIAL STRIPPING IN HIGHLY IONIZED NEON. (AUGER TRANSITI 2-0441
UORESCENCE YIELD. (AUGER TRANSITION) STRIPPING MECHANISM DEPENDENCE OF THE ARGON 2S FL 4-0957
LINE WIDTH AND AUGER EFFECT IN CALCIUM, STRONTIUM, AND BARIUM FLUORIDES. 4-1116
TION OF THE METAL-RICH SURFACE MODEL FOR THE OXIDATION OF STRONTIUM, BY AUGER ELECTRON SPECTROSCOPY. /IFICA 6-2045
ACTION INVESTIGATION OF OXYGEN AND HYDROGEN ADSORPTION ON STRONTIUM SURFACES. /ND REFLECTION ELECTRON DIFFR 5-1498
/ ELECTRON SPECTROSCOPY. A NEW TECHNIQUE FOR MATERIAL AND STRUCTURAL ANALYSIS AND A REVIEW OF ITS POSSIBLE 1-0084
AUGER ELECTRONIC SPECTROSCOPY AND LOW ENERG/ CHEMICAL AND STRUCTURAL ANALYSIS OF IRON SURFACES BY MEANS OF 5-1940
RON AFFINITY ON THE SILICON- CESIUM- OXYGEN SURFACE. (AU/ STRUCTURAL AND ELECTRONIC MODEL OF NEGATIVE ELECT 2-0426
/PTION OF WATER VAPOR ON CLEAN CLEAVED GERMANIUM. PART-1: STRUCTURAL AND KINETIC PROPERTIES. (AUGER ELECTR/ 5-1432
TROSCOPY) CORROSION AND STRUCTURAL CHANGES IN GLASS. (AUGER ELECTRON SPEC 5-1425
EALING OF (100) SURFACES OF GALLIUM AR/ COMPOSITIONAL AND STRUCTURAL CHANGES THAT ACCOMPANY THE THERMAL ANN 5-1163
AR OXYGEN ON SILICON. PART-1: ADSORPTION STUDIES. PART-2: STRUCTURAL EFFECTS. / CARBON MONOXIDE AND MOLECUL 5-1543
. PART-1: CARBON MONOXIDE ON (100) PALLADIUM. STRUCTURAL INFLUENCES ON ADSORBATE BINDING ENERGY 5-1913
2: CARBON MONOXIDE ON NICKEL (100). PART-3: CARBON MONOX/ STRUCTURAL INFLUENCES ON ADSORPTION ENERGY. PART- 5-1911
ORS. WORK FUNCTION AND STRUCTURAL STUDIES OF ALKALI- COVERED SEMICONDUCT 5-1958
D OXYGEN ON SILICON (100). STRUCTURAL STUDIES OF THE ADSORPTION OF CESIUM AN 5-1389
RGY ELECTRON DIFFRACTION AND AUGER ELECTRON SPECTROSCOPY/ STRUCTURAL STUDY OF OXYGEN- IRON (001) BY LOW ENE 5-1469
OXYGEN PHOTOSURFACE. AN ELECTRICAL, CHEMICAL, AND STRUCTURAL STUDY OF THE GERMANIUM (100)- CESIUM- 5-1319
TENT ANALYSIS OF CORE BAND-TO-BAND EXCITATION: AUGER FINE STRUCTURE. SELF-CONSIS 2-0509
ER ELECTRON SPECTROSCOPY. AN ORDERED BORON STRUCTURE AFTER DEPOSITION ON SILICON (111). (AUG 5-1168
/W ENERGY ELECTRON DIFFRACTION. PART-1: ELECTRODE SURFACE STRUCTURE AFTER EXPOSURE TO WATER AND AQUEOUS EL/ 5-1486
TECHNIQUE. (COMPARISON BETWEEN / SURFACE COMPOSITION AND STRUCTURE ANALYSIS WITH A ONE APPARATUS ION PROBE 5-1166
ROSCOPY AND AUGER ELECTRON SPECTROSCOPY. STRUCTURE ANALYSIS WITH X-RAY PHOTOELECTRON SPECT 1-0050
PRE- AND BACKSPUTTERING, STUDIED WITH AUGER SPECTROSCOP/ STRUCTURE AND CHEMISTRY OF SILICON SURFACES AFTER 5-1233
/COPY. AUGER ELECTRON SPECTROSCOPY, AND LEED STUDY OF THE STRUCTURE AND COMPOSITION OF ION BOMBARDED TUNGS/ 5-1951
VIEW OF RECENT LEED, AES, AND XPS RESULTS. STRUCTURE AND COMPOSITION OF METAL SURFACES: A RE 5-1341
D III-V SEMICONDUCTORS. SURFACE STRUCTURE AND COMPOSITION OF VARIOUS ELEMENTAL AN 5-1161
DE ON SILICON. (AUGER SPECTROSCOPY) STRUCTURE AND GROWTH KINETICS OF NICKEL(2) SILICI 5-1917
/IUM (100) CRYSTAL FACE. A STUDY OF ITS CLEANING, SURFACE STRUCTURE, AND INTERACTION WITH OXYGEN AND CARBO/ 5-1883
ICRORADIOGRAPHY USING AUGER ELECTRONS IN THE STUDY OF THE STRUCTURE AND PITTING CORROSION OF ALUMINUM. /W M 6-2066
DE LAYERS APPLIED ON A MOLYBDENUM SINGLE CRYSTAL. STRUCTURE AND SECONDARY EMISSION OF MAGNESIUM OXI 5-1191
ANALYSIS OF FRACTURE SURFACES. GRAIN BOUNDARY STRUCTURE AND SEGREGATION. AUGER SPECTROGRAPH 6-2118
HIDE SURFACES. EFFECT OF TEMPERATURE ON THE SURFACE STRUCTURE AND STOICHIOMETRY OF (100) INDIUM PHOSP 5-1164
X-RAY PHOTOEMISSION AND SURFACE STRUCTURE. (AUGER DISPLACEMENTS) 5-1811
SURFACE DEBYE TEMPERATURE OF THE SILICON (001) - 2X2 STRUCTURE. (AUGER EFFECT) 5-1477
OXIDE CRYSTALLINE FILMS. PART-1: PREPARATION AND PHYSICAL STRUCTURE. (AUGER SPECTROSCOPY) /D SULFIDE- LEAD 5-1790
RETRIEVAL OF ELECTRON EXCITED AUGER STRUCTURE BY DYNAMIC BACKGROUND SUBTRACTION. 1-0090
CONDUCTORS. STRUCTURE DEPENDENT PROPERTIES OF THIN FILM SUPER 5-1300
MANIUM. FINE STRUCTURE DETAILS IN THE LMM AUGER SPECTRA OF GER 4-0937
ON CHARACTERISTICS (GERMANIUM AND BERYLLIUM). FILM STRUCTURE EFFECT ON THE SECONDARY ELECTRON EMISSI 4-0923
ON EMISSION. SURFACE STRUCTURE FROM ANGULAR DEPENDENCE OF AUGER ELECTR 5-1977
L-SHELL IONIZATION THRESHOLD STRUCTURE FROM AUGER ELECTRON SPECTRA. 2-0360
SOLID SURFACES. NEW FINE STRUCTURE IN ELECTRON EXCITED AUGER SPECTRA FROM 2-0473
OF VANADIUM PENTOXIDE (010) AND VA/ INELASTIC EFFECTS AND STRUCTURE IN THE AUGER ELECTRON EMISSION SPECTRA 5-1340
ICON CARBIDE. HIGH ENERGY FINE STRUCTURE IN THE AUGER SPECTRA OF SILICON AND SIL 4-1036
D INDIUM. QUASIATOMIC FINE STRUCTURE IN THE AUGER SPECTRA OF SOLID SILVER AN 4-0771
VALENCE BAND STRUCTURE IN THE AUGER SPECTRUM OF ALUMINUM. 4-0889
CKEL SURFACE. ORIGIN OF FINE STRUCTURE IN THE AUGER SPECTRUM OF SULFUR ON A NI 5-1271
PER. AUGER VACANCY SATELLITE STRUCTURE IN THE L3,M4,5M4,5 AUGER SPECTRA OF COP 4-1034
D AND THE APPLICATION TO SILIC/ METHOD FOR DETECTING FINE STRUCTURE IN THE SECONDARY ELECTRON EMISSION YIEL 4-0852
/OLUTION MEASUREMENTS OF AUGER ELECTRON AND PHOTOELECTRON STRUCTURE IN THE SECONDARY ELECTRON ENERGY DISTR/ 4-1023
E. (AUGER SURFACE SPECTRA) SILICON (111)-7 STRUCTURE OBTAINED BY CLEAVAGE AT ROOM TEMPERATUR 5-1767
DENUM (001). ATOMIC ARRANGEMENT IN THE 1 K 1 STRUCTURE OF A SILICON ORDERED MONOLAYER ON MOLYB 5-1478
CON SUBSTRATE STUDIED BY AUGER ELECTRON AND X/ ELECTRONIC STRUCTURE OF A THIN GOLD FILM DEPOSITED ON A SILI 5-1435
LEED AND AES STUDIES OF THE STRUCTURE OF ADSORBED GASES ON SOLID SURFACES. 5-1835
/GRAPHY BY BETA-RAY AUGER ELECTRONS TO THE STUDIES OF THE STRUCTURE OF ALUMINUM ALLOYS AND CORROSION OF TH/ 6-2051
STAL SURFACES. SURFACE STRUCTURE OF AND CATALYSIS BY PLATINUM SINGLE CRY 5-1837
ELECTRONIC AND LATTICE STRUCTURE OF CESIUM FILMS ADSORBED ON TUNGSTEN. 5-1604
SURFACE STOICHIOMETRY AND STRUCTURE OF GALLIUM ARSENIDE. 5-1143
ON EMISSION. BAND STRUCTURE OF GRAPHITE STUDIED BY SECONDARY ELECTR 2-0569
STRUCTURE OF LOW ENERGY ELECTRON SPECTRUM. 4-0948
TION. CHANGE IN THE SURFACE ELECTRONIC STRUCTURE OF PLATINUM (100) INDUCED BY RECONSTRUC 5-1186
ECTRON SPECTROSCOPY) GROWTH AND STRUCTURE OF SEMICONDUCTING THIN FILMS. (AUGER EL 5-1509
TRUM ANALYSIS. BAND STRUCTURE OF SILICON BY CHARACTERISTIC AUGER SPEC 4-0740
USE OF AUGER ELECTRON SPECTROSCOPY TO DETERMINE THE STRUCTURE OF SILICON OXIDE FILMS. 5-1501
THIN REACTION LAYERS AND SURFACE STRUCTURE OF SILICON (111). 5-1879
N TECHNIQUES. (AUGER EFFECT) ELECTRONIC STRUCTURE OF SOLID SURFACES: CORE LEVEL EXCITATIO 2-0468
OBSERVATION OF A PLASMON GAIN IN THE FINE STRUCTURE OF THE ALUMINUM AUGER SPECTRUM. 4-1085
ARY ELECTRON EMISSION COEFFICIENT OF/ CHANGES IN THE FINE STRUCTURE OF THE ANGULAR DEPENDENCE OF THE SECOND 4-1061
SEC/ USE OF X-RAY SPECTROSCOPIC DATA TO EXPLAIN THE FINE STRUCTURE OF THE DEPENDENCE OF THE COEFFICIENT OF 4-0891
STRUCTURE OF THE PLATINUM (100) SURFACE. 5-1374
OF NICKEL. GROWTH AND STRUCTURE OF THIN COPPER FILMS ON (001) SURFACES 5-1227
OMINE AND CHLORINE. (AUGER SPECTROSCOPY) SURFACE STRUCTURE OF TITANIUM AND ITS INTERACTION WITH BR 5-1538
AUGER SPECTRA) GROWTH AND STRUCTURE OF TITANIUM OXIDE THIN FILMS. PART-1. (5-1980
ALS STU/ EFFECT OF ION BOMBARDMENT ON THE COMPOSITION AND STRUCTURE OF TUNGSTEN AND MOLYBDENUM SINGLE CRYST 5-1215
ON. INFLUENCE OF THE BAND STRUCTURE ON THE AUGER ELECTRON SPECTRUM OF SILIC 4-0944
PLATINUM. EFFECT OF STRUCTURE ON THE AUGER SPECTRA OF BARIUM FILMS ON 5-1359
UGER PEAK FROM SOLID ALUMINUM: INTERNAL PHOTOEMISSION. STRUCTURE ON THE HIGH ENERGY SIDE OF THE KL2,3M A 4-1021
A CARBON STRUCTURE ON THE RHENIUM (0001) SURFACE. 5-1989
GERMANIUM (111) SURFACES. STRUCTURE TRANSFORMATIONS ON CLEAVED AND ANNEALED 5-1713
NEW MICROAUTORADIOGRAPHY OF METAL STRUCTURE USING AUGER ELECTRONS. 6-2052
N (001) BY LEED- AES. ANALYSIS OF THE SURFACE STRUCTURE WITH ANTIPHASE DOMAINS OF SULFUR ON IRO 5-1470
STRUCTURES. ATOMIC ARRA 5-1716
NGEMENT OF GOLD (100) AND RELATED METAL OVERLAYER STRUCTURES. AUGER ELECTRON SPECTROSCOPY STU 5-1774
DIES OF GALLIUM ARSENIDE AND SILICON METAL- SEMICONDUCTOR STRUCTURES. /FUSED IMPURITY PROFILES IN SILICON D 5-1481
IOXIDE- SILICON AND POLYSILICON- SILICON DIOXIDE- SILICON STRUCTURES. (AES) /TANTALUM, SILICON DIOXIDE- TAN 5-1260
TALUM, AND SILICON- PLATINUM SILICIDE- TANTALUM THIN FILM STRUCTURES. (AES) /TANTALUM, SILICON DIOXIDE- TAN 5-1260

LECTRON EMISSION FROM SILICON (111). (AUGER ELECTRO/ FINE STRUCTURES AND ENERGY DISTRIBUTION OF SECONDARY E 4-0853
STUDY OF THE CHEMICAL COMPOSITION OF MOS AND MNOS STRUCTURES BY AUGER ELECTRON SPECTROSCOPY. 5-1503
URFACE. DETERMINATION OF SURFACE STRUCTURES BY LEED (AND AES): THE SILICON (111) S 5-1342
Y. DETERMINATION OF SURFACE STRUCTURES BY LEED AND AUGER ELECTRON SPECTROSCOP 5-1957
UGER EFFECT) PLATINUM (100) STRUCTURES: IMPURITY CAUSATION AND PROPERTIES. (A 5-1311
FILM) SYSTEMS STUDIED BY AUGER ELECTRO/ DIFFUSE INTERFACE STRUCTURES IN SOME (SUBSTRATE- VACUUM EVAPORATED 5-1436
OWTH OF METAL ON THEM OBSERVED BY LOW ENERGY ELECTRON DI/ STRUCTURES OF SUBSTRATE CRYSTALS AND EPITAXIAL GR 5-1906
SURFACE STRUCTURES ON IRON (110) ULTRAHIGH VACUUM. 5-1346
ION MICROSCOPY, LEED/ A CLOSE LOOK AT SURFACES. (SURFACE STRUCTURES, SPECIES, TOPOGRAPHY, PHENOMENA, FIELD 5-1492
/ IN-PROCESS CONTROL TECHNIQUES FOR COMPLEX SEMICONDUCTOR STRUCTURES. TASK-II. APPLICATION OF SECONDARY ELE 5-1411
LYSIS OF GRAIN BOUNDARY DIFFUSION IN BIMETALLIC THIN FILM STRUCTURES USING AUGER ELECTRON SPECTROSCOPY. ANA 6-2146
CHARACTERIZATION OF SOME VANADIUM (100) SURFACE STRUCTURES USING LEED AND AES. 5-1864
TERED ELECTRONS. DETERMINATION OF SURFACE STRUCTURES USING LEED AND ENERGY ANALYSIS OF SCAT 5-1956
NEW TECHNIQUE FOR AUGER ANALYSIS OF SURFACE SPECIES SUBJECT TO ELECTRON INDUCED DESORPTION. 3-0675
ROPERTIES OF FILMS. SILICON ON SAPPHIRE EPITAXY BY VACUUM SUBLIMATION: LEED- AUGER STUDIES AND ELECTRONIC P 5-1230
AUGER ELECTRON SPECTRUM OF CARBON SUBOXIDE. 5-1529
PENDENCE OF ATOMIC PHOTOEFFECT ON BOUND ELECTRON MAGNETIC SUBSTATE (AUGER EFFECT). DE 2-0524
NDENCE OF ATOMIC PHOTOEFFECT ON A BOUND ELECTRON MAGNETIC SUBSTATE. (AUGER ELECTRON DISTRIBUTION) DEPE 2-0463
MENTS FOR EPITAXIAL GROWTH OF THORIUM ON A TUNGSTEN (100) SUBSTRATE. /SCOPY, LEED AND WORK FUNCTION MEASURE 5-1739
ELECTRON SPECTROSCOPY) DIFFUSE INTERFACE IN SILICON (SUBSTRATE)- GOLD (EVAPORATED FILM) SYSTEM. (AUGER 5-1681
EMPERATURE DIFFUSION OF GOLD INTO SILICON IN THE SILICON (SUBSTRATE)- GOLD (FILM) SYSTEM. LOW T 5-1673
OF EPITAXIAL SILICON FILMS. INFLUENCE OF SUBSTRATE CONDITIONS ON THE NUCLEATION AND GROWTH 5-1514
ON THEM OBSERVED BY LOW ENERGY ELECTRON DI/ STRUCTURES OF SUBSTRATE CRYSTALS AND EPITAXIAL GROWTH OF METAL 5-1906
AHIGH VACUUM. A VERSATILE SUBSTRATE DESIGN FOR LEED AND AES STUDIES IN ULTR 3-0631
/NIC STRUCTURE OF A THIN GOLD FILM DEPOSITED ON A SILICON SUBSTRATE STUDIED BY AUGER ELECTRON AND X-RAY PH/ 5-1435
FILMS. (AUGER ELECTRON SPECTROSCOP/ RELATIONSHIPS BETWEEN SUBSTRATE SURFACE CHEMISTRY AND ADHESION OF THIN 5-1851
/EED AND AUGER ELECTRON SPECTROSCOPIC OBSERVATIONS OF THE SUBSTRATE SURFACE DURING SPUTTERING AND VACUUM V/ 5-1678
CHANISM OF SURFACE ALIGNMENT IN NEMATIC LIQUID CRYSTALS. (SUBSTRATE SURFACE EFFECTS, AUGER SPECTROMETRY) ME 5-1281
ELECTRON SPECTROSCOPY) SPONTANEOUS ALLOYING OF A GOLD SUBSTRATE WITH LEAD MONOLAYERS. (INTERFACE AUGER 5-1178
TRON SPECTROSCOPY TO SURFACE CHEMISTRY STUDIES OF CERAMIC SUBSTRATES. APPLICATION OF AUGER ELEC 5-1275
HYDROGEN CHLORIDE VAPOR ETCHED N+ GALLIUM ARSENIDE (100) SUBSTRATES. AUGER SPECTRA OF 4-0992
ON OF THIN FILMS. (AUGER ELECTRON SPECTROSCOPY OF ALUMINA SUBSTRATES) /BSTRATE SURFACE CHEMISTRY AND ADHESI 5-1851
SURFACE COMPOSITION OF MICA SUBSTRATES. 5-1742
HETEROGENEOUS NUCLEATION AND GROWTH OF GOLD ON SUBSTRATES. (AUGER SPECTRA) 5-1137
QUID PHASE EPITAXY OF III-V COMPOUNDS ON INDIUM PHOSPHIDE SUBSTRATES. (AUGER SPECTROSCOPY) LI 5-1590
/ OF THE TRANSITION DENSITY FUNCTION OF CARBON ON VARIOUS SUBSTRATES FROM THE KLL AUGER SPECTRUM. APPLICAT/ 5-1651
N. (AUGER SP/ GROWTH OF THIN METALLIC FILMS ON INSULATING SUBSTRATES IN THE PRESENCE OF ELECTRON IRRADIATIO 5-1317
ITAXIAL STUDIES USING VACUUM DEPOSITION ON HEATED SILICON SUBSTRATES. (SCANNING AUGER SPECTROSCOPY) /ASE EP 5-1287
EPITAXY OF ULTRATHIN METAL FILMS ON BODY CENTERED CUBIC SUBSTRATES USING LEED- AUGER TECHNIQUES. 5-1488
D BY AUGER ELECTRO/ DIFFUSE INTERFACE STRUCTURES IN SOME (SUBSTRATE- VACUUM EVAPORATED FILM) SYSTEMS STUDIE 5-1436
OF ELECTRON EXCITED AUGER STRUCTURE BY DYNAMIC BACKGROUND SUBTRACTION. RETRIEVAL 1-0090
ROSCOPY./ USE OF ANALOG INTEGRATION IN DYNAMIC BACKGROUND SUBTRACTION FOR QUANTITATIVE AUGER ELECTRON SPECT 1-0069
COPY. SPECTRUM SUBTRACTION TECHNIQUES IN AUGER ELECTRON SPECTROS 1-0068
ND SOFT X-RAY APPEARAN/ APPLICATION OF DYNAMIC BACKGROUND SUBTRACTION TO QUANTIFICATION OF AUGER ELECTRON A 1-0067
LECTRONS CREATED AS THE RESULT OF ELECTRON SHAKE-OFF AND (SUGAR) VACANCY CASCADES. NATURE OF SECONDARY E 2-0307
LMM AUGER SPECTRA OF SULFUR IN SODIUM SULFATE AND SODIUM PYROSULFITE. 4-0829
ROSCOPICAL STUDY OF FRICTIONAL PROPERTIES OF MOLYBDENUM DISULFIDE. (AUGER) MIC 5-1872
LEED AND AES STUDY OF THE INTERACTION OF HYDROGEN SULFIDE AND MOLYBDENUM (100). 5-1973
EN- COPPER LUMINESCENCE IN COPPER AND ALUMINUM DOPED ZINC SULFIDE. (AUGER EFFECT, COLOR CENTERS) /CE OF GRE 2-0402
ECTS OF LOW ENERGY ARGON- ION BOMBARDMENT ON MOLYBDENUM DISULFIDE. (AUGER MEASUREMENTS) EFF 2-0348
OBSERVATION OF EXTRINSIC SURFACE STATES ON (1120) CADMIUM SULFIDE BY AUGER SPECTROSCOPY. 5-1202
CONCENTRATION-DEPENDENT NONRADIATIVE PROCESSES IN ZINC SULFIDE PHOSPHORS. (AUGER RECOMBINATIONS) 2-0388
RECOMBINATION IN DONOR ACCEPTOR COMPLEXES IN CADMIUM SULFIDE SINGLE CRYSTALS. (AUGER) 2-0369
EXCITON SPECTRA OF HEAVILY DOPED CADMIUM SULFIDE SINGLE CRYSTALS. (AUGER PROCESS) 2-0372
SULFUR KLL AUGER SPECTRA IN HYDROGEN SULFIDE, SULFUR DIOXIDE, AND SULFUR HEXAFLUORIDE. 4-0910
OVOLTAGE AND AUGER SPECTROSCOPY STUDIES OF (1120) CADMIUM SULFIDE SURFACE. SURFACE PHOT 5-1201
ON. (USING LEED A/ EFFECT OF SULFUR, OXYGEN, AND HYDROGEN SULFIDE SURFACE FILMS ON THE ADHESION OF CLEAN IR 5-1211
ELECTRONIC PROPERTIES OF CLEAVED MOLYBDENUM DISULFIDE SURFACES. 5-1634
CHEMISORPTION REACTIONS ON HIGH INDEX ZINC SULFIDE SURFACES. (AUGER EFFECT) 4-0788
TWO DIMENSIONAL CHEMISORBED LAYERS. (AUGER SPECTROSCOPY, SULFIDE SYSTEMS) PHASE EQUILIBRIA IN 5-1170
COPPER, COPPER(II) OXIDE, COPPER(I) OXIDE, AND COPPER(I) SULFIDE THIN FILMS. /LECTRON AND AUGER SPECTRA OF 4-0932
INTERACTION OF HYDROGEN SULFIDE WITH COPPER (001). 5-1516
/DUCTIVITY OF HIGHLY EXCITED CADMIUM SELENIDE AND CADMIUM SULFIDE(0.48) SELENIDE(0.52) CRYSTALS. (AUGER RE/ 2-0432
EPARATION AND PHYSICAL STRUCTURE./ PROPERTIES OF THE LEAD SULFIDE- LEAD OXIDE CRYSTALLINE FILMS. PART-1: PR 5-1790
INATION BY AUGER SPECTROSCOPY BETWEEN ADSORBED SULFUR AND SULFIDES OF COPPER AND NICKEL. DISCRIM 5-1734
Y STUDY OF NICKEL (110) SURFACE EFFECTS DUE TO CARBON AND SULFUR. LEED AND AUGER ELECTRON SPECTROSCOP 5-1814
ERING WITH AUGER ELECTRON SPECTROSCOPY IN THE ANALYSIS OF SULFUR ABSORBED ON NICKEL. / ENERGY ION BACKSCATT 5-1869
NG. PART-1: AUGER ELECTRONS. (KLL TRANSITION ENERGIES FOR SULFUR, ALUMINUM, SILICON) /ON AND CHEMICAL BONDI 2-0327
/OREATE INTERACTIONS ON A PLATINUM (110) SURFACE. PART-1: SULFUR AND CARBON MONOXIDE LEED PATTERNS. (AUGER/ 5-1190
SULFUR AND CARBON ON (110) SURFACE OF NICKEL. 5-1813
/ARY EMBRITTLEMENT IN IRON- PHOSPHORUS, IRON- PHOSPHORUS- SULFUR AND IRON- ANTIMONY- SULFUR ALLOYS. (AUGER/ 6-2098
/ COATING ON SORPTION CHARACTERISTICS OF CARBON MONOXIDE, SULFUR, AND OTHER GASES ON STAINLESS STEEL SHEET/ 5-1551
R ELECT/ STUDIES OF THE MECHANISM OF THE REACTION BETWEEN SULFUR AND OXYGEN ON A MOLYBDENUM SURFACE BY AUGE 5-1530
/NAMIC INVESTIGATION OF THE MECHANISM OF REACTION BETWEEN SULFUR AND OXYGEN ON A MOLYBDENUM SURFACE BY MEA/ 5-1534
N SURFACE. DIFFUSION OF SULFUR AND PHOSPHORUS IMPURITY ATOMS TO A TUNGSTE 5-1226
DISCRIMINATION BY AUGER SPECTROSCOPY BETWEEN ADSORBED SULFUR AND SULFIDES OF COPPER AND NICKEL. 5-1734
TION OF BENZENE ON THE (100) FACE OF NICKEL. INFLUENCE OF SULFUR AT THE SURFACE. STUDY OF THE ADSORP 5-1284
K-SHELL FLUORESCENCE YIELDS OF ARGON, CHLORINE AND SULFUR. (AUGER TRANSITION) 4-1014
E SURFACES. (AES) EFFECTS OF SULFUR, BARIUM, AND CALCIUM ON IMPREGNATED CATHOD 5-1391
ER EFFECT) DETERMINATION OF SULFUR CONTAMINATION ON MOLYBDENUM SURFACES. (AUG 5-1647
ER IN ULTRAHIGH VACUUM. (AES) INTERACTION OF SULFUR DIOXIDE AND CARBON DIOXIDE WITH CLEAN SILV 5-1575
Y AUGER ELECTRON SPECTROSCOPY AND DEPTH PROF/ REACTION OF SULFUR DIOXIDE WITH MODIFIED 440C STEEL STUDIED B 6-2031
INTERACTION OF SULFUR DIOXIDE WITH TUNGSTEN. (AUGER ANALYSIS) 5-1367
BOUNDARY SEGREGATION) STUDY OF SULFUR EMBRITTLED OXYGEN FREE COPPER. (AES, GRAIN 6-2015
CTRON SPECTROSCOPY) CONFIRMATION OF SULFUR EMBRITTLEMENT IN NICKEL ALLOYS. (AUGER ELE 6-2053
LMM AUGER SPECTRA OF SULFUR IN SODIUM SULFATE AND SODIUM PYROSULFITE. 4-0829
AUGER LINE SHAPE COMPARISON OF NITROGEN AND SULFUR IN TWO DIFFERENT CHEMICAL ENVIRONMENTS. 5-1816
FFECT OF SLIDING ON THESE I/ AUGER ANALYSIS OF OXYGEN AND SULFUR INTERACTIONS WITH VARIOUS METALS AND THE E 5-1206
FUR DIOXIDE, AND SULFUR HEXAFLUORIDE. SULFUR KLL AUGER SPECTRA IN HYDROGEN SULFIDE, SUL 4-0910
AUGER SPEC/ PHASE TRANSITIONS IN TWO DIMENSIONAL ABSORBED SULFUR LAYERS ON COPPER INVESTIGATED BY LEED AND 5-1302

LECTRON SPECTROSCOPY.   SEGREGATION OF SULFUR ON A MOLYBDENUM SURFACE STUDIED BY AUGER E 5-1564
    ORIGIN OF FINE STRUCTURE IN THE AUGER SPECTRUM OF SULFUR ON A NICKEL SURFACE.   5-1271
D AND AES)   KINETICS OF THE REACTION BETWEEN OXYGEN AND SULFUR ON A NICKEL (111) SURFACE. (STUDIED BY LEE 5-1451
    AND AUGER SPECTROSCOPY.   ADSORPTION OF SULFUR ON GOLD BY LOW ENERGY ELECTRON DIFFRACTION 5-1558
ALYSIS OF THE SURFACE STRUCTURE WITH ANTIPHASE DOMAINS OF SULFUR ON IRON (001) BY LEED- AES.   AN 5-1470
HE MEASUREMENT OF AUGER ELECTRON CURRENTS. APPLICATION TO SULFUR ON NICKEL (110).   ANALOG TECHNIQUE FOR T 3-0635
N DIFFRACTION AND AUGER SPECTROSCOPY OF THE ADSORPTION OF SULFUR ON PLATINUM.   LOW ENERGY ELECTRO 5-1175
RGY ELECTRON DIFFRACTION AND AUGER INVESTIGATIONS.   SULFUR ON SINGLE CRYSTAL PLATINUM PLANES. LOW ENE 5-1420
LOW ENERGY ELECTRON DIFFRACTION AND AUGER / ADSORPTION OF SULFUR ON THE (100) AND (111) FACES OF PLATINUM. 5-1421
                                     ADSORPTION OF SULFUR ON THE (100) FACE OF MOLYBDENUM.   5-1732
S ON THE ADHESION OF CLEAN IRON. (USING LEED A/ EFFECT OF SULFUR, OXYGEN, AND HYDROGEN SULFIDE SURFACE FILM 5-1211
F PALLADIUM (/ LEED- AES STUDIES OF CHEMISORPTION INDUCED SULFUR SEGREGATION FROM THE BULK TO THE SURFACE O 5-1865
/XCITED AUGER ELECTRON SPECTRUM OF A NICKEL (110)- C(2X2) SULFUR SURFACE: LINE SHAPE ANALYSIS AND CORRELA/ 4-1065
                                     DIFFUSION OF SULFUR TOWARD THE (110) FACE OF NICKEL.   5-1771
CRYSTAL.   AUGER ELECTRON SPECTROSCOPY OF A SULFUR- OXYGEN SURFACE REACTION ON A COPPER (110) 5-1183
LLURGY ON THE PENETRATION DEPTH AND RF BREAKDOWN FIELD OF SUPERCONDUCTING NIOBIUM. (AUGER EFFECT) /ACE META 6-2137
/ HIGH TEMPERATURE HEAT TREATMENT ON PENETRATION DEPTH OF SUPERCONDUCTING NIOBIUM. (AUGER SPECTROSCOPIC ME/ 5-1938
VELY SPUTTERED FILMS OF TANTALUM NITRIDE AND TUNGSTEN NI/ SUPERCONDUCTING TRANSITION TEMPERATURES OF REACTI 5-1541
        STRUCTURE DEPENDENT PROPERTIES OF THIN FILM SUPERCONDUCTORS.   5-1300
EED- AES STUDIES OF / SOME PROPERTIES OF TRANSITION METAL SUPERCONDUCTORS AT SURFACES AND IN THIN FILMS. (L 5-1848
        NIOBIUM SURFACES FOR RF SUPERCONDUCTORS. (AUGER SURFACE ANALYSIS)   5-1472
N).   SECONDARY ELECTRON EMISSION AT SUPERHIGH FREQUENCIES. (COPPER- GERMANIUM JUNCTIO 2-0356
COPY.   GALLIUM(1-X) ALUMINUM(X) ARSENIDE SUPERLATTICES PROFILED BY AUGER ELECTRON SPECTROS 5-1596
CTRON IMAGES IN AUGER ELECTRON SPECTROSCOPY.   ELECTRONIC SUPERPOSITION OF SAMPLE CURRENT AND SECONDARY ELE 2-0450
SURFACE BY HIGH ENERGY ELECTRON DIFFRACT/ OBSERVATION OF SUPERSTRUCTURES ON CARBON COVERED GERMANIUM (100) 5-1805
RVED BY LEED AES.   GOLD-INDUCED SUPERSTRUCTURES ON SILICON (100) SURFACES AS OBSE 5-1708
OF ALUMINUM BY ELECTRON IMPACT.   AUGER TRANSITION AND SUPPLEMENTARY ELECTRON EJECTION IN THE IONIZATION 2-0230
A CARBON STRUCTURE ON THE RHENIUM (0001) SURFACE.   5-1989
    ADSORPTION OF HYDROGEN ON A PLATINUM (111) SURFACE.   5-1255
    ADSORPTION OF IONIC SALTS ON A TUNGSTEN (100) SURFACE.   5-1659
RESOLVED AUGER EMISSION SPECTRA FROM A CLEAN COPPER (100) SURFACE.   ANGULAR 5-1697
STUDIES OF OXYGEN AND CARBON MONOXIDE ON A PLATINUM (100) SURFACE.   (AUGER) ADSORPTION 5-1546
    (AUGER) ADSORPTION STUDIES ON THE PLATINUM (100) SURFACE.   5-1570
ER ELECTRON EMISSION FROM GOLD DEPOSITED ON SILICON (111) SURFACE.   AUG 5-1676
URING THE DIFFUSION OF FOREIGN ATOMS IN SOLIDS NEAR THEIR SURFACE. /LECTRON SPECTROSCOPY AS A TOOL FOR MEAS 1-0186
TRON SPECTROSCOPY OF NICKEL DEPOSITS ON THE SILICON (111) SURFACE. /LECTRON SPECTROSCOPY STUDY OF ELECTRON 5-1238
BEAM STIMULATED DESORPTION OF OXYGEN FROM A SEMICONDUCTOR SURFACE. /LECTRON SPECTROSCOPY STUDY OF THE ELECT 5-1617
RON STIMULATED DESORPTION OF OXYGEN FROM A TUNGSTEN (100) SURFACE.   5-1954
AUGER NEUTRALIZATION OF MULTIPLY CHARGED IONS ON A METAL SURFACE.   2-0247
D MOLECULAR OXYGEN AND CARBON DIOXIDE ON A TUNGSTEN (111) SURFACE. /OLECULAR OXYGEN AND CARBON MONOXIDE, AN 5-1465
OF OSMIUM, AND THE SEGREGATION OF PLATINUM TO THE OSMIUM SURFACE.   AUGER SPECTROSCOPY 5-1770
N OF THE DEPOSITION OF ALUMINUM ONTO THE MOLYBDENUM (110) SURFACE.   AUGER- LEED INVESTIGATIO 5-1489
BON MONOXIDE OXIDATION ON A PLATINUM (110) SINGLE CRYSTAL SURFACE.   CAR 5-1187
N BONDING. (REMOVAL OF OXIDE FILMS, AUGER SPECTROSCOPY OF SURFACE) CHEMICAL CLEANING FOR THERMOCOMPRESSIO 5-1454
    CHEMISORPTION OF OXYGEN ON THE SILVER (110) SURFACE.   5-1778
        CHLORINE REACTIONS ON THE SILICON (111) SURFACE.   5-1344
        CLEANING PROCEDURES OF MOLYBDENUM (110) SURFACE.   5-1649
COMBINATION OF AUGER SPECTROSCOPY WITH IMAGING OF STUDIED SURFACE.   3-0607
ED AUGER STUDY OF OXYGEN ADSORPTION ON THE TUNGSTEN (112) SURFACE.   COMBINED LEED- RHE 5-1464
ION OF SULFUR AND PHOSPHORUS IMPURITY ATOMS TO A TUNGSTEN SURFACE.   DIFFUS 5-1226
    ELECTRON BEAM ASSISTED ADSORPTION ON SILICON (111) SURFACE.   5-1269
COPIC STUDIES OF CLEAN AND OXIDIZED IRON. (AUGER STUDY OF SURFACE)   ELECTRON SPECTROS 5-1323
FURTHER STUDIES OF GAS ADSORPTION ON MOLYBDENUM (100) SURFACE.   5-1305
CTRON SPECTROSCOPY OF NITRIC OXIDE ADSORBED ON A TUNGSTEN SURFACE.   HIGH RESOLUTION AUGER ELE 5-1565
AUGER TRANSITIONS OF CHEMISORBED AMMONIA ON A MOLYBDENUM SURFACE. /LUTION MEASUREMENTS OF THE NITROGEN KLL 5-1532
    INTERACTION OF NITRIC OXIDE WITH A NICKEL (111) SURFACE.   5-1278
        LEED AND AES STUDIES OF LITHIUM HYDRIDE (100) SURFACE.   5-1446
ND AUGER ELECTRON SPECTROSCOPIC STUDIES ON TUNGSTEN (110) SURFACE.   LEED A 5-1677
    LEED AND AUGER INVESTIGATIONS OF COPPER (111) SURFACE.   5-1499
UGER) STUDY OF CHLORINE CHEMISORPTION ON THE SILVER (111) SURFACE.   LEED (A 5-1779
            LEED STUDY OF IRIDIUM (100) SURFACE.   5-1370
            LEED STUDY OF PLATINUM (100) SURFACE.   5-1377
MECHANISM OF DEPOSITED IRON FILM ON MAGNESIUM OXIDE (100) SURFACE.   LEED- AES OBSERVATION OF GROWTH 5-1525
    LEED- AUGER ANALYSIS OF THE BERYLLIUM (0001) SURFACE.   5-1581
IGATION OF A STABLE CARBIDE OVERLAYER ON A PLATINUM (111) SURFACE.   LEED- AUGER INVEST 5-1569
AUGER ELECTRON OBSERVATIONS OF THE SILICON CARBIDE (0001) SURFACE. LOW ENERGY ELECTRON DIFFRACTION AND 5-1932
SION STUDIES OF THE CESIUM COVERED GALLIUM ARSENIDE (110) SURFACE. /ER ELECTRON SPECTROSCOPY, AND PHOTOEMIS 5-1931
ON- AUGER ELECTRON SPECTROSCOPY STUDY OF CLEAN (001) IRON SURFACE.   LOW ENERGY ELECTRON DIFFRACTI 5-1527
INE STRUCTURE IN THE AUGER SPECTRUM OF SULFUR ON A NICKEL SURFACE.   ORIGIN OF F 5-1271
        PHASE TRANSFORMATIONS OF SILICON (111) SURFACE.   5-1343
IVE ESCAPE DEPTH OF AUGER AND PHOTOELECTRONS FROM A SOLID SURFACE.   RELAT 5-1903
SCATTERING OF LOW ENERGY ELECTRONS FROM A COPPER (111) SURFACE.   5-1621
    SOME PROPERTIES OF THE RHENIUM (0001) SURFACE.   5-1304
        STRUCTURE OF THE PLATINUM (100) SURFACE.   5-1374
E ON THE (100) FACE OF NICKEL. INFLUENCE OF SULFUR AT THE SURFACE.   STUDY OF THE ADSORPTION OF BENZEN 5-1284
THE CHEMICAL, PHYSICAL, AND CRYSTALLOGRAPHIC NATURE OF A SURFACE. /ICAL TECHNIQUES FOR CHARACTERIZATION OF 5-1433
INTERACTION OF CHLORINE WITH THE ZINC OXIDE (0001)- ZINC SURFACE: A LEED, AUGER AND WORK FUNCTION STUDY. 5-1463
ROSC/ EQUILIBRIUM SEGREGATION OF CARBON TO A NICKEL (111) SURFACE: A SURFACE PHASE TRANSITION. (AUGER SPECT 5-1803
BY AUGER EMISSION SPECTROSCOPY.   SURFACE ABSORPTION BY POLYPEPTIDE FILMS EXAMINED 5-1179
BSTRATE SURFACE EFFECTS, AUGER SPECTROMETRY) MECHANISM OF SURFACE ALIGNMENT IN NEMATIC LIQUID CRYSTALS. (SU 5-1281
(110) FACES OF GOLD. (AUGER EFFECT)   ADSORPTION AND SURFACE ALLOYING OF LEAD MONOLAYERS ON (111) AND 5-1733
SPECTROSCOPY AND AUGER ELECTRON EMISSION SPECTROSCOPY.   SURFACE ANALYSES BY COMBINED X-RAY PHOTOELECTRON 5-1272
    AUGER ELECTRON SPECTROSCOPY FOR SURFACE ANALYSIS.   1-0072
    AUGER ELECTRON SPECTROSCOPY FOR SURFACE ANALYSIS.   1-0200
    CHOOSING BETWEEN ESCA AND AUGER FOR SURFACE ANALYSIS.   5-1635
COMPARISON OF THE TECHNIQUES FOR SILICON SURFACE ANALYSIS.   5-1327
    ELECTRON SPECTROMETER FOR SURFACE ANALYSIS.   3-0715
        ELECTRONS OR IONS. (IN SURFACE ANALYSIS)   5-1765
NIOBIUM SURFACES FOR RF SUPERCONDUCTORS. (AUGER SURFACE ANALYSIS)   5-1472
        PHYSICAL METHODS OF SURFACE ANALYSIS.   1-0078

SOME TOPICS ON AES. (SURFACE ANALYSIS)   1-0106
USE OF SCANNING AUGER SPECTROSCOPY FOR SURFACE ANALYSIS.   5-1280
COMPARISON OF THE TECHNIQUES FOR SILICON SURFACE ANALYSIS. (AES, ESCA, SIMS, AND OTHERS)   1-0052
S ON IRON AND STEEL. (AES, ESCA, SIMS)   NEW METHODS IN SURFACE ANALYSIS AND THEIR APPLICATION FOR STUDIE 6-2035
F LEAD(0.8) TIN(0.2) TELLURIDE DETECTOR FABRICATION USING SURFACE ANALYSIS. (AUGER)   EVALUATION O 5-1593
COPY.   SURFACE ANALYSIS BY APPEARANCE POTENTIAL SPECTROS 5-1545
SURFACE ANALYSIS BY AUGER ELECTRON SPECTROSCOPY.   5-1402
GLASS SURFACE ANALYSIS BY AUGER ELECTRON SPECTROSCOPY.   5-1720
SURFACE ANALYSIS BY AUGER SPECTROSCOPY.   1-0103
RELATIVE SPUTTERING YIELDS AND QUANTITATIVE SURFACE ANALYSIS BY AUGER SPECTROSCOPY.   5-1964
APPARATUS FOR SURFACE ANALYSIS BY ELECTRON SPECTROSCOPY.   3-0664
SURFACE ANALYSIS BY LOW ENERGY ION REFLECTION.   5-1822
AUGER ELECTRON SPECTROMETER AS TOOL FOR SURFACE ANALYSIS. (CONTAMINATION MONITOR)   3-0706
DE)   COMPUTER-CONTROLLED MOLECULAR BEAM EPITAXY. (AUGER SURFACE ANALYSIS, GALLIUM(X) ALUMINUM(1-X) ARSENI 5-1324
) AND SECONDARY ION MASS SPECTROSCOPY. COMPARISON OF TWO SURFACE ANALYSIS METHODS. /RON SPECTROSCOPY (ESCA 1-0088
TOMIC LAYERS OF SOLIDS.   SURFACE ANALYSIS, METHODS OF STUDYING THE OUTER A 1-0016
OPY AND ESCA)   ASPECTS OF SPUTTERING IN SURFACE ANALYSIS METHODS. (REVIEW AUGER SPECTROSC 1-0215
INLESS STEEL.   AUGER FRACTURE SURFACE ANALYSIS OF A TEMPER EMBRITTLED 3340- STA 6-2085
AUGER ELECTRON EMISSI/ SECONDARY ELECTRON SPECTROSCOPY. (SURFACE ANALYSIS OF MATERIALS BY ELECTRON EXCITED 5-1403
IALS.   METHODS FOR SURFACE ANALYSIS OF PRECIOUS METALS CONTACT MATER 5-1785
NEW DEVELOPMENTS IN THE SURFACE ANALYSIS OF SOLIDS.   1-0011
GER ELECTRON SPECTROSCOPY.   FRACTURE SURFACE ANALYSIS OF TEMPER EMBRITTLED STEEL BY AU 6-2084
AES FOR SURFACE ANALYSIS. (REVIEW, 24 REFS)   1-0195
TECHNIQUE OF AUGER ELECTRON SPECTROSCOPY IN SURFACE ANALYSIS. (REVIEW, 76 REFS)   1-0196
DING AES)   SURFACE ANALYSIS TODAY. (REVIEW OF METHODS, INCLU 1-0038
ND CHEMICAL MEASUREMENTS. (AUGER)   SURFACE ANALYSIS USING COMPLEMENTARY ELECTRONIC A 5-1390
SURFACE ANALYSIS USING PROTON BEAMS.   1-0143
CTRONS. BASIC PROPERTIES OF A NEW METHOD FOR MATERIALS S/ SURFACE ANALYSIS WITH HEAVY ION INDUCED AUGER ELE 1-0070
CTRONS.   SURFACE ANALYSIS WITH HEAVY ION INDUCED AUGER ELE 1-0387
SURFACE ANALYSIS WITH LOW ENERGY ION SCATTERING.   5-1871
PECTROMETRY, AUGER ELECTRON SPECTROSCOPY AND SECONDARY I/ SURFACE ANALYTICAL STUDIES USING ION SCATTERING S 5-1423
/ DIRECT COMPARISON OF CORE ELECTRON BINDING ENERGIES OF SURFACE AND BULK ATOMS OF TITANIUM, CHROMIUM, AND 2-0392
ION OF ETHYLENE OXIDE AND VINYL CHLORIDE ON AN IRON (011) SURFACE AND EFFECT OF THESE FILMS ON ADHESION. /T 5-1207
MPURITIES. (TEMPER EMBRITTLEMENT BY AUGER SPECTROSCOPY)   SURFACE AND GRAIN BOUNDARY SEGREGATION OF STEEL I 6-2109
ERMANIUM (111). (AUG/ PHOTOEMISSION MEASUREMENTS OF BULK, SURFACE, AND HYDROGEN INDUCED STATES ON CLEAVED G 5-1781
/NG ELECTRON MICROSCOPY- AUGER SPECTROSCOPY FOR MICROSPOT SURFACE AND IN-DEPTH ANALYSIS OF SILICON AND TRA/ 5-1262
CHROMIUM THIN FILMS. (AES)   CORROSION, SURFACE, AND MAGNETIC PROPERTIES OF NICKEL- IRON- 5-1766
/HANE, ETHANE, ETHYLENE, AND ACETYLENE WITH AN IRON (001) SURFACE AND THEIR INFLUENCE ON ADHESION STUDIED / 5-1208
ERS)   SURFACE AND THIN FILM ANALYSIS. (AUGER SPECTROMET 3-0609
PY)   ESTABLISHING DETECTION AND DETERMINATION LIMITS IN SURFACE AND THIN FILM ANALYSIS. (AUGER SPECTROSCO 5-1572
ATERIALS. (SCANNING AUGER MICROPROBE)   SURFACE AND THIN FILM ANALYSIS OF SEMICONDUCTOR M 1-0089
CRIPTION AND COMPARISON OF TECHNIQUES. (REVIEW)   SURFACE AND THIN FILM COMPOSITIONAL ANALYSIS. DES 5-0053
ROLE OF COSTER-KRONIG TRANSITIONS IN THE IONIZATION OF SURFACE ATOMS.   2-0338
/ELECTRON IMPACT IONIZATION CROSS SECTIONS ON K-SHELLS OF SURFACE ATOMS. (AUGER INTENSITIES OF TUNGSTEN (1/ 4-0847
DIFFERENTIAL CROSS SECTIONS FOR K-SHELL IONIZATION OF SURFACE ATOMS BY ELECTRON IMPACT.   2-0358
INNER SHELL IONIZATION OF SURFACE ATOMS BY ELECTRON IMPACT.   2-0359
SM AND PURITY OF COPPER VAPOR DEPOSITED ON A COPPER (111) SURFACE. (AUGER)   ISOMORPHI 5-1171
CHEMISORPTION OF ATOMIC HYDROGEN ON THE SILICON (111) 7*7 SURFACE. (AUGER EFFECT)   5-1789
NEGATIVE ELECTRON AFFINITY ON THE SILICON- CESIUM- OXYGEN SURFACE. (AUGER EFFECT) /AND ELECTRONIC MODEL OF 2-0426
ADSORPTION OF COBALT ON A NICKEL (111) SURFACE. (AUGER EFFECT, AUGER SPECTROSCOPY)   5-1256
ADSORPTION OF OXYGEN ON A CLEAN SILICON SURFACE. (AUGER ELECTRON SPECTROSCOPY)   5-1638
DECOMPOSITION OF CARBON MONOXIDE ON A (110) NICKEL SURFACE. (AUGER ELECTRON SPECTROSCOPY)   5-1606
GATION OF CESIUM ADSORPTION ONTO A GALLIUM ARSENIDE (110) SURFACE. (AUGER ELECTRON SPECTROSCOPY)   INVESTI 5-1644
STIGATION OF HYDROGEN CHEMISORPTION ON THE TUNGSTEN (100) SURFACE. (AUGER ELECTRON SPECTROSCOPY)   INVE 5-1974
OEFFIC/ KINETICS OF OXYGEN ADSORPTION ON A PLATINUM (111) SURFACE. (AUGER ELECTRON SPECTROSCOPY, STICKING C 5-1189
OF OXYGEN AND OXIDATION OF COBALT ON THE RUTHENIUM (001) SURFACE. (AUGER SPECTRA)   ADSORPTION 5-1608
SODIUM SEGREGATION ONTO A LITHIUM METAL SURFACE. (AUGER SPECTRA)   5-1750
CRYSTAL FILMS AND THEIR USE IN STUDYING THE IRIDIUM (100) SURFACE. (AUGER SPECTRA OF IRIDIUM) /ENON SINGLE 5-1479
BINDING OF CARBON ATOMS AT A STEPPED NICKEL SURFACE. (AUGER SPECTROSCOPY)   5-1485
OWTH MODE OF IRON FILM DEPOSITED ON MAGNESIUM OXIDE (001) SURFACE. (AUGER SPECTROSCOPY)   GR 5-1528
SEGREGATION OF CADMIUM TO A (111) SILVER CHLORIDE SURFACE. (AUGER SPECTROSCOPY)   5-1918
/NTITATIVE ESTIMATES OF THE CHEMICAL COMPOSITION NEAR THE SURFACE BASED ON THE ELECTRON AUGER SPECTROSCOPY/ 5-1645
TROSCOPY)   SOME ASPECTS OF THE SURFACE BEHAVIOR OF SILICON. (AUGER ELECTRON SPEC 5-1510
/F THE REACTION BETWEEN SULFUR AND OXYGEN ON A MOLYBDENUM SURFACE BY AUGER ELECTRON SPECTROSCOPY: WORKING/ 5-1530
DETECTION OF IMPURITIES ON A SILICON (111) SURFACE BY AUGER- LEED ANALYSIS.   5-1360
DYNAMIC ELUCIDATION OF REACTION STEPS ON A SOLID SURFACE BY ELECTRON SPECTROSCOPY.   5-1535
TION OF SUPERSTRUCTURES ON CARBON COVERED GERMANIUM SURFACE BY HIGH ENERGY ELECTRON DIFFRACTION. /RVA 5-1805
GER E/ INVESTIGATION OF EUROPIUM(II) OXIDE (100) CLEAVAGE SURFACE BY LOW ENERGY ELECTRON DIFFRACTION AND AU 5-1157
/SM OF REACTION BETWEEN SULFUR AND OXYGEN ON A MOLYBDENUM SURFACE BY MEANS OF AUGER ELECTRON SPECTROSCOPY. 5-1534
/ INITIAL OXIDATION AND REDUCTION PROCESSES OF IRON SURFACE BY MEANS OF COMBINED LOW ENERGY ELECTRON/ 6-2141
/ SECONDARY ELECTRON YIELDS FROM COPPER BERYLLIUM OXYGEN SURFACE BY THERMAL CARBON OXYGEN NITROGEN(2), AND 4-0786
CHEMICAL EFFECTS ON THE KLL AUGER ELECTRON SPECTRUM FROM SURFACE CARBON.   4-0862
IN HEAT TREATED GALLIUM ARSENIDE. (AUGER SPECTROSCOPY)   SURFACE CARRIER CONCENTRATION AND AS EVAPORATION 5-1915
F LITHIUM NIOBATE AND POTASSIUM CHLORIDE SURFACES. (AUGE/ SURFACE CHARACTERISTICS RELATED TO LASER DAMAGE O 5-1746
OMETRY.   SURFACE CHARACTERIZATION BY AUGER ELECTRON SPECTR 5-1815
R ELECTRON SPECTROSCOPY.   SURFACE CHARACTERIZATION BY LOW ENERGY PHOTO AUGE 5-1429
AUGER SPECTROSCOPY)   SURFACE CHARACTERIZATION OF INDIUM PHOSPHIDE. (BY 5-1968
DISPLAYING POOR WETTABILITY DURING BRAZING. (MONEL, AUGE/ SURFACE CHARACTERIZATION OF NICKEL- COPPER ALLOY 5-1573
APPLICATIONS OF SURFACE CHARACTERIZATION TO GLASS SURFACES.   5-1705
N DIFFRACTION AND AUGER ELECTRON SPEC/ RECENT ADVANCES IN SURFACE CHARACTERIZATION USING LOW ENERGY ELECTRO 1-0201
ROSCOPY AND APPEARANCE POTENTIAL SPECTROSCOPY: A COMPAR/ SURFACE CHEMICAL ANALYSIS BY AUGER ELECTRON SPECT 5-1910
AND OXIDIZED VANADIUM/ AUGER ELECTRON SPECTROSCOPY OF THE SURFACE CHEMICAL COMPOSITION OF VANADIUM OXIDES, 4-1089
/ION OF CARBON MONOXIDE AND OXYGEN WITH A PALLADIUM (100) SURFACE. (CHEMISORPTION AND CATALYTIC REACTIONS,/ 5-1321
ING SPECTROMETRY AND AUGER ELECTRON SPECTR/ COMPARISON OF SURFACE CHEMISTRIES OBTAINED BY USING ION SCATTER 6-1999
GER ELECTRON SPECTROSCP/ RELATIONSHIPS BETWEEN SUBSTRATE SURFACE CHEMISTRY AND ADHESION OF THIN FILMS. (AU 5-1851
ELECTRON SPECTROSCOPY)   SURFACE CHEMISTRY OF CONTACT MATERIALS. (BY AUGER 5-1394
AUGER ELECTRON SPECTROSCOPY AND ITS APPLICATIONS IN SURFACE CHEMISTRY. (REVIEW)   1-0102
AUGER SPECTROMETRY AS A TOOL FOR SURFACE CHEMISTRY. (REVIEW)   1-0162
APPLICATION OF AUGER ELECTRON SPECTROSCOPY TO SURFACE CHEMISTRY STUDIES OF CERAMIC SUBSTRATES. 5-1275
GERMANIUM SURFACE CLEANING: AUGER ANALYSIS.   5-1540

(111), NICKEL (100), AND N/ EFFECTIVENESS OF THE STANDARD	SURFACE	CLEANING TECHNIQUES AS APPLIED TO NICKEL	5-1467
UDIES BY SECONDARY ION MASS SPECTROMETRY AND AUGER ELECT/	SURFACE	CLEANLINESS OF 316 L-N STAINLESS STEEL ST	3-0654
UDIED BY SECONDARY ION MASS SPECTROMETRY AND AUGER ELECT/	SURFACE	CLEANLINESS OF 316 L-N STAINLESS STEEL ST	6-2087
. PHOTOELECTRIC WORK FUNCTION STUDY ON IRON (100)	SURFACE	COMBINED WITH AUGER ELECTRON SPECTROSCOPY	5-1926
TIC AUGER ELECTRON EMISSION AS A TOOL FOR THE ANALYSIS OF	SURFACE	COMPOSITION. CHARACTERIS	1-0170
ERMALLOY FILMS OBSERVED BY X-RAY PHOTOEMISSION SPECTROSC/	SURFACE	COMPOSITION AND CHEMISTRY OF EVAPORATED P	5-1738
/ON OF HYDROGEN DESORBED FROM NICKEL SURFACES: EFFECTS OF	SURFACE	COMPOSITION AND CRYSTAL ORIENTATION. (SU/	5-1197
LE OF PLATINUM- TIN ALLOYS FROM COMBINED X-RAY PHOTOELEC/	SURFACE	COMPOSITION AND DEPTH CONCENTRATION PROFI	5-1195
ONE APPARATUS ION PROBE TECHNIQUE. (COMPARISON BETWEEN /	SURFACE	COMPOSITION AND STRUCTURE ANALYSIS WITH A	5-1166
EPETITIVE AUGER ELECTRON SPECTROSCOPY WITH APPLICATION TO	SURFACE	COMPOSITION CHANGES. /ANNEL MONITOR FOR R	3-0695
OYS DURING ARGON BOMBARD/ EFFECT OF TARGET TEMPERATURE ON	SURFACE	COMPOSITION CHANGES OF COPPER- NICKEL ALL	5-1808
BEAM ANALYSIS METHOD OF FILM AND	SURFACE	COMPOSITION DETERMINATION.	5-1810
/ INVESTIGATING THE CATALYTIC ACTIVITY OF ALLOY SYSTEMS. (SURFACE	SURFACE	COMPOSITION, INTERDIFFUSION, AUGER ELECT/	6-2003
CTROSCOPY. QUANTITATIVE STUDY OF THE	SURFACE	COMPOSITION OF BINARY ALLOYS BY AUGER SPE	6-2005
LICATION/ RELATION BETWEEN CATALYTIC ACTIVITY PATTERN AND	SURFACE	COMPOSITION OF COPPER- NICKEL ALLOY. APP	5-1875
GER SPECTRA IN THE LOWER ENERGY REGION AT/ OBSERVATION OF	SURFACE	COMPOSITION OF COPPER- NICKEL ALLOY BY AU	5-1982
R ANNEALING AND SPUTTERI/ AUGER SPECTROSCOPY STUDY ON THE	SURFACE	COMPOSITION OF COPPER- NICKEL ALLOYS AFTE	5-1675
CLEAN AND ADSORBATE COVERED SURFACE/ DETERMINATION OF THE	SURFACE	COMPOSITION OF COPPER- NICKEL ALLOYS FOR	5-1424
	SURFACE	COMPOSITION OF MICA SUBSTRATES.	5-1742
	SURFACE	COMPOSITION OF NICKEL- GOLD ALLOYS.	5-1966
RAY PHOTOELECTRON AND AUGER SPECTROSCOPY TO DETERMINE THE	SURFACE	COMPOSITION OF OXIDIZED VANADIUM. / OF X-	5-1205
/AY PHOTOELECTRON AND AUGER SPECTROSCOPY TO DETERMINE THE	SURFACE	COMPOSITION OF OXIDIZED VANADIUM. (COMME/	5-1863
IUM- SILVER CATALYSTS BY AUGER ELECTRON SPECTROSCOPY.	SURFACE	COMPOSITION OF PALLADIUM- GOLD AND PALLAD	5-1976
LM CATALYSTS USING AUGER ELECTRON S/ DETERMINATION OF THE	SURFACE	COMPOSITION OF PALLADIUM- NICKEL ALLOY FI	5-1846
IN EQUILIBRIUM.	SURFACE	COMPOSITION OF (PALLADIUM- SILVER) ALLOYS	5-1788
ICKEL ALLOYS BY AUGER ELECTRON SPECTROSC/ ANALYSIS OF THE	SURFACE	COMPOSITION OF PASSIVE FILMS ON COPPER- N	5-1874
(AUGER SPECTRA)	SURFACE	COMPOSITION OF PLATINUM HEATED IN OXYGEN.	5-1737
VACUUM AND IN THE PRES/ AUGER SPECTROSCOPIC STUDY OF THE	SURFACE	COMPOSITION OF PLATINUM- TIN IN ULTRAHIGH	5-1196
AUGER ELECTRON SPECTROSCOPY.	SURFACE	COMPOSITION OF PROMOTED IRON CATALYSTS BY	5-1817
	SURFACE	COMPOSITION OF SILVER- COPPER ALLOYS.	5-1199
AES STUDIES OF	SURFACE	COMPOSITION OF SILVER- COPPER ALLOYS. QUA	5-1200
NTITATIVE AUGER A/ AUGER ELECTRON SPECTROSCOPY STUDIES OF	SURFACE	COMPOSITION OF SILVER- COPPER ALLOYS. QUA	5-1740
NTITATIVE AUGER A/ AUGER ELECTRON SPECTROSCOPY STUDIES OF	SURFACE	COMPOSITION OF THE LEAD- INDIUM SYSTEM.	5-1172
AUGER ELECTRON SPECTROSCOPY STUDY OF THE	SURFACE	COMPOSITION OF THE SILVER- GOLD SYSTEM BY	5-1710
AUGER ELECTRON SPECTROSCOPY.	SURFACE	COMPOSITION OF TITANIUM(IV) OXIDE.	5-1893
AES ION SPUTTERING ANALYSIS AND THE	SURFACE	COMPOSITION PROBLEMS.	5-1812
AUGER ELECTRON SPECTROSCOPY APPLIED TO	SURFACE	COMPOSITION STUDIES OF THE (001) AND (110	5-1591
) FACES OF MONOCRYSTALLINE IRON(0.84) CHROMIUM(0.16). (A/	SURFACE	COMPOSITION WITH LOW ENERGY BACKSCATTERED	5-1823
IONS. ANALYSIS OF	SURFACE	COMPOSITIONAL CHANGES WITH ELECTRON BOMBA	5-1891
RDMENT OBSERVED BY AUGER ELECTRON SPECTROSCOPY.	SURFACE	CONCENTRATION OF MAGNESIUM IN 6061 ALUMIN	5-1824
UM BY AUGER ELECTRON SPECTROSCOPY.	SURFACE	CONCENTRATION OF MOLYBDENUM IN TYPES 316	6-1995
AND 304 STAINLESS STEEL BY AUGER ELECTRON SPECTROSCOPY.	SURFACE	CONDITION AND THERMIONIC EMISSION OF LANT	5-1471
HANUM HEXABORIDE. (AUGER SPECTROSCOPY)	SURFACE	CONSTITUTENTS AND SOME RESULTS.	3-0682
AUGER CYLINDRICAL MIRROR SPECTROMETER FOR THE ANALYSIS OF	SURFACE	CONTAINING SEVERAL KINDS OF ATOMS.	1-0039
COMPUTER PROGRAM TO CALCULATE AUGER TRANSITIONS FOR A	SURFACE	CONTAMINANTS) /C-80A ALLOY AT 800 DEGREES	6-2144
C IN LOW OXYGEN PRESSURES. (AUGER ELECTRON SPECTROSCOPY,	SURFACE	CONTAMINANTS. USE	5-1959
OF LEED APPARATUS FOR THE DETECTION AND IDENTIFICATION OF	SURFACE	CONTAMINANTS IN ELECTRON DEVICE TECHNOLOG	5-1758
Y. ROLE OF	SURFACE	CONTAMINANTS ON GALLIUM ARSENIDE- CESIUM/	5-1919
/ECTROSCOPY IN DETERMINING THE EFFECT OF CARBON AND OTHER	SURFACE	CONTAMINATION) /ASTABLE ATOMS. PART-3: EN	5-1133
ERGY AND ANGULAR DISTRIBUTIONS OF THE EJECTED ELECTRONS. (SURFACE	SURFACE	CONTAMINATION EFFECT ON THE GROWTH MODE O	5-1524
F IRON EPITAXIAL FI/ AUGER AND LCSS SPECTROSCOPY STUDY OF	SURFACE	CONTAMINATION IN SEMICONDUCTOR DEVICES.	5-1978
AUGER ELECTRON SPECTROSCOPY FOR THE STUDY OF	SURFACE	CONTROL PROGRAM IN THE MANUFACTURING OF /	5-1411
/NDARY ELECTRON SPECTROSCOPY (AUGER ELECTRON ANALYSIS) TO	SURFACE	DEBYE TEMPERATURE OF THE SILICON (001) -	5-1477
2X2 STRUCTURE. (AUGER EFFECT)	SURFACE	DEFORMATION CAUSED ON NATURAL MOLYBDENITE	6-2132
BY ABRASION. (AUGER ELECTRON SPECTROSCOPY)	SURFACE	DIFFUSION OF TIN OVER A GOLD SURFACE OBSE	6-2136
RVED WITH AUGER ELECTRON SPECTROSCOPY.	SURFACE	DIFFUSION ON MAGNESIUM OXIDE (001) BY AU/	5-1494
/CTRON STIMULATED DESORPTION OF SODIUM AND MEASUREMENT OF	SURFACE	DISSOCIATION IN POTASSIUM CHLORIDE AND LI	5-1279
THIUM FLUORIDE BY OBSERVATION OF PLASMON LOSSES/ STUDY OF	SURFACE	DISSOCIATION OF POTASSIUM CHLORIDE BY LOW	5-1719
ENERGY ELECTRON BOMBARDMENT.	SURFACE	DURING SLIDING USING AUGER SPECTROSCOPY.	6-2013
ELEMENTAL ANALYSIS OF FRICTION AND WEAR	SURFACE	DURING SPUTTERING AND VACUUM VAPOR DEPOS/	5-1678
/GER ELECTRON SPECTROSCOPIC OBSERVATIONS OF THE SUBSTRATE	SURFACE	DURING THERMAL TREATMENT. WORK FUNCTION C	6-2069
HANGES AND AU/ IMPURITIES SEGREGATION TO THE (001) NICKEL	SURFACE	EFFECTS DUE TO CARBON AND SULFUR.	5-1814
EED AND AUGER ELECTRON SPECTROSCOPY STUDY OF NICKEL (110)	SURFACE	EFFECTS ON THE L23 VV AUGER SPECTRA OF AL	4-0833
UMINUM. LIFETIME AND	SURFACE	ELECTRON CENTERS. (AUGER EFFECT, ZIRCONI/	4-0915
/ETHOD FOR DETERMINATION OF THE ENERGY DEPTH OF LEVELS OF	SURFACE	ELECTRONIC STRUCTURE OF PLATINUM (100) IN	5-1186
DUCED BY RECONSTRUCTION. CHANGE IN THE	SURFACE	ELEMENTAL ANALYSIS.	1-0194
ROLE OF AUGER ELECTRON SPECTROSCOPY IN	SURFACE	ELEMENTAL COMPOSITON) /MINATION IN THE SC	3-0596
ANNING ELECTRON MICROSCOPE. (AUGER ELECTRON SPECTROSCOPY,	SURFACE	ENRICHMENT IN COPPER(3) GOLD AND GOLD(3)	5-1934
COPPER ALLOYS.	SURFACE	ENRICHMENT OF INDIUM IN EVAPORATED GOLD-	5-1892
INDIUM FILMS. (AUGER ELECTRON SPECTROSCOPY)	SURFACE	EQUILIBRIA.	5-1314
IMPURITIES ON THORIUM METAL: AUGER STUDY OF BULK TO	SURFACE	FILMS. /GER CHARACTERISTICS OF MAGNESIUM	5-1220
SINGLE CRYSTALS AND MAGNESIUM OXIDE AND MAGNESIUM NITRIDE	SURFACE	FILMS ON SILVER- TIN.	5-1386
AUGER ANALYSIS OF	SURFACE	FILMS ON THE ADHESION OF CLEAN IRON. (USI	5-1211
NG LEED A/ EFFECT OF SULFUR, OXYGEN, AND HYDROGEN SULFIDE	SURFACE	FINISHING AND COATING TECHNOLOGY. (AUGER	1-0015
ELECTRON SPECTROSCOPY)	SURFACE	IMPURITIES IN SILVER HALIDE FILMS.	5-1149
LASER WINDOW	SURFACE	INTERACTIONS AT THE (100) CRYSTAL FACE OF	5-1963
PLATINUM BY MOLECULAR BEAM SCATTERING, / STUDIES OF GAS-	SURFACE	LAYER ANALYSIS TECHNIQUES. (AUGER ELECTRO	5-1627
N SPECTROSCOPY, SECONDARY ION MASS SPECTRO/ COMPARISON OF	SURFACE	LAYER OF COPPER- TIN ALLOY SYSTEM DUE TO	5-1867
FRICTION. (AUGER SPEC/ CONCENTRATION CHANGE OF TIN IN THE	SURFACE	LAYERS BY LIGHT ION BACKSCATTERING AND SP	5-1167
UTTERING COMBINED WITH AUGER ELECTRON SPECTR/ ANALYSIS OF	SURFACE	LAYERS OF SILICON DIOXIDE DURING ELECTRO/	4-1044
/AL OBSERVATION OF CHEMICAL SHIFTS IN AUGER SPECTRUM FROM	SURFACE	LAYER: LINE SHAPE ANALYSIS AND CORRELATION WI/	4-1065
/AUGER ELECTRON SPECTRUM OF A NICKEL (110)- C(2X2) SULFUR	SURFACE	MADELUNG POTENTIALS IN ELECTRON SPECTROSC	2-0511
OPY.	SURFACE	MATERIAL STATES.	5-1521
MEASUREMENTS OF	SURFACE	METALLURGY ON THE PENETRATION DEPTH AND R	6-2137
F BREAKDOWN FIELD OF SUPERCONDUCTING NIOBIUM. / EFFECT OF	SURFACE	MODEL FOR THE OXIDATION OF STRONTIUM, BY	6-2045
AUGER ELECTRON SPECTROSCO/ VERIFICATION OF THE METAL-RICH	SURFACE	MONOLAYERS. (BY AUGER EMISSION)	5-1589
CHEMICAL IDENTIFICATION OF	SURFACE	NITROGEN IN HIGH CRITICAL TEMPERATURE NIO	5-1804
BIUM(3) GERMANIUM FILMS BY AUGER ANALYSIS. DETECTION OF	SURFACE	OBSERVED BY AES AND LEED.	5-1650
INITIAL OXIDATION OF THE MOLYBDENUM (110)	SURFACE	OBSERVED BY AUGER EMISSION SPECTROSCOPY A	5-1180
ND BY LEED. SEGREGATION OF GOLD TO SILICON (111)	SURFACE	OBSERVED BY LOW ENERGY ELECTRON DIFFRACTI	5-1696
O/ EPITAXIAL GROWTH OF SILVER ON POTASSIUM CHLORIDE (100)	SURFACE		

RECORDING THE COMPONENTS ON THE SURFACE OF A BINARY ALLOY. (AUGER SPECTROSCOPY) 5-1358
AUGER E/ OBSERVATION OF THE SEGREGATION CF COPPER TO THE SURFACE OF A CLEAN ANNEALED 50-50 NICKEL ALLOY BY 6-2044
EED, AUGER ELECTRON / EPITAXIAL GROWTH CF COPPER ON (110) SURFACE OF A TUNGSTEN SINGLE CRYSTAL STUDIED BY L 5-1660
REQUENCY OF AUGER NEUTRALIZATION OF INERT GAS IONS ON THE SURFACE OF ALKALI HALIDE CRYSTALS. /TION OF THE F 5-1544
AUGER ELECTRON SPECTROSCOPY STUDIES ON ABRADED SURFACE OF ALLOYS. 6-2131
. ORDER- DISORDER PHENOMENA AT THE SURFACE OF ALPHA TITANIUM- OXYGEN SOLID SOLUTIONS 5-1819
LEED- AUGER INVESTIGATION ON A (100) SURFACE OF AUSTENITIC CHROMIUM- NICKEL STEEL. 6-1991
ELECTRON AUGER SPECTROSCOPIC STUDY OF THE SURFACE OF CARBIDE CATHODES. 5-1242
ELECTRON AUGER SPECTROSCOPIC STUDY OF THE SURFACE OF CARBIDE CATHODES. 5-1561
/LECTRON DIFFRACTION AND AUGER INVESTIGATION OF THE (100) SURFACE OF COPPER(3) GOLD. TEMPERATURE DEPENDENC/ 5-1854
ORDER- DISORDER TRANSFORMATION AT A (100) SURFACE OF COPPER(3) GOLD, THEORY, EXPERIMENT. 5-1855
IED BY AUGER ELECTRON SPECTROSCOPY. COMPOSITION OF THE SURFACE OF COPPER- MAGNESIUM OXIDE CATALYSTS STUD 5-1361
ECTRON SPECTROSCOPY) WORK FUNCTION OF WELL-DEFINED SURFACE OF COPPER- NICKEL ALLOY PLATES. (AUGER EL 5-1873
RECORDING OF COMPONENTS ON THE SURFACE OF FUSED EMITTERS. (AUGER SPECTRA) 4-0736
RACTIO/ INVESTIGATION OF THE PURE AND CESIUM COATED (100) SURFACE OF GALLIUM ARSENIDE BY SLOW ELECTRON DIFF 5-1639
CTRUM DURING THE INTERACTION BETWEEN OXYGEN AND THE (111) SURFACE OF GALLIUM PHOSPHIDE. AUGER ELECTRON SPE 5-1642
SPECTRUM IN THE INTERACTION BETWEEN OXYGEN AND THE (111) SURFACE OF GALLIUM PHOSPHIDE. /THE AUGER ELECTRON 5-1643
CTROSCOPY AND DIFFRACTION OF SLOW ELECTRONS. (111) SURFACE OF GALLIUM PHOSPHIDE STUDIED BY AUGER SPE 5-1640
INTERACTION BETWEEN CHLORINE AND THE (100) SURFACE OF GOLD. 5-1335
ER ELECTRON EMISSION MICROGRAPHIC STUDIES OF THE CLEAVAGE SURFACE OF GRAPHITE SINGLE CRYSTAL. AUG 5-1413
XIDATION. AUGER SPECTROSCOPIC STUDY OF THE SURFACE OF IRON AFTER IONIC BOMBARDMENT AND AIR O 5-1939
ECTRON SPECTRAL DATA. CHEMICAL COMPOSITION OF THE SURFACE OF MOLYBDENUM AND STEEL BASED ON AUGER EL 5-1985
RDING TO AUGER E/ DECREASE OF CARBON CONCENTRATION ON THE SURFACE OF MOLYBDENUM CARBIDE DURING HEATING ACCO 5-1759
THE (111) SURFACE OF N-TYPE SILICON. 5-1349
SEGREGATION OF CARBON TO (100) SURFACE OF NICKEL. 6-2000
SULFUR AND CARBON ON (110) SURFACE OF NICKEL. 5-1813
ISORPTION INDUCED SULFUR SEGREGATION FROM THE BULK TO THE SURFACE OF PALLADIUM (100). / AES STUDIES OF CHEM 5-1865
F THE CHEMISORPTION OF UNSATURATED MOLECULES BY THE (100) SURFACE OF PLATINUM. /CTRON DIFFRACTION STUDIES O 5-1267
CTRON EMISSION MICROGRAPHY AND MICROANALYSIS OF A STAINED SURFACE OF PURE IRON PLATE. AUGER ELE 5-1415
ON. AUGER SPECTRA FROM THE (100) SURFACE OF SILICON DURING OXIDATION AND NITRIDATI 5-1310
RELATION TO ITS FILM SYSTEM. (AUGER E/ CHARACTERISTICS OF SURFACE OF THE PALLADIUM BARIUM ALLOY CATHODE IN 5-1129
MPOUNDS STUDIED BY AN ELECTRONIC AUGER SPECTROSCOPIC MET/ SURFACE OF THERMIONIC EMITTERS MADE OF CARBIDE CO 5-1562
ER EFFECT) PREPARATION AND SURFACE OXIDATION OF A VANADIUM (100) PLANE. (AUG 6-1992
PREPARATION AND SURFACE OXIDATION OF A (110) FACE OF VANADIUM. 5-1683
DROGEN-FREE PROTECTIVE GASES. (USING/ OBSERVATIONS ON THE SURFACE OXIDATION OF STEEL DURING ANNEALING IN HY 6-2040
BY COMBINATION AUGER SPECT/ QUANTITATIVE DETERMINATION OF SURFACE OXIDE THICKNESS ON DEPOSITED METAL FILMS 6-2048
GAIN FATIGUE MECHANISM IN CHANNEL ELECTRON MULTIPLIERS. (SURFACE OXYGEN ON LEAD OXIDE BY AES) 3-0681
/ETTING BEHAVIOR OF SODIUM ON NIMONIC-PE16 ALLOY WITH THE SURFACE OXYGEN- CHROMIUM RATIO DETERMINED BY AUG/ 5-1441
PATTERNS. (AU/ ADSORBATE INTERACTIONS ON A PLATINUM (110) SURFACE. PART-1: SULFUR AND CARBON MONOXIDE LEED 5-1190
/EAN NICKEL SINGLE CRYSTAL SURFACES. PART-1: NICKEL (100) SURFACE. PART-2: NICKEL (111) SURFACE. (AUGER EL/ 5-1452
/NTERACTION OF OXYGEN AND NITROGEN WITH THE NIOBIUM (100) SURFACE. PART-2: REACTION KINETICS. (AUGER EFFEC/ 5-1333
ELECTRON AUGER SPECTROSCOPY: A METHOD OF INVESTIGATING SURFACE PHENOMENA (REVIEW). 1-0132
ES OF (1120) CADMIUM SULFIDE SURFACE. SURFACE PHOTOVOLTAGE AND AUGER SPECTROSCOPY STUDI 5-1201
LTRAVACUUM. (AUGER SPECTROSCOPY, SURFACE PREPARATION AND/ SURFACE PHYSICS OF SOLIDS FROM THE VIEWPOINT OF U 5-1419
EMICONDUCTORS BY AUGER ELECTRON SPECTROSCOPY. STUDIES ON SURFACE PLASMA OXIDIZED FILMS OF III-V COMPOUND S 5-1699
ISSION. EFFECT OF SURFACE PLASMA OSCILLATIONS ON AUGER ELECTRON EM 2-0236
MINUM (001) SINGLE CRYSTAL SURFACES USING INELASTIC LOW / SURFACE PLASMON STUDIES ON ALUMINUM (111) AND ALU 5-1745
AUGER ELECTRON SPECTROSCOPIC STUDY OF THE SURFACE POISONING OF COPPER CATALYSTS. 5-1176
ORPTION ON TUNGSTEN: ELLIPSOMETRY, AUGER SPECTROSCOPY AND SURFACE POTENTIAL DIFFERENCE STUDIES. CESIUM ADS 5-1830
NON ADSORBED ON COPPER (100). LEED AND SURFACE POTENTIAL STUDY OF CARBON MONOXIDE AND XE 5-1246
/PPLICATION OF AUGER ELECTRON SPECTROSCOPY TO THE BULK TO SURFACE PRECIPITATION AND SURFACE DIFFUSION OF C/ 5-1651
ES (AUGER, CLEANING) FOR QUARTZ RESONATORS. SURFACE PREPARATION AND CHARACTERIZATION TECHNIQU 5-1943
D SECONDARY ION MASS SPECTROMETRY AND AUGER ELE/ STUDY OF SURFACE PROCESSES ON COPPER- BERYLLIUM BY COMBINE 5-1213
SSING AND DECORATION STUDIES, AUGER SPECTRA) CHANGE OF SURFACE PROPERTIES OF MICA AFTER CLEAVAGE. (OUTGA 5-1743
ES) INTERFERENCE OF AN ELECTRON BEAM WITH THE SURFACE REACTION BETWEEN OXYGEN AND GERMANIUM. (A 5-1620
AUGER ELECTRON SPECTROSCOPY OF A SULFUR- OXYGEN SURFACE REACTION ON A COPPER (110) CRYSTAL. 5-1183
AES COMPOSITIONAL PROFILES OF MOBILE IONS IN THE SURFACE REGION OF GLASS. 5-1722
GY ELECTRON DIFFRACTIONS AND AUGE/ DEVELOPMENT OF STEPPED SURFACE REGIONS ON POLYCRYSTALLINE GOLD. LOW ENER 5-1483
EFFECT OF SURFACE ROUGHNESS OF AUGER ELECTRON SPECTROSCOPY. 5-1448
AUGER EFFECT IN SURFACE SCIENCE. 5-1626
/SION, FRICTION, WEAR, AND LUBRICATION RESEARCH BY MODERN SURFACE SCIENCE TECHNIQUES. (AES, ELLIPSOMETRY, / 1-0104
ES PERFORMED ON COPPER- ALUMINUM, COPPER- TI/ A REVIEW OF SURFACE SEGREGATION, ADHESION, AND FRICTION STUDI 6-2033
NITROGEN WITH ZIRCONIUM IN NIOBIUM- 1% ZIRCONIUM. SURFACE SEGREGATION AND INTERACTION OF OXYGEN AND / 5-1508
NDARY SEGREGATION (AUGER ELECTRON SPECTROSCOPY). SURFACE SEGREGATION AND ITS RELATION TO GRAIN BOU 6-2106
SEGREGATION. (AUGER SPECTROSCOPY) SURFACE SEGREGATION AS A GUIDE TO GRAIN BOUNDARY 6-2073
SITE COMPETITION IN SURFACE SEGREGATION. (AUGER EFFECT) 6-2074
SOME OBSERVATIONS OF SURFACE SEGREGATION BY AUGER ELECTRON EMISSION. 5-1401
D EXPERIMENT. SURFACE SEGREGATION: COMPARISON BETWEEN THEORY AN 5-1566
CTION WITH CHLORINE GA/ CHLORINE ADSORPTION AT A TITANIUM SURFACE. SEGREGATION FROM THE BULK CRYSTAL INTERA 5-1833
T ALUMINUM ALLOY USING AU/ THERMAL EFFECTS IN EQUILIBRIUM SURFACE SEGREGATION IN A COPPER- 10 ATOMIC PERCEN 6-2030
ATION OF GOLD TO THE (111) SURFACE OF NICKEL. (AES) SURFACE SEGREGATION IN ALLOYS: EQUILIBRIUM SEGREG 5-1216
BY AUGER EMISSION SPECTROSCOPY. SURFACE SEGREGATION IN BORON DOPED IRON OBSERVED 5-1181
AUGER ELECTRON SPECTROSCOPY STUDY OF SURFACE SEGREGATION IN COPPER- ALUMINUM ALLOYS. 6-2012
AUGER SPECTROSCOPY AND LEED STUDY OF EQUILIBRIUM SURFACE SEGREGATION IN COPPER- ALUMINUM ALLOYS. 6-2025
1 ATOMIC PERCENT IN/ AUGER ELECTRON SPECTROSCOPY STUDY OF SURFACE SEGREGATION IN THE BINARY ALLOYS COPPER- 6-2027
UGER ELECTRON SPECTROSCOPY. SURFACE SEGREGATION IN TITANIUM AS MONITORED BY A 6-2067
. (AUGER EFFECT) EQUILIBRIUM SURFACE SEGREGATION OF CARBON ON IRON (100) FACES 5-1369
SURFACE SEGREGATION OF OXYGEN IN NIOBIUM. (AES) 5-1443
AND TANTALUM- OXYGEN ALLOYS. (AUGER ELECTRON SPECTROSCOP/ SURFACE SEGREGATION OF OXYGEN IN NIOBIUM- OXYGEN 6-2065
C PERCENT, AND PALLADIUM- GOLD 70 ATOMIC PERCENT CONTACT/ SURFACE SEGREGATION ON PALLADIUM- SILVER 40 ATOMI 6-2041
ELECTRON SPECTROSCOPY. SURFACE SEGREGATION STUDIES IN ALLOYS USING AUGER 6-2019
COPY. MEASUREMENT OF EQUILIBRIUM SURFACE SEGREGATION USING AUGER ELECTRON SPECTROS 5-1184
COPY. (REPLY TO COMMENTS) MEASUREMENT OF EQUILIBRIUM SURFACE SEGREGATION USING AUGER ELECTRON SPECTROS 6-2026
LEED- AUGER ELECTRON SPECTROSCOPY AND THEIR INFLUENCE ON SURFACE SELF DIFFUSION OF COPPER. /ER SURFACES BY 5-1185
SECONDARY ELECTRON SPECTROSCOPY. SURFACE SENSITIVE TOOL. (AES) 1-0095
SECONDARY ELECTRON SPECTROSCOPY. SURFACE SENSITIVE TOOL. (REVIEW, 22 REFS) 1-0094
PTION. NEW TECHNIQUE FOR AUGER ANALYSIS OF SURFACE SPECIES SUBJECT TO ELECTRON INDUCED DESOR 3-0675
TRUCTURE OBTAINED BY CLEAVAGE AT ROOM TEMPERATURE. (AUGER SURFACE SPECTRA) SILICON (111)-7 S 5-1767
COMPARISON OF MANY BODY EFFECTS IN CORE LEVEL SURFACE SPECTROSCOPIES. 5-1723
LY RESONANT CURRENT-TO-VOLTAGE CONVERTER PREAMPLIFIER FOR SURFACE SPECTROSCOPY. DOUB 3-0677

SEMICONDUCTOR SURFACE SPECTROSCOPY. (REVIEW, 21 REFS) 1-0165
AUGER OBSERVATION OF A SURFACE STATE ON TUNGSTEN. 5-1146
ENERGY LOSS SPECTRA. (AUGER EFFECT) SURFACE STATE TRANSITIONS OF SILICON IN ELECTRON 4-1039
SPECTROSCOPY. OBSERVATION OF EXTRINSIC SURFACE STATES ON (1120) CADMIUM SULFIDE BY AUGER 5-1202
SENIDE. SURFACE STOICHIOMETRY AND STRUCTURE OF GALLIUM AR 5-1143
/Y AND LOW ENERGY ELECTRON DIFFRACTION. PART-1: ELECTRODE SURFACE STRUCTURE AFTER EXPOSURE TO WATER AND AQ/ 5-1486
ENTAL AND III-V SEMICONDUCTORS. SURFACE STRUCTURE AND COMPOSITION OF VARIOUS ELEM 5-1161
D C/ THORIUM (100) CRYSTAL FACE. A STUDY OF ITS CLEANING. SURFACE STRUCTURE, AND INTERACTION WITH OXYGEN AN 5-1883
UM PHOSPHIDE SURFACES. EFFECT OF TEMPERATURE ON THE SURFACE STRUCTURE AND STOICHIOMETRY OF (100) INDI 5-1164
X-RAY PHOTOEMISSION AND SURFACE STRUCTURE. (AUGER DISPLACEMENTS) 5-1811
R ELECTRON EMISSION. SURFACE STRUCTURE FROM ANGULAR DEPENDENCE OF AUGE 5-1977
NGLE CRYSTAL SURFACES. SURFACE STRUCTURE OF AND CATALYSIS BY PLATINUM SI 5-1837
THIN REACTION LAYERS AND SURFACE STRUCTURE OF SILICON (111). 5-1879
WITH BROMINE AND CHLORINE. (AUGER SPECTROSCOPY) SURFACE STRUCTURE OF TITANIUM AND ITS INTERACTION 5-1538
R ON IRON (001) BY LEED- AES. ANALYSIS OF THE SURFACE STRUCTURE WITH ANTIPHASE DOMAINS OF SULFU 5-1470
MIC ARRANGEMENT OF GOLD (100) AND RELATED METAL OVERLAYER SURFACE STRUCTURES. ATO 5-1716
(111) SURFACE. DETERMINATION OF SURFACE STRUCTURES BY LEED (AND AES): THE SILICON 5-1342
CTROSCOPY. DETERMINATION OF SURFACE STRUCTURES BY LEED AND AUGER ELECTRON SPE 5-1957
SURFACE STRUCTURES ON IRON (110) ULTRAHIGH VACUUM 5-1346
A, FIELD ION MICROSCOPY, LEED/ A CLOSE LOOK AT SURFACES. (SURFACE STRUCTURES, SPECIES, TOPOGRAPHY, PHENOMEN 5-1492
CHARACTERIZATION OF SOME VANADIUM (100) SURFACE STRUCTURES USING LEED AND AES. 5-1864
OF SCATTERED ELECTRONS. DETERMINATION OF SURFACE STRUCTURES USING LEED AND ENERGY ANALYSIS 5-1956
ELECTRON SPECTROSCOPY AND SURFACE STUDIES. 5-1194
ELLIPSOMETRY AS A TOOL IN SURFACE STUDIES. 1-0128
MODERN TECHNIQUES FOR SURFACE STUDIES. 1-0063
TECTION OF CONVERSION AND AUGER ELECTRONS. APPLICATION TO SURFACE STUDIES. /PY OF IRON-57 AND TIN-119 BY DE 5-1796
SCANNING AUGER ELECTRON MICROSCOPE FOR SURFACE STUDIES. 3-0679
REVIEW, 44 REFS) SURFACE STUDIES BY AUGER ELECTRON SPECTROSCOPY. (1-0144
OF ELECTRONS. (AUGER ELECTRON SPECTROSCOPY) METHODS OF SURFACE STUDIES DEPENDING ON INELASTIC SCATTERING 5-1159
AUGER SPECTROSCOPY) SURFACE STUDIES FOR QUARTZ RESONATORS. (CLEANING, 5-1942
OSCOPY, REVIEW) RECENT ADVANCES IN SURFACE STUDIES. ION BEAM ANALYSIS. (AUGER SPECTR 1-0137
OGENIDES. SURFACE STUDIES OF SOME TRANSITION METAL DICHALOC 5-1969
AUGER ELECTRON SPECTROSCOPY AND ITS APPLICATION TO SURFACE STUDIES. (REVIEW, 130 REFS) 1-0059
APPLICATION OF ELECTRON SPECTROSCOPY TO SURFACE STUDIES. (REVIEW, 85 REFS) 1-0018
S. (AUGER EFFECT) SURFACE STUDIES WITH EPITAXIALLY GROWN METAL FILM 5-1253
SEGREGATION OF SULFUR ON A MOLYBDENUM SURFACE STUDIED BY AUGER ELECTRON SPECTROSCOPY. 5-1564
THE REACTION BETWEEN OXYGEN AND SULFUR ON A NICKEL (111) SURFACE. (STUDIED BY LEED AND AES) KINETICS OF 5-1451
TROSCOPY. ADHESION OF METALS TO A CLEAN IRON SURFACE STUDIED WITH LEED AND AUGER EMISSION SPEC 6-2008
LOW ENERGY ELECTRON DIFFRACTION AND SURFACE TOPOGRAPHY. (REVIEW, AUGER SPECTROSCOPY) 1-0185
S) HOW TO ASSURE ONESELF OF THE CLEANLINESS OF A SURFACE. (USE OF AES TO DETERMINE TRACE IMPURITIE 5-1154
IDENTIFICATION OF THE FORM OF CARBON AT A SILICON (100) SURFACE USING AUGER ELECTRON SPECTROSCOPY. 5-1378
Y. STUDY OF IRIDIUM (100) SURFACE USING LEED AND AUGER ELECTRON SPECTROSCOP 5-1371
Y. STUDIES ON THE IRIDIUM (111) SURFACE USING LEED AND AUGER ELECTRON SPECTROSCOP 5-1372
Y. A STUDY OF THE IRON (111) SURFACE USING LEED AND AUGER EMISSION SPECTROSCOP 5-1670
N CLEAVED SILICON. SURFACE VALENCE BAND AND PLASMON FEATURES ON CLEA 5-1141
UDY OF THE ADSORPTION OF OXYGEN ON COPPER (100) AND (111) SURFACES. /ACTION- AUGER ELECTRON SPECTROSCOPY ST 6-2089
ION AND CONDENSATION OF COPPER ON TUNGSTEN SINGLE CRYSTAL SURFACES. ADSORPT 5-1160
ORPTION OF OXYGEN ON CLEAN CLEAVED (110) GALLIUM ARSENIDE SURFACES. ADS 5-1306
ANGULAR DEPENDENCE OF AUGER ELECTRON EMISSION FROM SOLID SURFACES. 5-1447
CE OF AUGER ELECTRON EMISSION FROM COPPER (111) AND (100) SURFACES. ANGULAR DEPENDEN 5-1632
R ELECTRON EMISSION FROM CLEAN AND GAS-COVERED IRON (100) SURFACES. ANGULAR DISTRIBUTION OF AUGE 5-1623
APPLICATIONS OF SURFACE CHARACTERIZATION TO GLASS SURFACES. 5-1705
100), GOLD-COVERED (100), AND SILVER-COVERED COPPER (100) SURFACES. ATOMIC ARRANGEMENT OF GOLD (5-1718
AUGER ANALYSIS OF TANTALUM SURFACES. 5-1480
AUGER AND IONIZATION LOSS SPECTRA FROM BERYLLIUM SURFACES. 5-1987
STIC ENERGIES IN SECONDARY ELECTRON SPECTRA FROM ALUMINUM SURFACES. AUGER AND OTHER CHARACTERI 4-0896
ESTIGATION OF OXYGEN AND HYDROGEN ADSORPTION ON STRONTIUM SURFACES. /ND REFLECTION ELECTRON DIFFRACTION INV 5-1498
AUGER ELECTRON EMISSION FROM BERYLLIUM SURFACES. 5-1613
ELECTRON EMISSION MICROGRAPHY AND MICROANALYSIS OF SOLID SURFACES. AUGER 5-1416
LECTRON DIFFRACTION (LEED) OF THE ZINC OXIDE (0001) POLAR SURFACES. /ON SPECTROSCOPY (AES) AND LOW ENERGY E 5-1338
N SPECTROSCOPY ANALYSIS OF VANADIUM AND VANADIUM COMPOUND SURFACES. AUGER ELECTRO 5-1862
AND INELASTIC ELECTRON SCATTERING IN THE STUDIES OF METAL SURFACES. AUGER ELECTRON SPECTROSCOPY 5-1818
ECTROSCOPY AND ITS APPLICATION IN STUDIES OF SOME SILICON SURFACES. AUGER ELECTRON SP 5-1418
LLIUM ARSENIDE- CESIUM- OXYGEN NEGATIVE ELECTRON AFFINITY SURFACES. /FRACTION STUDY OF THE ACTIVATION OF GA 5-1845
AUGER ELECTRON SPECTROSCOPY AND MICROANALYSIS OF SOLID SURFACES. 5-1414
AUGER ELECTRON SPECTROSCOPY FOR THE ANALYSIS OF SOLID SURFACES. 5-1193
ROSCOPY: NEW TOOL IN THE CHARACTERIZATION OF GLASS FIBER SURFACES. AUGER ELECTRON SPECT 1-0175
SPECTROSCOPY OF CLEANUP-RELATED CONTAMINATION ON SILICON SURFACES. AUGER ELECTRON 5-1983
ER ELECTRON SPECTROSCOPY OF CONTAMINATED GALLIUM ARSENIDE SURFACES. AUG 5-1920
ON SPECTROSCOPY OF GALLIUM ARSENIDE AND ALUMINUM ARSENIDE SURFACES. (AUGER) ELECTR 5-1595
AUGER ELECTRON SPECTROSCOPY OF GRAPHITE FIBER SURFACES. 5-1276
AUGER ELECTRON SPECTROSCOPY OF METAL SURFACES. 6-2122
AUGER ELECTRON SPECTROSCOPY OF SILICON SURFACES. 5-1397
AUGER ELECTRON SPECTROSCOPY OF SILICON SURFACES. 5-1945
RON SPECTROSCOPY OF (SILVER- GOLD AND INDIUM- LEAD) ALLOY SURFACES. AUGER ELECT 5-1839
(AUGER) ELECTRON SPECTROSCOPY OF TRANSITION METAL OXIDE SURFACES. 5-1339
AUGER ELECTRON SPECTROSCOPY ON SURFACES. 5-1840
ECTRON SPECTROSCOPY STUDIES OF CARBON OVERLAYERS ON METAL SURFACES. AUGER EL 5-1376
F ARSENIC VACANCIES IN (-1-1-1) GALLIUM ARSENIDE ANNEALED SURFACES. AUGER ELECTRON SPECTROSCOPY STUDIES O 5-1914
AUGER ELECTRON SPECTROSCOPY STUDY OF FLOAT GLASS SURFACES. 5-1236
AUGER ELECTRON STUDIES OF URANIUM DIOXIDE SURFACES. 5-1218
AUGER ELECTRONS, INDICATORS IN ANALYSIS OF SOLID BODY SURFACES. 5-1662
SPECTROSCOPY VANADIUM PENTOXIDE (010) AND VANADIUM (100) SURFACES. AUGER EMISSION 5-1941
TION BEHAVIOR IN SCATTERING OF LOW ENERGY IONS FROM SOLID SURFACES. (AUGER) NEUTRALIZA 2-0294
AUGER SPECTROSCOPIC STUDY OF PLATINUM, NICKEL, AND IRON SURFACES. 5-1791
AUGER SPECTROSCOPY AND LEED OF SILICON SURFACES. 5-1396
AUGER SPECTROSCOPY FOR THE CHEMICAL ANALYSIS OF SURFACES. 1-0150
AUGER SPECTROSCOPY OF SEMICONDUCTOR AND METAL SURFACES. 5-1404
AUGER STUDIES OF CLEAVED SILICON (111) SURFACES. 5-1382
AUGER STUDIES OF (111) SILICON SURFACES. 5-1375

ALENCY AND CHARGE TRANSFER IN CHEMICAL REACTIONS ON METAL SURFACES.   BINDING, V 5-1763
     CARBON CONTAMINATION OF SILICON (111) SURFACES.   5-1237
ENERGIES IN SECONDARY ELECTRON SPECTRA FROM SILICON (111) SURFACES.   CHARACTERISTIC 5-1263
OPY APPLIED TO THE STUDY OF ADSORPTION PHENOMENA AT METAL SURFACES. /ERGY LOSS AND AUGER ELECTRON SPECTROSC 5-1776
OSS SPECTRA FROM DIAMOND, GRAPHITE, AND EVAPORATED CARBON SURFACES. /TERISTIC K-SHELL LOW ENERGY ELECTRON L 4-0945
     CHARACTERIZATION OF COPPER- GOLD ALLOY SINGLE SURFACES.   5-1747
ISSION DURING OXIDATION OF NICKEL (100) AND (111) CRYSTAL SURFACES. /MICAL CHANGES IN SECONDARY ELECTRON EM 5-1453
ER SPECTRA OF CHEMISORBED NITROGEN ON IRON AND MOLYBDENUM SURFACES. /EMICAL EFFECTS OF THE NITROGEN KLL AUG 4-0925
     CHEMISORPTION ON SINGLE CRYSTAL MOLYBDENUM (112) SURFACES.   5-1303
     CLEAN SURFACES.   5-1625
ORPTION STUDIES OF THE INTERACTIONS OF GASES WITH SILICON SURFACES. /N SPECTROSCOPY AND ELECTRON IMPACT DES 5-1691
ROSCOPY AND ELECTRON IMPACT DESORPTION STUDIES OF SILICON SURFACES.   COMBINED AUGER ELECTRON SPECT 5-1694
ROSCOPY AND ELECTRON IMPACT DESORPTION STUDIES OF SILICON SURFACES.   COMBINED AUGER ELECTRON SPECT 5-1695
F THE STRUCTURE AND COMPOSITION OF ION BOMBARDED TUNGSTEN SURFACES. /LECTRON SPECTROSCOPY, AND LEED STUDY O 5-1951
     COMBINED LEED AND AES STUDIES OF INDIUM ARSENIDE SURFACES.   5-1373
LICON, GERMANIUM, GALLIUM ARSENIDE, AND INDIUM ANTIMONIDE SURFACES. /ER ELECTRON SPECTROSCOPY STUDIES OF SI 5-1379
FLASH DESORPTION SPECTROSCOPY OF METALS ON SINGLE CRYSTAL SURFACES. /TRON DIFFRACTION, AUGER ELECTRON, AND 1-0008
LECTRON SPECTROSCOPY STUDY OF THE ZINC OXIDE (0001) POLAR SURFACES. /LECTRON SPECTROSCOPY, AND X-RAY PHOTOE 5-1337
RING SPECTROMETRY AND AUGER ELECTRON SPECTROSCOPY. (STEEL SURFACES) /EMISTRIES OBTAINED BY USING ION SCATTE 6-1999
GER ELECTRON SPECTROSCOPY OF AIR OXIDIZED STAINLESS STEEL SURFACES. /VERSUS DEPTH PROFILES OBTAINED WITH AU 6-1998
OF ELECTRONIC, LEED, AND AUGER DIAGNOSTICS ON ZINC OXIDE SURFACES.   CORRELATION 5-1584
NS OBTAINED BY PROTON EXCITED X-RAY, AES ANALYSIS OF IRON SURFACES. /ACTIONAL MONOLAYER OXYGEN DETERMINATIO 5-1682
OF COPPER- NICKEL ALLOYS FOR CLEAN AND ADSORBATE COVERED SURFACES. /TERMINATION OF THE SURFACE COMPOSITION 5-1424
OPY FOR STUDYING SMALL QUANTITIES OF ELEMENTS ON METALLIC SURFACES.   DIFFERENCE AUGER SPECTROSC 5-1937
FECTS IN BACKSCATTERING AND AUGER PRODUCTION NEAR CRYSTAL SURFACES.   DIFFRACTION EF 2-0243
LS. (IMPURITY GRAIN BOUNDARY SEGREGATION, AES OF FRACTURE SURFACES) /TEMPER EMBRITTLEMENT OF LOW ALLOY STEE 6-2083
ACE STRUCTURE AND STOICHIOMETRY OF (100) INDIUM PHOSPHIDE SURFACES.   EFFECT OF TEMPERATURE ON THE SURF 5-1164
     ELECTRON EMISSION STUDIES FROM GALLIUM SELENIDE SURFACES.   5-1967
CTRON INDUCED SECONDARY ELECTRON EMISSION SPECTROSCOPY OF SURFACES.   ELE 5-1406
     ELECTRON SPECTROSCOPY OF SURFACES.   1-0017
     ELECTRONIC PROPERTIES OF CLEAVED MOLYBDENUM DISULFIDE SURFACES.   5-1634
D SEGREGATION.   AUGER SPECTROGRAPHIC ANALYSIS OF FRACTURE SURFACES.   GRAIN BOUNDARY STRUCTURE AN 6-2118
IGH RESOLUTION LMM AUGER SPECTRA OF LOW ENERGY FROM SOLID SURFACES.   H 4-0732
HOW MUCH CAN AUGER ELECTRONS TELL US ABOUT SOLID SURFACES.   5-1663
ON SPECTRA OF VANADIUM PENTOXIDE (010) AND VANADIUM (100) SURFACES. /STRUCTURE IN THE AUGER ELECTRON EMISSI 5-1340
INELASTIC SCATTERING OF LOW ENERGY ELECTRONS FROM SURFACES.   2-0495
RA OF ELECTRON INELASTIC SCATTERING FROM RARE EARTH METAL SURFACES.   IONIZATION LOSSES IN (AUGER) SPECT 2-0538
IONIZATION SPECTROMETER FOR ELEMENTAL ANALYSIS OF SURFACES.   3-0615
IONIZATION SPECTROSCOPY OF CONTAMINATED METAL SURFACES.   5-1351
IONIZATION SPECTROSCOPY OF SURFACES.   5-1352
D AES STUDIES OF THE STRUCTURE OF ADSORBED GASES ON SOLID SURFACES.   LEED AN 5-1835
FFECT OF OXYGEN ON ADHESION OF CLEAN IRON (001) AND (011) SURFACES.   LEED AND AUGER STUDIES OF E 6-2010
CT POTENTIAL STUDIES OF COPPER- GOLD ALLOY SINGLE CRYSTAL SURFACES.   LEED, AUGER SPECTROSCOPY, AND CONTA 5-1748
AUGER ELECTRON SPECTROSCOPY STUDY OF IRON SINGLE CRYSTAL SURFACES.   LOW ENERGY ELECTRON DIFFRACTION- 5-1866
MICROSCOPIC AUGER ELECTRON ANALYSIS OF FRACTURE SURFACES.   5-1598
NATURE OF SILICON (111) SURFACES.   5-1383
NE STRUCTURE IN ELECTRON EXCITED AUGER SPECTRA FROM SOLID SURFACES.   NEW FI 2-0473
BETA- SILICON CARBIDE FORMATION ON RECONSTRUCTED SILICON SURFACES.   OBSERVATIONS OF 5-1560
OXYGEN ADSORPTION ON CLEAN MOLYBDENUM (100) SURFACES.   5-1772
OEMISSION STUDIES OF CHEMISORPTION ON NICKEL AND TUNGSTEN SURFACES.   PHOT 5-1308
AND ELECTRON EXCITED AUGER ELECTRON SPECTRA OF MOLYBDENUM SURFACES.   PROTON 5-1667
RECENT ADVANCES IN THE STUDY OF SOLID SURFACES.   1-0051
OCESSES INVOLVING HELIUM(+)(2S) AND HELIUM(2+) NEAR SOLID SURFACES. RESONANCE, AUGER, AND AUTOIONIZATION PR 2-0375
SECONDARY ELECTRON EMISSION IN THE STUDY OF SOLID SURFACES.   5-1555
Y ELECTRON ENERGY DISTRIBUTION STUDIES OF URANIUM DIOXIDE SURFACES.   SECONDAR 5-1316
SECONDARY EMISSION AND CONTAMINATION OF METAL SURFACES.   6-2071
ALL MOLECULE REACTIONS ON STEPPED SINGLE CRYSTAL PLATINUM SURFACES.   SM 5-1173
SPUTTER ETCHING AND AUGER ELECTRON SPECTROSCOPY OF METAL SURFACES.   5-1552
SPUTTER ETCHING AND AUGER ELECTRON SPECTROSCOPY OF METAL SURFACES.   5-1554
STABILITY OF CESIUM COVERED ZINC OXIDE SURFACES.   5-1882
E TRANSFORMATIONS ON CLEAVED AND ANNEALED GERMANIUM (111) SURFACES.   STRUCTUR 5-1713
ACE STRUCTURE OF AND CATALYSIS BY PLATINUM SINGLE CRYSTAL SURFACES.   SURF 5-1837
EMPER EMBRITTLEMENT OF LOW ALLOY STEELS. (AES OF FRACTURE SURFACES)   T 6-2062
D AUGER ELECTRON SPECTRA OF OXIDIZED STATES ON SOME METAL SURFACES.   X-RAY EXCITED PHOTOELECTRON AN 5-1690
Y PHOTOELECTRON AND AUGER SPECTROSCOPIC ANALYSES OF STEEL SURFACES.   X-RA 5-1268
ESULTS.   STRUCTURE AND COMPOSITION OF METAL SURFACES: A REVIEW OF RECENT LEED, AES, AND XPS R 5-1341
AND PHOTOELECTRON SPECTRA OF PARTIALLY OXIDIZED MAGNESIUM SURFACES.   ABNORMAL CHEMICAL SHIFTS. /ITED AUGER 5-1949
CTS OF SULFUR, BARIUM, AND CALCIUM ON IMPREGNATED CATHODE SURFACES. (AES)   EFFE 5-1391
QUANTITATIVE AUGER ANALYSIS OF COPPER- NICKEL ALLOY SURFACES AFTER ARGON ION BOMBARDMENT.   5-1807
LOYS. QUANTITATIVE AUGER ANALYSIS OF COPPER- NICKEL ALLOY SURFACES AFTER ARGON ION BOMBARDMENT. COMMENTS. / 5-1740
/OYS. QUANTITATIVE AUGER ANALYSIS OF COPPER- NICKEL ALLOY SURFACES AFTER ARGON ION BOMBARDMENT. REPLY TO C/ 5-1200
MMEN/ QUANTITATIVE AUGER ANALYSIS OF COPPER- NICKEL ALLOY SURFACES AFTER ARGON ION BOMBARDMENT. REPLY TO CO 5-1702
ITH AUGER SPECTROSC/ STRUCTURE AND CHEMISTRY OF SILICON SURFACES AFTER PRE- AND BACKSPUTTERING, STUDIED W 5-1233
G AND AUGER ELECTRON SPECTROSCOPY STUDIES OF CLEAN NICKEL SURFACES AND ADSORBED LAYERS. /ERGY ION SCATTERIN 5-1870
/ SOME PROPERTIES OF TRANSITION METAL SUPERCONDUCTORS AT SURFACES AND IN THIN FILMS. (LEED- AES STUDIES OF 5-1848
UM / MANY PARTICLE RECOMBINATION PROCESSES (AUGER) ON THE SURFACES AND IN THIN FILMS OF SILICON AND GERMANI 5-1990
ECTRON MICROANALYT/ THREE-DIMENSIONAL ELEMENT ANALYSIS OF SURFACES AND INTERFACES OF SOLIDS BY THE AUGER EL 5-1412
ELECTRON ORBITAL ENERGIES OF OXYGEN ADSORBED ON SILICON SURFACES AND OF SILICON DIOXIDE.   5-1475
/NSATION COEFFICIENTS OF GOLD ON ROCK SALT (100) CLEAVAGE SURFACES AND ON AMORPHOUS CARBON BY MEANS OF QUA/ 5-1135
REALIZATION OF A MOSSBAUER SPECTROMETER FOR SURFACES AND THIN FILMS STUDY. (AUGER) 3-0690
GOLD-INDUCED SUPERSTRUCTURES ON SILICON (100) SURFACES AS OBSERVED BY LEED AES.   5-1708
ORBAS: AN APPARATUS FOR INVESTIGATING ION SCATTERING FROM SURFACES AT ENERGIES 100-2000 EV. (AES)   S 3-0704
NATURE OF ANNEALED SEMICONDUCTOR SURFACES. (AUGER)   2-0263
OTODESORPTION AND CONDUCTIVITY MEASUREMENTS ON ZINC OXIDE SURFACES. (AUGER ANALYSIS)   CHEMISORPTION, PH 5-1801
ANGULAR DEPENDENCE OF ELECTRON EMISSION FROM SURFACES. (AUGER EFFECT)   5-1726
CHEMISORPTION REACTIONS ON HIGH INDEX ZINC SULFIDE SURFACES. (AUGER EFFECT)   4-0788
DETERMINATION OF SULFUR CONTAMINATION ON MOLYBDENUM SURFACES. (AUGER EFFECT)   5-1647
INTERACTION OF CARBON MONOXIDE WITH (110) NICKEL SURFACES. (AUGER EFFECT)   5-1607
ION SELECTIVE SPUTTERING OF III-V COMPOUND SEMICONDUCTOR SURFACES. (AUGER EFFECT)   5-1637
ADSORPTION OF HYDROGEN ON NICKEL SINGLE CRYSTAL SURFACES. (AUGER ELECTRON SPECTROSCOPY)   5-1257

AMORPHOUS TO CRYSTALLINE TRANSITION OF SILICON (111) SURFACES. (AUGER ELECTRON SPECTROSCOPY) 5-1780
INTERACTION OF OXYGEN WITH SILICON (111) SURFACES. (AUGER ELECTRON SPECTROSCOPY) 5-1693
PLASMA CLEANING OF METAL SURFACES. (AUGER ELECTRON SPECTROSCOPY) 6-2093
BED FROM IRON, PLATINUM, COPPER, NIOBIUM, STAINLESS STEEL SURFACES. (AUGER ELECTRON SPECTROSCOPY) /EN DESOR 6-2006
SCOPIC STUDIES. ARSENIC (0001) SURFACES: AUGER, LOSS, AND PHOTOELECTRON SPECTRO 5-1315
ROOM TEMPERATURE ADSORPTION OF OXYGEN ON TUNGSTEN SURFACES. (AUGER MEASUREMENT METHODS) 5-1664
/TERACTION OF LOW ENERGY ATMOSPHERIC IONS WITH CONTROLLED SURFACES. (AUGER NEUTRALIZATION AT (100) FACE OF/ 5-1736
RADIATIONLESS DEEXCITATION OF EXCITED HELIUM ATOMS AT SURFACES. (AUGER PROCESS) 2-0255
CHEMISORPTION OF OXYGEN ON PLATINUM (111) SURFACES. (AUGER SPECTRA) 5-1500
CHEMISORPTION ON (001), (110) AND (111) NICKEL SURFACES. (AUGER SPECTRA) 5-1764
T-3: OXIDE NUCLEATION RATES ON (100) AND (110) MOLYBDENUM SURFACES. (AUGER SPECTRA) /ION OF MOLYBDENUM. PAR 5-1536
SECONDARY ELECTRON EMISSION FROM PLATINUM BLACK COATED SURFACES. (AUGER SPECTRA) 5-1979
T PHOTOELECTRON STUDIES OF ADSORPTION ON CLEAN METAL FILM SURFACES. (AUGER SPECTRA) / AND VACUUM ULTRAVIOLE 5-1203
CHARACTERIZATION OF CERAMIC SURFACES. (AUGER SPECTROSCOPY) 5-1174
GY STUDIES OF CLEAN (001), (110) AND ELASTIC (111) NICKEL SURFACES. (AUGER SPECTROSCOPY) /ED INTENSITY ENER 5-1292
ELECTRON BEAM ADSORBATE INTERACTIONS ON SILICON SURFACES. (AUGER SPECTROSCOPY) 5-1513
ELECTRON STIMULATED PROCESSES AT SOLID SURFACES. (AUGER SPECTROSCOPY) 5-1605
TATES OF OXYGEN ADSORBED ON CLEAN SILICON (111) AND (100) SURFACES. (AUGER SPECTROSCOPY) ELECTRONIC S 5-1476
INTERACTION OF CESIUM WITH CLEAN ZINC OXIDE SURFACES. (AUGER SPECTROSCOPY) 5-1588
RACTIONS OF CARBON MONOXIDE AND OXYGEN WITH IRIDIUM (110) SURFACES. (AUGER SPECTROSCOPY) INTE 5-1252
OTON INDUCED ELECTRON EMISSION FROM CHARACTERIZED NIOBIUM SURFACES. (AUGER SPECTROSCOPY) PR 5-1666
SCATTERING OF MOLECULAR BEAMS FROM SURFACES. (AUGER SPECTROSCOPY) 5-1905
SEGREGATION AT COPPER- GOLD ALLOY SURFACES. (AUGER SPECTROSCOPY) 5-1631
BE ANALYSIS) CRYSTALLOGRAPHY AND CHEMISTRY OF CERAMIC SURFACES. (AUGER SPECTROSCOPY, LEED, ION MICROPRO 5-1852
TO LASER DAMAGE OF LITHIUM NIOBATE AND POTASSIUM CHLORIDE SURFACES. (AUGER SURFACE PROFILE) /STICS RELATED 5-1746
M, NICKEL, TANTAL/ SECONDARY ELECTRON EMISSION FROM SOLID SURFACES BOMBARDED BY MEDIUM ENERGY IONS (PLATINU 2-0309
STUDY OF MULBERRY SURFACES BY AUGER AND IEE SPECTROSCOPY. 5-1274
ANALYSIS OF IMPURITIES ON SILICON SURFACES BY AUGER ELECTRON SPECTROSCOPY. 5-1674
CHARACTERIZATION OF ALUMINUM ADHEREND SURFACES (BY AUGER ELECTRON SPECTROSCOPY). 5-1725
CHEMICAL ANALYSIS OF PLATINUM (100) AND GOLD (100) SURFACES BY AUGER ELECTRON SPECTROSCOPY. 5-1712
DETECTION OF ORGANIC CONTAMINANTS ON ALUMINUM SURFACES BY AUGER ELECTRON SPECTROSCOPY. 5-1487
IMPURITY CONCENTRATIONS ON METAL SURFACES BY AUGER ELECTRON SPECTROSCOPY. 5-1923
QUANTITATIVE ANALYSIS OF SOLID SURFACES BY AUGER ELECTRON SPECTROSCOPY. 5-1715
STUDY OF THE PREPARATION OF ATOMICALLY CLEAN TUNGSTEN SURFACES BY AUGER ELECTRON SPECTROSCOPY. 5-1517
DARY ION/ SIMULTANEOUS OBSERVATIONS OF PARTIALLY OXIDIZED SURFACES BY AUGER ELECTRON SPECTROSCOPY AND SECON 5-1553
COMPOSITION OF PLATINUM ANODE SURFACES BY AUGER SPECTROSCOPY. 5-1504
DETECTION OF THIN CONTAMINATED LAYERS ON CONTACT SURFACES. (BY AUGER SPECTROSCOPY) 5-1165
DETERMINATION OF THE COMPOSITION OF GLASS SURFACES BY AUGER SPECTROSCOPY. 5-1365
DETECTION OF SODIUM ON SILICON AND SILICON DIOXIDE SURFACES BY CYLINDRICAL MIRROR AUGER ANALYZER. 3-0647
CHARACTERIZATION OF GLASS SURFACES BY ELECTRON SPECTROSCOPY. 5-1787
H AUGER ELECTRON SP/ STUDY OF THE OXIDATION OF MOLYBDENUM SURFACES BY ENERGY LOSS SPECTROSCOPY COMBINED WIT 5-1531
METRY. STUDY OF METAL SURFACES BY IONIC MICROANALYSIS AND AUGER SPECTRO 5-1582
HORUS ADSORPTION AND DESORPTION KINETICS ON SILICON (111) SURFACES. (BY LEED AND AES) A STUDY OF PHOSP 5-1288
THEIR INFLUENCE ON SU/ DETECTION OF IMPURITIES ON COPPER SURFACES BY LEED- AUGER ELECTRON SPECTROSCOPY AND 5-1185
/F THE INTERACTION OF OXYGEN WITH PLATINUM SINGLE CRYSTAL SURFACES BY LOW ENERGY ELECTRON DIFFRACTION AND / 5-1223
UGER ELECTRON SPECTR/ STUDY OF THE NITRIDATION OF SILICON SURFACES BY LOW ENERGY ELECTRON DIFFRACTION AND A 5-1417
UGER ELECTRON SPECTROSCOPY. STUDIES OF CRYSTAL SURFACES BY LOW ENERGY ELECTRON DIFFRACTION AND A 5-1836
Y AND LOW ENERG/ CHEMICAL AND STRUCTURAL ANALYSIS OF IRON SURFACES BY MEANS OF AUGER ELECTRONIC SPECTROSCOP 5-1940
SURVEYING SILICON SURFACES BY MEANS OF THE LEED- AES METHOD. 5-1809
S OF SECONDARY EM/ SECONDARY ELECTRON EJECTION FROM METAL SURFACES BY METASTABLE ATOMS. PART-1: MEASUREMENT 2-0340
ANGULAR DISTRIBUT/ SECONDARY ELECTRON EJECTION FROM METAL SURFACES BY METASTABLE ATOMS. PART-3: ENERGY AND 5-1133
LEED AND AES. INVESTIGATION OF REORDERED (001) GOLD SURFACES BY POSITIVE ION CHANNELING SPECTROSCOPY, 5-1139
LOW ENERGY ELE/ CHARACTERIZATION OF REORDERED (001) GOLD SURFACES BY POSITIVE ION CHANNELING SPECTROSCOPY, 5-1986
ON/ MEASUREMENT OF MOMENTUM ACCOMMODATION COEFFICIENTS ON SURFACES CHARACTERIZED BY AUGER SPECTROSCOPY, SEC 2-0501
TION AND SPUTTER REMOVAL OF MOLYBDENUM FROM VARIOUS METAL SURFACES (COPPER, GOLD, ALUMINUM, TUNGSTEN). /OSI 5-1878
R EFFECT) ELECTRONIC STRUCTURE OF SOLID SURFACES: CORE LEVEL EXCITATION TECHNIQUES. (AUGE 2-0468
RON DIFF/ CHEMISORPTION ON (001), (110), AND (111) NICKEL SURFACES. CORRELATED STUDY USING LOW ENERGY ELECT 5-1291
BARIUM ON / AUGER ELECTRON SPECTROSCOPY STUDY OF CATHODE SURFACES DURING ACTIVATION AND POISONING. PART-1: 5-1841
ON. CARBIDE FORMATION ON NIOBIUM SURFACES DURING HIGH TEMPERATURE PROTON IRRADIATI 5-1886
ANALYSIS, STAIN/ SEGREGATION OF ALLOYING ELEMENTS TO FREE SURFACES DURING IRRADIATION. (AUGER SPECTROSCOPY 6-2091
CTRA) METHODS OF CLEANING CONTAMINANTS FROM QUARTZ SURFACES DURING RESONATOR FABRICATION. (AUGER SPE 5-1410
/HE SPATIAL DISTRIBUTION OF HYDROGEN DESORBED FROM NICKEL SURFACES: EFFECTS OF SURFACE COMPOSITION AND CRY/ 5-1197
N, AUGER, AND CONTAC/ SODIUM ADSORPTION ON ALUMINUM (111) SURFACES: ELECTRIC LOW ENERGY ELECTRON DIFFRACTIO 5-1744
A STUDY OF ACTIVATED SILICON SURFACES FOR ELECTROLESS NICKEL PLATING BY AES. 5-1539
ANALYSIS AND CHARACTERIZATION OF CERAMIC SURFACES FOR ELECTRONIC APPLICATIONS. 5-1853
NALYSIS) NIOBIUM SURFACES FOR RF SUPERCONDUCTORS. (AUGER SURFACE A 5-1472
AUGER ELECTRON SPECTROSCOPY OF FACE CENTERED CUBIC METAL SURFACES. (GOLD, SILVER, PALLADIUM, COPPER, NICK/ 5-1717
AUGER ELECTRON SPECTROSCOPY OF ALLOY SURFACES. (GOLD- SILVER, LEAD- INDIUM) 5-1709
ELECTRON MICROSCOPE. AUGER SPECTROSCOPY OF SOLID SURFACES IN A DRY PUMPED HIGH RESOLUTION SCANNING 5-1258
TRON SPECTROSCOPY TO THE STUDY OF DISCOLORING OF ALUMINUM SURFACES IN BOILING WATER. /ICATION OF AUGER ELEC 5-1127
R ELECTRON SPECTROSCOPY) OF GRAIN BOUNDARY FRACTURE SURFACES IN DOPED TUNGSTEN- RHENIUM ALLOYS. (AUGE 6-2112
AUGER ELECTRON SPECTROSCOPY OF FRACTURE SURFACES IN IRRADIATED TYPE 304 STAINLESS STEEL. 6-2001
/ORPTION ISOTHERMS OF OXYGEN ON TUNGSTEN (100), AND (111) SURFACES IN THE PRESSURE RANGE 10(-9) TO 10(-4) / 5-1148
AND AUGER ELECTRON SPECTROS/ BEHAVIOR OF REFRACTORY METAL SURFACES IN ULTRAHIGH VACUUM AS OBSERVED BY LEED 6-2020
TH TECHNIQUES INVOLVING/ ELECTRON SPECTROSCOPY FROM SOLID SURFACES IN ULTRAHIGH VACUUM. CURRENT PROGRESS WI 5-1901
/Y OF FIELD ELECTRON EMISSION AND VACUUM BREAKDOWN. (FROM SURFACES) INVESTIGATIONS WITH FIELD EMISSION MIC/ 5-1520
TECHNIQUES FOR THE CHEMICAL CHARACTERIZATION OF CERAMIC SURFACES. (ION SCATTERING SPECTROMETRY, AUGER EL/ 5-1326
E PROPERTIES OF CLEAN AND ALKALI METAL COVERED ZINC OXIDE SURFACES. (LEED, AES) SOM 5-1881
CONDENSATION OF GOLD ONTO TANTALUM (100) SINGLE CRYSTAL SURFACES. LEED AND AES ANALYSIS. 5-1312
DEX ((111) AND (100)) AND STEPPED PLATINUM SINGLE CRYSTAL SURFACES. (LEED, AUGER ELECTRON SPECTROSCOPY) /IN 5-1838
CONTAMINANTS ON CHEMICALLY ETCHED SILICON SURFACES. LEED- AUGER METHOD. 5-1229
DEPTH ANALYSIS OF CLEAVED MICA SURFACES MONITORED BY AUGER SPECTROSCOPY. 5-1842
BSERVED BY LEED AND AES. CRYSTAL SURFACES OF ALKALI HALIDES AND THIN FILM GROWTH O 5-1907
M CLEANED SURFACES. (100) SURFACES OF ALKALI HALIDES. PART-1: AIR AND VACUU 5-1347
ULATED DISSOCIATION. (100) SURFACES OF ALKALI HALIDES. PART-2: ELECTRON STIM 5-1908
AUGER SPECTROSCOPIC STUDY OF THE SURFACES OF ALLOY EMITTERS. 5-1192
/ DIFFRACTION AND AUGER SPECTROSCOPY STUDIES OF THE (100) SURFACES OF ALUMINUM AND ALUMINUM- COPPER ALLOYS. 5-1755
AUGER SPECTROSCOPY OF FRACTURE SURFACES OF CERAMICS. 5-1616
ADSORPTION ON SINGLE CRYSTAL SURFACES OF COPPER- NICKEL ALLOYS. PART-1. 5-1322
D METHOD. MEASUREMENTS ON SURFACES OF ELECTRICAL SOLID MATERIALS BY THE LEE 5-1981

R, AND WORK FUNCTION STUDIES OF CLEAN AND SODIUM- COVERED SURFACES OF GALLIUM ARSENIDE.   LEED, AUGE 5-1240
LOW ENERGY ELECTRON DIFFRACTION STUDY OF THE POLAR (111) SURFACES OF GALLIUM ARSENIDE AND GALLIUM ANTIMON/ 5-1602
/AL CHANGES THAT ACCOMPANY THE THERMAL ANNEALING OF (100) SURFACES OF GALLIUM ARSENIDE, INDIUM PHOSPHIDE, / 5-1163
MENT WITH 2 KV EL/ DEPARTURES FROM STOICHIOMETRY IN (100) SURFACES OF GALLIUM PHOSPHIDE, INDUCED BY BOMBARD 5-1162
COMPOSITION PROFILES OF SEVERAL CONTAMINATED AND CLEANED SURFACES OF GOLD THICK FILMS ON COPPER PLATES BY/ 5-1550
ELECTRON SPECTROSCOPY.   STUDY OF FRACTURE SURFACES OF HOT PRESSED SILICON NITRIDE BY AUGER 5-1444
AUGER ELECTRON SPECTROSCOPY ANALYSIS OF THE FRACTURE SURFACES OF IRRADIATED PRESSURE VESSEL STEELS. 6-2113
AUGER SPECTROSCOPY OF FRACTURE SURFACES OF IRRADIATED STAINLESS STEEL. 6-2002
RON SPECTROSCOPY TO DETERMINE THE COMPOSITION OF FRACTURE SURFACES OF MINERALS.   USE OF AUGER ELECT 5-1307
GROWTH AND STRUCTURE OF THIN COPPER FILMS ON (001) SURFACES OF NICKEL. 5-1227
EQUILIBRATION OF HYDROGEN AND DEUTERIUM ON SINGLE CRYSTAL SURFACES OF PLATINUM. (AUGER SPECTROSCOPY) 5-1594
R ELECTRON SPECTROS/ INVESTIGATIONS OF CARBON RESIDUES ON SURFACES OF SILICON INTEGRATED CIRCUITS. (BY AUGE 5-1345
ON EMISSIO/ DETECTION OF SILICON OXYNITRIDE LAYERS ON THE SURFACES OF SILICON NITRIDE FILMS BY AUGER ELECTR 5-1611
(DETERMINATION OF SURFACES/ SEGREGATION OF IMPURITIES ON SURFACES OF THORIUM. THERMODYNAMICS AND KINETICS. 5-1313
LEED AND AES OBSERVATIONS ON THE POLAR SURFACES OF ZINC OXIDE. 5-1619
OR ABRADED B/ AUGER ELECTRON SPECTROSCOPIC STUDIES OF THE SURFACES OF 18-8 STAINLESS STEEL ELECTROPOLISHED 6-2130
CALLY AND ELECTROLYTICALLY.   AES STUDIES ON SURFACES OF 18-8 STAINLESS STEEL FINISHED MECHANI 6-2129
ADSORPTION OF OXYGEN ON SILICON (111) SURFACES. PART-1. (AUGER ELECTRON SPECTROSCOPY) 5-1474
/ THE REACTION OF OXYGEN WITH CLEAN NICKEL SINGLE CRYSTAL SURFACES. PART-1: NICKEL (100) SURFACE. PART-2: / 5-1452
/UE FOR DETERMINING THE ELEMENTAL COMPOSITION OF FRACTURE SURFACES PRODUCED BY CRACK GROWTH IN HYDROGEN AN/ 6-2142
AUGER SPECTROSCOPY ON COPPER- NICKEL ALLOY SURFACES RELATED TO CATALYSIS. 5-1703
INTERACTION OF OXYGEN WITH SILICON (111) SURFACES. (REPLY TO COMMENT) 5-1512
CHARACTERIZATION OF SOLID SURFACES. (REVIEW, AUGER SPECTROSCOPY, ESCA) 1-0134
INSTRUMENTATION FOR THE CHEMICAL ANALYSIS OF MANUFACTURED SURFACES. (REVIEW, SPECTROSCOPY) 3-0719
PY STUDIES OF ADSORPTION AND OXIDATION PROCESSES AT METAL SURFACES. (REVIEW, 50 REFS)   ELECTRON SPECTROSCO 1-0019
ON MASS SPECTROSCOPY ON WELL CHARACTERIZED SINGLE CRYSTAL SURFACES. (SILVER AND COPPER ON TUNGSTEN) /DARY I 5-1689
CLEAN TELLURIUM SURFACES STUDIED BY LEED. 5-1134
Y, PHENOMENA, FIELD ION MICROSCOPY, LEED/ A CLOSE LOOK AT SURFACES. (SURFACE STRUCTURES, SPECIES, TOPOGRAPH 5-1492
ONTAINING IMPURITIES. (AUGER ELECTRON SPECTROSCOPY) SURFACES TEXTURES: AN INVESTIGATION OF SURFACES C 5-1707
BY AUGER SPECTRO/ SPUTTER CLEANING AND ETCHING OF CRYSTAL SURFACES (TITANIUM, TUNGSTEN, SILICON) MONITORED 5-1828
W)   THE RELATIONSHIP OF SOLID SURFACES TO VACUUM SCIENCE AND TECHNOLOGY. (REVIE 1-0086
SERIES STAINLESS STEEL, AND N/ AUGER ELECTRON STUDIES OF SURFACES: URANIUM DIOXIDE, URANIUM, GRAPHITE, 300 5-1217
ION SPECTROSCOPY.   STUDY OF CARBON FIBER SURFACES USING AUGER AND SECONDARY ELECTRON EMISS 5-1970
/UANTITATIVE COMPARISON OF TITANIUM AND TITANIUM MONOXIDE SURFACES USING AUGER ELECTRON AND SOFT X-RAY APP/ 5-1384
E OXIDATION AND REDUCTION BY WATER VAPOR OF SILICON (111) SURFACES USING AUGER ELECTRON EMISSION. /DY OF TH 5-1612
INVESTIGATION OF SILVER CESIUM OXIDE PHOTOEMISSIVE SURFACES USING AUGER ELECTRON SPECTROSCOPY. 5-1782
PHOTOELECTRON SPECTROS/ A PRELIMINARY STUDY OF PURE METAL SURFACES USING AUGER ELECTRON SPECTROSCOPY X-RAY 5-1353
/DIES ON ALUMINUM (111) AND ALUMINUM (001) SINGLE CRYSTAL SURFACES USING INELASTIC LOW ENERGY ELECTRON DIF/ 5-1745
PY.   STUDY OF RUTHENIUM (0001) AND RHODIUM (111) SURFACES USING LEED AND AUGER ELECTRON SPECTROSCO 5-1380
PY.   STUDY OF INDIUM ARSENIDE (111) AND (-1-1-1) SURFACES USING LEED AND AUGER ELECTRON SPECTROSCO 5-1381
GER, AND LOW ENERGY ELEC/ CHEMISORPTION STUDIES OF NICKEL SURFACES UTILIZING WORK FUNCTION MEASUREMENTS, AU 5-1290
AES METHOD.   SURVEYING SILICON SURFACES BY MEANS OF THE LEED- 5-1809
PHENOMENOLOGICAL SYSTEMATICS OF L- AND M-AUGER SPECTRA. 4-0875

T

.   AUGER MEASUREMENTS ON T-027 SAMPLES EXPOSED DURING THE SKYLAB-2 MISSION 5-1735
TRON EMISSION DUE TO PRIMARY AND BACKSCATTERED ELECTRONS (TABLE FOR 60 ELEMENTS).   SECONDARY ELEC 4-0907
INTEGRAL AUGER INFORMATION VIA TAILORED MODULATION TECHNIQUES. 1-0187
DEUTERIUM, NITROGEN, OXYGEN AND CARBON MONOXIDE BY (100) TANTALUM. (AUGER EFFECT) /D SOLUTION OF HYDROGEN, 5-1549
DSORPTION ON TANTALUM AND POLYC/ AUGER SPECTRA OF CARBON. TANTALUM CARBIDES, ETHYLENE AND CARBON MONOXIDE A 4-0820
CARBON AND OXYGEN ON THE PHYSICAL PROPERTIES OF SPUTTERED TANTALUM FILMS. (AUGER ELECTRON SPECTROSCOPY) /, 5-1473
/GHT ELEMENTS (NITROGEN, CARBON, AND OXYGEN) IN SPUTTERED TANTALUM FILMS BY AUGER ELECTRON SPECTROSCOPY AN/ 5-1652
RFACES BOMBARDED BY MEDIUM ENERGY IONS (PLATINUM, NICKEL, TANTALUM, GRAPHITE). /TRON EMISSION FROM SOLID SU 2-0309
/TRANSITION TEMPERATURES OF REACTIVELY SPUTTERED FILMS OF TANTALUM NITRIDE AND TUNGSTEN NITRIDE. (AUGER EF/ 5-1541
/ILICIDE FORMATION AND INTERDIFFUSION EFFECTS IN SILICON- TANTALUM, SILICON DIOXIDE- TANTALUM, AND SILICON/ 5-1260
AUGER ANALYSIS OF TANTALUM SURFACES. 5-1480
CHEMISORPTION ON TANTALUM (100). (AUGER SPECTROSCOPY) 5-1245
AES ANALYSIS.   CONDENSATION OF GOLD ONTO TANTALUM (100) SINGLE CRYSTAL SURFACES. LEED AND 5-1312
IN AUGER ELECTRON SPECTROSCOPY FROM INITIAL OXIDATION OF TANTALUM (110).   CHEMICAL SHIFTS 6-2038
COP/ SURFACE SEGREGATION OF OXYGEN IN NIOBIUM- OXYGEN AND TANTALUM- OXYGEN ALLOYS. (AUGER ELECTRON SPECTROS 6-2065
ASYMMETRIC ION- ATOM COLLISIONS. (AUGER EFFECT) GAS TARGET EXPERIMENTS OF K-SHELL VACANCY SHARING IN 2-0523
EMISSION OF AUGER ELECTRONS BY ATOMS IN A METALLIC TARGET SUBJECTED TO AN IONIC BOMBARDMENT. 2-0383
OF COPPER- NICKEL ALLOYS DURING ARGON BOMBARD/ EFFECT OF TARGET TEMPERATURE ON SURFACE COMPOSITION CHANGES 5-1808
CONDARY ELECTRON EMISSION OF SOLID ALUMINUM AND MAGNESIUM TARGETS.   SE 0-0373
ENERGETIC AND SPATIAL DI/ ELECTRONIC EMISSION FROM SOLID TARGETS BOMBARDED BY NOBLE GAS IONS (10-100 KEV): 4-0776
SECONDARY ELECTRON SPECTRA OF SOME SOLID TARGETS UNDER IONIC BOMBARDMENT. 4-0942
ND CHEMICA/ STRESS CORROSION CRACKING OF ALPHA BRASS IN A TARNISHING AMMONIACAL ENVIRONMENT: FRACTOGRAPHY A 6-2095
ROSCOPY. TECHNOLOGICAL APPLICATION OF AUGER ELECTRON SPECT 1-0064
CITATION AND AUGER DECAY OF 5D CORE HOLES IN BISMUTH(III) TELLURIDE. /RIX ELEMENTS DEPENDENCE OF OPTICAL EX 2-0398
/M ELECTRON EFFECTS IN NARROW GAP MERCURY(1-X) CADMIUM(X) TELLURIDE AMBIENT TEMPERATURES. (AUGER RECOMBINA/ 2-0343
AUGER LIMITED CARRIER LIFETIMES IN MERCURY CADMIUM TELLURIDE AT HIGH EXCESS CARRIER CONCENTRATIONS. 2-0262
CARRIER DECAY AND TRANSPORT PROPERTIES IN MERCURY CADMIUM TELLURIDE. (AUGER RECOMBINATION)   TRANSIENT 2-0461
INATION MECHANISMS IN 8-14 MICRON MERCURY(1-X) CADMIUM(X) TELLURIDE. (AUGER-LIMITED)   RECOMB 2-0406
YSIS. (AUGER)   EVALUATION OF LEAD(0.8) TIN(0.2) TELLURIDE DETECTOR FABRICATION USING SURFACE ANAL 5-1593
RECOMBINATION IN A SEMICONDUCTOR (MERCURY(1-X) CADMIUM(X) TELLURIDE) UNDER A MAGNETIC FIELD.   AUGER 4-1092
N MNN AUGER SPECTRA OF SILVER, CADMIUM, INDIUM, ANTIMONY, TELLURIUM AND IODINE.   HIGH RESOLUTIO 4-0730
SPECTRUM OF AUGER KLL ELECTRONS OF TELLURIUM AND XENON. 4-0755
ETASTABLE GRAIN BOUNDARY SEGREGATION IN COPPER CONTAINING TELLURIUM. (AUGER SPECTROSCOPY)   M 6-2086
UGER RECOMBINATION OF EXCITONS BOUND TO NEUTRAL DONORS IN TELLURIUM DOPED GALLIUM PHOSPHIDE. /CIATED WITH A 2-0453
EFFECT OF TELLURIUM ON INTERGRANULAR COHESION OF IRON. 6-2099
CLEAN TELLURIUM SURFACES STUDIED BY LEED. 5-1134
EFFECT OF PRIOR AUSTENITIC GRAIN BOUNDARY COMPOSITION ON TEMPER BRITTLENESS IN A NICKEL- CHROMIUM- ANTIMO/ 6-2114
/ION OF MANGANESE AND ANTIMONY TO THE GRAIN BOUNDARIES OF TEMPER EMBRITTLED STEEL. (AUGER ELECTRON SPECTRO/ 6-2070
SCOPY.   FRACTURE SURFACE ANALYSIS OF TEMPER EMBRITTLED STEEL BY AUGER ELECTRON SPECTRO 6-2084
AUGER FRACTURE SURFACE ANALYSIS OF A TEMPER EMBRITTLED 3340- STAINLESS STEEL. 6-2085
FACE AND GRAIN BOUNDARY SEGREGATION OF STEEL IMPURITIES. (TEMPER EMBRITTLEMENT BY AUGER SPECTROSCOPY) SUR 6-2109
CONTAINING PHOSPHORUS AS AN IMPURITY. (AUGER EMISSION AN/ TEMPER EMBRITTLEMENT IN A NICKEL- CHROMIUM STEEL 6-2138
STUDIES USING AUGER ELECTRON EMISSION SPECTROSCOPY ON TEMPER EMBRITTLEMENT IN LOW ALLOY STEELS. 6-2119
NUM- VANADIUM.   LONG TIME ISOTHERMAL TEMPER EMBRITTLEMENT IN NICKEL- CHROMIUM- MOLYBDE 6-2140

FRACTURE SURFACES)
/EGREGATION AT SELECTIVE GRAIN BOUNDARIES AND ITS ROLE IN
SPECTROSCOPY)
TY GRAIN BOUNDARY SEGREGATI/ EFFECT OF SOLUTE ELEMENTS ON
CKEL LOW ALLOY STEEL/ EFFECTS OF COPPER AND PHOSPHORUS ON

ELECTRON SPECTROSCOPIC METHOD.
CTRON SPECTROSCOPY TO THE PLATINUM- OXYGEN SYSTEM AT HIGH
N ISOBARS FOR OXYGEN ON TUNGSTEN AT LOW PRESSURE AND HIGH
ACES. (AUGER MEASUREMENT METHODS)
APPLICATION OF AN "AUGER ELECTRON SPECTROMETER" FOR HIGH

IC PROPERTIES IN THIN INTRINSIC INDIUM ANTIMONIDE AT ROOM
SILICON (111)-7 STRUCTURE OBTAINED BY CLEAVAGE AT ROOM
/ER INVESTIGATION OF THE (100) SURFACE OF COPPER(3) GOLD.
LLIUM SELENIDE SINGLE CRYSTALS AT HIGH EXCITATION LEVELS/
ECTRUM OF NICKEL(II) OXIDE (100).
RON ENERGY DISTRIBUTIONS.   A
SILICON (SUBSTRATE)- GOLD (FILM) SYSTEM.   LOW
DIATIVE (AUGER) RECOMBINATION.   CARRIER
DEPOSITION. (HEED, LEED, AUGER ELECTRON SPECTROSCOP/ LOW
F SUPERCONDUCTING NIOBIUM. (AUGER SPECTRO/ EFFECT OF HIGH
NALYSIS.   DETECTION OF SURFACE NITROGEN IN HIGH CRITICAL
(AUGER EFFECT)   SURFACE DEBYE
PER- NICKEL ALLOYS DURING ARGON BOMBARD/ EFFECT OF TARGET
METRY OF (100) INDIUM PHOSPHIDE SURFACES.   EFFECT OF
SENSITIVE ESR AND AUGER ELECTRON SPECTROSCOPY.   LOW
AUGER ELECTRON SPECTROSCOPY OF HIGH
CARBIDE FORMATION ON NIOBIUM SURFACES DURING HIGH
/SURFACES IN THE PRESSURE RANGE 10(-9) TO 10(-4) TORR AND
YSTEMS. (AUGER SPECTROSCOPY)   POSSIBLE ORIGIN OF LOW
CHARACTERIZATION. (LEED- AUGER SYS/ ULTRAHIGH VACUUM LOW
AES ANALYSIS OF SODIUM IN A CORRODED BIOGLASS USING A LOW
NALYSIS OF SILICON THIN FILMS DEPOSITED ON CARBON AT HIGH
OBSERVATION OF SILICON- GOLD MIXED PHASE FORMATION AT HIGH
S IN NARROW GAP MERCURY(1-X) CADMIUM(X) TELLURIDE AMBIENT
ADSORPTION OF COBALT ON (001) RUTHENIUM AT
VACUUM ANNEALING OF STAINLESS STEEL AT
IGATIONS OF THE GAS- SOLID INTERFACE INTERACTIONS AT HIGH
INVESTIGATION OF OXYGEN ADSORPTION ON TUNGSTEN AT HIGH
TALUM NITRIDE AND TUNGSTEN NI/ SUPERCONDUCTING TRANSITION
OXYGEN EXPOSURE OF SAMARIUM, GADOLINIUM, AND
STRESS CORROSION CRACKING OF AN ALUMINUM- 5% MAGNESIUM
ARGON KLL AUGER SPECTRUM.
N TTL LOGIC FOR THE AUTOMATIC EXECUTION OF AUGER AND ESCA
TION OF ION, ELECTRON, AND X-RAY BEAMS IN MODERN MATERIAL
ON SPECTROSCOPY.   MECHANICAL

CTRONICS.   ELECTRONIC AUGER SPECTROSCOPY:
IMPURITIES. (AUGER ELECTRON SPECTROSCOPY)   SURFACES
ANE.
RATE)
IES, AND FLUORESCENCE YIELDS FOR MULTIPLY IONIZED NEON. /
IC RADIATION TRANSITION PROBABILITIES TO THE 1S STATE AND
RONIG TRANSITION PROBABILITIES TO THE ATOMIC 2S STATE AND
S AND L2-L3(X) COSTER-KRONIG TRANSITION PROBABILITIES.
OSTER-KRONIG TRANSITION PROBABILITIES F23 BASED ON THE G/
T TO MEDIUM RANGE OF ATOMIC NUMBER.
S OXIDE.   EXPERIMENTAL AND
ARGON KLL AUGER SPECTRUM. TEST OF
UCTORS. (SILICON, GERMANIUM, GRAPHITE, ELECTRON DIFFUSION
L2,3M4,5M4,5 AUGER SPECTRA OF METALLIC COPPER AND ZINC:
SURFACE SEGREGATION: COMPARISON BETWEEN
S.   MANY BODY PERTURBATION
O BOUND PARTICLES IN GALLIUM PHOSPHIDE.   AUGER
RDER TRANSFORMATION AT A (100) SURFACE OF COPPER(3) GOLD,
RONS.
DYNAMIC
BY IONS.

OMBINATION IN SIMPLE SEMICONDUCTORS.

AXATION.
S.

TION ENERGIES, AND/ APPLICATION OF MANY BODY PERTURBATION
/COMPOSITIONAL PROFILE AND ELECTRICAL CONDUCTIVITY OF THE
/ COMPOSITIONAL AND STRUCTURAL CHANGES THAT ACCOMPANY THE
/ ELECTRON YIELDS FROM COPPER BERYLLIUM OXYGEN SURFACE BY
ISM OF A CATALYTIC REAC/ USE OF PHOTOEMISSION, AUGER, AND
OPY. APPLICATION TO CESIUM-COVERED (110) GAL/ KINETICS OF
N IN A COPPER- 10 ATOMIC PERCENT ALUMINUM ALLOY USING AU/
/MPURITIES SEGREGATION TO THE (001) NICKEL SURFACE DURING
RON SPECTROSCOPY)   STUDY OF GRAIN BOUNDARY SEGREGANTS IN
NIDE INTERFACES: AN ASSESSMENT BY RUTHERFORD BACKSCATTER/
LLIUM OXIDE LAYERS. AUGER ELECTRON SPECTROSCOPY INVESTIG/
ER SPECTROSCOPY)   SURFACE CONDITION AND
DIED BY AN ELECTRONIC AUGER SPECTROSCOPIC MET/ SURFACE OF
S, AUGER SPECTROSCOPY OF SURFACE)   CHEMICAL CLEANING FOR

TEMPER EMBRITTLEMENT OF LOW ALLOY STEELS. (AES OF   6-2062
TEMPER EMBRITTLEMENT OF LOW ALLOY STEELS. (AUGER/   6-2057
TEMPER EMBRITTLEMENT OF LOW ALLOY STEELS. (AUGER   6-2059
TEMPER EMBRITTLEMENT OF LOW ALLOY STEELS. (IMPURI   6-2083
TEMPER EMBRITTLEMENT OF MANGANESE- MOLYBDENUM- NI   6-2043
TEMPER EMBRITTLEMENT OF STEELS.   6-2081
TEMPER EMBRITTLEMENT OF TIN DOPED STEELS BY AUGER   6-2128
TEMPERATURE.   APPLICATION OF AUGER ELE   5-1955
TEMPERATURE. /CTROSCOPY MEASUREMENTS OF ADSORPTIO   5-1282
TEMPERATURE ADSORPTION OF OXYGEN ON TUNGSTEN SURF   5-1664
TEMPERATURE ADSORPTION STUDIES.   DEVELOPMENT AND 3-0586
TEMPERATURE AUGER ELECTRON SPECTROSCOPY.   1-0005
TEMPERATURE. (AUGER PROCESS) /NT ON GALVANOMAGNET   2-0354
TEMPERATURE. (AUGER SURFACE SPECTRA)   5-1767
TEMPERATURE DEPENDENCE OF LONG RANGE ORDER AT A /   5-1854
TEMPERATURE DEPENDENCE OF THE EDGE EMISSION OF GA   2-0258
TEMPERATURE DEPENDENCE OF THE LOW ENERGY AUGER SP   4-0999
TEMPERATURE DEPENDENT CONTRIBUTION TO AUGER ELECT   2-0434
TEMPERATURE DIFFUSION OF GOLD INTO SILICON IN THE   5-1673
TEMPERATURE EFFECTS AND ENERGY TRANSFERS IN NONRA   2-0470
TEMPERATURE EPITAXY OF GERMANIUM FILMS BY SPUTTER   5-1537
TEMPERATURE HEAT TREATMENT ON PENETRATION DEPTH O   5-1938
TEMPERATURE NIOBIUM(3) GERMANIUM FILMS BY AUGER A   5-1804
TEMPERATURE OF THE SILICON (001) - 2X2 STRUCTURE.   5-1477
TEMPERATURE ON SURFACE COMPOSITION CHANGES OF COP   5-1808
TEMPERATURE ON THE SURFACE STRUCTURE AND STOICHIO   5-1164
TEMPERATURE OXIDATION OF SILICON STUDIED BY PHOTO   5-1786
TEMPERATURE OXYGEN- RHENIUM INTERACTIONS.   4-1111
TEMPERATURE PROTON IRRADIATION.   5-1886
TEMPERATURE RANGE 1500 TO 2600 DEGREES K AS MEAS/   5-1148
TEMPERATURE SILICON MIGRATION IN SILICON- METAL S   5-1437
TEMPERATURE STAGE FOR IN-SITU FILM DEPOSITION AND   3-0657
TEMPERATURE TECHNIQUE.   AUGER A   5-1721
TEMPERATURES.   5-1228
TEMPERATURES.   AUGER SPECTROSCOPIC 5-1680
TEMPERATURES. (AUGER RECOMBINATION) /CTRON EFFECT   2-0343
TEMPERATURES BELOW 300 DEGREES K.   5-1609
TEMPERATURES BETWEEN 770 AND 1470 K.   6-2145
TEMPERATURES BY AUGER ELECTRON SPECTROSCOPY. /EST   5-1147
TEMPERATURES BY AUGER ELECTRON SPECTROSCOPY.   5-1156
TEMPERATURES OF REACTIVELY SPUTTERED FILMS OF TAN   5-1541
TERBIUM STUDIED BY AUGER ELECTRON SPECTROSCOPY.   5-1332
TERNARY AND VARIOUS QUATERNARY ALLOYS. (AES)   6-2036
TEST OF THEORY.   2-0414
TEST PROGRAMS.   DIGITAL CONTROL UNIT I   3-0722
TESTING.   APPLICA   1-0034
TESTING: IN-SITU FRACTURE DEVICE FOR AUGER ELECTR   3-0659
TETRA..., SEE THE PARENT WORD.
TEXTBOOK FOR A COURSE OF LECTURES ON PHYSICAL ELE   1-0107
TEXTURES: AN INVESTIGATION OF SURFACES CONTAINING   5-1707
THEORETICAL ANALYSIS OF THE AUGER SPECTRA OF METH   4-1012
THEORETICAL FLUORESCENCE YIELDS FOR NEON. (AUGER   2-0283
THEORETICAL K-SHELL AUGER RATES, TRANSITION ENERG   2-0284
THEORETICAL K-SHELL FLUORESCENCE YIELDS.   ATOM   4-0918
THEORETICAL L1 FLUORESCENCE YIELDS. /AND COSTER-K   2-0329
THEORETICAL L2 AND L3 SUBSHELL FLUORESCENCE YIELD   2-0316
THEORETICAL L2 SUBSHELL FLUORESCENCE YIELDS AND C   2-0313
THEORETICAL STUDY OF THE AUGER EFFECT IN THE LIGH   2-0486
THEORETICAL STUDY OF THE AUGER SPECTRUM OF NITROU   4-0814
THEORY.   2-0414
THEORY) /SCOPY AND SECONDARY EMISSION IN SEMICOND   2-0240
THEORY AND EXPERIMENT.   4-0919
THEORY AND EXPERIMENT.   5-1566
THEORY APPLIED TO ATOMIC RADIATIONLESS TRANSITION   2-0386
THEORY AT DEFECTS.   APPLICATION TO STATES WITH TW 2-0457
THEORY, EXPERIMENT.   ORDER- DISO   5-1855
THEORY FOR ENERGY DISTRIBUTION OF SECONDARY ELECT   2-0241
THEORY OF AUGER EFFECT.   2-0381
THEORY OF AUGER EJECTION OF ELECTRONS FROM METALS   2-0563
THEORY OF AUGER SATELLITE ENERGY SHIFTS.   2-0507
THEORY OF AUGER TRANSITIONS.   1-0033
THEORY OF DONOR- ACCEPTOR RADIATIVE AND AUGER REC   2-0420
THEORY OF KLL AUGER ENERGIES.   4-1000
THEORY OF KLL AUGER ENERGIES INCLUDING STATIC REL   2-0506
THEORY OF THE AUGER EFFECT IN III-V SEMICONDUCTOR   2-0498
THEORY OF THE RADIATIVE AUGER EFFECT.   2-0224
THEORY TO THE CALCULATION OF AUGER RATES, CORRELA   2-0311
THERMAL AND ANODIC OXIDES OF INDIUM ANTIMONIDE. /   5-1971
THERMAL ANNEALING OF (100) SURFACES OF GALLIUM AR   5-1163
THERMAL CARBON OXYGEN NITROGEN(2), AND NOBLE GAS/   4-0786
THERMAL DESORPTION TECHNIQUES TO STUDY THE MECHAN   1-0042
THERMAL DESORPTION USING AUGER ELECTRON SPECTROSC   5-1293
THERMAL EFFECTS IN EQUILIBRIUM SURFACE SEGREGATIO   6-2030
THERMAL TREATMENT. WORK FUNCTION CHANGES AND AUG/   6-2069
THERMALLY EMBRITTLED MARAGING STEEL. (AUGER ELECT   6-2055
THERMALLY INDUCED PROCESSES AT GOLD- GALLIUM ARSE   5-1902
THERMALLY STIMULATED EXOELECTRON EMISSION OF BERY   5-1325
THERMIONIC EMISSION OF LANTHANUM HEXABORIDE. (AUG   5-1471
THERMIONIC EMITTERS MADE OF CARBIDE COMPOUNDS STU   5-1562
THERMOCOMPRESSION BONDING. (REMOVAL OF OXIDE FILM   5-1454

RFACES/ SEGREGATION OF IMPURITIES ON SURFACES OF THORIUM. THERMODYNAMICS AND KINETICS. (DETERMINATION OF SU 5-1313
R ADSORPTION OF XENON ON THE (0001) GRAPHITE FACE. THERMODYNAMICS AND KINETICS OF THE FIRST MONOLAYE 5-1859
RON SPECTROSCOPY.   SPUTTERING MEASUREMENTS ON CONTROLLED THERMONUCLEAR REACTOR MATERIALS USING AUGER ELECT 5-1826
ROLE OF AUGER EFFECT IN THERMOSTIMULATED PHENOMENA IN IONIC CRYSTALS.   2-0539
/LES OF SEVERAL CONTAMINATED AND CLEANED SURFACES OF GOLD THICK FILMS ON COPPER PLATES BY AUGER ELECTRON A/ 5-1550
F GOLD THROUGH PLATINUM FILMS ABOUT 2000 AND 600 ANGTROMS THICK STUDIED WITH AUGER SPECTROSCOPY. /INETICS O 5-1235
DISTINCTION BETWEEN ADSORBED MONOLAYERS AND THICKER LAYERS IN AUGER ELECTRON SPECTROSCOPY.   5-1797
PLE MODEL FOR DEPENDENCE OF AUGER INTENSITIES ON SPECIMEN THICKNESS.   SIM 2-0355
ER ELECTRON SPECTROSCOPY. THICKNESS DETERMINATION OF ULTRATHIN FILMS BY AUG 5-1449
ECTRONS.   NEW METHOD OF THIN FILM THICKNESS MEASUREMENT USING RADIOISOTOPE AUGER EL 5-1658
R ELECTRON SPECTROSCOPY. THICKNESS MEASUREMENTS OF ADSORBED LAYERS BY AUGE 5-1618
AUGER SPECT/ QUANTITATIVE DETERMINATION OF SURFACE OXIDE THICKNESS ON DEPOSITED METAL FILMS BY COMBINATION 6-2048
PECTROSCOPY.   DETERMINATION OF FILM THICKNESS OR ION ETCH RATE USING AUGER ELECTRON S 5-1984
CTRON ATTENUATION LENGTHS IN METALS: TRANSMISSION THROUGH THIN ADSORBED FILMS. (AUGER EFFECT) /ATION OF ELE 4-1088
AUGER ELECTRON EMISSION OF THIN CARBON FOILS IN REFLECTION AND TRANSMISSION. 4-0893
AUGER SPECTROSCOPY) DETECTION OF THIN CONTAMINATED LAYERS ON CONTACT SURFACES. (BY 5-1165
GROWTH AND STRUCTURE OF THIN COPPER FILMS ON (001) SURFACES OF NICKEL. 5-1227
TROSCOPY.   STUDY OF DIFFUSION IN THIN COPPER- LEAD FILMS USING AUGER ELECTRON SPEC 5-1972
ATIGUE CRACKS. (AUGER ELECTRON SPECTROSCOPY) EFFECT OF A THIN DIFFUSED SOLUTE LAYER ON THE NUCLEATION OF F 6-2037
HARACTERISTIC ENERGIES IN SECONDARY ELECTRON SPECTRA FROM THIN EVAPORATED PHOSPHORUS FILMS. /ER AND OTHER C 4-0986
AUGER ELECTRON SPECTROSCOPY FOR THIN FILM ANALYSIS.   1-0211
AUGER ELECTRON SPECTROSCOPY FOR THIN FILM ANALYSIS.   1-0212
SCANNING AUGER MICROSCOPE FOR THIN FILM ANALYSIS.   3-0645
IN SOLAR ENERGY CONVERSION.   AUGER THIN FILM ANALYSIS AS APPLIED TO VARIOUS ASPECTS 5-1960
SURFACE AND THIN FILM ANALYSIS. (AUGER SPECTROMETERS)   3-0609
LISHING DETECTION AND DETERMINATION LIMITS IN SURFACE AND THIN FILM ANALYSIS. (AUGER SPECTROSCOPY)   ESTAB 5-1572
. (REVIEW) THIN FILM ANALYSIS BY AUGER ELECTRON SPECTROSCOPY 1-0210
THIN FILM ANALYSIS FOR PROCESS EVALUATION. (AES) 1-0118
COPY) SPUTTERING IN THIN FILM ANALYSIS METHODS. (ESCA, AUGER SPECTROS 1-0214
CANNING AUGER MICROPROBE) SURFACE AND THIN FILM ANALYSIS OF SEMICONDUCTOR MATERIALS. (S 1-0089
SPECTROMETER.   DIGITAL METHODS FOR THIN FILM ANALYSIS USING A COMPUTER COUPLED AUGER 3-0640
ECHNIQUES. (AUGER SPECTROSCOPY) THIN FILM COMPOSITIONAL ANALYSIS. COMPARISON OF T 1-0054
COMPARISON OF TECHNIQUES. (REVIEW) SURFACE AND THIN FILM COMPOSITIONAL ANALYSIS. DESCRIPTION AND 1-0053
CHARACTERIZATION OF A CHROMIUM- GOLD THIN FILM DEPOSITION PROCESS BY AUGER ANALYSIS. 1-0037
SECONDARY ELECTRON EMISSION FROM THIN FILM GOLD METAL- OXIDE CERMETS. 4-1015
00) STUDIED WITH AUGER SPECTROSCOPY.   INTERDIFFUSIONS IN THIN FILM GOLD ON PLATINUM ON GALLIUM ARSENIDE (1 5-1232
CRYSTAL SURFACES OF ALKALI HALIDES AND THIN FILM GROWTH OBSERVED BY LEED AND AES. 5-1907
AUGER EXAMINATION OF CONTAMINANTS IN THIN FILM METALLIZATIONS. 5-1890
EFFECT) EXPERIMENTAL PROBLEMS OF STUDYING THIN FILM NUCLEATION. (ELECTRON MICROSCOPY, AUGER 5-1408
OXIDE- TANTALUM, AND SILICON- PLATINUM SILICIDE- TANTALUM THIN FILM STRUCTURES. (AES) /TANTALUM, SILICON DI 5-1260
SCOPY. ANALYSIS OF GRAIN BOUNDARY DIFFUSION IN BIMETALLIC THIN FILM STRUCTURES USING AUGER ELECTRON SPECTRO 6-2146
STRUCTURE DEPENDENT PROPERTIES OF THIN FILM SUPERCONDUCTORS. 5-1300
TY CHANGES W/ DIFFUSION MECHANISMS IN THE PALLADIUM- GOLD THIN FILM SYSTEM AND THE CORRELATION OF RESISTIVI 5-1393
E AUGER ELECTRONS.   NEW METHOD OF THIN FILM THICKNESS MEASUREMENT USING RADIOISOTOP 5-1658
PPLICATION OF AES TO THE STUDY OF SELECTIVE SPUTTERING OF THIN FILMS.   A 5-1933
AUGER ELECTRON SPECTRA OF MAGNESIUM THIN FILMS.   4-0873
ETRY AND AUGER ELECTRON SPECTROSCOPY, FOR THE ANALYSIS OF THIN FILMS. /ATION OF SECONDARY ION MASS SPECTROM 5-1214
ND EUTECTIC FORMATION IN TUNGSTEN PLATINUM- TUNGSTEN GOLD THIN FILMS. /ECTROSCOPY STUDY OF INTERDIFFUSION A 5-1259
SECONDARY ELECTRON EMISSION FROM THIN FILMS.   4-0894
SECONDARY ELECTRON EMISSION OF AMORPHOUS SEMICONDUCTOR THIN FILMS.   5-1239
SELECTED AREA AND IN-DEPTH AUGER ANALYSIS OF THIN FILMS.   5-1653
COPPER(II) OXIDE, COPPER(I) OXIDE, AND COPPER(I) SULFIDE THIN FILMS. /LECTRON AND AUGER SPECTRA OF COPPER, 4-0932
RAY PHOTOELECTRON AND AUGER ELECTRON SPECTRA OF MAGNESIUM THIN FILMS.   X- 5-1392
URFACE, AND MAGNETIC PROPERTIES OF NICKEL- IRON- CHROMIUM THIN FILMS. (AES)   CORROSION, S 5-1766
ECTRICAL PROPERTIES OF RF SPUTTERED YTTRIA- DOPED ZIRCONIA THIN FILMS. (AUGER EFFECT)   MORPHOLOGICAL AND EL 5-1385
GROWTH AND STRUCTURE OF SEMICONDUCTING THIN FILMS. (AUGER ELECTRON SPECTROSCOPY) 5-1509
/HIPS BETWEEN SUBSTRATE SURFACE CHEMISTRY AND ADHESION OF THIN FILMS. (AUGER ELECTRON SPECTROSCOPY OF ALUM/ 5-1851
TECHNIQUES FOR ELEMENTAL COMPOSITION PROFILING IN THIN FILMS. (AUGER SPECTROSCOPY) 5-1273
DIFFUSION STUDIES IN GOLD- PLATINUM THIN FILMS BY AUGER ELECTRON SPECTROSCOPY. 5-1636
GROWTH MODE DISCRIMINATION IN THIN FILMS BY AUGER ELECTRON SPECTROSCOPY. 5-1624
DIFFUSION MEASUREMENT IN THIN FILMS BY AUGER SPUTTER ETCHING TECHNIQUE. 5-1439
ES.   AUGER ANALYSIS OF SILICON THIN FILMS DEPOSITED ON CARBON AT HIGH TEMPERATUR 5-1228
TRANSPORT OF OXYGEN IN AMORPHOUS GERMANIUM THIN FILMS DURING ANNEALING. (AUGER SPECTROSCOPY) 5-1548
/S OF TRANSITION METAL SUPERCONDUCTORS AT SURFACES AND IN THIN FILMS. (LEED- AES STUDIES OF NIOBIUM SURFAC/ 5-1848
UGER ELECTRON SPECTROSCOPY. THIN FILMS. LOW ENERGY ELECTRON DIFFRACTION AND A 1-0100
SECONDARY ELECTRON EMISSION FROM THIN FILMS. (OF GOLD, APPARATUS) 3-0638
/E RECOMBINATION PROCESSES (AUGER) ON THE SURFACES AND IN THIN FILMS OF SILICON AND GERMANIUM SUBJECTED TO/ 5-1990
GROWTH AND STRUCTURE OF TITANIUM OXIDE THIN FILMS. PART-1. (AUGER SPECTRA) 5-1980
ELECTRON SPECTROSCOPY ON THIN FILMS. (SCANNING AUGER MICROSCOPE) 5-1405
REALIZATION OF A MOSSBAUER SPECTROMETER FOR SURFACES AND THIN FILMS STUDY. (AUGER) 3-0690
LOW INTRINSIC STRESS AND GOOD ADHESION ALLOY THIN FILMS. (TITANIUM AND CHROMIUM ALLOYS, AES) 5-1909
DIFFUSION STUDIES IN CHROMIUM- PLATINUM THIN FILMS USING AUGER ELECTRON SPECTROSCOPY. 5-1286
IUM INTO GOLD. (AUGER ELECTRON/ DIFFUSION MEASUREMENTS IN THIN FILMS UTILIZING WORK FUNCTION CHANGES: CHROM 5-1889
LECTRON SPECTROSCOPY STUDY.   SILICON HOMOEPITAXIAL THIN FILMS VIA SILANE PYROLYSIS: HEED AND AUGER E 5-1428
MATERIAL ANALYSIS OF THIN FOILS BY AUGER ELECTRON SPECTROSCOPY. 5-1798
TUDIED BY AUGER ELECTRON AND X/ ELECTRONIC STRUCTURE OF A THIN GOLD FILM DEPOSITED ON A SILICON SUBSTRATE S 5-1435
/TS OF DIFFUSION CURRENT ON GALVANOMAGNETIC PROPERTIES IN THIN INTRINSIC INDIUM ANTIMONIDE AT ROOM TEMPERA/ 2-0354
ON DIFFRACTION./ STUDY OF PLATINUM ELECTRODES BY MEANS OF THIN LAYER ELECTROCHEMISTRY AND LOW ENERGY ELECTR 5-1486
HE PRESENCE OF ELECTRON IRRADIATION. (AUGER SP/ GROWTH OF THIN METALLIC FILMS ON INSULATING SUBSTRATES IN T 5-1317
EFFECT OF IMPURITIES ON INTRINSIC STRESS IN THIN NICKEL FILMS. (AUGER EFFECT) 5-1131
EFFECT OF IMPURITIES ON INTRINSIC STRESS IN THIN NICKEL FILMS. (AUGER SPECTROSCOPY) 5-1130
AUGER ANALYSIS OF THIN OXIDE FILMS ON LEAD- INDIUM ALLOYS. 5-1249
ICON (111).   THIN REACTION LAYERS AND SURFACE STRUCTURE OF SIL 5-1879
SOME RECENT STUDIES OF VERY THIN SILICON DIOXIDE FILMS. 5-1935
ELECTROCHEMISTRY AT THIN SOLID FILMS. (AUGER SPECTRA) 4-1026
R SPECTROSCOPY.   SILICON DIFFUSION THROUGH THIN TUNGSTEN FILMS ON SILICON, STUDIED WITH AUGE 5-1234
SCANNING AUGER MICROSCOPY. THIRD DIMENSION IN SCANNING ELECTRON MICROSCOPY. 1-0117
, AND RADIATIVE RATES AND FLUORESCENCE YIELDS FOR SODIUM- THORIUM.   ATOMIC L-SHELL COSTER-KRONIG, AUGER 4-0963
D RADIATIVE RATES, AND FLUORESCENCE YIELDS FOR CALCIUM TO THORIUM. ATOMIC M-SHELL COSTER-KRONIG, AUGER, AN 4-0964
RON EMISSION.   CARBON EVAPORATION FROM A THORIUM DISPENSER CATHODE OBSERVED BY AUGER ELECT 5-1400

CED MULTILAYER DIELECTRIC MIRROR DESIGNS. (ZINC SELENIDE, THORIUM FLUORIDE) AUGER ANALYSIS OF THREE ENHAN 5-1328
ILIBRIA. IMPURITIES ON THORIUM METAL: AUGER STUDY OF BULK TO SURFACE EQU 5-1314
ED AND WORK FUNCTION MEASUREMENTS FOR EPITAXIAL GROWTH OF THORIUM ON A TUNGSTEN (100) SUBSTRATE. /SCOPY, LE 5-1739
ION OF SURFACES/ SEGREGATION OF IMPURITIES ON SURFACES OF THORIUM. THERMODYNAMICS AND KINETICS. (DETERMINAT 5-1313
NG, SURFACE STRUCTURE, AND INTERACTION WITH OXYGEN AND C/ THORIUM (100) CRYSTAL FACE. A STUDY OF ITS CLEANI 5-1883
GER DECAY) METASTABLE AUTOIONIZING THREE- AND FOUR ELECTRON STATES IN BERYLLIUM. (AU 2-0295
D INTERFACES OF SOLIDS BY THE AUGER ELECTRON MICROANALYT/ THREE-DIMENSIONAL ELEMENT ANALYSIS OF SURFACES AN 5-1412
TIES. DIGITIZING AUGER DATA. THREEFOLD APPROACH TO ENHANCE ANALYTICAL CAPABILI 1-0183
,500 AND KRYPTON M4,5NN PROCESSES BY ELECTRON IMPACT NEAR THRESHOLD. AUGER ELECTRON EJECTION FROM XENON N4 4-1010
LL IONIZATION OF THE N6,7 LEVEL IN GOLD, BISMU/ ANOMALOUS THRESHOLD BEHAVIOR FOR ELECTRON INDUCED INNER SHE 4-1067
AUGER SPECTRA OF GERMANIUM AND SELENIUM. THRESHOLD EFFECTS ON RELATIVE INTENSITIES IN LMM 4-0767
L-SHELL IONIZATION THRESHOLD STRUCTURE FROM AUGER ELECTRON SPECTRA. 2-0360
) INTERPRETATION OF KL2,3L2,3 RADIATIVE AUGER THRESHOLDS IN METALS AND SEMICONDUCTORS. (SILICON 4-0723
KLL AUGER ELECTRON SPECTRA OF THULIUM AND LUTETIUM. 4-0753
DLE REGION. AUGER RECOMBINATION IN SILICON RECTIFIERS AND THYRISTORS LOADED IN FORWARD DIRECTION IN THE MID 2-0415
CE AUGER LINE SHAPE: SILICON (111). TIGHT BINDING CALCULATION OF A CORE VALENCE VALEN 4-0831
HROMIUM- MOLYBDENUM- VANADIUM. TIME ISOTHERMAL TEMPER EMBRITTLEMENT IN NICKEL- C 6-2140
LONG TIN. 5-1386
AUGER ANALYSIS OF SURFACE FILMS ON SILVER- TIN ALLOYS FROM COMBINED X-RAY PHOTOELECTRON AND/ 5-1195
/COMPOSITION AND DEPTH CONCENTRATION PROFILE OF PLATINUM- TIN, AND IRON- ALUMINUM ALLOYS. (BY LEED AND AES) 6-2033
/ FRICTION STUDIES PERFORMED ON COPPER- ALUMINUM, COPPER- TIN, AND IRON- 6.55 ATOMIC PERCENT SILICON. /OYS 6-2027
COPPER- 1 ATOMIC PERCENT INDIUM, COPPER- 2 ATOMIC PERCENT TIN DOPED STEELS BY AUGER ELECTRON SPECTROSCOPIC 6-2128
METHOD. TEMPER EMBRITTLEMENT OF TIN IN THE SURFACE LAYER OF COPPER- TIN ALLOY SYS 5-1867
TEM DUE TO FRICTION. (AUGER SPEC/ CONCENTRATION CHANGE OF TIN IN ULTRAHIGH VACUUM AND IN THE PRESENCE OF O/ 5-1196
/CTROSCOPIC STUDY OF THE SURFACE COMPOSITION OF PLATINUM- TIN ON IRON. QUANTITATIVE AUGER ELE 1-0178
CTRON SPECTROSCOPY: COMPARISON OF TECHNIQUES FOR ADSORBED TIN ON MOLYBDENUM (100). 5-1490
AUGER- LEED INVESTIGATION OF TIN OVER A GOLD SURFACE OBSERVED WITH AUGER ELECT 6-2136
RON SPECTROSCOPY. SURFACE DIFFUSION OF TIN(0.2) TELLURIDE DETECTOR FABRICATION USING SUR 5-1593
FACE ANALYSIS. (AUGER) EVALUATION OF LEAD(0.8) TIN- PHOSPHORUS ALLOYS. (AUGER SPECTROSCOPY) /OSP 5-1868
HORUS ON THE FRICTION AND WEAR CHARACTERISTICS OF COPPER- TIN-119. 2-0576
BACKSCATTER MOSSBAUER STUDIES USING CONVERSION ELECTRONS. TIN-119 BY DETECTION OF CONVERSION AND AUGER ELEC 5-1796
TRONS. APPLICATION/ MOSSBAUER SPECTROSCOPY OF IRON-57 AND TISSUE: USE OF SOFT X-RAYS. (AUGER ELECTRON IRR/ 2-0531
/ON EFFECT ON INCLUSION OF MODERATELY HEAVY NUCLEI IN THE TITANIUM. 4-0783
AUGER SPECTROSCOPY OF TITANIUM. EFFECT OF 2-0367
MODULATION AMPLITUDE ON ELECTRON EXCITED AUGER DATA FROM TITANIUM. EFF 5-1158
ECT OF OXIDATION ON THE HIGH RESOLUTION AUGER SPECTRUM OF TITANIUM AND CHROMIUM ALLOYS, AES) 5-1909
LOW INTRINSIC STRESS AND GOOD ADHESION ALLOY THIN FILMS. TITANIUM AND IRON. IN 4-1060
TERFACIAL AND PLASMON DECAY PEAKS IN THE AUGER SPECTRA OF TITANIUM AND ITS INTERACTION WITH BROMINE AND CHL 5-1538
ORINE. (AUGER SPECTROSCOPY) SURFACE STRUCTURE OF TITANIUM AND TITANIUM MONOXIDE SURFACES USING AUG 5-1384
ER ELECTRON AND SOFT X-RAY AP/ QUANTITATIVE COMPARISON OF TITANIUM AS MONITORED BY AUGER ELECTRON SPECTROSC 6-2067
OPY. SURFACE SEGREGATION IN TITANIUM BETWEEN 25 DEGREES C AND 400 DEGREES C. 6-2115
OXIDATION OF TITANIUM, CHROMIUM, AND NICKEL. (AUGER EFFECT) /O 2-0392
RE ELECTRON BINDING ENERGIES OF SURFACE AND BULK ATOMS OF TITANIUM DIOXIDE AUGER AND APPEARANCE POTENTIAL S 5-1834
PECTRA INDUCED BY INCIDENT ELECTRON BEAM / CHANGES IN THE TITANIUM(II) AND TITANIUM(IV) OXIDES. MOLEC 4-1069
ULAR ORBITAL EFFECTS ON THE TITANIUM LMV AUGER SPECTRA OF TITANIUM(IV) OXIDE. A 5-1893
ES ION SPUTTERING ANALYSIS AND THE SURFACE COMPOSITION OF TITANIUM(IV) OXIDES. MOLECULAR ORBITAL EFFE 4-1069
CTS ON THE TITANIUM LMV AUGER SPECTRA OF TITANIUM(II) AND TITANIUM LMV AUGER SPECTRA OF TITANIUM(II) AND TI 4-1069
TANIUM(IV) OXIDES. MOLECULAR ORBITAL EFFECTS ON THE TITANIUM ON TUNGSTEN. 5-1140
AUGER ELECTRON SPECTROSCOPY OF LAYERED GROWTH OF TITANIUM OXIDE THIN FILMS. PART-1. (AUGER SPECTRA 5-1980
/PECTROSCOPY, AND WORK FUNCTION OF CESIUM ON TUNGSTEN AND TITANIUM SINGLE CRYSTALS IN THE SUBMONOLAYER REG/ 5-1829
AL INTERACTION WITH CHLORINE GA/ CHLORINE ADSORPTION AT A TITANIUM SURFACE. SEGREGATION FROM THE BULK CRYST 5-1833
PECTRO/ SPUTTER CLEANING AND ETCHING OF CRYSTAL SURFACES. (TITANIUM, TUNGSTEN, SILICON) MONITORED BY AUGER S 5-1828
ND SECONDARY ION MASS SPECTROMETRY FOR ALUMINUM, SILICON, TITANIUM, VANADIUM, AND CHROMIUM. /SPECTROSCOPY A 5-1553
ORDER- DISORDER PHENOMENA AT THE SURFACE OF ALPHA TITANIUM- OXYGEN SOLID SOLUTIONS. 5-1819
SECONDARY ELECTRON SPECTROSCOPY. SURFACE SENSITIVE TOOL. (AES) 1-0095
IN SOLIDS NEAR THEIR S/ AUGER ELECTRON SPECTROSCOPY AS A TOOL FOR MEASURING THE DIFFUSION OF FOREIGN ATOMS 1-0186
) AUGER ELECTRON SPECTROMETER AS TOOL FOR SURFACE ANALYSIS. (CONTAMINATION MONITOR 3-0706
AUGER SPECTROMETRY AS A TOOL FOR SURFACE CHEMISTRY. (REVIEW) 1-0162
CHARACTERISTIC AUGER ELECTRON EMISSION AS A TOOL FOR THE ANALYSIS OF SURFACE COMPOSITION. 1-0170
ELLIPSOMETRY AS A TOOL IN SURFACE STUDIES. 1-0128
CES. AUGER ELECTRON SPECTROSCOPY: NEW TOOL IN THE CHARACTERIZATION OF GLASS FIBER SURFA 1-0175
SECONDARY ELECTRON SPECTROSCOPY. SURFACE SENSITIVE TOOL. (REVIEW, 22 REFS) 1-0094
/ A CLOSE LOOK AT SURFACES. (SURFACE STRUCTURES, SPECIES, TOPOGRAPHY, PHENOMENA, FIELD ION MICROSCOPY, LEED 5-1492
LOW ENERGY ELECTRON DIFFRACTION AND SURFACE TOPOGRAPHY. (REVIEW, AUGER SPECTROSCOPY) 1-0185
TOTAL AND PARTIAL ATOMIC LEVEL WIDTHS. 4-0911
N K-SHELLS OF SURFACE ATOMS. (AUGER INTENSITIES OF TUNGS/ TOTAL ELECTRON IMPACT IONIZATION CROSS SECTIONS O 4-0847
N ARGON BY THE IMPACT/ CHARGE DEPENDENCE OF DEPENDENCE OF TOTAL K-SHELL VACANCY PRODUCTION CROSS SECTIONS I 2-0446
ROBE ANALYSIS, AUGER EFFECT, REVIEW) TRACE ANALYSIS TECHNIQUES FOR SOLIDS. (ELECTRON P 1-0093
ADSORPTION OF OXYGEN ON MOLYBDENUM (111): EFFECT OF TRACE IMPURITIES. 5-1568
OF THE CLEANLINESS OF A SURFACE. (USE OF AES TO DETERMINE TRACE IMPURITIES) HOW TO ASSURE ONESELF 5-1154
ED BY AUGER EMISSION SPECTROSCOPY. METALLIC TRANSFER BETWEEN METALS IN SLIDING CONTACT EXAMIN 5-1729
BINDING, VALENCY AND CHARGE TRANSFER IN CHEMICAL REACTIONS ON METAL SURFACES. 5-1763
N AUGER SPECTRA (MONOLAYER ADSORPTION, ESCA SIMPLE CHARGE TRANSFER MODEL). CHEMICAL SHIFTS I 4-1115
GER EMISSION SPECTROSCOPY. ADHESION AND TRANSFER OF PLATINUM IRON TO METALS STUDIED BY AU 5-1730
UDIED BY AUGER EMISSION SPECTROSCOPY. ADHESION AND TRANSFER OF POLYTETRAFLUORO ETHYLENE TO METALS ST 5-1731
CARRIER TEMPERATURE EFFECTS AND ENERGY TRANSFERS IN NONRADIATIVE (AUGER) RECOMBINATION. 2-0470
LD, THEORY, EXPERIMENT. ORDER- DISORDER TRANSFORMATION AT A (100) SURFACE OF COPPER(3) GO 5-1855
AFTEREFFECTS) CHEMICAL EFFECTS OF NUCLEAR TRANSFORMATIONS IN MOSSBAUER SPECTROSCOPY. (AUGER 2-0564
PHASE TRANSFORMATIONS OF SILICON (111) SURFACE. 5-1343
(111) SURFACES. STRUCTURE TRANSFORMATIONS ON CLEAVED AND ANNEALED GERMANIUM 5-1713
IN MERCURY CADMIUM TELLURIDE. (AUGER RECOMBINATION) TRANSIENT CARRIER DECAY AND TRANSPORT PROPERTIES 2-0461
OR MICROSPOT SURFACE AND IN-DEPTH ANALYSIS OF SILICON AND TRANSISTOR METALLIZATIONS. / AUGER SPECTROSCOPY F 5-1262
FLUORESCENCE YIELDS OF ARGON, CHLORINE AND SULFUR. (AUGER TRANSITION) K-SHELL 4-1014
ECTRONS IN SOFT X-RAY DOUBLE IONIZATION SATELLITE (AUGER) TRANSITION. /SSIBLE LOCALISATION OF CONDUCTION EL 4-0437
ISM DEPENDENCE OF THE ARGON 2S FLUORESCENCE YIELD. (AUGER TRANSITION) STRIPPING MECHAN 4-0957
S 30-94) KLL AUGER TRANSITION ABSOLUTE PROBABILITIES. (ATOMIC NUMBER 2-0251
THE IONIZATION OF ALUMINUM BY ELECTRON IMPACT. AUGER TRANSITION AND SUPPLEMENTARY ELECTRON EJECTION IN 2-0230
TION OF CARBON TO A NICKEL (111) SURFACE: A SURFACE PHASE TRANSITION. (AUGER SPECTROSCOPY) /LIBRIUM SEGREGA 5-1803
SUBSTRATES FROM THE KLL AUGER SPECTRUM/ EVALUATION OF THE TRANSITION DENSITY FUNCTION OF CARBON ON VARIOUS 5-1651
L1,L2,3M COSTER-KRONIG TRANSITION ENERGIES. 4-0929

MULTIPLY IONIZED NEON. / THEORETICAL K-SHELL AUGER RATES, TRANSITION ENERGIES, AND FLUORESCENCE YIELDS FOR 2-0284
ARIOUSLY IONIZED STATES CF ARGON.   K-SHELL AUGER RATES, TRANSITION ENERGIES, AND FLUORESCENCE YIELDS OF V 2-0280
/TION AND CHEMICAL BONDING. PART-1: AUGER ELECTRONS. (KLL TRANSITION ENERGIES FOR SULFUR, ALUMINUM, SILICO/ 2-0327
ATOM.   CALCULATED K-AUGER ELECTRON AND K X-RAY TRANSITION ENERGIES FOR THE MULTIPLY IONIZED NEON 4-0955
            AUGER CATALOG CALCULATED TRANSITION ENERGIES LISTED BY ENERGY AND ELEMENT. 1-0041
                    (AUGER) TRANSITION ENERGIES. (REVIEW) 1-0109
CALCULATI/ DETAILED ANALYSIS OF THE L3M4,5M4,5: 1G4 AUGER TRANSITION IN ATOMIC ZINC: IMPROVED AUGER ENERGY 4-0884
            NEW RADIATIVE TRANSITION IN GERMANIUM. 4-1002
MICROWAVE METHOD. (PRODUCTION OF AUGER/ METAL- DIELECTRIC TRANSITION IN GERMANIUM AND SILICON STUDIED BY A 4-0752
S SECTION M/ A STUDY OF THE INTENSITY OF THE L2,3VV TRANSITION IN MAGNESIUM AND SILICON FROM THE CROS 4-0765
POLARIZATION OF THE ELECTRON GAS ON THE ENERGIES CF AUGER TRANSITION IN METALS.   EFFECT OF 2-0435
TERMINATION OF OSCILLATCR STRENGTH FOR THE FIRST DISCRETE TRANSITION IN NITROGEN FROM THE AUGER DECAY.   DE 2-0575
            L1,L2,3V AUGER TRANSITION IN SILICON. 4-0779
N (0001) GRAPHITE. (AUGER SPECTRCS/ TWO-DIMENSIONAL PHASE TRANSITION IN XENON SUBMONOLAYER FILMS ADSORBED O 5-1858
LOWING INTERNAL CONVERSICN AND ELECTRON CAPTURE. (K-AUGER TRANSITION INTENSITIES) /CLOUDS REARRANGEMENT FOL 2-0334
RONS.   MULTIPLE EXCITATION PROCESSES IN TRANSITION METAL ATOMS. PART-2: INNER AUGER ELECT 2-0362
            SURFACE STUDIES OF SOME TRANSITION METAL DICHALOGOGENIDES. 5-1969
            (AUGER) ELECTRON SPECTROSCOPY OF TRANSITION METAL OXIDE SURFACES. 5-1339
IN THIN FILMS. (LEED- AES STUDIES OF / SCME FROPERTIES OF TRANSITION METAL SUPERCONDUCTORS AT SURFACES AND 5-1848
            AUGER ELECTRON SPECTROSCOPY OF TRANSITION METALS. 4-0864
INTERATOMIC AUGER PROCESSES IN SOME ADSORBATES ON TRANSITION METALS. 4-1045
ION EFFECTS IN AUGER SPECTRA OF SEVERAL GAS ALSORBATES ON TRANSITION METALS. /TOMIC TRANSITIONS AND RELAXAT 4-1042
L-SHELL AUGER ENERGIES AND INTENSITIES IN 3D TRANSITION METALS. 4-1018
            RADIATIVE KM2 AUGER EFFECT IN SOME TRANSITION METALS. 4-1057
            THE K-FAR-M2 RADIATIVE AUGER EFFECT IN TRANSITION METALS. 2-0502
RAATOMIC RELAXATION CN AUGER AND BINDING ENERGY SHIFTS IN TRANSITION METALS AND SALTS. /ATIVE EFFECT OF EXT 2-0412
            LMM AUGER SPECTRA OF SOME TRANSITION METALS OF THE FIRST SERIES. 4-0810
TRON SPECTROSCOPY)   AMORPHOUS TC CRYSTALLINE TRANSITION OF SILICON (111) SURFACES. (AUGER ELEC 5-1780
K X-RAY AND AUGER ELECTRCN ENERGIES FOR NEON BY A TRANSITION OPERATOR METHOD. 2-0393
ECTRA.   TRANSITION POTENTIAL MODELS FOR AUGER ELECTRON SP 2-0510
            AUGER AND RADIATIVE TRANSITION PROBABILITIES. 2-0328
            KLL AUGER TRANSITION PROBABILITIES. 2-0304
            RELATIVISTIC STUDY OF KLL AUGER TRANSITION PROBABILITIES. 4-0802
3 SUBSHELL FLUORESCENCE YIELDS AND L2-L3(X) COSTER-KRONIG TRANSITION PROBABILITIES.   THEORETICAL L2 AND L 2-0316
X-RAY FLUORESCENCE YIELDS, AUGER, AND COSTER-KRONIG TRANSITION PROBABILITIES. 2-0259
D INTERMEDIATE ATOMIC NUMBERS.   KLL AUGER TRANSITION PROBABILITIES FOR ELEMENTS WITH LOW AN 4-0803
CTRUM OF SELENIUM.   TRANSITION PROBABILITIES FOR THE L2,3MM AUGER SPE 4-1033
/ETICAL L2 SUBSHELL FLUORESCENCE YIELDS AND CCSTER-KRONIG TRANSITION PROBABILITIES F23 BASED ON THE GREEN-/ 2-0313
TES IN NEON.   TRANSITION PROBABILITIES OF KL- LLL AUGER SATELLI 4-1049
ND THEORETICAL L1 FLUORESCENCE Y/ AUGER AND COSTER-KRONIG TRANSITION PROBABILITIES TO THE ATOMIC 2S STATE A 2-0329
RETICAL K-SHELL FLUCRESCENCE YIELDS.   ATOMIC RADIATION TRANSITION PROBABILITIES TO THE 1S STATE AND THEO 4-0918
RA.   AUGER TRANSITION RATE FOR THE NOBLE GAS SATELLITE SPECT 4-0972
            AUGER TRANSITION RATE IN MIXED COUPLING SCHEME. 2-0342
RELATIVISTIC CALCULATIONS OF ATOMIC X-RAY AND AUGER TRANSITION RATES. 2-0485
            RELATIVISTIC KLL AUGER TRANSITION RATES. 4-0782
NTS ARGON TO XENON.   K-SHELL AUGER TRANSITION RATES AND FLUORESCENT YIELDS FOR ELEME 4-0961
NTS BERYLLIUM TO ARGON.   K-SHELL AUGER TRANSITION RATES AND FLUORESCENT YIELDS FOR ELEME 4-0962
            TRANSITION RATES FOR INTERATOMIC AUGER PROCESSES. 2-0438
ND RADIATIVE MATRIX ELEMENTS, AND AUGER AND COSTER-KRONIG TRANSITION RATES IN J-J COUPLING. /STER-KRONIG, A 4-0967
F ELECTRONS.   AUGER TRANSITION RATES IN J-J COUPLING FOR D HOLES AND 2-0444
ND SCREENING PA/ SENSITIVITY OF L1,L2,3M4,5 COSTER-KRONIG TRANSITION RATES TO VARIATIONS IN ENERGY LEVELS A 2-0303
/ UNEXPLAINED PHENOMENA IN THE AUGER SPECTRA OF THE FIRST TRANSITION SERIES. (COMMENTS ON PAPER BY CHUNG AN 4-0812
ILMS OF TANTALUM NITRIDE AND TUNGSTEN NI/ SUPERCONDUCTING TRANSITION TEMPERATURES OF REACTIVELY SPUTTERED F 5-1541
GRAPHITE. (AUGER ELECTRON SP/ FIRST CRDER TWO DIMENSIONAL TRANSITION: XENON ADSORBED ON THE (0001) FACE OF 5-1857
TIONS OF ATOMIC NITROGEN, ATOMIC OXYGEN, AND NEON. (AUGER TRANSITIONS) /LECTRON IMPACT IONIZATION CROSS SEC 2-0363
S PRODUCED BY INNER SHELL IONIZATION (AUGER AND RADIATIVE TRANSITIONS).   ATOMIC VACANCY DISTRIBUTION 2-0544
            AUGER TRANSITIONS. 1-0218
            CHEMICAL SHIFTS IN URANIUM AUGER TRANSITIONS. 4-0737
            DCUBLE PHOTOIONIZATION OF NEON. (AUGER TRANSITIONS) 2-0310
            EFFECT OF ATOMIC CORRELATIONS CN KLL AUGER TRANSITIONS. 4-0801
D IN ARGON(+)- ARGON COLLISIONS. (AUGER AND COSTER-KRONIG TRANSITIONS)   ENERGY SPECTRA OF ELECTRONS EJECTE 4-1007
LOW ENERGY K-ALPHA SPECTRUM OF SILICON. (RADIATIVE AUGER TRANSITIONS) INFLUENCE OF CHEMICAL BONDING ON THE 2-0226
BODY PERTURBATION THEORY APPLIED TO ATOMIC RADIATIONLESS TRANSITIONS.   MANY 2-0386
STATES BY THE L X-RAY- L X-RAY COINCIDENCE METHOD. (AUGER TRANSITIONS) /NT OF THE PRODUCTION OF LL VACANCY 2-0482
NATURE OF SCME X-RAY SATELLITES (RADIATIVE AUGER TRANSITIONS). 2-0290
THE RARE GASES BY 30 MEV OXYGEN IONS. (INNER SHELL AUGER TRANSITIONS)   OUTER SHELL EXCITATION OF 2-0298
AXATION PROCESSES ACCOMPANYING CADMIUM M4,5N4,5N4,5 AUGER TRANSITIONS.   REL 2-0559
            SEMI-AUGER ELECTRON TRANSITIONS. 2-0424
SEQUENTIAL STRIPPING IN HIGHLY IONIZED NEON. (AUGER TRANSITIONS) 2-0441
THE AUGER EFFECT AND OTHER RADIATIONLESS TRANSITIONS. 1-0022
THE L1,L2,3M4,5 COSTER-KRONIG TRANSITIONS. 2-0248
THEORY OF AUGER TRANSITIONS. 1-0033
NON(III) LEVELS EXCITED BY ELECTRON IMPACT VIA N4,5 AUGER TRANSITIONS.   VACUUM ULTRAVIOLET EMISSION OF XE 4-0878
S, AND IONIZATION EQUILIBRIUM EQUATIONS MODIFIED BY AUGER TRANSITIONS.   X-RAY IONIZATION CROSS SECTION 2-0562
X-RAY PHOTOELECTRON SPECTRUM OF DIAMOND. (AUGER TRANSITIONS) 4-0851
DISTINCTION BETWEEN WEAK AUGER TRANSITIONS AND INTERNAL X-RAY PHOTOEMISSION. 2-0436
RA OF SEVERAL GAS ADSORBATES ON TRANSITION M/ INTERATOMIC TRANSITIONS AND RELAXATION EFFECTS IN AUGER SPECT 4-1042
            AUGER TRANSITIONS AND SHAPE OF X-RAY SPECTRUM. 4-1104
PENETRATION EFFECTS IN CONVERTED MUONIC TRANSITIONS. (AUGER COEFFICIENTS) 2-0478
TO Z EQUALS 88.   AUGER TRANSITIONS BETWEEN 0 AND 1500 EV FOR ELEMENTS UP 4-1013
PY.   INVESTIGATION OF LOW-Z COSTER-KRONIG TRANSITIONS BY AUGER AND PHOTOELECTRON SPECTROSCO 2-0578
4. PART-2.   ABSOLUTE PROBABILITIES FOR AUGER- KLL TRANSITIONS DEPENDENT ON THE ATOMIC NUMBER Z=30-9 2-0252
94.   RELATIVE PRCBABILITIES FOR AUGER KLL TRANSITIONS DEPENDENT ON THE ATOMIC NUMBER 29 TO 2-0253
S OF ATOMS.   COMPUTER PROGRAM TO CALCULATE AUGER TRANSITIONS FOR A SURFACE CONTAINING SEVERAL KIND 1-0039
            LLM VERSUS LMM AUGER TRANSITIONS FOR LIGHT ELEMENTS. 4-0811
CATALOG OF CALCULATED AUGER TRANSITIONS FOR THE ELEMENTS. (REVIEW) 1-0040
LLM VERSUS LMM AUGER TRANSITIONS FOR THE LIGHT ELEMENTS. 4-0804
            4D AUGER TRANSITIONS FROM EXCITED STATES IN SAMARIUM METAL 4-0823
OBSERVATION OF THREE ELECTRON AUGER TRANSITIONS IN ATOMS WITH TWO INNER VACANCIES. 2-0235
DETECTION OF A NEW TYPE OF AUGER TRANSITIONS IN ATOMS WITH TWO INTERNAL VACANCIES. 2-0234

CT.   MULTIPLE IONIZATION AND AUGER TRANSITIONS IN INDIUM AND SILVER BY ELECTRON IMPA 2-0229
M.   AUGER RATES FOR SOFT X-RAY TRANSITIONS IN IONIC, ATOMIC, AND METALLIC LITHIU 4-0835
                                                    AUGER TRANSITIONS IN LITHIUM-LIKE BERYLLIUM-LIKE IONS.   2-0296
      PARITY NONCONSERVATION IN RADIATIONLESS (AUGER) TRANSITIONS IN MU MESIC ATOMS.   2-0364
                     BEAM FOIL EXCITED AUGER TRANSITIONS IN NEON.   4-0860
                         INTERFACIAL AUGER TRANSITIONS IN OXIDIZED SODIUM AND MAGNESIUM.   5-1497
ON)   RECOMBINATION-INDUCED NONEQUILIBRIUM PHASE TRANSITIONS IN SEMICONDUCTORS. (AUGER RECOMBINATI 2-0421
                          RADIATIVE AUGER TRANSITIONS IN SOFT X-RAY PLASMA EMISSION.   2-0344
L VACANCY STATES AND QUASIATOMIC ASPECTS OF RADIATIONLESS TRANSITIONS IN SOLIDS.   WIDTHS OF ATOMIC M-SHEL 3-0718
LICON)   TWO ELECTRON BAND-TO-BAND TRANSITIONS IN SOLIDS. (AUGER RECOMBINATION IN SI 2-0275
              EVIDENCE OF RADIATIVE AUGER TRANSITIONS IN THE FREE ARGON ATOM.   2-0403
              EVIDENCE OF RADIATIVE AUGER TRANSITIONS IN THE FREE ARGON ATOM.   2-0404
                    ROLE OF COSTER-KRONIG TRANSITIONS IN THE IONIZATION OF SURFACE ATOMS.   2-0338
YERS ON COPPER INVESTIGATED BY LEED AND AUGER SPEC/ PHASE TRANSITIONS IN TWO DIMENSIONAL ABSORBED SULFUR LA 5-1302
                                  KMM AUGER TRANSITIONS IN XENON-131.   4-0758
ZED STATES IN LITHIUM METAL.   AUGER TRANSITIONS INITIATED FROM SINGLY AND DOUBLY IONI 2-0581
S BY A COINCIDENCE TECHNIQUE.   AUGER TRANSITIONS LEADING TO DEFINED FINAL CHARGE STATE 2-0371
M/ HIGH RESOLUTION MEASUREMENTS OF THE NITROGEN KLL AUGER TRANSITIONS OF CHEMISORBED AMMONIA ON A MOLYBDENU 5-1532
ECTRA. (AUGER EFFECT)   SURFACE STATE TRANSITIONS OF SILICON IN ELECTRON ENERGY LOSS SP 4-1039
                                K-AUGER TRANSITIONS OF THE FREE SODIUM ATOM.   4-0880
S IN THE 3-21 EV REGION OF THE SPECTRUM. (INTERBAND AUGER TRANSITIONS, RADIATIONLESS RECOMBINATIONS) /POUND 4-1048
              AUGER AND COSTER-KRONIG TRANSITIONS. (REVIEW)   1-0123
                     INTERATOMIC AUGER TRANSITIONS. SURVEY. (36 REFS)   2-0318
      HIGH RESOLUTION MEASUREMENTS OF THE L3M2,3M4,5 AUGER TRANSITONS IN NICKEL AND COPPER.   4-1024
ELECTRON EMISSION OF THIN CARBON FOILS IN REFLECTION AND TRANSMISSION.   AUGER 4-0893
/F ELECTRON PROBE ANALYZER, SCANNING ELECTRON MICROSCOPE, TRANSMISSION ELECTRON MICROSCOPE, ENERGY DISPERS/ 1-0010
      DETERMINATION OF ELECTRON ATTENUATION LENGTHS IN METALS: TRANSMISSION THROUGH THIN ADSORBED FILMS. (AUGER/ 4-1088
AMORPHOUS GERMANIUM. (DEPTH PROFILES OF OXYGEN A/ IN-SITU TRANSPORT MEASUREMENTS IN ULTRAHIGH VACUUM GROWN 5-1547
                           WEAR AND INTERFACIAL TRANSPORT OF MATERIAL. (AES)   5-1210
ILMS DURING ANNEALING. (AUGER SPECTROSCOPY)   TRANSPORT OF OXYGEN IN AMORPHOUS GERMANIUM THIN F 5-1548
N FLOWING LIQUID SODIUM. (SCANNING ELECTRON MICROSC/ MASS TRANSPORT OF STAINLESS STEEL CORROSION PRODUCTS I 6-2014
. (AUGER RECOMBINATION)   TRANSIENT CARRIER DECAY AND TRANSPORT PROPERTIES IN MERCURY CADMIUM TELLURIDE 2-0461
L- AND M-SHELL YIELDS AND ELECTRON SPECTRA FOR THE TRANSURANIC ELEMENTS.   4-0968
                                RECENT TRENDS IN AUGER ELECTRON SPECTROSCOPY.   1-0126
/FFECTS OF AUGER IONIZATION FOLLOWING ELECTRON CAPTURE IN TRIS (1,10-PHENANTHROLINE) COBALT(III)-57 PERCHL/ 2-0516
S.   APPLICATION OF TRIPLE GRID LEED SYSTEM TO AUGER SPECTRUM ANALYSE 3-0661
IUM BY ELECTRON IMPACT. (AUGER SPECTRA)   DOUBLE AND TRIPLE IONIZATION IN ALUMINUM, GALLIUM, AND SCAND 2-0232
                                     TRUE SECONDARY ELECTRON ENERGY DISTRIBUTIONS.   2-0242
/LOW ELECTRON SCATTERING FROM METALS. PART-1: EMISSION OF TRUE SECONDARY ELECTRONS. PART-2: INELASTICALLY / 2-0500
D ESCA TEST PROGRAMS.   DIGITAL CONTROL UNIT IN TTL LOGIC FOR THE AUTOMATIC EXECUTION OF AUGER AN 3-0722
              ADSORPTION OF NITRIC OXIDE ON TUNGSTEN.   5-1929
AN ELECTRON DIFFRACTION STUDY OF CESIUM ADSORPTION ON TUNGSTEN.   5-1603
ENERGY SPECTRA OF SECONDARY ELECTRONS FROM MOLYBDENUM AND TUNGSTEN.   AUGER ELECTRON EMISSION IN THE 4-0874
OPY OF CESIUM ADSORBED ON CLEAN AND OXYGEN- COVERED (100) TUNGSTEN.   AUGER ELECTRON SPECTROSC 5-1295
ER ELECTRON SPECTROSCOPY OF LAYERED GROWTH OF TITANIUM ON TUNGSTEN.   AUG 5-1140
                  (AUGER) ELECTRON SPECTROSCOPY OF TUNGSTEN.   4-0845
ENUM FROM VARIOUS METAL SURFACES (COPPER, GOLD, ALUMINUM, TUNGSTEN). /OSITION AND SPUTTER REMOVAL OF MOLYBD 5-1878
              AUGER OBSERVATION OF A SURFACE STATE ON TUNGSTEN.   5-1146
                CESIUM- OXYGEN COADSORPTION ON (100) TUNGSTEN.   5-1296
RACTERIZED SINGLE CRYSTAL SURFACES. (SILVER AND COPPER ON TUNGSTEN) /DARY ION MASS SPECTROSCOPY ON WELL CHA 5-1689
CTRONIC AND LATTICE STRUCTURE OF CESIUM FILMS ADSORBED ON TUNGSTEN.   ELE 5-1604
OF INELASTICALLY SCATTERED ELECTRONS AND LEED STUDIES OF TUNGSTEN.   ENERGY SPECTRA 4-1094
DY OF OXYGEN- CESIUM ADSORPTION AND COADSORPTION ON (100) TUNGSTEN. /N DIFFRACTION AUGER, WORK FUNCTION STU 5-1297
ICS OF OXYGEN ADSORPTION ON THE (112) AND (110) PLANES OF TUNGSTEN.   THE KINET 5-1912
                SECONDARY ELECTRON EMISSION FROM TUNGSTEN AND MOLYBDENUM.   4-0850
OXYGEN.   CHEMICAL SHIFT IN AUGER SPECTRA OF TUNGSTEN AND MOLYBDENUM DURING THE ADSORPTION OF 4-0816
                SECONDARY ELECTRON EMISSION FROM TUNGSTEN AND MOLYBDENUM SINGLE CRYSTALS.   4-0742
/T OF ION BOMBARDMENT ON THE COMPOSITION AND STRUCTURE OF TUNGSTEN AND MOLYBDENUM SINGLE CRYSTALS STUDIED / 5-1215
/ETRY, AUGER SPECTROSCOPY, AND WORK FUNCTION OF CESIUM ON TUNGSTEN AND TITANIUM SINGLE CRYSTALS IN THE SUB/ 5-1829
PECTROSCOPY.   INVESTIGATION OF OXYGEN ADSORPTION ON TUNGSTEN AT HIGH TEMPERATURES BY AUGER ELECTRON S 5-1156
TROSCOPY MEASUREMENTS OF ADSORPTION ISOBARS FOR OXYGEN ON TUNGSTEN AT LOW PRESSURE AND HIGH TEMPERATURE. /C 5-1282
BY METASTABLE ATOMS: HELIUM AND ARGON ON (111) AND (110) TUNGSTEN. (AUGER) / FUNCTION IN ELECTRON EJECTION 5-1600
              INTERACTION OF SULFUR DIOXIDE WITH TUNGSTEN. (AUGER ANALYSIS)   5-1367
YLINDRICAL MIRROR ANALYZER INCORPORATING PRERETARDATION. (TUNGSTEN AUGER ELECTRON)   C 3-0590
              ADSORPTION OF OXYGEN ON THE (110) PLANE OF TUNGSTEN. (AUGER ELECTRON SPECTROSCOPY)   5-1318
      IDENTIFICATION OF BUBBLE FORMING IMPURITIES IN DOPED TUNGSTEN. (AUGER SPECTROSCOPY)   6-2107
E GRAIN BOUNDARY IMPURITIES ON THE PROPERTIES OF SINTERED TUNGSTEN. (BY AUGER SPECTROSCOPY) /F THE INFLUENC 6-1997
              AUGER EJECTION OF ELECTRONS FROM TUNGSTEN BY OXYGEN CHEMISORPTION.   5-1599
              DITUNGSTEN CARBIDE OVERLAYER ON TUNGSTEN (112).   5-1241
/NS BETWEEN INTERFACIAL OXIDES AND RESISTANCE CHANGES FOR TUNGSTEN CONTACTS ON HEAVILY DOPED SILICON- GERM/ 5-1861
MENT IN SOME METALS AND ALLOYS. (AUGER EMISSION ANALYSIS, TUNGSTEN, COPPER, STEELS) /AIN BOUNDARY EMBRITTLE 6-2058
/ED STRENGTH OF METALS AND ALLOYS. (SEGREGATION IN STEEL, TUNGSTEN, COPPER- BISMUTH ALLOYS STUDIED BY AUGE/ 6-2120
ING ACTIVATION AND POISONING. PART-1: BARIUM OR OXYGEN ON TUNGSTEN DISPENSER CATHODE. /CATHODE SURFACES DUR 5-1841
RFACE POTENTIAL DIFFERENCE STUDIES. CESIUM ADSORPTION ON TUNGSTEN: ELLIPSOMETRY, AUGER SPECTROSCOPY AND SU 5-1830
S AND AUGER SPECTROSCOP/ STUDY OF BARIUM FILMS ON A (110) TUNGSTEN FACE BY THE DIFFRACTION OF SLOW ELECTRON 5-1368
ER ELECTRON SPECTROSCOPIC TECHNIQUE/ RECRYSTALLIZATION OF TUNGSTEN FIBERS IN NICKEL MATRIX COMPOSITES. (AUG 6-2046
CTROSCOPY.   SILICON DIFFUSION THROUGH THIN TUNGSTEN FILMS ON SILICON, STUDIED WITH AUGER SPE 5-1234
              CHEMICAL SHIFT IN THE AUGER SPECTRUM OF TUNGSTEN FROM OXYGEN ADSORPTION.   4-0922
AUGER ELECTRON SPECTROSCOPY ON THE FRACTURE OF NICKEL TUNGSTEN MATERIALS AND SILICON NITRIDE.   5-1445
RES OF REACTIVELY SPUTTERED FILMS OF TANTALUM NITRIDE AND TUNGSTEN NITRIDE. (AUGER EFFECT) /ITION TEMPERATU 5-1541
/FECTS OF INTERFACIAL OXIDES ON THE CONTACT RESISTANCE OF TUNGSTEN ON HEAVILY DOPED SILICON- GERMANIUM ALL/ 5-1456
ROSCOPY STUDY OF INTERDIFFUSION AND EUTECTIC FORMATION IN TUNGSTEN PLATINUM- TUNGSTEN GOLD THIN FILMS. /ECT 5-1259
NTROLLED SURFACES. (AUGER NEUTRALIZATION AT (100) FACE OF TUNGSTEN, POLYCRYSTALLINE MOLYBDENUM). /S WITH CO 5-1736
/TTER CLEANING AND ETCHING OF CRYSTAL SURFACES (TITANIUM, TUNGSTEN, SILICON) MONITORED BY AUGER SPECTROSCO/ 5-1828
ECTRON / EPITAXIAL GROWTH OF COPPER ON (110) SURFACE OF A TUNGSTEN SINGLE CRYSTAL STUDIED BY LEED, AUGER EL 5-1660
      ADSORPTION AND CONDENSATION OF COPPER ON TUNGSTEN SINGLE CRYSTAL SURFACES.   5-1160
              CHEMISORPTION OF NITROGEN ON TUNGSTEN STUDIED BY AUGER ELECTRON SPECTROSCOPY.   5-1518
PECTROSCOPY.   CESIUM ADSORPTION ON TUNGSTEN STUDIED BY LEED AND SECONDARY ELECTRON S 5-1661
      DIFFUSION OF SULFUR AND PHOSPHORUS IMPURITY ATOMS TO A TUNGSTEN SURFACE.   5-1226
AUGER ELECTRON SPECTROSCOPY OF NITRIC OXIDE ADSORBED ON A TUNGSTEN SURFACE.   HIGH RESOLUTION 5-1565

D STUDY OF THE STRUCTURE AND COMPOSITION OF ION BOMBARDED TUNGSTEN SURFACES. /LECTRON SPECTROSCOPY, AND LEE 5-1951
PHOTOEMISSION STUDIES OF CHEMISORPTION ON NICKEL AND TUNGSTEN SURFACES. 5-1308
ROOM TEMPERATURE ADSORPTION OF OXYGEN ON TUNGSTEN SURFACES. (AUGER MEASUREMENT METHODS) 5-1664
STUDY OF THE PREPARATION OF ATOMICALLY CLEAN TUNGSTEN SURFACES BY AUGER ELECTRON SPECTROSCOPY. 5-1517
INTERANGULAR BRITTLENESS STUDIES IN TUNGSTEN USING AUGER SPECTROSCOPY. 6-2061
USING AUGER PEAK SHAPES. ETHYLENE AND CARBON MONOXIDE ON TUNGSTEN (100). /ACTERIZATION OF ADSORBED SPECIES 5-1243
ELECTRON SPECTROSCOPY. DISCREPANCY FOR CARBON DIOXIDE ON TUNGSTEN (100). FLASH DESORPTION- AUGER 5-1461
ELECTRON DIFFRACTION STUDY OF ALKALI METAL OVERLAYERS ON TUNGSTEN (100) /TRON SPECTROSCOPY AND LOW ENERGY 5-1895
TIONS ON K-SHELLS OF SURFACE ATOMS. (AUGER INTENSITIES OF TUNGSTEN (100) /TRON IMPACT IONIZATION CROSS SEC 4-0847
RON SPECTROSCOPY) OXIDATION OF CARBON ON TUNGSTEN (100) AND MOLYBDENUM (100). (AUGER ELECT 5-1944
E RANGE 10 (-9) TO 10 (-/ ADSORPTION ISOTHERMS OF OXYGEN ON TUNGSTEN (100), AND (111) SURFACES IN THE PRESSUR 5-1148
, AND/ ELECTRON BEAM EFFECTS ON CARBON MONOXIDE SATURATED TUNGSTEN (100), ETHYLENE SATURATED TUNGSTEN (100) 5-1244
UNCTION MEASUREMENTS FOR EPITAXIAL GROWTH OF THORIUM ON A TUNGSTEN (100) SUBSTRATE. /SCOPY, LEED AND WORK F 5-1739
ADSORPTION OF IONIC SALTS ON A TUNGSTEN (100) SURFACE. 5-1659
DY OF THE ELECTRON STIMULATED DESORPTION OF OXYGEN FROM A TUNGSTEN (100) SURFACE. /LECTRON SPECTROSCOPY STU 5-1954
OPY) INVESTIGATION OF HYDROGEN CHEMISORPTION ON THE TUNGSTEN (100) SURFACE. (AUGER ELECTRON SPECTROSC 5-1974
AUGER ELECTRON SPECTROSCOPY STUDY OF OXYGEN ADSORPTION ON TUNGSTEN (110). 5-1668
LEED AND AES STUDY OF OXYGEN ABSORPTION ON TUNGSTEN (110). 5-1669
EED- AUGER, AND WORK FUNCTION STUDY OF IODINE ADSORBED ON TUNGSTEN (110). L 5-1145
00). (BY L/ PHOTOEMISSION SPECTRA FROM ADSORBED OXYGEN ON TUNGSTEN (110) AND CARBON MONOXIDE ON TUNGSTEN (1 5-1152
IDE FORMATION AT 650-1100 DEGREES K. OXIDATION OF TUNGSTEN (110). PART-2: LEED- AUGER STUDY OF OX 6-1994
LEED AND AUGER ELECTRON SPECTROSCOPIC STUDIES ON TUNGSTEN (110) SURFACE. 5-1677
ON MONOXIDE, AND MOLECULAR OXYGEN AND CARBON DIOXIDE ON A TUNGSTEN (111) SURFACE. /OLECULAR OXYGEN AND CARB 5-1465
DITUNGSTEN CARBIDE OVERLAYER ON TUNGSTEN (112). 5-1241
BINED LEED- RHEED AUGER STUDY OF OXYGEN ADSORPTION ON THE TUNGSTEN (112) SURFACE. COM 5-1464
/ (RHENIUM DUCTILIZING EFFECT AND BUBBLE STRENGTHENING IN TUNGSTEN- RHENIUM ALLOYS, AUGER ELECTRON MICROSC/ 6-2111
SCOPY) STUDY OF GRAIN BOUNDARY FRACTURE SURFACES IN DOPED TUNGSTEN- RHENIUM ALLOYS. (AUGER ELECTRON SPECTRO 6-2112
PECTROSCOPY ANALYSIS OF A COMMERCIAL COBALT BASE AIRCRAFT TURBINE SHROUD ALLOY. /SION, FRICTION AND AUGER S 6-2009
LAYER FILMS ADSORBED ON (0001) GRAPHITE. (AUGER SPECTROS/ TWO-DIMENSIONAL PHASE TRANSITION IN XENON SUBMONO 5-1858
. AUGER ELECTRON SPECTROSCOPY USING THE TWO-GRID LEED SYSTEM AND PHOTOELECTRON MULTIPLIER 3-0656
AND THE DOUBLE AUGER PROCESS. TWO-PHOTON EMISSION, THE RADIATIVE AUGER EFFECT, 2-0225

U

A MULTIPLE SAMPLE HOLDER FOR USE IN ULTRAHIGH VACUUM. 3-0603
A VERSATILE SUBSTRATE DESIGN FOR LEED AND AES STUDIES IN ULTRAHIGH VACUUM. 3-0631
SURFACE STRUCTURES ON IRON (110) ULTRAHIGH VACUUM. 5-1346
ANALYTICAL SYSTEM FOR SECONDARY ION MASS SPECTROMETRY IN ULTRAHIGH VACUUM. (AES) AN 3-0632
OF SULFUR DIOXIDE AND CARBON DIOXIDE WITH CLEAN SILVER IN ULTRAHIGH VACUUM. (AES) INTERACTION 5-1575
ROSCOPY. ULTRAHIGH VACUUM AIRLOCK FOR LEED AND AUGER SPECT 3-0669
/PIC STUDY OF THE SURFACE COMPOSITION OF PLATINUM- TIN IN ULTRAHIGH VACUUM AND IN THE PRESENCE OF OXYGEN A/ 5-1196
ECTRON SPECTROS/ BEHAVIOR OF REFRACTORY METAL SURFACES IN ULTRAHIGH VACUUM AS OBSERVED BY LEED AND AUGER EL 6-2020
NEW SEM AUGER MICROANALYZER WITH ULTRAHIGH VACUUM CAPABILITY. 3-0636
S INVOLVING/ ELECTRON SPECTROSCOPY FROM SOLID SURFACES IN ULTRAHIGH VACUUM. CURRENT PROGRESS WITH TECHNIQUE 5-1901
DEVICE FOR LOW ENERGY ELECTRON DIFFRACTION AND AUGER EM/ ULTRAHIGH VACUUM EVAPORATOR AND MULTIPLE CLEAVAGE 3-0707
ISSION STUDIES. ULTRAHIGH VACUUM EVAPORATOR FOR LEED AND AUGER EM 3-0683
H PROFILES OF OXYGEN A/ IN-SITU TRANSPORT MEASUREMENTS IN ULTRAHIGH VACUUM GROWN AMORPHOUS GERMANIUM. (DEPT 5-1547
U FILM DEPOSITION AND CHARACTERIZATION. (LEED- AUGER SYS/ ULTRAHIGH VACUUM LOW TEMPERATURE STAGE FOR IN-SIT 3-0657
H A FIELD EMISSION CATHODE AND AUGER ANALYZER. ULTRAHIGH VACUUM SCANNING ELECTRON MICROSCOPE WIT 3-0621
H AUGER ANALYSIS FACILITIES. ULTRAHIGH VACUUM SCANNING ELECTRON MICROSCOPE WIT 3-0622
EVIEW, AUGER, MASS SPECTROMETRY) ROLE OF ULTRAHIGH VACUUM TECHNIQUES IN BASIC RESEARCH. (R 1-0065
CONTACT ANGLES. PART-4: WATER ON GRAPHITE (0001). (AES) ULTRAHIGH VACUUM TECHNIQUES IN THE MEASUREMENT OF 5-1795
ELECTRON GUN FOR AUGER AND ULTRASOFT X-RAY APPARATUS. 3-0676
/N OF MECHANISM OF S IONIZATION OF NOBLE GAS ATOMS BY THE ULTRASOFT X-RAY SPECTROSCOPY TECHNIQUE. (AUGER D/ 2-0580
THICKNESS DETERMINATION OF ULTRATHIN FILMS BY AUGER ELECTRON SPECTROSCOPY. 5-1449
AUGER AND ELLIPSOMETRIC STUDIES OF ULTRATHIN LEAD (II) OXIDE GROWTH ON LEAD. 5-1248
TRATES USING LEED- AUGER TECHNIQUES. EPITAXY OF ULTRATHIN METAL FILMS ON BODY CENTERED CUBIC SUBS 5-1488
TION AND/ SURFACE PHYSICS OF SOLIDS FROM THE VIEWPOINT OF ULTRAVACUUM. (AUGER SPECTROSCOPY, SURFACE PREPARA 5-1419
BY ELECTRON IMPACT VIA N4,5 AUGER TRANSITIONS. VACUUM ULTRAVIOLET EMISSION OF XENON (III) LEVELS EXCITED 4-0878
ULTRAVIOLET EMISSIONS IN CALCITE. (AUGER EFFECT) 4-0798
D MAGNESIUM (2) STANNIDE. PART-1: X-RAY EXCITATION PART-2: ULTRAVIOLET EXCITATION. (AUGER ELECTRONS) /UM, AN 2-0306
N CLEAN METAL FILM SURFAC/ X-RAY PHOTOELECTRON AND VACUUM ULTRAVIOLET PHOTOELECTRON STUDIES OF ADSORPTION O 5-1203
ODIUM, POTASSIUM, RUBIDIUM, AND CESIUM IONS IN THE VACUUM ULTRAVIOLET SPECTRAL RANGE. (AUGER DECAYS) /FOR S 4-0735
FIRST TRANSITION SERIES. (COMMENTS ON PAPER BY CHUNG AN/ UNEXPLAINED PHENOMENA IN THE AUGER SPECTRA OF THE 5-0812
/RGY ELECTRON DIFFRACTION STUDIES OF THE CHEMISORPTION OF UNSATURATED MOLECULES BY THE (100) SURFACE OF PL/ 5-1267
PHIDE AND GALLIUM ARSENIDE (111) FACES STUDIED BY AES AND UPS. OXIDATION AND ANNEALING OF GALLIUM PHOS 5-1493
DEPOSITED SILVER HALIDE FILMS. UPS, ESCA, AND AUGER SPECTROSCOPY STUDY OF VACUUM 5-1947
AUGER SPECTRA OF URANIUM. 4-1125
AUGER SPECTROSCOPY OF URANIUM. 4-0738
CHEMICAL SHIFTS IN URANIUM AUGER TRANSITIONS. 4-0737
AUGER ELECTRON STUDIES OF URANIUM DIOXIDE SURFACES. 5-1218
SECONDARY ELECTRON ENERGY DISTRIBUTION STUDIES OF URANIUM DIOXIDE SURFACES. 5-1316
AINLESS STEEL, AND N/ AUGER ELECTRON STUDIES OF SURFACES: URANIUM DIOXIDE, URANIUM, GRAPHITE, 300 SERIES ST 5-1217
AUGER ELECTRON SPECTROSCOPY. TOWARD OPTIMUM UTILIZATION OF THE CYLINDRICAL MIRROR ANALYZER IN 3-0678

V

NEW TYPE OF AUGER TRANSITIONS IN ATOMS WITH TWO INTERNAL VACANCIES. DETECTION OF A 2-0234
THREE ELECTRON AUGER TRANSITIONS IN ATOMS WITH TWO INNER VACANCIES. OBSERVATION OF 2-0235
URFACES. AUGER ELECTRON SPECTROSCOPY STUDIES OF ARSENIC VACANCIES IN (-1-1-1) GALLIUM ARSENIDE ANNEALED S 5-1914
PRODUCTION AND DECAY OF DOUBLE L VACANCIES IN ARGON AND PHOSPHORUS. 2-0489
FFECT) PRODUCTION OF INNER SHELL VACANCIES IN HEAVY ION- ATOM COLLISIONS. (AUGER E 2-0405
LATION ENERGIES AND AUGER RATES IN ATOMS WITH INNER SHELL VACANCIES (NEON WITH 1S OR 2S VACANCY). CORRE 2-0312
S IN ATOMS WITH INNER SHELL VACANCIES (NEON WITH 1S OR 2S VACANCY). CORRELATION ENERGIES AND AUGER RATE 2-0312
S CREATED AS THE RESULT OF ELECTRON SHAKE-OFF AND (SUGAR) VACANCY CASCADES. NATURE OF SECONDARY ELECTRON 2-0307
IZATION (AUGER AND RADIATIVE TRANSITIONS). ATOMIC VACANCY DISTRIBUTIONS PRODUCED BY INNER SHELL ION 2-0544
E IONS. (AUGER ELECTRON EMISSION) NEON K-SHELL VACANCY PRODUCTION BY INTERMEDIATE ENERGY FLUORIN 2-0572
IMPACT/ CHARGE DEPENDENCE OF DEPENDENCE OF TOTAL K-SHELL VACANCY PRODUCTION CROSS SECTIONS IN ARGON BY THE 2-0446
IONS AT KEV ENE/ DYNAMIC CHARGE STATE DEPENDENCE OF THE K-VACANCY PRODUCTION YIELD IN HEAVY ION ATOM COLLIS 2-0346
GER SPECTRA OF COPPER. AUGER VACANCY SATELLITE STRUCTURE IN THE L3,M4,5M4,5 AU 4-1034

S. (AUGER EFFECT)   GAS TARGET EXPERIMENTS OF K-SHELL VACANCY SHARING IN ASYMMETRIC ION- ATOM COLLISION 2-0523
COLLISIONS. (QUASIMOLECULE AUGER SPECTRA)   K-SHELL VACANCY SHARING IN 300 TO 500 KEV NEON(+)- SODIUM 2-0582
S.   K-VACANCY SHARING STUDIED FROM AUGER ELECTRON YIELD 2-0345
ONLESS TRANSITIONS IN SOLIDS.   WIDTHS OF ATOMIC M-SHELL VACANCY STATES AND QUASIATOMIC ASPECTS OF RADIATI 3-0718
E METHOD. (AUGER TRA/ MEASUREMENT OF THE PRODUCTION OF LL VACANCY STATES BY THE L X-RAY- L X-RAY COINCIDENC 2-0482
L-SHELL FLUORESCENCE YIELDS OF DOUBLE VACANCY STATES IN LEAD.   4-1105
AUGER ELECTRONS FROM DOUBLE K-SHELL VACANCY STATES IN NEON.   2-0573
WIDTH OF ATOMIC L2 AND L3 VACANCY STATES NEAR Z = 30. (AUGER EFFECT)   4-0939
ATOMIC N-SHELL VACANCY WIDTHS IN SOLIDS.   2-0577
A MULTIPLE SAMPLE HOLDER FOR USE IN ULTRAHIGH VACUUM.   3-0603
LE SUBSTRATE DESIGN FOR LEED AND AES STUDIES IN ULTRAHIGH VACUUM.   A VERSATI 3-0631
(AUGER STUDIES OF) ADHESION, FRICTION, LUBRICATION IN VACUUM.   5-1209
SURFACE STRUCTURES ON IRON (110) ULTRAHIGH VACUUM.   5-1346
L SYSTEM FOR SECONDARY ION MASS SPECTROMETRY IN ULTRAHIGH VACUUM. (AES)   AN ANALYTICA 3-0632
DIOXIDE AND CARBON DIOXIDE WITH CLEAN SILVER IN ULTRAHIGH VACUUM. (AES)   INTERACTION OF SULFUR 5-1575
ULTRAHIGH VACUUM AIRLOCK FOR LEED AND AUGER SPECTROSCOPY.   3-0669
/OF THE SURFACE COMPOSITION OF PLATINUM- TIN IN ULTRAHIGH VACUUM AND IN THE PRESENCE OF OXYGEN AND HYDROGE/ 5-1196
/IGN OF AN X-RAY DOUBLE CRYSTAL SPECTROMETER INSTALLED IN VACUUM AND THE MEASUREMENT OF EXCITED ELECTRONS / 3-0641
ES BETWEEN 770 AND 1470 K.   VACUUM ANNEALING OF STAINLESS STEEL AT TEMPERATUR 6-2145
CTROS/ BEHAVIOR OF REFRACTORY METAL SURFACES IN ULTRAHIGH VACUUM AS OBSERVED BY LEED AND AUGER ELECTRON SPE 6-2020
/IN BOUNDARY SEGREGATION IN MAGNESIA DOPED ALUMINA. (HIGH VACUUM AUGER ELECTRON SPECTROMETER, FRACTURED AL/ 5-1615
LIUM ARSENIDE, INDIUM PHOSPHIDE, AND GALLIUM PHOSPHIDE IN VACUUM. (AUGER SPECTROSCOPY) /00) SURFACES OF GAL 5-1163
WITH FIELD EMIS/ STABILITY OF FIELD ELECTRON EMISSION AND VACUUM BREAKDOWN. (FROM SURFACES) INVESTIGATIONS 5-1520
INVESTIGATION OF VACUUM BREAKDOWN USING AUGER SPECTROSCOPY.   5-1936
NEW SEM AUGER MICROANALYZER WITH ULTRAHIGH VACUUM CAPABILITY.   3-0636
EPOSITS OF IRON.   AUGER SPECTRA AND LEED PATTERNS FROM VACUUM CLEANED SILICON CRYSTALS WITH CALIBRATED D 5-1769
(100) SURFACES OF ALKALI HALIDES. PART-1: AIR AND VACUUM CLEANED SURFACES.   5-1347
G/ ELECTRON SPECTROSCOPY FROM SOLID SURFACES IN ULTRAHIGH VACUUM. CURRENT PROGRESS WITH TECHNIQUES INVOLVIN 5-1901
UPS, ESCA, AND AUGER SPECTROSCOPY STUDY OF VACUUM DEPOSITED SILVER HALIDE FILMS.   5-1947
SCANNING AUGER SPECT/ SOLID PHASE EPITAXIAL STUDIES USING VACUUM DEPOSITION ON HEATED SILICON SUBSTRATES. (5-1287
ELECTRO/ DIFFUSE INTERFACE STRUCTURES IN SOME (SUBSTRATE- VACUUM EVAPORATED FILM) SYSTEMS STUDIED BY AUGER 5-1436
R LOW ENERGY ELECTRON DIFFRACTION AND AUGER EM/ ULTRAHIGH VACUUM EVAPORATOR AND MULTIPLE CLEAVAGE DEVICE FO 3-0707
DIES.   ULTRAHIGH VACUUM EVAPORATOR FOR LEED AND AUGER EMISSION STU 3-0683
OF OXYGEN A/ IN-SITU TRANSPORT MEASUREMENTS IN ULTRAHIGH VACUUM GROWN AMORPHOUS GERMANIUM. (DEPTH PROFILES 5-1547
OSITION AND CHARACTERIZATION. (LEED- AUGER SYS/ ULTRAHIGH VACUUM LOW TEMPERATURE STAGE FOR IN-SITU FILM DEP 3-0657
EMISSION CATHODE AND AUGER ANALYZER.   ULTRAHIGH VACUUM SCANNING ELECTRON MICROSCOPE WITH A FIELD 3-0621
ALYSIS FACILITIES.   ULTRAHIGH VACUUM SCANNING ELECTRON MICROSCOPE WITH AUGER AN 3-0622
THE RELATIONSHIP OF SOLID SURFACES TO VACUUM SCIENCE AND TECHNOLOGY. (REVIEW)   1-0086
RONIC PROPERTIES OF FILMS. SILICON ON SAPPHIRE EPITAXY BY VACUUM SUBLIMATION: LEED- AUGER STUDIES AND ELECT 5-1230
ER, MASS SPECTROMETRY)   ROLE OF ULTRAHIGH VACUUM TECHNIQUES IN BASIC RESEARCH. (REVIEW, AUG 1-0065
NGLES. PART-4: WATER ON GRAPHITE (0001). (AES)   ULTRAHIGH VACUUM TECHNIQUES IN THE MEASUREMENT OF CONTACT A 5-1795
EXCITED BY ELECTRON IMPACT VIA N4,5 AUGER TRANSITIONS. VACUUM ULTRAVIOLET EMISSION OF XENON(III) LEVELS 4-0878
PTION ON CLEAN METAL FILM SURFAC/ X-RAY PHOTOELECTRON AND VACUUM ULTRAVIOLET PHOTOELECTRON STUDIES OF ADSOR 5-1203
/ FOR SODIUM, POTASSIUM, RUBIDIUM, AND CESIUM IONS IN THE VACUUM ULTRAVIOLET SPECTRAL RANGE. (AUGER DECAYS) 4-0735
SERVATIONS OF THE SUBSTRATE SURFACE DURING SPUTTERING AND VACUUM VAPOR DEPOSITION. /ECTRON SPECTROSCOPIC OB 5-1678
CAY OF CORE EXCITONS IN ALKALI HALIDES. (AUGER PROCESS IN VALENCE BAND)   PHOTOEMISSION STUDIES OF THE DE 2-0289
IMONIDE VIA PHOTOEMISSION MEASUREMENTS IN THE 20-70 EV R/ VALENCE BAND AND CORE LEVEL STUDIES OF INDIUM ANT 2-0341
D SILICON.   SURFACE VALENCE BAND AND PLASMON FEATURES ON CLEAN CLEAVE 5-1141
COMMENTS.   VALENCE BAND AUGER ELECTRON SPECTRA FOR ALUMINUM: 4-0839
D ALUMINUM.   CONTRASTING VALENCE BAND AUGER ELECTRON SPECTRA FOR SILVER AN 4-1022
AUGER AND PHOTOEMISSION SPECTROSCOPY.   VALENCE BAND OF MAGNESIUM STANNIDE DETERMINED BY 5-1885
LUMINUM.   VALENCE BAND STRUCTURE IN THE AUGER SPECTRUM OF A 4-0889
/T X-RAY PHOTOEMISSION, AND AUGER ELECTRON SPECTRA OF THE VALENCE BANDS OF MAGNESIUM, MAGNESIUM(2) COPPER / 4-0838
) USING AUGER ELECTRON SPECTROSCOPY: CHEMICAL SHIFTS AND VALENCE SPECTRA. /ORPTION OF OXYGEN ON NICKEL(III 5-1466
TIGHT BINDING CALCULATION OF A CORE VALENCE VALENCE AUGER LINE SHAPE: SILICON (111).   4-0831
ON METAL SURFACES.   BINDING, VALENCY AND CHARGE TRANSFER IN CHEMICAL REACTIONS 5-1763
ATION OF THE KLL AUGER/ APPLICATION OF THE ZETA EFFECTIVE VALUES OF HYDROGENIC WAVE FUNCTIONS IN THE CALCUL 2-0526
AL SHIFTS IN THE AUGER SPECTRA FROM OXIDIZED CHROMIUM AND VANADIUM.   CHEMIC 4-0809
MAL TEMPER EMBRITTLEMENT IN NICKEL- CHROMIUM- MOLYBDENUM- VANADIUM.   LONG TIME ISOTHER 6-2140
PREPARATION AND SURFACE OXIDATION OF A (110) FACE OF VANADIUM.   5-1683
TROSCOPY TO DETERMINE THE SURFACE COMPOSITION OF OXIDIZED VANADIUM. / OF X-RAY PHOTOELECTRON AND AUGER SPEC 5-1205
RY ION MASS SPECTROMETRY FOR ALUMINUM, SILICON, TITANIUM, VANADIUM, AND CHROMIUM. /SPECTROSCOPY AND SECONDA 5-1553
AUGER ELECTRON SPECTROSCOPY ANALYSIS OF VANADIUM AND VANADIUM COMPOUND SURFACES.   5-1862
TROSCOPY TO DETERMINE THE SURFACE COMPOSITION OF OXIDIZED VANADIUM. (COMMENTS) /HOTOELECTRON AND AUGER SPEC 5-1863
/ SPECTROSCOPIC INVESTIGATIONS OF CHEMICAL SHIFTS IN SOME VANADIUM COMPOUNDS. (VANADIUM OXIDE, NITRIDE, CA/ 5-1090
DISAPPEARANCE POTENTIAL SPECTROSCOPY (VANADIUM OXIDATION).   6-2068
/TRON SPECTROSCOPY OF THE SURFACE CHEMICAL COMPOSITION OF VANADIUM OXIDES, AND OXIDIZED VANADIUM: CHEMICAL/ 4-1089
/ AND STRUCTURE IN THE AUGER ELECTRON EMISSION SPECTRA OF VANADIUM PENTOXIDE (010) AND VANADIUM (100) SURF/ 5-1340
CES.   AUGER EMISSION SPECTROSCOPY VANADIUM PENTOXIDE (010) AND VANADIUM (100) SURFA 5-1941
DARY SEGREGATION OF PHOSPHORUS IN A CHROMIUM- MOLYBDENUM- VANADIUM STEEL. (AUGER SPECTROSCOPY) / GRAIN BOUN 6-2139
OPY AND CHARACTERISTIC LOSS SPECTROSCOPY FOR THE ELEMENTS VANADIUM TO COBALT. /MBINATION OF AUGER SPECTROSC 4-0813
PREPARATION AND SURFACE OXIDATION OF A VANADIUM (100) PLANE. (AUGER EFFECT)   6-1992
AES.   CHARACTERIZATION OF SOME VANADIUM (100) SURFACE STRUCTURES USING LEED AND 5-1864
AUGER ELECTRON SPECTRUM OF WATER VAPOR.   4-1066
,5N4,5N4,5 AND M4,5O1 AUGER ELECTRON SPECTRA FROM CADMIUM VAPOR.   M4 4-0728
URFACES PRODUCED BY CRACK GROWTH IN HYDROGEN AND IN WATER VAPOR. (AUGER SPECTROMETRY, STEEL) /OF FRACTURE S 6-2142
)   ISOMORPHISM AND PURITY OF COPPER VAPOR DEPOSITED ON A COPPER (111) SURFACE. (AUGER 5-1171
AUGER ELECTRON SPECT/ COMPOSITION PROFILES OF CHEMICALLY VAPOR DEPOSITED PLATINUM AND PLATINUM SILICIDE BY 5-1656
ONS OF THE SUBSTRATE SURFACE DURING SPUTTERING AND VACUUM VAPOR DEPOSITION. /ECTRON SPECTROSCOPIC OBSERVATI 5-1678
.   AUGER SPECTRA OF HYDROGEN CHLORIDE VAPOR ETCHED N+ GALLIUM ARSENIDE (100) SUBSTRATES 4-0992
RON EMISSI/ STUDY OF THE OXIDATION AND REDUCTION BY WATER VAPOR OF SILICON (111) SURFACES USING AUGER ELECT 5-1612
RAL AND KINETIC PROPERTIES. (AUGER E/ ADSORPTION OF WATER VAPOR ON CLEAN CLEAVED GERMANIUM. PART-1: STRUCTU 5-1432
BINDING STATES OF WATER VAPOR ON CLEAVED GERMANIUM. (LEED, AES)   5-1820
PECTRO/ METALLIC STATE OF SILICON IN SILICON- NOBLE METAL VAPOR QUENCHED ALLOYS STUDIED BY AUGER ELECTRON S 5-1438
AUGER ELECTRON STUDIES OF METAL VAPORS.   4-0729
TIPLE IONIZATION INDUCED BY ELECTRON IMPACT IN SOME METAL VAPORS. (AUGER PROCESS) /ANISMS OF SIMPLE AND MUL 2-0231
VARIAN CHART OF AUGER ELECTRON ENERGIES.   4-1082
ER RATES, TRANSITION ENERGIES, AND FLUORESCENCE YIELDS OF VARIOUSLY IONIZED STATES OF ARGON.   K-SHELL AUG 2-0280
INTERATOMIC AUGER EFFECT AT HIGH VELOCITIES.   2-0442
SPECTRA OF CARBON AND ARGON FOLLOWING IONIZATION BY EQUAL VELOCITY ALPHA PARTICLES AND DEUTERONS.   AUGER 4-1109

SCINTILLATION EFFICIENCY FOR LOW VELOCITY HEAVY IONS. (AUGER EMISSION) 2-0431
THE OXIDATION OF STRONTIUM, BY AUGER ELECTRON SPECTROSCO/ VERIFICATION OF THE METAL-RICH SURFACE MODEL FOR 6-2045
ANALYSIS OF THE FRACTURE SURFACES OF IRRADIATED PRESSURE VESSEL STEELS.   AUGER ELECTRON SPECTROSCOPY 6-2113
RFACE PREPARATION AND/ SURFACE PHYSICS OF SOLIDS FROM THE VIEWPOINT OF ULTRAVACUUM. (AUGER SPECTROSCOPY, SU 5-1419
T OF THESE FILMS ON ADH/ ABSORPTION OF ETHYLENE OXIDE AND VINYL CHLORIDE ON AN IRON (011) SURFACE AND EFFEC 5-1207
EXTENDED ION- ELECTRON COINCIDENCE MEASUREMENTS OF THE VIOLENT ARGON(+)- ARGON COLLISION. 2-0530
GER EL/ FAST ELECTRON- SLOW ELECTRON COINCIDENCE STUDY OF VIOLENT ARGON(+)- ARGON COLLISIONS. COMMENTS. (AU 2-0423
LIFETIME AND SURFACE EFFECTS ON THE L23 VV AUGER SPECTRA OF ALUMINUM. 4-0833

W

CADMIUM(X) TELLURIDE AMBIENT TEMPERATURES. (AUGER RECOMB/ WARM ELECTRON EFFECTS IN NARROW GAP MERCURY(1-X) 2-0343
THE STUDY OF DISCOLORING OF ALUMINUM SURFACES IN BOILING WATER. /ICATION OF AUGER ELECTRON SPECTROSCOPY TO 5-1127
CTRUM AND OF THE SATELLITES IN THE SOFT X-RAY SPECTRUM OF WATER. /CULATIONS FOR ASSIGNMENT OF THE AUGER SPE 4-0724
ON. PART-1: ELECTRODE SURFACE STRUCTURE AFTER EXPOSURE TO WATER AND AQUEOUS ELECTROLYTES. /ECTRON DIFFRACTI 5-1486
IONS OF THE AUGER SPECTRUM OF METHANE, HYDROGEN FLUORIDE, WATER, AND CARBON MONOXIDE. /INTERACTION CALCULAT 4-0879
PECTRA OF NITROGEN, OXYGEN, CARBON DIOXIDE, NITRIC OXIDE, WATER, AND CARBON MONOXIDE. /ATION OF KLL AUGER S 4-0990
TECHNIQUES IN THE MEASUREMENT OF CONTACT ANGLES. PART-4: WATER ON GRAPHITE (0001). (AES)   ULTRAHIGH VACUUM 5-1795
AUGER ELECTRON SPECTRUM OF WATER VAPOR. 4-1066
TURE SURFACES PRODUCED BY CRACK GROWTH IN HYDROGEN AND IN WATER VAPOR. (AUGER SPECTROMETRY, STEEL) /OF FRAC 6-2142
ELECTRON EMISSI/ STUDY OF THE OXIDATION AND REDUCTION BY WATER VAPOR OF SILICON (111) SURFACES USING AUGER 5-1612
TRUCTURAL AND KINETIC PROPERTIES. (AUGER E/ ADSORPTION OF WATER VAPOR ON CLEAN CLEAVED GERMANIUM. PART-1: S 5-1432
R/ APPLICATION OF THE ZETA EFFECTIVE VALUES OF HYDROGENIC WATER VAPOR ON CLEAVED GERMANIUM. (LEED, AES) 5-1820
METHOD FOR PACKET OF WAVE FUNCTIONS IN THE CALCULATION OF THE KLL AUGE 2-0526
ISSION.   DISTINCTION BETWEEN WAVES IN AUGER ELECTRON-ELECTRON PROCESSES. 2-0380
OF POLYMER- METAL INTERACTIONS IN ADHESION, FRICTION, AND WEAK AUGER TRANSITIONS AND INTERNAL X-RAY PHOTOEM 2-0436
WEAR.   ATOMIC NATURE 5-1212
SCIENCE TECHNIQUES. (AES, ELLIPSOMET/ ADHESION, FRICTION, WEAR AND INTERFACIAL TRANSPORT OF MATERIAL. (AES) 5-1210
LLOYS. (AUGER S/ EFFECT OF PHOSPHORUS ON THE FRICTION AND WEAR, AND LUBRICATION RESEARCH BY MODERN SURFACE 1-0104
INFLUENCE OF SILICON ON FRICTION AND WEAR CHARACTERISTICS OF COPPER- TIN- PHOSPHORUS A 5-1868
OPY.   ELEMENTAL ANALYSIS OF FRICTION AND WEAR OF IRON- COBALT ALLOYS. (AUGER ANALYSIS) 6-2011
ES. (AUGER ELECTRON SPECTROSCOPY)   WORK FUNCTION OF WEAR SURFACE DURING SLIDING USING AUGER SPECTROSC 6-2013
/CHARACTERIZATION OF NICKEL- COPPER ALLOY DISPLAYING POOR WELL-DEFINED SURFACE OF COPPER- NICKEL ALLOY PLAT 5-1873
WITH THE SURFACE OXYGEN- CHROMIUM RAT/ CORRELATION OF THE WETTABILITY DURING BRAZING. (MONEL, AUGER SPECTR/ 5-1573
BARIUM FLUORIDES.   LINE WETTING BEHAVIOR OF SODIUM ON NIMONIC-PE16 ALLOY 5-1441
H A PHOTOELECTRON SPEC/ CHEMICAL SHIFTS AND VARIATIONS OF WIDTH AND AUGER EFFECT IN CALCIUM, STRONTIUM, AND 4-1116
30. (AUGER EFFECT) WIDTH AND INTENSITY OF AUGER SIGNALS OBTAINED WIT 2-0273
OPY FOR CHEMICAL ANALYSIS.   INFORMATION OBTAINED FROM THE WIDTH OF ATOMIC L2 AND L3 VACANCY STATES NEAR Z = 4-0939
TOTAL AND PARTIAL ATOMIC LEVEL WIDTH OF SPECTRAL LINES IN THE ELECTRON SPECTROSC 4-1009
N ARGON L2,3 AUGER SPECTRA. WIDTHS. 4-0911
ATOMIC N-SHELL VACANCY WIDTHS AND INTENSITIES OF HIGHLY RESOLVED LINES I 4-1032
ATOMIC ASPECTS OF RADIATIONLESS TRANSITIONS IN SOLIDS. WIDTHS IN SOLIDS. 2-0057
MULTIPLET EFFECTS ON THE WIDTHS OF ATOMIC M-SHELL VACANCY STATES AND QUASI 3-0718
ROSCOPY.   ENERGY WIDTHS OF PHOTOELECTRON PEAKS. (AUGER EFFECT) 2-0445
ARE EARTHS. WIDTHS OF ROENTGEN LEVELS BY AUGER ELECTRON SPECT 4-0982
(AUGER ELECTRON SPECTROSCOPY)   LASER WIDTHS OF 4S AND 5S MULTIPLET COMPONENTS IN THE R 4-0970
NITIAL OXIDATION ON SILICON- IRON ALLOY (100) BY MEANS OF WINDOW SURFACE FINISHING AND COATING TECHNOLOGY. 1-0015
5.59 ATOMIC PERCENT SILICON- IRON ALLOY (1/ PHOTOELECTRIC WORK FUNCTION AND AUGER ELECTRON SPECTROSCOPY. /I 5-1924
OVERED SEMICONDUCTORS. WORK FUNCTION AND AUGER ELECTRON SPECTROSCOPY OF 5-1925
ILICON) MONITORED BY AUGER SPECTROSCOPY, ELLIPSOMETRY AND WORK FUNCTION AND STRUCTURAL STUDIES OF ALKALI- C 5-1958
ELECTRON DIFFRACTION AUGER APPARATUS.   MEASUREMENT OF WORK FUNCTION CHANGE. /CES (TITANIUM, TUNGSTEN, S 5-1828
W ENERGY ELECTRON DIFFRACTION SPECTRA, AUGER SPECTRA, AND WORK FUNCTION CHANGE IN A DISPLAY TYPE LOW ENERGY 3-0611
/ON TO THE (001) NICKEL SURFACE DURING THERMAL TREATMENT. WORK FUNCTION CHANGE MEASUREMENTS. /TUDY USING LO 5-1291
ELECTRON/ DIFFUSION MEASUREMENTS IN THIN FILMS UTILIZING WORK FUNCTION CHANGES AND AUGER ELECTRON SPECTRO/ 6-2069
R ELECTRON SPECTROSCOPY) WORK FUNCTION CHANGES: CHROMIUM INTO GOLD. (AUGER 5-1889
BY NOBLE GAS METASTABLE ATOMS. WORK FUNCTION CHANGES ON CONTACT MATERIALS. (AUGE 5-1395
ATOMS: HELIUM AND ARGON ON (111) AND (110) TUNGS/ ROLE OF WORK FUNCTION EFFECTS ON AUGER ELECTRON EJECTION 5-1601
AL TECHNIQUE FOR THE CONTINUOUS MEASUREMENT OF CHANGES IN WORK FUNCTION IN ELECTRON EJECTION BY METASTABLE 5-1600
TION OF CONTAMINANTS ON POLYCRYSTALLINE SILVER BY AES AND WORK FUNCTION. (LEED- AUGER SYSTEM) /DING POTENTI 3-0663
CTRO/ INITIAL OXIDATION STUDIES BY MEANS OF PHOTOELECTRIC WORK FUNCTION MEASUREMENTS.   DETEC 5-1574
ELEC/ CHEMISORPTION STUDIES OF NICKEL SURFACES UTILIZING WORK FUNCTION MEASUREMENTS AND AUGER ELECTRON SPE 5-1928
/XIDATION STUDIES ON IRON (100) BY MEANS OF PHOTOELECTRIC WORK FUNCTION MEASUREMENTS, AUGER, AND LOW ENERGY 5-1290
F THORIUM / A CORRELATION OF AUGER SPECTROSCOPY, LEED AND WORK FUNCTION MEASUREMENTS COMBINED WITH AUGER E/ 5-1927
/Y ELECTRON DIFFRACTION, AUGER ELECTRON SPECTROSCOPY, AND WORK FUNCTION MEASUREMENTS FOR EPITAXIAL GROWTH O 5-1739
SINGLE CRYSTALS IN/ ELLIPSOMETRY, AUGER SPECTROSCOPY, AND WORK FUNCTION MEASUREMENTS ON CLEAN AND CESIUM C/ 5-1294
NICKEL ALLOY PLATES. (AUGER ELECTRON SPECTROSCOPY) WORK FUNCTION OF CESIUM ON TUNGSTEN AND TITANIUM 5-1829
D SURFACES OF GALLIUM ARSENIDE.   LEED, AUGER, AND WORK FUNCTION OF WELL-DEFINED SURFACE OF COPPER- 5-1873
TH THE ZINC OXIDE (0001)- ZINC SURFACE: A LEED, AUGER AND WORK FUNCTION STUDIES OF CLEAN AND SODIUM- COVERE 5-1240
N (110).   LEED- AUGER, AND WORK FUNCTION STUDY.   INTERACTION OF CHLORINE WI 5-1463
AND COADSORPTION / LOW ENERGY ELECTRON DIFFRACTION AUGER, WORK FUNCTION STUDY OF IODINE ADSORBED ON TUNGSTE 5-1145
D WITH AUGER ELECTRON SPECTROSCOPY.   PHOTOELECTRIC WORK FUNCTION STUDY OF OXYGEN- CESIUM ADSORPTION 5-1297
NGSTEN SINGLE CRYSTAL STUDIED BY LEED, AUGER ELECTRON AND WORK FUNCTION STUDY ON IRON (100) SURFACE COMBINE 5-1926
OPPER- GOLD. (AUGER SPECTROSCOPY) WORK FUNCTION TECHNIQUES. / (110) SURFACE OF A TU 5-1660
LVER- GOLD. (AUGER ELECTRON SPECTROSCOPY) WORK FUNCTION VARIATION WITH ALLOY COMPOSITION: C 5-1330
ON A MOLYBDENUM SURFACE BY AUGER ELECTRON SPECTROSCOPY: WORK FUNCTION VARIATION WITH ALLOY COMPOSITION SI 4-0828
WORKING STATE OF THE CATALYST SURFACE. /ND OXYGEN 5-1530

X

ONOLAYER OXYGEN DETERMINATIONS OBTAINED BY PROTON EXCITED X-RAY, AES ANALYSIS OF IRON SURFACES. /ACTIONAL M 5-1682
R EFFECT.   SOFT X-RAY AMPLIFIED SPONTANEOUS EMISSION VIA THE AUGE 2-0443
/TROSCOPY, SCANNING ELECTRON MICROSCOPY AND NONDISPERSIVE X-RAY ANALYSIS TECHNIQUES FOR EXAMINATION OF SIL/ 5-1799
CTROSCOPY.   COMPARISON OF SOFT X-RAY AND AUGER ELECTRON APPEARANCE POTENTIAL SPE 1-0092
/E IN THE CHROMIUM L3/L2 INTENSITY RATIO MEASURED BY SOFT X-RAY AND AUGER ELECTRON APPEARANCE POTENTIAL SP/ 4-0890
E GAS ATOMS. PART-1. THE ARGO/ CHARGE STATE DEPENDENCE OF X-RAY AND AUGER ELECTRON EMISSION SPECTRA FOR RAR 4-0927
RANSITION OPERATOR METHOD.   K X-RAY AND AUGER ELECTRON ENERGIES FOR NEON BY A T 2-0393
ON COLLISIONS. COMMENTS.   SIMPLE MODEL FOR K X-RAY AND AUGER ELECTRON ENERGY SHIFTS IN HEAVY I 2-0301
RELATIVISTIC CALCULATIONS OF ATOMIC X-RAY AND AUGER TRANSITION RATES. 2-0485
AUGER ELECTRON SPECTROSCOPY)   INTRODUCTION TO LOW ENERGY X-RAY AND ELECTRON ANALYSIS. (ANALYZER FOR PHOTO 3-0627
ELECTRON SPECTROSCOPY. X-RAY AND ELECTRON EXCITATION. 1-0082

```
)                                              ELECTRON SPECTROSCOPY: X-RAY AND ELECTRON EXCITATION. (REVIEW, ESCA, AES  1-0081
CTROSCOPY.                                              COMPARISON OF X-RAY AND ELECTRON IMPACT IONIZATION IN AUGER SPE  1-0058
                            ELECTRON GUN FOR AUGER AND ULTRASOFT X-RAY APPARATUS.                                       3-0676
          SUBTRACTION TO QUANTIFICATION OF AUGER ELECTRON AND SOFT X-RAY APPEARANCE POTENTIAL SPECTROSCOPIES. /ROUND    1-0067
          TITANIUM MONOXIDE SURFACES USING AUGER ELECTRON AND SOFT X-RAY APPEARANCE POTENTIAL SPECTROSCOPIES. /M AND    5-1384
                                                           SOFT X-RAY APPEARANCE POTENTIAL SPECTROSCOPY IN A DISP       5-1952
          ITH COPPER- NICKEL AND IRON- NICKEL ALLOYS. (AES)    SOFT X-RAY APPEARANCE POTENTIAL SPECTROSCOPY STUDIES W   5-1952
                              APPLICATION OF ION, ELECTRON, AND X-RAY BEAMS IN MODERN MATERIAL TESTING.                 1-0034
          OF THE PRODUCTION OF LL VACANCY STATES BY THE L X-RAY- L X-RAY COINCIDENCE METHOD. (AUGER TRANSITIONS) /NT    2-0482
          CUUM AND THE MEASUREMENT OF EXCITED ELECTRO/ DESIGN OF AN X-RAY DOUBLE CRYSTAL SPECTROMETER INSTALLED IN VA   3-0641
          IO/ POSSIBLE LOCALISATION OF CONDUCTION ELECTRONS IN SOFT X-RAY DOUBLE IONIZATION SATELLITE (AUGER) TRANSIT   2-0437
                                                                X-RAY EDGE ANOMALIES IN LITHIUM.                        2-0270
                                                                X-RAY ELECTRON SPECTROMETERS AND AUGER ANALYZERS.       3-0599
          ELDS FOR LIGHT ATOMS AND MOLECULES BY ELECTRON IMPACT. (/ X-RAY EMISSION CROSS SECTIONS AND FLUORESCENCE YI   2-0528
          S. (AUGER BROADENING)                               K X-RAY EMISSION EDGE SHAPES OF FREE ELECTRON METAL       2-0337
          .                                           SCANDIUM L X-RAY EMISSION FROM PURE METAL AND SCANDIUM OXIDE      4-1102
          /GNESIUM(2) GERMANIUM, AND MAGNESIUM(2) STANNIDE. PART-1: X-RAY EXCITATION PART-2: ULTRAVIOLET EXCITATION./   2-0306
          /ARISON OF IONIZATION CROSS SECTIONS AND OTHER FACTORS IN X-RAY EXCITED AND ELECTRON EXCITED AUGER SPECTRO/   2-0261
          MAGNESIUM, SOME ALLOYS OF MAGNESIUM, AND ITS OXIDE.    X-RAY EXCITED AUGER AND PHOTOELECTRON SPECTRA OF       4-0837
          PARTIALLY OXIDIZED MAGNESIUM SURFACES.  ABNORMAL CHEMICA/ X-RAY EXCITED AUGER AND PHOTOELECTRON SPECTRA OF    5-1949
                                            ESCAPE DEPTHS OF X-RAY EXCITED (AUGER) ELECTRONS.                           2-0409
          AND IRON.                                             X-RAY EXCITED LMM AUGER SPECTRA OF COPPER, NICKEL       4-1123
          PHOTOELECTRON CHEMICAL SHIFTS IN COORDINATION COMPOUNDS. X-RAY EXCITED LMM AUGER SPECTRA OF GERMANIUM.        4-0924
          ECTRA OF OXIDIZED STATES ON SOME METAL SURFACES.     X-RAY EXCITED PHOTOELECTRON AND AUGER ELECTRON SP        5-1690
          M METAL AND SOME OF ITS COMPOUNDS.                   X-RAY FLUORESCENCE L SPECTRA OF NIOBIUM IN NIOBIU        4-0762
          IG TRANSITION PROBABILITIES.                         X-RAY FLUORESCENCE YIELDS, AUGER, AND COSTER-KRON        2-0259
          ECTRON AND AUGER SPECTRA. CARBON MONOXIDE / COMPARISON OF X-RAY INDUCED AND ELECTRON IMPACT INDUCED PHOTOEL   4-0768
          COPPER, COPPER(II) OXIDE, COPPER(I) OXIDE, AND COPPER(I)/ X-RAY INDUCED PHOTOELECTRON AND AUGER SPECTRA OF    4-0932
          R SPECTRA OF THE ZINC CHALCOGENID/ LIFETIME BROADENING IN X-RAY INDUCED PHOTOEMISSION SPECTROSCOPY AND AUGE   4-0832
          QUILIBRIUM EQUATIONS MODIFIED BY AUGER TRANSITIONS.  X-RAY IONIZATION CROSS SECTIONS, AND IONIZATION E        2-0562
          INES EXCITED BY ELECTRON BOMBARDMENT AND ALUMINUM K-ALPHA X-RAY IRRADIATION. /UDY OF INTENSITIES OF AUGER L   4-1017
                               QUANTITATIVE ASPECTS OF A SOFT X-RAY LASER. (AUGER EFFECT)                               2-0425
          AUGER-TYPE PROCESSES)    CLUSTER CALCULATIONS OF THE SOFT X-RAY L3 EMISSION SPECTRA OF NICKEL AND COPPER. (   4-0888
          LYSIS AND MICROSCOPY. PRINCIPLES AND EXPECTED PERFORMANC/ X-RAY PHOTOELECTRON (AND AUGER ELECTRON) MICROANA   1-0027
          MAGNESIUM THIN FILMS.                                X-RAY PHOTOELECTRON AND AUGER ELECTRON SPECTRA OF        5-1392
          / SPECTROSCOPY AND ADSORPTION STUDIES. CHEMICAL SHIFTS IN X-RAY PHOTOELECTRON AND AUGER SPECTRA FOR MONOLA/   5-1204
          SES OF STEEL SURFACES.                               X-RAY PHOTOELECTRON AND AUGER SPECTROSCOPIC ANALY        5-1268
          /CENTRATION PROFILE OF PLATINUM- TIN ALLOYS FROM COMBINED X-RAY PHOTOELECTRON AND AUGER SPECTROSCOPIC DATA.   5-1195
          NUATION OF LOW ENERGY ELECTRONS FROM SOLIDS: RESULTS FROM X-RAY PHOTOELECTRON (AND AUGER) SPECTROSCOPY. /TE   5-1844
          ERMINE THE SURFACE COMPOSITION OF OXIDIZED VANADI/ USE OF X-RAY PHOTOELECTRON AND AUGER SPECTROSCOPY TO DET   5-1205
          ERMINE THE SURFACE COMPOSITION OF OXIDIZED VANADI/ USE OF X-RAY PHOTOELECTRON AND AUGER SPECTROSCOPY TO DET   5-1863
          LECTRON STUDIES OF ADSORPTION ON CLEAN METAL FILM SURFAC/ X-RAY PHOTOELECTRON AND VACUUM ULTRAVIOLET PHOTOE   5-1203
                                    AUGER LINES IN X-RAY PHOTOELECTRON SPECTROMETRY.                                    1-0208
          METAL AND SODIUM OXIDE.                      AUGER AND X-RAY PHOTOELECTRON SPECTROSCOPIC STUDY OF SODIUM       4-0770
          ES OF OXIDE FILMS ON CHROMIUM STEEL BY AUGER EMISSION AND X-RAY PHOTOELECTRON SPECTROSCOPIES. /LYSIS PROFIL   6-2016
          ITED ON A SILICON SUBSTRATE STUDIED BY AUGER ELECTRON AND X-RAY PHOTOELECTRON SPECTROSCOPIES. /D FILM DEPOS   5-1435
          ON KINETICS OF CESIUM ON GALLIUM ARSENIDE. (USING AES AND X-RAY PHOTOELECTRON SPECTROSCOPY)        ADSORPTI   5-1821
          COPPER- NICKEL ALLOYS BY AUGER ELECTRON SPECTROSCOPY AND X-RAY PHOTOELECTRON SPECTROSCOPY. /SSIVE FILMS ON   5-1874
                             AUGER SPECTRA IN X-RAY PHOTOELECTRON SPECTROSCOPY.                                         1-0209
          BEAM EFFECTS IN AUGER ELECTRON SPECTROSCOPY REVEALED BY X-RAY PHOTOELECTRON SPECTROSCOPY.                     3-0601
          TIAL SPECTROSCOPY. COMPARISON WITH AUGER SPECTROSCOPY AND X-RAY PHOTOELECTRON SPECTROSCOPY. /PEARANCE POTEN   1-0173
          ANALYZER. APPLICATION TO AUGER ELECTRON SPECTROSCOPY AND X-RAY PHOTOELECTRON SPECTROSCOPY. /DING POTENTIAL    3-0634
          ON SPECTROSCOPY.                STRUCTURE ANALYSIS WITH X-RAY PHOTOELECTRON SPECTROSCOPY AND AUGER ELECTR      1-0050
          ON EMISSION SPECTROSCOPY.      SURFACE ANALYSES BY COMBINED X-RAY PHOTOELECTRON SPECTROSCOPY AND AUGER ELECTR  5-1272
          /OF PURE METAL SURFACES USING AUGER ELECTRON SPECTROSCOPY X-RAY PHOTOELECTRON SPECTROSCOPY AND SECONDARY I/   5-1353
          /ON AND DEPOSITION IN LIQUID METALS. (AUGER SPECTROSCOPY, X-RAY PHOTOELECTRON SPECTROSCOPY, MOSSBAUER SPEC/   6-2050
          /Y ELECTRON DIFFRACTION, AUGER ELECTRON SPECTROSCOPY, AND X-RAY PHOTOELECTRON SPECTROSCOPY STUDY OF THE ZI/   5-1337
          RANSITIONS)                                          X-RAY PHOTOELECTRON SPECTRUM OF DIAMOND. (AUGER T        4-0851
          GALLIUM, AND GALLIU/ AUGER AND DIRECT ELECTRON SPECTRA IN X-RAY PHOTOELECTRON STUDIES OF ZINC, ZINC OXIDE,    4-1054
          DISTINCTION BETWEEN WEAK AUGER TRANSITIONS AND INTERNAL X-RAY PHOTOEMISSION.                                  2-0436
          F THE VALENCE BANDS OF MAGNESIUM,/ COMPARISON OF THE SOFT X-RAY PHOTOEMISSION, AND AUGER ELECTRON SPECTRA O   4-0838
          DISPLACEMENTS)                                       X-RAY PHOTOEMISSION AND SURFACE STRUCTURE. (AUGER        5-1811
                    A SIMPLE ANALYTICAL CALCULATION FOR X-RAY PHOTOEMISSION. (AUGER PROCESS)                            2-0429
          )                                 MANY BODY EFFECTS IN X-RAY PHOTOEMISSION FROM MAGNESIUM. (AUGER EFFECT      2-0427
          / AND CHEMISTRY OF EVAPORATED PERMALLOY FILMS OBSERVED BY X-RAY PHOTOEMISSION SPECTROSCOPY AND BY AUGER EL/   5-1738
          /CES AND RECOIL ENERGIES OF FRAGMENT IONS FORMED FROM THE X-RAY PHOTOIONIZATION OF NITROGEN, OXYGEN, CARBO/   2-0308
                   EVIDENCE FOR A RADIATIVE AUGER EFFECT IN X-RAY PHOTON EMISSION.                                      2-0223
                 RADIATIVE AUGER TRANSITIONS IN SOFT X-RAY PLASMA EMISSION.                                             2-0344
          NT, AUGER SPECTRUM)                                  X-RAY PRODUCTION BY HEAVY IONS. (OXYGEN BOMBARDME        2-0479
          LISIONS.                                AUGER ELECTRON AND X-RAY PRODUCTION IN 50- TO 200 KEV NEON- NEON COL  4-1079
          ELL.                           NONRELATIVISTIC AUGER RATES, X-RAY RATES, AND FLUORESCENCE YIELDS FOR THE K-SH 2-0556
          HELL.                          NONRELATIVISTIC AUGER RATES, X-RAY RATES, AND FLUORESCENCE YIELDS FOR THE 2P S 2-0558
          , MAGNESIUM, ALUMINUM, AND/ INTERPRETATION OF HIGH ENERGY X-RAY SATELLITES OF L2,3 EMISSION BANDS OF SODIUM   2-0239
                     NATURE OF SOME X-RAY SATELLITES (RADIATIVE AUGER TRANSITIONS).                                     2-0290
          TER: COMBINATION OF CYLINDRICAL MIRROR ANALYZER WITH SOFT X-RAY SOURCE.     A NEW PHOTOELECTRON SPECTROME     3-0652
          -KRONIG YIELDS IN BISMUTH AND NEPTUNIUM.             X-RAY SPECTRA, L-SUBSHELL FLUORESCENCE AND COSTER       4-1112
          / INDIUM ANTIMONIDE ON THE EXTERNAL PHOTOEFFECT IN THE X-RAY SPECTRAL REGION. (EFFECT OF IMPURITIES ON /      4-1001
          CTURE OF THE DEPENDENCE OF THE COEFFICIENT OF SEC/ USE OF X-RAY SPECTROSCOPIC DATA TO EXPLAIN THE FINE STRU   4-0891
          DING ENERGIES AND AUGER ENERGIES IN FREE ATOMS FOR USE IN X-RAY SPECTROSCOPY.           INNER ELECTRON BIN    4-0774
          RON SPECTROSCOPY)                                    X-RAY SPECTROSCOPY. (REVIEW, AUGER AND PHOTOELECT        1-0203
          ANISM OF S IONIZATION OF NOBLE GAS ATOMS BY THE ULTRASOFT X-RAY SPECTROSCOPY TECHNIQUE. (AUGER DECAYS) /ECH   2-0580
          A NEW TYPE OF AUGER EFFECT AND ITS INFLUENCE ON THE X-RAY SPECTRUM.                                           1-0045
                             AUGER TRANSITIONS AND SHAPE OF X-RAY SPECTRUM.                                             4-1104
          T OF THE AUGER SPECTRUM AND OF THE SATELLITES IN THE SOFT X-RAY SPECTRUM OF WATER. /CULATIONS FOR ASSIGNMEN   4-0724
          D NEON ATOM.          CALCULATED K-AUGER ELECTRON AND K X-RAY TRANSITION ENERGIES FOR THE MULTIPLY IONIZE     4-0955
          LITHIUM.                             AUGER RATES FOR SOFT X-RAY TRANSITIONS IN IONIC, ATOMIC, AND METALLIC    4-0835
          IZATION ON THE FLUORESCENCE YIELD OF NEON. (K-AUGER AND K X-RAY YIELDS)       EFFECT OF MULTIPLE ION          2-0299
          EXCITATION CROSS SECTIONS AND RANGES USING PROTON EXCITED X-RAYS.          DETERMINATION OF AUGER ELECTRON    2-0456
                 SHIFT ENERGIES OF CHARACTERISTIC X-RAYS AND AUGER ELECTRONS FOR IONIZED ATOMS.                         4-1041
          EMPHASIS ON INSTRUMENTATIION.                        X-RAYS AND ELECTRONS IN ANALYTICAL CHEMISTRY WITH        3-0646
```

ON OF MODERATELY HEAVY NUCLEI IN THE TISSUE: USE OF SOFT X-RAYS. (AUGER ELECTRON IRRADIATION) / ON INCLUSI 2-0531
LECTRON SPECTRA, REVIEW, 6 REFS)   ELECTRONS AND X-RAYS IN NONDESTRUCTIVE ANALYSIS. (AUGER, PHOTOE 1-0083
/SUREMENT OF THE PRODUCTION OF LL VACANCY STATES BY THE L X-RAY- L X-RAY COINCIDENCE METHOD. (AUGER TRANSI/ 2-0482
ELECTRON PHOTOEMISSION INDUCED BY LASER-PRODUCED PLASMA X-RAYS. (PLASMA ELECTRONS, AUGER PROCESS)   2-0276
ELECTRON BEAM EFFECTS IN AUGER ANALYSIS OF PHYSISORBED XENON.   5-1151
RA FROM KRYPTON AND M4, 5NN AND N4,5OO AUGER SPECTRA FROM XENON. / RESOLUTION L2,3MM AND M4,5NN AUGER SPECT 4-1113
SITION RATES AND FLUORESCENT YIELDS FOR ELEMENTS ARGON TO XENON.   K-SHELL AUGER TRAN 4-0961
CORRELATION EFFECTS IN THE M4,5N4,5N4,5 AUGER SPECTRUM OF XENON.   SEARCH FOR 4-0872
SOLID STATE BROADENING EFFECTS IN THE AUGER SPECTRUM OF XENON.   4-1005
NING IN THE AUGER SPECTRA OF RARE GASES. (ARGON, KRYPTON, XENON)   SOLID STATE BROADE 4-1006
SPECTRUM OF AUGER KLL ELECTRONS OF TELLURIUM AND XENON.   4-0755
LEED AND SURFACE POTENTIAL STUDY OF CARBON MONOXIDE AND XENON ADSORBED ON COPPER (100).   5-1246
UGER ELECTRON SP/ FIRST ORDER TWO DIMENSIONAL TRANSITION: XENON ADSORBED ON THE (0001) FACE OF GRAPHITE. (A 5-1857
ADSORPTION OF XENON AND COBALT ON SILVER (111).   5-1633
CT. (AUGER EMISSION)   MULTIIONIZATION OF NEON, ARGON, XENON AND THEIR IONS BY HIGH ENERGY ELECTRON IMPA 2-0492
ISTRIBUTION IN ENERGY AND ANGLE OF ELECTRONS EJECTED FROM L XENON BY 0.3 TO 2.0 MEV PROTONS. (AUGER PEAKS) 2-0534
MENT OF COSTER-KRONIG RATES IN SOLID STATE ENVIRONMENTS. (XENON, CADMIUM, ZINC)   ENHANCE 2-0439
N4,5 AUGER TRANSITIONS.   VACUUM ULTRAVIOLET EMISSION OF XENON(III) LEVELS EXCITED BY ELECTRON IMPACT VIA 4-0878
(AUGER SPECTROSCOPY)   ENTROPY OF ADSORPTION OF XENON IN EPITAXY ON THE (0001) FACE OF GRAPHITE.   4-1860
PHOTOELECTRON AND AUGER SPECTRA OF XENON IN SOLID PERXENATES.   4-0902
ECTRON SPECTROSCOPY AND SECONDARY ION MASS SPECTRA DURING XENON ION BOMBARDMENT. /PPER ANALYZED BY AUGER EL 5-1679
TRON IMPACT NEAR THRESHOLD.   AUGER ELECTRON EJECTION FROM XENON N4,5OO AND KRYPTON M4,5NN PROCESSES BY ELEC 4-1010
ELLIPSOMETRIC STUDY OF ADSORPTION: ADSORPTION OF XENON ON GRAPHITE (001).   5-1754
QUANTITATIVE AUGER SPECTROSCOPY OF PHYSICALLY ADSORBED XENON ON NICKEL.   5-1150
PHYSICAL ADSORPTION OF XENON ON PALLADIUM (100).   5-1711
YNAMICS AND KINETICS OF THE FIRST MONOLAYER ADSORPTION OF XENON ON THE (0001) GRAPHITE FACE.   THERMOD 5-1859
GY ELECTRON DIFFRACTION STUDIES OF ADSORPTION ISOTHERMS: XENON ON (0001) GRAPHITE. /CTROSCOPY AND LOW ENER 5-1856
ING THE IRIDIUM (100) SURFACE. (A/ LEED INVESTIGATIONS OF XENON SINGLE CRYSTAL FILMS AND THEIR USE IN STUDY 5-1479
ITE. (AUGER SPECTROS/ TWO-DIMENSIONAL PHASE TRANSITION IN XENON SUBMONOLAYER FILMS ADSORBED ON (0001) GRAPH 5-1858
KMM AUGER TRANSITIONS IN XENON-131.   4-0758
TION OF METAL SURFACES: A REVIEW OF RECENT LEED, AES, AND XPS RESULTS.   STRUCTURE AND COMPOSI 5-1341
OXIDATION OF CUPRONICKEL ALLOYS. PART-1: XPS STUDY OF INTERDIFFUSION. (AND AES)   5-1224
SOME ASPECTS OF AN AES AND XPS STUDY OF THE ADSORPTION OF OXYGEN ON NICKEL.   5-1459

Y

NCIDENCE ANGLE OF PRIMARY ELECTRONS ON THE AUGER EMISSION YIELD.   EFFECT OF THE I 2-0237
TECTING FINE STRUCTURE IN THE SECONDARY ELECTRON EMISSION YIELD AND THE APPLICATION TO SILICON (111). /R DE 4-0852
RIPPING MECHANISM DEPENDENCE OF THE ARGON 2S FLUORESCENCE YIELD. (AUGER TRANSITION)   ST 4-0957
R) SP/ CHARGE STATE DEPENDENCE OF THE NEON K FLUORESCENCE YIELD DEDUCED FROM HIGH RESOLUTION EMISSION (AUGE 2-0520
/AMIC CHARGE STATE DEPENDENCE OF THE K-VACANCY PRODUCTION YIELD IN HEAVY ION ATOM COLLISIONS AT KEV ENERGI/ 2-0346
GER ELECTRON PRODUCT/ K-SHELL IONIZATION AND FLUORESCENCE YIELD IN 50 MEV CHLORINE(+)- NEON COLLISIONS. (AU 2-0518
/E DEPENDENCE OF NEON K-SHELL IONIZATION AND FLUORESCENCE YIELD IN 50 MEV CHLORINE(N+) ION- NEON COLLISION/ 2-0300
CORRECTIONS FOR POTENTIAL MODULATION DISTORTION IN AUGER YIELD MEASUREMENTS.   EXACT 2-0389
SPECTROMETERS FOR RETARDING POTENTIAL SECONDARY ELECTRON YIELD MEASUREMENTS.   USE OF CYLINDRICAL AUGER 3-0628
- AUGER SYSTEM)   FAST, ACCURATE SECONDARY ELECTRON YIELD MEASUREMENTS AT LOW PRIMARY ENERGIES. (LEED 3-0629
FLUORESCENCE YIELD OF CARBON. (AUGER MEASUREMENTS)   5-1571
K-SHELL FLUORESCENCE YIELD OF IONIZED ALUMINUM. (AUGER RATE)   4-0975
MULTIPLET EFFECTS ON THE L2,3 FLUORESCENCE YIELD OF MULTIPLY IONIZED ARGON. (AUGER EFFECT)   4-0956
EFFECT OF MULTIPLE IONIZATION ON THE FLUORESCENCE YIELD OF NEON. (K-AUGER AND K X-RAY YIELDS)   2-0299
NG INCIDENCE.   DEPENDENCE OF AUGER ELECTRON YIELD ON PRIMARY BEAM ENERGY AT NORMAL AND GLANCI 4-0455
/IUM OXIDE- GOLD CERMET FILMS AND ITS APPLICATION TO HIGH YIELD SECONDARY ELECTRON EMITTERS. (AUGER ELECTR/ 5-1430
ON BINDING ENERGIES OF BARIUM FROM THE SECONDARY ELECTRON YIELD SPECTRUM.   ELECTR 4-0846
/HARGE STATE DEPENDENCE OF THE ARGON L-SHELL FLUORESCENCE YIELD STUDIED BY ATOMIC HYDROGEN(+), MOLECULAR H/ 2-0517
TIES TO THE 1S STATE AND THEORETICAL K-SHELL FLUORESCENCE YIELDS.   ATOMIC RADIATION TRANSITION PROBABILI 4-0918
ES TO THE ATOMIC 2S STATE AND THEORETICAL L1 FLUORESCENCE YIELDS. /AND COSTER-KRONIG TRANSITION PROBABILITI 2-0329
EFFECTS ON RADIATIVE RATES, AUGER RATES, AND FLUORESCENCE YIELDS.   DOUBLE IONIZATION 4-0976
N ON THE FLUORESCENCE YIELD OF NEON. (K-AUGER AND K X-RAY YIELDS)   EFFECT OF MULTIPLE IONIZATIO 2-0299
INFLUENCE OF COSTER-KRONIG PROCESSES ON L-AUGER YIELDS.   4-1098
K-VACANCY SHARING STUDIED FROM AUGER ELECTRON YIELDS.   2-0345
- AND M-SHELL AND SUBSHELL FLUORESCENCE AND COSTER-KRONIG YIELDS.   L 4-0959
REVIEW OF FLUORESCENCE YIELDS AND AUGER ELECTRON MISSION.   1-0121
K- AND L-SHELL FLUORESCENCE AND AUGER YIELDS AND AUGER ELECTRON SPECTROSCOPY.   4-0966
GEN.   FLUORESCENCE YIELDS AND AUGER RATES FOR MULTIPLY IONIZED NITRO 4-0781
F23 BASED ON THE G/ THEORETICAL L2 SUBSHELL FLUORESCENCE YIELDS AND COSTER-KRONIG TRANSITION PROBABILITIES 2-0313
ELEMENTS.   L- AND M-SHELL YIELDS AND ELECTRON SPECTRA FOR THE TRANSURANIC E 4-0968
ABILITIES.   THEORETICAL L2 AND L3 SUBSHELL FLUORESCENCE YIELDS AND L2-L3(X) COSTER-KRONIG TRANSITION PROB 2-0316
SPECTROSCOPY.   RELATIVE SPUTTERING YIELDS AND QUANTITATIVE SURFACE ANALYSIS BY AUGER 5-1964
BILITIES.   X-RAY FLUORESCENCE YIELDS, AUGER, AND COSTER-KRONIG TRANSITION PROBA 2-0259
STER-KRONIG, AUGER, AND RADIATIVE RATES, AND FLUORESCENCE YIELDS FOR CALCIUM TO THORIUM.   ATOMIC M-SHELL CO 4-0964
K-SHELL AUGER TRANSITION RATES AND FLUORESCENT YIELDS FOR ELEMENTS ARGON TO XENON.   4-0961
K-SHELL AUGER TRANSITION RATES AND FLUORESCENT YIELDS FOR ELEMENTS BERYLLIUM TO ARGON.   4-0962
IMPACT. (/ X-RAY EMISSION CROSS SECTIONS AND FLUORESCENCE YIELDS FOR LIGHT ATOMS AND MOLECULES BY ELECTRON 2-0528
K-SHELL FLUORESCENCE YIELDS FOR LIGHT ELEMENTS.   4-0818
K-SHELL (AUGER) FLUORESCENT YIELDS FOR MULTIPLY IONIZED NEON.   4-0994
/SHELL AUGER RATES, TRANSITION ENERGIES, AND FLUORESCENCE YIELDS FOR MULTIPLY IONIZED NEON. (AUGER EFFECT) 2-0283
THEORETICAL FLUORESCENCE YIELDS FOR NEON. (AUGER RATE)   2-0283
OSTER-KRONIG, AUGER, AND RADIATIVE RATES AND FLUORESCENCE YIELDS FOR SODIUM- THORIUM.   ATOMIC L-SHELL C 4-0963
ONRELATIVISTIC AUGER RATES, X-RAY RATES, AND FLUORESCENCE YIELDS FOR THE K-SHELL.   N 2-0556
ONRELATIVISTIC AUGER RATES, X-RAY RATES, AND FLUORESCENCE YIELDS FOR THE 2P SHELL.   N 2-0558
AUGER SPECTRA AND FLUORESCENCE YIELDS FOR THE 3P AND THE 3D SHELL.   2-0278
NONRELATIVISTIC FLUORESCENCE YIELDS FOR THE 3P AND THE 3D SHELLS.   2-0277
STER-KRONIG, AUGER, AND RADIATIVE RATES, AND FLUORESCENCE YIELDS FOR Z BETWEEN 38 AND 103. /OMIC N-SHELL CO 4-0969
ERMAL CARBON OXYGEN NITROGEN(2), AND / SECONDARY ELECTRON YIELDS FROM COPPER BERYLLIUM OXYGEN SURFACE BY TH 4-0786
X-RAY SPECTRA, L-SUBSHELL FLUORESCENCE AND COSTER-KRONIG YIELDS IN BISMUTH AND NEPTUNIUM.   4-1112
D M-SHELL FLUORESCENCE YIELDS)   ATOMIC FLUORESCENCE YIELDS. (L-SHELL COSTER-KRONIG YIELDS, K-, L-, AN 2-0351
(AUGER EMISSION) MEASUREMENTS OF THE L-SHELL FLUORESCENCE YIELDS OF ARGON AND CHLORINE BY ELECTRON IMPACT. 2-0387
SITION)   K-SHELL FLUORESCENCE YIELDS OF ARGON, CHLORINE AND SULFUR. (AUGER TRAN 4-1014
L-SHELL FLUORESCENCE YIELDS OF DOUBLE VACANCY STATES IN LEAD.   4-1105
THE L(II) SUBSHELL ATOMIC YIELDS OF ELEMENTS WITH Z APPROXIMATELY 90.   2-0349

ELL IONIZATION OF CARBON BY FAST PROTONS. (AUGER ELECTRON YIELDS OF METHANE, ETHANE, ETHYLENE, ACETYLENE) / 2-0532
MULTIPLET EFFECT IN FLUORESCENCE YIELDS OF MULTIPLY IONIZED ATOMS. (AUGER EFFECT) 2-0314
-SHELL AUGER RATES, TRANSITION ENERGIES, AND FLUORESCENCE YIELDS OF VARIOUSLY IONIZED STATES OF ARGON. K 2-0280
NTHANUM, PRASEODYMIUM, NEODYMIUM, GADOLINIUM, DYSPROSIUM, YTTERBIUM, HAFNIUM). /SOME RARE EARTH METALS. (LA 4-1100
ELECTRONIC SPECTRA ACCOMPANYING THE DECAY OF ERBIUM-160 YTTERBIUM-166, YTTERBIUM-169, AND LUTETIUM-171. / 4-0743
CTRA ACCOMPANYING THE DECAY OF ERBIUM-160, YTTERBIUM-166, YTTERBIUM-169, AND LUTETIUM-171. / ELECTRONIC SPE 4-0743
(AUGER ANALYSIS) EFFECT OF YTTRIA ADDITIONS ON HOT PRESSED SILICON NITRIDE. 5-1350
MORPHOLOGICAL AND ELECTRICAL PROPERTIES OF RF SPUTTERED YTTRIA-DOPED ZIRCONIA THIN FILMS. (AUGER EFFECT) 5-1385
CHEMICAL SHIFTS IN THE AUGER SPECTRUM OF YTTRIUM ON OXYGEN ADSORPTION. 5-1153

Z

THE L(II) SUBSHELL ATOMIC YIELDS OF ELEMENTS WITH Z APPROXIMATELY 90. 2-0349
, AUGER, AND RADIATIVE RATES, AND FLUORESCENCE YIELDS FOR Z BETWEEN 38 AND 103. /OMIC N-SHELL COSTER-KRONIG 4-0969
IN THE KLL AUGER SPECTRUM. Z DEPENDANCE OF THE KLL AUGER RATES. 2-0557
NERGIES. EMISSION OF KLL AUGER ELECTRONS PRODUCED IN Z DEPENDENCE OF ENERGIES AND RELATIVE INTENSITIES 4-0886
WIDTH OF ATOMIC L2 AND L3 VACANCY STATES NEAR Z (1)- NITROGEN AND Z (1)- NEON COLLISIONS AT KEV E 4-0830
S IN THE CALCULATION OF THE KLL AUGER/ APPLICATION OF THE Z = 30. (AUGER EFFECT) 4-0939
RONIG RATES IN SOLID STATE ENVIRONMENTS. (XENON, CADMIUM, ZETA EFFECTIVE VALUES OF HYDROGENIC WAVE FUNCTION 2-0526
LEVEL AND AUGER SPECTRA OF NICKEL AND NICKEL ALLOYED WITH ZINC) ENHANCEMENT OF COSTER-K 2-0439
TATION OF HIGH RESOLUTION LMM AUGER SPECTRA OF COPPER AND ZINC. /L STATE EFFECTS IN THE PHOTOEMISSION CORE 5-1806
QUASIATOMIC LMM AUGER SPECTRA OF SOLID COPPER AND ZINC. L-S COUPLING INTERPRE 4-1122
HIGH RESOLUTION L2,3M4,5M4,5 AUGER SPECTRUM OF FREE ZINC. 4-1121
L2,3MM AUGER ELECTRON SPECTRUM FROM FREE ZINC ATOMS. 4-0733
DUCED PHOTOEMISSION SPECTROSCOPY AND AUGER SPECTRA OF THE ZINC ATOMS. 4-0731
NALYSIS OF THE L3M4,5M4,5: 1G4 AUGER TRANSITION IN ATOMIC ZINC CHALCOGENIDES. /ETIME BROADENING IN X-RAY IN 4-0832
ENA INVOLVED IN THE L3M4,5M4,.5M4,.5:1G4 AUGER PROCESS IN ZINC: IMPROVED AUGER ENERGY CALCULATIONS. /ILED A 4-0884
ADSORPTION OF ZINC METAL. RELAXATION PHENOM 4-0882
LEED AND AES OBSERVATIONS ON THE POLAR SURFACES OF ZINC ON GALLIUM ARSENIDE. (AUGER SPECTROSCOPY) 5-1144
RCLE OF IMPURITIES IN THE STABILITY OF ZINC OXIDE. 5-1619
TRON SPECTROSCOPY) EXOELECTRON EMISSION FROM ZINC OXIDE. 5-1462
CORRELATION OF ELECTRONIC, LEED, AND AUGER DIAGNOSTICS ON ZINC OXIDE. (OXYGEN ADSORPTION EFFECT, AUGER ELEC 4-0921
STABILITY OF CESIUM COVERED ZINC OXIDE SURFACES. 5-1584
ORPTION, PHOTODESORPTION AND CONDUCTIVITY MEASUREMENTS ON ZINC OXIDE SURFACES. 5-1882
INTERACTION OF CESIUM WITH CLEAN ZINC OXIDE SURFACES. (AUGER ANALYSIS) CHEMIS 5-1801
SOME PROPERTIES OF CLEAN AND ALKALI METAL COVERED ZINC OXIDE SURFACES. (AUGER SPECTROSCOPY) 5-1588
D WORK FUNCTION STUDY. INTERACTION OF CHLORINE WITH THE ZINC OXIDE SURFACES. (LEED, AES) 5-1881
Y (AES) AND LOW ENERGY ELECTRON DIFFRACTION (LEED) OF THE ZINC OXIDE (0001)- ZINC SURFACE: A LEED, AUGER AN 5-1463
OSCOPY, AND X-RAY PHOTOELECTRON SPECTROSCOPY STUDY OF THE ZINC OXIDE (0001) POLAR SURFACES. /ON SPECTROSCOP 5-1338
OF THREE ENHANCED MULTILAYER DIELECTRIC MIRROR DESIGNS. ZINC OXIDE (0001) POLAR SURFACES. /LECTRON SPECTR 5-1337
F GREEN- COPPER LUMINESCENCE IN COPPER AND ALUMINUM DOPED (ZINC SELENIDE, THORIUM FLUORIDE) AUGER ANALYSIS 5-1328
CONCENTRATION-DEPENDENT NONRADIATIVE PROCESSES IN ZINC SULFIDE. (AUGER EFFECT, COLOR CENTERS) /CE O 2-0402
CHEMISORPTION REACTIONS ON HIGH INDEX ZINC SULFIDE PHOSPHORS. (AUGER RECOMBINATIONS) 2-0388
L2,3M4,5M4,5 AUGER SPECTRA OF METALLIC COPPER AND ZINC SULFIDE SURFACES. (AUGER EFFECT) 4-0788
DIRECT ELECTRON SPECTRA IN X-RAY PHOTOELECTRON STUDIES OF ZINC: THEORY AND EXPERIMENT. 4-0919
PTH OF LEVELS OF SURFACE ELECTRON CENTERS. (AUGER EFFECT, ZINC, ZINC OXIDE, GALLIUM, AND GALLIUM OXIDE. /D 4-1054
AL AND ELECTRICAL PROPERTIES OF RF SPUTTERED YTTRIA-DOPED ZIRCONIA, SILICA) /DETERMINATION OF THE ENERGY DE 4-0915
DIFFERENTIAL AUGER SPECTROMETRY. (NIOBIUM- ZIRCONIA THIN FILMS. (AUGER EFFECT) MORPHOLOGIC 5-1385
EXOEMISSION PROPERTIES OF ZIRCONIUM ALLOY ANALYSIS 5-1849
E SEGREGATION AND INTERACTION OF OXYGEN AND NITROGEN WITH ZIRCONIUM DIOXIDE. (AUGER EFFECT) 2-0410
OPY) SECONDARY ELECTRON EMISSION OF A ZIRCONIUM IN NIOBIUM- 1% ZIRCONIUM. SURFAC 5-1508
ZIRCONIUM- ALUMINUM BULK GETTER. (AUGER SPECTROSC 5-1407

ELECTRON SPECTROSCOPY. KIKUCHI CORRELATIONS FOR ARSENIC (00001). HIGH ANGULAR RESOLUTION SECONDARY 5-1784
ION STUDY. INTERACTION OF CHLORINE WITH THE ZINC OXIDE (0001)- ZINC SURFACE: A LEED, AUGER AND WORK FUNCT 5-1463
MEASUREMENT OF CONTACT ANGLES. PART-4: WATER ON GRAPHITE (0001). (AES) ULTRAHIGH VACUUM TECHNIQUES IN THE 5-1795
UGER ELECTRON SPECTROSCOPY. STUDY OF RUTHENIUM (0001) AND RHODIUM (111) SURFACES USING LEED AND A 5-1380
/ORDER TWO DIMENSIONAL TRANSITION: XENON ADSORBED ON THE (0001) FACE OF GRAPHITE. (AUGER ELECTRON SPECTROS/ 5-1857
ENTROPY OF ADSORPTION OF XENON IN EPITAXY ON THE (0001) FACE OF GRAPHITE. (AUGER SPECTROSCOPY) 5-1860
N DIFFRACTION STUDIES OF ADSORPTION ISOTHERMS: XENON ON (0001) GRAPHITE. /CTROSCOPY AND LOW ENERGY ELECTRO 5-1856
/HASE TRANSITION IN XENON SUBMONOLAYER FILMS ADSORBED ON (0001) GRAPHITE. (AUGER SPECTROSCOPY, LOW ENERGY / 5-1858
NETICS OF THE FIRST MONOLAYER ADSORPTION OF XENON ON THE (0001) GRAPHITE FACE. THERMODYNAMICS AND KI 5-1859
LOW ENERGY ELECTRON DIFFRACTION (LEED) OF THE ZINC OXIDE (0001) POLAR SURFACES. /ON SPECTROSCOPY (AES) AND 5-1338
X-RAY PHOTOELECTRON SPECTROSCOPY STUDY OF THE ZINC OXIDE (0001) POLAR SURFACES. /LECTRON SPECTROSCOPY, AND 5-1337
A CARBON STRUCTURE ON THE RHENIUM (0001) SURFACE. 5-1989
LEED- AUGER ANALYSIS OF THE BERYLLIUM (0001) SURFACE. 5-1581
N AND AUGER ELECTRON OBSERVATIONS OF THE SILICON CARBIDE (0001) SURFACE. LOW ENERGY ELECTRON DIFFRACTIO 5-1932
SOME PROPERTIES OF THE RHENIUM (0001) SURFACE. 5-1304
PECTROSCOPIC STUDIES. ARSENIC (0001) SURFACES: AUGER, LOSS, AND PHOTOELECTRON S 5-1315
1 STRUCTURE OF A SILICON ORDERED MONOLAYER ON MOLYBDENUM (001). ATOMIC ARRANGEMENT IN THE 1 K 5-1478
E GROWTH MODE OF IRON EPITAXIAL FILMS ON MAGNESIUM OXIDE (001). /TUDY OF SURFACE CONTAMINATION EFFECT ON TH 5-1524
RIC STUDY OF ADSORPTION: ADSORPTION OF XENON ON GRAPHITE (001). ELLIPSOMET 5-1754
INTERACTION OF HYDROGEN SULFIDE WITH COPPER (001). 5-1516
ROSCOPY STUDY OF THE INITIAL STAGES OF OXIDATION OF IRON (001). /ELECTRON DIFFRACTION- AUGER ELECTRON SPECT 6-2110
SURFACE DEBYE TEMPERATURE OF THE SILICON (001) - 2X2 STRUCTURE. (AUGER EFFECT) 5-1754
ER STUDIES OF EFFECT OF OXYGEN ON ADHESION OF CLEAN IRON (001) AND (011) SURFACES. LEED AND AUG 6-2010
) CHROMIUM(0.16). (A/ SURFACE COMPOSITION STUDIES OF THE (001) AND (110) FACES OF MONOCRYSTALLINE IRON(0.84 5-1591
AND MEASUREMENT OF SURFACE DIFFUSION ON MAGNESIUM OXIDE (001) BY AUGER SPECTROSCOPY. /DESORPTION OF SODIUM 5-1494
RFACE STRUCTURE WITH ANTIPHASE DOMAINS OF SULFUR ON IRON (001) BY LEED- AES. ANALYSIS OF THE SU 5-1470
ELECTRON SPECTROSCOPY/ STRUCTURAL STUDY OF OXYGEN- IRON (001) BY LOW ENERGY ELECTRON DIFFRACTION AND AUGER 5-1469
/VIOR OF IMPURITY ATOMS AND ADSORBED OXYGEN ATOMS ON THE (001) FACE OF IRON EPITAXIAL FILM. (AUGER EFFECT) 5-1523
CTROSCOPY, LEED AND AES. INVESTIGATION OF REORDERED (001) GOLD SURFACES BY POSITIVE ION CHANNELING SPE 5-1139
CTROSCOPY, LOW ENERGY ELE/ CHARACTERIZATION OF REORDERED (001) GOLD SURFACES BY POSITIVE ION CHANNELING SPE 5-1986
DIFFRACTION- AUGER ELECTRON SPECTROSCOPY STUDY OF CLEAN (001) IRON SURFACE. LOW ENERGY ELECTRON 5-1527
ADSORPTION OF CARBON MONOXIDE ON THE (001) LEED AND AUGER EMISSION STUDIES. 5-1515
K. ADSORPTION OF COBALT ON (001) RUTHENIUM AT TEMPERATURES BELOW 300 DEGREES 5-1609
/ SURFACE PLASMON STUDIES ON ALUMINUM (111) AND ALUMINUM (001) SINGLE CRYSTAL SURFACES USING INELASTIC LOW 5-1745
OWTH MECHANISM OF DEPOSITED IRON FILM ON MAGNESIUM OXIDE (001) SURFACE. LEED- AES OBSERVATION OF GR 5-1525

/F METHANE, ETHANE, ETHYLENE, AND ACETYLENE WITH AN IRON (001) SURFACE AND THEIR INFLUENCE ON ADHESION STU/ 5-1208
PTION OF OXYGEN AND OXIDATION OF COBALT ON THE RUTHENIUM (001) SURFACE. (AUGER SPECTRA)   ADSOR 5-1608
GROWTH MODE OF IRON FILM DEPOSITED ON MAGNESIUM OXIDE (001) SURFACE. (AUGER SPECTROSCOPY)   5-1528
GROWTH AND STRUCTURE OF THIN COPPER FILMS CN (001) SURFACES OF NICKEL.   5-1227
/OF THE EPITAXIAL GRCWTH OF MAGNESIUM ON MAGNESIUM OXIDE (001) USING REFLECTION DIFFRACTION, LEED AND AUGE/   5-1496
UGER SPE/ ELASTIC LEED INTENSITY ENERGY STUDIES OF CLEAN (001), (110) AND ELASTIC (111) NICKEL SURFACES. (A   5-1292
CTRA)   CHEMISORPTION ON (001), (110) AND (111) NICKEL SURFACES. (AUGER SPE   5-1764
D STUDY USING LOW ENERGY ELECTRON DIFF/ CHEMISORPTION ON (001), (110), AND (111) NICKEL SURFACES. CORRELATE   5-1291
AUGER EMISSION SPECTROSCOPY VANADIUM PENTOXIDE (010) AND VANADIUM (100) SURFACES.   5-1941
HE AUGER ELECTRON EMISSION SPECTRA OF VANADIUM PENTOXIDE (010) AND VANADIUM (100) SURFACES. /STRUCTURE IN T   5-1340
/ORPTION OF ETHYLENE OXIDE AND VINYL CHLORIDE ON AN IRON (011) SURFACE AND EFFECT OF THESE FILMS ON ADHESI/   5-1207
OF EFFECT OF OXYGEN ON ADHESION OF CLEAN IRON (001) AND (011) SURFACES.   LEED AND AUGER STUDIES 6-2010
MOLYBDENUM (001).   ATOMIC ARRANGEMENT IN THE 1 K 1 STRUCTURE OF A SILICON ORDERED MONOLAYER ON   5-1478
ER ENERGY CALCULATI/ DETAILED ANALYSIS OF THE L3M4,5M4,5: 1G4 AUGER TRANSITION IN ATOMIC ZINC: IMPROVED AUG   4-0884
UGER RATES IN ATOMS WITH INNER SHELL VACANCIES (NEON WITH 1S OR 2S VACANCY).   CORRELATION ENERGIES AND A   2-0312
LDS.   ATOMIC RADIATION TRANSITION PROBABILITIES TO THE 1S STATE AND THEORETICAL K-SHELL FLUORESCENCE YIE   4-0918
ADSORPTION OF CYANOGEN ON PLATINUM (100).   5-1684
OSCOPY STUDY OF THE ADSORPTION OF RUBIDIUM ON MOLYBDENUM (100).   AUGER SPECTR   5-1894
AUGER- LEED INVESTIGATION OF TIN ON MOLYBDENUM (100).   5-1490
ER PEAK SHAPES. ETHYLENE AND CARBON MONOXIDE ON TUNGSTEN (100). /ACTERIZATION OF ADSORBED SPECIES USING AUG   5-1243
TION PEAKS IN SECONDARY ELECTRON ENERGY SPECTRA. (NICKEL (100).   DIFFRAC   4-0775
SPECTROSCOPY. DISCREPANCY FOR CARBON DIOXIDE ON TUNGSTEN (100).   FLASH DESORPTION- AUGER ELECTRON   5-1461
GH RESOLUTION AUGER ELECTRON SPECTRUM OF MAGNESIUM OXIDE (100).   HI   4-0772
DY OF THE INTERACTION OF HYDROGEN SULFIDE AND MOLYBDENUM (100).   LEED AND AES STU   5-1973
D ELECTRON SPECTROSCCPIC OBSERVATIONS ON NICKEL MONOXIDE (100).   LEED AN   5-1687
AL STUDY OF CARBON MONOXIDE AND XENON ADSORBED ON COPPER (100).   LEED AND SURFACE POTENTI   5-1246
UR SEGREGATION FROM THE BULK TO THE SURFACE OF PALLADIUM (100). / AES STUDIES OF CHEMISORPTION INDUCED SULF   5-1865
ECTROSCOPY STUDY OF THE OXIDATION OF IRON (110) AND IRON (100). /GY ELECTRON DIFFRACTION- AUGER ELECTRON SP   5-1585
PY STUDY OF THE OXIDATION OF CHROMIUM (110) AND CHROMIUM (100). /RON DIFFRACTION- AUGER ELECTRON SPECTROSCO   6-2022
PHYSICAL ADSORPTION OF XENON ON PALLADIUM (100).   5-1711
DIFFRACTION STUDY OF ALKALI METAL OVERLAYERS CN TUNGSTEN (100). /TRON SPECTROSCOPY AND LOW ENERGY ELECTRON   5-1895
SLOW ELECTRON SPECTROSCOPY ON MOLYBDENUM (100).   5-1578
TUDIES OF THE ADSORPTION OF CESIUM AND OXYGEN ON SILICON (100).   STRUCTURAL S   5-1389
NCE OF THE LOW ENERGY AUGER SPECTRUM OF NICKEL(II) OXIDE (100).   TEMPERATURE DEPENDE   4-0999
-SHELLS OF SURFACE ATOMS. (AUGER INTENSITIES OF TUNGSTEN (100)) /TRON IMPACT IONIZATION CROSS SECTIONS ON K   4-0847
RGY ELECTRON SPECTROSCOPY OF PLATINUM (100) AND PLATINUM (100)- CARBON MONOXIDE.   LOW ENE   5-1686
CTRICAL, CHEMICAL, AND STRUCTURAL STUDY OF THE GERMANIUM (100)- CESIUM- OXYGEN PHOTOSURFACE.   AN ELE   5-1319
ECTROSCOPY.   CHEMICAL ANALYSIS OF PLATINUM (100) AND GOLD (100) SURFACES BY AUGER ELECTRON SP   5-1712
GER ELECTRON SPECTROSCCPY STUDY OF THE OXIDATION OF IRON (100) AND IRON (110). /GY ELECTRON DIFFRACTION- AU   5-1586
OSCOPY)   OXIDATION OF CARBON ON TUNGSTEN (100) AND MOLYBDENUM (100). (AUGER ELECTRON SPECTR   5-1944
/ CLEANING TECHNIQUES AS APPLIED TO NICKEL (111), NICKEL (100), AND NICKEL SHEET USING AUGER ELECTRON SPEC/   5-1467
LOW ENERGY ELECTRON SPECTROSCOPY OF PLATINUM (100) AND PLATINUM (100)- CARBON MONOXIDE.   5-1686
ES.   ATOMIC ARRANGEMENT OF GOLD (100) AND RELATED METAL OVERLAYER SURFACE STRUCTUR   5-1716
S. (LEED, AUGER ELEC/ REACTIVITY OF LOW INDEX ((111) AND (100)) AND STEPPED PLATINUM SINGLE CRYSTAL SURFACE   5-1838
LEED- AES STUDY OF THE OXIDATION OF IRCN- CHROMIUM (100) AND (110).   5-1587
/DATION OF MOLYBDENUM. PART-3: OXIDE NUCLEATION RATES ON (100) AND (110) MOLYBDENUM SURFACES. (AUGER SPECT/   5-1536
N SECONDARY ELECTRON EMISSION DURING OXIDATION OF NICKEL (100) AND (111) CRYSTAL SURFACES. /MICAL CHANGES I   5-1453
TRON DIFFRACTION AND AUGER / ADSORPTION OF SULFUR ON THE (100) AND (111) FACES OF PLATINUM. LOW ENERGY ELEC   5-1421
SPECTROSCOPY STUDY OF THE ADSORPTION OF OXYGEN ON COPPER (100) AND (111) SURFACES. /ACTION- AUGER ELECTRON   6-2089
(-9) TO 10(-/ ADSORPTICN ISOTHERMS OF OXYGEN ON TUNGSTEN (100), AND (111) SURFACES IN THE PRESSURE RANGE 10   5-1148
GEN AND THE REACTION OF HYDROGEN WITH OXYGEN ON PLATINUM (100). (AUGER SPECTROSCOPY)   ADSORPTION OF HYDRO 5-1685
CARBON MONOXIDE ADSORPTION KINETICS ON MOLYBDENUM (100). (AUGER SPECTROSCOPY)   5-1579
CHEMISORPTION ON TANTALUM (100). (AUGER SPECTROSCOPY)   5-1245
OXYGEN ON TUNGSTEN (110) AND CARBON MONOXIDE ON TUNGSTEN (100). (BY LEED AND AES) /N SPECTRA FROM ADSORBED   5-1152
UREMENTS COMBINED WIT/ INITIAL OXIDATION STUDIES ON IRON (100) BY MEANS OF PHOTOELECTRIC WORK FUNCTION MEAS   5-1927
SP/ STUDIES OF INITIAL OXIDATION ON SILICON- IRON ALLOY (100) BY MEANS OF WORK FUNCTION AND AUGER ELECTRON   5-1924
RACTION AND AUGER E/ INVESTIGATION OF EUROPIUM(II) OXIDE (100) CLEAVAGE SURFACE BY LOW ENERGY ELECTRON DIFF   5-1157
CE STRUCTURE, AND INTERACTION WITH OXYGEN AND C/ THORIUM (100) CRYSTAL FACE. A STUDY OF ITS CLEANING, SURFA   5-1883
CATTERING, / STUDIES OF GAS- SURFACE INTERACTIONS AT THE (100) CRYSTAL FACE OF PLATINUM BY MOLECULAR BEAM S   5-1963
/TRON BEAM EFFECTS ON CARBON MONOXIDE SATURATED TUNGSTEN (100), ETHYLENE SATURATED TUNGSTEN (100), AND TUN/   5-1244
/ERGY HELIUM(+) IONS (LESS THAN 10 KEV) SCATTERED FROM A (100) FACE OF A COPPER SINGLE CRYSTAL. (AUGER NEU/   2-0546
ADSORPTION OF SULFUR ON THE (100) FACE OF MOLYBDENUM.   5-1732
URFACE.   STUDY OF THE ADSORPTION OF BENZENE ON THE (100) FACE OF NICKEL. INFLUENCE OF SULFUR AT THE S   5-1284
/IONS WITH CONTROLLED SURFACES. (AUGER NEUTRALIZATION AT (100) FACE OF TUNGSTEN, POLYCRYSTALLINE MOLYBDENU/   5-1736
EQUILIBRIUM SURFACE SEGREGATION OF CARBON ON IRON (100) FACES. (AUGER EFFECT)   5-1369
EFFECT)   BINDING ENERGIES OF CARBON TO NICKEL (100) FROM EQUILIBRIUM SEGREGATION STUDIES. (AUGER   5-1484
FRACTION AND AUGER SPECTR/ GROWTH OF GALLIUM ARSENIDE ON (100) GERMANIUM STUDIED BY LOW ENERGY ELECTRON DIF   5-1646
ER (100) SURFACES.   ATOMIC ARRANGEMENT OF GOLD (100) GOLD- COVERED (100), AND SILVER- COVERED COPP 5-1718
EMPERATURE ON THE SURFACE STRUCTURE AND STOICHIOMETRY OF (100) INDIUM PHOSPHIDE SURFACES.   EFFECT OF T   5-1164
CHANGE IN THE SURFACE ELECTRONIC STRUCTURE OF PLATINUM (100) INDUCED BY RECONSTRUCTION.   5-1186
ON ADSORBATE BINDING ENERGY. PART-1: CARBON MONOXIDE ON (100) PALLADIUM.   STRUCTURAL INFLUENCES   5-1913
A LEED STUDY OF MAGNESIUM OXIDE (100). PART-1: EXPERIMENT.   5-1580
/ON ADSORPTION ENERGY. PART-2: CARBON MONOXIDE ON NICKEL (100). PART-3: CARBON MONOXIDE ON COPPER (100). (/   5-1911
PREPARATION AND SURFACE OXIDATION OF A VANADIUM (100) PLANE. (AUGER EFFECT)   6-1992
/HE ADSORPTION OF HYDROGEN ON CLEAN NICKEL (111), NICKEL (100), SHEET AND EVAPORATED NICKEL FILMS. (AUGER /   5-1468
IS.   CONDENSATION OF GOLD ONTO TANTALUM (100) SINGLE CRYSTAL SURFACES. LEED AND AES ANALYS   5-1312
S AND AUGER ELECTRON SPECTROSCOPY ON 5.59% SILICON- IRON (100) SINGLE CRYSTALS. / WORK FUNCTION MEASUREMENT   5-1928
SPECTROSCOPY OF 5.59 ATOMIC PERCENT SILICON- IRON ALLOY (100) SINGLE CRYSTALS. /UNCTION AND AUGER ELECTRON   5-1925
S. (AUGER EFFECT)   PLATINUM (100) STRUCTURES: IMPURITY CAUSATION AND PROPERTIE   5-1311
USIONS IN THIN FILM GOLD ON PLATINUM ON GALLIUM ARSENIDE (100) STUDIED WITH AUGER SPECTROSCOPY.   INTERDIFF   5-1232
ASUREMENTS FOR EPITAXIAL GROWTH OF THORIUM ON A TUNGSTEN (100) SUBSTRATE. /SCOPY, LEED AND WORK FUNCTION ME   5-1739
RA OF HYDROGEN CHLORIDE VAPOR ETCHED N+ GALLIUM ARSENIDE (100) SUBSTRATES.   AUGER SPECT   4-0992
ADSORPTION OF IONIC SALTS ON A TUNGSTEN (100) SURFACE.   5-1659
ULAR RESOLVED AUGER EMISSION SPECTRA FROM A CLEAN COPPER (100) SURFACE.   ANG   5-1697
TION STUDIES OF OXYGEN AND CARBON MONOXIDE ON A PLATINUM (100) SURFACE.   (AUGER) ADSORP   5-1546
(AUGER) ADSORPTION STUDIES ON THE PLATINUM (100) SURFACE.   5-1570
ELECTRON STIMULATED DESORPTION OF OXYGEN FROM A TUNGSTEN (100) SURFACE. /LECTRON SPECTROSCOPY STUDY OF THE   5-1954
FURTHER STUDIES OF GAS ADSORPTION ON MOLYBDENUM (100) SURFACE.   5-1305
LEED AND AES STUDIES OF LITHIUM HYDRIDE (100) SURFACE.   5-1446
LEED STUDY OF IRIDIUM (100) SURFACE.   5-1370

```
                      LEED STUDY OF PLATINUM (100) SURFACE.                                       5-1377
                STRUCTURE OF THE PLATINUM (100) SURFACE.                                          5-1374
      INVESTIGATION OF HYDROGEN CHEMISORPTION ON THE TUNGSTEN (100) SURFACE. (AUGER ELECTRON SPECTROSCOPY)     5-1974
   NGLE CRYSTAL FILMS AND THEIR USE IN STUDYING THE IRIDIUM (100) SURFACE. (AUGER SPECTRA OF IRIDIUM) /ENON SI  5-1479
   /ERVATION OF SUPERSTRUCTURES ON CARBON COVERED GERMANIUM (100) SURFACE BY HIGH ENERGY ELECTRON DIFFRACTION.  5-1805
          /N THE INITIAL OXIDATION AND REDUCTION PROCESSES OF IRON (100) SURFACE BY MEANS OF COMBINED LOW ENERGY ELE/  6-2141
         /ERACTION OF CARBON MONOXIDE AND OXYGEN WITH A PALLADIUM (100) SURFACE. (CHEMISORPTION AND CATALYTIC REACT/  5-1321
SCOPY.    PHOTOELECTRIC WORK FUNCTION STUDY ON IRON (100) SURFACE COMBINED WITH AUGER ELECTRON SPECTRO  6-1936
   RACTIO/ EPITAXIAL GROWTH OF SILVER ON POTASSIUM CHLORIDE (100) SURFACE OBSERVED BY LOW ENERGY ELECTRON DIFF  5-1696
.                        LEED- AUGER INVESTIGATION ON A (100) SURFACE OF AUSTENITIC CHROMIUM- NICKEL STEEL  6-1991
      /RGY ELECTRON DIFFRACTION AND AUGER INVESTIGATION OF THE (100) SURFACE OF COPPER(3) GOLD. TEMPERATURE DEPE/  5-1854
T.              ORDER- DISORDER TRANSFORMATION AT A (100) SURFACE OF COPPER(3) GOLD, THEORY, EXPERIMEN  5-1855
    DIFFRACTIO/ INVESTIGATION OF THE PURE AND CESIUM COATED (100) SURFACE OF GALLIUM ARSENIDE BY SLOW ELECTRON  5-1639
                  INTERACTION BETWEEN CHLORINE AND THE (100) SURFACE OF GOLD.                          5-1335
                      SEGREGATION OF CARBON TO (100) SURFACE OF NICKEL.                           6-2000
   IES OF THE CHEMISORPTION OF UNSATURATED MOLECULES BY THE (100) SURFACE OF PLATINUM. /CTRON DIFFRACTION STUD  5-1267
   IDATION.                AUGER SPECTRA FROM THE (100) SURFACE OF SILICON DURING OXIDATION AND NITR  5-1310
   /TH CLEAN NICKEL SINGLE CRYSTAL SURFACES. PART-1: NICKEL (100) SURFACE. PART-2: NICKEL (111) SURFACE. (AUG/  5-1452
   FFE/ INTERACTION OF OXYGEN AND NITROGEN WITH THE NIOBIUM (100) SURFACE. PART-2: REACTION KINETICS. (AUGER E  5-1333
              CHARACTERIZATION OF SOME VANADIUM (100) SURFACE STRUCTURES USING LEED AND AES.      5-1864
       IDENTIFICATION OF THE FORM OF CARBON AT A SILICON (100) SURFACE USING AUGER ELECTRON SPECTROSCOPY.  5-1378
   OSCOPY.                STUDY OF IRIDIUM (100) SURFACE USING LEED AND AUGER ELECTRON SPECTR  5-1371
   ENDENCE OF AUGER ELECTRON EMISSION FROM COPPER (111) AND (100) SURFACES.               ANGULAR DEP  5-1632
       AUGER ELECTRON EMISSION FROM CLEAN AND GAS-COVERED IRON (100) SURFACES.        ANGULAR DISTRIBUTION OF  5-1623
   SSION SPECTROSCOPY VANADIUM PENTOXIDE (010) AND VANADIUM (100) SURFACES.               AUGER EMI  5-1941
   MISSION SPECTRA OF VANADIUM PENTOXIDE (010) AND VANADIUM (100) SURFACES. /STRUCTURE IN THE AUGER ELECTRON E  5-1340
                  OXYGEN ADSORPTION ON CLEAN MOLYBDENUM (100) SURFACES.                           5-1772
                GOLD-INDUCED SUPERSTRUCTURES ON SILICON (100) SURFACES AS OBSERVED BY LEED AES.       5-1708
   NIC STATES OF OXYGEN ADSORBED ON CLEAN SILICON (111) AND (100) SURFACES. (AUGER SPECTROSCOPY)    ELECTRO  5-1476
   VACUUM CLEANED SURFACES.                          (100) SURFACES OF ALKALI HALIDES. PART-1: AIR AND  5-1347
       STIMULATED DISSOCIATION.                     (100) SURFACES OF ALKALI HALIDES. PART-2: ELECTRON  5-1908
   /CTRON DIFFRACTION AND AUGER SPECTROSCOPY STUDIES OF THE (100) SURFACES OF ALUMINUM AND ALUMINUM- COPPER A/  5-1755
   /UCTURAL CHANGES THAT ACCOMPANY THE THERMAL ANNEALING OF (100) SURFACES OF GALLIUM ARSENIDE, INDIUM PHOSPH/  5-1163
   MBARDMENT WITH 2 KV. EL/ DEPARTURES FROM STOICHIOMETRY IN (100) SURFACES OF GALLIUM PHOSPHIDE, INDUCED BY BO  5-1162
   OGEN, DEUTERIUM, NITROGEN, OXYGEN AND CARBON MONOXIDE BY (100) TANTALUM. (AUGER EFFECT) /D SOLUTION OF HYDR  5-1549
   TROSCOPY OF CESIUM ADSORBED ON CLEAN AND OXYGEN- COVERED (100) TUNGSTEN.           AUGER ELECTRON SPEC  5-1295
                  CESIUM- OXYGEN CCADSORPTION ON (100) TUNGSTEN.                           5-1296
   N STUDY OF OXYGEN- CESIUM ADSORPTION AND COALSORPTION ON (100) TUNGSTEN. /N DIFFRACTION AUGER, WORK FUNCTIO  5-1297
   (AU/ ANGULAR DISTRIBUTIONS OF HYDROGEN DESORBED FROM THE (100), (110), AND (111) FACES OF COPPER CRYSTALS.  5-1155
   ATIVE RATES, AND FLUORESCENCE YIELDS FOR Z BETWEEN 38 AND 103. /OMIC N-SHELL COSTER-KRONIG, AUGER, AND RADI  4-0969
   ECTROSCOPY STUDY OF THE OXIDATION CF IRON (100) AND IRON (110). /GY ELECTRON DIFFRACTION- AUGER ELECTRON SP  5-1586
   AUGER ELECTRON CURRENTS. APPLICATION TO SULFUR ON NICKEL (110).   ANALOG TECHNIQUE FOR THE MEASUREMENT OF  3-0635
   ANGULAR EFFECTS IN AUGER ELECTRON EMISSION FROM COPPER (110).                           5-1988
   TRON SPECTROSCOPY STUDY OF OXYGEN ADSORPTION ON TUNGSTEN (110).                   AUGER ELEC  5-1668
   RGY LOSS AND AUGER ELECTRON SPECTRA OF GALLIUM PHOSPHIDE (110).        CHARACTERISTIC ENE  5-1657
   ELECTRON SPECTROSCOPY FROM INITIAL OXIDATION OF TANTALUM (110).         CHEMICAL SHIFTS IN AUGER  6-2038
     LEED AND AES OF OXYGEN ABSORPTION ON TUNGSTEN (110).                           5-1669
       LEED AUGER STUDY CF GROWTH OF ALUMINUM ON NIOBIUM (110).                           5-1491
   - AES STUDY OF THE OXIDATION OF IRON- CHROMIUM (100) AND (110).                   LEED  5-1587
   , AND WORK FUNCTION STUDY OF IODINE ADSORBED ON TUNGSTEN (110).             LEED- AUGER  5-1145
   Y STUDY OF THE ADSORPTION OF ALKALI METALS ON MOLYBDENUM (110).       LEED- AUGER SPECTROSCOP  5-1896
   S OF THE CATALYTIC CARBON MONOXIDE OXIDATION ON PLATINUM (110).               MECHANISM  5-1188
   ON SPECTROSCOPY. (STUDY OF CARBON MONOXIDE ON MOLYBDENUM (110)) / SUBTRACTION FOR QUANTITATIVE AUGER ELECTR  1-0069
   A/ ELECTRON EXCITED AUGER ELECTRON SPECTRUM OF A NICKEL (110)- C(2X2) SULFUR SURFACE:  LINE SHAPE ANALYSIS  4-1065
   / PHOTOEMISSION SPECTRA FROM ADSORBED OXYGEN ON TUNGSTEN (110) AND CARBON MONOXIDE ON TUNGSTEN (100). (BY L  5-1152
   ELECTRON SPECTROSCOPY STUDY OF THE OXIDATION CF CHROMIUM (110) AND CHROMIUM (100). /RON DIFFRACTION- AUGER  6-2022
   E/ ELASTIC LEED INTENSITY ENERGY STUDIES OF CLEAN (001), (110) AND ELASTIC (111) NICKEL SURFACES. (AUGER SP  5-1292
   GER ELECTRON SPECTROSCOPY STUDY OF THE OXIDATION OF IRON (110) AND IRON (100). /GY ELECTRON DIFFRACTION- AU  5-1585
   /ULAR DISTRIBUTIONS OF HYDROGEN DESORBED FROM THE (100), (110), AND (111) FACES OF COPPER CRYSTALS. (AUGER/  5-1155
   CHEMISORPTION ON (001), (110), AND (111) NICKEL SURFACES. (AUGER SPECTRA)      5-1764
   USING LOW ENERGY ELECTRON DIFF/ CHEMISORPTICN ON (001), (110), AND (111) NICKEL SURFACES. CORRELATED STUDY  5-1291
   CARBON MONOXIDE ADSORPTICN ON NICKEL (110). (AUGER EFFECT)                        5-1884
   ROSCOPY OF A SULFUR- OXYGEN SURFACE REACTION ON A COPPER (110) CRYSTAL.           AUGER ELECTRON SPECT  5-1183
   /ICAL EFFECTS IN THE M4,5NN AUGER SPECTRUM OF MOLYBDENUM (110) DUE TO ADSORPTION OF MOLECULAR OXYGEN AND C/  4-0885
                  DIFFUSION OF SULFUR TOWARD THE (110) FACE OF NICKEL.                       5-1771
       PREPARATION AND SURFACE OXIDATION OF A (110) FACE OF VANADIUM.                     5-1683
   ION AND SURFACE ALLOYING OF LEAD MONOLAYERS ON (111) AND (110) FACES OF GOLD. (AUGER EFFECT)      ADSORPT  5-1733
   (0.16). (A/ SURFACE COMPOSITION STUDIES OF THE (001) AND (110) FACES OF MONOCRYSTALLINE IRON(0.84) CHROMIUM  5-1591
   GER ELECTRON SPECTROSCOPY. APPLICATION TO CESIUM-COVERED (110) GALLIUM ARSENIDE. /ERMAL DESORPTION USING AU  5-1293
       ADSORPTION OF OXYGEN ON CLEAN CLEAVED (110) GALLIUM ARSENIDE SURFACES.              5-1306
       FORMIC ACID DESORPTION FROM GRAPHITIZED NICKEL (110). (LEED- AES)                    5-1630
   MOLYBDENUM. PART-3: OXIDE NUCLEATION RATES ON (100) AND (110) MOLYBDENUM SURFACES. (AUGER SPECTRA) /ION OF  5-1536
   )          DECOMPOSITION OF CARBON MONOXIDE ON A (110) NICKEL SURFACE. (AUGER ELECTRON SPECTROSCOPY  5-1606
       INTERACTION OF CARBON MONOXIDE WITH (110) NICKEL SURFACES. (AUGER EFFECT)       5-1607
   AND AUGER ELECT/ GROWTH OF GOLD FILM ON GALLIUM ARSENIDE (110) OBSERVED BY LOW ENERGY ELECTRON DIFFRACTION  5-1876
   ION AT 650-1100 DEGREES K.          OXIDATION OF (110) PART-2: LEED- AUGER STUDY OF OXIDE FORMAT  6-1994
   OPY)           ADSORPTION OF OXYGEN ON THE (110) PLANE OF TUNGSTEN. (AUGER ELECTRON SPECTROSC  5-1318
       THE KINETICS OF OXYGEN ADSORPTION ON THE (112) AND (110) PLANES OF TUNGSTEN.                 5-1912
       OXYGEN ADSORPTION ON (110) SILVER.                           5-1422
       CARBON MONOXIDE OXIDATION ON A PLATINUM (110) SINGLE CRYSTAL SURFACE.                5-1187
   GATION OF THE DEPOSITION OF ALUMINUM ONTO THE MOLYBDENUM (110) SURFACE.           AUGER- LEED INVESTI  5-1489
   CHEMISORPTION OF OXYGEN ON THE SILVER (110) SURFACE.                           5-1778
   CLEANING PROCEDURES OF MOLYBDENUM (110) SURFACE.                           5-1649
   EED AND AUGER ELECTRON SPECTROSCCPIC STUDIES ON TUNGSTEN (110) SURFACE.               L  5-1677
   OEMISSION STUDIES OF THE CESIUM COVERED GALLIUM ARSENIDE (110) SURFACE. /ER ELECTRON SPECTROSCOPY, AND PHOT  5-1931
   VESTIGATION OF CESIUM ADSORPTION ONTO A GALLIUM ARSENIDE (110) SURFACE. (AUGER ELECTRON SPECTROSCOPY)    IN  5-1644
   LEED AND AUGER ELECTRON SPECTROSCOPY STUDY OF NICKEL (110) SURFACE EFFECTS DUE TO CARBON AND SULFUR.  5-1814
       INITIAL OXIDATION OF THE MOLYBDENUM (110) SURFACE OBSERVED BY AES AND LEED.      5-1650
   BY LEED, AUGER ELECTRON / EPITAXIAL GROWTH OF COPPER ON (110) SURFACE OF A TUNGSTEN SINGLE CRYSTAL STUDIED  5-1660
   SULFUR AND CARBON ON (110) SURFACE OF NICKEL.                       5-1813
```

LEED PATTERNS. (AU/ ADSORBATE INTERACTIONS ON A PLATINUM (110)   SURFACE. PART-1: SULFUR AND CARBON MONOXIDE   5-1190
INTERACTIONS OF CARBON MONOXIDE AND OXYGEN WITH IRIDIUM (110)   SURFACES. (AUGER SPECTROSCOPY)   5-1252
CTION BY METASTABLE ATOMS: HELIUM AND ARGON ON (111) AND (110)   TUNGSTEN. (AUGER) / FUNCTION IN ELECTRON EJE   5-1600
CTRONS AND AUGER SPECTROSCOP/ STUDY OF BARIUM FILMS ON A (110)   TUNGSTEN FACE BY THE DIFFRACTION OF SLOW ELE   5-1368
SURFACE STRUCTURES ON IRON (110)   ULTRAHIGH VACUUM.   5-1346
TION OF FORMIC ACID ON CARBURIZED AND GRAPHITIZED NICKEL (110)   USING AES, LEED, AND FLASH DESORPTION. /POSI   5-1629
OSCOPY.   REEXAMINATION OF ALUMINUM (110)   WITH LOW ENERGY DIFFRACTION AND AUGER SPECTR   5-1507
ADSORPTION OF XENON AND COBALT ON SILVER (111).   5-1633
GEN CONTAMINANT ON THE ADSORPTION OF ALUMINUM ON SILICON (111).   5-1169
ONS AND CHARACTERISTIC LOSS AND GAIN PHENOMENA OF COPPER (111).   EFFECTS OF PREADSORBED OXY   5-1169
ON OF THE CHARACTERISTIC AUGER ENERGY LOSSES OF PLATINUM (111).   ENERGY SPECTRUM OF BACK SCATTERED ELECTR   4-0897
CTROSCOPY STUDY OF PHOSPHINE ADSORPTION ON CLEAN SILICON (111).   INVESTIGATI   2-0499
Y ELECTRON EMISSION YIELD AND THE APPLICATION TO SILICON (111). /LECTRON DIFFRACTION AND AUGER ELECTRON SPE   5-1930
THIN REACTION LAYERS AND SURFACE STRUCTURE OF SILICON (111). /R DETECTING FINE STRUCTURE IN THE SECONDAR   4-0852
TION OF A CORE VALENCE VALENCE AUGER LINE SHAPE: SILICON (111).   5-1879
PERATURE. (AUGER SURFACE SPECTRA)   SILICON (111)-   TIGHT BINDING CALCULA   4-0831
LECTRON SPECTROSCOPY.   STUDY OF INDIUM ARSENIDE (111)-7 STRUCTURE OBTAINED BY CLEAVAGE AT ROOM TEM   5-1767
SING INELASTIC LOW / SURFACE PLASMON STUDIES ON ALUMINUM (111) AND (-1-1-1) SURFACES USING LEED AND AUGER E   5-1381
AL SURFACES. (LEED, AUGER ELEC/ REACTIVITY OF LOW INDEX ((111) AND ALUMINUM (001) SINGLE CRYSTAL SURFACES U   5-1745
NGULAR DEPENDENCE OF AUGER ELECTRON EMISSION FROM COPPER (111) AND (100)) AND STEPPED PLATINUM SINGLE CRYST   5-1838
ELECTRONIC STATES OF OXYGEN ADSORBED ON CLEAN SILICON (111) AND (100) SURFACES.   A   5-1632
ADSORPTION AND SURFACE ALLOYING OF LEAD MONOLAYERS ON (111) AND (100) SURFACES. (AUGER SPECTROSCOPY)   5-1476
ECTRON EJECTION BY METASTABLE ATOMS: HELIUM AND ARGON ON (111) AND (110) FACES OF GOLD. (AUGER EFFECT)   5-1733
DISTRIBUTION OF SECONDARY ELECTRON EMISSION FROM SILICON (111) AND (110) TUNGSTEN. (AUGER) / FUNCTION IN EL   5-1600
AN ORDERED BORON STRUCTURE AFTER DEPOSITION ON SILICON (111). (AUGER ELECTRON EXCITATION) /ES AND ENERGY   4-0853
ORPTION AND DECOMPOSITION OF CARBON MONOXIDE ON PLATINUM (111). (AUGER ELECTRON SPECTROSCOPY)   5-1168
URFACE, AND HYDROGEN INDUCED STATES ON CLEAVED GERMANIUM (111). (AUGER SPECTROMETRY)   ADS   5-1622
DESORPTION AND DISSOCIATION: CARBON MONOXIDE ON PLATINUM (111). (AUGER SPECTROSCOPY) /ASUREMENTS OF BULK, S   5-1781
Y ELECTRON EMISSION DURING OXIDATION OF NICKEL (100) AND (111). (AUGER SPECTRUM) /RONS IN ELECTRON INDUCED   5-1567
ADSORPTION OF OXYGEN ON MOLYBDENUM (111): EFFECT OF TRACE IMPURITIES.   5-1568
RON SPECTROSCOPY)   ADSORPTION OF KRYPTON ON THE (111) FACE OF COPPER SINGLE CRYSTALS. (AUGER ELECT   5-1221
/BUTIONS OF HYDROGEN DESORBED FROM THE (100), (110), AND (111) FACES OF COPPER CRYSTALS. (AUGER ELECTRON S/   5-1155
ACTION AND AUGER / ADSORPTION OF SULFUR ON THE (100) AND (111) FACES OF PLATINUM. LOW ENERGY ELECTRON DIFFR   5-1421
AND ANNEALING OF GALLIUM PHOSPHIDE AND GALLIUM ARSENIDE (111) FACES STUDIED BY AES AND UPS.   OXIDATION   5-1493
/ THROUGH EPITAXIAL FILMS CF COPPER AND NICKEL ON SILVER (111): FORMATION OF AN EQUILIBRIUM LAYER OF SILVE/   5-1356
CHEMISORPTION ON (001), (110) AND (111) NICKEL SURFACES. (AUGER SPECTRA)   5-1764
TENSITY ENERGY STUDIES OF CLEAN (001), (110) AND ELASTIC (111) NICKEL SURFACES. (AUGER SPECTROSCOPY) /ED IN   5-1292
ENERGY ELECTRON DIFF/ CHEMISORPTION ON (001), (110), AND (111) NICKEL SURFACES. CORRELATED STUDY USING LOW   5-1291
/ANDARD SURFACE CLEANING TECHNIQUES AS APPLIED TO NICKEL (111), NICKEL (100), AND NICKEL SHEET USING AUGER/   5-1467
/BILITIES FOR THE ADSORPTION OF HYDROGEN ON CLEAN NICKEL (111), NICKEL (100), SHEET AND EVAPORATED NICKEL /   5-1468
AUGER STUDIES OF (111) SILICON SURFACES.   5-1375
RVATIONS OF THE EPITAXIAL GROWTH OF NICKEL AND COPPER ON (111) SILVER. (BY AUGER SPECTROSCOPY)   OBSE   5-1354
)   SEGREGATION OF CADMIUM TO A (111) SILVER CHLORIDE SURFACE. (AUGER SPECTROSCOPY   5-1918
ADSORPTION OF HYDROGEN ON A PLATINUM (111) SURFACE.   5-1255
AUGER ELECTRON EMISSION FROM GOLD DEPOSITED ON SILICON (111) SURFACE.   5-1676
ELECTRON SPECTROSCOPY OF NICKEL DEPOSITS ON THE SILICON (111) SURFACE.   AUGER   5-1238
E. AND MOLECULAR OXYGEN AND CARBON DIOXIDE ON A TUNGSTEN (111) SURFACE. /OLECULAR OXYGEN AND CARBON MONOXID   5-1465
CHLORINE REACTIONS ON THE SILICON (111) SURFACE.   5-1344
ION OF SURFACE STRUCTURES BY LEED (AND AES): THE SILICON (111) SURFACE.   DETERMINAT   5-1342
ELECTRON BEAM ASSISTED ADSORPTION ON SILICON (111) SURFACE.   5-1269
INTERACTION OF NITRIC OXIDE WITH A NICKEL (111) SURFACE.   5-1278
LEED AND AUGER INVESTIGATIONS CF COPPER (111) SURFACE.   5-1499
ED (AUGER) STUDY OF CHLORINE CHEMISORPTION ON THE SILVER (111) SURFACE.   LE   5-1779
NVESTIGATION OF A STABLE CARBIDE OVERLAYER ON A PLATINUM (111) SURFACE.   LEED- AUGER I   5-1569
PHASE TRANSFCRMATIONS OF SILICON (111) SURFACE.   5-1343
SCATTERING OF LOW ENERGY ELECTRONS FRCM A COPPER (111) SURFACE.   5-1621
SPECTROSC/ EQUILIBRIUM SEGREGATION OF CARBON TO A NICKEL (111) SURFACE: A SURFACE PHASE TRANSITION. (AUGER   5-1803
ORPHISM AND PURITY OF COPPER VAPOR DEPOSITED CN A COPPER (111) SURFACE. (AUGER)   ISOM   5-1171
ADSORPTION OF COBALT ON A NICKEL (111) SURFACE. (AUGER EFFECT, AUGER SPECTROSCOPY)   5-1256
L SURFACES. PART-1: NICKEL (100) SURFACE. PART-2: (111) SURFACE. (AUGER ELECTRON SPECTROSCOPY) /YSTA   5-1452
ING COEFFIC/ KINETICS CF OXYGEN ADSORPTION ON A PLATINUM (111) SURFACE. (AUGER ELECTRON SPECTROSCOPY, STICK   5-1189
DETECTION OF IMPURITIES ON A SILICON (111) SURFACE BY AUGER- LEED ANALYSIS.   5-1360
OPY AND BY LEED.   SEGREGATION OF GOLD TO SILICON (111) SURFACE OBSERVED BY AUGER EMISSION SPECTROSC   5-1180
N SPECTRUM DURING THE INTERACTION BETWEEN OXYGEN AND THE (111) SURFACE OF GALLIUM PHOSPHIDE. AUGER ELECTRO   5-1642
CTRON SPECTRUM IN THE INTERACTION BETWEEN OXYGEN AND THE (111) SURFACE OF GALLIUM PHOSPHIDE. /THE AUGER ELE   5-1643
R SPECTROSCOPY AND DIFFRACTION OF SLOW ELECTRONS. (111) SURFACE OF GALLIUM PHOSPHIDE STUDIED BY AUGE   5-1640
GATION IN ALLOYS: EQUILIBRIUM SEGREGATION OF GOLD TO THE (111) SURFACE OF NICKEL. (AES)   SURFACE SEGRE   5-1216
CS OF THE REACTION BETWEEN OXYGEN AND SULFUR ON A NICKEL (111) SURFACE. (STUDIED BY LEED AND AES)   KINETI   5-1451
OSCOPY.   STUDIES ON THE IRIDIUM (111) SURFACE USING LEED AND AUGER ELECTRON SPECTR   5-1372
OSCOPY.   A STUDY OF THE IRON (111) SURFACE USING LEED AND AUGER EMISSION SPECTR   5-1670
PY STUDY OF THE ADSORPTION OF OXYGEN ON COPPER (100) AND (111) SURFACES. /ACTION- AUGER ELECTRON SPECTROSCO   6-2089
AUGER STUDIES OF CLEAVED SILICON (111) SURFACES.   5-1382
CARBON CONTAMINATION OF SILICON (111) SURFACES.   5-1237
STIC ENERGIES IN SECONDARY ELECTRON SPECTRA FROM SILICON (111) SURFACES.   CHARACTERI   5-1263
NATURE OF SILICON (111) SURFACES.   5-1383
UCTURE TRANSFORMATIONS ON CLEAVED AND ANNEALED GERMANIUM (111) SURFACES.   STR   5-1713
AMORPHOUS TO CRYSTALLINE TRANSITION OF SILICON (111) SURFACES. (AUGER ELECTRON SPECTROSCOPY)   5-1780
INTERACTION OF OXYGEN WITH SILICON (111) SURFACES. (AUGER ELECTRON SPECTROSCOPY)   5-1693
CHEMISORPTION OF OXYGEN ON PLATINUM (111) SURFACES. (AUGER SPECTRA)   5-1500
PHOSPHORUS ADSORPTION AND DESORPTION KINETICS ON SILICON (111) SURFACES. (BY LEED AND AES)   A STUDY OF   5-1288
ACTION, AUGER, AND CONTAC/ SODIUM ADSORPTION ON ALUMINUM (111) SURFACES: ELECTRIC LOW ENERGY ELECTRON DIFFR   5-1744
-/ ADSORPTION ISOTHERMS OF OXYGEN ON TUNGSTEN (100), AND (111) SURFACES IN THE PRESSURE RANGE 10 (-9) TO 10(5-1148
IMON/ LOW ENERGY ELECTRON DIFFRACTION STUDY OF THE POLAR (111) SURFACES OF GALLIUM ARSENIDE AND GALLIUM ANT   5-1602
PY/   ADSORPTION CF OXYGEN ON SILICON (111) SURFACES. PART-1. (AUGER ELECTRON SPECTROSCO   5-1474
INTERACTION OF OXYGEN WITH SILICON (111) SURFACES. (REPLY TO COMMENT)   5-1512
OF THE OXIDATION AND REDUCTION BY WATER VAPOR OF SILICON (111) SURFACES USING AUGER ELECTRON EMISSION. /DY   5-1612
ROSCOPY.   STUDY OF RUTHENIUM (0001) AND RHODIUM (111) SURFACES USING LEED AND AUGER ELECTRON SPECT   5-1380
CHEMISORPTION OF ATOMIC HYDROGEN ON THE SILICON (111) 7*7 SURFACE. (AUGER EFFECT)   5-1789
DITUNGSTEN CARBIDE OVERLAYER ON TUNGSTEN (112).   5-1241
THE KINETICS OF OXYGEN ADSORPTION ON THE (112) AND (110) PLANES OF TUNGSTEN.   5-1912

- RHEED AUGER STUDY OF OXYGEN ADSORPTION ON THE TUNGSTEN (112) SURFACE.   COMBINED LEED 5-1464
                    CHEMISORPTION ON SINGLE CRYSTAL MOLYBDENUM (112) SURFACES.   5-1303
             OBSERVATION OF EXTRINSIC SURFACE STATES ON (1120) CADMIUM SULFIDE BY AUGER SPECTROSCOPY.   5-1202
   SURFACE PHOTOVOLTAGE AND AUGER SPECTROSCOPY STUDIES OF (1120) CADMIUM SULFIDE SURFACE.   5-1201
/ AUGER ELECTRON SPECTROSCOPIC STUDIES OF THE SURFACES OF 18-8 STAINLESS STEEL ELECTROPOLISHED OR ABRADED B 6-2130
ECTROLYTICALLY.   AES STUDIES ON SURFACES OF 18-8 STAINLESS STEEL FINISHED MECHANICALLY AND EL 6-2129
                AUGER ELECTRON SPECTROSCOPY STUDY OF 18-8 STAINLESS STEEL OXIDATION.   6-1993
AUGER EFFECT)   COMPARISON OF SILICON L3,2 ELECTRON LOSS SPECTRA AND OPTICAL ABSORPTION. (4-1038
AUGER RATES, X-RAY RATES, AND FLUORESCENCE YIELDS FOR THE 2P SHELL.   NONRELATIVISTIC 2-0558
         STRIPPING MECHANISM DEPENDENCE OF THE ARGON 2S FLUORESCENCE YIELD. (AUGER TRANSITION)   4-0957
/AND COSTER-KRONIG TRANSITION PROBABILITIES TO THE ATOMIC 2S STATE AND THEORETICAL L1 FLUORESCENCE YIELDS. 2-0329
ATES IN ATOMS WITH INNER SHELL VACANCIES (NEON WITH 1S OR 2S VACANCY).   CORRELATION ENERGIES AND AUGER R 2-0312
         SURFACE DEBYE TEMPERATURE OF THE SILICON (001) - 2X2 STRUCTURE. (AUGER EFFECT)   5-1477
             HIGH RESOLUTION L2,3M2,3M2,3 AUGER ELECTRON SPECTRA OF ARGON.   4-0912
PHOSPHORUS.   AUGER SATELLITES OF THE L2,3 AUGER EMISSION BANDS OF ALUMINUM, SILICON, AND 4-0985
OLLISIONS.   CROSS SECTIONS FOR L2,3 AUGER EMISSION IN 22 TO 300 KEV PROTON- ARGON C 4-1040
DTHS AND INTENSITIES OF HIGHLY RESOLVED LINES IN ARGON L2,3 AUGER SPECTRA.   WI 4-1032
MEANS OF THE DECOMPOSI/ DETECTION OF EXCITONS NEAR THE L2,3 EDGE OF THE CHLORIDE ION IN SODIUM CHLORIDE BY 4-1126
AND/ INTERPRETATION OF HIGH ENERGY X-RAY SATELLITES OF L2,3 EMISSION BANDS OF SODIUM, MAGNESIUM, ALUMINUM, 2-0239
                    "SEMI-AUGER" PROCESSES IN ARGON AND POTASSIUM CHLORIDE.   2-0326
AUGER EFFECT)   MULTIPLET EFFECTS ON THE L2,3 FLUORESCENCE YIELD OF MULTIPLY IONIZED ARGON. (4-0956
THE PEAK SHAPES IN THE AUGER SPECTRUM OF OXYGEN (KL2,3L2,3) ON MOLYBDENUM.   ORIGIN OF 4-1004
IUM OX/ ELECTRON- HOLE INTERACTION EFFECTS IN THE KL2,3L2,3 RADIATIVE AUGER SPECTRUM OF INSULATORS. (MAGNES 2-0540
NDUCTORS. (SILICON)   INTERPRETATION OF KL2,3L2,3 RADIATIVE AUGER THRESHOLDS IN METALS AND SEMICO 4-0723
ION.   IONIZATION CROSS SECTION OF THE L2,3 SHELL OF ALUMINUM IN A NEON(+)- ALUMINUM COLLIS 2-0267
CO/ MEASUREMENT OF THE IONIZATION CROSS SECTION OF THE L2,3 SHELL OF CHLORINE USING AUGER ELECTRON SPECTROS 4-0987
             ANGULAR DISTRIBUTION OF L3M2,3M2,3(1SO) AUGER ELECTRONS OF ARGON.   2-0322
   AUGER SPECTRA AND FLUORESCENCE YIELDS FOR THE 3P AND THE 3D SHELL.   2-0278
       NONRELATIVISTIC FLUORESCENCE YIELDS FOR THE 3P AND THE 3D SHELLS.   2-0277
         L-SHELL AUGER ENERGIES AND INTENSITIES IN 3D TRANSITION METALS.   4-1018
N OF THE PEAK SHAPES IN THE AUGER SPECTRUM OF OXYGEN (KL2,3L2,3) ON MOLYBDENUM.   ORIGI 4-1004
GNESIUM OX/ ELECTRON- HOLE INTERACTION EFFECTS IN THE KL2,3L2,3 RADIATIVE AUGER SPECTRUM OF INSULATORS. (MA 2-0540
MICONDUCTORS. (SILICON)   INTERPRETATION OF KL2,3L2,3 RADIATIVE AUGER THRESHOLDS IN METALS AND SE 4-0723
EMISSION. STRUCTURE ON THE HIGH ENERGY SIDE OF THE KL2,3M AUGER PEAK FROM SOLID ALUMINUM: INTERNAL PHOTO 4-1021
             L1,L2,3M COSTER-KRONIG TRANSITION ENERGIES.   4-0929
5NN AND N4,5OO AUGER SPECTRA FROM XE/ HIGH RESOLUTION L2,3MM AND M4,5NN AUGER SPECTRA FROM KRYPTON AND M4, 4-1113
ION.   CCMPARISON OF MAGNESIUM L2,3MM AUGER CURRENTS USING ELECTRON AND ION EXCITAT 4-0855
         PHOTOELECTRON AND L2,3MM AUGER ELECTRON ENERGIES FOR ARSENIC.   4-1035
             L2,3MM AUGER ELECTRON SPECTRUM FROM FREE ZINC ATOMS.   4-0731
             L2,3MM AUGER PROCESSES IN SELENIUM.   2-0560
             L2,3MM AUGER SPECTRUM OF ARGON.   4-1114
       TRANSITION PROBABILITIES FOR THE L2,3MM AUGER SPECTRUM OF SELENIUM.   4-1033
         ANGULAR DISTRIBUTION OF L3M2,3M2,3(1SO) AUGER ELECTRONS OF ARGON.   2-0322
             HIGH RESOLUTION L2,3M2,3M2,3 AUGER ELECTRON SPECTRA OF ARGON.   4-0912
   HIGH RESOLUTION MEASUREMENTS OF THE L3M2,3M4,5 AUGER TRANSITONS IN NICKEL AND COPPER.   4-1024
S IN ENERGY LEVELS AND SCREENING PA/ SENSITIVITY OF L1,L2,3M4,5 COSTER-KRONIG TRANSITION RATES TO VARIATION 2-0303
             THE L1,L2,3M4,5 COSTER-KRONIG TRANSITIONS.   2-0248
NC: THEORY AND EXPERIMENT.   L2,3M4,5M4,5 AUGER SPECTRA OF METALLIC COPPER AND ZI 4-0919
         HIGH RESOLUTION L2,3M4,5M4,5 AUGER SPECTRUM OF FREE ZINC ATOMS.   4-0733
   AUGER SPECTRA AND FLUORESCENCE YIELDS FOR THE 3P AND THE 3D SHELL.   2-0278
       NONRELATIVISTIC FLUORESCENCE YIELDS FOR THE 3P AND THE 3D SHELLS.   2-0277
             L1,L2,3V AUGER TRANSITION IN SILICON.   4-0779
         HIGH ENERGY SATELLITES OF THE M2,3VV AND M1VV AUGER PEAKS OF COPPER.   4-1043
M THE CROSS SECTION M/ A STUDY OF THE INTENSITY OF THE L2,3VV AUGER TRANSITION IN MAGNESIUM AND SILICON FRO 4-0765
TRON SPECTROSCOPY OF FRACTURE SURFACES IN IRRADIATED TYPE 304 STAINLESS STEEL.   AUGER ELEC 6-2001
Y. SURFACE CONCENTRATION OF MOLYBDENUM IN TYPES 316 AND 304 STAINLESS STEEL BY AUGER ELECTRON SPECTROSCOP 6-1995
CTROSCOPY. SURFACE CONCENTRATION OF MOLYBDENUM IN TYPES 316 AND 304 STAINLESS STEEL BY AUGER ELECTRON SPE 6-1995
MASS SPECTROMETRY AND AUGER ELECT/ SURFACE CLEANLINESS OF 316 L-N STAINLESS STEEL STUDIES BY SECONDARY ION 3-0654
MASS SPECTROMETRY AND AUGER ELECT/ SURFACE CLEANLINESS OF 316 L-N STAINLESS STEEL STUDIED BY SECONDARY ION 6-2087
DETERMINATION OF THE COMPOSITION OF PASSIVE FILMS ON TYPE 316 STAINLESS STEEL. /ECTRON SPECTROSCOPY TO THE 6-2080
       CHEMICAL PROFILE ANALYSIS OF THE 316 STAINLESS STEEL SHEET BY AUGER SPECTROSCOPY.   6-1996
ND RADIATIVE RATES, AND FLUORESCENCE YIELDS FOR Z BETWEEN 38 AND 103. /OMIC N-SHELL COSTER-KRONIG, AUGER, A 4-0969
EODYMIUM.   N5,4 AND M5,4 APPEARANCE POTENTIAL SPECTRA FROM PRAS 4-0908
       HIGH RESOLUTION NEON AUGER SPECTRUM PRODUCED IN 4.2 MEV ATOMIC HYDROGEN(+)- NEON COLLISIONS.   4-1078
IUM METAL.   4D AUGER TRANSITIONS FROM EXCITED STATES IN SAMAR 4-0823
.   WIDTHS OF 4S AND 5S MULTIPLET COMPONENTS IN THE RARE EARTHS 4-0970
 AND DEPTH PROF/ REACTION OF SULFUR DIOXIDE WITH MODIFIED 440C STEEL STUDIED BY AUGER ELECTRON SPECTROSCOPY 6-2031
PECTROSCOPY AND DEPTH PROFILE STUDY OF OXIDATION MODIFIED 440C STEEL.   AUGER ELECTRON S 6-2029
TROSCOPY AND DEPTH PROFILE STUDY OF OXIDATION OF MODIFIED 440C STEEL.   AUGER ELECTRON SPEC 6-2032
VAPOR.   M4,5N4,5N4,5 AND M4,5O1 AUGER ELECTRON SPECTRA FROM CADMIUM 4-0728
ND EXTRAATOMIC STATIC RELAXATION ENERGIES IN THE L3M4,5M4,5 AUGER PROCESS IN ARSENIC AND SELENIUM. ATOMIC A 4-0883
   AUGER VACANCY SATELLITE STRUCTURE IN THE L3,M4,5M4,5 AUGER SPECTRA OF COPPER.   4-1034
             N4,5 AUGER SPECTRA OF GADOLINIUM.   4-0821
RY AND EXPERIMENT.   L2,3M4,5M4,5 AUGER SPECTRA OF METALLIC COPPER AND ZINC: THEO 4-0919
       HIGH RESOLUTION L2,3M4,5M4,5 AUGER SPECTRUM OF FREE ZINC ATOMS.   4-0733
   SEARCH FOR CORRELATION EFFECTS IN THE M4,5N4,5N4,5 AUGER SPECTRUM OF XENON.   4-0872
   RELAXATION PROCESSES ACCOMPANYING CADMIUM M4,5N4,5N4,5 AUGER TRANSITIONS.   2-0559
ON OF XENON(III) LEVELS EXCITED BY ELECTRON IMPACT VIA N4,5 AUGER TRANSITIONS. VACUUM ULTRAVIOLET EMISSI 4-0878
       HIGH RESOLUTION MEASUREMENTS OF THE L3M2,3M4,5 AUGER TRANSITONS IN NICKEL AND COPPER.   4-1024
ENERGY LEVELS AND SCREENING PA/ SENSITIVITY OF L1,L2,3M4,5 COSTER-KRONIG TRANSITION RATES TO VARIATIONS IN 2-0303
             THE L1,L2,3M4,5 COSTER-KRONIG TRANSITIONS.   2-0248
   RELAXATION PHENOMENA INVOLVED IN THE L3M4,5M4,.5M4,5:1G4 AUGER PROCESS IN ZINC METAL.   4-0882
AUGER ENERGY CALCULATI/ DETAILED ANALYSIS OF THE L3M4,5M4,5: 1G4 AUGER TRANSITION IN ATOMIC ZINC: IMPROVED 4-0884
ECTRON ENERGIES (0-2000 EV) FOR ELEMENTS OF ATOMIC NUMBER 5-103.   AUGER EL 4-1119
MENTS DEPENDENCE OF OPTICAL EXCITATION AND AUGER DECAY OF 5D CORE HOLES IN BISMUTH(III) TELLURIDE. /RIX ELE 2-0398
IC AND EXTRAATOMIC STATIC RELAXATION ENERGIES IN THE L3M4,5M4,5 AUGER PROCESS IN ARSENIC AND SELENIUM. ATOM 4-0883
       AUGER VACANCY SATELLITE STRUCTURE IN THE L3,M4,5M4,5 AUGER SPECTRA OF COPPER.   4-1034
THEORY AND EXPERIMENT.   L2,3M4,5M4,5 AUGER SPECTRA OF METALLIC COPPER AND ZINC: 4-0919
       HIGH RESOLUTION L2,3M4,5M4,5 AUGER SPECTRUM OF FREE ZINC ATOMS.   4-0733
VED AUGER ENERGY CALCULATI/ DETAILED ANALYSIS OF THE L3M4,5M4,5: 1G4 AUGER TRANSITION IN ATOMIC ZINC: IMPRO 4-0884

RELAXATION PHENOMENA INVOLVED IN THE L3M4,5M4,.5M4,.5:1G4 AUGER PROCESS IN ZINC METAL.   4-0882
RON IMPACT.   ANGULAR DISTRIBUTION OF M4,5NN AUGER ELECTRONS EJECTED FROM KRYPTON BY ELECT 4-1003
,5OO AUGER SPECTRA FROM XE/ HIGH RESOLUTION L2,3MM AND M4,5NN AUGER SPECTRA FROM KRYPTON AND M4, 5NN AND N4 4-1113
ORPTION OF MOLECULAR OXYGEN A/ CHEMICAL EFFECTS IN THE M4,5NN AUGER SPECTRUM OF MOLYBDENUM (110) DUE TO ADS 4-0885
 AUGER ELECTRON EJECTION FROM XENON N4,5OO AND KRYPTON M4,5NN PROCESSES BY ELECTRON IMPACT NEAR THRESHOLD.   4-1010
CADMIUM VAPOR.   M4,5N4,5N4,5 AND M4,5O1 AUGER ELECTRON SPECTRA FROM 4-0728
          SEARCH FOR CORRELATION EFFECTS IN THE M4,5N4,5N4,5 AUGER SPECTRUM OF XENON.   4-0872
          RELAXATION PROCESSES ACCOMPANYING CADMIUM M4,5N4,5N4,5 AUGER TRANSITIONS.   2-0559
CT NEAR THRESHOLD. AUGER ELECTRON EJECTION FROM XENON N4,5OO AND KRYPTON M4,5NN PROCESSES BY ELECTRON IMPA 4-1010
AND M4,5NN AUGER SPECTRA FROM KRYPTON AND M4, 5NN AND N4,5OO AUGER SPECTRA FROM XENON. / RESOLUTION L2,3MM 4-1113
          M4,5N4,5N4,5 AND M4,5O1 AUGER ELECTRON SPECTRA FROM CADMIUM VAPOR.   4-0728
          WIDTHS OF 4S AND 5S MULTIPLET COMPONENTS IN THE RARE EARTHS.   4-0970
          SURFACE CONCENTRATION OF MAGNESIUM IN 6061 ALUMINUM BY AUGER ELECTRON SPECTROSCOPY.   5-1824
/OR FOR ELECTRON INDUCED INNER SHELL IONIZATION OF THE N6,7 LEVEL IN GOLD, BISMUTH, AND LEAD MONITORED BY / 4-1067
    CHEMISORPTION OF ATOMIC HYDROGEN ON THE SILICON (111) 7*7 SURFACE. (AUGER EFFECT)   5-1789
) SUBSHELL ATOMIC YIELDS OF ELEMENTS WITH Z APPROXIMATELY 90.   THE L(II 2-0349

Author Index

AAGAARD B	2-0345	K-VACANCY SHARING STUDIED FROM AUGER ELECTRON YIELDS.
	4-1079	AUGER ELECTRON AND X-RAY PRODUCTION IN 50- TO 200 KEV NEON- NEON COLLISIONS.
AARON HB	5-1184	MEASUREMENT OF EQUILIBRIUM SURFACE SEGREGATION USING AUGER ELECTRON SPECTROSCOPY
ABBATI A	3-0583	IMPROVED SPHERICAL GRID RETARDING POTENTIAL ANALYZER. (FOR USE IN AUGER ELECTRON
ABBATI I	3-0584	SECONDARY EMISSION ANALOG FOR IMPROVED AUGER SPECTROSCOPY WITH RETARDING POTENTI
ABE T	5-1127	APPLICATION CF AUGER ELECTRON SPECTROSCOPY TO THE STUDY OF DISCOLORING OF ALUMIN
	5-1487	DETECTION OF ORGANIC CONTAMINANTS ON ALUMINUM SURFACES BY AUGER ELECTRON SPECTRO
ABERG T	2-0223	EVIDENCE FOR A RADIATIVE AUGER EFFECT IN X-RAY PHOTON EMISSION.
	2-0224	THECRY OF THE RADIATIVE AUGER EFFECT.
	2-0225	TWO-PHOTON EMISSION, THE RADIATIVE AUGER EFFECT, AND THE DOUBLE AUGER PROCESS.
	2-0226	INFLUENCE OF CHEMICAL BONDING ON THE LOW ENERGY K-ALPHA SPECTRUM OF SILICON. (RA
	2-0393	K X-RAY AND AUGER ELECTRON ENERGIES FOR NEON BY A TRANSITION OPERATOR METHOD.
	2-0540	ELECTRON- HOLE INTERACTION EFFECTS IN THE $KL_{2,3}L_{2,3}$ RADIATIVE AUGER SPECTRUM OF
	4-0723	INTERPRETATION OF $KL_{2,3}L_2$ RADIATIVE AUGER THRESHOLDS IN METALS AND SEMICONDUCT
	4-1101	INTERPRETATION OF THE LOW ENERGY K-BETA SPECTRUM IN SOLIDS.
ABLOVA LA	2-0227	RADIATIVE AUGER RECOMBINATION OF TWO INDIRECT EXCITONS.
ABOUAF R	2-0228	AUGER EFFECT IN THE MULTIPLE IONIZATION OF MANGANESE AND CADMIUM BY ELECTRON IMP
	2-0229	MULTIPLE IONIZATION AND AUGER TRANSITIONS IN INDIUM AND SILVER BY ELECTRON IMPAC
	2-0230	AUGER TRANSITION AND SUPPLEMENTARY ELECTRON EJECTION IN THE IONIZATION OF ALUMIN
	2-0231	MECHANISMS CF SIMPLE AND MULTIPLE IONIZATION INDUCED BY ELECTRON IMPACT IN SOME
	2-0232	DOUBLE AND TRIPLE IONIZATION IN ALUMINUM, GALLIUM, AND SCANDIUM BY ELECTRON IMPA
ADAMCZYK B	2-0376	CROSS SECTIONS FOR THE PRODUCTION OF NITROGEN(2)(X), NITROGEN(X), AND NITROGEN(2
ADAMS AC	5-1231	FHOSPHORUS CONCENTRATION PROFILES IN P-DOPED SILICON DIOXIDE MEASURED USING AUGE
ADAMS MJ	2-0419	RADIATIVE AND AUGER PROCESSES IN SEMICONDUCTORS.
	2-0420	THEORY OF DONOR- ACCEPTOR RADIATIVE AND AUGER RECOMBINATION IN SIMPLE SEMICONDUC
ADAMS RO	4-0931	AUGER ELECTRON AND CHARACTERISTIC ENERGY LOSS SPECTRA OF PLUTONIUM.
ADLER I	1-0222	SOME PARAMETERS AFFECTING AUGER AND PHOTOELECTRON SPECTROSCOPY AS AN ANALYTICAL
	2-0577	ATOMIC N-SHELL VACANCY WIDTHS IN SOLIDS.
	2-0578	INVESTIGATICN OF LOW-Z COSTER-KRONIG TRANSITIONS BY AUGER AND PHOTOELECTRON SPEC
	2-0579	FREE ATOM BEHAVIOR OF AUGER ELECTRONS IN SOLIDS.
	3-0718	WIDTHS OF ATOMIC M-SHELL VACANCY STATES AND QUASIATOMIC ASPECTS OF RADIATIONLESS
	4-0939	WIDTH OF ATOMIC L2 AND L3 VACANCY STATES NEAR Z = 30. (AUGER EFFECT)
	4-1121	QUASIATOMIC LMM AUGER SPECTRA OF SOLID COPPER AND ZINC.
	4-1122	L-S COUPLING INTERPRETATION OF HIGH RESOLUTION LMM AUGER SPECTRA OF COPPER AND Z
	4-1123	X-RAY EXCITED LMM AUGER SPECTRA OF COPPER, NICKEL AND IRON.
ADOLPHI B	6-1991	LEED- AUGER INVESTIGATION ON A (100) SURFACE OF AUSTENITIC CHROMIUM- NICKEL STEE
AFFLECK JH	5-1411	IN-PROCESS CONTROL TECHNIQUES FOR COMPLEX SEMICONDUCTOR STRUCTURES. TASK-II. APP
AFFROSSMAN S	4-0837	X-RAY EXCITED AUGER AND PHOTOELECTRON SPECTRA OF MAGNESIUM, SOME ALLOYS OF MAGNE
AFROSIMOV VV	2-0233	ENERGY LOSS AND IONIZATION IN FAST ION- ATOM COLLISIONS.
	2-0234	DETECTION OF A NEW TYPE OF AUGER TRANSITIONS IN ATOMS WITH TWO INTERNAL VACANCIE
	2-0235	OBSERVATION CF THREE ELECTRON AUGER TRANSITIONS IN ATOMS WITH TWO INNER VACANCIE
AGREN H	4-0724	SCF AND LIMITED CI CALCULATIONS FOR ASSIGNMENT OF THE AUGER SPECTRUM AND OF THE
AGUSTI J	4-0776	ELECTRONIC EMISSION FROM SOLID TARGETS BOMBARDED BY NOBLE GAS IONS (10-100 KEV).
AHMAD I	4-1019	PROTON STATES, ANOMALCUS CONVERSION, AND K-AUGER SPECTRA IN AMERICIUM-241 FROM C
AHN J	5-1128	ELECTRON BEAM EFFECTS IN DEPTH PROFILING MEASUREMENTS WITH AUGER ELECTRON SPECTR
AIZAWA K	5-1127	APPLICATION OF AUGER ELECTRON SPECTROSCOPY TO THE STUDY OF DISCOLORING OF ALUMIN
AKHMETOV KM	4-0725	KLL AUGER ELECTRON SPECTRUM OF SAMARIUM.
	4-0726	SPECTRUM OF AUGER KLL ELECTRONS OF SAMARIUM.
AKIMOTO K	4-0727	CHEMICAL EFFECTS IN THE AUGER SPECTRA OF IRON DUE TO OXYGEN ADSORPTION.
AKSELA H	4-0728	M4,5N4,5N4,5 AND M4,5O1 AUGER ELECTRON SPECTRA FROM CADMIUM VAPOR.
	4-0729	AUGER ELECTRON STUDIES OF METAL VAPORS.
	4-0731	L2,3MM AUGER ELECTRON SPECTRUM FROM FREE ZINC ATOMS.
	4-0733	HIGH RESOLUTION L2,3M4,5M4,5 AUGER SPECTRUM OF FREE ZINC ATOMS.
AKSELA S	1-0001	ELECTRON SPECTROSCOPY. (PHOTOELECTRON, AUGER, AND IONIZATION SPECTROSCOPY, REVIE
	2-0473	NEW FINE STRUCTURE IN ELECTRON EXCITED AUGER SPECTRA FROM SOLID SURFACES.
	4-0728	M4,5N4,5N4,5 AND M4,5O1 AUGER ELECTRON SPECTRA FROM CADMIUM VAPOR.
	4-0729	AUGER ELECTRON STUDIES OF METAL VAPORS.
	4-0730	HIGH RESOLUTION MNN AUGER SPECTRA OF SILVER, CADMIUM, INDIUM, ANTIMONY, TELLURIU
	4-0731	L2,3MM AUGER ELECTRON SPECTRUM FROM FREE ZINC ATOMS.
	4-0732	HIGH RESOLUTION LMM AUGER SPECTRA OF LOW ENERGY FROM SOLID SURFACES.
	4-0733	HIGH RESOLUTION L2,3M4,5M4,5 AUGER SPECTRUM OF FREE ZINC ATOMS.
ALBEN RS	5-1855	ORDER- DISORDER TRANSFORMATION AT A (100) SURFACE OF COPPER(3) GOLD, THEORY, EXP
ALBRIDGE RG	4-0734	KLL AUGER SPECTRUM OF FLUORINE. (IN FLUORIDE SALTS)
	4-0938	KLL AUGER SPECTRUM OF MANGANESE.
	4-1097	K-, L-, AND M-AUGER AND L- COSTER-KRONIG SPECTRA OF PLATINUM.
ALDAG AW	6-1995	SURFACE CONCENTRATION OF MOLYBDENUM IN TYPES 316 AND 304 STAINLESS STEEL BY AUGE
	6-1996	CHEMICAL PROFILE ANALYSIS OF TYPE 316 STAINLESS STEEL SHEET BY AUGER SPECTROSCOP
ALDER K	2-0478	PENETRATION EFFECTS IN CONVERTED MUONIC TRANSITIONS. (AUGER COEFFICIENTS)
ALEKSAKHIN IS	4-0735	SPECTRAL LINE EXCITATION FUNCTIONS FOR SODIUM, POTASSIUM, RUBIDIUM, AND CESIUM I
ALEKSANDROV AL	3-0658	AUGER ELECTRON SPECTROMETER INTEGRATED WITH A LOW ENERGY ELECTRON DIFFRACTION SY
ALEKSEEV YUV	4-0736	RECORDING OF COMPONENTS ON THE SURFACE OF FUSED EMITTERS. (AUGER SPECTRA)
	5-1129	CHARACTERISTICS OF SURFACE OF THE PALLADIUM BARIUM ALLOY CATHODE IN RELATION TO
ALEXANDER PM	5-1130	EFFECT OF IMPURITIES CN INTRINSIC STRESS IN THIN NICKEL FILMS. (AUGER SPECTROSCO
	5-1131	EFFECT OF IMPURITIES ON INTRINSIC STRESS IN THIN NICKEL FILMS. (AUGER EFFECT)
ALIEV BZ	4-0850	SECONDARY ELECTRON EMISSION FROM TUNGSTEN AND MOLYBDENUM.
ALKHOURY NE	6-1992	PREPARATION AND SURFACE OXIDATION OF A VANADIUM (100) PLANE. (AUGER EFFECT)
ALLEN GC	4-0737	CHEMICAL SHIFTS IN URANIUM AUGER TRANSITIONS.
	4-0738	AUGER SPECTRCSCOPY OF URANIUM.
	6-1993	AUGER ELECTRON SPECTROSCOPY STUDY OF 18-8 STAINLESS STEEL OXIDATION.
ALLEN R	2-0262	AUGER LIMITED CARRIER LIFETIMES IN MERCURY CADMIUM TELLURIDE AT HIGH EXCESS CARR
ALLEN SD	1-0015	LASER WINDOW SURFACE FINISHING AND COATING TECHNOLOGY. (AUGER ELECTRON SPECTROSC
ALLEN TH	5-1132	STUDY OF ALUMINUM WITH A COMBINED AUGER ELECTRON SPECTROMETER ELLIPSOMETER SYSTE
ALLIE G	2-0236	EFFECT OF SURFACE PLASMON OSCILLATIONS ON AUGER ELECTRON EMISSION.
	2-0237	EFFECT OF THE INCIDENCE ANGLE OF PRIMARY ELECTRONS ON THE AUGER EMISSION YIELD.
	4-0739	ENERGY ANALYSIS OF BACKSCATTERED ELECTRONS FROM A SILICON SINGLE CRYSTAL UNDER A
ALLISON W	5-1133	SECONDARY ELECTRON EJECTION FROM METAL SURFACES BY METASTABLE ATOMS. PART-3: ENE
ALMBLADH CO	2-0238	INCOMPLETE RELAXATION IN A MODEL PROBLEM FOR THE AUGER PROCESS.
	2-0239	INTERPRETATION OF HIGH ENERGY X-RAY SATELLITES OF L2,3 EMISSION BANDS OF SODIUM,
ALONSO J	4-0873	AUGER ELECTRON SPECTRA OF MAGNESIUM THIN FILMS.
	5-1392	X-RAY PHOTOELECTRON AND AUGER ELECTRON SPECTRA OF MAGNESIUM THIN FILMS.

ALTRICHTER B   5-1520   STABILITY OF FIELD ELECTRON EMISSION AND VACUUM BREAKDOWN. (FROM SURFACES) INVES
AMELIO GF   2-0240   AUGER ELECTRON SPECTROSCOPY AND SECONDARY EMISSION IN SEMICONDUCTORS. (SILICON,
                      2-0241   THEORY FOR ENERGY DISTRIBUTION OF SECONDARY ELECTRONS.
                      2-0242   TRUE SECONDARY ELECTRON ENERGY DISTRIBUTIONS.
                      4-0740   BAND STRUCTURE OF SILICON BY CHARACTERISTIC AUGER SPECTRUM ANALYSIS.
                      4-0741   AUGER SPECTRCSCCPY CF GRAPHITE SINGLE CRYSTALS WITH LOW ENERGY ELECTRONS.
ANAND Y   5-1961   HIGH SPATIAL RESOLUTION SCANNING AUGER SPECTROSCOPY APPLIED TO ANALYSIS OF X-BAN
ANDERSEN SK   2-0243   DIFFRACTION EFFECTS IN BACKSCATTERING AND AUGER PRODUCTION NEAR CRYSTAL SURFACES
                      2-0254   CRYSTALLINE EFFECTS IN BACKSCATTERING AND AUGER PRODUCTION. (ALUMINUM)
ANDERSON J   2-0422   PHOTOEMISSION STUDIES OF CORE EXCITON DECAY IN POTASSIUM IODIDE.
ANDERSON MW   5-1429   SURFACE CHARACTERIZATION BY LOW ENERGY PHOTO AUGER ELECTRON SPECTROSCOPY.
ANDERSON N   2-0244   FORMATION OF A PLASMA SHEATH WITH SECONDARY ELECTRON EMISSION.
ANDERSON WA   3-0585   SPECTROMETER FOR INDUCED ELECTRON EMISSION. (AUGER ELECTRONS)
ANDERSSON D   5-1134   CLEAN TELLURIUM SURFACES STUDIED BY LEED.
ANDERSSON S   5-1134   CLEAN TELLURIUM SURFACES STUDIED BY LEED.
                      5-1621   SCATTERING CF LOW ENERGY ELECTRONS FROM A COPPER (111) SURFACE.
ANDRAE J   2-0295   METASTABLE AUTOIONIZING THREE- AND FOUR ELECTRON STATES IN BERYLLIUM. (AUGER DEC
                      2-0296   AUGER TRANSITIONS IN LITHIUM-LIKE BERYLLIUM-LIKE IONS.
ANTON R   5-1135   MEASUREMENTS OF CONDENSATION COEFFICIENTS OF GOLD ON ROCK SALT (100) CLEAVAGE SU
                      5-1136   QUANTITATIVE AUGER ELECTRON SPECTROSCOPY OF GOLD CONDENSED ON ROCK SALT AND ON A
                      5-1137   HETEROGENEOUS NUCLEATION AND GROWTH OF GOLD ON SUBSTRATES. (AUGER SPECTRA)
ANTONIEWICZ J   5-1138   INCREASING THE ACCURACY OF MATERIAL TESTS BY THE APPLICATION OF ELECTRON SPECTRC
ANTYPAS G   5-1962   DEPOSITION AND AUGER ANALYSIS OF DEPOSITED SILICON DIOXIDE ON ALUMINUM (X) GALLIU
APPLETON BR   5-1139   INVESTIGATION OF REORDERED (001) GOLD SURFACES BY POSITIVE ION CHANNELING SPECTR
                      5-1986   CHARACTERIZATION OF REORDERED (001) GOLD SURFACES BY POSITIVE ION CHANNELING SPE
ARAMATI VS   3-0586   DEVELOPMENT AND APPLICATION OF AN "AUGER ELECTRON SPECTROMETER" FOR HIGH TEMPERA
                      5-1282   AUGER ELECTRCN SPECTROSCOPY MEASUREMENTS OF ADSORPTION ISOBARS FOR OXYGEN ON TUN
ARGILE C   1-0002   QUANTITATIVE AUGER ELECTRON SPECTROSCOPY USING COADSORPTION.
                      3-0587   CALIBRATION IN AUGER ELECTRON SPECTROSCOPY BY MEANS OF COADSORPTION.
ARIFOV UA   2-0245   ANISOTROPY OF SECONDARY ELECTRON EMISSION FROM PASSAGE OF LITHIUM(+) IONS THROUG
                      2-0246   ENERGY SPECTRUM OF ELECTRONS EMITTED BY ALKALI HALIDE CRYSTALS UNDER IMPACT OF I
                      2-0247   AUGER NEUTRALIZATION OF MULTIPLY CHARGED IONS ON A METAL SURFACE.
                      4-0742   SECONDARY ELECTRON EMISSICN FROM TUNGSTEN AND MOLYBDENUM SINGLE CRYSTALS.
ARJIS E   5-1337   COMBINED LOW ENERGY ELECTRON DIFFRACTION, AUGER ELECTRON SPECTROSCOPY, AND X-RAY
                      5-1338   AUGER ELECTRON EMISSION SPECTROSCOPY (AES) AND LOW ENERGY ELECTRON DIFFRACTION (
ARMSTRONG NR   4-1026   ELECTROCHEMISTRY AT THIN SOLID FILMS. (AUGER SPECTRA)
ARMSTRONG RA   3-0618   SENSITIVITY VARIATIONS IN BAYARD- ALPERT GAUGES CAUSED BY AUGER EMISSION AT COLL
                      5-1140   AUGER ELECTRCN SPECTROSCOPY OF LAYERED GROWTH OF TITANIUM ON TUNGSTEN.
ARNOTT DR   3-0588   CONSTRUCTION AND OPERATION OF A CYLINDRICAL MIRROR TYPE OF AUGER SPECTROMETER.
                      5-1141   SURFACE VALENCE BAND AND PLASMON FEATURES ON CLEAN CLEAVED SILICON.
ARNS T   3-0657   ULTRAHIGH VACUUM LOW TEMPERATURE STAGE FOR IN-SITU FILM DEPOSITION AND CHARACTER
ARSENIO TP   5-1299   PURITY AND MORPHOLOGY OF ALUMINUM FILMS. (AUGER SPECTROSCOPY)
ARTAMONOVA KP   4-0743   AUGER L-SERIES ELECTRONIC SPECTRA ACCOMPANYING THE DECAY OF ERBIUM-160, YTTERBIU
                      4-0744   L-SERIES AUGER ELECTRON SPECTRA FOR ELEMENTS WITH Z = 66-70.
ARTHUR JR   5-1142   USE OF SCANNING AUGER MICROSCOPY IN MOLECULAR BEAM EPITAXY OF GALLIUM ARSENIDE A
                      5-1143   SURFACE STOICHIOMETRY AND STRUCTURE OF GALLIUM ARSENIDE.
                      5-1144   ADSORPTION OF ZINC ON GALLIUM ARSENIDE. (AUGER SPECTROSCOPY)
ASAAD WN   1-0023   AUGER EFFECT.
                      2-0248   THE L1,L2,3M4,5 COSTER-KRONIG TRANSITIONS.
                      2-0342   AUGER TRANSITION RATE IN MIXED COUPLING SCHEME.
                      4-0745   A NEW TREATMENT OF THE KLL AUGER SPECTRUM.
                      4-0746   ENERGIES OF THE KLL AUGER LINES.
                      4-0747   POSITIONS OF THE KLL AUGER LINES.
                      4-0748   INTENSITIES OF KLL AUGER LINES.
                      4-0749   CALCULATION OF AUGER SPECTRA IN J-J COUPLING.
                      4-0750   RELATIVISTIC CALCULATION OF THE KLL AUGER SPECTRUM. (OF MERCURY)
                      4-0751   THE K-AUGER SPECTRUM.
                      4-0980   THE KLL AUGER SPECTRUM FOR ATOMIC NUMBERS BETWEEN 10 AND 36.
ASANO K   1-0133   SAMPLES HEATED DIRECTLY BY ELECTRIC CURRENT OBSERVED BY AUGER ELECTRON SPECTROSC
                      3-0639   ELECTRODES AND CIRCUITS OF A LEED- AES SYSTEM.
                      5-1522   LOW ENERGY ELECTRON DIFFRACTION- AUGER ELECTRON SPECTROSCOPY STUDY OF IRON FILM
                      5-1523   BEHAVIOR OF IMPURITY ATOMS AND ADSORBED OXYGEN ATOMS ON THE (001) FACE OF IRON E
ASHKINADZE BM   4-0752   METAL- DIELECTRIC TRANSITION IN GERMANIUM AND SILICON STUDIED BY A MICROWAVE MET
ASHLEY EJ   5-1700   RESIDUAL GAS AND THE OPTICAL PROPERTIES OF SILVER FILMS. (AUGER SPECTROSCOPY)
ASHWELL GWB   1-0003   COMBINED AUGER ELECTRON SPECTROSCOPY AND SCANNING ELECTRON MICROSCOPY.
                      5-1902   THERMALLY INDUCED PROCESSES AT GOLD- GALLIUM ARSENIDE INTERFACES: AN ASSESSMENT
ASPLUND L   4-1066   AUGER ELECTRON SPECTRUM OF WATER VAPOR.
AUGER P   1-0004   AUGER EFFECT. (REVIEW, 16 REFS)
                      2-0249   ON THE COMPOUND PHOTOELECTRIC EFFECT.
AUGUSTUS PD   4-0944   INFLUENCE OF THE BAND STRUCTURE ON THE AUGER ELECTRON SPECTRUM OF SILICON.
                      5-1611   DETECTION OF SILICON OXYNITRIDE LAYERS ON THE SURFACES OF SILICON NITRIDE FILMS
                      5-1612   STUDY OF THE OXIDATION AND REDUCTION BY WATER VAPOR OF SILICON (111) SURFACES US
                      5-1613   AUGER ELECTRON EMISSICN FROM BERYLLIUM SURFACES.
AUSTON DH   2-0250   PICOSECOND OPTICAL MEASUREMENTS OF BAND-TO-BAND AUGER RECOMBINATION OF HIGH DENS
AVERY NR   5-1145   LEED- AUGER, AND WORK FUNCTION STUDY OF IODINE ADSORBED ON TUNGSTEN (110).
                      5-1146   AUGER OBSERVATION OF A SURFACE STATE ON TUNGSTEN.
                      6-1994   OXIDATION OF TUNGSTEN (110). PART-2: LEED- AUGER STUDY OF OXIDE FORMATION AT 6
BABENKOV MI   2-0251   KLL AUGER TRANSITION ABSOLUTE PROBABILITIES. (ATOMIC NUMBERS 30-94)
                      2-0252   ABSOLUTE PRCBABILITIES FOR AUGER- KLL TRANSITIONS DEPENDENT ON THE ATOMIC NUMBER
                      2-0253   RELATIVE PROBABILITIES FOR AUGER KLL TRANSITIONS DEPENDENT ON THE ATOMIC NUMBER
                      4-0753   KLL AUGER ELECTRON SPECTRA OF THULIUM AND LUTETIUM.
                      4-0754   KLL AUGER ELECTRONS OF CADMIUM AND ANTIMONY.
                      4-0755   SPECTRUM OF AUGER KLL ELECTRONS OF TELLURIUM AND XENON.
                      4-0756   SPECTRUM OF AUGER K,L,M,N,O ELECTRONS OF SAMARIUM.
                      4-0757   KLM SPECTRUM OF SAMARIUM AUGER ELECTRONS.
                      4-0758   KMM AUGER TRANSITIONS IN XENON-131.
BABOUX JC   4-0759   SECONDARY ELECTRON EMISSION FROM SODIUM CHLORINE SINGLE CRYSTAL BOMBARDED BY RAR
BADDE HG   4-1031   ORIENTATION ANISOTROPY OF BACKSCATTERING COEFFICIENT AND SECONDARY ELECTRON EMIS
BAENNINGER U   1-0005   HIGH TEMPERATURE AUGER ELECTRON SPECTROSCOPY.
                      5-1147   INVESTIGATICNS OF THE GAS- SOLID INTERFACE INTERACTIONS AT HIGH TEMPERATURES BY

BAENNINGER U   5-1148   ADSORPTION ISOTHERMS OF OXYGEN ON TUNGSTEN (100), AND (111) SURFACES IN THE PRES
   5-1156   INVESTIGATION OF OXYGEN ADSORPTION ON TUNGSTEN AT HIGH TEMPERATURES BY AUGER ELE
   5-1157   INVESTIGATION OF EUROPIUM(II) OXIDE (100) CLEAVAGE SURFACE BY LOW ENERGY ELECTRO
BAER AD   2-0422   PHOTOEMISSION STUDIES OF CORE EXCITON DECAY IN POTASSIUM IODIDE.
BAETZOLD RC   4-0760   AUGER STUDIES OF ELECTROLESS NICKEL FILMS.
   5-1149   SURFACE IMPURITIES IN SILVER HALIDE FILMS.
   5-1947   UPS, ESCA, AND AUGER SPECTROSCOPY STUDY OF VACUUM DEPOSITED SILVER HALIDE FILMS.
BAEV IA   4-0761   CARRIER RECOMBINATION IN N-INDIUM ANTIMONIDE.
BAGUS PS   4-1012   THEORETICAL ANALYSIS OF THE AUGER SPECTRA OF METHANE.
BAHL MK   4-0762   X-RAY FLUORESCENCE L SPECTRA OF NIOBIUM IN NIOBIUM METAL AND SOME OF ITS COMPOUN
   4-0763   RELAXATION DURING PHOTOEMISSION AND LMM AUGER DECAY IN ARSENIC AND SOME OF ITS C
BAINES M   2-0254   CRYSTALLINE EFFECTS IN BACKSCATTERING AND AUGER PRODUCTION. (ALUMINUM)
   3-0589   ELECTRON BEAM CURRENT CONTROLLER FOR LEED- AUGER STUDIES.
BAIRD WE   2-0255   RADIATIONLESS DEEXCITATION OF EXCITED HELIUM ATOMS AT SURFACES. (AUGER PROCESS)
BAJOREK CH   5-1738   SURFACE COMPOSITION AND CHEMISTRY OF EVAPORATED PERMALLOY FILMS OBSERVED BY X-RA
BAKER AD   1-0021   TECHNIQUES RELATED TO PHOTOELECTRON SPECTROSCOPY. (REVIEW)
BAKER BG   5-1150   QUANTITATIVE AUGER SPECTROSCOPY OF PHYSICALLY ADSORBED XENON ON NICKEL.
   5-1151   ELECTRON BEAM EFFECTS IN AUGER ANALYSIS OF PHYSISORBED XENON.
   5-1761   PASSIVE FILM ON IRON. APPLICATION OF AUGER ELECTRON SPECTROSCOPY.
   5-1762   DEGRADED PASSIVE FILM ON IRON. APPLICATION OF AUGER ELECTRON SPECTROSCOPY.
BAKER JH   3-0590   CYLINDRICAL MIRROR ANALYZER INCORPORATING PRERETARDATION. (TUNGSTEN AUGER ELECTR
BAKER JM   5-1152   PHOTOEMISSION SPECTRA FROM ADSORBED OXYGEN ON TUNGSTEN (110) AND CARBON MONOXIDE
   5-1153   CHEMICAL SHIFTS IN THE AUGER SPECTRUM OF YTTRIUM ON OXYGEN ADSORPTION.
   5-1308   PHOTOEMISSION STUDIES OF CHEMISORPTION ON NICKEL AND TUNGSTEN SURFACES.
BAKU J   3-0668   PRODUCTION OF A SPHERICAL GRID FOR LOW ENERGY ELECTRON DIFFRACTION- AUGER ELECTR
BALAGUROV LA   4-0785   RECOMBINATION OF CARRIERS IN N-TYPE INDIUM ARSENIDE AT 77 DEGREES K. (AUGER)
BALANDIN GD   3-0658   AUGER ELECTRON SPECTROMETER INTEGRATED WITH A LOW ENERGY ELECTRON DIFFRACTION SY
BALASHOV VV   2-0256   RESONANCE IONIZATION OF ATOMIC HELIUM BY FAST ELECTRONS. (ANGULAR DISTRIBUTION O
   2-0257   POSSIBILITY OF INVESTIGATING ATOMIC AUTOIONIZING STATES IN COINCIDENCE EXPERIMEN
BALASHOVA AP   4-1120   CHARACTERISTICS OF SHAPING SECONDARY ELECTRON EMISSION FROM POROUS POTASSIUM CHL
BALES M   5-1904   AUGER ELECTRON SPECTROSCOPY AT HIGH SPATIAL RESOLUTION AND PRIMARY BEAM CURRENTS
BALIBAR F   5-1154   HOW TO ASSURE ONESELF OF THE CLEANLINESS OF A SURFACE. (USE OF AES TO DETERMINE
BALLA CP   2-0556   NONRELATIVISTIC AUGER RATES, X-RAY RATES, AND FLUORESCENCE YIELDS FOR THE K-SHEL
   2-0557   Z DEPENDANCE OF THE KLL AUGER RATES.
BALOOCH M   5-1155   ANGULAR DISTRIBUTIONS OF HYDROGEN DESORBED FROM THE (100), (110), AND (111) FACE
BALTRAMIEJUNAS R   2-0258   TEMPERATURE DEPENDENCE OF THE EDGE EMISSION OF GALLIUM SELENIDE SINGLE CRYSTALS
BAMBYNEK W   2-0259   X-RAY FLUORESCENCE YIELDS, AUGER, AND COSTER-KRONIG TRANSITION PROBABILITIES.
BARBER M   4-0814   EXPERIMENTAL AND THEORETICAL STUDY OF THE AUGER SPECTRUM OF NITROUS OXIDE.
BARBULESCO N   5-1987   AUGER AND IONIZATION LOSS SPECTRA FROM BERYLLIUM SURFACES.
BARDSLEY JN   4-0906   AUGER SPECTRA OF HIGHLY IONIZED OXYGEN AND FLUORINE.
BARNES GJ   6-1995   SURFACE CONCENTRATION OF MOLYBDENUM IN TYPES 316 AND 304 STAINLESS STEEL BY AUGE
   6-1996   CHEMICAL PROFILE ANALYSIS OF TYPE 316 STAINLESS STEEL SHEET BY AUGER SPECTROSCOP
BARNETT CF   4-1029   SECONDARY ELECTRON EMISSION OF METALS BOMBARDED WITH 120-EV TO 5-KEV PROTONS.
BARO AM   2-0491   INTERATOMIC AUGER PROCESSES AND THE DENSITY OF STATES.
   3-0688   EXPERIMENTAL ASSEMBLY FOR OBTAINING THE SPECTRUM OF SECONDARY ELECTRON EMISSION
   4-0765   SECONDARY ELECTRON EMISSION ROCKING CURVE FROM COPPER.
   4-0765   A STUDY OF THE INTENSITY OF THE L2,3VV AUGER TRANSITION IN MAGNESIUM AND SILICON
   4-0766   MVV AUGER SPECTRA OF COPPER AND NICKEL.
   4-0779   L1,L2,3V AUGER TRANSITION IN SILICON.
   4-1042   INTERATOMIC TRANSITIONS AND RELAXATION EFFECTS IN AUGER SPECTRA OF SEVERAL GAS A
   4-1044   EXPERIMENTAL OBSERVATION OF CHEMICAL SHIFTS IN AUGER SPECTRUM FROM SURFACE LAYER
   4-1045   INTERATOMIC AUGER PROCESSES IN SOME ADSORBATES ON TRANSITION METALS.
   4-1046   HIGH ENERGY SATELLITES IN AUGER PEAKS OF MAGNESIUM AND SILICON.
   4-1047   ENERGY SPECTRA OF SECONDARY ELECTRONS EMITTED BY MAGNESIUM AND SILICON WHEN BOMB
BARRAU J   2-0260   EVIDENCES FOR AUGER RECOMBINATION IN ELECTRON HOLE DROPLETS. (SILICON)
BARRETT JH   5-1139   INVESTIGATION OF REORDERED (001) GOLD SURFACES BY POSITIVE ION CHANNELING SPECTR
   5-1986   CHARACTERIZATION OF REORDERED (001) GOLD SURFACES BY POSITIVE ION CHANNELING SPE
BARRIE A   2-0261   COMPARISON OF IONIZATION CROSS SECTIONS AND OTHER FACTORS IN X-RAY EXCITED AND E
   4-0767   THRESHOLD EFFECTS ON RELATIVE INTENSITIES IN LMM AUGER SPECTRA OF GERMANIUM AND
   4-0768   COMPARISON OF X-RAY INDUCED AND ELECTRON IMPACT INDUCED PHOTOELECTRON AND AUGER
   4-0769   EXPERIMENTAL DETERMINATION OF THE INTERNAL PHOTOEMISSION CONTRIBUTION TO KLM AUG
   4-0770   AUGER AND X-RAY PHOTOELECTRON SPECTROSCOPIC STUDY OF SODIUM METAL AND SODIUM OXI
   4-0814   EXPERIMENTAL AND THEORETICAL STUDY OF THE AUGER SPECTRUM OF NITROUS OXIDE.
BARRY TL   1-0197   TECHNIQUE OF PREPARING POWDER SAMPLES FOR AES AND/OR ESCA ANALYSIS.
BARTOLI F   2-0262   AUGER LIMITED CARRIER LIFETIMES IN MERCURY CADMIUM TELLURIDE AT HIGH EXCESS CARR
BAS EB   1-0005   HIGH TEMPERATURE AUGER ELECTRON SPECTROSCOPY.
   5-1147   INVESTIGATIONS OF THE GAS- SOLID INTERFACE INTERACTIONS AT HIGH TEMPERATURES BY
   5-1148   ADSORPTION ISOTHERMS OF OXYGEN ON TUNGSTEN (100), AND (111) SURFACES IN THE PRES
   5-1156   INVESTIGATION OF OXYGEN ADSORPTION ON TUNGSTEN AT HIGH TEMPERATURES BY AUGER ELE
   5-1157   INVESTIGATION OF EUROPIUM(II) OXIDE (100) CLEAVAGE SURFACE BY LOW ENERGY ELECTRO
BASIN LA   4-0725   KLL AUGER ELECTRON SPECTRUM OF SAMARIUM.
BASSETT PB   4-0771   QUASIATOMIC FINE STRUCTURE IN THE AUGER SPECTRA OF SOLID SILVER AND INDIUM.
BASSETT PJ   3-0591   IMPROVED AUGER ELECTRON SPECTROMETER USING CONCENTRIC HEMISPHERES.
   3-0592   HIGH ENERGY RESOLUTION AUGER ELECTRON SPECTROMETER USING CONCENTRIC HEMISPHERES.
   3-0624   SOME PROPERTIES OF RETARDING FIELD SPECTROMETERS USING POST MONOCHROMATORS. (FOR
   4-0772   HIGH RESOLUTION AUGER SPECTRUM OF MAGNESIUM OXIDE (100).
   4-0892   THE KVV AUGER SPECTRUM OF LITHIUM METAL.
   5-1158   EFFECT OF OXIDATION ON THE HIGH RESOLUTION AUGER SPECTRUM OF TITANIUM.
BATTYE FL   4-0773   PHOTOELECTRON DETERMINATION OF THE ATTENUATION OF THE LOW ENERGY ELECTRONS IN AL
BAUER E   1-0006   PRESENT STATE OF AUGER ELECTRON SPECTROSCOPY AND LEED.
   1-0007   LOW ENERGY ELECTRON DIFFRACTION AND AUGER METHODS. (REVIEW)
   1-0008   COMBINED LOW ENERGY ELECTRON DIFFRACTION, AUGER ELECTRON, AND FLASH DESORPTION S
   2-0263   NATURE OF ANNEALED SEMICONDUCTOR SURFACES. (AUGER)
   3-0665   LOW ENERGY ION BACKSCATTERING SPECTROSCOPY WITH A COMMERCIAL AUGER CYLINDRICAL M
   3-0666   QUANTITATIVE ASPECTS OF ION SCATTERING SPECTROSCOPY. (CYLINDRICAL MIRROR ANALYZE
   5-1159   METHODS OF SURFACE STUDIES DEPENDING ON INELASTIC SCATTERING OF ELECTRONS. (AUGE
   5-1160   ADSORPTION AND CONDENSATION OF COPPER ON TUNGSTEN SINGLE CRYSTAL SURFACES.
   5-1318   ADSORPTION OF OXYGEN ON THE (110) PLANE OF TUNGSTEN. (AUGER ELECTRON SPECTROSCOP
   5-1689   COMPARISON BETWEEN AUGER ELECTRON SPECTROSCOPY AND SECONDARY ION MASS SPECTROSCO
BAUER G   2-0461   TRANSIENT CARRIER DECAY AND TRANSPORT PROPERTIES IN MERCURY CADMIUM TELLURIDE. (

BAUER H	3-0692	ELECTRON CHANNELING PATTERNS IN AN AUGER ELECTRON SPECTROMETER WITH SCANNING SAM
BAUER W	1-0143	SURFACE ANALYSIS USING PROTON BEAMS.
	5-1667	PROTON AND ELECTRON EXCITED AUGER ELECTRON SPECTRA OF MOLYBDENUM SURFACES.
	5-1886	CARBIDE FORMATION ON NIOBIUM SURFACES DURING HIGH TEMPERATURE PROTON IRRADIATION
BAUN WL	1-0009	SECONDARY ION MASS SPECTROMETRY (AND AUGER ELECTRON SPECTROSCOPY): A USEFUL COMP
	1-0183	DIGITIZING AUGER DATA. THREEFOLD APPROACH TO ENHANCE ANALYTICAL CAPABILITIES.
	4-1069	MOLECULAR ORBITAL EFFECTS ON THE TITANIUM LMV AUGER SPECTRA OF TITANIUM(II) AND
	5-1834	CHANGES IN THE TITANIUM DIOXIDE AUGER AND APPEARANCE POTENTIAL SPECTRA INDUCED B
BAYLISS CR	5-1161	SURFACE STRUCTURE AND COMPOSITION OF VARIOUS ELEMENTAL AND III-V SEMICONDUCTORS.
	5-1162	DEPARTURES FROM STOICHIOMETRY IN (100) SURFACES OF GALLIUM PHOSPHIDE, INDUCED BY
	5-1163	COMPOSITIONAL AND STRUCTURAL CHANGES THAT ACCOMPANY THE THERMAL ANNEALING OF (10
	5-1164	EFFECT OF TEMPERATURE ON THE SURFACE STRUCTURE AND STOICHIOMETRY OF (100) INDIUM
BEAMAN DR	1-0010	ELECTRON BEAM MICROANALYSIS. PART-1. (REVIEW OF ELECTRON PROBE ANALYZER, SCANNIN
BECHTOLD E	5-1420	SULFUR ON SINGLE CRYSTAL PLATINUM PLANES. LOW ENERGY ELECTRON DIFFRACTION AND AU
	5-1421	ADSORPTION OF SULFUR ON THE (100) AND (111) FACES OF PLATINUM. LOW ENERGY ELECTR
BECK DE	2-0264	MOLECULAR ADSORPTION: INCLUSION OF BULK DIFFUSION AND ITS EFFECT ON AUGER INTEN
BECK DR	2-0265	ACCURATE ONE ELECTRON BINDING ENERGIES AND AUGER ENERGIES IN MANY ELECTRON ATOMS
	4-0774	INNER ELECTRON BINDING ENERGIES AND AUGER ENERGIES IN FREE ATOMS FOR USE IN X-RA
	4-1000	THEORY OF KLL AUGER ENERGIES.
BECK JD	2-0266	AUGER RECOMBINATION IN SILICON.
BECK K	5-1165	DETECTION OF THIN CONTAMINATED LAYERS ON CONTACT SURFACES. (BY AUGER SPECTROSCOP
BECKER GE	2-0375	RESONANCE, AUGER, AND AUTOIONIZATION PROCESSES INVOLVING HELIUM(+)(2S) AND HELIU
	4-0775	DIFFRACTION PEAKS IN SECONDARY ELECTRON ENERGY SPECTRA. (NICKEL (100))
BEGEMANN SHA	5-1166	SURFACE COMPOSITION AND STRUCTURE ANALYSIS WITH A ONE APPARATUS ION PROBE TECHNI
BEHRISCH R	5-1167	ANALYSIS OF SURFACE LAYERS BY LIGHT ION BACKSCATTERING AND SPUTTERING COMBINED W
BELLINA JJ	5-1168	AN ORDERED BORON STRUCTURE AFTER DEPOSITION ON SILICON (111). (AUGER ELECTRON SP
	5-1169	EFFECTS OF PREADSORBED OXYGEN CONTAMINANT ON THE ADSORPTION OF ALUMINUM ON SILIC
BENARD J	5-1170	PHASE EQUILIBRIA IN TWO DIMENSIONAL CHEMISORBED LAYERS. (AUGER SPECTROSCOPY, SUL
BENAZETH C	2-0267	IONIZATION CROSS SECTION OF THE L2,3 SHELL OF ALUMINUM IN A NEON(+)- ALUMINUM CO
	2-0268	INTERPRETATION OF ENERGY LOSS BY AUGER ELECTRONS EMITTED DURING ION BOMBARDMENT.
	2-0269	ENERGY DISTRIBUTION OF SECONDARY ELECTRONS EMITTED BY INDIUM PHOSPHIDE AND GALLI
	2-0550	EMISSION OF AUGER ELECTRONS BY ION BOMBARDMENT.
	4-0776	ELECTRONIC EMISSION FROM SOLID TARGETS BOMBARDED BY NOBLE GAS IONS (10-100 KEV):
	4-0942	SECONDARY ELECTRON SPECTRA OF SOME SOLID TARGETS UNDER IONIC BOMBARDMENT.
	4-0988	ENERGY SPECTRUM AND ANGULAR DISTRIBUTION OF SECONDARY ELECTRONS EMITTED BY IONIC
	4-1106	AUGER SPECTRA OF SOME FREE ELEMENTS AND BINARY COMPOUNDS BOMBARDED BY 60 KEV ARG
BENAZETH N	4-0776	ELECTRONIC EMISSION FROM SOLID TARGETS BOMBARDED BY NOBLE GAS IONS (10-100 KEV):
	4-1106	AUGER SPECTRA OF SOME FREE ELEMENTS AND BINARY COMPOUNDS BOMBARDED BY 60 KEV ARG
BENESOVSKY F	6-1997	DETERMINATICN OF THE INFLUENCE GRAIN BOUNDARY IMPURITIES ON THE PROPERTIES OF SI
BENNDORF C	5-1171	ISOMORPHISM AND PURITY CF COPPER VAPOR DEPOSITED ON A COPPER (111) SURFACE. (AUG
BENNINGHOVEN A	1-0011	NEW DEVELOPMENTS IN THE SURFACE ANALYSIS OF SOLIDS.
	3-0632	AN ANALYTICAL SYSTEM FOR SECONDARY ION MASS SPECTROMETRY IN ULTRAHIGH VACUUM. (A
BENOITALAGUILLAUME C	4-0777	ELECTRON HOLE DROPS IN GERMANIUM- SILICON ALLOYS. (AUGER RECOMBINATION)
BENZ G	4-0778	AUGER RECOMBINATION IN GALLIUM ANTIMONIDE.
BERGERSEN B	2-0270	X-RAY EDGE ANOMALIES IN LITHIUM.
	2-0447	PLASMON SATELLITES IN AUGER SPECTRA.
BERGLUND S	5-1172	AUGER ELECTRON SPECTROSCOPY STUDY OF THE SURFACE COMPOSITION OF THE LEAD- INDIUM
BERGMARK T	2-0515	ELECTRON SPECTROSCOPIC INVESTIGATION OF AUGER PROCESSES IN BROMINE-SUBSTITUTED M
	4-1113	HIGH RESOLUTION L2,3MM AND M4,5NN AUGER SPECTRA FROM KRYPTON AND M4, 5NN AND N4,
	4-1114	L2,3MM AUGER SPECTRUM OF ARGON.
	5-1529	AUGER ELECTRON SPECTRUM OF CARBON SUBOXIDE.
BERGSTROM I	2-0271	AUGER ELECTRCNS FROM THE L-SHELL IN MERCURY.
BERGSTROM J	1-0012	THE AUGER EFFECT.
BERKEY E	6-2107	IDENTIFICATION OF BUBBLE FORMING IMPURITIES IN DOPED TUNGSTEN. (AUGER SPECTROSCO
BERNASEK SL	5-1173	SMALL MOLECULE REACTIONS ON STEPPED SINGLE CRYSTAL PLATINUM SURFACES.
BERNDTSSON A	2-0408	ELECTRON ESCAPE DEPTH IN SILICON. (AUGER ENERGIES)
	2-0409	ESCAPE DEPTHS OF X-RAY EXCITED (AUGER) ELECTRONS.
BERRER S	4-0779	L1,L2,3V AUGER TRANSITION IN SILICON.
BERRIN L	5-1174	CHARACTERIZATION OF CERAMIC SURFACES. (AUGER SPECTROSCOPY)
	5-1853	ANALYSIS AND CHARACTERIZATION OF CERAMIC SURFACES FOR ELECTRONIC APPLICATIONS.
BERTHIER Y	5-1175	LOW ENERGY ELECTRON DIFFRACTION AND AUGER SPECTROSCOPY OF THE ADSORPTION OF SULF
	5-1732	ADSORPTION OF SULFUR ON THE (100) FACE OF MOLYBDENUM.
BERTHOU H	2-0272	AUGER SIGNALS OBTAINED WITH A PHOTOELECTRON SPECTROMETER.
	2-0273	CHEMICAL SHIFTS AND VARIATIONS OF WIDTH AND INTENSITY OF AUGER SIGNALS OBTAINED
	4-0902	PHOTOELECTRON AND AUGER SPECTRA OF XENON IN SOLID PERXENATES.
	4-0905	LINEAR LIGAND FIELDS INFLUENCING AUGER SPECTRA OF FLUORIDES, COPPER(I) AND SILVE
BERTINO JP	5-1783	KIKUCHI CORRELATIONS IN AUGER ELECTRON SPECTROSCOPY.
BERTOLINI JC	5-1283	AUGER SPECTRCSCOPY OF CARBON ON NICKEL. COMMENTS.
BERTOLINI K	5-1284	STUDY OF THE ADSORPTION OF BENZENE ON THE (100) FACE OF NICKEL. INFLUENCE OF SUL
BETZ G	1-0014	CHEMICAL EFFECTS IN AUGER ELECTRON SPECTROSCOPY.
	5-1331	SPUTTERING OF THE ALLOY SYSTEMS SILVER- GOLD, GOLD- COPPER, AND SILVER- COPPER S
	6-1998	COMPOSITION VERSUS DEPTH PROFILES OBTAINED WITH AUGER ELECTRON SPECTROSCOPY OF A
BETZLER K	2-0274	RELAXATION OF AUGER EXCITED CARRIERS IN SILICON.
	2-0275	TWO ELECTRON BAND-TO-BAND TRANSITIONS IN SOLIDS. (AUGER RECOMBINATION IN SILICON
	4-0780	MAGNETIC FIELD DEPENDENT INTENSITY OSCILLATIONS OF THE ELECTRON HOLE LUMINESCENC
BEVERLY RE	2-0276	ELECTRON PHOTOEMISSION INDUCED BY LASER-PRODUCED PLASMA X-RAYS. (PLASMA ELECTRON
BHALLA CP	2-0277	NONRELATIVISTIC FLUORESCENCE YIELDS FOR THE 3P AND THE 3D SHELLS.
	2-0278	AUGER SPECTRA AND FLUORESCENCE YIELDS FOR THE 3P AND THE 3D SHELL.
	2-0279	EFFECTS OF THE CONFIGURATION INTERACTION ON THE K-SHELL AUGER SPECTRUM OF NEON.
	2-0280	K-SHELL AUGER RATES, TRANSITION ENERGIES, AND FLUORESCENCE YIELDS OF VARIOUSLY I
	2-0281	DEEXCITATION OF MULTIPLY IONIZED ATOMS. (AUGER EFFECT)
	2-0282	K-SHELL AUGER RATES FOR MULTIPLY IONIZED ATOMS. PART-1: NEON.
	2-0283	THEORETICAL FLUORESCENCE YIELDS FOR NEON. (AUGER RATE)
	2-0284	THEORETICAL K-SHELL AUGER RATES, TRANSITION ENERGIES, AND FLUORESCENCE YIELDS FO
	2-0285	EXCITATION OF MULTIPLET STATES IN NEON.
	2-0558	NONRELATIVISTIC AUGER RATES, X-RAY RATES, AND FLUORESCENCE YIELDS FOR THE 2P SHE
	4-0781	FLUORESCENCE YIELDS AND AUGER RATES FOR MULTIPLY IONIZED NITROGEN.
	4-0782	RELATIVISTIC KLL AUGER TRANSITION RATES.
BHASIN MM	5-1176	AUGER ELECTRON SPECTROSCOPIC STUDY OF THE SURFACE POISONING OF COPPER CATALYSTS.
	5-1177	AUGER SPECTROSCOPIC STUDY OF THE POISONING OF A COMMERCIAL PALLADIUM ALUMINA HYD

BIBERIAN JP	5-1178	SPONTANEOUS ALLOYING OF A GOLD SUBSTRATE WITH LEAD MONOLAYERS. (INTERFACE AUGER
	5-1733	ADSORPTION AND SURFACE ALLOYING OF LEAD MONOLAYERS ON (111) AND (110) FACES OF G
BICKEL WS	2-0298	OUTER SHELL EXCITATION OF THE RARE GASES BY 30 MEV OXYGEN IONS. (INNER SHELL AUG
BIENFAIT M	5-1856	AUGER ELECTRCN SPECTROSCOPY AND LOW ENERGY ELECTRON DIFFRACTION STUDIES OF ADSOR
	5-1857	FIRST ORDER TWO DIMENSIONAL TRANSITION: XENON ADSORBED ON THE (0001) FACE OF GRA
	5-1858	TWO-DIMENSIONAL PHASE TRANSITION IN XENON SUBMONOLAYER FILMS ADSORBED ON (0001)
	5-1859	THERMODYNAMICS AND KINETICS OF THE FIRST MONOLAYER ADSORPTION OF XENON ON THE (0
	5-1860	ENTROPY OF ADSORPTION OF XENON IN EPITAXY ON THE (0001) FACE OF GRAPHITE. (AUGER
BIGEARD B	4-1111	AUGER ELECTRON SPECTROSCOPY OF HIGH TEMPERATURE OXYGEN- RHENIUM INTERACTIONS.
BILOEN P	1-0013	(AUGER) ELECTRON SPECTROSCOPY, A METHOD FOR STUDYING THE SOLID GAS INTERFACE.
	5-1195	SURFACE COMPOSITION AND DEPTH CONCENTRATION PROFILE OF PLATINUM- TIN ALLOYS FROM
	5-1949	X-RAY EXCITED AUGER AND PHOTOELECTRON SPECTRA OF PARTIALLY OXIDIZED MAGNESIUM SU
BINGLE WD	6-1999	COMPARISON OF SURFACE CHEMISTRIES OBTAINED BY USING ION SCATTERING SPECTROMETRY
BISHOP HE	2-0286	ANGULAR DEPENDENCIES IN ELECTRON EXCITED AUGER EMISSION. (COMMENTS)
	2-0287	CHARACTERISTIC IONIZATION LOSSES OBSERVED IN AUGER EMISSION SPECTROSCOPY.
	2-0288	ESTIMATES OF EFFICIENCIES OF PRODUCTION AND DETECTION OF ELECTRON EXCITED AUGER
	3-0593	A SCANNING ELECTRON MICROSCOPE WITH FACILITIES FOR AUGER EMISSION SPECTROSCOPY.
	3-0594	DESIGN FOR A CYLINDRICAL MIRROR ANALYZER FOR USE IN AUGER SPECTROSCOPY.
	4-0783	AUGER SPECTROSCOPY OF TITANIUM.
	4-0784	AUGER SPECTROSCOPY OF SILICON.
	5-1179	SURFACE ABSORPTION BY POLYPEPTIDE FILMS EXAMINED BY AUGER EMISSION SPECTROSCOPY.
	5-1180	SEGREGATION OF GOLD TO SILICON (111) SURFACE OBSERVED BY AUGER EMISSION SPECTROS
	5-1181	SURFACE SEGREGATION IN BORON DOPED IRON OBSERVED BY AUGER EMISSION SPECTROSCOPY.
	5-1269	ELECTRON BEAM ASSISTED ADSORPTION ON SILICON (111) SURFACE.
BLACK JN	1-0134	CHARACTERIZATION OF SOLID SURFACES. (REVIEW, AUGER SPECTROSCOPY, ESCA)
BLAKELY JM	3-0635	ANALOG TECHNIQUE FOR THE MEASUREMENT OF AUGER ELECTRON CURRENTS. APPLICATION TO
	5-1484	BINDING ENERGIES OF CARBON TO NICKEL (100) FROM EQUILIBRIUM SEGREGATION STUDIES.
	5-1485	BINDING OF CARBON ATOMS AT A STEPPED NICKEL SURFACE. (AUGER SPECTROSCOPY)
	5-1748	LEED, AUGER SPECTROSCOPY, AND CONTACT POTENTIAL STUDIES OF COPPER- GOLD ALLOY SI
	5-1803	EQUILIBRIUM SEGREGATION OF CARBON TO A NICKEL (111) SURFACE: A SURFACE PHASE TRA
	5-1912	THE KINETICS OF OXYGEN ADSORPTION ON THE (112) AND (110) PLANES OF TUNGSTEN.
	5-1918	SEGREGATION OF CADMIUM TO A (111) SILVER CHLORIDE SURFACE. (AUGER SPECTROSCOPY)
	6-2000	SEGREGATION OF CARBON TO (100) SURFACE OF NICKEL.
BLANC E	2-0236	EFFECT OF SURFACE PLASMON OSCILLATIONS ON AUGER ELECTRON EMISSION.
	2-0237	EFFECT OF THE INCIDENCE ANGLE OF PRIMARY ELECTRONS ON THE AUGER EMISSION YIELD.
BLANK G	5-1443	SURFACE SEGREGATION OF OXYGEN IN NIOBIUM. (AES)
BLATTNER RJ	5-1182	EFFECT OF OXIDIZING AMBIENTS ON PLATINUM SILICIDE FORMATION. PART-2: AUGER AND B
	5-1577	ANTIMONY DOPING OF SILICON LAYERS GROWN BY SOLID PHASE EPITAXY. (AES MEASUREMENT
	5-1672	INTERACTION OF ALUMINUM LAYERS WITH POLYCRYSTALLINE SILICON. (AUGER SPECTROSCOPY
BLAUTBLACHEV AN	4-0785	RECOMBINATION OF CARRIERS IN N-TYPE INDIUM ARSENIDE AT 77 DEGREES K. (AUGER)
BLECHSCHMIDT D	2-0289	PHOTOEMISSION STUDIES OF THE DECAY OF CORE EXCITONS IN ALKALI HALIDES. (AUGER PR
BLOKHIN MA	2-0290	NATURE OF SOME X-RAY SATELLITES (RADIATIVE AUGER TRANSITIONS).
BLOOM EE	6-2001	AUGER ELECTRON SPECTROSCOPY OF FRACTURE SURFACES IN IRRADIATED TYPE 304 STAINLES
	6-2002	AUGER SPECTROSCOPY OF FRACTURE SURFACES OF IRRADIATED STAINLESS STEEL.
BLOTT BH	5-1660	EPITAXIAL GROWTH OF COPPER ON (110) SURFACE OF A TUNGSTEN SINGLE CRYSTAL STUDIED
BOBYKIN BV	2-0251	KLL AUGER TRANSITION ABSOLUTE PROBABILITIES. (ATOMIC NUMBERS 30-94)
	2-0252	ABSOLUTE PRCBABILITIES FOR AUGER- KLL TRANSITIONS DEPENDENT ON THE ATOMIC NUMBER
	2-0253	RELATIVE PRCBABILITIES FOR AUGER KLL TRANSITIONS DEPENDENT ON THE ATOMIC NUMBER
	4-0725	KLL AUGER ELECTRON SPECTRUM OF SAMARIUM.
	4-0726	SPECTRUM OF AUGER KLL ELECTRONS OF SAMARIUM.
	4-0753	KLL AUGER ELECTRON SPECTRA OF THULIUM AND LUTETIUM.
	4-0754	KLL AUGER ELECTRONS OF CADMIUM AND ANTIMONY.
	4-0755	SPECTRUM OF AUGER KLL ELECTRONS OF TELLURIUM AND XENON.
	4-0756	SPECTRUM OF AUGER K,L,M,N,O ELECTRONS OF SAMARIUM.
	4-0757	KLM SPECTRUM OF SAMARIUM AUGER ELECTRONS.
BOCKRIS JOM	5-1761	PASSIVE FILM ON IRON. APPLICATION OF AUGER ELECTRON SPECTROSCOPY.
	5-1762	DEGRADED PASSIVE FILM ON IRON. APPLICATION OF AUGER ELECTRON SPECTROSCOPY.
BOERS AL	2-0545	CHARGE EXCHANGE OF LOW ENERGY HELIUM IONS AND ATOMS SCATTERED FROM A COPPER SING
	2-0546	CHARGE EXCHANGE OF LOW ENERGY HELIUM(+) IONS (LESS THAN 10 KEV) SCATTERED FROM A
	5-1166	SURFACE COMPOSITION AND STRUCTURE ANALYSIS WITH A ONE APPARATUS ION PROBE TECHNI
BOEVING E	4-1079	AUGER ELECTRON AND X-RAY PRODUCTION IN 50- TO 200 KEV NEON- NEON COLLISIONS.
BOHN GK	3-0595	APPARATUS FOR ANALYSIS BY AUGER ELECTRON SPECTROSCOPY.
	3-0672	HIGH SENSITIVITY AUGER ELECTRON SPECTROMETER.
	3-0708	AUGER ELECTRON SPECTROMETER PREAMPLIFIER.
BOHOUN A	2-0539	ROLE OF AUGER EFFECT IN THERMOSTIMULATED PHENOMENA IN IONIC CRYSTALS.
BOLGER JE	4-0904	HIGH RESOLUTION BEAM GAS AUGER ELECTRON SPECTRA FOR OXYGEN IONS EXCITED BY COLLI
	4-0993	OBSERVATION OF A PROJECTILE CHARGE DEPENDENCE FOR THE NEON K-AUGER ELECTRON SPEC
BONCZEK F	5-1160	ADSORPTION AND CONDENSATION OF COPPER ON TUNGSTEN SINGLE CRYSTAL SURFACES.
BOND VP	2-0531	BIOLOGICAL RADIATION EFFECT ON INCLUSION OF MODERATELY HEAVY NUCLEI IN THE TISSU
BONNELLE C	2-0373	SECONDARY ELECTRON EMISSION OF SOLID ALUMINUM AND MAGNESIUM TARGETS.
	4-0821	N4,5 AUGER SPECTRA OF GADOLINIUM.
	4-0822	AUGER SPECTRA OF MAGNESIUM AND ALUMINUM.
	4-0823	4D AUGER TRANSITIONS FROM EXCITED STATES IN SAMARIUM METAL.
BONSAAN R	5-1934	SURFACE ENRICHMENT IN COPPER(3) GOLD AND GOLD(3) COPPER ALLOYS.
BONZEL HP	1-0063	MODERN TECHNIQUES FOR SURFACE STUDIES.
	5-1183	AUGER ELECTRON SPECTROSCOPY OF A SULFUR- OXYGEN SURFACE REACTION ON A COPPER (11
	5-1184	MEASUREMENT OF EQUILIBRIUM SURFACE SEGREGATION USING AUGER ELECTRON SPECTROSCOPY
	5-1185	DETECTION OF IMPURITIES ON COPPER SURFACES BY LEED- AUGER ELECTRON SPECTROSCOPY
	5-1186	CHANGE IN THE SURFACE ELECTRONIC STRUCTURE OF PLATINUM (100) INDUCED BY RECONSTR
	5-1187	CARBON MONOXIDE OXIDATION ON A PLATINUM (110) SINGLE CRYSTAL SURFACE.
	5-1188	MECHANISMS CF THE CATALYTIC CARBON MONOXIDE OXIDATION ON PLATINUM (110).
	5-1189	KINETICS OF OXYGEN ADSORPTION ON A PLATINUM (111) SURFACE. (AUGER ELECTRON SPECT
	5-1190	ADSORBATE INTERACTIONS ON A PLATINUM (110) SURFACE. PART-1: SULFUR AND CARBON MO
	6-2003	A TECHNIQUE FOR INVESTIGATING THE CATALYTIC ACTIVITY OF ALLOY SYSTEMS. (SURFACE
BORISOV VL	4-0736	RECORDING OF COMPONENTS ON THE SURFACE OF FUSED EMITTERS. (AUGER SPECTRA)
	5-1191	STRUCTURE AND SECONDARY EMISSION OF MAGNESIUM OXIDE LAYERS APPLIED ON A MOLYBDEN
	5-1192	AUGER SPECTROSCOPIC STUDY OF THE SURFACES OF ALLOY EMITTERS.
	6-2004	AUGER SPECTRA OF ACTIVATED SILVER- MAGNESIUM AND SILVER- LITHIUM ALLOYS.
BORST WL	4-0786	SECONDARY ELECTRON YIELDS FROM COPPER BERYLLIUM OXYGEN SURFACE BY THERMAL CARBON
BOSE SM	4-0787	LIFETIME EFFECT ON THE AUGER SPECTRA OF METALS. (SODIUM)

BOSE SM   4-0833 LIFETIME AND SURFACE EFFECTS ON THE L23 VV AUGER SPECTRA OF ALUMINUM.
BOTTOMS WR   3-0596 CONTAMINATION IN THE SCANNING ELECTRON MICROSCOPE. (AUGER ELECTRON SPECTROSCOPY,
                         4-0788 CHEMISORPTION REACTIONS ON HIGH INDEX ZINC SULFIDE SURFACES. (AUGER EFFECT)
                         5-1584 CORRELATION OF ELECTRONIC, LEED, AND AUGER DIAGNOSTICS ON ZINC OXIDE SURFACES.
BOUDART M   5-1966 SURFACE COMPOSITION OF NICKEL- GOLD ALLOYS.
BOUDRY MR   3-0664 APPARATUS FOR SURFACE ANALYSIS BY ELECTRON SPECTROSCOPY.
BOUWMAN R   1-0013 (AUGER) ELECTRON SPECTROSCOPY, A METHOD FOR STUDYING THE SOLID GAS INTERFACE.
                         5-1193 AUGER ELECTRON SPECTROSCOPY FOR THE ANALYSIS OF SOLID SURFACES.
                         5-1194 ELECTRON SPECTROSCOPY AND SURFACE STUDIES.
                         5-1195 SURFACE COMPOSITION AND DEPTH CONCENTRATION PROFILE OF PLATINUM- TIN ALLOYS FROM
                         5-1196 AUGER SPECTROSCOPIC STUDY OF THE SURFACE COMPOSITION OF PLATINUM- TIN IN ULTRAHI
                         6-2005 QUANTITATIVE STUDY OF THE SURFACE COMPOSITION OF BINARY ALLOYS BY AUGER SPECTROS
BOVING E   2-0345 K-VACANCY SHARING STUDIED FROM AUGER ELECTRON YIELDS.
                         2-0346 DYNAMIC CHARGE STATE DEPENDENCE OF THE K-VACANCY PRODUCTION YIELD IN HEAVY ION A
BRABANT JC   2-0260 EVIDENCES FOR AUGER RECOMBINATION IN ELECTRON HOLE DROPLETS. (SILICON)
BRADLEY TL   5-1197 MEASUREMENTS OF THE SPATIAL DISTRIBUTION OF HYDROGEN DESORBED FROM NICKEL SURFAC
                         6-2006 SPATIAL DISTRIBUTIONS OF HYDROGEN DESORBED FROM IRON, PLATINUM, COPPER, NIOBIUM,
BRADSHAW AM   2-0291 PLASMON COUPLING TO CORE HOLE EXCITATIONS IN CARBON. (AUGER APPEARANCE SPECTROSC
BRAICOVICH L   3-0583 IMPROVED SPHERICAL GRID RETARDING POTENTIAL ANALYZER. (FOR USE IN AUGER ELECTRON
                         3-0584 SECONDARY EMISSION ANALOG FOR IMPROVED AUGER SPECTROSCOPY WITH RETARDING POTENTI
BRAINARD WA   1-0057 USE OF LOW ENERGY ELECTRON DIFFRACTION, AUGER EMISSION SPECTROSCOPY AND FIELD IO
                         5-1198 ADHESION AND FRICTION OF POLYTETRAFLUORO ETHYLENE IN CONTACT WITH METALS AS STUD
                         5-1212 ATOMIC NATURE OF POLYMER- METAL INTERACTIONS IN ADHESION, FRICTION, AND WEAR.
                         6-2011 INFLUENCE OF SILICON ON FRICTION AND WEAR OF IRON- COBALT ALLOYS. (AUGER ANALYSI
BRANDIS EK   3-0597 HIGH SPATIAL RESOLUTION AUGER ELECTRON SPECTROSCOPY IN AN ORDINARY DIFFUSION PUM
                         3-0598 AUGER ELECTRON SPECTROSCOPY IN A DIFFUSION PUMPED SCANNING ELECTRON MICROSCOPE.
BRAU MJ   2-0406 RECOMBINATION MECHANISMS IN 8-14 MICRON MERCURY(1-X) CADMIUM(X) TELLURIDE. (AUGE
BRAUN P   1-0014 CHEMICAL EFFECTS IN AUGER ELECTRON SPECTROSCOPY.
                         5-1199 AES STUDIES OF SURFACE COMPOSITION OF SILVER- COPPER ALLOYS.
                         5-1200 AUGER ELECTRON SPECTROSCOPY STUDIES OF SURFACE COMPOSITION OF SILVER- COPPER ALL
                         5-1329 INVESTIGATIONS OF THE OXIDE FORMATION ON METALS WITH AUGER ELECTRON SPECTROSCOPY
                         5-1331 SPUTTERING OF THE ALLOY SYSTEMS SILVER- GOLD, GOLD- COPPER, AND SILVER- COPPER S
                         5-1332 OXYGEN EXPOSURE OF SAMARIUM, GADOLINIUM, AND TERBIUM STUDIED BY AUGER ELECTRON S
                         6-1997 DETERMINATION OF THE INFLUENCE GRAIN BOUNDARY IMPURITIES ON THE PROPERTIES OF SI
                         6-2007 AUGER ELECTRON SPECTROSCOPY OF METALS DURING OXIDATION.
                         6-2024 AUGER ELECTRON SPECTROSCOPY OF SILVER- GOLD ALLOYS.
BRAUNSTEIN AI   1-0015 LASER WINDOW SURFACE FINISHING AND COATING TECHNOLOGY. (AUGER ELECTRON SPECTROSC
BRAUNSTEIN M   1-0015 LASER WINDOW SURFACE FINISHING AND COATING TECHNOLOGY. (AUGER ELECTRON SPECTROSC
BREITZER D   2-0292 G-R NOISE FOR AUGER BAND-TO-BAND PROCESSES.
BREUCKMAN B   4-0789 THE K-AUGER SPECTRUM OF THE FREE MAGNESIUM ATOM.
BRIAND JP   2-0293 NEW SATELLITE BANDS CORRESPONDING TO DOUBLE L IONIZATION. (OF AUGER ELECTRONS)
BRILLSON LJ   5-1201 SURFACE PHOTOVOLTAGE AND AUGER SPECTROSCOPY STUDIES OF (1120) CADMIUM SULFIDE SU
                         5-1202 OBSERVATION OF EXTRINSIC SURFACE STATES ON (1120) CADMIUM SULFIDE BY AUGER SPECT
BRONGERSMA HH   1-0016 SURFACE ANALYSIS, METHODS OF STUDYING THE OUTER ATOMIC LAYERS OF SOLIDS.
                         2-0294 (AUGER) NEUTRALIZATION BEHAVIOR IN SCATTERING OF LOW ENERGY IONS FROM SOLID SURF
BROOK RJ   6-2134 GRAIN BOUNDARY SEGREGATION IN ALUMINUM OXIDE. (AUGER SPECTROSCOPY)
BROUSSEAU M   2-0260 EVIDENCES FOR AUGER RECOMBINATION IN ELECTRON HOLE DROPLETS. (SILICON)
BROWN MD   2-0472 ELECTRON DECAY-IN-FLIGHT SPECTRA FROM AUTOIONIZING STATES OF HIGHLY STRIPPED OXY
BRUCH R   2-0295 METASTABLE AUTOIONIZING THREE- AND FOUR ELECTRON STATES IN BERYLLIUM. (AUGER DEC
                         2-0296 AUGER TRANSITIONS IN LITHIUM-LIKE BERYLLIUM-LIKE IONS.
BRUNDLE CR   1-0017 ELECTRON SPECTROSCOPY OF SURFACES.
                         1-0018 APPLICATION OF ELECTRON SPECTROSCOPY TO SURFACE STUDIES. (REVIEW, 85 REFS)
                         1-0019 ELECTRON SPECTROSCOPY STUDIES OF ADSORPTION AND OXIDATION PROCESSES AT METAL SUR
                         1-0020 AUGER SPECTROSCOPY. (REVIEW)
                         1-0021 TECHNIQUES RELATED TO PHOTOELECTRON SPECTROSCOPY. (REVIEW)
                         4-0768 COMPARISON OF X-RAY INDUCED AND ELECTRON IMPACT INDUCED PHOTOELECTRON AND AUGER
                         5-1203 X-RAY PHOTOELECTRON AND VACUUM ULTRAVIOLET PHOTOELECTRON STUDIES OF ADSORPTION O
                         5-1204 ELECTRON SPECTROSCOPY AND ADSORPTION STUDIES. CHEMICAL SHIFTS IN X-RAY PHOTOELEC
                         5-1205 USE OF X-RAY PHOTOELECTRON AND AUGER SPECTROSCOPY TO DETERMINE THE SURFACE COMPO
BRUNNER J   2-0297 ANGLE DISTRIBUTION OF SODIUM CHLORIDE SINGLE CRYSTALS. (PHOTO- AND AUGER ELECTRO
BRYSON CE   3-0603 A MULTIPLE SAMPLE HOLDER FOR USE IN ULTRAHIGH VACUUM.
                         5-1583 QUANTITATIVE STUDY OF AUGER ELECTRON SIGNALS OF PHOSPHORUS ON SILICON USING A QU
BRYTOV IA   3-0599 X-RAY ELECTRON SPECTROMETERS AND AUGER ANALYZERS.
                         4-0790 EMISSION LINES OF THE M SERIES OF RARE EARTH METALS. (AUGER EFFECT)
BUCHANAN DN   2-0565 SIZE EFFECT IN IONIC CHARGE RELAXATION FOLLOWING AUGER EFFECT.
BUCK TM   2-0294 (AUGER) NEUTRALIZATION BEHAVIOR IN SCATTERING OF LOW ENERGY IONS FROM SOLID SURF
BUCKLEY DH   1-0057 USE OF LOW ENERGY ELECTRON DIFFRACTION, AUGER EMISSION SPECTROSCOPY AND FIELD IO
                         5-1198 ADHESION AND FRICTION OF POLYTETRAFLUORO ETHYLENE IN CONTACT WITH METALS AS STUD
                         5-1206 AUGER ANALYSIS OF OXYGEN AND SULFUR INTERACTIONS WITH VARIOUS METALS AND THE EFF
                         5-1207 ABSORPTION OF ETHYLENE OXIDE AND VINYL CHLORIDE ON AN IRON (011) SURFACE AND EFF
                         5-1208 INTERACTION OF METHANE, ETHANE, ETHYLENE, AND ACETYLENE WITH AN IRON (001) SURFA
                         5-1209 (AUGER STUDIES OF) ADHESION, FRICTION, LUBRICATION IN VACUUM.
                         5-1210 WEAR AND INTERFACIAL TRANSPORT OF MATERIAL. (AES)
                         5-1211 EFFECT OF SULFUR, OXYGEN, AND HYDROGEN SULFIDE SURFACE FILMS ON THE ADHESION OF
                         5-1212 ATOMIC NATURE OF POLYMER- METAL INTERACTIONS IN ADHESION, FRICTION, AND WEAR.
                         5-1730 ADHESION AND TRANSFER OF PLATINUM IRON TO METALS STUDIED BY AUGER EMISSION SPECT
                         5-1731 ADHESION AND TRANSFER OF POLYTETRAFLUORO ETHYLENE TO METALS STUDIED BY AUGER EMI
                         6-2008 ADHESION OF METALS TO A CLEAN IRON SURFACE STUDIED WITH LEED AND AUGER EMISSION
                         6-2009 ADHESION, FRICTION AND AUGER SPECTROSCOPY ANALYSIS OF A COMMERCIAL COBALT BASE A
                         6-2010 LEED AND AUGER STUDIES OF EFFECT OF OXYGEN ON ADHESION OF CLEAN IRON (001) AND (
                         6-2011 INFLUENCE OF SILICON ON FRICTION AND WEAR OF IRON- COBALT ALLOYS. (AUGER ANALYSI
                         6-2012 AUGER ELECTRON SPECTROSCOPY STUDY OF SURFACE SEGREGATION IN COPPER- ALUMINUM ALL
                         6-2013 ELEMENTAL ANALYSIS OF FRICTION AND WEAR SURFACE DURING SLIDING USING AUGER SPECT
                         6-2033 A REVIEW OF SURFACE SEGREGATION, ADHESION, AND FRICTION STUDIES PERFORMED ON COP
BUHL R   5-1213 STUDY OF SURFACE PROCESSES ON COPPER- BERYLLIUM BY COMBINED SECONDARY ION MASS S
                         5-1214 COMBINATION OF SECONDARY ION MASS SPECTROMETRY AND AUGER ELECTRON SPECTROSCOPY,
BUHRING W   4-1031 ORIENTATION ANISOTROPY OF BACKSCATTERING COEFFICIENT AND SECONDARY ELECTRON EMIS
BULL WE   4-0990 DETERMINATION OF KLL AUGER SPECTRA OF NITROGEN, OXYGEN, CARBON DIOXIDE, NITRIC O
BURCH D   2-0298 OUTER SHELL EXCITATION OF THE RARE GASES BY 30 MEV OXYGEN IONS. (INNER SHELL AUG
                         2-0299 EFFECT OF MULTIPLE IONIZATION ON THE FLUORESCENCE YIELD OF NEON. (K-AUGER AND K

BURCH D	2-0300	PROJECTILE CHARGE STATE DEPENDENCE OF NEON K-SHELL IONIZATION AND FLUORESCENCE Y
	2-0301	SIMPLE MODEL FOR K X-RAY AND AUGER ELECTRON ENERGY SHIFTS IN HEAVY ION COLLISION
	2-0518	K-SHELL IONIZATION AND FLUORESCENCE YIELD IN 50 MEV CHLORINE(+)- NEON COLLISIONS
	4-1079	AUGER ELECTRON AND X-RAY PRODUCTION IN 50- TO 200 KEV NEON- NEON COLLISIONS.
BURGE DK	5-1700	RESIDUAL GAS AND THE OPTICAL PROPERTIES OF SILVER FILMS. (AUGER SPECTROSCOPY)
BURHOP EHS	1-0022	THE AUGER EFFECT AND OTHER RADIATIONLESS TRANSITIONS.
	1-0023	AUGER EFFECT.
	4-0751	THE K-AUGER SPECTRUM.
BURIBAEV I	2-0302	OPTICALLY STIMULATED SECONDARY ELECTRON EMISSION AND PHOTOELECTRON EMISSION STUD
BURKSTRAND JM	1-0201	RECENT ADVANCES IN SURFACE CHARACTERIZATION USING LOW ENERGY ELECTRON DIFFRACTIO
BURMAKA DS	5-1215	EFFECT OF ICN BCMBARDMENT ON THE COMPOSITION AND STRUCTURE OF TUNGSTEN AND MOLYB
BURMAKE LS	5-1939	AUGER SPECTROSCOPIC STUDY OF THE SURFACE OF IRON AFTER IONIC BOMBARDMENT AND AIR
BURTON JJ	5-1216	SURFACE SEGREGATION IN ALLOYS: EQUILIBRIUM SEGREGATION OF GOLD TO THE (111) SURF
BUSHMIRE DW	5-1450	DETECTION BY AUGER ELECTRON SPECTROSCOPY AND REMOVAL BY OZONIZATION OF PHOTORESI
BYRUM BW	1-0024	GATED ELECTRCN BEAM TECHNIQUE FOR AUGER ELECTRON SPECTROSCOPY.
CAHILL J	5-1250	AUGER AND ELLIPSOMETRIC STUDY OF PHOSPHORUS SEGREGATION IN OXIDIZED DEGENERATE S
CAIN EFC	4-0947	CHEMICAL ANALYSIS OF ELECTRODEPOSITED NICKEL- NICKEL BONDS BY AES.
CALLAN EJ	2-0303	SENSITIVITY CF L1,L2,3M4,5 COSTER-KRONIG TRANSITION RATES TO VARIATIONS IN ENERG
	2-0304	KLL AUGER TRANSITION PROBABILITIES.
CAMPBELL BD	5-1217	AUGER ELECTRON STUDIES OF SURFACES: URANIUM DIOXIDE, URANIUM, GRAPHITE, 300 SERI
	5-1218	AUGER ELECTRON STUDIES OF URANIUM DIOXIDE SURFACES.
	5-1316	SECONDARY ELECTRON ENERGY DISTRIBUTION STUDIES OF URANIUM DIOXIDE SURFACES.
CANALI C	5-1577	ANTIMONY DOPING OF SILICON LAYERS GROWN BY SOLID PHASE EPITAXY. (AES MEASUREMENT
CARBONI G	2-0305	MEASUREMENT OF THE AUGER EFFECT IN THE MU-4 HELIUM(2S)(+) IONIC SYSTEM.
CARD HC	5-1219	IN-DEPTH AUGER ANALYSIS OF ALUMINUM- SILICON INTERFACIAL REACTIONS.
CARDEN JL	5-1220	OPTICAL PROPERTIES AND AUGER CHARACTERISTICS OF MAGNESIUM SINGLE CRYSTALS AND MA
	5-1221	ADSORPTION CF KRYPTON ON THE (111) FACE OF COPPER SINGLE CRYSTALS. (AUGER ELECTR
	5-1386	AUGER ANALYSIS OF SURFACE FILMS ON SILVER- TIN.
CARDONA M	2-0306	PHOTOELECTRIC PROPERTIES OF MAGNESIUM(2) SILICIDE, MAGNESIUM(2) GERMANIUM, AND M
	5-1885	VALENCE BAND OF MAGNESIUM STANNIDE DETERMINED BY AUGER AND PHOTOEMISSION SPECTRO
CARLEY AF	5-1204	ELECTRON SPECTROSCOPY AND ADSORPTION STUDIES. CHEMICAL SHIFTS IN X-RAY PHOTOELEC
CARLSON DE	5-1222	INJECTION OF IONS INTO GLASS FROM A GLOW DISCHARGE. (AUGER SPECTROSCOPY)
	5-1365	DETERMINATION OF THE COMPOSITION OF GLASS SURFACES BY AUGER SPECTROSCOPY.
CARLSON TA	1-0025	PRESENT AND FUTURE APPLICATIONS OF AUGER SPECTROSCOPY.
	1-0026	PHOTOELECTRON AND AUGER SPECTROSCOPY.
	2-0307	NATURE OF SECONDARY ELECTRONS CREATED AS THE RESULT OF ELECTRON SHAKE-OFF AND (S
	2-0308	RELATIVE ABUNDANCES AND RECOIL ENERGIES OF FRAGMENT IONS FORMED FROM THE X-RAY P
	3-0600	HIGH RESOLUTION ELECTRON SPECTROMETER FOR PHOTOELECTRON AND AUGER ELECTRON STUDI
	4-0791	IDENTIFICATICN OF HIGH ENERGY LINES IN THE KLL AUGER SPECTRUM OF NITROGEN.
	4-0920	MANIFESTATION OF ATOMIC DYNAMICS THROUGH THE AUGER EFFECT. (NEON AUGER SPECTRUM)
	4-0990	DETERMINATION OF KLL AUGER SPECTRA OF NITROGEN, OXYGEN, CARBON DIOXIDE, NITRIC O
	4-0991	AUGER SPECTRA OF SIMPLE MOLECULES.
CARRIERE B	4-0792	STUDY OF MUSCOVITE AND SILICON BY AUGER ELECTRON SPECTROSCOPY.
	4-0793	AUGER ELECTRCN SPECTROSCOPY STUDY OF SOME OXYGENATED SILICON COMPOUNDS.
	4-0794	AUGER SPECTROSCOPY OF SILICON IN SILICATES.
	4-0795	AUGER ELECTRON SPECTROSCOPY OF INSULATING SILICON COMPOUNDS.
	4-0796	STUDY OF AMCRPHOUS AND CRYSTALLINE SILICATES BY AUGER ELECTRON SPECTROSCOPY.
	5-1223	STUDY OF THE INTERACTION OF OXYGEN WITH PLATINUM SINGLE CRYSTAL SURFACES BY LOW
CARVER JC	1-0082	ELECTRON SPECTROSCOPY. X-RAY AND ELECTRON EXCITATION.
CASSUTO A	4-0820	AUGER SPECTRA OF CARBON. TANTALUM CARBIDES, ETHYLENE AND CARBON MONOXIDE ADSORPT
	4-1111	AUGER ELECTRON SPECTROSCOPY OF HIGH TEMPERATURE OXYGEN- RHENIUM INTERACTIONS.
	5-1688	ADSORPTION AND DECOMPOSITION OF ACETYLENE ON POLYCRYSTALLINE RHENIUM FILMS AND R
	5-1955	APPLICATION OF AUGER ELECTRON SPECTROSCOPY TO THE PLATINUM- OXYGEN SYSTEM AT HIG
CASTLE JE	4-0797	CHEMICAL SHIFTS IN PHOTOEXCITED AUGER SPECTRA.
	5-1224	OXIDATION OF CUPRONICKEL ALLOYS. PART-1: XPS STUDY OF INTERDIFFUSION. (AND AES)
CAVALERU A	5-1225	AUGER ELECTRON SPECTROSCOPY ANALYSIS OF ION PLATED GOLD FILMS ON SILICON.
CAWTHRON ER	2-0309	SECCNDARY ELECTRON EMISSION FROM SOLID SURFACES BOMBARDED BY MEDIUM ENERGY IONS
CAZAUX J	1-0027	X-RAY PHOTOELECTRON (AND AUGER ELECTRON) MICROANALYSIS AND MICROSCOPY. PRINCIPLE
CEASAR GP	5-1947	UPS, ESCA, AND AUGER SPECTROSCOPY STUDY OF VACUUM DEPOSITED SILVER HALIDE FILMS.
CEDERBAUM SL	2-0291	PLASMON COUPLING TO CORE HOLE EXCITATIONS IN CARBON. (AUGER APPEARANCE SPECTROSC
CEVA T	4-0798	ULTRAVIOLET EMISSIONS IN CALCITE. (AUGER EFFECT)
CHAIKOVSKII EF	4-1124	PLASMA SATELLITES IN THE AUGER ELECTRON SPECTRUM OF MOLYBDENUM.
	5-1226	DIFFUSION OF SULFUR AND PHOSPHORUS IMPURITY ATOMS TO A TUNGSTEN SURFACE.
	5-1759	DECREASE OF CARBON CONCENTRATION ON THE SURFACE OF MOLYBDENUM CARBIDE DURING HEA
CHAMBERLAIN MB	5-1834	CHANGES IN THE TITANIUM DIOXIDE AUGER AND APPEARANCE POTENTIAL SPECTRA INDUCED B
CHAMBERS A	2-0399	MODEL FOR THE AUGER ELECTRON SPECTROSCOPY OF SYSTEMS EXHIBITING LAYER GROWTH, AN
	3-0637	A CRYSTAL CLEAVAGE DEVICE FOR USE IN LOW ENERGY ELECTRON DIFFRACTION- AUGER STUD
	4-0895	LOW ENERGY AUGER AND LOSS ELECTRON SPECTRA FROM MAGNESIUM AND ITS OXIDE.
	5-1227	GROWTH AND STRUCTURE OF THIN COPPER FILMS ON (001) SURFACES OF NICKEL.
	5-1495	COMPARATIVE STUDY OF SINGLE CRYSTAL MAGNESIUM OXIDE AND OXIDIZED MAGNESIUM BY AU
	5-1496	STUDY OF THE EPITAXIAL GROWTH OF MAGNESIUM ON MAGNESIUM OXIDE (001) USING REFLEC
	5-1497	INTERFACIAL AUGER TRANSITIONS IN OXIDIZED SODIUM AND MAGNESIUM.
CHAMBERS WJ	1-0028	ELECTRON PRCBE MICROANALYSIS AND ELECTRON SPECTROSCOPY.
CHANG CA	5-1228	AUGER ANALYSIS OF SILICON THIN FILMS DEPOSITED ON CARBON AT HIGH TEMPERATURES.
CHANG CC	1-0029	AUGER ELECTRON SPECTROSCOPY FOR CHEMICAL ANALYSIS.
	1-0030	AUGER ELECTRCN SPECTROSCOPY. (REVIEW)
	1-0031	ANALYTICAL AUGER ELECTRON SPECTROSCOPY.
	1-0032	GENERAL FORMALISM FOR QUANTITATIVE AUGER ANALYSIS.
	5-1229	CONTAMINANTS ON CHEMICALLY ETCHED SILICON SURFACES. LEED- AUGER METHOD.
	5-1230	SILICON ON SAPPHIRE EPITAXY BY VACUUM SUBLIMATION: LEED- AUGER STUDIES AND ELECT
	5-1231	PHOSPHORUS CONCENTRATION PROFILES IN P-DOPED SILICON DIOXIDE MEASURED USING AUGE
	5-1232	INTERDIFFUSICNS IN THIN FILM GOLD ON PLATINUM ON GALLIUM ARSENIDE (100) STUDIED
	5-1233	STRUCTURE ANC CHEMISTRY OF SILICON SURFACES AFTER PRE- AND BACKSPUTTERING, STUDI
	5-1234	SILICON DIFFUSION THROUGH THIN TUNGSTEN FILMS ON SILICON, STUDIED WITH AUGER SPE
	5-1235	DIFFUSION KINETICS OF GOLD THROUGH PLATINUM FILMS ABOUT 2000 AND 600 ANGSTROMS TH
CHANG LL	5-1596	GALLIUM(1-X) ALUMINUM(X) ARSENIDE SUPERLATTICES PROFILED BY AUGER ELECTRON SPECT
CHANG TN	2-0310	DOUBLE PHOTOIONIZATION OF NEON. (AUGER TRANSITIONS)
CHAPPELL RA	5-1236	AUGER ELECTRCN SPECTROSCOPY STUDY OF FLOAT GLASS SURFACES.
CHARIG JM	5-1237	CARBON CONTAMINATION OF SILICON (111) SURFACES.
	5-1238	AUGER ELECTRON SPECTROSCOPY OF NICKEL DEPOSITS ON THE SILICON (111) SURFACE.

CHARLTON DE	4-0799	FLUORESCENCE PHOTON AND AUGER ELECTRON SPECTRA.
CHASE RL	2-0311	APPLICATION OF MANY BODY PERTURBATION THEORY TO THE CALCULATION OF AUGER RATES,
	2-0312	CORRELATION ENERGIES AND AUGER RATES IN ATOMS WITH INNER SHELL VACANCIES (NEON W
CHATTARJI D	1-0033	THEORY OF AUGER TRANSITIONS.
	4-0800	EFFECT OF ELECTRONIC CORRELATIONS ON KLL AUGER LINES IN ELEMENTS MAGNESIUM TO SC
	4-0801	EFFECT OF ATOMIC CORRELATIONS ON KLL AUGER TRANSITIONS.
	4-0802	RELATIVISTIC STUDY OF KLL AUGER TRANSITION PROBABILITIES.
CHAUKA NG	1-0063	MODERN TECHNIQUES FOR SURFACE STUDIES.
	1-0064	TECHNOLOGICAL APPLICATION OF AUGER ELECTRON SPECTROSCOPY.
CHEBOTINKOV VA	4-0725	KLL AUGER ELECTRON SPECTRUM OF SAMARIUM.
CHEN ACM	5-1239	SECONDARY ELECTRON EMISSION OF AMORPHOUS SEMICONDUCTOR THIN FILMS.
CHEN JM	2-0348	EFFECTS OF LOW ENERGY ARGON- ION BOMBARDMENT ON MOLYBDENUM DISULFIDE. (AUGER MEA
	5-1240	LEED, AUGER, AND WORK FUNCTION STUDIES OF CLEAN AND SODIUM- COVERED SURFACES OF
	5-1241	DITUNGSTEN CARBIDE OVERLAYER ON TUNGSTEN (112).
CHEN MH	2-0313	THEORETICAL L2 SUBSHELL FLUORESCENCE YIELDS AND COSTER-KRONIG TRANSITION PROBABI
	2-0314	MULTIPLET EFFECT IN FLUORESCENCE YIELDS OF MULTIPLY IONIZED ATOMS. (AUGER EFFECT
	2-0315	AUGER AND RADIATIVE DEEXCITATION OF MULTIPLY IONIZED NEON.
	2-0316	THEORETICAL L2 AND L3 SUBSHELL FLUORESCENCE YIELDS AND L2-L3(X) COSTER-KRONIG TR
	2-0329	AUGER AND COSTER-KRONIG TRANSITION PROBABILITIES TO THE ATOMIC 2S STATE AND THEO
	2-0441	SEQUENTIAL STRIPPING IN HIGHLY IONIZED NEON. (AUGER TRANSITIONS)
	2-0544	ATOMIC VACANCY DISTRIBUTIONS PRODUCED BY INNER SHELL IONIZATION (AUGER AND RADIA
	3-0718	WIDTHS OF ATOMIC M-SHELL VACANCY STATES AND QUASIATOMIC ASPECTS OF RADIATIONLESS
	4-0803	KLL AUGER TRANSITION PROBABILITIES FOR ELEMENTS WITH LOW AND INTERMEDIATE ATOMIC
	4-0918	ATOMIC RADIATION TRANSITION PROBABILITIES TO THE 1S STATE AND THEORETICAL K-SHEL
	4-1121	QUASIATOMIC LMM AUGER SPECTRA OF SOLID COPPER AND ZINC.
CHEREPIN VT	5-1215	EFFECT OF ION BOMBARDMENT ON THE COMPOSITION AND STRUCTURE OF TUNGSTEN AND MOLYB
	5-1939	AUGER SPECTROSCOPIC STUDY OF THE SURFACE OF IRON AFTER IONIC BOMBARDMENT AND AIR
CHEREVATSKII NYA	1-0132	ELECTRON AUGER SPECTROSCOPY: A METHOD OF INVESTIGATING SURFACE PHENOMENA (REVIEW
	2-0449	EFFECT OF THE ELECTRON EJECTION DEPTH ON QUANTITATIVE ANALYSIS BY AUGER SPECTROS
	3-0658	AUGER ELECTRON SPECTROMETER INTEGRATED WITH A LOW ENERGY ELECTRON DIFFRACTION SY
	5-1242	ELECTRON AUGER SPECTROSCOPIC STUDY OF THE SURFACE OF CARBIDE CATHODES.
	5-1561	ELECTRON AUGER SPECTROSCOPIC STUDY OF THE SURFACE OF CARBIDE CATHODES.
	5-1562	SURFACE OF THERMIONIC EMITTERS MADE OF CARBIDE COMPOUNDS STUDIED BY AN ELECTRONI
	5-1640	(111) SURFACE OF GALLIUM PHOSPHIDE STUDIED BY AUGER SPECTROSCOPY AND DIFFRACTION
	5-1642	AUGER ELECTRON SPECTRUM DURING THE INTERACTION BETWEEN OXYGEN AND THE (111) SURF
	5-1643	SOME FEATURES OF THE AUGER ELECTRON SPECTRUM IN THE INTERACTION BETWEEN OXYGEN A
	5-1644	INVESTIGATION OF CESIUM ADSORPTION ONTO A GALLIUM ARSENIDE (110) SURFACE. (AUGER
CHESTERS MA	5-1243	CHARACTERIZATION OF ADSORBED SPECIES USING AUGER PEAK SHAPES. ETHYLENE AND CARBO
	5-1244	ELECTRON BEAM EFFECTS ON CARBON MONOXIDE SATURATED TUNGSTEN (100), ETHYLENE SATU
	5-1245	CHEMISORPTION ON TANTALUM (100). (AUGER SPECTROSCOPY)
	5-1246	LEED AND SURFACE POTENTIAL STUDY OF CARBON MONOXIDE AND XENON ADSORBED ON COPPER
CHEVALLIER P	2-0293	NEW SATELLITE BANDS CORRESPONDING TO DOUBLE L IONIZATION. (OF AUGER ELECTRONS)
CHIARENA JC	5-1355	AUGER SPECTROMETRIC AND MASS SPECTROMETRIC STUDY OF THE CHEMISORPTION OF CARBON
CHO AY	5-1247	MAGNESIUM DOPED GALLIUM ARSENIDE AND ALUMINUM(X) GALLIUM(1-X) ARSENIDE BY MOLECU
CHOU NJ	5-1248	AUGER AND ELLIPSOMETRIC STUDIES OF ULTRATHIN LEAD(II) OXIDE GROWTH ON LEAD.
	5-1249	AUGER ANALYSIS OF THIN OXIDE FILMS ON LEAD- INDIUM ALLOYS.
	5-1250	AUGER AND ELLIPSOMETRIC STUDY OF PHOSPHORUS SEGREGATION IN OXIDIZED DEGENERATE S
	5-1251	AUGER ANALYSIS OF CHLORINE IN HYDROGEN CHLORIDE OR CHLORINE GROWN SILICON DIOXID
CHRISTIAN H	1-0034	APPLICATION OF ION, ELECTRON, AND X-RAY BEAMS IN MODERN MATERIAL TESTING.
CHRISTMAN SB	4-1036	HIGH ENERGY FINE STRUCTURE IN THE AUGER SPECTRA OF SILICON AND SILICON CARBIDE.
	4-1037	ABSENCE OF PLASMON GAIN SATELLITES IN THE AUGER SPECTRUM OF SILICON.
	4-1038	COMPARISON OF SILICON L3,2 ELECTRON LOSS SPECTRA AND OPTICAL ABSORPTION. (AUGER
CHRISTMANN K	5-1252	INTERACTIONS OF CARBON MONOXIDE AND OXYGEN WITH IRIDIUM (110) SURFACES. (AUGER S
	5-1253	SURFACE STUDIES WITH EPITAXIALLY GROWN METAL FILMS. (AUGER EFFECT)
	5-1254	ADSORPTION OF CARBON MONOXIDE ON SILVER- PALLADIUM ALLOYS.
	5-1255	ADSORPTION OF HYDROGEN ON A PLATINUM (111) SURFACE.
	5-1256	ADSORPTION OF COBALT ON A NICKEL (111) SURFACE. (AUGER EFFECT, AUGER SPECTROSCOP
	5-1257	ADSORPTION OF HYDROGEN ON NICKEL SINGLE CRYSTAL SURFACES. (AUGER ELECTRON SPECTR
CHRISTOU A	5-1258	AUGER SPECTROSCOPY OF SOLID SURFACES IN A DRY PUMPED HIGH RESOLUTION SCANNING EL
	5-1259	SCANNING MICROSPOT AUGER SPECTROSCOPY STUDY OF INTERDIFFUSION AND EUTECTIC FORMA
	5-1260	SILICIDE FORMATION AND INTERDIFFUSION EFFECTS IN SILICON- TANTALUM, SILICON DIOX
	5-1261	SCANNING ELECTRON MICROSCOPY, AUGER SPECTROSCOPY, AND ION BACKSCATTERING TECHNIQ
	5-1262	COMBINED SCANNING ELECTRON MICROSCOPY- AUGER SPECTROSCOPY FOR MICROSPOT SURFACE
	5-1287	SOLID PHASE EPITAXIAL STUDIES USING VACUUM DEPOSITION ON HEATED SILICON SUBSTRAT
	5-1961	HIGH SPATIAL RESOLUTION SCANNING AUGER SPECTROSCOPY APPLIED TO ANALYSIS OF X-BAN
CHU WK	5-1917	STRUCTURE AND GROWTH KINETICS OF NICKEL(2) SILICIDE ON SILICON. (AUGER SPECTROSC
CHUNG MF	4-0804	LLM VERSUS LMM AUGER TRANSITIONS FOR THE LIGHT ELEMENTS.
	4-0805	AUGER ELECTRON SPECTROSCOPY OF THE OUTER SHELL ELECTRONS.
	4-0896	AUGER AND OTHER CHARACTERISTIC ENERGIES IN SECONDARY ELECTRON SPECTRA FROM ALUMI
	4-0897	ENERGY SPECTRUM OF BACK SCATTERED ELECTRONS AND CHARACTERISTIC LOSS AND GAIN PHE
	4-0898	AUGER SATELLITE AND OTHER CHARACTERISTIC EVENTS IN MAGNESIUM SECONDARY ELECTRON
	4-0901	CHARACTERISTIC ENERGY GAIN AND LOSS, DOUBLE IONIZATION, AND IONIZATION LOSS EVEN
	5-1263	CHARACTERISTIC ENERGIES IN SECONDARY ELECTRON SPECTRA FROM SILICON (111) SURFACE
	5-1499	LEED AND AUGER INVESTIGATIONS OF COPPER (111) SURFACE.
CILLIE GG	4-0945	CHARACTERISTIC K-SHELL LOW ENERGY ELECTRON LOSS SPECTRA FROM DIAMOND, GRAPHITE,
CINTI RC	5-1683	PREPARATION AND SURFACE OXIDATION OF A (110) FACE OF VANADIUM.
	6-1992	PREPARATION AND SURFACE OXIDATION OF A VANADIUM (100) PLANE. (AUGER EFFECT)
CIOMARTAN D	1-0035	ELECTRON SPECTROMETRY. (REVIEW)
CITRIN PH	2-0317	INTERATOMIC AUGER PROCESS: EFFECTS ON LIFETIMES OF CORE HOLE STATES.
	2-0318	INTERATOMIC AUGER TRANSITIONS. SURVEY. (36 REFS)
CLARK AE	5-1264	AUGER SPECTROSCOPIC ANALYSIS OF BIOGLASS CORROSION FILMS.
CLARK DE	5-1265	AQUEOUS CORROSION OF SODA- SILICA AND SODA- LIME- SILICA GLASS. (AES)
	5-1266	AUGER ANALYSIS OF BRASS- ENAMEL AND STAINLESS STEEL- ENAMEL INTERFACES.
CLARKE TA	2-0319	AUGER ELECTRON EMISSION SPECTRA: A SIMPLE TREATMENT OF ENERGIES AND INTENSITIES
	5-1267	AUGER SPECTROSCOPY AND LOW ENERGY ELECTRON DIFFRACTION STUDIES OF THE CHEMISORPT
CLAUSING RE	1-0036	APPLICATION OF AUGER ELECTRON SPECTROSCOPY TO REACTOR MATERIALS RESEARCH.
	1-0040	CATALOG OF CALCULATED AUGER TRANSITIONS FOR THE ELEMENTS. (REVIEW)
	1-0041	AUGER CATALOG CALCULATED TRANSITION ENERGIES LISTED BY ENERGY AND ELEMENT.
	2-0581	AUGER TRANSITIONS INITIATED FROM SINGLY AND DOUBLY IONIZED STATES IN LITHIUM MET
	4-0806	AUGER SPECTRA OF LITHIUM METAL AND LITHIUM OXIDE.

CLAUSING RE	4-0807	LEED- AES ELECTRON INDUCED EMISSION STUDIES OF LITHIUM HYDRIDE.
	4-1025	AUGER SPECTRA OF LITHIUM HYDRIDE.
	5-1446	LEED AND AES STUDIES OF LITHIUM HYDRIDE (100) SURFACE.
	5-1750	SODIUM SEGREGATION ONTO A LITHIUM METAL SURFACE. (AUGER SPECTRA)
	6-2001	AUGER ELECTRON SPECTROSCOPY OF FRACTURE SURFACES IN IRRADIATED TYPE 304 STAINLES
	6-2002	AUGER SPECTRCSCOPY CF FRACTURE SURFACES OF IRRADIATED STAINLESS STEEL.
CLAXTON KT	6-2014	MASS TRANSPORT OF STAINLESS STEEL CORROSION PRODUCTS IN FLOWING LIQUID SODIUM. (
CLAY FA	1-0037	CHARACTERIZATION OF A CHROMIUM- GOLD THIN FILM DEPOSITION PROCESS BY AUGER ANALY
CLEFF B	2-0320	ANGULAR DISTRIBUTION OF AUGER ELECTRONS.
	2-0321	ANGULAR DISTRIBUTION OF AUGER ELECTRONS FOLLOWING IMPACT IONIZATION.
	2-0322	ANGULAR DISTRIBUTION OF L3M2,3M2,3(1SO) AUGER ELECTRONS OF ARGON.
	4-0808	KLL AUGER SPECTRUM OF CHLORINE. (IN CARBON TETRACHLORIDE)
	4-0880	K-AUGER TRANSITIONS OF THE FREE SODIUM ATOM.
	4-1071	ENERGIES OF EXCITED STATES OF DOUBLY IONIZED MOLECULES BY MEANS OF AUGER ELECTRO
CLOUGH SP	6-2015	STUDY OF SULFUR EMBRITTLED OXYGEN FREE COPPER. (AES, GRAIN BOUNDARY SEGREGATION)
	6-2095	STRESS CORROSION CRACKING OF ALPHA BRASS IN A TARNISHING AMMONIACAL ENVIRONMENT:
COAD JP	3-0594	DESIGN FOR A CYLINDRICAL MIRROR ANALYZER FOR USE IN AUGER SPECTROSCOPY.
	3-0601	BEAM EFFECTS IN AUGER ELECTRON SPECTROSCOPY REVEALED BY X-RAY PHOTOELECTRON SPEC
	4-0809	CHEMICAL SHIFTS IN THE AUGER SPECTRA FROM OXIDIZED CHROMIUM AND VANADIUM.
	4-0810	LMM AUGER SPECTRA OF SOME TRANSITION METALS OF THE FIRST SERIES.
	4-0811	LLM VERSUS LMM AUGER TRANSITIONS FOR LIGHT ELEMENTS.
	4-0812	UNEXPLAINED PHENOMENA IN THE AUGER SPECTRA OF THE FIRST TRANSITION SERIES. (COMM
	4-0813	COMBINATION OF AUGER SPECTROSCOPY AND CHARACTERISTIC LOSS SPECTROSCOPY FOR THE E
	5-1268	X-RAY PHOTOELECTRON AND AUGER SPECTROSCOPIC ANALYSES OF STEEL SURFACES.
	5-1269	ELECTRON BEAM ASSISTED ADSORPTION ON SILICON (111) SURFACE.
	5-1270	AUGER SPECTROSCOPY OF CARBON ON NICKEL.
	5-1271	ORIGIN OF FINE STRUCTURE IN THE AUGER SPECTRUM OF SULFUR ON A NICKEL SURFACE.
	5-1272	SURFACE ANALYSES BY COMBINED X-RAY PHOTOELECTRON SPECTROSCOPY AND AUGER ELECTRON
	5-1353	A PRELIMINAFY STUDY OF PURE METAL SURFACES USING AUGER ELECTRON SPECTROSCOPY X-R
	6-2016	ANALYSIS PRCFILES OF OXIDE FILMS ON CHROMIUM STEEL BY AUGER EMISSION AND X-RAY P
	6-2134	GRAIN BCUNDARY SEGREGATION IN ALUMINUM OXIDE. (AUGER SPECTROSCOPY)
COBURN JW	1-0038	SURFACE ANALYSIS TODAY. (REVIEW OF METHODS, INCLUDING AES)
	5-1128	ELECTRON BEAM EFFECTS IN DEPTH PROFILING MEASUREMENTS WITH AUGER ELECTRON SPECTR
	5-1273	TECHNIQUES FOR ELEMENTAL COMPOSITION PROFILING IN THIN FILMS. (AUGER SPECTROSCOP
COCKE CL	2-0323	ARGON K-AUGER CROSS SECTIONS IN COLLISION WITH 30 MEV FLUORINE IONS.
	2-0571	NEON K-AUGER CROSS SECTIONS FROM 1.5 MEV PER AMU AND ATOMIC HYDROGEN, OXYGEN, AN
COGHLAN WA	1-0039	COMPUTER PROGRAM TO CALCULATE AUGER TRANSITIONS FOR A SURFACE CONTAINING SEVERAL
	1-0040	CATALOG OF CALCULATED AUGER TRANSITIONS FOR THE ELEMENTS. (REVIEW)
	1-0041	AUGER CATALOG CALCULATED TRANSITION ENERGIES LISTED BY ENERGY AND ELEMENT.
COHEN SA	6-2017	SPUTTERING OF IONS FROM STAINLESS STEEL BY HYDROGEN BOMBARDMENT. (AES)
COLDREN AP	6-2018	IDENTIFICATION OF ATOMIC BORON IN STEEL BY AUGER ELECTRON SPECTROSCOPY.
COLLET J	2-0260	EVIDENCES FOR AUGER RECOMBINATION IN ELECTRON HOLE DROPLETS. (SILICON)
COLLIER JG	6-2014	MASS TRANSPCRT OF STAINLESS STEEL CORROSION PRODUCTS IN FLOWING LIQUID SODIUM. (
COLLINS DA	2-0567	MINORITY CARRIER LIFETIME IN INDIUM ARSENIDE EPILAYERS. (AUGER RECOMBINATION PRO
COLLINS DM	1-0042	USE OF PHOTOEMISSION, AUGER, AND THERMAL DESORPTION TECHNIQUES TO STUDY THE MECH
COLLINS RL	2-0576	BACKSCATTER MOSSBAUER STUDIES USING CONVERSION ELECTRONS. TIN-119.
COLMENARES CA	5-1274	STUDY OF MULBERRY SURFACES BY AUGER AND IEE SPECTROSCOPY.
	5-1883	THORIUM (100) CRYSTAL FACE. A STUDY OF ITS CLEANING, SURFACE STRUCTURE, AND INTE
COLOMBIE N	2-0268	INTERPRETATION OF ENERGY LOSS BY AUGER ELECTRONS EMITTED DURING ION BOMBARDMENT.
	2-0550	EMISSION OF AUGER ELECTRONS BY ION BOMBARDMENT.
	2-0551	OBSERVATION OF AUGER ELECTRONS IN THE ENERGY SPECTRUM OF SECONDARY ELECTRONS EMI
	4-0942	SECONDARY ELECTRON SPECTRA OF SOME SOLID TARGETS UNDER IONIC BOMBARDMENT.
	4-0988	ENERGY SPECTRUM AND ANGULAR DISTRIBUTION OF SECONDARY ELECTRONS EMITTED BY IONIC
COLVIN AD	3-0695	A MULTICHANNEL MONITOR FOR REPETITIVE AUGER ELECTRON SPECTROSCOPY WITH APPLICATI
COMAS J	1-0216	DIRECT COMPARISON OF AUGER, SECONDARY ION MASS SPECTROSCOPY, AND PROTON RESONANC
COMRIE CM	5-1567	ROLE OF PRIMARY AND SECONDARY ELECTRONS IN ELECTRON INDUCED DESORPTION AND DISSO
	5-1569	LEED- AUGER INVESTIGATION OF A STABLE CARBIDE OVERLAYER ON A PLATINUM (111) SURF
CONLEY DK	5-1275	APPLICATION OF AUGER ELECTRON SPECTROSCOPY TO SURFACE CHEMISTRY STUDIES OF CERAM
CONNELL GL	1-0043	AUGER ELECTRON SPECTROSCOPY.
	5-1276	AUGER ELECTRON SPECTROSCOPY OF GRAPHITE FIBER SURFACES.
	5-1277	AUGER ELECTRON SPECTROSCOPY OF LUNAR MATERIAL.
CONNOR JA	4-0814	EXPERIMENTAL AND THEORETICAL STUDY OF THE AUGER SPECTRUM OF NITROUS OXIDE.
CONRAD H	5-1278	INTERACTION OF NITRIC OXIDE WITH A NICKEL (111) SURFACE.
CONRADT R	1-0044	AUGER RECOMBINATION IN SEMICONDUCTORS. (REVIEW, 26 REFS)
	2-0266	AUGER RECOMBINATION IN SILICON.
	2-0275	TWO ELECTRON BAND-TO-BAND TRANSITIONS IN SOLIDS. (AUGER RECOMBINATION IN SILICON
	4-0778	AUGER RECOMBINATION IN GALLIUM ANTIMONIDE.
CONSTANTINESCU C	2-0324	AUGER EFFECT AND RADIATIVE RECOMBINATION IN N-GALLIUM ARSENIDE- P-ALUMINUM GALLI
CONSTANTINESCU L	2-0325	EMISSION SPECTRA OF SILICON COMPENSATED ALUMINUM(X) GALLIUM(1-X) ARSENIDE ELECTR
COOK CF	5-1942	SURFACE STUDIES FOR QUARTZ RESONATORS. (CLEANING, AUGER SPECTROSCOPY)
	5-1943	SURFACE PREPARATION AND CHARACTERIZATION TECHNIQUES (AUGER, CLEANING) FOR QUARTZ
COOPER JW	2-0326	"SEMI-AUGER" PROCESSES IN L2,3 EMISSION IN ARGON AND POTASSIUM CHLORIDE.
CORVIN I	3-0596	CONTAMINATION IN THE SCANNING ELECTRON MICROSCOPE. (AUGER ELECTRON SPECTROSCOPY,
COSTER D	1-0045	A NEW TYPE OF AUGER EFFECT AND ITS INFLUENCE ON THE X-RAY SPECTRUM.
COTAARAIZA LS	5-1279	STUDY OF SURFACE DISSOCIATION IN POTASSIUM CHLORIDE AND LITHIUM FLUORIDE BY OBSE
COTHERN CR	1-0134	CHARACTERIZATION OF SOLID SURFACES. (REVIEW, AUGER SPECTROSCOPY, ESCA)
COTOMBIE N	2-0269	ENERGY DISTRIBUTION OF SECONDARY ELECTRONS EMITTED BY INDIUM PHOSPHIDE AND GALLI
COULOMB JP	5-1856	AUGER ELECTRON SPECTROSCOPY AND LOW ENERGY ELECTRON DIFFRACTION STUDIES OF ADSOR
	5-1857	FIRST ORDER TWO DIMENSIONAL TRANSITION: XENON ADSORBED ON THE (0001) FACE OF GRA
	5-1858	TWO-DIMENSICNAL PHASE TRANSITION IN XENON SUBMONOLAYER FILMS ADSORBED ON (0001)
	5-1859	THERMODYNAMICS AND KINETICS OF THE FIRST MONOLAYER ADSORPTION OF XENON ON THE (0
COULSON CA	2-0327	INNER SHELL DOUBLE IONIZATION AND CHEMICAL BONDING. PART-1: AUGER ELECTRONS. (KL
COUNTS RL	5-1280	USE OF SCANNING AUGER SPECTROSCOPY FOR SURFACE ANALYSIS.
COX SM	5-1801	CHEMISORPTION, PHOTODESORPTION AND CONDUCTIVITY MEASUREMENTS ON ZINC OXIDE SURFA
CRASEMANN B	2-0259	X-RAY FLUORESCENCE YIELDS, AUGER, AND COSTER-KRONIG TRANSITION PROBABILITIES.
	2-0313	THEORETICAL L2 SUBSHELL FLUORESCENCE YIELDS AND COSTER-KRONIG TRANSITION PROBABI
	2-0314	MULTIPLET EFFECT IN FLUORESCENCE YIELDS OF MULTIPLY IONIZED ATOMS. (AUGER EFFECT
	2-0315	AUGER AND RADIATIVE DEEXCITATION OF MULTIPLY IONIZED NEON.
	2-0316	THEORETICAL L2 AND L3 SUBSHELL FLUORESCENCE YIELDS AND L2-L3(X) COSTER-KRONIG TR
	2-0328	AUGER AND RADIATIVE TRANSITION PROBABILITIES.

CRASEMANN B	2-0329	AUGER AND CCSTER-KRONIG TRANSITION PROBABILITIES TO THE ATOMIC 2S STATE AND THEO
	2-0441	SEQUENTIAL STRIPPING IN HIGHLY IONIZED NEON. (AUGER TRANSITIONS)
	2-0544	ATOMIC VACANCY DISTRIBUTIONS PRODUCED BY INNER SHELL IONIZATION (AUGER AND RADIA
	3-0718	WIDTHS CF ATCMIC M-SHELL VACANCY STATES AND QUASIATOMIC ASPECTS OF RADIATIONLESS
	4-0803	KLL AUGER TRANSITION PROBABILITIES FOR ELEMENTS WITH LOW AND INTERMEDIATE ATOMIC
	4-0918	ATOMIC RADIATION TRANSITION PROBABILITIES TO THE 1S STATE AND THEORETICAL K-SHEL
	4-0939	WIDTH OF ATOMIC L2 AND L3 VACANCY STATES NEAR Z = 30. (AUGER EFFECT)
	4-0956	MULTIPLET EFFECTS ON THE L2,3 FLUORESCENCE YIELD OF MULTIPLY IONIZED ARGON. (AUG
	4-0957	STRIPPING MECHANISM DEPENDENCE OF THE ARGON 2S FLUORESCENCE YIELD. (AUGER TRANSI
	4-1121	QUASIATOMIC LMM AUGER SPECTRA OF SOLID COPPER AND ZINC.
CREAGH LT	5-1281	MECHANISM OF SURFACE ALIGNMENT IN NEMATIC LIQUID CRYSTALS. (SUBSTRATE SURFACE EF
CROMBEEN JE	5-1930	LOW ENERGY ELECTRON DIFFRACTION AND AUGER ELECTRON SPECTROSCOPY STUDY OF PHOSPHI
	5-1931	LOW ENERGY ELECTRON DIFFRACTION, AUGER ELECTRON SPECTROSCOPY, AND PHOTOEMISSION
	5-1932	LOW ENERGY ELECTRON DIFFRACTION AND AUGER ELECTRON OBSERVATIONS OF THE SILICON C
CROSS JA	3-0681	GAIN FATIGUE MECHANISM IN CHANNEL ELECTRON MULTIPLIERS. (SURFACE OXYGEN ON LEAD
CROTTY JM	4-0815	SEMIEMPIRICAL AUGER ELECTRON ENERGIES. PART-2: THE ARGON ATOM.
CRUSET A	2-0330	EFFICIENCY OF THE AUGER CASCADE IN SOLID STATE.
CUNNINGHAM JG	6-2016	ANALYSIS PROFILES OF OXIDE FILMS ON CHROMIUM STEEL BY AUGER EMISSION AND X-RAY P
CURNUTTE B	2-0323	ARGON K-AUGER CROSS SECTIONS IN COLLISION WITH 30 MEV FLUORINE IONS.
DABIRI AE	5-1197	MEASUREMENTS OF THE SPATIAL DISTRIBUTION OF HYDROGEN DESORBED FROM NICKEL SURFAC
	5-1282	AUGER ELECTRON SPECTROSCOPY MEASUREMENTS OF ADSORPTION ISOBARS FOR OXYGEN ON TUN
DACUNHABELO M	6-2102	AUGER SPECTROMETRY STUDY OF PASSIVE FILMS FORMED ON STAINLESS STEELS.
DADAYAN KA	4-0816	CHEMICAL SHIFT IN AUGER SPECTRA OF TUNGSTEN AND MOLYBDENUM DURING THE ADSORPTION
	5-1310	AUGER SPECTRA FROM THE (100) SURFACE OF SILICON DURING OXIDATION AND NITRIDATION
DAHL P	2-0331	ENERGY DISTRIBUTION OF (AUGER) ELECTRONS RESULTING FROM ION- ATOM COLLISIONS WIT
	2-0346	DYNAMIC CHARGE STATE DEPENDENCE OF THE K-VACANCY PRODUCTION YIELD IN HEAVY ION A
	2-0489	PRODUCTION AND DECAY OF DOUBLE L VACANCIES IN ARGON AND PHOSPHORUS.
DALAL VL	2-0332	CARRIER LIFETIMES IN EPITAXIAL INDIUM ARSENIDE. (AUGER RECOMBINATION)
DALINS I	5-1466	STUDY OF THE ADSORPTION OF OXYGEN ON NICKEL(III) USING AUGER ELECTRON SPECTROSCO
	5-1467	EFFECTIVENESS OF THE STANDARD SURFACE CLEANING TECHNIQUES AS APPLIED TO NICKEL (
	5-1468	STICKING PROBABILITIES FOR THE ADSORPTION OF HYDROGEN ON CLEAN NICKEL (111), NIC
DALMAI-IMELIK G	5-1283	AUGER SPECTROSCOPY OF CARBON ON NICKEL. COMMENTS.
	5-1284	STUDY OF THE ADSORPTION ON THE (100) FACE OF NICKEL. INFLUENCE OF SUL
DANCY JH	5-1746	SURFACE CHARACTERISTICS RELATED TO LASER DAMAGE OF LITHIUM NIOBATE AND POTASSIUM
DANYLUK S	5-1285	PLATINUM SILICIDE FORMATION. ELECTRON SPECTROSCOPY OF THE PLATINUM- PLATINUM SIL
	5-1286	DIFFUSION STUDIES IN CHROMIUM- PLATINUM THIN FILMS USING AUGER ELECTRON SPECTROS
DARLING RL	3-0602	ANALYSIS OF HIGH CHARGE STATE OUTPUT FROM A PENNING ION SOURCE.
DAVEY JE	5-1287	SOLID PHASE EPITAXIAL STUDIES USING VACUUM DEPOSITION ON HEATED SILICON SUBSTRAT
DAVIS DL	2-0398	MATRIX ELEMENTS DEPENDENCE OF OPTICAL EXCITATION AND AUGER DECAY OF 5D CORE HOLE
DAVIS LE	1-0101	AUGER ELECTRCN SPECTROSCOPY. (REVIEW, 154 REFS)
	3-0603	A MULTIPLE SAMPLE HCLDER FOR USE IN ULTRAHIGH VACUUM.
	4-0985	AUGER SATELLITES OF THE L2,3 AUGER EMISSION BANDS OF ALUMINUM, SILICON, AND PHOS
	4-0986	AUGER AND CTHER CHARACTERISTIC ENERGIES IN SECONDARY ELECTRON SPECTRA FROM THIN
	5-1288	A STUDY OF PHOSPHORUS ADSORPTION AND DESORPTION KINETICS ON SILICON (111) SURFAC
	5-1289	SILICON DEVICE APPLICATIONS USING A COMBINED ESCA- AUGER ANALYSIS SYSTEM.
	5-1583	QUANTITATIVE STUDY OF AUGER ELECTRON SIGNALS OF PHOSPHORUS ON SILICON USING A QU
DAVIS RH	3-0602	ANALYSIS OF HIGH CHARGE STATE OUTPUT FROM A PENNING ION SOURCE.
DAVITAYA FA	5-1293	KINETICS OF THERMAL DESORPTION USING AUGER ELECTRON SPECTROSCOPY. APPLICATION TO
DAY HM	5-1259	SCANNING MICROSPOT AUGER SPECTROSCOPY STUDY OF INTERDIFFUSION AND EUTECTIC FORMA
	5-1260	SILICIDE FORMATION AND INTERDIFFUSION EFFECTS IN SILICON- TANTALUM, SILICON DIOX
	5-1262	COMBINED SCANNING ELECTRON MICROSCOPY- AUGER SPECTROSCOPY FOR MICROSPOT SURFACE
	5-1287	SOLID PHASE EPITAXIAL STUDIES USING VACUUM DEPOSITION ON HEATED SILICON SUBSTRAT
DE HEER FJ	2-0517	CHARGE STATE DEPENDENCE OF THE ARGON L-SHELL FLUORESCENCE YIELD STUDIED BY ATOMI
DE LESEGNO PV	4-0817	AUGER SPECTRA OF LITHIUM AND SODIUM UNDER ION BOMBARDMENT.
	4-0876	ENERGY DISTRIBUTION OF AUGER ELECTRONS EMITTED FROM BERYLLIUM, MAGNESIUM, ALUMIN
DE PINHO AG	4-1112	X-RAY SPECTRA, L-SUBSHELL FLUORESCENCE AND COSTER-KRONIG YIELDS IN BISMUTH AND N
DEAN EM	3-0589	ELECTRON BEAM CURRENT CONTROLLER FOR LEED- AUGER STUDIES.
DECOSTERD JP	5-1345	INVESTIGATIONS OF CARBON RESIDUES ON SURFACES OF SILICON INTEGRATED CIRCUITS. (B
DEHEER FJ	2-0387	MEASUREMENTS OF THE L-SHELL FLUORESCENCE YIELDS OF ARGON AND CHLORINE BY ELECTRO
	2-0528	X-RAY EMISSION CROSS SECTIONS AND FLUORESCENCE YIELDS FOR LIGHT ATOMS AND MOLECU
DELCEA B	5-1225	AUGER ELECTRCN SPECTROSCOPY ANALYSIS OF ION PLATED GOLD FILMS ON SILICON.
DELCHAR TA	2-0333	ADSORBATE EFFECTS IN ELECTRON EJECTION BY RARE GAS METASTABLE ATOMS. (AUGER PROC
	5-1600	ROLE OF WORK FUNCTION IN ELECTRON EJECTION BY METASTABLE ATOMS: HELIUM AND ARGON
	5-1601	WORK FUNCTION EFFECTS ON AUGER ELECTRON EJECTION BY NOBLE GAS METASTABLE ATOMS.
DELONG A	3-0604	AUGER ELECTRCN SPECTROSCOPY IN THE ELECTRON EMISSION MICROSCOPE.
	3-0605	POSSIBILITY OF AUGER ELECTRON SPECTROSCOPY IN THE ELECTRON EMISSION MICROSCOPE.
DEMICHELIS B	3-0583	IMPROVED SPHERICAL GRID RETARDING POTENTIAL ANALYZER. (FOR USE IN AUGER ELECTRON
DEMJOKHIN VF	2-0290	NATURE OF SOME X-RAY SATELLITES (RADIATIVE AUGER TRANSITIONS).
DEMUTH JE	5-1290	CHEMISORPTION STUDIES OF NICKEL SURFACES UTILIZING WORK FUNCTION MEASUREMENTS, A
	5-1291	CHEMISORPTION ON (001), (110), AND (111) NICKEL SURFACES. CORRELATED STUDY USING
	5-1292	ELASTIC LEED INTENSITY ENERGY STUDIES OF CLEAN (001), (110) AND ELASTIC (111) NI
	5-1308	PHOTOEMISSION STUDIES OF CHEMISORPTION ON NICKEL AND TUNGSTEN SURFACES.
	5-1764	CHEMISORPTICN ON (001), (110) AND (111) NICKEL SURFACES. (AUGER SPECTRA)
DENISOV EP	3-0717	HIGH RESOLUTION ELECTROSTATIC SPECTROMETER USED IN ELECTRON SPECTROMETRY FOR CHE
DENNE DR	3-0673	A NOVEL DETECTION CIRCUIT FOR AUGER SPECTROSCOPY.
DERRIEN J	5-1293	KINETICS OF THERMAL DESORPTION USING AUGER ELECTRON SPECTROSCOPY. APPLICATION TO
	5-1294	LOW ENERGY ELECTRON DIFFRACTION, AUGER ELECTRON SPECTROSCOPY, AND WORK FUNCTION
DESPLAT JL	5-1295	AUGER ELECTRON SPECTROSCOPY OF CESIUM ADSORBED ON CLEAN AND OXYGEN- COVERED (100
	5-1296	CESIUM- OXYGEN COADSORPTION ON (100) TUNGSTEN.
	5-1297	LOW ENERGY ELECTRON DIFFRACTION AUGER, WORK FUNCTION STUDY OF OXYGEN- CESIUM ADS
DEVILLE JP	1-0046	SPECTROSCOPY OF AUGER ELECTRONS. (REVIEW, 24 REFS)
	1-0047	SPECTROGRAPHY BY AUGER ELECTRONS.
	4-0792	STUDY OF MUSCOVITE AND SILICON BY AUGER ELECTRON SPECTROSCOPY.
	4-0793	AUGER ELECTRON SPECTROSCOPY STUDY OF SOME OXYGENATED SILICON COMPOUNDS.
	4-0794	AUGER SPECTROSCOPY OF SILICON IN SILICATES.
	4-0795	AUGER ELECTRON SPECTROSCOPY OF INSULATING SILICON COMPOUNDS.
	4-0796	STUDY OF AMORPHOUS AND CRYSTALLINE SILICATES BY AUGER ELECTRON SPECTROSCOPY.
	5-1298	MUSCOVITE CLEAVAGE STUDIED BY AUGER ELECTRON SPECTROSCOPY.
DHERE NG	5-1299	PURITY AND MORPHOLOGY OF ALUMINUM FILMS. (AUGER SPECTROSCOPY)
DICK CE	4-0818	K-SHELL FLUCRESCENCE YIELDS FOR LIGHT ELEMENTS.

DICKEY JM   5-1300   STRUCTURE DEPENDENT PROPERTIES OF THIN FILM SUPERCONDUCTORS.
                     5-1301   INTERACTION OF NITROGEN WITH NIOBIUM. (AUGER EFFECT)
DIEBALL JW   5-1543   ELECTRON BEAM INDUCED EFFECTS ON GAS ADSORPTION UTILIZING AUGER ELECTRON SPECTRO
                     5-1799   AUGER ELECTRON SPECTROSCOPY, SCANNING ELECTRON MICROSCOPY AND NONDISPERSIVE X-RA
DIERINGER J   4-1032   WIDTHS AND INTENSITIES OF HIGHLY RESOLVED LINES IN ARGON L2,3 AUGER SPECTRA.
DILMORE MF   5-1265   AQUEOUS CORROSION OF SODA- SILICA AND SODA- LIME- SILICA GLASS. (AES)
DINCA G   5-1225   AUGER ELECTRON SPECTROSCOPY ANALYSIS OF ION PLATED GOLD FILMS ON SILICON.
DINI JOW   1-0048   APPLICATIONS OF AUGER ELECTRON SPECTROSCOPY IN ELECTROPLATING.
DIONISIO JS   2-0334   MERCURY AND GOLD ATOMIC CLOUDS REARRANGEMENT FOLLOWING INTERNAL CONVERSION AND E
DISTEFANO TH   3-0606   ENERGY RESOLUTION OF THE PHOTOEMISSION ANALYZER.
DIXON RD   4-0935   INTERPRETATION OF THE AUGER ELECTRON SPECTRUM FROM OXIDIZED BERYLLIUM.
DOBELIN E   2-0335   INNER SHELL ALIGNMENT OF ATOMS IN ELECTRON IMPACT IONIZATION. (AUGER ELECTRONS)
DOBROTT RD   5-1572   ESTABLISHING DETECTION AND DETERMINATION LIMITS IN SURFACE AND THIN FILM ANALYSI
DOBSON PJ   5-1354   OBSERVATIONS OF THE EPITAXIAL GROWTH OF NICKEL AND COPPER ON (111) SILVER. (BY A
                     5-1356   DIFFUSION OF SILVER THROUGH EPITAXIAL FILMS OF COPPER AND NICKEL ON SILVER (111)
                     5-1757   AUGER SPECTROSCOPY OF OXYGEN AND OTHER "CONTAMINANTS" ON SILVER.
DOHERTY JE   6-2053   CONFIRMATION OF SULFUR EMBRITTLEMENT IN NICKEL ALLOYS. (AUGER ELECTRON SPECTROSC
DOLL P   2-0336   DECAY OF HOLE STATES IN OXYGEN-16 VIA A "NUCLEAR AUGER EFFECT".
DOMANGE JL   5-1302   PHASE TRANSITIONS IN TWO DIMENSIONAL ABSORBED SULFUR LAYERS ON COPPER INVESTIGAT
                     5-1558   ADSORPTION OF SULFUR ON GOLD BY LOW ENERGY ELECTRON DIFFRACTION AND AUGER SPECTR
                     5-1566   SURFACE SEGREGATION: COMPARISON BETWEEN THEORY AND EXPERIMENT.
DOMCKE W   2-0291   PLASMON COUPLING TO CORE HOLE EXCITATIONS IN CARBON. (AUGER APPEARANCE SPECTROSC
DONG D   5-1248   AUGER AND ELLIPSOMETRIC STUDIES OF ULTRATHIN LEAD(II) OXIDE GROWTH ON LEAD.
DOOLEY GJ   1-0049   RECENT ADVANCES IN AES.
                     1-0071   A BIBLIOGRAPHY OF LOW ENERGY ELECTRON DIFFRACTION AND AUGER ELECTRON SPECTROSCOP
                     4-0819   CHEMICAL EFFECTS ON KLL AUGER ELECTRON SPECTRUM FROM OXYGEN.
                     4-0863   CHEMICAL SHIFTS IN AUGER ELECTRON SPECTROSCOPY.
                     4-0864   AUGER ELECTRON SPECTROSCOPY OF TRANSITION METALS.
                     4-0865   AUGER ELECTRON SPECTROSCOPY OF SOME REFRACTORY METALS.
                     4-0866   SOME PROBLEMS IN ANALYSIS OF AUGER ELECTRON SPECTRA.
                     4-0867   CHEMICAL EFFECTS IN AUGER ELECTRON SPECTROSCOPY (STATES OF CARBON).
                     4-0868   IDENTIFICATION OF THE CHEMICAL STATE OF ADSORBED SPECIES WITH AUGER ELECTRON SPE
                     4-0870   SOFT X-RAY APPEARANCE POTENTIAL SPECTROSCOPY IN A DISPLAY LEED SYSTEM.
                     5-1303   CHEMISORPTION ON SINGLE CRYSTAL MOLYBDENUM (112) SURFACES.
                     5-1304   SOME PROPERTIES OF THE RHENIUM (0001) SURFACE.
                     5-1305   FURTHER STUDIES OF GAS ADSORPTION ON MOLYBDENUM (100) SURFACE.
                     6-2019   SURFACE SEGREGATION STUDIES IN ALLOYS USING AUGER ELECTRON SPECTROSCOPY.
                     6-2020   BEHAVIOR OF REFRACTORY METAL SURFACES IN ULTRAHIGH VACUUM AS OBSERVED BY LEED AN
                     6-2021   AUGER ELECTRON SPECTROSCOPY: METALLURGICAL APPLICATIONS.
                     6-2112   STUDY OF GRAIN BOUNDARY FRACTURE SURFACES IN DOPED TUNGSTEN- RHENIUM ALLOYS. (AU
DORN R   5-1306   ADSORPTION OF OXYGEN ON CLEAN CLEAVED (110) GALLIUM ARSENIDE SURFACES.
                     5-1474   ADSORPTION OF OXYGEN ON SILICON (111) SURFACES. PART-1. (AUGER ELECTRON SPECTROS
DORNHAUS R   2-0461   TRANSIENT CARRIER DECAY AND TRANSPORT PROPERTIES IN MERCURY CADMIUM TELLURIDE. (
DOSYAGAEV AV   5-1191   STRUCTURE AND SECONDARY EMISSION OF MAGNESIUM OXIDE LAYERS APPLIED ON A MOLYBDEN
DOVE DB   5-1704   ANALYSIS OF GLASS CONTAINER COATINGS BY AUGER ELECTRON SPECTROSCOPY.
                     5-1705   APPLICATIONS OF SURFACE CHARACTERIZATION TO GLASS SURFACES.
                     5-1720   GLASS SURFACE ANALYSIS BY AUGER ELECTRON SPECTROSCOPY.
                     5-1721   AES ANALYSIS OF SODIUM IN A CORRODED BIOGLASS USING A LOW TEMPERATURE TECHNIQUE.
                     5-1722   AES COMPOSITIONAL PROFILES OF MOBILE IONS IN THE SURFACE REGION OF GLASS.
DOW JD   2-0337   K X-RAY EMISSION EDGE SHAPES OF FREE ELECTRON METALS. (AUGER BROADENING)
                     4-0835   AUGER RATES FOR SOFT X-RAY TRANSITIONS IN IONIC, ATOMIC, AND METALLIC LITHIUM.
DOYLE WD   5-1752   LOW ENERGY ELECTRON DIFFRACTION FROM CLEAN (100) NICKEL OXIDE.
DRAHOS V   3-0604   AUGER ELECTRON SPECTROSCOPY IN THE ELECTRON EMISSION MICROSCOPE.
                     3-0605   POSSIBILITY OF AUGER ELECTRON SPECTROSCOPY IN THE ELECTRON EMISSION MICROSCOPE.
                     3-0607   COMBINATION OF AUGER SPECTROSCOPY WITH IMAGING OF STUDIED SURFACE.
DREW P   5-1749   IDENTIFICATION OF A GRAIN BOUNDARY PHASE IN HOT PRESSED SILICON NITRIDE BY AUGER
DRISCOLL TJ   2-0456   DETERMINATION OF AUGER ELECTRON EXCITATION CROSS SECTIONS AND RANGES USING PROTO
                     5-1682   CORRELATION OF FRACTIONAL MONOLAYER OXYGEN DETERMINATIONS OBTAINED BY PROTON EXC
D'AVITAYA FA   5-1294   LOW ENERGY ELECTRON DIFFRACTION, AUGER ELECTRON SPECTROSCOPY, AND WORK FUNCTION
D'HAENENS IJ   5-1540   GERMANIUM SURFACE CLEANING: AUGER ANALYSIS.
DUBININA EM   5-1317   GROWTH OF THIN METALLIC FILMS ON INSULATING SUBSTRATES IN THE PRESENCE OF ELECTR
DUCHARME AR   2-0338   ROLE OF COSTER-KRONIG TRANSITIONS IN THE IONIZATION OF SURFACE ATOMS.
                     2-0339   L-SHELL IONIZATION CROSS SECTIONS IN AUGER ELECTRON SPECTROSCOPY.
                     2-0358   DIFFERENTIAL CROSS SECTIONS FOR K-SHELL IONIZATION OF SURFACE ATOMS BY ELECTRON
                     2-0359   INNER SHELL IONIZATION OF SURFACE ATOMS BY ELECTRON IMPACT.
                     2-0360   L-SHELL IONIZATION THRESHOLD STRUCTURE FROM AUGER ELECTRON SPECTRA.
                     4-0847   TOTAL ELECTRON IMPACT IONIZATION CROSS SECTIONS ON K-SHELLS OF SURFACE ATOMS. (A
DUCLOS J   2-0305   MEASUREMENT OF THE AUGER EFFECT IN THE MU-4 HELIUM(2S)(+) IONIC SYSTEM.
DUCROS R   4-0820   AUGER SPECTRA OF CARBON. TANTALUM CARBIDES, ETHYLENE AND CARBON MONOXIDE ADSORPT
                     5-1688   ADSORPTION AND DECOMPOSITION OF ACETYLENE ON POLYCRYSTALLINE RHENIUM FILMS AND R
DUFAYARD D   2-0236   EFFECT OF SURFACE PLASMON OSCILLATIONS ON AUGER ELECTRON EMISSION.
                     2-0237   EFFECT OF THE INCIDENCE ANGLE OF PRIMARY ELECTRONS ON THE AUGER EMISSION YIELD.
DUFOUR G   2-0373   SECONDARY ELECTRON EMISSION OF SOLID ALUMINUM AND MAGNESIUM TARGETS.
                     4-0821   N4,5 AUGER SPECTRA OF GADOLINIUM.
                     4-0822   AUGER SPECTRA OF MAGNESIUM AND ALUMINUM.
                     4-0823   4D AUGER TRANSITIONS FROM EXCITED STATES IN SAMARIUM METAL.
DUNN CG   4-0824   AUGER ELECTRON ANALYSIS OF ELECTROPOLISHED HIGH PURITY ALUMINUM.
DUNNING FB   2-0340   SECONDARY ELECTRON EJECTION FROM METAL SURFACES BY METASTABLE ATOMS. PART-1: MEA
                     5-1133   SECONDARY ELECTRON EJECTION FROM METAL SURFACES BY METASTABLE ATOMS. PART-3: ENE
DUNNING KL   1-0216   DIRECT COMPARISON OF AUGER, SECONDARY ION MASS SPECTROSCOPY, AND PROTON RESONANC
DURAUD JP   4-0948   STRUCTURE OF LOW ENERGY ELECTRON SPECTRUM.
DVORYANKIN VF   1-0132   ELECTRON SPECTROSCOPY: A METHOD OF INVESTIGATING SURFACE PHENOMENA (REVIEW
DWIGHT DJ   3-0608   IRON-55 AS AN AUGER ELECTRON EMITTER. NOVEL SOURCE FOR GAS CHROMATOGRAPHY DETECT
DWYER DJ   6-2110   LOW ENERGY ELECTRON DIFFRACTION- AUGER ELECTRON SPECTROSCOPY STUDY OF THE INITIA
DZHEMILEV NKH   2-0245   ANISOTROPY OF SECONDARY ELECTRON EMISSION FROM PASSAGE OF LITHIUM(+) IONS THROUG
EADINGTON P   5-1307   USE OF AUGER ELECTRON SPECTROSCOPY TO DETERMINE THE COMPOSITION OF FRACTURE SURF
EASTMAN DE   2-0341   VALENCE BAND AND CORE LEVEL STUDIES OF INDIUM ANTIMONIDE VIA PHOTOEMISSION MEASU
                     5-1152   PHOTOEMISSION SPECTRA FROM ADSORBED OXYGEN ON TUNGSTEN (110) AND CARBON MONOXIDE
                     5-1308   PHOTOEMISSION STUDIES OF CHEMISORPTION ON NICKEL AND TUNGSTEN SURFACES.
EASTON DS   4-0806   AUGER SPECTRA OF LITHIUM METAL AND LITHIUM OXIDE.

EASTON DS	4-1025	AUGER SPECTRA OF LITHIUM HYDRIDE.
EBEL H	1-0050	STRUCTURE ANALYSIS WITH X-RAY PHOTOELECTRON SPECTROSCOPY AND AUGER ELECTRON SPEC
EBIKO H	5-1309	(AUGER ELECTRON AND MOSSBAUER) SPECTROSCOPICAL STUDIES OF THE COMPOSITION OF PAS
ECKERTORA L	3-0638	SECONDARY ELECTRON EMISSION FROM THIN FILMS. (OF GOLD, APPARATUS)
EDELMAN FL	5-1310	AUGER SPECTRA FROM THE (100) SURFACE OF SILICON DURING OXIDATION AND NITRIDATION
EDMONDS T	5-1311	PLATINUM (100) STRUCTURES: IMPURITY CAUSATION AND PROPERTIES. (AUGER EFFECT)
EHRHARDT JJ	5-1688	ADSORPTION AND DECOMPOSITION OF ACETYLENE ON POLYCRYSTALLINE RHENIUM FILMS AND R
EKARDT W	2-0379	INFLUENCE OF SCREENING EFFECTS ON THE AUGER RECOMBINATION IN SEMICONDUCTORS. (SI
EKELUND S	5-1585	LOW ENERGY ELECTRON DIFFRACTION- AUGER ELECTRON SPECTROSCOPY STUDY OF THE OXIDAT
	5-1586	A LOW ENERGY ELECTRON DIFFRACTION- AUGER ELECTRON SPECTROSCOPY STUDY OF THE OXID
	5-1587	LEED- AES STUDY OF THE OXIDATION OF IRON- CHROMIUM (100) AND (110).
	5-1591	SURFACE COMPOSITION STUDIES OF THE (001) AND (110) FACES OF MONOCRYSTALLINE IRON
	6-2022	LOW ENERGY ELECTRON DIFFRACTION- AUGER ELECTRON SPECTROSCOPY STUDY OF THE OXIDAT
EL IBYARI SN	2-0342	AUGER TRANSITION RATE IN MIXED COUPLING SCHEME.
ELANGO M	4-1126	DETECTION OF EXCITONS NEAR THE L2,3 EDGE OF THE CHLORIDE ION IN SODIUM CHLORIDE
ELDRIDGE JM	5-1248	AUGER AND ELLIPSOMETRIC STUDIES OF ULTRATHIN LEAD(II) OXIDE GROWTH ON LEAD.
ELLIOT AG	5-1312	CONDENSATION OF GOLD ONTO TANTALUM (100) SINGLE CRYSTAL SURFACES. LEED AND AES A
	5-1742	SURFACE COMPOSITION OF MICA SUBSTRATES.
ELLIOTT CT	2-0343	WARM ELECTRON EFFECTS IN NARROW GAP MERCURY(1-X) CADMIUM(X) TELLURIDE AMBIENT TE
ELLIS WP	1-0051	RECENT ADVANCES IN THE STUDY OF SOLID SURFACES.
	2-0512	SPREADING FUNCTIONS IN SECONDARY ELECTRON ANALYSIS.
	5-1218	AUGER ELECTRON STUDIES OF URANIUM DIOXIDE SURFACES.
	5-1313	SEGREGATION OF IMPURITIES ON SURFACES OF THORIUM. THERMODYNAMICS AND KINETICS. (
	5-1314	IMPURITIES ON THORIUM METAL: AUGER STUDY OF BULK TO SURFACE EQUILIBRIA.
	5-1315	ARSENIC (0001) SURFACES: AUGER, LOSS, AND PHOTOELECTRON SPECTROSCOPIC STUDIES.
	5-1316	SECONDARY ELECTRON ENERGY DISTRIBUTION STUDIES OF URANIUM DIOXIDE SURFACES.
	5-1783	KIKUCHI CORRELATIONS IN AUGER ELECTRON SPECTROSCOPY.
	5-1784	HIGH ANGULAR RESOLUTION SECONDARY ELECTRON SPECTROSCOPY. KIKUCHI CORRELATIONS FO
ELOVICOV SS	5-1317	GROWTH OF THIN METALLIC FILMS ON INSULATING SUBSTRATES IN THE PRESENCE OF ELECTR
ELTON RC	2-0344	RADIATIVE AUGER TRANSITIONS IN SOFT X-RAY PLASMA EMISSION.
ENGEL T	5-1318	ADSORPTION OF OXYGEN ON THE (110) PLANE OF TUNGSTEN. (AUGER ELECTRON SPECTROSCOP
ENGELHARDT HA	5-1608	ADSORPTION OF OXYGEN ON COBALT ON THE RUTHENIUM (001) SURFACE. (AU
EPLER D	4-0797	CHEMICAL SHIFTS IN PHOTOEXCITED AUGER SPECTRA.
ERICSON RE	5-1319	AN ELECTRICAL, CHEMICAL, AND STRUCTURAL STUDY OF THE GERMANIUM (100)- CESIUM- OX
ERIKSSON JC	5-1591	SURFACE COMPOSITION STUDIES OF THE (001) AND (110) FACES OF MONOCRYSTALLINE IRON
ERTL G	5-1252	INTERACTIONS OF CARBON MONOXIDE AND OXYGEN WITH IRIDIUM (110) SURFACES. (AUGER S
	5-1253	SURFACE STUDIES WITH EPITAXIALLY GROWN METAL FILMS. (AUGER EFFECT)
	5-1254	ADSORPTION OF CARBON MONOXIDE ON SILVER- PALLADIUM ALLOYS.
	5-1255	ADSORPTION OF HYDROGEN ON A PLATINUM (111) SURFACE.
	5-1256	ADSORPTION OF COBALT ON A NICKEL (111) SURFACE. (AUGER EFFECT, AUGER SPECTROSCOP
	5-1257	ADSORPTION OF HYDROGEN ON NICKEL SINGLE CRYSTAL SURFACES. (AUGER ELECTRON SPECTR
	5-1278	INTERACTION OF NITRIC OXIDE WITH A NICKEL (111) SURFACE.
	5-1320	ELECTRON SPECTROSCOPY AND LEED STUDIES OF GAS- METAL INTERFACES. (AUGER EFFECT)
	5-1321	INTERACTION OF CARBON MONOXIDE AND OXYGEN WITH A PALLADIUM (100) SURFACE. (CHEMI
	5-1322	ADSORPTION ON SINGLE CRYSTAL SURFACES OF COPPER- NICKEL ALLOYS. PART-1.
	5-1323	ELECTRON SPECTROSCOPIC STUDIES OF CLEAN AND OXIDIZED IRON. (AUGER STUDY OF SURFA
	5-1606	DECOMPOSITION OF CARBON MONOXIDE ON A (110) NICKEL SURFACE. (AUGER ELECTRON SPEC
	5-1607	INTERACTION OF CARBON MONOXIDE WITH (110) NICKEL SURFACES. (AUGER EFFECT)
	5-1952	SOFT X-RAY APPEARANCE POTENTIAL SPECTROSCOPY STUDIES WITH COPPER- NICKEL AND IRO
	5-1953	ELECTRON SPECTROSCOPIC STUDIES OF THE OXIDATION OF NICKEL- PALLADIUM ALLOYS.
ESAKI L	5-1324	COMPUTER-CONTROLLED MOLECULAR BEAM EPITAXY. (AUGER SURFACE ANALYSIS, GALLIUM(X)
	5-1595	(AUGER) ELECTRON SPECTROSCOPY OF GALLIUM ARSENIDE AND ALUMINUM ARSENIDE SURFACES
	5-1596	GALLIUM(1-X) ALUMINUM(X) ARSENIDE SUPERLATTICES PROFILED BY AUGER ELECTRON SPECT
ESTEROWITZ L	2-0262	AUGER LIMITED CARRIER LIFETIMES IN MERCURY CADMIUM TELLURIDE AT HIGH EXCESS CARR
ESTRUP PJ	5-1884	CARBON MONOXIDE ADSORPTION ON NICKEL (110). (AUGER EFFECT)
ETHRIDGE EC	5-1265	AQUEOUS CORROSION OF SODA- SILICA AND SODA- LIME- SILICA GLASS. (AES)
EULER M	5-1325	THERMALLY STIMULATED EXOELECTRON EMISSION OF BERYLLIUM OXIDE LAYERS. AUGER ELECT
EVANS CA	1-0052	COMPARISON OF THE TECHNIQUES FOR SILICON SURFACE ANALYSIS. (AES, ESCA, SIMS, AND
	1-0053	SURFACE AND THIN FILM COMPOSITIONAL ANALYSIS. DESCRIPTION AND COMPARISON OF TECH
	1-0054	THIN FILM COMPOSITIONAL ANALYSIS. COMPARISON OF TECHNIQUES. (AUGER SPECTROSCOPY)
	3-0609	SURFACE AND THIN FILM ANALYSIS. (AUGER SPECTROMETERS)
	5-1182	EFFECT OF OXIDIZING AMBIENTS ON PLATINUM SILICIDE FORMATION. PART-2: AUGER AND B
	5-1326	TECHNIQUES FOR THE CHEMICAL CHARACTERIZATIONS OF CERAMIC SURFACES. (ION SCATTERI
	5-1327	COMPARISON OF THE TECHNIQUES FOR SILICON SURFACE ANALYSIS.
	5-1577	ANTIMONY DOPING OF SILICON LAYERS GROWN BY SOLID PHASE EPITAXY. (AES MEASUREMENT
	5-1672	INTERACTION OF ALUMINUM LAYERS WITH POLYCRYSTALLINE SILICON. (AUGER SPECTROSCOPY
EWING WS	5-1328	AUGER ANALYSIS OF THREE ENHANCED MULTILAYER DIELECTRIC MIRROR DESIGNS. (ZINC SEL
EXNER HE	6-2047	AUGER ELECTRON SPECTROSCOPY OF CONTRAST FORMING LAYERS ON METALS.
FABIAN DJ	2-0337	K X-RAY EMISSION EDGE SHAPES OF FREE ELECTRON METALS. (AUGER BROADENING)
	4-0837	X-RAY EXCITED AUGER AND PHOTOELECTRON SPECTRA OF MAGNESIUM, SOME ALLOYS OF MAGNE
	4-0838	COMPARISON OF THE SOFT X-RAY PHOTOEMISSION, AND AUGER ELECTRON SPECTRA OF THE VA
FAERBER W	1-0014	CHEMICAL EFFECTS IN AUGER ELECTRON SPECTROSCOPY.
	5-1199	AES STUDIES OF SURFACE COMPOSITION OF SILVER- COPPER ALLOYS.
	5-1200	AUGER ELECTRON SPECTROSCOPY STUDIES OF SURFACE COMPOSITION OF SILVER- COPPER ALL
	5-1329	INVESTIGATIONS OF THE OXIDE FORMATION ON METALS WITH AUGER ELECTRON SPECTROSCOPY
	6-2007	AUGER ELECTRON SPECTROSCOPY OF METALS DURING OXIDATION.
FAGOT B	2-0550	EMISSION OF AUGER ELECTRONS BY ION BOMBARDMENT.
	2-0551	OBSERVATION OF AUGER ELECTRONS IN THE ENERGY SPECTRUM OF SECONDARY ELECTRONS EMI
	4-0942	SECONDARY ELECTRON SPECTRA OF SOME SOLID TARGETS UNDER IONIC BOMBARDMENT.
FAHLINAN A	4-0887	AUGER SPECTRA IN THE INTERMEDIATE COUPLING REGION.
FAHLMAN A	4-0734	KLL AUGER SPECTRUM OF FLUORINE. (IN FLUORIDE SALTS)
	4-0825	CHEMICAL SHIFTS IN AUGER ELECTRON SPECTRA.
	4-0826	CHEMICAL SHIFT IN AUGER SPECTRA.
	4-0827	AUGER SPECTRA FOR ELEMENTS OF LOW ATOMIC NUMBER.
FAIN SC	4-0828	WORK FUNCTION VARIATION WITH ALLOY COMPOSITION SILVER- GOLD. (AUGER ELECTRON SPE
	5-1330	WORK FUNCTION VARIATION WITH ALLOY COMPOSITION: COPPER- GOLD. (AUGER SPECTROSCOP
	5-1631	SEGREGATION AT COPPER- GOLD ALLOY SURFACES. (AUGER SPECTROSCOPY)
FAITH WN	5-1745	SURFACE PLASMON STUDIES ON ALUMINUM (111) AND ALUMINUM (001) SINGLE CRYSTAL SURF
FALK H	1-0050	STRUCTURE ANALYSIS WITH X-RAY PHOTOELECTRON SPECTROSCOPY AND AUGER ELECTRON SPEC
FAN JCC	2-0384	EFFECTS OF CESIATION ON SECONDARY ELECTRON EMISSION FROM MAGNESIUM OXIDE- GOLD C

FAN JCC	5-1430	DIFFERENTIAL SPUTTERING OF MAGNESIUM OXIDE- GOLD CERMET FILMS AND ITS APPLICATIO
	5-1431	AUGER SPECTROSCOPY STUDIES OF THE OXIDATION OF AMORPHOUS AND CRYSTALLINE GERMANI
	6-2023	OXIDATION STUDIES OF AMORPHOUS AND CRYSTALLINE GERMANIUM FILMS BY AUGER SPECTROS
FARBER W	5-1331	SPUTTERING OF THE ALLOY SYSTEMS SILVER- GOLD, GOLD- COPPER, AND SILVER- COPPER S
	5-1332	OXYGEN EXPOSURE OF SAMARIUM, GADOLINIUM, AND TERBIUM STUDIED BY AUGER ELECTRON S
	6-2024	AUGER ELECTRON SPECTROSCOPY OF SILVER- GOLD ALLOYS.
FARRELL HH	4-0829	LMM AUGER SPECTRA OF SULFUR IN SODIUM SULFATE AND SODIUM PYROSULFITE.
	5-1333	INTERACTION OF OXYGEN AND NITROGEN WITH THE NIOBIUM (100) SURFACE. PART-2: REACT
FARROW RFC	5-1334	MOLECULAR BEAM STUDIES OF INDIUM PHOSPHIDE EVAPORATION AND GROWTH. (AES)
FARTNER RJ	4-0997	OBSERVATION AND INTERPRETATION OF THE AUGER ELECTRON SPECTRUM FROM CLEAN BERYLLI
FASTRUP B	2-0345	K-VACANCY SHARING STUDIED FROM AUGER ELECTRON YIELDS.
	2-0346	DYNAMIC CHARGE STATE DEPENDENCE OF THE K-VACANCY PRODUCTION YIELD IN HEAVY ION A
	2-0405	PRODUCTION OF INNER SHELL VACANCIES IN HEAVY ION- ATOM COLLISIONS. (AUGER EFFECT
	2-0489	PRODUCTION AND DECAY OF DOUBLE L VACANCIES IN ARGON AND PHOSPHORUS.
	4-0830	EMISSION OF KLL AUGER ELECTRONS PRODUCED IN Z(1)- NITROGEN AND Z(1)- NEON COLLIS
FAULK JD	2-0477	DIFFERENTIAL CROSS SECTION FOR K-SHELL IONIZATION OF COPPER AND SILVER BY ELECTR
FEDAK DG	5-1335	INTERACTION BETWEEN CHLORINE AND THE (100) SURFACE OF GOLD.
	5-1367	INTERACTION OF SULFUR DIOXIDE WITH TUNGSTEN. (AUGER ANALYSIS)
FEDER R	5-1346	SURFACE STRUCTURES ON IRON (110) ULTRAHIGH VACUUM.
FEIBELMAN PJ	2-0347	AUGER PLASMON SATELLITE INTENSITIES VERSUS DEPTH: A MEANS FOR DETERMINING ADATOM
	4-0831	TIGHT BINDING CALCULATION OF A CORE VALENCE VALENCE AUGER LINE SHAPE: SILICON (1
FEIGE Y	1-0055	MICRODOSIMETRY OF AUGER ELECTRONS. (REVIEW)
FEINENDEGEN LE	1-0056	BIOLOGICAL DAMAGE FROM THE AUGER EFFECT, POSSIBLE BENEFITS. (DNA)
	2-0531	BIOLOGICAL RADIATION EFFECT ON INCLUSION OF MODERATELY HEAVY NUCLEI IN THE TISSU
FENG HC	2-0348	EFFECTS OF LOW ENERGY ARGON- ION BOMBARDMENT ON MOLYBDENUM DISULFIDE. (AUGER MEA
FERRANDINO F	5-1220	OPTICAL PROPERTIES AND AUGER CHARACTERISTICS OF MAGNESIUM SINGLE CRYSTALS AND MA
FERRANTE J	1-0057	USE OF LOW ENERGY ELECTRON DIFFRACTION, AUGER EMISSION SPECTROSCOPY AND FIELD IO
	2-0452	AUGER ELECTRON SPECTROSCOPY STUDY OF ELECTRON IMPACT DESORPTION.
	5-1668	AUGER ELECTRON SPECTROSCOPY STUDY OF OXYGEN ADSORPTION ON TUNGSTEN (110).
	5-1669	LEED AND AES STUDY OF OXYGEN ABSORPTION ON TUNGSTEN (110).
	6-2012	AUGER ELECTRON SPECTROSCOPY STUDY OF SURFACE SEGREGATION IN COPPER- ALUMINUM ALL
	6-2025	AUGER SPECTROSCOPY AND LEED STUDY OF EQUILIBRIUM SURFACE SEGREGATION IN COPPER-
	6-2026	MEASUREMENT OF EQUILIBRIUM SURFACE SEGREGATION USING AUGER ELECTRON SPECTROSCOPY
	6-2027	AUGER ELECTRON SPECTROSCOPY STUDY OF SURFACE SEGREGATION IN THE BINARY ALLOYS CO
	6-2028	AUGER ELECTRON SPECTROSCOPY STUDY OF INITIAL STAGES OF OXIDATION IN A COPPER- 19
	6-2029	AUGER ELECTRON SPECTROSCOPY AND DEPTH PROFILE STUDY OF OXIDATION MODIFIED 440C S
	6-2030	THERMAL EFFECTS IN EQUILIBRIUM SURFACE SEGREGATION IN A COPPER- 10 ATOMIC PERCEN
	6-2031	REACTION OF SULFUR DIOXIDE WITH MODIFIED 440C STEEL STUDIED BY AUGER ELECTRON SP
	6-2032	AUGER ELECTRON SPECTROSCOPY AND DEPTH PROFILE STUDY OF OXIDATION OF MODIFIED 440
	6-2033	A REVIEW OF SURFACE SEGREGATION, ADHESION, AND FRICTION STUDIES PERFORMED ON COP
FERREIRA JG	2-0349	THE L(II) SUBSHELL ATOMIC YIELDS OF ELEMENTS WITH Z APPROXIMATELY 90.
FERRONI E	5-1779	LEED (AUGER) STUDY OF CHLORINE CHEMISORPTION ON THE SILVER (111) SURFACE.
FERTIG DJ	5-1336	STUDY OF PALLADIUM SILICIDE FILMS ON SILICON USING AUGER ELECTRON SPECTROSCOPY.
	5-1774	AUGER ELECTRON SPECTROSCOPY STUDIES OF GALLIUM ARSENIDE AND SILICON METAL- SEMIC
FEUERBACHER B	2-0568	GRAPHITE CONDUCTION BAND STATES FROM SECONDARY ELECTRON EMISSION SPECTRA.
FIERMANS L	1-0204	AUGER ELECTRON SPECTROSCOPY.
	2-0350	USE OF CHARACTERISTIC IONIZATION LOSSES IN AUGER ELECTRON EMISSION SPECTROSCOPY.
	3-0610	AVERAGING, CURVE FITTING, AND RELATED TECHNIQUES IN AUGER ELECTRON EMISSION SPEC
	4-0832	LIFETIME BROADENING IN X-RAY INDUCED PHOTOEMISSION SPECTROSCOPY AND AUGER SPECTR
	4-0882	RELAXATION PHENOMENA INVOLVED IN THE L3M4,5M4,.5M4,.5:1G4 AUGER PROCESS IN ZINC
	4-0883	ATOMIC AND EXTRAATOMIC STATIC RELAXATION ENERGIES IN THE L3M4,5M4,5 AUGER PROCES
	4-0884	DETAILED ANALYSIS OF THE L3M4,5M4,5: 1G4 AUGER TRANSITION IN ATOMIC ZINC: IMPROV
	5-1337	COMBINED LOW ENERGY ELECTRON DIFFRACTION, AUGER ELECTRON SPECTROSCOPY, AND X-RAY
	5-1338	AUGER ELECTRON EMISSION SPECTROSCOPY (AES) AND LOW ENERGY ELECTRON DIFFRACTION (
	5-1339	(AUGER) ELECTRON SPECTROSCOPY OF TRANSITION METAL OXIDE SURFACES.
	5-1340	INELASTIC EFFECTS AND STRUCTURE IN THE AUGER ELECTRON EMISSION SPECTRA OF VANADI
	5-1341	STRUCTURE AND COMPOSITION OF METAL SURFACES: A REVIEW OF RECENT LEED, AES, AND X
FIGUEROA J	2-0431	SCINTILLATION EFFICIENCY FOR LOW VELOCITY HEAVY IONS. (AUGER EMISSION)
FILINOV GP	3-0617	DRYING OF BUTADIENE- NITRILE RUBBERS IN AN AUGER MACHINE.
FINE ME	5-1615	GRAIN BOUNDARY SEGREGATION IN MAGNESIA DOPED ALUMINA. (HIGH VACUUM AUGER ELECTRO
FINK RW	2-0259	X-RAY FLUORESCENCE YIELDS, AUGER, AND COSTER-KRONIG TRANSITION PROBABILITIES.
	2-0351	ATOMIC FLUORESCENCE YIELDS. (L-SHELL COSTER-KRONIG YIELDS, K-, L-, AND M-SHELL F
FITCHEK J	4-0787	LIFETIME EFFECT ON THE AUGER SPECTRA OF METALS. (SODIUM)
	4-0833	LIFETIME AND SURFACE EFFECTS ON THE L23 VV AUGER SPECTRA OF ALUMINUM.
FITTON B	2-0568	GRAPHITE CONDUCTION BAND STATES FROM SECONDARY ELECTRON EMISSION SPECTRA.
	2-0569	BAND STRUCTURE OF GRAPHITE STUDIED BY SECONDARY ELECTRON EMISSION.
	5-1970	STUDY OF CARBON FIBER SURFACES USING AUGER AND SECONDARY ELECTRON EMISSION SPECT
FLORIO JV	5-1335	INTERACTION BETWEEN CHLORINE AND THE (100) SURFACE OF GOLD.
	5-1342	DETERMINATION OF SURFACE STRUCTURES BY LEED (AND AES): THE SILICON (111) SURFACE
	5-1343	PHASE TRANSFORMATIONS OF SILICON (111) SURFACE.
	5-1344	CHLORINE REACTIONS ON THE SILICON (111) SURFACE.
FLUGGE S	2-0352	ANGULAR DISTRIBUTION OF AUGER ELECTRONS FOLLOWING PHOTOIONIZATION.
FOKON CT	2-0455	DEPENDENCE OF AUGER ELECTRON YIELD ON PRIMARY BEAM ENERGY AT NORMAL AND GLANCING
FOLLAND NO	2-0284	THEORETICAL K-SHELL AUGER RATES, TRANSITION ENERGIES, AND FLUORESCENCE YIELDS FO
FONTANA PV	5-1345	INVESTIGATIONS OF CARBON RESIDUES ON SURFACES OF SILICON INTEGRATED CIRCUITS. (B
FORD JLC	2-0336	DECAY OF HOLE STATES IN OXYGEN-16 VIA A "NUCLEAR AUGER EFFECT".
FORTNER RJ	2-0441	SEQUENTIAL STRIPPING IN HIGHLY IONIZED NEON. (AUGER TRANSITIONS)
	4-0834	CHEMICAL EFFECTS ON THE AUGER ELECTRON SPECTRA OF BERYLLIUM.
FORTY AJ	3-0691	FARADAY CUP LEED APPARATUS WITH FACILITY FOR INVESTIGATING ENERGY AND ANGULAR DI
FRABER W	6-1997	DETERMINATION OF THE INFLUENCE GRAIN BOUNDARY IMPURITIES ON THE PROPERTIES OF SI
FRANCESCHETTI DR	2-0353	AUGER EFFECT AND CORE BROADENING IN LOW-Z ATOMS, IONS, AND METALS.
	4-0835	AUGER RATES FOR SOFT X-RAY TRANSITIONS IN IONIC, ATOMIC, AND METALLIC LITHIUM.
FREEDMAN MS	4-0836	THE L-AUGER SPECTRA OF PLUTONIUM, AMERICIUM, CALIFORNIUM, AND FERMIUM.
	4-1019	PROTON STATES, ANOMALOUS CONVERSION, AND K-AUGER SPECTRA IN AMERICIUM-241 FROM C
FREEOUF J	2-0341	VALENCE BAND AND CORE LEVEL STUDIES OF INDIUM ANTIMONIDE VIA PHOTOEMISSION MEASU
FRENCH JB	5-1736	INTERACTION OF LOW ENERGY ATMOSPHERIC IONS WITH CONTROLLED SURFACES. (AUGER NEUT
FREUND HU	2-0259	X-RAY FLUORESCENCE YIELDS, AUGER, AND COSTER-KRONIG TRANSITION PROBABILITIES.
FRICKE B	2-0296	AUGER TRANSITIONS IN LITHIUM-LIKE BERYLLIUM-LIKE IONS.
	4-0860	BEAM FOIL EXCITED AUGER TRANSITIONS IN NEON.
FRIDRIKHOV SA	2-0356	SECONDARY ELECTRON EMISSION AT SUPERHIGH FREQUENCIES. (COPPER- GERMANIUM JUNCTIO

FRIED LJ	5-1984	DETERMINATION OF FILM THICKNESS OR ION ETCH RATE USING AUGER ELECTRON SPECTROSCO
FRIEDMAN AM	4-1019	PROTON STATES, ANOMALOUS CONVERSION, AND K-AUGER SPECTRA IN AMERICIUM-241 FROM C
FRIEDT JM	5-1796	MOSSBAUER SPECTROSCOPY OF IRON-57 AND TIN-119 BY DETECTION OF CONVERSION AND AUG
FRITZ JH	3-0611	MEASUREMENT OF WORK FUNCTION CHANGE IN A DISPLAY TYPE LOW ENERGY ELECTRON DIFFRA
	5-1395	WORK FUNCTION CHANGES ON CONTACT MATERIALS. (AUGER ELECTRON SPECTROSCOPY)
FUGGLE JC	4-0837	X-RAY EXCITED AUGER AND PHOTOELECTRON SPECTRA OF MAGNESIUM, SOME ALLOYS OF MAGNE
	4-0838	COMPARISON OF THE SOFT X-RAY PHOTOEMISSION, AND AUGER ELECTRON SPECTRA OF THE VA
FUJISADA H	2-0354	EFFECTS OF DIFFUSION CURRENT ON GALVANOMAGNETIC PROPERTIES IN THIN INTRINSIC IND
FUJIWARA K	1-0141	AUGER ELECTRON SPECTROSCOPY AND ITS APPLICATION.
	5-1691	COMBINED AUGER ELECTRON SPECTROSCOPY AND ELECTRON IMPACT DESORPTION STUDIES OF T
FUKUDA Y	5-1982	OBSERVATION OF SURFACE COMPOSITION OF COPPER- NICKEL ALLOY BY AUGER SPECTRA IN T
FUKUDOME R	3-0641	DESIGN OF AN X-RAY DOUBLE CRYSTAL SPECTROMETER INSTALLED IN VACUUM AND THE MEASU
FUSY J	5-1955	APPLICATION OF AUGER ELECTRON SPECTROSCOPY TO THE PLATINUM- OXYGEN SYSTEM AT HIG
GABLER H	4-1078	HIGH RESOLUTION NEON AUGER SPECTRUM PRODUCED IN 4.2 MEV ATOMIC HYDROGEN(+)- NEON
	4-1080	ARGON L-SHELL AUGER SPECTRA PRODUCED IN ARGON(+) ION- ARGON COLLISIONS.
GADZUK JW	4-0839	VALENCE BAND AUGER ELECTRON SPECTRA FOR ALUMINUM: COMMENTS.
	4-1088	DETERMINATION OF ELECTRON ATTENUATION LENGTHS IN METALS: TRANSMISSION THROUGH TH
GAFNER G	5-1346	SURFACE STRUCTURES ON IRON (110) ULTRAHIGH VACUUM.
GAIPOV S	2-0246	ENERGY SPECTRUM OF ELECTRONS EMITTED BY ALKALI HALIDE CRYSTALS UNDER IMPACT OF I
GALKIN GN	4-0840	(AUGER) RECOMBINATION OF NONEQUILIBRIUM CHARGE CARRIERS OF INDIUM ARSENIDE AT HI
GALLON TE	1-0058	COMPARISON OF X-RAY AND ELECTRON IMPACT IONIZATION IN AUGER SPECTROSCOPY.
	1-0059	AUGER ELECTRON SPECTROSCOPY AND ITS APPLICATION TO SURFACE STUDIES. (REVIEW, 130
	2-0355	SIMPLE MODEL FOR DEPENDENCE OF AUGER INTENSITIES ON SPECIMEN THICKNESS.
	2-0399	MODEL FOR THE AUGER ELECTRON SPECTROSCOPY OF SYSTEMS EXHIBITING LAYER GROWTH, AN
	2-0430	AUGER EMISSION AND COLOR CENTERS IN LITHIUM FLUORIDE.
	2-0439	ENHANCEMENT OF COSTER-KRONIG RATES IN SOLID STATE ENVIRONMENTS. (XENON, CADMIUM,
	2-0513	AUGER EMISSION FROM SOLIDS. ESTIMATION OF BACKSCATTERING EFFECTS AND IONIZATION
	3-0592	HIGH ENERGY RESOLUTION AUGER ELECTRON SPECTROMETER USING CONCENTRIC HEMISPHERES.
	3-0612	IMPROVED APPARATUS FOR MEASUREMENT OF AUGER ELECTRON SPECTRA.
	4-0771	QUASIATOMIC FINE STRUCTURE IN THE AUGER SPECTRA OF SOLID SILVER AND INDIUM.
	4-0772	HIGH RESOLUTION AUGER ELECTRON SPECTRUM OF MAGNESIUM OXIDE (100).
	4-0841	ESTIMATION OF BACKSCATTERING EFFECTS IN ELECTRON INDUCED AUGER SPECTRA (SILVER A
	4-0842	LOW ENERGY AUGER EMISSION FROM LITHIUM FLUORIDE.
	4-0843	SOLID STATE BROADENING IN AUGER SPECTRA. (CADMIUM, RARE GASES)
	4-0892	THE KVV AUGER SPECTRUM OF LITHIUM METAL.
	4-1005	SOLID STATE BROADENING EFFECTS IN THE AUGER SPECTRUM OF XENON.
	4-1006	SOLID STATE BROADENING IN THE AUGER SPECTRA OF RARE GASES. (ARGON, KRYPTON, XENO
	4-1067	ANOMALOUS THRESHOLD BEHAVIOR FOR ELECTRON INDUCED INNER SHELL IONIZATION OF THE
	5-1158	EFFECT OF OXIDATION ON THE HIGH RESOLUTION AUGER SPECTRUM OF TITANIUM.
	5-1347	(100) SURFACES OF ALKALI HALIDES. PART-1: AIR AND VACUUM CLEANED SURFACES.
	5-1348	GROWTH OF SILVER ON POTASSIUM CHLORIDE OBSERVED BY LEED AND AUGER EMISSION SPECT
	5-1349	THE (111) SURFACE OF N-TYPE SILICON.
	5-1908	(100) SURFACES OF ALKALI HALIDES. PART-2: ELECTRON STIMULATED DISSOCIATION.
GANICHEV DA	2-0356	SECONDARY ELECTRON EMISSION AT SUPERHIGH FREQUENCIES. (COPPER- GERMANIUM JUNCTIO
GASTALDI U	2-0305	MEASUREMENT OF THE AUGER EFFECT IN THE MU-4 HELIUM(2S) (+) IONIC SYSTEM.
GAUCKLER LJ	5-1444	STUDY OF FRACTURE SURFACES OF HOT PRESSED SILICON NITRIDE BY AUGER ELECTRON SPEC
	5-1445	AUGER ELECTRON SPECTROSCOPY ON THE FRACTURE OF NICKEL TUNGSTEN MATERIALS AND SIL
GAVRON A	1-0055	MICRODOSIMETERY OF AUGER ELECTRONS. (REVIEW)
GAZZA GE	5-1350	EFFECT OF YTTRIA ADDITIONS ON HOT PRESSED SILICON NITRIDE. (AUGER ANALYSIS)
GEBALLE R	2-0482	MEASUREMENT OF THE PRODUCTION OF LL VACANCY STATES BY THE L X-RAY- L X-RAY COINC
GEBRANZIG U	2-0357	INFLUENCE OF AN ELECTRIC FIELD ON THE AUGER RECOMBINATION IN SEMICONDUCTORS.
	4-0844	INFLUENCE OF AN ELECTRIC FIELD ON THE AUGER RECOMBINATION IN SEMICONDUCTORS.
GEIGER JS	1-0060	THE K-SHELL AUGER ELECTRON SPECTRA. REVIEW.
GENTILE AL	1-0015	LASER WINDOW SURFACE FINISHING AND COATING TECHNOLOGY. (AUGER ELECTRON SPECTROSC
GEORGESCU S	2-0325	EMISSION SPECTRA OF SILICON COMPENSATED ALUMINUM(X) GALLIUM(1-X) ARSENIDE ELECTR
GERGELY GY	1-0061	AUGER ELECTRON SPECTROSCOPY.
	4-0845	(AUGER) ELECTRON SPECTROSCOPY OF TUNGSTEN.
GERLACH RL	1-0062	APPLICATIONS OF ELECTRON SPECTROSCOPY. (RELATIONSHIP OF IONIZATION SPECTROSCOPY
	1-0119	APPLICATIONS OF SCANNING AUGER MICROANALYSIS.
	2-0339	L-SHELL IONIZATION CROSS SECTIONS IN AUGER ELECTRON SPECTROSCOPY.
	2-0358	DIFFERENTIAL CROSS SECTIONS FOR K-SHELL IONIZATION OF SURFACE ATOMS BY ELECTRON
	2-0359	INNER SHELL IONIZATION OF SURFACE ATOMS BY ELECTRON IMPACT.
	2-0360	L-SHELL IONIZATION THRESHOLD STRUCTURE FROM AUGER ELECTRON SPECTRA.
	3-0613	DISTORTION OF DIFFERENTIATED DEFLECTION ANALYZER CURRENT. (USED IN AES)
	3-0614	RETARDING FIELD CYLINDRICAL MIRROR ANALYZER. (FOR AUGER SPECTROSCOPY)
	3-0615	IONIZATION SPECTROMETER FOR ELEMENTAL ANALYSIS OF SURFACES.
	4-0846	ELECTRON BINDING ENERGIES OF BARIUM FROM THE SECONDARY ELECTRON YIELD SPECTRUM.
	4-0847	TOTAL ELECTRON IMPACT IONIZATION CROSS SECTIONS ON K-SHELLS OF SURFACE ATOMS. (A
	5-1351	IONIZATION SPECTROSCOPY OF CONTAMINATED METAL SURFACES.
	5-1352	IONIZATION SPECTROSCOPY OF SURFACES.
	5-1962	DEPOSITION AND AUGER ANALYSIS OF DEPOSITED SILICON DIOXIDE ON ALUMINUM(X) GALLIU
GERSTEN JI	2-0361	MANY BODY EFFECTS IN AUGER DEEXCITATION OF ATOMS NEAR SOLIDS.
GERSTENBERG D	5-1473	EFFECT OF LIGHT ELEMENTS NITROGEN, CARBON AND OXYGEN ON THE PHYSICAL PROPERTIES
GERVAIS A	4-0739	ENERGY ANALYSIS OF BACKSCATTERED ELECTRONS FROM A SILICON SINGLE CRYSTAL UNDER A
GETTINGS M	3-0601	BEAM EFFECTS IN AUGER ELECTRON SPECTROSCOPY REVEALED BY X-RAY PHOTOELECTRON SPEC
	5-1353	A PRELIMINARY STUDY OF PURE METAL SURFACES USING AUGER ELECTRON SPECTROSCOPY X-R
GIAMEI AF	6-2053	CONFIRMATION OF SULFUR EMBRITTLEMENT IN NICKEL ALLOYS. (AUGER ELECTRON SPECTROSC
GIANTURCO FA	2-0327	INNER SHELL DOUBLE IONIZATION AND CHEMICAL BONDING. PART-1: AUGER ELECTRONS. (KL
	2-0362	MULTIPLE EXCITATION PROCESSES IN TRANSITION METAL ATOMS. PART-2: INNER AUGER ELE
GIBSON MJ	5-1354	OBSERVATIONS OF THE EPITAXIAL GROWTH OF NICKEL AND COPPER ON (111) SILVER. (BY A
GILLET E	5-1355	AUGER SPECTROMETRIC AND MASS SPECTROMETRIC STUDY OF THE CHEMISORPTION OF CARBON
GILLET M	5-1355	AUGER SPECTROMETRIC AND MASS SPECTROMETRIC STUDY OF THE CHEMISORPTION OF CARBON
GIORDANO S	6-2137	EFFECT OF SURFACE METALLURGY ON THE PENETRATION DEPTH AND RF BREAKDOWN FIELD OF
GISON MJ	5-1356	DIFFUSION OF SILVER THROUGH EPITAXIAL FILMS OF COPPER AND NICKEL ON SILVER (111)
GIULIANO CR	1-0015	LASER WINDOW SURFACE FINISHING AND COATING TECHNOLOGY. (AUGER ELECTRON SPECTROSC
GJOSTEIN NA	1-0063	MODERN TECHNIQUES FOR SURFACE STUDIES.
	1-0064	TECHNOLOGICAL APPLICATION OF AUGER ELECTRON SPECTROSCOPY.
	5-1185	DETECTION OF IMPURITIES ON COPPER SURFACES BY LEED- AUGER ELECTRON SPECTROSCOPY
	6-2081	AES AND TEMPER EMBRITTLEMENT OF STEELS.
GLACHANT A	5-1294	LOW ENERGY ELECTRON DIFFRACTION, AUGER ELECTRON SPECTROSCOPY, AND WORK FUNCTION

GLICK AJ	4-0871	PLASMON SATELLITE IN THE AUGER EMISSION SPECTRUM OF METALS. (ALUMINUM)
GLUPE G	2-0363	ABSOLUTE ELECTRON IMPACT IONIZATION CROSS SECTIONS OF ATOMIC NITROGEN, ATOMIC OX
	4-0848	A NEW METHOD FOR MEASURING ELECTRON IMPACT IONIZATION CROSS SECTIONS OF INNER SH
GNUCHEV NM	4-0736	RECORDING OF COMPONENTS ON THE SURFACE OF FUSED EMITTERS. (AUGER SPECTRA)
	4-0849	AUGER SPECTROSCOPY OF A BARIUM- METAL FILM SYSTEM.
	5-1357	AUGER SPECTROSCOPY OF THE BARIUM- PLATINUM FILM SYSTEM.
	5-1358	RECORDING THE COMPONENTS ON THE SURFACE OF A BINARY ALLOY. (AUGER SPECTROSCOPY)
	5-1359	EFFECT OF STRUCTURE ON THE AUGER SPECTRA OF BARIUM FILMS ON PLATINUM
GOBBY PL	2-0398	MATRIX ELEMENTS DEPENDENCE OF OPTICAL EXCITATION AND AUGER DECAY OF 5D CORE HOLE
	2-0422	PHOTOEMISSION STUDIES OF CORE EXCITON DECAY IN POTASSIUM IODIDE.
GODGE W	4-1050	AUGER ELECTRON EMISSION SPECTRA FROM FOIL AND GAS EXCITED CARBON BEAMS.
GOFF RF	5-1360	DETECTION OF IMPURITIES ON A SILICON (111) SURFACE BY AUGER- LEED ANALYSIS.
	5-1765	ELECTRONS OR IONS. (IN SURFACE ANALYSIS)
GOLDEN DE	3-0616	AN ENERGY MODULATED (CYLINDRICAL) HIGH ENERGY RESOLUTION ELECTRON SPECTROMETER.
GOLDENBLUM A	2-0324	AUGER EFFECT AND RADIATIVE RECOMBINATION IN N-GALLIUM ARSENIDE- P-ALUMINUM GALLI
	2-0325	EMISSION SPECTRA OF SILICON COMPENSATED ALUMINUM(X) GALLIUM(1-X) ARSENIDE ELECTR
GOLDOBIN AN	5-1361	COMPOSITION OF THE SURFACE OF COPPER- MAGNESIUM OXIDE CATALYSTS STUDIED BY AUGER
GOLDSCHMIDT M	2-0336	DECAY OF HOLE STATES IN OXYGEN-16 VIA A "NUCLEAR AUGER EFFECT".
GOLDSTEIN B	5-1362	ELECTRON BEAM INDUCED REDUCTION OF CARBON CONCENTRATION ON BERYLLIUM OXIDE AND I
	5-1363	LEED, AUGER AND PLASMON STUDIES OF NEGATIVE ELECTRON AFFINITY ON SILICON PRODUCE
	5-1364	LOW ENERGY ELECTRON DIFFRACTION- AUGER CHARACTERIZATION OF GALLIUM ARSENIDE DURI
	5-1365	DETERMINATION OF THE COMPOSITION OF GLASS SURFACES BY AUGER SPECTROSCOPY.
	5-1366	PREFERENTIAL EVAPORATION OF INDIUM FROM GALLIUM(X) INDIUM(1-X) ARSENIDE. (AUGER
GOLDSZTAUB S	4-0792	STUDY OF MUSCOVITE AND SILICON BY AUGER ELECTRON SPECTROSCOPY.
	4-0793	AUGER ELECTRON SPECTROSCOPY STUDY OF SOME OXYGENATED SILICON COMPOUNDS.
	4-0794	AUGER SPECTROSCOPY OF SILICON IN SILICATES.
	4-0795	AUGER ELECTRON SPECTROSCOPY OF INSULATING SILICON COMPOUNDS.
	4-0796	STUDY OF AMORPHOUS AND CRYSTALLINE SILICATES BY AUGER ELECTRON SPECTROSCOPY.
	5-1298	MUSCOVITE CLEAVAGE STUDIED BY AUGER ELECTRON SPECTROSCOPY.
GOLOVANOVA AN	3-0617	DRYING OF BUTADIENE- NITRILE RUBBERS IN AN AUGER MACHINE.
GOLUB S	5-1367	INTERACTION OF SULFUR DIOXIDE WITH TUNGSTEN. (AUGER ANALYSIS)
GOMOYUNOVA MV	4-0850	SECONDARY ELECTRON EMISSION FROM TUNGSTEN AND MOLYBDENUM.
GONZALES AJ	6-2034	FAILURE ANALYSIS APPLICATIONS OF AUGER ELECTRON SPECTROSCOPY.
GOPALARAMAN CP	1-0065	ROLE OF ULTRAHIGH VACUUM TECHNIQUES IN BASIC RESEARCH. (REVIEW, AUGER, MASS SPEC
	3-0618	SENSITIVITY VARIATIONS IN BAYARD- ALPERT GAUGES CAUSED BY AUGER EMISSION AT COLL
GORA T	4-0851	X-RAY PHOTOELECTRON SPECTRUM OF DIAMOND. (AUGER TRANSITIONS)
GORDEEV YS	2-0234	DETECTION OF A NEW TYPE OF AUGER TRANSITIONS IN ATOMS WITH TWO INTERNAL VACANCIE
	2-0235	OBSERVATION OF THREE ELECTRON AUGER TRANSITIONS IN ATOMS WITH TWO INNER VACANCIE
GORINI G	2-0305	MEASUREMENT OF THE AUGER EFFECT IN THE MU-4 HELIUM(2S) (+) IONIC SYSTEM.
GORODETSKII DA	5-1368	STUDY OF BARIUM FILMS ON A (110) TUNGSTEN FACE BY THE DIFFRACTION OF SLOW ELECTR
GORSHKOV VG	2-0364	PARITY NONCONSERVATION IN RADIATIONLESS (AUGER) TRANSITIONS IN MU MESIC ATOMS.
GOSCINSKI O	2-0393	K X-RAY AND AUGER ELECTRON ENERGIES FOR NEON BY A TRANSITION OPERATOR METHOD.
	2-0510	TRANSITION POTENTIAL MODELS FOR AUGER ELECTRON SPECTRA.
GOTO K	2-0365	AUGER AND SECONDARY ELECTRONS EXCITED BY BACKSCATTERED ELECTRONS.
	2-0366	AUGER AND SECONDARY ELECTRONS EXCITED BY BACKSCATTERED ELECTRONS. APPROACH TO QU
	3-0619	COMPACT, INEXPENSIVE HIGH RESOLUTION RETARDING FIELD ENERGY ANALYZER. (AUGER ELE
	4-0852	METHOD FOR DETECTING FINE STRUCTURE IN THE SECONDARY ELECTRON EMISSION YIELD AND
	4-0853	FINE STRUCTURES AND ENERGY DISTRIBUTION OF SECONDARY ELECTRON EMISSION FROM SILI
GOTOH Y	5-1414	AUGER ELECTRON SPECTROSCOPY AND MICROANALYSIS OF SOLID SURFACES.
GOULD RW	5-1425	CORROSION AND STRUCTURAL CHANGES IN GLASS. (AUGER ELECTRON SPECTROSCOPY)
GOUTTE R	4-0759	SECONDARY ELECTRON EMISSION FROM SODIUM CHLORINE SINGLE CRYSTAL BOMBARDED BY RAR
GRAAS H	5-1582	STUDY OF METAL SURFACES BY IONIC MICROANALYSIS AND AUGER SPECTROMETRY.
GRABKE HJ	5-1369	EQUILIBRIUM SURFACE SEGREGATION OF CARBON ON IRON (100) FACES. (AUGER EFFECT)
	5-1940	CHEMICAL AND STRUCTURAL ANALYSIS OF IRON SURFACES BY MEANS OF AUGER ELECTRONIC S
	6-2035	NEW METHODS IN SURFACE ANALYSIS AND THEIR APPLICATION FOR STUDIES ON IRON AND ST
GRANT JT	1-0066	QUANTITATIVE AUGER ANALYSIS USING INTEGRATION TECHNIQUES.
	1-0067	APPLICATION OF DYNAMIC BACKGROUND SUBTRACTION TO QUANTIFICATION OF AUGER ELECTRO
	1-0068	SPECTRUM SUBTRACTION TECHNIQUES IN AUGER ELECTRON SPECTROSCOPY.
	1-0069	USE OF ANALOG INTEGRATION IN DYNAMIC BACKGROUND SUBTRACTION FOR QUANTITATIVE AUG
	1-0071	A BIBLIOGRAPHY OF LOW ENERGY ELECTRON DIFFRACTION AND AUGER ELECTRON SPECTROSCOP
	2-0367	EFFECT OF MODULATION AMPLITUDE ON ELECTRON EXCITED AUGER DATA FROM TITANIUM.
	3-0620	CORRECTIONS OF AUGER ELECTRON SIGNAL STRENGTHS FOR MODULATION AMPLITUDE DISTORTI
	3-0700	EASY METHOD TO ACCURATELY ALIGN ION BOMBARDMENT GUNS FOR DEPTH PROFILING IN AUGE
	4-0819	CHEMICAL EFFECTS IN KLL AUGER ELECTRON SPECTRUM FROM OXYGEN.
	4-0854	AUGER ELECTRON SPECTROSCOPY OF SILICON.
	4-0855	COMPARISON OF MAGNESIUM L2,3MM AUGER CURRENTS USING ELECTRON AND ION EXCITATION.
	4-0856	AUGER CURRENT MEASUREMENTS FOR QUANTITATIVE AUGER ELECTRON SPECTROSCOPY OF SOLID
	4-0857	COMPARISON OF AUGER SPECTRA OF MAGNESIUM, ALUMINUM, AND SILICON EXCITED BY LOW E
	4-0862	CHEMICAL EFFECTS ON THE KLL AUGER ELECTRON SPECTRUM FROM SURFACE CARBON.
	4-0863	CHEMICAL SHIFTS IN AUGER ELECTRON SPECTROSCOPY.
	4-0864	AUGER ELECTRON SPECTROSCOPY OF TRANSITION METALS.
	4-0865	AUGER ELECTRON SPECTROSCOPY OF SOME REFRACTORY METALS.
	4-0866	SOME PROBLEMS IN ANALYSIS OF AUGER ELECTRON SPECTRA.
	4-0867	CHEMICAL EFFECTS IN AUGER ELECTRON SPECTROSCOPY (STATES OF CARBON).
	4-0868	IDENTIFICATION OF THE CHEMICAL STATE OF ADSORBED SPECIES WITH AUGER ELECTRON SPE
	4-0869	ION EXCITED AUGER SPECTRA OF ALUMINUM.
	4-0885	CHEMICAL EFFECTS IN THE M4,5NN AUGER SPECTRUM OF MOLYBDENUM (110) DUE TO ADSORPT
	5-1370	LEED STUDY OF IRIDIUM (100) SURFACE.
	5-1371	STUDY OF IRIDIUM (100) SURFACE USING LEED AND AUGER ELECTRON SPECTROSCOPY.
	5-1372	STUDIES ON THE IRIDIUM (111) SURFACE USING LEED AND AUGER ELECTRON SPECTROSCOPY.
	5-1373	COMBINED LEED AND AES STUDIES OF INDIUM ARSENIDE SURFACES.
	5-1374	STRUCTURE OF THE PLATINUM (100) SURFACE.
	5-1375	AUGER STUDIES OF (111) SILICON SURFACES.
	5-1376	AUGER ELECTRON SPECTROSCOPY STUDIES OF CARBON OVERLAYERS ON METAL SURFACES.
	5-1377	LEED STUDY OF PLATINUM (100) SURFACE.
	5-1378	IDENTIFICATION OF THE FORM OF CARBON AT A SILICON (100) SURFACE USING AUGER ELEC
	5-1379	COMBINED LEED AND AUGER ELECTRON SPECTROSCOPY STUDIES OF SILICON, GERMANIUM, GAL
	5-1380	STUDY OF RUTHENIUM (0001) AND RHODIUM (111) SURFACES USING LEED AND AUGER ELECTR
	5-1381	STUDY OF INDIUM ARSENIDE (111) AND (-1-1-1) SURFACES USING LEED AND AUGER ELECTR
	5-1382	AUGER STUDIES OF CLEAVED SILICON (111) SURFACES.

GRANT JT	5-1383	NATURE OF SILICON (111) SURFACES.
	5-1384	QUANTITATIVE COMPARISON OF TITANIUM AND TITANIUM MONOXIDE SURFACES USING AUGER E
	5-1457	FACTORS AFFECTING DEPTH PROFILING MEASUREMENTS USING AUGER ELECTRON SPECTROSCOPY
	5-1458	AUGER ELECTRON SPECTROSCOPY STUDIES OF CARBON MONOXIDE ON NICKEL. SPECTRAL LINE
	5-1459	SOME ASPECTS OF AN AES AND XPS STUDY OF THE ADSORPTION OF OXYGEN ON NICKEL.
	6-2038	CHEMICAL SHIFTS IN AUGER ELECTRON SPECTROSCOPY FROM INITIAL OXIDATION OF TANTALU
GRANT RW	4-0858	AUGER ELECTRON SPECTROSCOPY OF LUNAR SAMPLES.
GRAY H	1-0216	DIRECT COMPARISON OF AUGER, SECONDARY ION MASS SPECTROSCOPY, AND PROTON RESONANC
GRAY HF	5-1391	EFFECTS OF SULFUR, BARIUM, AND CALCIUM ON IMPREGNATED CATHODE SURFACES. (AES)
GREEN JAS	6-2036	STRESS CORROSION CRACKING OF AN ALUMINUM- 5% MAGNESIUM TERNARY AND VARIOUS QUATE
GREENE JE	5-1385	MORPHOLOGICAL AND ELECTRICAL PROPERTIES OF RF SPUTTERED YTTRIA-DOPED ZIRCONIA TH
GREENFIELD IG	6-2037	EFFECT OF A THIN DIFFUSED SOLUTE LAYER ON THE NUCLEATION OF FATIGUE CRACKS. (AUG
GRENGA HE	5-1386	AUGER ANALYSIS OF SURFACE FILMS ON SILVER- TIN.
GRIBKOVSKII VP	2-0368	AUGER RECOMBINATION IN SEMICONDUCTORS DURING INTENSE EXCITATION.
GRIFFITHS BW	3-0621	ULTRAHIGH VACUUM SCANNING ELECTRON MICROSCOPE WITH A FIELD EMISSION CATHODE AND
	3-0622	ULTRAHIGH VACUUM SCANNING ELECTRON MICROSCOPE WITH AUGER ANALYSIS FACILITIES.
	3-0679	SCANNING AUGER ELECTRON MICROSCOPE FOR SURFACE STUDIES.
GRIGOREV EP	4-0743	AUGER L-SERIES ELECTRONIC SPECTRA ACCOMPANYING THE DECAY OF ERBIUM-160, YTTERBIU
	4-0744	L-SERIES AUGER ELECTRON SPECTRA FOR ELEMENTS WITH Z = 66-70.
GRIGOREV VB	3-0617	DRYING OF BUTADIENE- NITRILE RUBBERS IN AN AUGER MACHINE.
GRIN VF	2-0369	RECOMBINATION IN DONOR ACCEPTOR COMPLEXES IN CADMIUM SULFIDE SINGLE CRYSTALS. (A
GRIVICKAS V	4-1103	ELECTRON HOLE SCATTERING AND (AUGER) RECOMBINATION OF NONEQUILIBRIUM CHARGE CARR
GROENEVELD KO	1-0070	SURFACE ANALYSIS WITH HEAVY ION INDUCED AUGER ELECTRONS. BASIC PROPERTIES OF A N
	2-0370	AUGER ELECTRONS FROM FOIL EXCITED HEAVY ION BEAMS. (OF NEON)
	2-0371	AUGER TRANSITIONS LEADING TO DEFINED FINAL CHARGE STATES BY A COINCIDENCE TECHNI
	4-0859	DOPPLER SHIFTED AUGER ELECTRONS FROM FOIL EXCITED HEAVY ION BEAMS.
	4-0860	BEAM FOIL EXCITED AUGER TRANSITIONS IN NEON.
	5-1387	SURFACE ANALYSIS WITH HEAVY ION INDUCED AUGER ELECTRONS.
GROSS EF	2-0372	EXCITON SPECTRA OF HEAVILY DOPED CADMIUM SULFIDE SINGLE CRYSTALS. (AUGER PROCESS
GUENNOU H	2-0373	SECONDARY ELECTRON EMISSION OF SOLID ALUMINUM AND MAGNESIUM TARGETS.
	4-0822	AUGER SPECTRA OF MAGNESIUM AND ALUMINUM.
GUENTHER AH	2-0425	QUANTITATIVE ASPECTS OF A SOFT X-RAY LASER. (AUGER EFFECT)
GUENTHNER CS	5-1388	ANODIC OXIDATION OF GALLIUM ARSENIDE PHOSPHIDE. (AES)
GUGGENHEIM HJ	2-0565	SIZE EFFECT IN IONIC CHARGE RELAXATION FOLLOWING AUGER EFFECT.
GUILLAUD C	4-0759	SECONDARY ELECTRON EMISSION FROM SODIUM CHLORINE SINGLE CRYSTAL BOMBARDED BY RAR
GUILLOT C	5-1579	CARBON MONOXIDE ADSORPTION KINETICS ON MOLYBDENUM (100). (AUGER SPECTROSCOPY)
	5-1772	OXYGEN ADSORPTION ON CLEAN MOLYBDENUM (100) SURFACES.
GUNDRY PM	5-1389	STRUCTURAL STUDIES OF THE ADSORPTION OF CESIUM AND OXYGEN ON SILICON (100).
GUPTA YP	1-0043	AUGER ELECTRON SPECTROSCOPY.
	5-1277	AUGER ELECTRON SPECTROSCOPY OF LUNAR MATERIAL.
GUSEINOV G	2-0258	TEMPERATURE DEPENDENCE OF THE EDGE EMISSION OF GALLIUM SELENIDE SINGLE CRYSTALS
GUTIERREZ WA	4-0861	SECONDARY ELECTRON EMISSION FROM GALLIUM ARSENIDE.
GUTTMAN M	6-2070	SEGREGATION OF MANGANESE AND ANTIMONY TO THE GRAIN BOUNDARIES OF TEMPER EMBRITTL
GYORFFY BL	4-0888	CLUSTER CALCULATIONS OF THE SOFT X-RAY L3 EMISSION SPECTRA OF NICKEL AND COPPER.
HAAS GA	5-1390	SURFACE ANALYSIS USING COMPLEMENTARY ELECTRONIC AND CHEMICAL MEASUREMENTS. (AUGE
	5-1391	EFFECTS OF SULFUR, BARIUM, AND CALCIUM ON IMPREGNATED CATHODE SURFACES. (AES)
	5-1889	DIFFUSION MEASUREMENTS IN THIN FILMS UTILIZING WORK FUNCTION CHANGES: CHROMIUM I
	5-1890	AUGER EXAMINATION OF CONTAMINANTS IN THIN FILM METALLIZATIONS.
HAAS TW	1-0049	RECENT ADVANCES IN AES.
	1-0066	QUANTITATIVE AUGER ANALYSIS USING INTEGRATION TECHNIQUES.
	1-0067	APPLICATION OF DYNAMIC BACKGROUND SUBTRACTION TO QUANTIFICATION OF AUGER ELECTRO
	1-0068	SPECTRUM SUBTRACTION TECHNIQUES IN AUGER ELECTRON SPECTROSCOPY.
	1-0069	USE OF ANALOG INTEGRATION IN DYNAMIC BACKGROUND SUBTRACTION FOR QUANTITATIVE AUG
	1-0071	A BIBLIOGRAPHY OF LOW ENERGY ELECTRON DIFFRACTION AND AUGER ELECTRON SPECTROSCOP
	1-0166	HIGH SPATIAL RESOLUTION AUGER SPECTROSCOPY AND AUGER INTEGRATION APPLICATIONS.
	1-0187	INTEGRAL AUGER INFORMATION VIA TAILORED MODULATION TECHNIQUES.
	2-0367	EFFECT OF MODULATION AMPLITUDE ON ELECTRON EXCITED AUGER DATA FROM TITANIUM.
	3-0620	CORRECTIONS OF AUGER ELECTRON SIGNAL STRENGTHS FOR MODULATION AMPLITUDE DISTORTI
	3-0700	EASY METHOD TO ACCURATELY ALIGN ION BOMBARDMENT GUNS FOR DEPTH PROFILING IN AUGE
	4-0819	CHEMICAL EFFECTS ON KLL AUGER ELECTRON SPECTRUM FROM OXYGEN.
	4-0854	AUGER ELECTRON SPECTROSCOPY OF SILICON.
	4-0856	AUGER CURRENT MEASUREMENTS FOR QUANTITATIVE AUGER ELECTRON SPECTROSCOPY OF SOLID
	4-0857	COMPARISON OF AUGER SPECTRA OF MAGNESIUM, ALUMINUM, AND SILICON EXCITED BY LOW E
	4-0862	CHEMICAL EFFECTS ON THE KLL AUGER ELECTRON SPECTRUM FROM SURFACE CARBON.
	4-0863	CHEMICAL SHIFTS IN AUGER ELECTRON SPECTROSCOPY.
	4-0864	AUGER ELECTRON SPECTROSCOPY OF TRANSITION METALS.
	4-0865	AUGER ELECTRON SPECTROSCOPY OF SOME REFRACTORY METALS.
	4-0866	SOME PROBLEMS IN ANALYSIS OF AUGER ELECTRON SPECTRA.
	4-0867	CHEMICAL EFFECTS IN AUGER ELECTRON SPECTROSCOPY (STATES OF CARBON).
	4-0868	IDENTIFICATION OF THE CHEMICAL STATE OF ADSORBED SPECIES WITH AUGER ELECTRON SPE
	4-0869	ION EXCITED AUGER SPECTRA OF ALUMINUM.
	4-0870	SOFT X-RAY APPEARANCE POTENTIAL SPECTROSCOPY IN A DISPLAY LEED SYSTEM.
	4-0885	CHEMICAL EFFECTS IN THE M4,5NN AUGER SPECTRUM OF MOLYBDENUM (110) DUE TO ADSORPT
	5-1303	CHEMISORPTION ON SINGLE CRYSTAL MOLYBDENUM (112) SURFACES.
	5-1304	SOME PROPERTIES OF THE RHENIUM (0001) SURFACE.
	5-1305	FURTHER STUDIES OF GAS ADSORPTION ON MOLYBDENUM (100) SURFACE.
	5-1374	STRUCTURE OF THE PLATINUM (100) SURFACE.
	5-1375	AUGER STUDIES OF (111) SILICON SURFACES.
	5-1376	AUGER ELECTRON SPECTROSCOPY STUDIES OF CARBON OVERLAYERS ON METAL SURFACES.
	5-1377	LEED STUDY OF PLATINUM (100) SURFACE.
	5-1378	IDENTIFICATION OF THE FORM OF CARBON AT A SILICON (100) SURFACE USING AUGER ELEC
	5-1379	COMBINED LEED AND AUGER ELECTRON SPECTROSCOPY STUDIES OF SILICON, GERMANIUM, GAL
	5-1380	STUDY OF RUTHENIUM (0001) AND RHODIUM (111) SURFACES USING LEED AND AUGER ELECTR
	5-1381	STUDY OF INDIUM ARSENIDE (111) AND (-1-1-1) SURFACES USING LEED AND AUGER ELECTR
	5-1382	AUGER STUDIES OF CLEAVED SILICON (111) SURFACES.
	5-1383	NATURE OF SILICON (111) SURFACES.
	5-1384	QUANTITATIVE COMPARISON OF TITANIUM AND TITANIUM MONOXIDE SURFACES USING AUGER E
	5-1459	SOME ASPECTS OF AN AES AND XPS STUDY OF THE ADSORPTION OF OXYGEN ON NICKEL.
	5-1841	AUGER ELECTRON SPECTROSCOPY STUDY OF CATHODE SURFACES DURING ACTIVATION AND POIS
	5-1894	AUGER SPECTROSCOPY STUDY OF THE ADSORPTION OF RUBIDIUM ON MOLYBDENUM (100).

HAAS TW	5-1895	QUANTITATIVE AUGER ELECTRON SPECTROSCOPY AND LOW ENERGY ELECTRON DIFFRACTION STU
	5-1896	LEED- AUGER SPECTROSCOPY STUDY OF THE ADSORPTION OF ALKALI METALS ON MOLYBDENUM
	6-2020	BEHAVIOR OF REFRACTORY METAL SURFACES IN ULTRAHIGH VACUUM AS OBSERVED BY LEED AN
	6-2021	AUGER ELECTRON SPECTROSCOPY: METALLURGICAL APPLICATIONS.
	6-2038	CHEMICAL SHIFTS IN AUGER ELECTRON SPECTROSCOPY FROM INITIAL OXIDATION OF TANTALU
	6-2039	APPLICATIONS OF AUGER ELECTRON SPECTROSCOPY TO METALLURGICAL PROBLEMS.
	6-2112	STUDY OF GRAIN BOUNDARY FRACTURE SURFACES IN DOPED TUNGSTEN- RHENIUM ALLOYS. (AU
HABEL L	6-2040	OBSERVATIONS ON THE SURFACE OXIDATION OF STEEL DURING ANNEALING IN HYDROGEN-FREE
HABIG PS	5-1541	SUPERCONDUCTING TRANSITION TEMPERATURES OF REACTIVELY SPUTTERED FILMS OF TANTALU
HABRAKEN L	5-1582	STUDY OF METAL SURFACES BY IONIC MICROANALYSIS AND AUGER SPECTROMETRY.
	6-2075	POSSIBILITIES OF THE ION MICROPROBE MASS ANALYZER AND AUGER SPECTROMETER IN PHYS
HACKETT LH	6-2083	EFFECT OF SOLUTE ELEMENTS ON TEMPER EMBRITTLEMENT OF LOW ALLOY STEELS. (IMPURITY
HAFNER E	5-1942	SURFACE STUDIES FOR QUARTZ RESONATORS. (CLEANING, AUGER SPECTROSCOPY)
	5-1943	SURFACE PREPARATION AND CHARACTERIZATION TECHNIQUES (AUGER, CLEANING) FOR QUARTZ
HAFNER H	3-0623	COMPARISON OF SPHERICAL DEFLECTOR AND CYLINDRICAL MIRROR ANALYZERS.
HAGEN AL	4-0871	PLASMON SATELLITE IN THE AUGER EMISSION SPECTRUM OF METALS. (ALUMINUM)
HAGMANN S	4-0872	SEARCH FOR CORRELATION EFFECTS IN THE M4,5N4,5N4,5 AUGER SPECTRUM OF XENON.
HAGSTRUM HD	2-0374	ELECTRON EMISSION FROM SOLIDS BY PARTICLES CARRYING POTENTIAL ENERGY. (AUGER NEU
	2-0375	RESONANCE, AUGER, AND AUTOIONIZATION PROCESSES INVOLVING HELIUM(+) (2S) AND HELIU
	4-0775	DIFFRACTION PEAKS IN SECONDARY ELECTRON ENERGY SPECTRA. (NICKEL (100))
	5-1789	CHEMISORPTION OF ATOMIC HYDROGEN ON THE SILICON (111) 7*7 SURFACE. (AUGER EFFECT
HAIRR LM	1-0219	AUGER ELECTRON DOSIMETRY.
HALAS ST	2-0376	CROSS SECTIONS FOR THE PRODUCTION OF NITROGEN(2) (X), NITROGEN(X), AND NITROGEN(2
HALDER NC	4-0873	AUGER ELECTRON SPECTRA OF MAGNESIUM THIN FILMS.
	5-1392	X-RAY PHOTOELECTRON AND AUGER ELECTRON SPECTRA OF MAGNESIUM THIN FILMS.
HALL JM	2-0481	RADIATIVE AUGER EFFECT IN ION- ATOM COLLISIONS.
HALL PM	5-1393	DIFFUSION MECHANISMS IN THE PALLADIUM- GOLD THIN FILM SYSTEM AND THE CORRELATION
HALPERN A	2-0377	RADIATION CHEMICAL RESONANCE EFFECT IN SOLID 5-BROMO DEOXYURIDINE: CHEMICAL CONS
HAMMER R	5-1248	AUGER AND ELLIPSOMETRIC STUDIES OF ULTRATHIN LEAD(II) OXIDE GROWTH ON LEAD.
	5-1249	AUGER ANALYSIS OF THIN OXIDE FILMS ON LEAD- INDIUM ALLOYS.
	5-1250	AUGER AND ELLIPSOMETRIC STUDY OF PHOSPHORUS SEGREGATION IN OXIDIZED DEGENERATE S
	5-1251	AUGER ANALYSIS OF CHLORINE IN HYDROGEN CHLORIDE OR CHLORINE GROWN SILICON DIOXID
HAMRIN K	4-0734	KLL AUGER SPECTRUM OF FLUORINE. (IN FLUORIDE SALTS)
	4-0826	CHEMICAL SHIFT IN AUGER SPECTRA.
HANAWA T	5-1708	GOLD-INDUCED SUPERSTRUCTURES ON SILICON (100) SURFACES AS OBSERVED BY LEED AES.
	5-1876	GROWTH OF GOLD FILM ON GALLIUM ARSENIDE (110) OBSERVED BY LOW ENERGY ELECTRON DI
HANEMAN D	5-1141	SURFACE VALENCE BAND AND PLASMON FEATURES ON CLEAN CLEAVED SILICON.
	5-1767	SILICON (111)-7 STRUCTURE OBTAINED BY CLEAVAGE AT ROOM TEMPERATURE. (AUGER SURFA
	5-1768	AUGER SPECTRA AND LEED PATTERNS FROM NICKEL DEPOSITS ON CLEAVED SILICON.
	5-1769	AUGER SPECTRA AND LEED PATTERNS FROM VACUUM CLEANED SILICON CRYSTALS WITH CALIBR
HAQUE CA	3-0611	MEASUREMENT OF WORK FUNCTION CHANGE IN A DISPLAY TYPE LOW ENERGY ELECTRON DIFFRA
	5-1394	SURFACE CHEMISTRY OF CONTACT MATERIALS. (BY AUGER ELECTRON SPECTROSCOPY)
	5-1395	WORK FUNCTION CHANGES ON CONTACT MATERIALS. (AUGER ELECTRON SPECTROSCOPY)
	6-2041	SURFACE SEGREGATION ON PALLADIUM- SILVER 40 ATOMIC PERCENT, AND PALLADIUM- GOLD
	6-2042	COMBINED MASS SPECTROMETRIC AND AUGER ELECTRON SPECTROSCOPIC TECHNIQUES FOR META
HARMAN R	5-1396	AUGER SPECTROSCOPY AND LEED OF SILICON SURFACES.
	5-1397	AUGER ELECTRON SPECTROSCOPY OF SILICON SURFACES.
HARRIS FM	3-0624	SOME PROPERTIES OF RETARDING FIELD SPECTROMETERS USING POST MONOCHROMATORS. (FOR
HARRIS JM	5-1398	FLUORINE ION IMPLANTATION PROFILES IN GALLIUM ARSENIDE AS DETERMINED BY AES.
	5-1616	AUGER SPECTROSCOPY OF FRACTURE SURFACES OF CERAMICS.
HARRIS JS	2-0555	POTENTIAL PROFILING ACROSS SEMICONDUCTOR JUNCTIONS BY AUGER ELECTRON SPECTROSCOP
	5-1398	FLUORINE ION IMPLANTATION PROFILES IN GALLIUM ARSENIDE AS DETERMINED BY AES.
HARRIS LA	1-0072	AUGER ELECTRON SPECTROSCOPY FOR SURFACE ANALYSIS.
	1-0073	AUGER ELECTRON EMISSION ANALYSIS.
	1-0074	SECONDARY ELECTRON SPECTROSCOPY.
	1-0075	MISCELLANEOUS TOPICS IN AUGER ELECTRON SPECTROSCOPY.
	2-0378	ANGULAR DEPENDENCIES OF ELECTRON EXCITED AUGER EMISSION.
	4-0824	AUGER ELECTRON ANALYSIS OF ELECTROPOLISHED HIGH PURITY ALUMINUM.
	5-1399	ANALYSIS OF MATERIALS BY ELECTRON EXCITED AUGER ELECTRONS.
	5-1400	CARBON EVAPORATION FROM A THORIUM DISPENSER CATHODE OBSERVED BY AUGER ELECTRON E
	5-1401	SOME OBSERVATIONS OF SURFACE SEGREGATION BY AUGER ELECTRON EMISSION.
	5-1402	SURFACE ANALYSIS BY AUGER ELECTRON SPECTROSCOPY.
	5-1403	SECONDARY ELECTRON SPECTROSCOPY. (SURFACE ANALYSIS OF MATERIALS BY ELECTRON EXCI
	5-1404	AUGER SPECTROSCOPY OF SEMICONDUCTOR AND METAL SURFACES.
	5-1405	ELECTRON SPECTROSCOPY ON THIN FILMS. (SCANNING AUGER MICROSCOPE)
	5-1411	IN-PROCESS CONTROL TECHNIQUES FOR COMPLEX SEMICONDUCTOR STRUCTURES. TASK-II. APP
HARRIS PR	4-1016	SECONDARY ELECTRON SPECTROSCOPY OF POLYCRYSTALLINE SILVER.
	5-1406	ELECTRON INDUCED SECONDARY ELECTRON EMISSION SPECTROSCOPY OF SURFACES.
	5-1407	SECONDARY ELECTRON EMISSION OF A ZIRCONIUM- ALUMINUM BULK GETTER. (AUGER SPECTRO
	5-1724	CHARACTERISTIC GAIN PHENOMENA IN POLYCRYSTALLINE COPPER, SILVER, AND ALUMINUM. (
HARRISON KG	1-0112	BEAM FOIL AUGER SPECTROSCOPY.
	2-0519	K-SHELL IONIZATION OF MOLECULAR NITROGEN AND METHANE BY 50- TO 600 KEV HYDROGEN(
	2-0528	X-RAY EMISSION CROSS SECTIONS AND FLUORESCENCE YIELDS FOR LIGHT ATOMS AND MOLECU
HARROWER GA	4-0874	AUGER ELECTRON EMISSION IN THE ENERGY SPECTRA OF SECONDARY ELECTRONS FROM MOLYBD
HARSDORFF M	5-1136	QUANTITATIVE AUGER ELECTRON SPECTROSCOPY OF GOLD CONDENSED ON ROCK SALT AND ON A
	5-1137	HETEROGENEOUS NUCLEATION AND GROWTH OF GOLD ON SUBSTRATES. (AUGER SPECTRA)
	5-1408	EXPERIMENTAL PROBLEMS OF STUDYING THIN FILM NUCLEATION. (ELECTRON MICROSCOPY, AU
HART RK	3-0625	INVESTIGATION OF HIGH RESOLUTION SCANNING TECHNIQUES FOR AUGER ELECTRON SPECTROS
	5-1409	(AUGER ELECTRON SPECTROSCOPY OF) PRECISION SINGLE SIDEBAND CRYSTAL UNITS.
	5-1410	METHODS OF CLEANING CONTAMINANTS FROM QUARTZ SURFACES DURING RESONATOR FABRICATI
	5-1411	IN-PROCESS CONTROL TECHNIQUES FOR COMPLEX SEMICONDUCTOR STRUCTURES. TASK-II. APP
HARTMANN DK	6-2043	EFFECTS OF COPPER AND PHOSPHORUS ON TEMPER EMBRITTLEMENT OF MANGANESE- MOLYBDENU
HASEGAWA M	5-1521	MEASUREMENTS OF SURFACE MATERIAL STATES.
HASEGAWA Y	5-1982	OBSERVATION OF SURFACE COMPOSITION OF COPPER- NICKEL ALLOY BY AUGER SPECTRA IN T
HASHIBA M	3-0636	NEW SEM AUGER MICROANALYZER WITH ULTRAHIGH VACUUM CAPABILITY.
HASHIMOTO H	5-1928	INITIAL OXIDATION STUDIES BY MEANS OF PHOTOELECTRIC WORK FUNCTION MEASUREMENTS A
HAUG A	2-0357	INFLUENCE OF AN ELECTRIC FIELD ON THE AUGER RECOMBINATION IN SEMICONDUCTORS.
	2-0379	INFLUENCE OF SCREENING EFFECTS ON THE AUGER RECOMBINATION IN SEMICONDUCTORS. (SI
	4-0844	INFLUENCE OF AN ELECTRIC FIELD ON THE AUGER RECOMBINATION IN SEMICONDUCTORS.
HAYAKAWA K	1-0076	MICROANALYSIS BY AUGER ELECTRON SPECTROSCOPY. (REVIEW, 15 REFS)

HAYAKAWA K	1-0077	THREE DIMENSIONAL CHARACTERIZATION OF METALS AND SEMICONDUCTORS BY THE AUGER ELE
	5-1412	THREE-DIMENSIONAL ELEMENT ANALYSIS OF SURFACES AND INTERFACES OF SOLIDS BY THE A
	5-1413	AUGER ELECTRON EMISSION MICROGRAPHIC STUDIES OF THE CLEAVAGE SURFACE OF GRAPHITE
	5-1414	AUGER ELECTRON SPECTROSCOPY AND MICROANALYSIS OF SOLID SURFACES.
	5-1415	AUGER ELECTRON EMISSION MICROGRAPHY AND MICROANALYSIS OF A STAINED SURFACE OF PU
	5-1416	AUGER ELECTRON EMISSION MICROGRAPHY AND MICROANALYSIS OF SOLID SURFACES.
	5-1471	SURFACE CONDITION AND THERMIONIC EMISSION OF LANTHANUM HEXABORIDE. (AUGER SPECTR
	5-1677	LEED AND AUGER ELECTRON SPECTROSCOPIC STUDIES ON TUNGSTEN (110) SURFACE.
HAYASHI C	5-1551	EFFECT OF GOLD COATING ON SORPTION CHARACTERISTICS OF CARBON MONOXIDE, SULFUR, A
	5-1552	SPUTTER ETCHING AND AUGER ELECTRON SPECTROSCOPY OF METAL SURFACES.
	5-1554	SPUTTER ETCHING AND AUGER ELECTRON SPECTROSCOPY OF METAL SURFACES.
HAYASHI Y	5-1487	DETECTION OF ORGANIC CONTAMINANTS ON ALUMINUM SURFACES BY AUGER ELECTRON SPECTRO
HAYEK K	5-1420	SULFUR ON SINGLE CRYSTAL PLATINUM PLANES. LOW ENERGY ELECTRON DIFFRACTION AND AU
	5-1421	ADSORPTION OF SULFUR ON THE (100) AND (111) FACES OF PLATINUM. LOW ENERGY ELECTR
HAYMANN P	2-0236	EFFECT OF SURFACE PLASMON OSCILLATIONS ON AUGER ELECTRON EMISSION.
	2-0380	METHOD FOR PACKET OF WAVES IN AUGER ELECTRON-ELECTRON PROCESSES.
	2-0381	DYNAMIC THEORY OF AUGER EFFECT.
HAYNES SK	4-0875	PHENOMENOLOGICAL SYSTEMATICS OF L- AND M-AUGER SPECTRA.
HECKINGBOTTOM R	1-0003	COMBINED AUGER ELECTRON SPECTROSCOPY AND SCANNING ELECTRON MICROSCOPY.
	1-0078	PHYSICAL METHODS OF SURFACE ANALYSIS.
	5-1417	STUDY OF THE NITRIDATION OF SILICON SURFACES BY LOW ENERGY ELECTRON DIFFRACTION
	5-1418	AUGER ELECTRON SPECTROSCOPY AND ITS APPLICATION IN STUDIES OF SOME SILICON SURFA
	5-1902	THERMALLY INDUCED PROCESSES AT GOLD- GALLIUM ARSENIDE INTERFACES: AN ASSESSMENT
	5-1903	RELATIVE ESCAPE DEPTH OF AUGER AND PHOTOELECTRONS FROM A SOLID SURFACE.
HECKMANN M	2-0260	EVIDENCES FOR AUGER RECOMBINATION IN ELECTRON HOLE DROPLETS. (SILICON)
HEDBAVNY P	5-1419	SURFACE PHYSICS OF SOLIDS FROM THE VIEWPOINT OF ULTRAVACUUM. (AUGER SPECTROSCOPY
HEDDLE DWO	3-0626	A COMPARISON OF THE EXTENDURE OF ELECTRON SPECTROMETERS. (USED IN AUGER SPECTROS
HEDMAN J	2-0408	ELECTRON ESCAPE DEPTH IN SILICON. (AUGER ENERGIES)
	2-0409	ESCAPE DEPTHS OF X-RAY EXCITED (AUGER) ELECTRONS.
HEEGEMANN W	5-1420	SULFUR ON SINGLE CRYSTAL PLATINUM PLANES. LOW ENERGY ELECTRON DIFFRACTION AND AU
	5-1421	ADSORPTION OF SULFUR ON THE (100) AND (111) FACES OF PLATINUM. LOW ENERGY ELECTR
HEFFNER R	2-0299	EFFECT OF MULTIPLE IONIZATION ON THE FLUORESCENCE YIELD OF NEON. (K-AUGER AND K
HEILAND W	2-0382	LOW ENERGY ION SCATTERING: ELASTIC AND INELASTIC EFFECTS. (AUGER NEUTRALIZATION)
	3-0704	SORBAS: AN APPARATUS FOR INVESTIGATING ION SCATTERING FROM SURFACES AT ENERGIES
	5-1422	OXYGEN ADSORPTION ON (110) SILVER.
	5-1423	SURFACE ANALYTICAL STUDIES USING ION SCATTERING SPECTROMETRY, AUGER ELECTRON SPE
	5-1869	DIRECT COMPARISON OF LOW ENERGY ION BACKSCATTERING WITH AUGER ELECTRON SPECTROSC
	5-1870	LOW ENERGY ION SCATTERING AND AUGER ELECTRON SPECTROSCOPY STUDIES OF CLEAN NICKE
	5-1871	SURFACE ANALYSIS WITH LOW ENERGY ION SCATTERING.
HEIN MA	2-0284	THEORETICAL K-SHELL AUGER RATES, TRANSITION ENERGIES, AND FLUORESCENCE YIELDS FO
HEINRICH KJF	1-0079	SECONDARY ION MASS SPECTROMETRY.
HELDT LA	5-1504	COMPOSITION OF PLATINUM ANODE SURFACES BY AUGER SPECTROSCOPY.
	6-2095	STRESS CORROSION CRACKING OF ALPHA BRASS IN A TARNISHING AMMONIACAL ENVIRONMENT:
HELM RF	5-1427	HEED AND AES STUDY OF SILICON EPITAXY BY SILANE PYROLYSIS.
	5-1428	SILICON HOMOEPITAXIAL THIN FILMS VIA SILANE PYROLYSIS: HEED AND AUGER ELECTRON S
HELMS CR	5-1186	CHANGE IN THE SURFACE ELECTRONIC STRUCTURE OF PLATINUM (100) INDUCED BY RECONSTR
	5-1216	SURFACE SEGREGATION IN ALLOYS: EQUILIBRIUM SEGREGATION OF GOLD TO THE (111) SURF
	5-1424	DETERMINATION OF THE SURFACE COMPOSITION OF COPPER- NICKEL ALLOYS FOR CLEAN AND
	6-2044	OBSERVATION OF THE SEGREGATION OF COPPER TO THE SURFACE OF A CLEAN ANNEALED 50-5
	6-2045	VERIFICATION OF THE METAL-RICH SURFACE MODEL FOR THE OXIDATION OF STRONTIUM, BY
HENCH LL	5-1264	AUGER SPECTROSCOPIC ANALYSIS OF BIOGLASS CORROSION FILMS.
	5-1265	AQUEOUS CORROSION OF SODA- SILICA AND SODA- LIME- SILICA GLASS. (AES)
	5-1425	CORROSION AND STRUCTURAL CHANGES IN GLASS. (AUGER ELECTRON SPECTROSCOPY)
	5-1704	ANALYSIS OF GLASS CONTAINER COATINGS BY AUGER ELECTRON SPECTROSCOPY.
HENDERSON RC	5-1426	SILICON CLEANING WITH HYDROGEN PEROXIDE SOLUTIONS: HIGH ENERGY ELECTRON DIFFRA
	5-1427	HEED AND AES STUDY OF SILICON EPITAXY BY SILANE PYROLYSIS.
	5-1428	SILICON HOMOEPITAXIAL THIN FILMS VIA SILANE PYROLYSIS: HEED AND AUGER ELECTRON S
HENKE BL	1-0080	LOW ENERGY PHOTO- AUGER ELECTRON SPECTROSCOPY.
	3-0627	INTRODUCTION TO LOW ENERGY X-RAY AND ELECTRON ANALYSIS. (ANALYZER FOR PHOTO AUGE
	5-1429	SURFACE CHARACTERIZATION BY LOW ENERGY PHOTO AUGER ELECTRON SPECTROSCOPY.
HENNEQUIN JF	2-0383	EMISSION OF AUGER ELECTRONS BY ATOMS IN A METALLIC TARGET SUBJECTED TO AN IONIC
	2-0547	ARGON AUGER EMISSION BY ARGON(+) ICN BOMBARDMENT OF SOME METALS.
	2-0548	RELATIVE EMISSION INTENSITIES OF AUGER ELECTRONS FROM LIGHT METALS (LITHIUM, SOD
	2-0549	ORIGIN OF ICN INDUCED AUGER ELECTRON EMISSION FROM METALS.
	4-0817	AUGER SPECTRA OF LITHIUM AND SODIUM UNDER ION BOMBARDMENT.
	4-0876	ENERGY DISTRIBUTION OF AUGER ELECTRONS EMITTED FROM BERYLLIUM, MAGNESIUM, ALUMIN
	4-0877	AUGER ELECTRON EMISSION BY ION BOMBARDMENT OF LIGHT METALS.
HENRICH K	3-0621	ULTRAHIGH VACUUM SCANNING ELECTRON MICROSCOPE WITH A FIELD EMISSION CATHODE AND
HENRICH VE	2-0384	EFFECTS OF CESIATION ON SECONDARY ELECTRON EMISSION FROM MAGNESIUM OXIDE- GOLD C
	3-0628	USE OF CYLINDRICAL AUGER SPECTROMETERS FOR RETARDING POTENTIAL SECONDARY ELECTRO
	3-0629	FAST, ACCURATE SECONDARY ELECTRON YIELD MEASUREMENTS AT LOW PRIMARY ENERGIES. (L
	5-1430	DIFFERENTIAL SPUTTERING OF MAGNESIUM OXIDE- GOLD CERMET FILMS AND ITS APPLICATIO
	5-1431	AUGER SPECTROSCOPY STUDIES OF THE OXIDATION OF AMORPHOUS AND CRYSTALLINE GERMANI
	6-2023	OXIDATION STUDIES OF AMORPHOUS AND CRYSTALLINE GERMANIUM FILMS BY AUGER SPECTROS
HENSON JE	3-0603	A MULTIPLE SAMPLE HOLDER FOR USE IN ULTRAHIGH VACUUM.
HENZLER M	5-1432	ADSORPTION OF WATER VAPOR ON CLEAN CLEAVED GERMANIUM. PART-1: STRUCTURAL AND KIN
	5-1820	BINDING STATES OF WATER VAPOR ON CLEAVED GERMANIUM. (LEED, AES)
HERCULES DM	1-0081	ELECTRON SPECTROSCOPY: X-RAY AND ELECTRON EXCITATION. (REVIEW, ESCA, AES)
	1-0082	ELECTRON SPECTROSCOPY. X-RAY AND ELECTRON EXCITATION.
HERFERT RE	5-1433	USE OF ADVANCED ANALYTICAL TECHNIQUES FOR CHARACTERIZATION OF THE CHEMICAL, PHYS
HERMANN G	4-0872	SEARCH FOR CORRELATION EFFECTS IN THE M4,5N4,5N4,5 AUGER SPECTRUM OF XENON.
HERMANSON J	2-0422	PHOTOEMISSION STUDIES OF CORE EXCITON DECAY IN POTASSIUM IODIDE.
HERRMANN KH	1-0083	ELECTRONS AND X-RAYS IN NONDESTRUCTIVE ANALYSIS. (AUGER, PHOTOELECTRON SPECTRA,
HERTZ H	4-0878	VACUUM ULTRAVIOLET EMISSION OF XENON(III) LEVELS EXCITED BY ELECTRON IMPACT VIA
HICINBOTHEM WA	2-0332	CARRIER LIFETIMES IN EPITAXIAL INDIUM ARSENIDE. (AUGER RECOMBINATION)
HICKLIN WH	5-1409	(AUGER ELECTRON SPECTROSCOPY OF) PRECISION SINGLE SIDEBAND CRYSTAL UNITS.
	5-1410	METHODS OF CLEANING CONTAMINANTS FROM QUARTZ SURFACES DURING RESONATOR FABRICATI
HICKSON K	1-0084	ELECTRON SPECTROSCOPY. A NEW TECHNIQUE FOR MATERIAL AND STRUCTURAL ANALYSIS AND
HIGGINBOTHAM IG	3-0612	IMPROVED APPARATUS FOR MEASUREMENT OF AUGER ELECTRON SPECTRA.
	5-1347	(100) SURFACES OF ALKALI HALIDES. PART-1: AIR AND VACUUM CLEANED SURFACES.

HIGGINBOTHAM IG	5-1348	GROWTH OF SILVER ON POTASSIUM CHLORIDE OBSERVED BY LEED AND AUGER EMISSION SPECT
	5-1434	ADHESION AT GOLD MICA INTERFACES. (LEED, AES)
	5-1908	(100) SURFACES OF ALKALI HALIDES. PART-2: ELECTRON STIMULATED DISSOCIATION.
HILL D	2-0385	A FORMALISM FOR THE INDIRECT AUGER EFFECT. PART-1. PART-2.
HILL RD	2-0271	AUGER ELECTRONS FROM THE L-SHELL IN MERCURY.
HILLIER IH	4-0814	EXPERIMENTAL AND THEORETICAL STUDY OF THE AUGER SPECTRUM OF NITROUS OXIDE.
HILLIER IN	4-0879	CONFIGURATION INTERACTION CALCULATIONS OF THE AUGER SPECTRUM OF METHANE, HYDROGE
HILLIG H	4-0880	K-AUGER TRANSITIONS OF THE FREE SODIUM ATOM.
	4-1071	ENERGIES OF EXCITED STATES OF DOUBLY IONIZED MOLECULES BY MEANS OF AUGER ELECTRO
HIRAKI A	5-1435	ELECTRONIC STRUCTURE OF A THIN GOLD FILM DEPOSITED ON A SILICON SUBSTRATE STUDIE
	5-1436	DIFFUSE INTERFACE STRUCTURES IN SOME (SUBSTRATE- VACUUM EVAPORATED FILM) SYSTEMS
	5-1437	POSSIBLE ORIGIN OF LOW TEMPERATURE SILICON MIGRATION IN SILICON- METAL SYSTEMS.
	5-1438	METALLIC STATE OF SILICON IN SILICON- NOBLE METAL VAPOR QUENCHED ALLOYS STUDIED
	5-1673	LOW TEMPERATURE DIFFUSION OF GOLD INTO SILICON IN THE SILICON (SUBSTRATE)- GOLD
	5-1680	AUGER SPECTROSCOPIC OBSERVATION OF SILICON- GOLD MIXED PHASE FORMATION AT LOW TE
	5-1681	DIFFUSE INTERFACE IN SILICON (SUBSTRATE)- GOLD (EVAPORATED FILM) SYSTEM. (AUGER
HIRVONEN JK	5-1261	SCANNING ELECTRON MICROSCOPY, AUGER SPECTROSCOPY, AND ION BACKSCATTERING TECHNIQ
HO PS	1-0085	QUANTITATIVE DEPTH PROFILING BY AUGER ION SPUTTERING TECHNIQUE.
	3-0630	DECONVOLUTION METHOD FOR COMPOSITION PROFILING BY AUGER SPUTTERING TECHNIQUE.
	5-1439	DIFFUSION MEASUREMENT IN THIN FILMS BY AUGER SPUTTER ETCHING TECHNIQUE.
	5-1440	INTERDIFFUSION IN BIMETALLIC POLYCRYSTALLINE FILMS MEASURED BY AUGER ION SPUTTER
	6-2146	ANALYSIS OF GRAIN BOUNDARY DIFFUSION IN BIMETALLIC THIN FILM STRUCTURES USING AU
HOBSON JP	1-0086	THE RELATIONSHIP OF SOLID SURFACES TO VACUUM SCIENCE AND TECHNOLOGY. (REVIEW)
HOCHMAN RF	5-1386	AUGER ANALYSIS OF SURFACE FILMS ON SILVER- TIN.
HODGE W	2-0496	HIGH RESOLUTION ARGON K-AUGER SPECTRUM PRODUCED IN 4 MEV PROTON ON ARGON COLLISI
	2-0497	RELATIVE SATELLITE LINE INTENSITIES IN ARGON K AUGER ELECTRON SPECTRA PRODUCED B
	4-0954	AUGER DECAY OF NEON FOLLOWING ENERGETIC ION BOMBARDMENT.
HODKIN EN	5-1441	CORRELATION OF THE WETTING BEHAVIOR OF SODIUM ON NIMONIC-PE16 ALLOY WITH THE SUR
HOENERLAGE B	4-1116	LINE WIDTH AND AUGER EFFECT IN CALCIUM, STRONTIUM, AND BARIUM FLUORIDES.
HOFFMAN RW	5-1131	EFFECT OF IMPURITIES ON INTRINSIC STRESS IN THIN NICKEL FILMS. (AUGER EFFECT)
HOFFMAN V	5-1962	DEPOSITION AND AUGER ANALYSIS OF DEPOSITED SILICON DIOXIDE ON ALUMINUM(X) GALLIU
HOFFMANN J	6-2046	RECRYSTALLIZATION OF TUNGSTEN FIBERS IN NICKEL MATRIX COMPOSITES. (AUGER ELECTRO
HOFMANN S	5-1442	EVALUATION OF CONCENTRATION DEPTH PROFILES BY SPUTTERING IN SIMS AND AES.
	5-1443	SURFACE SEGREGATION OF OXYGEN IN NIOBIUM. (AES)
	5-1444	STUDY OF FRACTURE SURFACES OF HOT PRESSED SILICON NITRIDE BY AUGER ELECTRON SPEC
	5-1445	AUGER ELECTRON SPECTROSCOPY ON THE FRACTURE OF NICKEL TUNGSTEN MATERIALS AND SIL
	6-2046	RECRYSTALLIZATION OF TUNGSTEN FIBERS IN NICKEL MATRIX COMPOSITES. (AUGER ELECTRO
	6-2047	AUGER ELECTRON SPECTROSCOPY OF CONTRAST FORMING LAYERS ON METALS.
HOLCOMBE CE	4-0807	LEED- AES ELECTRON INDUCED EMISSION STUDIES OF LITHIUM HYDRIDE.
	5-1446	LEED AND AES STUDIES OF LITHIUM HYDRIDE (100) SURFACE.
HOLLAND BW	5-1447	ANGULAR DEPENDENCE OF AUGER ELECTRON EMISSION FROM SOLID SURFACES.
	5-1632	ANGULAR DEPENDENCE OF AUGER ELECTRON EMISSION FROM COPPER (111) AND (100) SURFAC
HOLLOWAY DM	1-0087	APPLICATIONS OF DEPTH PROFILING BY AUGER- SPUTTER TECHNIQUES.
	3-0696	SPECIMEN POSITION EFFECTS ON ENERGY SHIFTS AND SIGNAL INTENSITY IN A SINGLE STAG
	6-2048	QUANTITATIVE DETERMINATION OF SURFACE OXIDE THICKNESS ON DEPOSITED METAL FILMS B
HOLLOWAY PH	3-0631	A VERSATILE SUBSTRATE DESIGN FOR LEED AND AES STUDIES IN ULTRAHIGH VACUUM.
	4-0881	CHEMICAL SHIFTS IN AUGER ELECTRON SPECTRA FROM SILICON IN SILICON NITRIDE.
	5-1448	EFFECT OF SURFACE ROUGHNESS OF AUGER ELECTRON SPECTROSCOPY.
	5-1449	THICKNESS DETERMINATION OF ULTRATHIN FILMS BY AUGER ELECTRON SPECTROSCOPY.
	5-1450	DETECTION BY AUGER ELECTRON SPECTROSCOPY AND REMOVAL BY OZONIZATION OF PHOTORESI
	5-1451	KINETICS OF THE REACTION BETWEEN OXYGEN AND SULFUR ON A NICKEL (111) SURFACE. (S
	5-1452	KINETICS OF THE REACTION OF OXYGEN WITH CLEAN NICKEL SINGLE CRYSTAL SURFACES. PA
	5-1453	CHEMICAL CHANGES IN SECONDARY ELECTRON EMISSION DURING OXIDATION OF NICKEL (100)
	5-1454	CHEMICAL CLEANING FOR THERMOCOMPRESSION BONDING. (REMOVAL OF OXIDE FILMS, AUGER
	5-1455	AUGER ELECTRON SPECTROSCOPIC ANALYSIS OF SILICON NITRIDE ON SILICON.
	5-1456	EFFECTS OF INTERFACIAL OXIDES ON THE CONTACT RESISTANCE OF TUNGSTEN ON HEAVILY D
	5-1861	CORRELATIONS BETWEEN INTERFACIAL OXIDES AND RESISTANCE CHANGES FOR TUNGSTEN CONT
HOLM R	1-0088	PHOTO- AND AUGER ELECTRON SPECTROSCOPY (ESCA) AND SECONDARY ION MASS SPECTROSCOP
	5-1701	MATERIAL ANALYSIS USING THE SCANNING ELECTRON MICROSCOPE. (AUGER SPECTROSCOPY)
HOLMES CP	2-0386	MANY BODY PERTURBATION THEORY APPLIED TO ATOMIC RADIATIONLESS TRANSITIONS.
HOLSCHER AA	5-1196	AUGER SPECTROSCOPIC STUDY OF THE SURFACE COMPOSITION OF PLATINUM- TIN IN ULTRAHI
	6-2005	QUANTITATIVE STUDY OF THE SURFACE COMPOSITION OF BINARY ALLOYS BY AUGER SPECTROS
HOLT SL	4-0861	SECONDARY ELECTRON EMISSION FROM GALLIUM ARSENIDE.
HOLTON R	5-1389	STRUCTURAL STUDIES OF THE ADSORPTION OF CESIUM AND OXYGEN ON SILICON (100).
HOLZI J	2-0499	INVESTIGATION OF THE CHARACTERISTIC AUGER ENERGY LOSSES OF PLATINUM (111).
HOMBROUCK JP	5-1590	LIQUID PHASE EPITAXY OF III-V COMPOUNDS ON INDIUM PHOSPHIDE SUBSTRATES. (AUGER S
HONDROS ED	4-1073	OXIDE FILMS: ELEMENT PROFILES BY AUGER ELECTRON SPECTROSCOPY.
	4-1074	AUGER ELECTRON SPECTROSCOPY STUDY OF BETA ALUMINA ELECTROLYTE.
	6-2049	GRAIN BOUNDARY ACTIVITY MEASUREMENTS BY AUGER ELECTRON SPECTROSCOPY.
	6-2103	GRAIN BOUNDARY EMBRITTLEMENT OF STEELS STUDIED BY AUGER ELECTRON SPECTROSCOPY.
	6-2104	GRAIN BOUNDARY SEGREGATION. (AUGER ELECTRON SPECTROSCOPY)
HONIG RE	1-0089	SURFACE AND THIN FILM ANALYSIS OF SEMICONDUCTOR MATERIALS. (SCANNING AUGER MICRO
HONJO G	5-1624	GROWTH MODE DISCRIMINATION OF THIN FILMS BY AUGER ELECTRON SPECTROSCOPY.
HOOGEWIJS R	4-0882	RELAXATION PHENOMENA INVOLVED IN THE L3M4,5M4,.5M4,.5:1G4 AUGER PROCESS IN ZINC
	4-0883	ATOMIC AND EXTRAATOMIC STATIC RELAXATION ENERGIES IN THE L3M4,5M4,5 AUGER PROCES
	4-0884	DETAILED ANALYSIS OF THE L3M4,5M4,5: 1G4 AUGER TRANSITION IN ATOMIC ZINC: IMPROV
	5-1339	(AUGER) ELECTRON SPECTROSCOPY OF TRANSITION METAL OXIDE SURFACES.
HOOGKAMER TP	2-0387	MEASUREMENTS OF THE L-SHELL FLUORESCENCE YIELDS OF ARGON AND CHLORINE BY ELECTRO
HOOKER MP	1-0068	SPECTRUM SUBTRACTION TECHNIQUES IN AUGER ELECTRON SPECTROSCOPY.
	1-0069	USE OF ANALOG INTEGRATION IN DYNAMIC BACKGROUND SUBTRACTION FOR QUANTITATIVE AUG
	1-0071	A BIBLIOGRAPHY OF LOW ENERGY ELECTRON DIFFRACTION AND AUGER ELECTRON SPECTROSCOP
	3-0700	EASY METHOD TO ACCURATELY ALIGN ION BOMBARDMENT GUNS FOR DEPTH PROFILING IN AUGE
	4-0855	COMPARISON OF MAGNESIUM L2,3MM AUGER CURRENTS USING ELECTRON AND ION EXCITATION.
	4-0856	AUGER CURRENT MEASUREMENTS FOR QUANTITATIVE AUGER ELECTRON SPECTROSCOPY OF SOLID
	4-0857	COMPARISON OF AUGER SPECTRA OF MAGNESIUM, ALUMINUM, AND SILICON EXCITED BY LOW E
	4-0869	ION EXCITED AUGER SPECTRA OF ALUMINUM.
	4-0885	CHEMICAL EFFECTS IN THE M4,5NN AUGER SPECTRUM OF MOLYBDENUM (110) DUE TO ADSORPT
	5-1457	FACTORS AFFECTING DEPTH PROFILING MEASUREMENTS USING AUGER ELECTRON SPECTROSCOPY
	5-1458	AUGER ELECTRON SPECTROSCOPY STUDIES OF CARBON MONOXIDE ON NICKEL. SPECTRAL LINE
	5-1459	SOME ASPECTS OF AN AES AND XPS STUDY OF THE ADSORPTION OF OXYGEN ON NICKEL.

HOOKER MP	5-1489	AUGER- LEED INVESTIGATION OF THE DEPOSITION OF ALUMINUM ONTO THE MOLYBDENUM (110
	5-1490	AUGER- LEED INVESTIGATION OF TIN ON MOLYBDENUM (100).
	5-1491	LEED AUGER STUDY OF GROWTH OF ALUMINUM ON NIOBIUM (110).
HOOPER AJ	5-1460	AUGER ELECTRON SPECTROSCOPY OF MATERIALS EXPOSED TO FLOWING SODIUM.
	6-2050	APPLICATION OF PHYSICAL EXAMINATION TECHNIQUES IN THE STUDY OF CORROSION AND DEP
HOOVER RA	3-0598	AUGER ELECTRON SPECTROSCOPY IN A DIFFUSION PUMPED SCANNING ELECTRON MICROSCOPE.
HOPKINS BJ	3-0662	LOW NOISE DETECTION SYSTEM FOR THE MEASUREMENT OF AUGER SPECTRA.
	3-0663	AN AC RETARDING POTENTIAL TECHNIQUE FOR THE CONTINUOUS MEASUREMENT OF CHANGES IN
	5-1243	CHARACTERIZATION OF ADSORBED SPECIES USING AUGER PEAK SHAPES. ETHYLENE AND CARBO
	5-1244	ELECTRON BEAM EFFECTS ON CARBON MONOXIDE SATURATED TUNGSTEN (100), ETHYLENE SATU
	5-1245	CHEMISORPTION ON TANTALUM (100). (AUGER SPECTROSCOPY)
	5-1461	FLASH DESORPTION- AUGER ELECTRON SPECTROSCOPY. DISCREPANCY FOR CARBON DIOXIDE ON
	5-1462	ROLE OF IMPURITIES IN THE STABILITY OF ZINC OXIDE.
	5-1463	INTERACTION OF CHLORINE WITH THE ZINC OXIDE (0001)- ZINC SURFACE: A LEED, AUGER
	5-1464	COMBINED LEED- RHEED AUGER STUDY OF OXYGEN ADSORPTION ON THE TUNGSTEN (112) SURF
	5-1465	AUGER SPECTROSCOPIC STUDY OF THE COADSORPTION OF MOLECULAR OXYGEN AND CARBON MON
	5-1588	INTERACTION OF CESIUM WITH CLEAN ZINC OXIDE SURFACES. (AUGER SPECTROSCOPY)
	5-1881	SOME PROPERTIES OF CLEAN AND ALKALI METAL COVERED ZINC OXIDE SURFACES. (LEED, AE
	5-1882	STABILITY OF CESIUM COVERED ZINC OXIDE SURFACES.
	5-1954	AUGER ELECTRON SPECTROSCOPY STUDY OF THE ELECTRON STIMULATED DESORPTION OF OXYGE
HORGAN AM	5-1466	STUDY OF THE ADSORPTION OF OXYGEN ON NICKEL(III) USING AUGER ELECTRON SPECTROSCO
	5-1467	EFFECTIVENESS OF THE STANDARD SURFACE CLEANING TECHNIQUES AS APPLIED TO NICKEL (
	5-1468	STICKING PROBABILITIES FOR THE ADSORPTION OF HYDROGEN ON CLEAN NICKEL (111), NIC
HORI H	4-1041	SHIFT ENERGIES OF CHARACTERISTIC X-RAYS AND AUGER ELECTRONS FOR IONIZED ATOMS.
HORIGUCHI T	5-1469	STRUCTURAL STUDY OF OXYGEN- IRON (001) BY LOW ENERGY ELECTRON DIFFRACTION AND AU
	5-1470	ANALYSIS OF THE SURFACE STRUCTURE WITH ANTIPHASE DOMAINS OF SULFUR ON IRON (001)
HORIUCHI S	5-1481	AUGER STUDIES OF DIFFUSED IMPURITY PROFILES IN SILICON DIOXIDE- SILICON AND POLY
HORN K	5-1474	ADSORPTION OF OXYGEN ON SILICON (111) SURFACES. PART-1. (AUGER ELECTRON SPECTROS
HORNFELDT O	4-0886	Z DEPENDENCE OF ENERGIES AND RELATIVE INTENSITIES IN THE KLL AUGER SPECTRUM.
	4-0887	AUGER SPECTRA IN THE INTERMEDIATE COUPLING REGION.
HOSHINA T	2-0388	CONCENTRATION-DEPENDENT NONRADIATIVE PROCESSES IN ZINC SULFIDE PHOSPHORS. (AUGER
	2-0402	CONCENTRATION DEPENDENCE OF GREEN- COPPER LUMINESCENCE IN COPPER AND ALUMINUM DO
HOSOKI S	5-1471	SURFACE CONDITION AND THERMIONIC EMISSION OF LANTHANUM HEXABORIDE. (AUGER SPECTR
HOTTA M	6-2133	STUDY OF THE INTERMITTENT MOTION OF FRICTION IN STAINLESS STEEL BY AN AUGER MICR
HOUSE D	4-0888	CLUSTER CALCULATIONS OF THE SOFT X-RAY L3 EMISSION SPECTRA OF NICKEL AND COPPER.
HOUSLEY RM	4-0858	AUGER ELECTRON SPECTROSCOPY OF LUNAR SAMPLES.
HOUSTON JE	1-0066	QUANTITATIVE AUGER ANALYSIS USING INTEGRATION TECHNIQUES.
	1-0067	APPLICATION OF DYNAMIC BACKGROUND SUBTRACTION TO QUANTIFICATION OF AUGER ELECTRO
	1-0090	RETRIEVAL OF ELECTRON EXCITED AUGER STRUCTURE BY DYNAMIC BACKGROUND SUBTRACTION.
	1-0091	THE SECONDARY ELECTRON BACKGROUND PROBLEM IN ELECTRON EXCITED AUGER ELECTRON SPE
	1-0092	COMPARISON OF SOFT X-RAY AND AUGER ELECTRON APPEARANCE POTENTIAL SPECTROSCOPY.
	1-0161	CHARACTERIZATION OF CHEMISORPTION BY LEED.
	2-0367	EFFECT OF MODULATION AMPLITUDE ON ELECTRON EXCITED AUGER DATA FROM TITANIUM.
	2-0389	EXACT CORRECTIONS FOR POTENTIAL MODULATION DISTORTION IN AUGER YIELD MEASUREMENT
	2-0390	AUGER EXCITATION BY INTERNAL SECONDARY ELECTRONS.
	2-0391	CROSS CORRELATION TECHNIQUES IN AUGER SPECTROSCOPY.
	2-0392	DIRECT COMPARISON OF CORE ELECTRON BINDING ENERGIES OF SURFACE AND BULK ATOMS OF
	2-0468	ELECTRONIC STRUCTURE OF SOLID SURFACES: CORE LEVEL EXCITATION TECHNIQUES. (AUGER
	4-0889	VALENCE BAND STRUCTURE IN THE AUGER SPECTRUM OF ALUMINUM.
	4-0890	DIFFERENCE IN THE CHROMIUM L3/L2 INTENSITY RATIO MEASURED BY SOFT X-RAY AND AUGE
	5-1352	IONIZATION SPECTROSCOPY OF SURFACES.
	5-1384	QUANTITATIVE COMPARISON OF TITANIUM AND TITANIUM MONOXIDE SURFACES USING AUGER E
	5-1723	COMPARISON OF MANY BODY EFFECTS IN CORE LEVEL SURFACE SPECTROSCOPIES.
HOWARD JK	1-0085	QUANTITATIVE DEPTH PROFILING BY AUGER ION SPUTTERING TECHNIQUE.
	5-1439	DIFFUSION MEASUREMENT IN THIN FILMS BY AUGER SPUTTER ETCHING TECHNIQUE.
	5-1440	INTERDIFFUSION IN BIMETALLIC POLYCRYSTALLINE FILMS MEASURED BY AUGER ION SPUTTER
	6-2146	ANALYSIS OF GRAIN BOUNDARY DIFFUSION IN BIMETALLIC THIN FILM STRUCTURES USING AU
HOWAT G	2-0393	K X-RAY AND AUGER ELECTRON ENERGIES FOR NEON BY A TRANSITION OPERATOR METHOD.
HOWELLS GP	1-0219	AUGER ELECTRON DOSIMETRY.
HOWIE A	2-0254	CRYSTALLINE EFFECTS IN BACKSCATTERING AND AUGER PRODUCTION. (ALUMINUM)
HOWSMAN AJ	2-0563	THEORY OF AUGER EJECTION OF ELECTRONS FROM METALS BY IONS.
HOYT EW	5-1472	NIOBIUM SURFACES FOR RF SUPERCONDUCTORS. (AUGER SURFACE ANALYSIS)
HRABAK G	3-0657	ULTRAHIGH VACUUM LOW TEMPERATURE STAGE FOR IN- SITU FILM DEPOSITION AND CHARACTER
HRIBAR M	4-1014	K-SHELL FLUORESCENCE YIELDS OF ARGON, CHLORINE AND SULFUR. (AUGER TRANSITION)
HUBBARD AT	5-1486	STUDY OF PLATINUM ELECTRODES BY MEANS OF THIN LAYER ELECTROCHEMISTRY AND LOW ENE
HUBER WK	3-0632	AN ANALYTICAL SYSTEM FOR SECONDARY ION MASS SPECTROMETRY IN ULTRAHIGH VACUUM. (A
	5-1213	STUDY OF SURFACE PROCESSES ON COPPER- BERYLLIUM BY COMBINED SECONDARY ION MASS S
	5-1214	COMBINATION OF SECONDARY ION MASS SPECTROMETRY AND AUGER ELECTRON SPECTROSCOPY,
HUBLER GK	1-0216	DIRECT COMPARISON OF AUGER, SECONDARY ION MASS SPECTROSCOPY, AND PROTON RESONANC
HUCHITAL DA	3-0633	HIGH SENSITIVITY ELECTRON SPECTROMETER.
	3-0634	RESOLUTION AND SENSITIVITY OF THE SPHERICAL GRID RETARDING POTENTIAL ANALYZER. A
	3-0686	RETARDING FIELD ELECTRON SPECTROMETER.
	5-1821	ADSORPTION KINETICS OF CESIUM ON GALLIUM ARSENIDE. (USING AES AND X-RAY PHOTOELE
HUDIS J	5-1844	ATTENUATION OF LOW ENERGY ELECTRONS FROM SOLIDS: RESULTS FROM X-RAY PHOTOELECTRO
HUDSON JB	3-0631	A VERSATILE SUBSTRATE DESIGN FOR LEED AND AES STUDIES IN ULTRAHIGH VACUUM.
	5-1451	KINETICS OF THE REACTION BETWEEN OXYGEN AND SULFUR ON A NICKEL (111) SURFACE. (S
	5-1452	KINETICS OF THE REACTION OF OXYGEN WITH CLEAN NICKEL SINGLE CRYSTAL SURFACES. PA
	5-1453	CHEMICAL CHANGES IN SECONDARY ELECTRON EMISSION DURING OXIDATION OF NICKEL (100)
	5-1622	ADSORPTION AND DECOMPOSITION OF CARBON MONOXIDE ON PLATINUM (111). (AUGER SPECTR
HUGGINS RA	1-0093	TRACE ANALYSIS TECHNIQUES FOR SOLIDS. (ELECTRON PROBE ANALYSIS, AUGER EFFECT, RE
HULDT L	2-0394	BAND-TO-BAND AUGER RECOMBINATION IN INDIRECT GAP SEMICONDUCTORS.
	2-0395	AUGER RECOMBINATION IN GERMANIUM.
	2-0396	AUGER RECOMBINATION AND IMPACT IONIZATION IN INDIRECT GAP SEMICONDUCTORS.
	2-0397	PHONON ASSISTED AUGER RECOMBINATION IN GERMANIUM.
HULTQUIST G	5-1587	LEED- AES STUDY OF THE OXIDATION OF IRON- CHROMIUM (100) AND (110).
	5-1591	SURFACE COMPOSITION STUDIES OF THE (001) AND (110) FACES OF MONOCRYSTALLINE IRON
HURD J	5-1741	INTEGRATED CIRCUIT PROCESS CHARACTERIZATION USING AES.
HURYCH Z	2-0398	MATRIX ELEMENTS DEPENDENCE OF OPTICAL EXCITATION AND AUGER DECAY OF 5D CORE HOLE
HUTTEMANN RD	5-1473	EFFECT OF LIGHT ELEMENTS NITROGEN, CARBON AND OXYGEN ON THE PHYSICAL PROPERTIES
IBACH H	4-1039	SURFACE STATE TRANSITIONS OF SILICON IN ELECTRON ENERGY LOSS SPECTRA. (AUGER EFF

IBACH H	5-1474	ADSORPTION OF OXYGEN ON SILICON (111) SURFACES. PART-1. (AUGER ELECTRON SPECTROS
	5-1475	ELECTRON ORBITAL ENERGIES OF OXYGEN ADSORBED ON SILICON SURFACES AND OF SILICON
	5-1476	ELECTRONIC STATES OF OXYGEN ADSORBED ON CLEAN SILICON (111) AND (100) SURFACES.
IBERL F	5-1422	OXYGEN ADSORPTION ON (110) SILVER.
ICHIKAWA M	5-1412	THREE-DIMENSIONAL ELEMENT ANALYSIS OF SURFACES AND INTERFACES OF SOLIDS BY THE A
	5-1414	AUGER ELECTRON SPECTROSCOPY AND MICROANALYSIS OF SOLID SURFACES.
IGNATIEV A	1-0094	SECONDARY ELECTRON SPECTROSCOPY. SURFACE SENSITIVE TOOL. (REVIEW, 22 REFS)
	1-0095	SECONDARY ELECTRON SPECTROSCOPY. SURFACE SENSITIVE TOOL. (AES)
	5-1477	SURFACE DEBYE TEMPERATURE OF THE SILICON (001) - 2X2 STRUCTURE. (AUGER EFFECT)
	5-1478	ATOMIC ARRANGEMENT IN THE 1 K 1 STRUCTURE OF A SILICON ORDERED MONOLAYER ON MOLY
	5-1479	LEED INVESTIGATIONS OF XENON SINGLE CRYSTAL FILMS AND THEIR USE IN STUDYING THE
IHARA T	3-0652	A NEW PHOTOELECTRON SPECTROMETER: COMBINATION OF CYLINDRICAL MIRROR ANALYZER WIT
IL'IN AI	3-0653	SPECTROMETER OF AUGER ELECTRONS.
INGALLS WB	2-0299	EFFECT OF MULTIPLE IONIZATION ON THE FLUORESCENCE YIELD OF NEON. (K-AUGER AND K
INGREY SJ	5-1480	AUGER ANALYSIS OF TANTALUM SURFACES.
INO S	1-0096	LEED AND AES. (REVIEW, 53 REFS)
INOUE T	5-1481	AUGER STUDIES OF DIFFUSED IMPURITY PROFILES IN SILICON DIOXIDE- SILICON AND POLY
	5-1482	OBSERVATION OF IMPURITY PROFILE IN ION IMPLANTED SILICON BY AUGER ELECTRON SPECT
	6-2051	APPLICATION OF MICROAUTORADIOGRAPHY BY BETA-RAY AUGER ELECTRONS TO THE STUDIES O
	6-2052	NEW MICROAUTORADIOGRAPHY OF METAL SURFACE STRUCTURE USING AUGER ELECTRONS.
	6-2066	NEW MICRORADIOGRAPHY USING AUGER ELECTRONS IN THE STUDY OF THE STRUCTURE AND PIT
IRGOLIC KJ	4-0763	RELAXATION DURING PHOTOEMISSION AND LMM AUGER DECAY IN ARSENIC AND SOME OF ITS C
ISA SA	5-1483	DEVELOPMENT OF STEPPED SURFACE REGIONS ON POLYCRYSTALLINE GOLD. LOW ENERGY ELECT
ISAACS HS	5-1333	INTERACTION OF OXYGEN AND NITROGEN WITH THE NIOBIUM (100) SURFACE. PART-2: REACT
ISASI JA	1-0010	ELECTRON BEAM MICROANALYSIS. PART-1. (REVIEW OF ELECTRON PROBE ANALYZER, SCANNIN
ISETT LC	3-0635	ANALOG TECHNIQUE FOR THE MEASUREMENT OF AUGER ELECTRON CURRENTS. APPLICATION TO
	5-1484	BINDING ENERGIES OF CARBON TO NICKEL (100) FROM EQUILIBRIUM SEGREGATION STUDIES.
	5-1485	BINDING OF CARBON ATOMS AT A STEPPED NICKEL SURFACE. (AUGER SPECTROSCOPY)
ISHIDA T	3-0636	NEW SEM AUGER MICROANALYZER WITH ULTRAHIGH VACUUM CAPABILITY.
ISHIHARA N	5-1696	EPITAXIAL GROWTH OF SILVER ON POTASSIUM CHLORIDE (100) SURFACE OBSERVED BY LOW E
ISHIKAWA K	2-0365	AUGER AND SECONDARY ELECTRONS EXCITED BY BACKSCATTERED ELECTRONS.
	2-0366	AUGER AND SECONDARY ELECTRONS EXCITED BY BACKSCATTERED ELECTRONS. APPROACH TO QU
	3-0619	COMPACT, INEXPENSIVE HIGH RESOLUTION RETARDING FIELD ENERGY ANALYZER. (AUGER ELE
	4-0852	METHOD FOR DETECTING FINE STRUCTURE IN THE SECONDARY ELECTRON EMISSION YIELD AND
	4-0853	FINE STRUCTURES AND ENERGY DISTRIBUTION OF SECONDARY ELECTRON EMISSION FROM SILI
ISHIKAWA RM	5-1486	STUDY OF PLATINUM ELECTRODES BY MEANS OF THIN LAYER ELECTROCHEMISTRY AND LOW ENE
ISOGAI A	5-1670	A STUDY OF THE IRON (111) SURFACE USING LEED AND AUGER EMISSION SPECTROSCOPY.
	5-1866	LOW ENERGY ELECTRON DIFFRACTION- AUGER ELECTRON SPECTROSCOPY STUDY OF IRON SINGL
	5-1867	CONCENTRATION CHANGE OF TIN IN THE SURFACE LAYER OF COPPER- TIN ALLOY SYSTEM DUE
ISOYAMA E	5-1127	APPLICATION OF AUGER ELECTRON SPECTROSCOPY TO THE STUDY OF DISCOLORING OF ALUMIN
	5-1487	DETECTION OF ORGANIC CONTAMINANTS ON ALUMINUM SURFACES BY AUGER ELECTRON SPECTRO
IVAN E	2-0325	EMISSION SPECTRA OF SILICON COMPENSATED ALUMINUM(X) GALLIUM(1-X) ARSENIDE ELECTR
IVANOV AV	4-0891	USE OF X-RAY SPECTROSCOPIC DATA TO EXPLAIN THE FINE STRUCTURE OF THE DEPENDENCE
IVASHCHENKO YN	5-1215	EFFECT OF ION BOMBARDMENT ON THE COMPOSITION AND STRUCTURE OF TUNGSTEN AND MOLYB
	5-1939	AUGER SPECTROSCOPIC STUDY OF THE SURFACE OF IRON AFTER IONIC BOMBARDMENT AND AIR
IWAMI M	5-1435	ELECTRONIC STRUCTURE OF A THIN GOLD FILM DEPOSITED ON A SILICON SUBSTRATE STUDIE
	5-1436	DIFFUSE INTERFACE STRUCTURES IN SOME (SUBSTRATE- VACUUM EVAPORATED FILM) SYSTEMS
	5-1437	POSSIBLE ORIGIN OF LOW TEMPERATURE SILICON MIGRATION IN SILICON- METAL SYSTEMS.
	5-1438	METALLIC STATE OF SILICON IN SILICON- NOBLE METAL VAPOR QUENCHED ALLOYS STUDIED
	5-1673	LOW TEMPERATURE DIFFUSION OF GOLD INTO SILICON IN THE SILICON (SUBSTRATE)- GOLD
JACKSON AG	1-0071	A BIBLIOGRAPHY OF LOW ENERGY ELECTRON DIFFRACTION AND AUGER ELECTRON SPECTROSCOP
	5-1488	EPITAXY OF ULTRATHIN METAL FILMS ON BODY CENTERED CUBIC SUBSTRATES USING LEED- A
	5-1489	AUGER- LEED INVESTIGATION OF THE DEPOSITION OF ALUMINUM ONTO THE MOLYBDENUM (110
	5-1490	AUGER- LEED INVESTIGATION OF TIN ON MOLYBDENUM (100).
	5-1491	LEED AUGER STUDY OF GROWTH OF ALUMINUM ON NIOBIUM (110).
JACKSON AJ	4-0892	THE KVV AUGER SPECTRUM OF LITHIUM METAL.
JACKSON DC	2-0399	MODEL FOR THE AUGER ELECTRON SPECTROSCOPY OF SYSTEMS EXHIBITING LAYER GROWTH, AN
	5-1227	GROWTH AND STRUCTURE OF THIN COPPER FILMS ON (001) SURFACES OF NICKEL.
JACKSON KA	5-1492	A CLOSE LOOK AT SURFACES. (SURFACE STRUCTURES, SPECIES, TOPOGRAPHY, PHENOMENA, F
JACOBI K	4-0893	AUGER ELECTRON EMISSION OF THIN CARBON FOILS IN REFLECTION AND TRANSMISSION.
	5-1493	OXIDATION AND ANNEALING OF GALLIUM PHOSPHIDE AND GALLIUM ARSENIDE (111) FACES ST
	5-1756	ELECTRON STIMULATED OXIDATION OF GALLIUM ARSENIDE, STUDIED BY QUANTITATIVE AUGER
JACOBSON RL	5-1360	DETECTION OF IMPURITIES ON A SILICON (111) SURFACE BY AUGER- LEED ANALYSIS.
JAHRREISS H	4-0894	SECONDARY ELECTRON EMISSION FROM THIN FILMS.
JAMES LW	4-0992	AUGER SPECTRA OF HYDROGEN CHLORIDE VAPOR ETCHED N+ GALLIUM ARSENIDE (100) SUBSTR
	5-1920	AUGER ELECTRON SPECTROSCOPY OF CONTAMINATED GALLIUM ARSENIDE SURFACES.
JAMISON KA	2-0481	RADIATIVE AUGER EFFECT IN ION- ATOM COLLISIONS.
	2-0571	NEON K-AUGER CROSS SECTIONS FROM 1.5 MEV PER AMU AND ATOMIC HYDROGEN, OXYGEN, AN
	2-0572	NEON K-SHELL VACANCY PRODUCTION BY INTERMEDIATE ENERGY FLUORINE IONS. (AUGER ELE
	2-0573	AUGER ELECTRONS FROM DOUBLE K-SHELL VACANCY STATES IN NEON.
	2-0574	K-SHELL AUGER ELECTRON PRODUCTION CROSS SECTIONS FROM ION BOMBARDMENT. (NEON, FL
	4-1117	K-SHELL AUGER ELECTRON HYPERSATELLITES OF NEON.
JANSSEN AP	3-0637	A CRYSTAL CLEAVAGE DEVICE FOR USE IN LOW ENERGY ELECTRON DIFFRACTION- AUGER STUD
	4-0895	LOW ENERGY AUGER AND LOSS ELECTRON SPECTRA FROM MAGNESIUM AND ITS OXIDE.
	5-1494	ELECTRON STIMULATED DESORPTION OF SODIUM AND MEASUREMENT OF SURFACE DIFFUSION ON
	5-1495	COMPARATIVE STUDY OF SINGLE CRYSTAL MAGNESIUM OXIDE AND OXIDIZED MAGNESIUM BY AU
	5-1496	STUDY OF THE EPITAXIAL GROWTH OF MAGNESIUM ON MAGNESIUM OXIDE (001) USING REFLEC
	5-1497	INTERFACIAL AUGER TRANSITIONS IN OXIDIZED SODIUM AND MAGNESIUM.
	5-1498	AUGER AND REFLECTION ELECTRON DIFFRACTION INVESTIGATION OF OXYGEN AND HYDROGEN A
JARVIS L	5-1261	SCANNING ELECTRON MICROSCOPY, AUGER SPECTROSCOPY, AND ION BACKSCATTERING TECHNIQ
JARVIS NL	5-1737	SURFACE COMPOSITION OF PLATINUM HEATED IN OXYGEN. (AUGER SPECTRA)
	5-1775	AES AND SPUTTER ETCHING OF NICKEL- GOLD- GERMANIUM ON N-GALLIUM ARSENIDE.
JASPER M	5-1593	EVALUATION OF LEAD(0.8) TIN(0.2) TELLURIDE DETECTOR FABRICATION USING SURFACE AN
JASPERSON SN	5-1700	RESIDUAL GAS AND THE OPTICAL PROPERTIES OF SILVER FILMS. (AUGER SPECTROSCOPY)
JENA P	2-0270	X-RAY EDGE ANOMALIES IN LITHIUM.
JENKIN JG	2-0427	MANY BODY EFFECTS IN X-RAY PHOTOEMISSION FROM MAGNESIUM. (AUGER EFFECT)
	4-0773	PHOTOELECTRON DETERMINATION OF THE ATTENUATION OF THE LOW ENERGY ELECTRONS IN AL
JENKINS LH	2-0581	AUGER TRANSITIONS INITIATED FROM SINGLY AND DOUBLY IONIZED STATES IN LITHIUM MET
	4-0804	LLM VERSUS LMM AUGER TRANSITIONS FOR THE LIGHT ELEMENTS.
	4-0805	AUGER ELECTRON SPECTROSCOPY OF THE OUTER SHELL ELECTRONS.

JENKINS LH	4-0896	AUGER AND OTHER CHARACTERISTIC ENERGIES IN SECONDARY ELECTRON SPECTRA FROM ALUMI
	4-0897	ENERGY SPECTRUM OF BACK SCATTERED ELECTRONS AND CHARACTERISTIC LOSS AND GAIN PHE
	4-0898	AUGER SATELLITE AND OTHER CHARACTERISTIC EVENTS IN MAGNESIUM SECONDARY ELECTRON
	4-0899	PLASMON ENERGY GAIN AND DOUBLE IONIZATION SATELLITES IN THE BERYLLIUM AUGER SPEC
	4-0900	HIGHER ENERGY SATELLITES IN AUGER SPECTRA FROM METALS.
	4-0901	CHARACTERISTIC ENERGY GAIN AND LOSS, DOUBLE IONIZATION, AND IONIZATION LOSS EVEN
	5-1139	INVESTIGATION OF REORDERED (001) GOLD SURFACES BY POSITIVE ION CHANNELING SPECTR
	5-1263	CHARACTERISTIC ENERGIES IN SECONDARY ELECTRON SPECTRA FROM SILICON (111) SURFACE
	5-1499	LEEL AND AUGER INVESTIGATIONS OF COPPER (111) SURFACE.
	5-1697	ANGULAR RESOLVED AUGER EMISSION SPECTRA FROM A CLEAN COPPER (100) SURFACE.
	5-1986	CHARACTERIZATION OF REORDERED (001) GOLD SURFACES BY POSITIVE ION CHANNELING SPE
	5-1988	ANGULAR EFFECTS IN AUGER ELECTRON EMISSION FROM COPPER (110).
JEPSEN DW	5-1478	ATOMIC ARRANGEMENT IN THE 1 K 1 STRUCTURE OF A SILICON ORDERED MONOLAYER ON MOLY
JERNER RC	6-1995	SURFACE CONCENTRATION OF MOLYBDENUM IN TYPES 316 AND 304 STAINLESS STEEL BY AUGE
	6-1996	CHEMICAL PROFILE ANALYSIS OF TYPE 316 STAINLESS STEEL SHEET BY AUGER SPECTROSCOP
JOEBSTL JA	5-1500	CHEMISORPTION OF OXYGEN ON PLATINUM (111) SURFACES. (AUGER SPECTRA)
JOERGENSEN CK	2-0272	AUGER SIGNALS OBTAINED WITH A PHOTOELECTRON SPECTROMETER.
	4-0902	PHOTOELECTRON AND AUGER SPECTRA OF XENON IN SOLID PERXENATES.
JOHANNESSEN JS	4-1081	AUGER ELECTRON SPECTROSCOPY STUDY OF SILICON SPECTRA FROM SILICON MONOXIDE, SILI
	5-1501	USE OF AUGER ELECTRON SPECTROSCOPY TO DETERMINE THE STRUCTURE OF SILICON OXIDE F
	5-1502	PHASE SEPARATION IN SILICON OXIDES AS SEEN BY AUGER ELECTRON SPECTROSCOPY.
	5-1503	STUDY OF THE CHEMICAL COMPOSITION OF MOS AND MNOS STRUCTURES BY AUGER ELECTRON S
	5-1847	AES CHARACTERIZATION OF OXIDIZED FILMS OF MAGNESIUM, ALUMINUM, AND SILICON.
JOHANSSON G	4-0734	KLL AUGER SPECTRUM OF FLUORINE. (IN FLUORIDE SALTS)
JOHARI O	1-0097	CHARACTERIZING MATERIALS WITH THE SEM TODAY. (AUGER SPECTROSCOPY)
	3-0596	CONTAMINATION IN THE SCANNING ELECTRON MICROSCOPE. (AUGER ELECTRON SPECTROSCOPY,
JOHNSON AL	5-1956	DETERMINATION OF SURFACE STRUCTURES USING LEED AND ENERGY ANALYSIS OF SCATTERED
	5-1957	DETERMINATION OF SURFACE STRUCTURES BY LEED AND AUGER ELECTRON SPECTROSCOPY.
JOHNSON BM	2-0400	OXYGEN BEAM GAS AUGER ELECTRON EMISSION.
	4-0903	HIGH RESOLUTION STUDY OF ARGON LMM AUGER ELECTRONS PRODUCED BY ION BOMBARDMENT.
	4-0904	HIGH RESOLUTION BEAM GAS AUGER ELECTRON SPECTRA FOR OXYGEN IONS EXCITED BY COLLI
	4-0943	HIGH RESOLUTION MEASUREMENTS OF P(+) ARGON KLL AND KLM AUGER ELECTRONS.
	4-0952	HIGH RESOLUTION K-AUGER SPECTRA FOR MULTIPLY IONIZED NEON.
	4-0953	HIGH RESOLUTION NEON KLL AUGER SPECTRA PRODUCED BY HYDROGEN(+) AND HELIUM(+) ION
	4-0954	AUGER DECAY OF NEON FOLLOWING ENERGETIC ION BOMBARDMENT.
	4-0955	CALCULATED K-AUGER ELECTRON AND K X-RAY TRANSITION ENERGIES FOR THE MULTIPLY ION
	4-0993	OBSERVATION OF A PROJECTILE CHARGE DEPENDENCE FOR THE NEON K-AUGER ELECTRON SPEC
	4-1050	AUGER ELECTRON EMISSION SPECTRA FROM FOIL AND GAS EXCITED CARBON BEAMS.
	4-1051	HIGH RESOLUTION NEON K-AUGER ELECTRON SPECTRUM PRODUCED BY 45 MEV CHLORINE(12+)
JOHNSON CE	2-0560	L2,3MM AUGER PROCESSES IN SELENIUM.
	4-1033	TRANSITION PROBABILITIES FOR THE L2,3MM AUGER SPECTRUM OF SELENIUM.
	4-1034	AUGER VACANCY SATELLITE STRUCTURE IN THE L3,M4,5M4,5 AUGER SPECTRA OF COPPER.
	4-1035	PHOTOELECTRON AND L2,3MM AUGER ELECTRON ENERGIES FOR ARSENIC.
JOHNSON WC	1-0098	AUGER ELECTRON SPECTROSCOPY. (REVIEW, 18 REFS)
	5-1504	COMPOSITION OF PLATINUM ANODE SURFACES BY AUGER SPECTROSCOPY.
	5-1505	ADDITIVE AND IMPURITY DISTRIBUTIONS AT GRAIN BOUNDARIES IN SINTERED ALUMINA. (SC
	5-1506	ANALYSIS OF GRAIN BOUNDARY IMPURITIES AND FLUORIDE ADDITIVES IN HOT PRESSED OXID
	6-2053	CONFIRMATION OF SULFUR EMBRITTLEMENT IN NICKEL ALLOYS. (AUGER ELECTRON SPECTROSC
	6-2054	CONFIRMATION OF IMPURITY ADSORPTION AT FLAKE IRON INTERFACES IN GRAY CAST IRON.
	6-2055	STUDY OF GRAIN BOUNDARY SEGREGANTS IN THERMALLY EMBRITTLED MARAGING STEEL. (AUGE
	6-2056	GRAIN BOUNDARY SEGREGATION IN SINTERED ALUMINA. (AUGER ELECTRON SPECTROSCOPY)
JOLLY WL	2-0471	STATIC RELAXATION OF GERMANE AND THE ESTIMATION OF RELAXATION ENERGY DIFFERENCES
JONA F	1-0099	LOW ENERGY ELECTRON DIFFRACTION, CIRCA 1968. (REVIEW OF LEED AND AUGER)
	1-0100	THIN FILMS. LOW ENERGY ELECTRON DIFFRACTION AND AUGER ELECTRON SPECTROSCOPY.
	4-1060	INTERFACIAL AND PLASMON DECAY PEAKS IN THE AUGER SPECTRA OF TITANIUM AND IRON.
	5-1477	SURFACE DEBYE TEMPERATURE OF THE SILICON (001) - 2X2 STRUCTURE. (AUGER EFFECT)
	5-1478	ATOMIC ARRANGEMENT IN THE 1 K 1 STRUCTURE OF A SILICON ORDERED MONOLAYER ON MOLY
	5-1507	REEXAMINATION OF ALUMINUM (110) WITH LOW ENERGY DIFFRACTION AND AUGER SPECTROSCO
JONES AR	5-1243	CHARACTERIZATION OF ADSORBED SPECIES USING AUGER PEAK SHAPES. ETHYLENE AND CARBO
	5-1244	ELECTRON BEAM EFFECTS ON CARBON MONOXIDE SATURATED TUNGSTEN (100), ETHYLENE SATU
	5-1461	FLASH DESORPTION- AUGER ELECTRON SPECTROSCOPY. DISCREPANCY FOR CARBON DIOXIDE ON
	5-1954	AUGER ELECTRON SPECTROSCOPY STUDY OF THE ELECTRON STIMULATED DESORPTION OF OXYGE
JONES AV	3-0622	ULTRAHIGH VACUUM SCANNING ELECTRON MICROSCOPE WITH AUGER ANALYSIS FACILITIES.
	5-1479	LEED INVESTIGATIONS OF XENON SINGLE CRYSTAL FILMS AND THEIR USE IN STUDYING THE
JOPSON RC	2-0351	ATOMIC FLUORESCENCE YIELDS. (L-SHELL COSTER-KRONIG YIELDS, K-, L-, AND M-SHELL F
JORGENSEN CK	2-0273	CHEMICAL SHIFTS AND VARIATIONS OF WIDTH AND INTENSITY OF AUGER SIGNALS OBTAINED
	4-0905	LINEAR LIGAND FIELDS INFLUENCING AUGER SPECTRA OF FLUORIDES, COPPER(I) AND SILVE
JOSHI A	1-0098	AUGER ELECTRON SPECTROSCOPY. (REVIEW, 18 REFS)
	1-0101	AUGER ELECTRON SPECTROSCOPY. (REVIEW, 154 REFS)
	4-1108	AUGER ELECTRON SPECTROSCOPY OF A STABLE GERMANIUM OXIDE.
	5-1508	SURFACE SEGREGATION AND INTERACTION OF OXYGEN AND NITROGEN WITH ZIRCONIUM IN NIO
	5-1849	DIFFERENTIAL AUGER SPECTROMETRY. (NIOBIUM- ZIRCONIUM ALLOY ANALYSIS)
	5-1937	DIFFERENCE AUGER SPECTROSCOPY FOR STUDYING SMALL QUANTITIES OF ELEMENTS ON METAL
	5-1938	EFFECT OF HIGH TEMPERATURE HEAT TREATMENT ON PENETRATION DEPTH OF SUPERCONDUCTIN
	6-1998	COMPOSITION VERSUS DEPTH PROFILES OBTAINED WITH AUGER ELECTRON SPECTROSCOPY OF A
	6-2018	IDENTIFICATION OF ATOMIC BORON IN STEEL BY AUGER ELECTRON SPECTROSCOPY.
	6-2057	SEGREGATION AT SELECTIVE GRAIN BOUNDARIES AND ITS ROLE IN TEMPER EMBRITTLEMENT O
	6-2058	GRAIN BOUNDARY EMBRITTLEMENT IN SOME METALS AND ALLOYS. (AUGER EMISSION ANALYSIS
	6-2059	ROLE OF MANGANESE AND SILICON IN TEMPER EMBRITTLEMENT OF LOW ALLOY STEELS. (AUGE
	6-2060	AUGER SPECTROSCOPIC ANALYSIS OF BISMUTH SEGREGATED TO GRAIN BOUNDARIES IN COPPER
	6-2061	INTERANGULAR BRITTLENESS STUDIES IN TUNGSTEN USING AUGER SPECTROSCOPY.
	6-2062	TEMPER EMBRITTLEMENT OF LOW ALLOY STEELS. (AES OF FRACTURE SURFACES)
	6-2063	IMPURITY SEGREGATION TO GRAIN BOUNDARIES.
	6-2064	CHEMISTRY OF GRAIN BOUNDARIES AND ITS RELATION TO INTERGRANULAR CORROSION OF AUS
	6-2065	SURFACE SEGREGATION OF OXYGEN IN NIOBIUM- OXYGEN AND TANTALUM- OXYGEN ALLOYS. (A
	6-2107	IDENTIFICATION OF BUBBLE FORMING IMPURITIES IN DOPED TUNGSTEN. (AUGER SPECTROSCO
	6-2113	AUGER ELECTRON SPECTROSCOPY ANALYSIS OF THE FRACTURE SURFACES OF IRRADIATED PRES
	6-2118	GRAIN BOUNDARY STRUCTURE AND SEGREGATION. AUGER SPECTROGRAPHIC ANALYSIS OF FRAC
	6-2119	STUDIES USING AUGER ELECTRON EMISSION SPECTROSCOPY ON TEMPER EMBRITTLEMENT IN LO
	6-2120	RELATION OF GRAIN BOUNDARY SEGREGATION TO REDUCED STRENGTH OF METALS AND ALLOYS.

JOSHI A   6-2137 EFFECT OF SURFACE METALLURGY ON THE PENETRATION DEPTH AND RF BREAKDOWN FIELD OF
                 6-2139 EFFECT OF MICROSTRUCTURE ON GRAIN BOUNDARY SEGREGATION OF PHOSPHORUS IN A CHROMI
JOYCE BA   2-0455 DEPENDENCE OF AUGER ELECTRON YIELD ON PRIMARY BEAM ENERGY AT NORMAL AND GLANCING
                 5-1509 GROWTH AND STRUCTURE OF SEMICONDUCTING THIN FILMS. (AUGER ELECTRON SPECTROSCOPY)
                 5-1510 SOME ASPECTS OF THE SURFACE BEHAVIOR OF SILICON. (AUGER ELECTRON SPECTROSCOPY)
                 5-1511 SILICON- OXYGEN INTERACTIONS USING AUGER ELECTRON SPECTROSCOPY.
                 5-1512 INTERACTION OF OXYGEN WITH SILICON (111) SURFACES. (REPLY TO COMMENT)
                 5-1513 ELECTRON BEAM ADSORBATE INTERACTIONS ON SILICON SURFACES. (AUGER SPECTROSCOPY)
                 5-1514 INFLUENCE OF SUBSTRATE CONDITIONS ON THE NUCLEATION AND GROWTH OF EPITAXIAL SILI
JOYES P   2-0548 RELATIVE EMISSION INTENSITIES OF AUGER ELECTRONS FROM LIGHT METALS (LITHIUM, SOD
                 4-0817 AUGER SPECTRA OF LITHIUM AND SODIUM UNDER ION BOMBARDMENT.
                 4-0998 SINGULARITIES IN AUGER EMISSION SPECTRUM OF METALS.
JOYNER RW   1-0102 AUGER ELECTRON SPECTROSCOPY AND ITS APPLICATIONS IN SURFACE CHEMISTRY. (REVIEW)
                 5-1483 DEVELOPMENT OF STEPPED SURFACE REGIONS ON POLYCRYSTALLINE GOLD. LOW ENERGY ELECT
                 5-1515 ADSORPTION OF CARBON MONOXIDE ON COPPER (001). LEED AND AUGER EMISSION STUDIES.
                 5-1516 INTERACTION OF HYDROGEN SULFIDE WITH COPPER (001).
                 5-1517 STUDY OF THE PREPARATION OF ATOMICALLY CLEAN TUNGSTEN SURFACES BY AUGER ELECTRON
                 5-1518 CHEMISORPTION OF NITROGEN ON TUNGSTEN STUDIED BY AUGER ELECTRON SPECTROSCOPY.
                 5-1519 AUGER ELECTRON SPECTROSCOPY STUDIES OF CLEAN POLYCRYSTALLINE GOLD AND OF THE ADS
                 5-1838 REACTIVITY OF LOW INDEX ((111) AND (100)) AND STEPPED PLATINUM SINGLE CRYSTAL SU
JUNKER BR   4-0906 AUGER SPECTRA OF HIGHLY IONIZED OXYGEN AND FLUORINE.
JUTTNER B   5-1520 STABILITY OF FIELD ELECTRON EMISSION AND VACUUM BREAKDOWN. (FROM SURFACES) INVES
KADCHENKO VN   4-1126 DETECTION OF EXCITONS NEAR THE L2,3 EDGE OF THE CHLORIDE ION IN SODIUM CHLORIDE
KADLEC J   3-0638 SECONDARY ELECTRON EMISSION FROM THIN FILMS. (OF GOLD, APPARATUS)
KAGOTANI T   2-0401 DETERMINATION OF THE ESCAPE LENGTH OF LOW ENERGY ELECTRONS BY AUGER AND LOSS SPE
                 5-1525 LEED- AES OBSERVATION OF GROWTH MECHANISM OF DEPOSITED IRON FILM ON MAGNESIUM OX
KAMADA H   5-1521 MEASUREMENTS OF SURFACE MATERIAL STATES.
                 5-1690 X-RAY EXCITED PHOTOELECTRON AND AUGER ELECTRON SPECTRA OF OXIDIZED STATES ON SOM
KAMMERER OF   6-2137 EFFECT OF SURFACE METALLURGY ON THE PENETRATION DEPTH AND RF BREAKDOWN FIELD OF
KANAJI T   2-0401 DETERMINATION OF THE ESCAPE LENGTH OF LOW ENERGY ELECTRONS BY AUGER AND LOSS SPE
                 3-0639 ELECTRODES AND CIRCUITS OF A LEED- AES SYSTEM.
                 5-1522 LOW ENERGY ELECTRON DIFFRACTION- AUGER ELECTRON SPECTROSCOPY STUDY OF IRON FILM
                 5-1523 BEHAVIOR OF IMPURITY ATOMS AND ADSORBED OXYGEN ATOMS ON THE (001) FACE OF IRON E
                 5-1524 AUGER AND LOSS SPECTROSCOPY STUDY OF SURFACE CONTAMINATION EFFECT ON THE GROWTH
                 5-1525 LEED- AES OBSERVATION OF GROWTH MECHANISM OF DEPOSITED IRON FILM ON MAGNESIUM OX
                 5-1526 LOW ENERGY ELECTRON DIFFRACTION- AUGER ELECTRON SPECTROSCOPY STUDY OF AN EPITAXI
                 5-1527 LOW ENERGY ELECTRON DIFFRACTION- AUGER ELECTRON SPECTROSCOPY STUDY OF CLEAN (001
                 5-1528 GROWTH MODE OF IRON FILM DEPOSITED ON MAGNESIUM OXIDE (001) SURFACE. (AUGER SPEC
KANAYA K   4-0907 SECONDARY ELECTRON EMISSION DUE TO PRIMARY AND BACKSCATTERED ELECTRONS (TABLE FO
KANICHEVA I   4-0736 RECORDING OF COMPONENTS ON THE SURFACE OF FUSED EMITTERS. (AUGER SPECTRA)
                 5-1129 CHARACTERISTICS OF SURFACE OF THE PALLADIUM BARIUM ALLOY CATHODE IN RELATION TO
KANOTANI T   5-1524 AUGER AND LOSS SPECTROSCOPY STUDY OF SURFACE CONTAMINATION EFFECT ON THE GROWTH
KANSKI J   4-0908 N5,4 AND M5,4 APPEARANCE POTENTIAL SPECTRA FROM PRASEODYMIUM.
KARASEK FW   1-0103 SURFACE ANALYSIS BY AUGER SPECTROSCOPY.
KARATAEV VV   4-0785 RECOMBINATION OF CARRIERS IN N-TYPE INDIUM ARSENIDE AT 77 DEGREES K. (AUGER)
KARLSSON L   5-1529 AUGER ELECTRON SPECTRUM OF CARBON SUBOXIDE.
KARNATAK RC   4-0823 4D AUGER TRANSITIONS FROM EXCITED STATES IN SAMARIUM METAL.
KARRAS M   2-0473 NEW FINE STRUCTURE IN ELECTRON EXCITED AUGER SPECTRA FROM SOLID SURFACES.
                 4-0732 HIGH RESOLUTION LMM AUGER SPECTRA OF LOW ENERGY FROM SOLID SURFACES.
KARUZSKII AL   4-0780 MAGNETIC FIELD DEPENDENT INTENSITY OSCILLATIONS OF THE ELECTRON HOLE LUMINESCENC
KASYMOV AKH   4-0742 SECONDARY ELECTRON EMISSION FROM TUNGSTEN AND MOLYBDENUM SINGLE CRYSTALS.
KATO M   6-2052 NEW MICROAUTORADIOGRAPHY OF METAL STRUCTURE USING AUGER ELECTRONS.
                 6-2066 NEW MICRORADIOGRAPHY USING AUGER ELECTRONS IN THE STUDY OF THE STRUCTURE AND PIT
KATZ M   5-1943 SURFACE PREPARATION AND CHARACTERIZATION TECHNIQUES (AUGER, CLEANING) FOR QUARTZ
KAUFFMAN RL   2-0481 RADIATIVE AUGER EFFECT IN ION- ATOM COLLISIONS.
                 2-0520 CHARGE STATE DEPENDENCE OF THE NEON K FLUORESCENCE YIELD DEDUCED FROM HIGH RESOL
                 2-0571 NEON K-AUGER CROSS SECTIONS FROM 1.5 MEV PER AMU AND ATOMIC HYDROGEN, OXYGEN, AN
                 2-0572 NEON K-SHELL VACANCY PRODUCTION BY INTERMEDIATE ENERGY FLUORINE IONS. (AUGER ELE
                 2-0573 AUGER ELECTRONS FROM DOUBLE K-SHELL VACANCY STATES IN NEON.
                 2-0574 K-SHELL AUGER ELECTRON PRODUCTION CROSS SECTIONS FROM ION BOMBARDMENT. (NEON, FL
                 4-1117 K-SHELL AUGER ELECTRON HYPERSATELLITES OF NEON.
KAVROV VP   3-0599 X-RAY ELECTRON SPECTROMETERS AND AUGER ANALYZERS.
KAWAI H   2-0388 CONCENTRATION-DEPENDENT NONRADIATIVE PROCESSES IN ZINC SULFIDE PHOSPHORS. (AUGER
                 2-0402 CONCENTRATION DEPENDENCE OF GREEN- COPPER LUMINESCENCE IN COPPER AND ALUMINUM DO
KAWAI S   5-1706 AUGER ELECTRON SPECTROSCOPY STUDY OF OXIDATION ON LANTHANUM HEXABORIDE.
KAWAI T   4-0912 HIGH RESOLUTION L2,3M2,3M2,3 AUGER ELECTRON SPECTRA OF ARGON.
                 4-0925 CHEMICAL EFFECTS OF THE NITROGEN KLL AUGER SPECTRA OF CHEMISORBED NITROGEN ON IR
                 5-1530 STUDIES OF THE MECHANISM OF THE REACTION BETWEEN SULFUR AND OXYGEN ON A MOLYBDEN
                 5-1531 STUDY OF THE OXIDATION OF MOLYBDENUM SURFACES BY ENERGY LOSS SPECTROSCOPY COMBIN
                 5-1532 HIGH RESOLUTION MEASUREMENTS OF THE NITROGEN KLL AUGER TRANSITIONS OF CHEMISORBE
                 5-1533 APPLICATIONS OF AUGER ELECTRON AND ENERGY LOSS SPECTROSCOPIES TO CATALYTIC PROCE
                 5-1534 DYNAMIC INVESTIGATION OF THE MECHANISM OF REACTION BETWEEN SULFUR AND OXYGEN ON
                 5-1535 DYNAMIC ELUCIDATION OF REACTION STEPS ON A SOLID SURFACE BY ELECTRON SPECTROSCOP
                 5-1564 SEGREGATION OF SULFUR ON A MOLYBDENUM SURFACE STUDIED BY AUGER ELECTRON SPECTROS
                 5-1565 HIGH RESOLUTION AUGER ELECTRON SPECTROSCOPY OF NITRIC OXIDE ADSORBED ON A TUNGST
KAWAKATSU H   4-0907 SECONDARY ELECTRON EMISSION DUE TO PRIMARY AND BACKSCATTERED ELECTRONS (TABLE FO
KAWASE S   1-0076 MICROANALYSIS BY AUGER ELECTRON SPECTROSCOPY. (REVIEW, 15 REFS)
                 1-0077 THREE DIMENSIONAL CHARACTERIZATION OF METALS AND SEMICONDUCTORS BY THE AUGER ELE
                 5-1412 THREE-DIMENSIONAL ELEMENT ANALYSIS OF SURFACES AND INTERFACES OF SOLIDS BY THE A
                 5-1413 AUGER ELECTRON EMISSION MICROGRAPHIC STUDIES OF THE CLEAVAGE SURFACE OF GRAPHITE
                 5-1414 AUGER ELECTRON SPECTROSCOPY AND MICROANALYSIS OF SOLID SURFACES.
                 5-1415 AUGER ELECTRON EMISSION MICROGRAPHY AND MICROANALYSIS OF A STAINED SURFACE OF PU
                 5-1416 AUGER ELECTRON EMISSION MICROGRAPHY AND MICROANALYSIS OF SOLID SURFACES.
KAY E   1-0038 SURFACE ANALYSIS TODAY. (REVIEW OF METHODS, INCLUDING AES)
                 5-1273 TECHNIQUES FOR ELEMENTAL COMPOSITION PROFILING IN THIN FILMS. (AUGER SPECTROSCOP
KEAR BH   6-2053 CONFIRMATION OF SULFUR EMBRITTLEMENT IN NICKEL ALLOYS. (AUGER ELECTRON SPECTROSC
KEENAN JA   3-0640 DIGITAL METHODS FOR THIN FILM ANALYSIS USING A COMPUTER COUPLED AUGER SPECTROMET
                 5-1572 ESTABLISHING DETECTION AND DETERMINATION LIMITS IN SURFACE AND THIN FILM ANALYSI
KELEMEN S   5-1186 CHANGE IN THE SURFACE ELECTRONIC STRUCTURE OF PLATINUM (100) INDUCED BY RECONSTR
KELFVE P   4-1066 AUGER ELECTRON SPECTRUM OF WATER VAPOR.

KELLER DV	1-0104	ADHESION, FRICTION, WEAR, AND LUBRICATION RESEARCH BY MODERN SURFACE SCIENCE TEC
KELLY HP	2-0312	CORRELATION ENERGIES AND AUGER RATES IN ATOMS WITH INNER SHELL VACANCIES (NEON W
	4-0909	K-AUGER RATES CALCULATED FOR NEON(+).
KENDRICK J	4-0814	EXPERIMENTAL AND THEORETICAL STUDY OF THE AUGER SPECTRUM OF NITROUS OXIDE.
	4-0879	CONFIGURATION INTERACTION CALCULATIONS OF THE AUGER SPECTRUM OF METHANE, HYDROGE
KENNETT HM	5-1536	INITIAL OXIDATION OF MOLYBDENUM. PART-3: OXIDE NUCLEATION RATES ON (100) AND (11
KERN R	5-1754	ELLIPSOMETRIC STUDY OF ADSORPTION: ADSORPTION OF XENON ON GRAPHITE (001).
KESKI-RAHKONEN O	2-0403	EVIDENCE OF RADIATIVE AUGER TRANSITIONS IN THE FREE ARGON ATOM.
	2-0404	EVIDENCE OF RADIATIVE AUGER TRANSITIONS IN THE FREE ARGON ATOM.
	2-0502	THE K-FAR-M2 RADIATIVE AUGER EFFECT IN TRANSITION METALS.
	4-0910	SULFUR KLL AUGER SPECTRA IN HYDROGEN SULFIDE, SULFUR DIOXIDE, AND SULFUR HEXAFLU
	4-0911	TOTAL AND PARTIAL ATOMIC LEVEL WIDTHS.
	4-1057	RADIATIVE KM2 AUGER EFFECT IN SOME TRANSITION METALS.
KESSEL QC	2-0405	PRODUCTION OF INNER SHELL VACANCIES IN HEAVY ION- ATOM COLLISIONS. (AUGER EFFECT
KESSLER J	5-1171	ISOMORPHISM AND PURITY OF COPPER VAPOR DEPOSITED ON A COPPER (111) SURFACE. (AUG
KHADZHI PI	2-0227	RADIATIVE AUGER RECOMBINATION OF TWO INDIRECT EXCITONS.
KHAN IH	5-1537	LOW TEMPERATURE EPITAXY OF GERMANIUM FILMS BY SPUTTER DEPOSITION. (HEED, LEED, A
	5-1538	SURFACE STRUCTURE OF TITANIUM AND ITS INTERACTION WITH BROMINE AND CHLORINE. (AU
	6-2067	SURFACE SEGREGATION IN TITANIUM AS MONITORED BY AUGER ELECTRON SPECTROSCOPY.
KHARAKHORIN FF	4-0840	(AUGER) RECOMBINATION OF NONEQUILIBRIUM CHARGE CARRIERS OF INDIUM ARSENIDE AT HI
KHARE PL	2-0526	APPLICATION OF THE ZETA EFFECTIVE VALUES OF HYDROGENIC WAVE FUNCTIONS IN THE CAL
KHERA RP	5-1539	A STUDY OF ACTIVATED SILICON SURFACES FOR ELECTROLESS NICKEL PLATING BY AES.
KHRONOPULO KA	5-1641	PROCESSES OF ADSORPTION OF OXYGEN ON GALLIUM ARSENIDE METHOD OF LOW ENERGY ELECT
	5-1642	AUGER ELECTRON SPECTRUM DURING THE INTERACTION BETWEEN OXYGEN AND THE (111) SURF
	5-1643	SOME FEATURES OF THE AUGER ELECTRON SPECTRUM IN THE INTERACTION BETWEEN OXYGEN A
	5-1644	INVESTIGATION OF CESIUM ADSORPTION ONTO A GALLIUM ARSENIDE (110) SURFACE. (AUGER
KIEWIT DA	5-1540	GERMANIUM SURFACE CLEANING: AUGER ANALYSIS.
KIKUTA S	3-0641	DESIGN OF AN X-RAY DOUBLE CRYSTAL SPECTROMETER INSTALLED IN VACUUM AND THE MEASU
KILBANE FM	5-1541	SUPERCONDUCTING TRANSITION TEMPERATURES OF REACTIVELY SPUTTERED FILMS OF TANTALU
KIM JS	6-2000	SEGREGATION OF CARBON TO (100) SURFACE OF NICKEL.
KINCH MA	2-0406	RECOMBINATION MECHANISMS IN 8-14 MICRON MERCURY(1-X) CADMIUM(X) TELLURIDE. (AUGE
KINNIBURGH C	5-1580	A LEED STUDY OF MAGNESIUM OXIDE (100). PART-1: EXPERIMENT.
KIRBY RE	5-1542	ELECTRON BEAM INDUCED EFFECTS ON GAS ADSORPTION STUDIES UTILIZING AUGER ELECTRON
	5-1543	ELECTRON BEAM INDUCED EFFECTS ON GAS ADSORPTION UTILIZING AUGER ELECTRON SPECTRO
	5-1619	LEED AND AES OBSERVATIONS ON THE POLAR SURFACES OF ZINC OXIDE.
	5-1620	INTERFERENCE OF AN ELECTRON BEAM WITH THE SURFACE REACTION BETWEEN OXYGEN AND GE
KIRK DL	5-1162	DEPARTURES FROM STOICHIOMETRY IN (100) SURFACES OF GALLIUM PHOSPHIDE, INDUCED BY
	5-1163	COMPOSITIONAL AND STRUCTURAL CHANGES THAT ACCOMPANY THE THERMAL ANNEALING OF (10
	5-1164	EFFECT OF TEMPERATURE ON THE SURFACE STRUCTURE AND STOICHIOMETRY OF (100) INDIUM
KIRSANOVA TS	4-0736	RECORDING OF COMPONENTS ON THE SURFACE OF FUSED EMITTERS. (AUGER SPECTRA)
	4-0849	AUGER SPECTROSCOPY OF A BARIUM- METAL FILM SYSTEM.
	5-1357	AUGER SPECTROSCOPY OF THE BARIUM- PLATINUM FILM SYSTEM.
	5-1358	RECORDING THE COMPONENTS ON THE SURFACE OF A BINARY ALLOY. (AUGER SPECTROSCOPY)
	5-1359	EFFECT OF STRUCTURE ON THE AUGER SPECTRA OF BARIUM FILMS ON PLATINUM.
KIRSCHNER J	5-1843	ABSOLUTE ATOMIC DENSITIES DETERMINED BY AUGER ELECTRON SPECTROSCOPY.
	6-2068	DISAPPEARANCE POTENTIAL SPECTROSCOPY (VANADIUM OXIDATION).
KISHINEVSKII LM	2-0247	AUGER NEUTRALIZATION OF MULTIPLY CHARGED IONS ON A METAL SURFACE.
	2-0407	AUGER IONIZATION OF ATOMS BY MULTIPLY CHARGED IONS. MODEL OF TWO COULOMB CENTERS
	5-1544	EVALUATION OF THE FREQUENCY OF AUGER NEUTRALIZATION OF INERT GAS IONS ON THE SUR
KJAER HA	2-0243	DIFFRACTION EFFECTS IN BACKSCATTERING AND AUGER PRODUCTION NEAR CRYSTAL SURFACES
KLASSON M	2-0408	ELECTRON ESCAPE DEPTH IN SILICON. (AUGER ENERGIES)
	2-0409	ESCAPE DEPTHS OF X-RAY EXCITED (AUGER) ELECTRONS.
KLEIN W	3-0642	SIMPLIFIED RETARDING FIELD PHOTOELECTRON AND AUGER ELECTRON SPECTROGRAPH.
KLUGE A	3-0643	HIGH SENSITIVITY APPEARANCE POTENTIAL SPECTROMETER FOR THE DETECTION OF PHOTONS,
	5-1545	SURFACE ANALYSIS BY APPEARANCE POTENTIAL SPECTROSCOPY.
KMETZ AR	5-1281	MECHANISM OF SURFACE ALIGNMENT IN NEMATIC LIQUID CRYSTALS. (SUBSTRATE SURFACE EF
KNAPP JA	2-0398	MATRIX ELEMENTS DEPENDENCE OF OPTICAL EXCITATION AND AUGER DECAY OF 5D CORE HOLE
	2-0422	PHOTOEMISSION STUDIES OF CORE EXCITON DECAY IN POTASSIUM IODIDE.
KNECHT TA	2-0398	MATRIX ELEMENTS DEPENDENCE OF OPTICAL EXCITATION AND AUGER DECAY OF 5D CORE HOLE
KNERINGER G	5-1546	(AUGER) ADSORPTION STUDIES OF OXYGEN AND CARBON MONOXIDE ON A PLATINUM (100) SUR
	5-1685	ADSORPTION OF HYDROGEN AND THE REACTION OF HYDROGEN WITH OXYGEN ON PLATINUM (100
KNOPFLE KT	2-0336	DECAY OF HOLE STATES IN OXYGEN-16 VIA A "NUCLEAR AUGER EFFECT".
KNOTEK ML	5-1547	IN-SITU TRANSPORT MEASUREMENTS IN ULTRAHIGH VACUUM GROWN AMORPHOUS GERMANIUM. (D
	5-1548	TRANSPORT OF OXYGEN IN AMORPHOUS GERMANIUM THIN FILMS DURING ANNEALING. (AUGER S
KO SM	5-1549	ADSORPTION AND SOLUTION OF HYDROGEN, DEUTERIUM, NITROGEN, OXYGEN AND CARBON MONO
KOBAYASHI N	4-1041	SHIFT ENERGIES OF CHARACTERISTIC X-RAYS AND AUGER ELECTRONS FOR IONIZED ATOMS.
KOCH J	3-0644	SMALL RETARDING FIELD ANALYZER FOR AUGER ELECTRON SPECTROSCOPY. (PALLADIUM- NICK
	5-1321	INTERACTION OF CARBON MONOXIDE AND OXYGEN WITH A PALLADIUM (100) SURFACE. (CHEMI
KODRE A	4-1014	K-SHELL FLUORESCENCE YIELDS OF ARGON, CHLORINE AND SULFUR. (AUGER TRANSITION)
KOHARA R	5-1671	CHEMICAL ANALYSIS OF SEMICONDUCTORS. (IMPURITIES AND COMPOSITION)
KOHLER HS	2-0312	CORRELATION ENERGIES AND AUGER RATES IN ATOMS WITH INNER SHELL VACANCIES (NEON W
KOIKE J	5-1658	NEW METHOD OF THIN FILM THICKNESS MEASUREMENT USING RADIOISOTOPE AUGER ELECTRONS
KOIKE K	5-1482	OBSERVATION OF IMPURITY PROFILE IN ION IMPLANTED SILICON BY AUGER ELECTRON SPECT
KOKTA L	2-0474	COINCIDENCE METHODS OF STANDARDIZATION FOR CESIUM-131 AND MEASUREMENT OF DECAY P
KOLACZKIEWICZ J	6-2069	IMPURITIES SEGREGATION TO THE (001) NICKEL SURFACE DURING THERMAL TREATMENT. WOR
KOLARIK V	3-0604	AUGER ELECTRON SPECTROSCOPY IN THE ELECTRON EMISSION MICROSCOPE.
	3-0605	POSSIBILITY OF AUGER ELECTRON SPECTROSCOPY IN THE ELECTRON EMISSION MICROSCOPE.
KOLESNIKOV VV	4-1104	AUGER TRANSITIONS AND SHAPE OF X-RAY SPECTRUM.
KOLIWAD KM	5-1286	DIFFUSION STUDIES IN CHROMIUM- PLATINUM THIN FILMS USING AUGER ELECTRON SPECTROS
	5-1983	AUGER ELECTRON SPECTROSCOPY OF CLEANUP-RELATED CONTAMINATION ON SILICON SURFACES
KOMAREK KL	5-1249	AUGER ANALYSIS OF THIN OXIDE FILMS ON LEAD- INDIUM ALLOYS.
KOMIYA A	5-1550	COMPOSITION PROFILES OF SEVERAL CONTAMINATED AND CLEANED SURFACES OF GOLD THICK
KOMIYA S	1-0105	AUGER ELECTRON SPECTROSCOPY. (REVIEW, 39 REFS)
	1-0106	SOME TOPICS ON AES. (SURFACE ANALYSIS)
	3-0647	DETECTION OF SODIUM ON SILICON AND SILICON DIOXIDE SURFACES BY CYLINDRICAL MIRRO
	5-1436	DIFFUSE INTERFACE STRUCTURES IN SOME (SUBSTRATE- VACUUM EVAPORATED FILM) SYSTEMS
	5-1438	METALLIC STATE OF SILICON IN SILICON- NOBLE METAL VAPOR QUENCHED ALLOYS STUDIED
	5-1551	EFFECT OF GOLD COATING ON SORPTION CHARACTERISTICS OF CARBON MONOXIDE, SULFUR, A
	5-1552	SPUTTER ETCHING AND AUGER ELECTRON SPECTROSCOPY OF METAL SURFACES.
	5-1553	SIMULTANEOUS OBSERVATIONS OF PARTIALLY OXIDIZED SURFACES BY AUGER ELECTRON SPECT

KOMIYA S	5-1554	SPUTTER ETCHING AND AUGER ELECTRON SPECTROSCOPY OF METAL SURFACES.
	5-1678	LEED AND AUGER ELECTRON SPECTROSCOPIC OBSERVATIONS OF THE SUBSTRATE SURFACE DURI
	5-1679	COMPOSITION PROFILE OF ION PLATED GOLD FILM ON COPPER ANALYZED BY AUGER ELECTRON
	5-1680	AUGER SPECTROSCOPIC OBSERVATION OF SILICON- GOLD MIXED PHASE FORMATION AT LOW TE
	5-1681	DIFFUSE INTERFACE IN SILICON (SUBSTRATE)- GOLD (EVAPORATED FILM) SYSTEM. (AUGER
	6-2090	COMPOSITION OF BINARY ALLOYS BY SIMULTANEOUS SIMS AND AES MEASUREMENTS.
KOMNINOS Y	2-0438	TRANSITION RATES FOR INTERATOMIC AUGER PROCESSES.
KOMYAK NI	3-0599	X-RAY ELECTRON SPECTROMETERS AND AUGER ANALYZERS.
KONDOW T	4-0912	HIGH RESOLUTION L2,3M2,3M2,3 AUGER ELECTRON SPECTRA OF ARGON.
	4-0925	CHEMICAL EFFECTS OF THE NITROGEN KLL AUGER SPECTRA OF CHEMISORBED NITROGEN ON IR
	5-1530	STUDIES OF THE MECHANISM OF THE REACTION BETWEEN SULFUR AND OXYGEN ON A MOLYBDE
	5-1531	STUDY OF THE OXIDATION OF MOLYBDENUM SURFACES BY ENERGY LOSS SPECTROSCOPY COMBIN
	5-1532	HIGH RESOLUTION MEASUREMENTS OF THE NITROGEN KLL AUGER TRANSITIONS OF CHEMISORBE
	5-1533	APPLICATIONS OF AUGER ELECTRON AND ENERGY LOSS SPECTROSCOPIES TO CATALYTIC PROCE
	5-1534	DYNAMIC INVESTIGATION OF THE MECHANISM OF REACTION BETWEEN SULFUR AND OXYGEN ON
	5-1564	SEGREGATION OF SULFUR ON A MOLYBDENUM SURFACE STUDIED BY AUGER ELECTRON SPECTROS
	5-1565	HIGH RESOLUTION AUGER ELECTRON SPECTROSCOPY OF NITRIC OXIDE ADSORBED ON A TUNGST
KONNO H	5-1873	WORK FUNCTION OF WELL-DEFINED SURFACE OF COPPER- NICKEL ALLOY PLATES. (AUGER ELE
KONRAD GT	4-0913	SECONDARY ELECTRON EMISSION FROM ION IMPLANTED SILICON (CESIUM).
KONSOW T	4-1004	ORIGIN OF THE PEAK SHAPES IN THE AUGER SPECTRUM OF OXYGEN (KL2,3L2,3) ON MOLYBDE
KORABLEV VV	1-0107	ELECTRONIC AUGER SPECTROSCOPY: TEXTBOOK FOR A COURSE OF LECTURES ON PHYSICAL EL
	2-0508	SECONDARY EMISSION AND ELASTIC REFLECTION OF ELECTRONS FROM MONOCRYSTALS.
	4-1061	CHANGES IN THE FINE STRUCTURE OF THE ANGULAR DEPENDENCE OF THE SECONDARY ELECTRO
	4-1062	SECONDARY ELECTRON EMISSION OF SILICON DIOXIDE SINGLE CRYSTALS.
	4-1063	SECONDARY ELECTRON EMISSION OF MOLYBDENUM CRYSTALS.
	4-1064	ENERGY SPECTRA OF INELASTICALLY SCATTERED AND AUGER ELECTRONS FROM SINGLE CRYSTA
KORBER H	4-0914	K-AUGER SPECTRUM OF NEON.
KORBITZ FW	5-1573	SURFACE CHARACTERIZATION OF NICKEL- COPPER ALLOY DISPLAYING POOR WETTABILITY DUR
KOROLKOV NS	3-0658	AUGER ELECTRON SPECTROMETER INTEGRATED WITH A LOW ENERGY ELECTRON DIFFRACTION SY
KORSUNSKII MI	2-0538	IONIZATION LOSSES IN (AUGER) SPECTRA OF ELECTRON INELASTIC SCATTERING FROM RARE
	3-0721	FOCUSING PROPERTIES OF AN ELECTROSTATIC MIRROR WITH CYLINDRICAL FIELD.
	4-1030	AUGER ELECTRON SPECTRUM OF OSMIUM AT ENERGIES UP TO 300 EV.
	4-1100	AUGER ELECTRON SPECTRA OF SOME RARE EARTH METALS. (LANTHANUM, PRASEODYMIUM, NEOD
	4-1124	PLASMA SATELLITES IN THE AUGER ELECTRON SPECTRUM OF MOLYBDENUM.
	5-1759	DECREASE OF CARBON CONCENTRATION ON THE SURFACE OF MOLYBDENUM CARBIDE DURING HEA
	5-1985	CHEMICAL COMPOSITION OF THE SURFACE OF MOLYBDENUM AND STEEL BASED ON AUGER ELECT
KORTOV VS	2-0410	EXOEMISSION PROPERTIES OF ZIRCONIUM DIOXIDE. (AUGER EFFECT)
	4-0915	EXOELECTRON SPECTROMETRY AS A METHOD FOR DETERMINATION OF THE ENERGY DEPTH OF LE
KOSHIKAWA T	2-0365	AUGER AND SECONDARY ELECTRONS EXCITED BY BACKSCATTERED ELECTRONS.
	2-0366	AUGER AND SECONDARY ELECTRONS EXCITED BY BACKSCATTERED ELECTRONS. APPROACH TO QU
	2-0411	BACKGROUND INFORMATION IN AUGER ELECTRON SPECTROSCOPY AND SECONDARY ELECTRON EMI
	4-0916	SECONDARY AND AUGER ELECTRON SPECTRA OF A COPPER- BERYLLIUM ALLOY.
	4-0917	SECONDARY ELECTRON EMISSION FROM IRON SINGLE AND POLYCRYSTALS.
	5-1555	SECONDARY ELECTRON EMISSION IN THE STUDY OF SOLID SURFACES.
KOSMACHEV OS	3-0721	FOCUSING PROPERTIES OF AN ELECTROSTATIC MIRROR WITH CYLINDRICAL FIELD.
KOSSOWSKY R	5-1556	MICROSTRUCTURE OF HOT PRESSED SILICON NITRIDE. (AUGER ANALYSIS, MICROPROBE ANALY
	5-1557	IMPURITIES AND INCLUSIONS IN HOT PRESSED SILICON NITRIDE. (AUGER SPECTROGRAPHIC
KOSTELITZ M	5-1558	ADSORPTION OF SULFUR ON GOLD BY LOW ENERGY ELECTRON DIFFRACTION AND AUGER SPECTR
KOSTER AS	4-0762	X-RAY FLUORESCENCE L SPECTRA OF NIOBIUM IN NIOBIUM METAL AND SOME OF ITS COMPOUN
KOSTROUN VO	2-0316	THEORETICAL L2 AND L3 SUBSHELL FLUORESCENCE YIELDS AND L2-L3(X) COSTER-KRONIG TR
	2-0329	AUGER AND COSTER-KRONIG TRANSITION PROBABILITIES TO THE ATOMIC 2S STATE AND THEO
	2-0386	MANY BODY PERTURBATION THEORY APPLIED TO ATOMIC RADIATIONLESS TRANSITIONS.
	4-0918	ATOMIC RADIATION TRANSITION PROBABILITIES TO THE 1S STATE AND THEORETICAL K-SHEL
KOSYACHKOV AA	5-1939	AUGER SPECTROSCOPIC STUDY OF THE SURFACE OF IRON AFTER IONIC BOMBARDMENT AND AIR
KOU WH	5-1583	QUANTITATIVE STUDY OF AUGER ELECTRON SIGNALS OF PHOSPHORUS ON SILICON USING A QU
KOVAL IF	4-0923	FILM STRUCTURE EFFECT ON THE SECONDARY ELECTRON EMISSION CHARACTERISTICS (GERMAN
KOWALCZYK SP	2-0412	RELATIVE EFFECT OF EXTRAATOMIC RELAXATION ON AUGER AND BINDING ENERGY SHIFTS IN
	2-0427	MANY BODY EFFECTS IN X-RAY PHOTOEMISSION FROM MAGNESIUM. (AUGER EFFECT)
	4-0919	L2,3M4,5M4,5 AUGER SPECTRA OF METALLIC COPPER AND ZINC: THEORY AND EXPERIMENT.
KOZIOL C	6-2069	IMPURITIES SEGREGATION TO THE (001) NICKEL SURFACE DURING THERMAL TREATMENT. WOR
KRAHE PR	6-2070	SEGREGATION OF MANGANESE AND ANTIMONY TO THE GRAIN BOUNDARIES OF TEMPER EMBRITTL
KRAMER HM	5-1559	ADSORPTION OF KRYPTON ON THE BASAL PLANE OF GRAPHITE: LEED AND AUGER MEASUREMENT
KRAUSE MO	2-0308	RELATIVE ABUNDANCES AND RECOIL ENERGIES OF FRAGMENT IONS FORMED FROM THE X-RAY P
	2-0413	REARRANGEMENT OF INNER SHELL IONIZED ATOMS. (AUGER SPECTRA)
	2-0414	ARGON KLL AUGER SPECTRUM. TEST OF THEORY.
	2-0575	DETERMINATION OF OSCILLATOR STRENGTH FOR THE FIRST DISCRETE TRANSITION IN NITROG
	3-0600	HIGH RESOLUTION ELECTRON SPECTROMETER FOR PHOTOELECTRON AND AUGER ELECTRON STUDI
	4-0791	IDENTIFICATION OF HIGH ENERGY LINES IN THE KLL AUGER SPECTRUM OF NITROGEN.
	4-0911	TOTAL AND PARTIAL ATOMIC LEVEL WIDTHS.
	4-0920	MANIFESTATION OF ATOMIC DYNAMICS THROUGH THE AUGER EFFECT. (NEON AUGER SPECTRUM)
	4-0990	DETERMINATION OF KLL AUGER SPECTRA OF NITROGEN, OXYGEN, CARBON DIOXIDE, NITRIC O
	4-0991	AUGER SPECTRA OF SIMPLE MOLECULES.
	5-1560	OBSERVATIONS OF BETA- SILICON CARBIDE FORMATION ON RECONSTRUCTED SILICON SURFACE
KRAUSE U	2-0291	PLASMON COUPLING TO CORE HOLE EXCITATIONS IN CARBON. (AUGER APPEARANCE SPECTROSC
KRAUSSE J	2-0415	AUGER RECOMBINATION IN SILICON RECTIFIERS AND THYRISTORS LOADED IN FORWARD DIREC
KRESSEL H	2-0332	CARRIER LIFETIMES IN EPITAXIAL INDIUM ARSENIDE. (AUGER RECOMBINATION)
KRIEGSEIS W	4-0921	EXOELECTRON EMISSION FROM ZINC OXIDE. (OXYGEN ADSORPTION EFFECT, AUGER ELECTRON
KRIGER YUG	4-0816	CHEMICAL SHIFT IN AUGER SPECTRA OF TUNGSTEN AND MOLYBDENUM DURING THE ADSORPTION
	4-0922	CHEMICAL SHIFT IN THE AUGER SPECTRUM OF TUNGSTEN FROM OXYGEN ADSORPTION.
KRONIG RL	1-0045	A NEW TYPE OF AUGER EFFECT AND ITS INFLUENCE ON THE X-RAY SPECTRUM.
KROYER JM	3-0645	SCANNING AUGER MICROSCOPE FOR THIN FILM ANALYSIS.
KRUER M	2-0262	AUGER LIMITED CARRIER LIFETIMES IN MERCURY CADMIUM TELLURIDE AT HIGH EXCESS CARR
KRYNKO YUN	4-0923	FILM STRUCTURE EFFECT ON THE SECONDARY ELECTRON EMISSION CHARACTERISTICS (GERMAN
KU R	5-1187	CARBON MONOXIDE OXIDATION ON A PLATINUM (110) SINGLE CRYSTAL SURFACE.
	5-1188	MECHANISMS OF THE CATALYTIC CARBON MONOXIDE OXIDATION ON PLATINUM (110).
	5-1189	KINETICS OF OXYGEN ADSORPTION ON A PLATINUM (111) SURFACE. (AUGER ELECTRON SPECT
	5-1190	ADSORBATE INTERACTIONS ON A PLATINUM (110) SURFACE. PART-1: SULFUR AND CARBON MO
KUBONIWA S	2-0402	CONCENTRATION DEPENDENCE OF GREEN- COPPER LUMINESCENCE IN COPPER AND ALUMINUM DO
KUCHITSU K	1-0108	ELECTRON SPECTROMETRY.
KUDO M	5-1690	X-RAY EXCITED PHOTOELECTRON AND AUGER ELECTRON SPECTRA OF OXIDIZED STATES ON SOM

KUHNLE G	3-0692	ELECTRON CHANNELING PATTERNS IN AN AUGER ELECTRON SPECTROMETER WITH SCANNING SAM
KULOV SK	6-2071	SECONDARY EMISSION AND CONTAMINATION OF METAL SURFACES.
KUL'VARSKAYA BS	5-1242	ELECTRON AUGER SPECTROSCOPIC STUDY OF THE SURFACE OF CARBIDE CATHODES.
	5-1561	ELECTRON AUGER SPECTROSCOPIC STUDY OF THE SURFACE OF CARBIDE CATHODES.
	5-1562	SURFACE OF THERMIONIC EMITTERS MADE OF CARBIDE COMPOUNDS STUDIED BY AN ELECTRONI
KUMAR G	4-0924	PHOTOELECTRON CHEMICAL SHIFTS IN COORDINATION COMPOUNDS. X-RAY EXCITED LMM AUGER
KUMAR V	5-1232	INTERDIFFUSIONS IN THIN FILM GOLD ON PLATINUM ON GALLIUM ARSENIDE (100) STUDIED
	5-1563	REACTION OF SPUTTERED PLATINUM FILMS ON GALLIUM ARSENIDE. (AUGER SPECTROSCOPY)
KUNIMORI K	4-0912	HIGH RESOLUTION L2,3M2,3M2,3 AUGER ELECTRON SPECTRA OF ARGON.
	4-0925	CHEMICAL EFFECTS OF THE NITROGEN KLL AUGER SPECTRA OF CHEMISORBED NITROGEN ON IR
	5-1530	STUDIES OF THE MECHANISM OF THE REACTION BETWEEN SULFUR AND OXYGEN ON A MOLYBDEN
	5-1531	STUDY OF THE OXIDATION OF MOLYBDENUM SURFACES BY ENERGY LOSS SPECTROSCOPY COMBIN
	5-1532	HIGH RESOLUTION MEASUREMENTS OF THE NITROGEN KLL AUGER TRANSITIONS OF CHEMISORBE
	5-1533	APPLICATIONS OF AUGER ELECTRON AND ENERGY LOSS SPECTROSCOPIES TO CATALYTIC PROCE
	5-1534	DYNAMIC INVESTIGATION OF THE MECHANISM OF REACTION BETWEEN SULFUR AND OXYGEN ON
	5-1564	SEGREGATION OF SULFUR ON A MOLYBDENUM SURFACE STUDIED BY AUGER ELECTRON SPECTROS
	5-1565	HIGH RESOLUTION AUGER ELECTRON SPECTROSCOPY OF NITRIC OXIDE ADSORBED ON A TUNGST
KUPPERS J	5-1278	INTERACTION OF NITRIC OXIDE WITH A NICKEL (111) SURFACE.
	5-1322	ADSORPTION ON SINGLE CRYSTAL SURFACES OF COPPER- NICKEL ALLOYS. PART-1.
	5-1607	INTERACTION OF CARBON MONOXIDE WITH (110) NICKEL SURFACES. (AUGER EFFECT)
KURODA N	2-0416	STIMULATED PHOTOLUMINESCENCE IN GALLIUM SELENIDE. (AUGER PROCESS)
KUSHIDA T	2-0451	LUMINESCENCE DUE TO EXCITON-EXCITON COLLISION IN GALLIUM ARSENIDE.
KUSUNOKI N	6-2043	EFFECTS OF COPPER AND PHOSPHORUS ON TEMPER EMBRITTLEMENT OF MANGANESE- MOLYBDENU
KUYATT CE	3-0623	COMPARISON OF SPHERICAL DEFLECTOR AND CYLINDRICAL MIRROR ANALYZERS.
LAFORCE RP	6-2119	STUDIES USING AUGER ELECTRON EMISSION SPECTROSCOPY ON TEMPER EMBRITTLEMENT IN LO
LAGUES M	5-1566	SURFACE SEGREGATION: COMPARISON BETWEEN THEORY AND EXPERIMENT.
LAHIRI SK	5-1249	AUGER ANALYSIS OF THIN OXIDE FILMS ON LEAD- INDIUM ALLOYS.
LAHTEENKORVA E	4-0729	AUGER ELECTRON STUDIES OF METAL VAPORS.
LAMBERT RM	5-1567	ROLE OF PRIMARY AND SECONDARY ELECTRONS IN ELECTRON INDUCED DESORPTION AND DISSO
	5-1568	ADSORPTION OF OXYGEN ON MOLYBDENUM (111): EFFECT OF TRACE IMPURITIES.
	5-1569	LEED- AUGER INVESTIGATION OF A STABLE CARBIDE OVERLAYER ON A PLATINUM (111) SURF
LAMOTHE R	1-0222	SOME PARAMETERS AFFECTING AUGER AND PHOTOELECTRON SPECTROSCOPY AS AN ANALYTICAL
LANDER JJ	3-0661	APPLICATION OF TRIPLE GRID LEED SYSTEM TO AUGER SPECTRUM ANALYSES.
	4-0926	AUGER PEAKS IN THE ENERGY SPECTRA OF SECONDARY ELECTRONS FROM VARIOUS MATERIALS.
	5-1603	AN ELECTRON DIFFRACTION STUDY OF CESIUM ADSORPTION ON TUNGSTEN.
	5-1604	ELECTRONIC AND LATTICE STRUCTURE OF CESIUM FILMS ADSORBED ON TUNGSTEN.
	5-1659	ADSORPTION OF IONIC SALTS ON A TUNGSTEN (100) SURFACE.
LANDERS AM	2-0333	ADSORBATE EFFECTS IN ELECTRON EJECTION BY RARE GAS METASTABLE ATOMS. (AUGER PROC
LANDOLT D	6-2088	QUANTITATIVE AUGER ELECTRON SPECTROSCOPY ANALYSIS OF SILVER- PALLADIUM AND NICKE
LANDSBERG PT	2-0385	A FORMALISM FOR THE INDIRECT AUGER EFFECT. PART-1. PART-2.
	2-0417	DETAILED BALANCE BETWEEN AUGER RECOMBINATION AND IMPACT IONIZATION IN SEMICONDUC
	2-0418	IMPACT IONIZATION AND AUGER RECOMBINATION IN BANDS.
	2-0419	RADIATIVE AND AUGER PROCESSES IN SEMICONDUCTORS.
	2-0420	THEORY OF DONOR- ACCEPTOR RADIATIVE AND AUGER RECOMBINATION IN SIMPLE SEMICONDUC
	2-0421	RECOMBINATION-INDUCED NONEQUILIBRIUM PHASE TRANSITIONS IN SEMICONDUCTORS. (AUGER
LANG B	5-1570	(AUGER) ADSORPTION STUDIES ON THE PLATINUM (100) SURFACE.
	5-1838	REACTIVITY OF LOW INDEX ((111) AND (100)) AND STEPPED PLATINUM SINGLE CRYSTAL SU
LANGENBERG A	5-1571	FLUORESCENCE YIELD OF CARBON. (AUGER MEASUREMENTS)
LANGER DW	2-0306	PHOTOELECTRIC PROPERTIES OF MAGNESIUM(2) SILICIDE, MAGNESIUM(2) GERMANIUM, AND M
	5-1885	VALENCE BAND OF MAGNESIUM STANNIDE DETERMINED BY AUGER AND PHOTOEMISSION SPECTRO
LANGERON JP	5-1740	AUGER ELECTRON SPECTROSCOPY STUDIES OF SURFACE COMPOSITION OF SILVER- COPPER ALL
	6-2102	AUGER SPECTROMETRY STUDY OF PASSIVE FILMS FORMED ON STAINLESS STEELS.
LAPEYRE GJ	2-0398	MATRIX ELEMENTS DEPENDENCE OF OPTICAL EXCITATION AND AUGER DECAY OF 5D CORE HOLE
	2-0422	PHOTOEMISSION STUDIES OF CORE EXCITON DECAY IN POTASSIUM IODIDE.
LARAMORE GE	2-0392	DIRECT COMPARISON OF CORE ELECTRON BINDING ENERGIES OF SURFACE AND BULK ATOMS OF
	5-1723	COMPARISON OF MANY BODY EFFECTS IN CORE LEVEL SURFACE SPECTROSCOPIES.
LARKINS FP	1-0109	(AUGER) TRANSITION ENERGIES. (REVIEW)
	2-0423	FAST ELECTRON- SLOW ELECTRON COINCIDENCE STUDY OF VIOLENT ARGON(+)- ARGON COLLIS
	4-0815	SEMIEMPIRICAL AUGER ELECTRON ENERGIES. PART-2: THE ARGON ATOM.
	4-0927	CHARGE STATE DEPENDENCE OF X-RAY AND AUGER ELECTRON EMISSION SPECTRA FOR RARE GA
	4-0928	KRYPTON M-SHELL AUGER AND COSTER-KRONIG ELECTRON ENERGIES.
	4-0929	L1,L2,3M COSTER-KRONIG TRANSITION ENERGIES.
	4-0930	SEMI-EMPIRICAL AUGER ELECTRON ENERGIES. PART-1: GENERAL METHOD AND KLL LINE ENER
LARRABEE GB	5-1572	ESTABLISHING DETECTION AND DETERMINATION LIMITS IN SURFACE AND THIN FILM ANALYSI
LARSEN GA	2-0346	DYNAMIC CHARGE STATE DEPENDENCE OF THE K-VACANCY PRODUCTION YIELD IN HEAVY ION A
	4-0830	EMISSION OF KLL AUGER ELECTRONS PRODUCED IN Z(1)- NITROGEN AND Z(1)- NEON COLLIS
LARSEN J	2-0255	RADIATIONLESS DEEXCITATION OF EXCITED HELIUM ATOMS AT SURFACES. (AUGER PROCESS)
	5-1220	OPTICAL PROPERTIES AND AUGER CHARACTERISTICS OF MAGNESIUM SINGLE CRYSTALS AND MA
LARSON DT	4-0931	AUGER ELECTRON AND CHARACTERISTIC ENERGY LOSS SPECTRA OF PLUTONIUM.
LARSON PE	4-0932	SURFACE CHARACTERIZATION OF NICKEL- COPPER ALLOY DISPLAYING POOR WETTABILITY DUR
	4-0932	X-RAY INDUCED PHOTOELECTRON AND AUGER SPECTRA OF COPPER, COPPER(II) OXIDE, COPPE
LASKIN A	3-0617	DRYING OF BUTADIENE- NITRILE RUBBERS IN AN AUGER MACHINE.
LASSITER WS	5-1574	DETECTION OF CONTAMINANTS ON POLYCRYSTALLINE SILVER BY AES AND WORK FUNCTION MEA
	5-1575	INTERACTION OF SULFUR DIOXIDE AND CARBON DIOXIDE WITH CLEAN SILVER IN ULTRAHIGH
	5-1576	AUGER SPECTROSCOPY OF HYDROCHLORIC ACID INTERACTION WITH ALUMINUM AT LOW PRESSUR
LASSNER E	6-1997	DETERMINATION OF THE INFLUENCE GRAIN BOUNDARY IMPURITIES ON THE PROPERTIES OF SI
LATANISION RM	6-2072	INTERGRANULAR EMBRITTLEMENT OF NICKEL BY HYDROGEN: EFFECT OF GRAIN BOUNDARY SEGR
LATTA EE	5-1278	INTERACTION OF NITRIC OXIDE WITH A NICKEL (111) SURFACE.
LAU SS	5-1182	EFFECT OF OXIDIZING AMBIENTS ON PLATINUM SILICIDE FORMATION. PART-2: AUGER AND B
	5-1577	ANTIMONY DOPING OF SILICON LAYERS GROWN BY SOLID PHASE EPITAXY. (AES MEASUREMENT
LAVILLA RE	2-0326	"SEMI-AUGER" PROCESSES IN L2,3 EMISSION IN ARGON AND POTASSIUM CHLORIDE.
	2-0424	SEMI-AUGER ELECTRON TRANSITIONS.
LAX B	2-0425	QUANTITATIVE ASPECTS OF A SOFT X-RAY LASER. (AUGER EFFECT)
LAYTON JK	1-0182	SPUTTERING MEASUREMENTS ON CTR MATERIALS USING AUGER ELECTRON SPECTROSCOPY
	5-1825	AUGER ELECTRON SPECTROSCOPY IN SPUTTERING MEASUREMENTS. APPLICATION TO LOW ENER
	5-1826	SPUTTERING MEASUREMENTS ON CONTROLLED THERMONUCLEAR REACTOR MATERIALS USING AUGE
LE GRESSUS C	4-0948	STRUCTURE OF LOW ENERGY ELECTRON SPECTRUM.
LE HERICY J	5-1740	AUGER ELECTRON SPECTROSCOPY STUDIES OF SURFACE COMPOSITION OF SILVER- COPPER ALL
LEA C	6-2073	SURFACE SEGREGATION AS A GUIDE TO GRAIN BOUNDARY SEGREGATION. (AUGER SPECTROSCOP
	6-2074	SITE COMPETITION IN SURFACE SEGREGATION. (AUGER EFFECT)

LEA C	6-2105	SCANNING ELECTRON AUGER MICROSCOPY IN METALLURGY.
	6-2106	SURFACE SEGREGATION AND ITS RELATION TO GRAIN BOUNDARY SEGREGATION (AUGER ELECTR
LEBUS JW	5-1942	SURFACE STUDIES FOR QUARTZ RESONATORS. (CLEANING, AUGER SPECTROSCOPY)
LECANTE J	5-1578	SLOW ELECTRON SPECTROSCOPY ON MOLYBDENUM (100).
	5-1579	CARBON MONOXIDE ADSORPTION KINETICS ON MOLYBDENUM (100). (AUGER SPECTROSCOPY)
LECKEY RCG	3-0673	A NOVEL DETECTION CIRCUIT FOR AUGER SPECTROSCOPY.
	4-0773	PHOTOELECTRON DETERMINATION OF THE ATTENUATION OF THE LOW ENERGY ELECTRONS IN AL
LEE AB	5-1536	INITIAL OXIDATION OF MOLYBDENUM. PART 3: OXIDE NUCLEATION RATES ON (100) AND (11
LEE EH	5-1743	CHANGE OF SURFACE PROPERTIES OF MICA AFTER CLEAVAGE. (OUTGASSING AND DECORATION
LEE JB	1-0042	USE OF PHOTOEMISSION, AUGER, AND THERMAL DESORPTION TECHNIQUES TO STUDY THE MECH
LEFUR P	2-0250	PICOSECOND OPTICAL MEASUREMENTS OF BAND-TO-BAND AUGER RECOMBINATION OF HIGH DENS
LEGARE P	5-1223	STUDY OF THE INTERACTION OF OXYGEN WITH PLATINUM SINGLE CRYSTAL SURFACES BY LOW
	5-1570	(AUGER) ADSORPTION STUDIES ON THE PLATINUM (100) SURFACE.
LEGG K	5-1580	A LEED STUDY OF MAGNESIUM OXIDE (100). PART-1: EXPERIMENT.
	5-1752	LOW ENERGY ELECTRON DIFFRACTION FROM CLEAN (100) NICKEL OXIDE.
LEGG KO	4-1060	INTERFACIAL AND PLASMON DECAY PEAKS IN THE AUGER SPECTRA OF TITANIUM AND IRON.
LEGGETT MR	5-1245	CHEMISORPTION ON TANTALUM (100). (AUGER SPECTROSCOPY)
LEGRESSUS C	4-0933	ABSORPTION BANDS IN SECONDARY ELECTRON CONTINUOUS SPECTRUM.
	4-0934	NEW METHOD FOR ANALYZING THE SECONDARY ELECTRON SPECTRUM.
	5-1753	AUGER SCANNING ELECTRON MICROSCOPIC STUDY OF THE PASSIVE FILM FORMED ON AN IRON-
LEHENY RF	2-0503	OPTICAL GAIN IN LIGHTLY DOPED GALLIUM ARSENIDE. (AUGER RECOMBINATION)
LEHERICY J	6-2102	AUGER SPECTROMETRY STUDY OF PASSIVE FILMS FORMED ON STAINLESS STEELS.
LEITHAEUSER U	4-1078	HIGH RESOLUTION NEON AUGER SPECTRUM PRODUCED IN 4.2 MEV ATOMIC HYDROGEN(+)- NEON
LEJEUNE EJ	4-0935	INTERPRETATION OF THE AUGER ELECTRON SPECTRUM FROM OXIDIZED BERYLLIUM.
	5-1581	LEED- AUGER ANALYSIS OF THE BERYLLIUM (0001) SURFACE.
LENC M	3-0604	AUGER ELECTRON SPECTROSCOPY IN THE ELECTRON EMISSION MICROSCOPE.
	3-0605	POSSIBILITY OF AUGER ELECTRON SPECTROSCOPY IN THE ELECTRON EMISSION MICROSCOPE.
LEPESHINSKAYA VN	4-0736	RECORDING OF COMPONENTS ON THE SURFACE OF FUSED EMITTERS. (AUGER SPECTRA)
	5-1192	AUGER SPECTROSCOPIC STUDY OF THE SURFACES OF ALLOY EMITTERS.
LEROY V	5-1582	STUDY OF METAL SURFACES BY IONIC MICROANALYSIS AND AUGER SPECTROMETRY.
	6-2075	POSSIBILITIES OF THE ION MICROPROBE MASS ANALYZER AND AUGER SPECTROMETER IN PHYS
LEVENSON LL	3-0603	A MULTIPLE SAMPLE HOLDER FOR USE IN ULTRAHIGH VACUUM.
	4-0985	AUGER SATELLITES OF THE L2,3 AUGER EMISSION BANDS OF ALUMINUM, SILICON, AND PHOS
	4-0986	AUGER AND OTHER CHARACTERISTIC ENERGIES IN SECONDARY ELECTRON SPECTRA FROM THIN
	5-1288	A STUDY OF PHOSPHORUS ADSORPTION AND DESORPTION KINETICS ON SILICON (111) SURFAC
	5-1583	QUANTITATIVE STUDY OF AUGER ELECTRON SIGNALS OF PHOSPHORUS ON SILICON USING A QU
LEVERETT V	5-1389	STRUCTURAL STUDIES OF THE ADSORPTION OF CESIUM AND OXYGEN ON SILICON (100).
LEVINE JD	2-0426	STRUCTURAL AND ELECTRONIC MODEL OF NEGATIVE ELECTRON AFFINITY ON THE SILICON- CE
	5-1584	CORRELATION OF ELECTRONIC, LEED, AND AUGER DIAGNOSTICS ON ZINC OXIDE SURFACES.
LEWIS JE	1-0085	QUANTITATIVE DEPTH PROFILING BY AUGER ION SPUTTERING TECHNIQUE.
	3-0630	DECONVOLUTION METHOD FOR COMPOSITION PROFILING BY AUGER SPUTTERING TECHNIQUE.
	5-1439	DIFFUSION MEASUREMENT IN THIN FILMS BY AUGER SPUTTER ETCHING TECHNIQUE.
	5-1440	INTERDIFFUSION IN BIMETALLIC POLYCRYSTALLINE FILMS MEASURED BY AUGER ION SPUTTER
LEWIS RK	1-0136	USE OF AUGER ELECTRON SPECTROSCOPY AND SECONDARY ION MASS SPECTROMETRY IN THE MI
LEWIS SJ	5-1766	CORROSION, SURFACE, AND MAGNETIC PROPERTIES OF NICKEL- IRON- CHROMIUM THIN FILMS
LEY L	2-0412	RELATIVE EFFECT OF EXTRAATOMIC RELAXATION ON AUGER AND BINDING ENERGY SHIFTS IN
	2-0427	MANY BODY EFFECTS IN X-RAY PHOTOEMISSION FROM MAGNESIUM. (AUGER EFFECT)
	4-0919	L2,3M4,5M4,5 AUGER SPECTRA OF METALLIC COPPER AND ZINC: THEORY AND EXPERIMENT.
LEYGRAF C	5-1585	LOW ENERGY ELECTRON DIFFRACTION- AUGER ELECTRON SPECTROSCOPY STUDY OF THE OXIDAT
	5-1586	A LOW ENERGY ELECTRON DIFFRACTION- AUGER ELECTRON SPECTROSCOPY STUDY OF THE OXID
	5-1587	LEED- AES STUDY OF THE OXIDATION OF IRON- CHROMIUM (100) AND (110).
	6-2022	LOW ENERGY ELECTRON DIFFRACTION- AUGER ELECTRON SPECTROSCOPY STUDY OF THE OXIDAT
LEYSEN R	5-1462	ROLE OF IMPURITIES IN THE STABILITY OF ZINC OXIDE.
	5-1588	INTERACTION OF CESIUM WITH CLEAN ZINC OXIDE SURFACES. (AUGER SPECTROSCOPY)
	5-1881	SOME PROPERTIES OF CLEAN AND ALKALI METAL COVERED ZINC OXIDE SURFACES. (LEED, AE
	5-1882	STABILITY OF CESIUM COVERED ZINC OXIDE SURFACES.
LIAU ZL	5-1577	ANTIMONY DOPING OF SILICON LAYERS GROWN BY SOLID PHASE EPITAXY. (AES MEASUREMENT
LICHTMAN D	5-1543	ELECTRON BEAM INDUCED EFFECTS ON GAS ADSORPTION UTILIZING AUGER ELECTRON SPECTRO
	5-1589	CHEMICAL IDENTIFICATION OF SURFACE MONOLAYERS. (BY AUGER EMISSION)
	5-1799	AUGER ELECTRON SPECTROSCOPY, SCANNING ELECTRON MICROSCOPY AND NONDISPERSIVE X-RA
	5-1801	CHEMISORPTION, PHOTODESORPTION AND CONDUCTIVITY MEASUREMENTS ON ZINC OXIDE SURFA
LIDAY J	5-1396	AUGER SPECTROSCOPY AND LEED OF SILICON SURFACES.
	5-1397	AUGER ELECTRON SPECTROSCOPY OF SILICON SURFACES.
LIDOW DB	4-0788	CHEMISORPTION REACTIONS ON HIGH INDEX ZINC SULFIDE SURFACES. (AUGER EFFECT)
LIEBHAFSKY HA	3-0646	X-RAYS AND ELECTRONS IN ANALYTICAL CHEMISTRY WITH EMPHASIS ON INSTRUMENTATIION.
LIESEGANG JL	4-0773	PHOTOELECTRON DETERMINATION OF THE ATTENUATION OF THE LOW ENERGY ELECTRONS IN AL
LINDAU I	2-0428	PROBING DEPTH IN PHOTOEMISSION AND AUGER ELECTRON SPECTROSCOPY.
	4-0936	HIGH RESOLUTION LMM AUGER SPECTRA OF GERMANIUM.
	4-0937	FINE STRUCTURE DETAILS IN THE LMM AUGER SPECTRA OF GERMANIUM.
LINH NT	5-1590	LIQUID PHASE EPITAXY OF III-V COMPOUNDS ON INDIUM PHOSPHIDE SUBSTRATES. (AUGER S
LINKOAHO M	4-1101	INTERPRETATION OF THE LOW ENERGY K-BETA SPECTRUM IN SOLIDS.
	4-1102	SCANDIUM L X-RAY EMISSION FROM PURE METAL AND SCANDIUM OXIDE.
LINNETT JW	5-1568	ADSORPTION OF OXYGEN ON MOLYBDENUM (111): EFFECT OF TRACE IMPURITIES.
	5-1569	LEED- AUGER INVESTIGATION OF A STABLE CARBIDE OVERLAYER ON A PLATINUM (111) SURF
LIPOVETSKII SS	2-0256	RESONANCE IONIZATION OF ATOMIC HELIUM BY FAST ELECTRONS. (ANGULAR DISTRIBUTION O
	2-0257	POSSIBILITY OF INVESTIGATING ATOMIC AUTOIONIZING STATES IN COINCIDENCE EXPERIMEN
LIPSKI M	5-1865	LEED- AES STUDIES OF CHEMISORPTION INDUCED SULFUR SEGREGATION FROM THE BULK TO T
LISTENGARTEN MA	1-0110	THE AUGER EFFECT (REVIEW, 116 REFS)
LITOVCHENKO VG	5-1990	MANY PARTICLE RECOMBINATION PROCESSES (AUGER) ON THE SURFACES AND IN THIN FILMS
LITVINOV OK	5-1357	AUGER SPECTROSCOPY OF THE BARIUM- PLATINUM FILM SYSTEM.
LIU YY	4-0938	KLL AUGER SPECTRUM OF MANGANESE.
LLABADOR Y	5-1796	MOSSBAUER SPECTROSCOPY OF IRON-57 AND TIN-119 BY DETECTION OF CONVERSION AND AUG
LLYGRAF C	5-1591	SURFACE COMPOSITION STUDIES OF THE (001) AND (110) FACES OF MONOCRYSTALLINE IRON
LO CM	5-1909	LOW INTRINSIC STRESS AND GOOD ADHESION ALLOY THIN FILMS. (TITANIUM AND CHROMIUM
LO IY	4-0939	WIDTH OF ATOMIC L2 AND L3 VACANCY STATES NEAR Z = 30. (AUGER EFFECT)
LOEBACH E	5-1213	STUDY OF SURFACE PROCESSES ON COPPER- BERYLLIUM BY COMBINED SECONDARY ION MASS S
	5-1214	COMBINATION OF SECONDARY ION MASS SPECTROMETRY AND AUGER ELECTRON SPECTROSCOPY,
LOFGREN H	4-0940	HIGH ENERGY SATELLITES IN AUGER SPECTRA FROM METALS.
LOHMEYER S	5-1592	NEW METHODS FOR THE DETECTION AND ELIMINATION OF COATING FLAWS.
LOHS KP	4-0941	NUCLEAR AUGER EFFECTS IN MUONIC ATOMS. (LEAD, BISMUTH)

LONG RL	5-1454	CHEMICAL CLEANING FOR THERMOCOMPRESSION BONDING. (REMOVAL OF OXIDE FILMS, AUGER
LONGSHORE R	5-1593	EVALUATION OF LEAD(0.8) TIN(0.2) TELLURIDE DETECTOR FABRICATION USING SURFACE AN
LONGUA KJ	6-2108	EFFECTS OF TEN YEARS EXPERIMENTAL BREEDER REACTOR II SERVICE ON CHROMIUM- MOLYBD
LOPEZ O	2-0429	A SIMPLE ANALYTICAL CALCULATION FOR X-RAY PHOTOEMISSION. (AUGER PROCESS)
LORCH EA	3-0608	IRON-55 AS AN AUGER ELECTRON EMITTER. NOVEL SOURCE FOR GAS CHROMATOGRAPHY DETECT
LORD DG	2-0430	AUGER EMISSION AND COLOR CENTERS IN LITHIUM FLUORIDE.
LORENTS DC	2-0331	ENERGY DISTRIBUTION OF (AUGER) ELECTRONS RESULTING FROM ION- ATOM COLLISIONS WIT
LOSCH WHP	1-0111	PERTURBING EFFECTS OF SPUTTERING DURING THE AUGER STUDY OF INSULATORS.
	6-2076	AUGER ELECTRON SPECTROSCOPY AND ITS APPLICATION IN METALLURGY.
	6-2077	SPECTROSCOPY OF AUGER ELECTRONS AND ITS USE IN METALLURGY.
LOSIEVSKAYA NV	3-0617	DRYING OF BUTADIENE- NITRILE RUBBERS IN AN AUGER MACHINE.
LOUCHET F	2-0550	EMISSION OF AUGER ELECTRONS BY ION BOMBARDMENT.
	4-0942	SECONDARY ELECTRON SPECTRA OF SOME SOLID TARGETS UNDER IONIC BOMBARDMENT.
LOVECCHIO P	5-1593	EVALUATION OF LEAD(0.8) TIN(0.2) TELLURIDE DETECTOR FABRICATION USING SURFACE AN
LOVELOCK JE	3-0608	IRON-55 AS AN AUGER ELECTRON EMITTER. NOVEL SOURCE FOR GAS CHROMATOGRAPHY DETECT
LOW JR	6-2078	IMPURITIES, INTERFACES, AND BRITTLE FRACTURE. (REVIEW OF AUGER ELECTRON SPECTROS
	6-2079	GRAIN BOUNDARY SEGREGATION OF IMPURITIES IN METALS AND INTERGRANULAR BRITTLE FRA
	6-2114	EFFECT OF PRIOR AUSTENITIC GRAIN BOUNDARY COMPOSITION ON TEMPER BRITTLENESS IN A
LU KE	5-1594	EQUILIBRATION OF HYDROGEN AND DEUTERIUM ON SINGLE CRYSTAL SURFACES OF PLATINUM.
LUCAS AC	4-0818	K-SHELL FLUORESCENCE YIELDS FOR LIGHT ELEMENTS.
LUCAS MW	1-0112	BEAM FOIL AUGER SPECTROSCOPY.
LUDEKE R	5-1595	(AUGER) ELECTRON SPECTROSCOPY OF GALLIUM ARSENIDE AND ALUMINUM ARSENIDE SURFACES
	5-1596	GALLIUM(1-X) ALUMINUM(X) ARSENIDE SUPERLATTICES PROFILED BY AUGER ELECTRON SPECT
LUHMAN T	5-1938	EFFECT OF HIGH TEMPERATURE HEAT TREATMENT ON PENETRATION DEPTH OF SUPERCONDUCTIN
LUHMAN TS	6-2137	EFFECT OF SURFACE METALLURGY ON THE PENETRATION DEPTH AND RF BREAKDOWN FIELD OF
LUMSDEN JB	4-1056	CHEMICAL SHIFTS IN THE AUGER SPECTRA OF PASSIVE FILMS.
	5-1800	AES ANALYSIS OF OXIDE FILMS ON IRON.
	6-2080	APPLICATON OF AUGER ELECTRON SPECTROSCOPY TO THE DETERMINATION OF THE COMPOSITIO
LUNTZ M	2-0431	SCINTILLATION EFFICIENCY FOR LOW VELOCITY HEAVY IONS. (AUGER EMISSION)
LUSS D	5-1793	PHYSICAL AND CHEMICAL CHARACTERIZATION OF PLATINUM- RHODIUM GAUZE CATALYSTS. (AU
LUTH H	5-1306	ADSORPTION OF OXYGEN ON CLEAN CLEAVED (110) GALLIUM ARSENIDE SURFACES.
	5-1474	ADSORPTION OF OXYGEN ON SILICON (111) SURFACES. PART-1. (AUGER ELECTRON SPECTROS
LYO K	1-0105	AUGER ELECTRON SPECTROSCOPY. (REVIEW, 39 REFS)
	1-0106	SOME TOPICS ON AES. (SURFACE ANALYSIS)
	3-0647	DETECTION OF SODIUM ON SILICON AND SILICON DIOXIDE SURFACES BY CYLINDRICAL MIRRO
LYUBCHENKO AV	2-0369	RECOMBINATION IN DONOR ACCEPTOR COMPLEXES IN CADMIUM SULFIDE SINGLE CRYSTALS. (A
MAAREFF H	2-0260	EVIDENCES FOR AUGER RECOMBINATION IN ELECTRON HOLE DROPLETS. (SILICON)
MACDONALD JR	2-0472	ELECTRON DECAY-IN-FLIGHT SPECTRA FROM AUTOIONIZING STATES OF HIGHLY STRIPPED OXY
MACDONALD NC	1-0113	POTENTIAL MAPPING USING AUGER ELECTRON SPECTROSCOPY.
	1-0114	AUGER ELECTRON SPECTROSCOPY FOR SCANNING ELECTRON MICROSCOPY.
	1-0115	HIGH SPATIAL RESOLUTION AES.
	1-0116	SCANNING AUGER MICROANALYSIS.
	1-0117	THIRD DIMENSION IN SCANNING ELECTRON MICROSCOPY. SCANNING AUGER MICROSCOPY.
	1-0118	THIN FILM ANALYSIS FOR PROCESS EVALUATION. (AES)
	1-0119	APPLICATIONS OF SCANNING AUGER MICROANALYSIS.
	1-0159	HANDBOOK OF AUGER ELECTRON SPECTROSCOPY. (EXPERIMENTAL SPECTRA)
	3-0648	AUGER ELECTRON SPECTROSCOPY IN SCANNING ELECTRON MICROSCOPY: POTENTIAL MEASUREME
	3-0649	AUGER ELECTRON IMAGES IN A SCANNING ELECTRON MICROSCOPE.
	3-0650	AUGER ELECTRON SPECTROSCOPY IN THE SCANNING ELECTRON MICROSCOPE: AUGER ELECTRON
	3-0651	LABORATORY COMPUTER SCANNING ELECTRON MICROSCOPE SYSTEM.
	5-1597	THREE DIMENSIONAL ELEMENTAL ANALYSIS OF MATERIALS BY AUGER ELECTRON SPECTROSCOPY
	5-1598	MICROSCOPIC AUGER ELECTRON ANALYSIS OF FRACTURE SURFACES.
MACEK JH	2-0490	MECHANISMS OF ELECTRON PRODUCTION IN ION- ATOM COLLISIONS. (AUGER EFFECT, REVIEW
MACK RE	4-0760	AUGER STUDIES OF ELECTROLESS NICKEL FILMS.
MACKEY JJ	4-0943	HIGH RESOLUTION MEASUREMENTS OF P(+) ARGON KLL AND KLM AUGER ELECTRONS.
	4-0952	HIGH RESOLUTION K-AUGER SPECTRA FOR MULTIPLY IONIZED NEON.
	4-0953	HIGH RESOLUTION NEON KLL AUGER SPECTRA PRODUCED BY HYDROGEN(+) AND HELIUM(+) ION
	4-0954	AUGER DECAY OF NEON FOLLOWING ENERGETIC ION BOMBARDMENT.
	4-0993	OBSERVATION OF A PROJECTILE CHARGE DEPENDENCE FOR THE NEON K-AUGER ELECTRON SPEC
MACLENNAN DA	2-0333	ADSORBATE EFFECTS IN ELECTRON EJECTION BY RARE GAS METASTABLE ATOMS. (AUGER PROC
	5-1599	AUGER EJECTION OF ELECTRONS FROM TUNGSTEN BY OXYGEN CHEMISORPTION.
	5-1600	ROLE OF WORK FUNCTION IN ELECTRON EJECTION BY METASTABLE ATOMS: HELIUM AND ARGON
	5-1601	WORK FUNCTION EFFECTS ON AUGER ELECTRON EJECTION BY NOBLE GAS METASTABLE ATOMS.
MACRAE AU	5-1602	LOW ENERGY ELECTRON DIFFRACTION STUDY OF THE POLAR (111) SURFACES OF GALLIUM ARS
	5-1603	AN ELECTRON DIFFRACTION STUDY OF CESIUM ADSORPTION ON TUNGSTEN.
	5-1604	ELECTRONIC AND LATTICE STRUCTURE OF CESIUM FILMS ADSORBED ON TUNGSTEN.
MADDEN HH	5-1605	ELECTRON STIMULATED PROCESSES AT SOLID SURFACES. (AUGER SPECTROSCOPY)
	5-1606	DECOMPOSITION OF CARBON MONOXIDE ON A (110) NICKEL SURFACE. (AUGER ELECTRON SPEC
	5-1607	INTERACTION OF CARBON MONOXIDE WITH (110) NICKEL SURFACES. (AUGER EFFECT)
MADEY TE	5-1608	ADSORPTION OF OXYGEN AND OXIDATION OF COBALT ON THE RUTHENIUM (001) SURFACE. (AU
	5-1609	ADSORPTION OF COBALT ON (001) RUTHENIUM AT TEMPERATURES BELOW 300 DEGREES K.
MADIX RJ	5-1629	STUDY OF THE KINETICS AND MECHANISM OF THE DECOMPOSITION OF FORMIC ACID ON CARBU
	5-1630	FORMIC ACID DESORPTION FROM GRAPHITIZED NICKEL (110). (LEED- AES)
MAEDA H	5-1550	COMPOSITION PROFILES OF SEVERAL CONTAMINATED AND CLEANED SURFACES OF GOLD THICK
MAEDA K	3-0652	A NEW PHOTOELECTRON SPECTROMETER: COMBINATION OF CYLINDRICAL MIRROR ANALYZER WIT
MAEDA N	4-1041	SHIFT ENERGIES OF CHARACTERISTIC X-RAYS AND AUGER ELECTRONS FOR IONIZED ATOMS.
MAENHOUT-VAN DER VORST W	5-1337	COMBINED LOW ENERGY ELECTRON DIFFRACTION, AUGER ELECTRON SPECTROSCOPY, AND X-RAY
	5-1338	AUGER ELECTRON EMISSION SPECTROSCOPY (AES) AND LOW ENERGY ELECTRON DIFFRACTION (
MAGEE CL	6-2081	AES AND TEMPER EMBRITTLEMENT OF STEELS.
MAGLIETTA M	5-1610	QUANTITATIVE DETERMINATION OF CARBON ON SILVER BY AUGER SPECTROSCOPY.
	5-1779	LEED (AUGER) STUDY OF CHLORINE CHEMISORPTION ON THE SILVER (111) SURFACE.
MAGNUSSON N	2-0515	ELECTRON SPECTROSCOPIC INVESTIGATION OF AUGER PROCESSES IN BROMINE-SUBSTITUTED M
MAGUIRE HG	4-0944	INFLUENCE OF THE BAND STRUCTURE ON THE AUGER ELECTRON SPECTRUM OF SILICON.
	4-0945	CHARACTERISTIC K-SHELL LOW ENERGY ELECTRON LOSS SPECTRA FROM DIAMOND, GRAPHITE,
	5-1611	DETECTION OF SILICON OXYNITRIDE LAYERS ON THE SURFACES OF SILICON NITRIDE FILMS
	5-1612	STUDY OF THE OXIDATION AND REDUCTION BY WATER VAPOR OF SILICON (111) SURFACES US
	5-1613	AUGER ELECTRON EMISSION FROM BERYLLIUM SURFACES.
MAIRE G	5-1223	STUDY OF THE INTERACTION OF OXYGEN WITH PLATINUM SINGLE CRYSTAL SURFACES BY LOW
	5-1570	(AUGER) ADSORPTION STUDIES ON THE PLATINUM (100) SURFACE.
MAIRLE G	2-0336	DECAY OF HOLE STATES IN OXYGEN-16 VIA A "NUCLEAR AUGER EFFECT".

MAITRA S	5-1614	AUGER ELECTRCN SPECTROSCOPIC ANALYSIS OF ANODIC FILMS ON ALUMINUM.
MAITREPIERRE PH	1-0120	AUGER ELECTRON SPECTROSCOPY. (REVIEW)
MAKAMURA K	5-1577	ANTIMONY DOFING OF SILICON LAYERS GROWN BY SOLID PHASE EPITAXY. (AES MEASUREMENT
MAKINO Y	5-1708	GOLD-INDUCED SUPERSTRUCTURES ON SILICON (100) SURFACES AS OBSERVED BY LEED AES.
MAKSIMENKO VI	5-1368	STUDY OF BARIUM FILMS ON A (110) TUNGSTEN FACE BY THE DIFFRACTION OF SLOW ELECTR
MAMEDA J	5-1525	LEED- AES OBSERVATION CF GROWTH MECHANISM OF DEPOSITED IRON FILM ON MAGNESIUM OX
	5-1526	LOW ENERGY ELECTRON DIFFRACTION- AUGER ELECTRON SPECTROSCOPY STUDY OF AN EPITAXI
	5-1527	LOW ENERGY ELECTRON DIFFRACTION- AUGER ELECTRON SPECTROSCOPY STUDY OF CLEAN (001
	5-1528	GROWTH MODE OF IRON FILM DEPOSITED ON MAGNESIUM OXIDE (001) SURFACE. (AUGER SPEC
MANDE C	2-0526	APPLICATION OF THE ZETA EFFECTIVE VALUES OF HYDROGENIC WAVE FUNCTIONS IN THE CAL
MANDL A	4-1023	HIGH RESOLUTION MEASUREMENTS OF AUGER ELECTRON AND PHOTOELECTRON STRUCTURE IN TH
	4-1024	HIGH RESOLUTION MEASUREMENTS OF THE L3M2,3M4,5 AUGER TRANSITONS IN NICKEL AND CO
MANENKOV AA	4-0946	HIGH FREQUENCY EXCITON BREAKDOWN AND FREE CARRIER AND EXCITON KINETICS IN GERMAN
MANN R	1-0070	SURFACE ANALYSIS WITH HEAVY ION INDUCED AUGER ELECTRONS. BASIC PROPERTIES OF A N
	2-0370	AUGER ELECTRCNS FROM FOIL EXCITED HEAVY ION BEAMS. (OF NEON)
	4-0859	DOPPLER SHIFTED AUGER ELECTRONS FROM FOIL EXCITED HEAVY ION BEAMS.
	4-0860	BEAM FOIL EXCITED AUGER TRANSITIONS IN NEON.
MARCUS HL	3-0710	AUGER ELECTRON SPECTROSCOPY IN THE SCANNING ELECTRON MICROSCOPE.
	4-0858	AUGER ELECTRON SPECTROSCOPY OF LUNAR SAMPLES.
	4-0947	CHEMICAL ANALYSIS OF ELECTRODEPOSITED NICKEL- NICKEL BONDS BY AES.
	5-1398	FLUORINE ION IMPLANTATION PROFILES IN GALLIUM ARSENIDE AS DETERMINED BY AES.
	5-1598	MICROSCOPIC AUGER ELECTRON ANALYSIS OF FRACTURE SURFACES.
	5-1615	GRAIN BOUNDARY SEGREGATION IN MAGNESIA DOPED ALUMINA. (HIGH VACUUM AUGER ELECTRO
	5-1616	AUGER SPECTROSCOPY OF FRACTURE SURFACES OF CERAMICS.
	6-2082	GRAIN BOUNDARY SEGREGATION RELATING TO INTERGRANULAR FRACTURE. (AUGER ELECTRON S
	6-2083	EFFECT OF SCLUTE ELEMENTS ON TEMPER EMBRITTLEMENT OF LOW ALLOY STEELS. (IMPURITY
	6-2084	FRACTURE SURFACE ANALYSIS OF TEMPER EMBRITTLED STEEL BY AUGER ELECTRON SPECTROSC
	6-2085	AUGER FRACTURE SURFACE ANALYSIS OF A TEMPER EMBRITTLED 3340- STAINLESS STEEL.
	6-2086	METASTABLE GRAIN BOUNDARY SEGREGATION IN COPPER CONTAINING TELLURIUM. (AUGER SPE
	6-2094	AUGER SPECTROSCOPIC ANALYSIS OF THE EXTENT OF GRAIN BOUNDARY SEGREGATION.
	6-2099	EFFECT OF TELLURIUM ON INTERGRANULAR COHESION OF IRON.
MARCUS PM	5-1478	ATOMIC ARRANGEMENT IN THE 1 K 1 STRUCTURE OF A SILICON ORDERED MONOLAYER ON MOLY
MARGONINSKI Y	5-1617	AUGER ELECTRON SPECTROSCOPY STUDY OF ELECTRON BEAM STIMULATED DESORPTION OF OXYG
	5-1618	THICKNESS MEASUREMENTS OF ADSORBED LAYERS BY AUGER ELECTRON SPECTROSCOPY.
	5-1619	LEED AND AES OBSERVATIONS ON THE POLAR SURFACES OF ZINC OXIDE.
	5-1620	INTERFERENCE OF AN ELECTRON BEAM WITH THE SURFACE REACTION BETWEEN OXYGEN AND GE
MARINOVA K	2-0432	PHOTOCONDUCTIVITY OF HIGHLY EXCITED CADMIUM SELENIDE AND CADMIUM SULFIDE(0.48) S
MARIOT JM	4-0823	4D AUGER TRANSITIONS FROM EXCITED STATES IN SAMARIUM METAL.
MARK H	1-0121	REVIEW OF FLUCRESCENCE YIELDS AND AUGER ELECTRON MISSION.
	2-0259	X-RAY FLUORESCENCE YIELDS, AUGER, AND COSTER-KRONIG TRANSITION PROBABILITIES.
	2-0351	ATOMIC FLUORESCENCE YIELDS. (L-SHELL COSTER-KRONIG YIELDS, K-, L-, AND M-SHELL F
MARK P	5-1584	CORRELATION CF ELECTRONIC, LEED, AND AUGER DIAGNOSTICS ON ZINC OXIDE SURFACES.
MARKLUND I	5-1134	CLEAN TELLURIUM SURFACES STUDIED BY LEED.
	5-1621	SCATTERING OF LOW ENERGY ELECTRONS FROM A COPPER (111) SURFACE.
MARTI C	4-0798	ULTRAVIOLET EMISSIONS IN CALCITE. (AUGER EFFECT)
MARTIN AD	2-0433	INELASTIC SCATTERING EFFECTS IN QUANTITATIVE AUGER ELECTRON SPECTROSCOPY.
	5-1951	COMBINED FIELD ION MICROSCOPY. AUGER ELECTRON SPECTROSCOPY, AND LEED STUDY OF TH
MARTINEZ JM	5-1622	ADSORPTION AND DECOMPOSITION OF CARBON MONOXIDE ON PLATINUM (111). (AUGER SPECTR
MARTINSEN WE	5-1824	SURFACE CONCENTRATION OF MAGNESIUM IN 6061 ALUMINUM BY AUGER ELECTRON SPECTROSCO
MARTINSON J	5-1621	SCATTERING OF LOW ENERGY ELECTRONS FROM A COPPER (111) SURFACE.
MARU S	5-1874	ANALYSIS OF THE SURFACE COMPOSITION OF PASSIVE FILMS ON COPPER- NICKEL ALLOYS BY
MASON R	2-0319	AUGER EMISSION SPECTRA: A SIMPLE TREATMENT OF ENERGIES AND INTENSITIES
	5-1267	AUGER SPECTROSCOPY AND LOW ENERGY ELECTRON DIFFRACTION STUDIES OF THE CHEMISORPT
MASRI P	5-1860	ENTROPY OF ADSORPTION OF XENON IN EPITAXY ON THE (0001) FACE OF GRAPHITE. (AUGER
MASSIGNON D	4-0933	ABSORPTION BANDS IN SECONDARY ELECTRON CONTINUOUS SPECTRUM.
	4-0934	NEW METHOD FOR ANALYZING THE SECONDARY ELECTRON SPECTRUM.
	4-0948	STRUCTURE OF LOW ENERGY ELECTRON SPECTRUM.
MASYAGIN VE	3-0653	SPECTROMETER OF AUGER ELECTRONS.
MATHEWSON AG	3-0654	SURFACE CLEANLINESS OF 316 L-N STAINLESS STEEL STUDIES BY SECONDARY ION MASS SPE
	6-2087	SURFACE CLEANLINESS OF 316 L-N STAINLESS STEEL STUDIED BY SECONDARY ION MASS SPE
MATHIEU HJ	6-2088	QUANTITATIVE AUGER ELECTRON SPECTROSCOPY ANALYSIS OF SILVER- PALLADIUM AND NICKE
MATSUDA Y	5-1874	ANALYSIS OF THE SURFACE COMPOSITION OF PASSIVE FILMS ON COPPER- NICKEL ALLOYS BY
MATSUDAIRA T	3-0655	A SMALL RETARDING FIELD ENERGY ANALYZER WITH CEM. (AUGER SPECTROMETER)
	3-0656	AUGER ELECTRON SPECTROSCOPY USING THE TWO-GRID LEED SYSTEM AND PHOTOELECTRON MUL
	5-1623	ANGULAR DISTRIBUTION OF AUGER ELECTRON EMISSION FROM CLEAN AND GAS-COVERED IRON
	6-2141	STUDY ON THE INITIAL OXIDATION AND REDUCTION PROCESSES OF IRON (100) SURFACE BY
MATSUMOTO Y	4-1004	ORIGIN OF THE PEAK SHAPES IN THE AUGER SPECTRUM OF OXYGEN (KL2,3L2,3) ON MOLYBDE
	5-1698	INITIAL OXIDATION OF MOLYBDENUM STUDIED BY A HIGH RESOLUTION AUGER PHOTOELECTRON
MATSUSHITA Y	5-1624	GROWTH MODE DISCRIMINATION OF THIN FILMS BY AUGER ELECTRON SPECTROSCOPY.
MATTHEW JAD	1-0058	COMPARISON CF X-RAY AND ELECTRON IMPACT IONIZATION IN AUGER SPECTROSCOPY.
	1-0059	AUGER ELECTRON SPECTROSCOPY AND ITS APPLICATION TO SURFACE STUDIES. (REVIEW, 130
	2-0434	A TEMPERATURE DEPENDENT CONTRIBUTION TO AUGER ELECTRON ENERGY DISTRIBUTIONS.
	2-0435	EFFECT OF POLARIZATION OF THE ELECTRON GAS ON THE ENERGIES OF AUGER TRANSITION I
	2-0436	DISTINCTION BETWEEN WEAK AUGER TRANSITIONS AND INTERNAL X-RAY PHOTOEMISSION.
	2-0437	POSSIBLE LOCALISATION OF CONDUCTION ELECTRONS IN SOFT X-RAY DOUBLE IONIZATION SA
	2-0438	TRANSITION RATES FOR INTERATOMIC AUGER PROCESSES.
	2-0439	ENHANCEMENT OF COSTER-KRONIG RATES IN SOLID STATE ENVIRONMENTS. (XENON, CADMIUM,
	2-0440	MECHANISM OF PLASMON GAIN IN AUGER SPECTRA.
	4-0771	QUASIATOMIC FINE STRUCTURE IN THE AUGER SPECTRA OF SOLID SILVER AND INDIUM.
	4-0772	HIGH RESOLUTION AUGER ELECTRON SPECTRUM OF MAGNESIUM OXIDE (100).
	4-0842	LOW ENERGY AUGER EMISSION FROM LITHIUM FLUORIDE.
	4-0892	THE KVV AUGER SPECTRUM OF LITHIUM METAL.
	4-0949	AUGER ELECTRON FINE PROFILES IN IONIC CRYSTALS.
	4-0950	EFFECTS OF EXTRA CLOSED CORE RELAXATION ON AUGER AND BINDING ENERGY SHIFTS.
	4-1067	ANOMALOUS THRESHOLD BEHAVIOR FOR ELECTRON INDUCED INNER SHELL IONIZATION OF THE
	5-1497	INTERFACIAL AUGER TRANSITIONS IN OXIDIZED SODIUM AND MAGNESIUM.
	5-1625	CLEAN SURFACES.
	5-1626	AUGER EFFECT IN SURFACE SCIENCE.
	5-1686	LOW ENERGY ELECTRON SPECTROSCOPY OF PLATINUM (100) AND PLATINUM (100)- CARBON MO
MATTHEWS DL	2-0285	EXCITATION OF MULTIPLET STATES IN NEON.

MATTHEWS DL	2-0441	SEQUENTIAL STRIPPING IN HIGHLY IONIZED NEON. (AUGER TRANSITIONS)
	4-0903	HIGH RESOLUTION STUDY OF ARGON LMM AUGER ELECTRONS PRODUCED BY ION BOMBARDMENT.
	4-0943	HIGH RESOLUTION MEASUREMENTS OF P(+) ARGON KLL AND KLM AUGER ELECTRONS.
	4-0951	MEASUREMENTS OF NEON K-AUGER ELECTRONS PRODUCED BY ENERGETIC HEAVY ION BOMBARDME
	4-0952	HIGH RESOLUTION K-AUGER SPECTRA FOR MULTIPLY IONIZED NEON.
	4-0953	HIGH RESOLUTION NEON KLL AUGER SPECTRA PRODUCED BY HYDROGEN(+) AND HELIUM(+) ION
	4-0954	AUGER DECAY OF NEON FOLLOWING ENERGETIC ION BOMBARDMENT.
	4-0955	CALCULATED K-AUGER ELECTRON AND K X-RAY TRANSITION ENERGIES FOR THE MULTIPLY ION
	4-0993	OBSERVATION OF A PROJECTILE CHARGE DEPENDENCE FOR THE NEON K-AUGER ELECTRON SPEC
	4-0994	K-SHELL (AUGER) FLUORESCENT YIELDS FOR MULTIPLY IONIZED NEON.
MATVEEV VI	2-0442	INTERATOMIC AUGER EFFECT AT HIGH VELOCITIES.
MAU HC	4-0939	WIDTH OF ATOMIC L2 AND L3 VACANCY STATES NEAR Z = 30. (AUGER EFFECT)
	4-0956	MULTIPLET EFFECTS ON THE L2,3 FLUORESCENCE YIELD OF MULTIPLY IONIZED ARGON. (AUG
MAUHSIUNG C	4-0957	STRIPPING MECHANISM DEPENDENCE OF THE ARGON 2S FLUORESCENCE YIELD. (AUGER TRANSI
MAYER JW	5-1182	EFFECT OF OXIDIZING AMBIENTS ON PLATINUM SILICIDE FORMATION. PART-2: AUGER AND B
	5-1577	ANTIMONY DOPING OF SILICON LAYERS GROWN BY SOLID PHASE EPITAXY. (AES MEASUREMENT
	5-1627	COMPARISON OF SURFACE LAYER ANALYSIS TECHNIQUES. (AUGER ELECTRON SPECTROSCOPY, S
	5-1672	INTERACTION OF ALUMINUM LAYERS WITH POLYCRYSTALLINE SILICON. (AUGER SPECTROSCOPY
	5-1917	STRUCTURE AND GROWTH KINETICS OF NICKEL(2) SILICIDE ON SILICON. (AUGER SPECTROSC
MAYNE JEO	5-1628	DETECTION OF LEAD, BY AUGER SPECTROSCOPY, ON IRON INHIBITED IN LEAD AZELATE SOLU
MAZAK RA	2-0576	BACKSCATTER MOSSBAUER STUDIES USING CONVERSION ELECTRONS. TIN-119.
MCCARROLL JJ	5-1311	PLATINUM (100) STRUCTURES: IMPURITY CAUSATION AND PROPERTIES. (AUGER EFFECT)
MCCARTY JG	5-1629	STUDY OF THE KINETICS AND MECHANISM OF THE DECOMPOSITION OF FORMIC ACID ON CARBU
	5-1630	FORMIC ACID DESORPTION FROM GRAPHITIZED NICKEL (110). (LEED- AES)
MCDAVID JM	4-0828	WORK FUNCTION VARIATION WITH ALLOY COMPOSITION SILVER- GOLD. (AUGER ELECTRON SPE
	5-1330	WORK FUNCTION VARIATION WITH ALLOY COMPOSITION: COPPER- GOLD. (AUGER SPECTROSCOP
	5-1631	SEGREGATION AT COPPER- GOLD ALLOY SURFACES. (AUGER SPECTROSCOPY)
MCDEVITT NT	1-0009	SECONDARY ION MASS SPECTROMETRY (AND AUGER ELECTRON SPECTROSCOPY): A USEFUL COMP
MCDONNELL L	4-0958	ANGULAR DEPENDENCE OF AUGER ELECTRON EMISSION FROM A SINGLE CRYSTAL SPECIMEN.
	5-1447	ANGULAR DEPENDENCE OF AUGER ELECTRON EMISSION FROM SOLID SURFACES.
	5-1632	ANGULAR DEPENDENCE OF AUGER ELECTRON EMISSION FROM COPPER (111) AND (100) SURFAC
	6-2089	A LOW ENERGY ELECTRON DIFFRACTION- AUGER ELECTRON SPECTROSCOPY STUDY OF THE ADSO
MCELHINEY G	5-1633	ADSORPTION OF XENON AND COBALT ON SILVER (111).
MCELROY J	2-0292	G-R NOISE FOR AUGER BAND-TO-BAND PROCESSES.
MCEVOY AJ	5-1434	ADHESION AT GOLD MICA INTERFACES. (LEED, AES)
	5-1967	ELECTRON EMISSION STUDIES FROM GALLIUM SELENIDE SURFACES.
MCFARLANE SC	1-0122	DIRECTIONAL CORRELATION BETWEEN PHOTOELECTRONS AND AUGER ELECTRONS.
MCFEELY FR	2-0412	RELATIVE EFFECT OF EXTRAATOMIC RELAXATION ON AUGER AND BINDING ENERGY SHIFTS IN
	2-0427	MANY BODY EFFECTS IN X-RAY PHOTOEMISSION FROM MAGNESIUM. (AUGER EFFECT)
	4-0919	L2,3M4,5M4,5 AUGER SPECTRA OF METALLIC COPPER AND ZINC: THEORY AND EXPERIMENT.
MCGEORGE JC	4-0959	L- AND M-SHELL AND SUBSHELL FLUORESCENCE AND COSTER-KRONIG YIELDS.
MCGOVERN IT	5-1634	ELECTRONIC PROPERTIES OF CLEAVED MOLYBDENUM DISULFIDE SURFACES.
	5-1968	SURFACE CHARACTERIZATION OF INDIUM PHOSPHIDE. (BY AUGER SPECTROSCOPY)
	5-1969	SURFACE STUDIES OF SOME TRANSITION METAL DICHALCOGENIDES.
MCGUIRE EJ	1-0123	AUGER AND COSTER-KRONIG TRANSITIONS. (REVIEW)
	2-0342	AUGER TRANSITION RATE IN MIXED COUPLING SCHEME.
	2-0443	SOFT X-RAY AMPLIFIED SPONTANEOUS EMISSION VIA THE AUGER EFFECT.
	2-0444	AUGER TRANSITION RATES IN J-J COUPLING FOR D HOLES AND F ELECTRONS.
	4-0831	TIGHT BINDING CALCULATION OF A CORE VALENCE VALENCE AUGER LINE SHAPE: SILICON (1
	4-0960	L-SHELL AUGER AND COSTER-KRONIG ELECTRON SPECTRA.
	4-0961	K-SHELL AUGER TRANSITION RATES AND FLUORESCENT YIELDS FOR ELEMENTS ARGON TO XENO
	4-0962	K-SHELL AUGER TRANSITION RATES AND FLUORESCENT YIELDS FOR ELEMENTS BERYLLIUM TO
	4-0963	ATOMIC L-SHELL COSTER-KRONIG, AUGER, AND RADIATIVE RATES AND FLUORESCENCE YIELDS
	4-0964	ATOMIC M-SHELL COSTER-KRONIG, AUGER, AND RADIATIVE RATES, AND FLUORESCENCE YIELD
	4-0965	M-SHELL AUGER AND COSTER-KRONIG ELECTRON SPECTRA.
	4-0966	K- AND L-SHELL FLUORESCENCE AND AUGER YIELDS AND AUGER ELECTRON SPECTROSCOPY.
	4-0967	M-SHELL AUGER, COSTER-KRONIG, AND RADIATIVE MATRIX ELEMENTS, AND AUGER AND COSTE
	4-0968	L- AND M-SHELL YIELDS AND ELECTRON SPECTRA FOR THE TRANSURANIC ELEMENTS.
	4-0969	ATOMIC N-SHELL COSTER-KRONIG, AUGER, AND RADIATIVE RATES, AND FLUORESCENCE YIELD
	4-0970	WIDTHS OF 4S AND 5S MULTIPLET COMPONENTS IN THE RARE EARTHS.
	4-0971	AUGER SPECTRA OF THE NOBLE GASES. KRYPTON.
	4-0972	AUGER TRANSITION RATE FOR THE NOBLE GAS SATELLITE SPECTRA.
	4-0973	AUGER CASCADES IN ALUMINUM.
	4-0974	AUGER SPECTRUM OF THE NOBLE GASES. PART-2: ARGON.
	4-0975	K-SHELL FLUORESCENCE YIELD OF IONIZED ALUMINUM. (AUGER RATE)
MCGUIRE GE	2-0445	MULTIPLET EFFECTS ON THE WIDTHS OF PHOTOELECTRON PEAKS. (AUGER EFFECT)
	2-0581	AUGER TRANSITIONS INITIATED FROM SINGLY AND DOUBLY IONIZED STATES IN LITHIUM MET
	3-0640	DIGITAL METHODS FOR THIN FILM ANALYSIS USING A COMPUTER COUPLED AUGER SPECTROMET
	4-1025	AUGER SPECTRA OF LITHIUM HYDRIDE.
	5-1285	PLATINUM SILICIDE FORMATION. ELECTRON SPECTROSCOPY OF THE PLATINUM- PLATINUM SIL
	5-1286	DIFFUSION STUDIES IN CHROMIUM- PLATINUM THIN FILMS USING AUGER ELECTRON SPECTROS
	5-1572	ESTABLISHING DETECTION AND DETERMINATION LIMITS IN SURFACE AND THIN FILM ANALYSI
	5-1635	CHOOSING BETWEEN ESCA AND AUGER FOR SURFACE ANALYSIS.
	5-1636	DIFFUSION STUDIES IN GOLD- PLATINUM THIN FILMS BY AUGER ELECTRON SPECTROSCOPY.
	5-1637	ION SELECTIVE SPUTTERING OF III-V COMPOUND SEMICONDUCTOR SURFACES. (AUGER EFFECT
	5-1750	SODIUM SEGREGATION ONTO A LITHIUM METAL SURFACE. (AUGER SPECTRA)
	5-1983	AUGER ELECTRON SPECTROSCOPY OF CLEANUP-RELATED CONTAMINATION ON SILICON SURFACES
	6-2002	AUGER SPECTROSCOPY OF FRACTURE SURFACES OF IRRADIATED STAINLESS STEEL.
MCGUIRE JH	2-0446	CHARGE DEPENDENCE OF DEPENDENCE OF TOTAL K-SHELL VACANCY PRODUCTION CROSS SECTIO
	4-0976	DOUBLE IONIZATION EFFECTS ON RADIATIVE RATES, AUGER RATES, AND FLUORESCENCE YIEL
MCKEE CS	1-0124	LEED AND AUGER ELECTRON SPECTROSCOPY. (REVIEW)
	5-1515	ADSORPTION OF CARBON MONOXIDE ON COPPER (001). LEED AND AUGER EMISSION STUDIES.
	5-1516	INTERACTION OF HYDROGEN SULFIDE WITH COPPER (001).
MCLEAN M	2-6124	LEAD MONOLAYER LUBRICATION IN STEEL MACHINING STUDIED BY AUGER ELECTRON SPECTROS
MCMAHON CJ	6-2099	EFFECT OF TELLURIUM ON INTERGRANULAR COHESION OF IRON.
MCMULLEN T	2-0270	X-RAY EDGE ANOMALIES IN LITHIUM.
	2-0447	PLASMON SATELLITES IN AUGER SPECTRA.
MCNATT JL	5-1153	CHEMICAL SHIFTS IN THE AUGER SPECTRUM OF YTTRIUM ON OXYGEN ADSORPTION.
MEAKIN JD	5-1725	CHARACTERIZATION OF ALUMINUM ADHEREND SURFACES (BY AUGER ELECTRON SPECTROSCOPY).
MECKBACH W	1-0070	SURFACE ANALYSIS WITH HEAVY ION INDUCED AUGER ELECTRONS. BASIC PROPERTIES OF A N

MEE CHB	5-1634	ELECTRONIC PROPERTIES OF CLEAVED MOLYBDENUM DISULFIDE SURFACES.
MEHLHORN W	1-0125	AUGER ELECTRON SPECTROSCOPY. (REVIEW, 44 REFS)
	1-0126	RECENT TRENDS IN AUGER ELECTRON SPECTROSCOPY.
	2-0320	ANGULAR DISTRIBUTION OF AUGER ELECTRONS.
	2-0321	ANGULAR DISTRIBUTION OF AUGER ELECTRONS FOLLOWING IMPACT IONIZATION.
	2-0322	ANGULAR DISTRIBUTION OF L3M2,3M2,3(1S0) AUGER ELECTRONS OF ARGON.
	2-0335	INNER SHELL ALIGNMENT OF ATOMS IN ELECTRON IMPACT IONIZATION. (AUGER ELECTRONS)
	2-0352	ANGULAR DISTRIBUTION OF AUGER ELECTRONS FOLLOWING PHOTOIONIZATION.
	2-0363	ABSOLUTE ELECTRON IMPACT IONIZATION CROSS SECTIONS OF ATOMIC NITROGEN, ATOMIC OX
	4-0808	KLL AUGER SPECTRUM OF CHLORINE. (IN CARBON TETRACHLORIDE)
	4-0848	A NEW METHOD FOR MEASURING ELECTRON IMPACT IONIZATION CROSS SECTIONS OF INNER SH
	4-0872	SEARCH FOR CORRELATION EFFECTS IN THE M4,5N4,5N4,5 AUGER SPECTRUM OF XENON.
	4-0880	K-AUGER TRANSITIONS OF THE FREE SODIUM ATOM.
	4-0914	K-AUGER SPECTRUM OF NEON.
	4-0977	COSTER-KRONIG AND AUGER SPECTRUM OF THE L1 SHELL OF ARGON.
	4-0978	AUGER ELECTRON SPECTROMETRY OF SINGLY AND MULTIPLY IONIZED ATOMS.
	4-0979	AUGER AND COSTER-KRONIG STUDIES OF THE M-SHELL OF KRYPTON.
	4-0980	THE KLL AUGER SPECTRUM FOR ATOMIC NUMBERS BETWEEN 10 AND 36.
	4-0981	CORRELATION EFFECTS IN THE AUGER SPECTRA OF NOBLE GASES.
	4-0982	ENERGY WIDTHS OF ROENTGEN LEVELS BY AUGER ELECTRON SPECTROSCOPY.
	4-0983	AUGER SPECTRUM OF L2 AND L3 SHELLS OF ARGON.
	4-1071	ENERGIES OF EXCITED STATES OF DOUBLY IONIZED MOLECULES BY MEANS OF AUGER ELECTRO
MEI JY	1-0127	APPLICATION OF (AUGER) ELECTRON SPECTROSCOPY TO CHEMISTRY. (REVIEW)
MEIJER F	1-0016	SURFACE ANALYSIS, METHODS OF STUDYING THE OUTER ATOMIC LAYERS OF SOLIDS.
MEISTER KH	5-1421	ADSORPTION OF SULFUR ON THE (100) AND (111) FACES OF PLATINUM. LOW ENERGY ELECTR
MELCHIOR W	3-0704	SOREAS: AN APPARATUS FOR INVESTIGATING ION SCATTERING FROM SURFACES AT ENERGIES
MELLES JJ	3-0603	A MULTIPLE SAMPLE HOLDER FOR USE IN ULTRAHIGH VACUUM.
	4-0984	AUGER SATELLITES ON THE L23 AUGER EMISSION BANDS OF ALUMINUM, SILICON, AND PHOSP
	4-0985	AUGER SATELLITES OF THE L2,3 AUGER EMISSION BANDS OF ALUMINUM, SILICON, AND PHOS
	4-0986	AUGER AND OTHER CHARACTERISTIC ENERGIES IN SECONDARY ELECTRON SPECTRA FROM THIN
	5-1288	A STUDY OF PHOSPHORUS ADSORPTION AND DESORPTION KINETICS ON SILICON (111) SURFAC
	5-1583	QUANTITATIVE STUDY OF AUGER ELECTRON SIGNALS OF PHOSPHORUS ON SILICON USING A QU
MELNIK PV	2-0409	ESCAPE DEPTHS OF X-RAY EXCITED (AUGER) ELECTRONS.
	4-0923	FILM STRUCTURE EFFECT ON THE SECONDARY ELECTRON EMISSION CHARACTERISTICS (GERMAN
MELNIK YP	5-1368	STUDY OF BARIUM FILMS ON A (110) TUNGSTEN FACE BY THE DIFFRACTION OF SLOW ELECTR
MENZEL D	2-0448	INELASTIC INTERACTIONS OF SLOW ELECTRONS WITH ABSORBED PARTICLES. (AUGER SPECTRO
	5-1422	OXYGEN ADSORPTION ON (110) SILVER.
	5-1608	ADSORPTION OF OXYGEN AND OXIDATION OF COBALT ON THE RUTHENIUM (001) SURFACE. (AU
	5-1609	ADSORPTION OF COBALT ON (001) RUTHENIUM AT TEMPERATURES BELOW 300 DEGREES K.
MEYER CH	1-0182	SPUTTERING MEASUREMENTS ON CTR MATERIALS USING AUGER ELECTRON SPECTROSCOPY
	5-1825	AUGER ELECTRON SPECTROSCOPY IN SPUTTERING MEASUREMENTS. APPLICATION TO LOW ENER
	5-1826	SPUTTERING MEASUREMENTS ON CONTROLLED THERMONUCLEAR REACTOR MATERIALS USING AUGE
MEYER F	1-0128	ELLIPSOMETRY AS A TOOL IN SURFACE STUDIES.
	1-0129	QUANTITATIVE ASPECTS OF AUGER ELECTRON SPECTROSCOPY.
	1-0130	IN DEPTH INFORMATION FROM AUGER ELECTRON SPECTROSCOPY. (SILICON MONOLAYERS)
	1-0205	AUGER ELECTRON SPECTROSCOPY MADE QUANTITATIVE BY ELLIPSOMETRIC CALIBRATION.
	1-0206	QUANTITATIVE ASPECTS OF AUGER ELECTRON SPECTROSCOPY. IMPORTANCE OF ATTENUATION O
	1-0207	QUANTITATIVE AUGER ELECTRON SPECTROSCOPY.
	2-0553	ELECTRON IMPACT IONIZATION CROSS SECTIONS OF INNER SHELLS MEASURED BY AUGER ELEC
	2-0554	MEASUREMENT OF IONIZATION CROSS SECTIONS AND BACKSCATTERING FACTORS FOR USE IN Q
	4-0987	MEASUREMENT OF THE IONIZATION CROSS SECTION OF THE L2,3 SHELL OF CHLORINE USING
	5-1638	ADSORPTION OF OXYGEN ON A CLEAN SILICON SURFACE. (AUGER ELECTRON SPECTROSCOPY)
MEYERHOF WE	2-0301	SIMPLE MODEL FOR K X-RAY AND AUGER ELECTRON ENERGY SHIFTS IN HEAVY ION COLLISION
MEYERS V	1-0184	AUGER PROFILE ANOMALIES. (REVIEW)
MIHE JP	4-1111	AUGER ELECTRON SPECTROSCOPY OF HIGH TEMPERATURE OXYGEN- RHENIUM INTERACTIONS.
MIKHAILOV AI	2-0364	PARITY NONCONSERVATION IN RADIATIONLESS (AUGER) TRANSITIONS IN MU MESIC ATOMS.
MIKHAILOVA GN	4-0946	HIGH FREQUENCY EXCITON BREAKDOWN AND FREE CARRIER AND EXCITON KINETICS IN GERMAN
MILLER DL	3-0657	ULTRAHIGH VACUUM LOW TEMPERATURE STAGE FOR IN-SITU FILM DEPOSITION AND CHARACTER
MILLER JW	5-1139	INVESTIGATION OF REORDERED (001) GOLD SURFACES BY POSITIVE ION CHANNELING SPECTR
	5-1986	CHARACTERIZATION OF REORDERED (001) GOLD SURFACES BY POSITIVE ION CHANNELING SPE
MILLS B	6-2124	LEAD MONOLAYER LUBRICATION IN STEEL MACHINING STUDIED BY AUGER ELECTRON SPECTROS
MILSTED J	4-1019	PROTON STATES, ANOMALOUS CONVERSION, AND K-AUGER SPECTRA IN AMERICIUM-241 FROM C
MILYAEV VA	4-0946	HIGH FREQUENCY EXCITON BREAKDOWN AND FREE CARRIER AND EXCITON KINETICS IN GERMAN
MISCHLER J	4-0776	ELECTRONIC EMISSION FROM SOLID TARGETS BOMBARDED BY NOBLE GAS IONS (10-100 KEV):
	4-0988	ENERGY SPECTRUM AND ANGULAR DISTRIBUTION OF SECONDARY ELECTRONS EMITTED BY IONIC
MITCHELL KAR	1-0131	LOW ENERGY ELECTRON DIFFRACTION.
MITTAL KL	6-2093	PLASMA CLEANING OF METAL SURFACES. (AUGER ELECTRON SPECTROSCOPY)
MITTLEMAN MH	4-0976	DOUBLE IONIZATION EFFECTS ON RADIATIVE RATES, AUGER RATES, AND FLUORESCENCE YIEL
MITYAGIN AYU	1-0132	ELECTRON AUGER SPECTROSCOPY: A METHOD OF INVESTIGATING SURFACE PHENOMENA (REVIEW
	2-0449	EFFECT OF THE ELECTRON EJECTION DEPTH ON QUANTITATIVE ANALYSIS BY AUGER SPECTROS
	3-0658	AUGER ELECTRON SPECTROMETER INTEGRATED WITH A LOW ENERGY ELECTRON DIFFRACTION SY
	5-1639	INVESTIGATION OF THE PURE AND CESIUM COATED (100) SURFACE OF GALLIUM ARSENIDE BY
	5-1640	(111) SURFACE OF GALLIUM PHOSPHIDE STUDIED BY AUGER SPECTROSCOPY AND DIFFRACTION
	5-1641	PROCESSES OF ADSORPTION OF OXYGEN ON GALLIUM ARSENIDE METHOD OF LOW ENERGY ELECT
	5-1642	AUGER ELECTRON SPECTRUM DURING THE INTERACTION BETWEEN OXYGEN AND THE (111) SURF
	5-1643	SOME FEATURES OF THE AUGER ELECTRON SPECTRUM IN THE INTERACTION BETWEEN OXYGEN A
	5-1644	INVESTIGATION OF CESIUM ADSORPTION ONTO A GALLIUM ARSENIDE (110) SURFACE. (AUGER
	5-1645	QUANTITATIVE ESTIMATES OF THE CHEMICAL COMPOSITION NEAR THE SURFACE BASED ON THE
	5-1646	GROWTH OF GALLIUM ARSENIDE ON (100) GERMANIUM STUDIED BY LOW ENERGY ELECTRON DIF
MIURA T	1-0133	SAMPLES HEATED DIRECTLY BY ELECTRIC CURRENT OBSERVED BY AUGER ELECTRON SPECTROSC
	5-1647	DETERMINATION OF SULFUR CONTAMINATION ON MOLYBDENUM SURFACES. (AUGER EFFECT)
	5-1648	CARBON CONTAMINATION PRODUCED BY ELECTRON BOMBARDMENT. POLYCRYSTALLINE MOLYBDENU
	5-1649	CLEANING PROCEDURES OF MOLYBDENUM (110) SURFACE.
	5-1650	INITIAL OXIDATION OF THE MOLYBDENUM (110) SURFACE OBSERVED BY AES AND LEED.
MIYAMOTO N	5-1786	LOW TEMPERATURE OXIDATION OF SILICON STUDIED BY PHOTOSENSITIVE ESR AND AUGER ELE
MIYAMURA M	6-2141	STUDY ON THE INITIAL OXIDATION AND REDUCTION PROCESSES OF IRON (100) SURFACE BY
MIYAZAKI E	2-0264	MOLECULAR ADSORPTION: INCLUSION OF BULK DIFFUSION AND ITS EFFECT ON AUGER INTEN
MIZUNO M	5-1550	COMPOSITION PROFILES OF SEVERAL CONTAMINATED AND CLEANED SURFACES OF GOLD THICK
MODDEMAN WE	1-0134	CHARACTERIZATION OF SOLID SURFACES. (REVIEW, AUGER SPECTROSCOPY, ESCA)
	3-0600	HIGH RESOLUTION ELECTRON SPECTROMETER FOR PHOTOELECTRON AND AUGER ELECTRON STUDI

MODDEMAN WE   4-0791 IDENTIFICATION OF HIGH ENERGY LINES IN THE KLL AUGER SPECTRUM OF NITROGEN.
                 4-0920 MANIFESTATION OF ATOMIC DYNAMICS THROUGH THE AUGER EFFECT. (NEON AUGER SPECTRUM)
                 4-0989 AUGER SPECTROSCOPY OF SIMPLE GASEOUS MOLECULES.
                 4-0990 DETERMINATION OF KLL AUGER SPECTRA OF NITROGEN, OXYGEN, CARBON DIOXIDE, NITRIC O
                 4-0991 AUGER SPECTRA OF SIMPLE MOLECULES.
MOJICA JF   5-1651 EVALUATION OF THE TRANSITION DENSITY FUNCTION OF CARBON ON VARIOUS SUBSTRATES FR
MOLJK A   4-1014 K-SHELL FLUORESCENCE YIELDS OF ARGON, CHLORINE AND SULFUR. (AUGER TRANSITION)
MONTAGUE WG   6-2036 STRESS CORROSION CRACKING OF AN ALUMINUM- 5% MAGNESIUM TERNARY AND VARIOUS QUATE
MOON RL   4-0992 AUGER SPECTRA OF HYDROGEN CHLORIDE VAPOR ETCHED N+ GALLIUM ARSENIDE (100) SUBSTR
MOORE CF   2-0285 EXCITATION OF MULTIPLET STATES IN NEON.
                 2-0496 HIGH RESOLUTION ARGON K-AUGER SPECTRUM PRODUCED IN 4 MEV PROTON ON ARGON COLLISI
                 2-0497 RELATIVE SATELLITE LINE INTENSITIES IN ARGON K AUGER ELECTRON SPECTRA PRODUCED B
                 4-0903 HIGH RESOLUTION STUDY OF ARGON LMM AUGER ELECTRONS PRODUCED BY ION BOMBARDMENT.
                 4-0904 HIGH RESOLUTION BEAM GAS AUGER ELECTRON SPECTRA FOR OXYGEN IONS EXCITED BY COLLI
                 4-0943 HIGH RESOLUTION MEASUREMENTS OF P(+) ARGON KLL AND KLM AUGER ELECTRONS.
                 4-0952 HIGH RESOLUTION K-AUGER SPECTRA FOR MULTIPLY IONIZED NEON.
                 4-0953 HIGH RESOLUTION NEON KLL AUGER SPECTRA PRODUCED BY HYDROGEN(+) AND HELIUM(+) ION
                 4-0954 AUGER DECAY OF NEON FOLLOWING ENERGETIC ION BOMBARDMENT.
                 4-0955 CALCULATED K-AUGER ELECTRON AND K X-RAY TRANSITION ENERGIES FOR THE MULTIPLY ION
                 4-0993 OBSERVATION OF A PROJECTILE CHARGE DEPENDENCE FOR THE NEON K-AUGER ELECTRON SPEC
                 4-0994 K-SHELL (AUGER) FLUORESCENT YIELDS FOR MULTIPLY IONIZED NEON.
                 4-1050 AUGER ELECTRON EMISSION SPECTRA FROM FOIL AND GAS EXCITED CARBON BEAMS.
                 4-1051 HIGH RESOLUTION NEON K-AUGER ELECTRON SPECTRUM PRODUCED BY 45 MEV CHLORINE(12+)
MOORHEAD D   5-1904 AUGER ELECTRON SPECTROSCOPY AT HIGH SPATIAL RESOLUTION AND PRIMARY BEAM CURRENTS
MOORHEAD RD   3-0659 MECHANICAL TESTING: IN-SITU FRACTURE DEVICE FOR AUGER ELECTRON SPECTROSCOPY.
MORABITO JM   1-0135 COMPARISON OF AUGER ELECTRON SPECTROSCOPY AND SECONDARY ION MASS SPECTROMETRY.
                 1-0136 USE OF AUGER ELECTRON SPECTROSCOPY AND SECONDARY ION MASS SPECTROMETRY IN THE MI
                 2-0450 ELECTRONIC SUPERPOSITION OF SAMPLE CURRENT AND SECONDARY ELECTRON IMAGES IN AUGE
                 3-0660 FIRST ORDER APPROXIMATION TO QUANTITATIVE AUGER ANALYSIS IN THE RANGE 100 TO 100
                 4-0995 IN-DEPTH PROFILES BY AUGER SPECTROSCOPY AND SECONDARY ION EMISSION
                 4-0996 IN-DEPTH PROFILES OF PHOSPHORUS ION IMPLANTED SILICON BY AUGER SPECTROSCOPY AND
                 5-1393 DIFFUSION MECHANISMS IN THE PALLADIUM- GOLD THIN FILM SYSTEM AND THE CORRELATION
                 5-1473 EFFECT OF LIGHT ELEMENTS NITROGEN, CARBON AND OXYGEN ON THE PHYSICAL PROPERTIES
                 5-1652 QUANTITATIVE ANALYSIS OF LIGHT ELEMENTS (NITROGEN, CARBON, AND OXYGEN) IN SPUTTE
                 5-1653 SELECTED AREA AND IN-DEPTH AUGER ANALYSIS OF THIN FILMS.
                 5-1654 ELEMENTAL ANALYSIS WITH AUGER ELECTRON SPECTROSCOPY AND SECONDARY ION MASS SPECT
                 5-1655 APPLICATIONS OF SCANNING AUGER SPECTROSCOPY TO THE SILICON INTEGRATED CIRCUIT TE
                 5-1656 COMPOSITION PROFILES OF CHEMICALLY VAPOR DEPOSITED PLATINUM AND PLATINUM SILICID
                 5-1887 DETECTABILITY LIMITS FOR BORON AND PHOSPHORUS IN SILICON BY AUGER ELECTRON SPECT
                 5-1916 IN-DEPTH PROFILE DETECTION LIMITS OF NITROGEN IN GALLIUM PHOSPHIDE (AND) NITROGE
MORGAN AE   5-1657 CHARACTERISTIC ENERGY LOSS AND AUGER ELECTRON SPECTRA OF GALLIUM PHOSPHIDE (110)
MORGAN DV   1-0137 RECENT ADVANCES IN SURFACE STUDIES. ION BEAM ANALYSIS. (AUGER SPECTROSCOPY, REVI
MORI C   5-1658 NEW METHOD OF THIN FILM THICKNESS MEASUREMENT USING RADIOISOTOPE AUGER ELECTRONS
MORIGAKI K   2-0453 PHOTOCONDUCTIVITY ASSOCIATED WITH AUGER RECOMBINATION OF EXCITONS BOUND TO NEUTR
                 2-0454 EXCITON LUMINESCENCE AND PHOTOCONDUCTIVITY IN HIGHLY EXCITED GALLIUM PHOSPHIDE.
MORITA I   5-1809 SURVEYING SILICON SURFACES BY MEANS OF THE LEED- AES METHOD.
MORIYA T   2-0451 LUMINESCENCE DUE TO EXCITON-EXCITON COLLISION IN GALLIUM ARSENIDE.
MOROZOV YA   2-0508 SECONDARY EMISSION AND ELASTIC REFLECTION OF ELECTRONS FROM MONOCRYSTALS.
                 4-1062 SECONDARY ELECTRON EMISSION OF SILICON DIOXIDE SINGLE CRYSTALS.
                 4-1063 SECONDARY ELECTRON EMISSION OF MOLYBDENUM CRYSTALS.
                 4-1064 ENERGY SPECTRA OF INELASTICALLY SCATTERED AND AUGER ELECTRONS FROM SINGLE CRYSTA
MORRISON J   3-0661 APPLICATION OF TRIPLE GRID LEED SYSTEM TO AUGER SPECTRUM ANALYSES.
                 5-1603 AN ELECTRON DIFFRACTION STUDY OF CESIUM ADSORPTION ON TUNGSTEN.
                 5-1604 ELECTRONIC AND LATTICE STRUCTURE OF CESIUM FILMS ADSORBED ON TUNGSTEN.
                 5-1659 ADSORPTION OF IONIC SALTS ON A TUNGSTEN (100) SURFACE.
MORTIMER DA   5-1441 CORRELATION OF THE WETTING BEHAVIOR OF SODIUM ON NIMONIC-PE16 ALLOY WITH THE SUR
MOSKALEV AN   2-0364 PARITY NONCONSERVATION IN RADIATIONLESS (AUGER) TRANSITIONS IN MU MESIC ATOMS.
MOSS ARL   5-1660 EPITAXIAL GROWTH OF COPPER ON (110) SURFACE OF A TUNGSTEN SINGLE CRYSTAL STUDIED
MOSS RL   5-1846 DETERMINATION OF THE SURFACE COMPOSITION OF PALLADIUM- NICKEL ALLOY FILM CATALYS
MOWAT JR   2-0472 ELECTRON DECAY-IN-FLIGHT SPECTRA FROM AUTOIONIZING STATES OF HIGHLY STRIPPED OXY
MROZ S   6-2069 IMPURITIES SEGREGATION TO THE (001) NICKEL SURFACE DURING THERMAL TREATMENT. WOR
MSTIBOVSKAYA LE 4-0790 EMISSION LINES OF THE M SERIES OF RARE EARTH METALS. (AUGER EFFECT)
MUEHLETHALER H   5-1157 INVESTIGATION OF EUROPIUM(II) OXIDE (100) CLEAVAGE SURFACE BY LOW ENERGY ELECTRO
MUELLER KH   2-0461 TRANSIENT CARRIER DECAY AND TRANSPORT PROPERTIES IN MERCURY CADMIUM TELLURIDE. (
MUESSIG HJ   6-1991 LEED- AUGER INVESTIGATION ON A (100) SURFACE OF AUSTENITIC CHROMIUM- NICKEL STEE
MUKHAMADIEV ES   2-0247 AUGER NEUTRALIZATION OF MULTIPLY CHARGED IONS ON A METAL SURFACE.
MULARIE WM   1-0138 DECONVOLUTION TECHNIQUES IN AUGER ELECTRON SPECTROSCOPY.
                 1-0139 INELASTIC EFFECTS IN AUGER ELECTRON SPECTROSCOPY.
MULLER K   5-1603 AN ELECTRON DIFFRACTION STUDY OF CESIUM ADSORPTION ON TUNGSTEN.
                 5-1604 ELECTRONIC AND LATTICE STRUCTURE OF CESIUM FILMS ADSORBED ON TUNGSTEN.
                 5-1661 CESIUM ADSORPTION ON TUNGSTEN STUDIED BY LEED AND SECONDARY ELECTRON SPECTROSCOP
                 5-1662 AUGER ELECTRONS, INDICATORS IN ANALYSIS OF SOLID BODY SURFACES.
                 5-1663 HOW MUCH CAN AUGER ELECTRONS TELL US ABOUT SOLID SURFACES.
MUNOZ E   5-1914 AUGER ELECTRON SPECTROSCOPY STUDIES OF ARSENIC VACANCIES IN (-1-1-1) GALLIUM ARS
                 5-1915 SURFACE CARRIER CONCENTRATION AND AS EVAPORATION IN HEAT TREATED GALLIUM ARSENID
MUNRO DF   2-0450 ELECTRONIC SUPERPOSITION OF SAMPLE CURRENT AND SECONDARY ELECTRON IMAGES IN AUGE
MURAKAWA T   5-1487 DETECTION OF ORGANIC CONTAMINANTS ON ALUMINUM SURFACES BY AUGER ELECTRON SPECTRO
MURARKA SP   5-1232 INTERDIFFUSIONS IN THIN FILM GOLD ON PLATINUM ON GALLIUM ARSENIDE (100) STUDIED
MURATA Y   1-0140 AUGER ELECTRON SPECTROSCOPY.
MUROTANI T   1-0141 AUGER ELECTRON SPECTROSCOPY AND ITS APPLICATION.
                 5-1691 COMBINED AUGER ELECTRON SPECTROSCOPY AND ELECTRON IMPACT DESORPTION STUDIES OF T
                 5-1692 IDENTIFICATION OF ION SPECIES IN THE COMBINED AUGER ELECTRON SPECTROSCOPY AND EL
                 5-1693 INTERACTION OF OXYGEN WITH SILICON (111) SURFACES. (AUGER ELECTRON SPECTROSCOPY)
                 5-1694 COMBINED AUGER ELECTRON SPECTROSCOPY AND ELECTRON IMPACT DESORPTION STUDIES OF S
                 5-1695 COMBINED AUGER ELECTRON SPECTROSCOPY AND ELECTRON IMPACT DESORPTION STUDIES OF S
MUSIKHIN VA   4-0736 RECORDING OF COMPONENTS ON THE SURFACE OF FUSED EMITTERS. (AUGER SPECTRA)
                 5-1192 AUGER SPECTROSCOPIC STUDY OF THE SURFACES OF ALLOY EMITTERS.
MUSKET RG   1-0048 APPLICATIONS OF AUGER ELECTRON SPECTROSCOPY IN ELECTROPLATING.
                 1-0142 DIRECT COMPARISON OF AUGER ELECTRON SPECTROSCOPY WITH APPEARANCE POTENTIAL SPECT
                 1-0143 SURFACE ANALYSIS USING PROTON BEAMS.

MUSKET RG   2-0452 AUGER ELECTRON SPECTROSCOPY STUDY OF ELECTRON IMPACT DESORPTION.
                   4-0834 CHEMICAL EFFECTS ON THE AUGER ELECTRON SPECTRA OF BERYLLIUM.
                   4-0997 OBSERVATION AND INTERPRETATION OF THE AUGER ELECTRON SPECTRUM FROM CLEAN BERYLLI
                   5-1664 ROOM TEMPERATURE ADSORPTION OF OXYGEN ON TUNGSTEN SURFACES. (AUGER MEASUREMENT M
                   5-1665 ENERGY DEPENDENCE OF THE SENSITIVITY OF IONIZATION SPECTROSCOPY: SILICON ON BERY
                   5-1666 PROTON INDUCED ELECTRON EMISSION FROM CHARACTERIZED NIOBIUM SURFACES. (AUGER SPE
                   5-1667 PROTON AND ELECTRON EXCITED AUGER ELECTRON SPECTRA OF MOLYBDENUM SURFACES.
                   5-1668 AUGER ELECTRON SPECTROSCOPY STUDY OF OXYGEN ADSORPTION ON TUNGSTEN (110).
                   5-1669 LEED AND AES STUDY OF OXYGEN ABSORPTION ON TUNGSTEN (110).
MYKURA H   6-2096 SEGREGATION OF BISMUTH TO GRAIN BOUNDARIES IN COPPER- BISMUTH ALLOYS.
MYKURA M   6-2097 STUDY OF INTERGRANULAR FRACTURE IN IRON USING AUGER SPECTROSCOPY.
NAGATA S   2-0401 DETERMINATION OF THE ESCAPE LENGTH OF LOW ENERGY ELECTRONS BY AUGER AND LOSS SPE
                   3-0639 ELECTRODES AND CIRCUITS OF A LEED- AES SYSTEM.
                   5-1522 LOW ENERGY ELECTRON DIFFRACTION- AUGER ELECTRON SPECTROSCOPY STUDY OF IRON FILM
                   5-1523 BEHAVIOR OF IMPURITY ATOMS AND ADSORBED OXYGEN ATOMS ON THE (001) FACE OF IRON E
                   5-1524 AUGER AND LOSS SPECTROSCOPY STUDY OF SURFACE CONTAMINATION EFFECT ON THE GROWTH
                   5-1525 LEED- AES OBSERVATION OF GROWTH MECHANISM OF DEPOSITED IRON FILM ON MAGNESIUM OX
                   5-1526 LOW ENERGY ELECTRON DIFFRACTION- AUGER ELECTRON SPECTROSCOPY STUDY OF AN EPITAXI
                   5-1527 LOW ENERGY ELECTRON DIFFRACTION- AUGER ELECTRON SPECTROSCOPY STUDY OF CLEAN (001
                   5-1528 GROWTH MODE OF IRON FILM DEPOSITED ON MAGNESIUM OXIDE (001) SURFACE. (AUGER SPEC
NAHORY RE   2-0503 OPTICAL GAIN IN LIGHTLY DOPED GALLIUM ARSENIDE. (AUGER RECOMBINATION)
NAKAGIMA T   5-1929 ADSORPTION OF NITRIC OXIDE ON TUNGSTEN.
NAKAJIMA K   5-1670 A STUDY OF THE IRON (111) SURFACE USING LEED AND AUGER EMISSION SPECTROSCOPY.
                   5-1866 LOW ENERGY ELECTRON DIFFRACTION- AUGER ELECTRON SPECTROSCOPY STUDY OF IRON SINGL
                   5-1867 CONCENTRATION CHANGE OF TIN IN THE SURFACE LAYER OF COPPER- TIN ALLOY SYSTEM DUE
                   5-1868 EFFECT OF PHOSPHORUS ON THE FRICTION AND WEAR CHARACTERISTICS OF COPPER- TIN- PH
NAKAJIMA N   6-2043 EFFECTS OF COPPER AND PHOSPHORUS ON TEMPER EMBRITTLEMENT OF MANGANESE- MOLYBDENU
NAKAJIMA S   5-1671 CHEMICAL ANALYSIS OF SEMICONDUCTORS. (IMPURITIES AND COMPOSITION)
NAKAMURA A   2-0453 PHOTOCONDUCTIVITY ASSOCIATED WITH AUGER RECOMBINATION OF EXCITONS BOUND TO NEUTR
                   2-0454 EXCITON LUMINESCENCE AND PHOTOCONDUCTIVITY IN HIGHLY EXCITED GALLIUM PHOSPHIDE.
NAKAMURA H   2-0494 RADIATIVE AUGER EFFECT WITH AND WITHOUT ELECTRON EMISSION.
NAKAMURA K   5-1672 INTERACTION OF ALUMINUM LAYERS WITH POLYCRYSTALLINE SILICON. (AUGER SPECTROSCOPY
NAKANISHI S   5-1469 STRUCTURAL STUDY OF OXYGEN- IRON (001) BY LOW ENERGY ELECTRON DIFFRACTION AND AU
                   5-1470 ANALYSIS OF THE SURFACE STRUCTURE WITH ANTIPHASE DOMAINS OF SULFUR ON IRON (001)
NAKASHIMA K   5-1673 LOW TEMPERATURE DIFFUSION OF GOLD INTO SILICON IN THE SILICON (SUBSTRATE)- GOLD
NAKAYAMA K   1-0144 SURFACE STUDIES BY AUGER ELECTRON SPECTROSCOPY. (REVIEW, 44 REFS)
                   1-0145 AUGER ELECTRON SPECTROSCOPY.
                   3-0669 ULTRAHIGH VACUUM AIRLOCK FOR LEED AND AUGER SPECTROSCOPY.
                   3-0670 INSTRUMENTAL PROBLEMS FOR ELECTRON MICROPROBE AUGER SPECTROSCOPY.
                   5-1674 ANALYSIS OF IMPURITIES ON SILICON SURFACES BY AUGER ELECTRON SPECTROSCOPY.
                   5-1675 AUGER SPECTROSCOPY STUDY ON THE SURFACE COMPOSITION OF COPPER- NICKEL ALLOYS AFT
                   5-1702 QUANTITATIVE AUGER ANALYSIS OF COPPER- NICKEL ALLOY SURFACES AFTER ARGON ION BOM
                   5-1703 AUGER SPECTROSCOPY ON COPPER- NICKEL ALLOY SURFACES RELATED TO CATALYSIS.
                   5-1807 QUANTITATIVE AUGER ANALYSIS OF COPPER- NICKEL ALLOY SURFACES AFTER ARGON ION BOM
                   5-1808 EFFECT OF TARGET TEMPERATURE ON SURFACE COMPOSITION CHANGES OF COPPER- NICKEL AL
                   6-2109 SURFACE AND GRAIN BOUNDARY SEGREGATION OF STEEL IMPURITIES. (TEMPER EMBRITTLEMEN
NARKEVICIUS V   2-0258 TEMPERATURE DEPENDENCE OF THE EDGE EMISSION OF GALLIUM SELENIDE SINGLE CRYSTALS
NARUSAWA T   5-1436 DIFFUSE INTERFACE STRUCTURES IN SOME (SUBSTRATE- VACUUM EVAPORATED FILM) SYSTEMS
                   5-1438 METALLIC STATE OF SILICON IN SILICON- NOBLE METAL VAPOR QUENCHED ALLOYS STUDIED
                   5-1550 COMPOSITION PROFILES OF SEVERAL CONTAMINATED AND CLEANED SURFACES OF GOLD THICK
                   5-1551 EFFECT OF GOLD COATING ON SORPTION CHARACTERISTICS OF CARBON MONOXIDE, SULFUR, A
                   5-1552 SPUTTER ETCHING AND AUGER ELECTRON SPECTROSCOPY OF METAL SURFACES.
                   5-1553 SIMULTANEOUS OBSERVATIONS OF PARTIALLY OXIDIZED SURFACES BY AUGER ELECTRON SPECT
                   5-1554 SPUTTER ETCHING AND AUGER ELECTRON SPECTROSCOPY OF METAL SURFACES.
                   5-1624 GROWTH MODE DISCRIMINATION OF THIN FILMS BY AUGER ELECTRON SPECTROSCOPY.
                   5-1676 AUGER ELECTRON EMISSION FROM GOLD DEPOSITED ON SILICON (111) SURFACE.
                   5-1677 LEED AND AUGER ELECTRON SPECTROSCOPIC STUDIES ON TUNGSTEN (110) SURFACE.
                   5-1678 LEED AND AUGER ELECTRON SPECTROSCOPIC OBSERVATIONS OF THE SUBSTRATE SURFACE DURI
                   5-1679 COMPOSITION PROFILE OF ION PLATED GOLD FILM ON COPPER ANALYZED BY AUGER ELECTRON
                   5-1680 AUGER SPECTROSCOPIC OBSERVATION OF SILICON- GOLD MIXED PHASE FORMATION AT LOW TE
                   5-1681 DIFFUSE INTERFACE IN SILICON (SUBSTRATE)- GOLD (EVAPORATED FILM) SYSTEM. (AUGER
                   6-2090 COMPOSITION OF BINARY ALLOYS BY SIMULTANEOUS SIMS AND AES MEASUREMENTS.
NASSERIAN-RIABI M 5-1224 OXIDATION OF CUPRONICKEL ALLOYS. PART-1: XPS STUDY OF INTERDIFFUSION. (AND AES)
NATH A   2-0516 MOSSBAUER STUDIES OF THE AFTEREFFECTS OF AUGER IONIZATION FOLLOWING ELECTRON CAP
NATHAN R   3-0662 LOW NOISE DETECTION SYSTEM FOR THE MEASUREMENT OF AUGER SPECTRA.
                   3-0663 AN AC RETARDING POTENTIAL TECHNIQUE FOR THE CONTINUOUS MEASUREMENT OF CHANGES IN
                   5-1243 CHARACTERIZATION OF ADSORBED SPECIES USING AUGER PEAK SHAPES. ETHYLENE AND CARBO
                   5-1244 ELECTRON BEAM EFFECTS ON CARBON MONOXIDE SATURATED TUNGSTEN (100), ETHYLENE SATU
NATTA M   4-0998 SINGULARITIES IN AUGER EMISSION SPECTRUM OF METALS.
NEAVE JH   2-0455 DEPENDENCE OF AUGER ELECTRON YIELD ON PRIMARY BEAM ENERGY AT NORMAL AND GLANCING
                   3-0664 APPARATUS FOR SURFACE ANALYSIS BY ELECTRON SPECTROSCOPY.
                   5-1511 SILICON- OXYGEN INTERACTIONS USING AUGER ELECTRON SPECTROSCOPY.
                   5-1512 INTERACTION OF OXYGEN WITH SILICON (111) SURFACES. (REPLY TO COMMENT)
                   5-1513 ELECTRON BEAM ADSORBATE INTERACTIONS ON SILICON SURFACES. (AUGER SPECTROSCOPY)
                   5-1514 INFLUENCE OF SUBSTRATE CONDITIONS ON THE NUCLEATION AND GROWTH OF EPITAXIAL SILI
NEEDHAM PB   2-0456 DETERMINATION OF AUGER ELECTRON EXCITATION CROSS SECTIONS AND RANGES USING PROTO
                   5-1682 CORRELATION OF FRACTIONAL MONOLAYER OXYGEN DETERMINATIONS OBTAINED BY PROTON EXC
NEKRASHEVICH IG   4-1001 EFFECT OF THE POLARITY OF INDIUM ANTIMONIDE ON THE EXTERNAL PHOTOEFFECT IN THE X
NEMEH EAK   5-1683 PREPARATION AND SURFACE OXIDATION OF A (110) FACE OF VANADIUM.
NERI G   2-0305 MEASUREMENT OF THE AUGER EFFECT IN THE MU-4 HELIUM(2S)(+) IONIC SYSTEM.
NETZER FP   4-0999 TEMPERATURE DEPENDENCE OF THE LOW ENERGY AUGER SPECTRUM OF NICKEL(II) OXIDE (100
                   5-1546 (AUGER) ADSORPTION STUDIES OF OXYGEN AND CARBON MONOXIDE ON A PLATINUM (100) SUR
                   5-1684 ADSORPTION OF CYANOGEN ON PLATINUM (100).
                   5-1685 ADSORPTION OF HYDROGEN AND THE REACTION OF HYDROGEN WITH OXYGEN ON PLATINUM (100
                   5-1686 LOW ENERGY ELECTRON SPECTROSCOPY OF PLATINUM (100) AND PLATINUM (100)- CARBON MO
                   5-1687 LEED AND ELECTRON SPECTROSCOPIC OBSERVATIONS ON NICKEL MONOXIDE (100).
NEUMANN M   5-1257 ADSORPTION OF HYDROGEN ON NICKEL SINGLE CRYSTAL SURFACES. (AUGER ELECTRON SPECTR
NEUMARK GF   2-0457 AUGER THEORY AT DEFECTS. APPLICATION TO STATES WITH TWO BOUND PARTICLES IN GALL
NEWBURY DE   1-0079 SECONDARY ION MASS SPECTROMETRY.
NICHOLAS MG   5-1441 CORRELATION OF THE WETTING BEHAVIOR OF SODIUM ON NIMONIC-PE16 ALLOY WITH THE SUR

NICLOU Y	5-1688	ADSORPTION AND DECOMPOSITION OF ACETYLENE ON POLYCRYSTALLINE RHENIUM FILMS AND R
NICOLAIDES CA	2-0265	ACCURATE ONE ELECTRON BINDING ENERGIES AND AUGER ENERGIES IN MANY ELECTRON ATOMS
	4-0774	INNER ELECTRON BINDING ENERGIES AND AUGER ENERGIES IN FREE ATOMS FOR USE IN X-RA
	4-1000	THEORY OF KLL AUGER ENERGIES.
NICOLET MA	5-1577	ANTIMONY DOPING OF SILICON LAYERS GROWN BY SOLID PHASE EPITAXY. (AES MEASUREMENT
	5-1672	INTERACTION OF ALUMINUM LAYERS WITH POLYCRYSTALLINE SILICON. (AUGER SPECTROSCOPY
NIEHUS H	3-0665	LOW ENERGY ION BACKSCATTERING SPECTROSCOPY WITH A COMMERCIAL AUGER CYLINDRICAL M
	3-0666	QUANTITATIVE ASPECTS OF ION SCATTERING SPECTROSCOPY. (CYLINDRICAL MIRROR ANALYZE
	5-1318	ADSORPTION OF OXYGEN ON THE (110) PLANE OF TUNGSTEN. (AUGER ELECTRON SPECTROSCOP
	5-1689	COMPARISON BETWEEN AUGER ELECTRON SPECTROSCOPY AND SECONDARY ION MASS SPECTROSCO
NIHEI Y	5-1690	X-RAY EXCITED PHOTOELECTRON AND AUGER ELECTRON SPECTRA OF OXIDIZED STATES ON SOM
NIKEI Y	5-1521	MEASUREMENTS OF SURFACE MATERIAL STATES.
NIKOLAENYA AZ	4-1001	EFFECT OF THE POLARITY OF INDIUM ANTIMONIDE ON THE EXTERNAL PHOTOEFFECT IN THE X
NIKOLSHII AP	4-1070	INTENSITIES OF SPECTRA- AUGER AND PHOTOELECTRON.
NILSSON NG	2-0458	BAND-TO-BAND AUGER RECOMBINATION AND CARRIER-CARRIER SCATTERING IN POWER RECTIFI
	2-0459	BAND-TO-BAND AUGER RECOMBINATION IN SILICON AND GERMANIUM.
	2-0460	SPECTRUM AND DECAY OF THE RECOMBINATION RADIATION FROM STRONGLY EXCITED SILICON.
	4-1002	NEW RADIATIVE TRANSITION IN GERMANIUM.
NILSSON PO	4-0908	N5,4 AND M5,4 APPEARANCE POTENTIAL SPECTRA FROM PRASEODYMIUM.
NILSSON R	2-0408	ELECTRON ESCAPE DEPTH IN SILICON. (AUGER ENERGIES)
	2-0409	ESCAPE DEPTHS OF X-RAY EXCITED (AUGER) ELECTRONS.
NIMTZ G	2-0461	TRANSIENT CARRIER DECAY AND TRANSPORT PROPERTIES IN MERCURY CADMIUM TELLURIDE. (
NISHIJIMA A	5-1521	MEASUREMENTS OF SURFACE MATERIAL STATES.
	5-1690	X-RAY EXCITED PHOTOELECTRON AND AUGER ELECTRON SPECTRA OF OXIDIZED STATES ON SOM
NISHIJIMA M	1-0141	AUGER ELECTRON SPECTROSCOPY AND ITS APPLICATION.
	5-1691	COMBINED AUGER ELECTRON SPECTROSCOPY AND ELECTRON IMPACT DESORPTION STUDIES OF T
	5-1692	IDENTIFICATION OF ION SPECIES IN THE COMBINED AUGER ELECTRON SPECTROSCOPY AND EL
	5-1693	INTERACTION OF OXYGEN WITH SILICON (111) SURFACES. (AUGER ELECTRON SPECTROSCOPY)
	5-1694	COMBINED AUGER ELECTRON SPECTROSCOPY AND ELECTRON IMPACT DESORPTION STUDIES OF S
	5-1695	COMBINED AUGER ELECTRON SPECTROSCOPY AND ELECTRON IMPACT DESORPTION STUDIES OF S
NISHIMORI K	1-0199	QUANTIFICATION OF AUGER ELECTRON SPECTROSCOPY.
	5-1696	EPITAXIAL GROWTH OF SILVER ON POTASSIUM CHLORIDE (100) SURFACE OBSERVED BY LOW E
NISHIMURA F	4-1003	ANGULAR DISTRIBUTION OF M4,5NN AUGER ELECTRONS EJECTED FROM KRYPTON BY ELECTRON
NISHIMURA H	4-1010	AUGER ELECTRON EJECTION FROM XENON N4,5OO AND KRYPTON M4,5NN PROCESSES BY ELECTR
NISHINA Y	2-0416	STIMULATED PHOTOLUMINESCENCE IN GALLIUM SELENIDE. (AUGER PROCESS)
NISHIO M	1-0146	AUGER ELECTRON SPECTROSCOPY.
NISHIZAWA J	5-1786	LOW TEMPERATURE OXIDATION OF SILICON STUDIED BY PHOTOSENSITIVE ESR AND AUGER ELE
NIUNKA V	2-0258	TEMPERATURE DEPENDENCE OF THE EDGE EMISSION OF GALLIUM SELENIDE SINGLE CRYSTALS
NOELLER HG	3-0667	HIGH RESOLUTION ELECTRON SPECTROMETER FOR AUGER ELECTRON SPECTROSCOPY.
NOGGLE TS	5-1139	INVESTIGATICN OF REORDERED (001) GOLD SURFACES BY POSITIVE ION CHANNELING SPECTR
	5-1986	CHARACTERIZATION OF RECRDERED (001) GCLD SURFACES BY POSITIVE ION CHANNELING SPE
NOLLE EL	2-0462	STIMULATED EMISSION OF LIGHT FROM A NONIDEAL EXCITON GAS IN A SEMICONDUCTOR. (AU
NOLTE G	2-0370	AUGER ELECTRONS FROM FOIL EXCITED HEAVY ION BEAMS. (OF NEON)
	2-0371	AUGER TRANSITIONS LEADING TO DEFINED FINAL CHARGE STATES BY A COINCIDENCE TECHNI
	4-0859	DOPPLER SHIFTED AUGER ELECTRONS FROM FOIL EXCITED HEAVY ION BEAMS.
	4-0860	BEAM FOIL EXCITED AUGER TRANSITIONS IN NEON.
NOONAN JR	5-1697	ANGULAR RESOLVED AUGER EMISSION SPECTRA FROM A CLEAN COPPER (100) SURFACE.
	5-1988	ANGULAR EFFECTS IN AUGER ELECTRON EMISSION FROM COPPER (110).
NORDBERG R	4-0826	CHEMICAL SHIFT IN AUGER SPECTRA.
	4-0827	AUGER SPECTRA FOR ELEMENTS OF LOW ATOMIC NUMBER.
NORDLING C	1-0012	THE AUGER EFFECT.
	2-0408	ELECTRON ESCAPE DEPTH IN SILICON. (AUGER ENERGIES)
	2-0409	ESCAPE DEPTHS OF X-RAY EXCITED (AUGER) ELECTRONS.
	2-0515	ELECTRON SPECTROSCOPIC INVESTIGATION OF AUGER PROCESSES IN BROMINE-SUBSTITUTED M
	4-0826	CHEMICAL SHIFT IN AUGER SPECTRA.
	4-0827	AUGER SPECTRA FOR ELEMENTS OF LOW ATOMIC NUMBER.
	4-0887	AUGER SPECTRA IN THE INTERMEDIATE COUPLING REGION.
NORGORODOV AF	4-0754	KLL AUGER ELECTRONS OF CADMIUM AND ANTIMONY.
NORRIS PR	4-0838	COMPARISON OF THE SOFT X-RAY PHOTOEMISSION, AND AUGER ELECTRON SPECTRA OF THE VA
NORTON JF	5-1239	SECONDARY ELECTRON EMISSION OF AMORPHOUS SEMICONDUCTOR THIN FILMS.
NOVOJILOV VP	5-1317	GROWTH OF THIN METALLIC FILMS ON INSULATING SUBSTRATES IN THE PRESENCE OF ELECTR
NOZOYE H	4-1004	ORIGIN OF THE PEAK SHAPES IN THE AUGER SPECTRUM OF OXYGEN (KL2,3L2,3) ON MOLYBDE
	5-1698	INITIAL OXIDATION OF MOLYBDENUM STUDIED BY A HIGH RESOLUTION AUGER PHOTOELECTRON
NUTTALL JD	2-0439	ENHANCEMENT OF COSTER-KRONIG RATES IN SOLID STATE ENVIRONMENTS. (XENON, CADMIUM,
	4-0843	SOLID STATE BROADENING IN AUGER SPECTRA. (CADMIUM, RARE GASES)
	4-1005	SOLID STATE BROADENING EFFECTS IN THE AUGER SPECTRUM OF XENON.
	4-1006	SOLID STATE BROADENING IN THE AUGER SPECTRA OF RARE GASES. (ARGON, KRYPTON, XENO
NYHOLM R	2-0408	ELECTRON ESCAPE DEPTH IN SILICON. (AUGER ENERGIES)
OBBYKIN BV	4-0758	KMM AUGER TRANSITIONS IN XENON-131.
ODA N	4-1003	ANGULAR DISTRIBUTION OF M4,5NN AUGER ELECTRONS EJECTED FROM KRYPTON BY ELECTRON
ODA T	5-1699	STUDIES ON SURFACE PLASMA OXIDIZED FILMS OF III-V COMPOUND SEMICONDUCTORS BY AUG
ODA Z	3-0636	NEW SEM AUGER MICRCANALYZER WITH ULTRAHIGH VACUUM CAPABILITY.
OGAWA H	1-0147	PRESENT STATUS AND PROBLEMS ON THE APPLICATION OF AUGER ELECTRON SPECTROSCOPY.
OGURTSOV GN	4-1007	ENERGY SPECTRA OF ELECTRONS EJECTED IN ARGON(+)- ARGON COLLISIONS. (AUGER AND CO
	4-1008	ENERGY DISTRIBUTION OF ELECTRONS EMITTED BY ARGON ATOMS UNDER ELECTRON IMPACT.
OH SD	2-0463	DEPENDENCE OF ATOMIC PHOTOEFFECT ON A BOUND ELECTRON MAGNETIC SUBSTATE. (AUGER E
OHANDLEY RC	5-1700	RESIDUAL GAS AND THE OPTICAL PROPERTIES OF SILVER FILMS. (AUGER SPECTROSCOPY)
OHMURA Y	5-1482	OBSERVATION CF IMPURITY PROFILE IN ION IMPLANTED SILICON BY AUGER ELECTRON SPECT
OHNSORGE J	5-1701	MATERIAL ANALYSIS USING THE SCANNING ELECTRON MICROSCOPE. (AUGER SPECTROSCOPY)
OHTA T	4-1009	INFORMATION CBTAINED FROM THE WIDTH OF SPECTRAL LINES IN THE ELECTRON SPECTROSCO
OHTANI S	4-1010	AUGER ELECTRON EJECTION FROM XENON N4,5OO AND KRYPTON M4,5NN PROCESSES BY ELECTR
OKABE T	5-1386	AUGER ANALYSIS OF SURFACE FILMS ON SILVER- TIN.
OKADA H	1-0147	PRESENT STATUS AND PROBLEMS ON THE APPLICATION OF AUGER ELECTRON SPECTROSCOPY.
OKADA K	5-1872	(AUGER) MICROSCOPICAL STUDY OF FRICTIONAL PROPERTIES OF MOLYBDENUM DISULFIDE.
	6-2129	AES STUDIES ON SURFACES OF 18-8 STAINLESS STEEL FINISHED MECHANICALLY AND ELECTR
	6-2130	AUGER ELECTRON SPECTROSCOPIC STUDIES OF THE SURFACES OF 18-8 STAINLESS STEEL ELE
	6-2131	AUGER ELECTRON SPECTROSCOPY STUDIES ON ABRADED SURFACE OF ALLOYS.
	6-2132	SURFACE DEFORMATION CAUSED ON NATURAL MOLYBDENITE BY ABRASION. (AUGER ELECTRON S
	6-2133	STUDY OF THE INTERMITTENT MOTION OF FRICTION IN STAINLESS STEEL BY AN AUGER MICR
OKAMOTO PR	6-2091	SEGREGATION OF ALLOYING ELEMENTS TO FREE SURFACES DURING IRRADIATION. (AUGER SPE

OKAMOTO PR	6-2092	IRRADIATION INDUCED SOLUTE SEGREGATION (IN STEEL, AUGER SPECTROSCOPY).
OKANE DF	6-2093	PLASMA CLEANING OF METAL SURFACES. (AUGER ELECTRON SPECTROSCOPY)
OKANO H	1-0076	MICROANALYSIS BY AUGER ELECTRON SPECTROSCOPY. (REVIEW, 15 REFS)
	1-0077	THREE DIMENSIONAL CHARACTERIZATION OF METALS AND SEMICONDUCTORS BY THE AUGER ELE
	5-1412	THREE-DIMENSIONAL ELEMENT ANALYSIS OF SURFACES AND INTERFACES OF SOLIDS BY THE A
	5-1413	AUGER ELECTRON EMISSION MICROGRAPHIC STUDIES OF THE CLEAVAGE SURFACE OF GRAPHITE
	5-1414	AUGER ELECTRON SPECTROSCOPY AND MICROANALYSIS OF SOLID SURFACES.
	5-1415	AUGER ELECTRON EMISSION MICROGRAPHY AND MICROANALYSIS OF A STAINED SURFACE OF PU
	5-1416	AUGER ELECTRON EMISSION MICROGRAPHY AND MICROANALYSIS OF SOLID SURFACES.
	5-1471	SURFACE CONDITION AND THERMIONIC EMISSION OF LANTHANUM HEXABORIDE. (AUGER SPECTR
OLSON C	2-0398	MATRIX ELEMENTS DEPENDENCE OF OPTICAL EXCITATION AND AUGER DECAY OF 5D CORE HOLE
OLTJEN J	2-0481	RADIATIVE AUGER EFFECT IN ION- ATOM COLLISIONS.
OMATA H	1-0147	PRESENT STATUS AND PROBLEMS ON THE APPLICATION OF AUGER ELECTRON SPECTROSCOPY.
OMELYANOVSKII EM	4-0785	RECOMBINATION OF CARRIERS IN N-TYPE INDIUM ARSENIDE AT 77 DEGREES K. (AUGER)
ONCHI M	3-0655	A SMALL RETARDING FIELD ENERGY ANALYZER WITH CEM. (AUGER SPECTROMETER)
	3-0656	AUGER ELECTRON SPECTROSCOPY USING THE TWO-GRID LEED SYSTEM AND PHOTOELECTRON MUL
	4-0727	CHEMICAL EFFECTS IN THE AUGER SPECTRA OF IRON DUE TO OXYGEN ADSORPTION.
	5-1623	ANGULAR DISTRIBUTION OF AUGER ELECTRON EMISSION FROM CLEAN AND GAS-COVERED IRON
	6-2141	STUDY ON THE INITIAL OXIDATION AND REDUCTION PROCESSES OF IRON (100) SURFACE BY
ONISHI T	1-0108	ELECTRON SPECTROMETRY.
	4-0912	HIGH RESOLUTION L2,3M2,3M2,3 AUGER ELECTRON SPECTRA OF ARGON.
	4-0925	CHEMICAL EFFECTS OF THE NITROGEN KLL AUGER SPECTRA OF CHEMISORBED NITROGEN ON IR
	4-1004	ORIGIN OF THE PEAK SHAPES IN THE AUGER SPECTRUM OF OXYGEN (KL2,3L2,3) ON MOLYBDE
	5-1530	STUDIES OF THE MECHANISM OF THE REACTION BETWEEN SULFUR AND OXYGEN ON A MOLYBDEN
	5-1531	STUDY OF THE OXIDATION OF MOLYBDENUM SURFACES BY ENERGY LOSS SPECTROSCOPY COMBIN
	5-1532	HIGH RESOLUTION MEASUREMENTS OF THE NITROGEN KLL AUGER TRANSITIONS OF CHEMISORBE
	5-1533	APPLICATIONS OF AUGER ELECTRON AND ENERGY LOSS SPECTROSCOPIES TO CATALYTIC PROCE
	5-1534	DYNAMIC INVESTIGATION OF THE MECHANISM OF REACTION BETWEEN SULFUR AND OXYGEN ON
	5-1564	SEGREGATION OF SULFUR ON A MOLYBDENUM SURFACE STUDIED BY AUGER ELECTRON SPECTROS
	5-1565	HIGH RESOLUTION AUGER ELECTRON SPECTROSCOPY OF NITRIC OXIDE ADSORBED ON A TUNGST
	5-1698	INITIAL OXIDATION OF MOLYBDENUM STUDIED BY A HIGH RESOLUTION AUGER PHOTOELECTRON
ONO M	1-0148	BIBLIOGRAPHY ON AUGER ELECTRON SPECTROSCOPY.
	3-0668	PRODUCTION OF A SPHERICAL GRID FOR LOW ENERGY ELECTRON DIFFRACTION- AUGER ELECTR
	3-0669	ULTRAHIGH VACUUM AIRLOCK FOR LEED AND AUGER SPECTROSCOPY.
	3-0670	INSTRUMENTAL PROBLEMS FOR ELECTRON MICROPROBE AUGER SPECTROSCOPY.
	5-1675	AUGER SPECTROSCOPY STUDY ON THE SURFACE COMPOSITION OF COPPER- NICKEL ALLOYS AFT
	5-1702	QUANTITATIVE AUGER ANALYSIS OF COPPER- NICKEL ALLOY SURFACES AFTER ARGON ION BOM
	5-1703	AUGER SPECTROSCOPY ON COPPER- NICKEL ALLOY SURFACES RELATED TO CATALYSIS.
	5-1807	QUANTITATIVE AUGER ANALYSIS OF COPPER- NICKEL ALLOY SURFACES AFTER ARGON ION BOM
	5-1808	EFFECT OF TARGET TEMPERATURE ON SURFACE COMPOSITION CHANGES OF COPPER- NICKEL AL
	6-2109	SURFACE AND GRAIN BOUNDARY SEGREGATION OF STEEL IMPURITIES. (TEMPER EMBRITTLEMEN
	6-2128	TEMPER EMBRITTLEMENT OF TIN DOPED STEELS BY AUGER ELECTRON SPECTROSCOPIC METHOD.
ONODA BY	5-1722	AES COMPOSITIONAL PROFILES OF MOBILE IONS IN THE SURFACE REGION OF GLASS.
ONODA GY	5-1266	AUGER ANALYSIS OF BRASS- ENAMEL AND STAINLESS STEEL- ENAMEL INTERFACES.
	5-1704	ANALYSIS OF GLASS CONTAINER COATINGS BY AUGER ELECTRON SPECTROSCOPY.
	5-1705	APPLICATIONS OF SURFACE CHARACTERIZATION TO GLASS SURFACES.
	5-1720	GLASS SURFACE ANALYSIS BY AUGER ELECTRON SPECTROSCOPY.
	5-1721	AES ANALYSIS OF SODIUM IN A CORRODED BIOGLASS USING A LOW TEMPERATURE TECHNIQUE.
OONA H	4-1011	ARGON CHARGE STATES DUE TO IMPACT OF HYDROGEN(+) AND HELIUM(+). (AUGER SPECTRA)
OPANOWICZ A	2-0432	PHOTOCONDUCTIVITY OF HIGHLY EXCITED CADMIUM SELENIDE AND CADMIUM SULFIDE(0.48) S
OPPERHAUSER H	6-2072	INTERGRANULAR EMBRITTLEMENT OF NICKEL BY HYDROGEN: EFFECT OF GRAIN BOUNDARY SEGR
ORCHARD AF	1-0149	PHOTOELECTRON SPECTROSCOPY AND ALLIED TECHNIQUES: GENERAL INTRODUCTION.
ORE A	2-0464	ELECTRON SPRAYING ATOMS. (AUGER CASCADES)
ORLOV VP	5-1639	INVESTIGATION OF THE PURE AND CESIUM COATED (100) SURFACE OF GALLIUM ARSENIDE BY
	5-1640	(111) SURFACE OF GALLIUM PHOSPHIDE STUDIED BY AUGER SPECTROSCOPY AND DIFFRACTION
	5-1641	PROCESSES OF ADSORPTION OF OXYGEN ON GALLIUM ARSENIDE METHOD OF LOW ENERGY ELECT
	5-1642	AUGER ELECTRON SPECTRUM DURING THE INTERACTION BETWEEN OXYGEN AND THE (111) SURF
	5-1643	SOME FEATURES OF THE AUGER ELECTRON SPECTRUM IN THE INTERACTION BETWEEN OXYGEN A
	5-1644	INVESTIGATION OF CESIUM ADSORPTION ONTO A GALLIUM ARSENIDE (110) SURFACE. (AUGER
	5-1645	QUANTITATIVE ESTIMATES OF THE CHEMICAL COMPOSITION NEAR THE SURFACE BASED ON THE
ORLOVA GA	4-1012	THEORETICAL ANALYSIS OF THE AUGER SPECTRA OF METHANE.
ORTENBURGER IB	5-1706	AUGER ELECTRON SPECTROSCOPY STUDY OF OXIDATION ON LANTHANUM HEXABORIDE.
OSHIMA C	1-0150	AUGER SPECTROSCOPY FOR THE CHEMICAL ANALYSIS OF SURFACES.
OSWALD RC	1-0141	AUGER ELECTRON SPECTROSCOPY AND ITS APPLICATION.
OTANI M	5-1175	LOW ENERGY ELECTRON DIFFRACTION AND AUGER SPECTROSCOPY OF THE ADSORPTION OF SULF
OUDAR J	5-1558	ADSORPTION OF SULFUR ON GOLD BY LOW ENERGY ELECTRON DIFFRACTION AND AUGER SPECTR
	5-1707	SURFACES TEXTURES: AN INVESTIGATION OF SURFACES CONTAINING IMPURITIES. (AUGER EL
	5-1732	ADSORPTION OF SULFUR ON THE (100) FACE OF MOLYBDENUM.
OURA K	5-1708	GOLD-INDUCED SUPERSTRUCTURES ON SILICON (100) SURFACES AS OBSERVED BY LEED AES.
OVERBURY SH	5-1709	AUGER ELECTRON SPECTROSCOPY OF ALLOY SURFACES. (GOLD- SILVER, LEAD- INDIUM)
	5-1710	SURFACE COMPOSITION OF THE SILVER- GOLD SYSTEM BY AUGER ELECTRON SPECTROSCOPY.
	5-1839	AUGER ELECTRON SPECTROSCOPY OF (SILVER- GOLD AND INDIUM- LEAD) ALLOY SURFACES.
OXENGENDLER BL	2-0465	AUGER DISSOCIATION OF MOLECULES AS A RESULT OF INNER SHELL IONIZATION.
PACKER ME	1-0218	AUGER TRANSITIONS.
	4-1013	AUGER TRANSITIONS BETWEEN 0 AND 1500 EV FOR ELEMENTS UP TO Z EQUALS 88.
PAHOR J	4-1014	K-SHELL FLUORESCENCE YIELDS OF ARGON, CHLORINE AND SULFUR. (AUGER TRANSITION)
PAIGNE J	5-1772	OXYGEN ADSORPTION ON CLEAN MOLYBDENUM (100) SURFACES.
PAKSWER S	4-1015	SECONDARY ELECTRON EMISSION FROM THIN FILM GOLD METAL- OXIDE CERMETS.
PALMBERG PW	1-0101	AUGER ELECTRON SPECTROSCOPY. (REVIEW, 154 REFS)
	1-0151	TECHNIQUE AND APPLICATIONS OF AUGER ELECTRON SPECTROSCOPY.
	1-0152	OPTIMIZATION OF AUGER ELECTRON SPECTROSCOPY IN LEED SYSTEMS.
	1-0153	AUGER ELECTRON SPECTROSCOPY IN LEED SYSTEMS.
	1-0154	AUGER ELECTRON SPECTROSCOPY. (REVIEW, 37 REFS)
	1-0155	RECENT ADVANCES IN COMPOSITION PROFILING BY SIMULTANEOUS SPUTTERING AND AUGER AN
	1-0156	USE OF AUGER ELECTRON SPECTROSCOPY AND INERT GAS SPUTTERING FOR OBTAINING CHEMIC
	1-0157	QUANTITATIVE AUGER ELECTRON SPECTROSCOPY USING ELEMENTAL SENSITIVITY FACTORS.
	1-0158	AUGER ELECTRON SPECTROSCOPY IN LEED SYSTEMS.
	1-0159	HANDBOOK OF AUGER ELECTRON SPECTROSCOPY. (EXPERIMENTAL SPECTRA)
	3-0654	SURFACE CLEANLINESS OF 316 L-N STAINLESS STEEL STUDIES BY SECONDARY ION MASS SPE
	3-0671	COMBINED ESCA AND AUGER SPECTROMETER.

PALMBERG PW
3-0672 HIGH SENSITIVITY AUGER ELECTRON SPECTROMETER.
5-1598 MICROSCOPIC AUGER ELECTRON ANALYSIS OF FRACTURE SURFACES.
5-1711 PHYSICAL ADSORPTION OF XENON ON PALLADIUM (100).
5-1712 CHEMICAL ANALYSIS OF PLATINUM (100) AND GOLD (100) SURFACES BY AUGER ELECTRON SP
5-1713 STRUCTURE TRANSFORMATIONS ON CLEAVED AND ANNEALED GERMANIUM (111) SURFACES.
5-1714 SECONDARY EMISSION STUDIES ON GERMANIUM AND SODIUM- COVERED GERMANIUM.
5-1715 QUANTITATIVE ANALYSIS OF SOLID SURFACES BY AUGER ELECTRON SPECTROSCOPY.
5-1716 ATOMIC ARRANGEMENT OF GOLD (100) AND RELATED METAL OVERLAYER SURFACE STRUCTURES.
5-1717 AUGER ELECTRON SPECTROSCOPY OF FACE CENTERED CUBIC METAL SURFACES. (GOLD, SILVER
5-1718 ATOMIC ARRANGEMENT OF GOLD (100), GOLD-COVERED (100), AND SILVER-COVERED COPPER
5-1719 SURFACE DISSOCIATION OF POTASSIUM CHLORIDE BY LOW ENERGY ELECTRON BOMBARDMENT.
5-1913 STRUCTURAL INFLUENCES ON ADSORBATE BINDING ENERGY. PART-1: CARBON MONOXIDE ON (1
6-2059 ROLE OF MANGANESE AND SILICON IN TEMPER EMBRITTLEMENT OF LOW ALLOY STEELS. (AUGE
6-2083 EFFECT OF SOLUTE ELEMENTS ON TEMPER EMBRITTLEMENT OF LOW ALLOY STEELS. (IMPURITY
6-2084 FRACTURE SURFACE ANALYSIS OF TEMPER EMBRITTLED STEEL BY AUGER ELECTRON SPECTROSC
6-2085 AUGER FRACTURE SURFACE ANALYSIS OF A TEMPER EMBRITTLED 3340- STAINLESS STEEL.
6-2094 AUGER SPECTROSCOPIC ANALYSIS OF THE EXTENT OF GRAIN BOUNDARY SEGREGATION.
6-2099 EFFECT OF TELLURIUM ON INTERGRANULAR COHESION OF IRON.
6-2122 AUGER ELECTRON SPECTROSCOPY OF METAL SURFACES.
6-2123 METALLURGICAL APPLICATION OF AUGER ELECTRON SPECTROSCOPY.
PALMIER JF
2-0466 IMPURITY IONIZATION IN N-TYPE GERMANIUM. (AUGER RECOMBINATION PARAMETER)
PALMS JM
4-1105 L-SHELL FLUORESCENCE YIELDS OF DOUBLE VACANCY STATES IN LEAD.
PALUMBO LJ
2-0344 RADIATIVE AUGER TRANSITIONS IN SOFT X-RAY PLASMA EMISSION.
PANDEY KC
4-0831 TIGHT BINDING CALCULATION OF A CORE VALENCE VALENCE AUGER LINE SHAPE: SILICON (1
PANISH MB
5-1247 MAGNESIUM DOPED GALLIUM ARSENIDE AND ALUMINUM(X) GALLIUM(1-X) ARSENIDE BY MOLECU
PANKOV JI
2-0467 IDENTIFICATION OF AUGER ELECTRONS IN GALLIUM ARSENIDE.
PANTANO CG
5-1264 AUGER SPECTROSCOPIC ANALYSIS OF BIOGLASS CORROSION FILMS.
5-1266 AUGER ANALYSIS OF BRASS- ENAMEL AND STAINLESS STEEL- ENAMEL INTERFACES.
5-1705 APPLICATIONS OF SURFACE CHARACTERIZATION TO GLASS SURFACES.
5-1720 GLASS SURFACE ANALYSIS BY AUGER ELECTRON SPECTROSCOPY.
5-1721 AES ANALYSIS OF SODIUM IN A CORRODED BIOGLASS USING A LOW TEMPERATURE TECHNIQUE.
5-1722 AES COMPOSITIONAL PROFILES OF MOBILE IONS IN THE SURFACE REGION OF GLASS.
PANTELEEV VV
2-0449 EFFECT OF THE ELECTRON EJECTION DEPTH ON QUANTITATIVE ANALYSIS BY AUGER SPECTROS
5-1644 INVESTIGATION OF CESIUM ADSORPTION ONTO A GALLIUM ARSENIDE (110) SURFACE. (AUGER
5-1646 GROWTH OF GALLIUM ARSENIDE ON (100) GERMANIUM STUDIED BY LOW ENERGY ELECTRON DIF
PAPGEORGOPOULOS CA
5-1241 DITUNGSTEN CARBIDE OVERLAYER ON TUNGSTEN (112).
PAPP H
5-1633 ADSORPTION OF XENON AND COBALT ON SILVER (111).
PARILIS ES
2-0247 AUGER NEUTRALIZATION OF MULTIPLY CHARGED IONS ON A METAL SURFACE.
2-0407 AUGER IONIZATION OF ATOMS BY MULTIPLY CHARGED IONS. MODEL OF TWO COULOMB CENTERS
2-0465 AUGER DISSOCIATION OF MOLECULES AS A RESULT OF INNER SHELL IONIZATION.
2-0469 THE AUGER EFFECT.
PARK RL
1-0092 COMPARISON OF SOFT X-RAY AND AUGER ELECTRON APPEARANCE POTENTIAL SPECTROSCOPY.
1-0160 INNER SHELL SPECTROSCOPY. (REVIEW)
1-0161 CHARACTERIZATION OF CHEMISORPTION BY LEED.
2-0390 AUGER EXCITATION BY INTERNAL SECONDARY ELECTRONS.
2-0391 CROSS CORRELATION TECHNIQUES IN AUGER SPECTROSCOPY.
2-0392 DIRECT COMPARISON OF CORE ELECTRON BINDING ENERGIES OF SURFACE AND BULK ATOMS OF
2-0468 ELECTRONIC STRUCTURE OF SOLID SURFACES: CORE LEVEL EXCITATION TECHNIQUES. (AUGER
4-0890 DIFFERENCE IN THE CHROMIUM L3/L2 INTENSITY RATIO MEASURED BY SOFT X-RAY AND AUGE
5-1352 IONIZATION SPECTROSCOPY OF SURFACES.
5-1723 COMPARISON OF MANY BODY EFFECTS IN CORE LEVEL SURFACE SPECTROSCOPIES.
PARROTT JE
2-0470 CARRIER TEMPERATURE EFFECTS AND ENERGY TRANSFERS IN NONRADIATIVE (AUGER) RECOMBI
PASCHOA AS
1-0219 AUGER ELECTRON DOSIMETRY.
PATERSON PJK
3-0673 A NOVEL DETECTION CIRCUIT FOR AUGER SPECTROSCOPY.
PATIL HR
5-1803 EQUILIBRIUM SEGREGATION OF CARBON TO A NICKEL (111) SURFACE: A SURFACE PHASE TRA
PATNAIK BK
5-1299 PURITY AND MORPHOLOGY OF ALUMINUM FILMS. (AUGER SPECTROSCOPY)
PATON NE
6-2086 METASTABLE GRAIN BOUNDARY SEGREGATION IN COPPER CONTAINING TELLURIUM. (AUGER SPE
PATTINSON EB
4-1016 SECONDARY ELECTRON SPECTROSCOPY OF POLYCRYSTALLINE SILVER.
4-1084 AUGER SPECTROSCOPY OF BERYLLIUM.
4-1085 OBSERVATION OF A PLASMON GAIN IN THE FINE STRUCTURE OF THE ALUMINUM AUGER SPECTR
4-1086 CHANGES IN AUGER SPECTRA OF MAGNESIUM AND IRON DUE TO OXIDATION.
4-1087 INTERPRETATION OF THE AUGER SPECTRUM OF CLEAN AND OXIDIZED BERYLLIUM.
4-1118 OXIDATION EFFECTS ON THE SPECTRUM OF THE SLOW SECONDARY ELECTRON EMISSION PEAK I
5-1407 SECONDARY ELECTRON EMISSION OF A ZIRCONIUM- ALUMINUM BULK GETTER. (AUGER SPECTRO
5-1724 CHARACTERISTIC GAIN PHENOMENA IN POLYCRYSTALLINE COPPER, SILVER, AND ALUMINUM. (
5-1897 ELECTRON SPECTROSCOPY OF METAL BLACKS.
5-1979 SECONDARY ELECTRON EMISSION FROM PLATINUM BLACK COATED SURFACES. (AUGER SPECTRA)
PATTNAIK A
5-1725 CHARACTERIZATION OF ALUMINUM ADHEREND SURFACES (BY AUGER ELECTRON SPECTROSCOPY).
PAUL G
2-0295 METASTABLE AUTOIONIZING THREE- AND FOUR ELECTRON STATES IN BERYLLIUM. (AUGER DEC
2-0296 AUGER TRANSITIONS IN LITHIUM-LIKE BERYLLIUM-LIKE IONS.
PAULSON W
5-1388 ANODIC OXIDATION OF GALLIUM ARSENIDE PHOSPHIDE. (AES)
PAUNOV M
5-1137 HETEROGENEOUS NUCLEATION AND GROWTH OF GOLD ON SUBSTRATES. (AUGER SPECTRA)
PAVLICHENKOV AV
2-0256 RESONANCE IONIZATION OF ATOMIC HELIUM BY FAST ELECTRONS. (ANGULAR DISTRIBUTION O
PEASE DE
3-0674 SPECTROSCOPY IN THE SCANNING ELECTRON MICROSCOPE NOW INCLUDES SECONDARY ION MASS
PENCHINA CM
5-1806 FINAL STATE EFFECTS IN THE PHOTOEMISSION CORE LEVEL AND AUGER SPECTRA OF NICKEL
PENDTRY JB
5-1726 ANGULAR DEPENDENCE OF ELECTRON EMISSION FROM SURFACES. (AUGER EFFECT)
PENTENERO A
1-0162 AUGER SPECTROMETRY AS A TOOL FOR SURFACE CHEMISTRY. (REVIEW)
PEPPER SV
1-0057 USE OF LOW ENERGY ELECTRON DIFFRACTION, AUGER EMISSION SPECTROSCOPY AND FIELD IO
3-0675 NEW TECHNIQUE FOR AUGER ANALYSIS OF SURFACE SPECIES SUBJECT TO ELECTRON INDUCED
5-1727 SLIDING OF POLY(VINYL CHLORIDE) ON METALS STUDIED BY AUGER ELECTRON SPECTROSCOPY
5-1728 AUGER ANALYSIS OF FILMS FORMED ON METALS IN SLIDING CONTACT WITH HALOGENATED POL
5-1729 METALLIC TRANSFER BETWEEN METALS IN SLIDING CONTACT EXAMINED BY AUGER EMISSION S
5-1730 ADHESION AND TRANSFER OF PLATINUM IRON TO METALS STUDIED BY AUGER EMISSION SPECT
5-1731 ADHESION AND TRANSFER OF POLYTETRAFLUORO ETHYLENE TO METALS STUDIED BY AUGER EMI
6-2013 ELEMENTAL ANALYSIS OF FRICTION AND WEAR SURFACE DURING SLIDING USING AUGER SPECT
PERALTA L
5-1732 ADSORPTION OF SULFUR ON THE (100) FACE OF MOLYBDENUM.
PERDEREAU J
5-1733 ADSORPTION AND SURFACE ALLOYING OF LEAD MONOLAYERS ON (111) AND (110) FACES OF G
PERDEREAU M
1-0163 QUANTITATIVE USE OF AUGER SPECTROSCOPY: STANDARDIZATION OF THE METHOD.
5-1175 LOW ENERGY ELECTRON DIFFRACTION AND AUGER SPECTROSCOPY OF THE ADSORPTION OF SULF
5-1734 DISCRIMINATION BY AUGER SPECTROSCOPY BETWEEN ADSORBED SULFUR AND SULFIDES OF COP

PERDRIX M	4-0759	SECONDARY ELECTRON EMISSION FROM SODIUM CHLORINE SINGLE CRYSTAL BOMBARDED BY RAR
PERELYGINA SA	3-0617	DRYING OF BUTADIENE- NITRILE RUBBERS IN AN AUGER MACHINE.
PERIA WT	1-0138	DECONVOLUTICN TECHNIQUES IN AUGER ELECTRON SPECTROSCOPY.
	5-1958	WORK FUNCTICN AND STRUCTURAL STUDIES OF ALKALI- COVERED SEMICONDUCTORS.
	5-1959	USE OF LEED APPARATUS FOR THE DETECTION AND IDENTIFICATION OF SURFACE CONTAMINAN
PERLEBERG CR	5-1128	ELECTRON BEAM EFFECTS IN DEPTH PROFILING MEASUREMENTS WITH AUGER ELECTRON SPECTR
PERLMAN ML	5-1844	ATTENUATION CF LOW ENERGY ELECTRONS FROM SOLIDS: RESULTS FROM X-RAY PHOTOELECTRO
PERMOGOROV SA	2-0372	EXCITON SPECTRA OF HEAVILY DOPED CADMIUM SULFIDE SINGLE CRYSTALS. (AUGER PROCESS
PERRY WB	2-0471	STATIC RELAXATION OF GERMANE AND THE ESTIMATION OF RELAXATION ENERGY DIFFERENCES
PERSON WB	5-1425	CORROSION AND STRUCTURAL CHANGES IN GLASS. (AUGER ELECTRON SPECTROSCOPY)
PESS DJ	2-0472	ELECTRON DECAY-IN-FLIGHT SPECTRA FROM AUTOIONIZING STATES OF HIGHLY STRIPPED OXY
PESSA M	1-0164	LOW ENERGY ELECTRON SPECTROSCOPY. (REVIEW, 16 REFS)
	2-0473	NEW FINE STRUCTURE IN ELECTRON EXCITED AUGER SPECTRA FROM SOLID SURFACES.
	3-0676	ELECTRON GUN FOR AUGER AND ULTRASOFT X-RAY APPARATUS.
	4-0732	HIGH RESOLUTION LMM AUGER SPECTRA OF LOW ENERGY FROM SOLID SURFACES.
	4-1017	STUDY OF INTENSITIES OF AUGER LINES EXCITED BY ELECTRON BOMBARDMENT AND ALUMINUM
PETERS E	2-0499	INVESTIGATION OF THE CHARACTERISTIC AUGER ENERGY LOSSES OF PLATINUM (111).
PETERS PN	5-1735	AUGER MEASUREMENTS ON T-027 SAMPLES EXPOSED DURING THE SKYLAB-2 MISSION.
PETERSON R	2-0472	ELECTRON DECAY-IN-FLIGHT SPECTRA FROM AUTOIONIZING STATES OF HIGHLY STRIPPED OXY
PETROFF P	5-1233	STRUCTURE AND CHEMISTRY OF SILICON SURFACES AFTER PRE- AND BACKSPUTTERING, STUDI
PETROFF Y	4-0777	ELECTRON HOLE DROPS IN GERMANIUM- SILICON ALLOYS. (AUGER RECOMBINATION)
PETROV IA	3-0717	HIGH RESOLUTION ELECTROSTATIC SPECTROMETER USED IN ELECTRON SPECTROMETRY FOR CHE
PETTER H	6-1997	DETERMINATION OF THE INFLUENCE GRAIN BOUNDARY IMPURITIES ON THE PROPERTIES OF SI
PETUKHOV VK	2-0251	KLL AUGER TRANSITION ABSOLUTE PROBABILITIES. (ATOMIC NUMBERS 30-94)
	2-0252	ABSOLUTE PROBABILITIES FOR AUGER- KLL TRANSITIONS DEPENDENT ON THE ATOMIC NUMBER
	2-0253	RELATIVE PROBABILITIES FOR AUGER KLL TRANSITIONS DEPENDENT ON THE ATOMIC NUMBER
	4-0725	KLL AUGER ELECTRON SPECTRUM OF SAMARIUM.
	4-0726	SPECTRUM OF AUGER KLL ELECTRONS OF SAMARIUM.
	4-0755	SPECTRUM OF AUGER KLL ELECTRONS OF TELLURIUM AND XENON.
	4-0756	SPECTRUM OF AUGER K,L,M,N,O ELECTRONS OF SAMARIUM.
	4-0757	KLM SPECTRUM OF SAMARIUM AUGER ELECTRONS.
	4-0758	KMM AUGER TRANSITIONS IN XENON-131.
PEYTON B	2-0292	G-R NOISE FOR AUGER BAND-TO-BAND PROCESSES.
PFEIFFER HG	3-0646	X-RAYS AND ELECTRONS IN ANALYTICAL CHEMISTRY WITH EMPHASIS ON INSTRUMENTATIION.
PHILLIPS JC	1-0165	SEMICONDUCTOR SURFACE SPECTROSCOPY. (REVIEW, 21 REFS)
	5-1604	ELECTRONIC AND LATTICE STRUCTURE OF CESIUM FILMS ADSORBED ON TUNGSTEN.
PHILLIPS JD	4-1018	L-SHELL AUGER ENERGIES AND INTENSITIES IN 3D TRANSITION METALS.
PHILLIPS LA	5-1410	METHODS OF CLEANING CONTAMINANTS FROM QUARTZ SURFACES DURING RESONATOR FABRICATI
PIASTOGI AK	5-1787	CHARACTERIZATION OF GLASS SURFACES BY ELECTRON SPECTROSCOPY.
PICARD J	2-0305	MEASUREMENT OF THE AUGER EFFECT IN THE MU-4 HELIUM(2S)(+) IONIC SYSTEM.
PICH J	2-0474	COINCIDENCE METHODS OF STANDARDIZATION FOR CESIUM-131 AND MEASUREMENT OF DECAY P
PIERCE DT	3-0606	ENERGY RESOLUTION OF THE PHOTOEMISSION ANALYZER.
PIERCE RH	5-1736	INTERACTION OF LOW ENERGY ATMOSPHERIC IONS WITH CONTROLLED SURFACES. (AUGER NEUT
PIEROTTI RA	5-1221	ADSORPTION OF KRYPTON ON THE (111) FACE OF COPPER SINGLE CRYSTALS. (AUGER ELECTR
PIGNET TP	5-1255	ADSORPTION OF HYDROGEN ON A PLATINUM (111) SURFACE.
	5-1737	SURFACE COMPOSITION OF PLATINUM HEATED IN OXYGEN. (AUGER SPECTRA)
PIMPALE A	2-0421	RECOMBINATION-INDUCED NONEQUILIBRIUM PHASE TRANSITIONS IN SEMICONDUCTORS. (AUGER
PINCHBACK TR	6-2095	STRESS CORROSION CRACKING OF ALPHA BRASS IN A TARNISHING AMMONIACAL ENVIRONMENT:
PIQUARD G	4-0820	AUGER SPECTRA OF CARBON. TANTALUM CARBIDES, ETHYLENE AND CARBON MONOXIDE ADSORPT
PIQUERAS J	5-1914	AUGER ELECTRON SPECTROSCOPY STUDIES OF ARSENIC VACANCIES IN (-1-1-1) GALLIUM ARS
	5-1915	SURFACE CARRIER CONCENTRATION AND AS EVAPORATION IN HEAT TREATED GALLIUM ARSENID
PITKETHLY RC	5-1311	PLATINUM (100) STRUCTURES: IMPURITY CAUSATION AND PROPERTIES. (AUGER EFFECT)
PITZURRA O	2-0305	MEASUREMENT OF THE AUGER EFFECT IN THE MU-4 HELIUM(2S)(+) IONIC SYSTEM.
PLACCI A	2-0305	MEASUREMENT OF THE AUGER EFFECT IN THE MU-4 HELIUM(2S)(+) IONIC SYSTEM.
POATE JM	5-1393	DIFFUSION MECHANISMS IN THE PALLADIUM- GOLD THIN FILM SYSTEM AND THE CORRELATION
POCKER DJ	1-0166	HIGH SPATIAL RESOLUTION AUGER SPECTROSCOPY AND AUGER INTEGRATION APPLICATIONS.
	1-0187	INTEGRAL AUGER INFORMATION VIA TAILORED MODULATION TECHNIQUES.
	3-0677	DOUBLY RESONANT CURRENT-TO-VOLTAGE CONVERTER PREAMPLIFIER FOR SURFACE SPECTROSCO
	3-0678	TOWARD OPTIMUM UTILIZATION OF THE CYLINDRICAL MIRROR ANALYZER IN AUGER ELECTRON
	6-2039	APPLICATIONS OF AUGER ELECTRON SPECTROSCOPY TO METALLURGICAL PROBLEMS.
POE RT	2-0310	DOUBLE PHOTOIONIZATION OF NEON. (AUGER TRANSITIONS)
POELSEMA B	2-0545	CHARGE EXCHANGE OF LOW ENERGY HELIUM IONS AND ATOMS SCATTERED FROM A COPPER SING
	2-0546	CHARGE EXCHANGE OF LOW ENERGY HELIUM(+) IONS (LESS THAN 10 KEV) SCATTERED FROM A
POLACCO E	2-0305	MEASUREMENT OF THE AUGER EFFECT IN THE MU-4 HELIUM(2S)(+) IONIC SYSTEM.
POLASCHEGG HD	3-0667	HIGH RESOLUTION ELECTRON SPECTROMETER FOR AUGER SPECTROSCOPY.
POLIZZOTTI RS	5-1216	SURFACE SEGREGATION IN ALLOYS: EQUILIBRIUM SEGREGATION OF GOLD TO THE (111) SURF
POLLAK RA	2-0412	RELATIVE EFFECT OF EXTRAATOMIC RELAXATION ON AUGER AND BINDING ENERGY SHIFTS IN
	4-0919	L2,3M4,5M4,5 AUGER SPECTRA OF METALLIC COPPER AND ZINC: THEORY AND EXPERIMENT.
	5-1738	SURFACE COMPOSITION AND CHEMISTRY OF EVAPORATED PERMALLOY FILMS OBSERVED BY X-RA
POLLARD JH	5-1739	A CORRELATION OF AUGER SPECTROSCOPY, LEED AND WORK FUNCTION MEASUREMENTS FOR EPI
POLYUDOV AN	2-0256	RESONANCE IONIZATION OF ATOMIC HELIUM BY FAST ELECTRONS. (ANGULAR DISTRIBUTION O
POMMERRENIG HD	4-0861	SECONDARY ELECTRON EMISSION FROM GALLIUM ARSENIDE.
PONS F	5-1740	AUGER ELECTRON SPECTROSCOPY STUDIES OF SURFACE COMPOSITION OF SILVER- COPPER ALL
	6-2102	AUGER SPECTROMETRY STUDY OF PASSIVE FILMS FORMED ON STAINLESS STEELS.
POPE D	5-1846	DETERMINATION OF THE SURFACE COMPOSITION OF PALLADIUM- NICKEL ALLOY FILM CATALYS
POPE M	5-1741	INTEGRATED CIRCUIT PROCESS CHARACTERIZATION USING AES.
POPPA H	1-0008	COMBINED LOW ENERGY ELECTRON DIFFRACTION, AUGER ELECTRON, AND FLASH DESORPTION S
	5-1160	ADSORPTION AND CONDENSATION OF COPPER ON TUNGSTEN SINGLE CRYSTAL SURFACES.
	5-1742	SURFACE COMPOSITION OF MICA SUBSTRATES.
	5-1743	CHANGE OF SURFACE PROPERTIES OF MICA AFTER CLEAVAGE. (OUTGASSING AND DECORATION
	5-1904	AUGER ELECTRON SPECTROSCOPY AT HIGH SPATIAL RESOLUTION AND PRIMARY BEAM CURRENTS
PORTER FT	4-0836	THE L-AUGER SPECTRA OF PLUTONIUM, AMERICIUM, CALIFORNIUM, AND FERMIUM.
	4-1019	PROTON STATES, ANOMALOUS CONVERSION, AND K-AUGER SPECTRA IN AMERICIUM-241 FROM C
PORTEUS JO	5-1744	SODIUM ADSORPTION ON ALUMINUM (111) SURFACES: ELECTRIC LOW ENERGY ELECTRON DIFFR
	5-1745	SURFACE PLASMON STUDIES ON ALUMINUM (111) AND ALUMINUM (001) SINGLE CRYSTAL SURF
	5-1746	SURFACE CHARACTERISTICS RELATED TO LASER DAMAGE OF LITHIUM NIOBATE AND POTASSIUM
POTTER HC	5-1747	CHARACTERIZATION OF COPPER- GOLD ALLOY SINGLE SURFACES.
	5-1748	LEED, AUGER SPECTROSCOPY, AND CONTACT POTENTIAL STUDIES OF COPPER- GOLD ALLOY SI
	6-2000	SEGREGATION OF CARBON TO (100) SURFACE OF NICKEL.
POWELL BD	3-0621	ULTRAHIGH VACUUM SCANNING ELECTRON MICROSCOPE WITH A FIELD EMISSION CATHODE AND

POWELL BD	3-0679	SCANNING AUGER ELECTRON MICROSCOPE FOR SURFACE STUDIES.
	4-1020	PLASMON EFFECTS IN ELECTRON ENERGY LOSS AND GAIN SPECTRA IN ALUMINUM.
	5-1279	STUDY OF SURFACE DISSOCIATION IN POTASSIUM CHLORIDE AND LITHIUM FLUORIDE BY OBSE
	5-1749	IDENTIFICATICN OF A GRAIN BOUNDARY PHASE IN HOT PRESSED SILICON NITRIDE BY AUGER
	6-2096	SEGREGATION OF BISMUTH TO GRAIN BOUNDARIES IN COPPER- BISMUTH ALLOYS.
	6-2097	STUDY OF INTERGRANULAR FRACTURE IN IRON USING AUGER SPECTROSCOPY.
POWELL CJ	2-0475	CROSS SECTICNS FOR IONIZATION OF INNER SHELL ELECTRONS BY ELECTRON IMPACT. (AUGE
	3-0680	SEMIAUTOMATED DATA RECORDING AND CONTROL SYSTEM FOR AN ELECTRON ENERGY ANALYZER.
	4-1021	STRUCTURE ON THE HIGH ENERGY SIDE OF THE KL2,3M AUGER PEAK FROM SOLID ALUMINUM:
	4-1022	CONTRASTING VALENCE BAND AUGER ELECTRON SPECTRA FOR SILVER AND ALUMINUM.
	4-1023	HIGH RESOLUTION MEASUREMENTS OF AUGER ELECTRON AND PHOTOELECTRON STRUCTURE IN TH
	4-1024	HIGH RESOLUTION MEASUREMENTS OF THE L3M2,3M4,5 AUGER TRANSITONS IN NICKEL AND CO
POWELL GL	4-0806	AUGER SPECTRA OF LITHIUM METAL AND LITHIUM OXIDE.
	4-0807	LEED- AES ELECTRON INDUCED EMISSION STUDIES OF LITHIUM HYDRIDE.
	4-1025	AUGER SPECTRA OF LITHIUM HYDRIDE.
	5-1446	LEED AND AES STUDIES OF LITHIUM HYDRIDE (100) SURFACE.
	5-1750	SODIUM SEGREGATION ONTO A LITHIUM METAL SURFACE. (AUGER SPECTRA)
PRADAL F	3-0688	EXPERIMENTAL ASSEMBLY FOR OBTAINING THE SPECTRUM OF SECONDARY ELECTRON EMISSION
PRALIAND H	5-1776	CHARACTERISTIC ENERGY LOSS AND AUGER ELECTRON SPECTROSCOPY APPLIED TO THE STUDY
PRATESI F	5-1610	QUANTITATIVE DETERMINATION OF CARBON ON SILVER BY AUGER SPECTROSCOPY.
	5-1777	CHLCRINE MONOLAYERS ON THE LOW INDEX FACES OF SILVER. (AUGER EFFECT)
	5-1778	CHEMISORPTION OF OXYGEN ON THE SILVER (110) SURFACE.
	5-1779	LEED (AUGER) STUDY OF CHLORINE CHEMISORPTION ON THE SILVER (111) SURFACE.
PRATT RH	2-0463	DEPENDENCE CF ATOMIC PHOTOEFFECT ON A BOUND ELECTRON MAGNETIC SUBSTATE. (AUGER E
	2-0524	DEPENDENCE CF ATOMIC PHOTOEFFECT ON BOUND ELECTRON MAGNETIC SUBSTATE (AUGER EFFE
PRESSER G	2-0476	RADIATIVE AUGER EFFECT IN HEAVY ATOMS.
PRINCE RH	3-0681	GAIN FATIGUE MECHANISM IN CHANNEL ELECTRON MULTIPLIERS. (SURFACE OXYGEN ON LEAD
PRITCHARD J	5-1246	LEED ANC SURFACE POTENTIAL STUDY OF CARBON MONOXIDE AND XENON ADSORBED ON COPPER
	5-1633	ADSCRPTION CF XENON AND COBALT ON SILVER (111).
	5-1751	SPECTROSCOPY OF A METAL- GAS INTERFACE.
PROJAHN H	3-0682	AUGER CYLINDRICAL MIRROR SPECTROMETER FOR THE ANALYSIS OF SURFACE CONSTITUTENTS
PRUTTON M	3-0592	HIGH ENERGY RESOLUTION AUGER ELECTRON SPECTROMETER USING CONCENTRIC HEMISPHERES.
	3-0612	IMPROVED APPARATUS FOR MEASUREMENT OF AUGER ELECTRON SPECTRA.
	3-0624	SOME PROPERTIES OF RETARDING FIELD SPECTROMETERS USING POST MONOCHROMATORS. (FOR
	3-0683	ULTRAHIGH VACUUM EVAPORATOR FOR LEED AND AUGER EMISSION STUDIES.
	3-0707	ULTRAHIGH VACUUM EVAPORATOR AND MULTIPLE CLEAVAGE DEVICE FOR LOW ENERGY ELECTRON
	4-0771	QUASIATOMIC FINE STRUCTURE IN THE AUGER SPECTRA OF SOLID SILVER AND INDIUM.
	4-0772	HIGH RESOLUTION AUGER ELECTRON SPECTRUM OF MAGNESIUM OXIDE (100).
	4-0895	LOW ENERGY AUGER AND LOSS ELECTRON SPECTRA FROM MAGNESIUM AND ITS OXIDE.
	4-0999	TEMPERATURE DEPENDENCE OF THE LOW ENERGY AUGER SPECTRUM OF NICKEL(II) OXIDE (100
	5-1347	(100) SURFACES OF AIKALI HALIDES. PART-1: AIR AND VACUUM CLEANED SURFACES.
	5-1348	GROWTH OF SILVER ON POTASSIUM CHLORIDE OBSERVED BY LEED AND AUGER EMISSION SPECT
	5-1349	THE (111) SURFACE OF N-TYPE SILICON.
	5-1580	A LEED STUDY OF MAGNESIUM OXIDE (100). PART-1: EXPERIMENT.
	5-1687	LEED AND ELECTRON SPECTROSCOPIC OBSERVATIONS ON NICKEL MONOXIDE (100).
	5-1752	LOW ENERGY ELECTRON DIFFRACTION FROM CLEAN (100) NICKEL OXIDE.
	5-1908	(100) SURFACES OF ALKALI HALIDES. PART-2: ELECTRON STIMULATED DISSOCIATION.
PULLEN BP	3-0600	HIGH RESOLUTION ELECTRON SPECTROMETER FOR PHOTOELECTRON AND AUGER ELECTRON STUDI
	4-0791	IDENTIFICATION OF HIGH ENERGY LINES IN THE KLL AUGER SPECTRUM OF NITROGEN.
	4-0990	DETERMINATION OF KLL AUGER SPECTRA OF NITROGEN, OXYGEN, CARBON DIOXIDE, NITRIC O
	4-0991	AUGER SPECTRA OF SIMPLE MOLECULES.
PURCELL EM	3-0684	FOCUSING OF CHARGED PARTICLES BY A SPHERICAL CONDENSER.
QUARLES CA	2-0477	DIFFERENTIAL CROSS SECTION FOR K-SHELL IONIZATION OF COPPER AND SILVER BY ELECTR
QUDAR J	5-1753	AUGER SCANNING ELECTRON MICROSCOPIC STUDY OF THE PASSIVE FILM FORMED ON AN IRON-
QUENTEL G	5-1754	ELLIPSOMETRIC STUDY OF ADSORPTION: ADSORPTION OF XENON ON GRAPHITE (001).
QUINN RK	4-1026	ELECTRCCHEMISTRY AT THIN SOLID FILMS. (AUGER SPECTRA)
QUINTANA G	5-1231	PHOSPHORUS CONCENTRATION PROFILES IN P-LOPED SILICON DIOXIDE MEASURED USING AUGE
	5-1232	INTERDIFFUSIONS IN THIN FILM GOLD ON PLATINUM ON GALLIUM ARSENIDE (100) STUDIED
	5-1233	STRUCTURE AND CHEMISTRY OF SILICON SURFACES AFTER PRE- AND BACKSPUTTERING, STUDI
	5-1234	SILICON DIFFUSION THROUGH THIN TUNGSTEN FILMS ON SILICON, STUDIED WITH AUGER SPE
	5-1235	DIFFUSION KINETICS OF GOLD THROUGH PLATINUM FILMS ABOUT 2000 AND 600 ANGTROMS TH
QUINTO DT	4-1027	IDENTIFICATION OF AUGER SPECTRA FROM ALUMINUM.
	4-1028	AUGER SPECTRA CF COPPER- NICKEL ALLOYS.
	5-1755	LOW ENERGY ELECTRON DIFFRACTION AND AUGER SPECTROSCOPY STUDIES OF THE (100) SURF
RADEKA V	5-1937	DIFFERENCE AUGER SPECTROSCOPY FOR STUDYING SMALL QUANTITIES OF ELEMENTS ON METAL
RADZHABOV TD	2-0245	ANISOTROPY CF SECONDARY ELECTRON EMISSION FROM PASSAGE OF LITHIUM(+) IONS THROUG
RAFF U	2-0246	PENETRATION EFFECTS IN CONVERTED MUONIC TRANSITIONS. (AUGER COEFFICIENTS)
RAKHIMOV RR	2-0478	ENERGY SPECTRUM OF ELECTRONS EMITTED BY ALKALI HALIDE CRYSTALS UNDER IMPACT OF I
RAMASUBRAMANIAN PV	6-2098	GRAIN BOUNDARY EMBRITTLEMENT IN IRON- PHOSPHORUS, IRON- PHOSPHORUS- SULFUR AND I
	6-2120	RELATION OF GRAIN BOUNDARY SEGREGATION TO REDUCED STRENGTH OF METALS AND ALLOYS.
	6-2121	STUDY OF GRAIN BOUNDARY SEGREGATION USING AUGER EMISSION SPECTRA.
RAMSDALE DJ	4-0782	RELATIVISTIC KLL AUGER TRANSITION RATES.
RAMSEY JA	1-0167	SECONDARY (AUGER) ELECTRON SPECTROSCOPY.
	1-0168	AUGER ELECTRON SPECTROSCOPY. (REVIEW)
	3-0588	CONSTRUCTION AND OPERATION OF A CYLINDRICAL MIRROR TYPE OF AUGER SPECTROMETER.
RAND MJ	5-1656	COMPOSITION PROFILES OF CHEMICALLY VAPOR DEPOSITED PLATINUM AND PLATINUM SILICID
RANDACCIO L	2-0319	AUGER ELECTRON EMISSION SPECTRA: A SIMPLE TREATMENT OF ENERGIES AND INTENSITIES
RANDALL RR	2-0323	ARGON K-AUGER CROSS SECTIONS IN COLLISION WITH 30 MEV FLUORINE IONS.
RANKE W	5-1493	OXIDATION AND ANNEALING OF GALLIUM PHOSPHIDE AND GALLIUM ARSENIDE (111) FACES ST
	5-1756	ELECTRON STIMULATED OXIDATION OF GALLIUM ARSENIDE, STUDIED BY QUANTITATIVE AUGER
RAO NG	5-1682	CORRELATION OF FRACTIONAL MONOLAYER OXYGEN DETERMINATIONS OBTAINED BY PROTON EXC
RAO PV	4-1105	L-SHELL FLUORESCENCE YIELDS OF DOUBLE VACANCY STATES IN LEAD.
RAO PVS	5-1388	ANODIC OXIDATION OF GALLIUM ARSENIDE PHOSPHIDE. (AES)
RASTOGI AK	1-0175	AUGER ELECTRON SPECTROSCOPY: NEW TOOL IN THE CHARACTERIZATION OF GLASS FIBER SU
RASULOV DK	2-0234	DETECTION OF A NEW TYPE OF AUGER TRANSITIONS IN ATOMS WITH TWO INTERNAL VACANCIE
	2-0235	OBSERVATION OF THREE ELECTRON AUGER TRANSITIONS IN ATOMS WITH TWO INNER VACANCIE
RAWLINGS KJ	5-1757	AUGER SPECTROSCOPY OF OXYGEN AND OTHER "CONTAMINANTS" ON SILVER.
RAY JA	4-1029	SECONDARY ELECTRON EMISSION OF METALS BOMBARDED WITH 120-EV TO 5-KEV PROTONS.
REDHEAD PA	3-0618	SENSITIVITY VARIATIONS IN BAYARD- ALPERT GAUGES CAUSED BY AUGER EMISSION AT COLL
	5-1758	ROLE OF SURFACE CONTAMINANTS IN ELECTRON DEVICE TECHNOLOGY.

REDKIN VS	2-0538	IONIZATION LOSSES IN (AUGER) SPECTRA OF ELECTRON INELASTIC SCATTERING FROM RARE
	3-0653	SPECTROMETER OF AUGER ELECTRONS.
	4-1030	AUGER ELECTRON SPECTRUM OF OSMIUM AT ENERGIES UP TO 300 EV.
	4-1100	AUGER ELECTRON SPECTRA OF SOME RARE EARTH METALS. (LANTHANUM, PRASEODYMIUM, NEOD
	4-1124	PLASMA SATELLITES IN THE AUGER ELECTRON SPECTRUM OF MOLYBDENUM.
	5-1759	DECREASE OF CARBON CONCENTRATION ON THE SURFACE OF MOLYBDENUM CARBIDE DURING HEA
	5-1985	CHEMICAL COMPOSITION OF THE SURFACE OF MOLYBDENUM AND STEEL BASED ON AUGER ELECT
REDOZUBOV VD	6-2004	AUGER SPECTRA OF ACTIVATED SILVER- MAGNESIUM AND SILVER- LITHIUM ALLOYS.
REED-HILL RE	5-1425	CORROSION AND STRUCTURAL CHANGES IN GLASS. (AUGER ELECTRON SPECTROSCOPY)
REHME H	3-0685	SPECTROSCOPIC METHODS OF MICROANALYSIS USING THE SCANNING ELECTRON MICROSCOPE. (
REID RJ	5-1760	AUGER ELECTRON SPECTROSCOPY OF CLEAN POLYCRYSTALLINE GOLD FILMS.
REIMER L	4-1031	ORIENTATION ANISOTROPY OF BACKSCATTERING COEFFICIENT AND SECONDARY ELECTRON EMIS
RELLICK JR	6-2099	EFFECT OF TELLURIUM ON INTERGRANULAR COHESION OF IRON.
REVIE RW	5-1761	PASSIVE FILM ON IRON. APPLICATION OF AUGER ELECTRON SPECTROSCOPY.
	5-1762	DEGRADED PASSIVE FILM ON IRON. APPLICATION OF AUGER ELECTRON SPECTROSCOPY.
REZNITSKII AN	2-0372	EXCITON SPECTRA OF HEAVILY DOPED CADMIUM SULFIDE SINGLE CRYSTALS. (AUGER PROCESS
RHEAD GE	1-0002	QUANTITATIVE AUGER ELECTRON SPECTROSCOPY USING COADSORPTION.
	3-0587	CALIBRATION IN AUGER ELECTRON SPECTROSCOPY BY MEANS OF COADSORPTION.
	5-1178	SPONTANEOUS ALLOYING CF A GOLD SUBSTRATE WITH LEAD MONOLAYERS. (INTERFACE AUGER
	5-1733	ADSORPTION AND SURFACE ALLOYING OF LEAD MONOLAYERS ON (111) AND (110) FACES OF G
RHODIN TN	1-0094	SECONDARY ELECTRON SPECTROSCOPY. SURFACE SENSITIVE TOOL. (REVIEW, 22 REFS)
	1-0095	SECONDARY ELECTRON SPECTROSCOPY. SURFACE SENSITIVE TOOL. (AES)
	5-1291	CHEMISORPTION ON (001), (110), AND (111) NICKEL SURFACES. CORRELATED STUDY USING
	5-1292	ELASTIC LEED INTENSITY ENERGY STUDIES OF CLEAN (001), (110) AND ELASTIC (111) NI
	5-1479	LEED INVESTIGATIONS OF XENON SINGLE CRYSTAL FILMS AND THEIR USE IN STUDYING THE
	5-1716	ATOMIC ARRANGEMENT OF GOLD (100) AND RELATED METAL OVERLAYER SURFACE STRUCTURES.
	5-1717	AUGER ELECTRON SPECTROSCOPY OF FACE CENTERED CUBIC METAL SURFACES. (GOLD, SILVER
	5-1718	ATOMIC ARRANGEMENT OF GOLD (100), GOLD-COVERED (100), AND SILVER-COVERED COPPER
	5-1719	SURFACE DISSOCIATION OF POTASSIUM CHLORIDE BY LOW ENERGY ELECTRON BOMBARDMENT.
	5-1763	BINDING, VALENCY AND CHARGE TRANSFER IN CHEMICAL REACTIONS ON METAL SURFACES.
	5-1764	CHEMISORPTION ON (001), (110) AND (111) NICKEL SURFACES. (AUGER SPECTRA)
RIACH GE	1-0118	THIN FILM ANALYSIS FOR PROCESS EVALUATION. (AES)
	1-0119	APPLICATIONS OF SCANNING AUGER MICROANALYSIS.
	1-0159	HANDBOOK OF AUGER ELECTRON SPECTROSCOPY. (EXPERIMENTAL SPECTRA)
	5-1289	SILICON DEVICE APPLICATIONS USING A COMBINED ESCA- AUGER ANALYSIS SYSTEM.
	5-1765	ELECTRONS OR IONS. (IN SURFACE ANALYSIS)
RIBBING CG	4-0936	HIGH RESOLUTION LMM AUGER SPECTRA OF GERMANIUM.
	4-0937	FINE STRUCTURE DETAILS IN THE LMM AUGER SPECTRA OF GERMANIUM.
RICE DW	5-1766	CORROSION, SURFACE, AND MAGNETIC PROPERTIES OF NICKEL- IRON- CHROMIUM THIN FILMS
RICE RW	5-1506	ANALYSIS OF GRAIN BOUNDARY IMPURITIES AND FLUORIDE ADDITIVES IN HOT PRESSED OXID
RICHARD P	1-0169	ION- ATOM COLLISIONS. (AUGER EMISSION, REVIEW)
	2-0479	X-RAY PRODUCTION BY HEAVY IONS. (OXYGEN BOMBARDMENT, AUGER SPECTRUM)
	2-0480	ION- ATOM COLLISIONS AT HIGH ENERGIES. (AUGER EFFECT)
	2-0481	RADIATIVE AUGER EFFECT IN ION- ATOM COLLISIONS.
	2-0520	CHARGE STATE DEPENDENCE CF THE NEON K FLUORESCENCE YIELD DEDUCED FROM HIGH RESOL
	2-0571	NEON K-AUGER CROSS SECTIONS FROM 1.5 MEV PER AMU AND ATOMIC HYDROGEN, OXYGEN, AN
	2-0572	NEON K-SHELL VACANCY PRODUCTION BY INTERMEDIATE ENERGY FLUORINE IONS. (AUGER ELE
	2-0573	AUGER ELECTRONS FRCM DOUBLE K-SHELL VACANCY STATES IN NEON.
	2-0574	K-SHELL AUGER ELECTRON PRODUCTION CROSS SECTIONS FROM ION BOMBARDMENT. (NEON, FL
	4-1117	K-SHELL AUGER ELECTRON HYPERSATELLITES OF NEON.
RICKARD JM	5-1754	ELLIPSOMETRIC STUDY OF ADSORPTION: ADSORPTION OF XENON ON GRAPHITE (001).
RICKMAN J	5-1517	STUDY OF THE PREPARATION OF ATOMICALLY CLEAN TUNGSTEN SURFACES BY AUGER ELECTRON
	5-1518	CHEMISORPTION OF NITROGEN ON TUNGSTEN STUDIED BY AUGER ELECTRON SPECTROSCOPY.
RIDDER D	2-0522	MOLECULAR EFFECTS ON K-VAANCY SHARING IN ASYMMETRIC ION- ATOM COLLISIONS. (AUGER
	2-0523	GAS TARGET EXPERIMENTS OF K-SHELL VACANCY SHARING IN ASYMMETRIC ION- ATOM COLLIS
	2-0582	K-SHELL VACANCY SHARING IN 300 TO 500 KEV NEON (+)- SODIUM COLLISIONS. (QUASIMOLE
	4-1032	WIDTHS AND INTENSITIES OF HIGHLY RESOLVED LINES IN ARGON L2,3 AUGER SPECTRA.
RIDGWAY JWT	5-1767	SILICON (111)-7 STRUCTURE OBTAINED BY CLEAVAGE AT ROOM TEMPERATURE. (AUGER SURFA
	5-1768	AUGER SPECTRA AND LEED PATTERNS FROM NICKEL DEPOSITS ON CLEAVED SILICON.
	5-1769	AUGER SPECTRA AND LEED PATTERNS FROM VACUUM CLEANED SILICON CRYSTALS WITH CALIBR
RIGDEN JD	3-0633	HIGH SENSITIVITY ELECTRON SPECTROMETER.
	3-0634	RESOLUTION AND SENSITIVITY OF THE SPHERICAL GRID RETARDING POTENTIAL ANALYZER. A
	3-0686	RETARDING FIELD ELECTRON SPECTROMETER.
RIMSTIDT JD	4-0851	X-RAY PHOTOELECTRON SPECTRUM OF DIAMOND. (AUGER TRANSITIONS)
RINGERS DA	3-0718	WIDTHS OF ATOMIC M-SHELL VACANCY STATES AND QUASIATOMIC ASPECTS OF RADIATIONLESS
RISLEY JS	2-0299	EFFECT OF MULTIPLE IONIZATION ON THE FLUORESCENCE YIELD OF NEON. (K-AUGER AND K
	2-0300	PROJECTILE CHARGE STATE DEPENDENCE OF NEON K-SHELL IONIZATION AND FLUORESCENCE Y
	2-0482	MEASUREMENT OF THE PRODUCTION OF LL VACANCY STATES BY THE L X-RAY- L X-RAY COINC
	2-0518	K-SHELL IONIZATION AND FLUORESCENCE YIELD IN 50 MEV CHLORINE(+)- NEON COLLISIONS
	3-0687	DESIGN PARAMETERS FOR THE CYLINDRICAL MIRROR ENERGY ANALYZER.
RITCHIE K	5-1827	AUGER ELECTRON SPECTROSCOPY DETERMINATION OF THE OXYGEN- SILICON RATIO IN SPIN O
RIVAIS G	2-0549	ORIGIN OF ION INDUCED AUGER ELECTRON EMISSION FROM METALS.
RIVIERE JC	1-0170	CHARACTERISTIC AUGER ELECTRON EMISSION AS A TOOL FOR THE ANALYSIS OF SURFACE COM
	1-0171	AUGER ELECTRON SPECTROSCOPY.
	1-0172	AUGER SPECTROSCOPY. (REVIEW)
	1-0173	IONIZATION (CHARACTERISTIC LOSS) SPECTROSCOPY AND APPEARANCE POTENTIAL SPECTROSC
	2-0286	ANGULAR DEPENDENCIES IN ELECTRON EXCITED AUGER EMISSION. (COMMENTS)
	2-0287	CHARACTERISTIC IONIZATION LOSSES OBSERVED IN AUGER EMISSION SPECTROSCOPY.
	2-0288	ESTIMATES OF EFFICIENCIES OF PRODUCTION AND DETECTION OF ELECTRON EXCITED AUGER
	3-0594	DESIGN FOR A CYLINDRICAL MIRROR ANALYZER FOR USE IN AUGER SPECTROSCOPY.
	3-0601	BEAM EFFECTS IN AUGER ELECTRON SPECTROSCOPY REVEALED BY X-RAY PHOTOELECTRON SPEC
	4-0783	AUGER SPECTROSCOPY OF TITANIUM.
	4-0784	AUGER SPECTRCSCOPY OF SILICON.
	4-0813	COMBINATION OF AUGER SPECTROSCOPY AND CHARACTERISTIC LOSS SPECTROSCOPY FOR THE E
	5-1179	SURFACE ABSORPTION BY POLYPEPTIDE FILMS EXAMINED BY AUGER EMISSION SPECTROSCOPY.
	5-1180	SEGREGATION OF GOLD TO SILICON (111) SURFACE OBSERVED BY AUGER EMISSION SPECTROS
	5-1181	SURFACE SEGREGATION IN BORON DOPED IRON OBSERVED BY AUGER EMISSION SPECTROSCOPY.
	5-1269	ELECTRON BEAM ASSISTED ADSORPTION ON SILICON (111) SURFACE.
	5-1270	AUGER SPECTROSCOPY OF CARBON ON NICKEL.
	5-1271	ORIGIN OF FINE STRUCTURE IN THE AUGER SPECTRUM OF SULFUR ON A NICKEL SURFACE.

RIVIERE JC	5-1272	SURFACE ANALYSES BY COMBINED X-RAY PHOTOELECTRON SPECTROSCOPY AND AUGER ELECTRON
	5-1441	CORRELATION OF THE WETTING BEHAVIOR OF SODIUM ON NIMONIC-PE16 ALLOY WITH THE SUR
	5-1770	AUGER SPECTROSCOPY OF OSMIUM, AND THE SEGREGATION OF PLATINUM TO THE OSMIUM SURF
	5-1802	QUANTITATIVE AUGER SPECTROSCOPIC ANALYSIS OF SEGREGATION OF PHOSPHORUS IN IRON.
	6-2100	AUGER SPECTROSCOPY AND ITS APPLICATIONS TO SOME TECHNICAL PROBLEMS. (EMBRITTLEME
RIWAN R	1-0174	AUGER ELECTRON SPECTROSCOPY. IONIZATION LOSS SPECTROSCOPY. APPEARANCE POTENTIAL
	5-1579	CARBON MONOXIDE ADSORPTION KINETICS ON MOLYBDENUM (100). (AUGER SPECTROSCOPY)
	5-1771	DIFFUSION OF SULFUR TOWARD THE (110) FACE OF NICKEL.
	5-1772	OXYGEN ADSORPTION ON CLEAN MOLYBDENUM (100) SURFACES.
ROBERTS ED	2-0560	L2,3MM AUGER PROCESSES IN SELENIUM.
	4-1033	TRANSITION PROBABILITIES FOR THE L2,3MM AUGER SPECTRUM OF SELENIUM.
	4-1034	AUGER VACANCY SATELLITE STRUCTURE IN THE L3,M4,5M4,5 AUGER SPECTRA OF COPPER.
	4-1035	PHOTOELECTRON AND L2,3MM AUGER ELECTRON ENERGIES FOR ARSENIC.
ROBERTS K	2-0496	HIGH RESOLUTION ARGON K-AUGER SPECTRUM PRODUCED IN 4 MEV PROTON ON ARGON COLLISI
	2-0497	RELATIVE SATELLITE LINE INTENSITIES IN ARGON K AUGER SPECTRA PRODUCED B
ROBERTS KS	4-0904	HIGH RESOLUTION BEAM GAS AUGER ELECTRON SPECTRA FOR OXYGEN IONS EXCITED BY COLLI
ROBERTS MW	1-0102	AUGER ELECTRON SPECTROSCOPY AND ITS APPLICATIONS IN SURFACE CHEMISTRY. (REVIEW)
	1-0124	LEED AND AUGER ELECTRON SPECTROSCOPY. (REVIEW)
	5-1483	DEVELOPMENT OF STEPPED SURFACE REGIONS ON POLYCRYSTALLINE GOLD. LOW ENERGY ELECT
	5-1515	ADSORPTION OF CARBON MONOXIDE ON COPPER (001). LEED AND AUGER EMISSION STUDIES.
	5-1516	INTERACTION OF HYDROGEN SULFIDE WITH COPPER (001).
	5-1517	STUDY OF THE PREPARATION OF ATOMICALLY CLEAN TUNGSTEN SURFACES BY AUGER ELECTRON
	5-1518	CHEMISORPTION OF NITROGEN ON TUNGSTEN STUDIED BY AUGER ELECTRON SPECTROSCOPY.
	5-1519	AUGER ELECTRON SPECTROSCOPY STUDIES OF CLEAN POLYCRYSTALLINE GOLD AND OF THE ADS
ROBERTSON WD	4-1027	IDENTIFICATION OF AUGER SPECTRA FROM ALUMINUM.
	4-1028	AUGER SPECTRA OF COPPER- NICKEL ALLOYS.
	5-1335	INTERACTION BETWEEN CHLORINE AND THE (100) SURFACE OF GOLD.
	5-1343	PHASE TRANSFORMATIONS OF SILICON (111) SURFACE.
	5-1344	CHLORINE REACTIONS ON THE SILICON (111) SURFACE.
	5-1855	ORDER- DISORDER TRANSFORMATION AT A (100) SURFACE OF COPPER(3) GOLD, THEORY, EXP
	5-1989	A CARBON STRUCTURE ON THE RHENIUM (0001) SURFACE.
ROBINSON GY	5-1336	STUDY OF PALLADIUM SILICIDE FILMS ON SILICON USING AUGER ELECTRON SPECTROSCOPY.
	5-1773	PALLADIUM SILICIDE FORMATION OBSERVED BY AUGER ELECTRON SPECTROSCOPY.
	5-1774	AUGER ELECTRON SPECTROSCOPY STUDIES OF GALLIUM ARSENIDE AND SILICON METAL- SEMIC
	5-1775	AES AND SPUTTER ETCHING OF NICKEL- GOLD- GERMANIUM ON N-GALLIUM ARSENIDE.
	6-2101	METALLURGICAL AND ELECTRICAL PROPERTIES OF ALLOYED NICKEL- GOLD- GERMANIUM FILMS
ROJO JM	2-0491	INTERATOMIC AUGER PROCESSES AND THE DENSITY OF STATES.
	4-0766	MVV AUGER SPECTRA OF COPPER AND NICKEL.
	4-1042	INTERATOMIC TRANSITIONS AND RELAXATION EFFECTS IN AUGER SPECTRA OF SEVERAL GAS A
	4-1046	HIGH ENERGY SATELLITES IN AUGER PEAKS OF MAGNESIUM AND SILICON.
	4-1047	ENERGY SPECTRA OF SECONDARY ELECTRONS EMITTED BY MAGNESIUM AND SILICON WHEN BOMB
RONDOT B	6-2102	AUGER SPECTROMETRY STUDY OF PASSIVE FILMS FORMED ON STAINLESS STEELS.
ROSENTHAL W	2-0357	INFLUENCE OF AN ELECTRIC FIELD ON THE AUGER RECOMBINATION IN SEMICONDUCTORS.
	2-0483	BAND-TO-BAND AUGER PROCESSES AT HIGH CARRIER CONCENTRATION.
	2-0484	PHONON ASSISTED AUGER PROCESSES IN GROUP III-V COMPOUNDS.
	4-0844	INFLUENCE OF AN ELECTRIC FIELD ON THE AUGER RECOMBINATION IN SEMICONDUCTORS.
ROSNER HR	2-0485	RELATIVISTIC CALCULATIONS OF ATOMIC X-RAY AND AUGER TRANSITION RATES.
ROTH JA	5-1540	GERMANIUM SURFACE CLEANING: AUGER ANALYSIS.
ROUSSEAU J	5-1283	AUGER SPECTROSCOPY OF CARBON ON NICKEL. COMMENTS.
	5-1284	STUDY OF THE ADSORPTION OF BENZENE ON THE (100) FACE OF NICKEL. INFLUENCE OF SUL
	5-1776	CHARACTERISTIC ENERGY LOSS AND AUGER ELECTRON SPECTROSCOPY APPLIED TO THE STUDY
ROVIDA G	5-1610	QUANTITATIVE DETERMINATION OF CARBON ON SILVER BY AUGER SPECTROSCOPY.
	5-1777	CHLORINE MONOLAYERS ON THE LOW INDEX FACES OF SILVER. (AUGER EFFECT)
	5-1778	CHEMISORPTION OF OXYGEN ON THE SILVER (110) SURFACE.
	5-1779	LEED (AUGER) STUDY OF CHLORINE CHEMISORPTION ON THE SILVER (111) SURFACE.
ROWE JE	4-1036	HIGH ENERGY FINE STRUCTURE IN THE AUGER SPECTRA OF SILICON AND SILICON CARBIDE.
	4-1037	ABSENCE OF PLASMON GAIN SATELLITES IN THE AUGER SPECTRUM OF SILICON.
	4-1038	COMPARISON OF SILICON L3,2 ELECTRON LOSS SPECTRA AND OPTICAL ABSORPTION. (AUGER
	4-1039	SURFACE STATE TRANSITIONS OF SILICON IN ELECTRON ENERGY LOSS SPECTRA. (AUGER EFF
	5-1475	ELECTRON ORBITAL ENERGIES OF OXYGEN ADSORBED ON SILICON SURFACES AND OF SILICON
	5-1476	ELECTRONIC STATES OF OXYGEN ADSORBED ON CLEAN SILICON (111) AND (100) SURFACES.
	5-1780	AMORPHOUS TO CRYSTALLINE TRANSITION OF SILICON (111) SURFACES. (AUGER ELECTRON S
	5-1781	PHOTOEMISSION MEASUREMENTS OF BULK, SURFACE, AND HYDROGEN INDUCED STATES ON CLEA
RUBENSTEIN RA	2-0486	THEORETICAL STUDY OF THE AUGER EFFECT IN THE LIGHT TO MEDIUM RANGE OF ATOMIC NUM
RUDD ME	2-0487	MECHANISMS OF INNER SHELL EXCITATION AND DEEXCITATION IN MULTIPLY IONIZED ATOMS.
	2-0488	MECHANISMS OF ELECTRON PRODUCTION IN ION- ATOM COLLISIONS. (AUGER EFFECT)
	2-0489	PRODUCTION AND DECAY OF DOUBLE L VACANCIES IN ARGON AND PHOSPHORUS.
	2-0490	MECHANISMS OF ELECTRON PRODUCTION IN ION- ATOM COLLISIONS. (AUGER EFFECT, REVIEW
	2-0552	AUGER ELECTRONS FROM ARGON WITH ENERGIES 150-210 EV PRODUCED BY HYDROGEN(+) IMPA
	4-1040	CROSS SECTIONS FOR L2,3 AUGER EMISSION IN 22 TO 300 KEV PROTON- ARGON COLLISIONS
RUPNIK T	4-1014	K-SHELL FLUORESCENCE YIELDS OF ARGON, CHLORINE AND SULFUR. (AUGER TRANSITION)
RUPPEL W	2-0432	PHOTOCONDUCTIVITY OF HIGHLY EXCITED CADMIUM SELENIDE AND CADMIUM SULFIDE(0.48) S
RUSCH TW	1-0139	INELASTIC EFFECTS IN AUGER ELECTRON SPECTROSCOPY.
	5-1782	INVESTIGATION OF SILVER CESIUM OXIDE PHOTOEMISSIVE SURFACES USING AUGER ELECTRON
	5-1783	KIKUCHI CORRELATIONS IN AUGER ELECTRON SPECTROSCOPY.
	5-1784	HIGH ANGULAR RESOLUTION SECONDARY ELECTRON SPECTROSCOPY. KIKUCHI CORRELATIONS FO
RUSSEK A	2-0530	EXTENDED ION- ELECTRON COINCIDENCE MEASUREMENTS OF THE VIOLENT ARGON(+)- ARGON C
RUSSELL GJ	5-1306	ADSORPTION OF OXYGEN ON CLEAN CLEAVED (110) GALLIUM ARSENIDE SURFACES.
RUTHARDT R	5-1785	METHODS FOR SURFACE ANALYSIS OF PRECIOUS METALS CONTACT MATERIALS.
RUZYLLO J	5-1786	LOW TEMPERATURE OXIDATION OF SILICON STUDIED BY PHOTOSENSITIVE ESR AND AUGER ELE
RYE RR	5-1594	EQUILIBRATION OF HYDROGEN AND DEUTERIUM ON SINGLE CRYSTAL SURFACES OF PLATINUM.
RYND JP	1-0175	AUGER ELECTRON SPECTROSCOPY: NEW TOOL IN THE CHARACTERIZATION OF GLASS FIBER SU
	5-1787	CHARACTERIZATION OF GLASS SURFACES BY ELECTRON SPECTROSCOPY.
SAAR A	4-1126	DETECTION OF EXCITONS NEAR THE L2,3 EDGE OF THE CHLORIDE ION IN SODIUM CHLORIDE
SACHENKO VP	2-0290	NATURE OF SOME X-RAY SATELLITES (RADIATIVE AUGER TRANSITIONS).
SACHTLER WMH	5-1788	SURFACE COMPOSITION OF (PALLADIUM- SILVER) ALLOYS IN EQUILIBRIUM.
SAITO S	5-1552	SPUTTER ETCHING AND AUGER ELECTRON SPECTROSCOPY OF METAL SURFACES.
	5-1554	SPUTTER ETCHING AND AUGER ELECTRON SPECTROSCOPY OF METAL SURFACES.
SAKISAKA M	4-1041	SHIFT ENERGIES OF CHARACTERISTIC X-RAYS AND AUGER ELECTRONS FOR IONIZED ATOMS.
SAKURAI T	5-1789	CHEMISORPTION OF ATOMIC HYDROGEN ON THE SILICON (111) 7*7 SURFACE. (AUGER EFFECT

284

SALMERON M   2-0491 INTERATOMIC AUGER PROCESSES AND THE DENSITY OF STATES.
                    3-0688 EXPERIMENTAL ASSEMBLY FOR OBTAINING THE SPECTRUM OF SECONDARY ELECTRON EMISSION
                    4-0764 SECONDARY ELECTRON EMISSION ROCKING CURVE FROM COPPER.
                    4-0765 A STUDY OF THE INTENSITY OF THE L2,3VV AUGER TRANSITION IN MAGNESIUM AND SILICON
                    4-0766 MVV AUGER SPECTRA OF COPPER AND NICKEL.
                    4-0779 L1,L2,3V AUGER TRANSITION IN SILICON.
                    4-1042 INTERATOMIC TRANSITIONS AND RELAXATION EFFECTS IN AUGER SPECTRA OF SEVERAL GAS A
                    4-1043 HIGH ENERGY SATELLITES OF THE M2,3VV AND M1VV AUGER PEAKS OF COPPER.
                    4-1044 EXPERIMENTAL OBSERVATION OF CHEMICAL SHIFTS IN AUGER SPECTRUM FROM SURFACE LAYER
                    4-1045 INTERATOMIC AUGER PROCESSES IN SOME ADSORBATES ON TRANSITION METALS.
                    4-1046 HIGH ENERGY SATELLITES IN AUGER PEAKS OF MAGNESIUM AND SILICON.
                    4-1047 ENERGY SPECTRA OF SECONDARY ELECTRONS EMITTED BY MAGNESIUM AND SILICON WHEN BOMB
SALOP A   2-0492 MULTIIONIZATION OF NEON, ARGON, XENON AND THEIR IONS BY HIGH ENERGY ELECTRON IMP
                    2-0493 MULTIIONIZATION OF KRYPTON AND ITS IONS BY HIGH ENERGY ELECTRON IMPACT. (AUGER E
SAL'KOV EA   2-0369 RECOMBINATION IN DONOR ACCEPTOR COMPLEXES IN CADMIUM SULFIDE SINGLE CRYSTALS. (A
SALTSBURG HM   5-1947 UPS, ESCA, AND AUGER SPECTROSCOPY STUDY OF VACUUM DEPOSITED SILVER HALIDE FILMS.
SAMEH-SAID M   5-1790 PROPERTIES OF THE LEAD SULFIDE- LEAD OXIDE CRYSTALLINE FILMS. PART-1: PREPARATIO
SANDERS DM   5-1425 CORROSION AND STRUCTURAL CHANGES IN GLASS. (AUGER ELECTRON SPECTROSCOPY)
SANDNER W   2-0335 INNER SHELL ALIGNMENT OF ATOMS IN ELECTRON IMPACT IONIZATION. (AUGER ELECTRONS)
SANINA VA   4-0946 HIGH FREQUENCY EXCITON BREAKDOWN AND FREE CARRIER AND EXCITON KINETICS IN GERMAN
SARD E   2-0292 G-R NOISE FOR AUGER BAND-TO-BAND PROCESSES.
SARIS FW   2-0387 MEASUREMENTS OF THE L-SHELL FLUORESCENCE YIELDS OF ARGON AND CHLORINE BY ELECTRO
SAR-EL HZ   3-0689 CYLINDRICAL CAPACITOR AS AN ANALYZER. PART-1: NONRELATIVISTIC PART.
SATAKE T   5-1553 SIMULTANEOUS OBSERVATIONS OF PARTIALLY OXIDIZED SURFACES BY AUGER ELECTRON SPECT
                    6-2090 COMPOSITION OF BINARY ALLOYS BY SIMULTANEOUS SIMS AND AES MEASUREMENTS.
SATO O   6-2052 NEW MICROAUTORADIOGRAPHY OF METAL STRUCTURE USING AUGER ELECTRONS.
                    6-2066 NEW MICRORADIOGRAPHY USING AUGER ELECTRONS IN THE STUDY OF THE STRUCTURE AND PIT
SAVCHENKO VI   4-0816 CHEMICAL SHIFT IN AUGER SPECTRA OF TUNGSTEN AND MOLYBDENUM DURING THE ADSORPTION
                    5-1310 AUGER SPECTRA FROM THE (100) SURFACE OF SILICON DURING OXIDATION AND NITRIDATION
                    5-1361 COMPOSITION OF THE SURFACE OF COPPER- MAGNESIUM OXIDE CATALYSTS STUDIED BY AUGER
                    5-1791 AUGER SPECTROSCOPIC STUDY OF PLATINUM, NICKEL, AND IRON SURFACES.
                    5-1792 CHEMICAL SHIFT IN THE AUGER SPECTRUM OF IRON DURING OXYGEN ADSORPTION.
SAVIKHINA TI   4-1048 EXCITATION SPECTRUM OF OXYGEN DOMINATED COMPOUNDS IN THE 3-21 EV REGION OF THE S
SAWADA M   2-0494 RADIATIVE AUGER EFFECT WITH AND WITHOUT ELECTRON EMISSION.
SCHARMANN A   4-0921 EXOELECTRON EMISSION FROM ZINC OXIDE. (OXYGEN ADSORPTION EFFECT, AUGER ELECTRON
                    5-1325 THERMALLY STIMULATED EXOELECTRON EMISSION OF BERYLLIUM OXIDE LAYERS. AUGER ELECT
SCHEIBNER EJ   1-0176 INTERFACE PHENOMENA IN ENGINEERING MATERIALS.
                    2-0242 TRUE SECONDARY ELECTRON ENERGY DISTRIBUTIONS.
                    2-0495 INELASTIC SCATTERING OF LOW ENERGY ELECTRONS FROM SURFACES.
                    4-0741 AUGER SPECTROSCOPY OF GRAPHITE SINGLE CRYSTALS WITH LOW ENERGY ELECTRONS.
                    4-1094 ENERGY SPECTRA OF INELASTICALLY SCATTERED ELECTRONS AND LEED STUDIES OF TUNGSTEN
                    5-1819 ORDER- DISORDER PHENOMENA AT THE SURFACE OF ALPHA TITANIUM- OXYGEN SOLID SOLUTIO
SCHERZER BMU   5-1167 ANALYSIS OF SURFACE LAYERS BY LIGHT ION BACKSCATTERING AND SPUTTERING COMBINED W
SCHIFF KL   5-1165 DETECTION OF THIN CONTAMINATED LAYERS ON CONTACT SURFACES. (BY AUGER SPECTROSCOP
SCHILLALIES H   3-0667 HIGH RESOLUTION ELECTRON SPECTROMETER FOR AUGER ELECTRON SPECTROSCOPY.
SCHMEISSER H   5-1137 HETEROGENEOUS NUCLEATION AND GROWTH OF GOLD ON SUBSTRATES. (AUGER SPECTRA)
SCHMIDT LD   5-1549 ADSORPTION AND SOLUTION OF HYDROGEN, DEUTERIUM, NITROGEN, OXYGEN AND CARBON MONO
                    5-1737 SURFACE COMPOSITION OF PLATINUM HEATED IN OXYGEN. (AUGER SPECTRA)
                    5-1793 PHYSICAL AND CHEMICAL CHARACTERIZATION OF PLATINUM- RHODIUM GAUZE CATALYSTS. (AU
                    5-1944 OXIDATION OF CARBON ON TUNGSTEN (100) AND MOLYBDENUM (100). (AUGER ELECTRON SPEC
SCHMIDT V   2-0352 ANGULAR DISTRIBUTION OF AUGER ELECTRONS FOLLOWING PHOTOIONIZATION.
                    4-0789 THE K-AUGER SPECTRUM OF THE FREE MAGNESIUM ATOM.
                    4-1049 TRANSITION PROBABILITIES OF KL- LLL AUGER SATELLITES IN NEON.
SCHMITZ W   4-0880 K-AUGER TRANSITIONS OF THE FREE SODIUM ATOM.
                    4-0981 CORRELATION EFFECTS IN THE AUGER SPECTRA OF NOBLE GASES.
SCHNEIDER D   2-0300 PROJECTILE CHARGE STATE DEPENDENCE OF NEON K-SHELL IONIZATION AND FLUORESCENCE Y
                    2-0345 K-VACANCY SHARING STUDIED FROM AUGER ELECTRON YIELDS.
                    2-0496 HIGH RESOLUTION ARGON K-AUGER SPECTRUM PRODUCED IN 4 MEV PROTON ON ARGON COLLISI
                    2-0497 RELATIVE SATELLITE LINE INTENSITIES IN ARGON K AUGER ELECTRON SPECTRA PRODUCED B
                    2-0518 K-SHELL IONIZATION AND FLUORESCENCE YIELD IN 50 MEV CHLORINE(+)- NEON COLLISIONS
                    2-0519 K-SHELL IONIZATION OF MOLECULAR HYDROGEN AND METHANE BY 50- 600 KEV HYDROGEN(
                    2-0520 CHARGE STATE DEPENDENCE OF THE NEON K FLUORESCENCE YIELD DEDUCED FROM HIGH RESOL
                    2-0521 ARGON L-SHELL IONIZATION BY 50 TO 600 KEV PROTON, DEUTERON, MOLECULAR HYDROGEN,
                    2-0522 MOLECULAR EFFECTS ON K-VAANCY SHARING IN ASYMMETRIC ION- ATOM COLLISIONS. (AUGER
                    4-0904 HIGH RESOLUTION BEAM GAS AUGER ELECTRON SPECTRA FOR OXYGEN IONS EXCITED BY COLLI
                    4-1050 AUGER ELECTRON EMISSION SPECTRA FROM FOIL AND GAS EXCITED CARBON BEAMS.
                    4-1051 HIGH RESOLUTION NEON K-AUGER ELECTRON SPECTRUM PRODUCED BY 45 MEV CHLORINE(12+)
                    4-1079 AUGER ELECTRON AND X-RAY PRODUCTION IN 50- TO 200 KEV NEON- NEON COLLISIONS.
                    4-1080 ARGON L-SHELL AUGER SPECTRA PRODUCED IN ARGON(+) ION- ARGON COLLISIONS.
SCHOBER O   5-1256 ADSORPTION OF COBALT ON A NICKEL (111) SURFACE. (AUGER EFFECT, AUGER SPECTROSCOP
                    5-1257 ADSORPTION OF HYDROGEN ON NICKEL SINGLE CRYSTAL SURFACES. (AUGER ELECTRON SPECTR
SCHOEPE H   6-1991 LEED- AUGER INVESTIGATION ON A (100) SURFACE OF AUSTENITIC CHROMIUM- NICKEL STEE
SCHON G   4-1052 HIGH RESOLUTION AUGER ELECTRON SPECTROSCOPY OF METALLIC COPPER.
                    4-1053 SPIN SPLITTING IN COPPER AUGER ELECTRON SPECTRUM.
                    4-1054 AUGER AND DIRECT ELECTRON SPECTRA IN X-RAY PHOTOELECTRON STUDIES OF ZINC, ZINC O
                    4-1055 ESCA STUDIES OF SILVER, DISILVER OXIDE AND SILVER OXIDE.
                    5-1794 ESCA (AND AUGER) STUDIES OF COPPER, COPPER(I), AND COPPER(II) OXIDE.
SCHOONMAKER RC   4-0895 LOW ENERGY AUGER AND LOSS ELECTRON SPECTRA FROM MAGNESIUM AND ITS OXIDE.
                    5-1495 COMPARATIVE STUDY OF SINGLE CRYSTAL MAGNESIUM OXIDE AND OXIDIZED MAGNESIUM BY AU
                    5-1496 STUDY OF THE EPITAXIAL GROWTH OF MAGNESIUM ON MAGNESIUM OXIDE (001) USING REFLEC
                    5-1497 INTERFACIAL AUGER TRANSITIONS IN OXIDIZED SODIUM AND MAGNESIUM.
                    5-1498 AUGER AND REFLECTION ELECTRON DIFFRACTION INVESTIGATION OF OXYGEN AND HYDROGEN A
SCHOW OE   5-1139 INVESTIGATION OF REORDERED (001) GOLD SURFACES BY POSITIVE ION CHANNELING SPECTR
                    5-1986 CHARACTERIZATION OF REORDERED (001) GOLD SURFACES BY POSITIVE ION CHANNELING SPE
SCHOWENGERDT FD   2-0489 PRODUCTION AND DECAY OF DOUBLE L VACANCIES IN ARGON AND PHOSPHORUS.
SCHRADER ME   5-1795 ULTRAHIGH VACUUM TECHNIQUES IN THE MEASUREMENT OF CONTACT ANGLES. PART-4: WATER
SCHRENE D   2-0498 THEORY OF THE AUGER EFFECT IN III-V SEMICONDUCTORS.
SCHRODER W   2-0499 INVESTIGATION OF THE CHARACTERISTIC AUGER ENERGY LOSSES OF PLATINUM (111).
SCHULER FT   4-0947 CHEMICAL ANALYSIS OF ELECTRODEPOSITED NICKEL- NICKEL BONDS BY AES.
SCHULTZ H   5-1443 SURFACE SEGREGATION OF OXYGEN IN NIOBIUM. (AES)

SCHUMANN S   2-0370 AUGER ELECTRONS FROM FOIL EXCITED HEAVY ION BEAMS. (OF NEON)
                      2-0371 AUGER TRANSITIONS LEADING TO DEFINED FINAL CHARGE STATES BY A COINCIDENCE TECHNI
                      4-0859 DOPPLER SHIFTED AUGER ELECTRONS FROM FOIL EXCITED HEAVY ION BEAMS.
                      4-0860 BEAM FOIL EXCITED AUGER TRANSITIONS IN NEON.
SCHUNCK JP   3-0690 REALIZATION OF A MOSSBAUER SPECTROMETER FOR SURFACES AND THIN FILMS STUDY. (AUGE
                      5-1796 MOSSBAUER SPECTROSCOPY OF IRON-57 AND TIN-119 BY DETECTION OF CONVERSION AND AUG
SCHUSTER F   3-0704 SORBAS: AN APPARATUS FOR INVESTIGATING ION SCATTERING FROM SURFACES AT ENERGIES
SCHWARZ JA   5-1568 ADSORPTION OF OXYGEN ON MOLYBDENUM (111): EFFECT OF TRACE IMPURITIES.
SCHWEITZ G   4-0990 DETERMINATION OF KLL AUGER SPECTRA OF NITROGEN, OXYGEN, CARBON DIOXIDE, NITRIC O
SCHWIDTAL K   5-1942 SURFACE STUDIES FOR QUARTZ RESONATORS. (CLEANING, AUGER SPECTROSCOPY)
SEAH MP   1-0177 QUANTITATIVE AUGER ELECTRON SPECTROSCOPY AND ELECTRON RANGES.
                      1-0178 QUANTITATIVE AUGER ELECTRON SPECTROSCOPY: COMPARISON OF TECHNIQUES FOR ADSORBED
                      2-0500 SLOW ELECTRON SCATTERING FROM METALS. PART-1: EMISSION OF TRUE SECONDARY ELECTRO
                      3-0691 FARADAY CUP LEED APPARATUS WITH FACILITY FOR INVESTIGATING ENERGY AND ANGULAR DI
                      5-1797 DISTINCTION BETWEEN ADSORBED MONOLAYERS AND THICKER LAYERS IN AUGER ELECTRON SPE
                      6-2049 GRAIN BOUNDARY ACTIVITY MEASUREMENTS BY AUGER ELECTRON SPECTROSCOPY.
                      6-2073 SURFACE SEGREGATION AS A GUIDE TO GRAIN BOUNDARY SEGREGATION. (AUGER SPECTROSCOP
                      6-2074 SITE COMPETITION IN SURFACE SEGREGATION. (AUGER EFFECT)
                      6-2103 GRAIN BOUNDARY EMBRITTLEMENT OF STEELS STUDIED BY AUGER ELECTRON SPECTROSCOPY.
                      6-2104 GRAIN BOUNDARY SEGREGATION. (AUGER ELECTRON SPECTROSCOPY)
                      6-2105 SCANNING ELECTRON AUGER MICROSCOPY IN METALLURGY.
                      6-2106 SURFACE SEGREGATION AND ITS RELATION TO GRAIN BOUNDARY SEGREGATION (AUGER ELECTR
                      6-2124 LEAD MONOLAYER LUBRICATION IN STEEL MACHINING STUDIED BY AUGER ELECTRON SPECTROS
SEFEROV AS   4-0946 HIGH FREQUENCY EXCITON BREAKDOWN AND FREE CARRIER AND EXCITON KINETICS IN GERMAN
SEGAL D   5-1620 INTERFERENCE OF AN ELECTRON BEAM WITH THE SURFACE REACTION BETWEEN OXYGEN AND GE
SEIDEL H   4-1031 ORIENTATION ANISOTROPY OF BACKSCATTERING COEFFICIENT AND SECONDARY ELECTRON EMIS
SEIDL M   2-0501 MEASUREMENT OF MOMENTUM ACCOMMODATION COEFFICIENTS ON SURFACES CHARACTERIZED BY
SEILER H   1-0179 SECONDARY ELECTRON EMISSION. (REVIEW, 86 REFS)
                      3-0692 ELECTRON CHANNELING PATTERNS IN AN AUGER ELECTRON SPECTROMETER WITH SCANNING SAM
                      5-1798 MATERIAL ANALYSIS OF THIN FOILS BY AUGER ELECTRON SPECTROSCOPY.
SELHOFER H   3-0632 AN ANALYTICAL SYSTEM FOR SECONDARY ION MASS SPECTROMETRY IN ULTRAHIGH VACUUM. (A
SELL HG   6-2107 IDENTIFICATION OF BUBBLE FORMING IMPURITIES IN DOPED TUNGSTEN. (AUGER SPECTROSCO
SELLIN IA   2-0472 ELECTRON DECAY-IN-FLIGHT SPECTRA FROM AUTOIONIZING STATES OF HIGHLY STRIPPED OXY
SELZER A   5-1799 AUGER ELECTRON SPECTROSCOPY, SCANNING ELECTRON MICROSCOPY AND NONDISPERSIVE X-RA
SEMERENKO VV   4-1001 EFFECT OF THE POLARITY OF INDIUM ANTIMONIDE ON THE EXTERNAL PHOTOEFFECT IN THE X
SEMPRINI E   2-0362 MULTIPLE EXCITATION PROCESSES IN TRANSITION METAL ATOMS. PART-2: INNER AUGER ELE
SEN AP   4-0736 RECORDING OF COMPONENTS ON THE SURFACE OF FUSED EMITTERS. (AUGER SPECTRA)
                      5-1358 RECORDING THE COMPONENTS ON THE SURFACE OF A BINARY ALLOY. (AUGER SPECTROSCOPY)
SENASHENKO VS   2-0257 POSSIBILITY OF INVESTIGATING ATOMIC AUTOIONIZING STATES IN COINCIDENCE EXPERIMEN
SEO M   4-1056 CHEMICAL SHIFTS IN THE AUGER SPECTRA OF PASSIVE FILMS.
                      5-1800 AES ANALYSIS OF OXIDE FILMS ON IRON.
SERGEEV VO   4-0743 AUGER L-SERIES ELECTRONIC SPECTRA ACCOMPANYING THE DECAY OF ERBIUM-160, YTTERBIU
                      4-0744 L-SERIES AUGER ELECTRON SPECTRA FOR ELEMENTS WITH Z = 66-70.
SERVAIS JP   5-1582 STUDY OF METAL SURFACES BY IONIC MICROANALYSIS AND AUGER SPECTROMETRY.
                      6-2075 POSSIBILITIES OF THE ION MICROPROBE MASS ANALYZER AND AUGER SPECTROMETER IN PHYS
SERVOMAA A   2-0502 THE K-FAR-M2 RADIATIVE AUGER EFFECT IN TRANSITION METALS.
                      4-1057 RADIATIVE KM2 AUGER EFFECT IN SOME TRANSITION METALS.
SEVIER KD   1-0180 THE AUGER, COSTER-KRONIG, AND AUTOIONIZATION EFFECTS. (REVIEW, 477 REFS)
                      2-0371 AUGER TRANSITIONS LEADING TO DEFINED FINAL CHARGE STATES BY A COINCIDENCE TECHNI
SEXTON BA   5-1150 QUANTITATIVE AUGER SPECTROSCOPY OF PHYSICALLY ADSORBED XENON ON NICKEL.
                      5-1151 ELECTRON BEAM EFFECTS IN AUGER ANALYSIS OF PHYSISORBED XENON.
SHAFFER JC   2-0398 MATRIX ELEMENTS DEPENDENCE OF OPTICAL EXCITATION AND AUGER DECAY OF 5D CORE HOLE
SHAKLESS KL   2-0503 OPTICAL GAIN IN LIGHTLY DOPED GALLIUM ARSENIDE. (AUGER RECOMBINATION)
SHANK CV   2-0250 PICOSECOND OPTICAL MEASUREMENTS OF BAND-TO-BAND AUGER RECOMBINATION OF HIGH DENS
SHAPIRA Y   5-1801 CHEMISORPTION, PHOTODESORPTION AND CONDUCTIVITY MEASUREMENTS ON ZINC OXIDE SURFA
SHARMA J   4-0851 X-RAY PHOTOELECTRON SPECTRUM OF DIAMOND. (AUGER TRANSITIONS)
SHATKOVSKII EV   4-0840 (AUGER) RECOMBINATION OF NONEQUILIBRIUM CHARGE CARRIERS OF INDIUM ARSENIDE AT HI
SHAW RW   4-1058 AUGER ELECTRON SPECTRUM AND IONIZATION POTENTIALS OF THE HYDROGEN FLUORIDE MOLEC
SHEINKMAN MK   2-0369 RECOMBINATION IN DONOR ACCEPTOR COMPLEXES IN CADMIUM SULFIDE SINGLE CRYSTALS. (A
                      2-0536 NONRADIATIVE AUGER RECOMBINATION OF ELECTRONS ON DONOR- ACCEPTOR PAIRS.
SHELL CA   5-1802 QUANTITATIVE AUGER SPECTROSCOPIC ANALYSIS OF SEGREGATION OF PHOSPHORUS IN IRON.
SHELTON JC   3-0693 ORIENTATION DEPENDENCE OF OVERLAYER ATTENUATION OF ELECTRONS FOR THE CYLINDRICAL
                      4-1059 INELASTIC MEAN FREE PATH FOR ELECTRONS IN BULK JELLIUM. (AUGER, LEED, ENERGY LOS
                      5-1803 EQUILIBRIUM SEGREGATION OF CARBON TO A NICKEL (111) SURFACE: A SURFACE PHASE TRA
SHEN LYL   5-1804 DETECTION OF SURFACE NITROGEN IN HIGH CRITICAL TEMPERATURE NIOBIUM(3) GERMANIUM
                      5-1805 OBSERVATION OF SUPERSTRUCTURES ON CARBON COVERED GERMANIUM (100) SURFACE BY HIGH
SHENG TT   5-1231 PHOSPHORUS CONCENTRATION PROFILES IN P-DOPED SILICON DIOXIDE MEASURED USING AUGE
SHERGIN AP   2-0234 DETECTION OF A NEW TYPE OF AUGER TRANSITIONS IN ATOMS WITH TWO INTERNAL VACANCIE
                      2-0235 OBSERVATION OF THREE ELECTRON AUGER TRANSITIONS IN ATOMS WITH TWO INNER VACANCIE
SHERLOCK TP   6-2140 LONG TIME ISOTHERMAL TEMPER EMBRITTLEMENT IN NICKEL- CHROMIUM- MOLYBDENUM- VANAD
SHERTNEV LG   6-2071 SECONDARY EMISSION AND CONTAMINATION OF METAL SURFACES.
SHEVCHIK NJ   2-0306 PHOTOELECTRIC PROPERTIES OF MAGNESIUM(2) SILICIDE, MAGNESIUM(2) GERMANIUM, AND M
                      5-1806 FINAL STATE EFFECTS IN THE PHOTOEMISSION CORE LEVEL AND AUGER SPECTRA OF NICKEL
                      5-1885 VALENCE BAND OF MAGNESIUM STANNIDE DETERMINED BY AUGER AND PHOTOEMISSION SPECTRO
SHEVTSOVA IN   2-0504 ROLE OF THE AUGER EFFECT IN THE FORMATION OF AN ELECTRON SHOWER SPECTRUM.
SHIELDS JA   6-2108 EFFECTS OF TEN YEARS EXPERIMENTAL BREEDER REACTOR II SERVICE ON CHROMIUM- MCLYBD
SHIFRIN VP   2-0410 EXOEMISSION PROPERTIES OF ZIRCONIUM DIOXIDE. (AUGER EFFECT)
                      4-0915 EXOELECTRON SPECTROMETRY AS A METHOD FOR DETERMINATION OF THE ENERGY DEPTH OF LE
SHIH HD   4-1060 INTERFACIAL AND PLASMON DECAY PEAKS IN THE AUGER SPECTRA OF TITANIUM AND IRON.
SHIMIZU A   5-1437 POSSIBLE ORIGIN OF LOW TEMPERATURE SILICON MIGRATION IN SILICON- METAL SYSTEMS.
                      5-1438 METALLIC STATE OF SILICON IN SILICON- NOBLE METAL VAPOR QUENCHED ALLOYS STUDIED
SHIMIZU H   3-0669 ULTRAHIGH VACUUM AIRLOCK FOR LEED AND AUGER SPECTROSCOPY.
                      3-0670 INSTRUMENTAL PROBLEMS FOR ELECTRON MICROPROBE AUGER SPECTROSCOPY.
                      5-1675 AUGER SPECTROSCOPY STUDY ON THE SURFACE COMPOSITION OF COPPER- NICKEL ALLOYS AFT
                      5-1702 QUANTITATIVE AUGER ANALYSIS OF COPPER- NICKEL ALLOY SURFACES AFTER ARGON ION BOM
                      5-1807 QUANTITATIVE AUGER ANALYSIS OF COPPER- NICKEL ALLOY SURFACES AFTER ARGON ION BOM
                      5-1808 EFFECT OF TARGET TEMPERATURE ON SURFACE COMPOSITION CHANGES OF COPPER- NICKEL AL
                      5-1874 ANALYSIS OF THE SURFACE COMPOSITION OF PASSIVE FILMS ON COPPER- NICKEL ALLOYS BY
                      5-1875 RELATION BETWEEN CATALYTIC ACTIVITY PATTERN AND SURFACE COMPOSITION OF COPPER- N
                      6-2109 SURFACE AND GRAIN BOUNDARY SEGREGATION OF STEEL IMPURITIES. (TEMPER EMBRITTLEMEN

SHIMIZU R	2-0365	AUGER AND SECONDARY ELECTRONS EXCITED BY BACKSCATTERED ELECTRONS.
	2-0366	AUGER AND SECONDARY ELECTRONS EXCITED BY BACKSCATTERED ELECTRONS. APPROACH TO QU
	2-0411	BACKGROUND INFORMATION IN AUGER ELECTRON SPECTROSCOPY AND SECONDARY ELECTRON EMI
	4-0916	SECONDARY AND AUGER ELECTRON SPECTRA OF A COPPER- BERYLLIUM ALLOY.
	4-0917	SECONDARY ELECTRON EMISSION FROM IRON SINGLE AND POLYCRYSTALS.
	5-1555	SECONDARY ELECTRON EMISSION IN THE STUDY OF SOLID SURFACES.
	5-1923	IMPURITY CONCENTRATIONS ON METAL SURFACES BY AUGER ELECTRON SPECTROSCOPY.
	5-1924	STUDIES OF INITIAL OXIDATION ON SILICON IRON ALLOY (100) BY MEANS OF WORK FUNCT
	5-1925	PHOTOELECTRIC WORK FUNCTION AND AUGER ELECTRON SPECTROSCOPY OF 5.59 ATOMIC PERCE
	5-1926	PHOTOELECTRIC WORK FUNCTION STUDY ON IRON (100) SURFACE COMBINED WITH AUGER ELEC
	5-1927	INITIAL OXIDATION STUDIES ON IRON (100) BY MEANS OF PHOTOELECTRIC WORK FUNCTION
	5-1928	INITIAL OXIDATION STUDIES BY MEANS OF PHOTOELECTRIC WORK FUNCTION MEASUREMENTS A
SHIMOJO T	5-1809	SURVEYING SILICON SURFACES BY MEANS OF THE LEED- AES METHOD.
	5-1876	GROWTH OF GOLD FILM ON GALLIUM ARSENIDE (110) OBSERVED BY LOW ENERGY ELECTRON DI
SHINODA G	1-0181	PRESENT STATUS OF BEAM ANALYSIS AND ITS FUTURE. (REVIEW, AUGER, MASS SPECTROSCOP
	5-1810	BEAM ANALYSIS METHOD OF FILM AND SURFACE COMPOSITION DETERMINATION.
SHIOTA I	5-1786	LOW TEMPERATURE OXIDATION OF SILICON STUDIED BY PHOTOSENSITIVE ESR AND AUGER ELE
SHIRLEY DA	2-0412	RELATIVE EFFECT OF EXTRAATOMIC RELAXATION ON AUGER AND BINDING ENERGY SHIFTS IN
	2-0427	MANY BODY EFFECTS IN X-RAY PHOTOEMISSION FROM MAGNESIUM. (AUGER EFFECT)
	2-0505	RELAXATION EFFECTS ON AUGER ENERGIES.
	2-0506	THEORY OF KLL AUGER ENERGIES INCLUDING STATIC RELAXATION.
	2-0507	THEORY OF AUGER SATELLITE ENERGY SHIFTS.
	4-0919	L2,3M4,5M4,5 AUGER SPECTRA OF METALLIC COPPER AND ZINC: THEORY AND EXPERIMENT.
	5-1811	X-RAY PHOTOEMISSION AND SURFACE STRUCTURE. (AUGER DISPLACEMENTS)
SHISHKIN BB	2-0302	OPTICALLY STIMULATED SECONDARY ELECTRON EMISSION AND PHOTOELECTRON EMISSION STUD
SHULMAN AR	2-0508	SECONDARY EMISSION AND ELASTIC REFLECTION OF ELECTRONS FROM MONOCRYSTALS.
	4-1061	CHANGES IN THE FINE STRUCTURE OF THE ANGULAR DEPENDENCE OF THE SECONDARY ELECTRO
	4-1062	SECONDARY ELECTRON EMISSION OF SILICON DIOXIDE SINGLE CRYSTALS.
	4-1063	SECONDARY ELECTRON EMISSION OF MOLYBDENUM CRYSTALS.
	4-1064	ENERGY SPECTRA OF INELASTICALLY SCATTERED AND AUGER ELECTRONS FROM SINGLE CRYSTA
SICKAFUS EN	2-0509	SELF-CONSISTENT ANALYSIS OF CORE BAND-TO-BAND EXCITATION: AUGER FINE STRUCTURE.
	3-0694	SECONDARY EMISSION ANALOG FOR IMPROVED AUGER SPECTROSCOPY WITH RETARDING POTENTI
	3-0695	A MULTICHANNEL MONITOR FOR REPETITIVE AUGER ELECTRON SPECTROSCOPY WITH APPLICATI
	3-0696	SPECIMEN POSITION EFFECTS ON ENERGY SHIFTS AND SIGNAL INTENSITY IN A SINGLE STAG
	4-1065	ELECTRON EXCITED AUGER ELECTRON SPECTRUM OF A NICKEL (110)- C(2X2) SULFUR SURFAC
	5-1812	AUGER ELECTRON SPECTROSCOPY APPLIED TO SURFACE COMPOSITION PROBLEMS.
	5-1813	SULFUR AND CARBON ON (110) SURFACE OF NICKEL.
	5-1814	LEED AND AUGER ELECTRON SPECTROSCOPY STUDY OF NICKEL (110) SURFACE EFFECTS DUE T
	5-1815	SURFACE CHARACTERIZATION BY AUGER ELECTRON SPECTROMETRY.
	5-1816	AUGER LINE SHAPE COMPARISON OF NITROGEN AND SULFUR IN TWO DIFFERENT CHEMICAL ENV
SIEGBAHN H	2-0510	TRANSITION POTENTIAL MODELS FOR AUGER ELECTRON SPECTRA.
	4-1066	AUGER ELECTRON SPECTRUM OF WATER VAPOR.
SIEGBAHN K	2-0515	ELECTRON SPECTROSCOPIC INVESTIGATION OF AUGER PROCESSES IN BROMINE-SUBSTITUTED M
	4-0826	CHEMICAL SHIFT IN AUGER SPECTRA.
	4-0827	AUGER SPECTRA FOR ELEMENTS OF LOW ATOMIC NUMBER.
	4-1113	HIGH RESOLUTION L2,3MM AND M4,5NN AUGER SPECTRA FROM KRYPTON AND M4, 5NN AND N4,
	4-1114	L2,3MM AUGER SPECTRUM OF ARGON.
	5-1529	AUGER ELECTRON SPECTRUM OF CARBON SUBOXIDE.
SIEGRIST M	2-0297	ANGLE DISTRIBUTION OF SODIUM CHLORIDE SINGLE CRYSTALS. (PHOTO- AND AUGER ELECTRO
SIEKHAUS WJ	5-1228	AUGER ANALYSIS OF SILICON THIN FILMS DEPOSITED ON CARBON AT HIGH TEMPERATURES.
SILVERMAN DC	5-1817	SURFACE COMPOSITION OF PROMOTED IRON CATALYSTS BY AUGER ELECTRON SPECTROSCOPY.
SIMMONS A	2-0406	RECOMBINATION MECHANISMS IN 8-14 MICRON MERCURY(1-X) CADMIUM(X) TELLURIDE. (AUGE
SIMMONS GW	5-1818	AUGER ELECTRON SPECTROSCOPY AND INELASTIC ELECTRON SCATTERING IN THE STUDIES OF
	5-1819	ORDER- DISORDER PHENOMENA AT THE SURFACE OF ALPHA TITANIUM- OXYGEN SOLID SOLUTIO
	6-2110	LOW ENERGY ELECTRON DIFFRACTION- AUGER ELECTRON SPECTROSCOPY STUDY OF THE INITIA
	6-2142	A TECHNIQUE FOR DETERMINING THE ELEMENTAL COMPOSITION OF FRACTURE SURFACES PRODU
SIMPSON JA	3-0623	COMPARISON OF SPHERICAL DEFLECTOR AND CYLINDRICAL MIRROR ANALYZERS.
	3-0697	HIGH RESOLUTION, LOW ENERGY ELECTRON SPECTROMETER.
SIMPSON RP	6-2111	SOLID SOLUTION SOFTENING (RHENIUM DUCTILIZING EFFECT AND BUBBLE STRENGTHENING IN
	6-2112	STUDY OF GRAIN BOUNDARY FRACTURE SURFACES IN DOPED TUNGSTEN- RHENIUM ALLOYS. (AU
SINGER KE	5-1219	IN-DEPTH AUGER ANALYSIS OF ALUMINUM- SILICON INTERFACIAL REACTIONS.
SINHAROY S	5-1820	BINDING STATES OF WATER VAPOR ON CLEAVED GERMANIUM. (LEED, AES)
SKIBOWSKI M	2-0289	PHOTOEMISSION STUDIES OF THE DECAY OF CORE EXCITONS IN ALKALI HALIDES. (AUGER PR
SKINNER DK	3-0698	REFINEMENTS TO A STANDARD LEED- AUGER SYSTEM FOR THE ANALYSIS OF ELECTRON EMISSI
	5-1237	CARBON CONTAMINATION OF SILICON (111) SURFACES.
	5-1238	AUGER ELECTRON SPECTROSCOPY OF NICKEL DEPOSITS ON THE SILICON (111) SURFACE.
	5-1970	STUDY OF CARBON FIBER SURFACES USING AUGER AND SECONDARY ELECTRON EMISSION SPECT
SLATER RR	2-0511	SURFACE MADELUNG POTENTIALS IN ELECTRON SPECTROSCOPY.
SLATIS H	3-0699	NEW, STRIATED, HIGH EFFICIENCY PHOTOGRAPHIC PLATES FOR RECORDING CONVERSION AND/
SLOCOMB CA	2-0512	SPREADING FUNCTIONS IN SECONDARY ELECTRON ANALYSIS.
SMARTT HB	6-2054	CONFIRMATION OF IMPURITY ADSORPTION AT FLAKE IRON INTERFACES IN GRAY CAST IRON.
SMIDT FA	6-2113	AUGER ELECTRON SPECTROSCOPY ANALYSIS OF THE FRACTURE SURFACES OF IRRADIATED PRES
SMITH ACH	2-0340	SECONDARY ELECTRON EJECTION FROM METAL SURFACES BY METASTABLE ATOMS. PART-1: MEA
	5-1133	SECONDARY ELECTRON EJECTION FROM METAL SURFACES BY METASTABLE ATOMS. PART-3: ENE
SMITH AW	3-0715	ELECTRON SPECTROMETER FOR SURFACE ANALYSIS.
SMITH CL	6-2079	GRAIN BOUNDARY SEGREGATION OF IMPURITIES IN METALS AND INTERGRANULAR BRITTLE FRA
	6-2114	EFFECT OF PRIOR AUSTENITIC GRAIN BOUNDARY COMPOSITION ON TEMPER BRITTLENESS IN A
SMITH DL	5-1821	ADSORPTION KINETICS OF CESIUM ON GALLIUM ARSENIDE. (USING AES AND X-RAY PHOTOELE
SMITH DM	2-0513	AUGER EMISSION FROM SOLIDS. ESTIMATION OF BACKSCATTERING EFFECTS AND IONIZATION
	4-1067	ANOMALOUS THRESHOLD BEHAVIOR FOR ELECTRON INDUCED INNER SHELL IONIZATION OF THE
SMITH DP	5-1822	SURFACE ANALYSIS BY LOW ENERGY ION REFLECTION.
	5-1823	ANALYSIS OF SURFACE COMPOSITION WITH LOW ENERGY BACKSCATTERED IONS.
SMITH JA	5-1824	SURFACE CONCENTRATION OF MAGNESIUM IN 6061 ALUMINUM BY AUGER ELECTRON SPECTROSCO
SMITH JN	1-0182	SPUTTERING MEASUREMENTS ON CTR MATERIALS USING AUGER ELECTRON SPECTROSCOPY
	5-1825	AUGER ELECTRON SPECTROSCOPY IN SPUTTERING MEASUREMENTS. APPLICATION TO LOW ENER
	5-1826	SPUTTERING MEASUREMENTS ON CONTROLLED THERMONUCLEAR REACTOR MATERIALS USING AUGE
	5-1827	AUGER ELECTRON SPECTROSCOPY DETERMINATION OF THE OXYGEN- SILICON RATIO IN SPIN O
SMITH LE	2-0497	RELATIVE SATELLITE LINE INTENSITIES IN ARGON K AUGER ELECTRON SPECTRA PRODUCED B
	4-0903	HIGH RESOLUTION STUDY OF ARGON LMM AUGER ELECTRONS PRODUCED BY ION BOMBARDMENT.
	4-0943	HIGH RESOLUTION MEASUREMENTS OF P(+) ARGON KLL AND KLM AUGER ELECTRONS.

SMITH LE	4-0954	AUGER DECAY OF NEON FOLLOWING ENERGETIC ION BOMBARDMENT.
	4-0993	OBSERVATION OF A PROJECTILE CHARGE DEPENDENCE FOR THE NEON K-AUGER ELECTRON SPEC
	4-1050	AUGER ELECTRON EMISSION SPECTRA FROM FOIL AND GAS EXCITED CARBON BEAMS.
SMITH PV	4-0888	CLUSTER CALCULATIONS OF THE SOFT X-RAY L3 EMISSION SPECTRA OF NICKEL AND COPPER.
SMITH RL	5-1883	THORIUM (100) CRYSTAL FACE. A STUDY OF ITS CLEANING, SURFACE STRUCTURE, AND INTE
SMITH T	5-1828	SPUTTER CLEANING AND ETCHING OF CRYSTAL SURFACES (TITANIUM, TUNGSTEN, SILICON) M
	5-1829	ELLIPSOMETRY, AUGER SPECTROSCOPY, AND WORK FUNCTION OF CESIUM ON TUNGSTEN AND TI
	5-1830	CESIUM ADSORPTION ON TUNGSTEN: ELLIPSOMETRY, AUGER SPECTROSCOPY AND SURFACE POTE
	5-1831	CHLORINE ADSORPTION ON CLEAN ALUMINUM. (AUGER SPECTOSCOPY)
	5-1832	MARKER EXPERIMENTS WITH AUGER SPECTROSCOPY AND ELLIPSOMETRY.
	5-1833	CHLORINE ADSORPTION AT A TITANIUM SURFACE. SEGREGATION FROM THE BULK CRYSTAL INT
	6-2115	OXIDATION OF TITANIUM BETWEEN 25 DEGREES C AND 400 DEGREES C.
	6-2116	CHARACTERIZING OF ALUMINUM FATIGUE DAMAGE BY PHOTOEMISSION, ELLIPSOMETRY, AND AU
SMITH WW	2-0472	ELECTRON DECAY-IN-FLIGHT SPECTRA FROM AUTOIONIZING STATES OF HIGHLY STRIPPED OXY
	2-0530	EXTENDED ION- ELECTRON COINCIDENCE MEASUREMENTS OF THE VIOLENT ARGON(+)- ARGON C
SMOLYAKOV A	4-0736	RECORDING OF COMPONENTS ON THE SURFACE OF FUSED EMITTERS. (AUGER SPECTRA)
	5-1129	CHARACTERISTICS OF SURFACE OF THE PALLADIUM BARIUM ALLOY CATHODE IN RELATION TO
SNYDER DM	6-2037	EFFECT OF A THIN DIFFUSED SOLUTE LAYER ON THE NUCLEATION OF FATIGUE CRACKS. (AUG
SOKCEVIC D	4-1088	DETERMINATION OF ELECTRON ATTENUATION LENGTHS IN METALS: TRANSMISSION THROUGH TH
SOL N	5-1590	LIQUID PHASE EPITAXY OF III-V COMPOUNDS ON INDIUM PHOSPHIDE SUBSTRATES. (AUGER S
SOLOMON JS	1-0183	DIGITIZING AUGER DATA. THREEFOLD APPROACH TO ENHANCE ANALYTICAL CAPABILITIES.
	1-0184	AUGER PROFILE ANOMALIES. (REVIEW)
	4-1068	A1 KLL AUGER PROFILE ARTIFACT AT THE OXIDE- METAL INTERFACE.
	4-1069	MOLECULAR ORBITAL EFFECTS ON THE TITANIUM LMV AUGER SPECTRA OF TITANIUM(II) AND
	5-1834	CHANGES IN THE TITANIUM DIOXIDE AUGER AND APPEARANCE POTENTIAL SPECTRA INDUCED B
SOMORJAI GA	1-0185	LOW ENERGY ELECTRON DIFFRACTION AND SURFACE TOPOGRAPHY. (REVIEW, AUGER SPECTROSC
	2-0514	EMISSION AND RECOMBINATION INVOLVING INNER SHELLS. (AUGER PROCESS, AES, PHOTOELE
	4-1089	AUGER ELECTRON SPECTROSCOPY OF THE SURFACE CHEMICAL COMPOSITION OF VANADIUM OXID
	4-1090	AUGER ELECTRON SPECTROSCOPIC INVESTIGATIONS OF CHEMICAL SHIFTS IN SOME VANADIUM
	5-1172	AUGER ELECTRON SPECTROSCOPY STUDY OF THE SURFACE COMPOSITION OF THE LEAD- INDIUM
	5-1173	SMALL MOLECULE REACTIONS ON STEPPED SINGLE CRYSTAL PLATINUM SURFACES.
	5-1205	USE OF X-RAY PHOTOELECTRON AND AUGER SPECTROSCOPY TO DETERMINE THE SURFACE COMPO
	5-1709	AUGER ELECTRON SPECTROSCOPY OF ALLOY SURFACES. (GOLD- SILVER, LEAD- INDIUM)
	5-1710	SURFACE COMPOSITION OF THE SILVER- GOLD SYSTEM BY AUGER ELECTRON SPECTROSCOPY.
	5-1835	LEED AND AES STUDIES OF THE STRUCTURE OF ADSORBED GASES ON SOLID SURFACES.
	5-1836	STUDIES OF CRYSTAL SURFACES BY LOW ENERGY ELECTRON DIFFRACTION AND AUGER ELECTRO
	5-1837	SURFACE STRUCTURE OF AND CATALYSIS BY PLATINUM SINGLE CRYSTAL SURFACES.
	5-1838	REACTIVITY OF LOW INDEX ((111) AND (100)) AND STEPPED PLATINUM SINGLE CRYSTAL SU
	5-1839	AUGER ELECTRON SPECTROSCOPY OF (SILVER- GOLD AND INDIUM- LEAD) ALLOY SURFACES.
	5-1840	AUGER ELECTRON SPECTROSCOPY ON SURFACES.
	5-1863	USE OF X-RAY PHOTOELECTRON AND AUGER SPECTROSCOPY TO DETERMINE THE SURFACE COMPO
	5-1864	CHARACTERIZATION OF SOME VANADIUM (100) SURFACE STRUCTURES USING LEED AND AES.
	5-1883	THORIUM (100) CRYSTAL FACE. A STUDY OF ITS CLEANING, SURFACE STRUCTURE, AND INTE
SOPIZET R	4-0934	NEW METHOD FOR ANALYZING THE SECONDARY ELECTRON SPECTRUM.
	4-0948	STRUCTURE OF LOW ENERGY ELECTRON SPECTRUM.
SOSNIAK J	5-1233	STRUCTURE AND CHEMISTRY OF SILICON SURFACES AFTER PRE- AND BACKSPUTTERING, STUDI
SOSTARICH M	2-0324	AUGER EFFECT AND RADIATIVE RECOMBINATION IN N-GALLIUM ARSENIDE- P-ALUMINUM GALLI
SOTNIKOV VT	4-1124	PLASMA SATELLITES IN THE AUGER ELECTRON SPECTRUM OF MOLYBDENUM.
	5-1226	DIFFUSION OF SULFUR AND PHOSPHORUS IMPURITY ATOMS TO A TUNGSTEN SURFACE.
	5-1759	DECREASE OF CARBON CONCENTRATION ON THE SURFACE OF MOLYBDENUM CARBIDE DURING HEA
SOUTHWORTH HN	5-1951	COMBINED FIELD ION MICROSCOPY, AUGER ELECTRON SPECTROSCOPY, AND LEED STUDY OF TH
SPAIN IL	2-0343	WARM ELECTRON EFFECTS IN NARROW GAP MERCURY(1-X) CADMIUM(X) TELLURIDE AMBIENT TE
SPARNAAY MJ	1-0186	AUGER ELECTRON SPECTROSCOPY AS A TOOL FOR MEASURING THE DIFFUSION OF FOREIGN ATO
SPEIGHT JD	5-1902	THERMALLY INDUCED PROCESSES AT GOLD- GALLIUM ARSENIDE INTERFACES: AN ASSESSMENT
SPICER WE	1-0042	USE OF PHOTOEMISSION, AUGER, AND THERMAL DESORPTION TECHNIQUES TO STUDY THE MECH
	2-0428	PROBING DEPTH IN PHOTOEMISSION AND AUGER ELECTRON SPECTROSCOPY.
	5-1501	USE OF AUGER ELECTRON SPECTROSCOPY TO DETERMINE THE STRUCTURE OF SILICON OXIDE F
	5-1502	PHASE SEPARATION IN SILICON OXIDES AS SEEN BY AUGER ELECTRON SPECTROSCOPY.
	5-1503	STUDY OF THE CHEMICAL COMPOSITION OF MOS AND MNOS STRUCTURES BY AUGER ELECTRON S
	6-2045	VERIFICATION OF THE METAL-RICH SURFACE MODEL FOR THE OXIDATION OF STRONTIUM, BY
SPOHR R	1-0070	SURFACE ANALYSIS WITH HEAVY ION INDUCED AUGER ELECTRONS. BASIC PROPERTIES OF A N
	2-0370	AUGER ELECTRONS FROM FOIL EXCITED HEAVY ION BEAMS. (OF NEON)
	2-0515	ELECTRON SPECTROSCOPIC INVESTIGATION OF AUGER PROCESSES IN BROMINE-SUBSTITUTED M
	4-0859	DOPPLER SHIFTED AUGER ELECTRONS FROM FOIL EXCITED HEAVY ION BEAMS.
	4-0860	BEAM FOIL EXCITED AUGER TRANSITIONS IN NEON.
	5-1387	SURFACE ANALYSIS WITH HEAVY ION INDUCED AUGER ELECTRONS.
SPRINGER RW	1-0187	INTEGRAL AUGER INFORMATION VIA TAILORED MODULATION TECHNIQUES.
	3-0700	EASY METHOD TO ACCURATELY ALIGN ION BOMBARDMENT GUNS FOR DEPTH PROFILING IN AUGE
	4-0857	COMPARISON OF AUGER SPECTRA OF MAGNESIUM, ALUMINUM, AND SILICON EXCITED BY LOW E
	4-0869	ION EXCITED AUGER SPECTRA OF ALUMINUM.
	5-1841	AUGER ELECTRON SPECTROSCOPY STUDY OF CATHODE SURFACES DURING ACTIVATION AND POIS
SRIVASTAVA TS	2-0516	MOSSBAUER STUDIES OF THE AFTEREFFECTS OF AUGER IONIZATION FOLLOWING ELECTRON CAP
STADLER K	1-0188	A MEASURING METHOD TO DISTINGUISH AUGER FROM ENERGY LOSS PEAKS OF THE PRIMARY BE
STADNIKOV CG	4-1070	INTENSITIES OF SPECTRA- AUGER AND PHOTOELECTRON.
STAEHLE RW	4-1056	CHEMICAL SHIFTS IN THE AUGER SPECTRA OF PASSIVE FILMS.
	5-1800	AES ANALYSIS OF OXIDE FILMS ON IRON.
	6-2080	APPLICATON OF AUGER ELECTRON SPECTROSCOPY TO THE DETERMINATION OF THE COMPOSITIO
STAHLHERM D	4-0981	CORRELATION EFFECTS IN THE AUGER SPECTRA OF NOBLE GASES.
	4-0982	ENERGY WIDTHS OF ROENTGEN LEVELS BY AUGER ELECTRON SPECTROSCOPY.
	4-0983	AUGER SPECTRUM OF L2 AND L3 SHELLS OF ARGON.
	4-1071	ENERGIES OF EXCITED STATES OF DOUBLY IONIZED MOLECULES BY MEANS OF AUGER ELECTRO
STAIB P	1-0189	HIGHLY SENSITIVE AUGER ELECTRON SPECTROSCOPY.
	3-0701	IMPROVED RETARDING FIELD ANALYZER. (FOR AUGER AND PHOTOELECTRON SPECTROSCOPY).
	3-0702	NEW COMPACT ENERGY ANALYZER FOR AUGER AND ESCA SPECTROSCOPY.
	4-1072	PREDICTION OF THE RELATIVE INTENSITIES OF AUGER LINES IN MICA.
	5-1167	ANALYSIS OF SURFACE LAYERS BY LIGHT ION BACKSCATTERING AND SPUTTERING COMBINED W
	5-1842	DEPTH ANALYSIS OF CLEAVED MICA SURFACES MONITORED BY AUGER SPECTROSCOPY.
	5-1843	ABSOLUTE ATOMIC DENSITIES DETERMINED BY AUGER ELECTRON SPECTROSCOPY.
	6-2068	DISAPPEARANCE POTENTIAL SPECTROSCOPY (VANADIUM OXIDATION).
STALEY R	4-0851	X-RAY PHOTOELECTRON SPECTRUM OF DIAMOND. (AUGER TRANSITIONS)

STANSKII VA	2-0356	SECONDARY ELECTRON EMISSION AT SUPERHIGH FREQUENCIES. (COPPER- GERMANIUM JUNCTIO
STEBBINGS RF	2-0340	SECONDARY ELECTRON EJECTION FROM METAL SURFACES BY METASTABLE ATOMS. PART-1: MEA
STEFANI F	2-0362	MULTIPLE EXCITATION PROCESSES IN TRANSITION METAL ATOMS. PART-2: INNER AUGER ELE
STEIDEL CA	5-1473	EFFECT OF LIGHT ELEMENTS NITROGEN, CARBON AND OXYGEN ON THE PHYSICAL PROPERTIES
STEIGER J	6-2040	OBSERVATIONS ON THE SURFACE OXIDATION OF STEEL DURING ANNEALING IN HYDROGEN-FREE
STEIN DF	1-0098	AUGER ELECTRON SPECTROSCOPY. (REVIEW, 18 REFS)
	1-0190	APPLICATIONS OF AUGER SPECTROSCOPY TO MATERIALS RESEARCH. (REVIEW, 19 REFS)
	5-1505	ADDITIVE AND IMPURITY DISTRIBUTIONS AT GRAIN BOUNDARIES IN SINTERED ALUMINA. (SC
	5-1506	ANALYSIS OF GRAIN BOUNDARY IMPURITIES AND FLUORIDE ADDITIVES IN HOT PRESSED OXID
	6-2015	STUDY OF SULFUR EMBRITTLED OXYGEN FREE COPPER. (AES, GRAIN BOUNDARY SEGREGATION)
	6-2018	IDENTIFICATION OF ATOMIC BORON IN STEEL BY AUGER ELECTRON SPECTROSCOPY.
	6-2055	STUDY OF GRAIN BOUNDARY SEGREGANTS IN THERMALLY EMBRITTLED MARAGING STEEL. (AUGE
	6-2056	GRAIN BOUNDARY SEGREGATION IN SINTERED ALUMINA. (AUGER ELECTRON SPECTROSCOPY)
	6-2059	ROLE OF MANGANESE AND SILICON IN TEMPER EMBRITTLEMENT OF LOW ALLOY STEELS. (AUGE
	6-2060	AUGER SPECTROSCOPIC ANALYSIS OF BISMUTH SEGREGATED TO GRAIN BOUNDARIES IN COPPER
	6-2061	INTERANGULAR BRITTLENESS STUDIES IN TUNGSTEN USING AUGER SPECTROSCOPY.
	6-2062	TEMPER EMBRITTLEMENT OF LOW ALLOY STEELS. (AES OF FRACTURE SURFACES)
	6-2063	IMPURITY SEGREGATION TO GRAIN BOUNDARIES.
	6-2064	CHEMISTRY OF GRAIN BOUNDARIES AND ITS RELATION TO INTERGRANULAR CORROSION OF AUS
	6-2098	GRAIN BOUNDARY EMBRITTLEMENT IN IRON- PHOSPHORUS, IRON- PHOSPHORUS- SULFUR AND I
	6-2107	IDENTIFICATION OF BUBBLE FORMING IMPURITIES IN DOPED TUNGSTEN. (AUGER SPECTROSCO
	6-2113	AUGER ELECTRON SPECTROSCOPY ANALYSIS OF THE FRACTURE SURFACES OF IRRADIATED PRES
	6-2117	STUDY OF GRAIN BOUNDARY SEGREGATION USING AUGER ELECTRON EMISSION SPECTROSCOPY.
	6-2118	GRAIN BOUNDARY STRUCTURE AND SEGREGATION. AUGER SPECTROGRAPHIC ANALYSIS OF FRAC
	6-2119	STUDIES USING AUGER ELECTRON EMISSION SPECTROSCOPY ON TEMPER EMBRITTLEMENT IN LO
	6-2120	RELATION OF GRAIN BOUNDARY SEGREGATION TO REDUCED STRENGTH OF METALS AND ALLOYS.
	6-2121	STUDY OF GRAIN BOUNDARY SEGREGATION USING AUGER EMISSION SPECTRA.
	6-2122	AUGER ELECTRON SPECTROSCOPY OF METAL SURFACES.
	6-2123	METALLURGICAL APPLICATION OF AUGER ELECTRON SPECTROSCOPY.
STEIN HJ	5-1455	AUGER ELECTRON SPECTROSCOPIC ANALYSIS OF SILICON NITRIDE ON SILICON.
STEINHARDT RG	5-1844	ATTENUATION OF LOW ENERGY ELECTRONS FROM SOLIDS: RESULTS FROM X-RAY PHOTOELECTRO
STEINHEIL E	2-0501	MEASUREMENT OF MOMENTUM ACCOMMODATION COEFFICIENTS ON SURFACES CHARACTERIZED BY
STEINMANN W	2-0289	PHOTOEMISSION STUDIES OF THE DECAY OF CORE EXCITONS IN ALKALI HALIDES. (AUGER PR
STEINRISSER F	5-1816	AUGER LINE SHAPE COMPARISON OF NITROGEN AND SULFUR IN TWO DIFFERENT CHEMICAL ENV
STERN RM	2-0237	EFFECT OF THE INCIDENCE ANGLE OF PRIMARY ELECTRONS ON THE AUGER EMISSION YIELD.
STEVENS MT	4-1015	SECONDARY ELECTRON EMISSION FROM THIN FILM GOLD METAL- OXIDE CERMETS.
STEVENSON JR	5-1220	OPTICAL PROPERTIES AND AUGER CHARACTERISTICS OF MAGNESIUM SINGLE CRYSTALS AND MA
STICKLER R	6-2107	IDENTIFICATION OF BUBBLE FORMING IMPURITIES IN DOPED TUNGSTEN. (AUGER SPECTROSCO
STICKNEY RE	5-1155	ANGULAR DISTRIBUTIONS OF HYDROGEN DESORBED FROM THE (100), (110), AND (111) FACE
	5-1197	MEASUREMENTS OF THE SPATIAL DISTRIBUTION OF HYDROGEN DESORBED FROM NICKEL SURFAC
	5-1282	AUGER ELECTRON SPECTROSCOPY MEASUREMENTS OF ADSORPTION ISOBARS FOR OXYGEN ON TUN
	6-2006	SPATIAL DISTRIBUTIONS OF HYDROGEN DESORBED FROM IRON, PLATINUM, COPPER, NIOBIUM,
STOCKER BJ	5-1845	AUGER ELECTRON SPECTROSCOPY AND LOW ENERGY ELECTRON DIFFRACTION STUDY OF THE ACT
STODDART CTH	4-1073	OXIDE FILMS: ELEMENT PROFILES BY AUGER ELECTRON SPECTROSCOPY.
	4-1074	AUGER ELECTRON SPECTROSCOPY STUDY OF BETA ALUMINA ELECTROLYTE.
	5-1236	AUGER ELECTRON SPECTROSCOPY STUDY OF FLOAT GLASS SURFACES.
	5-1846	DETERMINATION OF THE SURFACE COMPOSITION OF PALLADIUM- NICKEL ALLOY FILM CATALYS
	6-2124	LEAD MONOLAYER LUBRICATION IN STEEL MACHINING STUDIED BY AUGER ELECTRON SPECTROS
STOECKLIN G	2-0377	RADIATION CHEMICAL RESONANCE EFFECT IN SOLID 5-BROMO DEOXYURIDINE: CHEMICAL CONS
STOLTERFOHT N	2-0300	PROJECTILE CHARGE STATE DEPENDENCE OF NEON K-SHELL IONIZATION AND FLUORESCENCE Y
	2-0345	K-VACANCY SHARING STUDIED FROM AUGER ELECTRON YIELDS.
	2-0517	CHARGE STATE DEPENDENCE OF THE ARGON L-SHELL FLUORESCENCE YIELD STUDIED BY ATOMI
	2-0518	K-SHELL IONIZATION AND FLUORESCENCE YIELD IN 50 MEV CHLORINE(+)- NEON COLLISIONS
	2-0519	K-SHELL IONIZATION OF MOLECULAR NITROGEN AND METHANE BY 50- TO 600 KEV HYDROGEN(
	2-0520	CHARGE STATE DEPENDENCE OF THE NEON K FLUORESCENCE YIELD DEDUCED FROM HIGH RESOL
	2-0521	ARGON L-SHELL IONIZATION BY 50 TO 600 KEV PROTON, DEUTERON, MOLECULAR HYDROGEN,
	2-0522	MOLECULAR EFFECTS ON K-VAANCY SHARING IN ASYMMETRIC ION- ATOM COLLISIONS. (AUGER
	2-0523	GAS TARGET EXPERIMENTS OF K-SHELL VACANCY SHARING IN ASYMMETRIC ION- ATOM COLLIS
	2-0572	NEON K-SHELL VACANCY PRODUCTION BY INTERMEDIATE ENERGY FLUORINE IONS. (AUGER ELE
	2-0573	AUGER ELECTRONS FROM DOUBLE K-SHELL VACANCY STATES IN NEON.
	2-0574	K-SHELL AUGER ELECTRON PRODUCTION CROSS SECTIONS FROM ION BOMBARDMENT. (NEON, FL
	2-0582	K-SHELL VACANCY SHARING IN 300 TO 500 KEV NEON(+)- SODIUM COLLISIONS. (QUASIMOLE
	4-1032	WIDTHS AND INTENSITIES OF HIGHLY RESOLVED LINES IN ARGON L2,3 AUGER SPECTRA.
	4-1075	ANGULAR AND ENERGY DISTRIBUTION OF ELECTRONS PRODUCED BY 200-500 KEV PROTONS IN
	4-1076	PRODUCTION OF INTENSE SATELLITE LINES IN AUGER SPECTRA OF NITROGEN BY SLOW PROTO
	4-1077	CROSS SECTIONS FOR INNER SHELL IONIZATION OF GASEOUS MOLECULES BY 50-500 KEV PRO
	4-1078	HIGH RESOLUTION NEON AUGER SPECTRUM PRODUCED IN 4.2 MEV ATOMIC HYDROGEN(+)- NEON
	4-1079	AUGER ELECTRON AND X-RAY PRODUCTION IN 50- TO 200 KEV NEON- NEON COLLISIONS.
	4-1080	ARGON L-SHELL AUGER SPECTRA PRODUCED IN ARGON(+) ION- ARGON COLLISIONS.
	4-1117	K-SHELL AUGER ELECTRON HYPERSATELLITES OF NEON.
STORASTA J	4-1103	ELECTRON HOLE SCATTERING AND (AUGER) RECOMBINATION OF NONEQUILIBRIUM CHARGE CARR
STRAUSSER YE	4-1081	AUGER ELECTRON SPECTROSCOPY STUDY OF SILICON SPECTRA FROM SILICON MONOXIDE, SILI
	4-1082	VARIAN CHART OF AUGER ELECTRON ENERGIES.
	5-1502	PHASE SEPARATION IN SILICON OXIDES AS SEEN BY AUGER ELECTRON SPECTROSCOPY.
	5-1503	STUDY OF THE CHEMICAL COMPOSITION OF MOS AND MNOS STRUCTURES BY AUGER ELECTRON S
	5-1847	AES CHARACTERIZATION OF OXIDIZED FILMS OF MAGNESIUM, ALUMINUM, AND SILICON.
STREET FJ	4-0769	EXPERIMENTAL DETERMINATION OF THE INTERNAL PHOTOEMISSION CONTRIBUTION TO KLM AUG
	4-0770	AUGER AND X-RAY PHOTOELECTRON SPECTROSCOPIC STUDY OF SODIUM METAL AND SODIUM OXI
STRONGIN M	5-1333	INTERACTION OF OXYGEN AND NITROGEN WITH THE NIOBIUM (100) SURFACE. PART-2: REACT
	5-1508	SURFACE SEGREGATION AND INTERACTION OF OXYGEN AND NITROGEN WITH ZIRCONIUM IN NIO
	5-1848	SOME PROPERTIES OF TRANSITION METAL SUPERCONDUCTORS AT SURFACES AND IN THIN FILM
	5-1849	DIFFERENTIAL AUGER SPECTROMETRY. (NIOBIUM- ZIRCONIUM ALLOY ANALYSIS)
	5-1937	DIFFERENCE AUGER SPECTROSCOPY FOR STUDYING SMALL QUANTITIES OF ELEMENTS ON METAL
	5-1938	EFFECT OF HIGH TEMPERATURE HEAT TREATMENT ON PENETRATION DEPTH OF SUPERCONDUCTIN
	6-2065	SURFACE SEGREGATION OF OXYGEN IN NIOBIUM- OXYGEN AND TANTALUM- OXYGEN ALLOYS. (A
	6-2137	EFFECT OF SURFACE METALLURGY ON THE PENETRATION DEPTH AND RF BREAKDOWN FIELD OF
STROZIER JA	1-0100	THIN FILMS. LOW ENERGY ELECTRON DIFFRACTION AND AUGER ELECTRON SPECTROSCOPY.
	5-1507	REEXAMINATION OF ALUMINUM (110) WITH LOW ENERGY DIFFRACTION AND AUGER SPECTROSCO
STUPIAN GW	1-0191	AUGER ELECTRON SPECTROMETRY.
	5-1850	AUGER SPECTROSCOPY OF SILICONES.

STYRIS DL	5-1888	AUGER SPECTRCSCOPY OF SUBMONOLAYER GOLD DEPOSITIONS ON SILICON.
SUGANO T	5-1699	STUDIES ON SURFACE PLASMA OXIDIZED FILMS OF III-V COMPOUND SEMICONDUCTORS BY AUG
SUITS JC	5-1766	CORROSION, SURFACE, AND MAGNETIC PROPERTIES OF NICKEL- IRON- CHROMIUM THIN FILMS
SUKACH GA	5-1990	MANY PARTICLE RECOMBINATION PROCESSES (AUGER) ON THE SURFACES AND IN THIN FILMS
SULEMAN M	4-1083	AUGER ELECTRON AND CHARACTERISTIC ENERGY LOSS SPECTROSCOPY OF SOME METALS.
	4-1084	AUGER SPECTRCSCOPY OF BERYLLIUM.
	4-1085	OBSERVATION CF A PLASMON GAIN IN THE FINE STRUCTURE OF THE ALUMINUM AUGER SPECTR
	4-1086	CHANGES IN AUGER SPECTRA OF MAGNESIUM AND IRON DUE TO OXIDATION.
	4-1087	INTERPRETATION OF THE AUGER SPECTRUM OF CLEAN AND OXIDIZED BERYLLIUM.
	5-1897	ELECTRON SPECTROSCOPY OF METAL BLACKS.
SULTANOV FK	4-0752	METAL- DIELECTRIC TRANSITION IN GERMANIUM AND SILICON STUDIED BY A MICROWAVE MET
SUMNER B	5-1593	EVALUATION OF LEAD(0.8) TIN(0.2) TELLURIDE DETECTOR FABRICATION USING SURFACE AN
SUNDAHL RC	5-1174	CHARACTERIZATION OF CERAMIC SURFACES. (AUGER SPECTROSCOPY)
	5-1851	RELATIONSHIPS BETWEEN SUBSTRATE SURFACE CHEMISTRY AND ADHESION OF THIN FILMS. (A
	5-1852	CRYSTALLOGRAPHY AND CHEMISTRY OF CERAMIC SURFACES. (AUGER SPECTROSCOPY, LEED, IO
	5-1853	ANALYSIS ANC CHARACTERIZATION OF CERAMIC SURFACES FOR ELECTRONIC APPLICATIONS.
SUNDARAM VS	4-1028	AUGER SPECTRA OF COPPER- NICKEL ALLOYS.
	5-1854	LOW ENERGY ELECTRON DIFFRACTION AND AUGER INVESTIGATION OF THE (100) SURFACE OF
	5-1855	ORDER- DISORDER TRANSFORMATION AT A (100) SURFACE OF COPPER(3) GOLD, THEORY, EXP
SUNDERLAND RJ	6-2125	(AUGER) ANALYSIS OF CHROMATE CONVERSION COATINGS ON ALUMINUM.
SUNG DO	2-0524	DEPENDENCE OF ATOMIC PHOTOEFFECT ON BOUND ELECTRON MAGNETIC SUBSTATE (AUGER EFFE
SUNJIC M	4-1088	DETERMINATION OF ELECTRON ATTENUATION LENGTHS IN METALS: TRANSMISSION THROUGH TH
SUONINEN EJ	3-0703	HIGH RESOLUTION ELECTRON SPECTROGRAPHS FOR SOLID STATE RESEARCH. (REVIEW, 12 REF
SUZANNE J	5-1559	ADSORPTION OF KRYPTON ON THE BASAL PLANE OF GRAPHITE: LEED AND AUGER MEASUREMENT
	5-1856	AUGER ELECTRCN SPECTROSCOPY AND LOW ENERGY ELECTRON DIFFRACTION STUDIES OF ADSOR
	5-1857	FIRST ORDER TWO DIMENSIONAL TRANSITION: XENON ADSORBED ON THE (0001) FACE OF GRA
	5-1858	TWO-DIMENSIONAL PHASE TRANSITION IN XENON SUBMONOLAYER FILMS ADSORBED ON (0001)
	5-1859	THERMODYNAMICS AND KINETICS OF THE FIRST MONOLAYER ADSORPTION OF XENON ON THE (0
	5-1860	ENTROPY OF ADSORPTION OF XENON IN EPITAXY ON THE (0001) FACE OF GRAPHITE. (AUGER
SUZUKI HG	4-1010	AUGER ELECTRON EJECTION FROM XENON N4,5OO AND KRYPTON M4,5NN PROCESSES BY ELECTR
	6-2126	APPLICATION OF AUGER ELECTRON SPECTROSCOPY TO THE STUDY OF INTERFACE IN STEELS.
	6-2127	INTERGRANULAR FRACTURE AND IMPURITY SEGREGATION IN STEEL. STUDIES BY AUGER ELEC
	6-2128	TEMPER EMBRITTLEMENT OF TIN DOPED STEELS BY AUGER ELECTRON SPECTROSCOPY METHOD.
SUZUKI K	6-2043	EFFECTS OF CCPPER AND PHOSPHORUS ON TEMPER EMBRITTLEMENT OF MANGANESE- MOLYBDENU
SVANTESSON KG	2-0460	SPECTRUM AND DECAY OF THE RECOMBINATION RADIATION FROM STRONGLY EXCITED SILICON.
	4-1002	NEW RADIATIVE TRANSITION IN GERMANIUM.
SVENSSON S	4-0724	SCF AND LIMITED CI CALCULATIONS FOR ASSIGNMENT OF THE AUGER SPECTRUM AND OF THE
SWAIN M	6-2037	EFFECT OF A THIN DIFFUSED SOLUTE LAYER ON THE NUCLEATION OF FATIGUE CRACKS. (AUG
SWARTZ WE	4-0873	AUGER ELECTRON SPECTRA OF MAGNESIUM THIN FILMS.
	5-1392	X-RAY PHOTOELECTRON AND AUGER ELECTRON SPECTRA OF MAGNESIUM THIN FILMS.
SWEET JN	5-1456	EFFECTS OF INTERFACIAL OXIDES ON THE CONTACT RESISTANCE OF TUNGSTEN ON HEAVILY D
	5-1861	CORRELATIONS BETWEEN INTERFACIAL OXIDES AND RESISTANCE CHANGES FOR TUNGSTEN CONT
SWIFT CD	2-0259	X-RAY FLUORESCENCE YIELDS, AUGER, AND COSTER-KRONIG TRANSITION PROBABILITIES.
	2-0351	ATOMIC FLUORESCENCE YIELDS. (L-SHELL COSTER-KRONIG YIELDS, K-, L-, AND M-SHELL F
SZALKOWSKI FJ	4-0858	AUGER ELECTRON SPECTROSCOPY OF LUNAR SAMPLES.
	4-1089	AUGER ELECTRCN SPECTROSCOPY OF THE SURFACE CHEMICAL COMPOSITION OF VANADIUM OXID
	4-1090	AUGER ELECTRON SPECTROSCOPIC INVESTIGATIONS OF CHEMICAL SHIFTS IN SOME VANADIUM
	5-1205	USE OF X-RAY PHOTOELECTRON AND AUGER SPECTROSCOPY TO DETERMINE THE SURFACE COMPO
	5-1616	AUGER SPECTRCSCOPY OF FRACTURE SURFACES OF CERAMICS.
	5-1840	AUGER ELECTRON SPECTROSCOPY ON SURFACES.
	5-1862	AUGER ELECTRON SPECTROSCOPY ANALYSIS OF VANADIUM AND VANADIUM COMPOUND SURFACES.
	5-1863	USE OF X-RAY PHOTOELECTRON AND AUGER SPECTROSCOPY TO DETERMINE THE SURFACE COMPO
	5-1864	CHARACTERIZATION OF SOME VANADIUM (100) SURFACE STRUCTURES USING LEED AND AES.
SZOFRAN FR	5-1385	MORPHOLOGICAL AND ELECTRICAL PROPERTIES OF RF SPUTTERED YTTRIA-DOPED ZIRCONIA TH
SZOSTAK D	5-1366	PREFERENTIAL EVAPORATION OF INDIUM FROM GALLIUM(X) INDIUM(1-X) ARSENIDE. (AUGER
SZYMERSKA I	1-0192	AUGER ELECTRON SPECTROSCOPY.
	5-1865	LEED- AES STUDIES OF CHEMISORPTION INDUCED SULFUR SEGREGATION FROM THE BULK TO T
TAGA Y	5-1670	A STUDY OF THE IRON (111) SURFACE USING LEED AND AUGER EMISSION SPECTROSCOPY.
	5-1866	LOW ENERGY ELECTRON DIFFRACTION- AUGER ELECTRON SPECTROSCOPY STUDY OF IRON SINGL
	5-1867	CONCENTRATION CHANGE OF TIN IN THE SURFACE LAYER OF COPPER- TIN ALLOY SYSTEM DUE
	5-1868	EFFECT OF PHOSPHORUS ON THE FRICTION AND WEAR CHARACTERISTICS OF COPPER- TIN- PH
TAGLAUER E	2-0382	LOW ENERGY ION SCATTERING: ELASTIC AND INELASTIC EFFECTS. (AUGER NEUTRALIZATION)
	3-0704	SORBAS: AN APPARATUS FOR INVESTIGATING ION SCATTERING FROM SURFACES AT ENERGIES
	5-1422	OXYGEN ADSORPTION ON (110) SILVER.
	5-1423	SURFACE ANALYTICAL STUDIES USING ION SCATTERING SPECTROMETRY, AUGER ELECTRON SPE
	5-1869	DIRECT COMPARISON OF LOW ENERGY ION BACKSCATTERING WITH AUGER ELECTRON SPECTROSC
	5-1870	LOW ENERGY ICN SCATTERING AND AUGER ELECTRON SPECTROSCOPY STUDIES OF CLEAN NICKE
	5-1871	SURFACE ANALYSIS WITH LOW ENERGY ION SCATTERING.
TAHIRA S	2-0525	ENERGY AND ANGULAR DISTRIBUTION OF ELECTRONS RESULTING FROM DIRECT IONIZING COLL
	4-1003	ANGULAR DISTRIBUTION OF M4,5NN AUGER ELECTRONS EJECTED FROM KRYPTON BY ELECTRON
TAKAHASHI N	1-0193	MECHANISM OF FRICTION BY AES. (REVIEW)
	5-1872	(AUGER) MICROSCOPICAL STUDY OF FRICTIONAL PROPERTIES OF MOLYBDENUM DISULFIDE.
	6-2129	AES STUDIES ON SURFACES OF 18-8 STAINLESS STEEL FINISHED MECHANICALLY AND ELECTR
	6-2130	AUGER ELECTRON SPECTROSCOPIC STUDIES OF THE SURFACES OF 18-8 STAINLESS STEEL ELE
	6-2131	AUGER ELECTRCN SPECTRCSCOPY STUDIES ON ABRADED SURFACE OF ALLOYS.
	6-2132	SURFACE DEFORMATION CAUSED ON NATURAL MOLYBDENITE BY ABRASION. (AUGER ELECTRON S
	6-2133	STUDY OF THE INTERMITTENT MOTION OF FRICTION IN STAINLESS STEEL BY AN AUGER MICR
TAKAHASHI T	3-0641	DESIGN OF AN X-RAY DOUBLE CRYSTAL SPECTROMETER INSTALLED IN VACUUM AND THE MEASU
TAKASHIMA K	1-0199	QUANTIFICATION OF AUGER ELECTRON SPECTROSCOPY.
	5-1696	EPITAXIAL GRCWTH OF SILVER ON POTASSIUM CHLORIDE (100) SURFACE OBSERVED BY LOW E
TAKASU Y	5-1703	AUGER SPECTROSCOPY ON COPPER- NICKEL ALLOY SURFACES RELATED TO CATALYSIS.
	5-1873	WORK FUNCTION OF WELL-DEFINED SURFACE OF COPPER- NICKEL ALLOY PLATES. (AUGER ELE
	5-1874	ANALYSIS OF THE SURFACE COMPOSITION OF PASSIVE FILMS ON COPPER- NICKEL ALLOYS BY
	5-1875	RELATION BETWEEN CATALYTIC ACTIVITY PATTERN AND SURFACE COMPOSITION OF COPPER- N
TAKEDA K	5-1876	GROWTH OF GOLD FILM ON GALLIUM ARSENIDE (110) OBSERVED BY LOW ENERGY ELECTRON DI
TAKESHIMA M	4-1091	AUGER RECOMBINATION IN INDIUM ARSENIDE, GALLIUM ANTIMONIDE, INDIUM PHOSPHIDE AND
	4-1092	AUGER RECOMBINATION IN A SEMICONDUCTOR (MERCURY(1-X) CADMIUM(X) TELLURIDE) UNDER
	4-1093	EFFECT OF ELECTRON HOLE INTERACTION ON THE AUGER RECOMBINATION PROCESS IN A SEMI
TALUKDAR B	4-0800	EFFECT OF ELECTRONIC CORRELATIONS ON KLL AUGER LINES IN ELEMENTS MAGNESIUM TO SC
	4-0801	EFFECT OF ATOMIC CORRELATIONS ON KLL AUGER TRANSITIONS.

TALUKDAR B   4-0802 RELATIVISTIC STUDY OF KLL AUGER TRANSITION PROBABILITIES.
TAMARU K   4-0912 HIGH RESOLUTION L2,3M2,3M2,3 AUGER ELECTRON SPECTRA OF ARGON.
                  4-0925 CHEMICAL EFFECTS OF THE NITROGEN KLL AUGER SPECTRA OF CHEMISORBED NITROGEN ON IR
                  4-1004 ORIGIN OF THE PEAK SHAPES IN THE AUGER SPECTRUM OF OXYGEN (KL2,3L2,3) ON MOLYBDE
                  5-1530 STUDIES OF THE MECHANISM OF THE REACTION BETWEEN SULFUR AND OXYGEN ON A MOLYBDEN
                  5-1531 STUDY OF THE OXIDATION OF MOLYBDENUM SURFACES BY ENERGY LOSS SPECTROSCOPY COMBIN
                  5-1532 HIGH RESOLUTION MEASUREMENTS OF THE NITROGEN KLL AUGER TRANSITIONS OF CHEMISORBE
                  5-1533 APPLICATIONS OF AUGER ELECTRON AND ENERGY LOSS SPECTROSCOPIES TO CATALYTIC PROCE
                  5-1534 DYNAMIC INVESTIGATION OF THE MECHANISM OF REACTION BETWEEN SULFUR AND OXYGEN ON
                  5-1535 DYNAMIC ELUCIDATION OF REACTION STEPS ON A SOLID SURFACE BY ELECTRON SPECTROSCOP
                  5-1564 SEGREGATION OF SULFUR ON A MOLYBDENUM SURFACE STUDIED BY AUGER ELECTRON SPECTROS
                  5-1565 HIGH RESOLUTION AUGER ELECTRON SPECTROSCOPY OF NITRIC OXIDE ADSORBED ON A TUNGST
                  5-1698 INITIAL OXIDATION OF MOLYBDENUM STUDIED BY A HIGH RESOLUTION AUGER PHOTOELECTRON
TANIGUCHI K   2-0494 RADIATIVE AUGER EFFECT WITH AND WITHOUT ELECTRON EMISSION.
TANKHIWALE AV   2-0526 APPLICATION OF THE ZETA EFFECTIVE VALUES OF HYDROGENIC WAVE FUNCTIONS IN THE CAL
TANUMA Y   5-1481 AUGER STUDIES OF DIFFUSED IMPURITY PROFILES IN SILICON DIOXIDE- SILICON AND POLY
TAPILIN VM   4-0816 CHEMICAL SHIFT IN AUGER SPECTRA OF TUNGSTEN AND MOLYBDENUM DURING THE ADSORPTION
                  4-0922 CHEMICAL SHIFT IN THE AUGER SPECTRUM OF TUNGSTEN FROM OXYGEN ADSORPTION.
TAPLIN DMR   6-2097 STUDY OF INTERGRANULAR FRACTURE IN IRON USING AUGER SPECTROSCOPY.
TARNG ML   2-0527 ESCAPE LENGTH OF AUGER ELECTRONS.
                  5-1877 ALLCY SPUTTERING STUDIES WITH IN-SITU AUGER ELECTRON SPECTROSCOPY.
                  5-1878 AUGER ELECTRON SPECTROSCOPY STUDIES OF SPUTTER DEPOSITION AND SPUTTER REMOVAL OF
TASKARIN BT   4-0754 KLL AUGER ELECTRONS OF CADMIUM AND ANTIMONY.
TATE C   4-0892 THE KVV AUGER SPECTRUM OF LITHIUM METAL.
TAUBER G   5-1369 EQUILIBRIUM SURFACE SEGREGATION OF CARBON ON IRON (100) FACES. (AUGER EFFECT)
                  5-1940 CHEMICAL AND STRUCTURAL ANALYSIS OF IRON SURFACES BY MEANS OF AUGER ELECTRONIC S
TAVERNIER M   2-0293 NEW SATELLITE BANDS CORRESPONDING TO DOUBLE L IONIZATION. (OF AUGER ELECTRONS)
TAWARA H   2-0528 X-RAY EMISSION CROSS SECTIONS AND FLUORESCENCE YIELDS FOR LIGHT ATOMS AND MOLECU
TAYLOR NJ   1-0194 ROLE OF AUGER ELECTRON SPECTROSCOPY IN SURFACE ELEMENTAL ANALYSIS.
                  1-0195 AES FOR SURFACE ANALYSIS. (REVIEW, 24 REFS)
                  1-0196 TECHNIQUE OF AUGER ELECTRON SPECTROSCOPY IN SURFACE ANALYSIS. (REVIEW, 76 REFS)
                  3-0705 RESOLUTICN AND SENSITIVITY CONSIDERATIONS OF AN AUGER ELECTRON SPECTROMETER BASE
                  3-0706 AUGER ELECTRCN SPECTROMETER AS TOOL FOR SURFACE ANALYSIS. (CONTAMINATION MONITOR
                  4-0784 AUGER SPECTROSCOPY OF SILICON.
                  5-1879 THIN REACTION LAYERS AND SURFACE STRUCTURE OF SILICON (111).
                  5-1880 AUGER SPECTRCSCOPY OF SILICON. (REPLY TO COMMENTS)
                  5-1921 AUGER ELECTRON SPECTROSCOPY OF CLEAN GALLIUM ARSENIDE.
                  5-1922 AUGER ELECTRON SPECTROSCOPY OF GALLIUM ARSENIDE PHOTOSURFACES.
                  5-1962 DEPOSITION AND AUGER ANALYSIS OF DEPOSITED SILICON DIOXIDE ON ALUMINUM(X) GALLIU
TAYLOR PA   5-1462 ROLE OF IMPURITIES IN THE STABILITY OF ZINC OXIDE.
                  5-1463 INTERACTION CF CHLORINE WITH THE ZINC OXIDE (0001)- ZINC SURFACE: A LEED, AUGER
                  5-1588 INTERACTION OF CESIUM WITH CLEAN ZINC OXIDE SURFACES. (AUGER SPECTROSCOPY)
                  5-1881 SOME PROPERTIES OF CLEAN AND ALKALI METAL COVERED ZINC OXIDE SURFACES. (LEED, AE
                  5-1882 STABILITY OF CESIUM COVERED ZINC OXIDE SURFACES.
TAYLOR RI   6-2134 GRAIN BOUNDARY SEGREGATION IN ALUMINUM OXIDE. (AUGER SPECTROSCOPY)
TAYLOR TN   5-1883 THORIUM (100) CRYSTAL FACE. A STUDY OF ITS CLEANING, SURFACE STRUCTURE, AND INTE
                  5-1884 CARBON MONOXIDE ADSORPTION ON NICKEL (110). (AUGER EFFECT)
TAZHIBAEVA S   2-0538 IONIZATION LOSSES IN (AUGER) SPECTRA OF ELECTRON INELASTIC SCATTERING FROM RARE
TEJEDA J   2-0306 PHOTOELECTRIC PROPERTIES OF MAGNESIUM(2) SILICIDE, MAGNESIUM(2) GERMANIUM, AND M
                  5-1885 VALENCE BAND OF MAGNESIUM STANNIDE DETERMINED BY AUGER AND PHOTOEMISSION SPECTRO
TEPPO EA   5-1746 SURFACE CHARACTERISTICS RELATED TO LASER DAMAGE OF LITHIUM NIOBATE AND POTASSIUM
TERRY LE   5-1898 AUGER SPECTROSCOPY ANALYSIS OF PALLADIUM SILICIDE FILMS.
                  5-1899 AES STUDY OF THE SILICIDE FORMATION IN 85% NICKEL- 15% PLATINUM ALLOY FILMS.
                  5-1900 COMPOSITION PROFILES AND SCHOTTKY BARRIER HEIGHTS OF SILICIDES FORMED IN NICKEL-
TESCARI M   5-1267 AUGER SPECTROSCOPY AND LOW ENERGY ELECTRON DIFFRACTION STUDIES OF THE CHEMISORPT
THARP LN   2-0495 INELASTIC SCATTERING OF LOW ENERGY ELECTRONS FROM SURFACES.
                  3-0625 INVESTIGATION OF HIGH RESOLUTION SCANNING TECHNIQUES FOR AUGER ELECTRON SPECTROS
                  4-1094 ENERGY SPECTRA OF INELASTICALLY SCATTERED ELECTRONS AND LEED STUDIES OF TUNGSTEN
THERIAULT GE   1-0197 TECHNIQUE OF PREPARING POWDER SAMPLES FOR AES AND/OR ESCA ANALYSIS.
THIEME F   5-1171 ISOMORPHISM AND PURITY OF COPPER VAPOR LEPOSITED ON A COPPER (111) SURFACE. (AUG
THOMAS EW   2-0255 RADIATIONLESS DEEXCITATION OF EXCITED HELIUM ATOMS AT SURFACES. (AUGER PROCESS)
THOMAS GJ   5-1886 CARBIDE FORMATION ON NIOBIUM SURFACES DURING HIGH TEMPERATURE PROTON IRRADIATION
THOMAS JH   5-1887 DETECTABILITY LIMITS FOR BORON AND PHOSPHORUS IN SILICON BY AUGER ELECTRON SPECT
THOMAS JM   2-0319 AUGER ELECTRON EMISSION SPECTRA: A SIMPLE TREATMENT OF ENERGIES AND INTENSITIES
THOMAS MJB   1-0197 TECHNIQUE OF PREPARING POWDER SAMPLES FOR AES AND/OR ESCA ANALYSIS.
THOMAS MT   5-1888 AUGER SPECTROSCOPY OF SUBMONOLAYER GOLD DEPOSITIONS ON SILICON.
THOMAS RE   5-1391 EFFECTS OF SULFUR, BARIUM, AND CALCIUM ON IMPREGNATED CATHODE SURFACES. (AES)
                  5-1889 DIFFUSION MEASUREMENTS IN THIN FILMS UTILIZING WORK FUNCTION CHANGES: CHROMIUM I
                  5-1890 AUGER EXAMINATION OF CONTAMINANTS IN THIN FILM METALLIZATIONS.
THOMAS S   4-0870 SOFT X-RAY APPEARANCE POTENTIAL SPECTROSCOPY IN A DISPLAY LEED SYSTEM.
                  4-1095 PLASMON ENERGY GAIN IN THE BERYLLIUM AUGER SPECTRUM.
                  4-1096 ELECTRON IRRADIATION EFFECT IN THE AUGER ANALYSIS OF SILICON DIOXIDE.
                  5-1388 ANODIC OXIDATION OF GALLIUM ARSENIDE PHOSPHIDE. (AES)
                  5-1539 A STUDY OF ACTIVATED SILICON SURFACES FOR ELECTROLESS NICKEL PLATING BY AES.
                  5-1827 AUGER ELECTRON SPECTROSCOPY DETERMINATION OF THE OXYGEN- SILICON RATIO IN SPIN O
                  5-1891 SURFACE COMPOSITIONAL CHANGES WITH ELECTRON BOMBARDMENT OBSERVED BY AUGER ELECTR
                  5-1892 SURFACE ENRICHMENT OF INDIUM IN EVAPORATED GOLD- INDIUM FILMS. (AUGER ELECTRON S
                  5-1893 AES ION SPUTTERING ANALYSIS AND THE SURFACE COMPOSITION OF TITANIUM(IV) OXIDE.
                  5-1894 AUGER SPECTROSCOPY STUDY OF THE ADSORPTION OF RUBIDIUM ON MOLYBDENUM (100).
                  5-1895 QUANTITATIVE AUGER ELECTRON SPECTROSCOPY AND LOW ENERGY ELECTRON DIFFRACTION STU
                  5-1896 LEED- AUGER SPECTROSCOPY STUDY OF THE ADSORPTION OF ALKALI METALS ON MOLYBDENUM
                  5-1897 ELECTRON SPECTROSCOPY OF METAL BLACKS.
                  5-1898 AUGER SPECTROSCOPY ANALYSIS OF PALLADIUM SILICIDE FILMS.
                  5-1899 AES STUDY OF THE SILICIDE FORMATION IN 85% NICKEL- 15% PLATINUM ALLOY FILMS.
                  5-1900 COMPOSITION PROFILES AND SCHOTTKY BARRIER HEIGHTS OF SILICIDES FORMED IN NICKEL-
THOMAS TD   4-1058 AUGER ELECTRON SPECTRUM AND IONIZATION POTENTIALS OF THE HYDROGEN FLUORIDE MOLEC
THOMPSON DO   6-2116 CHARACTERIZING OF ALUMINUM FATIGUE DAMAGE BY PHOTOEMISSION, ELLIPSOMETRY, AND AU
THOMPSON M   1-0198 ANALYTICAL ASPECTS OF AUGER SPECTROSCOPY. (REVIEW)
THOMSON GM   2-0529 ELECTRON ION COINCIDENCE STUDIES OF HIGHLY IONIZING ARGON(+)- ARGON COLLISIONS.
                  2-0530 EXTENDED ION- ELECTRON COINCIDENCE MEASUREMENTS OF THE VIOLENT ARGON(+)- ARGON C

THORNTON ST	2-0336	DECAY OF HOLE STATES IN OXYGEN-16 VIA A "NUCLEAR AUGER EFFECT".
TILES B	6-1997	DETERMINATION OF THE INFLUENCE GRAIN BOUNDARY IMPURITIES ON THE PROPERTIES OF SI
TILLMANN L	5-1445	AUGER ELECTRON SPECTROSCOPY ON THE FRACTURE OF NICKEL TUNGSTEN MATERIALS AND SIL
	6-2046	RECRYSTALLIZATION OF TUNGSTEN FIBERS IN NICKEL MATRIX COMPOSITES. (AUGER ELECTRO
TIPPING DW	3-0615	IONIZATION SPECTROMETER FOR ELEMENTAL ANALYSIS OF SURFACES.
TISLJAR-LENTULIS G	2-0531	BIOLOGICAL RADIATION EFFECT ON INCLUSION OF MODERATELY HEAVY NUCLEI IN THE TISSU
TITOV AP	3-0617	DRYING OF BUTADIENE- NITRILE RUBBERS IN AN AUGER MACHINE.
TOBUREN LH	2-0532	K-SHELL IONIZATION OF CARBON BY FAST PROTONS. (AUGER ELECTRON YIELDS OF METHANE,
	2-0533	K-SHELL IONIZATION BY FAST PROTONS.
	2-0534	DISTRIBUTION IN ENERGY AND ANGLE OF ELECTRONS EJECTED FROM XENON BY 0.3 TO 2.0 M
	2-0535	AUGER ELECTRON SPECTROSCOPY APPLIED TO INNER SHELL IONIZATION BY FAST CHARGED PA
	4-1097	K-, L-, AND M-AUGER AND L- COSTER-KRONIG SPECTRA OF PLATINUM.
	4-1109	AUGER SPECTRA OF CARBON AND ARGON FOLLOWING IONIZATION BY EQUAL VELOCITY ALPHA P
TODD CJ	1-0003	COMBINED AUGER ELECTRON SPECTROSCOPY AND SCANNING ELECTRON MICROSCOPY.
	5-1901	ELECTRON SPECTROSCOPY FROM SOLID SURFACES IN ULTRAHIGH VACUUM. CURRENT PROGRESS
	5-1902	THERMALLY INDUCED PROCESSES AT GOLD- GALLIUM ARSENIDE INTERFACES: AN ASSESSMENT
	5-1903	RELATIVE ESCAPE DEPTH OF AUGER AND PHOTOELECTRONS FROM A SOLID SURFACE.
TODD G	5-1160	ADSORPTION AND CONDENSATION OF COPPER ON TUNGSTEN SINGLE CRYSTAL SURFACES.
	5-1904	AUGER ELECTRON SPECTROSCOPY AT HIGH SPATIAL RESOLUTION AND PRIMARY BEAM CURRENTS
TOENNIES JP	5-1905	SCATTERING OF MOLECULAR BEAMS FROM SURFACES. (AUGER SPECTROSCOPY)
TOKUTAKA H	1-0199	QUANTIFICATION OF AUGER ELECTRON SPECTROSCOPY.
	3-0683	ULTRAHIGH VACUUM EVAPORATOR FOR LEED AND AUGER EMISSION STUDIES.
	3-0707	ULTRAHIGH VACUUM EVAPORATOR AND MULTIPLE CLEAVAGE DEVICE FOR LOW ENERGY ELECTRON
	5-1347	(100) SURFACES OF ALKALI HALIDES. PART-1: AIR AND VACUUM CLEANED SURFACES.
	5-1348	GROWTH OF SILVER ON POTASSIUM CHLORIDE OBSERVED BY LEED AND AUGER EMISSION SPECT
	5-1696	EPITAXIAL GROWTH OF SILVER ON POTASSIUM CHLORIDE (100) SURFACE OBSERVED BY LOW E
	5-1906	STRUCTURES OF SUBSTRATE CRYSTALS AND EPITAXIAL GROWTH OF METAL ON THEM OBSERVED
	5-1907	CRYSTAL SURFACES OF ALKALI HALIDES AND THIN FILM GROWTH OBSERVED BY LEED AND AES
	5-1908	(100) SURFACES OF ALKALI HALIDES. PART-2: ELECTRON STIMULATED DISSOCIATION.
TOLPYGO EI	2-0536	NONRADIATIVE AUGER RECOMBINATION OF ELECTRONS ON DONOR- ACCEPTOR PAIRS.
TOLPYGO KB	2-0536	NONRADIATIVE AUGER RECOMBINATION OF ELECTRONS ON DONOR- ACCEPTOR PAIRS.
TOMASETTA L	2-0467	IDENTIFICATION OF AUGER ELECTRONS IN GALLIUM ARSENIDE.
TOMPKINS HG	6-2135	DIFFUSION OF COBALT OUT OF COBALT HARDENED GOLD MEASURED WITH AUGER ELECTRON SPE
	6-2136	SURFACE DIFFUSION OF TIN OVER A GOLD SURFACE OBSERVED WITH AUGER ELECTRON SPECTR
TONEMAN LH	5-1196	AUGER SPECTROSCOPIC STUDY OF THE SURFACE COMPOSITION OF PLATINUM- TIN IN ULTRAHI
	5-1934	SURFACE ENRICHMENT IN COPPER(3) GOLD AND GOLD(3) COPPER ALLOYS.
	6-2005	QUANTITATIVE STUDY OF THE SURFACE COMPOSITION OF BINARY ALLOYS BY AUGER SPECTROS
TONG HC	5-1909	LOW INTRINSIC STRESS AND GOOD ADHESION ALLOY THIN FILMS. (TITANIUM AND CHROMIUM
TOPLER J	5-1432	ADSORPTION OF WATER VAPOR ON CLEAN CLEAVED GERMANIUM. PART-1: STRUCTURAL AND KIN
TORELLI G	2-0305	MEASUREMENT OF THE AUGER EFFECT IN THE MU-4 HELIUM(2S)(+) IONIC SYSTEM.
TOTH L	6-1998	COMPOSITION VERSUS DEPTH PROFILES OBTAINED WITH AUGER ELECTRON SPECTROSCOPY OF A
TRABER WF	5-1909	LOW INTRINSIC STRESS AND GOOD ADHESION ALLOY THIN FILMS. (TITANIUM AND CHROMIUM
TRACY CE	5-1222	INJECTION OF IONS INTO GLASS FROM A GLOW DISCHARGE. (AUGER SPECTROSCOPY)
TRACY JC	1-0200	AUGER ELECTRON SPECTROSCOPY FOR SURFACE ANALYSIS.
	1-0201	RECENT ADVANCES IN SURFACE CHARACTERIZATION USING LOW ENERGY ELECTRON DIFFRACTIO
	2-0537	ELECTRON ESCAPE DEPTHS IN ALUMINUM. (AUGER ELECTRONS)
	3-0672	HIGH SENSITIVITY AUGER ELECTRON SPECTROMETER.
	3-0708	AUGER ELECTRON SPECTROMETER PREAMPLIFIER.
	4-1098	INFLUENCE OF COSTER-KRONIG PROCESSES ON L-AUGER YIELDS.
	5-1910	SURFACE CHEMICAL ANALYSIS BY AUGER ELECTRON SPECTROSCOPY AND APPEARANCE POTENTIA
	5-1911	STRUCTURAL INFLUENCES ON ADSORPTION ENERGY. PART-2: CARBON MONOXIDE ON NICKEL (1
	5-1912	THE KINETICS OF OXYGEN ADSORPTION ON THE (112) AND (110) PLANES OF TUNGSTEN.
	5-1913	STRUCTURAL INFLUENCES ON ADSORBATE BINDING ENERGY. PART-1: CARBON MONOXIDE ON (1
	4-1099	SEPARATION OF THE SPECTRA OF PHOTO AND AUGER ELECTRONS.
TRAPEZNIKOV VA	5-1914	AUGER ELECTRON SPECTROSCOPY STUDIES OF ARSENIC VACANCIES IN (-1-1-1) GALLIUM ARS
TRUEBA A	5-1915	SURFACE CARRIER CONCENTRATION AND AS EVAPORATION IN HEAT TREATED GALLIUM ARSENID
TSAI JC	4-0996	IN-DEPTH PROFILES OF PHOSPHORUS ION IMPLANTED SILICON BY AUGER SPECTROSCOPY AND
	5-1916	IN-DEPTH PROFILE DETECTION LIMITS OF NITROGEN IN GALLIUM PHOSPHIDE (AND) NITROGE
TSANG T	2-0577	ATOMIC N-SHELL VACANCY WIDTHS IN SOLIDS.
	2-0578	INVESTIGATION OF LOW-Z COSTER-KRONIG TRANSITIONS BY AUGER AND PHOTOELECTRON SPEC
	2-0579	FREE ATOM BEHAVIOR OF AUGER ELECTRONS IN SOLIDS.
	3-0718	WIDTHS OF ATOMIC M-SHELL VACANCY STATES AND QUASIATOMIC ASPECTS OF RADIATIONLESS
	4-1121	QUASIATOMIC LMM AUGER SPECTRA OF SOLID COPPER AND ZINC.
	4-1122	L-S COUPLING INTERPRETATION OF HIGH RESOLUTION LMM AUGER SPECTRA OF COPPER AND Z
TSUJI Y	1-0202	AUGER ELECTRON ANALYSIS AND ITS APPLICATIONS.
TSVEIMAN EV	2-0538	IONIZATION LOSSES IN (AUGER) SPECTRA OF ELECTRON INELASTIC SCATTERING FROM RARE
	4-1030	AUGER ELECTRON SPECTRUM OF OSMIUM AT ENERGIES UP TO 300 EV.
	4-1100	AUGER ELECTRON SPECTRA OF SOME RARE EARTH METALS. (LANTHANUM, PRASEODYMIUM, NEOD
TU KN	5-1917	STRUCTURE AND GROWTH KINETICS OF NICKEL(2) SILICIDE ON SILICON. (AUGER SPECTROSC
TU Y	5-1918	SEGREGATION OF CADMIUM TO A (111) SILVER CHLORIDE SURFACE. (AUGER SPECTROSCOPY)
TUAN BT	2-0539	ROLE OF AUGER EFFECT IN THERMOSTIMULATED PHENOMENA IN IONIC CRYSTALS.
TURGOOSE S	5-1628	DETECTION OF LEAD, BY AUGER SPECTROSCOPY, ON IRON INHIBITED IN LEAD AZELATE SOLU
TUROS A	5-1627	COMPARISON OF SURFACE LAYER ANALYSIS TECHNIQUES. (AUGER ELECTRON SPECTROSCOPY, S
TUZI Y	1-0133	SAMPLES HEATED DIRECTLY BY ELECTRIC CURRENT OBSERVED BY AUGER ELECTRON SPECTROSC
	1-0148	BIBLIOGRAPHY ON AUGER ELECTRON SPECTROSCOPY.
	3-0641	DESIGN OF AN X-RAY DOUBLE CRYSTAL SPECTROMETER INSTALLED IN VACUUM AND THE MEASU
	5-1648	CARBON CONTAMINATION PRODUCED BY ELECTRON BOMBARDMENT. POLYCRYSTALLINE MOLYBDENU
	5-1649	CLEANING PROCEDURES OF MOLYBDENUM (110) SURFACE.
	5-1650	INITIAL OXIDATION OF THE MOLYBDENUM (110) SURFACE OBSERVED BY AES AND LEED.
TYUTIKOV AM	4-0891	USE OF X-RAY SPECTROSCOPIC DATA TO EXPLAIN THE FINE STRUCTURE OF THE DEPENDENCE
TZOAR N	2-0361	MANY BODY EFFECTS IN AUGER DEEXCITATION OF ATOMS NEAR SOLIDS.
UCHIYAMA M	3-0636	NEW SEM AUGER MICROANALYZER WITH ULTRAHIGH VACUUM CAPABILITY.
UCHIYAMA T	5-1127	APPLICATION OF AUGER ELECTRON SPECTROSCOPY TO THE STUDY OF DISCOLORING OF ALUMIN
	5-1487	DETECTION OF ORGANIC CONTAMINANTS ON ALUMINUM SURFACES BY AUGER ELECTRON SPECTRO
UEBBING JJ	4-1082	VARIAN CHART OF AUGER ELECTRON ENERGIES.
	5-1919	USE OF AUGER ELECTRON SPECTROSCOPY IN DETERMINING THE EFFECT OF CARBON AND OTHER
	5-1920	AUGER ELECTRON SPECTROSCOPY OF CONTAMINATED GALLIUM ARSENIDE SURFACES.
	5-1921	AUGER ELECTRON SPECTROSCOPY OF CLEAN GALLIUM ARSENIDE.
	5-1922	AUGER ELECTRON SPECTROSCOPY OF GALLIUM ARSENIDE PHOTOSURFACES.
UEDA K	4-0917	SECONDARY ELECTRON EMISSION FROM IRON SINGLE AND POLYCRYSTALS.

UEDA K	5-1923	IMPURITY CONCENTRATIONS ON METAL SURFACES BY AUGER ELECTRON SPECTROSCOPY.
	5-1924	STUDIES OF INITIAL OXIDATION ON SILICON- IRON ALLOY (100) BY MEANS OF WORK FUNCT
	5-1925	PHOTOELECTRIC WORK FUNCTION AND AUGER ELECTRON SPECTROSCOPY OF 5.59 ATOMIC PERCE
	5-1926	PHOTOELECTRIC WORK FUNCTION STUDY ON IRON (100) SURFACE COMBINED WITH AUGER ELEC
	5-1927	INITIAL OXIDATION STUDIES ON IRON (100) BY MEANS OF PHOTOELECTRIC WORK FUNCTION
	5-1928	INITIAL OXIDATION STUDIES BY MEANS OF PHOTOELECTRIC WORK FUNCTION MEASUREMENTS A
ULLRICH BM	5-1182	EFFECT OF OXIDIZING AMBIENTS ON PLATINUM SILICIDE FORMATION. PART-2: AUGER AND B
URCH DS	1-0203	X-RAY SPECTROSCOPY. (REVIEW, AUGER AND PHOTOELECTRON SPECTROSCOPY)
USAMI S	5-1929	ADSORPTION OF NITRIC OXIDE ON TUNGSTEN.
USAROV EN	2-0372	EXCITON SPECTRA OF HEAVILY DOPED CADMIUM SULFIDE SINGLE CRYSTALS. (AUGER PROCESS
UTRIAINEN J	2-0226	INFLUENCE OF CHEMICAL BONDING ON THE LOW ENERGY K-ALPHA SPECTRUM OF SILICON. (RA
	2-0403	EVIDENCE OF RADIATIVE AUGER TRANSITIONS IN THE FREE ARGON ATOM.
	2-0404	EVIDENCE OF RADIATIVE AUGER TRANSITIONS IN THE FREE ARGON ATOM.
	2-0540	ELECTRON- HCLE INTERACTION EFFECTS IN THE KL2,3L2,3 RADIATIVE AUGER SPECTRUM OF
	4-0723	INTERPRETATION OF KL2,3L2,3 RADIATIVE AUGER THRESHOLDS IN METALS AND SEMICONDUCT
	4-1101	INTERPRETATION OF THE LOW ENERGY K-BETA SPECTRUM IN SOLIDS.
	4-1102	SCANDIUM L X-RAY EMISSION FROM PURE METAL AND SCANDIUM OXIDE.
VAITKUS J	2-0258	TEMPERATURE DEPENDENCE OF THE EDGE EMISSION OF GALLIUM SELENIDE SINGLE CRYSTALS
	4-1103	ELECTRON HOLE SCATTERING AND (AUGER) RECOMBINATION OF NONEQUILIBRIUM CHARGE CARR
VAKHIDOV UV	4-0742	SECONDARY ELECTRON EMISSION FROM TUNGSTEN AND MOLYBDENUM SINGLE CRYSTALS.
VALJAKKA J	4-1102	SCANDIUM L X-RAY EMISSION FROM PURE METAL AND SCANDIUM OXIDE.
VAN BOMMEL AJ	1-0186	AUGER ELECTRON SPECTROSCOPY AS A TOOL FOR MEASURING THE DIFFUSION OF FOREIGN ATO
	5-1930	LOW ENERGY ELECTRON DIFFRACTION AND AUGER ELECTRON SPECTROSCOPY STUDY OF PHOSPHI
	5-1931	LOW ENERGY ELECTRON DIFFRACTION, AUGER ELECTRON SPECTROSCOPY, AND PHOTOEMISSION
	5-1932	LOW ENERGY ELECTRON DIFFRACTION AND AUGER ELECTRON OBSERVATIONS OF THE SILICON C
VAN DER BORGH NE	4-1026	ELECTROCHEMISTRY AT THIN SOLID FILMS. (AUGER SPECTRA)
VAN DER MEULEN YJ	5-1250	AUGER AND ELLIPSOMETRIC STUDY OF PHOSPHORUS SEGREGATION IN OXIDIZED DEGENERATE S
	5-1251	AUGER ANALYSIS OF CHLORINE IN HYDROGEN CHLORIDE OR CHLORINE GROWN SILICON DIOXID
	5-1935	SOME RECENT STUDIES OF VERY THIN SILICON DIOXIDE FILMS.
VAN DER WIEL MJ	2-0541	EXCITATION OF INNER SHELL ELECTRONS BY PHOTONS AND FAST ELECTRONS.
VAN ECK J	2-0517	CHARGE STATE DEPENDENCE OF THE ARGON L-SHELL FLUORESCENCE YIELD STUDIED BY ATOMI
VAN OOSTROM A	5-1933	APPLICATION OF AES TO THE STUDY OF SELECTIVE SPUTTERING OF THIN FILMS.
	5-1936	INVESTIGATION OF VACUUM BREAKDOWN USING AUGER SPECTROSCOPY.
VAN SANTEN RA	5-1934	SURFACE ENRICHMENT IN COPPER(3) GOLD AND GOLD(3) COPPER ALLOYS.
VAN TOOREN A	1-0186	AUGER ELECTRON SPECTROSCOPY AS A TOOL FOR MEASURING THE DIFFUSION OF FOREIGN ATO
VAN VELZEN WJM	1-1657	CHARACTERISTIC ENERGY LOSS AND AUGER SPECTRUM OF GALLIUM PHOSPHIDE (110)
VANCE DW	2-0542	ANGLE OF INCIDENCE EFFECTS IN AUGER ELECTRON EMISSION FROM CLEAN MOLYBDENUM.
	2-0543	AUGER ELECTRON EMISSION FROM CLEAN AND CARBON CONTAMINATED MOLYBDENUM BOMBARDED
VANECK J	1-1571	FLUORESCENCE YIELD OF CARBON. (AUGER MEASUREMENTS)
VARMA MN	5-1508	SURFACE SEGREGATION AND INTERACTION OF OXYGEN AND NITROGEN WITH ZIRCONIUM IN NIO
	5-1849	DIFFERENTIAL AUGER SPECTROMETRY. (NIOBIUM- ZIRCONIUM ALLOY ANALYSIS)
	5-1937	DIFFERENCE AUGER SPECTROSCOPY FOR STUDYING SMALL QUANTITIES OF ELEMENTS ON METAL
VARMAZIS C	5-1938	EFFECT OF HIGH TEMPERATURE HEAT TREATMENT ON PENETRATION DEPTH OF SUPERCONDUCTIN
	6-2137	EFFECT OF SURFACE METALLURGY ON THE PENETRATION DEPTH AND RF BREAKDOWN FIELD OF
VASILEV MA	5-1939	AUGER SPECTROSCOPIC STUDY OF THE SURFACE OF IRON AFTER IONIC BOMBARDMENT AND AIR
VASINA P	3-0709	SECONDARY ELECTRON SPECTROMETER.
VAYRYNEN J	4-0729	AUGER ELECTRON STUDIES OF METAL VAPORS.
	4-0733	HIGH RESOLUTION L2,3M4,5M4,5 AUGER SPECTRUM OF FREE ZINC ATOMS.
VEDRINSKII RV	4-1104	AUGER TRANSITIONS AND SHAPE OF X-RAY SPECTRUM.
VELICKY B	2-0539	ROLE OF AUGER EFFECT IN THERMOSTIMULATED PHENOMENA IN IONIC CRYSTALS.
VELURI VR	4-1105	L-SHELL FLUORESCENCE YIELDS OF DOUBLE VACANCY STATES IN LEAD.
VENNIK J	1-0204	AUGER ELECTRON SPECTROSCOPY.
	2-0350	USE OF CHARACTERISTIC IONIZATION LOSSES IN AUGER ELECTRON EMISSION SPECTROSCOPY.
	3-0610	AVERAGING, CURVE FITTING, AND RELATED TECHNIQUES IN AUGER ELECTRON EMISSION SPEC
	4-0832	LIFETIME BROADENING IN X-RAY INDUCED PHOTOEMISSION SPECTROSCOPY AND AUGER SPECTR
	4-0882	RELAXATION PHENOMENA INVOLVED IN THE L3M4,5M4,.5M4,.5:1G4 AUGER PROCESS IN ZINC
	4-0883	ATOMIC AND EXTRAATOMIC STATIC RELAXATION ENERGIES IN THE L3M4,5M4,5 AUGER PROCES
	4-0884	DETAILED ANALYSIS OF THE L3M4,5M4,5: 1G4 AUGER TRANSITION IN ATOMIC ZINC: IMPROV
	5-1337	COMBINED LOW ENERGY ELECTRON DIFFRACTION, AUGER ELECTRON SPECTROSCOPY, AND X-RAY
	5-1338	AUGER ELECTRON EMISSION SPECTROSCOPY (AES) AND LOW ENERGY ELECTRON DIFFRACTION (
	5-1339	(AUGER) ELECTRON SPECTROSCOPY OF TRANSITION METAL OXIDE SURFACES.
	5-1340	INELASTIC EFFECTS AND STRUCTURE IN THE AUGER ELECTRON EMISSION SPECTRA OF VANADI
	5-1341	STRUCTURE AND COMPOSITION OF METAL SURFACES: A REVIEW OF RECENT LEED, AES, AND X
	5-1941	AUGER EMISSION SPECTROSCOPY VANADIUM PENTOXIDE (010) AND VANADIUM (100) SURFACES
VENTGOPALA RP	2-0544	ATOMIC VACANCY DISTRIBUTIONS PRODUCED BY INNER SHELL IONIZATION (AUGER AND RADIA
VERBEEK H	4-0982	ENERGY WIDTHS OF ROENTGEN LEVELS BY AUGER ELECTRON SPECTROSCOPY.
VERDI FW	6-2079	GRAIN BOUNDARY SEGREGATION OF IMPURITIES IN METALS AND INTERGRANULAR BRITTLE FRA
VERHEY LK	2-0545	CHARGE EXCHANGE OF LOW ENERGY HELIUM IONS AND ATOMS SCATTERED FROM A COPPER SING
	2-0546	CHARGE EXCHANGE OF LOW ENERGY HELIUM(+) IONS (LESS THAN 10 KEV) SCATTERED FROM A
VERINK ED	5-1614	AUGER ELECTRON SPECTROSCOPIC ANALYSIS OF ANODIC FILMS ON ALUMINUM.
VESELY M	5-1396	AUGER SPECTROSCOPY AND LEED OF SILICON SURFACES.
	5-1397	AUGER ELECTRON SPECTROSCOPY OF SILICON SURFACES.
VIARIS DE LESEGNO P	2-0547	ARGON AUGER EMISSION BY ARGON(+) ION BOMBARDMENT OF SOME METALS.
	2-0548	RELATIVE EMISSION INTENSITIES OF AUGER ELECTRONS FROM LIGHT METALS (LITHIUM, SOD
	2-0549	ORIGIN OF ICN INDUCED AUGER ELECTRON EMISSION FROM METALS.
	4-0877	AUGER ELECTRON EMISSION BY ION BOMBARDMENT OF LIGHT METALS.
VIEFHAUS H	5-1369	EQUILIBRIUM SURFACE SEGREGATION OF CARBON ON IRON (100) FACES. (AUGER EFFECT)
	5-1940	CHEMICAL AND STRUCTURAL ANALYSIS OF IRON SURFACES BY MEANS OF AUGER ELECTRONIC S
	6-2035	NEW METHODS IN SURFACE ANALYSIS AND THEIR APPLICATION FOR STUDIES ON IRON AND ST
VIEHBOCK FP	6-1997	DETERMINATION OF THE INFLUENCE GRAIN BOUNDARY IMPURITIES ON THE PROPERTIES OF SI
VIEL L	2-0267	IONIZATION CROSS SECTION OF THE L2,3 SHELL OF ALUMINUM IN A NEON(+)- ALUMINUM CO
	2-0268	INTERPRETATION OF ENERGY LOSS BY AUGER ELECTRONS EMITTED DURING ION BOMBARDMENT.
	2-0269	ENERGY DISTRIBUTION OF SECONDARY ELECTRONS EMITTED BY INDIUM PHOSPHIDE AND GALLI
	2-0550	EMISSION OF AUGER ELECTRONS BY ION BOMBARDMENT.
	2-0551	OBSERVATION OF AUGER ELECTRONS IN THE ENERGY SPECTRUM OF SECONDARY ELECTRONS EMI
	4-0776	ELECTRONIC EMISSION FROM SOLID TARGETS BOMBARDED BY NOBLE GAS IONS (10-100 KEV):
	4-0942	SECONDARY ELECTRON SPECTRA OF SOME SOLID TARGETS UNDER IONIC BOMBARDMENT.
	4-0988	ENERGY SPECTRUM AND ANGULAR DISTRIBUTION OF SECONDARY ELECTRONS EMITTED BY IONIC
	4-1106	AUGER SPECTRA OF SOME FREE ELEMENTS AND BINARY COMPOUNDS BOMBARDED BY 60 KEV ARG
VIERMANS L	5-1941	AUGER EMISSION SPECTROSCOPY VANADIUM PENTOXIDE (010) AND VANADIUM (100) SURFACES

VIEW C	2-0334	MERCURY AND GOLD ATOMIC CLOUDS REARRANGEMENT FOLLOWING INTERNAL CONVERSION AND E
VIG JR	5-1942	SURFACE STUDIES FOR QUARTZ RESONATORS. (CLEANING, AUGER SPECTROSCOPY)
	5-1943	SURFACE PREPARATION AND CHARACTERIZATION TECHNIQUES (AUGER, CLEANING) FOR QUARTZ
VIOLLIER RD	2-0478	PENETRATION EFFECTS IN CONVERTED MUONIC TRANSITIONS. (AUGER COEFFICIENTS)
VISCAKAS J	2-0258	TEMPERATURE DEPENDENCE OF THE EDGE EMISSION OF GALLIUM SELENIDE SINGLE CRYSTALS
VISWANATH Y	5-1944	OXIDATION OF CARBON ON TUNGSTEN (100) AND MOLYBDENUM (100). (AUGER ELECTRON SPEC
VISWANATHAN R	6-2138	TEMPER EMBRITTLEMENT IN A NICKEL- CHROMIUM STEEL CONTAINING PHOSPHORUS AS AN IMP
	6-2139	EFFECT OF MICROSTRUCTURE ON GRAIN BOUNDARY SEGREGATION OF PHOSPHORUS IN A CHROMI
	6-2140	LONG TIME ISOTHERMAL TEMPER EMBRITTLEMENT IN NICKEL- CHROMIUM- MOLYBDENUM- VANAD
VITALE A	2-0305	MEASUREMENT OF THE AUGER EFFECT IN THE MU-4 HELIUM(2S)(+) IONIC SYSTEM.
VLACHOVA B	5-1945	AUGER ELECTRON SPECTROSCOPY OF SILICON SURFACES.
	5-1946	AUGER SPECTROSCOPY OF SILICON.
VLASENKO VA	5-1759	DECREASE OF CARBON CONCENTRATION ON THE SURFACE OF MOLYBDENUM CARBIDE DURING HEA
VOLZ DJ	2-0552	AUGER ELECTRONS FROM ARGON WITH ENERGIES 150-210 EV PRODUCED BY HYDROGEN(+) IMPA
VON BACHO PS	5-1947	UPS, ESCA, AND AUGER SPECTROSCOPY STUDY OF VACUUM DEPOSITED SILVER HALIDE FILMS.
VON BARTH ULF	2-0239	INTERPRETATION OF HIGH ENERGY X-RAY SATELLITES OF L2,3 EMISSION BANDS OF SODIUM,
VOOS M	4-0777	ELECTRON HOLE DROPS IN GERMANIUM- SILICON ALLOYS. (AUGER RECOMBINATION)
VORONKOV AA	4-0743	AUGER L-SERIES ELECTRONIC SPECTRA ACCOMPANYING THE DECAY OF ERBIUM-160, YTTERBIU
	4-0744	L-SERIES AUGER ELECTRON SPECTRA FOR ELEMENTS WITH Z = 66-70.
VRAKKING JJ	1-0129	QUANTITATIVE ASPECTS OF AUGER ELECTRON SPECTROSCOPY.
	1-0130	IN DEPTH INFORMATION FROM AUGER ELECTRON SPECTROSCOPY. (SILICON MONOLAYERS)
	1-0205	AUGER ELECTRON SPECTROSCOPY MADE QUANTITATIVE BY ELLIPSOMETRIC CALIBRATION.
	1-0206	QUANTITATIVE ASPECTS OF AUGER ELECTRON SPECTROSCOPY. IMPORTANCE OF ATTENUATION O
	1-0207	QUANTITATIVE AUGER ELECTRON SPECTROSCOPY.
	2-0553	ELECTRON IMPACT IONIZATION CROSS SECTIONS OF INNER SHELLS MEASURED BY AUGER ELEC
	2-0554	MEASUREMENT OF IONIZATION CROSS SECTIONS AND BACKSCATTERING FACTORS FOR USE IN Q
	4-0987	MEASUREMENT OF THE IONIZATION CROSS SECTION OF THE L2,3 SHELL OF CHLORINE USING
	5-1638	ADSORPTION OF OXYGEN ON A CLEAN SILICON SURFACE. (AUGER ELECTRON SPECTROSCOPY)
VUKSTICH VS	4-0735	SPECTRAL LINE EXCITATION FUNCTIONS FOR SODIUM, POTASSIUM, RUBIDIUM, AND CESIUM I
WAGNER CD	1-0208	AUGER LINES IN X-RAY PHOTOELECTRON SPECTROMETRY.
	1-0209	AUGER SPECTRA IN X-RAY PHOTOELECTRON SPECTROSCOPY.
	4-1107	CHEMICAL SHIFTS OF AUGER LINES, AND THE AUGER PARAMETER.
	5-1948	AUGER PARAMETER IN ELECTRON SPECTROSCOPY FOR THE IDENTIFICATION OF CHEMICAL SPEC
	5-1949	X-RAY EXCITED AUGER AND PHOTOELECTRON SPECTRA OF PARTIALLY OXIDIZED MAGNESIUM SU
WAGNER GJ	2-0336	DECAY OF HOLE STATES IN OXYGEN-16 VIA A "NUCLEAR AUGER EFFECT".
WAHID AA	1-0217	USE OF AUGER AND CONVERSION ELECTRONS FOR THE DETECTION OF GAMMA EMITTING RADION
WAHLGREN UI	4-0724	SCF AND LIMITED CI CALCULATIONS FOR ASSIGNMENT OF THE AUGER SPECTRUM AND OF THE
WAKIYA K	4-1010	AUGER ELECTRON EJECTION FROM XENON N4,5OO AND KRYPTON M4,5NN PROCESSES BY ELECTR
WALDROP JR	2-0555	POTENTIAL PROFILING ACROSS SEMICONDUCTOR JUNCTIONS BY AUGER ELECTRON SPECTROSCOP
	3-0650	AUGER ELECTRON SPECTROSCOPY IN THE SCANNING ELECTRON MICROSCOPE: AUGER ELECTRON
	3-0651	LABORATORY COMPUTER SCANNING ELECTRON MICROSCOPE SYSTEM.
	3-0710	AUGER ELECTRON SPECTROSCOPY IN THE SCANNING ELECTRON MICROSCOPE.
	4-0947	CHEMICAL ANALYSIS OF ELECTRODEPOSITED NICKEL- NICKEL BONDS BY AES.
	5-1950	DEPLETION EFFECTS IN SEMIINSULATING GALLIUM ARSENIDE. (AUGER SPECTROSCOPY)
WALL W	5-1220	OPTICAL PROPERTIES AND AUGER CHARACTERISTICS OF MAGNESIUM SINGLE CRYSTALS AND MA
WALLDEN L	4-0940	HIGH ENERGY SATELLITES IN AUGER SPECTRA FROM METALS.
WALLS JM	5-1951	COMBINED FIELD ION MICROSCOPY. AUGER ELECTRON SPECTROSCOPY, AND LEED STUDY OF TH
WALTERS DL	2-0556	NONRELATIVISTIC AUGER RATES, X-RAY RATES, AND FLUORESCENCE YIELDS FOR THE K-SHEL
	2-0557	Z DEPANDANCE OF THE KLL AUGER RATES.
	2-0558	NONRELATIVISTIC AUGER RATES, X-RAY RATES, AND FLUORESCENCE YIELDS FOR THE 2P SHE
WANDELT K	5-1323	ELECTRON SPECTROSCOPIC STUDIES OF CLEAN AND OXIDIZED IRON. (AUGER STUDY OF SURFA
	5-1952	SOFT X-RAY APPEARANCE POTENTIAL SPECTROSCOPY STUDIES WITH COPPER- NICKEL AND IRO
	5-1953	ELECTRON SPECTROSCOPIC STUDIES OF THE OXIDATION OF NICKEL- PALLADIUM ALLOYS.
WANG JM	5-1239	SECONDARY ELECTRON EMISSION OF AMORPHOUS SEMICONDUCTOR THIN FILMS.
WANG KL	4-1108	AUGER ELECTRON SPECTROSCOPY OF A STABLE GERMANIUM OXIDE.
WARD FW	3-0600	HIGH RESOLUTION ELECTRON SPECTROMETER FOR PHOTOELECTRON AND AUGER ELECTRON STUDI
WARDELL IRM	3-0622	ULTRAHIGH VACUUM SCANNING ELECTRON MICROSCOPE WITH AUGER ANALYSIS FACILITIES.
WASSHAUSEN H	5-1943	SURFACE PREPARATION AND CHARACTERIZATION TECHNIQUES (AUGER, CLEANING) FOR QUARTZ
WATANABE K	5-1982	OBSERVATION OF SURFACE COMPOSITION OF COPPER- NICKEL ALLOY BY AUGER SPECTRA IN T
WATANABE M	3-0656	AUGER ELECTRON SPECTROSCOPY USING THE TWO-GRID LEED SYSTEM AND PHOTOELECTRON MUL
	5-1623	ANGULAR DISTRIBUTION OF AUGER ELECTRON EMISSION FROM CLEAN AND GAS-COVERED IRON
	6-2141	STUDY ON THE INITIAL OXIDATION AND REDUCTION PROCESSES OF IRON (100) SURFACE BY
WATANABE T	5-1658	NEW METHOD OF THIN FILM THICKNESS MEASUREMENT USING RADIOISOTOPE AUGER ELECTRONS
WATSON LM	4-0837	X-RAY EXCITED AUGER AND PHOTOELECTRON SPECTRA OF MAGNESIUM, SOME ALLOYS OF MAGNE
	4-0838	COMPARISON OF THE SOFT X-RAY PHOTOEMISSION, AND AUGER ELECTRON SPECTRA OF THE VA
WATSON LN	2-0337	K X-RAY EMISSION EDGE SHAPES OF FREE ELECTRON METALS. (AUGER BROADENING)
WATSON RL	4-0763	RELAXATION DURING PHOTOEMISSION AND LMM AUGER DECAY IN ARSENIC AND SOME OF ITS C
	4-1109	AUGER SPECTRA OF CARBON AND ARGON FOLLOWING IONIZATION BY EQUAL VELOCITY ALPHA P
WATTS BE	5-1514	INFLUENCE OF SUBSTRATE CONDITIONS ON THE NUCLEATION AND GROWTH OF EPITAXIAL SILI
WATTS CMK	2-0440	MECHANISM OF PLASMON GAIN IN AUGER SPECTRA.
	4-1110	DYNAMICAL SCREENING EFFECTS IN THE AUGER SPECTRUM OF METALS.
WATTS GD	5-1464	COMBINED LEED- RHEED AUGER STUDY OF OXYGEN ADSORPTION ON THE TUNGSTEN (112) SURF
	5-1465	AUGER SPECTROSCOPIC STUDY OF THE COADSORPTION OF MOLECULAR OXYGEN AND CARBON MON
	5-1954	AUGER ELECTRON SPECTROSCOPY STUDY OF THE ELECTRON STIMULATED DESORPTION OF OXYGE
WEBER B	4-0820	AUGER SPECTRA OF CARBON. TANTALUM CARBIDES, ETHYLENE AND CARBON MONOXIDE ADSORPT
	4-1111	AUGER ELECTRON SPECTROSCOPY OF HIGH TEMPERATURE OXYGEN- RHENIUM INTERACTIONS.
	5-1955	APPLICATION OF AUGER ELECTRON SPECTROSCOPY TO THE PLATINUM- OXYGEN SYSTEM AT HIG
WEBER RE	1-0159	HANDBOOK OF AUGER ELECTRON SPECTROSCOPY. (EXPERIMENTAL SPECTRA)
	1-0210	THIN FILM ANALYSIS BY AUGER ELECTRON SPECTROSCOPY. (REVIEW)
	1-0211	AUGER ELECTRON SPECTROSCOPY FOR THIN FILM ANALYSIS.
	1-0212	AUGER ELECTRON SPECTROSCOPY FOR THIN FILM ANALYSIS.
	3-0595	APPARATUS FOR ANALYSIS BY AUGER ELECTRON SPECTROSCOPY.
	5-1956	DETERMINATION OF SURFACE STRUCTURES USING LEED AND ENERGY ANALYSIS OF SCATTERED
	5-1957	DETERMINATION OF SURFACE STRUCTURES BY LEED AND AUGER ELECTRON SPECTROSCOPY.
	5-1958	WORK FUNCTION AND STRUCTURAL STUDIES OF ALKALI- COVERED SEMICONDUCTORS.
	5-1959	USE OF LEED APPARATUS FOR THE DETECTION AND IDENTIFICATION OF SURFACE CONTAMINAN
	6-2122	AUGER ELECTRON SPECTROSCOPY OF METAL SURFACES.
	6-2123	METALLURGICAL APPLICATION OF AUGER ELECTRON SPECTROSCOPY.
WEBER U	1-0213	BASIC INVESTIGATION IN THE AREA OF SCANNING ELECTRON MICROSCOPY. PART-1: APPLICA
	3-0711	ELECTRON BEAM MICROANALYZER WITH AUGER ELECTRON DETECTION.

WEBER U	3-0712	ELECTRON BEAM ANALYZER FOR DETECTING AUGER ELECTRONS.
	3-0713	ANALYSIS APPARATUS FOR INVESTIGATING A SAMPLE BY MEANS OF EMITTED ELECTRONS.
	3-0714	ELECTRON BEAM MICROPROBES USED IN THE ANALYSIS AND DEVELOPMENT OF MATERIALS. (AU
WEGMANN L	5-1345	INVESTIGATIONS OF CARBON RESIDUES ON SURFACES OF SILICON INTEGRATED CIRCUITS. (B
WEHNER GK	1-0214	SPUTTERING IN THIN FILM ANALYSIS METHODS. (ESCA, AUGER SPECTROSCOPY)
	1-0215	ASPECTS OF SPUTTERING IN SURFACE ANALYSIS METHODS. (REVIEW AUGER SPECTROSCOPY AN
	2-0527	ESCAPE LENGTH OF AUGER ELECTRONS.
	5-1877	ALLOY SPUTTERING STUDIES WITH IN-SITU AUGER ELECTRON SPECTROSCOPY.
	5-1878	AUGER ELECTRON SPECTROSCOPY STUDIES OF SPUTTER DEPOSITION AND SPUTTER REMOVAL OF
	5-1960	AUGER THIN FILM ANALYSIS AS APPLIED TO VARIOUS ASPECTS IN SOLAR ENERGY CONVERSIO
	6-1998	COMPOSITION VERSUS DEPTH PROFILES OBTAINED WITH AUGER ELECTRON SPECTROSCOPY OF A
WEI PSP	3-0715	ELECTRON SPECTROMETER FOR SURFACE ANALYSIS.
WEI RP	6-2142	A TECHNIQUE FOR DETERMINING THE ELEMENTAL COMPOSITION OF FRACTURE SURFACES PRODU
WEIGHTMAN P	2-0559	RELAXATION PROCESSES ACCOMPANYING CADMIUM M4,5N4,5N4,5 AUGER TRANSITIONS.
	2-0560	L2,3MM AUGER PROCESSES IN SELENIUM.
	4-1033	TRANSITION PROBABILITIES FOR THE L2,3MM AUGER SPECTRUM OF SELENIUM.
	4-1034	AUGER VACANCY SATELLITE STRUCTURE IN THE L3,M4,5M4,5 AUGER SPECTRA OF COPPER.
	4-1035	PHOTOELECTRON AND L2,3MM AUGER ELECTRON ENERGIES FOR ARSENIC.
WEINBERG WH	5-1569	LEED- AUGER INVESTIGATION OF A STABLE CARBIDE OVERLAYER ON A PLATINUM (111) SURF
WEISBERG LR	2-0561	AUGER RECOMBINATION IN GALLIUM ARSENIDE.
WEISENBERGER W	5-1262	COMBINED SCANNING ELECTRON MICROSCOPY- AUGER SPECTROSCOPY FOR MICROSPOT SURFACE
WEISENBERGER WH	1-0216	DIRECT COMPARISON OF AUGER, SECONDARY ION MASS SPECTROSCOPY, AND PROTON RESONANC
	5-1261	SCANNING ELECTRON MICROSCOPY, AUGER SPECTROSCOPY, AND ION BACKSCATTERING TECHNIQ
	5-1961	HIGH SPATIAL RESOLUTION SCANNING AUGER SPECTROSCOPY APPLIED TO ANALYSIS OF X-BAN
WEISHEIT JC	2-0562	X-RAY IONIZATION CROSS SECTIONS, AND IONIZATION EQUILIBRIUM EQUATIONS MODIFIED B
WEISSMAN I	5-1962	DEPOSITION AND AUGER ANALYSIS OF DEPOSITED SILICON DIOXIDE ON ALUMINUM(X) GALLIU
WEKSLER M	4-1112	X-RAY SPECTRA, L-SUBSHELL FLUORESCENCE AND COSTER-KRONIG YIELDS IN BISMUTH AND N
WELLER T	2-0275	TWO ELECTRON BAND-TO-BAND TRANSITIONS IN SOLIDS. (AUGER RECOMBINATION IN SILICON
WELSH LB	5-1385	MORPHOLOGICAL AND ELECTRICAL PROPERTIES OF RF SPUTTERED YTTRIA-DOPED ZIRCONIA TH
WEN CP	5-1950	DEPLETION EFFECTS IN SEMIINSULATING GALLIUM ARSENIDE. (AUGER SPECTROSCOPY)
WENAAS EP	2-0563	THEORY OF AUGER EJECTION OF ELECTRONS FROM METALS BY IONS.
WENZEL M	1-0217	USE OF AUGER AND CONVERSION ELECTRONS FOR THE DETECTION OF GAMMA EMITTING RADION
WERME LO	2-0515	ELECTRON SPECTROSCOPIC INVESTIGATION OF AUGER PROCESSES IN BROMINE-SUBSTITUTED M
	4-1113	HIGH RESOLUTION L2,3MM AND M4,5NN AUGER SPECTRA FROM KRYPTON AND M4, 5NN AND N4,
	4-1114	L2,3MM AUGER SPECTRUM OF ARGON.
	5-1529	AUGER ELECTRON SPECTRUM OF CARBON SUBOXIDE.
WERNER HW	1-0016	SURFACE ANALYSIS, METHODS OF STUDYING THE OUTER ATOMIC LAYERS OF SOLIDS.
WERTHEIM GK	2-0564	CHEMICAL EFFECTS OF NUCLEAR TRANSFORMATIONS IN MOSSBAUER SPECTROSCOPY. (AUGER AF
	2-0565	SIZE EFFECT IN IONIC CHARGE RELAXATION FOLLOWING AUGER EFFECT.
WEST LA	5-1963	STUDIES OF GAS- SURFACE INTERACTIONS AT THE (100) CRYSTAL FACE OF PLATINUM BY MO
	5-1964	RELATIVE SPUTTERING YIELDS AND QUANTITATIVE SURFACE ANALYSIS BY AUGER SPECTROSCO
WESTON WF	2-0566	IDENTIFICATION AND MEASUREMENT OF LOW ENERGY ELECTRONS BY CLOUD CHAMBER TECHNIQU
WESTWOOD HJ	6-2097	STUDY OF INTERGRANULAR FRACTURE IN IRON USING AUGER SPECTROSCOPY.
WESTWOOD WD	5-1480	AUGER ANALYSIS OF TANTALUM SURFACES.
	5-1965	POROSITY IN SPUTTERED PLATINUM FILMS. (AUGER ELECTRON SPECTROSCOPY)
WHEELER DR	6-2143	EFFECT OF ADSORBED CHLORINE AND OXYGEN ON SHEAR STRENGTH OF IRON AND COPPER JUNC
WHITAKER MAB	4-1115	CHEMICAL SHIFTS IN AUGER SPECTRA (MONOLAYER ADSORPTION, ESCA SIMPLE CHARGE TRANS
WHITENTON J	2-0497	RELATIVE SATELLITE LINE INTENSITIES IN ARGON K AUGER ELECTRON SPECTRA PRODUCED B
WHITMOYER RD	4-1119	AUGER ELECTRON ENERGIES (0-2000 EV) FOR ELEMENTS OF ATOMIC NUMBER 5-103.
WICKERSHAM CE	5-1385	MORPHOLOGICAL AND ELECTRICAL PROPERTIES OF RF SPUTTERED YTTRIA-DOPED ZIRCONIA TH
WIEDER HH	2-0567	MINORITY CARRIER LIFETIME IN INDIUM ARSENIDE EPILAYERS. (AUGER RECOMBINATION PRO
WIEDERSICH H	6-2091	SEGREGATION OF ALLOYING ELEMENTS TO FREE SURFACES DURING IRRADIATION. (AUGER SPE
	6-2092	IRRADIATION INDUCED SOLUTE SEGREGATION (IN STEEL, AUGER SPECTROSCOPY).
WIEDMANN P	5-1798	MATERIAL ANALYSIS OF THIN FOILS BY AUGER ELECTRON SPECTROSCOPY.
WIEDNER CA	2-0336	DECAY OF HOLE STATES IN OXYGEN-16 VIA A "NUCLEAR AUGER EFFECT".
WIEMAN H	2-0298	OUTER SHELL EXCITATION OF THE RARE GASES BY 30 MEV OXYGEN IONS. (INNER SHELL AUG
	2-0300	PROJECTILE CHARGE STATE DEPENDENCE OF NEON K-SHELL IONIZATION AND FLUORESCENCE Y
	2-0518	K-SHELL IONIZATION AND FLUORESCENCE YIELD IN 50 MEV CHLORINE(+)- NEON COLLISIONS
WIESNER H	4-1116	LINE WIDTH AND AUGER EFFECT IN CALCIUM, STRONTIUM, AND BARIUM FLUORIDES.
WILCOX DL	5-1128	ELECTRON BEAM EFFECTS IN DEPTH PROFILING MEASUREMENTS WITH AUGER ELECTRON SPECTR
WILD RK	4-0737	CHEMICAL SHIFTS IN URANIUM AUGER TRANSITIONS.
	4-0738	AUGER SPECTROSCOPY OF URANIUM.
	5-1460	AUGER ELECTRON SPECTROSCOPY OF MATERIALS EXPOSED TO FLOWING SODIUM.
	6-1993	AUGER ELECTRON SPECTROSCOPY STUDY OF 18-8 STAINLESS STEEL OXIDATION.
	6-2144	OXIDATION OF NIMONIC-80A ALLOY AT 800 DEGREES C IN LOW OXYGEN PRESSURES. (AUGER
	6-2145	VACUUM ANNEALING OF STAINLESS STEEL AT TEMPERATURES BETWEEN 770 AND 1470 K.
WILDMAN HS	1-0085	QUANTITATIVE DEPTH PROFILING BY AUGER ION SPUTTERING TECHNIQUE.
	5-1439	DIFFUSION MEASUREMENT IN THIN FILMS BY AUGER SPUTTER ETCHING TECHNIQUE.
	5-1440	INTERDIFFUSION IN BIMETALLIC POLYCRYSTALLINE FILMS MEASURED BY AUGER ION SPUTTER
	6-2146	ANALYSIS OF GRAIN BOUNDARY DIFFUSION IN BIMETALLIC THIN FILM STRUCTURES USING AU
WILETS L	2-0301	SIMPLE MODEL FOR K X-RAY AND AUGER ELECTRON ENERGY SHIFTS IN HEAVY ION COLLISION
WILLIAMS BF	2-0467	IDENTIFICATION OF AUGER ELECTRONS IN GALLIUM ARSENIDE.
WILLIAMS EM	3-0590	CYLINDRICAL MIRROR ANALYZER INCORPORATING PRERETARDATION. (TUNGSTEN AUGER ELECTR
WILLIAMS FL	5-1966	SURFACE COMPOSITION OF NICKEL- GOLD ALLOYS.
WILLIAMS RH	5-1434	ADHESION AT GOLD MICA INTERFACES. (LEED, AES)
	5-1634	ELECTRONIC PROPERTIES OF CLEAVED MOLYBDENUM DISULFIDE SURFACES.
	5-1967	ELECTRON EMISSION STUDIES FROM GALLIUM SELENIDE SURFACES.
	5-1968	SURFACE CHARACTERIZATION OF INDIUM PHOSPHIDE. (BY AUGER SPECTROSCOPY)
	5-1969	SURFACE STUDIES OF SOME TRANSITION METAL DICHALCOGENIDES.
WILLIS A	5-1584	CORRELATION OF ELECTRONIC, LEED, AND AUGER DIAGNOSTICS ON ZINC OXIDE SURFACES.
WILLIS RF	2-0568	GRAPHITE CONDUCTION BAND STATES FROM SECONDARY ELECTRON EMISSION SPECTRA.
	2-0569	BAND STRUCTURE OF GRAPHITE STUDIED BY SECONDARY ELECTRON EMISSION.
	3-0698	REFINEMENTS TO A STANDARD LEED- AUGER SYSTEM FOR THE ANALYSIS OF ELECTRON EMISSI
	5-1970	STUDY OF CARBON FIBER SURFACES USING AUGER AND SECONDARY ELECTRON EMISSION SPECT
WILMSEN CW	5-1971	CORRELATION BETWEEN THE COMPOSITIONAL PROFILE AND ELECTRICAL CONDUCTIVITY OF THE
WILSON JM	1-0218	AUGER TRANSITIONS.
	3-0589	ELECTRON BEAM CURRENT CONTROLLER FOR LEED- AUGER STUDIES.
	4-1013	AUGER TRANSITIONS BETWEEN 0 AND 1500 EV FOR ELEMENTS UP TO Z EQUALS 88.
	5-1628	DETECTION OF LEAD, BY AUGER SPECTROSCOPY, ON IRON INHIBITED IN LEAD AZELATE SOLU
	5-1972	STUDY OF DIFFUSION IN THIN COPPER- LEAD FILMS USING AUGER ELECTRON SPECTROSCOPY.

WILSON JM   5-1973   LEED AND AES STUDY OF THE INTERACTION OF HYDROGEN SULFIDE AND MOLYBDENUM (100).
WINTERS HF   5-1128   ELECTRON BEAM EFFECTS IN DEPTH PROFILING MEASUREMENTS WITH AUGER ELECTRON SPECTR
WINTON RI   5-1461   FLASH DESORPTION- AUGER ELECTRON SPECTROSCOPY. DISCREPANCY FOR CARBON DIOXIDE ON
WISE H   5-1976   SURFACE COMPOSITION OF PALLADIUM- GOLD AND PALLADIUM- SILVER CATALYSTS BY AUGER
WISSEMAN WR   5-1636   DIFFUSION STUDIES IN GOLD- PLATINUM THIN FILMS BY AUGER ELECTRON SPECTROSCOPY.
WITHROW SP   5-1974   INVESTIGATICN OF HYDROGEN CHEMISORPTION ON THE TUNGSTEN (100) SURFACE. (AUGER EL
WITTRY DB   3-0716   RECENT ADVANCES IN INSTRUMENTATION FOR MICROPROBE ANALYSIS. (AUGER ELECTRON SPEC
WOITSCHECK A   1-0034   APPLICATION OF ION, ELECTRON, AND X-RAY BEAMS IN MODERN MATERIAL TESTING.
WOLFF H   5-1520   STABILITY OF FIELD ELECTRON EMISSION AND VACUUM BREAKDOWN. (FROM SURFACES) INVES
WOLFGANG E   3-0685   SPECTROSCOPIC METHODS OF MICROANALYSIS USING THE SCANNING ELECTRON MICROSCOPE. (
WOLSCHIN GW   4-0941   NUCLEAR AUGER EFFECTS IN MUONIC ATOMS. (LEAD, BISMUTH)
WONG C   5-1507   REEXAMINATION OF ALUMINUM (110) WITH LOW ENERGY DIFFRACTION AND AUGER SPECTROSCO
WONG R   5-1975   RECENT ASPECTS OF GLASS FIBER- RESIN INTERFACES. (AUGER SPECTROSCOPY OF GLASS FI
WOOD BJ   5-1976   SURFACE COMPOSITION OF PALLADIUM- GOLD AND PALLADIUM- SILVER CATALYSTS BY AUGER
WOOD PR   5-1417   STUDY OF THE NITRIDATION OF SILICON SURFACES BY LOW ENERGY ELECTRON DIFFRACTION
                 5-1418   AUGER ELECTRON SPECTROSCOPY AND ITS APPLICATION IN STUDIES OF SOME SILICON SURFA
WOOD RE   4-1105   L-SHELL FLUORESCENCE YIELDS OF DOUBLE VACANCY STATES IN LEAD.
WOODALL RO   4-0763   RELAXATION DURING PHOTOEMISSION AND LMM AUGER DECAY IN ARSENIC AND SOME OF ITS C
WOODRUFF DP   3-0621   ULTRAHIGH VACUUM SCANNING ELECTRON MICROSCOPE WITH A FIELD EMISSION CATHODE AND
                 3-0679   SCANNING AUGER ELECTRON MICROSCOPE FOR SURFACE STUDIES.
                 4-0958   ANGULAR DEPENDENCE OF AUGER ELECTRON EMISSION FROM A SINGLE CRYSTAL SPECIMEN.
                 4-1020   PLASMON EFFECTS IN ELECTRON ENERGY LOSS AND GAIN SPECTRA IN ALUMINUM.
                 5-1447   ANGULAR DEPENDENCE OF AUGER ELECTRON EMISSION FROM SOLID SURFACES.
                 5-1632   ANGULAR DEPENDENCE OF AUGER ELECTRON EMISSION FROM COPPER (111) AND (100) SURFAC
                 5-1977   SURFACE STRUCTURE FROM ANGULAR DEPENDENCE OF AUGER ELECTRON EMISSION.
                 6-2089   A LOW ENERGY EIECTRON DIFFRACTION- AUGER ELECTRON SPECTROSCOPY STUDY OF THE ADSO
WOODS CW   2-0481   RADIATIVE AUGER EFFECT IN ION- ATOM COLLISIONS.
                 2-0570   K-SHELL AUGER ELECTRON PRODUCTION CROSS SECTIONS FROM ION BOMBARDMENT.
                 2-0571   NEON K-AUGER CROSS SECTIONS FROM 1.5 MEV PER AMU AND ATOMIC HYDROGEN, OXYGEN, AN
                 2-0572   NEON K-SHELL VACANCY PRODUCTION BY INTERMEDIATE ENERGY FLUORINE IONS. (AUGER ELE
                 2-0573   AUGER ELECTRONS FROM DOUBLE K-SHELL VACANCY STATES IN NEON.
                 2-0574   K-SHELL AUGER ELECTRON PRODUCTION CROSS SECTIONS FROM ION BOMBARDMENT. (NEON, FL
                 4-1117   K-SHELL AUGER ELECTRON HYPERSATELLITES OF NEON.
WOODWARD RP   3-0625   INVESTIGATICN OF HIGH RESOLUTION SCANNING TECHNIQUES FOR AUGER ELECTRON SPECTROS
                 5-1978   AUGER ELECTRON SPECTROSCOPY FOR THE STUDY OF SURFACE CONTAMINATION IN SEMICONDUC
WRAY L   5-1349   THE (111) SURFACE OF N-TYPE SILICON.
WRENN ME   1-0219   AUGER ELECTRON DOSIMETRY.
WRIGHT B   4-1118   OXIDATION EFFECTS ON THE SPECTRUM OF THE SLOW SECONDARY ELECTRON EMISSION PEAK I
                 5-1979   SECONDARY ELECTRON EMISSION FROM PLATINUM BLACK COATED SURFACES. (AUGER SPECTRA)
WUILLEUMIER F   2-0575   DETERMINATION OF OSCILLATOR STRENGTH FOR THE FIRST DISCRETE TRANSITION IN NITROG
WYNBLATT P   6-2003   A TECHNIQUE FOR INVESTIGATING THE CATALYTIC ACTIVITY OF ALLOY SYSTEMS. (SURFACE
YADREEV FI   3-0617   DRYING OF BUTADIENE- NITRILE RUBBERS IN AN AUGER MACHINE.
YAGI K   5-1624   GROWTH MODE DISCRIMINATION OF THIN FILMS BY AUGER ELECTRON SPECTROSCOPY.
YAGNIK CM   2-0576   BACKSCATTER MOSSBAUER STUDIES USING CCNVERSION ELECTRONS. TIN-119.
YAKOWITZ H   1-0220   FUTURE GROWTH OF SEM: TOWARD SUPER EFFICIENT MICROSCOPY. (AUGER SPECTROSCOPY)
YAMADA M   6-2109   SURFACE AND GRAIN BOUNDARY SEGREGATION OF STEEL IMPURITIES. (TEMPER EMBRITTLEMEN
YAMADA Y   5-1980   GROWTH AND STRUCTURE OF TITANIUM OXIDE THIN FILMS. PART-1. (AUGER SPECTRA)
YAMAGUCHI K   5-1981   MEASUREMENTS ON SURFACES OF ELECTRICAL SOLID MATERIALS BY THE LEED METHOD.
YAMAMOTO S   5-1413   AUGER ELECTRON EMISSION MICROGRAPHIC STUDIES OF THE CLEAVAGE SURFACE OF GRAPHITE
                 5-1415   AUGER ELECTRCN EMISSION MICROGRAPHY AND MICROANALYSIS OF A STAINED SURFACE OF PU
                 5-1416   AUGER ELECTRON EMISSION MICROGRAPHY AND MICROANALYSIS OF SOLID SURFACES.
                 5-1471   SURFACE CONDITION AND THERMIONIC EMISSION OF LANTHANUM HEXABORIDE. (AUGER SPECTR
YAMASHINA T   1-0221   AUGER ELECTRCN SPECTROSCOPY. (REVIEW, 8 REFS)
                 5-1703   AUGER SPECTROSCOPY ON COPPER- NICKEL ALLOY SURFACES RELATED TO CATALYSIS.
                 5-1873   WORK FUNCTICN OF WELL-DEFINED SURFACE OF COPPER- NICKEL ALLOY PLATES. (AUGER ELE
                 5-1982   OBSERVATION CF SURFACE COMPOSITION OF COPPER- NICKEL ALLOY BY AUGER SPECTRA IN T
YANG MG   5-1286   DIFFUSION STUDIES IN CHROMIUM- PLATINUM THIN FILMS USING AUGER ELECTRON SPECTROS
                 5-1637   ION SELECTIVE SPUTTERING OF III-V COMPOUND SEMICONDUCTOR SURFACES. (AUGER EFFECT
                 5-1983   AUGER ELECTRON SPECTROSCOPY OF CLEANUP-RELATED CONTAMINATION ON SILICON SURFACES
YASKEVICH GN   2-0368   AUGER RECOMBINATION IN SEMICONDUCTORS DURING INTENSE EXCITATION.
YASKO RN   4-1119   AUGER ELECTRON ENERGIES (0-2000 EV) FOR ELEMENTS OF ATOMIC NUMBER 5-103.
                 5-1984   DETERMINATION OF FILM THICKNESS OR ION ETCH RATE USING AUGER ELECTRON SPECTROSCO
YASNOPOLSKII NL 4-1120   CHARACTERISTICS OF SHAPING SECONDARY ELECTRON EMISSION FROM POROUS POTASSIUM CHL
YAVOR SY   3-0717   HIGH RESOLUTION ELECTROSTATIC SPECTROMETER USED IN ELECTRON SPECTROMETRY FOR CHE
YELLIN E   4-1122   L-S COUPLING INTERPRETATION OF HIGH RESOLUTION LMM AUGER SPECTRA OF COPPER AND Z
                 4-1123   X-RAY EXCITED LMM AUGER SPECTRA OF COPPER, NICKEL AND IRON.
YIN LI   1-0222   SOME PARAMETERS AFFECTING AUGER AND PHOTOELECTRON SPECTROSCOPY AS AN ANALYTICAL
                 2-0577   ATOMIC N-SHELL VACANCY WIDTHS IN SOLIDS.
                 2-0578   INVESTIGATION OF LOW-Z COSTER-KRONIG TRANSITIONS BY AUGER AND PHOTOELECTRON SPEC
                 2-0579   FREE ATOM BEHAVIOR OF AUGER ELECTRONS IN SOLIDS.
                 3-0718   WIDTHS OF ATOMIC M-SHELL VACANCY STATES AND QUASIATOMIC ASPECTS OF RADIATIONLESS
                 4-1121   QUASIATOMIC LMM AUGER SPECTRA OF SOLID COPPER AND ZINC.
                 4-1122   L-S COUPLING INTERPRETATION OF HIGH RESOLUTION LMM AUGER SPECTRA OF COPPER AND Z
                 4-1123   X-RAY EXCITED LMM AUGER SPECTRA OF COPPER, NICKEL AND IRON.
YOSHIDA K   5-1980   GROWTH AND STRUCTURE OF TITANIUM OXIDE THIN FILMS. PART-1. (AUGER SPECTRA)
YOSHII S   5-1481   AUGER STUDIES OF DIFFUSED IMPURITY PROFILES IN SILICON DIOXIDE- SILICON AND POLY
YOSHIKAWA M   5-1550   COMPOSITION PROFILES OF SEVERAL CONTAMINATED AND CLEANED SURFACES OF GOLD THICK
YOSHIKOKA T   5-1482   OBSERVATION OF IMPURITY PROFILE IN ION IMPLANTED SILICON BY AUGER ELECTRON SPECT
YOUNG RD   3-0719   INSTRUMENTATION FOR THE CHEMICAL ANALYSIS OF MANUFACTURED SURFACES. (REVIEW, SPE
YU KY   5-1424   DETERMINATICN OF THE SURFACE COMPOSITION OF COPPER- NICKEL ALLOYS FOR CLEAN AND
ZAPESOCHNYI IP   2-0580   INVESTIGATION OF MECHANISM OF S IONIZATION OF NOBLE GAS ATOMS BY THE ULTRASOFT X
                 4-0735   SPECTRAL LINE EXCITATION FUNCTIONS FOR SODIUM, POTASSIUM, RUBIDIUM, AND CESIUM I
ZASHKVARA VV   2-0538   IONIZATION LCSSES IN (AUGER) SPECTRA OF ELECTRON INELASTIC SCATTERING FROM RARE
                 3-0653   SPECTROMETER OF AUGER ELECTRONS.
                 3-0720   FOCUSING A CHARGED PARTICLE BEAM OF FINITE ANGULAR SPREAD IN AN ELECTROSTATIC AN
                 3-0721   FOCUSING PROPERTIES OF AN ELECTROSTATIC MIRROR WITH CYLINDRICAL FIELD.
                 4-1030   AUGER ELECTRON SPECTRUM OF OSMIUM AT ENERGIES UP TO 300 EV.
                 4-1100   AUGER ELECTRON SPECTRA OF SOME RARE EARTH METALS. (LANTHANUM, PRASEODYMIUM, NEOD
                 4-1124   PLASMA SATELLITES IN THE AUGER ELECTRON SPECTRUM OF MOLYBDENUM.
                 5-1759   DECREASE OF CARBON CONCENTRATION ON THE SURFACE OF MOLYBDENUM CARBIDE DURING HEA

ZASHKVARA VV	5-1985	CHEMICAL COMPOSITION OF THE SURFACE OF MOLYBDENUM AND STEEL BASED ON AUGER ELECT
ZAVATTINI E	2-0305	MEASUREMENT OF THE AUGER EFFECT IN THE MU-4 HELIUM(2S)(+) IONIC SYSTEM.
ZDERADICKA J	2-0474	COINCIDENCE METHODS OF STANDARDIZATION FOR CESIUM-131 AND MEASUREMENT OF DECAY P
ZECCA A	3-0616	AN ENERGY MODULATED (CYLINDRICAL) HIGH ENERGY RESOLUTION ELECTRON SPECTROMETER.
ZEHNER DM	2-0581	AUGER TRANSITIONS INITIATED FROM SINGLY AND DOUBLY IONIZED STATES IN LITHIUM MET
	4-0899	PLASMON ENERGY GAIN AND DOUBLE IONIZATION SATELLITES IN THE BERYLLIUM AUGER SPEC
	4-0900	HIGHER ENERGY SATELLITES IN AUGER SPECTRA FROM METALS.
	4-0901	CHARACTERISTIC ENERGY GAIN AND LOSS, DOUBLE IONIZATION, AND IONIZATION LOSS EVEN
	5-1139	INVESTIGATION OF REORDERED (001) GOLD SURFACES BY POSITIVE ION CHANNELING SPECTR
	5-1697	ANGULAR RESOLVED AUGER EMISSION SPECTRA FROM A CLEAN COPPER (100) SURFACE.
	5-1986	CHARACTERIZATION OF REORDERED (001) GOLD SURFACES BY POSITIVE ION CHANNELING SPE
	5-1987	AUGER AND IONIZATION LOSS SPECTRA FROM BERYLLIUM SURFACES.
	5-1988	ANGULAR EFFECTS IN AUGER ELECTRON EMISSION FROM COPPER (110).
ZEMEL JN	5-1790	PROPERTIES OF THE LEAD SULFIDE- LEAD OXIDE CRYSTALLINE FILMS. PART-1: PREPARATIO
ZENDER MJ	4-1125	AUGER SPECTRA OF URANIUM.
ZHDANOV VS	4-0756	SPECTRUM OF AUGER K,L,M,N,O ELECTRONS OF SAMARIUM.
	4-0757	KLM SPECTRUM OF SAMARIUM AUGER ELECTRONS.
	4-0758	KMM AUGER TRANSITIONS IN XENON-131.
ZHOW LD	1-0127	APPLICATION OF (AUGER) ELECTRON SPECTROSCOPY TO CHEMISTRY. (REVIEW)
ZHUKOV IG	2-0580	INVESTIGATION OF MECHANISM OF S IONIZATION OF NOBLE GAS ATOMS BY THE ULTRASOFT X
ZHURAKOVSKII AP	4-1126	DETECTION OF EXCITONS NEAR THE L2,3 EDGE OF THE CHLORIDE ION IN SODIUM CHLORIDE
ZHURKIN BG	4-0780	MAGNETIC FIELD DEPENDENT INTENSITY OSCILLATIONS OF THE ELECTRON HOLE LUMINESCENC
ZIEM P	2-0345	K-VACANCY SHARING STUDIED FROM AUGER ELECTRON YIELDS.
	2-0521	ARGON L-SHELL IONIZATION BY 50 TO 600 KEV PROTON, DEUTERON, MOLECULAR HYDROGEN,
	2-0523	GAS TARGET EXPERIMENTS OF K-SHELL VACANCY SHARING IN ASYMMETRIC ION- ATOM COLLIS
	2-0582	K-SHELL VACANCY SHARING IN 300 TO 500 KEV NEON(+)- SODIUM COLLISIONS. (QUASIMOLE
ZILDO JL	5-1385	MORPHOLOGICAL AND ELECTRICAL PROPERTIES OF RF SPUTTERED YTTRIA-DOPED ZIRCONIA TH
ZIMMER RS	5-1989	A CARBON STRUCTURE ON THE RHENIUM (0001) SURFACE.
ZINOVEV AN	2-0234	DETECTION OF A NEW TYPE OF AUGER TRANSITIONS IN ATOMS WITH TWO INTERNAL VACANCIE
	2-0235	OBSERVATION OF THREE ELECTRON AUGER TRANSITIONS IN ATOMS WITH TWO INNER VACANCIE
ZINSER J	3-0722	DIGITAL CONTROL UNIT IN TTL LOGIC FOR THE AUTOMATIC EXECUTION OF AUGER AND ESCA
ZIVITZ M	2-0255	RADIATIONLESS DEEXCITATION OF EXCITED HELIUM ATOMS AT SURFACES. (AUGER PROCESS)
ZOGG H	2-0297	ANGLE DISTRIBUTION OF SODIUM CHLORIDE SINGLE CRYSTALS. (PHOTO- AND AUGER ELECTRO
ZOLOTAVIN AV	4-0743	AUGER L-SERIES ELECTRONIC SPECTRA ACCOMPANYING THE DECAY OF ERBIUM-160, YTTERBIU
	4-0744	L-SERIES AUGER ELECTRON SPECTRA FOR ELEMENTS WITH Z = 66-70.
ZUCCA R	5-1950	DEPLETION EFFECTS IN SEMIINSULATING GALLIUM ARSENIDE. (AUGER SPECTROSCOPY)
ZUEV VA	5-1990	MANY PARTICLE RECOMBINATION PROCESSES (AUGER) ON THE SURFACES AND IN THIN FILMS